KU-004-361

BIOCHEMISTRY
The Chemical Reactions of Living Cells

David E. Metzler
Distinguished Professor Emeritus
Iowa State University

Volume 2

in association with
Carol M. Metzler

designed and illustrated by
David J. Sauke

ACADEMIC
PRESS

An imprint of Elsevier Science

Amsterdam London New York Oxford Paris Tokyo
Boston San Diego San Francisco Singapore Sydney

Second Edition

Feedback of typographical and scientific errors can be posted at <**http://www.academicpressbooks.com**>. Brief explanations of important alternative or controversial interpretations and important new information are also welcome.

Study question answers are posted at <**http://www.academicpressbooks.com**>.

Senior Publishing Editor	Jeremy Hayhurst
Editorial Coordinator	Nora Donaghy
Senior Project Manager	Julio Esperas
Production Services	Graphic World
Copyeditor	Kristin Landon
Printer	Courier
Permissions	The Permissions Company

This book is printed on acid-free paper. ∞

Academic Press
An imprint of Elsevier Science
525 B Street, Suite 1900, San Diego, California 92101-4495, USA
http://www.academicpress.com

Academic Press
84 Theobald's Road, London WC1X 8RR, UK
http://www.academicpress.com

Academic Press
200 Wheeler Road, Burlington, Massachusetts 01803, USA
www.academicpressbooks.com

Library of Congress Catalog Card Number: 2002117798

International Standard Book Number: 0-12-492543-X (Set)
International Standard Book Number: 0-12-492540-5 (Volume 1)
International Standard Book Number: 0-12-492541-3 (Volume 2)

PRINTED IN THE UNITED STATES OF AMERICA
03 04 05 06 07 08 9 8 7 6 5 4 3 2 1

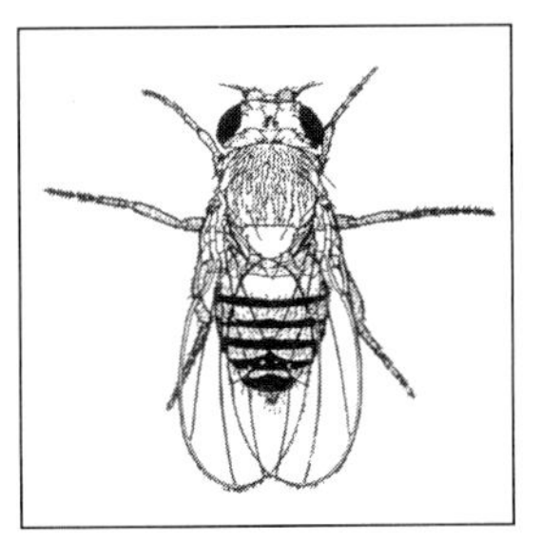

A Brief Table of Contents *Volume 2*

A Brief Table of Contents *Volume 1*

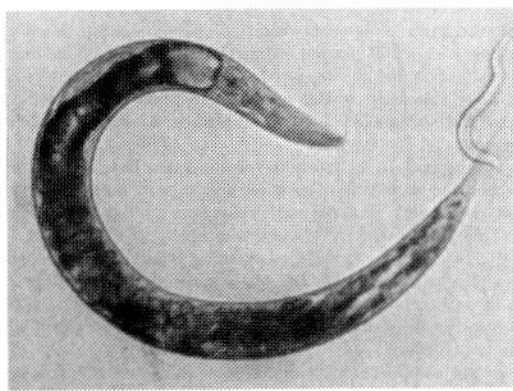

Volume 1

Chapter 1
The Scene of Action

Chapter 2
Amino Acids, Peptides, and Proteins

Chapter 3
Determining Structures and Analyzing Cells

Chapter 8
Lipids, Membranes, and Cell Coats

Chapter 9
Enzymes: The Catalysts of Cells

Chapter 10
An Introduction to Metabolism

Chapter 11
The Regulation of Enzymatic Activity and Metabolism

Chapter 12

Transferring Groups by Displacment Reactions

Chapter 13

Enolate Anions in Enzymatic Addition, Elimination, Isomerization, and Condensation Reactions

Chapter 14

Coenzymes: Nature's Special Reagents

Chapter 15
Coenzymes of Oxidation–Reduction Reactions

Chapter 16
Transition Metals in Catalysis and Electron Transport

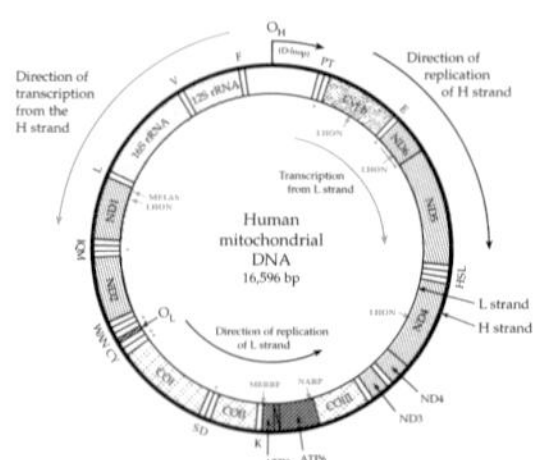

Volume 2

Chapter 17
The Organization of Metabolism

Chapter 18
Electron Transport, Oxidative Phosphorylation, and Hydroxylation

Chapter 19
The Chemistry of Movement

Chapter 20
Some Pathways of Carbohydrate Metabolism

Chapter 21
Specific Aspects of Lipid Metabolism

Chapter 22
Polyprenyl (Isoprenoid) Compounds

Chapter 23
Light and Life

Chapter 30
Chemical Communication Between Cells

Chapter 31
Biochemical Defense Mechanisms

Chapter 32
Growth and Development

Preface

When Volume I was published early in 2001 I thought that Volume II would be ready by the end of the same year. However, updating the first edition to cover the major aspects of the rapidly expanding scope of Biochemistry was more difficult than I had imagined. This second volume brings many topics up-to-date. The first chapter (Chapter 17), which has been available on the Web, is an introduction to metabolism, more extensive than that in Chapter 10. This is followed by specialized chapters on oxidative metabolism, muscle chemistry, and on selected aspects of the metabolism of carbohydrates, lipids, and polyprenyl compounds. Chapter 23 deals with light, photochemistry, and photobiology. Chapter 24 describes metabolism of nitrogen fixation, amino acid synthesis and degradation and chemistry of other nitrogenous compounds. Chapter 25 provides a summary of biosynthesis and catabolism of aromatic compounds including the nucleic acid bases. Chapters 26–29 deal with molecular genetics and the biosynthesis and metabolism of DNA, RNA, and proteins. Chapters 30–32 are introductions to the chemistry of intercellular signaling, brain chemistry, immunochemistry, and development.

Volume II contains a huge amount of information. However, it represents only a tiny fraction of the biochemical knowledge presently available. The many references given may help the reader to get started in consulting this literature. However, the references provide neither a comprehensive review of literature nor an accurate historical record. I hope that my fellow biochemists will forgive me where I have failed to cite their favorite articles. The main purpose of the references is to document the material that I used in preparing the text. The coverage is limited. I used a selection of journals that I could scan quickly, hoping to provide a broad view. However, there are now hundreds of journals that contain biochemical information. An important entrance to this literature is via the World Wide Web, which by now may contain ~10^{11} pages of information. (See D. Butler, *Nature*, **405**, 112-115, 2000). I especially recommend *http://highwire.stanford.edu* which provides free of charges both abstracts and full articles from many journals. I hope that you will read many original papers and not only reviews. Above all I hope that this book will help you to find excitement in the many scientific discoveries that are reported week after week. I am especially appreciative of the efforts made by our artist David J. Sauke, who died suddenly on November 6, 2002. He was designer, artist, compositor, and close friend. I am indebted to Robert R. Louden, Emily L. Osam, and Kim McDermott for taking over David's responsibilities and allowing us to complete the volume in a timely fashion.

David E. Metzler

Acknowledgments

I wish to express my appreciation to the following reviewers, each of whom read parts of the manuscript, for their generous assistance: George M. Carman, Rutgers University; Guillaume Chanfreau, University of California-Los Angeles; Eric Conn, University of California-Davis; Gloria Culver, Iowa State University; David Daleke, Indiana University; John Golbeck, University of Pennsylvania; Richard Gumport, University of Illinois, Champaign-Urbana; Jack Horowitz, Iowa State University; Craig Jackson, Hemosaga Diagnostics Corporation; Jørgen Johansen, Iowa State University; L. Andrew Lyon, Georgia Institute of Technology; Eric C. Niederhoffer, Southern Illinois University; Jim Olson, Iowa State University; Ivan Rayment, University of Wisconsin; John Robyt, Iowa State University; Kevin Sarge, University of Kentucky; Andrzej Stasiak, Universite de Lausanne; Laurey Steinke, University of Nebraska Medical Center; Marvin Stromer, Iowa State University; Robert W. Thornburg, Iowa State University; Skip Waechter, University of Kentucky.

Special thanks are due Jack Kirsch, University of California, Berkeley and to my colleagues at Iowa State University for providing many of the study questions.

Journal Acknowledgments

The cooperation of the publishers of books and journals in permitting inclusion of material in this book is gratefully acknowledged. The figures listed have been taken directly from the following journals or have been redrawn and adapted from published drawings. The literature citations appear in the chapter bibliographies.

Biochemistry
Copyright by the American Chemical Society
Fig. B in Box 17C, Fig. 23-3A, Fig. 23-12, Fig. 23-40D, Fig. 23-47, Fig. 24-3A, Fig. 24-3B,C, Fig. 24-3A,B,C, Fig. 29-9A,C,D, Fig. 30-6E, Fig. 31-2C

Journal of the American Chemical Society
Copyright by the American Chemical Society
Fig. 23-16

Journal of Molecular Biology
Copyright by Academic Press, London
Fig. 18-7, Fig. 19-1A,B,E, Fig. 19-9A,B, Fig. 19-11A, Fig. 19-12A,B, Fig. 19-13, Fig. 20-4F, Fig. 23-24A,B, Fig. 23-25B,C, Fig. 23-31C, Fig. 23-33B,C,D, Fig. 23-41C, Fig. 26-5A, Fig. 26-5A, Banner Chapter 27, Fig. 27-8, Fig. 28-3A, Fig. 28-27, Fig. 29-4C,D,E,H,I, Fig. 29-7B, Fig. 30-16E,F, Fig. 31-4E

Proceedings of the National Academy of Sciences, U. S. A.
Copyright by The National Academy of Sciences
Fig. 19-21C,D, Fig. 19-22, Fig. 20-4A–E, Fig. 23-24C, Fig. 26-13A–F, Fig. 26-13A, Fig. 27-1, Fig. 27-5B, Fig. 27-23, Fig. 27-28, Fig. 28-6, Fig. 28-13B, Fig. 28-19A,B, Fig. 29-4B, Fig. 30-5, Fig. 30-7, Fig. 30-31, Fig. 31-5

Scientific American
Copyright by Scientific American, Inc.
Fig. 30-15A, Fig. 30-20C, Banner Chaper 31, Fig. 31-2A,B, Fig. 31-11, Fig. 32-6B

Journal of Cell Biology
Copyright by The Rockefeller University Press
Fig. B in Box 19C, Fig. 27-5A,B

Nature (London)
Copyright by Macmillan Magazines Limited
Fig. 18-8A,B,C, Fig. 18-14C,D, Fig, 18-16A,B,C, Fig. 19-19, Fig. 19-20, Fig. 23-27, Fig. 23-28B, Fig. 23-30A,B, Fig. 23-34, Fig. 23-45A, Fig. 24-2, Banner Chapter 27, Fig. 27-4B, Fig. 27-7, Fig. 27-16A, Fig. 27-24, Fig. 27-26A, Fig. 28-9D, Fig. 28-13A,C, Fig. 28-14, Fig. 28-15, Fig. 29-2D,E,F,H, Fig. in Box 29B, Fig. 29-17, Fig. 30-15B, Fig. 30-17, Fig. 30-18, Fig. 30-22, Fig. 31-6, Fig. 31-12, Fig. 31-13A,B,C, Fig. 31-14A,B,C,D, Fig. 31-15, Fig. 31-17, Fig. 32-2, Fig. 32-5, Fig. 32-8

Nature New Biology
Copyright by MacMillan Magazines Limited
Fig. 30-22

Science
Copyright by the American Association for the Advancement of Science
Fig. in Box 18C, Fig. 18-10A,B,D, Fig. 21-3, Fig. 23-15, Fig. 23-26A,B,C, Fig. 23-29A, Fig. 23-41D, Fig. 23-49, Fig. in Box 23-A, Fig. 26-4B, Fig. 26-5B, Fig. 27-5C, Fig. 27-18B, Fig. 27-26B, Fig. 27-34, Banner Chapter 28, Fig. 28-3B, Fig. 28-4, Table 28-2, Banner Chapter 29, Fig. 29-1E,F,I,J, Fig. 29-2C,G, Fig. 29-4A,B,G, Fig. 29-5, Fig. 29-6A,C, Fig. 29-14C,F,G,H, Fig. 31-7B, Fig. 32-1

Journal of Biological Chemistry
Copyright by the American Society for Biochemistry and Molecular Biology, Inc.
Fig. 18-14A,B, Fig. 18-23A,B, Fig. 19-8C, Fig. 19-10B, Fig. 19-21B, Fig. 23-4A,B, Fig. 23-8A,B, Fig. 23-33A, Fig. 23-42, Fig. 24-6, Fig. 27-15, Fig. 27-16C, Fig. 28-9A,B,C, Fig. 29-1K, Fig. 29-6F,G

Journal of Biochemistry
Copyright by The Japanese Biochemical Society
Fig. 23-13A

EMBO Journal
Copyright by the European Molecular Biology Organization
Fig. 19-8B, Fig. in Box 21A, Fig. 27-28A,B

Trends in Biochemical Science
Copyright by Elsevier Science Ltd.
Fig. in Box 17C, Fig. 18-4, Fig. 19-11B,C, Fig. 23-3B, Fig. 23-5A,B, Fig. 23-40C, Fig. 26-14, Fig. 27-2, Fig. 27-9, Fig. 28-20, Fig. 29-3, Fig. in Box 30A, Fig. 31-4C, Fig. 31-9, Fig. 31-15

FASEB Journal
Copyright by The Federation of American Societies for Experimental Biology
Fig. 30-6A,B

Biophysical Journal
Copyright by the Biophysical Society
Fig. 23-25A, Fig. 30-21

Annual Reviews of Biochemistry
Copyright by Annual Reviews, Inc.
Fig. 27-20B, Fig. 28-26,

Annual Reviews of Biophysics and Biophysical Chemistry
Copyright by Annual Reviews, Inc.
Fig. 23-31A,B

Protein Science
Copyright by The Protein Society
Fig. 18-10C, Fig. 23-51, Fig. 30-6C,D, Fig. 31-7

Cell
Copyright by Cell Press
Fig. 19-2B, Fig. A in Box 19C, Fig. 26-13G, Fig. 26-13G, Fig. 27-12D

Chromosoma
Copyright by Springer, Berlin, Heidelberg
Fig. 16-13A

Current Topics in Bioenergetics
Copyright by Academic Press
Fig. 23-40A

New England Journal of Medicine
Copyright by The Massachusetts Medical Society
Fig. 18-17

Journal of Bacteriology
Copyright by the American Society for Microbiology
Fig. 19-2A, Fig. 24-4

Genome
Copyright by National Research Council of Canada
Fig. 26-13B–F

Book Acknowledgments

The cooperation of the publishers of books in permitting inclusion of material in this book is gratefully acknowledged.

Academic Press
Reproduced by permission of the publisher.

Fig. 18-2 from *Mechanisms in Bioenergetics* by L. Packer, (G. F. Azzone, L. Ernster and S. Papa, eds.), © 1973

Fig. 19-23A from *Cilia and Flagella* by M. A. Sleigh, ed., © 1974

Harcourt/Academic Press
Reproduced by permission of the publisher.

Fig. 31-18 from *Medical Biochemistry*, 4th ed., by N. V. Bhagavan © 2002

American Society for Microbiology
Reproduced by permission of the publisher.

Fig. 19-1C,D from *Escherichia coli and Salmonella typhimurium* by R. M. Macnab, (F. C. Niedhardt, ed.), © 1987

Garland Science Publishing
Reproduced by permission of the publisher.

Fig. 19-21A, Banner Chapter 26, Fig. 26-15A from *Molecular Biology of the Cell*, 3rd ed., by B. Alberts, D. Bray, J. Lewis, M. Raff, K. Roberts and J. D. Watson, © 1994

Wiley-Liss
Reproduced by permission of the publisher.

Fig. 19-23B,C from *Microtubules* by E. F. Smith and W. S. Sale, (J. S. Hyams and C. W. Lloyd, eds.), © 1994

Birkhäuser Verlag Basel
Reproduced by permission of the publisher.

Fig. 20-8 from *The Golgi Apparatus* by R. Rambourg and Y. Clermont, (E. G. Berger and J. Roth, eds.), © 1997

Fig. 23-38 from *Bioenergetics* by E. Haag and G. Renger, (P. Gräber and G. Milazzo, eds.), © 1997

John Wiley & Sons, Inc.
Reproduced by permission of the publisher.

Fig. 23-2, Fig. 23-6, Fig. 23-14 from *Photochemistry* by J. G. Calvert and J. N. Pitts, Jr. © 1966

Plenum Publishing Corporation
Reproduced by permission of the publisher.

Fig. 23-13B from *Fluorescence and Phosphorescence of Proteins and Nucleic Acids* by S. V. Konev © 1967

Cambridge University Press
Reproduced by permission of the publisher.

Fig. 23-19B from *Photosynthesis*, 5th ed., by D. O. Hall and K. K. Rao © 1994

Junk, The Hague
Reproduced by permission of the publisher.

Fig. 23-21 from *Photosynthesis. Two Centuries after Its Discovery by Joseph Priestly*, 2nd ed., by C. S. French and J. S. Brown, (G. Forti, M. Avron, A. Melandri, eds.), © 1972

Saunders College Publishing
Reproduced by permission of the publisher.

Fig. 26-9 from *The Science of Genetics* by A.G. Atherly, J. R. Girton and J. F. McDonald, © 1999

The McGraw-Hill Companies
Reproduced by permission of the publisher.

Fig. 26-15B from *The Metabolic and Molecular Bases of Inherited Disease*, 7th ed., by V. A. McKusick and J. S. Amberger, (C. R. Scriver, A. L. Beaudet, W. S. Sly and D. Valle, eds.), © 1995

Macmillan Publishing USA
Reproduced by permission of the publisher.

Fig. 27-6A from *Genetics, A Molecular Approach*, by L. M. Mays, © 1981

Brookhaven Laboratories
Reproduced by permission of the publisher.

Figure 27-6B,C,D from *Brookhaven Symposium Biology*, by J. Gall, © 1955

W. H. Freeman and Company
Reproduced by permission of the publisher.

Fig. 27-20A from *DNA Replication*, 2nd ed., by A. Kornberg and T. A. Baker, © 1992

Mosby Publishing Company
Reproduced by permission of the publisher.

Fig. 28-25 from *The Biology of Viruses* by B. A. Voyles © 1993

Charles C. Thomas, Publisher, LTD
Reproduced by permission of the publisher.

Fig. 30-11 from *Conduction of Nervous Impulses* by A. Hodgkin, © 1964

Blackwell Scientific Publications
Reproduced by permission of the publisher.

Fig.30-13 from *Proteins, Transmitters and Synapses* by D. G. Nicholls, © 1994

Many drawings in this book have been made using Dr. Per J. Kraulis' program MolScript. (Kraulis, P. J. (1991) *Journal of Applied Crystallography*, **24**, 946–950. MOLSCRIPT, a program to produce both detailed and schematic plots of protein structures.) These include Figures 2-6, 5-35B, 7-8C, 7-18, 7-20B, 7-30A, 8-19B, 8-20A,B,C, 8-24A, Box 11-A (fig. A), 12-3, 12-18, 15-19A,B, 18-10C, 19-15, 23-25, 23-30, 27-12A, and 28-6.

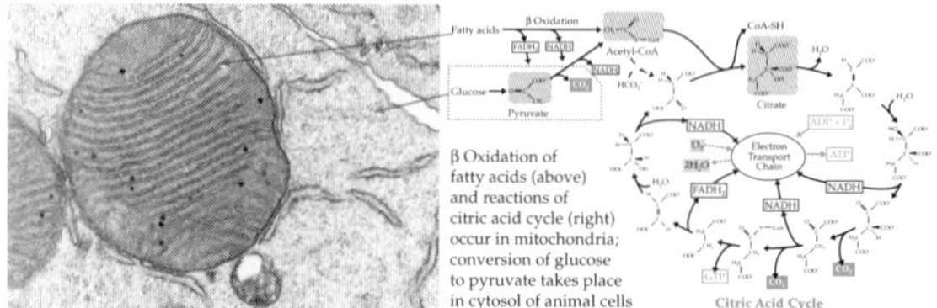

β Oxidation of fatty acids (above) and reactions of citric acid cycle (right) occur in mitochondria; conversion of glucose to pyruvate takes place in cytosol of animal cells

Metabolism, a complex network of chemical reactions, occurs in several different compartments in eukaryotic cells. Fatty acids, a major source of energy for many human cells, are oxidized in the mitochondria via β oxidation and the citric acid cycle. Glucose, a primary source of energy, is converted to pyruvate in the cytosol. Biosynthetic reactions occurring in both compartments form proteins, nucleic acids, storage polymers such as glycogen, and sparingly soluble lipid materials which aggregate to form membranes. Hydrophobic groups in proteins and other polymers also promote self-assembly of the cell. At the same time, oxidative processes, initiated by O_2, increase the water solubility of molecules, leading to metabolic turnover. Micrograph courtesy of Kenneth Moore.

Contents

17 The Organization of Metabolism

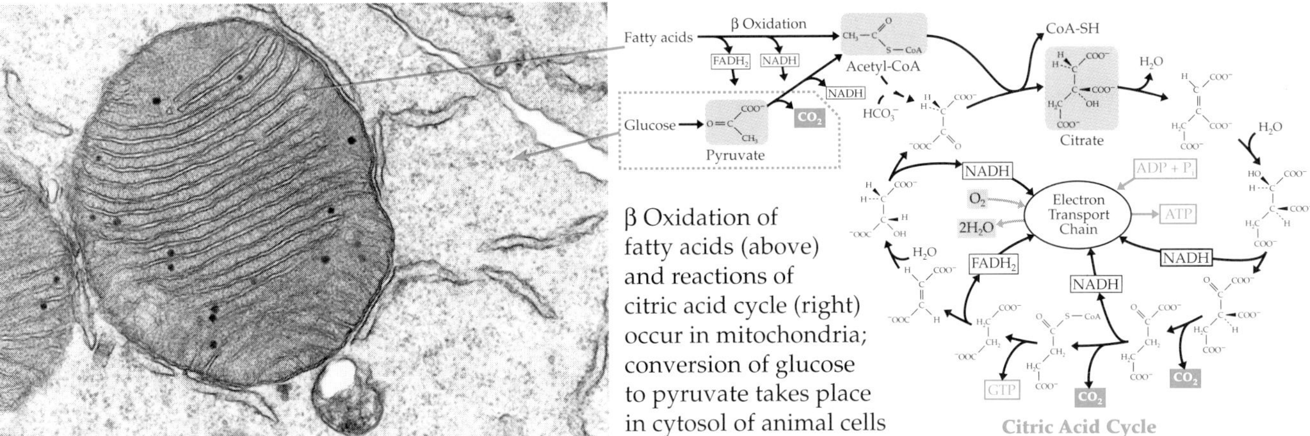

β Oxidation of fatty acids (above) and reactions of citric acid cycle (right) occur in mitochondria; conversion of glucose to pyruvate takes place in cytosol of animal cells

Metabolism involves a bewildering array of chemical reactions, many of them organized as complex cycles which may appear difficult to understand. Yet, there is logic and orderliness. With few exceptions, metabolic pathways can be regarded as sequences of the reactions considered in Chapters 12–16 (and summarized in the table inside the back cover) which are organized to accomplish specific chemical goals. In this chapter we will examine the chemical logic of the major pathways of catabolism of foods and of cell constituents as well as some reactions of biosynthesis (anabolism). A few of the sequences have already been discussed briefly in Chapter 10.

A. The Oxidation of Fatty Acids

Hydrocarbons yield more energy upon combustion than do most other organic compounds, and it is, therefore, not surprising that one important type of food reserve, the fats, is essentially hydrocarbon in nature. In terms of energy content the component fatty acids are the most important. Most aerobic cells can oxidize fatty acids completely to CO_2 and water, a process that takes place within many bacteria, in the matrix space of animal mitochondria, in the peroxisomes of most eukaryotic cells, and to a lesser extent in the endoplasmic reticulum.

The carboxyl group of a fatty acid provides a point for chemical attack. The first step is a priming reaction in which the fatty acid is converted to a water-soluble acyl-CoA derivative in which the α hydrogens of the fatty acyl radicals are "activated" (step *a*, Fig. 17-1). This synthetic reaction is catalyzed by **acyl-CoA synthetases** (fatty acid:CoA ligases). It is driven by the hydrolysis of ATP to AMP and two inorganic phosphate ions using the sequence shown in Eq. 10-1 (p. 508). There are isoenzymes that act on short-, medium-, and long-chain fatty acids. Yeast contains at least five of these.[1] In every case the acyl group is activated through formation of an intermediate acyl adenylate; hydrolysis of the released pyrophosphate helps to carry the reaction to completion (see discussion in Section H).

1. Beta Oxidation

The reaction steps in the oxidation of long-chain acyl-CoA molecules to acetyl-CoA were outlined in Fig. 10-4. Because of the great importance of this β oxidation sequence in metabolism the steps are shown again in Fig. 17-1 (steps *b–e*). The chemical logic becomes clear if we examine the structure of the acyl-CoA molecule and consider the types of biochemical reactions available. If the direct use of O_2 is to be avoided, the only reasonable mode of attack on an acyl-CoA molecule is dehydrogenation. Removal of the α hydrogen as a proton is made possible by the activating effect of the carbonyl group of the thioester. The β hydrogen can be transferred from the intermediate enolate, as a hydride ion, to the bound FAD present in the **acyl-CoA dehydrogenases** that catalyze this reaction[2–5] (step *b*, Fig. 17-1; see also Eq. 15-23). These enzymes contain FAD, and the reduced coenzyme $FADH_2$ that is formed is reoxidized by an **electron transferring flavoprotein** (Chapter 15), which also contains FAD. This protein carries the electrons abstracted in the oxidation process to the inner membrane of the mitochondrion where they enter the mitochondrial electron transport system,[5a] as depicted in Fig. 10-5 and as discussed in detail in

Chapter 18.

The product of step *b* is always a **trans-Δ²-enoyl-CoA**. One of the few possible reactions of this unsaturated compound is nucleophilic addition at the β position. The reacting nucleophile is an HO^- ion from water. This reaction step (step *c*, Fig. 17-1) is completed by addition of H^+ at the α position. The resulting **β-hydroxyacyl-CoA** (3-hydroxyacyl-CoA) is dehydrogenated to a ketone by NAD^+ (step *d*).[5b] This series of three reactions is the β oxidation sequence.

At the end of this sequence, the β-oxoacyl-CoA derivative is cleaved (Fig. 17-1, step *e*) by a **thiolase** (see also Eq. 13-35). One of the products is acetyl-CoA, which can be catabolized to CO_2 through the citric acid cycle. The other product of the thiolytic cleavage is an acyl-CoA derivative that is *two carbon atoms shorter than the original acyl-CoA*. This molecule is recycled through the β oxidation process, a two-carbon acetyl unit being removed as acetyl-CoA during each turn of the cycle (Fig. 17-1). The process continues until the fatty acid chain is completely degraded. If the original fatty acid contained an *even* number of carbon atoms in a straight chain, acetyl-CoA is the only product. However, if the original fatty acid contained an *odd* number of carbon atoms, **propionyl-CoA** is formed at the end.

For every step of the β oxidation sequence there is a small family of enzymes with differing chain length preferences.[6,7] For example, in liver mitochondria one acyl-CoA dehydrogenase acts most rapidly on *n*-butyryl and other short-chain acyl-CoA; a second prefers a substrate of medium chain length such as *n*-octanoyl-CoA; a third prefers long-chain substrates such as palmitoyl-CoA; and a fourth, substrates with 2-methyl branches. A fifth enzyme acts specifically on isovaleryl-CoA. Similar preferences exist for the other enzymes of the β oxidation pathway. In *Escherichia coli*

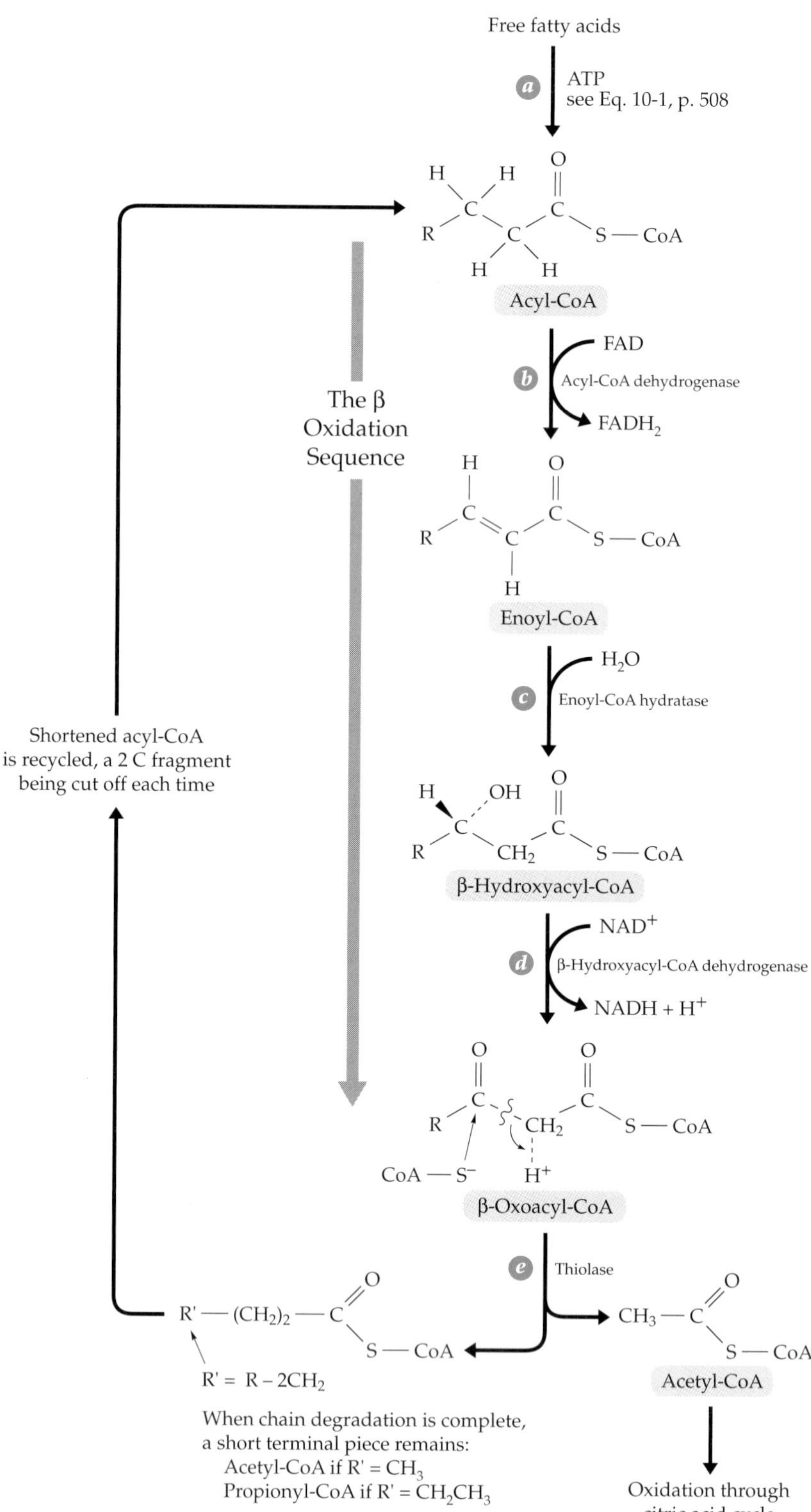

Figure 17-1 The β oxidation cycle for fatty acids. Fatty acids are converted to acyl-CoA derivatives from which 2-carbon atoms are cut off as acetyl-CoA to give a shortened chain which is repeatedly sent back through the cycle until only a 2- or 3-carbon acyl-CoA remains. The sequence of steps *b*, *c*, and *d* also occurs in many other places in metabolism.

most of these enzymes are present as a complex of multifunctional proteins[8] while the mitochondrial enzymes may be organized as a multiprotein complex.[9,10]

Peroxisomal beta oxidation. In animal cells β oxidation is primarily a mitochondrial process,[5] but it also takes place to a limited extent within peroxisomes and within the endoplasmic reticulum.[11–14] This "division of labor" is still not understood well. Straight-chain fatty acids up to 18 carbons in length appear to be metabolized primarily in mitochondria, but in the liver fatty acids with very long chains are processed largely in peroxisomes.[13] There, a very long-chain acyl-CoA synthetase acts on fatty acids that contain 22 or more carbon atoms.[15] In yeast all β oxidation takes place in peroxisomes,[15,16] and in most organisms, including green plants,[17–18a] the peroxisomes are the most active sites of fatty acid oxidation. However, animal peroxisomes cannot oxidize short-chain acyl-CoA molecules; they must be returned to the mitochondria.[16] The activity of peroxisomes in β oxidation is greatly increased by the presence of a variety of compounds known as **peroxisome proliferators**. Among them are drugs such as aspirin and clofibrate and environmental xenobiotics such as the plasticizer bis-(2-ethyl-hexyl)-phthalate. They may induce as much as a tenfold increase in peroxisomal β oxidation.[11,12,19,19a]

Several other features also distinguish β oxidation in peroxisomes. The peroxisomal flavoproteins that catalyze the dehydrogenation of acyl-CoA molecules to unsaturated enoyl-CoAs (step *b* of Fig. 17-1) are **oxidases** in which the $FADH_2$ that is formed is reoxidized by O_2 to form H_2O_2.[13,20] In peroxisomes the enoyl-hydratase and the NAD^+-dependent dehydrogenase catalyzing steps *c* and *d* of Fig. 17-1 are present together with an enoyl-CoA isomerase (next section) as a trifunctional enzyme consisting of a single polypeptide chain.[21] As in mitochondrial β oxidation the 3-hydroxyacyl-CoA intermediates formed in both animal peroxisomes and plant peroxisomes (glyoxysomes) have the L configuration. However, in fungal peroxisomes as well as in *E. coli* they have the D configuration.[22,23] Further metabolism in these organisms requires an epimerase that converts the D-hydroxyacyl-CoA molecules to L.[24] In the past it has often been assumed that peroxisomal membranes are freely permeable to NAD^+, NADH, and acyl-CoA molecules. However, genetic experiments with yeast and other recent evidence indicate that they are impermeable *in vivo* and that carrier and shuttle mechanisms similar to those in mitochondria may be required.[14,25]

Unsaturated fatty acids. Mitochondrial β oxidation of such unsaturated acids as the Δ^9-oleic acid begins with removal of two molecules of acetyl-CoA to form a Δ^5-acyl-CoA. However, further metabolism is slow. Two pathways have been identified (Eq. 17-1).[26–29b] The first step for both is a normal dehydrogenation to a 2-*trans*-5-*cis*-dienoyl-CoA. In pathway I this intermediate reacts slowly by the normal β oxidation sequence to form a 3-*cis*-enoyl-CoA intermediate which must then be acted upon by an auxiliary enzyme, a *cis*-Δ^3-*trans*-Δ^2-enoyl-CoA isomerase (Eq. 17-1, step *c*), before β oxidation can continue.

The alternative reductase pathway (II in Eq. 17-1) is often faster. It makes use of an additional isomerase which converts 3-*trans*, 5-*cis*-dienoyl-CoA into the 2-*trans*, 4-*trans* isomer in which the double bonds are conjugated with the carbonyl group.[29] This permits removal of one double bond by reduction with NADPH as shown (Eq. 17-1, step *f*).[29a,29b] The peroxisomal

a: Acyl-CoA dehydrogenase
I: β Oxidation continued
b: Acetyl-CoA
c: Δ^3,Δ^2-Enoyl-CoA isomerase
II, d: Enoyl-CoA isomerase
3-*trans*-5-*cis*-Dienoyl-CoA
e: From NADPH → H^-; H^+
f: 2,4-Dienoyl-CoA reductase
g: Enoyl-CoA isomerase
Complete β oxidation

(17-1)

pathway is similar.[21] However, the intermediate formed in step *e* of Eq. 17-1 may sometimes have the 2-*trans*, 4-*cis* configuration.[17] The NADH for the reductive step *f* may be supplied by an NADP-dependent isocitrate dehydrogenase.[29c] Repetition of steps *a*, *d*, *e*, and *f* of Eq. 17-1 will lead to β oxidation of the entire chain of polyunsaturated fatty acids such as linoleoyl-CoA or arachidonoyl-CoA. Important additional metabolic routes for polyunsaturated fatty acid derivatives are described in Chapter 21.

Branched-chain fatty acids. Most of the fatty acids in animal and plant fats have straight unbranched chains. However, branches, usually consisting of methyl groups, are present in lipids of some microorganisms, in waxes of plant surfaces, and also in polyprenyl chains. As long as there are not too many branches and if they occur only in the even-numbered positions (i.e., on carbons 2, 4, etc.) β oxidation proceeds normally. Propionyl-CoA is formed in addition to acetyl-CoA as a product of the chain degradation. On the other hand, if methyl groups occur in positions 3, 5, etc., β oxidation is blocked at step *d* of Fig. 17-1. A striking example of the effect of such blockage was

A CH_3 branch here blocks β oxidation

$$\overset{4}{CH_2}-\overset{3}{CH_2}-\overset{2}{CH_2}-\overset{1}{C}(=O)-S-CoA$$

A CH_3 branch here leads to formation of propionyl-CoA

provided by the synthetic detergents in common use until about 1966. These detergents contained a hydrocarbon chain with methyl groups distributed more or less randomly along the chain. Beta oxidation was blocked at many points and the result was a foamy pollution crisis in sewage plants in the United States and in some other countries. Since 1966, only biodegradable detergents having straight hydrocarbon chains have been sold.

In fact, cells *are* able to deal with small amounts of these hard-to-oxidize substrates. The O_2-dependent reactions called α oxidation and ω oxidation are used. These are related also to the oxidation of hydrocarbons which we will consider next.

Oxidation of saturated hydrocarbons. Although the initial oxidation step is chemically difficult, the tissues of our bodies are able to metabolize saturated hydrocarbons such as *n*-heptane slowly, and some microorganisms oxidize straight-chain hydrocarbons rapidly.[30,31] Strains of *Pseudomonas* and of the yeast *Candida* have been used to convert petroleum into edible proteins.[9]

The first step in oxidation of alkanes is usually an O_2-requiring **hydroxylation** (Chapter 18) to a primary alcohol. Further oxidation of the alcohol to an acyl-CoA derivative, presumably via the aldehyde (Eq. 17-2), is a frequently encountered biochemical oxidation sequence.

$$n\text{-Octane} \xrightarrow[\text{Hydroxylation}]{O_2} \underset{\text{Octanol}}{C_7H_{15}-CH_2-OH} \rightarrow C_7H_{15}-CHO \rightarrow \underset{\text{Acyl-CoA}}{C_7H_{15}-C(=O)-S-CoA} \quad (17\text{-}2)$$

Alpha oxidation and omega oxidation. Animal tissues degrade such straight-chain fatty acids as palmitic acid, stearic acid, and oleic acid almost entirely by β oxidation, but plant cells often oxidize fatty acids one carbon at a time. The initial attack may involve hydroxylation on the α-carbon atom (Eq. 17-3) to form either the D- or the L-2-hydroxy acid.[17,18,32,32a] The L-hydroxy acids are oxidized rapidly, perhaps by dehydrogenation to the oxo acids (Eq. 17-3, step *b*) and oxidative decarboxylation, possibly utilizing H_2O_2 (see Eq. 15-36). The D-hydroxy acids tend to accumulate

R—CH₂—COO⁻

Left branch: (a) CoA-SH → R—CH₂—C(=O)—S—CoA; (b) O₂, Fe²⁺, 2-Oxoglutarate, Ascorbate → R—CH(OH)—C(=O)—S—CoA [(2S), L-]; (c) Thiamin diphosphate → H—C(=O)—S—CoA + R—C(=O)H → H–COOH → CO₂

Right branch: (d) → R—CH(O—OH)—COO⁻ [(2R), D-]; (e) → R—C(=O)H → R—C(=O)O⁻ → R—C(=O)—S—CoA → β Oxidation

(17-3)

and are normally present in green leaves. However, they too are oxidized further, with retention of the α hydrogen as indicated by the shaded squares in Eq. 17-3, step *e*. This suggests a new type of dehydrogenation with concurrent decarboxylation. Alpha oxidation also occurs to some extent in animal tissues. For example, when β oxidation is blocked by the presence of a methyl side chain, the body may use α oxidation to get past the block (see **Refsum disease**, Box 17-A). As in plants, this occurs principally in the peroxisomes[33–35] and is important for degradation not only of polyprenyl chains but also bile acids. In the brain some of the fatty acyl groups of sphingolipids are hydroxylated to α-hydroxyacyl groups.[36] Alpha oxidation in animal cells occurs after conversion of free fatty acids to their acyl-CoA derivatives (Eq. 7-3, step *a*). This is followed by a 2-oxoglutarate-dependent hydroxylation (step *b*, see also Eq. 18-51) to form the 2-hydroxyacyl-CoA, which is cleaved in a standard thiamin diphosphate-requiring α cleavage (step *c*). The products are formyl-CoA, which is hydrolyzed and oxidized to CO_2, and a fatty aldehyde which is metabolized further by β oxidation.[34a]

In plants α-dioxygenases (Chapter 18) convert free fatty acids into 2(*R*)-hydroperoxy derivatives (Eq. 7-3, step *d*).[32a] These may be decarboxylated to fatty aldehydes (step *e*, see also Eq. 15-36) but may also give rise to a variety of other products. Compounds arising from linoleic and linolenic acids are numerous and include epoxides, epoxy alcohols, dihydroxy acids, short-chain aldehydes, divinyl ethers, and jasmonic acid (Eq. 21-18).[32a]

On other occasions, **omega (ω) oxidation** occurs at the opposite end of the chain to yield a dicarboxylic acid. Within the human body 3,6-dimethyloctanoic acid and other branched-chain acids are degraded largely via ω oxidation. The initial oxidative attack is by a hydroxylase of the cytochrome P450 group (Chapter 18). These enzymes act not only on fatty acids but also on prostaglandins, sterols, and many other lipids. In the animal body fatty acids are sometimes hydroxylated both at the terminal (ω) position and at the next (ω-2 or ω2) carbon. In plants hydroxylation may occur at the ω2, ω3, and ω4 positions as well.[17,37] Dicarboxylates resulting from ω oxidation of straight-chain fatty acids

BOX 17-A REFSUM DISEASE

Phytol

(a)

β Oxidation is blocked here

$H\text{-}(CH_2\text{-}CH(CH_3)\text{-}CH_2\text{-}CH_2)_3\text{-}CH_2\text{-}CH(CH_3)\text{-}CH_2\text{-}C(=O)OH$

Phytanic acid

(b) α Oxidation (Eq. 17-3)

$R\text{-}CH_2\text{-}C(CH_3)(H)\text{-}C(=O)\text{-}S\text{-}CoA$

(c) Degradation via β oxidation

$3\ CH_3\text{-}CH_2\text{-}C(=O)\text{-}S\text{-}CoA$

$+\ 3\ CH_3\text{-}C(=O)\text{-}S\text{-}CoA$

$+\ CH_3\text{-}CH(CH_3)\text{-}C(=O)\text{-}S\text{-}CoA$

In this autosomally inherited disorder of lipid metabolism the 20-carbon branched-chain fatty acid **phytanic acid** accumulates in tissues. Phytanic acid is normally formed in the body (step *a* in the accompanying scheme) from the polyprenyl plant alcohol **phytol**, which is found as an ester in the chlorophyll present in the diet (Fig. 23-20). Although only a small fraction of the ingested phytol is oxidized to phytanic acid, this acid accumulates to a certain extent in animal fats and is present in dairy products. Because β oxidation is blocked, the first step (step *b*) in degradation of phytanic acid is α oxidation in peroxisomes.[a] The remainder of the molecule undergoes β oxidation (step *c*) to three molecules of propionyl-CoA, three of acetyl-CoA, and one of isobutyryl-CoA. The disease, which was described by Refsum in 1946, causes severe damage to nerves and brain as well as lipid accumulation and early death.[b–d] This rare disorder apparently results from a defect in the initial hydroxylation. The causes of the neurological symptoms of Refsum disease are not clear, but it is possible that the isoprenoid phytanic acid interferes with prenylation of membrane proteins.[b]

[a] Singh, I., Pahan, K., Dhaunsi, G. S., Lazo, O., and Ozand, P. (1993) *J. Biol. Chem.* **268**, 9972–9979

[b] Steinberg, D. (1995) in *The Metabolic and Molecular Bases of Inherited Disease*, 7th ed., Vol. 2 (Scriver, C. R., Beaudet, A. L., Sly, W. S., and Valle, D., eds), pp. 2351–2369, McGraw-Hill, New York

[c] Steinberg, D., Herndon, J. H., Jr., Uhlendorf, B. W., Mize, C. E., Avigan, J., and Milne, G. W. A. (1967) *Science* **156**, 1740–1742

[d] Muralidharan, V. B., and Kishimoto, Y. (1984) *J. Biol. Chem.* **259**, 13021–13026

can undergo β oxidation from both ends. The resulting short-chain dicarboxylates, which appear to be formed primarily in the peroxisomes,[38] may be converted by further β oxidation into succinyl-CoA and free succinate.[39] Incomplete β oxidation in mitochondria (Fig. 17-1) releases small amounts of 3(β)-hydroxy fatty acids, which also undergo ω oxidation and give rise to free 3-hydroxydicarboxylic acids which may be excreted in the urine.[40]

2. Carnitine and Mitochondrial Permeability

A major factor controlling the oxidation of fatty acids is the rate of entry into the mitochondria. While some long-chain fatty acids (perhaps 30% of the total) enter mitochondria as such and are converted to CoA derivatives in the matrix, the majority are "activated" to acyl-CoA derivatives on the inner surface of the outer membranes of the mitochondria. Penetration of these acyl-CoA derivatives through the mitochondrial inner membrane is facilitated by **L-carnitine**.[41–44]

$(H_3C)_3N^+–CH_2–CH(OH)–CH_2–COO^-$

L-Carnitine

Carnitine is present in nearly all organisms and in all animal tissues. The highest concentration is found in muscle where it accounts for almost 0.1% of the dry matter. Carnitine was first isolated from meat extracts in 1905 but the first clue to its biological action was obtained in 1948 when Fraenkel and associates described a new dietary factor required by the mealworm, *Tenebrio molitor*. At first designated **vitamin B_t**, it was identified in 1952 as carnitine. Most organisms synthesize their own carnitine from lysine side chains (Eq. 24-30). The inner membrane of mitochondria contains a long-chain acyltransferase (carnitine palmitoyltransferase I) that catalyzes transfer of the fatty acyl group from CoA to the hydroxyl group of carnitine (Eq. 17-4).[45–47a] Perhaps acyl carnitine derivatives pass through the membrane more easily than do acyl-CoA derivatives because the positive and negative charges can swing together and neutralize each other as shown in Eq. 17-4. Inside the mitochondrion the acyl group is transferred back from carnitine onto CoA (Eq. 17-4, reverse) by carnitine palmitoyltransferase II prior to initiation of β oxidation.

Tissues contain not only long-chain acylcarnitines but also **acetylcarnitine** and other short-chain acylcarnitines, some with branched chains.[41] By accepting acetyl groups from acetyl-CoA, carnitine causes the release of free coenzyme A which can then be reused.

$R–C(=O)–S–CoA$

Acyl-CoA

Acyltransferase: Carnitine → CoA—SH

$R–C(=O)–O–CH(CH_2–N^+(CH_3)_3)(CH_2–COO^-)$

Acyl carnitine (17-4)

Thus, carnitine may have a regulatory function. In flight muscles of insects acetylcarnitine serves as a reservoir for acetyl groups. Carnitine acyltransferases that act on short-chain acyl-CoA molecules are also present in peroxisomes and microsomes, suggesting that carnitine may assist in transferring acetyl groups and other short acyl groups between cell compartments. For example, acetyl groups from peroxisomal β oxidation can be transferred into mitochondria where they can be oxidized in the citric acid cycle.[41]

3. Human Disorders of Fatty Acid Oxidation

Mitochondrial β oxidation of fatty acids is the principal source of energy for the heart. Consequently, inherited defects of fatty acid oxidation or of carnitine-assisted transport often appear as serious heart disease (inherited cardiomyopathy). These may involve heart failure, pulmonary edema, or sudden infant death. As many as 1 in 10,000 persons may inherit such problems.[48–50a] The proteins that may be defective include a plasma membrane carnitine transporter; carnitine palmitoyltransferases; carnitine/acylcarnitine translocase; long-chain, medium-chain, and short-chain acyl-CoA dehydrogenases; 2,4-dienoyl-CoA reductase (Eq. 17-1); and long-chain 3-hydroxyacyl-CoA dehydrogenase. Some of these are indicated in Fig. 17-2.

Several cases of genetically transmitted carnitine deficiency in children have been recorded. These children have weak muscles and their mitochondria oxidize long-chain fatty acids slowly. If the inner mitochondrial membrane carnitine palmitoyltransferase II is lacking, long-chain acylcarnitines accumulate in the mitochondria and appear to have damaging effects on membranes. In the unrelated condition of **acute myocardial ischemia** (lack of oxygen, e.g., during a heart attack) there is also a large accumulation of long-chain acylcarnitines.[51,52] These compounds may induce cardiac arrhythmia and may also account for sudden death from deficiency of carnitine palmitoyltransferase II. Treatment of disorders of carnitine metabolism with daily oral ingestion of several grams of carnitine is helpful, especially for deficiency of the plasma membrane transporter.[50a,53] Metabolic abnormalities may be corrected completely.[50a]

One of the most frequent defects of fatty acid oxidation is deficiency of a mitochondrial acyl-CoA dehydrogenase.[50] If the long-chain-specific enzyme is lacking, the rate of β oxidation of such substrates as octanoate is much less than normal and afflicted individuals excrete in their urine hexanedioic (adipic), octanedioic, and decanedioic acids, all products of ω oxidation.[54] Much more common is the lack of the mitochondrial *medium-chain* acyl-CoA dehydrogenase. Again, dicarboxylic acids, which are presumably generated by ω oxidation in the peroxisomes, are present in blood and urine. Patients must avoid fasting and may benefit from extra carnitine.

A deficiency of very long-chain fatty acid oxidation in peroxisomes is apparently caused by a defective transporter of the ABC type (Chapter 8).[55] The disease, **X-linked adrenoleukodystrophy (ALD)**, has received considerable publicity because of attempts to treat it with "Lorenzo's oil," a mixture of triglycerides of oleic and the C_{22} monoenoic **erucic acid**. The hope has been that these acids would flush out the very long-chain fatty acids that accumulate in the myelin sheath of neurons in the central nervous system and may be responsible for the worst consequences of the disease. However, there has been only limited success.[56,57]

Several genetic diseases involve the development of peroxisomes.[14,35,58,59] Most serious is the **Zellweger syndrome** in which there are no functional peroxisomes. Only "ghosts" of peroxisomes are present and they fail to take up proteins containing the C-terminal peroxisome-targeting sequence SKL.[60,60a] There are many symptoms and death occurs within the first year. Less serious disorders include the presence of catalaseless peroxisomes.[60a]

4. Ketone Bodies

When a fatty acid with an even number of carbon atoms is broken down through β oxidation the last intermediate before complete conversion to acetyl-CoA is the four-carbon **acetoacetyl-CoA**:

$$H_3C-C(=O)-CH_2-C(=O)-S-CoA$$

Acetoacetyl-CoA

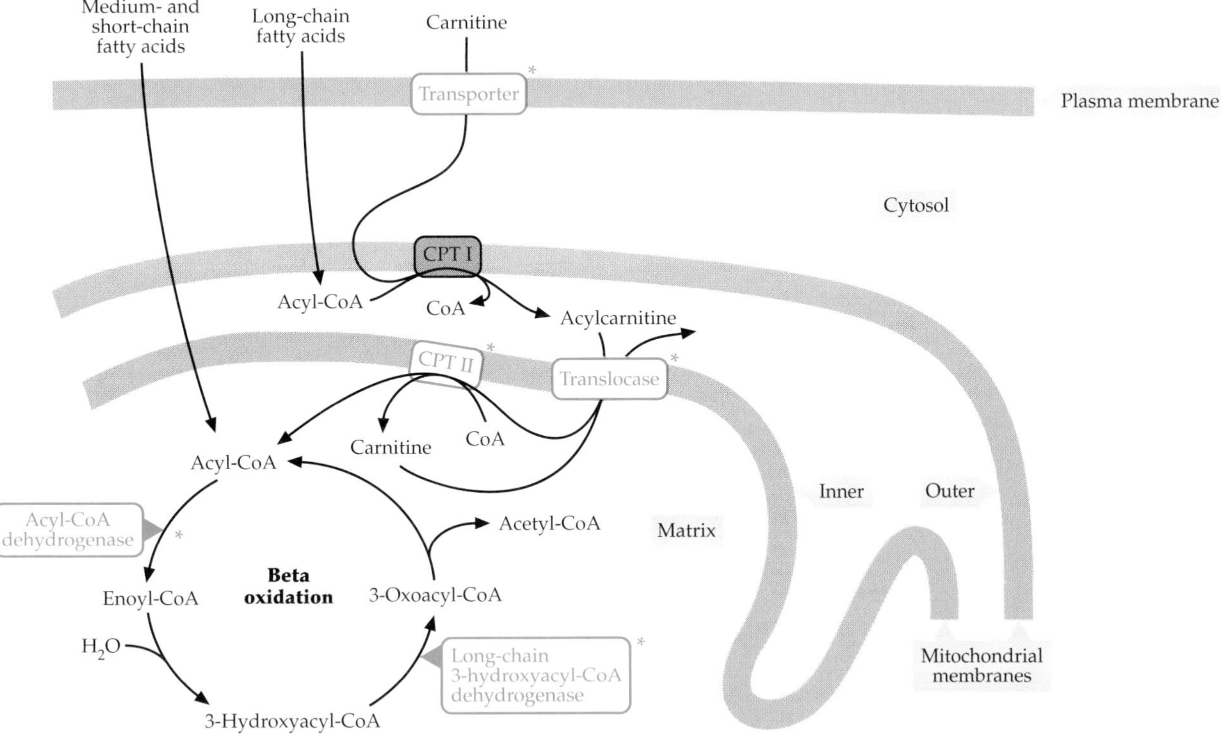

Figure 17-2 Some specific defects in proteins of β oxidation and acyl-carnitine transport causing cardiomyopathy are indicated by the green asterisks. CPT I and CPT II are carnitine palmitoyltransferases I and II. After Kelly and Strauss.[48]

Acetoacetyl-CoA appears to be in equilibrium with acetyl-CoA within the body and is an important metabolic intermediate. It can be cleaved to two molecules of acetyl-CoA which can enter the citric acid cycle. It is also a precursor for synthesis of polyprenyl (isoprenoid) compounds, and it can give rise to free **acetoacetate**, an important constituent of blood. Acetoacetate is a β-oxoacid that can undergo decarboxylation to acetone or can be reduced by an NADH-dependent dehydrogenase to D-3-hydroxybutyrate. Notice that the configuration of this compound is opposite to that of L-3-hydroxybutyryl-CoA which is

$CH_3—C(OH)(H)—CH_2—COO^-$

D-3-Hydroxybutyrate

formed during β oxidation of fatty acids (Fig. 17-1). D-3-Hydroxybutyrate is sometimes stored as a polymer in bacteria (Box 21-D).

The three compounds, acetoacetate, acetone, and 3-hydroxybutyrate, are known as **ketone bodies**.[60b] The inability of the animal body to form the glucose precursors, pyruvate or oxaloacetate, from acetyl units sometimes causes severe metabolic problems. The condition known as **ketosis**, in which excessive amounts of ketone bodies are present in the blood, develops when too much acetyl-CoA is produced and its combustion in the critic acid cycle is slow. Ketosis often develops in patients with Type I **diabetes mellitus** (Box 17-G), in anyone with high fevers, and during starvation. Ketosis is dangerous, if severe, because formation of ketone bodies produces hydrogen ions (Eq. 17-5) and acidifies the blood. Thousands of young persons with insulin-dependent diabetes die annually from ketoacidosis.

Rat blood normally contains about 0.07 mM acetoacetate, 0.18 mM hydroxybutyrate, and a variable amount of acetone. These amounts increase to 0.5 mM acetoacetate and 1.6 mM hydroxybutyrate after 48 h of starvation. On the other hand, the blood glucose concentration falls from 6 to 4 mM after 48 h starvation.[61] Under these conditions acetoacetate and hydroxybutyrate are an important alternative energy source for muscle and other tissues.[62,63] Acetoacetate can be thought of as a transport form of acetyl units, which can be reconverted to acetyl-CoA and oxidized in the citric acid cycle.

Some free acetoacetate is formed by direct hydrolysis of acetoacetyl-CoA. In rats, ~11% of the hydroxybutyrate that is excreted in the urine comes from acetoacetate generated in this way.[64] However, most acetoacetate arises in the liver indirectly in a two-step process (Eq. 17-5) that is closely related to the synthesis

2 $CH_3—C(=O)—S—CoA$

Acetyl-CoA

(a) → CoA—SH

$H_3C—C(=O)—CH_2—C(=O)—S—CoA$

Acetoacetyl-CoA

$CoA—S—C(=O)—CH_3$

Acetyl-CoA

(b) HMG-CoA synthase; H_2O → CoA—SH

$^-OOC—CH_2—C(H_3C)(HO)—CH_2—C(=O)—S—CoA$

S-3-Hydroxy-3-methylglutaryl-CoA (HMG-CoA)

(c) HMG-CoA lyase

$H_3C—C(=O)—CH_2—COO^- + H^+$

Acetoacetate

(17-5)

of cholesterol and other polyprenyl compounds. Step *a* of this sequence is a Claisen condensation, catalyzed by 3-hydroxy-3-methyl-glutaryl CoA synthase (HMG-CoA synthase)[64a–c] and followed by hydrolysis of one thioester linkage. It is therefore similar to the citrate synthase reaction (Eq. 13-38). Step *c* is a simple aldol cleavage. The overall reaction has the stoichiometry of a direct hydrolysis of acetoacetyl-CoA. Liver mitochondria contain most of the body's HMG-CoA synthase and are the major site of ketone body formation (**ketogenesis**). Cholesterol is synthesized from HMG-CoA that is formed in the cytoplasm (Chapter 22).

Utilization of 3-hydroxybutyrate or acetoacetate for energy requires their reconversion to acetyl-CoA as indicated in Eq. 17-6. All of the reactions of this sequence may be nearly at equilibrium in tissues that use ketone bodies for energy.[61]

Acetone, in the small amounts normally present in the body, is metabolized by hydroxylation to acetol (Eq. 17-7, step *a*), hydroxylation and dehydration to methylglyoxal (step *b*), and conversion to D-lactate and pyruvate. A second pathway via 1,2-propanediol and L-lactate is also shown in Eq. 17-7. During fasting the acetone content of human blood may rise to as much as 1.6 mM. As much as two-thirds of this may be converted to glucose.[65–69] Accumulation of acetone

3-Hydroxybutyrate → (NAD$^+$ → NADH) → Acetoacetate

Acetoacetate → (Succinyl-CoA → Succinate; see Eq. 12-50) → Acetoacetyl-CoA

Acetoacetyl-CoA → (Thiolase; CoA—SH) → 2 Acetyl-CoA

2 Acetyl-CoA → Oxidation via citric acid cycle (17-6)

appears to induce the synthesis of the hydroxylases needed for methylglyoxal formation,[68] and the pyruvate formed by Eq. 17-7 may give rise to glucose by the gluconeogenic pathway. However, at high acetone concentrations most metabolism may take place through a poorly understood conversion of 1,2-propanediol to acetate and formate or CO_2.[69] No net conversion of acetate into glucose can occur in animals, but isotopic labels from acetate can enter glucose via acetyl-CoA and the citric acid cycle.

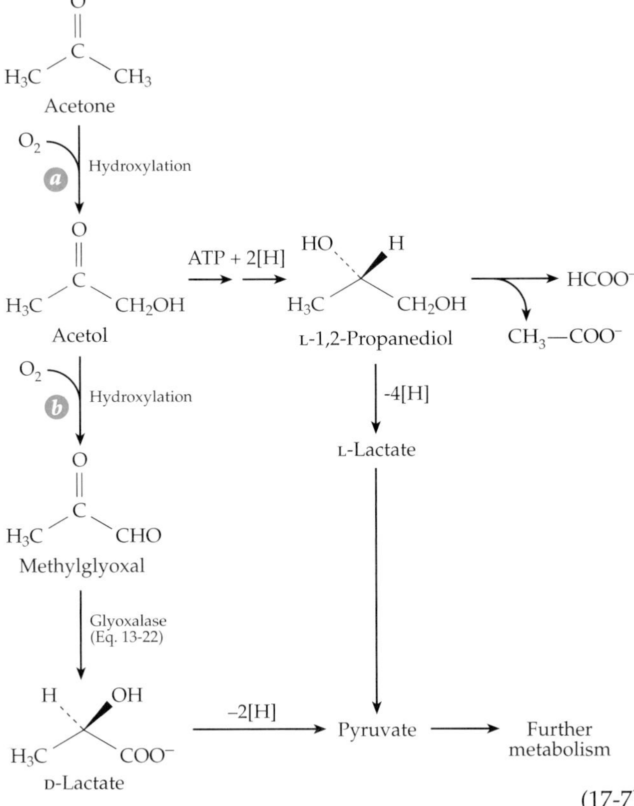

B. Catabolism of Propionyl Coenzyme A and Propionate

Beta oxidation of fatty acids with an odd number of carbon atoms leads to the formation of propionyl-CoA as well as acetyl-CoA. The three-carbon propionyl unit is also produced by degradation of cholesterol and other isoprenoid compounds and of isoleucine, valine, threonine, and methionine. Human beings ingest small amounts of free propionic acid, e.g., from Swiss cheese (which is cultivated with propionic acid-producing bacteria) and from propionate added to bread as a fungicide. In **ruminant** animals, such as cattle and sheep, the ingested food undergoes extensive fermentation in the **rumen**, a large digestive organ containing cellulose-digesting bacteria and protozoa. Major products of the rumen fermentations include acetate, propionate, and butyrate. Propionate is an important source of energy for these animals.

1. The Malonic Semialdehyde Pathways

The most obvious route of metabolism of propionyl-CoA is further β oxidation which leads to 3-hydroxypropionyl-CoA (Fig. 17-3, step *a*). This appears to be the major pathway in green plants.[17] Continuation of the β oxidation via steps *a–c* of Fig. 17-3 produces the CoA derivative of malonic semialdehyde. The latter can, in turn, be oxidized to malonyl-CoA, a β-oxoacid which can be decarboxylated to acetyl-CoA. The necessary enzymes have been found in *Clostridium kluyveri*,[70] but the pathway appears to be little used.

Nevertheless, malonyl-CoA is a major metabolite. It is an intermediate in fatty acid synthesis (see Fig. 17-12) and is formed in the peroxisomal β oxidation of odd chain-length dicarboxylic acids.[70a] Excess malonyl-CoA is decarboxylated in peroxisomes, and lack of the decarboxylase enzyme in mammals causes the lethal **malonic aciduria**.[70a] Some propionyl-CoA may also be metabolized by this pathway. The modified β oxidation sequence indicated on the left side of Fig. 17-3 is used in green plants and in many microorganisms. 3-Hydroxypropionyl-CoA is hydrolyzed to *free* β-hydroxypropionate, which is then oxidized to malonic semialdehyde and converted to acetyl-CoA by reactions that have not been completely described. Another possible pathway of propionate metabolism is the direct conversion to pyruvate via α oxidation into lactate, a mechanism that may be employed by some bacteria. Another route to lactate is through addition of water to acrylyl-CoA, the product of step *a* of Fig. 17-3. The water molecule adds in the "wrong way," the OH^- ion going to the α carbon instead of the β (Eq. 17-8). An enzyme with an active site similar to that of histidine ammonia-lyase (Eq. 14-48) could

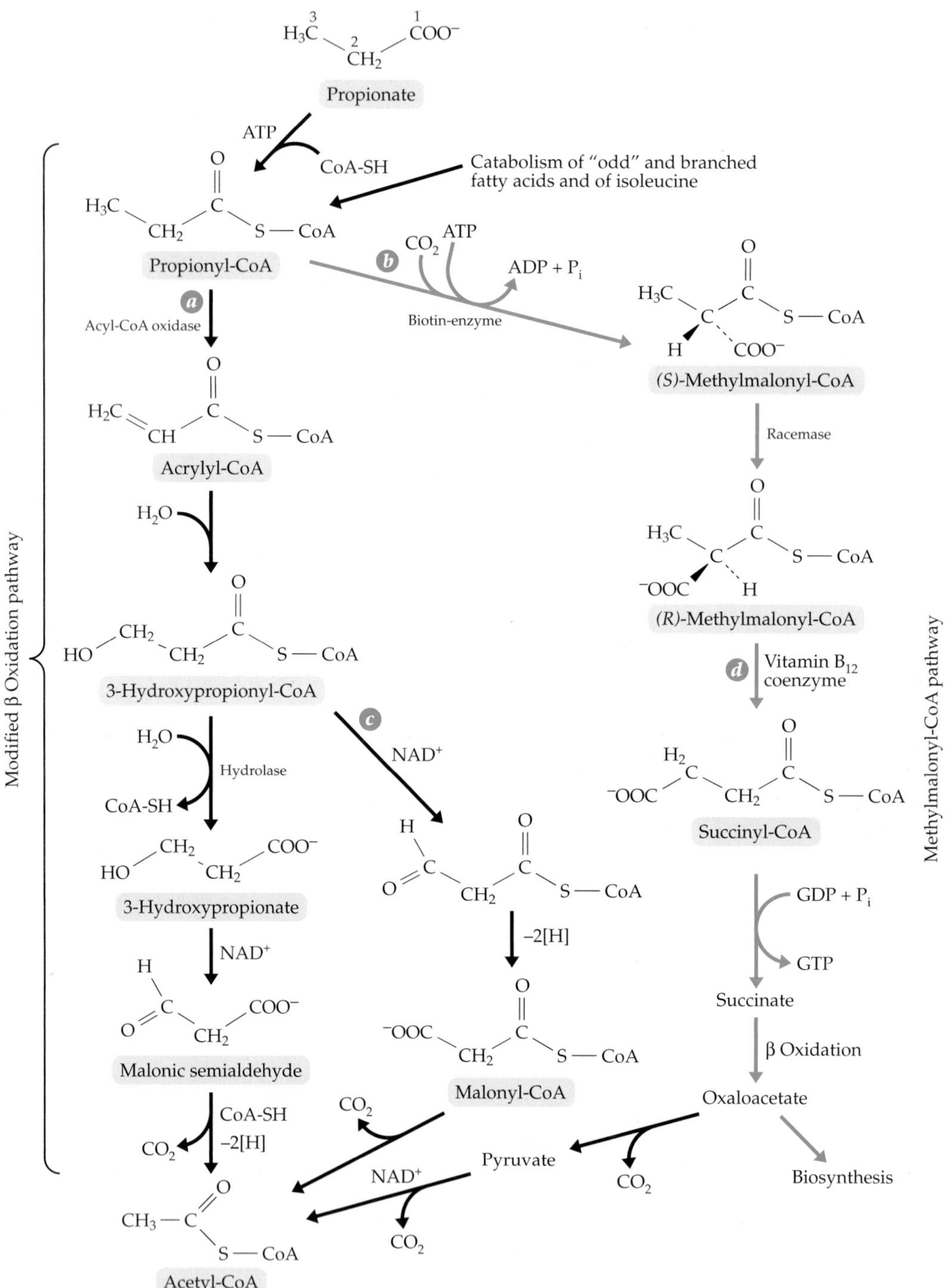

Figure 17-3 Catabolism of propionate and propionyl-CoA. In the names for methylmalonyl-CoA the *R* and *S* refer to the methylmalonyl part of the structure. Coenzyme A is also chiral.

presumably catalyze such a reaction. Lactyl-CoA could be converted to pyruvate readily. *Clostridium propionicum* does interconvert propionate, lactate, and pyruvate via acrylyl-CoA and lactyl-CoA as part of a fermentation of alanine (Fig. 24-19).[71–74] The enzyme that catalyzes hydration of acrylyl-CoA in this case is a complex flavoprotein that may function via a free radical mechanism.[71,72,74]

$$H_2C{=}CH{-}C({=}O){-}S{-}CoA \xrightarrow{H_2O} H_3C{-}CH(OH){-}C({=}O){-}S{-}CoA \quad \text{(17-8)}$$

D-Lactyl-CoA

BOX 17-B METHYLMALONIC ACIDURIA

In this hereditary disease up to 1–2 g of methylmalonic acid per day (compared to a normal of <5 mg/day) is excreted in the urine, and a high level of the compound is present in blood. Two causes of the rare disease are known.[a–d] One is the lack of functional vitamin B_{12}-containing coenzyme. This can be a result of a mutation in any one of several different genes involved in the synthesis and transport of the cobalamin coenzyme.[e] Cultured fibroblasts from patients with this form of the disease contain a very low level of the vitamin B_{12} coenzyme (Chapter 16), and addition of excess vitamin B_{12} to the diet may restore coenzyme synthesis to normal. Among elderly patients a smaller increase in methylmalonic acid excretion is a good indicator of vitamin B_{12} deficiency. A second form of the disease, which does not respond to vitamin B_{12}, arises from a defect in the methylmalonyl mutase protein. Methylmalonic aciduria is often a very severe disease, frequently resulting in death in infancy. Surprisingly, some children with the condition are healthy and develop normally.[a,f]

A closely related disease is caused by a deficiency of propionyl-CoA carboxylase.[a] This may be a result of a defective structural gene for one of the two subunits of the enzyme, of a defect in the enzyme that attaches biotin to carboxylases, or of biotinitase, the enzyme that hydrolytically releases biotin from linkage with lysine (Chapter 14). The latter two defects lead to a multiple carboxylase deficiency and to methylmalonyl aciduria as well as ketoacidosis and propionic acidemia.[g]

Both methylmalonic aciduria and propionyl-CoA decarboxylase deficiency are usually accompanied by severe ketosis, hypoglycemia, and hyperglycinemia. The cause of these conditions is not entirely clear. However, methylmalonyl-CoA, which accumulates in methylmalonic aciduria, is a known inhibitor of pyruvate carboxylase. Therefore, ketosis may develop because of impaired conversion of pyruvate to oxaloacetate.

Patients with propionic or methylmalonic acidemia also secrete 2,3-butanediols (D-,L- or meso) and usually also 1,2-propandiol in their urine. Secretion of 1,2-propanediol is also observed during starvation and in diabetic ketoacidosis. Propanediol may be formed from acetone (Eq. 17-7), and butanediols may originate from acetoin, which is a side reaction product of pyruvate dehydrogenase. However, in the metabolic defects under consideration here, acetoin may be formed by hydroxylation of methylethyl ketone which can arise, as does acetone, by decarboxylation of an oxoacid precursor formed by β oxidation.[h]

$$H_3C{-}CH_2{-}C(=O){-}CH_2{-}COO^- \xrightarrow{\;-CO_2\;} H_3C{-}CH_2{-}C(=O){-}CH_3$$

Methylmalonic aciduria is rare and can be diagnosed incorrectly. In 1989 a woman in St. Louis, Missouri, was convicted and sentenced to life in prison for murdering her 5-month-old son by poisoning with ethylene glycol. While in prison she gave birth to another son who soon fell ill of methylmalonyl aciduria and was successfully treated. Reexamination of the evidence revealed that the first boy had died of the same disease and the mother was released.[i]

---

[a] Fenton, W. A., and Rosenberg, L. E. (1995) in *The Metabolic and Molecular Bases of Inherited Disease*, 7th ed., Vol. 1 (Scriver, C. R., Beaudet, A. L., Sly, W. S., and Valle, D., eds), pp. 1423–1449, McGraw-Hill, New York

[b] Matsui, S. M., Mahoney, M. J., and Rosenberg, L. E. (1983) *N. Engl. J. Med.* **308**, 857–861

[c] Hubbard, S. R., Wei, L., Ellis, L., and Hendrickson, W. A. (1994) *Nature (London)* **372**, 746–754

[d] Luschinsky Drennan, C., Matthews, R. G., Rosenblatt, D. S., Ledley, F. D., Fenton, W. A., and Ludwig, M. L. (1996) *Proc. Natl. Acad. Sci. U.S.A.* **93**, 5550–5555

[e] Fenton, W. A., and Rosenberg, L. E. (1995) in *The Metabolic and Molecular Bases of Inherited Disease*, 7th ed., Vol. 2 (Scriver, C. R., Beaudet, A. L., Sly, W. S., and Valle, D., eds), pp. 3129–3149, McGraw-Hill, New York

[f] Ledley, F. D., Levy, H. L., Shih, V. E., Benjamin, R., and Mahoney, M. J. (1984) *N. Engl. J. Med.* **311**, 1015–1018

[g] Wolf, B. (1995) in *The Metabolic and Molecular Bases of Inherited Disease*, 7th ed., Vol. 2 (Scriver, C. R., Beaudet, A. L., Sly, W. S., and Valle, D., eds), pp. 3151–3177, McGraw-Hill, New York

[h] Casazza, J. P., Song, B. J., and Veech, R. L. (1990) *Trends Biochem. Sci.* **15**, 26–30

[i] Zurer, P. (1991) *Chem. Eng. News* **69** Sep 30, 7–8

2. The Methylmalonyl-CoA Pathway of Propionate Utilization

Despite the simplicity and logic of the β oxidation pathway of propionate metabolism, higher animals use primarily the more complex methylmalonyl-CoA pathway (Fig. 17-3, step *b*). This is one of the two processes in higher animals presently known to depend upon vitamin B_{12}. This vitamin has never been found in higher plants, nor does the methylmalonyl pathway occur in plants. The pathway (Fig. 17-3) begins with the biotin- and ATP-dependent carboxylation of propionyl-CoA. The *S*-methylmalonyl-CoA so formed is isomerized to *R*-methylmalonyl-CoA, after which the methylmalonyl-CoA is converted to succinyl-CoA in a vitamin B_{12} coenzyme-requiring reaction step *d* (Table 16-1). The succinyl-CoA is converted to free succinate (with the formation of GTP compensating for the ATP used initially). The succinate, by β oxidation, is converted to oxaloacetate which is decarboxylated to pyruvate. This, in effect, removes the carboxyl group that was put on at the beginning of the sequence in the ATP-dependent step. Pyruvate is converted by oxidative decarboxylation to acetyl-CoA.

A natural question is "Why has this complex pathway evolved to do something that could have been done much more directly?" One possibility is that the presence of too much malonyl-CoA, the product of the β oxidation pathway of propionate metabolism (Fig. 17-3, pathways *a* and *c*), would interfere with lipid metabolism. Malonyl-CoA is formed in the cytosol during fatty acid biosynthesis and retards mitochondrial β oxidation by inhibiting carnitine palmitoyltransferase I.[46,70a,75] However, a relationship to mitochondrial propionate catabolism is not clear. On the other hand, the tacking on of an extra CO_2 and the use of ATP at the beginning suggests that the *methylmalonyl-CoA pathway* (Fig. 17-3) *is a biosynthetic rather than a catabolic route* (see Section H,4). The methylmalonyl pathway provides a means for converting propionate to oxaloacetate, a transformation that is chemically difficult.

In this context it is of interest that cows, whose metabolism is based much more on acetate than is ours, often develop a severe ketosis spontaneously. A standard treatment is the administration of a large dose of propionate which is presumably effective because of the ease of its conversion to oxaloacetate via the methylmalonyl-CoA pathway. It is possible that this pathway was developed by animals as a means of capturing propionyl units, scanty though they may be, for conversion to oxaloacetate and use in biosynthesis. In ruminant animals, the pathway is especially important. Whereas we have 5.5 mM glucose in our blood, the cow has only half as much, and a substantial fraction of this glucose is derived, in the liver, from the propionate provided by rumen microorganisms.[76] The need for vitamin B_{12} in the formation of propionate by these organisms also accounts for the high requirement for cobalt in the ruminant diet (Chapter 16).

C. The Citric Acid Cycle

To complete the oxidation of fatty acids the acetyl units of acetyl-CoA generated in the β oxidation sequence must be oxidized to carbon dioxide and water.[77] The citric acid (or tricarboxylic acid) cycle by which this oxidation is accomplished is a vital part of the metabolism of almost all aerobic creatures. It occupies a central position in metabolism because of the fact that acetyl-CoA is also an intermediate in the catabolism of carbohydrates and of many amino acids and other compounds. The cycle is depicted in detail in Fig. 10-6 and in an abbreviated form, but with more context, in Fig. 17-4.

1. A Clever Way to Cleave a Reluctant Bond

Oxidation of the chemically resistant two-carbon acetyl group to CO_2 presents a chemical problem. As we have seen (Chapter 13), cleavage of a C–C bond occurs most frequently between atoms that are α and β to a carbonyl group. Such β cleavage is clearly impossible within the acetyl group. The only other common type of cleavage is that of a C–C bond adjacent to a carbonyl group (α cleavage), a thiamin-dependent process (Chapter 14). However, α cleavage would require the prior oxidation (hydroxylation) of the methyl group of acetate. Although many biological hydroxylation reactions occur, they are rarely used in the major pathways of rapid catabolism. Perhaps this is because the overall yield of energy obtainable via hydroxylation is less than that gained from dehydrogenation and use of an electron transport chain.[78]

The solution to the chemical problem of oxidizing acetyl groups efficiently is one very commonly found in nature; a catalytic cycle. Although direct cleavage is impossible, the two-carbon acetyl group of acetyl-CoA *can* undergo a Claisen condensation with a second compound that contains a carbonyl group. The condensation product has more than two carbon atoms, and a β cleavage to yield CO_2 is now possible. Since the cycle is designed to oxidize acetyl units we can regard acetyl-CoA as the **primary substrate** for the cycle. The carbonyl compound with which it condenses can be described as the **regenerating substrate**. To complete the catalytic cycle it is necessary that two carbon atoms be removed as CO_2 from the compound formed by condensation of the two substrates and that the remaining molecule be reconvertible to the original regenerating substrate. The reader may wish to play a

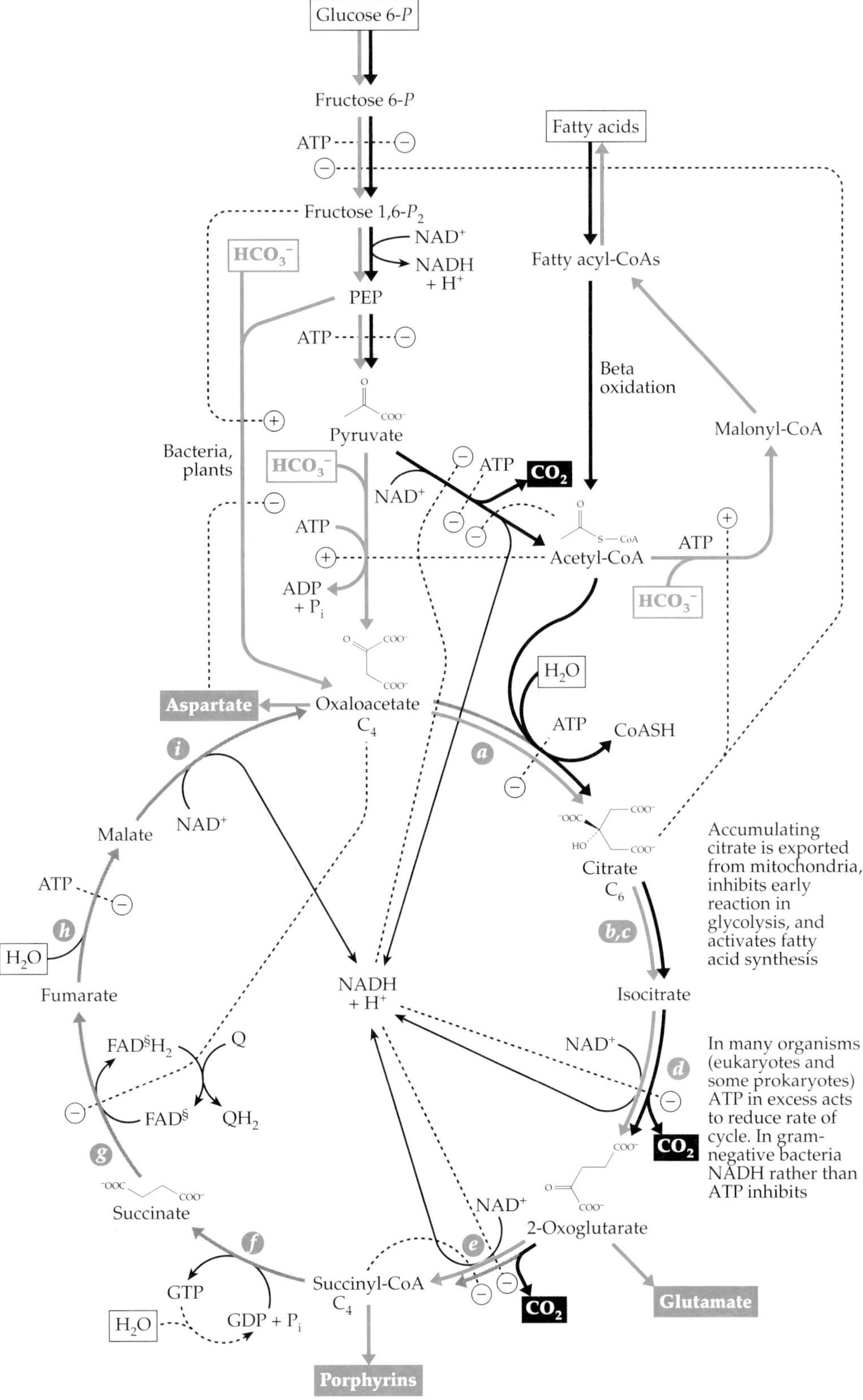

Figure 17-4 The Krebs citric acid cycle. Some of its controlling interactions and its relationship to glycolysis. See also Figure 10-6. Positive and negative regulatory influences, whether arising by allosteric effects or via covalent modification, are indicated by ⊕ or ⊖. Some biosynthetic reaction pathways related to the cycle are shown in green. Steps are lettered to correspond to the numbering in Fig. 10-6, which shows more complete structural formulas. Three molecules of H_2O (boxed) enter the cycle at each turn, providing hydrogen atoms for generation of NADH + H^+ and reduced ubiquinone (QH_2). The covalently attached FAD is designated $FAD^§$.

game by devising suitable sequences of reactions for an acetyl-oxidizing cycle and finding the simplest possible regenerating substrate. Ask yourself whether nature could have used anything simpler than **oxaloacetate**, the molecule actually employed in the citric acid cycle.

The first step in the citric acid cycle (step *a*, Fig. 17-4) is the condensation of acetyl-CoA with oxaloacetate to form citrate. The synthase that catalyzes this condensation also removes the CoA by hydrolysis after it has served its function of activating a methyl hydrogen. This hydrolysis also helps to drive the cycle by virtue of the high group transfer potential of the thioester linkage that is cleaved. Before the citrate can be degraded through β cleavage, the hydroxyl group must be moved from its tertiary position to an adjacent carbon where, as a secondary alcohol, it can be oxidized to a carbonyl group. This is accomplished through steps *b* and *c*, both catalyzed by the enzyme aconitase (Eq. 13-17). Isocitrate can then be oxidized to the β-oxoacid **oxalosuccinate**, which does not leave the enzyme surface but undergoes decarboxylation while still bound (step *d*; also see Eq. 13-45).

The second carbon to be removed from citrate is released as CO_2 through catalysis by the thiamin diphosphate dependent **oxidative decarboxylation** of **2-oxoglutarate** (α-ketoglutarate; Chapter 15). To complete the cycle the four-carbon succinyl unit of succinyl-CoA must be converted back to oxaloacetate through a pathway requiring two more oxidation steps: Succinyl-CoA is converted to free succinate (step *f*) followed by a β oxidation sequence (steps *g–i*; Figs. 10-6 and 17-4). Steps *e* and *f* accomplish a substrate-level phosphorylation (Fig. 15-16). Succinyl-CoA is an unstable thioester with a high group transfer potential. Therefore, step *f* could be accomplished by simple hydrolysis. However, this would be energetically wasteful. The cleavage of succinyl-CoA is coupled to synthesis of ATP in *E. coli* and higher plants and to GTP in mammals. Some of the succinyl-CoA formed in mitochondria is used in other ways, e.g., as in Eq. 17-6 and for biosynthesis of porphyrins.

2. Synthesis of the Regenerating Substrate Oxaloacetate

The primary substrate of the citric acid cycle is acetyl-CoA. Despite many references in the biochemical literature to substrates "entering" the cycle as oxaloacetate (or as one of the immediate precursors succinate, fumarate, or malate), *these compounds are not consumed* by the cycle but are completely regenerated; hence the term *regenerating substrate*, which can be applied to any of these four substances. A prerequisite for the operation of a catalytic cycle is that a regenerating substrate be readily available and that its concentration be increased if necessary to accommodate a more rapid rate of reaction of the cycle. Oxaloacetate can normally be formed in any amount needed for operation of the citric acid cycle from **PEP** or from **pyruvate**, both of these compounds being available from metabolism of sugars.

In bacteria and green plants **PEP carboxylase** (Eq. 13-53), a highly regulated enzyme, is responsible for synthesizing oxaloacetate. In animal tissues **pyruvate carboxylase** (Eq. 14-3) plays the same role. The latter enzyme is almost inactive in the absence of the allosteric effector acetyl-CoA. For this reason, it went undetected for many years. In the presence of high concentrations of acetyl-CoA the enzyme is fully activated and provides for synthesis of a high enough concentration of oxaloacetate to permit the cycle to function. Even so, the oxaloacetate concentration in mitochondria is low, only 0.1 to 0.4×10^{-6} M (10–40 molecules per mitochondrion), and is relatively constant.[65,79]

3. Common Features of Catalytic Cycles

The citric acid cycle is not only one of the most important metabolic cycles in aerobic organisms, including bacteria, protozoa, fungi, higher plants, and animals, but also *a typical catalytic cycle*. Other cycles also have one or more primary substrates and at least one regenerating substrate. Associated with every catalytic cycle there must be a metabolic pathway that provides for synthesis of the regenerating substrate. Although it usually needs to operate only slowly to replenish regenerating substrate lost in side reactions, the pathway also provides *a mechanism for the net biosynthesis of any desired quantity of any intermediate in the cycle*. Cells draw off from the citric acid cycle considerable amounts of oxaloacetate, 2-oxoglutarate, and succinyl-CoA for synthesis of other compounds. For example, aspartate and glutamate are formed directly from oxaloacetate and 2-oxoglutarate, respectively, by transamination (Eq. 14-25).[79a,b] Citrate itself is exported from mitochondria and used for synthesis of fatty acids. It is often stated that the citric acid cycle functions in biosynthesis, but when intermediates in the cycle are drawn off for synthesis the complete cycle does not operate. Rather, *the pathway for synthesis of the regenerating substrate, together with some of the enzymes of the cycle, is used to construct a biosynthetic pathway.*

The word **amphibolic** is often applied to those metabolic sequences that are part of a catabolic cycle and at the same time are involved in a biosynthetic (anabolic) pathway. Another term, **anaplerotic**, is sometimes used to describe pathways for the synthesis of regenerating substrates. This word, which was suggested by H. L. Kornberg, comes from a Greek root meaning "filling up."[80]

4. Control of the Cycle

What factors determine the rate of oxidation by the citric acid cycle? As with most other important pathways of metabolism, several control mechanisms operate and different steps may become rate limiting under different conditions.[81] Factors influencing the flux through the cycle include (1) the rate of generation of acetyl groups, (2) the availability of oxaloacetate, and (3) the rate of reoxidation of NADH to NAD^+ in the electron transport chain. As indicated in Fig. 17-4, acetyl-CoA is a positive effector for conversion of pyruvate to oxaloacetate. Thus, acetyl-CoA "turns on" the formation of a substance required for its own further metabolism. However, when no pyruvate is available operation of the cycle may be impaired by lack of oxaloacetate. This may be the case when liver metabolizes high concentrations of ethanol. The latter is oxidized to acetate but it cannot provide oxaloacetate. Accumulating NADH reduces pyruvate to lactate, further interfering with formation of oxaloacetate.[82] In some individuals the accumulating acetyl units cannot all be oxidized in the cycle and instead are converted to the ketone bodies (Section A,4). A similar problem arises during metabolism of fatty acids by diabetic individuals with inadequate insulin. The accelerated breakdown of fatty acids in the liver overwhelms the system and results in ketosis, even though the oxaloacetate concentration remains normal.[83]

The rates of the oxidative steps in the citric acid cycle are limited by the rate of reoxidation of NADH and reduced ubiquinone in the electron transport chain which may sometimes be restricted by the availability of O_2. However, in aerobic organisms this rate is usually determined by the concentration of ADP and/or P_i available for conversion to ATP in the oxidative phosphorylation process (Chapter 18). If catabolism supplies an excess of ATP over that needed to meet the cell's energy needs, the concentration of ADP falls to a low level, cutting off phosphorylation. At the same time, ATP is present in high concentration and acts as a feedback inhibitor for the catabolism of carbohydrates and fats. This inhibition is exerted at many points, a few of which are indicated in Fig. 17-4. Important sites of inhibition are the **pyruvate dehydrogenase complex**,[84–85a] which converts pyruvate into acetyl-CoA; **isocitrate dehydrogenase**,[86,86a] which converts isocitrate into 2-oxoglutarate; and **2-oxoglutarate dehydrogenase**.[87] The enzyme **citrate synthase**, which catalyzes the first reaction of the cycle, is also inhibited by ATP.[88,89]

Mitochondrial pyruvate dehydrogenase, which contains a 60-subunit icosohedral core of dihydrolipoyl-transacylase (Fig. 15-14), is associated with three molecules of a two-subunit kinase as well as six molecules of a structural **binding protein** which contains a lipoyl group that can be reduced and acetylated by other subunits of the core protein. The binding protein is apparently essential to the functioning of the dehydrogenase complex but not through its lipoyl group.[90,91] The specific pyruvate dehydrogenase kinase is thought to be one of the most important regulatory proteins involved in controlling energy metabolism in most organisms.[92–92b] Phosphorylation of up to three specific serine hydroxyl groups in the thiamin-containing decarboxylase subunit (designated E1) converts the enzyme into an inactive form (Eq. 17-9). A specific phosphatase reverses the inhibition. The kinase is most active on enzymes whose core lipoyl (E2) subunits are reduced and acetylated, a condition favored by high ratios of [acetyl-CoA] to free [CoASH] and of [NADH] to [NAD^+]. Since the kinase inactivates the enzyme the effect is to decrease the pyruvate dehydrogenase action when the system becomes saturated and NADH and acetyl-CoA accumulate. Conversely, a high [pyruvate] inhibits the kinase and increases the action of the dehydrogenase complex. This system also permits various external signals to be felt. For example, insulin has a pronounced stimulatory effect on mitochondrial energy.[65,92,93] One way in which this may be accomplished is through stimulation of the pyruvate dehydrogenase phosphatase, as indicated in Eq. 17-9. A **kinase activator protein** (Eq. 17-9) may also respond to various external stimuli and may be inhibited by insulin.[92]

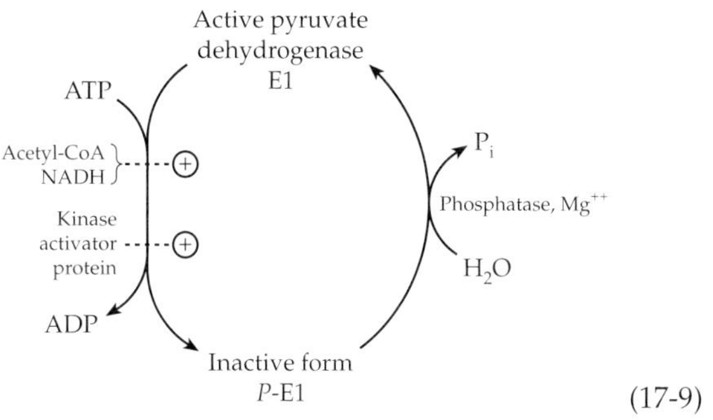

(17-9)

The activities of 2-oxoglutarate dehydrogenase,[94] and to a lesser extent of pyruvate and isocitrate dehydrogenases, are increased by increases in the free Ca^{2+} concentration.[87] Calcium ions stimulate the phosphatase that dephosphorylates the deactivated phosphorylated pyruvate dehydrogenase and activate the other two dehydrogenases allosterically, increasing the affinities for the substrates.[87] Phosphorylation of the NAD^+-dependent isocitrate dehydrogenase also decreases its activity. In *E. coli* the isocitrate dehydrogenase kinase and a protein phosphatase exist as a bifunctional protein able to both deactivate the dehydrogenase and restore its activity.[86] For this organism, the decrease in activity forces substrate into the glyoxylate pathway (Section J,4) instead of the citric acid cycle.

BOX 17-C USE OF ISOTOPIC TRACERS IN STUDY OF THE TRICARBOXYLIC ACID CYCLE

The first use of isotopic labeling in the study of the citric acid cycle and one of the first in the history of biochemistry was carried out by Harland G. Wood and C. H. Werkman in 1941.[a,b] The aim was to study the fermentation of glycerol by propionic acid bacteria, a process that was not obviously related to the citric acid cycle. Some succinate was also formed in

$$\text{Glycerol} \rightarrow \text{propionate}^- + H^+ + H_2O$$
$$\Delta G' = -69 \text{ kJ/mol}$$

the fermentation, and on the basis of simple measurements of the fermentation balance reported in 1938 it was suggested that CO_2 was incorporated into oxaloacetate, which was then reduced to succinate. As we now know, this is indeed an essential step in the propionic acid fermentation (Section F,3). At the time ^{14}C was not available but the mass spectrometer, newly developed by A. O. Nier, permitted the use of the stable ^{13}C as a tracer. Wood and Werkman constructed a thermal diffusion column and used it to prepare bicarbonate enriched in ^{13}C and also built a mass spectrometer. By 1941 it was established unequivocally that carbon dioxide was incorporated into succinate by the bacteria.[c]

To test the idea that animal tissues could also incorporate CO_2 into succinate Wood examined the metabolism of a pigeon liver preparation to which malonate had been added to block succinate dehydrogenase (Box 10-B). Surprisingly, the accumulating succinate, which arose from oxalocetate via citrate, isocitrate, and 2-oxoglutarate (traced by green arrows in accompanying scheme), contained no ^{13}C. Soon, however, it was shown that CO_2 was incorporated into the carboxyl group of 2-oxoglutarate that is adjacent to the carbonyl group. That carboxyl is lost in conversion to succinate (Fig. 10-6) explains the lack of ^{13}C in succinate. It is of historical interest that these observations were incorrectly interpreted by many of the biochemists of the time. They agreed that *citrate could not be a member of the tricarboxylic acid cycle.* Since citrate is a symmetric compound it was thought that any ^{13}C incorporated into citrate would be present in equal amounts in both terminal carboxyl groups. This would necessarily result in incorporation of ^{13}C into succinate. It was not until 1948 that Ogston popularized the concept that by binding with substrates at three points, enzymes were capable of asymmetric attack upon symmetric substrates.[d] In other words, an enzyme could synthesize citrate with the carbon atoms from acetyl-CoA occupying one of the two $-CH_2COOH$ groups surrounding the prochiral center. Later, the complete stereochemistry of the citric acid cycle was elucidated through the use of a variety of isotopic labels (see p. 704). Some of the results are indicated by the asterisks and daggers in the structures in Fig. 10-6.

The operation of the citric acid cycle in living cells, organs, and whole animals has also been observed using NMR and mass spectroscopy with ^{13}C-containing compounds. For example, a heart can be perfused with a suitable oxygenated perfusion fluid[e] containing various ^{13}C-enriched substrates such as [U-^{13}C]fatty acids, [2-^{13}C]acetate, [3-^{13}C]L-lactate, or [2,3-^{13}C]propionate.[e–k] NMR spectroscopy allows direct and repeated observation of the ^{13}C nuclei from a given substrate and its entry into a variety of metabolic pathways. Because of the high dispersion of chemical shift values for ^{13}C the NMR resonance for the isotope can be seen at each position within a single compound.

A compound that is especially easy to observe is glutamate. This amino acid, most of which is found in the cytoplasm, is nevertheless in relatively rapid equilibrium with 2-oxoglutarate of the citric acid cycle in the mitochondria. The accompanying scheme shows where isotopic carbon from certain compounds will be located *when it first enters* the citric acid cycle and traces some of the labels into glutamate. For example, uniformly enriched fatty acids will introduce label into the two atoms of the *pro-S* arm of citrate and into 4- and 5-positions of glutamate whereas [2-^{13}C]acetate will introduce label only into the C4 position as marked by ● in the scheme. In the NMR spectrum a singlet resonance at 32.4 ppm will be observed. However, as successive turns of the citric acid cycle occur the isotope will appear in increasing amounts in the adjacent 3-position of glutamate. They will be recognized readily by the appearance of a multiplet. The initial singlet will be flanked by a pair of peaks that arise from spin–spin coupling with the adjacent 3-^{13}C of the [2,3-^{13}C]isotopomer (see accompanying figure). After longer periods of time the central resonance will weaken and the outer pair strengthen as the recycling occurs.

Metabolism with [U-^{13}C]fatty acids gives a labeling pattern similar to that with [2,3-^{13}C]acetate and it has been deduced that heart muscle normally metabolizes principally fatty acids for energy. What will happen to the glutamate C4 resonance if [3-^{13}C]lactate is added to the perfusion solution? It will enter both acetyl-CoA and oxaloacetate as indicated by ● in the following scheme. That will also intro-duce ^{13}C at C3 of glutamate. By looking at spectra at short times the relative amounts of lactate being oxidized via the cycle and that being converted

BOX 17-C (continued)

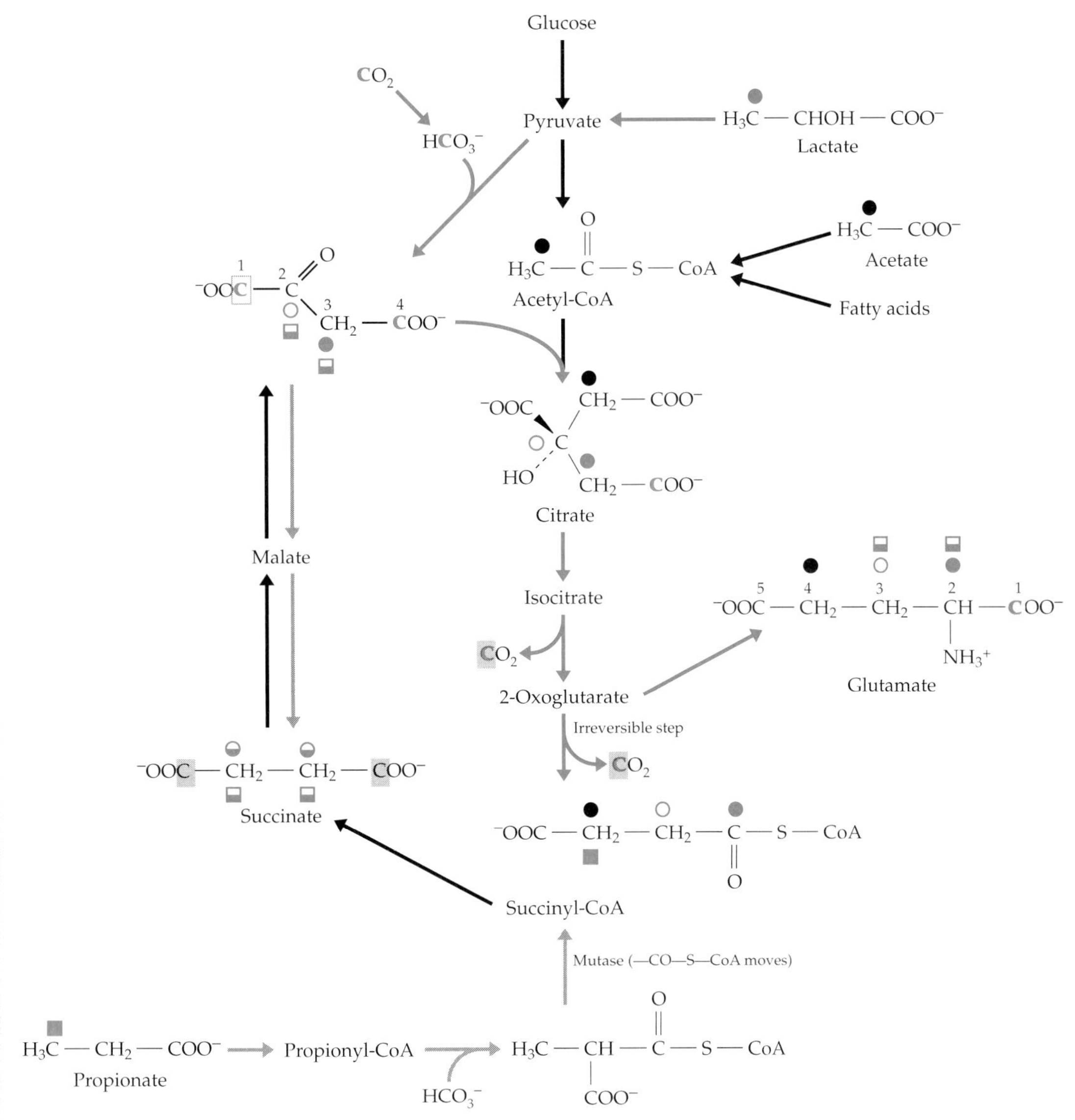

biosynthetically (anaplerotically) to glutamate can be estimated. There is a complication that has long been recognized. Oxaloacetate can be converted by exchange processes to succinate. Since succinate is symmetric the effect is to put 50% of the label into each of the central atoms of succinate (◒ in scheme). The exchange will then transfer label back into the C2 position of oxaloacetate (○) and through citric acid cycle reactions into C3 of glutamate. Now the C4 NMR resonance will contain an additional pair of peaks arising from spin–spin coupling with C2 but which will have a different coupling constant than that for coupling to C3.

If uniformly labeled [U-^{13}C]acetate is introduced the additional isotopomers, [3,4 - ^{13}C]glutamate and [3,4,5 - ^{13}C]glutamate, will be formed as will others with ^{13}C in the C1 and C5 positions but which will not affect the C4 resonance. A total of nine lines will be seen as illustrated in curve *a* of the accompanying figure. We see that the multiplet patterns arising from mass isotopomers are complex, but they can be predicted accurately with a computer program.[f] Isotopomers of succinate have also been analyzed.[g]

It is also of interest to introduce ^{13}C from propionate labeled in various positions. One of these is illustrated in the scheme. In this case the appearance of multiplets arising from [3,4-^{13}C] glutamate verifies the existence of end-to-end scrambling of the isotope in succinate. However, is the scrambling complete or are some molecules

BOX 17-C USE OF ISOTOPIC TRACERS IN STUDY OF THE TRICARBOXYLIC ACID CYCLE (cont.)

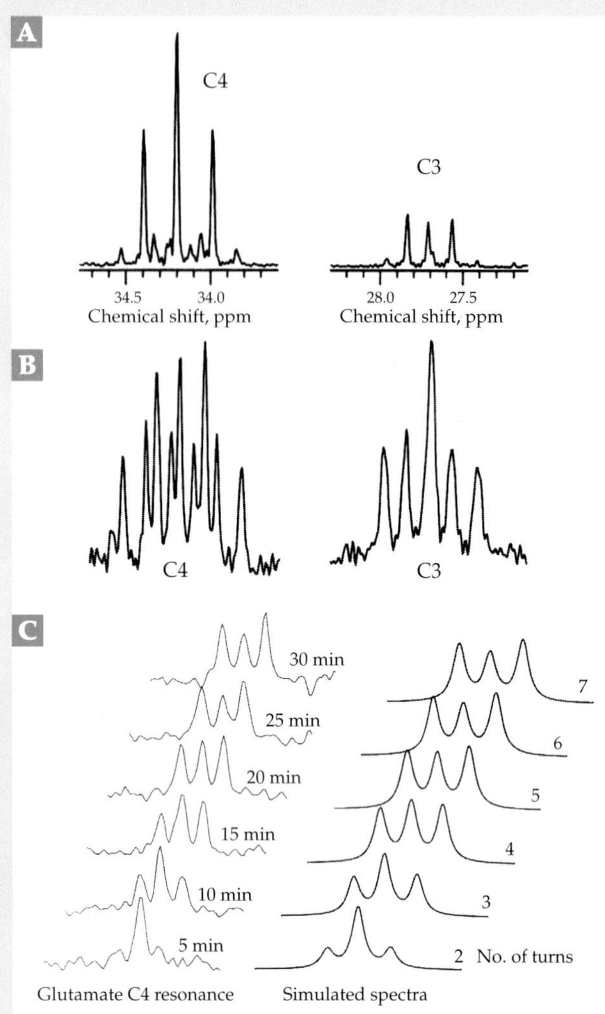

A. ^{13}C-NMR spectrum of extracts of Langendorff-perfused rat hearts perfused for 5 min with [1,2^{13}C] acetate, [3^{13}C]-lactate and glucose. Only the glutamate C4 (left) and C3 (right) resonances are shown. B. Spectrum after perfusion for 30 min. From Malloy *et al.*[r] C. The glutamate C4 resonance of an intact Langendorff-perfused rat heart supplied with 2 mM [2-^{13}C]acetate showing evolution of the multiplet as a function of time after introducing the label. The right panel shows glutamate C4 resonances generated by a computer simulation after turnover of citric acid cycle pools the indicated number of time. From Jeffrey *et al.*[f]

efficiently "channeled" through enzyme–enzyme complexes in such a way as to avoid scrambling? As shown in the scheme, full scrambling would give equal labeling of C2 and C3 of oxaloacetate and of glutamate. Experimentally greater labeling was seen at C3 than at C2 during the first few turns of the cycle suggesting that some channeling does occur.[e]

Isotopomer analysis can also be conducted by mass spectroscopy, which is more sensitive than NMR, using ^{13}C[h,k,l] or ^{2}H labeling.[j] Making use of a technique like that employed by Knoop (Box 10-A), a "chemical biopsy" can be performed on animals or on human beings, who may ingest gram quantities of sodium phenylacetate without harm. The phenylacetate is converted to an amide with glutamine (phenylacetylglutamine) which is excreted in the urine, from which it can easily be recovered for analysis.[l–n] This provides a non-invasive way of studying the operation of the citric acid cycle in the human body. Direct measurement on animal brains[o,p] and on human limbs or brain has also been accomplished by NMR spectroscopy[q] and may become more routine as instrumentation is improved.

[a] Wood, H. G. (1972) in *The Molecular Basis of Biological Transport* (Woessner, J. F., and Huijing, F., eds), pp. 1–54, Academic Press, New York
[b] Krampitz, L. O. (1988) *Trends Biochem. Sci.* **13**, 152–155
[c] Wood, H. G., Werkman, C. H., Hemingway, A., and Nier, A. O. (1941) *J. Biol. Chem.* **139**, 377–381
[d] Ogston, A. G. (1948) *Nature (London)* **162**, 936
[e] Sherry, A. D., Sumegi, B., Miller, B., Cottam, G. L., Gavva, S., Jones, J. G., and Malloy, C. R. (1994) *Biochemistry* **33**, 6268–6275
[f] Jeffrey, F. M. H., Rajagopal, A., Malloy, C. R., and Sherry, A. D. (1991) *Trends Biochem. Sci.* **16**, 5–10
[g] Jones, J. G., Sherry, A. D., Jeffrey, F. M. H., Storey, C. J., and Malloy, C. R. (1993) *Biochemistry* **32**, 12240–12244
[h] Des Rosiers, C., Di Donato, L., Comte, B., Laplante, A., Marcoux, C., David, F., Fernandez, C. A., and Brunengraber, H. (1995) *J. Biol. Chem.* **270**, 10027–10036
[i] Sherry, A. D., and Malloy, C. R. (1996) *Cell Biochem. Funct.* **14**, 259–268
[j] Yudkoff, M., Nelson, D., Daikhin, Y., and Erecinska, M. (1994) *J. Biol. Chem.* **269**, 27414–27420
[k] Beylot, M., Soloviev, M. V., David, F., Landau, B. R., and Brunengraber, H. (1995) *J. Biol. Chem.* **270**, 1509–1514
[l] Di Donato, L., Des Rosiers, C., Montgomery, J. A., David, F., Garneau, M., and Brunengraber, H. (1993) *J. Biol. Chem.* **268**, 4170–4180
[m] Magnusson, I., Schumann, W. C., Bartsch, G. E., Chandramouli, V., Kumaran, K., Wahren, J., and Landau, B. R. (1991) *J. Biol. Chem.* **266**, 6975–6984
[n] Chervitz, S. A., and Falke, J. J. (1995) *J. Biol. Chem.* **270**, 24043–24053
[o] Hyder, F., Chase, J. R., Behar, K. L., Mason, G. F., Siddeek, M., Rothman, D. L., and Shulman, R. G. (1996) *Proc. Natl. Acad. Sci. U.S.A.* **93**, 7612–7617
[p] Cerdan, S., Künnecke, B., and Seelig, J. (1990) *J. Biol. Chem.* **265**, 12916–12926
[q] Rothman, D. L., Novotny, E. J., Shulman, G. I., Howseman, A. M., Petroff, O. A. C., Mason, G., Nixon, T., Hanstock, C. C., Prichard, J. W., and Shulman, R. G. (1992) *Proc. Natl. Acad. Sci. U.S.A.* **89**, 9603–9606
[r] Malloy, C. R., Thompson, J. R., Jeffery, F. M. H., and Sherry, A. D. (1990) *Biochemistry* **29**, 6756–6761

Acting to counteract any drop in ATP level, accumulating ADP acts as a positive effector for isocitrate dehydrogenases.

Another way in which the phosphorylation state of the adenylate system can regulate the cycle depends upon the need for GDP in step *f* of the cycle (Fig. 17-4). Within mitochondria, GTP is used largely to reconvert AMP to ADP. Consequently, formation of GDP is promoted by AMP, a compound that arises in mitochondria from the utilization of ATP for activation of fatty acids (Eq. 13-44) and activation of amino acids for protein synthesis (Eq. 17-36).

In *E. coli* and some other bacteria ATP does not inhibit citrate synthase but NADH does; the control is via the redox potential of the NAD^+ system rather than by the level of phosphorylation of the adenine nucleotide system.[95] Succinic dehydrogenase may be regulated by the redox state of ubiquinone (Chapter 15). Another mechanism of regulation may be the formation of specific protein–protein complexes between enzymes catalyzing reactions of the cycle.[96–97a] This may permit one enzyme to efficiently have a product of its action transferred to the enzyme catalyzing the next step in the cycle.

5. Catabolism of Intermediates of the Citric Acid Cycle

Acetyl-CoA is the only substrate that can be completely oxidized to CO_2 by the reaction of the citric acid cycle alone. Nevertheless, cells must sometimes

BOX 17-D FLUOROACETATE AND "LETHAL SYNTHESIS"

Among the most deadly of simple compounds is sodium fluoroacetate. The LD_{50} (the dose lethal for 50% of animals receiving it) is only 0.2 mg/kg for rats, over tenfold less than that of the nerve poison diisopropylphosphofluoridate (Chapter 12).[a,b] Popular, but controversial, as the rodent poison "1080," fluoroacetate is also found in the leaves of several poisonous plants in Africa, Australia, and South America. Surprisingly, difluoroacetate $HCF_2–COO^-$ is nontoxic and biochemical studies reveal that monofluoroacetate has no toxic effect on cells until it is converted metabolically in a "lethal synthesis" to 2*R*,3*R*-2-fluorocitrate, which is a competitive inhibitor of aconitase (aconitate hydratase, Eq. 13-17).[b–g] This fact was difficult to understand since citrate formed by the reaction of fluorooxaloacetate and acetyl-CoA has only weak inhibitory activity toward the same enzyme. Yet, it is the fluorocitrate formed from fluorooxaloacetate that contains a fluorine atom at a site that is attacked by aconitase in the citric acid cycle.

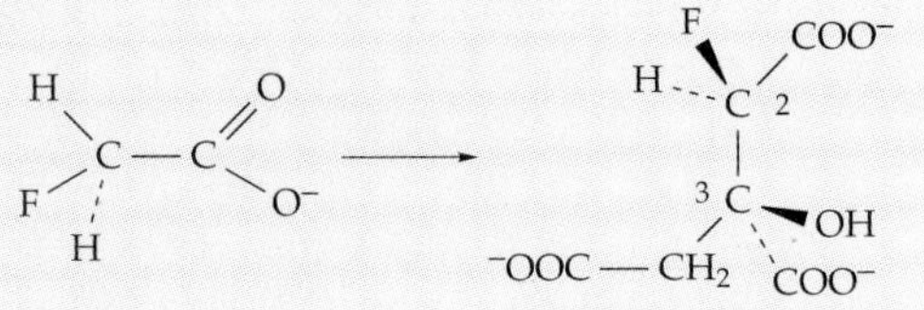

2*R*,3*R* 2-Fluoroacetate

The small van der Waals radius of fluorine (0.135 nm), comparable to that of hydrogen (0.12 nm), is often cited as the basis for the ability of fluoro compounds to "deceive" enzymes. However, the high electronegativity and ability to enter into hydrogen bonds may make F more comparable to –OH in metabolic effects. In the case of fluorocitrate it was proposed that the inhibitory isomer binds in the "wrong way" to aconitase in such a manner that the fluorine atom is coordinated with the ferric ion at the catalytic center.[c] However, 2*R*,3*R*-2-fluorocitrate is a simple competitive inhibitor of aconitase but an irreversible poison. It is especially toxic to nerves and also appears to affect mitochondrial membranes. Therefore, this poison may affect some other target, such as a citrate transporter.[d] Fluoroacetate is only one of many known naturally occurring fluorine compounds.[c]

Another example of lethal synthesis is seen in the use of 5-fluorouracil in cancer therapy (Box 15-E). In this compound and in many other fluorine-containing inhibitors the F atom replaces the H atom that is normally removed as H^+ in the enzymatic reaction. The corresponding F^+ cannot be formed.[h] Because of the high electronegativity of fluorine a C–F bond is polarized: $C^{\delta+}–F^{\delta-}$. This may have very large effects on reactivity at adjacent positions. For example, the reactivity of 2-fluoroglycosyl groups toward glycosyl transfer is decreased by several orders of magnitude (p. 597).

[a] Gibble, G. W. (1973) *J. Chem. Educ.* **50**, 460–462

[b] Elliott, K., and Birch, J., eds. (1972) *Carbon–Fluorine Compounds*, Elsevier, Amsterdam

[c] Glusker, J. P. (1971) in *The Enzymes*, 3rd ed., Vol. 5 (Boyer, P. D., ed), pp. 413–439, Academic Press, New York

[d] Kun, E. (1976) in *Biochemistry Involving Carbon–Fluorine Bonds* (Filler, R., ed), pp. 1–22, American Chemical Society, Washington, DC

[e] Marletta, M. A., Srere, P. A., and Walsh, C. (1981) *Biochemistry* **20**, 3719–3723

[f] Rokita, S. E., and Walsh, C. T. (1983) *Biochemistry* **22**, 2821–2828

[g] Peters, R. A. (1957) *Adv. Enzymol.* **18**, 113–159

[h] Abeles, R. H., and Alston, T. A. (1990) *J. Biol. Chem.* **265**, 16705–16708

oxidize large amounts of one of the compounds found in the citric acid cycle to CO_2.[98,99] For example, bacteria subsisting on succinate as a carbon source must oxidize it for energy as well as convert some of it to carbohydrates, lipids, and other materials. Complete combustion of *any citric acid cycle intermediate* can be accomplished by conversion to malate followed by oxidation of malate to oxaloacetate (Eq. 17-10, step *a*) and decarboxylation (β cleavage) to pyruvate, or (Eq. 17-10, step *b*) oxidation and decarboxylation of malate by the **malic enzyme** (Eq. 13-45) without free oxaloacetate as an intermediate. Pathway *b* is probably the most important. It is catalyzed by two different malic enzymes present in animal mitochondria. One is specific for $NADP^+$ while the other reacts with NAD^+ as well.[100,101] They both have complex regulatory properties. For example, the less specific NAD^+-utilizing enzyme is allosterically inhibited by ATP but is activated by fumarate, succinate, or isocitrate.[100] Thus, accumulation of citric acid cycle intermediates "turns on" the malic enzyme, allowing the excess to leave the cycle and reenter as acetyl groups. Since the Michaelis constant for malate is high, this will not happen unless malate accumulates, signaling a need for acetyl-CoA. The $NADP^+$-dependent enzyme is activated by a high concentration of free CoA and is inhibited by NADH. Perhaps when glycolysis becomes slow the free CoA level rises and turns on malate oxidation.[101] On the other hand, rapid glycolysis increases the NADH concentration which inhibits the malic enzyme. The result is a buildup of the oxaloacetate concentration and an increase in activity of the citric acid cycle. The malic enzymes are also present in the cytoplasm, where one of them functions as part of an NADPH-generating cycle (Eq. 17-46).

L-Malate
Malate dehydrogenase (a): NAD^+ → NADH
"Malic enzyme" (b): NAD^+ or $NADP^+$ → NADH or NADPH, CO_2
Oxaloacetate
Oxaloacetate decarboxylase (also spontaneous): CO_2
PEPCK (Eq 13-46) (c): GTP → GDP
Phosphoenolpyruvate
Pyruvate kinase: ADP → ATP
Pyruvate
Oxidative decarboxylation
Acetyl-CoA

(17-10)

D. Oxidative Pathways Related to the Citric Acid Cycle

In this section we will consider some other catalytic cycles as well as some noncyclic pathways of oxidation of one- and two-carbon substrates that are utilized by microorganisms.

1. The γ-Aminobutyrate Cycle

A modification of the citric acid cycle which involves glutamate and gamma (γ) aminobutyrate (GABA) has an important function in the brain (Fig. 17-5). Both glutamate and γ-aminobutyrate occur in high concentrations in brain (10 and 0.8 mM, respectively). Both are important neurotransmitters, γ-aminobutyrate being a principal neuronal inhibitory substance[102,103] (Chapter 30). In the γ-aminobutyrate cycle acetyl-CoA and oxaloacetate are converted into citrate (step *a*) in the usual way and the citrate is then converted into 2-oxoglutarate. The latter is transformed to L-glutamate either by direct amination (*b*) or by transamination (*c*), the amino donor being γ-aminobutyrate.

γ-Aminobutyrate is formed by decarboxylation of glutamate (Fig. 17-5, step *d*)[104] and is catabolized via transamination (step *e*)[105] to succinic semialdehyde, which is oxidized to succinate[106] and oxaloacetate. The two transamination steps in the pathways may be linked, as indicated in Fig. 17-5, to form a complete cycle that parallels the citric acid cycle but in which 2-oxoglutarate is oxidized to succinate via glutamate and γ-aminobutyrate. No thiamin diphosphate is required, but 2-oxoglutarate is reductively aminated to glutamate. The cycle is sometimes called the **γ-aminobutyrate shunt**, and it plays a significant role in the overall oxidative processes of brain tissue. This pathway is also prominent in green plants.[107–109] For example, under anaerobic conditions the radish *Raphanus sativus* accumulates large amounts of γ-aminobutyrate.[110] Most animal tissues contain very little γ-aminobutyrate, although it has been found in the oviducts of rats at concentrations that exceed those in the brain.[111]

2. The Dicarboxylic Acid Cycle

Some bacteria can subsist solely on glycolate, glycine, or oxalate, all of which

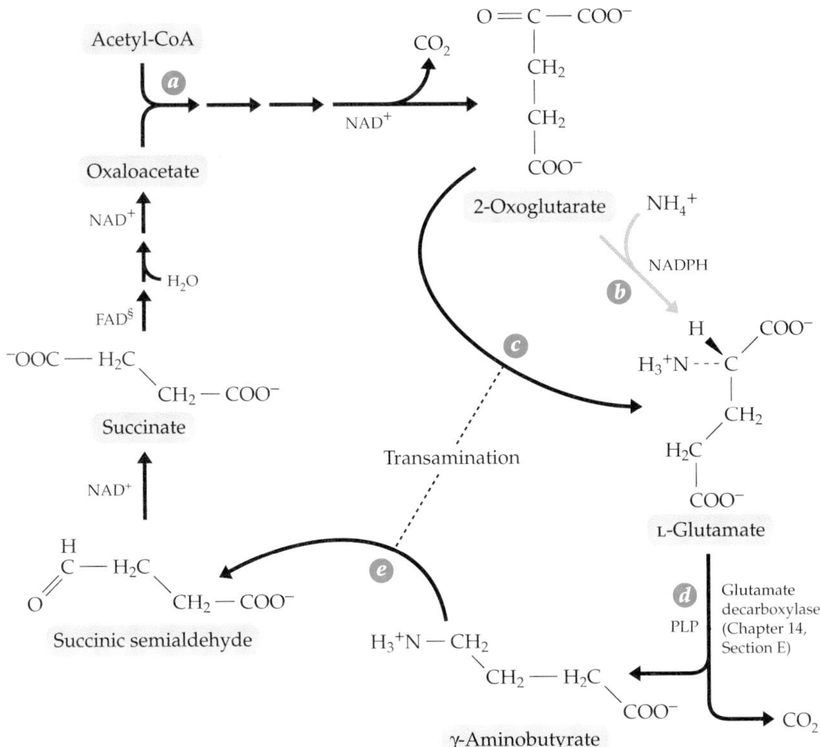

Figure 17-5 Reactions of the γ-aminobutyrate (GABA) cycle.

are converted to glyoxylate (Eq. 17-11). Glyoxylate is oxidized to CO_2 and water to provide energy to the bacteria and is also utilized for biosynthetic purposes. The energy-yielding process is found in the **dicarboxylic acid cycle** (Fig. 17-6), which catalyzes the complete oxidation of glyoxylate. Four hydrogen atoms are removed with generation of two molecules of NADH which can be oxidized by the respiratory chain to provide energy.[112,113] In the dicarboxylic acid cycle glyoxylate is the principal substrate and acetyl-CoA is the regenerating substrate rather than the principal substrate as it is for the citric acid cycle.

The logic of the dicarboxylic acid cycle is simple. Acetyl-CoA contains a potentially free carboxyl group. After the acetyl group of acetyl-CoA has been condensed with glyoxylate and the resulting hydroxyl group has been oxidized, the free carboxyl group appears in oxaloacetate in a position β to the carbonyl group. The carboxyl

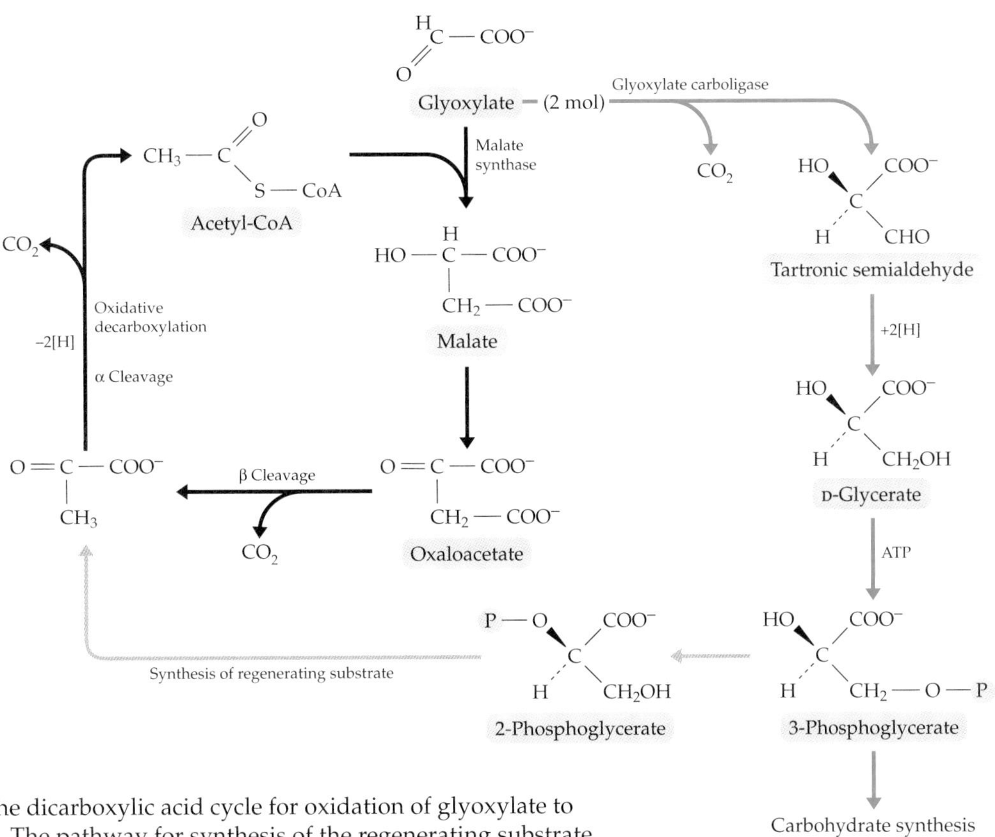

Figure 17-6 The dicarboxylic acid cycle for oxidation of glyoxylate to carbon dioxide. The pathway for synthesis of the regenerating substrate is indicated by green lines. This pathway is also needed for synthesis of carbohydrates and all other cell constituents.

$HO—CH_2—COO^-$ (Glycolate) —(−2[H])→ Glyoxylate

$H_3^+N—CH_2—COO^-$ (Glycine) —(Transamination)→ Glyoxylate ($H—C(=O)—COO^-$)

$HO—CH_2—COO^-$ (Oxalate) → Oxalyl-CoA —(2[H])→ Glyoxylate

(17-11)

donated by the glyoxylate is still in the α position. A consecutive β cleavage and an oxidative α cleavage release the two carboxyl groups as carbon dioxide to reform the regenerating substrate. The cycle is simple and efficient. Like the citric acid cycle, it depends upon thiamin diphosphate, without which the α cleavage would be impossible. Comparing the citric acid cycle (Fig. 17-2) with the simpler dicarboxylic acid cycle, we see that in the former the initial condensation product citrate contains a hydroxyl group attached to a tertiary carbon atom. With no adjacent hydrogen it is impossible to oxidize it directly to the carbonyl group which is essential for subsequent chain cleavage; hence the dependence on aconitase to shift the OH to an adjacent carbon. Both cycles involve oxidation of a hydroxy acid to a ketone followed by β cleavage and oxidative α cleavage. In the citric acid cycle additional oxidation steps are needed to convert succinate back to oxaloacetate, corresponding to the fact that the citric acid cycle deals with a more reduced substrate than does the dicarboxylic acid cycle.

The synthetic pathway for the regenerating substrate of the dicarboxylic acid cycle is quite complex. Two molecules of glyoxylate undergo α condensation with decarboxylation by glyoxylate carboligase[114] (see also Chapter 14, Section D) to form **tartronic semialdehyde**. The latter is reduced to D-glycerate, which is phosphorylated to 3-phosphoglycerate and 2-phosphoglycerate. Since the phosphoglycerates are carbohydrate precursors, this **glycerate pathway** provides the organisms with a means for synthesis of carbohydrates and other complex materials from glyoxylate alone. At the same time, 2-phosphoglycerate can be converted to pyruvate and the pyruvate, by oxidative decarboxylation, to the regenerating substrate acetyl-CoA.

E. Catabolism of Sugars

In most sugars each carbon atom bears an oxygen atom which facilitates chemical attack by oxidation at any point in the carbon chain. Every sugar contains a potentially free aldehyde or ketone group, and the carbonyl function can be moved readily to adjacent positions by isomerases. Consequently, aldol cleavage is also possible at many points. For these reasons, the metabolism of carbohydrates is complex and varied.

A sugar chain can be cut in several places giving rise to a variety of metabolic pathways. However, in the energy economy of most organisms, including human beings, the **Embden–Meyerhof–Parnas** or **glycolysis pathway** by which hexoses are converted to pyruvate (Fig. 17-7) stands out above all others. We have already considered this pathway, which is also outlined in Figs. 10-2 and 10-3. Some history and additional important details follow.

1. The Glycolysis Pathway

The discovery of glycolysis followed directly the early observations of Buchner and of Harden and Young on fermentation of sugar by yeast juice (p. 767). Another line of research, the study of muscle, soon converged with the investigations of alcoholic fermentation. Physiologists were interested in the process by which an isolated muscle could obtain energy for contraction in the absence of oxygen. It was shown by A. V. Hill that glycogen was converted to lactate to supply the energy, and Otto Meyerhof later demonstrated that the chemical reactions were related to those of alcoholic fermentation. The establishment of the structures and functions of the pyridine nucleotides in 1934 (Chapter 15) coincided with important studies by G. Embden in Frankfurt and of J. K. Parnas in Poland. The sequence of reactions in glycolysis soon became clear. All of the 15 enzymes catalyzing the individual steps in the sequence have been isolated and crystallized and are being studied in detail.[115]

Formation of pyruvate. The conversion of glucose to pyruvate requires ten enzymes (Fig. 17-7), and the sequence can be divided into four stages: preparation for chain cleavage (reactions *1–3*), cleavage and equilibration of triose phosphates (reactions *4* and *5*), oxidative generation of ATP (reactions *6* and *7*), and conversion of 3-phosphoglycerate to pyruvate (reactions *8–10*).

In preparation for chain cleavage, free glucose is phosphorylated to glucose 6-phosphate by ATP under the action of hexokinase (reaction *1*). Glucose 6-phosphate can also arise by cleavage of a glucosyl unit from glycogen by the consecutive action of glycogen phosphorylase (reaction 1*a*) and phosphoglucomutase, which transfers a phospho group from the oxygen at C-1 to that at C-6 (reaction 1*b*) (see also Eq. 12-39 and associated discussion of the mechanism of this enzyme). Why do cells attach phospho groups to sugars to initiate metabolism of the sugars? Four reasons can be given:

(a) The phospho group constitutes an electrically charged handle for binding the sugar phosphate to enzymes.

(b) There is a kinetic advantage in initiating a reaction sequence with a highly irreversible reaction

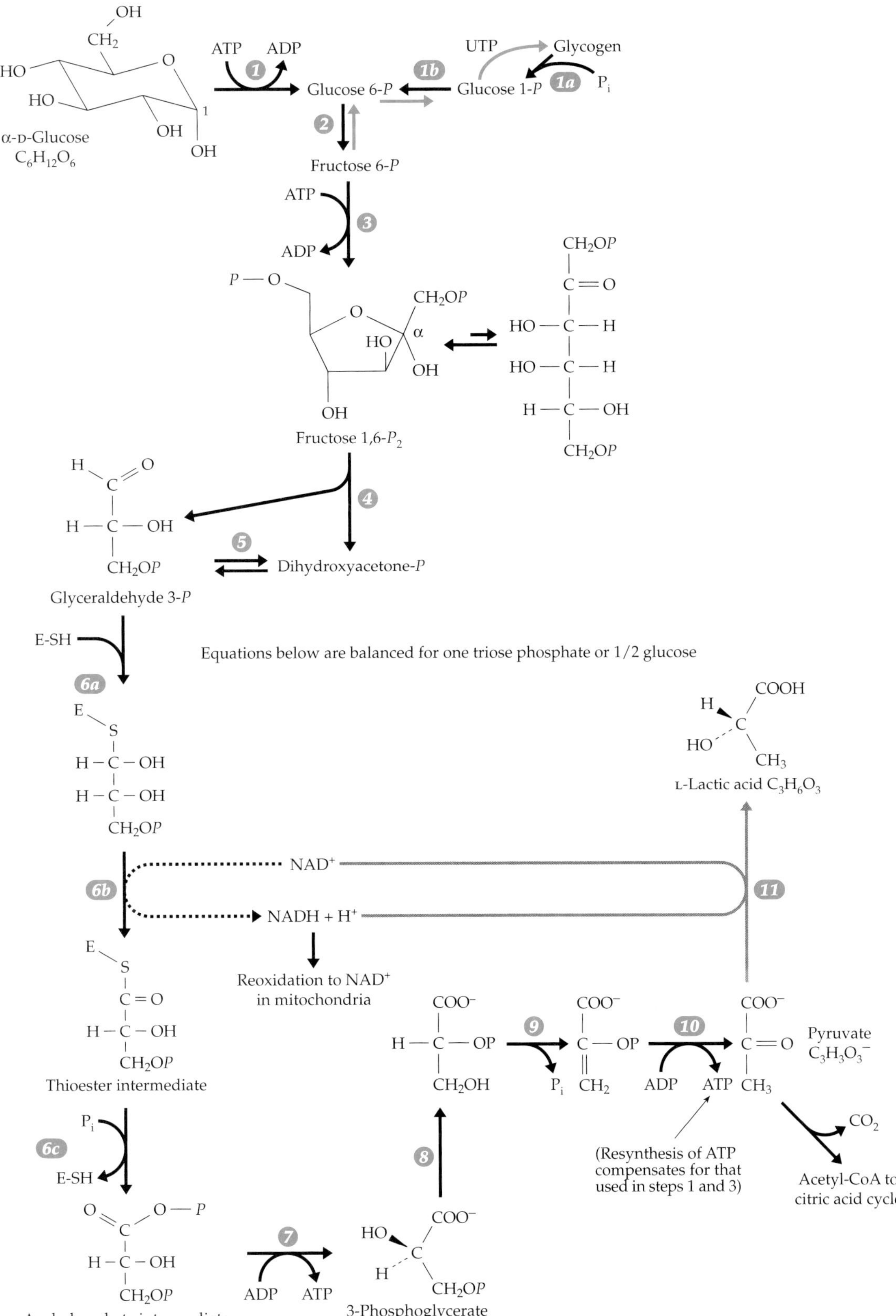

Figure 17-7 Outline of the glycolysis pathway by which hexoses are broken down to pyruvate. The ten enzymes needed to convert D-glucose to pyruvate are numbered. The pathway from glycogen using glycogen phosphorylase is also included, as is the reduction of pyruvate to lactate (step *11*). Steps *6a–7*, which are involved in ATP synthesis via thioester and acyl phosphate intermediates, are emphasized. See also Figures 10-2 and 10-3, which contain some additional information.

such as the phosphorylation of glucose.

(c) Phosphate esters are unable to diffuse out of cells easily and be lost.

(d) There is at least a possibility that the phospho groups may function in catalysis.

Reaction *2* of Fig. 17-7 is a simple isomerization that moves the carbonyl group to C-2 so that β cleavage to two three-carbon fragments can occur. Before cleavage a second phosphorylation (reaction *3*) takes place to form fructose 1,6-bisphosphate. This ensures that when fructose bisphosphate is cleaved by aldolase each of the two halves will have a phosphate handle. This second priming reaction (reaction *3*) is the first step in the series that is unique to glycolysis. The catalyst for the reaction, **phosphofructokinase**, is carefully controlled, as discussed in Chapter 11 (see Fig. 11-2).

Fructose bisphosphate is cleaved by action of an aldolase (reaction *4*) to give glyceraldehyde 3-phosphate and dihydroxyacetone phosphate. These two triose phosphates are then equilibrated by triose phosphate isomerase (reaction *5*; see also Chapter 13). As a result, both halves of the hexose can be metabolized further via glyceraldehyde 3-*P* to pyruvate. The oxidation of glyceraldehyde 3-*P* to the corresponding carboxylic acid, 3-phosphoglyceric acid (Fig. 17-7, reactions *6* and *7*), is coupled to synthesis of a molecule of ATP from ADP and P_i. This means that two molecules of ATP are formed per hexose cleaved, and that two molecules of NAD^+ are converted to NADH in the process.

The conversion of 3-phosphoglycerate to pyruvate begins with transfer of a phospho group from the C-3 to the C-2 oxygen (reaction *8*) and is followed by dehydration through an α, β elimination catalyzed by **enolase** (reaction *9*). The product, phosphoenolpyruvate (PEP), has a high group transfer potential. Its phospho group can be transferred easily to ADP via the action of the enzyme **pyruvate kinase**, to leave the enol of pyruvic acid which is spontaneously converted to the much more stable pyruvate ion (see Eq. 7-59). Because two molecules of PEP are formed from each glucose molecule, the process provides for the recovery of the two molecules of ATP that were expended in the initial formation of fructose 1,6-bisphosphate from glucose. Several isoenzyme forms exist in mammals. Most of these are allosterically activated by fructose 1,6-bisphosphate.[115a,b] However, the enzyme from trypanosomes is activated by fructose 2,6-P_2.[115c]

The further metabolism of pyruvate. In the aerobic metabolism that is characteristic of most tissues of our bodies, pyruvate is oxidatively decarboxylated to acetyl-CoA, which can then be completely oxidized in the citric acid cycle (Fig. 17-4). The NADH produced in reaction *6* of Fig. 17-7, as well as in the oxidative decarboxylation of pyruvate and in subsequent reactions of the citric acid cycle, is reoxidized in the electron transport chain of the mitochondria as described in detail in Chapter 18 (see Fig. 18-5). An important alternative fate of pyruvate is to enter into fermentation reactions. For example, the enzyme lactate dehydrogenase (Fig. 17-7, reaction *11*) catalyzes reduction by NADH of pyruvate to L-lactate, or, for some bacteria, to D-lactate. This reaction can be coupled to the NADH-producing reaction *6* to give a balanced process by which glucose is fermented to lactic acid in the absence of oxygen (see also Eq. 10-3). In a similar process, yeast cells decarboxylate pyruvate (α cleavage) to acetaldehyde which is reduced to ethanol using the NADH produced in reaction *6*. These fermentation reactions are summarized in Fig. 10-3 and, along with many others, are discussed further in Section F of this chapter.

2. Generation of ATP by Substrate Oxidation

The formation of ATP from ADP and P_i is a vital process for all cells. It is usually referred to as "phosphorylation" and includes **oxidative phosphorylation** associated with the passage of electrons through an electron transport chain—usually in mitochondria; **photosynthetic phosphorylation**, a similar process occurring in chloroplasts under the influence of light; and **substrate-level phosphorylation**. Only the last is fully understood chemically. The dehydrogenation of glyceraldehyde 3-*P* and the accompanying ATP formation (reactions *6* and *7*, Fig. 17-7; Fig. 15-6) is the best known example of substrate-level phosphorylation and is tremendously important for yeasts and other microorganisms that live anaerobically. They depend upon this one reaction for their entire supply of energy. The conversion of glucose either to lactate or to ethanol and CO_2 is accompanied by a net synthesis of only two molecules of ATP and it is most logical to view these as arising from oxidation of glyceraldehyde 3-*P*. The formation of ATP from PEP and ADP in reaction *10* of Fig. 17-7 can be regarded as the recapturing of ATP "spent" in the priming reactions of steps *1* and *3*. With a gain of only two molecules of ATP for each molecule of hexose fermented, it is not surprising that yeast must ferment enormous amounts of sugar to meet its energy needs.

Each glucose unit of glycogen stored in our bodies can be converted to pyruvate with an apparent net gain of *three* molecules of ATP. However, two molecules of ATP were needed for the initial synthesis of each hexose unit of glycogen (Fig. 12-2). Therefore, the overall net yield for fermentation of stored polysaccharide is still only two ATP per hexose. The fermentation of glycogen accounts for the very rapid generation of lactic acid during intense muscular activity. However, in most circumstances within aerobic tissues reoxidation of NADH occurs via the electron transport chain of mitochondria with a much higher yield of ATP. Substrate-level phosphorylation can also follow oxidative decarboxylation of an α-oxoacid. For example, in the citric acid cycle GTP is formed following oxidative decarboxylation of 2-oxoglutarate (Fig. 17-4, steps *e* and *f*).

3. The Pentose Phosphate Pathways

A second way of cleaving glucose 6-phosphate utilizes sequences involving the five-carbon pentose sugars. They are referred to as **pentose phosphate pathways**, the phosphogluconate pathway, or the hexose monophosphate shunt. Historically, the evidence for such routes dates from the experiments of Warburg on the oxidation of glucose 6-*P* to 6-phosphogluconate (Chapter 15). For many years the oxidation remained an enzymatic reaction without a defined pathway. However, it was assumed to be part of an alternative method of degradation of glucose. Supporting evidence was found in the observation that tissues continue to respire in the presence of a high concentration of fluoride ion, a known inhibitor of the enolase reaction and capable of almost completely blocking glycolysis. Some tissues, e.g., liver, are especially active in respiration through this alternative pathway, whose details were elucidated by Horecker and associates.[116,117] We now know that the pentose phosphate pathways are multiple as well as multipurpose. They function in catabolism and also, when operating in the reverse direction, as a **reductive pentose phosphate pathway** that lies at the heart of the sugar-forming reactions of photosynthesis.

The oxidative pentose pathway provides *a means for cutting the chain of a sugar molecule one carbon at a time*, with the carbon removed appearing as CO_2. The enzymes required can be grouped into three distinct systems, all of which are found in the cytosol of animal cells: (i) a dehydrogenation–decarboxylation system, (ii) an isomerizing system, and (iii) a sugar rearrangement system. The dehydrogenation–decarboxylation system cleaves glucose 6-*P* to CO_2 and the pentose phosphate, ribulose 5-*P* (Eq. 17-12). Three enzymes are required, the first being glucose 6-*P* dehydrogenase[117a] (Eq. 17-12, step *a*; see also Eq. 15-10). The immediate product, a lactone, undergoes spontaneous hydrolysis. However, the action of **gluconolactonase** (Eq. 17-12, step *b*) causes a more rapid ring opening. A second dehydrogenation is catalyzed by **6-phosphogluconate dehydrogenase** (Eq. 17-12, step *c*),[117b] and this reaction is immediately followed by a β decarboxylation catalyzed by the same enzyme (as in Eq. 13-45). The value of $\Delta G°$ for oxidation of glucose 6-*P* to ribulose 5-*P* by $NADP^+$ according to Eq. 17-12 is −30.8 kJ mol^{-1}, a negative enough value to drive the [NADPH]/[$NADP^+$] ratio to an equilibrium value of over 2000 at a CO_2 tension of 0.05 atm.

The isomerizing system, consisting of two enzymes,

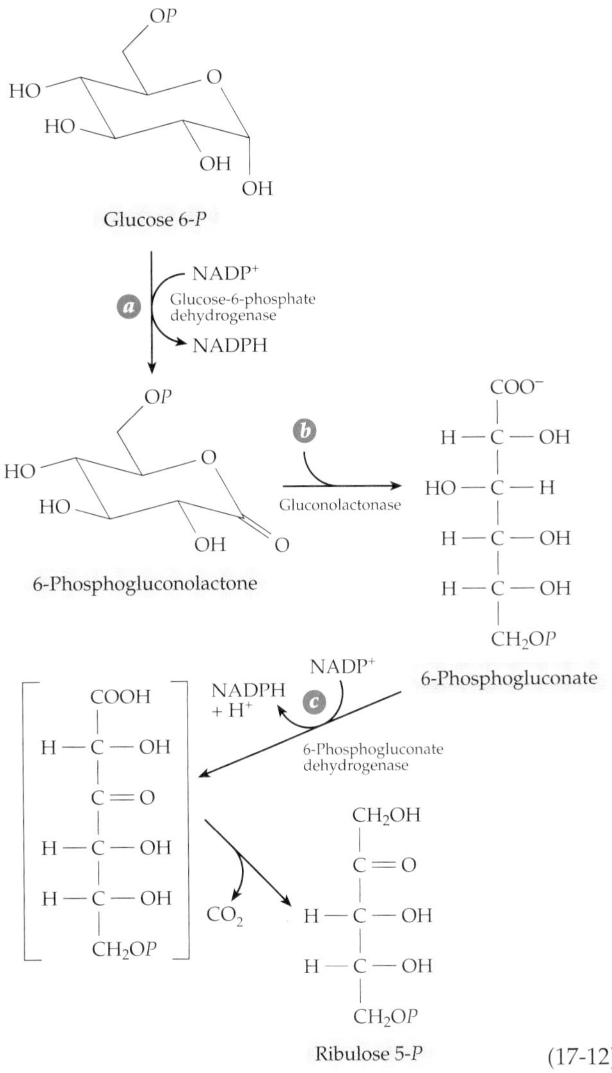

Glucose 6-*P*

(a) $NADP^+$ → NADPH; Glucose-6-phosphate dehydrogenase

6-Phosphogluconolactone

(b) Gluconolactonase

6-Phosphogluconate

(c) $NADP^+$ → NADPH + H^+; 6-Phosphogluconate dehydrogenase; CO_2

Ribulose 5-*P* (17-12)

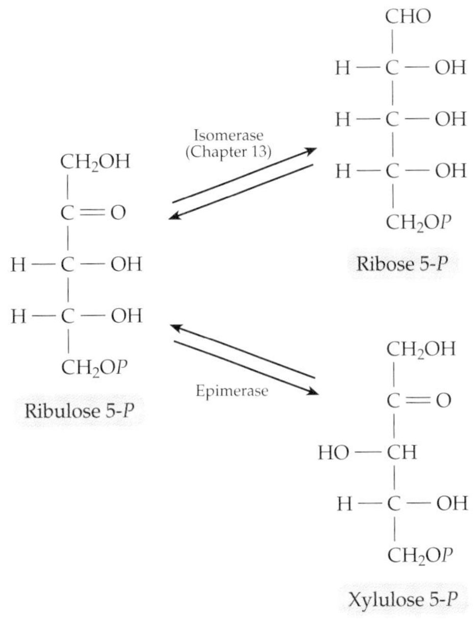

Ribulose 5-*P* ⇌ Ribose 5-*P* (Isomerase (Chapter 13))

Ribulose 5-*P* ⇌ Xylulose 5-*P* (Epimerase) (17-13)

CH_2OH–C=O–HO–C–H–R (Xylulose 5-*P*, fructose 6-*P*, or sedoheptulose 7-*P*) + CHO–R′ (Ribose 5-*P*, erythrose 4-*P*, or glyceraldehyde 3-*P*)

Transketolase (TK), a thiamin diphosphate enzyme (Chapter 14)

→ H–C(=O)–R + CH_2OH–C=O–HO–C–H–R′ (17-14)

CH_2OH–C=O–HO–C–H–H–C–OH–R″ + CHO–R′

Transaldolase (TA) (Chapter 13)

→ CHO–R″ + CH_2OH–C=O–HO–C–H–H–C–OH–R′ (17-15)

interconverts three pentose phosphates (Eq. 17-13). As a consequence the three compounds exist as an equilibrium mixture. Both xylulose 5-*P* and ribose 5-*P* are needed for further reactions in the pathways.

The ingenious sugar rearrangement system uses two enzymes, **transketolase** and **transaldolase**. Both catalyze chain cleavage and transfer reactions (Eqs. 17-14 and 17-15) that involve the same group of substrates. These enzymes use the two basic types of C–C bond cleavage, adjacent to a carbonyl group (α) and one carbon removed from a carbonyl group (β). Both types are needed in the pentose phosphate pathways just as they are in the citric acid cycle. The enzymes of the pentose phosphate pathway are found in the cytoplasm of both animal and plant cells.[117c] Mammalian cells appear to have an additional set that is active in the endoplasmic reticulum and plants have another set in the chloroplasts.[117c]

An oxidative pentose phosphate cycle. Putting the three enzyme systems together, we can form a cycle that oxidizes hexose phosphates. Three carbon atoms are chopped off one at a time (Fig. 17-8A) leaving a three-carbon triose phosphate as the product. Since the dehydrogenation system works only on glucose 6-*P*, a part of the sugar rearrangement system must be utilized between each of the three oxidation steps. Notice that a C_5 unit (ribose 5-*P*) is used in the first reaction with transketolase but is regenerated at the end of the sequence. This C_5 unit is the regenerating substrate for the cycle. As indicated by the dashed arrows, it is formed readily in any quantity needed by oxidation of glucose 6-*P*. Before the C_5 unit that is formed in each oxidation step can be processed by the sugar rearrangement reactions, it must be isomerized[117c,118,118a,b] from ribulose 5-*P* to xylulose 5-*P*; before the C_5 unit, produced at the end of the sequence in Fig. 17-8, can be reutilized as a regenerating substrate, it must be isomerized to ribose 5-*P*. Thus, the cycle is quite complex. The same C_5 substrates appear at several points in Fig. 17-8A and substrates from different parts of the cycle become scrambled and the pathway does not degrade all the hexose molecules in a uniform manner. For this reason, Zubay described the pentose phosphate pathways as a "swamp."[119]

The oxidative pentose phosphate cycle is often presented as a means for complete oxidation of hexoses to CO_2. For this to happen the C_3 unit indicated as the product in Fig. 17-8A must be converted (through the action of aldolase, a phosphatase, and hexose phosphate isomerase) back to one-half of a molecule of glucose-6-*P* which can enter the cycle at the beginning. On the other hand, alternative ways of degrading the C_3 product glyceraldehyde-*P* are available. For example, using glycolytic enzymes, it can be oxidized to pyruvate and to CO_2 via the citric acid cycle.

As a general rule, NAD^+ is associated with catabolic reactions and it is somewhat unusual to find $NADP^+$ acting as an oxidant. However, in mammals the enzymes of the pentose phosphate pathway are specific for $NADP^+$. The reason is thought to lie in the need of NADPH for biosynthesis (Section I). On this basis, the occurrence of the pentose phosphate pathway in tissues having an unusually active biosynthetic function (liver and mammary gland) is understandable. In these tissues the cycle may operate as indicated in Fig. 17-8A with the C_3 product also being used in biosynthesis. Furthermore, any of the products from C_4 to C_7 may be withdrawn in any desired amounts without disrupting the smooth operation of the cycle. For example, the C_4 intermediate **erythrose 4-*P*** is required in synthesis of aromatic amino acids by bacteria and plants (but not in animals). **Ribose 5-*P*** is needed for formation of several amino acids and of nucleic acids by all organisms. In some circumstances the formation of ribose 5-*P* may be the only essential function for the pentose phosphate pathway.[120]

Several studies of the metabolism of isotopically labeled glucose[121–122a] have been in accord with

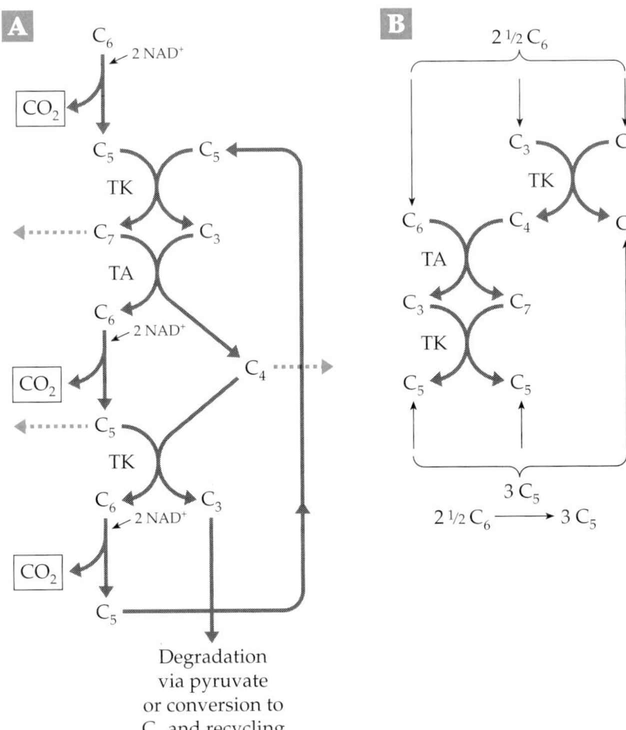

Figure 17-8 The pentose phosphate pathways. (A) Oxidation of a hexose (C_6) to three molecules of CO_2 and a three-carbon fragment with the option of removing C_3, C_4, C_5, and C_7 units for biosynthesis (dashed arrows). (B) Nonoxidative pentose pathways: $2\,^1\!/_2\,C_6 \rightarrow 3\,C_5$ or $2\,C_6 \rightarrow 3\,C_4$ or $3\,^1\!/_2\,C_6 \rightarrow 3\,C_7$.

operation of the pentose phosphate pathway as is depicted in Fig. 17-8. However, Williams and associates proposed a modification in the sugar rearrangement sequence in liver[123–126] to include the formation of arabinose 5-*P* (from ribose 5-*P*), an octulose bisphosphate, and an octulose 8-monophosphate. Many investigators argue that these additional reactions are of minor significance.[121,122,127] The measured concentrations of pentose phosphate pathway intermediates in rat livers are close to those predicated for a near-equilibrium state from equilibrium constants measured for the individual steps of Fig. 17-8.[128] Most of the concentrations are in the 4- to 10-μM range but the level of erythrose 4-*P*, which is predicted to be ~ 0.2 μM, is too low to measure.

In contrast to animals, the resurrection plant *Craterostigma plantaginenm* accumulates large amounts of a 2-oxo-octulose. This plant is one of a small group of angiosperms that can withstand severe dehydration and are able to rehydrate and resume normal metabolism within a few hours. During desiccation much of the octulose is converted into sucrose. The plant has extra transketolase genes which may be essential for this rapid interconversion of sugars.[129]

Nonoxidative pentose phosphate pathways. The sugar rearrangement system together with the glycolytic enzymes that convert glucose 6-*P* to glyceraldehyde 3-*P* can function to transform hexose phosphates into pentose phosphates (Fig. 17-8B; Eq. 17-16) which may be utilized for nucleic acid synthesis in erythrocytes and other cells.[130,131]

$$2\tfrac{1}{2}\ C_6 \rightarrow 3\,C_5 \qquad (17\text{-}16)$$

The reader can easily show that the same enzymes will catalyze the net conversion of hexose phosphate to erythrose 4-phosphate or to sedoheptulose 7-phosphate (Eq. 17-17):

$$2\,C_6 \rightarrow 3\,C_4;\ \ 3\tfrac{1}{2}C_6 \rightarrow 3\,C_7 \qquad (17\text{-}17)$$

An investigation of metabolism of the red lipid-forming yeast *Rhodotorula gracilis* (which lacks phosphofructokinase and is thus unable to break down sugars through the glycolytic pathway) indicated that 20% of the glucose is *oxidized* through the pentose phosphate pathways while 80% is *altered* by the nonoxidative pentose phosphate pathway.[100] However, it is not clear how the C_3 unit used in the nonoxidative pathway (Fig. 17-8B) is formed if glycolysis is blocked. A number of fermentations are also based on the pentose phosphate pathways (Section F,5).

4. The Entner–Doudoroff Pathway

An additional way of cleaving a six-carbon sugar chain provides the basis for the **Entner–Doudoroff pathway** which is used by *Zymomonas lindneri* and many other species of bacteria. Glucose 6-*P* is oxidized first to 6-phosphogluconate, which is converted by dehydration to a 2-oxo-3-deoxy derivative (Eq. 17-18,

6-Phosphogluconate —(a), $-H_2O$→

COO^-
|
C=O
|
CH_2
|
HCOH
|
HCOH
|
CH_2O–*P*

2-Oxo-3-deoxy-6-phosphogluconate

(b) An aldolase ↓

Pyruvate		Glyceraldehyde 3-*P*
COO^-		HC=O
C=O	+	HCOH
CH_3		CH_2O–*P*

(17-18)

step *a*). The resulting 2-oxo-3-deoxy sugar is cleaved by an aldolase (Eq. 17-18, step *b*) to pyruvate and glyceraldehyde 3-*P*, which are then metabolized in standard ways.

F. Fermentation: "Life without Oxygen"

Pasteur recognized in 1860 that fermentation was not a spontaneous process but a result of life in the near absence of air.[132] He realized that yeasts decompose much more sugar under anaerobic conditions than they do aerobically, and that the anaerobic fermentation was essential to the life of these organisms. In addition to the alcoholic fermentation of yeast, there are many other fermentations which have been attractive subjects for biochemical study. If life evolved at a time when no oxygen was available, the most primitive organisms must have used fermentations. They may be the oldest as well as the simplest ways in which cells obtain energy. The enzymes of the glycolysis pathway are found in the small genomes of *Mycoplasma, Haemophila,* and *Methanococcus*.[133,134]

Fermentation is also a vital process in the human body. Our muscles usually receive enough oxygen to oxidize pyruvate and to obtain ATP through aerobic metabolism, but there are circumstances in which the oxygen supply is inadequate. During extreme exertion, after most oxygen is consumed, muscle cells produce lactate by fermentation. White muscle of fish and fowl has little aerobic metabolism and normally yields L-lactic acid as a principal end product. Likewise, a variety of tissues within the human body, including the transparent lens and cornea of our eyes, are poorly supplied with blood and depend upon fermentation of glucose to lactic acid. Red blood cells and skin and sometimes adipose tissue are also major producers of lactic acid.[135] Of the ~ 115 g of lactic acid present in a 70-kg human body, about 29% comes from erythrocytes, 29% from skin, 17% from the brain, and 16% from skeletal muscle.[136] Because lactic acid lowers the pH of cells it must be removed efficiently.

Some of the lactic acid formed in muscle and most of the lactate formed in less aerobic tissues (e.g., adipose tissue)[136a] enters the bloodstream, which normally contains 1–2 mM lactate,[136] and is carried to the liver where it is reoxidized to pyruvate. Part of the pyruvate is then oxidized via the citric acid cycle while a larger part is reconverted to glucose (Section J,5). This glucose may be released into the bloodstream and returned to the muscles. The overall process is known as the **Cori cycle**. Lactic acid accumulates in muscle after vigorous exercise. It is exported to the liver slowly, but if mild exercise continues the lactate may be largely oxidized within muscle via the tricarboxylic acid cycle. Recent NMR studies have shown that lactic acid is formed rapidly during muscular contraction, even when exercise is mild.[136b] During the initial 15 ms of contraction the ATP utilized is regenerated from creatine phosphate (Eq. 6-67). During the remainder of the contraction (up to ~100 ms) glycogen is converted to lactic acid to provide ATP and to replenish the creatine phosphate. In the resting period following contraction most of the lactate is either dehydrogenated to pyruvate and oxidized in mitochondria or exported to other tissues. The glycogen stores in muscle are renewed by synthesis from blood glucose. Lactic acid is a convenient energy carrier and a precursor for gluconeogenesis which can be transferred between tissues easily.[136c] Cancer cells often take advantage of this opportunity to grow rapidly using fermentation of glucose to lactic acid as a source of energy.[136d]

Alcoholic fermentation allows roots of some plants to survive short periods of flooding. Ethanol does not acidify the tissues as does lactic acid, avoiding possible damage from low pH.[137,138] Goldfish can also use the ethanolic fermentation for short times, excreting the ethanol.[139]

1. Fermentations Based on the Embden–Meyerhof Pathway

Homolactic and alcoholic fermentations. The reactions by which glucose can be converted to lactate and, by yeast cells, to ethanol and CO_2 (Figs. 10-3 and 17-7) illustrate several features common to all fermentations. The NADH produced in the oxidation step is reoxidized in a reaction by which substrate is reduced to the final end product. The NAD alternates between oxidized and reduced forms. This coupling of oxidation steps with reduction steps in exact equivalence is characteristic of all true (anaerobic) fermentations. The formation of ATP from ADP and P_i by substrate-level phosphorylation is also common to all fermentations. The stoichiometry is often nearly exact and simple. For example, according to the reaction of Eq. 17-19, which is outlined step-by-step in Fig. 17-7, a net total of two moles of ATP is formed per mole of glucose fermented.

Energy relationships. If we disregard the synthesis of ATP, the equations for the lactic acid and ethanol fermentations are given by Eqs. 17-19 and 17-20.

$$\text{Glucose} \rightarrow 2\ \text{lactate}^- + 2\ \text{H}^+$$
$$\Delta G'\ (\text{pH } 7) = -196\ \text{kJ mol}^{-1}\ (-46.8\ \text{kcal mol}^{-1}) \quad (17\text{-}19)$$

$$\text{Glucose} \rightarrow 2\ \text{CO}_2 + 2\ \text{ethanol}$$
$$\Delta G^\circ = -235\ \text{kJ mol}^{-1} \quad (17\text{-}20)$$

The Gibbs energy changes are negative and of sufficient magnitude that the reactions will unquestionably go to completion. However, the synthesis of two molecules of ATP from inorganic phosphate and ADP, a reaction

(Eq. 17-21) for which $\Delta G'$ is substantially positive, is coupled to the fermentation.

$$ADP^{3-} + HPO_4^{2-} + H^+ \rightarrow ATP^{4-} + H_2O$$
$$\Delta G' \text{ (pH 7)} = +34.5 \text{ kJ mol}^{-1} \text{ (Table 6-5)} \quad (17\text{-}21)$$

To obtain the net Gibbs energy change for the complete reaction we must add 2 x 34.5 = +69.0 kJ to the values of $\Delta G'$ for Eqs. 17-19 and 17-20. When this is done we see that the net Gibbs energy changes are still highly negative, that the reactions will proceed to completion, and that these fermentations can serve as an usable energy source for organisms.

Biochemists sometimes divide ΔG for the ATP synthesis in a coupled reaction sequence (in this case +69 kJ) by the overall Gibbs energy decrease for the coupled process (196 or 235 kJ mol^{-1}) to obtain an "efficiency." In the present case the efficiency would be 35% and 29% for coupling of Eq. 17-21 (for 2 mol of ATP) to Eqs. 17-19 and 17-20, respectively. According to this calculation, nature is approximately one-third efficient in the utilization of available metabolic Gibbs energy for ATP synthesis. However, it is important to realize that this calculation of efficiency has no exact thermodynamic meaning. Furthermore, the utilization of ATP formed by a cell for various purposes is far from 100% efficient.

Why are the Gibbs energy decreases for Eqs. 17-19 and 17-20 so large? No overall oxidation takes place; there is only a rearrangement of the existing bonds between atoms of the substrate. Why does this rearrangement of bonds lead to a substantial negative ΔG? An answer is suggested by an examination of the numbers of each type of bond in the substrate and in the products. During the conversion from glucose to two molecules of lactate one C–C bond, one C–O bond, and one O–H bond are lost and one C–H bond and one C=O are gained. If we add up the bond energies for these bonds (Table 6-6) we find that the difference (ΔH) between substrate and products amounts to only about 20 kJ/mol. However, lactic acid contains a carboxyl group, and carboxyl groups have a special stability as a result of resonance. The extra resonance energy of a carboxyl group (Table 6-6) is ~117 kJ (28 kcal) per mole or 234 kJ/mol for two carboxyl groups. This is approximately the same as the Gibbs energy change (Eq. 17-19) for fermentation of glucose to lactate. Thus, the energy available results largely from the rearrangement of bonds by which the carboxyl groups of lactate are formed. Likewise, the resonance stabilization of CO_2 is given by Pauling as 151 kJ/mol, again of just the right magnitude to explain ΔG in alcoholic fermentations (Eq. 17-20).

On this basis we can state as a general rule that fermentations can occur when substrates consisting of largely singly bonded atoms and groups, such as the carbonyl groups that are not highly stabilized by resonance, are converted to products containing carboxyl groups or to CO_2. If we assume an efficiency of ~30%, the energy available will be about sufficient for synthesis of one ATP molecule for each carboxyl group or CO_2 created. Bear in mind that generation of ATP also depends upon availability of a mechanism. It is of interest that most synthesis of ATP is linked directly to the chemical processes by which carboxyl groups or CO_2 molecules are created in a fermentation process. The most important single reaction is the oxidation of the aldehyde group of glyceraldehyde 3-*P* to the carboxyl group of 3-phosphoglycerate (steps *6a*–*6c* and *7* in Fig. 17-7; see also Fig. 15-6).

Compare the fermentation of glucose with the complete oxidation of the sugar to carbon dioxide and water (Eq. 17-22), a process which yeast cells (as well as our own cells) carry out in the presence of air. The overall Gibbs energy change is over 10 times greater than that for fermentation, a fact that permits the cell

$$\text{Glucose} + 6\ O_2 \rightarrow 6\ CO_2 + 6\ H_2O$$
$$\Delta G' = -2872 \text{ kJ } (-686.5 \text{ kcal}) \text{ mol}^{-1} \quad (17\text{-}22)$$

to form an enormously greater quantity of ATP. The net gain in ATP synthesis, accompanying Eq. 17-22, is usually about 38 mol of ATP—19 times more than is available from fermentation of glucose. Thus, the explanation of Pasteur's observation that yeast decomposes much less sugar in the presence of air than in its absence is clear. Also, we can understand why a cell, living anaerobically, must metabolize a very large amount of substrate to grow. (Recall from Chapter 6 that ~1 mol of ATP energy is needed to produce 10 g of cells.)

Variations of the alcoholic and homolactic fermentations. The course of a fermentation is often affected drastically by changes in conditions. Many variations can be visualized by reference to Fig. 17-9, which shows a number of available metabolic sequences. We have already discussed the conversion of glucose to triose phosphate and via reaction pathway *a* to pyruvate, via reaction *c* to lactate, and via reaction *d* to ethanol.

If bisulfite is added to a fermenting culture of yeast, the acetaldehyde formed through reaction *d* is trapped as the bisulfite adduct blocking the reduction of acetaldehyde to ethanol, an essential part of the fermentation. Yeast cells accommodate this change by using the accumulating NADH to reduce half of the triose-*P* to glycerol through pathway *b*. Two enzymes are needed, a dehydrogenase and a phosphatase, to hydrolytically cleave off the phosphate. The balanced reaction is given by Eq. 17-23:

$$\text{Glucose} \rightarrow \text{glycerol} + \text{acetaldehyde (trapped)} + CO_2$$
$$\Delta G' \text{ (pH 7)} = -105 \text{ kJ mol}^{-1} \quad (17\text{-}23)$$

In this reaction only one molecule of CO_2 is produced

but the overall Gibbs energy change is still adequate to make the reaction highly spontaneous. However (referring to Fig. 17-9), we see that the net synthesis of ATP is now apparently zero. The fermentation apparently does not permit cell growth. Nevertheless, it has been used industrially for production of glycerol.

Reduction of dihydroxyacetone phosphate to glycerol phosphate also occurs in insect flight muscle and apparently operates as an alternative to lactic acid formation in that tissue. There is no net gain of ATP in the conversion of free glucose to glycerol phosphate and pyruvate, but using stored glycogen in muscle as the starting material, the dismutation of triose-*P* to glycerol-*P* and pyruvate provides one ATP per glucose unit rapidly during the vigorous contraction of the powerful insect flight muscle. During the slower recovery phase, glycerol-*P* is thought to be reoxidized after entering the mitochondria of these highly aerobic cells. Thus, the transport of glycerol-*P* into mitochondria serves as a means for transporting reducing equivalents derived from reoxidation of NADH into the mitochondria. Indeed, the significance of glycerol-*P* to muscle metabolism may be more related to this function than to the rapid formation of ATP (see Chapter 18).

Why does the glycolysis sequence begin with phosphorylation of glucose by ATP? The phospho groups probably provide convenient handles and doubtless assist in substrate recognition. There may be a kinetic advantage but also a danger. Unless there is adequate regulation the "turbo design," in which ATP is used at the outset to drive glycolysis, may lead to accumulation of phosphorylated intermediates and to inadequate concentrations of ATP and inorganic phosphate.[139a,b] Yeast cells guard against this problem by synthesizing trehalose 6-phosphate, which acts as a feedback inhibitor of hexokinase.[139a] Trypanosomes utilize a different type of control. The enzymes that convert glucose into 3-phosphoglycerate are present in membrane-bounded organelles called **glycosomes**. Phosphoglycerate is exported from them into the cytosol where glycolysis is completed.[139b] Since inorganic phosphate is essential for ATP formation, if the P_i concentration falls too low the rate of fermentation by yeast juice is greatly decreased, an observation made by Harden and Young[139c] in 1906.

2. The Mixed Acid Fermentation

Enterobacteria, including *E. coli*, convert glucose to ethanol and acetic acid and either formic acid or CO_2 and H_2 derived from it. The stoichiometry is variable but the fermentation can be described in an idealized form as follows:

$$\text{Glucose} + H_2O \rightarrow \text{ethanol} + \text{acetate}^- + H^+ + 2\,H_2 + 2\,CO_2$$

$$\Delta G' \text{ (pH 7)} = -225 \text{ kJ mol}^{-1} \quad (17\text{-}24)$$

The details of the process and the oxidation–reduction balance can be pictured as in Eq. 17-25. Pyruvate is cleaved by the pyruvate formate–lyase reaction (Eq. 15-37) to acetyl-CoA and formic acid. Half of the acetyl-CoA is cleaved to acetate via acetyl-*P* with generation of ATP, while the other half is reduced in two steps to ethanol using the two molecules of NADH produced in the initial oxidation of triose phosphate (Eq. 17-25). The overall energy yield is three molecules of ATP per glucose. The "efficiency" is thus (3 x 34.5) ÷ 225 = 46%. Some of the glucose is converted to D-lactic and to succinic acids (pathway *f*, Fig. 17-9); hence the name **mixed acid fermentation**. Table 17-1 gives typical yields of the mixed acid fermentation of *E. coli*. Among the four major products are acetate, ethanol, H_2, and CO_2, as shown in Eq. 17-25. However, at high pH formate accumulated instead of CO_2.

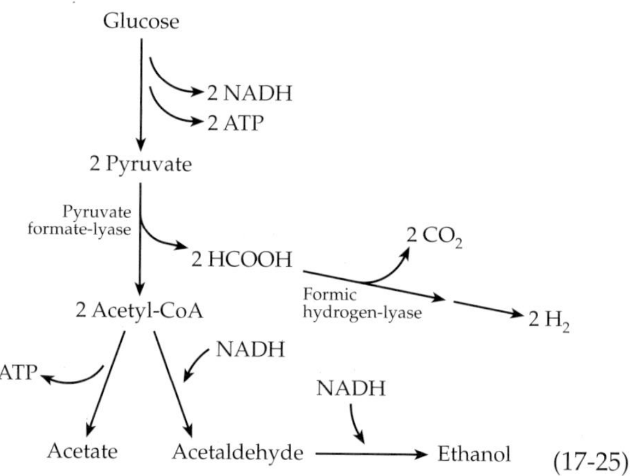

(17-25)

TABLE 17-1
Products of the Mixed Acid Fermentation by *E. coli* at Low and High Values of pH[a]

Product (Millimole formed from 100 mmol of glucose)	pH 6.2	pH 7.8
Acetate	36	39
Ethanol	50	51
H_2	70	0.3
CO_2	88	1.7
Formate	2.4	86
Lactate	79	70
Succinate	11	15
Glycerol	1.4	0.3
Acetoin	0.1	0.2
Butanediol	0.3	0.2

[a] From Tempest, D. W. and Neijssel, O. M.[140] Based on data of Blackwood.[141]

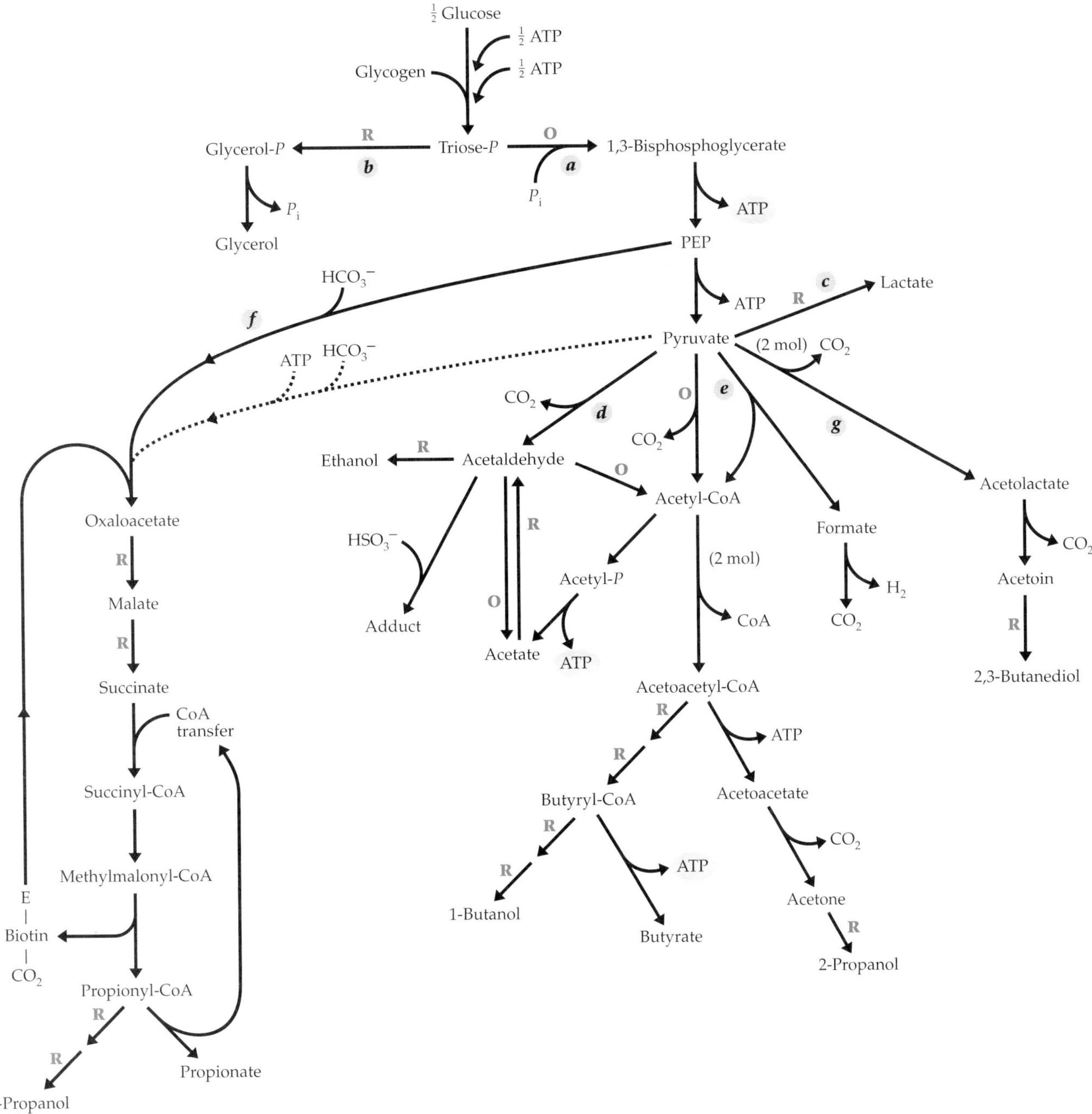

Figure 17-9 Reaction sequences in fermentation based on the Embden–Meyerhof–Parnas pathway. Oxidation steps (producing NADH + H^+) are marked "O"; reduction steps (using NADH + H^+) are marked "R."

Over one-third of the glucose was fermented to lactate in both cases.

In some mixed acid fermentations (e.g., that of *Shigella*) formic acid accumulates, but in other cases (e.g., with *E. coli* at pH 6) it is converted to CO_2 and H_2 (Eq. 17-25). The equilibration of formic acid with CO_2 and hydrogen is catalyzed by the **formic hydrogen–lyase** system which consists of two iron–sulfur enzymes. The selenium-containing **formate dehydrogenase** (Eq. 16-63) catalyzes oxidation of formate to CO_2 by NAD^+, while a membrane-bound **hydrogenase** (Eq. 16-48) equilibrates NADH + H^+ with NAD^+ + H_2. Hydrogenase also serves to release H_2 from excess NADH. Krebs pointed out that an excess of NADH may arise because growth of cells requires biosynthesis of many components such as amino acids. When glucose is the sole source of carbon, biosynthetic reactions involve an excess of oxidation steps over reduction steps.[142] The excess of reducing equivalents may be released as H_2 or may be used to form highly reduced products such as succinate.

Among such genera as *Aerobacter* and *Serratia* part of the pyruvate formed is condensed with decarboxylation to form **S acetolactate**,[143] which is decarboxylated to acetoin (Eq. 17-26; pathway *g* of Fig. 17-9). The acetoin is reduced with NADH to **2,3-butanediol**, while a third molecule of pyruvate is converted to ethanol, hydrogen, and CO_2 (Eq. 17-26). The reaction provides the basis for industrial production of butanediol, which can be dehydrated nonenzymatically to butadiene.

1 ½ Glucose
→ 3 ATP, 3 NADH
3 Pyruvate
two molecules, − CO_2 → Acetolactate (H_3C–C(=O)–C(OH)(CH_3)–COO^-)
→ $H_2 + CO_2$ → Acetyl-CoA → (2 NADH) → Ethanol
Acetolactate → − CO_2 → Acetoin (H_3C–C(=O)–C(OH)(CH_3)–H)
Acetoin → NADH → 2,3-Butanediol

(17-26)

Mixed acid fermentations are not limited to bacteria. For example, trichomonads, parasitic flagellated protozoa, have no mitochondria. They export pyruvate into the bloodstreams of their hosts and also contain particles called **hydrogenosomes** which can convert pyruvate to acetate, succinate, CO_2, and H_2.[144] Hydrogenosomes are bounded by double membranes and have a common evolutionary relationship with both mitochondria and bacteria. The enzyme that catalyzes pyruvate cleavage in hydrogenosomes apparently does not contain lipoate and may be related to the pyruvate–ferredoxin oxidoreductase of clostridia (Eq. 15-35). The hydrogenosomes also contain an active hydrogenase.

Many invertebrate animals are true facultative anaerobes, able to survive for long periods, sometimes indefinitely, without oxygen.[145–147] Among these are *Ascaris* (Fig. 1-14), oysters, and other molluscs. Succinate and alanine are among the main end products of anaerobic metabolism. The former may arise by a mixed acid fermentation that also produces pyruvate. The pyruvate is converted to acetate to balance the fermentation in *Ascaris lumbricoides*, which is in effect an obligate anaerobe. However, in molluscs the pyruvate may undergo transamination with glutamate to form alanine and 2-oxoglutarate; the oxoglutarate may be oxidatively decarboxylated to succinate. The reactions depend upon the availability of a store of glutamate or of other amino acids, such as arginine, that can give rise to glutamate.

3. The Propionic Acid Fermentation

Propionic (propanoic) acid-producing bacteria are numerous in the digestive tract of ruminants. Within the rumen some bacteria digest cellulose to form glucose, which is then converted to lactate and other products. The propionic acid bacteria can convert either glucose or lactate into propionic and acetic acids which are absorbed into the bloodstream of the host. Usually some succinic acid is also formed.

The basis of the propionic acid fermentation is conversion of pyruvate to oxaloacetate by carboxylation and the further conversion through succinate and succinyl-CoA to methylmalonyl-CoA and propionyl-CoA, reactions which are almost the exact reverse of those for the oxidation of propionate in the animal body (Fig. 17-3, pathway *d*). However, whereas the carboxylation of pyruvate to oxaloacetate in the animal body requires ATP, the propionic acid bacteria save one equivalent of ATP by using a carboxyltransferase (p. 725). This enzyme donates a carboxyl group from a preformed carboxybiotin compound generated in the decarboxylation of methylmalonyl-CoA in the next to final step of the reaction sequence (Fig. 17-10). A second molecule of ATP is saved by linking directly the conversion of succinate to succinyl-CoA to the cleavage of propionyl-CoA to propionate through the use of a CoA transferase (Eq. 12-50). To provide for oxidation–reduction balance, two-thirds of the glucose goes to propionate and one-third to acetate (Eq. 17-27):

$$1\tfrac{1}{2}\,\text{Glucose} \rightarrow 2\ \text{propionate}^- + \text{acetate}^- + 3\ H^+ + CO_2 + H_2O$$
$$\Delta G'\ (\text{pH 7}) = -465\ \text{kJ per } 1\tfrac{1}{2}\ \text{mol of glucose} \quad (17\text{-}27)$$

More carboxyl groups and CO_2 molecules are formed in this fermentation ($2\frac{2}{3}$ per glucose molecule) than in the regular lactic acid fermentation. The yield of ATP (also $2\ \frac{2}{3}$ mol/mol of glucose fermented) is correspondingly greater and $\Delta G'$ is more negative.

Using the same mechanism (Fig. 17-10), propionic acid bacteria are also able to ferment lactate, the product of fermentation by other bacteria, to propionate and acetate (Eq. 17-28). The net gain is one molecule of ATP. This reaction probably accounts for the niche

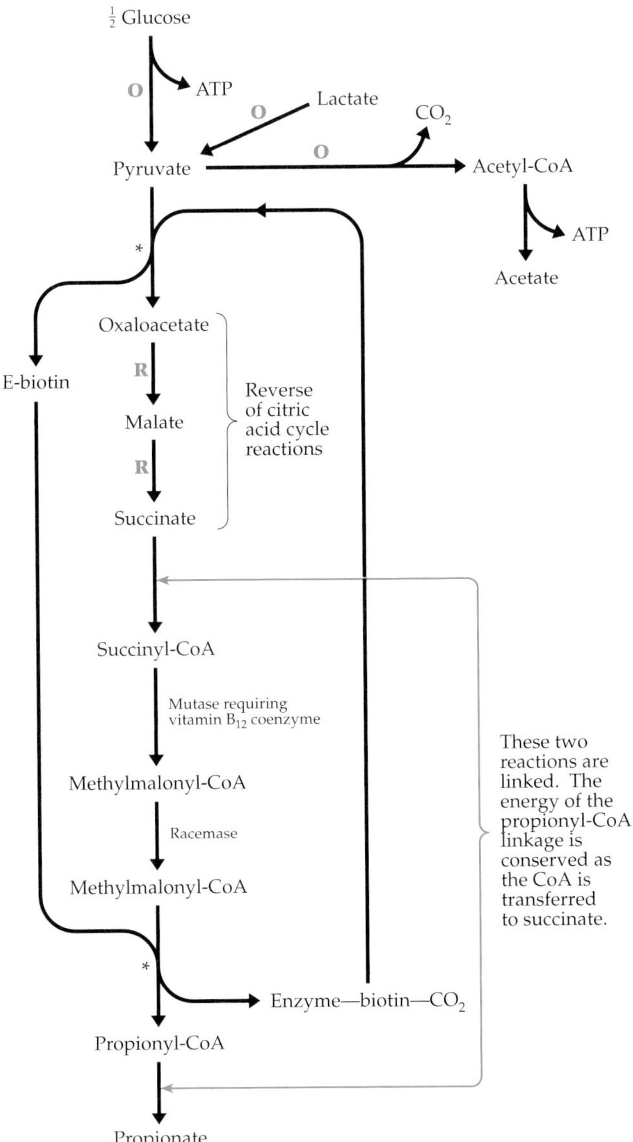

Figure 17-10 Propionic acid fermentation of *Propionobacteria* and *Veillonella.* Oxidation steps are designated by the symbol "O" and reduction steps by "R." The two coupled reactions marked by asterisks are catalyzed by carboxyltransferase.

$$3\ \text{Lactate}^- \rightarrow 2\ \text{propionate}^- + \text{acetate}^- + H_2O + CO_2$$
$$\Delta G'\ (\text{pH 7}) = -171\ \text{kJ per 3 mol of lactate} \qquad (17\text{-}28)$$

in the ecology of the animal rumen that is occupied by propionic acid bacteria.

4. Butyric Acid and Butanol-Forming Fermentations

A variety of fermentations are carried out by bacteria of the genus *Clostridium* and by the rumen organisms *Eubacterium (Butyribacterium)* and *Butyrivibrio.* For example, glucose may be converted to butyric and acetic acids together with CO_2 and H_2 (Eqs. 17-29 and 17-30).

$$2\ \text{Glucose} + 2\ H_2O \rightarrow \text{butyrate}^- + 2\ \text{acetate}^- + 4\ CO_2 + 6\ H_2 + 3\ H^+$$
$$\Delta G'\ (\text{pH 7}) = -479\ \text{kJ per 2 mol of glucose} \qquad (17\text{-}29)$$

The yield of ATP (3½ mol/mol of glucose) is the highest we have discussed giving an efficiency of 50%. Another fermentation yields butanol, isopropanol, ethanol, and acetone.

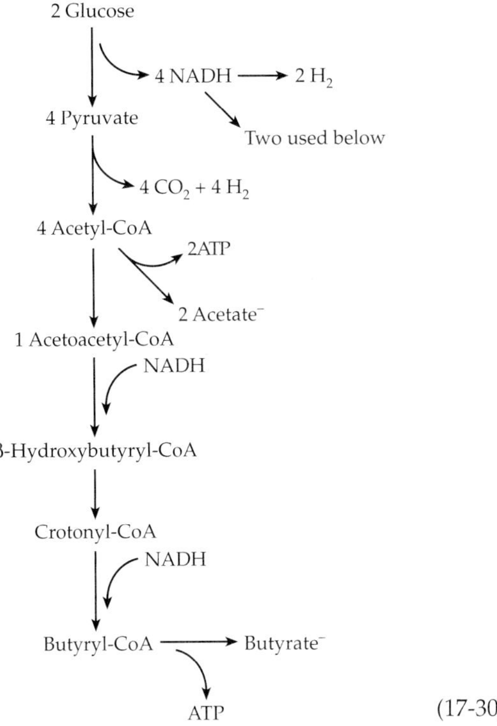

(17-30)

The fermentation of Eq. 17-31 is catalyzed by *Clostridium kluyveri.* The value of $-\Delta G'$ is one of the lowest that we have considered but is still enough to provide easily for the synthesis of one molecule of ATP.

$$2\ CH_3CH{=}CH{-}COO^- + 2\ H_2O \rightarrow \text{butyrate}^- + 2\ \text{acetate}^- + H^+$$
Crotonate
$$\Delta G'\ (\text{pH 7}) = -105\ \text{kJ mol}^{-1} \qquad (17\text{-}31)$$

The energy of the butyryl-CoA linkage and of one of the acetyl-CoA linkages is conserved and utilized in the initial formation of crotonyl-CoA (Eq. 17-32). That leaves one acetyl-CoA which can be converted via acetyl-*P* to acetate with formation of ATP.

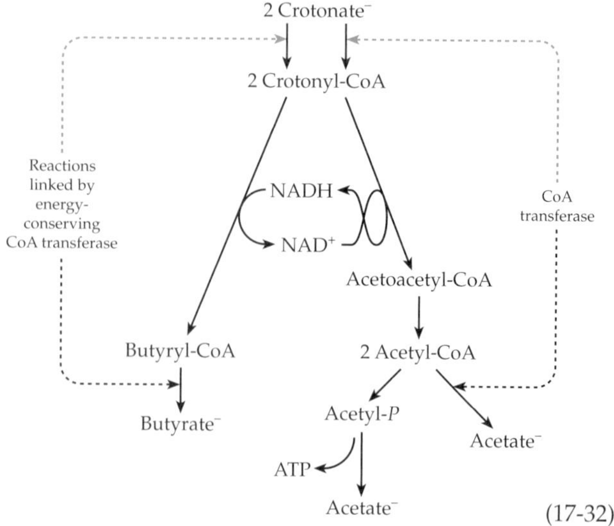

(17-32)

5. Fermentations Based on the Phosphogluconate and Pentose Phosphate Pathways

Some lactic acid bacteria of the genus *Lactobacillus,* as well as *Leuconostoc mesenteroides* and *Zymomonas mobilis,* carry out the **heterolactic** fermentation (Eq. 17-33) which is based on the reactions of the pentose phosphate pathway. These organisms lack aldolase, the key enzyme necessary for cleavage of fructose 1,6-bisphosphate to the triose phosphates. Glucose is converted to ribulose 5-*P* using the oxidative reactions of the pentose phosphate pathway. The ribulose-phosphate is cleaved by phosphoketolase (Eq. 14-23) to acetyl-phosphate and glyceraldehyde 3-phosphate, which are converted to ethanol and lactate, respectively. The overall yield is only one ATP per glucose fermented.

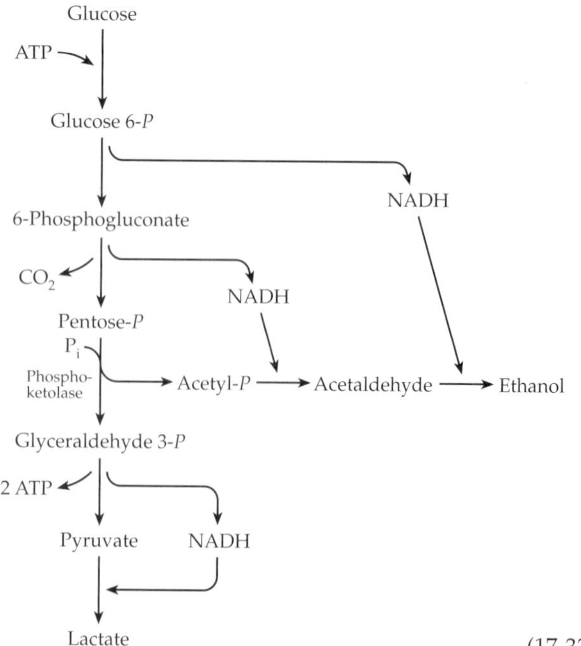

(17-33)

This is generated in the substrate level oxidative phosphorylation catalyzed by phosphoketolase. Metabolic engineering of *Zymomonas* was accomplished by transferring from other bacteria two operons that provide for assimilation of xylose and a complete set of enzymes for the pentose phosphate pathway. The engineered bacteria are able to convert pentose phosphates nonoxidatively (see Fig. 17-8) into glyceraldehyde 3-phosphate, which is converted to ethanol in high yield and with much greater synthesis of ATP than according to Eq. 17-33.[148]

A variation of the heterolactic fermentation is used by *Bifidobacterium* (Eq. 17-34).[149] Phosphoketolase and a **phosphohexoketolase**, which cleaves fructose 6-*P* to erythrose 4-*P* and acetyl-*P*, are required, as are the enzymes of the sugar rearrangement system (Section E,3). The net yield of ATP is 2 ½ molecules per molecule of glucose.

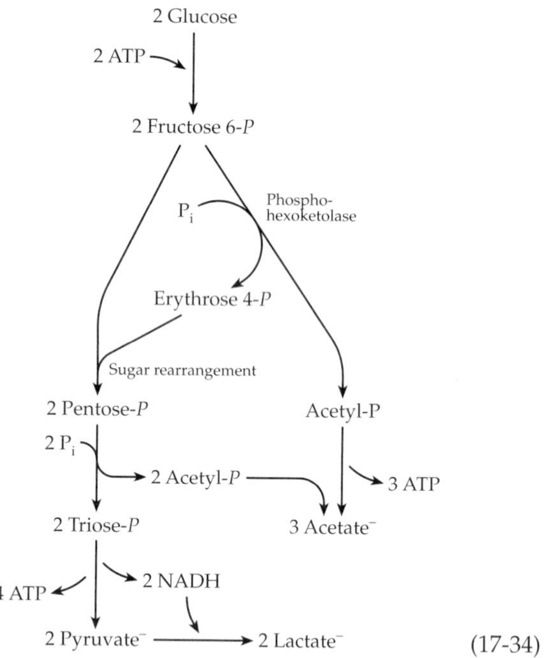

(17-34)

G. Biosynthesis

In this section and sections H – K the general principles and strategy of synthesis of the many carbon compounds found in living things will be considered. Since green plants and autotrophic bacteria are able to assemble all of their needed carbon compounds from CO_2, let us first examine the mechanisms by which this is accomplished. We will also need to ask how some organisms are able to subsist on such simple compounds as methane, formate, or acetate.

1. Metabolic Loops and Biosynthetic Families

As was pointed out in Chapter 10, routes of biosynthesis (anabolism) often closely parallel pathways of biodegradation (catabolism). Thus, catabolism begins with hydrolytic breakdown of polymeric molecules; the resulting monomers are then cleaved into small two- and three-carbon fragments. Biosynthesis begins with formation of monomeric units from small pieces followed by assembly of the monomers into polymers. The mechanisms of the individual reactions of biosynthesis and biodegradation are also often closely parallel. However, in most instances, there are clear-cut differences. A first principle of biosynthesis is that *biosynthetic pathways, although related to catabolic pathways, differ from them in distinct ways and are often catalyzed by completely different sets of enzymes.*

The sum of the pathways of biosynthesis and biodegradation form a continuous loop – a series of reactions that take place concurrently and often within the same part of a cell. Metabolic loops often begin in the central pathways of carbohydrate metabolism with three- or four-carbon compounds such as phosphoglycerate, pyruvate, and oxaloacetate. After loss of some atoms as CO_2 the remainder of the compound rejoins the "mainstream" of metabolism by entering a catabolic pathway leading to acetyl-CoA and oxidation in the citric acid cycle. Not all of the loops are closed within a given species. For example, *human beings are unable to synthesize the vitamins and the "essential amino acids."* We depend upon other organisms to make these compounds, but we do degrade them. Some metabolites, such as uric acid, are excreted by humans and are further catabolized by bacteria. From a chemical viewpoint the whole of nature can be regarded as an enormously complex set of branching and interconnecting metabolic cycles. Thus, the synthetic pathways used by autotrophs are all parts of metabolic loops terminating in oxidation back to CO_2.

It is often not possible to state at what point in a metabolic loop biosynthesis has been completed and biodegradation begins. An end product X that serves one need of a cell may be a precursor to another cell component Y which is then degraded to complete the loop. The reactions that convert X to Y can be regarded as either biosynthetic (for Y) or catabolic (for X).

2. Key Intermediates and Biosynthetic Families

In examining routes of biosynthesis it is helpful to identify some key intermediates. One of these is **3-phosphoglycerate**. This compound is a primary product of photosynthesis and may reasonably be regarded as the starting material from which all other carbon compounds in nature are formed. Phosphoglycerate, in most organisms, is readily interconvertible with both **glucose** and **phosphoenolpyruvate (PEP)**. Any of these three compounds can serve as the precursor for synthesis of other organic materials. A first stage in biosynthesis consists of those reactions by which 3-phosphoglycerate or PEP arise, whether it be from CO_2, formate, acetate, lipids, or polysaccharides.

The further biosynthetic pathways from 3-phosphoglycerate to the myriad amino acids, nucleotides, lipids, and miscellaneous compounds found in cells are complex and numerous. However, the basic features are relatively simple. Figure 17-11 indicates the origins of many substances including the 20 amino acids present in proteins, nucleotides, and lipids. Among the additional key biosynthetic precursors that can be identified from this chart are **glucose 6-phosphate**, **pyruvate**, **oxaloacetate**, **acetyl-CoA**, **2-oxoglutarate**, and **succinyl-CoA**.

The amino acid **serine** originates almost directly from 3-phosphoglycerate. **Aspartate** arises from oxaloacetate and **glutamate** from 2-oxoglutarate. These three amino acids each are converted to "families" of other compounds.[150] A little attention paid to establishing correct family relationships will make the study of biochemistry easier. Besides the serine, aspartate, and glutamate–oxoglutarate families, a fourth large family originates directly from pyruvate and a fifth (mostly lipids) from acetyl-CoA. The aromatic amino acids are formed from erythrose 4-*P* and PEP via the key intermediate **chorismic acid** (Box 9-E; Fig. 25-1). Other families of compounds arise from glucose 6-*P* and from the **pentose phosphates**. These groups have been set off roughly by the boxes outlined in green in Fig. 17-11.

H. Harnessing the Energy of ATP for Biosynthesis

In the past it seemed reasonable to think that some biosynthetic pathways involved exact reversal of catabolic pathways. For example, it was observed that glycogen phosphorylase catalyzed elongation of glycogen branches by transfer of glycosyl groups from glucose 1-phosphate. Likewise, the enzymes needed for the β oxidation of fatty acid derivatives, when isolated from mitochondria, catalyze formation of fatty acyl-CoA derivatives from acetyl-CoA and a reducing agent such as NADH. However, reactant concentrations within cells are rarely appropriate for reversal of a catabolic sequence. For catabolic sequences the Gibbs energy change is usually distinctly negative and reversal requires high concentrations of end products. However, the latter are often removed promptly from the cells. For example, NADH produced in degradation of fatty acids is oxidized to NAD^+ and is therefore never available in sufficient concentrations to reverse the β oxidation sequence.

Nature's answer to the problem of reversing a catabolic pathway lies in the coupling of cleavage of ATP to the biosynthetic reaction. The concept was introduced in Chapter 10, in which one sequence for linking hydrolysis of ATP to biosynthesis was discussed. However, living cells employ several different methods of harnessing the Gibbs energy of hydrolysis of ATP to drive biosynthetic processes. Many otherwise strange aspects of metabolism become clear if it is recognized that they provide a means for coupling ATP cleavage to biosynthesis. A few of the most important of these coupling mechanisms are summarized in this section.

1. Group Activation

Consider the formation of an ester (or of an amide) from a free carboxylic acid and an alcohol (or amine) by elimination of a molecule of water (Eq. 17-35). The reaction is thermodynamically unfavorable with values of $\Delta G'$ (pH 7) ranging from ~ +10 to 30 kJ mol^{-1} depending on conditions and structures of the specific compounds. Long ago, organic chemists learned that such reactions can be made to proceed by careful removal of the water that is generated (Eq. 17-35).

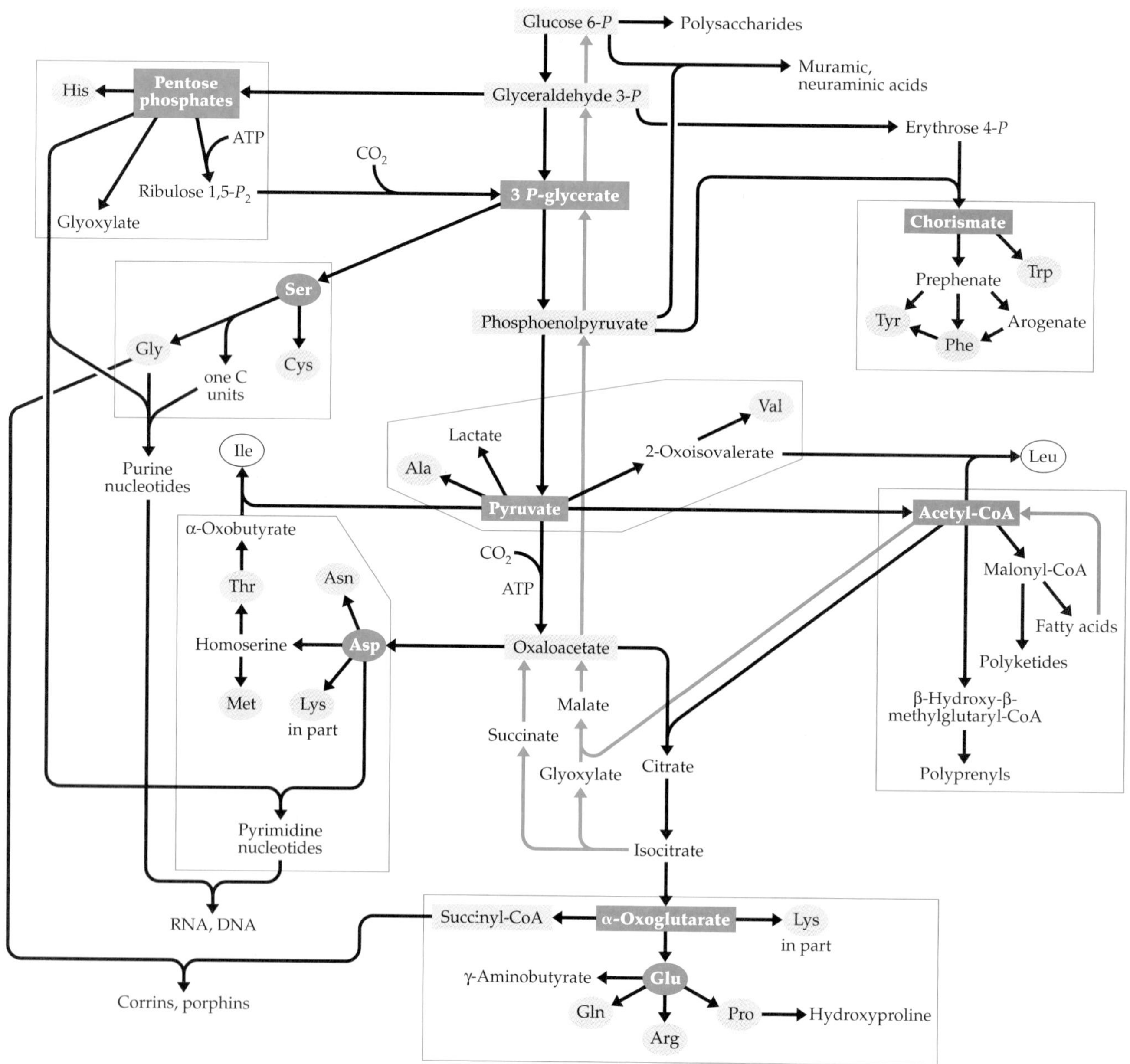

Figure 17-11 Some major biosynthetic pathways. Some key intermediates are enclosed in boxes and the 20 common amino acids of proteins are encircled. Key intermediates for each family are in shaded boxes or elipses. Green lines trace the reactions of the glyoxylate pathway and of glucogenesis.

$$R{-}C({=}O){-}\mathbf{OH} \;\xrightarrow[\;-\,H_2O\;]{\mathbf{H}{-}O{-}R' \text{ or } \mathbf{H}{-}NH{-}R'}\; R{-}C({=}O){-}O{-}R' \text{ or } R{-}C({=}O){-}NH{-}R' \qquad (17\text{-}35)$$

However, it is often better to "activate" the carboxylic acid by conversion to an acyl chloride or an anhydride:

$R{-}C({=}O){-}Cl$ Acyl chloride $\qquad R{-}C({=}O){-}O{-}C({=}O){-}R$ Acid anhydride

Nucleophilic attack on the carbonyl group of such a compound results in displacement of a good leaving group, Cl^- or R–COO⁻. Nature has followed the same approach in forming from carboxylic acids **acyl phosphates** or **acyl-CoA** derivatives.

The virtue of these "activated" acyl compounds in biosynthetic reactions was considered in Chapter 12 and Table 15-1. Just as a carboxylic acid can be converted to an active acyl derivative, so other groups can be activated. ATP and other compounds with phospho groups of high group transfer potential are **active phospho** compounds. Sulfate is converted to a phosphosulfate anhydride, an **active sulfo** derivative. Sugars are converted to compounds such as glucose 1-*P* or sucrose, which contain **active glycosyl** groups. The group transfer potentials of the latter, though not as great as that of the phospho groups of ATP, are still high enough to make glucose 1-phosphate and sucrose effective glycosylating reagents. Table 17-2 lists several of the more important activated groups.

Group activation usually takes place at the expense of ATP cleavage.

TABLE 17-2
"Activated" Groups Used in Biosynthesis

Group	Typical activated forms
$-P(=O)(O^-){-}O^-$ **Phospho**	$R{-}O{-}P(=O)(O^-){-}O{-}P(=O)(O^-){-}O^-$ Pyrophosphate; $R{-}NH{-}C(=NH_2^+){-}NH{-}P(=O)(O^-){-}O^-$ Guanidine phosphate; $R{-}C(=CH_2){-}O{-}P(=O)(O^-){-}O^-$ Enol phosphate; $R{-}C(=O){-}O{-}P(=O)(O^-){-}O^-$ Acyl phosphate
$-S(=O)_2{-}O^-$ **Sulfo**	$R{-}O{-}P(=O)(O^-){-}O{-}S(=O)_2{-}O^-$ in PAPS (3′-Phosphoadenosine-5′-phosphosulfate)
$-C(=O){-}R$ **Acyl**	$R{-}C(=O){-}S{-}R'$ Thioester; $R{-}C(=O){-}O{-}P(=O)(O^-){-}O{-}R'$ Acyl phosphate
Glycosyl	Glycosyl–$O{-}PO_3^{2-}$ Glycosyl phosphate; Glucosyl–O–Fructosyl Sucrose (HOCH₂, HO, OH, HOCH₂, HO, CH₂OH, OH)
$R{-}C(=CH_2){-}$ **Enoyl**	$R{-}C(=CH_2){-}O{-}P(=O)(O^-){-}O^-$ Enol phosphate
$H_2N{-}C(=O){-}$ **Carbamoyl**	$H_2N{-}C(=O){-}O{-}P(O)(O^-){-}O^-$ Carbamoyl phosphate
$R{-}$ **Alkyl**	$H_3C{-}S^+(R){-}R'$ *S*-Adenosylmethionine

As pointed out in Chapter 12, acyl phosphates play a central role in metabolism by virtue of the fact that they contain both an activated acyl group and an activated phospho group. The high group transfer potential can be conserved in subsequent reactions *in either one group or the other* (but not in both). Thus, displacement on P by an oxygen of ADP will regenerate ATP and attack on C by an – SH will give a thioester.

$$R-\overset{O}{\overset{\|}{C}}-O-\overset{O}{\overset{\|}{\underset{O^-}{\underset{|}{P}}}}-O^-$$

Acyl phosphate

Activated acyl group — Activated phospho group

Several of the other compounds in Table 17-2 can also be split in two ways to yield different activated groups, e.g., the phosphosulfate anhydride, enoyl phosphate, and carbamoyl phosphate. It is probably only through intermediates of this type that cleavage of ATP can be coupled to synthesis of activated groups. Such **common intermediates** are essential to the synthesis of ATP by substrate-level phosphorylation (Fig. 15-6).

2. Hydrolysis of Pyrophosphate

The splitting of inorganic pyrophosphate (PP_i) into two inorganic phosphate ions is catalyzed by **pyrophosphatases** (p. 636)[150a,b] that apparently occur universally. Their function appears to be simply to remove the product PP_i from reactions that produce it, shifting the equilibrium toward formation of a desired compound. An example is the formation of **aminoacyl-tRNA** molecules needed for protein synthesis. As shown in Eq. 17-36, the process requires the use of two ATP molecules to activate one amino acid. While the "spending" of two ATPs for the addition of one monomer unit to a polymer does not appear necessary from a thermodynamic viewpoint, it is frequently observed, and there is no doubt that hydrolysis of PP_i ensures that the reaction will go virtually to completion. Transfer RNAs tend to become saturated with amino acids according to Eq. 17-36 even if the concentration of free amino acid in the cytoplasm is low. On the other hand, kinetic considerations may be involved. Perhaps the biosynthetic sequence would move too slowly if it were not for the extra boost given by the removal of PP_i. Part of the explanation for the complexity may depend on control mechanisms which are only incompletely understood.

In some metabolic reactions pyrophosphate esters are formed by consecutive transfer of the terminal phospho groups of two ATP molecules onto a hydroxyl

Amino acid ($^-O-C(=O)-CH(R)-NH_3^+$)

ATP → PP_i (Pyrophosphatase: PP_i + H_2O → 2 P_i)

Aminoacyl adenylate (activated amino acid): Adenosine — O — P(=O) — O — C(=O) — CH(R) — NH_3^+

tRNA (with free —OH) → AMP (Adenylate kinase: AMP + ATP → 2 ADP)

tRNA — O — C(=O) — CH(R) — NH_3^+

Activated amino acid in final form ready for protein synthesis

$$\text{Net: 2 ATP + amino acid + tRNA} \longrightarrow \text{aminoacyl tRNA + 2 ADP + 2 } P_i \qquad (17\text{-}36)$$

group. Such esters often react with elimination of PP_i, e.g., in polymerization of prenyl units (reaction type 6B, Table 10-1; Fig. 22-1). Again, hydrolysis to P_i follows. Thus, *cleavage of pyrophosphate is a second very general method for coupling ATP cleavage to synthetic reactions.*

Although pyrophosphatases are ubiquitous, there are organisms in which PP_i is conserved by the cell and replaces ATP in several glycolytic reactions. These include *Propionibacterium*,[151,152] sulfate-reducing bacteria,[153] the photosynthetic *Rhodospirillum*, and the parasitic *Entamoeba histolytica*.[152,154] In the latter the internal concentration of PP_i is about 0.2 mM. Green plants also accumulate PP_i at concentrations of up to 0.2 mM.[155] Apparently, pyrophosphate is not always hydrolyzed immediately. Another mystery of metabolism is the accumulation of inorganic **polyphosphate** in chains of tens to many hundreds of phospho groups linked, as in pyrophosphate, by phosphoanhydride bonds. These polyphosphates are present in many bacteria, including *E. coli*, and also in fungi, plants, and animals.[156–156b] They constitute a store of energy as well as of phosphate. Various other functions have also been proposed. A polyphosphate kinase transfers a terminal phospho group from polyphosphate chains onto ADP to form ATP. This source of metabolic energy is evidently essential to the ability of *Pseudomonas aeruginosa* to form biofilms.[156a] Both endophosphatases and exophosphatases, of uncertain function, can degrade the chains hydrolytically. An exophosphatase from *E. coli* can completely hydrolyze polyphosphate chains of 1000 units processively without release of intermediates.[156b]

Methionine

ATP → $P_i + PP_i$

S configuration around sulfur

Adenine

S-Adenosylmethionine (AdoMet or SAM - older) See also p. 581

(17-37)

ATP^{3-} + SO_4^{2-} — $\Delta G^0 = +45$ kJ mol^{-1} — *a* ATP sulfurylase

Adenylyl sulfate + PP_i^{3-}

H_2O — *b* — $\Delta G'$ (pH 7) = −33 kJ mol^{-1} → $2P_i + H^+$

c ATP → ADP

3′-Phosphoadenosine-5′-Phosphosulfate (PAPS)

Acceptor — *d* → Acceptor-sulfate + Adenosine 3′,5′-bisphosphate

H_2O — *e* (Phosphatase) → AMP + P_i

(17-38)

In a few instances group activation is coupled to cleavage of ATP at C-5' presumably with formation of bound tripolyphosphate (PPP_i). The latter is hydrolyzed to P_i and PP_i and ultimately to *three* molecules of P_i. An example is the formation of *S*-adenosylmethionine[157] shown in Eq. 17-37. The reaction is a displacement on the 5'-methylene group of ATP by the sulfur atom of methionine. While the initial product may be enzyme-bound PPP_i, it is P_i and PP_i that are released from the enzyme, the P_i arising from the terminal phosphorus (P_γ) of ATP.[157] The *S*-adenosylmethionine formed has the *S* configuration around the sulfur.[158]

3. Coupling by Phosphorylation and Subsequent Cleavage by a Phosphatase

A third general method for coupling the hydrolysis of ATP to drive a synthetic sequence is to transfer the terminal phospho group from ATP to a hydroxyl group *somewhere* on a substrate. Then, after the substrate has undergone a synthetic reaction, the phosphate is removed by action of a phosphatase. For example, in the activation of sulfate (Eq. 17-38),[159] the overall standard Gibbs energy change for steps *a* (catalyzed by **ATP sulfurylase**[160,161]) and *b* is distinctly positive (+12 kJ mol^{-1}). The equilibrium concentration of adenylyl sulfate formed in this group activation process is extremely low. Nature's solution to this problem is to spend another molecule of ATP to phosphorylate the 3' – OH of adenosine phosphosulfate. As the latter is formed, it is converted to 3'-phosphoadenosine-5'-phosphosulfate (Eq. 17-38, step *c*) by a kinase, which is often part of a bifunctional enzyme that also contains the active site of ATP sulfurylase.[162–163a] Since the equilibrium in this step lies far toward the right, the product accumulates in a substantial concentration (up to 1 mM in cell-free systems) and serves as the active sulfo group donor in formation of sulfate esters. The reaction cycle is completed by two more reactions. In Eq. 17-38, step *d*, the sulfo group is transferred to an acceptor, and in step *e* the extra phosphate group is removed from adenosine 3',5'-bisphosphate by a specific phosphatase. Since the reconversion of AMP to ADP requires expenditure of still a third high-energy linkage of ATP, the overall process makes use of three high-energy phosphate linkages for formation of one sulfate ester.

An analogous use of ATP is found in photosynthetic reduction of carbon dioxide in which ATP phosphorylates ribulose 5-*P* to ribulose bisphosphate and the phosphate groups are removed later by phosphatase action on fructose bisphosphate and sedoheptulose bisphosphate (Section J,2). Phosphatases involved in synthetic pathways usually have a high substrate specificity and are to be distinguished from nonspecific phosphatases which are essentially digestive enzymes (Chapter 12).

4. Carboxylation and Decarboxylation: Synthesis of Fatty Acids

A fourth way in which cleavage of ATP can be coupled to biosynthesis was recognized in about 1958 when Wakil and coworkers discovered that synthesis of fatty acids in animal cytoplasm is stimulated by carbon dioxide. However, when $^{14}CO_2$ was used in the experiment no radioactivity appeared in the fatty acids formed. Rather, it was found that acetyl-CoA was carboxylated to **malonyl-CoA** in an ATP- and biotin-requiring process (Eq. 17-39; see also Chapter 13). The carboxyl group formed in this reaction is later converted back to CO_2 in a decarboxylation (Fig. 17-12).

Acetyl-CoA

HCO_3^- ATP ADP + P_i

Malonyl-CoA

Activated hydrogens

Group to be removed later by decarboxylation

(17-39)

We know now that in most bacteria and green plants both an acetyl group of acetyl-CoA and a malonyl group of malonyl-CoA are transferred (steps *a* and *d* of Fig. 17-12) to the sulfur atoms of the phosphopantetheine groups of a low-molecular-weight **acyl carrier protein** (ACP; Chapter 14). The malonyl group of the malonyl-ACP is then condensed (step *f* of Fig. 17-12) with an acetyl group, which has been transferred from acetyl-ACP onto a thiol group of the enzyme (E in Eq. 17-40). The enolate anion indicated in this equation is generated by decarboxylation of the malonyl-ACP. It is this decarboxylation that drives the reaction to completion and, in effect, links C–C bond formation to the cleavage of the ATP required for the carboxylation step. A related sequence involving multifunctional proteins is used by animals and fungi[164] (Section J,6).

β-Oxoacyl-ACP

(17-40)

Carboxylation followed by a later decarboxylation is an important pattern in other biosynthetic pathways, too. Sometimes the decarboxylation follows the carboxylation by many steps. For example, pyruvate (or PEP) is converted to uridylic acid (Eq. 17-41; details are shown in Fig. 25-14):

Pyruvate

ATP HCO_3^- ADP + P_i

Oxaloacetate

Transamination

Aspartate

Orotidine 5'-*P*

CO_2

5'-Uridylic acid (UMP)

(17-41)

I. Reducing Agents for Biosynthesis

Still another difference between biosynthesis of fatty acids and oxidation (in mammals) is that the former has an absolute requirement for NADPH (Fig. 17-12) while the latter requires NAD^+ and flavoproteins (Fig. 17-1). This fact, together with many other observations, has led to the generalization that *biosynthetic reduction reactions usually require NADPH rather than NADH.* Many measurements have shown that in the cytosol of eukaryotic cells the ratio [NADPH]/[$NADP^+$] is high, whereas the ratio [NADH]/[NAD^+] is low. Thus, the NAD^+/NADH system is kept highly oxidized, in line with the role of NAD^+ as a principal biochemical oxidant, while the $NADP^+$/NADPH system is kept reduced.

The use of NADPH in step *g* of Fig. 17-12 ensures that significant amounts of the β-oxoacyl-ACP derivative are reduced to the alcohol. Another difference between β oxidation and biosynthesis is that the alcohol formed in this reduction step in the biosynthetic process

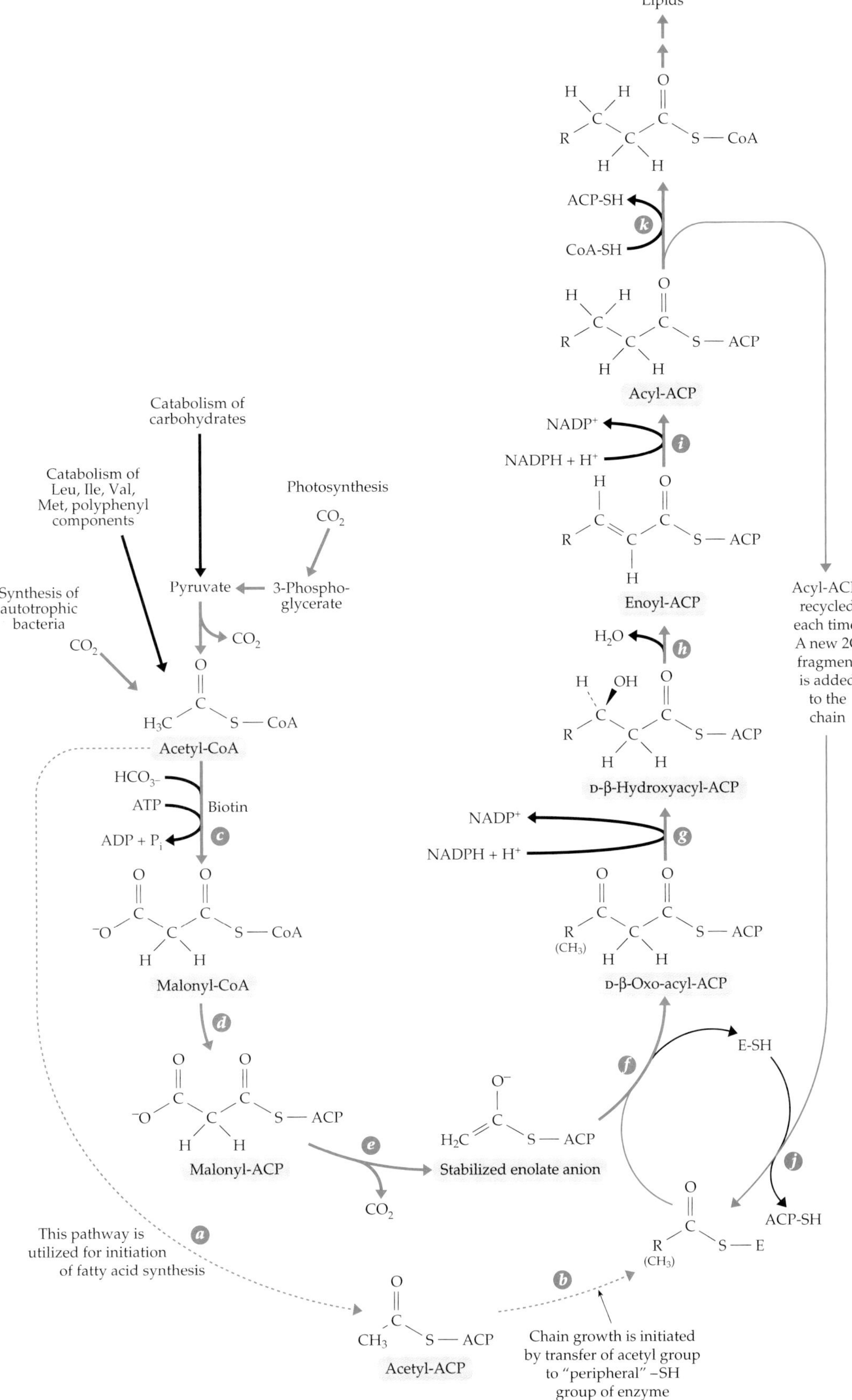

Figure 17-12 The reactions of cytoplasmic biosynthesis of saturated fatty acids. Compare with pathway of β oxidation (Fig. 17-1).

has the D configuration while the corresponding alcohol in β oxidation has the L configuration.

1. Reversing an Oxidative Step with a Strong Reducing Agent

The second reduction step in biosynthesis of fatty acids in the rat liver (step *i*) also required NADPH. The corresponding step in β oxidation utilizes FAD, but NADPH is a stronger reducing agent than $FADH_2$. Therefore, use of a reduced pyridine nucleotide again provides a thermodynamic advantage in pushing the reaction in the biosynthetic direction. Interesting variations have been observed among different species. For example, fatty acid synthesis in the rat requires only NADPH, but the multienzyme complexes from *Mycobacterium phlei*, *Euglena gracilis*, and the yeast *Saccharomyces cerevisiae* all give much better synthesis with a mixture of NADPH and NADH than with NADPH alone.[165] Apparently, NADPH is required in step *g* and NADH in step *i*. This seems reasonable because the equilibrium in step *i* lies far toward the product formation, and NADH at a very low concentration could carry out the reduction.

2. Regulation of the State of Reduction of the NAD and NADP Systems

The ratio $[NAD^+]/[NADH]$ appears to be maintained at a relatively constant value and in equilibrium with a series of different reduced and oxidized substrate pairs. Thus, it was observed that in the cytoplasm of rat liver cells, the dehydrogenations catalyzed by lactate dehydrogenase, *sn*-glycerol 3-phosphate dehydrogenase, and malate dehydrogenase are all at equilibrium with the same ratio of $[NAD^+]/[NADH]$.[166] In one experiment rat livers were removed and frozen in less than 8 s by "freeze-clamping" (Section L,2) and the concentrations of different components of the cytoplasm determined[167]; the ratio $[NAD^+]/[NADH]$ was found to be 634, while the ratio of [lactate]/[pyruvate] was 14.2. From these values an

Glyceraldehyde 3-*P* | Lactate
P_i | NAD^+ | $NADH + H^+$ | *c*
ADP | *a* | ATP
3-Phosphoglycerate —*b* (Slower)→ Pyruvate
Further metabolism

(17-42)

apparent equilibrium constant for reaction *c* of Eq. 17-42 was calculated as $K_c' = 9.0 \times 10^3$. The known equilibrium constant for the reaction (from *in vitro* experiments) is 8.8×10^3 (Eq. 17-43). In a similar way it was shown that several other dehydrogenation reactions are nearly at equilibrium. This conclusion has been confirmed more recently by NMR observations.[168]

$$K_c'\,(\text{pH 7, 38°C}) = \frac{[\text{lactate}]}{[\text{pyruvate}]} \times \frac{[\text{NAD}^+]}{[\text{NADH}]} = 8.8 \times 10^3 \qquad (17\text{-}43)$$

Now consider Eq. 17-42, step *a*, the ADP- and P_i-requiring oxidation of glyceraldehyde 3-phosphate (Fig. 15-6). Experimental measurements indicated that this reaction is also at equilibrium in the cytoplasm. In one series of experiments the measured phosphorylation state ratio $[ATP]/[ADP][P_i]$ was 709, while the ratio [3-phosphoglycerate]/[glyceraldehyde 3-phosphate] was 55.5. The overall equilibrium constant for Eq. 17-42*a* is given by Eq. 17-44. That calculated from known equilibrium constants is 60.

$$K_a'\,(\text{pH 7, 38°C}) = \frac{[\text{ATP}]}{[\text{ADP}][\text{P}_i]} \times \frac{[\text{3-phosphoglycerate}]}{[\text{glyceraldehyde phosphate}]} \times \frac{[\text{NADH}]}{[\text{NAD}^+]} = 709 \times 55.5 \times 1/634 = 62 \qquad (17\text{-}44)$$

From these data Krebs and Veech concluded that the oxidation state of the NAD system is determined largely by the phosphorylation state ratio of the adenylate system.[169] If the ATP level is high the equilibrium in Eq. 17-42*a* will be reached at a higher $[NAD^+]/[NADH]$ ratio and lactate may be oxidized to pyruvate to adjust the [lactate]/[pyruvate] ratio.

It is important not to confuse the reactions of Eq. 17-42 as they occur in an aerobic cell with the tightly coupled pair of redox reactions in the homolactate fermentation (Fig. 10-3; Eq. 17-19). The reactions of steps *a* and *c* of Eq. 17-42 are essentially at equilibrium, but the reaction of step *b* may be relatively slow. Furthermore, pyruvate is utilized in many other metabolic pathways and ATP is hydrolyzed and converted to ADP through innumerable processes taking place within the cell. Reduced NAD does not cycle between the two enzymes in a stoichiometric way and the "reducing equivalents" of NADH formed are, in large measure, transferred to the mitochondria. The proper view of the reactions of Eq. 17-42 is that the redox pairs represent a kind of **redox buffer system** that poises the $NAD^+/NADH$ couple at a ratio appropriate for its metabolic function.

Somewhat surprisingly, within the mitochondria the ratio [NAD^+]/[NADH] is 100 times lower than in the cytoplasm. Even though mitochondria are the site of oxidation of NADH to NAD^+, the intense catabolic activity occurring in the β oxidation pathway and the citric acid cycle ensure extremely rapid production of NADH. Furthermore, the reduction state of NAD is apparently buffered by the low potential of the β-hydroxybutyrate–acetoacetate couple (Chapter 18, Section C,2). Mitochondrial pyridine nucleotides also appear to be at equilibrium with glutamate dehydrogenase.[169]

How is the cytoplasmic [NADPH]/[$NADP^+$] ratio maintained at a value higher than that of [NADH]/[NAD^+]? Part of the answer is from operation of the pentose phosphate pathway (Section E,3). The reactions of Eq. 17-12, if they attained equilibrium, would give a ratio of cytosolic [NADPH]/[$NADP^+$] > 2000 at 0.05 atm CO_2. Compare this with the ratio 1/634 for [NADH/NAD^+] deduced from the observation on the reactions of Eq. 17-42.

Consider also the following **transhydrogenation** reaction (Eq. 17-45):

$$NADH + NADP^+ \rightarrow NAD^+ + NADPH \quad (17\text{-}45)$$

There are soluble enzymes that catalyze this reaction for which *K* equals ~ 1. Within mitochondria an energy-linked system (Chapter 18) involving the membrane shifts the equilibrium to favor NADPH. However, within the cytoplasm, the reaction of Eq. 17-45 is driven by coupling ATP cleavage to the transhydrogenation via carboxylation followed by eventual decarboxylation. One cycle that accomplishes this is given in Eq. 17-46. The first step (step *a*) is ATP-dependent carboxylation of pyruvate to oxaloacetate, a reaction that occurs only within mitochondria (Eq. 14-3). Oxaloacetate can be reduced by malate dehydrogenase using NADH (Eq. 17-46, step *b*), and the resulting malate can be exported from the mitochondria. In the cytoplasm the malate is oxidized to pyruvate, with decarboxylation, by the

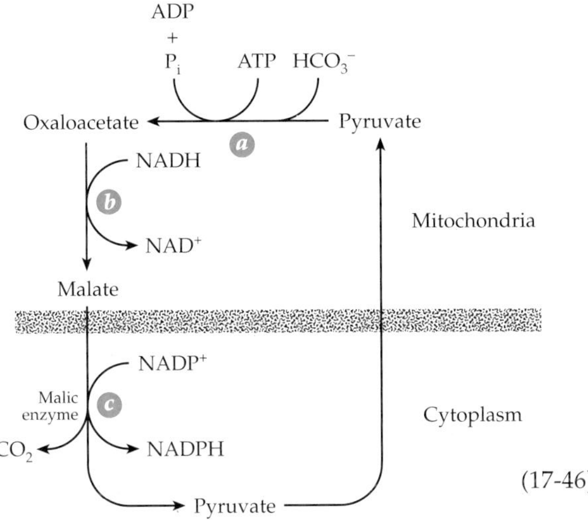

(17-46)

malic enzyme (Eq. 13-45). The malic enzyme (Eq. 17-46, step *c*) is specific for $NADP^+$, is very active, and also appears to operate at or near equilibrium within the cytoplasm. On this basis, using known equilibrium constants, it is easy to show that the ratio [NADPH]/[$NADP^+$] will be ~ 10^5 times higher at equilibrium than the ratio [NADH]/[NAD^+].[169,170]

Since NADPH is continuously used in biosynthetic reactions, and is thereby reconverted to $NADP^+$, the cycle of Eq. 17-46 must operate continuously. As in Eq. 17-42, a true equilibrium does not exist but steps *b* and *c* are both essentially at equilibrium. These equilibria, together with those of Eq. 17-42 for the NAD system, ensure the correct redox potential of both pyridine nucleotide coenzymes in the cytoplasm.

Malate is not the only form in which C_4 compounds are exported from mitochondria. Much oxaloacetate is combined with acetyl-CoA to form citrate; the latter leaves the mitochondria and is cleaved by the ATP-dependent citrate-cleaving enzymes (Eq. 13-39). This, in effect, exports both acetyl-CoA (needed for lipid synthesis) and oxaloacetate which is reduced to malate within the cytoplasm. Alternatively, oxaloacetate may be transaminated to aspartate. The aspartate, after leaving the mitochondria, may be converted in another transamination reaction back to oxaloacetate. All of these are part of the nonequilibrium process by which C_4 compounds diffuse out of the mitochondria before completing the reaction sequence of Eq. 17-46 and entering into other metabolic processes. Note that the reaction of Eq. 17-46 leads to the *export* of reducing equivalents from mitochondria, the opposite of the process catalyzed by the malate–aspartate shuttle which is discussed in Chapter 18 (Fig. 18-18). The two processes are presumably active under different conditions.

While the difference in the redox potential of the two pyridine nucleotide systems is clear-cut in mammalian tissues, in *E. coli* the apparent potentials of the two systems are more nearly the same.[171]

3. Reduced Ferredoxin in Reductive Biosynthesis

Both the NAD^+ and $NADP^+$ systems have standard electrode potentials $E°'$ (pH 7) of −0.32 V. However, because of the differences in concentration ratios, the NAD^+ system operates at a less negative potential (−0.24 V) and the $NADP^+$ system at a more negative potential (−0.38 V) within the cytoplasm of eukaryotes. In green plants and in many bacteria a still more powerful reducing agent is available in the form of reduced ferredoxin. The value of $E°'$ (pH 7) for clostridial ferredoxin is −0.41 V, corresponding to a Gibbs energy change for the two-electron reduction of a substrate ~ 18 kJ mol^{-1} more negative than the corresponding value of $\Delta G'$ for reduction by NADPH. Using reduced ferredoxin (Fd) some photosynthetic bacteria and anaerobic bacteria are able to carry out reductions that are virtually impossible with the pyridine nucleotide system. For example, pyruvate and 2-oxoglutarate can be formed from acetyl-CoA (Eq. 15-35) and succinyl-CoA, respectively (Eq. 17-47).[172–173a] In our bodies the reaction of Eq. 17-47, with NAD^+ as the oxidant, goes only in the opposite direction and is essentially irreversible.

$$\text{Acetyl-CoA} + CO_2 + Fd_{red} \longrightarrow \text{Pyruvate}^- + \text{CoA} + Fd_{ox}$$

$$\Delta G' \text{ (pH 7)} = +17 \text{ kJ mol}^{-1} \qquad (17\text{-}47)$$

J. Constructing the Monomer Units

Now let us consider the synthesis of the monomeric units from which biopolymers are made. How can simple one-carbon compounds such as CO_2 and formic acid be incorporated into complex carbon compounds? How can carbon chains grow in length or be shortened? How are branched chains and rings formed?

1. Carbonyl Groups in Chain Formation and Cleavage

Except for some vitamin B_{12}-dependent reactions, the cleavage or formation of carbon–carbon bonds usually depends upon the participation of carbonyl groups. For this reason, carbonyl groups have a central mechanistic role in biosynthesis. The activation of hydrogen atoms β to carbonyl groups permits β condensations to occur during biosynthesis. Aldol or Claisen condensations require the participation of two carbonyl compounds. Carbonyl compounds are also essential to thiamin diphosphate-dependent condensations and the aldehyde pyridoxal phosphate is needed for most C–C bond cleavage or formation within amino acids.

Because of the importance of carbonyl groups to the mechanism of condensation reactions, much of the assembly of either straight-chain or branched-carbon skeletons takes place between compounds in which the average oxidation state of the carbon atoms is similar to that in carbohydrates (or in formaldehyde, H_2CO). The diversity of chemical reactions possible with compounds at this state of oxidation is a maximum, a fact that may explain why carbohydrates and closely related substances are major biosynthetic precursors and why the average state of oxidation of the carbon in most living things is similar to that in carbohydrates.[174] This fact may also be related to the presumed occurrence of formaldehyde as a principal component of the earth's atmosphere in the past and to the ability of formaldehyde to condense to form carbohydrates.

In Fig. 17-13 several biochemicals have been arranged according to the oxidation state of carbon. Most of the important biosynthetic intermediates lie within ± 2 electrons per carbon atom of the oxidation state of carbohydrates. As the chain length grows, they tend to fall even closer. It is extremely difficult to move through enzymatic processes between 2C, 3C, and 4C compounds (i.e., vertically in Fig. 17-13) except at the oxidation level of carbohydrates or somewhat to its right, at a slightly higher oxidation level. On the other hand, it is often possible to move horizontally with ease using oxidation–reduction reactions. Thus, fatty acids are assembled from acetate units, which lie at the same oxidation state as carbohydrates and, after assembly, are reduced.

Among compounds of the same overall oxidation state, e.g., acetic acid and sugars, the oxidation states of individual carbon atoms can be quite different. Thus, in a sugar every carbon atom can be regarded as immediately derived from formaldehyde, but in acetic acid one end has been oxidized to a carboxyl group and the other has been reduced to a methyl group. Such internal oxidation–reduction reactions play an important role in the chemical manipulations necessary to assemble the carbon skeletons needed by a cell. Decarboxylation is a feature of many biosynthetic routes. Referring again to Fig. 17-13, notice that many of the biosynthetic intermediates such as pyruvate, oxoglutarate, and oxaloacetate are more oxidized than the carbohydrate level. However, their decarboxylation products, which become incorporated into the compounds being synthesized, are closer to the oxidation level of carbohydrates.

2. Starting with CO_2

There are three known pathways by which autotrophic organisms can use CO_2 to synthesize triose phosphates or 3-phosphoglycerate, three-carbon compounds from which all other biochemical substances can be formed.[175–177] The first of these is the **reductive tricarboxylic cycle.** This is a reversal of the oxidative citric acid cycle in which reduced ferredoxin is used as a reductant in the reaction of Eq. 17-47 to incorporate CO_2 into pyruvate. Succinyl-CoA can react with CO_2 in the same type of reaction to form 2-oxoglutarate, accomplishing the reversal of the only irreversible step in the citric acid cycle. Using these reactions photosynthetic bacteria and some anaerobes that can generate a high ratio of reduced to oxidized ferredoxin carry out the reductive tricarboxylic acid cycle. Together with

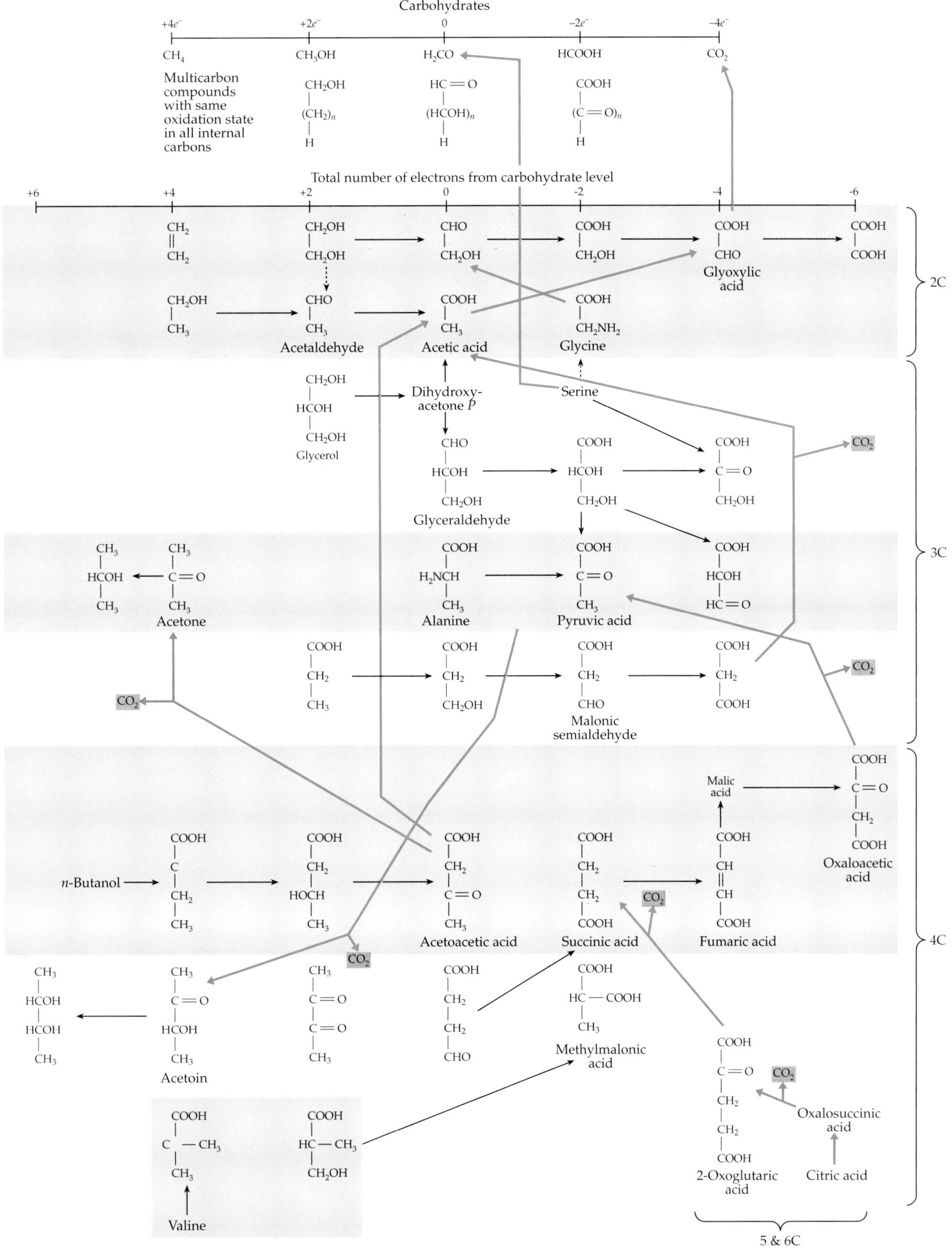

Figure 17-13 Some biochemical compounds arranged in order of average oxidation state of the carbon atoms and by carbon-chain lengths. Black horizontal arrows mark some biological interconversions among compounds with the same chain length, while green lines show changes in chain length and are often accompanied by decarboxylation.

Eq. 17-47, the cycle provides for the complete synthesis of pyruvate from CO_2.[178,179]

A quantitatively much more important pathway of CO_2 fixation is the **reductive pentose phosphate pathway** (ribulose bisphosphate cycle or **Calvin–Benson cycle**; Fig. 17-14). This sequence of reactions, which takes place in the chloroplasts of green plants and also in many chemiautotrophic bacteria, is essentially a way of reversing the oxidative pentose phosphate pathway (Fig. 17-8). The latter accomplishes the complete oxidation of glucose or of glucose 1-phosphate by $NADP^+$ (Eq. 17-48):

$$\text{Glucose 1-}P^{2-} + ATP^{4-} + 8\ H_2O + 12\ NADP^+ \rightarrow$$
$$6\ CO_2 + 12\ NADPH + ADP^{3-} + 2HPO_4^{2-} + 13\ H^+$$
$$\Delta G'\ (\text{pH 7}) = -299\ \text{kJ mol}^{-1} \qquad (17\text{-}48)$$

It would be almost impossible for a green plant to fix CO_2 using photochemically generated NADPH by an exact reversal of Eq. 17-48 because of the high positive Gibbs energy change. To solve this thermodynamic problem the reductive pentose phosphate pathway has been modified in a way that couples ATP cleavage to the synthesis.

The **reductive carboxylation** system is shown within the green shaded box of Fig. 17-14. Ribulose 5-phosphate is the starting compound and in the first step one molecule of ATP is expended to form **ribulose 1,5-bisphosphate**. The latter is carboxylated and cleaved to two molecules of 3-phosphoglycerate. This reaction was discussed in Chapter 13. The reductive step (step *c*) of the system employs NADPH together with ATP. Except for the use of the NADP system instead of the NAD system, it is exactly the reverse of the glyceraldehyde phosphate dehydrogenase reaction of glycolysis. Looking at the first three steps of Fig. 17-14 it is clear that in the reductive pentose phosphate pathway three molecules of ATP are utilized for each CO_2 incorporated. On the other hand, in the oxidative direction *no* ATP is generated by the operation of the pentose phosphate pathway.

The reactions enclosed within the shaded box of Fig. 17-14 do not give the whole story about the coupling mechanism. A phospho group was transferred from ATP in step *a* and to complete the hydrolysis it must be removed in some future step. This is indicated in a general way in Fig. 17-14 by the reaction steps *d*, *e*, and *f*. Step *f* represents the action of specific phosphatases that remove phospho groups from the seven-carbon sedoheptulose bisphosphate and from fructose bisphosphate. In either case the resulting ketose monophosphate reacts with an aldose (via transketolase, step *g*) to regenerate ribulose 5-phosphate, the CO_2 acceptor. The overall reductive pentose phosphate cycle (Fig. 17-14B) is easy to understand as a reversal of the oxidative pentose phosphate pathway in which the oxidative decarboxylation system of Eq. 17-12 is replaced by the reductive carboxylation system of Fig. 17-14A. The scheme as written in Fig. 17-14B shows the incorporation of three molecules of CO_2. The reductive carboxylation system operates three times with a net production of one molecule of triose phosphate. As with other biosynthetic cycles, any amount of any of the intermediate metabolites may be withdrawn into various biosynthetic pathways without disruption of the flow through the cycle.

The overall reaction of carbon dioxide reduction in the Calvin–Benson cycle (Fig. 17-14) becomes

$$6\ CO_2 + 12\ NADPH + 18\ ATP^{4-} + 11\ H_2O \rightarrow$$
$$\text{glucose-1-}P^{2-} + 12\ NADP^+ + 18\ ADP^{3-}$$
$$17\ HPO_4^{2-} + 6\ H^+$$
$$(17\text{-}49)$$

The Gibbs energy change $\Delta G'$ (pH 7) is now -357 kJ mol^{-1} instead of the $+299$ kJ mol^{-1} required to reverse the reaction of Eq. 17-48.

The third pathway for reduction of CO_2 to acetyl-CoA is utilized by acetogenic bacteria, by methanogens, and probably by sulfate-reducing bacteria.[179–181] This **acetyl-CoA pathway** (or **Wood–Ljungdahl pathway**) involves reduction by H_2 of one of the two molecules of CO_2 to the methyl group of methyl-tetrahydromethanopterin in methanogens and of methyltetra-hydrofolate in acetogens. The pathway utilized by methanogens is illustrated in Fig. 15-22.[182–184] A similar process utilizing H_2 as the reductant is employed by acetogens.[179,185–188a] In both cases a methyl corrinoid is formed and its methyl group is condensed with a molecule of carbon monoxide bound to a copper ion in a Ni–Cu cluster.[189a,b] The resulting acetyl group is transferred to a molecule of coenzyme A as illustrated in Eq. 16-52.[189] The bound CO is formed by reduction of CO_2, again using H_2 as the reductant.[190] The overall reaction for acetyl-CoA synthesis is given by Eq. 17-50. Conversion of acetyl-CoA to pyruvate via Eq. 17-47 leads into the glucogenic pathway.

$$CO_2 \xrightarrow[\text{(Fig. 15-22)}]{\text{6-Electron reduction}} CH_3-\text{Pterin} \rightarrow CH_3-\text{Co (Corrin)}$$

$$CO_2 \xrightarrow{2e^-} CO \rightarrow \text{CO–Cu (Ni)}$$

CH₃–Co (Corrin) + CO–Cu, with CoA-SH (Eq. 16-52), Carbon monoxide dehydrogenase:

$$\rightarrow CH_3-\overset{\displaystyle O}{\overset{\|}{C}}-S-\text{CoA}$$

Overall:

$$2\ CO_2 + 4\ H_2 + \text{CoASH} \longrightarrow CH_3-\overset{\displaystyle O}{\overset{\|}{C}}-S-\text{CoA} + 3\ H_2O \qquad (17\text{-}50)$$

An alternative pathway by which some acetogenic bacteria form acetate is via reversal of the glycine decarboxylase reaction of Fig. 15-20. Methylene-THF is formed by reduction of CO_2, and together with NH_3 and CO_2 a lipoamide group of the enzyme and PLP forms glycine. The latter reacts with a second methylene-THF to form serine, which can be deaminated to pyruvate and assimilated. Methanogens may use similar pathways but ones that involve methanopterin (Fig. 15-17).[191]

3. Biosynthesis from Other Single-Carbon Compounds

Various bacteria and fungi are able to subsist on such one-carbon compounds as methane, methanol, methylamine, formaldehyde, and formate.[192–197] Energy is obtained by oxidation to CO_2. **Methylotrophic bacteria** initiate oxidation of methane by hydroxylation (Chapter 18) and dehydrogenate the resulting methanol or exogenous methanol using the PPQ cofactor (Eq. 15-51).[198] Further dehydrogenation to formate and of formate to CO_2 via formate dehydrogenase (Eq. 16-63) completes the process. Some methylotrophic bacteria incorporate CO_2 for biosynthetic purposes via the ribulose bisphosphate (Calvin–Benson) cycle but many use pathways that begin with formaldehyde (or methylene-THF). Others employ variations of the reductive pentose phosphate pathway to convert formaldehyde to triose phosphate. In one of these, the **ribulose monophosphate cycle** or Quayle cycle,[192,193] ribulose 5-*P* undergoes an aldol condensation with formaldehyde to give a 3-oxo-hexulose 6-phosphate (Eq. 17-51, step *a*). The latter is isomerized to fructose 6-*P* (Eq. 17-51, step *b*). If this equation is applied to the

BOX 17-E ^{14}C AND THE CALVIN–BENSON CYCLE

The chemical nature of photosynthesis had intrigued chemists for decades but little was learned about the details until radioactive ^{14}C became available. Discovered in 1940 by Ruben and Kamen, the isotope was available in quantity by 1946 as a product of nuclear reactors. Initial studies of photosynthesis had been conducted by Ruben and Kamen using ^{11}C but ^{14}C made rapid progress possible. In 1946 Melvin Calvin and Andrew A. Benson began their studies that elucidated the mechanism of incorporation of CO_2 into organic materials.

A key development was two-dimensional paper chromatography with radioautography (Box 3-C). A suspension of the alga *Chlorella* (Fig. 1-11) was allowed to photosynthesize in air. At a certain time, a portion of $H^{14}CO_3$ was injected into the system, and after a few seconds of photosynthesis with ^{14}C present the suspension of algae was run into hot methanol to denature proteins and to stop the reaction. The soluble materials extracted from the algal cells were concentrated and chromatographed; radioautographs were then prepared. It was found that after 10 s of photosynthesis in the presence of $^{14}CO_2$, the algae contained a dozen or more ^{14}C labeled compounds. These included malic acid, aspartic acid, phosphoenolpyruvate, alanine, triose phosphates, and other sugar phosphates and diphosphates. However, *during the first five seconds a single compound, 3-phosphoglycerate, contained most of the radioactivity.*[a,b] This finding suggested that a two-carbon regenerating substrate might be carboxylated by $^{14}CO_2$ to phosphoglycerate. Search for this two-carbon compound was unsuccessful, but Benson, in Calvin's laboratory, soon identified ribulose

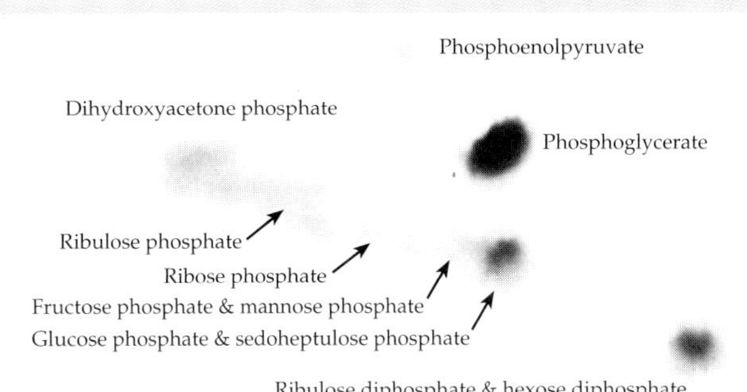

Chromatogram of extract of the alga *Scenedesmus* after photosynthesis in the presence of $^{14}CO_2$ for 10 s. Courtesy of J. A. Bassham. The origin of the chromatogram is at the lower right corner.

bisphosphate,[c] which kinetic studies proved to be the true regenerating substrate.[c,d] Its carboxylation and cleavage[e] represent the first step in what has come to be known as the Calvin–Benson cycle (Fig. 17-14).[f]

---

[a] Benson, A. A., Bassham, J. A., Calvin, M., Goodale, T. C., Haas, V. A., and Stepka, W. (1950) *J. Am. Chem. Soc.* **72**, 1710–1718
[b] Benson, A. A. (1951) *J. Am. Chem. Soc.* **73**, 2971–2972
[c] Benson, A. A., Kawaguchi, S., Hayes, P., and Calvin, M. (1952) *J. Am. Chem. Soc.* **74**, 4477–4482
[d] Bassham, J. A., Benson, A. A., Kay, L. D., Harris, A. Z., Wilson, A. T., and Calvin, M. (1954) *J. Am. Chem. Soc.* **76**, 1760–1770
[e] Calvin, M., and Bassham, J. A. (1962) *The Photosynthesis of Carbon Compounds*, Benjamin, New York
[f] Fuller, R. C. (1999) *Photosynth Res.* **62**, 1–29

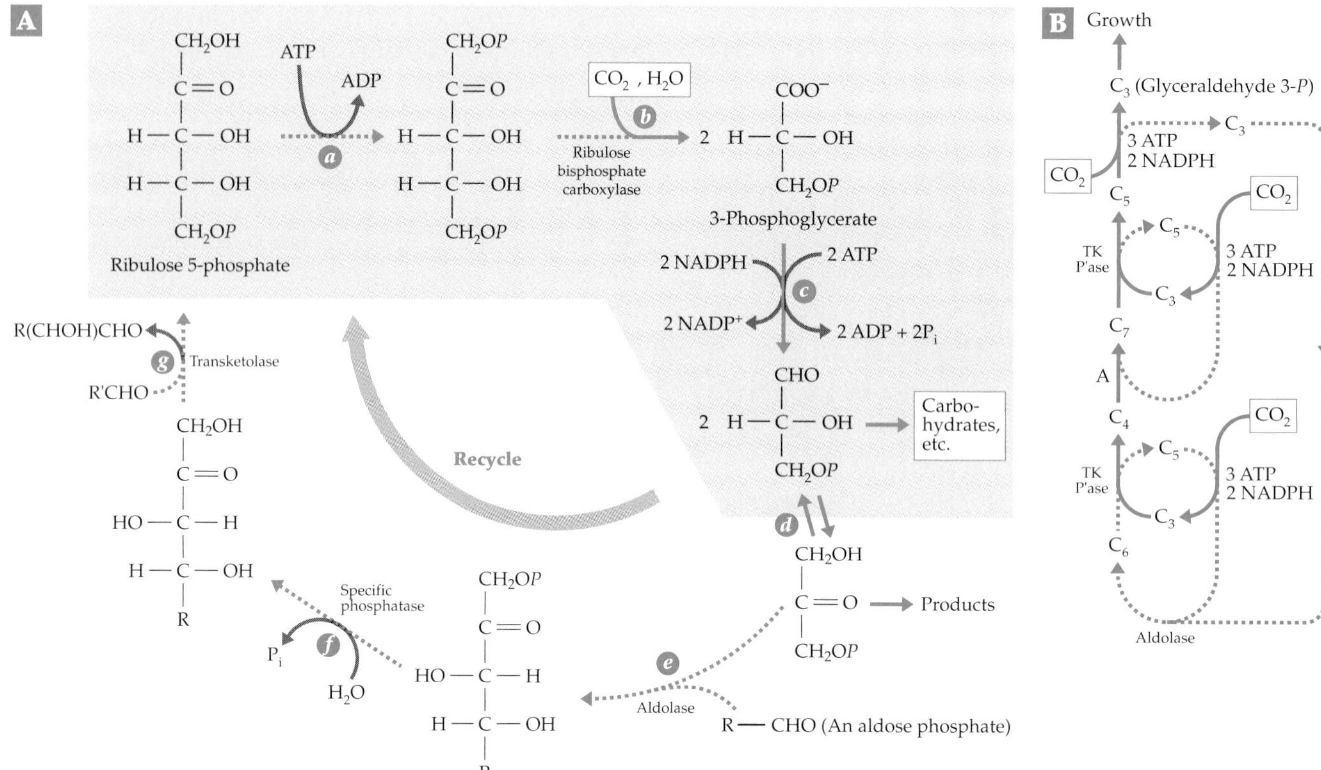

Figure 17-14 (A) The reductive carboxylation system used in reductive pentose phosphate pathway (Calvin–Benson cycle). The essential reactions of this system are enclosed within the dashed box. Typical subsequent reactions follow. The phosphatase action completes the phosphorylation–dephosphorylation cycle. (B) The reductive pentose phosphate cycle arranged to show the combining of three CO_2 molecules to form one molecule of triose phosphate. Abbreviations are RCS, reductive carboxylation system (from above); A, aldolase, Pase, specific phosphatase; and TK, transketolase.

three C_5 sugars three molecules of fructose 6-phosphate will be formed. One of these can be phosphorylated by ATP to fructose 1,6-bisphosphate, which can be cleaved by aldolase. One of the resulting triose phosphates can then be removed for biosynthesis and the second, together with the other two molecules of fructose 6-*P*, can be recycled through the sugar rearrangement sequence of Fig. 17-8B to regenerate the three ribulose 5-*P* molecules that serve as the regenerating substrate.

In bacteria, which lack formate dehydrogenase, formaldehyde can be oxidized to CO_2 to provide energy beginning with the reactions of Eq. 17-51. The resulting fructose 6-*P* is isomerized to glucose 6-*P*, which is then dehydrogenated via Eq. 17-12 to form CO_2 and the regenerating substrate ribulose 5-phosphate.

A number of pseudomonads and other bacteria convert C_1 compounds to acetate via tetrahydrofolic acid-bound intermediates and CO_2 using the **serine pathway**[179,192,193] shown in Fig. 17-15. This is a cyclic process for converting one molecule of formaldehyde (bound to tetrahydrofolate) plus one of CO_2 into acetate. The regenerating substrate is **glyoxylate**. Before condensation with the "active formaldehyde" of methylene THF, the glyoxylate undergoes transamination to glycine (Fig. 17-15, step *a*). The glycine plus formaldehyde forms serine (step *b*), which is then transaminated to hydroxypyruvate, again using step *a*. Glyoxylate plus formaldehyde could have been joined in a thiamin-dependent condensation. However, as in the γ-aminobutyrate shunt (Fig. 17-5), the coupled transamination step of Fig 17-15 permits use of PLP-dependent C–C bond formation.

Conversion of hydroxypyruvate to PEP (Fig. 17-15) involves reduction by NADH and phosphorylation by ATP to form 3-phosphoglycerate, which is converted to PEP as in glycolysis. The conversion of malate to acetate and glyoxylate via malyl-CoA and isocitrate lyase (Eq. 13-40) forms the product acetate and regenerates glyoxylate. As with other metabolic cycles, various intermediates, such as PEP, can be withdrawn for biosynthesis. However, there must be an independent route of synthesis of the regenerating substrate glyoxylate. One way in which this can be accomplished is to form glycine via the reversal of the glycine decarboxylase pathway as is indicated by the shaded green lines in Fig. 17-15.

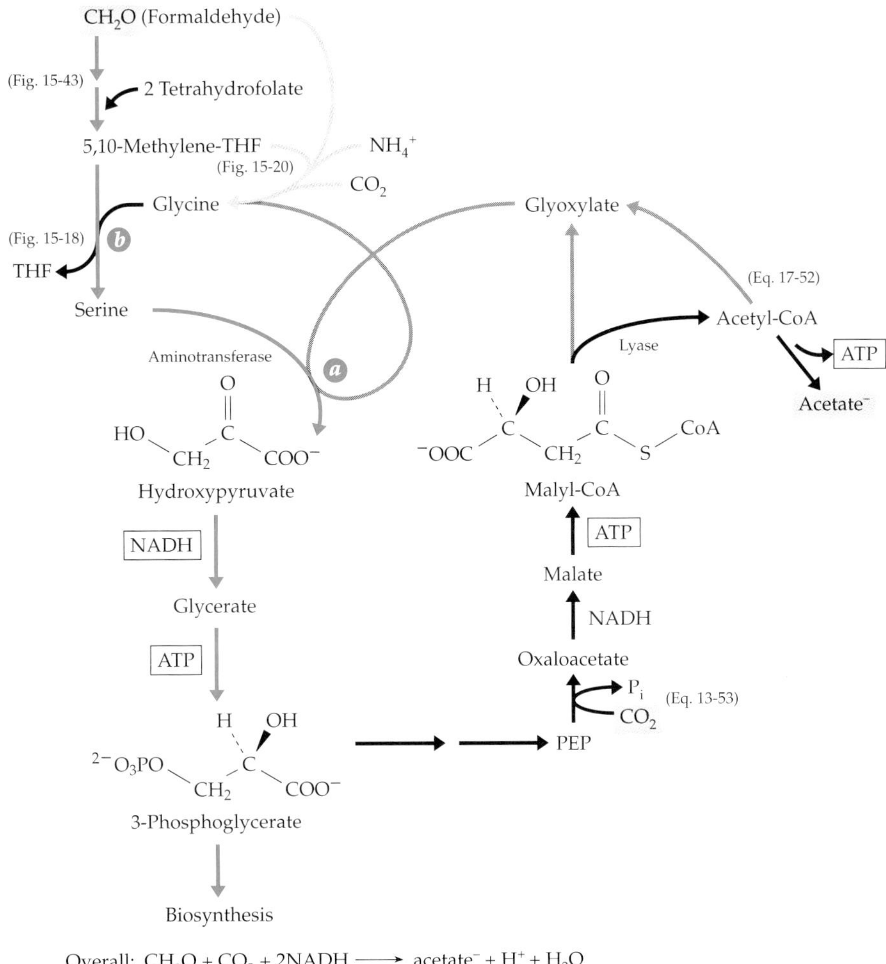

Overall: $CH_2O + CO_2 + 2NADH \longrightarrow acetate^- + H^+ + H_2O$

Figure 17-15 One of the serine pathways for assimilation of one-carbon compounds.

Ribulose 5-*P* $\xrightarrow[a]{H_2C{=}O}$ Hexulose 6-*P* $\xrightarrow[\text{Isomerase}]{b}$ Fructose 6-*P* (17-51)

4. The Glyoxylate Pathways

The reductive carboxylation of acetyl-CoA to pyruvate (Eq. 17-47) occurs only in a few types of bacteria. For most species, from microorganisms to animals, the oxidative decarboxylation of pyruvate to acetyl-CoA is irreversible. This fact has many important consequences. For example, carbohydrate is readily converted to fat; because of the irreversibility of this process, excess calories lead to the deposition of fats. However, in animals fat cannot be used to generate most of the biosynthetic intermediates needed for formation of carbohydrates and proteins because those intermediates originate largely from C_3 units.

This limitation on the conversion of C_2 acetyl units to C_3 metabolites is overcome in many organisms by the **glyoxylate cycle** (Fig. 17-16), which converts *two* acetyl units into one C_4 unit. The cycle provides a way for organisms, such as *E. coli*,[111,199] *Saccharomyces*,[200] *Tetrahymena*, and the nematode *Caenorhabditis*,[201] to subsist on acetate as a sole or major carbon source. It is especially prominent in plants that store large amounts of fat in their seeds (**oil seeds**). In the germinating oil seed the glyoxylate cycle allows fat to be converted rapidly to sucrose, cellulose, and other carbohydrates needed for growth.

A key enzyme in the glyoxylate cycle is **isocitrate lyase**, which cleaves isocitrate (Eq. 13-40) to succinate and glyoxylate. The latter is condensed with a second acetyl group by the action of **malate synthase** (Eq. 13-38). The L-malate formed in this reaction is dehydrogenated to the regenerating substrate oxaloacetate. Some of the reaction steps are those of the citric acid cycle and it appears that in bacteria there is no spatial separation of the citric acid cycle and glyoxylate pathway. However, in plants the enzymes of the glyoxylate cycle are present in specialized peroxisomes known as **glyoxysomes**.[60] The glyoxysomes also contain the enzymes for β oxidation of fatty acids, allowing for efficient conversion of fatty acids to **succinate**. This compound is exported from the glyoxysomes and enters the mitochondria where it undergoes β oxidation to oxaloacetate. The latter can be converted by PEP carboxylase (Eq. 13-53) or by PEP carboxykinase (Eq. 13-46) to PEP.

An **acetyl-CoA-glyoxylate** cycle, which catalyzes oxidation of acetyl groups to glyoxylate, can also be constructed from isocitrate lyase and citric acid cycle

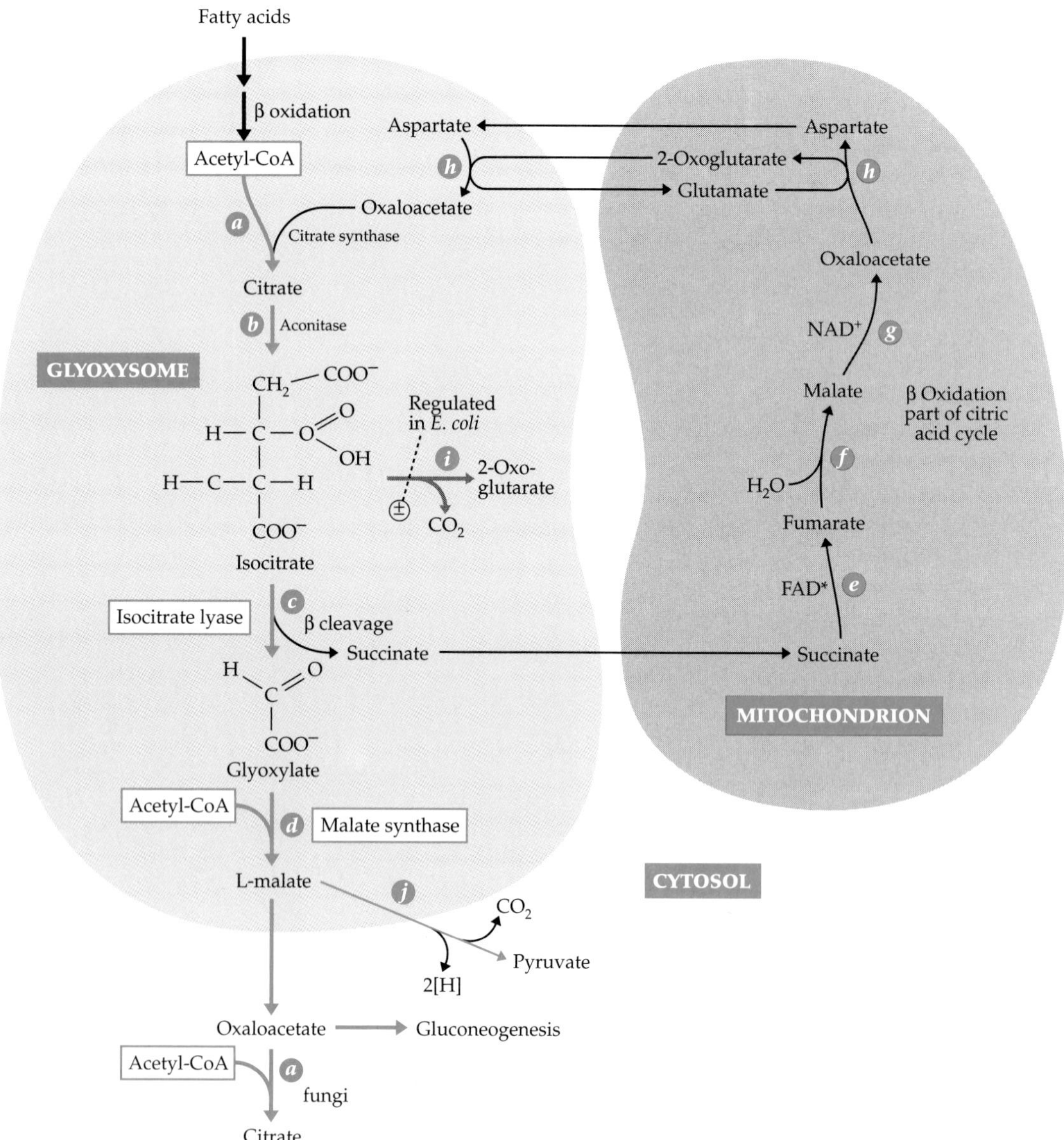

Figure 17-16 The glyoxylate pathway. The green line traces the pathway of labeled carbon from fatty acids or acetyl-CoA into malate and other products.

enzymes. Glyoxylate is taken out of the cycle as the product and succinate is recycled (Eq. 17-52). The independent pathway for synthesis of the regenerating substrate oxaloacetate is condensation of glyoxylate with acetyl-CoA (malate synthetase) to form malate and oxidation of the latter to oxaloacetate as in the main cycle of Fig. 17-16.

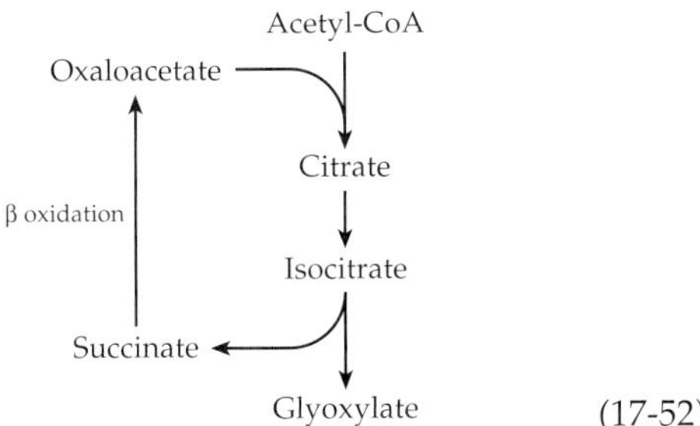

(17-52)

5. Biosynthesis of Glucose from Three-Carbon Compounds

Now let us consider the further conversion of PEP and of the triose phosphates to **glucose 1-phosphate**, the key intermediate in biosynthesis of other sugars and polysaccharides. The conversion of PEP to glucose 1-*P* represents a reversal of part of the glycolysis sequence. It is convenient to discuss this along with **gluconeogenesis**, the reversal of the complete glycolysis sequence from lactic acid. This is an essential part of the Cori cycle (Section F) in our own bodies, and the same process may be used to convert pyruvate derived from deamination of alanine or serine (Chapter 24) into carbohydrates.

Just as with the pentose phosphate cycle, an exact reversal of the glycolysis sequence (Eq. 17-53) is precluded on thermodynamic grounds. Even at very high values of the phosphorylation state ratio R_p, the reaction:

$$2\ \text{Lactate}^- + 3\ \text{ATP}^{4-} + 2\ H_2O \rightarrow C_6H_{10}O_5\ (\text{glycogen}) + 3\ \text{ADP}^{3-} + 3\ HPO_4^{2-} + H^+$$
$$\Delta G'\ (\text{pH 7}) = +\ 107\ \text{kJ per glycosyl unit} \quad (17\text{-}53)$$

would be unlikely to go to completion. The actual pathways used for gluconeogenesis (Fig. 17-17, green lines) differ from those of glycolysis (black lines) in three significant ways. First, while glycogen breakdown is initiated by the reaction with inorganic phosphate catalyzed by phosphorylase (Fig. 17-17, step *a*), the biosynthetic sequence from glucose 1-*P*, via uridine diphosphate glucose (Fig. 17-17, step *b*; see also Eq. 17-56), is coupled to cleavage of ATP. Second, in the catabolic process (glycolysis) fructose 6-*P* is converted to fructose 1,6-P_2 through the action of a kinase (Fig. 17-17, step *c*), which is then cleaved by aldolase. The resulting triose phosphate is degraded to PEP. In glucogenesis a phosphatase is used to form fructose *P* from fructose P_2 (Fig. 17-17, step *d*). Third, during gly-colysis PEP is converted to pyruvate by a kinase with generation of ATP (Fig. 17-17, step *e*). During glucogenesis pyruvate is converted to PEP indirectly via oxaloacetate (Fig. 17-17, steps *f* and *g*) using pyruvate carboxylase (Eq. 14-3) and PEP carboxykinase (Eq. 13-46). This is another example of the coupling of ATP cleavage through a carboxylation–decarboxylation sequence. The net effect is to use two molecules of ATP (actually one ATP and one GTP) rather than *one* to convert pyruvate to PEP.

The overall reaction for reversal of glycolysis to form glycogen (Eq. 17-54) now has a comfortably negative standard Gibbs energy change as a result of coupling the cleavage of 7 ATP to the reaction.

$$2\ \text{Lactate}^- + 7\ \text{ATP}^{4-} + 6\ H_2O \rightarrow \text{glycogen} + 7\ \text{ADP}^{3-} + 7\ HPO_4^{2-} + 5\ H^+$$
$$\Delta G'\ (\text{pH 7}) = -\ 31\ \text{kJ mol}^{-1}\ \text{per glycosyl unit} \quad (17\text{-}54)$$

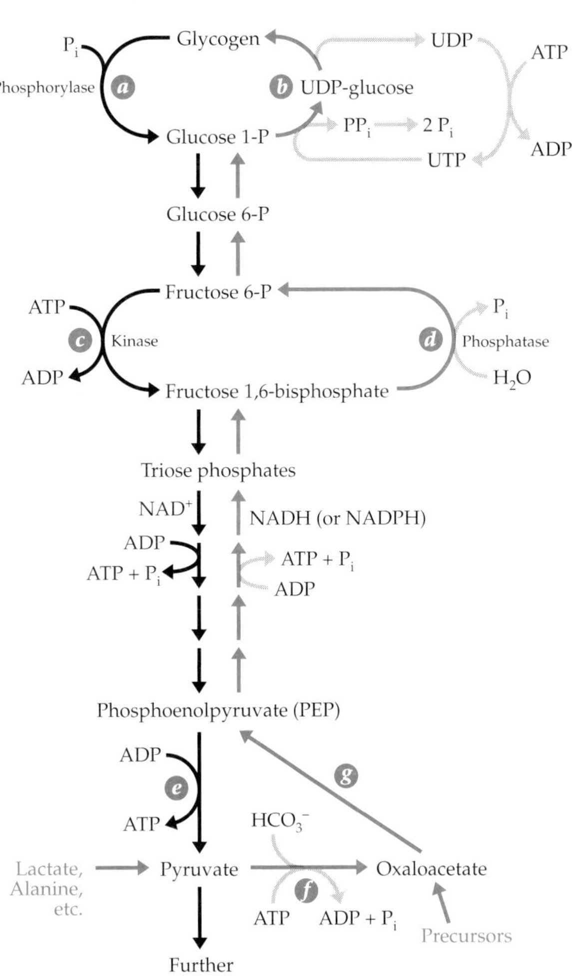

Figure 17-17 Comparison of glycolytic pathway (left) with pathway of gluconeogenesis (right, green arrows).

Two enzymes that are able to convert pyruvate directly to PEP are found in some bacteria and plants. In each case, as in the animal enzyme system discussed in the preceding paragraph, the conversion involves expenditure of two high-energy linkages of ATP. The **PEP synthase** of *E. coli* first transfers a pyrophospho group from ATP onto an imidazole group of histidine in the enzyme (Eq. 17-55). A phospho group is hydrolyzed from this intermediate (dashed green line in Eq. 17-55, step *b*), ensuring that sufficient intermediate E–His–*P* is present. The latter reacts with pyruvate to form PEP.[202,203] **Pyruvate–phosphate dikinase** is a similar enzyme first identified in tropical grasses and known to play an important role in the CO_2 concentrating system of the so-called "C_4 plants" (Chapter 23).[204] The same enzyme participates in gluconeogenesis in *Acetobacter*. The reaction cycle for this enzyme is also portrayed in Eq. 17-55. In this case P_i, rather than water, is the attacking nucleophile in Eq. 17-55 and PP_i is a product. The latter is probably hydrolyzed by pyrophosphatase action, the end result being an overall reaction that is the same as with PEP synthase. Kinetic and positional isotope exchange studies suggest that the P_i must be bound to pyruvate–phosphate dikinase before the bound ATP can react with the imidazole group.[202] Likewise, AMP doesn't dissociate until P_i has reacted to form PP_i.

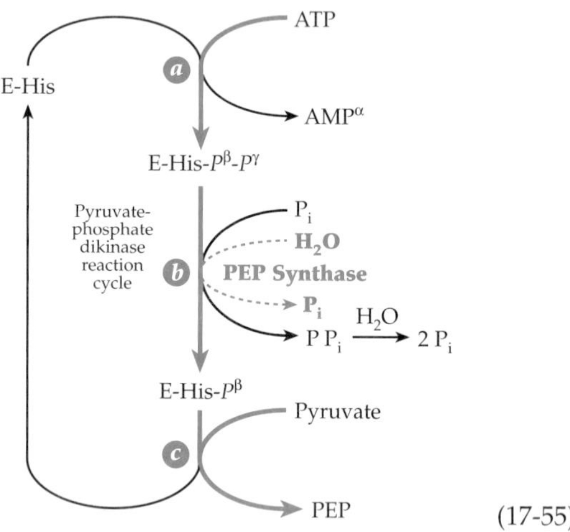

(17-55)

6. Building Hydrocarbon Chains with Two-Carbon Units

Fatty acid chains are taken apart two carbon atoms at a time by β oxidation. Biosynthesis of fatty acids reverses this process by using the two-carbon acetyl unit of acetyl-CoA as a starting material. The coupling of ATP cleavage to this process by a carboxylation–decarboxylation sequence, the role of acyl carrier protein (Section H,4), and the use of NADPH as a reductant (Section I) have been discussed and are summarized in Fig. 17-12, which gives the complete sequence of reactions for fatty acid biosynthesis. Why does β oxidation require CoA derivatives while biosynthesis requires the more complex acyl carrier protein (ACP)? The reason may involve control. ACP is a complex handle able to hold the growing fatty acid chain and to guide it from one enzyme to the next. In *E. coli* the various enzymes catalyzing the reactions of Fig. 17-12 are found in the cytosol and behave as independent proteins. The same is true for fatty acid synthases of higher plants which resemble those of bacteria.[205,205a]

It is thought that the ACP molecule lies at the center of the complex and that the growing fatty acid chain on the end of the phosphopantetheine prosthetic group moves from one subunit to the other.[164,206] The process is started by a **primer** which is usually acetyl-CoA in *E. coli*. Its acyl group is transferred first to the central molecule of ACP (step *a*, Fig. 17-12) and then to a "peripheral" thiol group, probably that of a cysteine side chain on a separate protein subunit (step *b*, Fig. 17-12). Next, a malonyl group is transferred (step *d*) from malonyl-CoA to the free thiol group on the ACP. The condensation (steps *e* and *f*) occurs with the freeing of the peripheral thiol group. The latter does not come into use again until the β-oxoacyl group formed has undergone the complete sequence of reduction reactions (steps *g*–*i*). Then the growing chain is again transferred to the peripheral –SH (step *j*) and a new malonyl unit is introduced on the central ACP.

After the chain reaches a length of 12 carbon atoms, the acyl group tends to be transferred off to a CoA molecule (step *k*) rather than to pass around the cycle again. Thus, chain growth is terminated. This tendency systematically increases as the chain grows longer.

In higher animals as well as in *Mycobacterium*,[207] yeast,[208] and *Euglena*, the **fatty acid synthase** consists of only one or two multifunctional proteins. The synthase from animal tissues has seven catalytic activities in a single 263-kDa 2500-residue protein.[209] The protein consists of a series of domains that contain the various catalytic activities needed for the entire synthetic sequence. One domain contains an ACP-like site with a bound 4'-phosphopantetheine as well as a cysteine side chain in the second acylation site. This synthase produces free fatty acids, principally the C_{16} palmitate. The final step is cleavage of the acyl-CoA by a thioesterase, one of the seven enzymatic activities of the synthase. See Chapter 21 for further discussion.

7. The Oxoacid Chain Elongation Process

As mentioned in Section 4, glyoxylate can be converted to oxaloacetate by condensation with acetyl-CoA (Fig. 17-16) and the oxaloacetate can be decarboxylated to pyruvate. This sequence of reactions resembles that of the conversion of oxaloacetate to 2-oxoglutarate in the citric acid cycle (Fig. 17-4). *Both are examples of a frequently used general chain elongation process for α-oxo acids.* This sequence, which is illustrated in Fig. 17-18, has four steps: (1) condensation of the α-oxo acid with an acetyl group, (2) isomerization by dehydration and rehydration (catalyzed by aconitase in the case of the citric acid cycle), (3) dehydrogenation, and (4) β decarboxylation. In many cases steps 3 and 4 are combined as a single enzymatic reaction. The isomerization of the intermediate hydroxy acid in step 2 is required because the hydroxyl group, which is attached to a tertiary carbon bearing no hydrogen, must be moved to the adjacent carbon atom before oxidation to a ketone can take place. However, in the case of glyoxylate, isomerization is not necessary because R = H.

It may be protested that the reaction of the citric acid cycle by which oxaloacetate is converted to oxoglutarate does not follow exactly the pattern of Fig. 17-18. The carbon dioxide removed in the decarboxylation step does not come from the part of the molecule donated by the acetyl group but from that formed from oxaloacetate. However, the end result is the same. Furthermore, there are two known citrate-forming enzymes with different stereospecificities (Chapter 13), one of which leads to a biosynthetic pathway strictly according to the sequence of Fig. 17-18.

At the bottom of Fig. 17-18 several stages of the α-oxo acid elongation process are arranged in tandem. We see that glyoxylate (a product of the acetyl-CoA–glyoxylate cycle) can be built up systematically to

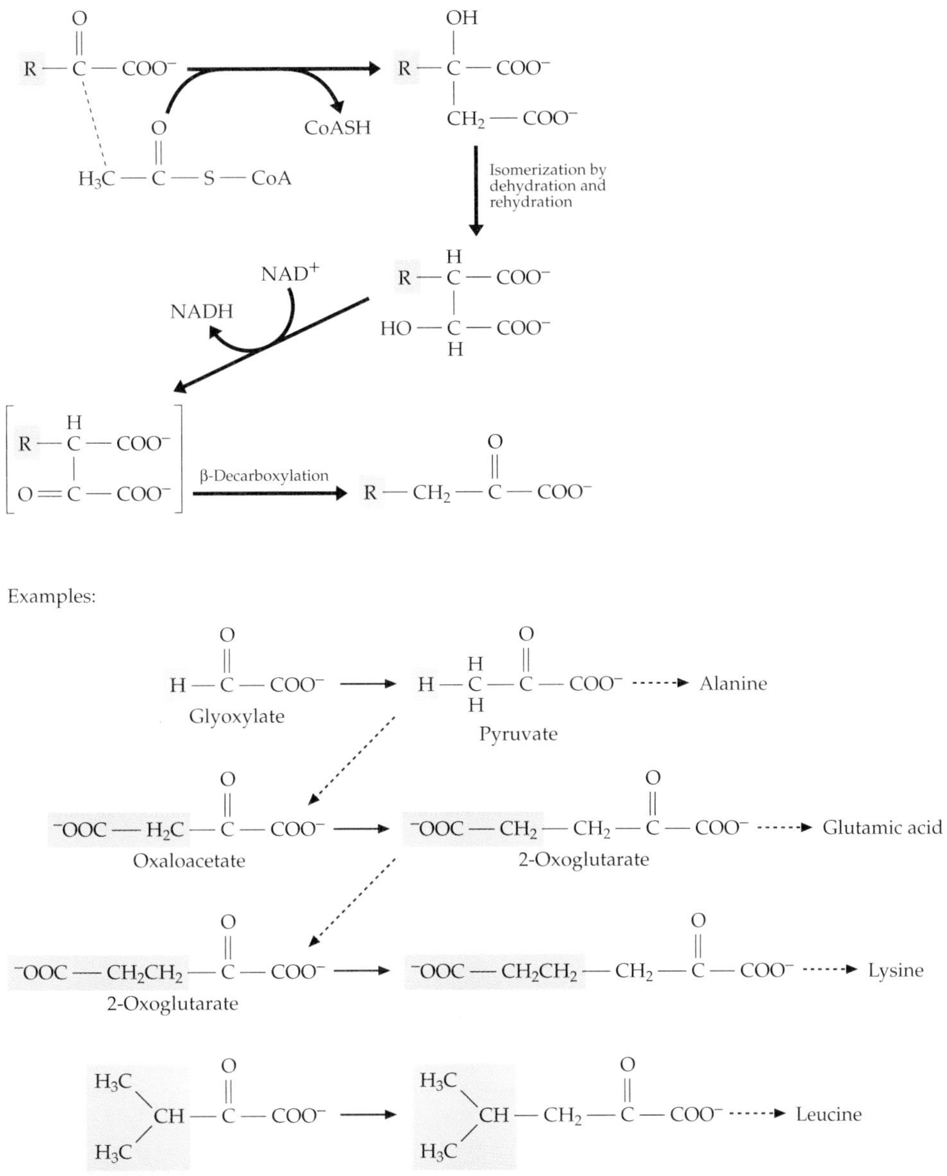

CoASH

Isomerization by dehydration and rehydration

NAD^+

NADH

β-Decarboxylation

Examples:

Glyoxylate → Pyruvate ······▸ Alanine

Oxaloacetate → 2-Oxoglutarate ······▸ Glutamic acid

2-Oxoglutarate → ······▸ Lysine

······▸ Leucine

Figure 17-18 The oxoacid chain elongation process.

pyruvate, oxaloacetate, 2-oxoglutarate, and 2-oxoadipate (a precursor of lysine) using this one reaction sequence. Methanogens elongate 2-oxoadipate by one and two carbon atoms using the same sequence to give 7- and 8-carbon dicarboxylates.[210]

8. Decarboxylation as a Driving Force in Biosynthesis

Consider the relationship of the following prominent biosynthetic intermediates one to another:

CH_3–C(=O)–S-CoA
Acetyl-CoA

HOOC–H_2C–C(=O)–S-CoA
Malonyl-CoA

CH_3–C(=O)–COO^-
Pyruvate

CH_2=C(O–P)–COO^-
PEP

HOOC–H_2C–C(=O)–COO^-
Oxaloacetate (OAA)

Utilization of acetyl-CoA for the synthesis of long-chain fatty acids occurs via carboxylation to malonyl-CoA. *We can think of the malonyl group as a β-carboxylated acetyl group.* During synthesis of a fatty acid the carboxyl group is lost, and only the acetyl group is ultimately incorporated into the fatty acid. In a similar way *pyruvate can be thought of as an α-carboxylated acetaldehyde and oxaloacetate as an α- and β-dicarboxylated acetaldehyde.* During biosynthetic reactions these three- and four-carbon compounds also often undergo decarboxylation. Thus, they both can be regarded as "activated acetaldehyde units." Phosphoenolpyruvate is an α-carboxylated phosphoenol form of acetaldehyde and undergoes both decarboxylation and dephosphorylation before contributing a two-carbon unit to the final product.

It is of interest to compare two chain elongation processes by which two-carbon units are combined. In the synthesis of fatty acids the acetyl units are condensed and then are reduced to form straight hydrocarbon chains. In the oxo-acid chain elongation mechanism, the acetyl unit is introduced but is later decarboxylated. Thus, the chain is increased in length by one carbon atom at a time. These two mechanisms account for a great deal of the biosynthesis by chain extension. However, there are other variations. For example, glycine (a carboxylated methylamine), under the influence of pyridoxal phosphate and with accompanying decarboxylation, condenses with succinyl-CoA (Eq. 14-32) to extend the carbon chain and at the same time to introduce an amino group. Likewise, serine (a carboxylated ethanolamine) condenses with palmitoyl-CoA in biosynthesis of sphingosine (as in Eq. 14-32). Phosphatidylserine is decarboxylated to phosphatidylethanolamine in the final synthetic step for that phospholipid (Fig. 21-5).

9. Stabilization and Termination of Chain Growth by Ring Formation

Biochemical substances frequently undergo cyclization to form stable five- and six-atom ring structures. The three-carbon glyceraldehyde phosphate exists in solution primarily as the free aldehyde (and its covalent hydrate) but glucose 6-phosphate exists largely as the cyclic hemiacetal. In this ring form no carbonyl group is present and further chain elongation is inhibited. When the hemiacetal of glucose 6-*P* is enzymatically isomerized to glucose 1-*P* the ring is firmly locked. Glucose 1-*P*, in turn, serves as the biosynthetic precursor of polysaccharides and related compounds, in all of which the sugar rings are stable. Ring formation can occur in lipid biosynthesis, too. Among the **polyketides** (Chapter 21), polyprenyl compounds (Chapter 22), and aromatic amino acids (Chapter 25) are many substances in which ring formation has occurred by ester or aldol condensations followed by reduction and elimination processes. This is a typical sequence for biosynthesis of highly stable aromatic rings.

10. Branched Carbon Chains

Branched carbon skeletons are formed by standard reaction types but sometimes with addition of rearrangement steps. Compare the biosynthetic routes to three different branched five-carbon units (Fig. 17-19) The first is the use of a **propionyl group** to initiate formation of a branched-chain fatty acid. Propionyl-CoA is carboxylated to methylmalonyl-CoA, whose acyl group is transferred to the acyl carrier protein before condensation. Decarboxylation and reduction yields an acyl-CoA derivative with a methyl group in the 3-position.

The second five-carbon branched unit, in which the branch is one carbon further down the chain, is an intermediate in the biosynthesis of **polyprenyl** (isoprenoid) compounds and steroids. Three two-carbon units are used as the starting material with decarboxylation of one unit. Two acetyl units are first condensed to form acetoacetyl-CoA. Then a third acetyl unit, which has been transferred from acetyl-CoA onto an SH group of the enzyme, is combined with the acetoacetyl-CoA through an ester condensation. The thioester linkage to the enzyme is hydrolyzed to free the product **3-hydroxy-3-methylglutaryl-CoA** (HMG-CoA). This sequence is illustrated in Eq. 17-5. The thioester group of HMG-CoA is reduced to the alcohol **mevalonic acid**, a direct precursor to isopentenyl pyrophosphate, from which the polyprenyl compounds are formed (Fig. 22-1).

The third type of carbon-branched unit is 2-oxoisovalerate, from which valine is formed by transamination. The starting units are two molecules of pyruvate which combine in a thiamin diphosphate-dependent α condensation with decarboxylation. The resulting α-acetolactate contains a branched chain but is quite unsuitable for formation of an α amino acid. A rearrangement moves the methyl group to the β position (Fig. 24-17), and elimination of water from the diol forms the enol of the desired α-oxo acid (Fig. 17-19). The precursor of isoleucine is formed in an analogous way by condensation, with decarboxylation of one molecule of pyruvate with one of 2-oxobutyrate.

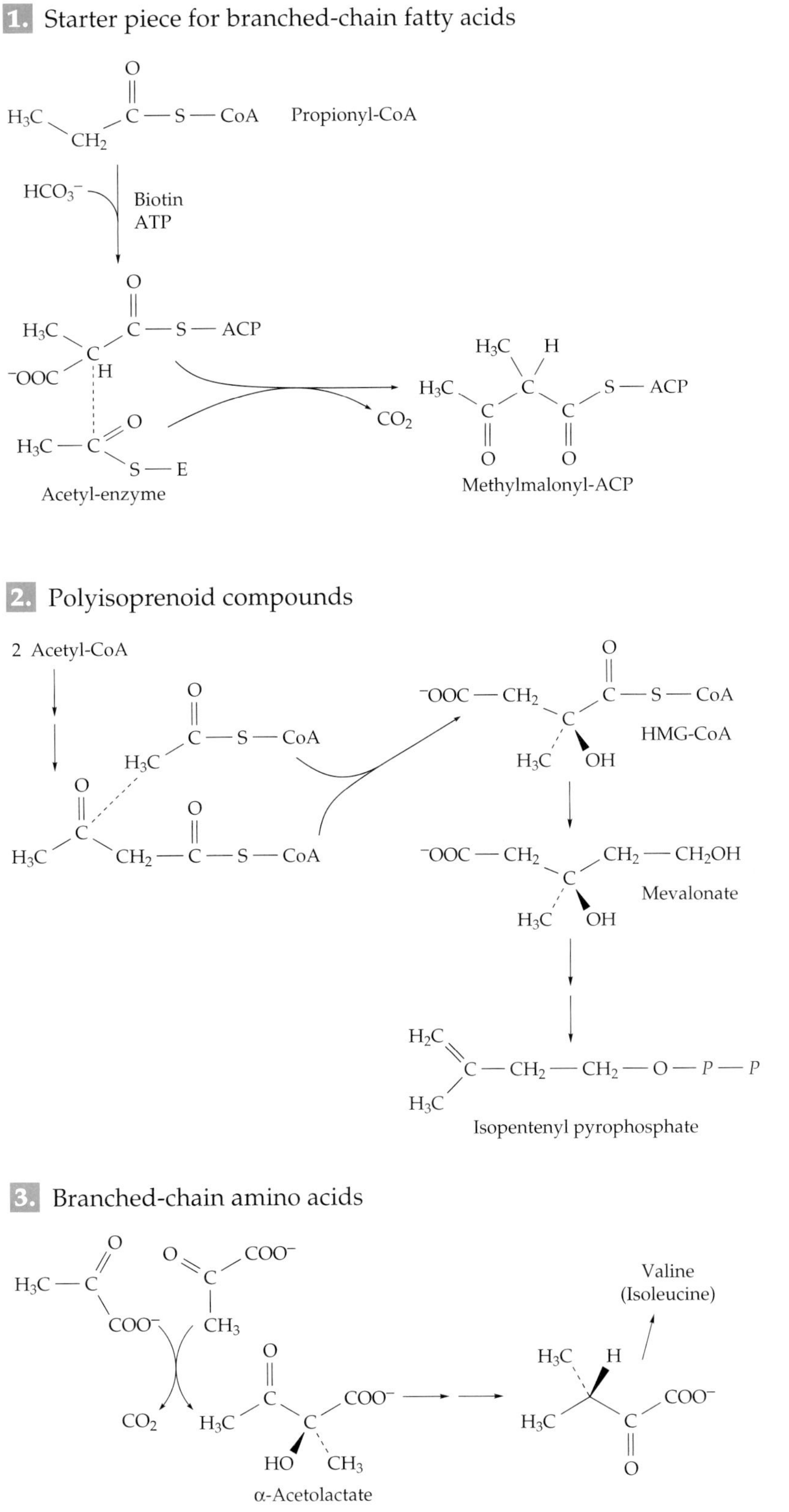

1. Starter piece for branched-chain fatty acids

Propionyl-CoA

HCO_3^- Biotin ATP

Acetyl-enzyme

CO_2

Methylmalonyl-ACP

2. Polyisoprenoid compounds

2 Acetyl-CoA

HMG-CoA

Mevalonate

Isopentenyl pyrophosphate

3. Branched-chain amino acids

CO_2

α-Acetolactate

Valine (Isoleucine)

Figure 17-19 Biosynthetic origins of three five-carbon branched structural units. Notice that decarboxylation is involved in driving each sequence.

K. Biosynthesis and Modification of Polymers

There are three chemical problems associated with the assembly of a protein, nucleic acid, or other biopolymer. The first is *to overcome thermodynamic barriers*. The second is *to control the rate of synthesis*, and the third is *to establish the pattern or sequence in which the monomer units are linked together*. Let us look briefly at how these three problems are dealt with by living cells.

1. Peptides and Proteins

Activation of amino acids for incorporation into oligopeptides and proteins can occur via two routes of acyl activation. In the first of these an **acyl phosphate** (or acyl adenylate) is formed and reacts with an amino group to form a peptide linkage (Eq. 13-4). The tripeptide **glutathione** is formed in two steps of this type (Box 11-B). In the second method of activation **aminoacyl**

adenylates are formed. They transfer their activated aminoacyl groups onto specific tRNA molecules during synthesis of proteins (Eq. 17-36). In other cases activated aminoacyl groups are transferred onto –SH groups to form intermediate **thioesters**. An example is the synthesis of the antibiotic **gramicidin S** formed by *Bacillus brevis*. The antibiotic is a cyclic decapeptide with the following five-amino-acid sequence repeated twice in the ringlike molecule[211]:

$$(\text{-D-Phe-L-Pro-L-Val-L-Orn-L-Leu-})_2$$

The soluble enzyme system responsible for its synthesis contains a large 280-kDa protein that not only activates the amino acids as aminoacyl adenylates and transfers them to thiol groups of 4'-phosphopantetheine groups covalently attached to the enzyme but also serves as a template for joining the amino acids in proper sequence.[211–214] Four amino acids—proline, valine, ornithine (Orn), and leucine—are all bound. A second enzyme (of mass 100 kDa) is needed for activation of phenylalanine. It is apparently the activated phenylalanine (which at some point in the process is isomerized from L- to D-phenylalanine) that initiates polymer formation in a manner analogous to that of fatty acid elongation (Fig. 17-12). Initiation occurs when the amino group of the activated phenylalanine (on the second enzyme) attacks the acyl group of the aminoacyl thioester by which the activated proline is held. Next, the freed imino group of proline attacks the activated valine, etc., to form the pentapeptide. Then two pentapeptides are joined and cyclized to give the antibiotic. The sequence is absolutely specific, and it is remarkable that this relatively small enzyme system is able to carry out each step in the proper sequence. Many other peptide antibiotics, such as the bacitracins, tyrocidines,[215] and enniatins, are synthesized in a similar way,[213,216,217] as are depsipeptides and the immunosuppresant cyclosporin. A virtually identical pattern is observed for formation of **polyketides**,[218,219] whose chemistry is considered in Chapter 21.

While peptide antibiotics are synthesized according to enzyme-controlled polymerization patterns, both proteins and nucleic acids are made by **template mechanisms**. The sequence of their monomer units is determined by genetically encoded information. A key reaction in the formation of proteins is the transfer of activated aminoacyl groups to molecules of tRNA (Eq. 17-36). The tRNAs act as carriers or adapters as explained in detail in Chapter 29. Each **aminoacyl-tRNA synthetase** must recognize the correct tRNA and attach the correct amino acid to it. The tRNA then carries the activated amino acid to a ribosome, where it is placed, at the correct moment, in the active site. **Peptidyltransferase**, using a transacylation reaction, in an *insertion mechanism* transfers the C terminus of the growing peptide chain onto the amino group of the new amino acid to give a tRNA-bound peptide one unit longer than before.

2. Polysaccharides

Incorporation of a sugar monomer into a polysaccharide also involves cleavage of two high-energy phosphate linkages of ATP. However, the activation process has its own distinctive pattern (Eq. 17-56). Usually a sugar is first phosphorylated by a kinase or a kinase plus a phosphomutase (Eq. 17-56, step *a*). Then a nucleoside triphosphate (NuTP) reacts under the influence of a second enzyme with elimination of pyrophosphate and formation of a **glycopyranosyl ester** of the nucleoside diphosphate, more often known as a **sugar nucleotide** (Eq. 17-56, step *b*). The inorganic pyrophosphate is hydrolyzed by pyrophosphatase while the sugar nucleotide donates the activated glycosyl group for polymerization (Eq. 17-56, step *c*). In this step the glycosyl group is transferred with displacement of the nucleoside diphosphate. Thus, the overall process involves first the cleavage of ATP to ADP and P_i, and then the cleavage of a nucleoside triphosphate to a nucleoside diphosphate plus P_i. The nucleoside triphosphate in Eq. 17-56, step *b* is sometimes ATP, in which case the overall result is the splitting of two molecules of ATP to ADP. However, as detailed in Chapter 20, the whole series of nucleotide "handles" serve to carry various activated glycosyl units.

What determines the pattern of incorporation of sugar units into polysaccharides? Homopolysaccharides, like cellulose and the linear amylose form of starch, contain only one monosaccharide component in only one type of linkage. A single synthetase enzyme can add unit after unit of an activated sugar (UDP glucose or other sugar nucleotide) to the growing end. However, at least two enzymes are needed to assemble a branched molecule such as that of the glycogen molecule. One is the synthetase; the second is a **branching enzyme**, a transglycosylase. After the chain ends attain a length of about ten monosaccharide units the branching enzyme attacks a glycosidic linkage somewhere in the chain. Acting much like a hydrolase, it forms a glycosyl enzyme (or a stabilized carbocation) intermediate. The enzyme does not release the severed chain fragment but transfers it to another nearby site on the branched polymer. In the synthesis of glycogen, the chain fragment is joined to a free 6-hydroxyl group of the glycogen, creating a new branch attached by an α-1,6-linkage.

Other carbohydrate polymers consist of **repeating oligosaccharide units**. Thus, in hyaluronan units of glucuronic acid and N-acetyl-D-glucosamine alternate (Fig. 4-11). The "O antigens" of bacterial cell coats (p. 180) contain repeating subunits made up of a "block" of four or five different sugars. In these and

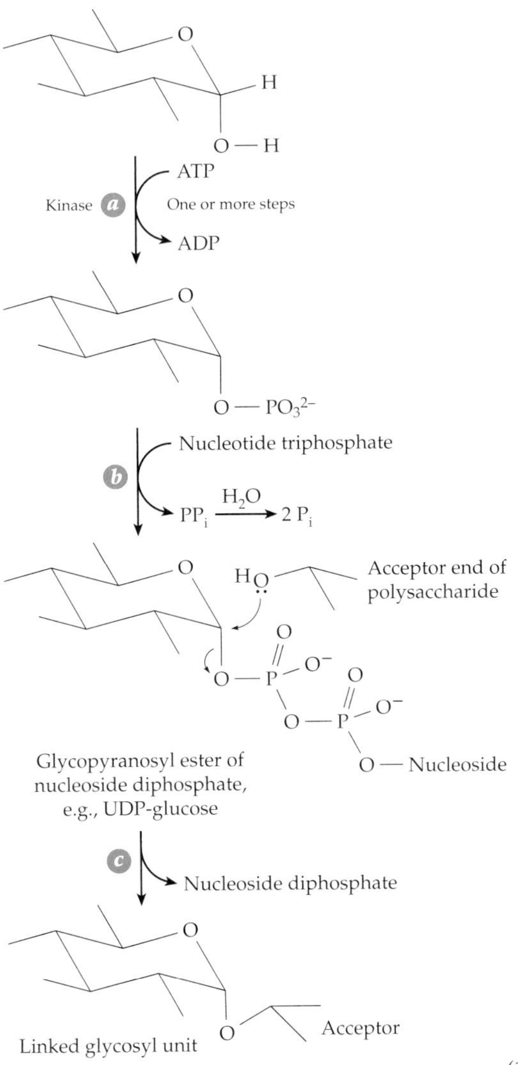

(17-56)

(17-57)

many other cases the pattern of polymerization is established by the specificities of individual enzymes. An enzyme capable of joining an activated glucosyl unit to a growing polysaccharide will do so only if the proper structure has been built up to that point. In cases where a block of sugar units is transferred it is usually *inserted* at the nonreducing end of the polymer, which may be covalently attached to a protein. Notice that the insertion mode of chain growth exists for lipids, polysaccharides, and proteins.

3. Nucleic Acids

The activated nucleotides are the nucleoside 5'-triphosphates. The ribonucleotides ATP, GTP, UTP, and CTP are needed for RNA synthesis and the 2'-deoxyribonucleotide triphosphates, dATP, dTTP, dGTP, and dCTP for DNA synthesis. In every case, the addition of activated monomer units to a growing polynucleotide chain is catalyzed by an enzyme that binds to the template nucleic acid. The choice of the proper nucleotide unit to place next in the growing strand is determined by the nucleotide already in place in the complementary strand, a matter that is dealt with in Chapters 27 and 28. The chemistry is a simple displacement of pyrophosphate (Eq. 17-57). The 3'-hydroxyl of the polynucleotide attacks the phosphorus atom of the activated nucleoside triphosphate. Thus, *nucleotide chains always grow from the 5' end, with new units being added at the 3' end.*

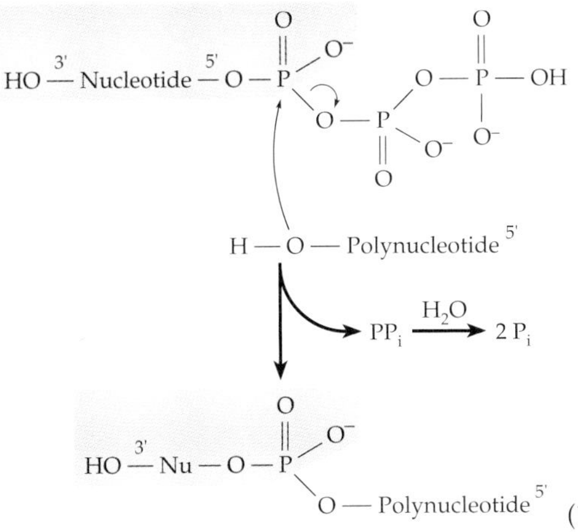

4. Phospholipids and Phosphate–Sugar Alcohol Polymers

Choline and ethanolamine are activated in much the same way as are sugars. For example, choline can be phosphorylated using ATP (Eq. 17-58, step *a*) and the phosphocholine formed can be further converted (Eq. 17-58, step *b*) to **cytidine diphosphate choline**. Phosphocholine is transferred from the latter onto a suitable acceptor to form the final product (Eq. 17-58, step *c*). The polymerization pattern differs from that for polysaccharide synthesis. When the sugar nucleotides react, the entire nucleoside diphosphate is eliminated (Eq. 17-56), but CDP-choline and CDP-ethanolamine react with elimination of CMP (Eq. 17-58, step *c*), leaving one phospho group in the final product. The same thing is true in the synthesis of the bacterial teichoic acids (Chapter 8). Either CDP-glycerol or CDP-ribitol is formed first and polymerization takes place with elimination of CMP to form the alternating phosphate–sugar alcohol polymer.[220]

5. Irreversible Modification and Catabolism of Polymers

While polymers are being synthesized continuously by cells, they are also being modified and torn down. Nothing within a cell is static. As discussed in Chapters

$(H_3C)_3N^+-CH_2-CH_2-OH$

Choline

↓ (a) ATP → ADP

$(CH_3)_3N^+-CH_2CH_2-O-PO_3^{2-}$

↓ (b) CTP → PP_i $\xrightarrow{H_2O}$ 2 P_i

$(CH_3)_3N^+-CH_2CH_2-O-P(=O)(O^-)-O-P(=O)(O^-)-O-$cytidine

Cytidine diphosphate choline

↓ (c) Y—OH (Phosphocholine acceptor such as the hydroxyl group of a diglyceride) → CMP

$(CH_3)_3N^+-CH_2-CH_2-O-P(=O)(O^-)-O-Y$

Phospholipid
(phosphatidylcholine if Y is a diglyceride) (17-58)

10 and 29, everything turns over at a slower or faster rate. Hydrolases attack all of the polymers of which cells are composed, and active catabolic reactions degrade the monomers formed. Membrane surfaces are also altered, for example, by hydroxylation and glycosylation of both glycoproteins and lipid head groups. It is impossible to list all of the known modification reactions of biopolymers. They include hydrolysis, methylation, acylation, isopentenylation, phosphorylation, sulfation, and hydroxylation. Precursor molecules are cut and trimmed and often modified further to form functional proteins or nucleic acids. Phosphotransferase reactions splice RNA transcripts to form mRNA and a host of alterations convert precursors into mature tRNA molecules (Chapter 28). Even DNA, which remains relatively unaltered, undergoes a barrage of chemical attacks. Only because of the presence of an array of repair enzymes (Chapter 27) does our DNA remain nearly unchanged so that faithful copies can be provided to each cell in our bodies and can be passed on to new generations.

L. Regulation of Biosynthesis

A simplified view of metabolism is to consider a cell as a "bag of enzymes." Indeed, much of metabolism can be explained by the action of several thousand enzymes promoting specific reactions of their substrates. These reactions are based upon the natural chemical reactivities of the substrates. However, the enzymes, through the specificity of their actions and through association with each other,[96,221–223] channel the reactions into a selected series of metabolic pathways. The reactions are often organized as cycles which are inherently stable. We have seen that biosynthesis often involves ATP-dependent reductive reactions. *It is these reductive processes that produce the less reactive nonpolar lipid groupings and amino acid side chains so essential to the assembly of insoluble intracellular structures.* Oligomeric proteins, membranes, microtubules, and filaments are all the natural result of aggregation caused largely by hydrophobic interactions with electrostatic forces and hydrogen bonding providing specificity. A major part of metabolism is the creation of complex molecules that aggregate spontaneously to generate structure. This structure includes the lipid-rich cytoplasmic membranes which, together with embedded carrier proteins, control the entry of substances into cells. Clearly, the cell is now much more than a bag of enzymes, containing several compartments, each of which contains its own array of enzymes and other components. Metabolite concentrations may vary greatly from one compartment to another.

The reactions that modify lipids and glycoproteins provide a driving force that assists in moving membrane materials generated internally into the outer surface of cells. Other processes, including the breakdown by lysosomal enzymes, help to recycle membrane materials. Oxidative attack on hydrophobic materials such as the sterols and the fatty acids of membrane lipids results in their conversion into more soluble substances which can be degraded and completely oxidized. The flow of matter within cells tends to occur in metabolic loops and some of these loops lead to formation of membranes and organelles and to their turnover. This flow of matter, which is responsible for growth and development of cells, is driven both by hydrolysis of ATP coupled to biosynthesis and by irreversible degradative alterations of polymers and lipid materials. It also provides for transient formation and breakup of complexes of macromolecules, which may be very large, in response to varying metabolic needs.

Anything that affects the rate of a reaction involved in either biosynthesis or degradation of any component of the cell will affect the overall picture in some way. Thus, every chemical reaction that contributes to a quantitatively significant extent to metabolism has some controlling influence. Since molecules interact with each other in so many ways, reactions of metabolic control are innumerable. Small molecules act on macromolecules as effectors that influence conformation and reactivity. Enzymes act on each other to break covalent bonds, to oxidize, and to crosslink. Transferases add phospho, glycosyl, methyl, and other groups to various sites. The resulting alterations often affect catalytic activities. The number of such interactions significant to metabolic control within an organism may be in the millions. Small wonder that biochemical journals are filled with a confusing number of postulated control mechanisms.

Despite this complexity, some regulatory mechanisms stand out clearly. The control of enzyme synthesis through feedback repression and the rapid control of activity by feedback inhibition (Chapter 11) have been considered previously. Under some circumstances, in which there is a constant growth rate, these controls may be sufficient to ensure the harmonious and proportional increase of all constitutents of a cell. Such may be the case for bacteria during logarithmic growth (Box 9-B) or for a mammalian embryo growing rapidly and drawing all its nutrients from the relatively constant supply in the maternal blood.

Contrast the situation in an adult. Little growth takes place, but the metabolism must vary with time and physiological state. The body must make drastic readjustments from normal feeding to a starvation situation and from resting to heavy exercise. The metabolism needed for rapid exertion is different from that needed for sustained work. A fatty diet requires different metabolism than a high-carbohydrate diet. The necessary control mechanisms must be rapid and sensitive.

1. Glycogen and Blood Glucose

Two special features of glucose metabolism in animals are dominant.[224] The first is the storage of glycogen for use in providing muscular energy rapidly. This is a relatively short-term matter but the rate of glycolysis can be intense: The entire glycogen content of muscle could be exhausted in only 20 s of anaerobic fermentation or in 3.5 min of oxidative metabolism.[225] There must be a way to turn on glycolysis quickly and to turn it off when it is no longer needed. At the same time, it must be possible to reconvert lactate to glucose or glycogen (gluconeogenesis). The glycogen stores of the muscle must be repleted from glucose of the blood. If insufficient glucose is available from the diet or from the glycogen stores of the liver, it must be synthesized from amino acids.

The second special feature of glucose metabolism is that certain tissues, including brain, blood cells, kidney medulla, and testis, ordinarily obtain most of their energy through oxidation of glucose.[226,227] For this reason, the glucose level of blood cannot be allowed to drop much below the normal 5 mM. The mechanism of regulation of the blood glucose level is complex and incompletely understood. A series of hormones are involved.

Insulin. This 51-residue cross-linked polypeptide (Fig. 7-17) is synthesized in the pancreatic islets of Langerhans, a tissue specialized for synthesis and secretion into the bloodstream of a series of small peptide hormones. One type of islet cells, the β cells, forms primarily insulin which is secreted in response to high (> 5mm) blood [glucose].[228] Insulin has a wide range of effects on metabolism,[228a] which are discussed in Chapter 11, Section G. Most of these effects are thought to arise from binding to insulin receptors (Figs. 11-11 and 11-12) and are mediated by cascades such as that pictured in Fig. 11-13.[229–232c] The end result is to increase or decrease activities of a large number of enzymes as is indicated in Table 17-3. Some of those are also shown in Fig. 17-20, which indicates interactions with the tricarboxylic acid cycle and lipid metabolism. Binding of insulin to the extracellular domain of its dimeric receptor induces a conformational change that activates the intracellular tyrosine kinase domains of the two subunits. Recent studies suggest that in the activated receptor the two transmembrane helices and the internal tyrosine kinase domains move closer together, inducing the essential autophosphosphorylation.[232b] The kinase domain of the phosphorylated receptor, in turn, phosphorylates several additional proteins, the most important of which seem to be the insulin receptor substrates IRS-1 and IRS-2. Both appear to be essential in different tissues.[232d,e] Phosphorylated forms of these proteins initiate a confusing variety of signaling cascades.[232f–i]

One of the most immediate effects of insulin is to stimulate an increased rate of uptake of glucose by muscle and adipocytes (fat cells) and other insulin-sensitive tissues. This uptake is accomplished largely by movement of the glucose transporter GLUT4 (Chapter 8) from internal "sequestered" storage vesicles located near the cell membrane into functioning positions in the membranes.[232f,j–m] Activation of this translocation process apparently involves IRS-1 and phosphatidylinositol (PI) 3-kinase, which generates PI-3,4,5-P_3 (Fig. 11-9).[232c,n,o] The latter induces the translocation. However, the mechanism remains obscure. The process may also require a second signaling pathway which involves action of the insulin receptor kinase on an adapter protein known as **CAP**, a transmembrane caveolar protein **flotillin**, and a third protein **Cbl**, a known cellular protooncogene. Phosphorylated Cbl forms a complex with CAP and flotillin in a "lipid raft" which induces the exocytosis of the sequestered GLUT4 molecules.[232o,p]

TABLE 17-3
Some Effects of Insulin on Enzymes

Name of Enzyme	Type of Regulation
A. Activity increased	
Enzymes of glycolysis	
Glucokinase	Transcription induced
Phosphofructokinase	via 2,6-fructose P_2
Pyruvate kinase	Dephosphorylation
6-Phosphofructo-2-kinase	Dephosphorylation
Enzymes of glycogen synthesis	
Glucokinase	Transcription
Glycogen synthase (muscle)	Dephosphorylation
Enzymes of lipid synthesis	
Pyruvate dehydrogenase (adipose)	Dephosphorylation (Eq. 17-9)
Acetyl-CoA carboxylase	Dephosphorylation
ATP-citrate lyase	Phosphorylation
Fatty acid synthase	
Lipoprotein lipase	
Hydroxymethylglutaryl-CoA reductase	
B. Activity decreased	
Enzymes of gluconeogenesis	
Pyruvate carboxylase	
PEP carboxykinase	Transcription inhibited
Fructose 1,6-bisphosphate	
Glucose 6-phosphatase	
Enzymes of lipolysis	
Triglyceride lipase	
(hormone-sensitive lipase)	Dephosphorylation
Enzymes of glycogenolysis	
Glycogen phosphorylase	
C. Other proteins affected by insulin	
Glucose transporter GLUT4	Redistribution
Ribosomal protein S6	Phosphorylation by p90^{rsk}
IGF-II receptor	Redistribution
Transferrin receptor	Redistribution
Calmodulin	Phosphorylation

D-*chiro*-Inositol

myo-Inositol
(numbered as D)

A clue to another possible unrecognized mechanism of action for insulin comes from the observation that urine of patients with non-insulin-dependent diabetes contains an unusual isomer of inositol, D-*chiro*-inositol.[233,234] Plasma of such individuals contains an antagonist of insulin action, an inositol phosphoglycan containing *myo*-inositol as a cyclic 1,2-phosphate ester and galactosamine and mannose in a 1:1:3 ratio.[235] This appears to be related to the glycosyl phosphatidylinositol (GPI) membrane anchors (Fig. 8-13). It has been suggested that such a glycan, perhaps containing *chiro*-inositol, is released in response to insulin and serves as a second messenger for insulin.[235–236a] This hypothesis remains unproved.[237] However, insulin does greatly stimulate a GPI-specific phospholipase C, at least in yeast.[237a] Another uncertainty surrounds the possible cooperation of chromium (Chapter 16) in the action of insulin.

How do the insulin-secreting pancreatic β cells sense a high blood glucose concentration? Two specialized proteins appear to be involved. The sugar transporter **GLUT2** allows the glucose in blood to equilibrate with the free glucose in the β cells,[237b] while **glucokinase** (hexokinase IV or D) apparently serves as the glucose sensor.[228,238] Despite the fact that glucokinase is a monomer, it displays a cooperative behavior toward glucose binding, having a low affinity at low [glucose] and a high affinity at high [glucose]. Mutant mice lacking the glucokinase gene develop early onset diabetes which is mild in heterozygotes but severe and fatal within a week of birth for homozygotes.[239,240] These facts alone do not explain how the sensor works and there are doubtless other components to the signaling system.

A current theory is that the increased rate of glucose catabolism in the β cells when blood [glucose] is high leads to a high ratio of [ATP]/[ADP] which induces closure of ATP-sensitive K^+ channels and opening of voltage-gated Ca^{2+} channels.[241] This could explain the increase in [Ca^{2+}] within β cells which has been associated with secretion of insulin[242,243] and which is thought to induce the exocytosis in insulin storage granules. The internal [Ca^{2+}] in pancreatic islet cells is observed to oscillate in a characteristic way that is synchronized with insulin secretion.[243]

Glucagon. This 29-residue peptide is the principal hormone that counteracts the action of insulin. Glucagon acts primarily on liver cells (hepatocytes) and adipose tissue and is secreted by the α cells of the islets of Langerhans in the pancreas, the same tissue whose β cells produce insulin, if the blood glucose concentration falls much below 2 mM.[244–250] Like the insulin-secreting β cells, the pancreatic α cells contain glucokinase, which may be involved in sensing the drop in glucose concentration. However, the carrier GLUT2 is not present and there is scant information on the sensing mechanism.[248]

Glucagon promotes an increase in the blood glucose level by stimulating breakdown of liver glycogen, by inhibiting its synthesis, and by stimulating gluconeogenesis. All of these effects are mediated by cyclic AMP through cAMP-activated protein kinase (Fig. 11-4) and through fructose 2,6-P_2 (Fig. 11-4 and next section). Glucagon also has a strong effect in promoting the release of glucose into the bloodstream. **Adrenaline** has similar effects, again mediated by cAMP. However, glucagon affects the liver, while adrenaline affects many tissues. **Glucocorticoids** such as cortisol (Chapter 22) also promote gluconeogenesis and the accumulation of glycogen in the liver through their action on gene transcription.

The release of glucose from the glycogen stores in the liver is mediated by **glucose 6-phosphatase**, which is apparently embedded within the membranes of the endoplasmic reticulum. A labile enzyme, it consists of a 357-residue catalytic subunit,[251,252] which may be associated with other subunits that participate in transport.[252,253] A deficiency of this enzyme causes the very severe type 1a **glycogen storage disease** (see Box 20-D).[251,253] Only hepatocytes have significant glucose 6-phosphatase activity.

2. Phosphofructo-1-Kinase in the Regulation of Glycolysis

The metabolic interconversions of glucose 1-*P*, glucose 6-*P*, and fructose 6-*P* are thought to be at or near equilibrium within most cells. However, the phosphorylation by ATP of fructose 6-*P* to fructose 1,6-P_2 catalyzed by phosphofructose-1-kinase (Fig. 11-2, step *b*; Fig. 17-17, top center) is usually far from equilibrium. This fact was established by comparing the mass action ratio [fructose 1,6-P_2] [ADP]/[fructose 6-P] [ATP] measured within tissues with the known equilibrium constant for the reaction. At equilibrium this mass action ratio should be equal to the equilibrium constant (Section I,2). The experimental techniques for determining the four metabolite concentrations that are needed for evaluation of the mass action ratio in tissues are of interest. The tissues must be frozen very rapidly. This can be done by compressing them between large liquid nitrogen-cooled aluminum clamps. For details see Newsholme and Start,[225] pp. 30 – 32. Tissues can be cooled to – 80°C in less than 0.1 s in this manner. The frozen tissue is then powdered, treated with a frozen protein denaturant such as perchloric acid, and analyzed. From data obtained in this way, a mass action ratio of 0.03 was found for the phosphofructo-1-kinase reaction in heart muscle.[225] This is much lower than the equilibrium constant of over 3000 calculated from the value of $\Delta G'$ (pH 7) = –20.1 kJ mol^{-1}. Thus, like other biochemical reactions that are nearly irreversible thermodynamically, this reaction is far from equilibrium in tissues.

The effects of ATP, AMP, and fructose 2,6-bisphosphate on phosphofructokinase have been discussed in Chapter 11, Section C. Fructose 2,6-P_2 is a potent allosteric activator of phosphofructokinase and a strong competitive inhibitor of fructose 1,6-bisphosphatase (Fig. 11-2). It is formed from fructose 6-*P* and ATP by the 90-kDa bifunctional phosphofructo-2-kinase/fructose 2,6-bisphosphatase. Thus, the same protein forms and destroys this allosteric effector. Since the bifunctional enzyme is present in very small amounts, the rate of ATP destruction from the substrate cycling is small.

Glucagon causes the concentration of *liver* fructose 2,6-P_2 to drop precipitously from its normal value. This, in turn, causes a rapid drop in glycolysis rate and shifts metabolism toward gluconeogenesis. At the same time, liver glycogen breakdown is inceased and glucose is released into the bloodstream more rapidly. The effect on fructose 2,6-P_2 is mediated by a cAMP-dependent protein kinase which phosphorylates the bifunctional kinase/phosphatase in the liver.[254] This modification greatly reduces the kinase activity and strongly activates the phosphatase, thereby destroying the fructose 2,6-P_2. The changes in activity appear to be largely a result of changes in the appropriate K_m values which are increased for fructose 6-*P* and decreased for fructose 2,6-P_2.[255]

3. Gluconeogenesis

If a large amount of lactate enters the liver, it is oxidized to pyruvate which enters the mitochondria. There, part of it is oxidized through the tricarboxylic acid cycle. However, if [ATP] is high, pyruvate dehydrogenase is inactivated by phosphorylation (Eq. 17-9) and the amount of pyruvate converted to oxaloacetate and malate (Eq. 17-46) may increase. Malate may leave the mitochondrion to be reoxidized to oxaloacetate, which is then converted to PEP and on to glycogen (heavy green arrows in Fig. 17-20). When [ATP] is high, phosphofructokinase is also blocked, but the fructose 1,6-bisphosphatase, which hydrolyzes one phosphate group from fructose 1,6-P_2 (Fig. 11-2, step *d*), is active. If the glucose content of blood is low, the glucose 6-*P* in the liver is hydrolyzed and free glucose is secreted. Otherwise, most of the glucose 6-*P* is converted to glycogen. Muscle is almost devoid of glucose 6-phosphatase, the export of glucose not being a normal activity of that tissue.

Gluconeogenesis in liver is strongly promoted by glucagon and adrenaline. The effects, mediated by cAMP, include stimulation of fructose 1,6-bisphosphatase and inhibition of phosphofructo-1-kinase, both caused by the drop in the level of fructose 2,6-P_2.[254,256] The conversion of pyruvate to PEP via oxaloacetate is also promoted by glucagon. This occurs primarily by stimulation of pyruvate carboxylase (Eq. 14-3).[257,258] However, it has been suggested that the most important mechanism by which glucagon enhances gluconeogenesis is through stimulation of mitochondrial respiration, which in turn may promote gluconeogenesis.[257]

The conversion of oxaloacetate to PEP by PEP-carboxykinase (PEPCK, Eq. 14-43; Fig. 17-20) is another control point in gluconeogenesis. Insulin inhibits gluconeogenesis by decreasing transcription of the mRNA for this enzyme.[259–261a] Glucagon and cAMP stimulate its transcription. The activity of PEP carboxykinase[262] is also enhanced by Mn^{2+} and by very low concentrations of Fe^{2+}. However, the enzyme is readily inactivated by Fe^{2+} and oxygen.[263] Any regulatory significance is uncertain.

Although the regulation of gluconeogenesis in the liver may appear to be well understood, some data indicate that the process can occur efficiently in the presence of high average concentrations of fructose 2,6-P_2. A possible explanation is that liver consists of several types of cells, which may contain differing concentrations of this inhibitor of gluconeogenesis.[264] However, mass spectroscopic studies suggest that glucose metabolism is similar throughout the liver.[265]

4. Substrate Cycles

The joint actions of phosphofructokinase and fructose 1,6-bisphosphatase (Fig. 11-2, steps *b* and *c*; see also Fig. 17-20) create a substrate cycle of the type discussed in Chapter 11, Section F. Such cycles apparently accomplish nothing but the cleavage of ATP to ADP and P_i (ATPase activity). There are many cycles of this type in metabolism and the fact that they do not ordinarily cause a disastrously rapid loss of ATP is a consequence of the tight control of the metabolic pathways involved. In general, only one of the two enzymes of Fig. 11-2, steps *b* and *c*, is fully activated at any time. Depending upon the metabolic state of the cell, degradation may occur with little biosynthesis or biosynthesis with little degradation. Other obvious substrate cycles involve the conversion of glucose to glucose 6-*P* and hydrolysis of the latter back to glucose (Fig. 17-20, upper left-hand corner), the synthesis and breakdown of glycogen (upper right), and the conversion of PEP to pyruvate and the reconversion of the latter to PEP via oxaloacetate and malate (partially within the mitochondria).

While one might suppose that cells always keep substrate cycling to a bare minimum, experimental measurements on tissues *in vivo* have indicated surprisingly high rates for the fructose 1,6-bisphosphatase–phosphofructokinase cycle in mammalian tissues when glycolytic flux rates are low and also for the pyruvate → oxaloacetate → PEP → pyruvate cycle.[266] As pointed out in Chapter 11, by maintaining a low rate of substrate cycling under conditions in which the carbon flux is low (in either the glycolytic or glucogenic direction) the system is more sensitive to allosteric effectors than it would be otherwise. However, when the flux through the glycolysis pathway is high the relative amount of cycling is much less and the amount of ATP formed approaches the theoretical 2.0 per glucose.[267]

Substrate cycles generate heat, a property that is apparently put to good use by cold bumblebees whose thoracic temperature must reach at least 30°C before they can fly. The insects apparently use the fructose bisphosphatase–phosphofructokinase substrate cycle (Fig. 11-2, steps *b* and *c*) to warm their flight muscles.[268] It probably helps to keep us warm, too.

5. Nuclear Magnetic Resonance, Isotopomer Analysis, and Modeling of Metabolism

As as been pointed out in Boxes 3-C and 17-C, the use of ^{13}C and other isotopic tracers together with NMR and mass spectroscopy have provided powerful tools for understanding the complex interrelationships among the various interlocking pathways of metabolism. In Box 17-C the application of ^{13}C NMR to the citric acid cycle was described. Similar approaches have been used to provide direct measurement of the glucose concentration in human brain (1.0 ± 0.1 mM; 4.7 ± 0.3 mM in plasma)[226] and to study gluconeogenesis[269–271] as well as fermentation.[271a] Similar investigations have been made using mass spectroscopy.[272] The metabolism of acetate through the glycoylate pathway in yeast has been observed by ^{13}C NMR.[200] Data obtained from such experiments are being used in attempts to model metabolism and to understand how flux rates through the various pathways are altered in response to varying conditions.[65,273–276]

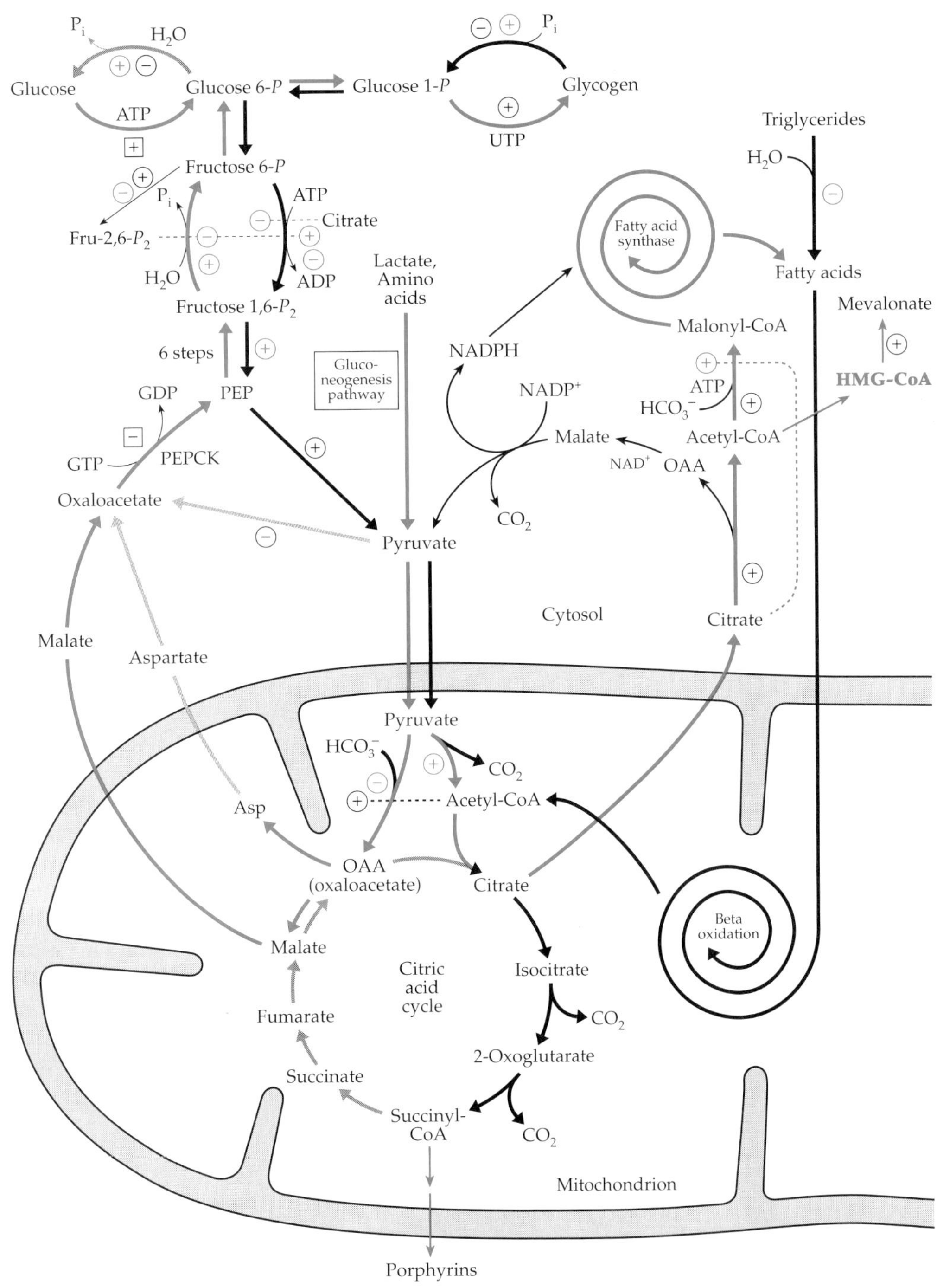

Figure 17-20 The interlocking pathways of glycolysis, gluconeogenesis, and fatty acid oxidation and synthesis with indications of some aspects of control in hepatic tissues. (⟶) Reactions of glycolysis, fatty acid degradation, and oxidation by the citric acid cycle. (⟶) Biosynthetic pathways. Some effects of insulin via indirect action on enzymes ⊕, ⊖, or on transcription ⊞, ⊟. Effects of glucagon ⊕, ⊖.

6. The Fasting State

During prolonged fasting, glycogen supplies are depleted throughout the body and fats become the principal fuels. Both glucose and pyruvate are in short supply. While the hydrolysis of lipids provides some glycerol (which is phosphorylated and oxidized to dihydroxyacetone-*P*), the quantity of glucose precursors formed in this way is limited. Since the animal body cannot reconvert acetyl-CoA to pyruvate, there is a continuing need for both glucose and pyruvate. The former is needed for biosynthetic processes, and the latter is a precursor of oxaloacetate, the regenerating substrate of the citric acid cycle. For this reason, during fasting the body readjusts its metabolism. As much as 75% of the glucose need of the brain can gradually be replaced by ketone bodies derived from the breakdown of fats (Section A,4).[277] Glucocorticoids (e.g., cortisol; Chapter 22) are released from the adrenal glands. By inducing enzyme synthesis, these hormones increase the amounts of a variety of enzymes within the cells of target organs such as the liver. Glucocorticoids also appear to increase the sensitivity of cell responses to cAMP and hence to hormones such as glucagon.[268]

The overall effects of glucocorticoids include an increased release of glucose from the liver (increased activity of glucose 6-phosphatase), an elevated blood glucose and liver glycogen, and a decreased synthesis of mucopolysaccharides. The reincorporation of amino acids released by protein degradation is inhibited and synthesis of enzymes degrading amino acids is enhanced. Among these enzymes are tyrosine and alanine aminotransferases, enzymes that initiate amino acid degradation which gives rise to the glucogenic precursors fumarate and pyruvate.

The inability of the animal body to form the glucose

BOX 17-F LACTIC ACIDEMIA AND OTHER DEFICIENCIES IN CARBOHYDRATE METABOLISM

The lactate concentration in blood can rise from its normal value of 1–2 mM to as much as 22 mM after very severe exercise such as sprinting, but it gradually returns to normal, requiring up to 6–8 h, less if mild exercise is continued. However, continuously high lactic acid levels are observed when enzymes of the gluconeogenic pathway are deficient or when oxidation of pyruvate is partially blocked.[a,b] Severe and often lethal deficiencies of the four key gluconeogenic enzymes pyruvate carboxylase, PEP carboxykinase, fructose 1,6-bisphosphatase, and glucose 6-phosphatase are known.[b] Pyruvate carboxylase deficiency may be caused by a defective carboxylase protein, by an absence of the enzyme that attaches biotin covalently to the three mitochondrial biotin-containing carboxylases (Chapter 14, Section C), or by defective transport of biotin from the gut into the blood. The latter types of deficiency can be treated successfully with 10 mg biotin per day.

Deficiency of pyruvate dehydrogenase is the most frequent cause of lactic acidemia.[a,c] Since this enzyme has several components (Fig. 15-15), a number of forms of the disease have been observed. Patients are benefitted somewhat by a high-fat, low-carbohydrate diet. Transient lactic acidemia may result from infections or from heart failure. One treatment is to administer dichloroacetate, which stimulates increased activity of pyruvate dehydrogenase, while action is also taken to correct the underlying illness.[d] Another problem arises if a lactate transporter is defective so that lactic acid accumulates in muscles.[e]

A different problem results from deficiency of enzymes of glycolysis such as phosphofructokinase (see Box 20-D), phosphoglycerate mutase, and pyruvate kinase. Lack of one isoenzyme of phosphoglycerate mutase in muscle leads to intolerance to strenuous exercise.[f] A deficiency in pyruvate kinase is one of the most common defects of glycolysis in erythrocytes and leads to a shortened erythrocyte lifetime and hereditary hemolytic anemia.[g]

Deficiency of the first enzyme of the pentose phosphate pathway, glucose 6-phosphate dehydrogenase, is widespread.[h] Its geographical distribution suggests that, like the sickle-cell trait, it confers some resistance to malaria. A partial deficiency of 6-phosphogluconolactonase (Eq. 17-12, step *b*) has also been detected within a family and may have contributed to the observed hemolytic anemia.[i]

---

[a] Robinson, B. H. (1995) in *The Metabolic and Molecular Bases of Inherited Disease*, 7th ed., Vol. 1 (Scriver, C. R., Beaudet, A. L., Sly, W. S., and Valle, D., eds), pp. 1479–1499, McGraw-Hill, New York

[b] Robinson, B. H. (1982) *Trends Biochem. Sci.* **7**, 151–153

[c] McCartney, R. G., Sanderson, S. J., and Lindsay, J. G. (1997) *Biochemistry* **36**, 6819–6826

[d] Stacpoole, P. W., 17 other authors, and Dichloroacetate-Lactic Acidosis Study Group (1992) *N. Engl. J. Med.* **327**, 1564–1569

[e] Fishbein, W. N. (1986) *Science* **234**, 1254–1256

[f] DiMauro, S., Mirando, A. F., Khan, S., Gitlin, K., and Friedman, R. (1981) *Science* **212**, 1277–1279

[g] Tanaka, K. R., and Paglia, D. E. (1995) in *The Metabolic and Molecular Bases of Inherited Disease* (Scriver, C. R., Beaudet, A. L., Sly, W. S., and Valle, D., eds), pp. 3485–3511, McGraw-Hill, New York

[h] Pandolfi, P. P., Sonati, F., Rivi, R., Mason, P., Grosveld, F., and Luzzatto, L. (1995) *EMBO J.* **14**, 5209–5215

[i] Beutler, E., Kuhl, W., and Gelbart, J. (1985) *Proc. Natl. Acad. Sci. U.S.A.* **82**, 3876–3878

precursors pyruvate or oxaloacetate from acetyl units is sometimes a cause of severe metabolic problems. Ketosis, which was discussed in Section A,4, develops when too much acetyl-CoA is produced and not efficiently oxidized in the citric acid cycle. Ketosis occurs during starvation, with fevers, and in insulin-dependent diabetes (see also Box 17-G). In cattle, whose metabolism is based much more on acetate than is ours, spontaneously developing ketosis is a frequent problem.

7. Lipogenesis

A high-carbohydrate meal leads to an elevated blood glucose concentration. The glycogen reserves within cells are filled. The ATP level rises, blocking the citric acid cycle, and citrate is exported from mitochondria (Fig. 17-20). Outside the mitochondria citrate is cleaved by the ATP-requiring citrate lyase (Eq. 14-37) to acetyl-CoA and oxaloacetate. The oxaloacetate can be reduced to malate and the latter oxidized with $NADP^+$ to pyruvate (Eq. 17-46), which can again enter the mitochondrion. In this manner acetyl groups are exported from the mitochondrion as acetyl-CoA which can be carboxylated, under the activating influence of citrate, to form malonyl-CoA, the precursor of fatty acids. The NADPH formed from oxidation of the malate provides part of the reducing equivalents needed for fatty acid synthesis. Additional NADPH is available from the pentose phosphate pathway. Thus, excess carbohydrate is readily converted into fat by our bodies. These reactions doubtless occur to some extent in most cells, but they are quantitatively

BOX 17-G DIABETES MELLITUS

The most prevalent metabolic problem affecting human beings is diabetes mellitus.[a–c] Out of a million people about 400 develop **type I** (or juvenile-onset) **insulin-dependent diabetes mellitus** (IDDM) between the ages of 8 and 12. Another 33,000 (over 3%) develop diabetes by age 40–50, and by the late 70s over 7% are affected. A propensity toward diabetes is partially hereditary, and recessive susceptibility genes are present in a high proportion of the population. The severity of the disease varies greatly. About half of the type I patients can be treated by diet alone, while the other half must receive regular insulin injections because of the atrophy of the insulin-producing cells of the pancreas. Type I diabetes sometimes develops very rapidly with only a few days of ravenous hunger and unquenchable thirst before the onset of ketoacidosis. Without proper care death can follow quickly. This suggested that a virus infection might cause the observed death of the insulin-secreting β cells of the pancreatic islets. However, the disease appears to be a direct result of an autoimmune response (Chapter 31). Antibodies directed against such proteins as insulin, glutamate decarboxylase,[d,e] and a tyrosine phosphatase[f] of the patient's own body are present in the blood. There may also be a direct attack on the β cells by T cells of the immune system (see Chapter 31).[g,h] The events that trigger such autoimmune attacks are not clear, but there is a strong correlation with susceptibility genes, in both human beings[i,j] and mice.[k,l]

Adults seldom develop type I diabetes but often suffer from **type II** or **non-insulin-dependent diabetes mellitus** (NIDDM). This is not a single disease but a syndrome with many causes. There is usually a marked decrease in sensitivity to insulin (referred to as **insulin resistance**) and poor uptake and utilization of glucose by muscles.[m] In rare cases this is a result of a mutation in the gene for the insulin molecule precursors (Eq. 10-8) or in the gene regulatory regions of the DNA.[n,o] Splicing of the mRNA[p] may be faulty or there may be defects in the structure or in the mechanisms of activation of the insulin receptors (Figs. 11-11 and 11-12).[q] The number of receptors may be too low or they may be degraded too fast to be effective. About 15% of persons with NIDDM have mutations in the insulin substrate protein IRS-1 (Chapter 11, Section G) but the significance is not clear.[m,r] Likewise, the causes of the loss of sensitivity of insulin receptors as well as other aspects of insulin resistance are still poorly understood.[s] In addition, prolonged high glucose concentrations result in decreased insulin synthesis or secretion, both of which are also complex processes. After synthesis the insulin hexamer is stored as granules of the hexamer $(insulin)_6Zn_2$ (Fig. 7-18) in vesicles at low pH. For secretion to occur the vesicles must first dock at membrane sites and undergo exocytosis. The insulin dissolves, releasing the Zn^{2+}, and acts in the monomeric form.[t] Because the mechanisms of action of insulin are still not fully understood, it is difficult to interpret the results of the many studies of diabetes mellitus.

A striking symptom of diabetes is the high blood glucose level which may range from 8 to 60 mM. Lower values are more typical for mild diabetes because when the glucose concentration exceeds the renal threshold of ~8 mM the excess is secreted into the urine. Defective utilization of glucose seems to be tied to a failure of glucose to exert proper

BOX 17-G DIABETES MELLITUS (continued)

feedback control. The result is that gluconeogenesis is increased with corresponding breakdown of proteins and amino acids. The liver glycogen is depleted and excess nitrogen from protein degradation appears in the urine. In IDDM diabetes the products of fatty acid degradation accumulate, leading to ketosis. The volume of urine is excessive and tissues are dehydrated.

Although the acute problems of diabetes, such as coma induced by ketoacidosis, can usually be avoided, it has not been possible to prevent long-term complications that include cataract formation and damage to the retina and kidneys. Most diabetics eventually become blind and half die within 15–20 years. Many individuals with NIDDM develop insulin-dependent diabetes in later life as a result of damage to the pancreatic β cells. The high glucose level in blood appears to be a major cause of these problems. The aldehyde form of glucose reacts with amino groups of proteins to form Schiff bases which undergo the Amadori rearrangement to form ketoamines (Eq. 4-8). The resulting modified proteins tend to form abnormal disulfide crosslinkages. Crosslinked aggregates of lens proteins may be a cause of cataract. The accumulating glucose-modified proteins may also induce autoimmune responses that lead to the long-term damage to kidneys and other organs. Another problem results from reduction of glucose to sorbitol (Box 20-A). Accumulation of sorbitol in the lens may cause osmotic swelling, another factor in the development of cataracts.[w,x] Excessive secretion of the 37-residue polypeptide **amylin**, which is synthesized in the β cells along with insulin, is another frequent complication of diabetes.[u,v] Amylin precipitates readily within islet cells to form **amyloid deposits** which are characteristic of NIDDM.

For many persons with diabetes regular injections of insulin are essential. Insulin was discovered in 1921 in Toronto by Banting and Best, with a controversial role being played by Professor J. J. R. Macleod, who shared the Nobel prize with Banting in 1923.[y,z] In 1922 the first young patients received pancreatic extracts and a new, prolonged life.[z-bb] Persons with IDDM are still dependent upon daily injections of insulin, but attempts are being made to treat the condition with transplanted cells from human cadavers.[cc] Animal insulins are suitable for most patients, but allergic reactions sometimes make **human insulin** essential. The human hormone, which differs from bovine insulin in three positions (Thr in human vs Ala in bovine at positions 8 of the A chain and 30 of the B chain and Ile vs Val at position 10 of the A chain), is now produced in bacteria using recombinant DNA. Nonenzymatic laboratory synthesis of insulin has also been achieved, but it is difficult to place the disulfide crosslinks in the proper positions. New approaches mimic the natural synthesis, in which the crosslinking takes place in proinsulin (Fig. 10-7).

NIDDM is strongly associated with obesity,[dd] and dieting and exercise often provide adequate control of blood glucose. Sulfonylurea drugs such as the following induce an increase in the number of insulin receptors formed and are also widely used in treatment of the condition.[ee,ff] These drugs bind to and inhibit ATP-sensitive K^+ channels in the β cell membranes. A defect in this sulfonylurea receptor has been associated with excessive insulin secretion

H_3C—C_6H_4—SO_2—NH—C(=O)—NH—$(CH_2)_3CH_3$

1-Butyl-*p*-tolylsulfonylurea

in infants.[17] New types of drugs are being tested.[gg-kk] These include inhibitors of aldose reductase,[ii] which forms sorbitol; compounds such as aminoguanidine, which inhibit formation of advanced products of glycation and newly discovered fungal metabolites that activate insulin receptors.[jj]

H_2N—C(=NH)—NH—NH_2

Aminoguanidine

[a] Atkinson, M. A., and Maclaren, N. K. (1990) *Sci. Am.* **263**(Jul), 62–71

[b] Taylor, S. I. (1995) in *The Metabolic and Molecular Bases of Inherited Disease*, 7th ed., Vol. 1 (Scriver, C. R., Beaudet, A. L., Sly, W. S., and Valle, D., eds), pp. 843–896, McGraw-Hill, New York

[c] Draznin, B., and LeRoith, D., eds. (1994) *Molecular Biology of Diabetes, Parts I and II*, Humana Press, Totowa, New Jersey

[d] Baekkeskov, S., Aanstoot, H.-J., Christgau, S., Reetz, A., Solimena, M., Cascalho, M., Folli, F., Richter-Olesen, H., and Camilli, P.-D. (1990) *Nature (London)* **347**, 151–156

[e] Nathan, B., Bao, J., Hsu, C.-C., Aguilar, P., Wu, R., Yarom, M., Kuo, C.-Y., and Wu, J.-Y. (1994) *Proc. Natl. Acad. Sci. U.S.A.* **91**, 242–246

[f] Lu, J., Li, Q., Xie, H., Chen, Z.-J., Borovitskaya, A. E., Maclaren, N. K., Notkins, A. L., and Lan, M. S. (1996) *Proc. Natl. Acad. Sci. U.S.A.* **93**, 2307–2311

[g] Solimena, M., Dirkx, R., Jr., Hermel, J.-M., Pleasic-Williams, S., Shapiro, J. A., Caron, L., and Rabin, D. U. (1996) *EMBO J.* **15**, 2102–2114

[h] MacDonald, H. R., and Acha-Orbea, H. (1994) *Nature (London)* **371**, 283–284

[i] Todd, J. A. (1995) *Proc. Natl. Acad. Sci. U.S.A.* **92**, 8560–8565

BOX 17-G (continued)

[j] Davies, J. L., Kawaguchi, Y., Bennett, S. T., Copeman, J. B., Cordell, H. J., Pritchard, L. E., Reed, P. W., Gough, S. C. L., Jenkins, S. C., Palmer, S. M., Balfour, K. M., Rowe, B. R., Farrall, M., Barnett, A. H., Bain, S. C., and Todd, J. A. (1994) *Nature (London)* **371**, 130–136

[k] Todd, J. A., Aitman, T. J., Cornall, R. J., Ghosh, S., Hall, J. R. S., Hearne, C. M., Knight, A. M., Love, J. M., McAleer, M. A., Prins, J.-B., Rodrigues, N., Lathrop, M., Pressey, A., DeLarato, N. H., Peterson, L. B., and Wicker, L. S. (1991) *Nature (London)* **351**, 542–547

[l] Leiter, E. H. (1989) *FASEB J.* **3**, 2231–2241

[m] Kim, J. K., Gavrilova, O., Chen, Y., Reitman, M. L., and Shulman, G. I. (2000) *J. Biol. Chem.* **275**, 8456–8460

[n] Zhao, L., Cissell, M. A., Henderson, E., Colbran, R., and Stein, R. (2000) *J. Biol. Chem.* **275**, 10532–10537

[o] Catasti, P., Chen, X., Moyzis, R. K., Bradbury, E. M., and Gupta, G. (1996) *J. Mol. Biol.* **264**, 534–545

[p] Wang, J., Shen, L., Najafi, H., Kolberg, J., Matschinsky, F. M., Urdea, M., and German, M. (1997) *Proc. Natl. Acad. Sci. U.S.A.* **94**, 4360–4365

[q] Schmid, E., Hotz-Wagenblatt, A., Hack, V., and Dröge, W. (1999) *FASEB J.* **13**, 1491–1500

[r] Thomas, P. M., Cote, G. J., Wohllk, N., Haddad, B., Mathew, P. M., Rabl, W., Aguilar-Bryan, L., Gagel, R. F., and Bryan, J. (1995) *Science* **268**, 426–429

[s] Nakajima, K., Yamauchi, K., Shigematsu, S., Ikeo, S., Komatsu, M., Aizawa, T., and Hashizume, K. (2000) *J. Biol. Chem.* **275**, 20880–20886

[t] Aspinwall, C. A., Brooks, S. A., Kennedy, R. T., and Lakey, J. R. T. (1997) *J. Biol. Chem.* **272**, 31308–31314

[u] Leighton, B., and Cooper, G. J. S. (1990) *Trends Biochem. Sci.* **15**, 295–299

[v] Lorenzi, A., Razzaboni, B., Weir, G. C., and Yankner, B. A. (1994) *Nature (London)* **368**, 756–760

[w] De Winter, H. L., and von Itzstein, M. (1995) *Biochemistry* **34**, 8299–8308

[x] Wilson, D. K., Bohren, K. M., Gabbay, K. H., and Quiocho, F. A. (1992) *Science* **257**, 81–84

[y] Stevenson, L. G. (1979) *Trends Biochem. Sci.* **4**, N158–N160

[z] Broad, W. J. (1982) *Science* **217**, 1120–1122

[aa] Orci, L., Vassalli, J.-D., and Perrelet, A. (1988) *Sci. Am.* **259**(Sep), 85–94

[bb] Marliss, E. B. (1982) *N. Engl. J. Med.* **306**, 362–364

[cc] Lacy, P. E. (1995) *Sci. Am.* **273**(Jul), 50–58

[dd] Simoneau, J.-A., Colberg, S. R., Thaete, F. L., and Kelley, D. E. (1995) *FASEB J.* **9**, 273–278

[ee] Aguilar-Bryan, L., Nichols, C. G., Wechsler, S. W., Clement, J. P., IV, Boyd, A. E., III, González, G., Herrera-Sosa, H., Nguy, K., Bryan, J., and Nelson, D. A. (1995) *Science* **268**, 423–426

[ff] Eliasson, L., Renström, E., Ämmälä, C., Berggren, P.-O., Bertorello, A. M., Bokvist, K., Chibalin, A., Deeney, J. T., Flatt, P. R., Gäbel, J., Bromada, J., Larsson, O., Lindström, P., Rhodes, C. J., and Rorsman, P. (1996) *Science* **271**, 813–815

[gg] Keen, H. (1994) *N. Engl. J. Med.* **331**, 1226–1227

[hh] Clark, C. M., Jr., and Lee, D. A. (1995) *N. Engl. J. Med.* **332**, 1210–1216

[ii] Bohren, K. M., Grimshaw, C. E., Lai, C.-J., Harrison, D. H., Ringe, D., Petsko, G. A., and Gabbay, K. H. (1994) *Biochemistry* **33**, 2021–2032

[jj] Qureshi, S. A., Ding, V., Li, Z., Szalkowski, D., Biazzo-Ashnault, D. E., Xie, D., Saperstein, R., Brady, E., Huskey, S., Shen, X., Liu, K., Xu, L., Salituro, G. M., Heck, J. V., Moller, D. E., Jones, A. B., and Zhang, B. B. (2000) *J. Biol. Chem.* **275**, 36590–36595

[kk] Moler, D. E. (2001) *Nature (London)* **414**, 821–827

most important in the liver, in fat cells of adipose tissue, and in mammary glands. The process is also facilitated by insulin, which promotes the activation of pyruvate dehydrogenase (Eq. 17-9) and of fatty acid synthase of adipocytes.[277a] Activity of fatty acid synthase seems to be regulated by the rate of transcription of its gene, which is controlled by a transcription factor designated either as **adipocyte determination and differentiation factor-1** (**ADD-1**) or **sterol regulatory element-binding protein-1c** (**SREBP-1c**). This protein (ADD-1/SREBP-1c) may be a general mediator of insulin action.[277b] The nuclear DNA-binding protein known as **peroxisome proliferator-activated receptor gamma** (**PPAR$_\gamma$**) is also involved in the control of insulin action, a conclusion based directly on discovery of mutations in persons with type II diabetes.[277c] A newly discovered hormone **resistin**, secreted by adipocytes, may also play a role.[277d] Another adipocyte hormone, **leptin**, impairs insulin action.[277e] Recent evidence suggests that both insulin and leptin may have direct effects on the brain which also influence blood glucose levels.[277f] Malonyl-CoA, which may also play a role in insulin secretion,[278,279] inhibits carnitine palmitoyltransferase I (CPT I; Fig. 17-2), slowing fatty acid catabolism.[280]

References

1. Knoll, L. J., Schall, O. F., Suzuki, I., Gokel, G. W., and Gordon, J. I. (1995) *J. Biol. Chem.* **270**, 20090–20097
2. Thorpe, C., and Kim, J.-J. P. (1995) *FASEB J.* **9**, 718–725
3. Vock, P., Engst, S., Eder, M., and Ghisla, S. (1998) *Biochemistry* **37**, 1848–1860
4. Aoyama, T., Ueno, I., Kamijo, T., and Hashimoto, T. (1994) *J. Biol. Chem.* **269**, 19088–19094
5. Eaton, S., Bartlett, K., and Pourfarzan, M. (1996) *Biochem. J.* **320**, 345–357

5a. Dwyer, T. M., Mortl, S., Kemter, K., Bacher, A., Fauq, A., and Frerman, F. E. (1999) *Biochemistry* **38**, 9735–9745

5b. Barycki, J. J., O'Brien, L. I., Strauss, A. W., and Banaszak, L. J. (2001) *J. Biol. Chem.* **276**, 36718–36726

6. Ikeda, Y., Okamure-Ikeda, K., and Tanaka, K. (1985) *J. Biol. Chem.* **260**, 1311–1325
7. Yang, S.-Y., Bittman, R., and Schultz, H. (1985) *J. Biol. Chem.* **260**, 2862–2888
8. Yang, S.-Y., Yang, X.-Y. H., Healy-Louie, G., Schulz, H., and Elzinga, M. (1990) *J. Biol. Chem.* **265**, 10424–10429
9. Wyatt, J. M. (1984) *Trends Biochem. Sci.* **9**, 20–23
10. Nada, M. A., Rhead, W. J., Sprecher, H., Schulz, H., and Roe, C. R. (1995) *J. Biol. Chem.* **270**, 530–535
11. Osumi, T., and Hashimoto, T. (1984) *Trends Biochem. Sci.* **9**, 317–319
12. Kaikaus, R. M., Sui, Z., Lysenko, N., Wu, N. Y., Ortiz de Montellano, P. R., Ockner, R. K., and Bass, N. M. (1993) *J. Biol. Chem.* **268**, 26866–26871
13. Chu, R., Varanasi, U., Chu, S., Lin, Y., Usuda, N., Rao, M. S., and Reddy, J. K. (1995) *J. Biol. Chem.* **270**, 4908–4915
14. Masters, C., and Crane, D. (1995) *The Peroxisome: A Vital Organelle*, Cambridge Univ. Press, London
15. Luo, Y., Karpichev, I. V., Kohanski, R. A., and Small, G. M. (1996) *J. Biol. Chem.* **271**, 12068–12075
16. Elgersma, Y., van Roermund, C. W. T., Wanders, R. J. A., and Tabak, H. F. (1995) *EMBO J.* **14**, 3472–3479
17. Gerhardt, B. (1993) in *Lipid Metabolism in Plants* (Moore, T. S., Jr., ed), pp. 528–565, CRC Press, Boca Raton, Florida
18. Moore, T. S., Jr., ed. (1993) *Lipid Metabolism in Plants*, CRC Press, Boca Raton, Florida

18a. Geisbrecht, B. V., Zhang, D., Schulz, H., and Gould, S. J. (1999) *J. Biol. Chem.* **274**, 21797–21803

19. Novikov, D. K., Vanhove, G. F., Carchon, H., Asselberghs, S., Eyssen, H. J., Van Veldhoven, P. P., and Mannaerts, G. P. (1994) *J. Biol. Chem.* **269**, 27125–27135

19a. Lopez-Huertas, E., Charlton, W. L., Johnson, B., Graham, I. A., and Baker, A. (2000) *EMBO J.* **19**, 6770–6777

20. Orci, L., Vassalli, J.-D., and Perrelet, A. (1988) *Sci. Am.* **259**(Sep), 85–94
21. Luthria, D. L., Baykousheva, S. P., and Sprecher, H. (1995) *J. Biol. Chem.* **270**, 13771–13776
22. Filppula, S. A., Sormunen, R. T., Hartig, A., Kunau, W.-H., and Hiltunen, J. K. (1995) *J. Biol. Chem.* **270**, 27453–27457
23. Smeland, T. E., Cuebas, D., and Schulz, H. (1991) *J. Biol. Chem.* **266**, 23904–23908
24. Preisig-Müller, R., Gühnemann-Schäfer, K., and Kindl, H. (1994) *J. Biol. Chem.* **269**, 20475–20481
25. van Roermund, C. W. T., Elgersma, Y., Singh, N., Wanders, R. J. A., and Tabak, H. F. (1995) *EMBO J.* **14**, 3480–3486
26. Malila, L. H., Siivari, K. M., Mäkelä, M. J., Jalonen, J. E., Latipää, P. M., Kunau, W.-H., and Hiltunen, J. K. (1993) *J. Biol. Chem.* **268**, 21578–21585
27. Chen, L.-S., Jin, S.-J., and Tserng, K.-Y. (1994) *Biochemistry* **33**, 10527–10534
28. Chen, L.-S., Jin, S.-J., Dejak, I., and Tserng, K.-Y. (1995) *Biochemistry* **34**, 442–450
29. Zhang, D., Liang, X., He, X.-Y., Alipui, O. D., Yang, S.-Y., and Schulz, H. (2001) *J. Biol. Chem.* **276**, 13622–13627

29a. Mursula, A. M., van Aalten, D. M. F., Hiltunen, J. K., and Wierenga, R. K. (2001) *J. Mol. Biol.* **309**, 845–853

29b. Fillgrove, K. L., and Anderson, V. E. (2001) *Biochemistry* **40**, 12412–12421

29c. Henke, B., Girzalsky, W., Berteaux-Lecellier, V., and Erdmann, R. (1998) *J. Biol. Chem.* **273**, 3702–3711

30. Johnson, M. J. (1967) *Science* **155**, 1515–1519
31. Eggink, G., Engel, H., Meijer, W. G., Otten, J., Kingma, J., and Witholt, B. (1988) *J. Biol. Chem.* **263**, 13400–13405
32. Stumpf, P. K., ed. (1987) *The Biochemistry of Plants; A Comprehensive Treatise*, Vol. 9, Academic Press, Orlando, Florida

32a. Hamberg, M., Sanz, A., and Castresana, C. (1999) *J. Biol. Chem.* **274**, 24503–24513

33. Singh, H., Beckman, K., and Poulos, A. (1994) *J. Biol. Chem.* **269**, 9514–9520
34. Vanhove, G. F., Van Veldhoven, P. P., Fransen, M., Denis, S., Eyssen, H. J., Wanders, R. J. A., and Mannaerts, G. P. (1993) *J. Biol. Chem.* **268**, 10335–10344

34a. Foulon, V., Antonenkov, V. D., Croes, K., Waelkens, E., Mannaerts, G. P., Van Veldhoven, P. P., and Casteels, M. (1999) *Proc. Natl. Acad. Sci. U.S.A.* **96**, 10039–10044

35. Lazarow, P. B., and Moser, H. W. (1995) in *The Metabolic and Molecular Bases of Inherited Disease*, 7th ed., Vol. 2 (Scriver, C. R., Beaudet, A. L., Sly, W. S., and Valle, D., eds), pp. 2287–2324, McGraw-Hill, New York
36. Kaya, K., Ramesha, C. S., and Thompson, G. A., Jr. (1984) *J. Biol. Chem.* **259**, 3548–3553
37. Hardwick, J. P., Song, B.-J., Huberman, E., and Gonzalez, F. J. (1987) *J. Biol. Chem.* **262**, 801–810
38. Cerdan, S., Künnecke, B., Dölle, A., and Seelig, J. (1988) *J. Biol. Chem.* **263**, 11664 –11674
39. Tserng, K.-Y., and Jin, S.-J. (1991) *J. Biol. Chem.* **266**, 2924–2929
40. Jin, S.-J., Hoppel, C. L., and Tserng, K.-Y. (1992) *J. Biol. Chem.* **267**, 119–125
41. Bieber, L. L., Emaus, R., Valkner, K., and Farrell, S. (1982) *Fed. Proc.* **41**, 2858–2862
42. Bieber, L. L. (1988) *Ann. Rev. Biochem.* **57**, 261–283
43. Rebouche, C. J. (1992) *FASEB J.* **6**, 3379–3386
44. Ramsay, R. R. (1994) in *Essays in Biochemistry*, Vol. 28 (Tipton, K. F., ed), pp. 47–62, Portland Press, London and Chapel Hill, North Carolina
45. Farrell, S. O., Fiol, C. J., Reddy, J. K., and Bieber, L. L. (1984) *J. Biol. Chem.* **259**, 13089–13095
46. Brady, P. S., Ramsay, R. R., and Brady, L. J. (1993) *FASEB J.* **7**, 1039–1044

46a. Schmalix, W., and Bandlow, W. (1993) *J. Biol. Chem.* **268**, 27428–27439

47. Brown, N. F., Weis, B. C., Husti, J. E., Foster, D. W., and McGarry, J. D. (1995) *J. Biol. Chem.* **270**, 8952–8957

47a. Yamazaki, N., Shinohara, Y., Kajimoto, K., Shindo, M., and Terada, H. (2000) *J. Biol. Chem.* **275**, 31739–31746

48. Kelly, D. P., and Strauss, A. W. (1994) *N. Engl. J. Med.* **330**, 913–919
49. Marinetti, G. V. (1990) *Disorders of Lipid Metabolism*, Plenum, New York

49a. Tamai, I., Ohashi, R., Nezu, J.-i, Sai, Y., Kobayashi, D., Oku, A., Shimane, M., and Tsuji, A. (2000) *J. Biol. Chem.* **275**, 40064–40072

50. Roe, C. R., and Coates, P. M. (1995) in *The Metabolic and Molecular Bases of Inherited Disease*, 7th ed., Vol. 1 (Scriver, C. R., Beaudet, A. L., Sly, W. S., and Valle, D., eds), pp. 1501–1533, McGraw-Hill, New York

50a. Wang, Y., Ye, J., Ganapathy, V., and Longo, N. (1999) *Proc. Natl. Acad. Sci. U.S.A.* **96**, 2356–2360

51. Ford, D. A., Han, X., Horner, C. C., and Gross, R. W. (1996) *Biochemistry* **35**, 7903–7909
52. Requero, M. A., Goñi, F. M., and Alonso, A. (1995) *Biochemistry* **34**, 10400–10405
53. Treem, W. R., Stanley, C. A., Finegold, D. N., Hale, D. E., and Coates, P. M. (1988) *N. Engl. J. Med.* **319**, 1331–1336
54. Rhead, W. J., Amendt, B. A., Fritchman, K. S., and Felts, S. J. (1983) *Science* **221**, 73–75
55. Mosser, J., Douar, A.-M., Sarde, C.-O., Kioschis, P., Feil, R., Moser, H., Poustka, A.-M., Mandel, J.-L., and Aubourg, P. (1993) *Nature (London)* **361**, 726–730
56. Aubourg, P., Adamsbaum, C., Lavallard-Rousseau, M.-C., Rocchiccioli, F., Cartier, N., Jambaqué, I., Jakobezak, C., Lemaitre, A., Boureau, F., Wolf, C., and Bougnères, P.-F. (1993) *N. Engl. J. Med.* **329**, 745–752
57. Moser, H. W., Smith, K. D., and Moser, A. B. (1995) in *The Metabolic and Molecular Bases of Inherited Disease*, 7th ed., Vol. 2 (Scriver, C. R., Beaudet, A. L., Sly, W. S., and Valle, D., eds), pp. 2325–2349, McGraw-Hill, New York
58. Yahraus, T., Braverman, N., Dodt, G., Kalish, J. E., Morrell, J. C., Moser, H. W., Valle, D., and Gould, S. J. (1996) *EMBO J.* **15**, 2914–2923
59. Street, J. M., Evans, J. E., and Natowicz, M. R. (1996) *J. Biol. Chem.* **271**, 3507–3516
60. Wolins, N. E., and Donaldson, R. P. (1994) *J. Biol. Chem.* **269**, 1149–1153

60a. Fujiwara, C., Imamura, A., Hashiguchi, N., Shimozawa, N., Suzuki, Y., Kondo, N., Imanaka, T., Tsukamoto, T., and Osumi, T. (2000) *J. Biol. Chem.* **275**, 37271–37277

60b. Quant, P. A. (1994) in *Essays in Biochemistry*, Vol. 28 (Tipton, K. F., ed), Portland Press, Chapel Hill, North Carolina

61. Krebs, H. A., Williamson, D. H., Bates, M. W., Page, M. A., and Hawkins, R. A. (1971) *Adv. Enzyme Regul.* **9**, 387–409
62. McGarry, J. D., and Foster, D. W. (1980) *Ann. Rev. Biochem.* **49**, 395–420
63. Endemann, G., Goetz, P. G., Edmond, J., and Brunengraber, H. (1982) *J. Biol. Chem.* **257**, 3434–3440
64. Ohgaku, S., Brady, P. S., Schumann, W. C., Bartsch, G. E., Margolis, J. M., Kumaran, K., Landau, S. B., and Landau, B. R. (1982) *J. Biol. Chem.* **257**, 9283–9289

64a. Misra, I., and Miziorko, H. M. (1996) *Biochemistry* **35**, 9610–9616

64b. Chun, K. Y., Vinarov, D. A., and Miziorko, H. M. (2000) *Biochemistry* **39**, 14670–14681

64c. Chun, K. Y., Vinarov, D. A., Zajicek, J., and Miziorko, H. M. (2000) *J. Biol. Chem.* **275**, 17946–17953

65. Sato, K., Kashiwaya, Y., Keon, C. A., Tsuchiya, N., King, M. T., Radda, G. K., Chance, B., Clarke, K., and Veech, R. L. (1995) *FASEB J.* **9**, 651–658
66. Casazza, J. P., Felver, M. E., and Veech, R. L. (1984) *J. Biol. Chem.* **259**, 231–236
67. Argilés, J. M. (1986) *Trends Biochem. Sci.* **11**, 61–63
68. Landau, B. R., and Brunengraber, H. (1987) *Trends Biochem. Sci.* **12**, 113–114

References

69. Gavino, V. C., Somma, J., Philbert, L., David, F., Garneau, M., Bélair, J., and Brunengraber, H. (1987) *J. Biol. Chem.* **262**, 6735–6740
70. Vagelos, P. R. (1960) *J. Biol. Chem.* **235**, 346–350
70a. Sacksteder, K. A., Morrell, J. C., Wanders, R. J. A., Matalon, R., and Gould, S. J. (1999) *J. Biol. Chem.* **274**, 24461–24468
71. Hofmeister, A. E. M., and Buckel, W. (1992) *Eur. J. Biochem.* **206**, 547–552
72. Buckel, W. (1992) *FEMS Microbiol. Rev.* **88**, 211–232
73. Vagelos, P. R., Earl, J. M., and Stadtman, E. R. (1959) *J. Biol. Chem.* **234**, 490–497, 765–769
74. Kuchta, R. D., and Abeles, R. H. (1985) *J. Biol. Chem.* **260**, 13181–13189
75. Taylor, S. I. (1995) in *The Metabolic and Molecular Bases of Inherited Disease*, 7th ed., Vol. 1 (Scriver, C. R., Beaudet, A. L., Sly, W. S., and Valle, D., eds), pp. 843–896, McGraw–Hill, New York
76. Weigand, E., Young, J. W., and McGilliard, A. D. (1972) *Biochem. J.* **126**, 201–209
77. Baldwin, J. E., and Krebs, H. (1981) *Nature (London)* **291**, 381–385
78. Nishimura, J. S., and Grinnell, F. (1972) *Adv. Enzymol.* **36**, 183–202
79. Rolleson, F. S. (1972) *Curr. Top. Cell. Regul.* **5**, 47–75
79a. Jucker, B. M., Lee, J. Y., and Shulman, R. G. (1998) *J. Biol. Chem.* **273**, 12187–12194
79b. Petersen, S., de Graaf, A. A., Eggeling, L., Möllney, M., Wiechert, W., and Sahm, H. (2000) *J. Biol. Chem.* **275**, 35932–35941
80. Kornberg, H. L. (1966) *Essays Biochem.* **2**, 1–31
81. Newsholme, E. A., and Start, C. (1973) *Regulation in Metabolism*, Wiley, New York (pp. 124–145)
82. Lieber, C. S. (1976) *Sci. Am.* **234**(Mar), 25–33
83. Kalnitsky, G., and Tapley, D. F. (1958) *Biochem. J.* **70**, 28–34
84. Rahmatullah, M., and Roche, T. E. (1987) *J. Biol. Chem.* **262**, 10265–10271
85. Yudkoff, M., Nelson, D., Daikhin, Y., and Erecinska, M. (1994) *J. Biol. Chem.* **269**, 27414–27420
85a. Reed, L. J. (2001) *J. Biol. Chem.* **276**, 38329–38336
86. Stueland, C. S., Eck, K. R., Stieglbauer, K. T., and LaPorte, D. C. (1987) *J. Biol. Chem.* **262**, 16095–16099
86a. Panisko, E. A., and McAlister-Henn, L. (2001) *J. Biol. Chem.* **276**, 1204–1210
87. Rashed, H. M., Waller, F. M., and Patel, T. B. (1988) *J. Biol. Chem.* **263**, 5700–5706
88. Chiang, P. K., and Sacktor, B. (1975) *J. Biol. Chem.* **250**, 3399–3408
89. Krebs, H. A. (1970) *Adv. Enzyme Regul.* **8**, 335–353
90. Maeng, C.-Y., Yazdi, M. A., and Reed, L. J. (1996) *Biochemistry* **35**, 5879–5882
91. Harris, R. A., Bowker-Kinley, M. M., Wu, P., Jeng, J., and Popov, K. M. (1997) *J. Biol. Chem.* **272**, 19746–19751
92. Patel, M. S., and Roche, T. E. (1990) *FASEB J.* **4**, 3224–3233
92a. Yang, D., Gong, X., Yakhnin, A., and Roche, T. E. (1998) *J. Biol. Chem.* **273**, 14130–14137
92b. McCartney, R. G., Sanderson, S. J., and Lindsay, J. G. (1997) *Biochemistry* **36**, 6819–6826
93. Bessman, S. P., Mohan, C., and Zaidise, I. (1986) *Proc. Natl. Acad. Sci. U.S.A.* **83**, 5067–5070
94. Wan, B., LaNoue, K. F., Cheung, J. Y., and Scaduto, R. C., Jr. (1989) *J. Biol. Chem.* **264**, 13430–13439
95. Srere, P. A. (1971) *Adv. Enzyme Regul.* **9**, 221–233
96. Ovádi, J., and Srere, P. A. (1996) *Cell Biochem. Funct.* **14**, 249–258
97. Sherry, A. D., and Malloy, C. R. (1996) *Cell Biochem. Funct.* **14**, 259–268
97a. Vélot, C., Mixon, M. B., Teige, M., and Srere, P. A. (1997) *Biochemistry* **36**, 14271–14276
98. Palmer, T. N., and Sugden, M. C. (1983) *Trends Biochem. Sci.* **8**, 161–162
99. Loeber, G., Infante, A. A., Maurer-Fogy, I., Krystek, E., and Dworkin, M. B. (1991) *J. Biol. Chem.* **266**, 3016–3021
100. Mandella, R. D., and Sauer, L. A. (1975) *J. Biol. Chem.* **250**, 5877–5884
101. Macrae, A. R. (1971) *Biochem. J.* **122**, 495–501
102. Huh, T.-L., Casazza, J. P., Huh, J.-W., Chi, Y.-T., and Song, B. J. (1990) *J. Biol. Chem.* **265**, 13320–13326
103. Nathan, B., Bao, J., Hsu, C.-C., Aguilar, P., Wu, R., Yarom, M., Kuo, C.-Y., and Wu, J.-Y. (1994) *Proc. Natl. Acad. Sci. U.S.A.* **91**, 242–246
104. Nathan, B., Hsu, C.-C., Bao, J., Wu, R., and Wu, J.-Y. (1994) *J. Biol. Chem.* **269**, 7249–7254
105. Toney, M. D., Pascarella, S., and De Biase, D. (1995) *Protein Sci.* **4**, 2366–2374
106. Hearl, W. G., and Churchich, J. E. (1985) *J. Biol. Chem.* **260**, 16361–16366
107. Baum, G., Lev-Yadum, S., Fridmann, Y., Arazi, T., Katsnelson, H., Zik, M., and Fromm, H. (1996) *EMBO J.* **15**, 2988–2996
108. Baum, G., Chen, Y., Arazi, T., Takatsuji, H., and Fromm, H. (1993) *J. Biol. Chem.* **268**, 19610–19617
109. Satyanarayan, V., and Nair, P. M. (1990) *Phytochemistry* **29**, 367–375
110. Streeter, J. G., and Thompson, J. F. (1972) *Plant Physiol.* **49**, 579–584
111. del Rio, R. M. (1981) *J. Biol. Chem.* **256**, 9816–9819
112. Bartley, W., Kornberg, H. L., and Quayle, J. R. (1970) *Essays in Cell Metabolism*, Wiley-Interscience, New York (p. 125)
113. Quayle, J. R. (1963) *Biochem. J.* **89**, 492–503
114. Chang, Y.-Y., Wang, A.-Y., and Cronan, J. E., Jr. (1993) *J. Biol. Chem.* **268**, 3911–3919
115. Fothergill-Gilmore, L. A. (1986) *Trends Biochem. Sci.* **11**, 47–51
115a. Bond, C. J., Jurica, M. S., Mesecar, A., and Stoddard, B. L. (2000) *Biochemistry* **39**, 15333–15343
115b. Valentini, G., Chiarelli, L., Fortin, R., Speranza, M. L., Galizzi, A., and Mattevi, A. (2000) *J. Biol. Chem.* **275**, 18145–18152
115c. Rigden, D. J., Phillips, S. E. V., Michels, P. A. M., and Fothergill-Gilmore, L. A. (1999) *J. Mol. Biol.* **291**, 615–635
116. Horecker, B. L., Gibbs, M., Klenow, H., and Smyrniotis, P. Z. (1954) *J. Biol. Chem.* **207**, 393–403
117. Wood, T. (1985) *The Pentose Phosphate Pathway*, Academic Press, Orlando, Florida
117a. Vought, V., Ciccone, T., Davino, M. H., Fairbairn, L., Lin, Y., Cosgrove, M. S., Adams, M. J., and Levy, H. R. (2000) *Biochemistry* **39**, 15012–15021
117b. Karsten, W. E., Chooback, L., and Cook, P. F. (1998) *Biochemistry* **37**, 15691–15697
117c. Kopriva, S., Koprivova, A., and Süss, K.-H. (2000) *J. Biol. Chem.* **275**, 1294–1299
118. Bublitz, C., and Steavenson, S. (1988) *J. Biol. Chem.* **263**, 12849–12853
118a. Johnson, A. E., and Tanner, M. E. (1998) *Biochemistry* **37**, 5746–5754
118b. Chen, Y.-R., Larimer, F. W., Serpersu, E. H., and Hartman, F. C. (1999) *J. Biol. Chem.* **274**, 2132–2136
119. Zubay, G. L., Parsons, W. W., and Vance, D. E. (1995) *Principles of Biochemistry*, W.C.Brown, Dubuque, Iowa (p. 276)
120. Reitzer, L. J., Wice, B. M., and Kennell, D. (1980) *J. Biol. Chem.* **255**, 5616–5626
121. Scofield, R. E., Kosugi, K., Chandramouli, V., Kumaran, K., Schumann, W. C., and Landau, B. R. (1985) *J. Biol. Chem.* **260**, 15439–15444
122. Magnusson, I., Chandramouli, V., Schumann, W. C., Kumaran, K., Wahren, J., and Landau, B. R. (1988) *Proc. Natl. Acad. Sci. U.S.A.* **85**, 4682–4685
122a. Kurland, I. J., Alcivar, A., Bassilian, S., and Lee, W.-N. P. (2000) *J. Biol. Chem.* **275**, 36787–36793
123. Longenecker, J. P., and Williams, J. F. (1980) *Biochem. J.* **188**, 847–857
124. Williams, J. F. (1980) *Trends Biochem. Sci.* **5**, 315–320
125. Williams, J. F. (1983) *Trends Biochem. Sci.* **8**, 275–277
126. Flanigan, I., Collins, J. G., Arora, K. K., Macleod, J. K., and Williams, J. F. (1993) *Eur. J. Biochem.* **213**, 477–485
127. Landau, B. R., and Wood, H. G. (1983) *Trends Biochem. Sci.* **8**, 292–296; 312–313
128. Casazza, J. P., and Veech, R. L. (1986) *J. Biol. Chem.* **261**, 690–698
129. Bernacchia, G., Schwall, G., Lottspeich, F., Salamini, F., and Bartels, D. (1995) *EMBO J.* **14**, 610–618
130. Berthon, H. A., Kuchel, P. W., and Nixon, P. F. (1992) *Biochemistry* **31**, 12792–12798
131. Horecker, B. L. (1965) *J. Chem. Educ.* **42**, 244–253
132. Decker, K., Jungermann, K., and Thauer, R. K. (1970) *Angew. Chem. Int. Ed. Engl.* **9**, 138–158
133. Mushegian, A. R., and Koonin, E. V. (1996) *Proc. Natl. Acad. Sci. U.S.A.* **93**, 10268–10273
134. Bult, C. J., and 39 other authors. (1996) *Science* **273**, 1058–1073
135. DiGirolamo, M., Newby, F. D., and Lovejoy, J. (1992) *FASEB J.* **6**, 2405–2412
136. Robinson, B. H. (1982) *Trends Biochem. Sci.* **7**, 151–153
136a. DiGirolamo, M., Newby, F. D., and Lovejoy, J. (1992) *FASEB J.* **6**, 2405–2412
136b. Shulman, R. G., and Rothman, D. L. (2001) *Proc. Natl. Acad. Sci. U.S.A.* **98**, 457–461
136c. Gladden, L. B. (2001) *Proc. Natl. Acad. Sci. U.S.A.* **98**, 395–397
136d. Bouzier, A.-K., Goodwin, R., Macouillard-Poulletier de Gannes, F., Valeins, H., Voisin, P., Canioni, P., and Merle, M. (1998) *J. Biol. Chem.* **273**, 27162–27169
137. Roberts, J. K. M., Callis, J., Jardetzky, O., Walbot, V., and Freeling, M. (1984) *Proc. Natl. Acad. Sci. U.S.A.* **81**, 6029–6033
138. Hake, S., Kelley, P. M., Taylor, W. C., and Freeling, M. (1985) *J. Biol. Chem.* **260**, 5050–5054
139. Shoubridge, E. A., and Hochachka, P. W. (1980) *Science* **209**, 308–309
139a. Teusink, B., Walsh, M. C., van Dam, K., and Westerhoff, H. V. (1998) *Trends Biochem. Sci.* **23**, 162–169
139b. Bakker, B. M., Mensonides, F. I. C., Teusink, B., van Hoek, P., Michels, P. A. M., and Westerhoff, H. V. (2000) *Proc. Natl. Acad. Sci. U.S.A.* **97**, 2087–2092
139c. Manchester, K. L. (2000) *Trends Biochem. Sci.* **25**, 89–92
140. Tempest, D. W., and Neijssel, O. M. (1987) in *Escherichia coli and Salmonella typhimurium* (Neidhardt, F. C., ed), pp. 797–806, Am. Soc. for Microbiology, Washington, D.C.
141. Blackwood, A. C., Neish, A. C., and Ledingham, G. A. (1956) *J. Bacteriol.* **72**, 497–499
142. Krebs, H. A. (1972) *Essays Biochem.* **8**, 1–34
143. Tse, J. M.-T., and Schloss, J. V. (1993) *Biochemistry* **32**, 10398–10403

References

144. Bui, E. T. N., Bradley, P. J., and Johnson, P. J. (1996) *Proc. Natl. Acad. Sci. U.S.A.* **93**, 9651–9656
145. Hochachka, P. W., and Mustafa, T. (1972) *Science* **178**, 1056–1060
146. Hochachka, P. W., and Somero, G. N. (1973) *Strategies of Biochemical Adaptation*, Saunders, Philadelphia, Pennsylvania (pp. 46–61)
147. Hochachka, P. W. (1980) *Living Without Oxygen*, Harvard Univ. Press, Cambridge, Massachusetts
148. Zhang, M., Eddy, C., Deanda, K., Finkelstein, M., and Picataggio, S. (1995) *Science* **267**, 240–243
149. Stanier, R. Y., Doudoroff, M., and Adelberg, E. A. (1970) *Microbial World*, 3rd ed., Prentice-Hall, Englewood Cliffs, New Jersey (p. 191)
150. Switzer, R. L. (1974) in *The Enzymes*, 3rd ed., Vol. 10 (Boyer, P. D., ed), pp. 607–629, Academic Press, New York
150a. Baykov, A. A., Fabrichniy, I. P., Pohjanjoki, P., Zyryanov, A. B., and Lahti, R. (2000) *Biochemistry* **39**, 11939–11947
150b. Ahn, S., Milner, A. J., Fütterer, K., Konopka, M., Ilias, M., Young, T. W., and White, S. A. (2001) *J. Mol. Biol.* **313**, 797–811
151. Wood, H. G., and Goss, N. H. (1985) *Proc. Natl. Acad. Sci. U.S.A.* **82**, 312–315
152. Wood, H. G. (1977) *Fed. Proc.* **36**, 2197–2205
153. Liu, C.-L., Hart, N., and Peck, H. D., Jr. (1982) *Science* **217**, 363–364
154. Weinbach, E. C. (1981) *Trends Biochem. Sci.* **6**, 254–257
155. Takeshige, K., and Tazawa, M. (1989) *J. Biol. Chem.* **264**, 3262–3266
156. Kornberg, A., Rao, N. N., and Ault-Riché, D. (1999) *Ann. Rev. Biochem.* **68**, 89–125
156a. Rashid, M. H., Rumbaugh, K., Passador, L., Davies, D. G., Hamood, A. N., Iglewski, B. H., and Kornberg, A. (2000) *Proc. Natl. Acad. Sci. U.S.A.* **97**, 9636–9641
156b. Bolesch, D. G., and Keasling, J. D. (2000) *J. Biol. Chem.* **275**, 33814–33819
157. Mudd, S. H. (1973) *The Enzymes*, 3rd ed., Vol. 8, Academic Press, New York (pp. 121–154)
158. Cornforth, J. W., Reichard, S. A., Talalay, P., Carrell, H. L., and Glusker, J. P. (1977) *J. Am. Chem. Soc.* **99**, 7292–7301
159. Peck, H. D., Jr. (1974) in *The Enzymes*, 3rd ed., Vol. 10 (Boyer, P. D., ed), pp. 651–669, Academic Press, New York
160. Geller, D. H., Henry, J. G., Belch, J., and Schwartz, N. B. (1987) *J. Biol. Chem.* **262**, 7374–7382
161. Liu, C., Martin, E., and Leyh, T. S. (1994) *Biochemistry* **33**, 2042–2047
162. Li, H., Deyrup, A., Mensch, J. R., Jr., Domowicz, M., Konstantinidis, A. K., and Schwartz, N. B. (1995) *J. Biol. Chem.* **270**, 29453–29459
163. Moréra, S., Chiadmi, M., LeBras, G., Lascu, I., and Janin, J. (1995) *Biochemistry* **34**, 11062–11070
163a. MacRae, I. J., Segel, I. H., and Fisher, A. J. (2000) *Biochemistry* **39**, 1613–1621
164. Wakil, S. J., Stoops, J. K., and Joshi, V. C. (1983) *Ann. Rev. Biochem.* **52**, 537–579
165. White, H. B., III, Mitsuhashi, O., and Bloch, K. (1971) *J. Biol. Chem.* **246**, 4751–4754
166. Williamson, D. H., Lund, P., and Krebs, H. A. (1967) *Biophys. J.* **103**, 514–527
167. Stubbs, M., Veech, R. L., and Krebs, H. A. (1972) *Biophys. J.* **126**, 59–65
168. Chung, Y., and Jue, T. (1992) *Biochemistry* **31**, 11159–11165
169. Krebs, H. A., and Veech, R. L. (1969) *Mitochondria: Structure & Function*, Academic Press, New York (pp. 101–109)
170. Veech, R. L., Guynn, R., and Veloso, D. (1972) *Biochem. J.* **127**, 387–397
171. Lundquist, R., and Olivera, B. M. (1971) *J. Biol. Chem.* **246**, 1107–1116
172. Buchanan, B. B. (1969) *J. Biol. Chem.* **244**, 4218–4223
173. Gehring, U., and Arnon, D. I. (1972) *J. Biol. Chem.* **247**, 6963–6969
173a. Furdui, C., and Ragsdale, S. W. (2002) *Biochemistry* **41**, 9921–9937
174. Stanier, R. Y., Doudoroff, M., and Adelberg, E. A. (1970) *The Microbial World*, 3rd ed., Prentice-Hall, Englewood Cliffs, New Jersey
175. Stanier, R. Y., Ingraham, J. L., Wheelis, M. L., and Painter, P. R. (1986) *The Microbial World*, 5th ed., Prentice-Hall, Englewood Cliffs, New Jersey
176. Brock, T. D., and Madigan, M. T. (1988) *Microbiology*, Prentice-Hall, Englewood Cliffs, New Jersey
177. Gottschalk, G. (1979) *Bacterial Metabolism*, Springer-Verlag, New York
178. Buchanan, B. B., Schürmann, P., and Shanmugam, K. T. (1972) *Biochim. Biophys. Acta.* **283**, 136–145
179. Wood, H. G., Ragsdale, S. W., and Pezacka, E. (1986) *Trends Biochem. Sci.* **11**, 14–17
180. Ferry, J. G., ed. (1993) *Methanogenesis: Ecology, Physiology, Biochemistry and Genetics*, Chapman & Hall, New York
181. Thauer, R. K., Hedderich, R., and Fischer, R. (1993) in *Methanogenesis: Ecology, Physiology, Biochemistry and Genetics* (Ferry, J. G., ed), pp. 209–252, Chapman & Hall, New York
182. Thauer, R. K., Schwörer, B., and Zirngibl, C. (1993) in *Microbial Growth on C1 Compounds* (Murrell, J. C., and Kelly, D. P., eds), pp. 151–162, Intercept Ltd., Andover, UK
183. Müller, V., Blaut, M., and Gottschalk, G. (1993) in *Methanogenesis: Ecology, Physiology, Biochemistry and Genetics* (Ferry, J. G., ed), pp. 360–406, Chapman & Hall, New York
184. Simpson, P. G., and Whitman, W. B. (1993) in *Methanogenesis: Ecology, Physiology, Biochemistry and Genetics* (Ferry, J. G., ed), pp. 445–472, Chapman & Hall, New York
185. Ljungdahl, L., Irion, E., and Wood, H. G. (1965) *Biochemistry* **4**, 2771–2780
186. Wood, H. G. (1991) *FASEB J.* **5**, 156–163
187. Drake, H. L. (1993) in *Microbial Growth on C1 Compounds* (Murrell, J. C., and Kelly, D. P., eds), pp. 493–507, Intercept Ltd., Andover, UK
188. Wood, H. G., and Ljungdahl, L. G. (1991) in *Variations in Autotrophic Life* (Shively, J. M., and Barton, L. L., eds), pp. 201–250, Academic Press, San Diego, California
188a. Tan, X. S., Sewell, C., and Lindahl, P. A. (2002) *J. Am. Chem. Soc.* **124**, 6277–6284
189. Qiu, D., Kumar, M., Ragsdale, S. W., and Spiro, T. G. (1996) *J. Am. Chem. Soc.* **118**, 10429–10435
189a. Doukov, T. I., Iverson, T. M., Seravalli, J., Ragsdale, S. W., and Drennan, C. L. (2002) *Science* **298**, 567–572
189b. Peters, J. W. (2002) *Science* **298**, 552–553
190. Menon, S., and Ragsdale, S. W. (1996) *Biochemistry* **35**, 12119–12125
191. Hemming, A., and Blotevogel, K. H. (1985) *Trends Biochem. Sci.* **10**, 198–200
192. Salem, A. R., Hacking, A. J., and Quayle, J. R. (1973) *Biochem. J.* **136**, 89–96
193. Quayle, J. R. (1980) *Biochem. Soc. Trans.* **8**, 1–10
194. Anthony, C. (1982) *The Biochemistry of Methylotrophs*, Academic Press, New York
195. Colby, J., Dalton, H., and Whittenbury, R. (1979) *Ann. Rev. Microbiol.* **33**, 481–517
196. Haber, C. L., Allen, L. N., Zhao, S., and Hanson, R. S. (1983) *Science* **221**, 1147–1153
197. Higgins, I. J., Best, D. J., and Hammond, R. C. (1980) *Nature (London)* **286**, 561–567
198. Anthony, C. (1996) *Biochem. J.* **320**, 697–711
199. Stueland, C. S., Gorden, K., and LaPorte, D. C. (1988) *J. Biol. Chem.* **263**, 19475–19479
200. Dickinson, J. R., Dawes, I. W., Boyd, A. S. F., and Baxter, R. L. (1983) *Proc. Natl. Acad. Sci. U.S.A.* **80**, 5847–5851
201. Liu, F., Thatcher, J. D., and Epstein, H. F. (1997) *Biochemistry* **36**, 255–260
202. Cook, A. G., and Knowles, J. R. (1985) *Biochemistry* **24**, 51–58
203. Pocalyko, D. J., Carroll, L. J., Martin, B. M., Babbitt, P. C., and Dunaway-Mariano, D. (1990) *Biochemistry* **29**, 10757–10765
204. Xu, Y., Yankie, L., Shen, L., Jung, Y.-S., Mariano, P. S., and Dunaway-Mariano, D. (1995) *Biochemistry* **34**, 2181–2187
205. Ohlrogge, J. B. (1982) *Trends Biochem. Sci.* **7**, 386–387
205a. Heath, R. J., Su, N., Murphy, C. K., Rock, C. O. (2000) *J. Biol. Chem.* **275**, 40128–40133
206. Holak, T. A., Kearsley, S. K., Kim, Y., and Prestegard, J. H. (1988) *Biochemistry* **27**, 6135–6142
207. Wood, W. I., Peterson, D. O., and Bloch, K. (1978) *J. Biol. Chem.* **253**, 2650–2656
208. Kolodziej, S. J., Penczek, P. A., Schroeter, J. P., and Stoops, J. K. (1996) *J. Biol. Chem.* **271**, 28422–28429
209. Smith, S. (1994) *FASEB J.* **8**, 1248–1259
210. White, R. H. (1989) *Arch. Biochem. Biophys.* **270**, 691–697
211. Lipmann, F. (1971) *Science* **173**, 875–884
212. Kurahashi, K. (1974) *Ann. Rev. Biochem.* **43**, 445–459
213. Kleinkauf, H., and von Döhren, H. (1983) *Trends Biochem. Sci.* **8**, 281–283
214. Stein, T., Vater, J., Kruft, V., Otto, A., Wittmann-Liebold, B., Franke, P., Panico, M., McDowell, R., and Morris, H. R. (1996) *J. Biol. Chem.* **271**, 15428–15435
215. Pfeifer, E., Pavela-Vrancic, M., von Döhren, H., and Kleinkauf, H. (1995) *Biochemistry* **34**, 7450–7459
216. Pavela-Vrancic, M., Pfeifer, E., Schröder, W., von Döhren, H., and Kleinkauf, H. (1994) *J. Biol. Chem.* **269**, 14962–14966
217. Haese, A., Pieper, R., von Ostrowski, T., and Zocher, R. (1994) *J. Mol. Biol.* **243**, 116–122
218. Kao, C. M., Pieper, R., Cane, D. E., and Khosla, C. (1996) *Biochemistry* **35**, 12363–12368
219. Cortes, J., Wiesmann, K. E. H., Roberts, G. A., Brown, M. J. B., Staunton, J., and Leadlay, P. F. (1995) *Science* **268**, 1487–1489
220. Pooley, H. M., and Karamata, D. (1994) in *Bacterial Cell Wall (New Comprehensive Biochemistry)*, Vol. 27 (Ghuysen, J.-M., and Hakenbeck, R., eds), pp. 187–198, Elsevier, Amsterdam
221. Knull, H., and Minton, A. P. (1996) *Cell Biochem. Funct.* **14**, 237–248
222. Low, P. S., Rathinavelu, P., and Harrison, M. L. (1993) *J. Biol. Chem.* **268**, 14627–14631
223. Hardin, C. D., and Roberts, T. M. (1995) *Biochemistry* **34**, 1323–1331
224. Milner, Y., and Wood, H. G. (1972) *Proc. Natl. Acad. Sci. U.S.A.* **69**, 2463–2468
225. Newsholme, E. A., and Start, C. (1973) *Regulation in Metabolism*, Wiley, New York
226. Gruetter, R., Novotny, E. J., Boulware, S. D., Rothman, D. L., Mason, G. F., Shulman, G. I., Shulman, R. G., and Tamborlane, W. V. (1992) *Proc. Natl. Acad. Sci. U.S.A.* **89**, 1109–1112
227. Lapidot, A., and Gopher, A. (1994) *J. Biol. Chem.* **269**, 27198–27208
228. Efrat, S., Tal, M., and Lodish, H. F. (1994) *Trends Biochem. Sci.* **19**, 535–538

References

228a. Saltiel, A. R., and Kahn, C. R. (2001) *Nature (London)* **414**, 799–806

229. Paz, K., Voliovitch, H., Hadari, Y. R., Roberts, C. T., Jr., LeRoith, D., and Zick, Y. (1996) *J. Biol. Chem.* **271**, 6998–7003

230. Moxham, C. M., Tabrizchi, A., Davis, R. J., and Malbon, C. C. (1996) *J. Biol. Chem.* **271**, 30765–30773

231. Guilherme, A., Klarlund, J. K., Krystal, G., and Czech, M. P. (1996) *J. Biol. Chem.* **271**, 29533–29536

232. Tsakiridis, T., Taha, C., Grinstein, S., and Klip, A. (1996) *J. Biol. Chem.* **271**, 19664–19667

232a. Krook, A., Whitehead, J. P., Dobson, S. P., Griffiths, M. R., Ouwens, M., Baker, C., Hayward, A. C., Sen, S. K., Maassen, J. A., Siddle, K., Tavaré, J. M., and O'Rahilly, S. (1997) *J. Biol. Chem.* **272**, 30208–30214

232b. Ottensmeyer, F. P., Beniac, D. R., Luo, R. Z.-T., and Yip, C. C. (2000) *Biochemistry* **39**, 12103–12112

232c. Aguirre, V., Uchida, T., Yenush, L., Davis, R., and White, M. F. (2000) *J. Biol. Chem.* **275**, 9047–9054

232d. Alper, J. (2000) *Science* **289**, 37,39

232e. Withers, D. J., Gutierrez, J. S., Towery, H., Burks, D. J., Ren, J.-M., Previs, S., Zhang, Y., Bernal, D., Pons, S., Shulman, G. I., Bonner-Weir, S., and White, M. F. (1998) *Nature (London)* **391**, 900–904

232f. Czech, M. P., and Corvera, S. (1999) *J. Biol. Chem.* **274**, 1865–1868

232g. Lavan, B. E., Fantin, V. R., Chang, E. T., Lane, W. S., Keller, S. R., and Lienhard, G. E. (1997) *J. Biol. Chem.* **272**, 21403–21407

232h. Inoue, G., Cheatham, B., Emkey, R., and Kahn, C. R. (1998) *J. Biol. Chem.* **273**, 11548–11555

232i. Qiao, L.-y, Goldberg, J. L., Russell, J. C., and Sun, X. J. (1999) *J. Biol. Chem.* **274**, 10625–10632

232j. Pessin, J. E., Thurmond, D. C., Elmendorf, J. S., Coker, K. J., and Okada, S. (1999) *J. Biol. Chem.* **274**, 2593–2596

232k. Shepherd, P. R., and Kahn, B. B. (1999) *N. Engl. J. Med.* **341**, 248–257

232l. Rea, S., Martin, L. B., McIntosh, S., Macaulay, S. L., Ramsdale, T., Baldini, G., and James, D. E. (1998) *J. Biol. Chem.* **273**, 18784–18792

232m. Lee, W., Ryu, J., Souto, R. P., Pilch, P. F., and Jung, C. Y. (1999) *J. Biol. Chem.* **274**, 37755–37762

232n. Kanzaki, M., Watson, R. T., Artemyev, N. O., and Pessin, J. E. (2000) *J. Biol. Chem.* **275**, 7167–7175

232o. Czech, M. P. (2000) *Nature (London)* **407**, 147–148

232p. Baumann, C. A., Ribon, V., Kanzaki, M., Thurmond, D. C., Mora, S., Shigematsu, S., Bickel, P. E., Pessin, J. E., and Saltiel, A. R. (2000) *Nature (London)* **407**, 202–207

233. Ostlund, R. E., Jr., Seemayer, R., Gupta, S., Kimmel, R., Ostlund, E. L., and Sherman, W. R. (1996) *J. Biol. Chem.* **271**, 10073–10078

234. Ostlund, R. E., Jr., McGill, J. B., Herskowitz, I., Kipnis, D. M., Santiago, J. V., and Sherman, W. R. (1993) *Proc. Natl. Acad. Sci. U.S.A.* **90**, 9988–9992

235. Galasko, G. T. F., Abe, S., Lilley, K., Zhang, C., and Larner, J. (1996) *J. Clin. Endocrinol. Metab.* **81**, 1051–1057

236. Romero, G., Luttrell, L., Rogol, A., Zeller, K., Hewlett, E., and Larner, J. (1988) *Science* **240**, 509–511

236a. Frick, W., Bauer, A., Bauer, J., Wied, S., and Müller, G. (1998) *Biochemistry* **37**, 13421–13436

237. Saltiel, A. R. (1994) *FASEB J.* **8**, 1034–1040

237a. Müller, G., Grey, S., Jung, C., and Bandlow, W. (2000) *Biochemistry* **39**, 1475–1488

237b. Thorens, B., Guillam, M.-T., Beermann, F., Burcelin, R., and Jaquet, M. (2000) *J. Biol. Chem.* **275**, 23751–23758

238. Van Schaftingen, E., Detheux, M., and Veiga da Cunha, M. (1994) *FASEB J.* **8**, 414–419

239. Terauchi, Y., Sakura, H., Yasuda, K., Iwamoto, K., Takahashi, N., Ito, K., Kasai, H., Suzuki, H., Ueda, O., Kamada, N., Jishage, K., Komeda, K., Noda, M., Kanazawa, Y., Taniguchi, S., Miwa, I., Akanuma, Y., Kodama, T., Yazaki, Y., and Kadowaki, T. (1995) *J. Biol. Chem.* **270**, 30253–30256

240. Agius, L., Peak, M., Newgard, C. B., Gomez-Foix, A. M., and Guinovart, J. J. (1996) *J. Biol. Chem.* **271**, 30479–30486

241. Webb, D.-L., Islam, M. S., Efanov, A. M., Brown, G., Köhler, M., Larsson, O., and Berggren, P.-O. (1996) *J. Biol. Chem.* **271**, 19074–19079

242. Deeney, J. T., Cunningham, B. A., Chheda, S., Bokvist, K., Juntti-Berggren, L., Lam, K., Korchak, H. M., Corkey, B. E., and Berggren, P.-O. (1996) *J. Biol. Chem.* **271**, 18154–18160

243. Bergsten, P., Grapengiesser, E., Gylfe, E., Tangholm, A., and Hellman, B. (1994) *J. Biol. Chem.* **269**, 8749–8753

244. Buggy, J. J., Livingston, J. N., Rabin, D. U., and Yoo-Warren, H. (1995) *J. Biol. Chem.* **270**, 7474–7478

245. Walajtys-Rhode, E., Zapatero, J., Moehren, G., and Hoek, J. B. (1992) *J. Biol. Chem.* **267**, 370–379

246. de Duve, C. (1994) *FASEB J.* **8**, 979–981

247. Rubanyi, G. M., and Botelho, L. H. P. (1991) *FASEB J.* **5**, 2713–2720

248. Heimberg, H., De Vos, A., Moens, K., Quartier, E., Bouwens, L., Pipeleers, D., Van Schaftingen, E., Madsen, O., and Schuit, F. (1996) *Proc. Natl. Acad. Sci. U.S.A.* **93**, 7036–7041

249. Granner, D., and Pilkis, S. (1990) *J. Biol. Chem.* **265**, 10173–10176

250. Unson, C. G., Wu, C.-R., and Merrifield, R. B. (1994) *Biochemistry* **33**, 6884–6887

251. Lei, K.-J., Pan, C.-J., Liu, J.-L., Shelly, L. L., and Chou, J. Y. (1995) *J. Biol. Chem.* **270**, 11882–11886

252. Berteloo, A., St-Denis, J.-F., and van de Werve, G. (1995) *J. Biol. Chem.* **270**, 21098–21102

253. Chen, Y.-T., and Burchell, A. (1995) in *The Metabolic and Molecular Bases of Inherited Disease*, 7th ed., Vol. 1 (Scriver, C. R., Beaudet, A. L., Sly, W. S., and Valle, D., eds), pp. 935–965, McGraw-Hill, New York

254. Pilkis, S. J., El-Maghrabi, M. R., and Claus, T. C. (1988) *Ann. Rev. Biochem.* **57**, 755–783

255. Tornheim, K. (1988) *J. Biol. Chem.* **263**, 2619–2624

256. Scheffler, J. E., and Fromm, H. J. (1986) *Biochemistry* **25**, 6659–6665

257. Kraus-Friedmann, N. (1986) *Trends Biochem. Sci.* **11**, 276–279

258. Blackard, W. G., and Clore, J. N. (1988) *J. Biol. Chem.* **263**, 16725–16730

259. Liu, J., Park, E. A., Gurney, A. L., Roesler, W. J., and Hanson, R. W. (1991) *J. Biol. Chem.* **266**, 19095–19102

260. Beale, E. G., Chrapkiewicz, N. B., Scoble, H. A., Metz, R. J., Quick, D. P., Noble, R. L., Donelson, J. E., Biemann, K., and Granner, D. K. (1985) *J. Biol. Chem.* **260**, 10748–10760

261. Scott, D. K., Mitchell, J. A., and Granner, D. K. (1996) *J. Biol. Chem.* **271**, 31909–31914

261a. Yeagley, D., Moll, J., Vinson, C. A., and Quinn, P. G. (2000) *J. Biol. Chem.* **275**, 17814–17820

262. Höppner, W., Beckert, L., Buck, F., and Seitz, H.-J. (1991) *J. Biol. Chem.* **266**, 17257–17260

263. Punekar, N. S., and Lardy, H. A. (1987) *J. Biol. Chem.* **262**, 6714–6719

264. Kuwajima, M., Golden, S., Katz, J., Unger, R. H., Foster, D. W., and McGarry, J. D. (1986) *J. Biol. Chem.* **261**, 2632–2637

265. Cline, G. W., and Shulman, G. I. (1995) *J. Biol. Chem.* **270**, 28062–28067

266. Rognstad, R., and Katz, J. (1972) *J. Biol. Chem.* **247**, 6047–6054

267. Clark, M. G., Bloxham, D. P., Holland, P. C., and Lardy, H. A. (1973) *Biochem. J.* **134**, 589–597

268. Exton, J. H., Friedmann, N., Wong, E. H.-A., Brineaux, J. P., Corbin, J. D., and Park, C. R. (1972) *J. Biol. Chem.* **247**, 3579–3588

269. Katz, J., Wals, P., and Lee, W.-N. P. (1993) *J. Biol. Chem.* **268**, 25509–25521

270. Des Rosiers, C., Di Donato, L., Comte, B., Laplante, A., Marcoux, C., David, F., Fernandez, C. A., and Brunengraber, H. (1995) *J. Biol. Chem.* **270**, 10027–10036

270a. Previs, S. F., Hallowell, P. T., Neimanis, K. D., David, F., and Brunengraber, H. (1998) *J. Biol. Chem.* **273**, 16853–16859

271. Previs, S. F., Fernandez, C. A., Yang, D., Soloviev, M. V., David, F., and Brunengraber, H. (1995) *J. Biol. Chem.* **270**, 19806–19815

271a. Zhang, B. L., Yunianta, and Martin, M. L. (1995) *J. Biol. Chem.* **270**, 16023–16029

272. Neese, R. A., Schwarz, J.-M., Faix, D., Turner, S., Letscher, A., Vu, D., and Hellerstein, M. K. (1995) *J. Biol. Chem.* **270**, 14452–14463

273. Martin, G., Chauvin, M.-F., Dugelay, S., and Baverel, G. (1994) *J. Biol. Chem.* **269**, 26034–26039

274. Fernandez, C. A., and Des Rosiers, C. (1995) *J. Biol. Chem.* **270**, 10037–10042

275. Ni, T.-C., and Savageau, M. A. (1996) *J. Biol. Chem.* **271**, 7927–7941

276. Shiraishi, F., and Savageau, M. A. (1993) *J. Biol. Chem.* **268**, 16917–16928

277. Wang, H.-C., Ciskanik, L., Dunaway-Mariano, D., van der Saal, W., and Villafranca, J. J. (1988) *Biochemistry* **27**, 625–633

277a. Moon, Y. S., Latasa, M.-J., Kim, K.-H., Wang, D., and Sul, H. S. (2000) *J. Biol. Chem.* **275**, 10121–10127

277b. Flier, J. S., and Hollenberg, A. N. (1999) *Proc. Natl. Acad. Sci. U.S.A.* **96**, 14191–14192

277c. Barroso, I., Gurnell, M., Crowley, V. E. F., Agostini, M., Schwabe, J. W., Soos, M. A., Li Maslen, G., Williams, T. D. M., Lewis, H., Schafer, A. J., Chatterjee, V. K. K., and O'Rahilly, S. (1999) *Nature (London)* **402**, 880–883

277d. Steppan, C. M., Bailey, S. T., Bhat, S., Brown, E. J., Banerjee, R. R., Wright, C. M., Patel, H. R., Ahima, R. S., and Lazar, M. A. (2001) *Nature (London)* **409**, 307–312

277e. Müller, G., Ertl, J., Gerl, M., and Preibisch, G. (1997) *J. Biol. Chem.* **272**, 10585–10593

277f. Schwartz, M. W. (2000) *Science* **289**, 2066–2067

278. Newgard, C. B., and McGarry, J. D. (1995) *Ann. Rev. Biochem.* **64**, 689–719

279. Corkey, B. E., Glennon, M. C., Chen, K. S., Deeney, J. T., Matschinsky, F. M., and Prentki, M. (1989) *J. Biol. Chem.* **264**, 21608–21612

280. Kudo, N., Barr, A. J., Barr, R. L., Desai, S., and Lopaschuk, G. D. (1995) *J. Biol. Chem.* **270**, 17513–17520

281. Munir, E., Yoon, J. J., Tokimatsu, T., Hattori, T., and Shimada, M. (2001) *Proc. Natl. Acad. Sci. U.S.A.* **98**, 11126–11130

Study Questions

1. Write out a complete step-by-step mechanism for the reactions by which citrate can be synthesized from pyruvate and then exported from mitochondria for use in the biosynthesis of fatty acids. Include a chemically reasonable mechanism for the action of ATP–citrate lyase, which catalyzes the following reaction:

   ATP + citrate + CoA-SH → Acetyl-CoA + oxaloacetate + ADP + P_i

   Show how this reaction can be incorporated into an ATP-driven cyclic pathway for generating NADPH from NADH.

2. Show which parts (if any) of the citric acid cycle are utilized in each of the following reactions and what, if any, additional enzymes are needed in each case.
   a. Oxidation of acetyl-CoA to CO_2
   b. Catabolism of glutamate to CO_2
   c. Biosynthesis of glutamate from pyruvate
   d. Formation of propionate from pyruvate

3. Here is a possible metabolic reaction for a fungus.

   L-Leucine + 2-oxoglutarate^{2-} + 2 ½ O_2 → L-glutamate$^-$ + citrate^{3-} + H_2O + $2H^+$
   $\Delta G°'$ (pH 7) = –1026 kJ/mol

   Suggest a metabolic pathway for this reaction. Is it thermodynamically feasible?

4. It has been suggested that in *Escherichia coli* pyruvate may act as the regenerating substrate for a catalytic cycle by which glyoxylate, OHC–COO$^-$, is oxidized to CO_2. Key enzymes in this cycle are thought to be a 2-oxo-4-hydroxyglutarate aldolase and 2-oxoglutarate dehydrogenase. Propose a detailed pathway for this cycle.

5. Some bacteria use a "dicarboxylic acid cycle" to oxidize glyoxylate OHC–COO$^-$ to CO_2. The regenerating substrate for this cycle is acetyl-CoA. It is synthesized from glyoxylate by a complex pathway that begins with conversion of two molecules of glyoxylate to tartronic semialdehyde: $^-$OOC–CHOH–CHO. The latter is then dehydrogenated to D-glycerate.

   Write out a detailed scheme for the dicarboxylate cycle. Also indicate how glucose and other cell constituents can be formed from intermediates created in this biosynthetic pathway.

6. Some bacteria that lack the usual aldolase produce ethanol and lactic acid in a 1:1 molar ratio via the "heterolactic fermentation." Glucose is converted to ribulose 5-phosphate via the pentose phosphate pathway enzymes. A thiamin diphosphate-dependent "phosphoketolase" cleaves xylulose 5-phosphate in the presence of inorganic phosphate to acetyl phosphate and glyceraldehyde 3-phosphate.

   Propose a mechanism for the phosphoketolase reaction and write a balanced set of equations for the fermentation.

7. Bacteria of the genera *Aerobacter* and *Serratia* ferment glucose according to the following equation:

   1 ½ Glucose → 2,3-butanediol + 3 CO_2 + H_2 + ethanol

   Write out a detailed pathway for the reactions. Use the pyruvate formate–lyase reaction. What yield of ATP do you expect per molecule of glucose fermented?

8. Some Clostridia ferment glucose as follows:

   2 Glucose + 2 H_2O → butyrate$^-$ + 2 acetate$^-$ + 4 CO_2 + 6 H_2 + 3 H^+
   $\Delta G°'$ (pH 7) = –479 kJ/mol of butyrate formed

   Write out detailed pathways. How much ATP do you think can be formed per glucose molecule fermented?

9. Propionic acid bacteria use the following fermentation:

   1 ½ Glucose → 2 propionate + acetate$^-$ + 3 H^+ + CO_2 + H_2O
   $\Delta G°'$ = –464 kJ/mol of acetate formed

   Write out a detailed pathway for the reactions. How much ATP can be formed per molecule of glucose?

10. Consider the following reaction which can occur in the animal body:

    2 Lactate$^-$ ($C_3H_5O_3^-$) + CoASH + 2 NAD^+ → butyryl-CoA (C_4H_7O–S–CoA) + 2 CO_2 + 2 NADH + H_2O
    $\Delta G°'$ (pH 7) ~ 70 kJ/mol of butyryl-CoA formed

    Outline the sequence of reactions involved in this

transformation. Do you think that any ATP will either be used or generated in the reaction? Explain.

11. Leucine, an essential dietary constituent for human beings, is synthesized in many bacteria and plants using pyruvate as a starting material. Outline this pathway of metabolism and illustrate the chemical reaction mechanisms involved in each step.

12. Write a step-by-step sequence for all of the chemical reactions involved in the biosynthesis of L-leucine from 2-oxoisovalerate:

```
       CH3
       ||
H3C – CH – C – COO⁻
            ||
            O
```

Notice that this compound contains one carbon atom less than leucine. Start by condensing 2-oxoisovalerate with acetyl-CoA in a reaction similar to that of citrate synthase. Use structural formulas. Show all intermediate structures and indicate what coenzymes are needed. Use curved arrows to indicate the flow of electrons in each step.

13. Some fungi synthesize lysine from 2-oxoglutarate by elongating the chain using a carbon atom derived from acetyl-CoA to form the 6-carbon 2-oxoadipate. The latter is converted by an ATP-dependent reduction to the ε-aldehyde. Write out reasonable mechanisms for the conversion of 2-oxoglutarate to the aldehyde. The latter is converted on to L-lysine by a non–PLP-dependent transamination via saccharopine (Chapter 24).

14. Outline the pathway for biosynthesis of L-leucine from glucose and NH_4^+ in autotrophic organisms. In addition, outline the pathways for degradation of leucine to CO_2, water, and NH_4^+ in the human body. For this overall pathway or "metabolic loop," mark the locations (one or more) at which each of the following processes occurs.
    a. Synthesis of a thioester by dehydrogenation
    b. Substrate-level phosphorylation
    c. Thiamin-dependent α condensation
    d. Oxoacid chain-elongation process
    e. Transamination
    f. Oxidative decarboxylation of an α-oxoacid
    g. Partial β oxidation
    h. Thiolytic cleavage
    i. Claisen condensation
    j. Biotin-dependent carboxylation

15. A photosynthesizing plant is exposed to $^{14}CO_2$. On which carbon atoms will the label first appear in glucose?

16. The Calvin–Benson cycle and the pentose phosphate pathway (Eq. 17-12) have many features in common but run in opposite directions. Since the synthesis of glucose from CO_2 requires energy, the energy expenditure for the two processes will obviously differ. Describe the points in each pathway where a Gibbs energy difference is used to drive the reaction in the desired direction.

17. Draw the structure of ribulose 1(^{14}C), 5-bisphosphate. Enter an asterisk (*) next to carbon 1 to show that this position is ^{14}C-labeled.

    Draw the structures of the products of the ribulose bisphosphate carboxylase reaction, indicating the radioactive carbon position with an asterisk.

18. A wood-rotting fungus is able to convert glucose to oxalate approximately according to the following equation:

    $$\text{Glucose} + 5\ O_2 \rightarrow 2\ \text{Oxalic acid}\ (C_2O_4H_2) + 2\ CO_2 + 4\ H_2O$$

    Propose a mechanism. See Munir *et al.*[281] for details.

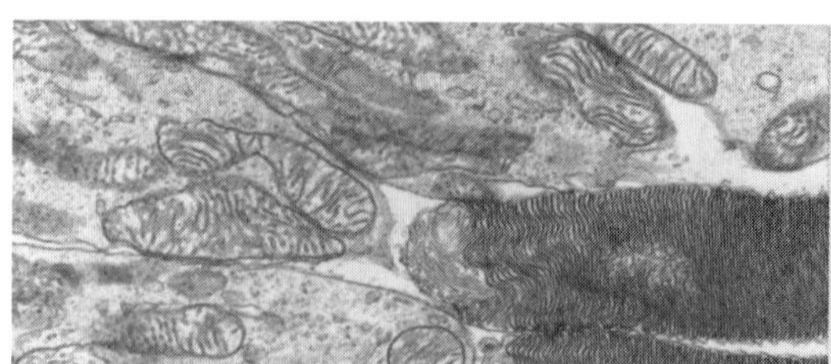

Electrons flowing through the electron transport chains in the membranes of the mitochondria, (at the left) in this thin section through the retina of a kangaroo rat (*Dipodomys ordi*) generates ATP. The ATP provides power for the synthesis and functioning of the stacked photoreceptor disks seen at the right. The outer segment of each rod cell (See Fig. 23-40), which may be 15–20 mm in length, consists of these disks, whose membranes contain the photosensitive protein pigment rhodopsin. Absorption of light initiates an electrical excitation which is sent to the brain. Micrograph from Porter and Bonneville (1973) Fine Structure of Cells and Tissues, Lea and Febiger, Philadelphia, Pennsylvania. Courtesy of Mary Bonneville.

Contents

18 Electron Transport, Oxidative Phosphorylation, and Hydroxylation

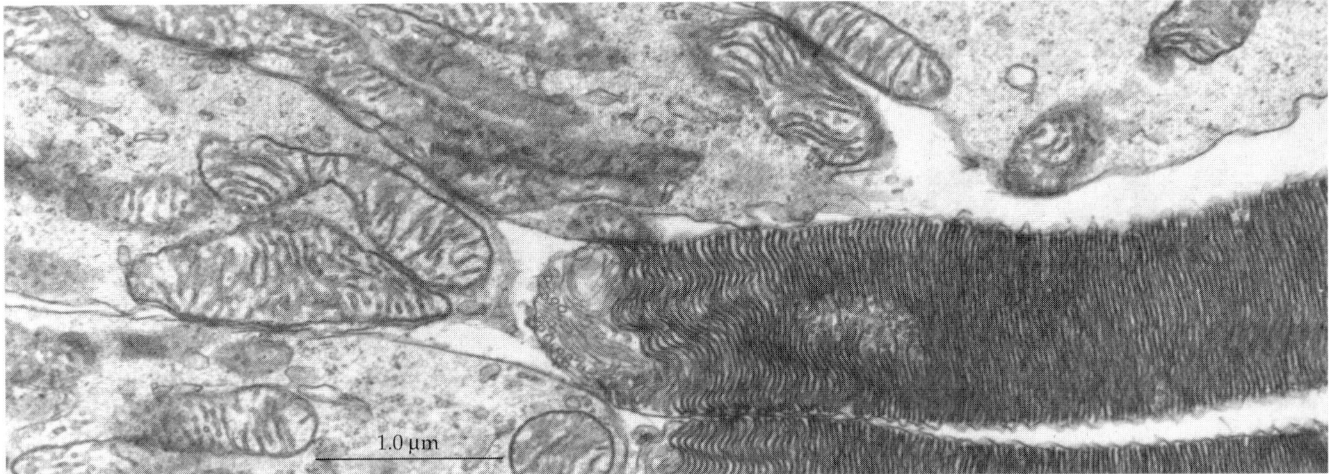

In this chapter we will look at the processes by which reduced carriers such as NADH and $FADH_2$ are oxidized within cells. Most familiar to us, because it is used in the human body, is **aerobic respiration**. Hydrogen atoms of NADH, $FADH_2$, and other reduced carriers appear to be transferred through a chain of additional carriers of increasingly positive reduction potential and are finally combined with O_2 to form H_2O. In fact, the hydrogen nuclei move freely as protons (or sometimes as H^- ions); it is the *electrons* that are deliberately transferred. For this reason, the chain of carriers is often called the **electron transport chain**. It is also referred to as the **respiratory chain**.

Because far more energy is available to cells from oxidation of NADH and $FADH_2$ than can be obtained by fermentation, the chemistry of the electron transport chain and of the associated reactions of ATP synthesis assumes great importance. A central question becomes "How is ATP generated by flow of electrons through this series of carriers"? Not only is most of the ATP formed in aerobic and in some anaerobic organisms made by this process of **oxidative phosphorylation**, but the solar energy captured during photosynthesis is used to form ATP in a similar manner. The mechanism of ATP generation may also be intimately tied to the function of membranes in the transport of ions. In a converse manner, the mechanism of oxidative phosphorylation may be related to the utilization of ATP in providing energy for the contraction of muscles.

In some organisms, especially bacteria, energy may be obtained through oxidation of H_2, H_2S, CO, or Fe^{2+} rather than of the hydrogen atoms removed from organic substrates. Furthermore, some bacteria use **anaerobic respiration** in which NO_3^-, SO_4^{2-}, or CO_2 act as oxidants either of reduced carriers or of reduced inorganic substances. In the present chapter, we will consider these energy-yielding processes as well as the chemistry of reactions of oxygen that lead to incorporation of atoms from O_2 into organic compounds.

The oxidative processes of cells have been hard to study, largely because the enzymes responsible are located in or on cell membranes. In bacteria the sites of electron transport and oxidative phosphorylation are on the inside of the plasma membrane or on membranes of mesosomes. In eukaryotes they are found in the inner membranes of the mitochondria and, to a lesser extent, in the endoplasmic reticulum. For this reason we should probably start with a closer look at mitochondria, the "power plants of the cell."

A. The Architecture of the Mitochondrion

Mitochondria are present in all eukaryotic cells that use oxygen in respiration, but the number per cell and the form and size vary.[1–4] Certain tiny trypanosomes have just *one* mitochondrion but some oocytes have as many as 3×10^5. Mammalian cells typically contain several hundred mitochondria and liver cells[5] more than 1000. Mammalian sperm cells may contain 50–75 mitochondria,[6] but in some organisms only one very large helical mitochondrion, formed by the fusion of many individual mitochondria, wraps around the base of the tail. Typical mitochondria appear to be about the size of cells of *E. coli*. However, study of ultrathin serial sections of a single yeast cell by electron microscopy has shown that, under some growth conditions, all of the mitochondria are interconnected.[7]

In every case a mitochondrion is enclosed by two

concentric membranes, an *outer* and an *inner* membrane, each ~5–7 nm thick (Figs. 18-1, 18-2). The inner membrane is folded to form the **cristae**. The number of cristae, the form of the cristae, and the relative amount of the internal **matrix** space are variable. In liver there is little inner membrane and a large matrix space, while in heart mitochondria there are more folds and a higher rate of oxidative phosphorylation. The enzymes catalyzing the tricarboxylic acid cycle are also unusually active in heart mitochondria. A typical heart mitochondrion has a volume of 0.55 μm^3; for every cubic micrometer of mitochondrial volume there are 89 μm^2 of inner mitochondrial membranes.[9]

Mitchondria can swell and contract, and forms other than that usually seen in osmium-fixed electron micrographs have been described. In some mitochondria the cristae are swollen, the matrix volume is much reduced, and the **intermembrane space** between the membranes is increased. Rapidly respiring mitochondria fixed for electron microscopy exhibit forms that have been referred to as "energized" and "energized-twisted."[10] The micrograph (Fig. 18-1) and drawing (Fig. 18-2) both show a significant amount of intermembrane space. However, electron micrographs of mitochondria from rapidly frozen aerobic tissues show almost none.[11] Recent studies by electron microscopic tomography show cristae with complex tubular structures. The accepted simple picture of mitochondrial structure (Fig. 18-1) is undergoing revision.[12–12b] The isolated mitochondria that biochemists have studied may be fragments of an interlinked **mitochondrial reticulum** that weaves its way through the cell.[12b] However, this reticulum may not be static but may break and reform. The accepted view that the mitochondrial matrix space is continuous with the internal space in the cristae is also the subject of doubts. Perhaps they are two different compartments.[12a]

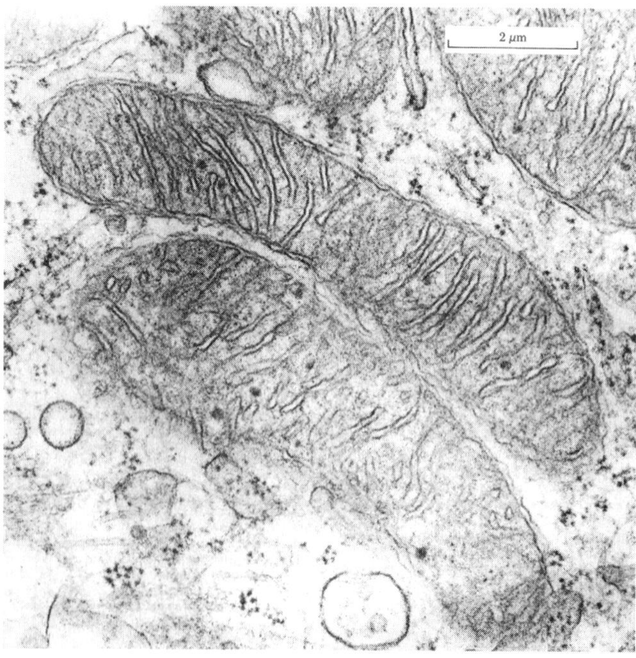

Figure 18-1 Thin section of mitochondria of a cultured kidney cell from a chicken embryo. The small, dark, dense granules within the mitochondria are probably calcium phosphate. Courtesy of Judie Walton.

1. The Mitochondrial Membranes and Compartments

The outer membranes of mitochondria can be removed from the inner membranes by osmotic rupture.[13] Analyses on separated membrane fractions show that *the outer membrane* is less dense (density ~1.1 g / cm^3) than the inner (density ~1.2 g / cm^3). It is highly permeable to most substances of molecular mass 10 kDa or less because of the presence of pores of ~2 nm diameter. These are formed by **mitochondrial porins**,[14–17] which are similar to the outer membrane porins of gram-negative bacteria (Fig. 8-20). The ratio of phospholipid to protein (~0.82 on a weight basis) is much higher than in the inner membrane. Extraction of the phospholipids by acetone destroys the membrane. Of the lipids present, there is a low content of cardiolipin, a high content of phosphatidylinositol and cholesterol, and no ubiquinone.

The inner membrane is impermeable to many substances. Neutral molecules of <150 Da can penetrate the membranes, but the permeability for all other materials including small ions such as H^+, K^+, Na^+, and Cl^- is tightly controlled. The ratio of phospholipid to protein in the inner membrane is ~0.27, and cardiolipin makes up ~20% of the phospholipid present. Cholesterol is absent. Ubiquinone and other components of the respiratory chain are all found in the inner membrane. Proteins account for 75% of the mass of the membrane.

Another characteristic of the inner mitochondrial membrane is the presence of projections on the inside surface, which faces the mitochondrial matrix. See Fig. 18-14. These spherical 85-kDa particles, discovered by Fernandez Moran in 1962 and attached to the membrane through a "stalk", display ATP-hydrolyzing (ATPase) activity. The latter was a clue that the knobs might participate in the *synthesis* of ATP during oxidative phosphorylation. In fact, they are now recognized as a complex of proteins called **coupling factor 1** (F_1) or **ATP synthase**.

In addition to bacterialike mitochondrial ribosomes and small circular molecules of DNA, mitochondria may contain variable numbers of dense granules of calcium phosphate, either $Ca_3(PO_4)_2$ or hydroxylapatite (Fig. 8-34),[4,18] as well as of phospholipoprotein.[4]

Quantitatively the major constituents of the matrix space are a large number of proteins. These account for about 56% by weight of the matrix material and exist in a state closer to that in a protein crystal than in a true solution.[19–20a] Mitochondrial membranes also contain proteins, both tightly bound relatively nonpolar intrinsic proteins and extrinsic proteins bound to the membrane surfaces. The other mitochondrial compartment, the intermembrane space, may normally be very small but it is still "home" for a few enzymes.

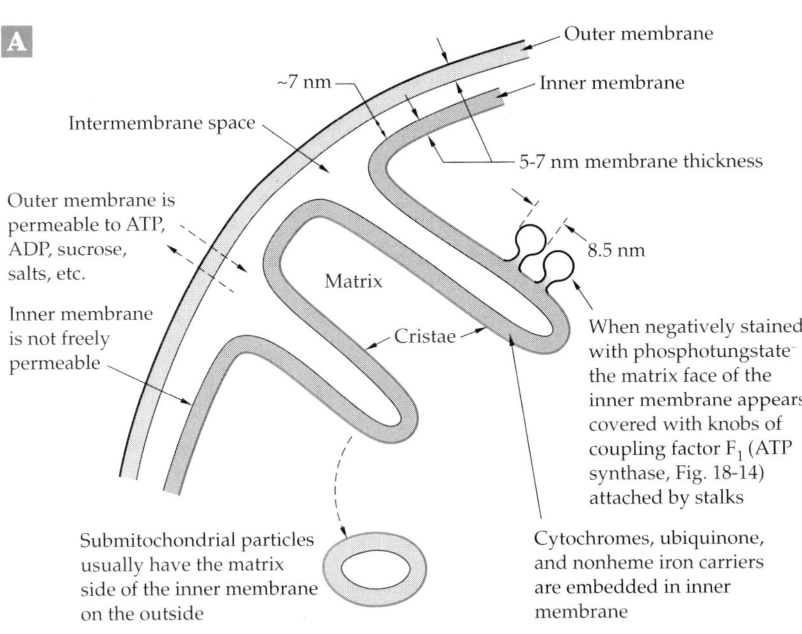

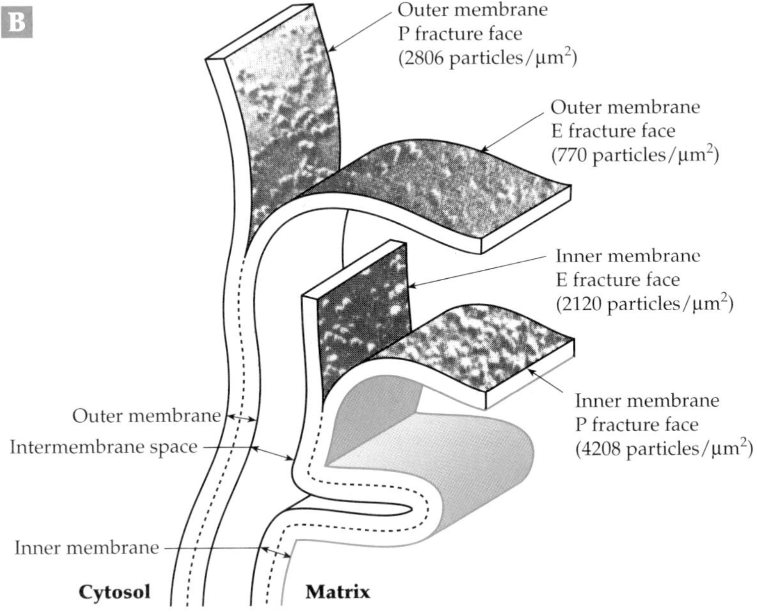

Figure 18-2 (A) Schematic diagram of mitochondrial structure. (B) Model showing organization of particles in mitochondrial membranes revealed by freeze-fracture electron microscopy. The characteristic structural features seen in the four half-membrane faces (EF and PF) that arise as a result of fracturing of the outer and inner membranes are shown. The four smooth membrane surfaces (ES and PS) are revealed by etching. From Packer.[8]

2. The Chemical Activities of Mitochondria

Mention of mitochondria usually brings to the mind of the biochemist the **citric acid cycle,** the **β oxidation pathway** of fatty acid metabolism, and **oxidative phosphorylation**. In addition to these major processes, many other chemical events also occur. Mitochondria concentrate Ca^{2+} ions and control the entrance and exit of Na^+, K^+, dicarboxylates, amino acids, ADP, P_i and ATP, and many other substances.[16] Thus, they exert regulatory functions both on catabolic and biosynthetic sequences. The glycine decarboxylase system (Fig. 15-20) is found in the mitochondrial matrix and is especially active in plant mitochondria (Fig. 23-37). Several cytochrome P450-dependent hydroxylation reactions, important to the biosynthesis and catabolism of steroid hormones and to the metabolism of vitamin D, take place within mitochondria. Mitochondria make only a few of their own proteins but take in several hundred other proteins from the cytoplasm as they grow and multiply.

Where within the mitochondria are specific enzymes localized? One approach to this question is to see how easily the enzymes can be dissociated from mitochondria. Some enzymes come out readily under hypotonic conditions. Some are released only upon sonic oscillation, suggesting that they are inside the matrix space. Others, including the cytochromes and the flavoproteins that act upon succinate and NADH, are so firmly embedded in the inner mitochondrial membranes that they can be dissociated only through the use of non-denaturing detergents.

Because the enzymes of the citric acid cycle[20a] (with the exception of succinate dehydrogenase) and β oxidation are present in the matrix, the reduced electron carriers must approach the inner membrane from the matrix side (the M side). Thus, the embedded enzymes designed to oxidize NADH, succinate, and other reduced substrates must be accessible from the matrix side. However, *sn*-glycerol 3-phosphate dehydrogenase, a flavoprotein, is accessible from the "outside" of the cytoplasmic (C side) of the inner membrane.[21] Fluorescent antibodies to cytochrome *c* bind only to the cytoplasmic (intermembrane) side of the inner membrane, but antibodies to cytochrome oxidase label both sides, which suggested that this protein complex spans the membrane.[22] However, oxidation of cytochrome *c* by cytochrome oxidase occurs only on the cytoplasmic surface.[22] Antibodies to the ATP synthase that makes up the knobs bind strictly to the matrix side.

The outer mitochondrial membrane contains monoamine oxidase, cytochrome b_5, fatty acyl-CoA synthase, and enzymes of cardiolipin synthesis[22a] as well as other proteins. Cardiolipin (diphosphatidylglycerol; Fig. 21-4) is found only in the inner mitochondrial membrane and in bacteria. It is functionally important for several mitochondrial enzymes including cytochrome oxidase and cytochome bc_1.[22a–c] It is also essential to photosynthetic membranes for which an exact role in an interaction between the lipid membrane and the associated protein has been revealed by crystallography.[22d] In other respects the composition of the inner mitochondrial membrane resembles that of the membranes of the endoplasmic reticulum. Isoenzyme III of adenylate kinase, a key enzyme involved in equilibrating ATP and AMP with ADP (Eq. 6-65), is one of the enzymes present in the intermembrane space. A number of other kinases, as well as sulfite oxidase, are also present between the membranes.[4]

As mentioned in Box 6-D, mitochondria sometimes take up calcium ions. The normal total concentration of Ca^{2+} is ~1 mM and that of free Ca^{2+} may be only ~0.1 µM.[22e,f] However, under some circumstances mitochondria accumulate large amounts of calcium, perhaps acting as a Ca^{2+} buffer.[22g,f] The so called ryanodine receptors (Fig. 19-21), prominent in the endoplasmic reticulum, have also been found in heart mitochondria, suggesting a function in control of calcium oscillations.[22i,j] On the other hand, accumulation of calcium by mitochondria may be pathological and the activation of Ca^{2+}-dependent proteases may be an initial step in apoptosis.[22h,22k]

TABLE 18-1
Catalog of Mitochondrial Genes[a]

Name and symbol	*Homo sapiens*	*Reclinomonas americana*	*Saccharomyces cerevisiae*	*Arabidopsis thaliana*
Ribosomal RNA				
s rRNA (small, 12s)	1	1	1	1
l rRNA (large, 16s)	1	1	1	1
5 S RNA		1		1
Transfer RNAs	22[b]	26	24	22
NADH dehydrogenase				
Subunits ND1–6, ND4L	7	12	0	9
Cytochrome *b*	1	1	1	1
Cytochrome oxidase				
Subunits I, II, III	3	3	3	3
ATP synthase				
Subunits 6, 8, others	2	5	3	4
Total protein coding genes	13	62	8	27
Total genes	37	92	35	53
Size of DNA (kbp)	16.596	69	75	367

[a] Data from Palmer, J. D. (1997) *Nature (London)* **387**, 454–455.[1]
[b] One for each amino acid of the genetic code but two each for serine and leucine.

3. The Mitochondrial Genome

Each mitochondrion contains several molecules of DNA (**mtDNA**), usually in a closed, circular form, as well as the ribosomes, tRNA molecules, and enzymes needed for protein synthesis.[1,23–26] With rare exceptions almost all of the mitochondrial DNA in a human cell is inherited from the mother.[6,26a] The size of the DNA circles varies from 16–19 kb in animals[27] to over 200 kb in many higher plants. Complete sequences of many mitochondrial DNAs are known.[28,28a] Among these are the 16,569 bp human mtDNA,[29] the 16,338 bp bovine mtDNA, the 16,896 bp mtDNA of the wallaroo *Macropus robustus*,[30] and the 17,533 bp mtDNA of the amphibian *Xenopus laevis*.[31,32] The sea urchin *Paracentotus lividus* has a smaller 15,697 bp genome. However, the order of the genes in this and other invertebrate mtDNA is different from that in mammalian mitochondria.[33] Protozoal mtDNAs vary in size from ~5900 bp for the parasitic malaria organism *Plasmodium falciparum*[34,35] to 41,591 bp for *Acanthamoeba castellani*[36] and 69,034 bp for the fresh water flagellate *Reclinomonas americana*.[26,37]

All of the mammalian mtDNAs are organized as shown in Fig. 18-3. The two strands of the DNA can be separated by virtue of their differing densities. The heavy (H) strand has a 5'→3' polarity in a counterclockwise direction in the map of Fig. 18-3, while the light (L) strand has a clockwise polarity. From the sequences 13 genes for specific proteins, 2 genes for ribosomal RNA molecules, and 22 genes for transfer RNAs have been identified. The genes are listed in Table 18-1 and have also been marked on the map in Fig. 18-3. The map also shows the tRNA genes, labeled with standard one-letter amino acid abbreviations, and the directions of transcription. Most of the protein genes are on the H-strand. One small region, the D-loop, contains an origin of replication and control signals for transcription (see Chapters 27 and 28).

The genes in mammalian mtDNA are closely packed with almost no nucleotides between them. However, the 19.5-kb mtDNA of *Drosophila* contains an (A+T)-rich region, which varies among species.[38] In the much larger 78-kb genome of yeast *Saccharomyces cerevisiae* many (A+T)-rich spacer regions are present.[39] The yeast mitochondrial genome also contains genes for several additional proteins. Mitochondria of *Reclinomonas americana* contain 97 genes, including those for 5S RNA, the RNA of ribonuclease P as well as a variety of protein coding genes. Perhaps this organism is primitive, resembling the original progenitor of eukaryotic life.[26] The mtDNA of trypanosomes is present in the large mitochondrion or kinetoplast as 40–50 "maxicircles" ~20–35 kb in size, together with 5000–10,000 "minicircles", each of 645–2500 bp (see Fig. 5-16). The latter encode **guide RNA** for use in RNA editing (Chapter 28). The large mitochondrial DNAs of higher plants, e.g., *Arabidopsis* (Table 18-1), also contain additional protein genes as well as large segments of DNA between the genes. The genome of the turnip (*Brassica campestris*) exists both as a 218-kb **master chromosome** and smaller 83- and 135-kb incomplete chromosomes, a pattern existing for most land plants.[40–42] The muskmelon contains 2500 kb of mitochondrial DNA (mtDNA). On the other hand, most mtDNA of the liverwort *Marchantia polymorpha* consists of 186-kb linear duplexes.[42a]

The compact size of the mammalian genome is dependent, in part, on alterations in the genetic code, as shown in Table 18-2, and a modification of tRNA structures that permits mitochondria to function with a maximum of 22 tRNAs (Chapter 28).[43–45] However, the more primitive *Reclinomonas* utilizes the standard genetic code in its mitochondria.[26] The mammalian mitochondrial genes contain no introns, but some yeast mitochondrial genes do. Furthermore, some of the introns contain long open reading frames. At least two of these genes-within-genes encode enzymes that excise the introns.

Why does mtDNA contain *any* protein genes, or why does mtDNA even exist? It seems remarkable that the cells of our bodies make the 100 or so extra proteins (encoded in the nucleus) needed for replication, transcription, amino acid activation, and mitochondrial ribosome formation and bring these into the mitochondria for the sole purpose of permitting the synthesis there of 13 proteins. The explanation is not evident. What are the 13 proteins?

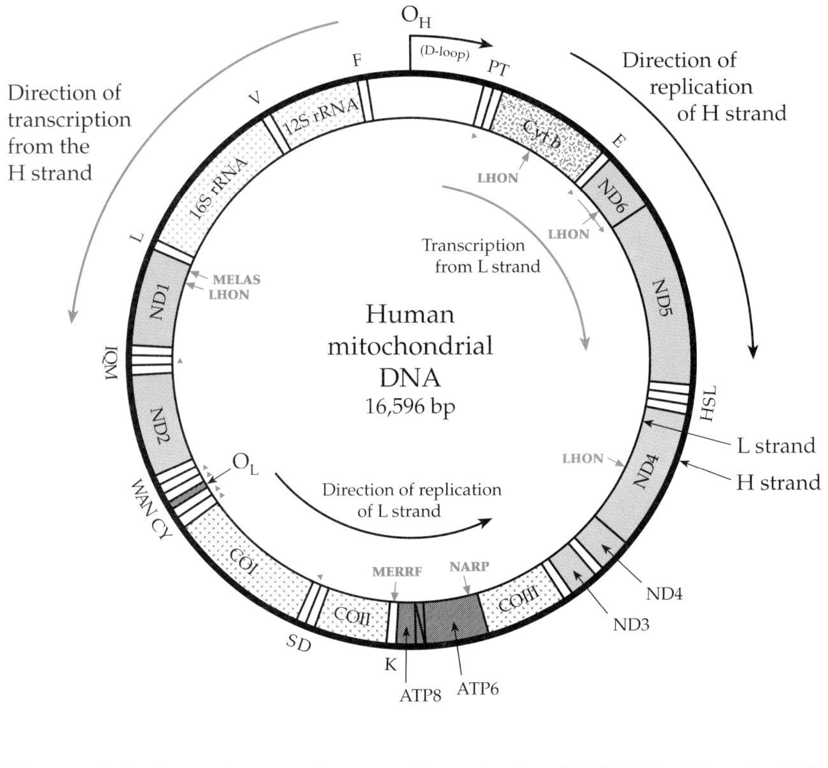

Figure 18-3 Genomic map of mammalian mitochondrial DNA. The stippled areas represent tRNA genes which are designated by the single-letter amino acid code; polarity is counterclockwise except for those marked by green arrow heads. All protein-coding genes are encoded on the H strand (with counterclockwise polarity), with the exception of ND6, which is encoded on the L strand. COI, COII, and COIII: cytochrome oxidase subunits I, II, and III; Cyt *b*: cytochrome *b*; ND: NADH dehydrogenase; ATP:ATP synthase. O_H and O_L: the origins of H and L strand replication, respectively. After Wallace[46] and Shoffner and Wallace.[45] Positions of a few of many known mutations that cause serious diseases are marked using abbrevations defined in Box 18-B.

TABLE 18-2
Alterations in the Genetic Code in the DNA of Animal Mitochondria

Codons	Nuclear DNA[a]	Mitochondrial DNA
AGA, ACG	Arg	Termination
AUA	Ile	Met
UGA	Termination	Trp

[a] See Table 5-5 for the other "standard" codons.

Three are the large functional subunits of cytochrome oxidase, one is cytochrome *b*, and seven are subunits of the NADH dehydrogenase system (Complex I). Two are subunits of ATP synthase. These are all vitally involved in the processes of electron transport and oxidative phosphorylation, but so are other proteins that are imported from the cytoplasm.

One gene in yeast mtDNA is especially puzzling. The *var 1* gene encodes a mitochondrial ribosomal protein, whose sequence varies with the strain of yeast. The gene is also involved in unusual recombinational events.[47] Another unusual aspect of yeast mitochondrial genetics is the frequent appearance of "petite" mutants, which grow on an agar surface as very small colonies. These have lost a large fraction of their mitochondrial DNA and, therefore, the ability to make ATP by oxidative phosphorylation. The remaining petite mtDNA may sometimes become integrated into nuclear DNA.[48] A few eukaryotes that have no aerobic metabolism also have no mitochondria.[49]

4. Growth and Development

Mitochondria arise by division and growth of preexisting mitochondria. Because they synthesize only a few proteins and RNA molecules, they must import many proteins and other materials from the cytoplasm. A mitochondrion contains at least 100 proteins that are encoded by nuclear genes.[50,50a] The mechanisms by which proteins are taken up by mitochondria are complex and varied. Many of the newly synthesized proteins carry, at the N terminus, presequences that contain **mitochondrial targeting signals**[51–53] (Chapter 10). These amino acid sequences often lead the protein to associate with receptor proteins on the outer mitochondrial membrane and subsequently to be taken up by the mitochondria. While the targeting sequences are usually at the N terminus of a polypeptide, they are quite often internal. The N-terminal sequences are usually removed by action of the **mitochondrial processing peptidase** (MPP) in the matrix, but internal targeting sequences are not removed.[52] Targeting of proteins to mitochondria may be assisted by **mRNA binding proteins** that guide appropriate mRNAs into the vicinity of mitochondria or other organelles.[53]

In addition to targeting signals, polypeptides destined for the inner mitochondrial membrane contain additional **topogenic signals** that direct the polypeptide to its destination. These topogenic signals are distinct from the targeting signals, which they sometimes follow. Topogenic signals are usually hydrophobic sequences, which may become transmembrane segments of the protein in its final location.[52,54] The uptake of many proteins by mitochondria requires the electrical potential that is usually present across the inner membrane (Section E). The fact that mitochondrial proteins usually have higher isoelectric points and carry more positive charges at neutral pH favors uptake.[55] In addition, chaperonins assist in

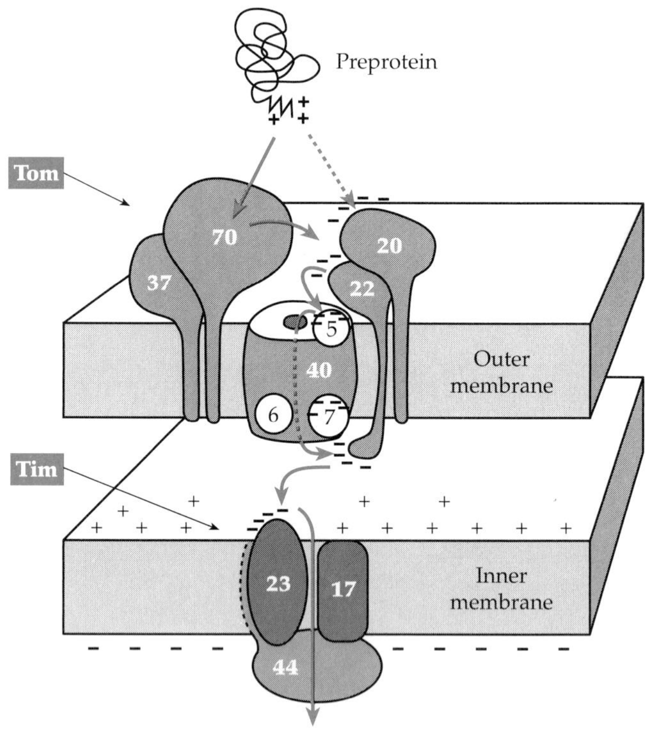

Figure 18-4 Schematic diagram of the protein transport machinery of mitochondrial membranes labeled according to the uniform nomenclature.[57] Subunits of outer membrane receptors and translocase are labeled Tom (translocase of outer membrane) and those of the inner membrane Tim (translocase of inner membrane). They are designated Tom70, etc., according to their sizes in kilodaltons (kDa). Preproteins are recognized by receptor Tom70•Tom 37 and / or by Tom22•Tom20. Clusters of negative charges on many components help guide the preprotein through the uptake pores in one or both membranes.[50,58] See Pfanner *et al.*,[57] Schatz,[50,50a] and Gabriel *et al.*[50b]

unfolding the protein to be taken up, assist in transport of some proteins,[50a] and may help the imported proteins to assemble into oligomeric structures.[51–53,56]

Protein uptake also requires a set of special proteins described as the **translocase of the outer mitochondrial membrane (Tom)** and **translocase of the mitochondrial inner membrane (Tim)**. Subunits that form the receptor targets and transport pores are designated, according to their approximate molecular masses in kD as Tom70, Tim23, etc. (Fig. 18-4).[57] Preproteins are recognized by the receptor complexes Tom70 • Tom37 and / or Tom22 • Tom20 on the mitochondrial surface. They then enter the **general import pore** formed by Tom40, Tom6, and Tom7 with the assistance of a small integral membrane protein Tom5, which has a positively charged C-terminal membrane anchor segment and a negatively charged N-terminal portion that may bind to the positively charged mitochondrial targeting sequences.[50,59] A number of other translocase components, including Tom20 and Tom22 of the outer membrane and Tim23 of the inner membrane, also have acidic extramembranous domains.[58] This suggested an "acid chain" hypothesis according to which the targeting signal interacts consecutively with a series of acidic protein domains that help to guide it across the two membranes.[50a,58,59] A series of small proteins, Tim8, 9, 10, 12, 13, function in yeast mitochondria to mediate the uptake of metabolite transporters. A defect in the human nuclear gene for a protein that resembles Tim8 causes **deafness dystonia**, a recessive X-linked neurodegenerative disorder.[59a,b]

B. Electron Transport Chains

During the 1940s, when it had become clear that formation of ATP in mitochondria was coupled to electron transport, the first attempts to pick the system apart and understand the molecular mechanism began. This effort led to the identification and at least partial characterization of several flavoproteins, iron-sulfur centers, ubiquinones, and cytochromes, most of which have been described in Chapters 15 and 16. It also led to the picture of mitochondrial electron transport shown in Fig. 10-5 and which has been drawn in a modern form in Fig. 18-5.

1. The Composition of the Mitochondrial Electron Transport System

Because of the difficulty of isolating the electron transport chain from the rest of the mitochondrion, it is easiest to measure ratios of components (Table 18-3). Cytochromes a, a_3, b, c_1, and c vary from a 1:1 to a 3:1 ratio while flavins, ubiquinone, and nonheme iron occur in relatively larger amounts. The much larger amount of pyridine nucleotides is involved in carrying electrons from the various soluble dehydrogenases of the matrix to the immobile carriers in the inner membrane, while ubiquinone has a similar function within the lipid bilayer of mitochondrial membranes.

TABLE 18-3
Ratios of Components in the Electron Transport Chain of Mitochondria[a,b]

Electron carrier	Rat liver mitochondria	Beef heart mitochondria
Cytochrome a_3	1.0	1.1
Cytochrome a	1.0	1.1
Cytochrome b	1.0	1.0
Cytochrome c_1	0.63	0.33–0.51
Cytochrome c	0.78	0.66–0.85
Pyridine nucleotides	24	
Flavins	3	1
Ubiquinone	3–6	7
Copper		2.2
Nonheme iron		5.5

[a] From Wainio, W. W. (1970) *The Mammalian Mitochondrial Respiratory Chain*, Academic Press, New York, and references cited therein.
[b] Molecular ratios are given. Those for the cytochromes refer to the relative numbers of heme groups.

What are the molar concentrations of the electron carriers in mitochondrial membranes? In one experiment, cytochrome b was found in rat liver mitochondria to the extent of 0.28 μmol/g of protein. If we take a total mitochondrion as about 22% protein, the average concentration of the cytochrome would be ~0.06 mM. Since all the cytochromes are concentrated in the inner membranes, which may account for 10% or less of the volume of the mitochondrion, the concentration of cytochromes may approach 1 mM in these membranes. This is sufficient to ensure rapid reactions with substrates.

2. The Sequence of Electron Carriers

Many approaches have been used to deduce the sequence of carriers through which electron flow takes place (Fig. 18-5). In the first place, it seemed reasonable to suppose that the carriers should lie in order of increasing oxidation–reduction potential going from left to right of the figure. However, since the redox potentials existing in the mitochondria may be somewhat different from those in isolated enzyme preparations, this need not be strictly true.

The development by Chance of a dual wavelength spectrophotometer permitted easy observation of the state of oxidation or reduction of a given carrier within mitochondria.[60] This technique, together with the study of specific inhibitors (some of which are indicated in Fig. 18-5 and Table 18-4), allowed some electron transport sequences to be assigned. For example, blockage with **rotenone** and **amytal** prevented reduction of the cytochrome system by NADH but allowed reduction by succinate and by other substrates having their own flavoprotein components in the chain. Artificial electron acceptors, some of which are shown in Table 18-5, were used to bypass parts of the chain as indicated in Fig. 18-5.

Submitochondrial particles and complexes. Many methods have been employed to break mitochondrial membranes into submitochondrial particles that retain an ability to catalyze some of the reactions of the chain.[61] For example, the Keilin–Hartree preparation of heart muscle is obtained by homogenizing mitochondria and precipitation at low pH.[62] The resulting particles have a low cytochrome *c* content and do not carry out oxidative phosphorylation.

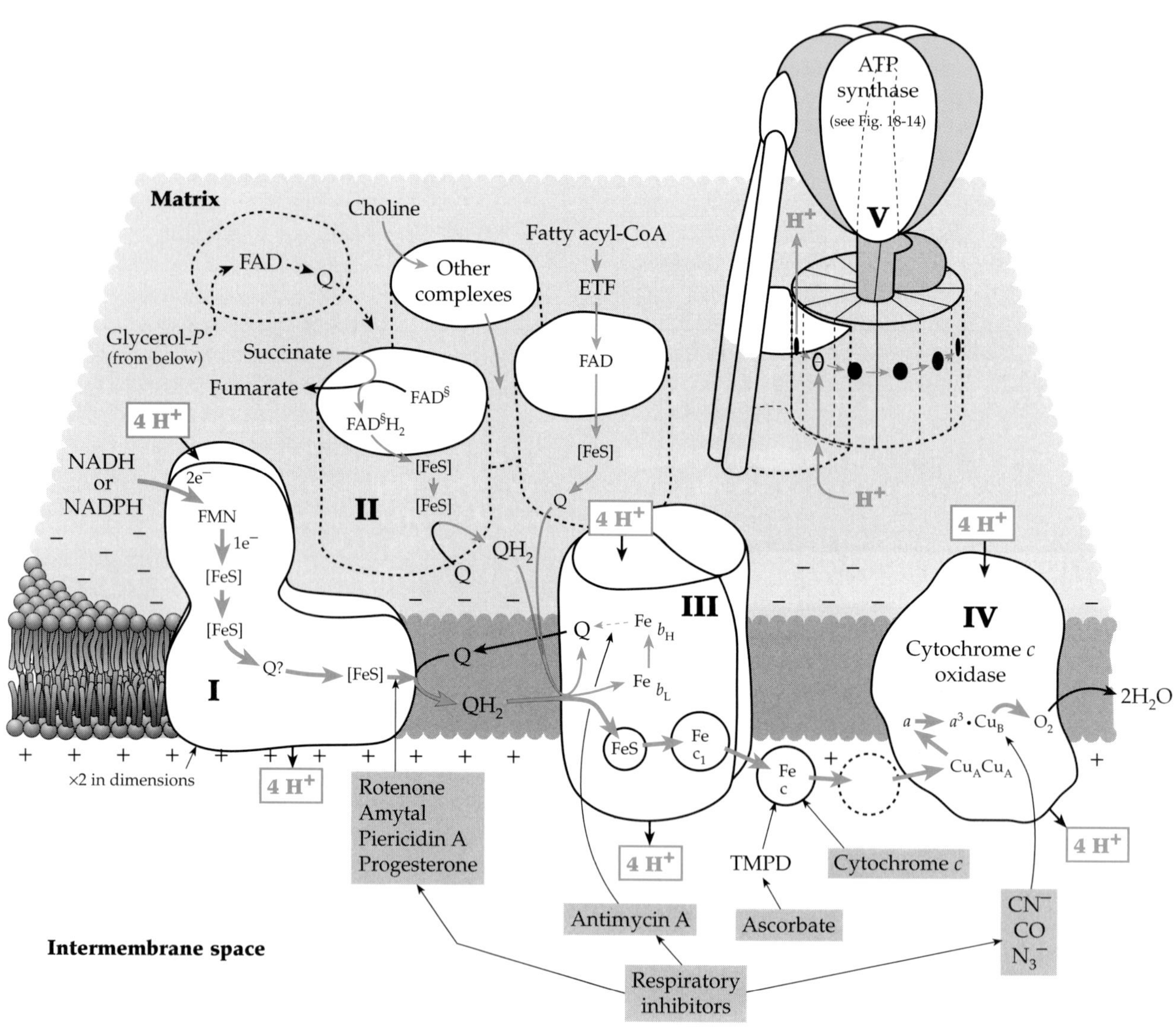

Figure 18-5 A current concept of the electron transport chain of mitochondria. Complexes I, III, and IV pass electrons from NADH or NADPH to O_2, one NADH or two electrons reducing one O to H_2O. This electron transport is coupled to the transfer of about 12 H^+ from the mitochondrial matrix to the intermembrane space. These protons flow back into the matrix through ATP synthase (V), four H^+ driving the synthesis of one ATP. Succinate, fatty acyl-CoA molecules, and other substrates are oxidized via complex II and similar complexes that reduce ubiquinone Q, the reduced form QH_2 carrying electrons to complex III. In some tissues of some organisms, glycerol phosphate is dehydrogenated by a complex that is accessible from the intermembrane space.

However, they do transport electrons and react with O_2. Other electron transport particles have been prepared by sonic oscillation. Under the electron microscope such particles appear to be small membranous vesicles resembling mitochondrial cristae.

Many detergents are strong denaturants of proteins, but some of them disrupt mitochondrial membranes without destroying enzymatic activity. A favorite is **digitonin** (Fig. 22-12), which causes disintegration of the outer membrane. The remaining fragments of inner membrane retain activity for oxidative phosphorylation. Such submitochondrial particles can be fractionated further by chemical treatments. Separate complexes can be obtained by treating the inner membranes with the nondenaturing detergent cholate (Fig. 22-10) and isolating the complexes by differential salt fractionation using ammonium sulfate. The isolated complexes I – IV catalyze reactions of four different portions of the electron transport process[63–65] as indicated in Eq. 18-1:

$$\text{Succinate} \xrightarrow{\text{II}} \text{Q}; \quad \text{NADH} \xrightarrow{\text{I}} \text{Q} \xrightarrow{\text{III}} \text{cytochrome } c \xrightarrow{\text{IV}} O_2 \qquad (18\text{-}1)$$

These complexes are usually named as follows: I, **NADH-ubiquinone oxidoreductase**; II, **succinate-ubiquinone oxidoreductase**; III, **ubiquinol-cytochrome *c* oxidoreductase**; IV, **cytochrome *c* oxidase**. The designation **complex V** is sometimes applied to ATP synthase (Fig. 18-14). Chemical analysis of the electron transport complexes verified the probable location of some components in the intact chain. For example, a high iron content was found in both complexes I and II and copper in complex IV.

We now recognize not only that these complexes are discrete structural units but also that they are functional units. Complete X-ray crystallographic structures are available for complexes III and IV and for much of the ATP synthase complex. As is indicated in Fig. 18-5, complexes I – IV are linked by two soluble electron carriers, ubiquinone and cytochrome *c*.

The lipid-soluble ubiquinone (Q) is present in both bacterial and mitochondrial membranes in relatively large amounts compared to other electron carriers (Table 18-2). It seems to be located at a point of convergence of the NADH, succinate, glycerol phosphate, and choline branches of the electron transport chain. Ubiquinone plays a role somewhat like that of NADH, which carries electrons between dehydrogenases in the cytoplasm and from soluble dehydrogenases in the aqueous mitochondrial matrix to flavoproteins embedded in the membrane. Ubiquinone transfers electrons plus protons between proteins within the

TABLE 18-4
Some Well-Known Respiratory Inhibitors[a]

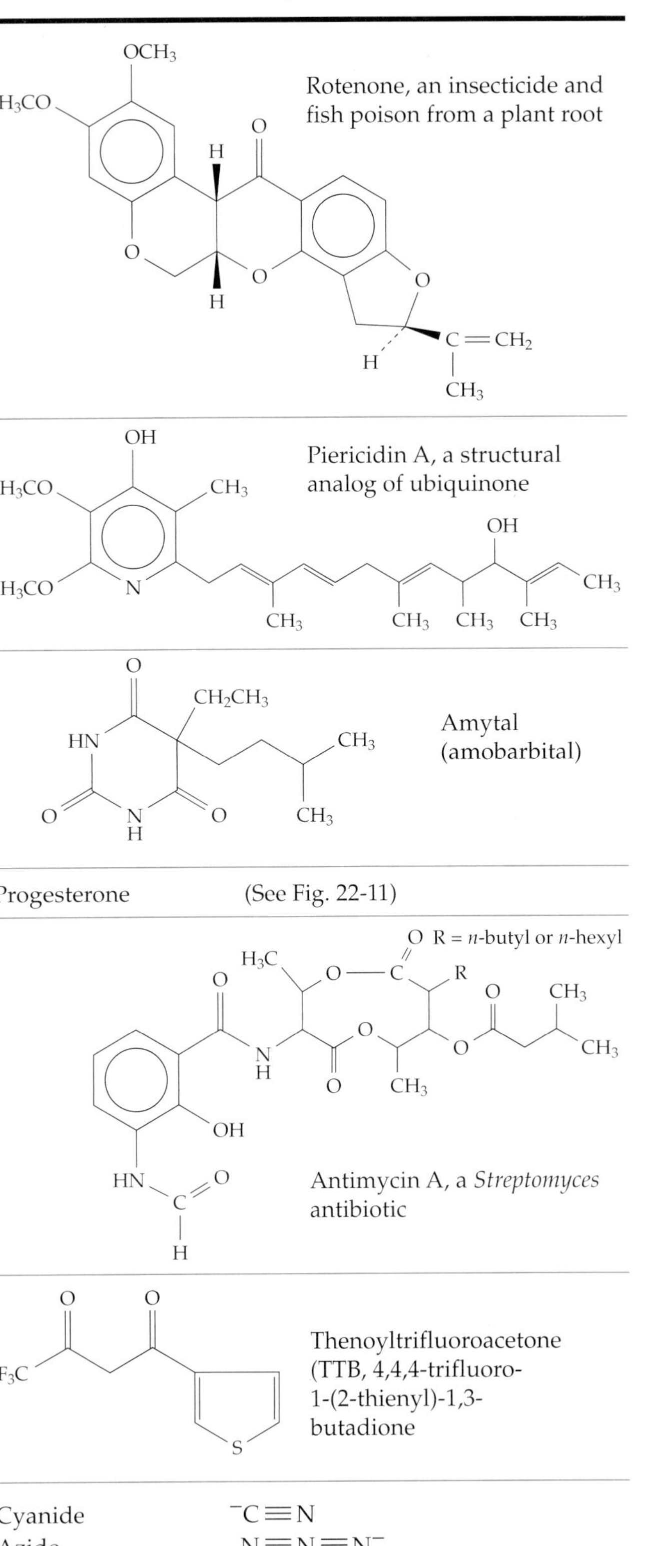

[a] See Fig. 18-5 for sites of inhibition.

BOX 18-A HISTORICAL NOTES ON RESPIRATION

Animal respiration has been of serious interest to scientists since 1777, when Lavoisier concluded that foods undergo slow combustion within the body, supposedly in the blood. In 1803–1807, Spallanzani established for the first time that the tissues were the actual site of respiration, but his conclusions were largely ignored. In 1884, MacMunn discovered that cells contain the heme pigments, which are now known as **cytochromes**. However, the leading biochemists of the day dismissed the observations as experimental error, and it was not until the present century that serious study of the chemistry of biological oxidations began.[a,b]

Recognition that substrates are oxidized by **dehydrogenation** is usually attributed to H. Wieland. During the years 1912–1922 he showed that synthetic dyes, such as methylene blue, could be substituted for oxygen and would allow respiration of cells in the absence of O_2. Subsequent experiments (see Chapter 15) led to isolation of the soluble pyridine nucleotides and flavoproteins and to development of the concept of an electron transport chain.

Looking at the other end of the respiratory chain, Otto Warburg[c,d] noted in 1908 that all aerobic cells contain iron. Moreover, iron-containing charcoal prepared from blood catalyzed nonenzymatic oxidation of many substances, but iron-free charcoal prepared from cane sugar did not. Cyanide was found to inhibit tissue respiration at low concentrations similar to those needed to inhibit nonenzymatic catalysis by iron salts. On the basis of these investigations, Warburg proposed in 1925 that aerobic cells contain an iron-based *Atmungsferment* (respiration enzyme), which was later called **cytochrome oxidase**. It was inhibited by carbon monoxide.

Knowing that carbon monoxide complexes of hemes are dissociated by light, Warburg and Negelein, in 1928, determined the photochemical **action spectrum** (see Chapter 23) for reversal of the carbon monoxide inhibition of respiration of the yeast *Torula utilis*. The spectrum closely resembled the absorption spectrum of known heme derivatives (Fig. 16-7). Thus, it was proposed that O_2, as well as CO, combines with the iron of the heme group in the Atmungsferment.

Meanwhile, during 1919–1925, David Keilin, while peering through a microscope equipped with a spectroscopic ocular at thoracic muscles of flies and other insects, observed a pigment with four distinct absorption bands. At first he thought it was derived by some modification of hemoglobin, but when he found the same pigment in fresh baker's yeast, he recognized it as an important new substance. In his words:[e]

One day while I was examing a suspension of yeast freshly prepared from a few bits of compressed yeast shaken vigorously with a little water in a test-tube, I failed to find the characteristic four-banded absorption spectrum, but before I had time to remove the suspension from the field of vision of the microspectroscope the four absorption bands suddenly reappeared. This experiment was repeated time after time and always with the same result: the absorption bands disappeared on shaking the suspension with air and reappeared within a few seconds on standing.

I must admit that this first visual perception of an intracellular respiratory process was one of the most impressive spectacles I have witnessed in the course of my work. Now I have no doubt that cytochrome is not only widely distributed in nature and completely independent of haemoglobin, but that it is an intracellular respiratory pigment which is much more important than haemoglobin.

Keilin soon realized that three of the absorption bands, those at 604, 564, and 550 nm (*a*, *b*, and *c*), represented different pigments, while the one at 521 nm was common to all three. Keilin proposed the names cytochromes *a*, *b*, and *c*. The idea of an electron transport or respiratory chain followed[e] quickly as the flavin and pyridine nucleotide coenzymes were recognized to play their role at the dehydrogenase level. Hydrogen removed from substrates by these carriers could be used to oxidize reduced cytochromes. The latter would be oxidized by oxygen under the influence of cytochrome oxidase.

In 1929, Fiske and Subbarow,[d,f–h] curious about the occurrence of purine compounds in muscle extracts, discovered and characterized ATP. It was soon shown (largely through the work of Lundsgaard and Lohman)[f] that hydrolysis of ATP provided energy for muscular contraction. At about the same time, it was learned that synthesis of ATP accompanied glycolysis. That ATP could also be formed as a result of electron transport became clear following an observation of Engelhardt[h,i] in 1930, that methylene blue stimulated ATP synthesis by tissues.

The study of electron transport chains and of oxidative phosphorylation began in earnest after Kennedy and Lehninger,[j] in 1949, showed that mitochondria were the site not only of ATP synthesis but also of the operation of the citric acid cycle and fatty acid oxidation pathways. By 1959, Chance had introduced elegant new techniques of spectrophotometry that led to formulation of the electron

BOX 18-A (continued)

transport chain as follows:

Substrate → pyridine nucleotides → flavoprotein → cyt b → cyt c → cyt a → cyt a_3 → O_2

Since that time, some new components have been added, notably the ubiquinones and iron-sulfur proteins, but the basic form proposed for the chain was correct.

[a] Kalckar, H. M. (1969) *Biological Phosphorylations*, Prentice-Hall, Englewood Cliffs, New Jersey
[b] Kalckar, H. M. (1991) *Ann. Rev. Biochem.* **60**, 1–37
[c] Edsall, J. T. (1979) *Science* **205**, 384–385
[d] Fiske, C. H., and Subbarow, Y. (1929) *Science* **70**, 381–382
[e] Keilin, D. (1966) *The History of Cell Respiration and Cytochrome*, Cambridge Univ. Press, London and New York
[f] Kalckar, H. (1980) *Trends Biochem. Sci.* **5**, 56–57
[g] Schlenk, F. (1987) *Trends Biochem. Sci.* **12**, 367–368
[h] Saraste, M. (1998) *Science* **283**, 1488–1493
[i] Slater, E. C. (1981) *Trends Biochem. Sci.* **6**, 226–227
[j] Talalay, P., and Lane, M. D. (1986) *Trends Biochem. Sci.* **11**, 356–358

membrane bilayer. Membranes also contain **ubiquinone-binding proteins**,[66,67] which probably hold the ubiquinone that is actively involved in electron transport. Perhaps some ubiquinone molecules function as fixed carriers. There is also uncertainty about the number of sites at which ubiquinone functions in the chain.

Mitochondrial electron transport in plants and fungi. Plant mitochondria resemble those of mammals in many ways, but they contain additional dehydrogenases and sometimes utilize alternative pathways of electron transport,[68–73] as do fungi.[74] Mitochondria are impermeable to NADH and NAD$^+$. Animal mitochondria have shuttle systems (see Fig. 18-16) for bringing the reducing equivalents of NADH into mitochondria and to the NADH dehydrogenase that faces the matrix side of the inner membrane. However, plant mitochondria also have an NADH dehydrogenase on the *outer* surface of the inner membrane (Fig. 18-6). This enzyme transfers electrons to ubiquinone, is not inhibited by rotenone (see Fig. 18-5), and also acts on NADPH. Inside the mitochondria a high-affinity NADH dehydrogenase resembles complex I of animal mitochondria and is inhibited by rotenone.[73] There is also a low-affinity NADH dehydrogenase, which is insensitive to rotenone. Some plant mitochondria respire slowly in the presence of cyanide. They utilize an **alternative oxidase** that replaces complex III and cytochome c oxidase and which is not inhibited by antimycin or by cyanide (Fig. 18-6).[68,71,75] It is especially active in thermogenic plant tissues (Box 18-C). A

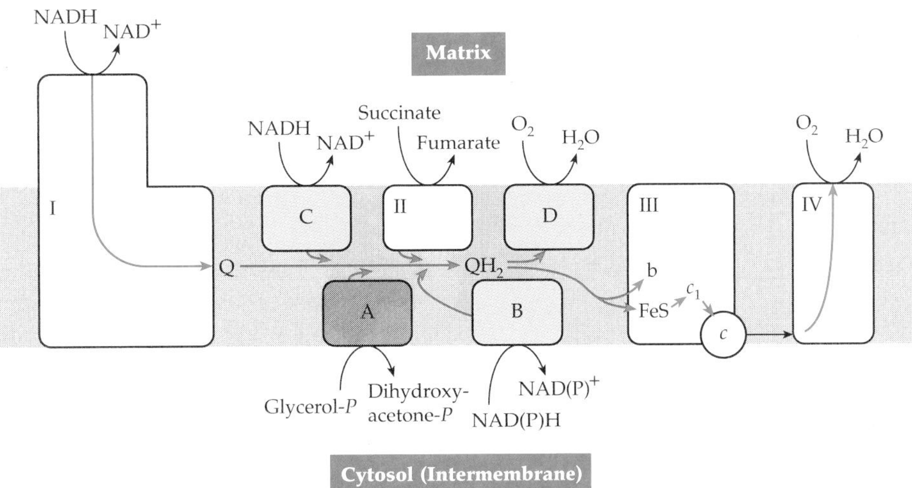

Figure 18-6 Schematic diagram of some mitochondrial dehydrogenase and oxidase complexes of plants and also the glycerol phosphate dehydrogenase of animals, which is embedded in the inner membrane. Complexes I–IV are also shown. (A) The glycerol phosphate dehydrogenase of some animal tissues. It is accessible from the intermembrane space on the cytosolic side. (B) The rotenone-insensitive NAD(P)H dehydrogenase of the external membrane surface of plants. (C) The rotenone-insensitive NADH dehydrogenase facing the matrix side in some plants. (D) The plant alternative oxidase. Ubiquinone, Q. The three green stippled dehydrogenases are not coupled to proton pumps or ATP synthesis. After Hoefnagel *et al.*[73]

BOX 18-B DEFECTS OF MITOCHONDRIAL DNA

A mutation in any of the 13 protein subunits, the 22 tRNAs, or the two rRNAs whose genes are carried in mitochondrial DNA may possibly cause disease. The 13 protein subunits are all involved in electron transport or oxidative phosphorylation. The syndromes resulting from mutations in mtDNA frequently affect oxidative phosphorylation (OXPHOS) causing what are often called "OXPHOS diseases."[a–g] Mitochondrial oxidative phosphorylation also depends upon ~100 proteins encoded in the nucleus. Therefore, OXPHOS diseases may result from defects in either mitochondrial or nuclear genes. The former are distinguished by the fact that they are inherited almost exclusively maternally. Most mitochondrial diseases are rare. However, mtDNA is subject to rapid mutation, and it is possible that accumulating mutants in mtDNA may be an important component of aging.[h–k]

The first recognition of mitochondrial disease came in 1959. A 30-year old Swedish woman was found to have an extremely high basal metabolic rate (180% of normal), a high caloric intake (>3000 kcal / day), and an enormous perspiration rate. She had developed these symptoms at age seven. Examination of her mitochondria revealed that electron transport and oxidative phosphorylation were very loosely coupled. This explains the symptoms. However, the disease (Luft disease) is extremely rare and the underlying cause isn't known.[i] Its recognition did focus attention on mitochondria, and by 1988, 120 different mtDNA defects had been described.[e,i]

Some OXPHOS disorders, including Luft disease, result from mutations in nuclear DNA. A second group arise from point mutations in mtDNA and a third group involve deletions, often very large, in mtDNA. Persons with these deletions survive because they have both mutated and normal mtDNA, a condition of **heteroplasmy** of mtDNA. As these persons age their disease may become more severe because they lose many normal mitochondria.[d,e]

The names of mitochondrial diseases are often complex and usually are described by abbreviations. Here are a few of them: **LHON**, Lebers hereditary optical neuropathy; **MERRF**, myoclonic epilepsy and ragged-red-fiber disease; **MELAS**, mitochondrial encephalomyopathy, lactic acidosis, and strokelike episodes; **NARP**, neurological muscle weakness, ataxia, and retinitis pigmentosa; **Leigh disease — SNE**, subacute necrotizing encephalomyelopathy; **KSS**, Kearns–Sayre syndrome; **CPEO**, chronic progressive external ophthalmoplegia. LHON is a hereditary disease that often leads to sudden blindness from death of the optic nerve especially among males. Any one of several point mutations in subunits ND1, 2, 4, 5, and 6 of NADH dehydrogenase (complex I; Figs. 18-3 and 18-5), cytochrome *b* of complex II, or subunit I of cytochrome oxidase may cause this syndrome. Most frequent is an R340H mutation of the ND4 gene at position 11,778 of mtDNA (Fig. 18-3).[e,l,m] It may interfere with reduction of ubiquinone.[n] Mutations in the ND1 gene at position 3460 and in the ND6 gene at position 14484 or in the cytochrome *b* gene at position 15257 cause the same disease.[l] The most frequent (80–90%) cause of MERRF, which is characterized by epilepsy and by the appearance of ragged red fibers in stained sections of muscle, is an A → G substitution at position 8344 of mtDNA in the TψC loop (Fig. 5-30) of mitochondrial $tRNA^{Lys}$. A similar disease, MELAS, is accompanied by strokes (not seen in MERRF) and is caused in 80% of cases by an A → G substitution in the dihydrouridine loop (Fig. 5-30) of mitochondrial $tRNA^{Leu}$.[o] CPEO, Leigh disease, and KSS often result from large deletions of mtDNA.[p] NARP and related conditions have been associated with an L156R substitution in the ATPase 6 gene of ATP synthase.[q]

Can mitochondrial diseases be treated? Attempts are being made to improve the function of impaired mitochondria by adding large amounts of ubiquinone, vitamin K, thiamin, riboflavin, and succinate to the diet.[e] One report suggests that mitochondrial decay during aging can be reversed by administration of *N*-acetylcarnitine.[k]

---

[a] Palca, J. (1990) *Science* **249**, 1104–1105
[b] Capaldi, R. A. (1988) *Trends Biochem. Sci.* **13**, 144–148
[c] Darley-Usmar, V., ed. (1994) *Mitochondria: DNA, Proteins and Disease*, Portland Press, London
[d] Wallace, D. C. (1999) *Science* **283**, 1482–1488
[e] Shoffner, J. M., and Wallace, D. C. (1995) in *The Metabolic and Molecular Bases of Inherited Disease*, 7th ed., Vol. 1 (Scriver, C. R., Beaudet, A. L., Sly, W. S., and Valle, D., eds), pp. 1535–1609, McGraw-Hill, New York
[f] Schon, E. A. (2000) *Trends Biochem. Sci.* **25**, 555–560
[g] Tyler, D. D. (1992) *The Mitochondrion in Health and Disease*, VCH Publ., New York
[h] Wallace, D. C. (1992) *Science* **256**, 628–632
[i] Luft, R. (1994) *Proc. Natl. Acad. Sci. U.S.A.* **91**, 8731–8738
[j] Tanhauser, S. M., and Laipis, P. J. (1995) *J. Biol. Chem.* **270**, 24769–24775
[k] Shigenaga, M. K., Hagen, T. M., and Ames, B. N. (1994) *Proc. Natl. Acad. Sci. U.S.A.* **91**, 10771–10778
[l] Hofhaus, G., Johns, D. R., Hurko, O., Attardi, G., and Chomyn, A. (1996) *J. Biol. Chem.* **271**, 13155–13161
[m] Brown, M. D., Trounce, I. A., Jun, A. S., Allen, J. C., and Wallace, D. C. (2000) *J. Biol. Chem.* **275**, 39831–39836
[n] Zickermann, V., Barquera, B., Wikström, M., and Finel, M. (1998) *Biochemistry* **37**, 11792–11796
[o] Hayashi, J.-I., Ohta, S., Kagawa, Y., Takai, D., Miyabayashi, S., Tada, K., Fukushima, H., Inui, K., Okada, S., Goto, Y., and Nonaka, I. (1994) *J. Biol. Chem.* **269**, 19060–19066
[p] Moraes, C. T., and 19 other authors. (1989) *N. Engl. J. Med.* **320**, 1293–1299
[q] Hartzog, P. E., and Cain, B. D. (1993) *J. Biol. Chem.* **268**, 12250–12252

TABLE 18-5
Some Artificial Electron Acceptors[a,b]

Compound	Structure	E°′(pH 7) 30°C
Ferricyanide	$Fe(CN)_6^{3-}$	+0.36 V (25°C)
Oxidized form of tetramethyl-*p*-phenylenediamine	$(H_3C)_2\overset{+}{N}{=}C_6H_4{=}\overset{+}{N}(CH_3)_2$	+0.260 V
2,6-Dichlorophenol-indophenol (DCIP)	Cl, Cl; O=…=N–C₆H₄–OH	+0.217 V
Phenazine methosulfate (PMS)	N, $\overset{+}{N}$–CH_3; 1/2 SO_4^{2-}	+0.080 V
Ascorbate	(See Box 18-D)	+0.058 V
Methylene blue	H_3C–N–CH_3; N; S; $\overset{+}{N}$(CH_3)CH_3	+0.011 V
Menadione	O; CH_3; O	+0.008 V (°25C)
Tetrazolium salts, e.g., "neotetrazolium chloride"	(C, N, N, N^+, N)$_2$	−0.125 V

[a] From Wainio, W. W. (1970) *The Mammalian Mitochondrial Respiratory Chain*, Academic Press, New York, pp. 106–111.
[b] See Fig. 18-5 for sites of action.

similar oxidase is present in trypanosomes.[72] Neither the rotenone-insensitive dehydrogenases nor the alternative oxidases are coupled to synthesis of ATP.

Electron transport chains of bacteria. The bacterial electron transport systems are similar to that of mitochondria but simpler. Bacteria also have a variety of alternative pathways that allow them to adapt to various food sources and environmental conditions.[76,77] The gram-negative soil bacterium *Paracoccus denitrificans*, which has been called "a free-living mitochondrion," has a mammalian-type respiratory system. Its complexes I–IV resemble those of animals and of fungi,[78–79] but *Paracoccus* has fewer subunits in each complex. Complex I of *E. coli* is also similar to that of our own bodies.[79–80] However, other major flavoprotein dehydrogenases in *E. coli* act on D-lactate and *sn*-3-glycerol phosphate.[81] Pyruvate is oxidized by a membrane-bound flavoprotein (Fig. 14-2). All of these enzymes pass electrons to ubiquinone-8 (Q_8).[82] Succinate dehydrogenase of *E. coli* resembles that of mitochondria,[83] and the ubiquinol oxidase of *Paracoccus* resembles complexes III + IV of mitochondria. It can be resolved into a three-subunit bc_1 complex, a three-subunit c_1aa_3 complex, and another 57-kDa peptide.[84] The last contains a 22-kDa cytochrome c_{552}, which is considerably larger than mitochondrial cytochrome *c*.

The cytochrome aa_3 terminal oxidase is produced constitutively, i.e., under all conditions. However, when cells are grown on succinate or H_2 another set of enzymes is produced with the *b*-type cytochrome o_3 as the terminal **quinol oxidase** (Eq. 18-2).[85]

$$\begin{array}{l} \qquad\quad H_2 \\ \qquad\quad \downarrow \\ \text{Succinate} \rightarrow Q \rightarrow \text{cytochrome } b \\ \qquad\qquad\qquad\qquad \downarrow \\ \qquad\qquad\qquad \text{cytochrome } o_3 \\ \qquad\qquad\qquad\qquad \downarrow \\ \qquad\qquad\qquad\qquad O_2 \end{array} \qquad (18\text{-}2)$$

Two terminal quinol oxidase systems, both related to cytochrome *c* oxidase, are utilized by *E. coli* to oxidize ubiquinol-8. When cultured at high oxygen tensions, cytochrome bo_3 (also called cytochrome *bo*) is the major oxidase. It utilizes heme *o* (Fig. 16-5) instead of heme *a*. However, at low oxygen tension, e.g., in the late logarithmic stage of growth, the second oxidase, cytochrome *bd*, is formed.[76,86–88a] It contains two molecules of the chlorin heme *d* (Fig. 16-5), which appear to be involved directly in binding O_2. This terminal oxidase system is present in many bacteria and can utilize either O_2 or nitrite as the oxidant. A simpler electron transport chain appears to be involved in the oxidation of pyruvate by *E. coli*. The flavoprotein pyruvate oxidase passes electrons to Q_8, whose reduced form can pass electrons directly to cytochrome *d*. Incorporation of these two pure protein complexes and ubiquinone-8 into phospholipid vesicles has given an active reconstituted chain.[82] Other bacteria utilize a variety of quinol oxidase systems, which contain various combinations of cytochromes: aa_3, caa_3, *cao*, bo_3, and ba_3.[88b,c]

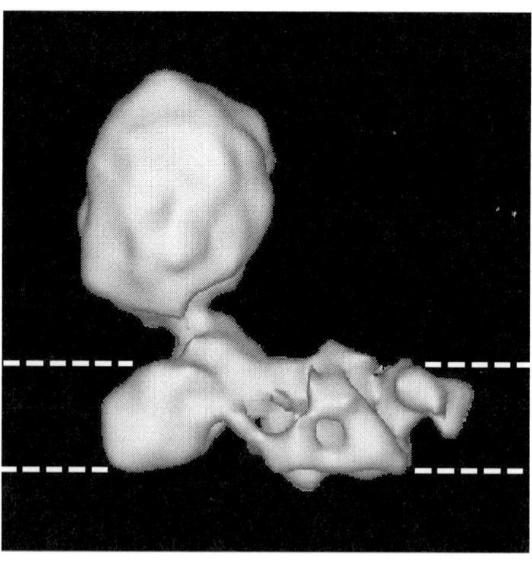

Figure 18-7 Three-dimensional image of bovine NADH-Ubiquinone oxidoreductase (complex I) reconstructed from individual images obtained by electron cyro-microscopy. The resolution is 2.2 nm. The upper portion projects into the mitochondrial matrix while the horizontal part lies within the membrane as indicated. Courtesy of N. Grigorieff.[90]

3. Structures and Functions of the Individual Complexes I – IV and Related Bacterial Assemblies

What are the structures of the individual electron transport complexes? What are the subunit compositions? What cofactors are present? How are electrons transferred? How are protons pumped? We will consider these questions for each of complexes I–IV, as found in both prokaryotes and eukaryotes.[88d,e]

Complex I, NADH-ubiquinone oxidoreductase. Complex I oxidizes NADH, which is generated within the mitochondrial matrix by many dehydrogenases. Among these are the pyruvate, 2-oxoglutarate, malate, and isocitrate dehydrogenases, which function in the tricarboxylic acid cycle; the β-oxoacyl-CoA dehydrogenase of the β oxidation system for fatty acids; and 2-hydroxybutyrate, glutamate, and proline dehydrogenases. All produce NADH, which reacts with the flavoprotein component of complex I. Whether from bacteria,[79] fungal mitochondria,[89] or mammalian mitochondria[89a,90] complex I exists as an L-shaped object, of which each of the two arms is ~23 nm long. One arm projects into the matrix while the other lies largely within the inner mitochondrial membrane (Fig. 18-7). The mitochondrial complex, which has a mass of ~1 MDa, has the same basic structure as the 530-kDa bacterial complex. However, the arms are thicker in the mitochondrial complex. Analysis of the denatured proteins by gel electrophoresis revealed at least 43 peptides.[78,90] Bound to some of these are the electron carriers FMN, Fe_2S_2, and Fe_4S_4 clusters, ubiquinone or other quinones, and perhaps additional unidentified cofactors.[79] Complex I from *E. coli* is smaller, containing only 14 subunits. These are encoded by a cluster of 14 genes, which can be directly related by their sequences to subunits of mitochondrial complex I and also to the corresponding genes of *Paracoccus denitrificans*.[80,91] Complex I of *Neurospora* contains at least 35 subunits.[89] The 14 subunits that are present both in bacteria and in mitochondria probably form the structural core of the complex. The other subunits thicken, strengthen, and rigidify the arms. Some of the "extra" subunits have enzymatic activities that are not directly related to electron transport. Among these are a 10-kDa prokaryotic type acyl carrier protein (ACP), which may be a relic of a bacterial fatty acid synthase, reflecting the endosymbiotic origin of mitochondria.[92] Also present is a 40-kDa NAD(P)H dependent reductase / isomerase, which may be involved in a biosynthetic process, e.g., synthesis of a yet unknown redox group.[79,92]

In all cases, FMN is apparently the immediate acceptor of electrons from NADH. From the results of extrusion of the Fe–S cores (Chapter 16) and EPR measurements it was concluded that there are three tetranuclear (Fe_4S_4) iron–sulfur centers and at least two binuclear (Fe_2S_2) centers[93,94] as well as bound ubiquinone.[95] Chemical analysis of iron and sulfide suggested up to eight Fe–S clusters per FMN, while gene sequences reveal potential sites for formation of six Fe_4S_4 clusters and two Fe_2S_2 clusters.[78] Treatment of complex I with such "chaotropic agents" as 2.5 M urea or 4 M sodium trichloroacetate followed by fractionation with ammonium sulfate[95] gave three fractions:

(1) A soluble NADH dehydrogenase consisting of a 51-kDa peptide that binds both the FMN and also one tetranuclear Fe – S cluster (designated N3) and a 24-kDa peptide that carries a binuclear Fe–S center designated N1b. (2) A 75-kDa peptide bearing two binuclear Fe–S centers, one of which is called N1a and also 47-, 30-, and 13-kDa peptides. One of these carries tetranuclear center N4. (3) A group of insoluble, relatively nonpolar proteins, one of which carries tetranuclear cluster N2. It may be the immediate donor of electrons to a ubiquinone held by a ubiquinone-binding protein designated QP-N. In bacteria seven of these are homologs of the seven NADH dehydrogenase subunits encoded by mtDNA (Fig. 18-3). A 49-kDa subunit of complex I in the yeast *Yarrowia lipolytica* is strikingly similar to the hydrogen reactive subunit of NiFe hydrogenases (Fig. 16-26).[95a] These proteins are thought to lie within the membrane arm and to form ~55 transmembrane α helices. Ubiquinone may also function as a carrier within complex I,[96,97] and there may be a new redox cofactor as well.[79] The following tentative sequence (Eq. 18-3) for electron transfer within complex I (with apparent $E^{\circ\prime}$ values of carriers) has been suggested. By equilibration with external redox systems, the redox potentials of these centers within the mitochondria have been estimated and are given (in V) in Eq. 18-3. The presence of a

$$\begin{array}{cccccccccccc} & & & & & & & & & & \text{rotenone} \\ \text{NADH} \rightarrow & \text{FMN} \rightarrow & \text{N1a} \rightarrow & (\text{N3}, & \text{N1b}, & \text{N4}) & \rightarrow \text{N2} & \rightarrow \text{QP-N} & \dashrightarrow \text{Q} \\ & & & & & & & & \ominus \\ & & Fe_2S_2 & Fe_4S_4 & Fe_2S_2 & Fe_4S_4 & Fe_4S_4 & & \\ -0.32 & & -0.38 & & -0.24 & & -0.21\ \text{V} & & \end{array} \quad (18\text{-}3)$$

large fraction of the bound ubiquinone as a free radical suggests that the quinone functions as a one-electron acceptor rather than a two-electron acceptor. A characteristic of complex I is inhibition by rotenone or piericidin, both of which block electron transport at the site indicated in Fig. 18-5.

Complex II, succinate-ubiquinone oxidoreductase. Complex II, which carries electrons from succinate to ubiquinone, contains covalently linked 8^{α}-(*N*-histidyl)-FAD (Chapter 15) as well as Fe–S centers and one or more ubiquinone-binding sites. There are four subunits whose structures and properties have been highly conserved among mitochondria and bacteria and also in **fumarate reductases**. The latter function in the opposite direction during anaerobic respiration with fumarate as the terminal oxidant, both in bacteria[98–99a] and in parasitic helminths and other eukaryotes that can survive prolonged anaerobic conditions (Chapter 17, Section F,2).[100] Complex II from *E. coli* consists of 64-, 27-, 14-, and 13-kDa subunits, which are encoded by genes *sdhCDAB* of a single operon.[101–103] The two larger hydrophilic subunits associate to form the readily soluble succinate dehydrogenase. The 64-kDa subunit carries the covalently bound FAD while the 27-kDa subunit carries three Fe–S centers. The two small 13- and 14-kDa subunits form a hydrophobic anchor and contain a ubiquinol-binding site (QD-S)[103] as well as a heme that may bridge the two subunits[102] to form cytochrome b_{556}. The functions of the heme is uncertain. The soluble mammalian succinate dehydrogenase resembles closely that of *E. coli* and contains three Fe – S centers: binuclear S1 of $E^{\circ\prime}$ 0 V, and tetranuclear S2 and S3 of −0.25 to −0.40 and + 0.065 V, respectively. Center S3 appears to operate between the −2 and −1 states of Eq. 16-17 just as does the cluster in the *Chromatium* high potential iron protein. The function of the very low potential S2 is not certain, but the following sequence of electron transport involving S1 and S3 and the bound ubiquinone QD–S[66] has been proposed (Eq. 18-4).

$$\text{Succinate} \rightarrow (\text{FAD*, S1}) \rightarrow \text{S3} \rightarrow \text{QP–S} \rightarrow \text{Q} \qquad (18\text{-}4)$$

In addition to complexes I and II several other membrane-associated FAD-containing dehydrogenase systems also send electrons to soluble ubiquinone. These include dehydrogenases for choline, *sn*-glycerol 3-phosphate, and the electron-transfer protein (ETF) of the fatty acyl-CoA β oxidation system (Fig. 18-5). The last also accepts electrons from dehydrogenases for sarcosine (*N*-methylglycine), dimethylglycine, and other substrates. The *sn*-glycerol 3-phosphate dehydrogenase is distinguished by its accessibility from the intermembrane (cytosolic) face of the inner mitochondrial membrane (Fig. 18-6).

Complex III (ubiquinol-cytochrome c oxidoreductase or cytochrome bc_1 complex). Mitochondrial complex III is a dimeric complex, each subunit of which contains 11 different subunits with a total molecular mass of ~240 kDa per monomer.[104–107] However, in many bacteria the complex consists of only three subunits, cytochrome *b*, cytochrome c_1, and the high potential (~0.3 V) Rieske iron-sulfur protein, which is discussed in Chapter 16, Section A,7. These three proteins are present in all bc_1 complexes. In eukaryotes the 379-residue cytochrome *b* is mitochondrially encoded. Although there is only one cytochrome *b* gene in the mtDNA, two forms of cytochrome *b* can be seen in absorption spectra: b_H (also called b_{562} or b_K) and lower potential b_L (also called b_{566} or b_T).[107a,b]

X-ray diffraction studies have revealed the complete 11-subunit structure of bovine bc_1 complex[104,106–107] as well as a nearly complete structure of the chicken bc_1 complex (Fig. 18-8).[105] The bovine complex contains 2166 amino acid residues per 248-kDa monomer and exists in crystals as a 496-kDa dimer and probably functions as a dimer.[106-107] The two hemes of cytochrome *b* are near the two sides of the membrane, and the Fe–S and cytochrome c_1 subunits are on the surface next to the intermembrane space (Fig. 18-8). On the matrix side (bottom in Fig. 18-8A) are two large ~440 residue "core" subunits that resemble subunits of the mitochondrial processing protease. They may be evolutionary relics of that enzyme.[106,108,108a] Mitochondrial cytochrome b_H has an $E°'$ value of +0.050 V, while that of b_L is −0.090 V at pH 7.[109] That of the Rieske Fe–S protein is + 0.28 V.[110]

The sequence of electron transport within complex III has been hard to determine in detail. For reasons discussed in Section C, the "Q-cycle" shown in Fig. 18-9 has been proposed.[111–114a] As is indicated in Fig. 18-9, complex II accepts electrons from QH_2 and passes them consecutively to the Fe–S protein, cytochrome c_1, and the external cytochrome *c*. However, half of the electrons are recycled through the two heme groups of cytochrome *b*, as is indicated in the figure and explained in the legend. The X-ray structure (Fig. 18-8) is consistent with this interpretation. Especially intriguing is the fact that the Fe_2S_2 cluster of the Rieske protein subunit has been observed in two or three different conformations.[105–107,114a-c] In Fig. 18-8C the structures of two conformations are superimposed. The position of the long helix at the right side is unchanged but the globular domain at the top can be tilted up to bring the Fe_2S_2 cluster close to the heme of cytochrome c_1, or down to bring the cluster close to heme b_L. Movement between these two positions is probably part of the catalytic cycle.[115]

The simpler cytochrome bc_1 complexes of bacteria such as *E. coli*,[102] *Paracoccus dentrificans*,[116] and the photosynthetic *Rhodobacter capsulatus*[117] all appear to function in a manner similar to that of the large mitochondrial complex. The bc_1 complex of *Bacillus subtilis* oxidizes reduced **menaquinone** (Fig. 15-24) rather than ubiquinol.[118] In chloroplasts of green plants photochemically reduced **plastoquinone** is oxidized by a similar complex of cytochrome *b*, *c*-type cytochrome *f*, and a Rieske Fe–S protein.[119–120a] This cytochrome b_6f complex delivers electrons to the copper protein plastocyanin (Fig. 23-18).

The electron acceptor for complex III is cytochrome *c*, which, unlike the other cytochromes, is water soluble and easily released from mitochondrial membranes. Nevertheless, it is usually present in a roughly 1:1 ratio with the fixed cytochromes, and it seems unlikely that it is as free to diffuse as are ubiquinone and NAD^+.[121,122] However, a small fraction of the cytochrome *c* may diffuse through the intermembrane space and accept electrons from cytochrome b_5, which is located in the outer membrane.[123] Cytochrome *c* forms a complex with cardiolipin (diphosphatidylglycerol), a characteristic component of the inner mitochondrial membrane.[124]

Complex IV. Cytochrome c oxidase (ubiquinol-cytochrome c oxidoreductase). Complex IV from mammalian mitochondria contains 13 subunits. All of them have been sequenced, and the three-dimensional structure of the complete complex is known (Fig. 18-10).[125–127] The simpler cytochrome *c* oxidase from *Paracoccus denitrificans* is similar but consists of only three subunits. These are homologous in sequence to those of the large subunits I, II, and III of the mitochondrial complex. The three-dimensional structure of the *Paracoccus* complex is also known. Its basic structure is nearly identical to that of the catalytic core of subunits I, II, and III of the mitochondrial complex (Fig. 18-10,A).[128] All three subunits have transmembrane helices. Subunit III seems to be structural in function, while subunits I and II contain the oxidoreductase centers: two hemes *a* (*a* and a_3) and two different copper centers, Cu_A (which contains two Cu^{2+}) and a third Cu^{2+} (Cu_B) which exists in an EPR-silent exchange coupled pair with a_3. Bound Mg^{2+} and Zn^{2+} are also present in the locations indicated in Fig. 18-10.

The Cu_A center has an unusual structure.[130–132] It was thought to be a single atom of copper until the three-dimensional structure revealed a dimetal center, whose structure follows. The Cu_B-cytochrome a_3 center is also unusual. A histidine ring is covalently attached to tyrosine.[133–135a] Like the tyrosine in the active site of galactose oxidase (Figs. 16-29, 16-30), which carries a covalently joined cysteine, that of cytochrome oxidase may be a site of tyrosyl radical formation.[135]

Cytochrome *c* oxidase accepts four electrons, one at a time from cytochrome *c*, and uses them to reduce O_2 to two H_2O. Electrons enter the oxidase via the Cu_A center and from there pass to the cytochrome *a* and on to the cytochrome a_3 – Cu_B center where the reduction of O_2 takes place. A possible sequence of steps in the catalytic cycle is given in Fig. 18-11. Reduction of O_2 to two H_2O requires four electrons and also four protons. An additional four protons are evidently pumped across the membrane for each catalytic cycle.[136–138] The overall reaction is:

$$4\,e^- + 8\,H^+_{IN} + O_2 \longrightarrow 2\,H_2O + 4\,H^+_{OUT} \quad (18\text{-}5)$$

The reaction of O_2 with cytochrome *c* oxidase to form the oxygenated species A (Fig. 18-11) is very rapid, occurring with apparent lifetime τ (Eq. 9-5) of ~8–10 μs.[139] Study of such rapid reactions has depended upon a flow-flash technique developed by Greenwood and Gibson.[136,140,141] Fully reduced cytochrome oxidase is allowed to react with carbon monoxide, which binds to the iron in cytochrome a_3 just as does O_2. In fact, it was the spectroscopic observation that only half of the

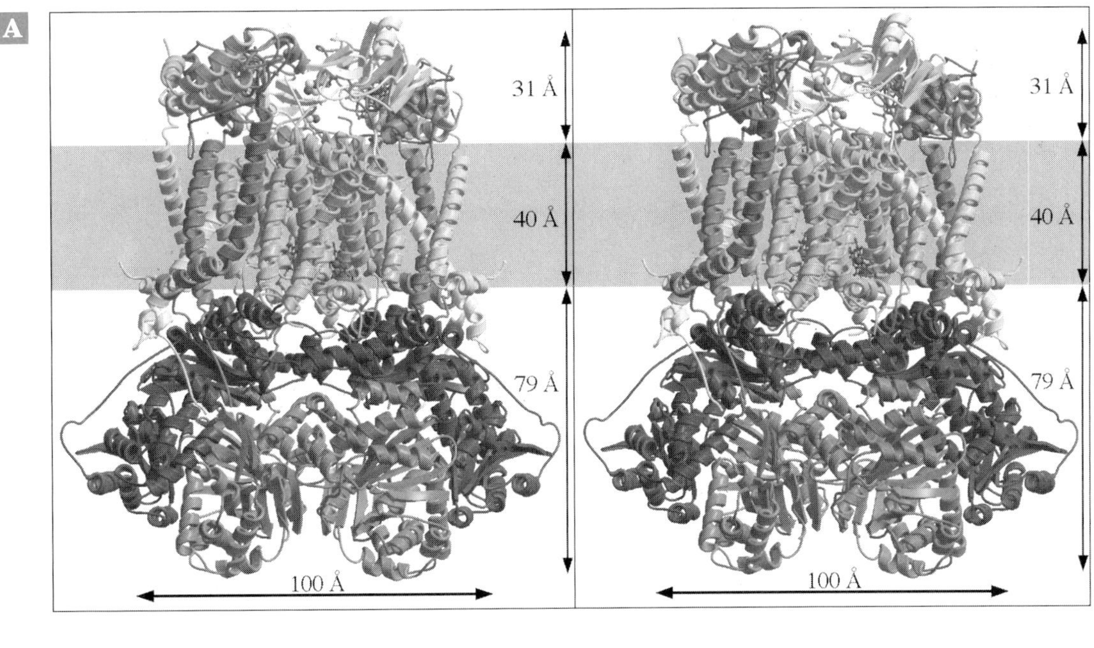

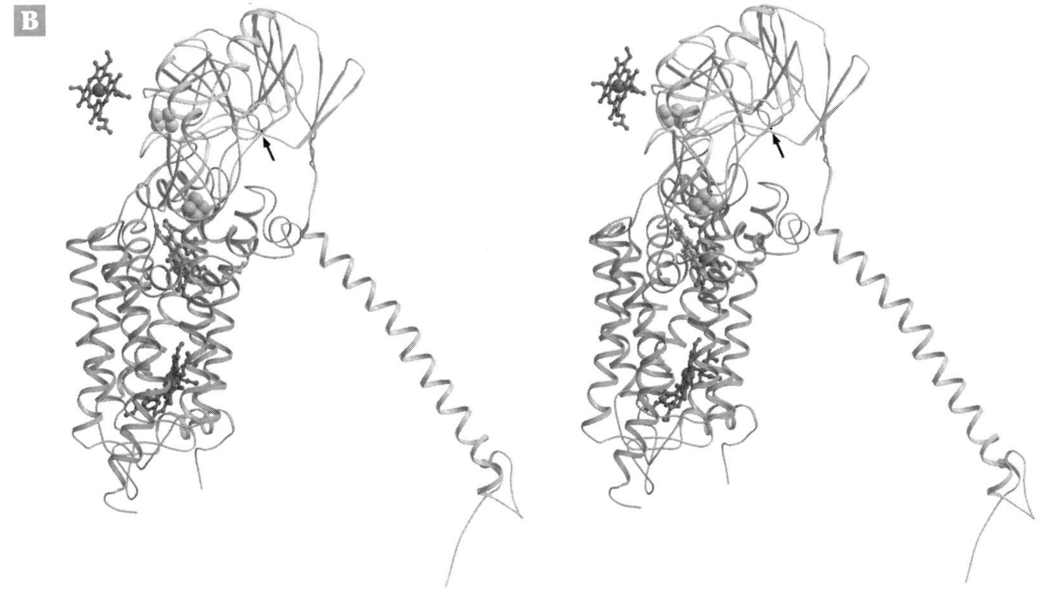

Figure 18-8 Stereoscopic ribbon diagrams of the chicken bc_1 complex (A) The native dimer. The molecular twofold axis runs vertically between the two monomers. Quinones, phospholipids, and detergent molecules are not shown for clarity. The presumed membrane bilayer is represented by a gray band. (B) Isolated close-up view of the two conformations of the Rieske protein (top and long helix at right) in contact with cytochrome *b* (below), with associated heme groups and bound inhibitors, stigmatellin, and antimycin. The isolated heme of cytochrome c_1 (left, above) is also shown. (C) Structure of the intermembrane (external surface) domains of the chicken bc_1 complex. This is viewed from within the membrane, with the transmembrane helices truncated at roughly the membrane surface. Ball-and-stick models represent the heme group of cytochrome c_1, the Rieske iron–sulfur cluster, and the disulfide cysteines of subunit 8. SU, subunit; cyt, cytochrome. From Zhang *et al.*[105]

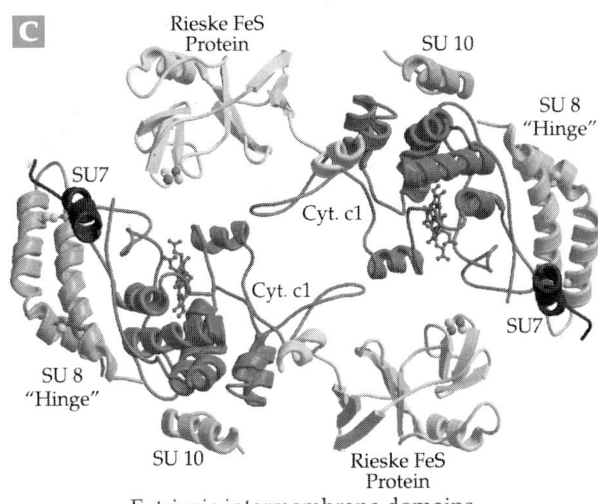

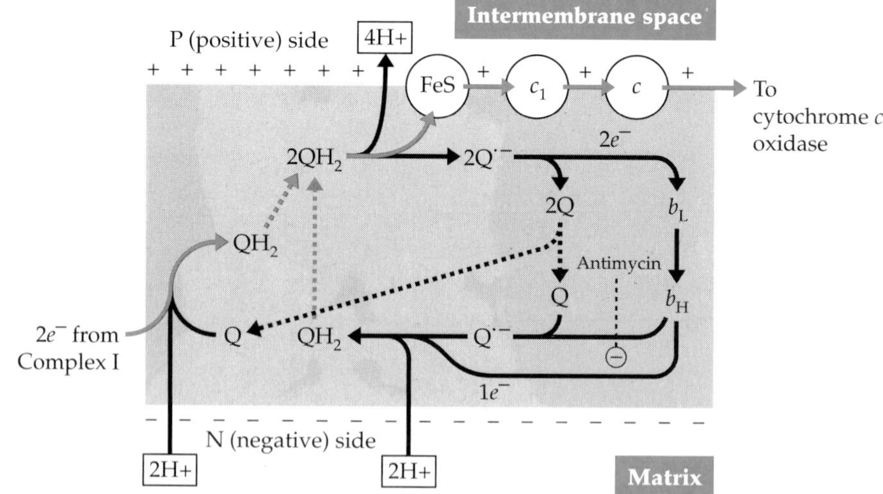

Figure 18-9 Proposed routes of electron transfer in mitochondrial complex III according to Peter Mitchell's Q cycle. Ubiquinone (Q) is reduced to QH_2 by complex I (left side of diagram) using two H^+ taken up from the matrix (leaving negative charges on the inner membrane surface). After diffusing across the bilayer (dashed line) the QH_2 is oxidized in the two steps with release of the two protons per QH_2 on the positive (P) side of the membrane. In the two-step oxidation via anionic radical Q^- one electron flows via the Rieske Fe–S protein and the cytochrome c_1 heme to external cytochrome c. The other electron is transferred to heme b_L of cytochrome b, then across the membrane to heme b_H which now reduces Q to Q^-. A second QH_2 is dehydrogenated in the same fashion and the electron passed through the cytochrome b centers is used to reduce Q^- to QH_2 with uptake from the matrix of 2 H^+. The resulting QH_2 diffuses back across the membrane to function again while the other Q diffuses back to complex I. The net result is pumping of 4 H^+ per 2 e^- passed through the complex. Notice that in the orientation used in this figure the matrix is at the bottom, not the top as in Figs. 18-4 and 18-5.

The Cu_A center of cytochrome oxidase

The $Cu_B \bullet A_3$ center of cytochrome oxidase

cytochrome a combined with CO that led Keilin to designate the reactive component a_3. This CO complex is mixed with O_2-containing buffer and irradiated with a laser pulse to release the CO and allow O_2 to react. The first rapid reaction observed is the binding of O_2 (step c in Fig. 18-11). Formation of a peroxy intermediate from the initial oxygenated form (A in Fig. 18-11) is very fast. The O–O bond of O_2 has already been cleaved in form P (Fig. 18-11), which has until recently been thought to be the peroxy intermediate. In fact, spectroscopic measurements indicate that form P contains an oxo-ferryl ion with the second oxygen of the original O_2 converted to an OH ion and probably coordinated with Cu_B.[136a,136c,142,142a–c] P may also contain an organic radical, perhaps formed from tyrosine 244 as indicated in Fig. 18-11.

A second relaxation time of τ = 32–45 μs has been assigned[139] to the conversion of the peroxide intermediate P to P′. A third relaxation time (τ = 100–140 μs) is associated with the oxidation of Cu_A by a (not shown in Fig. 18-11).[143] This electron transfer step limits the rate of step f of Fig. 18-11. Another reduction step with $\tau \sim 1.2$ ms is apparently associated with electron transfer in step h. This slowest step still allows a first-order reaction rate of ~800 s^{-1}.

When O_2 reacts with cytochrome c oxidase, it may be bound initially to either the a_3 iron or to Cu_B, but in the peroxy intermediate P it may bind to both atoms. Oxyferryl compound F (Fig. 8-11) as well as radical species, can also be formed by treatment of the oxidized

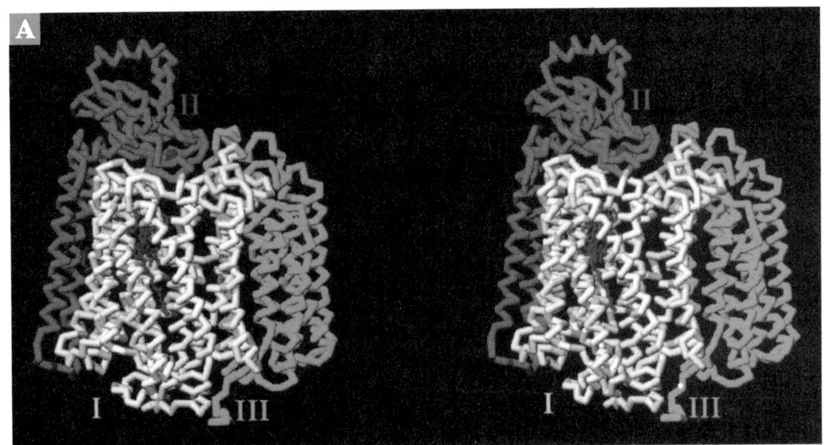

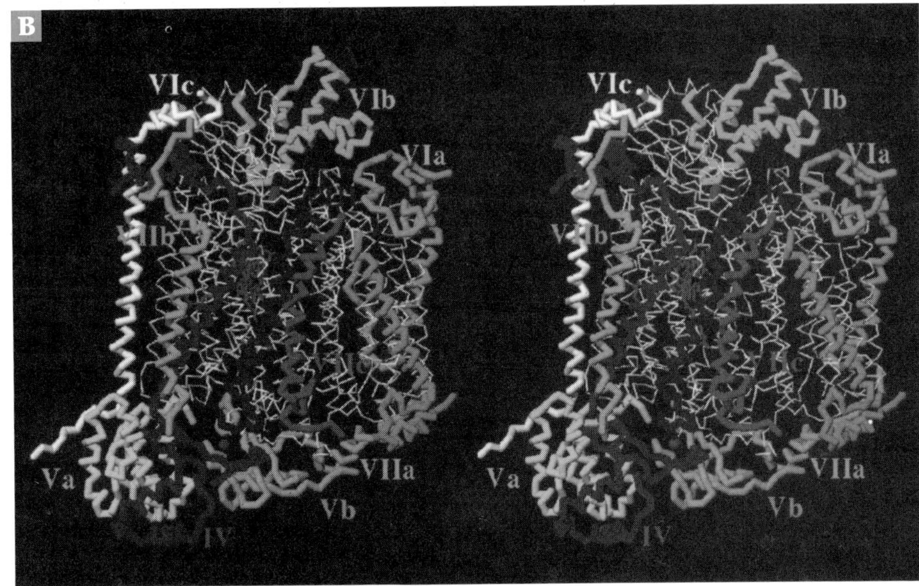

Figure 18-10 Structure of mitochondrial cytochrome *c* oxidase. (A) Stereoscopic C_α backbone trace for one monomeric complex of the core subunits I, II, and III. (B) Stereoscopic view showing all 13 subunits. The complete complex is a dimer of this structure. From Tsukihara *et al.*[125] (C) MolScript ribbon drawing of one monomeric unit. The horizontal lines are drawn at distances of ±1.0 and ± 2.0 nm from the center of the membrane bilayer as estimated from eight phospholipid molecules bound in the structure. From Wallin *et al.*[127] Courtesy of Arne Elofsson. (D) Schematic drawing of the same complex showing positions of the Cu_A dimetal center, bound Mg^{2+}, heme *a*, the bimetal heme a_3 - Cu_B center, and bound Zn^{2+}. The location of an 0.48-nm membrane bilayer is marked. From Tsukihara *et al.*[129] (A), (B), and (D) courtesy of Shinya Yoshikawa.

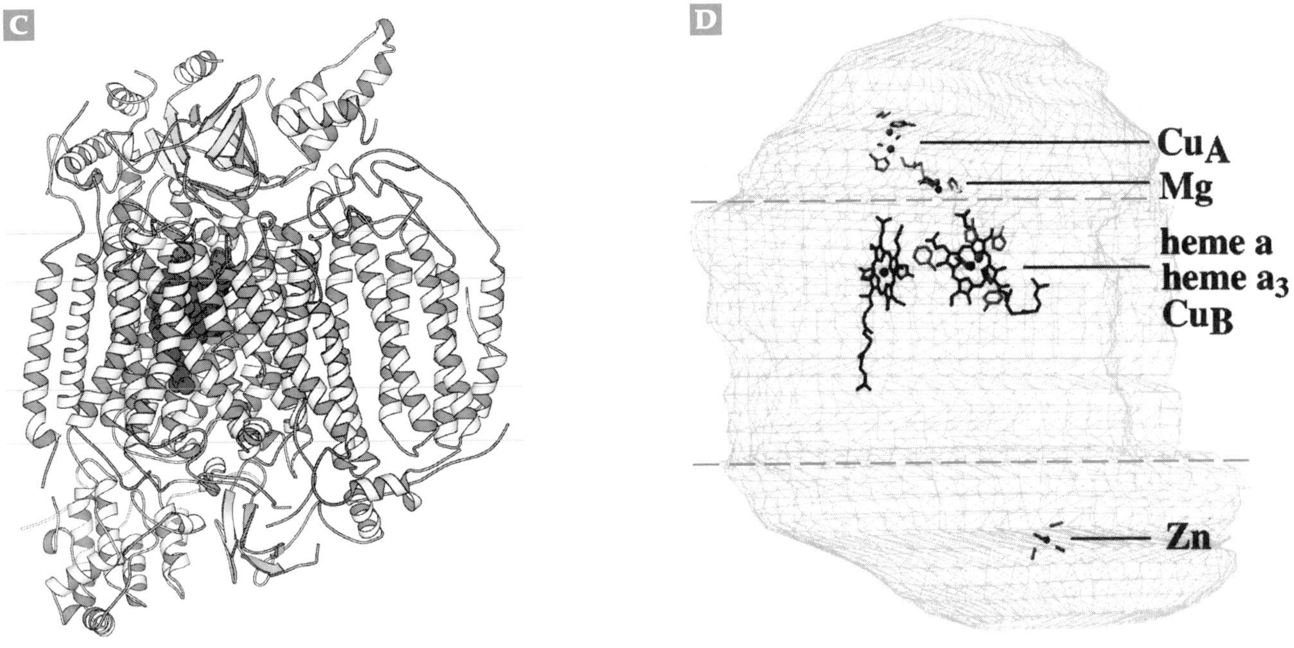

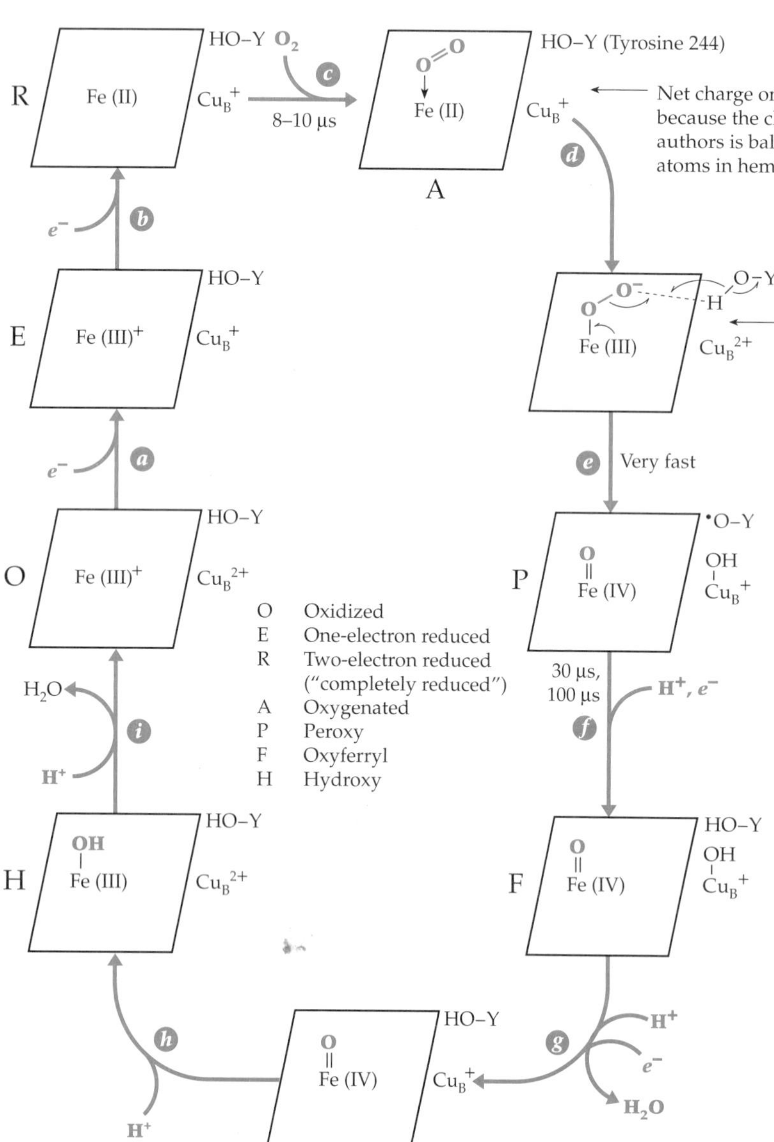

Figure 18-11 Possible catalytic cycle of cytochrome *c* oxidase at the cytochrome $a_3 - Cu_B$ site. The fully oxidized enzyme (O; left center) receives four electrons consecutively from the cyt $c \rightarrow Cu_A \rightarrow$ cyt *a* chain. In steps *a* and *b* both heme *a* and Cu_B, as well as the Cu_A center and cyt a_3, are reduced to give the fully reduced enzyme (R). In the very fast step *c* the cyt a_3 heme becomes oxygenated and in step *d* is converted to a peroxide with oxidation of both the Fe and Cu. Intermediate P was formerly thought to be a peroxide but is now thought to contain ferryl iron and an organic radical. This radical is reduced by the third electron in step *f* to give the ferryl form F, with Cu^{2+} participating in the oxidation. The fourth electron reduces Cu_B again (step *g*) allowing reduction to the hydroxy form H in step *h*. Protonation to form H_2O (step *i*) completes the cycle which utilizes $4 e^- + 4 H^+ + O_2$ to form $2 H_2O$. Not shown is the additional pumping of $4 H^+$ across the membrane from the matrix to the intermembrane space.

enzyme O with hydrogen peroxide.[143a-144] Use of various inhibitors has also been important in studying this enzyme. Cyanide, azide, and sulfide ions, as well as carbon monoxide, are powerful inhibitors. Cyanide specifically binds to the Fe^{3+} form of cytochrome a_3 preventing its reduction,[145] while CO competes with O_2 for its binding site. A much-used reagent that modifies carboxyl groups in proteins, and which inhibits many proton translocating proteins, is dicyclohexyl carbodiimide (Eq. 3-10).[146] The step-by-step flow of electrons through cytochrome *c* oxidase seems quite well defined. However, one of the most important aspects is unclear. How is the pumping of protons across the membrane coupled to electron transport?[137,138,142,147,147a] Many recent studies have employed directed mutation of residues in all four subunits to locate possible proton pathways or channels.[148–152] Most ideas involve movement through hydrogen bonded chains (Eq. 9-94), which may include the carboxylate groups of the bound hemes.[153] Conformational changes may be essential to the gating of proton flow by electron transfers.[143]

The surface of the matrix side of cytochrome oxidase contains histidine and aspartate side chains close together. It has been suggested that they form a proton collecting antenna that contains groups basic enough to extract protons from the buffered matrix and guide them to a proton conduction pathway.[154] Calcium ions also affect proton flow.[153a,b] We will return to this topic in Section C,3 (p. 1040).

C. Oxidative Phosphorylation

During the 1940s when it had become clear that formation of ATP from ADP and inorganic phosphate was coupled to electron transport in mitochondria, intensive efforts were made to discover the molecular mechanisms. However, nature sometimes strongly resists attempts to pry out her secrets, and the situation which prevailed was aptly summarized by Ephraim Racker: "Anyone who is not confused about oxidative phosphorylation just doesn't understand the situation."[155] The confusion is only now being resolved.

1. The Stoichiometry (P/O Ratio) and Sites of Oxidative Phosphorylation

Synthesis of ATP *in vitro* by tissue homogenates was demonstrated in 1937 by Kalckar, who has written a historical account.[156] In 1941, Ochoa[157] obtained the first reliable measurement of the P/O ratio, the *number of moles of ATP generated per atom of oxygen utilized* in respiration. The P/O ratio is also equal to the number of moles of ATP formed for each pair of electrons passing through an electron transport chain. Ochoa established that for the oxidation of pyruvate to acetyl-CoA and CO_2, with two electrons passed down the mitochondrial electron transport chain, the P/O ratio was ~3. This value has since been confirmed many times.[158–160] However, experimental difficulties in measuring the P/O ratio are numerous.[161] Many errors have been made, even in recent years, and some investigators[162] have contended that this ratio is closer to 2.5 than to 3. One method for measuring the P/O ratio is based on the method of determining the amount of ATP used that is described in the legend to Fig. 15-2.

The experimental observation of a P/O ratio of ~3 for oxidation of pyruvate and other substrates that donate NADH to the electron transport chain led to the concept that there are *three sites for generation of ATP*. It was soon shown that the P/O ratio was only 2 for oxidation of succinate. This suggested that one of the sites (site I) is located between NADH and ubiquinone and precedes the diffusion of QH_2 formed in the succinate pathway to complex III.

In 1949, Lehninger used ascorbate plus tetramethylphenylene-diamine (TMPD, Table 18-4) to introduce electrons into the chain at cytochrome *c*. The sequence ascorbate → TMPD → cytochrome *c* was shown to occur nonenzymatically. Later, it became possible to use cytochrome *c* as an electron donor directly. In either case only one ATP was generated, as would be anticipated if only site III were found to the right of cytochrome *c*. Site I was further localized by Lardy, who used hexacyanoferrate (III) (ferricyanide) as an artificial oxidant to oxidize NADH in the presence of antimycin *a*. Again a P/O ratio of one was observed. Finally, in 1955, Slater showed that passage of electrons from succinate to cytochrome c also gave only one ATP, the one generated at site II. The concept of three sites of ATP formation became generally accepted. However, as we shall see, these sites are actually proton-pumping sites, and there may be more than three of them.

Respiratory control and uncoupling. With proper care relatively undamaged mitochondria can be isolated. Such mitochondria are said to be **tightly coupled**. By this we mean that electrons cannot pass through the electron transport chain without generation of ATP. If the concentration of ADP or of P_i becomes too low, both phosphorylation and respiration cease. This **respiratory control** by ADP and P_i is a property of undamaged mitochondria. It may seem surprising that damaged mitochondria or submitochondrial particles are often able to transfer electrons at a faster rate than do undamaged mitochondria. However, electron transfer in damaged mitochondria occurs *without synthesis of ATP* and with no slowdown as the ADP concentration drops. A related kind of **uncoupling** of electron transport from ATP synthesis is brought about by various lipophilic anions called **uncouplers**, the best known of which is **2,4-dinitrophenol**. Even before the phenomenon of uncoupling was discovered, it had been known that dinitrophenol substantially increased the respiration rates of animals. The compound had even been used (with some fatal results) in weight control pills. The chemical basis of uncoupling will be considered in Section D.

"States" of mitochondria and spectrophotometric observation. Chance and Williams defined five **states** of tightly coupled mitochondria[60,163]; of these, states 3 and 4 are most often mentioned. If no oxidizable substrate or ADP is added the mitochondria have a very low rate of oxygen uptake and are in state 1. If oxidizable substrate and ADP are added rapid O_2 uptake is observed, the rate depending upon the rate of flow of electrons through the electron transport chain. This is state 3. As respiration occurs the coupled phosphorylation converts ADP into ATP, exhausting the ADP. Respiration slows to a very low value and the mitochondria are in state 4. If the substrate is present in excess, addition of more ADP will return the mitochondria to state 3.

Chance and associates employed spectrophotometry on intact mitochondria or submitochondrial particles to investigate both the sequence of carriers and the sites of phosphorylation. Using the dual wavelength spectrophotometer, the light absorption at the absorption maximum (λ_{max}) of a particular component was followed relative to the absorption at some other reference wavelength (λ_{ref}). The principal wavelengths used are given in Table 18-6. From these measurements the state of oxidation or reduction of each one of the carriers could be observed in the various states and in the presence of inhibitors. The

TABLE 18-6
Wavelengths of Light Used to Measure States of Oxidation of Carriers in the Electron Transport Chain of Mitochondria[a]

Carrier	λ_{max} (nm)[b]	λ_{ref} (nm)
NADH	340	374
Flavins	465	510
Cytochromes		
b^{2+}	564(α)	575
	530(β)	
	430(γ)	
c_1^{2+}	534(α)	
	523(β)	
	418(γ)	
c^{2+}	550(α)	540
	521(β)	
	416(γ)	
a^{2+}	605(α)	630(590)
	450(γ)[a]	
a_3^{2+}	600(α)[a]	
	445(γ)	455

[a] After Chance, B. and Williams, G. R. (1955) *J. Biol. Chem*, **217**, 409–427; (1956) Adv. Enzymol. **17**, 409–427.
[b] The wavelengths used for each carrier in dual wavelength spectroscopy appear opposite each other in the two columns. Some positions of other absorption bands of cytochromes are also given.

experiments served to establish that electrons passing down the chain do indeed reside for a certain length of time on particular carriers. That is, in a given state each carrier exists in a defined ratio of oxidized to reduced forms ([ox] / [red]). Such a result would not be seen if the entire chain functioned in a cooperative manner with electrons passing from the beginning to the end in a single reaction. By observing changes in the ratio [ox] / [red] under different conditions, some localization of the three phosphorylation sites could be made. In one experiment antimycin *a* was added to block the chain ahead of cytochrome c_1. Then tightly coupled mitochondria were allowed to go into state 4 by depletion of ADP. Since the concentration of oxygen was high and cytochrome a_3 has a low K_m for O_2 (~3 μM) cytochrome a_3 was in a highly oxidized state. Cytochrome *a* was also observed to be oxidized, while cytochrome c_1 and *c* remained reduced. The presence of this **crossover point** suggested at the time that cytochrome *c* might be at or near one of the "energy conservation sites." Accounts of more recent experiments using the same approach are given by Wilson *et al.*[164]

2. Thermodynamics and Reverse Electron Flow

From Table 6-8 the value of $\Delta G'$ for oxidation of one mole of NADH by oxygen (1 atm) is –219 kJ. At a pressure of ~10^{-2} atm O_2 in tissues the value is –213 kJ. However, when the reaction is coupled to the synthesis of three molecules of ATP ($\Delta G'$ = +34.5 kJ mol^{-1}) the net Gibbs energy change for the overall reaction becomes $\Delta G'$ = –110 kJ mol^{-1}. This is still very negative. However, we must remember that the concentrations of ATP, ADP, and P_i can depart greatly from the 1:1:1 ratio implied by the $\Delta G'$ value.

An interesting experiment is to allow oxidative phosphorylation to proceed until the mitochondria reach state 4 and to measure the **phosphorylation state ratio R_p**, which equals the value of [ATP] / [ADP][P_i] that is attained. This mass action ratio, which has also been called the "phosphorylation ratio" or "phosphorylation potential" (see Chapter 6 and Eq. 6-29), often reaches values greater than 10^4–10^5 M^{-1} in the cytosol.[164] An extrapolated value for a zero rate of ATP hydrolysis of log R_p = 6.9 was estimated. This corresponds (Eq. 6-29) to an increase in group transfer potential (ΔG of hydrolysis of ATP) of 39 kJ/mol. It follows that the overall value of ΔG for oxidation of NADH in the coupled electron transport chain is less negative than is $\Delta G'$. If synthesis of three molecules of ATP is coupled to electron transport, the system should reach an equilibrium when $R_p = 10^{6.4}$ at 25°C, the difference in ΔG and $\Delta G'$ being 3RT ln Rp = 3 x 5.708 x 6.4 = 110 kJ mol^{-1}. This value of Rp is, within experimental error, the same as the maximum value observed.[165] There apparently is an almost true equilibrium among NADH, O_2, and the adenylate system if the P/O ratio is 3.

Within more restricted parts of the chain it is possible to have *reversed electron flow*. Consider the passage of electrons from NADH, partway through the chain, and back out to fumarate, the oxidized form of the succinate–fumarate couple. The Gibbs energy change $\Delta G'$ (pH 7) for oxidation of NADH by fumarate is –67.7 kJ mol^{-1}. In uncoupled mitochondria electron flow would always be from NADH to fumarate. However, in tightly coupled mitochondria, in which ATP is being generated at site I, the overall value of $\Delta G'$ becomes much less negative. If $R_p = 10^4$ M^{-1}, $\Delta G'$ for the coupled process becomes approximately zero (–67.7 + 68 kJ mol^{-1}). Electron flow can easily be reversed so that succinate reduces NAD^+. Such ATP-driven reverse flow occurs under some physiological conditions within mitochondria of living cells, and some anaerobic bacteria generate all of their NADH by reversed electron flow (see Section E).

Another experiment involving equilibration with the electron transport chain is to measure the "observed potential" of a carrier in the chain as a function of the concentrations of ATP, ADP, and P_i. The observed potential E is obtained by measuring log([ox] / [red]) and applying Eq. 18-6 in which $E°'$ is the known mid-point potential of the couple (Table 6-8) and n is the number of electrons required to reduce one molecule of the carrier. If the system is equilibrated with a

$$E = \frac{-\Delta G}{nF} = E°' + \frac{0.0592}{n}\log\frac{[ox]}{[red]}$$
$$= \text{observed potential of carrier} \quad (18\text{-}6)$$

"redox buffer" (Chapter 6), E can be fixed at a pre-selected value. For example, a 1:1 mixture of succinate and fumarate would fix E at +0.03 V while the couple 3-hydroxybutyrate-acetoacetate in a 1:1 ratio would fix it at $E°' = -0.266$ V. Consider the potential of cytochrome b_{562} (b_H), which has an $E°'$ value of 0.030 V. Substituting this in Eq. 18-7 and using $E = -0.266$ V (as obtained by equilibration with 3-hydroxybutyrate–acetoacetate), it is easy to calculate that at equilibrium the ratio [ox] / [red] for cytochrome b_{562} is about 10^{-5}. In other words, in the absence of O_2 this cytochrome will be kept almost completely in the reduced form in an uncoupled mitochondrion.

However, if the electron transport between 3-hydroxybutyrate and cytochrome b_{562} is tightly coupled to the synthesis of one molecule of ATP, the observed potential of the carrier will be determined not only by the imposed potential E_i of the equilibrating system but also by the phosphorylation state ratio of the adenylate system (Eq. 18-7). Here $\Delta G'_{ATP}$ is the group transfer potential ($-\Delta G'$ of hydrolysis) of ATP at pH 7 (Table 6-6), and n' is the number of electrons passing through the chain required to synthesize one ATP. In the upper part of the equation n is the number of electrons required to reduce the carrier, namely one in the case of cytochrome b_{562}.

From Eq. 18-7 it is clear that in the presence of a high phosphorylation state ratio a significant fraction of cytochrome b_{562} may remain in the reduced form at equilibrium. Thus, if $R_p = 10^4$, if $E°'$ for cytochrome b_{562} is 0.030 V, if $n' = 2$, and the potential E is fixed at -0.25 V using the hydroxybutyrate–acetoacetate couple, we calculate, from Eq. 18-7, that the ratio [ox] / [red] for cytochrome b_{562} will be 1.75. Now, if R_p is varied the observed potential of the carrier should change as predicted by Eq. 18-7. This variation has been observed.[164] For a tenfold change in R_p the observed potential of cytochrome b_{562} changed by 0.030 V, just that predicted if $n' = 2$. On the other hand, the observed potential of cytochrome c varied by 0.059 V for every tenfold change in the ratio. This is just as expected if $n' = 2$, and if synthesis of two molecules of

TABLE 18-7
Electrode Potentials of Mitochondrial Electron Carriers and Gibbs Energy Changes Associated with Passage of Electrons[a]

	Electron carrier	$E°'$ (pH 7) isolated	$E°'$ (pH 7.2) in mitochondria	ΔG (kJ mol^{-1}) for 2 e^- flow to O_2 at 10^{-2} atm, carriers at pH 7
	NADH / NAD+	–0.320		–213
Group I	Flavoprotein		~ –0.30	
~ –0.30 V	Fe–S protein		~ –0.305	
	β-Hydroxybutyrate–acetoacetate	–0.266		–203
	Lactate–pyruvate	–0.185		–187
	Succinate-fumarate	0.031		–146
Group II	Flavoprotein		~ –0.045	
~ 0 V	Cytochrome b_T		–0.030	
	Cu		0.001	
	Fe–S protein		0.030	
	Cytochrome b_K		0.030	
	Ubiquinone	0.10	0.045	–132
	Cytochrome a_3 + ATP		0.155	
Group III	Cytochrome c_1		0.215	
	Cytochrome c	0.254	0.235	–102
	Cytochrome b_T + ATP		0.245	
	Cytochrome a	0.29	0.210	
	Cu		0.245	
	Fe–S protein		0.28	
Group IV	Cytochrome a_3		0.385	–77
	O_2 (10^{-2} atm)	0.785		0.00
	1 atm	0.815		

[a] Data from Wilson, D. F., Dutton, P. L., Erecinska, M., Lindsay, J. G., and Soto, N. (1972) *Acc. Chem. Res.* **5**, 234–241 and Wilson, D. F., Erecinska, M., and Sutoon, P. L. (1974) *Ann. Rev. Biophys. Bioeng.* **3**, 203–230.

$$E(\text{observed}) = E^{\circ\prime} + \frac{0.0592}{n}\log_{10}\frac{[\text{ox}]}{[\text{red}]}$$
$$= E_i + \frac{\Delta G'_{\text{ATP}}}{96.5n'} + \frac{RT}{n'F}\ln\frac{[\text{ATP}]}{[\text{ADP}][\text{P}_i]}$$
$$= E_i + \frac{0.358}{n'} + \frac{0.0592}{n'}\log_{10} R_P \qquad (18\text{-}7)$$

ATP is coupled to the electron transport to cytochrome *c*. Thus, we have experimental evidence that when one-electron carriers such as the cytochromes are involved, the passage of *two electrons* is required to synthesize one molecule of ATP. Furthermore, from experiments of this type it was concluded that the sites of phosphorylation were localized in or related to complexes I, III, and IV.

Another kind of experiment is to equilibrate the electron transport chain with an external redox pair of known potential using *uncoupled* mitochondria. The value of $E^{\circ\prime}$ of a particular carrier can then be measured by observation of the ratio [ox] / [red] and applying Eq. 18-7. While changes in the equilibrating potential E will be reflected by changes in [ox] / [red] the value of $E^{\circ\prime}$ will remain constant. The $E^{\circ\prime}$ values of Fe–S proteins and copper atoms in the electron transport chain have been obtained by equilibrating mitochondria, then rapidly freezing them in liquid nitrogen, and observing the ratios [ox] / [red] by EPR at 77K (Table 18-7).

The values of $E^{\circ\prime}$ of the mitochondrial carriers fall into four **isopotential groups** at ~–0.30, ~0, ~+0.22, and ~+0.39 V (Table 18-7). When tightly coupled mitochondria are allowed to go into state 4 (low ADP, high ATP, O_2 present but low respiration rate), the observed potentials change. That of the lowest isopotential group (which includes NAD^+/ NADH) falls to ~–0.38 V, corresponding to a high state of reduction of the carriers to the left of the first phosphorylation site in Fig. 18-4. Groups 2 and 3 remain close to their midpoint potentials at ~–0.05 and +0.26 V. In this condition the potential difference between each successive group of carriers amounts to ~0.32 V, just enough to balance the formation of one molecule of ATP for each two electrons passed at a ratio $R_p \approx 10^4$ M^{-1} (Eq. 18-7).

Two cytochromes show exceptional behavior and appear twice in Table 18-7. The midpoint potential $E^{\circ\prime}$ of cytochrome b_{566} (b_L) changes from –0.030 V in the absence of ATP to +0.245 V in the presence of a high concentration of ATP. On the other hand, $E^{\circ\prime}$ for cytochrome a_3 drops from +0.385 to 0.155 V in the presence of ATP. These shifts in potential must be related to the coupling of electron transport to phosphorylation.

3. The Mechanism of Oxidative Phosphorylation

It was natural to compare mitochondrial ATP synthesis with substrate-level phosphorylations, in which high-energy intermediates are generated *by the passage of electrons through the substrates*. The best known example is oxidation of the aldehyde group of glyceraldehyde 3-phosphate to an acyl phosphate, which, after transfer of the phospho group to ADP, becomes a carboxylate group (Fig. 15-6). The Gibbs energy of oxidation of the aldehyde to the carboxylate group provides the energy for the synthesis of ATP. However, this reaction differs from mitochondrial electron transport in that *the product, 3-phosphoglycerate, is not reconverted to glyceraldehyde 3-phosphate*. Electron carriers of the respiratory chain must be regenerated in some cyclic process. Because of this, it was difficult to imagine practical mechanisms for oxidative phosphorylation that could be related to those of substrate level phosphorylation. Nevertheless, many efforts were made over a period of several decades to find such high-energy intermediates.

Search for chemical intermediates. An early hypothetical model, proposed by Lipmann,[166] is shown in Fig. 18-12. Here A, B, and C are three electron carriers in the electron transport chain. Carrier C is a better oxidizing agent than B or A. Carrier B has some special chemistry that permits it, in the reduced state, to react with group Y of a protein (step *b*) to form Y–BH_2. The latter, an unidentified adduct, is converted by oxidation with carrier C (step *c*) to a "high energy" oxidized form indicated as Y ~ B. Once the possibility of generating such an intermediate is conceded, it is easy to imagine plausible ways in which the energy of this intermediate could be transferred into forms with which we are already familiar. For example, another protein X could react (step *d*) to form X ~ Y in which the X ~ Y linkage could be a thioester, an acyl phosphate, or other high-energy form. Furthermore, it might not be necessary to have two proteins; X and Y could be different functional groups of the same protein. They might be nonprotein components, e.g., Y might be a phospholipid.

Generation of ATP by the remaining reactions (steps *e* and *f* of Fig. 18-12) is straightforward. For example, if X ~ Y were a thioester the reactions would be the reverse of Eq. 12-48. These reaction steps would also be responsible for observed exchange reactions, for example, the mitochondrially catalyzed exchange of inorganic phosphate ($H^{32}PO_4^{2-}$) into the terminal position of ATP. Mitochondria and submitochondrial particles also contain ATP-hydrolyzing **(ATPase)** activity, which is thought to depend upon the same machinery that synthesizes ATP in tightly coupled mitochondria. In the scheme of Fig. 18-12, ATPase

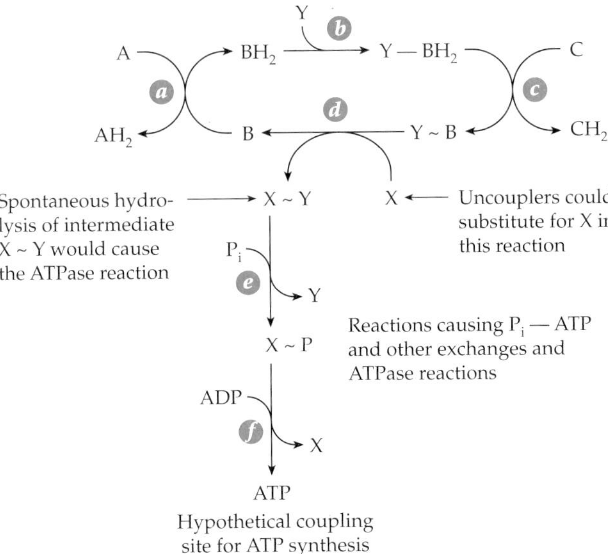

Figure 18-12 An early proposal for formation of ATP via "high-energy" chemical intermediates.

activity would be observed if hydrolysis of X ~ Y were to occur. Partial disruption of the system would lead to increased ATPase reactivity, as is observed. Uncouplers such as the dinitrophenolate ion or arsenate ion, acting as nucleophilic displacing groups, could substitute for a group such as X. Spontaneous breakdown of labile intermediates would permit oxidation to proceed unimpaired. Since there are three different sites of phosphorylation, we might expect to have three different enzymes of the type Y in the scheme of Fig. 18-12, but it would be necessary to have only one X.

In Lipmann's original scheme group Y was visualized as adding to a carbon–carbon double bond to initiate the sequence. Isotopic exchange reactions ruled out the possibility that either ADP or P_i might serve as Y, but it was attractive to think that a bound phosphate ion, e.g., in a phospholipid or coenzyme, could be involved. Y ~ B of Fig. 18-12 would be similar in reactivity to an acyl phosphate or thioester. However, whatever the nature of Y ~ B, part of group Y would be left attached to B after the transfer of Y to X. For example, if Y were Y'OH compound X ~ OY' would be formed, and the carrier would be left in step *d* in the form of B–OH. Elimination of a hydroxyl group would be required to regenerate B. Perhaps nature has shunned this mechanism because there is no easy way to accomplish such an elimination. Many variations on the scheme of Fig. 18-12 were proposed,[166] and some were discussed in the first edition of this textbook.[167] However, as attractive as these ideas may have seemed, *all attempts to identify discrete intermediates that might represent X ~ Y failed. Furthermore, most claims to have seen Y ~ B by any means have been disproved.*

Peter Mitchell's chemiosmotic theory. To account for the inability to identify high energy intermediates as well as the apparent necessity for an intact membrane, Peter Mitchell, in 1961, offered his **chemiosmotic theory** of oxidative phosphorylation.[168–175a] aThis theory also accounts for the existence of **energy-linked processes** such as the accumulation of cations by mitochondria. The principal features of the Mitchell theory are illustrated in Fig. 18-13. Mitchell proposed

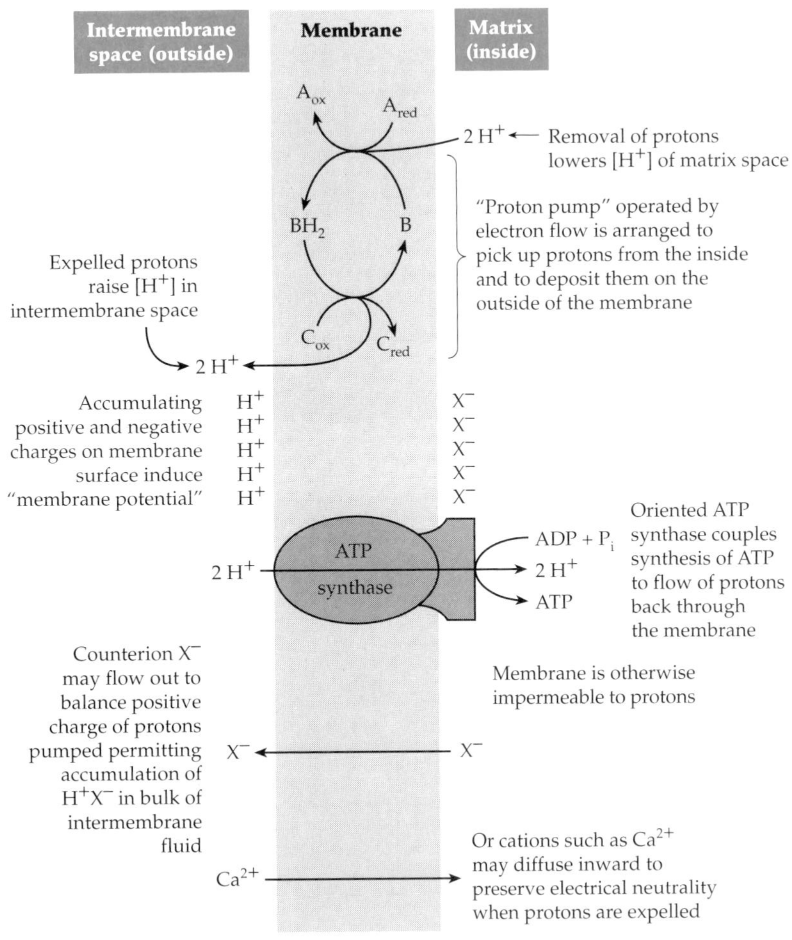

Figure 18-13 Principal features of Mitchell's chemiosmotic theory of oxidative phosphorylation.

that the inner membrane of the mitochondrion is a closed, proton-impermeable **coupling membrane**, which contains **proton pumps** operated by electron flow and which cause protons to be expelled through the membrane from the matrix space. As indicated in Fig. 18-13, an oxidized carrier B, upon reduction to BH_2, acquires two protons. These protons do not necessarily come from reduced carrier AH_2, and Mitchell proposed that they are picked up from the solvent on the matrix side of the membrane. Then, when BH_2 is reoxidized by carrier C, protons are released on the outside of the membrane. On the basis of existing data, *Mitchell assumed a stoichiometry of two protons expelled for each ATP synthesized.* It followed that there should be three different proton pumps in the electron transport chain corresponding to the three phosphorylation sites.

The postulated proton pumps would lead either to bulk accumulation of protons in the intermembrane space and cytoplasm, with a corresponding drop in pH, *or to an accumulation of protons along the membrane itself.* The latter would be expected if counterions X^- do not pass through the membrane with the protons. The result in such a case would be the development of a **membrane potential**, a phenomenon already well documented for nerve membranes (Chapter 8).

A fundamental postulate of the chemiosmotic theory is the presence of an oriented ATP synthase that utilizes the Gibbs energy difference of the proton gradient to drive the synthesis of ATP (Fig. 18-9). Since $\Delta G'$ (pH 7) for ATP synthesis is +34.5 kJ mol^{-1} and, if as was assumed by Mitchell, the passage of two protons through the ATP synthase is required to form one ATP, the necessary pH gradient (given by Eq. 6-25 or Eq. 18-9 with $E_m = 0$) would be 34.5/(2 x 5.708) = 3.0 pH units at 25°C. On the other hand, if the phosphorylation state ratio is ~10^4 M^{-1}, the pH difference would have to be 5 units. Most investigators now think that 4 H^+ per ATP are needed by the synthase. If so, a pH difference of 2.5 units would be adequate. Various experiments have shown that passage of electrons does induce a pH difference, and that an artifically induced pH difference across mitochondrial membranes leads to ATP synthesis. However, pH gradients of the required size have not been observed. Nevertheless, if the membrane were charged as indicated in Fig. 18-13, without accumulation of protons in the bulk medium, a membrane potential would be developed, and this could drive the ATP synthase, just as would a proton gradient.

The mitochondrial membrane potential E_m (or $\Delta\psi$) is the potential difference measured across a membrane relative to a reference electrode present in the surrounding solution.[176] For both mitochondria and bacteria E_m normally has a negative value. The Gibbs energy change $\Delta\psi_H+$ for transfer of one mole of H^+ from the inside of the mitochondrion to the outside, against the concentration and potential gradients, is given by Eq. 18-8. This equation follows directly from Eqs. 6-25

$$\begin{aligned}\Delta G_H+ &= 2.303\ RT\ \Delta pH - E_m F\\ &= 5.708\ \Delta pH - 96.5\ E_m\ kJ/mol\ at\ 25°C\\ \text{where } \Delta pH &= pH\ (inside) - pH\ (outside)\end{aligned} \tag{18-8}$$

and 6-63 with n = 1. The same information is conveyed in Eq. 18-9, which was proposed by Mitchell for what he calls the **total protonic potential difference Δ*p*.**

$$\begin{aligned}\Delta p\ (volts) &= E_m(volts) - 2.303\ \frac{RT}{F}\ \Delta\ pH\\ \Delta p\ (mV) &= E_m(mV) - 59.2\ \Delta\ pH\ at\ 25°C\\ E_m &= \Delta\psi\end{aligned} \tag{18-9}$$

Mitchell was struck by the parallel between the force and flow of electrons, which we call electricity, and the force and flow of protons, which he named **proticity**.[174] This led one headline writer in *Nature*[177] to describe Mitchell as "a man driven by proticity," but if Mitchell is right, as seems to be the case, we are all driven by proticity! Mitchell also talked about **protonmotive** processes and referred to Δp as the **protonmotive force**. Although it is a potential rather than a force, this latter name is a popular designation for Δp.

The reader should be aware that considerable confusion exists with respect to names and definitions.[176] For example, the ΔG_H+ of Eq. 18-8 can also be called the **proton electrochemical potential** $\Delta\mu_H+$, which is analogous to the chemical potential μ of an ion (Eq. 6-24) and has units of kJ/mol (Eq. 18-10).

$$\begin{aligned}-\Delta G_H+ = \Delta\mu_H+ &= F\ \Delta p\\ &= 96.5\ \Delta p\ kJ/mol\ at\ 25°C\end{aligned} \tag{18-10}$$

However, many authors use $\Delta\mu_H+$ as identical to the protonmotive force Δp.

From Eq. 18-9 or Eq. 18-10 it can be seen that a membrane potential E_m of −296 mV at 25°C would be equivalent to a 5.0 unit change in pH and would be sufficient, if coupled to ATP synthesis via 2 H^+, to raise R_p to 10^4 M^{-1}. Any combination of ΔpH and E_m providing Δp of −296 mV would also suffice. If the ratio H^+/ATP = 4, Δp of −148 mV would suffice.

The chemiosmotic hypothesis had the great virtue of predicting the following consequences which could be tested: (1) electron-transport driven proton pumps with defined stoichiometries and (2) a separate ATP synthase, which could be driven by a pH gradient or membrane potential. Mitchell's hypothesis was initially greeted with skepticism but it encouraged many people, including Mitchell and his associate Jennifer Moyle, to test these predictions, which were soon found to be correct.[178]

Observed values of E_m and pH. One of the problems[179] in testing Mitchell's ideas has been the difficulty of reliably measuring Δp. To evaluate the pH term in Eq. 18-10 measurements have been made with microelectrodes and indicator dyes. However, the most reliable approach has been to observe the distribution of weak acids and bases across the mitochondrial membrane.[180] This is usually done with a suspension of freshly isolated active mitochondria. The method has been applied widely using, for example, methylamine. A newer method employs an isotope exchange procedure to measure the pH-sensitive carbonic anhydrase activity naturally present in mitochondria.[181]

The measurement of E_m ($\Delta\psi$) is also difficult.[179] Three methods have been used: (1) measurement with microelectrodes; (2) observation of fluorescent probes; (3) distribution of permeant ions. Microelectrodes inserted into mitochondria[182] have failed to detect a significant value for E_m. Fluorescent probes are not very reliable,[179,183] leaving the distribution of permeant ions the method of choice. In this method a mitochondrial suspension is exposed to an ion that can cross the membrane but which is not pumped or subject to other influences that would affect its distribution. Under such conditions the ion will be distributed according to Eq. 18-11. The most commonly used ions are K^+, the same ion that is thought to reflect the membrane potential of nerve axons (Chapter 30), or Rb^+. To make the inner mitochondrial membrane permeable to K^+, valinomycin (Fig. 8-22) is added. The membrane potential, with n = 1 in Eq. 9-1, becomes:

$$E_m = -59.2\ ([K^+]_{inside} / [K^+]_{outside})\ \text{volts} \tag{18-11}$$

In these experiments respiring mitochondria are observed to take up the K^+ or Rb^+ to give a high ratio of K^+ inside to that outside and consequently a negative E_m. There are problems inherent in the method. The introduction of a high concentration of ion perturbs the membrane potential, and there are uncertainties concerning the contribution of the Donnan equilibrium (Eq. 8-5) to the observed ion distribution.[184]

In most instances, either for mitochondrial suspensions or whole bacteria, ΔpH is less negative than –0.5 unit making a contribution of, at most, –30 mV to Δp. The exception is found in the thylakoid membranes of chloroplasts (Chapter 23) in which protons are pumped into the thylakoid vesicles and in which the internal pH falls dramatically upon illumination of the chloroplasts.[185] The ΔpH reaches a value of –3.0 or more units and Δp is ~180 mV, while E_m remains ~0. Reported values of E_m for mitochondria and bacteria range from –100 to –168 mV and Δp from –140 to –230 mV.[172,179] Wilson concluded that E_m for actively respiring mitochondria, using malate or glutamate as substrates, attains maximum (negative) values of E_m= –130 mV and Δp = –160 mV.[179] However, Tedeschi and associates[183,184] argued that E_m is nearly zero for liver mitochondria and seldom becomes more negative than –60 mV for any mitochondria.

A crucially important finding is that submitochondrial particles or vesicles from broken chloroplasts will synthesize ATP from ADP and P_i, when an artificial pH gradient is imposed.[172,186] Isolated purified F_1F_0 ATPase from a thermophilic *Bacillus* has been coreconstituted into liposomes with the light-driven proton pump **bacteriorhodopsin** (Chapter 23). Illumination induced ATP synthesis.[187] These observations support Mitchell's proposal that the ATP synthase is both spatially separate from the electron carriers in the membrane and utilizes the protonmotive force to make ATP. Thus, the passage of protons from the outside of the mitochondria back in through the ATP synthase induces the formation of ATP. What is the stoichiometry of this process?

It is very difficult to measure the flux of protons across the membrane either out of the mitochondria into the cytoplasm or from the cytoplasm through the ATP synthase into the mitochondria. Therefore, estimates of the stoichiometry have often been indirect. One argument is based on thermodynamics. If Δp attains values no more negative than –160 mV and R_p within mitochondria reaches at least 10^4 M^{-1}, we must couple ΔG_H of –15.4 kJ/mol to ΔG of formation of ATP of +57.3 kJ/mol. To do this four H^+ must be translocated per ATP formed. Recent experimental measurements with chloroplast ATP synthase[188] also favor four H^+. It is often proposed that one of these protons is used to pump ADP into the mitochondria via the ATP-ADP exchange carrier (Section D). Furthermore, if R_p reaches 10^6 M^{-1} in the cytoplasm, it must exceed 10^4 M^{-1} in the mitochondrial matrix.

Proton pumps driven by electron transport. What is the nature of the proton-translocating pumps that link Δp with electron transport? In his earliest proposals Mitchell suggested that electron carriers, such as flavins and ubiquinones, each of which accepts two protons as well as two electrons upon reduction, could serve as the proton carriers. Each pump would consist of a pair of oxidoreductases. One, on the inside (matrix side) of the coupling membrane, would deliver two electrons (but no protons) to the carrier (B in Fig. 18-13). The two protons needed for the reduction would be taken from the solvent in the matrix. The second oxidoreductase would be located on the outside of the membrane and would accept two electrons from the reduced carrier (BH_2 in Fig. 18-13) leaving the two released protons on the outside of the membrane. To complete a "loop" that would allow the next carrier to be reduced, electrons would have to be transferred through fixed electron carriers embedded in the membrane from the reduced electron acceptor (C_{red} in Fig. 18-13) to the oxidized form of the oxidoreductase to be used as reductant for the next loop. These loops, located in complexes I and III of Fig. 18-5, would pump three protons per electron or six H^+/O. With a P/O ratio of three this would provide two H^+ per ATP formed. Mitchell regarded this stoichiometry as appropriate.

The flavin of NAD dehydrogenase was an obvious candidate for a carrier, as was ubiquinone. However, the third loop presented a problem. Mitchell's solution was the previously discussed **Q cycle**, which is shown in Fig. 18-9. This accomplishes the pumping in complex III of 2 H^+/e^-, the equivalent of two loops.[111] However, as we have seen, the magnitude of Δp suggests that 4 H^+, rather than 2 H^+, may be coupled to synthesis of one ATP. If this is true, mitochondria must pump 12 H^+/O rather than six when dehydrogenating NADH, or eight H^+/O when dehydrogenating succinate.

The stoichiometry of proton pumping was measured by Lehninger and associates using a fast-responding O_2 electrode and a glass pH electrode.[189,190] They observed an export of eight H^+/O for oxidation of succinate rat liver mitochondria in the presence of a permeant cation that would prevent the buildup of E_m, and four H^+/O (2 H^+/e^-) for the cytochrome oxidase system. These are equivalent to two H^+/e^- at each of sites II and III as is indicated in Fig. 18-4. Some others have found lower H^+/e^- ratios.

If two H^+/e^- are pumped out of mitochondria, where do we find the pumping sites? One possibility is that protons are pumped through the membrane by a **membrane Bohr effect**, so named for its similarity to the Bohr effect observed upon oxygenation of hemoglobin. In the latter case (Chapter 7), the pK_a values of certain imidazole and terminal amino groups are decreased when O_2 binds. This may result, in part, from an electrostatic effect of O_2 in inducing a partial positive charge in the heme. This partial charge may then cause a decrease in the pK_a values of nearby groups. Similarly, complete loss of an electronic charge from a heme group or an iron–sulfur protein in the electron transport chain would leave a positive charge, an electron "hole," which could induce a large change in the pK_a of a neighboring group. One manifestation of this phenomenon may be a strong pH dependence of the reduction potential (Eq. 16-19).

Protons that could logically be involved in a membrane Bohr effect are those present on imidazole rings coordinated to Fe or Cu in redox proteins. Removal of an electron from the metal ion could be accompanied by displacement of electrons within the imidazole, within a peptide group that is hydrogen-bonded to an imidazole, or within some other acidic group. A hypothetical example is illustrated in Eq. 18-12 in which a carboxyl group loses a proton when "handed" a second. If the transiently enolized peptide linkage formed in

Fe---N ... N—H---O=C ... N—H---O—C=O (e−; Peptide)

→ H^+ (released)

Fe—N ... N---H—O—C ... N---H—O—C=O (Enolized peptide)

(18-12)

this process is tautomerized back to its original state before the iron is reduced again, the proton originally present on the carboxyl group will be released. It is easy to imagine that a proton could then be "ferried" in (as in Eq. 9-96) from the opposite side of the membrane to reprotonate the imidazole group and complete the pumping process.

In view of the large number of metal-containing electron carriers in the mitochondrial chain, there are many possible locations for proton pumps. However, the presence of the three isopotential groups of Table 18-7 suggests that the pumps are clustered in complexes I-III as pictured in Fig. 18-5. One site of pumping is known to be in the cytochrome *c* oxidase complex. When reconstituted into phospholipid, the purified complex does pump protons in response to electron transport, H^+/e^- ratios of ~1 being observed.[136,137,147,191] As mentioned in Section B,3 a large amount of experimental effort has been devoted to identifying proton transport pathways in cytochrome *c* oxidase and also in the cytochrome bc_1 (complex II).[192] Proton pumping appears to be coupled to chemical changes occurring between intermediates P and F of Fig. 18-11, between F and O,[136,193] and possibly between O and R.[137,138] Mechanisms involving direct coupling of chemical changes at the A_3Cu_B center and at the Cu_A dimetal center have been proposed.[147,194]

How do protons move from the pumping sites to ATP synthase molecules? Since protons, as H_3O^+, are sufficiently mobile, ordinary diffusion may suffice. Because of the membrane potential they will tend to stay close to the membrane surface, perhaps being transported on phosphatidylethanolamine head groups (see Chapter 8). According to the view of R. J. P. Williams protons are not translocated across the entire membrane by the proton pumps, but flow through the proteins of the membrane to the ATP synthase.[195] There the protons induce the necessary conformational changes to cause ATP synthesis. A related idea is that transient high-energy intermediates generated by electron transport within membranes are proton-carrying conformational isomers. When an electron is removed from an electron-transporting metalloprotein, the resulting positively charged "hole" could be stable for some short time, while the protein diffused within the membrane until it encountered an F_0 protein of an ATP synthase. Then it might undergo an induced conformational change at the same time that it "handed" the Bohr effect proton of Eq. 18-14 to the F_0 protein and simultaneously induced a conformational change in that protein. The coupling of proton transport to conformational changes seems plausible, when we recall that the induction of conformational changes within proteins almost certainly involves rearrangement of hydrogen bonds.

A consequence of the chemiosmotic theory is that there is no need for an integral stoichiometry between protons pumped and ATP formed or for an integral P/O ratio. There are bound to be inefficiencies in coupling, and Δp is also used in ways other than synthesis of ATP.

4. ATP Synthase

In 1960, Racker and associates[196,197] discovered that the "knobs" or "little mushrooms" visible in negatively stained mitochondrial fragments or fragments of bacterial membranes possess ATP-hydrolyzing **(ATPase)** activity. Earlier the knob protein had been recognized as one of several **coupling factors** required for reconstitution of oxidative phosphorylation by submitochondrial particles.[197] Electron micrographs showed that the submitochondrial particles consist of closed vesicles derived from the mitochondrial cristae, and that the knobs (Fig. 18-14A) are on the *outside* of the vesicles. They can be shaken loose by ultrasonic oscillation with loss of phosphorylation and can be added back with restoration of phosphorylation. The knob protein became known as **coupling factor F_1**. Similar knobs present on the outside of the thylakoids became **CF_1** and those inside thermophilic bacteria **TF_1**. The ATPase activity of F_1 was a clue that *the knobs were really ATP synthase.* It also became clear that a portion of the ATP synthase is firmly embedded in the membranes. This part became known as **F_0**. Both the names F_1F_0 ATP synthase and F_1F_0 ATPase are applied to the complex, the two names describing different catalytic activities. The ATPase activity is usually not coupled to proton pumping but is a readily measurable property of the F_1 portion. In a well-coupled submitochondrial particle the ATPase activity will be coupled to proton transport and will represent a reversal of the ATP synthase activity.

The synthase structure. The F_1 complex has been isolated from *E. coli*,[204,205] other bacteria,[206,207] yeast,[208a,b] animal tissues,[199,209–211] and chloroplasts.[212–214] In every case it consists of five kinds of subunits with the stoichiometry $\alpha_3\beta_3\gamma\delta\varepsilon$.[214a,b] The F_0 complex of *E. coli* contains three subunits designated a,b, and c. All of these proteins are encoded in one gene cluster, the *unc* operon (named for uncoupled mutants), with the following order:

I	*B*	*E*	*F*	*H*	*A*	*G*	*D*	*C*	--- Gene symbol
i	a	c	b	δ	α	γ	β	ε	--- Subunit symbol

Here I is the regulatory gene (as in Fig. 28-1). The *E. coli* F_0 appears to have approximately the unusual stoichiometry ab_2c_{9-11}. This suggested the possibility that 12 c subunits form a ring with D_6 or D_{12} symmetry, the latter being illustrated in the structural proposal shown in Fig. 18-14E. However, crystallographic evidence suggests that there may be 10, not 12 subunits.[214c]

Mitochondrial ATP synthase of yeast contains at least 13 different kinds of subunits[208] and that of animals[215] 16, twice as many as in *E. coli.* Subunits α, β, γ, a, b, and c of the mitochondrial synthase correspond to those of *E. coli.* However, the mitochondrial homolog of *E. coli* δ is called the **oligomycin-sensitivity-conferring protein** (OSCP).[216–218] It makes the ATPase activity sensitive to oligomycin. The mitochondrial δ subunit corresponds to ε of *E. coli* or of chloroplasts.[217,219] Mitochondrial ε has no counterpart in bacteria.[209,220] In addition,[209,215] mitochondria contain subunits called d, e, f, g, A6L, F6, and IF_1 the last being an 84-residue inhibitor, a regulatory subunit.[221] The subunits of yeast ATP synthase correspond to those of the animal mitochondrial synthase but include one additional protein (**h**).[208a]

	F_1							F_0								
E. coli	α	β	γ	δ	ε			a	b	c						
Mitochondria	α	β	γ	OSCP	δ	ε	IF_1	a	b	c	d	e	f	g	A6L	F6

Six of the relatively large (50–57 kDa) α and β subunits associate to an $\alpha_3\beta_3$ complex that constitutes the knobs.[202,210] Chemical crosslinking, directed mutation, electron cryomicroscopy,[222,222a] and high-resolution X-ray diffraction measurements[199,207,211,223,224] have established that the α and β subunits alternate in a quasisymmetric cyclic head that contains active sites for ATP formation in the three β subunits (Fig. 18-14C–E). The α subunits also contain ATP-binding sites, but they are catalytically inactive, and their bound MgATP does not exchange readily with external ATP and can be replaced by the nonhydrolyzable AMP-PNP (Fig. 12-31) with retention of activity. The $\alpha_3\beta_3$ complex is associated with the F_0 part by a slender **central stalk**

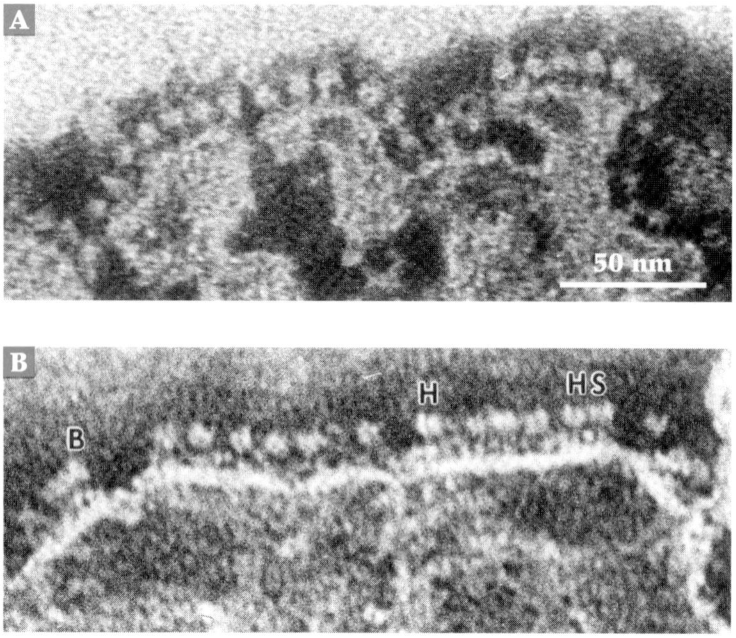

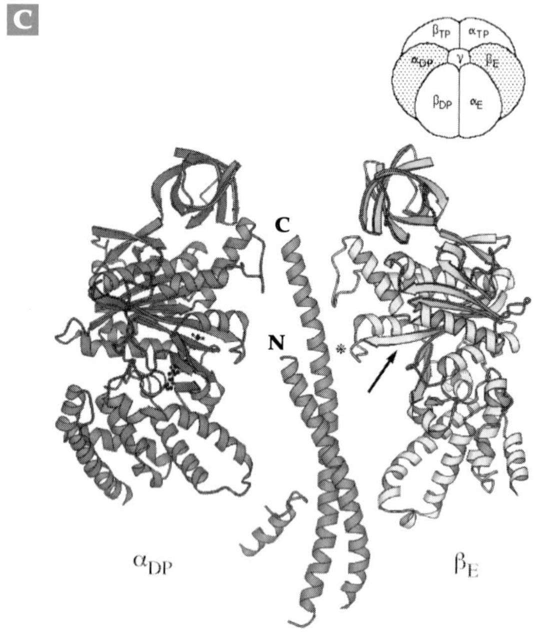

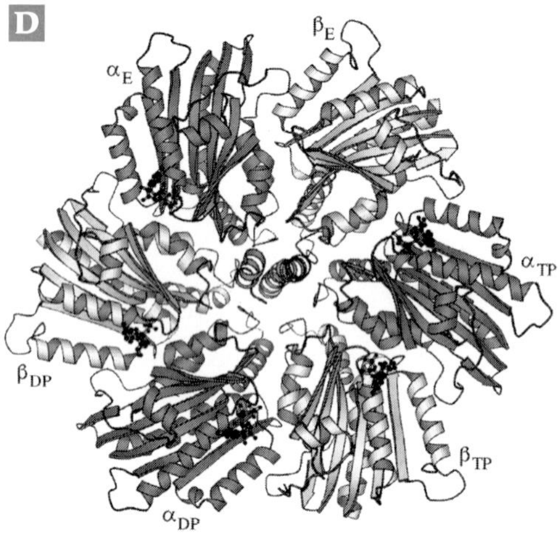

Figure 18-14 ATP synthase and vacuolar ATPase. (A) "Knobs" of ATP synthase on mitochondrial membranes negatively stained with phosphotungstate. (B) Vacuolar proton-pumping ATPase from an intact vacuolar membrane stained in the same way. Some images have been marked to indicate well-resolved head groups (H), stalks (S), and basal components (B). (A) and (B) are from Dschida and Bowman.[198] Courtesy of Barry J. Bowman. (C) Ribbon model of the atomic structure of the F_1 part of bovine heart mitochondrial ATP synthase. This section through the knob, which is drawn schematially at the upper right, shows one α subunit, containing bound ADP, α_{DP}, to the left and one empty β subunit, β_E, to the right. In the center are the N and C termini of subunit γ. The arrow points to a disruption in the β sheet of the subunit structure in the β_E conformation. The asterisk marks a loop that would collide with the C-terminal part of subunit γ if the latter were rotated. (D) View of the F_1 ATP synthase from the membrane side. The section shown contains the nucleotide binding sites. Subunits with empty sites are labeled α_E and β_E. Those with bound ADP are labeled α_{DP} and β_{DP}, and those with bound ATP α_{TP} and β_{TP}. Deviation from perfect threefold symmetry can be seen in this view. (C) and (D) are from Abrahams *et al.*[199] Courtesy of John E. Walker. (E) A recent model of the *E. coli* ATP synthase. The $\alpha_3\beta_3$ head contains the ATP-synthesizing sites in the β subunits. The $\alpha_3\beta_3$ complex and also subunits a, b, and δ remain stationary and form the stator for a protic engine, whose rotor consists of 12 c subunits and attached γ and ε subunits. Rotation is induced by the membrane potential or difference in H^+ activity on the two sides of the membrane. The carboxylate group of Asp 61 in each c subunit must be protonated to –COOH in order for it to move away from the entrance channel in subunit a. The presence of the positively charged Arg 210 near the exit channel in subunit a induces release of the proton when the c subunit has rotated almost 360°. According to this picture, 12 protons are required for one rotation with synthesis of three molecules of ATP. After Elston *et al.*[200] and Zhou *et al.*[201] See also Junge *et al.*[202] and Engelbrecht and Junge.[203]

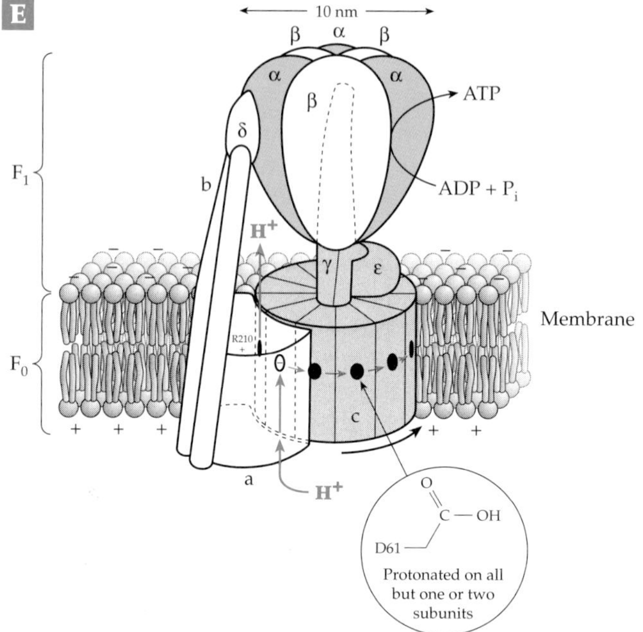

or **shaft**. Much effort has gone into establishing the subunit composition of the shaft and the F_0 parts of the structure. As is indicated in Fig. 18-14E, subunits γ and ε of the *E. coli* enzyme are both part of the central shaft.[219,225,226] The same is true for the mitochondrial complex, in which the δ subunit corresponds to bacterial ε.[227] The role of this subunit is uncertain. It is part of the shaft but is able to undergo conformational alterations that can permit its C-terminal portion to interact either with F_0 or with the $\alpha_3\beta_3$ head.[227,227a] The unique ε subunit of mitochondrial ATPase appears also to be part of the shaft.[220]

The most prominent component of the central shaft is the 270-residue subunit γ, which associates loosely with the $\alpha_3\beta_3$ head complex but more tightly with F_0. About 40 residues at the N terminus and 60 at the C terminus form an α-helical coiled coil, which is visible in Fig. 18-14E[199,211] and which protrudes into the central cavity of the $\alpha_3\beta_3$ complex. Because it is asymmetric, the γ subunit apparently acts as a rotating camshaft to physically alter the α and β subunits in a cyclic manner. Asymmetries are visible in Fig. 18-14D.[211] The central part of subunit γ forms a more globular structure, which bonds with the c subunits of F_0.[205] Exact structures are not yet clear.

The δ subunit of *E. coli* ATP synthase (OSCP of mitochondria) was long regarded as part of the central stalk. However, more recent results indicate that it is found in a **second stalk**, which joins the $\alpha_3\beta_3$ complex to F_0. The central stalk rotates, relative to the second stalk. The second stalk may be regarded as stationary and part of a **stator** for a protic engine.[228,229] This stalk has been identified[230] in electron micrographs of chloroplast F_1F_0 and by crosslinking studies. As is depicted in Fig. 18-14E, a major portion of the second stalk is formed by two molecules of subunit b. Recent results indicate that bacterial subunit δ (mitochondrial OSCP) extends further up than is shown in Fig. 18-14, and together with subunit F6 may form a cap at the top of the $\alpha_3\beta_3$ head.[230a, 230b]

The F_0 portion of bacterial ATP synthase, which is embedded in the membrane, consists of one 271-residue subunit a, an integral membrane protein probably with five transmembrane helices,[231,232] two 156-residue b subunits, and ~twelve 79-residue c subunits. The c subunit is a proteolipid, insoluble in water but soluble in some organic solvents. The structure of monomeric c in chloroform:methanol:H_2O (4:4:1) solution has been determined by NMR spectroscopy. It is a hairpin consisting of two antiparallel α helices.[233] Twelve of the c subunits are thought to assemble into a ring with both the N and C termini of the subunit chains in the periplasmic (or intermembrane) face of the membrane.[234,235] The ratio of c to a subunits has been difficult to measure but has been estimated as 9–12. The fact that both genetically fused c_2 dimers and c_3 trimers form function F_0 suggested that they assemble to a c_{12} ring as shown in Fig. 18-14E.[236] However, the recent crystallographic results that revealed a C_{10} ring[214c] raise questions about stoichiometry.

Since ATP synthesis takes place in F_1, it has long been thought that the F_0 part of the ATP synthase contains a "proton channel," which leads from the inside of the mitochondria to the F_1 assemblage.[146] Such a channel would probably not be an open pore but a chain of hydrogen-bonded groups, perhaps leading through the interior of the protein and able to transfer protons via icelike conduction. One residue in the c subunit, Asp 61, which lies in the center of the second of the predicted transmembrane helices, is critical for proton transport.[236a] Natural or artificial mutants at this position (e.g., D61G or D61N) do not transport protons. This carboxyl group also has an unusually high reactivity and specificity toward the protein-modifying reagent dicyclohexylcarbodiimide (DCCD; see Eq. 3-10).[146,237] Modification of a single c subunit with DCCD blocks the proton conductance.

An interesting mutation is replacement of alanine 62 of the c subunit with serine. This mutant will support ATP synthase using Li^+ instead of H^+.[237] Certain alkylophilic bacteria, such as *Propionigenium modestum*, have an ATP synthase that utilizes the membrane potential and a flow of Na^+ ions rather than protons through the c subunits.[238–240c] The sodium transport requires glutamate 65, which fulfills the same role as D61 in *E. coli*, and also Q32 and S66. Study of mutants revealed that the polar side chains of all three of these residues bind Na^+, that E65 and S66 are needed to bind Li^+, and that only E65 is needed for function with H^+.

The a subunit is also essential for proton translocation.[231,232,241] Structural work on this extremely hydrophobic protein has been difficult, but many mutant forms have been studied. Arginine 210 is essential as are E219 and H245. However, if Q252 is mutated to glutamate, E219 is no longer essential.[241] One of the OXPHOS diseases (NARP; Box 18-B) is a result of a leucine-to-arginine mutation in human subunit a.[241a] The b subunit is an elongated dimer, largely of α-helical structure.[242,243] Its hydrophobic N terminus is embedded in the membrane,[229,244] while the hydrophilic C-terminal region interacts with subunit σ of F_1, in the stator structure (Fig. 18-14E). Some of the F_0 subunits (d, e, f, g, A6L) may form a collar around the lower end of the central stalk.[210a,b]

How is ATP made? No covalent intermediates have been identified, and isotopic exchange studies indicate a direct dehydration of ADP and P_i to form bound ATP.[245] For the nucleophilic terminal phospho group of ADP to generate a high-energy linkage directly by attack on the phosphorus atom of P_i an OH^- ion must be eliminated (Eq. 18-13). This is not a probable reaction at pH 7, but it would be reasonable at low pH. Thus, one function of the oriented ATP synthase might be to deliver one or more protons flowing in from F_0 specifically to the oxygen that is to be eliminated (Eq. 18-13). As we have seen (Section 2), on thermodynamic grounds 3–4 protons would probably be needed. Perhaps they could be positioned nearby to exert a large electrostatic effect, or they could assist in releasing the ATP formed from the synthase by inducing a conformational change. However, it isn't clear how protons could be directed to the proper spots.

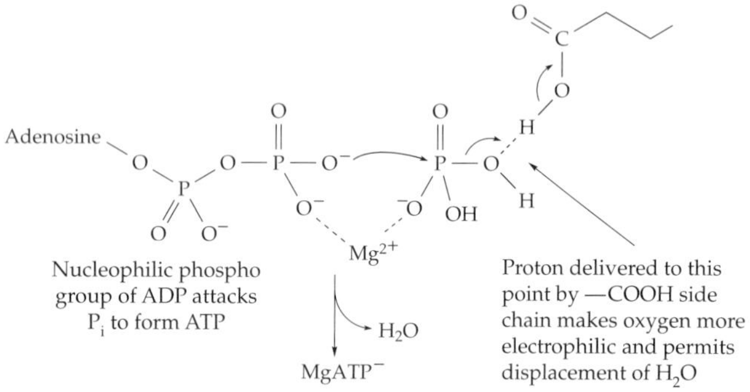

Adenosine

Nucleophilic phospho group of ADP attacks P_i to form ATP

Mg^{2+}

H_2O

$MgATP^-$

Proton delivered to this point by —COOH side chain makes oxygen more electrophilic and permits displacement of H_2O

(18-13)

Paul Boyer's binding change mechanism. Boyer and associates suggested that ATP synthesis occurs rapidly and reversibly in a closed active site of the ATP synthase in an environment that is essentially anhydrous. ATP would then be released by an energy-dependent conformational change in the protein.[245–249] Oxygen isotope exchange studies verified that a rapid interconversion of bound ADP, P_i, and ATP does occur. Studies of soluble ATP synthase, which is necessarily uncoupled from electron transport or proton flow, shows that ATP is exceedingly tightly bound to F_1 as expected by Boyer's mechanism.[248] According to his **conformational coupling** idea, protons flowing across the membrane into the ATP synthase would in some way induce the conformational change necessary for release of ATP.

The idea of conformational coupling of ATP synthesis and electron transport is especially attractive when we recall that ATP is used in muscle to carry out mechanical work. Here we have the hydrolysis of ATP coupled to motion in the protein components of the muscle. It seems reasonable that ATP should be formed as a result of motion induced in the protein components of the ATPase. Support for this analogy has come from close structural similarities of the F_1 ATPase β subunits and of the active site of ATP cleavage in the muscle protein myosin (Chapter 19).

A simple version of Boyer's binding change mechanism is shown in Figure 18-15. The three F_1 β subunits are depicted in three different conformations. In O the active site is open, in T it is closed, and if ATP is present in the active site it is tightly bound. In the low affinity L conformation ligands are bound weakly. In step *a* MgADP and P_i enter the L site while MgATP is still present in the T site. In step *b* a protonic-energy-dependent step causes synchronous conformational changes in all of the subunits. The tight site opens and MgATP is free to leave. At the same time MgADP and P_i in the T site are converted spontaneously to tightly bound ATP. The MgATP is in reversible equilibrium with MgADP + P_i, which must be bound less tightly than is MgATP. That is, the high positive value of $\Delta G'$ for formation of ATP must be balanced by a corresponding negative $\Delta G'$ for a conformational or electronic reorganization of the protein in the T conformation. Opening of the active site in step *b* of Fig. 18-15 will have a high positive $\Delta G'$ unless it is coupled to proton flow throught F_0. Of three sites in the subunits, one binds MgATP very tightly (K_d ~ 0.1 μM) while the other sites bind less tightly (K_d ~ 20 μM).[250,251] However, it has been very difficult to establish binding constants or K_m values for the ATPase reaction.[248] Each of the three β sites probably, in turn, becomes the high-affinity site, consistent with an ATP synthase mechanism involving protein conformational changes.

Rotational catalysis. Boyer suggested that there is a cyclic rotation in the conformations of the three β subunits of the ATP synthase, and that this might involve rotation about the stalk. By 1984, it had been shown that bacterial flagella are rotated by a protonic motor (Chapter 19), and a protic rotor for ATP synthase had been proposed by Cox *et al.*[252] and others.[245] However, the b subunits were thought to be in the central stalk.[222] More recently chemical crosslinking experiments,[201,253] as well as electron microscopy, confirmed the conclusion that an intact stator structure must also be present as in Fig. 18-14E.[202] The necessary second stalk is visible in CF_1F_0 ATPase of chloroplasts[230] and also in the related vacuolar ATPase, a proton or Na^+ pump from a clostridium.[254] See also Section 5. Another technique, **polarized absorption recovery after photobleaching**, was applied after labeling of Cys 322, the penultimate residue at the C terminus of the γ subunit with the dye eosin. After photobleaching with a laser beam the polarization of the light absorption by the dye molecule relaxed because of rotation. Relaxation was observed when ATP was added but not with ADPPNP.[202,255,256]

The most compelling experiments were performed by Noji *et al.*[202,257–260] They prepared the $\alpha_3\beta_3\gamma$ subcomplex of ATPase from a thermophilic bacterium. The complex was produced in *E. coli* cells from the cloned genes allowing for some "engineering" of the proteins. A ten-histidine "tag" was added at the N termini of the β subunits so that the complex could be "glued" to a microscope coverslip coated with a nickel complex with a high affinity for the His tags. The γ subunit shafts protrude upward as shown in Fig. 18-16. The γ subunit was mutated to replace its

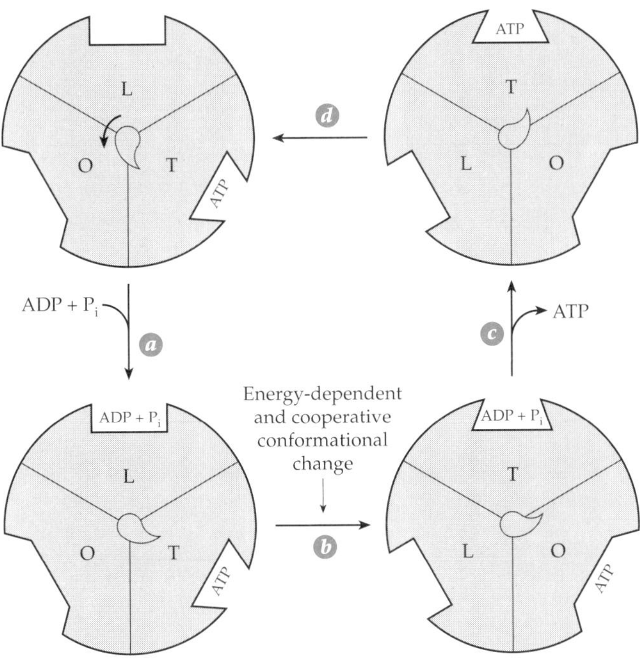

Figure 18-15 Boyer's binding change mechanism for ATP synthase in a simple form. After Boyer[245] but modified to include a central camshaft which may drive a cyclic alteration in conformations of the subunits. The small "pointer" on this shaft is not to be imagined as real but is only an indicator of rotation with induced conformational changes. The rotation could occur in 120° steps rather than the smaller steps suggested here.

only cysteine by serine and to introduce a cysteine in place of Ser 107 of the stalk region of γ. The new cysteine was biotinylated and attached to streptavidin (see Box 14-B) which was also attached to a fluorescently labeled actin filament (Fig. 7-10) ~1–3 µm in length as shown in Fig. 18-16. The actin fiber rotated in a counterclockwise direction when ATP was added but did not rotate with AMPPNP. At low ATP concentrations the rotation could be seen to occur in discrete 120° steps.[258,261,262] Each 120° step seems to consist of ~90° and ~30° substeps, each requiring a fraction of a millisecond.[262a] The ATPase appears to be acting as a **stepper motor**, hydrolysis of a single ATP turning the shaft 120°. Rotation at a rate of ~14 revolutions per second would require the hydrolysis of ~42 ATP per second. If the motor were attached to the F_0 part it would presumably pump four (or perhaps three) H^+ across a membrane for each ATP hydrolyzed. Acting in reverse, it would make ATP. A modification of the experiment of Fig. 18-16 was used to demonstrate that the c subunits also rotate with respect to the $\alpha_3\beta_3$ head.[262b] Other experiments support rotation of the c ring relative to subunit a.[262c,d]

Still to be answered are important questions. How does ATP hydrolysis turn the shaft? Are four H^+ pumped for each step, or are there smaller single proton substeps? Is the simple picture in Fig. 18-15 correct or, as proposed by some investigators,[263–265] must all three β subunits be occupied for maximum catalytic activity?[266] How is the coupling of H^+ transport to mechanical motion accomplished?[267,267a–d]

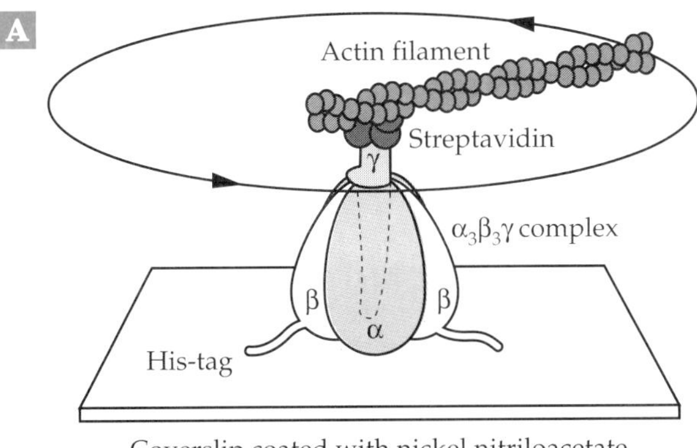

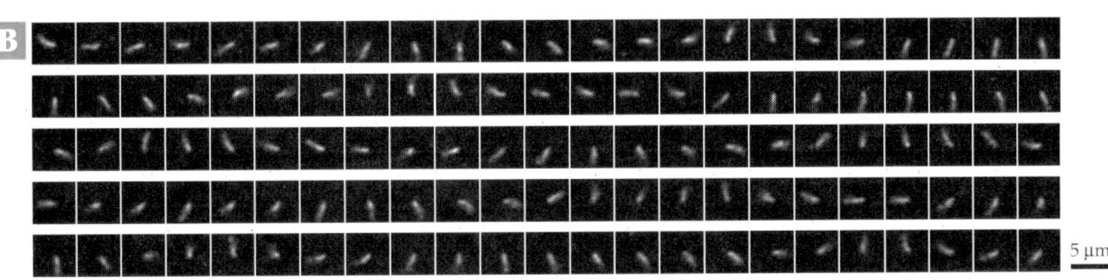

Figure 18-16 (A) The system used for observation of the rotation of the γ subunit in the $\alpha_3\beta_3\gamma$ subcomplex. The observed direction of the rotation of the γ subunit is indicated by the arrows. (B) Sequential images of a rotating actin filament attached as in (A). (C) Similar images obtained with the axis of rotation near the middle of the filament. The images correspond to the view from the top in (A). Total length of the filament, 2.4 µm; rotary rate, 1.3 revolutions per second; time interval between images, 33 ms. From Noji *et al.*[258]

5. ATP-driven Proton Pumps

Not all proton pumps are driven by electron transport. ATP synthase is reversible, and if Δp is low, hydrolysis of ATP can pump protons out of mitochondria or across bacterial plasma membranes.[268] Cells of *Streptococcus faecalis*, which have no respiratory chain and form ATP by glycolysis, use an F_1F_0 ATP synthase complex to pump protons out to help regulate cytoplasmic pH.[268] Similar **vacuolar (V-type) H⁺-ATPases** or V_1V_0 ATPases pump protons into vacuoles, Golgi and secretory vesicles, coated vesicles, and lysosomes[198,269-270] in every known type of eukaryotic cell.[271,272] These proton pumps are similar in appearance (Fig. 18-14,B) and in structure to F_1F_0 ATPases.[272a–c] The 65- to 77-kDa A subunits and 55- to 60-kDa B subunits are larger than the corresponding F_1F_0 α and β subunits. Accessory 40-, 39-, and 33-kDa subunits are also present in V_1. The V_0 portion appears to contain a hexamer of a 16-kDa proteolipid together with 110- and 21-kDa subunits.[271] V-type ATPases are also found in archaebacteria[271,273] and also in some clostridia[254] and other eubacteria.[273a] A type of proton pump, the **V-PPase**, uses hydrolysis of inorganic pyrophosphate as a source of energy.[274] It has been found in plants, in some phototrophic bacteria, and in acidic calcium storage vesicles (acidocalcisomes) of trypanosomes.[274a]

Other ATP-dependent proton pumps are present in the plasma membranes of yeast and other fungi[274b] and also in the acid-secreting parietal cells of the stomach (Fig. 18-17). The H^+-ATPase of *Neurospora* pumps H^+ from the cytoplasm without a counterion. It is electrogenic.[275,275a] However, the gastric H^+,K^+-ATPase exchanges H_3O^+ for K^+ and cleaves ATP with formation of a phosphoenzyme.[276] It belongs to the family of P-type ion pumps that includes the mammalian Na^+,K^+-ATPase (Fig. 8-25) and Ca^{2+}-ATPase (Fig. 8-26). These are discussed in Chapter 8. The H^+,K^+-ATPases, which are widely distributed within eukaryotes, are also similar, both in sequence and in the fact that a phospho group is transferred from ATP onto a carbox-ylate group of an aspartic acid residue in the protein. All of them, including a Mg-ATPase of *Salmonella*, are two-subunit proteins. A large catalytic α subunit contains the site of phosphorylation as well as the ATP- and ion-binding sites. It associates noncovalently with the smaller heavily glycosylated β subunit.[276–278] For example, the rabbit H^+,K^+-ATPase consists of a 1035-residue α chain which has ten transmembrane segments and a 290-residue β chain with a single transmembrane helix and seven N-linked glycosylation sites.[278]

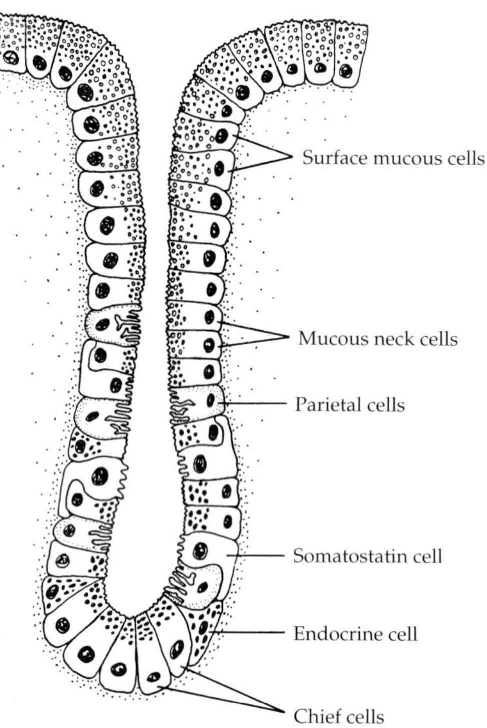

Figure 18-17 Schematic diagram of an acid-producing oxyntic gland of the stomach. The normal human stomach contains about 10^9 parietal (oxyntic) cells located in the walls of these glands. From Wolfe and Soll.[279] Modified from Ito. These glands also produce mucus, whose role in protecting the stomach lining from the high acidity is uncertain.[280]

6. Uncouplers and Energy-linked Processes in Mitochondria

Many compounds that uncouple electron transport from phosphorylation, like 2,4-dinitrophenol, are weak acids. Their anions are nucleophiles. According to the scheme of Fig. 18-12, they could degrade a high energy intermediate, such as Y ~ B, by a nucleophilic attack on Y to give an inactive but rapidly hydrolyzed derivative of Y. On the other hand, according to Mitchell's hypothesis uncouplers facilitate the transport of protons back into the mitochondria thereby destroying Δp. The fact that the anions of the uncouplers are large, often aromatic, and therefore soluble in the lipid bilayer supports this interpretation; the protonated uncouplers can diffuse into the mitochondria and the anion can diffuse back out. Mitochondria can also be uncoupled by a combination of ionophores, e.g., a mixture of valinomycin (Fig. 8-22), which carries K^+ into the mitochondria, plus nigericin, which catalyzes an exchange of K^+ (out) for H^+ (in).[172]

The uncoupler carbonyl cyanide *p*-(trifluoromethoxy)phenylhydrazone (FCCP) and related compounds are widely used in biochemical studies. Their action can be explained only partially by increased proton conduction.

$$F_3C—O—C_6H_4—NH—N=C(CN)_2$$

Carbonyl cyanide *p*-(trifluoromethoxy)phenylhydrazone (FCCP)

Uncoupling is sometimes important to an organism. The generation of heat by uncoupling is discussed in Box 18-C. The fungus *Bipolaris maydis* caused a crisis in maize production when it induced pore formation in mitochondrial membranes of a special strain used in production of hybrid seeds.[281,282]

Synthesis of ATP by mitochondria is inhibited by oligomycin, which binds to the OSCP subunit of ATP synthase. On the other hand, there are processes that require energy from electron transport and that are not inhibited by oligomycin. These **energy-linked processes** include the transport of many ions across the mitochondrial membrane (Section E) and reverse electron flow from succinate to NAD^+ (Section C,2). Dinitrophenol and many other uncouplers block the reactions, but oligomycin has no effect. This fact can be rationalized by the Mitchell hypothesis if we assume that Δp can drive these processes.

Another energy-linked process is the **transhydrogenase** reaction by which NADH reduces $NADP^+$ to form NADPH. In the cytoplasm various other reactions are used to generate NADPH (Chapter 17, Section I), but within mitochondria a membrane-bound transhydrogenase has this function.[283–286a] It couples the transhydrogenation reaction to the transport of one (or possibly more than one) proton back into the mitochondria (Eq. 18-14). A value of Δp of –180 mV could increase the ratio of [NADPH] / [$NADP^+$] within mitochondria by a factor of as much as 1000.

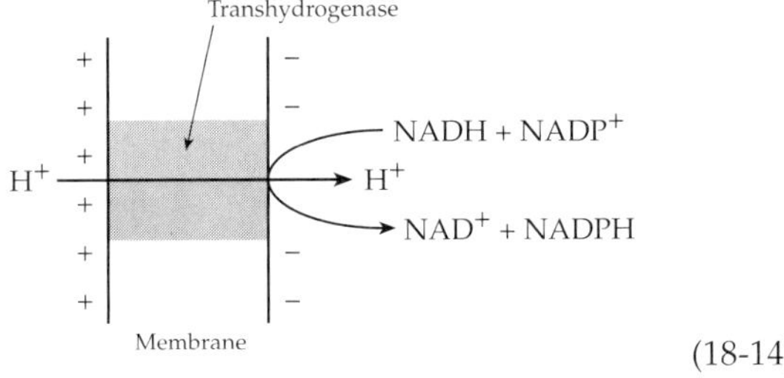

(18-14)

Transhydrogenases function in a similar way within bacteria. Whether from *E. coli*, photosynthetic bacteria, or bovine mitochondria, transhydrogenases have similar structures.[285] Two 510-residue α subunits associate with two 462-residue β subunits to form an $\alpha_2\beta_2$ tetramer with 10–14 predicted transmembrane helices. The α subunits contain separate NAD(H) and NADP(H) binding sites. A conformational change appears to be associated with the binding or release of the $NADP^+$ or NADPH.[287]

D. Transport and Exchange across Mitochondrial Membranes

Like the external plasma membrane of cells, the inner mitochondrial membrane is selective. Some nonionized materials pass through readily but the transport of ionic substances, including the anions of the dicarboxylic and tricarboxylic acids, is restricted. In some cases energy-dependent active transport is involved but in others one anion passes inward in exchange for another anion passing outward. In either case specific translocating carrier proteins are needed.

Solutes enter mitochondria through pores in thousands of molecules of the **voltage-gated anion-selective channel VDAC**, also known as mitochondrial porin.[15,16,288,289] In the absence of a membrane potential these pores allow free diffusion to molecules up to ~1.2 kDa in mass and may selectively permit passage of anions of 3- to 5-kDa mass. However, a membrane potential greater than ~20 mV causes the pores to close. NADH also decreases permeability. In the closed state the outer membrane becomes almost impermeant to ATP.[289,290]

An example of energy-dependent transport is the uptake of Ca^{2+} by mitochondria. As indicated in the lower part of Fig. 18-13, there are two possibilities for preservation of electrical neutrality according to the chemiosmotic theory. Counterions X^- may flow out to balance the protons discharged on the outside. On the other hand, if a cation such as Ca^{2+} flows inward to balance the two protons flowing outward, neutrality will be preserved and the mitochondrion will accumulate calcium ions. Experimentally such accumulations via a **calcium uniporter**[4] are observed to accompany electron transport. In the presence of a suitable ionophore energy-dependent accumulation of potassium ions also takes place.[291] In contrast, an **electroneutral exchange** of one Ca^{2+} for two Na^+ is mediated by a Na^+–Ca^{2+} exchanger.[292,293] It permits Ca^{2+} to leave mitochondria. A controversial role of mitochondria in accumulating Ca^{2+} postulates a special **rapid uptake mode** of exchange (see p. 1049).[294]

It is thought that glutamate enters mitochondria as the monoanion Glu^- in exchange for the dianion of aspartate Asp^{2-}. Like the uptake of Ca^{2+} this exchange is driven by Δp. Since a membrane potential can be created by this exchange in the absence of Δp, the process is electrogenic.[4] In contrast, an **electroneutral** exchange of $malate^{2-}$ and 2-$oxoglutarate^{2-}$ occurs by means of carriers that are not energy-linked.[295,296] This dicarboxylate transporter is only one of 35 structurally related mitochondrial carriers identified in the complete genome of yeast.[296,297] Another is the **tricarboxylic transporter** (citrate transport protein) which exchanges the dianionic form of citrate for malate, succinate, isocitrate, phosphoenolpyruvate, etc.[298,298a,b]

The important **adenine nucleotide carrier** takes ADP into the mitochondrial matrix for phosphorylation in a 1:1 ratio with ATP that is exported into the cytoplasm.[299-300b] This is one of the major rate-determining processes in respiration. It has been widely accepted that the carrier is electrogenic,[300]

BOX 18-C USING METABOLISM TO GENERATE HEAT: THERMOGENIC TISSUES

A secondary but important role of metabolism in warm-blooded animals is to generate heat. The heat evolved from ordinary metabolism is often sufficient, and an animal can control its temperature by regulating the heat exchange with the environment. Shivering also generates heat and is used from birth by pigs.[a] However, this is insufficient for many newborn animals, for most small mammals of all ages, and for animals warming up after hibernation. The need for additional heat appears to be met by **brown fat**, a tissue which contrasts strikingly with the more abundant white adipose tissue. Brown fat contains an unusually high concentration of blood vessels, many mitochondria with densely packed cristae, and a high ratio of cytochrome *c* oxidase to ATP synthase. Also present are a large number of sympathetic nerve connections, which are also related to efficient generation of heat. Newborn humans have a small amount of brown fat, and in newborn rabbits it accounts for 5–6% of the body weight.[b–d] It is especially abundant in species born without fur and in hibernating animals. Swordfish also have a large mass of brown adipose tissue that protects their brains from rapid cooling when traveling into cold water.[a]

The properties of brown fat pose an interesting biochemical question. Is the energy available from electron transport in the mitochondria dissipated as heat because ATP synthesis is uncoupled from electron transport? Or does ATP synthesis take place but the resulting ATP is hydrolyzed wastefully through the action of ATPases? Part of the answer came from the discovery that mitochondria of brown fat cells synthesize a 32-kDa **uncoupling protein** (UCP1 or thermogenin). It is incorporated into the inner mitochondrial membranes where it may account for 10–15% of total protein.[d–f] This protein, which is a member of a family of mitochondrial membrane metabolic carriers (Table 18-8), provides a "short-circuit" that allows the protonmotive force to be dissipated rapidly, perhaps by a flow of protons out through the uncoupling protein.[g–i] Synthesis of the uncoupling protein is induced by exposure to cold, but when an animal is warm the uncoupling action is inhibited.

The uncoupling protein resembles the ATP/ADP and phosphate *anion* carriers (Table 18-8),[g,i] which all have similar sizes and function as homodimers. Each monomeric subunit has a triply repeated ~100-residue sequence, each repeat forming two transmembrane helices. Most mitochondrial transporters carry anions, and UCP1 will transport Cl^-.[h,i] However, the relationship of chloride transport to its real function is unclear. Does the protein transport H^+ into, or does it carry HO^- out from, the mitochondrial matrix?[g,h] Another possibility is that a fatty acid anion binds H^+ on the intermembrane surface and carries it across into the matrix as an unionized fatty acid. The fatty acid anion could then pass back out using the anion transporter function and assisted by the membrane potential.[h,i]

The uncoupling protein is affected by several control mechanisms. It is inhibited by nucleotides such as GDP, GTP, ADP, and ATP which may bind at a site corresponding to that occupied by ATP or ADP in the ADP/ATP carrier.[i] Uncoupling is stimulated by noradrenaline,[f] which causes a rapid increase in heat production by brown fat tissues, apparently via activation of adenylate cyclase. Uncoupling is also stimulated by fatty acids.[j] Recently UCP1 and related uncoupling proteins have been found to require both fatty acids and **ubiquinone** for activity.[jj,jk]

It has been suggested that brown adipose tissue may also function to convert excess dietary fat into heat and thereby to resist obesity.[k–m] Mice lacking the gene for the mitochondrial uncoupling protein are cold-sensitive but not obese. However, other proteins, homologous to UCP1, have been discovered. They may partially compensate for the loss.[m,h]

The bombardier beetle generates a hot, quinone-containing defensive discharge, which is sprayed in a pulsed jet from a special reaction chamber at a temperature of 100°C.[n–p] The reaction mixture of 25% hydrogen peroxide and 10% hydroquinone plus methylhydroquinone is stored in a reservoir as shown in the accompanying figure and reacts with explosive force when it comes into contact with catalase and peroxidases in the reaction chamber. The synthesis and storage of 25% H_2O_2 poses interesting biochemical questions!

Some plant tissues are thermogenic. For example, the spadix (or inflorescence, a sheathed floral spike) of the skunk cabbage *Symplocarpus foetidus* can maintain a 15–35°C higher temperature than that of the surrounding air.[q] The voodoo lily in a single day heats the upper end of its long spadex to a temperature 22°C above ambient, volatilizing a foul smelling mixture of indoles and amines.[r,s] This is accomplished using the alternative oxidase system[s] (Box D in Fig. 18-6). The lotus flower maintains a temperature of 30–35°C, while the ambient temperature may vary from 10–30°C.[t] While the volatilization of insect attractants may be the primary role for thermogenesis in plants, the warm flowers may also offer an important reward to insect pollinators. Beetles and bees require thoracic temperatures above 30°C to initiate flight and, therefore,

BOX 18-C (continued)

benefit from the warm flowers.[t] While in flight bees vary their metabolic heat production by altering their rate of flight, hovering, and other changes in physical activity.[u]

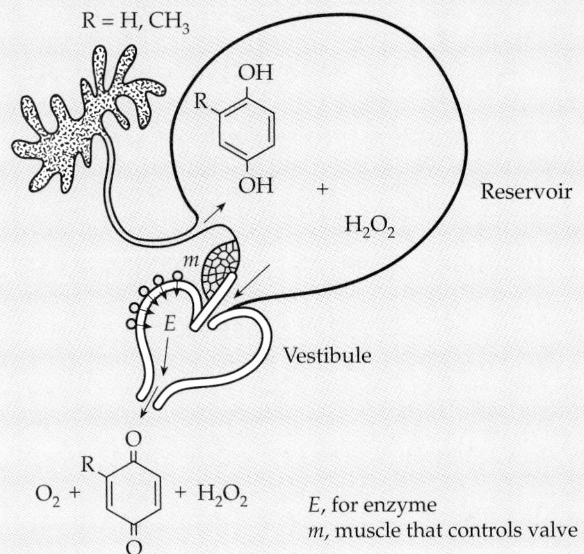

Reservoir and reaction vessel of the bombardier beetle. From D. J. Aneshansley, *et al.*[n]

[a] Tyler, D. D. (1992) *The Mitochondrion in Health and Disease*, VCH Publ., New York

[b] Dawkins, M. J. R., and Hull, D. (1965) *Sci. Am.* **213**(Aug), 62–67

[c] Lindberg, O., ed. (1970) *Brown Adipose Tissue*, Am. Elsevier, New York

[d] Nicholls, D. G., and Rial, E. (1984) *Trends Biochem. Sci.* **9**, 489–491

[e] Cooney, G. J., and Newsholme, E. A. (1984) *Trends Biochem. Sci.* **9**, 303–305

[f] Ricquier, D., Casteilla, L., and Bouillaud, F. (1991) *FASEB J.* **5**, 2237–2242

[g] Klingenberg, M. (1990) *Trends Biochem. Sci.* **15**, 108–112

[h] Jaburek, M., Varecha, M., Gimeno, R. E., Dembski, M., Jezek, P., Zhang, M., Burn, P., Tartaglia, L. A., and Garlid, K. D. (1999) *J. Biol. Chem.* **274**, 26003–26007

[i] González-Barroso, M. M., Fleury, C., Levi-Meyrueis, C., Zaragoza, P., Bouillaud, F., and Rial, E. (1997) *Biochemistry* **36**, 10930–10935

[j] Hermesh, O., Kalderon, B., and Bar-Tana, J. (1998) *J. Biol. Chem.* **273**, 3937–3942

[jj] Echtay, K. S., Winkler, E., and Klingenberg, M. (2000) *Nature (London)* **408**, 609–613

[jk] Echtay, K. S., Winkler, E., Frischmuth, K., and Klingenberg, M. (2001) *Proc. Natl. Acad. Sci. U.S.A.* **98**, 1416–1421

[k] Rothwell, N. J., and Stock, M. J. (1979) *Nature (London)* **281**, 31–35

[l] Tai, T.-A. C., Jennermann, C., Brown, K. K., Oliver, B. B., MacGinnitie, M. A., Wilkison, W. O., Brown, H. R., Lehmann, J. M., Kliewer, S. A., Morris, D. C., and Graves, R. A. (1996) *J. Biol. Chem.* **271**, 29909–29914

[m] Enerbäck, S., Jacobsson, A., Simpson, E. M., Guerra, C., Yamashita, H., Harper, M.-E., and Kozak, L. P. (1997) *Nature (London)* **387**, 90–94

[n] Aneshansley, D. J., Eisner, T., Widom, J. M., and Widom, B. (1969) *Science* **165**, 61–63

[o] Eisner, T., and Aneshansley, D. J. (1999) *Proc. Natl. Acad. Sci. U.S.A.* **96**, 9705–9709

[p] Dean, J., Aneshansley, D. J., Edgerton, H. E., and Eisner, T. (1990) *Science* **248**, 1219–1221

[q] Knutson, R. M. (1974) *Science* **186**, 746–747

[r] Diamond, J. M. (1989) *Nature (London)* **339**, 258–259

[s] Rhoads, D. M., and McIntosh, L. (1991) *Proc. Natl. Acad. Sci. U.S.A.* **88**, 2122–2126

[t] Seymour, R. S., and Schultze-Motel, P. (1996) *Nature (London)* **383**, 305

[u] Harrison, J. F., Fewell, J. H., Roberts, S. P., and Hall, H. G. (1996) *Science* **274**, 88 – 90

bringing in ADP^{3-} and exporting ATP^{4-} in an exchange driven by Δp. However, an electroneutral exchange, e.g., of ADP^{3-} for ATP^{3-}, may also be possible. The carrier is an ~300-residue 32-kDa protein, which is specifically inhibited by the plant glycoside **atractyloside** or the fungal antibiotic **bongkrekate**. The carrier is associated with bound cardiolipin.[301] This one transporter accounts for ~10% of all of the mitochondrial protein.[302,303]

A separate dimeric carrier allows P_i to enter, probably as $H_2PO_4^-$.[304-305a] This ion enters mitochondria in an electroneutral fashion, either in exchange for OH^- or by cotransport with H^+. A less important carrier[306] exchanges HPO_4^{2-} for malate^{2-}. Several other transporters help to exchange organic and inorganic ions. One of them allows pyruvate to enter mitochondria in exchange for OH^- or by cotransport with H^+. Some of the identified carriers are listed in Table 18-8. As discussed in Chapter 8, exchange carriers are also important in plasma membranes of organisms from bacteria to human beings. For example, many metabolites enter cells by cotransport with Na^+ using the energy of the Na^+ gradient set up across the membrane by the Na^+,K^+ pump.

Under some circumstances the inner membrane develops one or more types of large-permeability pore. An increase in Ca^{2+} may induce opening of an unselective pore which allows rapid uptake of Ca^{2+}.[294,307,307a] A general anion-specific channel may be involved in volume homeostasis of mitochondria.[308]

Mitochondria are not permeable to NADH. However, reactions of glycolysis and other dehydrogenations in the cytoplasm quickly reduce available NAD^+ to NADH. For aerobic metabolism to occur, the "reducing equivalents" from the NADH must be transferred into the mitochondria. Fungi and green plants have solved the problem by providing *two* NADH dehydrogenases embedded in the inner mitochondrial membranes (Fig. 18-6). One faces the matrix space and oxidizes the NADH produced in the matrix while the second faces outward to the intermembrane space and is able to oxidize the NADH formed in the cytoplasm. In animals the reducing equivalents from NADH enter the mitochondria indirectly. There are several mechanisms, and more than one may function simultaneously in a tissue.

In insect flight muscle, as well as in many mammalian tissues, NADH reduces dihydroxyacetone phosphate. The resulting *sn*-3-glycerol *P* passes through the permeable outer membrane of the mitochondria, where it is reoxidized to dihydroxyacetone phosphate by the FAD-containing glycerol-phosphate dehydrogenase embedded in the outer surface of the inner membrane (Figs. 18-5, 18-6). The dihydroxyacetone can then be returned to the cytoplasm. The overall effect of this **glycerol-phosphate shuttle** (Fig. 18-18A) is to provide for mitochondrial oxidation of NADH produced in the cytoplasm. In heart and liver the same function is served by a more complex **malate–aspartate shuttle** (Fig. 18-18B).[309] Reduction of oxaloacetate to malate by NADH, transfer of malate into mitochondria, and reoxidation with NAD^+ accomplishes the transfer of reduction equivalents into the mitochondria. Mitochondrial membranes are not very permeable to oxaloacetate. It returns to the cytoplasm mainly via transamination to aspartate, which leaves the mitochondria together with 2-oxoglutarate. At the same time glutamate enters the mitochondria in exchange for aspartate. The 2-oxoglutarate presumably exchanges with the entering malate as is indicated in Fig. 18-18B. The export of aspartate may be energy-linked as a result of the use of an electrogenic carrier that exchanges glutamate$^-$ + H^+ entering mitochondria for aspartate$^-$ leaving the mitochondria. Thus, Δp may help to expel aspartate from mitochondria and to drive the shuttle.

The glycerol-phosphate shuttle, because it depends upon a mitochondrial flavoprotein, provides ~ 2 ATP per electron pair (P/O = 2), whereas the malate–aspartate shuttle may provide a higher yield of ATP. The glycerol-phosphate shuttle is essentially irreversible, but the reactions of the malate–aspartate shuttle can be reversed and utilized in gluconeogenesis (Chapter 17).

TABLE 18-8
Some Mitochondrial Membrane Transporters[a]

Ion Diffusing In	Ion Diffusing Out	Comment[b]
ADP^{3-}	ATP^{3-}	
or ADP^{3-}	ATP^{4-}	Electrogenic symport
$H_2PO_4^-$ + H^+		Electroneutral symport
or $H_2PO_4^-$	OH^-	
HPO_4^{2-}	Malate^{2-}	
Malate^{2-}	2-Oxoglutarate^{2-}	
Glutamate^{2-} + H^+	Aspartate^{2-}	
Glutamate$^-$	OH^-	
Pyruvate$^-$	OH^-	
or Pyruvate$^-$ + H^+		Electroneutral symport
Citrate^{3-} + H^+	Malate^{2-}	
Ornithine$^+$	H^+	
Acylcarnitine	Carnitine	
2 Na^+	Ca^{2+}	
H^+	K^+	
H^+	Na^+	
General transporters		
VDAC (porin)		Outer membrane
Large anion pores		Inner membrane

[a] From Nicholls and Ferguson[172] and Tyler[4].
[b] Unless indicated otherwise the transporters are *antiporters* that catalyze an electroneutral ion exchange.

E. Energy from Inorganic Reactions

Some bacteria obtain all of their energy from inorganic reactions. These **chemolithotrophs** usually have a metabolism that is similar to that of heterotrophic organisms, but they also have the capacity to obtain all of their energy from an inorganic reaction. In order to synthesize carbon compounds they must be able to fix CO_2 either via the reductive pentose phosphate cycle or in some other way. The chloroplasts of green plants, using energy from sunlight, supply the organism with both ATP and the reducing agent NADPH (Chapter 17). In a similar way the lithotrophic bacteria obtain both energy and reducing materials from inorganic reactions.

Chemolithotrophic organisms often grow slowly, making study of their metabolism difficult.[310] Nevertheless, these bacteria usually use electron transport chains similar to those of mitochondria. ATP is formed by oxidative phosphorylation, the amount formed per electron pair depending upon the number of proton-pumping sites in the chain. This, in turn, depends upon the electrode potentials of the reactions involved. For example, H_2, when oxidized by O_2, leads to passage of electrons through the entire electron transport chain with synthesis of ~3 molecules of ATP per electron pair. On the other hand, oxidation by O_2 of nitrite, for which $E^{\circ\prime}$ (pH 7) = +0.42 V, can make use only of the site III part of the chain. Not only is the yield of ATP less than in the oxidation of H_2 but also there is another problem. Whereas reduced pyridine

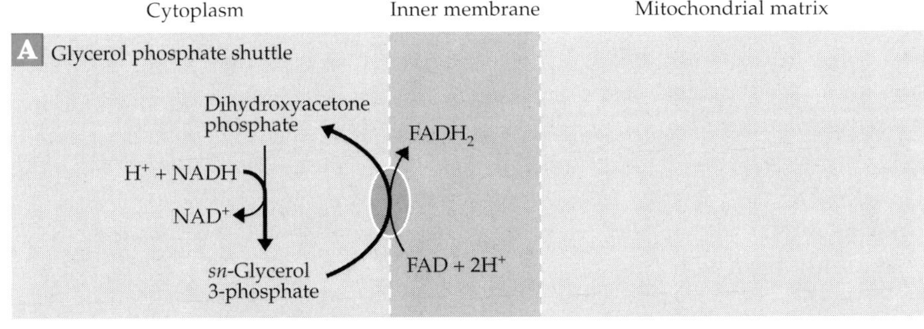

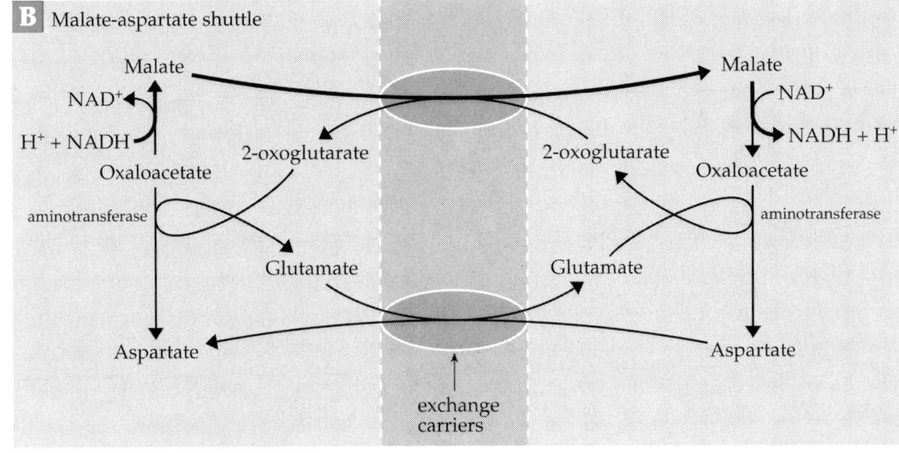

Figure 18-18 (A) The glycerol-phosphate shuttle and (B) the malate–aspartate shuttle for transport from cytoplasmic NADH into mitochondria. The heavy arrows trace the pathway of the electrons (as 2H) transported.

nucleotides needed for biosynthesis can be generated readily from H_2, nitrite is not a strong enough reducing agent to reduce NAD^+ to NADH. The only way that reducing agents can be formed in cells utilizing oxidation of nitrite as an energy source is via *reverse electron flow* driven by hydrolysis of ATP or by Δp. Such reverse electron flow is a common process for many chemolithotrophic organisms.

Let us consider the inorganic reactions in two groups: (1) oxidation of reduced inorganic compounds by O_2 and (2) oxidation reactions in which an inorganic oxidant, such as nitrate or sulfate, substitutes for O_2. The latter reactions are often referred to as **anaerobic respiration**.

1. Reduced Inorganic Compounds as Substrates for Respiration

The hydrogen-oxidizing bacteria. Species from several genera including *Hydrogenomonas*, *Pseudomonas*, and *Alcaligenes* oxidize H_2 with oxygen:

$$H_2 + 1/2\ O_2 \rightarrow H_2O \qquad \Delta G^\circ = -237.2\ \text{kJ mol}^{-1} \qquad (18\text{-}15)$$

Some can also oxidize carbon monoxide:

$$CO + 1/2\ O_2 \rightarrow CO_2 \qquad \Delta G^\circ = -257.1\ \text{kJ mol}^{-1} \qquad (18\text{-}16)$$

The hydrogen bacteria can also oxidize organic compounds using straightforward metabolic pathways. The key enzyme is a membrane-bound nickel-containing hydrogenase (Fig. 16-26), which delivers electrons from H_2 into the electron transport chain.[310a] A second soluble hydrogenase (sometimes called **hydrogen dehydrogenase**) transfers electrons to $NADP^+$ to form NADPH for use in the reductive pentose phosphate cycle and for other biosynthetic purposes.

Nitrifying bacteria. Two genera of soil bacteria oxidize ammonium ion to nitrite and nitrate (Eqs. 18-17 and 18-18).[311]

$$NH_4^+ + 3/2\ O_2 \rightarrow NO_2^- + 2\ H^+ + H_2O\ (\textit{Nitrosomonas}) \qquad \Delta G'\ (\text{pH } 7) = -272\ \text{kJ mol}^{-1} \qquad (18\text{-}17)$$

$$NO_2^- + 1/2\ O_2 \rightarrow NO_3^-\ (\textit{Nitrobacter}) \qquad \Delta G'\ (\text{pH } 7) = -76\ \text{kJ mol}^{-1} \qquad (18\text{-}18)$$

The importance of these reactions to the energy metabolism of the bacteria was recognized in 1895 by Winogradsky, who first proposed the concept of chemiautotrophy. Because the nitrifying bacteria grow slowly (generation time ~10–12 h) it has been hard to get enough cells for biochemical studies and progress has been slow. The reaction catalyzed by *Nitrosomonas* (Eq. 18-17) is the more complex; it occurs in two or more stages and is catalyzed by two enzymes as illustrated in Fig. 18-19. The presence of hydrazine blocks oxidation of hydroxylamine (NH_2OH) in step *b* and permits that intermediate to accumulate. The oxidation of ammonium ion by O_2 to hydroxylamine (step *a*) is endergonic with $\Delta G'$ (pH 7) = 16 kJ mol^{-1} and is incapable of providing energy to the cell. It occurs by a hydroxylation mechanism (see Section G). On the other hand, the oxidation of hydroxylamine to nitrite by O_2 in step *b* is highly exergonic with $\Delta G'$ (pH 7) = –228 kJ mol^{-1}. The hydroxylamine oxidoreductase that catalyzes this reaction is a trimer of 63-kDa subunits, each containing seven *c*-type hemes and an unusual heme P450, which is critical to the enzyme's function[312–314a] and which is covalently linked to a tyrosine as well as to two cysteines.

The electrode potentials for the two- and four-electron oxidation steps are indicated in Fig. 18-19. It is apparent that step *b* can feed four electrons into the electron transport system at about the potential of ubiquinone. Two electrons are needed to provide a cosubstrate (Section G) for the ammonia monooxygenase and two could be passed on to the terminal cytochrome aa_3 oxidase. The stoichiometry of proton pumping in complexes III and IV is uncertain, but if it is assumed to be as shown in Fig. 18-19 and similar to that in Fig. 18-5, there will be ~13 protons available to be passed through ATP synthase to generate ~3 ATP per NH_3 oxidized. However, to generate NADH for reductive biosynthesis *Nitrosomonas* must send some electrons to NADH dehydrogenase (complex I) using reverse electron transport, a process that depends upon Δp to drive the reaction via a flow of protons through the NADH dehydrogenase from the periplasm back into the bacterial cytoplasm (Fig. 18-19).

Nitrobacter depends upon a simpler energy-yielding reaction (Eq. 18-18) with a relatively small Gibbs energy decrease. The two-electron oxidation delivers electrons to the electron transport chain at $E^{\circ\prime}$ = +0.42 V. The third oxygen in NO_3^- originates from H_2O, rather than from O_2 as might be suggested by Eq. 18-18.[316,317] It is reasonable to anticipate that a single molecule of ATP should be formed for each pair of electrons reacting with O_2. However, *Nitrobacter* contains a confusing array of different cytochromes in its membranes.[311] Some of the ATP generated by passage of electrons from nitrite to oxygen must be used to drive a reverse flow of electrons through both a bc_1-type complex and NADH dehydrogenases. This generates reduced pyridine nucleotides required for biosynthetic reactions (Fig. 18-20).

An interesting feature of the structure of *Nitrobacter* is the presence of several double-layered membranes which completely envelop the interior of the cell. Nitrite entering the cell is oxidized on these membranes and cannot penetrate to the interior, where it might have toxic effects.

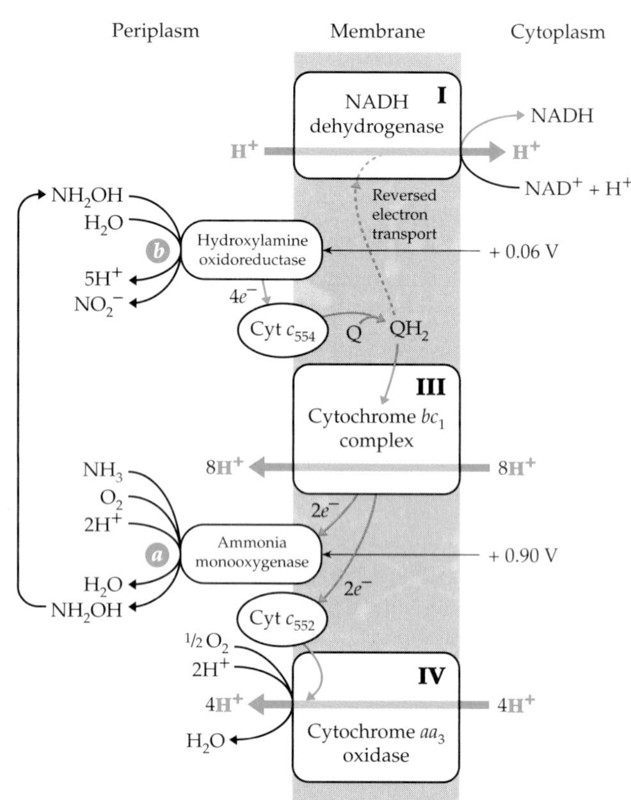

Figure 18-19 The ammonia oxidation system of the bacterium *Nitrosomonas*. Oxidation of ammonium ion (as free NH_3) according to Eq. 18-17 is catalyzed by two enzymes. The location of ammonia monooxygenase (step *a*) is uncertain but hydroxylamine oxidoreductase (step *b*) is periplasmic. The membrane components resemble complexes I, III, and IV of the mitochondrial respiratory chain (Fig. 18-5) and are assumed to have similar proton pumps. Solid green lines trace the flow of electrons in the energy-producing reactions. This includes flow of electrons to the ammonia monoxygenase. Complexes III and IV pump protons out but complex I catalyzes reverse electron transport for a fraction of the electrons from hydroxylamine oxidoreductase to NAD^+. Modified from Blaut and Gottschalk.[315]

The sulfur-oxidizing bacteria. Anaerobic conditions prevail in marine sediments, in poorly stirred swamps, and around hydrothermal vents at the bottom of the sea. Sulfate-reducing bacteria form high concentrations (up to mM) of H_2S (in equilibrium with HS^- and S^{2-})[318–320] This provides the substrate for bacteria of the genus *Thiobacillus*, which are able to oxidize sulfide, elemental sulfur, thiosulfate, and sulfite to sulfate and live where the aerobic and anaerobic regions meet.[311,321–323] Most of these small gram-negative

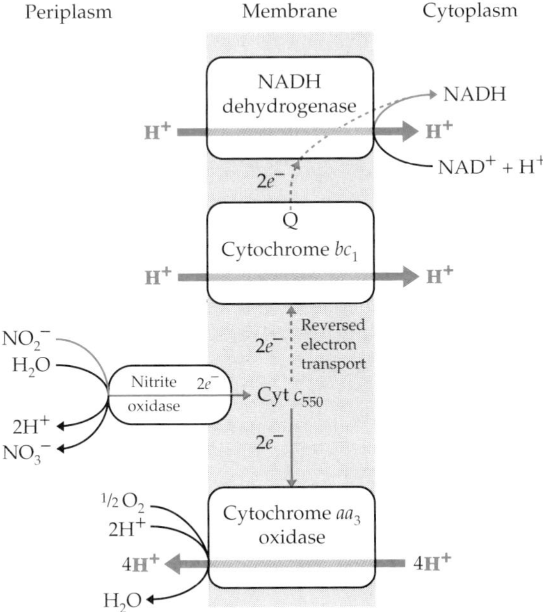

Figure 18-20 Electron transport system for oxidation of the nitrite ion to the nitrate ion by *Nitrobacter*. Only one site of proton pumping for oxidative phosphorylation is available. Generation of NADH for biosynthesis requires two stages of reverse electron transport.

organisms, found in water and soil, are able to grow in a simple salt medium containing an oxidizable sulfur compound and CO_2. One complexity in understanding their energy-yielding reactions is the tendency of sulfur to form chain molecules. Thus, when sulfide is oxidized, it is not clear that it is necessarily converted to monoatomic elemental sulfur as indicated in Eq. 18-19. Elemental sulfur ($S_8{}^0$) often precipitates. In *Beggiotoa*, another sulfide-oxidizing bacterium, sulfur is often seen as small globules within the cells. Fibrous sulfur precipitates are often abundant in the sulfide-rich layers of ponds, lakes, and oceans.[318]

$$\begin{aligned} &H_2S \xrightarrow{a} S^0 + 2H^+ + 2e^- \\ &S^0 + 3H_2O \xrightarrow{b} SO_3{}^{2-} + 6H^+ + 4e^- \\ &SO_3{}^{2-} + H_2O \xrightarrow[\text{Sulfite oxidase}]{c} SO_4{}^{2-} + 2H^+ + 2e^- \end{aligned} \qquad (18\text{-}19)$$

The reactions of Eq. 18-19 occur in the periplasmic space of some species.[315,323a,324] Steps *a* and *b* of Eq. 18-19 are catalyzed by a 67-kDa sulfide dehydrogenase in the periplasm of a purple photosynthetic bacterium.[324] The enzyme consists of a 21-kDa subunit containing two cytochrome *c*-like hemes, presumably the site of binding of S^{2-}, and a larger 46-kDa FAD-containing flavoprotein resembling glutathione reductase.[324] The molybdenum-containing sulfite oxidase (Fig. 16-32), which is found in the intermembrane space of mitochondria, may be present in the periplasmic space of these bacteria. However, there is also an intracellular pathway for sulfite oxidation (see Eq. 18-22).

The sulfide-rich layers inhabited by the sulfur oxidizers also contain thiosulfate, $S_2O_3{}^{2-}$. It may arise, in part, from reaction of glutathione with elemental sulfur:

$$GSH + S_8{}^0 \rightarrow G{-}S{-}S_8{-}H \qquad (18\text{-}20)$$

The linear polysulfide obtained by this reaction may be oxidized, the sulfur atoms being removed from the chain either one at a time to form sulfite or two at a time to form thiosulfate.[322,322a] Thiosulfate is oxidized by all species, the major pathway beginning with cleavage to S^0 and $SO_3{}^{2-}$ (Eq. 18-21, step *a*). At high thiosulfate concentrations some may be oxidized to tetrathionate (step *b*), which is hydrolyzed to sulfate (step *c*).

$$\begin{aligned} &\underset{\text{Thiosulfate}}{S{-}SO_3{}^-} \xrightarrow{a} S^0 + SO_3{}^{2-} \\ &\underset{\text{Thiosulfate}}{S{-}SO_3{}^-} \xrightarrow{b} 2e^- \rightarrow \text{cyt } c \;;\; \underset{\text{Tetrathionite}}{{}^-O_3S{-}S{-}S{-}SO_3{}^-} \\ &\underset{\text{Tetrathionite}}{{}^-O_3S{-}S{-}S{-}SO_3{}^-} + H_2O \xrightarrow{c} SO_4{}^{2-} + 2H^+ + S{-}SO_3{}^- + S^0 \end{aligned} \qquad (18\text{-}21)$$

Oxidation of sulfite to sulfate within cells occurs by a pathway through **adenosine 5'-phosphosulfate** (**APS**, adenylyl sulfate). Oxidation via APS (Eq. 18-22) provides a means of substrate-level phosphorylation,

$$\begin{aligned} &SO_3{}^{2-} + \text{AMP} \rightarrow \text{Adenosine 5'-phosphosulfate (APS)} + 2e^- \\ &\text{APS} + P_i{}^{2-} \rightarrow \text{ADP} + SO_4{}^{2-} \\ &\text{ADP} \rightarrow \text{AMP} \quad (\text{ADP} \rightarrow \text{ATP}) \end{aligned} \qquad (18\text{-}22)$$

the only one known among chemolithotrophic bacteria. No matter which of the two pathways of sulfite oxidation is used, thiobacilli also obtain energy via electron transport. With a value of $E°'$ (pH 7) of −0.454 V [$E°'$ (pH 2) = −0.158 V] for the sulfate–sulfite couple an abundance of energy may be obtained. Oxidation of sulfite to sulfate produces hydrogen ions. Indeed, pH 2 is optimal for the growth of *Thiobacillus thiooxidans*, and the bacterium withstands 5% sulfuric acid.[322]

The "iron bacterium" *Thiobacillus ferrooxidans* obtains energy from the oxidation of Fe^{2+} to Fe^{3+} with subsequent precipitation of ferric hydroxide (Eq. 18-23). However, it has been recognized recently that a previously unknown species of Archaea is much more important than *T. ferrooxydans* in catalysis of this reaction.[324a]

$$2\,Fe^{2+} \xrightarrow[2\,H^+]{1/2\,O_2 \;\to\; H_2O} 2\,Fe^{3+} \xrightarrow[\;\to\; 6\,H^+]{6\,H_2O} 2\,Fe(OH)_3 \quad (18\text{-}23)$$

Since the reduction potential for the Fe(II) / Fe(III) couple is + 0.77 V at pH 7, the energy obtainable in this reaction is small. These bacteria always oxidize reduced sulfur compounds, too. Especially interesting is their oxidation of **pyrite**, ferrous sulfide (Eq. 18-24). The Gibbs energy change was calculated from published data[325] using a value of $G_f°$ for Fe $(OH)_3$ of

$$2\,FeS + 4\tfrac{1}{2}\,O_2 + 5\,H_2O \rightarrow 2\,Fe(OH)_3 + 2\,SO_4^{2-} + 4\,H^+$$
$$\Delta G'\ (pH\ 2) = -1340\ kJ \quad (18\text{-}24)$$

−688 kJ mol^{-1} estimated from its solubility product. Because sulfuric acid is generated in this reaction, a serious water pollution problem is created by the bacteria living in abandoned mines. Water running out of the mines often has a pH of 2.3 or less.[326]

Various invertebrates live in S^{2-}-containing waters. Among these is a clam that has symbiotic sulfur-oxidizing bacteria living in its gills. The clam tissues apparently carry out the first step in oxidation of the sulfide.[327] Among the animals living near sulfide-rich thermal vents in the ocean floor are giant 1-meter-long tube worms. Both a protective outer tube and symbiotic sulfide-oxidizing bacteria protect them from toxic sulfides.[319,320]

2. Anaerobic Respiration

Nitrate as an electron acceptor. The use of nitrate as an alternative oxidant to O_2 is widespread among bacteria. For example, *E. coli* can subsist anaerobically by reducing nitrate to nitrite (Eq. 18-25).[311,328] The respiratory (dissimilatory) nitrate reductase that

$$NO_3^- + 2\,H^+ + 2e^- \rightarrow NO_2^- + H_2O$$
$$E°'\ (pH7) = +0.421\ V \quad (18\text{-}25)$$

catalyzes the reaction is a large three-subunit molybdenum-containing protein. The enzyme is present in the plasma membrane, and electrons flow from ubiquinone through as many as six heme and Fe–S centers to the molybdenum atom.[328-329] A second molybdoenzyme, formate dehydrogenase (discussed in Chapter 16), appears to be closely associated with nitrate reductase. Formate is about as strong a reducing agent as NADH (Table 6-8) and is a preferred electron donor for the reduction of NO_3^- (Eq. 18-26).[329a,b] Since cytochrome *c* oxidase of the electron transport chain is bypassed, one less ATP is formed than when O_2 is the oxidant. Nitrate is the oxidant preferred by bacteria grown under anaerobic conditions. The

$$HCOO^- + NO_3^- + H^+ \rightarrow NO_2^- + CO_2 + H_2O$$
$$\Delta G'\ (pH\ 7) = 165\ kJ\ mol^{-1} \quad (18\text{-}26)$$

presence of NO_3^- induces the synthesis of nitrate reductase and represses the synthesis of alternative enzymes such as **fumarate reductase**,[330,331] which reduces fumarate to succinate (see also p. 1027). On the other hand, if NO_3^- is absent and fumarate, which can be formed from pyruvate, is present, synthesis of fumarate reductase is induced. Although it is a much weaker oxidant than is nitrate ($E°'$ = 0.031 V), fumarate is still able to oxidize H_2 or NADH with oxidative phosphorylation. Like the related succinate dehydrogenase, fumarate reductase of *E. coli* is a flavoprotein with associated Fe / S centers. It contains covalently linked FAD and Fe_2S_2, Fe_4S_4, and three-Fe iron–sulfur centers.[332] In some bacteria a soluble periplasmic cytochrome c_3 carried out the fumarate reduction step.[332a] Trimethylamine *N*-oxide[330,333] or dimethylsulfoxide (DMSO; Eq. 16-62)[334,335] can also serve as alternative oxidants for anaerobic respiration using appropriate molybdemum-containing reductases (Chapter 16).

Reduction of nitrite: denitrification. The nitrite formed in Eq. 18-25 is usually reduced further to ammonium ions (Eq. 18-27). The reaction may not be important to the energy metabolism of the bacteria, but it provides NH_4^+ for biosynthesis. This six-electron reduction is catalyzed by a hexaheme protein containing six *c*-type hemes bound to a single 63-kDa polypeptide chain.[336,337]

$$NO_2^- + 8\,H^+ + 6\,e^- \rightarrow 2\,H_2O + NH_4^+ \quad (18\text{-}27)$$

Several types of **denitrifying bacteria**[315,338–340]

use either nitrate or nitrite ions as oxidants and reduce nitrite to N_2. A typical reaction for *Micrococcus denitrificans* is oxidation of H_2 by nitrate (Eq. 18-28). *Thiobacillus denitrificans*, like other thiobacilli, can oxidize

$$5\,H_2 + 2\,NO_3^- + 2\,H^+ \rightarrow N_2 + 6\,H_2O$$
$$\Delta G'\ (\text{pH } 7) = -561\ \text{kJ mol}^{-1}\ \text{of nitrate reduced or}\ -224\ \text{kJ mol}^{-1}\ \text{of } H_2\ \text{oxidized} \qquad (18\text{-}28)$$

sulfur as well as H_2S or thiosulfate using nitrate as the oxidant (Eq. 18-29):

$$5\,S + 6\,NO_3^- + 2\,H_2O \rightarrow 5\,SO_4^{2-} + 3\,N_2 + 4\,H^+$$
$$\Delta G'\ (\text{pH } 7) = -455\ \text{kJ mol}^{-1}\ \text{of nitrate reduced or}\ -546\ \text{kJ mol}^{-1}\ \text{of S oxidized} \qquad (18\text{-}29)$$

The reactions begin with reduction of nitrate to nitrite (Eq. 18-25) and continue with further reduction of nitrite to nitric oxide, **NO**; nitrous oxide, **N_2O**; and dinitrogen, **N_2**. A probable arrangement of the four enzymes needed for the reactions of Eq. 18-30 in *Paracoccus denitrificans* is shown in Fig. 18-21. See also pp. 884, 885.

Two types of dissimilatory nitrite reductases catalyze step *b* of Eq. 18-30. Some bacteria use a copper-containing enzyme, which contains a type 1 (blue) copper bound to a β barrel domain of one subunit and a type 2 copper at the catalytic center. The type 1 copper is thought to receive electrons from the small copper-containing carrier pseudoazurin (Chapter 16).[341–342b]

$$NO_3^{2-} \xrightarrow[\text{(a) } H_2O]{2e^-,\ 2H^+} NO_2^- \quad \text{Nitrate reductase (Eq. 18-25)}$$
$$NO_2^- \xrightarrow[\text{(b) } H_2O]{e^-,\ 2H^+} NO \quad \text{Nitrite reductase}$$
$$2\,NO \xrightarrow{\text{(c) } 2e^-} N_2O \quad \text{Nitric oxide reductase}$$
$$N_2O \xrightarrow{\text{(d) } 2e^-} N_2 \quad \text{Nitrous oxide reductase} \qquad (18\text{-}30)$$

More prevalent is **cytochrome *cd*$_1$** nitrite reductase.[340,343–346] The water-soluble periplasmic enzyme is a homodimer of ~60-kDa subunits, each containing a *c*-type heme in a small N-terminal domain and **heme *d*$_1$**, a ferric dioxoisobacteriochlorin (Fig. 16-6). The latter is present in the central channel of an eight-bladed β-sheet propeller[345-346g] similar to that in Fig. 15-23A. The heme d_1 is unusual in having its Fe atom ligated by a tyrosine hydroxyl oxygen, which may be displaced to allow binding of NO_3^-. The electron required for the reduction is presumably transferred from the electron transfer chain in the membrane to the heme *d*, via the heme *c* group.[347] Cytochrome cd_1 nitrite reductases have an unexpected second enzymatic activity. They catalyze the four-electron reduction of O_2 to H_2O, as does cytochrome *c* oxidase. However, the rate is much slower than that of nitrite reduction.[340,348]

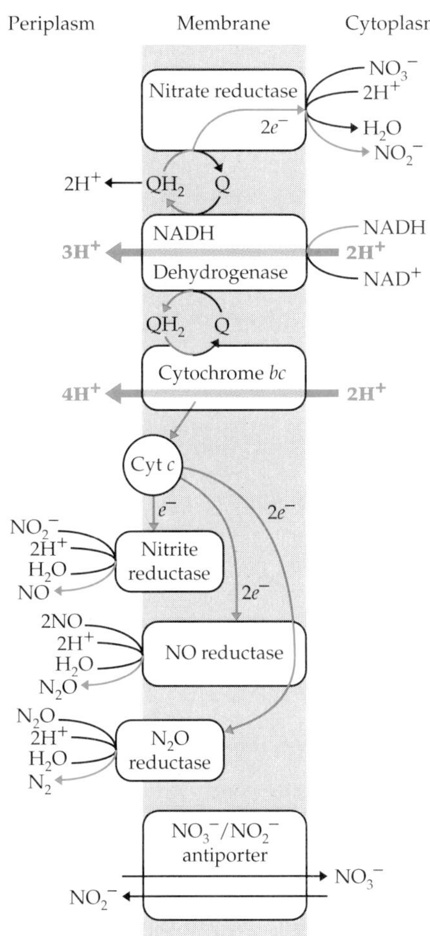

Figure 18-21 Organization of the nitrate reduction system in the outer membrane of the bacterium *Paracoccus dinitrificans* as outlined by Blaut and Gottschalk.[315] The equations are not balanced as shown but will be balanced if two NO_3^- ions are reduced to N_2 by five molecules of NADH (see also Eq. 18-28). Although this will also require seven protons, about 20 additional H^+ will be pumped to provide for ATP synthesis.

The enzyme catalyzing the third step of Eq. 18-30 (step *c*), **nitric oxide reductase**, is an unstable membrane-bound protein cytochrome *bc* complex.[349,350]

It has been isolated as a two-subunit protein, but genetic evidence suggests the presence of additional subunits.[350] The small subunit is a cytochrome *c*, while the larger subunit is predicted to bind two protohemes as well as a nonheme iron center. This protein also shows sequence homology with cytochrome *c* oxidase. It contains no copper, but it has been suggested that a heme *b*–nonheme Fe center similar to the heme a–Cu_B center of cytochrome *c* oxidase may be present. It may be the site at which the nitrogen atoms of two molecules of NO are joined to form N_2O.[350,351] A different kind of NO reductase is utilized by the denitrifying fungus *Fusarium oxysporum*. It is a cytochrome P450 but with an unusually low redox potential (–0.307 V). This **cytochrome P450$_{nor}$** does not react with O_2 (as in Eq. 18-57) but binds NO to its heme Fe^{3+}, reduces the complex with two electrons from NADH, then reacts with a second molecule of NO to give N_2O and H_2O.[352]

Reduction of N_2O to N_2 by bacteria (Eq. 18-30, step *d*) is catalyzed by the copper-containing nitrous oxide reductase. The purple enzyme is a dimer of 66-kDa subunits, each containing four atoms of Cu.[353] It has spectroscopic properties similar to those of cytochome *c* oxidase and a dinuclear copper–thiolate center similar to that of Cu_A in cytochrome *c* oxidase (p. 1030). The nature of the active site is uncertain.[354]

Sulfate-reducing and sulfur-reducing bacteria. A few obligate anaerobes obtain energy by using sulfate ion as an oxidant.[355–356a] For example, *Desulfovibrio desulfuricans* catalyzes a rapid oxidation of H_2 with reduction of sulfate to H_2S (Eq. 18-31).

$$4\ H_2 + SO_4^{2-} + 2\ H^+ \rightarrow H_2S + 4\ H_2O$$
$$\Delta G'\ (\text{pH } 7) = -154\ \text{kJ mol}^{-1} \text{ of sulfate reduced} \quad (18\text{-}31)$$

While this may seem like an esoteric biological process, the reaction is quantitatively significant. For example, it has been estimated that within the Great Salt Lake basin bacteria release sulfur as H_2S in an amount of 10^4 metric tons (10^7 kg) per year.[357]

The reduction potential for sulfate is extremely low ($E°'$, pH 7 = –0.454 V), and organisms are not known to reduce it directly to sulfite. Rather, a molecule of ATP is utilized to form adenosine 5'-phosphosulfate (APS) through the action of **ATP sulfurylase** (ATP:sulfate adenylyltransferase, Eq. 17-38).[358,359] APS is then reduced by cytochrome c_3 (Eq. 18-32, step *b*). The 13-kDa low-potential ($E°'$, pH 7 = 0.21 V) cytochrome c_3 contains four heme groups (Figure 16-8C) and is found in high concentration in sulfate-reducing bacteria.[360,361] Some of these bacteria have larger polyheme cytochromes *c*.[361a] For example, *Desulfovibrio vulgaris* forms a 514-residue protein carrying 16 hemes organized as four cytochrome c_3-like domains.[362] Each heme in cytochrome c_3 has a distinct redox potential within the range –0.20 to –0.40 V.[361–363]

APS is reduced (Eq. 18-32, step *b*) by **APS reductase**, a 220-kDa iron–sulfur protein containing FAD and several Fe–S clusters. An intermediate in the reaction may be the adduct of sulfite with FAD, which may be formed as in Eq. 18-32. The initial step in this hypothetical mechanism is displacement on sulfur by a strong nucleophile generated by transfer of electrons from reduced ferredoxin to cytochrome c_3 to the flavin.[364]

Cyt c_3
SO_4^{2-}
ATP
a Eq. 17-38
2 P_i
FAD
Adenosine 5′-phosphosulfate (APS) or adenylyl sulfate
b APS reductase
HSO_3^-

(18-32)

Bisulfite produced according to Eq. 18-32 is reduced further by a **sulfite reductase,** which is thought to receive electrons from flavodoxin, cyt c_3, and a hydrogenase. ATP synthesis is coupled to the reduction. Sulfite reductases generally contain both siroheme and Fe_4S_4 clusters (Fig. 16-19). They appear to be able to carry out the 6-electron reduction to sulfide without accumulation of intermediates.[365,366] However, in contrast to the assimilatory sulfite reductases present in many organisms, the dissimilatory nitrite reductases of sulfur-reducing bacteria may also release some thiosulfate $S_2O_3^{2-}$.[367] A possible role of menaquinone (vitamin K_2), present in large amounts in *Desulfovibrio,* has been suggested.[311] Although *Desulfovibrio* can obtain their energy from Eq. 18-31, they are not true autotrophs and must utilize compounds such as acetate together with CO_2 as a carbon source.

Some thermophilic archaeobacteria are able to live with CO_2 as their sole source of carbon and reduction of elemental sulfur with H_2 (Eq. 18-33) as their sole source of energy.[368,369]

$$H_2 + S^0 \rightarrow H_2S$$
$$\Delta G° = -27.4 \text{ kJ mol}^{-1} \quad (18\text{-}33)$$

This is remarkable in view of the small standard Gibbs energy decrease. Some species of the archaeobacterium *Sulfolobus* are able either to live aerobically oxidizing sulfide to sulfate with O_2 (Eq. 18-22) or to live anaerobically using reduction of sulfur by Eq. 18-33 as their source of energy.[369]

The sulfate-reducing bacterium *Desulfovibrio sulfodismutans* carries out what can be described as "inorganic fermentations" which combine the oxidation of compounds such as sulfite or thiosulfate (as observed for sulfur-oxidizing bacteria; Eq. 18-22), with reduction of the same compounds (Eq. 18-34).[370,370a] Dismutation of $S_2O_3^{2-}$ plus H_2O to form SO_4^{2-} and H_2S also occurs but with a less negative Gibbs energy change.

$$4\ SO_3^{2-} + H^+ \rightarrow 3\ SO_4^{2-} + HS^-$$
$$\Delta G' \text{ (pH 7)} = -185 \text{ kJ/mol} \quad (18\text{-}34)$$

A strain of *Pseudomonas* obtains all of its energy by reducing sulfate using phosphite, which is oxidized to phosphate.[370b]

Methane bacteria. The methane-producing bacteria (Chapter 15) are also classified as chemiautotrophic organisms. While they can utilize substances such as methanol and acetic acid, they can also reduce CO_2 to methane and water using H_2 (Fig. 15-22). The electron transport is from hydrogenase, perhaps through ferredoxin to formate dehydrogenase and via the deazaflavin F_{420} and $NADP^+$ to the methanopterin-dependent dehydrogenases that carry out the stepwise reduction of formate to methyl groups (Fig. 16-28). Generation of ATP probably involves proton pumps, perhaps in internal coupling membranes .[315,371]

F. Oxygenases and Hydroxylases

For many years the idea of dehydrogenation dominated thinking about biological oxidation. Many scientists assumed that the oxygen found in organic substances always came from water, e.g., by addition of water to a double bond followed by dehydrogenation of the resulting alcohol. Nevertheless, it was observed that small amounts of O_2 were essential, even to anaerobically growing cells.[372] In 1955, Hayaishi and Mason independently demonstrated that ^{18}O was sometimes incorporated into organic compounds directly from $^{18}O_2$ as in Eq. 18-35. Today a bewildering variety of **oxygenases** are

Catechol $+ {}^{18}O_2$ —(Catechol 1,2-dioxygenase)→ *Cis,cis*-muconic acid (two ^{18}COOH groups)

(18-35)

known to function in forming such essential metabolites as sterols, prostaglandins, and active derivatives of vitamin D. Oxygenases are also needed in the catabolism of many substances, often acting on nonpolar groups that cannot be attacked readily by other types of enzyme.[372]

Oxygenases are classified either as **dioxygenases** or as **monooxygenases**. The monooxygenases are also called mixed function oxidases or **hydroxylases**. Dioxygenases catalyze incorporation of two atoms of oxygen as in Eq. 18-35, but monooxygenases incorporate only one atom. The other oxygen atom from the O_2 is converted to water. A typical monooxygenase-catalyzed reaction is the hydroxylation of an alkane to an alcohol (Eq. 18-36).

$$H_3C-(CH_2)_n-CH_3 \xrightarrow[BH_2 \text{ (cosubstrate)} \rightarrow B]{O_2 \rightarrow H_2O} H_3C-(CH_2)_n-CH_2OH \quad (18\text{-}36)$$

A characteristic of the monooxygenases is that an additional reduced substrate, a **cosubstrate** (BH_2 in Eq. 18-36), is usually required to reduce the second atom of the O_2 molecule to H_2O.

Since O_2 exists in a "triplet" state with two unpaired electrons, it reacts rapidly only with transition metal ions or with organic radicals (Chapter 16). For this reason, most oxygenases contain a transition metal ion, usually of iron or copper, or contain a cofactor, such as FAD, that can easily form a radical or act on a cosubstrate or substrate to form a free radical.

1. Dioxygenases

Among the best known of the oxygenases that incorporate both atoms of O_2 into the product are those that participate in the biological degradation of aromatic compounds by cleaving double bonds at positions between two OH groups as in Eq. 18-35 or adjacent to one OH group of an *ortho* or *para* hydroxyl pair.[373] A much studied example is **protocatechuate 3,4-dioxygenase**,[373–375] which cleaves its substrate between the two OH groups (*intradiol cleavage*) as in Eq. 18-35. A different enzyme, **protocatechuate 4,5-dioxygenase**,[357] cleaves the same substrate next to just one of the two OH groups (*extradiol cleavage*; Eq. 18-37) to form the aldehyde α-hydroxy-δ-carboxymuconic semialdehyde. Another extradiol cleaving enzyme, **protocatechuate 2,3-dioxygenase**, acts on the same substrate. Many other dioxygenases attack related substrates.[376–380] Intradiol-cleaving enzymes are usually iron-tyrosinate proteins (Chapter 16) in which the

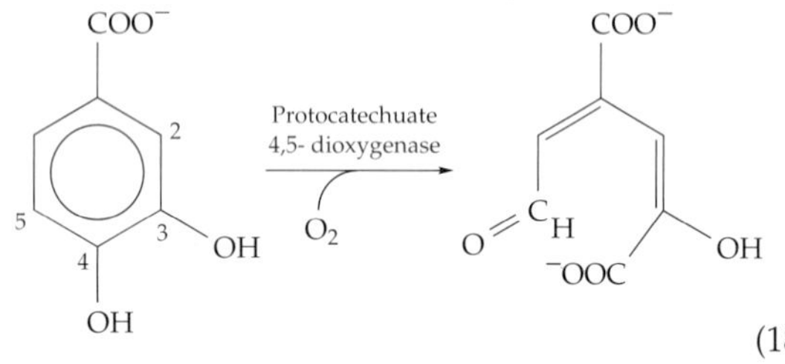

(18-37)

iron is present in the Fe(III) oxidation state and remains in this state throughout the catalytic cycle.[375] The enzymes usually have two subunits and no organic prosthetic groups. For example, a protocatechuate 3,4-dioxygenase from *Pseudomonas* has the composition $(\alpha\beta Fe)_{12}$ with subunit masses of 23 (α) and 26.5 (β) kDa. The iron is held in the active site cleft between the α and β subunits by Tyr 408, Tyr 447, His 460, and His 462 of the β subunit and a water molecule.[375] These enzymes and many other oxygenases probably assist the substrate in forming radicals that can react with O_2 to form organic peroxides. Some plausible intermediate species are pictured in Fig. 18-22. The reactions are depicted as occurring in two-electron steps. However, O_2 is a diradical, and it is likely that the Fe^{3+}, which is initially coordinated to both phenolate groups of the ionized substrate, assists in forming an organic free radical that reacts with O_2.

Extradiol dioxygenases have single Fe^{2+} ions in their active sites. The O_2 probably binds to the Fe^{2+} and may be converted transiently to an Fe^{3+}-superoxide complex which adds to the substrate. Some extradiol dioxygenases require an Fe_2S_2 ferredoxin to reduce any Fe^{3+}-enzyme that is formed as a side reaction back to the Fe^{2+} state.[381] Possible intermediates are given in Fig. 18-22 (left side) with two-electron steps used to save space and to avoid giving uncertain details about free radical intermediates. Formation of the organic radical is facilitated by the iron atom, which may be coordinated initially to both phenolate groups of the ionized substrate. The peroxide intermediates, for both types of dioxygenases, may react and be converted to various final products by several mechanisms.[382]

Tryptophan dioxygenase (indoleamine 2,3-dioxygenase)[383] is a heme protein which catalyzes the reaction of Eq. 18-38. The oxygen atoms designated by

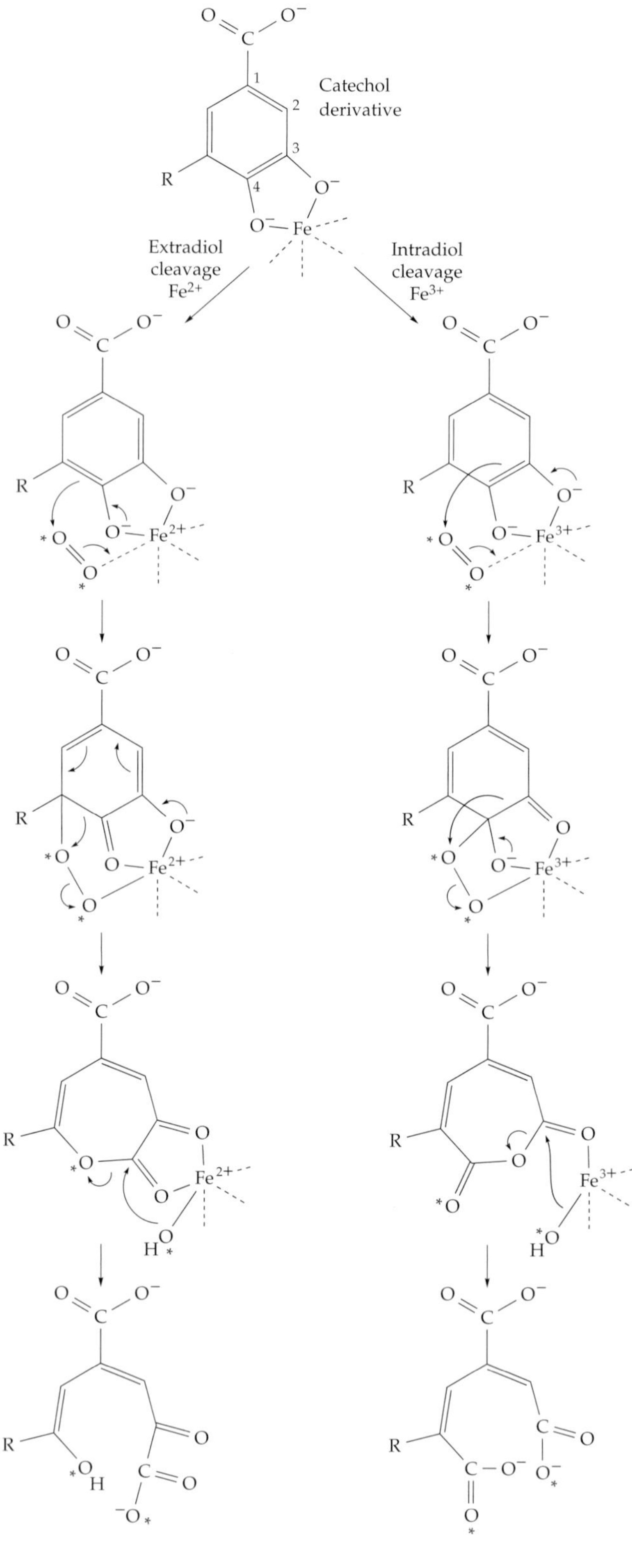

Figure 18-22 Some possible intermediates in the action of extradiol (left) and intradiol (right) aromatic dioxygenases. Although the steps depict the flow of pairs of electrons during the formation and reaction of peroxide intermediates, the mechanisms probably involve free radicals whose formation is initiated by O_2. The asterisks show how two atoms of labeled oxygen can be incorporated into final products. After Ohlendorf *et al.*[374]

the asterisks are derived from O_2. Again, the first step is probably the formation of a complex between Fe(II) and O_2, but tryptophan must also be present before this can occur. At 5°C the enzyme, tryptophan, and O_2 combine to give an altered spectrum reminiscent of that of compound III of peroxidase (Fig. 16-14). This oxygenated complex may, perhaps, then be converted to a complex of Fe(II) and superoxide ion.

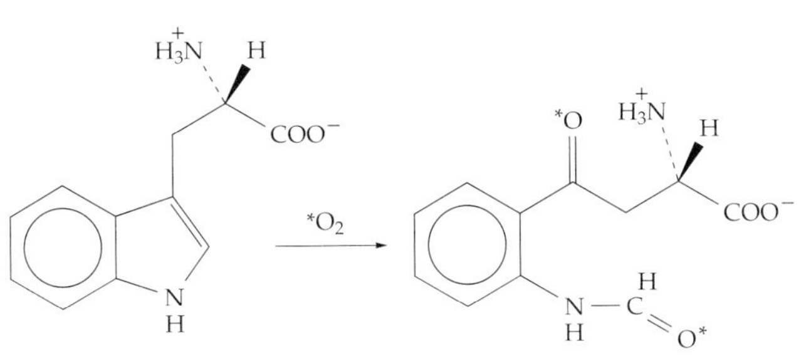

Tryptophan → (*O_2) → Formylkynurenine

(18-38)

There is much evidence, including inhibition by superoxide dismutase and stimulation by added potassium superoxide,[384] that the superoxide anion radical is the species that attacks the substrate (Eq. 18-39). In this reaction one electron is returned to the Fe(III) form of the enzyme to regenerate the original Fe(II) form. Subsequent reaction of the hydroperoxide anion would give the observed products.

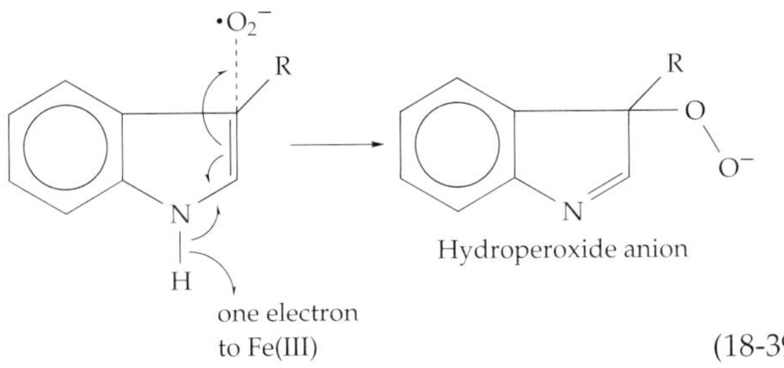

$\cdot O_2^-$ ... one electron to Fe(III) → Hydroperoxide anion

(18-39)

Some dioxygenases require a cosubstrate. For example, **phthalate dioxygenase**[385] converts phthalate to a *cis*-dihydroxy derivative with NADH as the cosubstrate (Eq. 18-40). Similar double hydroxylation reactions catalyzed by soil bacteria are known for benzene, benzoate,[386] toluene, naphthalene, and several other aromatic compounds.[386a] The formation of the *cis*-glycols is usually followed by dehydrogenation or oxidative decarboxylation by NAD^+ to give a catechol, whose ring is then opened by another dioxygenase reaction (Chapter 25). An elimination of Cl^- follows dioxygenase action on *p*-chlorophenylacetate and produces 3,4-dihydroxyphenylacetate as a product. Pthalate dioxygenases consist of two subunits. The 50-kDa dioxygenase subunits receive electrons from reductase subunits that contain a Rieske-type Fe–S

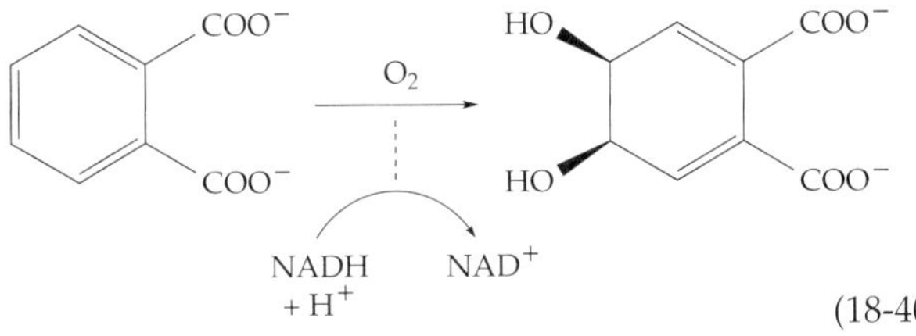

O_2; NADH + H^+ → NAD^+

(18-40)

centers and bound FMN.[387] The dioxygenase also contains an Fe_2S_2 center, and electrons flow from NADH to FMN and through the two Fe–S centers to the Fe^{2+} of the active site.[387–388]

Lipoxygenases catalyze oxidation of polyunsaturated fatty acids in plant lipids. Within animal tissues the lipoxygenase-catalyzed reaction of arachidonic acid with O_2 is the first step in formation of **leukotrienes** and other mediators of inflammation. These reactions are discussed in Chapter 21.

2. Monooxygenases

Two classes of monooxygenases are known. Those requiring a cosubstrate (BH_2 of Eq. 18-36) in addition to the substrate to be hydroxylated are known as **external monooxygenases**. In the other group, the **internal monooxygenases**, some portion of the substrate being hydroxylated also serves as the cosubstrate. Many internal monooxygenases contain flavin cofactors and are devoid of metal ions.

Flavin-containing monooxygenases. One group of flavin-dependent monooxygenases form H_2O_2 by reaction of O_2 with the reduced flavin and use the H_2O_2 to hydroxylate 2-oxoacids. An example is **lactate monooxygenase,** which apparently dehydrogenates lactate to pyruvate and then oxidatively decarboxylates the pyruvate to acetate with H_2O_2 (Eq. 15-36). One atom of oxygen from O_2 is incorporated into the acetate formed.[389,390] In a similar manner, the FAD-containing bacterial **lysine monooxygenase** probably catalyzes the sequence of reactions shown in Eq. 18-41.[391] When native lysine monooxygenase was treated with sulfhydryl-blocking reagents the resulting modified enzyme produced a 2-oxoacid, ammonia, and H_2O_2, just the products predicted from the hydrolytic decomposition of the bracketed intermediate of Eq. 18-41. Similar bacterial enzymes act on tryptophan and phenylalanine.[392]

NADPH can serve as a cosubstrate of flavoprotein monooxygenase by first reducing the flavin, after which the reduced flavin can react with O_2 to generate the hydroxylating reagent.[393] An example is the bacterial **4-hydroxybenzoate hydroxylase** which forms 3,4-dihydroxybenzoate.[394] The 43-kDa protein consists of three domains, the large FAD-binding domain being folded in nearly the same way as that of glutathione reductase (Fig. 15-10). The 4-hydroxybenzoate binds

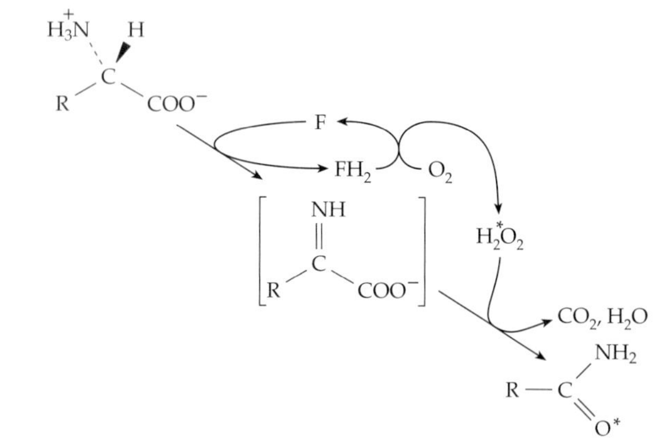

(18-41)

first into a deep cleft below the N-5 edge of the isoalloxazine ring of the FAD; then the NADH binds. Spectroscopic studies have shown the existence of at least three intermediates. The first of these has been identified as the 4a-peroxide whose formation (Eq. 15-31) is discussed in Chapter 15. The third intermediate is the corresponding 4a-hydroxyl compound. The substrate hydroxylation must occur in a reaction with the flavin peroxide, presumably with the phenolate anion form of the substrate (Eq. 18-42).[395] The initial hydroxylation product is tautomerized to form the product 3,4-dihydroxybenzoate.

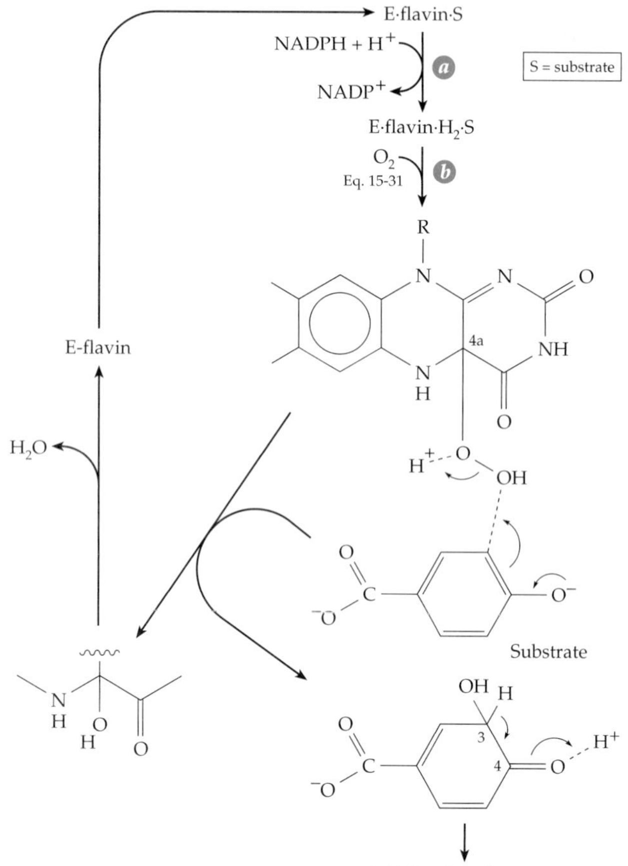

(18-42)

According to this mechanism, one of the two oxygen atoms in the hydroperoxide reacts with the aromatic substrate, perhaps as OH^+ or as a superoxide radical. A variety of mechanisms for activating the flavin peroxide to give a more potent hydroxylating reagent have been proposed. These include opening of the central ring of the flavin to give a carbonyl oxide intermediate which could transfer an oxygen atom to the substrate,[396] elimination of H_2O to form an **oxaziridine,**[397] or rearrangment to a **nitroxyl radical**.[398] Any of these might be an active electrophilic hydroxylating reagent. However, X-ray structural studies suggest that conformational changes isolate the substrate–FAD-enzyme complex from the medium stabilizing the 4a peroxide via hydrogen bonding[399–400] in close proximity to the substrate. Reaction could occur by the simple mechanism of Eq. 18-42, a mechanism also supported by ^{19}F NMR studies with fluorinated substrate analogs[401] and other investigations.[401a,b]

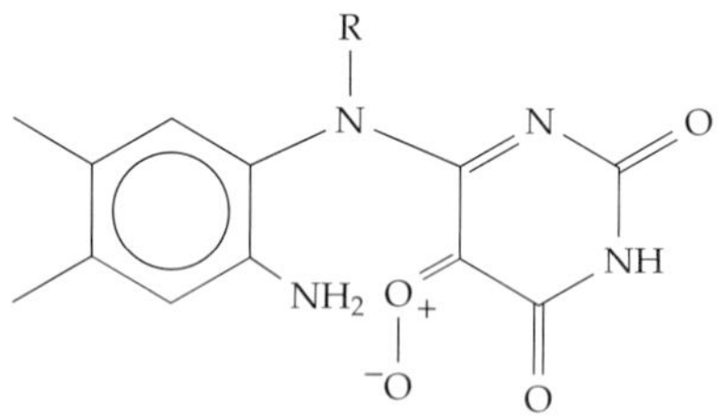

Carbonyl oxide

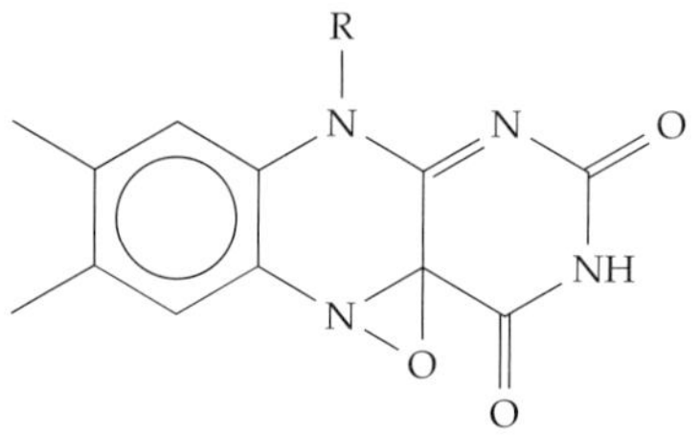

An oxaziridine

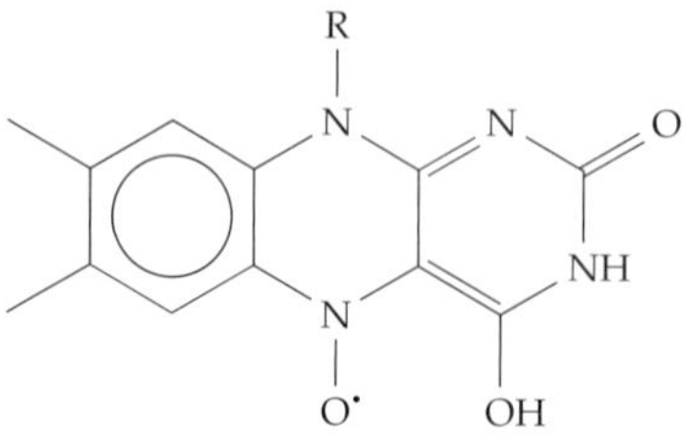

A nitroxyl radical

Related flavin hydroxylases act at nucleophilic positions on a variety of molecules[393,402] including phenol,[403] salicylate,[404] anthranilate,[405] *p*-cresol,[406] 4-hydroxyphenylacetate,[407,408] and 4-aminobenzoate.[409] Various microsomal flavin hydroxylases are also known.[410] Flavin peroxide intermediates are also able to hydroxylate some electrophiles.[411] For example, the bacterial **cyclohexanone oxygenase** catalyzes the ketone to lactone conversion of Eq. 18-43.[411a] The mechanism presumably involves the nucleophilic attack of the flavin hydroperoxide on the carbonyl group of the substrate followed by rearrangement. This parallels the Baeyer–Villiger rearrangement that results from treatment of ketones with peracids.[393] Cyclohexanone oxygenase also catalyzes a variety of other reactions,[412] including conversion of sulfides to sulfoxides.

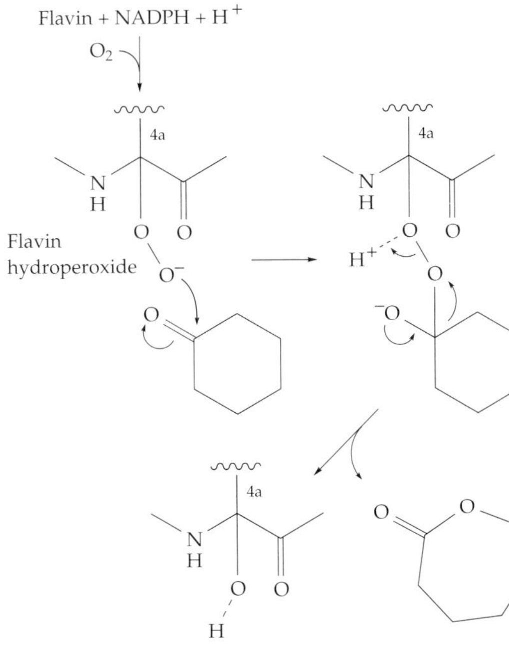

Flavin + NADPH + H^+, O_2; Flavin hydroperoxide

(18-43)

Reduced pteridines as cosubstrates. A dihydro form of biopterin (Fig. 15-17) serves as a cosubstrate, that is reduced by NADPH (Eq. 18-44) in hydroxylases that act on phenylalanine, tyrosine, and tryptophan.

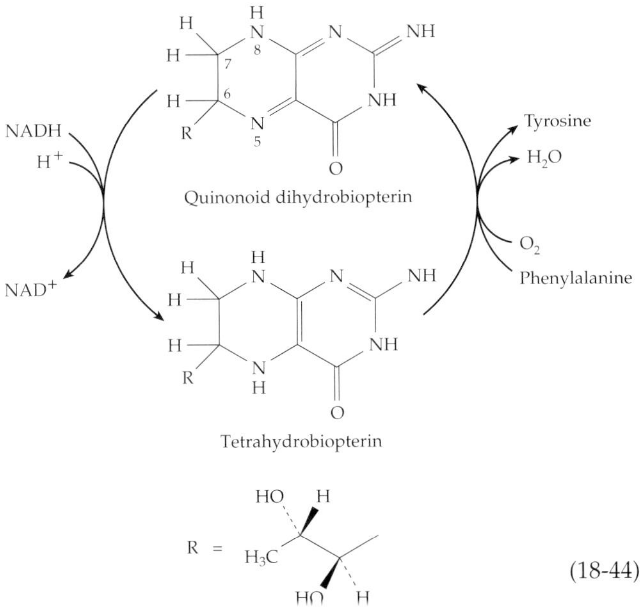

Quinonoid dihydrobiopterin; Tetrahydrobiopterin; NADH, H^+, NAD^+; Tyrosine, H_2O; O_2, Phenylalanine

R = H_3C–CH(OH)–CH(OH)–

(18-44)

The tetrahydrobiopterin formed in this reaction is similar in structure to a reduced flavin. The mechanism of its interaction with O_2 could reasonably be the same as that of 4-hydroxybenzoate hydroxylase. However, **phenylalanine hydroxylase**, which catalyzes the formation of tyrosine (Eq. 18-45), a dimer of 451-residue subunits, contains one Fe per subunit,[413–415a] whereas flavin monooxygenases are devoid of iron. **Tyrosine hydroxylase**[416–419a] and **tryptophan hydroxylase**[420] have very similar properties. All three enzymes contain regulatory, catalytic, and tetramerization domains as well as a common Fe-binding motif in their active sites.[413,421,421a]

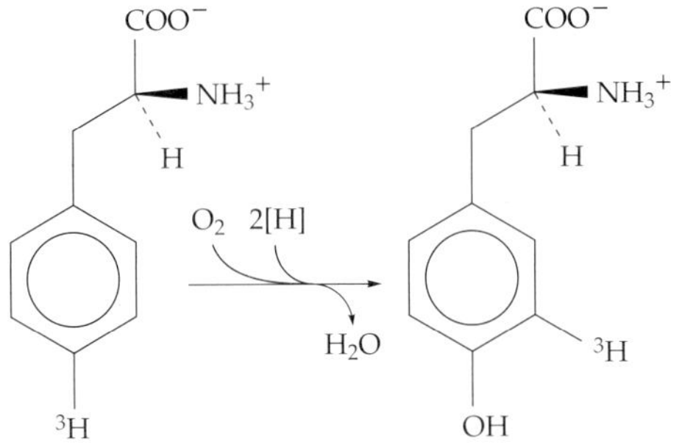

COO^-, NH_3^+, H; O_2 2[H]; H_2O; ^3H; OH

(18-45)

The role of the iron atom in these enzymes must be to accept an oxygen atom from the flavin peroxide, perhaps forming a reactive ferryl ion and transferring the oxygen atom to the substrate, e.g., as do cytochromes P450 (see Eq. 18-57). The 4*a*-hydroxytetrahydrobiopterin, expected as an intermediate if the mechanism parallels that of Eq. 18-42, has been identified by its ultraviolet absorption spectrum.[422] A ring-opened intermediate has also been ruled out for phenylalanine hydroxylase.[423] However, the 4*a* –OH adduct has been observed by ^{13}C-NMR spectroscopy. Its absolute configuration is 4*a*(S) and the observation of an ^{18}O-induced shift in the ^{13}C resonance of the 4*a*-carbon atom[424] confirms the origin of this oxygen from $^{18}O_2$ (see Eq. 18-42). A "stimulator protein" needed for rapid reaction of phenylalanine hydroxylase has been identified as a **4*a*-carbinolamine dehydratase** (Eq. 18-46).[425–426] This protein also has an unexpected function as part of a complex with transcription factor HNF1 which is found in nuclei of liver cells.[425a,426]

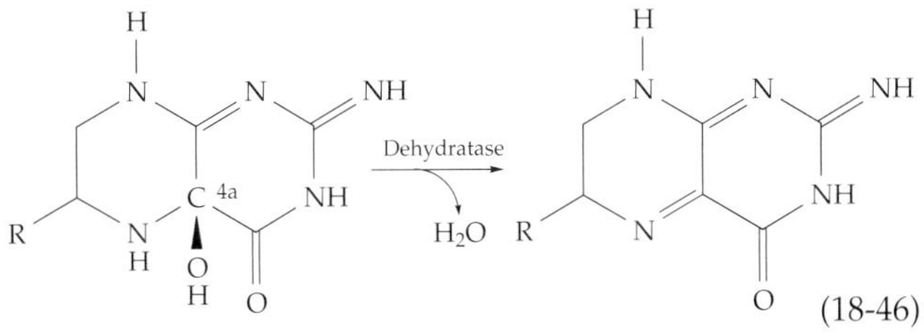

Dehydratase; H_2O

(18-46)

Dihydrobiopterin can exist as a number of isomers. The quinonoid form shown in Eqs. 18-44 and 18-46 is a tautomer of 7,8-dihydrobiopterin, the form generated by dihydrofolate reductase (Chapter 15). A pyridine nucleotide-dependent **dihydropteridine reductase**[427–429] catalyzes the left-hand reaction of Eq. 18-44.

The hereditary absence of phenylalanine hydroxylase, which is found principally in the liver, is the cause of the biochemical defect **phenylketonuria** (Chapter 25, Section B).[430,430a] Especially important in the metabolism of the brain are tyrosine hydroxylase, which converts tyrosine to 3,4-dihydroxyphenylalanine, the rate-limiting step in biosynthesis of the catecholamines (Chapter 25), and tryptophan hydroxylase, which catalyzes formation of 5-hydroxytryptophan, the first step in synthesis of the neurotransmitter 5-hydroxytryptamine (Chapter 25). All three of the pterin-dependent hydroxylases are under complex regulatory control.[431,432] For example, tyrosine hydroxylase is acted on by at least four kinases with phosphorylation occurring at several sites.[431,433,433a] The kinases are responsive to nerve growth factor and epidermal growth factor,[434] cAMP,[435] Ca^{2+} + calmodulin, and Ca^{2+} + phospholipid (protein kinase C).[436] The hydroxylase is inhibited by its endproducts, the catecholamines,[435] and its activity is also affected by the availability of tetrahydrobiopterin.[436]

Hydroxylation-induced migration. A general result of enzymatic hydroxylation of aromatic compounds is the intramolecular migration of a hydrogen atom or of a substituent atom or group as is shown for the ^3H atom in Eq. 18-45.[437] Dubbed the NIH shift (because the workers discovering it were in a National Institutes of Health laboratory), the migration tells us something about possible mechanisms of hydroxylation. In Eq. 18-45 a tritium atom has shifted in response to the entering of the hydroxyl group. The migration can be visualized as resulting from electrophilic attack on the aromatic system, e.g., by an oxygen atom from Fe(N)=O or by OH^+ (Eq. 18-47).

Such an attack could lead in step *a* either to an **epoxide** (**arene oxide**) or directly to a carbocation as shown in Eq. 18-47. Arene oxides can be converted, via the carbocation step *b*, to end products in which the NIH shift has occurred.[438] The fact that phenylalanine hydroxylase also catalyzes the conversion of the special substrate shown in Eq. 18-48 to a stable epoxide, which cannot readily undergo ring opening, also supports this mechanism.

Operation of the NIH shift can cause migration of a large substituent as is illustrated by the hydroxylation of 4-hydroxyphenylpyruvate (Eq. 18-49), a key step in the catabolism of tyrosine (Chapter 25). Human 4-hydroxyphenylpyruvate dioxygenase is a dimer of 43-kDa subunits.[439] A similar enzyme from *Pseudomonas* is a 150-kDa tetrameric iron-tyrosinate protein, which must be maintained in the reduced Fe(II) state for catalytic activity.[440] Although this enzyme is a

H+
a
[HO+]
H
3H
H
H
O
Arene oxide
b
c
+
3H
O—H
3H
H
O
d
3H
OH

(18-47)

dioxygenase, it is probably related in its mechanism of action to the 2-oxoglutarate-dependent monooxygenases discussed in the next section (Eqs. 18-51, 18-52). It probably uses the oxoacid side chain of the substrate

COO^-
NH_3^+
H
O_2
COO^-
NH_3^+
H
O

(18-48)

to generate a reactive oxygen intermediate such as Fe(IV)=O by the decarboxylative mechanism of Eqs. 18-50 and 18-51. The iron-bound oxygen attacks C1 of

OH
4
1
COO^-
O
p-Hydroxyphenylpyruvate
$^*O_2$
CO_2
OH
$\overset{*}{C}OO^-$
HO^*
Attack here by a dioxygenase opens ring
Homogentisate

(18-49)

the aromatic ring, the electron-donating *p*-hydroxyl group assisting. This generates a hydroxylated carbocation of the type shown in Eq. 18-47 in which the whole two-carbon side chain undergoes the NIH shift.

2-Oxoglutarate as a decarboxylating cosubstrate. Several oxygenases accept hydrogen atoms from 2-oxoglutarate, which is decarboxylated in the process to form succinate. Among these are enzymes catalyzing hydroxylation of residues of proline in both the 3- and 4-positions (Eq. 8-6)[441–444] and of lysine in the 5-position (Eq. 8-7)[445,446] in the collagen precursor **procollagen**. The hydroxylation of prolyl residues also takes place within the cell walls of plants.[447] Similar enzymes hydroxylate the β-carbon of aspartyl or asparaginyl side chains in EGF domains (Table 7-3) of proteins.[441] Thymine[448] and taurine[449,449a] are acted on by related dioxygenases. A bacterial oxygenase initiates the degradation of the herbicide 2,4-dichlorophenoxyacetic acid (2,4-D) using another 2-oxoglutarate-dependent hydroxylase.[450,451] In the human body a similar enzyme hydroxylates γ-butyrobetaine to form carnitine (Eq. 18-50).[452] All of

γ-Butyrobetaine → L-Carnitine (2-Oxoglutarate, O_2^* → Succinate*, CO_2) (18-50)

these enzymes contain iron and require ascorbate, whose function is apparently to prevent the oxidation of the iron to the Fe(III) state.

When $^{18}O_2$ is used for the hydroxylation of γ-butyrobetaine (Eq. 18-51), one atom of ^{18}O is found in the carnitine and one in succinate. The reaction is stereospecific and occurs with retention of configuration at C-3, the *pro*-R hydrogen being replaced by OH while the *pro*-S hydrogen stays.[453] Under some conditions these enzymes decarboxylate 2-oxoglutarate in the absence of a hydroxylatable substrate, the iron being oxidized to Fe^{3+} and ascorbate being consumed stoichiometrically.[454] A plausible mechanism (Eq. 18-51) involves formation of an Fe(II)–O_2 complex, conversion to $Fe(III)^+ \cdot O_2^-$, and addition of the superoxide ion to 2-oxoglutarate to form an adduct.[451,455] Decarboxylation of this adduct could generate the oxidizing reagent, perhaps Fe(IV)=O. In the absence of substrate S the ferryl iron could be reconverted to Fe(II) by a suitable reductant such as ascorbate. In the absence of ascorbate the Fe(IV) might be reduced to a catalytically inert Fe(III) form.

(18-51)

An unusual oxygenase with a single Fe^{2+} ion in its active site closes the four-membered ring in the biosynthesis of penicillins (Eq. 18-52). It transfers four

L-δ-(α-Aminoadipoyl)-L-cysteinyl-D-valine

O_2 → 2 H_2O (Isopenicillin N synthase)

Isopenicillin N (18-52)

hydrogen atoms from its dipeptide substrate to form two molecules of water and the product isopenicillin N.[456,457] Sequence comparison revealed several regions including the Fe-binding sites that are homologous with the oxoacid-dependent oxygenases. A postulated mechanism for **isopenicillin N synthase** involves formation of an Fe^{3+} superoxide anion complex as in Eq. 18-51. However, instead of attack on an oxoacid as in Eq. 18-51, it removes a hydrogen from the substrate to initiate the reaction sequence.[457] Other related oxygenases include **aminocyclopropane-1-carboxylate oxidase** (Eq. 24-35); **deacetoxycephalosporin C synthase**,[457a] an enzyme that converts penicillins to cephalosporins (Box 20-G); and **clavaminate synthase**,[458,459] an enzyme needed for synthesis of the β-lactamase inhibitor clavulanic acid, and **clavaminate synthase**.[458,459] This 2-oxoglutarate-dependent oxygenase catalyzes three separate reactions in the synthesis of the clinically important β-lactamase inhibitor clavulanic acid. The first step is similar to that in Eq. 18-50. The second is an oxidative cyclization and the third a desaturation reaction.

Copper-containing hydroxylases. Many Fe(II)-containing hydroxylases require a reducing agent to maintain the iron in the reduced state, and ascorbate is often especially effective. In addition, ascorbate is apparently a true cosubstrate for the copper-containing **dopamine β-hydroxylase**, an enzyme required in the synthesis of noradrenaline according to Eq. 18-53. This reaction takes place in neurons of the brain and in the adrenal gland, a tissue long known as especially rich in ascorbic acid. The reaction requires two molecules of ascorbate, which are converted in two one-electron steps to **semidehydroascorbate**.[460] Both the structure of this free radical and that of the fully oxidized form of vitamin C, **dehydroascorbic acid**, are shown in Box 18-D. Dopamine β-hydroxylase is a 290-kDa tetramer, consisting of a pair of identical disulfide-crosslinked homodimers, which contains two Cu ions per subunit.[461]

HO, HO, NH_2
Dopamine
Ascorbate
O_2
H_2O
Dehydroascorbate (Box 18-D)
HO, HO, NH_2, H, OH
Noradrenaline

(18-53)

A similar copper-dependent hydroxylase constitutes the N-terminal domain of the **peptidylglycine α-amidating enzyme** (Eq. 10-11). This bifunctional enzyme hydroxylates C-terminal glycines in a group of neuropeptide hormones and other secreted peptides. The second functional domain of the enzyme cleaves the hydroxylated glycine to form a C-terminal amide group and glyoxylate.[462–464b] The three-dimensional structure of a 314-residue catalytic core of the hydroxylase domain is known.[463] Because of similar sequences and other properties, the structures of this enzyme and of dopamine β-hydroxylase are thought to be similar. The hydroxylase domain of the α-amidating enzyme is folded into two eight-stranded antiparallel jelly-roll motifs, each of which binds one of the two copper ions. Both coppers can exist in a Cu(II) state and be reduced by ascorbate to Cu(I). One Cu (Cu_A) is held by three imidazole groups and is thought to be the site of interaction with ascorbate. The other copper, Cu_B, which is 1.1 nm away from Cu_A, is held by two imidazoles. The substrate binds adjacent to Cu_B.[463]

The reaction cycle of these enzymes begins with reduction of both coppers from Cu(II) to Cu(I) (Eq. 18-54, step *a*). Both O_2 and substrate bind (steps *b* and *c*, but not necessarily in this order). The O_2 bound to Cu_B is reduced to a peroxide anion that remains bound to Cu_B. Both Cu_A and Cu_B donate one electron, both being oxidized to Cu(II). These changes are also included in step *c* of Eq. 18-54. One proposal is that the resulting peroxide is cleaved homolytically while removing the *pro-S* hydrogen of the glycyl residue.

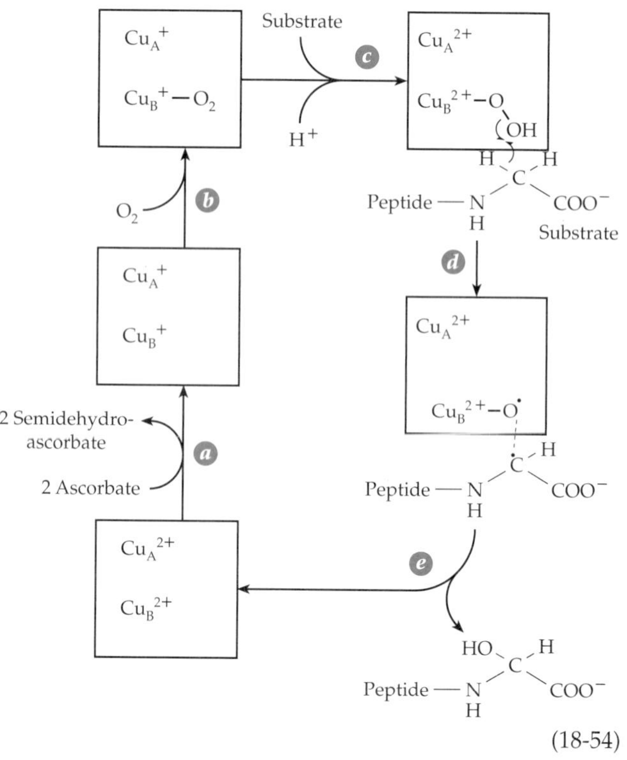

(18-54)

The resulting glycyl radical couples with the oxygen radical that is bound to Cu_B (step *e*).

A variety of other copper hydroxylases are known. For example, **tyrosinase**, which contains a binuclear copper center, catalyzes both hydroxylation of phenols and aromatic amines and dehydrogenation of the resulting catechols or *o*-aminophenols (Eq. 16-57). As in hemocyanin, the O_2 is thought to be reduced to a peroxide which bridges between the two copper atoms. Methane-oxidizing bacteria, such as *Methylococcus capsulatus*, oxidize methane to methanol to initiate its metabolism. They do this with a copper-containing membrane-embedded monooxygenase whose active site is thought to contain a trinuclear copper center. Again a bridging peroxide may be formed and may insert an oxygen atom into the substrate.[465,466] The same bacteria produce a soluble methane monooxygenase containing a binuclear iron center.

Hydroxylation with cytochrome P450. An important family of heme-containing hydroxylases, found in most organisms from bacteria to human beings, are the cytochromes P450. The name comes from the fact that in their reduced forms these enzymes form a complex with CO that absorbs at 450 nm. In soil bacteria cytochromes P450 attack compounds of almost any structure. In the adrenal gland they participate in steroid metabolism,[467,468] and in the liver microsomal cytochromes P450 attack drugs, carcinogens, and other xenobiotics (foreign compounds).[469–471] They convert cholesterol to bile acids[472] and convert vitamin D,[473] prostaglandins, and many other metabolites to more soluble and often biologically more active forms. In plants cytochromes P450 participate in hydroxylation of fatty acids at many positions.[474] They play a major role in the biosynthetic phenylpropanoid pathway (Fig. 25-8) and in lignin synthesis.[475] More than 700 distinct isoenzyme forms have been described.[476,476a]

Cytochromes P450 are monooxygenases whose cosubstrates, often NADH or NADPH, deliver electrons to the active center heme via a separate flavoprotein and often via an iron-sulfur protein as well.[476a,b] A typical reaction (Eq. 18-55) is the 11 β-hydroxylation of a steroid, an essential step in the biosynthesis of steroid hormones (Fig. 22-11). The hydroxyl group is introduced without inversion of configuration. The same enzyme converts unsaturated derivatives to epoxides (Eq. 18-56), while other cytochromes P450

H, H, 11, CH_3 —[O_2; $2e^-$ + $2H^+$ → H_2O]→ HO, H, CH_3 (18-55)

epoxidize olefins.[477] Epoxide hydrolases, which act by a mechanism related to haloalkane dehalogenase (Fig. 12-1), convert the epoxides to diols.[478] Cytochromes P450 are able to catalyze a bewildering array of other

CH_3 —[O_2; $2e^-$ + $2H^+$ → H_2O]→ O, CH_3 (18-56)

reactions[479–481] as well. Most of these, such as conversion of amines and thioesters to *N*- or *S*-oxides, also involve transfer of an oxygen atom to the substrate. Others, such as the reduction of epoxides, *N*-oxides, or nitro compounds, are electron-transfer reactions.

Several different cytochromes P450 are present in mammalian livers.[470] All are bound to membranes of the endoplasmic reticulum and are difficult to solubilize. Biosynthesis of additional forms is induced by such agents as phenobarbital,[470] 3-methylcholanthrene,[469] dioxin,[482] and ethanol.[483] These substances may cause as much as a 20-fold increase in P450 activity. Another family of cytochrome P450 enzymes is present in mitochondria.[483a] A large number of cytochrome P450 genes have been cloned and sequenced. Although they are closely related, each cytochrome P450 has its own gene. There are at least ten families of known P450 genes and the total number of these enzymes in mammals may be as high as 200. Microorganisms, from bacteria to yeast, produce many other cytochromes P450.

Microsomal cytochromes P450 receive electrons from an **NADPH-cytochrome P450** reductase, a large 77-kDa protein that contains one molecule each of FAD and FMN.[484–485a] It is probably the FAD which accepts electrons from NADPH and the FMN which passes them on to the heme of cytochrome P450. Cytochrome b_5 is also reduced by this enzyme,[486] and some cytochromes P450 may accept one electron directly from the flavin of the reductase and the second electron via cytochrome b_5. However, most bacterial and mitochondrial cytochromes P450 accept electrons only from small iron–sulfur proteins. Those of the adrenal gland receive electrons from the 12-kDa **adrenodoxin**.[487,488] This small protein of the ferredoxin class contains one Fe_2S_2 cluster and is, therefore, able to transfer electrons one at a time from the FAD-containing NADPH-adrenodoxin reductase[489] to the cytochrome P450. The camphor 5-monoxygenase from *Pseudomonas putida* consists of three components: an FAD-containing reductase, the Fe_2S_2 cluster-containin **putidaredoxin**,[489a] and cytochrome $P450_{cam}$.[490] Some other bacterial

BOX 18-D VITAMIN C: ASCORBIC ACID

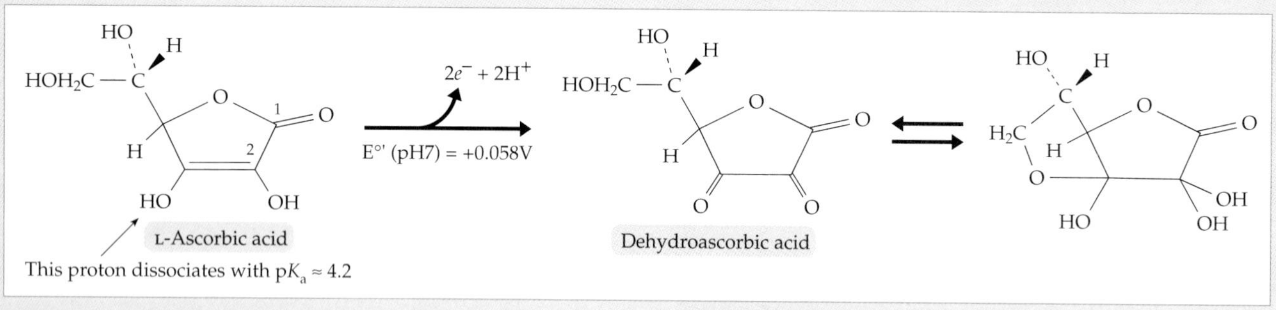

L-Ascorbic acid

$2e^- + 2H^+$

$E°'$ (pH7) = +0.058V

Dehydroascorbic acid

This proton dissociates with $pK_a \approx 4.2$

Hemorrhages of skin, gums, and joints were warnings that death was near for ancient sea voyagers stricken with **scurvy**. It was recognized by the year 1700 that the disease could be prevented by eating citrus fruit, but it was 200 years before efforts to isolate vitamin C were made. Ascorbic acid was obtained in crystalline form in 1927,[a–e] and by 1933 the structure had been established. Only a few vertebrates, among them human beings, monkeys, guinea pigs, and some fishes, require ascorbic acid in the diet; most species are able to make it themselves. Compared to that of other vitamins, the nutritional requirement is large.[e] Ten milligrams per day prevents scurvy, but subclinical deficiency, as judged by fragility of small capillaries in the skin, is present at that level of intake. "Official" recommendations for vitamin C intake have ranged from 30 to 70 mg / day. A more recent study[f] suggests 200 mg / day, a recommendation that is controversial.[g]

The biological functions of vitamin C appear to be related principally to its well-established reducing properties and easy one-electron oxidation to a free radical or two-electron reduction to **dehydroascorbic acid**. The latter is in equilibrium with the hydrated hemiacetal shown at the beginning of this box as well as with other chemical species.[h–j] Vitamin C is a weak acid which also has metal complexing properties.

Ascorbate, the anion of ascorbic acid, tends to be concentrated in certain types of animal tissues and may reach 3 mM or more in leukocytes, in tissues of eyes and lungs, in pituitary, adrenal, and parotid glands,[k,l] and in gametes.[m] Uptake into vesicles of the endoplasmic reticulum may occur via glucose transporters.[n] Ascorbate concentrations are even higher in plants and may exceed 10 mM in chloroplasts.[o] In animals the blood plasma ascorbate level of 20–100 µM is tightly controlled.[p,q] Cells take up ascorbate but any excess is excreted rapidly in the urine.[q] Both in plasma and within cells most vitamin C exists as the reduced form, ascorbate. When it is formed, the oxidized dehydroascorbate is reduced back to ascorbate or is degraded. The lactone ring is readily hydrolyzed to 2,3-dioxogulonic acid, which can undergo decarboxylation and oxidative degradation, one product being oxalate (see Fig. 20-2).[r] Tissues may also contain smaller amounts of L-ascorbic acid 2-sulfate, a compound originally discovered in brine shrimp. It is more stable than free ascorbate and may be hydrolyzed to ascorbate in tissues.[s]

In the chromaffin cells of the adrenal glands and in the neurons that synthesize catecholamines as neurotransmitters, ascorbate functions as a cosubstrate for dopamine β-hydroxylase (Eq. 18-53).[t,u] In fibroblasts it is required by the prolyl and lysyl hydroxylases and in hepatocytes by homogentisate dioxygenase (Eq. 18-49). Any effect of ascorbic acid in preventing colds may be a result of increased hydroxylation of procollagen and an associated stimulation of procollagen secretion.[v] High levels of ascorbate in guinea pigs lead to more rapid healing of wounds.[w] An important function of ascorbate in the pituitary and probably in other endocrine glands is in the α-amidation of peptides (Eq. 10-11).[x,y] Together with Fe(II) and O_2 ascorbate is a powerful nonenzymatic hydroxylating reagent for aromatic compounds. Like hydroxylases, the reagent attacks nucleophilic sites, e.g., converting phenylalanine to tyrosine. Oxygen atoms from $^{18}O_2$ are incorporated into the hydroxylated products. While H_2O_2 is formed in the reaction mixture, it cannot replace ascorbate. The relationship of this system to biochemical functions of ascorbate is not clear. An unusual function for vitamin C has been proposed for certain sponges that are able to etch crystalline quartz (SiO_2) particles from sand or rocks.[z]

Ascorbate is a major antioxidant, protecting cells and tissues from damage by free radicals, peroxides, and other metabolites of O_2.[p,r,aa,bb] It is chemically suited to react with many biologically important radicals and is present in high enough concentrations to be effective. It probably functions in cooperation with glutathione (Box 11-B),[cc] α-tocopherol (Fig. 15-24),[dd] and lipoic acid.[ee] Ascorbate can react with radicals in one-election transfer reactions to give the monodehydroascorbate radical[aa]:

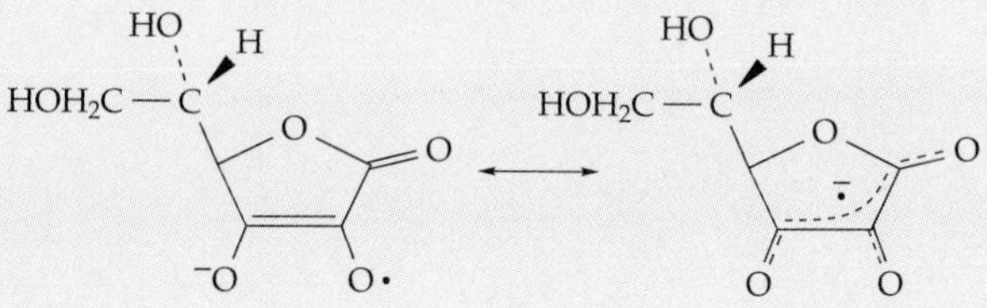

BOX 18-D (continued)

Two ascorbate radicals can react with each other in a disproportionation reaction to give ascorbate plus dehydroascorbate. However, most cells can reduce the radicals more directly. In many plants this is accomplished by NADH + H^+ using a flavoprotein **monodehydroascorbate reductase**.[o] Animal cells may also utilize NADH or may reduce dehydroascorbate with reduced glutathione.[cc,ff] Plant cells also contain a very active blue copper ascorbate oxidase (Chapter 16, Section D,5), which catalyzes the opposite reaction, formation of dehydroascorbate.[gg] A heme ascorbate oxidase has been purified from a fungus.[hh] Action of these enzymes initiates an oxidative degradation of ascorbate, perhaps through the pathway of Fig. 20-2.

Ascorbate can also serve as a signal. In cultured cells, which are usually deficient in vitamin C, addition of ascorbate causes an enhanced response to added iron, inducing synthesis of the iron storage protein ferritin.[ii] Ascorbate indirectly stimulates transcription of procollagen genes[v] and decreases secretion of insulin by the pancreas.[jj] However, since its concentration in blood is quite constant this effect is not likely to cause a problem for a person taking an excess of vitamin C.

Should we take extra vitamin C to protect us from oxygen radicals and slow down aging? Linus Pauling, who recommended an intake of 0.25–10 g / day, maintained that ascorbic acid also has a specific beneficial effect in preventing or ameliorating symptoms of the common cold.[j,kk] However, critics point out that unrecognized hazards may exist in high doses of this seemingly innocuous compound. Ascorbic acid has antioxidant properties, but it also promotes the generation of free radicals in the presence of Fe(III) ions, and it is conceivable that too much may be a bad thing.[ll] Catabolism to oxalate may promote formation of calcium oxalate kidney stones. Under some conditions products of dehydroascorbic acid breakdown may accumulate in the lens and contribute to cataract formation.[i,mm,nn] However, dehydroascorbate, or its decomposition products, apparently *protects* low-density lipoproteins against oxidative damage.[bb] Pauling pointed out that nonhuman primates synthesize within their bodies many grams of ascorbic acid daily, and that there is little evidence for toxicity. Pauling's claim that advanced cancer patients are benefited by very high (10 g daily) doses of vitamin C has been controversial, and some studies have failed to substantiate the claim.[oo,pp]

[a] Hughes, R. E. (1983) *Trends Biochem. Sci.* **8**, 146–147
[b] Staudinger, H. J. (1978) *Trends Biochem. Sci.* **3**, 211–212
[c] Szent-Györgyi, A. (1963) *Ann. Rev. Biochem.* **32**, 1–14
[d] Bradford, H. F. (1987) *Trends Biochem. Sci.* **12**, 344–347
[e] Packer, L., and Fuchs, J., eds. (1997) *Vitamin C in Health and Disease*, Dekker, New York
[f] Levine, M., Conry-Cantilena, C., Wang, Y., Welch, R. W., Washko.PW, Dhariwal, K. R., Park, J. B., Lazarev, A., Graumlich, J. F., King, J., and Cantilena, L. R. (1996) *Proc. Natl. Acad. Sci. U.S.A.* **93**, 3704–3709
[g] Young, V. R. (1996) *Proc. Natl. Acad. Sci. U.S.A.* **93**, 14344–14348
[h] Hroslif, J., and Pederson, B. (1979) *Acta Chem. Scand.* **B33**, 503–511
[i] Counsell, J. N., and Horing, D. H., eds. (1981) *Vitamin C*, Applied Sciences Publ., London
[j] Burns, J. J., Rivers, J. M., and Machlin, L. J., eds. (1987) *Third Conference on Vitamin C*, Vol. 498, New York Academy of Sciences, New York
[k] Washko, P. W., Wang, Y., and Levine, M. (1993) *J. Biol. Chem.* **268**, 15531–15535
[l] Rosen, G. M., Pou, S., Ramos, C. L., Cohen, M. S., and Britigan, B. E. (1995) *FASEB J.* **9**, 200–209
[m] Moreau, R., and Dabrowski, K. (1998) *Proc. Natl. Acad. Sci. U.S.A.* **95**, 10279–10282
[n] Bánhegyi, G., Marcolongo, P., Puskás, F., Fulceri, R., Mandl, J., and Benedetti, A. (1998) *J. Biol. Chem.* **273**, 2758–2762
[o] Sano, S., Miyake, C., Mikami, B., and Asada, K. (1995) *J. Biol. Chem.* **270**, 21354–21361
[p] May, J. M., Qu, Z.-c, and Whitesell, R. R. (1995) *Biochemistry* **34**, 12721–12728
[q] Vera, J. C., Rivas, C. I., Velásquez, F. V., Zhang, R. H., Concha, I. I., and Golde, D. W. (1995) *J. Biol. Chem.* **270**, 23706–23712
[r] Rose, R. C., and Bode, A. M. (1993) *FASEB J.* **7**, 1135 – 1142
[s] Benitez, L. V., and Halver, J. E. (1982) *Proc. Natl. Acad. Sci. U.S.A.* **79**, 5445–5449
[t] Dhariwal, K. R., Shirvan, M., and Levine, M. (1991) *J. Biol. Chem.* **266**, 5384–5387
[u] Tian, G., Berry, J. A., and Klinman, J. P. (1994) *Biochemistry* **33**, 226–234
[v] Chojkier, M., Houglum, K., Solis-Herruzo, J., and Brenner, D. A. (1989) *J. Biol. Chem.* **264**, 16957–16962
[w] Harwood, R., Grant, M. E., and Jackson, D. S. (1974) *Biochem. J.* **142**, 641–651
[x] Bradbury, A. F., and Smyth, D. G. (1991) *Trends Biochem. Sci.* **16**, 112–115
[y] Eipper, B. A., Milgram, S. L., Husten, E. J., Yun, H.-Y., and Mains, R. E. (1993) *Protein Sci.* **2**, 489–497
[z] Bavestrello, G., Arillo, A., Benatti, U., Cerrano, C., Cattaneo-Vietti, R., Cortesogno, L., Gaggero, L., Giovine, M., Tonetti, M., and Sarà, M. (1995) *Nature (London)* **378**, 374–376
[aa] Kobayashi, K., Harada, Y., and Hayashi, K. (1991) *Biochemistry* **30**, 8310–8315
[bb] Retsky, K. L., Freeman, M. W., and Frei, B. (1993) *J. Biol. Chem.* **268**, 1304–1309
[cc] Ishikawa, T., Casini, A. F., and Nishikimi, M. (1998) *J. Biol. Chem.* **273**, 28708–28712
[dd] May, J. M., Qu, Z.-c, and Morrow, J. D. (1996) *J. Biol. Chem.* **271**, 10577–10582
[ee] Lykkesfeldt, J., Hagen, T. M., Vinarsky, V., and Ames, B. N. (1998) *FASEB J.* **12**, 1183–1189
[ff] May, J. M., Cobb, C. E., Mendiratta, S., Hill, K. E., and Burk, R. F. (1998) *J. Biol. Chem.* **273**, 23039–23045
[gg] Gaspard, S., Monzani, E., Casella, L., Gullotti, M., Maritano, S., and Marchesini, A. (1997) *Biochemistry* **36**, 4852–4859
[hh] Kim, Y.-R., Yu, S.-W., Lee, S.-R., Hwang, Y.-Y., and Kang, S.-O. (1996) *J. Biol. Chem.* **271**, 3105–3111
[ii] Toth, I., and Bridges, K. R. (1995) *J. Biol. Chem.* **270**, 19540–19544
[jj] Bergsten, P., Sanchez Moura, A., Atwater, I., and Levine, M. (1994) *J. Biol. Chem.* **269**, 1041–1045
[kk] Pauling, L. (1970) *Vitamin C and the Common Cold*, Freeman, San Francisco, California
[ll] Halliwell, B. (1999) *Trends Biochem. Sci.* **24**, 255–259
[mm] Russell, P., Garland, D., Zigler, J. S., Jr., Meakin, S. O., Tsui, L.-C., and Breitman, M. L. (1987) *FASEB J.* **1**, 32–35
[nn] Nagaraj, R. H., Sell, D. R., Prabhakaram, M., Ortwerth, B. J., and Monnier, V. M. (1991) *Proc. Natl. Acad. Sci. U.S.A.* **88**, 10257–10261
[oo] Moertel, C. G., Fleming, T. R., Creagan, E. T., Rubin, J., O'Connell, M. J., and Ames, M. M. (1985) *N. Engl. J. Med.* **312**, 142–146
[pp] Lee, S. H., Oe, T., and Blair, I. A. (2001) *Science* **292**, 2083–2086

cytochrome P450s, such as a soluble fatty acid hydroxylase from *Bacillus megaterium*, have reductase domains with tightly bound FMN and FAD bound to the same polypeptide chain as is the heme.[491]

All cytochromes P450 appear to have at their active sites a molecule of heme with a thiolate anion as an axial ligand in the fifth position (Fig. 18-23). These relatively large heme proteins of ~ 45- to 55-kDa mass may consist of as many as 490 residues. Only a few three-dimensional structures are known,[490,492–494] and among these there are significant differences. However, on the basis of a large amount of experimental effort[487,495,496] it appears that all cytochromes P450 act by basically similar mechanisms.[474,496a,b,497] As indicated in Eq. 18-57, the substrate AH binds to the protein near the heme, which must be in the Fe(III) form. An electron delivered from the reductase then reduces the iron to the Fe(II) state (Eq. 18-57, step *b*). Then O_2 combines with the iron, the initial oxygenated complex formed in step *c* being converted to an Fe(III)-superoxide complex (Eq. 18-57, step *d*). Subsequent events are less certain.[497a] Most often a second electron is transferred in from the reductase (Eq. 18-57, step *e*) to give a peroxide complex of Fe(III), which is then converted in step *f* to a ferryl iron form, as in the action of peroxidases (Fig. 16-14). This requires transfer of two H^+ into the active site. The ferryl Fe(IV)=O donates its oxygen atom to the substrate regenerating the Fe(III) form of the heme (step *g*) and releasing the product (step *h*).

Microsomal cytochromes P450 often form hydrogen peroxide as a side product. This may arise directly from the Fe–O–O⁻ intermediate shown in Eq. 18-57. Some cytochromes P450 use this reaction in reverse to carry out hydroxylation utilizing peroxides instead of O_2 (Eq. 18-58).

$$AH + ROOH \rightarrow AOH + ROH \qquad (18\text{-}58)$$

Cytochromes P450 often convert drugs or other foreign compounds to forms that are more readily excreted.[499] However, the result is not always beneficial. For example, 3-methylcholanthrene, a strong inducer of cytochrome P450, is converted to a powerful carcinogen by the hydroxylation reaction.[500] See also Box 18-E.

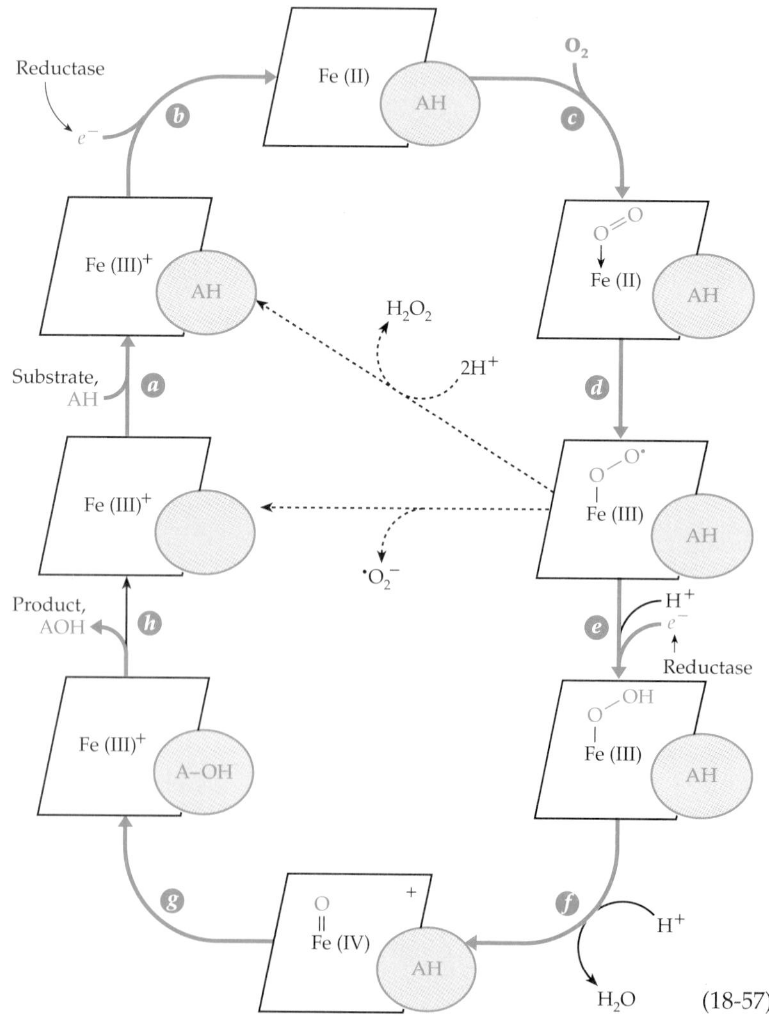

(18-57)

Other iron-containing oxygenases. Hydroxylases with properties similar to those of cytochrome P450 but containing nonheme iron catalyze ω-oxidation of alkanes and fatty acids in certain bacteria, e.g., *Pseudomonas oleovarans*. A flavoprotein rubredoxin reductase, is also required.[501] The methylotrophs *Methylococcus* and *Methylosinus* hydroxylate methane using as cosubstrate NADH or NADPH (Eq. 18-59). A soluble complex consists of 38-kDa reductase containing FAD and an Fe_2S_2

$$CH_4 + O_2 + NADH + H^+ \rightarrow CH_3OH + H_2O + NAD^+ \qquad (18\text{-}59)$$

center, a small 15-kDa component, and a 245-kDa hydroxylase with an $(\alpha\beta\gamma)_2$ composition and a three-dimensional structure[502–503] similar to that of ribonucleotide reductase (Chapter 16, Section A,9). Each large α subunit contains a diiron center similar to that shown in Fig. 16-20C. It is likely that O_2 binds between the two iron atoms in the Fe(II) oxidation state and, oxidizing both irons to Fe(III), is converted to a bridging peroxide group as shown in Eq. 18-60. In this intermediate, in which the two metals are held rigidly by the surrounding ligands including a bridging carboxylate side chain, the O–O bond may be broken as in Eq. 18-60, steps *a* and *b*, to generate an Fe(IV)–O• radical that may

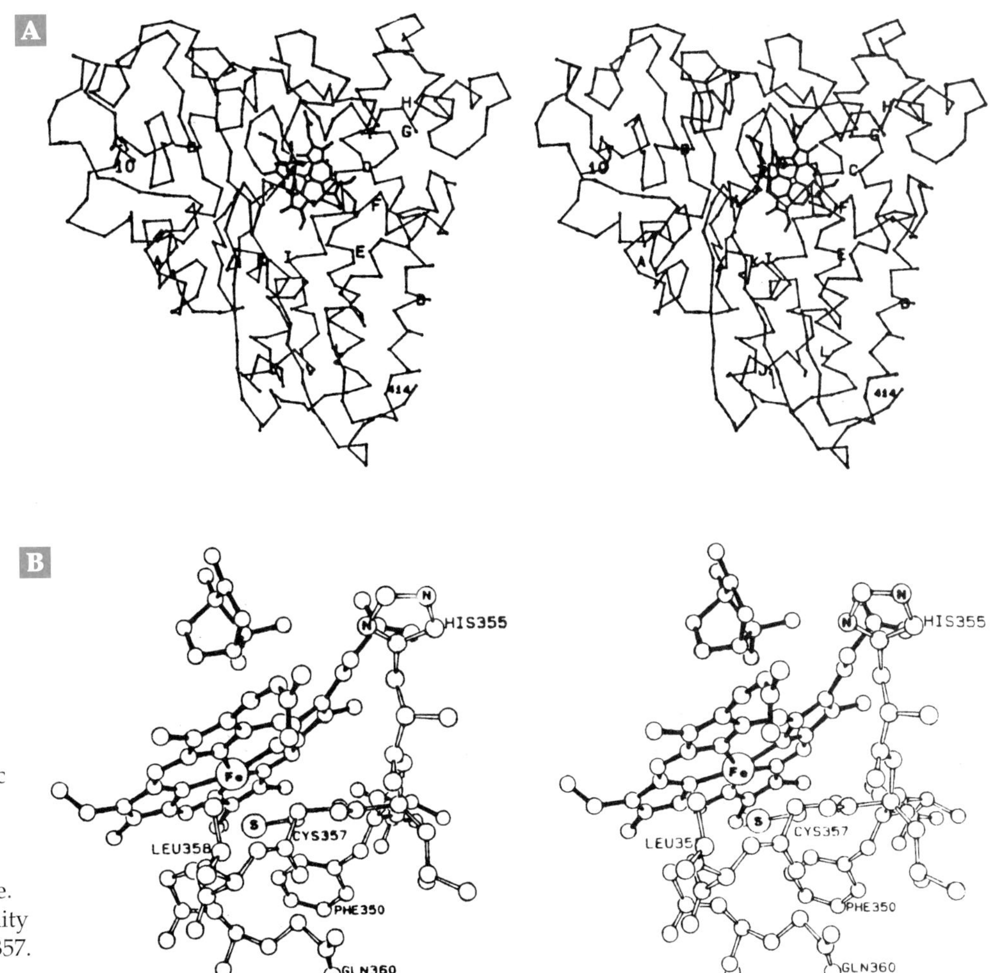

Figure 18-23 (A) Stereoscopic α-carbon backbone model of cytochrome P450cam showing the locations of the heme and of the bound camphor molecule. (B) View in the immediate vicinity of the thiolate ligand from Cys 357. From Poulos *et al.*[498]

remove a hydrogen atom from the substrate (step *c*) and undergo subsequent reaction steps analogous to those in the cytochrome P450 reaction cycle.[504–506]

$$\text{Fe(III)}-\text{O}-\text{O}-\text{Fe(III)} \xrightarrow{a} \text{Fe(IV)}(=\text{O})_2\text{Fe(IV)} \xrightarrow{b} \text{Fe(III)}-\text{O}-\text{Fe(IV)}-\text{O}\cdot \;+\; \text{H}-\text{CH}_3 \xrightarrow{c} \text{H}_3\text{C}-\text{OH} \quad (18\text{-}60)$$

A group of related bacterial enzymes hydroxylate alkanes,[507] toluene,[508] phenol,[509] and other substrates.[509a] Eukaryotic fatty acid desaturases (Fig. 16-20B) belong to the same family.[508] Some bacteria use cytochrome P450 or other oxygenase to add an oxygen atom to an alkene to form an epoxide. For example, propylene may be converted to either *R* or *S* epoxypropane which may be hydrolyzed, rearranged by a coenzyme M-dependent reaction, and converted to acetoacetate, which can be used as an energy source.[509a,b]

Nitric oxide and NO synthases. Nitric oxide (NO) is a reactive free radical whose formula is often written as $^{\bullet}$NO to recognize this characteristic. However, NO is not only a toxic and sometimes dangerous metabolite but also an important hormone with functions in the circulatory system, the immune system, and the brain.[510–512] The hormonal effects of NO are discussed in Chapter 30, but it is appropriate here to mention a few reactions. Nitric oxide reacts rapidly with O_2 to form nitrite (Eq. 18-61).

$$4\ \text{NO} + \text{O}_2 + 2\ \text{H}_2\text{O} \rightarrow 4\ \text{NO}_2^- + 4\ \text{H}^+ \quad (18\text{-}61)$$

It also combines very rapidly with superoxide anion radical to form **peroxynitrite** (Eq. 18-62).[513] This is another reactive oxidant which, because of its relatively high pK_a of 6.8, is partially protonated and able to diffuse through phospholipids within cells.[514,515]

BOX 18-E THE TOXICITY OF ACETAMINOPHEN

Most drugs, as well as toxins and other xenobiotic compounds, enter the body through membranes of the gastrointestinal tract, lungs, or skin. Drugs are frequently toxic if they accumulate in the body. They are often rather hydrophobic and are normally converted to more polar, water-soluble substances before elimination from the body. Two major types of reaction take place, usually in the liver. These are illustrated in the accompanying scheme for acetaminophen (*N*-acetyl-*p*-aminophenol), a widely used analgesic and antipyretic (fever relieving) nonprescription drug sold under a variety of trade names: (1) A large water-soluble group such as sulfate[a] or glucuronate is transferred onto the drug by a nucleophilic displacement reaction (steps *a* and *b* of scheme). (2) Oxidation, demethylation, and other alterations are catalyzed by one or more of the nearly 300 cytochrome P450 monooxygenases present in the liver (step *c*). Oxidation products may be detoxified by glutathione S-transferases, step *d* (see also Box 11-B).[b,c,cc]

These reactions protect the body from the accumulation of many compounds but in some cases can cause serious problems. The best known of these involves acetaminophen. Its oxidation by cytochrome P450 2E1 or by prostaglandin H synthase[d] yields a highly reactive quinone imine which reacts with cell proteins.[e] Since the cytochrome P450 oxidation can occur in two steps, a reactive intermediate radical is also created.[c,f] At least 20 drug-labeled proteins arising in this way have been identified.[c] Both addition of thiol groups of proteins to the quinone imine (step *e* of scheme) and oxidation of protein thiols occur.[g] Mitochondria suffer severe damage,[h] some of which is related to induction of Ca^{2+} release.[i]

Acetaminophen is ordinarily safe at the recommended dosages, but large amounts exhaust the reserve of glutathione and may cause fatal liver damage. By 1989, more than 1000 cases of accidental or intentional (suicide) overdoses had been reported with many deaths. Prompt oral or intravenous administration of *N*-acetylcysteine over a 72-hour period promotes synthesis of glutathione and is an effective antidote.[j]

Similar problems exist for many other drugs. Both acetaminophen and phenacetin, its ethyl ether derivative, may cause kidney damage after many years of use.[k,l] Metabolism of phenacetin and several other drugs varies among individuals. Effective detoxification may not occur in individuals lacking certain isoenzyme forms of cytochrome P450.[m] Use of the anticancer drugs daunomycin (daunorubicin; Figs. 5-22 and 5-23) and adriamycin is limited by severe cardiac toxicity arising from free radicals generated during oxidation of the drugs.[n] These are only a few examples of the problems with drugs, pesticides, plasticizers, etc.

Acetaminophen (*N*-Acetyl-*p*-aminophenol)
PAPS (Eq. 17-38)
Sulfotransferase *a*
UDP-glucuronosyl-transferase *b*
Cytochrome P450 2E1 *c*
Quinone imine
Glutathione
Glutathione transferase *d*
S-glutathione
Mercapturic acid
Cell proteins *e*
S-protein

[a] Klaassen, C. D., and Boles, J. W. (1997) *FASEB J.* **11**, 404–418

[b] Lee, W. M. (1995) *N. Engl. J. Med.* **333**, 1118–1127

[c] Qiu, Y., Benet, L. Z., and Burlingame, A. L. (1998) *J. Biol. Chem.* **273**, 17940–17953

[cc] Chen, W., Shockcor, J. P., Tonge, R., Hunter, A., Gartner, C., and Nelson, S. D. (1999) *Biochemistry* **38**, 8159 – 8166

[d] Potter, D. W., and Hinson, J. A. (1987) *J. Biol. Chem.* **262**, 974–980

[e] Lee, S. S. T., Buters, J. T. M., Pineau, T., Fernandez-Salguero, P., and Gonzalez, F. J. (1996) *J. Biol. Chem.* **271**, 12063–12067

[f] Rao, D. N. R., Fischer, V., and Mason, R. P. (1990) *J. Biol. Chem.* **265**, 844–847

[g] Tirmenstein, M. A., and Nelson, S. D. (1990) *J. Biol. Chem.* **265**, 3059–3065

[h] Burcham, P. C., and Harman, A. W. (1991) *J. Biol. Chem.* **266**, 5049–5054

[i] Weis, M., Kass, G. E. N., Orrenius, S., and Moldéus, P. (1992) *J. Biol. Chem.* **267**, 804–809

[j] Smilkstein, M. J., Knapp, G. L., Kulig, K. W., and Rumack, B. H. (1988) *N. Engl. J. Med.* **319**, 1557–1562

[k] Stolley, P. D. (1991) *N. Engl. J. Med.* **324**, 191–193

[l] Rocha, G. M., Michea, L. F., Peters, E. M., Kirby, M., Xu, Y., Ferguson, D. R., and Burg, M. B. (2001) *Proc. Natl. Acad. Sci. U.S.A.* **98**, 5317–5322

[m] Distlerath, L. M., Reilly, P. E. B., Martin, M. V., Davis, G. G., Wilkinson, G. R., and Guengerich, F. P. (1985) *J. Biol. Chem.* **260**, 9057–9067

[n] Davies, K. J. A., and Doroshow, J. H. (1986) *J. Biol. Chem.* **261**, 3060–3067

$$\cdot NO + \cdot O_2^- \rightarrow O{=}N{-}O{-}O^- \quad (18\text{-}62)$$

NO binds to the iron atoms in accessible heme groups such as those of hemoglobin[516] and of guanylate cylases,[517,518] and in some Fe – S proteins such as aconitase. Some blood-sucking insects utilize salivary heme proteins called **nitrophorins** to carry NO into host tissues where it activates guanylate cyclase causing vasodilation. Nitrophorins also bind histamine and inhibit blood coagulation, which assists feeding.[518a,b] In the presence of a suitable oxidant such as O_2, nitric oxide reacts with thiol groups of proteins and small molecules to give **S-nitrosothiols** (Eq. 18-63).[511,519-520]

$$\cdot NO + R{-}SH \longrightarrow R{-}S{-}\dot{N}{-}OH \xrightarrow[\cdot O_2^-]{O_2} R{-}S{-}NO \quad (18\text{-}63)$$

S-Nitrosothiol

However, the physiological mechanisms of formation of these *S*-nitroso compounds is not clear.[520–521] One mechanism may involve conversion by O_2 to nitrous anhydride.[511,522]

$$2\ NO + O_2 \rightarrow ON{-}O{-}NO \quad (18\text{-}64)$$

Nucleophilic attack on this compound by –SH, $-NH_2$, and other nucleophiles would yield *S*-nitroso and *N*- nitroso compounds with release of nitrous oxide N_2O.

The relatively stable *S*-nitrosothiols derived from glutathione, cysteine, and proteins such as hemoglobin[516] may be important storage and transport forms of NO. If so, mechanisms of release of NO are important. A simple homolytic cleavage of R–S–NO to NO plus a thiyl radical R–S• has often been assumed. However, rapid cleavage requires catalysis by a transition metal ion or reaction with reducing agents such as ascorbate or other thiols.[523,524] *S*-Nitrosothiols may also give rise to nitrosonium (NO^+) or nitroxy (NO^-) ions.[524]

NO synthases are oxygenases that carry out a two-step oxidation of L-arginine to L-citrulline with production of NO. In the first step, a normal monooxygenase reaction, L-N^γ-hydroxyarginine is formed (Eq. 18-65, step *a*). In the second step (Eq. 18-65, step *b*) NO is formed in a three-electron oxidation. In this equation the symbols * and † indicate positions of incorporation of labeled O_2 atoms in the intermediate and final products.

The human body contains three types of nitric oxide synthase known as **neuronal** (nNOS or NOS1), **inducible** (iNOS or NOS2), and **endothelial** (eNOS or NOS3).[511a–f] These enzymes have a broad distribution within various tissues, but NOS1 is especially active in neurons and NOS3 in endothelial cells. The inducible NOS2 originally discovered in macrophages is transcriptionally regulated. When these phagocytic cells are at rest the activity of NOS2 is very low, but it becomes highly active after induction by cytokines or by the lipopolysaccharides of bacterial cell membranes.[525,526] Both NOS1 and NOS2 are constitutively expressed but require calcium ions, which bind to a calmodulin domain of the protein. The inducible NOS2 doesn't require added Ca^{2+} but does contain the calmodulin domain. NOS3 carries an N-terminal myristoyl group as well as cysteines that may be palmitoylated. It is located in caveolae of plasma membranes and in Golgi complexes.[527,528]

Nucleotide sequences revealed a close similarity of NO synthases to cytochrome P450 reductase.[525,529,530] Study of spectra suggested that NO synthases, in which the heme is held by a thiolate sulfur of a cysteine side chain, might be specialized cytochromes P450.[531] However, although the heme in NO synthases is bound by a thiolate group the protein fold is unlike that of cytochromes P450.[532,533] The NO synthases all share a three-component structure.[534] In the NOS2 from macrophages residues 1–489 form the catalytic oxygenase domain, residues 499–530 bind calmodulin, and residues 531–1144 form the reductase domain.[535] The last contains a binding site for NADPH as well as bound FMN and FAD. The reductase domain structure can be modeled after that of cytochrome P450 reductase.[533] Electrons from NADPH are apparently transferred to FAD, then to FMN, and then to the catalytic site.[525]

The least understood aspect of NO synthases is the requirement for tetrahydrobiopterin, BH_4, the same coenzyme required by the other pterin-dependent monooxygenases (Eq. 18-44). The presence of this coenzyme in the reduced BH_4 form is essential for step *a* of Eq. 18-65 but not for step *b*. This suggests that in step *a* an organic peroxide might be generated by BH_4 and used to form an oxo-iron hydroxylating reagent.

L-Arginine $\xrightarrow[\text{NADH}]{O_2^*\ (a)}$ L-N^γ-hydroxyarginine $\xrightarrow[\text{½ NADH}]{O_2^\dagger,\ \cdot NO^*\ (b)}$ L-Citrulline

(18-65)

However, there is no evidence for the expected quinonoid dihydropterin, and the three-dimensional structure suggested that BH_4 plays a structural role in mediating essential conformational changes.[532,535a,b] Nevertheless, newer data indicate a role in electron transfer.[535c]

Step *b* of Eq. 18-65 is an unusual three-electron oxidation, which requires only one electron to be delivered from NADPH by the reductase domain. Hydrogen peroxide can replace O_2 in this step.[536] A good possibility is that a peroxo or superoxide complex of the heme in the Fe(III) state adds to the hydroxyguanidine group. For example, the following structure could arise from addition of Fe(III)–O–O⁻:

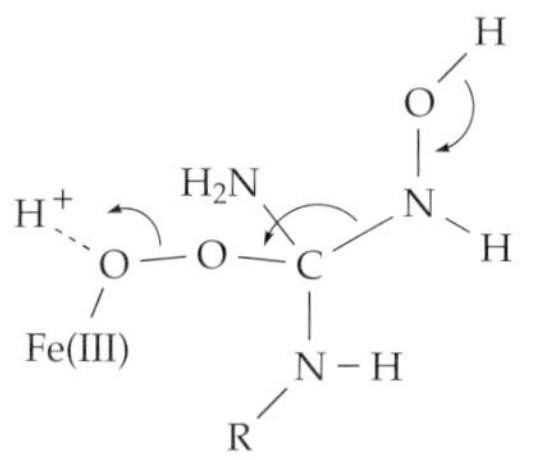

H
O
H₂N
H⁺
N
O—O—C
H
Fe(III)
N–H
R

Breakup as indicated by the arrows on this structure would give Fe(III)–OH, citrulline, and O=N–H, **nitroxyl**. This is one electron ($e^- + H^+$) more reduced than •NO. Perhaps the adduct forms from Fe(III)–O–O•. On the other hand, there is evidence that NO synthases may produce nitroxyl or nitroxyl ion NO^- as the initial product.[537–538] NO and other products such as N_2O and NO_2^- may arise rapidly in subsequent reactions. Nitrite is a major oxidation product of NO in tissues.[538a] The chemistry of NO in biological systems is complex and not yet fully understood. See also pp. 1754, 1755.

G. Biological Effects of Reduced Oxygen Compounds

Although molecular oxygen is essential to the aerobic mode of life, it is toxic at high pressures. Oxidative damage from O_2 appears to be an important cause of aging and also contributes to the development of cancer. Reduced forms of oxygen such as superoxide, hydrogen peroxide, and hydroxyl radicals are apparently involved in this toxicity.[539,540] The same agents are deliberately used by phagocytic cells such as the neutrophils (polymorphonuclear leukocytes) to kill invading bacteria or fungi and to destroy malignant cells.[541]

The reactions shown with vertical arrows in Eq. 18-66 can give rise to the reduced oxygen compounds. The corresponding standard redox potential at pH 7 for each is also given.[539,542–544] As indicated by the low value of the redox potential for the $O_2 / O_2^{\bullet-}$ couple, the formation of superoxide by reduction of O_2 is spontaneous only for strongly reducing one-electron donors. Superoxide ion is a strong reductant, but at the same time a powerful one-electron oxidant, as is indicated by the high electrode potential of the $O_2^{\bullet-}/H_2O_2$ couple.

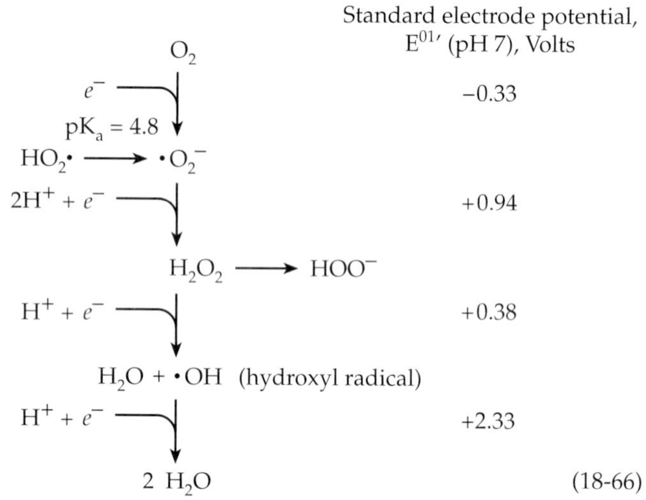

	Standard electrode potential, $E^{01'}$ (pH 7), Volts
O_2	
e^- ↓	−0.33
$HO_2^\bullet$ —(pK_a = 4.8)→ $^\bullet O_2^-$	
$2H^+ + e^-$ ↓	+0.94
H_2O_2 ⟶ HOO^-	
$H^+ + e^-$ ↓	+0.38
$H_2O + {}^\bullet OH$ (hydroxyl radical)	
$H^+ + e^-$ ↓	+2.33
2 H_2O	

(18-66)

1. The Respiratory Burst of Neutrophils

Some 25 x 10^9 neutrophils circulate in an individual's blood, and an equal number move along the surfaces of red blood cells. Invading microorganisms are engulfed after they are identified by the immune system as foreign. Phagocytosis is accompanied by a rapid many-fold rise in the rate of oxygen uptake as well as an increased glucose metabolism. One purpose of this **respiratory burst**[545–548] is the production of reduced oxygen compounds that kill the ingested microorganisms. In the very serious **chronic granulomatous disease** the normal respiratory burst does not occur, and bacteria are not killed.[549] The respiratory burst seems to be triggered not by phagocytosis itself, but by stimulation of the neutrophil by chemotactic formylated peptides such as formyl-Met-Leu-Phe[550] and less rapidly by other agonists such as phorbol esters.

The initial product of the respiratory burst appears to be superoxide ion O_2^-. It is formed by an **NADPH oxidase,** which transports electrons from NADPH to O_2 probably via a flavoprotein and cytochrome b_{558}. Either the flavin or the cytochrome b_{558} must donate one electron to O_2 to form the superoxide anion.

$$\text{NADPH} \rightarrow \text{FAD} \rightarrow \text{cyt } b_{558} \rightarrow O_2 \quad (18\text{-}67)$$

Flavocytochrome b_{558} (also called b_{-245}) has the unusually low redox potential of −0.245 V. It exists in phagocytic cells as a heterodimer of membrane-associated subunits p22-*phox* and gp91-*phox* where *phox* indicates phagocytic oxidase. The larger 91-kDa subunit contains two heme groups as well as one FAD and the presumed NADPH binding site.[548,551–552a] The mechanism of interaction with O_2 is unclear. Unlike hemoglobin but like other cytochromes *b*, cytochrome b_{558} does not form a complex with CO.[553] NADPH oxidase also requires two cytosolic components p47-*phox* and p67-*phox*. In resting phagocytes they reside in the cytosol as a 240-kDa complex with a third component, p40-*phox*, which may serve as an inhibitor.[548,554] Upon activation of the phagocyte in response to chemotactic signals the cytosolic components undergo phosphorylation at several sites,[554] and protein p47-*phox* and p67-*phox* move to the membrane and bind to and with the assistance of the small G protein Rac[552a,554a] activate flavocytochrome b_{558}. Phosphorylation of p47-*phox* may be especially important.[555]

In the X-chromosome-linked type of chronic granulomatous disease flavocytochrome b_{558} is absent or deficient, usually because of mutation in gp91-*phox*.[556,557] In an autosomal recessive form the superoxide-forming oxidase system is not activated properly. In some patients protein kinase C fails to phosphorylate p47-*phox*.[556,558] Less severe symptoms arise from deficiences in myeloperoxidase, chloroperoxidase (Chapter 16), glucose 6-phosphate dehydrogenase, glutathione synthetase, and glutathione reductase. The importance of these enzymes can be appreciated by examination of Fig. 18-24, which illustrates the relationship of several enzymatic reactions to the formation of superoxide anion and related compounds. Not only neutrophils but monocytes, macrophages, **natural killer cells** (NK cells), and other phagocytes apparently use similar chemistry in attacking ingested cells (Chapter 31).[559] Superoxide-producing NADH oxidases have also been found in nonphagocytic cells in various tissues.[559a]

What kills the ingested bacteria and other microorganisms? Although superoxide anion is relatively unreactive, its protonated form $HO_2\bullet$, is very reactive. Since its pK_a is 4.8, there will be small amounts present even at neutral pH. Some of the $^\bullet O_2^-$ may react with NO to form peroxynitrite (Eq. 18-62).[559b] Peroxynitrite, in turn, can react with the ubitquitous CO_2 to give $^\bullet CO_3^-$ and $^\bullet NO_2$ radicals.[559c] Peroxynitrite anion also reacts with metalloenzyme centers[559d] and causes nitration and oxidation of aromatic residues in proteins.[559d,e] However, neutrophils contain active superoxide dismutases, and most of the superoxide that is formed is converted quickly to O_2 and H_2O_2. The latter may diffuse into the phagosomes as well as into the extracellular space. The H_2O_2 itself is toxic, but longer lived, more toxic oxidants are also formed. Reaction of H_2O_2 with **myeloperoxidase** (Chapter 16) produces hypochlorous acid, **(HOCl;** Eqs. 16-12, Fig. 18-24) and **chloramines** such as NH_2Cl, RNHCl, and $RNCl_2$. An important intracellular chloramine may be that of taurine.

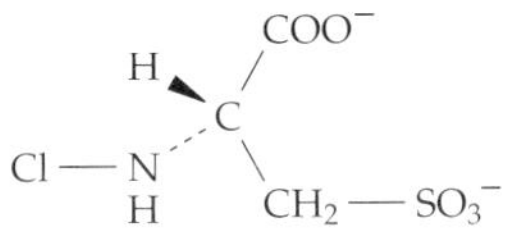

Chloramine formed from taurine

Human neutrophils use HOCl formed by myeloperoxidase to oxidize α-amino acids such as tyrosine to reactive aldehydes that form adducts with –SH, $-NH_2$, imidazole, and other nucleophilic groups.[560] They also contain NO synthases, which form NO, peroxynitrite (Fig. 18-24), and nitrite.[561,562]

Hydroxyl radicals •OH, which attack proteins, nucleic acids, and a large variety of other cellular constituents, may also be formed. Although too reactive to diffuse far, they can be generated from H_2O_2 by Eq. 18-68. This reaction involves catalysis by Fe ions as shown.[562a]

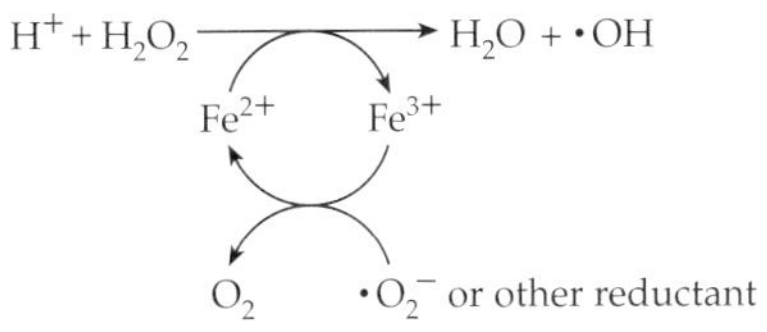

$$H^+ + H_2O_2 \longrightarrow H_2O + \bullet OH \quad (Fe^{2+} \rightarrow Fe^{3+});\quad \bullet O_2^- \text{ or other reductant} \rightarrow O_2 \quad (Fe^{3+} \rightarrow Fe^{2+}) \tag{18-68}$$

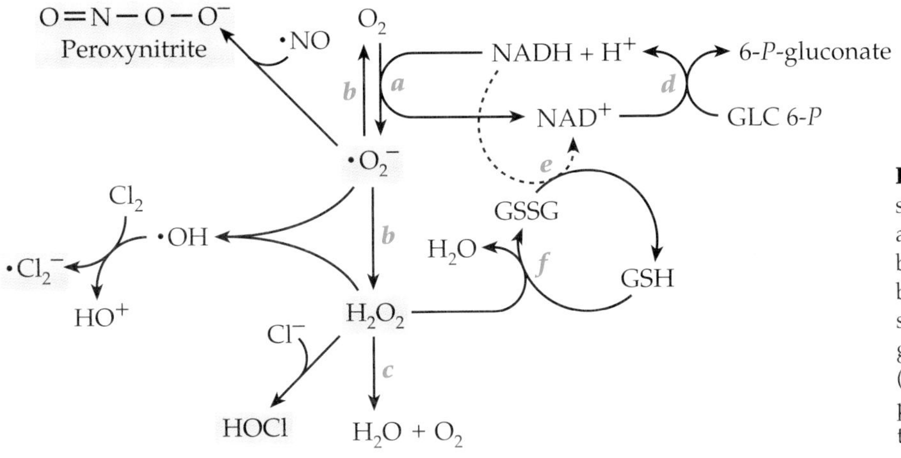

Figure 18-24 Some reactions by which superoxide anions, hydrogen peroxide and related compounds are generated by neutrophils and to a lesser extent by other cells: (*a*) NADPH oxidase, (*b*) superoxide dismutase, (*c*) catalase, (*d*) glucose-6-phosphate dehydrogenase, (*e*) glutathione reductase, (*f*) glutathione peroxidase. Abbrevations: GSH, glutathione; GSSB, oxidized glutathione.

Because Fe^{3+} is present in such low concentrations, there is uncertainty as to the biological significance of this reaction.[563] However, other iron compounds may function in place of Fe^{3+} and Fe^{2+} in Eq. 18-68.[564] A mixture of ferrous salts and H_2O_2 (Fenton reagent) has long been recognized as a powerful oxidizing mixture, which generates •OH or compounds of similar reactivity.[564–567] Ascorbate, and various other compounds, can also serve as the reductant in Eq. 18-68.[568]

Eosinophils, whose presence is stimulated by parasitic infections, have a peroxidase which acts preferentially on Br^- to form HOBr.[569] This compound can react with H_2O_2 more efficiently than does HOCl (Eq. 16-16) to form the very reactive **singlet oxygen.**[570] Singlet oxygen can also be generated from H_2O_2 and $•O_2$ by Eq. 18-69[571] and also photochemically.[572]

$$•O_2^- + H_2O_2 \rightarrow {}^1O_2({}^1\Delta_g) + OH + OH^- \quad (18\text{-}69)$$

Additional killing mechanisms used by phagocytes include acidification of the phagocytic lysosomes with the aid of a proton pump[573] and formation of toxic peptides. For example, bovine neutrophils produce the bacteriocidal peptide RLCRIVVIRVCR which has a disulfide crosslinkage between the two cysteine residues.[574] Microorganisms have their own defenses against the oxidative attack by phagocytes. Some bacteria have very active superoxide dismutases. The protozoan *Leishmania* produces an acid phosphatase that shuts down the production of superoxide of the host cells in response to activating peptides.[575]

A respiratory burst accompanies fertilization of sea urchin eggs.[576,577] In this case, the burst appears to produce H_2O_2 as the major or sole product and is accompanied by release of **ovoperoxidase** from cortical granules. This enzyme uses H_2O_2 to generate **dityrosine crosslinkages** between tyrosine side chains during formation of the fertilization membrane. Defensive respiratory bursts are also employed by plant cells.[578,579] See also Box 18-B.

2. Oxidative Damage to Tissues

Superoxide anion radicals are formed not only in phagocytes but also as an accidental by-product of the action of many flavoproteins,[580,581] heme enzymes, and other transition metal-containing proteins. An example is xanthine oxidase. It is synthesized as xanthine dehydrogenase which is able to use NAD^+ as an oxidant, but upon aging, some is converted into the $^{\bullet}O_2$-utilizing xanthine oxidase (Chapter 16). This occurs extensively during ischemia. When oxygen is readmitted to a tissue in which this conversion of xanthine dehydrogenase to xanthine oxidase has occurred, severe oxidative injury may occur.[582] In animals the intravenous administration of superoxide dismutase or pretreatment with the xanthine oxidase inhibitor **allopurinol** (Chapter 25) prevents much of the damage, suggesting that superoxide is the culprit.

Hydrogen peroxide is also generated within cells[583] by flavoproteins and metalloenzymes and by the action of superoxide dismutase on $^{\bullet}O_2^-$. Since H_2O_2 is a small uncharged molecule, it can diffuse out of cells and into other cells readily. If it reacts with Fe(II), it can be converted within cells to $^{\bullet}OH$ radicals according to Eq. 18-68. Such radicals and others have been detected upon readmission of oxygen to ischemic animal hearts.[584,585] It has also been suggested that NADH may react with Fe(III) compounds in the same way as does O_2^- in Eq. 18-68 to provide a mechanism for producing hydroxyl radicals from H_2O_2.[539] Nitric oxide, formed by the various NO synthases in the cytosol and in mitochondria[586] and by some cytochromes P450,[587] is almost ubiquitous and can also lead to formation of peroxynitrate (Eq. 18-62). Thus, the whole range of reduced oxygen compounds depicted in Eq. 18-24 are present in small amounts throughout cells.

There is little doubt that these compounds cause extensive damage to DNA, proteins, lipids, and other cell constituents.[539,540,563,588] For example, one base in 150,000 in nuclear DNA is apparently converted from guanine to 8-hydroxyguanine presumably as a result of attack by oxygen radicals.[589] In mitochondrial DNA one base in 8000 undergoes this alteration. This may be a result of the high rate of oxygen metabolism in mitochondria and may also reflect the lack of histones and the relatively inefficient repair of DNA within mitochondria. Proteins undergo chain cleavage, crosslinking, and numerous side chain modification reactions.[588] Dissolved O_2 can react directly with exposed glycyl residues in protein backbones to create glycyl radicals which may lead to chain cleavage as in Eq. 15-39.[588a] Iron–sulfur clusters, such as the Fe_4S_4 center of aconitase (Fig. 13-4), are especially sensitive to attack by superoxide anions.[588a–c] "Free iron" released from the Fe–S cluster may catalyze formation of additional damaging radicals.

Antioxidant systems. Cells have numerous defenses against oxidative damage.[563,590,591] Both within cells and in extracellular fluids superoxide dismutase (Eq. 16-27) decomposes superoxide to O_2 and H_2O_2. The H_2O_2 is then broken down by catalase (Eq. 16-8) to O_2 and H_2O. In higher animals the selenoenzyme glutathione peroxidase (Chapter 16) provides another route for decomposition of H_2O_2 and lipid peroxides of membranes. The oxidized glutathione formed is reduced by NADPH. The system has a critical role within erythrocytes (Box 15-H). In chloroplasts an analogous system utilizes ascorbate peroxidase, ascorbate, and glutathione to break down peroxides.[592]

$$\text{NADPH} + \text{H}^+ \rightarrow \text{glutathione} \rightarrow \text{ascorbate} \rightarrow \text{H}_2\text{O}_2 \quad (18\text{-}70)$$

Ascorbate,[593–594] glutathione, NADPH,[594a] and tocopherols (Box 15-G)[595] all act as scavengers of free radicals such as O_2^-, $^{\bullet}OH$ and $ROO^{\bullet}$, $^{\bullet}CO_3^-$, and of singlet oxygen. Antioxidant protection is needed in extracellular fluids as well as within cells. In addition to glutathione and ascorbate, bilirubin,[596] uric acid,[597] melatonin,[598,598a] circulating superoxide dismutase, and the copper protein ceruloplasmin (Chapter 16) all act as antioxidants. Methionine residues of proteins may have a similar function.[599] Various proteins and small chelating compounds such as citrate tie up Fe^{3+} preventing it from promoting radical formation. Tocopherols, ubiquinols, and lipoic acid[600,600a–c] protect membranes. Beta carotene (Fig. 22-5), another lipid-soluble antioxidant, is the most effective quencher of singlet O_2 that is known. Even nitric oxide, usually regarded as toxic, sometimes acts as an antioxidant.[601] Trehalose protects plants against oxidative damage.[601a]

An increasing number of proteins are being recognized as protectants against oxidative damage. The exposed –SH and $-SCH_3$ groups of cysteine and methionine residues in proteins may function as appropriately located scavengers which may donate electrons to destroy free radicals or react with superoxide ions to become sulfonated. The thioredoxin (Box 15-C) and glutathione (Eq. 18-70) systems, in turn, reduce the protein radicals formed in this way.[601b–d] Methionine sulfoxide, both free and in polypeptides, is reduced by **methionine sulfoxide reductase** in organisms from bacteria to humans.[601e–g] Biotin, together with biotin sulfoxide reductase,[601h] may provide another antioxidant system. Some bacteria utilize glutathione-independent **alkylperoxide reductases** to scavenge organic peroxides.[601i] while mammals accomplish the same result with **peroxiredoxins** and with thioredoxin.[601j] Many other proteins will doubtless be found to participate in defense against oxidative damage. Oxygen is always present and its reactions in our bodies are essential. Generation of damaging reduced oxygen compounds and radicals is inevitable. Evolution will select in favor of many proteins that have been modified to minimize the damage.

Antioxidant enzymes do not always protect us. There was great excitement when it was found that victims of a hereditary form of the terrible neurological disease **amyotrophic lateral sclerosis** (**ALS**; see also Chapter 30) carry a defective gene for Cu / Zn-superoxide dismutase (SOD; Eq. 16-27).[602–603b] This discovery seemed to support the idea that superoxide anions in the brain were killing neurons. However, it now appears that in some cases of ALS the defective SOD is *too active*, producing an excess of H_2O_2, which damages neurons.

Transcriptional regulation of antioxidant proteins. Certain proteins with easily accessible Fe–S clusters, e.g., aconitase, are readily inactivated by oxidants such as peroxynitrite.[540,604] At least two proteins of this type function as transcription factors in *E. coli*. These are known as **SoxR** and **OxyR**. The SoxR protein is sensitive to superoxide anion, which carries out a one-electron oxidation on its Fe_2S_2 centers.[540,605–607] In its oxidized form SoxR is a transcriptional regulator that controls 30–40 genes, among them several that are directly related to "**oxidative stress.**"[608] These include genes for manganese SOD, glucose-6-phosphate dehydrogenase, a DNA repair nuclease, and aconitase (to replace the inactivated enzyme). The OxyR protein, which responds to elevated $[H_2O_2]$, is activated upon oxidation of a pair of nearby –SH groups to form a disulfide bridge.[607,609] It controls genes for catalase, glutathione reductase, an alkyl hydroperoxide reductase,[610,610a] and many others. Similar transcriptional controls in yeast result in responses to low doses of H_2O_2 by at least 167 different proteins.[608] Animal mitochondria also participate in sensing oxidant levels.[611,612] (See also Chapter 28, Section C,6.)

References

1. Tzagoloff, A. (1982) *Mitochondria*, Plenum, New York
1a. Scheffler, I. E. (1999) *Mitochondria*, Wiley-Liss, New York
2. Chappell, J. B. (1979) *The Energetics of Mitochondria*, 2nd ed., Oxford Univ. Press, London (Carolina Biology Reader No. 19)
3. Lee, C. P., Schatz, G., and Dallner, G., eds. (1981) *Mitochondria and Microsomes*, Addison-Wesley, Reading, Massachusetts
4. Tyler, D. D. (1992) *The Mitochondrion in Health and Disease*, VCH Publ., New York
5. Noble, R. W., and Gibson, Q. H. (1970) *J. Biol. Chem.* **245**, 2409–2413
6. Ankel-Simons, F., and Cummins, J. M. (1996) *Proc. Natl. Acad. Sci. U.S.A.* **93**, 13859–13863
7. Champion, P. M., Münck, E., Debrunner, P. G., Hollenberg, P. F., and Hager, L. P. (1973) *Biochemistry* **12**, 426–435
8. Packer, L. (1973) in *Mechanisms in Bioenergetics* (Azzone, G. F., Ernster, L., Papa, S., Quagliariello, E., and Siliprandi, N., eds), pp. 33–52, Academic Press, New York
9. Smith, L. D., and Gholson, R. K. (1969) *J. Biol. Chem.* **244**, 68–71
10. Harris, R. A., Williams, C. H., Caldwell, M., Green, D. E., and Valdivia, E. (1969) *Science* **165**, 700–703
11. Malhotra, S. K., and Sikewar, S. S. (1983) *Trends Biochem. Sci.* **8**, 358–359
12. Frey, T. G., and Mannella, C. A. (2000) *Trends Biochem. Sci.* **25**, 319–324
12a. Varmus, H. E. (1985) *Nature (London)* **314**, 583–584
12b. Rutter, G. A., and Rizzuto, R. (2000) *Trends Biochem. Sci.* **25**, 215–221
13. Ernster, L., and Drahota, Z., eds. (1969) *Mitochondria: Structure and Function*, Vol. 17, FEBS Symposium, (pp. 5–31)
14. Pfaller, R., Freitag, H., Harmey, M. A., Benz, R., and Neupert, W. (1985) *J. Biol. Chem.* **260**, 8188–8193
15. Heins, L., Mentzel, H., Schmid, A., Benz, R., and Schmitz, U. K. (1994) *J. Biol. Chem.* **269**, 26402–26410
15a. Bölter, B., and Soll, J. (2001) *EMBO J.* **20**, 935–940
16. Mannella, C. A. (1992) *Trends Biochem. Sci.* **17**, 315–320
17. Ha, H., Hajek, P., Bedwell, D. M., and Burrows, P. D. (1993) *J. Biol. Chem.* **268**, 12143–12149
18. Sutfin, L. V., Holtrop, M. E., and Ogilvie, R. E. (1971) *Science* **174**, 947–949
19. Srere, P. A. (1982) *Trends Biochem. Sci.* **7**, 375–377
20. Srere, P. A. (1981) *Trends Biochem. Sci.* **6**, 4–7
20a. Haggie, P. M., and Brindle, K. M. (1999) *J. Biol. Chem.* **274**, 3941–3945
21. McCabe, E. R. B. (1995) in *The Metabolic and Molecular Bases of Inherited Disease*, 7th ed., Vol. 1 (Scriver, C. R., Beaudet, A. L., Sly, W. S., and Valle, D., eds), pp. 1631–1652, McGraw-Hill, New York
22. Hackenbrock, C. R., and Hammon, K. M. (1975) *J. Biol. Chem.* **250**, 9185–9197
22a. Jiang, F., Ryan, M. T., Schlame, M., Zhao, M., Gu, Z., Klingenberg, M., Pfanner, N., and Greenberg, M. L. (2000) *J. Biol. Chem.* **275**, 22387–22394
22b. Sedlák, E., and Robinson, N. C. (1999) *Biochemistry* **38**, 14966–14972
22c. Gomez, B., Jr., and Robinson, N. C. (1999) *Biochemistry* **38**, 9031–9038
22d. McAuley, K. E., Fyfe, P. K., Ridge, J. P., Isaacs, N. W., Cogdell, R. J., and Jones, M. R. (1999) *Proc. Natl. Acad. Sci. U.S.A.* **96**, 14706–14711
22e. Bazhenova, E. N., Deryabina, Y. I., Eriksson, O., Zvyagilskaya, R. A., and Saris, N.-E. L. (1998) *J. Biol. Chem.* **273**, 4372–4377
22f. Horikawa, Y., Goel, A., Somlyo, A. P., and Somlyo, A. V. (1998) *Biophys. J.* **74**, 1579–1590
22g. Territo, P. R., French, S. A., Dunleavy, M. C., Evans, F. J., and Balaban, R. S. (2001) *J. Biol. Chem.* **276**, 2586–2599
22h. Smaili, S. S., Stellato, K. A., Burnett, P., Thomas, A. P., and Gaspers, L. D. (2001) *J. Biol. Chem.* **276**, 23329–23340
22i. Beutner, G., Sharma, V. K., Giovannucci, D. R., Yule, D. I., and Sheu, S.-S. (2001) *J. Biol. Chem.* **276**, 21482–21488
22j. Arnaudeau, S., Kelley, W. L., Walsh, J. V., Jr., and Demaurex, N. (2001) *J. Biol. Chem.* **276**, 29430–29439
22k. Berridge, M. J., Bootman, M. D., and Lipp, P. (1998) *Nature (London)* **395**, 645–648
23. Clayton, D. A. (1984) *Ann. Rev. Biochem.* **53**, 573–594
24. Slonirnski, P., Borst, P., and Attardi, G., eds. (1982) *Mitochondrial Genes*, Cold Spring Harbor Lab. Press, Cold Spring Harbor, New York
25. Attardi, G. (1981) *Trends Biochem. Sci.* **6**, 86–89; 100–103
26. Palmer, J. D. (1997) *Nature (London)* **387**, 454–455
26a. Schwartz, M., and Vissing, J. (2002) *N. Engl. J. Med.* **347**, 576–580
27. Wolstenholme, D. R. (1992) in *Mitochondrial Genomes—International Review of Cytology*, Vol. 141 (Wolstenholme, D. R., and Jeon, K. W., eds), pp. 173–216, Academic Press, San Diego, California
28. Wolstenholme, D. R., and Jeon, K. W., eds. (1992) *Mitochondrial Genomes*, Vol. 141, Academic Press, San Diego, California
28a. Wolfsberg, T. G., Schafer, S., Tatusov, R. L., and Tatusova, T. A. (2001) *Trends Biochem. Sci.* **26**, 199–203
29. Borst, P., and Grivell, L. A. (1981) *Nature (London)* **290**, 443–444
30. Janke, A., Xu, X., and Arnason, U. (1997) *Proc. Natl. Acad. Sci. U.S.A.* **94**, 1276–1281
31. Anderson, S., Bankier, A. T., Barrell, B. G., deBruijn, M. H. L., Coulson, A. R., Drouin, J., Eperon, I. C., Nierlich, D. P., Roe, B. A., Sanger, F., Schreier, P. H., Smith, A. J. H., Staden, R., and Young, I. G. (1981) *Nature (London)* **290**, 457–470
32. Anderson, S., de Bruijn, M. H. L., Coulson, A. R., Eperon, I. C., Sanger, F., and Young, I. G. (1984) *J. Mol. Biol.* **156**, 683–717
33. Cantatore, P., Roberti, M., Rainaldi, G., Gadaleta, M. N., and Saccone, C. (1989) *J. Biol. Chem.* **264**, 10965–10975
34. Gardner, M. J., and 26 other authors (1998) *Science* **282**, 1126–1132
35. Köhler, S., Delwiche, C. F., Denny, P. W., Tilney, L. G., Webster, P., Wilson, R. J. M., Palmer, J. D., and Roos, D. S. (1997) *Science* **275**, 1485–1489
36. Burger, G., Plante, I., Lonergan, K. M., and Gray, M. W. (1995) *J. Mol. Biol.* **245**, 522–537
37. Lang, B. F., Burger, G., O'Kelly, C. J., Cedergren, R., Golding, G. B., Lemieux, C., Sankoff, D., Turmel, M., and Gray, M. W. (1997) *Nature (London)* **387**, 493–497
38. de Bruijn, M. H. L. (1983) *Nature (London)* **304**, 234–241
39. Bernardi, G. (1982) *Trends Biochem. Sci.* **7**, 404–408
40. Palmer, J. D., and Shields, C. R. (1984) *Nature (London)* **307**, 437–440
41. Leblanc, C., Boyen, C., Richard, O., Bonnard, G., Grienenberger, J.-M., and Kloareg, B. (1995) *J. Mol. Biol.* **250**, 484–495
42. Oldenburg, D. J., and Bendich, A. J. (1998) *J. Mol. Biol.* **276**, 745–758
42a. Oldenburg, D. J., and Bendich, A. J. (2001) *J. Mol. Biol.* **310**, 549–562
43. Barrell, B. G., Bankier, A. T., and Drouin, J. (1979) *Nature (London)* **282**, 189–194
44. Wallace, D. C. (1982) *Microbiol. Rev.* **46**, 208–240
45. Shoffner, J. M., and Wallace, D. C. (1995) in *The Metabolic and Molecular Bases of Inherited Disease*, 7th ed., Vol. 1 (Scriver, C. R., Beaudet, A. L., Sly, W. S., and Valle, D., eds), pp. 1535–1609, McGraw-Hill, New York
46. Wallace, D. C. (1994) *Proc. Natl. Acad. Sci. U.S.A.* **91**, 8739–8746
47. Butow, R. A., Perlman, P. S., and Grossman, L. I. (1985) *Science* **228**, 1496–1501
48. Douglas, M., and Takeda, M. (1985) *Trends Biochem. Sci.* **10**, 192–194
49. Cavalier-Smith, T. (1987) *Nature (London)* **326**, 332–333
50. Schatz, G. (1997) *Nature (London)* **388**, 121–122
50a. Tokatlidis, K., and Schatz, G. (1999) *J. Biol. Chem.* **274**, 35285–35288
50b. Gabriel, K., Buchanan, S. K., and Lithgow, T. (2001) *Trends Biochem. Sci.* **26**, 36–40
51. Schatz, G., and Dobberstein, B. (1996) *Science* **271**, 1519–1526
52. Stuart, R. A., and Neupert, W. (1996) *Trends Biochem. Sci.* **21**, 261–267
53. Lithgow, T., Cuezva, J. M., and Silver, P. A. (1997) *Trends Biochem. Sci.* **22**, 110–113
54. Rojo, E. E., Guiard, B., Neupert, W., and Stuart, R. A. (1998) *J. Biol. Chem.* **273**, 8040–8047
55. Hartmann, C., Christen, P., and Jaussi, R. (1991) *Nature (London)* **352**, 762–763
56. Gärtner, F., Voos, W., Querol, A., Miller, B. R., Craig, E. A., Cumsky, M. G., and Pfanner, N. (1995) *J. Biol. Chem.* **270**, 3788–3795
57. Pfanner, N., Douglas, M. G., Endo, T., Hoogenraad, N. J., Jensen, R. E., Meijer, M., Neupert, W., Schatz, G., Schmitz, U. K., and Shore, G. C. (1996) *Trends Biochem. Sci.* **21**, 51–52
58. Komiya, T., Rospert, S., Koehler, C., Looser, R., Schatz, G., and Mihara, K. (1998) *EMBO J.* **17**, 3886–3898
59. Dietmeier, K., Hönlinger, A., Bömer, U., Dekker, P. J. T., Eckerskorn, C., Lottspeich, F., Kübrich, M., and Pfanner, N. (1997) *Nature (London)* **388**, 195–200
59a. Koehler, C. M., Leuenberger, D., Merchant, S., Renold, A., Junne, T., and Schatz, G. (1999) *Proc. Natl. Acad. Sci. U.S.A.* **96**, 2141–2146
59b. Wallace, D. C., and Murdock, D. G. (1999) *Proc. Natl. Acad. Sci. U.S.A.* **96**, 1817–1819
60. Chance, B., and Williams, G. R. (1955) *J. Biol. Chem.* **217**, 409–427
61. Prebble, J. N. (1981) *Mitochondria Chloroplasts and Bacterial Membranes*, Longman, London and New York
62. King, T. E. (1967) *Methods Enzymol.* **10**, 202–208
63. Ragen, C. I., and Racker, E. (1973) *J. Biol. Chem.* **248**, 2563–2569
64. Hafeti, Y. (1985) *Ann. Rev. Biochem.* **54**, 1015–1019
65. Yu, L., and Yu, C.-A. (1982) *J. Biol. Chem.* **257**, 2016–2021
66. Yu, C.-A., and Yu, L. (1982) *J. Biol. Chem.* **257**, 6127–6131
67. Wakabayashi, S., Takao, T., Shimonishi, Y., Kuramitsu, S., Matsubara, H., Wang, T., Zhang, Z., and King, T. E. (1985) *J. Biol. Chem.* **260**, 337–343
68. Moore, A. L., and Rich, P. R. (1980) *Trends Biochem. Sci.* **5**, 284–288
69. Palmer, J. M., and Moller, I. M. (1982) *Trends Biochem. Sci.* **7**, 258–261

References

70. Douce, R., and Day, D. A., eds. (1985) *Higher Plant Cell Respiration*, Vol. 18, Springer-Verlag, New York
71. Rhoads, D. M., and McIntosh, L. (1991) *Proc. Natl. Acad. Sci. U.S.A.* **88**, 2122–2126
72. Clarkson, A. B., JR, Bienen, E. J., Pollakis, G., and Grady, R. W. (1989) *J. Biol. Chem.* **264**, 17770–17776
73. Hoefnagel, M. H. N., Atkin, O. K., and Wiskich, J. T. (1998) *Biochim. Biophys. Acta.* **1366**, 235–255
74. Albury, M. S., Affourtit, C., and Moore, A. L. (1998) *J. Biol. Chem.* **273**, 30301–30305
75. Rhoads, D. M., Umbach, A. L., Sweet, C. R., Lennon, A. M., Rauch, G. S., and Siedow, J. N. (1998) *J. Biol. Chem.* **273**, 30750–30756
76. Anraku, Y. (1988) *Ann. Rev. Biochem.* **57**, 101–132
77. Trumpower, B. L., and Gennis, R. B. (1994) *Ann. Rev. Biochem.* **63**, 675–716
78. Yamaguchi, M., Belogrudov, G. I., and Hatefi, Y. (1998) *J. Biol. Chem.* **273**, 8094–8098
78a. Grivennikova, V. G., Kapustin, A. N., and Vinogradov, A. D. (2001) *J. Biol. Chem.* **276**, 9038–9044
79. Guénebaut, V., Schlitt, A., Weiss, H., Leonard, K., and Friedrich, T. (1998) *J. Mol. Biol.* **276**, 105–112
79a. Hellwig, P., Scheide, D., Bungert, S., Mäntele, W., and Friedrich, T. (2000) *Biochemistry* **39**, 10884–10891
80. Weidner, U., Geier, S., Ptock, A., Friedrich, T., Leif, H., and Weiss, H. (1993) *J. Mol. Biol.* **233**, 109–122
81. Schryvers, A., Lohmeier, E., and Weiner, J. H. (1978) *J. Biol. Chem.* **253**, 783–788
82. Koland, J. G., Miller, M. J., and Gennis, R. B. (1984) *Biochemistry* **23**, 445–453
83. Condon, C., Cammack, R., Patil, D. S., and Owen, P. (1985) *J. Biol. Chem.* **260**, 9427–9434
84. Berry, E. A., and Trumpower, B. L. (1985) *J. Biol. Chem.* **260**, 2458–2467
85. John, P. (1981) *Trends Biochem. Sci.* **6**, 8–10
86. Kaysser, T. M., Ghaim, J. B., Georgiou, C., and Gennis, R. B. (1995) *Biochemistry* **34**, 13491–13501
87. Anraku, Y., and Gennis, R. B. (1987) *Trends Biochem. Sci.* **12**, 262–266
88. Sun, J., Kahlow, M. A., Kaysser, T. M., Osborne, J. P., Hill, J. J., Rohlfs, R. J., Hille, R., Gennis, R. B., and Loehr, T. M. (1996) *Biochemistry* **35**, 2403–2412
88a. Zhang, J., Hellwig, P., Osborne, J. P., Huang, H.-w, Moënne-Loccoz, P., Konstantinov, A. A., and Gennis, R. B. (2001) *Biochemistry* **40**, 8548–8556
88b. Gerscher, S., Döpner, S., Hildebrandt, P., Gleissner, M., and Schäfer, G. (1996) *Biochemistry* **35**, 12796–12803
88c. Das, T. K., Gomes, C. M., Teixeira, M., and Rousseau, D. L. (1999) *Proc. Natl. Acad. Sci. U.S.A.* **96**, 9591–9596
88d. Saraste, M. (1999) *Science* **283**, 1488–1493
88e. Villani, G., Capitanio, N., Bizzoca, A., Palese, L. L., Carlino, V., Tattoli, M., Glaser, P., Danchin, A., and Papa, S. (1999) *Biochemistry* **38**, 2287–2294
89. Guénebaut, V., Vincentelli, R., Mills, D., Weiss, H., and Leonard, K. R. (1997) *J. Mol. Biol.* **265**, 409–418
89a. Schuler, F., Yano, T., Di Bernardo, S., Yagi, T., Yankovskaya, V., Singer, T. P., and Casida, J. E. (1999) *Proc. Natl. Acad. Sci. U.S.A.* **96**, 4149–4153
90. Grigorieff, N. (1998) *J. Mol. Biol.* **277**, 1033–1046
91. Di Bernardo, S., Yano, T., and Yagi, T. (2000) *Biochemistry* **39**, 9411–9418
92. Schneider, R., Brors, B., Massow, M., and Weiss, H. (1997) *FEBS Lett.* **407**, 249–252
93. Ohnishi, T., Ragan, C. I., and Hatefi, Y. (1985) *J. Biol. Chem.* **260**, 2782–2788
94. Yano, T., Sled, V. D., Ohnishi, T., and Yagi, Y. (1996) *J. Biol. Chem.* **271**, 5907–5913
95. Ragan, C. I., Galante, Y. M., Hatefi, Y., and Ohnishi, T. (1984) *Biochemistry* **21**, 590–594
95a. Kashani-Poor, N., Zwicker, K., Kerscher, S., and Brandt, U. (2001) *J. Biol. Chem.* **276**, 24082–24087
96. Heinrich, H., Azevedo, J. E., and Werner, S. (1992) *Biochemistry* **31**, 11420–11424
97. Ohshima, M., Miyoshi, H., Sakamoto, K., Takegami, K., Iwata, J., Kuwabara, K., Iwamura, H., and Yagi, T. (1998) *Biochemistry* **37**, 6436–6445
98. Kowal, A. T., Werth, M. T., Manodori, A., Cecchini, G., Schröder, I., Gunsalus, R. P., and Johnson, M. K. (1995) *Biochemistry* **34**, 12284–12293
98a. Iverson, T. M., Luna-Chavez, C., Cecchini, G., and Rees, D. C. (1999) *Science* **284**, 1961–1966
98b. Doherty, M. K., Pealing, S. L., Miles, C. S., Moysey, R., Taylor, P., Walkinshaw, M. D., Reid, G. A., and Chapman, S. K. (2000) *Biochemistry* **39**, 10695–10701
98c. Lancaster, C. R. D., Kröger, A., Auer, M., and Michel, H. (1999) *Nature (London)* **402**, 377–385
98d. Lancaster, C. R. D., Gross, R., Haas, A., Ritter, M., Mäntele, W., Simon, J., and Kröger, A. (2000) *Proc. Natl. Acad. Sci. U.S.A.* **97**, 13051–13056
98e. Matsson, M., Tolstoy, D., Aasa, R., and Hederstedt, L. (2000) *Biochemistry* **39**, 8617–8624
99. Westenberg, D. J., Gunsalus, R. P., Ackrell, B. A. C., Sices, H., and Cecchini, G. (1993) *J. Biol. Chem.* **268**, 815–822
99a. Mowat, C. G., Pankhurst, K. L., Miles, C. S., Leys, D., Walkinshaw, M. D., Reid, G. A., and Chapman, S. K. (2002) *Biochemistry* **41**, 11990–11996
100. Van Hellemond, J. J., Klockiewicz, M., Gaasenbeek, C. P. H., Roos, M. H., and Tielens, A. G. M. (1995) *J. Biol. Chem.* **270**, 31065–31070
101. Nakamura, K., Yamaki, M., Sarada, M., Nakayama, S., Vibat, C. R. T., Gennis, R. B., Nakayashiki, T., Inokuchi, H., Kojima, S., and Kita, K. (1996) *J. Biol. Chem.* **271**, 521–527
102. Vibat, C. R. T., Cecchini, G., Nakamura, K., Kita, K., and Gennis, R. B. (1998) *Biochemistry* **37**, 4148–4159
103. Yang, X., Yu, L., and Yu, C.-A. (1997) *J. Biol. Chem.* **272**, 9683–9689
104. Xia, D., Yu, C.-A., Kim, H., Xia, J.-Z., Kachurin, A. M., Zhang, L., Yu, L., and Deisenhofer, J. (1997) *Science* **277**, 60–66
105. Zhang, Z., Huang, L., Shulmeister, V. M., Chi, Y.-I., Kim, K. K., Hung, L.-W., Crofts, A. R., Berry, E. A., and Kim, S.-H. (1998) *Nature (London)* **392**, 677–684
105a. Crofts, A. R., Hong, S., Zhang, Z., and Berry, E. A. (1999) *Biochemistry* **38**, 15827–15839
106. Iwata, S., Lee, J. W., Okada, K., Lee, J. K., Iwata, M., Rasmussen, B., Link, T. A., Ramaswamy, S., and Jap, B. K. (1998) *Science* **281**, 64–71
107. Smith, J. L. (1998) *Science* **281**, 58–59
107a. Baymann, F., Robertson, D. E., Dutton, P. L., and Mäntele, W. (1999) *Biochemistry* **38**, 13188–13199
107b. Lange, C., and Hunte, C. (2002) *Proc. Natl. Acad. Sci. U.S.A.* **99**, 2800–2805
108. Braun, H.-P., and Schmitz, U. K. (1995) *Trends Biochem. Sci.* **20**, 171–175
108a. Deng, K., Shenoy, S. K., Tso, S.-C., Yu, L., and Yu, C.-A. (2001) *J. Biol. Chem.* **276**, 6499–6505
109. Saribas, A. S., Ding, H., Dutton, P. L., and Daldal, F. (1995) *Biochemistry* **34**, 16004–16012
110. Denke, E., Merbitz-Zahradnik, T., Hatzfeld, O. M., Snyder, C. H., Link, T. A., and Trumpower, B. L. (1998) *J. Biol. Chem.* **273**, 9085–9093
111. Mitchell, P. (1976) *J. Theor. Biol.* **62**, 327–367
112. Trumpower, B. L. (1990) *J. Biol. Chem.* **265**, 11409–11412
113. Orii, Y., and Miki, T. (1997) *J. Biol. Chem.* **272**, 17594–17604
113a. Bartoschek, S., Johansson, M., Geierstanger, B. H., Okun, J. G., Lancaster, C. R. D., Humpfer, E., Yu, L., Yu, C.-A., Griesinger, C., and Brandt, U. (2001) *J. Biol. Chem.* **276**, 35231–35234
114. Jünemann, S., Heathcote, P., and Rich, P. R. (1998) *J. Biol. Chem.* **273**, 21603–21607
114a. Darrouzet, E., Moser, C. C., Dutton, P. L., and Daldal, F. (2001) *Trends Biochem. Sci.* **26**, 445–451
114b. Darrouzet, E., Valkova-Valchanova, M., Moser, C. C., Dutton, P. L., and Daldal, F. (2000) *Proc. Natl. Acad. Sci. U.S.A.* **97**, 4567–4572
114c. Crofts, A. R., Barquera, B., Gennis, R. B., Kuras, R., Guergova-Kuras, M., and Berry, E. A. (1999) *Biochemistry* **38**, 15807–15826
115. Tian, H., Yu, L., Mather, M. W., and Yu, C.-A. (1998) *J. Biol. Chem.* **273**, 27953–27959
116. Meinharat, S. M., Yang, X., Trumpower, B. L., and Ohnishi, T. (1987) *J. Biol. Chem.* **262**, 8702–8706
117. Saribas, A. M., Valkova-Valchanova, M., Tokito, M. K., Zhang, Z., Berry, E. A., and Daldal, F. (1998) *Biochemistry* **37**, 8105–8114
118. Yu, J., and Le Brun, N. E. (1998) *J. Biol. Chem.* **273**, 8860–8866
119. Barber, J. (1984) *Trends Biochem. Sci.* **9**, 209–211
120. Joliot, P., and Joliot, A. (1998) *Biochem. Soc. Trans.* **37**, 10404–10410
120a. Finazzi, G. (2002) *Biochemistry* **41**, 7475–7482
121. Hochman, J., Ferguson-Miller, S., and Scindler, M. (1985) *Biochemistry* **24**, 2509–2516
122. Hackenbrock, C. R. (1981) *Trends Biochem. Sci.* **6**, 151–154
123. Bernardi, P., and Azzone, G. F. (1981) *J. Biol. Chem.* **256**, 7187–7192
124. Vincent, J. S., and Levin, I. W. (1986) *J. Am. Chem. Soc.* **108**, 3551–3554
125. Tsukihara, T., Aoyama, H., Yamashita, E., Tomizaki, T., Yamaguchi, H., Shinzawa-Itoh, K., Nakashima, R., Yaono, R., and Yoshikawa, S. (1996) *Science* **272**, 1136–1144
126. Gennis, R., and Ferguson-Miller, S. (1995) *Science* **269**, 1063–1064
127. Wallin, E., Tsukihara, T., Yoshikawa, S., Von Heijne, G., and Elofsson, A. (1997) *Protein Sci.* **6**, 808–815
128. Iwata, S., Ostermeier, C., Ludwig, B., and Michel, H. (1995) *Nature (London)* **376**, 660–669
129. Tsukihara, T., Aoyama, H., Yamashita, E., Tomizaki, T., Yamaguchi, H., Shinzawa-Itoh, K., Nakashima, R., Yaono, R., and Yoshikawa, S. (1995) *Science* **269**, 1069–1074
130. Luchinat, C., Soriano, A., Djinovic-Carugo, K., Saraste, M., Malmström, B. G., and Bertini, I. (1997) *J. Am. Chem. Soc.* **119**, 11023–11027
131. Salgado, J., Warmerdam, G., Bubacco, L., and Canters, G. W. (1998) *Biochemistry* **37**, 7378–7389
132. Blackburn, N. J., de Vries, S., Barr, M. E., Houser, R. P., Tolman, W. B., Sanders, D., and Fee, J. A. (1997) *J. Am. Chem. Soc.* **119**, 6135–6143
133. Yoshikawa, S., Shinzawa-Itoh, K., Nakashima, R., Yaono, R., Yamashita, E., Inoue, N., Yao, M., Fei, M. J., Libeu, C. P., Mizushima, T., Yamaguchi, H., Tomizaki, T., and Tsukihara, T. (1998) *Science* **280**, 1724–1729

References

134. Ostermeier, C., Harrenga, A., Ermler, U., and Michel, H. (1997) *Proc. Natl. Acad. Sci. U.S.A.* **94**, 10547–10553
135. Stubbe, J., and Riggs-Gelasco, P. (1998) *Trends Biochem. Sci.* **23**, 438–443
135a. Buse, G., Soulimane, T., Dewor, M., Meyer, H. E., and Blüggel, M. (1999) *Protein Sci.* **8**, 985–990
136. Babcock, G. T., and Wikström, M. (1992) *Nature (London)* **356**, 301–309
136a. Babcock, G. T. (1999) *Proc. Natl. Acad. Sci. U.S.A.* **96**, 12971–12973
136b. Wikström, M. (2000) *Biochemistry* **39**, 3515–3519
136c. Morgan, J. E., Verkhovsky, M. I., Palmer, G., and Wilström, M. (2001) *Biochemistry* **40**, 6882–6892
137. Gennis, R. B. (1998) *Proc. Natl. Acad. Sci. U.S.A.* **95**, 12747–12749
138. Michel, H. (1999) *Biochemistry* **38**, 15129–15140
139. Sucheta, A., Georgiadis, K. E., and Einarsdóttir, O. (1997) *Biochemistry* **36**, 554–565
140. Gibson, Q., and Greenwood, C. (1963) *Biochem. J.* **86**, 541–554
141. Greenwood, C., and Gibson, Q. H. (1967) *J. Biol. Chem.* **242**, 1782–1787
142. Fabian, M., and Palmer, G. (2001) *Biochemistry* **40**, 1867–1874
142a. Han, S., Takahashi, S., and Rousseau, D. L. (2000) *J. Biol. Chem.* **275**, 1910–1919
142b. Karpefors, M., Ädelroth, P., Namslauer, A., Zhen, Y., and Brzezinski, P. (2000) *Biochemistry* **39**, 14664–14669
142c. Capitanio, N., Capitanio, G., Minuto, M., De Nitto, E., Palese, L. L., Nicholls, P., and Papa, S. (2000) *Biochemistry* **39**, 6373–6379
143. Karpefors, M., Ädelroth, P., Zhen, Y., Ferguson-Miller, S., and Brzezinski, P. (1998) *Proc. Natl. Acad. Sci. U.S.A.* **95**, 13606–13611
143a. Rigby, S. E. J., Jünemann, S., Rich, P. R., and Heathcote, P. (2000) *Biochemistry* **39**, 5921–5928
143b. Pecoraro, C., Gennis, R. B., Vygodina, T. V., and Konstantinov, A. A. (2001) *Biochemistry* **40**, 9695–9708
144. Zaslavsky, D., Sadoski, R. C., Wang, K., Durham, B., Gennis, R. B., and Millett, F. (1998) *Biochemistry* **37**, 14910–14916
145. Nicholls, P. (1983) *Trends Biochem. Sci.* **8**, 353
146. Solioz, M. (1984) *Trends Biochem. Sci.* **9**, 309–312
147. Musser, S. M., and Chan, S. I. (1995) *Biophys. J.* **68**, 2543–2555
147a. Siletsky, S., Kaulen, A. D., and Konstantinov, A. A. (1999) *Biochemistry* **38**, 4853–4861
148. Aagaard, A., Gilderson, G., Mills, D. A., Ferguson-Miller, S., and Brzezinski, P. (2000) *Biochemistry* **39**, 15847–15850
148a. Brändén, M., Sigurdson, H., Namslauer, A., Gennis, R. B., Ädelroth, P., and Brzezinski, P. (2001) *Proc. Natl. Acad. Sci. U.S.A.* **98**, 5013–5018
149. Mitchell, D. M., Fetter, J. R., Mills, D. A., Adelroth, P., Pressler, M. A., Kim, Y., Aasa, R., Brzezinski, P., Malmström, B. G., Alben, J. O., Babcock, G. T., Ferguson-Miller, S., and Gennis, R. B. (1996) *Biochemistry* **35**, 13089–13093
150. Qian, J., Shi, W., Pressler, M., Hoganson, C., Mills, D., Babcock, G. T., and Ferguson-Miller, S. (1997) *Biochemistry* **36**, 2539–2543
151. Ädelroth, P., Gennis, R. B., and Brzezinski, P. (1998) *Biochemistry* **37**, 2470–2476
152. Konstantinov, A. A., Siletsky, S., Mitchell, D., Kaulen, A., and Gennis, R. B. (1997) *Proc. Natl. Acad. Sci. U.S.A.* **94**, 9085–9090
153. Behr, J., Hellwig, P., Mäntele, W., and Michel, H. (1998) *Biochemistry* **37**, 7400–7406
153a. Riistama, S., Laakkonen, L., Wikström, M., Verkhovsky, M. I., and Puustinen, A. (1999) *Biochemistry* **38**, 10670–10677
153b. Jünemann, S., Meunier, B., Fisher, N., and Rich, P. R. (1999) *Biochemistry* **38**, 5248–5255
154. Marantz, Y., Nachliel, E., Aagaard, A., Brzezinski, P., and Gutman, M. (1998) *Proc. Natl. Acad. Sci. U.S.A.* **95**, 8590–8595
155. Lehninger, L. A. (1972) in *International Symposium of Biochemistry and Biophysics of Mitochondrial Membranes* (Azzone, G. F., ed), p. 1, Academic Press, New York
156. Kalckar, H. M. (1969) *Biological Phosphorylations*, Prentice-Hall, Englewood Cliffs, New Jersey
157. Ernster, L. (1993) *FASEB J.* **7**, 1520–1524
158. Lemasters, J. J. (1984) *J. Biol. Chem.* **259**, 13123–13130
159. Flatt, J. P., Pahud, P., Ravussin, E., and Jéquier, E. (1984) *Trends Biochem. Sci.* **9**, 466–468
160. Hinkle, P. C., Kumar, M. A., Resetar, A., and Harris, D. L. (1991) *Biochemistry* **30**, 3576–3582
161. Ferguson, S. J. (1986) *Trends Biochem. Sci.* **11**, 351–353
162. Berry, E. A., and Hinkle, P. C. (1983) *J. Biol. Chem.* **258**, 1474–1486
163. Chance, B., and Williams, G. R. (1956) *Adv. Enzymol.* **17**, 65–134
164. Wilson, D. F., Erecinska, M., and Dutton, P. L. (1974) *Annu Rev Biophys Bioeng.* **3**, 203–230
165. Veech, R. L., Lawson, J. W. R., Cornell, N. W., and Krebs, H. A. (1979) *J. Biol. Chem.* **254**, 6538–6547
166. Lardy, H. A., and Ferguson, S. M. (1969) *Ann. Rev. Biochem.* **38**, 991–1034
167. Metzler, D. E. (1977) *Biochemistry; The Chemical Reactions of Living Cells*, Academic Press, New York (pp. 598-600)
168. Mitchell, P. (1961) *Nature (London)* **191**, 144–148
169. Mitchell, P. (1968) *Chemiosmotic Coupling and Energy Transduction*, Glynn Res., Bodmin., Cornwall, England
170. Hinkle, P. C., and McCarty, R. E. (1978) *Sci. Am.* **238**(Mar), 104–123
171. Skulachev, V. P., and Hinckle, P. C., eds. (1981) *Chemiosmotic Protein Circuits in Biological Membranes*, Addison-Wesley, Reading, Massachusetts
172. Nicholls, D. G., and Ferguson, S. J. (1992) *Bioenergetics 2*, Academic Press, London
173. Mitchell, P. (1966) *Biol. Rev. Cambridge Phil. Soc.* **41**, 445–502
174. Mitchell, P. (1979) *Science* **206**, 1148–1159
175. Mitchell, P. (1978) *Trends Biochem. Sci.* **3**, N58–N61
175a. Williams, R. J. P., and Prebble, J. (2002) *Trends Biochem. Sci.* **27**, 393–395
176. Lowe, A. G., and Jones, M. N. (1984) *Trends Biochem. Sci.* **9**, 11–13
177. Garland, P. (1978) *Nature (London)* **276**, 8–9
178. Mitchell, P., and Moyle, J. (1969) *Eur. J. Biochem.* **7**, 471–484
179. Wilson, D. F. (1980) in *Membrane Structure and Function*, Vol. 1 (Bittar, E. E., ed), Wiley, New York (pp. 153–195)
180. Waddell, W. J., and Butler, T. C. (1959) *J. Clin. Invest.* **38**, 770–
181. Dodgson, S. J., Forster, R. E., II, and Storey, B. T. (1982) *J. Biol. Chem.* **257**, 1705–1711
182. Bowman, C., and Tedeschi, H. (1979) *Nature (London)* **280**, 597–598
183. Kinnally, K. W., Tedeschi, H., and Maloff, B. L. (1978) *Biochemistry* **17**, 3419–3427
184. Tedeschi, H. (1980) *Trends Biochem. Sci.* **5**, VIII–IX
185. Jagendorf, A. T. (1975) in *Bioenergetics of Photosynthesis* (Govindjee, ed), Academic Press, New York (pp. 413–492)
186. Jagendorf, A. T., and Uribe, E. (1966) *Proc. Natl. Acad. Sci. U.S.A.* **55**, 170–177
187. Richard, P., Pitard, B., and Rigaud, J.-L. (1995) *J. Biol. Chem.* **270**, 21571–21578
188. van Walraven, H. S., Strotmann, H., Schwarz, O., and Rumberg, B. (1996) *FEBS Lett.* **379**, 309–313
189. Costa, L. E., Reynafarje, B., and Lehninger, A. L. (1984) *J. Biol. Chem.* **259**, 4802–4811
190. Reynafarje, B., Costa, L. E., and Lehninger, A. L. (1986) *J. Biol. Chem.* **261**, 8254–8262
191. Capitanio, N., Capitanio, G., Demarinis, D. A., De Nitto, E., Massari, S., and Papa, S. (1996) *Biochemistry* **35**, 10800–10806
192. Wang, Y., Howton, M. M., and Beattie, D. S. (1995) *Biochemistry* **34**, 7476–7482
193. Wikström, M. (1989) *Nature (London)* **338**, 776–778
194. Wikström, M., Bogacher, A., Finel, M., Morgan, J. E., Puustinen, M., Raitio, M. L., Verkhovskaya, M. L., and Verkhovsky, M. I. (1994) *Biochim. Biophys. Acta.* **1187**, 106–111
195. Williams, R. J. P. (1988) *Ann. Rev. Biophys. Biophys. Chem.* **17**, 71–97
196. Penefsky, H. S., Pullman, M. E., Datta, A., and Racker, E. (1960) *J. Biol. Chem.* **235**, 3330–3336
197. Racker, E. (1976) *A New Look at Mechanisms in Bioenergetics*, Academic Press, New York
198. Dschida, W. J., and Bowman, B. J. (1992) *J. Biol. Chem.* **267**, 18783–18789
199. Abrahams, J. P., Leslie, A. G. W., Lutter, R., and Walker, J. E. (1994) *Nature (London)* **370**, 621–628
200. Elston, T., Wang, H., and Oster, G. (1998) *Nature (London)* **391**, 510–513
201. Zhou, Y., Duncan, T. M., and Cross, R. L. (1997) *Proc. Natl. Acad. Sci. U.S.A.* **94**, 10583–10587
202. Junge, W., Lill, H., and Engelbrecht, S. (1997) *Trends Biochem. Sci.* **22**, 420–423
203. Engelbrecht, S., and Junge, W. (1997) *FEBS Letters* **414**, 485–491
204. Amzel, L. M., and Pedersen, P. L. (1983) *Ann. Rev. Biochem.* **52**, 801–824
205. Hausrath, A. C., Grüber, G., Matthews, B. W., and Capaldi, R. A. (1999) *Proc. Natl. Acad. Sci. U.S.A.* **96**, 13697–13702
206. Matsui, T., Muneyuki, E., Honda, M., Allison, W. S., Dou, C., and Yoshida, M. (1997) *J. Biol. Chem.* **272**, 8215–8221
207. Shirakihara, Y., Leslie, A. G. W., Abrahams, J. P., Walker, J. E., Ueda, T., Sekimoto, Y., Kambara, M., Saika, K., Kagawa, Y., and Yoshida, M. (1997) *Structure* **5**, 825–836
208. Spannagel, C., Vaillier, J., Chaignepain, S., and Velours, J. (1998) *Biochemistry* **37**, 615–621
208a. Arnold, I., Pfeiffer, K., Neupert, W., Stuart, R. A., and Schägger, H. (1999) *J. Biol. Chem.* **274**, 36–40
208b. Soubannier, V., Rusconi, F., Vaillier, J., Arselin, G., Chaignepain, S., Graves, P.-V., Schmitter, J.-M., Zhang, J. L., Mueller, D., and Velours, J. (1999) *Biochemistry* **38**, 15017–15024
209. Collinson, I. R., Runswick, M. J., Buchanan, S. K., Fearnley, I. M., Skehel, J. M., van Raaij, M. J., Griffiths, D. E., and Walker, J. E. (1994) *Biochemistry* **33**, 7971–7978
210. Pedersen, P. L., and Amzel, L. M. (1993) *J. Biol. Chem.* **268**, 9937–9940
210a. Hee Ko, Y., Hullihen, J., Hong, S., and Pedersen, P. L. (2000) *J. Biol. Chem.* **275**, 32931–32939
210b. Karrasch, S., and Walker, J. E. (1999) *J. Mol. Biol.* **290**, 379 – 384
211. Bianchet, M. A., Hullihen, J., Pedersen, P. L., and Amzel, L. M. (1998) *Proc. Natl. Acad. Sci. U.S.A.* **95**, 11065–11070

References

212. Possmayer, F. E., and Gräber, P. (1994) *J. Biol. Chem.* **269**, 1896–1904
213. Sokolov, M., and Gromet-Elhanan, Z. (1996) *Biochemistry* **35**, 1242–1248
213a. Groth, G., and Pohl, E. (2001) *J. Biol. Chem.* **276**, 1345–1352
214. Hammes, G. G. (1983) *Trends Biochem. Sci.* **8**, 131–134
214a. Noji, H., and Yoshida, M. (2001) *J. Biol. Chem.* **276**, 1665–1668
214b. Schnitzer, M. J. (2001) *Nature (London)* **410**, 878–881
214c. Stock, D., Leslie, A. G. W., and Walker, J. E. (1999) *Science* **286**, 1700–1705
215. Belogrudov, G. I., Tomich, J. M., and Hatefi, Y. (1996) *J. Biol. Chem.* **271**, 20340–20345
216. Collinson, I. R., van Raaij, M. J., Runswick, M. J., Fearnley, I. M., Skehel, J. M., Orriss, G. L., Miroux, B., and Walker, J. E. (1994) *J. Mol. Biol.* **242**, 408–421
217. Golden, T. R., and Pedersen, P. L. (1998) *Biochemistry* **37**, 13871–13881
218. Joshi, S., Cao, G. J., Nath, C., and Shah, J. (1997) *Biochemistry* **36**, 10936–10943
219. Uhlin, U., Cox, G. B., and Guss, J. M. (1997) *Structure* **5**, 1219–1230
220. Gabellieri, E., Strambini, G. B., Baracca, A., and Solaini, G. (1997) *Biophys. J.* **72**, 1818–1827
221. Gordon-Smith, D. J., Carbajo, R. J., Yang, J.-C., Videler, H., Runswick, M. J., Walker, J. E., and Neuhaus, D. (2001) *J. Mol. Biol.* **308**, 325–339
222. Capaldi, R. A., Aggeler, R., Turina, P., and Wilkens, S. (1994) *Trends Biochem. Sci.* **19**, 284–289
222a. Capaldi, R. A., and Aggeler, R. (2002) *Trends Biochem. Sci.* **27**, 154–160
223. van Raaij, M. J., Abrahams, J. P., Leslie, A. G. W., and Walker, J. E. (1996) *Proc. Natl. Acad. Sci. U.S.A.* **93**, 6913–6917
224. Abrahams, J. P., Buchanan, S. K., van Raaij, M. J., Fearnley, I. M., Leslie, A. G. W., and Walker, J. E. (1996) *Proc. Natl. Acad. Sci. U.S.A.* **93**, 9420–9424
225. Wilkens, S., and Capaldi, R. A. (1998) *J. Biol. Chem.* **273**, 26645–26651
226. Bulygin, V. V., Duncan, T. M., and Cross, R. L. (1998) *J. Biol. Chem.* **273**, 31765–31769
227. Pan, W., Ko, Y. H., and Pedersen, P. L. (1998) *Biochemistry* **37**, 6911–6923
227a. Tsunoda, S. P., Rodgers, A. J. W., Aggeler, R., Wilce, M. C. J., Yoshida, M., and Capaldi, R. A. (2001) *Proc. Natl. Acad. Sci. U.S.A.* **98**, 6560–6564
228. Rodgers, A. J. W., Wilkens, S., Aggeler, R., Morris, M. B., Howitt, S. M., and Capaldi, R. A. (1997) *J. Biol. Chem.* **272**, 31058–31064
229. Sawada, K., Kuroda, N., Watanabe, H., Moritani-Otsuka, C., and Kanazawa, H. (1997) *J. Biol. Chem.* **272**, 30047–30053
230. Böttcher, B., Schwarz, L., and Gräber, P. (1998) *J. Mol. Biol.* **281**, 757–762
230a. McLachlin, D. T., Coveny, A. M., Clark, S. M., and Dunn, S. D. (2000) *J. Biol. Chem.* **275**, 17571–17577
230b. Wilkens, S., Zhou, J., Nakayama, R., Dunn, S. D., and Capaldi, R. A. (2000) *J. Mol. Biol.* **295**, 387–391
231. Long, J. C., Wang, S., and Vik, S. B. (1998) *J. Biol. Chem.* **273**, 16235–16240
232. Jiang, W., and Fillingame, R. H. (1998) *Proc. Natl. Acad. Sci. U.S.A.* **95**, 6607–6612
233. Girvin, M. E., Rastogi, V. K., Abildgaard, F., Markley, J. L., and Fillingame, R. H. (1998) *Biochemistry* **37**, 8817–8824
234. Jones, P. C., Jiang, W., and Fillingame, R. H. (1998) *J. Biol. Chem.* **273**, 17178–17185
235. Groth, G., Tilg, Y., and Schirwitz, K. (1998) *J. Mol. Biol.* **281**, 49–59
236. Jones, P. C., and Fillingame, R. H. (1998) *J. Biol. Chem.* **273**, 29701–29705
236a. Dmitriev, O. Y., Abildgaard, F., Markley, J. L., and Fillingame, R. H. (2002) *Biochemistry* **41**, 5537–5547
237. Zhang, Y., and Fillingame, R. H. (1995) *J. Biol. Chem.* **270**, 87–93
238. Kaim, G., Matthey, U., and Dimroth, P. (1998) *EMBO J.* **17**, 688–695
239. Kaim, G., Wehrle, F., Gerike, U., and Dimroth, P. (1997) *Biochemistry* **36**, 9185–9194
240. Kaim, G., and Dimroth, P. (1998) *EMBO J.* **17**, 5887–5895
240a. Kaim, G., and Dimroth, P. (1999) *EMBO J.* **18**, 4118–4127
240b. Dimroth, P., Wang, H., Grabe, M., and Oster, G. (1999) *Proc. Natl. Acad. Sci. U.S.A.* **96**, 4924–4929
240c. Aufurth, S., Schägger, H., and Müller, V. (2000) *J. Biol. Chem.* **275**, 33297–33301
241. Vik, S. B., and Antonio, B. J. (1994) *J. Biol. Chem.* **269**, 30364–30369
241a. Nijtmans, L. G. J., Henderson, N. S., Attardi, G., and Holt, I. J. (2001) *J. Biol. Chem.* **276**, 6755–6762
242. Sorgen, P. L., Bubb, M. R., McCormick, K. A., Edison, A. S., and Cain, B. D. (1998) *Biochemistry* **37**, 923–932
243. Dunn, S. D., and Chandler, J. (1998) *J. Biol. Chem.* **273**, 8646–8651
244. McLachlin, D. T., Bestard, J. A., and Dunn, S. D. (1998) *J. Biol. Chem.* **273**, 15162–15168
245. Boyer, P. D. (1993) *Biochim. Biophys. Acta.* **1140**, 215–250
246. Zhou, J.-M., and Boyer, P. D. (1993) *J. Biol. Chem.* **268**, 1531–1538
247. Boyer, P. D. (1995) *FASEB J.* **9**, 559–561
248. Boyer, P. D. (1997) *Ann. Rev. Biochem.* **66**, 717–749
249. Smith, L. T., Rosen, G., and Boyer, P. D. (1983) *J. Biol. Chem.* **258**, 10887–10894
250. Souid, A.-K., and Penefsky, H. S. (1995) *J. Biol. Chem.* **270**, 9074–9082
251. Penefsky, H. S., and Cross, R. L. (1991) *Adv. Enzymol.* **64**, 173–214
252. Cox, G. B., Jans, D. A., Fimmel, A. L., Gibson, F., and Hatch, L. (1984) *Biochim. Biophys. Acta.* **768**, 201–208
253. Aggeler, R., Ogilvie, I., and Capaldi, R. A. (1997) *J. Biol. Chem.* **272**, 19621–19624
254. Boekema, E. J., Ubbink-Kok, T., Lolkema, J. S., Brisson, A., and Konings, W. N. (1997) *Proc. Natl. Acad. Sci. U.S.A.* **94**, 14291–14293
255. Sabbert, D., Engelbrecht, S., and Junge, W. (1996) *Nature (London)* **381**, 623–625
256. Sabbert, D., Engelbrecht, S., and Junge, W. (1997) *Proc. Natl. Acad. Sci. U.S.A.* **94**, 4401–4405
257. Noji, H. (1998) *Science* **282**, 1844–1845
258. Noji, H., Yasuda, R., Yoshida, M., and Kinosita, K., Jr. (1997) *Nature (London)* **386**, 299–302
259. Kato-Yamada, Y., Noji, H., Yasuda, R., Kinosita, K., Jr., and Yoshida, M. (1998) *J. Biol. Chem.* **273**, 19375–19377
260. Berg, H. C. (1998) *Nature (London)* **394**, 324–325
261. Yasuda, R., Noji, H., Kinosita, K. J., Jr., and Yoshida, M. (1998) *Cell* **93**, 1117–1124
262. Kinosita, K. J., Jr., Yasuda, R., Noji, H., Ishiwata, S., and Yoshida, S. (1998) *Cell* **93**, 21–24
262a. Yasuda, R., Noji, H., Yoshida, M., Kinosita, K., Jr., and Itoh, H. (2001) *Nature (London)* **410**, 898–904
262b. Sambongi, Y., Iko, Y., Tanabe, M., Omote, H., Iwamoto-Kihara, A., Ueda, I., Yanagida, T., Wada, Y., and Futai, M. (1999) *Science* **286**, 1722–1724
262c. Hutcheon, M. L., Duncan, T. M., Ngai, H., and Cross, R. L. (2001) *Proc. Natl. Acad. Sci. U.S.A.* **98**, 8519–8524
262d. Nishio, K., Iwamoto-Kihara, A., Yamamoto, A., Wada, Y., and Futai, M. (2002) *Proc. Natl. Acad. Sci. U.S.A.* **99**, 13448–13452
263. Ko, Y. H., Bianchet, M., Amzel, L. M., and Pedersen, P. L. (1997) *J. Biol. Chem.* **272**, 18875–18881
264. Dou, C., Grodsky, N. B., Matsui, T., Yoshida, M., and Allison, W. S. (1997) *Biochemistry* **36**, 3719–3727
265. Grodsky, N. B., Dou, C., and Allison, W. S. (1998) *Biochemistry* **37**, 1007–1014
266. Dou, C., Fortes, P. A. G., and Allison, W. S. (1998) *Biochemistry* **37**, 16757–16764
267. Wang, H., and Oster, G. (1998) *Nature (London)* **396**, 279–282
267a. Rastogi, V. K., and Girvin, M. E. (1999) *Nature (London)* **402**, 263–268
267b. Jones, P. C., Hermolin, J., Jiang, W., and Fillingame, R. H. (2000) *J. Biol. Chem.* **275**, 31340–31346
267c. Le, N. P., Omote, H., Wada, Y., Al-Shawi, M. K., Nakamoto, R. K., and Futai, M. (2000) *Biochemistry* **39**, 2778–2783
267d. Ko, Y. H., Hong, S., and Pedersen, P. L. (1999) *J. Biol. Chem.* **274**, 28853–28856
268. Kobayashi, H., Suzuki, T., and Unemoto, T. (1986) *J. Biol. Chem.* **261**, 627–630
269. Forgac, M. (1999) *J. Biol. Chem.* **274**, 12951–12954
269a. Oka, T., Toyomura, T., Honjo, K., Wada, Y., and Futai, M. (2001) *J. Biol. Chem.* **276**, 33079–33085
269b. Müller, M. L., Jensen, M., and Taiz, L. (1999) *J. Biol. Chem.* **274**, 10706–10716
269c. Kawamura, Y., Arakawa, K., Maeshima, M., and Yoshida, S. (2000) *J. Biol. Chem.* **275**, 6515–6522
270. Oka, T., Yamamoto, R., and Futai, M. (1998) *J. Biol. Chem.* **273**, 22570–22576
271. van Hille, B., Richener, H., Evans, D. B., Green, J. R., and Bilbe, G. (1993) *J. Biol. Chem.* **268**, 7075–7080
272. Nelson, H., Mandiyan, S., and Nelson, N. (1994) *J. Biol. Chem.* **269**, 24150–24155
272a. Wilkens, S., Vasilyeva, E., and Forgac, M. (1999) *J. Biol. Chem.* **274**, 31804–31810
272b. Boekema, E. J., van Breemen, J. F. L., Brisson, A., Ubbink-Kok, T., Konings, W. N., and Lolkema, J. S. (1999) *Nature (London)* **401**, 37–38
272c. Sagermann, M., Stevens, T. H., and Matthews, B. W. (2001) *Proc. Natl. Acad. Sci. U.S.A.* **98**, 7134–7139
273. Steinert, K., Wagner, V., Kroth-Pancic, P. G., and Bickel-Sandkötter, S. (1997) *J. Biol. Chem.* **272**, 6261–6269
273a. Yokoyama, K., Muneyuki, E., Amano, T., Mizutani, S., Yoshida, M., Ishida, M., and Ohkuma, S. (1998) *J. Biol. Chem.* **273**, 20504–20510
274. Zhen, R.-G., Kim, E. J., and Rea, P. A. (1997) *J. Biol. Chem.* **272**, 22340–22348
274a. Scott, D. A., and Docampo, R. (2000) *J. Biol. Chem.* **275**, 24215–24221
274b. Miranda, M., Allen, K. E., Pardo, J. P., and Slayman, C. W. (2001) *J. Biol. Chem.* **276**, 22485–22490
275. Pederson, P. L., and Carafoli, E. (1987) *Trends Biochem. Sci.* **12**, 146–150
275a. Kühlbrandt, W., Zeelen, J., and Dietrich, J. (2002) *Science* **297**, 1692–1696
276. Lambrecht, N., Corbett, Z., Bayle, D., Karlish, S. J. D., and Sachs, G. (1998) *J. Biol. Chem.* **273**, 13719–13728
277. Asano, S., Tega, Y., Konishi, K., Fujioka, M., and Takeguchi, N. (1996) *J. Biol. Chem.* **271**, 2740–2745
278. Melle-Milovanovic, D., Milovanovic, M., Nagpal, S., Sachs, G., and Shin, J. M. (1998) *J. Biol. Chem.* **273**, 11075–11081

References

279. Wolfe, M. M., and Soll, A. H. (1988) *N. Engl. J. Med.* **319**, 1707–1715
280. Priver, N. A., Rabon, E. C., and Zeidel, M. L. (1993) *Biochemistry* **32**, 2459–2468
281. Nicholls, D. G., and Ferguson, S. J. (1992) *Bioenergetics 2*, Academic Press, London (p. 139)
282. Levings, C. S., III. (1990) *Science* **250**, 942–947
283. Hatefi, Y., and Yamaguchi, M. (1996) *FASEB J.* **10**, 444–452
284. Glavas, N. A., Hou, C., and Bragg, P. D. (1995) *Biochemistry* **34**, 7694–7702
285. Meuller, J., and Rydström, J. (1999) *J. Biol. Chem.* **274**, 19072–19080
286. Grimley, R. L., Quirk, P. G., Bizouarn, T., Thomas, C. M., and Jackson, J. B. (1997) *Biochemistry* **36**, 14762–14770
286a. Venning, J. D., Rodrigues, D. J., Weston, C. J., Cotton, N. P. J., Quirk, P. G., Errington, N., Finet, S., White, S. A., and Jackson, J. B. (2001) *J. Biol. Chem.* **276**, 30678–30685
287. Fjellström, O., Axelsson, M., Bizouarn, T., Hu, X., Johansson, C., Meuller, J., and Rydström, J. (1999) *J. Biol. Chem.* **274**, 6350–6359
288. Lee, A.-C., Zizi, M., and Colombini, M. (1994) *J. Biol. Chem.* **269**, 30974–30980
289. Rostovtseva, T., and Colombini, M. (1996) *J. Biol. Chem.* **271**, 28006–28008
290. Rostovtseva, T., and Colombini, M. (1997) *Biophys. J.* **72**, 1954–1962
291. Beavis, A. D., Lu, Y., and Garlid, K. D. (1993) *J. Biol. Chem.* **268**, 997–1004
292. Li, W., Shariat-Madar, Z., Powers, M., Sun, X., Lane, R. D., and Garlid, K. D. (1992) *J. Biol. Chem.* **267**, 17983–17989
293. Cox, D. A., and Matlib, M. A. (1993) *J. Biol. Chem.* **268**, 938–947
294. Sparagna, G. C., Gunter, K. K., Sheu, S.-S., and Gunter, T. E. (1995) *J. Biol. Chem.* **270**, 27510–27515
295. Bisaccia, F., Capobianco, L., Brandolin, G., and Palmieri, F. (1994) *Biochemistry* **33**, 3705–3713
296. Fiermonte, G., Palmieri, L., Dolce, V., Lasorsa, F. M., Palmieri, F., Runswick, M. J., and Walker, J. E. (1998) *J. Biol. Chem.* **273**, 24754–24759
297. Nelson, D. R., Felix, C. M., and Swanson, J. M. (1998) *J. Mol. Biol.* **277**, 285–308
298. Kaplan, R. S., Mayor, J. A., and Wood, D. O. (1993) *J. Biol. Chem.* **268**, 13682–13690
298a. Xu, Y., Kakhniashvili, D. A., Gremse, D. A., Wood, D. O., Mayor, J. A., Walters, D. E., and Kaplan, R. S. (2000) *J. Biol. Chem.* **275**, 7117–7124
298b. Bandell, M., and Lolkema, J. S. (2000) *J. Biol. Chem.* **275**, 39130–39136
299. Müller, V., Basset, G., Nelson, D. R., and Klingenberg, M. (1996) *Biochemistry* **35**, 16132–16143
300. Broustovetsky, N., Bamberg, E., Gropp, T., and Klingenberg, M. (1997) *Biochemistry* **36**, 13865–13872
300a. Wiedemann, N., Pfanner, N., and Ryan, M. T. (2001) *EMBO J.* **20**, 951–960
300b. Heimpel, S., Basset, G., Odoy, S., and Klingenberg, M. (2001) *J. Biol. Chem.* **276**, 11499–11506
301. Beyer, K., and Klingenberg, M. (1985) *Biochemistry* **24**, 3821–3826
302. Muller, M., Krebs, J. J. R., Cherry, R. J., and Kawato, S. (1984) *J. Biol. Chem.* **259**, 3037–3043
303. Ziegler, M., and Penefsky, H. S. (1993) *J. Biol. Chem.* **268**, 25320 – 25328
304. Schroers, A., Burkovski, A., Wohlrab, H., and Krämer, R. (1998) *J. Biol. Chem.* **273**, 14269–14276
305. Briggs, C., Mincone, L., and Wohlrab, H. (1999) *Biochemistry* **38**, 5096–5102
305a. Majima, E., Ishida, M., Miki, S., Shinohara, Y., and Terada, H. (2001) *J. Biol. Chem.* **276**, 9792–9799
306. Kaplan, R. S., Pratt, R. D., and Pederson, P. L. (1986) *J. Biol. Chem.* **261**, 12767–12773
307. Nicolli, A., Petronilli, V., and Bernardi, P. (1993) *Biochemistry* **32**, 4461–4465
307a. Scorrano, L., Petronilli, V., Colonna, R., Di Lisa, F., and Bernardi, P. (1999) *J. Biol. Chem.* **274**, 24657–24663
308. Liu, G., Hinch, B., Davatol-Hag, H., Lu, Y., Powers, M., and Beavis, A. D. (1996) *J. Biol. Chem.* **271**, 19717–19723
309. Sugano, T., Handler, J. A., Yoshihara, H., Kizaki, Z., and Thurman, R. G. (1990) *J. Biol. Chem.* **265**, 21549–21553
310. Schlegel, H. G., and Eberhardt, U. (1972) *Adv. Microbiol. Physiol.* **7**, 205–242
310a. Pierik, A. J., Roseboom, W., Happe, R. P., Bagley, K. A., and Albracht, S. P. J. (1999) *J. Biol. Chem.* **274**, 3331–3337
311. Bothe, H., and Trebst, A., eds. (1981) *Biology of Inorganic Nitrogen and Sulfur*, Springer, Berlin
312. Arciero, D. M., Golombek, A., Hendrich, M. P., and Hooper, A. B. (1998) *Biochemistry* **37**, 523–529
313. Igarashi, N., Moriyama, H., Fujiwara, T., Fukumori, Y., and Tanaka, N. (1997) *Nature Struct. Biol.* **4**, 276–284
313a. Schalk, J., de Vries, S., Kuenen, J. G., and Jetten, M. S. M. (2000) *Biochemistry* **39**, 5405–5412
314. Hendrich, M. P., Petasis, D., Arciero, D. M., and Hooper, A. B. (2001) *J. Am. Chem. Soc.* **123**, 2997–3005
314a. Hendrich, M. P., Upadhyay, A. K., Riga, J., Arciero, D. M., and Hooper, A. B. (2002) *Biochemistry* **41**, 4603–4611
315. Blaut, M., and Gottschalk, G. (1997) in *Bioenergetics* (Gräber, P., and Milazzo, G., eds), pp. 139–211, Birkhäuser Verlag, Basel
316. DiSpirito, A. A., and Hooper, A. B. (1986) *J. Biol. Chem.* **261**, 10534–10537
317. Logan, M. S. P., and Hooper, A. B. (1995) *Biochemistry* **34**, 9257–9264
318. Taylor, C. D., and Wirsen, C. O. (1997) *Science* **277**, 1483–1485
319. Felbeck, H., and Somero, G. N. (1982) *Trends Biochem. Sci.* **7**, 201–204
320. Gaill, F. (1993) *FASEB J.* **7**, 558–565
321. Trudinger, P. A. (1969) *Ad. Microbiol. Physiol.* **3**, 111–158
322. Roy, A. B., and Trudinger, P. A. (1970) *The Biochemistry of Inorganic Compounds of Sulfur*, Cambridge Univ. Press, London and New York
322a. Cheesman, M. R., Little, P. J., and Berks, B. C. (2001) *Biochemistry* **40**, 10562–10569
323. Postgate, J. R., and Kelly, D. P., eds. (1982) *Sulphur Bacteria*, The Royal Society,
323a. Griesbeck, C., Schütz, M., Schödl, T., Bathe, S., Nausch, L., Mederer, N., Vielreicher, M., and Hauska, G. (2002) *Biochemistry* **41**, 11552–11565
324. Chen, Z.-w, Koh, M., Van Driessche, G., Van Beeumen, J. J., Bartsch, R. G., Meyer, T. E., Cusanovich, M. A., and Mathews, F. S. (1994) *Science* **266**, 430–432
324a. Edwards, K. J., Bond, P. L., Gihring, T. M., and Banfield, J. F. (2000) *Science* **287**, 1796–1799
325. Hodgman, C. D., ed. (1967–1968) *CRC Handbook of Chemistry and Physics*, 48th ed., Chem. Rubber Publ. Co., Cleveland, Ohio
326. Dugan, P. R. (1972) *Biochemical Ecology of Water Pollution*, Plenum, New York
327. Powell, M. A., and Somero, G. N. (1986) *Science* **233**, 563–566
328. Rothery, R. A., Blasco, F., and Weiner, J. H. (2001) *Biochemistry* **40**, 5260–5268
328a. Anderson, L. J., Richardson, D. J., and Butt, J. N. (2001) *Biochemistry* **40**, 11294–11307
329. Magalon, A., Asso, M., Guigliarelli, B., Rothery, R. A., Bertrand, P., Giordano, G., and Blasco, F. (1998) *Biochemistry* **37**, 7363–7370
329a. Richardson, D., and Sawers, G. (2002) *Science* **295**, 1842–1843
329b. Jormakka, M., Törnroth, S., Byrne, B., and Iwata, S. (2002) *Science* **295**, 1863–1868
330. Iuchi, S., and Lin, E. C. C. (1987) *Proc. Natl. Acad. Sci. U.S.A.* **84**, 3901–3905
331. Spiro, S., and Guest, J. R. (1991) *Trends Biochem. Sci.* **16**, 310–314
332. Cecchini, G., Thompson, C. R., Ackrell, B. A. C., Westenberg, D. J., Dean, N., and Gunsalus, R. P. (1986) *Proc. Natl. Acad. Sci. U.S.A.* **83**, 8898–8902
332a. Turner, K. L., Doherty, M. K., Heering, H. A., Armstrong, F. A., Reid, G. A., and Chapman, S. K. (1999) *Biochemistry* **38**, 3302–3309
333. Czjzek, M., Dos Santos, J.-P., Pommier, J., Giordano, G., Méjean, V., and Haser, R. (1998) *J. Mol. Biol.* **284**, 435–447
334. Trieber, C. A., Rothery, R. A., and Weiner, J. H. (1994) *J. Biol. Chem.* **269**, 7103–7109
335. Rothery, R. A., and Weiner, J. H. (1996) *Biochemistry* **35**, 3247–3257
336. Costa, C., Moura, J. J. G., Moura, I., Liu, M. Y., Peck, H. D., Jr., LeGall, J., Wang, Y., and Huynh, B. H. (1990) *J. Biol. Chem.* **265**, 14382–14387
337. Blackmore, R. S., Gibson, Q. H., and Greenwood, C. (1992) *J. Biol. Chem.* **267**, 10950–10955
338. Ferguson, S. J. (1987) *Trends Biochem. Sci.* **12**, 353–357
339. Wang, Y., and Averill, B. A. (1996) *J. Am. Chem. Soc.* **118**, 3972–3973
339a. Suharti, Strampraad, M. J. F., Schröder, I., and de Vries, S. (2001) *Biochemistry* **40**, 2632–2639
340. Nurizzo, D., Cutruzzolà, F., Arese, M., Bourgeois, D., Brunori, M., Cambillau, C., and Tegoni, M. (1998) *Biochemistry* **37**, 13987–13996
341. Kukimoto, M., Nishiyama, M., Tanokura, M., Adman, E. T., and Horinouchi, S. (1996) *J. Biol. Chem.* **271**, 13680–13683
342. Murphy, M. E. P., Turley, S., and Adman, E. T. (1997) *J. Biol. Chem.* **272**, 28455–28460
342a. Peters-Libeu, C. A., Kukimoto, M., Nishiyama, M., Horinouchi, S., Adman, E. T. (1997) *Biochemistry* **36**, 13160 – 13179
342b. Inoue, T., Nishio, N., Suzuki, S., Kataoka, K., Kohzuma, T., and Kai, Y. (1999) *J. Biol. Chem.* **274**, 17845 – 17852
343. Weeg-Aerssens, E., Wu, W., Ye, R. W., Tiedje, J. M., and Chang, C. K. (1991) *J. Biol. Chem.* **266**, 7496–7502
344. Williams, P. A., Fülöp, V., Garman, E. F., Saunders, N. F. W., Ferguson, S. J., and Hajdu, J. (1997) *Nature (London)* **389**, 406–412
345. Baker, S. C., Saunders, N. F. W., Willis, A. C., Ferguson, S. J., Hajdu, J., and Fülöp, V. (1997) *J. Mol. Biol.* **269**, 440–455
346. Nurizzo, D., Silvestrini, M.-C., Mathieu, M., Cutruzzolà, F., Bourgeois, D., Fülöp, V., Hajdu, J., Brunori, M., Tegoni, M., and Cambillau, C. (1997) *Structure* **5**, 1157–1171
346a. Cutruzzolà, F., Brown, K., Wilson, E. K., Bellelli, A., Arese, M., Tegoni, M., Cambillau, C., and Brunori, M. (2001) *Proc. Natl. Acad. Sci. U.S.A.* **98**, 2232–2237
346b. Sjögren, T., Svensson-EK, M., Hajdu, J., and Brzezinki, P. (2000) *Biochemistry* **39**, 10967–10974
346c. Einsle, O., Messerschmidt, A., Stach, P., Bourenkov, G. P., Bartunik, H. D., Huber, R., and Kroneck, P. M. H. (1999) *Nature (London)* **400**, 476–480
346d. Einsle, O., Stach, P., Messerschmidt, A., Simon, J., Kröger, A., Huber, R., and Kroneck, P. M. H. (2000) *J. Biol. Chem.* **275**, 39608–39616
346e. Sjögren, T., and Hajdu, J. (2001) *J. Biol. Chem.* **276**, 29450–29455
346f. Ranghino, G., Scorza, E., Sjögren, T., Williams, P. A., Ricci, M., and Hajdu, J. (2000) *Biochemistry* **39**, 10958–10966

References

346g. Kobayashi, K., Koppenhöfer, A., Ferguson, S. J., Watmough, N. J., and Tagawa, S. (2001) *Biochemistry* **40**, 8542–8547

347. Kobayashi, K., Koppenhöfer, A., Ferguson, S. J., and Tagawa, S. (1997) *Biochemistry* **36**, 13611–13616

348. Cheesman, M. R., Ferguson, S. J., Moir, J. W. B., Richardson, D. J., Zumft, W. G., and Thomson, A. J. (1997) *Biochemistry* **36**, 16267–16276

349. Sakurai, N., and Sakurai, T. (1997) *Biochemistry* **36**, 13809–13815

350. Hendriks, J., Warne, A., Gohlke, U., Haltia, T., Ludovici, C., Lübben, M., and Saraste, M. (1998) *Biochemistry* **37**, 13102–13109

351. Moënne-Loccoz, P., and de Vries, S. (1998) *J. Am. Chem. Soc.* **120**, 5147–5152

352. Shiro, Y., Fujii, M., Isogai, Y., Adachi, S.-i, Iizuka, T., Obayashi, E., Makino, R., Nakahara, K., and Shoun, H. (1995) *Biochemistry* **34**, 9052–9058

353. Neese, F., Zumft, W. G., Antholine, W. E., and Kroneck, P. M. H. (1996) *J. Am. Chem. Soc.* **118**, 8692–8699

354. Farrar, J. A., Zumft, W. G., and Thomson, A. J. (1998) *Proc. Natl. Acad. Sci. U.S.A.* **95**, 9891–9896

355. Postgate, J. R., FRS. (1984) in *The Sulfate-Reducing Bacteria*, 2nd ed., pp. 56–63, Cambridge University Press, Cambridge

356. Postgate, J. R. (1984) *The Sulfate-reducing Bacteria*, 2nd ed., Cambridge Univ. Press, London

356a. Michaelis, W., and 16 other authors. (2002) *Science* **297**, 1013–1015

357. Grey, D. C., and Jensen, M. L. (1972) *Science* **177**, 1099–1100

358. Gavel, O. Y., Bursakov, S. A., Calvete, J. J., George, G. N., Moura, J. J. G., and Moura, I. (1998) *Biochemistry* **37**, 16225–16232

359. Klenk, H.-P., and 50 other authors. (1997) *Nature (London)* **390**, 364–370

360. Morais, J., Palma, P. N., Frazao, C., Caldeira, J., LeGall, J., Moura, I., Moura, J. J. G., and Carrondo, M. A. (1995) *Biochemistry* **34**, 12830–12841

361. Louro, R. O., Catarino, T., Turner, D. L., Picarra-Pereira, M. A., Pacheco, I., LeGall, J., and Xavier, A. V. (1998) *Biochemistry* **37**, 15808–15815

361a. Fritz, G., Griesshaber, D., Seth, O., and Kroneck, P. M. H. (2001) *Biochemistry* **40**, 1317–1324

362. Florens, L., Ivanova, M., Dolla, A., Czjzek, M., Haser, R., Verger, R., and Bruschi, M. (1995) *Biochemistry* **34**, 11327–11334

363. Banci, L., Bertini, I., Bruschi, M., Sompornpisut, P., and Turano, P. (1996) *Proc. Natl. Acad. Sci. U.S.A.* **93**, 14396–14400

364. Feng, Y., and Swenson, R. P. (1997) *Biochemistry* **36**, 13617–13628

365. Lui, S. M., and Cowan, J. A. (1994) *J. Am. Chem. Soc.* **116**, 11538–11549

366. Crane, B. R., Siegel, L. M., and Getzoff, E. D. (1997) *Biochemistry* **36**, 12101–12119

367. Moura, I., LeGall, J., Lino, A. R., Peck, H. D., Jr., Fauque, G., Xavier, A. V., Der Vertanian, D. V., Moura, J. J. G., and Huynh, B. H. (1988) *J. Am. Chem. Soc.* **110**, 1075–1082

368. Fischer, F., Zillig, W., Stetter, K. O., and Schreiber, G. (1983) *Nature (London)* **301**, 511–515

369. Kelly, D. P. (1985) *Nature (London)* **313**, 734

370. Bak, F., and Cypionka, H. (1987) *Nature (London)* **326**, 891–892

370a. Morelli, X., Czjzek, M., Hatchikian, C. E., Bornet, O., Fontecilla-Camps, J. C., Palma, N. P., Moura, J. J. G., and Guerlesquin, F. (2000) *J. Biol. Chem.* **275**, 23204–23210

370b. Costas, A. M. C., White, A. K., and Metcalf, W. W. (2001) *J. Biol. Chem.* **276**, 17429–17436

371. Müller, V., Blaut, M., and Gottschalk, G. (1993) in *Methanogenesis: Ecology, Physiology, Biochemistry and Genetics* (Ferry, J. G., ed), pp. 360–406, Chapman and Hall, New York

372. Hayaishi, O., and Nozaki, M. (1969) *Science* **164**, 389–405

373. Orville, A. M., and Lipscomb, J. D. (1993) *J. Biol. Chem.* **268**, 8596–8607

374. Ohlendorf, D. H., Orville, A. M., and Lipscomb, J. D. (1994) *J. Mol. Biol.* **244**, 586–608

375. Orville, A. M., Lipscomb, J. D., and Ohlendorf, D. H. (1997) *Biochemistry* **36**, 10052–10066

376. Sanvoisin, J., Langley, G. J., and Bugg, T. D. H. (1995) *J. Am. Chem. Soc.* **117**, 7836–7837

377. Miller, M. A., and Lipscomb, J. D. (1996) *J. Biol. Chem.* **271**, 5524–5535

378. Han, S., Eltis, L. D., Timmis, K. N., Muchmore, S. W., and Bolin, J. T. (1995) *Science* **270**, 976–980

379. Shu, L., Chiou, Y.-M., Orville, A. M., Miller, M. A., Lipscomb, J. D., and Que, L., Jr. (1995) *Biochemistry* **34**, 6649–6659

380. Senda, T., Sugiyama, K., Narita, H., Yamamoto, T., Kimbara, K., Fukuda, M., Sato, M., Yano, K., and Mitsui, Y. (1996) *J. Mol. Biol.* **255**, 735–752

381. Hugo, N., Armengaud, J., Gaillard, J., Timmis, K. N., and Jouanneau, Y. (1998) *J. Biol. Chem.* **273**, 9622–9629

382. Fraser, M. S., and Hamilton, G. A. (1982) *J. Am. Chem. Soc.* **104**, 4203–4211

383. Leeds, J. M., Brown, P. J., McGeehan, G. M., Brown, F. K., and Wiseman, J. S. (1993) *J. Biol. Chem.* **268**, 17781–17786

384. Ohnishi, T., Hirata, F., and Hayaishi, O. (1977) *J. Biol. Chem.* **252**, 4643–4647

385. Gassner, G. T., Ballou, D. P., Landrum, G. A., and Whittaker, J. W. (1993) *Biochemistry* **32**, 4820–4825

386. Yamaguchi, M., and Fujisawa, H. (1982) *J. Biol. Chem.* **257**, 12497–12502

386a. Senda, T., Yamada, T., Sakurai, N., Kubota, M., Nishizaki, T., Masai, E., Fukuda, M., and Mitsui, Y. (2000) *J. Mol. Biol.* **304**, 397–410

387. Gassner, G. T., Ludwig, M. L., Gatti, D. L., Correll, C. C., and Ballou, D. P. (1995) *FASEB J.* **9**, 1411–1418

387a. Coulter, E. D., Moon, N., Batie, C. J., Dunham, W. R., and Ballou, D. P. (1999) *Biochemistry* **38**, 11062–11072

388. Bertrand, P., More, C., and Camensuli, P. (1995) *J. Am. Chem. Soc.* **117**, 1807–1809

389. Lockridge, O., Massey, V., and Sullivan, P. A. (1972) *J. Biol. Chem.* **247**, 8097–8106

390. Sanders, S. A., Williams, C. H., Jr., and Massey, V. (1999) *J. Biol. Chem.* **274**, 22289–22295

391. Flashner, M. I. S., and Massey, V. (1974) *J. Biol. Chem.* **249**, 2579–2586

392. Emanuele, J. J., and Fitzpatrick, P. F. (1995) *Biochemistry* **34**, 3710–3715

393. Massey, V. (1994) *J. Biol. Chem.* **269**, 22459–22462

394. Entsch, B., and van Berkel, W. J. H. (1995) *FASEB J.* **9**, 476–483

395. Gatti, D. L., Entsch, B., Ballou, D. P., and Ludwig, M. L. (1996) *Biochemistry* **35**, 567–578

396. Hamilton, G. A. (1971) *Prog. Bioorg. Chem.* **1**, 83

397. Orf, H. W., and Dolphin, D. (1974) *Proc. Natl. Acad. Sci. U.S.A.* **71**, 2646–2650

398. Wagner, W. R., Spero, D. M., and Rastetter, W. H. (1984) *J. Am. Chem. Soc.* **106**, 1476–1480

399. Moran, G. R., Entsch, B., Palfey, B. A., and Ballou, D. P. (1997) *Biochemistry* **36**, 7548–7556

399a. Frederick, K. K., Ballou, D. P., and Palfey, B. A. (2001) *Biochemistry* **40**, 3891–3899

400. Eppink, M. H. M., Schreuder, H. A., and van Berkel, W. J. H. (1998) *J. Biol. Chem.* **273**, 21031–21039

401. van der Bolt, F. J. T., van den Heuvel, R. H. H., Vervoort, J., and van Berkel, W. J. H. (1997) *Biochemistry* **36**, 14192–14201

401a. Eppink, M. H. M., Overkamp, K. M., Schreuder, H. A., and Van Berkel, W. J. H. (1999) *J. Mol. Biol.* **292**, 87–96

401b. Ortiz-Maldonado, M., Ballou, D. P., and Massey, V. (2001) *Biochemistry* **40**, 1091–1101

402. Eppink, M. H. M., Schreuder, H. A., and van Berkel, W. J. H. (1997) *Protein Sci.* **6**, 2454–2458

403. Maeda-Yorita, K., and Massey, V. (1993) *J. Biol. Chem.* **268**, 4134–4144

404. Einarsdottir, G. H., Stankovich, M. T., and Tu, S.-C. (1988) *Biochemistry* **27**, 3277–3285

405. Powlowski, J., Ballou, D. P., and Massey, V. (1990) *J. Biol. Chem.* **265**, 4969–4975

406. Mathews, F. S., Chen, Z.-w, and Bellamy, H. D. (1991) *Biochemistry* **30**, 238–247

407. Arunachalam, U., Massey, V., and Miller, S. M. (1994) *J. Biol. Chem.* **269**, 150–155

408. Prieto, M. A., and Garcia, J. L. (1994) *J. Biol. Chem.* **269**, 22823–22829

409. Ysuji, H., Ogawa, T., Bando, N., Kimoto, M., and Sasaoka, K. (1990) *J. Biol. Chem.* **265**, 16064–16067

410. Itagaki, K., Carver, G. T., and Philpot, R. M. (1996) *J. Biol. Chem.* **271**, 20102–20107

411. Jones, K. C., and Ballou, D. P. (1986) *J. Biol. Chem.* **261**, 2553–2559

411a. Sheng, D., Ballou, D. P., and Massey, V. (2001) *Biochemistry* **40**, 11156–11167

412. Branchaud, B. P., and Walsh, C. T. (1985) *J. Am. Chem. Soc.* **107**, 2153–2161

413. Erlandsen, H., Flatmark, T., Stevens, R. C., and Hough, E. (1998) *Biochemistry* **37**, 15638–15646

414. Loeb, K. E., Westre, T. E., Kappock, T. J., Mitic, N., Glasfeld, E., Caradonna, J. P., Hedman, B., Hodgson, K. O., and Solomon, E. I. (1997) *J. Am. Chem. Soc.* **119**, 1901–1915

415. Chen, D., and Frey, P. A. (1998) *J. Biol. Chem.* **273**, 25594–25601

415a. Teigen, K., Froystein, N. Å., and Martínez, A. (1999) *J. Mol. Biol.* **294**, 807–823

416. Goodwill, K. E., Sabatier, C., and Stevens, R. C. (1998) *Biochemistry* **37**, 13437–13445

417. Francisco, W. A., Tian, G., Fitzpatrick, P. F., and Klinman, J. P. (1998) *J. Am. Chem. Soc.* **120**, 4057–4062

418. Michaud-Soret, I., Andersson, K. K., and Que, L., Jr. (1995) *Biochemistry* **34**, 5504–5510

419. Hillas, P. J., and Fitzpatrick, P. F. (1996) *Biochemistry* **35**, 6969–6975

419a. Almås, B., Toska, K., Teigen, K., Groehn, V., Pfleiderer, W., Martínez, A., Flatmark, T., and Haavik, J. (2000) *Biochemistry* **39**, 13676–13686

420. Moran, G. R., Daubner, S. C., and Fitzpatrick, P. F. (1998) *J. Biol. Chem.* **273**, 12259–12266

421. Ramsey, A. J., Hillas, P. J., and Fitzpatrick, P. F. (1996) *J. Biol. Chem.* **271**, 24395–24400

421a. Fitzpatrick, P. F. (1999) *Ann. Rev. Biochem.* **68**, 355–381

422. Lazarus, R. A., DeBrosse, C. W., and Benkovic, S. J. (1982) *J. Am. Chem. Soc.* **104**, 6869–6871

423. Pike, D. C., Hora, M. T., Bailey, S. W., and Ayling, J. E. (1986) *Biochemistry* **25**, 4762–4771

424. Dix, T. A., Bollag, G. E., Domanico, P. L., and Benkovic, S. J. (1985) *Biochemistry* **24**, 2955–2958

425. Rebrin, I., Thöny, B., Bailey, S. W., and Ayling, J. E. (1998) *Biochemistry* **37**, 11246–11254

425a. Endrizzi, J. A., Cronk, J. D., Wang, W., Crabtree, G. R., and Alber, T. (1995) *Science* **268**, 556–559

426. Ficner, R., Sauer, U. H., Stier, G., and Suck, D. (1995) *EMBO J.* **14**, 2034–2042

References

427. Su, Y., Varughese, K. I., Xuong, N. H., Bray, T. L., Roche, D. J., and Whiteley, J. M. (1993) *J. Biol. Chem.* **268**, 26836–26841
428. Varughese, K. I., Xuong, N. H., Kiefer, P. M., Matthews, D. A., and Whiteley, J. M. (1994) *Proc. Natl. Acad. Sci. U.S.A.* **91**, 5582–5586
429. Kiefer, P. M., Grimshaw, C. E., and Whiteley, J. M. (1997) *Biochemistry* **36**, 9438–9445
430. Scriver, C. R., Kaufman, S., Eisensmith, R. C., and Woo, S. L. C. (1995) in *The Metabolic and Molecular Bases of Inherited Disease*, 7th ed., Vol. 1 (Scriver, C. R., Beaudet, A. L., Sly, W. S., and Valle, D., eds), pp. 1015–1075, McGraw-Hill, New York
430a. Kobe, B., Jennings, I. G., House, C. M., Michell, B. J., Goodwill, K. E., Santarsiero, B. D., Stevens, R. C., Cotton, R. G. H., and Kemp, B. E. (1999) *Nature Struct. Biol.* **6**, 442–448
431. Ginns, E. L., Rehavi, M., Martin, B. M., Weller, M., O'Malley, K. L., LeMarca, M. E., McAllister, C. G., and Paul, S. M. (1988) *J. Biol. Chem.* **263**, 7406–7410
432. Grenett, H. E., Ledley, F. D., Reed, L. L., and Woo, S. L. C. (1987) *Proc. Natl. Acad. Sci. U.S.A.* **84**, 5530–5534
433. Pigeon, D., Ferrara, P., Gros, F., and Thibault, J. (1987) *J. Biol. Chem.* **262**, 6155–6158
433a. Itagaki, C., Isobe, T., Taoka, M., Natsume, T., Nomura, N., Horigome, T., Omata, S., Ichinose, H., Nagatsu, T., Greene, L. A., and Ichimura, T. (1999) *Biochemistry* **38**, 15673–15680
434. McTigue, M., Cremins, J., and Halegoua, S. (1985) *J. Biol. Chem.* **260**, 9047–9056
435. Okuno, S., and Fujisawa, H. (1985) *J. Biol. Chem.* **260**, 2633–2635
436. Albert, K. A., Helmer-Matyjek, E., Nairn, A. C., Muller, T. H., Waycock, J. W., Greene, L. A., Goldstein, M., and Greengard, P. (1984) *Proc. Natl. Acad. Sci. U.S.A.* **81**, 7713–7717
437. Guroff, G., Daly, J. W., Jerina, D. M., Renson, J., Witkop, B., and Udenfriend, S. (1967) *Science* **157**, 1524–1530
438. Kasperek, G. J., Bruice, T. C., Yagi, H., Kaubisch, N., and Jerina, D. M. (1972) *J. Am. Chem. Soc.* **94**, 7876–7882
439. Endo, F., Awata, H., Tanoue, A., Ishiguro, M., Eda, Y., Titani, K., and Matsuda, I. (1992) *J. Biol. Chem.* **267**, 24235–24240
440. Bradley, F. C., Lindstedt, S., Lipscomb, J. D., Que, L., Jr., Roe, A. L., and Rundgren, M. (1986) *J. Biol. Chem.* **261**, 11693–11696
441. McGinnis, K., Ku, G. M., VanDusen, W. J., Fu, J., Garsky, V., Stern, A. M., and Friedman, P. A. (1996) *Biochemistry* **35**, 3957–3962
442. Lamberg, A., Pihlajaniemi, T., and Kivirikko, K. I. (1995) *J. Biol. Chem.* **270**, 9926–9931
443. Myllyharju, J., and Kivirikko, K. I. (1997) *EMBO J.* **16**, 1173–1180
444. Annunen, P., Autio-Harmainen, H., and Kivirikko, K. I. (1998) *J. Biol. Chem.* **273**, 5989–5992
445. Myllylä, R., Pihlajaniemi, T., Pajunen, L., Turpeenniemi-Hujanen, T., and Kivirikko, K. I. (1991) *J. Biol. Chem.* **266**, 2805–2810
446. Valtavaara, M., Szpirer, C., Szpirer, J., and Myllylä, R. (1998) *J. Biol. Chem.* **273**, 12881–12886
447. Bolwell, G. P., Robbins, M. P., and Dixon, R. A. (1985) *Biochem. J.* **229**, 693–699
448. Thornburg, L. D., and Stubbe, J. (1993) *Biochemistry* **32**, 14034–14042
449. Eichhorn, E., van der Ploeg, J. R., Kertesz, M. A., and Leisinger, T. (1997) *J. Biol. Chem.* **272**, 23031–23036
449a. Elkins, J. M., Ryle, M. J., Clifton, I. J., Hotopp, J. C. D., Lloyd, J. S., Burzlaff, N. I., Baldwin, J. E., Hausinger, R. P., and Roach, P. L. (2002) *Biochemistry* **41**, 5185–5192
450. Whiting, A. K., Que, L., Jr., Saari, R. E., Hausinger, R. P., Frederick, M. A., and McCracken, J. (1997) *J. Am. Chem. Soc.* **119**, 3413–3414
451. Hotopp, J. C. D., and Hausinger, R. P. (2002) *Biochemistry* **41**, 9787–9794
452. Henderson, L. M., Nelson, P. J., and Henderson, L. (1982) *Fed. Proc.* **41**, 2843–2847
453. Englard, S., Blanchard, J. S., and Midelfort, C. F. (1985) *Biochemistry* **24**, 1110–1116
454. Myllyla, R., Majamaa, K., Gunzler, V., Hanauske-Abel, H. M., and Kivirikko, K. I. (1984) *J. Biol. Chem.* **259**, 5403–5405
455. Ho, R. Y. N., Mehn, M. P., Hegg, E. L., Liu, A., Ryle, M. J., Hausinger, R. P., and Que, L., Jr. (2001) *J. Am. Chem. Soc.* **123**, 5022–5029
456. Roach, P. L., Clifton, I. J., Fülöp, V., Harlos, K., Barton, G. J., Hajdu, J., Andersson, I., Schofield, C. J., and Baldwin, J. E. (1995) *Nature (London)* **375**, 700–704
457. Roach, P. L., Clifton, I. J., Hensgens, C. M. H., Shibata, N., Schofield, C. J., Hajdu, J., and Baldwin, J. E. (1997) *Nature (London)* **387**, 827–830
457a. Lloyd, M. D., Lee, H.-J., Harlos, K., Zhang, Z.-H., Baldwin, J. E., Schofield, C. J., Charnock, J. M., Garner, C. D., Hara, T., Terwisscha van Scheltinga, A. C., Valegård, K., Viklund, J. A. C., Hajdu, J., Andersson, I., Danielsson, Å., and Bhikhabhai, R. (1999) *J. Mol. Biol.* **287**, 943–960
458. Busby, R. W., Chang, M. D.-T., Busby, R. C., Wimp, J., and Townsend, C. A. (1995) *J. Biol. Chem.* **270**, 4262–4269
459. Zhou, J., Kelly, W. L., Bachmann, B. O., Gunsior, M., Townsend, C. A., and Solomon, E. I. (2001) *J. Am. Chem. Soc.* **123**, 7388–7398
460. Tian, G., Berry, J. A., and Klinman, J. P. (1994) *Biochemistry* **33**, 226–234
461. Klinman, J. P., Krueger, M., Brenner, M., and Edmondson, D. E. (1984) *J. Biol. Chem.* **259**, 3399–3402
462. Merkler, D. J., Kulathila, R., Consalvo, A. P., Young, S. D., and Ash, D. E. (1992) *Biochemistry* **31**, 7282–7288
463. Prigge, S. T., Kolhekar, A. S., Eipper, B. A., Mains, R. E., and Amzel, L. M. (1997) *Science* **278**, 1300–1305
463a. Jaron, S., and Blackburn, N. J. (2001) *Biochemistry* **40**, 6867–6875
464. Francisco, W. A., Merkler, D. J., Blackburn, N. J., and Klinman, J. P. (1998) *Biochemistry* **37**, 8244–8252
464a. Jaron, S., and Blackburn, N. J. (1999) *Biochemistry* **38**, 15086–15096
464b. Kolhekar, A. S., Bell, J., Shiozaki, E. N., Jin, L., Keutmann, H. T., Hand, T. A., Mains, R. E., and Eipper, B. A. (2002) *Biochemistry* **41**, 12384–12394
465. Wilkinson, B., Zhu, M., Priestley, N. D., Nguyen, H.-H. T., Morimoto, H., Williams, P. G., Chan, S. I., and Floss, H. G. (1996) *J. Am. Chem. Soc.* **118**, 921–922
466. Elliott, S. J., Zhu, M., Tso, L., Nguyen, H.-H. T., Yip, J. H.-K., and Chan, S. I. (1997) *J. Am. Chem. Soc.* **119**, 9949–9955
467. Swinney, D. C., and Mak, A. Y. (1994) *Biochemistry* **33**, 2185–2190
468. Imai, T., Yamazaki, T., and Kominami, S. (1998) *Biochemistry* **37**, 8097–8104
469. Denison, M. S., and Whitlock, J. P., Jr. (1995) *J. Biol. Chem.* **270**, 18175–18178
470. Coon, M. J., Ding, X., Pernecky, S. J., and Vaz, A. D. N. (1992) *FASEB J.* **6**, 669–673
471. Gonzalez, F. J., and Lee, Y.-H. (1996) *FASEB J.* **10**, 1112–1117
472. Nishimoto, M., Gotoh, O., Okuda, K., and Noshiro, M. (1991) *J. Biol. Chem.* **266**, 6467–6471
473. Dilworth, F. J., Scott, I., Green, A., Strugnell, S., Guo, Y.-D., Roberts, E. A., Kremer, R., Calverley, M. J., Makin, H. L. J., and Jones, G. (1995) *J. Biol. Chem.* **270**, 16766–16774
474. Cabello-Hurtado, F., Batard, Y., Salaün, J.-P., Durst, F., Pinot, F., and Werck-Reichhart, D. (1998) *J. Biol. Chem.* **273**, 7260–7267
475. Meyer, K., Shirley, A. M., Cusumano, J. C., Bell-Lelong, D. A., and Chapple, C. (1998) *Proc. Natl. Acad. Sci. U.S.A.* **95**, 6619–6623
476. Harris, D. L., and Loew, G. H. (1998) *J. Am. Chem. Soc.* **120**, 8941–8948
476a. French, K. J., Strickler, M. D., Rock, D. A., Rock, D. A., Bennett, G. A., Wahlstrom, J. L., Goldstein, B. M., and Jones, J. P. (2001) *Biochemistry* **40**, 9532–9538
476b. Schlichting, I., Berendzen, J., Chu, K., Stock, A. M., Maves, S. A., Benson, D. E., Sweet, R. M., Ringe, D., Petsko, G. A., and Sligar, S. G. (2000) *Science* **287**, 1615–1622
477. Vaz, A. D. N., McGinnity, D. F., and Coon, M. J. (1998) *Proc. Natl. Acad. Sci. U.S.A.* **95**, 3555–3560
478. Tzeng, H.-F., Laughlin, L. T., and Armstrong, R. N. (1998) *Biochemistry* **37**, 2905–2911
479. De Montellano, P. R. O., ed. (1995) *Cytochrome P450. Structure, Mechanism, and Biochemistry*, 2nd ed., Plenum, New York
480. Coon, M. J., Vaz, A. D. N., and Bestervelt, L. L. (1996) *FASEB J.* **10**, 428–434
481. Negishi, M., Uno, T., Darden, T. A., Sueyoshi, T., and Pedersen, L. G. (1996) *FASEB J.* **10**, 683–689
482. Gilday, D., Gannon, M., Yutzey, K., Bader, D., and Rifkind, A. B. (1996) *J. Biol. Chem.* **271**, 33054–33059
483. Khani, S. C., Zophiropoulos, P. G., Fujita, V. S., Porter, T. D., Koop, D. R., and Coon, M. J. (1987) *Proc. Natl. Acad. Sci. U.S.A.* **84**, 638–642
483a. Pikuleva, I. A., Puchkaev, A., and Björkhem, I. (2001) *Biochemistry* **40**, 7621–7629
484. Porter, T. D. (1991) *Trends Biochem. Sci.* **16**, 154–158
485. Hubbard, P. A., Shen, A. L., Paschke, R., Kasper, C. B., and Kim, J.-J. P. (2001) *J. Biol. Chem.* **276**, 29163–29170
485a. Gutierrez, A., Lian, L.-Y., Wolf, C. R., Scrutton, N. S., and Roberts, G. C. K. (2001) *Biochemistry* **40**, 1964–1975
486. Perret, A., and Pompon, D. (1998) *Biochemistry* **37**, 11412–11424
487. Vidakovic, M., Sligar, S. G., Li, H., and Poulos, T. L. (1998) *Biochemistry* **37**, 9211–9219
488. Takemori, S., and Kominami, S. (1984) *Trends Biochem. Sci.* **9**, 393–396
489. Kobayashi, K., Miura, S., Miki, M., Ichikawa, Y., and Tagawa, S. (1995) *Biochemistry* **34**, 12932–12936
489a. Pochapsky, T. C., Jain, N. U., Kuti, M., Lyons, T. A., and Heymont, J. (1999) *Biochemistry* **38**, 4681–4690
490. Poulos, T. L., and Raag, R. (1992) *FASEB J.* **6**, 674–679
491. Murataliev, M. B., Klein, M., Fulco, A., and Feyereisen, R. (1997) *Biochemistry* **36**, 8401–8412
492. Ravichandran, K. G., Boddupalli, S. S., Hasemann, C. A., Peterson, J. A., and Deisenhofer, J. (1993) *Science* **261**, 731–736
493. Hasemann, C. A., Ravichandran, K. G., Peterson, J. A., and Deisenhofer, J. (1994) *J. Mol. Biol.* **236**, 1169–1185
494. Harris, D. L., and Loew, G. H. (1996) *J. Am. Chem. Soc.* **118**, 6377–6387

References

495. Higgins, L., Bennett, G. A., Shimoji, M., and Jones, J. P. (1998) *Biochemistry* **37**, 7039–7046
496. Toy, P. H., Newcomb, M., and Hollenberg, P. F. (1998) *J. Am. Chem. Soc.* **120**, 7719–7729
496a. French, K. J., Strickler, M. D., Rock, D. A., Rock, D. A., Bennett, G. A., Wahlstrom, J. L., Goldstein, B. M., and Jones, J. P. (2001) *Biochemistry* **40**, 9532–9538
496b. Schlichting, I., Berendzen, J., Chu, K., Stock, A. M., Maves, S. A., Benson, D. E., Sweet, R. M., Ringe, D., Petsko, G. A., and Sligar, S. G. (2000) *Science* **287**, 1615–1622
497. Harris, D., Loew, G., and Waskell, L. (1998) *J. Am. Chem. Soc.* **120**, 4308–4318
497a. Kupfer, R., Liu, S. Y., Allentoff, A. J., and Thompson, J. A. (2001) *Biochemistry* **40**, 11490–11501
498. Poulos, T. L., Finzel, B. C., Gunsalus, I. C., Wagner, G. C., and Kraut, J. (1985) *J. Biol. Chem.* **260**, 16122–16130
499. Wang, H., Dick, R., Yin, H., Licad-Coles, E., Kroetz, D. L., Szklarz, G., Harlow, G., Halpert, J. R., and Correia, M. A. (1998) *Biochemistry* **37**, 12536–12545
500. Custer, L., Zajc, B., Sayer, J. M., Cullinane, C., Phillips, D. R., Cheh, A. M., Jerina, D. M., Bohr, V. A., and Mazur, S. J. (1999) *Biochemistry* **38**, 569–581
501. Shanklin, J., Achim, C., Schmidt, H., Fox, B. G., and Münck, E. (1997) *Proc. Natl. Acad. Sci. U.S.A.* **94**, 2981–2986
502. Rosenzweig, A. C., Frederick, C. A., Lippard, S. J., and Nordlund, P. (1993) *Nature (London)* **366**, 537–543
502a. Whittington, D. A., Sazinsky, M. H., and Lippard, S. J. (2001) *J. Am. Chem. Soc.* **123**, 1794–1795
503. Elango, N., Radhakrishnan, R., Froland, W. A., Wallar, B. J., Earhart, C. A., Lipscomb, J. D., and Ohlendorf, D. H. (1997) *Protein Sci.* **6**, 556–568
504. Shu, L., Nesheim, J. C., Kauffmann, K., Münck, E., Lipscomb, J. D., and Que, L., Jr. (1997) *Science* **275**, 515–518
505. Siegbahn, P. E. M., and Crabtree, R. H. (1997) *J. Am. Chem. Soc.* **119**, 3103–3113
505a. Valentine, A. M., LeTadic-Biadatti, M.-H., Toy, P. H., Newcomb, M., and Lippard, S. J. (1999) *J. Biol. Chem.* **274**, 10771–10776
505b. Chang, S.-L., Wallar, B. J., Lipscomb, J. D., and Mayo, K. H. (2001) *Biochemistry* **40**, 9539–9551
506. Liu, Y., Nesheim, J. C., Paulsen, K. E., Stankovich, M. T., and Lipscomb, J. D. (1997) *Biochemistry* **36**, 5223–5233
507. Katopodis, A. G., Wimalasena, K., Lee, J., and May, S. W. (1984) *J. Am. Chem. Soc.* **106**, 7928–7935
508. Pikus, J. D., Studts, J. M., Achim, C., Kauffmann, K. E., Münck, E., Steffan, R. J., McClay, K., and Fox, B. G. (1996) *Biochemistry* **35**, 9106–9119
509. Qian, H., Edlund, U., Powlowski, J., Shingler, V., and Sethson, I. (1997) *Biochemistry* **36**, 495–504
509a. Eichhorn, E., van der Ploeg, J. R., and Leisinger, T. (1999) *J. Biol. Chem.* **274**, 26639–26646
509b. Allen, J. R., Clark, D. D., Krum, J. G., and Ensign, S. A. (1999) *Biochemistry* **96**, 8432–8437
510. Mayer, B., and Hemmens, B. (1997) *Trends Biochem. Sci.* **22**, 477–481
511. Stamler, J. S., Singel, D. J., and Loscalzo, J. (1992) *Science* **258**, 1898–1902
511a. Wolthers, K. R., and Schimerlik, M. I. (2001) *Biochemistry* **40**, 4722–4737
511b. Ledbetter, A. P., McMillan, K., Roman, L. J., Masters, BsS., Dawson, J. H., and Sono, M. (1999) *Biochemistry* **38**, 8014–8021
511c. Li, H., Raman, C. S., Martásek, P., Masters, B. S. S., and Poulos, T. L. (2001) *Biochemistry* **40**, 5399–5406
511d. Marletta, M. A. (2001) *Trends Biochem. Sci.* **26**, 519–521
511e. Rusche, K. M., and Marletta, M. A. (2001) *J. Biol. Chem.* **276**, 421–427
511f. Crane, B. R., Arvai, A. S., Ghosh, S., Getzoff, E. D., Stuehr, D. J., and Tainer, J. A. (2000) *Biochemistry* **39**, 4608–4621
512. Culotta, E., and Koshland, D. E., Jr. (1992) *Science* **258**, 1862–1863
513. Richeson, C. E., Mulder, P., Bowry, V. W., and Ingold, K. U. (1998) *J. Am. Chem. Soc.* **120**, 7211–7219
514. Pfeiffer, S., Gorren, A. C. F., Schmidt, K., Werner, E. R., Hansert, B., Bohle, D. S., and Mayer, B. (1997) *J. Biol. Chem.* **272**, 3465–3470
515. Marla, S. S., Lee, J., and Groves, J. T. (1997) *Proc. Natl. Acad. Sci. U.S.A.* **94**, 14243–14248
516. Chan, N.-L., Rogers, P. H., and Arnone, A. (1998) *Biochemistry* **37**, 16459–16464
517. Friebe, A., Schultz, G., and Koesling, D. (1996) *EMBO J.* **15**, 6863–6868
518. Schelvis, J. P. M., Zhao, Y., Marletta, M. A., and Babcock, G. T. (1998) *Biochemistry* **37**, 16289–16297
518a. Andersen, J. F., and Montfort, W. R. (2000) *J. Biol. Chem.* **275**, 30496–30503
518b. Roberts, S. A., Weichsel, A., Qiu, Y., Shelnutt, J. A., Walker, F. A., and Montfort, W. R. (2001) *Biochemistry* **40**, 11327–11337
519. DeMaster, E. G., Quast, B. J., Redfern, B., and Nagasawa, H. T. (1995) *Biochemistry* **34**, 11494–11499
519a. Lai, T. S., Hausladen, A., Slaughter, T. F., Eu, J. P., Stamler, J. S., and Greenberg, C. S. (2001) *Biochemistry* **40**, 4904–4910
520. Gow, A. J., Buerk, D. G., and Ischiropoulos, H. (1997) *J. Biol. Chem.* **272**, 2841–2845
520a. Nedospasov, A., Rafikov, R., Beda, N., and Nudler, E. (2000) *Proc. Natl. Acad. Sci. U.S.A.* **97**, 13543–13548
521. Goldstein, S., and Czapski, G. (1996) *J. Am. Chem. Soc.* **118**, 3419–3425
522. Caulfield, J. L., Wishnok, J. S., and Tannenbaum, S. R. (1998) *J. Biol. Chem.* **273**, 12689–12695
523. Singh, R. J., Hogg, N., Joseph, J., and Kalyanaraman, B. (1996) *J. Biol. Chem.* **271**, 18596–18603
524. Wong, P. S.-Y., Hyun, J., Fukuto, J. M., Shirota, F. N., DeMaster, E. G., Shoeman, D. W., and Nagasawa, H. T. (1998) *Biochemistry* **37**, 5362–5371
525. Ghosh, D. K., and Stuehr, D. J. (1995) *Biochemistry* **34**, 801–807
526. Nathan, C., and Xie, Q.-w. (1994) *J. Biol. Chem.* **269**, 13725–13728
527. Rodríguez-Crespo, I., Gerber, N. C., and Ortiz de Montellano, P. R. (1996) *J. Biol. Chem.* **271**, 11462–11467
528. Liu, J., García-Cardena, G., and Sessa, W. C. (1996) *Biochemistry* **35**, 13277–13281
529. Bredt, D. S., Hwang, P. M., Glatt, C. E., Lowenstein, C., Reed, R. R., and Snyder, S. H. (1991) *Nature (London)* **351**, 714–718
530. Masters, B. S. S., McMillan, K., Sheta, E. A., Nishimura, J. S., Roman, L. J., and Martasek, P. (1996) *FASEB J.* **10**, 552–558
531. Hurshman, A. R., and Marletta, M. A. (1995) *Biochemistry* **34**, 5627–5634
532. Crane, B. R., Arvai, A. S., Ghosh, D. K., Wu, C., Getzoff, E. D., Stuehr, D. J., and Tainer, J. A. (1998) *Science* **279**, 2121–2126
533. Poulos, T. L., Raman, C. S., and Li, H. (1998) *Structure* **6**, 255–258
534. Villeret, V., Huang, S., Zhang, Y., and Lipscomb, W. N. (1995) *Biochemistry* **34**, 4307–4315
535. González, D. H., and Andreo, C. S. (1989) *Trends Biochem. Sci.* **14**, 24–27
535a. Reif, A., Fröhlich, L. G., Kotsonis, P., Frey, A., Bömmel, H. M., Wink, D. A., Pfleiderer, W., and Schmidt, H. H. H. W. (1999) *J. Biol. Chem.* **274**, 24921–24929
535b. Heller, R., Unbehaun, A., Schellenberg, B., Mayer, B., Werner-Felmayer, G., and Werner, E. R. (2001) *J. Biol. Chem.* **276**, 40–47
535c. Hurshman, A. R., and Marletta, M. A. (2002) *Biochemistry* **41**, 3439–3456
536. Pufahl, R. A., Wishnok, J. S., and Marletta, M. A. (1995) *Biochemistry* **34**, 1930–1941
537. Rusche, K. M., Spiering, M. M., and Marletta, M. A. (1998) *Biochemistry* **37**, 15503–15512
537a. Miranda, K. M., Espey, M. G., Yamada, K., Krishna, M., Ludwick, N., Kim, S. M., Jourd'heuil, D., Grisham, M. B., Feelisch, M., Fukuto, J. M., and Wink, D. A. (2001) *J. Biol. Chem.* **276**, 1720–1727
538. Schmidt, H. H. H. W., Hofmann, H., Schindler, U., Shutenko, Z. S., Cunningham, D. D., and Feelisch, M. (1996) *Proc. Natl. Acad. Sci. U.S.A.* **93**, 14492–14497
538a. Burner, U., Furtmüller, P. G., Kettle, A. J., Koppenol, W. H., and Obinger, C. (2000) *J. Biol. Chem.* **275**, 20597–20601
539. Imlay, J. A., and Linn, S. (1988) *Science* **240**, 1302–1309
540. Fridovich, I. (1997) *J. Biol. Chem.* **272**, 18515–18517
541. Rosen, G. M., Pou, S., Ramos, C. L., Cohen, M. S., and Britigan, B. E. (1995) *FASEB J.* **9**, 200–209
542. Malmstrom, B. G. (1982) *Ann. Rev. Biochem.* **51**, 21–59
543. Naqui, A., and Chance, B. (1986) *Ann. Rev. Biochem.* **55**, 137–166
544. Wood, P. M. (1987) *Trends Biochem. Sci.* **12**, 250–251
545. Segal, A. W., and Abo, A. (1993) *Trends Biochem. Sci.* **18**, 43–47
546. Chanock, S. J., Benna, J. E., Smith, R. M., and Babior, B. M. (1994) *J. Biol. Chem.* **269**, 24519–24522
547. Baggiolini, M., Boulay, F., Badwey, J. A., and Curnutte, J. T. (1993) *FASEB J.* **7**, 1004–1010
548. Han, C.-H., Freeman, J. L. R., Lee, T., Motalebi, S. A., and Lambeth, J. D. (1998) *J. Biol. Chem.* **273**, 16663–16668
549. Forehand, J. R., Nauseef, W. M., Curnutte, J. T., and Johnston, R. B., Jr. (1995) in *The Metabolic and Molecular Bases of Inherited Disease*, 7th ed., Vol. 3 (Scriver, C. R., Beaudet, A. L., Sly, W. S., and Valle, D., eds), pp. 3995–4028, McGraw-Hill, New York
550. El Benna, J., Dang, P. M.-C., Gaudry, M., Fay, M., Morel, F., Hakim, J., and Gougerot-Pocidalo, M.-A. (1997) *J. Biol. Chem.* **272**, 17204–17208
551. Huang, J., Hitt, N. D., and Kleinberg, M. E. (1995) *Biochemistry* **34**, 16753–16757
552. Cross, A. R., Rae, J., and Curnutte, J. T. (1995) *J. Biol. Chem.* **270**, 17075–17077
552a. Dang, P. M.-C., Cross, A. R., and Babior, B. M. (2001) *Proc. Natl. Acad. Sci. U.S.A.* **98**, 3001–3005
553. Isogai, Y., Iizuka, T., and Shiro, Y. (1995) *J. Biol. Chem.* **270**, 7853–7857
554. Bouin, A.-P., Grandvaux, N., Vignais, P. V., and Fuchs, A. (1998) *J. Biol. Chem.* **273**, 30097–30103
554a. Di-Poi, N., Fauré, J., Grizot, S., Molnár, G., Pick, E., and Dagher, M.-C. (2001) *Biochemistry* **40**, 10014–10022
555. Park, J.-W., Hoyal, C. R., El Benna, J., and Babior, B. M. (1997) *J. Biol. Chem.* **272**, 11035–11043

References

556. Kramer, I. M., Verhoeven, A. J., van der Bend, R. L., Weening, R. S., and Roos, D. (1988) *J. Biol. Chem.* **263**, 2352–2357
557. Yoshida, L. S., Saruta, F., Yoshikawa, K., Tatsuzawa, O., and Tsunawaki, S. (1998) *J. Biol. Chem.* **273**, 27879–27886
558. Segal, A. W., Heyworth, P. G., Cockcroft, S., and Barrowman, M. M. (1985) *Nature (London)* **316**, 547–549
559. Cox, F. E. G. (1983) *Nature (London)* **302**, 19
559a. Shiose, A., Kuroda, J., Tsuruya, K., Hirai, M., Hirakata, H., Naito, S., Hattori, M., Sakaki, Y., and Sumimoto, H. (2001) *J. Biol. Chem.* **276**, 1417–1423
559b. Lee, C.-i, Miura, K., Liu, X., and Zweier, J. L. (2000) *J. Biol. Chem.* **275**, 38965–38972
559c. Bonini, M. G., and Augusto, O. (2001) *J. Biol. Chem.* **276**, 9749–9754
559d. Alvarez, B., Ferrer-Sueta, G., Freeman, B. A., and Radi, R. (1999) *J. Biol. Chem.* **274**, 842–848
559e. Zhang, H., Joseph, J., Feix, J., Hogg, N., and Kalyanaraman, B. (2001) *Biochemistry* **40**, 7675–7686
560. Hazen, S. L., Hsu, F. F., d'Avignon, A., and Heinecke, J. W. (1998) *Biochemistry* **37**, 6864–6873
561. Evans, T. J., Buttery, L. D. K., Carpenter, A., Springall, D. R., Polak, J. M., and Cohen, J. (1996) *Proc. Natl. Acad. Sci. U.S.A.* **93**, 9553–9558
562. Eiserich, J. P., Hristova, M., Cross, C. E., Jones, A. D., Freeman, B. A., Halliwell, B., and van der Vliet, A. (1998) *Nature (London)* **391**, 393–397
562a. Henle, E. S., Han, Z., Tang, N., Rai, P., Luo, Y., and Linn, S. (1999) *J. Biol. Chem.* **274**, 962–971
563. Halliwell, B., and Gutteridge, J. M. C. (1985) *Free Radicals in Biology and Medicine*, Clarendon Press, Oxford
564. Yamazaki, I., and Piette, L. H. (1990) *J. Biol. Chem.* **265**, 13589–13594
565. Wink, D. A., Nims, R. W., Saavedra, J. E., Utermahlen, W. E. J., and Ford, P. C. (1994) *Proc. Natl. Acad. Sci. U.S.A.* **91**, 6604–6608
566. Pogozelski, W. K., McNeese, T. J., and Tullius, T. D. (1995) *J. Am. Chem. Soc.* **117**, 6428–6433
567. Luo, Y., Henle, E. S., and Linn, S. (1996) *J. Biol. Chem.* **271**, 21167–21176
568. Hlavaty, J. J., and Nowak, T. (1997) *Biochemistry* **36**, 15514–15525
569. Weiss, S. J., Test, S. T., Eckmann, C. M., Roos, D., and Regiani, S. (1986) *Science* **234**, 200–203
570. Kanofsky, J. R., Hoogland, H., Wever, R., and Weiss, S. J. (1988) *J. Biol. Chem.* **263**, 9692–9696
571. Khan, A. U., and Kasha, M. (1994) *Proc. Natl. Acad. Sci. U.S.A.* **91**, 12365–12367
572. Stratton, S. P., and Liebler, D. C. (1997) *Biochemistry* **36**, 12911–12920
573. Mollinedo, F., Manara, F. S., and Scheider, D. L. (1986) *J. Biol. Chem.* **261**, 1077–1082
574. Romeo, D., Skerlavaj, B., Bolognesi, M., and Gennaro, R. (1988) *J. Biol. Chem.* **263**, 9573–9575
575. Remaley, A. T., Kuhns, D. B., Basford, R. E., Glew, R. H., and Kaplan, S. S. (1984) *J. Biol. Chem.* **259**, 11173–11175
576. Heinecke, J. W., Meier, K. E., Lorenzen, J. A., and Shapiro, B. M. (1990) *J. Biol. Chem.* **265**, 7717–7720
577. Shapiro, B. M. (1991) *Science* **252**, 533–536
578. Jabs, T., Tschöpe, M., Colling, C., Hahlbrock, K., and Scheel, D. (1997) *Proc. Natl. Acad. Sci. U.S.A.* **94**, 4800–4805
579. Chandra, S., and Low, P. S. (1997) *J. Biol. Chem.* **272**, 28274–28280
580. Imlay, J. A. (1995) *J. Biol. Chem.* **270**, 19767–19777
581. Zhang, L., Yu, L., and Yu, C.-A. (1998) *J. Biol. Chem.* **273**, 33972–33976
582. McCord, J. M. (1985) *N. Engl. J. Med.* **312**, 159–163
583. González-Flecha, B., and Demple, B. (1995) *J. Biol. Chem.* **270**, 13681–13687
584. Zweier, J. L. (1988) *J. Biol. Chem.* **263**, 1353–1357
585. Karoui, H., Hogg, N., Fréjaville, C., Tordo, P., and Kalyanaraman, B. (1996) *J. Biol. Chem.* **271**, 6000–6009
586. Giulivi, C., Poderoso, J. J., and Boveris, A. (1998) *J. Biol. Chem.* **273**, 11038–11043
587. Jousserandot, A., Boucher, J.-L., Henry, Y., Niklaus, B., Clement, B., and Mansuy, D. (1998) *Biochemistry* **37**, 17179–17191
588. Berlett, B. S., and Stadtman, E. R. (1997) *J. Biol. Chem.* **272**, 20313–20316
588a. Rauk, A., and Armstrong, D. A. (2000) *J. Am. Chem. Soc.* **122**, 4185–4192
588b. Messner, K. R., and Imlay, J. A. (1999) *J. Biol. Chem.* **274**, 10119–10128
588c. Srinivasan, C., Liba, A., Imlay, J. A., Valentine, J. S., and Gralla, E. B. (2000) *J. Biol. Chem.* **275**, 29187–29192
589. Beckman, K. B., and Ames, B. N. (1997) *J. Biol. Chem.* **272**, 19633–19636
590. Chow, C. K., ed. (1988) *Cellular Antioxidant Defense Mechanisms*, CRC Press, Boca Raton, Florida (3 volumes)
591. Halliwell, B., and Gutteridge, J. M. C. (1986) *Trends Biochem. Sci.* **11**, 372–375
592. Dalton, D. A., Russell, S. A., Hanus, F. J., Pascae, G. A., and Evans, H. J. (1986) *Proc. Natl. Acad. Sci. U.S.A.* **83**, 3811–3815
593. Berger, T. M., Polidori, M. C., Dabbagh, A., Evans, P. J., Halliwell, B., Morrow, J. D., Roberts, L. J., II, and Frei, B. (1997) *J. Biol. Chem.* **272**, 15656–15660
593a. Kirsch, M., and de Groot, H. (2000) *J. Biol. Chem.* **275**, 16702–16708
594. Conklin, P. L., Williams, E. H., and Last, R. L. (1996) *Proc. Natl. Acad. Sci. U.S.A.* **93**, 9970–9974
594a. Kirsch, M., and de Groot, H. (2001) *FASEB J.* **15**, 1569–1574
595. Christen, S., Woodall, A. A., Shigenaga, M. K., Southwell-Keely, P. T., Duncan, M. W., and Ames, B. N. (1997) *Proc. Natl. Acad. Sci. U.S.A.* **94**, 3217–3222
596. Stocker, R., Yamamoto, Y., McDonagh, A. F., Glazer, A. N., and Ames, B. N. (1987) *Science* **235**, 1043–1046
597. Peden, D. B., Hohman, R., Brown, M. E., Mason, R. T., Berkebile, C., Fales, H. M., and Kaliner, M. A. (1990) *Proc. Natl. Acad. Sci. U.S.A.* **87**, 7638–7642
598. Reiter, R. J. (1995) *FASEB J.* **9**, 526–533
598a. Martín, M., Macías, M., Escames, G., León, J., and Acuña-Castroviejo, D. (2000) *FASEB J.* **14**, 1677–1679
599. Levine, R. L., Mosoni, L., Berlett, B. S., and Stadtman, E. R. (1996) *Proc. Natl. Acad. Sci. U.S.A.* **93**, 15036–15040
600. Schultz, J. R., Ellerby, L. M., Gralla, E. B., Valentine, J. S., and Clarke, C. F. (1996) *Biochemistry* **35**, 6595–6603
600a. Beyer, R. E., Segura-Aguilar, J., Di Bernardo, S., Cavazzini, M., Fato, R., Fiorentini, D., Galli, M. C., Setti, M., Landi, L., and Lenaz, G. (1996) *Proc. Natl. Acad. Sci. U.S.A.* **93**, 2528–2532
600b. Lass, A., and Sohal, R. S. (2000) *FASEB J.* **14**, 87–94
600c. Suh, J. H., Shigeno, E. T., Morrow, J. D., Cox, B., Rocha, A. E., Frei, B., and Hagen, T. M. (2001) *FASEB J.* **15**, 700–706
601. Wink, D. A., Hanbauer, I., Krishna, M. C., DeGraff, W., Gamson, J., and Mitchell, J. B. (1993) *Proc. Natl. Acad. Sci. U.S.A.* **90**, 9813–9817
601a. Benaroudj, N., Lee, D. H., and Goldberg, A. L. (2001) *J. Biol. Chem.* **276**, 24261–24267
601b. Sun, Q.-A., Kirnarsky, L., Sherman, S., and Gladyshev, V. N. (2001) *Proc. Natl. Acad. Sci. U.S.A.* **98**, 3673–3678
601c. Kanzok, S. M., Fechner, A., Bauer, H., Ulschmid, J. K., Müller, H.-M., Botella-Munoz, J., Schneuwly, S., Schirmer, R. H., and Becker, K. (2001) *Science* **291**, 643–646
601d. Lee, S.-R., Bar-Noy, S., Kwon, J., Levine, R. L., Stadtman, T. C., and Rhee, S. G. (2000) *Proc. Natl. Acad. Sci. U.S.A.* **97**, 2521–2526
601e. Lowther, W. T., Brot, N., Weissbach, H., and Matthews, B. W. (2000) *Biochemistry* **39**, 13307–13312
601f. St. John, G., Brot, N., Ruan, J., Erdjument-Bromage, H., Tempst, P., Weissbach, H., and Nathan, C. (2001) *Proc. Natl. Acad. Sci. U.S.A.* **98**, 9901–9906
601g. Boschi-Muller, S., Azza, S., Sanglier-Cianferani, S., Talfournier, F., Van Dorsselear, A., and Branlant, G. (2000) *J. Biol. Chem.* **275**, 35908–35913
601h. Pollock, V. V., and Barber, M. J. (2001) *Biochemistry* **40**, 1430–1440
601i. Bieger, B., and Essen, L.-O. (2001) *J. Mol. Biol.* **307**, 1–8
601j. Seo, M. S., Kang, S. W., Kim, K., Baines, I. C., Lee, T. H., and Rhee, S. G. (2000) *J. Biol. Chem.* **275**, 20346–20354
602. Deng, H.-X., Hentati, A., Tainer, J. A., Iqbal, Z., Cayabyab, A., Hung, W.-Y., Getzoff, E. D., Hu, P., Herzfeldt, B., Roos, R. P., Warner, C., Deng, G., Soriano, E., Smyth, C., Parge, H. E., Ahmed, A., Roses, A. D., Hallewell, R. A., Pericak-Vance, M. A., and Siddique, T. (1993) *Science* **261**, 1047–1051
602a. Yim, M. B., Kang, J.-H., Yim, H.-S., Kwak, H.-S., Chock, P. B., and Stadtman, E. R. (1996) *Proc. Natl. Acad. Sci. U.S.A.* **93**, 5709–5714
603. Goto, J. J., Gralla, E. B., Valentine, J. S., and Cabelli, D. E. (1998) *J. Biol. Chem.* **273**, 30104–30109
603a. Estévez, A. G., Crow, J. P., Sampson, J. B., Reiter, C., Zhuang, Y., Richardson, G. J., Tarpey, M. M., Barbeito, L., and Beckman, J. S. (1999) *Science* **286**, 2498–2500
603b. Goto, J. J., Zhu, H., Sanchez, R. J., Nersissian, A., Gralla, E. B., Valentine, J. S., and Cabelli, D. E. (2000) *J. Biol. Chem.* **275**, 1007–1014
604. Gardner, P. R., Raineri, I., Epstein, L. B., and White, C. W. (1995) *J. Biol. Chem.* **270**, 13399–13405
605. Gaudu, P., and Weiss, B. (1996) *Proc. Roy. Soc. (London)* **93**, 10094–10098
606. Hidalgo, E., Bollinger, J. M., JR, Bradley, T. M., Walsh, C. T., and Demple, B. (1995) *J. Biol. Chem.* **270**, 20908–20914
607. Demple, B. (1998) *Science* **279**, 1655–1656
608. Godon, C., Lagniel, G., Lee, J., Buhler, J.-M., Kieffer, S., Perrot, M., Boucherie, H., Toledano, M. B., and Labarre, J. (1998) *J. Biol. Chem.* **273**, 22480–22489
609. Zheng, M., Åslund, F., and Storz, G. (1998) *Science* **279**, 1718–1721
610. Ellis, H. R., and Poole, L. B. (1997) *Biochemistry* **36**, 13349–13356
610a. Fuangthong, M., and Helmann, J. D. (2002) *Proc. Natl. Acad. Sci. U.S.A.* **99**, 6690–6695
611. Duranteau, J., Chandel, N. S., Kulisz, A., Shao, Z., and Schumacker, P. T. (1998) *J. Biol. Chem.* **273**, 11619–11624
612. Wenger, R. H. (2002) *FASEB J.* **16**, 1151–1162
613. DeLong, E. F. (2002) *Nature (London)* **419**, 676–677
614. Sinninghe Damsté, J. S., Strous, M., Rijpstra, W. I. C., Hopmans, E. C., Geenevasen, J. A. J., van Duin, A. C. T., van Niftrik, L. A., and Jetten, M. S. M. (2002) *Nature (London)* **419**, 708–712

Study Questions

1. Reticulocytes (immature red blood cells) contain mitochondria that are capable of both aerobic and anaerobic oxidation of glucose. In an experiment using these cells, incubated in oxygenated Krebs–Ringer solution with 10 mM glucose, the addition of antimycin A produced the following changes in metabolite concentration after 15 min (From Ghosh, A. K. and Sloviter, H. A. (1973) *J. Biol. Chem.* **248**, 3035–3040). Interpret the observed changes in ATP, ADP, and AMP concentrations (see tabulation). Express the concentration of each component after addition of antimycin as a percentage of that before addition. Then plot the resulting figures for each compound in the sequence found in glycolysis, i.e., label the *X* axis as follows:

G6P — F6P — FDP — TP — etc.

Metabolite	Abbreviation	Concentrations (mmol/1 of cells) Before addition of antimycin	After addition of antimycin
Glucose 6-phosphate	G6P	460	124
Fructose 6-phosphate	F6P	150	30
Fructose 1,6-bisphosphate	FBP	8	33
Triose phosphates	TP	18	59
3-Phosphoglycerate	3PGA	45	106
2-Phosphoglycerate	2PGA	26	19
Phosphoenolpyruvate	PEP	46	34
Pyruvate	Pyr	126	315
Lactate	Lac	1125	8750
ATP		2500	1720
ADP		280	855
AMP		36	206

2. The following problem can be solved using standard reduction potentials (Table 6-8). Use $E°'$ (pH 7) values for NAD^+, enzyme-bound FAD, and fumarate of −0.32, 0.0, and −0.03 volts, respectively. Values of numerical constants are given in Table 6-1.

   a) Derive an equation relating the equilibrium constant for a reaction, K_{eq}, to differences in E_0'.

   b) Calculate the numerical values of K_{eq} for the reactions

   $$\text{Succinate} + NAD^+ \rightarrow \text{Fumarate} + NADH + H^+$$

   $$\text{Succinate} + FAD \rightarrow \text{Fumarate} + FADH_2$$

   at pH 7 and 25°C. The values should be calculated for succinate and the oxidant in the numerator.

3. Compare the catalytic cycles of the following enzymes:

   Peroxidase
   Cytochrome *c* oxidase
   Cytochrome P450

4. What chemical properties are especially important for the following compounds in the electron transport complexes of mitochondria?

   FAD or FMN
   Ubiquinone (coenzyme Q)
   Cytochrome *c*

5. Describe the operation of the F_1F_0ATP synthase of mitochondrial membranes.

6. In studies of mitochondrial function the following stoichiometric ratios have been measured.

   a) The P/O ratio: number of molecules of ATP formed for each atom of oxygen (as O_2) taken up by isolated mitochondria under specified conditions.
   b) The ratio of H^+ ions translocated across a mitochondrial inner membrane to the molecules of ATP formed.
   c) The ratio of H^+ ions pumped out of a mitochondrion to the number of molecules of ATP formed.

   Discuss the experimental difficulties in such measurements. How do uncertainties affect conclusions about the mechanism of ATP synthase? Are the ratios in (b) and (c) above necessarily equal? Explain.

7. Compare P/O ratios observed for mitochondrial respiration with the following substrates and conditions:

Study Questions

a) Oxidation of NADH by O_2.
b) Oxidation of succinate by O_2.
c) Dehydrogenation of ascorbate by O_2.

How would the ratio of ATP formed to the number of electrons passing from NADH through the respiratory chain differ for these three oxidants: O_2, fumarate, nitrite?

8. What is the mitochondrial glycerol phosphate shuttle? Is it utilized by plant cells? Explain.

9. What chemical reactions are included in these two important components of the nitrogen cycle (see also Fig. 24-1)?

   Nitrification
   Denitrification

10. What is the difference between a dioxygenase and a monooxygenase? What is meant by a cosubstrate for a monooxygenase?

11. The enzyme *p*-hydroxybenzoate hydroxylase utilizes a cosubstrate together with O_2 to form 3,4-dihydroxybenzoate. Indicate the mechanisms by which the bound FAD cofactor participates in the reaction.

12. What pterin-dependent hydroxylation reactions are important to the human body? Point out similarities and differences between flavin and pterin hydroxylase mechanisms.

13. Describe the basic properties of nitric oxide synthases (NOSs) and their varied functions in the body. What are the three different types of NOS? In what ways do they differ?

14. List several compounds that cause oxidative stress in cells and describe some chemical and physiological characteristics of each.

15. Propylene glycol is metabolized by several aerobic bacteria to acetoacetate, which can be catabolized as an energy source (see references 509a and 509b). The first step is conversion to an epoxide which reacts further in coenzyme M-dependent and CO_2-dependent reactions to form acetoacetate. Can you propose chemical mechanisms?

16. A group of slow-growing denitrifying bacteria obtain energy by oxidizing ammonium ions anaerobically with nitrite ions.[613,614]

$$NO_2^- + NH_4^+ \rightarrow N_2 + 2\,H_2O$$

Intermediate metabolites are hydroxylamine (H_2NOH) and hydrazine (N_2H_4). The reaction takes place within internal vesicles known as **anammoxosomes**. Unusual cyclobutane- and cyclohexane-based lipids in their membranes are thought to partially prevent the escape of the toxic intermediates from the anammoxosomes.[614]

Four protons may move from the cytoplasm into the vesicles for each ammonium ion oxidized. Can you write a reaction sequence? What is the Gibbs energy change for the reaction? How is ATP generated? See p. 1052.

Notes

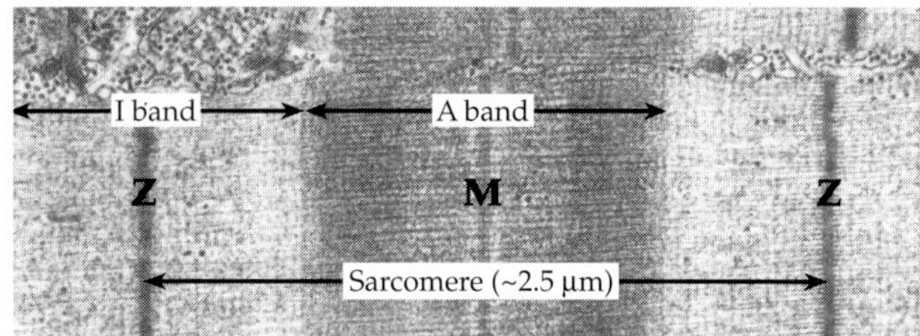

Electron micrograph of a thin longitudinal section of a myofibril from pig muscle. The basic contractile unit is the **sarcomere**, which extends from one Z line to the next. Thin **actin** filaments are anchored at the M lines and the thick **myosin** filaments at the Z lines. The (anisotropic) A bands are regions of overlap of interdigitated thick and thin filaments, while the I ("isotropic") bands are devoid of thick filaments. The ATP-driven contraction of muscle involves sliding of the interdigitated filaments and shortening of the sarcomere to ~1.8 μm. Micrograph courtesy of Marvin Stromer

Contents

The Chemistry of Movement

19

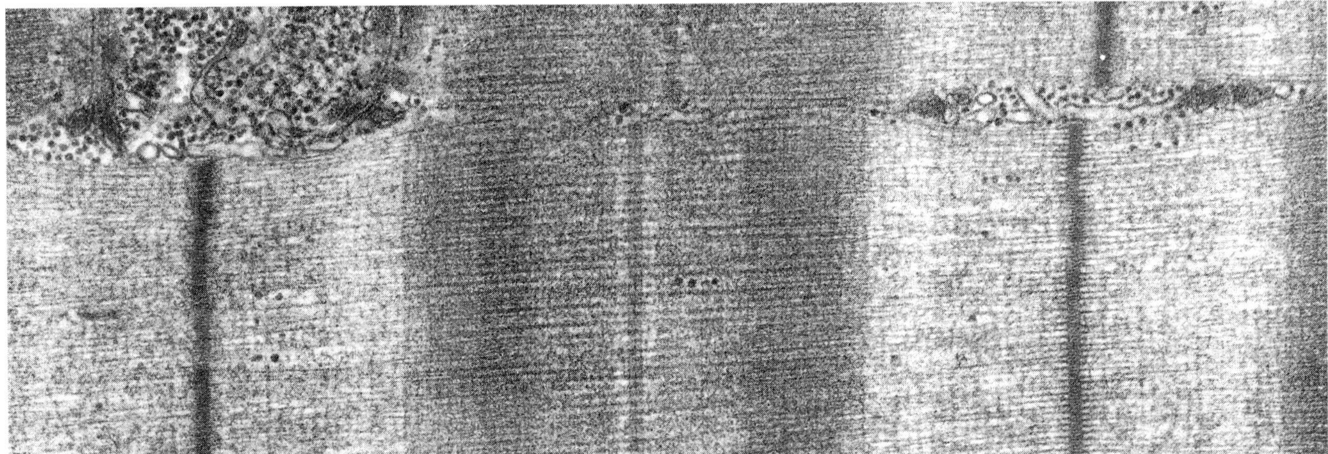

The swimming of bacteria, the flowing motion of the ameba, the rapid contraction of voluntary muscles, and the slower movements of organelles and cytoplasm within cells all depend upon transduction of chemical energy into mechanical work.

A. Motility of Bacteria

The smallest organs of propulsion are the bacterial flagella (Figs. 1-1, 1-3), and we have been able to unravel some of the mystery of movement by looking at them. When a cell of *E. coli* or *Salmonella* swims smoothly, each flagellum forms a left-handed superhelix with an ~2.3 µm pitch. Rotation of these "propellors" at rates of 100–200 revolutions / s (100–200 Hz) or more[1,2] in a counterclockwise direction, as viewed from the distal end of the flagellum, drives the bacterium forward in a straight line.[3–8] Several flagella rotate side-by-side as a bundle.[4] The observed velocities of 20–60 µm / s are remarkably high in comparison with the dimensions of the bacteria. Also remarkable is the fact that a cell may travel straight for a few seconds, but then tumble aimlessly for about 0.1 s before again moving in a straight line in a different direction. The tumbling occurs when the flagellum reverses its direction of rotation and also changes from a left-handed to a right-handed superhelix, which has just half the previous pitch.

Such behavior raised many questions. What causes reversal of direction of the propellor? Why do the bacteria tumble? How does a bacterium "decide" when to tumble? How is the flagellum changed from a left-handed to a right-handed superhelix? How does this behavior help the bacterium to find food? Most intriguing of all, what kind of motor powers the flagella? The answers are complex, more than 50 genes being needed to specify the proteins required for assembly and operation of the motility system of *E. coli* or *Salmonella typhimurium*.[9]

1. The Structure and Properties of Bacterial Flagella

Twenty or more structural proteins are present from the base to the tip of a complete bacterial flagellum. However, over most of their length the long thin shafts (Figs. 1-1, 19-1) are composed of subunits of single proteins called **flagellins**. Flagellin molecules have a high content of hydrophobic amino acids and, in *Salmonella*, contain one residue of the unusual N^{ε}-methyllysine. The subunits are arranged in a helix of outside diameter ~20 nm in which they also form 11 nearly longitudinal rows or **protofilaments**.[10–12a] Each subunit gives rise to one of the projections seen in the stereoscopic view in Fig. 19-1B. The flagella usually appear under the electron microscope to be supercoiled (Fig 19-1C–E) with a long "wavelength" (pitch) of ~2.5 µm. The supercoiled structure is essential for function, and mutant bacteria with straight flagella are nonmotile. Under some conditions and with some mutant flagellins, straight flagella, of the type shown in Fig. 19-1B, are formed. There is a central hole which is surrounded by what appears to be inner and outer tubes with interconnecting "spokes." However, all of the 494-residue flagellin subunits presumably have identical conformations, and each subunit contributes to both inner and outer tubes as well as to the outer projections. **Basal bodies** (Fig. 19-2) anchor the flagella to the cell wall and plasma membrane and contain the protic motors (Fig. 19-3) that drive the flagella.[14–16]

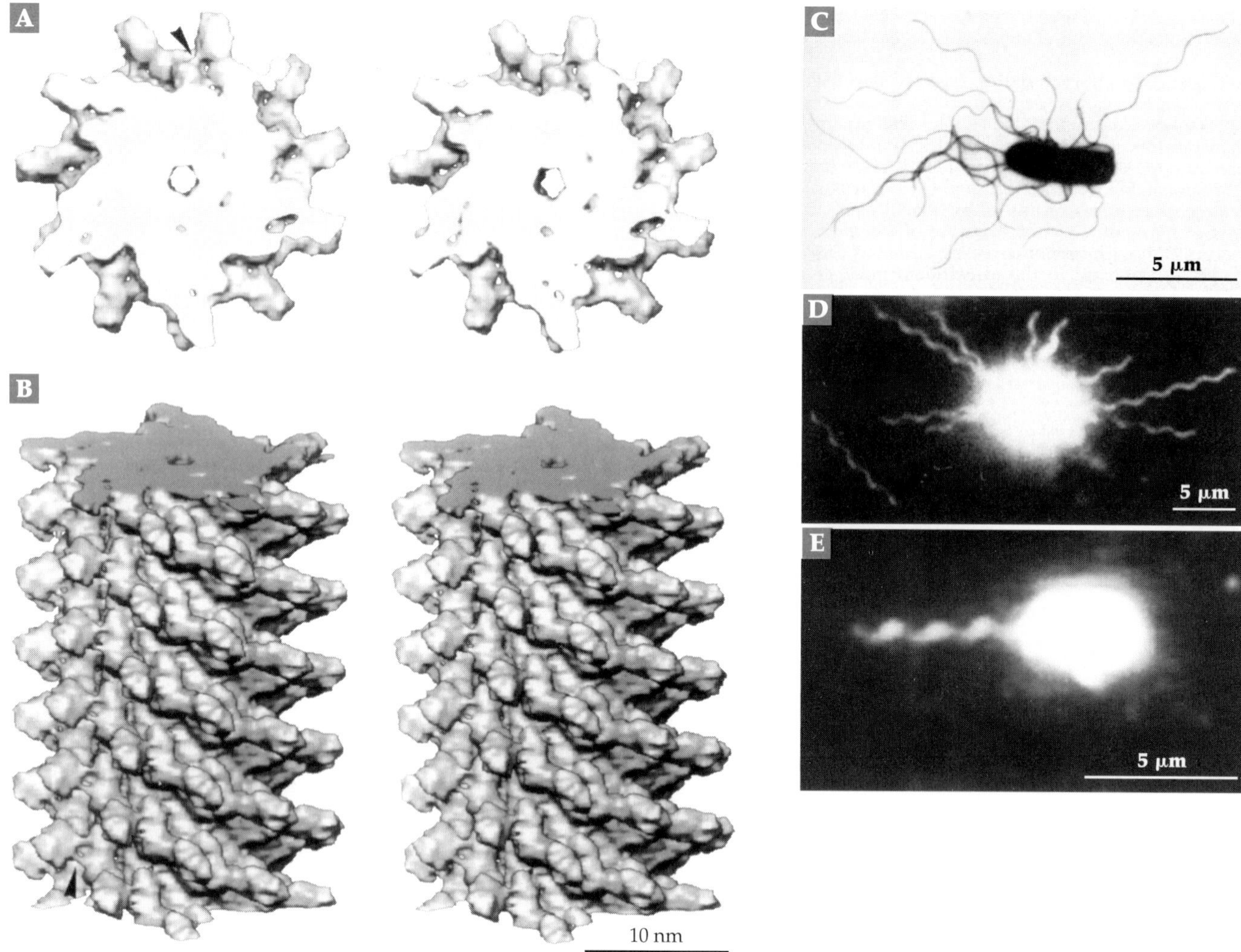

Figure 19-1 (A) Axial view of a 5-nm thick cross-section of the flagellar filament shown in (B). The 11 subunits form two turns of the one-start helix. (B) Stereoscopic oblique view of a 30-nm long section of a flagellum of *Salmonella typhimurium*. This is a straight flagellum from a nonmotile strain of bacteria. The structure was determined to a resolution of 0.9 nm by electron cryomicroscopy. From Mimori *et al.*[11] Courtesy of Keiichi Namba. (C) Electron micrograph of a cell of *S. typhimurium* showing peritrichous (all-around) distribution of flagella. Courtesy of S. Aizawa.[3] (D) Dark-field light micrograph of a flagellated cell of *S. typhimurium* with flagella dispersed during tumbling (see text). Courtesy of R. M. Macnab.[3] (E) Image of a cell of *Vibrio alginolyticus* obtained with dark-field illumination showing the single polar flagellum.[13] Because the illumination was strong, the size of the cell body and the thickness of the flagellum in the image appear large. Courtesy of Michio Homma.

Quasiequivalence. There are two distinct types of straight flagella: one (R) in which the protofilaments have a right-handed twist (as in Fig. 19-1) and the other (L) in which the protofilaments have a left-handed twist. These arise from two different conformations of the subunit proteins. Native supercoiled flagella contain a mixture of flagellins in the R- and L-states with all subunits in a given protofilament being in the same state. The supercoiling of the filament cannot be explained by stacking of identical subunits but is thought to arise because of an asymmetric distribution of protofilaments in a given state around the filament.[17–19a] Here, as with the icosahedral viruses (Chapter 7), quasiequivalence permits formation of a structure that would be impossible with full equivalence of subunits. The corkscrew shape of the flagellum is essential to the conversion of the motor's torque into a forward thrust.[18] Certain mutants of *Salmonella* have "curly" flagella with a superhelix of one-half the normal pitch. The presence of *p*-fluorophenylalanine in the growth medium also produces curly flagella, and normal flagella can be transformed to curly ones by a suitable change of pH. More important for biological function, the transformation from normal to curly also appears to take place during the tumbling of bacteria associated with chemotaxis.[17]

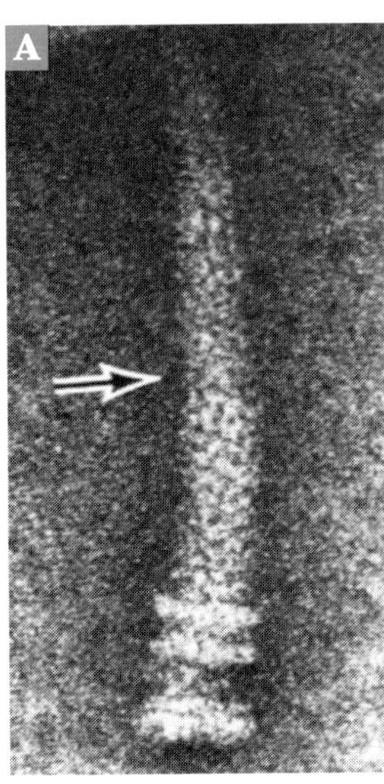

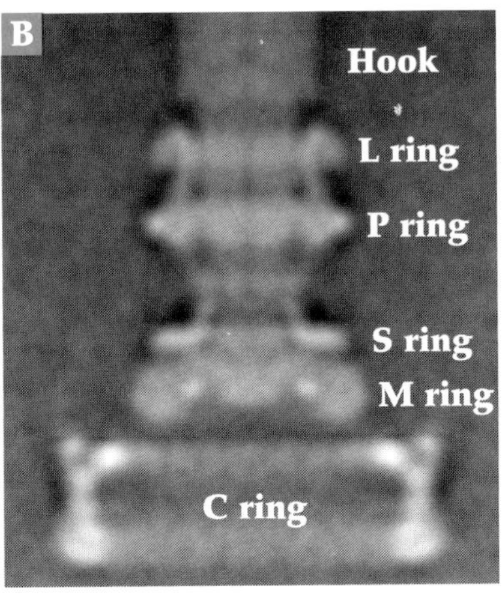

Figure 19-2 (A) Electron micrograph of a flagellum from *E. coli* stained with uranyl acetate. The M-and S-rings are seen at the end. Above them are the P-ring, thought to connect to the peptidoglycan layer, and the L-ring, thought to connect to the outer membrane or lipopolysaccharide layer (see Fig. 8-28). An arrow marks the junction between hook and thinner filament. From DePamphilis and Adler.[14] The hook is often bent to form an elbow. (B) Average of ~100 electron micrographs of frozen-hydrated preparations of basal bodies showing the cytoplasmic C-ring (see Fig. 19-3) extending from the thickened M-ring. From DeRosier.[16]

Growth of flagella. Iino added *p*-fluorophenylalanine to a suspension of bacteria, whose flagella had been broken off at various distances from the body.[20] Curly ends appeared as the flagella grew out. Unlike the growth of hairs on our bodies, the flagella grew from the outer ends. Because no free flagellin was found in the surrounding medium, it was concluded that the flagellin monomers are synthesized within the bacterium, then pass out, perhaps in a partially unfolded form, through the 2- to 3-nm diameter hole[10,12] in the flagella, and bind at the ends.[21] Flagella of *Salmonella* grow at the rate of 1 µm in 2–3 min initially, then more slowly until they attain a length of ~15 µm. More recent studies have provided details. The hook region (Fig. 19-3) grows first to a length of ~55 nm by addition to the basal-body rod of ~140 subunits of protein **FlgE**. During growth a **hook cap** formed from subunits **FlgD** prevents the FlgE subunits from passing out into the medium.[22,23] Hook subunits are added beneath the cap, moving the cap outward. Hook growth is terminated by protein **FlgK** (also called hook-associated protein Hap1). This protein displaces the hook cap and initiates growth of the main filament.[24] The first 10–20 subunits added are those of the FlgK (Fig. 19-3). These are followed by 10–20 subunits of **FlgL** (Hap3), a modified flagellin whose mechanical properties can accomodate the stress induced in the flagella by their rotation.[25] FlgJ is also needed for rod formation.[25a]

Growth of the flagellum to a length of up to 20 µm continues with subunits of **FliC** that are added at the tip, which is now covered by a dodecamer of the **cap protein FliD** (HAP2).[24,26,26a,b] Its 5-fold rotational symmetry means that this "star-cap" does not form a perfect plug for the 11-fold screw-symmetry of the flagellum, a fact that may be important in allowing new flagellin subunits to add at the growing tip. If the cap protein is missing, as in some *FliD* mutants, a large amount of flagellin leaks into the medium.[24]

Still unclear is how the protein synthesis that is taking place on the ribosomes in the bacterial cytoplasm is controlled and linked to "export machinery" at the base of the flagellin. As indicated in Fig. 19-3, the genetically identified proteins FlhA, FliH, and FliI are involved in the process that sends the correct flagellin subunits through the growing flagellum at the appropriate time. FliI contains an ATPase domain.[26c] FliS protein may be an export chaperone.[26d]

2. Rotation of Flagella

A variety of experiments showed that the flagellum is a rigid propellor that is rotated by a "motor" at the base. For example, a bacterium, artificially linked by means of antibodies to a short stub of a flagellum of another bacterium, can be rotated by the second bacterium. Rotation of cells tethered to a cover slip has also been observed. Although it is impossible to see individual flagella on live bacteria directly, bundles of flagella and even single filaments (Fig. 19-1C) can be viewed by dark-field light microscopy.[8,29] Normal flagella appear to have a left-handed helical form, but curly *Salmonella* flagella, which have a superhelix of one-half the normal pitch, form a right-handed helix.[5] Normal bacteria swim in straight lines but periodically "tumble" before swimming in a new random direction. This behavior is part of the system of **chemotaxis** by which the organism moves toward a food supply.[30] Curly mutants tumble continuously. When bacteria tumble the flagella change from normal to curly. The pitch is reversed and shortened. A proposed mechanism for the change of pitch involves propagation of cooperative conformational changes down additional

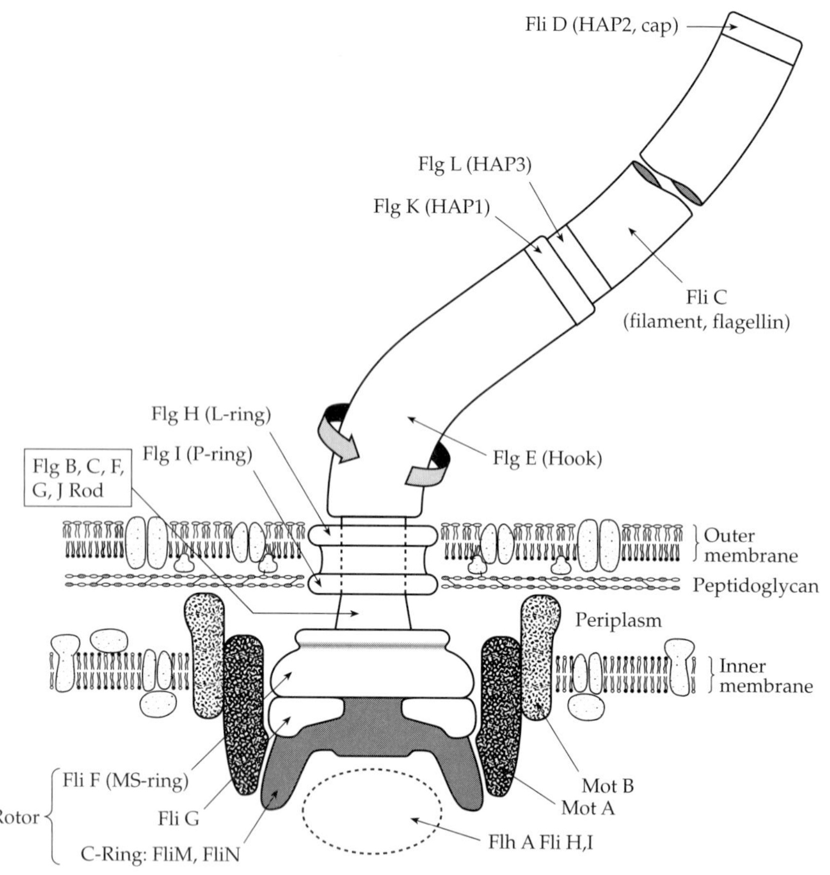

Figure 19-3 Schematic drawing of bacterial flagellar motor. Based on drawings of Berg,[27] Zhou and Blair,[28] and Elston and Oster.[1]

rows of flagellin subunits.[31]

There are no muscle-type proteins in the flagella. By incubating flagellated bacteria with penicillin and then lysing them osmotically, Eisenbach and Adler obtained cell envelopes whose flagella would rotate in a counterclockwise fashion if a suitable artificial electron donor was added.[32] This and other evidence showed that ATP is not needed. Rather, the torque developed is proportional to the **protonmotive force** and, under some circumstances, to ΔpH alone. It is the flow of protons from the external medium into the cytoplasm that drives the flagella.[8] Movement of *E. coli* cells in a capillary tube can also be powered by an external voltage.[33] In alkalophilic strains of *Bacillus* and some *Vibrio* species a sodium ion gradient will substitute.[13] Several hundred protons or Na^+ ions must pass through the motor per revolution.[8] Some estimates, based on energy balance,[29] are over 1000. However, Na^+-dependent rotation at velocities of up to 1700 Hz has been reported for the polar flagellum of *Vibrio alginolyticus*. It is difficult to understand how the bacterium could support the flow of 1000 Na^+ per revolution to drive the flagellum.[2]

What kind of protic motor can be imagined for bacterial flagella? Electron microscopy reveals that the flagellar hook is attached to a rod that passes through the cell wall and is, in turn, attached to a thin disc, the **M-ring** (or MS-ring), which is embedded in the cytoplasmic membrane both for gram-positive and gram-negative bacteria (Fig. 19-3). Two additional rings are present above the M-ring of flagella from gram-negative bacteria. The P-ring interacts with the peptidoglycan layer, and the L-ring contacts the outer membrane (lipopolysaccharide; Fig. 19-3). A logical possibility is that the M-ring, which lies within the plasma membrane, is the rotor, and a ring of surrounding protein subunits is the stator for the motor (Fig. 19-4).[34,35] Glagolev and Skulachev suggested in 1978 that attraction between $-COO^-$ and $-NH_3^+$ groups provides the force for movement.[34] Protons passing down an H^+-conducting pathway from the outer surface could convert $-NH_2$ groups to $-NH_3^+$, which would then be attracted to the $-COO^-$ groups on the stator subunits. When these two oppositely charged groups meet, a proton could be transferred from $-NH_3^+$ to $-COO^-$ destroying the electrostatic attraction. At the same time, movement of the M-ring would bring the next $-NH_2$ group to the H^+-conducting pathway from the outside. The $-COOH$ of the stator would now lose its proton through a conducting pathway to the inside of the bacterium, the proximity of the new $-NH_3^+$ assisting in this proton release. Since that time, other models based on electrostatic interactions have been advanced.[1,29,36]

Approximately 40 genes are required for assembly of the flagella, but mutations in only five motility genes have produced bacteria with intact flagella that do not rotate. Among these genes are *motA, mot B, FliG, FliM,* and *FliN*.[16,29,37,37a] Infection with a lambda transducing bacteriophage carrying functional *motB* genes restores motility to *motB* mutants by inducing synthesis of the motB protein. Block and Berg observed rotation of single bacteria tethered to a coverslip by their flagella. As the synthesis of the motB protein increased,the flagellar rotation rate increased in as many as 16 steps. This suggested that as many as 16 subunits of the *motB* gene product may contribute to the operation of the motor.[38] Later studies suggest eight subunits[39] rather than 16.

Both the M-ring and the thin S-ring, which lies directly above it and is now usually referred to as the MS-ring, are formed from ~20–25 subunits of the 61-kDa **FliF** protein.[39] Both the MotA and MotB proteins are embedded in the inner bacterial membrane and appear to form a circular array of "studs" around the M-ring.[16] MotA has a large cytosolic domain as well as four predicted trans-membrane helices[40] while MotB has a large periplasmic domain and probably binds to the peptidoglycan.[37,41,41a] The MotA and MotB proteins, which bind to each other, are thought to form the ~8 functional units in the stator of the motor.[37] Proteins FliG, FliM, and FliN are evidently parts of the rotor assembly. FliM and FliN form an additional ring, the cytoplasmic or **C-ring**, which had been difficult to see in early electron microscopy. As many as 40 of each of these subunits may be present in the ring.[42,43] A ring of FliG subunits joins the C-ring to the MS-ring (Fig. 19-3). FliE is also a part of the basal body.[25a]

From study of mutants it has been concluded that three charged residues of FliG, R279, D286, and D287 are directly involved in generation of torque by the motor.[44] Side chains of these residues may interact with the cytoplasmic domains of MotA and MotB. Residues R90 and E98 of MotA may be involved in controlling proton flow through the motor units.[28,44] The two prolines P173 and P122 are also essential for torque generation.[28]

There are obvious similarities between the flagellar motors and the protic turbines of ATP synthases (Fig. 18-14), but there are also substantial differences. It apparently takes about 12 protons for one revolution of the ATP synthase but about 1000, or ~125 per motor unit, for rotation of a bacterial flagellum. Elston and Oster propose an ion turbine more complex than that of ATP synthase. They suggest that the rotor might contain about 60 slanted rows of positively charged groups spaced as shown in Fig. 19-4. The motor is reversible, i.e., it can rotate in either direction. One possibility is that the subunits alter their conformations cooperatively in such a way that the slant of the rows of charged groups is reversed. Other possibilities for altering the constellation of charges via conformational changes can be imagined.[1] See also Thomas *et al.*[44a]

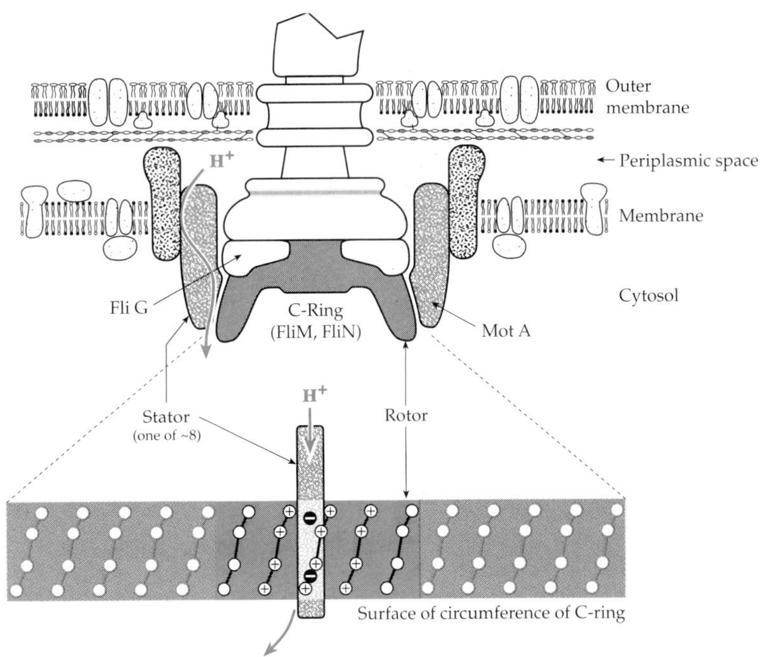

Figure 19-4 Schematic drawing of a hypothetical configuration of rotor and one stator unit in a flagellar motor as proposed by Elston and Oster.[1] The rotor can hold up to 60 positive charges provided by protons flowing from the periplasm through the stator motor units that surround the C-ring and hopping from one site to the next along the slanted lines. The rotor is composed of 15 repeating units, each able to accommodate four protons. Negative charges on the stator units are 0.5 nm from the rotor charges at their closest approach. For details see the original paper.

3. Chemotaxis

The flagellar motor is reversible, and in response to some signal from the bacterium it will turn in the opposite direction. At the same time, the flagellin subunits and those of the hook undergo conformational changes that change the superhelical twist. Perhaps synchronous conformational changes in the M-ring also are associated with the change in direction of rotation and are induced by interaction with a **switch complex** that lies below the M-ring. This consists of proteins FliG, FliM, and FliN.[44b] Mutations in any one of these proteins lead to the following four phenotypes: absence of flagella, paralyzed flagella, or flagella with the switch biased toward clockwise or toward counterclockwise rotation.[45]

What signals a change in direction of rotation? The answer lies in the attraction of bacteria to compounds that they can metabolize. Bacteria will swim toward such compounds but away from repellent substances, a response known as **chemotaxis**. Cells of *E. coli* swim toward higher concentrations of L-serine (but not of D-serine), of L-aspartate, or of D-ribose.

Phenol and Ni^{2+} ions are repellent.[46–48] By what mechanism can a minuscule prokaryotic cell sense a concentration gradient? It is known that the plasma membrane contains receptor proteins, whose response is linked to control of the flagella. Since the dimensions of a bacterium are so small, it would probably be impossible for them to sense the difference in concentration between one end and the other end of the cell. The chemotatic response apparently results from the fact that a bacterium swims for a relatively long time without tumbling when it senses that the concentration of the attractant is increasing with *time*. When it swims in the opposite direction and the concentration of attractant decreases, it tumbles sooner.[49]

Koshland[47] proposed that as the membrane receptors become increasingly occupied with the attractant molecule, the rate of formation v_f of some compound X, within the membrane or within the bacterium, is increased (Eq. 19-1). When [X] rises higher than a threshold level, tumbling is induced. At the same time, X is destroyed at a velocity of v_d.

$$\xrightarrow{v_f} [X] \xrightarrow{v_d} \qquad (19\text{-}1)$$

Subsequently, a readjustment of v_f and v_d occurs such that the concentration of X falls to its normal steady state level. X would act directly on the flagellar motor.

The receptors for L-serine[50–51a] and L-aspartate[52,53] are 60-kDa proteins encoded by genes *tsr* and *tar* in *Salmonella* or *E. coli*.[46,54] These proteins span the inner plasma membrane of the bacteria as shown in Figs. 11-8 and 19-5. The functioning of the receptor has been discussed in some detail in Chapter 11. However, there is still much that is not understood. The symmetric head, whose structure is known (Fig. 11-8), has two binding sites, but the aspartate receptor binds only one aspartate tightly. There is substantial evidence that suggests a piston-type sliding of one helix toward the cytoplasm as part of the signaling mechanism.[54a] While the flagella are distributed around the cell, the receptors appear to be clustered at the cell poles.[55]

Proteins encoded by genes *cheA*, *cheW*, *cheY* and *cheZ*, *cheB*, and *ChR* are all involved in controlling chemotaxis.[48,56] Their functions are indicated in the scheme of Fig. 19-5. All of the corresponding protein products have been isolated and purified, and the whole chemotaxis system has been reconstituted in phospholipid vesicles.[57] Gene *CheA* encodes a 73-kDa protein kinase, which binds as a dimer to the cytoplasmic domains of the related aspartate, serine, and ribose/galactose receptors with the aid of a coupling protein, cheW (Fig. 19-5). A great deal of effort has been expended in trying to understand how binding of an attractant molecule to the periplasmic domain of the receptor can affect the activity of the CheA kinase, but the explanation is unclear. There is a consensus that a small but distinct conformational alteration is transmitted through the receptor.[58–61a] An apparently α-helical region containing methylation sites (Fig. 19-5) appears to be critically involved in the signaling, responding not only to occupancy of the receptor site but also to intracellular pH and temperature and to methylation. Mutation of the buried Gly 278 found in this region to branched hydrophobic amino acids, such as Val or leucine, locks the receptor in state with a superactivated CheA kinase, while *substitution* of Gly 278 with aspartate leaves the kinase *inactive*.[61] Occupancy of the normal receptor site with ligand (aspartate, serine, etc.,) dramatically decreases the kinase activity.

The CheA protein is an autokinase which, upon activation by the receptor, becomes phosphorylated on N^{ε} of the imidazole ring of His 48. It then transfers this phospho group from His 48 to the carboxylate of Asp 57 of the 654-residue protein **CheY**, which is known as the **response regulator**.[62–65d] The unregulated flagellum rotates counterclockwise (CCW). *Phospho-CheY* (CheY-*P*, which qualifies as X in Eq. 19-1) *carries the message to the flagellar motor to turn clockwise (CW).* This is apparently accomplished through the binding of CheY-*P* to the N-terminal portion of protein FliM. This presumably induces a conformational change, which is propagated to FliG and to all of the proteins of the rotor and flagellar rod, hook, and filament.[45,65,66,66a] The flagella fly apart, and the bacterium tumbles and heads randomly in a new direction.

Tumbling occurs most often when receptors are unoccupied, and the bacteria change directions often, as if lost. However, if a receptor is occupied by an attractant, the activity of CheY is decreased and less CheY-*P* will be made. The carboxyl phosphate linkage in this compound is labile and readily hydrolyzed, a process hastened by the phosphatase **CheZ**.[67–69] Consequently, in the presence of a high enough attractant concentration the tumbling frequency is decreased, CCW flagellar rotation occurs, and the bacterium swims smoothly for a relatively long time.

There are still other important factors. Occupancy of the receptor by a ligand makes the receptor protein itself a substrate for the chemotaxis-specific methyltransferase encoded by the *cheR* gene.[62,70,71] This enzyme transfers methyl groups from *S*-adenosylmethionine to specific glutamate side chains of the receptor to form methyl esters. In the aspartate receptor there are four such glutamate residues in a large cytoplasmic domain that includes the C terminus. Two of these glutamates are initially glutamines and can undergo methylation only if they are deaminated first.[72] An esterase encoded by the *cheB* gene[72] removes the methyl ester groupings as methanol.

The action of the CheR methyltransferase is apparently unregulated, but the esterase activity of CheB is controlled by the phosphorylation state of the

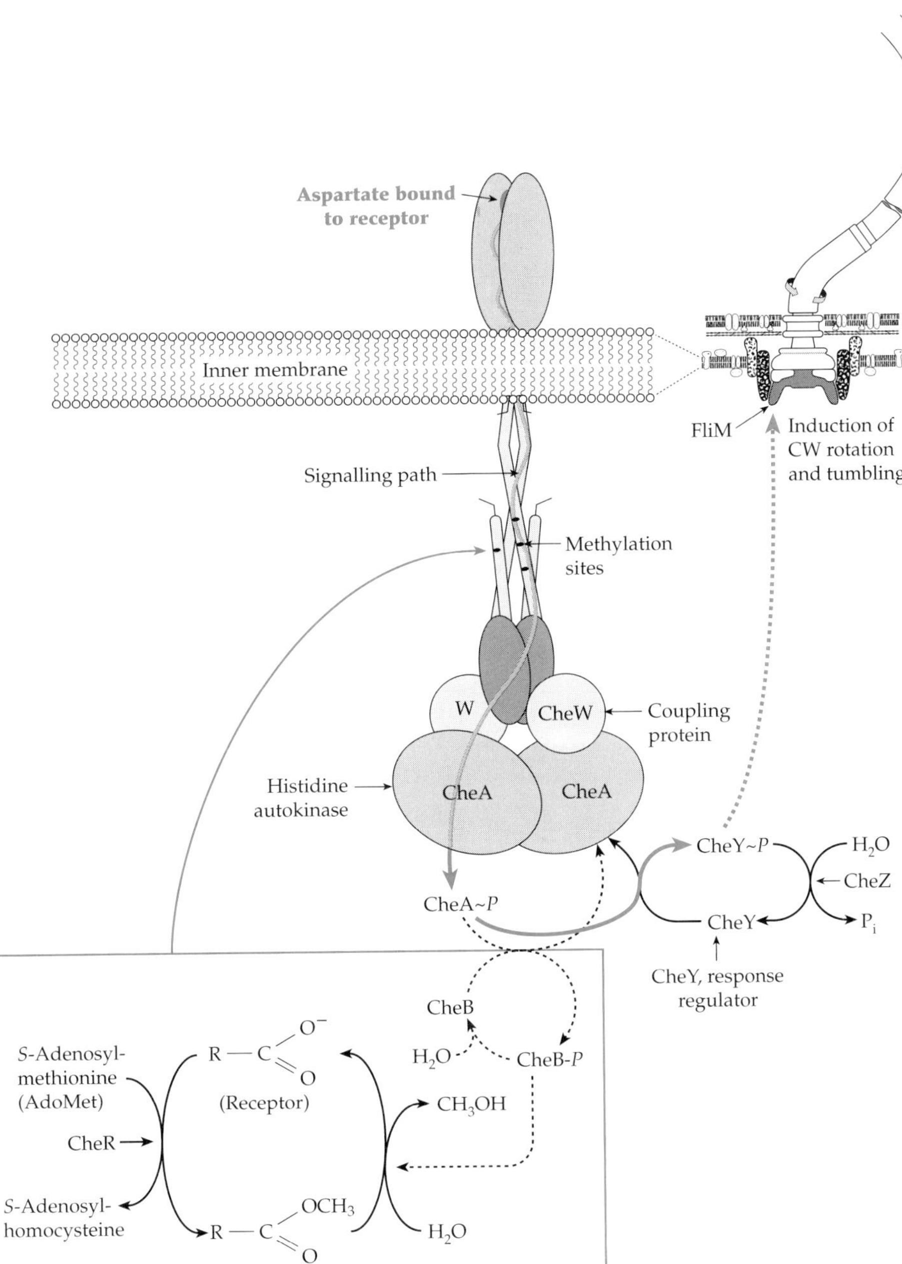

Figure 19-5 Schematic representation of an important chemotactic system of *E. coli*, *S. typhimurium*, and other bacteria. The transmembrane receptor activates the autokinase CheA, which transfers its phospho group to proteins CheY and CheB to form CheY-*P* and CheB-*P*. CheY-*P* regulates the direction of rotation of the flagella, which are distributed over the bacterial surface. CheR is a methyltransferase which methylates glutamate carboxyl groups in the receptor and modulates the CheA activity. CheZ is a phosphatase and CheB-*P* a methylesterase.

autokinase CheA. CheB competes with CheY (Fig. 19-5), and CheB-*P* is the active form of the esterase. After a chemotactic stimulus the level of CheA-*P* falls and so does the activity of the methylesterase. The number of methyl groups per receptor rises making the CheA kinase more active and opposing the decrease in kinase activity caused by receptor occupancy. The system is now less sensitive to the attractant; the bacterium has *adapted* to a higher attractant concentration.[62,73,73a] It tumbles more often unless the attractant concentration rises; if it is headed toward food tumbling is still inhibited. If it is headed away from the attractant the levels of both CheY-*P* and ChB-*P* rise. A high level of fumarate within the cell also acts on the switch–motor complex and favors CW rotation.[74]

For some bacterial attractants such as D-galactose, D-ribose, maltose, and dipeptides[75] the corresponding binding proteins,[38,76] which are required for the sugar uptake (e.g., Fig. 4-18A), are also necessary for chemotaxis. The occupied binding proteins apparently react with membrane-bound receptors to trigger the chemotactic response. The aspartate receptor (*tar* gene product) appears also to be the receptor for the maltose-binding protein complex,[47] and both the aspartate and the serine receptor (*tsr* gene product) also mediate thermotaxis and pH taxis.[77,77a] Clusters of identical receptors may function cooperatively to provide high sensitivity and dynamic range.[77b]

B. Muscle

There is probably no biological phenomenon that has excited more interest among biochemists than the movement caused by the contractile fibers of muscles. Unlike the motion of bacterial flagella, the movement of muscle is directly dependent on the hydrolysis of ATP as its source of energy. Several types of muscle exist within our bodies. **Striated** (striped) **skeletal muscles** act under voluntary control. Closely related are the **involuntary striated heart muscles,** while **smooth involuntary muscles** constitute a third type. Further distinctions are made between fast-twitch and slow-twitch **fibers**. **Fast-twitch fibers** have short isometric contraction times, high maximal velocities for shortening, and high rates of ATP hydrolysis. They occur predominately in white muscle. Because of the absence of the strong oxidative metabolism found in red muscles, fast-twitch fibers fatigue rapidly. Although red muscle sometimes contains fast-twitch fibers, it more often consists of **slow-twitch fibers,** which have a longer contraction time, low shortening velocity, and low ATPase. They are more resistant to fatigue[56] than fast-twitch fibers.[78] Embryonic muscle contains fast-twitch fibers as well as embryonic forms which contract slowly.[79] Some organisms contain specialized types of muscle. For example, the asynchronous flight muscles of certain insects cause the wings to beat at rates of 100–1000 Hz. In these muscles nerve impulses are used only to start and to stop the action; otherwise the cycle of contraction and relaxation continues automatically.[80] The adductor muscles, which close the shells of oysters and clams, can sustain large tensions for long periods of time with little expenditure of energy. This is accomplished by a **catch mechanism**.[81]

1. The Structural Organization of Striated Muscle

Skeletal muscles consist of bundles of long **muscle fibers,** which are *single cells* of diameter 10–100 µm formed by the fusion of many embryonic cells. The lengths are typically 2–3 cm in mammals but may sometimes be as great as 50 cm. Each fiber contains up to 100–200 nuclei. Typical cell organelles are present but are often given special names. Thus, the plasma membrane (plasmalemma) of muscle fibers is called the **sarcolemma**. The cytoplasm is **sarcoplasm,** and mitochondria may be called **sarcosomes**. The major characteristic of muscle is the presence of the contractile **myofibrils**, organized bundles of proteins 1–2 µm in diameter and not separated by membranes from the cytoplasm. Since they occupy most of the cytoplasm, a substantial number of myofibrils are present in each muscle fiber.

In the light microscope cross striations with a repeating distance of ~2.5 µm can be seen in the myofibrils (Figs. 19-6 and 19-7). The space between two of the dense **Z-discs** (Z lines) defines the **sarcomere**, the basic contractile unit. In the center of the sarcomere is a dense **A-band** (anisotropic band). The name refers to the intense birefringence of the band when viewed with plane polarized light. Straddling the Z-discs are less dense **I-bands** (the abbreviation stands for isotropic, a misnomer, for although the bands lack birefringence, they are not isotropic). Weakly staining **M-lines** (usually visible only with an electron microscope) mark the centers of the A-bands and of the sarcomeres.

The fine structure of the sarcomere was a mystery until 1953, when H. E. Huxley, examining thin sections of skeletal muscle with the electron microscope, discovered a remarkably regular array of interdigitated protein filaments.[82,83] **Thick filaments**, 12–16 nm in diameter and ~1.6 µm long, are packed in a hexagonal array on 40- to 50-nm centers throughout the A-bands (Fig. 19-6B). Between these thick filaments are **thin filaments** only 8 nm in diameter and extending from the Z-line for a length of ~1.0 µm. When contracted muscle was examined, it was found that the I-bands had become so thin that they had nearly disappeared and that the amount of overlap between the thick and the thin filaments had increased. This indicated that contraction had consisted of the sliding movement of the thick and thin filaments with respect to each other.[84] In skeletal muscle the sarcomere shortens to a length of ~2 µm, but in insect flight muscle a much smaller shortening occurs repetitively at a very high rate.

2. Muscle Proteins and Their Structures

The myofibrillar proteins make up 50–60% of the total protein of muscle cells. Insoluble at low ionic strengths, these proteins dissolve when the ionic strength exceeds ~0.3 and can be extracted with salt solutions. Analysis of isolated mammalian myofibrils[86] shows that nine proteins account for 96% or more of the protein; **myosin**, which constitutes the bulk of the thick filaments, accounts for 43% and **actin**, the principal component of the thin filaments, 22%.

Actin and the thin filaments. There are at least six forms of actin in adult mammalian tissues: α-cardiac, α-skeletal muscle, α- and γ-smooth muscle, β- and γ-cytoplasmic.[87–89] All of them are closely homologous, e.g., the 42-kDa α-skeletal muscle actin differs in only 4 of 375 residues from the α-cardiac form and only in 6 residues from the γ-smooth form. In almost all organisms actins contain one residue of N^{δ}-methylhistidine at position 73.[87,88,90] Actin is an unusual protein in that

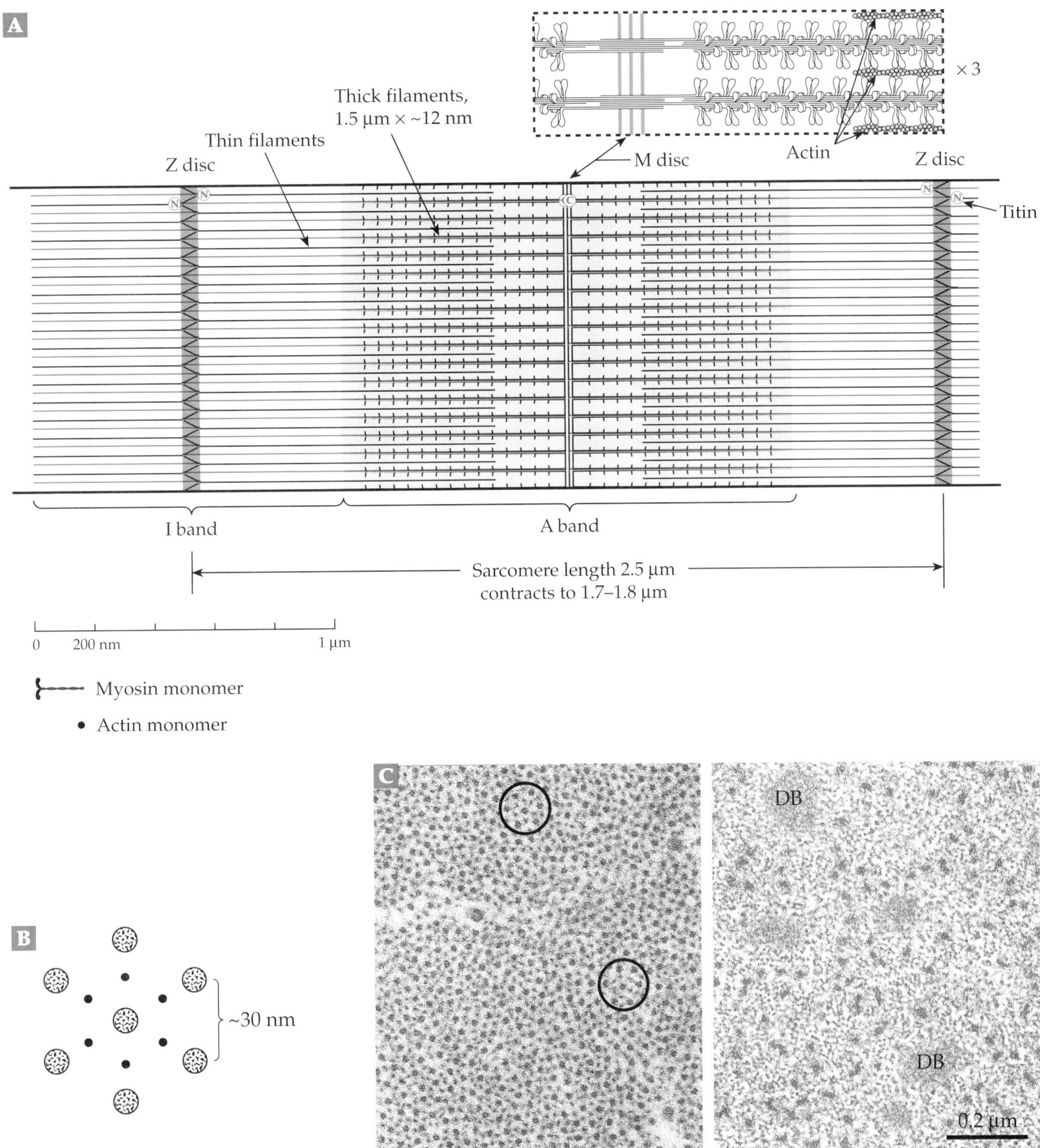

Figure 19-6 (A) The structure of a typical sarcomere of skeletal muscle. The longitudinal section depicted corresponds to that of the electron micrograph, Fig. 19-7A. The titin molecules in their probable positions are colored green. The heads of only a fraction of the myosin molecules are shown protruding toward the thin actin filaments with which they interact. A magnified section at the top is after Spudich.[85] It shows the interactions of the myosin heads with the thin filaments at the right-hand edge. (B) A sketch showing the arrangement of thick and thin filaments as seen in a transverse section of a striated muscle fiber. (C) Left: electron micrograph of a transverse section of a glycerated rabbit psoas muscle. The hexagonal arrangement of six thin filaments around one thick filament can be seen in the center of the circle. Six other thick filaments form a larger concentric circle as in (B). Right: transverse section of a smooth muscle fiber. Notice the irregular arrangement of thick and thin filaments. Filaments of intermediate diameter are also present, as are dense bodies (DB). The latter are characteristically present in smooth muscle.

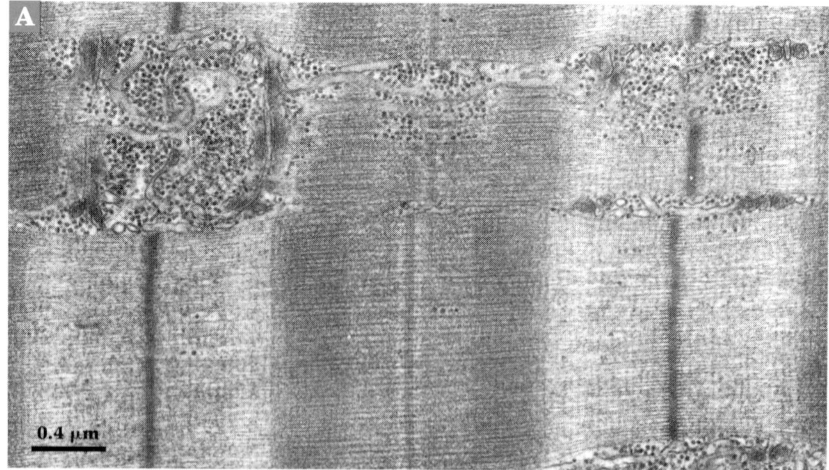

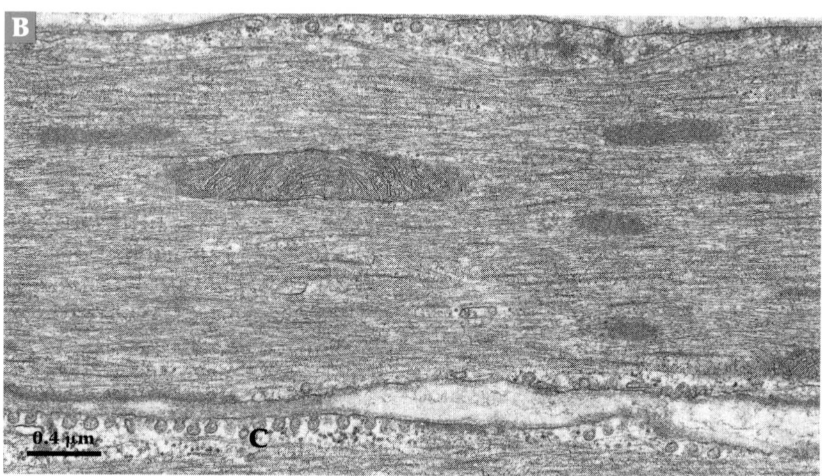

Figure 19-7 (A) Electron micrograph of a longitudinal section of a mammalian skeletal muscle (pig biceps muscle). The tissue was doubly fixed, first with formaldehyde and glutaraldehyde, then with osmium tetroxide. It was then stained with uranyl acetate and lead citrate. The section shows a white muscle fiber containing few mitochondria and narrow Z-lines. The Z-discs (marked Z), M-line, A- and I-bands, and thick and thin filaments can all be seen clearly. The periodicity of ~40 nm along the thin filaments corresponds to the length of the tropomyosin molecules, and the cross striation is thought to represent bound tropomyosin and troponin. The numerous dense particles in the upper part of the micrograph are glycogen granules, while the horizontal membranous structures are longitudinal tubules of the sarcoplasmic reticulum (endoplasmic reticulum). These come into close apposition to the T tubules leading from the surface of the muscle fiber. The T tubules (T) are visible in longitudinal section at the upper left of the micrograph on both sides of the Z-line and in cross-section in the upper right-hand corner. There a T tubule is seen lying between two lateral cisternae of the sarcoplasmic reticulum. (B) Longitudinal section of smooth muscle (chicken gizzard) fixed as in A. Thick filaments (Th), which are considerably thicker than those in striated muscle and less regular, can be seen throughout the section. They are surrounded by many thin filaments, which are often joined to dense bodies (DB). A mitochondrion (Mi) is seen in the center of the micrograph, and at the lower edge is a boundary between two adjacent cells. Notice the caveolae (C), which are present in large numbers in the plasma membrane and which are extremely active in smooth muscle. Micrographs courtesy of Marvin Stromer.

it can exist in both a filamentous and a soluble state. The interconversion between them is of great physiological importance. Actin filaments dissolve in a low ionic strength medium containing ATP to give the soluble, monomeric **G-actin**. Each G-actin monomer usually contains one molecule of bound ATP and a calcium ion.

Because of its tendency to polymerize, G-actin has been difficult to crystallize. However, it forms crystalline complexes with several other proteins, e.g., deoxyribonuclease I,[91] a fragment of gelsolin, and profilin,[92] which block polymerization and it has recently been crystallized as the free ADP complex.[92a] The three-dimensional structure of the actin is nearly the same in all cases. The molecule folds into four domains, the ATP binding site being buried in a deep cleft. The atomic structure (Fig. 7-10) resembles that of hexokinase, of glycerol kinase, and of an ATP-binding domain of a chaperonin of the Hsp 70 family.[90] As with the kinases, actin can exist in a closed and more open conformations, one of which is seen in the profilin complex. Addition of 1 mM Mg^{2+} or 0.1 M KCl to a solution of G-actin leads to spontaneous transformation into filaments of **F-actin** (Figs. 7-10 and 19-9) each containing 340–380 actin monomers and resembling the thin filaments of muscle.[93-94a] The ATP bound in the F-actin filament is hydrolyzed within ~100 s to ADP and P_i. However, the hydrolysis is not as rapid as polymerization so that a "cap" of ATP-containing monomers may be found at each end of the filament.[94,95,95a] There is a striking similarity to the binding of nucleotides to microtubule subunits (Fig. 7-33) and in the contractile tail sheath of bacteriophage (Box 7-C).

The two ends of the F-actin filaments have different surfaces of the monomer exposed and grow at different rates. This has been demonstrated by allowing the myosin fragment called heavy meromyosin (HMM; see Fig. 19-10) to bind to (or "decorate") an actin filament. The myosin heads bind at an angle, all pointed in one direction. This gives a "pointed" appearance to one end and a "barbed" appearance at the other. When monomeric actin is added to such an HMM-decorated F-actin filament the barbed ends grow much faster than the pointed ends.[94,96] In the intact sarcomere the ends that become pointed when decorated are free, while the opposite barbed ends of the filaments are attached at the Z-line (Fig. 19-6A). The existence in the cytoplasm of proteins that "cap" the fast-growing end of actin filaments thereby preventing further growth[96,97] suggests that cap proteins may be present at the ends of the thin filaments of the myofibrils.

Titin and nebulin. The third most abundant protein (10%), titin (also called **connectin**),[98–100a] is one of the largest of known proteins. Titin cDNA from human cardiac muscle encodes a 26,926-residue chain. Several tissue-specific isoforms of the protein are created by alternative mRNA splicing.[101] A single titin molecule stretches ~1200 nm from the Z-disc, where the N terminus is bound, to the M-line, where the C-terminal domain is attached (Fig. 19-8A). Thoughout much of the A-band titin binds to the thick filament and appears to be part of a scaffold for maintenance of the sarcomere structure. The I-band portion of titin has elastic properties that allow it to lengthen greatly or to shorten as the sarcomere changes length.[98,100,102]

Under the electron microscope titin appears as a flexible beaded string ~4 nm in diameter. Most of the molecule is made up of repetitive domains of two types. In human cardiac titin there are 132 folded domains that resemble type III fibronectin repeats and 112 immunoglobulin-like domains.[98] In a "PEVK region," between residues 163 and 2174, 70% of the residues are Pro, Glu, Val, or Lys. The titin molecule may be organized as polyproline helices in this elastic region.[102a] At the C terminus of titin 800 residues, including a Ser / Thr protein kinase domain, are found within the M-line.

Another very large protein, **nebulin** (3% of the total protein),[103] appears to be stretched alongside the thin filaments. In the electron microscope it appears as a flexible, beaded string ~4 nm in diameter. Ninety-seven percent of the 6669-residue human nebulin is organized as 185 copies of an ~ 35-residue module.[104,105] Nebulin has a proline residue at about every 35th position, possibly corresponding in length to the pitch of the actin helix (Fig. 7-10). At the C terminus is an SH3 domain (see Fig. 11-14), which is preceded by a 120-residue segment rich in potential phosphorylation sites.[106] This part of the peptide chain is anchored in the Z-discs (Fig. 19-8B, C). The three extreme N-terminal modules of nebulin bind to tropomodulin, which caps the pointed ends of thin filments.[106a] Avian cardiac muscle contains a much shorter 100-kDa protein called **nebulette**, which resembles the C-terminal parts of nebulin. Nebulin has been described as encoding a blueprint for thin filament architecture.[99,103]

Proteins of the M-line and Z-disc. The M-line region contains the structural protein **myomesin**, which binds to both titin and myosin and holds the two together.[107] Fast skeletal and cardiac fibers also contain another **M-protein**, which may bridge between myosin filaments. Both the C-terminal region of nebulin and the N termini of pairs of titin molecules meet in the Z-disc, where they are bound into a lattice containing **α-actinin**[98,108-109b] and other proteins (Fig. 19-8B). The dimeric α-actinin, a member of the spectrin family, has a subunit mass of ~97 kDa.[109a] Found primarily in the Z-discs, it is also present in nonmuscle cells in stress fibers and at other locations in the cytoskeleton (Chapter 7). It may anchor actin filaments to various structures outside of the sarcomere.[110] In the dense Z-disc of insect flight muscle a regular hexagonal lattice of α-actinin[111] and a large (500–700 kDa) modular protein called **kettin**[112,112a,b] bind the thin filaments of opposite polarity together.

The **C-protein** (thick filaments), myomesin (M-line protein), and α-actinin (Z-line protein)[110,113,114] each provide 2% of the protein in the myofibril. Less than 1% each of 11 or more other proteins may also be present within the sarcomere.[86,115] Several of these, including the cytoskeletal proteins **desmin** and **vimentin**, and **synemin** surround the Z-discs.[116,116a]

The regulatory proteins troponin and tropomyosin. These two proteins are also associated with the filaments, each one contributing ~5% to the total protein of myofibrils. Tropomyosin is an elongated α-helical coiled-coil molecule, each molecule of which associates with seven actin subunits of an actin filament. Troponin consists of three subunits known as troponins C, T, and I. The elongated troponin T binds to tropomyosin. Troponin I is an inhibitor of the interaction of myosin and actin necessary for muscle contraction. Troponin C, a member of the calmodulin family (Fig. 6-8), binds Ca^{2+} and induces conformational changes that relieve the inhibition and allow contraction to occur. Nebulin is also thought to bind to tropomyosin. A possible arrangement of one of the tropomyosin–troponin–nebulin complexes that lie along the length of the thin filaments is shown schematically and as a three-dimensional model in Figs. 19-8C and D. These proteins are discussed further in Section 4. Figure 19-9 shows a model of the thin filaments with tropomyosin coiled-coil molecules on each side. The troponin subunits are not shown.

Myosins. There are 15 distinct families of proteins within the myosin superfamily.[117–120] They vary greatly in size, but all of them bind and hydrolyze ATP, and all bind to actin. Most have C-terminal tails. At their N

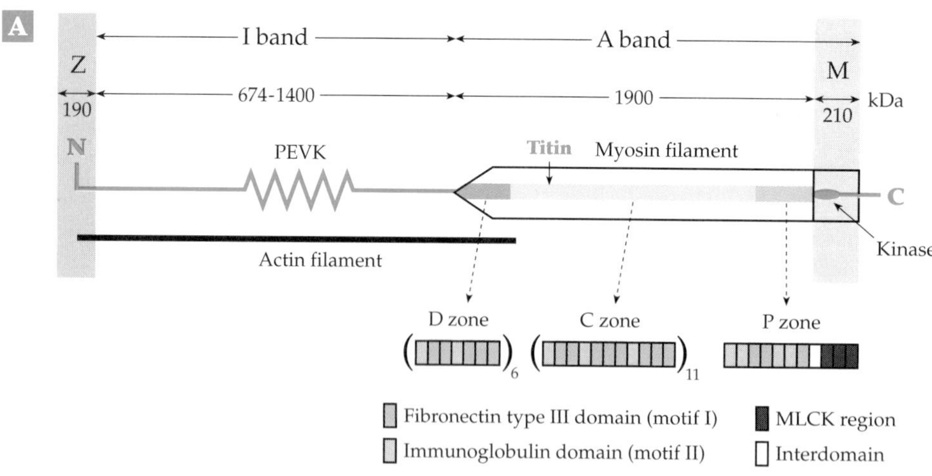

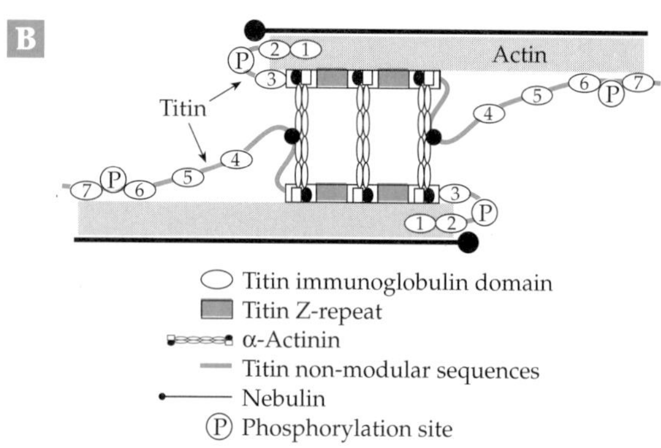

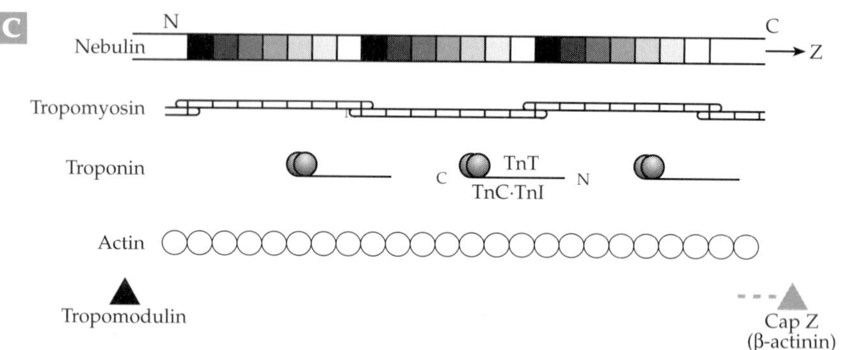

Figure 19-8 (A) Schematic drawing showing one molecule of titin (connectin) in a half sarcomere and its relationship to thick myosin filaments and thin actin filaments. The complex repeat patterns of fibronectin type III, immunoglobulin, in the three zones D, C, and P are also indicated.[98] See Maruyama.[98,98a] (B) Schematic drawing of the molecular structure of the sarcomere Z-disc. Titin, which is thought to parallel the thin filaments through the I-band, consists of various modules that are numbered from the N termini. In the Z-disc titin binds to α-actin, shown here as three vertical rods, and also to actin or actin-binding proteins. The SH3 domain (shown as a sphere) of nebulin and the N terminus of titin may interact. Regulatory phosphorylation sites are marked P. From Young *et al.*[108] Courtesy of Mathias Gautel. (C) Hypothetical model of a composite regulatory complex containing nebulin, tropomyosin, and troponin on the thin filaments of a skeletal muscle sarcomere. Each seven-module nebulin super-repeat (squares with graded shading) binds one tropomyosin, possible through the seven charge clusters along the length of each tropomyosin, and one troponin complex (shaded spheres with a tail). This complex consists of TnT, TnI, and TnC in orientations indicated by the N and C termini. Each nebulin super-repeat binds to seven actin monomers (open circles) along the thin filament. **Tropomodulin** caps the pointed ends of actin filaments and Cap Z, the "barbed ends." From Wang *et al.*[103]

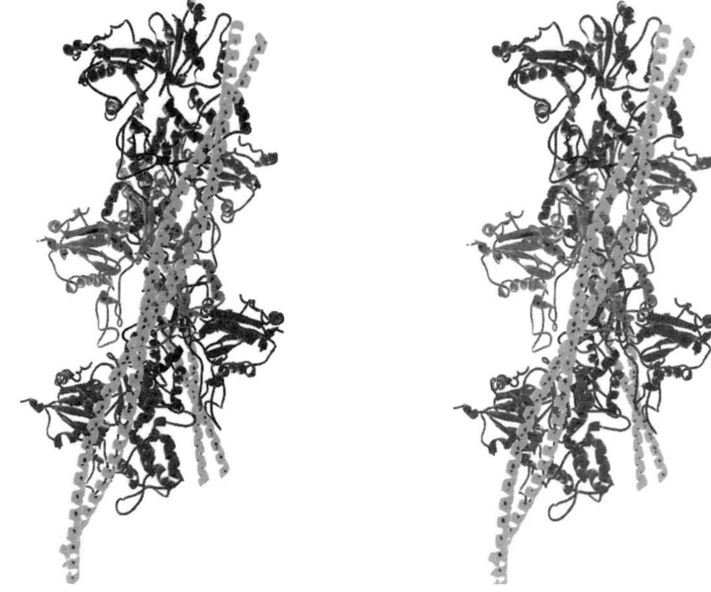

Figure 19-9 Stereoscopic ribbon drawing of the proposed structure of a thin actin filament with tropomyosin coiled-coils bound on opposing sides.[124] Five actin nomomers are assembled in the structure as is also illustrated in Fig. 7-10. From Lorenz *et al.*[125] Courtesy of Michael Lorenz.

termini are one or two globular heads, which contain the catalytic centers in which ATP hydrolysis occurs. Sizes vary from 93 kDa for a myosin with a very short tail from *Toxoplasma*[118] to over 300 kDa. Myosins I, found in ameboid organisms and also in our own bodies (for example in the microvilli of the brush border of intestinal epithelial cells), are small single-headed molecules.[117,121] Myosins II are the "conventional" myosins of myofibrils and are often referred to simply as myosin. However, each of the three muscle types (skeletal, cardiac, and smooth) has its own kind of myosin II.[121a–c] Likewise, at least six different genes have been identified for the light chains of the myosin heads.[122] Fast and slow muscle as well as embryonic muscle have their own light chains. Each myosin II molecule consists of two identical ~230-kDa **heavy chains,** which are largely α-helical, together with two pairs of smaller 16- to 21-kDa **light chains**. Human skeletal muscle heavy chains contain 1938 residues of which the first ~850 are folded into pear-shaped heads, which contain the catalytic sites involved in harnessing ATP cleavage to movement. Following proline 850 nearly all of the remaining 1088 residues form an α-helical coiled-coil rod of dimensions ~160 x 2 nm (Fig. 19-10) in which the two chains coil around each other. The two heavy chains are parallel, each having its N terminus in one of the two heads and its C terminus bound in the shaft of the thick filament.

Myosins II from other sources have similar structures. For example, analysis of the DNA sequence for a heavy chain gene from the nematode *Caenorhabditis* showed that the protein contains 1966 residues, 1095 of which contain an amino acid sequence appropriate for a 160-nm long coiled coil.[123] There are no prolines within this sequence, which lies between Pro 850 and Pro 1944. Although there are many bands containing positively and negatively charged side chains along the myosin rod, the interactions between the two coiled helices are largely nonpolar. In *Drosophila* 15 different heavy-chain isoforms are created by splicing of a single mRNA.[123a]

While the C-terminal portions of the two parallel myosin heavy chains form a rod, the N-terminal portions fold into two separate heads. Each head also contains two smaller 16- to 21-kDa peptide chains which belong to the calmodulin family. One of these, the **essential light chain**, is tightly bound to the heavy chain. The second, the **regulatory light chain**, is able to bind Ca^{2+} and is less tightly bound to the rest of the head. A short treatment with trypsin or papain cuts the myosin molecule into two pieces. The tail end gives rise to **light meromyosin (LMM)**, a molecule ~90 nm in length. The remainder of the molecule including the heads is designated **heavy meromysin (HMM)**. A longer trypsin treatment leads to cleavage of HMM into one ~62-kDa **S2** fragment 40 nm long, and two ~130-kDa **S1** fragments, each of the latter representing one of the two heads (Fig. 19-10).

The junction of the head and tail portions of myosin appears rigid in Fig. 19-10A. However, there must be considerable conformational flexibility and perhaps some uncoiling of the helices to allow the two heads to interact with a single thin filament as is observed by electron microscopy.[126,128] There also appears to be a hinge between the S2 and LMM segments (Figs. 19-10A and 19-14).

The thick filaments. Dissociated myosin molecules can be induced to aggregate into rods similar to the thick filaments of muscle.[129] Since the filaments

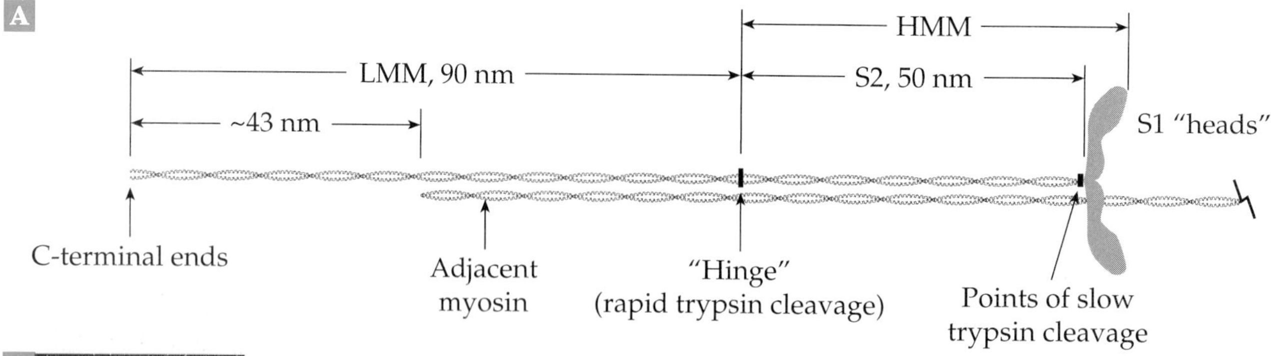

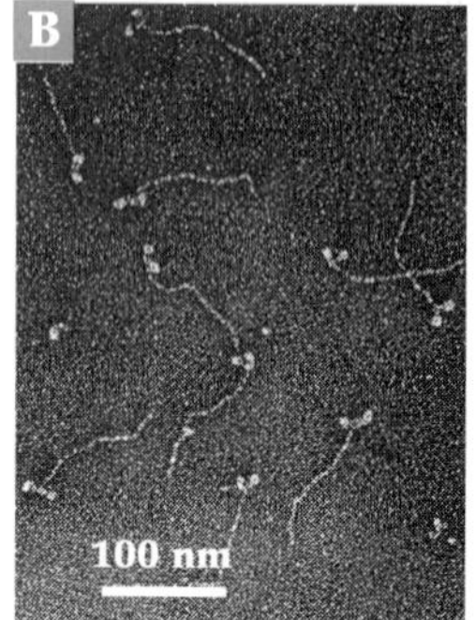

Figure 19-10 (A) An approximate scale drawing of the myosin molecule. The "hinge" is a region that is rapidly attacked by trypsin to yield the light and heavy meromyosins (LMM and HMM). Total length ~160 nm, molecular mass, 470 kDa; two ~200-kDa heavy chains; two pairs of 16- to 21-kDa light chains; heads: ~15 x 4 x 3 nm. (B) Electron micrograph of rabbit myosin monomers that became dissociated from thick filaments in the presence of ATP, fixed and shadowed with platinum.[127] Courtesy of Tsuyoshi Katoh.

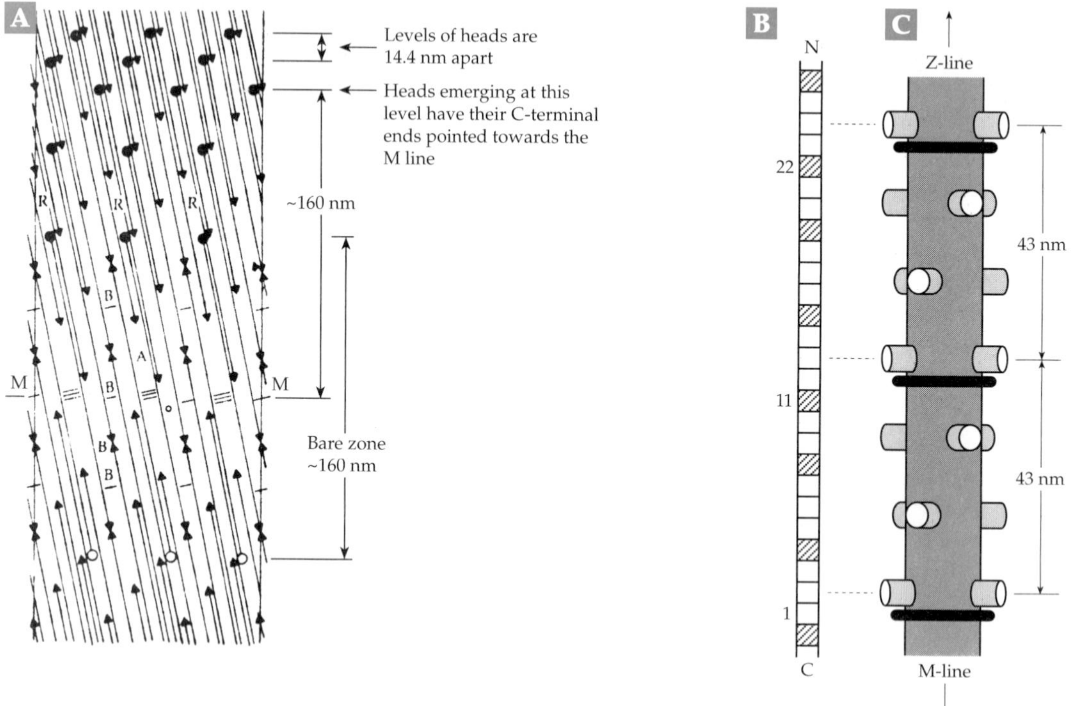

Figure 19-11 (A) Radial projection illustrating packing of myosin rods as suggested by Squire[130] for thick filaments of vertebrate skeletal muscle. The region of the bare zone at the M-line is shown. The filled circles represent the head ends of the myosin molecules and the arrowheads represent the other end of the rod, i.e., the end of the LMM portion. Antiparallel molecules interacting with overlaps of 43 and 130 nm are shown joined by single and triple cross-lines, respectively. Positions where two arrowheads meet are positions of end-to-end butting. O is an "up" molecule (thin lines) and A a "down" molecule (thick lines). The molecules move from the core at the C-terminal end to the filament surface at the head end. The levels marked B may be the levels of attachment of M-bridge material to the myosin filament. The level M-M is the center of the M-line and of the whole filament. The lateral scale is exaggerated more than threefold. (B) A segment of titin showing the 43-nm 11-domain super-repeat. (C) Model of a segment of a thick filament showing the 43-nm repeat, the C-protein, also bound at 43 nm intervals.[99] (B) and (C) Courtesy of John Trinick.

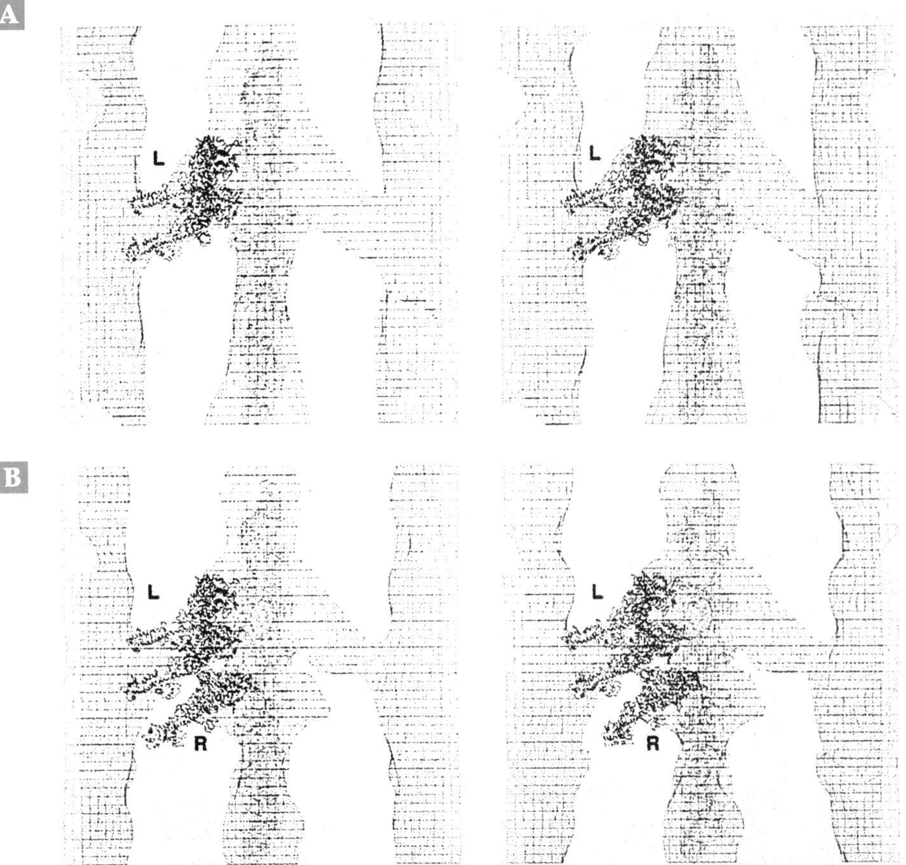

Figure 19-12 (A) Stereoscopic views of computer-assisted reconstructions of images of myosin heads attached to an F-actin filament centered between two thick filaments. Atomic structures of actin (Fig. 7-10) and of myosin heads (Fig. 19-15) have been built into the reconstructed images obtained by electron microscopy. (A) With the nonhydrolyzable ATP analog ATPPNP bound in the active sites. (B) Rigor. Two myosin heads are apparently bound to a single actin filament in (A). If they belong to the same myosin molecule the two C-terminal ends must be pulled together from the location shown here. In (B) a third head is attached, presumably from another myosin rod. This configuration is often seen in rigor. From Winkler *et al.*[134] Courtesy of K. A. Taylor.

have a diameter of ~14 to 20 nm, a large number of the thin 2-nm myosin molecules must be packed together. Electron microscopy reveals the presence of the heads projecting from the thick filaments at intervals of ~43.5 nm. However, there is a bare zone centered on the M-line, a fact that suggests tail-to-tail aggregation of the myosin monomers at the M-line in the centers of the thick filaments (see magnified section of Fig. 19-6, A). A helical packing arrangement involving about 300 myosin molecules (up to 30 rods in a single cross section) in close packing with a small central open core has been proposed for skeletal muscle myosin[130,131] and is illustrated in Fig. 19-11A,B. There are approximately three heads per turn of the helix, each group of three heads spaced 14.3 nm from the preceding one along the thick filament. It is apparently the zones of positive and negative charge, which are especially prevalent in the LMM segment toward the C termini, that lock the successive myosin molecules into this 14.3-nm spacing.[116,132] Titin also binds to the LMM segment of the myosin rod,[99,133] and its 11-domain super repeat of IgG-like and fibronectin-like modules are also 43 nm in length.[98,101] There are typically 47–49 of these super repeats in titin, and if each fits to a turn of the helix, as shown in Fig. 19-11B, there would be 147 myosin molecules in one-half of the thick filament.

Not all muscles have the thick filament structure of Fig. 19-11. In the tarantula muscle, which has a particularly well-ordered structure, there are four myosins per turn.[135,135a] Figure 19-13 shows a reconstruction of scallop myosin which has a 7-fold rotational symmetry. The thick filaments often contain

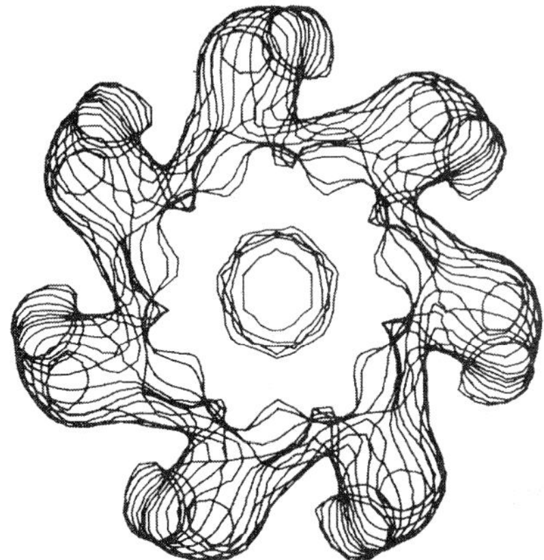

Figure 19-13 Superimposed sections for the 14 nm thickness of a computer-assisted reconstruction of the myosin filaments of the scallop adductor muscle. From Vibert and Craig.[137]

other proteins in addition to myosin. Thus, skeletal muscle contains the C-protein in a series of helical bands along the thick rod.[135b,c] In nematodes, molluscs, and insects the thick filament has a cylindrical core of **paramyosin**, another protein with a structure resembling that of the myosin rod. A minor component of *Drosophila* myosin, the **myosin rod protein**, lacks heads but is transcribed from the myosin heavy chain gene.[136]

3. Actomyosin and Muscle Contraction

That actin and myosin are jointly responsible for contraction was demonstrated long before the fine structure of the myofibril became known. In about 1929, ATP was recognized as the energy source for muscle contraction, but it was not until 10 years later that Engelhardt and Ljubimowa showed that isolated myosin preparations catalyzed the hydrolysis of ATP.[138] Szent-Györgi[139,140] showed that a combination of the two proteins actin (discovered by F. Straub[141]) and myosin was required for Mg^{2+}-stimulated ATP hydrolysis (ATPase activity). He called this combination **actomyosin**.

Under the electron microscope the myosin heads can sometimes be seen to be attached to the nearby thin actin filaments as **crossbridges**. When skeletal muscle is relaxed (not activated by a nerve impulse), the crossbridges are not attached, and the muscle can be stretched readily. The thin filaments are free to move past the thick filaments, and the muscle has some of the properties of a weak rubber band. However, when the muscle is activated and under tension, the crossbridges form more frequently. When ATP is exhausted (e.g., after death) muscle enters the state of **rigor** in which the crossbridges can be seen by electron microscopy to be almost all attached to thin filaments, accounting for the complete immobility of muscle in rigor (Figs. 19-12, 14).[134]

In rigor the crossbridges are almost all firmly attached to the thin actin filaments, making an approximately 45° angle to the actin filaments.[142–144] However, the addition of ATP causes their instantaneous release and the relaxation of the muscle fiber. In contrast, activation by a nerve impulse, with associated release of calcium ions (Section B,4), causes the thin filaments to slide between the thick filaments with shortening of the muscle. An activated muscle shortens if a low tension is applied to the muscle, but at a higher tension it maintains a constant length. Because the maximum tension developed is proportional to the length of overlap between the thick and thin filaments, it was natural to identify the individual crossbridges as the active centers for generation of the force needed for contraction.

The rowing hypothesis. H. E. Huxley[145,146] and A. F. Huxley and R. M. Simmons[147] independently proposed that during contraction the myosin heads attach themslves to the thin actin filaments. The hydrolysis of ATP is then coupled to the generation of a tension that causes the thick and thin filaments to be pulled past each other. The heads then release themselves and become attached at new locations on the actin filament. Repetition of this process leads to the sliding motion of the filaments (Fig. 19-14). The evidence in favor of this "rowing" or "swinging bridge" hypothesis was initially based largely on electron microscopy. For example, contracting muscle was frozen rapidly and fixed for microscopy in the frozen state.[148] Relaxed muscle shows no attached crossbridges, but contracting muscle has many. However,

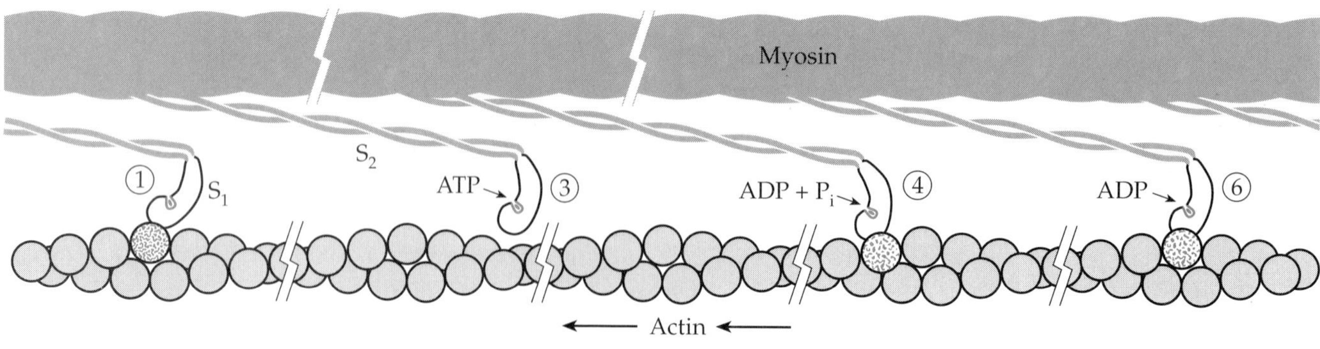

Figure 19-14 A model for the coupling of ATP hydrolysis to force production in muscle based on proposals of H. E. Huxley, and A. F. Huxley and Simmons. The power stroke is depicted here as a rotation of the crossbridge from a 90° to a 45° configuration. Four representative stages are shown: (1) the rigor complex, (3) the dissociated myosin ATP complex, (4) the actomyosin ADP pre-power stroke state in which the actin–myosin band has reformed but with a different actin subunit, which may be distant from that in (1), and (6) the actomyosin ADP post-power stroke state. Force production and contraction result from crossbridges passing cyclically through the steps depicted from left to right. Numbering of the stages corresponds approximately to that in Fig. 19-18. After H. Huxley.[146]

their appearance was distinct from that seen in rigor. The model was also supported by indirect physical methods.

An impressive demonstration that myosin heads do move along the actin filaments was provided by Sheetz and Spudich, who found that myosin-coated fluorescent beads ~0.7 μm in diameter will move along actin filaments from cells of the alga *Nitella* in an ATP-dependent fashion at velocities similar to those required in muscle.[149] The myosin heads literally glide along the thick cables of parallel actin filaments present in these algae.

Why two heads? The actin filament is a two-start helix, and it is natural to ask whether the two myosin heads bind to just one or simultaneously to both of the actin strands. Most evidence supports a 1:1 interaction of a single head with just one strand of actin. However, the other actin strand may associate with heads from a different thick filament. Another question concerns the role of the pairs of myosin heads. Could the two heads bind sequentially to the actin and exert their pull in a fixed sequence? In the reconstruction of the actomyosin complex in rigor (Fig. 19-12B) two different images are seen for the crossbridges. This suggests the existence of two different conformations for the attached myosin heads. Similar images for smooth muscle heavy meromyosin in its inactive (resting) dephosphorylated state (see p. 1116) show the two heads in very different orientations with one binding to the other of the pair and blocking its movement.[121b] Perhaps one head is tightly bound at the end of the power stroke while the other is at a different stage of the catalytic cycle. Nevertheless, single-headed myosin from *Acanthamoeba* will propell organelles along actin filaments,[150] and actin filaments will slide across a surface coated with single-headed myosin formed by controlled proteolysis.[151] The additional interactions seen in rigor may be peculiar to that state.

Structure of the myosin heads. Myosin and myosin fragments can be isolated in large quantities, but they have been difficult to crystallize. However, Rayment and coworkers purified S1 heads cleaved from chicken myosin by papain and subjected them to reductive methylation (using a dimethylamine–borane complex; see also Eq. 3-34). With most of the lysine side chain amino groups converted to dimethylamine groups, high-quality crystals were obtained, and a structure was determined by X-ray diffraction.[152] Since that time various forms of both modified and unmodified myosin heads from several species have been studied by X-ray crystallography.[153–160] Especially clear results were obtained with unmodified myosin from the ameba *Dictyostelium discoideum*. The head structure, shown in Fig. 19-11, includes a 95-kDa piece of the heavy chain and both light chains. A clearer picture of the neck region containing the light chains was provided by the structure of the "regulatory domain" of scallop myosin.[161] Unlike mammalian or avian myosins, molluscan myosins are regulated by binding of Ca^{2+} to a site in the essential light chain, but the structures are similar to those in Figs. 19-10 and 19-15.

Cleavage of the ~850-residue S1 heads with trypsin yields mainly three large fragments that correspond to structural domains of the intact protein as shown in Fig. 19-15. They are known as the 25-kDa (N-terminal), 50-kDa, and 20-kDa fragments, and for myosin from *D. discoideum* correspond to residues 1 to 204, 216 to 626, and 647 to 843, respectively. The ATP-binding site is in a deep cleft between the 20-kDa and 50-kDa

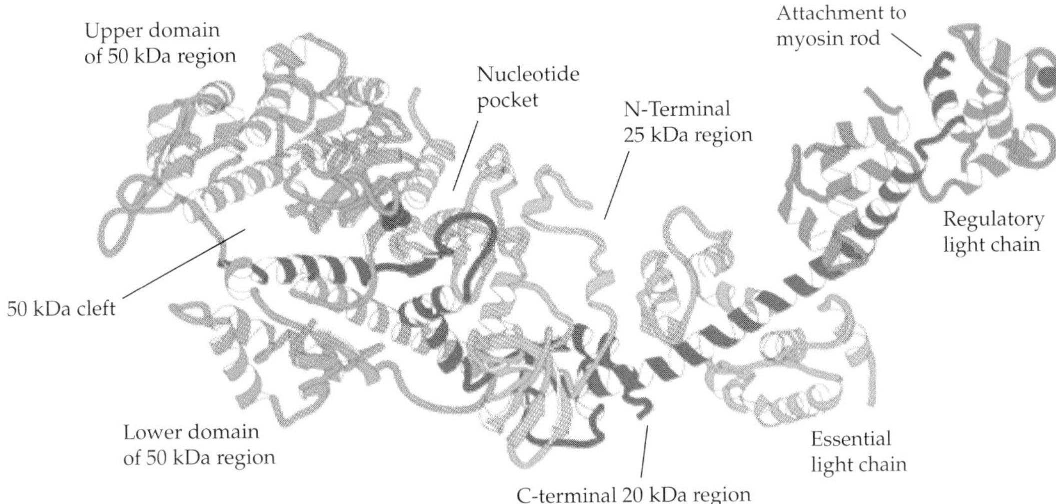

Figure 19-15 Ribbon representation of chicken skeletal myosin subfragment-1 showing the major domains and tryptic fragments. Prepared with the program MolScript. From Rayment.[157]

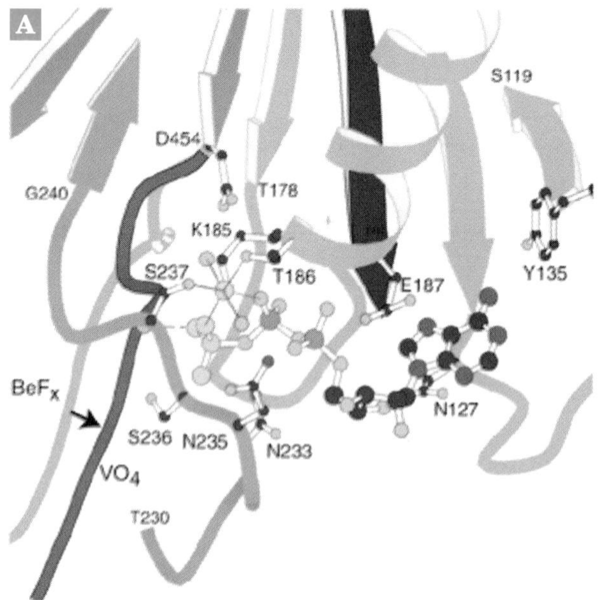

Figure 19-16 (A) The nucleotide binding site of myosin with MgADP•BeF$_x$ bound in a conformation thought to mimic that of ATP prior to hydrolysis. The β-sheet strands are contributed by both the 25-kDa and 50-kDa domains. The P-loop lies between T178 and E187. The conserved N233 to G240 loop, which also contributes important ATP-binding residues, comes from the 50-kDa region. (B) Stereoscopic view of the γ-phospho group binding pocket with the bound MgADP•VO$_4$ (vanadate) complex. The coordinated Mg^{2+} and associated water molecules are seen clearly. Courtesy of Ivan Rayment.[157]

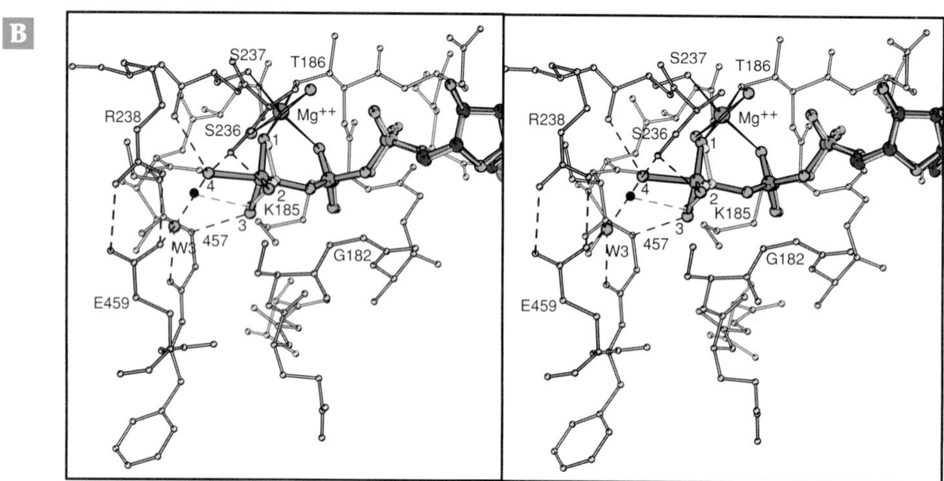

regions. Figure 19-16 illustrates the binding of an ATP analog, the beryllium fluoride complex of MgADP, in the active site. As can be seen, the ATP binds to loops at the C termini of the β strands of the 8-stranded β sheet from the 25-kDa domain. The conserved P-loop (Chapter 12, E), which lies between T178 and E187, curls around the α and β phospho groups, and has the sequence G(179)ESGAGKT. A second conserved loop N(233)SNSSR-G(240) from the 50-kDa domain contributes to the binding of ATP.

The actin-binding region of the myosin head is formed largely by the 50-kDa segment, which is split by a deep cleft into two separate domains (Fig. 19-15), both of which are thought to participate in binding to actin. A surface loop (loop 1) near the ATP-binding site at the junction of the 25- and 50-kDa regions affects the kinetic properties of myosin, probably by influencing product release. A second loop (loop 2, residues 626–647) at the junction of the 50- and 20-kDa regions interact with actin. Loop 2 contains a GKK sequence whose positive charges may interact with negative charges in the N-terminal part of actin.[162–164]

The C-terminal fragment of myosin contains a globular domain that interacts with both the 20-kDa and 50-kDa regions and contains an α-helical neck that connects to the helix of the coiled-coil myosin rod. This helical region is surrounded by the two myosin light chains (Fig. 19-15).[157] A pair of reactive thiol groups (from C697 and C707) in the globular domain are near the active site. Crosslinking of these cysteines by an –S–S– bridge has been utilized to trap nucleotide analogs in the active site.[165]

How does actin bind? The actin monomer consists of four subdomains, 1, 2, 3, and 4 numbered from the N terminus (Fig. 7-10). The negatively charged N-terminal region of actin contains the sequence

$$\overset{}{D}-E-D-\overset{4}{E}---\overset{24}{D}-\overset{25}{D}.$$

It may interact with loop 1 of myosin, which contains five lysines. However, to form a strong interaction with the myosin head a conformational change must occur in the myosin. A change may also occur in actin. Modeling suggests that a large nonpolar contact region involves actin residues A144, I341, I345, L349, and F352 and myosin residues P529, M530, I535, M541, F542, and P543. A conformational change in actin, which might involve largely the highly conserved actin subdomain 2, may also be required for tight interaction.[142,166–168]

Kinesins and other molecular motors. Before considering further how the myosin motor may work, we should look briefly at the **kinesins**, a different group of motor molecules,[168a] which transport various cellular materials along microtubule "rails." They also participate in organization of the mitotic spindle and other microtubule-dependent activities.[168a,b,c] See Section C,2 for further discussion. More than 90 members of the family have been identified. Kinesin heads have much shorter necks than do the myosin heads. A myosin head is made up of ~850 residues, but the motor domain of a kinesin contains only ~345. Like myosin, the 950- to 980-residue kinesins have a long coiled-coil C- terminal region that forms a "neck" of ~50 residues, a "stalk" of ~190 and ~330 residue segments with a Pro / Gly-rich hinge between them, and an ~45 residue "tail."[169–171]

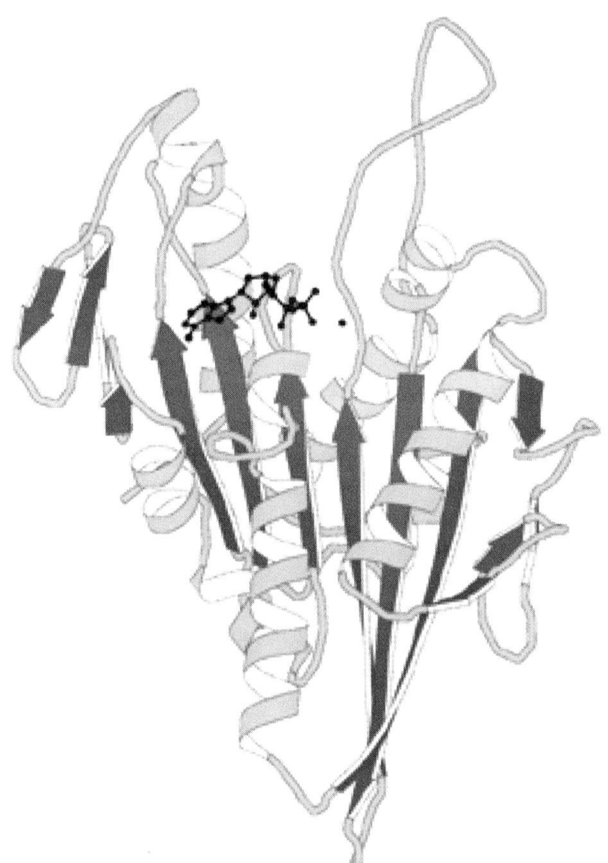

Figure 19-17 Ribbon drawing of human kinesin with bound Mg•ADP. From Gulick *et al.*[174] Courtesy of Ivan Rayment and Andy Gulick.

Crystal structures are known for motor domains of human kinesin[172] and of a kinesin from rat brain.[169,173] The structures of one of six yeast kinesins,[174] a protein called **Kar3**, and also of a *Drosophila* motor molecule designated **Ncd** have also been determined.[175] The last was identified through study of a *Drosophila* mutant called non-claret disjunctional (Ncd). The motor domains of various members of the kinesin family show ~40% sequence identity and very close structural identity (Fig. 19-17).[174] Although the sequences are different from those of the myosin heads or of G proteins, the folding pattern in the core structures is similar in all cases. An 8-stranded β sheet is flanked by three α helices on each side and a P-loop crosses over the ATP-binding site as in Fig. 19-16. Further similarity is found in the active site structures, which, for a monomeric kinesin KIF1A,[174a] have been determined both with bound ADP and with a nonhydrolyzable analog of ATP.[174b,174c] Although there is little similarity in amino acid sequences the structures in the catalytic core are clearly related to each other, to those of dimeric kinesins,[174d] to those of myosins, and to those of the GTP-hydrolyzing G proteins.

A puzzling discovery was that the motor domain of kinesin, which binds primarily to the β subunits of tubulin (Fig. 7-34) and moves toward the fast growing *plus* end of the microtubule,[176] is located at the N terminus of the kinesin molecule, just as is myosin. However, the Ncd and Kar3 motor domains are at the C-terminal ends of their peptide chains and move their "cargos" toward the *minus* ends of microtubules.[174] Nevertheless, the structures of all the kinesin heads are conserved as are the basic chemical mechanisms. The differences in directional preference are determined by a short length of peptide chain between the motor domain and the neck, which allows quite different geometric arrangements when bound to microtubules.[173,177] Like Ncd, myosin VI motor domains also move "backwards" toward the pointed (minus) ends of actin filaments.[178–179a]

Other major differences between kinesins and myosin II heads involve kinetics[180,181] and processivity.[173] Dimeric kinesin is a **processive** molecule. It moves rapidly along microtubules in 8-nm steps but remains attached.[182,182a] Myosins V and VI are also processive[183–183e] but myosin II is not. It binds, pulls on actin, and then releases it. The many myosin heads interacting with each actin filament accomplish muscle contraction with a high velocity in spite of the short time of attachment. Ncd and Kar3 are also nonprocessive and slower than the *plus* end-oriented kinesins.[184]

The ATPase cycles of actomyosin and of the kinesins. The properties of the protein assemblies found in muscle have been described in elegant details, but the most important question has not been fully answered. How can the muscle machinery use the Gibbs energy of hydrolysis of ATP to do mechanical work? Some insight has been obtained by studying the ATPase activity of isolated myosin heads (S1) alone or together with actin. Results of numerous studies of ATP binding, hydrolysis, and release of products using fast reaction techniques[185–191] and cryoenzymology[191a] are summarized in Fig. 19-18. In resting muscle the myosin heads swing freely in the ~20-nm space between the thick and thin filaments. However, in activated muscle some heads are bound tightly to actin as if in rigor (complex A•M in Fig. 19-18). When ATP is added MgATP binds into the active site of the myosin (Fig. 19-18, step *a*) inducing a conformational change to form A•M*•ATP in which the bond between actin and myosin is weakened greatly, while that between myosin and ATP is strengthened. The complex dissociates (step *b*) to give free actin and (M*•ATP), which accumulates at –15°C. However, at higher temperatures the bound ATP is hydrolyzed rapidly (step *c*) to a form M**•ADP•P_i in which the ATP has been cleaved to ADP + P_i but in which the split products remain bound at the active site.[116,192,192a,b] All of these reactions are reversible. That is, the split products can recombine to form ATP. This fact suggests that most of the Gibbs energy of hydrolysis of the ATP must be stored, possibly through a conformational change in the myosin head or through tighter bonding to ATP. As long as calcium ions are absent, there is only a slow release of the bound ADP and P_i and replacement with fresh ATP takes place. Thus, myosin alone shows a very weak ATPase activity.

On the other hand, in activated muscle the head with the split ATP products will bind to actin (step *d*), probably at a new position. The crossbridges that form appear to be attached almost at right angles to the thin filaments. In step *e*, P_i is released following a conformational alteration that is thought to open a "back door" to allow escape of the phosphate ion.[193] In the final two steps (*f* and *g*) the stored energy in the myosin head (or in the actin) is used to bring about another conformational change that alters the angle of attachment of myosin head to the thin filament from ~90° to ~45°. At least some indication of such a change can be observed directly by electron microscopy.[144] Such a change in angle is sufficient to cause the actin filament to move ~10 nm with respect to the thick filaments to complete the movement cycle (Fig. 19-18), if the head is hinged at the correct place. However, the existence of at least four different conformational states suggests a more complex sequence.[193a,193b] Examination of the three-dimensional structures available also suggest a complex sequence of alterations in structure and geometry. X-ray crystallographic structures of myosin heads, in states thought to correspond to states 1 and 3 of Figs. 19-14 and 19-18, are also in agreement on an ~10 (5–12) nm movement of the lever arm.[194,195] Six states of the actomyosin complex are depicted in Fig. 19-18, but a complete kinetic analysis requires at least eight and possibly 12 states.[196,197]

Observing single molecules. A major advance in the study of molecular motors has been the development of ways to observe and study single macromolecules. The methods make use of **optical traps** (optical "tweezers") that can hold a very small (~1 μm diameter) polystyrene or silica bead near the waist of a laser beam focused through a microscope objective.[198–202] In one experimental design an F-actin filament is stretched between two beads, held in a pair of optical traps. The filament is pulled taut and lowered onto a stationary silica bead to which a few myosin HMM fragments have been attached (Fig. 19-19). If ATP is present, short transient movements along the filament are detected by observation of displacements of one of the beads when the actin filament contacts HMM heads. An average lateral displacement of 11 nm was observed. Each HMM head exerted a force of 3–4 pN, a value consistent with expectations for the swinging bridge model.[200] From the duration of a single displacement (≤7 ms) and an estimated k_{cat} for ATP hydrolysis of 10 s^{-1}, the fraction of time that the head was attached during one catalytic cycle of the head was therefore only 0.07. This ratio, which is called the **duty ratio**, is low for actomyosin. However, many myosin heads bind to each actin filament in a muscle. Each head exerts its pull for a short time. but the actin is never totally unattached.[203] Similar measurements with smooth muscle revealed similar displacements but with a 10-fold slower sliding velocity and a 4-fold increase in the duty ratio. This may perhaps account for an observed 3-fold increase in force as compared with skeletal muscle.[204,204a]

Other single molecule techniques involve direct observation of motor molecules or of S1 myosin fragments tagged with highly fluorescent labels.[205,206] All measurements of single molecule movement are subject to many errors. Brownian motion of the beads makes measurements difficult.[207] Not all results are in agreement, and some are difficult to understand.[207a] Most investigators agree that there is a step size of ~4–10 nm. Kitamura *et al.* found 5.3 nm as the average.[206] However, they also reported the puzzling observation that some single S1 molecules moved 11–30 nm in two to five rapid successive steps during the time of hydrolysis of a single molecule at ATP. They suggested that some of the energy of ATP hydrolysis may be stored in S1 or in the actin filament and be released in multiple steps. Veigel *et al.*[208] observed that a brush border myosin I from chicks produced ~ 6 nm movements, each of which was followed by an additional ~ 5.5 nm step within ~5 ms. They attribute these steps to two stages in the power stroke, e.g., to steps *f* and *g* of Fig. 19-18. A value of ~10nm was reported recently by Piazzesi *et al.*[208a] Myosin V moves along actin filaments with very large 36-nm steps.[208b]

Motion of kinesin heads has been observed by movement of microtubules over biotinylated kinesin fixed to a steptavidin-coated surface,[209] by direct observation of fluorescent kinesin moving along microtubules,[171] and by optical trap interferometry.[210] Kinesin heads move by 8-nm steps, evidently the exact length

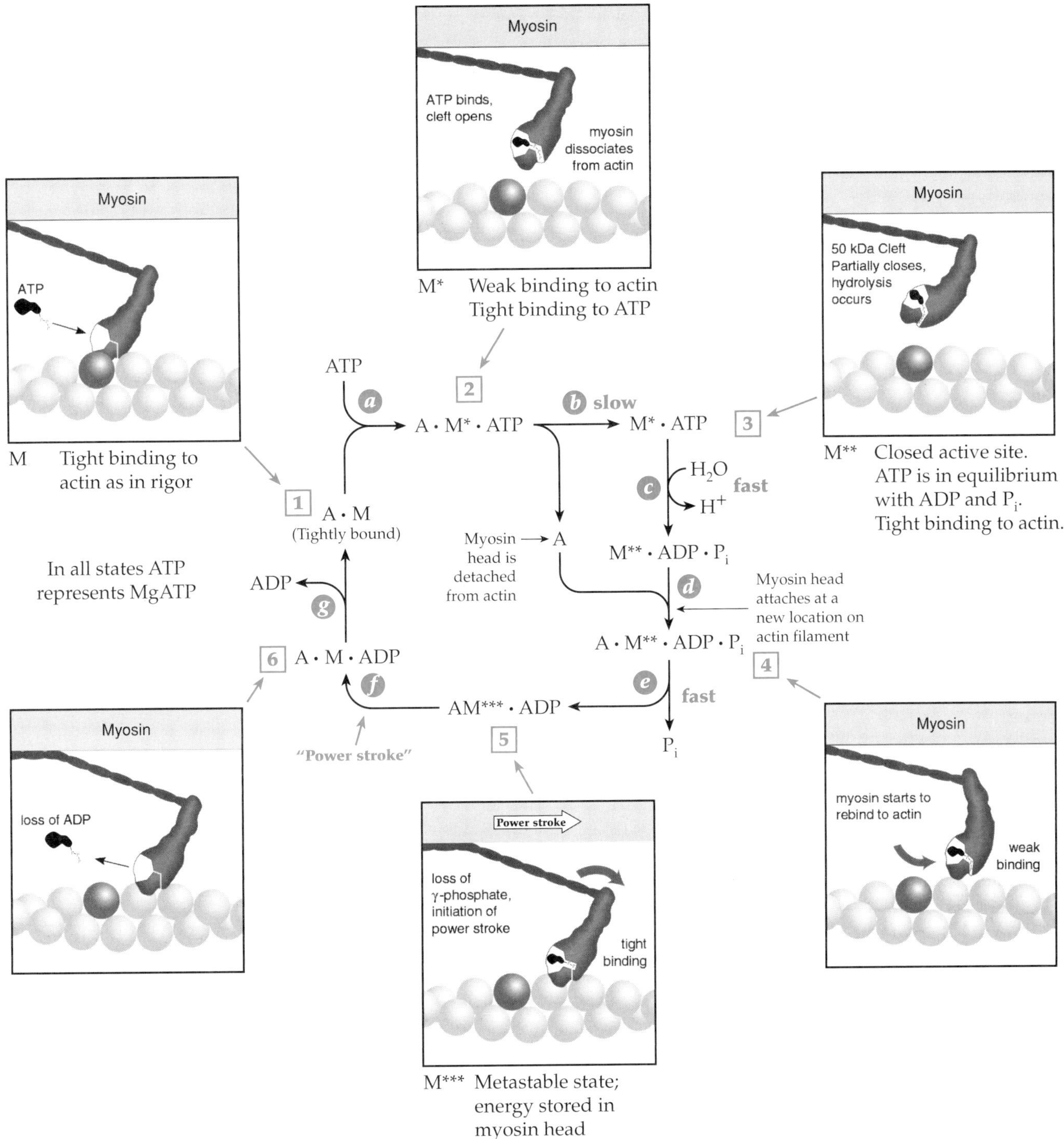

Figure 19-18 Simplified view of the ATP hydrolysis cycle for actomyosin. A similar cycle can be written for kinesins and dyneins. Here A stands for fibrous actin and M, M*, M**, and M*** for four different conformations of the myosin heads. As indicated by the numbers in squares, four of the six states of actomyosin shown here can also be correlated with those in Fig. 19-14.

of an αβ tubulin dimer in the microtubule structure (Fig. 7-33). One ATP is apparently hydrolyzed for each 8-nm step. However, shorter steps of ~5 nm have sometimes been seen.[211,212]

The movement is processive, kinesin motors typically taking 100 steps before dissociating from the microtubule.[201,212a] Kinesin is bound to the microtubule continuously. Its duty ratio is nearly 1.0 (the same is true for the bacterial flagellar motor; Fig. 19-4).[212b] However, single kinesin heads, which lack the coiled-coil neck region, have a duty ratio of <0.45. The movement is nonprocessive.[213] The Ncd motor is also nonprocessive.[214–216] As mentioned previously, the Ncd and kinesin motor domains are at opposite ends of the peptide chain, and the motors move in opposite directions along microtubules.[217,218] The critical difference between the two motor molecules was found in the neck domains, which gave rise to differing symmetrics in the two heads.[219] The latter are shown in Fig. 19-20, in which they have been docked onto the tubulin protofilament structure. One head, both of ncd and of kinesin, occupies a similar position on the microtubule, but the other head points toward the microtubular plus end for kinesin but toward the minus end for Ncd. Cryoelectron microscopy also supports this interpretation.[220]

Still not fully understood is the processive action of kinesin.[221–224] It is often assumed that this protein moves in a hand-over-hand fashion with the two heads binding alternatively to the microtubule. Some substantial reorganization of the peptide chain in the hinge region at the end of the neck is presumably involved.[173] An alternative "inchworm" mechanism has been suggested.[220a]

Thinking about chemical mechanisms. We have now examined the active sites of kinases that cleave ATP (Chapter 12), ATPases that pump ions by cleaving ATP, ATP synthases that form ATP from ADP and P_i (Chapter 18), and GTP hydrolyzing enzymes that cause movement and shape changes that control metabolic processes (Chapter 11). It is striking that the active site regions where the ATP or GTP bind have such a highly conserved structure.[225] This suggests that the secret of movement can be found in the very strong interactions of the nucleotides and their split products with the proteins. In every case there is at least one tight binding or closed conformation in which a large number of hydrogen bonds and ionic interactions bind the nucleotide. This is shown for a kinase in Fig. 12-32 and for myosin in Fig. 19-16. During the actomyosin reaction several conformational changes must occur. Not only does the affinity for the bound nucleotide vary, but also the binding of actin to myosin can be strong, as in the nucleotide-free state or

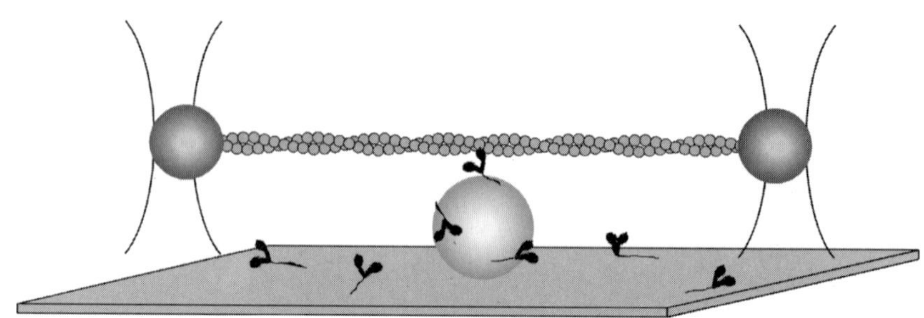

Figure 19-19 Schematic drawing (not to scale) illustrating the use of two optical traps that are focused on beads attached to a single actin filament. The filament is lowered onto a stationary silica bead sparsely coated with HMM fragments of myosin. In the presence of ATP the myosin heads bind transiently for a few milliseconds to the actin, moving it in one direction and displacing the beads from their positions in the optical traps. An image of one of the beads is projected onto photodiode detectors capable of measuring small displacements. The displacing force can also be recorded. For details of the experiments and of the optical traps and measuring devices see Finer *et al.*[200] Courtesy of J. A. Spudich.

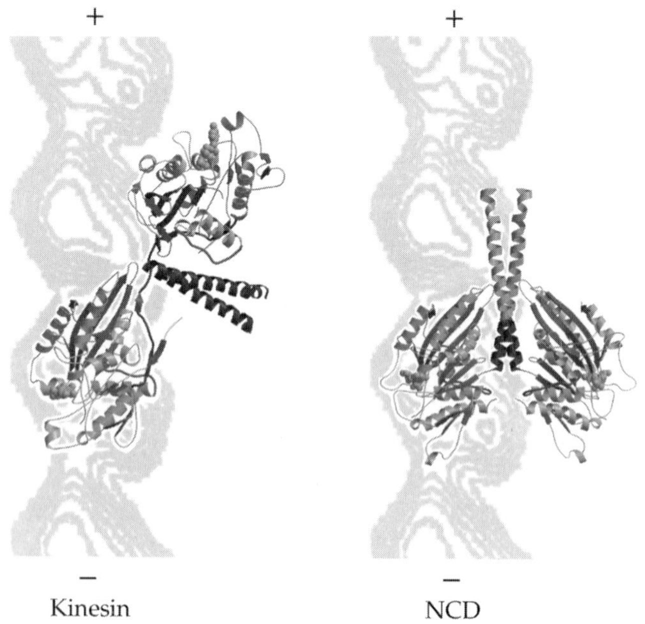

Figure 19-20 Model showing the ncd and kinesin dimer structures docked onto a tubulin protofilament. The bound ncd and kinesin heads are positioned similarly. Because of the distinct architectures of the kinesin and ncd necks, the unbound kinesin head points toward the plus end, whereas the unbound ncd head is tilted toward the minus end of the protofilament. From Sablin *et al.*[219] Courtesy of Ronald Vale.

in the presence of bound ADP. Binding is weak when ATP or the split product ADP + P_i are in the active site.

To understand these differences we should look at the structure of ATP itself. The triphosphate group has many negative charges repelling each other. What must happen to allow the binding of ATP to myosin to break the actin-myosin bond? The electrostatic attraction of these phospho groups for active site groups is doubtless one cause of the observed conformational changes. Could it be that electrostatic repulsion, via a proton shuttle mechanism, is also induced at the right point in the actin–myosin interface? Many studies with analogs of ATP have contributed to our understanding. Neither the purine nor the ribose ring of ATP is absolutely essential. The compound 2-[(4-azido-2-nitrophenyl) amino] diphosphate (NANDP) and related nonnucleotide analogs[165,196,226] support muscle contraction and relaxation in the same

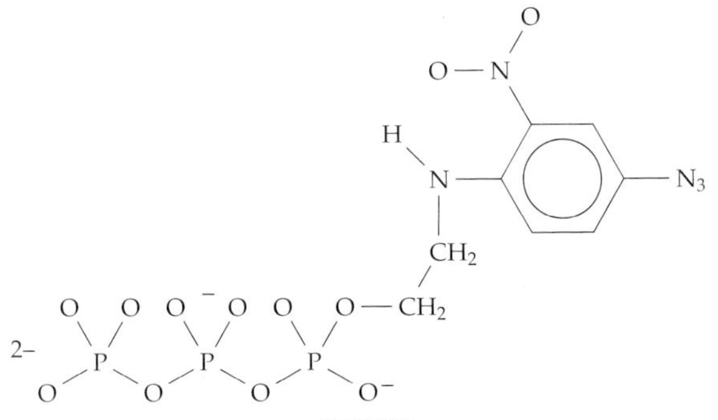

NANTP

way as does ATP. An analog with a rigid five-membered ring, 2',3'-dideoxydidehydro-ATP, is also active.[227] A comparison of kinetic data and X-ray structural data supports the proposal that the ATP must be bound in the conformation shown for MgADP•BeF_x in Fig. 19-16A.[196] When the two SH groups of C697 and C707 of the myosin head are crosslinked by various reagents,[227a] this NANDP analog can be trapped at the active site. Because of the presence of its azide ($-N_3$) group the trapped compound can serve as a photoaffinity label, attaching itself to a tryptophan side chain upon activation with visible light (Eq. 23-27).

How can cleavage of ATP to bound ADP + P_i create a metastable high-energy state of the myosin head ready to hold onto and pull the actin chain? This may be compared with the inverse problem of generating ATP in oxidative phosphorylation, in which ADP and P_i coexist in equilibrium with ATP in a closed active site (Fig. 18-14). Comparison should also be made with the GTP-hydrolyzing G proteins (Fig. 11-7).[227b,c] During hydrolysis of GTP by the Ras protein binding to the protein induces a shift of negative charge from the γ- to β-oxygens of GTP facilitating bond cleavage as in Eq. 19-2. G proteins also couple substrate hydrolysis to mechanical motion. We should also think about the fact that when ATP is cleaved within myosin there will necessarily be a flow of electrical charge from the water to the ADP (Eq. 19-2). This will be followed by some accomodation of the protein to the new charge constellation. As we have seen previously, movement of protons is often the key to conformational changes. In this case, the initial change must be to create a high-energy state of myosin which, following loss of the orthophosphate ion, can cause the major conformational change that swings the lever arm of the myosin. The conformational changes may occur in several steps in which the packing of groups within the myosin head is always tight in some places and rather loose in others. Movement within the head is being observed not only by X-ray crystallography but by **fluorescence resonance energy transfer** (FRET; Chapter 23)[227d] and by the newer **luminescence resonance energy transfer** (LRET). For example, a terbium chelate of azide-ATP was photochemically bound in the active site, and a fluorescent dye was attached to Cys 108 in the regulatory light chain. The terbium ion was irradiated, and fluorescence of the dye was observed. Distance changes, measured in the absence and presence of ATP, were consistent with the swinging arm model.[228] Dyes have been attached to –SH groups engineered into various locations in the myosin molecule to permit other distance measurements.[229,230] In another elegant application of the FRET technique the **green fluorescent protein** of *Aequoria* (Box 23-A) was fused to the C terminus of the motor domain of myosin giving a fluorescent lever arm. Energy transfer to blue fluorescent protein fused to the N terminus of the S1 head was measured. The distance between these was estimated by the FRET technique and was also consistent with expectations for the "rowing model."[231]

The "rowing model" is generally accepted, but other quite different processes have been proposed to account for the elementary cycle of muscle contraction. Muscle contracts nearly *isovolumetrically*; thus, anything that expands the sarcomere will cause a contraction. It has been suggested that the hydrolysis of ATP deposits negatively charged phospho groups on the actin filaments, and that the electrostatic repulsion is responsible for

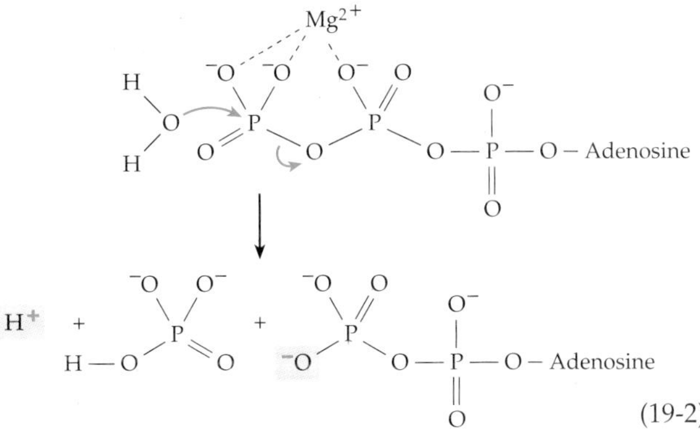

(19-2)

BOX 19-A HEREDITARY DISEASES OF MUSCLE

Considering the numerous specialized proteins in muscle it is not surprising that many rare hereditary muscle diseases are known. The most frequent and most studied of these is **Duchenne muscular dystrophy**. An X-linked disease of boys, it may not be recognized until two to three years of age, but victims are usually in a wheelchair by age 12 and die around age 20. Individual muscle fibers disintegrate, die, and are replaced by fibrous or fatty tissue.[a–d] The disease strikes about 1 out of 3500 boys born. The less serious **Becker muscular dystrophy** arises from defects in the same gene but affects only 1 in 30,000 males, some of whom have a normal life span. Because of its frequency and the knowledge that the gene must lie in the X-chromosome, an intensive search for the gene was made. It was found in 1986 after a five-year search.[a,e] This was the first attempt to find a faulty gene whose protein product was totally unknown. The project, which relied upon finding restriction fragment polymorphisms (Chapter 26) that could serve as markers in the genome, made use of the DNA from patients with a range of related diseases. The very rare female patients in whom the faulty gene had been translocated from the X-chromosome to an autosome also provided markers. DNA probes obtained from a young man with a large X-chromosome deletion that included genes related to retinitis pigmentosa and several other diseases provided additional markers. The result was a triumph of "reverse genetics" which has since been applied to the location of many other disease genes.[e]

The muscular dystrophy gene may be the largest human gene. It consists of 2.3 million base pairs, which include 79 exons which encode a huge 427 kDa protein named **dystrophin**. The protein consists of four main domains.[a,f,g] The N-terminal domain binds to actin and is homologous to β-actinin. The central domain is an elongated rod resembling spectrin. It contains repetitive coiled-coil segments and four hinge regions. The third domain is rich in cysteine and binds Ca^{2+}, while the fourth domain has a structure that is shared by several other proteins of the dystrophin family. Dystrophin is quantitatively a minor protein of muscle. It forms part of the cytoskeleton, lying adjacent to the sarcolemma (cell membrane) along with β-spectrin and vinculin (see Fig. 8-16).

While one end of the dystrophin molecule binds to actin filaments, the C-terminal domain associates with several additional proteins to form a **dystrophin–glycoprotein complex** (see figure).[f,h–k] Dystrophin is linked directly to the membrane-spanning protein **β-dystroglycan,** which in the outer membrane surfaces associates with a glycoprotein **α-dystroglycan**. The latter binds to laminin-2 (Fig. 8-33), a protein that binds the cell to the basal lamina. Four other membrane-spanning proteins, α-, β-, γ-, and **δ-sarcoglycans**, are among additional members of the complex.[k–m]

Patients with Duchenne muscular dystrophy are deficient not only in dystrophin but also in the dystroglycan and sarcoglycan proteins.[f,n] Evidently, dystrophin is needed for formation of the complex which plays an essential role in muscle. In both types of X-linked muscular dystrophy there are individuals with a wide range of point mutations, frame-shift mutations, and deletions in the dystrophin gene.[d] The essential function of dystrophin and associated proteins is uncertain but may be related to the linkage from actin filaments through the membrane to laminin. Individuals with Becker muscular dystrophy also have defects in dystrophin, but the protein is partially functional. Some other muscular dystrophies are caused by defects in the autosomal genes of any of the four sarcoglycan subunits.[j,k,o] or in that of laminin α1 chain.[p,q] The arrows in the accompanying drawing indicate chromosome locations of the sarcoglycan subunits, which are sites for mutations causing **limb girdle muscular dystrophy**.[k]

Dystrophin, shorter isoforms, and related proteins are found in many tissues including the brain.[s] One related protein, **utrophin** (dystrophin-related

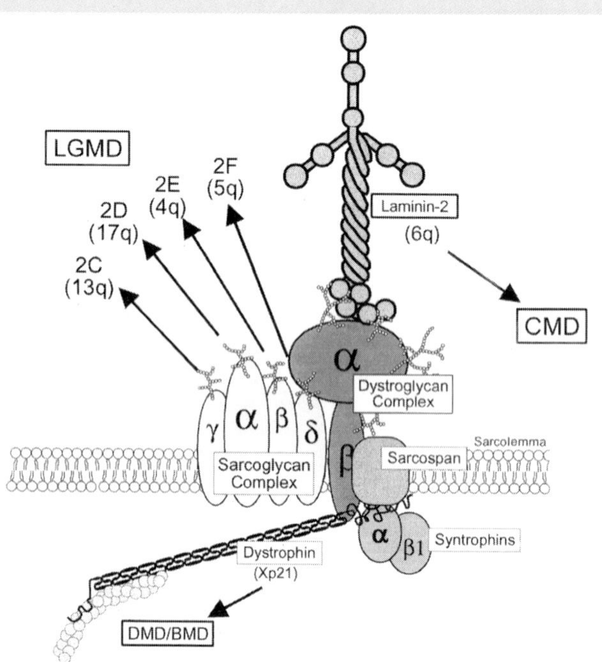

Schematic model of the dystrophin–glycoprotein complex. Courtesy of Kevin P. Campbell. See Lim and Campbell[r] Abbreviations: LGMD, Limb Girdle muscular dystrophy; CMD, congenital muscular dystrophy; DMD / BMD, Duchenne / Becker muscular dystrophies.

BOX 19-A (continued)

protein 1), is present in the neuromuscular junction of adult skeletal muscle. One approach to therapy of Duchenne muscular dystrophy is to stimulate a higher level of expression of the utrophin gene.[t] Because the dystrophin gene is so large treatment by gene transfer is not practical, but transfer of parts of the gene may be. Myoblast transfer has not been successful, but new approaches will be devised.[d]

Myotonic dystrophy is a generalized adult-onset disorder with muscular spasms, weakness, and many other symptoms.[u–y] It is one of the **triple-repeat diseases** (Table 26-4). The affected gene encodes a protein kinase of unknown function. The corresponding mRNA transcript has ~2400 nucleotides. The gene has a CTG repeat $(CTG)_n$ near the 3'-end with $n < 30$ normally. For persons with the mildest cases of myotonic dystrophy n may be over 50 while in severe cases it may be as high as 2000. As in other triple-repeat diseases the repeat number tends to increase in successive generations of people as does the severity of the disease.[x]

For some individuals, muscular dystrophy causes no obvious damage to skeletal muscle but affects the heart producing a severe **cardiomyopathy,** and persons with Duchenne muscular dystrophy often die from heart failure. Heart failure from other causes, some hereditary, is a major medical problem, especially among older persons. Hereditary forms are often autosomal dominant traits that may cause sudden death in young persons. At least seven genes for cardiac sarcomeric proteins including actin,[z] myosin, both heavy and light chains,[aa–dd] three subunits of troponin,[ee] tropomyosin, and protein C (p. 1104) may all carry mutations that cause cardiomyopathy.

A hereditary disease common in Japan results from a defect in migration of neurons and is associated with brain malformation as well as muscular dystrophy.[ff] In **nemaline myopathy** a defect in nebulin leads to progressive weakness and often to death in infancy. A characteristic is the appearance of "nemaline bodies" or thickened Z-discs containing Z-disc proteins.[gg] Some hereditary diseases involve nonmuscle myosins. Among these is **Usher syndrome**, the commonest cause of deaf-blindness. The disease, which results from a defect in the myosin VIA gene, typically causes impairment of hearing and retinitis pigmentosa (Chapter 23).[hh]

---

[a] Anderson, M. S., and Kunkel, L. M. (1992) *Trends Biochem. Sci.* **17**, 289–292

[b] Emery, A. E. H., and Emery, M. L. H. (1995) *The History of a Genetic Disease: Duchenne Muscular Dystrophy or Meryon's Disease*, Royal Society of Medicine, London

[c] Brown, S. C., and Lucy, J. A., eds. (1997) *Dystrophin: Gene, Protein and Cell Biology*, Cambridge Univ. Press, London

[d] Worton, R. G., and Brooke, M. H. (1995) in *The Metabolic and Molecular Bases of Inherited Disease*, 7th ed., Vol. 3 (Scriver, C. R., Beaudet, A. L., Sly, W. S., and Valle, D., eds), pp. 4195–4226, McGraw-Hill, New York

[e] Rowland, L. P. (1988) *N. Engl. J. Med.* **318**, 1392–1394

[f] Tinsley, J. M., Blake, D. J., Zuellig, R. A., and Davies, K. E. (1994) *Proc. Natl. Acad. Sci. U.S.A.* **91**, 8307–8313

[g] Fabbrizio, E., Bonet-Kerrache, A., Leger, J. J., and Mornet, D. (1993) *Biochemistry* **32**, 10457–10463

[h] Ervasti, J. M., and Campbell, K. P. (1991) *Cell* **66**, 1121–1131

[i] Madhavan, R., and Jarrett, H. W. (1995) *Biochemistry* **34**, 12204–12209

[j] Sweeney, H. L., and Barton, E. R. (2000) *Proc. Natl. Acad. Sci. U.S.A.* **97**, 13464–13466

[k] Jung, D., and 13 others. (1996) *J. Biol. Chem.* **271**, 32321–32329

[l] Winder, S. J. (2001) *Trends Biochem. Sci.* **26**, 118–124

[m] McDearmon, E. L., Combs, A. C., and Ervasti, J. M. (2001) *J. Biol. Chem.* **276**, 35078–35086

[n] Durbeej, M., and Campbell, K. P. (1999) *J. Biol. Chem.* **274**, 26609–26616

[o] Yang, B., Ibraghimov-Beskrovnaya, O., Moomaw, C. R., Slaughter, C. A., and Campbell, K. P. (1994) *J. Biol. Chem.* **269**, 6040–6044

[p] Noguchi, S., and 17 others. (1995) *Science* **270**, 819–822

[q] Tiger, C.-F., Champliaud, M.-F., Pedrosa-Domellof, F., Thornell, L.-E., Ekblom, P., and Gullberg, D. (1997) *J. Biol. Chem.* **272**, 28590–28595

[r] Lim, L. E., and Campbell, K. P. (1998) *Curr. Opin. Neurol.* **11**, 443–452

[s] Dixon, A. K., Tait, T.-M., Campbell, E. A., Bobrow, M., Roberts, R. G., and Freeman, T. C. (1997) *J. Mol. Biol.* **270**, 551–558

[t] Campbell, K. P., and Crosbie, R. H. (1996) *Nature (London)* **384**, 308–309

[u] Harper, P. S. (1995) in *The Metabolic and Molecular Bases of Inherited Disease*, 7th ed., Vol. 3 (Scriver, C. R., Beaudet, A. L., Sly, W. S., and Valle, D., eds), pp. 4227–4252, McGraw-Hill, New York

[v] Ptacek, L. J., Johnson, K. J., and Griggs, R. C. (1993) *N. Engl. J. Med.* **328**, 482–489

[w] Pearson, C. E., and Sinden, R. R. (1996) *Biochemistry* **35**, 5041–5053

[x] Tapscott, S. J., and Thornton, C. A. (2001) *Science* **293**, 816–817

[y] Miller, J. W., Urbinati, C. R., Teng-umnuay, P., Stenberg, M. G., Byrne, B. J., Thornton, C. A., and Swanson, M. S. (2000) *EMBO J.* **19**, 4439–4448

[z] Olson, T. M., Michels, V. V., Thibodeau, S. N., Tai, Y.-S., and Keating, M. T. (1998) *Science* **280**, 750–755

[aa] Rayment, I., Holden, H. M., Sellers, J. R., Fananapazir, L., and Epstein, N. D. (1995) *Proc. Natl. Acad. Sci. U.S.A.* **92**, 3864–3868

[bb] Roopnarine, O., and Leinwand, L. A. (1998) *Biophys. J.* **75**, 3023–3030

[cc] Yanaga, F., Morimoto, S., and Ohtsuki, I. (1999) *J. Biol. Chem.* **274**, 8806–8812

[dd] Martinsson, T., Oldfors, A., Darin, N., Berg, K., Tajsharghi, H., Kyllerman, M., and Wahlström, J. (2000) *Proc. Natl. Acad. Sci. U.S.A.* **97**, 14614–14619

[ee] Miller, T., Szczesna, D., Housmans, P. R., Zhao, J., de Freitas, F., Gomes, A. V., Culbreath, L., McCue, J., Wang, Y., Xu, Y., Kerrick, W. G. L., and Potter, J. D. (2001) *J. Biol. Chem.* **276**, 3743–3755

[ff] Kobayashi, K., and 17 others. (1998) *Nature (London)* **394**, 388–392

[gg] Pelin, K., and 19 others. (1999) *Proc. Natl. Acad. Sci. U.S.A.* **96**, 2305–2310

[hh] Weil, D., and 18 others. (1995) *Nature (London)* **374**, 60–61

lateral expansion of the sarcomere.[232] Still other ideas have been advanced.[233,234]

4. Control of Muscle Contraction

Skeletal muscle must be able to rest without excessive cleavage of ATP but able to act rapidly with a high expenditure of energy upon nervous excitation. Even a simple physical activity requires that a person's muscles individually contract and relax in rapid response to nerve impulses from the brain. To allow for this control the endoplasmic reticulum (**sarcoplasmic reticulum, SR**) of striated muscle fibers is organized in a striking regular manner.[235–237] Interconnecting tubules run longitudinally through the fibers among the bundles of contractile elements. At regular intervals they come in close contact with infoldings of the outer cell membrane (the **T system** of membranes, Fig. 19-21; see also Fig. 19-7A). A nerve impulse enters the muscle fiber through the neuromuscular junctions and travels along the sarcolemma and into the T tubules. At the points of close contact the signal is transmitted to the longitudinal tubules of the sarcoplasmic reticulum, which contain a high concentration of calcium ions.

Calcium ions in muscle. A nerve signal arriving at a muscle causes a sudden release of the calcium ions into the cytoplasm from cisternae of the sarcoplasmic reticulum (SR) that are located adjacent to the T-tubules. Diffusion of the Ca^{2+} into the myofibrils initiates contraction. In smooth muscle the signals do not come directly from the nervous system but involve hormonal regulation.[238] Again, calcium ions play a major role, which is also discussed in Chapter 6, Section E, and in Box 6-D. Muscle contains a large store of readily available Ca^{2+} in lateral cisternae of the SR. The free intracellular Ca^{2+} concentration is kept low by a very active ATP-dependent calcium ion pump (Fig. 8-26), which is embedded in the membranes of the SR.[238a] Within the vesicles Ca^{2+} is held loosely by the ~63-kDa protein **calsequestrin,** which binds as many as 50 calcium ions per molecule. When the cytoplasmic concentration of free Ca^{2+} falls below ~10^{-6} M, contraction ceases. In fast-contracting skeletal muscles the Ca^{2+}-binding protein **parvalbumin** (Fig. 6-7) may

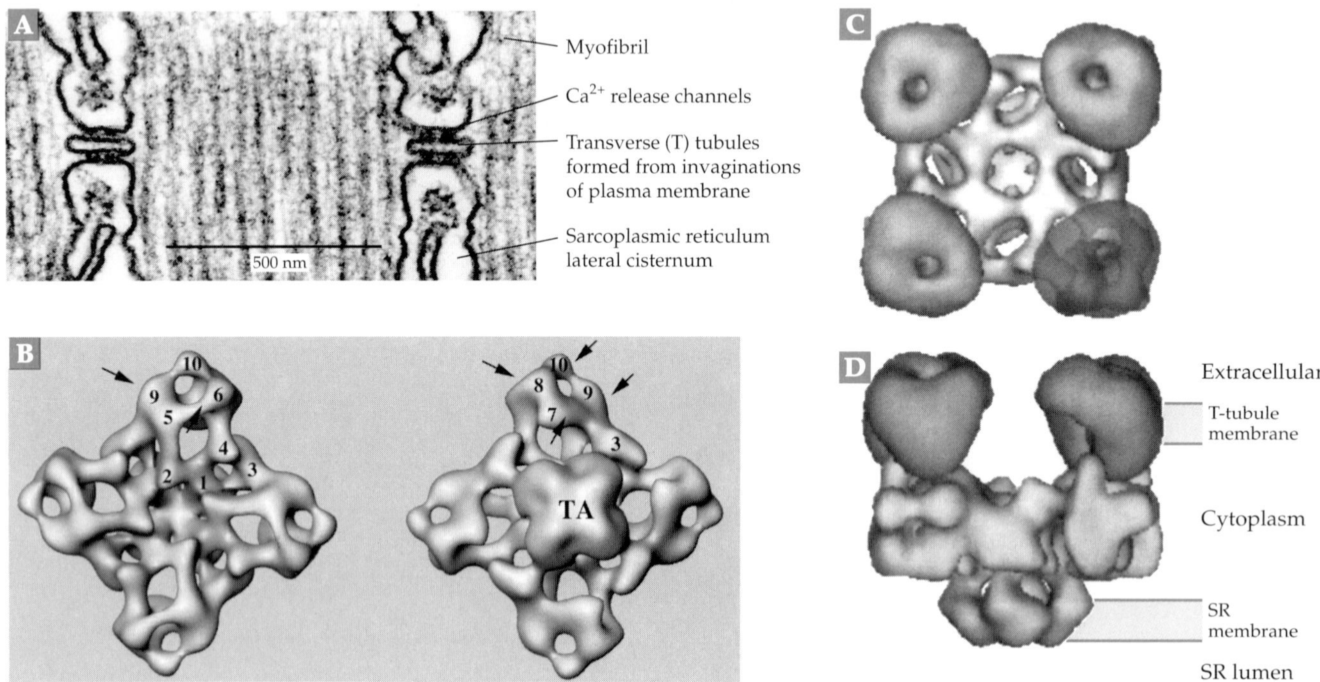

Figure 19-21 (A) Electron micrograph showing two transverse tubules (T-tubules) that are formed by infolding of the plasma membrane. They wrap around a skeletal muscle fiber and carry nerve impulses to all parts of the fiber. From Alberts *et al.*[237] Courtesy of Clara Franzini-Armstrong. (B) Three-dimensional surface representation of the calcium release channels known as ryanodine receptors, type RyR1 based on cryoelectron microscopy and image reconstruction at a resolution of 4.1 nm. The image to the left shows the surface that would face the cytoplasm while that to the right shows the surface that would interact with the sarcoplasmic reticulum, TA representing the transmembrane portion. Notice the fourfold symmetry of the particle, which is composed of four 565-kDa subunits. From Sharma *et al.*[239] Courtesy of Manjuli Rani Sharma. (C), (D) Model showing proposed arrangement of ryanodine receptors and dihydropyridine receptors (round) in the T tubule and SR membranes. From Serysheva *et al.*[245a]

assist in rapid removal of free Ca^{2+} from the cytoplasm. Contraction is activated when Ca^{2+} is released from the SR through the **calcium release channels**,[240–244] which are often called **ryanodine receptors**. The name arises from their sensitivity to the insecticidal plant alkaloid ryanodine, which at low concentration (≤60 μM) causes the channels to open, but a high concentration closes the channels.[243] These calcium release channels consist of tetramers of ~5000-residue proteins. The bulk of the 565-kDa polypeptides are on the cytosolic surface of the SR membranes, where they form a complex "foot" structure (Fig. 19-21B) that spans the ~12 nm gap between the SR vesicles and the T-tubule membrane.[239,240] Ryanodine receptor function is modulated by NO, which apparently binds to –SH groups within the Ca^{2+} channel.[243a,243b] Some ryanodine receptors are activated by cyclic ADP-ribose (cADPR, p. 564).[243c] Some have an oxidoreductase-like structural domain.[243d]

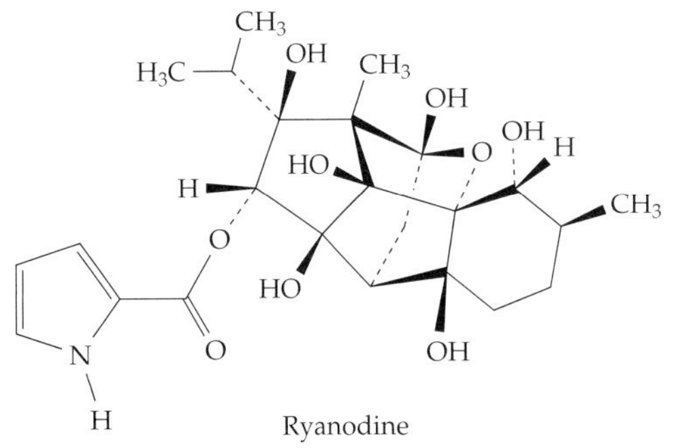

The release channels open in response to an incompletely characterized linkage to the **voltage sensor** that is present in the T-tubule membrane and is known as the **dihydropyridine receptor**.[240,245] This too, is a Ca^{2+} channel, which opens in response to arrival of an **action potential** (nerve impulse; see Chapter 30) that move along the T-tubule membrane. Because the

Ryanodine

action potential arrives almost simultaneously throughout the T-tubules of the muscles, the dihydropyridine receptors all open together. It isn't clear whether the linkage to the calcium channels is via stimulation from released Ca^{2+} passing from the dihydropyridine receptor to the surfaces of the feet of the release channels, or is a result of depolarization of the T-tubule membrane, or involves direct mechanically linked conformational changes.[240,245] The close cooperation of the Ca^{2+} release channel and the voltage sensor is reflected in their close proximity. In the sarcoplasmic reticulum every second release channel is adjacent to a voltage sensor in the opposing T-tubule membrane.[240,245a] The essential nature of the voltage

BOX 19-B MALIGNANT HYPERTHERMIA AND STRESS-PRONE PIGS

Very rarely during surgery the temperature of a patient suddenly starts to rise uncontrollably. Even when heroic measures are taken, sudden death may follow within minutes. This **malignant hyperthermia syndrome** is often associated with administration of halogenated anesthetics such as a widely used mixture of halothane (2-bromo-2-chloro-1,1,1-trifluoroethane) and succinylcholine.[a–d] There is often no warning that the patient is abnormally sensitive to anesthetic. However, development of an antidote together with increased alertness to the problem has greatly decreased the death rate. Nevertheless, severe damage to nerves and kidneys may still occur.[c] Biochemical investigation of the hyperthermia syndrome has been facilitated by the discovery of a similar condition that is prevalent among certain breeds of pigs. Such "stress-prone" pigs are likely to die suddenly of hyperthermia induced by some stress such as shipment to market. The sharp rise in temperature with muscles going into a state of rigor is accompanied by a dramatic lowering of the ATP content of the muscles.

The problem, both in pigs and in humans susceptible to malignant hyperthermia, was found in the Ca^{2+} release channels (ryanodine receptors). Study of inheritance in human families together with genetic studies in pigs led to the finding that the stress-prone pigs have cysteine replacing arginine 615 in the Ca^{2+} channel protein. This modification appears to facilitate opening of the channels but to inhibit their closing.[e] A similar mutation has been found in some human families in which the condition has been recognized. However, there is probably more than one site of mutation in humans.[c,f] Similar mutations in the nematode *C. elegans* are being investigated with the hope of shedding light both on the problem of hyperthermia and on the functioning of the Ca^{2+} release channels.[g]

[a] Gordon, R. A., Britt, B. A., and Kalow, W., eds. (1973) *International Symposium on Malignant Hyperthermia*, Thomas, Springfield, Illinois

[b] Clark, M. G., Williams, C. H., Pfeifer, W. F., Bloxham, D. P., Holland, P. C., Taylor, C. A., and Lardy, H. A. (1973) *Nature (London)* **245**, 99–101

[c] MacLennan, D. H., and Phillips, M. S. (1992) *Science* **256**, 789–794

[d] Simon, H. B. (1993) *N. Engl. J. Med.* **329**, 483–487

[e] Fujii, J., Otsu, K., Zorzato, F., de Leon, S., Khanna, V. K., Weiler, J. E., O'Brien, P. J., and MacLennan, D. H. (1991) *Science* **253**, 448–451

[f] MacLennan, D. H., Duff, C., Zorzato, F., Fujii, J., Phillips, M., Korneluk, R. G., Frodis, W., Britt, B. A., and Worton, R. G. (1990) *Nature (London)* **343**, 559–561

[g] Sakube, Y., Ando, H., and Kagawa, H. (1997) *J. Mol. Biol.* **267**, 849–864

sensor is revealed by a lethal mutation (**muscular disgenesis**) in mice. Animals with this autosomal recessive trait generate normal action potentials in the sarcolemma but Ca^{2+} is not released and no muscular contraction occurs. They lack a 170-kDa dihydropyridine-binding subunit of the sensor.[246]

Some aspects of regulation by calcium ions are poorly understood. The frequent observations of oscillations in [Ca^{2+}] in cells is described in Box 6-D. Another phenomenom is the observation of Ca^{2+} "sparks," detected with fluorescent dyes and observation by confocal microscopy.[247] These small puffs of Ca^{2+} have been seen in cardiac muscle[247] and in a somewhat different form in smooth muscle.[248] They may represent the release of Ca^{2+} from a single release channel or a small cluster of channels. When the calcium release channels open, Ca^{2+} ions flow from the cisternae of the SR into the cytoplasm, where they activate both the troponin–tropomyosin system and also the Ca^{2+}-calmodulin-dependent **light chain protein kinase,** which acts on the light chains of the myosin head. These light chains resemble calmodulin in their Ca^{2+}-binding properties. The function of light chain phosphorylation of skeletal muscle myosin is uncertain but it is very important in smooth muscle.[248a]

The regulatory complex of tropomyosin and troponin is attached to the actin filaments as indicated in Fig. 19-8D and also in Fig. 19-9. The latter shows a model at near atomic resolution but without side chains on the tropomyosin and without the troponin components. When the regulatory proteins are completely removed from the fibrils, contraction occurs until the ATP is exhausted. However, in the presence of the regulatory proteins and in the absence of calcium, both contraction and hydrolysis of ATP are blocked. Tropomyosin (Tm) is a helical coiled-coil dimer, a 40-nm rigid rod, in which the two 284-residue 33-kDa monomers have a parallel orientation (Fig. 19-9)[249] as in the myosin tail. However, an 8– to 9– residue overlap at the ends may permit end-to-end association of the Tm molecules bound to the actin filament. As with other muscle proteins there are several isoforms,[250,251] whose distribution differs in skeletal and smooth muscle and in platelets. The elongated Tm rods appear to fit into the grooves between the two strands of actin monomers in the actin filament.[252–254] In resting muscle the Tm is thought to bind to actin near the site at which the S1 portion of the myosin binds. As a consequence, the Tm rod may block the attachment of the myosin heads to actin and prevent actin-stimulated hydrolysis of ATP. The 40-nm Tm rod can contact about seven actin subunits at once (Fig. 19-9). Thus, one Tm–troponin complex controls seven actin subunits synchronously.

Troponin (Tn) consists of three polypeptides (TnC, TnI, TnT) that range in mass from 18 to 37 kDa. The complex binds both to Tm and to actin.[255,256] Peptide TnT binds tightly to Tm and is thought to link the TnI•TnC complex to Tm.[256,257] TnI interacts with actin and inhibits ATPase activity in the absence of Ca^{2+}.[258–261] It may work with the other two peptides to keep the Tm in the proper position to inhibit ATP hydrolysis. TnC binds calcium ions. This ~160-residue protein has a folding pattern almost identical to that of calmodulin (Fig. 6-8) with four Ca^{2+}-binding domains arranged in two pairs at the ends of a long 9-turn helix. When Ca^{2+} binds to TnC, a conformational change occurs[258,259,262-265] (p. 313). This induces changes in the troponin•tropomyosin•actin complex, releases the inhibition of actomyosin ATPase, and allows contraction to occur.[265a] In the heart additional effects are exerted by β-adrenergic stimulation, which induces phosphorylation of two sites on TnI by the action of the cAMP-dependent protein kinase PKA. Dephosphorylation by protein phosphatase 2A completes a regulatory cycle in which the doubly phosphorylated TnI has a decreased sensitivity to calcium ions.[266] Cardiac muscle also contains a specialized protein called **phospholamban**. An oligomer of 52-residue subunits, it controls the calcium ion pump in response to β-adrenergic stimulation. Unphosphorylated phospholamban inhibits the Ca ATPase, keeping [Ca^{2+}] high in the cytoplasm. Phosphorylation of phospholamban by cAMP and/or calmodulin-dependent protein kinase activates the Ca^{2+} pump,[267–268a] removing Ca^{2+} and ending contraction.

X-ray diffraction and electron microscopy in the 1970s suggested that when calcium binds to troponin the tropomyosin moves through an angle of ~ 20° away from S1, uncovering the active site for the myosin–ATP–actin interactions.[252,253] Tropomyosin could be envisioned as rolling along the surface of the actin, uncovering sites on seven actin molecules at once. Side-chain knobs protruding from the tropomyosin like teeth on a submicroscopic gear might engage complementary holes in the actin. At the same time a set of magnesium ion bridges between zones of negative charge on tropomyosin and actin could hold the two proteins together. This proposal has been difficult to test. Although the older image reconstruction is regarded as unreliable, recent work still supports this **steric blocking** model.[255,269-270c] Image reconstruction from electron micrographs of thin filaments shows that, in the presence of Ca^{2+}, the tropomyosin does move 25° away from the position in low [Ca^{2+}]. However, instead of two states of the thin filament, "on" and "off," there may be at least three, which have been called "blocked," "closed," and "open."[255,269,271,271a] The closed state may be attained in rigor.[269] In addition, the possibility that changes in the conformation of actin as well as of myosin occur during the contraction cycle must be considered.[255]

Smooth muscle. The primary regulation of smooth muscle contraction occurs via phosphorylation of the Ser 19 –OH group in the 20-kDa regulatory light chains of each myosin head.[121b,160,272–274] The phosphorylated form is active, participating in the contraction process. Removal of Ca^{2+} by the calcium pump and dephosphorylation of the light chains by a protein phosphatase[275] restores the muscle to a resting state. The N-terminal part of the myosin light chain kinase binds to actin, while the catalytic domain is in the center of the protein. The C-terminal part binds to myosin, and this binding also has an activating effect.[276]

Another protein, **caldesmon**, binds to smooth muscle actin and blocks actomyosin ATPase.[271,277–278a] It is present in smooth muscle in a ratio of actin:tropomyosin:caldesmon of ~14:2:1. Inhibition can be reversed by Ca^{2+}, but there is no agreement on the function of caldesmon.[277] It is an elongated ~756-residue protein with N-terminal domain, which binds to myosin, and a C-terminal domain, which binds to actin, separated by a long helix.[278] Caldesmon may be a substitute for troponin in a tropomyosin-type regulatory system, or it may promote actomyosin assembly. Another possibility is that it functions in a **latch state**, an energy-economic state of smooth muscle at low levels of ATP hydrolysis.[278,279] In molluscan muscles Ca^{2+} binds to a myosin light chain and activates contraction directly. Some molluscan smooth muscles (**catch muscles**) also have a latch state, which enables these animals to maintain muscular tension for long periods of time, e.g., holding their shells tightly closed, with little expenditure of energy.[279a] Catch muscles contain myosin plus a second protein, **catchin**, which is formed as a result of alternative mRNA splicing. Catchin contains an N-terminal sequence that may undergo phosphorylation as part of a regulatory mechanism.[280] However, recent experiments indicate that twitchin (see next paragraph), rather than catchin, is essential to the catch state and is regulated by phosphorylation.[280a] Regulation of the large groups of unconventional myosins is poorly understood. Phosphorylation of groups on the myosin heavy chains is involved in ameba myosins and others.[281]

An unexpected aspect of regulation was discovered from study of the 40 or more genes of *Caenorhabditis elegans* needed for assembly and function of muscle. The mutants designated *unc-22* showed a constant twitch arising from the muscles in the nematode's body. The gene was cloned using transposon tagging (Chapter 27) and was found to encode a mammoth 753-kDa 6839-residue protein which has been named **twitchin**.[282–285] Twitchin resembles titin (Fig. 19-8) and like titin has a protein kinase domain, which is normally inhibited by the end of its peptide chain, which folds over the active site of the kinase. Perhaps the protein kinase activities of twitchin, titin, and related proteins[285] are required in assembly of the sarcomere.

5. The Phosphagens

ATP provides the immediate source of energy for muscles but its concentration is only ~5 mM. As discussed in Chapter 6, Section D, **phosphagens**, such as **creatine phosphate**, are also present and may

$$\text{Creatine-}P + \text{ADP} \longrightarrow \text{Creatine} + \text{ATP}$$

attain a concentration of 20 mM in mammalian muscle. This provides a reserve of high-energy phospho groups and keeps the adenylate system buffered at a high phosphorylation state ratio[286] (see Eq. 6-67).

The concentration of both ATP and creatine phosphate as well as their rates of interconversion can be monitored by ^{31}P NMR within living muscles (Figs. 6-4 and 19-22). Phospho groups were observed to be

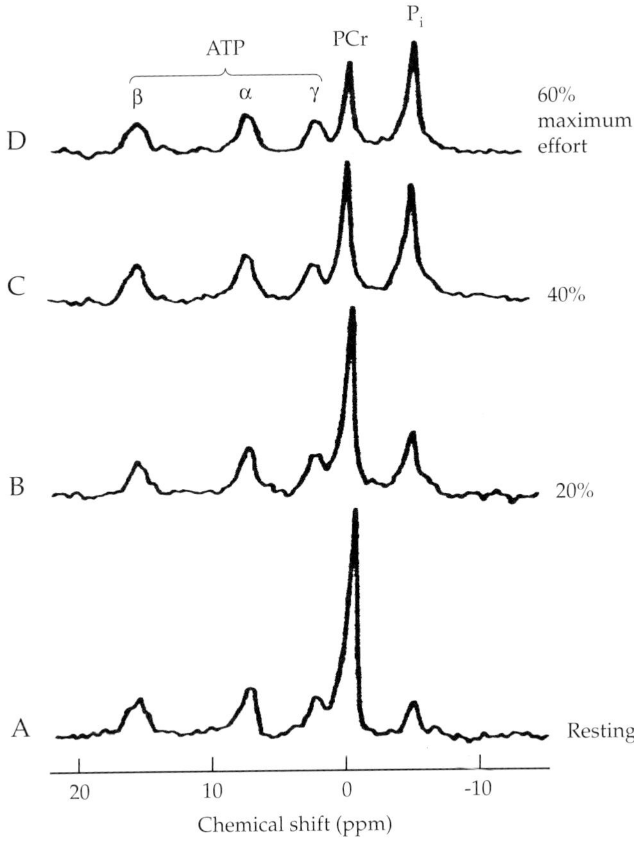

Figure 19-22 Phosphorus-31 magnetic resonance spectra of wrist flexor muscles of the forearm of a trained long-distance runner at rest and during contraction at three different levels of exercise. Ergometer measurements indicating the percent of initial maximum strength (% max) were recorded over each 6-min period. Spectra were obtained during the last 3 min of each period. Times of spectral data collection: A, resting; B, 4–6 min; C, 10–12 min; and D, 16–18 min. The pH ranged from 6.9 to 7.0. From Park *et al.*[288]

transferred from creatine phosphate to ADP to form ATP with a flux of 13 mmol $kg^{-1}s^{-1}$ in rat legs.[287] The reverse reaction must occur at about the same rate because little cleavage of ATP to P_i was observed in the anaesthetized rats. Use of surface coils has permitted direct observation of the operation of this shuttle system in human muscle (Fig. 19-22)[288] as well as in animal hearts (see Chapter 6). Only a fraction of the total creatine present within cells participates in the shuttle, however.[289]

C. Motion in Nonmuscle Cells

At one time actin and myosin were thought to be present only in muscles, but we know now that both actin proteins of the myosin family are present in all eukaryotic cells. Ameboid movement, the motion of cilia and flagella, and movement of materials along microtubules within cells also depend upon proteins of this group.

1. Actin-Based Motility

Ameboid movements of protozoa and of cells from higher organisms, the ruffling movements of cell membranes, phagocystosis, and the cytoplasmic streaming characteristic of many plant cells[289a] have all been traced to actin filaments or actin cables rather

TABLE 19-1
Some Actin-Binding Proteins

Function	Name
Bind and stabilize monomeric actin	Profilin[a,b,c,d,e,f,g,h]
	ADF/Cofilin[i,j,h,k]
	Thymosin[f,l]
Cap actin filament ends	CapZ[m,n]
barbed end	Fragmin[o,p]
pointed end	β-Actin[q]
	Tropomodulin[r]
	Arp2/3, a complex of seven polypeptides
Sever or dissociate actin filament	Gelsolin[s,t,u,v,w]
	Depactin
	Profilin[d,e]
	ADF/cofilin[h,k,x,y]

Function	Name
Crosslink actin filaments or monomers	
Tight bundles	Villin[c,z,aa]
Loose bundles	α-Actinin[bb]
Spectrin[bb]	
	Fascin[cc]
	MARCKS[a]
Network	Filamin[bb,c]
	Gelactins
Bind actin filaments to membrane	Talin[dd]
	"ERM" proteins[ee,ff]

[a] Aderem, A. (1992) *Trends Biochem. Sci.* **17**, 438–443
[b] Mannherz, H. G. (1992) *J. Biol. Chem.* **267**, 11661–11664
[c] Way, M., and Weeds, A. (1990) *Nature (London)* **344**, 292–293
[d] Eads, J. C., Mahoney, N. M., Vorobiev, S., Bresnick, A. R., Wen, K.-K., Rubenstein, P. A., Haarer, B. K., and Almo, S. C. (1998) *Biochemistry* **37**, 11171–11181
[e] Vinson, V. K., De La Cruz, E. M., Higgs, H. N., and Pollard, T. D. (1998) *Biochemistry* **37**, 10871–10880
[f] Kang, F., Purich, D. L., and Southwick, F. S. (1999) *J. Biol. Chem.* **274**, 36963–36972
[g] Gutsche-Perelroizen, I., Lepault, J., Ott, A., and Carlier, M.-F. (1999) *J. Biol. Chem.* **274**, 6234–6243
[h] Nodelman, I. M., Bowman, G. D., Lindberg, U., and Schutt, C. E. (1999) *J. Mol. Biol.* **294**, 1271–1285
[i] Lappalainen, P., Fedorov, E. V., Fedorov, A. A., Almo, S. C., and Drubin, D. G. (1997) *EMBO J.* **16**, 5520–5530
[j] Rosenblatt, J., and Mitchison, T. J. (1998) *Nature (London)* **393**, 739–740
[k] Chen, H., Bernstein, B. W., and Bamburg, J. R. (2000) *Trends Biochem. Sci.* **25**, 19–23
[l] Carlier, M.-F., Didry, D., Erk, I., Lepault, J., Van Troys, M. L., Vandekerckhove, J., Perelroizen, I., Yin, H., Doi, Y., and Pantaloni, D. (1996) *J. Biol. Chem.* **271**, 9231–9239
[m] Barron-Casella, E. A., Torres, M. A., Scherer, S. W., Heng, H. H. Q., Tsui, L.-C., and Casella, J. F. (1995) *J. Biol. Chem.* **270**, 21472–21479
[n] Kuhlman, P. A., and Fowler, V. M. (1997) *Biochemistry* **36**, 13461–13472
[o] Steinbacher, S., Hof, P., Eichinger, L., Schleicher, M., Gettemans, J., Vandekerckhove, J., Huber, R., and Benz, J. (1999) *EMBO J.* **18**, 2923–2929
[p] Khaitlina, S., and Hinssen, H. (1997) *Biophys. J.* **73**, 929–937
[q] See main text
[r] Gregorio, C. C., Weber, A., Bondad, M., Pennise, C. R., and Fowler, V. M. (1995) *Nature (London)* **377**, 83–86
[s] Azuma, T., Witke, W., Stossel, T. P., Hartwig, J. H., and Kwiatkowski, D. J. (1998) *EMBO J.* **17**, 1362–1370
[t] De Corte, V., Demol, H., Goethals, M., Van Damme, J., Gettemans, J., and Vandekerckhove, J. (1999) *Protein Sci.* **8**, 234–241
[u] McGough, A., Chiu, W., and Way, M. (1998) *Biophys. J.* **74**, 764–772
[v] Sun, H. Q., Yamamoto, M., Mejillano, M., and Yin, H. L. (1999) *J. Biol. Chem.* **274**, 33179–33182
[w] Robinson, R. C., Mejillano, M., Le, V. P., Burtnick, L. D., Yin, H. L., and Choe, S. (1999) *Science* **286**, 1939–1942
[x] Carlier, M.-F., Ressad, F., and Pantaloni, D. (1999) *J. Biol. Chem.* **274**, 33827–33830
[y] McGough, A., and Chiu, W. (1999) *J. Mol. Biol.* **291**, 513–519
[z] Markus, M. A., Matsudaira, P., and Wagner, G. (1997) *Protein Sci.* **6**, 1197–1209
[aa] Vardar, D., Buckley, D. A., Frank, B. S., and McKnight, C. J. (1999) *J. Mol. Biol.* **294**, 1299–1310
[bb] Matsudaira, P. (1991) *Trends Biochem. Sci.* **16**, 87–92
[cc] Ono, S., Yamakita, Y., Yamashiro, S., Matsudaira, P. T., Gnarra, J. R., Obinata, T., and Matsumura, F. (1997) *J. Biol. Chem.* **272**, 2527–2533
[dd] McLachlan, A. D., Stewart, M., Hynes, R. O., and Rees, D. J. G. (1994) *J. Mol. Biol.* **235**, 1278–1290
[ee] Tsukita, S., Yonemura, S., and Tsukita, S. (1997) *Trends Biochem. Sci.* **22**, 53–58
[ff] Tsukita, S., and Yonemura, S. (1999) *J. Biol. Chem.* **274**, 34507–34510

than to microtubules.[289b] Actin is one of the most abundant proteins in all eukaryotes. Its network of filaments is especially dense in the lamellipodia of cell edges, in microvilli, and in the specialized stereocilia and acrosomal processes (see also pp. 369-370).[289c] Actin filaments and cables are often formed rapidly and dissolve quickly. When actin filaments grow, the monomeric subunits with bound ATP are added most rapidly at the "barbed end" and dissociate from the filament at the "pointed end" (see Section B,2).[94,290] The rate of growth may be ~20–200 nm / s, which requires the addition of 10–100 subunits / s.[291] Various actin-binding proteins control the growth and stability of the filaments. The actin-related proteins **Arp2** and **Arp3**, as a complex Arp2/3, together with recently recognized **formins**.[291a], provide nuclei for rapid growth of new actin filaments as branches near the barbed ends.[290,292–293c]

Growth of the barbed ends of actin filaments is stimulated by phosphoinositides and by members of the Rho family of G proteins (p. 559)[293d] through interaction with proteins of the **WASp** group.[293b,d,e] The name WAS comes from the immune deficiency disorder Wiskott–Aldrich Syndrome. Yet another family, the **Ena/VASP** proteins, are also implicated in actin dynamics. They tend to localize at focal adhesions and edges of lamellipodia.[293e,f] Profilin (Table 19-1) stabilizes a pool of monomeric actin when the barbed ends of actin filaments are capped. However, it catalyzes both the addition of actin monomers to uncapped barbed ends and rapid dissociation of subunits from pointed ends, leading to increased **treadmilling**.[294,295] Actin-severing proteins such as the **actin depolymerizing factor** (**ADF** or **cofilin**, Table 19-1) promote breakdown of the filaments.[296–297a] Treadmilling in the actin filaments of the lamellipods of crawling cells or pseudopods of amebas provides a motive force for many cells[291,298–299] ranging from those of *Dictyostelium* to human leukocytes. A series of proteins known as the **ezrin**, **radixin**, **moesin** (**ERM**) group attach actin to integral membrane proteins (Fig. 8-17)[292,300,301] and may interact directly with membrane lipids.[301a,b] Bound ATP in the actin subunits is essential for polymerization, and excess ATP together with crosslinking proteins stabilize the filaments. However, when the bound ATP near the pointed ends is hydrolyzed to ADP the filaments become unstable and treadmilling is enhanced. Thus, as in skeletal muscle, ATP provides the energy for movement.

Bacteria also contain filamentous proteins that resemble F-actin and which may be utilized for cell-shape determination.[301c] Actin-based motility is used by some bacteria and other pathogens during invasion of host cells (Box 19-C). It is employed by sperm cells of *Ascaris* and of *C. elegans*, which crawl by an ameboid movement that utilizes treadmilling of filaments formed from a motile sperm protein, which does not resemble actin.[302,302a] Cells are propelled on a glass surface at rates up to ~1 µm / s.

Various nonmuscle forms of myosin also interact with actin without formation of the myofibrils of muscle.[299] In most higher organisms nonmuscle myosins often consist of two ~200-kDa subunits plus two pairs of light chains of ~17 and 24 kDa each. These may form bipolar aggregates, which may bind to pairs of actin filaments to cause relative movement of two parts of a cell.[303] Movement depending upon the cytoskeleton is complicated by the presence of a bewildering array of actin-binding proteins, some of which are listed in Table 19-1.

2. Transport along Microtubules by Kinesin and Dynein

Many materials are carried out from the cell bodies of neurons along microtubules in the axons, which in the human body may be as long as 1 m. The rates of this **fast axonal transport** in neurons may be as high as 5 µm / s or 0.43 m / day. The system depends upon ATP and kinesin (Fig. 19-17) and permits small vesicles to be moved along single microtubules.[304–305b] Movement is from the minus end toward the plus end of the microtubule as defined in Figs. 7-33, 7-34. Slower **retrograde axonal transport** carries vesicles from the synapses at the ends of the axons (Fig. 30-8) back toward the cell body. This retrograde transport depends upon the complex motor molecule **cytoplasmic dynein** which moves materials from the plus end of the microtubule toward the minus end.[305,305c] In addition to these movements, as mentioned in Chapter 7, microtubules often grow in length rapidly or dissociate into their tubulin subunits. Growth occurs at one end by addition of tubulin subunits with their bound GTP. The fast growing *plus*-ends of the microtubules are usually oriented toward the cell periphery, while the *minus*-ends are embedded in the **centrosome** or **microtubule-organizing center** (p. 372).[306] Just as with actin, in which bound ATP is hydrolyzed to ADP, the bound GTP in the β-tubulin subunits of microtubules is hydrolyzed to GDP[307–310] decreasing the stability of the microtubules, a phenomenon described as **dynamic instability**. Various **microtubule-associated proteins** (MAPs) have strong effects on this phenomenon.[311] The MAPs are often regulated by phosphorylation–dephosphorylation cycles involving serine / threonine kinases. Microtubules also undergo posttranslational alterations not seen in other proteins. These include addition or removal of tyrosine at the C terminus.[312] Polyglycyl groups containing 3–34 glycine residues may be bound covalently to γ-carboxyl groups of glutamate side chains in both α- and β-tubulins.[312,313] This stabilizes the microtubules and is important to the long-lived microtubules of the axonomes of flagella and

BOX 19-C ACTIN-BASED MOTILITY AND BACTERIAL INVASION

Listeria monocytogenes is a dangerous food-borne bacterium that has become a major problem in the United States. This is one of the best understood *intracellular* pathogens. It is able to enter cells, escape from phagocytic vesicles, spread from cell to cell, and cross intestinal, blood–brain, and placental barriers.[a–c] Within cells these bacteria move using actin-based motility. Actin subunits polymerize at one end of a bacterium leaving a "comet tail" of crosslinked fibrous actin behind (see micrographs). Actin polymerization occurs directly behind the bacteria with subunits of monomeric actin adding to the fast growing "barbed end" (see Section B,2) of the actin strands. Growth has been described as a "Brownian ratchet."[c,d] Continual Brownian movement opens up spaces behind the bacteria, spaces that are immediately filled by new actin subunits. This provides a propulsive force adequate to move the bacteria ahead at velocities of about 0.3 μm / s.

Polymerization of actin is induced by interaction of a dimer of a 610-residue bacterial protein **ActA** with proteins of the host cell.[a,e–h] ActA is a composite protein with an N-terminal region that protrudes from the bacterial cell, a central region of proline-rich repeats that appear to be essential for recognition by host cells, and a C-terminal hydrophobic membrane anchor. There are also regions of close sequence similarity to the human actin-binding proteins vinculin and zyxin. The number of host proteins needed in addition to monomeric actin are:[i,j] the two actin-related proteins, **Arp2** and **Arp3**, which stimulate actin polymerization and branching;[h] **ADF/ cofilin**, which increases the rates, both of growth at the barbed ends and dissociation from the pointed ends of the filaments; and **Cap Z**, which caps barbed ends (Table 19-1). The need for ADF/ cofilin and Cap Z seems paradoxical. Cap Z may cap mostly older and slower growing filaments, restricting rapid filament assembly to the region closest to the bacterium. The need for ADF/ cofilin is unclear.[i] Growth rates are also enhanced by the human protein called **VASP** (vasodilator-stimulated phosphoprotein). The proline-rich region of the bacterial ActA may bind to VASP to initiate polymerization.[g] Both profilin (Table 19-1) and the crosslinking protein α-actinin also stimulate comet tail growth. Myosin does not participate in actin-based motility, but the hydrolysis of ATP drives the process through its link to actin polymerization.[i]

Although *Listeria* has been studied most, actin-based motility is employed by other pathogens as well, e.g., *Shigella flexneri* (the dysentery bacterium),[k] *Rickettsia*,[l] and vaccinia virus.[l] Although enteropathogenic *E. coli* do not use this method of movement, they induce accumulation of actin beneath the bacteria. They also promote formation of actin-rich adherent pseudopods and highly organized cytoskeletal structures that presumably assist the bacteria in entering a cell.[m]

[a] Cossart, P., and Lecuit, M. (1998) *EMBO J.* **17**, 3797–3806

[b] Sechi, A. S., Wehland, J., and Small, J. V. (1997) *J. Cell Biol.* **137**, 155–167

[c] Pantaloni, D., Le Clainche, C., and Carlier, M.-F. (2001) *Science* **292**, 1502–1506

[d] Mogilner, A., and Oster, G. (1996) *Biophys. J.* **71**, 3030–3045

[e] Mourrain, P., Lasa, I., Gautreau, A., Gouin, E., Pugsley, A., and Cossart, P. (1997) *Proc. Natl. Acad. Sci. U.S.A.* **94**, 10034–10039

[f] Southwick, F. S., and Purich, D. L. (1994) *Proc. Natl. Acad. Sci. U.S.A.* **91**, 5168–5172

[g] Niebuhr, K., Ebel, F., Frank, R., Reinhard, M., Domann, E., Carl, U. D., Walter, U., Gertler, F. B., Wehland, J., and Chakraborty, T. (1997) *EMBO J.* **16**, 5433–5444

[h] Welch, M. D., Rosenblatt, J., Skoble, J., Portnoy, D. A., and Mitchison, T. J. (1998) *Science* **281**, 105–108

[i] Loisel, T. P., Boujemaa, R., Pantaloni, D., and Carlier, M.-F. (1999) *Nature (London)* **401**, 613–616

[j] Machesky, L. M., and Cooper, J. A. (1999) *Nature (London)* **401**, 542–543

[k] Bourdet-Sicard, R., Rüdiger, M., Jockusch, B. M., Gounon, P., Sansonetti, P. J., and Tran Van Nhieu, G. (1999) *EMBO J.* **18**, 5853–5862

[l] Cudmore, S., Cossart, P., Griffiths, G., and Way, M. (1995) *Nature (London)* **378**, 636–638

[m] Rosenshine, I., Ruschkowski, S., Stein, M., Reinscheid, D. J., Mills, S. D., and Finlay, B. B. (1996) *EMBO J.* **15**, 2613–2624

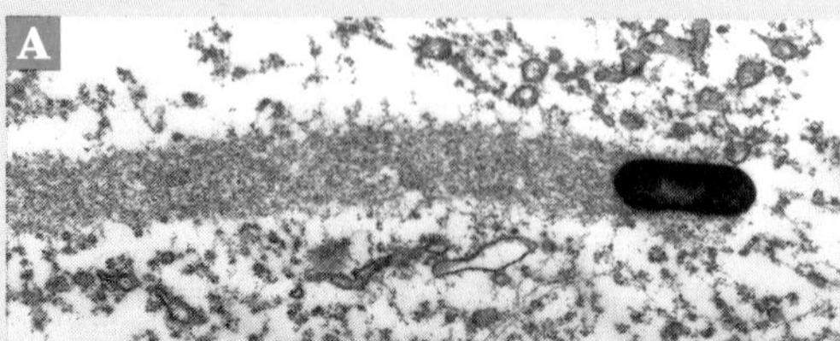

(A) *Listeria* cell with "comet tail" of cross-linked actin filaments. From Kocks *et al.* (1992) *Cell* **68**, 521–531. Courtesy of Pascale Cossart.

(B) Enlarged section of a thin comet tail of high resolution showing the actin filaments. From Sechi *et al.*[b] Courtesy of Antonio Sechi.

cilia. Polyglutamyl groups of 6–7 glutamates are also often added.[314]

Both dynein and several kinesins act as motors for formation of the spindle and for movement of chromosomes toward the minus ends of spindle microtubules during mitosis and meiosis (Fig. 26-11).[314a] In the genome of *Saccharomyces cerevisiae* there is only one dynein gene, but genes for six different kinesin-type motor molecules are present.[315] In higher organisms there may be even more genes for kinesins but there is apparently only one dynein in most species.[316] **Axonemal dyneins** drive the motion of eukaryotic flagella and cilia. As with the cytoplasmic dyneins a complete molecule consists of two or three heavy chains with molecular mass ~520 kDa, some localized in the dynein tail, and several lighter chains.[305a,317–321] Like myosin dynein is an ATPase.

3. Eukaryotic Cilia and Flagella

The motion of eukaryotic flagella (Fig. 1-8) involves a sliding of the microtubular filaments somewhat analogous to the sliding of muscle filaments.[305,322–325] Sliding between the outer doublet microtubules (Fig. 19-23) via their inner and outer arms (dynein compounds) is thought to provide the characteristic bending waves.[325a,b] The movement is powered by dynein and ATP hydrolysis. Force and displacement measurements made by optical trapping nanometry suggest that the characteristic rhythmic beating of flagella results from an oscillatory property of the dynein.[326] The extremely complex structure of flagella is illustrated in Fig. 19-23. About 250 individual axonemal proteins have been detected in flagella of the alga *Chlamydomonas* (Fig. 1-11),[327] and a large number of mutants with various defects in their flagella have been isolated. The radial spokes (Fig. 19-23) alone contain 17 different proteins. These spokes protrude at ~29-nm intervals while the dynein molecules lie between pairs of the outer microtubule doublets at ~24-nm intervals. The dynein "arms" protrude from the "A" microtubule of each outer doublet and make contact with the incomplete "B" microtubule of the next doublet (Fig. 19-12). Although the shapes of the molecules are quite different, the basic chemistry of the ATPase activity of the dynein–microtubule system resembles that of actomyosin. However, the complexity of the dynein arms,[328]

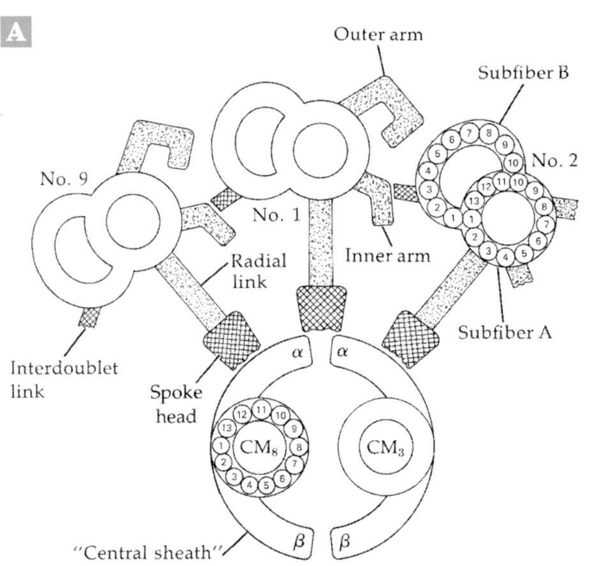

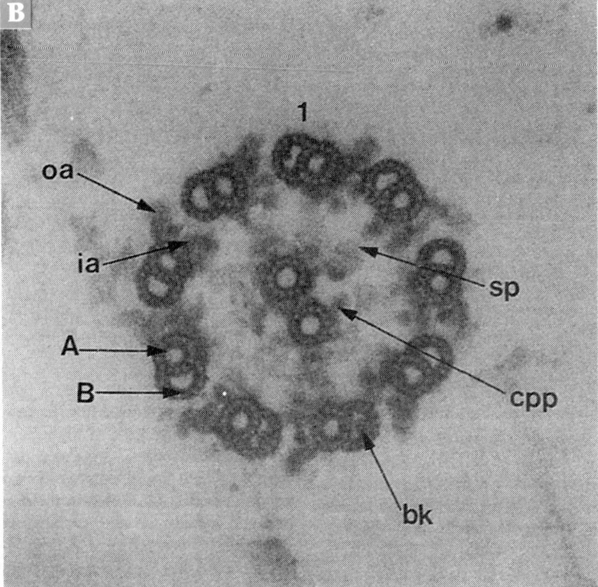

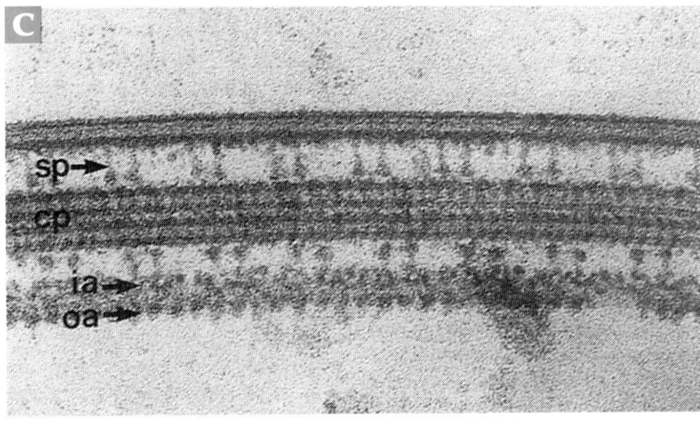

Figure 19-23 (A) Diagram of a cross-sectional view of the outer portion of a lamellibranch gill cilium. This has the 9+2 axoneme structure as shown in Fig. 1-8 and in (B). The viewing direction is from base to tip. From M. A. Sleigh.[329] (B, C) Thin-section electron micrographs of transverse (B) and longitudinal (C) sections of wild-type *Chlamydomonas* axonemes. In transverse section labels A and B mark A and B subtubules of microtubule doublets; oa, ia, outer and inner dynein arms, respectively; sp, spokes; cpp, central pair projections; bk, beaks. From Smith and Sale.[329a]

which exist in two types, inner and outer, suggests a complex contraction cycle.

4. Chemotaxis

As described in Box 11-C, the ameboid cells of the slime mold *Dictyostelium discoideum* are attracted to nutrients such as folic acid during their growth stage. Later, as the cells undergo developmental changes they become attracted by pulses of cyclic AMP.[330] Occupancy of 7-helix receptors for cAMP on the outer plasma membrane appears to induce methylation of both proteins and phospholipids and a rise in cytosolic Ca^{2+} and changes in the cytoskeleton that result in preferential extension of the actin-rich pseudopods toward the chemoattractant.[331]

In a similar manner, human ameboid leukocytes are attracted to sites of inflammation by various **chemotactic factors**.[332] These include the 74-residue cleavage product C5a formed from the fifth component of complement (Chapter 31),[333] various **lymphokines** (Chapter 31) secreted by lymphocytes, and peptides such as VGSE and AGSE, as well as larger peptides released by mast cells, basophils, or stimulated monocytes[334] and oxoicosenoids.[335] Polymorphonuclear leukocytes, upon engulfing sodium urate crystals in gouty joints, release an 8.4-kDa chemotactic protein which may cause a damaging response in this arthritic condition. Leukotriene B is a potent chemotactic agent as are a series of specific bacterial products, formylated peptides such as *N*-formyl-MLF.[332,336]

Neutrophils, monocytes, macrophages, eosinophils, basophils, and polymorphonuclear leukocytes are all affected by several or all of these factors. Binding to specific receptors results in a variety of changes in the cells. These include alterations in membrane potential, cyclic nucleotide levels, and ion fluxes (Na^+, K^+, Ca^{2+}) as well as increased methylation of specific proteins. A reorganization of microtubules and actin fibrils occurs, probably in response to an altered gradient of Ca^{2+}. The morphology of the cells changes, and they begin immediately to crawl toward the chemoattractants. It appears that these ameboid cells detect a gradient of attractant concentration between one end of the cell and the other, even though the anticipated difference may amount to only 0.1% of the total.[337,338]

5. Other Forms of Movement

Movement is characteristic of life and is caused not only by motor proteins but by various springs and ratchets which may be energized in a number of ways.[339] A striking example, which any one with a microscope and some fresh pond water can observe, is contraction of the stalk of protozoa of the genus *Vorticella*. Apparently first reported in 1676 by Leeuwenhoek the organism's 2–3 mm-long stalk contracts into a coiled spring (see p. 1 and also p. 281) when the animal is disturbed. Application of calcium ions causes contraction within a few milliseconds to ~40% of the original length. The process reverses slowly after a few seconds. Contraction is caused by a spring-like organ the **spasmoneme**, which is a bundle of short 2 nm-diameter fibrils inside the stalk. The fibrils are thought to be weakly cross-linked and held in the extended state by electrostatic repulsion between the negatively charged rods. Addition of Ca^{2+} neutralizes the charges permitting an entropy-driven collapse of the fibers.[339]

Another remarkable example is extension of the acrosomal process from a sperm cell of the horseshoe crab *Limulus polyphemus* at fertilization. A bundle of actin filaments in a crystalline state lies coiled around the base of the nucleus. At fertilization the bundle uncoils and slides through a tunnel in the nucleus to form a 60 μm-long acrosomal process within a few seconds. The uncoiled bundle is also crystalline. The coiled bundle is apparently overtwisted and an actin crosslinking protein **scruin** mediates the conformational alteration that takes place.[339] A somewhat related process may be involved in contraction of bacteriophage tails (pp. 363, 364)

Some bacteria glide with a twitching movement induced by rapid retraction of pili.[340] Another type of movement involves the pinching off of vesicles, e.g., of clathrin-coated pits (Fig. 8-27). This is a GTP-driven process that requires a mechanoenzyme called **dynamin**.[341,342]

References

1. Elston, T. C., and Oster, G. (1997) *Biophys. J.* **73**, 703–721
2. Magariyama, Y., Sugiyama, S., Muramoto, K., Maekawa, Y., Kawagishi, I., Imae, Y., and Kudo, S. (1994) *Nature (London)* **371**, 752
3. Macnab, R. M. (1987) in *Escherichia coli and Salmonella typhimurium* (Niedhardt, F. C., ed), pp. 70–83, Am. Soc. for Microbiology, Washington, D.C.
4. Berg, H. C. (1975) *Sci. Am.* **233**(Aug), 36–44
5. Shimada, K., Kamiya, R., and Asakura, S. (1975) *Nature (London)* **254**, 332–334
6. Berg, H. C. (1975) *Nature (London)* **254**, 389–392
7. Macnab, R. M. (1985) *Trends Biochem. Sci.* **10**, 185–188
8. Macnab, R. M., and Aizawa, S.-J. (1984) *Annu Rev Biophys Bioeng.* **13**, 51–83
9. Sharp, L. L., Zhou, J., and Blair, D. F. (1995) *Biochemistry* **34**, 9166–9171
10. Morgan, D. G., Owen, C., Melanson, L. A., and DeRosier, D. J. (1995) *J. Mol. Biol.* **249**, 88–110
11. Mimori, Y., Yamashita, I., Murata, K., Fujiyoshi, Y., Yonekura, K., Toyoshima, C., and Namba, K. (1995) *J. Mol. Biol.* **249**, 69–87
12. Yamashita, I., Vonderviszt, F., Mimori, Y., Suzuki, H., Oosawa, K., and Namba, K. (1995) *J. Mol. Biol.* **253**, 547–558

12a. Samatey, F. A., Imada, K., Nagashima, S., Vonderviszt, F., Kumasaka, T., Yamamoto, M., and Namba, K. (2001) *Nature (London)* **410**, 331–337

13. Muramoto, K., Kawagishi, I., Kudo, S., Magariyama, Y., Imae, Y., and Homma, M. (1995) *J. Mol. Biol.* **251**, 50–58
14. DePamphilis, M. L., and Adler, J. (1971) *J. Bacteriol.* **105**, 396–407

References

15. Stallmeyer, M. J. B., Aizawa, S.-I., Macnab, R. M., and DeRosier, D. J. (1989) *J. Mol. Biol.* **205**, 519–528
16. DeRosier, D. J. (1998) *Cell* **93**, 17–20
17. Trachtenberg, S., and DeRosier, D. J. (1992) *J. Mol. Biol.* **226**, 447–454
18. Trachtenberg, S., DeRosier, D. J., Zemlin, F., and Beckmann, E. (1998) *J. Mol. Biol.* **276**, 759–773
19. Hasegawa, K., Yamashita, I., and Namba, K. (1998) *Biophys. J.* **74**, 569–575
19a. Macnab, R. M. (2001) *Nature (London)* **410**, 321–322
20. Iino, T. (1969) *Bacteriol. Rev.* **33**, 454–475
21. Kubori, T., Shimamoto, N., Yamaguchi, S., Namba, K., and Aizawa, S.-I. (1992) *J. Mol. Biol.* **226**, 433–446
22. Vonderviszt, F., Závodszky, P., Ishimura, M., Uedaira, H., and Namba, K. (1995) *J. Mol. Biol.* **251**, 520–532
23. Muramoto, K., Makishima, S., Aizawa, S.-I., and Macnab, R. M. (1998) *J. Mol. Biol.* **277**, 871–882
24. Ikeda, T., Oosawa, K., and Hotani, H. (1996) *J. Mol. Biol.* **259**, 679–686
25. Fahrner, K. A., Block, S. M., Krishnaswamy, S., Parkinson, J. S., and Berg, H. C. (1994) *J. Mol. Biol.* **238**, 173–186
25a. Hirano, T., Minamino, T., and Macnab, R. M. (2001) *J. Mol. Biol.* **312**, 359–369
26. Maki, S., Vonderviszt, F., Furukawa, Y., Imada, K., and Namba, K. (1998) *J. Mol. Biol.* **277**, 771–777
26a. Yonekura, K., Maki, S., Morgan, D. G., DeRosier, D. J., Vonderviszt, F., Imada, K., and Namba, K. (2000) *Science* **290**, 2148–2152
26b. Macnab, R. M. (2000) *Science* **290**, 2086–2087
26c. Minamino, T., Tame, J. R. H., Namba, K., and Macnab, R. M. (2001) *J. Mol. Biol.* **312**, 1027–1036
26d. Auvray, F., Thomas, J., Fraser, G. M., and Hughes, C. (2001) *J. Mol. Biol.* **308**, 221–229
27. Berg, H. C. (1996) *Proc. Natl. Acad. Sci. U.S.A.* **93**, 14225–14228
28. Lloyd, S. A., Whitby, F. G., Blair, D. F., and Hill, C. P. (1999) *Nature (London)* **400**, 472–475
29. Schuster, S. C., and Khan, S. (1994) *Annu. Rev. Biophys. Biomol. Struct.* **23**, 509–539
30. Adler, J. (1976) *Sci. Am.* **234** (Apr), 40–47
31. Calladine, C. R. (1975) *Nature (London)* **255**, 121–124
32. Eisenbach, M., and Adler, J. (1981) *J. Biol. Chem.* **256**, 8807–8814
33. Fung, D. C., and Berg, H. C. (1995) *Nature (London)* **375**, 809–812
34. Glagolev, A. N., and Skulachev, V. P. (1978) *Nature (London)* **273**, 280–282
35. Chun, S. Y., and Parkinson, J. S. (1988) *Science* **239**, 276–278
36. Laüger, P. (1990) *Commun. Theor. Biol.* **2**, 99–123
37. Tang, H., Braun, T. F., and Blair, D. F. (1996) *J. Mol. Biol.* **261**, 209–221
37a. Ko, M., and Park, C. (2000) *J. Mol. Biol.* **303**, 371–382
38. Block, S. M., and Berg, H. C. (1984) *Nature (London)* **309**, 470–472
39. Ueno, T., Oosawa, K., and Aizawa, S.-I. (1994) *J. Mol. Biol.* **236**, 546–555
40. Zhou, J., Fazzio, R. T., and Blair, D. F. (1995) *J. Mol. Biol.* **251**, 237–242
41. Garza, A. G., Biran, R., Wohlschlegel, J. A., and Manson, M. D. (1996) *J. Mol. Biol.* **258**, 270–285
41a. Van Way, S. M., Hosking, E. R., Braun, T. F., and Manson, M. D. (2000) *J. Mol. Biol.* **297**, 7–24
42. Zhao, R., Pathak, N., Jaffe, H., Reese, T. S., and Khan, S. (1996) *J. Mol. Biol.* **261**, 195–208
43. Marykwas, D. L., Schmidt, S. A., and Berg, H. C. (1996) *J. Mol. Biol.* **256**, 564–576
44. Zhou, J., Lloyd, S. A., and Blair, D. F. (1998) *Proc. Natl. Acad. Sci. U.S.A.* **95**, 6436–6441
44a. Thomas, D. R., Morgan, D. G., and DeRosier, D. J. (1999) *Proc. Natl. Acad. Sci. U.S.A.* **96**, 10134–10139
44b. Lux, R., Kar, N., and Khan, S. (2000) *J. Mol. Biol.* **298**, 577–583
45. Toker, A. S., and Macnab, R. M. (1997) *J. Mol. Biol.* **273**, 623–634
46. Kehry, M. R., Doak, T. G., and Dahlquist, F. W. (1984) *J. Biol. Chem.* **259**, 11828–11835
47. Koshland, D. E., Jr. (1988) *Biochemistry* **27**, 5829–5834
48. Stock, J., and Stock, A. (1987) *Trends Biochem. Sci.* **12**, 371–375
49. Segall, J. E., Block, S. M., and Berg, H. C. (1986) *Proc. Natl. Acad. Sci. U.S.A.* **83**, 8987–8991
50. Jeffery, C. J., and Koshland, D. E., Jr. (1993) *Protein Sci.* **2**, 559–566
51. Li, J., Li, G., and Weis, R. M. (1997) *Biochemistry* **36**, 11851–11857
51a. Isaac, B., Gallagher, G. J., Balazs, Y. S., and Thompson, L. K. (2002) *Biochemistry* **41**, 3025–3036
52. Scott, W. G., Milligan, D. L., Milburn, M. V., Privé, G. G., Yeh, J., Koshland, D. E., Jr., and Kim, S.-H. (1993) *J. Mol. Biol.* **232**, 555–573
53. Yeh, J. I., Biemann, H.-P., Privé, G. G., Pandit, J., Koshland, D. E., Jr., and Kim, S.-H. (1996) *J. Mol. Biol.* **262**, 186–201
54. Foster, D. L., Mowbray, S. L., Jap, B. K., and Koshland, D. E., Jr. (1985) *J. Biol. Chem.* **260**, 11706–11710
54a. Falke, J. J., and Hazelbauer, G. L. (2001) *Trends Biochem. Sci.* **26**, 257–265
55. Maddock, J. R., and Shapiro, L. (1993) *Science* **259**, 1717–1723
56. Wolfe, A. J., Conley, M. P., and Berg, H. C. (1988) *Proc. Natl. Acad. Sci. U.S.A.* **85**, 6711–6715
57. Ninfa, E. G., Stock, A., Mowbray, S., and Stock, J. (1991) *J. Biol. Chem.* **266**, 9764–9770
58. Kim, S.-H. (1994) *Protein Sci.* **3**, 159–165
59. Chervitz, S. A., and Falke, J. J. (1995) *J. Biol. Chem.* **270**, 24043–24053
60. Ottemann, K. M., Thorgeirsson, T. E., Kolodziej, A. F., Shin, Y.-K., and Koshland, D. E., Jr. (1998) *Biochemistry* **37**, 7062–7069
61. Trammell, M. A., and Falke, J. J. (1999) *Biochemistry* **38**, 329–336
61a. Hirschman, A., Boukhvalova, M., VanBruggen, R., Wolfe, A. J., and Stewart, R. C. (2001) *Biochemistry* **40**, 13876–13887
62. Spiro, P. A., Parkinson, J. S., and Othmer, H. G. (1997) *Proc. Natl. Acad. Sci. U.S.A.* **94**, 7263–7268
63. Volz, K. (1993) *Biochemistry* **32**, 11741–11753
64. Bellsolell, L., Cronet, P., Majolero, M., Serrano, L., and Coll, M. (1996) *J. Mol. Biol.* **257**, 116–128
65. Jiang, M., Bourret, R. B., Simon, M. I., and Volz, K. (1997) *J. Biol. Chem.* **272**, 11850–11855
65a. Halkides, C. J., McEvoy, M. M., Casper, E., Matsumura, P., Volz, K., and Dahlquist, F. W. (2000) *Biochemistry* **39**, 5280–5286
65b. Lee, S.-Y., Cho, H. S., Pelton, J. G., Yan, D., Berry, E. A., and Wemmer, D. E. (2001) *J. Biol. Chem.* **276**, 16425–16431
65c. Cho, H. S., Lee, S.-Y., Yan, D., Pan, X., Parkinson, J. S., Kustu, S., Wemmer, D. E., and Pelton, J. G. (2000) *J. Mol. Biol.* **297**, 543–551
65d. Kim, C., Jackson, M., Lux, R., and Khan, S. (2001) *J. Mol. Biol.* **307**, 119–135
66. Bren, A., and Eisenbach, M. (1998) *J. Mol. Biol.* **278**, 507–514
66a. Bren, A., and Eisenbach, M. (2001) *J. Mol. Biol.* **312**, 699–709
67. Sanna, M. G., and Simon, M. I. (1996) *J. Biol. Chem.* **271**, 7357–7361
68. Blat, Y., and Eisenbach, M. (1996) *J. Biol. Chem.* **271**, 1226–1231
69. Silversmith, R. E., Appleby, J. L., and Bourret, R. B. (1997) *Biochemistry* **36**, 14965–14974
70. Weis, R. M., Chasalow, S., and Koshland, D. E., Jr. (1990) *J. Biol. Chem.* **265**, 6817–6826
71. Wu, J., Li, J., Li, G., Long, D. G., and Weis, R. M. (1996) *Biochemistry* **35**, 4984–4993
72. West, A. H., Martinez-Hackert, E., and Stock, A. M. (1995) *J. Mol. Biol.* **250**, 276–290
73. Alon, U., Surette, M. G., Barkai, N., and Leibler, S. (1999) *Nature (London)* **397**, 168–171
73a. Bray, D. (2002) *Proc. Natl. Acad. Sci. U.S.A.* **99**, 7–9
74. Prasad, K., Caplan, S. R., and Eisenbach, M. (1998) *J. Mol. Biol.* **280**, 821–828
75. Nickitenko, A. V., Trakhanov, S., and Quiocho, F. A. (1995) *Biochemistry* **34**, 16585–16595
76. Lux, R., Jahreis, K., Bettenbrock, K., Parkinson, J. S., and Lengeler, J. W. (1995) *Proc. Natl. Acad. Sci. U.S.A.* **92**, 11583–11587
77. Gebert, J. F., Overhoff, B., Manson, M. D., and Bods, W. (1988) *J. Biol. Chem.* **263**, 16652–16660
77a. Nishiyama, S.-i, Maruyama, I. N., Homma, M., and Kawagishi, I. (1999) *J. Mol. Biol.* **286**, 1275–1284
77b. Kim, S.-H., Wang, W., and Kim, K. K. (2002) *Proc. Natl. Acad. Sci. U.S.A.* **99**, 11611–11615
78. Unsworth, B. R., Witzmann, F. A., and Fitts, R. H. (1982) *J. Biol. Chem.* **257**, 15129–15136
79. Kelly, A. M., and Rubinstein, N. A. (1980) *Nature (London)* **288**, 267–269
80. Bullard, B. (1983) *Trends Biochem. Sci.* **8**, 68–70
81. Cohen, C. (1982) *Proc. Natl. Acad. Sci. U.S.A.* **79**, 3176–3178
82. Huxley, H., and Hanson, J. (1954) *Nature (London)* **173**, 973–976
83. Huxley, H. E. (1958) *Sci. Am.* **199**, 67–82
84. Huxley, H. E. (1990) *J. Biol. Chem.* **265**, 8347–8350
85. Glickson, J. D., Phillips, W. D., and Rupley, J. A. (1971) *J. Am. Chem. Soc.* **93**, 4031–4038
86. Yates, L. D., and Greaser, M. L. (1983) *J. Mol. Biol.* **168**, 123–141
87. Yao, X., Grade, S., Wriggers, W., and Rubenstein, P. A. (1999) *J. Biol. Chem.* **274**, 37443–37449
88. Rubenstein, P. A. (1990) *BioEssays* **12**, 309–315
89. Allen, P. G., Shuster, C. B., Käs, J., Chaponnier, C., Janmey, P. A., and Herman, I. M. (1996) *Biochemistry* **35**, 14062–14069
90. Kabsch, W., and Holmes, K. C. (1995) *FASEB J.* **9**, 167–174
91. Kabsch, W., Mannherz, H. G., Suck, D., Pai, E. F., and Holmes, K. C. (1990) *Nature (London)* **347**, 37–44
92. Chik, J. K., Lindberg, U., and Schutt, C. E. (1996) *J. Mol. Biol.* **263**, 607–623
92a. Otterbein, L. R., Graceffa, P., and Dominguez, R. (2001) *Science* **293**, 708–711
93. Pollard, T. D., and Craig, S. W. (1982) *Trends Biochem. Sci.* **7**, 55–58
94. Teubner, A., and Wegner, A. (1998) *Biochemistry* **37**, 7532–7538
94a. Orlova, A., Galkin, V. E., VanLoock, M. S., Kim, E., Shvetsov, A., Reisler, E., and Egelman, E. H. (2001) *J. Mol. Biol.* **312**, 95–106
95. Carlier, M.-F., and Pantaloni, D. (1988) *J. Biol. Chem.* **263**, 817–825
95a. Sablin, E. P., Dawson, J. F., VanLoock, M. S., Spudich, J. A., Egelman, E. H., and Fletterick, R. J. (2002) *Proc. Natl. Acad. Sci. U.S.A.* **99**, 10945–10947

References

96. Maruta, H., Knoerzer, W., Hinssen, H., and Isenberg, G. (1984) *Nature (London)* **312**, 424–427
97. Pollard, T. D., and Cooper, J. A. (1986) *Ann. Rev. Biochem.* **55**, 987–1035
98. Maruyama, K. (1997) *FASEB J.* **11**, 341–345
98a. Maruyama, K. (2002) *Trends Biochem. Sci.* **27**, 264–266
99. Trinick, J. (1994) *Trends Biochem. Sci.* **19**, 405–409
100. Labeit, S., and Kolmerer, B. (1995) *Science* **270**, 293–296
100a. Amodeo, P., Fraternali, F., Lesk, A. M., and Pastore, A. (2001) *J. Mol. Biol.* **311**, 283–296
101. Kolmerer, B., Olivieri, N., Witt, C. C., Herrmann, B. G., and Labeit, S. (1996) *J. Mol. Biol.* **256**, 556–563
102. Tskhovrebova, L., and Trinick, J. (2001) *J. Mol. Biol.* **310**, 755–771
102a. Ma, K., Kan, L.-s, and Wang, K. (2001) *Biochemistry* **40**, 3427–3438
103. Wang, K., Knipfer, M., Huang, Q.-Q., van Heerden, A., Hsu, L. C.-L., Gutierrez, G., Quian, X.-L., and Stedman, H. (1996) *J. Biol. Chem.* **271**, 4304–4314
104. Kalverda, A. P., Wymenga, S. S., Lomman, A., van de Ven, F. J. M., Hilbers, C. W., and Canters, G. W. (1994) *J. Mol. Biol.* **240**, 358–371
105. Millevoi, S., Trombitas, K., Kolmerer, B., Kostin, S., Schaper, J., Pelin, K., Granzier, H., and Labeit, S. (1998) *J. Mol. Biol.* **282**, 111–123
106. Politou, A. S., Millevoi, S., Gautel, M., Kolmerer, B., and Pastore, A. (1998) *J. Mol. Biol.* **276**, 189–202
106a. McElhinny, A. S., Kolmerer, B., Fowler, V. M., Labeit, S., and Gregorio, C. C. (2001) *J. Biol. Chem.* **276**, 583–592
107. Obermann, W. M. J., Gautel, M., Weber, K., and Fürst, D. O. (1997) *EMBO J.* **16**, 211–220
108. Young, P., Ferguson, C., Banuelos, S., and Gautel, M. (1998) *EMBO J.* **17**, 1614–1624
109. Sorimachi, H., Freiburg, A., Kolmerer, B., Ishiura, S., Stier, G., Gregorio, C. C., Labeit, D., Linke, W. A., Suzuki, K., and Labeit, S. (1997) *J. Mol. Biol.* **270**, 688–695
109a. Tang, J., Taylor, D. W., and Taylor, K. A. (2001) *J. Mol. Biol.* **310**, 845–858
109b. Joseph, C., Stier, G., O'Brien, R., Politou, A. S., Atkinson, R. A., Bianco, A., Ladbury, J. E., Martin, S. R., and Pastore, A. (2001) *Biochemistry* **40**, 4957–4965
110. Isobe, Y., Warner, F. D., and Lemanski, L. F. (1988) *Proc. Natl. Acad. Sci. U.S.A.* **85**, 6758–6762
111. Deatherage, J. F., Cheng, N., and Bullard, B. (1989) *J. Cell Biol.* **108**, 1775–1782
112. van Straaten, M., Goulding, D., Kolmerer, B., Labeit, S., Clayton, J., Leonard, K., and Bullard, B. (1999) *J. Mol. Biol.* **285**, 1549–1562
112a. Kolmerer, B., Clayton, J., Benes, V., Allen, T., Ferguson, C., Leonard, K., Weber, U., Knekt, M., Ansorge, W., Labeit, S., and Bullard, B. (2000) *J. Mol. Biol.* **296**, 435–448
112b. Fukuzawa, A., Shimamura, J., Takemori, S., Kanzawa, N., Yamaguchi, M., Sun, P., Maruyama, K., and Kimura, S. (2001) *EMBO J.* **20**, 4826–4835
113. Yamaguchi, M., Izumimoto, M., Robson, R. M., and Stromer, M. H. (1985) *J. Mol. Biol.* **184**, 621–644
114. Baron, M. D., Davison, M. D., Jones, P., Patel, B., and Critchley, D. R. (1987) *J. Biol. Chem.* **262**, 2558–2561
115. Pan, K.-M., Roelke, D. L., and Greaser, M. L. (1986) *J. Biol. Chem.* **261**, 9922–9928
116. McLachlan, A. D. (1984) *Ann. Rev. Biophys. Bioeng.* **13**, 167–189
116a. Bellin, R. M., Huiatt, T. W., Critchley, D. R., and Robson, R. M. (2001) *J. Biol. Chem.* **276**, 32330–32337
117. Carragher, B. O., Cheng, N., Wang, Z.-Y., Korn, E. D., Reilein, A., Belnap, D. M., Hammer, J. A., III, and Steven, A. C. (1998) *Proc. Natl. Acad. Sci. U.S.A.* **95**, 15206–15211
118. Heintzelman, M. B., and Schwartzman, J. D. (1997) *J. Mol. Biol.* **271**, 139–146
119. Baker, J. P., and Titus, M. A. (1997) *J. Mol. Biol.* **272**, 523–535
120. Hasson, T., and Mooseker, M. S. (1996) *J. Biol. Chem.* **271**, 16431–16434
121. Jontes, J. D., and Milligan, R. A. (1997) *J. Mol. Biol.* **266**, 331–342
121a. Ajtai, K., Garamszegi, S. P., Park, S., Dones, A. L. V., and Burghardt, T. P. (2001) *Biochemistry* **40**, 12078–12093
121b. Wendt, T., Taylor, D., Trybus, K. M., and Taylor, K. (2001) *Proc. Natl. Acad. Sci. U.S.A.* **98**, 4361–4366
121c. Andersen, J. L., Schjerling, P., and Saltin, B. (2000) *Sci. Am.* **283**(Sep), 48–55
122. Weiss, A., Schiaffino, S., and Leinwand, L. A. (1999) *J. Mol. Biol.* **290**, 61–75
123. Robillard, G., and Schulman, R. G. (1972) *J. Mol. Biol.* **71**, 507–511
123a. Swank, D. M., Bartoo, M. L., Knowles, A. F., Iliffe, C., Bernstein, S. I., Molloy, J. E., and Sparrow, J. C. (2001) *J. Biol. Chem.* **276**, 15117–15124
124. Ferrin, T. E., Huang, C. C., Jarvis, L. E., and Langridge, R. (1988) *J. Mol. Graphics* **6**, 13–27
125. Lorenz, M., Poole, K. J. V., Popp, D., Rosenbaum, G., and Holmes, K. C. (1995) *J. Mol. Biol.* **246**, 108–119
126. Knight, P. J. (1996) *J. Mol. Biol.* **255**, 269–274
127. Katoh, T., Konishi, K., and Yazawa, M. (1998) *J. Biol. Chem.* **273**, 11436–11439
128. Offer, G., and Knight, P. (1996) *J. Mol. Biol.* **256**, 407–416
129. Davis, J. S. (1988) *Ann. Rev. Biophys. Biophys. Chem.* **17**, 217–239
130. Squire, J. M. (1973) *J. Mol. Biol.* **77**, 291–323
131. Padrón, R., Alamo, L., Murgich, J., and Craig, R. (1998) *J. Mol. Biol.* **275**, 35–41
132. Sohn, R. L., Vikstrom, K. L., Strauss, M., Cohen, C., Szent-Gyorgyi, A. G., and Leinwand, L. A. (1997) *J. Mol. Biol.* **266**, 317–330
133. Bennett, P. M., and Gautel, M. (1996) *J. Mol. Biol.* **259**, 896–903
134. Winkler, H., Reedy, M. C., Reedy, M. K., Tregear, R., and Taylor, K. A. (1996) *J. Mol. Biol.* **264**, 302–322
135. Crowther, R. A., Padron, R., and Craig, R. (1985) *J. Mol. Biol.* **184**, 429–439
135a. Offer, G., Knight, P. J., Burgess, S. A., Alamo, L., and Padrón, R. (2000) *J. Mol. Biol.* **298**, 239–260
135b. Flavigny, J., Souchet, M., Sébillon, P., Berrebi-Bertrand, I., Hainque, B., Mallet, A., Bril, A., Schwartz, K., and Carrier, L. (1999) *J. Mol. Biol.* **294**, 443–456
135c. Witt, C. C., Gerull, B., Davies, M. J., Centner, T., Linke, W. A., and Thierfelder, L. (2001) *J. Biol. Chem.* **276**, 5353–5359
136. Standiford, D. M., Davis, M. B., Miedema, K., Franzini-Armstrong, C., and Emerson, C. P. J. (1997) *J. Mol. Biol.* **265**, 40–55
137. Vibert, P. and Craig, R. (1983) *J. Mol. Biol.* **165**, 303–320
138. Engelhardt, W. A., and Ljubimowa, M. N. (1939) *Nature (London)* **144**, 668–669
139. Szent-Gyorgyi, A. (1941) *Studies Inst. Med. Chem., Univ. Szeged* **1**, 17–26 (reprinted in H. M. Kalckar ed., *Biological Phosphorylations*, pp. 465–472, Prentice-Hall, Englewood Cliffs, New Jersey 1969)
140. Szent-Gyorgyi, A. (1947) *Chemistry of Muscular Contraction*, Academic Press, New York
141. Straub, F. B. (1969) in *Biological Phosphorylation* (Kalckar, H. M., ed), pp. 474–483, Prentice-Hall, Englewood Cliffs, New Jersey (reprinted from Studies Inst. Med. Chem. Univ. Szeged, 2, 3–15)
142. Milligan, R. A. (1996) *Proc. Natl. Acad. Sci. U.S.A.* **93**, 21–26
143. Schmitz, H., Reedy, M. C., Reedy, M. K., Tregear, R. T., Winkler, H., and Taylor, K. A. (1996) *J. Mol. Biol.* **264**, 279–301
144. Katayama, E. (1998) *J. Mol. Biol.* **278**, 349–367
145. Huxley, H. (1969) *Science* **114**, 1356–1366
146. Huxley, H. E. (1998) *Trends Biochem. Sci.* **23**, 84–87
147. Huxley, A. F., and Simmons, R. M. (1969) *Nature (London)* **233**, 533–538
148. Tsukita, S., and Yano, M. (1985) *Nature (London)* **317**, 182–184
149. Sheetz, M. P., and Spudich, J. A. (1983) *Nature (London)* **303**, 31–35
150. Adams, R. J., and Pollard, T. D. (1986) *Nature (London)* **322**, 754–756
151. Harada, Y., Noguchi, A., Kishino, A., and Yanagida, T. (1987) *Nature (London)* **326**, 805–808
152. Rayment, I., Rypniewski, W. R., Schmidt-Bäse, K., Smith, R., Tomchick, D. R., Benning, M. M., Winkelmann, D. A., Wesenberg, G., and Holden, H. M. (1993) *Science* **261**, 50–58
153. Fisher, A. J., Smith, C. A., Thoden, J. B., Smith, R., Sutoh, K., Holden, H. M., and Rayment, I. (1995) *Biochemistry* **34**, 8960–8972
154. Smith, C. A., and Rayment, I. (1995) *Biochemistry* **34**, 8973–8981
155. Fisher, A. J., Smith, C. A., Thoden, J., Smith, R., Sutoh, K., Holden, H. M., and Rayment, I. (1995) *Biophys. J.* **68**, 19s–29s
156. Rayment, I., Holden, H. M., Whittaker, M., Yohn, C. B., Lorenz, M., Holmes, K. C., and Milligan, R. A. (1993) *Science* **261**, 58–65
157. Rayment, I. (1996) *J. Biol. Chem.* **271**, 15850–15853
158. Smith, C. A., and Rayment, I. (1996) *Biochemistry* **35**, 5404–5417
159. Gulick, A. M., Bauer, C. B., Thoden, J. B., and Rayment, I. (1997) *Biochemistry* **36**, 11619–11628
160. Murphy, R. A. (1994) *FASEB J.* **8**, 311–318
161. Xie, X., Harrison, D. H., Schlichting, I., Sweet, R. M., Kalabokis, V. N., Szent-Györgyi, A. G., and Cohen, C. (1994) *Nature (London)* **368**, 306–312
162. Furch, M., Geeves, M. A., and Manstein, D. J. (1998) *Biochemistry* **37**, 6317–6326
163. Van Dijk, J., Fernandez, C., and Chaussepied, P. (1998) *Biochemistry* **37**, 8385–8394
164. Yengo, C. M., Chrin, L., Rovner, A. S., and Berger, C. L. (1999) *Biochemistry* **38**, 14515–14523
165. Grammer, J. C., Kuwayama, H., and Yount, R. G. (1993) *Biochemistry* **32**, 5725–5732
166. Miller, C. J., Wong, W. W., Bobkova, E., Rubenstein, P. A., and Reisler, E. (1996) *Biochemistry* **35**, 16557–16565
166a. Hansen, J. E., Marner, J., Pavlov, D., Rubenstein, P. A., and Reisler, E. (2000) *Biochemistry* **39**, 1792–1799
166b. Bertrand, R., Derancourt, J., and Kassab, R. (2000) *Biochemistry* **39**, 14626–14637
166c. Sasaki, N., Ohkura, R., and Sutoh, K. (2000) *J. Biol. Chem.* **275**, 38705–38709
167. Orlova, A., Chen, X., Rubenstein, P. A., and Egelman, E. H. (1997) *J. Mol. Biol.* **271**, 235–243
167a. Prochniewicz, E., and Thomas, D. D. (2001) *Biochemistry* **40**, 13933–13940

References

168. Belmont, L. D., Orlova, A., Drubin, D. G., and Egelman, E. H. (1999) *Proc. Natl. Acad. Sci. U.S.A.* **96**, 29–34

168a. Higgins, S. J., and Banting, G., eds. (2000) *Molecular Motors*, Vol. Essays in Biochemistry No. 35, Portland Press, London

168b. Vale, R. D., and Milligan, R. A. (2000) *Science* **288**, 88–95

168c. Verhey, K. J., and Rapoport, T. A. (2001) *Trends Biochem. Sci.* **26**, 545–550

169. Sack, S., Müller, J., Marx, A., Thormählen, M., Mandelkow, E.-M., Brady, S. T., and Mandelkow, E. (1997) *Biochemistry* **36**, 16155–16165

169a. Song, Y.-H., Marx, A., Müller, J., Woehlke, G., Schliwa, M., Krebs, A., Hoenger, A., and Mandelkow, E. (2001) *EMBO J.* **20**, 6213–6225

169b. Gilbert, S. P. (2001) *Nature (London)* **414**, 597–598

170. Stock, M. F., Guerrero, J., Cobb, B., Eggers, C. T., Huang, T.-G., Li, X., and Hackney, D. D. (1999) *J. Biol. Chem.* **274**, 14617–14623

171. Vale, R. D., Funatsu, T., Pierce, D. W., Romberg, L., Harada, Y., and Yanagida, T. (1996) *Nature (London)* **380**, 451–453

172. Kull, F. J., Sablin, E. P., Lau, R., Fletterick, R. J., and Vale, R. D. (1996) *Nature (London)* **380**, 550–555

173. Mandelkow, E., and Johnson, K. A. (1998) *Trends Biochem. Sci.* **23**, 429–433

174. Gulick, A. M., Song, H., Endow, S. A., and Rayment, I. (1998) *Biochemistry* **37**, 1769–1776

174a. Miki, H., Setou, M., Kaneshiro, K., and Hirokawa, N. (2001) *Proc. Natl. Acad. Sci. U.S.A.* **98**, 7004–7011

174b. Kikkawa, M., Sablin, E. P., Okada, Y., Yajima, H., Fletterick, R. J., and Hirokawa, N. (2001) *Nature (London)* **411**, 439–445

174c. Schliwa, M., and Woehlke, G. (2001) *Nature (London)* **411**, 424–425

174d. Yun, M., Zhang, X., Park, C.-G., Park, H.-W., and Endow, S. A. (2001) *EMBO J.* **20**, 2611–2618

175. Sablin, E. P., Kull, F. J., Cooke, R., Vale, R. D., and Fletterick, R. J. (1996) *Nature (London)* **380**, 555–559

176. Thormählen, M., Marx, A., Müller, S. A., Song, Y.-H., Mandelkow, E.-M., Aebi, U., and Mandelkow, E. (1998) *J. Mol. Biol.* **275**, 795–809

177. Block, S. M. (1998) *Cell* **93**, 5–8

178. Wells, A. L., Lin, A. W., Chen, L.-Q., Safer, D., Cain, S. M., Hasson, T., Carragher, B. O., Milligan, R. A., and Sweeney, H. L. (1999) *Nature (London)* **401**, 505–508

179. Schliwa, M. (1999) *Nature (London)* **401**, 431–432

179a. Homma, K., Yoshimura, M., Saito, J., Ikebe, R., and Ikebe, M. (2001) *Nature (London)* **412**, 831–834

180. Ma, Y.-Z., and Taylor, E. W. (1995) *Biochemistry* **34**, 13242–13251

181. Ma, Y.-Z., and Taylor, E. W. (1997) *J. Biol. Chem.* **272**, 717–723

182. Visscher, K., Schnitzer, M. J., and Block, S. M. (1999) *Nature (London)* **400**, 184–189

182a. Tomishige, M., Klopfenstein, D. R., and Vale, R. D. (2002) *Science* **297**, 2263–2267

183. Mehta, A. D., Rock, R. S., Rief, M., Spudich, J. A., Mooseker, M. S., and Cheney, R. E. (1999) *Nature (London)* **400**, 590–593

183a. Rief, M., Rock, R. S., Mehta, A. D., Mooseker, M. S., Cheney, R. E., and Spudich, J. A. (2000) *Proc. Natl. Acad. Sci. U.S.A.* **97**, 9482–9486

183b. Trybus, K. M., Krementsova, E., and Freyzon, Y. (1999) *J. Biol. Chem.* **274**, 27448–27456

183c. De La Cruz, E. M., Wells, A. L., Sweeney, H. L., and Ostap, E. M. (2000) *Biochemistry* **39**, 14196–14202

183d. Rock, R. S., Rice, S. E., Wells, A. L., Purcell, T. J., Spudich, J. A., and Sweeney, H. L. (2001) *Proc. Natl. Acad. Sci. U.S.A.* **98**, 13655–13659

183e. Karcher, R. L., Roland, J. T., Zappacosta, F., Huddleston, M. J., Annan, R. S., Carr, S. A., and Gelfand, V. I. (2001) *Science* **293**, 1317–1320

184. Goldman, Y. E. (1998) *Cell* **93**, 1–4

185. Taylor, E. W. (1991) *J. Biol. Chem.* **266**, 294–302

186. Houadjeto, M., Travers, F., and Barman, T. (1992) *Biochemistry* **31**, 1564–1569

187. White, H. D., Belknap, B., and Webb, M. R. (1997) *Biochemistry* **36**, 11828–11836

188. Spudich, J. A. (1994) *Nature (London)* **372**, 515–518

189. Cooke, R. (1993) *FASEB J.* **9**, 636–642

190. Ostap, E. M., Barnett, V. A., and Thomas, D. D. (1995) *Biophys. J.* **69**, 177–188

191. Kambara, T., Rhodes, T. E., Ikebe, R., Yamada, M., White, H. D., and Ikebe, M. (1999) *J. Biol. Chem.* **274**, 16400–16406

191a. Lionne, C., Stehle, R., Travers, F., and Barman, T. (1999) *Biochemistry* **38**, 8512–8520

192. Harrington, W. F., and Rodgers, M. E. (1984) *Ann. Rev. Biochem.* **53**, 35–73

192a. Uyeda, T. Q. P., Tokuraku, K., Kaseda, K., Webb, M. R., and Patterson, B. (2002) *Biochemistry* **41**, 9525–9534

192b. Himmel, D. M., Gourinath, S., Reshetnikova, L., Shen, Y., Szent-Györgyi, A. G., and Cohen, C. (2002) *Proc. Natl. Acad. Sci. U.S.A.* **99**, 12645–12650

193. Yount, R. G., Lawson, D., and Rayment, I. (1995) *Biophys. J.* **68**, 445–495

193a. Burghardt, T. P., Park, S., and Ajtai, K. (2001) *Biochemistry* **40**, 4834–4843

193b. Borejdo, J., Ushakov, D. S., Moreland, R., Akopova, I., Reshetnyak, Y., Saraswat, L. D., Kamm, K., and Lowey, S. (2001) *Biochemistry* **40**, 3796–3803

194. Dominguez, R., Freyzon, Y., Trybus, K. M., and Cohen, C. (1998) *Cell* **94**, 559–571

195. Highsmith, S. (1999) *Biochemistry* **38**, 9791–9797

196. Gulick, A. M., Bauer, C. B., Thoden, J. B., Pate, E., Yount, R. G., and Rayment, I. (2000) *J. Biol. Chem.* **275**, 398–408

197. Geeves, M. A., and Holmes, K. C. (1999) *Ann. Rev. Biochem.* **68**, 687–728

198. Block, S. M. (1992) *Nature (London)* **360**, 493–495

199. Mehta, A. D., Rief, M., Spudich, J. A., Smith, D. A., and Simmons, R. M. (1999) *Science* **283**, 1689–1695

200. Finer, J. T., Simmons, R. M., and Spudich, J. A. (1994) *Nature (London)* **368**, 113–119

201. Molloy, J. E., Burns, J. E., Kendrick-Jones, J., Tregear, R. T., and White, D. C. S. (1995) *Nature (London)* **378**, 209–212

202. Mehta, A. D., Rief, M., and Spudich, J. A. (1999) *J. Biol. Chem.* **274**, 14517–14520

203. Howard, J. (1997) *Nature (London)* **389**, 561–567

204. Guilford, W. H., Dupuis, D. E., Kennedy, G., Wu, J., Patlak, J. B., and Warshaw, D. M. (1997) *Biophys. J.* **72**, 1006–1021

204a. Rosenfeld, S. S., Xing, J., Whitaker, M., Cheung, H. C., Brown, F., Wells, A., Milligan, R. A., and Sweeney, H. L. (2000) *J. Biol. Chem.* **275**, 25418–25426

205. Sase, I., Miyata, H., Ishiwata, S., and Kinosita, K., Jr. (1997) *Proc. Natl. Acad. Sci. U.S.A.* **94**, 5646–5650

206. Kitamura, K., Tokunaga, M., Iwane, A. H., and Yanagida, T. (1999) *Nature (London)* **397**, 129–134

207. Smith, D. A. (1998) *Biophys. J.* **75**, 2996–3007

207a. Yanagida, T., and Iwane, A. H. (2000) *Proc. Natl. Acad. Sci. U.S.A.* **97**, 9357–9359

208. Veigel, C., Coluccio, L. M., Jontes, J. D., Sparrow, J. C., Milligan, R. A., and Molloy, J. E. (1999) *Nature (London)* **398**, 530–533

208a. Piazzesi, G., Reconditi, M., Linari, M., Lucii, L., Sun, Y.-B., Narayanan, T., Boesecke, P., Lombardi, V., and Irving, M. (2002) *Nature (London)* **415**, 659–662

208b. Tanaka, H., Homma, K., Iwane, A. H., Katayama, E., Ikebe, R., Saito, J., Yanagida, T., and Ikebe, M. (2002) *Nature (London)* **415**, 192–195

209. Berliner, E., Mahtani, H. K., Karki, S., Chu, L. F., Cronan, J. E., Jr., and Gelles, J. (1994) *J. Biol. Chem.* **269**, 8610–8615

210. Schnitzer, M. J., and Block, S. M. (1997) *Nature (London)* **388**, 386–390

211. Coppin, C. M., Finer, J. T., Spudich, J. A., and Vale, R. D. (1996) *Proc. Natl. Acad. Sci. U.S.A.* **93**, 1913–1917

212. Irving, M., and Goldman, Y. E. (1999) *Nature (London)* **398**, 463–465

212a. Shimizu, T., Thorn, K. S., Ruby, A., and Vale, R. D. (2000) *Biochemistry* **39**, 5265–5273

212b. Ryu, W. S., Berry, R. M., and Berg, H. C. (2000) *Nature (London)* **403**, 444–447

213. Young, E. C., Mahtani, H. K., and Gelles, J. (1998) *Biochemistry* **37**, 3467–3479

214. Foster, K. A., Correia, J. J., and Gilbert, S. P. (1998) *J. Biol. Chem.* **273**, 35307–35318

215. deCastro, M. J., Ho, C.-H., and Stewart, R. J. (1999) *Biochemistry* **38**, 5076–5081

216. Platts, J. A., Howard, S. T., and Bracke, B. R. F. (1996) *J. Am. Chem. Soc.* **118**, 2726–2733

217. Sosa, H., and Milligan, R. A. (1996) *J. Mol. Biol.* **260**, 743–755

218. Hirose, K., Cross, R. A., and Amos, L. A. (1998) *J. Mol. Biol.* **278**, 389–400

219. Sablin, E. P., Case, R. B., Dai, S. C., Hart, C. L., Ruby, A., Vale, R. D., and Fletterick, R. J. (1998) *Nature (London)* **395**, 813–816

220. Hirose, K., Lockhart, A., Cross, R. A., and Amos, L. A. (1996) *Proc. Natl. Acad. Sci. U.S.A.* **93**, 9539–9544

220a. Hua, W., Chung, J., and Gelles, J. (2002) *Science* **295**, 844–848

221. Rice, S., Lin, A. W., Safer, D., Hart, C. L., Naber, N., Carragher, B. O., Cain, S. M., Pechatnikova, E., Wilson-Kubalek, E. M., Whittaker, M., Pate, E., Cooke, R., Taylor, E. W., Milligan, R. A., and Vale, R. D. (1999) *Nature (London)* **402**, 778–784

222. Hancock, W. O., and Howard, J. (1999) *Proc. Natl. Acad. Sci. U.S.A.* **96**, 13147–13152

223. Crevel, I., Carter, N., Schliwa, M., and Cross, R. (1999) *EMBO J.* **18**, 5863–5872

224. Brendza, K. M., Rose, D. J., Gilbert, S. P., and Saxton, W. M. (1999) *J. Biol. Chem.* **274**, 31506–31514

225. Pate, E., Naber, N., Matuska, M., Franks-Skiba, K., and Cooke, R. (1997) *Biochemistry* **36**, 12155–12166

226. Moschcovich, L., Peyser, Y. M., Salomon, C., Burghardt, T. P., and Muhlrad, A. (1998) *Biochemistry* **37**, 15137–15143

227. Gopal, D., Pavlov, D. I., Levitsky, D. I., Ikebe, M., and Burke, M. (1996) *Biochemistry* **35**, 10149–10157

227a. Kliche, W., Pfannstiel, J., Tiepold, M., Stoeva, S., and Faulstich, H. (1999) *Biochemistry* **38**, 10307–10317

227b. Furch, M., Fujita-Becker, S., Geeves, M. A., Holmes, K. C., and Manstein, D. J. (1999) *J. Mol. Biol.* **290**, 797–809

227c. Allin, C., and Gerwert, K. (2001) *Biochemistry* **40**, 3037–3046

227d. Nyitrai, M., Hild, G., Lukács, A., Bódis, E., and Somogyi, B. (2000) *J. Biol. Chem.* **275**, 2404–2409

References

228. Xiao, M., Li, H., Snyder, G. E., Cooke, R., Yount, R. G., and Selvin, P. R. (1998) *Proc. Natl. Acad. Sci. U.S.A.* **95**, 15309–15314
229. Palm, T., Sale, K., Brown, L., Li, H., Hambly, B., and Fajer, P. G. (1999) *Biochemistry* **38**, 13026–13034
230. Corrie, J. E. T., Brandmeier, B. D., Ferguson, R. E., Trentham, D. R., Kendrick-Jones, J., Hopkins, S. C., van der Heide, U. A., Goldman, Y. E., Sabido-David, C., Dale, R. E., Criddle, S., and Irving, M. (1999) *Nature (London)* **400**, 425–430
231. Suzuki, Y., Yasunaga, T., Ohkura, R., Wakabayashi, T., and Sutoh, K. (1998) *Nature (London)* **396**, 380–383
232. Ashley, R. (1972) *J. Theor. Biol.* **36**, 339–354
233. McClare, C. W. F. (1972) *J. Theor. Biol.* **35**, 569–595
234. Davydov, A. S. (1973) *J. Theor. Biol.* **38**, 559–569
235. Porter, K. R., and Franzini-Armstrong, C. (1965) *Sci. Am.* **212**(Mar), 73–80
236. Hoyle, G. (1970) *Sci. Am.* **222**(Apr), 85–93
237. Alberts, B., Bray, D., Lewis, J., Raff, M., Roberts, K., and Watson, J. D. (1994) *Molecular Biology of the Cell*, 3rd ed., Garland, New York
238. Macrez-Leprêtre, N., Kalkbrenner, F., Schultz, G., and Mironneau, J. (1997) *J. Biol. Chem.* **272**, 5261–5268
238a. Toyoshima, C., and Nomura, H. (2002) *Nature (London)* **418**, 605–611
239. Sharma, M. R., Penczek, P., Grassucci, R., Xin, H.-B., Fleischer, S., and Wagenknecht, T. (1998) *J. Biol. Chem.* **273**, 18429–18434
240. Bers, D. M., and Fill, M. (1998) *Science* **281**, 790–791
241. Sonnleitner, A., Conti, A., Bertocchini, F., Schindler, H., and Sorrentino, V. (1998) *EMBO J.* **17**, 2790–2798
242. Du, G. G., and MacLennan, D. H. (1998) *J. Biol. Chem.* **273**, 31867–31872
243. Bidasee, K. R., and Besch, H. R., Jr. (1998) *J. Biol. Chem.* **273**, 12176–12186
243a. Sun, J., Xin, C., Eu, J. P., Stamler, J. S., and Meissner, G. (2001) *Proc. Natl. Acad. Sci. U.S.A.* **98**, 11158–11162
243b. Feng, W., Liu, G., Allen, P. D., and Pessah, I. N. (2000) *J. Biol. Chem.* **275**, 35902–35907
243c. Patel, S., Churchill, G. C., and Galione, A. (2001) *Trends Biochem. Sci.* **26**, 482–489
243d. Baker, M. L., Serysheva, I. I., Sencer, S., Wu, Y., Ludtke, S. J., Jiang, W., Hamilton, S. L., and Chiu, W. (2002) *Proc. Natl. Acad. Sci. U.S.A.* **99**, 12155–12160
244. Takeshima, H., Komazaki, S., Hirose, K., Nishi, M., Noda, T., and Lino, M. (1998) *EMBO J.* **17**, 3309–3316
245. Zhou, J., Cribbs, L., Yi, J., Shirokov, R., Perez-Reyes, E., and Ríos, E. (1998) *J. Biol. Chem.* **273**, 25503–25509
245a. Serysheva, I. I., Ludtke, S. J., Baker, M. R., Chiu, W., and Hamilton, S. L. (2002) *Proc. Natl. Acad. Sci. U.S.A.* **99**, 10370 –10375
246. Knudson, C. M., Chaudhari, N., Sharp, A. H., Powell, J. A., Beam, K. G., and Campbell, K. P. (1989) *J. Biol. Chem.* **264**, 1345–1348
247. Smith, G. D., Keizer, J. E., Stern, M. D., Lederer, W. J., and Cheng, H. (1998) *Biophys. J.* **75**, 15–32
248. Fay, F. S. (1995) *Science* **270**, 588–589
248a. Kamm, K. E., and Stull, J. T. (2001) *J. Biol. Chem.* **276**, 4527–4530
249. Xie, X., Rao, S., Walian, P., Hatch, V., Phillips, G. N., Jr., and Cohen, C. (1994) *J. Mol. Biol.* **236**, 1212–1226
250. Dufour, C., Weinberger, R. P., Schevzov, G., Jeffrey, P. L., and Gunning, P. (1998) *J. Biol. Chem.* **273**, 18547–18555
251. Kagawa, H., Sugimoto, K., Matsumoto, H., Inoue, T., Imadzu, H., Takuwa, K., and Sakube, Y. (1995) *J. Mol. Biol.* **251**, 603–613
252. Amos, L. A. (1985) *Ann. Rev. Biophys. Biophys. Chem.* **14**, 291–313
253. Zot, A. S., and Potter, J. D. (1987) *Ann. Rev. Biophys. Biophys. Chem.* **16**, 535–559
254. Saeki, K., Sutoh, K., and Wakabayashi, T. (1996) *Biochemistry* **35**, 14465–14472
255. Squire, J. M., and Morris, E. P. (1998) *FASEB J.* **12**, 761–771
256. Tripet, B., Van Eyk, J. E., and Hodges, R. S. (1997) *J. Mol. Biol.* **271**, 728–750
257. Malnic, B., Farah, C. S., and Reinach, F. C. (1998) *J. Biol. Chem.* **273**, 10594–10601
258. Hernández, G., Blumenthal, D. K., Kennedy, M. A., Unkefer, C. J., and Trewhella, J. (1999) *Biochemistry* **38**, 6911–6917
259. Vassylyev, D. G., Takeda, S., Wakatsuki, S., Maeda, K., and Meéda, Y. (1998) *Proc. Natl. Acad. Sci. U.S.A.* **95**, 4847–4852
260. McKay, R. T., Tripet, B. P., Pearlstone, J. R., Smillie, L. B., and Sykes, B. D. (1999) *Biochemistry* **38**, 5478–5489
260a. Mercier, P., Spyracopoulos, L., and Sykes, B. D. (2001) *Biochemistry* **40**, 10063–10077
261. Kobayashi, T., Kobayashi, M., Gryczynski, Z., Lakowicz, J. R., and Collins, J. H. (2000) *Biochemistry* **39**, 86–91
262. Strynadka, N. C. J., Cherney, M., Sielecki, A. R., Li, M. X., Smillie, L. B., and James, M. N. G. (1997) *J. Mol. Biol.* **273**, 238–255
263. da Silva, A. C. R., and Reinach, F. C. (1991) *Trends Biochem. Sci.* **16**, 53–57
264. Gagné, S. M., Li, M. X., and Sykes, B. D. (1997) *Biochemistry* **36**, 4386–4392
265. Li, H.-C., and Fajer, P. G. (1998) *Biochemistry* **37**, 6628–6635
265a. Lehman, W., Rosol, M., Tobacman, L. S., and Craig, R. (2001) *J. Mol. Biol.* **307**, 739–744
266. Reiffert, S. U., Jaquet, K., Heilmeyer, L. M. G., Jr., and Herberg, F. W. (1998) *Biochemistry* **37**, 13516–13525
267. Simmerman, H. K. B., Kobayashi, Y. M., Autry, J. M., and Jones, L. R. (1996) *J. Biol. Chem.* **271**, 5941–5946
268. Reddy, L. G., Jones, L. R., and Thomas, D. D. (1999) *Biochemistry* **38**, 3954–3962
268a. Asahi, M., Green, N. M., Kurzydlowski, K., Tada, M., and MacLennan, D. H. (2001) *Proc. Natl. Acad. Sci. U.S.A.* **98**, 10061–10066
269. Holmes, K. C. (1995) *Biophys. J.* **68**, 2s–7s
270. Vibert, P., Craig, R., and Lehman, W. (1997) *J. Mol. Biol.* **266**, 8–14
270a. Moraczewska, J., and Hitchcock-DeGregori, S. E. (2000) *Biochemistry* **39**, 6891–6897
270b. Narita, A., Yasunaga, T., Ishikawa, T., Mayanagi, K., and Wakabayashi, T. (2001) *J. Mol. Biol.* **308**, 241–261
270c. Craig, R., and Lehman, W. (2001) *J. Mol. Biol.* **311**, 1027–1036
271. Hnath, E. J., Wang, C.-L. A., Huber, P. A. J., Marston, S. B., and Phillips, G. N., Jr. (1996) *Biophys. J.* **71**, 1920–1933
271a. Gerson, J. H., Kim, E., Muhlrad, A., and Reisler, E. (2001) *J. Biol. Chem.* **276**, 18442–18449
272. Allen, B. G., and Walsh, M. P. (1994) *Trends Biochem. Sci.* **19**, 362–368
273. Ikura, M., Clore, G. M., Gronenborn, A. M., Zhu, G., Klee, C. B., and Bax, A. (1992) *Science* **256**, 632–638
274. Wu, X., Clack, B. A., Zhi, G., Stull, J. T., and Cremo, C. R. (1999) *J. Biol. Chem.* **274**, 20328–20335
275. Lee, M. R., Li, L., and Kitazawa, T. (1997) *J. Biol. Chem.* **272**, 5063–5068
276. Ye, L.-H., Kishi, H., Nakamura, A., Okagaki, T., Tanaka, T., Oiwa, K., and Kohama, K. (1999) *Proc. Natl. Acad. Sci. U.S.A.* **96**, 6666–6671
277. Lehman, W., Vibert, P., and Craig, R. (1997) *J. Mol. Biol.* **274**, 310–317
278. Graceffa, P. (1997) *Biochemistry* **36**, 3792–3801
278a. Wang, Z., and Yang, Z.-Q. (2000) *Biochemistry* **39**, 11114–11120
279. Gollub, J., Cremo, C. R., and Cooke, R. (1999) *Biochemistry* **38**, 10107–10118
279a. Stafford, W. F., Jacobsen, M. P., Woodhead, J., Craig, R., O'Neall-Hennessey, E. O., and Szent-Györgyi, A. G. (2001) *J. Mol. Biol.* **307**, 137–147
280. Yamada, A., Yoshio, M., Oiwa, K., and Nyitray, L. (2000) *J. Mol. Biol.* **295**, 169–178
280a. Yamada, A., Yoshio, M., Kojima, H., and Oiwa, K. (2001) *Proc. Natl. Acad. Sci. U.S.A.* **98**, 6635–6640
281. Brzeska, H., and Korn, E. D. (1996) *J. Biol. Chem.* **271**, 16983–16986
282. Benian, G. M., Kiff, J. E., Neckelmann, N., Moerman, D. G., and Waterston, R. H. (1989) *Nature (London)* **342**, 45–50
283. Heierhorst, J., Kobe, B., Feil, S. C., Parker, M. W., Benian, G. M., Weiss, K. R., and Kemp, B. E. (1996) *Nature (London)* **380**, 636–639
284. Lei, J., Tang, X., Chambers, T. C., Pohl, J., and Benian, G. M. (1994) *J. Biol. Chem.* **269**, 21078–21085
285. Johnson, K. A., and Quiocho, F. A. (1996) *Nature (London)* **380**, 585–587
286. Bessman, S. P., and Carpenter, C. L. (1985) *Ann. Rev. Biochem.* **54**, 831–862
287. Balaban, R. S., Kantor, H. L., and Ferretti, J. A. (1983) *J. Biol. Chem.* **258**, 12787–12789
288. Park, J. H., Brown, R. L., Park, C. R., McCully, K., Cohn, M., Haselgrove, J., and Chance, B. (1987) *Proc. Natl. Acad. Sci. U.S.A.* **84**, 8976–8980
289. Savabi, F. (1988) *Proc. Natl. Acad. Sci. U.S.A.* **85**, 7476–7480
289a. Kashiyama, T., Ito, K., and Yamamoto, K. (2001) *J. Mol. Biol.* **311**, 461–466
289b. Pantaloni, D., Le Clainche, C., and Carlier, M.-F. (2001) *Science* **292**, 1502–1506
289c. Pollard, T. D. (2000) *Trends Biochem. Sci.* **25**, 607–611
290. Machesky, L. M., and Way, M. (1998) *Nature (London)* **394**, 125–126
291. Carlier, M.-F., and Pantaloni, D. (1997) *J. Mol. Biol.* **269**, 459–467
291a. Pruyne, D., Evangelista, M., Yang, C., Bi, E., Zigmond, S., Bretscher, A., and Boone, C. (2002) *Science* **297**, 612–615
292. Yin, H. L., and Stull, J. T. (1999) *J. Biol. Chem.* **274**, 32529–32530
292a. Robinson, R. C., Turbedsky, K., Kaiser, D. A., Marchand, J.-B., Higgs, H. N., Choe, S., and Pollard, T. D. (2001) *Science* **294**, 1679–1684
292b. Dayel, M. J., Holleran, E. A., and Mullins, R. D. (2001) *Proc. Natl. Acad. Sci. U.S.A.* **98**, 14871–14876
293. Svitkina, T. M., and Borisy, G. G. (1999) *Trends Biochem. Sci.* **24**, 432–436
293a. Boldogh, I. R., Yang, H.-C., Nowakowski, W. D., Karmon, S. L., Hays, L. G., Yates, J. R., III, and Pon, L. A. (2001) *Proc. Natl. Acad. Sci. U.S.A.* **98**, 3162–3167
293b. Kim, A. S., Kakalis, L. T., Abdul-Manan, N., Liu, G. A., and Rosen, M. K. (2000) *Nature (London)* **404**, 151–158
293c. Higgs, H. (2001) *Trends Biochem. Sci.* **26**, 219
293d. Kaibuchi, K., Kuroda, S., and Amano, M. (1999) *Ann. Rev. Biochem.* **68**, 459–486
293e. Castellano, F., Le Clainche, C., Patin, D., Carlier, M.-F., and Chavrier, P. (2001) *EMBO J.* **20**, 5603–5614

References

293f. Reinhard, M., Jarchau, T., and Walter, U. (2001) *Trends Biochem. Sci.* **26**, 243–249

294. Gutsche-Perelroizen, I., Lepault, J., Ott, A., and Carlier, M.-F. (1999) *J. Biol. Chem.* **274**, 6234–6243

295. Kang, F., Purich, D. L., and Southwick, F. S. (1999) *J. Biol. Chem.* **274**, 36963–36972

296. Chen, H., Bernstein, B. W., and Bamburg, J. R. (2000) *Trends Biochem. Sci.* **25**, 19–23

297. Carlier, M.-F., Ressad, F., and Pantaloni, D. (1999) *J. Biol. Chem.* **274**, 33827–33830

297a. Ono, S., McGough, A., Pope, B. J., Tolbert, V. T., Bui, A., Pohl, J., Benian, G. M., Gernert, K. M., and Weeds, A. G. (2001) *J. Biol. Chem.* **276**, 5952–5958

298. Theriot, J. A., and Mitchison, T. J. (1991) *Nature (London)* **352**, 126–131

299. Stossel, T. P. (1993) *Science* **260**, 1086–1094

300. Tsukita, S., and Yonemura, S. (1999) *J. Biol. Chem.* **274**, 34507–34510

301. Hanakam, F., Gerisch, G., Lotz, S., Alt, T., and Seelig, A. (1996) *Biochemistry* **35**, 11036–11044

301a. Niggli, V. (2001) *Trends Biochem. Sci.* **26**, 604–611

301b. Caroni, P. (2001) *EMBO J.* **20**, 4332–4336

301c. van den Ent, F., Amos, L. A., and Löwe, J. (2001) *Nature (London)* **413**, 39–44

302. Bullock, T. L., Roberts, R. M., and Stewart, M. (1996) *J. Mol. Biol.* **263**, 284–296

302a. Villeneuve, A. M. (2001) *Science* **291**, 2099–2101

303. Spudich, J. D., and Lord, K. (1974) *J. Biol. Chem.* **249**, 6013–6020

304. Allen, R. D. (1987) *Sci. Am.* **256**(Feb), 42–47

305. Gibbons, I. R. (1988) *J. Biol. Chem.* **263**, 15837–15840

305a. Setou, M., Nakagawa, T., Seog, D.-H., and Hirokawa, N. (2000) *Science* **288**, 1796–1802

305b. Goldstein, L. S. B. (2001) *Proc. Natl. Acad. Sci. U.S.A.* **98**, 6999–7003

305c. Fan, J.-S., Zhang, Q., Tochio, H., Li, M., and Zhang, M. (2001) *J. Mol. Biol.* **306**, 97–108

306. Rodionov, V., Nadezhdina, E., and Borisy, G. (1999) *Proc. Natl. Acad. Sci. U.S.A.* **96**, 115–120

307. Burgess, S. A. (1995) *J. Mol. Biol.* **250**, 52–63

308. Caplow, M., and Shanks, J. (1998) *Biochemistry* **37**, 12994–13002

309. Dougherty, C. A., Himes, R. H., Wilson, L., and Farrell, K. W. (1998) *Biochemistry* **37**, 10861–10865

310. Hyams, J. S., and Lloyd, C. W., eds. (1994) *Microtubules*, Wiley-Liss, New York

311. Drewes, G., Ebneth, A., and Mandelkow, E.-M. (1998) *Trends Biochem. Sci.* **23**, 307–311

312. Vinh, J., Langridge, J. I., Bré, M.-H., Levilliers, N., Redeker, V., Loyaux, D., and Rossier, J. (1999) *Biochemistry* **38**, 3133–3139

313. Redeker, V., Levilliers, N., Schmitter, J.-M., Le Caer, J.-P., Rossier, J., Adoutte, A., and Bré, M.-H. (1994) *Science* **266**, 1688–1691

314. Regnard, C., Audebert, S., Desbruyères, E., Denoulet, P., and Eddé, B. (1998) *Biochemistry* **37**, 8395–8404

314a. Sharp, D. J., Rogers, G. C., and Scholey, J. M. (2000) *Nature (London)* **407**, 41–47

315. Hoyt, M. A., Hyman, A. A., and Bähler, M. (1997) *Proc. Natl. Acad. Sci. U.S.A.* **94**, 12747–12748

316. Barton, N. R., and Goldstein, L. S. B. (1996) *Proc. Natl. Acad. Sci. U.S.A.* **93**, 1735–1742

317. Habura, A., Tikhonenko, I., Chisholm, R. L., and Koonce, M. P. (1999) *J. Biol. Chem.* **274**, 15447–15453

318. Samsó, M., Radermacher, M., Frank, J., and Koonce, M. P. (1998) *J. Mol. Biol.* **276**, 927–937

319. Shimizu, T., Toyoshima, Y. Y., Edamatsu, M., and Vale, R. D. (1995) *Biochemistry* **34**, 1575–1582

320. King, S. M., Barbarese, E., Dillman, J. F., III, Patel-King, R. S., Carson, J. H., and Pfister, K. K. (1996) *J. Biol. Chem.* **271**, 19358–19366

321. Koonce, M. P., Köhler, J., Neujahr, R., Schwartz, J.-M., Tikhonenko, I., and Gerisch, G. (1999) *EMBO J.* **18**, 6786–6792

322. Sleigh, M. A., ed. (1974) *Cilia and Flagella*, Academic Press, New York

323. Smith, E. F., and Sale, W. S. (1994) in *Microtubules* (Hyams, J. S., and Lloyd, C. W., eds), pp. 381–392, Wiley-Liss, New York

324. Tanaka-Takiguchi, Y., Itoh, T. J., and Hotani, H. (1998) *J. Mol. Biol.* **280**, 365–373

325. Satir, P. (1999) *FASEB J.* **13**, S235–S237

325a. Fang, Y.-I., Yokota, E., Mabuchi, I., Nakamura, H., and Obizumi, Y. (1997) *Biochemistry* **36**, 15561–15567

325b. Sakakibara, H., Kojima, H., Sakai, Y., Katayama, E., and Oiwa, K. (1999) *Nature (London)* **400**, 586–590

326. Shingyoji, C., Higuchi, H., Yoshimura, M., Katayama, E., and Yanagida, T. (1998) *Nature (London)* **393**, 711–714

327. Hunt, A. J. (1998) *Nature (London)* **393**, 624–625

328. Benashski, S. E., Patel-King, R. S., and King, S. M. (1999) *Biochemistry* **38**, 7253–7264

329. Sleigh, M. A., ed. (1974) *Cilia and Flagella*, p. 14 Academic Press, New York

329a. Smith, E. F., and Sale, W. S. (1994) in *Microtubules* (Hyams, J. S., and Lloyd, C. W., eds), pp. 381–392, Wiley-Liss, New York

330. Klein, P. S., Sun, T. J., Saxe, C. L., III, Kimmel, A. R., Johnson, R. L., and Devreotes, P. N. (1988) *Science* **241**, 1467–1472

331. Caterina, M. J., and Devreotes, P. N. (1991) *FASEB J.* **5**, 3078–3085

332. Snyderman, R., and Goetzl, E. J. (1981) *Science* **213**, 830–837

333. Gerard, N. P., Bao, L., Xiao-Ping, H., Eddy, R. L., Jr., Shows, T. B., and Gerard, C. (1993) *Biochemistry* **32**, 1243–1250

334. Thelen, M., Peveri, P., Kernen, P., von Tscharner, V., Walz, A., and Baggiolini, M. (1988) *FASEB J.* **2**, 2702–2706

335. Schwenk, U., and Schröder, J.-M. (1995) *J. Biol. Chem.* **270**, 15029–15036

336. Hsu, M. H., Chiang, S. C., Ye, R. D., and Prossnitz, E. R. (1997) *J. Biol. Chem.* **272**, 29426–29429

337. Jin, T., Zhang, N., Long, Y., Parent, C. A., and Devreotes, P. N. (2000) *Science* **287**, 1034–1036

338. Chung, C. Y., Funamoto, S., and Firtel, R. A. (2001) *Trends Biochem. Sci.* **26**, 557–566

339. Mahadevan, L., and Matsudaira, P. (2000) *Science* **288**, 95–99

340. Merz, A. J., So, M., and Sheetz, M. P. (2000) *Nature (London)* **407**, 98–102

341. Kirchhausen, T. (1999) *Nature (London)* **398**, 470–471

342. McNiven, M. A., Cao, H., Pitts, K. R., and Yoon, Y. (2000) *Trends Biochem. Sci.* **25**, 115–120

Study Questions

1. Describe briefly major aspects of the structure, properties, locations, and functions of each of the following proteins of skeletal muscle.

Actin	Tropomyosin
Myosin	Troponin
Titin	Myomesin
Nebulin	Desmin
α-Actinin	Vimentin
C-protein	

2. Describe the generally accepted sliding filament model of muscle contraction. List some uncertainties in this description.

3. Compare mechanisms that regulate contraction in skeletal muscle and in smooth muscle.

4. Compare myosin with kinesins and dyneins. What features do they have in common? What differences can you describe?

5. Compare the properties of actin in skeletal muscle and in nonmuscle cells. What is meant by "treadmilling?" What is "actin-based motility?"

6. The human genome contains more than 100 genes for proteins of the kinesin superfamily. Why?

7. Describe some of the major diseases that involve muscle proteins.

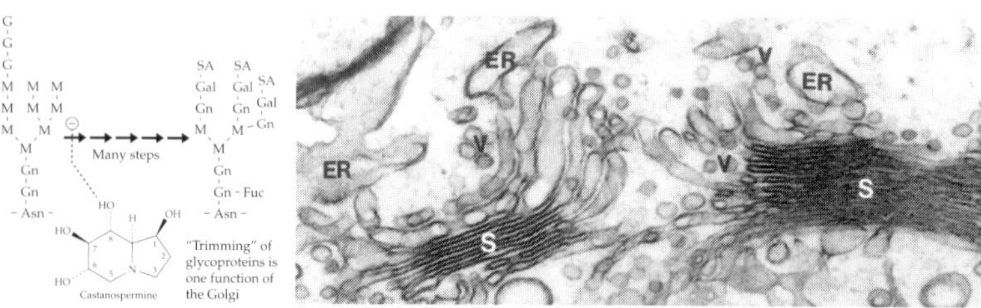

The branched oligosaccharides of glycoprotein surfaces are formed on asparagine side chains of selected cell surface proteins. The oligosaccharide at the left is formed in the ER and is transferred intact (Fig. 20-6) to an acceptor asparagine. It is then trimmed by removal of glucose and mannose units and residues of glucosamine, galactose, and fucose are added as in Fig. 20-7. These reactions begin in the ER and continue in the Golgi apparatus (right). See also Fig. 20-8.

Contents

Boxes

Tables

Some Pathways of Carbohydrate Metabolism

20

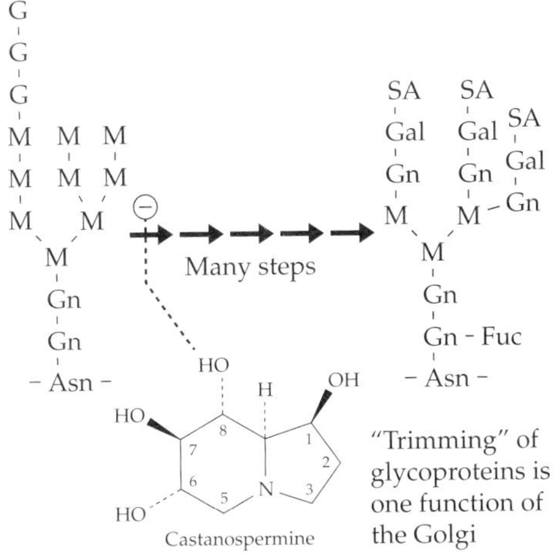

"Trimming" of glycoproteins is one function of the Golgi

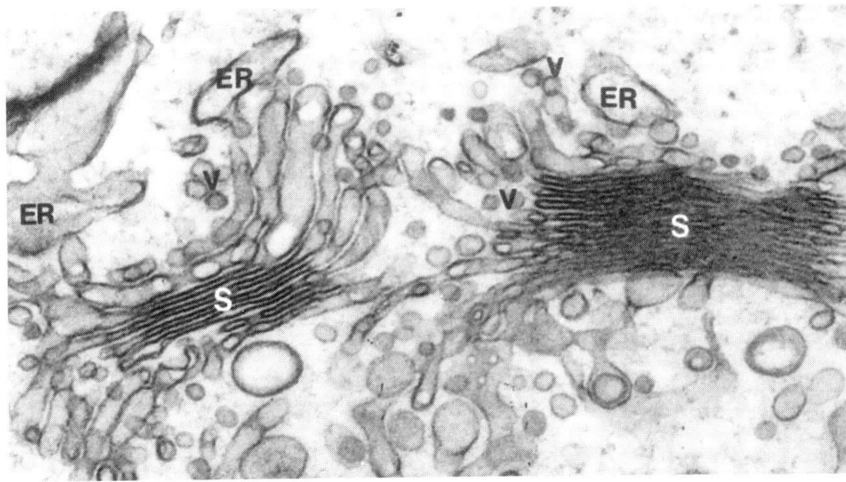

The general principles of biosynthesis, as well as the pathways of formation of major carbohydrate and lipid precursors, are considered in Chapter 17. Also described are the processes of gluconeogenesis, the synthesis of glucose 6-phosphate and fructose 6-phosphate from free glucose, and typical polymerization pathways for formation of polysaccharides. In this chapter, additional aspects of the metabolism of monosaccharides, oligosaccharides, polysaccharides, glycoproteins, and glycolipids are considered. These are metabolic transformations that affect the physical properties of cell surfaces and body fluids. They are essential to signaling between cells, to establishment of the immunological identity of individuals, and to the development of strong cell wall materials. Some of the differences in carbohydrates found in bacteria, fungi, green plants, and mammals are considered.

A. Interconversions of Monosaccharides

Chemical interconversions between compounds are easiest at the level of oxidation of carbohydrates. Consequently, many reactions by which one sugar can be changed into another are known. Most of the transformations take place in the "sugar nucleotide derivatives" (see also Eq. 17-56). The first of this group of compounds to be recognized was **uridine diphosphate glucose** (UDPG), which was discovered around 1950 by L. F. Leloir[1,2] during his investigation of the metabolism of galactose 1-*P*. The fact that interconversions of hexoses take place largely at the sugar nucleotide level was unknown at the time. Leloir's studies led to the characterization of both UDP-glucose and UDP-galactose.

Uridine diphosphate D-glucose

Figure 20-1 summarizes pathways by which glucose 6-phosphate or fructose 6-phosphate can be converted into many of the other sugars found in living things. Galactose and mannose can also be interconverted with the other sugars. A kinase forms **mannose 6-phosphate** which equilibrates with fructose 6-phosphate. Galactokinase converts free galactose to **galactose 1-phosphate**, which can be isomerized to glucose 1-phosphate by the reactions of Eq. 20-1. Fructose, an important human dietary constituent[3] derived largely by hydrolysis of sucrose, can also be formed in tissues via the **sorbitol pathway**[4] (Box 20-A). Fructose can be phosphorylated to fructose 1-phosphate by liver **fructokinase**. We have no mutase able to convert fructose 1-*P* to fructose 6-*P*, but a special aldolase cleaves fructose 1-*P* to dihydroxyacetone phosphate and free glyceraldehyde. Lack of this aldolase leads to occasionally observed cases of fructose intolerance.[5,6] The glyceraldehyde formed from fructose can be metabolized by reduction to glycerol followed by phosphorylation (glycerol kinase) and reoxidation to dihydroxyacetone phosphate. Some phosphorylation of fructose 1-*P* to fructose 1,6-P_2 apparently also occurs.[7] Interconversion of ribose 5-*P* and other sugar phosphates is a central part of the pentose phosphate pathway (Fig. 17-8). Free ribose can be phosphorylated by a **ribokinase**.[8]

Oxidation of UDP-glucose in two steps[9,9a] by NAD+ yields **UDP-glucuronic** acid, which can be epimerized to **UDP-galacturonic** acid. Likewise (see bottom of Fig. 20-1), **guanosine disphosphate-mannose** (GDP-mannose) is oxidized to **GDP-mannuronic acid**, which undergoes 4-epimerization to **GDP-guluronic acid**. Looking again at the top of the scheme, notice that UDP-D-glucuronic acid may be epimerized at the 5 position to **UDP-L-iduronic acid**. However, the iduronic acid residues in dermatan sulfate arise by inversion at C-5 of D-glucuronic acid residues in the polymer.[10] The mechanism of these reactions, like that of the decarboxylation of UDP-glucuronic acid to UDP-xylose (near the top of Fig. 20-1), apparently have not been well investigated.

Notice that glucuronic acid is abbreviated GlcA, in accord with IUB recommendations. However, many authors use GlcUA, ManUA, etc., for the uronic acids.

1. The Metabolism of Galactose

The reactions of galactose have attracted biochemists' interest because of the occurrence of the rare (30 cases / million births) hereditary disorder **galactosemia**. When this defect is present, the body cannot transform galactose into glucose metabolites but reduces it to the sugar alcohol **galactitol** or oxidizes it to **galactonate**, both products being excreted in the urine. Unfortunately, severe gastrointestinal troubles often appear within a few days or weeks of birth. Growth is slow and cataracts develop in the eyes, probably as a result of the accumulating galactitol. Death may come quickly from liver damage. Fortunately, galactose-free diets can be prepared for young infants, and if the disease is diagnosed promptly the most serious damage can be avoided. However, it has not been possible to prevent long-term effects that include speech difficulties, learning disabilities, and ovarian dysfunction.[1,11]

In some less seriously affected galactosemic patients **galactokinase** (Eq. 20-1, step *a*) is absent, but it is more often **galactose-1-phosphate uridyltransferase** (Eq. 20-1, step *b*) that is missing or inactive.[12–15a] This enzyme transforms galactose 1-*P* to UDP-galactose by displacing glucose 1-*P* from UDP-glucose. The UDP-galactose is then isomerized by the NAD+-dependent

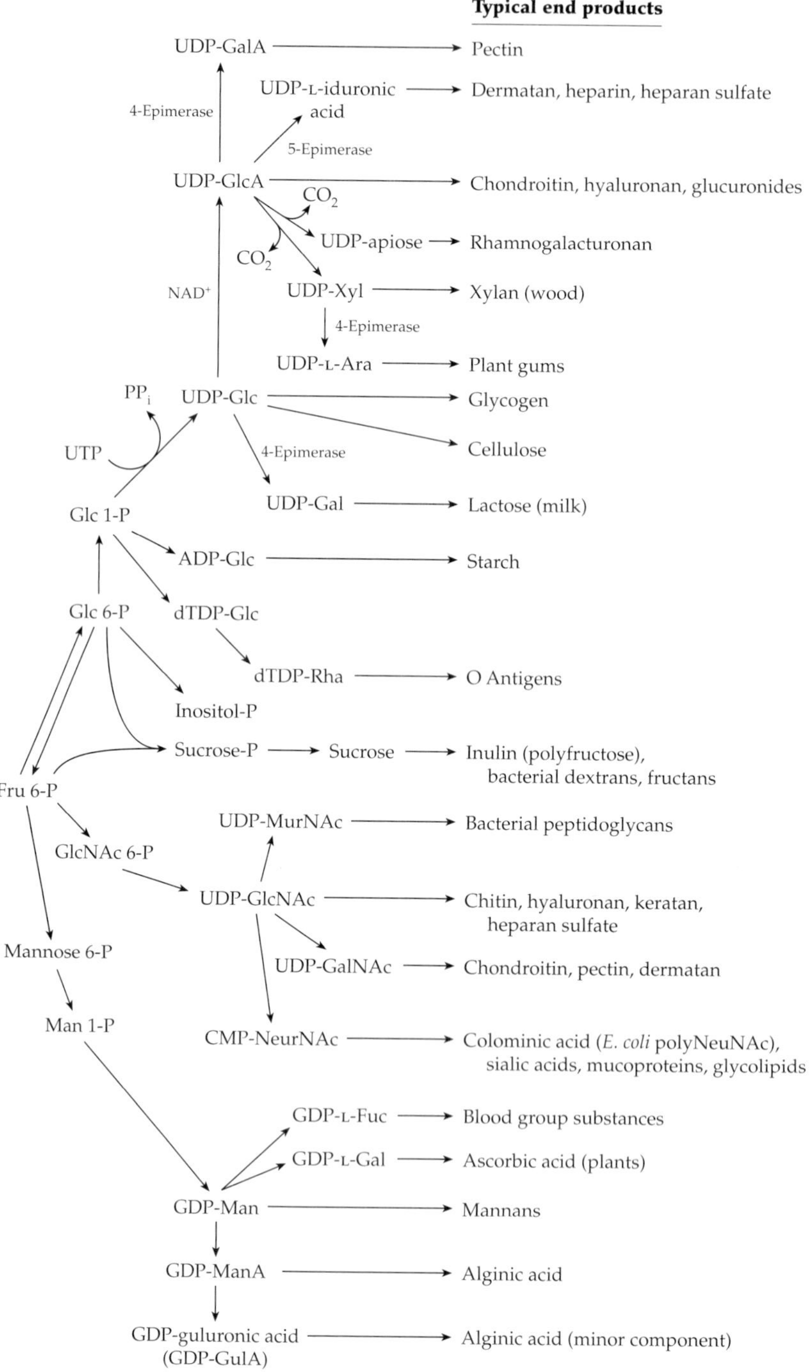

Figure 20-1 Some routes of interconversion of monosaccharides and of polymerization of the activated glycosyl units.

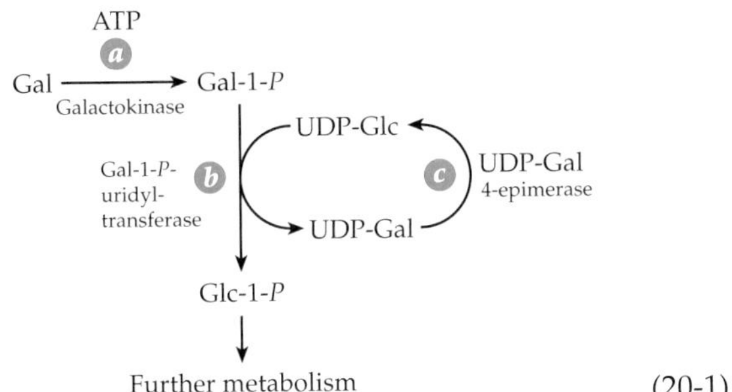

(20-1)

UDP-Gal 4-epimerase[16–16b] (Eq. 20-1, step *c*; see also Eq. 15-13 and accompanying discussion). Absence of this enzyme also causes galactosemia.[11] The overall effect of the reactions of Eq. 20-1 is to transform galactose into glucose 1-*P*. At the same time, the 4-epimerase can operate in the reverse direction to convert UDP-glucose to UDP-galactose, when the latter is needed for biosynthesis (Fig. 20-1).

Another enzyme important to galactose metabolism, at least in *E. coli*, is **galactose mutarotase**.[17] Cleavage of lactose by β-galactosidase produces β-D-galactose which must be converted to the α-anomer by the mutarotase before it can be acted upon by galactokinase. Galactose is present in most glycoproteins and glycolipids in the pyranose ring form. However, in bacterial O-antigens, in cell walls of mycobacteria and fungi, and in some protozoa glactose occurs in the furanose form. The precursor is UDP-Gal*f*, which is formed from UDP-Gal*p* by **UDP-Gal*p* mutase**.[17a]

2. Inositol

Related to the monosaccharides is the hexahydroxycyclohexane ***myo*-inositol** (Eq. 20-2). This **cyclitol**, which is apparently present universally within cells (Fig. 11-9), can be formed from glucose-6-*P* according to Eq. 20-2 using a synthase that contains bound

BOX 20-A FRUCTOSE FOR SPERM CELLS VIA THE POLYOL PATHWAY

An interesting example of the way in which the high [NADPH]/[NADP$^+$] and [NAD$^+$]/[NADH] ratios in cells can be used to advantage is found in the metabolism of sperm cells. Whereas D-glucose is the commonest sugar used as an energy source by mammalian cells, spermatozoa use principally D-fructose, a sugar that is not readily metabolized by cells of surrounding tissues.[a–c] Fructose, which is present in human semen at a concentration of 12 mM, is made from glucose by cells of the seminal vesicle by reduction with NADPH to the sugar alcohol D-sorbitol, which in turn is oxidized in the 2 position by NAD$^+$. The combination of high [NADPH]/[NADP$^+$] and high [NAD$^+$]/[NADH] ratio is sufficient to shift the equilibrium far toward fructose formation.[d]

D-Glucose (HC=O, H–C–OH, R) —[NADPH + H$^+$ → NADP$^+$]→ D-Sorbitol (D-Glucitol) (CH_2OH, H–C–OH, R) —[NAD$^+$ → NADH + H$^+$]→ D-Fructose (CH_2OH, C=O, R)

The polyol pathway is an active bypass of the dominant glycolysis pathway in many organisms.[e] Sorbitol and other polyols such as glycerol, erythritol, threitol, and ribitol serve as cryoprotectants in plants, insects, and other organisms.[f] Sorbitol is also an important osmolyte in some organisms (see Box 20-C). On the other hand, accumulation of sorbitol in lenses of diabetic individuals has often been blamed for development of cataract. However, doubts have been raised about this conclusion. The polyol pathway is more active than normal in diabetes, and there is evidence that the increased flow in this pathway may lead to an increase in oxidative damage to the lens. This may result, in part, from the depletion of NADPH needed for reduction of oxidized glutathione in the antioxidant system.[g] Aldose reductase inhibitors, which reduce the rate of sorbitol formation, decrease cataract formation. However, the reason for this is not yet clear.[h]

a McGilvery, R. W. (1970) *Biochemistry, A Functional Approach*, Saunders, Philadelphia, Pennsylvania (pp. 631–632)
b Hers, H. G. (1960) *Biochim. Biophys. Acta.* **37**, 127–
c Gitzelmann, R., Steinmann, B., and Van den Berghe, G. (1995) in *The Metabolic and Molecular Bases of Inherited Disease*, 7th ed., Vol. 1 (Scriver, C. R., Beaudet, A. L., Sly, W. S., and Valle, D., eds), pp. 905–935, McGraw-Hill, New York
d Prendergast, F. G., Veneziale, C. M., and Deering, N. G. (1975) *J. Biol. Chem.* **250**, 1282–1289
e Luque, T., Hjelmqvist, L., Marfany, G., Danielsson, O., El-Ahmad, M., Persson, B., Jörnvall, H., and Gonzàlez-Duarte, R. (1998) *J. Biol. Chem.* **273**, 34293–34301
f Podlasek, C. A., and Serianni, A. S. (1994) *J. Biol. Chem.* **269**, 2521–2528
g Lee, A. Y. W., and Chung, S. S. M. (1999) *FASEB J.* **13**, 23–30
h Srivastava, S., Watowich, S. J., Petrash, J. M., Srivastava, S. K., and Bhatnagar, A. (1999) *Biochemistry* **38**, 42–54

NAD^+. In addition to the two redox steps (Eq. 20-2, *b* and *d*), this enzyme catalyzes both the conversion of the β anomer of glucose 6-*P* to the open-chain aldehyde form and the internal aldol condensation of Eq. 20-2, step *c*.[18-19b] The pro-*R* hydrogen at C-6 of glucose 1-*P* is lost in step *b* while the pro-*S* hydrogen is retained.[20] The ring numbering system is different for glucose and for the inositols, C-5 of glucose 1-*P* becoming C-2 of L-*myo*-inositol. Since *myo*-inositol contains a plane of symmetry D- and L- forms are identical. However, they are numbered differently (Eq. 20-2). The phosphoinositides and inositol polyphosphates are customarily numbered as derivatives of D-*myo*-inositol.

Synthesis of inositol by animals is limited and *myo*-inositol is sometimes classified as a vitamin. Mice grow poorly and lose some of their hair if deprived of dietary inositol. Various phosphate esters of inositol occur in nature. For example, large amounts of the hexaphosphate (**phytic acid**) are present in grains, usually as the calcium or mixed Ca^{2+}–Mg^{2+} salts known as **phytin**. The two apical cells of the 28-cell larvae of mesozoa (Fig. 1-12A) contain enough magnesium phytate in granular form to account for up to half of the weight of the larvae.[21] Inositol pentaphosphate is an allosteric activator for hemoglobin in birds and turtles (p. 358). Di-*myo*-inositol-1,1'-phosphate is an osmolyte in some hypothermophilic archaea.[19a] Inositol is a component of **galactinol**, the β glycoside of D-galactose with inositol (Eq. 20-15). Galactinol, as well as free inositol, circulates in human blood and in plants and may be a precursor of cell wall polysaccharides. However, in our own bodies the greatest importance for inositol doubtless lies in the inositol-containing phospholipids known as **phosphoinositides** (Figs. 8-2, 11-9, 21-5). Their function in generation of "second messengers" for various hormones is dealt with in Chapters 11 and 21.

β-D-Glucopyranose 6-*P*

myo-Inositol

L-*myo*-Inositol 1-phosphate

D-*myo*-Inositol 3-phosphate

(20-2)

A person typically ingests daily about one gram of inositol, some in the free form, some as phosphoinositides, and some as phytin. As much as four grams of inositol per day may be synthesized in the kidneys.[22] Breast milk is rich in inositol and dietary supplementation with inositol has increased survival of premature infants with respiratory distress syndrome.[22] The action of insulin is reported to be improved by administration of D-*chiro*-inositol (p. 998) to women with polycystic ovary syndrome.[22a]

3. D-Glucuronic Acid, Ascorbic Acid, and Xylitol

In bacteria, as well as in animal kidneys,[23] inositol may be converted to D-glucuronic acid (Fig. 20-1) with the aid of an oxygenase. Free glucuronic acid may also be formed by animals from glucose or from UDP-glucose (Fig. 20-2). Within the animal body glucuronic acid can be reduced with NADH (Fig. 20-2, step *a*) to yield **L-gulonic acid**, an aldonic acid that could also be formed by oxidation at the aldehyde end of the sugar **gulose**. Because C-6 of the glucuronic acid has become C-1 of gulonic acid, the latter belongs to the L family of

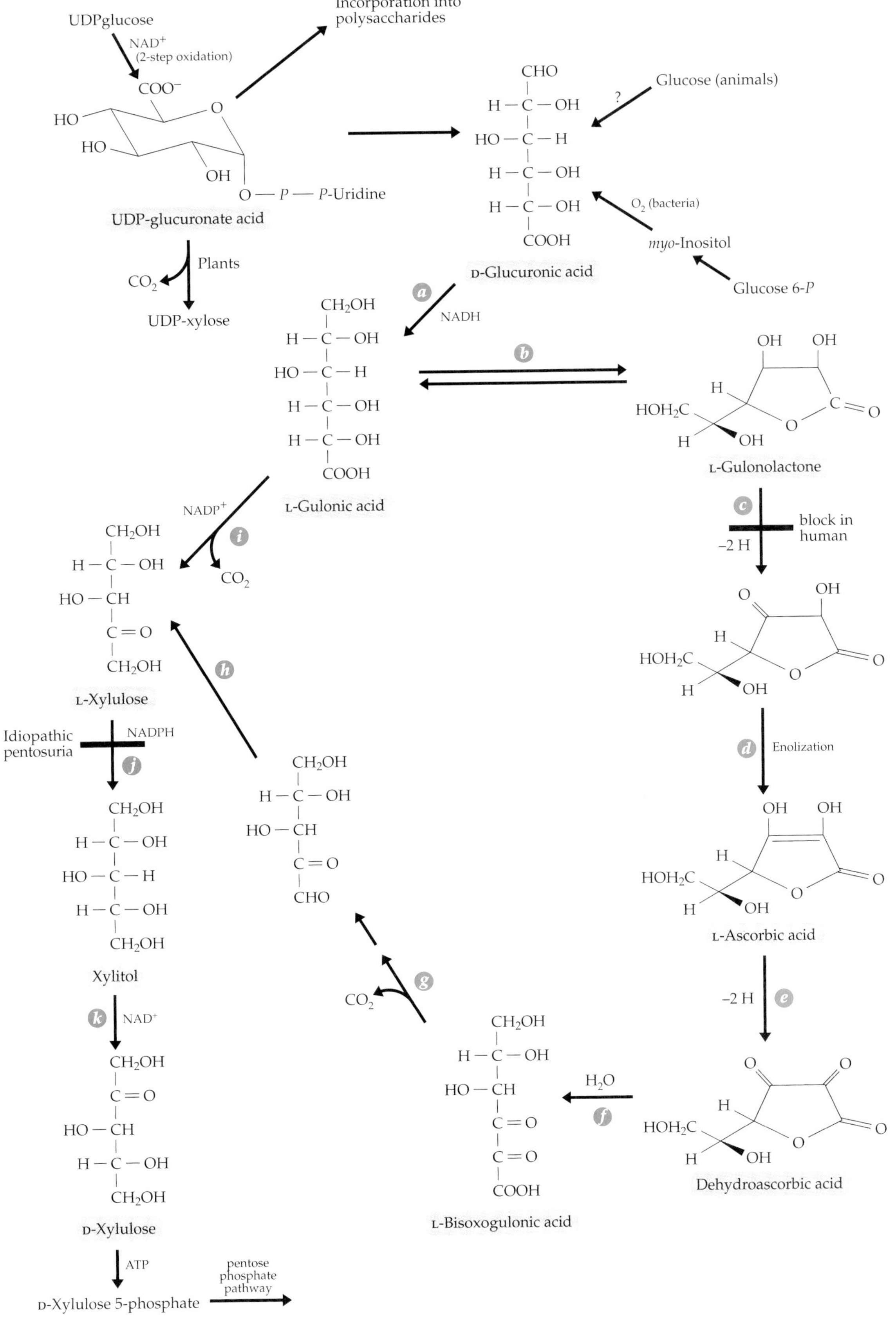

Figure 20-2 Some pathways of metabolism of D-glucuronic acid and of ascorbic acid, vitamin C.

sugars. Gulonic acid can be converted to a cyclic lactone (step *b*) which, in a two-step process involving dehydrogenation and enolization (steps *c* and *d*), is converted to **L-ascorbic acid**. This occurs in most higher animals.[24] However, the dehydrogenation step is lacking in human beings and other primates, in the guinea pig, and in a few other species. One might say that we and the guinea pig have a genetic defect at this point which obliges us to eat relatively large quantities of plant materials to satisfy our bodily needs for ascorbic acid (see Box 18-D). Gulonolactone oxidase is one of the enzymes containing covalently bound 8α-(N^1-histidyl)riboflavin.[25] The defective human gene for this enzyme has been identified, isolated, and sequenced. It is found to have accumulated a large number of mutations, which have rendered it inactive and now only a pseudogene.[26] Mice with an inactivated gulonolactone oxidase have a dietary requirement for vitamin C similar to that of humans. They suffer severe vascular damage on diets marginal in ascorbic acid.[26a] Even in rodents Na^+-dependent ascorbic acid transporters are present in metabolically active tissues to bring the vitamin from the blood into cells.[26b]

A clever bit of genetic engineering has permitted the conversion of D-glucose to 2-oxo-L-gulonate in the enzymatic sequence of Eq. 20-3, *a*, *b*.

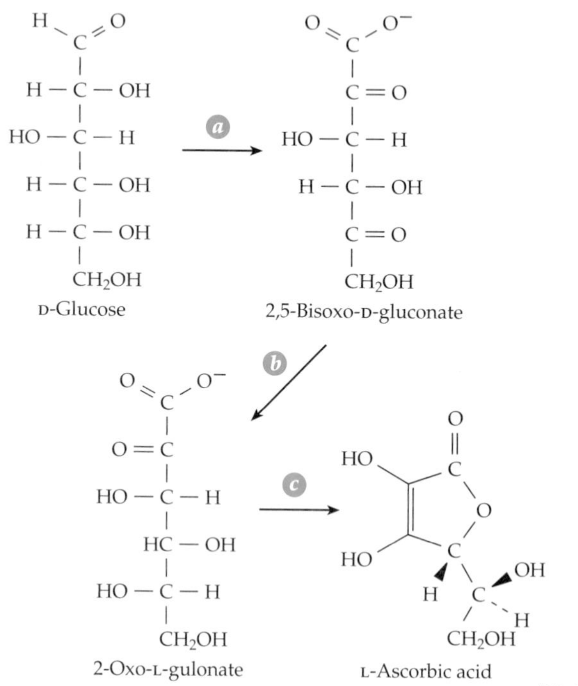

(20-3)

The bacterium *Erwinia herbicola* naturally has the ability to oxidize glucose to 2,5-bisoxo-D-gluconate (Eq. 20-3, step *a*) but cannot carry out the next step, the stereospecific reduction to 2-oxo-L-gulonate. However, a gene encoding a suitable reductase was isolated from a genomic library from *Corynebacterium*. The cloned gene was fused to an *E. coli trp* promoter (see Chapter 28) and was introduced in a multicopy plasmid into *E. herbicola*. The resultant organism can carry out both steps *a* and *b* of Eq. 20-3 leaving only step *c*, a nonenzymatic acid-catalyzed reaction, to complete an efficient synthesis of vitamin C from glucose.[27]

Higher plants make large amounts of L-ascorbate, which in leaves may account for 10% of the soluble carbohydrate content.[28] However, the pathway of synthesis differs from that in Fig. 20-2. Both D-mannose and L-galactose are efficient precursors. The pathway in Eq. 20-4, which starts with GDP-D-mannose and utilizes known enzymatic processes, has been suggested.[28,29] The GDP-D-mannose-3, 5-epimerase is a well documented but poorly understood enzyme. Multistep mechanisms related to that of UDP-glucose 4-epimerase (Eqs. 20-1, 15-14) can be envisioned.

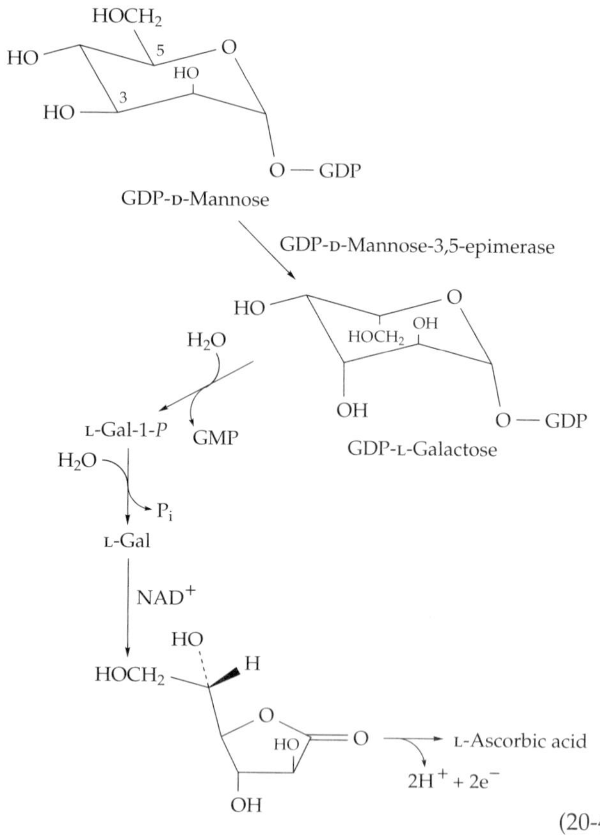

(20-4)

Ascorbic acid is readily oxidized to dehydroascorbic acid (Box 18-D; Fig. 20-2, step *e*), which may be hydrolyzed to L-bisoxogulonic acid (step *f*). The latter, after decarboxylation and reduction, is converted to L-xylulose (steps *g* and *h*), a compound that can also be formed by a standard oxidation and decarboxylation sequence on L-gulonic acid (step *i*). Reduction of xylulose to xylitol and oxidation of the latter with NAD^+ (steps *j* and *k*) produces D-xylulose, which can be phosphorylated with ATP and enter the pentose phosphate pathways. A metabolic variation produces a condition called **idiopathic pentosuria**. Affected individuals cannot reduce xylulose to xylitol and, hence, excrete large amounts of the pentose into the urine, especially if the diet is rich in glucuronic acid. The "defect" seems to be harmless, but the sugar in the urine can cause the condition to be mistaken for diabetes mellitus.[30]

Xylitol is as sweet as sucrose and has been used as a food additive. Because it does not induce formation of dental plaque, it is used as a replacement for sucrose in chewing gum. It appeared to be an ideal sugar substitute for diabetics. However, despite the fact that it is already naturally present in the body, ingestion of large amounts of xylitol causes bladder tumors as well as oxalate stones in rats and mice. Its use has, therefore, been largely discontinued. A possible source of the problem may lie in the conversion by fructokinase of some of the xylitol to D-xylulose 1-*P*, which can be cleaved by the xylulose 1-*P* aldolase to dihydroxyacetone *P* and glycolaldehyde.

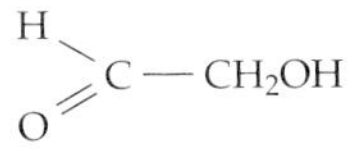

$$\mathrm{H(O{=})C-CH_2OH}$$

Glycolaldehyde

The latter can be oxidized to oxalate and may also be carcinogenic. As indicated in the upper left corner of Fig. 20-2, UDP-glucuronate can be decarboxylated to UDP-xylose.

4. Transformations of Fructose 6-Phosphate

Biosynthesis of **D-glucosamine 6-phosphate** is accomplished by reaction of fructose 6-*P* with glutamine (Eq. 20-5):

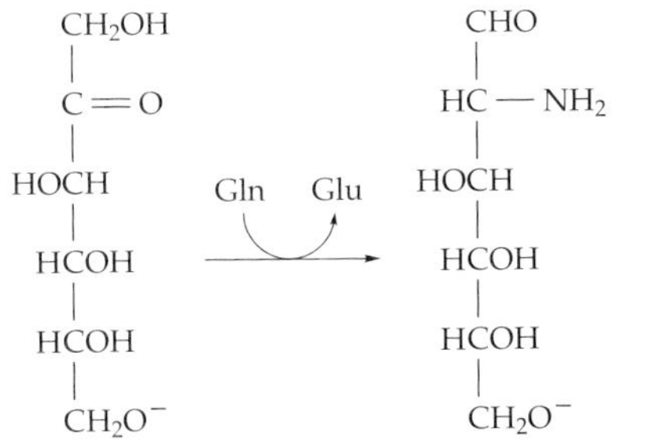

$$\begin{array}{ccc} \mathrm{CH_2OH} & & \mathrm{CHO} \\ \mathrm{C{=}O} & & \mathrm{HC-NH_2} \\ \mathrm{HOCH} & \xrightarrow{\text{Gln} \quad \text{Glu}} & \mathrm{HOCH} \\ \mathrm{HCOH} & & \mathrm{HCOH} \\ \mathrm{HCOH} & & \mathrm{HCOH} \\ \mathrm{CH_2O^-} & & \mathrm{CH_2O^-} \\ \text{Fructose 6-}P & & \text{D-Glucosamine 6-}P \end{array} \qquad (20\text{-}5)$$

Glutamine is one of the principal combined forms of ammonia that is transported throughout the body (Chapter 24). Glucosamine 6-phosphate synthase, which catalyzes the reaction of Eq. 20-5, is an amidotransferase of the N-terminal nucleophile hydrolase superfamily (Chapter 12).[31] It hydrolyzes the amide linkage of glutamine. The released ammonia presumably reacts with the carbonyl group of fructose 6-*P* to form an imine,[32–34a] which then undergoes a reaction analogous to that catalyzed by sugar isomerases.[35] The resulting D-glucosamine 6-*P* is acetylated on its amino group by transfer of an acetyl group,[36] and a mutase moves the phospho group to form *N*-acetylglucosamine 1-*P*. In *E. coli* acetylation occurs on GlcN 1-*P* and is catalyzed by a bifunctional enzyme that also has mutase activity.[37-37b] The resulting *N*-acetylglucosamine 1-*P* is converted to UDP-*N*-acetylglucosamine (UDP-GlcNAc) with cleavage of UTP to inorganic pyrophosphate as in the synthesis of UDP-glucose (Eq. 17-56). Cells of *E. coli* are also able to catabolize glucosamine 6-phosphate. A **deaminase,** with many properties similar to those of GlcN 6-*P* synthase, catalyzes a reaction resembling the reverse of Eq. 20-5 but releasing NH_3.[38,39]

One of the compounds formed from UDP-GlcNAc is **UDP-*N*-acetylmuramic acid**. The initial step in its synthesis is an unusual type of displacement reaction on the α-carbon of PEP by the 3-hydroxyl group of the sugar (Eq. 20-6, step *a*).[40-41c] Inorganic phosphate is

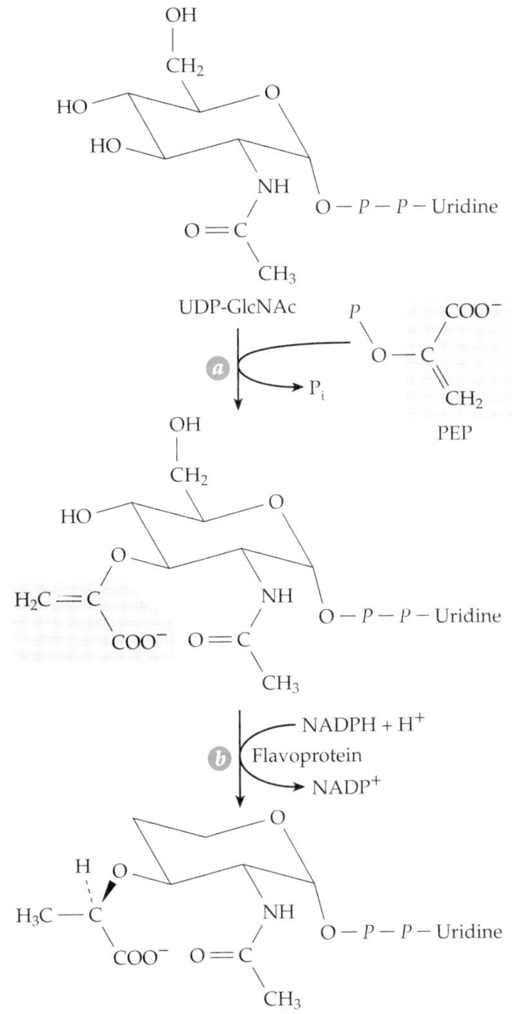

UDP-GlcNAc (OH, CH_2, HO, HO, O, NH, O=C, CH_3, O – *P* – *P* – Uridine)

(*a*) PEP (*P*–O–C(COO⁻)=CH₂) → P_i

(OH, CH_2, HO, O, $H_2C{=}C$, COO^-, NH, O=C, CH_3, O – *P* – *P* – Uridine)

(*b*) NADPH + H⁺ → Flavoprotein → NADP⁺

(H, O, H_3C-C, COO^-, NH, O=C, CH_3, O – *P* – *P* – Uridine)

UDP-*N*-acetylmuramic acid (UDP-MurNAc) (20-6)

displaced with formation of an enolpyruvyl derivative of UDP-GlcNAc. This derivative is then reduced by NADPH (Eq. 20-6, step *b*).[42-43a] A second sugar nucleotide formed from UDP-GlcNAc is **UDP-*N*-acetylgalactosamine** (UDP-GalNAc), which may be created by the same 4-epimerase that generates UDP-Gal (Eq. 20-1).[44] Some animal tissues such as kidney and liver also have a **GalNAc kinase** that may salvage, for reuse, GalNAc that arises from the degradation of complex polysaccharides.[44] Bacteria may dehydrogenate UDP-GalNAc to UDP-*N*-acetylgalactosaminuric acid (UDP-GalNAcA).[44a]

UDP-GlcNAc can be converted to UDP-*N*-acetylmannosamine (UDP-ManNAc) with concurrent elimination of UDP (Eq. 20-7).[45–47b] This unusual epimerization occurs without creation of an adjacent carbonyl group that would activate the 2-H for removal as a proton. As indicated by the small arrows in Eq. 20-7, step *a*′, the UDP is evidently eliminated. In a bacterial enzyme it remains in the E–S complex and is returned after a conformational change involving the acetamido group. This allows the transient C1–C2 double bond to be protonated from the opposite side (Eq. 20-7, step *a*).[47] In bacteria the UDP-ManNAc may be dehydrogenated to UDP-*N*-acetylmannos-aminuronic acid (ManNAcA). Both ManNAc and ManNAcA are components of bacterial capsules.[47]

In mammals the epimerase (Eq. 20-7, step *a*′) probably utilizes a similar chemical mechanism but eliminates UDP and replaces it with HO^- to give free *N*-acetylmannosamine, which is then phosphorylated on the 6-hydroxyl (Eq. 20-7, step *b*). ManNAc may also be formed from free GlcNAc by another 2-epimerase (step a″).[47c,d]

UDP-GlcNAC

N-acetyl-glucosamine (GlcNAc)

UDP-ManNAc

Polysaccharides

N-Acetylmannos-aminuronic acid (ManNAcA)

N-Acetylmannosamine (ManNAc)

N-Acetylmannosamine 6-*P* (ManNAc 6-*P*)

PEP

N-Acetylneur-aminic acid 9-*P* (Neu5Ac 9-*P*)

N-Acetylneuraminic acid (Neu5Ac)

CMP-Neu5Ac

(20-7)

5. Extending a Sugar Chain with Phosphoenolpyruvate (PEP)

The six-carbon chain of ManNAc 6-*P* can be extended by three carbon atoms using an aldol-type condensation with a three-carbon fragment from PEP (Eq. 20-7, step *c*) to give **N-acetylneuraminic acid** (sialic acid).[48] The nine-carbon chain of this molecule can cyclize to form a pair of anomers with 6-membered rings as shown in Eq. 20-7. In a similar manner, arabinose 5-*P* is converted to the 8-carbon **3-deoxy-D-*manno*-octulosonic acid** (**KDO**) (Fig. 4-15), a component of the lipopolysaccharide of gram-negative bacteria (Fig. 8-30), and D-Erythrose 4-*P* is converted to 3-deoxy-D-*arabino*-heptulosonate 7-*P*, the first metabolite in the shikimate pathway of aromatic synthesis (Fig. 25-1).[48a] The arabinose-*P* used for KDO synthesis is formed by isomerization of D-ribulose 5-*P* from the pentose phosphate pathway, and erythrose 4-*P* arises from the same pathway.

The mechanism of the aldol condenstion that forms these sugars is somewhat unexpected. A reactive enolate anion can be formed from PEP by hydrolytic attack on the phospho group with cleavage of the O–P bond. However, in reactions such as step *a* of

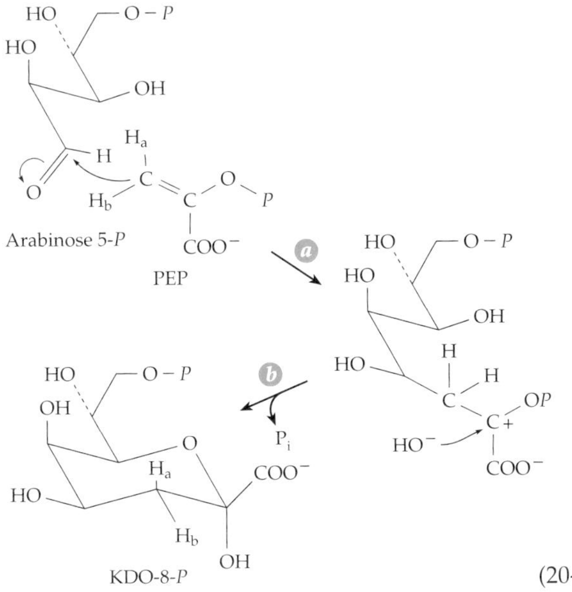

(20-8)

Eq. 20-6, step *c* of Eq. 20-7,and also in EPSP synthase (Eq. 25-4) the initial condensation does not involve O–P cleavage. NMR studies of the action of KDO synthase reveal that the C–O bond of PEP is cleaved as is indicated in Eq. 20-8.[49-52b] The *si* face of PEP faces the *re* face of the carbonyl group of the sugar phosphate. A carbanionic center is generated at C-3 of PEP with possible participation of the phosphate oxygen as well as electrostatic stabilization of the carbocation formed in step *a*. Ring closure (step *b*) occurs with loss of P_i. The immediate product of the aldol condensation, in Eq. 20-7, is *N*-acetylneuraminic acid 9-phosphate, which is cleaved through phosphatase action (step *d*) and is activated to the CMP derivative by reaction with CTP (Eq. 20-7,step *e*).[52c] Further alterations may occur. For example, CMP-Neu5Ac is hydroxylated to form CMP-*N*-glycolylneuraminic acid.[53] Furthermore, an additional type of sialic acid, 2-oxo-3-deoxy-D-*glycero*-D-*galacto*-nononic acid (**KDN**), has been found in human developmentally regulated glycoproteins and also in many other organisms.[54-55a] It has an –OH group in the 5-position rather than the acetamido group of the other sialic acids. Like NeuNAc it is activated by reaction with CTP forming CMP-KDN. These activated monosaccharides differ from most others in being derivatives of a CMP rather than of CDP. More than 40 different naturally occuring variations of sialic acid have been identified.[55b]

In a similar fashion, KDO is converted to the β-linked **CMP-KDO**,[56-56b] which is incorporated into lipid A as shown in Fig. 20-10. The ADP derivative of the **L-*glycero*-D-*manno*-heptose** (Fig. 4-15), which is also present in the lipopolysaccharide of gram-negative bacteria, is formed from sedoheptulose 7-*P* in a five-step process (Eq. 20-9).[57-58b]

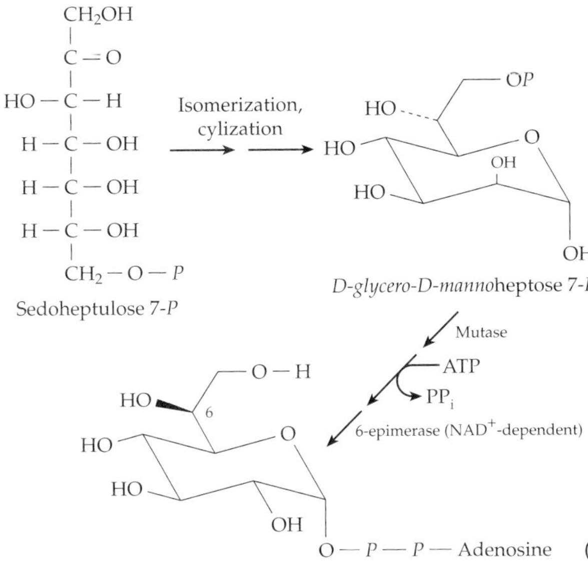

(20-9)

6. Synthesis of Deoxy Sugars

Metabolism of sugars often involves dehydration to α,β-unsaturated carbonyl compounds. An example is the formation of 2-oxo-3-deoxy derivatives of sugar acids (Eq. 14-36). Sometimes a carbonyl group is

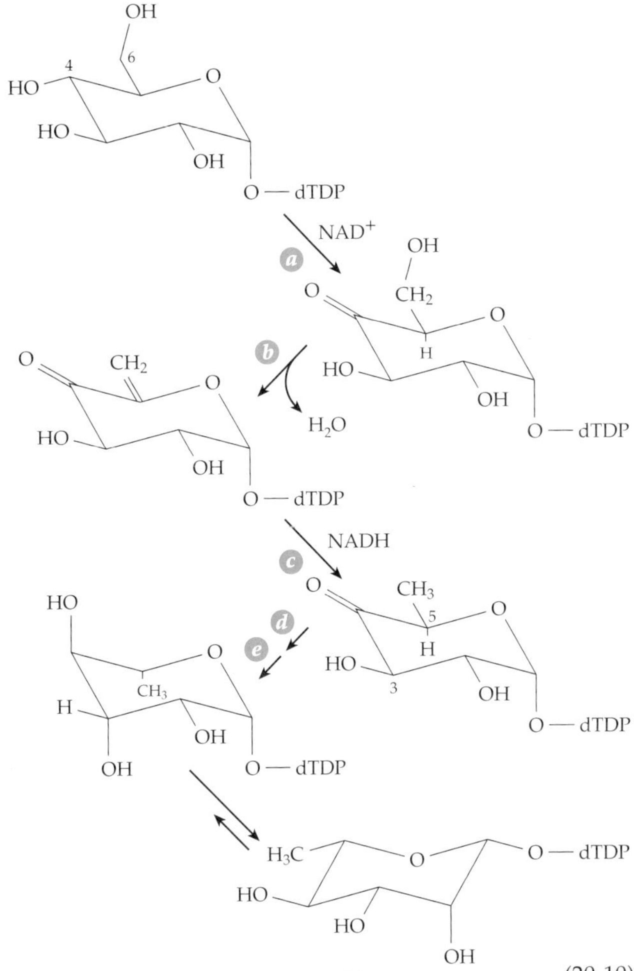

(20-10)

created by oxidation of an –OH group, apparently for the sole purpose of promoting dehydration. For example, the biosynthesis of L-rhamnose from D-glucose is a multistep process (Eq. 20-10) that takes place while the sugars are attached to deoxythymidine diphosphate.[59,59a,b] Introduction of the carbonyl group by dehydrogenation with tightly bound NAD^+ (Eq. 20-10, step *a*) is followed by dehydration (step *b*).[59c,d] To complete the sequence, the double bond formed by dehydration is reduced (step *c*) by the NADH produced in step *a*. A separate enzyme, a 3,5-epimerase catalyzes inversion at both C-3 and C-5 (step *d*).[59e] Finally, a third enzyme is needed for a second reduction (step *e*) using NADPH.[59f] The biosynthesis of **GDP-L-fucose** from GDP-D-mannose occurs by a parallel sequence.[60-61b]

The metabolism of free L-fucose (6-deoxy-L-galactose), which is present in the diet and is also generated by degradation of glycoproteins, resembles the Entner–Doudoroff pathway of glucose metabolism (Eq. 17-18). Similar degradative pathways act on D-arabinose and L-galactose.[60]

Bacterial surface polysaccharides contain a variety of dideoxy sugars. The four 3,6-dideoxy sugars **D-paratose** (3,6-dideoxy-D-glucose), **D-abequose** (3,6-dideoxy-D-galactose), **D-tyvelose** (3,6-dideoxy-D-mannose), and **L-ascarylose** (3,6-dideoxy-L-mannose), whose structures are shown in Fig. 4-15, arise from CDP-glucose.[60a] This substrate is first converted, in reactions parallel to the first three steps of Eq. 20-10, to 4-oxo-6-deoxy-CDP-glucose which reacts in two steps with pyridoxamine 5'-phosphate (PMP) and NADH (Eq. 20-11). This unusual reaction[62–65] is catalyzed by a two-enzyme complex. The first component, E1, catalyzes the formation of a Schiff base of the substrate with PMP and a transamination, which also accomplishes dehydration, to give an unsaturated sugar ring (Eq. 20-11, step *a*). The protein also contains an Fe_2S_2 center suggesting a possible one-electron transfer. The second component, E_3, contains both an Fe_2S_2 plant type ferredoxin center and bound FAD.[65] Observation by EPR spectroscopy revealed accumulation of an organic free radical[64] that may be an intermediate in step *b* of Eq. 20-11. Hydrolysis, epimerization at C-5, and reduction yields L-ascarylose. A similar reaction sequence without the last epimerization would yield D-abequose. CDP-D-tyvelose arises by C-2 epimerization of CDP-D-paratose.[65a] Other unusual sugars[66–68] are formed from intermediates in Eq. 20-11. One is a **3-amino-3,4,6-trideoxyhexose** in which the amino group has been provided by transamination[67] (see also Box 20-B).

The unusual sulfur-containing sugar **6-sulfoquinovose** is present in

CMP-D-glucose → (PMP) → E1 (*a*), H_2O → E3 (*b*), $2H^+ + 2e^-$ via (Fe_2S_2), FAD → H_2O → Epimerase, Reduction → L-Ascarylose; Reduction → D-Abequose

(20-11)

Cysteic acid → Transamination, Reduction (*a*) → Sulfolactaldehyde + Dihydroxyacetone-*P* → Aldol (H_2O, P_i) → 6-Sulfoquinovose

UDP-Glc → (*b*) (H_2O, NAD^+) → ... → NADH → 6-Sulfoquinovose

(20-12)

BOX 20-B THE BIOSYNTHESIS OF STREPTOMYCIN

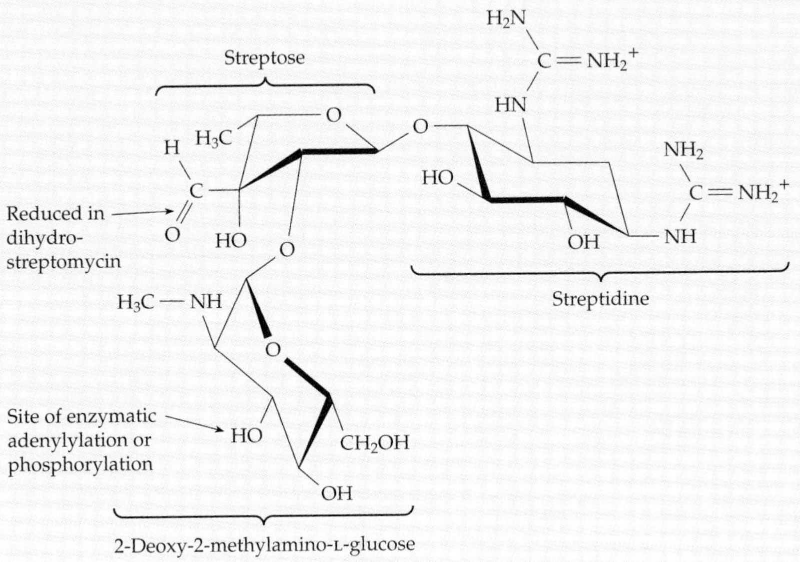

Streptose

Reduced in dihydro-streptomycin

Streptidine

Site of enzymatic adenylylation or phosphorylation

2-Deoxy-2-methylamino-L-glucose

Streptomycin

Streptomycin, the kanamycins, neomycins, and gentamycins form a family of medically important **aminoglycoside antibiotics**.[a] They are all water-soluble basic carbohydrates containing three or four unusual sugar rings. D-Glucose is a precursor of streptomycin, all three rings being derived from it. While the route of biosynthesis of 2-deoxy-2-methylamino-L-glucose is not entirely clear, the pathways to L-streptose and streptidine, the other two rings, have been characterized.[b–d] The starting material for streptidine synthesis is a nucleoside diphosphate sugar, which is an intermediate in the synthesis of L-rhamnose (Eq. 20-10). The carbon–carbon chain undergoes an aldol cleavage as shown in step *a* of the following equation:

Intermediate from Eq. 20-10

Streptosyl-TDP

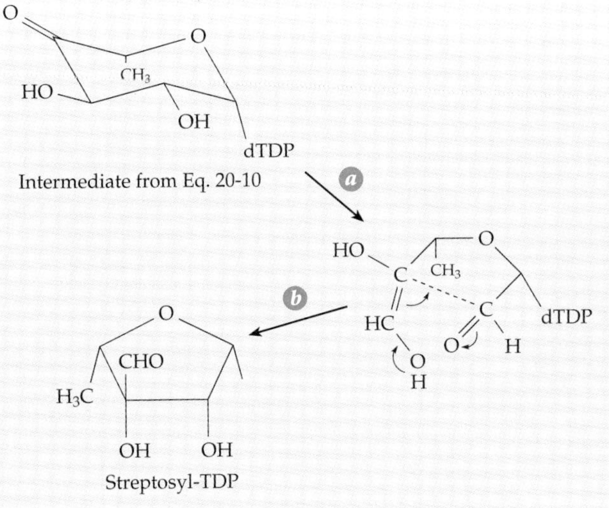

The ring-open product is written here as an enediol, which is able to recyclize in an aldol condensation (step *b*) to form a five-membered ring with a branch at C-3. The L-streptosyl nucleoside diphosphate formed in this way serves as the donor of streptose to streptomycin.

The basic cyclitol streptidine is derived from *myo*-inositol, which has been formed from glucose 6-*P* (Eq. 20-2). The guanidino groups are introduced by oxidation of the appropriate hydroxyl group to a carbonyl group followed by transamination from a specific amino donor. In the first step, illustrated by the following equation, glutamine is the amino donor for the transamination, the oxoacid product being α-oxoglutaramic acid.

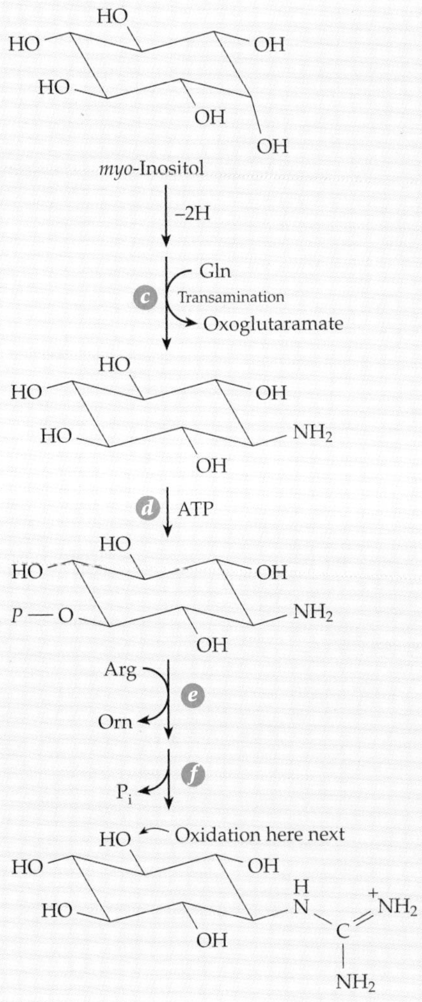

myo-Inositol

–2H

(c) Gln, Transamination, → Oxoglutaramate

(d) ATP

Arg (e) Orn

P_i (f)

Oxidation here next

BOX 20-B THE BIOSYNTHESIS OF STREPTOMYCIN (continued)

The amino group on the ring now receives an amidine group, which is transferred from arginine by nucleophilic displacement[e] in a reaction resembling that in the synthesis of urea (see Fig. 24-10, step *h*). However, there is first a phosphorylation at the 2 position. After the amidine transfer has occurred to form the guanidino group, the phospho group is hydrolyzed off by a phosphatase. This is another phosphorylation–dephosphorylation sequence (p. 977) designed to drive the reaction to completion in the desired direction. The second guanidino group is introduced in an analogous way by oxidation at the 3 position followed by transamination, this time with the amino group being donated by alanine. Again, a phosphorylation is followed by transfer of an amidine group from arginine. The final hydrolytic removal of the phospho group (which this time is added at C-6) does not occur until the two other sugar rings have been transferred on from nucleoside diphosphate precursors to form streptomycin phosphate.

As with other antibiotics,[f–i] streptomycin is subject to inactivation by enzymes encoded by genetic resistance factors (Chapter 26). Among these are enzymes that transfer phospho groups or adenylyl groups onto streptomycin at the site indicated by the arrow in the structure.[j,k] Thus, dephosphorylation at one site generates the active antibiotic as the final step in the biosynthesis, while phosphorylation at another site inactives the antibiotic.

[a] Benveniste, R., and Davies, J. (1973) *Ann. Rev. Biochem.* **42**, 471–506

[b] Luckner, M. (1972) *Secondary Metabolism in Plants and Animals*, Academic Press, New York (pp.78–80)

[c] Walker, J. B., and Skorvaga, M. (1973) *J. Biol. Chem.* **248**, 2441–2446

[d] Marquet, A., Frappier, F., Guillerm, G., Azoulay, M., Florentin, D., and Tabet, J.-C. (1993) *J. Am. Chem. Soc.* **115**, 2139–2145

[e] Fritsche, E., Bergner, A., Humm, A., Piepersberg, W., and Huber, R. (1998) *Biochemistry* **37**, 17664–17672

[f] Cox, J. R., and Serpersu, E. H. (1997) *Biochemistry* **36**, 2353–2359

[g] McKay, G. A., and Wright, G. D. (1996) *Biochemistry* **35**, 8680–8685

[h] Thompson, P. R., Hughes, D. W., Cianciotto, N. P., and Wright, G. D. (1998) *J. Biol. Chem.* **273**, 14788–14795

[i] Gerratana, G., Cleleand, W. W., and Reinhardt, L. A. (2001) *Biochemistry* **40**, 2964–2971

[j] Roestamadji, J., Grapsas, I., and Mobashery, S. (1995) *J. Am. Chem. Soc.* **117**, 80–84

[k] Thompson, P. R., Hughes, D. W., and Wright, G. D. (1996) *Biochemistry* **35**, 8686–8695

the sulfolipid of chloroplasts (p. 387).[69] A possible biosynthetic sequence begins with transamination of cysteic acid to 3-sulfopyruvate, reduction of the latter to sulfolactaldehyde, and aldol condensation with dihydroxyacetone-*P* as indicated in Eq. 20-12a.[70] See also Eq. 24-47 and Fig. 4-4. However, biosynthesis in chloroplasts appears to start with action of a 4,6-dehydratase on UDP-glucose followed by addition of sulfite and reduction (Eq. 20-12*b*).[70a,b] The sulfite is formed by reduction of sulfate via adenylyl sulfate (Fig. 24-25). However, biosynthesis in chloroplasts appears to start with action of a 4,6-dehydratase on UDP-glucose followed by addition of sulfite and reduction (Eq. 20-12*b*).[70a,b] The sulfite is formed by reduction of sulfate via adenylyl sulfate (Fig. 24-25).

B. Synthesis and Utilization of Oligosaccharides

Our most common food sugar **sucrose** is formed in all green plants and nowhere else. It is made both in the chloroplasts and in the vicinity of other starch deposits. It serves both as a transport sugar and, dissolved within vacuoles, as an energy store. Sucrose is very soluble in water and is chemically inert because the hemiacetal groups of both sugar rings are blocked. However, sucrose is thermodynamically reactive, the glucosyl group having a group transfer potential of 29.3 kJ mol^{-1}. It is extremely sensitive toward hydrolysis catalyzed by acid. Transport of sugar in the form of a disaccharide provides an advantage to plants in that the disaccharide has a lower osmotic pressure than would the same amount of sugar in monosaccharide form.

Biosynthesis of sucrose[71,71a] utilizes both UDP-glucose and fructose 6-*P* (Eq. 20-13). Reaction of UDP-glucose with fructose can also occur to give sucrose directly.[72] Because this reaction is reversible, sucrose serves as a source of UDP-glucose for synthesis of cellulose and other polysaccharides in plants. Metabolism of sucrose in the animal body begins with the action of **sucrase** (invertase), which hydrolyzes the disaccharide to fructose and glucose (Eq. 20-14, step *a*). The same enzyme is also found in higher plants and fungi. Mammalian sucrase is one of several carbohydrases that are anchored to the external surfaces of the microvilli of the small intestines. Sucrose is bound

$$\text{UDP–Glc} \xrightarrow{\text{Fru-6-}P} \text{Sucrose 6-}P \xrightarrow[P_i]{H_2O} \text{Sucrose} \qquad (20\text{-}13)$$

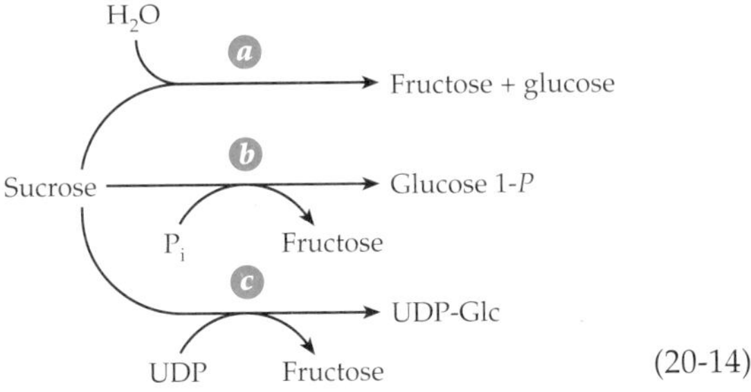

(20-14)

tightly but noncovalently to **isomaltase,** which hydrolyzes the α-1,6-linked isomaltose and related oligosaccharides. A nonpolar N-terminal segment of the isomaltase anchors the pair of enzymes to the microvillus membrane. The two-protein complex arises naturally because the two enzymes are synthesized as a single polypeptide, which is cleaved by intestinal proteases.[73,74]

Because of the relatively high group transfer potential of either the glucosyl or fructosyl parts, sucrose is a substrate for glucosyltransferases such as sucrose phosphorylase (Eq. 20-14, step *b*; see also Eq. 12-7 and associated discussion). In certain bacteria this reaction makes available the activated glucose 1-*P* which may enter catabolic pathways directly. Cleavage of sucrose for biosynthetic purposes can occur by reaction 20-14, step *c*, which yields UDP-glucose in a single step.

A disaccharide with many of the same properties as sucrose is **trehalose,** which consists of two α-glucopyranose units in 1,1 linkage (p. 168). The biosynthetic pathway from UDP-glucose and glucose 6-*P* parallels that for synthesis of sucrose (Eq. 20-13).[75,76] In *E. coli* the genes for the needed glucosyltransferase and phosphatase are part of a single operon. Its transcription is controlled in part by glucose-mediated catabolite repression (Chapter 28) and also by a repressor of the Lac family.[76,76a,77] The repressor is allosterically activated by trehalose 6-*P*, the intermediate in the synthesis. Trehalose formation in bacteria, fungi, plants, and microscopic animals is strongly induced during conditions of high osmolality (see Box 20-C).[77] Both trehalose and maltose can also be taken up via an ABC type transporter (p. 417).[77a,b]

Lactose, the characteristic sugar of milk, is formed by transfer of a galactosyl unit from UDP-galactose directly to glucose (Eq. 20-15, reaction *a*). The similar transfer of a galactosyl unit to *N*-acetylglucosamine to form *N*-acetyllactosamine (Eq. 20-15, reaction *b*) occurs in many animal tissues. An interesting regulatory mechanism is involved. The transferase catalyzing Eq. 20-15, reaction *b*, forms a complex with **α-lactalbumin** to become **lactose synthase**,[78–80b] the enzyme that catalyzes reaction *a*. Lactalbumin was identified as a milk constituent long before its role as a regulatory protein was recognized.

A very common biochemical problem is intolerance to lactose.[81] This results from the inability of the intestinal mucosa to make enough **lactase** to hydrolyze the sugar to its monosaccharide components galactose and glucose. Among most of the peoples of the earth only infants have a high lactase level, and the use of milk as a food for adults often leads to a severe diarrhea. The same is true for most animals. In fact, baby seals and walruses, which drink lactose-free milk, become very ill if fed cow's milk.

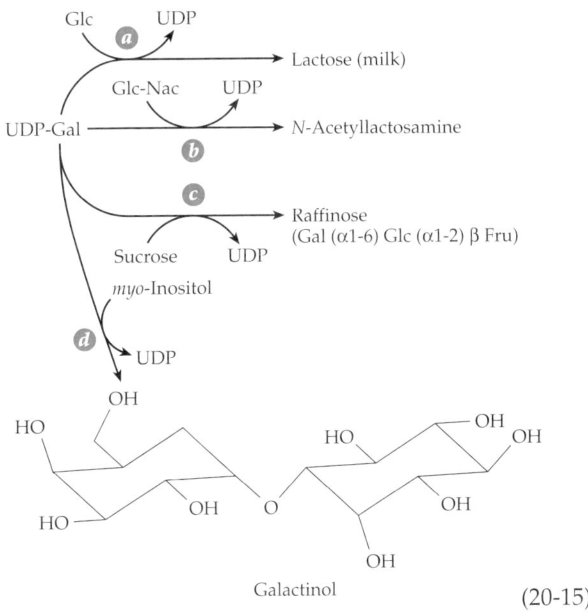

(20-15)

The plant trisaccharide **raffinose** arises from UDP-galactose by transfer of a galactosyl unit onto the 6-hydroxyl of the glucose ring of sucrose (Eq. 20-15, reaction *c*). Transfer of a galactosyl unit onto *myo*-inositol (Eq. 20-15, reaction *d*) produces **galactinol**, whose occurrence is widespread within the plant kingdom. Galactinol, in turn, can serve as a donor of activated galactosyl groups. Thus, many plants contain **stachyose** and higher homologs, all of which are formed by transfer of additional α-D-galactosyl units onto the 6-hydroxyl of the galactose unit of raffinose. These sugars appear to serve as antifreeze agents in the plants. The concentration of stachyose in soy beans can be as high as that of sucrose. Some seeds, e.g., those of maize, are coated with a glassy sugar mixture of sucrose and raffinose in a ratio of ~3:1.[81a]

Besides the oligosaccharides, living organisms form a great variety of glycosides that contain nonsugar components. Among these are the **glucuronides** (glucosiduronides), excretion products found in urine and derived by displacement of UDP from UDP-glucuronic acid by such compounds as phenol, benzoic acid, and sterols.[81b,c] Phenol is converted to phenyl glucuronide (Eq. 20-16), while benzoic acid (also excreted in part as hippuric acid, Box 10-A) yields an ester by the same type of displacement reaction. Many other aromatic or aliphatic compounds containing $-OH$, $-SH$, $-NH_2$, or $-COO^-$ groups also form glucuronides.[82]

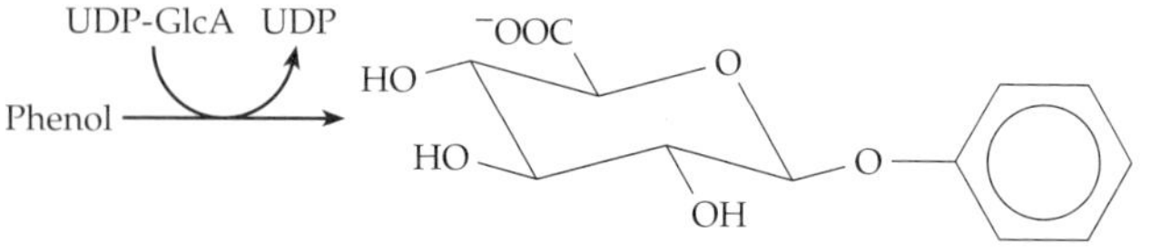

(20-16)

Among these is bilirubin (Fig. 24-24). UDP-glucuronosyltransferases responsible for their synthesis are present in liver microsomes.

Among the many glycosides and glycosylamines made by plants are the anthocyanin and flavonoid pigments of flowers (Box 21-E), cyanogenic glycosides such as amygdalin (Box 25-B), and antibiotics (e.g., see Box 20-B).[83,84] Some are characteristic of certain families of plants. For example, more than 100 β-thioglucosides known as **glucosinolides** are found in the Cruciferae (cabbages, mustard, rapeseed). The compounds impart the distinctive flavors and aromas of the plants. However, some are toxic and may cause goiter or liver damage. The enzyme **myrosinase** hydrolyzes these compounds releasing isothiocyanates, thiocyanates, and nitriles (Eq. 20-17).[85-86a] L-Ascorbate acts as a cofactor for this enzyme, evidently providing a catalytic base.[86a]

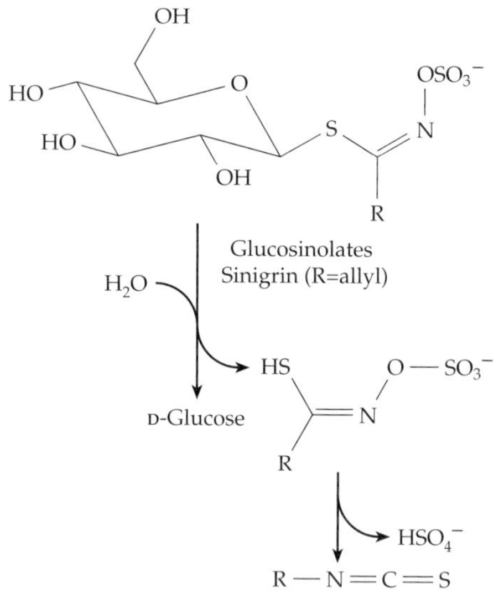

(20-17)

BOX 20-C OSMOTIC ADAPTATION

Bacteria, plants of many kinds, and a variety of other organisms are forced to adapt to conditions of variable osmotic pressure.[a,b] For example, plants must resist drought, and some must adapt to increased salinity. Some organisms live in saturated brine ~6 M in NaCl.[c] The **osmotic pressure** Π in dilute aqueous solutions is proportional to the total molar concentration of solute particles, c_s, as follows.

$$\Pi = RTc_s \quad \text{where } R \text{ is the gas constant and } T \text{ the Kelvin temperature}$$

At higher solute concentrations c_s must be replaced by the "effective molar concentration," which is called **osmolarity** (OsM) and has units of molarity (see Record *et al.* for discussion).[b] An osmolarity difference across a membrane of 0.04 OsM results in a turgor pressure of ~ 1 atmosphere. To adapt to changes in the environmental osmolarity organisms must alter their internal solute concentrations.

Cells of *E. coli* can adapt to at least 100-fold changes in osmolarity. Because of the porosity of the bacterial outer membrane the osmolarity of the periplasmic space is normally the same as that of the external medium. However, the inner membrane is freely permeable only to water and a few solutes such as glycerol.[b] The bacterial cells avoid loss of water when the external osmolarity is high by accumulating K^+ together with anions such as glutamate⁻ and nonionic **osmoprotectants** such as trehalose, sucrose, and oligosaccharides. *E. coli* cells will also take up other osmoprotectants such as **glycine betaine**, dimethylglycine, choline, proline, and proline betaine. Some methanogens accumulate N^{ε}-acetyl-β-lysine as well as glycine betaine.[d]

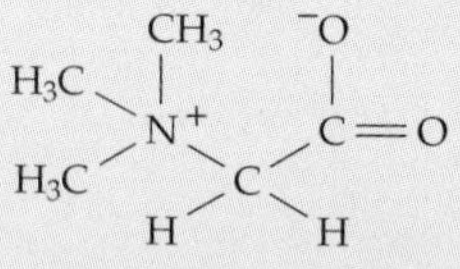

Betaine (glycine betaine) 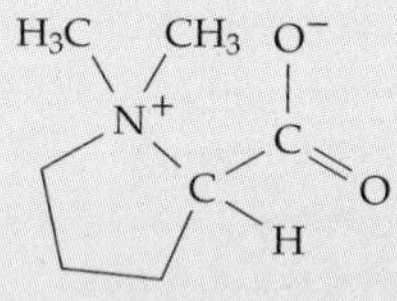 Proline betaine 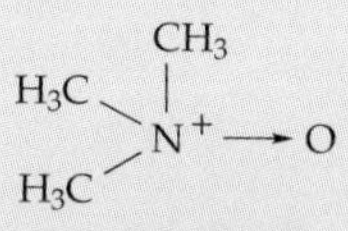 Trimethylamine N-oxide

Functioning in a somewhat different way in *E. coli* are 6- to 12-residue **periplasmic membrane-derived oligosaccharides**. These are β-1,2- and β-1,6-linked glucans covalently linked to *sn*-1-phosphoglycerol, phosphoethanolamine, or succinate (see Fig. 8-28).[b,e,f] They accumulate in the periplasm when cells are placed in a medium of low osmolarity. The resulting increased turgor in the periplasm is thought to buffer the cytoplasm against the loss of external osmolarity and to protect the periplasmic space from being eliminated by expansion of the plasma membrane. Related cyclic glucans, which are attached to *sn*-1-phosphoglycerol or *O*-succinyl ester residues, are accumulated by rhizobia.[g]

C. Synthesis and Degradation of Polysaccharides

Polysaccharides are all formed by transfer of glycosyl groups onto initiating molecules or onto growing polymer chains. The initiating molecule is usually a glycoprotein. However, let us direct our attention first to the growth of polysaccharide chains. The glycosyl are transferred by the action of glycosyltransferases from substrates such as UDP-glucose, other sugar nucleotides, and sometimes sucrose. The glycosyltransferases act by mechanisms discussed in Chapter 12 and are usually specific with respect both to substrate structure and to the type of linkage formed.

1. Glycogen and Starch

The bushlike glycogen molecules grow at their numerous nonreducing ends by the transfer of glucosyl units from UDP-glucose (Eq. 17-56)[87,87a] or in bacteria from ADP-glucose.[88–90] Utilization of glycogen by the cell involves removal of glucose units as glucose 1-*P* by the action of glycogen phosphorylase. The combination of growth and degradation from the same chain ends provides a means of rapidly storing and utilizing glucose units. The synthesis and breakdown of glycogen in mammalian muscle (Fig. 11-4) involves one of the first studied[91] and best known metabolic control systems. Various aspects have been discussed in Chapters 11, 12, and 17. The mechanism[92] and regulatory features[93–96b] have been described. An important recent development is the observation of glycogen concentrations in human muscles *in vivo* with ^{13}C NMR. This can be coupled with observation of glucose 6-*P* by ^{31}P NMR. The concentration of the latter is ~ 1 mM but increases after intense exercise.[94]

Glycogen phosphorylase and glycogen synthase alone are insufficient to synthesize and degrade glycogen. Synthesis also requires the action of the **branching enzyme** amylo-(1,4 → 1,6-transglycosylase,[97]

BOX 20-C (continued)

Fungi, green algae, and higher plants more often accumulate glycerol,[h,i] sorbitol, sucrose,[j] trehalose,[k] or proline.[a,l,m] These compounds are all "compatible solutes" which tend not to disrupt cellular structure.[n] Betaines and proline are especially widely used by a variety or organisms. How is it then that some desert rodents, some fishes, and other creatures accumulate **urea**, a well-known protein denaturant? The answer is that they also accumulate methylamine or trimethylamine *N*-oxide in an approximately 2:1 ratio of urea to amine. The mixture of compounds is compatible, the stabilizing effects of the amines offsetting the destabilizing effect of urea.[c,o]

Adaptation to changes in osmotic pressure involves sensing and signaling pathways that have been partially elucidated for *E. coli*[p] and yeasts.[i,q] Major changes in structure and metabolism may result. For example, in *E. coli* the outer membrane porin OmpF (Fig. 8-20) is replaced by OmpC (osmoporin), which has a smaller pore.[r]

A "resurrection plant" that normally contains an unusual 2-octulose converts this sugar almost entirely into sucrose when desiccated. This is one of a small group of plants that are able to withstand severe desiccation but can, within a few hours, reverse the changes when rehydrated.[j]

---

[a] Le Rudulier, D., Strom, A. R., Dandekar, A. M., Smith, L. T., and Valentine, R. C. (1984) *Science* **224**, 1064–1068

[b] Record, M. T., Jr., Courtenay, E. S., Cayley, D. S., and Guttman, H. J. (1998) *Trends Biochem. Sci.* **23**, 143–148

[c] Yancey, P. H., Clark, M. E., Hand, S. C., Bowlus, R. D., and Somero, G. N. (1982) *Science* **217**, 1214–1222

[d] Sowers, K. R., Robertson, D. E., Noll, D., Gunsalus, R. P., and Roberts, M. F. (1990) *Proc. Natl. Acad. Sci. U.S.A.* **87**, 9083–9087

[e] Kennedy, E. P. (1987) in *Escherichia coli and Salmonella typhimurium* (Neidhardt, F. C., ed), pp. 672–679, Am. Soc. for Microbiology, Washington,DC

[f] Fiedler, W., and Rotering, H. (1988) *J. Biol. Chem.* **263**, 14684–14689

[g] Weissborn, A. C., Rumley, M. K., and Kennedy, E. P. (1991) *J. Biol. Chem.* **266**, 8062–8067

[h] Ben-Amotz, A., and Avron, M. (1981) *Trends Biochem. Sci.* **6**, 297–299

[i] Davenport, K. R., Sohaskey, M., Kamada, Y., Levin, D. E., and Gustin, M. C. (1995) *J. Biol. Chem.* **270**, 30157–30161

[j] Bernacchia, G., Schwall, G., Lottspeich, F., Salamini, F., and Bartels, D. (1995) *EMBO J.* **14**, 610–618

[k] Dijkema, C., Kester, H. C. M., and Visser, J. (1985) *Proc. Natl. Acad. Sci. U.S.A.* **82**, 14–18

[l] García-Ríos, M., Fujita, T., LaRosa, P. C., Locy, R. D., Clithero, J. M., Bressan, R. A., and Csonka, L. N. (1997) *Proc. Natl. Acad. Sci. U.S.A.* **94**, 8249–8254

[m] Verbruggen, N., Hua, X.-J., May, M., and Van Montagu, M. (1996) *Proc. Natl. Acad. Sci. U.S.A.* **93**, 8787–8791

[n] Higgins, C. F., Cairney, J., Stirling, D. A., Sutherland, L., and Booth, I. R. (1987) *Trends Biochem. Sci.* **12**, 339–344

[o] Lin, T.-Y., and Timasheff, S. N. (1994) *Biochemistry* **33**, 12695–12701

[p] Racher, K. I., Voegele, R. T., Marchall, E. V., Culham, D. E., Wood, J. M., Jung, H., Bacon, M., Cairns, M. T., Ferguson, S. M., Liang, W.-J., Henderson, P. J. F., White, G., and Hallett, F. R. (1999) *Biochemistry* **38**, 1676–1684

[q] Shiozaki, K., and Russell, P. (1995) *EMBO J.* **14**, 492–502

[r] Kenney, L. J., Bauer, M. D., and Silhavy, T. J. (1995) *Proc. Natl. Acad. Sci. U.S.A.* **92**, 8866–8870

an enzyme with dual specificity. After the chain ends attain a length of about ten glucose units, the branching enzyme attacks a 1,4-glycosidic linkage somewhere in the chain. Acting much as does a hydrolase, it forms a glycosyl enzyme or a stabilized carbocation intermediate. The enzyme does not release the severed chain fragment but transfers it to another nearby site on the glycogen molecule. There the enzyme rejoins the bound oligosaccharide chain that it carries to a free 6-hydroxyl group of the glycogen creating a new branch attached in α-1,6-linkage. Degradation of glycogen requires **debranching** after the long nonreducing ends of the polysaccharide have been shortened until only four glycosyl residues remain at each branch point. This is accomplished by **amylo-1,6-glucosidase / 4- α-glucanotransferase**. This 165-kDa bifunctional enzyme transfers a trisaccharide unit from each branch end to the main chain and also removes hydrolytically the last glucosyl residue at each branch point.[98-99a]

How are new glycogen molecules made? There is some evidence that a 37-kDa protein primer **glycogenin** is needed to initiate their formation.[100-101a] Thus, glycogen synthesis may be analogous to that of the glycosaminoglycans considered in Section D,1. Muscle glycogenin is a self-glycosylating protein, which catalyzes attachment of ~7 to 11 glucose units in α–1,4 linkage to the hydroxyl group of Tyr 194. The glucose units are added one at a time and when the chain is long enough it becomes a substrate for glycogen synthase.[100,102] The role of glycogenin in liver has been harder to demonstrate,[103] but a second glycogenin gene, which is expressed in liver, has been identified.[104] Genes for several glycogenins or glycogenin-like proteins have been identified in yeast, *Caenorhabditis elegans*, and *Arabidopsis*.[101a,105]

In contrast to animals, bacteria such as *E. coli* synthesize glycogen via ADP-glucose rather than UDP-glucose.[88] ADP-glucose is also the glucosyl donor for synthesis of starch in plants. The first step in the biosynthesis (Eq. 20-18) is catalyzed by the enzyme ADP-glucose pyrophosphorylase (named for the reverse reaction).

$$\text{Glucose 1-}P + \text{ATP} \rightarrow \text{ADP-glucose} + \text{PP}_i \qquad (20\text{-}18)$$

In bacteria this enzyme is usually inhibited by AMP and ADP and activated by glycolytic intermediates such as fructose 1,6-P_2, fructose 6-P, or pyruvate. In higher plants, green algae, and cyanobacteria the enzyme is usually activated by 3-phosphoglycerate, a product of photosynthetic CO_2 fixation, and is inhibited by inorganic phosphate (P_i).[106–108].

In eukaryotic plants starch is deposited within chloroplasts or in the cytoplasm as granules (Fig. 4-6) in a specifically differentiated and physically fragile plastid, the **amyloplast**.[108–110] Within the granules the starch is deposited in layers ~ 9 nm in thickness. About two-thirds of the thickness consists of nearly crystalline arrays, probably of double helical amylopectin side chains (Figs. 4-7, 4-8, 20-3) with "amorphous" segments between the layers.[111–114] In maize there are at least five starch synthases, one of which forms the straight chain amylose.[115–117] There are also at least three branching enzymes[118] and two or three debranching enzymes.[119,120] As in the synthesis of glycogen the molecules of amylopectin may grow at the many nonreducing ends. A current model, which is related to the broom-like cluster model of French (Fig. 4-7), is shown in Fig. 20-3. The branches are thought to arise, in part, by transglycosylation within the double helical strands. After branching the two chains remain in a double helix but the cut chain can now grow. Only double helical parts of strands pack well in the crystalline layer. A recent suggestion is that debranching enzymes then trim the molecule, removing single-stranded regions.[112]

The location (within the granule) of amylose, which makes up 15–30% by weight of many starches,[121] is uncertain. It may fill in the amorphous layers. It may be cut and provide primer pieces for new amylopectin molecules.[122,122a] Another possibility is that it grows by an insertion mechanism such as that portrayed for cellulose in Fig. 20-5 and is extruded inward from the membrane of the amyloplast. This mechanism might explain a puzzling question about starch. The branched amylopectin presumably grows in much the same way as does glycogen. A branching enzyme transfers part of the growing glycan chain to the $-CH_2-OH$ group of

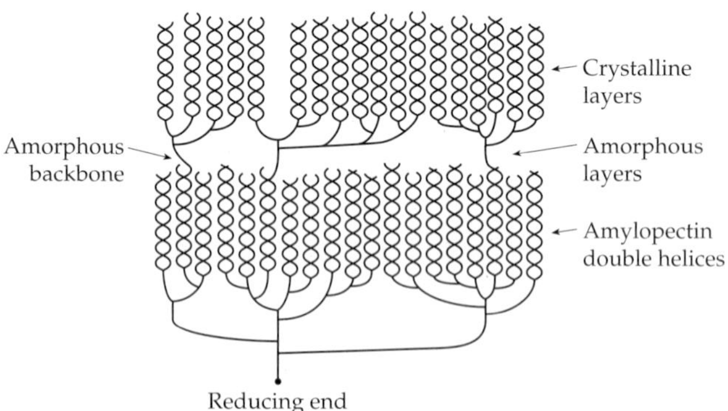

Figure 20-3 Proposed structure of a molecule of amylopectin in a starch granule. The highly branched molecule lies within 9 nm thick layers, about 2 / 3 of which contains parallel double helices of the kind shown in Fig. 4-8 in a semicrystalline array. The branches are concentrated in the amorphous region.[113,114,121] Some starch granules contain no amylose, but it may constitute up to 30% by weight of the starch. It may be found in part in the amorphous bands and in part intertwined with the amylopectin.[122]

BOX 20-D GENETIC DISEASES OF GLYCOGEN METABOLISM

In 1951, B. McArdle described a patient who developed pain and stiffness in muscles after moderate exericse.[a] Surprisingly, this person completely lacked muscle glycogen phosphorylase. Since that time several hundred others have been found with the same defect. Glycogen accumulates in muscle tissue in this disease, one of the several types of **glycogen storage disease**.[b] Severe exercise is damaging, but steady moderate exercise can be tolerated. Until the time of McArdle's discovery, it was assumed that glycogen was synthesized by reversal of the phosphorylase reaction. No hint of the UDP-glucose pathway had appeared, and it was, therefore, not obvious how glycogen could accumulate in the muscles of these patients. Leloir's discovery of UDP-glucose at about the same time provided the answer. Persons with McArdle syndrome are greatly benefitted by a high-protein diet, presumably because amino acids such as alanine and glutamine are converted efficiently to glucose and because branched-chain amino acids may serve as a direct source of muscle energy.[c,d]

Several other rare heritable diseases also lead to accumulation of glycogen because of some block in its breakdown through the glycolysis pathway. The enzyme deficiencies include those of muscle phosphofructokinase,[e] liver phosphorylase kinase, liver phosphorylase, and liver glucose-6-phosphatase. In the last case, glycogen accumulates because the liver stores cannot be released to the blood as free glucose.[b,f,g] This is a dangerous disease because blood glucose concentrations may fall too low at night. The prognosis improved greatly when methods were devised for providing the body with a continuous supply of glucose. The simplest treatment is ingestion of uncooked cornstarch which is digested slowly.[b,h] In one of the storage diseases the branching enzyme of glycogen synthesis is lacking, and glycogen is formed with unusually long outer branches. In another the debranching enzyme is lacking, and only the outer branches of glycogen can be removed readily.[i]

The most serious of the storage diseases involve none of the enzymes mentioned above. Pompe disease is a fatal generalized glycogen storage disease in which a lysosomal α-1,4-glucosidase is lacking. This observation suggested the existence of a new and essential pathway of degradation of glycogen to free glucose in the lysosomes. A few cases of glycogen synthase deficiency have been reported. Little or no glycogen is stored in muscle or liver, and patients must eat at regular intervals to prevent hypoglycemia. Severe diseases in which glycogen synthesis is impaired include deficiencies of the gluconeogenic enzymes pyruvate carboxylase and PEP carboxykinase.

The following tabulation includes deficiencies of glycogen metabolism, glycolysis, and gluconeogenesis.[a] Glycogen storage diseases are often designated as Types I–V and these terms are included.

Deficiency	Organ	Severity
Glycogen phosphorylase (Type V), McArdle disease	Muscle	Moderate, late onset
Glycogen phosphorylase	Liver	Very mild
Phosphorylase kinase	Liver	Very mild
Debranching enzyme (Type III)	Liver	Mild
Lysosomal α-glucosidase (Type II)		Lethal, infant and adult form
Phosphofructokinase	Muscle	Moderate, late onset
Phosphoglycerate mutase[j]	Muscle	Moderate
Pyruvate carboxylase		Lethal
PEP carboxykinase		Lethal
Fructose-1,6-bisphosphatase	Muscle	Severe
Glycogen synthase	Liver	Mild
Branching enzyme (Type IV)		Lethal, liver transplantation
Glucose-6-phosphatase (Type I)		Severe if untreated

[a] Huijing, F. (1979) *Trends Biochem. Sci.* **4**, 192

[b] Chen, Y.-T., and Burchell, A. (1995) in *The Metabolic and Molecular Bases of Inherited Disease*, 7th ed., Vol. 1 (Scriver, C. R., Beaudet, A. L., Sly, W. S., and Valle, D., eds), pp. 935–965, McGraw-Hill, New York

[c] Slonim, A. E., and Goans, P. J. (1985) *N. Engl. J. Med.* **312**, 355–359

[d] Goldberg, A. L., and Chang, T. W. (1987) *Fed. Proc.* **37**, 2301–2307

[e] Raben, N., Sherman, J., Miller, F., Mena, H., and Plotz, P. (1993) *J. Biol. Chem.* **268**, 4963–4967

[f] Nordlie, R. C., and Sukalski, K. A. (1986) *Trends Biochem. Sci.* **11**, 85–88

[g] Lei, K.-J., Shelly, L. L., Pan, C.-J., Sidbury, J. B., and Chou, J. Y. (1993) *Science* **262**, 580–583

[h] Chen, Y.-T., Cornblath, M., and Sidbury, J. B. (1984) *N. Engl. J. Med.* **310**, 171–175

[i] Thon, V. J., Khalil, M., and Cannon, J. F. (1993) *J. Biol. Chem.* **268**, 7509–7513

[j] Shanske, S., Sakoda, S., Hermodson, M. A., DiMauro, S., and Schon, E. A. (1987) *J. Biol. Chem.* **262**, 14612–14617

a glucose unit in an adjacent polysaccharide chain that lies parallel to the first, possibly in a double helix. Since amylose and amylopectin are intimately intermixed in the starch granules, it seems strange that the branching enzyme never transfers a branch to molecules of the straight-chain amylose. However, if the linear amylose chains are oriented in the opposite direction from the amylopectin chains, the nonreducing ends of the amylose molecules would be located toward the center of the starch granule. Growth could occur by an insertion mechanism at the reducing ends and the ends could move out continually with the amyloplast membrane as the granule grows.[123] Recent evidence from ^{14}C labeling indicates that both amylose and amylopectin too may grow by insertion at the reducing end of glucose units from ADP-glucose.[123a,b] Branching could occur to give the structure of Fig. 20-3. Starch synthesis in leaves occurs by day but at night the starch is degraded by amylases, α-glucosidases, and starch phosphorylase. Both the starch synthases and catabolic enzymes are present within the amyloplasts where they may be associated with regulatory proteins of the 14-3-3 class.[122a]

Digestion of dietary glycogen and starch in the human body begins with the salivary and pancreatic amylases, which cleave α-1,4 linkages at random. It continues with a **glucoamylase** found in the brush border membranes of the small intestine where it occurs as a complex with **maltase**.[74] Carbohydrases are discussed in Chapter 12, Section B.

2. Cellulose, Chitin, and Related Glycans

Cellulose synthases transfer glucosyl units from UDP-glucose, while chitin synthases utilize UDP-*N*-acetylglucosamine. Not only green plants but some fungi and a few bacteria form cellulose. The ameba *Dictyostelium discoideum* also coats its spores with cellulose.[124] Electron microscopic investigations suggest that both in bacteria[125] and in plants[126] multienzyme aggregates located at the plasma membrane synthesize many polymer chains side by side to generate hydrogen-bonded microfibrils which are extruded through the membrane. Both green plants and fungi also form important β-1,3-linked glycans.

The bacterial cellulose synthase from *Acetobacter xylinum* can be solubilized with detergents, and the resulting enzyme generates characteristic 1.7 nm cellulose fibrils (Fig. 20-4) from UDP-glucose.[125,127–129] These are similar, but not identical, to the fibrils of cellulose I produced by intact bacteria.[125,130] Each native fibril appears as a left-handed helix which may contain about nine parallel chains in a crystalline array. Three of these helices appear to coil together (Fig. 20-4) to form a larger 3.7-nm left-handed helical fibril. Similar fibrils are formed by plants. In both

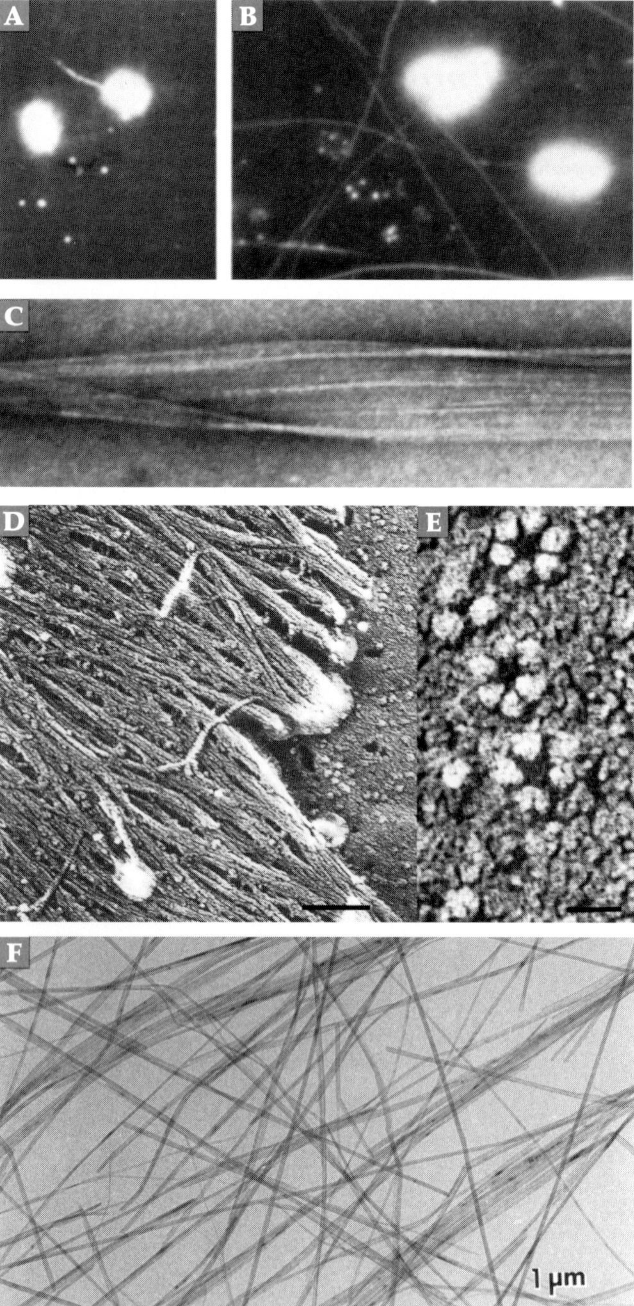

Figure 20-4 Cellulose microfibrils being formed by *Acetobacter xylinum*.[127] (A) Dark-field light micrograph after five minutes of cellulose production (x 1250). (B) After 15 minutes a pellicle of cellulose fibers is forming (x 2000) (C) Negatively stained cellulose ribbon. At the right the subdivision into microfibrils is visible. Courtesy of R. Malcolm Brown, Jr. (D) Cellulose microfibrils overlaying the plasma membrane in the secondary cell wall of a tracheary element of *Zinnia elegans*. Bar = 100 nm. (E) Rosettes in the plasma membrane underlying the cellulose-rich secondary cell wall thickening in *A. elegans*. Bar = 30 nm. (D) and (E) from Haigler and Blanton.[132] Courtesy of Candace H. Haigler. (F) Chitin microfibrils purified from protective tubes of the tube-worm *Lamellibrachia satsuma*.[137] Courtesy of Junji Sugiyama.

bacteria and plants the cellulose I fibrils that are formed are highly crystalline, contain parallel polysaccharide chains (Fig. 4-5), and have the tensile strength of steel. Electron micrographs show that the cell envelope of *A. xylinum* contains 5–80 pores, through which the cellulose is extruded, lying along the long axis of the cell.[129] The biosynthetic enzymes are probably bound to the plasma membrane. Similar, but more labile, cellulose synthases are present in green plants.[131] In *Arabidopsis* there are ten genes. The encoded cellulose synthases appear to be organized as rosettes on some cell surfaces (Fig. 20-4E).[131a-133a] The rosettes may be assembled to provide parallel synthesis of ~36 individual cellulose chains needed to form a fibril.[131a]

Because of the insolubility of cellulose fibrils it has been difficult to determine whether they grow from the reducing ends or the nonreducing ends of the chains. From silver staining of reducing ends and micro electron diffraction of cellulose fibrils attached to bacteria,[134] Koyama *et al.* concluded that the reducing ends are extruded from cells. New glucosyl rings would be added at the *nonreducing ends*, which remain attached noncovalently to the cells.[134] From amino acid sequence similarities it was also concluded that the same is true for *Arabidopsis*.[131,133] A single cellulose chain has a twofold screw axis, each residue being rotated 180° from the preceding residue (Fig. 4-5). It was postulated that two synthases act cooperatively to add cellobiose units. Another suggestion is that sitosterol β-glucoside acts in some fashion as a primer for cellulose synthesis in plants.[133b]

An insertion mechanism for synthesis of cellulose. Using ^{14}C "pulse and chase" labeling Han and Robyt found that new glucosyl units are added at the *reducing ends* of cellulose chains formed by cell membrane preparations from *A. xylinum*.[135] This conclusion is in accord with the generalization that extracellular polysaccharides made by bacteria usually grow from the reducing end by an insertion mechanism that depends upon a polyprenyl alcohol present in the cell membrane.[136] This lipid alcohol, often the C_{55}

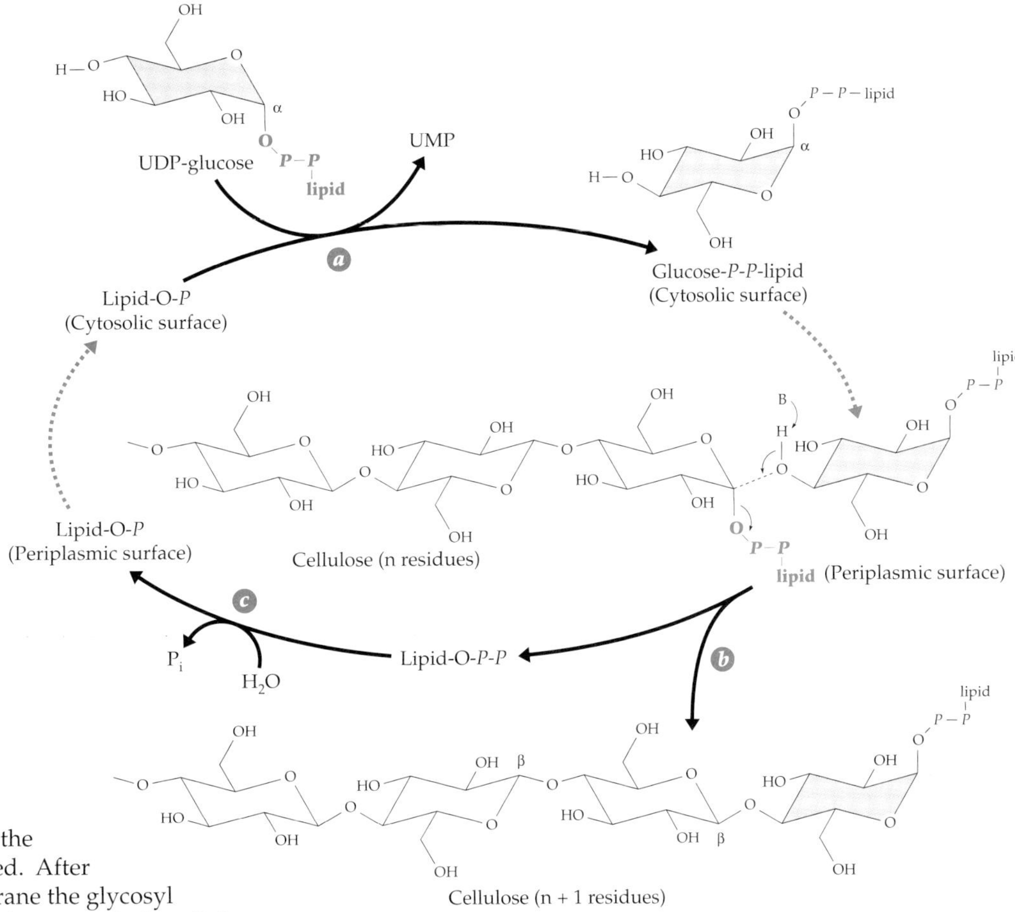

Figure 20-5 Proposed insertion mechanism for biosynthesis of cellulose. Three enzymatic steps are involved: a nucleophilic displacement reaction of a lipid phosphate on UDP-glucose yields a glucosyl diphosphate lipid in which the α-glycosyl linkage is retained. After passage through the membrane the glycosyl group is inserted into the reducing end of a cellulose chain, which is covalently attached by a pyrophosphate linkage to another lipid. The first lipid diphosphate is released and is hydrolyzed (step *c*) to the monophosphate, which crosses the membrane to complete the cycle. After Han and Robyt.[135] As throughout this book *P* represents the phospho group –PO_3H. The H may be replaced by groups which may contain oxygen atoms. This explains why an O is included in Lipid-O-*P* but no O is shown between the *P*'s in -O-*P*-*P*.

bactoprenol, reacts with UDP-glucose (or other glycosyl donor) to give a lipopyrophospho-glucose (step *a*, Fig. 20-5). The α linkage of the UDP-glucose is retained in this compound. The growing cellulose chain is attached at the reducing end by a similar linkage to a second lipid molecule. Then, in a displacement on the anomeric carbon of the first glucosyl residue of the cellulose chain, the new glucosyl unit is inserted with inversion of the α linkage to β. In step *c* the pyrophosphate linkage of the lipid diphosphate is hydrolyzed to regenerate the lipid monophosphate and to drive the reaction toward completion. Two of the steps in the cycle involve transport across the bacterial membrane. The first involves the lipid –O-*P*-*P*-glucose and the second the lipid monophosphate. This type of insertion mechanism is a common feature of polyprenol phosphate-dependent synthetic cycles for extracellular polysaccharides (Figs. 20-6, 20-9 and Eq. 20-20). However, further verification is needed for cellulose synthesis.

Regulation of cellulose synthesis in bacteria depends on allosteric activator of cellulose synthase, **cyclic diguanylate** (c-di-GMP), and a Ca^{2+}-activated phosphodiesterase that degrades the activator.[129,138–139a] Sucrose is the major transport form of glucose in plants. Synthesis of both cellulose and starch is reduced in mutant forms of maize deficient in sucrose synthase (Eq. 20-13). This synthase, acting in the reverse direction, forms UDP-Glc from sucrose.[140,141] The enzymatic degradation of cellulose is an important biological reaction, which is limited to certain bacteria, to fungi, and to organisms such as termites that obtain cellulases from symbiotic bacteria or by ingesting fungi.[142] These enzymes are discussed in Chapter 12, Section B,6. Genetic engineering methods now offer the prospect of designing efficient cellulose-digesting yeasts[143] that may be used to produce useful fermentation products from cellulose wastes.

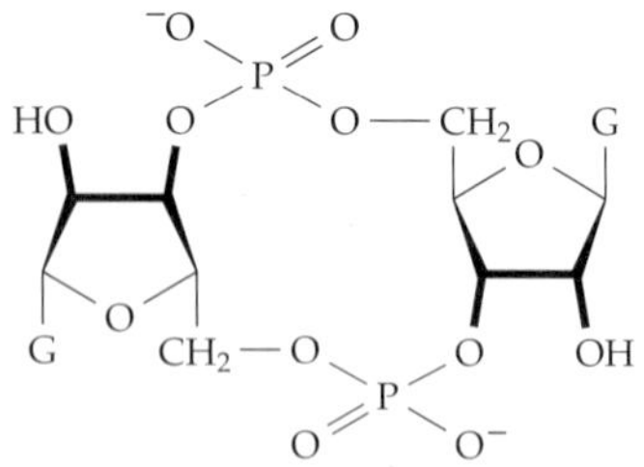

Cyclic diguanylate

Callose and other β-1,3-linked glycans. Attempts to produce cellulose from UDP-Glc using enzymes of isolated plasma membranes from higher plants have usually yielded the β-1,3-linked glucan (callose) instead. This is a characteristic polysaccharide of plant wounds which, as healing occurs, is degraded and replaced by cellulose.[140,144] Callose formation is induced by a specific activator β-furfuryl-β-glucoside, and callose synthase is virtually inactive unless both the activator and Ca^{2+} are present.[144]

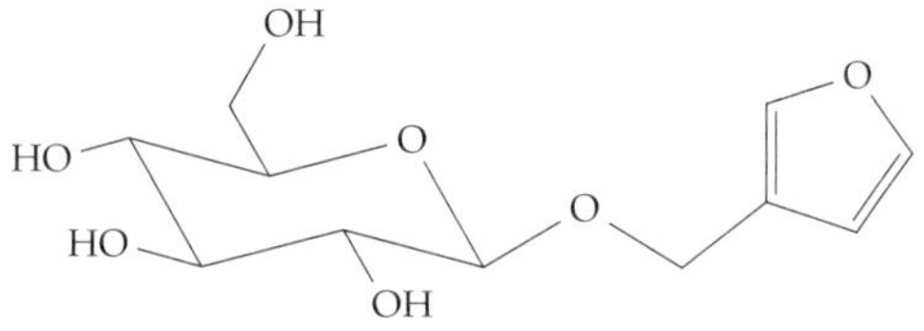

β-Furfuryl-β-glucoside

Beta-1,3-linked glycans are major components of the complex layered cell wall of yeasts and other fungi. In the fission yeast *Saccharomyces pombe* ~55% of the cell wall carbohydrate consists of β-1,3-linked glucan with some β-1,4-linked branches, ~ 28% is α-1,3-linked glucan, ~6% is α-1,6-linked glucan, and ~0.5% is chitin. There are two carbohydrate layers, the outer one appearing amorphous. The inner layer contains interwoven fibrils of both α-1,3-linked and α-1,4-linked glucans and holds the shape of the cell. The β-1,3 glucan synthase is localized on the inner side of the cell membrane and is activated by GTP and a small subunit of the Rho family of G proteins.[145]

Plants synthesize 1,3-β-glucanases that hydrolyze the glycans of fungal cell walls. Synthesis is induced by wounding as a defense reaction (see Box 20-E). These glycanases also function in the removal of callose.[146]

Chitin. Like cellulose synthase, fungal chitin synthases are present in the plasma membrane and extrude microfibrils of chitin to the outside.[147–150] In the fungus *Mucor* the majority of the chitin synthesized later has its *N*-acetyl groups removed hydrolytically to form the deacetylated polymer **chitosan**.[151,152] Chitin is also a major component of insect exoskeletons. For this reason, chitin synthase is an appropriate target enzyme for design of synthetic insecticides.[153]

Chitin hydrolyzing enzymes are formed by fungi and in marine bacteria.[154] Chitinases are also present in plant vacuoles, where they participate in defense against fungi and other pathogens[155] (Box 20-E). More recently a chitinase has been identified in human activated macrophages.[156] Another unanticipated discovery was that a developmental gene designated *DG42*, from *Xenopus*, has a sequence similar to that of the *NodC* gene. The latter encodes a synthase for chitin oligosaccharides (Nod factors) that serve as nodulation factors in *Rhizobia* (Chapter 24). The enzyme is synthesized for only a short time during early embryonic development.[157] The significance of this discovery is not yet clear. Synthesis of both the bacterial Nod factors and chitin oligosaccharides in zebrafish embryos occurs by transfer of GlcNAc residues from UDP-GlcNAc at the *nonreducing ends* of the chains.[158] Whether the same is true of chitin in fungi or arthropods remains uncertain.

Cell walls of plants. The thick walls of higher plant cells (Figs. 1-7, 4-14, and 20-4D) provide strength and rigidity to plants and, at the same time, allow rapid elongation during periods of growth.[159–163a] Northcote[164] likened the wall structure to glass fiber-reinforced plastic (fiber glass). Thus, the cell wall contains microfibrils of cellulose and other polysaccharides embedded in a matrix, also largely polysaccharide. The **primary cell wall** laid down in green plants during early stages of growth contains loosely interwoven cellulose fibrils ~10 nm in diameter and with an ~4 nm crystalline center. The cellulose in these fibrils has a degree of polymerization of 8000–12,000 glucose units. As the plant cell matures, a secondary cell wall is laid down on the inside of the primary wall. This contains many layers of closely packed microfibrils, alternate layers often being laid down at different angles to one another (Fig. 20-4D). The microfibrils in green plants are most often cellulose but may contain other polysaccharides as well. Some algae are rich in fibrils of xylan and mannan.

The materials present in the matrix phase vary with the growth period of the plant. During initial phases **pectin** (polygalacturonic acid derivatives) predominate but later xylans and a variety of other polysaccharides known as **crosslinking glycans** (or hemicelluloses) appear. Primary cell wall constituents of dicotyledons include **xyloglucans** (linear glucan chains with xylose, galactose, and fucose units in

BOX 20-E OLIGOSACCHARIDES IN DEFENSIVE AND OTHER RESPONSES OF PLANTS

Plants that are attacked by bacteria, fungi, or arthropods respond by synthesizing broad-spectrum antibiotics called **phytoalexins**,[a,b] by strengthening their cell walls with lignin and hydroxyproline-rich proteins called **extensins**,[c] and by making **protease inhibitors** and other proteins that help to block the chemical attack.[d] These plant responses seem to be initiated by the release from an invading organism of **elicitors**, which are often small oligosaccharide fragments, sometimes called **oligosaccharins**.[e] These include β-1,6-linked glucans that carry β-1,3-linked branches as well as chitin and chitosan oligomers, derived from fungal cell walls.[f] Other elicitors include galacturonic acid oligomers released from damaged plant cell walls,[g] metabolites such as arachidonic acid and glutathione,[h] and bacterial toxins.[i] Any of these may serve as signals to plants to take defensive measures.

Phytoalexins are often isoflavonoid derivatives (Box 21-E). Their synthesis, like that of lignin, occur via 4-coumarate (4-hydroxycinnamate, Fig. 25-8). The ligase which forms the thioester of 4-coumarate with coenzyme A is one of the **pathogenesis-related proteins** whose synthesis is induced.[j] A second induced enzyme is chalcone synthase, which condenses three acetyl units onto 4-coumaroyl-CoA as shown in Box 21-E. Its induction by elicitors acting on bean cells requires only five minutes.[h] Another rapidly induced gene is that of cinnamoyl alcohol dehydrogenase,[k] essential to lignin synthesis. Other proteins formed in response to infections include **chitinases** that are able to attack invading fungi[l,m] as well as the protease inhibitors. Their synthesis is induced via derivatives of **jasmonate**, a product of the octadecenoic acid pathway (Eq. 21-18).[a] As yet, little is known about the mechanism by which elicitors induce the defensive responses, but the presence of receptors, of phosphorylation, and of release of second messengers have been suggested.[d]

Lipooligosaccharides known as Nod factors (p. 1365) are another group of signaling molecules. These chitin-related *N*-acylated oligomers of *N*-acetylglucosamine (GlcNAc) do not defend against infection but invite infection of roots of legumes by appropriate species of *Rhizobia*[n–p] leading to formation of nitrogen-fixing root nodules.

---

a Blechert, S., Brodschelm, W., Hölder, S., Kammerer, L., Kutchan, T. M., Mueller, M. J., Xia, Z.-Q., and Zenk, M. H. (1995) *Proc. Natl. Acad. Sci. U.S.A.* **92**, 4099–4105

b Ebel, J., and Grisebach, H. (1988) *Trends Biochem. Sci.* **13**, 23–27

c Kieliszewski, M. J., O'Neill, M., Leykam, J., and Orlando, R. (1995) *J. Biol. Chem.* **270**, 2541–2549

d Ryan, C. A. (1988) *Biochemistry* **27**, 8879–8883

e Ryan, C. A. (1994) *Proc. Natl. Acad. Sci. U.S.A.* **91**, 1–2

f Baureithel, K., Felix, G., and Boller, T. (1994) *J. Biol. Chem.* **269**, 17931–17938

g Reymond, P., Grünberger, S., Paul, K., Müller, M., and Farmer, E. E. (1995) *Proc. Natl. Acad. Sci. U.S.A.* **92**, 4145–4149

h Dron, M., Clouse, S. D., Dixon, R. A., Lawton, M. A., and Lamb, C. J. (1988) *Proc. Natl. Acad. Sci. U.S.A.* **85**, 6738–6742

i Bidwai, A. P., and Takemoto, J. Y. (1987) *Proc. Natl. Acad. Sci. U.S.A.* **84**, 6755–6759

j Douglas, C., Hoffmann, H., Schulz, W., and Hahlbrock, K. (1987) *EMBO J.* **6**, 1189–1195

k Walter, M. H., Grima-Pettenati, J., Grand, C., Boudet, A. M., and Lamb, C. J. (1988) *Proc. Natl. Acad. Sci. U.S.A.* **85**, 5546–5550

l Legrand, M., Kauffmann, S., Geoffroy, P., and Fritig, B. (1987) *Proc. Natl. Acad. Sci. U.S.A.* **84**, 6750–6754

m Payne, G., Ahl, P., Moyer, M., Harper, A., Beck, J., Meins, F., Jr., and Ryals, J. (1990) *Proc. Natl. Acad. Sci. U.S.A.* **87**, 98–102

n Cedergren, R. A., Lee, J., Ross, K. L., and Hollingsworth, R. I. (1995) *Biochemistry* **34**, 4467–4477

o Jabbouri, S., Relic, B., Hanin, M., Kamalaprija, P., Burger, U., Promé, D., Promé, J. C., and Broughton, W. J. (1998) *J. Biol. Chem.* **273**, 12047–12055

p Dénarié, J., Debellé, F., and Promé, J.-C. (1996) *Ann. Rev. Biochem.* **65**, 503–535 branches),[164a] other crosslinking glycans, and galacturonic acid-rich pectic materials.[163a] The xyloglycans, which comprise 20% of the cell wall in some plants, have a backbone of α-1,4-linked glucose units with numerous α-1,6-linked xylose rings, some of which carry attached L-arabinose, galactose, or fucose. The structures, which vary from species to species, are organized as repeating blocks with a continuous glucan backbone. Another crosslinking glycan is **glucuronoarabinoxylan**. The backbone is β-1,4-linked xylose. Less abundant glucomannans, galactomannans, and galactoglucomannans, with β-1,4-linked mannan backbone structures, are also present in most angiosperms.[163a]

Pectins form a porous gel on the inside surface of plant cell walls.[163a,164a] A major component is a **homogalacturonan**, which consists of α-1,4-linked galacturonic acid (GalA). A second is rhamnogalacturonan I, an alternating polymer of (2-L-Rhaα1 → 4GalAα →) units. The most interesting pectin component is **rhamnogalacturonan II**, one of the less abundant constuents of pectin. It is obtained by hydrolytic cleavage of pectin by a polygalacturonidase. Before such release it forms parts (hairy regions) of pectin molecules that are largely homogalacturonans (in smooth regions). A rhamnogalacturonan II segment consists of 11 different monomer units.[164b–f] Attached to the polygalacturonic acid backbone are four oligosaccharides, consisting of rhamnose, galactose and fucose as well as some unusual sugars (see structure in Box 20-E). This polysaccharide is apparently present in all higher plants and is unusually stable, accumulating, for example, in red wine.[164e] It contains two residues of the branched chain sugar **apiose**, one of which is a site of crosslinking by boron (Box 20-F). A borate diol ester linkage binds two molecules of the pectin together as a dimer, perhaps controlling the porosity of the pectin gel. All of the complex cell wall polysaccharides bind, probably through multiple hydrogen bonds, to the cellulose microfibrils (Fig. 4-14). The resulting structures are illustrated in drawings of Carpita and McCann,[163a] which are more current than is Fig. 4-14. The cellulose plus crosslinking glycans form one network in the cell wall. The pectic substances form a second independent network. Some covalent crosslinking occurs, but most interactions are noncovalent.[163a] The site of biosynthesis of pectins and hemicelluloses is probably Golgi vesicles which pass to the outside via exocytosis. However, the cellulose fibrils as well as the chitin in fungi are apparently extruded from the plasma membrane.

Although the principal cell wall components of plants are carbohydrates, proteins account for 5–10% of the mass.[165] Predominant among these are glycoprotein **extensins**. Like collagen, they are rich in 4-hydroxyproline which is glycosylated with arabinose oligosaccharides and galactose (p. 181). Other hydroxyproline-containing proteins with the characteristic sequence (hydroxyproline)$_4$-Ser are also found, e.g., in soybean cell walls.[166] Some plant cell walls contain glycine-rich structural proteins. One in the petunia consists of 67% glycine residues.[167] During advanced stages of formation, as the walls harden into wood, large amounts of **lignins** are laid down in some plant cells. These chemically resistant phenylpropanoid polymers contain many crosslinked aromatic rings (Fig. 25-9).[163a]

A remarkable aspect of primary plant cell walls is their ability to be elongated extremely rapidly during growth. While the driving force for cell expansion is thought to be the development of pressure within the cell, the manner in which the wall expands is closely regulated. After a certain point in development, elongation occurs in one direction only and under the influence of plant hormones. Most striking is the effect of the **gibberellins** (Eq. 22-5), which cause very rapid elongation. Elongation of plant cell walls may depend to some extent upon chemical cleavage and reforming of crosslinking polysaccharides. However, the cellulose fibrils probably remain intact and slide past each other.[161] A curious effect, which is mediated by proteins called **expansins**,[133a,168] is the ability of plant tissues to extend rapidly when incubated in a mildly acidic buffer of pH <5.5. Expansins are also involved in ripening of fruit. They may disrupt noncovalent bonding between cellulose fibrils and the hemicelluloses.[169,170] The β-expansins of grasses are allergens found in grass pollens.[133a,168] The borate diol ester linkages in the pectin may also facilitate expansion.

3. Patterns in Polysaccharide Structures

How can the many complex polysaccharides found in nature be synthesized? Are there genetically determined patterns? How are these controlled? The answer can be found in the *specificities* of the hundreds of known *glycosyltransferases*[171,172] and in the *patterns of expression of the genes* for transferases and other proteins. As a consequence, a great variety of structurally varied polysaccharide structures arise, especially on cell surfaces. The structures are not random but depend upon the assortment of glycosyltransferases available at the particular stage of development in a tissue. The numerous possibilities can account for much of the variation observed between species, between tissues, and also among individuals.

The simplest pattern is the growth of straight-chain homopolysaccharides such as amylose, cellulose, and chitin. The glycosyltransferases must recognize both the glycosyl donor, e.g., ADP-glucose, UDP-glucose, and also the correct end of the growing polymer, always adding the same monomer unit.

BOX 20-F WHAT DOES BORON DO?

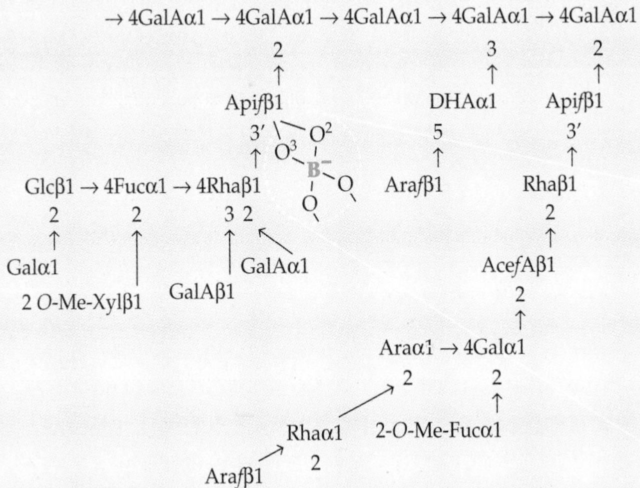

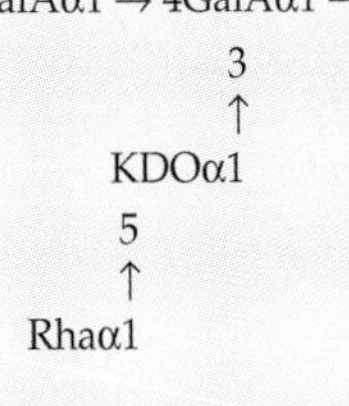

Rhamnogalacturonan II monomer showing position of the boron crosslinks. Most sugars are of the D-configuration with the exception of L-rhamnose and L-fucose. Except when designated *f* the sugar rings are pyranose. Unusual sugars include **apiose** (**Api**), aceric acid (**Ace**, 3-C′-carboxy-5-deoxyxylose), and 3-deoxy-D-*lyxo*-2-heptulosaric acid (**Dha**).

Borate diol ester linkage between apiose rings in oligosaccharide clusters of rhamnogalacturonan II in two pectin chains.

For 75 years or more it has been known that boron is essential for growth of green plants.[a,b] In its absence root tips fail to elongate normally, and synthesis of DNA and RNA is inhibited. Boron in the form of boric acid, $B(OH)_3$, is absorbed from soil. Although deficiency is rare it causes disintegration of tissues in such diseases as "heart rot" of beets and "drought spot" of apples. The biochemical role has been obscure, but is usually thought to involve formation of borate esters with sugar rings or other molecules with adjacent pairs of –OH groups (as in the accompanying structures). A regulatory role involving the plant hormones auxin, gibberelic acid, and cytokinin has also been suggested.

Diatoms also require boron, which is incorporated into the silicon-rich cell walls.[a] Some strains of *Streptomyces griseus* produce boron-containing macrolide antibiotics such as **aplasmomycin** (right).[c] Recently a function in plant cell walls has been identified (see also main text) as crosslinking of rhamnogalacturonan portions of pectin chains by borate diol ester linkages as illustrated.

It was long thought that boron was not required by human beings, but more recent studies suggest that we may need ~30 µg / day.[d] The possible functions are uncertain. Animals deprived of boron show effects on bone, kidney, and brain as well as a relationship to the metabolism of calcium, copper, and nitrogen. Nielson proposed a signaling function, perhaps via phosphoinositides, in animals.[b]

Aplasmomycin, a boron-containing antibiotic

[a] Salisbury, F. B., and Ross, C. W. (1992) *Plant Physiology*, 4th ed., Wadsworth, Belmont, California

[b] Nielsen, F. H. (1991) *FASEB J.* **5**, 2661–2667

[c] Lee, J. J., Dewick, P. M., Gorst-Allman, C. P., Spreafico, F., Kowal, C., Chang, C.-J., McInnes, A. G., Walter, J. A., Keller, P. J., and Floss, H. G. (1987) *J. Am. Chem. Soc.* **109**, 5426–5432

[d] Nielsen, F. H. (1999) in *Modern Nutrition in Health and Disease*, 9th ed. (Shils, M. E., Olson, J. A., Shike, M., and Ross, A. C., eds), pp. 283–303, Williams & Wilkins, Baltimore, Maryland

In contrast, hyaluronan and the polysaccharide chains of **glycosaminoglycans** (Fig. 4-11) have an alternating pattern. For a hyaluronan chain growing at the reducing end, one active site of hyaluronan synthase must be specific for UDP-GlcNAc and transfer the sugar unit only to the end of a glucuronic acid ring. A second active site must be specific for UDP-glucuronic acid but attach it only to the end of an acetylglucosamine unit.[172a,172b] There is still uncertainty about the direction of growth of hyaluronan.[173–175] Some hyaluronan synthases are lipid-dependent and their mechanism may resemble that proposed for cellulose synthesis (Fig. 20-5).

Dextrans. Some polysaccharides, such as the bacterial dextrans, are synthesized outside of cells by the action of secreted enzymes. An enzyme of this type, **dextran sucrase** of *Leuconostoc* and *Streptococcus*, adds glucosyl units at the *reducing* ends of the dextran chains (p. 174). Sucrose is the direct donor of the glucosyl groups, which are added by an insertion mechanism.[121,176–178] However, it is not dependent upon a membrane lipid as is that of Fig. 20-5. The glucosyl groups are transferred from sucrose to one of a pair of carboxylate groups of aspartate side chains in the active site.[179,180] If both carboxylates are glucosylated, a dextran chain can be initiated by insertion of one glucosyl group into the second (Eq. 20-19). The dextran grows alternating binding sites between the two carboxylates. Chain growth can be terminated by reaction with a sugar or oligosaccharide that fits into the active site and acts in place of the glucosyl group attached to Asp 1 as pictured in Eq. 20-19. The α-1,3-linked branches can be formed when a 3-OH group of

Sucrose — Fructose — Asp 1 — Asp 2 — Fructose — :Base — Asp 1

R = H or growing α–1,6-linked chain (20-19)

a second dextran chain enters the catalytic site, serving as the glycosyl acceptor. See Robyt for a detailed discussion of synthesis of dextrans and related polysaccharides such as **alternan** and the α-1,3-linked **mutan** (p. 175).[121,176] Some bacteria form β-2,6-linked **fructans** by a similar mechanism, with glucose being released by displacement on C2 of sucrose.[181] Fructans are also formed in green plants, apparently from reaction of two molecules of sucrose with release of glucose to form the trisaccharide Fru*f*β2 → 1Fru*f*β2–1α-Glc*p*, which then transfers a fructosyl group to the growing chain.

Lipid-dependent synthesis of polysaccharides. Insertion of monomer units at the base of a chain is a major mechanism of polymerization that is utilized for synthesis not only of polysaccharides but also of proteins (Chapter 29). For most carbohydrates the synthesis is dependent upon a polyprenyl lipid alcohol. In bacteria this is often the 55-carbon **undecaprenol** or **bactoprenol**,[136] which functions as a phosphate ester:

Undecaprenol phosphate

It serves as a membrane anchor for the growing polysaccharide. We have already discussed one example in the hypothetical cellulose synthase mechanism of Fig. 20-5. For some polysaccharides the mechanism is better established. The synthetic cycles all resemble that of Fig. 20-5 and can be generalized as in Eq. 20-20. Here NDP-Glx is a suitable nucleotide disphosphate derivative of sugar Glx, and Z-Glx is the repeating unit of the polysaccharide formed by the action of glycosyltransferases and other enzymes.

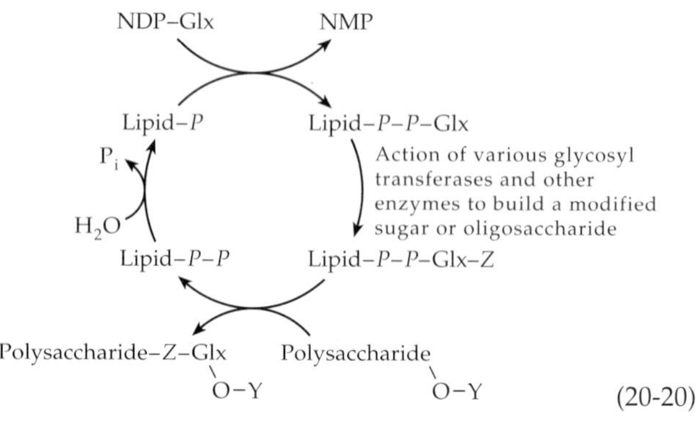

(20-20)

For example, the biosynthesis of alginate involves GDP-mannuronic acid (GDP-ManA) as NDP-Glx, bactoprenol as the lipid, and a glycosyltransferase that inserts a second mannuronate residue (as Z).

An additional transferase that uses acetyl-CoA as a substrate sometimes acetylates one mannuronate unit. The disaccharide units are then inserted into the growing chain. An additional modification, which occurs after polymerization, is random C5 epimerization of unacetylated D-mannuronate residues to L-guluronate.[136,182] Formation of alginate is of medical interest because infections by alginate-forming bacteria are a major cause of respiratory problems in cystic fibrosis.[182]

Sometimes an oligosaccharide assembled on the polyprenol phosphate represents a substantial block in assembly of a repeating polymer. For example, the xanthan gum (p. 179) produced by the bacterium *Xanthomonas campestris* is formed by several successive glycosyl transfers to bactoprenol-*P*-*P*-Glc. A second glucose is transferred onto the first from UDP-Glc, forming a pair of glucosyl groups in β-1,4 linkage. Mannose is then transferred from GDP-Man and joined in an α-1,3 linkage to the first GDP-Man to form a branch point. A glucuronate residue is then transferred from UDP-GlcA and another mannose from GDP-Man. The last mannose is modified by reaction with PEP to form a ketal (Eq. 4-9). The product of this assembly is the following lipid-bound oligosaccharide block.

$$\begin{array}{l} \text{Pyruvate} \qquad\qquad\qquad\qquad \overset{\text{Acetyl}}{\downarrow}\ \ \overset{\text{Glc}\beta 1}{\downarrow} \\ \quad \| \qquad\qquad\qquad\qquad\qquad\quad 6 \qquad\quad 4 \\ \text{Man}\beta 1 \rightarrow 4\ \text{GlcA}\beta 1 \rightarrow 2\ \text{Man}\alpha 1 \rightarrow 3\ \text{Glc}\beta 1 \rightarrow \text{O-}P\text{-}P\text{-lipid} \end{array}$$

This is inserted into the growing polysaccharide using the free 4-OH on the second glucose to link the units in a cellulose type chain. The twelve separate genes needed for synthesis of xanthan gum are contained in a 16-kb segment of the *X. campestris* genome.[136] Lipid-bound intermediates are also involved in synthesis of peptidoglycans (Fig. 20-9) and in the assembly of bacterial O-antigens (Fig. 8-30). Both of these also yield "block polymers."

D. Proteoglycans and Glycoproteins

The glycoproteins contain oligosaccharides attached to the protein either through *O*-glycosidic linkages with hydroxyl groups of side chains of serine, threonine, hydroxyproline, or hydroxylysine (*O*-linked) or via glycosylaminyl linkages to asparagine side chains (*N*-linked). The "core proteins" of the proteoglycans carry long polysaccharide chains, which are usually *O*-linked and are usually described as glycosaminoglycans.

1. Glycosaminoglycans

Synthesis of the alternating polysaccharide hyaluronan has been discussed in Section C,3 and may occur by an insertion mechanism. However, other glycosaminoglycans (sulfate esters of **chondroitin**, **dermatan**, **keratan**, **heparan**, and **heparin**) grow at their nonreducing ends.[183,184] Their synthesis is usually initiated by the hydroxyl group of serine or threonine side chains at special locations within several secreted proteins.[185] These proteins are synthesized in the rough ER and then move to the Golgi. Addition of the first sugar ring begins in the ER with transfer of single xylosyl residues to the initiating –OH groups.[186–190b] This reaction is catalyzed by the first of a group of special glycosyltransferases of high specificity that form the special terminal units (Chapter 4, Section D,1), that anchor the alternating polysaccharide represented here as $(X–Y)_n$:

$$(X \rightarrow Y)_n \rightarrow 4\ \text{GlcA}\beta 1 \rightarrow 3\ \text{Gal}\beta 1 \rightarrow 3\ \text{Gal}\beta 1 \rightarrow 4\ \text{Xyl}\beta 1 \rightarrow \text{O-Ser / Thr}$$

After transfer of the xylosyl residue from UDP-xylose to the –OH group in the protein,[190a] a second enzyme with proper specificity transfers a galactosyl group from UDP-galactose, joining it in β-1,4 linkage. A third enzyme transfers another galactosyl group onto the first one in β-1,3 linkage. A fourth enzyme, with a specificity different from that used in creating the main chain, then transfers a glucuronosyl group from UDP-glucuronic acid onto the chain terminus to complete the terminal unit.[190c] Then two more enzymes transfer the alternating units in sequence to form the repeating polymer with lengths of up to 100 or more monosaccharide residues. The sequence $(X–Y)_n$ in the preceding formula is:

$$(\rightarrow 4\ \text{GlcA}\beta 1 \rightarrow 3\ \text{GalNAc}\beta 1 \rightarrow)_n$$
for chondroitin[172a,191,192]

$$(\rightarrow 4\ \text{GlcA}\beta 1 \rightarrow 4\ \text{GlcNAc}\alpha 1 \rightarrow)_n$$
for heparan sulfate and heparin

$$(\rightarrow 3\ \text{Gal}\beta 1 \rightarrow 4\ \text{GlcNAc}\beta 1 \rightarrow)_n$$
for dermatan sulfate

Subsequent modifications of the polymers involve extensive formation of *O*-sulfate esters,[190a,193–197] *N*-deacetylation and *N*-sulfation,[198,199] and epimerization at C5.[10] In some tissues almost all GluA is epimerized.[200] The modifications are especially extensive in dermatan, heparan sulfates, and heparin (see also p. 177).[196,201–203b] The modifications are not random and follow a defined order. *N*-Deacetylation must precede *N*-sulfation, and *O*-sulfation is initiated only after *N*-sulfation of the entire chain is complete. The modifications occur within the Golgi (see Fig. 20-7) but not all of the glycosyltransferases, PAPS (3'-phosphoadenosine 5'-phosphosulfate)-dependent sulfotransferases, and epimerases are present within a single compartment. Nevertheless, an entire glycosaminoglycan chain can be synthesized within 1–3 min.[189]

The completed polymers are modified uniformly. There are clusters of sulfo groups with unusual structures in chondroitin from squid and shark cartilages[204,205] and fucosylated chondroitin from echinoderms.[206] Similar modifications are present less extensively in vertebrates. One of the best known modifications forms the unique pentasaccharide sequence shown in Fig. 4-13, which is essential to the anticoagulant activity of heparin. This sequence has been synthesized in the laboratory as have related longer heparin chains. A sequence about 17 residues in length containing an improved synthetic version of the unique pentasaccharide binds tightly to both thrombin and antithrombin (Chapter 12, Section C,9).[207,208] This opens the door to the development of improved substitutes for the medically important heparin. Heparan sulfate chains are found on proteoglycans throughout the body, but the highly modified heparin does not circulate in the blood. It is largely sequestered in cytoplasmic granules within mast cells and is released as needed.[208a] Heparin binds to many different proteins (p. 177). Among them is the glycoprotein selenoprotein P (p. 824), which may impart antioxidant properties to the extracellular matrix.[208b]

Although glycosaminoglycans are most often attached to *O*-linked terminal units, chondroitin sulfate chains can also be synthesized with *N*-linked oligosaccharides attached to various glycoproteins serving as initiators.[209] At least one form of keratan sulfate, found in the cornea, is linked to its initiator protein via GlcNAc-Man to *N*-linked oligosaccharides of the type present in many glycoproteins (Section D).

At least 25 different proteins that are secreted into the extracellular spaces of the mammalian body carry glycosaminoglycan chains.[183,210,211] Most of these proteins can be described as (1) **small leucine-rich proteoglycans** with 36- to 42-kDa protein cores and (2) **large modular proteoglycans** whose protein cores have molecular masses of 40 to 500 kDa.[210] The most studied of the second group is **aggrecan**, a major component of cartilage. This 220-kDa protein carries ~100 chondroitin chains, each averaging about 100 monosaccharide residues and ~100 negative charges from the carboxylate and sulfate groups. Aggrecan has three highly conserved globular domains near the N and C termini.[212–213a] The G1 domain near the N terminus is a **lectin** (p. 186), which, together with a small link protein that is structurally similar to the G1 domain, binds to a decasaccharide unit of hyaluronan. One hyaluronan molecule of 500- to1000-kDa mass (~2500–5000 residues) may bind 100 aggrecan and link molecules to form an ~200,000-kDa particle such as that shown in Fig. 4-16. These enormous highly negatively charged molecules, together with associated counterions, draw in water and preserve osmotic balance. It is these molecules that keep our joints mobile and which deteriorate by proteolytic degradation in the common **osteoarthritis**.[214,215] The keratan sulfate content of cartilage varies with age, and the level in serum and in synovial fluid is increased in osteoarthritis.[215] Keratan sulfate is also found in the cornea and the brain. Its content is dramatically decreased in the cerebral cortex of patients with Alzheimer disease.[216]

Other modular core proteins[210] include **versican** of blood vessels and skin,[210,213a,217,217a] **neurocam** and **brevican** of brain, **perlecan** of basement membranes,[218] **agrin** of neuromuscular junctions, and **testican** of seminal fluid. However, several of these have a broader distribution than is indicated in the foregoing description. The sizes vary from 44 kDa for testican to greater than 400 kDa for perlecan. The numbers of glycosaminoglycan chains are smaller than for aggrecan, varying from 1 to 30. Another of the chondroitin sulfate-bearing core protein is **appican**, a protein found in brain and one of the splicing variants of the amyloid precursor protein that gives rise to amyloid deposits in Alzheimer disease (Chapter 30).[217a,b]

The core proteins of the leucine-rich proteoglycans have characteristic horseshoe shapes and are constructed from ~28-residue repeats, each containing a β turn and an α helix. The three-dimensional structures are doubtless similar to that of a ribonuclease inhibitor of known structure which contains 15 tandem repeats.[219,220] A major function of these proteoglycans seems to be to interact with collagen fibrils, which have distinct proteoglyan-binding sites,[221] and also with fibronectin.[222] The small leucine-rich proteoglycans have names such as **biglycan**, **decorin**,[222a] **fibromodulin**, **lumican**, **keratoglycan**, **chondroadherin**, **osteoglycin**, and **osteoadherin**.[219,223–223c] The distribution varies with the tissue and the stage of development. For example, biglycan may function in early bone formation; decorin, which has a high affinity for type I collagen, disappears from bone tissue as mineralization takes place. Osteoadherin is found in mature osteoblasts.[223] **Phosphocan**, another brain proteoglycan, has an unusually high content (about one residue per mole) of L-isoaspartyl residues (see Box 12-A).[224]

Proteoglycans bind to a variety of different proteins and polysaccharides. For example, the large extracellular matrix protein **tenascin**, which is important to adhesion, cell migration, and proliferation, binds to chondroitin sulfate proteoglycans such as neurocan.[225] **Syndecan**, a transmembrane proteoglycan, carries both chondroitin and heparan sulfate chains, enabling it to interact with a variety of proteins that mediate cell–matrix adhesion.[185]

The ability of dissociated cells of sponges to aggregate with cells only of a like type (p. 29) depends upon large extracellular proteoglycans. That of *Microciona prolifera* appears to be an aggregate of about three hundred 35-kDa core protein molecules with equal masses of attached carbohydrate. This **aggregation factor** has a total mass of ~2×10^4 kDa.[226,227] It apparently interacts specifically, in the presence of Ca^{2+} ions, with a 210-kDa cell matrix protein to hold cells of the same species together.[227]

2. *O*-Linked Oligosaccharides

A variety of different oligosaccharides are attached to hydroxyl groups on appropriate residues of serine, threonine, hydroxylysine, and hydroxyproline in many different proteins (Chapter 4). Such oligosaccharides are present on external cell surfaces, on secreted proteins, and on some proteins of the cytosol and the nucleus.[228–231b] The rules that determine which –OH groups are to become glycosylated are not yet clear.[232] Glycosylation occurs in the ER, and, just as during synthesis of the long carbohydrate chains of proteoglycans, the sugar rings are added directly to an –OH group, either of the protein or of the growing oligosaccharide. The first glycosyl group transferred is most often **GalNAc** for external and secreted proteins[233] but more often **GlcNAc** for cytosolic and nuclear proteins.[228,231,233a c] Glycosylation of protein –OH groups can occur on either the luminal or cytosolic faces of the ER membranes.[234] The external *O*-linked glycoproteins often have large clusters of oligosaccharides attached to –OH groups of serine or threonine, but cytosolic proteins may carry only a small number of small oligosaccharides.

Of great importance are the **blood group determinants** which are discussed in Box 4-C. The ABO determinants are found at the nonreducing ends of *O*-linked oligosaccharides. Conserved Ser/Thr sites in the epidermal growth factor domains (Table 7-3) of various proteins carry ***O*-linked fucose**.[235]

The secreted **mucins** are unique in having clusters of large numbers of oligosaccharides linked by ***N*-acetylgalactosamine** to serine or threonine of the polypeptide.[236] The following core structures predominate.[237] These may be lengthened or further branched by the particular variety of glycosyltransferases present in a tissue and by their specificities.[233] The human genome contains at least nine mucin genes.[238] The large apomucins contain central domains with tandem repeats rich in Ser, Thr, Gly, and Ala and flanked at the ends by cysteine-rich domains.[239] For example, porcine submaxillary mucins are encoded by a gene with at least three alleles that encode 90, 125, or 133 repeats. The polypeptide may contain as many as 13,288 residues. N-terminal cysteine-rich regions are involved in dimer formation.[240]

```
core 1   Galβ1 → 3 GalNAc – Ser/Thr
core 2   Galβ1 → 3 GalNAc – Ser/Thr – Ser/Thr
                      6
                      ↑
                  GalNAcβ1

core 3   GlcNAcβ1 → 3 GalNAc – Ser/Thr
core 4   Galβ1 → 4 GlcNAcβ1  → 6 GalNAc– Ser/Thr
                                  3
                                  ↑
                              GlcNAcβ1
```

3. Assembly of *N*-Linked Oligosaccharides on Glycosyl Carrier Lipids

In eukaryotes the biosynthesis of the *N*-linked oligosaccharides of glycoproteins depends upon the polyprenyl alcohols known as **dolichols**, which are present in membranes of the endoplasmic reticulum. They contain 16–20 prenyl units, of which the one bearing the OH group is completely saturated as a result of the action of an NADPH-dependent reductase on the unsaturated precursor.[241] The predominant dolichol in mammalian cells contains 19 prenyl units. The structure of its mannosyl phosphate ester, one of the intermediates in the oligosaccharide synthesis, is illustrated below. The fully extended 95-carbon dolichol has a length of almost 10 nm, four times greater than that of oleic acid and twice the thickness of the nonpolar membrane bilayer core. The need for this great length is not clear nor is it clear why the first prenyl unit must be saturated for good acceptor activity.

The assembly of the oligosaccharides that will become linked to Asn residues in proteins occurs on the phosphate head of dolichol-*P*. The process begins on the cytoplasmic face of the membrane and within the lumen of the rough or smooth ER and continues within cisternae of the Golgi apparatus.[234,242–245] The initial transfer of GlcNAc-*P* to dolichol-*P* (Fig. 20-6, step *a*) appears to occur on the cytoplasmic face of the ER and is specifically inhibited by **tunicamycin**.[246,247] As the first "committed reaction" of *N*-glycosylation, it is regulated by a variety of hormonal and other factors.[248,249] The reaction takes place cotranslationally as the still unfolded peptide chain leaves the ribosome.[242] The oligosaccharide, still attached to the dolichol, continues to grow on the cytosolic surface of the ER membrane by transfer of GlcNAc and five residues of mannose from their sugar nucleotide forms (Fig. 20-6, steps *b* and *c*).[249a] The intermediate Dol-*P*-*P*-$GlcNAc_2Man_5$ crosses the membrane bilayer (Fig. 20-6, step *d*), after which mannosyl and glucosyl units are added (steps *e* and *f*). These sugars are carried across the membrane while attached to dolichol.

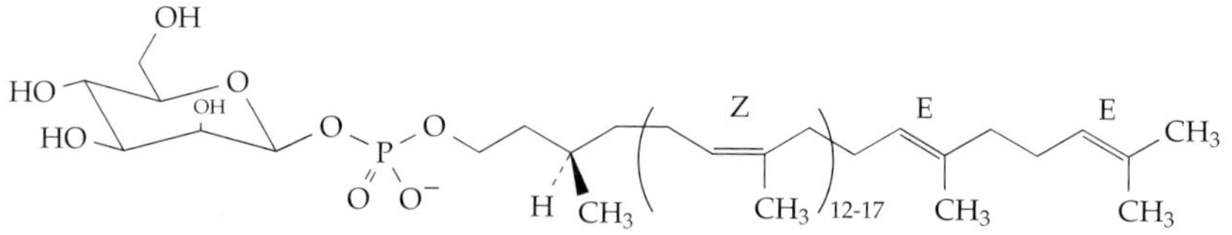

Man-*P*-dolichol

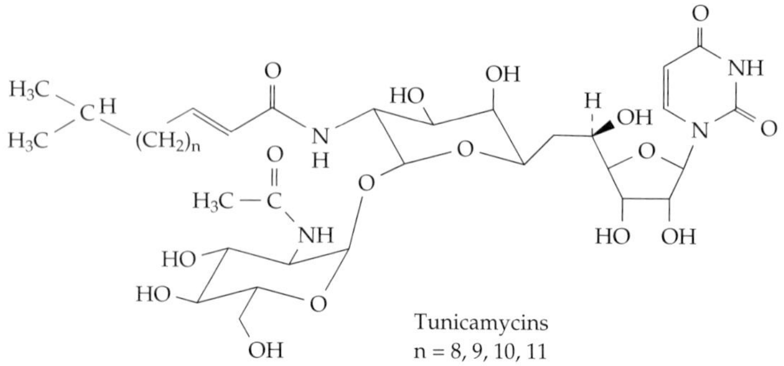

Tunicamycins
n = 8, 9, 10, 11

subunits have been reported for the enzyme in yeast. Genes for at least five of these are essential.[250–252b] One subunit serves to recognize suitable glycosylation sites.

Trimming of glycoprotein oligosaccharides. After transfer to glycoproteins the newly synthesized oligosaccharides undergo trimming, the hydrolytic removal of some of the sugar units, followed by addition of new sugar units to create the finished glycoproteins. The initial glycosylation process ensures that the glycoproteins remain in the lumen of the ER or within vesicles or cisternae separated from the cytoplasm. The subsequent processing appears to allow the cell to **sort** the proteins. Some remain attached to membranes and take up residence within ER, Golgi, or plasma membrane. Others are passed outward into transfer vesicles, Golgi, and secretion vesicles. A third group enter **lysosomes**. A series of specific inhibitors of trimming reactions, some of whose structures are shown in Fig. 20-7, has provided important insights.[253–255] Use of these inhibitors, together with immunochemical methods and study of yeast mutants,[250,252,256] is enabling us to learn many details of glycoprotein biosynthesis.

Whereas the formation of dolichol-linked oligosaccharides occurs in an identical manner in virtually all eukaryotic cells, trimming is highly variable as is the addition of new monosaccharide units.[257–257b] The major pathway for mammalian glycoproteins is shown in Fig. 20-7. Specific hydrolases in the ER remove all of the glucosyl units and one to three mannosyl units.[258] Removal of additional mannosyl residues occurs in the cis Golgi, to give the pentasaccharide core $Man_3GlcNAc_2$ which is common to all of the complex N-linked oligosaccharides. However, partial trimming without additional glycosylation produces some "high mannose" oligosaccharides.[258a] Removal of glucose may be necessary to permit some glycoproteins to leave the ER.

Sulfate groups and in some cases fatty acyl groups[259] may also be added. The exact composition of the oligosaccharides may depend upon the condition of the cell and may be altered in response to external influences.[257] Oligosaccharides attached to proteins that remain in the ER membranes may undergo

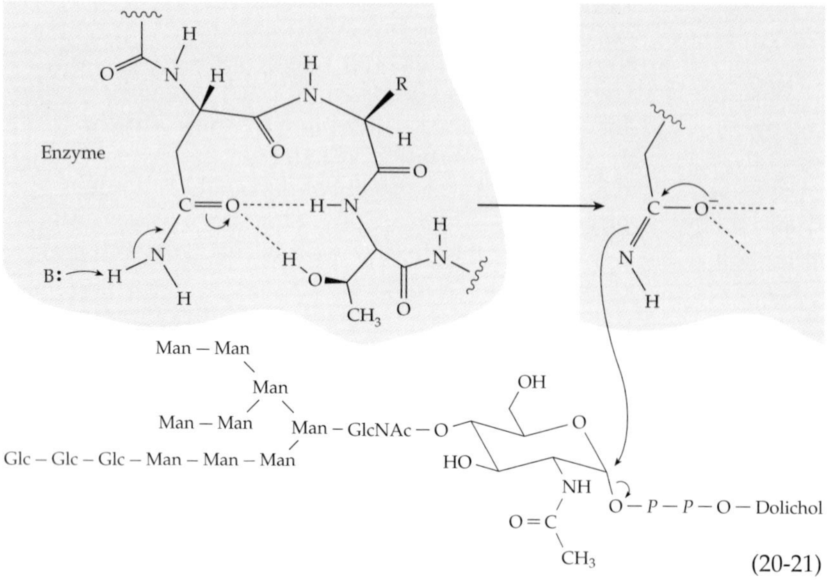

The completed 14-residue branched oligosaccharide $Glc_3Man_9GlcNAc_2$, with the structure indicated in Fig. 20-6, is then transferred to a suitable asparagine side chain (step *g*). This may be on a newly synthesized protein or on a still-growing polypeptide chain that is being extruded through the membrane into the lumenal space of the rough ER (Eq. 20-21; Fig. 20-6). The glycosylation site is often at the sequence Asn-X-Ser(Thr), which is likely to be present at a beta bend in the folded protein. Bends of the type illustrated in Eq. 20-21 and stabilized by the asparagine side chain are apparently favored.[250] In such a bend the –OH of the serine or threonine helps to polarize the amide group of the Asn side chain, perhaps enolizing it and generating a nucleophilic center that can participate in a displacement reaction[250,251] as indicated in Eq. 20-21. The **oligosaccharyltransferase** that catalyzes the reaction is a multisubunit protein. As many as eight different

Enzyme

Man – Man
Man
Man – Man Man – GlcNAc – O
Glc – Glc – Glc – Man – Man – Man
O – P – P – O – Dolichol

(20-21)

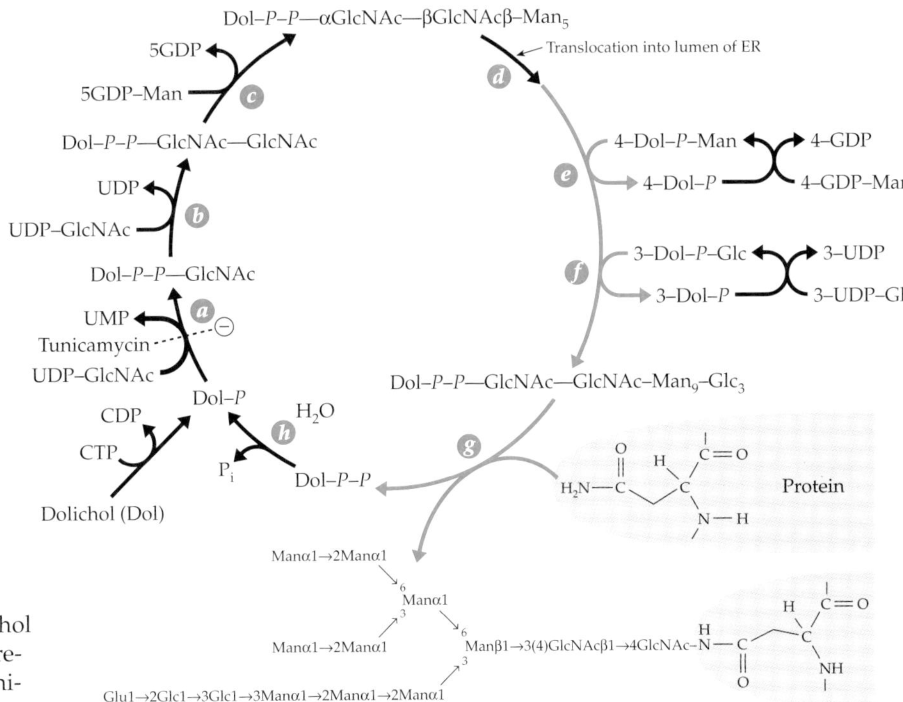

Figure 20-6 Biosynthesis of the dolichol diphosphate-linked oligosaccharide precursor to glycoproteins. The site of inhibition by tunicamycin is indicated.

very little trimming. However, the three glucosyl residues are usually removed by the glucosidases present in the rough ER. Some plant glycoproteins of the high-mannose type undergo no further processing.

Extensions and terminal elaborations. Even before trimming is completed, addition of new residues begins within the medial cisternae.[260] In mammalian Golgi, *N*-acetylglucosamine is added first. Galactose, sialic acid, and often fucose are then transferred from their activated forms to create such elaborate oligosaccharides as that shown in Fig. 4-17. As many as 500 glycosyltransferases, having different specificities for glycosyl donor and glucosyl acceptor, may be involved.[245] Extensions of the basic oligosaccharide structure often contain polylactosamine chains, branches with fucosyl residues, and sulfate groups.[245,261] More than 14 sialyltransferases place sialic acid residues, often in terminal positions, on these cell surface oligosaccharides.[262,263]

The cell wall of the yeast *Saccharomyces* is rich in **mannoproteins** that contain 50–90% mannose.[264] The ~250-residue mannan chains consist of an α-1,6-linked backbone with mono-, di-, and tri-mannosyl branches. These are attached to the same core structure as that of mammalian oligosaccharides. All of the core structures are formed in a similar way.[258,265] The mannoproteins may serve as a "filler" to occupy spaces in a cell wall constructed from β-1,3- and β-1,6-linked glycans and chitin. All of the four components, including the mannoproteins, are covalently linked together.[266] As was emphasized in Chapter 4 (pp. 186–188) glycoproteins serve many needs in biological recognition. The N-linked oligosaccharides play a major role in both animals and plants.[266a–c] Use of mass spectrometry, new automated methods of oligosaccharide synthesis,[266d] and development of new synthetic inhibitors[266c] are all contributing to current studies of what is commonly called "glycobiology."

The perplexing Golgi apparatus. First observed by Camillo Golgi[267,268] in 1898, the stacked membranes, now referred to simply as Golgi, remain somewhat mysterious.[268–271] There are at least three functionally distinct sets of Golgi cisternae, the **cis** (nearest the nucleus), **medial**, and **trans**. An additional series of tubules referred to as the **trans Golgi network** lies between the Golgi and the cell surface and may be the site at which lysosomal enzymes are sorted from proteins to be secreted.[260,272] Immunochemical staining directed toward specific glycosidases and glycosyl transferases suggested that the trimming reactions of glycoproteins start in the ER and continue as the proteins pass outward successively from one compartment of the Golgi to the next (Fig. 20-8). This has been the conventional view since the 1970s. The movement of the glycoproteins between compartments is thought to take place in small vesicles using a rather elaborate system of specialized proteins. Some of these coat the vesicles[273,274] while others target the vesicles to specific locations, e.g., the lysosomes[275] or the plasma membranes where they may be secreted.[273,277] A host of regulatory G-proteins assist these complex processes and drive them via hydolysis of GTP.[271,278]

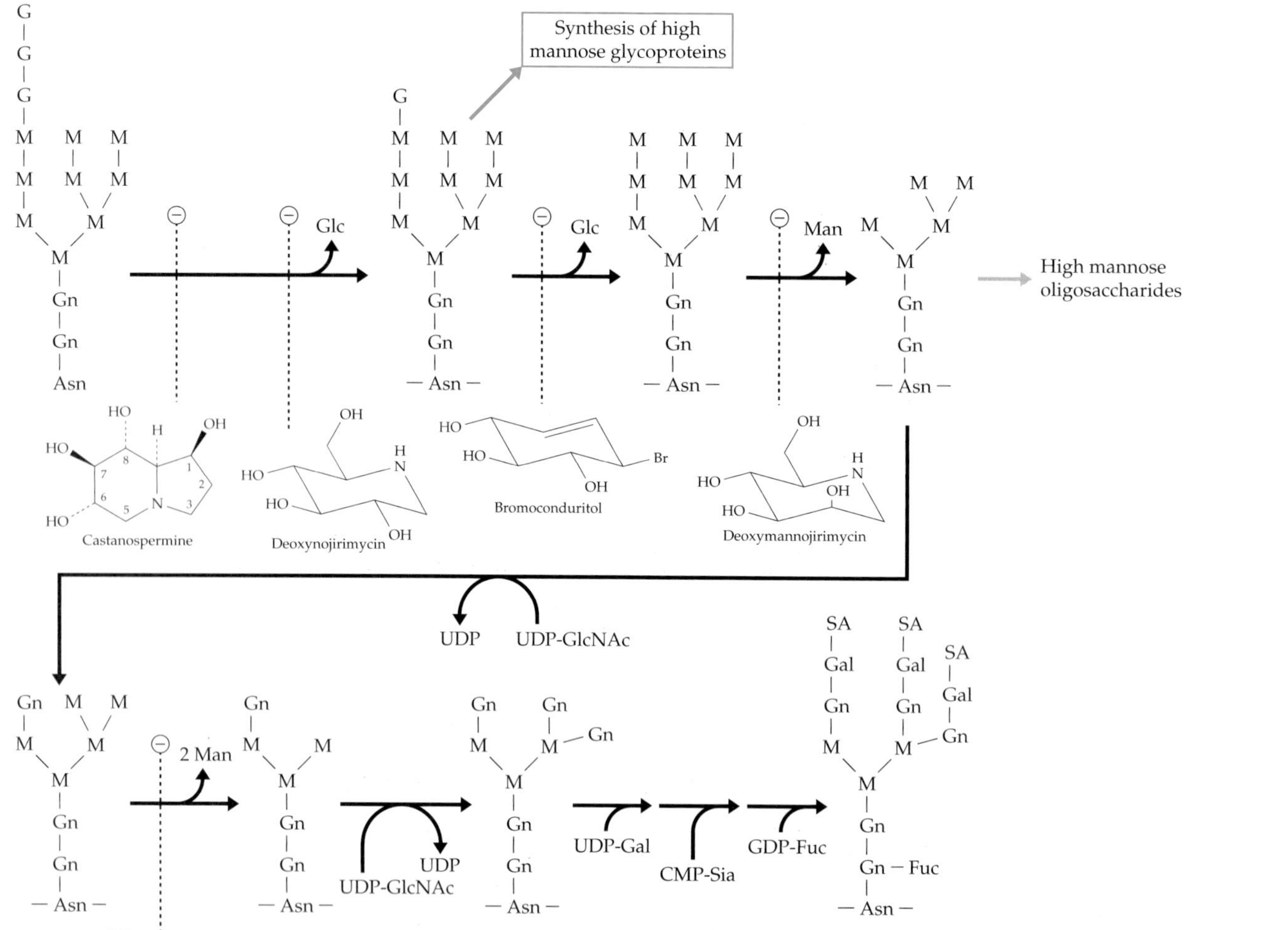

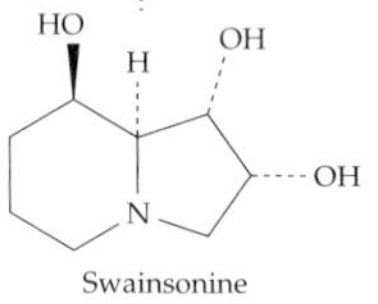

Figure 20-7 Steps in trimming of the oligosaccharide precursor and synthesis of complex N-linked oligosaccharides of mammalian glycoproteins. Sites of action of several inhibitors are shown. Abbrevations: M, mannosyl, Gn, *N*-acetylglucosaminyl, G, glucosyl residues. Structures of several inhibitors are shown.

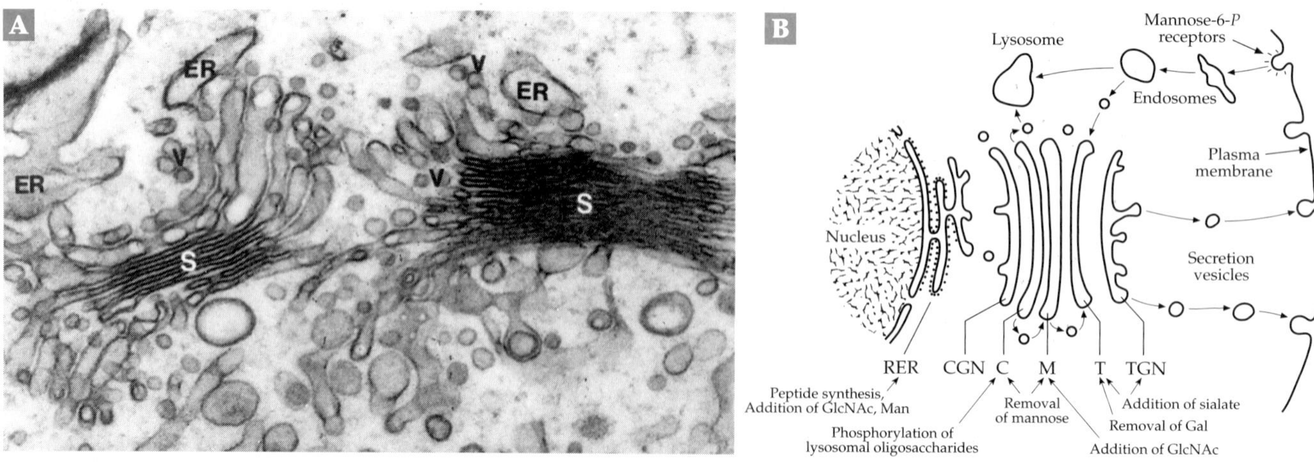

Figure 20-8 (A) Electron micrograph showing a transverse section through part of the Golgi apparatus of an early spermatid. Cisternae of the ER, Golgi stacks (S), and vesicles (V) can be seen. Curved arrows point to associated tubules. Magnification X45,000.[276] Courtesy of Y. Clermont. (B) Scheme showing functions of endoplasmic reticulum, transfer vesicles, Golgi apparatus, and secretion vesicles in the metabolism of glycoproteins.

While most proteins synthesized in the ER follow the exocytotic pathway through the Golgi, some are retained in the ER and some that pass on through the Golgi are returned to the ER.[279] In fact, such **retrograde transport** can carry some proteins taken up by endocytosis through the plasma membrane and through the Golgi to the ER where they undergo *N*-glycosylation. Retrograde transport is essential for recycling of plasma membrane proteins and lipids. The forward flow of glycoproteins and membrane components from the ER to the Golgi can be blocked by the fungal macrocyclic lactone **brefeldin A**. In cells treated with this drug, which inactivates a small

Brefeldin A

CTP-binding protein,[280] the Golgi apparatus is almost completely resorbed into the ER by retrograde transport. Proteins remaining in the ER undergo increased *O*-glycosylation as well as unusual types of *N*-glycosylation.

Although the conventional view of flow through the Golgi is generally accepted, it is difficult to distinguish it from an alternative explanation: *The Golgi compartments may move outward continuously while retrograde transport occurs via the observed vesicles.*[268,272] Some evidence for this **cisternal maturation model** has been known for many years but was widely regarded as reflecting unusual exceptions to the conventional model. In fact, both views could be partially correct; vesicular transport may function in both directions.[280a] High-resolution tomographic images are also altering our view of the Golgi.[280b]

The proteins of Golgi membranes are largely integral membrane proteins and peripheral proteins associated with the cytosolic face. Some of the integral membrane proteins are the oligosaccharide-modifying enzymes, which protrude into the Golgi lumen.[280c,280d] Many other proteins participate in transport,[280d–f] docking, membrane fusion,[280g,h] and acidification of Golgi compartments.[280i] Many of the first studies of vesicular transport were conducted with synaptic vesicles and are considered in Chapter 30 (see Fig. 30-20). Other aspects of membrane fusion and transport are discussed in Chapter 8. A group of specialized Golgi proteins, the **golgins**, are also present. They are designated golgin-84, -95, -160, -245, and -376 (giantin or macrogolgin) and were identified initially as human **autoantigens** (Chapter 31), appearing in the blood of persons with autoimmune disorders such as Sjögren's syndrome.[281,282] Another protein of molecular mass ~130 kDa, and which appears to be specific to the trans-Golgi network, has been found in human serum of patients with renal vasculitis.[283]

Lysosomal enzymes. Various **sorting signals** are encoded within proteins. These include the previously mentioned C-terminal KDEL (mammals) or HDEL (yeast) amino acid sequence, which serves as a retrieval signal for return of proteins from the Golgi to the ER (p. 521).[279,284] Other sorting signals are provided by the varied structures of the oligosaccharides attached to glycoproteins. These sugar clusters convey important biological information, which is "decoded" in the animal body by interaction with various **lectins** that serve as receptors.[285] This often leads to endocytosis of the glycoprotein. An example is provided by the more than 50 proteins that are destined to become lysosomal enzymes and which undergo phosphorylation of 6-OH groups of the mannosyl residue marked by asterisks on the following structure. This is an *N*-linked oligosaccharide that has been partially trimmed. The phosphorylation is accomplished in

```
Man — Man*
         \
          Man
     Man /   \
              Man — GlcNAc — GlcNAc — Asn
             /
Man — Man*— Man
```

two steps by enzymes present in the cis Golgi compartment (Eq. 20-22) An ***N*-acetylglucosaminylphosphotransferase** transfers phospho-GlcNAc units from UDP-GlcNAc onto the 6-OH groups of mannosyl residues. These must be recognized by the phosphotransferase as appropriate.[286,287] Then a hydrolase cleaves off GlcNAc.

The proteins carrying the mannose 6-phosphate groups bind to one of two different **mannose 6-*P* receptors** present in the Golgi membranes and are subsequently transported in clathrin-coated vesicles to endosomes where the low pH causes the proteins to dissociate from the receptors, which may be recycled.[288–290a] The hydrolytic enzymes are repackaged in lysosomes. The same mannose 6-*P* receptors also appear on the external surface of the plasma membrane allowing many types of cells to take up lysosomal enzymes that have "escaped" from the cell. These proteins, too, are transported to the lysosomes. The mannose 6-*P* receptors have a dual function, for they also remove insulinlike growth factor from the circulation, carrying it to the lysosomes for degradation.[287,290] Most Man 6-*P* groups are removed from proteins once they reach the lysosomes but this may not always be true.[291] Not all lysosomal proteins are recruited by the mannose 6-*P* receptors. Some lysosomal membrane proteins are sorted by other mechanisms.[292]

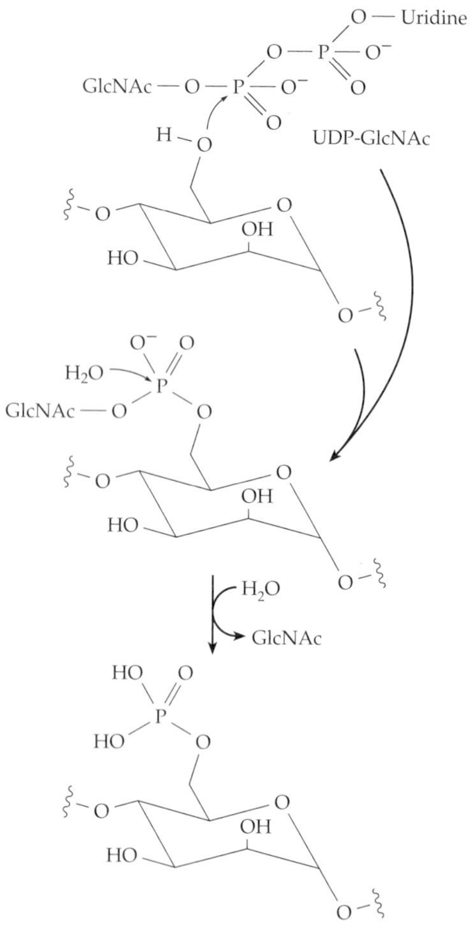

(20-22)

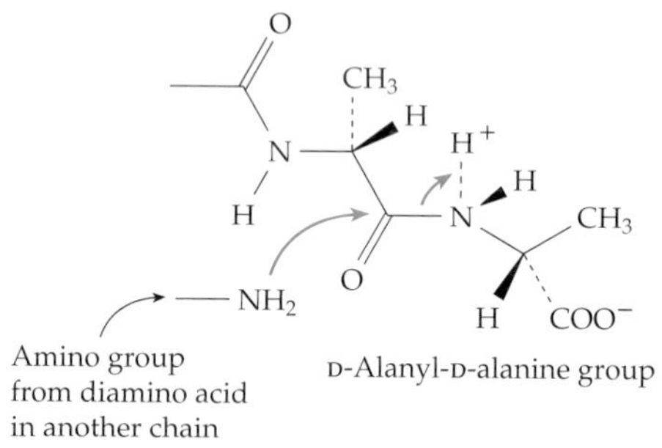

The hepatic asialoglycoprotein (Gal) receptor. A variety of proteins are taken out of circulation in the blood by the hepatocytes of the liver. Serum glycoproteins bearing sialic acid at the ends of their oligosaccharides (see Fig. 20-7) have relatively long lives, but if the sialic acid is removed by hydrolysis, the exposed galactosyl residues are recognized by the multisubunit **asialoglycoprotein receptor**.[293–295] The bound proteins are then internalized rapidly via the coated pit pathway and are degraded in the lysosomes. Other receptors, including those that recognize transferrin, low-density lipoprotein, α_2 macroglobulin, and T lymphocyte antigens, also depend upon interaction with oligosaccharides.[296]

E. Biosynthesis of Bacterial Cell Walls

The outer surfaces of bacteria are rich in specialized polysaccharides. These are often synthesized while attached to lipid membrane anchors as indicated in a general way in Eq. 20-20.[136,296a] One of the specific biosynthetic cycles (Fig. 20-9) that depends upon undecaprenol phosphate is the formation of the **peptidoglycan** (murein) layer (Fig. 8-29) of both gram-negative and gram-positive bacterial cell walls. Synthesis begins with attachment of L-alanine to the OH of the lactyl group of UDP-*N*-acetylmuramic acid in a typical ATP-requiring process (Fig. 20-9, step *a*).[297] Next D-glutamic acid, *meso*-diaminopimelate (Fig. 8-29), or L-lysine, and D-alanyl-D-alanine are joined in sequence, each in another ATP-requiring step.[298–301d] The entire unit assembled in this way is transferred to undecaprenol phosphate with creation of a pyrophosphate linkage (step *e*). An *N*-acetylglucosamine unit is added by action of another transferase (step *f*), and in an ATP-requiring process ammonia is sometimes added to cap the free α-carboxyl group of the D-glutamyl residue (step *g*). In *Staphylococcus aureus* and related gram-positive bacteria five glycyl units are also added, each from a molecule of glycyl-tRNA (green arrows in Fig. 20-9). The completely assembled repeating unit, together with the connecting peptide chain needed in the crosslinking reaction, is transferred onto the growing chain (step *h*). As in formation of dextrans, growth is by insertion of the repeating unit at the reducing end of the chain. The polyprenyl diphosphate is released, and the cycle is completed by the action of a pyrophosphatase (step *i*). This step is blocked by **bacitracin**, an antibiotic which forms an unreactive complex with the polyprenyl diphosphate carrier. Completion of the peptidoglycan requires crosslinking. This is accomplished by displacement of the terminal D-alanine of the pentapeptide by attack by the $-NH_2$ group of the diaminopimelate or lysine or other diamino acid (see also Fig. 8-29).[301e]

Amino group from diamino acid in another chain

D-Alanyl-D-alanine group

Because the peptidoglycan layer must resist swelling of the bacteria in media of low osmolarity, it must be strong and must enclose the entire bacterium. At the same time the bacterium must be able to grow in size and also to divide. For these reasons bacteria must continuously not only synthesize peptidoglycan but also degrade it.[302,303] The latter is accomplished by hydrolytic cleavage using cell wall enzymes to the ***N*-acetylglucosamine-anhydro-*N*-acetylmuramate-tripeptide** (GlcNAc-1,6-anhydro-MurNAc-L-Ala-D-Glu-A_2pm) fragment.[304,305] A hydrolase cuts the peptide bridge.[305a] This process is probably essential to formation of new growing points for expansion of the murein layer. Most of the peptide fragments that are released in the periplasm are transported back into the cytosol.

The anhydroMurNAc is removed, and new UDP-MurNAc and D-Ala-D-Ala units are added salvaging the tripeptide unit. The repaired UDP-MurNA-pentapeptide can then reenter the biosynthetic pathway (Fig. 20-9).

The O-antigens and lipid A. A cluster of sugar units of specific structure makes up the repeating unit of the "O-antigen" of *Salmonella*. The many structural variations in this surface polysaccharide account for the over two thousand serotypes of *Salmonella* (p. 180).[121,306] As is illustrated in Fig. 8-30, the O-antigen is a repeating block polymer that is attached to a complex lipopolysaccharide "core" and a hydrophobic membrane anchor known as lipid A (Figs. 8-28 and 8-30).[307–308a] Lipid A and the attached core and O-antigen are synthesized inside the bacterial cell by enzymes found in the cytoplasmic membrane.[309] The complete lipopolysaccharide units are then translocated from the inner membrane to the outer membrane of the bacteria. The synthesis of the O-antigen is understood best. Consider the following group E3 antigen, where Abe is abequose (Fig. 4-15) and Rha is rhamnose:

$$\begin{array}{l} \quad \text{Abe}\alpha 1 \\ \qquad \downarrow \\ \qquad 4 \\ (\rightarrow 6\,\text{Gal}\beta 1 \rightarrow 6\,\text{Man}1 \rightarrow 4\,\text{Rha}1 \rightarrow)_n \end{array}$$

Assembly of this repeating unit begins with the transfer of a *phosphogalactosyl* unit from UDP-Gal to the phospho group of the lipid carrier undecaprenol phosphate. The basic reaction cycle is much like that in Fig. 20-9 for assembly of a peptidoglycan.

The oligosaccharide repeating unit of the O-antigen is constructed by the consecutive transferring action of three more transferases. For the antigen shown above, one enzyme transfers a rhamnosyl unit, another a mannosyl unit, and another an abequosyl unit from the appropriate sugar nucleotides. Then the entire growing O-antigen chain, which is attached to a second molecule of undecaprenol diphosphate, is transferred onto the end of the newly assembled oligosaccharide unit. In effect, the newly formed oligosaccharide is inserted at the reducing end of the growing chain just as in Fig. 20-9. Elongation continues by the transfer of the entire chain onto yet another tetrasaccharide unit. As each oligosaccharide unit is added, an undecaprenol diphosphate unit is released and a phosphatase cleaves off the terminal phospho group to regenerate the original undecaprenol phosphate carrier. When the O-antigen is long enough, it is attached to the rest of the lipopolysaccharide.

The lipid A anchor is also based on a carbohydrate skeleton. Its assembly in *E. coli*,[307] which requires nine enzymes, is depicted in Fig. 20-10. *N*-Acetylglucosamine 6-*P* is acylated at the 3-position[310] and after deacetylation[311] at the 2-position. As shown in this figure, acylation is accomplished by transfer of hydroxymyristoyl groups from acyl carrier protein (ACP). Two molecules of the resulting UDP-2,3-diacyl-GlcN are then joined via the reactions shown to give the acylated disaccharide precursor to lipid A. Stepwise transfer of KDO, L-*glycero*-D-*manno*-heptose, and other monosaccharide units from the appropriate sugar nucleotides and further acylation follows (Fig. 20-10). The assembled O-antigen chain is transferred from undecaprenol diphosphate onto the lipopolysaccharide core. This apparently occurs on the periplasmic surface of the plasma membrane. If so, the core lipid domain must be flipped across the plasma membrane before the O-antigen chain is attached.[312] Less is known about the transport of the completed lipopolysaccharide across the periplasmic space and into the outer membrane.

The core structures of the lipopolysaccharides vary from one species to another or even from one strain of bacteria to another. All three domains (lipid A, core, and O-antigen) contribute to the antigenic properties of the bacterial surface[313] and to the virulence of the organism.[313,314] Nitrogen-fixing strains of *Rhizobium* require their own peculiar lipopolysaccharides for successful symbiosis with a host plant.[315] However, there are some features common to most lipopolysaccharides. Two to three residues of KDO are usually attached to the acylated diglucosamine anchor, and these are often followed by 3 – 4 heptose rings.[316–318]

The structure of the inner core regions of a typical lipopolysaccharide from *E. coli* is indicated in Fig. 8-30. The complete structure of the lipopolysaccharide from a strain of *Klebsiella* is shown at the top of the next page.[319] Here L, D-Hep*p* is D-*manno*-heptopyranose and D, D-Hep*p* is D-*glycero*-D-*manno*-heptose. As in this case, the outer core often contains several different hexoses. The lipopolysaccharide of *Neisseria meningititis* has sialic acid at the outer end.[320] However, the major virulence factor for this organism, which is a leading cause of bacterial meningitis in young children, is a capsule of poly(ribosyl)ribitol phosphate that surrounds the cell.[321] *Haemophilus influenzae*, a common cause of ear infections and meningitis in children, has no O-antigen but a more highly branched core oligosaccharide than is present in *E. coli*.[321a] *Legionella* has its own variations.[321b]

Gram-positive bacteria. Although their outer coatings are extremely varied, all gram-positive bacteria have a peptidoglycan similar to that of gram-negative bacteria but often containing the intercalated pentaglycine bridge indicated in Fig. 8-29. However, the peptidoglycan of gram-positive bacteria is 20–50 nm thick, as much as ten times thicker than that of *E. coli*. Furthermore, the peptidoglycan is intertwined with the anionic polymers known as teichoic acids and

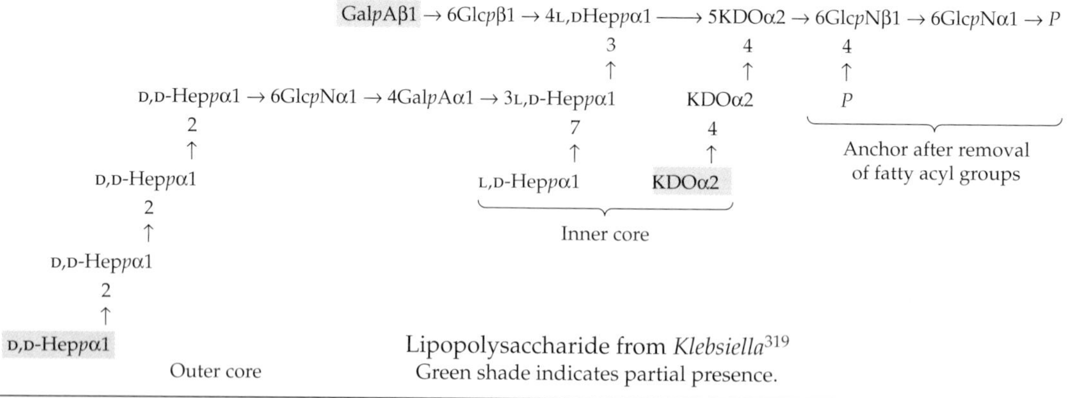

Lipopolysaccharide from *Klebsiella*[319]
Green shade indicates partial presence.

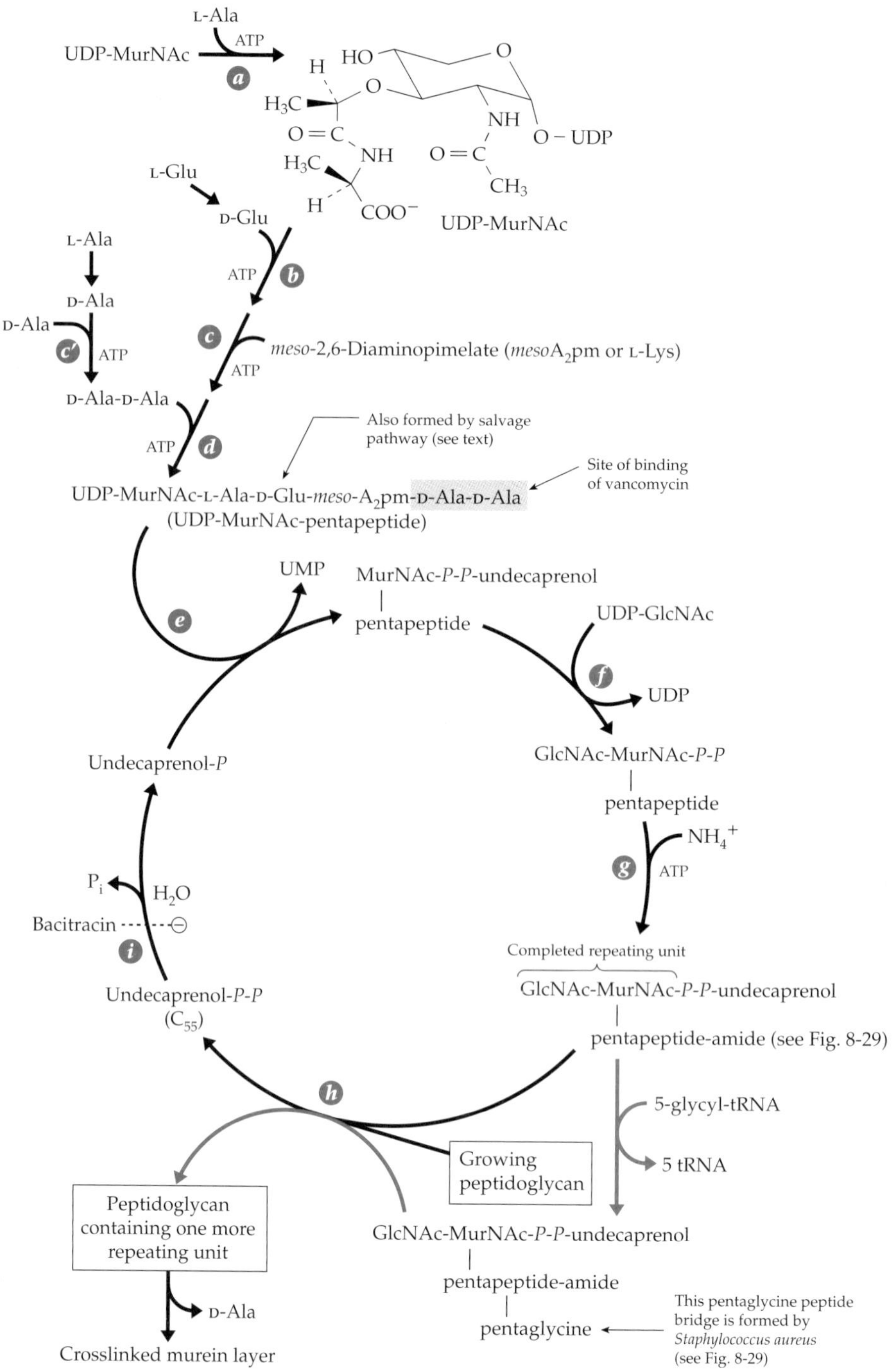

UDP-MurNAc → (L-Ala, ATP, *a*) → UDP-MurNAc (structure)

L-Glu → D-Glu (ATP, *b*)

meso-2,6-Diaminopimelate (*meso*A_2pm or L-Lys) (ATP, *c*)

L-Ala → D-Ala; D-Ala (ATP, *c'*) → D-Ala-D-Ala (ATP, *d*)

UDP-MurNAc-L-Ala-D-Glu-*meso*-A_2pm-D-Ala-D-Ala
(UDP-MurNAc-pentapeptide)

Also formed by salvage pathway (see text)

Site of binding of vancomycin

e: UMP; MurNAc-*P*-*P*-undecaprenol | pentapeptide

f: UDP-GlcNAc → UDP; GlcNAc-MurNAc-*P*-*P* | pentapeptide

g: NH_4^+, ATP

Completed repeating unit: GlcNAc-MurNAc-*P*-*P*-undecaprenol | pentapeptide-amide (see Fig. 8-29)

5-glycyl-tRNA → 5 tRNA

GlcNAc-MurNAc-*P*-*P*-undecaprenol | pentapeptide-amide | pentaglycine

This pentaglycine peptide bridge is formed by *Staphylococcus aureus* (see Fig. 8-29)

h: Growing peptidoglycan; Peptidoglycan containing one more repeating unit → D-Ala; Crosslinked murein layer

Undecaprenol-*P*-*P* (C_{55})

i: H_2O, P_i; Bacitracin ⊖

Undecaprenol-*P*

Figure 20-9 Biosynthesis of bacterial peptidoglycans. See Fig. 8-29 for details of the peptidoglycan structures. Green arrows show alternative route used by gram-positive bacteria.

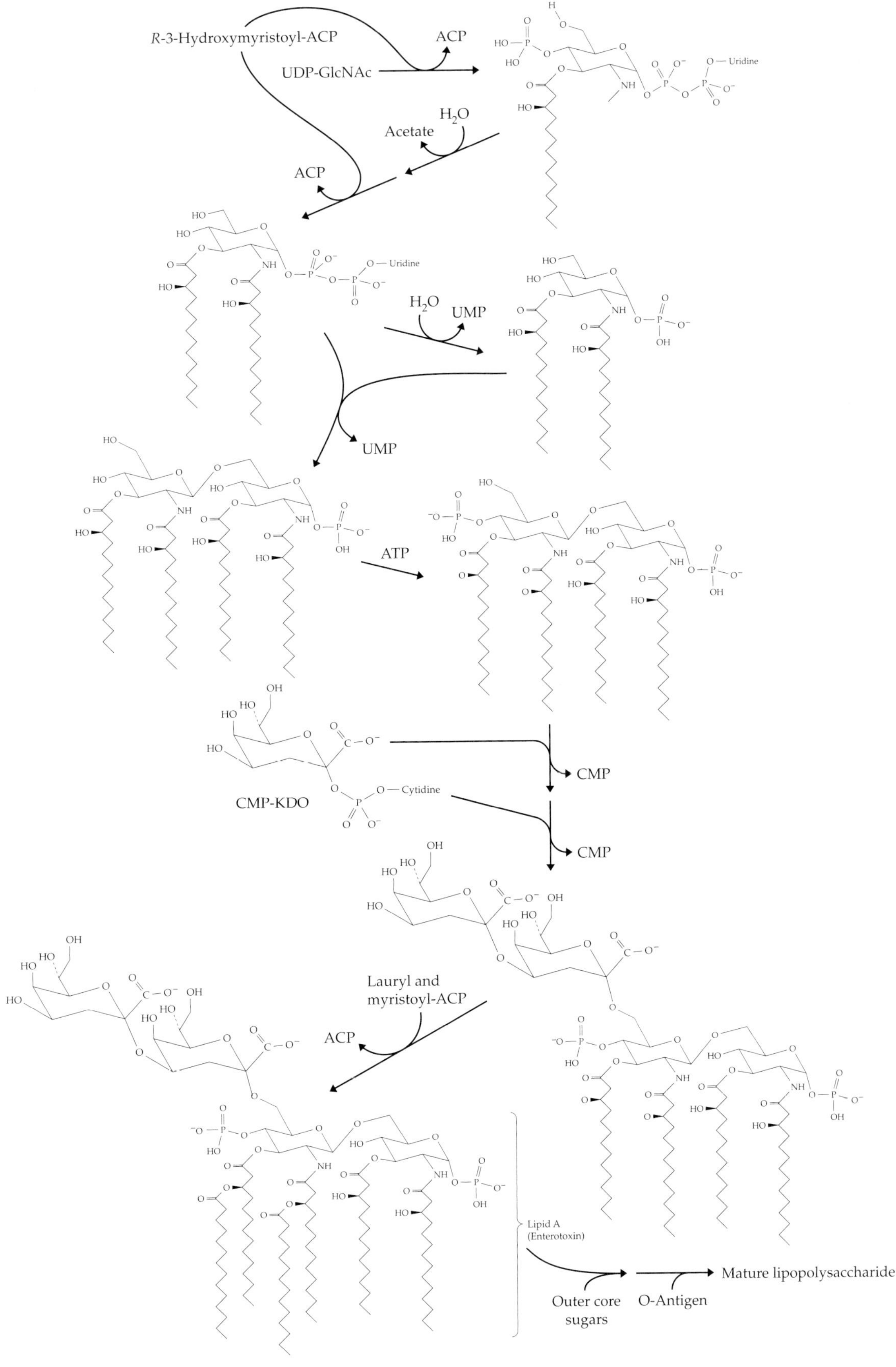

Figure 20-10 Proposed biosynthetic route for synthesis of lipid A and the mature lipopolysaccharide of the *E. coli* cell wall. After C. R. H. Raetz *et al.*[307]

BOX 20-G PENICILLINS AND RELATED ANTIBIOTICS

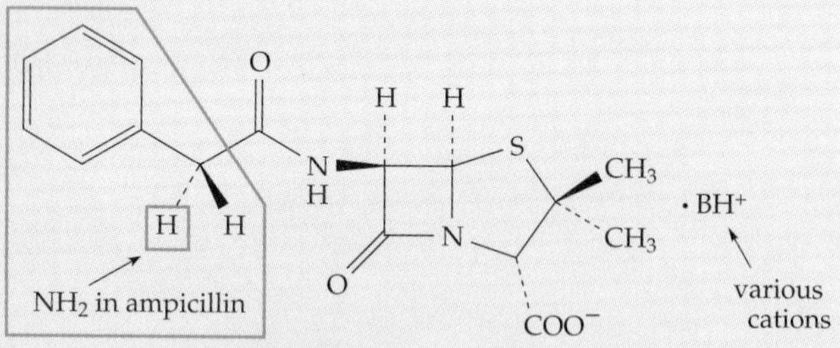

Penicillin G
(Benzylpenicillin)

Many organisms produce chemical substances that are toxic to other organisms. Some plants secrete from their roots or leaves compounds that block the growth of other plants. More familiar to us are the medicinal antibiotics produced by fungi and bacteria. The growth inhibition of one kind of organism upon another was well known in the last century, e.g., as reported by Tyndall[a,b] in 1876. The beginning of modern interest in the phenomenon is usually attributed to Alexander Fleming, who, in 1928, noticed the inhibition of growth of staphylococci by *Penicillium notatum*. His observation led directly to the isolation of penicillin, which was first used on a human patient in 1930. The early history as well as the subsequent purification, characterization, synthesis, and development as the first major antibiotic has been recorded in numerous books and articles.[c–i] During the same time period, Rene Dubos isolated the peptide antibiotics **gramicidin** and **tyrocidine**.[j] A few years later **actinomycin** (Box 28-A) and **streptomycin** (Box 20-B) were isolated from soil actinomycetes (streptomyces) by Waksman, who coined the name **antibiotic** for these compounds. Streptomycin was effective against tuberculosis, a finding that helped to stimulate an intensive search for additional antibacterial substances. Since that time, new antibiotics have been discovered at the rate of more than 50 a year. More than 100 are in commercial production.

Major classes of antibiotics include more than 200 peptides such as the gramicidins, bacitracin, tyrocidines and valinomycin (Fig. 8-22)[k]; more than 150 **penicillins, cephalosporins,** and related compounds; **tetracyclines** (Fig. 21-10); the **macrolides,** large ring lactones such as the **erythromycins** (Fig. 21-11); and the **polyene** antibiotics (Fig. 21-10).

Penicillin was the first antibiotic to find practical use in medicine. Commercial production began in the early 1940s and benzylpenicillin (penicillin G), one of several natural penicillins that differ in the R group boxed in the structure above, became one of the most important of all drugs. Most effective against gram-positive bacteria, at higher concentrations it also attacks gram-negative bacteria including *E. coli*. The widely used semisynthetic penicillin **ampicillin** (R = D-α-aminobenzyl) attacks both gram-negative and gram-positive organisms. It shares with penicillin extremely low toxicity but some danger of allergic reactions. Other semisynthetic penicillins are resistant to β-lactamases, enzymes produced by penicillin-resistant bacteria which cleave the four-membered β-lactam ring of natural penicillins and inactivate them.

Closely related to penicillin is the antibiotic **cephalosporin C**. It contains a D-α-aminoadipoyl side chain, which can be replaced to form various semisynthetic cephalosporins. **Carbapenems** have similar structures but with CH_2 replacing S and often a different chirality in the lactam ring.

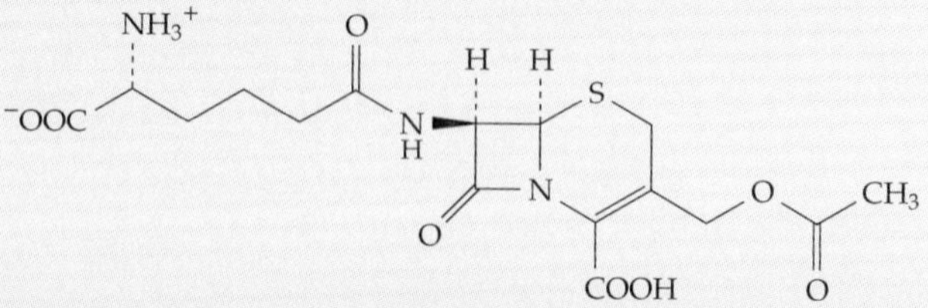

Cephalosporin C

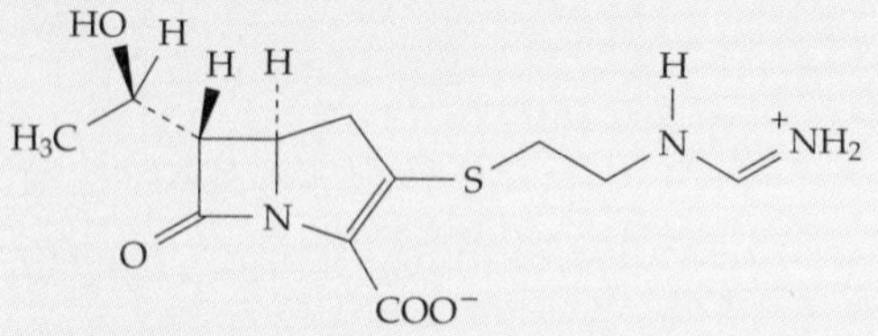

Imipenem, a carbapenem antibiotic of last resort

These and other related β-lactams are medically important antibacterial drugs whose numbers are increasing as a result of new isolations, synthetic modifications, and utilization of purified biosynthetic enzymes.[c,l,m]

How do antibiotics act? Some, like penicillin, block specific enzymes. Peptide antibiotics often form complexes with metal ions (Fig. 8-22) and disrupt the control of ion permeability in bacterial membranes. Polyene antibiotics interfere with proton and ion transport in fungal membranes. Tetracyclines and many other antibiotics interfere directly with protein synthesis (Box 29-B). Others intercalate into DNA molecules (Fig. 5-23; Box 28-A). There is no single mode of action. The search for suitable antibiotics for human use consists in finding compounds highly toxic to infective organisms but with low toxicity to human cells.

Penicillin kills only growing bacteria by preventing proper crosslinking of the peptidoglycan

BOX 20-G (continued)

layer of their cell walls. An amino group from a diamino acid in one peptide chain of the peptidoglycan displaces a D-alanine group in a transpeptidation (acyltransferase) reaction. The transpeptidase is also a hydrolase, a DD-carboxypeptidase. Penicillins are structural analogs of D-alanyl-D-alanine and bind to the active site of the transpeptidase.[l,n–p] The β-lactam ring of penicillins is unstable, making penicillins powerful acylating agents. The transpeptidase apparently acts by a double displacement mechanism, and the initial attack of a nucleophilic serine hydroxyl group of the enzyme on penicillin bound at the active site leads to formation of an inactivated, penicillinoylated enzyme.[q,r] More than one protein in a bacterium is derivatized by penicillin.[s] Therefore, more than one site of action may be involved in the killing of bacterial cells.

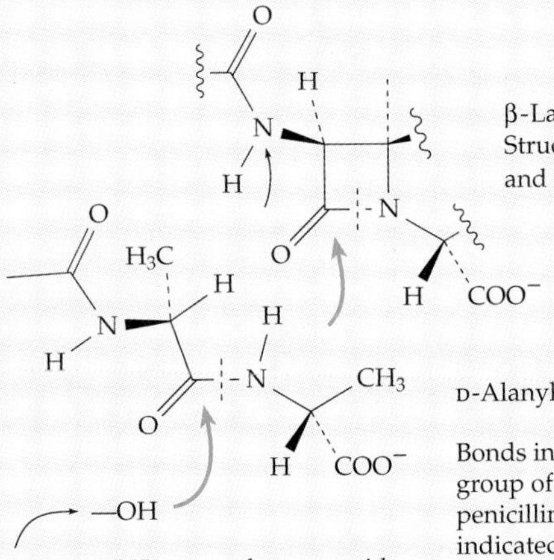

β-Lactam
Structure of penicillin and related antibiotics

D-Alanyl-D-alanine group

Bonds in D-alanyl-D-alanine group of peptidoglycan or in penicillins are cleaved as indicated by dashed lines

Serine —OH group of transpeptidase.

Several classes of **β-lactamases**, often encoded in transmissible plasmids, have spread worldwide rapidly among bacteria, seriously decreasing the effectivenss of penicillins and other β-lactam antibiotics.[t–y] Most β-lactamases (classes A and C) contain an active site serine and are thought to have evolved from the DD transpeptidases, but the B type[y] has a catalytic Zn^{2+}. The latter, as well as a recently discovered type A enzyme,[z] hydrolyze imipenem, currently one of the antibiotics of last resort used to treat infections by penicillin-resistant bacteria. Some β-lactam antibiotics are also powerful inhibitors of β-lactamases.[u,aa,bb] These antibiotics may also have uses in inhibition of serine proteases[cc,dd] such as elastase. Some antibiotic-resistant staphylococci produce an extra penicillin-binding protein that protects them from beta lactams.[ee] Because of antibiotic resistance the isolation of antibiotics from mixed populations of microbes from soil, swamps, and lakes continues. Renewed efforts are being made to find new targets for antibacterial drugs and to synthesize new compounds in what will evidently be a never-ending battle. We also need better antibiotics against fungi and protozoa.

[a] Tyndall, J. (1876) *Phil. Trans. Roy. Soc. London B* **166**, 27

[b] Reese, K. M. (1980) *Chem. Eng. News, Sept. 29*, p64

[c] Abraham, E. P. (1981) *Sci. Am.* **244** (Jun), 76–86

[d] Sheehan, J. C. (1982) *The Enchanted Ring: The Untold Story of Penicillin*, MIT Press, Cambridge, Massachusetts

[e] Hobby, G. L. (1985) *Penicillin: Meeting the Challenge*, Yale Univ. Press,

[f] Chain, B. (1991) *Nature (London)* **353**, 492–494

[g] Williams, T. I. (1984) *Howard Florey: Penicillin and After*, Oxford Univ. Press, London

[h] Moberg, C. L. (1991) *Science* **253**, 734–735

[i] Nayler, J. H. C. (1991) *Trends Biochem. Sci.* **16**, 195–197

[j] Crease, R. P. (1989) *Science* **246**, 883–884

[k] Gabay, J. E. (1994) *Science* **264**, 373–374

[l] Nayler, J. H. C. (1991) *Trends Biochem. Sci.* **16**, 234–237

[m] Mourey, L., Miyashita, K., Swarén, P., Bulychev, A., Samama, J.-P., and Mobashery, S. (1998) *J. Am. Chem. Soc.* **120**, 9382–9383

[n] Yocum, R. R., Waxman, D. J., and Strominger, J. L. (1980) *Trends Biochem. Sci.* **5**, 97–101

[o] Kelly, J. A., Moews, P. C., Knox, J. R., Frere, J.-M., and Ghuysen, J.-M. (1982) *Science* **218**, 479–480

[p] Kelly, J. A., and Kuzin, A. P. (1995) *J. Mol. Biol.* **254**, 223–236

[q] Englebert, S., Charlier, P., Fonzè, E., To'th, Y., Vermeire, M., Van Beeumen, J., Grandchamps, J., Hoffmann, K., Leyh-Bouille, M., Nguyen-Distèche, M., and Ghuysen, J.-M. (1994) *J. Mol. Biol.* **241**, 295–297

[r] Kuzin, A. P., Liu, H., Kelly, J. A., and Knox, J. R. (1995) *Biochemistry* **34**, 9532–9540

[s] Thunnissen, M. M. G. M., Fusetti, F., de Boer, B., and Dijkstra, B. W. (1995) *J. Mol. Biol.* **247**, 149–153

[t] Siemers, N. O., Yelton, D. E., Bajorath, J., and Senter, P. D. (1996) *Biochemistry* **35**, 2104–2111

[u] Swarén, P., Massova, I., Bellettini, J. R., Bulychev, A., Maveyraud, L., Kotra, L. P., Miller, M. J., Mobashery, S., and Samama, J.-P. (1999) *J. Am. Chem. Soc.* **121**, 5353–53595

[v] Guillaume, G., Vanhove, M., Lamotte-Brasseur, J., Ledent, P., Jamin, M., Joris, B., and Frère, J.-M. (1997) *J. Biol. Chem.* **272**, 5438–5444

[w] Adediran, S. A., Deraniyagala, S. A., Xu, Y., and Pratt, R. F. (1996) *Biochemistry* **35**, 3604–3613

[x] Brown, R. P. A., Aplin, R. T., and Schofield, C. J. (1996) *Biochemistry* **35**, 12421–12432

[y] Carfi, A., Pares, S., Duée, E., Galleni, M., Duez, C., Frère, J. M., and Dideberg, O. (1995) *EMBO J.* **14**, 4914–4921

[z] Swarén, P., Maveyraud, L., Raquet, X., Cabantous, S., Duez, C., Pédelacq, J.-D., Mariotte-Boyer, S., Mourey, L., Labia, R., Nicolas-Chanoine, M.-H., Nordmann, P., Frère, J.-M., and Samama, J.-P. (1998) *J. Biol. Chem.* **273**, 26714–26721

[aa] Trehan, I., Beadle, B. M., and Shoichet, B. K. (2001) *Biochemistry* **40**, 7992–7999

[bb] Swarén, P., Golemi, D., Cabantous, S., Bulychev, A., Maveyraud, L., Mobashery, S., and Samama, J.-P. (1999) *Biochemistry* **38**, 9570–9576

[cc] Wilmouth, R. C., Kassamally, S., Westwood, N. J., Sheppard, R. J., Claridge, T. D. W., Aplin, R. T., Wright, P. A., Pritchard, G. J., and Schofield, C. J. (1999) *Biochemistry* **38**, 7989–7998

[dd] Taylor, P., Anderson, V., Dowden, J., Flitsch, S. L., Turner, N. J., Loughran, K., and Walkinshaw, M. D. (1999) *J. Biol. Chem.* **274**, 24901–24905

[ee] Zhang, H. Z., Hackbarth, C. J., Chansky, K. M., and Chambers, H. F. (2001) *Science* **291**, 1962–1965

BOX 20-H ANTIBIOTIC RESISTANCE AND VANCOMYCIN

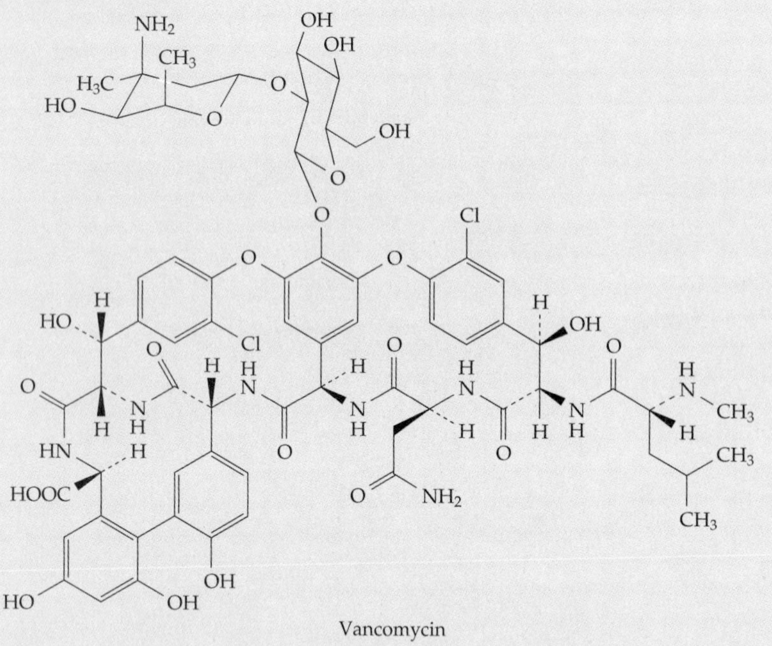

Vancomycin

As antibiotics came into widespread use, an unanticipated problem arose in the rapid development of resistance by bacteria. The problem was made acute by the fact that resistance genes are easily transferred from one bacterium to another by the infectious R-factor plasmids.[a–d] Since resistance genes for many different antibiotics may be carried on the same plasmid, "super bacteria," resistant to a large variety of antibiotics, have developed, often in hospitals.

The problem has reached the crisis stage, perhaps most acutely for tuberculosis. Drug-resistant *Mycobacteria tuberculosis* have emerged, especially, in patients being treated for HIV infection (see Box 21-C). Mechanisms of resistance often involve inactivation of the antibiotics. Aminoglycosides such as streptomycin, spectinomycin, and kanamycin (Box 20-B) are inactivated by en-zymes catalyzing phosphorylation or adenylylation of hydroxyl groups on the sugar rings.[e–g] Penicillin and related antibiotics are inactivated by β-lactamases (Box 20-G). Chloramphenicol (Fig. 25-10) is inactivated by acylation on one or both of the hydroxyl groups.

What is the origin of the drug resistance factors? Why do genes for inactivation of such unusual molecules as the antibiotics exist widely in nature? Apparently the precursor to drug resistance genes fulfill normal biosynthetic roles in nature. An antibiotic-containing environment, such as is found naturally in soil, leads to selection of mutants of such genes with drug-inactivating properties. Never-theless, it is not entirely clear why drug resistance factors have appeared so promptly in the population. Overuse of antibiotics in treating minor infections is one apparent cause.[h,i] Another is probably the widespread use on farms.[j] A nationwide effort to decrease the use of erythromycin in Finland had a very favorable effect in decreasing the incidence of erythromycin-resistant group A streptocci.[h]

Because of the rapid development of resistance, continuous efforts are made to alter antibiotics by semisynthesis (see Box 20-G) and to identify new targets for antibiotics or for synthetic antibacterial compounds.[d] An example is provided by the discovery of vancomycin. Like the penicillins, this antibiotic interferes with bacterial cell wall

teichuronic acids (p. 431). Both proteins and neutral polysaccharides, sometimes covalently bound, may also be present.[303] Like peptidoglycans, teichoic and teichuronic acids are assembled on undecaprenyl phosphates[303] or on molecules of diacylglycerol.[322] Either may serve as an anchor. A "linkage unit" may be formed by transfer of several glycosyl rings onto an anchor unit. For example, in synthesis of ribitol teichoic acid sugar rings are transferred from UDP-GlcNAc, UDP-ManNAC, and CDP-Gal to form the following linkage unit:

$$\text{Lipid}-P-P-\text{GlcNAc}-\text{ManNAc}-(P\text{-Gal})_{2,3}$$

Then many ribitol phosphate units are added by transfer from CDP-ribitol. Finally, the chain is capped by transfer of a glucose from UDP-Glc. Lipoteichoic acids often carry covalently linked D-alanine in ester linkage, altering the net electrical charge on the cell surface.[322a] The completed teichoic acid may then be transferred to a peptidoglycan, releasing the lipid phosphate for reuse.[303] Glycerol teichoic acid may be formed in a similar fashion.[322] Teichuronic acids arise by alternate transfers of *P*-GalNAc from UDP-GalNAc and of GlcA from UDP-GlcA.[303]

Gram-positive bacteria often carry surface proteins that interact with host tissues in establishing human infections. Protein A of *Staphylococcus* is a well-known

BOX 20-H (continued)

synthesis but does so by binding tightly to the D-alanyl-D-alanine termini of peptidoglycans that are involved in crosslinking (Fig. 8-29, Fig. 20-9).[k,l] Like penicillin, vancomycin prevents crosslinking but is unaffected by β-lactamases. Initially bacteria seemed unable to develop resistance to vancomycin, and this antibiotic was for 25 years the drug of choice for β-lactam resistant streptococci or staphylococci. However, during this period bacteria carrying a plasmid with nine genes on the transposon Tn1546 (see Fig. 27-30) developed resistance to vancomycin and were spread worldwide.[l] Vancomycin-resistant bacteria are able to sense the presence of the antibiotic and to synthesize an altered **D-alanine:D-alanine ligase**, the enzyme that joins two D-alanine molecules in an ATP-dependent reaction to form the D-alanyl-D-alanine needed to permit peptidoglycan crosslinking (Fig. 20-9, step *c'*). The altered enzyme adds D-lactate rather than D-alanine providing an – OH terminus in place of $-NH_3^+$. This prevents the binding of vancomycin[l,m] and allows crosslinking of the peptidoglycan via depsipeptide bonds. Another gene in the transposon encodes an oxoacid reductase which supplies the D-lactate.[l] A different resistant strain synthesizes a D-Ala-D-Ser ligase.[n] High-level vancomycin resistance is not attained unless the bacteria also synthesize a D-alanyl-D-alanine dipeptidase.[o]

The D-Ala-D-Ala ligase does provide yet another attractive target for drug design.[p,q] Still another is the D-Ala-D-Ala adding enzyme (Fig. 20-8, step *d*; encoded by the *E. coli MurF* gene).[r,s] Strategies for combatting vancomycin resistance include synthesis of new analogs of the antibiotic[t,u] and simultaneous administration of small molecules that catalyze cleavage of the D-Ala-D-lactate bond formed in cell wall precursors of resistant bacteria.[v]

Another possibility is to use **bacteriophages** directly as antibacterial medicines. This approach was introduced as early as 1919 and has enjoyed considerable success. It is now regarded as a promising alternative to the use of antibiotics in many instances.[w]

[a] Davies, J. (1994) *Science* **264**, 375–382
[b] Benveniste, R., and Davies, J. (1973) *Ann. Rev. Biochem.* **42**, 471–506
[c] Clowes, R. C. (1973) *Sci. Am.* **228**(Apr), 19–27
[d] Neu, H. C. (1992) *Science* **257**, 1064–1073
[e] McKay, G. A., and Wright, G. D. (1996) *Biochemistry* **35**, 8680–8685
[f] Thompson, P. R., Hughes, D. W., Cianciotto, N. P., and Wright, G. D. (1998) *J. Biol. Chem.* **273**, 14788–14795
[g] Cox, J. R., and Serpersu, E. H. (1997) *Biochemistry* **36**, 2353–2359
[h] Seppälä, H., Klaukka, T., Vuopio-Varkila, J., Muotiala, A., Helenius, H., Lager, K., and Huovinen, P. (1997) *N. Engl. J. Med.* **337**, 441–446
[i] Gorbach, S. L. (2001) *N. Engl. J. Med.* **345**, 1202–1203
[j] Witte, W. (1998) *Science* **279**, 996–997
[k] Sheldrick, G. M., Jones, P. G., Kennard, O., Williams, D. H., and Smith, G. A. (1978) *Nature (London)* **271**, 223–225
[l] Walsh, C. T. (1993) *Science* **261**, 308–309
[m] Sharman, G. J., Try, A. C., Dancer, R. J., Cho, Y. R., Staroske, T., Bardsley, B., Maguire, A. J., Cooper, M. A., O'Brien, D. P., and Williams, D. H. (1997) *J. Am. Chem. Soc.* **119**, 12041–12047
[n] Park, I.-S., Lin, C.-H., and Walsh, C. T. (1997) *Proc. Natl. Acad. Sci. U.S.A.* **94**, 10040–10044
[o] Aráoz, R., Anhalt, E., René, L., Badet-Denisot, M.-A., Courvalin, P., and Badet, B. (2000) *Biochemistry* **39**, 15971–15979
[p] Fan, C., Moews, P. C., Walsh, C. T., and Knox, J. R. (1994) *Science* **266**, 439–443
[q] Fan, C., Park, I.-S., Walsh, C. T., and Knox, J. R. (1997) *Biochemistry* **36**, 2531–2538
[r] Duncan, K., van Heijenoort, J., and Walsh, C. T. (1990) *Biochemistry* **29**, 2379–2386
[s] Anderson, M. S., Eveland, S. S., Onishi, H. R., and Pompliano, D. L. (1996) *Biochemistry* **35**, 16264–16269
[t] Walsh, C. (1999) *Science* **284**, 442–443
[u] Ge, M., Chen, Z., Onishi, H. R., Kohler, J., Silver, L. L., Kerns, R., Fukuzawa, S., Thompson, C., and Kahne, D. (1999) *Science* **284**, 507–511
[v] Chiosis, G., and Boneca, I. G. (2001) *Science* **293**, 1484–1487
[w] Stone, R. (2002) *Science* **298**, 728–731

example. After synthesis in the cytoplasm, it enters the secretory pathway. An N-terminal hydrophobic leader sequence and a 35-residue C-terminal sorting signal guide it to the correct destination. There a free amino group of an unlinked pentaglycyl group of the peptidoglycan carries out a transamidation reaction with an LPXTG sequence in the proteins, cutting the chain between the threonine and glycine residues, and anchoring the protein A to the peptidoglycan.[323]

Group A streptococci, which are serious human pathogens, form α-helical coiled-coil threads whose C termini are anchored in the cell membrane. They protrude through the peptidoglycan layers and provide a hairlike layer around the bacteria. A variable region at the N termini provides many antigens, some of which escape the host's immune system allowing infection to develop.[324] Group B streptococci form carbohydrate antigens linked to teichoic acid.[325] Streptococci, which are normally present in the mouth, utilize their carbohydrate surfaces as receptors for adhesion, allowing them to participate in formation of dental plaque.[326]

Cell walls of mycobacteria are composed of a peptidoglycan with covalently attached galactan chains. Branched chains of **arabinan**, a polymer of the furanose ring form of arabinose with covalently attached **mycolic acids**, are glycosidically linked to the galactan.[327] Shorter **glycopeptidolipids**, containing modified glucose and rhamnose rings as well as fatty acids, contribute to the complexity of mycobacterial surfaces.[328]

These examples describe only a small sample of the great diversity of cell coats found in the prokaryotic world. Some bacteria also provide themselves with additional protection in the form of external sheaths of crystalline arrays of proteins known as S-layers.[329]

F. Biosynthesis of Eukaryotic Glycolipids

Glycolipids may be thought of as membrane lipids bearing external oligosaccharides. In this sense, they are similar to glycoproteins both in location and in biological significance. Like the glycoproteins, glycolipids are synthesized in the ER, then transported into the Golgi apparatus and eventually outward to join the outer surface of the plasma membrane. Some glycolipids are attached to proteins by covalent linkage.

1. Glycophosphatidylinositol (GPI) Anchors

More than 100 different human proteins are attached to phosphatidylinositol anchors of the type shown in Fig. 8-13.[329a] Similar anchors are prevalent in yeast and in protozoa including *Leishmania* and *Trypanosoma*,[330–334b] and *Plasmodium*[334c] where they often bind major surface proteins to the plasma membrane.[335] They are also found in mycobacteria.[336] The structures of the hydrophobic anchor ends are all similar.[337] Two or three fatty acyl groups hold the molecule to the bilayer. Variations are found in the attached glycan portion, both in the number of sugar rings and in the structures of the covalently attached phosphoethanol-amine groups.[337–339] A typical assembly pathway is shown in Eq. 20-23. The first step (step *a* in Eq. 20-23), the transfer of an *N*-acetylglucosamine residue to phosphatidylinositol, is surprisingly complex, requiring at least three proteins.[340] The hydrolytic deacetylation (step *b*) helps to drive the synthetic process. Step *c* provides dolichol-*P*-mannose for the GPI anchors as well as for glycoproteins. The phosphoethanolamine part of the structure is added from phosphatidylethanolamine, apparently via direct nucleophilic displacement.[333] In this way the C terminus of the protein forms an amide linkage with the $-NH_2$ group of ethanolamine in the GPI anchor. Another unexpected finding was that this completed anchor unit undergoes "remodeling" during which the fatty acyl chains of the original phosphatidylinositol are replaced by other fatty acids.[339,339a]

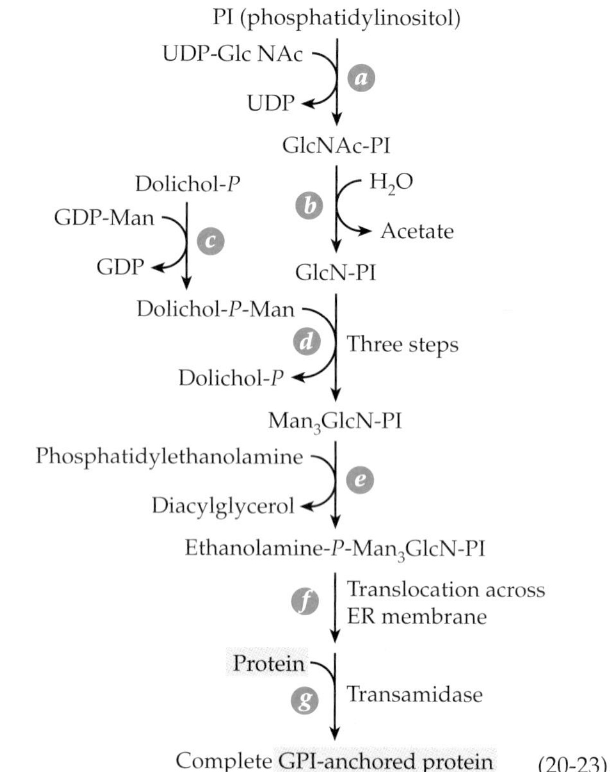

(20-23)

2. Cerebrosides and Gangliosides

These two groups of glycolipids are derived from the *N*-acylated sphingolipids known as **ceramides**. Some biosynthetic pathways from sphingosine to these substances are indicated in Fig. 20-11. Acyl, glycosyl, and sulfo groups are transferred from appropriate derivatives of CoA, CDP, UDP, CMP, and from PAPS to form more than 40 different gangliosides.[341,342] The biosynthesis of a sphingomyelin is also shown in this scheme but is discussed in Chapter 21.

Carbohydrate chain is added here in the glycolipids; phosphocholine in sphingomyelin

H OH

CH_2 — OH

H NH

Long-chain acyl group

R — C

O

A ceramide

Each biosynthetic step in Fig. 20-11 is catalyzed by a specific transferase. Most of these enzymes are present in membranes of the ER and the Golgi.[343–346] Furthermore, the sequence by which the transferases act may not always be fixed, and a complete biosynthetic scheme would be far more complex than is shown in the figure. For example, one alternative sequence is the synthesis of galactosyl ceramide by transfer of galactose to sphingosine followed by acylation (Eq. 20-24). However, the pathway shown in Fig. 20-11 is probably more important.

Sphingosine
↓ (UDP-Gal → UDP)
O-Galactosylsphingosine (psychosine) —(Acyl-CoA)→ Gal-Cer (20-24)

G. The Intracellular Breakdown of Polysaccharides and Glycolipids

The attention of biochemists has been drawn to the importance of pathways of degradation of complex polysaccharides through the existence of at least 35 inherited **lysosomal storage diseases**.[347–351] In many of these diseases one of the 40-odd lysosomal hydrolases is defective or absent.

1. Mucopolysaccharidoses

There are at least seven mucopolysaccharidoses (Table 20-1) in which glycosaminoglycans such as hyaluronic acid accumulate to abnormal levels in tissues and may be excreted in the urine. The diseases cause severe skeletal defects; varying degrees of mental retardation; and early death from liver, kidney, or cardiovascular problems. As in other lysosomal diseases, undegraded material is stored in intracellular inclusions lined by a single membrane. Various tissues are affected to different degrees, and the diseases tend to progress with time.

First described in 1919 by Hurler, **mucopolysaccharidosis I** (MPS I, the **Hurler syndrome**) leads to accumulation of partially degraded dermatan and heparan sulfates (Fig. 4-11).[347,352,353] A standard procedure in the study of diseases of this type is to culture fibroblasts from a skin biopsy. Such cells cultured from patients with the Hurler syndrome accumulate the polysaccharide, but when fibroblasts from a normal person are cultured in the same vessel the defect is "corrected." It was shown that a protein secreted by the normal fibroblasts is taken up by the defective fibroblasts, permitting them to complete the degradation of the stored polysaccharide.

This "Hurler corrective factor" was identified as an **α-L-iduronidase**. In the **Hunter syndrome** (MPS II) dermatan sulfate and heparan sulfate accumulate. The missing enzyme is a **sulfatase** for 2-sulfated iduronate residues.[354,355] The diagram at the bottom of the page illustrates the need for both of these enzymes as well as three others in the degradation of dermatan.[352,356,357]

The **Sanfilippo disease** type A (MPS III) corrective factor is a heparan *N*-sulfatase. However, as is true for many other metabolic diseases, the same symptoms

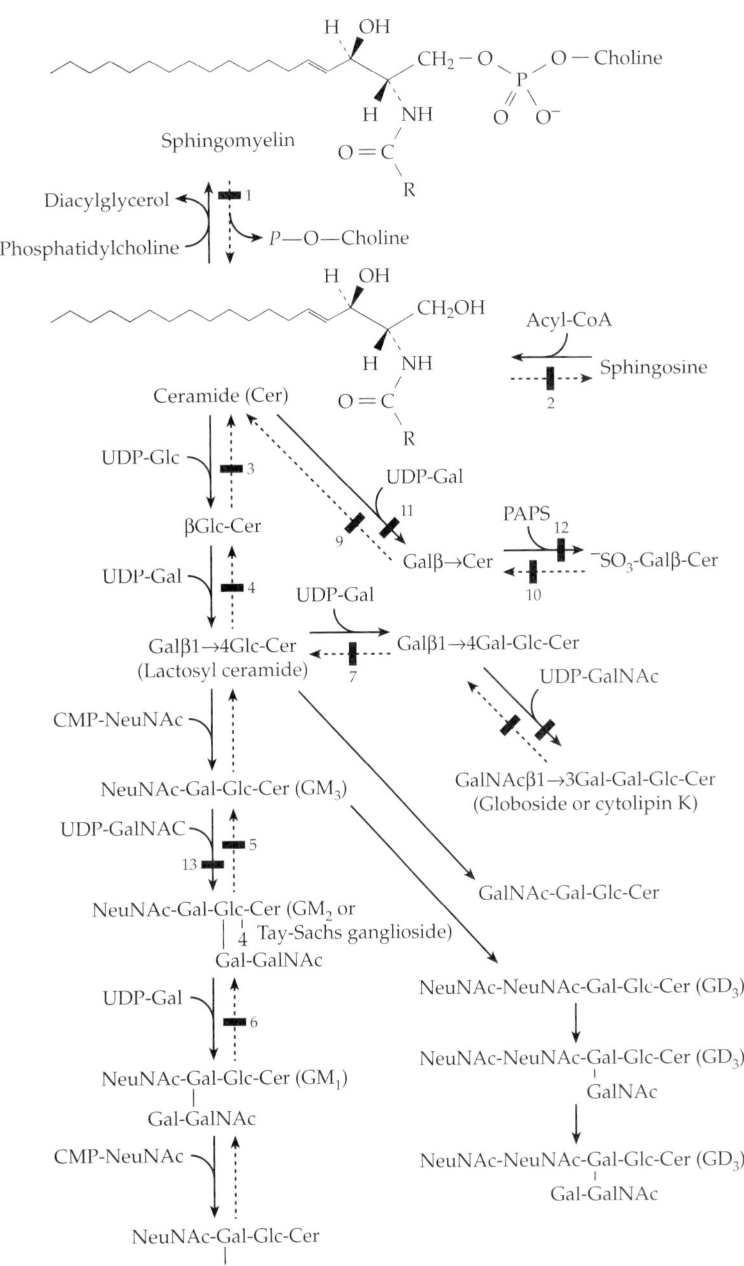

Figure 20-11 Biosynthesis and catabolism of glycosphingolipids. The heavy bars indicate metabolic blocks in known diseases.

may arise from several causes. Thus, Sanfilippo diseases B and D arise from lack of an *N*-acetylglucosaminidase and of a sulfatase for GlcNac-6-sulfate, respectively.[354] In Sanfilippo disease C the missing or defective enzyme is an acetyl transferase that transfers an acetyl group from acetyl-CoA onto the amino groups of glucosamine residues in heparan sulfate fragments. All four of these enzymes are needed to degrade the glucosamine-uronic acid pairs of heparan. The *N*-sulfate groups must be removed by the *N*-sulfatase. The free amino groups formed must then be acetylated before the *N*-acetylglucosaminidase can cut off the GlcNac groups. Remov-al of the 6-sulfate groups requires the fourth enzyme. Completion of the degradation also requires both β-glucuronidase and α-L-iduronidase. Another lysosomal enzyme deficiency, which is most prevalent in Finland, is the absence of **aspartylglucosaminidase**, an N-terminal nucleophile hydrolase (Chapter 12, Section C,3) that cleaves glucosamine from aspartate side chains to which oligosaccharides were attached in glycoproteins.[358–360]

Some hereditary diseases are characterized by lack of two or more lysosomal enzymes. In **I-cell disease** (mucolipidosis II), which resembles the Hurler syndrome, at least ten enzymes are absent or are present at much reduced levels.[350,361] The biochemical defect is the absence from the Golgi cisternae of the *N*-acetylglucosaminyl phosphotransferase that transfers *P*-GlcNAc units from UDP-GlcNAc onto mannose residues (Eq. 20-22) of glycoproteins marked for use in lysosomes.

2. Sphingolipidoses

There are at least ten lysosomal storage diseases, known as sphingolipidoses, that involve the metabolism of the glycolipids. Their biochemical bases are indicated in Fig. 20-11 and in Table 20-1. **Gaucher disease**[362–365] is a result of an autosomal recessive trait that permits glucosyl ceramide to accumulate in macrophages. The liver and spleen are seriously damaged, the latter becoming enlarged to four or five times normal size in the adult form of the disease. In the more severe juvenile form mental retardation occurs. By 1965, it was established that cerebroside is synthesized at a normal rate in the individuals affected, but that a lysosomal hydrolase was missing. This blocked the catabolic pathway indicated by dashed arrows in Fig. 20-11 (block No. 3 in the figure). In many patients a single base change causing a Leu → Pro substitution accounts for the defect. In **Fabry disease** an X-linked gene that provides for removal of galactosyl residues from cerebrosides is defective.[350] This leads to accumulation of the triglycosylceramide whose degradation is blocked at point 7 in Fig. 20-11.

The best known and the commonest sphingolipidosis is **Tay–Sachs disease**.[366–368] Several hundred cases have been reported since it was first described in 1881. A terrible disease, it is accompanied by mental deterioration, blindness, paralysis, dementia, and death by the age of three. About 15 children a year are born in North America with this condition, and the world figure must be 5–7 times this. The defect is in the α subunit of the β-hexosaminidase A (point 7 in Fig. 20-10)[366,366a] with accumulation of ganglioside G_{M2}. Somewhat less severe forms of the disease are caused by different mutations in the same gene[369] or in a protein activator. **Sandhoff disease**, which resembles Tay–Sachs disease, is caused by a defect in the β subunit, which is present in both β-hexosaminidases A and B.[368] Mutant "knockout" mice that produce only ganglioside GM_3 as the major ganglioside in their central nervous system die suddenly from seizures if they hear a loud sound. This provides further evidence of the essential nature of these components of nerve membranes.[369a]

3. Causes of Lysosomal Diseases

The descriptions given here have been simplified. For many lysosomal diseases there are mild and severe forms and infantile or juvenile forms to be contrasted with adult forms. Some of the enzymes exist as multiple isozymes. An enzyme may be completely lacking or

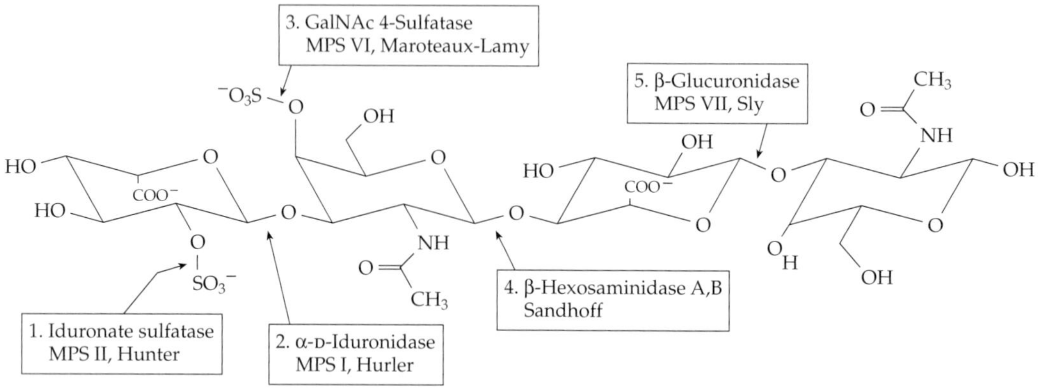

3. GalNAc 4-Sulfatase MPS VI, Maroteaux-Lamy
5. β-Glucuronidase MPS VII, Sly
4. β-Hexosaminidase A,B Sandhoff
1. Iduronate sulfatase MPS II, Hunter
2. α-D-Iduronidase MPS I, Hurler

TABLE 20-1
Lysosomal Storage Diseases: Sphingolipidoses and Mucopolysaccharidoses[a]

No. in Fig. 20-11	Name	Defective enzyme
1.	Niemann–Pick disease[b]	Sphingomyelinase
2.	Farber disease (lipogranulomatosis)	Ceramidase
3.	Gaucher disease[c]	β-Glucocerebrosidase
4.	Lactosyl ceramidosis	β-Galactosyl hydrolase
5.	Tay–Sachs disease[c]	β-Hexosaminidase A
6.	G_{M1} gangliosidosis[d]	β-Galactosidase
7.	Fabry disease[c]	α-Galactosidase
8.	Sandhoff disease[e]	β-Hexosaminidases A and B
9.	Globoid cell leukodystrophy	Galactocerebrosidase
10.	Metachromatic leukodystrophy	Arylsulfatase A
13.	Hematoside (G_{M3}) accumulation	G_{M3}-*N*-acetylgalactosaminyltransferase
	Pompe disease[f]	α-Glucosidase
	Hurler syndrome (MPS I)[c]	α-L-Iduronidase
	Hunter syndrome (MPS II)[c]	Iduronate 2-sulfate sulfatase
	Sanfilippo disease[c,g]	
	Type A (MPS III)	Heparan *N*-sulfatase
	Type B	*N*-Acetylglucosaminidase
	Type C	Acetyl-CoA: α-glucosaminide *N*-acetyltransferase
	Type D	GlcNAc-6-sulfate sulfatase
	Maroteaux–Lamy syndrome (MPS VI)[g]	Arylsulfatase B
	Sly syndrome (MPS VII)[g,h]	β-Glucuronidase
	Aspartylglycosaminuria[i]	Aspartylglucosaminidase
	Mannosidosis	β-Mannosidase
	Fucosidosis	α-L-Fucosidase
	Mucolipidosis	α-*N*-Acetylneuraminidase
	Sialidosis[j]	

[a] A general reference is Scriver, C. R., Beaudet, A. L., Sly, W. S., and Valle, D., eds. (1995) *The Metabolic and Molecular Bases of Inherited Disease*, 7th ed., Vol. 1, McGraw-Hill, New York (pp. 2427–2879)
[b] Wenger, D. A., Sattler, M., Kudoh, T., Snyder, S. P., and Kingston, R. S. (1980) *Science* **208**, 1471–1473
[c] See main text
[d] Hoogeveen, A. T., Reuser, A. J. J., Kroos, M., and Galjaard, H. (1986) *J. Biol. Chem.* **261**, 5702–5704
[e] Gravel, R. A., Clarke, J. T. R., Kaback, M. M., Mahuran, D., Sandhoff, K., and Suzuki, K. (1995) in *The Metabolic and Molecular Bases of Inherited Disease*, 7th ed., Vol. 1 (Scriver, C. R., Beaudet, A. L., Sly, W. S., and Valle, D., eds), pp. 2839–2879, McGraw-Hill, New York
[f] See Box 20-D
[g] Neufeld, E. F., and Muenzer, J. (1995) in *The Metabolic and Molecular Bases of Inherited Disease*, 7th ed., Vol. 2 (Scriver, C. R., Beaudet, A. L., Sly, W. S., and Valle, D., eds), pp. 2465 – 2494, McGraw-Hill, New York
[h] Wu, B. M., Tomatsu, S., Fukuda, S., Sukegawa, K., Orii, T., and Sly, W. S. (1994) *J. Biol. Chem.* **269**, 23681–23688
[i] Mononen, I., Fisher, K. J., Kaartinen, V., and Aronson, N. N., Jr. (1993) *FASEB J.* **7**, 1247–1256
[j] Seppala, R., Tietze, F., Krasnewich, D., Weiss, P., Ashwell, G., Barsh, G., Thomas, G. H., Packman, S., and Gahl, W. A. (1991) *J. Biol. Chem.* **266**, 7456–7461

may be low in concentration. The causes of the deficiencies may include total absence of the gene, absence of the appropriate mRNA, impaired conversion of a proenzyme to active enzyme, rapid degradation of a precursor or of the enzyme itself, incorrect transport of the enzyme precursor to its proper destination, presence of mutations that inactivate the enzyme, or absence of protective proteins. Several lysosomal hydrolases require auxiliary **activator proteins** that allow them to react with membrane-bound substrates.[350,351,365,370]

4. Can Lysosomal Diseases Be Treated?

There has been some success in using enzyme replacement therapy for lysosomal deficiency diseases.[347,371-371b] One approach makes use of the fact that the mannose-6-*P* receptors of the plasma membrane take up suitably marked proteins and transfer them into lysosomes (Section C,2). The missing enzyme might simply be injected into the patient's bloodstream from which it could be taken up into the lysosomes.[372] This carries a risk of allergic reaction, and it may be safer to attempt microencapsulation of the enzyme, perhaps in ghosts from the patient's own erythrocytes.[347] A second approach, which has had limited success, is transplantation of an organ[371] or of bone marrow[373] from a donor with a normal gene for the missing enzyme. This is dangerous and is little used at present. However, new hope is offered by the possibility of transferring a gene for the missing enzyme into some of the patient's cells. For example, the cloned gene for the transferase missing in Gaucher disease has been transferred into cultured cells from Gaucher disease patients with apparent correction of the defect.[374] Long-term correction of the Hurler syndrome in bone marrow cells also provides hope for an effective therapy involving gene transfer into a patient's own bone marrow cells[375] or transplantation of selected hematopoietic cells.[375a]

In the cases of Gaucher disease and Fabry disease, it is hoped that treatment of infants and young children may prevent brain damage. However, in Tay–Sachs disease the primary sites of accumulation of the ganglioside GM_2 are the ganglion and glial cells of the brain. Because of the "blood–brain barrier" and the severity of the damage it seems less likely that the disease can be treated successfully.

The approach presently used most often consists of identifying carriers of highly undesirable genetic traits and offering genetic counseling. For example, if both parents are carriers the risk of bearing a child with Tay–Sachs disease is one in four. Women who have borne a previous child with the disease usually have the genetic status of the fetus checked by **amniocentesis**. A sample of the amniotic fluid surrounding the fetus is withdrawn during the 16th to 18th week of pregnancy. The fluid contains fibroblasts that have become detached from the surface of the fetus. These cells are cultured for 2–3 weeks to provide enough cells for a reliable assay of the appropriate enzymes. Such tests for a variety of defects are becoming faster and more sensitive as new techniques are applied.[376] In the case of Tay–Sachs disease, most women who have one child with the disease choose abortion if a subsequent fetus has the disease.

The diseases considered here affect only a small fraction of the problems in the catabolism of body constituents. On the other hand, fewer cases are on record of deficiencies in biosynthetic pathways. These are more often absolutely lethal and lead to early spontaneous abortion. However, blockages in the biosynthesis of cerebrosides are known in the special strains of mice known as Jimpy, Quaking, and msd (myelin synthesis deficient).[377,378] The transferases (points 11 and 12 of Fig. 20-11) are not absent but are of low activity. The mice have distinct neurological defects and poor myelination of nerves in the brain. A human ailment involving impaired conversion of GM_3 to GM_2 (with accumulation of the former; point 13 of Fig. 20-11) has been reported. Excessive synthesis of sialic acid causes the rare human **sialuria.**[379] This is apparently a result of a failure in proper feedback inhibition.

Animals suffer many of the same metabolic diseases as humans. Among these are a large number of lysosomal deficiency diseases.[380] Their availability means that new methods of treating the diseases may, in many cases, be tried first on animals. For example, enzyme replacement therapy for the Hurler syndrome is being tested in dogs.[381] Bone marrow transplantation for human **α-mannosidosis** is being tested in cats with a similar disease.[382] Mice with a hereditary deficiency of β-glucuronidase are being treated by gene transfer from normal humans.[357]

References

1. Cardini, C. E., Paladini, A. C., Caputto, R., and Leloir, L. F. (1950) *Nature (London)* **165**, 191–192
2. Grisolía, S. (1988) *Nature (London)* **331**, 212
3. Hallfrisch, J. (1990) *FASEB J.* **4**, 2652–2660
4. Jefferey, J., and Jornvall, H. (1983) *Proc. Natl. Acad. Sci. U.S.A.* **80**, 901–905
5. Cox, T. M. (1994) *FASEB J.* **8**, 62–71
6. Gitzelmann, R., Steinmann, B., and Van den Berghe, G. (1995) in *The Metabolic and Molecular Bases of Inherited Disease*, 7th ed., Vol. 1 (Scriver, C. R., Beaudet, A. L., Sly, W. S., and Valle, D., eds), pp. 905–935, McGraw-Hill, New York
7. Gopher, A., Vaisman, N., Mandel, H., and Lapidot, A. (1990) *Proc. Natl. Acad. Sci. U.S.A.* **87**, 5449–5453
8. Hope, J. N., Bell, A. W., Hermodson, M. A., and Groarke, J. M. (1986) *J. Biol. Chem.* **261**, 7663–7668
9. Ridley, W. P., Houchins, J. P., and Kirkwood, S. (1975) *J. Biol. Chem.* **250**, 8761–8767

9a. Campbell, R. E., Mosimann, S. C., van de Rijn, I., Tanner, M. E., and Strynadka, N. C. J. (2000) *Biochemistry* **39**, 7012–7023

10. Li, J.-p, Hagner-McWhirter, Å., Kjellén, L., Palgi, J., Jalkanen, M., and Lindahl, U. (1997) *J. Biol. Chem.* **272**, 28158–28163
11. Segal, S., and Berry, G. T. (1995) in *The Metabolic and Molecular Bases of Inherited Disease*, 7th ed., Vol. 1 (Scriver, C. R., Beaudet, A. L., Sly, W. S., and Valle, D., eds), pp. 967–1000, McGraw-Hill, New York
12. Ruzicka, F. J., Wedekind, J. E., Kim, J., Rayment, I., and Frey, P. A. (1995) *Biochemistry* **34**, 5610–5617
13. Wedekind, J. E., Frey, P. A., and Rayment, I. (1996) *Biochemistry* **35**, 11560–11569
14. Thoden, J. B., Ruzicka, F. J., Frey, P. A., Rayment, I., and Holden, H. M. (1997) *Biochemistry* **36**, 1212–1222
15. Frey, P. A. (1996) *FASEB J.* **10**, 461–470

15a. Lai, K., Willis, A. C., and Elsas, L. J. (1999) *J. Biol. Chem.* **274**, 6559–6566

16. Thoden, J. B., and Holden, H. M. (1998) *Biochemistry* **37**, 11469–11477

16a. Thoden, J. B., Wohlers, T. M., Fridovich-Keil, J. L., and Holden, H. M. (2001) *J. Biol. Chem.* **276**, 15131–15136

16b. Thoden, J. B., Wohlers, T. M., Fridovich-Keil, J. L., and Holden, H. M. (2001) *J. Biol. Chem.* **276**, 20617–20623

17. Beebe, J. A., and Frey, P. A. (1998) *Biochemistry* **37**, 14989–14997

17a. Zhang, Q., and Liu, H.-w. (2001) *J. Am. Chem. Soc.* **123**, 6756–6766

18. Wong, Y.-H. H., and Sherman, W. R. (1985) *J. Biol. Chem.* **260**, 11083–11090
19. Migaud, M. E., and Frost, J. W. (1996) *J. Am. Chem. Soc.* **118**, 495–501

19a. Tian, F., Migaud, M. E., and Frost, J. W. (1999) *J. Am. Chem. Soc.* **121**, 5795–5796

19b. Chen, L., Zhou, C., Yang, H., and Roberts, M. F. (2000) *Biochemistry* **39**, 12415–12423

20. Loewus, M. W., Loewus, F. A., Brillinger, G. U., Otsuka, H., and Floss, H. G. (1980) *J. Biol. Chem.* **255**, 11710–11712
21. Lapan, E. A. (1975) *Exp. Cell. Res.* **94**, 277–282
22. Holub, B. J. (1992) *N. Engl. J. Med.* **326**, 1285–1287

22a. Nestler, J. E., Jakubowicz, D. J., Reamer, P., Gunn, R. D., and Allan, G. (1999) *N. Engl. J. Med.* **340**, 1314–1320

23. Reddy, C. C., Swan, J. S., and Hamilton, G. A. (1981) *J. Biol. Chem.* **256**, 8510–8518
24. Nishikimi, M., and Yagi, K. (1996) in *Subcellular Biochemistry*, Vol. 25 (Harris, J. R., ed), pp. 17–39, Plenum, New York
25. Koshizaka, T., Nishikimi, M., Ozawa, T., and Yagi, K. (1988) *J. Biol. Chem.* **263**, 1619–1621
26. Nishikimi, M., Fukuyama, R., Minoshima, S., Shimizu, N., and Yagi, K. (1994) *J. Biol. Chem.* **269**, 13685–13688

26a. Maeda, N., Hagihara, H., Nakata, Y., Hiller, S., Wilder, J., and Reddick, R. (2000) *Proc. Natl. Acad. Sci. U.S.A.* **97**, 841–846

26b. Tsukaguchi, H., Tokui, T., Mackenzie, B., Berger, U. V., Chen, X.-Z., Wang, Y., Brubaker, R. F., and Hediger, M. A. (1999) *Nature (London)* **399**, 70–75

27. Miller, J. V., Estell, D. A., and Lazarus, R. A. (1987) *J. Biol. Chem.* **262**, 9016–9020
28. Wheeler, G. L., Jones, M. A., and Smirnoff, N. (1998) *Nature (London)* **393**, 365–369
29. Conklin, P. L., Norris, S. R., Wheeler, G. L., Williams, E. H., Smirnoff, N., and Last, R. L. (1999) *Proc. Natl. Acad. Sci. U.S.A.* **96**, 4198–4203
30. Hiatt, H. H. (1995) in *The Metabolic and Molecular Bases of Inherited Disease*, 7th ed., Vol. 1 (Scriver, C. R., Beaudet, A. L., Sly, W. S., and Valle, D., eds), pp. 1001–1010, McGraw-Hill, New York
31. Massière, F., Badet-Denisot, M.-A., René, L., and Badet, B. (1997) *J. Am. Chem. Soc.* **119**, 5748–5749
32. Bearne, S. L., and Wolfenden, R. (1995) *Biochemistry* **34**, 11515–11520
33. Badet, B., Vermoote, P., Haumont, P. Y., Lederer, F., and Le Goffic, F. (1987) *Biochemistry* **26**, 1940–1948
34. Golinelli-Pimpaneau, B., Le Goffic, F., and Badet, B. (1989) *J. Am. Chem. Soc.* **111**, 3029–3034

34a. Bearne, S. L., and Blouin, C. (2000) *J. Biol. Chem.* **275**, 135–140

35. Leriche, C., Badet-Denisot, M.-A., and Badet, B. (1996) *J. Am. Chem. Soc.* **118**, 1797–1798
36. Mio, T., Yamada-Okabe, T., Arisawa, M., and Yamada-Okabe, H. (1999) *J. Biol. Chem.* **274**, 424–429
37. Mengin-Lecreulx, D., and van Heijenoort, J. (1996) *J. Biol. Chem.* **271**, 32–39

37a. Sulzenbacher, G., Gal, L., Peneff, C., Fassy, F., and Bourne, Y. (2001) *J. Biol. Chem.* **276**, 11844–11851

37b. Olsen, L. R., and Roderick, S. L. (2001) *Biochemistry* **40**, 1913–1921

38. Montero-Morán, G. M., Horjales, E., Calcagno, M. L., and Altamirano, M. M. (1998) *Biochemistry* **37**, 7844–7849
39. Wolosker, H., Kline, D., Bian, Y., Blackshaw, S., Cameron, A. M., Fralich, T. J., Schnaar, R. L., and Snyder, S. H. (1998) *FASEB J.* **12**, 91–99
40. Ramilo, C., Appleyard, R. J., Wanke, C., Krekel, F., Amrhein, N., and Evans, J. N. S. (1994) *Biochemistry* **33**, 15071–15079
41. Kim, D. H., Lees, W. J., Kempsell, K. E., Lane, W. S., Duncan, K., and Walsh, C. T. (1996) *Biochemistry* **35**, 4923–4928

41a. Krekel, F., Oecking, C., Amrhein, N., and Macheroux, P. (1999) *Biochemistry* **38**, 8864–8878

41b. Krekel, F., Samland, A. K., Macheroux, P., Amrhein, N., and Evans, J. N. S. (2000) *Biochemistry* **39**, 12671–12677

41c. Samland, A. K., Etezady-Esfarjani, T., Amrhein, N., and Macheroux, P. (2001) *Biochemistry* **40**, 1550–1559

42. Benson, T. E., Walsh, C. T., and Hogle, J. M. (1997) *Biochemistry* **36**, 806–811
43. Benson, T. E., Walsh, C. T., and Massey, V. (1997) *Biochemistry* **36**, 796–805

43a. Benson, T. E., Harris, M. S., Choi, G. H., Cialdella, J. I., Herberg, J. T., Martin, J. P., Jr., and Baldwin, E. T. (2001) *Biochemistry* **40**, 2340–2350

44. Pastuszak, I., O'Donnell, J., and Elbein, A. D. (1996) *J. Biol. Chem.* **271**, 23653–23656

44a. Zhao, X., Creuzenet, C., Bélanger, M., Egbosimba, E., Li, J., and Lam, J. S. (2000) *J. Biol. Chem.* **275**, 33252–33259

45. Schauer, R. (1985) *Trends Biochem. Sci.* **10**, 357–360
46. Simon, E. S., Bednarski, M. D., and Whitesides, G. M. (1988) *J. Am. Chem. Soc.* **110**, 7159–7163
47. Morgan, P. M., Sala, R. F., and Tanner, M. E. (1997) *J. Am. Chem. Soc.* **119**, 10269–10277

47a. Campbell, R. E., Mosimann, S. C., Tanner, M. E., and Strynadka, N. C. J. (2000) *Biochemistry* **39**, 14993–15001

47b. Keppler, O. T., Hinderlich, S., Langner, J., Schwartz-Albiez, R., Reutter, W., and Pawlita, M. (1999) *Science* **284**, 1372–1376

47c. Itoh, T., Mikami, B., Maru, I., Ohta, Y., Hashimoto, W., and Murata, K. (2000) *J. Mol. Biol.* **303**, 733–744

47d. Jacobs, C. L., Goon, S., Yarema, K. J., Hinderlich, S., Hang, H. C., Chai, D. H., and Bertozzi, C. R. (2001) *Biochemistry* **40**, 12864–12874

48. Stäsche, R., Hinderlich, S., Weise, C., Effertz, K., Lucka, L., Moormann, P., and Reutter, W. (1997) *J. Biol. Chem.* **272**, 24319–24324

48a. Jordan, P. A., Bohle, D. S., Ramilo, C. A., and Evans, J. N. S. (2001) *Biochemistry* **40**, 8387–8396

49. Dotson, G. D., Dua, R. K., Clemens, J. C., Wooten, E. W., and Woodard, R. W. (1995) *J. Biol. Chem.* **270**, 13698–13705
50. Sheflyan, G. Y., Howe, D. L., Wilson, T. L., and Woodard, R. W. (1998) *J. Am. Chem. Soc.* **120**, 11028–11032
51. Kaustov, L., Kababya, S., Du, S., Baasov, T., Gropper, S., Shoham, Y., and Schmidt, A. (2000) *Biochemistry* **39**, 14865–14876
52. Radaev, S., Dastidar, P., Patel, M., Woodard, R. W., and Gatti, D. L. (2000) *J. Biol. Chem.* **275**, 9476–9484

52a. Duewel, H. S., Radaev, S., Wang, J., Woodard, R. W., and Gatti, D. L. (2001) *J. Biol. Chem.* **276**, 8393–8402

52b. Wagner, T., Kretsinger, R. H., Bauerle, R., and Tolbert, W. D. (2000) *J. Mol. Biol.* **301**, 233–238

52c. Mosimann, S. C., Gilbert, M., Dombroswki, D., To, R., Wakarchuk, W., and Strynadka, N. C. J. (2001) *J. Biol. Chem.* **276**, 8190–8196

53. Kawano, T., Koyama, S., Takematsu, H., Kozutsumi, Y., Kawasaki, H., Kawashima, S., Kawasaki, T., and Suzuki, A. (1995) *J. Biol. Chem.* **270**, 16458–16463
54. Nishino, S., Kuroyanagi, H., Terada, T., Inoue, S., Inoue, Y., Troy, F. A., and Kitajima, K. (1996) *J. Biol. Chem.* **271**, 2909–2913
55. Inoue, S., Kitajima, K., and Inoue, Y. (1996) *J. Biol. Chem.* **271**, 24341–24344

55a. Angata, T., Nakata, D., Matsuda, T., Kitajima, K., and Troy, F. A., II. (1999) *J. Biol. Chem.* **274**, 22949–22956

55b. Lawrence, S. M., Huddleston, K. A., Pitts, L. R., Nguyen, N., Lee, Y. C., Vann, W. F., Coleman, T. A., and Betenbaugh, M. J. (2000) *J. Biol. Chem.* **275**, 17869–17877

56. Kohlbrenner, W. E., Nuss, M. M., and Fesik, S. W. (1987) *J. Biol. Chem.* **262**, 4534–4537

56a. Royo, J., Gómez, E., and Hueros, G. (2000) *J. Biol. Chem.* **275**, 24993–24999

56b. Jelakovic, S., and Schulz, G. E. (2001) *J. Mol. Biol.* **312**, 143–155

57. Kontrohr, T., and Kocsis, B. (1981) *J. Biol. Chem.* **256**, 7715–7718

References

58. Ding, L., Seto, B. L., Ahmed, S. A., and Coleman, W. G., Jr. (1994) *J. Biol. Chem.* **269**, 24384–24390
58a. Ni, Y., McPhie, P., Deacon, A., Ealick, S., and Coleman, W. G., Jr. (2001) *J. Biol. Chem.* **276**, 27329–27334
58b. Kneidinger, B., Graninger, M., Puchberger, M., Kosma, P., and Messner, P. (2001) *J. Biol. Chem.* **276**, 20935–20944
59. Sharon, N. (1975) *Complex Carbohydrates*, Addison-Wesley, Reading, Massachusetts (pp. 131–138)
59a. Blankenfeldt, W., Asuncion, M., Lam, J. S., and Naismith, J. H. (2000) *EMBO J.* **19**, 6652–6663
59b. Hegeman, A. D., Gross, J. W., and Frey, P. A. (2001) *Biochemistry* **40**, 6598–6610
59c. Gross, J. W., Hegeman, A. D., Gerratana, B., and Frey, P. A. (2001) *Biochemistry* **40**, 12497–12504
59d. Allard, S. T. M., Giraud, M.-F., Whitfield, C., Graninger, M., Messner, P., and Naismith, J. H. (2001) *J. Mol. Biol.* **307**, 283–295
59e. Christendat, D., Saridakis, V., Dharamsi, A., Bochkarev, A., Pai, E. F., Arrowsmith, C. H., and Edwards, A. M. (2000) *J. Biol. Chem.* **275**, 24608–24612
59f. Kneidinger, B., Graninger, M., Adam, G., Puchberger, M., Kosma, P., Zayni, S., and Messner, P. (2001) *J. Biol. Chem.* **276**, 5577–5583
60. Chan, J. Y., Nwokoro, N. A., and Schachter, H. (1979) *J. Biol. Chem.* **251**, 7060–7068
60a. He, X. M., and Liu, H.-w. (2002) *Ann. Rev. Biochem.* **71**, 701–754
61. Bonin, C. P., Potter, I., Vanzin, G. F., and Reiter, W.-D. (1997) *Proc. Natl. Acad. Sci. U.S.A.* **94**, 2085–2090
61a. Menon, S., Stahl, M., Kumar, R., Xu, G.-Y., and Sullivan, F. (1999) *J. Biol. Chem.* **274**, 26743–26750
61b. Rosano, C., Bisso, A., Izzo, G., Tonetti, M., Sturla, L., De Flora, A., and Bolognesi, M. (2000) *J. Mol. Biol.* **303**, 77–91
62. Rubenstein, P. A., and Strominger, J. L. (1974) *J. Biol. Chem.* **249**, 3776–3781
63. Pieper, P. A., Guo, Z., and Liu, H.-w. (1995) *J. Am. Chem. Soc.* **117**, 5158–5159
64. Johnson, D. A., Gassner, G. T., Bandarian, V., Ruzicka, F. J., Ballou, D. P., Reed, G. H., and Liu, H.-w. (1996) *Biochemistry* **35**, 15846–15856
65. Chen, X. M. H., Ploux, O., and Liu, H.-w. (1996) *Biochemistry* **35**, 16412–16420
65a. Hallis, T. M., Zhao, Z., and Liu, H.-w. (2000) *J. Am. Chem. Soc.* **122**, 10493–10503
66. Chang, C.-W. T., Chen, X. H., and Liu, H.-w. (1998) *J. Am. Chem. Soc.* **120**, 9698–9699
67. Zhao, L., Que, N. L. S., Xue, Y., Sherman, D. H., and Liu, H.-w. (1998) *J. Am. Chem. Soc.* **120**, 12159–12160
68. Toshima, K., Nozaki, Y., Mukaiyama, S., Tamai, T., Nakata, M., Tatsuta, K., and Kinoshita, M. (1995) *J. Am. Chem. Soc.* **117**, 3717–3727
69. Benning, C., Beatty, J. T., Prince, R. C., and Somerville, C. R. (1993) *Proc. Natl. Acad. Sci. U.S.A.* **90**, 1561–1565
70. Benson, A. A. (1963) *Adv. Lipid Res.* **1**, 387–394
70a. Mulichak, A. M., Theisen, M. J., Essigmann, B., Benning, C., and Garavito, R. M. (1999) *Proc. Natl. Acad. Sci. U.S.A.* **96**, 13097–13102
70b. Sanda, S., Leustek, T., Theisen, M. J., Garavito, R. M., and Benning, C. (2001) *J. Biol. Chem.* **276**, 3941–3946
71. Preiss, J. (1984) *Trends Biochem. Sci.* **9**, 24–27
71a. Lunn, J. E., Ashton, A. R., Hatch, M. D., and Heldt, H. W. (2000) *Proc. Natl. Acad. Sci. U.S.A.* **97**, 12914–12919
72. Singh, A. N., Hester, L. S., and Raushel, F. M. (1987) *J. Biol. Chem.* **262**, 2554–2557
73. Hunziker, W., Spiess, M., Semenza, G., and Lodish, H. F. (1986) *Cell* **46**, 227–234
74. Naim, H. Y., Sterchi, E. E., and Lentze, M. J. (1988) *J. Biol. Chem.* **263**, 19709–19717
75. Wolschek, M. F., and Kubicek, C. P. (1997) *J. Biol. Chem.* **272**, 2729–2735
76. Horlacher, R., and Boos, W. (1997) *J. Biol. Chem.* **272**, 13026–23032
76a. Hars, U., Horlacher, R., Boos, W., Welte, W., and Diederichs, K. (1998) *Protein Sci.* **7**, 2511–2521
77. Kandror, O., DeLeon, A., and Goldberg, A. L. (2002) *Proc. Natl. Acad. Sci. U.S.A.* **99**, 9727–9732
77a. Diederichs, K., Diez, J., Greller, G., Müller, C., Breed, J., Schnell, C., Vonrhein, C., Boos, W., and Welte, W. (2000) *EMBO J.* **19**, 5951–961
77b. Diez, J., Diederichs, K., Greller, G., Horlacher, R., Boos, W., and Welte, W. (2001) *J. Mol. Biol.* **305**, 905–915
78. Shaper, N. L., Hollis, G. F., Douglas, J. G., Kirsch, I. R., and Shaper, J. H. (1988) *J. Biol. Chem.* **263**, 10420–10428
79. Rajput, B., Shaper, N. L., and Shaper, J. H. (1996) *J. Biol. Chem.* **271**, 5131–5142
80. Yadav, S. P., and Brew, K. (1991) *J. Biol. Chem.* **266**, 698–703
80a. Gastinel, L. N., Cambillau, C., and Bourne, Y. (1999) *EMBO J.* **18**, 3546–3557
80b. Ramakrishnan, B., Shah, P. S., and Qasba, P. K. (2001) *J. Biol. Chem.* **276**, 37665–37671
81. Montgomery, R. K., Büller, H. A., Rings, E. H. H. M., and Grand, R. J. (1991) *FASEB J.* **5**, 2824–2832
81a. Potera, C. (1998) *Science* **281**, 1793
81b. Barbier, O., Girard, C., Breton, R., Bélanger, A., and Hum, D. W. (2000) *Biochemistry* **39**, 11540–11552
81c. Lévesque, E., Turgeon, D., Carrier, J.-S., Montminy, V., Beaulieu, M., and Bélanger, A. (2001) *Biochemistry* **40**, 3869–3881
82. Lewis, D. A., and Armstrong, R. N. (1983) *Biochemistry* **22**, 6297–6303
83. Hanessian, S., and Haskell, T. H. (1970) in *The Carbohydrates*, 2nd ed., Vol. 2A (Pigman, W., and Horton, D., eds), pp. 139–211, Academic Press, New York
84. Snell, J. F. (1966) *Biosynthesis of Antibiotics*, Vol. 1, Academic Press, New York
85. Botti, M. G., Taylor, M. G., and Botting, N. P. (1995) *J. Biol. Chem.* **270**, 20530–20535
86. Cottaz, S., Henrissat, B., and Driguez, H. (1996) *Biochemistry* **35**, 15256–15259
86a. Burmeister, W. P., Cottaz, S., Rollin, P., Vasella, A., and Henrissat, B. (2000) *J. Biol. Chem.* **275**, 39385–39393
87. Lomako, J., Lomako, W. M., and Whelan, W. J. (1988) *FASEB J.* **2**, 3097–3103
87a. Cid, E., Gomis, R. R., Geremia, R. A., Guinovart, J. J., and Ferrer, J. C. (2000) *J. Biol. Chem.* **275**, 33614–33621
88. Baecker, P. A., Greenberg, E., and Preiss, J. (1986) *J. Biol. Chem.* **261**, 8738–8743
89. Hill, M. A., Kaufmann, K., Otero, J., and Preiss, J. (1991) *J. Biol. Chem.* **266**, 12455–12460
90. Furukawa, K., Tagaya, M., Tanizawa, K., and Fukui, T. (1994) *J. Biol. Chem.* **269**, 868–871
91. Ochoa, S. (1985) *Trends Biochem. Sci.* **10**, 147–150
92. Kim, S. C., Singh, A. N., and Raushel, F. M. (1988) *J. Biol. Chem.* **263**, 10151–10154
93. Skurat, A. V., Wang, Y., and Roach, P. J. (1994) *J. Biol. Chem.* **269**, 25534–25542
94. Shulman, R. G., Bloch, G., and Rothman, D. L. (1995) *Proc. Natl. Acad. Sci. U.S.A.* **92**, 8535–8542
95. Villar-Palasi, C., and Guinovart, J. J. (1997) *FASEB J.* **11**, 544–558
96. Printen, J. A., Brady, M. J., and Saltiel, A. R. (1997) *Science* **275**, 1475–1478
96a. Halse, R., Rochford, J. J., McCormack, J. G., Vandenheede, J. R., Hemmings, B. A., and Yeaman, S. J. (1999) *J. Biol. Chem.* **274**, 776–780
96b. Oikonomakos, N. G., Schnier, J. B., Zographos, S. E., Skamnaki, V. T., Tsitsanou, K. E., and Johnson, L. N. (2000) *J. Biol. Chem.* **275**, 34566–34573
97. Thon, V. J., Khalil, M., and Cannon, J. F. (1993) *J. Biol. Chem.* **268**, 7509–7513
98. Takrama, J., and Madsen, N. B. (1988) *Biochemistry* **27**, 3308–3314
99. Yang, B.-Z., Ding, J.-H., Enghild, J. J., Bao, Y., and Chen, Y.-T. (1992) *J. Biol. Chem.* **267**, 9294–9299
99a. Nakayama, A., Yamamoto, K., and Tabata, S. (2001) *J. Biol. Chem.* **276**, 28824–28828
100. Alonso, M. D., Lomako, J., Lomako, W. M., and Whelan, W. J. (1995) *FASEB J.* **9**, 1126–1137
101. Blumenfeld, M. L., and Krisman, C. R. (1985) *J. Biol. Chem.* **260**, 11560–11566
101a. Pederson, B. A., Cheng, C., Wilson, W. A., and Roach, P. J. (2000) *J. Biol. Chem.* **275**, 27753–27761
102. Alonso, M. D., Lomako, J., Lomako, W. M., and Whelan, W. J. (1995) *J. Biol. Chem.* **270**, 15315–15319
103. Ercan, N., Gannon, M. C., and Nuttall, F. Q. (1994) *J. Biol. Chem.* **269**, 22328–22333
104. Mu, J., and Roach, P. J. (1998) *J. Biol. Chem.* **273**, 34850–34856
105. Mu, J., Skurat, A. V., and Roach, P. J. (1997) *J. Biol. Chem.* **272**, 27589–27597
106. Charng, Y.-y, Iglesias, A. A., and Preiss, J. (1994) *J. Biol. Chem.* **269**, 24107–24113
107. Nakata, P. A., Anderson, J. M., and Okita, T. W. (1994) *J. Biol. Chem.* **269**, 30798–30807
108. Van den Koornhuyse, N., Libessart, N., Delrue, B., Zabawinski, C., Decq, A., Iglesias, A., Carton, A., Preiss, J., and Ball, S. (1996) *J. Biol. Chem.* **271**, 16281–16287
109. Pozueta-Romero, J., Frehner, M., Viale, A. M., and Akazawa, T. (1991) *Proc. Natl. Acad. Sci. U.S.A.* **88**, 5769–5773
110. Yu, Y., Mu, H. H., Mu-Forster, C., and Wasserman, B. P. (1998) *Plant Physiol.* **116**, 1451–1460
111. Calvert, P. (1997) *Nature (London)* **389**, 338–339
112. Ball, S., Guan, H.-P., James, M., Myers, A., Keeling, P., Mouille, G., Buléon, A., Colonna, P., and Preiss, J. (1996) *Cell* **86**, 349–352
113. Waigh, T. A., Hopkinson, I., Donald, A. M., Butler, M. F., Heidelbach, F., and Riekel, C. (1997) *Macromolecules* **30**, 3813–3820
114. Gallant, D. J., Bouchet, B., and Baldwin, P. M. (1997) *Carbo. Polymers* **32**, 177–191
115. Fontaine, T., D'Hulst, C., Maddelein, M.-L., Routier, F., Pépin, T. M., Decq, A., Wieruszeski, J.-M., Delrue, B., Van den Koornhuyse, N., Bossu, J.-P., Fournet, B., and Ball, S. (1993) *J. Biol. Chem.* **268**, 16223–16230
116. Gao, M., Wanat, J., Stinard, P. S., James, M. G., and Myers, A. M. (1998) *Plant Cell* **10**, 399–412
117. Cao, H., Imparl-Radosevich, J., Guan, H. P., Keeling, P. L., James, M. G., and Myers, A. M. (1999) *Plant Physiol.* **120**, 1–11
118. Rahman, A., Wong, K.-s, Jane, J.-l, Myers, A. M., and James, M. G. (1998) *Plant Physiol.* **117**, 425–435
119. Guan, H., Kuriki, T., Sivak, M., and Preiss, J. (1995) *Proc. Natl. Acad. Sci. U.S.A.* **92**, 964–967
120. Beatty, M. K., Rahman, A., Cao, H., Woodman, W., Lee, M., Myers, A. M., and James, M. G. (1999) *Plant Physiol.* **119**, 255–266
121. Robyt, J. F. (1998) *Essentials of Carbohydrate Chemistry*, Springer, New York

References

122. Ball, S. G., van de Wal, M. H. B. J., and Visser, R. G. F. (1998) *Trends Plant Sci.* **3**, 462–467
122a. Sehnke, P. C., Chung, H.-J., Wu, K., and Ferl, R. J. (2001) *Proc. Natl. Acad. Sci. U.S.A.* **98**, 765–770
123. French, D. and Robyt, J. F. (1973) *Abstr., 166th Natl. Meet., Am. Chem. Soc.*, Abstract 65 BIOL
123a. Robyt, J. F. (2000) *Abstr., 220th Natl. Meet., Am. Chem. Soc.*, Abstract 84 CARB
123b. Mukerjea, R., Yu, L., and Robyt, J. F. (2002) *Carbohydr. Res.* **337**, 1015–1022
124. McGuire, V., and Alexander, S. (1996) *J. Biol. Chem.* **271**, 14596–14603
125. Bureau, T. E., and Brown, R. M., Jr. (1987) *Proc. Natl. Acad. Sci. U.S.A.* **84**, 6985–6989
126. Lloyd, C. (1980) *Nature (London)* **284**, 596–597
127. Brown, R. M., Jr., Willison, J. H. M., and Richardson, C. L. (1976) *Proc. Natl. Acad. Sci. U.S.A.* **73**, 4565–4569
128. Lin, F. C., Brown, R. M., Jr., Cooper, J. B., and Delmer, D. P. (1985) *Science* **230**, 822–825
129. Ross, P., Mayer, R., and Benziman, M. (1991) *Microbiol. Rev.* **55**, 35–58
130. Bokelman, G. H., Ruben, G. C., and Krakow, W. (1988) *J. Cell Biol.* **107**, 147a
131. Carpita, N., and Vergara, C. (1998) *Science* **279**, 672–673
131a. Scheible, W.-R., Eshed, R., Richmond, T., Delmer, D., and Somerville, C. (2001) *Proc. Natl. Acad. Sci. U.S.A.* **98**, 10079–10084
131b. Kurek, I., Kawagoe, Y., Jacob-Wilk, D., Doblin, M., and Delmer, D. (2002) *Proc. Natl. Acad. Sci. U.S.A.* **99**, 11109–11114
132. Haigler, C. H., and Blanton, R. L. (1996) *Proc. Natl. Acad. Sci. U.S.A.* **93**, 12082–12085
133. Arioli, T., Peng, L., Betzner, A. S., Burn, J., Wittke, W., Herth, W., Camilleri, C., Höfte, H., Plazinski, J., Birch, R., Cork, A., Glover, J., Redmond, J., and Williamson, R. E. (1998) *Science* **279**, 717–720
133a. Cosgrove, D. J. (2000) *Nature (London)* **407**, 321–326
133b. Peng, L., Kawagoe, Y., Hogan, P., and Delmer, D. (2002) *Science* **295**, 147–150
134. Koyama, M., Helbert, W., Imai, T., Sugiyama, J., and Henrissat, B. (1997) *Proc. Natl. Acad. Sci. U.S.A.* **94**, 9091–9095
135. Han, N. S., and Robyt, J. F. (1998) *Carbohydr. Res.* **313**, 125–133
136. Sutherland, I. W. (1993) in *Industrial Gums*, 3rd ed. (Whistler, R. L., and BeMiller, J. N., eds), pp. 69–85, Academic Press, San Diego, California
137. Sugiyama, J., Boisset, C., Hashimoto, M., and Watanabe, T. (1999) *J. Mol. Biol.* **286**, 247–255
138. Mayer, R., Ross, P., Weinhouse, H., Amikam, D., Volman, G., Ohana, P., Calhoon, R. D., Wong, H. C., Emerick, A. W., and Benziman, M. (1991) *Proc. Natl. Acad. Sci. U.S.A.* **88**, 5472–5476
139. Egli, M., Gessner, R. V., Williams, L. D., Quigley, G. J., van der Marel, G. A., van Boom, J. H., Rich, A., and Frederick, C. A. (1990) *Proc. Natl. Acad. Sci. U.S.A.* **87**, 3235–3239
139a. Chang, A. L., Tuckerman, J. R., Gonzalez, G., Mayer, R., Weinhouse, H., Volman, G., Amikam, D., Benziman, M., and Gilles-Gonzalez, M.-A. (2001) *Biochemistry* **40**, 3420–3426
140. Amor, Y., Haigler, C. H., Johnson, S., Wainscott, M., and Delmer, D. P. (1995) *Proc. Natl. Acad. Sci. U.S.A.* **92**, 9353–9357
141. Nakai, T., Tonouchi, N., Konishi, T., Kojima, Y., Tsuchida, T., Yoshinaga, F., Sakai, F., and Hayashi, T. (1999) *Proc. Natl. Acad. Sci. U.S.A.* **96**, 14–18
142. Martin, M. M., and Martin, J. S. (1978) *Science* **199**, 1453–1455
143. Skipper, N., Sutherland, M., Davies, R. W., Kilburn, D., Miller, R. C., Jr., Warren, A., and Wong, R. (1985) *Science* **230**, 958–960
144. Ohana, P., Delmer, D. P., Steffens, J. C., Matthews, D. E., Mayer, R., and Benziman, M. (1991) *J. Biol. Chem.* **266**, 13742–13745
145. Arellano, M., Durán, A., and Pérez, P. (1996) *EMBO J.* **15**, 4584–4591
146. Hrmova, M., Garrett, T. P. J., and Fincher, G. B. (1995) *J. Biol. Chem.* **270**, 14556–14563
147. Cabib, E. (1987) *Adv. Enzymol.* **59**, 59–101
148. Orlean, P. (1987) *J. Biol. Chem.* **262**, 5732–5739
149. Silverman, S. J., Sburlati, A., Slater, M. L., and Cabib, E. (1988) *Proc. Natl. Acad. Sci. U.S.A.* **85**, 4735–4739
150. Machida, S., and Saito, M. (1993) *J. Biol. Chem.* **268**, 1702–1707
151. Davis, L. L., and Bartnicki-Garcia, S. (1984) *Biochemistry* **23**, 1065–1073
152. Kafetzopoulos, D., Martinou, A., and Bouriotis, V. (1993) *Proc. Natl. Acad. Sci. U.S.A.* **90**, 2564–2568
153. Marx, J. L. (1977) *Science* **197**, 1170–1172
154. Bassler, B. L., Gibbons, P. J., Yu, C., and Roseman, S. (1991) *J. Biol. Chem.* **266**, 24268–24275
155. Sticher, L., Hofsteenge, J., Milani, A., Neuhaus, J.-M., and Meins, F., Jr. (1992) *Science* **257**, 655–657
156. Boot, R. G., Renkema, G. H., Verhoek, M., Strijland, A., Bliek, J., de Meulemeester, T. M. A. M. O., Mannens, M. M. A. M., and Aerts, J. M. F. G. (1998) *J. Biol. Chem.* **273**, 25680–25685
157. Semino, C. E., Specht, C. A., Raimondi, A., and Robbins, P. W. (1996) *Proc. Natl. Acad. Sci. U.S.A.* **93**, 4548–4553
158. Kamst, E., Bakkers, J., Quaedvlieg, N. E. M., Pilling, J., Kijne, J. W., Lugtenberg, B. J. J., and Spaink, H. P. (1999) *Biochemistry* **38**, 4045–4052
159. Gibeaut, D., and Carpita, N. C. (1994) *FASEB J.* **8**, 904–915
160. Darvill, S., McNeil, M., Albersheim, P., and Delmer, D. P. (1980) in *The Biochemisty of Plants*, Vol. I (Tolbert, N. E., ed), pp. 91–162, Academic Press, New York
161. Preston, R. D. (1979) *Ann. Rev. Plant Physiol.* **30**, 55
162. Brett, C. T., and Hillman, J. R., eds. (1985) *Biochemistry of Plant Cell Walls*, Cambridge Univ. Press, Cambridge
163. MacKay, A. L., Wallace, J. C., Sasaki, K., and Taylor, I. E. P. (1988) *Biochemistry* **27**, 1467–1473
163a. Carpita, N., and McCann, M. (2000) in *Biochemistry and Molecular Biology of Plants* (Buchanan, B. B., Gruissem, W., and Jones, R. L., eds), pp. 52–108, American Society of Plant Physiologists, Rockville, Maryland
164. Northcote, D. H. (1972) *Ann. Rev. Plant Physiol.* **23**, 113
164a. Perrin, R. M., DeRocher, A. E., Bar-Peled, M., Zeng, W., Norambuena, L., Orellana, A., Raikhel, N. V., and Keegstra, K. (1999) *Science* **284**, 1976–1979
164b. Kofod, L. V., Kauppinen, S., Christgau, S., Andersen, L. N., Heldt-Hansen, H. P., Dörreich, K., and Dalbøge, H. (1994) *J. Biol. Chem.* **269**, 29182–29189
164c. O'Neill, M. A., Warrenfeltz, D., Kates, K., Pellerin, P., Doco, T., Darvill, A. G., and Albersheim, P. (1996) *J. Biol. Chem.* **271**, 22923–22930
164d. Ishii, T., Matsunaga, T., Pellerin, P., O'Neill, M. A., Darvill, A., and Albersheim, P. (1999) *J. Biol. Chem.* **274**, 13098–13104
164e. Höfte, H. (2001) *Science* **294**, 795–797
164f. O'Neill, M. A., Eberhard, S., Albersheim, P., and Darvill, A. G. (2001) *Science* **294**, 846–849
165. Robertson, D., Mitchell, G. P., Gilroy, J. S., Gerrish, C., Bolwell, G. P., and Slabas, A. R. (1997) *J. Biol. Chem.* **272**, 15841–15848
166. Averyhart-Fullard, V., Datta, K., and Marcus, A. (1988) *Proc. Natl. Acad. Sci. U.S.A.* **85**, 1082–1085
167. Condit, C. M., and Meagher, R. B. (1986) *Nature (London)* **323**, 178–181
168. Cosgrove, D. J. (1997) *Proc. Natl. Acad. Sci. U.S.A.* **94**, 5504–5505
169. Rose, J. K. C., Lee, H. H., and Bennett, A. B. (1997) *Proc. Natl. Acad. Sci. U.S.A.* **94**, 5955–5960
170. Fleming, A. J., McQueen-Mason, S., Mandel, T., and Kuhlemeier, C. (1997) *Science* **276**, 1415–1418
171. Stanley, P., and Ioffe, E. (1995) *FASEB J.* **9**, 1436–1444
172. Baenziger, J. U. (1994) *FASEB J.* **8**, 1019–1025
172a. DeAngelis, P. L., and Padgett-McCue, A. J. (2000) *J. Biol. Chem.* **275**, 24124–24129
172b. Pummill, P. E., Kempner, E. S., and DeAngelis, P. L. (2001) *J. Biol. Chem.* **276**, 39832–39835
173. Laurent, T. C., and Fraser, J. R. E. (1992) *FASEB J.* **6**, 2397–2404
174. Tlapak-Simmons, V. L., Baggenstoss, B. A., Kumari, K., Heldermon, C., and Weigel, P. H. (1999) *J. Biol. Chem.* **274**, 4246–4253
175. Spicer, A. P., and McDonald, J. A. (1998) *J. Biol. Chem.* **273**, 1923–1932
176. Robyt, J. F. (1979) *Trends Biochem. Sci.* **4**, 47–49
177. Robyt, J. F. (1995) *Adv. Carbohydr. Chem. Biochem.* **51**, 133–168
178. Robyt, J. F., and Martin, P. J. (1983) *Carbohydr. Res.* **113**, 301–315
179. Mooser, G., Hefta, S. A., Paxton, R. J., Shively, J. E., and Lee, T. D. (1991) *J. Biol. Chem.* **266**, 8916–8922
180. Funane, K., Shiraiwa, M., Hashimoto, K., Ichishima, E., and Kobayashi, M. (1993) *Biochemistry* **32**, 13696–13702
181. Sprenger, N., Bortlik, K., Brandt, A., Boller, T., and Wiemken, A. (1995) *Proc. Natl. Acad. Sci. U.S.A.* **92**, 11652–11656
182. Beale, J. M., Jr., and Foster, J. L. (1996) *Biochemistry* **35**, 4492–4501
183. Hardingham, T. E., and Fosang, A. J. (1992) *FASEB J.* **6**, 861–870
184. Huang, S., Wang, Y.-X., and Draper, D. E. (1996) *J. Mol. Biol.* **258**, 308–321
185. Kokenyesi, R., and Bernfield, M. (1994) *J. Biol. Chem.* **269**, 12304–12309
186. Kearns, A. E., Campbell, S. C., Westley, J., and Schwartz, N. B. (1991) *Biochemistry* **30**, 7477–7483
187. Shworak, N. W., Shirakawa, M., Mulligan, R. C., and Rosenberg, R. D. (1994) *J. Biol. Chem.* **269**, 21204–21214
188. Vertel, B. M., Walters, L. M., Flay, N., Kearns, A. E., and Schwartz, N. B. (1993) *J. Biol. Chem.* **268**, 11105–11112
189. Fernández, C. J., and Warren, G. (1998) *J. Biol. Chem.* **273**, 19030–19039
190. Lugemwa, F. N., Sarkar, A. K., and Esko, J. D. (1996) *J. Biol. Chem.* **271**, 19159–19165
190a. Bai, X., Wei, G., Sinha, A., and Esko, J. D. (1999) *J. Biol. Chem.* **274**, 13017–13024
190b. Götting, C., Kuhn, J., Zahn, R., Brinkmann, T., and Kleesiek, K. (2000) *J. Mol. Biol.* **304**, 517–528
190c. Pedersen, L. C., Tsuchida, K., Kitagawa, H., Sugahara, K., Darden, T. A., and Negishi, M. (2000) *J. Biol. Chem.* **275**, 34580–34585
191. Sugahara, K., Ohkita, Y., Shibata, Y., Yoshida, K., and Ikegami, A. (1995) *J. Biol. Chem.* **270**, 7204–7212

References

192. Pavão, M. S. G., Aiello, K. R. M., Werneck, C. C., Silva, L. C. F., Valente, A.-P., Mulloy, B., Colwell, N. S., Tollefsen, D. M., and Mourão, P. A. S. (1998) *J. Biol. Chem.* **273**, 27848–27857
193. Sugumaran, G., Katsman, M., and Drake, R. R. (1995) *J. Biol. Chem.* **270**, 22483–22487
194. Yamauchi, S., Hirahara, Y., Usui, H., Takeda, Y., Hoshino, M., Fukuta, M., Kimura, J. H., and Habuchi, O. (1999) *J. Biol. Chem.* **274**, 2456–2463
195. Kobayashi, M., Habuchi, H., Habuchi, O., Saito, M., and Kimata, K. (1996) *J. Biol. Chem.* **271**, 7645–7653
196. Razi, N., and Lindahl, U. (1995) *J. Biol. Chem.* **270**, 11267–11275
197. Shworak, N. W., Liu, J., Petros, L. M., Zhang, L., Kobayashi, M., Copeland, N. G., Jenkins, N. A., and Rosenberg, R. D. (1999) *J. Biol. Chem.* **274**, 5170–5184
198. Toma, L., Berninsone, P., and Hirschberg, C. B. (1998) *J. Biol. Chem.* **273**, 22458–22465
199. Aikawa, J.-i, and Esko, J. D. (1999) *J. Biol. Chem.* **274**, 2690–2695
200. Scott, J. E., Heatley, F., and Wood, B. (1995) *Biochemistry* **34**, 15467–15474
201. Yanagashita, M., and Hascall, V. C. (1992) *J. Biol. Chem.* **267**, 9451–9454
202. Lane, D. A., and Björk, I., eds. (1992) *Heparin and Related Polysaccharides*, Plenum, New York
203. Lindahl, U., Kusche-Gullberg, M., and Kjellén, L. (1998) *J. Biol. Chem.* **273**, 24979–24982
203a. Safaiyan, F., Lindahl, U., and Salmivirta, M. (2000) *Biochemistry* **39**, 10823–10830
203b. Zhang, L., Lawrence, R., Schwartz, J. J., Bai, X., Wei, G., Esko, J. D., and Rosenberg, R. D. (2001) *J. Biol. Chem.* **276**, 28806–28813
204. Kinoshita, A., Yamada, S., Haslam, S. M., Morris, H. R., Dell, A., and Sugahara, K. (1997) *J. Biol. Chem.* **272**, 19656–19665
205. Nadanaka, S., Clement, A., Masayama, K., Faissner, A., and Sugahara, K. (1998) *J. Biol. Chem.* **273**, 3296–3307
206. Mourão, P. A. S., Pereira, M. S., Pavão, M. S. G., Mulloy, B., Tollefsen, D. M., Mowinckel, M.-C., and Abildgaard, U. (1996) *J. Biol. Chem.* **271**, 23973–23984
207. Petitou, M., Hérault, J.-P., Bernat, A., Driguez, P.-A., Duchaussoy, P., Lormeau, J.-C., and Herbert, J.-M. (1999) *Nature (London)* **398**, 417–422
208. Sinaÿ, P. (1999) *Nature (London)* **398**, 377–378
208a. Zehnder, J. L., and Galli, S. J. (1999) *Nature (London)* **400**, 714–715
208b. Hondal, R. J., Ma, S., Caprioli, R. M., Hill, K. E., and Burk, R. F. (2001) *J. Biol. Chem.* **276**, 15823–15831
209. Spiro, R. C., Casteel, H. E., Laufer, D. M., Reisfeld, R. A., and Harper, J. R. (1989) *J. Biol. Chem.* **264**, 1779–1786
210. Iozzo, R. V., and Murdoch, A. D. (1996) *FASEB J.* **10**, 598–614
211. Cheng, F., Heinegård, D., Fransson, L.-Å., Bayliss, M., Bielicki, J., Hopwood, J., and Yoshida, K. (1996) *J. Biol. Chem.* **271**, 28572–28580
212. Hauser, N., Paulsson, M., Heinegård, D., and Mörgelin, M. (1996) *J. Biol. Chem.* **271**, 32247–32252
213. Zheng, J., Luo, W., and Tanzer, M. L. (1998) *J. Biol. Chem.* **273**, 12999–13006
213a. Olin, A. I., Mörgelin, M., Sasaki, T., Timpl, R., Heinegård, D., and Aspberg, A. (2001) *J. Biol. Chem.* **276**, 1253–1261
214. Arner, E. C., Pratta, M. A., Trzaskos, J. M., Decicco, C. P., and Tortorella, M. D. (1999) *J. Biol. Chem.* **274**, 6594–6601
215. Brown, G. M., Huckerby, T. N., Bayliss, M. T., and Nieduszynski, I. A. (1998) *J. Biol. Chem.* **273**, 26408–26414
216. Lindahl, B., Eriksson, L., Spillmann, D., Caterson, B., and Lindahl, U. (1996) *J. Biol. Chem.* **271**, 16991–16994
217. Dours-Zimmermann, M. T., and Zimmermann, D. R. (1994) *J. Biol. Chem.* **269**, 32992–32998
217a. Yang, B. L., Cao, L., Kiani, C., Lee, V., Zhang, Y., Adams, M. E., and Yang, B. B. (2000) *J. Biol. Chem.* **275**, 21255–21261
217b. Pangalos, M. N., Efthimiopoulos, S., Shioi, J., and Robakis, N. K. (1995) *J. Biol. Chem.* **270**, 10388–10391
218. Halfter, W., Dong, S., Schurer, B., and Cole, G. J. (1998) *J. Biol. Chem.* **273**, 25404–25412
219. Scott, J. E. (1996) *Biochemistry* **35**, 8795–8799
220. Weber, I. T., Harrison, R. W., and Iozzo, R. V. (1996) *J. Biol. Chem.* **271**, 31767–31770
221. Tai, G.-H., Huckerby, T. N., and Nieduszynski, I. A. (1996) *J. Biol. Chem.* **271**, 23535–23546
222. Ungefroren, H., and Krull, N. B. (1996) *J. Biol. Chem.* **271**, 15787–15795
222a. Yang, V. W.-C., LaBrenz, S. R., Rosenberg, L. C., McQuillan, D., and Höök, M. (1999) *J. Biol. Chem.* **274**, 12454–12460
223. Sommarin, Y., Wendel, M., Shen, Z., Hellman, U., and Heinegård, D. (1998) *J. Biol. Chem.* **273**, 16723–16729
223a. Iozzo, R. V. (1999) *J. Biol. Chem.* **274**, 18843–18846
223b. Saika, S., Shiraishi, A., Saika, S., Liu, C.-Y., Funderburgh, J. L., Kao, C. W.-C., Converse, R. L., and Kao, W. W.-Y. (2000) *J. Biol. Chem.* **275**, 2607–2612
223c. Bengtsson, E., Aspberg, A., Heinegård, D., Sommarin, Y., and Spillmann, D. (2000) *J. Biol. Chem.* **275**, 40695–40702
224. David, C. L., Orpiszewski, J., Zhu, X.-C., Reissner, K. J., and Aswad, D. W. (1998) *J. Biol. Chem.* **273**, 32063–32070
225. Grumet, M., Milev, P., Sakurai, T., Karthikeyan, L., Bourdon, M., Margolis, R. K., and Margolis, R. U. (1994) *J. Biol. Chem.* **269**, 12142–12146
226. Fernàndez-Busquets, X., Kammerer, R. A., and Burger, M. M. (1996) *J. Biol. Chem.* **271**, 23558–23565
227. Varner, J. A. (1996) *J. Biol. Chem.* **271**, 16119–16125
228. Hart, G. W., Holt, G. D., and Haltiwanger, R. S. (1988) *Trends Biochem. Sci.* **13**, 380–384
229. Sadler, J. E. (1984) in *"Biology of Carbohydrates"*, Vol. 2 (Ginsburg, V., and Robbins, P. W., eds), pp. 199–288, Wiley, New York
230. Jentoft, N. (1990) *Trends Biochem. Sci.* **15**, 291–294
231. Dong, D. L.-Y., Xu, Z.-S., Chevrier, M. R., Cotter, R. J., Cleveland, D. W., and Hart, G. W. (1993) *J. Biol. Chem.* **268**, 16679–16687
231a. Varki, A., Cummings, R., Esko, J., Freeze, H., Hart, G., and Marth, J., eds. (1999) *Essentials of Glycobiology*, Cold Spring Harbor Lab. Press, Plainview, New York
231b. Roseman, S. (2001) *J. Biol. Chem.* **276**, 41527–41542
232. Hanisch, F.-G., Müller, S., Hassan, H., Clausen, H., Zachara, N., Gooley, A. A., Paulsen, H., Alving, K., and Peter-Katalinic, J. (1999) *J. Biol. Chem.* **274**, 9946–9954
233. Schwientek, T., Nomoto, M., Levery, S. B., Merkx, G., van Kessel, A. G., Bennett, E. P., Hollingsworth, M. A., and Clausen, H. (1999) *J. Biol. Chem.* **274**, 4504–4512
233a. Comer, F. I., and Hart, G. W. (2000) *J. Biol. Chem.* **275**, 29179–29182
233b. Wells, L., Vosseller, K., and Hart, G. W. (2001) *Science* **291**, 2376–2378
233c. Hanover, J. A. (2001) *FASEB J.* **15**, 1865–1876
234. Abeijon, C., and Hirschberg, C. B. (1992) *Trends Biochem. Sci.* **17**, 32–36
235. Moloney, D. J., Lin, A. I., and Haltiwanger, R. S. (1997) *J. Biol. Chem.* **272**, 19046–19050
236. Allen, A. (1983) *Trends Biochem. Sci.* **8**, 169–173
237. Yeh, J.-C., Ong, E., and Fukuda, M. (1999) *J. Biol. Chem.* **274**, 3215–3221
238. Gum, J. R., Jr., Ho, J. J. L., Pratt, W. S., Hicks, J. W., Hill, A. S., Vinall, L. E., Roberton, A. M., Swallow, D. M., and Kim, Y. S. (1997) *J. Biol. Chem.* **272**, 26678–6686
239. Desseyn, J.-L., Buisine, M.-P., Porchet, N., and Aubert, J.-P. (1998) *J. Biol. Chem.* **273**, 30157–30164
240. Perez-Vilar, J., and Hill, R. L. (1998) *J. Biol. Chem.* **273**, 34527–34534
241. Sagami, H., Kurisaki, A., and Ogura, K. (1993) *J. Biol. Chem.* **268**, 10109–10113
242. Opdenakker, G., Rudd, P. M., Ponting, C. P., and Dwek, R. A. (1993) *FASEB J.* **7**, 1330–1337
243. Gahmberg, C. G., and Tolvanen, M. (1996) *Trends Biochem. Sci.* **21**, 308–311
244. Manzella, S. M., Hooper, L. V., and Baenziger, J. U. (1996) *J. Biol. Chem.* **271**, 12117–12120
245. Drickamer, K., and Taylor, M. E. (1998) *Trends Biochem. Sci.* **23**, 321–324
246. Elbein, A. D. (1981) *Trends Biochem. Sci.* **6**, 219–221
247. Zhu, X., Zeng, Y., and Lehrman, M. A. (1992) *J. Biol. Chem.* **267**, 8895–8902
248. Rajput, B., Muniappa, N., and Vijay, I. K. (1994) *J. Biol. Chem.* **269**, 16054–16061
249. Zara, J., and Lehrman, M. A. (1994) *J. Biol. Chem.* **269**, 19108–19115
249a. Ünligil, U. M., Zhou, S., Yuwaraj, S., Sarkar, M., Schachter, H., and Rini, J. M. (2000) *EMBO J.* **19**, 5269–5280
250. Silberstein, S., and Gilmore, R. (1996) *FASEB J.* **10**, 849–858
251. Imperiali, B., and Shannon, K. L. (1991) *Biochemistry* **30**, 4374–4380
252. Yan, Q., Prestwich, G. D., and Lennarz, W. J. (1999) *J. Biol. Chem.* **274**, 5021–5025
252a. Chen, X., VanValkenburgh, C., Liang, H., Fang, H., and Green, N. (2001) *J. Biol. Chem.* **276**, 2411–2416
252b. Karaoglu, D., Kelleher, D. J., and Gilmore, R. (2001) *Biochemistry* **40**, 12193–12206
253. Schwarz, R. T., and Datema, R. (1984) *Trends Biochem. Sci.* **9**, 32–34
254. Elbein, A. D. (1987) *Ann. Rev. Biochem.* **56**, 497–534
255. Kaushal, G. P., Pan, Y. T., Tropea, J. E., Mitchell, M., Liu, P., and Elbein, A. D. (1988) *J. Biol. Chem.* **263**, 17278–17283
256. Kukuruzinska, M. A., Bergh, M. L. E., and Jackson, B. J. (1987) *Ann. Rev. Biochem.* **56**, 915–944
257. Hatton, M. W. C., Marz, L., and Regoeczi, E. (1983) *Trends Biochem. Sci.* **8**, 287–291
257a. Vallée, F., Karaveg, K., Herscovics, A., Moremen, K. W., and Howell, P. L. (2000) *J. Biol. Chem.* **275**, 41287–41298
257b. Van Petegem, F., Contreras, H., Contreras, R., and Van Beeumen, J. (2001) *J. Mol. Biol.* **312**, 157–165
258. Herscovics, A., and Orlean, P. (1993) *FASEB J.* **7**, 540–550
258a. Dell, A., and Morris, H. R. (2001) *Science* **291**, 2351–2356
259. Dolci, E. D., and Palade, G. E. (1985) *J. Biol. Chem.* **260**, 10728–10735
260. Gonatas, J. O., Mezitis, S. G. E., Stieber, A., Fleischer, B., and Gonatas, N. K. (1989) *J. Biol. Chem.* **264**, 646–653
261. Hooper, L. V., Hindsgaul, O., and Baenziger, J. U. (1995) *J. Biol. Chem.* **270**, 16327–16332
262. Okajima, T., Fukumoto, S., Miyazaki, H., Ishida, H., Kiso, M., Furukawa, K., Urano, T., and Furukawa, K. (1999) *J. Biol. Chem.* **274**, 11479–11486

References

263. Lee, Y.-C., Kaufmann, M., Kitazume-Kawaguchi, S., Kono, M., Takashima, S., Kurosawa, N., Liu, H., Pircher, H., and Tsuji, S. (1999) *J. Biol. Chem.* **274**, 11958–11967
264. Trimble, R. B., and Atkinson, P. H. (1986) *J. Biol. Chem.* **261**, 9815–9824
265. Chiba, Y., Suzuki, M., Yoshida, S., Yoshida, A., Ikenaga, H., Takeuchi, M., Jigami, Y., and Ichishima, E. (1998) *J. Biol. Chem.* **273**, 26298–26304
266. Kollár, R., Reinhold, B. B., Petrákova, E., Yeh, H. J. C., Ashwell, G., Drgonová, J., Kapteyn, J. C., Klis, F. M., and Cabib, E. (1997) *J. Biol. Chem.* **272**, 17762–17775
266a. Helenius, A., and Aebi, M. (2001) *Science* **291**, 2364–2369
266b. Lehrman, M. A. (2001) *J. Biol. Chem.* **276**, 8623–8626
266c. Bertozzi, C. R., and Kiessling, L. L. (2001) *Science* **291**, 2357–2364
266d. Plante, O. J., Palmacci, E. R., and Seeberger, P. H. (2001) *Science* **291**, 1523–1527
267. Mazzarello, P., and Bentivoglio, M. (1998) *Nature (London)* **392**, 543–544
268. Featherstone, C. (1998) *Science* **282**, 2172–2174
269. Berger, E. G., and Roth, J., eds. (1997) *The Golgi Apparatus*, Birkhäuser Verlag, Basel
270. Driouich, A., Faye, L., and Staehelin, L. A. (1993) *Trends Biochem. Sci.* **18**, 210–214
271. Lazar, T., Götte, M., and Gallwitz, D. (1997) *Trends Biochem. Sci.* **22**, 468–472
272. Farquhar, M. G., and Hauri, H.-P. (1997) in *The Golgi Apparatus* (Berger, E. G., and Roth, J., eds), pp. 63–129, Birkhäuser Verlag, Basel, Switzerland
273. Springer, S., and Schekman, R. (1998) *Science* **281**, 698–700
274. Walter, D. M., Paul, K. S., and Waters, M. G. (1998) *J. Biol. Chem.* **273**, 29565–29576
275. Dice, J. F. (1990) *Trends Biochem. Sci.* **15**, 305–309
276. Rambourg, A., and Clermont, Y. (1997) in *The Golgi Apparatus* (Berger, E. G., and Roth, J., eds), pp. 37–61, Birkhäuser Verlag, Basel, Switzerland
277. Ungermann, C., and Wickner, W. (1998) *EMBO J.* **17**, 3269–3276
278. Wu, S.-K., Zeng, K., Wilson, I. A., and Balch, W. E. (1996) *Trends Biochem. Sci.* **21**, 472–476
279. Webb, R. J., East, J. M., Sharma, R. P., and Lee, A. G. (1998) *Biochemistry* **37**, 673–679
280. Ivessa, N. E., De Lemos-Chiarandini, C., Gravotta, D., Sabatini, D. D., and Kreibich, G. (1995) *J. Biol. Chem.* **270**, 25960–25967
280a. Todorow, Z., Spang, A., Carmack, E., Yates, J., and Schekman, R. (2000) *Proc. Natl. Acad. Sci. U.S.A.* **97**, 13643–13648
280b. Marsh, B. J., Mastronarde, D. N., Buttle, K. F., Howell, K. E., and McIntosh, J. R. (2001) *Proc. Natl. Acad. Sci. U.S.A.* **98**, 2399–2406
280c. Jakymiw, A., Raharjo, E., Rattner, J. B., Eystathioy, T., Chan, E. K. L., and Fujita, D. J. (2000) *J. Biol. Chem.* **275**, 4137–4144
280d. Hirschberg, C. B., Robbins, P. W., and Abeijon, C. (1998) *Ann. Rev. Biochem.* **67**, 49–69
280e. Gao, X.-D., and Dean, N. (2000) *J. Biol. Chem.* **275**, 17718–17727
280f. Bell, A. W., and 16 other authors. (2001) *J. Biol. Chem.* **276**, 5152–5165
280g. Alvarez, C., Garcia-Mata, R., Hauri, H.-P., and Sztul, E. (2001) *J. Biol. Chem.* **276**, 2693–2700
280h. Charest, A., Lane, K., McMahon, K., and Housman, D. E. (2001) *J. Biol. Chem.* **276**, 29456–29465
280i. Demaurex, N., Furuya, W., D'Souza, S., Bonifacino, J. S., and Grinstein, S. (1998) *J. Biol. Chem.* **273**, 2044–2051
281. Fritzler, M. J., Lung, C.-C., Hamel, J. C., Griffith, K. J., and Chan, E. K. L. (1995) *J. Biol. Chem.* **270**, 31262–31268
282. Bascon, R. A., Srinivasan, S., and Nussbaum, R. L. (1999) *J. Biol. Chem.* **274**, 2953–2962
283. Kain, R., Angata, K., Kerjaschki, D., and Fukuda, M. (1998) *J. Biol. Chem.* **273**, 981–988
284. Johannes, L., Tenza, D., Antony, C., and Goud, B. (1997) *J. Biol. Chem.* **272**, 19554–19561
285. Drickamer, K. (1988) *J. Biol. Chem.* **263**, 9557–9560
286. Cantor, A. B., and Kornfeld, S. (1992) *J. Biol. Chem.* **267**, 23357–23363
287. Kornfeld, S. (1992) *Ann. Rev. Biochem.* **61**, 307–330
288. Pohlmann, R., Boeker, M. W. C., and von Figura, K. (1995) *J. Biol. Chem.* **270**, 27311–27318
289. Sleat, D. E., and Lobel, P. (1997) *J. Biol. Chem.* **272**, 731–738
290. York, S. J., Arneson, L. S., Gregory, W. T., Dahms, N. M., and Kornfeld, S. (1999) *J. Biol. Chem.* **274**, 1164–1171
290a. Zhu, Y., Doray, B., Poussu, A., Lehto, V.-P., and Kornfeld, S. (2001) *Science* **292**, 1716–1718
291. Sleat, D. E., Sohar, I., Lackland, H., Majercak, J., and Lobel, P. (1996) *J. Biol. Chem.* **271**, 19191–19198
292. Klionsky, D. J., and Emr, S. D. (1990) *J. Biol. Chem.* **265**, 5349–5352
293. Lodish, H. F. (1991) *Trends Biochem. Sci.* **16**, 374–377
294. Chao, W., Liu, H., Hanahan, D. J., and Olson, M. S. (1992) *J. Biol. Chem.* **267**, 6725–6735
295. Chiu, M. H., Thomas, V. H., Stubbs, H. J., and Rice, K. G. (1995) *J. Biol. Chem.* **270**, 24024–24031
296. Reichner, J. S., Whiteheart, S. W., and Hart, G. W. (1988) *J. Biol. Chem.* **263**, 16316–16326
296a. Filipe, S. R., Severina, E., and Tomasz, A. (2001) *J. Biol. Chem.* **276**, 39618–39628
297. Emanuele, J. J., Jr., Jin, H., Yanchunas, J., Jr., and Villafranca, J. J. (1997) *Biochemistry* **36**, 7264–7271
298. Anderson, M. S., Eveland, S. S., Onishi, H. R., and Pompliano, D. L. (1996) *Biochemistry* **35**, 16264–16269
299. Duncan, K., van Heijenoort, J., and Walsh, C. T. (1990) *Biochemistry* **29**, 2379–2386
300. Van Heijenoort, J. (1994) in *Bacterial Cell Wall (New Comprehensive Biochemistry)*, Vol. 27 (Ghuysen, J.-M., and Hakenbeck, R., eds), pp. 39–54, Elsevier, Amsterdam
301. Matsuhashi, M. (1994) in *Bacterial Cell Wall (New Comprehensive Biochemistry)*, Vol. 27 (Ghuysen, J.-M., and Hakenbeck, R., eds), pp. 55–71, Elsevier, Amsterdam
301a. Ha, S., Chang, E., Lo, M.-C., Men, H., Park, P., Ge, M., and Walker, S. (1999) *J. Am. Chem. Soc.* **121**, 8417–8426
301b. Marmor, S., Petersen, C. P., Reck, F., Yang, W., Gao, N., and Fisher, S. L. (2001) *Biochemistry* **40**, 12207–12214
301c. Gordon, E., Flouret, B., Chantalat, L., van Heijenoort, J., Mengin-Lecreulx, D., and Dideberg, O. (2001) *J. Biol. Chem.* **276**, 10999–11006
301d. Yan, Y., Munshi, S., Leiting, B., Anderson, M. S., Chrzas, J., and Chen, Z. (2000) *J. Mol. Biol.* **304**, 435–445
301e. Lee, W., McDonough, M. A., Kotra, L. P., Li, Z.-H., Silvaggi, N. R., Takeda, Y., Kelly, J. A., and Mobashery, S. (2001) *Proc. Natl. Acad. Sci. U.S.A.* **98**, 1427–1431
302. Koch, A. L. (1985) *Trends Biochem. Sci.* **10**, 11–14
303. Archibald, A. R., Hancock, I. C., and Harwood, C. R. (1993) in *Bacillus subtilis and Other Gram-Positive Bacteria* (Sonenshein, A. L., Hoch, J. A., and Losick, R., eds), pp. 381–410, American Society for Microbiology, Washington, D. C.
304. Jacobs, C. (1997) *Science* **278**, 1731–1732
305. Jacobs, C., Huang, L.-j, Bartowsky, E., Normark, S., and Park, J. T. (1994) *EMBO J.* **13**, 4684–4694
305a. Templin, M. F., Ursinus, A., and Höltje, J.-V. (1999) *EMBO J.* **18**, 4108–4117
306. Keenleyside, W. J., and Whitfield, C. (1996) *J. Biol. Chem.* **271**, 28581–28592
307. Raetz, C. R. H., Ulevitch, R. J., Wright, S. D., Sibley, C. H., Ding, A., and Nathan, C. F. (1991) *FASEB J.* **5**, 2652–2660
308. Odegaard, T. J., Kaltashov, I. A., Cotter, R. J., Steeghs, L., van der Ley, P., Khan, S., Maskell, D. J., and Raetz, C. R. H. (1997) *J. Biol. Chem.* **272**, 19688–19696
308a. Raetz, C. R. H., and Whitfield, C. (2002) *Ann. Rev. Biochem.* **71**, 635–700
309. Yethon, J. A., Heinrichs, D. E., Monteiro, M. A., Perry, M. B., and Whitfield, C. (1998) *J. Biol. Chem.* **273**, 26310–26316
310. Raetz, C. R. H., and Roderick, S. L. (1995) *Science* **270**, 997–1000
311. Jackman, J. E., Raetz, C. R. H., and Fierke, C. A. (1999) *Biochemistry* **38**, 1902–1911
312. Zhou, Z., White, K. A., Polissi, A., Georgopoulos, C., and Raetz, C. R. H. (1998) *J. Biol. Chem.* **273**, 12466–12475
313. Luk, J. M. C., Lind, S. M., Tsang, R. S. W., and Lindberg, A. A. (1991) *J. Biol. Chem.* **266**, 23215–23225
314. Kerwood, D. E., Schneider, H., and Yamasaki, R. (1992) *Biochemistry* **31**, 12760–12768
315. Basu, S. S., York, J. D., and Raetz, C. R. H. (1999) *J. Biol. Chem.* **274**, 11139–11149
316. Rietschel, E. T., Kirikae, T., Schade, F. U., Mamat, U., Schmidt, G., Loppnow, H., Ulmer, A. J., Zähringer, U., Seydel, U., Di Padova, F., Schreier, M., and Brade, H. (1994) *FASEB J.* **8**, 217–225
317. White, K. A., Kaltashov, I. A., Cotter, R. J., and Raetz, C. R. H. (1997) *J. Biol. Chem.* **272**, 16555–16563
318. Brooke, J. S., and Valvano, M. A. (1996) *J. Biol. Chem.* **271**, 3608–3614
319. Süsskind, M., Brade, L., Brade, H., and Holst, O. (1998) *J. Biol. Chem.* **273**, 7006–7017
320. Pavliak, V., Brisson, J.-R., Michon, F., Uhrín, D., and Jennings, H. J. (1993) *J. Biol. Chem.* **268**, 14146–14152
321. Phillips, N. J., Apicella, M. A., Griffiss, J. M., and Gibson, B. W. (1993) *Biochemistry* **32**, 2003–2012
321a. White, K. A., Lin, S., Cotter, R. J., and Raetz, C. R. H. (1999) *J. Biol. Chem.* **274**, 31391–31400
321b. Kooistra, O., Lüneberg, E., Lindner, B., Knirel, Y. A., Frosch, M., and Zähringer, U. (2001) *Biochemistry* **40**, 7630–7640
322. Ganfield, M.-C. W., and Pieringer, R. A. (1980) *J. Biol. Chem.* **255**, 5164–5169
322a. Volkman, B. F., Zhang, Q., Debabov, D. V., Rivera, E., Kresheck, G. C., and Neuhaus, F. C. (2001) *Biochemistry* **40**, 7964–7972
323. Ton-That, H., Labischinski, H., Berger-Bächi, B., and Schneewind, O. (1998) *J. Biol. Chem.* **273**, 29143–29149
324. Fischetti, V. A. (1991) *Sci. Am.* **264**(Jun), 58–65
325. Michon, F., Brisson, J.-R., Dell, A., Kasper, D. L., and Jennings, H. J. (1988) *Biochemistry* **27**, 5341–5351
326. Cassels, F. J., Fales, H. M., London, J., Carlson, R. W., and van Halbeek, H. (1990) *J. Biol. Chem.* **265**, 14127–14135
327. Wolucka, B. A., McNeil, M. R., de Hoffmann, E., Chonjnacki, T., and Brennan, P. J. (1994) *J. Biol. Chem.* **269**, 23328–23335
328. Besra, G. S., McNeil, M. R., Rivoire, B., Khoo, K.-H., Morris, H. R., Dell, A., and Brennan, P. J. (1993) *Biochemistry* **32**, 347–355

References

329. Beveridge, T. J., and Koval, S. F., eds. (1993) *Advances in Bacterial Paracrystalline Surface Layers*, Plenum, New York

329a. Maeda, Y., Watanabe, R., Harris, C. L., Hong, Y., Ohishi, K., Kinoshita, K., and Kinoshita, T. (2001) *EMBO J.* **20**, 250–261

330. Takeda, J., and Kinoshita, T. (1995) *Trends Biochem. Sci.* **20**, 367–371

331. Udenfriend, S., and Kodukula, K. (1995) *Ann. Rev. Biochem.* **64**, 563–591

332. Tarutani, M., Itami, S., Okabe, M., Ikawa, M., Tezuka, T., Yoshikawa, K., Kinoshita, T., and Takeda, J. (1997) *Proc. Natl. Acad. Sci. U.S.A.* **94**, 7400–7405

333. Menon, A. K., and Stevens, V. L. (1992) *J. Biol. Chem.* **267**, 15277–15280

334. Medof, M. E., Nagarajan, S., and Tykocinski, M. L. (1996) *FASEB J.* **10**, 574–586

334a. Ferguson, M. A. J. (2000) *Proc. Natl. Acad. Sci. U.S.A.* **97**, 10673–10675

334b. Mahoney, A. B., Sacks, D. L., Saraiva, E., Modi, G., and Turco, S. J. (1999) *Biochemistry* **38**, 9813–9823

334c. Smith, T. K., Gerold, P., Crossman, A., Paterson, M. J., Borissow, C. N., Brimacombe, J. S., Ferguson, M. A. J., and Schwarz, R. T. (2002) *Biochemistry* **41**, 12395–12406

335. Ralton, J. E., and McConville, M. J. (1998) *J. Biol. Chem.* **273**, 4245–4257

336. Gilleron, M., Nigou, J., Cahuzac, B., and Puzo, G. (1999) *J. Mol. Biol.* **285**, 2147–2160

337. Benghezal, M., Benachour, A., Rusconi, S., Aebi, M., and Conzelmann, A. (1996) *EMBO J.* **15**, 6575–6583

338. Zawadzki, J., Scholz, C., Currie, G., Coombs, G. H., and McConville, M. J. (1998) *J. Mol. Biol.* **282**, 287–299

339. Smith, T. K., Sharma, D. K., Crossman, A., Dix, A., Brimacombe, J. S., and Ferguson, M. A. J. (1997) *EMBO J.* **16**, 6667–6675

339a. Morita, Y. S., Acosta-Serrano, A., Buxbaum, L. U., and Englund, P. T. (2000) *J. Biol. Chem.* **275**, 14147–14154

340. Leidich, S. D., and Orlean, P. (1996) *J. Biol. Chem.* **271**, 27829–27837

341. Chien, J.-L., and Hogan, E. L. (1983) *J. Biol. Chem.* **258**, 10727–10730

342. van Echten, G., and Sandhoff, K. (1993) *J. Biol. Chem.* **268**, 5341–5344

343. Sprong, H., Kruithof, B., Leijendekker, R., Slot, J. W., van Meer, G., and van der Sluijs, P. (1998) *J. Biol. Chem.* **273**, 25880–25888

344. Schwientek, T., Almeida, R., Levery, S. B., Holmes, E. H., Bennett, E., and Clausen, H. (1998) *J. Biol. Chem.* **273**, 29331–29340

345. Lannert, H., Gorgas, K., Meissner, I., Wieland, F. T., and Jeckel, D. (1998) *J. Biol. Chem.* **273**, 2939–2946

346. Jaskiewicz, E., Zhu, G., Bassi, R., Darling, D. S., and Young, W. W., Jr. (1996) *J. Biol. Chem.* **271**, 26395–26403

347. Neufeld, E. F., Lim, T. W., and Shapiro, L. J. (1975) *Ann. Rev. Biochem.* **44**, 357–376

348. Neufeld, E. F. (1991) *Ann. Rev. Biochem.* **60**, 257–280

349. Tager, J. M. (1985) *Trends Biochem. Sci.* **10**, 324–326

350. Scriver, C. R., Beaudet, A. L., Sly, W. S., and Valle, D., eds (1995) *The Metabolic and Molecular Bases of Inherited Disease*, 7th ed., Vol. 1, McGraw-Hill, New York (pp. 2427–2879)

351. von Figura, K., and Hasilik, A. (1984) *Trends Biochem. Sci.* **9**, 29–31

352. Neufeld, E. F., and Muenzer, J. (1995) in *The Metabolic and Molecular Bases of Inherited Disease*, 7th ed., Vol. 2 (Scriver, C. R., Beaudet, A. L., Sly, W. S., and Valle, D., eds), pp. 2465–2494, McGraw-Hill, New York

353. Stoltzfus, L. J., Sosa-Pineda, B., Moskowitz, S. M., Menon, K. P., Dlott, B., Hooper, L., Teplow, D. B., Shull, R. M., and Neufeld, E. F. (1992) *J. Biol. Chem.* **267**, 6570–6575

354. Kresse, H., Paschke, E., von Figura, K., Gilberg, W., and Fuchs, W. (1980) *Proc. Natl. Acad. Sci. U.S.A.* **77**, 6822–6826

355. Wilson, P. J., Morris, C. P., Anson, D. S., Occhiodoro, T., Bielicki, J., Clements, P. R., and Hopwood, J. J. (1990) *Proc. Natl. Acad. Sci. U.S.A.* **87**, 8531–8535

356. Wicker, G., Prill, V., Brooks, D., Gibson, G., Hopwood, J., von Figura, K., and Peters, C. (1991) *J. Biol. Chem.* **266**, 21386–21391

357. Wu, B. M., Tomatsu, S., Fukuda, S., Sukegawa, K., Orii, T., and Sly, W. S. (1994) *J. Biol. Chem.* **269**, 23681–23688

358. McGovern, M. M., Aula, P., and Desnick, R. J. (1983) *J. Biol. Chem.* **258**, 10743–10747

359. Mononen, I., Fisher, K. J., Kaartinen, V., and Aronson, N. N., Jr. (1993) *FASEB J.* **7**, 1247–1256

360. Saarela, J., Laine, M., Tikkanen, R., Oinonen, C., Jalanko, A., Rouvinen, J., and Peltonen, L. (1998) *J. Biol. Chem.* **273**, 25320–25328

361. Rothman, J. E., and Lenard, J. (1984) *Trends Biochem. Sci.* **9**, 176–178

362. Beutler, E. (1992) *Science* **256**, 794–799

363. Beutler, E. (1993) *Proc. Natl. Acad. Sci. U.S.A.* **90**, 5384–5390

364. Beutler, E., and Grabowski, G. A. (1995) in *The Metabolic and Molecular Bases of Inherited Disease*, 7th ed., Vol. 2 (Scriver, C. R., Beaudet, A. L., Sly, W. S., and Valle, D., eds), pp. 2641–2670, McGraw-Hill, New York

365. Sano, A., Radin, N. S., Johnson, L. L., and Tarr, G. E. (1988) *J. Biol. Chem.* **263**, 19597–19601

366. Myerowitz, R., and Costigan, F. C. (1988) *J. Biol. Chem.* **263**, 18587–18589

366a. Mark, B. L., Vocadlo, D. J., Knapp, S., Triggs-Raine, B. L., Withers, S. G., and James, M. N. G. (2001) *J. Biol. Chem.* **276**, 10330–10337

367. Neufeld, E. F. (1989) *J. Biol. Chem.* **264**, 10927–10930

368. Gravel, R. A., Clarke, J. T. R., Kaback, M. M., Mahuran, D., Sandhoff, K., and Suzuki, K. (1995) in *The Metabolic and Molecular Bases of Inherited Disease*, 7th ed., Vol. 1 (Scriver, C. R., Beaudet, A. L., Sly, W. S., and Valle, D., eds), pp. 2839–2879, McGraw-Hill, New York

369. Paw, B. H., Moskowitz, S. M., Uhrhammer, N., Wright, N., Kaback, M. M., and Neufeld, E. F. (1990) *J. Biol. Chem.* **265**, 9452–9457

369a. Kawai, H., Allende, M. L., Wada, R., Kono, M., Sango, K., Deng, C., Miyakawa, T., Crawley, J. N., Werth, N., Bierfreund, U., Sandhoff, K., and Proia, R. L. (2001) *J. Biol. Chem.* **276**, 6885–6888

370. Hama, Y., Li, Y.-T., and Li, S.-C. (1997) *J. Biol. Chem.* **272**, 2828–2833

371. Beutler, E. (1981) *Trends Biochem. Sci.* **6**, 95–97

371a. Schiffmann, R., and 21 other authors (2000) *Proc. Natl. Acad. Sci. U.S.A.* **97**, 365–370

371b. Sly, W. S., Vogler, C., Grubb, J. H., Zhou, M., Jiang, J., Zhou, X. Y., Tomatsu, S., Bi, Y., and Snella, E. M. (2001) *Proc. Natl. Acad. Sci. U.S.A.* **98**, 2205–2210

372. Furbish, F. S., Steer, C. J., Barranger, J. A., Jones, E. A., and Brady, R. O. (1978) *Biochem. Biophys. Res. Commun.* **81**, 1047–1053

373. Krivit, W., Pierpont, M. E., Ayaz, K., Tsai, M., Ramsay, N. K. C., Kersey, J. H., Weisdorf, S., Sibley, R., Snover, D., McGovern, M. M., Schwartz, M. F., and Desnick, R. J. (1984) *N. Engl. J. Med.* **311**, 1606–1611

374. Sorge, J., Kuhl, W., West, C., and Beutler, E. (1987) *Proc. Natl. Acad. Sci. U.S.A.* **84**, 906–909

375. Fairbairn, L. J., Lashford, L. S., Spooncer, E., McDermott, R. H., Lebens, G., Arrand, J. E., Arrand, J. R., Bellantuono, I., Holt, R., Hatton, C. E., Cooper, A., Besley, G. T. N., Wraith, J. E., Anson, D. S., Hopwood, J. J., and Dexter, T. M. (1996) *Proc. Natl. Acad. Sci. U.S.A.* **93**, 2025–2030

375a. Qin, G., Takenaka, T., Telsch, K., Kelley, L., Howard, T., Levade, T., Deans, R., Howard, B. H., Malech, H. L., Brady, R. O., and Medin, J. A. (2001) *Proc. Natl. Acad. Sci. U.S.A.* **98**, 3428–3433

376. Beaudet, A. L., Scriver, C. R., Sly, W. S., and Valle, D. (1995) in *The Metabolic and Molecular Bases of Inherited Disease*, 7th ed., Vol. 1 (Scriver, C. R., Beaudet, A. L., Sly, W. S., and Valle, D., eds), pp. 53–118, McGraw-Hill, New York

377. Brenkert, A., Arora, R. C., Radin, N. S., Meier, H., and MacPike, A. D. (1972) *Brain Res.* **36**, 195–202

378. Morell, P., and Constantino-Ceccarini, E. (1972) *Lipids* **7**, 266–268

379. Seppala, R., Tietze, F., Krasnewich, D., Weiss, P., Ashwell, G., Barsh, G., Thomas, G. H., Packman, S., and Gahl, W. A. (1991) *J. Biol. Chem.* **266**, 7456–7461

380. Spellacy, E., Shull, R. M., Constantopoulos, G., and Neufeld, E. F. (1983) *Proc. Natl. Acad. Sci. U.S.A.* **80**, 6091–6095

381. Shull, R. M., Kakkis, E. D., McEntee, M. F., Kania, S. A., Jonas, A. J., and Neufeld, E. F. (1994) *Proc. Natl. Acad. Sci. U.S.A.* **91**, 12937–12941

382. Walkley, S. U., Thrall, M. A., Dobrenis, K., Huang, M., March, P. A., Siegel, D. A., and Wurzelmann, S. (1994) *Proc. Natl. Acad. Sci. U.S.A.* **91**, 2970–2974

Study Questions

1. Constitution of cell surface oligosaccharides or polysaccharides includes the following:
   D-Glucose
   D-Mannose
   D-Galactose
   L-Arabinose
   L-Fucose
   D-Glucuronic acid
   D-Neuraminic acid

   Outline pathways for biosynthesis of these compounds from glucose.

2. Decarboxylation steps are required for synthesis of UDP-xylulose and UDP-apiose (Fig. 21-1). Propose chemical mechanisms for these reactions.

3. How do animals and plants differ with respect to transport and storage of glucose?

4. Comment on unresolved questions about the biosynthesis of cellulose, amylose, and amylopectin. What glycosyl carrier groups are required?

5. Most 5- and 6-membered sugars are found in nature as pyranose ring forms. Why is ribose in RNA in the furanose ring form?

6. If the ratio $[NAD^+]/[NADH]$ in a cell were 500 and the ratio $[NADP^+]/[NADPH]$ were 0.002, what concentrations of fructose and sorbitol would be in equilibrium with 0.1 mM glucose? See Box 20-A and Table 6-4.

7. Write a balanced equation for reaction of boric acid (H_3BO_3) with two sugar rings to give a borate diol ester linkage (Box 20-E).

8. Describe in general terms the process by which *N*-linked oligosaccharides are synthesized and attached to proteins. What are the functions of the ER and the Golgi?

9. What, if any, restrictions do you think should be applied to the use of antibiotics on farms?

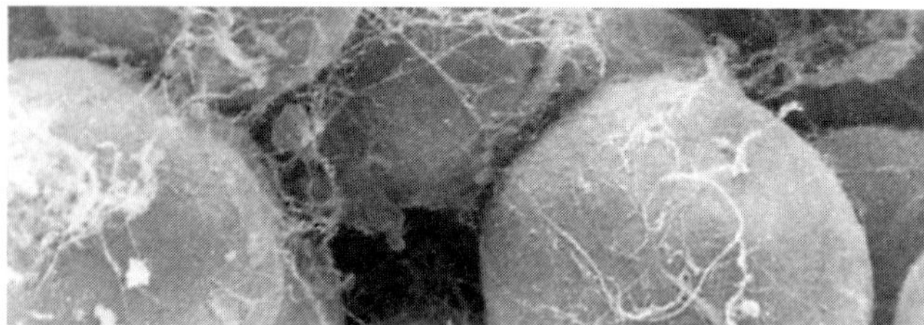

Two large fat-filled adipose cells are seen in the foreground of this scanning electron micrograph. They are part of a larger cluster of cells from rat tissue. Delicate strands of connective tissue fibers intertwine the cells and hold them together. While most of the connective tissue substance has been washed away during preparation of the specimen, the remnants give a realistic impression of the soft, loose nature of the intercellular material. From Porter and Bonneville (1973) *Fine Structure of Cells and Tissues*, Lea and Febiger, Philadelphia, Pennsylvania. Courtesy of Mary Bonneville.

Contents

Specific Aspects of Lipid Metabolism 21

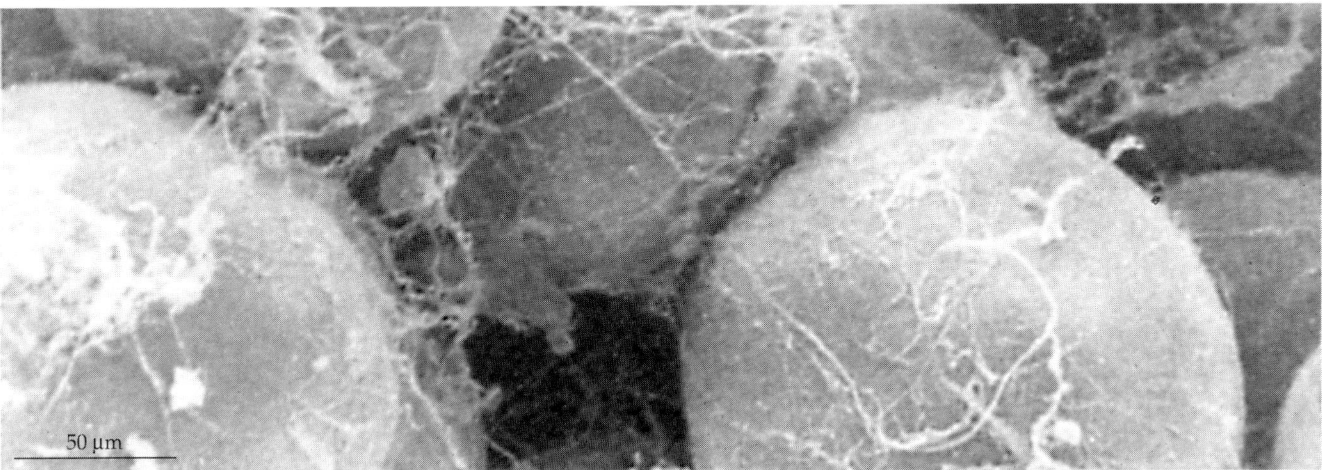

The basic pathways for both synthesis of fatty acids and for their β-oxidation (Fig. 17-1) have been described in Chapter 17. However, there are many variations to these pathways, and additional sets of enzymes are needed to synthesize the complex array of lipids present in most organisms. We will consider these details in this chapter. Like most other organisms, human beings are able to synthesize triacylglycerols (triglycerides), phospholipids, and glycolipids needed for cell membranes. Glucose can serve as the starting material. However, dietary lipids are also a major source. For this reason, we will start with a discussion of the digestion and uptake of lipids and of the distribution by way of the bloodstream of ingested lipids and of lipids synthesized in the liver or in other tissues.

A. Digestion, Synthesis, and Distribution of Triacylglycerols in the Human Body

Digestion of triglycerides begins in the stomach with emulsification and partial digestion by gastric lipase. Within the small intestine the ~100-residue protein called **colipase**[1–2a] binds to the surface of the fat droplets and provides an attachment site for the 449-residue **pancreatic lipase**. This Ca^{2+}-dependent serine esterase cleaves each triglyceride to two molecules of fatty acid and one of a 2-monoacylglycerol.[3–5] These products are emulsified by bile salts (Fig. 22-10) and are then taken up by the cells of the intestinal lining. The fatty acids are converted to acyl-CoA esters which transfer their acyl groups to the monoacylglycerols to resynthesize the triacylglycerols.[6] The latter are incorporated into the very large lipoprotein particles called **chylomicrons** (Table 21-1) and enter the bloodstream via the lymphatic system (Fig. 21-1).[7] Free fatty acids are also transported as complexes with serum albumin.

Synthesis of lipids from carbohydrates is an efficient process, which occurs largely in the liver and also in intestinal epithelial cells.[6] The newly synthesized triacylglycerols, together with smaller amounts of phospholipids and cholesterol, combine with specific **apolipoproteins**, which are also synthesized in the liver, to form **very low density lipoprotein (VLDL)** particles which are secreted into the blood stream. They transport the newly formed triacylglycerols from the liver to other body cells including the adipocytes, which store excess fat (Fig. 21-1).

1. Plasma Lipoproteins

The small particles of plasma lipoprotein, which carry triacylglycerols, can be separated according to their buoyant densities by centrifugation. They have been classified into five groups of increasing density but smaller size as **chylomicrons,** very low density lipoproteins **(VLDL)**, intermediate density lipoproteins (**IDL**), low density **(LDL),** and high density lipoproteins **(HDL)** (Table 21-1 and Fig. 21-2). Each lipoprotein particle contains one or more apolipoproteins (Table 21-2), whose sizes vary from the enormous 4536-residue apoB-100 to apoC-II and apoC-III, each of which contains just 79 residues[7a] and the 57-residue apoC-I.[7b]

The larger lipoproteins are spherical micelles containing a core of triacylglycerols and esters of cholesterol surrounded by a 2- to 3-nm-thick layer consisting of phospholipid, free cholesterol, and the apolipoprotein components.[8] The size of the lipoprotein particles also varies from a 200- to 500-nm diameter for chylomicrons to as little as 5 nm for the smallest HDL particles. The difference in volume is more impressive. If, as has been estimated,[9] a 22-nm diameter LDL particle contains about 2000 cholesterol and cholesteryl ester molecules and 800 phospholipids, a small HDL particle of 7-nm diameter will have room for only about 60 molecules of cholesterol and 90 of phospholipid, while a chylomicron may carry 10 million molecules of triacylglycerol. HDL particles are quite heterogeneous. As is indicated in Table 21-1, they are sometimes dividing into HDL2 and HDL3 density groups. In addition, there is a pre-HDL with lower phospholipid content and discoid forms low in cholesterol. Models of a reconstituted lipoprotein disc contain two molecules of apoA-I and ~160 phosphatidylcholines that form a bilayer core.[10–10b]

Each apolipoprotein has one or more distinct functions. The apoB proteins probably stabilize the lipoprotein micelles. In addition, apoB-100 is essential to recognition of LDL by its receptors. The 79-residue apoC-II has a specific function of activating the lipoprotein lipase that hydrolyses the triacylglycerols of chylomicrons and VLDL. Lack of either C-II or the lipase results in a very high level of triacylglycerols in the blood.[11]

The large apolipoprotein B-100 is synthesized in the liver and is a principal component of VLDL, IDL, and LDL. It is the sole protein in LDL, accounting for nearly 20% of the mass of LDL particles. Partly because of its insolubility in water, its detailed structure is uncertain. If it were all coiled into an α helix, it would be 680 nm long and could encircle the LDL particle nearly 10 times! While the true structure of apoB-100 is unknown, it is thought to be extended and to span at least a hemisphere of the LDL surface.[12] It consists of at least five domains. Sixteen cysteines are present in the first 25 residues at the N terminus, forming a crosslinked high-cysteine region. There are also 16 *N*-glycosylated sites. Domain IV (residues 3071–4011) is thought to contain the site that binds to its specific receptor, the LDL receptor.[12] Heterogeneity in the amide I band of the infrared absorption spectrum (Fig. 23-3) suggests that about 24% is α helix, 23% β sheet, and that a large fraction consists of turns, and unordered and extended peptide structures.[13]

In intestinal epithelial cells the same apoB gene that is used to synthesize apoB-100 in the liver is used to make the shorter **apoB-48** (48%) protein. This is accomplished in an unusual way that involves "editing" of the mRNA that is formed. Codon 2153 in the mRNA for the protein is CAA, encoding glutamine. However, the cytosine of the triplet is acted on by a deaminase, an editing enzyme, to form UAA, a chain termination codon.[14,15] A third form of apoB is found in **lipoprotein(a)** (Lpa). This LDL-like particle contains apoB-100 to which is covalently attached by a single disulfide linkage (probably to Cys 3734 of apoB-100) a second protein, **apo(a)**. The latter consists largely of a chain of from 11 to over 50 kringle domains resembling the 78-residue kringle-4 of plasminogen (see Fig. 7-30C)[16–19a] as well as a protease domain.[20] This additional chain may cause tighter binding to LDL receptors and may cause lipoprotein(a) to displace plasminogen from cell surface receptors.[21] The amount of Lp(a) varies over 1000-fold among individuals and is genetically determined. The number of kringle domains also varies.[17] Although the presence of high Lp(a) is associated with a high risk of atherosclerosis and stroke,[21a] many healthy 100-year olds also have high serum Lp(a).[22]

Apolipoprotein A-I is the primary protein component of HDL.[23–25b] Most of the 243 residues consist of a nearly continuous amphipathic α helix with kinks at regularly spaced proline residues.[26–28] Two disulfide-linked ApoA-I molecules may form a belt that encircles the discoid lipoprotein.[25b] ApoA-II is the second major HDL protein, but no clearly specialized function has been identified.[29,30] ApoA-I, II, and IV, apoC-I, II, and III, and apoE all have multiple repeats of 22 amino acids with sequences that suggest amphipathic helices. The 391-residue ApoA-IV has 13 tandem 22-residue repeats. Proline and glycine are present in intervening hinge regions.[23] This may enable these proteins to spread over and penetrate the surfaces of the lipoprotein micelles. Most of these proteins are encoded by a related multigene family.[7,30a]

The 299-residue apolipoprotein E plays a key role in metabolism of both triacylglycerols and cholesterol. Like apoB-100 it binds to cell surface receptors.[31–33a] Absence of functional apoE leads to elevated plasma triacylglycerol and cholesterol, a problem that is considered in Chapter 22, Section D. The N-terminal domain, from residues 23 to 164, forms a 6.5 nm-long four-helix bundle, which binds to the LDL receptors.[32] There are three common isoforms of alipoprotein E (apoE2, apoE3, and apoE4). ApoE3 is most common.[33b] The presence of apoE4 is associated with an increased risk of Alzheimer disease (Chapter 30).

The major lipoproteins of insect hemolymph, the **lipophorins**, transport diacylglycerols. The apolipophorins have molecular masses of ~250, 80, and sometimes 18 kDa.[34–37a] The three-dimensional structure of a small 166-residue lipophorin (apolipophorin-III) is that of a four-helix bundle. It has been suggested that it may partially unfold into an extended form, whose amphipathic helices may bind to a phospholipid surface of the lipid micelle of the lipophorin.[35] A similar behavior may be involved in binding of mammalian apolipoproteins. Four-helix lipid-binding proteins have also been isolated from plants.[38] See also Box 21-A. Specialized lipoproteins known as **lipovitellins**

TABLE 21-1
Classes of Lipoprotein Particles

Class			Composition (weight percent)[a]				
			Surface components			Core lipids	
	Diameter (nm)	Density (g/ml)	Protein	Phospholipid	Cholesterol	Cholesteryl esters	Triacylglycerol
Chylomicrons	75–1200	0.93	2	7	2	3	86
VLDL	30–80	0.93–1.006	8	18	7	12	55
IDL	25–35	1.006–1.019	19	19	9	29	23
LDL	18–25	1.019–1.063	22	22	8	42	6
HDL2	9–12	1.063–1.125	40	33	5	17	5
HDL3	5–9	1.125–1.21	45	35	4	13	3
Lp (a); slow pre-β	25–30	1.04–1.09					

[a] Data from Havel, R. J., and Kane, J. P. (1995) in *The Metabolic and Molecular Bases of Inherited Disease*, 7th ed., Vol. II (Scriver, C. R., Beaudet, A. L., Sly, W. S., and Valle, D., eds), pp. 1841–1852, McGraw-Hill, New York.

These are averages and there is considerable variation.

TABLE 21-2
Properties of Major Plasma Apolipoproteins

Designation	No. residues	Mass (kDa)	Source	Function
A-I	243	29		Major HDL protein
A-II	—	17.4	Liver and intestine	
A-IV	376	44.5		
B-100	4536	513	Liver	VLDL formation; ligand for LDL receptor
B-48	2152	241	Intestine	Chylomicron formation ligand for liver chylomicron receptor
C-I	57	6.6		
C-II	79	8.9	Liver	Cofactor for lipoprotein lipase
C-III	79	8.8		
D	—	31	Many tissues	A lipocalin
E	299	34	Liver, VLDL	Ligand for LDL receptor
(a)	Variable			Ligand for liver chylomicron receptor

store phospholipid in eggs whether from nematodes, frogs, or chickens.[39] There is some sequence similarity to that of human apolipoprotein B-100.

2. Movement of Lipid Materials Between Cells

After the synthesis and release of chylomicrons into the lymphatic circulation, various exchange processes occur by which apolipoproteins, as well as enzymes and other proteins, may be added or removed. These very complex and incompletely understood phenomena are presented in simplified form in Fig. 21-1. Chylomicrons donate apolipoproteins of the A and C families to HDL particles which, in turn, donate apoE and may also return some apoC protein to the chylomicrons.

Both chylomicrons and VLDL particles undergo similar processes in the capillary blood vessels, where their triacylglycerols are hydrolyzed to glycerol and free fatty acids by **lipoprotein lipase**.[40–42a] This enzyme requires for its activity the apolipoprotein C-II which is present in the chylomicrons and VLDL particles. Lipoprotein lipase is also known as the "clearing

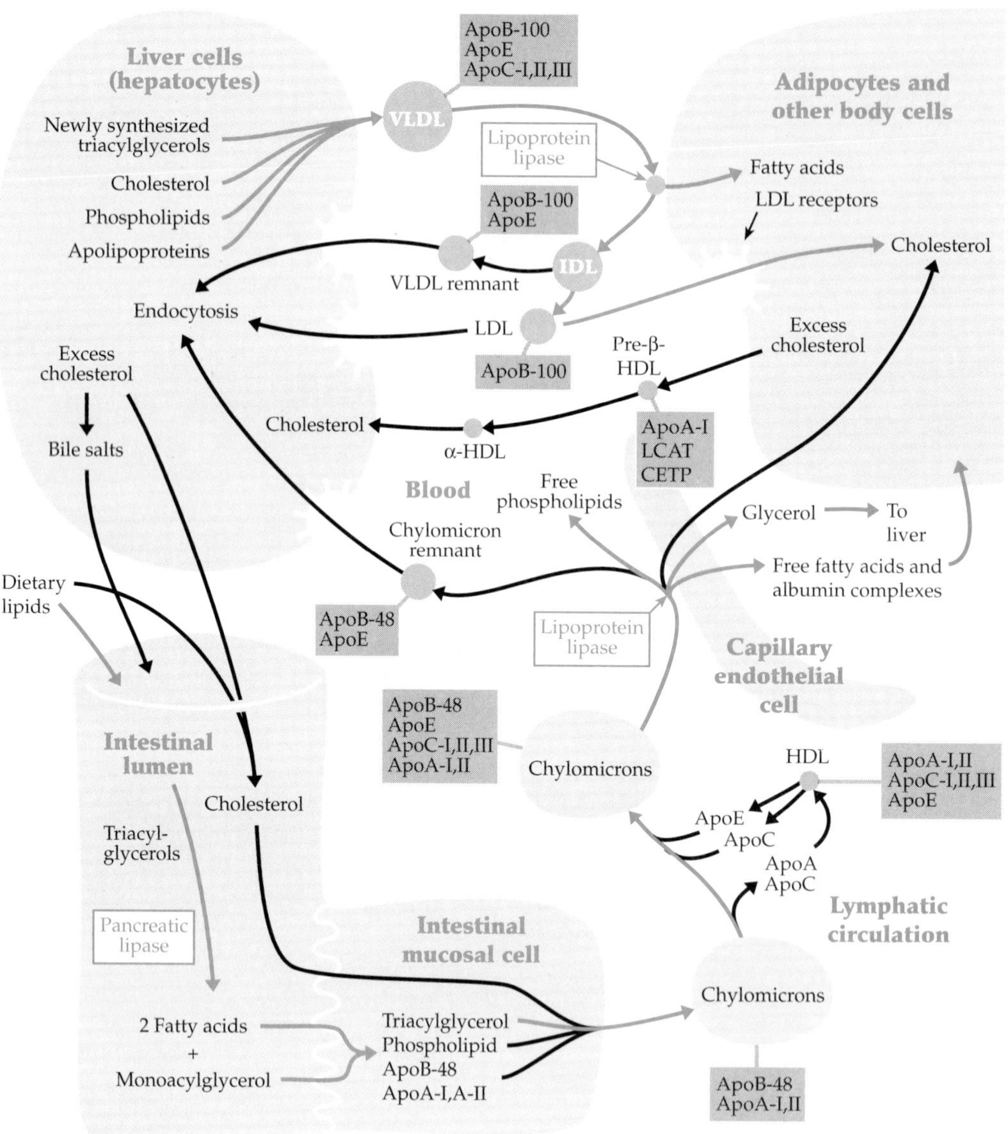

Figure 21-1 Movement of triacylglycerols from liver and intestine to body cells and lipid carriers of blood. VLDL; very low density lipoprotein which contains triacylglycerols, phospholipids, cholesterol, and apolipoproteins B, and C. IDL; intermediate density lipoproteins found in human plasma. LDL; low density lipoproteins which have lost most of their triacylglycerols. ApoB-100, etc., are apolipoproteins listed in Table 21-2. LCAT, lecithin: cholesterol acyltransferase; CETP, cholesteryl ester transfer protein (see Chapter 22).

factor" because it clears the milky chylomicron-containing lymph. It is secreted by adipocytes and other cells and becomes attached to heparan sulfate proteglycans on surfaces of capillary endothelial cells, a major site of its action.[43] Hereditary absence of functional lipoprotein lipase causes **chylomicronemia**, a massive buildup of chylomicrons in plasma.[41,44] The condition does not cause atherosclerosis but may lead to pancreatitis if not treated. Restriction of dietary fat to 20 g / day or less usually prevents problems. Naturally occurring mutations of lipoprotein lipase involving both the aspartate of the catalytic triad (p. 635)[45] and the flexible loop that covers the active site[46] have been discovered.

Both lipoprotein lipase and the less well understood **hepatic lipase** are related structurally to pancreatic lipase.[42,42b] In addition to hydrolysis of the triacylglycerols, the uptake of materials from lipoproteins probably involves shedding of intact phospholipids, perhaps as liposome-like particles.[40]

The free fatty acids and glycerol are taken up by mammalian tissue cells leaving the cholesterol and some of the phospholipids of the VLDL particles as LDL. In humans **intermediate density lipoproteins** (IDL) are formed initially, but some are converted to LDL later. Both LDL particles and the shrunken **chylomicron remnants** and **VLDL remnants** are taken up by endocytosis in coated pits and are degraded by body cells, principally of the liver.[47,48] The best known of these receptors is the 839-residue **LDL receptor,** which has a specific affinity for ApoB-100. The related **VLDL receptor** (apoE receptor) has a higher affinity for apoE[48–50] and may function in uptake of both VLDL and chylomicron remnants. The **LDL receptor-related protein** functions as a third lipoprotein receptor.[51] In addition, a series of **scavenger receptors**, found in abundance in macrophages, take up oxidized lipoproteins and other materials.[51,52] Scavenger receptor B1 (SR-B1), which is also found in liver cells, is involved in uptake of cholesterol from HDL particles by hepatocytes [53] (see also Chapter 22). Liver cells, and other cells as well, contain **lipocalins** and **fatty acid binding proteins** (Box 21-A) that help to carry these relatively insoluble acids to their destinations within the cells. Serum albumin (Box 2-A) is also a major carrier of free fatty acids.[53a] Within the adipocytes the fatty acids are reconverted to triacylglycerols. The low density (LDL) and high density (HDL) lipoproteins are involved primarily in transporting cholesterol to and from cells, a topic that is discussed in Chapter 22, Section D,2.

Fatty acids are carried to tissues for use in synthesis of triacylglycerols, phospholipids, and other membrane lipids. The mobilization of fatty acids from triacylglycerol stores and from cholesterol esters depends upon **hormone-sensitive lipase** (p. 635).[53b, 53c] This enzyme is activated by cAMP-dependent phosphorylation and moves from the cytoplasm to the surfaces of lipid droplets in response to catecholamines and other lipolytic hormones. Fatty acids are a major fuel for aerobic cells. Their conversion to acyl-CoA derivative and oxidation to CO_2 by beta oxidation (Fig. 17-1) and other pathways are discussed in Chapter 17 (pp. 939 – 950).

B. The Biosynthesis of Fatty Acids and Their Esters

The synthesis of fatty acids two carbon atoms at a time from acetyl-CoA has been considered in Chapter 17 and is outlined in Fig. 17-12. In this pathway, which resembles the β oxidation sequence in reverse, the products are saturated fatty acids with an even number of carbon atoms as shown in Fig. 21-2. In this section, we will consider some of the factors that lead to variations in the chain lengths and types of fatty acids.

1. Fatty Acid Synthases

Both bacteria and plants have separate enzymes that catalyze the individual steps in the biosynthetic sequence (Fig. 17-12). The fatty acyl group grows while attached to the small acyl carrier protein (ACP).[54–58] Control of the process is provided, in part, by the existence of isoenzyme forms. For example, in *E. coli* there are three different β-oxoacyl-ACP synthases. They carry out the transfer of any acyl primer from ACP to the enzyme, decarboxylate malonyl-ACP, and carry out the Claisen condensation (steps *b*, *e*, and *f* in Eq. 17-12)[58a – e] One of the isoenzymes is specialized for the initial elongation of acetyl-ACP and also provides feedback regulation.[58c] The other two function specifically in synthesis of unsaturated fatty acids.

In a few bacteria and protozoa and in higher animals the fatty acid synthase consists of only one or two multifunctional proteins. That from animal tissues contains six enzymes and an acyl carrier protein (ACP) domain as well. The human enzyme contains 2504 amino acid residues organized as a series of functional domains.[59–59b] Pairs of the 272-kDa chains associate to form 544-kDa dimers. The complex protein may have arisen via an evolutionary process involving fusion of formerly separate genes.[60] The enzyme contains an ACP-like site with a bound 4'-phosphopantetheine near the C terminus as well as a cysteine side chain near the N terminus in the second acylation site. Since the two –SH groups can be crosslinked by dibromopropanone,[61,62] an antiparallel linear arrangement of the two chains was proposed.[63–65] Locations of the six enzymatic activities in each chain are indicated on page 1187. According to this picture,

BOX 21-A LIPOCALINS, FATTY ACID-BINDING PROTEINS, AND LIPOPHORINS

Small hydrophobic molecules, which might easily bind in biologically undesirable ways, are chaperoned in animals, plants, and bacteria by binding proteins that provide hydrophobic cavities or crevices appropriate for holding these molecules in readily releasable forms. The **lipocalins**, most of which are extracellular proteins, have a conserved structural motif consisting of an 8-stranded β barrel arranged as two stacked orthogonal sheets with a C-terminal α helix that blocks one end. The other end is able to open and allow a small hydrophobic molecule to bind in the internal cavity[a–c] (see figure). Only three short amino acid sequences are conserved within a large family of lipocalins[b,d] which includes **plasma retinol-binding protein**,[e] mammalian **odorant-binding proteins**,[f] **α-lactalbumin**, **apolipoprotein D**,[a] and the blue biliverdin-binding protein **insecticyanin** of insect hemolymph.[g,h] Most lipocalins are soluble, but some such as the plasma **α1-microglobulin**,[i] which plays a role in the immune system (Chapter 31), have additional functions that require them to bind to other proteins or to cell surfaces.[j] The **gelatinase-associated lipocalin** of human neutrophils binds bacterially derived *N*-formylpeptides that act as chemotactic agents (Chapter 19) and induces release of materials from intracellular granules.[c] A few lipocalins have enzymatic activity. For example, **prostaglandin D synthase** is both an enzyme and a carrier of bile pigments and thyroid hormones.[k] Most lipocalins have been found in higher animals, but at least a few bacterial proteins belong to the family.[d] One is the 77-residue *E. coli* outer membrane lipoprotein.[a]

A related family of proteins are represented by **fatty acid-binding proteins**[l–o] and by the intracellular **retinol-** and **retinoic acid-binding proteins** (see also Box 22-A).[p] These are 10-stranded antiparallel β-barrels with two helices blocking an end (see Figure).

A third group of lipid-binding proteins have a four-helix bundle structure. They include the insect **lipophorins**, which transport diacylglycerols in the hemolymph (see main text), and nonspecific lipid carriers of green plants.[q] An 87-residue four-helix protein with a more open structure binds acyl-coenzyme A molecules in liver.[r]

A small 98-residue sterol-binding protein from

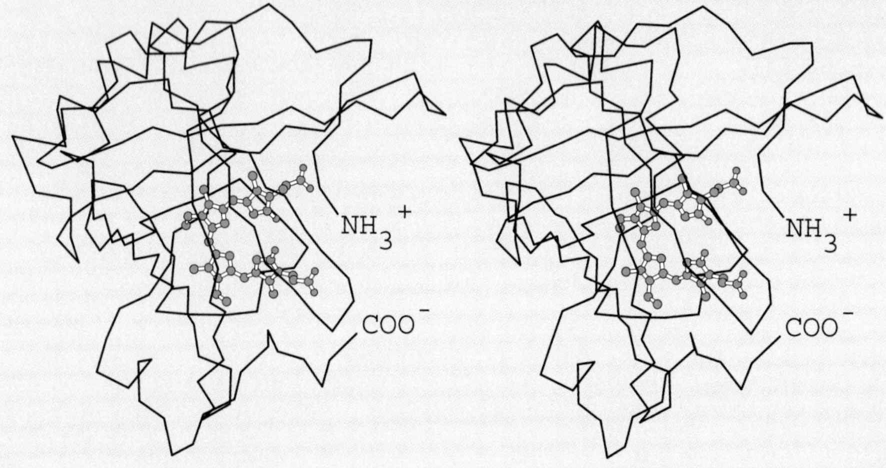

Stereoscopic view of an α-carbon model of an insecticyanin subunit with the bound biliverdin. The N and C termini are labeled NH_3^+ and COO^-, respectively. The positions of several amino acid residues are indicated. From Holden *et al.*[g] Courtesy of Hazel Holden.

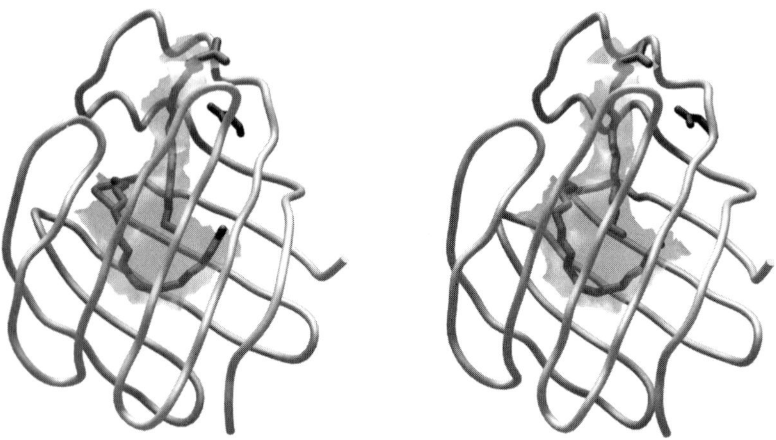

Structure of a crystalline fatty acid-binding protein from liver with two molecules of bound oleate (dark rods). The lower molecule is more deeply embedded in the protein and more tightly bound than the second molecule, which is closer to the outer surface of the protein. Semitransparent grey marks the solvent-accessible surface of the binding cavity. An unknown molecule, perhaps butanoic acid (as modeled), binds also at top of the protein. See Thompson *et al.*[n] Courtesy of Leonard Banaszak.

BOX 21-A (continued)

the fungus *Phytophthora*, an agriculturally important plant pathogen, has a very different folding pattern. The sterol binds into a cavity formed by six helices and two loops. The protein is not only a sterol carrier but an **elicitin**, which induces a defensive response in the invaded plant. The function of the protein for the invader may be to acquire sterols for the fungus, which is unable to synthesize them.[s]

Many larger lipid carrier proteins are known. The 476-residue plasma cholesteryl ester transfer protein is discussed briefly in Chapter 22. Plasma phospholipid transfer proteins are of similar size.[t,u] A 456-residue human phospholipid-binding protein interacts with the lipopolysaccharide of the surfaces of gram-negative bacteria (Fig. 8-30) and participates in the immune response to the bacteria. It has an elongated boomerang shape with two cavities, both of which bind a molecule of phosphatidylcholine. Other plasma lipid transfer proteins may have similar structures.[v]

[a] Bishop, R. E., Penfold, S. S., Frost, L. S., Höltje, J.-V., and Weiner, J. H. (1995) *J. Biol. Chem.* **270**, 23097–23103
[b] Flower, D. R., North, A. C. T., and Attwood, T. K. (1993) *Protein Sci.* **2**, 753–761
[c] Coles, M., Diercks, T., Muehlenweg, B., Bartsch, S., Zölzer, V., Tschesche, H., and Kessler, H. (1999) *J. Mol. Biol.* **289**, 139–157
[d] Flower, D. R., Sansom, C. E., Beck, M. E., and Attwood, T. K. (1995) *Trends Biochem. Sci.* **20**, 498–499
[e] Cowan, S. W., Newcomer, M. E., and Jones, T. A. (1990) *Proteins* **8**, 44–61
[f] Spinelli, S., Ramoni, R., Grolli, S., Bonicel, J., Cambillau, C., and Tagoni, M. (1998) *Biochemistry* **37**, 7913–7918
[g] Holden, H. M., Rypniewski, N. R., Law, J. H., and Rayment, I. (1987) *EMBO J.* **6**, 1565–1570
[h] Huber, R., Schneider, M., Mayr, I., Müller, R., Deutzmann, R., Suter, F., Zuber, H., Falk, H., and Kayser, H. (1987) *J. Mol. Biol.* **198**, 499–513
[i] Åkerström, B., and Lögdberg, L. (1990) *Trends Biochem. Sci.* **15**, 240–243
[j] Bishop, R. E., and Weiner, J. H. (1996) *Trends Biochem. Sci.* **21**, 127
[k] Beuckmann, C. T., Aoyagi, M., Okazaki, I., Hiroike, T., Toh, H., Hayaishi, O., and Urade, Y. (1999) *Biochemistry* **38**, 8006 – 8013
[l] Sacchettini, J. C., and Gordon, J. I. (1993) *J. Biol. Chem.* **268**, 18399–18402
[m] Wiesner, S., Kurian, E., Prendergast, F. G., and Halle, B. (1999) *J. Mol. Biol.* **286**, 233–246
[n] Thompson, J., Winter, N., Terwey, D., Bratt, J., and Banaszak, L. (1997) *J. Biol. Chem.* **272**, 7140–7150
[o] Hohoff, C., Börchers, T., Rüstow, B., Spener, F., and van Tilbeurgh, H. (1999) *Biochemistry* **38**, 12229–12239
[p] Thompson, J. R., Bratt, J. M., and Banaszak, L. J. (1995) *J. Mol. Biol.* **252**, 433–446
[q] Heinemann, B., Andersen, K. V., Nielsen, P. R., Bech, L. M., and Poulsen, F. M. (1996) *Protein Sci.* **5**, 13–23
[r] Andersen, K. V., and Poulsen, F. M. (1992) *J. Mol. Biol.* **226**, 1131–1141
[s] Boissy, G., O'Donohue, M., Gaudemer, O., Perez, V., Pernollet, J.-C., and Brunie, S. (1999) *Protein Sci.* **8**, 1191–1199
[t] Wirtz, K. W. A. (1991) *Ann. Rev. Biochem.* **60**, 73–99
[u] Tall, A. (1995) *Ann. Rev. Biochem.* **64**, 235–257
[v] Beamer, L. J., Carroll, S. F., and Eisenberg, D. (1997) *Science* **276**, 1861–1864

the ACP domain of one chain would cooperate with the β-oxacyl synthase (KS) domain of the second chain. However, more recent studies indicate greater flexibility with the ACP, MAS, and KS domains of a single chain also being able to function together.[58,62] Animal fatty acid synthases produce free fatty acids, principally the C_{16} palmitate. The final step is cleavage of the acyl-CoA by a thioesterase, one of the six enzymatic activities of the synthase.

Yeast fatty acid synthase[66,67] has an $\alpha_6\beta_6$ structure where the 208-kDa α subunit contains the ACP-like site, the active site –SH, and three catalytic activities. The 220-kDa β subunit has five catalytic activities. The yeast enzyme contains the FMN thought to act as $FMNH_2$ in the second reduction step. As in bacteria, the products of the complex are molecules of acyl-CoA of chain lengths C_{14}, C_{16}, and C_{18}.[68]

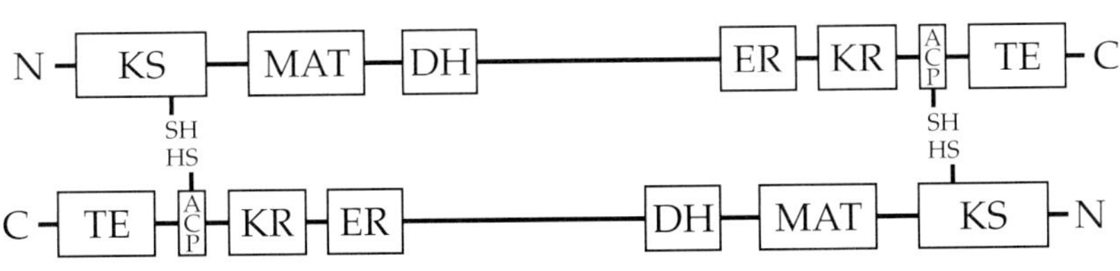

Abbreviation	Enzymatic activity	Residue numbers
KS	β-Oxoacyl (ketoacyl) synthase	1 – 406
MAT	Malonyl and acetyl transferase	428 – 815
DH	Dehydratase	829 – 969
Central region	Structural core (?)	970 – 1629
ER	Enoyl reductase	1630 – 1850
KR	β-Oxoacyl (ketoacyl) reductase	1870 – 2100
ACP	Acyl carrier protein	2114 – 2190
TE	Thioesterase	2200 – 2505

Organization of eukaryotic fatty acid synthase. From Joshi *et al.*[61]

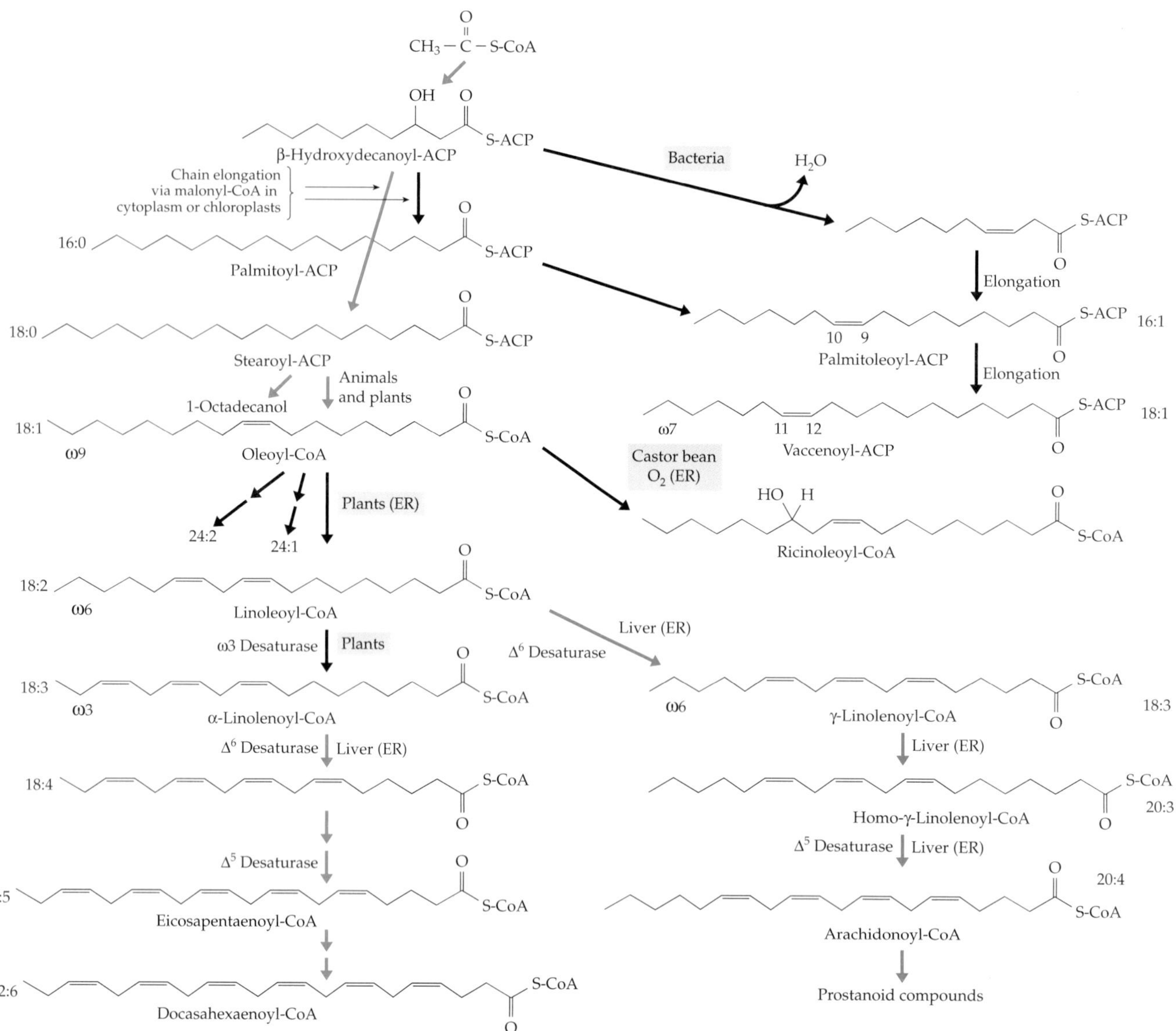

Figure 21-2 Some biosynthetic reactions of fatty acids. Green arrows indicate transformations carried out by the human body.

2. Control of Chain Length

The length of fatty acid chains is controlled largely by the enzymatic activity that releases the fatty acyl-CoA molecules or the free fatty acids from the synthase complex. In the animal enzymes the thioesterase, which is built into the synthase molecule, favors the release primarily of the 16-carbon saturated (16:0) palmitic acid. However, in mammary glands and in the uropygial glands (preen glands) of waterfowl shorter chain fatty acids predominate. These are released from the synthase by reaction with a second thioesterase, a 29-kDa protein[69,70] that catalyzes the otherwise premature release of shorter fatty acids. Cow's milk contains significant amounts of C_4-C_{14} acids as well as those with longer chains, whereas rabbit's milk contains largely C_8 to C_{10} fatty acids.[71]

In plants most biosynthesis occurs in the chloroplasts or in the protoplastids of seeds.[72–75] There are two different synthase systems in chloroplasts, one that forms primarily the 16:0 palmitoyl-ACP and the other the 18:0 stearoyl-ACP. Hydrolysis of the palmitoyl-ACP releases palmitate, one major product of chloroplasts. However, the stearoyl-ACP is desaturated to oleoyl-ACP[75a] before hydrolysis to free oleate or conversion to oleoyl-CoA. In many species oleic acid is almost the sole fatty acid exported by the chloroplasts. However, it undergoes a variety of modification reactions in the plant cytosol.

Plants, animals, and fungi all have fatty acid elongation systems in the endoplasmic reticulum. Using malonyl-CoA and NADPH,[76,77] chain lengths of fatty acids may be increased to C_{20} to C_{26}. Elongation of fatty acids can also occur in mitochondria by reactions that are essentially the reverse of β oxidation. The only deviation from an exact reversal of oxidation is the use of NADPH as the reductant for enoyl-CoA reductase. Elongation of fatty acids in the *outer* membrane of mitochondria, followed by transport of the elongated chains into the mitochondria, may even constitute another shuttle for transport of reducing equivalents from NADH into mitochondria (Chapter 18, Section D).[78] Elongation reactions may also occur in peroxisomes.[78a]

3. Starter Pieces and Branches

Acetyl-CoA is most often the primer or starter piece for fatty acid synthesis, but butyryl-CoA is a better primer for rabbits. Butyryl-CoA arises from acetyl-CoA by a reversal of β oxidation, the necessary enzymes occurring in significant amounts in the cytosol.[79] If either acetyl-CoA or butyryl-CoA is the starter piece, chain elongation via malonyl-CoA (Fig. 18-12) leads to fatty acids with an even number of carbon atoms. However, degradation of the branched chain amino acids valine, isoleucine, and leucine creates a series of branched starter pieces (Table 21-3), whose utilization leads to formation of branched fatty acids of the iso and anteiso series. These are found in bacteria, in the lipids of tobacco and wool, in the "sound lens" of echo-locating porpoises,[80] and in many other materials.[81] Propionyl-CoA serves as an intermediate for introduction (via methylmalonyl-CoA) of branches at various other points in a fatty acid chain.[82] For example, 2*R*- and 4*R*-methylhexanoic acids, 2,4,6,8-tetramethyldecanoic acid, and a variety of other branched chain acids are esterified with long- chain alcohols (mainly 1-octadecanol) to form the waxes of the preen glands of ducks and geese.[83] The C_{32} **mycocerosic acid** of *Mycobacterium tuberculosis* is also formed using both malonyl-CoA and methylmalonyl-CoA for chain elongation.[84] This acid is present in mycobacterial cell walls esterified with long-chain diols (Box 21-C).[85]

TABLE 21-3
Starter Pieces for Biosynthesis of Fatty Acids

Starter piece	Fatty acid products
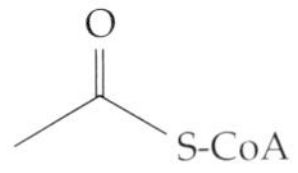 Acetyl-CoA (O, S-CoA)	Acid with even number of carbon atoms
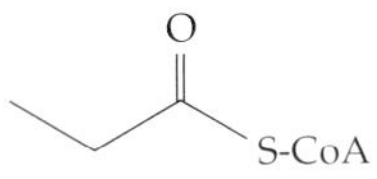 Propionyl-CoA (O, S-CoA)	Acid with odd number of carbon atoms
Valine 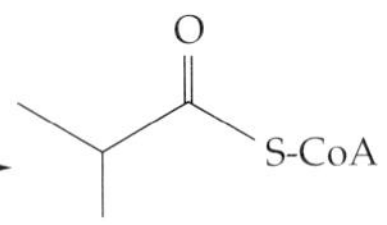 → Isobutyryl-CoA (O, S-CoA)	Iso series (even)
Leucine 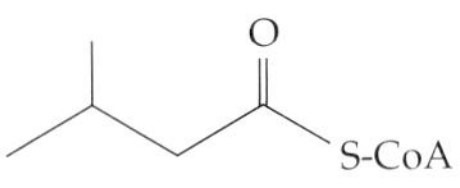 → Isovaleryl-CoA (O, S-CoA)	Iso series (odd)
Isoleucine 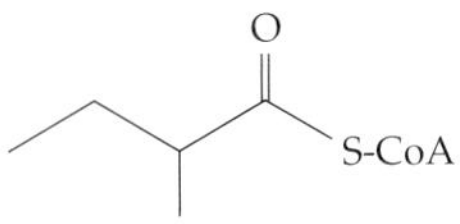 → (O, S-CoA)	Anteiso series (odd)

4. Synthesis by the Oxoacid Chain Elongation Process

The carbon skeleton of leucine is derived from that of valine by elongation of the corresponding oxoacid by a single carbon atom that is derived from acetyl-CoA (Fig. 17-18). At least in plants some branched-chain fatty acids of medium length are formed via the same oxoacid elongation process, which extends the chain one carbon atom at a time using 2-oxobutyrate as a starting compound. The same process can be used to form medium length straight-chain fatty acids (up to ~C_{12}) with either an odd or an even number of carbon atoms.[86]

2-Oxoglutarate can also serve as a starter piece for elongation by the oxoacid pathway. Extension by three carbon atoms yields 2-**oxosuberate** (Eq. 21-1). This dicarboxylate is converted by reactions shown in Eq. 24-39 into biotin and in archaebacteria into the coenzyme 7-mercaptoheptanoylthreonine phosphate (HTP), Eq. 21-1.[87] Lipoic acid is also synthesized from a fatty acid, the eight-carbon octanoate.[88,89] A fatty acid synthase system that utilizes a mitochondrial ACP may have as its primary function the synthesis of octanoate for lipoic acid formation.[90] The mechanism of insertion of the two sulfur atoms to form lipoate (Chapter 15) is uncertain. It requires an iron–sulfur protein[91,91a,b] and is probably similar to the corresponding process in the synthesis of biotin (Eq. 24-39)[92-93a] and in formation of HTP (Eq. 21-1). One component of the archaebacterial cofactor **methanofuran** (Chapter 15) is a tetracarboxylic acid that is formed from 2-oxoglutarate by successive condensations with two malonic acid units as in fatty acid synthesis.[94]

BOX 21-B THE ESSENTIAL FATTY ACIDS

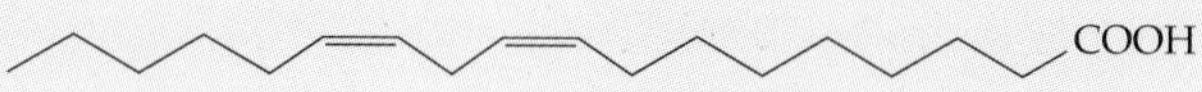

In 1930, George and Mildred Burr reported that the C18:2 ($\Delta^{9,12}$) **linoleic acid**, a fatty acid of exclusively plant origin, cured a disease condition observed in rats raised on a highly purified fat-free diet.[a,b] These animals grew poorly, developed a scaly dermatitis, and suffered kidney damage and impaired fertility. The symptoms could be prevented if 1% of the dietary energy was provided by linoleic acid. This C18:2 fatty acid can be converted in animals into a series of other fatty acids by chain elongation and desaturation. All of this series have a double bond six carbon atoms from the $-CH_3$ terminus and form an ω6 (or n-6) family.[c]

C18:2, ω6 — Linoleic acid — COOH

C18:3, ω3 — α-Linolenic acid — COOH

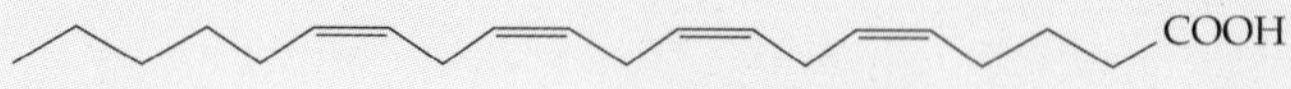

C20:4, ω6 — Arachidonic acid — COOH

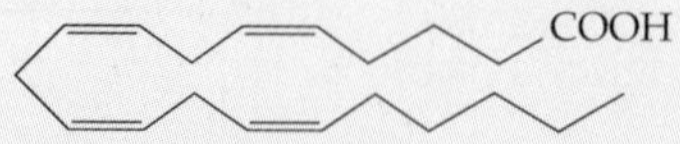

Arachidonic acid – folded — COOH

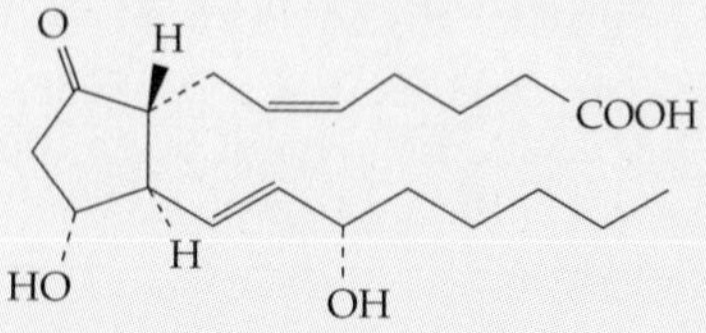

Prostaglandin E_2 — O, H, COOH, HO, H, OH

ω6: C18:2 → C18:3 → C20:3 → C20:4 → C22:4 → C22:5

Arachidonate

C20:3 ↓ PGE_1, $PGF_{1\alpha}$; C20:4 ↓ PGE_2, $PGF_{2\alpha}$

The major known essential function of linoleic acid is conversion to the C20:4 ($\Delta^{5,8,11,14}$) **arachidonic acid**, the major precursor to prostaglandins and other prostanoid compounds (Section D). This conversion occurs in infants as well as adults,[d] but the rate may not always be adequate, and arachidonic acid is usually classified as an essential fatty acid. It is not clear whether linoleic acid has any essential role of its own. However, while arachidonic acid can be converted into the prostaglandins designated PGE_2 and $PGF_{2\alpha}$, linoleic acid can also give rise, via the C20:30 **dihomolinolenic acid**, to PGE_1 and $PGF_{1\alpha}$ (see Eq. 21-16). The C18:3 ($\Delta^{9,12,15}$) **α-linolenic acid**, another plant acid, can partially replace linoleic acid and can be converted into PGE_3 and $PGF_{3\alpha}$.

ω3: C18:3 → C18:4 → C20:4 → C20:5 → C22:5 → C22:6

C20:5 ↓ PGE_3, $PGE_{3\alpha}$; C22:5 ↓ PG_4; C22:6 ↓ PG_5

Thus, it is not surprising that the three acids are not completely equivalent.[e–g]

Recent interest has focused on the C20:5 **eicosapentaenoic acid** (EPA) and the C22:6 **docosahexaenoic acids** (DHA). These ω3 (or n-3) polyunsaturated acids are formed from linolenic acid by marine algae and are found in fish oils.[h] The C22:5 and C22:6 acids can be converted to prostaglandins of the PG_4 and PG_5 series. DHA together with the ω6 C22:4 acid constitutes over 30% of the fatty acids in brain phospholipids. In the retina DHA accounts for over 60% of the total in the rod outer segments (Fig. 23-40). DHA may be formed in the human body from α-linolenic acid obtained from plant sources (Fig. 21-2). However, the rate of synthesis may be inadequate especially in infants and in persons of old age.[i–m] Deficiency of either EPA or DHA may lead to poor brain development during prenatal and infant life. Formation of new synapses between neurons as well as growth of new neurons in some areas of the brain is associated with thinking and with memory formation (Chapter 30). Therefore, a lack of the essential ω3 and ω6 fatty acids may contribute to mental deterioration in older adults. Eskimo populations, which consume large amounts of fish, have a very low incidence of coronary heart disease. An inverse relation between fish consumption and heart disease has also been demonstrated in other populations.[n] Inclusion of fish oils to 20–30% of total caloric intake in the diet causes a marked decrease in plasma triacylglycerols and very low density lipoproteins (VLDL).[o] This effect has been attributed to the altered composition of the prostanoid compounds known as thromboxanes and prostacyclins (PGI). For similar reasons the ω3 acids may have an antiinflammatory effect.[p,q] Administration of fish oil to patients with kidney disease has proved beneficial[r] and may decrease risk of some cancers.[s] However, a diet high in ω3 fatty acids has also been reported to increase cancer risk.[f] Long-chain ω3 fatty acids may protect against sudden death from heart disease.[t] They may promote lateral phase separation within membranes to form regions low in cholesterol (see references 95a and 119d).

BOX 21-B (continued)

[a] Burr, G. O., and Burr, M. M. (1930) *J. Biol. Chem.* **86**, 587–621
[b] Burr, G. (1980) *Trends Biochem. Sci.* **5**, 28
[c] Horrobin, D. F., ed. (1990) *Omega-6 Essential Fatty Acids*, Wiley-Liss, New York
[d] Salem, N., JR, Wegher, B., Mena, P., and Uauy, R. (1996) *Proc. Natl. Acad. Sci. U.S.A.* **93**, 49–54
[e] Leat, W. M. F. (1981) *Trends Biochem. Sci.* **6**, IX–X
[f] Cave, W. T., Jr. (1991) *FASEB J.* **5**, 2160–2166
[g] Lands, W. E. M. (1992) *FASEB J.* **6**, 2530–2536
[h] Lees, R. S., and Karel, M., eds. (1990) *Omega-3 Fatty Acids in Health and Disease*, Dekker, New York
[i] Cho, H. P., Nakamura, M., and Clarke, S. D. (1999) *J. Biol. Chem.* **274**, 37335–37339
[j] Farkas, T., Kitajka, K., Fodor, E., Csengeri, I., Lahdes, E., Yeo, Y. K., Krasznai, Z., and Halver, J. E. (2000) *Proc. Natl. Acad. Sci. U.S.A.* **97**, 6362–6366
[k] Qiu, X., Hong, H., and MacKenzie, S. L. (2001) *J. Biol. Chem.* **276**, 31561–31566
[l] Kang, Z. B., Ge, Y., Chen, Z., Cluette-Brown, J., Laposata, M., Leaf, A., and Kang, J. X. (2001) *Proc. Natl. Acad. Sci. U.S.A.* **98**, 4050–4054
[m] Carper, J. (2000) *Your Miracle Brain*, Harper Collins Publ., New York
[n] Kromhout, D., Bosschieter, E. B., and Coulander, C. L. (1985) *N. Engl. J. Med.* **312**, 1205–1209
[o] Phillipson, B. E., Rothrock, D. W., Connor, W. E., Harris, W. S., and Illingworth, D. R. (1985) *N. Engl. J. Med.* **312**, 1210–1216
[p] Lee, T. H., Hoover, R. L., Williams, J. D., Sperling, R. I., Ravalese, J., III, Spur, B. W., Robinson, D. W., Corey, E. J., Lewis, R. A., and Austen, K. F. (1985) *N. Engl. J. Med.* **312**, 1217–1224
[q] Hwang, D. (1989) *FASEB J.* **3**, 2052–2061
[r] Donadio, J. V., Jr., Bergstralh, E. J., Offord, K. P., Spencer, D. C., and Holley, K. E. (1994) *N. Engl. J. Med.* **331**, 1194–1199
[s] Hilakivi-Clarke, L., Clarke, R., Onojafe, I., Raygada, M., Cho, E., and Lippman, M. (1997) *Proc. Natl. Acad. Sci. U.S.A.* **94**, 9372–9377
[t] Albert, C. M., Campos, H., Stampfer, M. J., Ridker, P. M., Manson, J. E., Willett, W. C., Ma, J. (2002) *N. Engl. J. Med.* **346**, 1113 – 1118 See also **347**, 531 – 533

2-Oxoglutarate

Acetyl-CoA, NAD^+ → CO_2

Acetyl-CoA, NAD^+ → CO_2 — Oxoacid chain elongation process. Used three times

Acetyl-CoA, NAD^+ → CO_2

2-Oxosuberate → Biotin (Eq. 24-39)

NAD^+ → CO_2 — Oxidative decarboxylation

Sulfur

7-Mercaptoheptanoate

Threonine, ATP

7-Mercaptoheptanoylthreonine

ATP

7-Mercaptoheptanoylthreonine phosphate (HS-HTP) (see Fig. 15-22 for function)

(21-1)

5. Unsaturated Fatty Acids

Fatty acids containing one or more double bonds provide necessary fluidity to cell membranes[95,95a] and serve as precursors to other components of cells. Significant differences in the methods of introduction of double bonds into fatty acids are observed among various organisms. Bacteria such as *E. coli* that can live anaerobically often form **vaccenic acid** as the principal unsaturated fatty acid. It is formed by chain elongation after introduction of a cis double bond at the C_{10} stage of synthesis. The bacteria possess a **β-hydroxydecanoyl thioester dehydratase,** which catalyzes elimination of a β-hydroxyl group to yield primarily the *cis*-β,γ rather than the *trans*-α,β-unsaturated product (Eq. 21-2).[96] The mechanism may resem-

β-Hydroxydecanoyl-ACP (S–ACP: Acyl carrier protein)

→ H_2O

trans-α,β-Unsaturated product → Chain elongation → Longer saturated fatty acids

cis-β,γ-Unsaturated product

Chain elongation → Longer unsaturated fatty acids

(21-2)

ble that of enoyl hydratase (Eq. 13-7), the indicated *trans*-α,β-unsaturated intermediate (enzyme bound) being isomerized to the *cis*-β,γ-unsaturated product through an allylic rearrangement. The product can then be elongated to the C_{16} **palmitoleoyl-ACP** and C_{18} **vaccenoyl-ACP** derivatives (Fig. 21-2, right side). However, dehydration of β-hydroxydecanoyl-ACP lies at a branch point in the biosynthetic sequences. The *trans*-α,β-unsaturated fatty acyl compound lies on the usual route of chain elongation to palmitoyl-CoA (left side, Fig. 21-2).

In higher plants, animals, protozoa, and fungi, saturated fatty acids are acted upon by **desaturases** to introduce double bonds, usually of the cis (Z) configuration. The substrates may be fatty acyl-ACP, fatty acyl-CoA molecules, membrane phospholipids,[97] or glycolipids.[98] The Δ^9 desaturase, isolated from liver or from yeast, converts stearoyl-CoA to oleoyl-CoA (Eq. 21-3).[99–102] This membrane-associated enzyme system utilizes NADH as a reductant, passing electrons via cytochrome b_5 reductase and cytochrome b_5 itself to the desaturase.[103,104] The pro-*R* hydrogens are removed at both C-9 and C-10.

$$O_2 + 2\,H^+ + 2\,e^- + \text{stearoyl-CoA} \rightarrow \text{oleoyl-CoA} + 2\,H_2O \quad (21\text{-}3)$$

In plants a similar enzyme catalyzes formation of the first double bond in a fatty acyl group converting stearoyl-ACP into oleoyl-ACP in the chloroplasts.[72,75a,105–108] The soluble Δ^9 stearoyl-ACP desaturase has a diiron-oxo active site (Fig. 16-20, B, C).[109,110] Electrons are donated from light-generated reduced ferredoxin (see Chapter 23). In addition to the Δ^9 desaturase both plants and cyanobacteria usually desaturate C_{18} acids also at the Δ^{12} and Δ^{15} positions and C_{16} acids at the Δ^7, Δ^{10}, and Δ^{13} (ω3) positions.[111,112] Desaturation of oleate occurs primari-

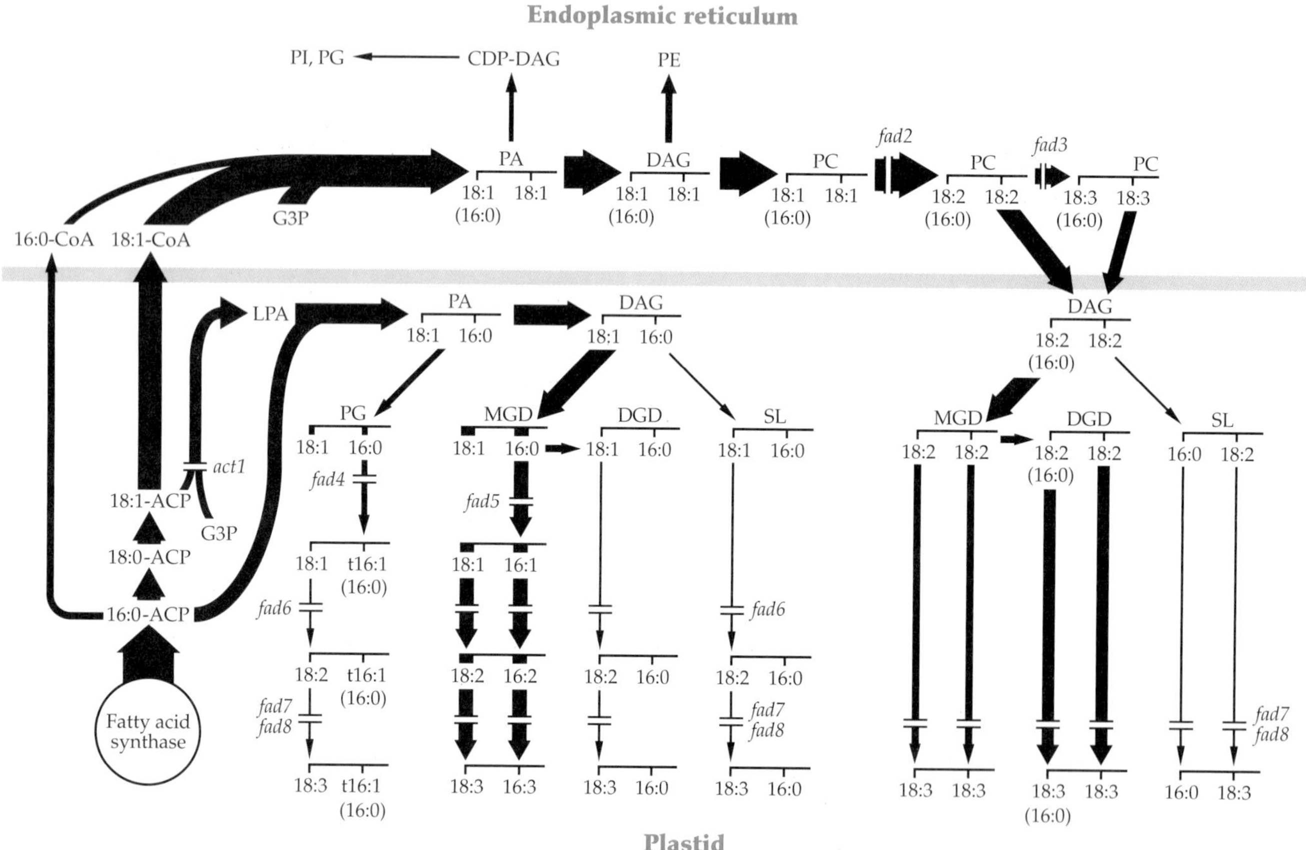

Figure 21-3 Major pathways of synthesis of fatty acids and glycerolipids in the green plant *Arabidopsis*. The major site of fatty acid synthesis is chloroplasts. Most is exported to the cytosol as oleic acid (18:1). After conversion to its coenzyme A derivative it is converted to phosphatidic acid (PA), diacylglycerol (DAG), and the phospholipids: phosphatidylcholine (PC), phosphatidylinositol (PI), phosphatidylglycerol (PG), and phosphatidylethanolamine (PE). Desaturation also occurs, and some linoleic and linolenic acids are returned to the chloroplasts. See text also. From Sommerville and Browse.[106] See also Figs. 21-4 and 21-5. Other abbreviations: monogalactosyldiacylglycerol (MGD), digalactosyldiacylglycerol (DGD), sulfolipid (SL), glycerol 3-phosphate (G3P), lysophosphatidic acid (LPA), acyl carrier protein (ACP), cytidine diphosphate-DAG (CDP-DAG).

ly in the ER after conversion of the free acid to its coenzyme A derivative or to phosphatidylcholine. Consecutive introduction of two double bonds forms linoleoyl-CoA (18:2, $\Delta^{9,12}$) and linolenoyl-CoA (18:3, $\Delta^{9,12,15}$; see Fig. 21-3).[108] All double bonds are cis. The membrane lipids of chloroplasts contain both linoleic and linolenic acids, which have apparently been returned to the chloroplasts from the cytosol[72] as indicated in Fig. 21-3. Plants grown at colder temperatures have a higher content of these trienoic acids than those grown at higher temperature.[72a]

The origin of ricinoleic acid, an abundant constituent of castor beans, is also shown in Fig. 21-2. It is formed by an **oleate hydroxylase** that has an amino acid sequence similar to those of oleate desaturases.[113] Both hydroxylation and desaturation are reactions catalyzed by diiron centers.[114] Other fatty acid hydroxylases act on the alpha[115] and the omega positions. The latter are members of the cytochrome P450 family.[116,117]

The conversion of oleoyl-CoA to linoleoyl-CoA is accomplished by some insects[118] but does not take place in most animals. As a result of this biosynthetic deficiency, polyunsaturated fatty acids such as linoleic, linolenic, and the C_{20} arachidonic acid are necessary in the diet (Box 21-B). One essential function of linoleic acid is to serve as a precursor of **prostaglandins** and related **prostanoids** (Section D). Dietary linoleate is converted to its CoA derivative and then by sequential Δ^6 desaturation,[119] elongation, and then Δ^5 desaturation, to the 20:4 ($\Delta^{5,8,11,14}$) arachidonoyl-CoA (Fig. 21-2, lower right). These acids are referred to as ω6 because of the position of the last double bond. Linolenic acid can be converted in an analogous fashion to the CoA derivative of the 20:5 ($\Delta^{5,8,11,14,17}$ ω6) eicosapentaenoic acid (**EPA**). The 22:6 docasahexaenoic acid (**DHA**; Fig. 21-2) is apparently formed by elongation of the 22:5 acyl-CoA to 24:5, desaturation, transfer to a peroxisome or mitochondrion, and β oxidation to shorten the chain.[95a]

These acids are very important in human nutrition.[119a–d] (See also Box 21-B.) In the absence of adequate essential fatty acids oleate is desaturated and elongated in a similar sequence to the unusual 20:3 ($\Delta^{5,8,11}$ ω9) acid. Vertebrate tissues also carry out desaturation at the Δ^4 and Δ^3 positions.[111] Lepidoptera, which synthesize a great diversity of pheromones, are rich in unusual desaturases such as the gland-specific acyl-CoA Δ^{11} desaturase of cabbage looper moths.[120]

Polyunsaturated fatty acids, containing 4, 5, or 6 double bonds and chain lengths of up to $C_{36,}$ are found in phosphatidylcholine of vertebrate retinas.[121] Although the double bonds are rarely in conjugated positions in food fats and in animal bodies, some plants convert oleic or linoleic acid into fatty acids with as many as three or four conjugated double bonds.[121a] **Conjugated linoleic acid** (9*c*, 11*t* 18:2) can be formed from 11-*trans*-octadecenoate in the human body.[121b] This compound is also found in meat and dairy products. It has been reported to have anticancer properties[121c] and may be another beneficial dietary constituent. An isomerase isolated from red algae converts polyunsaturated acids into forms with conjugated double bonds. For example, arachidonic acid (5*Z*, 8*Z*, 11*A*, 14*Z*) is converted to (5*Z*, 7*E*, 9*E*, 14*Z*)-eicosatetraenoic acid.[122,123]

6. Cyclopropane Fatty Acids and Mycolic Acids

Fatty acids containing one or more cyclopropane rings are present in many bacteria (p. 381).[124,125] The extra carbon of the cyclopropane ring is added from *S*-adenosylmethionine (AdoMet) at the site of a cis double bond in a fatty acyl group of a phosphatidylethanolamine molecule in a membrane (Eq. 21-4).[126,126a] The same type of intermediate carbocation can yield either a cyclopropane fatty acid (Eq. 21-4, step *a*) or a methenyl fatty acid (Eq. 21-4, step *b*). The latter can be reduced to a branched fatty acid. This is an alternative way of introducing methyl branches that is used by some bacteria.[127]

H H C=C
AdoMet
Adenosylhomocysteine
H CH_3 H C—C^+
H^+ *a*
H CH_2 H C—C
b H^+
CH_2 ‖ C CH_2
NADPH
H_3C H C CH_2

(21-4)

Mycobacteria are rich in cyclopropane-containing fatty acids. These **mycolic acids** are major components of the cell walls and may account for 30% of the dry weight of the cells.[128] The most abundant mycolic acid of *M. tuberculosis* consists of C_{52} fatty acid containing two cyclopropane rings joined via a Claisen-type condensation with a C_{26} carboxylate fatty acid (Eq. 21-5). A similar mycolic acid formed by *M. smegmatis* has double bonds instead of cyclopropane rings as indicated below Eq. 21-5.[128,129] There are other variations. In place of a double bond or cyclopropane group there may be –OH, $-OCH_3$, C=O, epoxide, or CH_3[127]

Cyclopropane fatty acids are catabolized via β oxidation,[130] which is modified as in Eq. 21-6 when the chain degradation reaches the cyclopropane ring. The

BOX 21-C TUBERCULOSIS

As many as one-third of the inhabitants of the earth are infected by *Mycobacterium tuberculosis*. For most the infection is dormant, but in some the slow-growing bacteria cause a progressive and deadly destruction of the lungs. There are still about three million deaths annually, and *M. tuberculosis* now, as in the past, kills more people then any other pathogen.[a,b] The development of drug-resistant strains of the bacterium and the threat of a world-wide resurgence of tuberculosis[c,d] has spurred new efforts to understand the unusual metabolism of mycobacteria and to develop new drugs. The complete 4.4 million base-pair sequence of the circular genome is known.[c] An unusually large fraction of the genes encode enzymes involved in synthesis and breakdown of lipids, including the synthesis of **mycolic acids** (Eq. 21-5) which are characteristic of mycobacteria.

The mycobacterial cell wall, discussed in Chapter 8, contains mycolic acids bound covalently at the nonreducing ends of arabinogalactans that are attached to the inner peptidoglycan layer,[e] as well as phosphatidylinositol-anchored lipoarabinomannans. Other unusual lipids that are also present and account for some of the difficulty of treatment with antibiotics include esters of **mycocerosic acid** with long-chain diols known as phenolphthiocerols and phthiocerols.[f,fg]

Streptomycin (Boxes 20-B, 20-H) was introduced into clinical use against tuberculosis in about 1943. However, resistant mutants always survived until newer drugs were developed. Isonicotinylhydrazide (**isoniazid**) is especially effective in combinations with suitable antibiotics and other drugs.[g] The four-drug combination isoniazid, rifampicin (Box 28-A), pyrazinamide, and ethambutol is often used. Nevertheless, bacteria resistant to all of these have developed.

Although isoniazid has been in use for about 45 years, the enzyme that it inhibits has been recognized only recently. It is a specific NADH-dependent **enoyl reductase** involved in synthesis of mycolic acids.[h,i] The isoniazid must be activated by action of a bacterial catalase-peroxidase.[j,k] This enzyme may convert the drug to a reactive radical that combines with a NADH-derived radical to form an adduct in the active site of the enzymes. One possible reaction sequence follows.[h] However, the mechanisms are not clear.

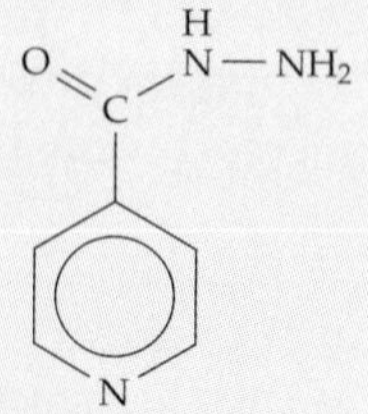

Isonicotinylhydrazide (Isoniazid)

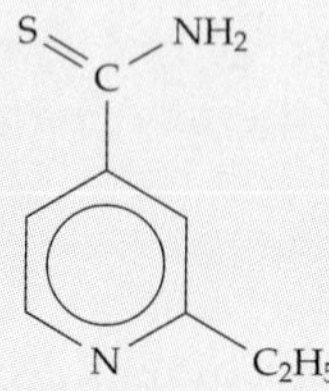

Ethionamide

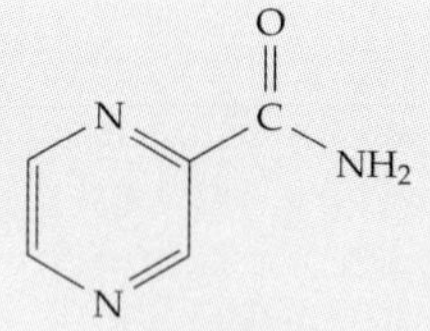

Pyrazinamide

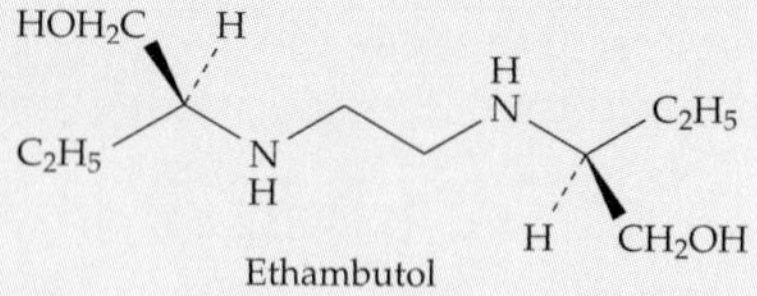

Ethambutol

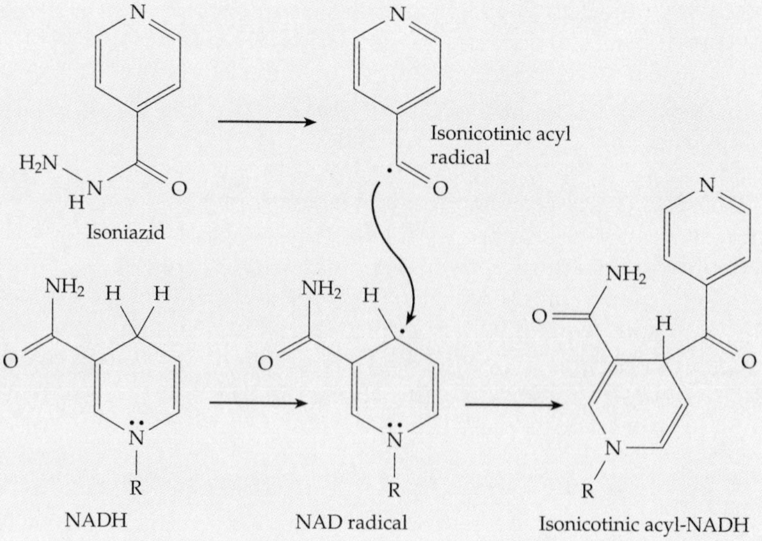

Isoniazid

Isonicotinic acyl radical

NADH

NAD radical

Isonicotinic acyl-NADH

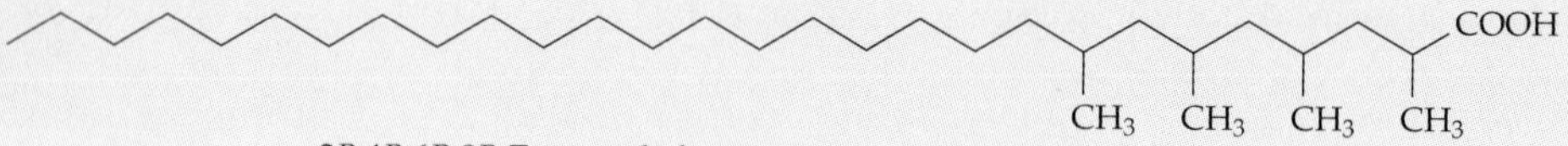

2*R*,4*R*,6*R*,8*R*-Tetramethyloctacosanoic acid (mycocerosic acid)

BOX 21-C (continued)

[a] Young, D. B. (1998) *Nature (London)* **393**, 515–516

[b] Venisse, A., Rivière, M., Vercauteren, J., and Puzo, G. (1995) *J. Biol. Chem.* **270**, 15012–15021

[c] Cole, S. T., and and 41 other authors. (1998) *Nature (London)* **393**, 537–544

[d] Iseman, M. D. (1994) *Proc. Natl. Acad. Sci. U.S.A.* **91**, 2428–2429

[e] Scherman, M. S., Kalbe-Bournonville, L., Bush, D., Xin, Y., Deng, L., and McNeil, M. (1996) *J. Biol. Chem.* **271**, 29652–29658

[f] Fitzmaurice, A. M., and Kolattukudy, P. E. (1998) *J. Biol. Chem.* **273**, 8033–8039

[fg] Patterson, J. H., McConville, M. J., Haites, R. E., Coppel, R. L., and Billman-Jacobe, H. (2000) *J. Biol. Chem.* **275**, 24900–24906

[g] Blanchard, J. S. (1996) *Ann. Rev. Biochem.* **65**, 215–239

[h] Rozwarski, D. A., Grant, G. A., Barton, D. H. R., Jacobs, W. R. J., and Sacchettini, J. C. (1998) *Science* **279**, 98–102

[i] Baldock, C., Rafferty, J. B., Stuitje, A. R., Slabas, A. R., and Rice, D. W. (1998) *J. Mol. Biol.* **284**, 1529–1546

[j] Sherman, D. R., Mdluli, K., Hickey, M. J., Arain, T. M., Morris, S. L., Barry, C. E., III, and Stover, C. K. (1996) *Science* **272**, 1641–1643

[k] Wengenack, N. L., Lopes, H., Kennedy, M. J., Tavares, P., Pereira, A. S., Moura, I., Moura, J. J. G., and Rusnak, F. (2000) *Biochemistry* **39**, 11508–11513

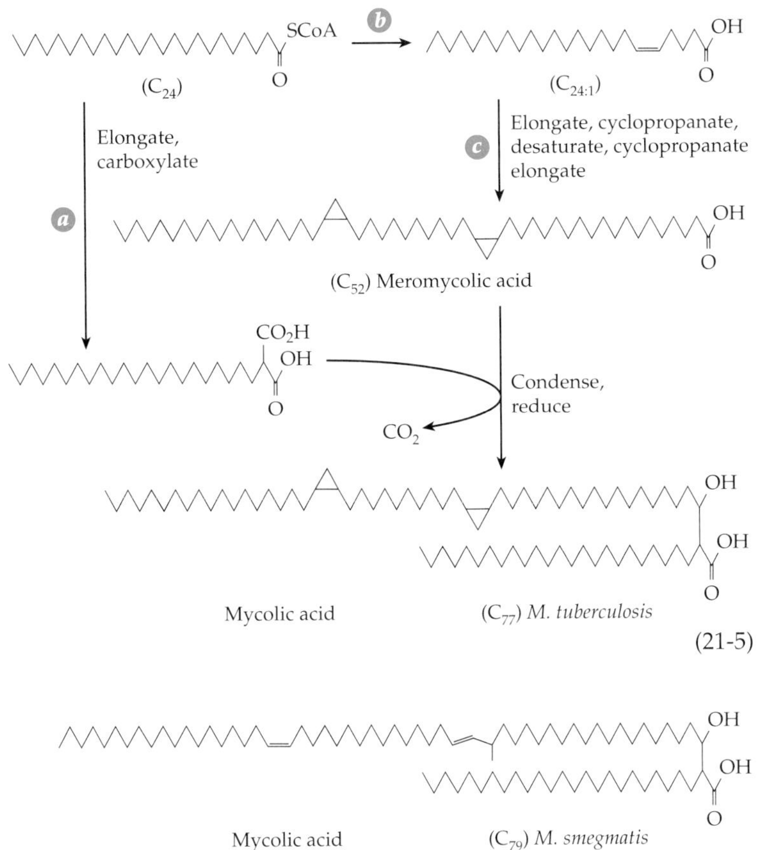

(21-5)

$^*CH_2$ / H / H / O / C—C / R / S—CoA

First two steps of β oxidation

$^*CH_2$ / H / OH / O / C—C / R / S—CoA

R / $^*CH_2$ / O / S—CoA

Futher β oxidation

(21-6)

ring opening of the cyclopropanol derivatives occurs readily, even with mild nonenzymatic acid–base catalysis.

Another alteration of unsaturated fatty acids is the formation of acetylenic groups (–C≡C–). This apparently occurs by dehydrogenation of –CH=CH–. Examples of naturally occurring acetylenes are **crepenynic acid** (p. 381), **alloxanthin** (p. 1240), and the following remarkable hydrocarbon from the common cornflower *Centaurea cyanus*[131]:

$$H_3C-C\equiv C-C\equiv C-C\equiv C-CH=CH-CH=CH-CH_2-CH_2-CH_2-CH_2-CH=CH_2$$

7. The Lipids of Skin and Other Surfaces

Special fatty materials are often secreted to form external surfaces of organisms.[81,132] An example already mentioned is the secretion of the uropygial glands (preen glands) of water fowl. In the goose 90% of this material is a wax consisting of monoesters of various acids with predominantly **1-octadecanol** as the long-chain fatty alcohol.[81] The latter is formed by reduction of stearoyl-CoA as indicated in Fig. 21-2. Waxes are also important constituents of marine environments, where they are not limited to surfaces. For example, copepods, which constitute a major component of marine zooplankton, may contain up to 70% of their dry weight as wax esters. Some marine animals, such as sperm whales, accumulate the same esters in major amounts as energy stores.[133]

Among the compounds present in the lipids of human skin are a variety of branched fatty acids, both free and combined. They may play a role in maintaining the ecological balance among microorganisms of the skin, and they also impart to each individual a distinct odor or "chemical fingerprint."[132] Some of the skin lipids are incorporated into the cornified outer skin surface (Box 8-F).[134] See also Section C, 3.

Surface lipids of plants. The thick cuticle (Fig. 1-6) that covers the outer surfaces of green plants consists largely of waxes and other lipids but also contains a complex polymeric matrix of **cutin** (stems and leaves) or **suberin** (roots and wound surfaces).[135,135a] Plant waxes commonly have $C_{10} - C_{30}$ chains in both acid and alcohol components. Methyl branches are frequently present. A major function of the waxes is to inhibit evaporation of water and to protect the outer cell layer. In addition, the methyl branched components may inhibit enzymatic breakdown by microbes. Free fatty acids, free alcohols, aldehydes, ketones, β-diketones, and alkanes are also present in plant surface waxes. Chain lengths are usually $C_{20} - C_{35}$.[136] Hydrocarbon formation can occur in other parts of a plant as well as in the cuticle. Thus, normal **heptane** constitutes up to 98% of the volatile portion of the turpentine of *Pinus jeffreyi*.[81]

Cutin is largely a polyester with a high content of ω-hydroxypalmitic acid and related fatty acids, which are also hydroxylated at a second position:

$$HO - CH_2 - (CH_2)_x - \underset{\displaystyle OH}{\underset{|}{CH}} - (CH_2)_y - COOH$$

Cutin monomers. C_{16} acids in which y = 8, 7, 6, or 5 and x + y = 13

This allows branching of the polymer. Monomers of other chain lengths as well as aromatic components related to lignin are also present and polymerized into a high molecular mass branched structure. Suberin is a more complex ligninlike polymer with a high content of phenolic constituents[135] such as vanillin (Fig. 25-8).

Formation of hydrocarbons. Alkanes and alkenes occur in plants, in preen gland secretions, and in insects. The alkanes of plant cuticle are thought to be formed by elongation of a C_{16} acid followed by loss of the carboxyl group. The mechanisms are not obvious. However, these hydrocarbons are often two carbon atoms shorter than the starting fatty acid. The pathway between them might begin by α-oxidation to form an α-peroxy acid which would decarboxylate to form an aldehyde, a reaction similar to that of Eq. 15-36. Alternatively, a long-chain acyl-CoA may be reduced directly to an aldehyde. In fact, when suitable inhibitors are present aldehydes do accumulate in tissues that are forming hydrocarbons.[137] Conversion of an aldehyde intermediate to an alkane may occur by **decarbonylation** (loss of CO). This has been demonstrated in pea (*Pisum sativum*) leaves,[138] in uropygial glands,[139] in flies, and in a colonial green alga, *Botyrococcus braunii*.[138] In the last case 32% of the dry weight of the cells is C_{27}, C_{29}, and C_{31} hydrocarbons. They appear to be formed by action of a decarbonylase that apparently contains a cobalt porphyrin.[137] Plants require cobalt for growth, but an enzymatic function has not previously been established.

In contrast, the sex pheromone of the female housefly is (Z)-9-tricosene, a hydrocarbon apparently formed by an oxidative decarboxylative process from a precursor aldehyde by an enzyme that requires NADPH and O_2 and is apparently a cytochrome P450.[140] Oxidative deformylation by a cytochrome P450 converts aldehydes to alkenes, presumably via a peroxo intermediate.[117] Formation of an alkene by decarboxylation has also been proposed,[141] but a mechanism is not obvious.

Insect waxes, hydrocarbons, and pheromones. Surface lipids of many insects contain esters of long-chain (as long as 66 carbon atoms) alcohols and long-chain acids.[142] On the other hand, waxes of the tobacco hornworm consist largely of 11- and 12-oxo derivatives of a C_{28} alcohol, which may be esterified to short-chain acids.[142] A major hydrocarbon of cockroaches is 6,9-heptacosadiene.[81]

Insects communicate through the use of a great variety of volatile pheromones. As mentioned in Chapter 8, Section A,1, some moths utilize acetate esters of various isomers of Δ^7 and Δ^{11} unsaturated C_{14} fatty acids as sex pheromones. Some other moths convert the *trans*-11-tetradecenyl acetate into the corresponding C_{14} aldehyde or alcohol, while others use similar compounds of shorter ($C_{11} - C_{12}$) chain length.[143] Some ants use ketones, such as 4-methyl-3-heptanone, as well as various isoprenoid compounds and pyrazines as volatile signaling compounds.[144] Other insects also utilize isoprenoids,[145] alkaloids,[146] and aromatic substances as pheromones.

C. Synthesis of Triacylglycerols, Glycolipids, and Phospholipids

Reduction of dihydroxyacetone phosphate yields *sn*-glycerol 3-phosphate, the starting compound for formation of the glycerol-containing lipids (Fig. 21-4 step *a*).[146a,b] Transfer of two acyl groups from ACP or CoA to the hydroxyl groups of this compound (steps *b* and *c*) yields 1,2-diacylglycerol 3-phosphate (phosphatidic acid). Two different acyltransferases are required.[147] Unsaturated fatty acids are incorporated preferentially into the 2-position. The intermediate 1-acyl-*sn*-glycero-3-phosphate, often called **lysophosphatidic acid** (LPA), is formed in excess in activated platelets and has a variety of signaling activities.[148,149] LPA for signaling is derived by turnover of existing phospholipids. An alternative route of LPA formation in liver is the transfer of one acyl group onto dihydroxyacetone phosphate and reduction prior to addition of the second acyl group (Eq. 21-7).

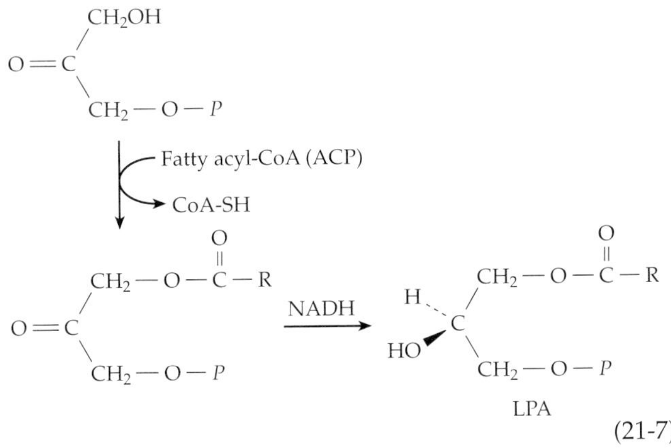

$$
\begin{array}{c}
O{=}C(CH_2OH)(CH_2{-}O{-}P) \\
\downarrow \text{Fatty acyl-CoA (ACP)} \to \text{CoA-SH} \\
O{=}C(CH_2{-}O{-}\overset{O}{\overset{\|}{C}}{-}R)(CH_2{-}O{-}P) \xrightarrow{NADH} H, HO{-}C(CH_2{-}O{-}\overset{O}{\overset{\|}{C}}{-}R)(CH_2{-}O{-}P) \\
\text{LPA}
\end{array}
\quad (21\text{-}7)
$$

Phosphatidic acid lies at a metabolic branch point. On the one hand, the phospho group can be removed by a specific phosphatase (step *d*)[150] and another acyl group (most often an unsaturated acyl group) may be transferred onto the resulting diacylglycerol (DAG, diglyceride, step *e*)[150a,b] to form a **triacylglycerol** (triglyceride). Alternatively, the phosphatidic acid may be converted to a **CDP-diacylglycerol** (step *g*), a key intermediate in phospholipid synthesis both in eukaryotes and in bacteria.[151] Not only can phosphatidic acid be hydrolyzed to 1,2-diacylglycerols, but the reverse process can occur by action of a kinase. This presumably permits recycling of the diacylglycerol formed by turnover of membrane phospholipids.[152]

Diacylglycerols can also be converted to a variety of glycolipids such as the **galactolipids** of chloroplasts (Eq. 21-8). See also Chapter 8. These are the major lipids of photosynthetic membranes.[98,153–155] Some bacteria, e.g., the mycoplasma *Acholeplasma laidlawii*, contain both monoglucosyl- and diglucosyl-DAG. Changes in the ratio of these two membrane components may regulate the phase equilibrium between bilayer and nonbilayer forms.[156] 1,2-Diacylglycerol can also react with UDP-sulfoquinovose (Eq. 20-12) to form the characteristic sulfolipid of chloroplasts.[154]

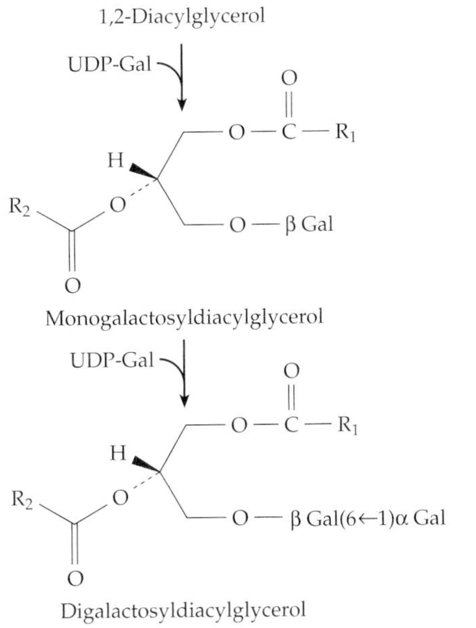

1,2-Diacylglycerol

UDP-Gal ↓

R_2–C(=O)–O–CH(H)(CH$_2$–O–C(=O)–R_1)(CH$_2$–O–β Gal)

Monogalactosyldiacylglycerol

UDP-Gal ↓

R_2–C(=O)–O–CH(H)(CH$_2$–O–C(=O)–R_1)(CH$_2$–O–β Gal(6←1)α Gal)

Digalactosyldiacylglycerol (21-8)

In animals a principal regulatory point for lipid synthesis is in the activation of acetyl-CoA carboxylase by citrate (Fig. 17-20).[156a,b] Beyond that, a complex hormonal control is exerted on both biosynthesis and the catabolism of triglycerides stored in liver and adipose tissues.[157] For example, adrenaline and glucagon, by stimulating production of cAMP, stimulate acetyl-CoA carboxylase,[158] activate lipases that cleave triacylglycerols, and mobilize depot fats.[159] Insulin, on the other hand, promotes lipid storage. It increases the activity of the enzymes of lipogenesis from the ATP-dependent citrate cleavage enzyme (Eq. 13-39) and inhibits cAMP production, thus blocking lipolysis within cells. At the transcriptional level sterols bind to activator proteins (**sterol regulatory element binding proteins,** SREBPs) and activate genes for acetyl-CoA carboxylase[158] and for stearoyl-CoA desaturases.[160] Fatty acid synthases, which play a central role in lipid formation, are controlled by both hormonal and nutritional factors at the transcriptional[161] and translational[162] levels. Environmental factors also have indirect effects. For example, the Δ^9 fatty acid desaturase activity of poikilothermic (cold-blooded) animals is increased at low temperatures. The resulting increased synthesis of unsaturated fats leads to increased fluidity of the membrane bilayer.[163] As mentioned on p. 1193; the same is true for green plants.

1. Phospholipids

Bacterial and also some eukaryotic phospholipids are formed following conversion (Fig. 21-4, step *g*) of phosphatidic acids to CDP-diacylglycerols, which are able to react with a variety of nucleophiles with displacement of CMP.[164–166] Reaction with L-serine (step *h*)[167] leads to **phosphatidylserine,** and reaction with glycerol 3-phosphate (step *i*),[168] which enters cells via a special transporter,[168a] produces **phosphatidyglycerol 3-*P*.** The enzyme catalyzing the formation of phosphatidylserine appears to occur naturally as an integral membrane protein of the ER. Some is also bound to ribosomes and to mitochondria.[169,169a] In contrast, most of the other enzymes of phospholipid formation are closely associated with or embedded in the cytoplasmic membrane. One of these, a pyruvoyl group dependent enzyme (Chapter 14, Section F), catalyzes decarboxylation of phosphatidylserine to **phosphatidylethanolamine** (PE, step *j*, Fig. 21-4).[170] This reaction had been thought unimportant in animals, but

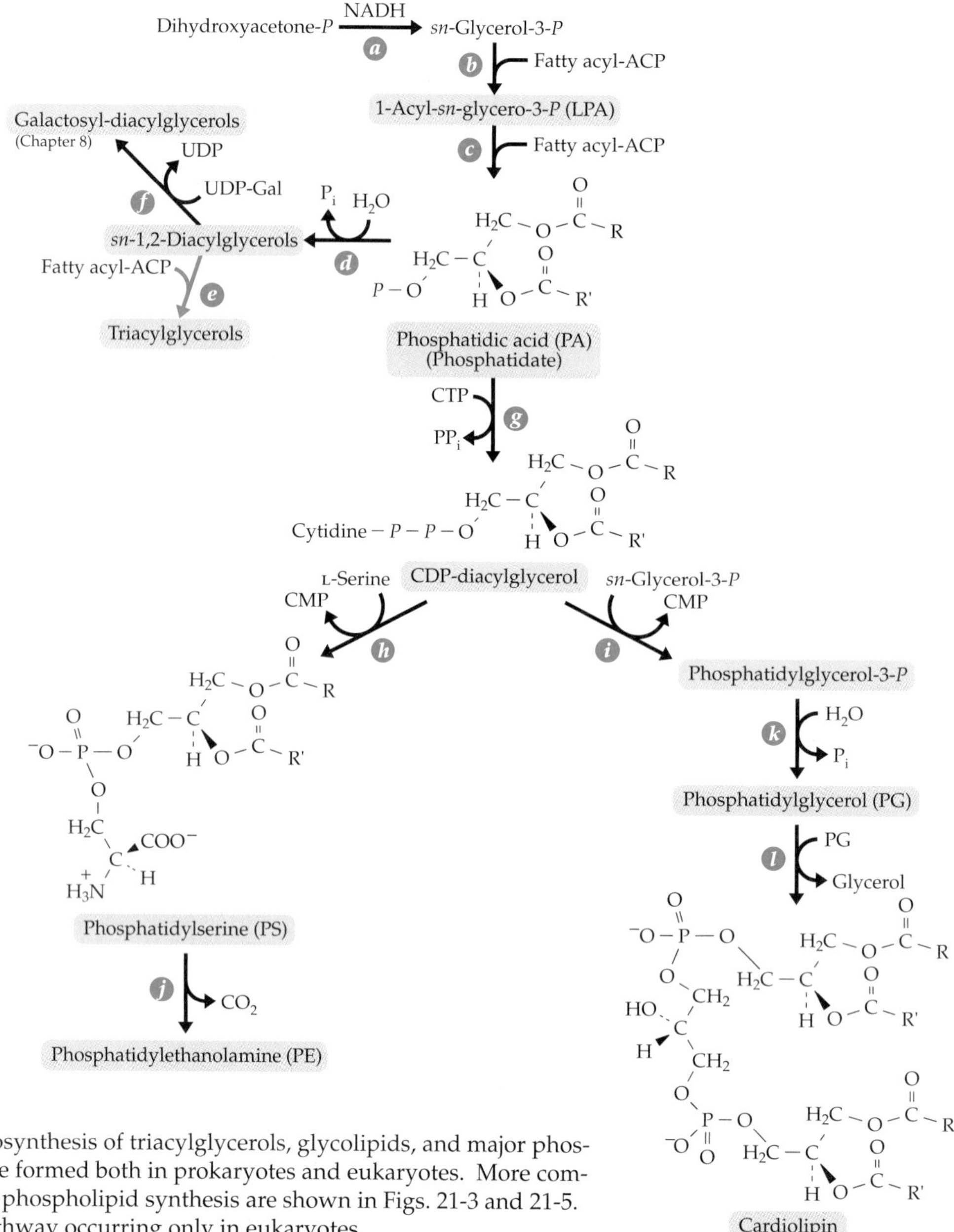

Figure 21-4 Biosynthesis of triacylglycerols, glycolipids, and major phospholipids that are formed both in prokaryotes and eukaryotes. More complete schemes of phospholipid synthesis are shown in Figs. 21-3 and 21-5. Green arrow: pathway occurring only in eukaryotes.

results with cultured cells show that decarboxylation of phosphatidylserine is often the major route of formation of phosphatidylethanolamine in mammalian cells.[171,172] This phospholipid also accounts for 75% of total phospholipid of the *E. coli* cell envelope. It is synthesized on the cytosolic side of the inner membrane, but it is also translocated to the outer membrane, where it is a major constituent of the inner bilayer leaflet.[173] PE is essential for viability of *E. coli* cells.[174] It provides dipolar ionic head groups and apparently serves as a chaperone for folding of some membrane proteins.[175]

After removal of a phosphate from phosphatidylglycerol 3-*P*, the resulting phosphatidylglycerol can be converted to **diphosphatidylglycerol** (known as **cardiolipin**). One manner in which this is accomplished in bacteria is indicated by step *l* of Fig. 21-4. One molecule of glycerol is displaced as two molecules of phosphatidylglycerol are coupled. The alternative pathway of Eq. 21-9 is followed in eukaryotic mitochondria and perhaps in some bacteria. The entire phosphatidic acid group is transferred from CDP-diacylglycerol to phosphatidylglycerol with displacement of CMP.[176–178] Gram-negative bacteria also synthesize a second set of membrane phospholipids, compounds such as **lipid A** (Figs. 8-30, 20-10) that are based on acylated glucosamine.[165]

Phosphatidylcholine, which is rarely present in bacteria, is formed in eukaryotes from phosphatidylethanolamine by three consecutive steps of methylation

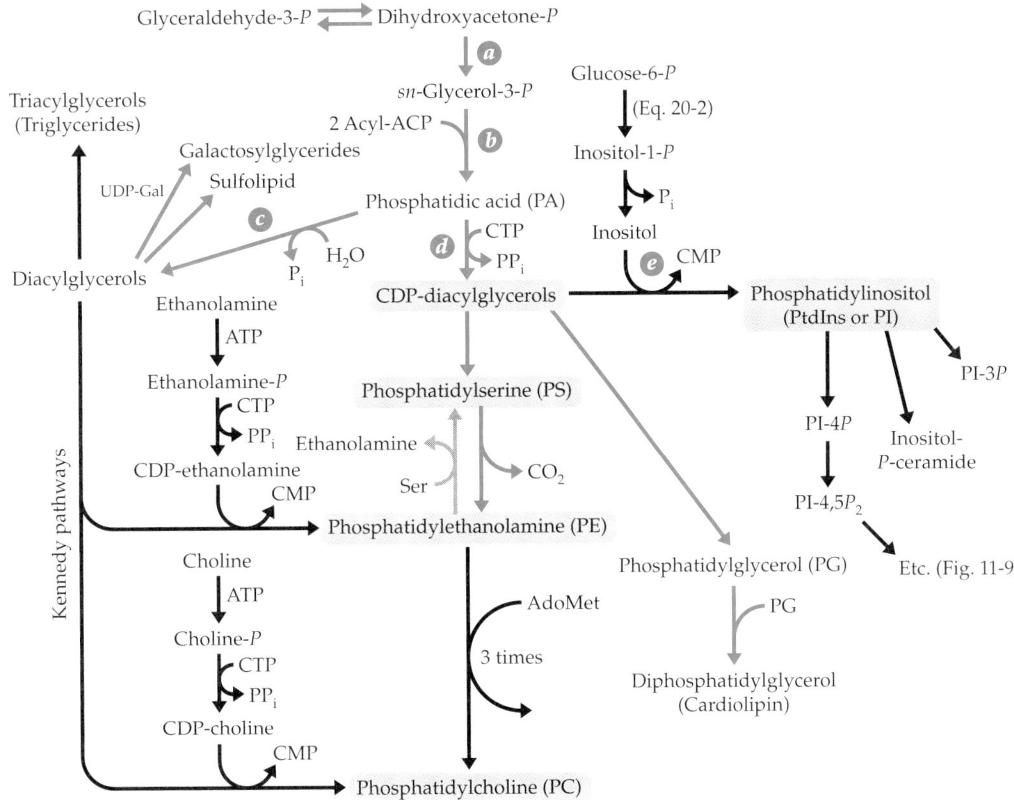

Figure 21-5 A more complete outline of the biosynthesis of triacylglycerols, glycolipids, and phospholipids including characteristic eukaryotic pathways. Green lines indicate pathways utilized by both bacteria and eukaryotes. Structures of some of the compounds are shown in Fig. 21-4. The gray arrows show the formation of phosphatidylserine by exchange with ethanolamine (Eq. 21-10).

CDP-Diacylglycerol

Phosphatidylglycerol

CMP

Diphosphatidylglycerol (21-9)

by *S*-adenosylmethionine (Fig. 21-5). This pathway is of major importance in eukaryotic cells.[179,180] However, alternative pathways (the Kennedy pathways),[166] which are represented by black lines on the left side of Fig. 21-5, are also used for formation of both phosphatidylcholine and phosphatidylethanolamine. In both cases, the free base, choline, or ethanolamine[180a,b] is phosphorylated with ATP. Choline phosphate formed in this manner is then converted by reaction with CTP to CDP-choline (Eq. 17-58).[181] Phosphatidylcholine is formed from this intermediate[181a,b] while CDP-ethanolamine is used to form phosphatidylethanolamine (Fig. 21-5). These synthetic reactions occur within cell nuclei as well as on surfaces of cytoplasmic membranes.[181c]

The formation of phosphatidylserine and possibly other phospholipids in animal tissues may also be accomplished by exchange reactions (Eq. 21-10, step *a*).[182,183] At the same time, decarboxylation of phosphatidylserine back to phosphatidylethanolamine (Eq. 21-10, step *b*) also takes place, the net effect being a catalytic cycle for decarboxylation of serine to ethanolamine. The latter can react with CTP to initiate synthesis of new phospholipid molecules or can be converted to phosphatidylcholine (step *c*). However, unless there is an excess of methionine and folate in the diet, choline is an essential human nutrient.[184]

BOX 21-D POLY-β-HYDROXYBUTYRATE AND BIODEGRADABLE PLASTICS

D-β-Hydroxybutyric acid

The important bacterial storage material poly-hydroxybutyric acid is related metabolically and structurally to the lipids. This highly reduced polymer is made up of D-β-hydroxybutyric acid units in ester linkage, about 1500 residues being present per chain. The structure is that of a compact right-handed coil with a twofold screw axis and a pitch of 0.60 nm.[a] Within bacteria it often occurs in thin lamellae ~5.0 nm thick. Since a chain of 1500 residues stretches to 440 nm, there must be ~88 folds in a single chain. Present in both cytoplasmic granules and in membranes,[b] polyhydroxybutyrate can account for as much as 50% of the total carbon of some bacteria.[c] In *E. coli* and many other bacteria polyhydroxybutyrate is present in a lower molecular mass form bound to calcium polyphosphates, proteins, or other macromolecules.[d,e] It has also been extracted from bovine serum albumin and may be ubiquitous in both eukaryotes and prokaryotes.[d,e] The polymer may function in formation of Ca^{2+} channels in membranes.[b,d]

Biosynthesis occurs from 3-hydroxybutyryl-CoA. Some bacteria incorporate other β-hydroxyacids into the polymer.[f] Apparently various hydroxyacyl-CoAs can be diverted from the β oxidation pathway to polymer synthesis,[g] and synthases that will accept a variety of β-hydroxyacyl-CoA substrates have been isolated.[h,i] More than 80 different hydroxyacyl groups can be incorporated into the polymer.[i] A bacterially produced copolymer of β-hydroxybutyrate and β-hydroxyvalerate resembles polypropylene but is biodegradeable. It not only can be used for sutures and other medical implants[j] but also could compete with petroleum-derived plastics[h] and be derived from renewable sources. To this end the synthase genes have been cloned, engineered, and transferred into other microorganisms and plants.[k–n] Transgenic cotton plants incorporate polyhydroxybutyrate granules into the cotton fibers altering the properties of the fibers.[m] The polyhydroxybutyrate synthases apper to be related mechanistically to bacterial lipases.[o]

[a] Okamura, K., and Marchessault, R. H. (1967) in *Conformation of Biopolymers*, Vol. 2 (Ramachandran, G. N., ed), pp. 709–720, Academic Press, New York

[b] Reusch, R. N., and Sadoff, H. L. (1988) *Proc. Natl. Acad. Sci. U.S.A.* **85**, 4176–4180

[c] Jacob, G. S., Garbow, J. R., and Schaefer, J. (1986) *J. Biol. Chem.* **261**, 16785–16787

[d] Reusch, R. N., Huang, R., and Bramble, L. L. (1995) *Biophys. J.* **69**, 754–766

[e] Huang, R., and Reusch, R. N. (1996) *J. Biol. Chem.* **271**, 22196–22202

[f] Peoples, O. P., and Sinskey, A. J. (1989) *J. Biol. Chem.* **264**, 15298–15303

[g] de Waard, P., van der Wal, H., Huijberts, G. N. M., and Eggink, G. (1993) *J. Biol. Chem.* **268**, 315–319

[h] Müh, U., Sinskey, A. J., Kirby, D. P., Lane, W. S., and Stubbe, J. (1999) *Biochemistry* **38**, 826–837

[i] Gerngross, T. U., and Martin, D. P. (1995) *Proc. Natl. Acad. Sci. U.S.A.* **92**, 6279–6283

[j] Pool, R. (1989) *Science* **245**, 1187–1189

[k] Poirier, Y., Dennis, D. E., Klomparens, K., and Somerville, C. (1992) *Science* **256**, 520–523

[l] Mittendorf, V., Robertson, E. J., Leech, R. M., Krüger, N., Steinbüchel, A., and Poirier, Y. (1998) *Proc. Natl. Acad. Sci. U.S.A.* **95**, 13397–13402

[m] John, M. E., and Keller, G. (1996) *Proc. Natl. Acad. Sci. U.S.A.* **93**, 12768–12773

[n] García, B., Olivera, E. R., Minambres, B., Fernández-Valverde, M., Canedo, L. M., Prieto, M. A., García, J. L., Martínez, M., and Luengo, J. M. (1999) *J. Biol. Chem.* **274**, 29228–29241.

[o] Crandall, W. V., and Lowe, M. E. (2001) *J. Biol. Chem.* **276**, 12505–12512

(21-10)

Apparently the synthesis via serine and phosphatidylserine cannot provide an adequate amount of choline, which is present in the body not only in phosphatidylcholine but in plasmalogens, sphingomyelins, and the neurotransmitter **acetylcholine**.[185]

Phosphatidylinositol (PI), a major component of membrane lipids, is formed by displacement of CMP from CMD-dialylglycerol by *myo*-inositol.[186] It is also converted into a variety of less abundant phosphorylated derivatives that engage in signaling activities (see Fig. 11-9). In addition, PI is a component of the glycosylphosphatidylinositol (GPI) membrane anchors for suface proteins (Fig. 8-13). Free GPI anchors, lacking bound proteins, are also present in membranes.

They are especially abundant in many parasitic protozoa and may carry additional glycosyl groups.[186a]

Regulation of phospholipid synthesis, which has been studied in detail in yeast,[187–191] is complex but highly coordinated. The committed step in the synthesis of PE and PC is the hydrolysis of phosphatidate (PA) by a phosphatase to generate diacylglycerols (Fig. 21-5, step *c*). Reaction of PA with CTP (step *d*) also affects synthesis of the other major phospholipids. Much of the coordinate regulation arises at the transcriptional level. For example, genes for the synthesis of PC or PI are repressed by inositol alone and in combination with choline.[187,188] Regulation of CTP synthetase controls the formation of CDP-diacylglycerols.[190] In mammalian cells PC synthesis appears to occur only during the S-phase of the cell cycle (Fig. 11-15).[191] The CTP: phosphocholine cytidyltransferase that catalyzes CDP-choline formation is controlled by storage in a resevoir in the nucleus from which it is transferred to ER membranes.[181b]

2. The Ether-Linked Lipids

Closely related to both the triacylglycerols and phospholipids, the ether-linked lipids contain in place of one ester group an *alkoxy* (–OR) or *alkenyl* (–O–CH =CH–R) group.[192] Phospholipids containing the alk-1-enyl group, the **plasmalogens**, were first recognized in 1924 by Feulgen and Voit, who were developing histological staining procedures. They observed that treatment of tissue slices with acid resulted in the liberation of aldehydes, which were later shown to be formed by breakdown of the alkenyl lipids (Eq. 21-11). Over 10% of the lipid in the human central nervous system is plasmalogen and about 1% is alkoxy lipid. Among the latter is the **platelet activating factor** (Box 8-A).[193-194a] In many mammalian cells the ethanolamine plasmalogen **plasmenylethanolamine** represents the major storage depot of arachidonic acid.[195]

$$\mathrm{Y{-}O{-}CH{=}CH{-}R} \xrightarrow[\mathrm{Y{-}OH}]{\mathrm{H_2O},\ \mathrm{H^+}} \mathrm{HC(=O){-}CH_2{-}R} \qquad (21\text{-}11)$$

Ether-linked lipids constitute up to 35% of the total phospholipid in molluscs. Although they are usually regarded as animal constituents, small amounts of ether-linked lipids have been identified in plants. The major phospholipids of archaebacterial membranes are ether-linked derivatives of the polyprenyl phytanyl group and of the dimeric biphytanyl group (Chapter 8).[196–198]

Biosynthesis of ether lipids begins with formation of fatty acyl derivatives of dihydroxyacetone phosphate. The acyl group is then displaced, along with the oxygen atom to which it is attached, by an alkoxy group of a long-chain fatty alcohol (Eq. 21-12, step *a*), which is formed by reduction of the corresponding acyl-CoA.[199] The oxygen of the alcohol (designated by an asterisk) is retained in the product.[200,201] The reaction differs significantly from displacements discussed in Chapter 12. The pro-*R* hydrogen atom (marked by the dagger, †) at C-1 exchanges with the medium during the reaction suggesting that enolization of the dihydroxyacetone phosphate takes place. A possible mechanism would be to add the incoming R–O⁻, generated as in serine proteases, to the double bond of

Alkylglycerol 3-*P*

Alkylacyl-glycerol

Alkyldiacylglycerol

Alkoxy-phospholipid (Glycerol-ether lipid) Y=choline or ethanolamine

Platelet activating factor (PAF)

Plasmalogen: Y=ethanolamine

(21-12)

the enol at C-1. This would generate a transient carbanion on C-2. It could then eliminate the carboxylate containing the original C-1 acyl group, and the enol could then ketonize.

Once an alkoxy derivative of dihydroxyacetone is formed, reduction to the 2-OH form, further acylation, and conversion to various alkyl phospholipids and neutral lipids can occur. The pathways (Eq. 21-12, steps *b–f*) are closely akin to those of Fig. 21-4. The conversion of alkoxy lipids to plasmalogens occurs by oxidative desaturation (Eq. 21-12, step *f*).[202] The initial steps in the synthesis of ether-linked lipids take place principally in the peroxisomes. Enzymes catalyzing both the acylation of dihydroxyacetone phosphate and the synthesis of alkyl-dihydroxyacetone-*P* (step *a*, Eq. 21-12) are found in high amounts in animal peroxisomes. In the rare autosomal recessive disorder known as the **Zellweger syndrome** peroxisomes are completely lacking.[203] Both the synthesis of ether-linked lipids[204] and the β oxidation of very-long-chain fatty acids are depressed. These acids, principally C26:0 and C26:1, accumulate in tissues[205,206] of patients with this severe disease, which is usually fatal during the first four months of life.

The platelet activating factor (PAF, Box 8-A) is formed in neutrophils and macrophages from alkylacyl-*sn*-glycero-3-phosphocholine by the action of phospholipase A_2. This enzyme removes the C2 acyl group, which is then replaced by an *acetyl group* transferred from acetyl-CoA to form PAF. Alternatively, a phosphocholine group may be transferred onto 1-alkyl-2-acyl-*sn*-glycerol from CDP-choline as in the formation of phosphatidylcholine (Fig. 21-5). PAF can undergo hydrolytic removal of its acetyl group in tissues but can also transfer it to such acceptors as lysoplasmalogens or sphingosine.[194] Hydrolytic loss of the acetyl group from PAF destroys biological activities including induction of allergic and inflammatory responses.[194a] The various signaling activities of PAF arise from binding to G-protein linked receptors in many cells and tissues.[207]

3. Sphingolipids

Sphinoglipids are phospholipids and glycolipids derived from **sphingosine** and other "long-chain bases."[208] At least 60 bases of this type have been identified.[209] They vary in chain length from C_{14} to C_{26} and include members of the iso and anteiso series. Up to two double bonds may be present. The C_{18} compound, usually called sphingosine, is derived from condensation of palmitoyl-CoA with serine.[209a] Carbon dioxide is lost from the serine during the condensation reaction (Fig. 21-6, step *a*; Chapter 14), and the resulting ketone is reduced with NADPH (step *b*) to form **sphinganine**, a common component of animal sphingolipids. It may be hydroxylated to phytosphingosine in plants and fungi (step *c*).[210,211] Sphinganine is converted to long-chain amides by acyl transfer from acyl-CoA (step *d*) and then undergoes desaturation (step *e*)[212–213a] to form **ceramides**, the precursors to more than 100 gangliosides (Fig. 20-11),[214,215] to the phospholipid **sphingomyelins** (Fig. 21-6, step *g*), and also to free sphingosine (step *h*). This last reaction is degradative and on the pathway of breakdown of gangliosides (Fig. 20-11). Further catabolism of sphingosine is thought to take place by a PLP-mediated chain cleavage to palmitaldehyde.[216]

The essential functions of sphingolipids, including the complex gangliosides, are only now being clarified.[215,217,218] The latter are abundant in brain and are thought to function in cell–cell recognition. On blood cell surfaces they carry blood group antigens (Box 4-C). They play an essential role in spermatogenesis[218] and may function in various signaling processes.[218a] In the outer cornified layers of skin, ceramides with very long chain (C_{28}–C_{36}) fatty acyl groups undergo ω hydroxylation (Fig. 21-6) and become esterified to glutamate side chains of specific skin proteins called **involucrins**. The long hydrocarbon chains are thought to pass entirely through the lipid bilayer to form rigid lamellae of a water-impermeable outer skin barrier.[134] An important hypothesis is that sphingolipids associate with cholesterol to form "lipid rafts," which float in a sea of glycerolipids and serve as bases for various signaling processes. The long hydrophobic acyl chains of the sphingolipids pack well with cholesterol to form a rigid lipid structure of high melting temperature.[218b,c]

4. Complex Lipids in Signaling

While pathways of synthesis of complex lipids have been described, we are far from understanding the dynamics of the synthesis and turnover of the membranous structures built from them. The fact that the lipid bilayer of a cell membrane is so thin means that any sudden changes in composition at a particular location will cause changes in physical properties and a wave of diffusion that will travel along the membrane. The membrane seems to be ideally structured to receive and propagate messages from outer surface or internal receptors, or messages sent along the bilayer.

One of the most studied examples of signaling with membrane lipids is provided by the **phosphoinositide cascade**, which is pictured in Fig. 11-9. Six or more phosphate esters of phosphatidylinositol (PI) are generated by the action of kinases.[219,220] More than 100 extracellular signaling molecules activate specific isozyme forms of **phospholipase C**,[221–224] releasing 20 or more different inositol phosphates from these

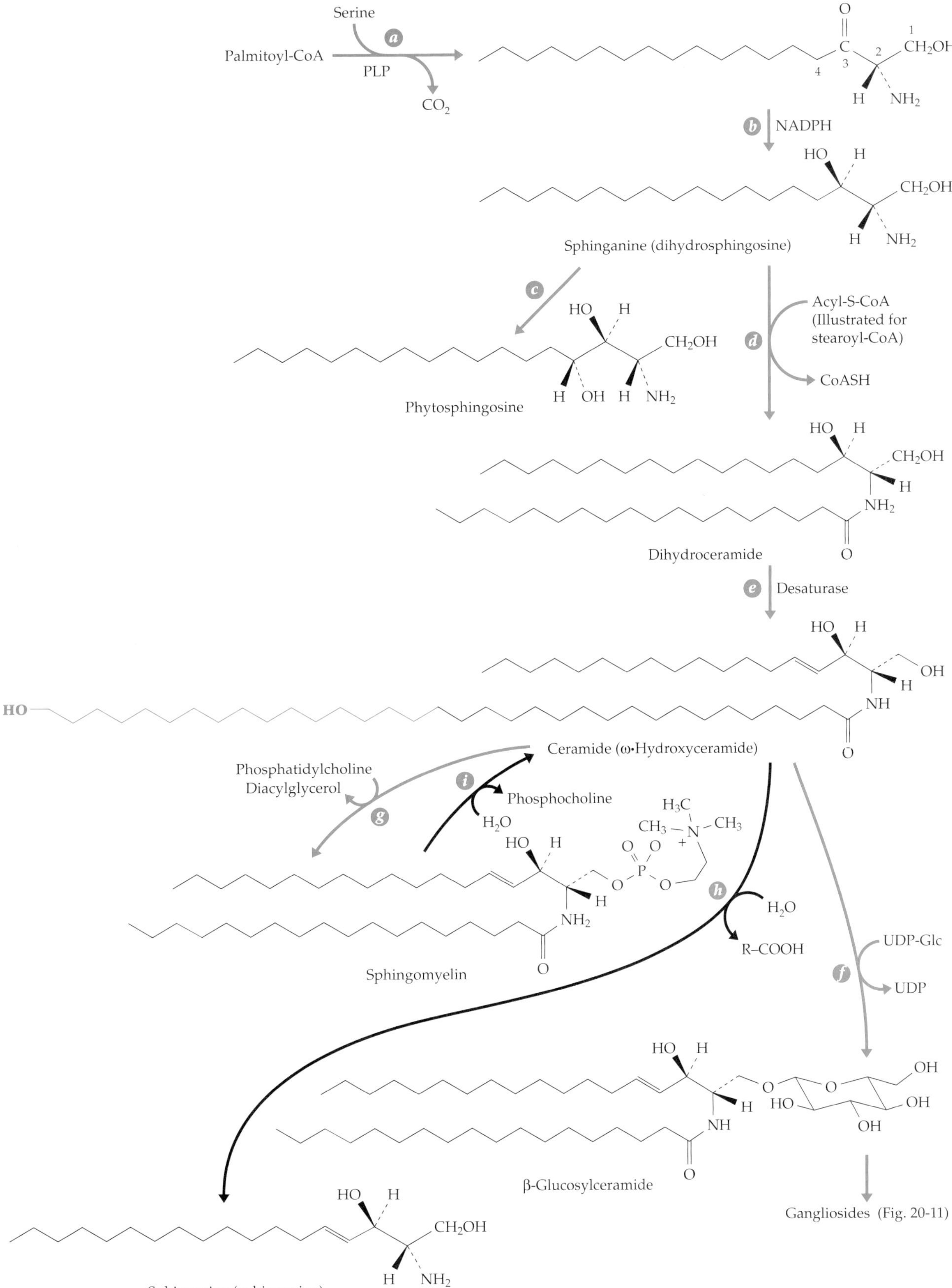

Figure 21-6 Pathways of synthesis and metabolism of sphingolipids. Gray arrows indicate catabolic pathways. See also Fig. 20-11. The green extension on the ceramide structure is that of a long-chain ω-hydroxyceramide that is covalently bound to protein in human skin.

phosphoinositide esters. The released inositol phosphates, which act as water-soluble messengers, are further modified by the action of several phosphatases (Fig. 11-9).[225] At the same time, **diacylglycerols** are left in the membrane. With loss of the negative charges of the PI phosphates there will be immediate electrostatic effects in the membrane, which may alter the ionic environment, open ion channels, etc. The diacylglycerols, which diffuse within the membrane, may lose arachidonic acid from the *sn*-2 position to supply substrate for the arachidonate cascades described in Section D (Eq. 21-16). Diacylglycerols also activate the 11 isozyme forms of **protein kinase C**.[226–228] Some of these enzymes not only are activated by diacylglycerols, but also require **phosphatidylserine** for activity.

Other lipid-based signaling cascades arise from reactions that modify phosphatidylcholine molecules. Unsaturated fatty acids in the *sn*-2 position are readily oxidized by free radicals with cleavage of the hydrocarbon chains to form alcohols, aldehydes, and carboxylic acids. These mimic PAF in their biological activities.[194a,228a] Phosphatidate molecules with saturated or monounsaturated fatty acids in the *sn*-2 position arise from breakdown of phosphatidylcholine catalyzed by **phospholipase D**.[229] Phosphatidate may also be formed by a family of lipid **diacylglycerol kinases**.[230,230a] Phosphatidates containing saturated or monounsaturated fatty acids also have a variety of signaling activities.[230a] An arachidonoyl-diacylglycerol kinase is thought to function in many processes. An example is a PI-mediated cycle in invertebrate vision.[231] Sphingomyelin breakdown (black arrows in Fig. 21-6) releases diffusible ceramides that have been implicated as signaling molecules in cell proliferation, differentiation, growth arrest, and apoptosis.[232–237a] Sphingosine and sphingosine 1-*P* also have signaling functions.[211,237b-e]

Phospholipids have been shown to exchange between different membranes, e.g., of mitochondria and the ER. Exchanges of phosphatidylcholine, phosphatidylinositol, and sphingomyelin are catalyzed by specific **exchange proteins** (Box 21-A).[238,238a] These proteins may also participate in signaling, but their major function may be to transport the phospholipids from their sites of synthesis to the various membranes of the cell.

5. Peroxidation of Lipids and Rancidity

Storage of fats and oils leads to **rancidity**, a largely oxidative deterioration that causes development of unpleasant tastes, odors, and toxic compounds.[239] Similar chemical changes account for the "drying" of oil-based paints and varnishes. These reactions occur most readily with polyunsaturated fatty acids, whether free or in ester linkage within triacylglycerols. The reactions are initiated by free radicals, which may be generated by oxidative enzymes within or outside of cells, or by nonenzymatic reactions catalyzed by traces of transition metals or by environmental pollutants. Characteristic of rancidity is an autocatalytic chain reaction (Eq. 21-13).[239–241]

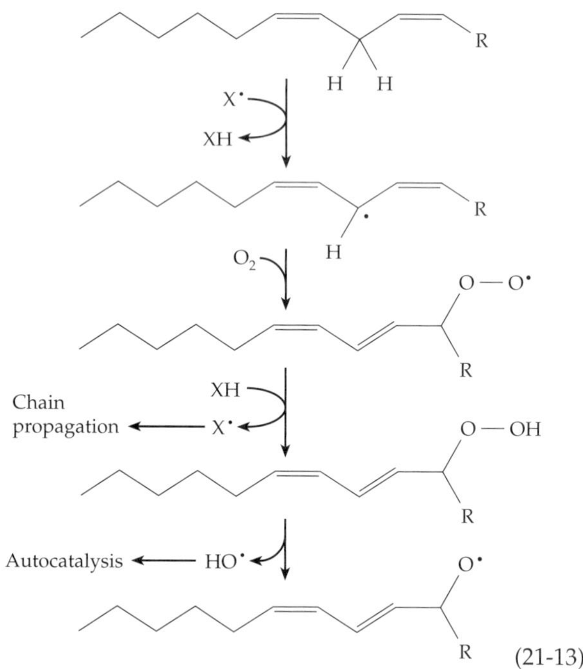

(21-13)

Radical X•, which initiates the reaction, is regenerated in a chain propagation sequence that, at the same time, produces an organic peroxide. The latter can be cleaved to form two additional radicals, which can also react with the unsaturated fatty acids to set up the autocatalytic process. Isomerization, chain cleavages, and radical coupling reactions also occur, especially with polyunsaturated fatty acids. For example, reactive unsaturated aldehydes can be formed (Eq. 21-14).

$$\text{R–CH=CH–}\overset{\bullet\text{O}}{\overset{|}{\text{CH}}}\text{–R}' \rightarrow \text{R–CH=CH–CHO} + {}^{\bullet}\text{R}' \qquad (21\text{-}14)$$

An intermediate in Eq. 21-13 may be converted to **4-hydroxy-2-nonenal**, a prominent product of the peroxidation of arachidonic or linoleic acids (Eq. 21-15).[242–243a] However, other biosynthetic pathways to this compound are possible.[244,244a] 4-Hydroxy-2-nonenal can react with side chains of lysine, cysteine, and histidine[245] to form fluorescent products such as the following cyclic compound generated by an oxidative reaction.[246]

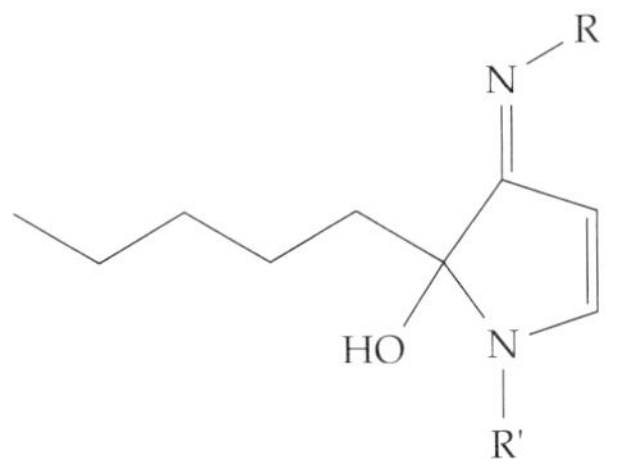

Polyenoic acids also give rise to malondialdehyde, a reactive mutagenic compound, which can be reduced

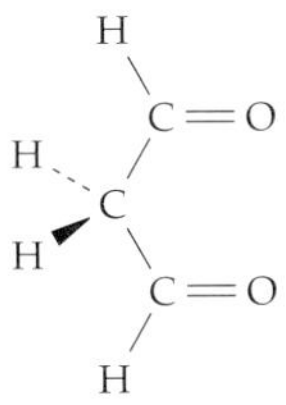

Malondialdehyde

and dehydrated to acrolein, a toxic compound which also reacts with both lysine and serine to

$$CH_2{=}CH{-}CHO$$

Acrolein

produce products excreted in the urine.[247,248] More dangerous are similar reactions of these aldehydes with proteins of the body in conditions such as diabetes or renal insufficiency.[242] The bifunctional malondialdehyde forms Schiff bases with protein amino groups and acts as a crosslinking agent.[249] **Age pigments** (also called **lipofuscin**), which tend to accumulate within neurons and other cells, are thought to represent precipitated lipid-protein complexes resulting from such reactions.[250] The reactions are similar to those of proteins with the products of sugar breakdown (glycation; p. 69).[250a] Organisms have developed multiple enzymatic mechanisms for detoxification of products of both glycation and oxidative degradation.[243a]

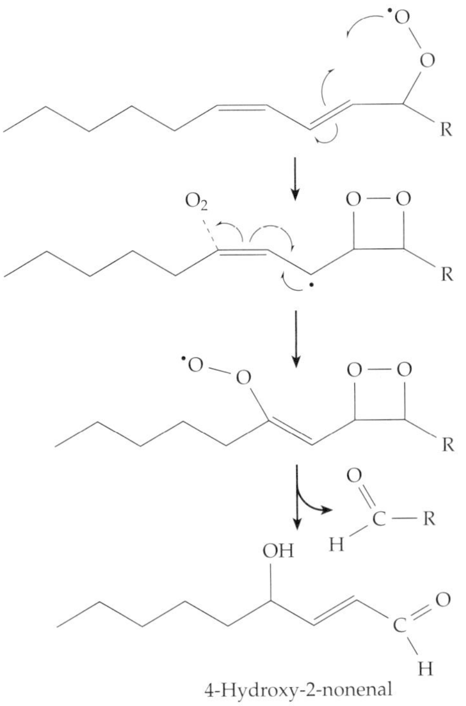

4-Hydroxy-2-nonenal (21-15)

The oxidative degradation represented by the foregoing reactions is referred to as peroxidation. Peroxidation can lead to rapid development of rancidity in fats and oils. However, the presence of a small amount of tocopherol inhibits this decomposition, presumably by trapping the intermediate radicals in the form of the more stable tocopherol radicals (Eq. 15-54), which may dimerize or react with other radicals to terminate the chain.

Catalytic hydrogenation of vegetable oils is widely used to form harder fats and to decrease the content of polyunsaturated fatty acyl groups. The products have a greatly increased resistance to rancidity. However, they also contain fats with trans double bonds as well as isomers with double bonds in unusual positions.[251–253] Such compounds may interfere with normal fatty acid metabolism and also appear to affect serum lipoprotein levels adversely. Trans fatty acids are present in some foods. One hundred grams of butter contain 4–8 g, but hydrogenated fats often contain much more. It has been estimated that in the United States trans fatty acids account for 6–8% of total dietary fat.[253]

6. Some Nutritional Questions

While many of the poorer people on earth starve to death the problems of atherosclerosis and obesity affect many in wealthier societies.[253a–c] The fat content of foods is often blamed, and, as discussed in Boxes 21-B and 22-B, the quality of fatty acids in the diet is very important. However, like fatty acids, carbohydrates are also metabolized via acetyl-CoA and can readily be converted to both fatty acids and cholesterol.[253d] Obesity is largely a problem of excessive total caloric intake.

Why do some people stay slim while others become obese? What are the regulatory mechanisms that affect appetite and body composition? The human body weight tends to be stable or to increase slowly during adult life.[253e] Is there a natural set point for each individual? No, an apparent set point is just a result of action of a multitude of factors including genetic variations[253f] and psychological factors that affect exercise levels, eating habits, etc.[253g] It is worthwhile to recognize that the basal metabolic rate, which is also affected by many factors, accounts for a very high fraction of a person's energy expenditure (p. 283).

The following are among specific biochemical

factors that act on the energy balance: the activity of acetyl-CoA carboxylase and the associated level of malonyl-CoA[253h,i]; the activity of mitochondrial uncoupling proteins (Box 18-C)[253j,k]; actions of the hormone leptin[253l] (which have been hard to interpret)[253m,n]; and other hormones including cholecystokinins and neuropeptide Y (Chapter 30).

D. Prostaglandins and Related Prostanoid Compounds

Lipid peroxidation has often been regarded simply as an undesirable side reaction, but it is also a normal part of metabolism. Initiated by enzymatically generated radicals, peroxidation occurs as specific metabolic pathways, such as the **arachidonate cascade**, which leads to a variety of local hormones and other substances (Eq. 21-16).[254–256a]

As early as 1930, it was recognized that seminal fluid contains materials that promote contraction of uterine muscles. The active compounds, the **prostaglandins**, were isolated and crystallized in 1960 and were identified shortly thereafter.[257,258] As many as 14 closely related compounds are found in human seminal fluid, one of the richest known sources. Prostaglandins are present in seminal fluid at a total concentration of ~1 mM, but their action on smooth muscles has been observed at a concentration as low as 10^{-9} M. The structures and biosynthetic pathways of several of the prostaglandins are indicated in Fig. 21-7. Prostaglandins are usually abbreviated PG with an additional letter and numerical subscript added to indicate the type. The E type are β-hydroxyketones, the F type 1,3-diols, and the A type α, β-unsaturated ketones. Series 2 prostaglandins arise from arachidonic acid, while series 1 and 3 arise from fatty acids containing one fewer or one more double bond, respectively (Fig. 21-7). Additional forms are known.[257,259]

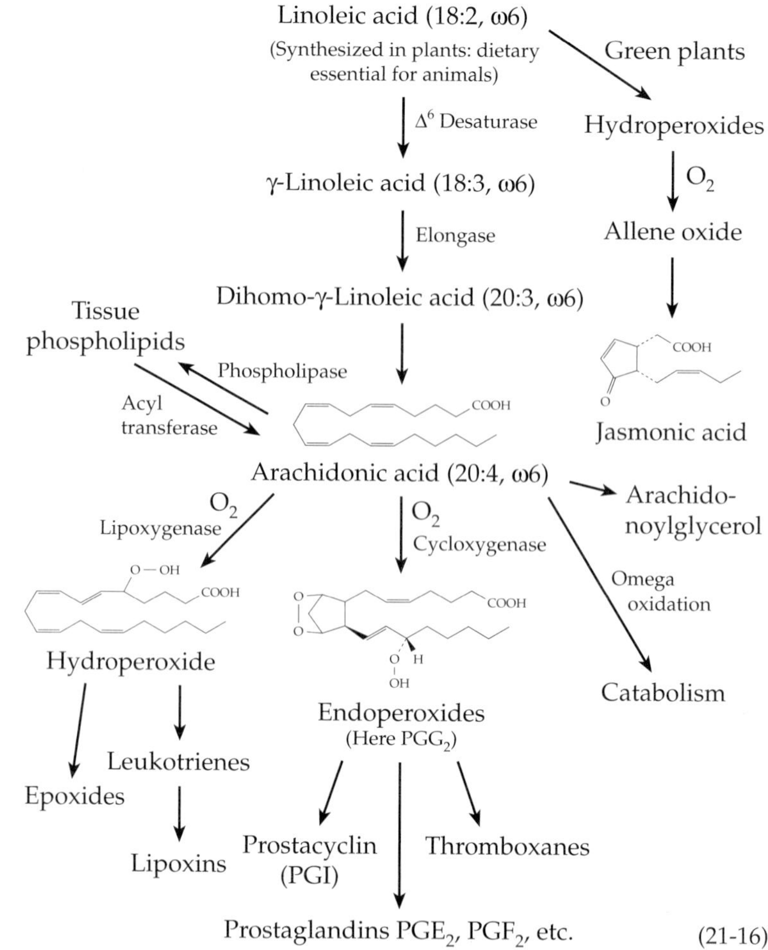

(21-16)

15 12 8 5

COOH

Arachidonic acid **C20 : 4 : ω6**

13

COOH

α-Linolenic acid **C18 : 3 : ω3**

13

COOH

H_S H_R

Linoleic acid **C18 : 2 : ω6**

E–Fe^{3+} H^+ E–Fe^{2+}

O_2 COOH H

H^+

•O–O COOH

E–Fe^{3+} H^+ E–Fe^{2+}

HO–O H COOH

(13S)-9,11 (Z,E)-Octadecadienoic

(21-17)

1. Metabolism of the Prostaglandins

Prostaglandins are not stored by cells but are synthesized in response to external stimuli. Arachidonic acid and other polyenoic acids are present in relatively small amounts (e.g., ~1% of total plasma fatty acids), but they are concentrated in the 2-position of phospholipids. This is in part a result of phospholipid "remodeling." Acyl groups are hydrolyzed from the *sn*-2 position by action of phospholipase A_2. An acyltransferase with a preference for arachidonoyl groups then transfers esterified arachidonic acid from

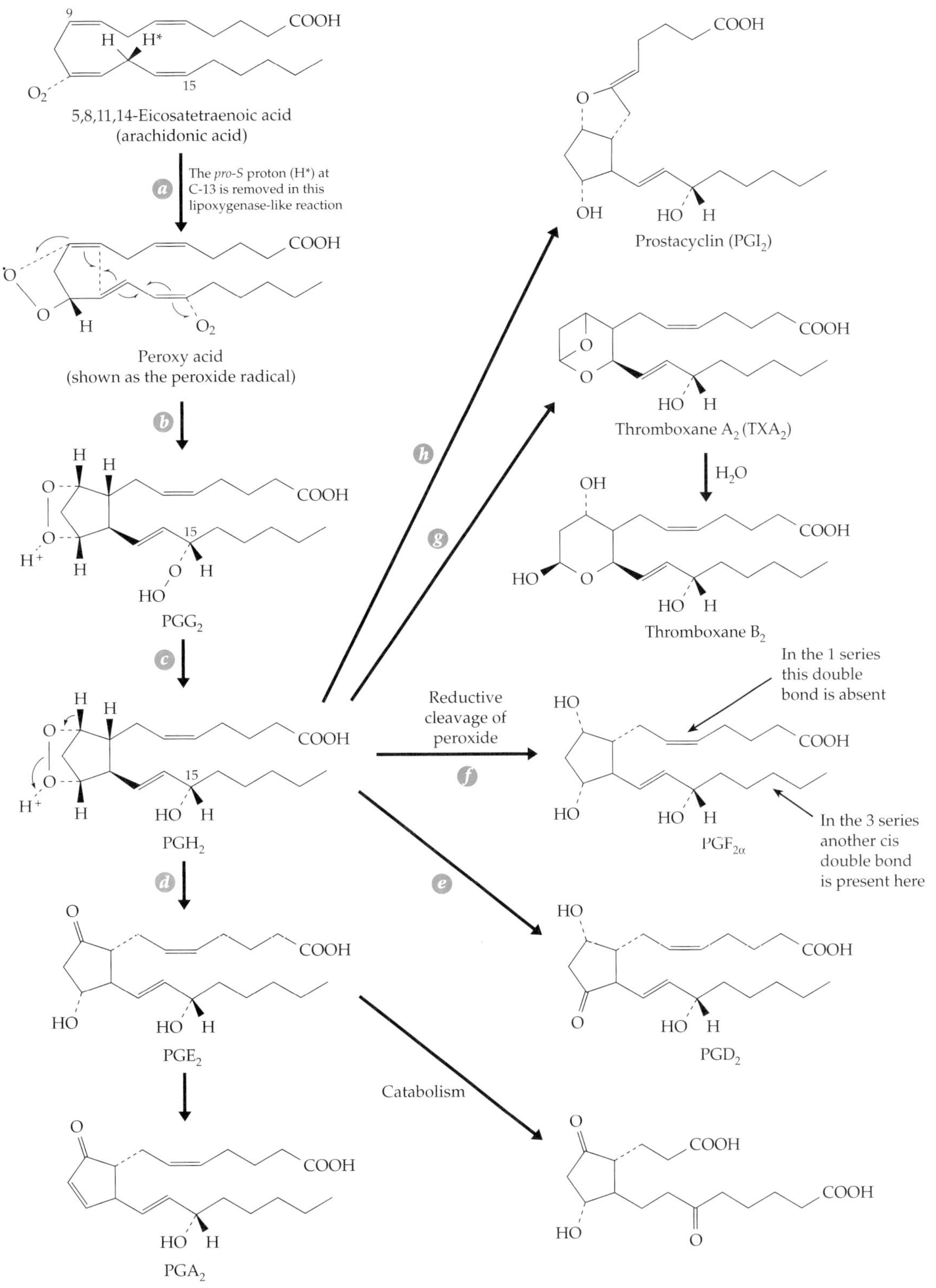

Figure 21-7 Pathway of synthesis and catabolism of the prostaglandins.

phosphatidylcholine and other phospholipids to the lyso forms of the phospholipids, which lack a 2-acyl group. The enzyme has a strong preference for the lyso-ethanolamine plasmalogens. As a consequence, in the plasmenylethanolamine of platelets arachidonoyl groups account for 66% of the acyl residues at the 2-position.[260] An arachidonate-specific acyl-CoA synthetase rapidly reconverts any free arachidonate that is not used for prostanoid synthesis back into phospholipids.[261]

The synthesis of prostaglandins, which was elucidated by Samuelsson,[258–260] begins with the release of arachidonate and other polyenoic acid precursors from phospholipids through the action of phospholipase A_2. The released arachidonate is then acted upon by **prostaglandin H synthases,** which catalyze two consecutive reactions at adjacent but distinct sites in a single protein.[262–262d] The first, **cyclooxygenase** or prostaglandin endoperoxide synthase reaction, forms PGG_2 from arachidonate and the second, a **peroxidase** reaction, generates PGH_2. There are two major mammalian isozymes of prostaglandin synthase, which are often called cyclooxygenase-1 (COX-1) and cyclooxygenase-2 (COX-2). From studies of stereospecifically synthesized ^{3}H-containing fatty acid precursors, it was established that the first step in cyclooxygenase action involves removal of the *pro*-S proton at C-13 of the fatty acid (step *a*, Fig. 21-7). The O_2-requiring cyclooxygenase resembles lipoxygenase (Eq. 21-17).[181] The product is a peroxy acid, possibly in the form of the peroxide radical shown in Fig. 21-7. This radical (or peroxide anion) undergoes cyclization with synchronous attack by a separate O_2 molecule at C-15 (Fig. 21-7, step *b*) to give the endoperoxide PGG. Reduction of the latter to an OH group by the NADPH-dependent peroxidase (step *c*) yields PGH. The entire sequence is catalyzed by the single 70-kDa PGH synthase, which contains a single heme prosthetic group.[263] During the cyclooxygenase reaction the enzyme appears to accept electrons from cytochrome b_5. During the peroxidase step the heme group undergoes formation of the characteristic peroxidase intermediate compounds I and II[263] (Fig. 16-14). It has been suggested that a tyrosyl radical is generated in the peroxidase active site (on Y385 of COX-1) and is used to form an arachidonate radical that reacts with O_2 in the cyclooxygenase reaction.[264–266b] Alternatively, a carbocation mechanism is also possible.[267]

PGH can break down in three ways to give the E and F series of prostaglandins.[268] In one the proton at C-9 is eliminated (step *d*) as indicated by the small arrows by the PGH_2 structure of Fig. 21-7. An alternative isomerization (step *e*) gives PGD_2. The F prostaglandins are formed by reductive cleavage of the endoperoxide (step *f*). The A series and other prostaglandins arise by secondary reactions, one of which is shown in Fig. 21-7.

A biochemical characteristic of the prostaglandins is rapid catabolism. The product shown in Fig. 21-7 (lower right) arises by oxidation of the 15-OH to a carbonyl group, permitting reduction of the adjacent trans double bond. Two steps of β oxidation as well as ω oxidation are also required[269] to produce the dicarboxylic acid product shown. However, a series of products appears, and the distribution varies among species. Catabolism of prostaglandins is especially active in the lungs, and any prostaglandins entering the bloodstream are removed by a single pass through the lungs. This observation has led to the conclusion that prostaglandins are not hormones in the classical sense but act on a more local basis.

2. Thromboxanes and Prostacyclins

In blood platelets and in some other tissues PGG is also transformed to another series of compounds, the **thromboxanes**,[270] which were identified in 1975. Labile hemiacetals, the thromboxanes A (TXA, Fig. 21-7), are derived by rearrangement of PGH (step *g*). Thromboxane synthase,[271–273] which catalyzes the reaction, has characteristics of a cytochrome P450. Cytochromes P450 are known to react with peroxides as well as with O_2, and the endoperoxide of PGH may be opened by the synthase prior to rearrangement to TXA.[273] Thromboxane A_2 is so unstable that its half-life at 37°C in water is ~36 s. It is spontaneously converted to TXB_2 (Fig. 21-7), which contains an –OH group at C-15. The thromboxanes B are much more stable than TXA but are not very active physiologically.

By 1976, Vane and associates had identified another prostanoid compound, **prostacyclin** (or PGI_2).[274–275a] This compound also arises from PGH_2 by action of a cytochrome P450-like prostacyclin synthase (Fig. 21-7).[273,275,276] It is thought to be an important vasoprotective molecule. As with the thromboxanes, prostacyclin undergoes rapid inactivation[277] by hydrolysis to the physiologically inactive 6-oxo-$PGF_1α$.

3. Lipoxygenases

Lipoxygenases, of which the enzyme from soy beans has been studied the most, also catalyze oxidation of polyunsaturated fatty acids in lipids as indicated in Eq. 21-17. Formation of the hydroperoxide product is accompanied by a shift of the double bond and conversion from cis to trans configuration. Soybean lipoxygenase is a member of a family of related lipoxygenases that are found in all eukaryotes. All appear to have similar iron- or manganese-containing active sites and to act by similar mechanisms.[278–280e] The major substrate in animals is arachidonic acid (probably as the arachidonate ion). As marked on the structures above Eq. 21-17, there are 5-, 8-, 12-, and 15-lipoxygenases, which catalyze reaction with dioxygen at the indicated places.[281,282] Linoleic and linolenic acids are the primary substrates in plants. Soybean lipoxygenase acts on the 13-position of linoleic acid as shown in Eq. 21-17. However, this enzyme is often referred to as a *15-lipoxygenase* because it acts on arachidonate at C15. The 100-kDa enzyme from

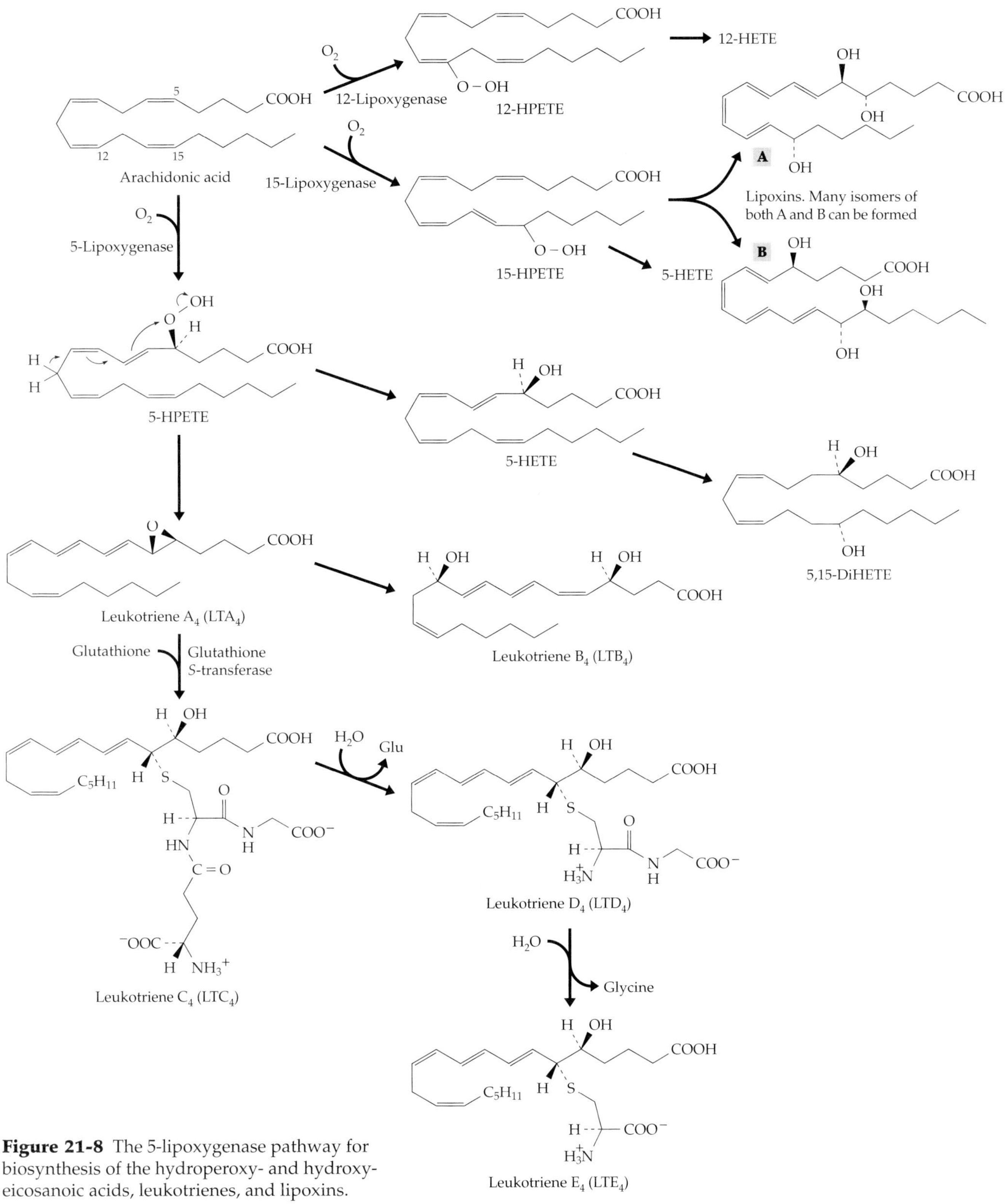

Figure 21-8 The 5-lipoxygenase pathway for biosynthesis of the hydroperoxy- and hydroxy-eicosanoic acids, leukotrienes, and lipoxins.

soybeans contains one atom of Fe(II), which is bound by a cluster of His and Tyr side chains.[283] It must be oxidized to the Fe(III) state before becoming active.[279,284] The initial reaction with O_2 may occur via an intermediate radical.

4. Leukotrienes, Lipoxins, and Related Compounds

Yet another series of products results from the action on arachidonate of tissue lipoxygenases, which compete with the prostaglandin-forming cyclooxygenases. The 5-lipoxygenase (Fig. 21-8) produces the unstable peroxide **5-hydroperoxy 6,8,11,14-eicosatetraenoic acid** (usually abbreviated **5-HPETE**). This enzyme requires ATP and Ca^{2+} and appears to be regulated by a series of other metabolites.[285] The 12- and 15-lipoxygenases, whose distribution varies among different mammalian organs and tissues, form the corresponding 12- and 15-HPETES as well as 5, 15-, 8,15-, 14,15-diHPETEs.[255,286] Some of these peroxides have physiological effects of their own, but they are largely transformed by peroxidases to more stable compounds such as the corresponding alcohols (hydroxy-icosatetraenoic acids or HETEs; Fig. 21-8).

The **leukotrienes** are formed from 5-HPETE (Fig. 21-8).[287,288] Dehydration of HPETE produces the unstable epoxide **leukotriene A_4** (LTA_4), which can be hydrolyzed enzymatically by leukocytes to the diol **leukotriene B_4** (LTB_4).[289,290] Alternatively leukotriene synthase, present in many cells, catalyzes the addition of glutathione (Box 11-B) to the LTA. This is a ring-opening reaction of the epoxide that can be visualized as a nucleophilic displacement by the thiolate anion of glutathione at C-6 (Fig. 21-8). The product is **leukotriene C_4** (LTC_4),[291] which can undergo consecutive removal of glutamate and glycine to form **leukotrienes D** and **E** (LTD_4, LTE_4), respectively. Removal of the glutamate occurs by the action of γ-glutamyl transpeptidase (Box 11-B), whereas removal of the glycine is hydrolytic. LTC_4 and the more potent LTD_4 have been identified as the **slow-reacting substance of anaphylaxis** (SRS-A), a long-sought mediator of bronchial asthma.[292,293] Leukotrienes can be formed from polyunsaturated acids other than arachidonic acid. Thus, eicosapentaenoic acid yields LTC_5 and LTD_5.[294] A lipoxygenase-derived product from the C18:2 linoleic acid is 13-hydroxylinoleic acid, which is made principally by endothelial cells that line blood vessels. It may contribute to resistance to blood clotting.[294]

Products of the 15-lipoxygenase pathway include a group of trihydroxytetraenes formed by leukocytes.[295] Several routes of biosynthesis, which may involve epoxide intermediates, are known.[296,297] The structures of two of these compounds, **lipoxin A** and **lipoxin B,** are shown in Fig. 21-8. Several stereoisomers and cis–trans isomers can be formed. These compounds can all arise from 15-HPETE, either by enzyme action or nonenzymatically. In fact, the entire series of prostanoid compounds arise by reactions related to but more specific than those that occur during nonenzymatic autoxidation of arachidonate.[255,298] Cytochrome P450-catalyzed reaction with O_2 can convert arachidonic acid into four different **epoxytrienoic acids** (**EETs**), which may also exist as stereoisomers. They are vasodilators which affect a variety of signaling pathways.[298a,b]

5. Physiological Effects of the Prostanoids

The release of arachidonate and initiation of the arachidonate cascade is induced by hormones, various inflammatory and immunological stimuli, and even mechanical agitation. Tissues do not all behave the same in response to the arachidonate cascade.[299] Blood platelets form largely thromboxane A_2, whereas tissues of the aorta form prostacyclin. Prostaglandin D_2 is a major prostanoid in the central nervous system.[300] Biological functions of prostanoids are also varied.[256a,301] The **primary prostaglandins** PGE and PGF were first recognized as mediators of inflammation. However, PGE_2 and PGF_2 sometimes have opposite effects. The unstable precursors PGG_2 and PGH_2, which have half-lives of only a few minutes, are much more powerful than the more stable PGEs and PGFs. Prostaglandin D_2, which is released in lungs during attacks of asthma, is thought to be a major bronchoconstrictor, but it may also serve as a neurotransmitter.[300] Thromboxanes released from platelets cause smooth muscle contraction and aggregation of the platelets, the first step in blood clot formation. Thromboxanes have half life-times of only seconds but are extremely potent not only in inducing platelet aggregation but also in causing contraction of blood vessels. Prostacyclin has the opposite effect, being a potent vasodilator that causes relaxation of smooth muscle. Upon release from blood vessel walls it acts to prevent clot formation.

The lipoxygenase pathway (Fig. 21-8) leading to the leukotrienes, lipoxins, and other products is especially active in leukocytes and in mast cells.[302] The leukotrienes promote inflammation, but lipoxins A_4 and B_4 are antiinflammatory.[303] The release of leukotrienes LTC_4, LTD_4, and LTE_4 in lung tissue is correlated with the long-lasting contractions of smooth muscle of the bronchi that are characteristic of asthma.[304] Leukotriene LTC_4 is ~1000 times more powerful than histamine in inducing such contraction. Leukotrienes have also been found in the central nervous system.[305]

Some effects of prostaglandins are mediated through cell surface G-protein coupled receptors (see Chapter 11).[306] Some other prostanoids bind to and activate nuclear peroxisome proliferator-activated receptors.[306] PGJ_2 may inhibit fatty acid synthesis and fat deposition in adipose tissue through these receptors. Some of the prostanoid derivatives enter membranes and may become incorporated into phospholipids and exert their effects there.

A number of medical uses of prostaglandins have been discovered and more will probably be developed. While prostaglandins may be required for conception, small amounts of PGE_2 or $PGF_{2\alpha}$ induce abortion. $PGF_{2\alpha}$ is also used to induce labor. Prostaglandins are widely employed to control breeding of farm animals, to synchronize their estrus cycles, and to improve the efficiency of artificial insemination.[307]

6. Inflammation

Special interest in the prostaglandins has focused on pain of inflammation and allergic responses. The medical significance is easy to see. Five million Americans have **rheumatoid arthritis**, an inflammatory disease. Bronchial asthma and other allergic diseases are equally important. Our most common medicine is **aspirin**, an anti-inflammatory drug. Both the inflammatory response and the immune response are normal parts of the defense mechanisms of the body, but both are potentially harmful, and it is their regulation that is probably faulty in rheumatoid arthritis and asthma. Overproduction of prostaglandins may be a cause of menstrual cramps.[308]

Prostaglandins have been implicated both in the induction of inflammation and in its relief. In inflammation small blood vessels become dilated, and fluid and proteins leak into the interstitial spaces to produce the characteristic swelling (edema). Many polymorphonuclear leukocytes attracted by chemotactic factors that include LTB_4[309] (Chapter 19) migrate into the inflamed area, engulfing dead tissue and bacteria. In this process lysosomes of the leukocytes release phospholipase A, which hydrolyzes phospholipids and initiates the arachidonate cascade. The leukotrienes that are formed promote the inflammatory response. However, cAMP can suppress inflammation, and PGE_2 has a similar effect. Indeed, E prostaglandins, when inhaled in small amounts, relieve asthma.

The synthesis of prostaglandins is inhibited by aspirin[310] and many other analgesic drugs. Aspirin is an acetylating reagent, and the inhibition has been traced to acetylation of the side chain –OH group of a single serine residue, Ser 530 of COX-1 or Ser 516 of COX-2 in the arachidonate binding channel.[311–313] Other nonsteroidal antiinflammatory drugs (NSAIDs),

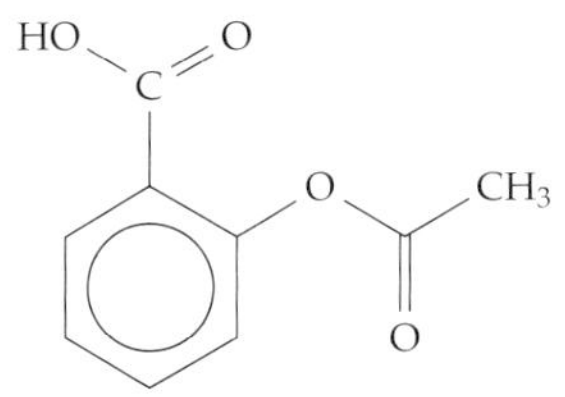

Aspirin (acetylsalicylic acid)

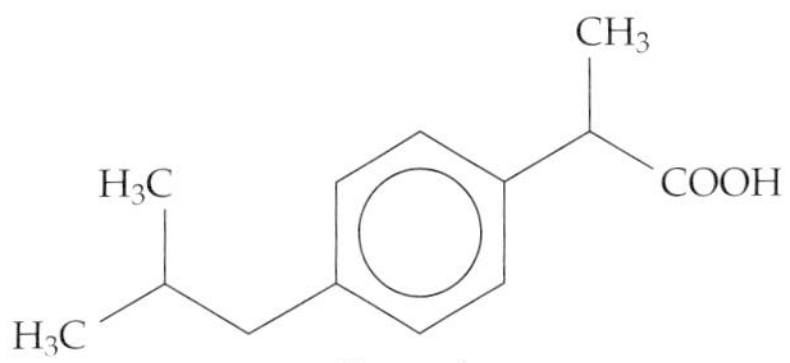

Ibuprofen

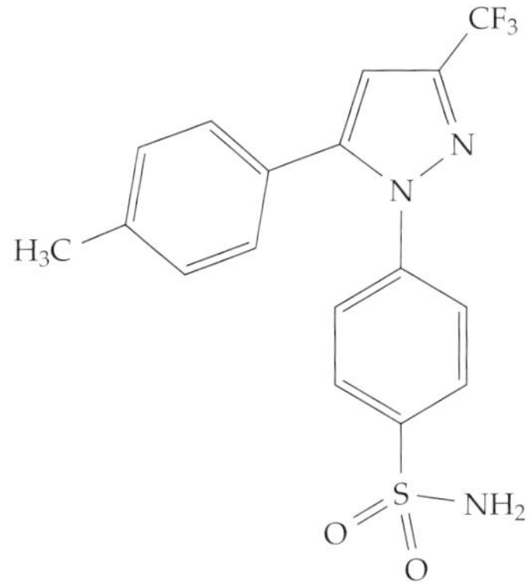

Celecoxib, a new COX-2 inhibitor

e.g., ibuprofen, are competitive inhibitors of the cyclooxygenases.[314] COX-3, a variant form of COX-1, may be the target for acetaminohen (Box 18-E).[314a] However, the same drugs also inhibit activation of neutrophils and may thus exercise their anti-inflammatory action in more than one way.[315] Since PGE_1 is a potent **pyrogen** (fever-inducing agent), a relationship to the ability of aspirin to reduce fever is also suggested. Unfortunately, all of these drugs inhibit COX-1 of platelets. Small regular doses of aspirin may be useful in preventing blood clots in persons with arterial disease, but they can be disastrous. Thousands of people die annually of hemorrhage caused by aspirin.[316,317]

Recently it was recognized that COX-1 provides eicosinoids for homeostatic purposes, while it is COX-2 that is inducible and generates prostoglandins for production of leukotrienes and induction of the inflammatory response. Now there is a major effort, with the first drugs already in use, to develop specific inhibitors for COX-2, which do not inhibit COX-1. It is hoped that these will be safer than aspirin.[256a,266a,311,317,317a-c] However, these drugs can also cause dangerous side effects.[317d] COX-2 of macrophages is also inhibited by γ-tocopherol, a major form of vitamin E.[317e]

(21-18)

7. Plant Lipoxygenases and Jasmonic Acid

An octadecenoid signaling pathway (Eq. 21-18), which resembles the arachidonate cascade in some respects, plays an important role in green plants.[318–320] Alpha-linolenic acid is acted upon by a lipoxgenase in plastids to form a 13-hydroperoxy derivative. This is dehydrated and cyclized by allene oxide synthase. Although this doesn't appear to be an oxidation–reduction process, the enzyme seems to be a cytochrome P450 and to initiate the cyclization to the unstable epoxide **allene oxide** by homolytic cleavage of the peroxy group.[321,321a] Allene oxide synthase and the cyclase that acts in the next step may be cytosolic, while the β oxidation that shortens the chain occurs in plants exclusively in peroxisomes or glyoxysomes.[319]

Jasmonic acid is a plant growth regulator that affects many aspects of plant development as well as responses to environmental signals. A very important function is mobilization of plant defenses in response to damage by herbivores, by bacterial or fungal pathogens, or by ultraviolet light.[322,323] The synthesis of protease inhibitors as well as phytoalexins is induced. A curious variant of the jasmonate pathway is the acquisition of α-linoleic acid from plants by chewing caterpillars. The linoleic acid is hydroxylated by the insect, and conjugated with glutamine to form **volicitin** (*N*-(17-hydroxylinoleoyl)-L-glutamine; Eq. 21-18). Some of this compound reenters the plant from material regurgitated into wounds by the caterpillars. Volicitin induces the plant to release volatile terpenes and other compounds. The value to the caterpillars is not clear, but not only does volicitin induce defensive reactions in plants but also the released volatile compounds attract wasps that parasitize the caterpillars.[324,325] Plants also form a group of **isoprostanes** E_1 from α-linolenic acid.[326]

Allene oxides are unusual biological products. However, they are formed from arachidonic acid by some corals and are evidently precursors to prostaglandin esters, which may be present in high concentrations.[327] Allene oxides are also present in starfish oocytes.[321,328]

E. The Polyketides

In 1907, Collie proposed that polymers of ketene ($CH_2=C=O$) might be precursors of such compounds as **orsellinic** acid, a common constituent of lichens. The hypothesis was modernized in 1953 by Birch and Donovan, who proposed that several molecules of acetyl-CoA are condensed (Eq. 21-19) but *without the two reduction steps required in biosynthesis of fatty acids* (Fig. 17-12).[329] As we now know they were correct in assuming that the condensation occurs via malonyl-CoA and an acyl carrier group of an enzyme. The resulting **β-polyketone** can react in various ways to give the large group of compounds known as polyketides.

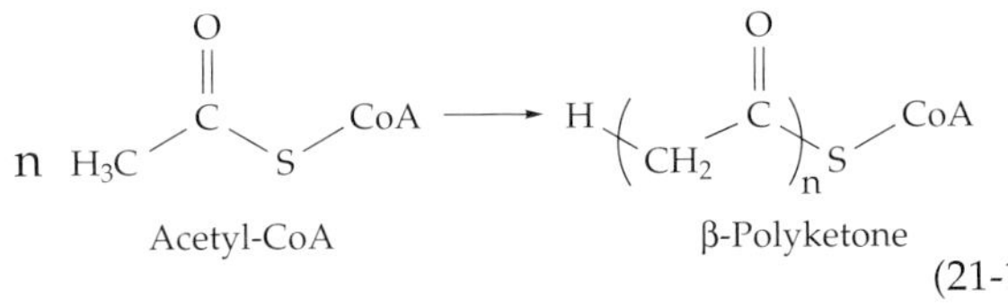

$$n\ H_3C{-}C({=}O){-}S{-}CoA \longrightarrow H{+}CH_2{-}C({=}O){+}_n S{-}CoA$$

Acetyl-CoA β-Polyketone (21-19)

β-Polyketones can be stabilized by ring formation through ester or aldol condensations. Remaining carbonyl groups can be reduced (prior to or after cyclization) to hydroxyl groups, and the latter can be eliminated as water to form benzene or other aromatic rings. Figure 21-9 illustrates two ways in which cyclization can occur. One involves a Claisen ester condensation during which the enzyme and its SH group are eliminated. Enolization of the product gives a trihydroxy-acetophenone. The second cyclization reaction is the aldol condensation. Following the condensation water is eliminated, and the product is hydrolyzed and enolized to form orsellinic acid. Another product of fungal metabolism is **6-methylsalicylic acid,** which lacks one OH group of orsellinic acid. This synthesis can be explained by assuming that the carbonyl group at C-5 of the original β-polyketone was reduced to an OH group at some point during the biosynthesis. Elimination of two molecules of water together with enolization of the remaining ring carbonyl gives the product (Fig. 21-9).[330]

By allowing a few variations in the basic polyketone structure, the biosynthesis of a large number of unusual compounds can be explained. Extra oxygen atoms can be added by hydroxylation, and methyl groups may be transferred from *S*-adenosylmethionine to form methoxyl groups.[331] Occasionally a methyl group may be transferred directly to the carbon chain. Glycosyl groups may also be attached.[332,333] Many starter pieces other than acetyl-CoA may initiate polyketide synthesis. These include the branched-chain acids of Table 21-3, nicotinic and benzoic acids, 4-coumaroyl-CoA, and a 14:1 Δ^9-ACP. The last of these starter pieces is formed by desaturation of the corresponding 14:0-ACP and is converted via polyketide synthesis to one of a family of **anacardic acids,** which provide pest resistance to a variety of dicotyledenous plants (Fig. 21-10, bottom).[334] The CoA derivative of malonic acid amide is the starter piece for synthesis of the antibiotic **tetracycline** as indicated in Fig. 21-10).[335] Polyketide origins of some other antibiotics are also indicated in this figure.

The cloning and sequencing of genes for enzymes involved in synthesis of polyketides of fungi and actinomycetes has shown that these enzymes are closely related to the fatty acid synthases and, like the latter, have a multidomain structure (Fig. 21-11). The possibilities of engineering these genes, together with the urgent need for new antibiotics, has led to an

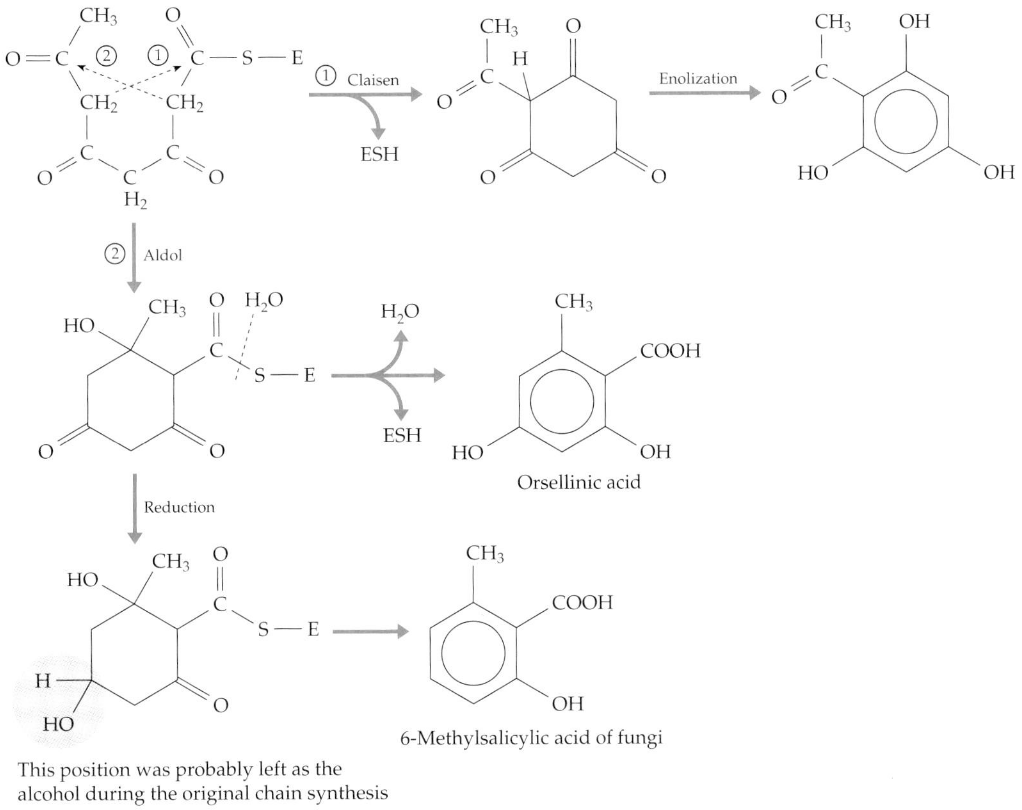

Figure 21-9 Postulated origin of orsellinic acid and other polyketides.

BOX 21-E HOW THE FLOWERS MAKE THEIR COLORS

Most of the pigments of flowers arise from a single polyketide precursor. Phenylalanine is converted to ***trans*-cinnamic acid** (Eq. 14-45) and then to cinnamoyl-CoA. The latter acts as the starter piece for chain elongation via malonyl-CoA (step *a* in the accompanying scheme). The resulting β-polyketone derivative can cyclize in two ways. The aldol condensation (step *b*) leads to **stilbenecarboxylic acid** and to such compounds as **pinosylvin** of pine trees. The Claisen condensation (step *c*) produces **chalcones**, **flavonones**, and **flavones**. These, in turn, can be converted to the yellow **flavonol pigments** and to the red, purple, and blue **anthocyanidins**.[a–c]

At the bottom of the synthetic scheme on the next page the structures and names of three common anthocyanidins are shown. The names are derived from those of flowers from which they have been isolated. The colors depend upon the number of hydroxyl groups and on the presence or absence of methylation and glycosylation. In addition to the three pigments indicated in the diagram, three other common anthocyanidins are formed by methylation. **Peonidin** is 3'-methylcyanidine. Methylation of delphinidin at position 3' yields **petunidin**, while methylation at both the 3' and 5' positions gives **malvidin**. There are many other anthocyanidins of more limited distribution. Anthocyanidins are nearly insoluble, but they exist in plants principally as glycosides known as **anthocyanins**. The number of different glycosides among the many species of flowering plants is large. Both the 3 and 5 –OH groups may be glycosylated with Glc, Gal, Rha, Ara, and by a large variety of oligosaccharides. The colors of the anthocyanins vary from red to violet and blue and are pH dependent. For example, **cyanin** (diglucosyl cyanidin) is red in acid solution and becomes violet upon dissociation of the 4'-hydroxyl group:

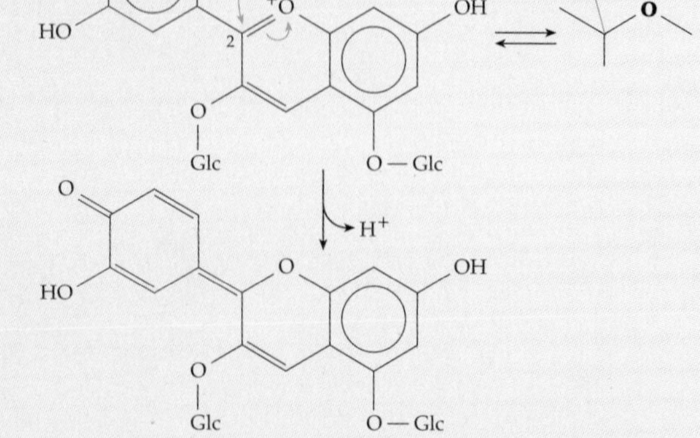

H–O 4' $^-$OH HO $^+$O 2 OH HO O Glc O–Glc

$\downarrow$ H$^+$

O HO O OH O Glc O–Glc

Dissociation of the 7–OH generates an anion with an extended conjugated π electron system, which will favor absorption of long-wavelength light (Chapter 23) and a blue color. Notice that a large number of resonance structures can be drawn for both the anthocyanin and the dissociated forms. Formation of complexes of Mg^{2+} or other metal ions with the 4' –O$^-$and adjacent OH groups may also stabilize blue colored forms.[d]

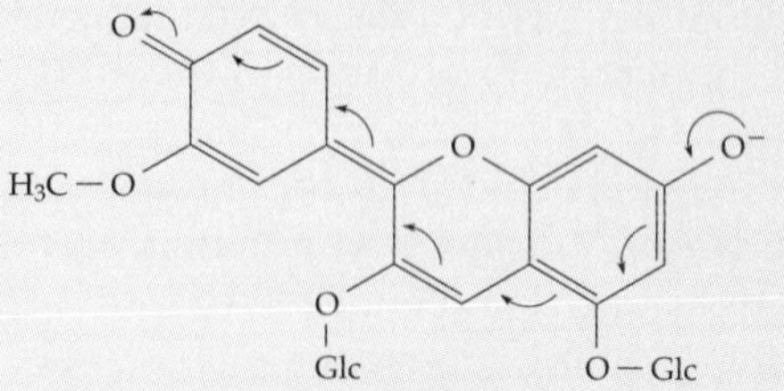

O H_3C–O O O$^-$ O Glc O–Glc

Most blue flower pigments are based on delphinidin,[b] but the "heavenly blue" of the morning glory is a peonidin with a complex caffeoylglucose-containing glycosyl group on the 3-position. Its blue color has been attributed to the relatively high pH of ~7.7 in vacuoles.[e] The aromatic rings within the glycosyl group of this and other complex anthocyanins may fold over the primary chromophore and stabilize the colored forms. A competing reaction, which is indicated in green on the first structure in this box, is the addition of a hydroxyl ion at C-2 to give a nearly colorless adduct.[f]

The yellow pigments of flowers are usually flavonols. The most common of all is **rutin**, the 3α-rhamnosyl-D-glucosyl derivative of **quercitin** (see diagram). An extraordinary number of other flavonols, flavones, and related compounds are found throughout the plant kingdom.[g] One of these is **phlorhizin**, a dihydrochalcone found in the root bark of pears, apples, and other plants of the rose family. Phlorhizin specifically blocks resorption of glucose by kidney tubules. As a result, the drug induces a strong glucosuria. The biochemical basis is uncertain, but the action on kidney tubules may be related to inhibition of mutarotase.[h]

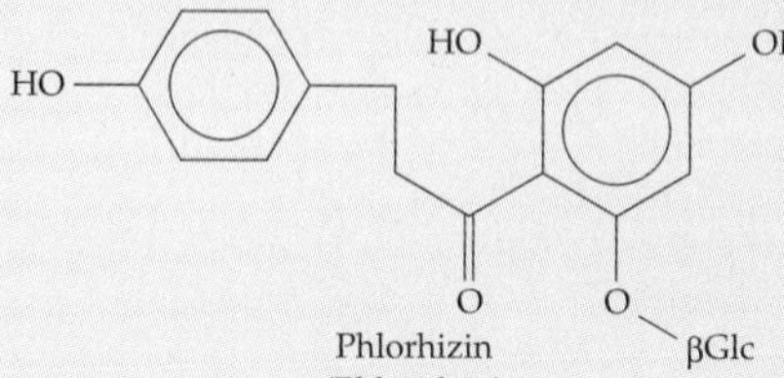

HO HO OH O O βGlc

Phlorhizin (Phloridzin)

BOX 21-E (continued)

Phenylalanine
Eq. 14-45
Cinnamoyl-CoA
(a) Chain elongation 3 malonyl-CoA
(b)
(c)
Chalcones
Flavonones
Stilbenecarboxylic acid
CO_2
Pinosylvin (present in most pines)
Rha-Glc
Flavonols (yellow and ivory)
This is the flavonol glycoside **rutin**
Flavones
Delphinidin contains one more —OH at this position
This —OH group is lacking in **pelargonidin** of the red geranium *Pelargonium*
Methylation of —OH groups at positions 3' and 5' yields other pigments
Anthocyanidins (red, blue, and violet)
Cyanidin, named after the blue cornflower *Centaurea cyanus*
Glycosylation at one or both of these points forms the water-soluble anthocyanins

The flavone glycoside **hesperidin** makes up 80% of the dry weight of orange peels. It has been claimed (but not proved) that this compound, also known as **vitamin P** and **citrus bioflavonoid**, is essential to good health. Another flavone, **maysin**, is a resistance factor for the corn earworm and is present in silks of resistant strains of *Zea mays*.[i]

Hesperidin

Maysin

[a] Clevenger, S. (1964) *Sci. Am.* **210**(Jun), 85–92

[b] Harborne, J. B. (1988) in *Plant Pigments* (Goodwin, T. W., ed), pp. 299–343, Academic Press, London

[c] Lloyd, A. M., Walbot, V., and Davis, R. W. (1992) *Science* **258**, 1773–1775

[d] Kondo, T., Yoshida, K., Nakagawa, A., Kawai, T., Tamura, H., and Goto, T. (1992) *Nature (London)* **358**, 515–518

[e] Yoshida, K., Kondo, T., Okazaki, Y., and Katou, K. (1995) *Nature (London)* **373**, 291

[f] Figueiredo, P., Elhabiri, M., Saito, N., and Brouillard, R. (1996) *J. Am. Chem. Soc.* **118**, 4788–4793

[g] Nakayama, T., Yonekura-Sakakibara, K., Sato, T., Kikuchi, S., Fukui, Y., Fukuchi-Mizutani, M., Ueda, T., Nakao, M., Tanaka, Y., Kusumi, T., and Nishino, T. (2000) *Science* **290**, 1163–1166.

[h] White, A., Handler, P., and Smith, E. L. (1973) *Principles of Biochemistry*, 5th ed., McGraw-Hill, New York (pp. 415–416)

[i] Byrne, P. F., McMullen, M. D., Snook, M. E., Musket, T. A., Theuri, J. M., Widstrom, N. W., Wiseman, B. R., and Coe, E. H. (1996) *Proc. Natl. Acad. Sci. U.S.A.* **93**, 8820–8825

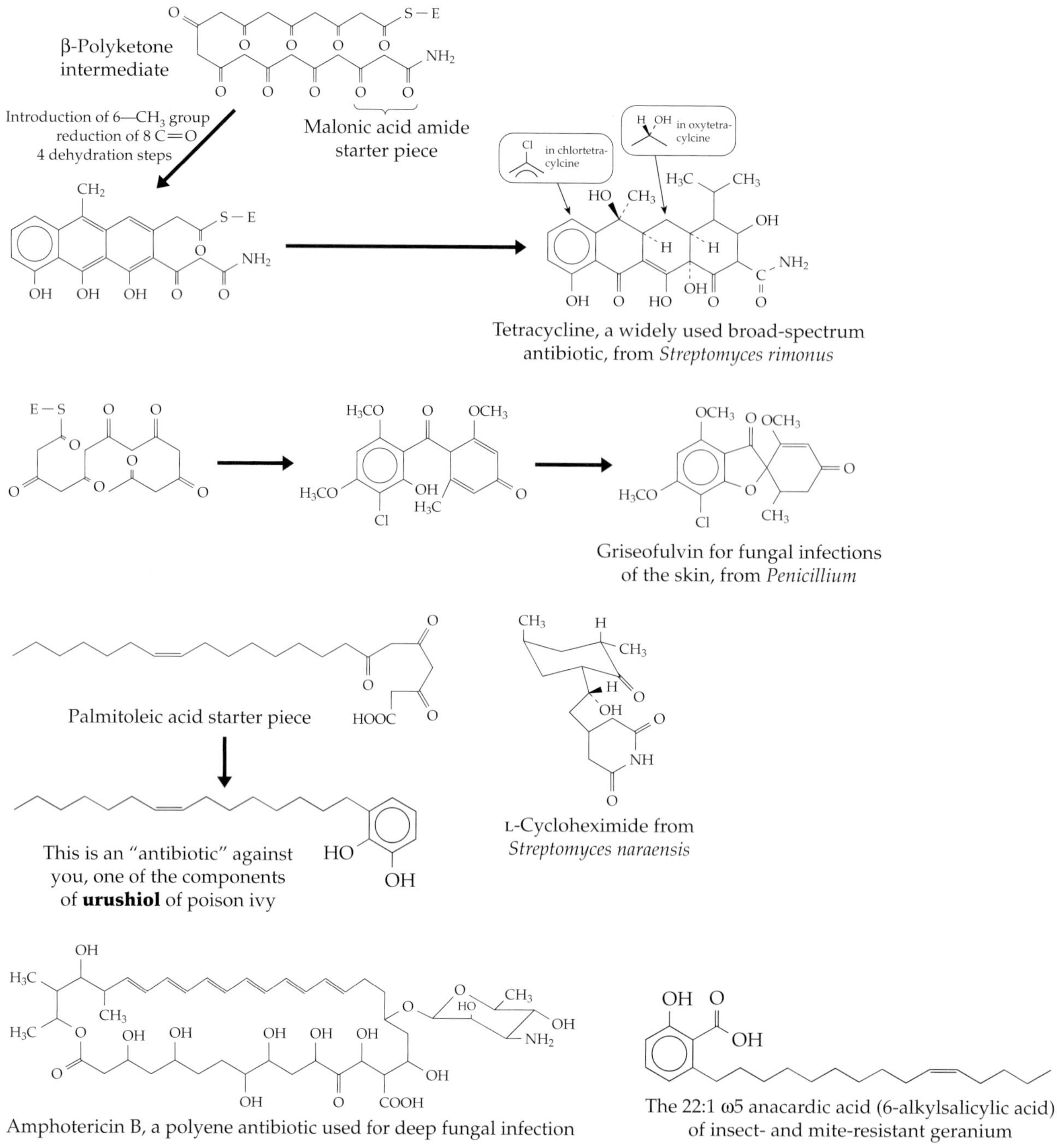

Figure 21-10 Some important polyketide antibiotics and plant defensive compounds.

explosion of information about polyketide synthases.[336,336a–c]

A 26-kb gene cluster encoding enzymes for synthesis of the blue antibiotic **actinorhodin** by *Streptomyces coelicolor* has been cloned and sequenced.[332,337] The three large ~10-kb genes required for formation of the broad-spectrum antibiotic **erythromycin** by *Saccharopolyspora erythraea* have also been cloned and sequenced.[337–339] In both cases, the genes encode large proteins with structures resembling those of the eukaryotic fatty acid synthases (Section B,1). However, a new feature is evident. As shown in Fig. 21-11, each of the three polypeptides of the deoxyerythronolide synthase, which synthesizes the aglycone of erythromycin, consists of two multidomain modules, each able to catalyze one round of reaction with a new molecule of malonyl-CoA. When reduction of an oxo group or dehydration and

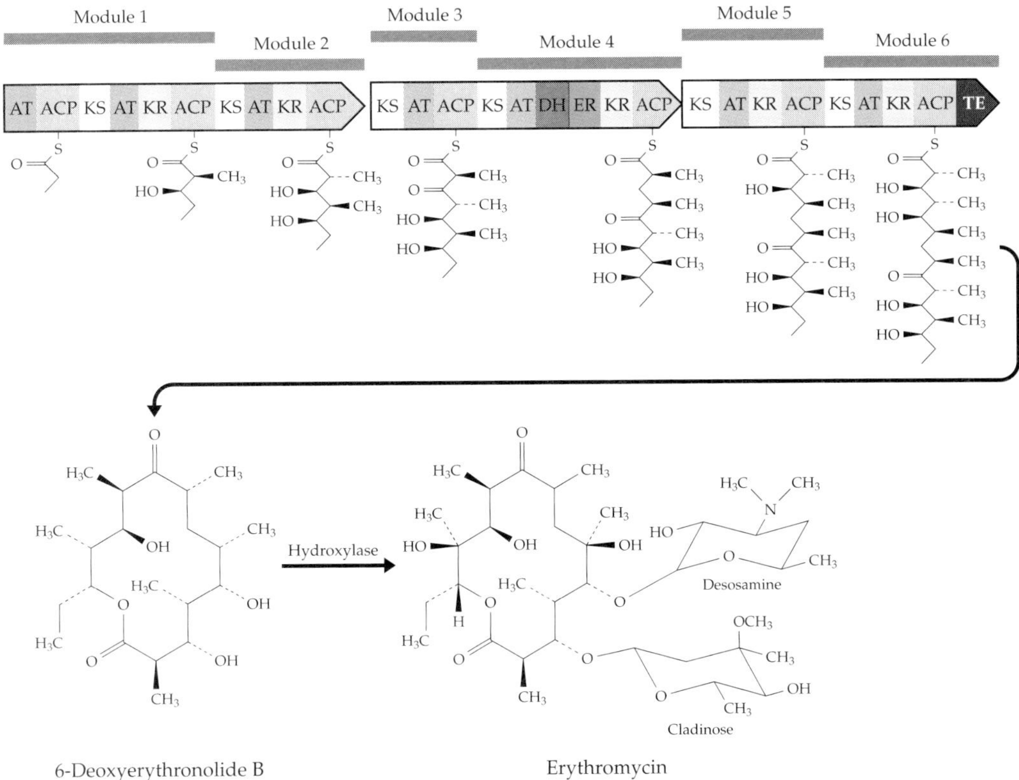

Figure 21-11 Catalytic domains within three polypeptide chains of the modular polyketide synthase that forms 6-deoxyerythronolide B, the aglycone of the widely used antibiotic erythromycin. The domains are labeled as for fatty acid synthases; AT, acyltransferase; ACP, acyl carrier protein; KS, β-ketoacyl-ACP synthase; KR, ketoreductase; DH, dehydrase; ER, enoylreductase; TE, thioesterase. After Pieper *et al.*[338] Courtesy of Chaitan Khosla.

reduction of an enoyl-CoA are not needed in a round, the KR, DH, and ER domains are absent (as in module 3 of Fig. 21-11). A final domain contains a thioesterase that releases and cyclizes the product. The "assembly line" sequence of synthetic steps, beginning with a propionyl group from propionyl-CoA, is pictured in Fig. 21-11. Two hydroxylation steps[340] and transfer of two unusual glycosyl groups complete the synthesis of the antibiotic.

Other medically important polyketides include the antibiotics **doxorubicin** (14-hydroxydaunomycin; Fig. 5-23),[341] rifamycin (Box 28-A),[342] and the antifungal **pimaricin**,[343] **griseofulvin**, and **amphotericin** (Fig. 21-10), the HMG-CoA reductase inhibitor **lovastatin**,[344] the 2-butanyl-4-methylthreonine of cyclosporin A (Box 9-F),[345] and other immunosuppressants such as **rapamycin**.[346] Many characteristic plant products, including **stilbenes**[347] and **chalcones**[348,348a] (Box 21-E), are polyketides. A variety of different polyketides serve as phytoalexins.[349] Some such as **aflatoxin**[350] are dangerous toxins. Ants and ladybird beetles make toxic polyamine alkaloids using a polyketide pathway.[350a]

Avermectin (Fig. 30–25), a widely used antibiotic against canine heartworms, is formed by a polyketide synthase with an unusually broad specificity for starter units. More than 40 alternative carboxylic acids are accepted. By grafting the first multidomain module of the erythromycin-forming synthase (of Fig. 21-11) onto the wide-specificity loading module of the avermectin-forming synthase, a whole new series of antibiotics have been created.[351] This is only one of many steps being taken to create new aliphatic and aromatic, linear and macrocyclic polyketides by genetic engineering.[336,352,353] Combinatorial biosynthesis (see Chapter 3) is also being developed[336,354] and has even been discovered in nature.[355]

References

1. Ayvazian, L., Crenon, I., Hermoso, J., Pignol, D., Chapus, C., and Kerfelec, B. (1998) *J. Biol. Chem.* **273**, 33604–33609
2. Bezzine, S., Ferrato, F., Ivanova, M. G., Lopez, V., Verger, R., and Carrière, F. (1999) *Biochemistry* **38**, 5499–5510

2a. Crandall, W. V., and Lowe, M. E. (2001) *J. Biol. Chem.* **276**, 12505–12512

3. Hermoso, J., Pignol, D., Kerfelec, B., Crenon, I., Chapus, C., and Fontecilla-Camps, J. C. (1996) *J. Biol. Chem.* **271**, 18007–18016
4. Kobayashi, J., Applebaum-Bowden, D., Dugi, K. A., Brown, D. R., Kashyap, V. S., Parrott, C., Duarte, C., Maeda, N., and Santamarina-Fojo, S. (1996) *J. Biol. Chem.* **271**, 26296–26301
5. Winkler, F. K., D'Arcy, A., and Hunziker, W. (1990) *Nature (London)* **343**, 771–774
6. Levy, E., Mehran, M., and Seidman, E. (1995) *FASEB J.* **9**, 626–635
7. Havel, R. J., and Kane, J. P. (1995) in *The Metabolic and Molecular Bases of Inherited Disease*, 7th ed., Vol. 2 (Scriver, C. R., Beaudet, A. L., Sly, W. S., and Valle, D., eds), pp. 1841–1852, McGraw-Hill, New York

7a. Liu, H., Talmud, P. J., Lins, L., Brasseur, R., Olivecrona, G., Peelman, F., Vandekerckhove, J., Rosseneu, M., and Labeur, C. (2000) *Biochemistry* **39**, 9201–9212

7b. Rozek, A., Sparrow, J. T., Weisgraber, K. H., and Cushley, R. J. (1999) *Biochemistry* **38**, 14475–14484

8. Orlova, E. V., Sherman, M. B., Chiu, W., Mowri, H., Smith, L. C., and Gotto, A. M., Jr. (1999) *Proc. Natl. Acad. Sci. U.S.A.* **96**, 8420–8425
9. Brown, M. S., and Goldstein, J. L. (1974) *Sci. Am.* **251**(Nov), 58–66
10. Phillips, J. C., Wriggers, W., Li, Z., Jonas, A., and Schulten, K. (1997) *Biophys. J.* **73**, 2337–2346

10a. Segrest, J. P., Jones, M. K., Klon, A. E., Sheldahl, C. J., Hellinger, M., De Loof, H., and Harvey, S. C. (1999) *J. Biol. Chem.* **274**, 31755–31758

10b. Tricerri, M. A., Agree, A. K. B., Sanchez, S. A., Bronski, J., and Jonas, A. (2001) *Biochemistry* **40**, 5065–5074

11. Rosseneu, M., and Labeur, C. (1995) *FASEB J.* **9**, 768–776
12. Chan, L. (1992) *J. Biol. Chem.* **267**, 25621–25624
13. Ferguson, S. J. (1987) *Trends Biochem. Sci.* **12**, 353–357
14. MacGinnitie, A. J., Anant, S., and Davidson, N. O. (1995) *J. Biol. Chem.* **270**, 14768–14775
15. Richardson, N., Navaratnam, N., and Scott, J. (1998) *J. Biol. Chem.* **273**, 31707–31717
16. Lawn, R. M. (1992) *Sci. Am.* **266**(Jun), 54–60
17. Utermann, G. (1995) in *The Metabolic and Molecular Bases of Inherited Disease*, 7th ed., Vol. 2 (Scriver, C. R., Beaudet, A. L., Sly, W. S., and Valle, D., eds), pp. 1887–1912, McGraw-Hill, New York
18. Trieu, V. N., and McConathy, W. J. (1995) *J. Biol. Chem.* **270**, 15471–15474
19. Mochalkin, I., Cheng, B., Klezovitch, O., Scanu, A. M., and Tulinsky, A. (1999) *Biochemistry* **38**, 1990–1998

19a. Wang, J., and White, A. L. (1999) *J. Biol. Chem.* **274**, 12883–12889

20. Fless, G. M., Santiago, J. Y., Furbee, J., Jr., and Meredith, S. C. (1997) *Biochemistry* **36**, 11304–11313
21. Williams, K. J., Fless, G. M., Petrie, K. A., Snyder, M. L., Brocia, R. W., and Swenson, T. L. (1992) *J. Biol. Chem.* **267**, 13284–13292

21a. Goldstein, J. L., and Brown, M. S. (2001) *Science* **292**, 1310–1312

22. Baggio, G., Donazzan, S., Monti, D., Mari, D., Martini, S., Gabelli, C., Dalla Vestra, M., Previato, L., Guido, M., Pigozzo, S., Cortella, I., Crepaldi, G., and Franceschi, C. (1998) *FASEB J.* **12**, 433–437
23. Bergeron, J., Frank, P. G., Scales, D., Meng, Q.-H., Castro, G., and Marcel, Y. L. (1995) *J. Biol. Chem.* **270**, 27429–27438
24. Deeb, S. S., Cheung, M. C., Peng, R., Wolf, A. C., Stern, R., Albers, J. J., and Knopp, R. H. (1991) *J. Biol. Chem.* **266**, 13654–13660
25. Laccotripe, M., Makrides, S. C., Jonas, A., and Zannis, V. I. (1997) *J. Biol. Chem.* **272**, 17511–17522

25a. Gorshkova, I. N., Liu, T., Zannis, V. I., and Atkinson, D. (2002) *Biochemistry* **41**, 10529–10539

25b. Klon, A. E., Segrest, J. P., and Harvey, S. C. (2002) *Biochemistry* **41**, 10895–10905

26. Borhani, D. W., Rogers, D. P., Engler, J. A., and Brouillette, C. G. (1997) *Proc. Natl. Acad. Sci. U.S.A.* **94**, 12291–12296
27. Wang, G., Sparrow, J. T., and Cushley, R. J. (1997) *Biochemistry* **36**, 13657–13666
28. Rogers, D. P., Roberts, L. M., Lebowitz, J., Engler, J. A., and Brouillette, C. G. (1998) *Biochemistry* **37**, 945–955
29. Lopez, J., Roghani, A., Bertrand, J., Zanni, E., Kalopissis, A., Zannis, V. I., and Chambaz, J. (1994) *Biochemistry* **33**, 4056–4064
30. Boisfer, E., Lambert, G., Atger, V., Tran, N. Q., Pastier, D., Benetollo, C., Trottier, J.-F., Beaucamps, I., Antonucci, M., Laplaud, M., Griglio, S., Chambaz, J., and Kalopissis, A.-D. (1999) *J. Biol. Chem.* **274**, 11564–11572

30a. Pennacchio, L. A., Olivier, M., Hubacek, J. A., Cohen, J. C., Cox, D. R., Fruchart, J.-C., Krauss, R. M., and Rubin, E. M. (2001) *Science* **294**, 169–173

31. Dong, L.-M., Wilson, C., Wardell, M. R., Simmons, T., Mahley, R. W., Weisgraber, K. H., and Agard, D. A. (1994) *J. Biol. Chem.* **269**, 22358–22365
32. Wilson, C., Wardell, M. R., Weisgraber, K. H., Mahley, R. W., and Agard, D. A. (1991) *Science* **252**, 1817–1822
33. Mahley, R. W., and Rall, S. C. J. (1995) in *The Metabolic and Molecular Bases of Inherited Disease*, 7th ed., Vol. 2 (Scriver, C. R., Beaudet, A. L., Sly, W. S., and Valle, D., eds), pp. 1953–1980, McGraw-Hill, New York

33a. Kypreos, K. E., Morani, P., van Dijk, K. W., Havekes, L. M., and Zannis, V. I. (2001) *Biochemistry* **40**, 6027–6035

33b. Morrow, J. A., Arnold, K. S., Dong, J., Balestra, M. E., Innerarity, T. L., and Weisgraber, K. H. (2000) *J. Biol. Chem.* **275**, 2576–2580

34. Law, J. H., and Wells, M. A. (1989) *J. Biol. Chem.* **264**, 16335–16338
35. Breiter, D. R., Kanost, M. R., Benning, M. M., Wesenberg, G., Law, J. H., Wells, M. A., Rayment, I., and Holden, H. M. (1991) *Biochemistry* **30**, 603–608
36. Zhang, Y., Lewis, R. N. A. H., McElhaney, R. N., and Ryan, R. O. (1993) *Biochemistry* **32**, 3942–3952
37. Soulages, J. L., Rivera, M., Walker, F. A., and Wells, M. A. (1994) *Biochemistry* **33**, 3245–3251

37a. Dettloff, M., Weers, P. M. M., Niere, M., Kay, C. M., Ryan, R. O., and Wiesner, A. (2001) *Biochemistry* **40**, 3150–3157

38. Heinemann, B., Andersen, K. V., Nielsen, P. R., Bech, L. M., and Poulsen, F. M. (1996) *Protein Sci.* **5**, 13–23
39. Banaszak, L., Sharrock, W., and Timmins, P. (1991) *Ann. Rev. Biophys. Biophys. Chem.* **20**, 221–246
40. Deckelbaum, R. J., Ramakrishnan, R., Eisenberg, S., Olivecrona, T., and Bengtsson-Olivecrona, G. (1992) *Biochemistry* **31**, 8544–8551
41. Brunzell, J. D. (1995) in *The Metabolic and Molecular Bases of Inherited Disease*, 7th ed., Vol. 2 (Scriver, C. R., Beaudet, A. L., Sly, W. S., and Valle, D., eds), pp. 1913–1932, McGraw-Hill, New York
42. van Tilbeurgh, H., Egloff, M.-P., Martinez, C., Rugani, N., Verger, R., and Cambillau, C. (1993) *Nature (London)* **362**, 814–820

42a. Borén, J., Lookene, A., Makoveichuk, E., Xiang, S., Gustafsson, M., Liu, H., Talmud, P., and Olivecrona, G. (2001) *J. Biol. Chem.* **276**, 26916–26922

42b. Hime, N. J., Barter, P. J., and Rye, K.-A. (1998) *J. Biol. Chem.* **273**, 27191–27198

43. Lookene, A., Savonen, R., and Olivecrona, G. (1997) *Biochemistry* **36**, 5267–5275
44. Dichek, H. L., Fojo, S. S., Beg, O. U., Skarlatos, S. I., Brunzell, J. D., Cutler, G. B., Jr., and Brewer, H. B., Jr. (1991) *J. Biol. Chem.* **266**, 473–477
45. Ma, Y., Bruin, T., Tuzgol, S., Wilson, B. I., Roederer, G., Liu, M.-S., Davignon, J., Kastelein, J. J. P., Brunzell, J. D., and Hayden, M. R. (1992) *J. Biol. Chem.* **267**, 1918–1923
46. Dugi, K. A., Dichek, H. L., Talley, G. D., Brewer, J., HB, and Santamarina-Fojo, S. (1992) *J. Biol. Chem.* **267**, 25086–25091
47. Mamo, J. C. L., Elsegood, C. L., Gennat, H. C., and Yu, K. (1996) *Biochemistry* **35**, 10210–10214
48. Strickland, D. K., Kounnas, M. Z., and Argraves, W. S. (1995) *FASEB J.* **9**, 890–898
49. Sakai, J., Hoshino, A., Takahashi, S., Miura, Y., Ishii, H., Suzuki, H., Kawarabayasi, Y., and Yamamoto, T. (1994) *J. Biol. Chem.* **269**, 2173–2182
50. Willnow, T. E., Sheng, Z., Ishibashi, S., and Herz, J. (1994) *Science* **264**, 1471–1474
51. Hajjar, D. P., and Haberland, M. E. (1997) *J. Biol. Chem.* **272**, 22975–22978
52. Suzuki, K., Yamada, T., and Tanaka, T. (1999) *Biochemistry* **38**, 1751–1756
53. Krieger, M. (1998) *Proc. Natl. Acad. Sci. U.S.A.* **95**, 4077–4080

53a. Bhattacharya, A. A., Grüne, T., and Curry, S. (2000) *J. Mol. Biol.* **303**, 721–732

53b. Saltiel, A. R. (2000) *Proc. Natl. Acad. Sci. U.S.A.* **97**, 535–537

53c. De Simone, G., Galdiero, S., Manco, G., Lang, D., Rossi, M., and Pedone, C. (2000) *J. Mol. Biol.* **303**, 761–771

54. Holak, T. A., Kearsley, S. K., Kim, Y., and Prestegard, J. H. (1988) *Biochemistry* **27**, 6135–6142
55. Heath, R. J., and Rock, C. O. (1995) *J. Biol. Chem.* **270**, 15531–15538
56. Huang, W., Jia, J., Edwards, P., Dehesh, K., Schneider, G., and Lindqvist, Y. 1998) *EMBO J.* **17**, 1183–1191
57. Heath, R. J., and Rock, C. O. (1995) *J. Biol. Chem.* **270**, 26538–26542
58. Joshi, A. K., Witkowski, A., and Smith, S. (1998) *Biochemistry* **37**, 2515–2523

58a. Zhang, Y.-M., Rao, M. S., Heath, R. J., Price, A. C., Olson, A. J., Rock, C. O., and White, S. W. (2001) *J. Biol. Chem.* **276**, 8231–8238

58b. Scarsdale, J. N., Kazanina, G., He, X., Reynolds, K. A., and Wright, H. T. (2001) *J. Biol. Chem.* **276**, 20516–20522

58c. Qiu, X., Janson, C. A., Smith, W. W., Head, M., Lonsdale, J., and Konstantinidis, A. K. (2001) *J. Mol. Biol.* **307**, 341–356

References

58d. McGuire, K. A., Siggaard-Andersen, M., Bangera, M. G., Olsen, J. G., and von Wettstein-Knowles, P. (2001) *Biochemistry* **40**, 9836–9845
58e. Moche, M., Dehesh, K., Edwards, P., and Lindqvist, Y. (2001) *J. Mol. Biol.* **305**, 491–503
59. Jayakumar, A., Tai, M.-H., Huang, W.-Y., Al-Feel, W., Hsu, M., Abu-Elheiga, L., Chirala, S. S., and Wakil, S. J. (1995) *Proc. Natl. Acad. Sci. U.S.A.* **92**, 8695–8699
59a. Chirala, S. S., Jayakumar, A., Gu, Z.-W., and Wakil, S. J. (2001) *Proc. Natl. Acad. Sci. U.S.A.* **98**, 3104–3108
59b. Witkowski, A., Joshi, A. K., and Smith, S. (2002) *Biochemistry* **41**, 10877–10887
60. McCarthy, A. D., and Hardie, D. G. (1984) *Trends Biochem. Sci.* **9**, 60–63
61. Joshi, A. K., Witkowski, A., and Smith, S. (1997) *Biochemistry* **36**, 2316–2322
62. Witkowski, A., Joshi, A. K., Rangan, V. S., Falick, A. M., Witkowska, H. E., and Smith, S. (1999) *J. Biol. Chem.* **274**, 11557–11563
63. Wakil, S. J. (1989) *Biochemistry* **28**, 4523–4530
64. Jayakumar, A., Chirala, S. S., and Wakil, S. J. (1997) *Proc. Natl. Acad. Sci. U.S.A.* **94**, 12326–12330
65. Smith, S. (1994) *FASEB J.* **8**, 1248–1259
66. Wakil, S. J., Stoops, J. K., and Joshi, V. C. (1983) *Ann. Rev. Biochem.* **52**, 537–579
67. Kolodziej, S. J., Penczek, P. A., Schroeter, J. P., and Stoops, J. K. (1996) *J. Biol. Chem.* **271**, 28422–28429
68. Orme, T. W., McIntyre, J., Lynen, F., Kühn, L., and Schweizer, E. (1972) *Eur. J. Biochem.* **24**, 407–415
69. Poulose, A. J., Rogers, L., Cheesbrough, T. M., and Kolattukudy, P. E. (1985) *J. Biol. Chem.* **260**, 15953–15958
70. Witkowska, H. F., Green, B. N., and Smith, S. (1990) *J. Biol. Chem.* **265**, 5662–5665
71. Smith, S., and Abraham, S. (1975) *Adv. Lipid Res.* **13**, 195–239
72. Stumpf, P. K. (1981) *Trends Biochem. Sci.* **6**, 173–176
72a. Murakami, Y., Tsuyama, M., Kobayashi, Y., Kodama, H., and Iba, K. (2000) *Science* **287**, 476–479
73. Ohlrogge, J. B., Jaworski, J. G., and Post-Beittenmiller, D. (1993) in *Lipid Metabolism in Plants* (Moore, T. S., Jr., ed), pp. 3–32, CRC Press, Boca Raton, Florida
74. Heinz, E. (1993) in *Lipid Metabolism in Plants* (Moore, T. S., Jr., ed), pp. 33–90, CRC Press, Boca Raton, Florida
75. Stumpf, P. K. (1984) in *Fatty Acid Metabolism and its Regulation: New Comprehensive Biochemistry*, Vol. 7 (Numa, S., ed.), pp. 155–179, Elsevier, Amsterdam
75a. Lyle, K. S., Möenne-Loccoz, P., Ai, J., Sanders-Loehr, J., Loehr, T. M., and Fox, B. G. (2000) *Biochemistry* **39**, 10507–10513
76. Choi, J.-Y., and Martin, C. E. (1999) *J. Biol. Chem.* **274**, 4671–4683
77. Watkins, P. A., Lu, J.-F., Steinberg, S. J., Gould, S. J., Smith, K. D., and Braiterman, L. T. (1998) *J. Biol. Chem.* **273**, 18210–18219
78. Whereat, A. F., Orishimo, M. W., Nelson, J., and Phillips, S. J. (1969) *J. Biol. Chem.* **244**, 6498–6506
78a. Das, A. K., Uhler, M. D., and Hajra, A. K. (2000) *J. Biol. Chem.* **275**, 24333–24340
79. Lin, C. Y., and Kumar, S. (1971) *J. Biol. Chem.* **246**, 3284–3290
80. Varanasi, U., Feldman, H. R., and Malins, D. C. (1975) *Nature (London)* **255**, 340–343
81. Kolattukudy, P. E. (1968) *Science* **159**, 498–505
82. Dillwith, J. W., Nelson, J. H., Pomonis, J. G., Nelson, D. R., and Blomquist, G. J. (1982) *J. Biol. Chem.* **257**, 11305–11314
83. Odham, G., and Stenhagen, E. (1971) *Acc. Chem. Res.* **4**, 121–128
84. Cole, S. T., and and 41 other authors. (1998) *Nature (London)* **393**, 537–544
85. Azad, A. K., Sirakova, T. D., Fernandes, N. D., and Kolattukudy, P. E. (1997) *J. Biol. Chem.* **272**, 16741–16745
86. Kroumova, A. B., Xie, Z., and Wagner, G. J. (1994) *Proc. Natl. Acad. Sci. U.S.A.* **91**, 11437–11441
87. Howell, D. M., Harich, K., Xu, H., and White, R. H. (1998) *Biochemistry* **37**, 10108–10117
88. Hayden, M. A., Huang, I. Y., Iliopoulos, G., Orozco, M., and Ashley, G. W. (1993) *Biochemistry* **32**, 3778–3782
89. Sulo, P., and Martin, N. C. (1993) *J. Biol. Chem.* **268**, 17634–17639
90. Wada, H., Shintani, D., and Ohlrogge, J. (1997) *Proc. Natl. Acad. Sci. U.S.A.* **94**, 1591–1596
91. Busby, R. W., Schelvis, J. P. M., Yu, D. S., Babcock, G. T., and Marletta, M. A. (1999) *J. Am. Chem. Soc.* **121**, 4706–4707
91a. Miller, J. R., Busby, R. W., Jordan, S. W., Cheek, J., Henshaw, T. F., Ashley, G. W., Broderick, J. B., Cronan, J. E., Jr., and Marletta, M. A. (2000) *Biochemistry* **39**, 15166–15178
91b. Gueguent, V., Macherel, D., Jaquinod, M., Douce, R., and Bourguignon, J. (2000) *J. Biol. Chem.* **275**, 5016–5025
92. Duin, E. C., Lafferty, M. E., Crouse, B. R., Allen, R. M., Sanyal, I., Flint, D. H., and Johnson, M. K. (1997) *Biochemistry* **36**, 11811–11820
93. Hayden, M. A., Huang, I., Bussiere, D. E., and Ashley, G. W. (1992) *J. Biol. Chem.* **267**, 9512–9515
93a. Ugulava, N. B., Sacanell, C. J., and Jarrett, J. T. (2001) *Biochemistry* **40**, 8352–8358
94. White, R. H. (1987) *Biochemistry* **26**, 3163–3167
95. deMendoza, D., and Cronan, J. E., Jr. (1983) *Trends Biochem. Sci.* **8**, 49–52
95a. Wallis, J. G., Watts, J. L., and Browse, J. (2002) *Trends Biochem. Sci.* **27**, 467–473
96. Henderson, B. S., Larsen, B. S., and Schwab, J. M. (1994) *J. Am. Chem. Soc.* **116**, 5025–5034
97. Roughan, G., and Slack, R. (1984) *Trends Biochem. Sci.* **9**, 383–386
98. Sperling, P., Linscheid, M., Stöcker, S., Mühlback, H.-P., and Heinz, E. (1993) *J. Biol. Chem.* **268**, 26935–26940
99. Jeffcoat, R., and James, A. T. (1984) in *Fatty Acid Metabolism and its Regulation: New Comprehensive Biochemistry*, Vol. 7 (Numa, S., ed), pp. 85–112, Elsevier, Amsterdam
100. Thiede, M. A., and Strittmatter, P. (1985) *J. Biol. Chem.* **260**, 14459–14463
101. Shanklin, J., Whittle, E., and Fox, B. G. (1994) *Biochemistry* **33**, 12787–12794
102. Buist, P. H., and Behrouzian, B. (1998) *J. Am. Chem. Soc.* **120**, 871–876
103. Vergères, G., Ramsden, J., and Waskell, L. (1995) *J. Biol. Chem.* **270**, 3414–3422
104. Mitchell, A. G., and Martin, C. E. (1995) *J. Biol. Chem.* **270**, 29766–29772
105. McKeon, T. A., and Stumpf, P. K. (1982) *J. Biol. Chem.* **257**, 12141–12147
106. Somerville, C., and Browse, J. (1991) *Science* **252**, 80–87
107. Töpfer, R., Martini, N., and Schell, J. (1995) *Science* **268**, 681–686
108. Browse, J., McConn, M., James, D., Jr., and Miquel, M. (1993) *J. Biol. Chem.* **268**, 16345–16351
109. Lindqvist, Y., Huang, W., Schneider, G., and Shanklin, J. (1996) *EMBO J.* **15**, 4081–4092
110. Fox, B. G., Shanklin, J., Ai, J., Loehr, T. M., and Sanders-Loehr, J. (1994) *Biochemistry* **33**, 12776–12786
111. Spychalla, J. P., Kinney, A. J., and Browse, J. (1997) *Proc. Natl. Acad. Sci. U.S.A.* **94**, 1142–1147
112. Mustardy, L., Los, D. A., Gombos, Z., and Murata, N. (1996) *Proc. Natl. Acad. Sci. U.S.A.* **93**, 10524–10527
113. Broun, P., Shanklin, J., Whittle, E., and Somerville, C. (1998) *Science* **282**, 1315–1317
114. van de Loo, F. J., Broun, P., Turner, S., and Somerville, C. (1995) *Proc. Natl. Acad. Sci. U.S.A.* **92**, 6743–6747
115. Kaya, K., Ramesha, C. S., and Thompson, G. A., Jr. (1984) *J. Biol. Chem.* **259**, 3548–3553
116. Ortiz de Montellano, P. R., Chan, W. K., Tuck, S. F., Kaikaus, R. M., Bass, N. M., and Peterson, J. A. (1992) *FASEB J.* **6**, 695–699
117. Vaz, A. D. N., McGinnity, D. F., and Coon, M. J. (1998) *Proc. Natl. Acad. Sci. U.S.A.* **95**, 3555–3560
118. de Renobales, M., Cripps, C., Stanley-Samuelson, D. W., Jurenka, R. A., and Blomquist, G. J. (1987) *Trends Biochem. Sci.* **12**, 364–366
119. Cho, H. P., Nakamura, M. T., and Clarke, S. D. (1999) *J. Biol. Chem.* **274**, 471–477
119a. Bernoud-Hubac, N., Davies, S. S., Boutaud, O., Montine, T. J., and Roberts, L. J., II. (2001) *J. Biol. Chem.* **276**, 30964–30970
119b. Mirnikjoo, B., Brown, S. E., Kim, H. F. S., Marangell, L. B., Sweatt, J. D., and Weeber, E. J. (2001) *J. Biol. Chem.* **276**, 10888–10896
119c. Brzustowicz, M. R., Cherezov, V., Zerouga, M., Caffrey, M., Stillwell, W., and Wassall, S. R. (2002) *Biochemistry* **41**, 12509–12519
119d. Albert, C. M., Campos, H., Stampfer, M. J., Ridker, P. M., Manson, J. E., Willett, W. C., and Ma, J. (2002) *M. Engl. J. Med.* **346**, 1113–1118 (and discussion in Vol. 347, pp. 531–533)
120. Knipple, D. C., Rosenfield, C.-L., Miller, S. J., Liu, W., Tang, J., Ma, P. W. K., and Roelofs, W. L. (1998) *Proc. Natl. Acad. Sci. U.S.A.* **95**, 15287–15292
121. Aveldaño, M. I., and Sprecher, H. (1997) *J. Biol. Chem.* **262**, 1180–1186
121a. Cahoon, E. B., Ripp, K. G., Hall, S. E., and Kinney, A. J. (2001) *J. Biol. Chem.* **276**, 2637–2643
121b. Adlof, R. O., Duval, S., and Emken, E. A. (2000) *Lipids* **35**, 131–135
121c. Majumder, B., Wahle, K. W. J., Moir, S., Schofield, A., Choe, S.-N., Farquharson, A., Grant, I., and Heys, S. D. (2002) *FASEB J.* **16**, 1447-1449
122. Wise, M. L., Hamberg, M., and Gerwick, W. H. (1994) *Biochemistry* **33**, 15223–15232
123. Wise, M. L., Rossi, J., and Gerwick, W. H. (1997) *Biochemistry* **36**, 2985–2992
124. Cox, G. S., Thomas, E., Kaback, H. R., and Weissbach, H. (1973) *Arch. Biochem. Biophys.* **158**, 667–676
125. Wang, A.-Y., Grogan, D. W., and Cronan, J. E., Jr. (1992) *Biochemistry* **31**, 11020–11028
126. Packter, N. M. (1973) *Biosynthesis of Acetate-Derived Compounds*, Wiley, New York
126a. Glickman, M. S., Cahill, S. M., and Jacobs, W. R., Jr. (2001) *J. Biol. Chem.* **276**, 2228–2233
127. Yuan, Y., Crane, D. C., Musser, J. M., Sreevatsan, S., and Barry, C. E., III. (1997) *J. Biol. Chem.* **272**, 10041–10049
128. Yuan, Y., Lee, R. E., Besra, G. S., Belisle, J. T., and Barry, C. E., III. (1995) *Proc. Natl. Acad. Sci. U.S.A.* **92**, 6630–6634
129. Liu, J., and Nikaido, H. (1999) *Proc. Natl. Acad. Sci. U.S.A.* **96**, 4011–4016
130. Tipton, C. L., and Al-Shathir, N. M. (1974) *J. Biol. Chem.* **249**, 886–889
131. Bohlmann, F., Burkhardt, T., and Zdero, C. (1973) *Naturally Occuring Acetylenes*, Academic Press, New York

References

132. Nicolaides, N. (1974) *Science* **186**, 19–26
133. Bauermeister, A., and Sargent, J. R. (1979) *Trends Biochem. Sci.* **4**, 209–211
134. Nemes, Z., Marekov, L. N., Fésüs, L., and Steinert, P. M. (1999) *Proc. Natl. Acad. Sci. U.S.A.* **96**, 8402–8407
135. Kolattukudy, P. E. (1980) *Science* **208**, 990–1000
135a. Wellesen, K., Durst, F., Pinot, F., Benveniste, I., Nettesheim, K., Wisman, E., Steiner-Lange, S., Saedler, H., and Yephremov, A. (2001) *Proc. Natl. Acad. Sci. U.S.A.* **98**, 9694–9699
136. von Wettstein-Knowles, P. M. (1993) in *Lipid Metabolism in Plants* (Moore, T. S., Jr., ed), pp. 127–166, CRC Press, Boca Raton, Florida
137. Dennis, M., and Kolattukudy, P. E. (1992) *Proc. Natl. Acad. Sci. U.S.A.* **89**, 5306–5310
138. Cheesbrough, T. M., and Kolattukudy, P. E. (1984) *Proc. Natl. Acad. Sci. U.S.A.* **81**, 6613–6617
139. Cheesbrough, T. M., and Kolattukudy, P. E. (1988) *J. Biol. Chem.* **263**, 2738–2743
140. Reed, J. R., Vanderwel, D., Choi, S., Pomonis, J. G., Reitz, R. C., and Blomquist, G. J. (1994) *Proc. Natl. Acad. Sci. U.S.A.* **91**, 10000–10004
141. Gorgen, G., and Boland, W. (1989) *Eur. J. Biochem.* **185**, 237–242
142. Buckner, J. S., Nelson, D. R., Fatland, C. L., Hakk, H., and Pomonis, J. G. (1984) *J. Biol. Chem.* **259**, 8461–8470
143. Morse, D., and Meighen, E. (1984) *Science* **226**, 1434–1436
144. Hölldobler, B. (1995) *Proc. Natl. Acad. Sci. U.S.A.* **92**, 19–22
145. Prestwich, G. D. (1987) *Science* **237**, 999 – 1006
146. Dussourd, D. E., Harvis, C. A., Meinwald, J., and Eisner, T. (1991) *Proc. Natl. Acad. Sci. U.S.A.* **88**, 9224–9227
146a. Athenstaedt, K., and Daum, G. (2000) *J. Biol. Chem.* **275**, 235–240
146b. Kalhan, S. C., Mahajan, S., Burkett, E., Reshef, L., and Hanson, R. W. (2001) *J. Biol. Chem.* **276**, 12928–12931
147. Rock, C. O., Goelz, S. E., and Cronan, J. E., Jr. (1981) *J. Biol. Chem.* **256**, 736–742
148. Eberhardt, C., Gray, P. W., and Tjoelker, L. W. (1997) *J. Biol. Chem.* **272**, 20299–20305
149. Guo, Z., Liliom, K., Fischer, D. J., Bathurst, I. C., Tomei, L. D., Kiefer, M. C., and Tigyi, A. (1996) *Proc. Natl. Acad. Sci. U.S.A.* **93**, 14367–14372
150. Kai, M., Wada, I., Imai, S.-i, Sakane, F., and Kanoh, H. (1997) *J. Biol. Chem.* **272**, 24572–24578
150a. Buhman, K. K., Chen, H. C., and Farese, R. V., Jr. (2001) *J. Biol. Chem.* **276**, 40369–40372
150b. Abo-Hashema, K. A. H., Cake, M. H., Power, G. W., and Clarke, D. (1999) *J. Biol. Chem.* **274**, 35577–35582
151. Shen, H., Heacock, P. N., Clancey, C. J., and Dowhan, W. (1996) *J. Biol. Chem.* **271**, 789–795
152. Badola, P., and Sanders, C. R., II. (1997) *J. Biol. Chem.* **272**, 24176–24182
153. Maréchal, E., Block, M. A., Joyard, J., and Douce, R. (1994) *J. Biol. Chem.* **269**, 5788–5798
154. Joyard, J., Block, M. A., Malherbe, A., Maréchal, E., and Douce, R. (1993) in *Lipid Metabolism in Plants* (Moore, T. S., Jr., ed), pp. 231–258, CRC Press, Boca Raton, Florida
155. Dörmann, P., Balbo, I., and Benning, C. (1999) *Science* **284**, 2181–2184
156. Karlsson, O. P., Rytömaa, M., Dahlqvist, A., Kinnunen, P. K. J., and Wieslander, A. (1996) *Biochemistry* **35**, 10094–10102
156a. Abu-Elheiga, L., Matzuk, M. M., Abo-Hashema, K. A. H., and Wakil, S. J. (2001) *Science* **291**, 2613–2616
156b. Jelenska, J., Crawford, M. J., Harb, O. S., Zuther, E., Haselkorn, R., Roos, D. S., and Gornicki, P. (2001) *Proc. Natl. Acad. Sci. U.S.A.* **98**, 2723–2728
157. Thompson, G. A., Jr. (1992) *The Regulation of Membrane Lipid Metabolism*, 2nd ed., CRC Press, Boca Raton, Florida
158. Lopez, J. M., Bennett, M. K., Sanchez, H. B., Rosenfeld, J. M., and Osborne, T. F. (1996) *Proc. Natl. Acad. Sci. U.S.A.* **93**, 1049–1053
159. Honnor, R. C., Dhillon, G. S., and Londos, C. (1985) *J. Biol. Chem.* **260**, 15130–15138
160. Tabor, D. E., Kim, J. B., Spiegelman, B. M., and Edwards, P. A. (1999) *J. Biol. Chem.* **274**, 20603–20610
161. Soncini, M., Yet, S.-F., Moon, Y., Chun, J.-Y., and Sul, H. S. (1995) *J. Biol. Chem.* **270**, 30339–30343
162. Dudek, S. M., and Semenkovich, C. F. (1995) *J. Biol. Chem.* **270**, 29323–29329
163. Tiku, P. E., Gracey, A. Y., Macartney, A. I., Beynon, R. J., and Cossins, A. R. (1996) *Science* **271**, 815–818
164. Jackson, B. J., Gennity, J. M., and Kennedy, E. P. (1986) *J. Biol. Chem.* **261**, 13464–13468
165. Raetz, C. R. H., and Dowhan, W. (1990) *J. Biol. Chem.* **265**, 1235–1238
166. Kennedy, E. P. (1992) *Ann. Rev. Biochem.* **61**, 1–28
167. Bae-Lee, M. S., and Carman, G. M. (1984) *J. Biol. Chem.* **259**, 10857–10862
168. Kawasaki, K., Kuge, O., Chang, S.-C., Heacock, P. N., Rho, M., Suzuki, K., Nishijima, M., and Dowhan, W. (1999) *J. Biol. Chem.* **274**, 1828–1834
168a. Auer, M., Kim, M. J., Lemieux, M. J., Villa, A., Song, J., Li, X.-D., and Wang, D.-N. (2001) *Biochemistry* **40**, 6628–6635
169. Louie, K., and Dowhan, W. (1980) *J. Biol. Chem.* **255**, 1124–1127
169a. Stone, S. J., and Vance, J. E. (2000) *J. Biol. Chem.* **275**, 34534–34540
170. Trotter, P. J., and Voelker, D. R. (1995) *J. Biol. Chem.* **270**, 6062–6070
171. Shiao, Y.-J., Lupo, G., and Vance, J. E. (1995) *J. Biol. Chem.* **270**, 11190–11198
172. Heikinheimo, L., and Somerharju, P. (1998) *J. Biol. Chem.* **273**, 3327–3335
173. Huijbregts, R. P. H., de Kroon, A. I. P. M., and de Kruijff, B. (1998) *J. Biol. Chem.* **273**, 18936–18942
174. Rietveld, A. G., Chupin, V. V., Koorengevel, M. C., Wienk, H. L. J., Dowhan, W., and de Kruijff, B. (1994) *J. Biol. Chem.* **269**, 28670–28675
175. Bogdanov, M., Umeda, M., and Dowhan, W. (1999) *J. Biol. Chem.* **274**, 12339–12345
176. Ohtsuka, T., Nishijima, M., and Akamatsu, Y. (1993) *J. Biol. Chem.* **268**, 22908–22911
177. Zhao, M., Schlame, M., Rua, D., and Greenberg, M. L. (1998) *J. Biol. Chem.* **273**, 2402–2408
178. Chang, S.-C., Heacock, P. N., Mileykovskaya, E., Voelker, D. R., and Dowhan, W. (1998) *J. Biol. Chem.* **273**, 14933–14941
179. McMaster, C. R., and Bell, R. M. (1994) *J. Biol. Chem.* **269**, 14776–14783
180. Kent, C., and Carman, G. M. (1999) *Trends Biochem. Sci.* **24**, 146–150
180a. Dowd, S. R., Bier, M. E., and Patton-Vogt, J. L. (2001) *J. Biol. Chem.* **276**, 3756–3763
180b. Lykidis, A., Wang, J., Karim, M. A., and Jackowski, S. (2001) *J. Biol. Chem.* **276**, 2174–2179
181. Vance, D. E., and Pelech, S. L. (1984) *Trends Biochem. Sci.* **9**, 17–20
181a. Henneberry, A. L., Wistow, G., and McMaster, C. R. (2000) *J. Biol. Chem.* **275**, 29808–29815
181b. Cornell, R. B., and Northwood, I. C. (2000) *Trends Biochem. Sci.* **25**, 441–447
181c. Hunt, A. N., Clark, G. T., Attard, G. S., and Postle, A. D. (2001) *J. Biol. Chem.* **276**, 8492–8499
182. Kuge, O., Saito, K., and Nishijima, M. (1997) *J. Biol. Chem.* **272**, 19133–19139
183. Vance, J. E. (1998) *Trends Biochem. Sci.* **23**, 423–428
184. Zeisel, S. H., da Costa, K.-A., Franklin, P. D., Alexander, E. A., LaMont, J. T., Sheard, N. F., and Beiser, A. (1991) *FASEB J.* **5**, 2093–2098
185. Lee, H.-C., Fellenz-Maloney, M.-P., Liscovitch, M., and Krzysztof Blusztajn, J. (1993) *Proc. Natl. Acad. Sci. U.S.A.* **90**, 10086–10090
186. Nikawa, J.-i, Kodaki, T., and Yamashita, S. (1987) *J. Biol. Chem.* **2 262**, 4876–4881
186a. Ilgoutz, S. C., Zawadzki, J. L., Ralton, J. E., and McConville, M. J. (1999) *EMBO J.* **18**, 2746–2755
187. Wu, W.-I., and Carman, G. M. (1996) *Biochemistry* **35**, 3790–3796
188. Griac, P., Swede, M. J., and Henry, S. A. (1996) *J. Biol. Chem.* **271**, 25692–25698
189. Kim, K.-H., Voelker, D. R., Flocco, M. T., and Carman, G. M. (1998) *J. Biol. Chem.* **273**, 6844–6852
190. Ostrander, D. B., O'Brien, D. J., Gorman, J. A., and Carman, G. M. (1998) *J. Biol. Chem.* **273**, 18992–19001
191. Wieprecht, M., Wieder, T., Paul, C., Geilen, C. C., and Orfanos, C. E. (1996) *J. Biol. Chem.* **271**, 9955–9961
192. Snyder, F., ed. (1972) *Ether Lipids: Chemistry and Biology*, Academic Press, New York
193. Lee, T.-c, Malone, B., and Snyder, F. (1988) *J. Biol. Chem.* **263**, 1755–1760
194. Karasawa, K., Qiu, X., and Lee, T.-c. (1999) *J. Biol. Chem.* **274**, 8655–8661
194a. Min, J.-H., Wilder, C., Aoki, J., Arai, H., Inoue, K., Paul, L., and Gelb, M. H. (2001) *Biochemistry* **40**, 4539–4549
195. Ford, D. A., Rosenbloom, K. B., and Gross, R. W. (1992) *JBC* **267**, 11222–11228
196. Langworthy, T. A. (1985) in *The Bacteria* (Woese, C. R., and Wolfe, R. S., eds), pp. 459–497, Academic Press, New York
197. Jones, W. J., Nagle, D. P., Jr., and Whitman, W. B. (1989) *Microbiol. Rev.* **51**, 135–177
198. DeRosa, M. A., Gambacorta, A., and Gliozzi, A. (1986) *Microbiol. Rev.* **50**, 70–80
199. Burdett, K., Larkins, L. K., Das, A. K., and Hajra, A. K. (1991) *J. Biol. Chem.* **266**, 12201–12206
200. Friedberg, S. J., Weintraub, S. T., Singer, M. R., and Greene, R. C. (1982) *J. Biol. Chem.* **258**, 136–142
201. Brown, A. J., and Snyder, F. (1983) *J. Biol. Chem.* **258**, 4184–4189
202. Paltauf, F., and Holasek, A. (1973) *J. Biol. Chem.* **248**, 1609–1615
203. Borst, P. (1983) *Trends Biochem. Sci.* **8**, 269–272
204. de Vet, E. C. J. M., Ijlst, L., Oostheim, W., Wanders, R. J. A., and van den Bosch, H. (1998) *J. Biol. Chem.* **273**, 10296–10301
205. Moser, A. E., Singh, I., Brown, F. R., Solish, G. I., Kelley, R. I., Benke, P. J., and Moser, H. W. (1984) *N. Engl. J. Med.* **310**, 1141–1146
206. Lazarow, P. B., and Moser, H. W. (1995) in *The Metabolic and Molecular Bases of Inherited Disease*, 7th ed., Vol. 2 (Scriver, C. R., Beaudet, A. L., Sly, W. S., and Valle, D., eds), pp. 2287–2324, McGraw-Hill, New York
207. Chao, W., Liu, H., Hanahan, D. J., and Olson, M. S. (1992) *J. Biol. Chem.* **267**, 6725–6735
208. Lynch, D. V. (1993) in *Lipid Metabolism in Plants* (Moore, T. S., Jr., ed), pp. 285–308, CRC Press, Boca Raton, Florida

References

209. Barenholz, Y., and Gatt, S. (1982) in *New Comprehensive Biochemistry*, Vol. 4 (Hawthorne, J. N., and Ansell, G. B., eds), pp. 129–177, Elsevier, Amsterdam

209a. Ikushiro, H., Hayashi, H., and Kagamiyama, H. (2001) *J. Biol. Chem.* **276**, 18249–18256

210. Stoffel, W., and Melzner, I. (1980) *Z. Physiol. Chem.* **361**, 755–771

211. Grilley, M. M., Stock, S. D., Dickson, R. C., Lester, R. L., and Takemoto, J. Y. (1998) *J. Biol. Chem.* **273**, 11062–11068

212. Sperling, P., Zähringer, U., and Heinz, E. (1998) *J. Biol. Chem.* **273**, 28590–28596

213. Toledo, M. S., Levery, S. B., Straus, A. H., Suzuki, E., Momany, M., Glushka, J., Moulton, J. M., and Takahashi, H. K. (1999) *Biochemistry* **38**, 7294–7306

213a. Savile, C. K., Fabriàs, G., and Buist, P. H. (2001) *J. Am. Chem. Soc.* **123**, 4382–4385

214. Yamakawa, T. (1988) *Trends Biochem. Sci.* **13**, 452–454

215. van Echten, G., and Sandhoff, K. (1993) *J. Biol. Chem.* **268**, 5341–5344

216. Wiegandt, H. (1971) *Adv. Lipid Res.* **9**, 249–289

217. Hakomori, S.-i. (1990) *J. Biol. Chem.* **265**, 18713–18716

218. Takamiya, K., Yamamoto, A., Furukawa, K., Zhao, J., Fukumoto, S., Yamashiro, S., Okada, M., Haraguchi, M., Shin, M., Kishikawa, M., Shiku, H., Aizawa, S., and Furukawa, K. (1998) *Proc. Natl. Acad. Sci. U.S.A.* **95**, 12147–12152

218a. Mylvaganam, M., and Lingwood, C. A. (1999) *J. Biol. Chem.* **274**, 20725–20732

218b. Uragami, M., Tokutake, N., Yan, X., and Regen, S. L. (2001) *J. Am. Chem. Soc.* **123**, 5124–5125

218c. Brown, D. A., and London, E. (2000) *J. Biol. Chem.* **275**, 17221–17224

219. Rameh, L. E., and Cantley, L. C. (1999) *J. Biol. Chem.* **274**, 8347–8350

220. Anderson, R. A., Boronenkov, I. V., Doughman, S. D., Kunz, J., and Loijens, J. C. (1999) *J. Biol. Chem.* **274**, 9907–9910

221. Rhee, S. G., and Bae, Y. S. (1997) *J. Biol. Chem.* **272**, 15045–15048

222. Heinz, D. W., Ryan, M., Bullock, T. L., and Griffith, O. H. (1995) *EMBO J.* **14**, 3855–3863

223. Essen, L.-O., Perisic, O., Katan, M., Wu, Y., Roberts, M. F., and Williams, R. L. (1997) *Biochemistry* **36**, 1704–1718

224. Hondal, R. J., Zhao, Z., Kravchuk, A. V., Liao, H., Riddle, S. R., Yue, X., Bruzik, K. S., and Tsai, M.-D. (1998) *Biochemistry* **37**, 4568–4580

225. Majerus, P. W., Kisseleva, M. V., and Norris, F. A. (1999) *J. Biol. Chem.* **274**, 10669–10672

226. Newton, A. C. (1995) *J. Biol. Chem.* **270**, 28495–28498

227. Newton, A. C., and Johnson, J. E. (1998) *Biochim. Biophys. Acta.* **1376**, 155–172

228. Epand, R. M., Stevenson, C., Bruins, R., Schram, V., and Glaser, M. (1998) *Biochemistry* **37**, 12068–12073

228a. McIntyre, T. M., Zimmerman, G. A., and Prescott, S. M. (1999) *J. Biol. Chem.* **274**, 25189–25192

229. Hodgkin, M. N., Pettitt, T. R., Martin, A., Michell, R. H., Pemberton, A. J., and Wakelam, M. J. O. (1998) *Trends Biochem. Sci.* **23**, 200–204

230. Topham, M. K., and Prescott, S. M. (1999) *J. Biol. Chem.* **274**, 11447–11450

230a. Jones, D. R., Pettitt, T. R., Sanjuán, M. A., Mérida, I., and Wakelam, M. J. O. (1999) *J. Biol. Chem.* **274**, 16846–16852

231. Walsh, J. P., Suen, R., and Glomset, J. A. (1995) *J. Biol. Chem.* **270**, 28647–28653

232. Schissel, S. L., Keesler, G. A., Schuchman, E. H., Williams, K. J., and Tabas, I. (1998) *J. Biol. Chem.* **273**, 18250–18259

233. Testi, R. (1996) *Trends Biochem. Sci.* **21**, 468–471

234. Spiegel, S., and Merrill, A. H., Jr. (1996) *FASEB J.* **10**, 1388–1397

235. Hannun, Y. A., and Obeid, L. M. (1995) *Trends Biochem. Sci.* **20**, 73–77

236. Usta, J., El Bawab, S., Roddy, P., Szulc, Z. M., Hannun, Y. A., and Bielawska, A. (2001) *Biochemistry* **40**, 9657–9668

236a. Hannun, Y. A., Luberto, C., and Argraves, K. M. (2001) *Biochemistry* **40**, 4893–4903

237. Mandala, S. M., Thornton, R., Tu, Z., Kurtz, M. B., Nickels, J., Broach, J., Menzeleev, R., and Spiegel, S. (1998) *Proc. Natl. Acad. Sci. U.S.A.* **95**, 150–155

237a. Liu, Y.-Y., Han, T.-Y., Giuliano, A. E., and Cabot, M. C. (2001) *FASEB J.* **15**, 719–730

237b. Windh, R. T., Lee, M.-J., Hla, T., An, S., Barr, A. J., and Manning, D. R. (1999) *J. Biol. Chem.* **274**, 27351–27358

237c. Ammit, A. J., Hastie, A. T., Edsall, L. C., Hoffman, R. K., Amrani, Y., Krymskaya, V. P., Kane, S. A., Peters, S. P., Penn, R. B., Spiegel, S., and Panettieri, R. A., Jr. (2001) *FASEB J.* **15**, 1212–1214

237d. Dickson, R. C. (1998) *Ann. Rev. Biochem.* **67**, 27–48

237e. Hla, T., Lee, M.-J., Ancellin, N., Paik, J. H., and Kluk, M. J. (2001) *Science* **294**, 1875–1878

238. Westerman, J., de Vries, K.-J., Somerharju, P., Timmermans-Hereijgers, J. L. P. M., Snoek, G. T., and Wirtz, K. W. A. (1995) *J. Biol. Chem.* **270**, 14263–14266

238a. Yoder, M. D., Thomas, L. M., Tremblay, J. M., Oliver, R. L., Yarbrough, L. R., and Helmkamp, G. M., Jr. (2001) *J. Biol. Chem.* **276**, 9246–9252

239. Gutteridge, J. M. C., and Halliwell, B. (1990) *Trends Biochem. Sci.* **15**, 129–135

240. Porter, N. A., and Wujek, D. G. (1984) *J. Am. Chem. Soc.* **106**, 2626–2629

241. Wagner, B. A., Buettner, G. R., and Burns, C. P. (1994) *Biochemistry* **33**, 4449–4453

242. Al-Abed, Y., Liebich, H., Voelter, W., and Bucala, R. (1996) *J. Biol. Chem.* **271**, 2892–2896

243. Uchida, K., Shiraishi, M., Naito, Y., Torii, Y., Nakamura, Y., and Osawa, T. (1999) *J. Biol. Chem.* **274**, 2234–2242

243a. Burczynski, M. E., Sridhar, G. R., Palackal, N. T., and Penning, T. M. (2001) *J. Biol. Chem.* **276**, 2890–2897

244. Gardner, H. W., and Hamberg, M. (1993) *J. Biol. Chem.* **268**, 6971–6977

244a. Schneider, C., Tallman, K. A., Porter, N. A., and Brash, A. R. (2001) *J. Biol. Chem.* **276**, 20831–20838

245. Kato, Y., Mori, Y., Makino, Y., Morimitsu, Y., Hiroi, S., Ishikawa, T., and Osawa, T. (1999) *J. Biol. Chem.* **274**, 20406–20414

246. Tsai, L., Szweda, P. A., Vinogradova, O., and Szweda, L. I. (1998) *Proc. Natl. Acad. Sci. U.S.A.* **95**, 7975–7980

247. McGirr, L. G., Hadley, M., and Draper, H. H. (1985) *J. Biol. Chem.* **260**, 15427–15431

248. Hadley, M., and Draper, H. H. (1988) *FASEB J.* **2**, 138–140

249. Slatter, D. A., Paul, R. G., Murray, M., and Bailey, A. J. (1999) *J. Biol. Chem.* **274**, 19661–19669

250. Armstrong, D., Dimmitt, S., Boehme, D. H., Leonberg, S. C., Jr., and Vogel, W. (1974) *Science* **186**, 155–156

250a. Onorato, J. M., Jenkins, A. J., Thorpe, S. R., and Baynes, J. W. (2000) *J. Biol. Chem.* **275**, 21177–21184

251. Hill, E. G., Johnson, S. B., Lawson, L. D., Mahfouz, M. M., and Holman, R. T. (1982) *Proc. Natl. Acad. Sci. U.S.A.* **79**, 953–957

252. Holman, R. T., Pusch, F., Svingen, B., and Dutton, H. J. (1991) *Proc. Natl. Acad. Sci. U.S.A.* **88**, 4830–4834

253. Mensink, R. P., and Katan, M. B. (1990) *N. Engl. J. Med.* **323**, 439–445

253a. Kopelman, P. G. (2000) *Nature (London)* **404**, 635–643

253b. Unger, R. H., and Orci, L. (2001) *FASEB J.* **15**, 312–321

253c. Taubes, G. (2001) *Science* **291**, 2536–2545

253d. Koo, S.-H., Dutcher, A. K., and Towle, H. C. (2001) *J. Biol. Chem.* **276**, 9437–9445

253e. Weigle, D. S. (1994) *FASEB J.* **8**, 302–310

253f. Barsh, G. S., Farooqi, I. S., and O'Rahilly, S. (2000) *Nature (London)* **404**, 644–651

253g. Harris, R. B. S. (1990) *FASEB J.* **4**, 3310–3318

253h. Ruderman, N., and Flier, J. S. (2001) *Science* **291**, 2558–2559

253i. Lee, J.-J., Moon, Y.-A., Ha, J.-H., Yoon, D.-J., Ahn, Y.-H., and Kim, K.-S. (2001) *J. Biol. Chem.* **276**, 2576–2585

253j. Gura, T. (1998) *Science* **280**, 1369–1370

253k. Jucker, B. M., Ren, J., Dufour, S., Cao, X., Previs, S. F., Cadman, K. S., and Shulman, G. I. (2000) *J. Biol. Chem.* **275**, 39279–39286

253l. Clément, K., Vaisse, C., Lahlou, N., Cabrol, S., Pelloux, V., Cassuto, D., Gourmelen, M., Dina, C., Chambaz, J., Lacorte, J.-M., Basdevant, A., Bougnères, P., Lebouc, Y., Froguel, P., and Guy-Grand, B. (1998) *Nature (London)* **392**, 398–401

253m. Gura, T. (2000) *Science* **287**, 1738–1741

253n. Lee, Y., Wang, M.-Y., Kakuma, T., Wang, Z.-W., Babcock, E., McCorkle, K., Higa, M., Zhou, Y.-T., and Unger, R. H. (2001) *J. Biol. Chem.* **276**, 5629–5635

254. Makita, K., Falck, J. R., and Capdevila, J. H. (1996) *FASEB J.* **10**, 1456–1463

255. Schewe, T., and Kühn, H. (1991) *Trends Biochem. Sci.* **16**, 369–73

256. Capdevila, J. H., Falck, J. R., and Estabrook, R. W. (1992) *FASEB J.* **6**, 731–736

256a. Funk, C. D. (2001) *Science* **294**, 1871–1875

257. Bergström, S. (1967) *Science* **157**, 382–391

258. Samuelsson, B., Granström, E., Green, K., Hamberg, M., and Hammarström, S. (1975) *Ann. Rev. Biochem.* **44**, 669–695

259. Newton, R. F., and Roberts, S. M., eds. (1982) *Prostaglandins and Thromboxanes*, Butterworth Scientific, London

260. Neufeld, E. J., Bross, T. E., and Majerus, P. W. (1984) *J. Biol. Chem.* **259**, 1986–1992

261. Kramer, R. M., and Deykin, D. (1983) *J. Biol. Chem.* **258**, 13806–13811

262. Picot, D., Loll, P. J., and Garavito, R. M. (1994) *Nature (London)* **367**, 243–249

262a. Seibold, S. A., Cerda, J. F., Mulichak, A. M., Song, I., Garavito, R. M., Arakawa, T., Smith, W. L., and Babcock, G. T. (2000) *Biochemistry* **39**, 6616–6624

262b. Klefer, J. R., Pawlitz, J. L., Moreland, K. T., Stegeman, R. A., Hood, W. F., Glerse, J. K., Stevens, A. M., Goodwin, D. C., Rowlinson, S. W., Marnett, L. J., Stallings, W. C., and Kurumball, R. G. (2000) *Nature (London)* **405**, 97–101

262c. Thuresson, E. D., Malkowski, M. G., Lakkides, K. M., Rieke, C. J., Mulichak, A. M., Ginell, S. L., Garavito, R. M., and Smith, W. L. (2001) *J. Biol. Chem.* **276**, 10358–10365

262d. Peng. S., Okeley, N. M., Tsai, A.-L., Wu, G., Kulmacz, R. J., and van der Donk. W. A. (2002) *J. Am. Chem. Soc.* **124**, 10785–10796

263. Lambeir, A.-M., Markey, C. M., Dunford, H. B., and Marnett, L. J. (1985) *J. Biol. Chem.* **260**, 14894–14896

264. Tsai, A.-l, Kulmacz, R. J., and Palmer, G. (1995) *J. Biol. Chem.* **270**, 10503–10508

References

265. Smith, W. L., Garavito, R. M., and DeWitt, D. L. (1996) *J. Biol. Chem.* **271**, 33157–33160
266. Tsai, A.-l, Palmer, G., Xiao, G., Swinney, D. C., and Kulmacz, R. J. (1998) *J. Biol. Chem.* **273**, 3888–3894
266a. Marnett, L. J., Rowlinson, S. W., Goodwin, D. C., Kalgutkar, A. S., and Lanzo, C. A. (1999) *J. Biol. Chem.* **274**, 22903–22906
266b. Malkowski, M. G., Ginell, S. L., Smith, W. L., and Garavito, R. M. (2000) *Science* **15**, 1933–1937
267. Dean, A. M., and Dean, F. M. (1999) *Protein Sci.* **8**, 1087–1098
268. Jakobsson, P.-J., Thorén, S., Morgenstern, R., and Samuelsson, B. (1999) *Proc. Natl. Acad. Sci. U.S.A.* **96**, 7220–7225
269. Williams, D. W., Hale, S. E., Okita, R. T., and Masters, B. S. S. (1984) *J. Biol. Chem.* **259**, 14600–14608
270. Hamberg, M., Svensson, J., and Samuelsson, B. (1975) *Proc. Natl. Acad. Sci. U.S.A.* **72**, 2994–2998
271. Hecker, M., and Ullrich, V. (1989) *J. Biol. Chem.* **264**, 141–150
272. Ruan, K.-H., Li, D., Ji, J., Lin, Y.-Z., and Gao, X. (1998) *Biochemistry* **37**, 822–830
273. Haurand, M., and Ullrich, V. (1985) *J. Biol. Chem.* **260**, 15059–15067
274. Moncada, S., Gryglewski, R., Bunting, S., and Vane, J. R. (1976) *Nature (London)* **263**, 663–665
275. Vane, J. R., and Bergström, S., eds. (1979) *Prostacyclin*, Raven Press, New York
275a. Cheng, Y., Austin, S. C., Rocca, B., Koller, B. H., Coffman, T. M., Grosser, T., Lawson, J. A., and FitzGerald, G. A. (2002) *Science* **296**, 539–541
276. Shyue, S.-K., Ruan, K.-H., Wang, L.-H., and Wu, K. K. (1997) *J. Biol. Chem.* **272**, 3657–3662
277. Wong, P. Y.-K., Malik, K. U., Taylor, B. M., Schneider, W. P., McGiff, J. C., and Sun, F. F. (1985) *J. Biol. Chem.* **260**, 9150–9153
278. Kramer, J. A., Johnson, K. R., Dunham, W. R., Sands, R. H., and Funk, M. O., Jr. (1994) *Biochemistry* **33**, 15017–15022
279. Glickman, M. H., and Klinman, J. P. (1996) *Biochemistry* **35**, 12882–12892
280. Nelson, M. J., Chase, D. B., and Seitz, S. P. (1995) *Biochemistry* **34**, 6159–6163
280a. Brash, A. R. (1999) *J. Biol. Chem.* **274**, 23679–23682
280b. Skrzypczak-Jankun, E., Bross, R. A., Carroll, R. T., Dunham, W. R., and Funk, M. O., Jr. (2001) *J. Am. Chem. Soc.* **123**, 10814–10820
280c. Knapp, M. J., Seebeck, F. P., and Klinman, J. P. (2001) *J. Am. Chem. Soc.* **123**, 2931–2932
280d. Mogul, R., and Holman, T. R. (2001) *Biochemistry* **40**, 4391–4397
280e. Tomchick, D. R., Phan, P., Cymborowski, M., Minor, W., and Holman, T. R. (2001) *Biochemistry* **40**, 7509–7517
281. Funk, C. D., Keeney, D. S., Oliw, E. H., Boeglin, W. E., and Brash, A. R. (1996) *J. Biol. Chem.* **271**, 23338–23344
282. Hamberg, M., Su, C., and Oliw, E. (1998) *J. Biol. Chem.* **273**, 13080–13088
283. Boyington, J. C., Gaffney, B. J., and Amzel, L. M. (1993) *Science* **260**, 1482–1486
284. Moiseyev, N., Rucker, J., and Glickman, M. H. (1997) *J. Am. Chem. Soc.* **119**, 3853–3860
285. Ford-Hutchinson, A. W., Gresser, M., and Young, R. N. (1994) *Ann. Rev. Biochem.* **63**, 383–417
286. Glasgow, W. C., Harris, T. M., and Brash, A. R. (1986) *J. Biol. Chem.* **261**, 200–204
287. Hammarström, S. (1983) *Ann. Rev. Biochem.* **52**, 355–377
288. Samuelsson, B., Dahlén, S.-E., Lindgren, J. Å., Rouzer, C. A., and Serhan, C. N. (1987) *Science* **237**, 1171–1176
289. Samuelsson, B., and Funk, C. D. (1989) *J. Biol. Chem.* **264**, 19469–19472
290. Blomster Andberg, M., Hamberg, M., and Haeggström, J. Z. (1997) *J. Biol. Chem.* **272**, 23057–23063
291. Carter, B. Z., Wiseman, A. L., Orkiszewski, R., Ballard, K. D., Ou, C.-N., and Lieberman, M. W. (1997) *J. Biol. Chem.* **272**, 12305–12310
292. Lewis, R. A., Austen, K. F., and Soberman, R. J. (1990) *N. Engl. J. Med.* **323**, 645–655
293. Lynch, K. R., O'Neill, G. P., Liu, Q., Im, D.-S., Sawyer, N., Metters, K. M., Coulombe, N., Abramovitz, M., Figueroa, D. J., Zeng, Z., Connolly, B. M., Bai, C., Austin, C. P., Chateauneuf, A., Stocco, R., Greig, G. M., Kargman, S., Hooks, S. B., Hosfield, E., Williams, D. L., Jr., Ford-Hutchinson, A. W., Caskey, C. T., and Evans, J. F. (1999) *Nature (London)* **399**, 789–793
294. Hammarström, S. (1983) *J. Biol. Chem.* **258**, 1427–1430
295. Serhan, C. N., Hamberg, M., and Samuelsson, B. (1984) *Proc. Natl. Acad. Sci. U.S.A.* **81**, 5335–5339
296. Rowley, A. F., Lloyd-Evans, P., Barrow, S. E., and Serhan, C. N. (1994) *Biochemistry* **33**, 856–863
297. Chang, M. S., Boeglin, W. E., Guengerich, F. P., and Brash, A. R. (1996) *Biochemistry* **35**, 464–471
298. Brash, A. R., Porter, A. T., and Maas, R. L. (1985) *J. Biol. Chem.* **260**, 4210–4216
298a. Zeldin, D. C. (2001) *J. Biol. Chem.* **276**, 36059–36062
298b. Node, K., Huo, Y., Ruan, X., Yang, B., Spiecker, M., Ley, K., Zeldin, D. C., and Liao, J. K. (1999) *Science* **285**, 1276–1279
299. Goetzl, E. J., An, S., and Smith, W. L. (1995) *FASEB J.* **9**, 1051–1058
300. Urade, Y., Fujimoto, N., and Hayaishi, O. (1985) *J. Biol. Chem.* **260**, 12410–12415
301. Brock, T. G., McNish, R. W., and Peters-Golden, M. (1999) *J. Biol. Chem.* **274**, 11660–11666
302. MacMillan, D. K., Hill, E., Sala.A, Sigal, E., Shuman, T., Henson, P. M., and Murphy, R. C. (1994) *J. Biol. Chem.* **269**, 26663–26668
303. Maddox, J. F., Colgan, S. P., Clish, C. B., Petasis, N. A., Fokin, V. V., and Serhan, C. N. (1998) *FASEB J.* **12**, 487–494
304. Dahlén, S.-E., Hansson, G., Hedqvist, P., Björck, T., Granström, E., and Dahlén, B. (1983) *Proc. Natl. Acad. Sci. U.S.A.* **80**, 1712–1716
305. Lindgren, J. Å., Hökfelt, T., Dahlén, S.-E., Patrono, C., and Samuelsson, B. (1984) *Proc. Natl. Acad. Sci. U.S.A.* **81**, 6212–6216
306. Reginato, M. J., Krakow, S. L., Bailey, S. T., and Lazar, M. A. (1998) *J. Biol. Chem.* **273**, 1855–1858
307. Nelson, N. A., Kelly, R. C., and Johnson, R. A. (1982) *Chem. Eng. News* **Aug 16**, 30–44
308. Marx, J. L. (1979) *Science* **205**, 175–176
309. Wheelan, P., and Murphy, R. C. (1995) *J. Biol. Chem.* **270**, 19845–19852
310. Weissmann, G. (1991) *Sci. Am.* **264**(Jan), 84–90
311. Kalgutkar, A. S., Crews, B. C., Rowlinson, S. W., Garner, C., Seibert, K., and Marnett, L. J. (1998) *Science* **280**, 1268–1270
312. Vane, J. (1994) *Nature (London)* **367**, 215–217
313. Lecomte, M., Laneuville, O., Ji, C., DeWitt, D. L., and Smith, W. L. (1994) *J. Biol. Chem.* **269**, 13207–13215
314. Loll, P. J., Picot, D., Ekabo, O., and Garavito, R. M. (1996) *Biochemistry* **35**, 7330–7340
314a. Warner, T. D., and Mitchell, J. A. (2002) *Proc. Natl. Acad. Sci. U.S.A.* **99**, 13371-13373
315. Abramson, S., Korchak, H., Ludewig, R., Edelson, H., Haines, K., Levin, R. I., Herman, R., Rider, L., Kimmel, S., and Weissman, G. (1985) *Proc. Natl. Acad. Sci. U.S.A.* **82**, 7227–7231
316. Patrono, C. (1994) *N. Engl. J. Med.* **330**, 1287–1294
317. DuBois, R. N., Abramson, S. B., Crofford, L., Gupta, R. A., Simon, L. S., van de Putte, L. B. A., and Lipsky, P. E. (1998) *FASEB J.* **12**, 1063–1073
317a. Selinsky, B. S., Gupta, K., Sharkey, C. T., and Loll, P. J. (2001) *Biochemistry* **40**, 5172 – 5180
317b. FitzGerald, G. A., and Patrono, C. (2001) *N. Engl. J. Med.* **345**, 433–442
317c. Killen, J. P., Nzerue, C. M., Rich, S. A., FitzGerald, G. A., and Patrono, C. (2001) *N. Engl. J. Med.* **345**, 1708–1709
317d. Vane, J. R. (2002) *Science* **296**, 474–475
317e. Jiang, Q., Elson-Schwab, I., Courtemanche, C., and Ames, B. N. (2000) *Proc. Natl. Acad. Sci. U.S.A.* **97**, 11494–11499
318. Vick, B. A. (1993) in *Lipid Metabolism in Plants* (Moore, T. S., Jr., ed), pp. 167–194, CRC Press, Boca Raton, Florida
319. Schaller, F., and Weiler, E. W. (1997) *J. Biol. Chem.* **272**, 28066–28072
320. Vijayan, P., Shockey, J., Lévesque, C. A., Cook, R. J., and Browse, J. (1998) *Proc. Natl. Acad. Sci. U.S.A.* **95**, 7209–7214
321. Song, W.-C., Baertschi, S. W., Boeglin, W. E., Harris, T. M., and Brash, A. R. (1993) *J. Biol. Chem.* **268**, 6293–6298
321a. Ziegler, J., Stenzel, I., Hause, B., Maucher, H., Hamberg, M., Grimm, R., Ganal, M., and Wasternack, C. (2000) *J. Biol. Chem.* **275**, 19132–19138
322. Conconi, A., Smerdon, M. J., Howe, G. A., and Ryan, C. A. (1996) *Nature (London)* **383**, 826–829
323. Xie, D.-X., Feys, B. F., James, S., Nieto-Rostro, M., and Turner, J. G. (1998) *Science* **280**, 1091–1094
324. Alborn, H. T., Turlings, T. C. J., Jones, T. H., Stenhagen, G., Loughrin, J. H., and Tumlinson, J. H. (1997) *Science* **276**, 945–949
325. Paré, P. W., Alborn, H. T., and Tumlinson, J. H. (1998) *Proc. Natl. Acad. Sci. U.S.A.* **95**, 13971–13975
326. Parchmann, S., and Mueller, M. J. (1998) *J. Biol. Chem.* **273**, 32650–32655
327. Koljak, R., Boutaud, O., Shieh, B.-H., Samel, N., and Brash, A. R. (1997) *Science* **277**, 1994–1996
328. Brash, A. R., Hughes, M. A., Hawkins, D. J., Boeglin, W. E., Song, W.-C., and Meijer, L. (1991) *J. Biol. Chem.* **266**, 22926–22931
329. Birch, A. J. (1967) *Science* **156**, 202–206
330. Spencer, J. B., and Jordan, P. M. (1992) *Biochemistry* **31**, 9107–9116
331. Fu, H., Alvarez, M. A., Khosla, C., and Bailey, J. E. (1996) *Biochemistry* **35**, 6527–6532
332. Kendrew, S. G., Harding, S. E., Hopwood, D. A., and Marsh, E. N. G. (1995) *J. Biol. Chem.* **270**, 17339–17343
333. Quirós, L. M., and Salas, J. A. (1995) *J. Biol. Chem.* **270**, 18234–18239
334. Schultz, D. J., Cahoon, E. B., Shanklin, J., Craig, R., Cox-Foster, D. L., Mumma, R. O., and Medford, J. I. (1996) *Proc. Natl. Acad. Sci. U.S.A.* **93**, 8771–8775
335. Michal, G., ed. (1999) *Biochemical Pathways*, Wiley - Spektrum Academischer Verlag, New York - Heidelberg, Germany
336. Hutchinson, C. R. (1999) *Proc. Natl. Acad. Sci. U.S.A.* **96**, 3336–3338
336a. Bao, W., Sheldon, P. J., and Hutchinson, C. R. (1999) *Biochemistry* **38**, 9752–9757

References

336b. Jez, J. M., Ferrer, J.-L., Bowman, M. E., Dixon, R. A., and Noel, J. P. (2000) *Biochemistry* **39**, 890–902

336c. Khosla, C., Gokhale, R. S., Jacobsen, J. R., and Cane, D. E. (1999) *Ann. Rev. Biochem.* **68**, 219–253

337. Cane, D. E. (1994) *Science* **263**, 338–340

338. Pieper, R., Gokhale, R. S., Luo, G., Cane, D. E., and Khosla, C. (1997) *Biochemistry* **36**, 1846–1851

339. Kao, C. M., Pieper, R., Cane, D. E., and Khosla, C. (1996) *Biochemistry* **35**, 12363–12368

340. Lambalot, R. H., Cane, D. E., Aparicio, J. J., and Katz, L. (1995) *Biochemistry* **34**, 1858–1866

341. Kendrew, S. G., Katayama, K., Deutsch, E., Madduri, K., and Hutchinson, C. R. (1999) *Biochemistry* **38**, 4794–4799

342. Kim, C.-G., Yu, T.-W., Fryhle, C. B., Handa, S., and Floss, H. G. (1998) *J. Biol. Chem.* **273**, 6030–6040

343. Aparicio, J. F., Colina, A. J., Ceballos, E., and Martín, J. F. (1999) *J. Biol. Chem.* **274**, 10133–10139

344. Kennedy, J., Auclair, K., Kendrew, S. G., Park, C., Vederas, J. C., and Hutchinson, C. R. (1999) *Science* **284**, 1368–1372

345. Offenzeller, M., Santer, G., Totschnig, K., Su, Z., Moser, H., Traber, R., and Schneider-Scherzer, E. (1996) *Biochemistry* **35**, 8401–8412

346. Schwecke, T., Aparicio, J. F., Molnár, I., König, A., Ee Khaw, L., Haydock, S. F., Oliynyk, M., Caffrey, P., Cortés, J., Lester, J. B., Böhm, G. A., Staunton, J., and Leadlay, P. F. (1995) *Proc. Natl. Acad. Sci. U.S.A.* **92**, 7839–7843

347. Preisig-Müller, R., Gehlert, R., Melchior, F., Stietz, U., and Kindl, H. (1997) *Biochemistry* **36**, 8349–8358

348. Schröder, J., Raiber, S., Berger, T., Schmidt, A., Schmidt, J., Soares-Sello, A. M., Bardshiri, E., Strack, D., Simpson, T. J., Veit, M., and Schröder, G. (1998) *Biochemistry* **37**, 8417–8425

348a. Jez, J. M., Bowman, M. E., and Noel, J. P. (2001) *Biochemistry* **40**, 14829–14838

349. Guo, L., Dixon, R. A., and Paiva, N. L. (1994) *J. Biol. Chem.* **269**, 22372–22378

350. Brown, D. W., Adams, T. H., and Keller, N. P. (1996) *Proc. Natl. Acad. Sci. U.S.A.* **93**, 14873–14877

350a. Schroeder, F. C., Farmer, J. J., Smedley, S. R., Attygalle, A. B., Eisner, T., and Meinwald, J. (2000) *J. Am. Chem. Soc.* **122**, 3628–3634

351. Marsden, A. F. A., Wilkinson, B., Cortés, J., Dunster, N. J., Staunton, J., and Leadlay, P. F. (1998) *Science* **279**, 199–202

352. Shen, Y., Yoon, P., Yu, T.-W., Floss, H. G., Hopwood, D., and Moore, B. S. (1999) *Proc. Natl. Acad. Sci. U.S.A.* **96**, 3622–3627

353. Gokhale, R. S., Tsuji, S. Y., Cane, D. E., and Khosla, C. (1999) *Science* **284**, 482–485

354. Reynolds, K. A. (1998) *Proc. Natl. Acad. Sci. U.S.A.* **95**, 12744–12746

355. Schröder, F. C., Farmer, J. J., Attygalle, A. B., Smedley, S. R., Eisner, T., and Meinwald, J. (1998) *Science* **281**, 428–431

356. Metz, J. G., Roessler, P., Facciotti, D., Levering, C., Dittrich, F., Lassner, M., Valentine, R., Lardizabal, K., Domergue, F., Yamada, A., Yazawa, K., Knauf, V., and Browse, J. (2001) *Science* **293**, 290–293

357. Nugent, C., Prins, J. B., Whitehead, J. P., Wentworth, J. M., Chatterjee, V. K. K., and O'Rahilly, S. (2001) *J. Biol. Chem.* **276**, 9149–9157

358. Gibbons, H. S., Lin, S., Cotter, R. J., and Raetz, C. R. H. (2000) *J. Biol. Chem.* **275**, 32940–32949

359. Sohlenkamp, C., de Rudder, K. E. E., Röhrs, V., López-Lara, I. M., and Geiger, O. (2000) *J. Biol. Chem.* **275**, 18919–18925

360. Moody, J. S., Kozak, K. R., Ji, C., and Marnett, L. J. (2001) *Biochemistry* **40**, 861–866

361. Nakajima, J.-i, Tanaka, Y., Yamazaki, M., and Saito, K. (2001) *J. Biol. Chem.* **276**, 25797–25803

Study Questions

1. Outline possible pathways of metabolism of dietary fats. Consider digestion, transport of fatty acids, storage, conversion to prostaglandins, steroids, etc. Will any of the fat be converted into glucose?

2. What are the functions in the human body of the following?

   Pancreatic lipase
   Lipocalins
   Lipoprotein lipase
   Very low density lipoprotein (VLDL)
   Hormone-sensitive lipase
   Chylomicrons
   Apolipoproteins

3. Describe two different types of fatty acid synthase. Compare the basic chemical reactions that are involved. Also, compare these with the reactions of fatty acid oxidation.

4. Discuss the different types of fatty acids found in the human body and the synthetic pathways by which they are formed.

5. What mechanisms are utilized for incorporation of double bonds into fatty acids? Propose a mechanism that makes use of polyketide synthase domains (Fig. 21-11) in the synthesis of polyunsaturated fatty acids. See Metz *et al.*[356]

6. In what locations would you expect to find the following?

   Tripalmitin
   Mycolic acids
   Arachidonic acid
   Propionic acid
   Docosahexaenoic acid

7. Substitution of a small percentage of ω6 fatty acids in the diet of insulin-resistant rodents with ω3 unsaturated fatty acids normalized insulin action.[357] Can you suggest possible mechanisms? Is this result significant to human nutrition?

8. Formation of the 3-hydroxymyristoyl groups of lipid A (Fig. 8-30) requires O_2. Comparisons of amino acid sequences suggest that an Fe^{2+}/2-oxoglutarate-dependent oxygenase is involved.[358] Write a balanced equation for this reaction.

9. Phosphatidylcholine can be formed by two pathways as described on pp. 1198–1199. A third pathway, used by some bacteria, involves a direct one-step reaction of choline with CDP-diacylglycerol.[359] Write a reasonable chemical mechanism.

Study Questions

10. ^{14}C-Carboxyl labeled palmitic acid is fed to a fasted rat. There is no increase in liver glycogen, but the glucose units of the glycogen contain ^{14}C.
    a) Outline, using appropriate equations, the reaction sequence by which the carbon atoms of glucose become labeled.
    b) Explain why there is no net synthesis of glycogen from the fatty acid.

11. a) Write the reactions that most *dietary* tripalmitin will undergo in the body of an adult human in order to be deposited in adipose tissue as tripalmitin.
    b) What is the minimum amount of ATP (high energy bonds) normally required to deposit the one mole of dietary tripalmitin in adipose tissue? Count only ATP involved in tripalmitin metabolism and consider the source of glycerol in the adipose tissue.

12. Describe the biochemical effects on lipid metabolism of injecting into a normal animal
    a) insulin
    b) glucagon
    c) epinephrine

13. Suggest a biosynthetic pathway for formation of the fungal metabolite **agaricic acid**:

$$\begin{array}{l} CH_3 — (CH_2)_{15} — CH — COOH \\ \qquad\qquad\qquad\quad | \\ \qquad\qquad\quad HOC — COOH \\ \qquad\qquad\qquad\quad | \\ \qquad\qquad\qquad CH_2 — COOH \end{array}$$

14. The ketone **palmiton** $CH_3(CH_2)14 – {}^*CO – (CH_2)_{14}CH_3$ is formed by mycobacteria. The carbon marked by an asterisk was found to be labeled after feeding of [1-^{14}C]palmitic acid. Suggest a biosynthetic pathway.

15. The following reaction occurs in the biosynthesis of fatty acids.

    Crotonyl-ACP + NADPH + H^+ → butyryl-ACP + $NADP^+$

    The reduction half-reactions for crotonyl-ACP and NADPH are

    Crotonyl-ACP + 2 H^+ + 2 e^- → butyryl-ACP
    $E^{o\prime} = -0.015$ V
    NADP + 2 H^+ + 2 e^- → NADPH + H
    $E^{o\prime} = -0.320$ V

    What is $\Delta G^{o\prime}$ for this reaction? What is the equilibrium constant for the reaction?

16. How does the inhibition of citrate synthase affect fatty acid synthesis?

17. Malonyl CoA is an allosteric effector of carnitine acyl transferase. What kind of effector is it, i.e., activator or inhibitor, and what is the logic behind the interaction?

18. Compound II is formed in a series of enzymatic reactions from compound I. Propose a mechanistically realistic sequence, showing by name any cofactors required.

$$\begin{array}{c} OH \\ | \\ R — CH — (CH_2)_3\,COOH \end{array}$$

I

$$\begin{array}{c} O \\ \| \\ R — C — CH{=}CH\,CH_3 \end{array}$$

II

19. Fatty acid biosynthesis requires NADPH. Where does the NADPH come from?

20. An individual has been found who is missing malic enzyme in his cytoplasm. He has instead an enzyme that converts the oxalacetate made from the citrate lyase reaction directly to pyruvate and CO_2. Discuss this patient in terms of the likely effect of these changes on his ability to synthesize fatty acids.

21. The $\Delta G^{o\prime}$ values for the hydrolysis of any P – O – P bond of ATP, inorganic pyrophosphate, or any acyl CoA thiolester are all about –34 kJ / mole, while the corresponding figure for the hydrolysis of a mixed carboxylic phosphate anhydride is about –55 kJ / mole. Calculate the value of $\Delta G^{o\prime}$ for the following reaction describing the activation of fatty acids to the fatty acyl adenylate.

$$RCOO^- + ATP \longrightarrow R — \overset{\overset{\displaystyle O}{\|}}{C} — O — AMP + PP_i$$

Study Questions

22. Fatty acid biosynthesis is made irreversible by the specific input of energy. Name the reactions or give equations for those steps in the pathway that require ATP. It is important that you consider both the mitochondrial and cytosolic components of the pathway.

23. The fatty acid biosynthesis pathway communicates with at least three other metabolic pathways either by sharing common intermediates or by regulatory mechanisms. Fill in the table below. List four molecules that have this function. You should name the additional pathway where each of these is found and briefly describe what it does in this second pathway. Do **not** use a redox cofactor as one of your choices.

Molecule	**Other Pathway**	**Role in Second Pathway**

24. The endogenous cannabinoid 2-arachidonoylglycerol is thought to play important roles both in the brain (Chapter 30) and in the immune system (Chapter 31). Leukocyte 12-oxygenase acts on this compound.[360] What products would be expected?

25. In the synthesis of anthocyanidins (Box 21-E) another Fe^{2+}/2-oxoglutarate-dependent oxygenase acts on the colorless leukoanthrocyanidin, which is then converted to the colored anthocyanidin:

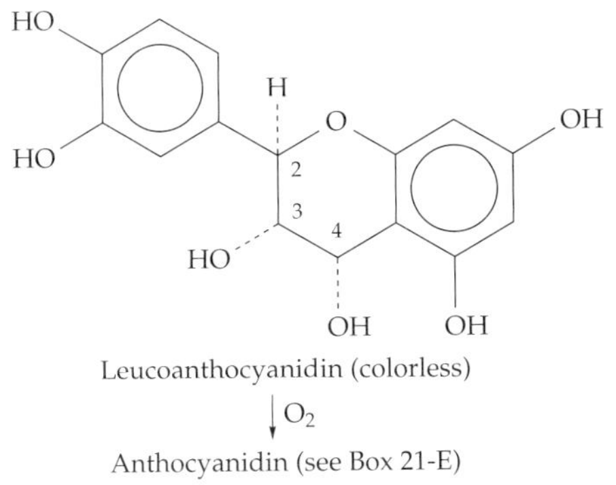

Leucoanthocyanidin (colorless)

$\downarrow O_2$

Anthocyanidin (see Box 21-E)

Propose a reasonable sequence for this reaction. See Nakajima *et al.*[361]

26. Why does a dietary deficiency of choline in humans induce a fatty liver, i.e., a liver in which the hepatocytes contain excess triglycerides?

27. Name two important functions of citrate in the conversion of acetate to palmitate by adipose tissues.

28. Outline the steps by which linoleic acid can be converted to prostaglandins in the human body.

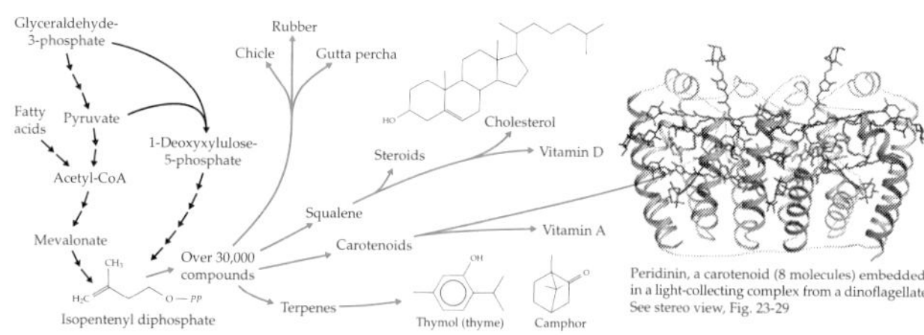

Peridinin, a carotenoid (8 molecules) embedded in a light-collecting complex from a dinoflagellate. See stereo view, Fig. 23-29

Starting with the simple compounds acetyl-CoA, glyceraldehyde-3-phosphate, and pyruvate, which arise via the central pathways of metabolism, the key intermediate **isopentenyl diphosphate** is formed by two independent routes. It is then converted by bacteria, fungi, plants, and animals into thousands of different naturally occurring products. These include high polymers, such as rubber, as well as vitamins, sterols, carotenoids, and over 30,000 different terpenes and related compounds. Many of the latter are found only in specific plants where they may function as defensive compounds or pheromones.

Contents

22 Polyprenyl (Isoprenoid) Compounds

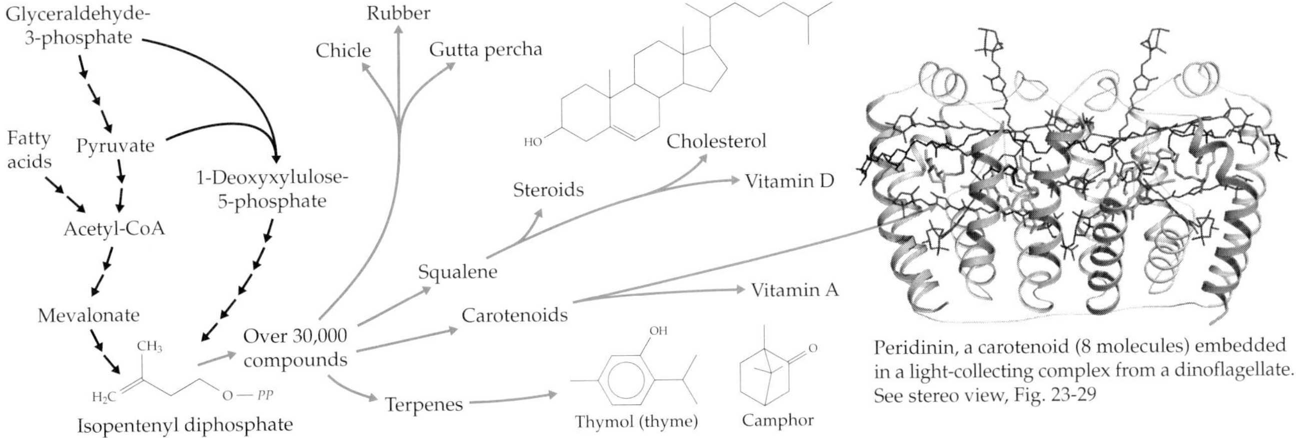

Peridinin, a carotenoid (8 molecules) embedded in a light-collecting complex from a dinoflagellate. See stereo view, Fig. 23-29

The **terpenes**, **carotenoids**, **steroids**, and many other compounds arise in a direct way from the prenyl group of **isopentenyl diphosphate** (Fig. 22-1).[1–6a] Biosynthesis of this five-carbon branched unit from **mevalonate** has been discussed previously (Chapter 17, Fig. 17-19) and is briefly recapitulated in Fig. 22-1. Distinct isoenzymes of 3-hydroxy-3-methylglutaryl-CoA synthase (HMG-CoA synthase) in the liver produce HMG-CoA destined for formation of ketone bodies (Eq. 17-5) or mevalonate.[7,8] A similar cytosolic enzyme is active in plants which, collectively, make more than 30,000 different isoprenoid compounds.[9,10] However, many of these are formed by an alternative pathway that does not utilize mevalonate but starts with a thiamin diphosphate-dependent condensation of glyceraldehyde 3-phosphate with pyruvate (Figs. 22-1, 22-2).

3(R) Mevalonate

Isopentenyl diphosphate

The two-step reduction of HMG-CoA to mevalonate (Fig. 22-1, step *a*)[11–15] is highly controlled, a major factor in regulating cholesterol synthesis in the human liver.[12,16,17] The N-terminal portion of the 97-kDa 888-residue mammalian HMG-CoA reductase is thought to be embedded in membranes of the ER, while the C-terminal portion is exposed in the cytoplasm.[16] The enzyme is sensitive to feedback inhibition by cholesterol (see Section D, 2). The regulatory mechanisms include a phosphorylation–dephosphorylation cycle and control of both the rates of synthesis and of proteolytic degradation of this key enzyme.[14,15,18–20]

A. Isopentenyl Diphosphate and Polyprenyl Synthases

In animals all isoprenoid compounds are apparently synthesized from mevalonate, which is converted by the consecutive action of two kinases[21–23] into mevalonate 5-diphosphate (Fig. 22-1, step *b*). Mevalonate kinase is found predominantly in peroxisomes, which are also active in other aspects of steroid synthesis in humans.[21,24] A deficiency of this enzyme is associated with mevalonic aciduria, a serious hereditary disease in which both blood and urine contain very high concentrations of mevalonate.[23] Mevalonate diphosphate kinase, which is also a decarboxylase, catalyzes phosphorylation of the 3-OH group of mevalonate (step *c*, Fig. 22-1) and decarboxylative elimination of phosphate (step *d*)[25] to form isopentenyl diphosphate.

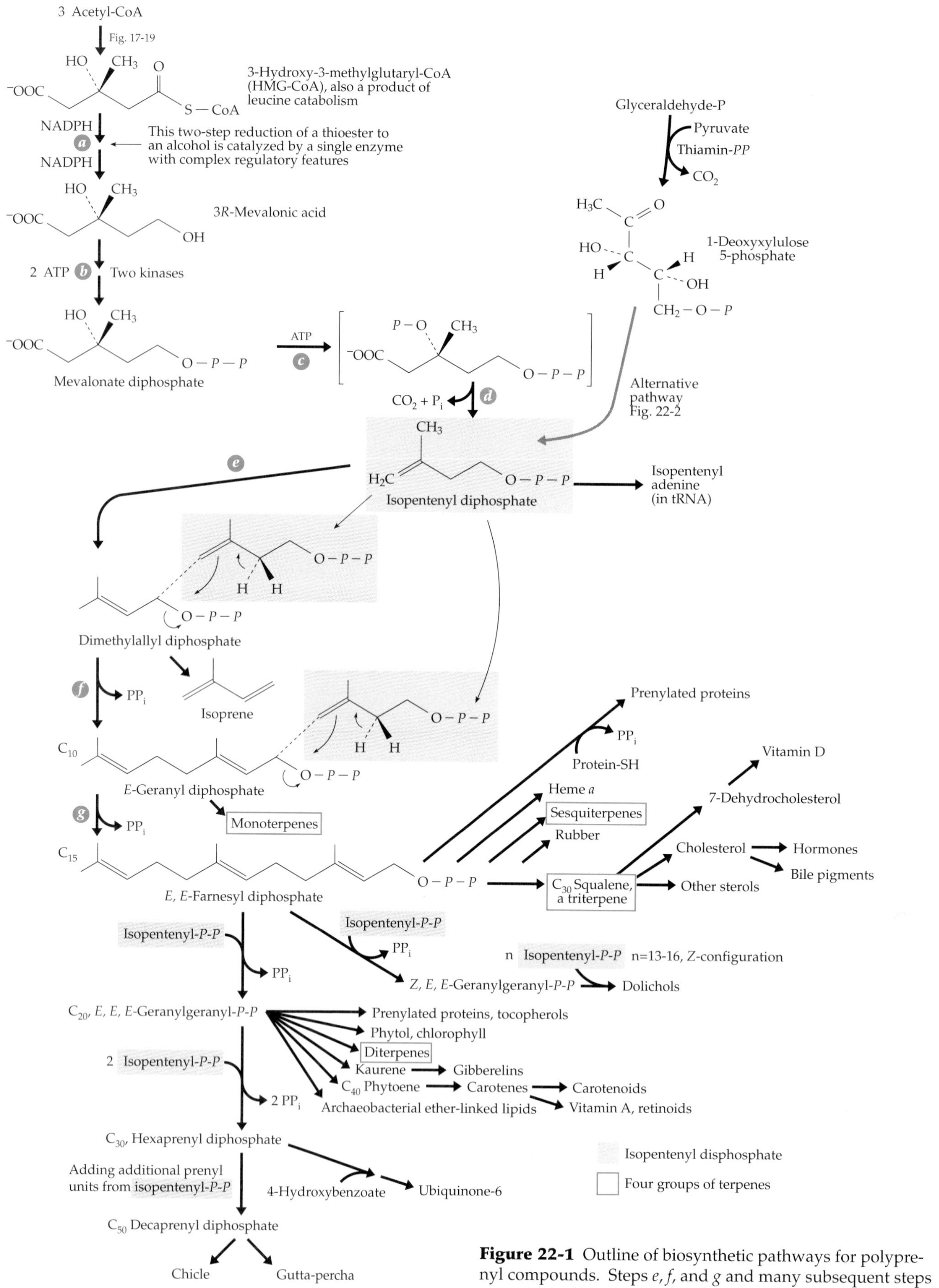

3 Acetyl-CoA
Fig. 17-19
3-Hydroxy-3-methylglutaryl-CoA (HMG-CoA), also a product of leucine catabolism
NADPH
a
NADPH
This two-step reduction of a thioester to an alcohol is catalyzed by a single enzyme with complex regulatory features
3*R*-Mevalonic acid
2 ATP b Two kinases
Mevalonate diphosphate
ATP
c
$CO_2 + P_i$ d
Glyceraldehyde-P
Pyruvate
Thiamin-*PP*
CO_2
1-Deoxyxylulose 5-phosphate
Alternative pathway Fig. 22-2
e
Isopentenyl diphosphate
Isopentenyl adenine (in tRNA)
Dimethylallyl diphosphate
f PP_i
Isoprene
C_{10}
E-Geranyl diphosphate
g PP_i
Monoterpenes
C_{15}
E, E-Farnesyl diphosphate
Prenylated proteins
PP_i
Protein-SH
Heme *a*
Sesquiterpenes
Rubber
C_{30} Squalene, a triterpene
Cholesterol
Other sterols
7-Dehydrocholesterol
Vitamin D
Hormones
Bile pigments
Isopentenyl-*P-P*
PP_i
Isopentenyl-*P-P*
PP_i
Z, E, E-Geranylgeranyl-*P-P*
n Isopentenyl-*P-P* n=13-16, *Z*-configuration
Dolichols
C_{20}, *E, E, E*-Geranylgeranyl-*P-P*
Prenylated proteins, tocopherols
Phytol, chlorophyll
Diterpenes
Kaurene → Gibberelins
C_{40} Phytoene → Carotenes → Carotenoids
Vitamin A, retinoids
Archaeobacterial ether-linked lipids
2 Isopentenyl-*P-P*
2 PP_i
C_{30}, Hexaprenyl diphosphate
4-Hydroxybenzoate
Ubiquinone-6
Adding additional prenyl units from isopentenyl-*P-P*
C_{50} Decaprenyl diphosphate
Chicle
Gutta-percha
Isopentenyl disphosphate
Four groups of terpenes

Figure 22-1 Outline of biosynthetic pathways for polyprenyl compounds. Steps *e, f,* and *g* and many subsequent steps are catalyzed by isoprenyl diphosphate synthases.

1. An Alternative Pathway for Isoprenoid Synthesis

It is generally agreed that mevalonate is the precursor to sterols in higher plants as well as in animals and is also the precursor to plant carotenoids. However, it is poorly incorporated into monoterpenes and into some diterpenes such as those of the taxane group.[26,27] The alternative **glyceraldehyde 3-phosphate:pyruvate pathway** explains this result. The pathway also operates in some bacteria and apparently is the sole source of isoprenoid compounds for the unicellular alga *Scenedesmus*.[28] The pathway is outlined in Fig. 22-2. Pyruvate is decarboxylated by a thiamin diphosphate-dependent enzyme,[29] and the resulting enamine is condensed with D-glyceraldehyde 3-phosphate to form 1-deoxyxylulose 5-phosphate.[28,30–31a] The latter undergoes an isomeroreductase rearrange-

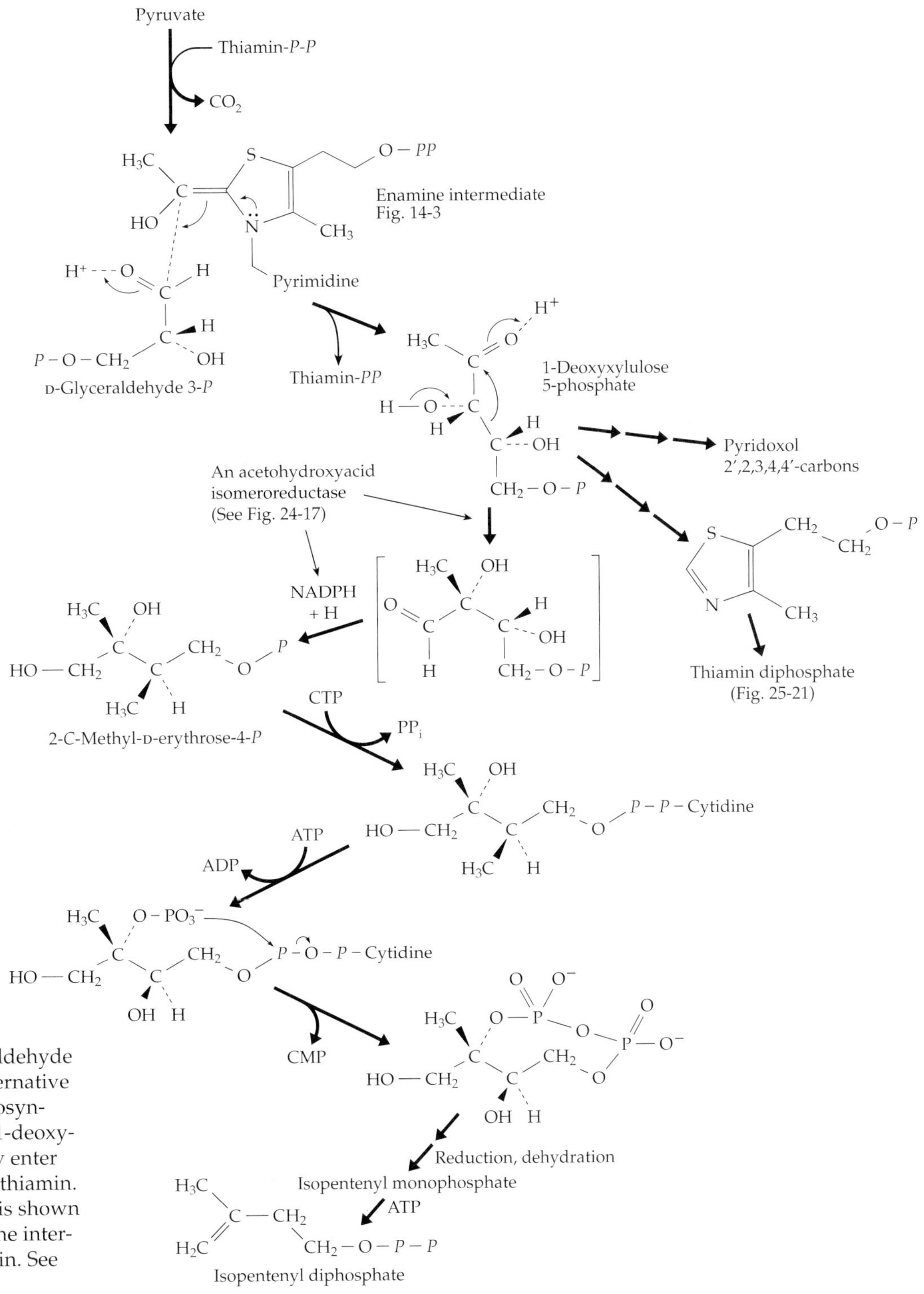

Figure 22-2 The glyceraldehyde 3-phosphate:pyruvate alternative pathway of isoprenoid biosynthesis. The intermediate 1-deoxyxylulose 5-phosphate may enter terpenes, vitamin B_6, and thiamin. Isopentenyl diphosphate is shown as the final product, but the intermediate steps are uncertain. See Lange *et al.*[32g]

ment of the type that occurs in the biosynthesis of valine and isoleucine (Fig. 24-17),[32] but the additional steps on to isopentenyl diphosphate are not obvious. However, some intermediate compounds have been identified as is indicated in Fig. 22-2.[32a–g] 1-Deoxyxylulose 5-phosphate has also been identified as an intermediate in the biosynthesis of vitamin B_6 in *E. coli*.[32a,h,33] It gives rise to the 2', 2, 3, 4, and 4' carbon atoms of pyridoxine and also to the pyrimidine ring of thiamin. See Chapter 25.

2. Isomerization and Isoprene Formation

Before polymer formation begins, one molecule of isopentenyl diphosphate must be isomerized to **dimethylallyl diphosphate** (Fig. 22-1, step *e*, Eq. 13-56).[10,33a,b] In this process the hydrogen that was in the 4-*pro-S* position of mevalonic acid (the *pro-R* position of isopentenyl pyrophosphate) is lost. Dimethylallyl diphosphate is unstable and can undergo acid-catalyzed elimination of PP_i to form isoprene, apparently by a carbocation mechanism.

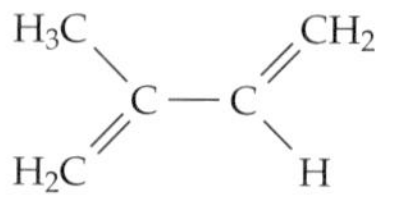

Isoprene

This evidently accounts for the presence of isoprene in the breath.[34] Isoprene is also formed by many plants and is released into the atmosphere in large amounts, which contribute to photochemical formation of haze. A Mg^{2+}-dependent enzyme catalyzes the elimination of pyrophosphate.[35] Isoprene emissions rise with increasing temperature, and it has been suggested that the isoprene may dissolve in chloroplast membranes and in some way confer increased heat resistance.[36,37] Hydrolytic dephosphorylation can lead to dimethylallyl alcohol, which is oxidized in the liver to dimethylacrylyl-CoA (Eq. 22-1).

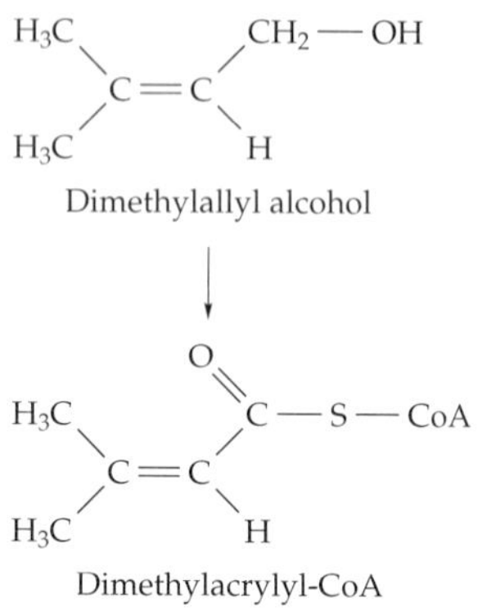

Dimethylallyl alcohol

Dimethylacrylyl-CoA (22-1)

The latter is also a catabolite of leucine and can be converted back to HMG-CoA via a biotin-dependent carboxylation (see Fig. 24-18). This provides a means of recycling the dimethylallyl alcohol back to the polyprenyl pathway.[38]

3. Polyprenyl Compounds

Dimethylallyl diphosphate serves as the starter piece for most polyprenyl compounds. Additional prenyl units are added, with elimination of pyrophosphate, by the action of **polyprenyl diphosphate synthases** as indicated in Fig. 22-1. Many of the products have all-trans (*E*) double bonds. A substantial number of these synthases are known and are distinguished by their chain length specificity and stereochemical properties.[39–46b] The most studied is farnesyl diphosphate synthase. The three-dimensional structure of an avian form is known.[47,47a] It catalyzes steps *f* and *g* of Fig. 22-1, joining three prenyl groups with the *E,E* (trans, trans) configuration. This protein, which consists almost entirely of packed α helices, has a large central cavity with conserved lysines and two aspartate-rich sequences (DDXXD) along its walls. These polar groups, together with magnesium ions, probably bind the pyrophosphate groups of the substrates.[48] Aspartates 224 and 225 of a bacterial form of the enzyme appear to be essential for catalytic activity.[49] The reaction is thought to be initiated by elimination of PP_i to form a carbocation to which the second prenyl unit adds as in Eq. 22-2.[50,51] For each prenyl unit a hydrogen atom that was originally the 4-*pro-S* hydrogen of mevalonic acid is lost as a proton.[52,53] Addition of another prenyl unit gives *E,E,E* geranylgeranyl diphosphate.[46a]

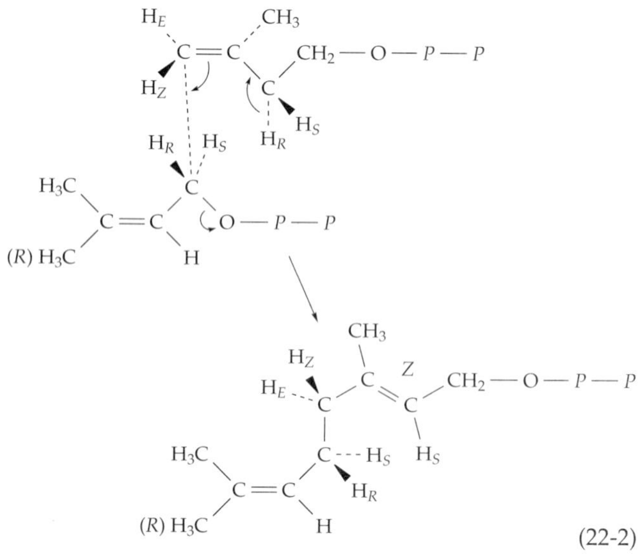

(22-2)

The chain length of the polyprenyl compounds appears to be determined by the protein structure of the synthase.[45,54,54a,b] Polymerization of prenyl units can continue with the formation of high molecular weight polyprenyl alcohols such as the **dolichols** and bacterial decaprenols (Chapter 20) or of the high polymers **rubber** (all *Z* configuration), chicle, and **gutta-percha**.[6a] Dolichols, which function in the biosynthesis of glycoproteins, consist of 16–21 prenyl units and are synthesized in the endoplasmic reticulum as the diphosphates.[55–57] Farnesyl diphosphate is elongated to *Z,E,E* geranylgeranyl diphosphate, and polymerization continues with addition of 13–18 more units, all with the *Z* (cis) configuration.[46b,57a,b] However, after dephosphorylation[58] the double bond of the last unit added becomes saturated.[59] Partial absence of the required reductase causes a serious human deficiency disease involving faulty glycoprotein synthesis.[60]

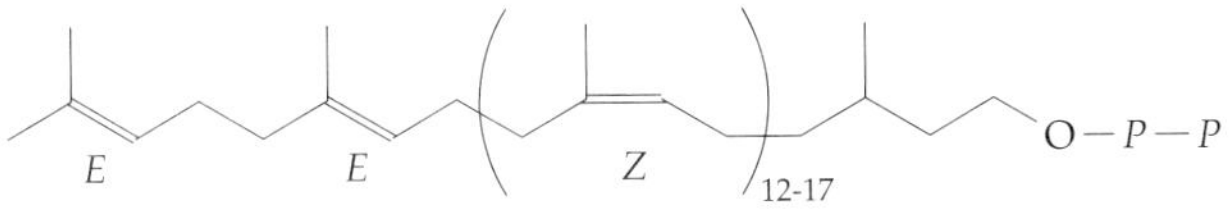

E E Z 12-17 O–P–P

A fully extended 19-unit dolichol (dolichol-19) would have a length of about 10 nm, twice that of the bilayer in which it is dissolved. It has been suggested that the central part of the molecule has a helical structure, while the ends are more flexible. Dolichols also appear to increase the fluidity of membrane bilayers.[61] Bacterial undecaprenyl diphosphate, which has a similar function, contains only one *E* and ten *Z* double bonds[62–63a] (see p. 1152).

Rubber also contains almost entirely *Z* double bonds. Consistent with this fact is the finding that the prenyltransferases catalyzing formation of rubber promote loss of the *pro-R* proton rather than the *pro-S* proton of mevalonic acid (see Eq. 22-2). There appear to be two types of prenyltransferase in animal mitochondria giving rise to *E* and *Z* double bonds, respectively.[64] In contrast, the rubber tree contains a 137-residue protein, the **rubber elongation factor**. This small protein binds to *E* prenyltransferases causing them to form *Z* double bonds.[65] The bacterium *Micrococcus luteus* synthesizes all *E* polyprenyl alcohol diphosphates up to the C_{45} nonaprenyl compound **solanesyl diphosphate**.[66]

Chain elongation during polymerization of prenyl units can be terminated in one of a number of ways. The pyrophosphate group may be hydrolyzed to a monophosphate or to a free alcohol. Alternatively, two polyprenyl compounds may join "head to head" to form a symmetric dimer. The C_{30} terpene **squalene**, the precursor to cholesterol, arises in this way from two molecules of farnesyl diphosphate as does **phytoene**, precursor of the C_{40} carotenoids, from *E,E,E* geranylgeranyl diphosphate. The phytanyl groups of archaebacterial lipids (p. 385) arise rather directly from geranylgeranyl diphosphate by transfer of the polyprenyl group to the $-CH_2OH$ group of *sn*-glycerol 3-phosphate.[67,67a] This is followed by hydrogenation of the double bonds. Formation of diphytanyl group (p. 388) must involve additional crosslinking reactions.

4. Prenylation of Proteins and Other Compounds

Polyprenyl groups are often transferred onto thiolate ions of cysteine side chains of certain proteins that bind to membranes (p. 559).[68,69] We have previously considered the Ras family (Chapter 11). Recoverin, an important protein in the visual cycle (Fig. 23-43), is another example of a prenylated protein. Both **farnesyltransferases**[70–76f] and **geranylgeranyltransferases**[72,77–78b] have been characterized, and the three- dimensional structure of the former has been established.[73,75–76] The two-domain protein contains a seven-helix crescent-shaped hairpin domain and an α,α-barrel similar to that in Fig. 2-29. A bound zinc ion in the active site may bind the $-S^-$ group of the substrate protein after the farnesyl diphosphate has been bound into the active site.[76,79] These enzymes are thought to function by a carbocation mechanism as shown in Eq. 22-3 and with the indicated inversion of configuration.[71]

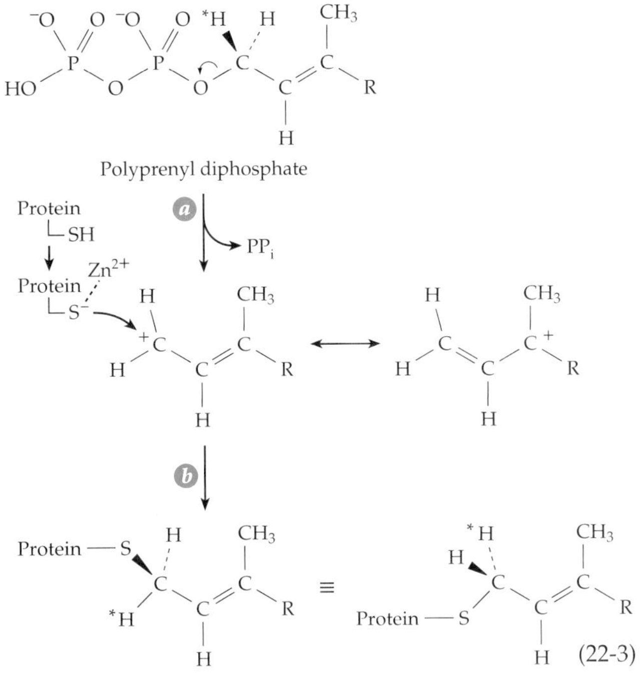

Polyprenyl diphosphate

Protein–SH → Protein–S⁻ (Zn^{2+})

(a) → PP_i

(b)

Protein — S

(22-3)

Inhibition of these prenyltransferases blocks growth of tumor cells. Many prenyltransferase inhibitors are apparently nontoxic to normal cells and are undergoing human clinical trials as anticancer drugs.[76a, 79a,b] Among other important polyprenyl compounds are the side chains of vitamin K, the ubiquinones, plastoquinones, tocopherols, and the phytyl group of chlorophyll. In all cases, a diphosphate of a polyprenyl alcohol serves as an alkyl group donor. Introduction of the polyprenyl chain into aromatic groups, such as those of the quinones (Fig. 15-24), occurs at a position ortho to a hydroxyl group in the reduced quinone (hydroquinone). The reader should be able to propose a reasonable prenyltransferase mechanism involving participation of the hydroxy group. The monoprenyl compound dimethylallyl diphosphate prenylates the N^6 position of adenine in a specific site in many tRNA molecules (Fig. 5-33)[80] as well as the C-4 position of L-tryptophan in the synthesis of ergot alkaloids.[81]

B. Terpenes

The number of small compounds that arise from isopentenyl diphosphate and are found in plants, animals, and bacteria is staggering. Just a few of these "terpenes" are shown in Figs. 22-3 and 22-4. The biosynthetic pathways have been worked out by "feeding" radioactively labeled acetate to plants and studying the characteristic labeling patterns in the terpenes. Many of the enzymes involved have been identified and studied. A given plant usually contains a large number of different terpenes, which are often concentrated in specialized "oil glands" or resinous duct tissues. Lesser amounts, often as glycosides of terpene alcohols, may be present within cells. Some terpenes occur in truly enormous amounts. For example, turpentine may contain 64% of **α-pinene** and juniper oil 65% **α-terpineol**.[82] The large quantities of α-pinene released into the air from pine trees are a major cause of photochemical smog.[83]

Terpenes have a variety of functions. Plant terpenes may deter herbivores and attract pollinators. They may participate in competition among plants and may act as antibiotics, called **phytoalexins**, to protect plants from bacteria and fungi.[84] In invertebrate animals terpenes serve as hormones, pheromones, and defensive repellants (Figs. 22-3, 22-4). The terpene squalene is the precursor to sterols. Some terpenes are toxic. For example, thujone (Fig. 22-3), which is present in the liqueur absinthe, causes serious chronic poisoning.[84]

1. Biosynthesis of Monoterpenes

The compounds of Figure 22-3 each contain ten carbon atoms and are called monoterpenes. They occur largely in plants, but some function in arthropods as pheromones. As with chain elongation, the cyclization of geranyl diphosphate to the various monoterpenes appears to occur through loss of pyrophosphate (as PP_i) with formation of an intermediate carbocation such as that depicted in Equation 22-3.[85–88a] Similar mechanisms initiate cyclization of sesquiterpenes and diterpenes. Numerous terpene cyclases have been isolated, and several have been studied carefully. A stereochemical view of the formation of borneol is illustrated in Eq. 22-4. Both linalyl-*PP* and bornyl-*PP* are intermediates. Croteau and associates suggested that a tight ion pair between carbocation and magnesium pyrophosphate is maintained at each stage.[86,89]

As is indicated in Fig. 22-3, the same intermediate cation can yield a variety of end products. For example, pure geranyl diphosphate: pinene cyclase catalyzes formation of several other terpenes in addition to α-pinene.[89] Another aspect of terpene synthesis is that insects may convert a plant terpene into new compounds for their own use. For example, **myrcene**, which is present in pine trees, is converted by bark beetles to **ipsenol** (Fig. 22-3), a compound that acts as an aggregation pheromone.[90]

Bound magnesium pyrophosphate

O–P–P

Geranyl diphosphate

Geranyl cation

$^-$OPP–Mg^{2+}

Bornyl diphosphate

α-Terpinyl cation

Linalyl diphosphate

H_2O

Borneol

(22-4)

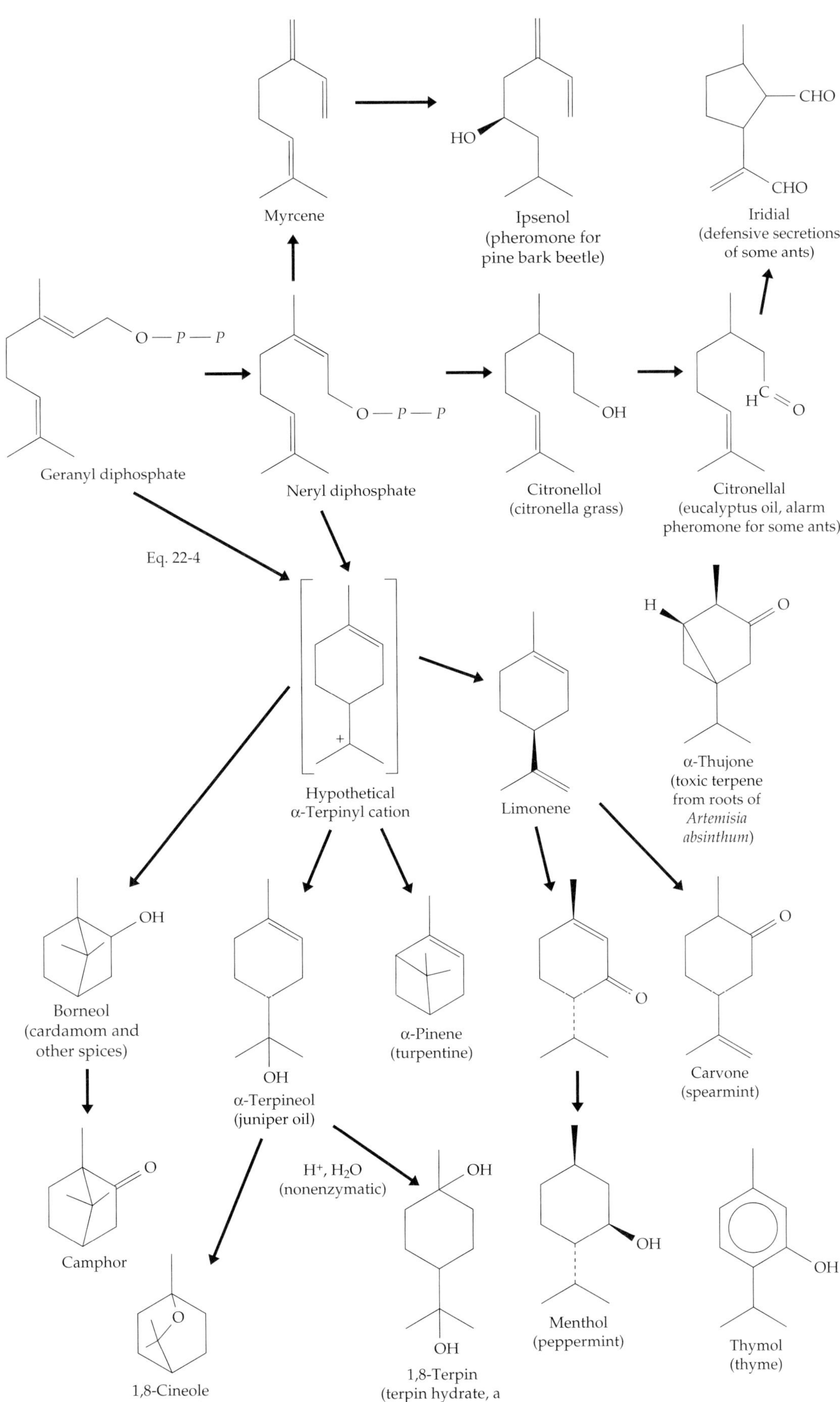

Figure 22-3 Probable biosynthetic pathways of some monoterpenes and related substances. Some of the natural sources are indicated.

2. Sesquiterpenes and Diterpenes

Most of the compounds shown in Figure 22-4 are derived from the C_{15} farnesyl diphosphate. There are more than 300 known cyclic structures among these **sesquiterpenes,** and many sesquiterpene synthases have been characterized.[91,91a] **Aristolochene** is formed by the action of a 38-kDa cyclase that has been isolated from species of *Penicillium* and *Aspergillus*.[92–94] Notice that the synthesis must involve two cyclization steps and migration of a methyl group. Three-dimensional structures are known for at least two terpene synthases,[95,96] and comparison of gene sequences suggests that many others have similar structures. The **5-epi-aristolochene synthase** of tobacco makes the 5-epimer of aristolochene (Fig. 22-4). It binds the diphosphate group within a central cavity using two Mg^{2+} ions held by carboxylates, some of which are in the DDXXD sequence found also in polyprenyl diphosphate synthases. The enzyme active sites of both the epi-aristolochene synthase and a **pentalenene synthase** from *Streptomyces*[96] are rich in polar groups that form hydrogen-bonded networks and which participate in proton abstraction and donation during the rearrangement reactions that must occur.

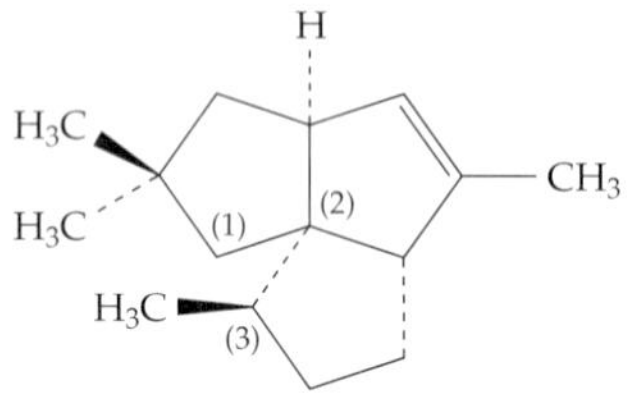

H, H_3C, H_3C, (1), (2), CH_3, H_3C, (3)

Pentalenene

Aromatic groups that are also present may assist in stabilizing intermediate carbocationic species. Deprotonation by an aspartate side chain in the epi-aristolochene synthase has been proposed to assist the cyclization; subsequent reprotonation by an Asp•Tyr•Asp triad would generate a new carbocation and promote the necessary methyl group migration. A detailed step-by-step mechanism has been proposed.[95] The fungal pentalenene synthase has an active site histidine, which is proposed to serve as proton acceptor and donor for the several steps of the reaction of farnesyl diphosphate.[96] The carbon atoms originating from C1 to C3 of the precursor have been marked on the pentalenene structure as well as on the aristolochene structure in Fig. 22-4. We now see how synthases can guide the terpene cyclase reactions to give specific products.

Another sesquiterpene synthase forms **trichodiene**, the parent compound for a family of mycotoxins and antibiotics.[97,97a,b] A different sesquiterpene synthase, present in peppermint and also in a wide range of other plants and animals, forms the acyclic **(E)-β-farnesene**.[98] It serves as alarm pheromone for aphids, has a variety of signaling functions in other insects, and is a urinary pheromone in mice.[98]

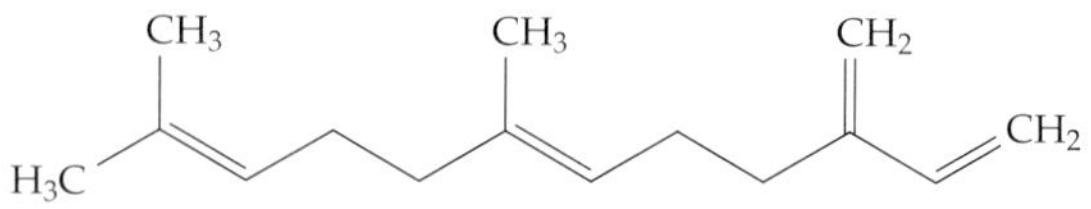

CH_3, CH_3, CH_2, CH_2, H_3C

(*E*)-β-Farnesene

Abscisic acid, one of five known types of plant hormone of general distribution throughout higher plants, is not regarded as a true terpene because it arises by degradation of a carotenoid.[99–99b] However, its structure (Fig. 22-4) is that of a sesquiterpene.

The C_{20} **diterpenes** are derived from geranylgeranyl-*PP*. Among the best known members are another group of plant hormones, the **gibberellins**.[100–102a] The first gibberellin was isolated as a product of plants infected with a *Fusarium* fungus. The rice plants grew in an abnormally tall, weak form. Subsequently, this multimembered class of over 50 highly modified diterpenes have been shown to have a variety of regulatory functions in all higher plants. For example, gibberellins are essential for stem elongation.

Equation 22-5 gives an abbreviated biosynthetic sequence for gibberellin A_1. The ring closure of step *a*, Eq. 22-5, may be initiated by protonation of the double bond at the left of the first structural formula. The resultant carbocation could initiate the consecutive closure of the two rings and the loss of a proton from a methyl group (step *b*) to yield copalyl-*PP*. Steps *c* and *d*, Eq. 22-5, each represent several reactions. In step *c*, pyrophosphate is eliminated, and the methyl group that becomes a methylene in *ent*-kaurene (enantiotopic kaurene) undergoes migration.[102b] Step *d* involves several hydroxylation and oxidation steps as well as a ring contraction through which one of the original methylene groups ends up as a carboxyl group in the final product.[102,103,103a] Deactivation of the hormone is initiated by a 2-oxoglutarate-dependent dioxygenase.[103b]

The **juvenile hormone** of insects (Fig. 22-4) is also of polyprenyl origin.[104,105] However, two of the methyl groups have been converted to (or replaced by) ethyl groups. The isolation and identification of the structure of the juvenile hormone was a difficult task. After its completion it was a surprise to researchers to discover that a large variety of synthetic compounds, sometimes with only a small amount of apparent structural similarity, also serve as juvenile hormones, keeping insects in the larval stage or preventing insect eggs from hatching. Furthermore, a number of plant products such as **juvebione** (Fig. 22-4),[106] which was originally isolated from paper, have the same effect. Thus, in nature products of plant metabolism have a profound effect upon the development of insects that

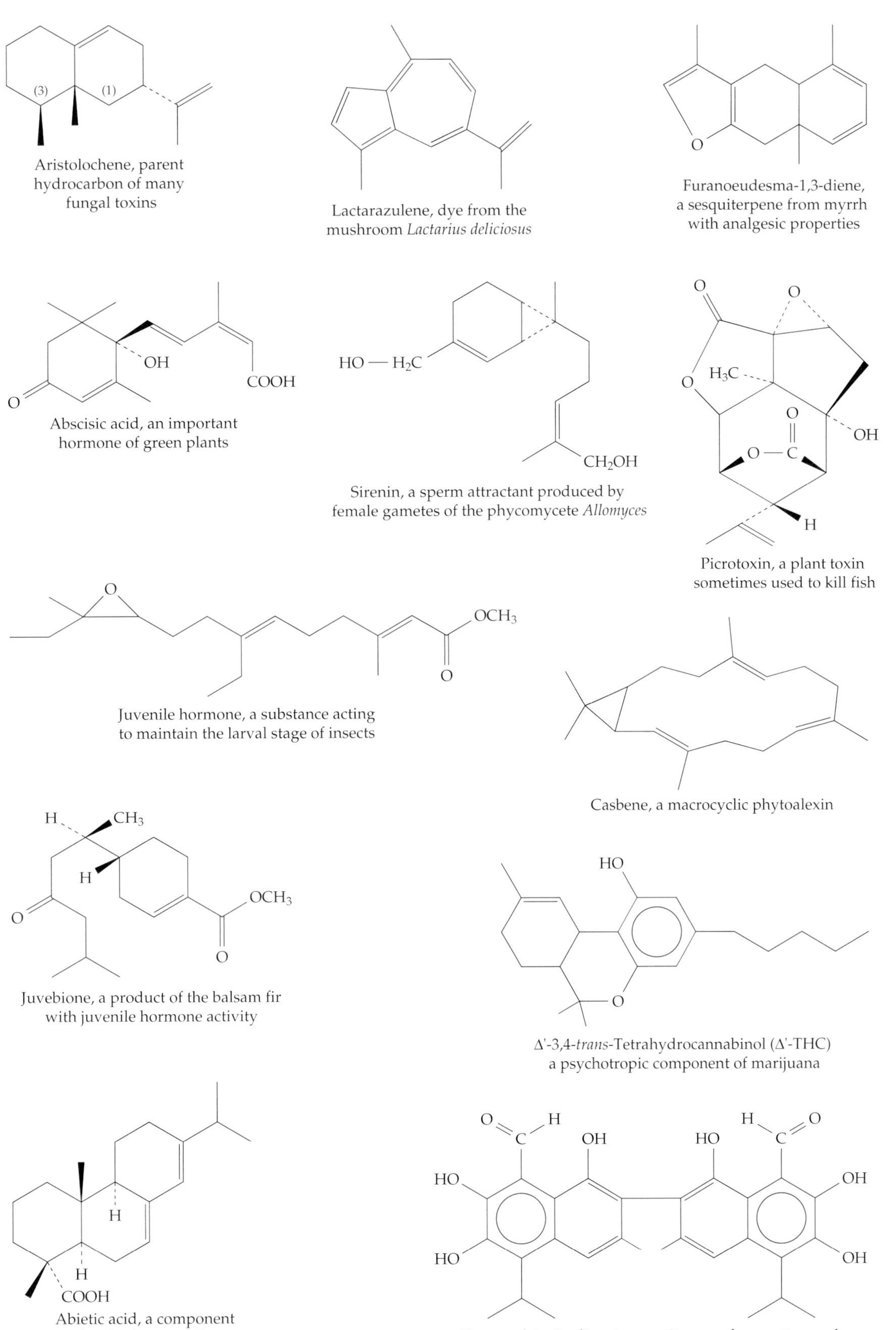

Figure 22-4 More terpenes and related substances. The numbers in parentheses on the aristolochene structure are those of atoms in the precursor farnesyl diphosphate.

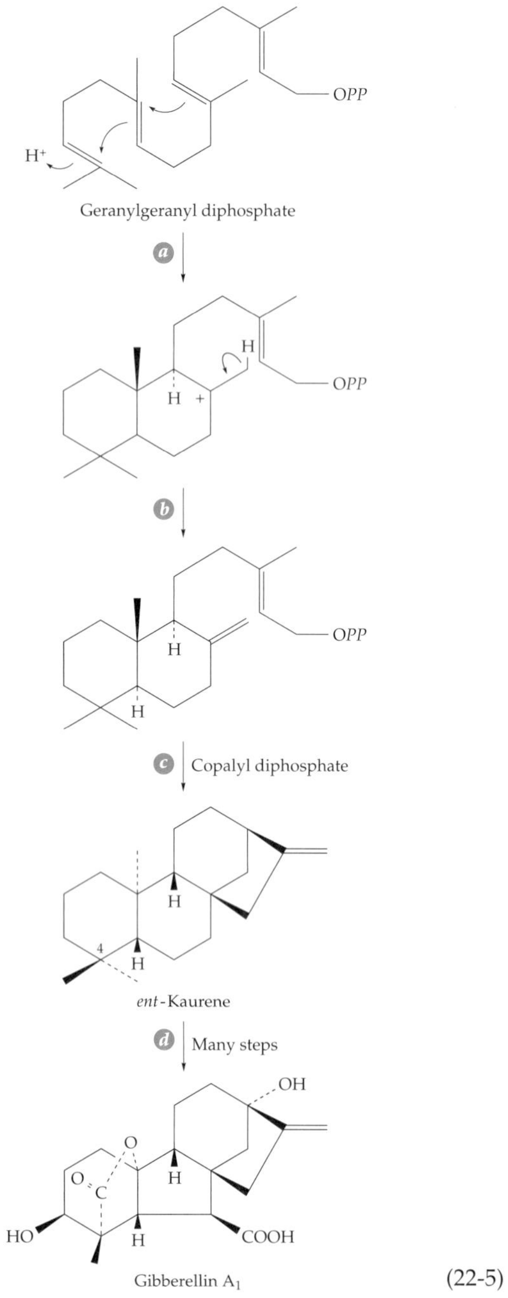

(22-5)

eat the plants. There is interest in the possible use of juvenile hormone, or of synthetic compounds mimicking its action, as insecticides.

Many conifers secrete **oleoresin** (pitch) in response to attack by bark beetles. The oleoresin contains approximately equal amounts of turpentine (a mixture of monoterpenes and sesquiterpenes) and diterpenoid rosin including **abietic acid** (Fig. 22-4).[91,107,107a] The oleoresin is toxic to beetles and, after evaporation of the turpentine, forms a hard rosin seal over the wounds. **Casbene** (Fig. 22-4) is a diterpene produced by castor beans as a phytoalexin (Chapter 31), an antifungal antibiotic.[108] The synthesis of the anticancer compound taxol (Box 7-D) from geranylgeranyl diphosphate involves extensive oxidative and other modification.[109]

3. Formation of the Symmetric Terpenes, Squalene and Phytoene

Two molecules of the C_{15} farnesyl diphosphate can be joined "head to head" to form the C_{30} squalene (Eq. 22-6). Similarly, two C_{20} **geranylgeranyl-*PP*** molecules can be joined to form the C_{40} phytoene (Fig. 22-5), a precursor of carotenoid pigments of plants.[110] In the synthesis of squalene both pyrophosphate groups are eliminated from the precursor molecules, and one proton from C-1 of one of the molecules of farnesyl-*PP* is lost. The other three C-1 hydrogens are retained. At the same time, one proton is introduced from the *pro-S* position of NADPH. Squalene synthase has been difficult to obtain,[111,112] and its mechanism has been uncertain. However, there is strong evidence in favor of carbocationic intermediates.[113–114b] The first step (Eq. 22-7, step *a*) involves reaction of the initial carbocation with the double bond of the second farnesyl-*PP* to form the cyclopropane derivative **presqualene alcohol-*PP***, which was first isolated from yeast as the free alcohol by Rilling and associates.[115] The loss of the second pyrophosphate (Eq. 22-7, step *b*) generates a cyclopropylcarbinyl cation, which can rearrange (step *c*) to a more stable tertiary cation.[112,113] The latter is reduced by NADPH with opening of the cyclopropane ring (Eq. 22-7, step *d*). Once formed squalene diffuses within and between membranes with the help of cytosolic protein carriers.[116]

Phytoene (Fig. 22-5) is apparently formed from geranylgeranyl-*PP* via **prephytoene-*PP*,** whose structure is entirely analogous to that of presqualene-*PP*.[44,117] However, no reduction by NADH is required (Eq. 22-8). It is known that the 5-*pro-R* hydrogen atoms of mevalonate are retained in the phytoene as indicated by a shaded box in Eq. 22-8. Elimination of the other (*pro-S*) hydrogen yields 15,15'–*Z* phytoene (*cis*-phytoene), while elimination of the *pro-R* hydrogen yields all-*E* (trans) phytoene. Higher plants and fungi form mostly *cis*-phytoene, but some bacteria produce the all-*E* isomer.[118]

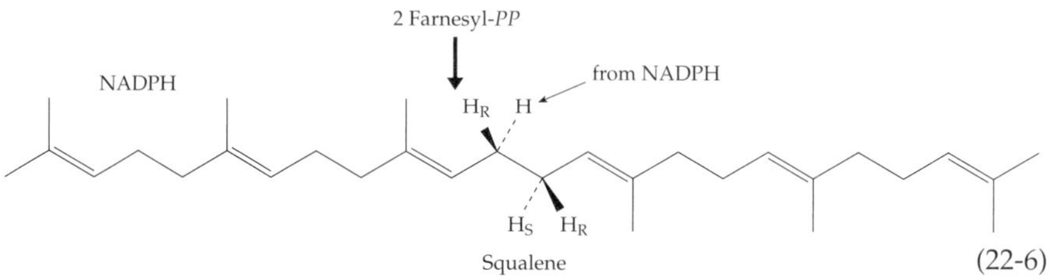

(22-6)

(22-7)

(22-8)

Another polyprenyl compound formed by a head to head condensation is **chrysanthemic acid**. This monoterpene component of the pyrethrum insecticides is formed by chrysanthemums from two molecules of dimethyallyl-*PP* via an intermediate analogous to presqualene alcohol-*PP*.[118a]

A quite different "tail to tail" condensation, whose chemistry is still obscure,[118b] must occur in archaebacteria whose lipids contain the C_{40} diphytanyl alcohol. An example is the diglyceryltetraether,[119,120] whose structure is shown on p. 388.

C. Carotenes and Their Derivatives

Phytoene can be converted to the carotenes by pathways indicated in part in Fig. 22-5 and Eq. 22-10. One of the first products is **lycopene**, the red pigment of tomatoes and watermelons, which is an all-trans compound. If 15-*Z* phytoene is formed, it must, at some point, be isomerized to an all-*E* isomer, and four additional double bonds must be introduced. The isomerization *may* be nonenzymatic. The double bonds are created by an oxygen-dependent **desaturation,** which occurs through the trans loss of hydrogen atoms.

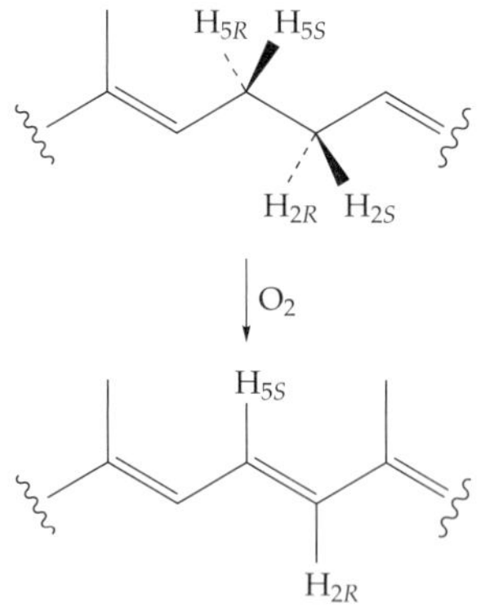

Desaturation takes place in a stepwise fashion, and many intermediate compounds with fewer double bonds are known (Eq. 22-10).[118,121–123] The enzymes required have not been characterized well until recently. Plant enzymes are present in small amounts, and isolation has been difficult. However, the genes for carotenoid biosynthesis in such bacteria as the purple photosynthetic *Rhodobacter*,[118,124] *Rhodospirillum*,[125] and *Rubrivarax*,[126] the cyanobacterium *Synechococcus*,[127] and the nonphotosynthetic *Erwinia*[44,118] have been cloned, sequenced, and used to produce enzymes in quantities that can be studied. Matching genes from higher plants have also been cloned and expressed in bacteria.[123]

Ring closure at the ends of the lycopene molecule to form the carotenes can be formulated most readily through an acid-catalyzed carbocation mechanism (Eq. 22-11). Loss of one or the other of two protons adjacent to the positive charge leads to the β ring of β-carotene or to the α ring of α-carotene.[110,123a] Compounds with only one ring may also be formed.[123b] In many bacteria

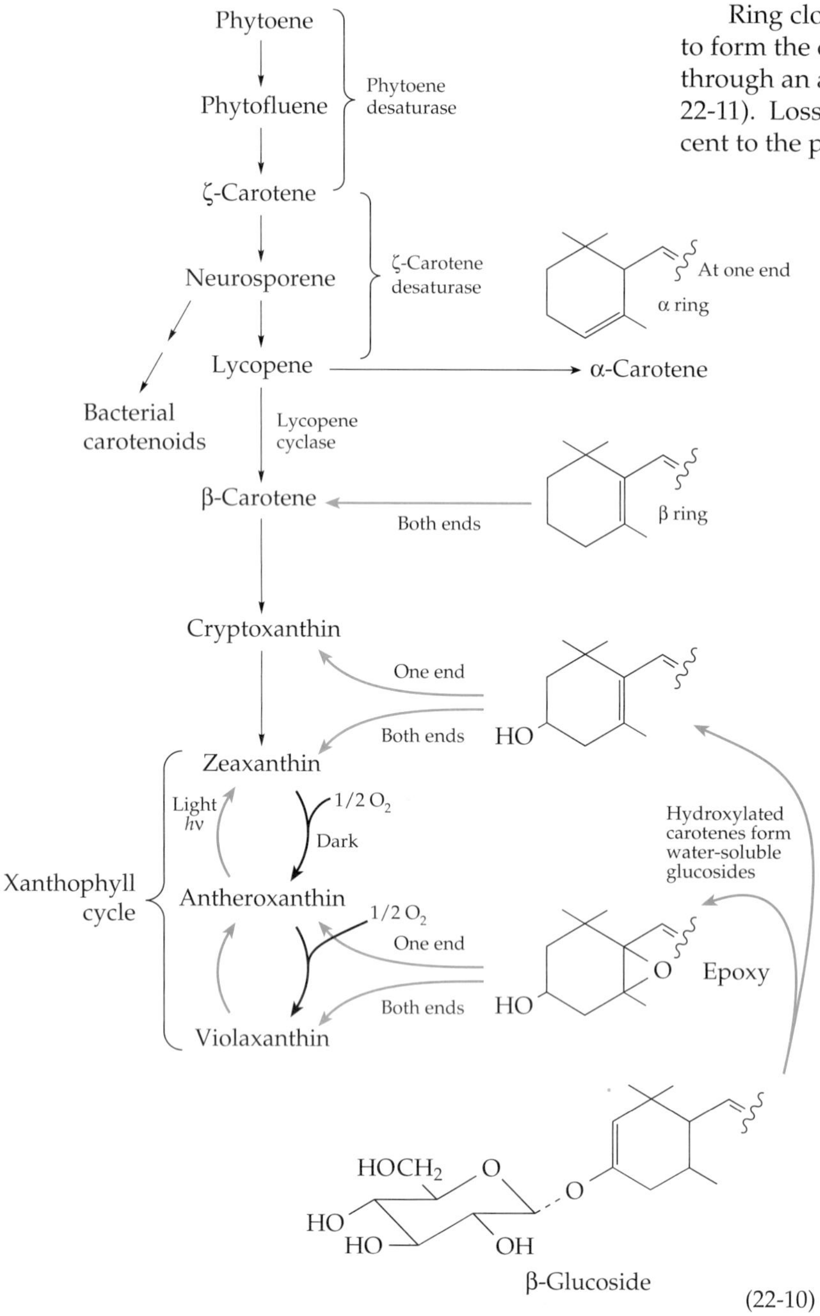

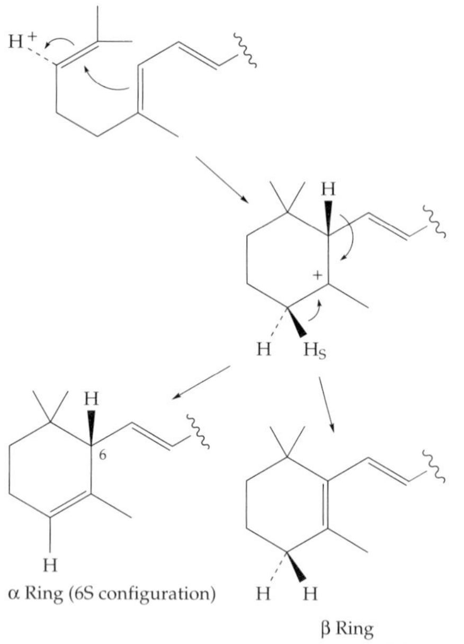

these rings are not formed at all, but the open-chain (acyclic) cartenoids may be modified in ways similar to those of higher plants.[118,124,125]

A genetic engineering success is the transfer of genes for synthesis of β-carotene into rice. The resulting "golden rice" contains enough carotene in its endosperm to make a significant contribution to the vitamin A needs of people for whom

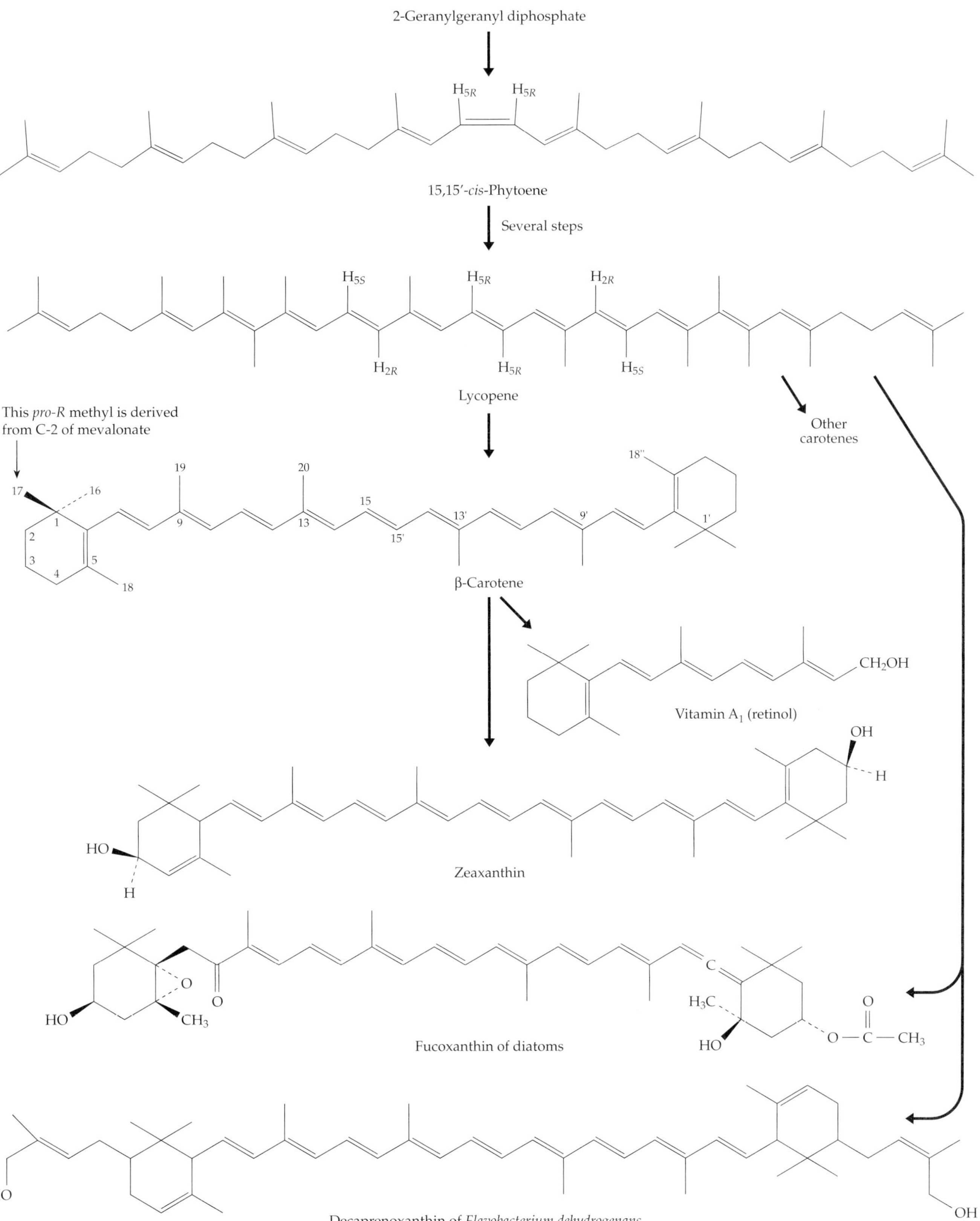

Figure 22-5 Structures and partial biosynthetic pathways for a few of the more than 600 known carotenoid compounds. The origin of some hydrogen atoms from mevalonate is shown, using the numbering for mevalonate. The numbering system for C_{40} carotenoids is also indicated.

rice is a major food. There are estimated to be over 100 million vitamin A-deficient children in the world. As many as one-quarter of a million of these go blind each year.[127a] It is hoped that the golden rice will help to alleviate the problem. At the same time, an ongoing program is supplying vitamin A, which is stored in the liver, at regular intervals of time to many children.[127b]

1. Xanthophylls and Other Oxidized Carotenes

Carotenes can be hydroxylated and otherwise modified in a number of ways.[110,128–131] The structure of zeaxanthin, one of the resulting **xanthophylls**, is indicated in Fig. 22-5. Some other xanthophylls are shown in Eq. 22-10. Lutein resembles zeaxanthin, but the ring at one end of the chain has been isomerized by a shift in double bond position to the accompanying structure. The photosynthetic bacterium *Rhodospirillum rubrum* has its own special carotenoid spirilloxanthin, which has the accompanying structure at both ends of the chain.

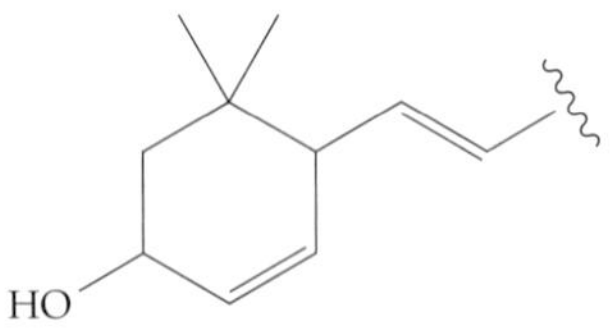

HO

Lutein (structure at one end of molecule)

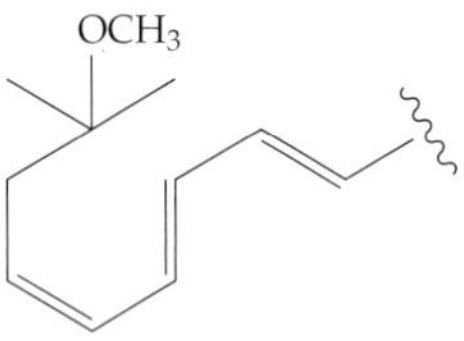

OCH3

Spirilloxanthin

Fucoxanthin (Fig. 22-5) is the characteristic brown pigment of diatoms. One end of the molecule has an epoxide, also formed by the action of O_2, while the other end contains an **allene** structure rare in nature. Even so, fucoxanthin may be the most abundant carotenoid of all. The structure of the allene-containing end of the fucoxanthin molecule (turned over from that shown in Fig. 22-5) is also given in Eq. 22-12. Figure 22-5 does not indicate the stereochemistry of the allene group correctly; the carotenoid chain protrudes behind the ring as drawn in the equation. **Violaxanthin** contains epoxide groups in the rings at both ends of the molecule (Eq. 22-12). An isomerase in algae converts violaxanthin into **neoxanthin** (Eq. 22-12), which contains the allene structure at one end. Subsequent acetylation yields fucoxanthin.

Other algal carotenoids contain acetylenic triple bonds. For example, **alloxanthin** has the following structure at both ends of the symmetric molecule. The symmetric carotenoids **canthaxanthin** and **astaxanthin** have oxo groups at both ends:

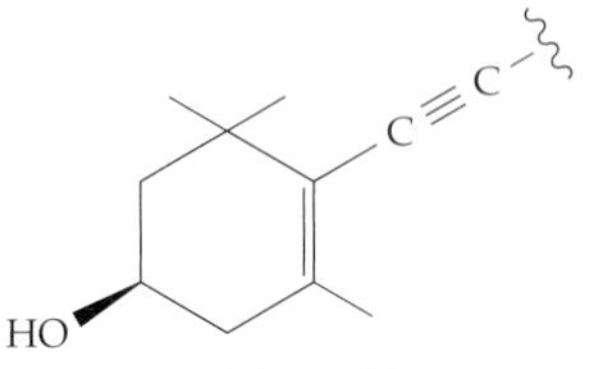

HO C≡C

Alloxanthin

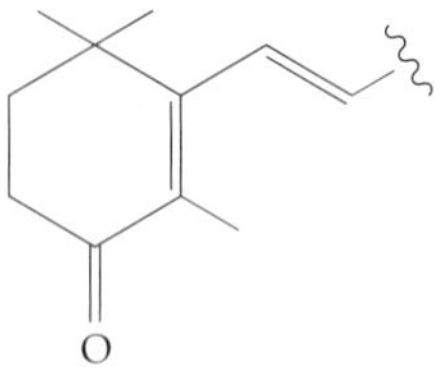

O

Canthaxanthin

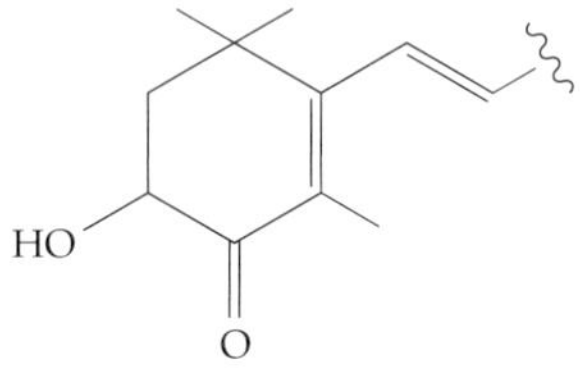

HO O

Astaxanthin

These carotenoids have a limited distribution and occur as complexes, perhaps in Schiff base linkage, with proteins. Astaxanthin-protein complexes with absorption maxima ranging from 410 nm to 625 nm or more provide the color to the lobster's exoskeleton.[130,132] Whereas most naturally occurring carotenoids have all-*E* double bonds, mono-*Z* isomers of canthaxanthin are found in the colored carotenoproteins of the brine shrimp *Artemesia*.[133]

Some bacteria synthesize C_{50} carotenoids such as decaprenoxanthin (Fig. 22-5), the extra carbon atoms at each end being donated from additional prenyl groups, apparently at the stage of cyclization of lycopene.[134] Thus, a carbocation derived by elimination of pyrophosphate from dimethylallyl-*PP* could replace the H^+ shown in the first step of Eq. 22-11. The foregoing descriptions deal with only a few of the many known structural modifications of carotenoids.[2,135,136]

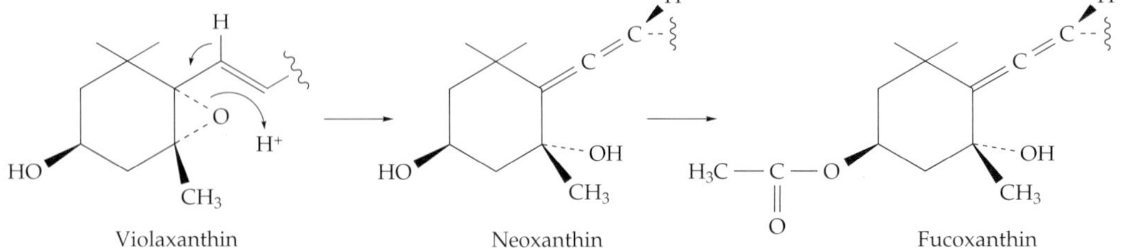

H O H^+ HO CH_3

Violaxanthin (ring structure present at both ends)

H C C OH HO CH_3

Neoxanthin (this structure occurs at one end only)

H C C OH H_3C—C—O O CH_3

Fucoxanthin

(22-12)

BOX 22-A VITAMIN A

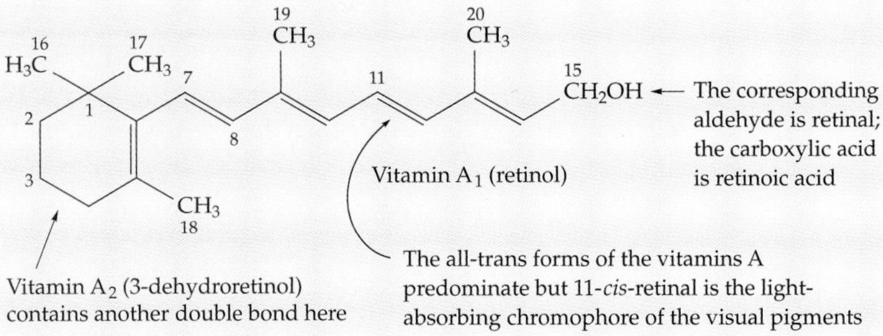

The recognition in 1913 of vitamin A (Box 14-A) was soon followed by its isolation from fish liver oils.[a–c] Both vitamin A_1 (**retinol**) and vitamin A_2 are 20-carbon polyprenyl alcohols. They are formed by cleavage of the 40-carbon β-carotene (Fig. 22-5) or other carotenoids containing a β-ionone ring. While the carotenes are plant products, vitamin A is produced only in animals, primarily within cells of the intestinal mucosa.[d–f] The carotene chains are cleaved in the center, and to some extent in other positions,[g] by oxygenases; β-carotene yields as many as two molecules of the vitamin A aldehyde **retinal**.[h–i] The retinal is reduced by NADH to retinol which is immediately esterified, usually with saturated fatty acids, by transfer of an acyl group from a fatty acyl-CoA or from phosphatidylcholine. The resulting retinyl esters are transported in chylomicrons. They remain in the chylomicron remnants (see Fig. 21-1), which are taken up by the liver where both hydrolysis and reesterification occur.[j] Vitamin A is one of the few vitamins that can be stored in animals in relatively large quantities. It accumulates in the liver, mainly as retinyl palmitate, in special storage cells termed stellate cells.[f] The human body usually contains enough vitamin A to last for several months.

Free retinol is released from the liver as a 1:1 complex of retinol with the 21-kDa **retinol-binding protein**.[k,l] This protein is normally almost saturated with retinol and is bound to another serum protein, the 127-residue **transthyretin** (prealbumin).[m,n] Some of the retinol is oxidized to **retinoic acid**. Both all-*trans* and 13-*cis*-retinoic acids as well as 5,6-epoxyretinoic acid are found in tissue.[f,o,p,pq]

Another metabolite, which may be very important, is **retinoyl β-glucuronide**.[q,r,rs]

Cell surfaces of body tissues appear to contain receptors for the retinol-binding protein. Many cells also contain cytoplasmic retinol-binding proteins[s–u] as well as proteins that bind retinoic acid.[u–y] These proteins are members of the large superfamily of hydrophobic transporter molecules described in Box 21-A. This includes the milk protein **β-lactoglobulin,** which also forms a complex with retinol.[z,aa]

A strikingly early symptom of vitamin A deficiency is **night blindness**. A variety of other symptoms include dry skin and hair, conjunctivitis of the eyes, retardation of growth, and low resistance to infection. The skin symptoms are particularly noticeable in the internal respiratory passages and alimentary canal lining. About 0.7 mg/day of vitamin A is required by an adult. The content of vitamin A in foods is often expressed in terms of international units: 1.0 mg of retinol equals 3333 I.U.

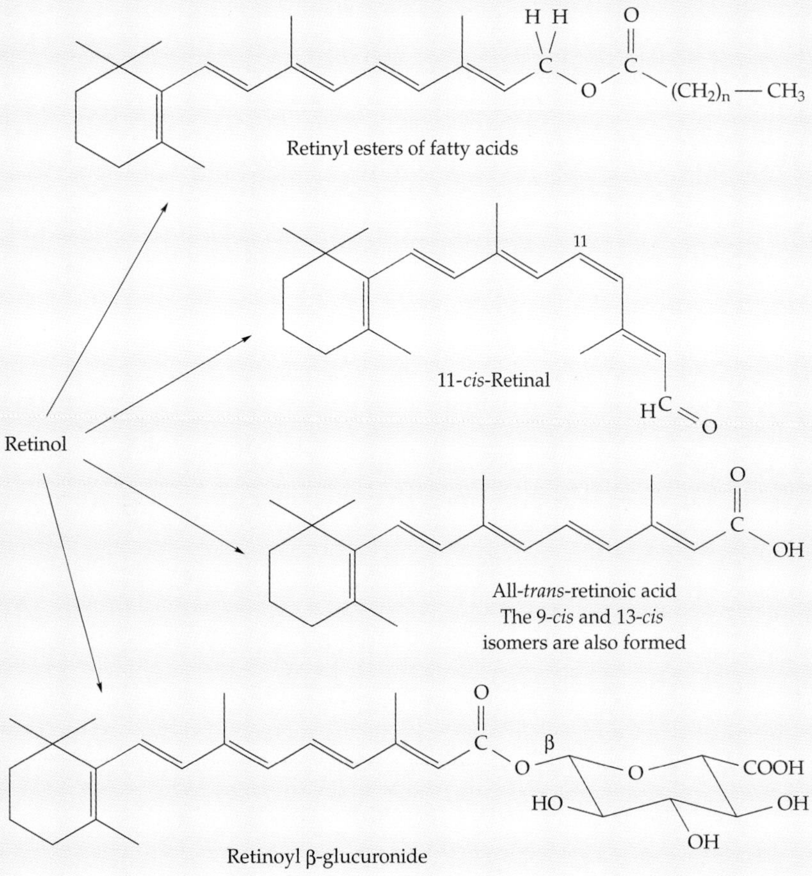

BOX 22-A VITAMIN A (continued)

Vitamin A, as retinal, has a clearly established role in vision (Chapter 23) and apparently has a specialized function in reproduction. In vitamin A deficiency no sperm cells are formed in males, and fetal resorption occurs in females. Rats deprived of vitamin A but fed retinoic acid become blind and sterile but otherwise appear healthy.[e,bb] Evidently either the alcohol or the aldehyde has an essential function in reproduction, whereas bone growth and maintenance of mucous secretions requires only retinoic acid. Indeed, retinoic acid is 100 to 1000 times more active than other forms of vitamin A in these differentiation functions.[r]

In vitamin A deficiency the internal epithelial surfaces of lungs and other tissues, which are usually rich in mucous secreting cells and in ciliated cells, develop thick layers of keratinizing squamous cells similar to those on the external surface of the body. The synthesis of some mannose- and glucosamine-containing glycoproteins consequently decreases.[cc] The major effects of retinoic acid is evidently through regulation of transcription (Chapter 28). In developing lungs retinoic acid promotes the transformation of undifferentiated epithelial cells into mucus-secreting cells.[dd]

Do we know all of the special chemistry of vitamin A that is involved in its functions? Retinal could form Schiff bases with protein groups as it does in the visual pigments. Redox reactions could occur. Conjugative elimination of water from retinol to form **anhydroretinol** is catalyzed nonenzymatically by HCl. Anhydroretinol occurs in nature and

CH_2

Anhydroretinol

may serve as an inhibitor of the action of 14-hydroxy-*retro*-retinol in lymphocyte differentiation.[ee,ff]

Much recent interest has been aroused by the fact that retinoid compounds, including both retinol and retinoic acid, reduce the incidence of experimentally induced cancer. In addition, 13-*cis*-retinoic acid taken orally is remarkably effective in treatment of severe cystic acne.[gg] However, both vitamin A and retinoic acid in large doses are **teratogenic**, i.e., they cause fetal abnormalities. The use of 13-*cis*-retinoic acid during early phases of pregnancy led to a high incidence of major malformations in infants born.[hh]

[a] Moore, T. (1981) *Trends Biochem. Sci.* **6**, 115–116
[b] Moore, T. (1957) *Vitamin A*, Elsevier, Amsterdam
[c] Olson, J. A. (1968) *Vitam. Horm.(N. Y.)* **26**, 1–63
[d] Blomhoff, R., ed. (1994) *Vitamin A in Health and Disease*, Dekker, New York
[e] Goodman, D. S. (1984) *N. Engl. J. Med.* **310**, 1023–1031
[f] Blomhoff, R., Green, M. H., Berg, T., and Norum, K. R. (1990) *Science* **250**, 399–404
[g] Tang, G., Wang, X.-D., Russell, R. M., and Krinsky, N. I. (1991) *Biochemistry* **30**, 9829–9834
[h] Redmond, T. M., Gentleman, S., Duncan, T., Yu, S., Wiggert, B., Gantt, E., and Cunningham, F. X., Jr. (2001) *J. Biol. Chem.* **276**, 6560–6565
[hi] Kiefer, C., Hessel, S., Lampert, J. M., Vogt, K., Lederer, M. O., Breithaupt, D. E., and von Lintig, J. (2001) *J. Biol. Chem.* **276**, 14110–14116
[i] Symons, M. (1996) *Trends Biochem. Sci.* **21**, 178–181
[j] Blomhoff, R., Rasmussen, M., Nilsson, A., Norum, K. R., Berg, T., Blaner, W. S., Kato, M., Mertz, J. R., Goodman, D. S., Eriksson, U., and Peterson, P. A. (1985) *J. Biol. Chem.* **260**, 13560–13565
[k] Zanotti, G., Ottonello, S., Berni, R., and Monaco, H. L. (1993) *J. Mol. Biol.* **230**, 613–624
[l] Zanotti, G., Marcello, M., Malpeli, G., Folli, C., Sartori, G., and Berni, R. (1994) *J. Biol. Chem.* **26**, 29613–29620
[m] Monaco, H. L., Rizzi, M., and Coda, A. (1995) *Science* **268**, 1039–1041
[n] Blake, C. C. F., Geisow, M. J., Swan, I. D. A., Rerat, C., and Rerat, B. (1974) *J. Mol. Biol.* **88**, 1–12
[o] Napoli, J. L., Khalil, H., and McCormick, A. M. (1982) *Biochemistry* **21**, 1942–1949
[p] Napoli, J. (1996) *FASEB J.* **10**, 993–1001
[pq] Tryggvason, K., Romert, A., and Eriksson, U. (2001) *J. Biol. Chem.* **276**, 19253–19258
[q] Barua, A. B., and Olson, J. A. (1986) *Am. J. Clin. Nutr.* **43**, 481–485
[r] Formelli, F., Barua, A. B., and Olson, J. A. (1996) *FASEB J.* **10**, 1014–1024
[rs] Cullum, M. E., and Zile, M. H. (1985) *J. Biol. Chem.* **260**, 10590–10596
[s] Rong, D., Lovey, A. J., Rosenberger, M., d'Avignon, A., Ponder, J., and Li, E. (1993) *J. Biol. Chem.* **268**, 7929–7934
[t] Quadro, L., Blaner, W. S., Salchow, D. J., Vogel, S., Piantedosi, R., Gouras, P., Freeman, S., Cosma, M. P., Colantuoni, V., and Gottesman, M. E. (1999) *EMBO J.* **18**, 4633–4644
[u] Newcomer, M. E. (1995) *FASEB J.* **9**, 229–239
[v] Ross, A. C. (1993) *FASEB J.* **7**, 317–327
[w] Thompson, J. R., Bratt, J. M., and Banaszak, L. J. (1995) *J. Mol. Biol.* **252**, 433–446
[x] Chen, X., Tordova, M., Gilliland, G. L., Wang, L., Li, Y., Yan, H., and Ji, X. (1998) *J. Mol. Biol.* **278**, 641–653
[y] Mansfield, S. G., Cammer, S., Alexander, S. C., Muehleisen, D. P., Gray, R. S., Tropsha, A., and Bollenbacher, W. E. (1998) *Proc. Natl. Acad. Sci. U.S.A.* **95**, 6825–6830
[z] Godovac-Zimmermann, J. (1988) *Trends Biochem. Sci.* **13**, 64–66
[aa] Cho, Y., Batt, C. A., and Sawyer, L. (1994) *J. Biol. Chem.* **269**, 11102–11107
[bb] Clamon, G. H., Sporn, M. B., Smith, J. M., and Saffiotti, V. (1974) *Nature (London)* **250**, 64–66
[cc] Rossa, G. C., Bendrick, C. J., and Wolf, G. (1981) *J. Biol. Chem.* **256**, 8341–8347
[dd] Chytil, F. (1996) *FASEB J.* **10**, 986–992
[ee] Buck, J., Derguini, F., Levi, E., Nakanishi, K., and Hammerling, U. (1991) *Science* **254**, 1654–1656
[ff] Grün, F., Noy, N., Hämmerling, U., and Buck, J. (1996) *J. Biol. Chem.* **271**, 16135–16138
[gg] Greenwald, D., DeWys, W., Black, G., Malone, W., Sporn, M., and Underwood, B. A. (1984) *Science* **224**, 338
[hh] Lammer, E. J., Chen, D. T., Hoar, R. M., Agnish, N. D., Benke, P. J., Braun, J. T., Curry, C. J., Fernhoff, P. M., Grix, A. W., Jr., Lott, I. T., Richard, J. M., and Sun, S. C. (1985) *N. Engl. J. Med.* **313**, 837–841

2. Properties and Functions of Carotenes

The most characteristic property of carotenoids is the striking color, most often yellow to red, which is used by birds as a sexual attractant and by plants to attract pollinators.[137,138] The associated light absorption fits these compounds for a role in photosynthetic light-harvesting,[139] in photoprotection, and in photoreception,[140] matters that are dealt with in Chapter 23. One aspect of photoprotection, which involves the **xanthophyll cycle,** is also indicated in Eq. 22-10. The cycle allows green plants to adjust to varying light intensity by altering the amount of zeaxanthin available for quenching excessive amounts of photoexcited chlorophyll (Chapter 23). Zeaxanthin undergoes epoxidation by O_2 to form antheroxanthin and violaxanthin as shown in Eq. 22-10. The process requires NADPH and reduced ferredoxin.[131] When light intensity is high the process is reversed by an ascorbate-dependent violaxanthin **de-epoxidase**.[99,128,141]

Violaxanthin also functions as a precursor to the plant hormone abscisic acid. Compare the structure of the latter (Fig. 22-4) with those of carotenoids. Oxidative cleavage of violaxanthin or related epoxy-carotenoids initiates the pathway of synthesis of this hormone.[142,143]

The system of conjugated double bonds responsible for carotenoid colors also helps to impart specific shapes to these largely hydrophobic molecules and ensures that they occupy the appropriate niches in the macromolecular complexes with which they associate. Information on stereochemistry is provided in a short review by Britton.[138]

β-Carotene, which can serve as an antioxidant at low oxygen pressures and can quench singlet oxygen,[144,145] has been associated with a reduced incidence of lung cancer.[137,146] While most animals do not synthesize carotenoids, they use them to make vitamin A and related retinoids and also as colorants. Yellow and red pigments of bird feathers[147] and the colors of tissues of salmon and of lobsters and other invertebrates are derived from dietary carotenoids, which are often modified further by the new host. The lobster accumulates astaxanthin, as a blue protein complex,[148] and the flamingo uses the astaxanthin of shrimp to color its feathers.[130]

Dietary carotenes and carotenoids are absorbed and transported in the plasma of humans and animals by lipoproteins.[149] The conversion of carotenes to vitamin A (Box 22-A) provides the aldehyde **retinal** for synthesis of visual pigments (Chapter 23) and **retinoic acid**, an important regulator of gene transcription and development (Chapter 32).[150–152c] See also Section E,5.

D. Steroid Compounds

The large class of **steroids** contains a characteristic four-ring nucleus consisting of three fused six-membered rings and one five-membered ring.[153] **Cholestanol** (dihydrocholesterol) may be taken as a representative steroid alcohol or **sterol**. Most sterols, including cholestanol, contain an 8- to 10-carbon side chain at position 17. The polyprenyl origin of the side chain is suggested by the structure. Steroid compounds usually contain an oxygen atom at C-3. This atom is present in an –OH group in the sterols and

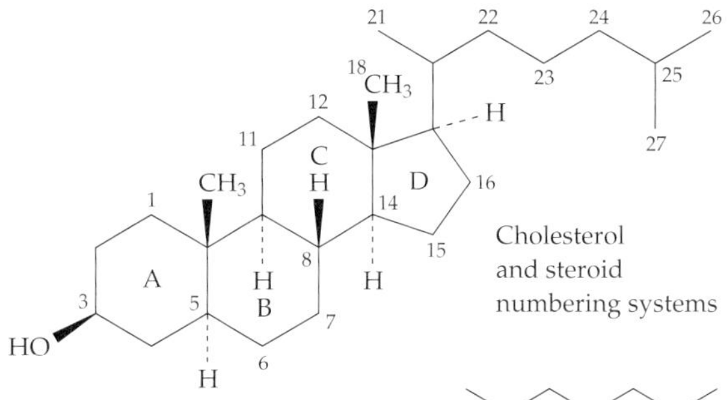

Cholesterol and steroid numbering systems

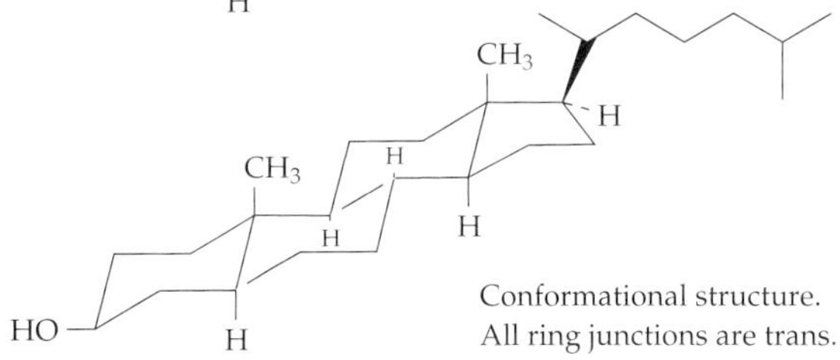

Conformational structure. All ring junctions are trans.

frequently in a carbonyl group in other steroids. Most steroids contain two axially oriented methyl groups, the "angular methyl groups,"which are attached to the ring system and numbered C-18 and C-19. In the customary projection formulas they are to be thought of as extending forward toward the viewer. In the same manner, the equatorially oriented 3-OH group of cholestanol and the side chain at C-17 also project forward toward the viewer in the projection formula.

The angular methyl groups, the 3-OH groups, and the side chain of cholestanol are all on the same side of the steroid ring in the projection formula and are all said to have a **β orientation**. Substituents projecting from the opposite side of the ring system are **α oriented**. While the methyl groups (C-18 and C-19) almost always have the β orientation, the 3-OH group has the α orientation in some sterols. Dashed lines are customarily used to connect α-oriented substituents, and solid lines are used for β-oriented substituents in structural formulas. Cholesterol is chiral and its enantiomer does not support life for *C. elegans* and presumably for other organisms.[153a]

In cholestanol the ring fusions between rings A and B, B and C, and C and D are all trans; that is, the hydrogen atoms or methyl groups attached to the bridgehead carbon atoms project on opposite sides of the ring system. This permits all three of the six-membered rings to assume relatively unstrained chair conformations. However, the introduction of a double bond alters the shape of the molecule significantly. Thus, in cholesterol the double bond between C-5 and C-6 (Δ^5) distorts both the A and B rings from the chair conformation. In some steroid compounds the junction between rings A and B is cis. This greatly alters the overall shape of the steroid from the relatively flat one of cholestanol to one that is distinctly bent. An example is **β-coprostanol**, a product of bacterial action on cholesterol and a compound occurring in large amounts in the feces. In some sterols, notably the estrogenic hormones, ring A is completely aromatic and the methyl group at C-19 is absent.

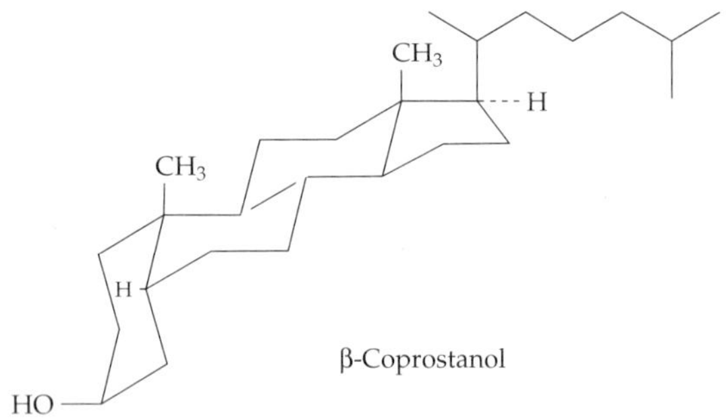

β-Coprostanol

1. Biosynthesis of Sterols

Most animal steroids arise from cholesterol, which in turn is derived from squalene. This C_{30} triterpene, whose biosynthesis is described in Section B, is named after the dogfish *Squalus* in whose liver it accumulates as a result of blockage in oxidation to cholesterol. Squalene is also a prominent constituent of human skin lipids. Its conversion to **cholesterol**, which takes place in most animal tissues,[117,154–156] is initiated by a microsomal enzyme system that utilized O_2 and NADPH to form **squalene 2,3-oxide** (Fig. 22-6, step *a*). The subsequent cyclization reaction, which probably takes place through a carbocation created by attack of a proton on the oxygen atom of the epoxide ring (Fig. 22-6, step *b*), is catalyzed by the large 70- to 80-kDa **oxidosqualene cyclase**.[157–159] The enzyme from rat liver consists of 733 residues and contains a highly conserved sequence with two consecutive aspartates that are thought to be at the active site. The sequence is somewhat similar to that of prenyltransferases and sesquiterpene cyclases.[157] The cyclization step appears to require that the enzyme hold the substrate in a rigid conformation as indicated in Fig. 22-6. The flow of electrons effects the closure of all four rings. The carbocation created at C-2 of squalene (C-4 of the sterol ring that is formed) by opening of the epoxide ring reacts with electrons from the 6,7 double bond to close ring A leaving a carbocation at C-6. This in turn reacts with the 10,11 double bond leaving a carbocation at C-10, etc. At the end of this cascade a carbocation is left on C-19 of squalene, which is numbered C-20 in the incipient sterol. The closures of rings A and B both follow the Markovnikov rule by generating relatively stable tertiary carbocations. Thus, the natural chemical reactivities of the substrates are followed in these enzymatic reaction steps. However, this is not the case in the closure of ring C to form a 6-membered ring instead of a 5-membered ring. This presumably happens because the enzyme imposes the correct geometry for a 6-membered ring on the squalene and the correct stereochemistry on the ring closure.[160]

The rearrangement of this initially created C-20 carbocation to **lanosterol** (Fig. 22-6, step *c*) is also a remarkable reaction that requires the shift of a hydride ion and of two methyl groups, as indicated by the arrows in the figure. In addition, a hydrogen at C-9 (sterol numbering) is lost as a proton. Lanosterol is named for its occurrence in lanolin, the waxy fat in wool. Although the principal component of lanolin is cholesterol, lanosterol is its precursor both in sheep and in all other animals. Cholesterol is in turn the precursor to other animal sterols. The cholesterol biosynthetic pathway also provides cells with a variety of important signaling molecules.[160a]

In green plants, which contain little or no cholesterol, **cycloartenol** is the key intermediate in sterol biosynthesis.[161–162a] As indicated in Fig. 22-6, step *c′*, cycloartenol can be formed if the proton at C-9 is shifted (as a hydride ion) to displace the methyl group from C-8. A proton is lost from the adjacent methyl group to close the cyclopropane ring. There are still other ways in which squalene is cyclized,[162,163,163a] including some that incorporate nitrogen atoms and form alkaloids.[163b] One pathway leads to the **hopanoids**. These triterpene derivatives function in bacterial membranes, probably much as cholesterol does in our membranes. The three-dimensional structure of a bacterial hopene synthase is known.[164,164a] Like glucoamylase (Fig. 2-29) and farnesyl transferase, the enzyme has an $(\alpha,\alpha)_6$-barrel structure in one domain and a somewhat similar barrel in a second domain. The active site lies in a large interior cavity. The properties of the hopene synthase are similar to those of oxidosqualene synthase, and it appears to function by a similar mechanism, which resembles that of Fig. 22-6 but does not depend upon O_2. Hopene lacks polar groups, but these are provided in the hopanoids by a polyol side chain. One of these compounds, **bacteriohopanetetrol**, may be one of the most abundant compounds on earth.[160,165,166] Hopanoids appear to originate from mevalonate synthesized via the 1-deoxyxylulose pathway (Fig. 22-2). The polyol side chain is probably formed from ribose.[166]

Hopene

OH OH OH OH

Bacteriohopanetetrol
(projection formula)

Formation of cholesterol. The conversion of lanosterol to cholesterol requires at least 19 steps,[167,168] which are catalyzed by enzymes bound to membranes of the ER. The removal of the three methyl groups of lanosterol, the migration of the double bond within the B ring, and the saturation of the double bond in the side chain may occur in more than one sequence, two of which are indicated in Fig. 22-7. The predominant pathway in many organisms including humans is the oxidative demethylation at the C / D ring junction (C-14) by a cytochrome P450 called **lanosterol 14α-demethylase**. This single enzyme catalyzes three consecutive O_2 and NADPH-dependent reactions that convert the methyl group to hydroxymethyl, formyl, and then free formate (right side of Fig. 22-8).[169–172b] Steps *a* and *b* are typical cytochrome P450 oxygenation reactions. In step *b* a geminal diol is formed and is dehydrated to the formyl derivative. The third step is atypical. Shyadehi *et al.* proposed the sequence depicted in steps *c-f* in which an Fe(III) peroxo intermediate reacts as shown.[172] This mechanism is supported by the fact that both ^{18}O present in the formyl group and ^{18}O from $^{18}O_2$ appear in the liberated formate.

The corresponding reactions of the methyl groups at C-4 on the A ring[167,168,173] are depicted on the left side of Fig. 22-8. The 4α methyl group is first hydroxylated by a microsomal (ER) system similar to cytochrome P450 but able to accept electrons from NADH and cytochrome b_5 rather than NADPH.[173] The two-step oxidation of the resulting alcohol to a carboxylic acid is catalyzed by the same enzyme. A second enzyme catalyzes the dehydrogenation of the 3-OH group to a ketone allowing for efficient β-decarboxylation (Fig. 22-8, steps *j* and *k*).[173a] Inversion of configuration at C-4, assisted by the 3-carbonyl group (step *l*), places the second 4-methyl group in the α orientation. After reduction of the 3-carbonyl by a third enzyme the sequence is repeated on this second methyl group.

In addition to the enzymes that are embedded in the membranes of the ER, conversion of lanosterol to cholesterol depends upon soluble cytoplasmic carrier proteins.[174] See also Box 21-A. Other sterols formed in

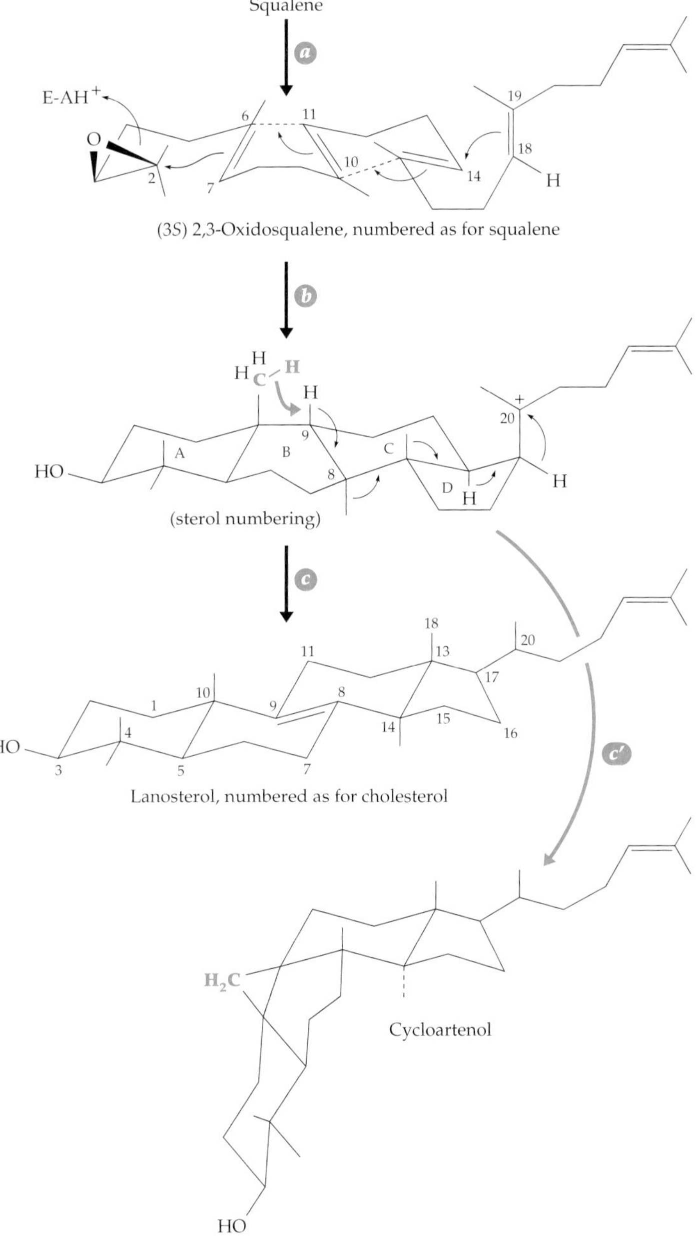

Figure 22-6 The cyclization of all-*trans* squalene to lanosterol and cycloartenol.

the animal body are **7-dehydrocholesterol**, prominent in skin and a precursor of vitamin D. Both β-cholestanol and its isomer β-coprostanol are formed by bacteria in the intestinal tract, and small amounts of cholesterol are converted to cholestanol within tissues. **Ergosterol**, the most common sterol in fungi, contains the $\Delta^{5,7}$ ring system of 7-dehydrocholesterol as well as an extra double bond in the side chain. It arises from zymosterol (Fig. 22-7).[173,174a]

As indicated previously, plant sterols are thought to be formed in most cases through cycloartenol which is often converted to **24-methylenecycloartenol**, a substance present in grapefruit peel and in many other plants. The methylene carbon is donated by *S*-adenosylmethionine (AdoMet) as shown in Eq. 22-13, which implies a transient intermediate carbocation. Saturation of the side chain and oxidative demethylation similar to that shown in Fig. 22-8[175] and introduction of a double bond[176] leads to **campesterol** (Fig. 22-9). It has the Δ^5-unsaturated ring of cholesterol but, like many other plant sterols, the side chain has one additional methyl group, which is also donated from *S*-adenosylmethionine.[161,177,178] Several more steps are required to convert campesterol into the plant steroid hormone **brassinolide**.[179–180] Among higher plants, **sitosterol** and **stigmasterol** are the most common sterols. Each contains an extra ethyl group in the side chain. Sitosterol is formed by the methylation (by AdoMet) of ergosterol. For the guinea pig stigmasterol is a vitamin, the"antistiffness factor" necessary to prevent stiffening of the joints. Some other plant steroids arise without addition of the extra carbons at C-23 or C-24 but usually via a different cyclization of squalene. Of these, the cucurbitacins (Fig. 22-9) are among the bitterest substances known.[181]

AdoMet
S^+
H_3C
20
CH_3
CH_3
Side chain of cycloartenol
→
CH_2
24
CH_3
CH_3
24-Methylenecycloartenol

(22-13)

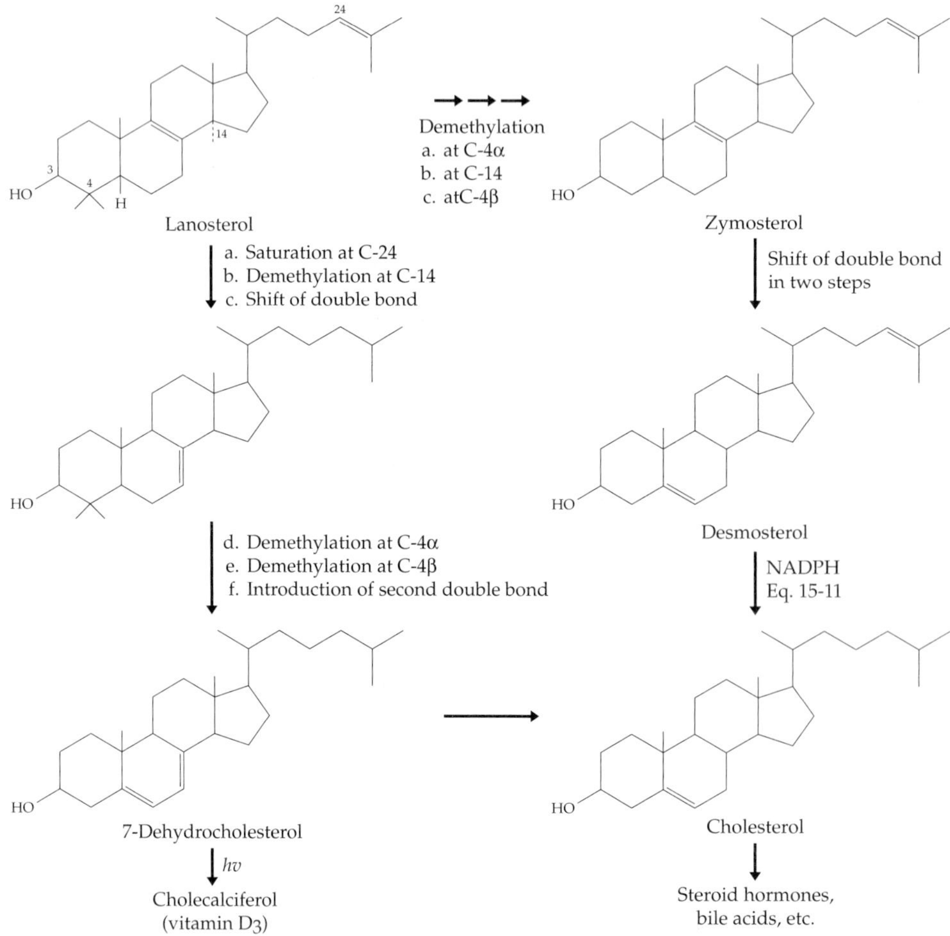

Figure 22-7 Conversion of lanosterol to cholesterol. Two of many possible sequences are shown.

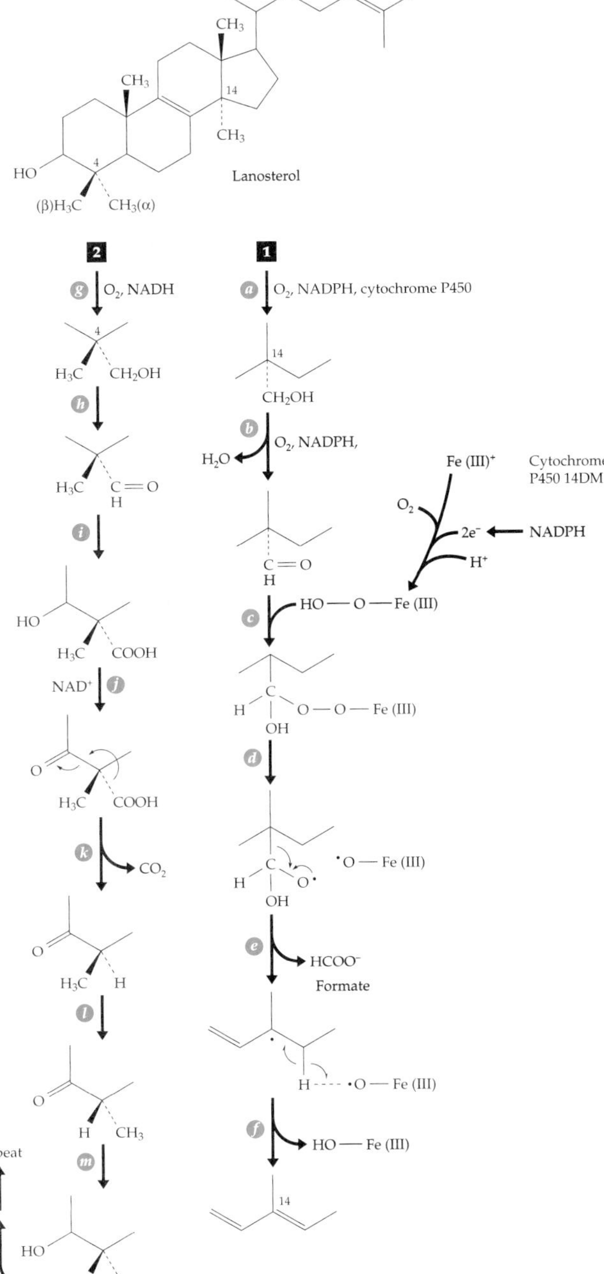

Figure 22-8 Steps in the demethylation of lanosterol. The most frequent sequence, labeled [1], begins with demethylation at C-14 by the action of a cytochrome P450 and is followed [2] by the successive demethylation of the α-CH_3 and β-CH_3 at C-4 by an NADH-dependent oxygenase.

2. Metabolism of Cholesterol in the Human Body

Cholesterol is both absorbed from the intestinal tract and synthesized from acetate via squalene, principally in the liver. The quantities produced are substantial. Daily biosynthesis is ~600 mg, and dietary uptake may supply another 300 mg.[182] Not only is there a large amount of cholesterol in the brain and other nervous tissues but also about 1.7 g of cholesterol per liter is present in blood plasma, about two-thirds of it being esterified principally to unsaturated fatty acids. The cholesterol content of plasma varies greatly with diet, age, and sex. By age 55 it averages 2.5 g / liter and may be considerably higher. Women up to the age of menopause have distinctly lower blood cholesterol than do men. Cholesterol regulates its own abundance by a variety of feedback mechanisms.[183,184] These include inhibition of the synthesis by means of reduced activities (step *a* of Fig. 22-1) of HMG-CoA reductase, farnesyl diphosphate synthase (step *g* of Fig. 22-1), and squalene synthase. All of these reactions are essential steps in cholesterol synthesis.[16,185] On the other hand, cholesterol induces an increase in acyl-CoA:cholesterol acyltransferase.

Dietary cholesterol, together with triacylglycerols, is absorbed from the intestinal tract and enters the large lipoprotein chylomicrons (see Fig. 21-1). Absorption of cholesterol is incomplete, usually amounting to less than 40% of that in the diet. Absorption requires bile salts and is influenced by other factors.[186] As it is needed cholesterol is taken from the plasma lipoproteins into cells by endocytosis. Much of the newly absorbed cholesterol is taken up by the liver. The liver also secretes cholesterol, in the form of esters with fatty acids, into the bloodstream.

Cholesterol is synthesized in the ER and other internal membranes by most cells of the body.[187,188] Newly formed cholesterol is sorted from the ER into the various membranes of the cell, the greatest abundance being in plasma membranes where cholesterol plays an essential role in decreasing fluidity. Cholesterol also aggregates with sphingolipids to form rigid lipid "rafts" floating in the plasma membrane. These rafts are thought to have important functions in signaling, in distribution of lipid materials,[188a] and in influencing protein translocation.[188b] Caveolae in cell surfaces may also arise from cholesterol-rich rafts.[188c] However, cholesterol must also be able to move out of the internal membranes back into the interior ER of the cell to provide for homeostasis and to allow formation of cholesteryl esters for transport, bile acids for excretion by liver, and the steroid hormones.[183,184,189] Movement between organelles occurs with the aid of sterol carrier proteins.[190–190c]

Liver and some intestinal cells export cholesterol into the bloodstream, together with triacylglycerols and phospholipids in the form of VLDL particles, for uptake by other tissues (see Fig. 21-1). Cholesteryl esters are formed in the ER by **lecithin:cholesterol acyltransferase** (LCAT), an enzyme that transfers the central acyl group from phosphatidylcholine to the hydroxyl group of cholesterol.[191,191a] This enzyme is also secreted by the liver and acts on free cholesterol in lipoproteins.[192] Tissue acyltransferases also form cholesteryl esters from fatty acyl-CoAs.[192a]

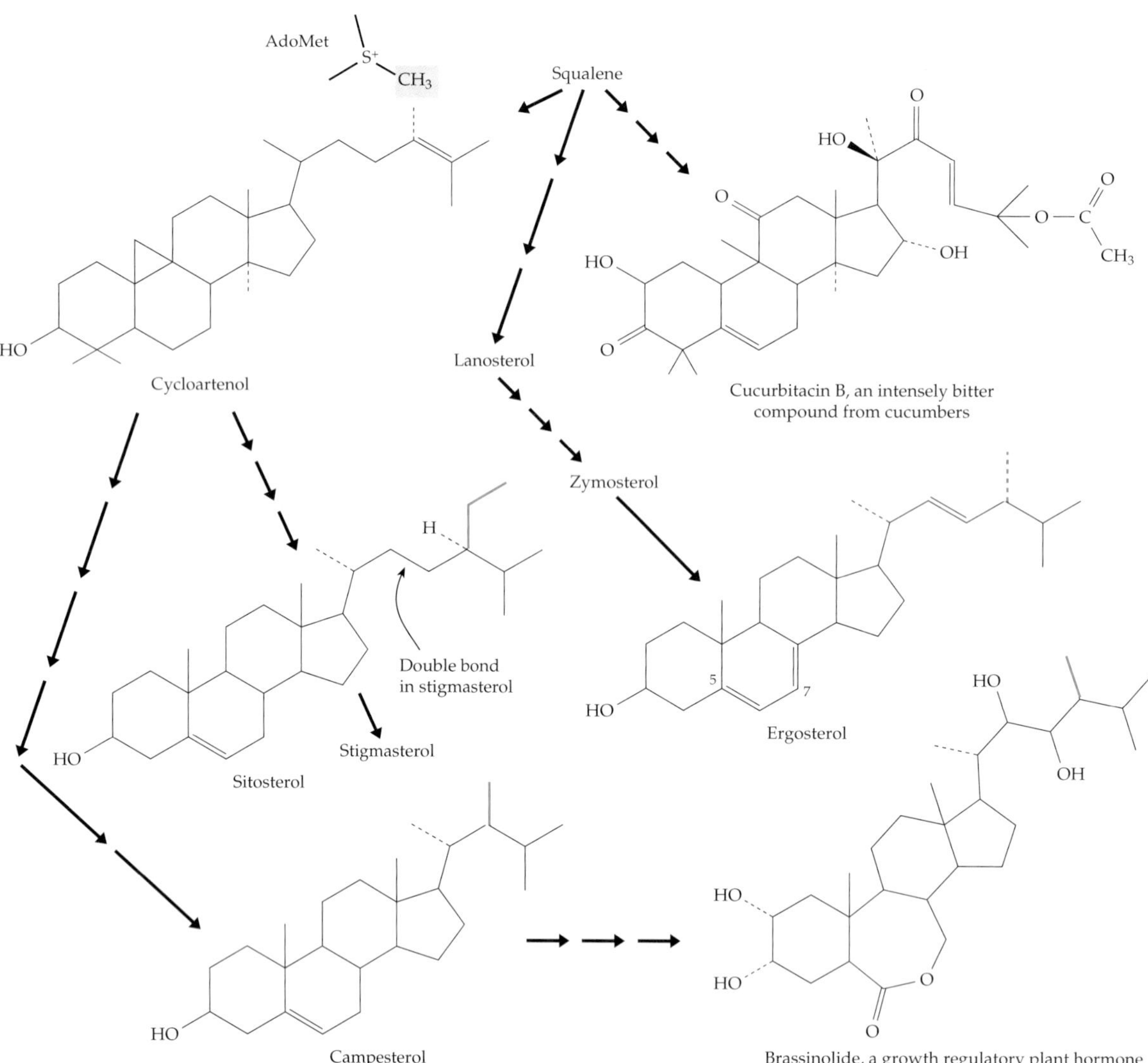

Figure 22-9 Structures and routes of biosynthesis for a few plant steroids.

In the brain a complex of cholesterol with apolipoprotein E (Table 21-2) promotes the formation of new synapses (Chapter 30). Synthesis of cholesterol for this purpose appears to occur within glial cells.[192b,c]

Serum cholesterol. Most cholesterol is carried in the blood by low density lipoprotein (LDL, Tables 21-1, 21-2), which delivers the cholesteryl esters directly to cells that need cholesterol. Both a 74-kDa **cholesteryl ester transfer protein**[193–195a] and a **phospholipid transfer protein**[196,196a] are also involved in this process. **Cholesterol esterases**, which release free cholesterol, may act both on lipoproteins and on pancreatic secretions.[197–199]

The LDL–cholesterol complex binds to LDL receptors on the cell surfaces.[167,168,200–202] These receptors are specific for apolipoprotein B-100 present in the LDL. The occupied LDL–receptor complexes are taken up by endocytosis through coated pits; the apolipoproteins are degraded in lysosomes, while the cholesteryl esters are released and cleaved by a specific **lysosomal acid lipase**[203,204] to form free cholesterol.

While the primary role of LDL appears to be the transport of esterified cholesterol to tissues, the high density lipoproteins (HDL) carry excess cholesterol away from most tissues to the liver.[205–207] The apoA-I present in the HDL particle not only binds lipid but activates LCAT, which catalyzes formation of cholesteryl esters which migrate into the interior of the HDL and are carried to the liver.

Unlike other lipoproteins, HDL particles are assembled outside of cells from lipids and proteins, some of which may be donated from chylomicrons (see Fig. 21-1) or other lipoprotein particles. HDL has a higher protein content than other lipoproteins and is more heterogeneous. The major HDL protein is apolipoprotein A-I, but many HDL particles also contain A-II,[205,208–210] and apolipoproteins A-IV, D, and E may also be present. A low plasma level of HDL cholesterol is associated with a high risk of atherosclerosis.[205,207]

BOX 22-B ATHEROSCLEROSIS

Our most common lethal disease is atherosclerosis, which causes constriction and blockage of arteries of the heart, brain, and other organs. In the United States, Europe, and Japan half of all deaths can be attributed to this ailment.[a,b] There seems to be a variety of causes. However, there is agreement that the disease begins with injury to the endothelial cells that form the inner lining of the arteries.[a,c,d] This is followed by the aggregation of blood platelets at the sites of injury and infiltration of smooth muscle cells, which may be attracted by 12-hydroxyeicosotetraenoic acid and other chemoattractants formed by activated platelets.[c] **"Foam cells"** laden with cholesterol and other lipids appear, and the lesions enlarge to become the characteristic plaques (**atheromas**).

The best understood cause of atherosclerosis is the genetic defect **familial hypercholesterolemia**, an autosomal dominant trait carried by one person in 500 all over the world.[e] Males with the defective gene tend to develop atherosclerosis when they are 35–50 years of age. The approximately one in a million persons *homozygous* for the trait develop coronary heart disease in their teens or earlier. Cultured fibroblasts from these patients have 40- to 60-fold higher levels of HMG-CoA reductase (Eq. 15-9) than are present normally, and the rate of cholesterol synthesis is increased greatly. The LDL level is very high and, as shown by Brown and Goldstein,[f–i] the LDL receptor gene is defective. Genetic defects associated with a low HDL level are also associated with atherosclerosis[b,j–l] as is a genetic variant of the metalloproteinase **stromelysin**.[m]

Other factors favoring development of atherosclerosis include hypertension and smoking. Chickens infected with a herpes virus (Marek disease virus) develop the disease after infection, and it is possible that artery damage in humans can also be caused by viruses[n] or bacterial infections.[o] In recent years it has been established that oxidative modification of the phospholipids in LDL induces the uptake of LDL by scavenger receptors of macrophages. This appears to trigger the development of foam cells and atherosclerotic plaques.[c,p,q] The initial damage is thought to be caused by lipid peroxides in the diet or generated by lipoxygenases in platelets and other cells.[c,p,r] Unsaturated fatty acids in lipoproteins can undergo oxidation (Chapter 21), especially in the presence of Cu^{2+} ions,[s,t] to yield malondialdehyde, 4-hydroxynonenal (Eq. 21-15), and other reactive compounds, which may damage the lipoproteins and cause them to have too high an affinity for their receptors in the smooth muscle cells of artery walls.[p,u] The 17β-hydroperoxy derivative of cholesterol has also been found in atherosclerotic lesions and may account for some of the toxicity of oxidized LDL.[r] Ascorbic acid may help to prevent formation of these oxidation products.[p,q,v] Chlorinated sterols may also be produced by the myeloperoxidase of the phagocytic macrophages that are abundant in atherosclerotic plaque.[w] Trans fatty acids, which are abundant in some margarines, and other hydrogenated fats raise both cholesterol and LDL levels.[x] Another cause of artery disease may be the presence of excessive homocysteine,[y,z] which can accumulate as a result of marginal deficiencies of folate, vitamin B_6, or vitamin B_{12}.

OH H C O

4-Hydroxy-2-nonenal

What can be done to prevent atherosclerosis? For persons with a high LDL level there is little doubt that a decreased dietary intake of cholesterol and a decrease in caloric intake are helpful. While such dietary restriction may be beneficial to the entire population, controlled studies of the effect of dietary modification on atherosclerosis have been disappointing and confusing.[aa] A diet that is unhealthy for some may be healthy for others. For example, an 88-year old man who ate 25 eggs a day for many years had a normal plasma cholesterol level of 150–200 mg / deciliter (3.9–5.2 mM)![bb] Comparisons of diets rich in unsaturated fatty acids, palmitic acid, or stearic acid have also been confusing.[cc,cd,dd] Can it be true that palmitic acid from tropical oils and other plant sources promotes atherogenesis, but that both unsaturated fatty acids and stearic acid from animal fats are less dangerous?

One of the best therapeutic approaches may be to prevent absorption of cholesterol from the intestines by inclusion of a higher fiber content in the diet.[ee] Supplementation with a cholesterol-binding resin may provide additional protection. Plant sterols also interfere with cholesterol absorption. Incorporation of esters of **sitostanol** into margarine provides an easy method of administration.[ff] Supplemental vitamin E may also be of value.[q] Another effective approach is to decrease the rate of cholesterol synthesis by administration of drugs that inhibit the synthesis of cholesterol. Inhibitors of HMG-CoA reductase,[gg,hh] (e.g., vaLostatin) isopentenyl-*PP* isomerase, squalene synthase (e.g.,

BOX 22-B ATHEROSCLEROSIS (continued)

squalestatin S1),[ii] and other enzymes in the biosynthetic pathway are targets for drug treatment.

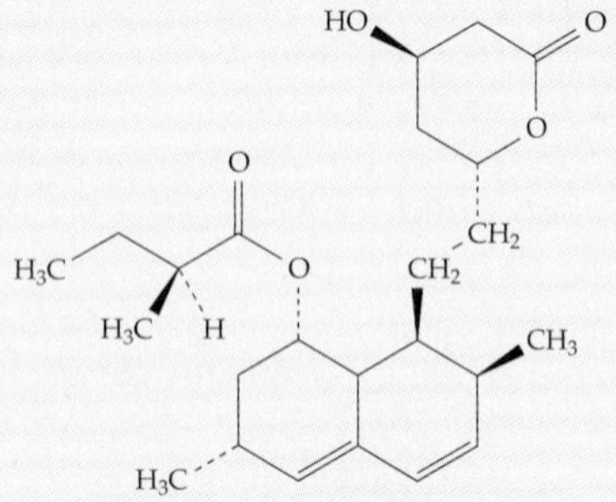

Lovastatin
(Mevinolin)

Questions of possible long-term toxicity remain. Since 1976 there has been a greater than 25% decrease in the incidence of ischemic heart disease in the United States.[jj] Increased exercise, a decreased severity of influenza epidemics, and fluoridation of water[kk] have been suggested as explanations.

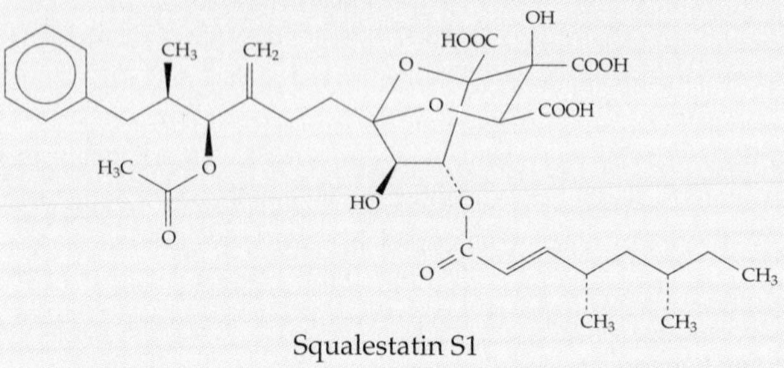

Squalestatin S1

[a] Ross, R. (1993) *Nature (London)* **362**, 801–809
[b] Krieger, M. (1998) *Proc. Natl. Acad. Sci. U.S.A.* **95**, 4077–4080
[c] Yagi, K. (1986) *Trends Biochem. Sci.* **11**, 18–19
[d] Breslow, J. L. (1996) *Science* **272**, 685–688
[e] Goldstein, J. L., Hobbs, H. H., and Brown, M. S. (1995) in *The Metabolic and Molecular Bases of Inherited Disease*, 7th ed., Vol. 2 (Scriver, C. R., Beaudet, A. L., Sly, W. S., and Valle, D., eds), pp. 1981–2030, McGraw-Hill, New York
[f] Brown, M. S., and Goldstein, J. L. (1974) *Sci. Am.* **251**(Nov), 58–66
[g] Brown, M. S., Kovanen, P. T., and Goldstein, J. L. (1981) *Science* **212**, 628–635
[h] Brown, M. S., and Goldstein, J. L. (1986) *Science* **232**, 34–47
[i] Motulsky, A. G. (1986) *Science* **231**, 126–128
[j] Breslow, J. L. (1993) *Proc. Natl. Acad. Sci. U.S.A.* **90**, 8314–8318
[k] Bergeron, J., Frank, P. G., Scales, D., Meng, Q.-H., Castro, G., and Marcel, Y. L. (1995) *J. Biol. Chem.* **270**, 27429–27438
[l] Breslow, J. L. (1995) in *The Metabolic and Molecular Bases of Inherited Disease*, 7th ed., Vol. 2 (Scriver, C. R., Beaudet, A. L., Sly, W. S., and Valle, D., eds), pp. 2031–2052, McGraw-Hill, New York
[m] Ye, S., Eriksson, P., Hamsten, A., Kurkinen, M., Humphries, S. E., and Henney, A. M. (1996) *J. Biol. Chem.* **271**, 13055–13060
[n] Benditt, E. P., Barrett, T., and McDougall, J. K. (1983) *Proc. Natl. Acad. Sci. U.S.A.* **80**, 6386–6389
[o] Gura, T. (1998) *Science* **281**, 35–37
[p] Holvoet, P., and Collen, D. (1994) *FASEB J.* **8**, 1279–1284
[q] Steinberg, D. (1997) *J. Biol. Chem.* **272**, 20963–20966
[r] Chisolm, G. M., Ma, G., Irwin, K. C., Martin, L. L., Gunderson, K. G., Linberg, L. F., Morel, D. W., and DiCorleto, P. E. (1994) *Proc. Natl. Acad. Sci. U.S.A.* **91**, 11452–11456
[s] Lynch, S. M., and Frei, B. (1995) *J. Biol. Chem.* **270**, 5158–5163
[t] Leeuwenburgh, C., Rasmussen, J. E., Hsu, F. F., Mueller, D. M., Pennathur, S., and Heinecke, J. W. (1997) *J. Biol. Chem.* **272**, 3520–3526
[u] Itabe, H., Yamamoto, H., Suzuki, M., Kawai, Y., Nakagawa, Y., Suzuki, A., Imanaka, T., and Takano, T. (1996) *J. Biol. Chem.* **271**, 33208–33217
[v] Retsky, K. L., Freeman, M. W., and Frei, B. (1993) *J. Biol. Chem.* **268**, 1304–1309
[w] Hazen, S. L., Hsu, F. F., Duffin, K., and Heinecke, J. W. (1996) *J. Biol. Chem.* **271**, 23080–23088
[x] Mensink, R. P., and Katan, M. B. (1990) *N. Engl. J. Med.* **323**, 439–445
[y] Chen, Z., Crippen, K., Gulati, S., and Banerjee, R. (1994) *J. Biol. Chem.* **269**, 27193–27197
[z] Kokame, K., Kato, H., and Miyata, T. (1996) *J. Biol. Chem.* **271**, 29659–29665
[aa] Kolata, G. (1985) *Science* **227**, 40–41
[bb] Kern, F., Jr. (1991) *N. Engl. J. Med.* **324**, 896–899
[cc] Bonanome, A., and Grundy, S. M. (1988) *N. Engl. J. Med.* **318**, 1244–1248
[cd] Merkel, M., Velez-Carrasco, W., Hudgins, L. C., and Breslow, J. L. (2001) *Proc. Natl. Acad. Sci. U.S.A.* **98**, 13294 – 13299
[dd] Hayes, K. C., and Khosla, P. (1992) *FASEB J.* **6**, 2600–2607
[ee] Jenkins, D. J. A., Wolever, T. M. S., Rao, A. V., Hegele, R. A., Mitchell, S. J., Ransom, T. P. P., Boctor, D. L., Spadafora, P. J., Jenkins, A. L., Mehling, C., Relle, L. K., Connelly, P. W., Story, J. A., Furumoto, E. J., Corey, P., and Würsch, P. (1993) *N. Engl. J. Med.* **329**, 21–26
[ff] Pedersen, T. R. (1995) *N. Engl. J. Med.* **333**, 1350–1351
[gg] Grundy, S. M. (1988) *N. Engl. J. Med.* **319**, 24–33
[hh] Brown, M. S., and Goldstein, J. L. (1996) *Science* **272**, 629
[ii] Kelly, M. J., and Roberts, S. M. (1995) *Nature (London)* **373**, 192–193
[jj] Stallones, R. A. (1980) *Sci. Am.* **243**(Nov), 53–59
[kk] Taves, D. R. (1978) *Nature (London)* **272**, 361–362

The LDL and related receptors. The LDL receptor gene extends over 50 kb of DNA and appears to be a mosaic of exons shared by several other genes that seem to have nothing to do with cholesterol metabolism.[201,211] The 839-residue receptor protein consists of five structural domains. The N-terminal domain that binds the LDL consists of seven repeated ~40-residue cysteine-rich modules.[212,213] This is followed by a large domain that is homologous to a precursor of the epidermal growth factor, a 48-residue domain containing many *O*-glycosylated serine and threonine residues, a 27-residue hydrophobic region that spans the membrane, and a 50-residue C-terminal cytoplasmic domain.[200,214–216] Synthesis of LDL receptors is regulated by a feedback mechanism, the cholesterol released within cells inhibiting the synthesis of new receptors.

As mentioned in Chapter 21, there are several related receptors with similar structures. Two of them have a specificity for apolipoprotein E and can accept remnants of VLDL particles and chylomicrons.[216–220] The LDL receptor-related protein is a longer-chain receptor.[216,221] LDL particles, especially when present in excess or when they contain oxidized lipoproteins, may be taken up by endocytosis into macrophages with the aid of the quite different **scavenger receptors**.[221–225] The uptake of oxidized lipoproteins by these receptors may be a major factor in promoting development of atherosclerosis (Box 22-B). On the other hand, **scavenger receptor SR-B1**, which is also present in liver cells, was recently identified as the receptor for HDL and essential to the "reverse cholesterol transport" that removes excess cholesterol for excretion in the bile.[213,213a]

Abnormalities of cholesterol metabolism. A variety of genetic problems have been identified, many of them being associated with atherosclerosis (Box 22-B).[218,226–230] In the commonest form of **familial hypercholesterolemia** a mutation in the LDL receptor protein prevents normal synthesis, binding, clustering into coated pits, or uptake of LDL and its cholesteryl esters. Over 600 mutations have been identified.[229,229a] Some of these are present in a Ca^{2+}-binding region of the 5th cysteine-rich module.[230] Other disorders that raise the plasma LDL level include a defective apoB-100 protein (see p. 1182)[230a] and deficiency of a protein that seems to be involved in incorporation of LDL receptors into clathrin-coated pits during endocytosis or in receptor recycling.[229a] In a **cholesteryl ester storage disease** the lysosomal lipase is lacking. Absence of lecithin:cholesterol acyltransferase from plasma causes corneal opacity and often kidney failure.[231]

In the very rare and fatal **Niemann–Pick C1** disease lysosomes in cells of the central nervous system and the viscera accumulate LDL-derived cholesterol. Study of the DNA of patients led to discovery of a 1278-residue integral membrane protein, which may be required for the Golgi-mediated transport of unesterified cholesterol from lysosomes to the ER.[189,232–234c]

Some people with elevated lipoprotein levels have VLDL that migrates on electrophoresis in the β band rather than the pre-β band (see Box 2-A). The presence of the β-VLDL is associated with a high incidence of artery disease,[218] which is most likely to develop in persons homozygous for a genetic variant of apolipoprotein E. The problem may arise because apo-E is required for receptor-mediated uptake of VLDL, which interacts both with tissue LDL receptors and with hepatic apo-E receptors. Genes for many of the apolipoproteins are polymorphic, and numerous alleles are present in a normal population.[218,235] In the rare **Tangier disease** apolipoprotein A-I is catabolized too rapidly, and the HDL level is depressed, resulting in accumulation of cholesterol esters in macrophages, Schwann cells, and smooth muscles. Orange-yellow enlarged tonsils are characteristic of the disease.[236] An ABC type transporter that allows cholesterol to leave cells is defective.[236a,b,c] The faulty component is known as the **cholesterol-efflux regulatory protein**. Another ABC transporter, apparently encoded by a pair of genes, which are expressed predominantly in liver and intestinal cells, prevents excessive accumulation of plant sterols such as sitosterol (Fig. 22-9).[236d,e]

Bile acids. Among the metabolites of cholesterol the bile acids (Fig. 22-10)[182,237,238] are quantitatively the most important (100–400 mg / day). These powerful emulsifying agents are formed in the liver and flow into the bile duct and the small intestine. A large fraction is later reabsorbed in the duodenum and is returned to the liver for reuse.[238a] Formation of the bile acids involves the removal of the double bond of cholesterol, inversion at C-3 to give a 3α-hydroxyl group, followed by hydroxylation and oxidation of the side chain.[238b–f] The principal human bile acids are **cholic acid** and **chenodeoxycholic acid** (Fig. 22-10). The free bile acids are then converted to CoA derivatives and conjugated with glycine and taurine to form **bile salts**, such as **glycocholic** and **taurocholic acids**.[238d] Several rare lipid-storage diseases are associated with defective bile acid formation.[239–241] In one of these, **cerebrotendinous xanthomatosis**, cholestanol is deposited both in tendons throughout the body and in the brain. Oxidation of the cholesterol side chain is incomplete with excretion, as glucuronides, of large amounts of bile alcohols (precursors to the bile acids). The synthesis of bile acids is regulated by feedback inhibition by the bile acids, but in this disease the inhibition is absent and the rates of both cholesterol biosynthesis and oxidation are increased. The problem is not one of storage of cholesterol but of the cholestanol that arises as a minor product of the pathway. A proper ratio of bile salt, phosphatidylcholine, and cholesterol in the bile is important to prevention of **cholesterol gallstones**.[207]

A variety of other oxidative modifications of cholesterol take place in tissues to give small amounts of diols.[242] Hydroperoxides of cholesterol may also be formed.[243] Some of the products are probably toxic, but others may be essential. One of these is 26-hydroxycholesterol, a minor component of plasma but a major neonatal excretion product.[244]

The body contains sulfate esters of cholesterol and other sterols,[245] sometimes in quite high concentrations relative to those of unesterified steroids. These esters are presumably soluble transport forms. They

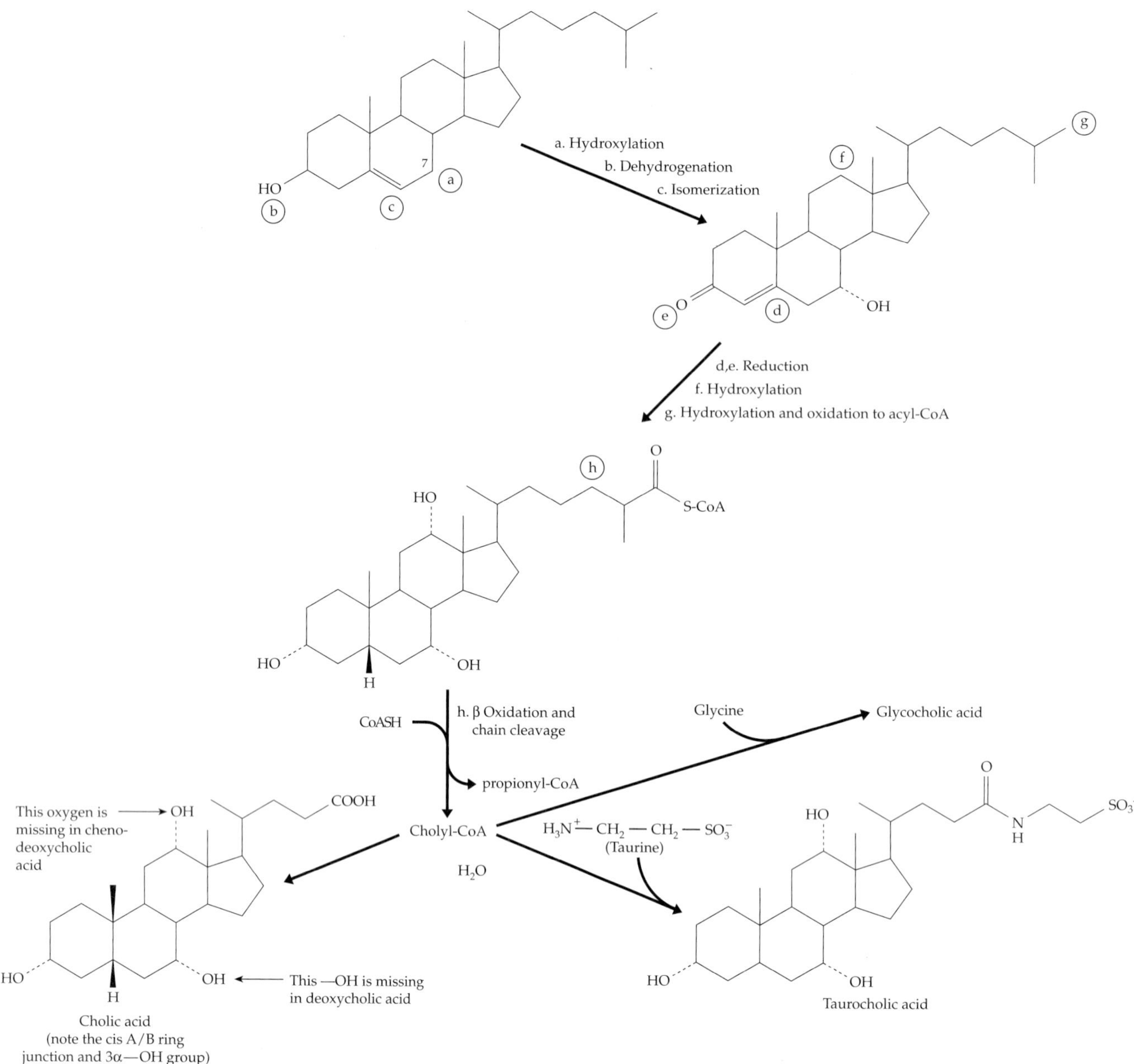

Figure 22-10 Formation of the bile acids.

are hydrolyzed by a steroid sulfatase present within cells and whose absence causes **X-linked ichthyosis**, one of the commoner forms of scaly skin. The defect may also lead to corneal opacity but appears not to cause any other problems.[246] Glucuronides of sterols are another group of water-soluble metabolites.[247] We tend to think of cholesterol as an inert structural component of cell membranes. However, it has been found in ester linkage to a signaling protein of the "hedgehog" family, important in embryo development.[248]

E. The Steroid Hormones

In the animal body three important groups of hormones are formed by the metabolism of cholesterol: the **progestins**, the **sex hormones**, and the **adrenal cortical hormones**.[249] Their synthesis occurs principally in mitochondria of the adrenal cortex and the gonads. Steroid hormone synthesis is regulated by hormones, such as **corticotropin** (ACTH), from the anterior pituitary[250] (see Chapter 30) and is also dependent upon the recently discovered **steroidogenic acute regulatory protein,** which in some way promotes the movement of stored cholesterol into mitochondrial membranes.[251,252] Some major pathways of biosynthesis are outlined in Fig. 22-11. The side chain is shortened to two carbon atoms through hydroxylation and oxidative cleavage to give the key intermediate **pregnenolone**. The reaction is initiated by the mitochondrial **cytochrome P450$_{ssc}$**, which receives electrons from NADPH and adrenodoxin (Chapter 16).[253] Hydroxylation occurs sequentially on C-22 and C-20 (Eq. 22-14). The chain cleavage is catalyzed by the same enzyme, an overall 6-electron oxidation occurring in three O_2-dependent steps. Dehydrogenation of the 3-OH group of pregnenolone to C=O is followed by a shift in the double bond, the oxosteroid isomerase reaction (Eq. 13-30, step *b*). In bacteria these two steps are catalyzed by different proteins, but a single human 3β-hydroxysteroid / Δ^5-Δ^4 isomerase catalyzes both reactions.[254,255] The product is the α,β-unsaturated ketone **progesterone**.

Most steroid hormones exist in part as sulfate esters and may also become esterified with fatty acids.[256] The fatty acid esters may have relatively long lives within tissues.[256] A special sex hormone-binding globulin transports sex hormones in the blood and regulates their access to target cells.[256a,b]

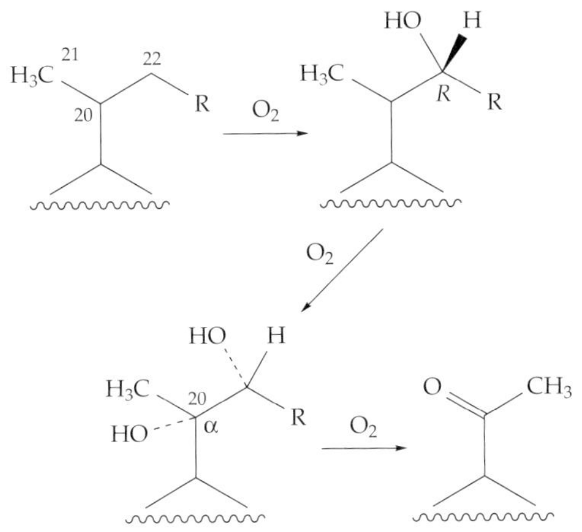

Pregnenolone

(22-14)

1. Progestins

Progesterone is the principal hormone of the **corpus luteum**, the endocrine gland that develops in the ovarian follicle after release of an ovum. Progesterone is also formed in the adrenals, testes, and placenta. It is metabolized rapidly, largely by reduction to alcohols, which may then be conjugated and excreted as glucuronides (see Eq. 20-16).[247] Reduction of the double bond within the A ring of progesterone leads to complete loss of activity, an indication that the α,β-unsaturated ketone group may play an essential role in the action of the hormone. Progesterone has a special role in the maintenance of pregnancy, and together with the estrogenic hormones it regulates the menstrual cycle. It is also essential for reproduction in lower animals such as birds and amphibians.[257,258]

The synthetic progesterone agonist **norethindrone** is widely used as one component of contraceptive pills. Having an opposite effect is **mifepristone** (also known as RU 486), a powerful antagonist of both progesterone and glucocorticoids. It is capable of inducing abortion and has other medical uses.[258–260] It is an effective emergency contraceptive agent that prevents implantation of an embryo.[261,262]

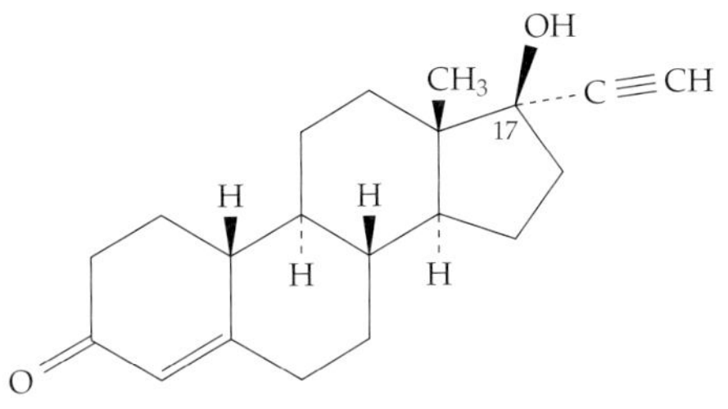

Norethindrone
(19-Nor-17α-ethynyltestosterone)

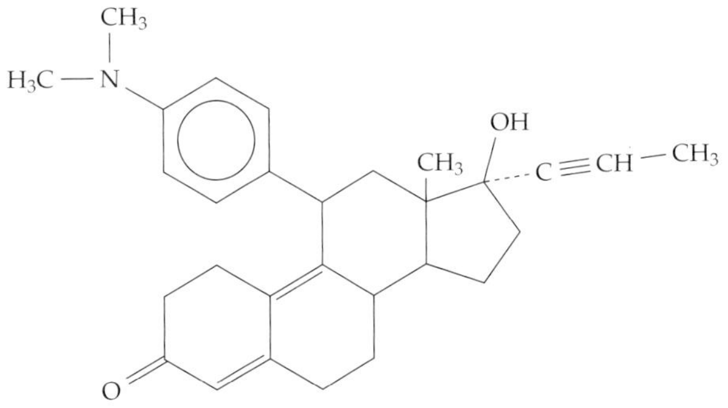

Mifepristone (RU 486)

2. Adrenal Cortical Steroids

Within the adrenal cortex (the outer portion of the adrenal glands) progesterone is converted into two groups of hormones of which **cortisol** and **aldosterone** are representative.[263] Two different cytochrome P450 hydroxylases, found in the ER and specific for C-21 and C-17α, respectively, together with a mitochondrial cytochrome P450 specific for C-11β (Eq. 18-55) participate in formation of cortisol.[264] Two of the same enzymes together with additional hydroxylases are required to form aldosterone.

Absence of the C-21 hydroxylase is one of the commonest of hereditary metabolic defects and is one of several enzymatic deficiencies that lead to **congenital adrenal hyperplasia**.[265–269] Cortisol, the synthesis of which is controlled by ACTH, is secreted by the adrenals in amounts of 15–30 mg daily in an adult. The hormone, which is essential to life, circulates in the blood, largely bound to the plasma protein **transcortin**. Cortisol, in turn, exerts feedback inhibition on ACTH production, and it is this feedback loop that fails when the C-21 hydroxylase is missing. Normally the circulating cortisol binds to receptors in both the pituitary and the hypothalamus of the brain to inhibit release of both ACTH and its hypothalamic releasing hormone (corticotropin-releasing hormone, CRF; see also Chapter 30). Girls are especially seriously affected by adrenal hyperplasia because, as ACTH production increases, the adrenal glands swell and produce an excessive amount of androgens. This occurs during the prenatal period of androgen release that initiates sexual differentiation. Girls with this deficiency are born with a masculine appearance of their external genitalia and continue to develop a masculine appearance. For reasons that are not clear the gene for the 21-hydroxylase is located within the HLA region (Chapter 31) of human chromosome 6.

Cortisol is a glucocorticoid which promotes gluconeogenesis and the accumulation of glycogen in the liver (Chapter 17). While it induces increased protein synthesis in the liver, it inhibits protein synthesis in muscle and many other tissues and leads to breakdown of fats to free fatty acids in adipose tissue.

Cortisol and its close relative **cortisone**, which was discovered by Kendall and Reichstein in the late 1940s, are probably best known for their anti-inflammatory effect in the body.[268,270] The effect depends upon several factors including inhibition of protein synthesis by fibroblasts, neutrophils, and antibody-forming cells. Migration of neutrophils into the inflamed area is also suppressed. Because of this action cortisone and synthetic analogs such as prednisolone and dexamethasone are among the modern "wonder drugs." They are used in controlling acute attacks of arthritis and of serious inflammations of the eyes and other organs. However, prolonged therapy can have serious side effects including decreased resistance to infections, wasting of muscle, and resorption of bone. The last results from a specific inhibition of calcium absorption from the gastrointestinal tract, glucocorticoids being antagonistic to the action of vitamin D (Box 22-C).

Aldosterone, which is classified as a **mineralocorticoid**, is produced under the control of the **renin–angiotensin** hormone system (Box 22-D), which is stimulated when sodium ion receptors in the kidneys detect an imbalance. It is synthesized in vascular cells of the body as well as in the adrenal cortex.[271] Aldosterone promotes the resorption of sodium ions in the kidney tubules and thus regulates water and electrolyte metabolism.[267,268] Glucocorticoids also have weak mineralocorticoid activity, and most patients with adrenocortical insufficiency (**Addison's disease**) can be maintained with glucocorticoids alone if their salt intake is adequate. Addison's disease develops when the adrenals are destroyed, most often by autoimmune disease or by tuberculosis.

3. Androgens

The principal **androgenic** or male sex hormone is **testosterone** formed from pregnenolone through removal of the side chain at C-17. The first step in the conversion is 17α-hydroxylation by a cytochrome P450 which may also mediate the further oxidative cleavage to **Δ⁴-androstenedione** (Fig. 22-11). Reduction of the 17-carbonyl group forms testosterone. About 6–10 mg are produced daily in men, and smaller amounts (~0.4 mg) are synthesized in women. Testosterone is carried in the blood as a complex with a β-globulin and affects a variety of target tissues including the reproductive organs. Another striking effect is stimulation of the growth of the beard. Testosterone also causes premature death of follicles of head hair in genetically susceptible individuals. However, a bald man can usually grow a full beard, and follicles of the beard type, when transplanted to the head, remain immune to the action of androgen. No one knows what regulatory differences explain this fact. Baldness might be cured by use of suitable antagonists of the androgenic hormones, but the beard might fall out and sexual interest could be lost.

Androgen synthesis in the human male fetus begins at the age of about 70 days when the testes enlarge and go through an important period of activity that begins the conversion of the infant body to a male type. Other bursts of testosterone synthesis occur during infancy, but there is little further synthesis until the onset of puberty.[268,272]

Within many target tissues testosterone is converted by an NADPH-dependent 5α-reductase into **5α-dihydrotestosterone**. That this transformation is important is shown by the fact that absence of one of the two 5α-reductase isomers causes a form of **pseudohermaphroditism** in which male children are often mistaken as female and raised as girls. However, at puberty they become unmistakably male.[268,273–275] Many other metabolites of testosterone are known. These include the

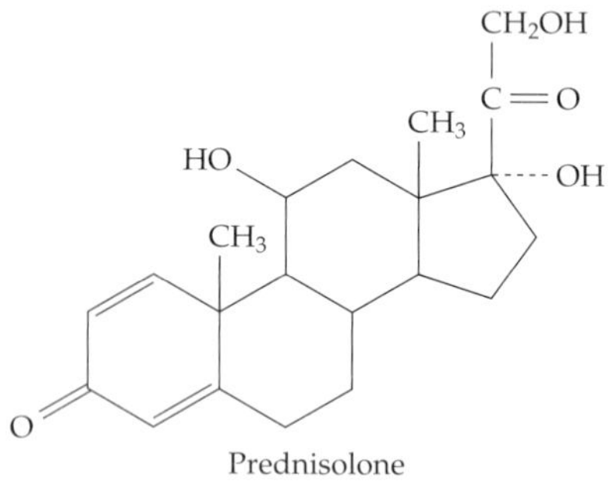

Prednisolone

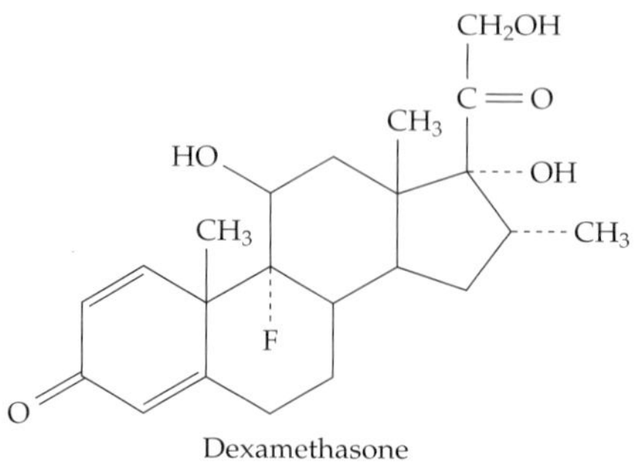

Dexamethasone

isomeric 5β-dihydrotestosterone and 5α-androstane-3β, 17β-diol which arises by reduction of the carbonyl group of 5α-dihydrotestosterone. Testosterone and dihydroxytestosterone have distinct roles in the body.

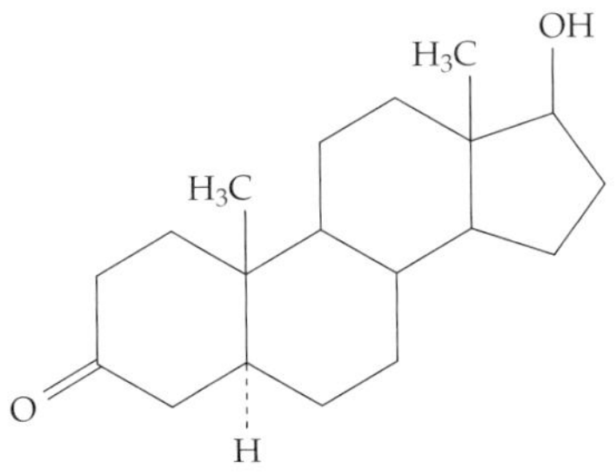

5α-Dihydrotestosterone

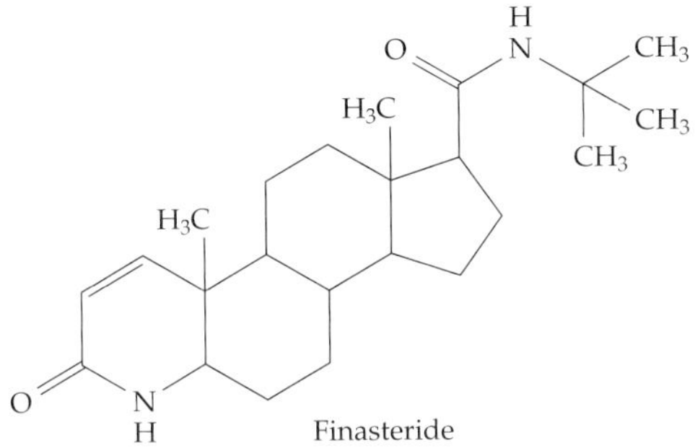

Finasteride

NADH NAD+ Finasteride Anion Adduct in inactivated enzyme

(22-15)

For example, testostosterone is required for sperm cell formation, voice deepening, and growth of pubic hair while dihydroxytestosterone stimulates development of the prostate gland and male pattern hair growth.[276]

Males with deficiency of the 5α-reductase isoenzyme do not develop acne, male pattern baldness, or enlarged prostates.[274] The last fact was some of the impetus for development of the steroid 5α-reductase inhibitor **finasteride,** which is widely used to treat benign prostate enlargement.[274,277,278] It is an enzyme-activated inhibitor in which the NADH reduces the C=C bond in the A ring, which is not in the same position as in the substrate. The resulting anion cannot become protonated but instead adds to the NAD^+ as shown in Eq. 22-15.

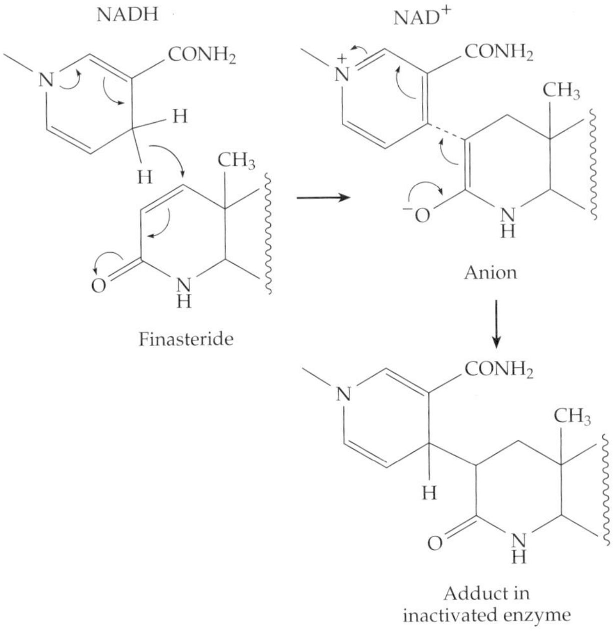

A number of other androgens are present in the body. The adrenal glands make **dehydroepiandrosterone** (DHEA; Fig. 22–11), which circulates in human blood as its sulfate ester in higher concentration than that of any other androgen.[279–283] However, this steroid is absent from most species. DHEA can be taken up by tissues and converted to testosterone, estrogens, or other steroids (Fig. 22–11).[280] Recent attention has been focused on this hormone because it reaches a peak plasma concentration at age 20–25 years and by age 70 has fallen to 1/5 this value or less.[283] Should older men supplement their circulating DHEA by oral ingestion of 25–50 mg per day of DHEA sulfate? The hormone depresses blood cholesterol and lowers blood glucose in diabetic individuals.[280,282] It seems to promote increased energy metabolism.[281] It may fight obesity and atherosclerosis,[282] increase levels of estrogen and other steroid hormones in the brain,[284] and enhance memory and immune function.[285–285c] However, the hormone may be metabolized differently in different tissues, and its pathway of biosynthesis in the brain is uncertain.[279,286] Will men synthesize more testosterone from DHEA or make more estrogens in their adipose tissues? Young women should not take DHEA. It may increase the testosterone and dihydrotestosterone levels in the blood manyfold, and the women may become hirsute and masculinized.[280] However, most DHEA is converted to estrogen which may be of value to older women.[285a]

In addition to their role in sexual development androgens have a generalized "anabolic" effect causing increased protein synthesis, especially in muscles.[287–289] They promote bone growth, and the adolescent growth spurt in both males and females is believed to result from androgens. The greater height attained by men results in part from the higher concentration of androgen than is present in women. Many synthetic steroids have been made in an attempt to find "anabolic hormones" with little or no androgenic activity. The effort has been at least partially successful, and the use of anabolic hormones by athletes has become both widespread and controversial.

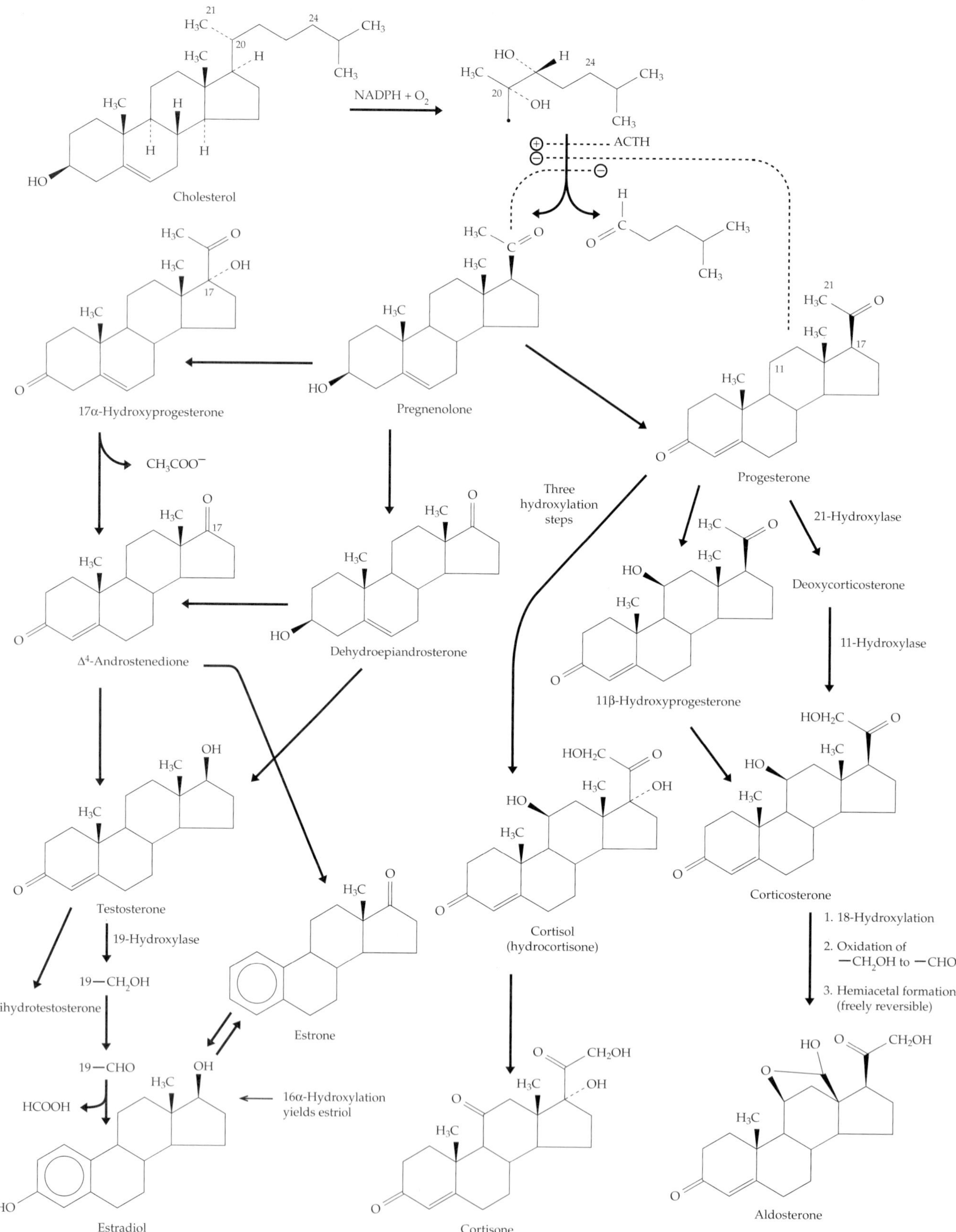

Figure 22-11 Biosynthesis of some steroid hormones.

BOX 22-C VITAMIN D

A lack of vitamin D causes **rickets**, a disease of humans and other animals in which the bones are soft, deformed, and poorly calcified. Rickets was recognized by some persons to result from a dietary deficiency well over a hundred years ago, and the use of cod liver oil to prevent the disease was introduced in about 1870. By 1890 an association of rickets with a lack of sunlight had been made.

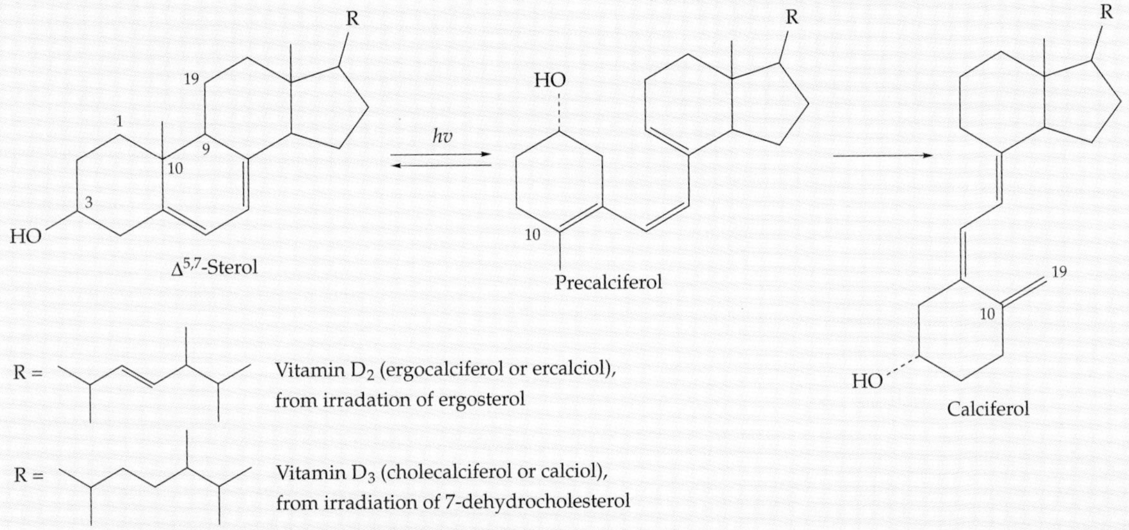

However, it was not until 1924, when Steenbock and Hess showed that irradiation of certain foods generated protective activity against the disease, that vitamin D (**calciferol**) was recognized as a second lipid-soluble vitamin. Vitamin D is a family of compounds formed by the irradiation of $\Delta^{5,7}$-unsaturated sterols such as ergosterol and 7-dehydrocholesterol.[a–e] The former yields **ergocalciferol** (vitamin D_2) and the latter **cholecalciferol** (vitamin D_3).

At low temperature the intermediate **precalciferol** can be isolated. Irradiation sets up a photochemical steady-state equilibrium between the $\Delta^{5,7}$-sterol and the precalciferol. At higher temperatures the latter is converted to calciferol.[f] Other products, including toxic ones, are produced in slower photochemical side reactions. Therefore, the irradiation of ergosterol for food fortification must be done with care.

With normal exposure to sunlight enough 7-dehydrocholesterol is converted to cholecalciferol in the skin that it was concluded that no dietary vitamin D is required by most adults except during pregnancy. However, recently it has been recognized that old and sick adults probably need 400–600 I.U. per day to maintain calcium absorption and to prevent osteoporesis and fractures.[g,gh] It is usually recommended that children receive ~20 µg (400 I.U.) of ergocalciferol per day in their diet. Larger amounts are undesirable, and at a tenfold higher level vitamin D is seriously toxic.[h]

The principal function of vitamin D is in the control of calcium metabolism. This is accomplished through the mediation of polar, hydroxylated metabolites, the most important of which is **1α,25-dihydroxyvitamin D3** (1,25-dihydroxycholecalciferol or calcitriol).[a,i,j] This compound may be properly described as a steroid hormone and vitamin D itself as a hormone precursor formed in the skin. The major hydroxylation reactions of vitamin D are summarized in the accompanying scheme. The first hydroxylation to 25-hydroxyvitamin D_3 occurs largely in the liver,[jk] but the subse-

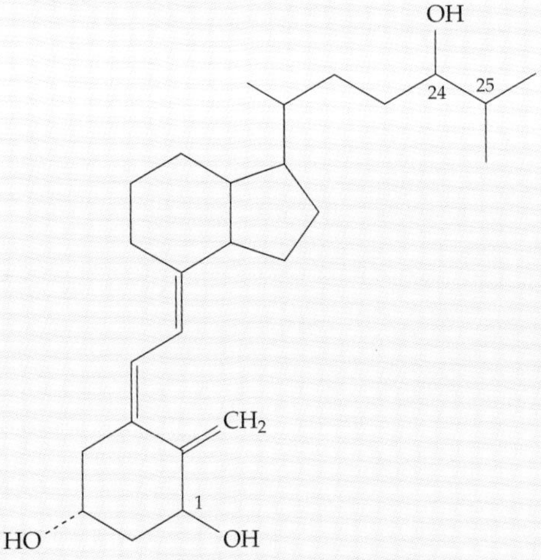

$\Delta^{5,7}$-Sterol — hv → Precalciferol → Calciferol

R = Vitamin D_2 (ergocalciferol or ercalciol), from irradation of ergosterol

R = Vitamin D_3 (cholecalciferol or calciol), from irradiation of 7-dehydrocholesterol

1 α,25-Dihydroxyvitamin D_3

BOX 22-C VITAMIN D (continued)

quent cytochrome P450 catalyzed 1α-hydroxylation takes place almost entirely in the kidneys.[k,l] Since it is 1,25-dihydroxy derivative that is essential for control of calcium ion metabolism, human patients with damaged kidneys often suffer severe demineralization of their bones (renal osteodystrophy). Administration of synthetic 1α,25-dihydroxyvitamin D_3 provides an effective treatment for these persons and also for children with an inherited defect in production of this hormone.[a,m] However, large doses, tested as an antileukemia drug, caused a severe hypercalcinemia limiting its use.[n]

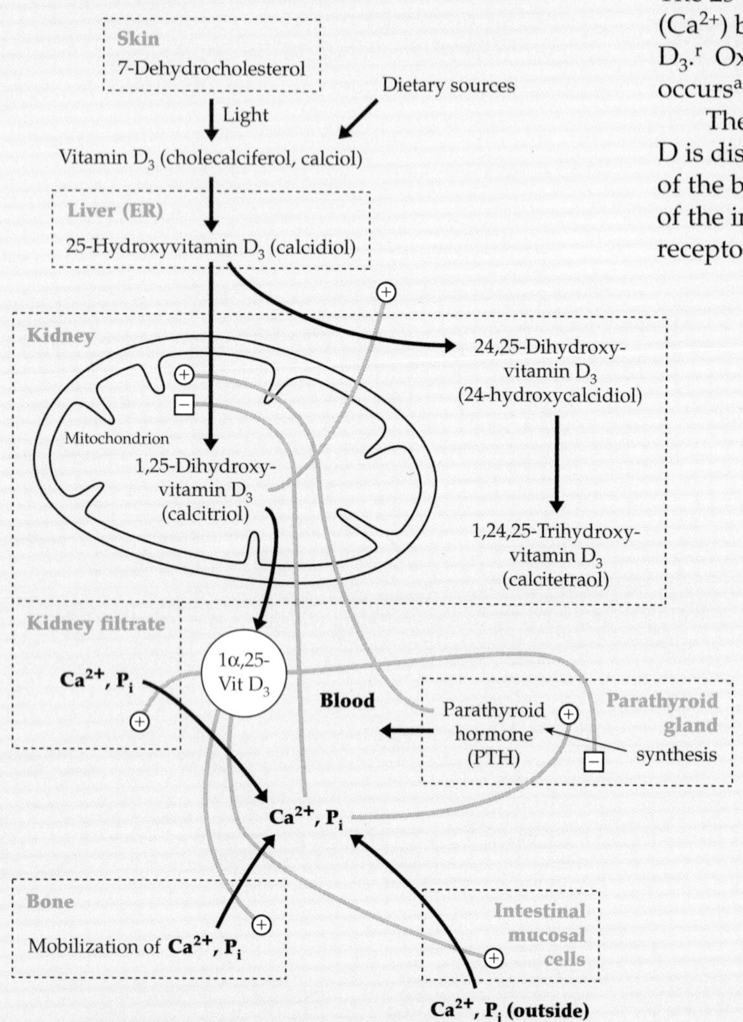

A second major vitamin D metabolite is 24*R*,25-dihydroxyvitamin D_3, a compound that circulates in the blood at a concentration 10 times higher than that of the 1,25-isomer.[a,b] However, no biological function has been discovered, and like a series of other polar metabolites (>30) it is probably on a pathway of inactivation and degradation of vitamin D. 1α,25-Dihydroxyvitamin D is also hydroxylated at C-24.[o] Additional hydroxylations occur at the 23- and 26-positions.[p,q] The 24-OH is often converted to an oxo group. Oxidation at C-26 together with cyclization yields 26,23 lactol and lactone species. The 25-hydroxy-26,23-lactone suppresses serum (Ca^{2+}) by competing with 1,25-dihydroxyvitamin D_3.[r] Oxidative cleavage of the side chains also occurs[a,s] as in the metabolism of cholesterol.

The hormonally active 1α,25-dihydroxyvitamin D is distributed through the bloodstream to all parts of the body. It is taken up rapidly by nuclei of cells of the intestinal lining where it binds to a 55-kDa receptor protein. In response, the cells synthesize **calbindins** (Chapter 6), Ca^{2+}-binding proteins which facilitate the uptake of calcium ions by the body.[t–v] (see also Fig. 6-7).

Other target organs for the action of 1,25-dihydroxyvitamin D include the kidneys, bone, muscle,[vw] and skin. The hormone promotes reabsorption of both Ca^{2+} and inorganic phosphate by kidney tubules. In bone it binds to a specific receptor where it promotes the mobilization of calcium ions. This effect may result in part from stimulation of calcium-activated ATPase of the outer membrane of bone cells. Dissolution of bone also requires the presence of **parathyroid hormone (PTH)**, the 83-residue hormone secreted by the parathyroid gland. In women past the age of menopause and in elderly men the production of 1,25-dihydroxyvitamin D decreases.[w] This may be a cause of the serious bone loss (**osteoporosis**) frequently observed. Treatment with 1,25-dihydroxyvitamin D_3 or a synthetic analog seems to be helpful to such individuals.[x,xy] See also Chapter 30, Section A,5.

There is another important member of the Ca^{2+} homeostatic system. While vitamin D and PTH act together to increase the calcium level of the blood, **calcitonin**, a hormone of the thyroid gland, lowers the level of Ca^{2+} by promoting deposition of calcium in bone by the osteoblasts. The overall effect is to hold the concentration of Ca^{2+} in the blood at 2.2–2.6 mM in most animals with bones serving as a mobile reserve. Another role for vitamin D is suggested by the observation that 1,25-dihydroxyvitamin D_3 inhibits the growth promoting effect of interleukin-2 on mitogen-activated lymphocytes (Chapter 31).

BOX 22-C (continued)

Like other steroid hormones 1,25-dihydroxyvitamin D_3 acts to regulate gene transcription. It binds to a specific receptor, a member of the v-*erb*-A superfamily of ligand-activated transcription factors and a relative of the steroid receptor family.[a,j,y] Like other hormone receptors this vitamin D_3 receptor (VDR) has some rapid "nongenomic" actions such as causing an increase in intracellular [Ca^{2+}] as well as slower effects on transcription.[y–bb] Like the steroid receptors it is found in both cytoplasm and nucleus.[y] When occupied by 1,25-dihydroxyvitamin D, the receptor binds tightly to specific dihydroxyvitamin D_3 response elements in the DNA. These are found in promoter sequences for genes such as that of bone protein **osteocalcin**.[cc] In some promoters the response element binds heterodimers of VDR with another receptor, e.g., the retinoid X receptor (Table 22-1).[cc,dd] One effect of this response element is to activate the gene for the 24-hydroxylase involved in degradation of 1,25-dihydroxyvitamin D.[dd] Heterodimers of VDR with the thyroid hormone receptor activate the transcription of genes for calbindins D_{28K} and D_{9K}.[u] VDR also binds to repressor sequences, e.g., in the parathyroid gland.[ee] See also Chapter 28 and Table 22-1.

[a] DeLuca, H. F. (1988) *FASEB J.* **2**, 224–236

[b] Reichel, H., Koeffler, H. P., and Norman, A. W. (1989) *N. Engl. J. Med.* **320**, 980–991

[c] Lawson, D. E. M. (1978) *Vitamin D*, Academic Press, New York

[d] Norman, A. W. (1979) *Vitamin D, the Calcium Homeostatic Steroid Hormone*, Academic Press, New York

[e] DeLuca, H. F. (1979) *Vitamin D – Metabolism and Function*, Springer, New York

[f] Holick, M. F., Tian, X. Q., and Allen, M. (1995) *Proc. Natl. Acad. Sci. U.S.A.* **92**, 3124–3126

[g] Utiger, R. D. (1998) *N. Engl. J. Med.* **338**, 828–829

[gh] Thomas, M. K., Lloyd-Jones, D. M., Thadhani, R. I., Shaw, A. C., Deraska, D. J., Kitch, B. T., Vamvakas, E. C., Dick, I. M., Prince, R. L., and Finkelstein, J. S. (1998) *N. Engl. J. Med.* **338**, 777–783

[h] Jacobus, C. H., Holick, M. F., Shao, Q., Chen, T. C., Holm, I. A., Kolodny, J. M., Fuleihan, G. E.-H., and Seely, E. W. (1992) *N. Engl. J. Med.* **326**, 1173–1177

[i] Holick, M. F., Schnoes, T. H. K., DeLuca, H. F., Suda, T., and Cousins, R. J. (1971) *Biochemistry* **10**, 2799–2804

[j] Minghetti, P. P., and Norman, A. W. (1988) *FASEB J.* **2**, 3043–3053

[jk] Hosseinpour, F., and Wikvall, K. (2000) *J. Biol. Chem.* **275**, 34650–34655

[k] Takeyama, K.-i, Kitanaka, S., Sato, T., Kobori, M., Yanagisawa, J., and Kato, S. (1997) *Science* **277**, 1827–1830

[l] Menaa, C., Vrtovsnik, F., Friedlander, G., Corvol, M., and Garabédian, M. (1995) *J. Biol. Chem.* **270**, 25461–25467

[m] Hughes, M. R., Malloy, P. J., Kieback, D. G., Kesterson, R. A., Pike, J. W., Feldman, D., and O'Malley, B. W. (1988) *Science* **242**, 1702–1705

[n] Lee, N. E., Reddy, G. S., Brown, A. J., and Williard, P. G. (1997) *Biochemistry* **36**, 9429–9437

[o] Miyamoto, Y., Shinki, T., Yamamoto, K., Ohyama, Y., Iwasaki, H., Hosotani, R., Kasama, T., Takayama, H., Yamada, S., and Suda, T. (1997) *J. Biol. Chem.* **272**, 14115–14119

[p] Mayer, E., Bishop, J. E., Chandraratna, R. A. S., Okamura, W. H., Kruse, J. R., Popjak, G., Ohnuma, N., and Norman, A. W. (1983) *J. Biol. Chem.* **258**, 13458–13465

[q] Koszewski, N. J., Reinhardt, T. A., Napoli, J. L., Beitz, D. C., and Horst, R. L. (1988) *Biochemistry* **27**, 5785–5790

[r] Yamada, S., Nakayama, K., Takayama, H., Shinki, T., Takasaki, Y., and Suda, T. (1984) *J. Biol. Chem.* **259**, 884–889

[s] Jones, G., Kano, K., Yamada, S., Furusawa, T., Takayama, H., and Suda, T. (1984) *Biochemistry* **23**, 3749–3754

[t] Szebenyi, D. M. E., and Moffat, K. (1986) *J. Biol. Chem.* **261**, 8761–8777

[u] Schräder, M., Müller, K. M., Nayeri, S., Kahlen, J.-P., and Carlberg, C. (1994) *Nature (London)* **370**, 382–386

[v] Heizmann, C. W., and Hunziker, W. (1991) *Trends Biochem. Sci.* **16**, 98–103

[vw] Morelli, S., Buitrago, C., Vazquez, G., De Boland, A. R., and Boland, R. (2000) *J. Biol. Chem.* **275**, 36021–36028

[w] Slovik, D. M., Adams, J. S., Neer, R. M., Holick, M. F., and Potts, J. T., Jr. (1981) *N. Engl. J. Med.* **305**, 372–374

[x] Tilyard, M. W., Spears, G. F. S., Thomson, J., and Dovey, S. (1992) *N. Engl. J. Med.* **326**, 357–362

[xy] Shevde, N. K., Plum, L. A., Clagett-Dame, M., Yamamoto, H., Pike, J. W., and Deluca, H. F. (2002) *Proc. Natl. Acad. Sci. U.S.A.* **99**, 13487–13491

[y] Barsony, J., Renyi, I., and McKoy, W. (1997) *J. Biol. Chem.* **272**, 5774–5782

[z] Yukihiro, S., Posner, G. H., and Guggino, S. E. (1994) *J. Biol. Chem.* **269**, 23889–23893

[aa] Lissoos, T. W., Beno, D. W. A., and Davis, B. H. (1993) *J. Biol. Chem.* **268**, 25132–25138

[bb] de Boland, A. R., Morelli, S., and Boland, R. (1994) *J. Biol. Chem.* **269**, 8675–8679

[cc] Schräder, M., Müller, K. M., and Carlberg, C. (1994) *J. Biol. Chem.* **269**, 5501–5504

[dd] Allegretto, E. A., Shevde, N., Zou, A., Howell, S. R., Boehm, M. F., Hollis, B. W., and Pike, J. W. (1995) *J. Biol. Chem.* **270**, 23906–23909

[ee] Kremer, R., Sebag, M., Champigny, C., Meerovitch, K., Hendy, G. N., White, J., and Goltzman, D. (1996) *J. Biol. Chem.* **271**, 16310–16316

O
‖
O—C—CH_2CH_2—(phenyl)

Norandrolone phenyl propionate, a synthetic drug having a five times higher ratio of anabolic to androgenic activities as does testosterone

Several sex-dependent differences have been observed in the action of cytochrome P450 isoenzymes on steroid hormones.[290,290a] Thus, androstenedione is hydroxylated by rat liver enzymes specific for the 6β, 7α, 16α, and 16β positions.[291] Of these the 16 hydroxylase is synthesized only in males, and synthesis of the 6 hydroxylase is also largely suppressed in females.

Δ^4-Androstenedione

A female-specific 15β hydroxylase acts on steroid sulfates such as corticosterone sulfate and forms the major urinary excretion product of that hormone in female rats.[292] These sex-specific differences in enzymes are thought to be related to secretions of growth hormone that are in turn controlled by the "programming" of the hypothalamus by androgen during the neonatal period in rats[272] or during human fetal development.

4. Estrogens

The principal **estrogenic** or female hormone is estradiol-17β. It is formed by oxidative removal of C-19 of testosterone followed by aromatization of the A ring.[268,293–295] All of the estrogenic hormones have this aromatic ring. Its formation involves three steps of hydroxylation followed by elimination of formate (Eq. 22-16). This **aromatase** appears to be a unique cytochrome P450, which catalyzes all of the steps of Eq. 22-16. It accepts electrons from NADPH via the flavoprotein NADH-cytochrome P450 reductase, which serves as the intermediate electron carrier.[296] It probably acts by a mechanism related to that illustrated for lanosterol 14α-demethylase on the right-hand side of Fig. 22-8.[294] This enzyme is the target for synthetic enzyme-activated inhibitors.[293] One of these is an androstenedione derivative with an acetylenic group attached to C-19. Passage through the first two steps of Eq. 22-16 generates a conjugated ketone to which a nucleophilic group of the enzyme can add irreversibly to inactivate the enzyme (Eq. 22-17). The C-17 acetylenic progesterone antagonist (norethyndrone) is also an enzyme activated inhibitor of the aromatase.

Estrogens are formed largely in the ovary and during pregnancy in the placenta. Estrogens are also synthesized in the testes, and the estrogen content of the horse testis is the highest of any endocrine organ. Target tissues for estrogens include the mammary glands, the uterus, and many other tissues throughout the body. Estrogens act on the growing ends of the long bones to stop growth and are therefore responsible, in part, for the shorter stature of females as compared to males. They are responsible for the overall higher fat content of the female body and for the smoother skin of the female. Recent attention has been focused on the effects of estrogens on brain neurons.

Estrogens stimulate sprouting of axons and dendrites in neurons in cell cultures, and there is preliminary evidence that the hormone improves memory in healthy women and in those with Alzheimer disease.[285]

The cooperative action of progesterone and estradiol regulate the menstrual cycle. At the beginning of the cycle the levels of both estrogen and progesterone are low. Estrogen synthesis increases as a result of release of **follicle-stimulating hormone** (FSH) from the anterior pituitary. This hormone stimulates growth of the graafian follicles of the ovary which in turn produce estrogen. At about the midpoint of the cycle, as a result of the action of the pituitary **luteinizing hormone** (LH), an ovum is released and progesterone secretion begins. The latter is essential to maintenance of pregnancy. If a blastocyst is not implanted, hormone production decreases and menstruation occurs.

(22-16)

(22-17)

BOX 22-D THE RENIN-ANGIOTENSIN-ALDOSTERONE SYSTEM AND THE REGULATION OF BLOOD PRESSURE

The mineralocorticoid **aldosterone** was isolated and identified in 1953. Although the function of adrenal cortical hormones in regulation of electrolytes had been known for many years, the special role of aldosterone had been overlooked.[a] Aldosterone works in concert with the aspartate protease **renin** and the octapeptide **angiotensin II** to regulate blood pressure. Angiotensin II, which is the most potent pressor substance known, is formed in the liver from the 477-residue (in the rat) **preangiotensinogen** as shown in the following cascade:

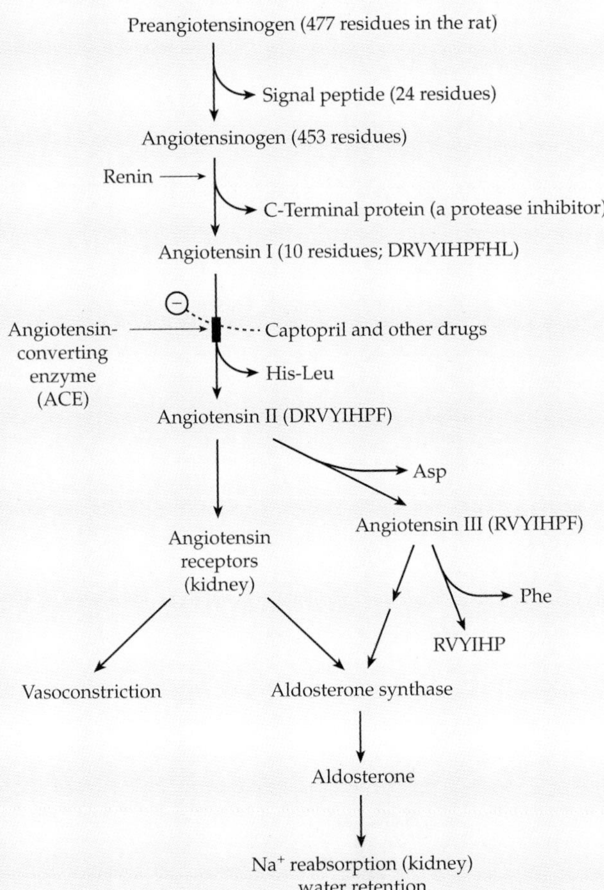

This **renin-angiotensin system** is a peptide hormone-generating system that operates in blood plasma rather than within tissues.[b,c]

Angiotensinogen, which is secreted by the liver and circulates in the blood, is converted to the physiologically inactive decapeptide angiotensin I by cleavage of a Leu-Leu peptide bond by the 328-residue renin.[d–f] Its precursor **preprorenin** is produced in the kidneys by the juxtaglomerular cells as well as in some other tissues and undergoes several steps of processing before the active enzyme is formed.[g–j] Active renin is released from the kidney cells into the bloodstream in response to various stimuli which include low arterial pressure resulting from constriction of the renal arteries or loss of blood.[k] Parathyroid hormone, glucagon, other adrenergic agonists, cAMP, some prostaglandins, low levels of aldosterone or Na^+, or high K^+ all induce secretion of renin. High blood pressure, α-adrenergic agonists, some prostaglandins, angiotensin, vasopressin, high Na^+, or low K^+ concentration decrease secretion of renin. It has been suggested that these diverse effects may be modified by a rise or fall in the Ca^{2+} concentration, high Ca^{2+} inhibiting secretion of renin, and low Ca^{2+} favoring secretion.

The only known physiological substrate for renin is angiotensinogen,[e] but it may also act on related proteins in the brain and other organs.[g] The inactive angiotensin I is converted to angiotensin II by the metal-containing carboxydipeptidase known as **angiotensin-converting enzyme**.[l,m] This enzyme is a target for drugs such as captopril, which is used to control high blood pressure (hypertension).[i] A zinc-dependent aminopeptidase may cut off the

Captopril

N-terminal aspartate to form angiotension III,[n] and degradation of angiotensions II and III can be initiated by removal of the C-terminal phenylalanine by a prolylcarboxypeptidase.[o]

Angiotensin II has a variety of effects. By constricting blood vessels it raises blood pressure, and by stimulating thirst centers in the brain it increases blood volume. Both angiotensins II and III also act on the adrenal gland to promote the synthesis and release of aldosterone. Most of the effects of angiotension II are mediated by 359-residue seven-helix G-protein linked receptors which activate phospholipase C.[p,q,qr] Like other steroid hormones aldosterone acts,via mineralocorticoid receptors, to control transcription of a certain set of proteins. The end effect is to increase the transport of Na^+ across the renal tubules and back into the blood. Thus, aldosterone acts to decrease the loss of Na^+ from the body. It promotes retention of water and raises

BOX 22-D THE RENIN–ANGIOTENSIN–ALDOSTERONE SYSTEM AND THE REGULATION OF BLOOD PRESSURE (continued)

blood pressure.[c,r,s] Its primary function is to provide adequate Na^+ to cells.[s] Dietary sodium appears to have little or no effect on blood pressure.[t]

The control of blood pressure is considerably more complex than it is described here. Another hormone system involving the peptide **bradykinin** and prostaglandins acts to lower blood pressure. Bradykin is also cleaved by the angiotensin-converting enzyme but is *inactivated* by the cleavage.[u,v] At least ten human genes have been shown to affect blood pressure.[c] One of these is the structural gene for angiotensin-converting enzyme, which has been linked to hypertension in both rats and humans.[w,x]

While several antagonists of angiotensin-converting enzyme are widely used to treat hypertension, they are not free of harmful side effects.[u,y,z] One alternative approach is to inhibit renin.[v,y]

While the angiotensins promote release of aldosterone, the **atrial natriuretic hormone**[r,aa–cc] inhibits release. This group of 21- to 33-residue polypeptides, secreted by cells of the atria (auricles) of the heart, also inhibits release of renin and promotes secretion of both Na^+ and water. Thus, they antagonize the action of aldosterone, which promotes Na^+ retention. However, there is uncertainty as to the significance of these peptides. The following metabolite of γ-tocopherol (Fig. 15-24) has been isolated from urine and is proposed as a new endogenous natriuretic factor.[dd]

HO, H_3C, CH_3, O, H, COOH

2, 7, 8-trimethyl-2-(β-carboxyethyl)-6-hydroxychroman

[a] Tait, J. F., and Tait, S. A. S. (1978) *Trends Biochem. Sci.* **3**, N273–N275

[b] Inagami, T. (1994) in *Essays in Biochemistry*, Vol. 28 (Tipton, K. F., ed), pp. 147– , Portland Press, Chapel Hill, North Carolina

[c] Lifton, R. P. (1996) *Science* **272**, 676–680

[d] Kem, D. C., and Brown, R. D. (1990) *N. Engl. J. Med.* **323**, 1136–1137

[e] Wang, W., and Liang, T. C. (1994) *Biochemistry* **33**, 14636–14641

[f] Tong, L., Pav, S., Lamarre, D., Pilote, L., LaPlante, S., Anderson, P. C., and Jung, G. (1995) *J. Mol. Biol.* **250**, 211–222

[g] Yanai, K., Saito, T., Kakinuma, Y., Kon, Y., Hirota, K., Taniguchi-Yanai, K., Nishijo, N., Shigematsu, Y., Horiguchi, H., Kasuya, Y., Sugiyama, F., Yagami, K.-i, Murakami, K., and Fukamizu, A. (2000) *J. Biol. Chem.* **275**, 5–8

[h] Smith, E. L., Hill, R. L., Lehman, I. R., Lefkowitz, R. J., Handler, P., and White, A. (1983) *Principles of Biochemistry, Mammalian Biochemistry*, 7th ed., McGraw-Hill, New York (pp. 157–163)

[i] Bull, H. G., Thornberry, N. A., Cordes, M. H. J., Patchett, A. A., and Cordes, E. H. (1985) *J. Biol. Chem.* **260**, 2952–2962

[j] Derkx, F. H. M., Schalekamp, M. P. A., and Schalekamp, M. A. D. H. (1987) *J. Biol. Chem.* **262**, 2472–2477

[k] Fray, J. C. S., Lush, D. J., and Valentine, A. N. D. (1983) *Fed. Proc.* **42**, 3150–3154

[l] Tipnis, S. R., Hooper, N. M., Hyde, R., Karran, E., Christie, G., and Turner, A. J. (2000) *J. Biol. Chem.* **275**, 33238–33243

[m] Ehlers, M. R. W., Schwager, S. L. U., Scholle, R. R., Manji, G. A., Brandt, W. F., and Riordan, J. F. (1996) *Biochemistry* **35**, 9549–9559

[n] Vazeux, G., Wang, J., Corvol, P., and Llorens-Cortès, C. (1996) *J. Biol. Chem.* **271**, 9069–9074

[o] Tan, F., Morris, P. W., Skidgel, R. A., and Erdös, E. G. (1993) *J. Biol. Chem.* **268**, 16631–16638

[p] Noda, K., Feng, Y.-H., Liu, X.-p, Saad, Y., Husain, A., and Karnik, S. S. (1996) *Biochemistry* **35**, 16435–16442

[q] Boucard, A. A., Wilkes, B. C., Laporte, S. A., Escher, E., Guillemette, G., and Leduc, R. (2000) *Biochemistry* **39**, 9662–9670

[qr] Heerding, J. N., Hines, J., Fluharty, S. J., and Yee, D. K. (2001) *Biochemistry* **40**, 8369–8377

[r] Flier, J. S., and Underhill, L. H. (1985) *N. Engl. J. Med.* **313**, 1330–1340

[s] Berger, S., Bleich, M., Schmid, W., Cole, T. J., Peters, J., Watanabe, H., Kriz, W., Warth, R., Greger, R., and Schütz, G. (1998) *Proc. Natl. Acad. Sci. U.S.A.* **95**, 9424–9429

[t] Taubes, G. (1998) *Science* **281**, 898–907

[u] Williams, G. H. (1988) *N. Engl. J. Med.* **319**, 1517–1525

[v] Dealwis, C. G., Frazao, C., Badasso, M., Cooper, J. B., Tickle, I. J., Driessen, H., Blundell, T. L., Murakami, K., Miyazaki, H., Sueiras-Diaz, J., Jones, D. M., and Szelke, M. (1994) *J. Mol. Biol.* **236**, 342–360

[w] Hilbert, P., Lindpaintner, K., Beckmann, J. S., Serikawa, T., Soubrier, F., Dubay, C., Cartwright, P., De Gouyon, B., Julier, C., Takahasi, S., Vincent, M., Ganten, D., Georges, M., and Lathrop, G. M. (1991) *Nature (London)* **353**, 521–529

[x] Kreutz, R., Hübner, N., James, M. R., Bihoreau, M.-T., Gauguier, D., Lathrop, G. M., Ganten, D., and Lindpaintner, K. (1995) *Proc. Natl. Acad. Sci. U.S.A.* **92**, 8778–8782

[y] Tong, L., Pav, S., Lamarre, D., Simoneau, B., Lavallée, P., and Jung, G. (1995) *J. Biol. Chem.* **270**, 29520–29524

[z] Warren, J. B., and Loi, R. K. (1995) *FASEB J.* **9**, 411–418

[aa] Sagnella, G. A., and MacGregor, G. A. (1986) *Trend Biochem. Sci.* II, 299–302

[bb] Cantin, M., and Genest, J. (1986) *Sci. Am.* **254**(February), 76–81

[cc] Lopez, M. J., Garbers, D. L., and Kuhn, M. (1997) *J. Biol. Chem.* **272**, 23064–23068

[dd] Wechter, W. J., Kantoci, D., Murray, E. D., Jr., D'Amico, D. C., Jung, M. E., and Wang, W.-H. (1996) *Proc. Natl. Acad. Sci. U.S.A.* **93**, 6002–6007

Administration of estrogens and progestins inhibits FSH and LH secretion from the pituitary (feedback inhibition) and hence ovulation. This effect is the action of contraceptive pills. A small amount of the synthetic estrogen 17-ethynylestradiol may be taken daily for 10–15 days followed by a combination of estrogen plus a progestin such as norethindrone for 0–15 days. Alternatively, a progestin alone may be

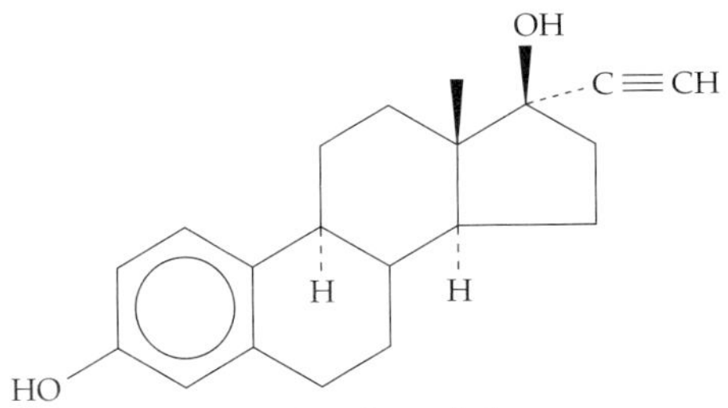

17-Ethynylestradiol

ingested over the entire period. Another synthetic compound with estrogenic activity is diethylstilbestrol. Its once widespread use in promoting growth of cattle and other animals has been discontinued because of carcinogenic action in rats fed large amounts of the compound.

Human cancers of the breast and endometrium are stimulated by estrogen. However, the mechanism is unknown.[297,298] It has usually been assumed that the proliferation of cells induced by estrogens leads to mutations and cancer. However, estrogens can form adducts with DNA after oxidation to 2- and 4-hydroxy derivatives and further oxidation to quinones.[297,298] In a similar manner prostate cancer is promoted by androgens. It has also been observed that in the United States the incidence of cancers of the ovary and endometrium has declined substantially during the past 35 years, perhaps as a result of the anti-estrogenic activity of the progestin in the widely used oral contraceptives.[299] This observation led to the idea of **hormonal chemoprevention**, the deliberate use of hormone antagonists to slow cancer formation.[299a] This may be especially attractive to persons carrying known cancer-susceptibility genes. The synthetic antiestrogen **tamoxifen** is being evaluated as a chemopreventive agent for breast cancer.[299,299b] However, a planned large-scale trial was postponed because of uncertainities about safety.[300]

In addition to steroids there are plant flavonoids (Box 21-E) that have estrogenic activity. These labile compounds are among the **"environmental estrogens."** In addition, there are many much more stable compounds, including the insecticides atrazine and DDT, PCBs, and phenolic softeners for plastics, that have weak estrogenic activity. Alarm has been sounded by some who maintain that these **xenoestrogens** are contributing to breast cancer, to reproductive difficulties in animals, and to low sperm counts in men.[301] It seems surprising that such small amounts of weakly estrogenic compounds could have such large effects. Some experiments suggest that two weak xenoestrogens may cooperate to give larger effects,[302,303] but this concept needs further evaluation.

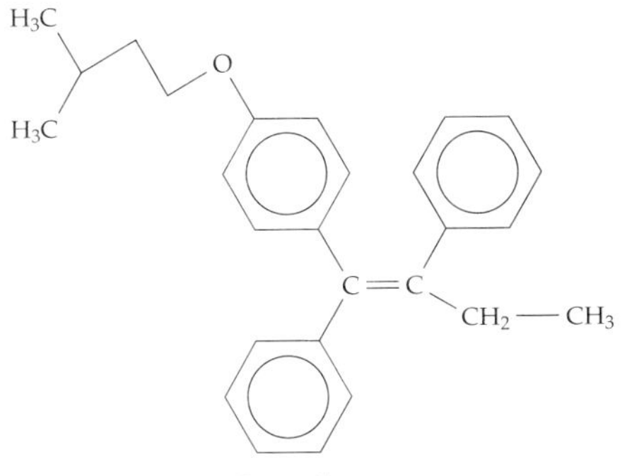

trans-Tamoxifen

Although estrogens are usually regarded as female hormones, they are present in small amounts in male blood and in high concentration in semen. Male reproductive tissues contain estrogen receptors, and mice deficient in these receptors are sterile and their testes degenerate.[304] Furthermore, as in females estrogen stops growth of long bones in late puberty. A few men lacking estrogen or estrogen receptors have grown very tall (>2.1 m and still growing).[305]

5. The Steroid Receptor Family

The principal mode of action of steroid hormones is to stimulate transcription of specific sets of genes. The plasma concentrations of these hormones are low, typically ~10^{-9} M, but they have a high affinity for their protein receptors, some of which are located initially in the cytosol but are found largely in the nucleus. The earliest identification of steroid receptors was accomplished with radioactive ^{3}H-labeled progesterone, estrogens, and glucocorticoids.[306,307] Autoradiographs of thin sections of tissue made after the uptake of the hormones revealed that the radioactivity was concentrated in the nuclei. The protein receptors were then isolated and were found to bind both to the hormone and to specific sequences in DNA, the hormone **response elements**.[308] The progesterone, estrogen, and glucocorticoid receptors are multidomain proteins with two of the domains having highly conserved sequences and structures. One of these domains binds to DNA and the other to the steroid hormone. Their amino acid sequences are also related to those of the v-*erb-A* oncogene (Table 11-3).[309] With this knowledge available and through use of methods of "reverse genetics," receptors for the other steroid hormones and also for vitamin D, retinoic acid, and thyroid hormone were identified as members of the family as were several "orphan" receptors of as yet unknown function (Table 22-1). Invertebrates have similar receptors. At least eight are present in *Drosophila*,[307] and the family is present in the nematode *Caenorhabditis*.[310]

Because of their hydrophobic character the steroid hormones or other nonpolar ligands diffuse through membranes into cells. There they may encounter a variety of binding proteins that affect their access to a receptor.[307,311] Some receptors, including glucocorticoid receptors, are found in the cytosol. After a hor-

TABLE 22-1
Known Members of the Steroid Receptor Family[a]

Glucocorticoid[b–d]	Thyroid α, β_1, β_2[r]
Mineralocorticoid[e,f]	Retinoic acid α, β, γ[s]
Progesterone[g,h]	Retinoid-X α,β[t–w]
Androgen[i,j,k]	Peroxisome proliferators[x,y]
Estrogen[l–o]	Farnesoid X: bile acids[z,aa]
Vitamin D_3[p,q]	Orphan receptors, 8 or more

[a] See Fuller (1991) *FASEB J.* **5**, 3092–3099
[b] Luisi, B. F., Xu, W. X., Otwinowski, Z., Freedman, L. P., Yamamoto, K. R., and Sigler, P. B. (1991) *Nature (London)* **352**, 497–505
[c] Eriksson, M. A. L., Härd, T., and Nilsson, L. (1995) *Biophys. J.* **68**, 402–426
[d] La Baer, J., and Yamamoto, K. R. (1994) *J. Mol. Biol.* **239**, 664–688
[e] Funder, J. W. (1993) *Science* **259**, 1132–1133
[f] Geller, D. S., Farhi, A., Pinkerton, N., Fradley, M., Moritz, M., Spitzer, A., Meinke, G., Tsai, F. T. F., Sigler, P. B., and Lifton, R. P. (2000) *Science* **289**, 119–123
[g] Williams, S. P., and Sigler, P. B. (1998) *Nature (London)* **393**, 392–396
[h] Knotts, T. A., Orkiszewski, R. S., Cook, R. G., Edwards, D. P., and Weigel, N. L. (2001) *J. Biol. Chem.* **276**, 8475–8483
[i] Grossmann, M. E., Lindzey, J., Blok, L., Perry, J. E., Kumar, M. V., and Tindall, D. J. (1994) *Biochemistry* **33**, 14594–14600
[j] Matias, P. M., and 13 other authors. (2000) *J. Biol. Chem.* **275**, 26164–26171
[k] Sack, J. S., Kish, K. F., Wang, C., Attar, R. M., Kiefer, S. E., An, Y., Wu, G. Y., Scheffler, J. E., Salvati, M. E., Krystek, S. R., Jr., Weinmann, R., and Einspahr, H. M. (2001) *Proc. Natl. Acad. Sci. U.S.A.* **98**, 4904–4909
[l] Ekena, K., Weis, K. E., Katzenellenbogen, J. A., and Katzenellenbogen, B. S. (1996) *J. Biol. Chem.* **271**, 20053–20059
[m] Tanenbaum, D. M., Wang, Y., Williams, S. P., and Sigler, P. B. (1998) *Proc. Natl. Acad. Sci. U.S.A.* **95**, 5998–6003
[n] Brzozowski, A. M., Pike, A. C. W., Dauter, Z., Hubbard, R. E., Bonn, T., Engström, O., Öhman, L., Greene, G. L., Gustafsson, J.-Å., and Carlquist, M. (1997) *Nature (London)* **389**, 753–758
[o] Kahlert, S., Nuedling, S., van Eickels, M., Vetter, H., Meyer, R., and Grohé, C. (2000) *J. Biol. Chem.* **275**, 18447–18453
[p] Craig, T. A., Veenstra, T. D., Naylor, S., Tomlinson, A. J., Johnson, K. L., Macura, S., Juranic, N., and Kumar, R. (1997) *Biochemistry* **36**, 10482–10491
[q] Hendick, M., and Carlberg, C. (2000) *J. Mol. Biol.* **304**, 793–801
[r] Kostrouch, Z., Kostrouchova, M., and Rall, J. E. (1995) *Proc. Natl. Acad. Sci. U.S.A.* **92**, 156–159
[s] Klaholz, B. P., Mitschler, A., and Moras, D. (2000) *J. Mol. Biol.* **302**, 155–170
[t] Bourguet, W., Ruff, M., Chambon, P., Gronemeyer, H., and Moras, D. (1995) *Nature (London)* **375**, 377–382
[u] Holmbeck, S. M. A., Foster, M. P., Casimiro, D. R., Sem, D. S., Dyson, J., and Wright, P. E. (1998) *J. Mol. Biol.* **281**, 271–284
[v] Kersten, S., Dong, D., Lee, W.-y, Reczek, P. R., and Noy, N. (1998) *J. Mol. Biol.* **284**, 21–32
[w] Chawla, A., Repa, J. J., Evans, R. M., and Mangelsdorf, D. J. (2001) *Science* **294**, 1866–1870
[x] Nolte, R. T., Wisely, G. B., Westin, S., Cobb, J. E., Lambert, M. H., Kurokawa, R., Rosenfeld, M. G., Willson, T. M., Glass, C. K., and Milburn, M. V. (1998) *Nature (London)* **395**, 137–143
[y] Rosen, E. D., and Spiegelman, B. M. (2001) *J. Biol. Chem.* **276**, 37731–37734
[z] Makishima, M., Okamoto, A. Y., Repa, J. J., Tu, H., Learned, R. M., Luk, A., Hull, M. V., Lustig, K. D., Mangelsdorf, D. J., and Shan, B. (1999) *Science* **284**, 1362–1365
[aa] Parks, D. J., Blanchard, S. G., Bledsoe, R. K., Chandra, G., Consler, T. G., Kliewer, S. A., Stimmel, J. B., Willson, T. M., Zavacki, A. M., Moore, D. D., and Lehmann, J. M. (1999) *Science* **284**, 1365–1368

mone binds to a cytosolic receptor, the complex apparently undergoes time and temperature dependent alterations that activate the receptor before it diffuses into the nucleus and binds to its proper response element in the DNA.[312] This process has been observed directly for glucocorticoid receptors labeled by fusion with green fluorescent protein (Box 23-A).[313]

The conserved 68-residue DNA binding domain of the glucocorticoid receptor contains two Zn^{2+} ions, each coordinated by four cysteine $-CH_2-S^-$ groups with tetrahedral geometry. These two consecutive motifs form structures somewhat similar to those of the "zinc fingers" shown in Fig. 5-38.[314–316] However, the overall folding pattern is different from those considered in Chapter 5. The two zinc-binding sites lie at the N termini of a pair of helices that cross at right angles near their centers. One of these is a DNA-recognition helix that fits into the major groove of DNA thereby allowing interaction of its amino acid side chains with the bases of the DNA response elements.

The response elements for glucocorticoids and estrogen receptors contain short palindromic sequences with various three-nucleotide "spacer" sequences in the center as follows.[308,314,316–318] Two receptor proteins bind to the palindromic DNA forming a homodimeric receptor pair. For the 9-*cis* retinoic acid receptor RXR-α the response element contains a pair of direct repeats of a 6-base consensus sequence with a two-base pair spacer:

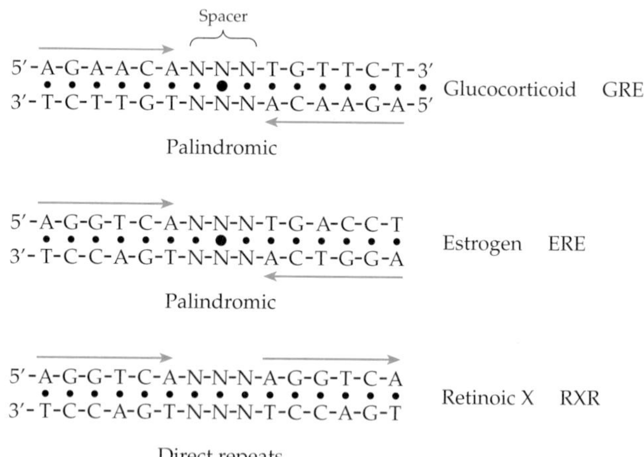

The RXR-α receptor binds differently and tends to form *heterodimeric* pairs with other receptors.[319,320] All of these receptors undergo conformational changes when agonists or antagonists bind.[320,321] Estrogens can also bind to androgen receptors, perhaps in playing their essential role in male reproductive physiology.[322] There are more than one type of receptor for each group of steroid hormones, and these may interact differently with the various response elements in DNA[323] making the effects of hormones complex and hard to analyze. Interactions with additional proteins also affect the response of a cell to hormones.[324] Furthermore, steroid hormones have **transcription-independent effects**. For example, progesterone binds to oxytocin receptors[325] as well as to other steroid receptors, which affect a broad range of biochemical processes.[326]

F. Other Steroids

The **saponins** are a series of steroid glycosides with detergent properties that are widespread among higher plants.[327] Some are toxic, and among these toxic materials are compounds of extraordinary medical importance. Best known are the steroid glycosides of *Digitalis*, among them **digitonin** (Fig. 22-12). The particular arrangement of sugar units in this molecule imparts a specificity toward heart muscle. The compound is extremely toxic; in small amounts it acts to increase the tone of heart muscle and is widely used in treatment of congestive heart failure.[328] The maintenance dose is only 0.1 mg / day. Another toxic glycoside and heart stimulant is **ouabain** (Fig. 22-12). Ouabain is a specific inhibitor of the membrane-bound ($Na^+ + K^+$)-ATPase believed to be the ion pump that keeps intracellular K^+ concentrations high and Na^+ concentrations low (Chapter 8). Similar glycosides account for the extreme toxicity of the leaves of the oleander and the roots of the lily of the valley. A steroid glycoside from red squill is used as a rat poison. A number of alkaloids (nitrogenous bases) are derived from steroids. An example is **solanidine** (Fig. 22-12), which is present in the skins and sprouts of potatoes, making both quite toxic.

Some animals also contain toxic steroids. **Batrachotoxin** of the Columbian poison arrow frog (Fig. 22-12) is present in amounts of only 50 μg per frog.[329,330]

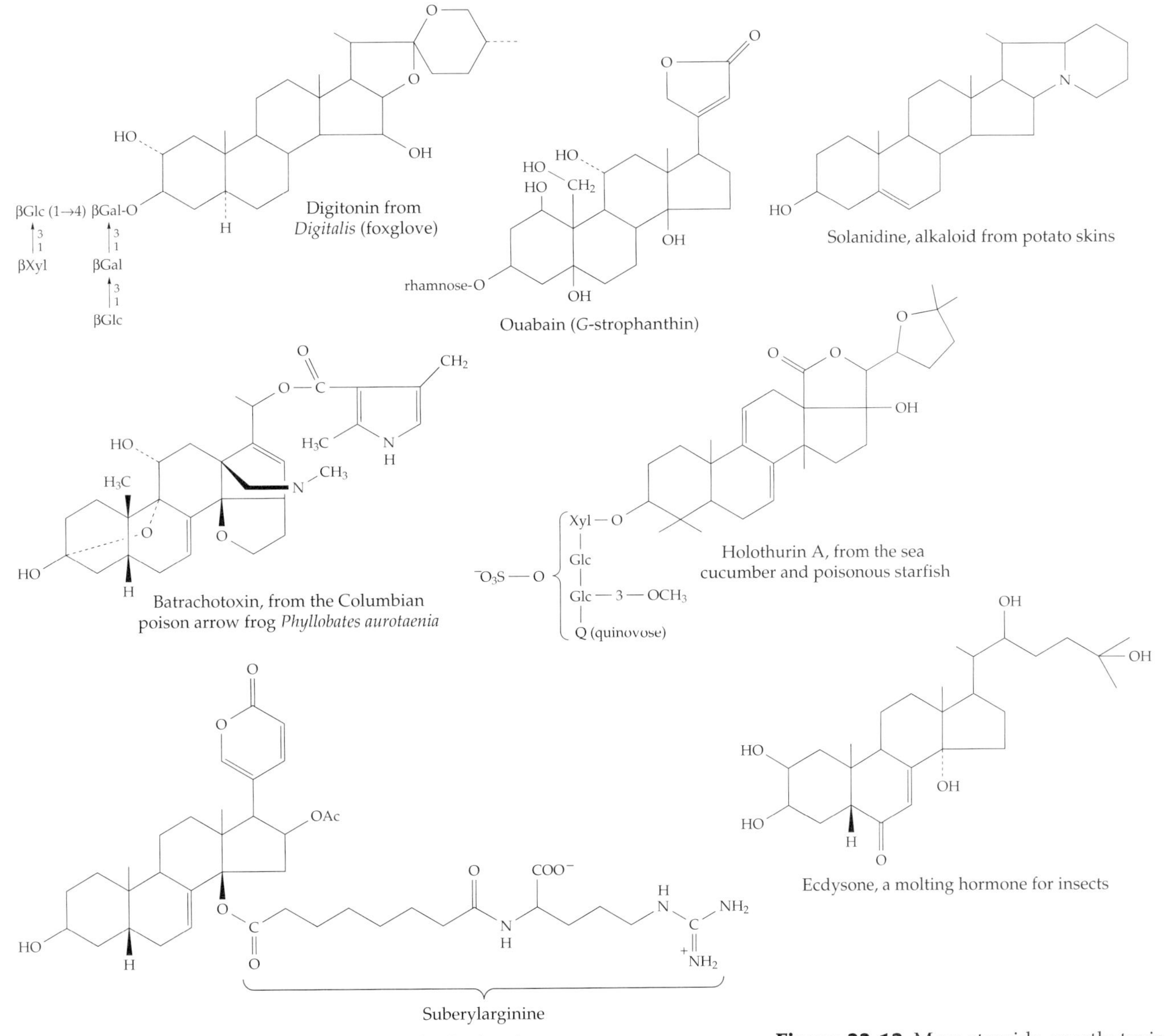

Digitonin from *Digitalis* (foxglove)

Ouabain (*G*-strophanthin)

Solanidine, alkaloid from potato skins

Batrachotoxin, from the Columbian poison arrow frog *Phyllobates aurotaenia*

Holothurin A, from the sea cucumber and poisonous starfish

Ecdysone, a molting hormone for insects

Suberylarginine

Bufotoxin, from the skin glands of toads

Figure 22-12 More steroids, mostly toxic.

The toxin acts on nerves to block transmissions to the muscle by increasing the permeability of membranes to sodium ions. It is specifically antagonized by **tetrodotoxin** (Fig. 30-16). Batrachotoxin alkaloids are present also in certain birds.[330a] Some echinoderms make powerful steroid toxins such as **holothurin A** (Fig. 22-12), a surface active agent that causes irreversible destruction of the excitability of neuromuscular tissues. The common toad produces in its skin steroid toxins such as **bufotoxin** (Fig. 22-12), which are sufficiently powerful to teach a dog to leave toads alone.

Ecdysone, a highly hydroxylated steroid (Fig. 22-12), is a molting hormone for insects.[331,332] Several molecules with ecdysone activity are known, and some of these are produced by certain plants. Although ecdysones are needed by insects for larval molting, they are toxic in excess. Perhaps plants protect themselves from insects by synthesizing these substances.

Among the antibiotics **fusidic acid** is a steroid. An inhibitor of chloramphenicol acetyltransferase[333] it is highly inhibitory to staphyococci but almost noninhibitory to *E. coli*. Note the boat conformation of the B ring.

CH_3 HO H_3C OH CH_3 CH_3 COO^- O C=O CH_3

Fusidic acid

Another steroid antibiotic, **squalamine**,[334] was isolated from the stomachs of sharks. It is effective against gram-positive and gram-negative bacteria, and some fungi as well.[335]

OSO_3^- $H_3\overset{+}{N}$ $\overset{+}{N}H_2$ OH

Squalamine

References

1. Luckner, M. (1984) *Secondary Metabolism in Plants and Animals*, 2nd ed., Academic Press, New York
2. Britton, G. (1976) in *Chemistry and Biochemistry of Plant Pigments*, 2nd ed., Vol. 1 (Goodwin, T. W., ed), pp. 262–327, Academic Press, New York
3. Haslam, E. (1985) *Metabolites and Metabolism*, Oxford Univ. Press, London
4. Herbert, R. B. (1989) *The Biosynthesis of Secondary Metabolites*, 2nd ed., Chapman and Hall, New York
5. Towers, G. H. N., and Stafford, H. A., eds. (1990) *Biochemistry of the Mevalonic Acid Pathway to Terpenoids*, Vol. 24, Plenum Press, New York and London
6. Nes, W. D., ed. (1994) *Isopentenoids and Other Natural Products*, American Chemical Society, Washington D.C.

6a. Wang, K., and Ohnuma, S.-i. (1999) *Trends Biochem. Sci.* **24**, 445–451

7. Misra, I., Narasimhan, C., and Miziorko, H. M. (1993) *J. Biol. Chem.* **268**, 12129–12135
8. Quant, P. A. (1994) in *Essays in Biochemistry*, Vol. 28 (Tipton, K. F., ed), Portland Press, Chapel Hill, North Carolina
9. Denbow, C. J., Lång, S., and Cramer, C. L. (1996) *J. Biol. Chem.* **271**, 9710–9715
10. Street, I. P., Coffman, H. R., Baker, J. A., and Poulter, C. D. (1994) *Biochemistry* **33**, 4212–4217
11. Frimpong, K., and Rodwell, V. W. (1994) *J. Biol. Chem.* **269**, 11478–11483
12. Bischoff, K. M., and Rodwell, V. W. (1997) *Protein Sci.* **6**, 156–161
13. Misra, I., and Miziorko, H. M. (1996) *Biochemistry* **35**, 9610–9616
14. Omkumar, R. V., and Rodwell, V. W. (1994) *J. Biol. Chem.* **269**, 16862–16866
15. Lawrence, C. M., Rodwell, V. W., and Stauffacher, C. V. (1995) *Science* **268**, 1758–1762
16. Luskey, K. L., and Stevens, B. (1985) *J. Biol. Chem.* **260**, 10271–10277
17. Goldstein, J. L., and Brown, M. S. (1990) *Nature (London)* **343**, 425–430
18. Meigs, T. E., Roseman, D. S., and Simoni, R. D. (1996) *J. Biol. Chem.* **271**, 7916–7922
19. Berkhout, T. A., Simon, H. M., Patel, D. D., Bentzen, C., Niesor, E., Jackson, B., and Suckling, K. E. (1996) *J. Biol. Chem.* **271**, 14376–14382
20. Correll, C. C., Ng, L., and Edwards, P. A. (1994) *J. Biol. Chem.* **269**, 17390–17393
21. Biardi, L., Sreedhar, A., Zokaei, A., Vartak, N. B., Bozeat, R. L., Shackelford, J. E., Keller, G.-A., and Krisans, S. K. (1994) *J. Biol. Chem.* **269**, 1197–1205
22. Potter, D., Wojnar, J. M., Narasimhan, C., and Miziorko, H. M. (1997) *J. Biol. Chem.* **272**, 5741–5746
23. Schafer, B. L., Bishop, R. W., Kratunis, V. J., Kalinowski, S. S., Mosley, S. T., Gibson, K. M., and Tanaka, R. D. (1992) *J. Biol. Chem.* **267**, 13229–13238
24. Paton, V. G., Shackelford, J. E., and Krisans, S. K. (1997) *J. Biol. Chem.* **272**, 18945–18950
25. Dhe-Paganon, S., Magrath, J., and Abeles, R. H. (1994) *Biochemistry* **33**, 13355 – 13362
26. Eisenreich, W., Menhard, B., Hylands, P. J., Zenk, M. H., and Bacher, A. (1996) *Proc. Natl. Acad. Sci. U.S.A.* **93**, 6431–6436
27. Kuo, S.-M., and Aronson, P. S. (1996) *J. Biol. Chem.* **271**, 15491–15497
28. Rohmer, M., Seemann, M., Horbach, S., Bringer-Meyer, S., and Sahm, H. (1996) *J. Am. Chem. Soc.* **118**, 2564–2566
29. Sprenger, G. A., Schörken, U., Wiegert, T., Grolle, S., de Graaf, A. A., Taylor, S. V., Begley, T. P., Bringer-Meyer, S., and Sahm, H. (1997) *Proc. Natl. Acad. Sci. U.S.A.* **94**, 12857–12862
30. Himmeldirk, K., Sayer, B. G., and Spenser, I. D. (1998) *J. Am. Chem. Soc.* **120**, 3581–3589
31. Taylor, S. V., Kelleher, N. L., Kinsland, C., Chiu, H.-J., Costello, C. A., Backstrom, A. D., McLafferty, F. W., and Begley, T. P. (1998) *J. Biol. Chem.* **273**, 16555–16560

31a. Rohdich, F., Hecht, S., Gärtner, K., Adam, P., Krieger, C., Amslinger, S., Arigoni, D., Bacher, A., and Eisenreich, W. (2002) *Proc. Natl. Acad. Sci. U.S.A.* **99**, 1158–1163

32. Arigoni, D., Sagner, S., Latzel, C., Eisenreich, W., Bacher, A., and Zenk, M. H. (1997) *Proc. Natl. Acad. Sci. U.S.A.* **94**, 10600–10605

32a. Cane, D. E. (2000) *Science* **287**, 818–819

32b. Arigoni, D., Eisenreich, W., Latzel, C., Sagner, S., Radykewicz, T., Zenk, M. H., and Bacher, A. (1999) *Proc. Natl. Acad. Sci. U.S.A.* **96**, 1309–1314

32c. Rohdich, F., Wungsintaweekul, J., Fellermeier, M., Sagner, S., Herz, S., Kis, K., Eisenreich, W., Bacher, A., and Zenk, M. H. (1999) *Proc. Natl. Acad. Sci. U.S.A.* **96**, 11758–11763

32d. Herz, S., and 10 other authors. (2000) *Proc. Natl. Acad. Sci. U.S.A.* **97**, 2486–2490

32e. Rohdich, F., Wungsintaweekul, J., Eisenreich, W., Richter, G., Schuhr, C. A., Hecht, S., Zenk, M. H., and Bacher, A. (2000) *Proc. Natl. Acad. Sci. U.S.A.* **97**, 6451–6456

32f. Lange, B. M., and Croteau, R. (1999) *Proc. Natl. Acad. Sci. U.S.A.* **96**, 13714–13719

32g. Lange, B. M., Rujan, T., Martin, W., and Croteau, R. (2000) *Proc. Natl. Acad. Sci. U.S.A.* **97**, 13172–13177

32h. Zeidler, J., Ullah, N., Gupta, R. N., Pauloski, R. M., Sayer, B. G., and Spenser, I. D. (2002) *J. Am. Chem. Soc.* **124**, 4542–4543

33. Hill, R. E., Himmeldirk, K., Kennedy, I. A., Pauloski, R. M., Sayer, B. G., Wolf, E., and Spenser, I. D. (1996) *J. Biol. Chem.* **271**, 30426–30435

33a. Paton, V. G., Shackelford, J. E., and Krisans, S. K. (1997) *J. Biol. Chem.* **272**, 18945–18950

References

33b. Kaneda, K., Kuzuyama, T., Takagi, M., Hayakawa, Y., and Seto, H. (2001) *Proc. Natl. Acad. Sci. U.S.A.* **98**, 932–937

34. Deneris, E. S., Stein, R. A., and Mead, J. F. (1985) *J. Biol. Chem.* **260**, 1382–1385

35. Silver, G. M., and Fall, R. (1995) *J. Biol. Chem.* **270**, 13010–13016

36. Sharkey, T. D., and Singsaas, E. L. (1995) *Nature (London)* **374**, 769

37. Mlot, C. (1995) *Science* **268**, 641–642

38. Brady, P. S., Scofield, R. F., Schumann, W. C., Ohgaku, S., Kumaran, K., Margolis, J. M., and Landau, B. R. (1982) *J. Biol. Chem.* **257**, 10742–10746

39. Chen, A., Kroon, P. A., and Poulter, C. D. (1994) *Protein Sci.* **3**, 600–607

40. Cunillera, N., Arró, M., Delourme, D., Karst, F., Boronat, A., and Ferrer, A. (1996) *J. Biol. Chem.* **271**, 7774–7780

41. Tarshis, L. C., Proteau, P. J., Kellogg, B. A., Sacchettini, J. C., and Poulter, C. D. (1996) *Proc. Natl. Acad. Sci. U.S.A.* **93**, 15018–15023

42. Ohnuma, S.-i, Narita, K., Nakazawa, T., Ishida, C., Takeuchi, Y., Ohto, C., and Nishino, T. (1996) *J. Biol. Chem.* **271**, 30748–30754

43. Ohnuma, S.-i, Nakazawa, T., Hemmi, H., Hallberg, A.-M., Koyama, T., Ogura, K., and Nishino, T. (1996) *J. Biol. Chem.* **271**, 10087–10095

44. Math, S. K., Hearst, J. E., and Poulter, C. D. (1992) *Proc. Natl. Acad. Sci. U.S.A.* **89**, 6761–6764

45. Ohnuma, S.-i, Hirooka, K., Hemmi, H., Ishida, C., Ohto, C., and Nishino, T. (1996) *J. Biol. Chem.* **271**, 18831–18837

46. Carattoli, A., Romano, N., Ballario, P., Morelli, G., and Macino, G. (1991) *J. Biol. Chem.* **266**, 5854–5859

46a. Kuzuguchi, T., Morita, Y., Sagami, I., Sagami, H., and Ogura, K. (1999) *J. Biol. Chem.* **274**, 5888–5894

46b. Kharel, Y., Zhang, Y.-W., Fujihashi, M., Miki, K., and Koyama, T. (2001) *J. Biol. Chem.* **276**, 28459–28464

47. Tarshis, L. C., Yan, M., Poulter, C. D., and Sacchettini, J. C. (1994) *Biochemistry* **33**, 10871–10877

47a. Fernandez, S. M. S., Kellogg, B. A., and Poulter, C. D. (2000) *Biochemistry* **39**, 15316–15321

48. Sacchettini, J. C., and Poulter, C. D. (1997) *Science* **277**, 1788–1789

49. Koyama, T., Tajima, M., Sano, H., Doi, T., Koike-Takeshita, A., Obata, S., Nishino, T., and Ogura, K. (1996) *Biochemistry* **35**, 9533–9538

50. Poulter, C. D., and Rilling, H. C. (1978) *Acc. Chem. Res.* **11**, 307–313

51. Poulter, C. D., Wiggins, D. L., and Le, A. T. (1981) *J. Am. Chem. Soc.* **103**, 3926–3927

52. Suga, T., Hirata, T., Aoki, T., and Shishibori, T. (1983) *J. Am. Chem. Soc.* **105**, 6178–6179

53. Davisson, V. J., Neal, T. R., and Poulter, C. D. (1993) *J. Am. Chem. Soc.* **115**, 1235–1245

54. Berdis, A. J., and Benkovic, S. J. (1996) *Biochemistry* **35**, 9253–9265

54a. Zhang, Y.-W., Li, X.-Y., and Koyama, T. (2000) *Biochemistry* **39**, 12717–12722

54b. Oh, S. K., Han, K. H., Ryu, S. B., and Kang, H. (2000) *J. Biol. Chem.* **275**, 18482–18488

55. Matsuoka, S., Sagami, H., Kurisaki, A., and Ogura, K. (1991) *J. Biol. Chem.* **266**, 3464–3468

56. Keller, R. K. (1987) *Trends Biochem. Sci.* **12**, 443–445

57. Ericsson, J., Scallen, T. J., Chojnacki, T., and Dallner, G. (1991) *J. Biol. Chem.* **266**, 10602–10607

57a. Pan, J.-J., Chiou, S.-T., and Liang, P.-H. (2000) *Biochemistry* **39**, 10936–10942

57b. Fujihashi, M., Zhang, Y.-W., Higuchi, Y., Li, X.-Y., Koyama, T., and Miki, K. (2001) *Proc. Natl. Acad. Sci. U.S.A.* **98**, 4337–4342

58. Frank, D. W., and Waechter, C. J. (1998) *J. Biol. Chem.* **273**, 11791–11798

59. Sagami, H., Igarashi, Y., Tateyama, S., Ogura, K., Roos, J., and Lennarz, W. J. (1996) *J. Biol. Chem.* **271**, 9560–9566

60. Ohkura, T., Fukushima, K., Kurisaki, A., Sagami, H., Ogura, K., Ohno, K., Hara-Kuge, S., and Yamashita, K. (1997) *J. Biol. Chem.* **272**, 6868–6875

61. Murgolo, N. J., Patel, A., Stivala, S. S., and Wong, T. K. (1989) *Biochemistry* **28**, 253–260

62. Takahashi, I., Ogura, K., and Seto, S. (1980) *J. Biol. Chem.* **255**, 4539–4543

63. Wolucka, B. A., and de Hoffmann, E. (1995) *J. Biol. Chem.* **270**, 20151–20155

63a. Schulbach, M. C., Mahapatra, S., Macchia, M., Barontini, S., Papi, C., Minutolo, F., Bertini, S., Brennan, P. J., and Crick, D. C. (2001) *J. Biol. Chem.* **276**, 11624–11630

64. Runquist, M., Ericsson, J., Thelin, A., Chojnacki, T., and Dallner, G. (1994) *J. Biol. Chem.* **269**, 5804–5809

65. Dennis, M. S., Henzel, W. J., Bell, J., Kohr, W., and Light, D. R. (1989) *J. Biol. Chem.* **264**, 18618–18626

66. Ohnuma, S.-i, Koyama, T., and Ogura, K. (1991) *J. Biol. Chem.* **266**, 23706–23713

67. Chen, A., Zhang, D., and Poulter, C. D. (1993) *J. Biol. Chem.* **268**, 21701–21705

67a. Morii, H., Nishihara, M., and Koga, Y. (2000) *J. Biol. Chem.* **275**, 36568–36574

68. Zhang, F. L., and Casey, P. J. (1996) *Ann. Rev. Biochem.* **65**, 241–269

69. Gelb, M. H. (1997) *Science* **275**, 1750–1751

70. Vogt, A., Sun, J., Qian, Y., Tan-Chiu, E., Hamilton, A. D., and Sebti, S. M. (1995) *Biochemistry* **34**, 12398–12403

71. Mu, Y. Q., Omer, C. A., and Gibbs, R. A. (1996) *J. Am. Chem. Soc.* **118**, 1819–1823

72. Del Villar, K., Mitsuzawa, H., Yang, W., Sattler, I., and Tamanoi, F. (1997) *J. Biol. Chem.* **272**, 680–687

73. Park, H.-W., Boduluri, S. R., Moomaw, J. F., Casey, P. J., and Beese, L. S. (1997) *Science* **275**, 1800–1804

74. Mathis, J. R., and Poulter, C. D. (1997) *Biochemistry* **36**, 6367–6376

75. Dunten, P., Kammlott, U., Crowther, R., Weber, D., Palermo, R., and Birktoft, J. (1998) *Biochemistry* **37**, 7907–7912

76. Long, S. B., Casey, P. J., and Beese, L. S. (1998) *Biochemistry* **37**, 9612–9618

76a. Crespo, N. C., Ohkanda, J., Yen, T. J., Hamilton, A. D., and Sebti, S. M. (2001) *J. Biol. Chem.* **276**, 16161–16167

76b. Long, S. B., Hancock, P. J., Kral, A. M., Hellinga, H. W., and Beese, L. S. (2001) *Proc. Natl. Acad. Sci. U.S.A.* **98**, 12948–12953

76c. Micali, E., Chehade, K. A. H., Isaacs, R. J., Andres, D. A., and Spielmann, H. P. (2001) *Biochemistry* **40**, 12254–12265

76d. Huang, C.-c, Hightower, K. E., and Fierke, C. A. (2000) *Biochemistry* **39**, 2593–2602

76e. Pang, Y.-P., Xu, K., El Yazal, J., and Prendergast, F. G. (2000) *Protein Sci.* **9**, 1857–1865

76f. Long, S. B., Casey, P. J., and Beese, L. S. (2002) *Nature (London)* **419**, 645–650

77. Yokoyama, K., McGeady, P., and Gelb, M. H. (1995) *Biochemistry* **34**, 1344–1354

78. Witter, D. J., and Poulter, C. D. (1996) *Biochemistry* **35**, 10454–10463

78a. Clausen, V. A., Edelstein, R. L., and Distefano, M. D. (2001) *Biochemistry* **40**, 3920–3930

78b. Dursina, B., Thomä, N. H., Sidorovitch, V., Niculae, A., Iakovenko, A., Rak, A., Albert, S., Ceacareanu, A.-C., Kölling, R., Herrmann, C., Goody, R. S., and Alexandrov, K. (2002) *Biochemistry* **41**, 6805–6816

79. Fu, H.-W., Beese, L. S., and Casey, P. J. (1998) *Biochemistry* **37**, 4465–4472

79a. Ashar, H. R., James, L., Gray, K., Carr, D., Black, S., Armstrong, L., Bishop, W. R., and Kirschmeier, P. (2000) *J. Biol. Chem.* **275**, 30451–30457

79b. Huber, H. E., Robinson, R. G., Watkins, A., Nahas, D. D., Abrams, M. T., Buser, C. A., Lobell, R. B., Patrick, D., Anthony, N. J., Dinsmore, C. J., Graham, S. L., Hartman, G. D., Lumma, W. C., Williams, T. M., and Heimbrook, D. C. (2001) *J. Biol. Chem.* **276**, 24457–24465

80. Leung, H.-C. E., Chen, Y., and Winkler, M. E. (1997) *J. Biol. Chem.* **272**, 13073–13083

81. Gebler, J. C., Woodside, A. B., and Poulter, C. D. (1992) *J. Am. Chem. Soc.* **114**, 7354–7360

82. Clause, E. P., Tyler, V. E., and Brady, L. R. (1970) *Pharmacognosy*, 6th ed., Lea & Febiger, Philadelphia, Pennsylvania

83. Loreto, F., Ciccioli, P., Brancaleoni, E., Cecinato, A., Frattoni, M., and Sharkey, T. D. (1996) *Proc. Natl. Acad. Sci. U.S.A.* **93**, 9966–9969

84. Harborne, J. B., and Tomas-Barberan, F. A., eds. (1991) *Ecological Chemistry and Biochemistry of Plant Terpenoids*, Clarendon Press, Oxford

85. Gershenzon, J., and Croteau, R. B. (1993) in *Lipid Metabolism in Plants* (Moore, T. S., Jr., ed), pp. 339–388, CRC Press, Boca Raton, Florida

86. Bohlmann, J., Meyer-Gauen, G., and Croteau, R. (1998) *Proc. Natl. Acad. Sci. U.S.A.* **95**, 4126–4133

87. Bohlmann, J., Steele, C. L., and Croteau, R. (1997) *J. Biol. Chem.* **272**, 21784–21792

88. Wise, M. L., Savage, T. J., Katahira, E., and Croteau, R. (1998) *J. Biol. Chem.* **273**, 14891–14899

88a. Williams, D. C., McGarvey, D. J., Katahira, E. J., and Croteau, R. (1998) *Biochemistry* **37**, 12213–12220

89. Croteau, R., and Satterwhite, D. M. (1989) *J. Biol. Chem.* **264**, 15309–15315

90. Seybold, S. J., Quilici, D. R., Tillman, J. A., Vanderwel, D., Wood, D. L., and Blomquist, G. J. (1995) *Proc. Natl. Acad. Sci. U.S.A.* **92**, 8393–8397

91. Steele, C. L., Crock, J., Bohlmann, J., and Croteau, R. (1998) *J. Biol. Chem.* **273**, 2078–2089

91a. Rising, K. A., Starks, C. M., Noel, J. P., and Chappell, J. (2000) *J. Am. Chem. Soc.* **122**, 1861–1866

92. Procter, R. H., and Hohn, T. M. (1993) *J. Biol. Chem.* **268**, 4543–4548

93. Cane, D. E., and Tsantrizos, Y. S. (1996) *J. Am. Chem. Soc.* **118**, 10037–10040

93a. Caruthers, J. M., Kang, I., Rynkiewicz, M. J., Cane, D. E., and Christianson, D. W. (2000) *J. Biol. Chem.* **275**, 25533–25539

94. Back, K., and Chappell, J. (1995) *J. Biol. Chem.* **270**, 7375–7381

95. Starks, C. M., Back, K., Chappell, J., and Noel, J. P. (1997) *Science* **277**, 1815–1820

96. Lesburg, C. A., Zhai, G., Cane, D. E., and Christianson, D. W. (1997) *Science* **277**, 1820–1824

97. Cane, D. E., and Xue, Q. (1996) *J. Am. Chem. Soc.* **118**, 1563–1564

97a. Rynkiewicz, M. J., Cane, D. E., and Christianson, D. W. (2002) *Biochemistry* **41**, 1732–1741

References

97b. Peters, R. J., and Croteau, R. B. (2002) *Biochemistry* **41**, 1836–1842

98. Crock, J., Wildung, M., and Croteau, R. (1997) *Proc. Natl. Acad. Sci. U.S.A.* **94**, 12833–12838

99. Marin, E., Nussaume, L., Quesada, A., Gonneau, M., Sotta, B., Hugueney, P., Frey, A., and Marion-Poll, A. (1996) *EMBO J.* **15**, 2331–2342

99a. Qin, X., and Zeevaart, J. A. D. (1999) *Proc. Natl. Acad. Sci. U.S.A.* **96**, 15354–15361

99b. Seo, M., Peeters, A. J. M., Koiwai, H., Oritani, T., Marion-Poll, A., Zeevaart, J. A. D., Koornneef, M., Kamiya, Y., and Koshiba, T. (2000) *Proc. Natl. Acad. Sci. U.S.A.* **97**, 12908–12913

100. He, X., Saint-Jeannet, J.-P., Woodgett, J. R., Varmus, H. E., and Dawid, I. B. (1995) *Nature (London)* **374**, 617–622

101. Spray, C. R., Kobayashi, M., Suzuki, Y., Phinney, B. O., Gaskin, P., and MacMillan, J. (1996) *Proc. Natl. Acad. Sci. U.S.A.* **93**, 10515–10518

102. Xu, Y.-L., Li, L., Wu, K., Peeters, A. J. M., Gage, D. A., and Zeevaart, J. A. D. (1995) *Proc. Natl. Acad. Sci. U.S.A.* **92**, 6640–6644

102a. Rojas, M. C., Hedden, P., Gaskin, P., and Tudzynski, B. (2001) *Proc. Natl. Acad. Sci. U.S.A.* **98**, 5838–5843

102b. Kawaide, H., Sassa, T., and Kamiya, Y. (2000) *J. Biol. Chem.* **275**, 2276–2280

103. Lange, T., Hedden, P., and Graebe, J. E. (1994) *Proc. Natl. Acad. Sci. U.S.A.* **91**, 8552–8556

103a. Helliwell, C. A., Chandler, P. M., Poole, A., Dennis, E. S., and Peacock, W. J. (2001) *Proc. Natl. Acad. Sci. U.S.A.* **98**, 2065–2070

103b. Thomas, S. G., Phillips, A. L., and Hedden, P. (1999) *Proc. Natl. Acad. Sci. U.S.A.* **96**, 4698–4703

104. Feyereisen, R., and Farnsworth, D. E. (1987) *J. Biol. Chem.* **262**, 2676–2681

105. Toong, Y. C., Schooley, D. A., and Baker, F. C. (1988) *Nature (London)* **333**, 170–171

106. Bohlmann, J., Crock, J., Jetter, R., and Croteau, R. (1998) *Proc. Natl. Acad. Sci. U.S.A.* **95**, 6756–6761

107. Vogel, B. S., Wildung, M. R., Vogel, G., and Croteau, R. (1996) *J. Biol. Chem.* **271**, 23262–23268

107a. Peters, R. J., and Croteau, R. B. (2002) *Proc. Natl. Acad. Sci. U.S.A.* **99**, 580–584

108. Mau, C. J. D., and West, C. A. (1994) *Proc. Natl. Acad. Sci. U.S.A.* **91**, 8497–8501

109. Lin, X., Hezari, M., Koepp, A. E., Floss, H. G., and Croteau, R. (1996) *Biochemistry* **35**, 2968–2977

110. Goodwin, T. W., ed. (1988) *Plant Pigments*, Academic Press, London

111. Mookhtiar, K. A., Kalinowski, S. S., Zhang, D., and Poulter, C. D. (1994) *J. Biol. Chem.* **269**, 11201–11207

112. Gu, P., Ishii, Y., Spencer, T. A., and Shechter, I. (1998) *J. Biol. Chem.* **273**, 12515–12525

113. Jarstfer, M. B., Blagg, B. S. J., Rogers, D. H., and Poulter, C. D. (1996) *J. Am. Chem. Soc.* **118**, 13089–13090

114. Zurer, P. (1997) *Chem. Eng. News* January 6, 6–7

114a. Radisky, E. S., and Poulter, C. D. (2000) *Biochemistry* **39**, 1748–1760

114b. Jarstfer, M. B., Zhang, D.-L., and Poulter, C. D. (2002) *J. Am. Chem. Soc.* **124**, 8834–8845

115. Musico, F., Carlson, J. P., Kuehl, L., and Rilling, H. C. (1974) *J. Biol. Chem.* **249**, 3746–3749

116. Kojima, Y., Friedlander, E. J., and Bloch, K. (1981) *J. Biol. Chem.* **256**, 7235–7239

117. Torssell, K. B. G. (1983) *Natural Product Chemistry*, Wiley, New York (pp. 181–215)

118. Armstrong, G. A., and Hearst, J. E. (1996) *FASEB J.* **10**, 228–237

118a. Rivera, S. B., Swedlund, B. D., King, G. J., Bell, R. N., Hussey, C. E., Jr., Shattuck-Eidens, D. M., Wrobel, W. M., Peiser, G. D., and Poulter, C. D. (2001) *Proc. Natl. Acad. Sci. U.S.A.* **98**, 4373–4378

118b. Soderberg, T., Chen, A., and Poulter, C. D. (2001) *Biochemistry* **40**, 14847–14854

119. Heathcock, D. H., Finkelstein, B. L., Aoki, T., and Poulter, C. D. (1985) *Science* **229**, 862–863

120. Poulter, C. D., Aoki, T., and Daniels, L. (1988) *J. Am. Chem. Soc.* **110**, 2620–2624

121. Britton, G. (1988) in *Plant Pigments* (Goodwin, T. W., ed), pp. 133–182, Academic Press, London

122. Chamovitz, D., Sandmann, G., and Hirschberg, J. (1993) *J. Biol. Chem.* **268**, 17348–17353

123. Bartley, G. E., Viitanen, P. V., Pecker, I., Chamovitz, D., Hirschberg, J., and Scolnik, P. A. (1991) *Proc. Natl. Acad. Sci. U.S.A.* **88**, 6532–6536

123a. Arrach, N., Fernández-Martín, R., Cerdá-Olmedo, E., and Avalos, J. (2001) *Proc. Natl. Acad. Sci. U.S.A.* **98**, 1687–1692

123b. Cunningham, F. X., Jr., and Gantt, E. (2001) *Proc. Natl. Acad. Sci. U.S.A.* **98**, 2905–2910

124. Armstrong, G. A., Schmidt, A., Sandmann, G., and Hearst, J. E. (1990) *J. Biol. Chem.* **265**, 8329–8338

125. Komori, M., Ghosh, R., Takaichi, S., Hu, Y., Mizoguchi, T., Koyama, Y., and Kuki, M. (1998) *Biochemistry* **37**, 8987–8994

126. Ouchane, S., Picaud, M., Vernotte, C., Reiss-Husson, F., and Astier, C. (1997) *J. Biol. Chem.* **272**, 1670–1676

127. Langer, M., Pauling, A., and Rétey, J. (1995) *Angew. Chem. Int. Ed. Engl.* **34**, 1464 – 1465

127a. Ye, X., Al-Babili, S., Klöti, A., Zhang, J., Lucca, P., Beyer, P., and Potrykus, I. (2000) *Science* **287**, 303–305

127b. A contribution of $40 to UNICEF (www.unicefusa.com) will provide vitamin A for a year to 1000 young children, protecting them from blindness.

128. Demmig-Adams, B., Gilmore, A. M., and Adams, W. W., III. (1996) *FASEB J.* **10**, 403–412

129. Sun, Z., Gantt, E., and Cunningham, F. X., Jr. (1996) *J. Biol. Chem.* **271**, 24349–24352

130. Fraser, P. D., Miura, Y., and Misawa, N. (1997) *J. Biol. Chem.* **272**, 6128–6135

131. Bouvier, F., d'Harlingue, A., Hugueney, P., Marin, E., Marion-Poll, A., and Camara, B. (1996) *J. Biol. Chem.* **271**, 28861–28867

132. Schroeder, W. A., and Johnson, E. A. (1995) *J. Biol. Chem.* **270**, 18374–18379

133. Nelis, H. J. C. F., Lavens, P., Moens, L., Sorgeloos, P., Jonckheere, J. A., Criel, G. R., and DeLeenheer, A. P. (1984) *J. Biol. Chem.* **259**, 6063–6066

134. Swift, I. E., and Milborrow, B. V. (1981) *J. Biol. Chem.* **256**, 11607–11611

135. Searcy, D. G. (1982) *Trends Biochem. Sci.* **7**, 183–185

136. Woese, C. R., Kandler, O., and Wheelis, M. L. (1990) *Proc. Natl. Acad. Sci. U.S.A.* **87**, 4576–4579

137. Olson, J. A., and Krinsky, N. I. (1995) *FASEB J.* **9**, 1547–1550

138. Britton, G. (1995) *FASEB J.* **9**, 1551–1558

139. Zuber, H. (1986) *Trends Biochem. Sci.* **11**, 414–419

140. Quiñones, M. A., Lu, Z., and Zeiger, E. (1996) *Proc. Natl. Acad. Sci. U.S.A.* **93**, 2224–2228

141. Bugos, R. C., and Yamamoto, H. Y. (1996) *Proc. Natl. Acad. Sci. U.S.A.* **93**, 6320–6325

142. Rock, C. D., and Zeevaart, J. A. D. (1991) *Proc. Natl. Acad. Sci. U.S.A.* **88**, 7496–7499

143. Schwartz, S. H., Tan, B. C., Gage, D. A., Zeevaart, J. A. D., and McCarty, D. R. (1997) *Science* **276**, 1872–1874

144. Kennedy, T. A., and Liebler, D. C. (1992) *J. Biol. Chem.* **267**, 4658–4663

145. Everett, S. A., Dennis, M. F., Patel, K. B., Maddix, S., Kundu, S. C., and Willson, R. L. (1996) *J. Biol. Chem.* **271**, 3988–3994

146. Burton, G. W., and Ingold, K. U. (1984) *Science* **224**, 569–573

147. Brush, A. H. (1990) *FASEB J.* **4**, 2969–2977

148. Weesie, R. J., Jansen, F. J. H. M., Merlin, J. C., Lugtenburg, J., Britton, G., and de Groot, H. J. M. (1997) *Biochemistry* **36**, 7288–7296

149. Parker, R. S. (1996) *FASEB J.* **10**, 542–551

150. Hoffman, M. (1990) *Science* **250**, 372–373

151. Gudas, L. J. (1994) *J. Biol. Chem.* **269**, 15399–15402

152. Means, A. L., and Gudas, L. J. (1995) *Ann. Rev. Biochem.* **64**, 201–233

152a. Mertz, J. R., Shang, E., Piantedosi, R., Wei, S., Wolgemuth, D. J., and Blaner, W. S. (1997) *J. Biol. Chem.* **272**, 11744–11749

152b. White, J. A., Guo, Y.-D., Baetz, K., Beckett-Jones, B., Bonasoro, J., Hsu, K. E., Dilworth, F. J., Jones, G., and Petkovich, M. (1996) *J. Biol. Chem.* **271**, 29922–29927

152c. Lamb, A. L., and Newcomer, M. E. (1999) *Biochemistry* **38**, 6003–6011

153. Parish, E. J., and Nes, W. D., eds. (1997) *Biochemistry and Function of Sterols*, CRC Press, Boca Raton, Florida

153a. Crowder, C. M., Westover, E. J., Kumar, A. S., Ostlund, R. E., Jr., and Covey, D. F. (2001) *J. Biol. Chem.* **276**, 44369–44372

154. van Tamelen, E. E. (1968) *Acc. Chem. Res.* **1**, 111–120

155. Heftmann, E. (1969) *Steroid Biochemistry*, Academic Press, New York

156. Templeton, W. (1969) *An Introduction to the Chemistry of Terpenoids and Steroids*, Butterworths, London

157. Abe, I., and Prestwich, G. D. (1994) *J. Biol. Chem.* **269**, 802–804

158. Abe, I., and Prestwich, G. D. (1995) *Proc. Natl. Acad. Sci. U.S.A.* **92**, 9274–9278

159. Shi, Z., Buntel, C. J., and Griffin, J. H. (1994) *Proc. Natl. Acad. Sci. U.S.A.* **91**, 7370–7374

160. Nes, W. D., and Venkatramesh, M. (1994) in *Isopentenoids and Other Natural Products* (Nes, W. D., ed), pp. 55–89, American Chemical Society, Washington D.C.

160a. Edwards, P. A., and Ericsson, J. (1999) *Ann. Rev. Biochem.* **68**, 157–185

161. Goad, L. J. (1991) in *Ecological Chemistry and Biochemistry of Plant Terpenoids* (Harborne, J. B., and Tomas-Barberan, F. A., eds), pp. 209–229, Clarendon Press, Oxford

162. Nes, W. D., Parker, S. R., Crumley, F. G., and Ross, S. A. (1993) in *Lipid Metabolism in Plants* (Moore, T. S., Jr., ed), pp. 389–426, CRC Press, Boca Raton, Florida

162a. Herrera, J. B. R., Wilson, W. K., and Matsuda, S. P. T. (2000) *J. Am. Chem. Soc.* **122**, 6765–6766

163. Nes, W. D. (1990) in *Biochemistry of the Mevalonic Acid Pathway to Terpenoids* (Towers, G. H. N., and Stafford, H. A., eds), pp. 283–327, Plenum, New York

163a. Kushiro, T., Shibuya, M., Masuda, K., and Ebizuka, Y. (2000) *J. Am. Chem. Soc.* **122**, 6816–6824

163b. Heathcock, C. H. (1996) *Proc. Natl. Acad. Sci. U.S.A.* **93**, 14323–14327

164. Wendt, K. U., Poralla, K., and Schulz, G. E. (1997) *Science* **277**, 1811–1815

164a. Wendt, K. U., Lenhart, A., and Schulz, G. E. (1999) *J. Mol. Biol.* **286**, 175–187

165. Prince, R. C. (1987) *Trends Biochem. Sci.* **12**, 455–456

References

166. Rohmer, M., and Bisseret, P. (1994) in *Isopentenoids and Other Natural Products* (Nes, W. D., ed), pp. 31–43, American Chemical Society, Washington D.C.
167. Fukushima, H., Grinstead, G. F., and Gaylor, J. L. (1981) *J. Biol. Chem.* **256**, 4822–4826
168. Krieger, M., Kingsley, D., Sege, R., Hobbie, L., and Kozarsky, K. (1985) *Trends Biochem. Sci.* **10**, 447–452
169. Williams, J. B., and Napoli, J. L. (1985) *Proc. Natl. Acad. Sci. U.S.A.* **82**, 4658–4662
170. Lamb, D. C., Kelly, D. E., Schunck, W.-H., Shyadehi, A. Z., Akhtar, M., Lowe, D. J., Baldwin, B. C., and Kelly, S. L. (1997) *J. Biol. Chem.* **272**, 5682–5688
171. Trzaskos, J. M., Ko, S. S., Magolda, R. L., Favata, M. F., Fischer, R. T., Stam, S. H., Johnson, P. R., and Gaylor, J. L. (1995) *Biochemistry* **34**, 9670–9676
172. Shyadehi, A. Z., Lamb, D. C., Kelly, S. L., Kelly, D. E., Schunck, W.-H., Wright, J. N., Corina, D., and Akhtar, M. (1996) *J. Biol. Chem.* **271**, 12445–12450
172a. Podust, L. M., Poulos, T. L., and Waterman, M. R. (2001) *Proc. Natl. Acad. Sci. U.S.A.* **98**, 3068–3073
172b. Bellamine, A., Mangla, A. T., Nes, W. D., and Waterman, M. R. (1999) *Proc. Natl. Acad. Sci. U.S.A.* **96**, 8937–8942
173. Bard, M., Bruner, D. A., Pierson, C. A., Lees, N. D., Biermann, B., Frye, L., and Koegel, C. (1996) *Proc. Natl. Acad. Sci. U.S.A.* **93**, 186–190
173a. Taton, M., Husselstein, T., Benveniste, P., and Rahier, A. (2000) *Biochemistry* **39**, 701–711
174. Seedorf, U., Brysch, P., Engel, T., Schrage, K., and Assmann, G. (1994) *J. Biol. Chem.* **269**, 21277–21283
174a. Gachotte, D., Barbuch, R., Gaylor, J., Nickel, E., and Bard, M. (1998) *Proc. Natl. Acad. Sci. U.S.A.* **95**, 13794–13799
175. Pascal, S., Taton, M., and Rahier, A. (1993) *J. Biol. Chem.* **268**, 11639–11654
176. Rahier, A., and Taton, M. (1996) *Biochemistry* **35**, 7069–7076
177. Nes, W. D., Janssen, G. G., and Bergenstrahle, A. (1991) *J. Biol. Chem.* **266**, 15202–15212
178. Shi, J., Gonzalas, R. A., and Bhattacharyya, M. K. (1996) *J. Biol. Chem.* **271**, 9384–9389
179. Russell, D. W. (1996) *Science* **272**, 370–371
179a. Rouleau, M., Marsolais, F., Richard, M., Nicolle, L., Voigt, B., Adam, G., and Varin, L. (1999) *J. Biol. Chem.* **274**, 20925–20930
180. Li, J., Biswas, M. G., Chao, A., Russell, D. W., and Chory, J. (1997) *Proc. Natl. Acad. Sci. U.S.A.* **94**, 3554–3559
181. Metcalf, R. L., Metcalf, R. A., and Rhodes, A. M. (1980) *Proc. Natl. Acad. Sci. U.S.A.* **77**, 3769–3772
182. Russell, D. W., and Setchell, K. D. R. (1992) *Biochemistry* **31**, 4737–4749
183. Lange, Y., and Steck, T. L. (1994) *J. Biol. Chem.* **269**, 29371–29374
184. Liscum, L., and Underwood, K. W. (1995) *J. Biol. Chem.* **270**, 15443–15446
185. Guan, G., Dai, P.-H., Osborne, T. F., Kim, J. B., and Shechter, I. (1997) *J. Biol. Chem.* **272**, 10295–10302
186. Mackay, K., Starr, J. R., Lawn, R. M., and Ellsworth, J. L. (1997) *J. Biol. Chem.* **272**, 13380–13389
187. Lange, Y., and Muraski, M. F. (1988) *J. Biol. Chem.* **263**, 9366–9373
188. Bretscher, M. S., and Munro, S. (1993) *Science* **261**, 1280–1281
188a. Simons, K., and Ikonen, E. (2000) *Science* **290**, 1721–1726
188b. Nilsson, I., Ohvo-Rekilä, H., Slotte, J. P., Johnson, A. E., and von Heijne, G. (2001) *J. Biol. Chem.* **276**, 41748–41754
188c. Parpal, S., Karlsson, M., Thorn, H., and Strålfors, P. (2001) *J. Biol. Chem.* **276**, 9670–9678
189. Lange, Y., Ye, J., and Steck, T. L. (1998) *J. Biol. Chem.* **273**, 18915–18922
190. Puglielli, L., Rigotti, A., Greco, A. V., Santos, M. J., and Nervi, F. (1995) *J. Biol. Chem.* **270**, 18723–18726
190a. López García, F., Szyperski, T., Dyer, J. H., Choinowski, T., Seedorf, U., Hauser, H., and Wüthrich, K. (2000) *J. Mol. Biol.* **295**, 595–603
190b. Schroeder, F., Frolov, A., Starodub, O., Atshaves, B. B., Russell, W., Petrescu, A., Huang, H., Gallegos, A. M., McIntosh, A., Tahotna, D., Russell, D. H., Billheimer, J. T., Baum, C. L., and Kier, A. B. (2000) *J. Biol. Chem.* **275**, 25547–25555
190c. Choinowski, T., Hauser, H., and Piontek, K. (2000) *Biochemistry* **39**, 1897–1902
191. Yang, H., Bard, M., Bruner, D. A., Gleeson, A., Deckelbaum, R. J., Aljinovic, G., Pohl, T. M., Rothstein, R., and Sturley, S. L. (1996) *Science* **272**, 1353–1356
191a. Spady, D. K., Willard, M. N., and Meidell, R. S. (2000) *J. Biol. Chem.* **275**, 27005–27012
192. Jauhiainen, M., and Dolphin, P. J. (1986) *J. Biol. Chem.* **261**, 7032–7043
192a. Seo, T., Oelkers, P. M., Giattina, M. R., Worgall, T. S., Sturley, S. L., and Deckelbaum, R. J. (2001) *Biochemistry* **40**, 4756–4762
192b. Mauch, D. H., Nägler, K., Schumacher, S., Göritz, C., Müller, E.-C., Otto, A., and Pfrieger, F. W. (2001) *Science* **294**, 1354–1357
192c. Barres, B. A., and Smith, S. J. (2001) *Science* **294**, 1296–1297
193. Jiang, X.-c, and Bruce, C. (1995) *J. Biol. Chem.* **270**, 17133–17138
194. Agellon, L. B., Quinet, E. M., Gillette, T. G., Drayna, D. T., Brown, M. L., and Tall, A. R. (1990) *Biochemistry* **29**, 1372–1376
195. Epps, D. E., Greenlee, K. A., Harris, J. S., Thomas, E. W., Castle, C. K., Fisher, J. F., Hozak, R. R., Marschke, C. K., Melchior, G. W., and Kézdy, F. J. (1995) *Biochemistry* **34**, 12560–12569
195a. Föger, B., Chase, M., Amar, M. J., Vaisman, B. L., Shamburek, R. D., Paigen, B., Fruchart-Najib, J., Paiz, J. A., Koch, C. A., Hoyt, R. F., Brewer, H. B., Jr., and Santamarina-Fojo, S. (1999) *J. Biol. Chem.* **274**, 36912–36920
196. Tall, A. (1995) *Ann. Rev. Biochem.* **64**, 235–257
196a. Desrumaux, C., Labeur, C., Verhee, A., Tavernier, J., Vandekerckhove, J., Rosseneu, M., and Peelman, F. (2001) *J. Biol. Chem.* **276**, 5908–5915
197. Feaster, S. R., Quinn, D. M., and Barnett, B. L. (1997) *Protein Sci.* **6**, 73–79
198. Lopez-Candales, A., Bosner, M. S., Spilburg, C. A., and Lange, L. G. (1993) *Biochemistry* **32**, 12085–12089
199. Sutton, L. D., Froelich, S., Hendrickson, H. S., and Quinn, D. M. (1991) *Biochemistry* **30**, 5888–5893
200. Brown, M. S., and Goldstein, J. L. (1986) *Science* **232**, 34–47
201. Motulsky, A. G. (1986) *Science* **231**, 126–128
202. Brown, M. S., and Goldstein, J. L. (1983) *Ann. Rev. Biochem.* **52**, 223–261
203. Griffiths, G., and Simons, K. (1986) *Science* **234**, 438–443
204. Golgi, C. (1898) *Arch. Ital. Biol.* **30**, 60 and 278
205. Breslow, J. L. (1995) in *The Metabolic and Molecular Bases of Inherited Disease*, 7th ed., Vol. 2 (Scriver, C. R., Beaudet, A. L., Sly, W. S., and Valle, D., eds), pp. 2031–2052, McGraw-Hill, New York
206. Dietschy, J. M. (1997) *Am. J. Clin. Nutr.* **65**, 1581S–1589S
207. Krieger, M. (1998) *Proc. Natl. Acad. Sci. U.S.A.* **95**, 4077–4080
208. Lopez, J., Roghani, A., Bertrand, J., Zanni, E., Kalopissis, A., Zannis, V. I., and Chambaz, J. (1994) *Biochemistry* **33**, 4056–4064
209. Bergeron, J., Frank, P. G., Scales, D., Meng, Q.-H., Castro, G., and Marcel, Y. L. (1995) *J. Biol. Chem.* **270**, 27429–27438
210. Gong, E. L., Stoltzfus, L. J., Brion, C. M., Murugesh, D., and Rubin, E. M. (1996) *J. Biol. Chem.* **271**, 5984–5987
211. Südhof, T. C., Goldstein, J. L., Brown, M. S., and Russell, D. W. (1985) *Science* **228**, 815–822
212. Daly, N. L., Scanlon, M. J., Djordjevic, J. T., Kroon, P. A., and Smith, R. (1995) *Proc. Natl. Acad. Sci. U.S.A.* **92**, 6334–6338
213. Kozarsky, K. F., Donahee, M. H., Rigotti, A., Iqbal, S. N., Edelman, E. R., and Krieger, M. (1997) *Nature (London)* **387**, 414–417
213a. Krieger, M. (1999) *Ann. Rev. Biochem.* **68**, 523–558
214. Brown, M. S., and Goldstein, J. L. (1974) *Sci. Am.* **251**(Nov), 58–66
215. Lehrman, M. A., Russell, D. W., Goldstein, J. L., and Brown, M. S. (1987) *J. Biol. Chem.* **262**, 3354–3361
216. Strickland, D. K., Kounnas, M. Z., and Argraves, W. S. (1995) *FASEB J.* **9**, 890–898
217. Chappell, D. A., Inoue, I., Fry, G. L., Pladet, M. W., Bowen, S. L., Iverius, P.-H., Lalouel, J.-M., and Strickland, D. K. (1994) *J. Biol. Chem.* **269**, 18001–18006
218. Rosseneu, M., and Labeur, C. (1995) *FASEB J.* **9**, 768–776
219. Sakai, J., Hoshino, A., Takahashi, S., Miura, Y., Ishii, H., Suzuki, H., Kawarabayasi, Y., and Yamamoto, T. (1994) *J. Biol. Chem.* **269**, 2173–2182
220. Takahashi, S., Suzuki, J., Kohno, M., Oida, K., Tamai, T., Miyabo, S., Yamamoto, T., and Nakai, T. (1995) *J. Biol. Chem.* **270**, 15747–15754
221. Krieger, M., and Herz, J. (1994) *Ann. Rev. Biochem.* **63**, 601–637
222. Doi, T., Kurasawa, M., Higashino, K.-i, Imanishi, T., Mori, T., Naito, M., Takahashi, K., Kawabe, Y., Wada, Y., Matsumoto, A., and Kodama, T. (1994) *J. Biol. Chem.* **269**, 25598–25604
223. Krieger, M., Acton, S., Ashkenas, J., Pearson, A., Penman, M., and Resnick, D. (1993) *J. Biol. Chem.* **268**, 4569–4572
224. Krieger, M. (1992) *Trends Biochem. Sci.* **17**, 141–146
225. Hajjar, D. P., and Haberland, M. E. (1997) *J. Biol. Chem.* **272**, 22975–22978
226. Schaefer, E. J., and Levy, R. I. (1985) *N. Engl. J. Med.* **312**, 1300–1310
227. Breslow, J. L. (1993) *Proc. Natl. Acad. Sci. U.S.A.* **90**, 8314–8318
228. Havel, R. J., and Kane, J. P. (1995) in *The Metabolic and Molecular Bases of Inherited Disease*, 7th ed., Vol. 2 (Scriver, C. R., Beaudet, A. L., Sly, W. S., and Valle, D., eds), pp. 1841–1852, McGraw-Hill, New York
229. Goldstein, J. L., Hobbs, H. H., and Brown, M. S. (1995) in *The Metabolic and Molecular Bases of Inherited Disease*, 7th ed., Vol. 2 (Scriver, C. R., Beaudet, A. L., Sly, W. S., and Valle, D., eds), pp. 1981–2030, McGraw-Hill, New York
229a. Goldstein, J. L., and Brown, M. S. (2001) *Science* **292**, 1310–1312
229b. Garcia, C. K., and 12 other authors. (2001) *Science* **292**, 1394–1398
230. Fass, D., Blacklow, S., Kim, P. S., and Berger, J. M. (1997) *Nature (London)* **388**, 691–693
230a. Borén, J., Ekström, U., Ågren, B., Nilsson-Ehle, P., and Innerarity, T. L. (2001) *J. Biol. Chem.* **276**, 9214–9218

References

231. Glomset, J. A., Assmann, G., Gjone, E., and Norum, K. R. (1995) in *The Metabolic and Molecular Bases of Inherited Disease*, 7th ed., Vol. 2 (Scriver, C. R., Beaudet, A. L., Sly, W. S., and Valle, D., eds), pp. 1933–1952, McGraw-Hill, New York
232. Pennisi, E. (1997) *Science* **277**, 180–181
233. Carstea, E. D., and 37 other authors. (1997) *Science* **277**, 228–231
234. Loftus, S. K., Morris, J. A., Carstea, E. D., Gu, J. Z., Cummings, C., Brown, A., Ellison, J., Ohno, K., Rosenfeld, M. A., Tagle, D. A., Pentchev, P. G., and Pavan, W. J. (1997) *Science* **277**, 232–235
234a. Davies, J. P., Ioannou,Y. A. (2000) *J. Biol. Chem.* **275**, 24367–24374
234b. Davies, J. P., Chen, F. W., and Ioannou, Y. A. (2000) *Science* **290**, 2295–2298
234c. Naureckiene, S., Sleat, D. E., Lackland, H., Fensom, A., Vanier, M. T., Wattiaux, R., Jadot, M., and Lobel, P. (2000) *Science* **290**, 2298–2301
235. Assmann, G., von Eckardstein, A., and Brewer, H. B., Jr. (1995) in *The Metabolic and Molecular Bases of Inherited Disease*, 7th ed., Vol. 2 (Scriver, C. R., Beaudet, A. L., Sly, W. S., and Valle, D., eds), pp. 2053–2072, McGraw-Hill, New York
236. Lohse, P., Kindt, M. R., Rader, D. J., and Brewer, H. B., Jr. (1991) *J. Biol. Chem.* **266**, 13513–13518
236a. Gura, T. (1999) *Science* **285**, 814–815
236b. Scott, J. (1999) *Nature (London)* **400**, 816–819
236c. Liscovitch, M., and Lavie, Y. (2000) *Trends Biochem. Sci.* **25**, 530–534
236d. Berge, K. E., Tian, H., Graf, G. A., Yu, L., Grishin, N. V., Schultz, J., Kwiterovich, P., Shan, B., Barnes, R., and Hobbs, H. H. (2000) *Science* **290**, 1771–1775
236e. Allayee, H., Laffitte, B. A., and Lusis, A. J. (2000) *Science* **290**, 1709–1711
237. Bloch, K. (1982) *Trends Biochem. Sci.* **7**, 334–336
238. Tanaka, N., Nonaka, T., Tanabe, T., Yoshimoto, T., Tsuru, D., and Mitsui, Y. (1996) *Biochemistry* **35**, 7715–7730
238a. Kramer, W., Sauber, K., Baringhaus, K.-H., Kurz, M., Stengelin, S., Lange, G., Corsiero, D., Girbig, F., König, W., and Weyland, C. (2001) *J. Biol. Chem.* **276**, 7291–7301
238b. Lundell, K., Hansson, R., and Wikvall, K. (2001) *J. Biol. Chem.* **276**, 9606–9612
238c. del Castillo-Olivares, A., and Gil, G. (2000) *J. Biol. Chem.* **275**, 17793–17799
238d. Steinberg, S. J., Mihalik, S. J., Kim, D. G., Cuebas, D. A., and Watkins, P. A. (2000) *J. Biol. Chem.* **275**, 15605–15608
238e. Kotti, T. J., Savolainen, K., Helander, H. M., Yagi, A., Novikov, D. K., Kalkkinen, N., Conzelmann, E., Hiltunen, J. K., and Schmitz, W. (2000) *J. Biol. Chem.* **275**, 20887–20895
238f. Pullinger, C. R., Eng, C., Salen, G., Shefer, S., Batta, A. K., Erickson, S. K., Verhagen, A., Rivera, C. R., Mulvihill, S. J., Malloy, M. J., and Kane, J. P. (2002) *J. Clin. Invest.* **110**, 109–117
239. Berginer, V. M., Salem, G., and Shefer, S. (1984) *N. Engl. J. Med.* **311**, 1649–1652
240. Björkhem, I., and Boberg, K. M. (1995) in *The Metabolic and Molecular Bases of Inherited Disease*, 7th ed., Vol. 2 (Scriver, C. R., Beaudet, A. L., Sly, W. S., and Valle, D., eds), pp. 2073–2099, McGraw-Hill, New York
241. de Vree, J. M. L., Jacquemin, E., Sturm, E., Cresteil, D., Bosma, P. J., Aten, J., Deleuze, J.-F., Desrochers, M., Burdelski, M., Bernard, O., Oude Elferink, R. P. J., and Hadchouel, M. (1998) *Proc. Natl. Acad. Sci. U.S.A.* **95**, 282–287
242. Breuer, O., and Björkhem, I. (1995) *J. Biol. Chem.* **270**, 20278–20284
243. Korytowski, W., Geiger, P. G., and Girotti, A. W. (1996) *Biochemistry* **35**, 8670–8679
244. Javitt, N. B., Kok, E., Burstein, S., Cohen, B., and Kutscher, J. (1981) *J. Biol. Chem.* **256**, 12644–12646
245. Anderson, C. J., Lucas, L. J. H., and Widlanski, T. S. (1995) *J. Am. Chem. Soc.* **117**, 3889–3890
246. Ballabio, A., Parenti, G., Carrozzo, R., Sebastio, G., Andria, G., Buckle, V., Fraser, N., Craig, I., Rocchi, M., Romeo, G., Jobsis, A. C., and Persico, M. G. (1987) *Proc. Natl. Acad. Sci. U.S.A.* **84**, 4519–4523
247. Beaulieu, M., Lévesque, E., Hum, D. W., and Bélanger, A. (1996) *J. Biol. Chem.* **271**, 22855–22862
248. Porter, J. A., Young, K. E., and Beachy, P. A. (1996) *Science* **274**, 255–258
249. Briggs, M. H., and Brotherton, J. (1970) *Steriod Biochemistry and Pharmacology*, Academic Press, New York
250. Waterman, M. R., and Bischof, L. J. (1997) *FASEB J.* **11**, 419–427
251. Waterman, M. R. (1995) *Science* **267**, 1780–1781
252. Lin, D., Sugawara, T., Strauss, J. F., III, Clark, B. J., Stocco, D. M., Saenger, P., Rogol, A., and Miller, W. L. (1995) *Science* **267**, 1828–1831
253. Heyl, B. L., Tyrrell, D. J., and Lambeth, J. D. (1986) *J. Biol. Chem.* **261**, 2743–2749
254. Rhéaume, E., Sanchez, R., Mébarki, F., Gagnon, E., Carel, J.-C., Chaussain, J.-L., Morel, Y., Labrie, F., and Simard, J. (1995) *Biochemistry* **34**, 2893–2900
255. Bain, P. A., Yoo, M., Clarke, T., Hammond, S. H., and Payne, A. H. (1991) *Proc. Natl. Acad. Sci. U.S.A.* **88**, 8870–8874
256. Borg, W., Shackleton, C. H. L., Pahuja, S. L., and Hochberg, R. B. (1995) *Proc. Natl. Acad. Sci. U.S.A.* **92**, 1545–1549
256a. Avvakumov, G. V., Muller, Y. A., and Hammond, G. L. (2000) *J. Biol. Chem.* **275**, 25920–25925
256b. Grishkovskaya, I., Avvakumov, G. V., Sklenar, G., Dales, D., Hammond, G. L., and Muller, Y. A. (2000) *EMBO J.* **19**, 504–512
257. Mester, J., and Baulieu, E.-E. (1984) *Trends Biochem. Sci.* **9**, 56–59
258. Baulieu, E.-E. (1989) *Science* **245**, 1351–1357
259. Ulmann, A., Teutsch, G., and Philibert, D. (1990) *Sci. Am.* **262**(Jun), 42–48
260. Beck, C. A., Weigel, N. L., Moyer, M. L., Nordeen, S. K., and Edwards, D. P. (1993) *Proc. Natl. Acad. Sci. U.S.A.* **90**, 4441–4445
261. Glasier, A., Thong, K. J., Dewar, M., Mackie, M., and Baird, D. T. (1992) *N. Engl. J. Med.* **327**, 1041–1044
262. Grimes, D. A., and Cook, R. J. (1992) *N. Engl. J. Med.* **327**, 1088–1089
263. Brann, D. W., and Mahesh, V. B. (1991) *FASEB J.* **5**, 2691–2698
264. Narasimhulu, S., Eddy, C. R., Dibartolomeis, M., Kowluru, R., and Jefcoate, C. R. (1985) *Biochemistry* **24**, 4287–4294
265. White, P. C. (1994) *N. Engl. J. Med.* **331**, 250–258
266. Phillips, I. R., and Shepard, E. A. (1985) *Nature (London)* **314**, 130–131
267. Goldsworthy, G. J., Robinson, J., and Mordue, W. (1981) *Endocrimology*, Blackie, Glascow
268. Makin, H. L. J., ed. (1984) *Biochemistry of the Steroid Hormones*, 2nd ed., Blackwell, Oxford
269. New, M. I., ed. (1985) *Congenital Adrenal Hyperplasia (Ann. N. York Acad. Sci.)*, Vol. 458, New York Acad. Sci., New York
270. Chrousos, G. P. (1995) *N. Engl. J. Med.* **332**, 1351–1362
271. Hatakeyama, H., Miyamori, I., Fujita, T., Takeda, Y., Takeda, R., and Yamamoto, H. (1994) *J. Biol. Chem.* **269**, 24316–24320
272. McEwen, B. (1981) *Nature (London)* **291**, 610
273. Moore, R. J., Griffin, J. E., and Wilson, J. D. (1975) *J. Biol. Chem.* **250**, 7168–7172
274. Bull, H. G., Garcia-Calvo, M., Andersson, S., Baginsky, W. F., Chan, H. K., Ellsworth, D. E., Miller, R. R., Stearns, R. A., Bakshi, R. K., Rasmusson, G. H., Tolman, R. L., Myers, R. W., Kozarich, J. W., and Harris, G. S. (1996) *J. Am. Chem. Soc.* **118**, 2359–2365
275. Russell, D. W., and Wilson, J. D. (1994) *Ann. Rev. Biochem.* **63**, 25–61
276. Lin, T.-M., and Chang, C. (1997) *Proc. Natl. Acad. Sci. U.S.A.* **94**, 4988–4993
277. Rittmaster, R. S. (1994) *N. Engl. J. Med.* **330**, 120–125
278. Gormley, G. J., Stoner, E., Bruskewitz, R. C., Imperato-McGinley, J., Walsh, P. C., McConnell, J. D., Andriole, G. L., Geller, J., Bracken, B. R., Tenover, J. S., Vaughan, E. D., Pappas, F., Taylor, A., Binkowitz, B., and Ng, J. (1992) *N. Engl. J. Med.* **327**, 1185–1191
279. Baulieu, E.-E., and Robel, P. (1998) *Proc. Natl. Acad. Sci. U.S.A.* **95**, 4089–4091
280. Lardy, H., Partridge, B., Kneer, N., and Wei, Y. (1995) *Proc. Natl. Acad. Sci. U.S.A.* **92**, 6617–6619
281. Berdanier, C. D., Parente, J. A., Jr., and Mcintosh, M. K. (1993) *FASEB J.* **7**, 414–419
282. Nestler, J. E., Clore, J. N., and Blackard, W. G. (1992) *FASEB J.* **6**, 3073–3075
283. Berr, C., Lafont, S., Debuire, B., Dartigues, J.-F., and Baulieu, E.-E. (1996) *Proc. Natl. Acad. Sci. U.S.A.* **93**, 13410–13415
284. Rose, K. A., Stapleton, G., Dott, K., Kieny, M. P., Best, R., Schwarz, M., Russell, D. W., Björkhem, I., Seckl, J., and Lathe, R. (1997) *Proc. Natl. Acad. Sci. U.S.A.* **94**, 4925–4930
285. Wickelgren, I. (1997) *Science* **276**, 675–678
285a. Yen, S. S. C. (2001) *Proc. Natl. Acad. Sci. U.S.A.* **98**, 8167–8169
285b. Baulieu, E.-E., Thomas, G., Legrain, S., Lahlou, N., Roger, M., Debuire, B., Faucounau, V., Girard, L., Hervy, M.-P., Latour, F., Leaud, M.-C., Mokrane, A., Pitti-Ferrandi, H., and 10 other authors. (2000) *Proc. Natl. Acad. Sci. U.S.A.* **97**, 4279–4284
285c. Rose, K., Allan, A., Gauldie, S., Stapleton, G., Dobbie, L., Dott, K., Martin, C., Wang, L., Hedlund, E., Seckl, J. R., Gustafsson, J.-Å., and Lathe, R. (2001) *J. Biol. Chem.* **276**, 23937–23944
286. Cascio, C., Prasad, V. V. K., Lin, Y. Y., Lieberman, S., and Papadopoulos, V. (1998) *Proc. Natl. Acad. Sci. U.S.A.* **95**, 2862–2867
287. Kochakian, C. D. (1989) *Foundations of Endocrinology*, Univ. of Alabama School of Medicine, Tuscaloosa
288. Kochakian, C. D. (1987) *Trends Biochem. Sci.* **12**, 446–450
289. Bardin, C. W. (1996) *N. Engl. J. Med.* **335**, 52
290. Matsunaga, T., Nomoto, M., Kozak, C. A., and Gonzalez, F. J. (1990) *Biochemistry* **29**, 1329–1341
290a. Yamada, A., Yamada, M., Fujita, Y., Nishigami, T., Nakasho, K., and Uematsu, K. (2001) *J. Biol. Chem.* **276**, 4604–4610
291. Waxman, D. J., Dannan, G. A., and Guengerich, F. P. (1985) *Biochemistry* **24**, 4409–4417
292. MacGeoch, C., Morgan, E. T., Halper, J., and Gustafsson, J.-A. (1984) *J. Biol. Chem.* **259**, 15433–15439
293. Covey, D. F., Hood, W. F., Beusen, D. D., and Carrell, H. L. (1984) *Biochemistry* **23**, 5398–5406
294. Vaz, A. D. N., Kessell, K. J., and Coon, M. J. (1994) *Biochemistry* **33**, 13651–13661

References

295. Harada, N., Ogawa, H., Shozu, M., Yamada, K., Suhara, K., Nishida, E., and Takagi, Y. (1992) *J. Biol. Chem.* **267**, 4781–4785
296. Graham-Lorence, S., Khalil, M. W., Lorence, M. C., Mendelson, C. R., and Simpson, E. R. (1991) *J. Biol. Chem.* **266**, 11939–11946
297. Terashima, I., Suzuki, N., Itoh, S., Yoshizawa, I., and Shibutani, S. (1998) *Biochemistry* **37**, 8803–8807
298. Service, R. F. (1998) *Science* **279**, 1631–1633
299. Henderson, B. E., Ross, R. K., and Pike, M. C. (1993) *Science* **259**, 633–638
299a. Hong, W. K., and Sporn, M. B. (1997) *Science* **278**, 1073–1077
299b. Shimotakahara, S., Gorin, A., Kolbanovskiy, A., Kettani, A., Hingerty, B. E., Amin, S., Broyde, S., Geacintov, N., and Patel, D. J. (2000) *J. Mol. Biol.* **302**, 377–393
300. Marshall, E. (1994) *Science* **264**, 1524–1527
301. Davis, D. L., and Bradlow, H. L. (1995) *Sci. Am.* **273**(Oct), 166–172
302. Kaiser, J. (1996) *Science* **272**, 1418
303. Arnold, S. F., Klotz, D. M., Collins, B. M., Vonier, P. M., Guillette, J., LJ, and McLachlan, J. A. (1996) *Science* **272**, 1489–1492
304. Hess, R. A., Bunick, D., Lee, K.-H., Bahr, J., Taylor, J. A., Korach, K. S., and Lubahn, D. B. (1997) *Nature (London)* **390**, 509–512
305. Sharpe, R. M. (1997) *Nature (London)* **390**, 447–448
306. O'Malley, B. W., and Schrader, W. T. (1976) *Sci. Am.* **234**(Feb), 32–43
307. Fuller, P. J. (1991) *FASEB J.* **5**, 3092–3099
308. La Baer, J., and Yamamoto, K. R. (1994) *J. Mol. Biol.* **239**, 664–688
309. Green, S., Walter, P., Kumar, V., Krust, A., Bornert, J.-M., Argos, P., and Chambon, P. (1986) *Nature (London)* **320**, 134–139
310. Kostrouch, Z., Kostrouchova, M., and Rall, J. E. (1995) *Proc. Natl. Acad. Sci. U.S.A.* **92**, 156–159
311. Scherrer, L. C., Dalman, F. C., Massa, E., Meshinchi, S., and Pratt, W. B. (1990) *J. Biol. Chem.* **265**, 21397–21400
312. Tsai, M.-J., and O'Malley, B. W. (1994) *Ann. Rev. Biochem.* **63**, 451–486
313. Htun, H., Barsony, J., Renyi, I., Gould, D. L., and Hager, G. L. (1996) *Proc. Natl. Acad. Sci. U.S.A.* **93**, 4845–4850
314. Luisi, B. F., Xu, W. X., Otwinowski, Z., Freedman, L. P., Yamamoto, K. R., and Sigler, P. B. (1991) *Nature (London)* **352**, 497–505
315. Eriksson, M. A. L., Härd, T., and Nilsson, L. (1995) *Biophys. J.* **68**, 402–426
316. Schwabe, J. W. R., and Rhodes, D. (1991) *Trends Biochem. Sci.* **16**, 291–296
317. Hyder, S. M., Nawaz, Z., Chiappetta, C., Yokoyama, K., and Stancel, G. M. (1995) *J. Biol. Chem.* **270**, 8506–8513
318. Rastinejad, F., Perlmann, T., Evans, R. M., and Sigler, P. B. (1995) *Nature (London)* **375**, 203–211
319. Lee, M.-O., Dawson, M. I., Picard, N., Hobbs, P. D., and Pfahl, M. (1996) *J. Biol. Chem.* **271**, 11897–11903
320. Vivat, V., Zechel, C., Wurtz, J.-M., Bourguet, W., Kagechika, H., Umemiya, H., Shudo, K., Moras, D., Gronemeyer, H., and Chambon, P. (1997) *EMBO J.* **16**, 5697–5709
321. Brzozowski, A. M., Pike, A. C. W., Dauter, Z., Hubbard, R. E., Bonn, T., Engström, O., Öhman, L., Greene, G. L., Gustafsson, J.-Å., and Carlquist, M. (1997) *Nature (London)* **389**, 753–758
322. Yeh, H., Miyamoto, H., Shima, H., and Chang, C. (1998) *Proc. Natl. Acad. Sci. U.S.A.* **95**, 5527–5532
323. Paech, K., Webb, P., Kuiper, G. G. J. M., Nilsson, S., Gustafsson, J.-Å., Kushner, P. J., and Scanlan, T. S. (1997) *Science* **277**, 1508–1510
324. Oñate, S. A., Tsai, S. Y., Tsai, M.-J., and O'Malley, B. W. O. (1995) *Science* **70**, 1354–1357
325. Grazzini, E., Guillon, G., Mouillac, B., and Zingg, H. H. (1998) *Nature (London)* **392**, 509–512
326. Kester, H. A., van der Leede, B.-jM., van der Saag, P. T., and van der Burg, B. (1997) *J. Biol. Chem.* **272**, 16637–16643
327. Bowyer, P., Clarke, B. R., Lunness, P., Daniels, M. J., and Osbourn, A. E. (1995) *Science* **267**, 371–374
328. Smith, T. W. (1988) *N. Engl. J. Med.* **318**, 358–365
329. Albuquerque, E. X., Daly, J. W., and Witkop, B. (1971) *Science* **172**, 995–1002
330. Daly, J. W. (1995) *Proc. Natl. Acad. Sci. U.S.A.* **92**, 9–13
330a. Dumbacher, J. P., Spande, T. F., and Daly, J. W. (2000) *Proc. Natl. Acad. Sci. U.S.A.* **97**, 12970–12975
331. Coudron, T. A., Law, J. H., and Koeppe, J. K. (1981) *Trends Biochem. Sci.* **6**, 248–252
332. Pongs, O. (1985) in *Interaction of Steroid Hormone Receptors with DNA* (Sluyser, M., ed), pp. 226–240, Ellis Horwood, Chichester
333. Murray, I. A., Cann, P. A., Day, P. J., Derrick, J. P., Sutcliffe, M. J., Shaw, W. V., and Leslie, A. G. W. (1995) *J. Mol. Biol.* **254**, 993–1005
334. Moore, K. S., Wehrli, S., Roder, H., Rogers, M., Forrest, J. N., Jr., McCrimmon, D., and Zasloff, M. (1993) *Proc. Natl. Acad. Sci. U.S.A.* **90**, 1354–1358
335. Stone, R. (1993) *Science* **259**, 1125

Study Questions

1. Outline the sequence and chemical mechanisms of the reactions involved in the conversion of acetyl-CoA into mevalonate.

2. a) Show the structures of the reactants for the hydroxymethylglutaryl CoA synthase reaction.
   b) Free coenzyme A is liberated in the above reaction. From which molecule did it come? Explain the metabolic purpose behind the liberation of free CoA.

3. List as many substances as you can that are of polyprenyl origin and are present
   a) in foods
   b) in various commercial products

4. What distinctly different functions do 3-hydroxy-3-methylglutaryl-CoA synthases serve in the cytosol and in mitochondria of the liver?

5. Outline the functions of mitochondrial enzymes in the conversion of fructose into cholesterol in the liver.

6. How do you think that hydroxycitrate, an inhibitor of ATP citrate lyase, would affect the ability of liver to convert dietary fructose into bile acids?

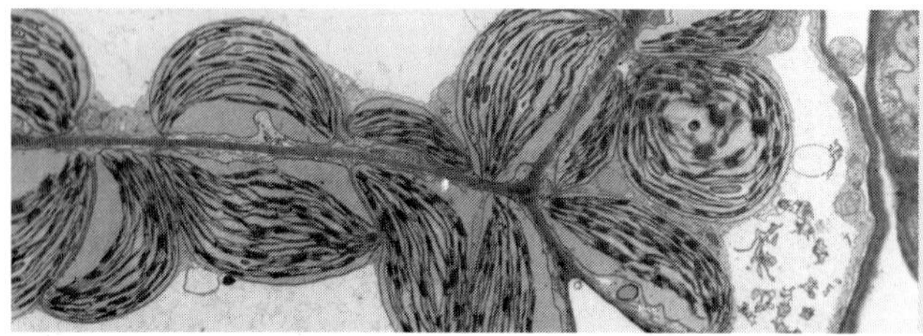

Chloroplasts fill most of the cytoplasm around the junction of three cells of *Arabidopsis thaliana* seen in this micrograph. Both grana stacks and stroma lamellae (pictured in more detail in Fig. 23-19) can be seen. Also present are several small mitochondria. Portions of the large vacuoles, characteristic of plant cells, are seen at top, right, and bottom. Micrograph courtesy of Kenneth Moore.

Contents

23 Light and Life

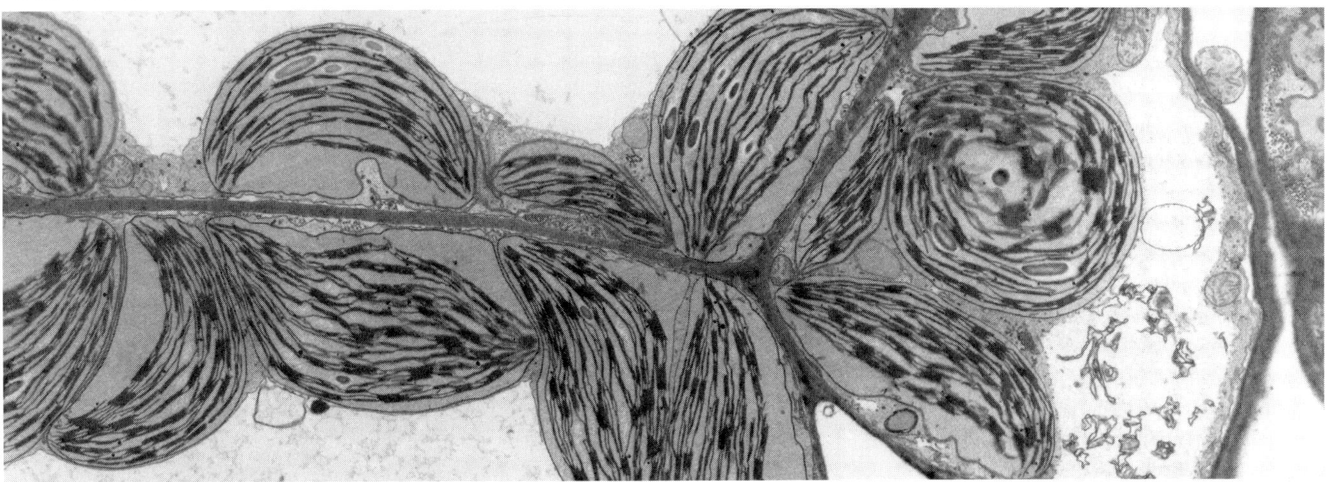

Light plays a pervasive role in human life. The earth is bathed in light from the sun, and from this light comes not only warmth but also energy for all living organisms. Of the 3×10^4 kJ m^{-2} of light energy falling on the earth each day,[1,2] ~30 kJ m^{-2} are captured by photosynthesis.[3] Light penetrating the atmosphere allows us to see and provides color to our environment. It controls the flowering of plants, the germination of seeds and spores, the greening of seedlings, and the daily cycles of many organisms. High in the stratosphere ultraviolet light reacts with oxygen to create a protective blanket of ozone. The ultraviolet light that is not screened out by the ozone layer kills bacteria, tans our skin, and often mutates our DNA, inducing many cancers.[4,5] Organisms, from bacteria to higher plants, display **phototaxis**, the ability to move toward a source of light or to orient themselves with respect to a source of light. In plants the chloroplasts assume an orientation that maximizes efficiency of light absorption. Plants grow toward light (**phototropism**), and some organisms avoid light. Many organisms emit light.

Many of our most important experimental techniques involve the use of light or of other forms of electromagnetic radiation of a wide range of energies. X-rays, ultraviolet light, infrared light, and microwaves all serve in the study of biomolecules.

A. Properties of Light

Light is a form of electromagnetic radiation and possesses characteristics of both waves and particles (**photons**). The energy of a photon is usually measured by the frequency (or by the wavelength in a vacuum to which it is inversely related, Table 23-1). A portion of the electromagnetic spectrum is shown on a logarithmic scale in Fig. 23-1.[1] At the high-energy end (off the scale of the figure to the right) are cosmic rays and gamma rays, while at the low-energy end radio waves extend to wavelengths of many kilometers. The narrow range of wavelengths from about 100 nm to a few micrometers, which is the subject of this chapter, includes the ultraviolet, visible, and near infrared

TABLE 23-1
Some Properties of Light

Velocity of light in a vacuum	$c = 2.998 \times 10^8$ m s^{-1}
Velocity of light in a medium	$c' = c/n$ where n = refractive index
Wave number (in nm)	$\bar{\nu} = 1/\lambda$; $\bar{\nu}$ (in cm^{-1}) $= 10^7/\lambda$
Frequency	$\nu = c/\lambda = c\bar{\nu}$ ν (in hertz) $= 2.998 \times 10^{10}\ \bar{\nu}$ (cm^{-1}) in a vacuum
Energy of quantum	$E = h\nu = hc\bar{\nu}$ E (joules) $= 1.986 \times 10^{-23}\ \bar{\nu}$ (cm^{-1}) E (eV) $= 1.240 \times 10^{-4}\ \bar{\nu}$ (cm^{-1})
Energy of einstein	$E = Nh\nu = Nhc\bar{\nu}$ $= 6.023 \times 10^{23}\ hc\bar{\nu}$ E (joules) $= 11.961\ \bar{\nu}$ (cm^{-1}) E (kcal) $= 2.859 \times 10^{-3}\ \bar{\nu}$ (cm^{-1})

ranges. The second line of Fig. 23-1 shows this region expanded. Note that the range of light reaching the earth's surface is narrow, largely being confined to wavelengths of 320–1100 nm. The human eye responds to an even more limited range of 380–760 nm, in which all of the colors of the rainbow can be found. The aromatic rings of proteins and nucleic acids absorb maximally at 280 and 260 nm, respectively. Even though these wavelengths are largely screened out by the ozone layer of the stratosphere, enough light penetrates to cause mutations and to damage the skin of the unwary sunbather.

The energy of a quantum of light is proportional to the **wave number** or **frequency**. The wave number $\bar{\nu}$ is the reciprocal of wavelength and is customarily given in units of cm^{-1} (**reciprocal centimeters**). Most of the absorption spectra in this book are plotted against wave number in cm^{-1}. The frequency ν in **hertz** is equal to $c'\nu$, where c' is the velocity of light in a medium. (The velocity of light in a vacuum is designated c and is equal to 3.00×10^8 m s^{-1}.) The energy of a quantum of light E is equal to $h\nu$, where h is Planck's constant, 6.626×10^{-34} J s^{-1}. From a chemical viewpoint, we are more interested in the energy of one **einstein**, i.e., one "mole" of light (6.023×10^{23} quanta). The energy in kJ per einstein is 11,960 $\bar{\nu}$ (in cm^{-1}, vacuum). Energy relationships are summarized in Table 23-1. The lower three scales of Fig. 23-1 also show the relationships of $\bar{\nu}$, ν, and E to wavelength.

The light wave is characterized by oscillating electrical and magnetic fields.[2,3,6] For propagation of light in the x direction the electric field vector **E**, which is customarily plotted in the y direction, is a function of the wavelength λ and the time (Eq. 23-1).

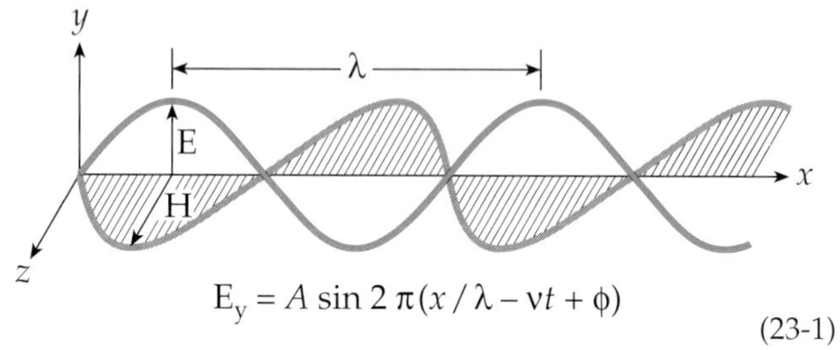

$$E_y = A \sin 2\pi(x/\lambda - \nu t + \phi) \tag{23-1}$$

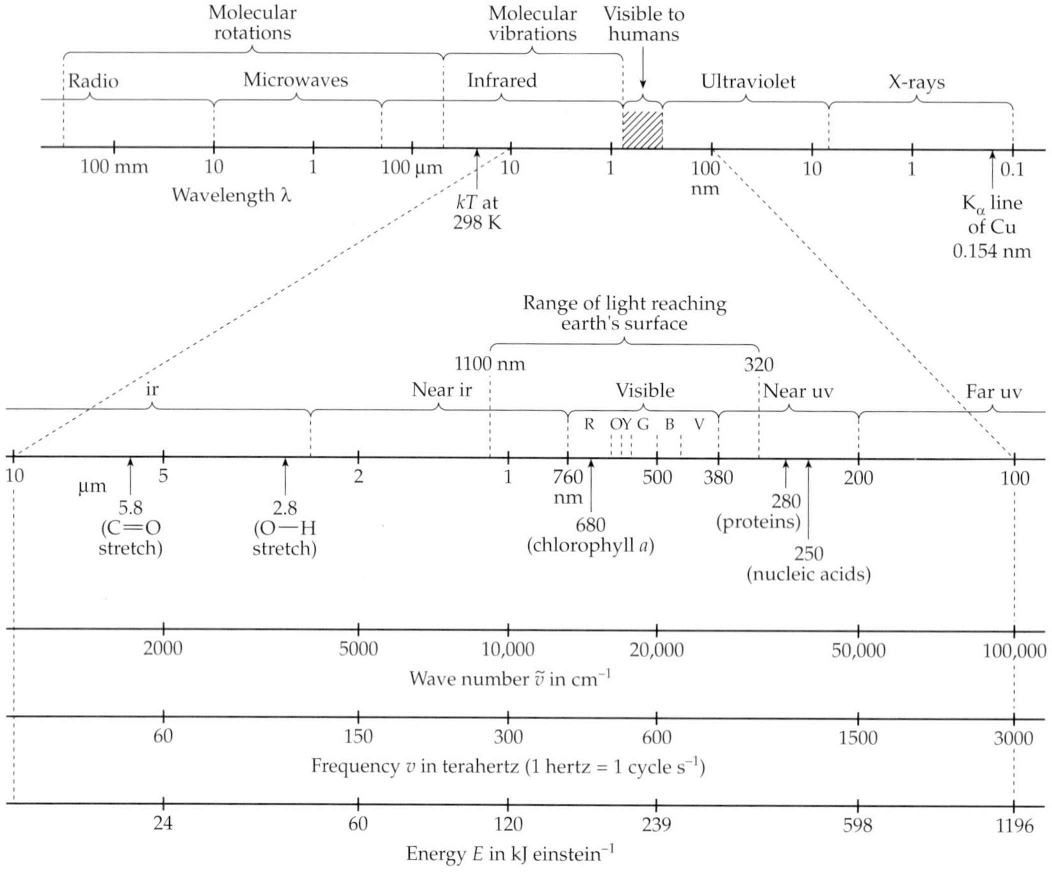

Figure 23-1 A part of the electromagnetic spectrum. The letters V, B, G, Y, O, R over the visible part of the spectrum refer to the colors of the light. The position marked "K_a line of Cu" is the wavelength of X-rays and most widely employed in X-ray diffraction studies of proteins and other organic materials.

The magnetic vector H is at right angles to the electric vector and is given by Eq. 23-2.

$$H_z = (\varepsilon / \mu) - A \sin 2\pi(x/\lambda - \upsilon t + \phi) \qquad (23\text{-}2)$$

The velocity of propagation of light in a medium, c' (Eq. 23-3), depends upon both ε, the **dielectric constant** of the medium, and μ, the **magnetic permeability**.

$$\text{Velocity in a medium: } c' = c / (\varepsilon/\mu) - = c / n \quad (23\text{-}3)$$

The **refractive index** of a medium relative to a vacuum is given the symbol n. It is the factor by which the velocity of light in a vacuum is diminished in a medium. It is a function of wavelength. For the 589 nm sodium line n is 1.00029 for air and 1.33 for water at 25°C.

The term ϕ in Eqs. 23-1 and 23-2 is a **phase factor**. Most light is called **incoherent** because ϕ varies for the many photons making up the beam. **Coherent light** produced by **lasers** contains photons all with the same phase relationship. If the electric vectors of all the photons in a beam of light are in the same plane (as will be the case for light emerging after passage through certain kinds of crystals), the light is called **plane polarized**. The direction of polarization is that of the electric vector E. **Circularly polarized light**, in which the electric vector rotates and traces out either a left-handed or right-handed helix, can also be generated. A beam of left-handed circularly polarized light, together with a comparable beam polarized in the right-handed direction, is equivalent to a beam of plane polarized light. Conversely, plane polarized light can be resolved mathematically into right- and left-handed circularly polarized components.

B. Absorption of Light

Absorption of light is fundamental to all aspects of photochemistry and provides the basis for absorption spectroscopy.[3,5–11] Light absorption is always **quantized**. It can take place only when the energy $h\nu$ of a quantum is equal to the difference in energy between two energy levels of the absorbing molecule (Eq. 23-4).

$$E_2 - E_1 = h\nu \qquad (23\text{-}4)$$

Not only must the difference $E_2 - E_1$ be correct for absorption but also there must always be a change in the dipole moment of the molecule in going from one energy level to another. Only when this is true can the electric field of the light wave interact with the molecule. A further limitation comes from the symmetry properties of the wave functions associated with each energy level. Quantum mechanical considerations show that transitions between certain energy levels are allowed, while others are forbidden. Consideration of such matters is beyond the scope of this book, but the student should be aware that the quantum mechanical selection rules that express this fact are an important determinant of light absorption.

Many types of light source are used in chemical measurements. Of great importance is the recent development of lasers that deliver very short pulses of light. Pulses as short as five femtoseconds (5 fs)[11a,11b] and even less[11c] are being utilized for very rapid spectroscopy and excitation of fluorescence. Structures are being determined by ultrafast electron diffraction[11d] or X-ray diffraction.[11e,11f] It takes 200 fs or more for a chemical bond to stretch and break during a reaction. The cleavage and formation of bonds during this time can be observed using 5-fs pulses. Lasers with pulses in the attosecond range may soon be used to observe movements of electrons.[11g]

1. Quantitative Measurement of Light Absorption, Spectroscopy

An absorption spectrum is a plot of some measure of the intensity of absorption as a function of wavelength or wave number. The **transmittance** of a sample held in a **cell** (or **cuvette**) is the fraction of incident light that is transmitted, i.e., transmittance = I/I_o where I_o is the intensity of light entering the sample and I is that of the emerging light. The transmittance is usually defined for a single wavelength, i.e., for **monochromatic** light. The absorbance (or optical density) is defined by Eq. 23-5, which also states the **Beer–Lambert law**. The length (in centimeters) of the

$$\text{Absorbance} = A = \log_{10}(I_o / I) = \varepsilon c l \qquad (23\text{-}5)$$

light path through the sample is l, c is the concentration in moles per liter, and ε is the **molar extinction coefficient** (molar absorptivity or molar absorption coefficient), whose units are liter mol^{-1} cm^{-1} (or $\mathbf{M^{-1}}$ $\mathbf{cm^{-1}}$). The reader can derive Eq. 23-5 by assuming that in a thin layer of thickness dx the number of light quanta absorbed is proportional to the number of absorbing molecules in the layer. Integration from x = 0 to l gives the Beer–Lambert law. Equation 23-5 generally holds very well for solutions containing single ionic or molecular forms. However, it is usually valid only for monochromatic light. Furthermore, making precise measurements of absorbance is not easy. At $A=1$ only 10% of the incident light is transmitted, and the utmost care is required to obtain a value of A good to within ±0.05. At $A=2$ only 1% of incident light is transmitted, and the value of A will be much less reliable. Very low absorbances are also difficult to measure. In view of the importance that spectrometry has played in biochemistry, it may seem surprising that the first reliable commercial laboratory ultraviolet–visible spectrophotometers became available in 1940 and the first commercial infrared spectrometer in 1942.[12]

2. The Energy Levels of Molecules

The energy of molecules consists of **kinetic** (translational), **rotational**, **vibrational**, and **electronic** components. The corresponding rotational, vibrational, and electronic energy levels are always quantized. Light quanta of wavelengths 0.2–20 mm (50–0.5 cm^{-1}; frequencies of 1.5 x 10^{12} to 1.5 x 10^{10} s^{-1}) with energies of 0.6–0.006 kJ/einstein are sufficient to excite molecules from a given rotational energy level to a higher one. Spectra in this "far infrared" or "microwave" region often consist of a closely spaced series of lines. For example, the rotational spectrum of gaseous HCl is a series of lines at 20.7 cm^{-1} intervals beginning at that wave number and reaching a maximum at about 186 cm^{-1} (54 μm). The energies involved in absorption of such light are far lower than energies of activation for common chemical reactions and lower than the average translational energy of molecules in solution at ordinary temperatures (3/2 k_BT or 3.7 kJ/mol at 25°C). However, they are still much higher than energies involved in the nuclear transitions of NMR spectra (Chapter 3). Compare 500 Hz for a proton resonating at 1 ppm in a 500 MHz NMR spectrometer with the 10^{10}–10^{12} s^{-1} frequencies of microwave spectra.

Vibrational energies range from about 6 to as much as 100 kJ mol^{-1} with corresponding wave numbers of ~500–8000 cm^{-1}. The resulting absorption spectra are in the infrared region. Excited electronic energy levels are ~120 – 200 kJ mol^{-1}, and the spectral transitions are at 10,000–100,000 cm^{-1} (1000–100 nm wavelengths) in the visible and ultraviolet region.

3. Infrared (IR) and Raman Spectra

Absorption in the near infrared region is dominated by changes in vibrational energy levels. A typical wave number is that of the "amide A" band at 3300 cm^{-1} (3.0 μm wavelength), approximately 10^{14} s^{-1}. First let us consider the stretching vibrations of a diatomic molecule. The two nuclei of the molecule can be thought of as connected with a spring. The energy of oscillation is approximately that of a harmonic oscillator. Application of quantum theory shows that the discrete energy levels that can be assumed by the oscillator are equally spaced. The difference between each pair of successive energy levels is $h\nu$, where ν is the frequency of light that must be absorbed to raise the energy from one level to the next. In the ground state (unexcited state) the molecule still possesses a **zero-point energy,** $E = 1/2\, h\nu_o$, equal to half the energy needed to induce a transition.

While the harmonic oscillator is a good approximation to the behavior of a molecule in the lower vibrational energy states, marked deviations occur at higher energies. At the lower energy levels the change in the distance between the atomic centers during the

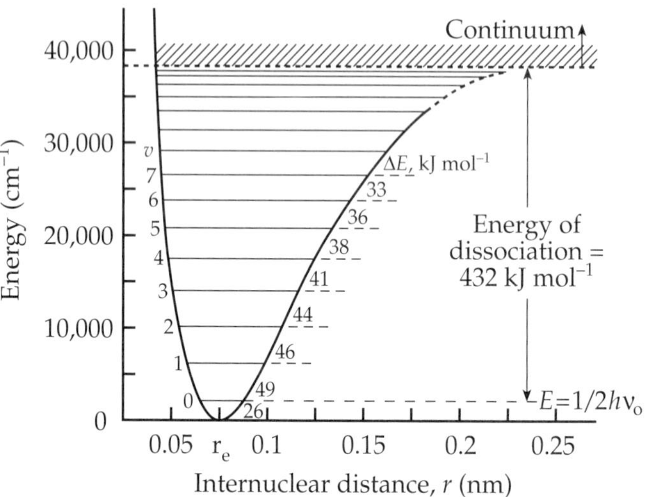

Figure 23-2 The potential energy of the hydrogen molecule as a function of internuclear distance, and the position of its vibrational energy levels. ΔE values are energy differences between successive levels; v designates vibrational quantum numbers. Adapted from Calvert and Pitts,[2] p. 135.

course of the vibration amounts to only ±10% or less, but as the energy becomes greater the bond stretches more and the motion becomes **anharmonic**. The energy states of molecules are often in the form of **Morse curves** in which energy is plotted against internuclear distance (Fig. 23-2). As the internuclear distance becomes very short, the energy rises steeply. As the bond is stretched, there comes a point at which addition of more energy ruptures the bond. A diatomic molecule will dissociate into atoms and more complex molecules into fragments. Vibrational energy levels can be portrayed as horizontal lines at appropriate heights on the Morse curve (Fig. 23-2).

Because there are many rotational energy levels corresponding to each vibrational level, IR spectra contain absorption bands resulting from simultaneous changes in both the vibrational and rotational energy levels of molecules. Instead of single peaks corresponding to single transitions in vibrational energy, progressions of sharp bands at closely spaced intervals are observed. An example is provided by the band corresponding to the stretching frequency of the H–Cl bond in gaseous HCl at 2886 cm^{-1} (3.46 µm). There is actually no band at this wave number but a series of almost equally spaced bands on either side of the fundamental frequency from ~2600 to ~3100 cm^{-1} at intervals of ~21 cm^{-1}, i.e., the wave number of the rotational frequency seen in the microwave spectrum (Herzberg,[8] p. 55). The effect is to broaden the band as seen in a low-resolution spectrum. This is only one cause of the broadening of IR bands in solution. Another cause is interaction with solvent to provide a heterogeneity in the environments of the absorbing molecules.

The IR spectra of diatomic molecules are relatively easy to interpret, but for more complex substances the infrared absorption bands often cannot be associated with individual chemical bonds. Instead, they correspond to the **fundamental vibrations** (normal vibrations) of the *molecule*. Fundamental vibrations are those in which the center of gravity does not change. For a molecule containing n atoms, there are $3n - 6$ such vibrations. They are sometimes dominated by a vibration of a single bond, but often involve synchronous motion of many atoms. The fundamental vibrations of a molecule are described by such words as *stretching*, *bending* (in-plane and out-of-plane), *twisting*, and *deformation*. Rarely are all $3n - 6$ bands seen in an infrared spectrum. Some of the vibrations, e.g., the symmetric stretching of the linear CO_2 molecule, are not accompanied by any change in dipole moment, while other bands may simply be too weak to be observed clearly.

Vibrations involving many atoms in a molecule, i.e., **skeletal vibrations**, are often found in the region of 700–1400 cm^{-1} (14–7 µm). Vibrational frequencies that are dominantly those of individual functional groups can often be identified in the range 1000–5000 cm^{-1} (10–2 µm). Examples of the latter are the stretching frequencies of C–H, N–H, and O–H bonds, which have wave numbers of ~2900, 3300, and 3600 cm^{-1}, respectively. The energy (and frequency) of the vibrations increases as the difference in electronegativity between the two atoms increases. When a bond connects two heavier atoms, the frequency is lower, e.g., the wave number for C–O in a primary alcohol is ~ 1053 cm^{-1}. For a double bond it increases; for C=O it is ~1700 cm^{-1}. This C=O stretching frequency usually gives rise to one of the strongest bands observed in IR spectra. Hydrogen bonding has a strong and characteristic effect. Thus, the O–H frequency at ~3600 cm^{-1} is decreased to ~3500 cm^{-1} by hydrogen bonding.

Theory predicts that for a harmonic oscillator only a change from one vibrational energy level to the next higher is allowed, but for anharmonic oscillators weaker transitions to higher vibrational energy levels can occur. The resulting "overtones" are found at approximate multiples of the frequency of the fundamental. Combination frequencies representing sums and differences of frequencies of individual IR bands may also be seen. The intensities of these bands are low, but their presence at relatively high energies in the near IR region (4000–12,500 cm^{-1}) means that they may be easier to observe than the fundamental frequencies in the more crowded IR region.[12a] Development of the very sensitive **Fourier-transform infrared spectroscopy** (FTIR) has made it possible to record the complex IR spectra of macromolecules in dilute aqueous solutions rapidly. The water spectrum is substracted digitally.[13–16] FTIR has been utilized to study amide groups in peptides, carboxyl groups in proteins,[17] conformations of sugar rings in DNA,[18] and the ionization state of phosphate groups.[19] New computational methods involving use of two-dimensional representations provide simplifications in interpretation of IR spectra.[19a] Another variant is **total reflection FTIR**, a technique that records spectra of thin films and has permitted the recording of transient changes in protein spectra with microsecond time resolution.[20,21]

Vibrational frequencies of amide groups. The IR absorption bands of amide groups, which are present in both proteins and in the purine and pyrimidine bases, have attracted a great deal of attention.[13,22–23a] The **amide I** band at ~1680 cm^{-1} is associated with an in-plane normal mode of vibration that involves primarily C=O stretching. (The band is designated I′ if the N–H has been exchanged to form $N–^2H$.)[16] The **amide II** band at ~1500 $^{-1}$ and the **amide III** band at ~1250 cm^{-1} both arise from in-plane modes that involve N–H bending, while the higher frequency **amide A** band at ~3450 cm^{-1} involves N–H stretching. It is shifted to ~3300 cm^{-1} when the N–H is hydrogen bonded. Examples of IR spectra of proteins, including the amide bands A, I, and II, are shown in Fig. 23-3. The band shapes are complex. Those of the amide bands I and III depend upon the conformation of the peptide chain. For example, (Fig 23-3B) amide groups in α helices have an amide I band about 20 cm^{-1} higher than do those in β structures.

If peptide chains can be oriented in a regular fashion, it may be useful to measure **infrared linear dichroism**.[24,25] Absorption spectra are recorded by passing plane polarized light through the protein in two mutually perpendicular directions, with the electric vector either parallel to the peptide chains or perpendicular to the chains. Such a pair of spectra is shown in Fig. 23-3A for oriented fibrils of insulin. In this instance, the insulin molecules are thought to assume a β conformation and to be stacked in such a way that they extend transverse to the fibril axis (a cross-β structure). When the electric vector is parallel to the fibril axis, it is perpendicular to the peptide chains. Since the amide I band is dominated by a carbonyl stretching motion that is perpendicular to the

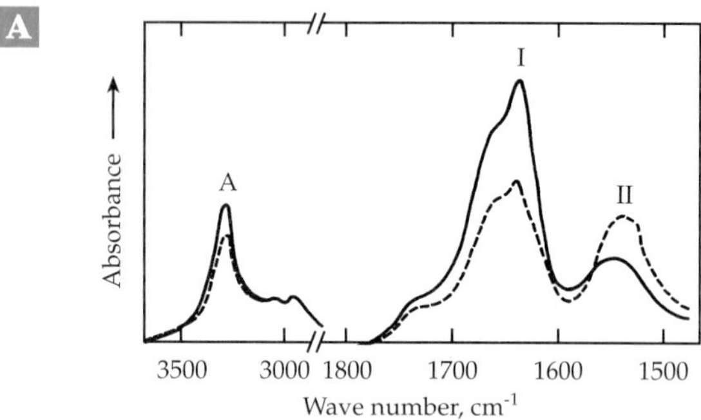

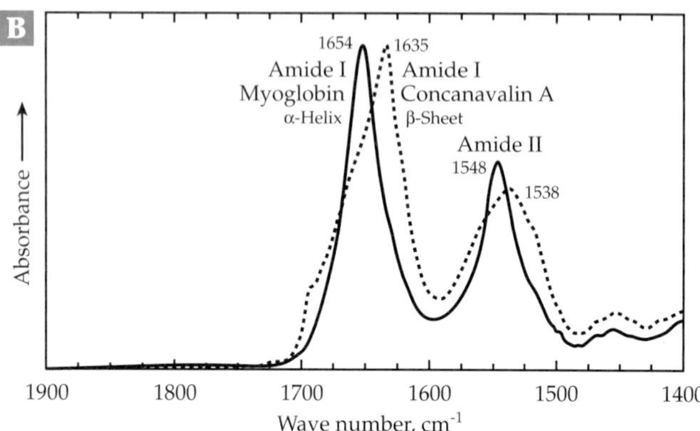

Figure 23-3 Infrared absorbance spectra of the amide regions of proteins. (A) Spectra of insulin fibrils illustrating dichroism. Solid line, electric vector parallel to fibril axis; broken line, electric vector perpendicular to fibril axis. From Burke and Rougvie.[24] Courtesy of Malcolm Rougvie. See also Box 29-E. (B) Fourier transform infrared (FTIR) spectra of two soluble proteins in aqueous solution obtained after subtraction of the background H_2O absorption. The spectrum of myoglobin, a predominantly α-helical protein, is shown as a continuous line. That of concanavalin A, a predominantly β-sheet containing protein, is shown as a broken line. From Haris and Chapman.[14] Courtesy of Dennis Chapman.

peptide chains in the β structure, this band is enhanced when the electric vector is also perpendicular to the peptide chains and is diminished when the electric vector is parallel to the peptide chains (perpendicular to the fibril axis, Fig. 23-3A). The same is true of the amide A band which is dominated by an N–H stretch. On the other hand, the dichroism of the amide II band is the opposite because it tends to be dominated by an N–H bending, which is in the plane of the peptide group but is longitudinal in direction. In **isotope-edited FTIR**, heavy atoms such as ^{13}C are introduced to shift IR bands and assist in their identification. The method can be combined with measurement of linear dichroism of oriented peptides.[25]

The loss of the amide II band in D_2O is one of the major tools for studying protein dynamics.[13,26] (see Chapter 3, Section I,5). In some cases the four main secondary structures, α helix, β sheet, β turn, and random coil, can be distinguished.[27] The amide bands of pyrimidines can also be observed in IR spectra of pyrimidines. Figure 23-4A shows the spectrum of 1-methyluracil in H_2O and also in D_2O. Notice that the amide II band is totally lacking in D_2O. The same figure also shows the IR spectrum of 1-methyluracil contaning ^{18}O in the 4 position. The shift of 9 cm^{-1} in the amide II position is part of the evidence that the NH bending vibration is extensively coupled to C=O and C=C stretching modes.

Raman spectra. In a collision between a photon and a molecule, the photon may undergo **elastic collision** in which the photon loses no energy but changes its direction of travel. Such scattering is known as **Rayleigh scattering** and forms the basis for a method of molecular mass determination. Sometimes **inelastic** collisions occur in which both the

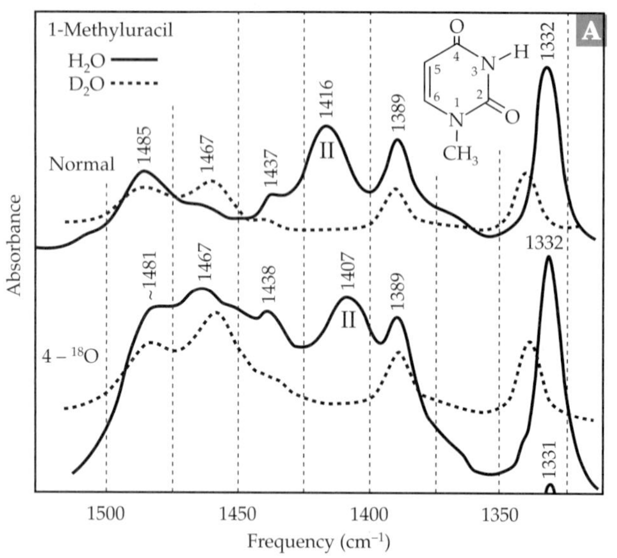

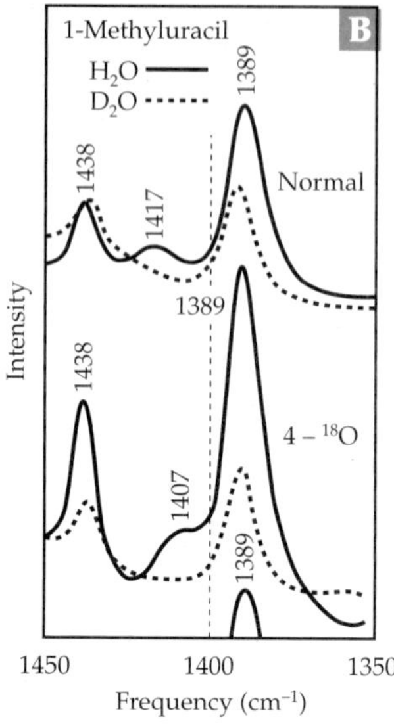

Figure 23-4 (A) Infrared and (B) Raman spectra of 1-methyluracil in H_2O and D_2O. Spectra for normal 1-methyluracil and for the specific isotopic derivatives with ^{18}O in the 4 position are shown. From Miles *et al.*[28]

molecule and the photon undergo changes in energy. Since such changes must be quantized and involve vibrational and rotational levels of the molecule, the spectrum of the scattered light (Raman spectrum) contains much of the same information as an ordinary IR spectrum. However, the selection rules are not the same. Some transitions are "infrared active" and others are "Raman active." IR-active transitions can occur only when the dipole moment varies with time as a molecule vibrates, whereas Raman-active transitions require that the polarizability change with time. For this reason, it, s useful to measure both IR and Raman spectra on the same sample. Until recently, Raman spectroscopy was not used much in biochemistry because of the low intensity of the scattered light. However, with laser excitation the technique is practical[7,13,23,28,29] and can be applied to aqueous or nonaqueous solutions, and to solid or dispersed forms of macromolecules.

Both amide I and amide III bands are seen in Raman spectra of proteins.[30] Lippert *et al.* devised the following method for estimating the fractions of α-helix, β sheet, and random coil conformations in proteins.[31] The amide I Raman bands are recorded at 1632 and 1660 cm^{-1} in D_2O (amide I′). The amide III band, which is weak in D_2O, is measured at 1240 cm^{-1} in H_2O. The intensities of the three bands relative to the intensity of an internal standard (the 1448 cm^{-1} CH_2 deformation) are related to those of standard poly-L-lysine in known conformations. See also Craig and Gaber.[32] The Raman spectrum of 2-methyluracil is shown in Fig. 23-4B. Note the low intensity of the amide II band relative to that of the amide I band, a characteristic of Raman spectra. Linear dichroism observed by polarized Raman microspectrophotometry has provided information about orientation of indole rings of tryptophan in filamentous virus particles.[33,33a]

In **resonance Raman spectroscopy**[34–37] a laser beam of a wavelength that is absorbed in an electronic transition is used. The scattered light is often strongly enhanced at frequencies differing from that of the laser by Raman frequencies of groups within the chromophore or of groups in another molecule adjacent to the chromophore. The resonance effect not only increases the sensitivity of Raman spectroscopy but also allows a person to study specifically the vibrational spectrum of a selected aromatic group or other structure within a macromolecule. Problems associated with the technique are fluorescence, which may be 10^6 times as strong as the Raman emission, and photochemical damage from the intense laser beam. Fluorescence is often quenched with KI (see Section C,1).

If the exciting laser has a frequency ν_0 and the frequency of a vibrationally excitation in a molecule is ν_1 the Raman spectrum will contain a pair of bands,

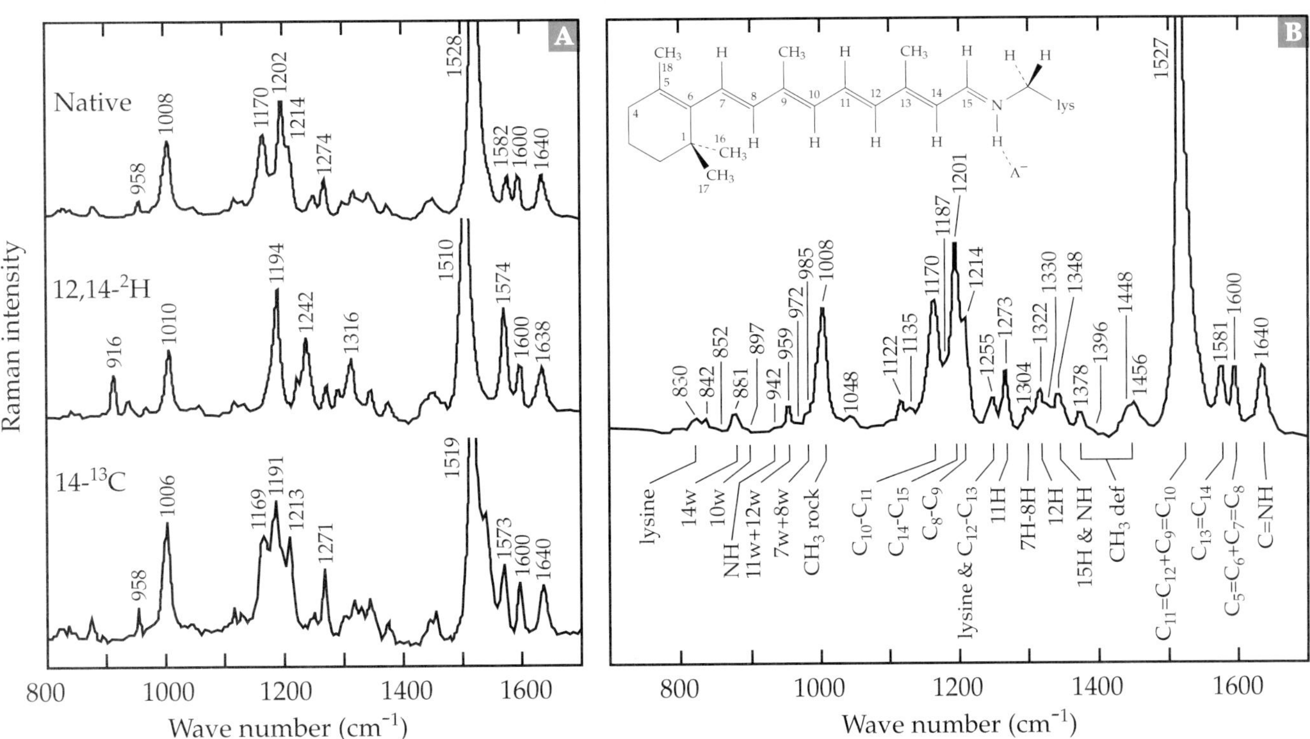

Figure 23-5 Resonance Raman spectra. (A) Of the retinaldehyde-containing bacteriorhodopsin bR_{568} (see Fig. 23-45) and its 12,14-^{2}H and 14-^{13}C isotopic derivatives. (B) Of bR_{568} labeled with the dominant internal coordinates that contribute to the normal modes. From Lugtenburg *et al.*[37]

the stronger one or "stokes" band of energy $h\ (\nu_0 - \nu_1)$ and a weaker one or "anti-stokes" band of $h\ (\nu_0 + \nu_1)$. Special techniques such as **coherent anti-stokes** Raman scattering provide a means of getting around the fluorescence problem.[13,36,38] Raman spectroscopy is also possible with excitation of ultraviolet absorption bands. It can be applied to peptides,[39–41] tryptophan or tyrosine residues of proteins,[42] nucleic acid bases,[43–45] heme proteins,[46–48] other metalloproteins,[49] flavin coenzymes,[50–52] pyridoxal phosphate,[53,54] flavoproteins,[54a] carotenoid-containing proteins,[55] and to substrates undergoing cleavage in the active site of papain.[56] Resonance Raman spectroscopy is very useful in the study of adducts of CO, NO, and O_2 with heme proteins because coupled vibrational modes of both the porphyrin rings and axial ligands can be observed.[56a,b] Resonance Raman spectra are strongly amplified for samples adsorbed to specially prepared colloidal silver particles.[57] Examples of resonance Raman spectra are shown in Fig. 23-5.

4. Electronic Spectra

Biochemists make extensive use of spectroscopy in the ultraviolet (UV) and visible range. Visible light begins at the red end at ~12,000 cm^{-1} (800 nm) and extends to 25,000 cm^{-1} (400 nm). The ultraviolet range begins at this point and extends upward, the upper limit accessible to laboratory spectrophotometers being ~55,000 cm^{-1} (180 nm). The energies covered in the visible–UV range are from ~140 to ~660 kJ/mol. The latter is greater than the bond energy of all but the strongest double and triple bonds (Table 6-7). It is understandable that UV light is effective in inducing photochemical reactions. Even the lower energy red light, which is used by plants in photosynthesis, contains enough energy per einstein to make it feasible to generate ATP, to reduce $NADP^+$, and to carry out other photochemical processes. Although the energies of light absorbed in electronic transitions are large, the geometry of molecules in the excited states is often only slightly altered from that in the ground state. The amount of vibration is increased, and the molecule usually expands moderately in one or more dimensions.

The significance of light absorption in biochemical studies lies in the great sensitivity of electronic energy levels of molecules to their immediate environment and to the fact that spectrophotometers are precise and sensitive. The related measurements of circular dichroism and fluorescence also have widespread utility for study of proteins, nucleic acids, coenzymes, and many other biochemical substances that contain intensely absorbing groups or **chromophores**.[58]

Shapes of absorption bands. Electronic absorption bands are usually quite broad, the width of the band at half-height often being 3000–4000 cm^{-1}. The breadth arises largely from the coupling of electronic excitation to changes in the vibrational and rotational energy levels. Inhomogeneity of environments in the solvent also contributes. Shapes of absorption bands are to a large extent determined by the **Franck–Condon principle**, which states that no significant change in the positions of the atomic nuclei of the molecule occurs during the time of the electronic transition. Since the frequency of light absorbed during these transitions is ~10^{15} to 10^{16} s^{-1}, the absorption of light energy occurs within 10^{-15} to 10^{-16} s, the time equivalent to the passage of one wavelength of light. During this period the vibrational motions of the nuclei are almost insignificant because of the much lower frequencies of vibration. Two types of potential energy curves for excited states of molecules are shown in Fig. 23-6.[2] In the first the geometry of the molecule is little changed between ground state and excited state. At room temperature most molecules are in the lowest energy states of at least the most energetic of the various vibrational modes of the molecule ($3/2\ k_BT$ ~300 cm^{-1}). Therefore, the most probable transitions occur from the lowest vibrational states of the ground electronic states. The most probable internuclear distance for a molecule in the ground state is the equilibrium distance r_e (Fig. 23-2). Since that distance is the same in all of the vibrational levels of the electronically excited state, transitions to any of these states may occur. The transition to the first vibrational level of the excited state is most likely. The result is an absorption spectrum in which the sharp band representing the "0–0 transition" is most intense and in which there are progressively weaker bands corresponding to the 0–1, 0–2, 0–3, etc., transitions (Fig. 23-6A). Many organic dyes with long series of conjugated double bonds have spectra of this type.[59,60]

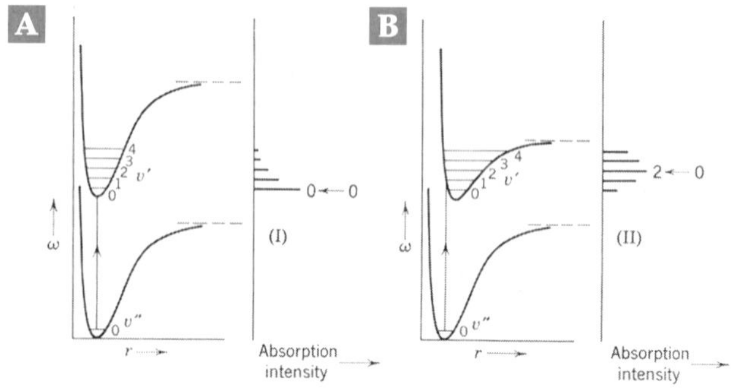

Figure 23-6 Typical potential energy curves for two types of band spectra: (A) For a transition in which the equilibrium internuclear distances r_e are about equal in the ground and excited states. (B) For a transition in which r_e' (excited state) > r_e (ground state). From Calvert and Pitts,[2] p. 179.

A second type of spectrum is illustrated in Fig. 23-6B. In this instance, the molecule has expanded in the excited state, and r_e is greater than in the ground state. The Franck–Condon principle suggests that a transition is likely only to those vibrational levels of the excited state in which the internuclear distance is compressed for a significant fraction of the time, approximately to that of r_e in the ground state. Examination of Fig. 23-6B explains why the resulting absorption spectra tend to have weak 0–0 bands and stronger bands corresponding to transitions to higher levels.

For real spectra of polyatomic molecules the situation is more complex. Some molecules in the ground state do occupy higher vibrational levels of the less energetic modes. Therefore, there will be weaker lines, some of which lie on the low-energy side of the 0–0 transition. Since in polyatomic molecules there are several normal modes of vibration, there will be other progressions of absorption bands paralleling those shown in Fig. 23-6 and filling in the valleys between them. All of the bands are broadened by rotational coupling and by interactions with solvent.

An example of a molecule giving a spectrum of the type shown in Fig. 23-6B is toluene. The vapor phase spectrum contains a large number of sharp lines, some of which can be seen in the low-resolution spectrum of Fig. 23-7. Several progressions can be identified.[61] One begins with the intense 0–0 line at 37.48×10^3 cm^{-1} and in which spacing of ~930 cm^{-1} between lines corresponds to a vibration causing symmetric expansion of the ring (ring breathing frequency), a frequency that can also be observed in the infrared spectrum. Other progressions beginning at the 0–0 line involve additional modes of vibration with frequencies (in the excited electronic state) of 460, 520, and 1190 cm^{-1}. Additional weaker bands are "buried" in the valleys in Fig. 23-6. When the spectrum of toluene is measured in solution, the sharp lines are broadened, but there are still indications of vibrational structure.

Gaussian curves (normal distribution functions) can sometimes be used to describe the shape of the overall envelope of the many vibrationally induced subbands that make up one electronic absorption band, e.g., for the absorption spectrum of the copper-containing blue protein of *Pseudomonas* (Fig. 23-8) Gaussian bands are appropriate. They permit resolution of the spectrum into components representing individual electronic transitions. Each transition is described by a **peak position**, **height** (molar extinction coefficient), and **width** (as measured at the half-height, in cm^{-1}). However, most absorption bands of organic compounds are not symmetric but are skewed

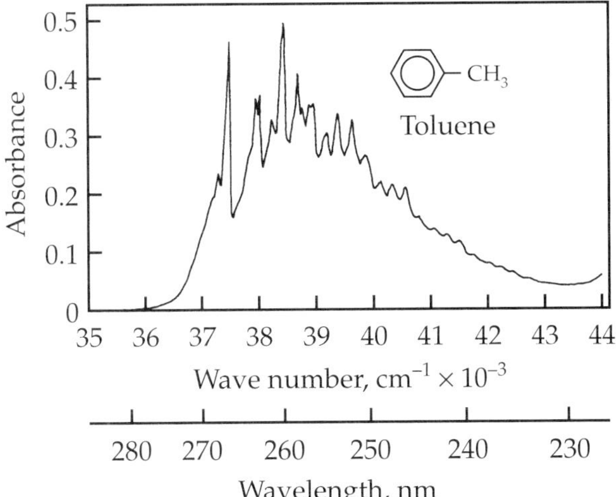

Figure 23-7 The spectrum of the first electronic transition of toluene vapor at low resolution. Cary 1501 spectrophotometer.

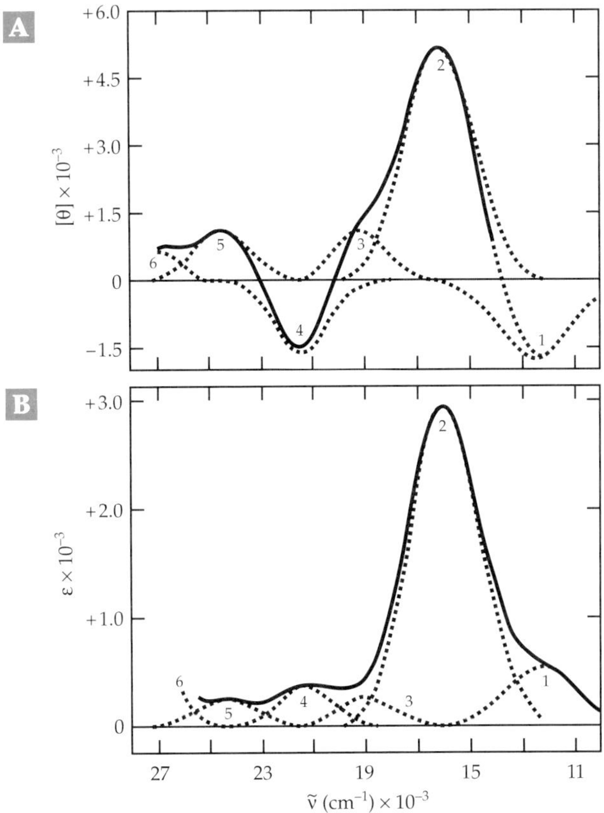

Figure 23-8 Resolution of the visible circular dichroism (ellipticity) spectrum (A) and absorption spectrum (B) of the *Pseudomonas* blue protein into series of overlapping Gaussian bands (—). The numbers 1 to 6 refer to bands of identical position and width in both spectra. Absorption envelopes resulting from the sum of the set of overlapping Gaussian bands (—) correspond within the error of the measurement to the experimental spectra. The *dashed part* of the CD envelope above 700 nm was completed by a curve fitter with the use of a band in the position of *band 1* of the absorption spectrum. From Tang *et al.*[68]

toward the high-energy side. It is best to fit such bands with a skewed function such as the **log normal** distribution curve.[62–64] In addition to position, height, and width, a fourth parameter provides a measure of **skewness**. Computer-assisted fitting with log normal curves gives precise values for the positions, widths, and intensities. In general, the peak position is somewhat to the high-energy side of the 0–0 transition.

Absorption spectra plotted as a linear function of wavelength are sometimes fitted with Gaussian curves. However, Gaussian curves only occasionally give a good fit for such spectra, and it is undesirable to measure bandwidths in nanometers. It is wave number that is proportional to energy. Spectral bands tend to have similar widths across the visible–ultraviolet range when plotted against wave number but not when ploted against wavelength. Another approach to the quantitative description of spectra is to fit the major progressions of vibrational subbands with series of narrow Gaussian curves.[65–67]

Classification of electronic transitions. The intense 600-nm absorption band of the copper blue protein in Fig. 23-8 is attributed to a *d–d* transition of an electron in the metal ion from one *d* orbital to another of higher energy.[68] The intensity is thought to arise from transfer of an electron from a cysteine thiolate to the copper (p. 883). The electronic transitions in most organic molecules are of a different type. Transitions lying at energies <55,000 cm^{-1} are classified as either $\pi-\pi^*$ or $n-\pi^*$. In the $\pi-\pi^*$ transitions an electron is moved from a bonding π molecular orbital to an antibonding (π^*) orbital. Such a transition is present in ethylene at 61,540 cm^{-1} (162.5 nm) with a maximum molar extinction coefficient ε_{max} of ~15,000 M^{-1} cm^{-1}. An $n-\pi^*$ transition results from the raising of an electron in an unshared pair of an oxygen or nitrogen atom into a π^* antibonding orbital. These transitions are invariably weak. For example, acetone in H_2O shows an $n-\pi^*$ transition at 37,740 cm^{-1} (265 nm). The value of ε_{max} is ~240 and the width is about 6400 cm^{-1}. A characteristic of $n-\pi^*$ transitions is a strong shift to lower energies as the compound is moved from water into less polar solvents. For example, the peak of the acetone band lies at 36,920 cm^{-1} in methanol and at 35,970 cm^{-1} (278 nm) in hexane. Such a solvent shift is often taken as diagnostic of an $n-\pi^*$ transition, and it is often stated that the $\pi-\pi^*$ bands shift in the opposite direction upon change of solvent character. However, the latter is not true for many of the polar chromophores found in biochemical substances. Thus, the $\pi-\pi^*$ bands of tyrosine also shift to lower energies when the molecule is moved from water into hexane. However, the magnitude of the shift is much less than for the $n-\pi^*$ band of acetone.

A molecule can have several different excited states of increasing energies. In benzene and its derivatives there are three easily detectable $\pi-\pi^*$ transitions (see Fig. 3-13). The first is a weak band centered at ~ 260 nm in toluene (Fig. 23-7) and ~275 nm in tyrosine (Fig. 3-13) with $\varepsilon = 10^2$ to 10^3. The second is a band at a higher frequency (at 1.35 0.10 times the frequency of the first band) with ε_{max} often as high as 10^4. The third band is found at still higher energies with ε_{max} reaching 5 x 10^4. The excited-state energy levels represented by these transitions were labeled ^1L_b, ^1L_a, and ^1B_a by Platt. Other authors described the levels in terms of the symmetries of the molecular orbitals, the ground state being $^1A_{1g}$, and the three excited states $^1B_{2u}$, $^1B_{1u}$, and $^1E_{1u}$. In these symbols the superscript 1 indicates that the excited states are **singlet** in nature; that is, the electrons remain paired in the excited states. Absorption of visible and ultraviolet light almost always leads to singlet excited states initially. For more complex ring systems the number of possible electronic transitions increases, but attempts are often made to relate these transitions back to those of benzene.

The intensities of electronic transitions vary greatly. The area ($\mathcal{A}$) under the absorption band, when ε is plotted against wave number $\bar{\nu}$, is directly proportional (Eq. 23-6) to a dimensionless quantity called the **oscillator strength** *f*.

$$f = \frac{2.303\, m_e c^2}{\pi N e^2} F\mathcal{A} = 4.32 \times 10^{-9}\, F\mathcal{A} \quad (23\text{-}6)$$

In this equation m_e and e are the mass and charge of the electron, c is the velocity of light, N is Avogadro's number, and $\mathcal{A}$ is the area in a plot of ε vs ν in cm^{-1}; F is a dimensionless correction factor that is related to the refractive index of the medium and is near unity for aqueous solutions. If the area is approximated as that of a triangle of height ε_{max} and width (at half-height) W, we find that for a typical absorption band of $\varepsilon_{max} = 10^4$ and $W = 3000$ cm^{-1}, $f = 0.13$.

The oscillator strength is related to the probability of a transition and can become approximately 1 only for the strongest electronic transitions. However, it is rarely this high. For example, the oscillator strength is ~10^{-4} for Cu^{2+} and ~2 x 10^{-3} for the toluene absorption band shown in Fig. 23-7. The low intensity of absorption bands of benzene derivatives is related to the fact that these transitions are quantum mechanically forbidden for a completely symmetric molecule. It is only because of coupling with asymmetric vibrations of the ring that the ^1L_b transition of benzene becomes weakly allowed. In the benzene spectrum the 0–0 transition is completely absent, and only those progressions involving uptake of an additional 520 cm^{-1} of a nonsymmetric vibrational energy are observed. In the case of toluene and phenylalanine, the asymmetry of the ring introduced by the substituents permits the 0–0 transition to occur and leads to a higher oscillator strength than that observed with benzene. The ^1L_a transition of benzenoid derivatives is also partially forbidden by selection rules, and only the third band begins to approach an oscillator strength of one.

Use of plane polarized light. The intensity of a spectral transition is directly related to the **transition dipole moment** (or simply the transition moment), a vector quantity that depends upon the dipole moments of the ground and excited states. For aromatic ring systems, the transition dipole moments of the $\pi-\pi^*$ transitions lie in the plane of the ring. However, both the directions and intensities for different $\pi-\pi^*$ transitions within a molecule vary.

The transition moment has a dimension of length (usually given in angstroms) and can be thought of as a measure of the extent of the charge migration during the transition. Light is absorbed best when the directions of polarization (i.e., of the electric vector of the light) and of the transition moment coincide. This fact can easily be verified by light absorption measurements on crystals. As with infrared spectra of oriented peptide chains (Fig. 23-3), the electronic spectra of crystals display a distinct dichroism. Crystals of coenzyme-containing proteins (Fig. 23-9) are very appropriate for spectroscopy with polarized light because the chromophores are spaced far enough apart to avoid electronic interaction and have absorbances low enough to record with crystals of the order of 0.1 mm thickness.[69,70]

In contrast to $\pi-\pi^*$ transitions, the $n-\pi^*$ transitions of heterocyclic compounds and carbonyl-containing rings are often polarized in a direction perpendicular to the plane of the ring. Linear dichroism of cytosine, adenine, and other nucleic acid bases has been measured on single crystals and in partially oriented polymer films.[71] Magnetically induced linear dichroism provides a new tool for study of metalloproteins.[72]

Relationship of absorption positions and intensity to structures. While quantum mechanical calculations permit prediction of the correct number and approximate positions of absorption bands, they are imprecise. For this reason, electronic spectroscopy also relies upon a combination of empirical rules and atlases of spectra that can be used for comparison purposes.[74–76] The following may help to orient the student. The position of an absorption band shifts **bathochromically** (to longer wavelength, lower energy) when the number of conjugated double bonds increases. Thus, **butadiene** absorbs at 46,100 cm^{-1} (217 nm) vs the 61,500 cm^{-1} of ethylene. As the number of double bonds increases further, the bathochromic shifts become progressively smaller (but remain more nearly constant in terms of wavelength than wave number). For **lycopene** (Fig. 23-10) with 11

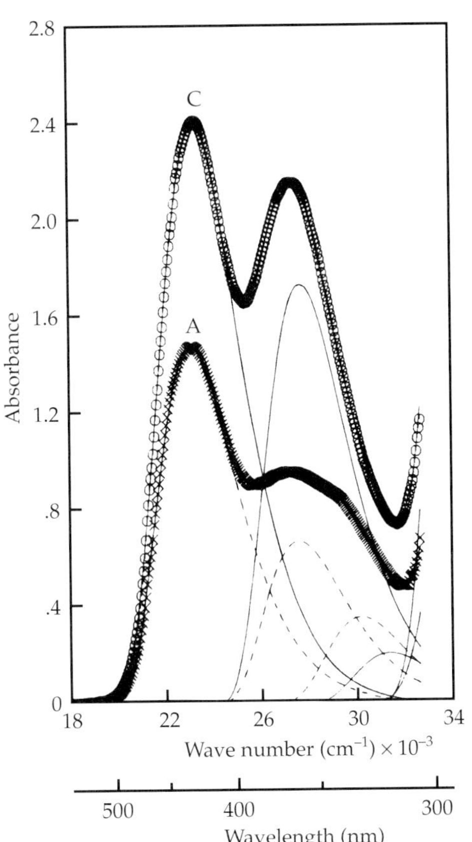

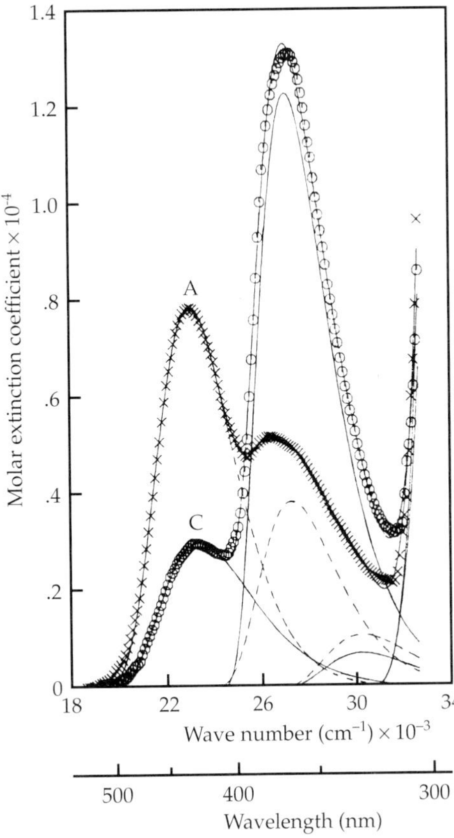

Figure 23-9 Polarized absorption spectra of orthorhombic crystals of cytosolic aspartate aminotransferase. The light beam passed through the crystals along the *b* axis with the plane of polarization parallel to the *a* axis (A) or the *c* axis (C). Left, native enzyme at pH 5.4; right, enzyme soaked with 300 mM 2-methylaspartate at pH 5.9. The band at ~430 nm represents the low pH protonated Schiff base form of the enzyme. Upon soaking with 2-methylaspartate the coenzyme rotates ~30° to form a Schiff base with this quasisubstrate. The result is a large change in the *c*/*a* polarization ratio. The ~364 nm band in the complex represents the free enzyme active site in the second subunit of the dimeric enzyme.[70,73] Courtesy of C. M. Metzler.

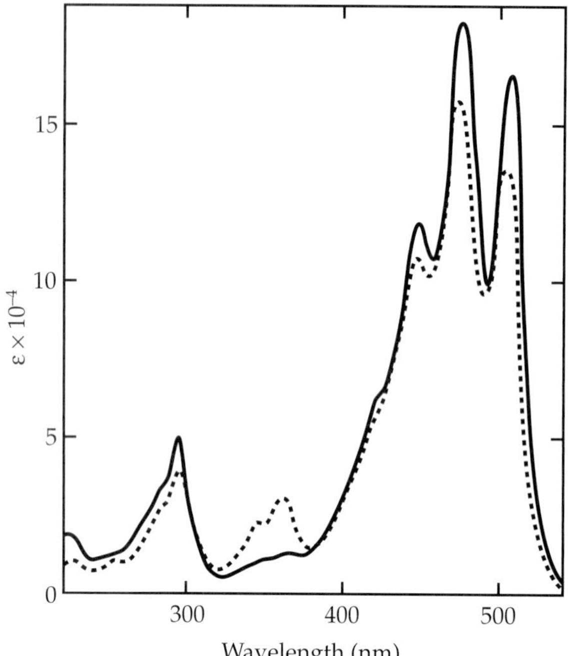

Figure 23-10 The absorption spectrum of lycopene (plotted vs wavelength). Note the vibrational structure, which has a spacing of ~1200–1500 cm^{-1}. The solid line is for all-translycopene while the dashed line is that of the sample after refluxing 45 min in the dark. The new peak at ~360 nm arises from isomers containing some cis double bonds.

conjugated double bonds the absorption band is located at 21,300 cm^{-1} and displays distinct vibrational structure (Fig. 23-10). Certain ring molecules such as the porphyrins and chlorophylls have spectra that can be related back to those of the linear polyenes. Note (Fig. 16-7) that the porphyrin α and β bands represent vibrational structure of a single electronic transition, whereas the intense Soret band results from a different transition.

Substituted benzenes almost invariably absorb at lower energies than the parent hydrocarbon. The stronger the electron withdrawing or donating ability of the substituent, the larger the bathochromic shift. The magnitude of the shift has been correlated with the Hammett σ constants. Thus, the first absorption band of tyrosine in water is shifted 2600 cm^{-1} toward the red from that of benzene, while that of the dissociated tyrosine anion is shifted 4700 cm^{-1}, very roughly in proportion to the σ_p values of Box 6-C. Especially large shifts are observed when functional groups of opposite types (that is, an electron donating group vs an electron accepting group) are both present in the same ring. The effects of ortho and meta substituent pairs are closely similar (in contrast to the differing electronic effects of ortho and meta pairs in chemical reactivity). Substituent pairs in para positions yield somewhat different spectral shifts. When there are more than two substituents, the two strongest groups often dominate in determining the character of the spectrum. Useful empirical rules have been developed.[77,78]

Spectra of proteins and nucleic acids. Most proteins have a strong light absorption band at 280 nm (35,700 cm^{-1}) which arises from the aromatic amino acids tryptophan, tyrosine, and phenylalanine (Fig. 3-14). The spectrum of phenylalanine resembles that of toluene (Fig. 23-7),[61] whose 0–0 band comes at 37.32 x 10^3 cm^{-1}. The vibrational structure of phenylalanine can be seen readily in the spectra of many proteins (e.g., see Fig. 23-11A). The spectrum of tyrosine is also similar (Fig. 3-13), but the 0–0 peak is shifted to a lower energy of ~35,500 cm^{-1} (in water). Progressions with spacings of 1200 and 800 cm^{-1} are prominent.[79] The low-energy band of tryptophan consists of two overlapping transitions ^1L_a and ^1L_b.[65] The ^1L_b transition has well-resolved vibrational subbands, whereas those of the ^1L_a transition are more diffuse. Tryptophan derivatives in hydrocarbon solvents show 0–0 bands for both of these transitions at approximately 289.5 nm (34,540 cm^{-1}). However, within proteins the ^1L_a band may be shifted 3–10 nm (up to 1100 cm^{-1}) toward lower energies, probably as a result of hydrogen bonding to other groups in the protein. The largest shifts can occur when the NH group of the indole ring is hydrogen bonded to COO^-, a ring nitrogen of histidine, or a carbonyl group of amides.[80] In an aqueous medium the ^1L_b band of tryptophan is shifted to higher energies and the ^1L_a band to lower energies than in a hydrocarbon solvent.

In addition to the three aromatic amino acids, disulfide bonds absorb in the near ultraviolet region as indicated in Fig. 3-14. Since the absorption characteristics depend upon the dihedral angles in the disulfide bridges, it is difficult to accurately evaluate the contribution of this chromophore to the 280-nm band.

Tyrosine, tryptophan, and phenylalanine all have additional transitions in the high-energy UV region of the spectrum (Fig. 3-13). Even more intense are the absorption bands of the amide groups, which become significant above 45,000 cm^{-1}.[81] These include a weak n–π^* transition at ~45,500 cm^{-1} (210 nm) overlapped by a strong π–π^* transition at ~52,000 cm^{-1} (192 nm).[82] Histidine also has absorption bands in this region.

As with polypeptides, the light absorption properties of polynucleotides reflect those of the individual components. The spectra of the purine and pyrimidine bases as ribonucleosides are shown in Fig. 5-5. The number of individual electronic transitions and their origins are not immediately obvious, but many measurements in solutions and in crystals, as well as theoretical computations,[7,83,84] have been made. Cytosine has π–π^* transitions at ~275, 230, 200, and

185 nm,[83] the two highest energy bands being overlapped. Adenine derivatives have seven $\pi-\pi^*$ transitions.[71] Spectra of flavins contain at least four intense transitions (Fig. 15-8).[85]

Whereas proteins have their low energy absorption band at ~280 nm, polynucleotides typically have maxima at ~260 nm (38,500 cm^{-1}). A phenomenon of particular importance in the study of nucleic acids is the **hypochromic effect**. In a denatured polynucleotide the absorption is approximately the sum of that of the individual components. However, when a double helical structure is formed and the bases are stacked together, there is as much as a 34% depression in the absorbance at 260 nm. This provides the basis for optical measurement of DNA melting curves (Fig. 5-45).[45,86] The physical basis for the hypochromic effect is found in dipole–dipole interactions between the closely stacked base pairs.[7,86,87]

Difference spectra and derivative spectra. Changes in light-absorbing properties of proteins and nucleic acids are often measured as a function of some quantity such as pH, temperature, ionic environment, or the presence or absence of another interacting molecule. The induced changes in the spectrum are small but can be seen if the *difference* between the two spectra, one "unperturbed" and the other in the presence of some "perturbant," is recorded. The perturbant might be an additional reagent, an altered solvent (e.g., with added glycerol, D_2O), a change in pH, or temperature. The difference spectrum shown in Fig. 23-11B arises from the binding of an inhibitor succinate together with a substrate carbamoyl phosphate to the catalytic subunit of aspartate carbamoyltransferase (Fig. 7-20).[88] The difference spectrum appears as a pair of peaks and a valley in the aromatic amino acid region. With proper interpretation (caution!) difference spectra can be used to infer something about the change in environment of aromatic amino acids in a protein.[89]

Difference spectra are usually recorded by placing the unperturbed spectrum in the *reference* light beam of a spectrophotometer and the perturbed solution in the *sample* beam in carefully matched cuvettes. However, the spectrum shown in Fig. 23-11B was obtained by recording the two spectra independently and subtracting them with the aid of a computer. The same data have been treated in another way by fitting two log normal curves (p. 1283) to the absorption bands and plotting the differences between the mathematically

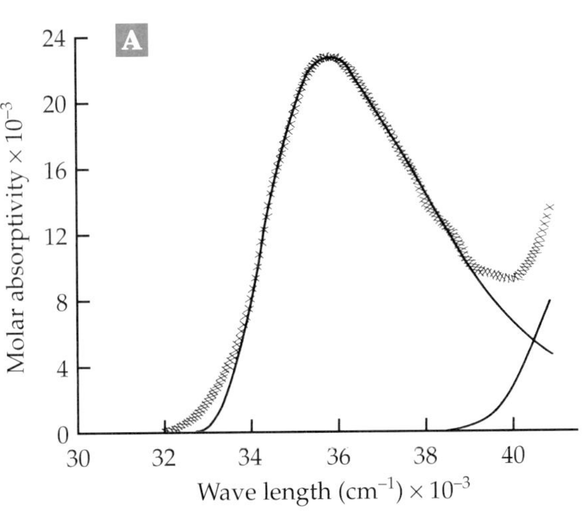

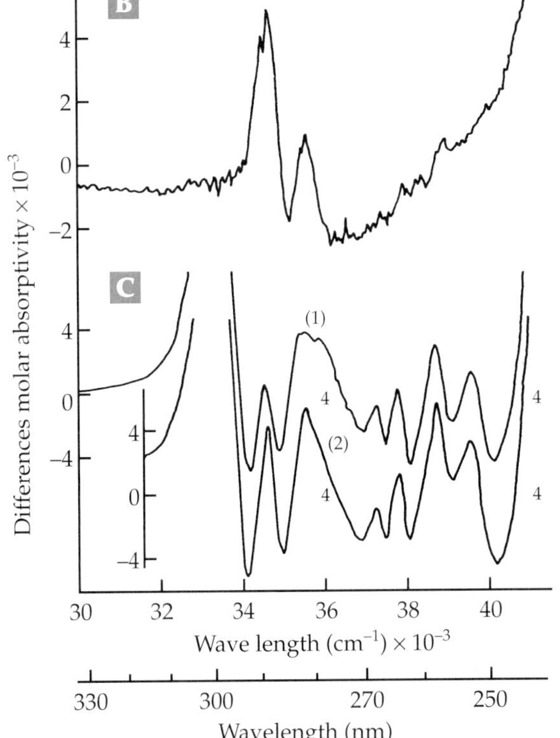

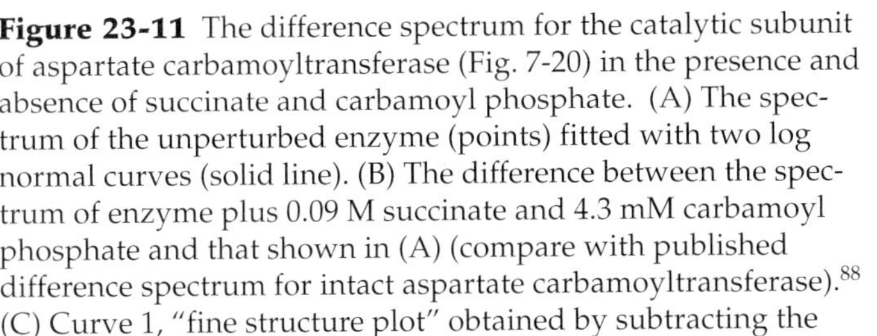

Figure 23-11 The difference spectrum for the catalytic subunit of aspartate carbamoyltransferase (Fig. 7-20) in the presence and absence of succinate and carbamoyl phosphate. (A) The spectrum of the unperturbed enzyme (points) fitted with two log normal curves (solid line). (B) The difference between the spectrum of enzyme plus 0.09 M succinate and 4.3 mM carbamoyl phosphate and that shown in (A) (compare with published difference spectrum for intact aspartate carbamoyltransferase).[88] (C) Curve 1, "fine structure plot" obtained by subtracting the spectrum in A from the smooth curve obtained by summing the two log normal curves. Curve 2, a similar plot for enzyme plus succinate and carbamoyl phosphate. The enzyme was supplied by G. Nagel and H. K. Schachman and the spectra were recorded by I.-Y. Yang.

smooth fitted curve and the experimental points taken at close intervals[90,91] as shown in Fig. 23-11C. The two "fine structure plots" obtained in this way are an alternative way of representing the same data that gave rise to the difference spectrum. The method has the advantage that information about the overall band shape is obtained from the computer-assisted curve fitting process. Thus, the binding of succinate and carbamoyl phosphate caused an almost insignificant shift (of 20 cm^{-1}) in the overall band position and a very slight broadening. The principal effect is an enhancement in the vibrational structure at 34,600 cm^{-1} in the 0–0 band of the two tryptophan residues present in the subunit. The cause of this change is not entirely obvious, a weakness of difference spectroscopy. Another way of examining a spectrum such as that in Fig. 23-11A is to plot a mathematical derivative of the curve. Both second[92] and fourth derivatives[91,93,94] yield curves similar to the difference plots of Fig. 23-11C.

5. Circular Dichroism and Optical Rotatory Dispersion

The circular dichroism of a sample is the difference between the molar extinction coefficients for left-handed and right-handed polarized light (Eq. 23-7) and is observed only for chiral molecules.[7,95–97]

$$\Delta\varepsilon = \varepsilon_L - \varepsilon_R \text{ (units are } M^{-1}\,cm^{-1}) \qquad (23\text{-}7)$$

The **dichrograph** gives a direct measure of $\Delta\varepsilon$. A circular dichroism (CD) spectrum often resembles an absorption spectrum, the peaks coming at the same positions as the peaks in the absorption spectrum of the same sample. However, the CD can be either positive or negative and may be positive for one transition and negative for another (Fig. 23-8). It is most convenient to plot $\Delta\varepsilon$ directly as a function of wavelength or wave number. However, much of the literature makes use of the **molar ellipticity** (Eq. 23-8):

$$\text{Molar ellipticity} = [\theta] = 3299\,\Delta\varepsilon \quad \text{(units are degrees } cm^2\, decimole^{-1}) \qquad (23\text{-}8)$$

The **rotational strength** may also be evaluated (Eq. 23-9):

$$\text{Rotational strength} = \int [(\Delta\epsilon)/\lambda]\,d\lambda \qquad (23\text{-}9)$$

The integration is carried out over the entire absorption band for a given transition.

Circular dichroism is closely related to **optical rotatory dispersion**, the variation of optical rotation with wavelength. Optical rotation depends upon the difference in refractive index ($\eta_L - \eta_R$) between left-handed and right-handed polarized light. Rotation α is measured as an angle in degrees or radians. Data are customarily reported in terms of **specific rotation**, that of a hypothetical solution containing 1 g / ml in a 1 dm (decimeter) tube. Specific rotation is calculated (Eq. 23-10) from the observed rotation, the concentration c' in g ml^{-1}, and the length of the tube l' in decimeters.

$$\text{Specific rotation} = [\alpha] = \alpha_{obs} / c'l' \qquad (23\text{-}10)$$

The **molecular rotation** is defined by Eq. 23-11 in which M_r is the molecular mass and c and l are in moles per liter and cm, respectively.

$$\text{Molecular rotation} = [\phi] = 100\,\alpha_{obs} / cl = [\alpha]\,M_r / 100 \qquad (23\text{-}11)$$

It is often multiplied by a factor of $3/(\eta^2 + 2)$ to correct for a minor effect of the polarizability of the field acting on the molecules. The rotation in the radians per centimeter of light path can be related (Eq. 23-12) directly to the wavelength of the light and the refractive indices η_L and η_R.

$$\alpha\,(\text{radians / cm}) = [\alpha]\,c' / 1800 = \pi / \lambda [\eta_L - \eta_R] \qquad (23\text{-}12)$$

In contrast to circular dichroism, optical rotary dispersion (ORD) extends far from absorption bands into spectral regions in which the compound is transparent. As an absorption band is approached, the optical rotation increases in either the positive or negative direction. Then, within the absorption band it drops abruptly through zero and assumes the opposite sign on the other side of the band (the Cotton effect). Although the occurrence of optical rotation in nonabsorbing regions of the spectrum provides an advantage to ORD measurements, the interpretation of ORD spectra is more complex than that of CD spectra. In principle, the two can be related mathematically and both are able to give the same kind of chemical information.[7]

The CD in the *d–d* bands of the blue copper protein (Fig. 23-8) arises in part from the fact that within the protein the copper ion is in an asymmetric environment. For a similar reason, the aromatic amino acids of proteins often give rise to circular dichroism. In the case of tyrosine, the sign of the CD bands can be either positive or negative but is the same throughout a given transition. The CD bands are similar in shape to the absorption bands.[36,98] The behavior of phenylalanine is more complex. The progression of vibrational subbands at 930 cm^{-1} intervals above the 0–0 band all have the same sign, and the intensities relative to that of the 0–0 band are similar to those in absorption. However, the vibrations of wave numbers equal to that of the 0–0 transition plus 180 and 520 cm^{-1} sometimes give rise to CD bands of the opposite sign, and the relative intensity relationships are variable.[61,98]

The binding of a symmetric chromophore to a protein or nucleic acid often induces CD in that chromophore. For example, the bands of enzyme-bound pyridoxal and pyridoxamine phosphates shown in Fig. 14-9 are positively dichroic in CD, but the band of the quinonoid intermediate at 20,400 cm^{-1} (490 nm) displays negative CD. When "transimination" occurs to form a substrate Schiff base (Eq. 14-26), the CD is greatly diminished. While the coenzyme ring is known to change its orientation (Eq. 14-39; Fig. 14-10), it is not obvious how the change in environment is related to the change in CD.

A series of octant rules make it possible to predict the sign and magnitude of CD to be expected for $n-\pi^*$ transitions of simple carbonyl compounds.[99] Theoretical approaches to the CD and ultraviolet absorption of proteins in the high-energy ultraviolet region have also been developed. In a regular β structure, in an α helix or in a crystalline array, the transitions of adjacent amide groups may be **coupled**, the excitation energy being delocalized. This **exciton** delocalization leads to a splitting (Davydov splitting) into two transitions of somewhat different energies and polarized in different directions.[7,9] The amide absorption band at 52,600 cm^{-1} is split in an α helix into components at ~48,500 and 52,600 cm^{-1}. Furthermore, low-energy $\pi-\pi^*$ and $n-\pi^*$ states are close together in energy, a fact that allows mixing of the two states and appearance of rotational strength in the $\pi-\pi^*$ band with a sign opposite to that in the $n-\pi^*$ band.[100]

Both the sign and intensity of the CD bands of peptides also depend upon conformation. Well-defined differences are observed among α helices, β structure, and random-coil conformations. Measurements may be extended into the "vacuum ultraviolet" region—up to 60,000 cm^{-1} in aqueous solutions.[101] A useful empirical approach is to deduce spectra of helices, β structures, and unordered peptide chains from measured spectra together with an examination of actual structures obtained by X-ray crystallography[7,95,97,102,103] (Fig. 23-12). Note that the CD curve for the α helix has a deep minimum at 222 nm, whereas the β form has a shallower minimum. The random structure has almost no CD at the same wavelength. The approximate helix content of a protein is often estimated from the depth of the trough at 222 nm in the CD spectrum. Better predictions can be made by using a computer-assisted comparison of an experimental CD spectrum with those of a series of proteins of known 3-D structure.[7]

The circular dichroism of polynucleotides at 275 nm is a linear function of both the helix winding angle and the base pair twist.[103] Measurement of CD spectra on large polynucleotides or large molecular aggregates is complicated by differential light scattering of right and left circularly polarized light. However, the phenomenon can also provide new structural information.[104] Progress has been made in attempts to predict the optical rotation of molecules from quantitative values for the polarizabilities of individual atoms.[105–109]

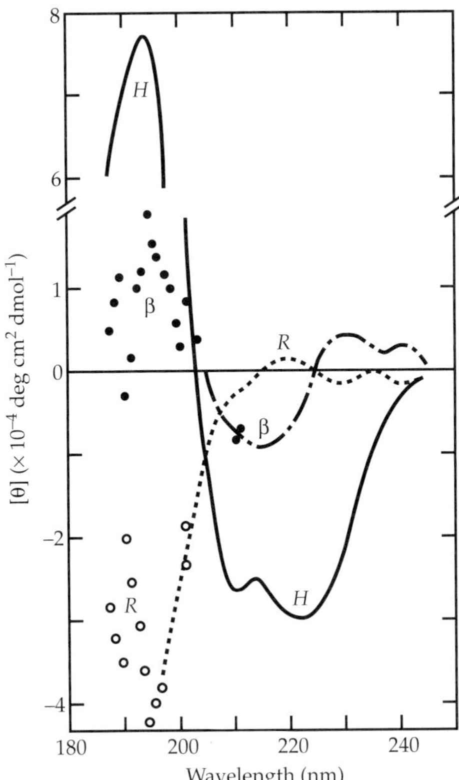

Figure 23-12 Circular dichroism of the helix (H), β, and unordered (R) form computed from the CD of five proteins. Points are plotted when a smooth curve could not be drawn. From Y.-H. Chen *et al.*[102]

Vibrational circular dichroism involves IR absorption bands. The technique has been applied to sugars,[110] oligosaccharides,[111] proteins,[112] and nucleic acids.[113] The related **vibrational Raman optical activity** has also been applied to polyribonucleotides.[114]

6. Photoacoustic Spectroscopy

Photoacoustic spectra are recordings of the energy emitted as heat after absorption of monochromatic light. The sample is placed in a closed photoacoustic cell. The light beam, which is chopped at an audiofrequency, induces a periodic heating and cooling of the gas in contact with the sample in the cell. This is sensed as sound by a sensitive microphone. The resulting electrical signals are sent to a computer for analysis. The output is an absorption spectrum resembling that measured optically. The samples do not have to be transparent.[5] A related technique is photoacoustic calorimetry.[115,116]

7. X-Ray Absorption and Mössbauer Spectroscopies

The importance of X-ray diffraction (discussed in Chapter 3) to biochemistry is obvious, but techniques related to absorption of X-rays and γ-rays have also come into widespread use.[117] Abbreviations such as **XANES** and **EXAFS** are common in the metalloprotein literature. The names arise from the sharp increase in the absorption coefficient for X-rays as their energy is increased to what is called the **K absorption edge**. At slightly lower energies absorption of an X-ray by an atom leads to expulsion of an electron or the raising of an electron to an excited state. Absorption of X-rays will expel all except the inner 1*s* electrons. As the energy is increased further, the stepwise increase in absorption that constitutes the edge is observed. At higher energies the absorption decreases. However, with a high-resolution instrument distinct oscillations are observed on the high-energy side of the edge, extending for ~20 eV. This is **X-ray absorption near-edge structure** (XANES). When an X-ray absorbing atom in a molecule is surrounded by other atoms, a fine structure that depends upon the nature of these atoms and their distances from the absorbing atom is observed over a range of several hundred electron volts above the edge. This is **extended X-ray absorption fine structure** (EXAFS).[117]

The EXAFS technique has been especially useful for metalloproteins. It has often provided the first clues as to the identity of atoms (O, N, S) surrounding a metal atom and either covalently bonded to it or coordinated with it (Chapter 16). Interpretations are often difficult, and a common approach is to try to simulate the observed spectrum by calculation from a proposed structure.[118] Tautomerism in crystalline Schiff bases (see Eq. 23-24) has been studied by near-edge X-ray absorption fine structure (NEXAFS) employing soft X-rays.[119]

Mössbauer spectroscopy, also called recoil-free nuclear resonance absorption, depends upon resonant absorption of γ-rays emitted by a radioactive source by atomic nuclei.[120] The phenomenon was initially difficult to observe, but the German physicist Mössbauer devised a way in which to record the absorption of a quantum of energy equal to the difference in two energy states of the atomic nucleus. The method depends upon a Doppler effect observed when the sample or source moves. Consequently, Mössbauer spectra, such as that in Fig. 16-18, are plots of absorption versus velocity. Mössbauer spectroscopy has been applied to numerous metalloproteins, especially those containing iron centers. It is a major tool in investigation of Fe–S proteins.[121,122] Since ^{56}Fe is "silent" in Mössbauer spectroscopy, proteins are often enriched with ^{57}Fe for observation.

C. Fluorescence and Phosphorescence

An electronically excited molecule is able to lose its excitation energy and return to the ground state in several ways. One of these is to reemit a quantum of light as fluorescence.[7,123–127] The intensity and spectral properties of fluorescent emission can be measured by illuminating a sample in a cuvette with four clear faces with the measuring photomultiplier set at right angles to the exciting light beam. In absorption spectrophotometry we measure a difference between the light intensity of the beam entering the sample and that emerging from the sample. In fluorescence spectroscopy we measure the absolute intensity of the light emitted. Although this intensity is small, the measurement can be made extremely sensitive, far more so than can light absorption. For this reason, fluorescence is widely used for detection and analysis, e.g., in DNA sequencers. Enzyme kinetics can be studied with fluorescent substrates at very low concentrations.[127a] Fluorescent antibodies, DNA chips, and numerous bioassay and imaging methods are dependent upon measurement of fluorescence. Fluorescence can also yield a wealth of information about the chemical and physical properties of electronically excited states of molecules.

1. Excitation and Emission Spectra

Measurements of the intensity of fluorescence at any wavelength vs the wavelength of monochromatic light used to excite the fluorescence give a fluorescence **excitation spectrum**. The excitation spectrum is an example of an **action spectrum**, which is a measure of any response to absorbed light. At very low concentrations of pure substances, action spectra tend to be identical to absorption spectra. However, since the observed response (fluorescence in this case) is proportional to light absorbed, action spectra should be compared to plots of $1-T$ (where T = transmittance, Section B,1) vs wavelength rather than to plots of ε vs λ. The two plots are proportional at low concentrations. For a discussion of action spectra see Clayton.[123]

A fluorescence **emission spectrum** is a record of fluorescence intensity vs wavelength for a constant intensity of exciting light. Excitation and emission spectra for a flavin and for the indole ring of tryptophan are both given in Fig. 23-13. The heights of the

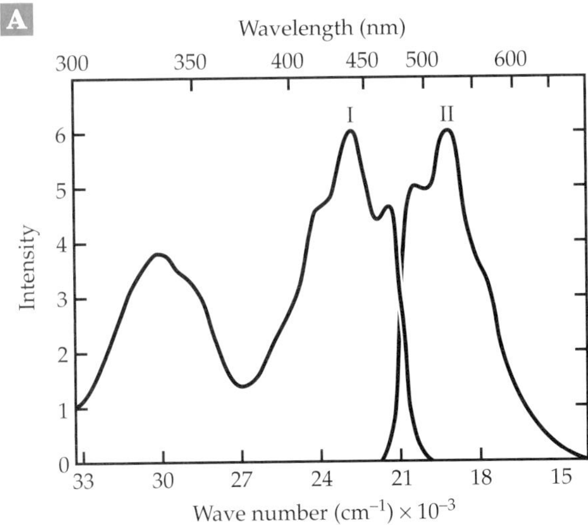

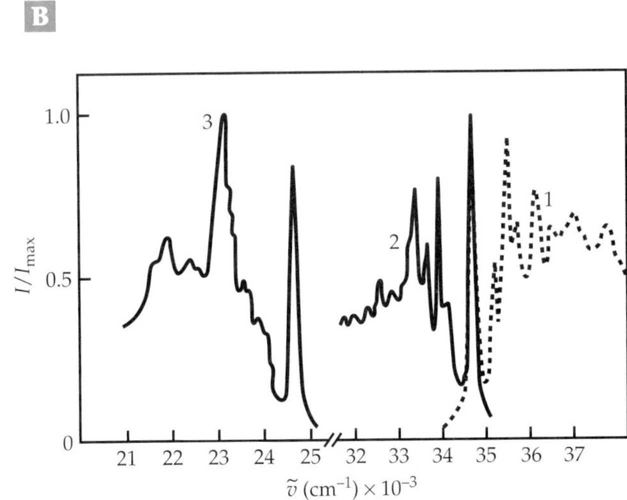

Figure 23-13 (A) Corrected emission and excitation spectra of riboflavin tetrabutyrate in *n*-heptane. Concentration, about 0.4 mg l^{-1}. Curve 1: excitation spectrum; emission at 525 nm. Curve 2: emission spectrum; excitation at 345 nm. From Kotaki and Yagi.[128] (B) Indole in cyclohexane, T = 196 K. 1, Fluorescence excitation spectrum; 2, fluorescence spectrum; and 3, phosphorescence spectrum. From Konev.[125]

emission spectra have been adjusted to the same scale as that of absorption. The fluorescent emission is always at a lower energy than that of the absorbed light. The excitation and emission spectra overlap only slightly, and the emission spectrum is an approximate mirror image of excitation spectrum. To understand this, refer to the diagram in Fig. 23-14. Absorption usually leads to a higher vibrational energy state after light absorption than before. However, most of the excess vibrational energy is dissipated before much fluorescent emission occurs. The excited molecule finds itself in the lowest vibrational state of the upper electronic state, and it is from this state that the bulk of the fluorescent emission takes place. Furthermore, whereas absorption usually occurs from the lowest vibrational state of the ground electronic level, fluorescence can populate many excited vibrational states of the ground electronic state (Fig. 23-14). Consequently, as indicated in the figure, the fluorescent emission spectrum consists of a series of subbands at lower energies than those observed in the absorption spectrum. The two spectra have only the 0–0 transition in common. As can be seen from Fig. 23-14, even the two 0–0 transitions do not coincide exactly. The peak of emission is shifted toward slightly lower energies than that of absorption because during or immediately following absorption of a photon there is some rearrangement of solvent molecules around the absorbing molecule to an energetically more stable arrangement. Just as excess vibrational energy is dissipated in the excited state, so relaxation of these solvent molecules around the excited chromophore leads to a small shift in energy. A similar relaxation

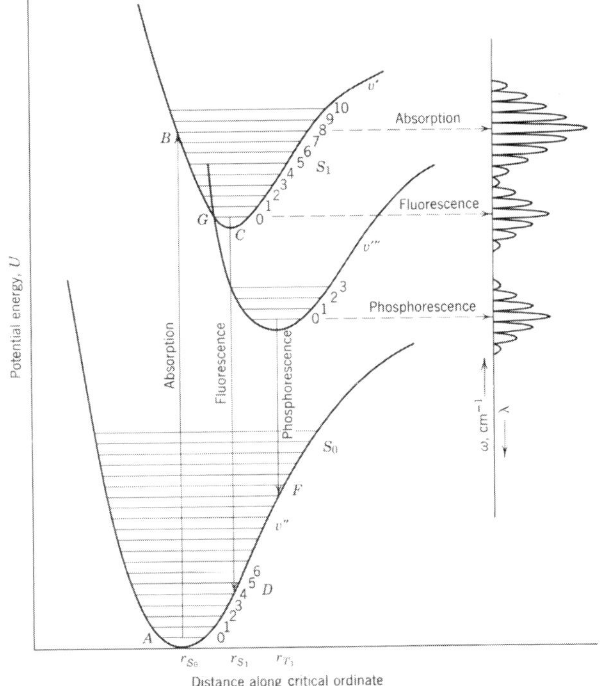

Figure 23-14 Potential energy diagram for the ground state S_0 and the first excited singlet S_1 and triplet T_1 states of a representative organic molecule in solution. G is a point of intersystem crossing $S_1 \rightarrow T_1$. For convenience in representation, the distances *r* were chosen $r_{S_0} < r_{S_1} < r_{T_1}$; thus, the spectra are spread out. Actually, in complex, fairly symmetric molecules, $r_{S_0} \sim r_{S_1} < r_{T_1}$ and the 0–0 absorption and fluorescence bands almost coincide, but phosphorescence bands are significantly displaced to the lower wavelengths. From Calvert and Pitts,[2] p. 274.

occurs in the ground state of a molecule that has just emitted a photon as fluorescence. This also contributes to the shift in position of the 0–0 band in fluorescence (see Parker,[124] p. 13).

Several molecular properties can be measured using emission and excitation spectra. These include fluorescence lifetime, efficiency, anisotropy of the emitted light, mobility of chromophores, rates of quenching, and energy transfer to other chromophores.

Fluorescence lifetimes. Why are some molecules fluorescent, while others are not? The possibility for fluorescent emission is limited by the radiative lifetime τ_r, which is related by Eq. 23-13 to the first-order rate constant k_f for exponential decay of the excited state by fluorescence.

$$\tau_r = 1/k_f \qquad (23\text{-}13)$$

The radiative lifetime is a function of the wavelength of the light and of the oscillator strength of the transition. For molecules absorbing in the near UV, the approximation of Eq. 23-14 is often made.

$$1/\tau_r \sim 10^4\varepsilon_{max}) \qquad (23\text{-}14)$$

Thus, if ε = 10,000, the radiative lifetime (the time in which the fluorescence decays to $1/e$, its initial value) is ~10^{-8} s (10 ns). If the absorption is more intense, the lifetime is shorter, and if it is less intense, it is longer. Other modes of deexcitation compete with fluorescence; therefore, the shorter the radiative lifetime the more likely that fluorescence will be observed.

The actual lifetime τ of an excited molecule is usually less than τ_r because of the competing nonradiative processes. The sum of their rate constants can be designated k_{nr}. The **fluorescence efficiency** (or **quantum yield**) ϕ_F is given by Eq. 23-15.

$$\phi_F = k_f / k_f + k_{nr}) = k_r\tau$$
$$\text{where } \tau^{-1} = k_f + k_{nr} \qquad (23\text{-}15)$$

For a highly fluorescent molecule such as riboflavin, ϕ_f may be 0.25 or more.[129] For tryptophan in water it is about 0.14, and in proteins it varies from near zero to 0.35.[130]

Time-resolved fluorescence spectroscopy. The fluorescence lifetime τ can be measured with either of two different types of fluorometer.[7,127,131–133] **Pulse fluorometers** use pulsed lasers that can deliver pulses of light lasting as little as one picosecond or less. This permits rapid excitation and permits the direct observation of emitted light, using photon countings, over the entire range of time from a few picoseconds to milliseconds required for decay of the fluorescence.[131] The observed value of τ for riboflavin 5'-phosphate (ε_{max} = 12,200 at 450 nm) at 25°C is ~5 ns.[134] That for tryptophan is 3 ns.

Phase fluorometers utilize continuous irradiation by a beam of light that is sinusoidally modulated. If the frequency of the modulation is set correctly, there will be a phase difference in the modulation of the fluorescent emission that will depend upon τ. Phase fluorometry can yield the same information as does pulse fluorometry.[127,132,133] By using two or more modulation frequencies the decay rates and fluorescence lifetimes for various chromophores in a protein can be observed. For example, the protein **colicin A** (Box 8-D) contains three tryptophans W86, W130, and W140. Their fluorescence decays with lifetimes τ_1, τ_2, τ_3 of ~0.6–0.9 ns, 2.0–2.2 ns, and 4.2–4.9 ns at pH 7. While τ_3 originates mainly from W140, both of the other tryptophans contribute to τ_1 and τ_2. Changes in fluorescence intensity with pH reflect a pK_a value of 5.2.[135] Tryptophan, which often occurs at only one or a few places in a protein, is a useful fluorescent probe for study of protein dynamics. The optical properties of 7-azatryptophan, 2-azotryptophan, and 5-hydroxytryptophan are even better because their absorption maxima occur at longer wavelengths. These amino acids can be biosynthetically introduced in place of tryptophan in proteins.[136–138] The maximum fluorescence of tryptophan in one protein is at 350 nm, but for 7-azatryptophan in the same protein it was shifted to 380 nm.[136]

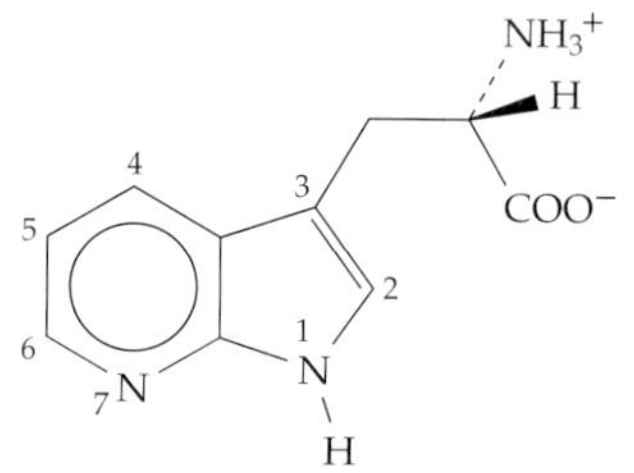

7-Azatryptophan

Triplet states, phosphoresence, and quenching. In addition to emitting fluorescent radiation, molecules can often pass from the excited singlet state to a lower energy **triplet state**, in which two electrons are now unpaired and the molecule assumes something of the character of a diradical (see Fig. 23-14). This process, known as **intersystem crossing**, competes directly with fluorescence and shortens the fluorescence lifetime. The triplet state is long-lived (e.g., for tryptophan in water at 20°C it is 1.2 ms[139]) and is responsible for much of the photochemical behavior of molecules. It also gives rise to the delayed light emission known as **phosphorescence**, as is illustrated in Figure 23-14. Other processes that compete with fluorescence are **photochemical reactions** of the singlet excited state and **internal conversion**. The latter is the process by which a molecule moves from the lowest vibrational state of the upper electronic level to some high vibrational state of the unexcited electronic level. This is the principal means of depopulating the electronic state and competes directly with fluorescence.

The rate of relaxation by nonradiative pathways can be increased by addition of **quenchers**. Quenching of fluorescence occurs by several mechanisms, many of which involve collision of the excited chromophore with the quenching molecule. Some substances such as iodide ion are especially effective quenchers. The fluorescence efficiency of a substance in the absence of a quencher can be expressed (Eq. 23-16) in terms of the rate constants for fluorescence (k_f), for nonradiative decay (k_{nr}), and for phosphorescence (k_p):

$$\phi_f = k_f / (k_f + k_{nr} + k_p) \qquad (23\text{-}16)$$

In the presence of a quencher, Q, there is an additional rate process for relaxation. The ratio of the fluorescence efficiency in the absence of (ϕ_f^0) and the presence of a quencher is given by the **Stern–Volmer equation**.[7,140]

$$\phi_f^0 / \phi_f = 1 + K[Q] = 1 + k_Q \tau_0 [Q] \qquad (23\text{-}17)$$

The constant K is known as the Stern–Volmer quenching constant; k_Q is the rate constant for the quenching reaction, and τ_0 the lifetime in the absence of quencher. Fluorescence quenching of tryptophan in proteins by acrylamide or O_2 has been used to determine whether tryptophan side chains are accessible to solvent or are "buried" in the protein.[141,142] The long-lived phosphorescence of tryptophan can be studied in a similar way.[143]

A recent application of fluorescence quenching is the development of "**molecular beacons**" for detection of viruses such as the AIDS viruses HIV-1 and HIV-2.[144] A single-stranded oligonucleotide is synthesized with a 25- or 33-nucleotide sequence complementary to a sequence in the target viral RNA. At the ends of this sequence are two 6-nucleotide arms with complementary sequences that will form a stable double-helical stem at the annealing temperature used for PCR amplification of the viral nucleic acid. The end of one arm carries a covalently bonded fluorescent dye, e.g., a fluorescein or rhodamine derivative. The other arm carries a potent covalently linked fluorescence quencher such as 4-(4'-dimethylaminophenylazo)benzoic acid. When the arms form a duplex, the quencher will be next to the fluorophore and no fluorescence will be seen upon irradiation with light of a suitable exciting wavelength. However, if viral DNA is present it will hydridize with the central polynucleotide, keeping the fluorophore and quenchers far apart

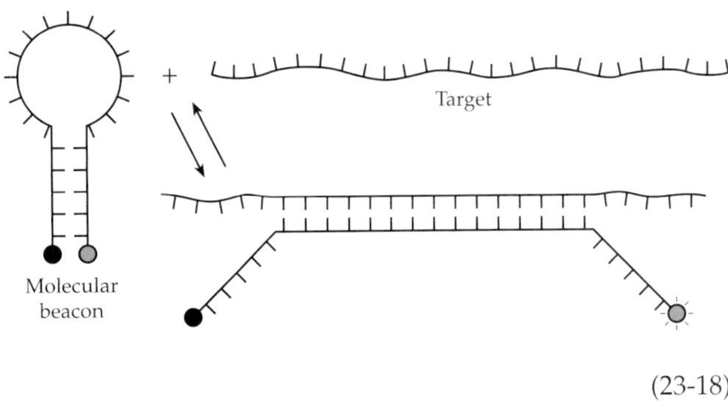

(23-18)

and allowing the beacon to signal the presence of a virus (Eq. 23-18). As few as ten retroviral genomes could be detected. By using a series of molecular beacons with different colored fluorescence and specific for different viruses, it is possible to test for more than one virus simultaneously.

Anisotropy. Light emitted from excited molecules immediately after absorption is always partially polarized, whether or not the exciting beam consists of plane polarized light. When light polarized in a vertical plane is used for excitation, part of the emitted light (of intensity I_v) will have its electric vector parallel to that of the exciting light. The remainder of intensity I_h, will be polarized in a horizontal plane. The **polarization *P*** of the emitted radiation is defined by Eq. 23-19 and the **anisotropy *R*** by Eq. 23-20. After excitation by a laser pulse both the fluorescence and its anisotropy decay with time and can be measured. The decay of R (but not of P) can usually be described as the sum

$$P = (I_v - I_h) / (I_v + I_h) \qquad (23\text{-}19)$$

$$R = (I_v - I_h) / (I_v + 2\, I_h) \qquad (23\text{-}20)$$

of simple exponential curves, which are readily obtained by phase fluorometry. These can in turn be related to specific types of motion, such as rotation of the emitting molecule or group.[7,145,146] Rotation of tryptophan rings, both free and restricted, has been studied in a variety of proteins.[145,147] However, interpretation is difficult.[130,148] The rotational rates obtained from anisotropy measurements are strongly affected by the viscosity of the medium (see Eq. 9-35).

2. Fluorescence Resonance Energy Transfer (FRET)

Electronic excitation of one chromophore sometimes elicits fluorescence from a different chromophore that is located nearby. For example, excitation of a monomolecular layer of dye can induce fluorescence in a layer of another dye spaced 5 nm away. Excitation of tyrosine residues in proteins can lead to fluorescence from tryptophan, and excitation of tryptophan can cause fluorescence in dyes attached to the surface of a protein or in an embedded coenzyme.[134] Such fluorescence resonance energy transfer (**FRET**) is expected for molecules, when the fluorescence spectrum of one overlaps the absorption spectrum of the other. The mechanism is not one of fluorescence emission and absorption but of nonradiative resonant transfer of energy. Resonant transfer of energy is of major biological significance in photosynthesis (Section E). Most of the chlorophyll molecules, which absorb light in a chloroplast, transfer the absorbed energy in a stepwise fashion to a **reaction center**.

Förster[149] calculated that the rate of energy transfer k_t should be proportional to the rate of fluorescence k_f, to an orientation factor K^2, to the spectral overlap interval J, to the inverse fourth power of the refractive index n, and to the inverse sixth power of the distance r separating the two chromophores.

$$k_t \propto k_f K^2 J n^{-4} r^{-6} \tag{23-21}$$

Besides predicting the inverse sixth power dependence of energy transfer, Förster provided a formula for calculating R_0, the distance between chromophores at which 50% efficient singlet-singlet energy transfer takes place. R_0 is commonly of the order of 2.0 nm. Making use of these relationships, Stryer proposed a method of measuring distances between chromophores. He calibrated the method by constructing a series of molecules containing various lengths of the rigid threefold polyproline helix to which dansyl groups were attached at one end and naphthyl groups

$(CH_3)_2N$ — Dansyl — L-Prolyl (n = 1-12) — Naphthyl

at the other.[150] By exciting the naphthyl group, which has the higher energy absorption band and is strongly fluorescent, the characteristic lower energy emission of the dansyl group could be observed if energy transfer took place. Since the fluorescent emission band of the naphthyl group overlaps the absorption band of the dansyl group, efficient transfer was expected. The results of a plot of transfer efficiency against distance is shown in Fig. 23-15. The inverse sixth power dependence was followed quite accurately with a value of R_0 ~3.4 nm. Having calibrated his "spectroscopic ruler," Stryer turned his attention to biochemical macromolecules. Attaching the same kinds of fluorescent probe to the visual light receptor rhodopsin, Wu and Stryer were able to estimate distances between specific parts of the molecule and to draw some conclusions about the overall shape.[151]

More recently the FRET technique has been widely applied to a broad range of biochemical problems. Sensitivity has been improved to the extent that fluorescence of single molecules can be detected.[152–154] Use of **terbium** (Tb^{3+}) or **europium** (Eu^{3+}) ions, which can provide luminous labels for metal-binding sites, has provided another advance. These ions absorb light poorly and are therefore only weakly fluorescent. However, they can be excited by resonance energy transfer and become brilliantly luminous. This **luminescence resonance energy transfer** (**LRET**) is a variant of FRET, which allows distances up to ~10 nm to be measured.[155–157] Another advance is the ability to graft into specific proteins fluorescent tags such as the intact **green fluorescent protein** (Section J)[158,159] or an amino acid sequence such as CCXXCC in which the four –SH groups of the cysteines serve to trap an arsenic derivative of fluorescein (see Box 12-B).[159] **Confocal laser scanning microscopy** (Chapter 3) is basic to many applications.[160]

Specific applications of FRET and LRET include observation of myosin movement (Fig. 19-14),[157] measurement of distances between binding sites on tubulin,[161] determining stoichiometry of subunit assembly in a γ-aminobutyrate receptor of brain,[162] association of proteins in peroxisomes,[160] study of hydridization of deoxyribonucleotides,[163] verifying the handedness of various forms of DNA,[164] and other studies of DNA and RNA.[164a,b]

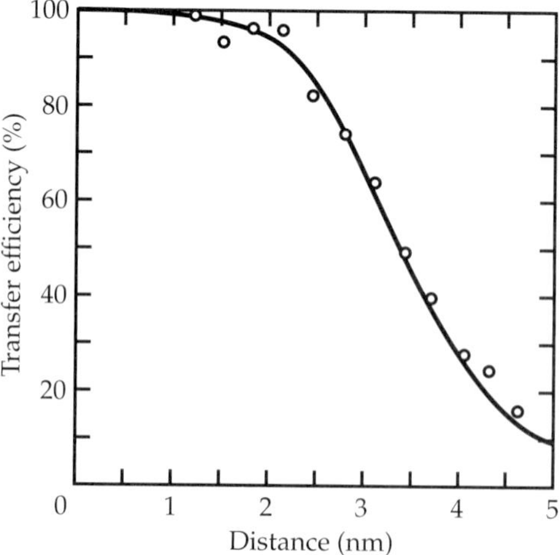

Figure 23-15 Efficiency of energy transfer as a function of distance between α-naphthyl and dansyl groups at the ends of a polyprolyl "rod" (L-prolyl)$_n$. The observed efficiencies of transfer for n = 1 to 12 are shown as points. The solid line corresponds to an r^{-6} distance dependence. From L. Stryer.[150]

Using the Förster equation the distance between the two calcium-binding sites in parvalbumin (Fig. 6-7) has been estimated by energy transfer from Eu(III) in one site to Tb(III) in the other[165] to within 10–15% of the distance of 1.18 nm based on X-ray crystallography.

3. Energy-Selective Spectroscopic Methods

At low enough temperatures vibrational fine structure of aromatic chromophores may be well resolved, especially if they are embedded in a suitable matrix such as argon or N_2, which is deposited on a transparent surface at 15 K. This **matrix isolation spectroscopy**[77,166] may reveal differences in spectra of conformers or, as in Fig. 23-16, of tautomers. In the latter example the IR spectra of the well-known amino-oxo and amino-hydroxy tautomers of cytosine can both be seen in the matrix isolation IR spectrum. Figure 23-16 is an IR spectrum, but at low temperatures electronic absorption spectra may display sharp vibrational structure. For example, aromatic hydrocarbons dissolved in *n*-heptane or *n*-octane and frozen often have absorption spectra, and therefore fluorescence excitation spectra, which often consist of very narrow lines. A laser can be tuned to excite only one line in the absorption spectrum. For example, in the spectrum of the carcinogen 11-methylbenz(*a*)anthrene in frozen octane three major transitions arise because there are three different environments for the molecule. Excitation of these lines separately yields distinctly different emission spectra.[77] Likewise, in complex mixtures of different hydrocarbons emission can be excited from each one at will and can be used for estimation of amounts. Other related methods of energy-selective laser spectroscopy include **fluorescence line narrowing**[167] and **spectral hole burning**.[167,168]

4. Analytical Applications of Fluorescence

Because of the high sensitivity with which fluorescence can be detected, its measurement is important as an analytical tool. As a result of improved techniques **fluorescence microscopy** has become one of the most important of all tools in biological studies.[168a] New types of microscopes (see also pp. 129–131) have increased resolution beyond what was thought possible.[168b–d] Studies such as those of lipid metabolism in the transparent zebrafish are possible using substrates that carry fluorescent labels.[168e] As mentioned in the preceding paragraph, many aromatic compounds can be detected by their fluorescence. The relatively weak fluorescence of proteins and nucleic acids can be greatly enhanced by the binding of a highly fluorescent dye to the macromolecules. Fluorescent antibodies are widely employed for this purpose. Fluorescent labels are rapidly replacing radioisotopes in analysis of nucleic acids. For example, biotin may be attached to a pyrimidine base of a nucleoside triphosphate by a long spacer arm. The modified base can then be incorporated enzymatically into polynucleotides, e.g., in the synthesis of probes used for hybridization. The attached biotin can be detected by binding to avidin or streptavidin (Box 14-B) and use of fluorescent antibodies to this

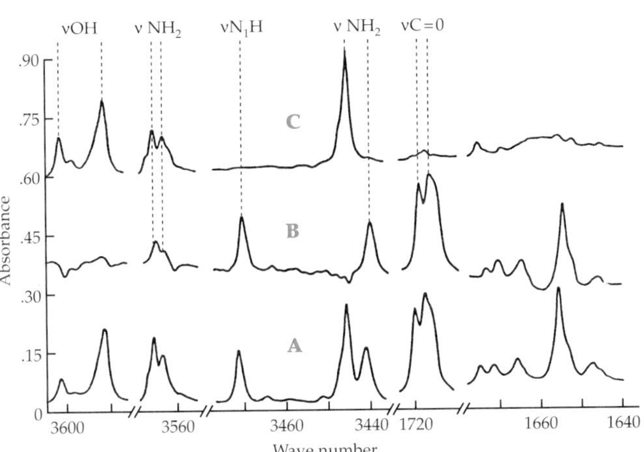

Fig. 23-16 Infrared spectrum of cytosine in the –OH, NH, and C=O stretching region in an argon matrix in 15 K. (A) Initial spectrum of a sample containing both tautomers. (B) Spectrum of the amino-oxo tautomer calculated by substraction of the spectrum of the amino-hydroxy tautomer. (C) Spectrum of the aminohydroxy tautomer obtained by irradiating the initial sample in order to photochemically convert it into the aminohydroxy tautomer. Spectra (B) and (C) were also corrected for residual amounts of the second tautomer. From Szczesniak *et al.*[166] Courtesy of W. B. Person.

protein.[169,170] Fluorescent dyes can also be covalently attached to nucleotides. Fluorescent dideoxynucleoside triphosphates are used as chain terminators in DNA sequencing (Chapter 5). Using a different dye that fluoresces at a different wavelength for each of the four dideoxynucleosides, polynucleotides can be sequenced automatically using a single column rather than four parallel lanes as in Fig. 5-49.

Fluorescent "probes" such as **1-anilinonapthalene-8-sulfonate** or 1,6-diphenyl-1,3,5-hexatriene embedded in membranes, contractile fibers, etc., can reveal changes in mobility that accompany alterations in physiological conditions. For example, molecular changes occurring in membranes during nerve conduction and in mitochondria during electron transport can be observed.[146,171] Another type of probe is exemplified by 2'-(*N*,*N*-dimethylamino)-6-4-*trans*-cyclohexamoic (DANCA). Its emission maximum shifts from 390 nm in cyclohexane to 520 nm in water, presumably as a consequence of increased polarization of the molecule in the excited state, as indicated by the green arrows in the accompanying structural formula.[172] DANCA can be used to obtain some idea of the polarity of sites within macromolecules to which it binds.

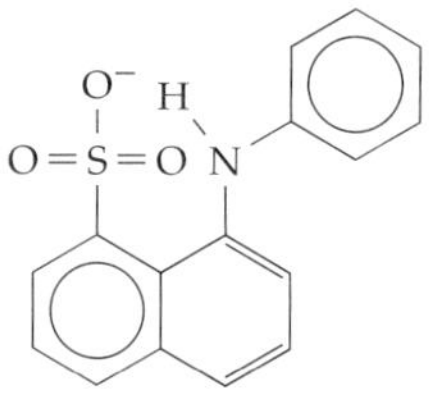

1-Anilinonaphthalene-8-sulfonate

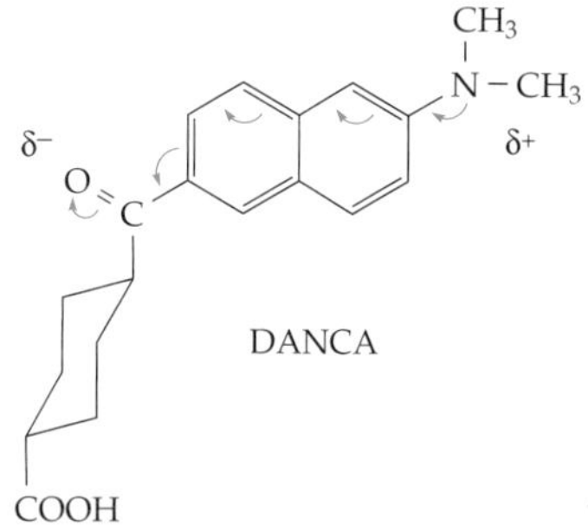

DANCA

Study of calcium ions in living cells has been immensely aided by calcium fluorophores[173] (Box 6-D), which are often derivatives of EDTA (Table 6-9). One of these, **calcein**,[174] is also very specific toward Fe^{3+}. The natural calcium-dependent luminous protein, aequorin (Section J), is also widely used.

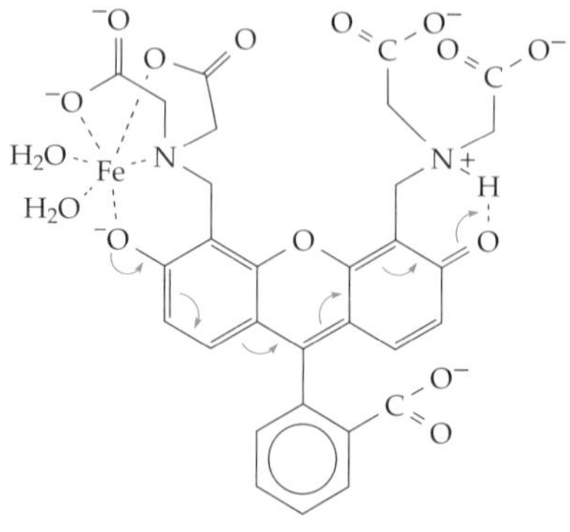

Fe^{III} - calcein • H^+

Nontoxic but highly fluorescent dyes are used to study diffusion within and between cells. The so called Lucifer dyes have been employed to trace shapes and branching patterns in neurons.[175]

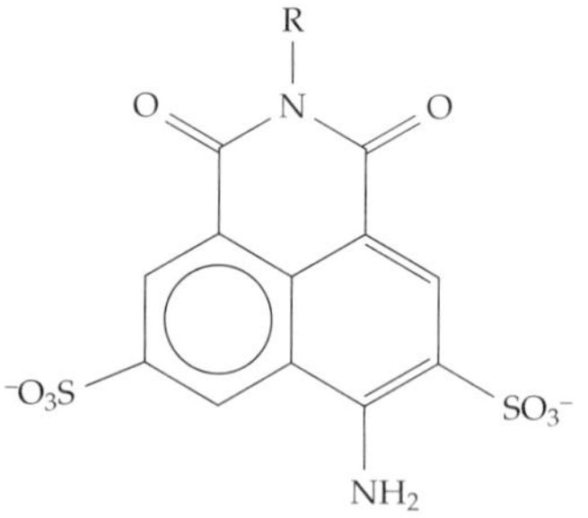

Structural formula for Lucifer dyes
For Lucifer yellow
R = —NH—CO—NH—NH_2

The recently developed **fluorescence correlation spectroscopy** permits studies of molecular association in one femtoliter of solution using a confocal or two-photon microscope. Two lasers are used to excite two fluorophores of different colors, each one on a different type of molecule. Fluorescence of single molecules can be detected, and molecular associations can be detected by changes in the distribution of the fluctuations in fluorescence intensity caused by Brownian motion.[176–178] A different type of advance is development of computer programs that analyze chromosomes stained with a mixture of dyes with overlapping spectra and display the result as if each chromosome were painted with a specific color.[179–180a] Yet another advance is development of **semiconductor nanocrystals** (or "quantum dots") with narrow absorption bands and intense fluorescence. The wavelength of absorption and fluorescence depends upon the size of the crystals. For example, CdSe crystals of diameter 2–5 nm coated with silica or with a surfactant fluoresce across the visible range. They have a variety of uses in biological staining.[181,182]

D. Photochemistry

Because of their high energy, molecules in either the singlet or triplet photoexcited state undergo a greater variety of chemical reactions than do molecules in the ground state.[5,183,184] Many of these photochemical reactions arise from the triplet state that is formed from the singlet by intersystem crossing. Selection rules forbid transitions between excited triplet state and ground state; therefore, the radiative lifetime of the triplet state is long. The diradical character of the triplet state also makes it unusually reactive. Despite its forbidden character, nonradiative deexcitation of the triplet state is possible, and phosphorescence is observed for most molecules at low temperatures if the solvent is immobilized as a glass. The intense light from lasers can also induce a variety of photochemical processes that arise from absorption of two or more photons.

One of the simplest of photochemical processes is the dissociation or uptake of a proton by an excited molecule as a result of a change in the pK_a value of a functional group. Various other bond cleavages may lead to dissociation into ions or radicals. Photoelimination and photoaddition reactions both occur. Molecules may be isomerized, a process of importance in visual receptors. Excited molecules may become strong oxidizing agents able to accept hydrogen atoms or electrons from other molecules. An example is the **photooxidation** of EDTA by riboflavin (which undergoes photoreduction as shown in Fig. 15-8). A biologically more important example is in photosynthesis, during which excited chlorophyll molecules carry out **photoreduction** of another molecule and are themselves transiently oxidized. A frustrating aspect of investigation of photochemical reactions is that the variety of reactions possible often leads to a superabundance of photochemical products, e.g., see the thin layer chromatogram of cleavage products of riboflavin in Fig. 3-5. However, biological photoprocesses are usually much more specific.

1. Chemical Equilibria in the Excited State

When pyridoxamine with a dipolar ionic ring structure (Fig. 14-9) and an absorption peak at 30,700 cm^{-1} (326 nm) is irradiated, fluorescence emission is observed at 25,000 cm^{-1} (400 nm). When basic pyridoxamine with an anionic ring structure and an absorption peak at 32,500 cm^{-1} (308 nm) is irradiated, fluorescence is observed at 27,000 cm^{-1} (370 nm), again shifted ~5500 cm^{-1} from the absorption peak. However, when the same molecule is irradiated in acidic solution, where the absorption peak is at 34,000 cm^{-1} (294 nm), the luminescent emission at 25,000 cm^{-1} is the same as from the neutral dipolar ionic form and abnormally far shifted (9000 cm^{-1}) from the 34,000 cm^{-1} absorption peak.[185,186] The phenomenon, which is observed for most phenols, results from rapid dissociation of a proton from the phenolic group in the photoexcited state. A phenolic group is more acidic in the excited state than in the ground state, and the excited pyridoxamine cation in acid solution is rapidly converted to a dipolar ion.

The variation of fluorescence intensity with pH can provide direct information about the pK_a in the excited state. Förster suggested the following indirect procedure for estimating excited-state pK_a values for phenols. Let E_1 represent the energy of the 0–0 transition (preferably measured as the mean of the observed 0–0 transition energies in absorption and fluorescent emission spectra); let E_2 represent the energy of the 0–0 transition in the dissociated (anionic in the case of a phenol) form, while ΔH and ΔH^* represent the enthalpies of dissociation in the ground and excited states, respectively. It is evident from the diagram that Eq. 23-22 holds.

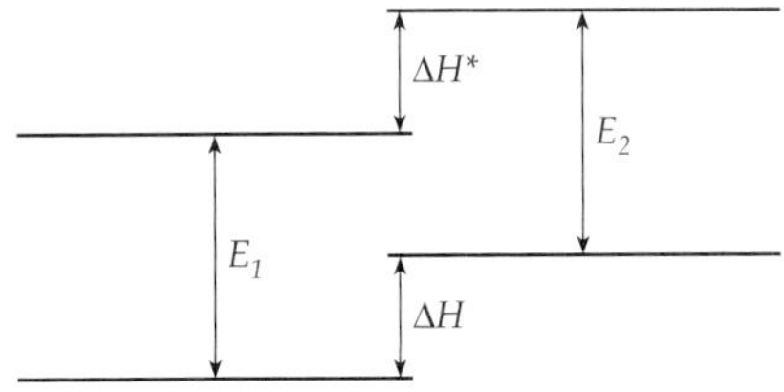

$$E_1 - E_2 = \Delta H - \Delta H^* \quad (23\text{-}22)$$

If we assume that the changes in entropy for the reaction are the same in the ground and excited state, Eq. 23-23 follows.

$$\log_{10}(K^*/K) = Nh\,(\Delta\,\bar{\nu})/(2.3RT)$$

or

$$pK^* = pK - (2.1 \times 10^3)\Delta\nu\ (cm^{-1})\ \text{at}\ 25°C \quad (23\text{-}23)$$

For example, a shift in the spectrum of the basic form by 1000 cm^{-1} to a lower wave number compared with the acid form corresponds to a decrease of 2.1 units in pK_a for dissociation of the acid form. Whereas it is desirable to use both absorption and fluorescent measurements to locate the approximate positions of 0–0 bands, absorption measurements alone are often used, and the positions of the band maxima are taken. Thus, for pyridoxamine the shift in absorption maximum from 34,000 cm^{-1} in the protonated form to 30,700 cm^{-1} in the dissociated form suggests that the pK_a of pyridoxamine of 3.4 in the ground state is shifted by 6.9 units to −3.5 in the excited state. Bridges *et al.*[186] evaluated this same pK_a from the pH dependence of fluorescence as pK^* ~ −4.1.

While phenols and amines are usually more acidic in the singlet excited state than in the ground state, some substances, e.g., aromatic ketones, may become more *basic* in the photoexcited state.

Observation of an abnormally large shift in the position of fluorescent emission of pyridoxal phosphate (PLP) in glycogen phosphorylase answered an interesting chemical question.[187,188] A 330 nm (30,300 cm^{-1}) absorption band could be interpreted either as arising from an adduct of some enzyme functional group with the Schiff base of PLP and a lysine side chain (structure A) or as a nonionic tautomer of a Schiff base in a hydrophobic environment (structure B, Eq. 23-24). For structure A, the fluorescent emission would be expected at a position similar to that of pyridoxamine. On the other hand, Schiff bases of the

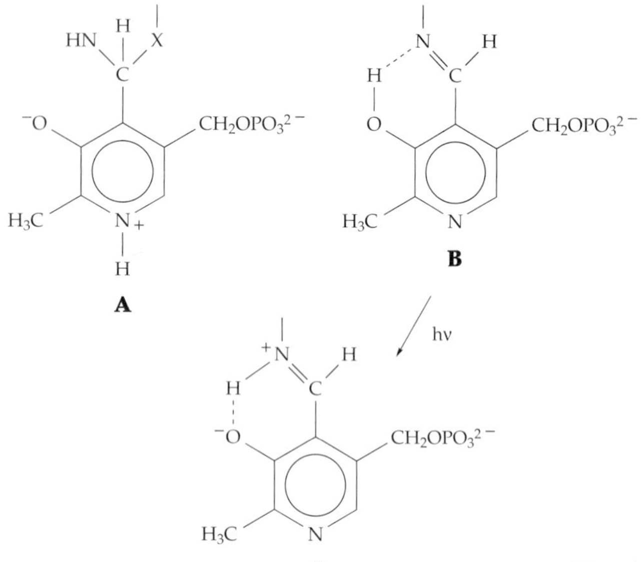

A B C (23-24)

type indicated by structure B would be expected to undergo a photoinduced proton shift (phototautomerization) to form structure C[124,187] with an absorption band at 430 nm (23,300 cm^{-1}) and fluorescent emission at a still lower energy. Since the observed fluorescence was at 530 nm, it was judged that the chromophore does have structure B.

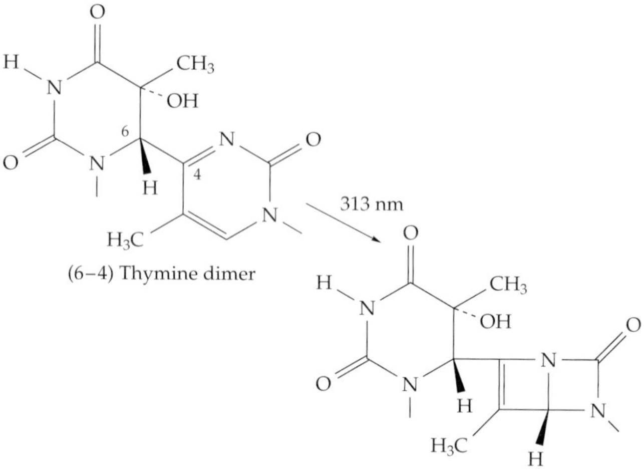

The rate of proton dissociations from the excited states of molecules can be measured directly by nanosecond fluorimetry.[189]

2. Photoreactions of Nucleic Acid Bases

Photochemical reactions of the purines and pyrimidines assume special significance because of the high molar extinction coefficients of the nucleic acids present in cells. Light is likely to be absorbed by nucleic acids and to induce photoreactions that lead to mutations.[190] Both pyrimidines and purines undergo photochemical alterations, but purines are only about one-tenth as sensitive as pyrimidines. **Photohydration** of cytidine (Eq. 23-25) is observed readily. The reaction is the photochemical analog of the hydration of α,β-unsaturated carboxylic acids. Uracil derivatives also undergo photohydration.

A more important reaction is the photodimerization of thymine (Eq. 23-26), a reaction also observed with uracil. A variety of stereoisomers of the resulting

(23-25)

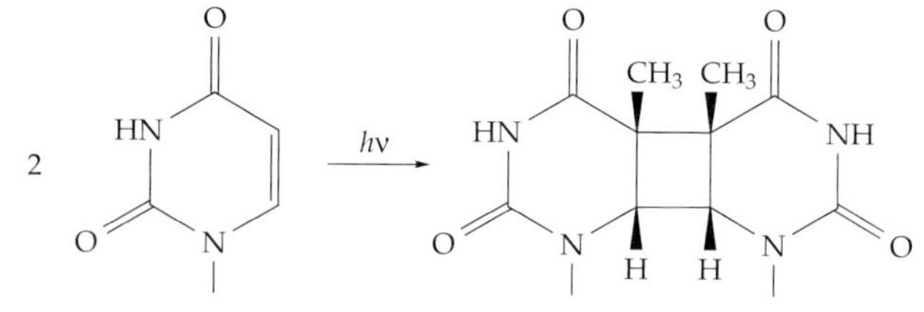

Thymine dimer (23-26)

(6–4) Thymine dimer

313 nm

Dewar product

cyclobutane-linked structure are formed. The one shown in the equation predominates after irradiation of frozen thymine. Another important isomer is the 6–4 thymine dimer.[191,191a] Both of these types of **cyclobutane dimers** block DNA replication. This accounts for much of the lethal and mutagenic effect of ultraviolet radiation on organisms. The matter is sufficiently important that a special "excision repair" process is used by cells to cut out the thymine dimers (Chapter 27, Section E). In addition, light-dependent **photolyases**, discussed in Section I, act to reverse the dimerization reactions.

Light can also cause addition and other crosslinking reactions between DNA and proteins or other cell constituents.[192,193] One use of such reactions in the laboratory is DNA "photo footprinting" (Fig. 5-50), a technique which reveals contact regions between DNA and associated proteins.[190] Another type of cellular damage is caused by photosensitization of DNA by a light-absorbing intercalating agent such as **8-methoxypsoralen**.[194,195] DNA as well as adjacent proteins can be damaged.

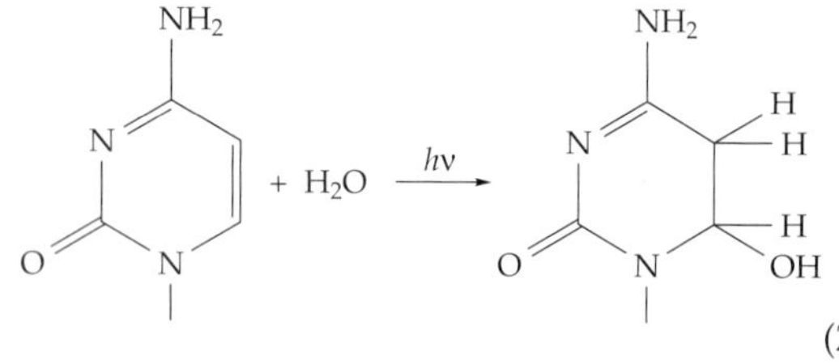

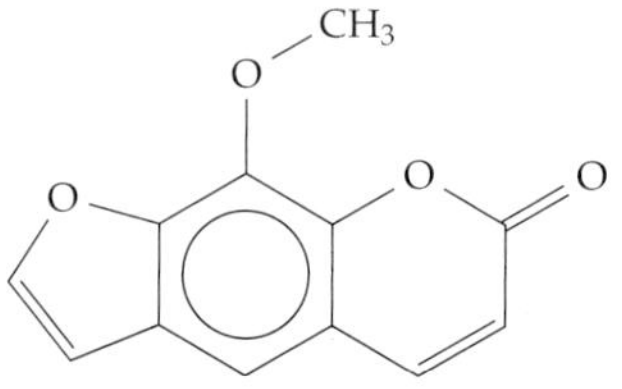

8-Methoxypsoralen

3. Sunburn, Cancer, and Phototherapy

Ultraviolet light is sometimes classified according to its energy and capacity for damaging cells as follows:[195a]

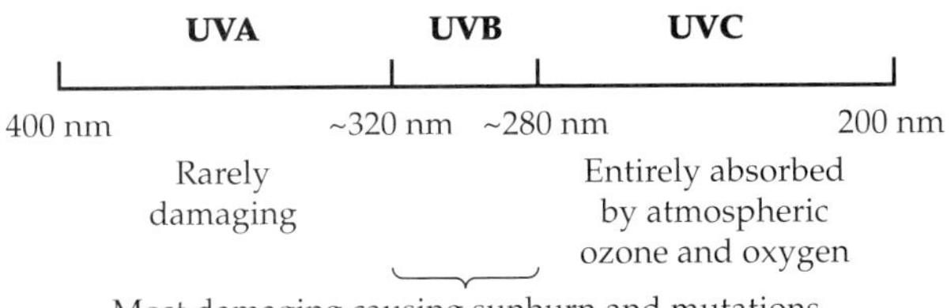

Most damaging causing sunburn and mutations.
Most is absorbed by the ozone layer.

The UVB radiation causes most damage to skin. UVA radiation is at least an order of magnitude less damaging and is usually harmless to human skin. The UVC solar radiation is all absorbed by ozone and dioxygen of the atmosphere. UVC radiation produced by ultraviolet lamps is usually all absorbed in the epidermis.[5]

Ultraviolet light damages proteins as well as DNA. Residues of Trp, Tyr, His, Cys, and Met are especially susceptible to photolysis, or photooxidation by O_2, or by singlet oxygen. Also damaged are unsaturated lipids, porphyrins, flavins, etc. Kynurenic acid (Fig. 25-11) and urocanic acid (Eq. 14-44), an important ultraviolet filter in skin,[196] are also decomposed by light.

Repeated sunburn ages skin and may induce cancer.[4] However, light also has beneficial effects. It allows us to see, provides a source of vitamin D (Box 22-C), induces enzymatic repair of some DNA damage, and provides all of our food energy, directly or indirectly, by photosynthesis.

Light is used in **phototherapy**. This happens most frequently in the irradiation of newborn babies with white light to isomerize bilirubin (Fig. 24-24) from the 4*Z*, 15*Z* form to more readily excreted forms such as the 4*E*, 15*Z* isomer. About half of newborns have some **jaundice** (elevated bilirubin), and if it is severe it must be treated promptly to avoid neurological damage (see Chapter 24). There are sometimes complications, but the treatment is usually effective.[5] **Psoriasis** is frequently treated by irradiation with UVB light, which is thought to inhibit growth of the abnormal skin cells. Ten to 35 treatments are usually required, and the condition may return after some years. An alternative treatment is irradiation with UVA light after ingestion of 8-methoxypsoralen or other psoralen derivative.[5] Another skin condition that may respond to phototherapy is **vitiligo** (Box 25-A). Treatment with UVA and psoralen derivatives may stimulate repigmentation. If depigmentation is extreme, the remaining pigmentation may be reduced by bleaching with the monobenzyl ester of hydroquinone.

Photodynamic therapy is a cancer treatment that involves intravenous injection of a light-absorbing molecule such as a porphyrin, which may be taken up preferentially by cancer cells. Laser irradiation by deeply penetrating red light (650–800 nm wavelength) causes oxygen-dependent photosensitization.[5,197,198] Improvements in lasers, in fiber optics, and in photosensitizers may lead to widespread use of this type of therapy both for cancer and for some other conditions.[5,199]

$$\text{Arylazide } (C_6H_5{-}N_3) \xrightarrow[-N_2]{h\nu} \text{Aryl nitrene } (C_6H_5{-}\ddot{N}:) \xrightarrow{R{-}C(R')(R'')H} R{-}C(R')(R''){-}\ddot{N}H{-}C_6H_5 \qquad (23\text{-}27)$$

4. Photoaffinity Labels

Photochemically reactive molecules have often been used as labels for specific sites in proteins and nucleic acids. Psoralen derivatives serve as relatively nonspecific photochemically activated crosslinking agents for DNA and double-stranded RNA.[195] **Aryl azides** are converted by light to aryl nitrenes, which react in a variety of ways including insertion into C–H bonds (Eq. 23-27).[200,201] In some cases UV irradiation can be used to join natural substrates to enzymes or hormones to receptors. For example, progesterone, testosterone, and other steroids have been used for direct photoaffinity labeling of their receptors.[202] Synthetic **benzophenones** have also been used widely as photoactivated probes.[203]

5. Microphotolysis and Ultrafast Light-Induced Reactions

Fluorescence microphotolysis, or photobleaching, has been widely used to study translational mobility of lipids and proteins in membranes. An attenuated laser beam may be focused down to the diameter of a cell or less. Then the intensity can be suddenly increased by several orders of magnitude, bleaching any fluorescent material present. The return of fluorescent material by free diffusion from a neighboring region (**fluorescence recovery after photobleaching**) or by diffusion through a membrane into a cell can then be observed.[204,205] Diffusion coefficients of labeled biopolymers or of components of cells can be evaluated, and translation and metabolism of lipids and other components can be followed.[206]

Laser-based techniques are being used for ultrafast observation of the results of a photochemical process, e.g., the light-induced dissociation of CO from the hemoglobin • CO complex. A dissociating laser pulse can be as short as 100 fs (0.1 ps) or less. This is shorter than the time of vibrational motion of nuclei in an electronically excited state (~0.3 ps). Using IR spectroscopy, events that follow can be observed at intervals as short as 0.1 ns.[207] X-ray diffraction measurements using 150 ps pulses have allowed direct observation of the CO dissociated from hemoglobin or myoglobin and its recombination with the same protein.[208–210] Femtosecond dynamics of electron transfer along a DNA helix is also being studied.[211,212]

6. Optical Tweezers, Light-Directed Synthesis, and Imaging

The radiation pressure exerted by light is very weak. A bright laser beam of several milliwatts of power can exert only a few piconewtons (pN) of force. However, a force of 10 pN is enough to pull a cell of *E. coli* through water ten times faster than it can swim.[213] In about 1986, it was found that a laser beam focused down to a spot of ~ one λ (~1 μm for an infrared laser) can trap and hold in its focus a refractile bead of ~1 μm diameter. This "optical tweezers" has become an important experimental tool with many uses.[213,214] For example, see Fig. 19-19. Not only are optical tweezers of utility in studying biological motors but also mechanical properties of all sorts of macromolecules can be examined. For example, DNA can be stretched and its extensibility measured.[215] Actin filaments have even been tied into knots![216]

Light-directed solid state synthesis of peptides[217] and oligonucleotides is another new tool.[218] Development of this technology may provide new advances in preparation of DNA chips (Chapter 26) with a higher density of components than are now available.

X-rays and more recently NMR (MRI)-imaging have become well known to us. It might seem impossible to use visible light for a similar type of imaging. However, a laser beam can pass through a person's head. Is it possible, using computer-based techniques, to create an image from the emergent light? Efforts are being made to do exactly this.[219,220] **Optical coherence tomography** using backscattered infrared light and related fast techniques have higher resolution than MRI, computerized tomography (CT), or ultrasound.[220–221a] Ultrabright synchrotron radiation is also being used in **infrared microspectrophotometry**.[222,223]

E. Photosynthesis

The photochemical reduction of CO_2 to organic materials[224–228] is the basic source of energy for the biosphere. Nevertheless, the process is limited to a few genera of photosynthetic bacteria (Table 1-1), eukaryotic algae, and higher green plants. Photosynthetic bacteria include the distinctly different purple, green, and bluegreen (cyanobacteria) groups, each of which has a different array of photosynthetic pigments. However, the basic mechanism of transduction of solar energy into chemical energy is the same in all of the bacteria and in green plants.

As discussed in Chapter 17, photosynthesis involves the incorporation of CO_2 into organic compounds by reduction with NADPH with coupled hydrolysis of ATP. This is most often via the Calvin–Benson cycle of Fig. 17-14. In a few organisms a reductive tricarboxylic acid cycle is employed. The idea that the chloroplasts of plants or the pigmented granules of photosynthetic bacteria generate NADPH or reduced ferredoxin plus ATP (Chapter 17) is now thoroughly accepted. However, it was not always obvious. Consider the overall equation (Eq. 23-28) for formation of glucose by photosynthesis in higher plants:

$$6\,CO_2 + 6\,H_2O \rightarrow 6\,O_2 + C_6H_{12}O_6 \quad (23\text{-}28)$$

$$6\,CO_2 + 12\,H_2O^* \rightarrow 6\,O_2^* + C_6H_{12}O_6 + 6\,H_2O \quad (23\text{-}29)$$

The stoichiometry of the reaction suggests that all 12 of the oxygen atoms of the evolved O_2 might come from CO_2 or that some might come from CO_2 and some from H_2O. In fact, water supplies both of the oxygen atoms needed for formation of O_2, as is indicated by the asterisks in Eq. 23-29. This possibility was suggested by van Niel[229,230] in 1931. He pointed out that in bacterial photosynthesis no O_2 is produced, and that bacteria must have access to a reducing agent to provide hydrogen for the reduction of CO_2 (Eq. 23-30).

$$H_2A \xrightarrow{h\nu} A + 2[H] \quad (23\text{-}30)$$

In this equation, H_2A might be H_2S (in the purple sulfur bacteria), elemental H_2, isopropanol, etc. From a consideration of these various reactions, van Niel concluded that in the O_2-producing cyanobacteria and eukaryotic plants water serves as the oxidizable substrate in Eq. 23-30 and is cleaved to form O_2 and to provide hydrogen atoms for reduction. This photochemical cleavage is the only known biological oxidation reaction of H_2O. No oxidizing agents present in living things are powerful enough to dehydrogenate water except for the photochemical **reaction centers** of photosynthetic organisms.

1. Two Photosystems, the Z Scheme, and Reaction Centers

It had long been known that for green plants light of wavelength 650 nm was much more efficient than that of 680 nm. However, Emerson and associates[34] in 1956 showed that a combination of light of 650 nm *plus* that of 680 nm gave a higher rate of photosynthesis than either kind of light alone. This result suggested that there might be two separate photosystems. What is now known as **photosystem I** (PSI) is excited by far red light (~700 nm), while **photosystem II** (PSII) depends upon the higher energy red light of 650 nm. Additional evidence supported the idea. Hill had shown many years before[231] that mild oxidizing agents such as ferricyanide and benzoquinone can serve as substrates for photoproduction of O_2, while Gaffron[232] found that some green algae could be adapted to photooxidize H_2 to protons (Eq. 23-30) and to use the electrons to reduce NADP. Thus, photosystem I could be disconnected from photosystem II. The powerful herbicide **dichlorophenyldimethylurea** (DCMU) was found to block electron transport between the two photosystems. In the presence of DCMU electrons from such artificial donors as ascorbic acid or an indophenol dye could be passed through photosystem I.

3-(3,4-Dichlorophenyl)-1,1-dimethylurea

Atrazine, another PSII inhibitor

The Z scheme. The result of these and other experiments was the development of the series formulation or zigzag scheme of photosynthesis[233] which is shown in Fig. 23-17. Passage of an electron through the system requires two quanta of light. Thus, four quanta are required for each NADPH formed and eight quanta for each CO_2 incorporated into carbohydrate.

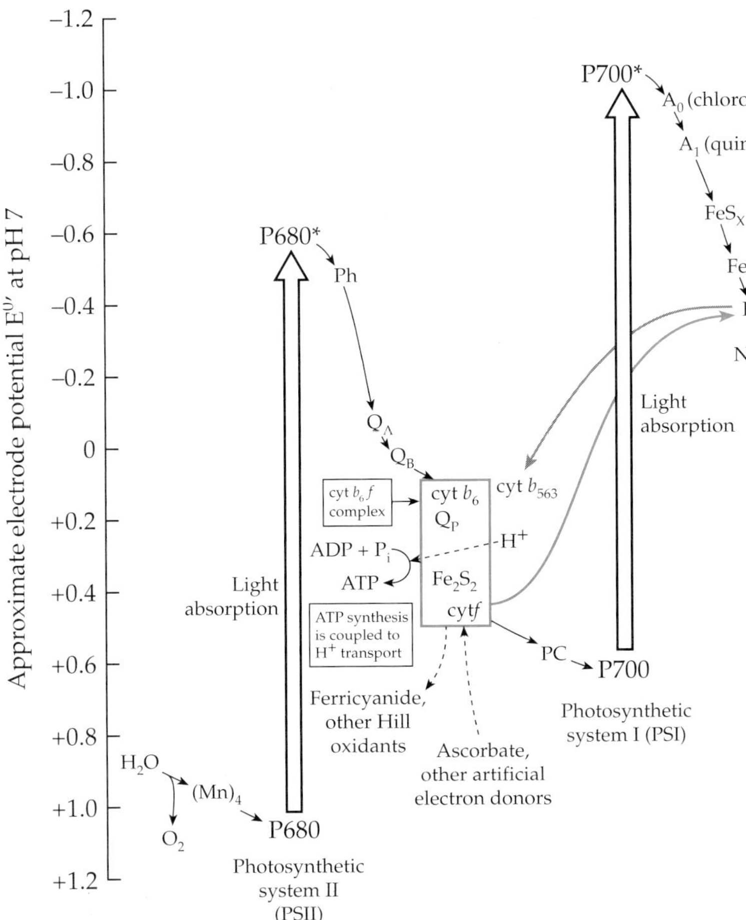

Figure 23-17 The zigzag scheme (Z scheme) for a two-quantum per electron photoreduction system of chloroplasts. Abbreviations are P680 and P700, reaction center chlorophylls; Ph, pheophytin acceptor of electrons from PSII; Q_A, Q_B, quinones bound to reaction center proteins; PQ, plastoquinone (mobile pool); Cyt, cytochromes; PC, plastocyanin; A_0 and A_1, early electron acceptors for PSI, possibly chlorophyll and quinone, respectively; F_X, Fe_2S_2 center bound to reaction center proteins; F_A, F_B, Fe_4S_4 centers; Fd, soluble ferredoxin; and DCMU, dichlorophenyldimethylurea. Note that the positions of P682, P700, Ph, Q_A, Q_B, A_0, and A_1 on the E°′ scale are uncertain. The E°′ values for P682 and P700 should be for the (chlorophyll / chlorophyll cation radical) pair in the reaction center environment. These may be lower than are shown.

An important experiment of Emerson and Arnold[35] employed very short flashes of light and measurement of the quantum efficiency of photosynthesis during those flashes. A striking fact was observed. At most, during a single turnover of the photosynthetic apparatus of the leaf, one molecule of O_2 would be released for each 3000 chlorophyll molecules. However, it could be calculated that for each O_2 released only about eight quanta of light had been absorbed. It followed that about 400 chlorophyll molecules were involved in the uptake of one quantum of light. This finding suggested that a large number of chlorophyll molecules act as a single light receiving unit (usually called a light-harvesting or antenna complex) able to feed energy to one **reaction center**. The concept is now fully accepted.

Electron transport and photophosphorylation. Two molecules of NADPH are required to reduce one molecule of CO_2 via the Calvin–Benson cycle (Fig. 17-14), and three molecules of ATP are also needed. How are these formed? The Z scheme provides part of the answer. There is enough drop in potential between the upper end of PSI and the lower end of PSII to permit synthesis of ATP by electron transport. It is likely that only one molecule of ATP is formed for each pair of electrons passing through this chain. Since, according to Fig. 17-14, one and a half molecules of ATP are needed per NADPH, some other mechanism must exist for the synthesis of additional ATP. Furthermore, many other processes in chloroplasts depend upon ATP so that the actual need for photogenerated ATP may be larger than this.

Arnon[234,235] demonstrated that additional ATP can be formed in chloroplasts by means of **cyclic photophosphorylation**: Electrons from the top of PSI can be recycled according to the dashed lines in Fig. 23-17. An electron transport system, probably that of the Z scheme, is used to synthesize ATP. As isolation of proteins and cloning of their genes progressed, it became clear that a complex of proteins known as **cytochrome b_6f** is closely related to the cytochrome bc_1 of mitochondria (Fig. 18-8).[236–237a] As in that complex, cytochrome b_6 carries two hemes, designated b_h and b_l with E_m of −84 and −158 mv, respectively. E_m values are for *Chlamydomonas* (see Fig. 1-11). Heme b_l is closer to the positively charged membrane surface (lumen side) and heme b_h is closer to the negative surface (stroma side). In maize cytochrome b_6 is a 23-kDa subunit, and the c-type cytochrome *f* is a larger ~ 34-kDa subunit whose heme is close to the lumenal side. Its E_m value is +330 mv. A 20-kDa Rieske Fe–S protein (Chapter 16) and an additional 17-kDa subunit complete the core four-subunit complex, which is found in green plants, green algae,[236] and cyanobacteria.[238,239] Other smaller subunits are also present. It is usually assumed that a Q cycle equivalent to that of mitochondrial complex III (Fig. 18-9) operates in the pumping of protons across the thylakoid membrane (Fig. 23-18).[237b,c] However, at high rates of photosynthesis the electron transfer may bypass cytochrome *f*.[237d] The associated ATP synthase is also subject to complex regulatory mechanisms.[237e] The Rieske protein is encoded by a nuclear gene, but genes for other subunits are chloroplastic. Electrons may be carried from the cytochrome *f* subunit to PSI by plastocyanin[240] or, in many algae and cyanobacteria, by the small **cytochrome c_6**.[241–242b] It is often synthesized when copper is inadequate for synthesis of plastocyanin. Figure 23-18 is a schematic view of PSI, PSII, and the intermediate cyt b_6f complex in a thylakoid membrane.

In spite of the close similarities in structures and function, there are distinct differences between cyto-

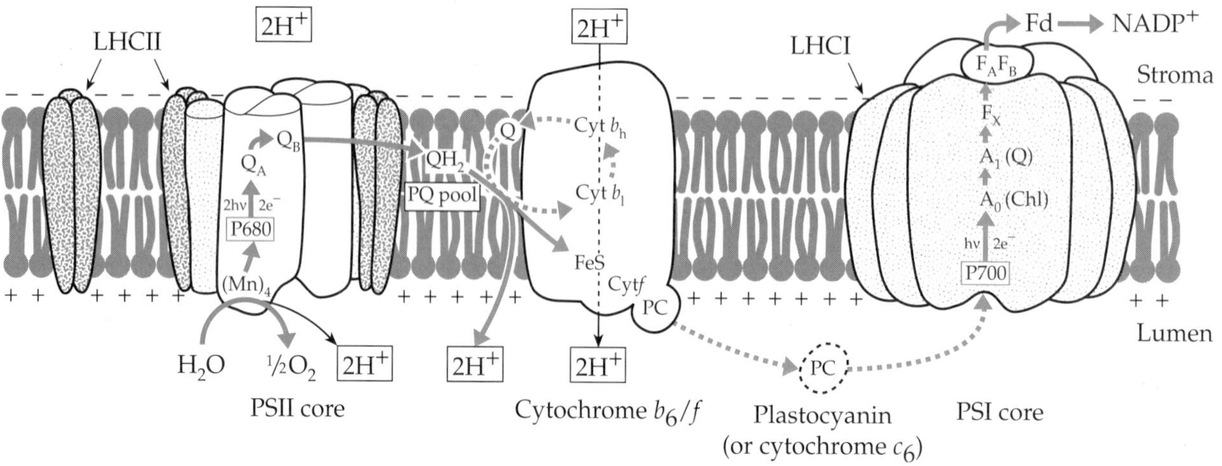

Figure 23-18 Schematic view of photosynthetic reaction centers and the cytochrome b_6f complex embedded in a thylakoid membrane. Plastocyanin (or cytochrome c_6 in some algae and cyanobacteria) carries electrons to the PSI core.

chrome b_6f and the cytochrome bc_1 of mitochondria.[243] Among these are the presence of stoichiometrically bound chlorophyll *a* and β-carotene[238,244] in the photosynthetic complex. The function of the chlorophyll is uncertain, but the carotene is probably there to quench the chlorophyll triplet state,[244] which would probably cause photodamage via formation of singlet O_2.

Bacterial photosynthesis. What is the relationship of the Z scheme of Fig. 23-17 to bacterial photosyntheses? In photoheterotrophs, such as the purple *Rhodospirillum*, organic compounds, e.g., succinate, serve as electron donors in Eq. 23-30. Because they can utilize organic compounds for growth, these bacteria have a relatively low requirement for NADPH or other photochemically generated reductants and a larger need for ATP. Their photosynthetic reaction centers receive electrons via cytochrome *c* from succinate ($E°'$ = + 0.03 V). The centers resemble PSII of chloroplasts and have a high midpoint electrode potential $E°'$ of 0.46 V. The initial electron acceptor is the Mg^{2+}-free bacteriopheophytin (see Fig. 23-20) whose midpoint potential is –0.7 V. Electrons flow from reduced bacteriopheophytin to menaquinone or ubiquinone or both via a cytochrome bc_1 complex, similar to that of mitochondria, then back to the reaction center P870. This is primarily a cyclic process coupled to ATP synthesis. Needed reducing equivalents can be formed by ATP-driven reverse electron transport involving electrons removed from succinate. Similarly, the purple sulfur bacteria can use electrons from H_2S.

In contrast, the reaction centers of green sulfur bacteria resemble PSI of chloroplasts. Their reaction centers also receive electrons from a reduced quinone via a cytochrome *bc* complex.[245] However, the reduced form of the reaction center bacteriochlorophyll donates electrons to iron–sulfur proteins as in PSI (Fig. 23-17). The latter can reduce a quinone to provide cyclic photophosphorylation. Cyanobacteria have a photosynthetic apparatus very similar to that of green algae and higher plants.

2. Chloroplast Structure

Chloroplasts come in various sizes and shapes, but all contain a small number of DNA molecules ranging in size from 120–160 kb. Complete sequences are known for DNA from chloroplasts of a liverwort (121,025 bp),[246] tobacco (155,844 bp),[247] maize (*Zea mays*),[248] and other plants. The 140,387 bp DNA from maize chloroplasts is a circular molecule containing the genes for 23S, 16S, 5S, and 4.5S RNA, for 30 species of tRNA, and for 70 different proteins. Among them are subunits of RNA polymerase, NADH dehydrogenase, subunits of both PSI and PSII, rubisco (large subunit), cytochromes *b* and *f*, six subunits of ATP synthase, and others. As with mitochondria, some subunits of the enzyme complexes that provide the cell with energy (e.g., ATP synthase) are encoded in the nucleus.

A characteristic of chloroplast genomes is the presence of a pair of large (in maize 22,348 bp) inverted repeat sequences. Since they can form a large hairpin structure with a very large loop, they may stabilize the gene sequence. The mutation rate within the repeat sequence is lower than in the single-stranded regions. The same genes are found in corresponding positions in both maize and rice. Both genomes have a number of sites of departure from the standard genetic code. These "editing sites" give rise to C→ U transitions in the RNA transcripts.

Most land plants have similar chloroplast DNA sequences, but considerable divergence is observed among algae.[249] For example, the red alga *Porphyra purpura* has 70 genes not found in chloroplasts of land plants. Each gene of the chloroplasts of the dinoflagellate *Heterocapsa triquetra* is carried on its own DNA minicircle.[250] However, ~2000 chloroplast proteins are encoded by nuclear DNA. The corresponding proteins are synthesized on cytoplasmic ribosomes and are transported into the chloroplasts.[249a] Some of these proteins must pass through both the double membrane of the envelope and the thylakoid membrane. As in mitochondria (Fig. 18-4) an array of different transport proteins are required. They are distinctly different from the mitochondrial transport proteins and involve their own unique targeting mechanisms.[251–253a]

Chloroplast membranes. Like the other energy-producing organelles, the mitochondria chloroplasts are surrounded by an outer double membrane or **envelope** and also contain an internal membrane system.[225–227,254–255a] Within the colorless **stroma** are stacks of flattened discs known as **grana** (Fig. 23-19). The discs themselves (the **thylakoids**) consist of pairs of closely spaced membranes 9 nm thick, each pair being separated by a thin internal space or loculus (Fig. 23-19). At least 75 different proteins are present in the isolated membranes. There is also a high content of **galactosyl diacylglycerol**, **digalactosyl diacylglycerol**, and **sulfolipid** (Chapter 8, Section A,4). Lipids account for half of the mass of thylakoid membranes.

Through the use of freeze-fracture and freeze-etching techniques of electron microscopy, it is possible to see, embeddded in the thylakoid membranes, particles which may represent individual **photosynthetic units** (also called quantosomes).[227,256–258] They are about 20 nm in diameter, and at least many of them presumably contain a reaction center surrounded by light-collecting chlorophyll–protein complexes. Others may represent the cytochrome b_6f complex and

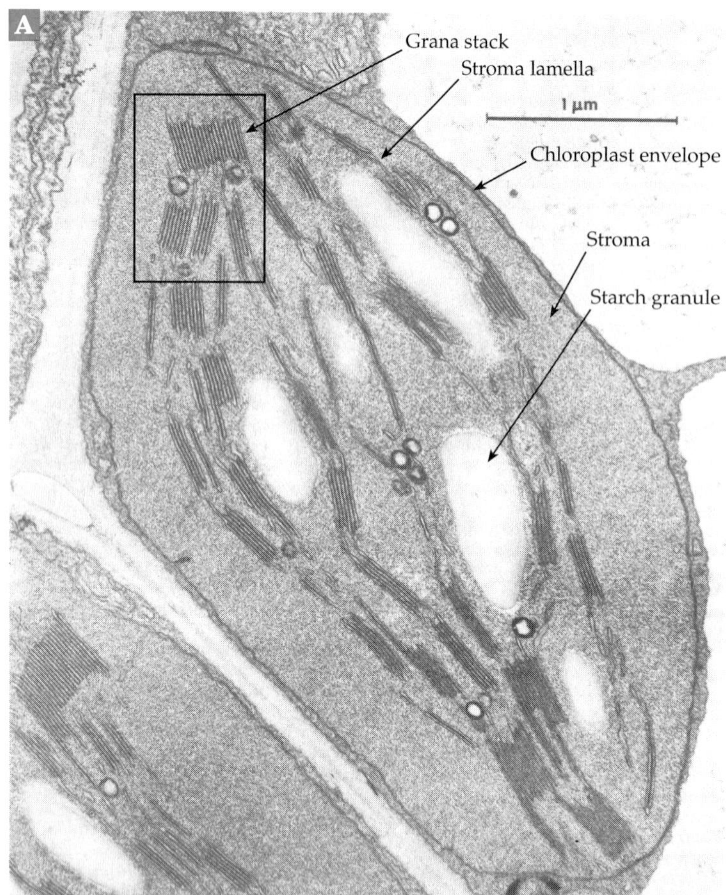

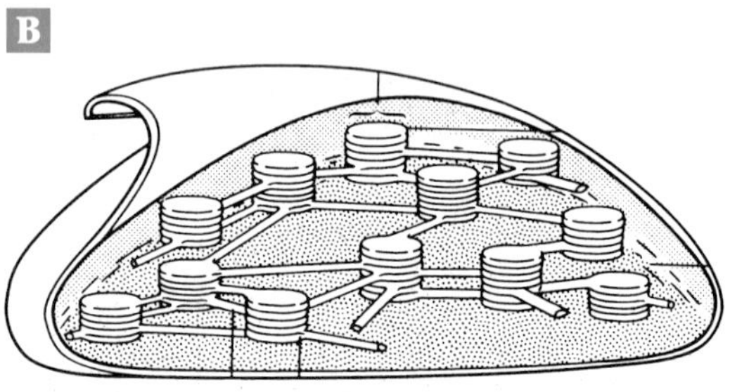

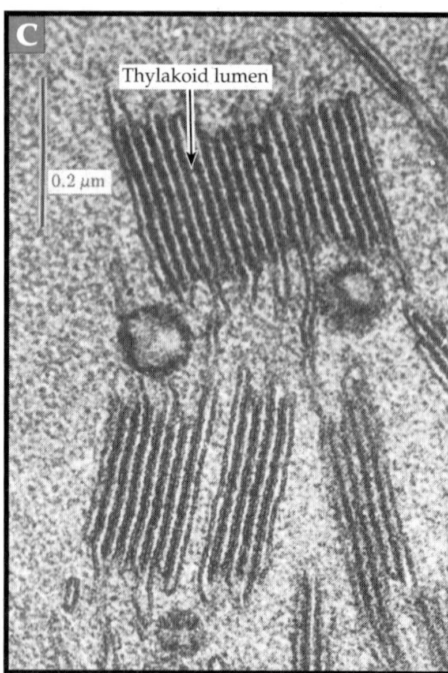

Figure 23-19 (A) Electron micrograph of alfalfa leaf chloroplast. Courtesy of Harry T. Horner, Jr., Iowa State University. (B) Schematic drawing of a chloroplast. From Hall and Rao[227] (C) Enlargement of a portion of (A) to show grana stacks more clearly.

ATP synthase, whose knobs also protrude into the stroma. A photosynthetic unit can also be defined chemically by the number of various types of molecules present in a chloroplast membrane for each four manganese atoms (Table 23-2). Separate units contain PSI and PSII. These reaction centers appear to have a different distribution within the thylakoids, the PSI units being located principally in the unstacked membranes and the PSII units in the grana stacks.[255,259]

Photosynthetic pigments and their environments. The chlorophylls (Fig. 23-20) are related in structure to the hemes (Figs. 16-5, 16-6), but ring IV (D) is not fully dehydrogenated as in the porphyrins. The **chlorin** ring system is further modified in chlorophyll by the addition of a fifth ring (V) containing an oxo group and a methyl ester. Ring V has been formed by crosslinking between the propionic acid side chain of ring III and a methine bridge carbon to give the parent compound **pheoporphyrin**. Chlorophylls contain constituents around the periphery that indicate a common origin with the porphyrins (Fig. 24-23). However, one of the carboxyethyl groups is esterified with the long-chain phytyl group in most of the chlorophylls. Chlorophyll *a* is the major pigment of chloroplasts and is a centrally important chromophore for photosynthesis in green plants. Most of the other chlorophylls, as well as the carotenoids and certain other pigments, are referred to as **accessory pigments**. Many of them have a light-receiving antenna function. Carotenoids are also photoprotectants. The relative numbers of pigment molecules in the photosynthetic units (average of PSI and PSII) of spinach chloroplasts are given in Table 23-2.

While the structure of chlorophyll *a* shown in Fig. 23-20 is the predominant one, other forms exist, e.g., with

$$-CH_2CH_3 \text{ or } -\underset{\displaystyle OH}{\underset{|}{C}H}-CH_3$$

replacing the vinyl group on ring I or with vinyl or hydroxyethyl replacing the ethyl group on ring II. The same kind of variation occurs for chlorophyll *b*.[260] In 80% acetone chlorophyll *a* has a sharp absorption band at 663 nm (15,100 cm^{-1}), but within chloroplasts the absorption maximum is shifted toward the red, the majority of the chlorophyll absorbing at 678 nm. Chlorophyll *b* (Fig. 23-20) is also nearly always present in green leaves.

The absorption peak in acetone is at 635 nm (15,800 cm^{-1}). Chlorophyll *c* found in diatoms, brown algae (Phaeophyta), and dinoflagellates (Fig. 1-9) lacks the phytyl group. Chlorophyll *d* contains a formyl group on ring I.[261]

Photosynthetic bacteria contain **bacteriochlorophylls** in which ring II is reduced (Fig. 23-20). The absorption band is shifted to the red from that of chlorophyll *a* to ~770 nm. The most abundant chlorophylls of green sulfur bacteria, **bacteriochlorophylls *c*, *d***, and ***e*** (or *Chlorobium* chlorophylls), contain a hydroxyethyl group on ring I; ethyl, *n*-propyl, or isobutyl groups on ring II; often an ethyl group instead of methyl on ring III; and a methyl group on the methine carbon linking rings I and IV. A variety of polyprenyl side chains can replace the phytyl group of the chlorophylls of higher plants.[262,263] The **pheophytins**, which are identical to the chlorophylls but lack the central magnesium ion, play an essential role in photosynthetic reaction centers. They can be formed in the laboratory by splitting the Mg^{2+} out from chlorophyll with a weak acid. Other derivatives are the **chlorophyllides** formed by hydrolysis of the methyl ester group and **chlorophyllins** formed by removal of both the methyl and phytyl groups.

Since chlorophyll can be removed readily from chloroplasts by mild solvent extraction, it might appear that it is simply dissolved in the lipid portion of the membranes. However, from measurements of dichroism (Gregory,[226] p. 111) it was concluded that the chlorophyll molecules within the membranes have a definite orientation with respect to the planes of the thylakoids and are probably bound to fixed structures. The absorption spectrum of chlorophyll in leaves has bands that are shifted to the red by up to 900 cm^{-1} from the position of chlorophyll *a* in acetone. Most green plants contain at least four major chlorophyll bands at ~662, 670, 677, and 683 nm as well as other minor bands[264] (Fig. 23-21). This fact suggested that

TABLE 23-2
Approximate Composition of Photosynthetic Units in a Spinach Chloroplast[a]

Component	Number of molecules[b]
Chlorophyll *a*	160
Chlorophyll *b*	70
Carotenoids	48
Plastoquinone A	16
Plastoquinone B	8
Plastoquinone C	4
α-Tocopherol	10
α-Tocopherylquinone	4
Vitamin K_2	4
Phospholipids	116
Sulfolipids	48
Galactosylglycerides	490
Iron	12 atoms
Ferredoxin	5
Cytochrome b_{563}	1
Cytochrome b_{559}	
Cytochrome *f*	1
Copper	6 atoms
Plastocyanin	1
Manganese	2 atoms
Protein	928 kDa

[a] Averaged for PSI and PSII. After Gregory, R. P. F. (1971) *Biochemistry of Photosynthesis*, Wiley, New York [data of Luchtenthaler, H. K., and Park, R. B. (1963) *Nature (London)* **198**, 1070] and White, A., Handler, P., and Smith, E. L. (1973) *Principles of Biochemistry*, 5th ed., p. 528, McGraw-Hill, New York.

[b] Numbers of molecules assuming 2 Mn^{2+} ions per unit (4 for PSII and 0 for PSI).

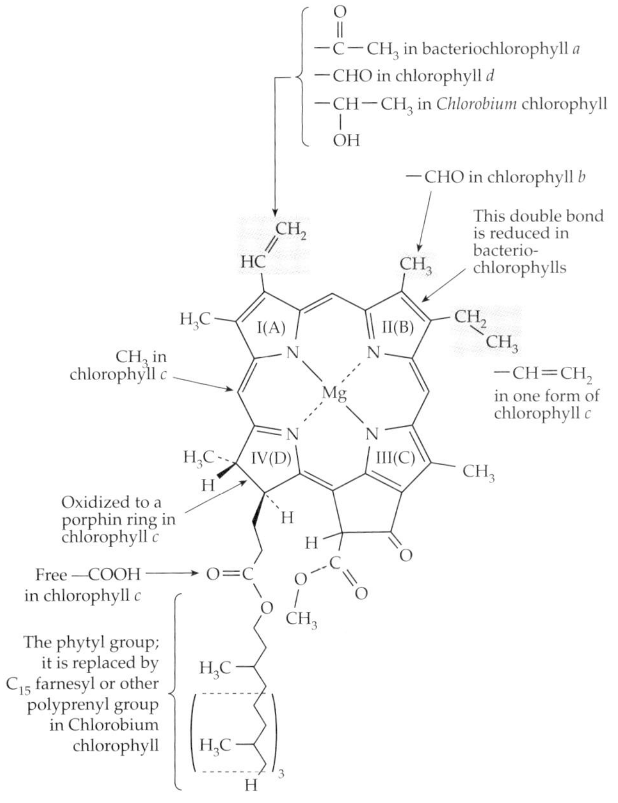

Name of parent ring system:

without ring V closed:	chlorin
with ring V closed:	pheoporphyrin
with free — COOH on ring V:	chlorophyllides
with free — COOH on both rings II and V:	chlorophyllins
with Mg_{2+} replaced by 2 H^+:	pheophytins

Figure 23-20 Structures of the chlorophylls.

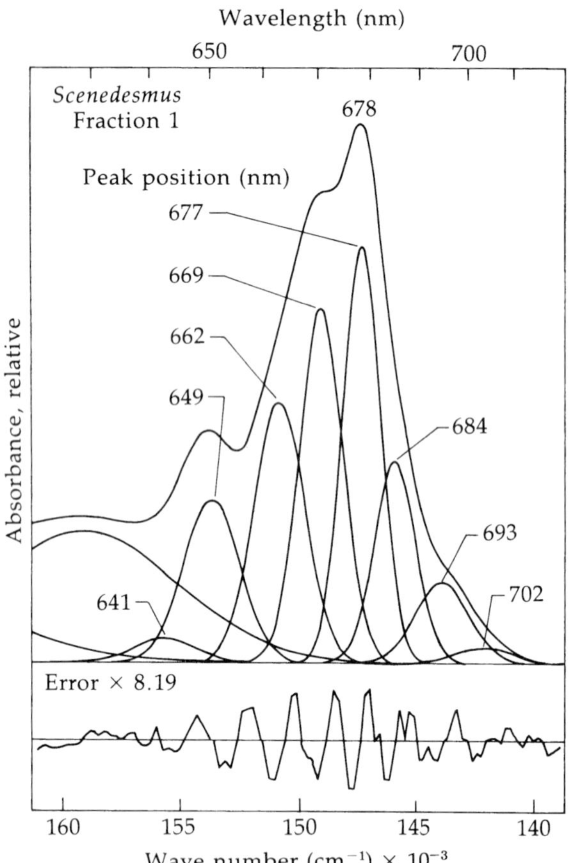

Figure 23-21 Absorption spectrum of chlorophyll in a suspension of chloroplast fragments from the green alga *Scenedesmus* showing the multicomponent nature of the chlorophyll environments. From French and Brown.[264]

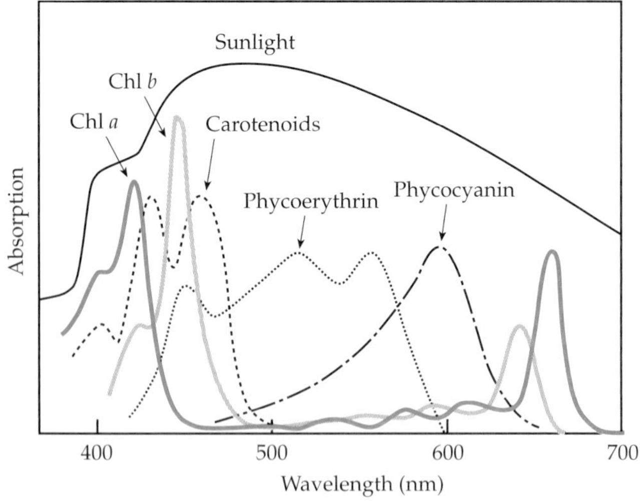

Figure 23-22 Absorption spectra of chlorophylls and accessory pigments compared. Redrawn from G. and R. Govindjee,[265] and from J. J. Wolken.[266]

the chlorophyll exists in a number of different environments. As a result, the absorption is spread over a broader region leading to more efficient capture of light. Only a small fraction of the total chlorophyll is in the reaction centers; that for PSI absorbs at ~700 nm and that for PSII at ~682 nm.

Bacteriochlorophyll in *Chromatium* has three absorption bands with peak positions at 800, 850, and 890 nm. The last includes the reaction center bacteriochlorophyll and is the only form that fluoresces. Recent studies have established that most if not all chlorophyll is bound to specific proteins, a fact that can account for the various overlapping absorption bands.

The **carotenes** and **carotenoids** are very important accessory pigments (Fig. 23-22). The major component in most green plants is β-carotene. Green sulfur bacteria contain γ-carotene in which one end of the molecule has not undergone cyclization and resembles lycopene (Fig. 22-5). Chloroplasts also contain a large variety of oxygenated carotenoids (xanthophylls). Of these, neoxanthin, violaxanthin (Eq. 22-12), and lutein (p. 1240) predominate in higher plants and green algae. *Euglena* and related microorganisms contain much antheraxanthin (Eq. 22-10). A light-collecting protein from marine dinoflagellates contains both chlorophyll *a* and several molecules of the carotenoid peridinin. See Fig. 23-29.[267,268] Brown algae and diatoms contain mostly fucoxanthin and zeaxanthin (Fig. 22-5), while the bacterium *Rhodospirillum rubrum* synthesizes spirilloxanthin (p. 1240).

It is a striking fact that there are no naturally occurring green plants that lack carotenoid pigments.[269] Carotenoidless mutants are used in photosynthesis research, but they apparently cannot survive under natural conditions. Carotenoids not only participate as members of the light-receiving complex but also confer protection to chlorophyll against light-induced destruction by singlet oxygen. This accounts for the fact that carotenoids are usually intimately associated with chlorophyll in the pigment complexes. For example, see Figs. 23-29 and 23-30 and discussion on pp. 1308–1310.

A third class of accessory pigment of more limited distribution are the **open tetrapyrroles**, sometimes called "plant bile pigments" because of their relationship to the pigments of animal bile (Fig. 24-24). Among these are the **phycocyanins**, which provide the characteristic color to cyanobacteria. They are conjugated proteins (biliproteins) containing covalently bound phycocyanobilin (Fig. 23-24).[270] The red **phycoerythrins** of the Rhodophyta contain bound phycoerythrobilin (Fig. 23-23), an isomer of phycoerythrobilin. There are four common isomeric **bilins**, each having a different number of conjugated double bonds.[272,273] Together, they provide for a broad range

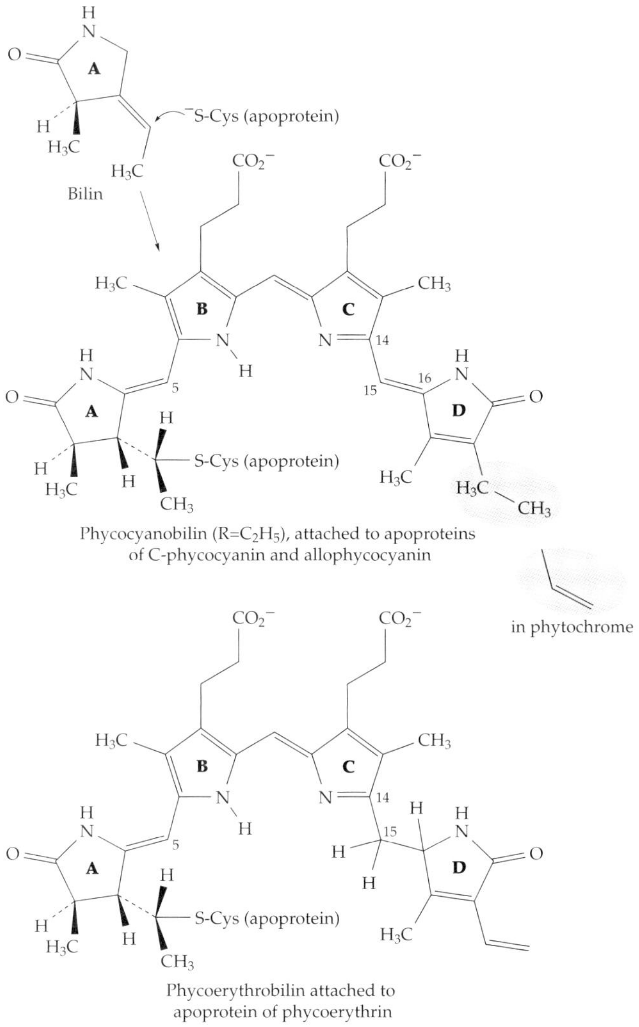

Phycocyanobilin (R=C_2H_5), attached to apoproteins of C-phycocyanin and allophycocyanin

Phycoerythrobilin attached to apoprotein of phycoerythrin

Figure 23-23 Structures of the open tetrapyrroles of plants. See also Fig. 24-24. After Szalontai *et al.*[271]

of colors from blue to red (Table 23-3). The bilins are attached to proteins via addition of a cysteine –SH group to the vinyl group of ring A of the bilin (Fig. 23-23). A double attachment may be formed by addition of –SH groups to both vinyl groups.[273–276] Isolated tetrapyrrole pigments tend to have a helical structure and to absorb light at lower wavelengths than do the protein-bound pigments which assume elongated conformations.

There are three major classes of conjugated phycobiliproteins,[273,277] all of which are $\alpha\beta$ heterodimers often associated as $(\alpha\beta)_6$ (Fig. 23-24). The **allophycocyanins** carry one bilin per subunit, the phycocyanins carry one on the α and two on the β subunit, and the phycoerythrins carry two or three on the α subunit and three on the β (Fig. 23-24). Cysteine α-84 is one of the frequent attachment sites.[273] Three-dimensional structures are known for several of these proteins[278–281]

TABLE 23-3
The Common Bilin Pigments Present in Phycobiliniproteins[a]

Isomer	Number of conjugated double bonds	Absorption maximum (nm) when conjugated to proteins
Phycocyanobilin	8	~640
Phycobiliviolin	7	~590
Phycoerythrobilin	6	~550
Phycourobilin	5	~490

[a] Wedemayer, G. J., Kidd, D. G., Wemmer, D. E., and Glazer, A. N. (1992) *J. Biol. Chem.* **267**, 7315–7331.

The bilins are derived biosynthetically rather directly[282] from biliverdin IXα, whose formation is described in Fig. 24-24. The addition of an apoprotein –SH group to a carbon–carbon double bond of the bilin is catalyzed by a specific lyase.[283] **Phytochrome** (Section H) arises in a similar way[284] as does the blue biliprotein **insecticyanin** (Box 21-A).[285]

3. The Light-Receiving Complexes

Irradiation of chloroplasts leads to easily measurable fluorescence from chlorophyll *a*, but no fluorescence is observed from chlorophyll *b* or from other forms of chlorophyll, carotenoids, or other pigments. It appears that the latter all serve as light-collecting or antenna pigments that efficiently transfer their energy to chlorophyll *a* at the reaction centers.[286] As is evident from Fig. 23-22, the light-collecting pigments generally have higher energy absorption bands than do the reaction centers. Thus, a broad range of wavelengths of light are absorbed by an organism, and energy from all of them is funneled into the reaction centers. The light-collecting pigments are bound to specific proteins, which are located close to the reaction centers and are arranged to provide efficient energy-transfer. Distances between adjacent pigment molecules vary from 1 to 7 nm.[287]

Phycobilosomes. Algal and cyanobacterial phycocyanins and phycoerythrins are aggregated in special granules that are on the outsides of the photosynthetic membranes. The granules in the cyanobacteria are known as phycobilosomes (Fig. 23-24).[272,286,286a,288] The $(\alpha\beta)_6$ hexamers form the disks of the phycobilosomes. These are held together by linker proteins,[281,289] which fit asymmetrically into the central cavities. As is indicated in Fig. 23-24C, the disks and linker proteins are assembled into rods which are joined to form the phycobilosomes. The latter are organized into closely packed parallel arrays on the surface of the photosynthetic membranes.

Purple photosynthetic bacteria. The reaction centers of *Rhodobacter spheroides, Rhodospirillum rubrum,* and related purple bacteria are embedded in the plasma membrane. Each center is surrounded by a ring of bacteriochlorophyll *a* molecules bound noncovalently to heterodimeric (αβ) protein subunits made up of ~52- to 54-residue chains. Each αβ dimer binds two molecules of BChl *a*, whose central Mg^{2+} ions are coordinated by conserved histidine imidazole groups, as well as a molecule of spirilloxanthin. About 15–17 αβ subunits form the ring, which is designated LH1 (Fig. 23-25A).[291–293] Most of these bacteria also have smaller rings, designated LH2, floating in the membrane near the LH1 complex. The LH2 rings (Figs. 23-25B, C) consist of about nine αβ subunits with associated BChl *a* and carotenoid.[294–298] Under some conditions a third complex LH3 may be formed.[298a] In *Rhodopseudomonas acidophila* nine of the 27 BChl *a* molecules absorb light maximally at ~800 nm and are designated **B800**. The other 18, designated **B850**, absorb maximally at ~860 nm.[294,299] The B850 BChl *a* molecules have direct contact with the chromophores of neighboring molecules, allowing for easy energy transfer. The B800 chromophores are more isolated.

Low-temperature (1.2 K) single-molecule spectroscopic techniques have been used to obtain the fluorescence-excitation spectra shown in Fig. 23-26C. For an ensemble of LH2 complexes (upper trace) the spectral absorption bands are broad, but for individual LH2 complexes structure can be seen clearly for the B800 chromophores but not for the B850 chromophores. This difference has been interpreted to mean that the excitation energy of an electronically excited B850 molecule is delocalized over the whole ring of 18 BChl *a* molecules as an **exciton**. This permits both fast and efficient energy transfers from B800 to B850 and from B850 of one LH2 ring to another or to an LH1 ring and to the reaction center (Fig. 23-27).[299–300a] Energy transfer may occur by the Förster dipole–dipole mechanism (Section C,2).[298] Many of the antenna are supported by binding through their Mg^{2+} ion to an imidazole group of a protein as can also be seen in reaction center chlorophylls (Fig. 23-31C). Hydrogen-bonding to the C13-oxo groups of the chlorophylls may also be possible.[300b] The orientations of the transition dipole moments of the chlorophyll molecules may be arranged to facilitate rapid energy transfer.[300c]

Green sulfur and nonsulfur bacteria. In these organisms chlorophylls are present in rodlike particles

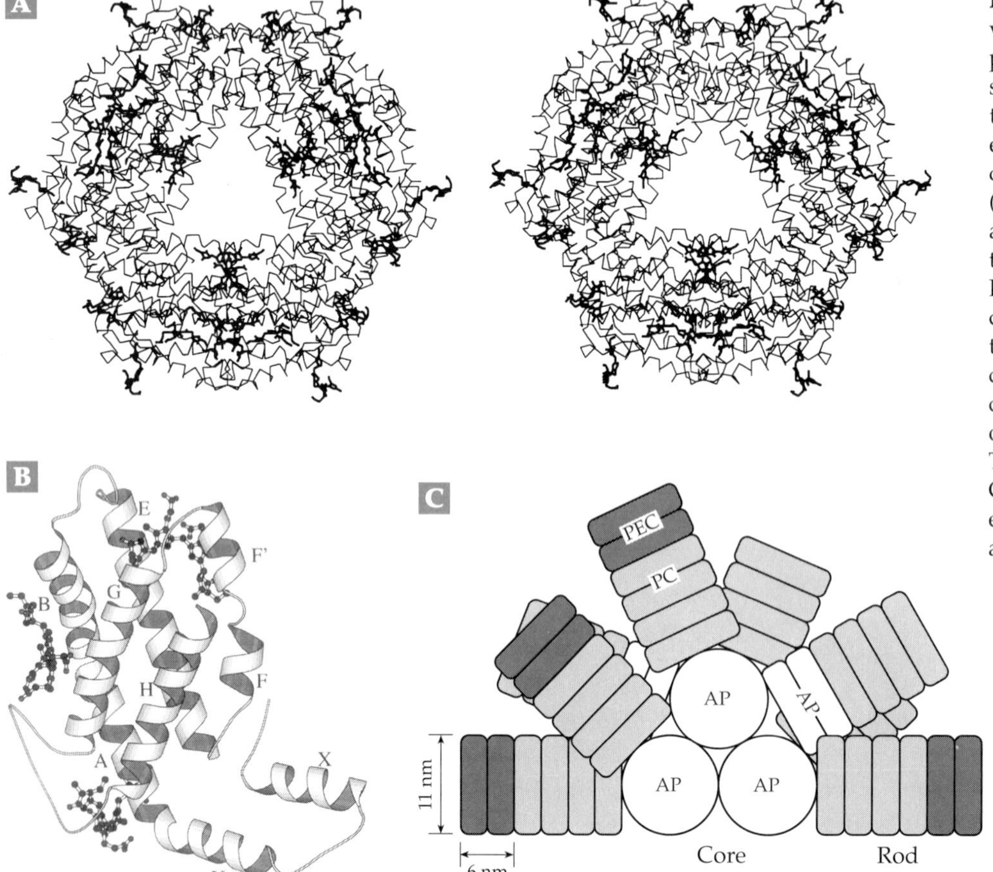

Figure 23-24 (A) Stereoscopic view of a hexameric $(\alpha\beta)_3$ phycobiliprotein. (B) The β subunit of the complex with two molecules of bound phycoerythrobilin and one of phycourobilin. From Chang *et al.*[279] (C) Schematic representation of a phycobilosome of a strain of the cyanobacterium *Anabaena*. Each disk in the structure contains an $(\alpha\beta)_3$ phycobiliprotein. The circles marked AP are cross-sections of rods, each one composed of about four disks of allophycocyanin (AP). The projecting rods contain C-phycocyanin (PC) and phycoerythrocyanin (PEC). From Lao and Glazer.[290]

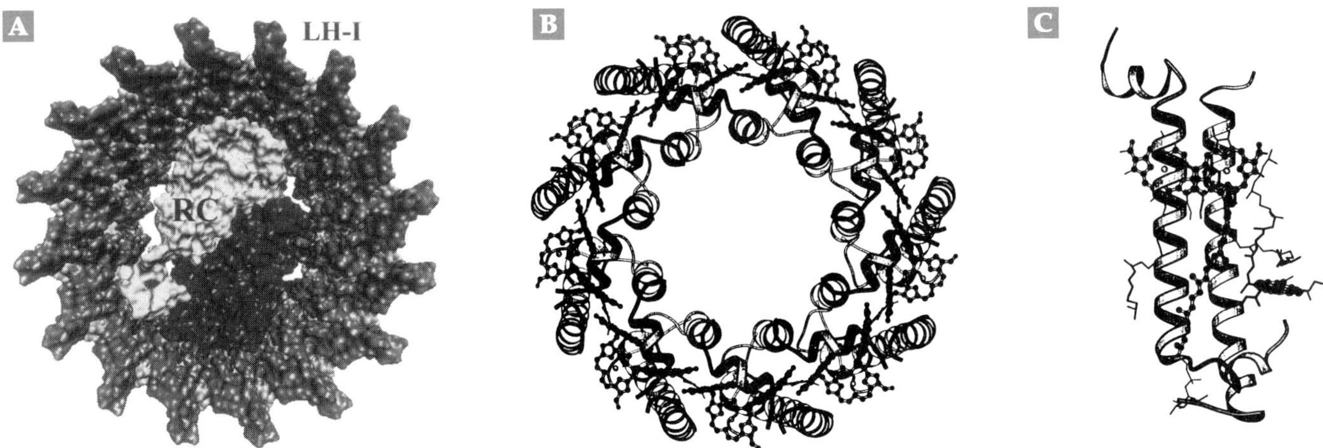

Figure 23-25 (A) The van der Waals contact surface of the periplasmic face of the reaction center and surrounding light-harvesting complex LH1 of *Rhodobacter sphaeroides*. Made with VMD by Theoretical Biophysics Group, UIUC. See also Hu and Schulten.[291] (B) Ribbon drawing of the structure of the circular light-harvesting complex LH2 of the purple photosynthetic bacterium *Rhodopseudomonas acidophila*. The tetrapyrrole rings of the 18 bacteriochlorophyll molecules are also shown. (C) Structure of one of the nine αβ protomers with three associated bacteriochlorophylls. One of these is near the top of the protein, and the other has its chromophoric group protruding at nearly a right angle on the right side. The complete phytyl side chains are also depicted in a stick representation. (B) and (C) are from Prince *et al.*[295] MolScript drawings courtesy of N. W. Isaacs.

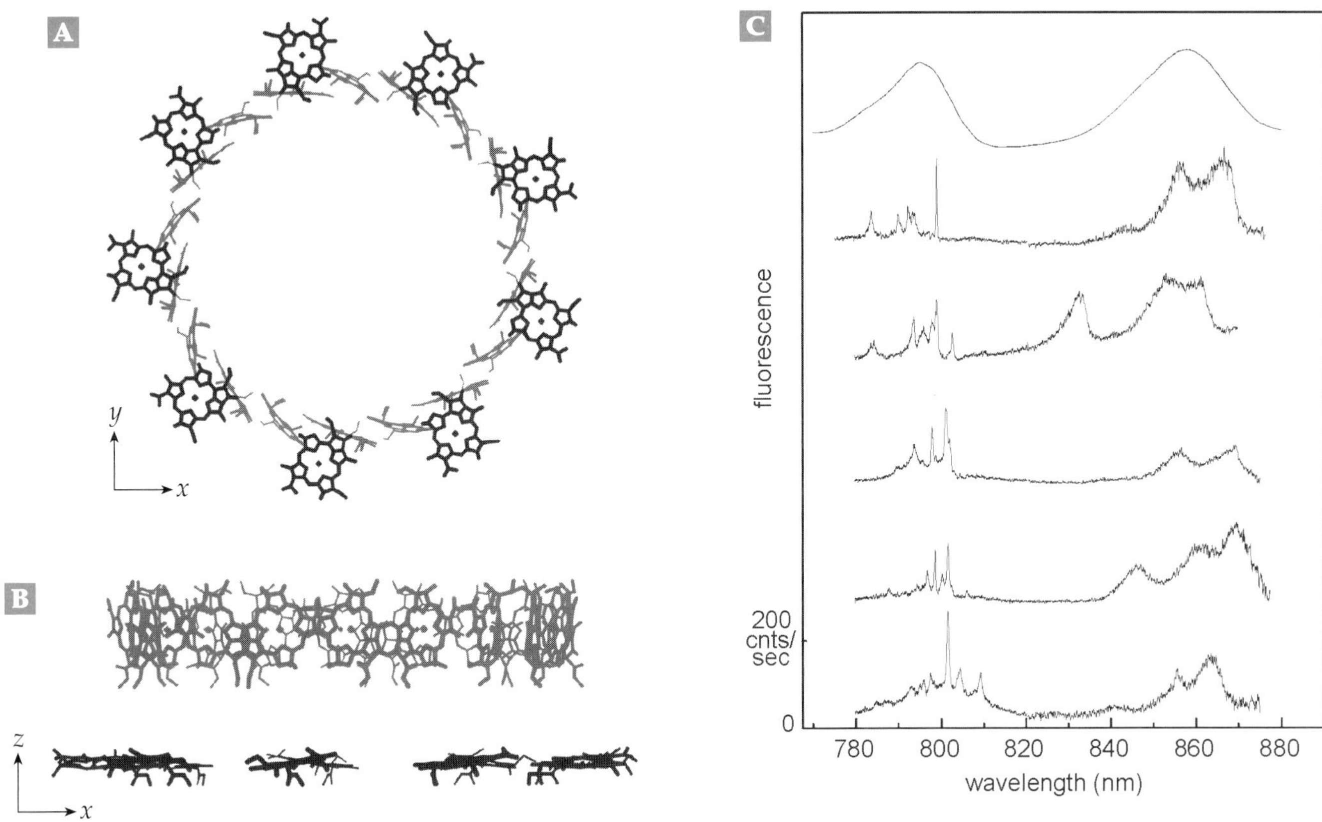

Figure 23-26 (A), (B) Arrangement of bacteriochlorophyll chromophores in the cyclic LH2 array of *Rhodopseudomonas acidophila*. The B850 subunits are gray while the B800 subunits are black. (C) Fluorescence-excitation spectra. Top trace, for an ensemble of LH2 complexes, other traces, for several individual LH2 complexes at 1.2K. Fine structure is evident for the B800 but not for the B850 chromophores. From van Oijen *et al.*[299] with permission.

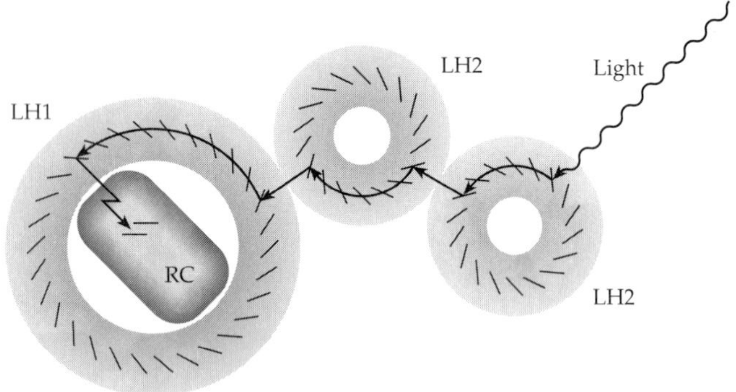

Figure 23-27 Illustration of proposed exciton transfer of the energy of light absorbed by bacteriochlorophyll *a* of purple bacteria. Energy absorbed by the light harvesting complex LH2 is transferred in steps to another LH2, to LH1 and to the reaction center. The short lines within the circles represent the edges of the BChl*a* chromophores. After Kühlbrandt[300] with permission.

of protein present within **chlorosomes**, baglike structures which may be as large as 100 x 260 nm and are attached to the inside of the cytoplasmic membranes,[301–302a] which contain the reaction centers (Fig. 23-28). The over 10,000 light-collecting chlorophylls *c*, *d*, or *e*, which may be present per reaction center, allow these bacteria to live in extremely weak light. The bacteria also contain a smaller "antenna" of ~5000 BChl *a* present as a complex with a water-soluble protein. The three-dimensional structure of this protein is also shown in Fig. 23-28. Each 45-kDa subunit of the trimeric protein contains seven embedded molecules of bacteriochlorophyll *a*.[302,303] Other light-collecting chlorophyll–protein complexes may contain an even higher ratio of chlorophyll to protein.[286]

Eukaryotic plants and cyanobacteria. Photosynthetic dinoflagellates, which make up much of the marine plankton, use both carotenoids and chlorophyll in light-harvesting complexes. The carotenoid **peridinin** (Fig. 23-29), which absorbs blue-green in the 470- to 550-nm range, predominates. The LH complex of *Amphidinium carterae* consists of a 30.2-kDA protein that forms a cavity into which eight molecules of peridinin but only two of chlorophyll a (Chl *a*) and two molecules of a galactolipid are bound (Fig. 23-29).[268]

The allenic carotenoid **fucoxanthin** (Fig. 22-5), which is absent in higher plants, predominates in brown algae, where it occurs in light-harvesting complexes along with Chl *a* and Chl *c*.[306,307]

A family of Chl *a*/*b* binding proteins are found in green plants.[308] These have apparently evolved inde-

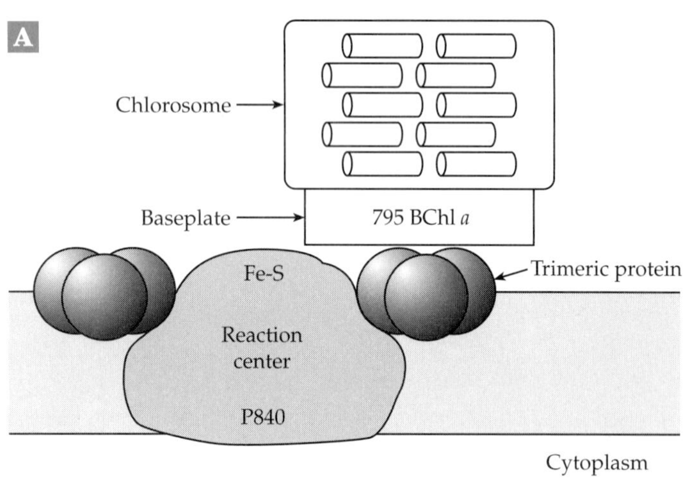

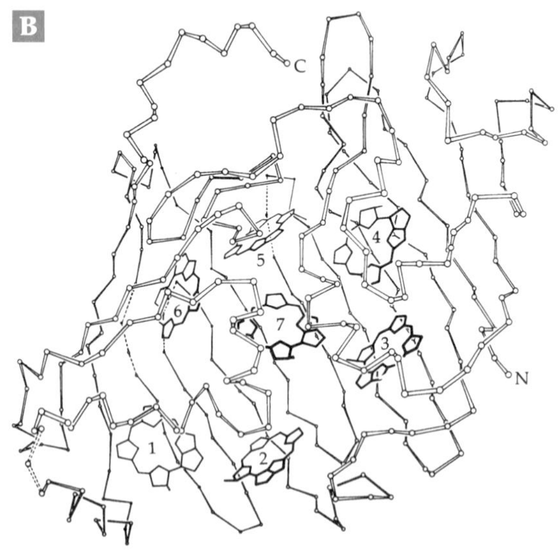

Figure 23-28 (A) Model of a light-harvesting chlorosome from green photosynthetic sulfur bacteria such as *Chlorobium tepidum* and species of *Prosthecochloris*. The chlorosome is attached to the cytoplasmic membrane via a baseplate, which contains the additional antenna bacteriochlorophylls (795 BChl *a*) and is adjacent to the trimeric BChl protein shown in (B) and near the reaction center. After Li *et al.*[302] and Rémigy *et al.*[304] (B) Alpha carbon diagram of the polypeptide backbone and seven bound BChl *a* molecules in one subunit of the trimeric protein from the green photosynthetic bacterium *Prosthecochloris*. For clarity, the magnesium atoms, the chlorophyll ring substituents, and the phytyl chains, except for the first bond, are omitted. The direction of view is from the three-fold axis, which is horizontal, toward the exterior of the molecule. From Fenna and Matthews.[305] See also Li *et al.*[302]

pendently of chlorophyll-binding proteins of green bacteria.[309] Quantitatively most important is the complex known as LHCII, the major Chl *a* / *b* protein associated with PSII and which may also provide energy to PSI. This one protein, whose structure is shown in Fig. 23-30,[310] is thought to bind half of all of the chlorophyll in green plants. The protein is organized as trimers.[311] Each 232-residue monomer binds 5–6 Chl *b*, 7–8 Chl *a*, ~ two molecules of lutein, and one of neoxanthin.[310,311] LHCII also carries all four characteristic thylakoid lipids: mono- and digalactosyl diacylglycerols, phosphatidylglycerol, and sulfoquinovosyl diacylglycerol.[312] The Chl *a* and Chl *b* molecules are in close contact (Fig. 23-30). Subpicosecond transient absorption spectroscopy[311] indicates that half of the Chl *b* to Chl *a* energy transfers occur in < 0.2 ps. Notice the close association of the two luteins in Fig. 23-30 with the chlorophyll rings. The carotenoids are thought to quench chlorophyll triplet states to prevent formation of singlet oxygen.

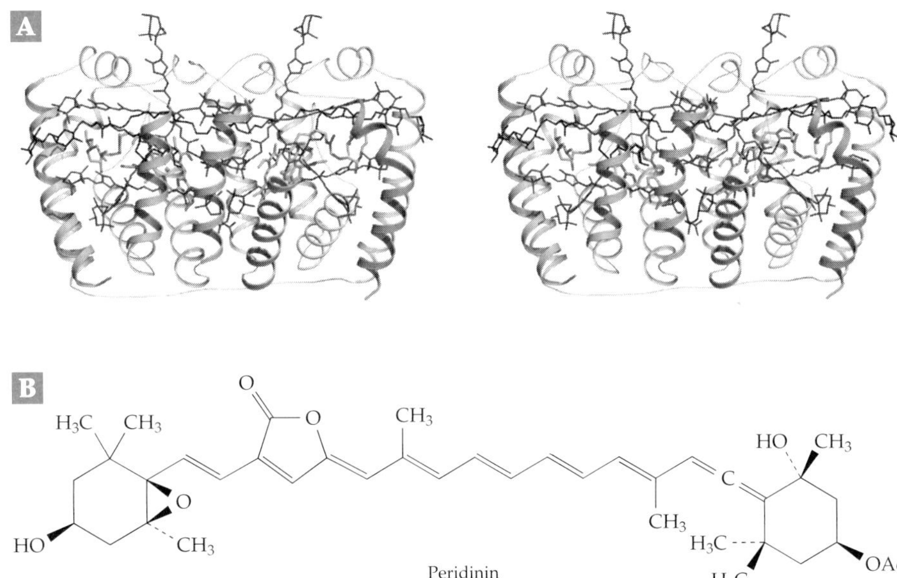

Figure 23-29 (A) Stereoscopic drawing of light-harvesting complex from the dinoflagellate protozoan *Amphidinium carterae*. The central cavity contains eight molecules of peridinin, two of which can be seen protruding from the top. Deeply buried toward the bottom are two molecules of Chl *a*. Also present are two molecules of digalactosyl diacylglycerol. From Hofmann *et al.*[268] Courtesy of Wolfram Welte. (B) Structure of peridinin.

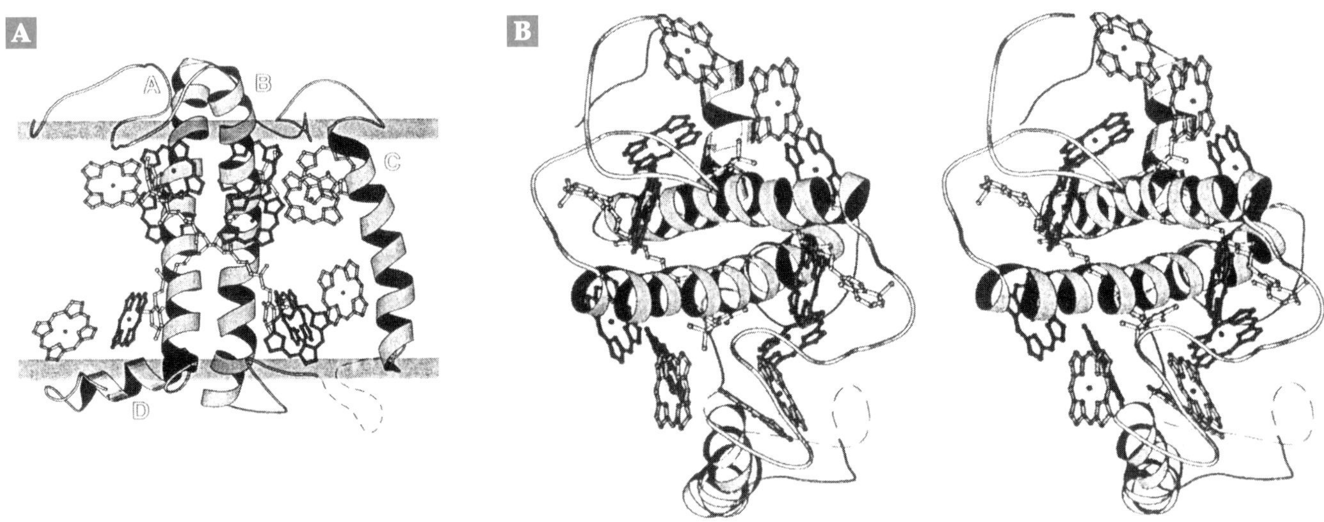

Figure 23-30 Views of light-harvesting protein LHCII of green plants. (A) Side view indicating the approximate position in the lipid bilayer of the thylakoid membrane. Helices are labeled A–D. (B) Stereoscopic top view from the stromal side of the membrane. The structure, at 0.34 nm resolution, was determined by electron crystallography on highly ordered two-dimensional crystals. MolScript drawings from Kühlbrandt *et al.*[310] Courtesy of Werner Kühlbrandt.

PSII also contains several additional chlorophyll-binding proteins, designated CP24, CP26, CP29, CP43, CP47, etc. These lie on both sides of a pair of reaction-center cores.[313–315a] A large fraction of the LHCII complexes are separate from the reaction centers and are mobile, while a smaller fraction are bound to the outer ends of the core complex.[308,314,315] The light-harvesting chlorophylls of PSI are arranged around the core. Some are bound in the N-terminal part of the major core subunits, the products of genes *PsaA* and *PsaB*.[316] They bind ~90 Chl *a* and 14 β-carotene molecules per reaction center. (See Fig. 23-33.) In addition, the peripheral LHCI, composed of four different proteins arranged around the core, binds ~110 Chl *a* and Chl *b* and ~70 molecules of xanthophyll.[317] Some species of cyanobacteria use antenna rings around their PSI trimers instead of phycobilinosomes.[317a]

In every case the light-harvesting complexes are arranged to allow rapid and very efficient transfer of electronic excitation energy from one chromophore to another and finally to the chromophores of the reaction center.[317b] The speed and efficiency appear to depend upon very rigid structures of the proteins and precise orientations of the bound chromophores to allow direct excitonic transfer of energy at distances of less than 2 nm or transfer by the Förster mechanism at distances not exceeding 10 nm.[302] An example of the precision of protein structures was observed when a posttranslationally modified asparagine N^5-methylasparagine at position 72 of the β subunit of many phycobiliniproteins was substituted by aspartate or glutamine. The fluorescence lifetime of the nearby bilin was reduced 7–10% in the mutants, an effect that could cut the >95% efficiency of energy transfer from the phycobilinisomes to the PSII reaction center.[318]

4. The Reaction Centers and Their Photochemistry

The initial or primary processes of photosynthesis occur in the reaction centers in which chlorophyll or bacteriochlorophyll absorbs a photon.[318a] Then, the chlorophyll, in its singlet excited state (Chl*), donates an electron to some acceptor A to form a radical A^- and to leave an oxidized chlorophyll Chl^+ radical (Eq. 23-31).

$$\text{Chl} \xrightarrow{h\nu} \text{Chl}^* \xrightarrow{\text{A}} \text{Chl}^+ + \text{A}^- \quad (23\text{-}31)$$

In the scheme of Fig. 23-18, acceptor A is Q_A for PSII and A_0 for PSI. The oxidized chlorophyll (Chl^+) quickly reacts further by receiving an electron from some donor.

The photooxidation of chlorophyll indicated in Eq. 23-31 is accompanied by bleaching in the principal light absorption band. However, since there is so much light-gathering chlorophyll for each reaction center, the effect is small. The study of the process has been aided greatly by preparation of isolated bacterial photochemical reaction centers.

Reaction centers of purple bacteria. The exact composition varies, but the properties of reaction centers from several genera of purple bacteria are similar. In *Rhodopseudomonas viridis* there are three peptide chains designated H, M, and L (for heavy, medium and light) with molecular masses of 33, 28, and 24 kDa, respectively. Together with a 38-kDa tetraheme cytochrome (which is absent from isolated reaction centers of other species) they form a 1:1:1:1 complex. This constitutes reaction center P870. The three-dimensional structure of this entire complex has been determined to 0.23-nm resolution[288,319–323] (Fig. 23-31). In addition to the 1182 amino acid residues there are four molecules of bacteriochlorophyll (BChl), two of bacteriopheophytin (BPh), a molecule of menaquinone-9, an atom of nonheme iron, and four molecules of heme in the *c* type cytochrome. In 1984, when the structure was determined by Deisenhofer and Michel, this was the largest and most complex object whose atomic structure had been described. It was also one of the first known structures for a membrane protein. The accomplishment spurred an enormous rush of new photosynthesis research, only a tiny fraction of which can be mentioned here.

The reaction centers are embedded in the cytoplasmic membranes of the bacteria, with the bottom of the structure, as shown in Fig. 23-31, protruding into the cytoplasm and the heme protein at the top projecting out into the periplasm which lies within infoldings of the plasma membrane. Subunits L and M each contain five ~4.0 nm long roughly parallel helices, which span the cytoplasmic membrane. Another membrane-spanning helix is contributed by subunit H, which is located mainly on the cytoplasmic side. An approximate twofold axis of symmetry relates subunits L and M and the molecules of bound chlorophyll and pheophytin.

Spectral measurements suggesting exciton splitting were among early observations that led to the conclusion that the bacteriochlorophyll involved in the initial photochemical process exists as a dimer or **special pair** (Fig. 23-31),[319,324] a conclusion verified by the structure determination. The special pair of BChl *b* lies in the center of the helical bundle that is embedded in the membrane. Nearly perpendicular to the rings of the special pair are two more molecules of BChl *b*. The central magnesium atoms of all four bacteriochlorophylls are held by imidazole groups of histidine side chains.[319,325] Below the chlorophylls are

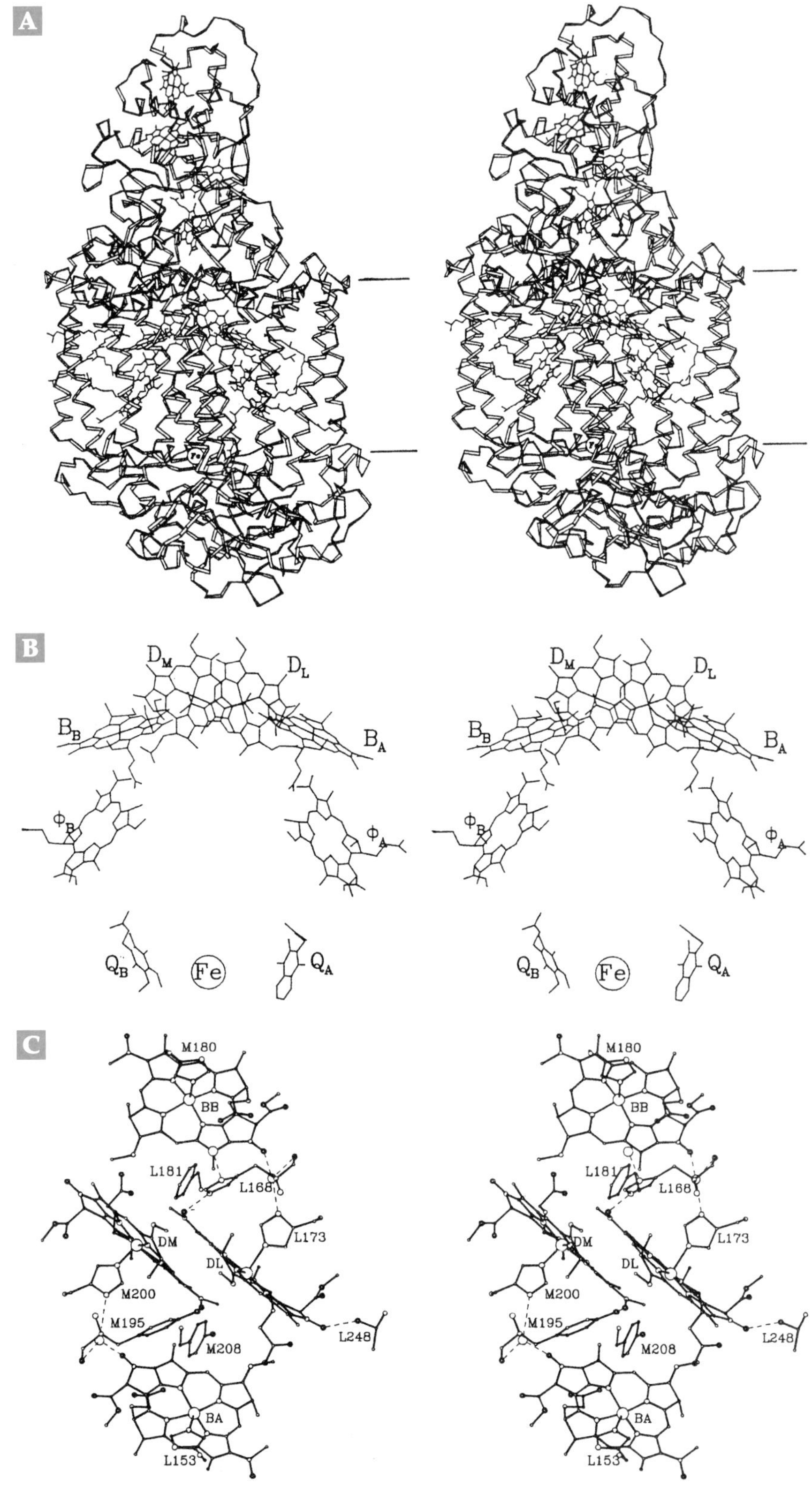

Figure 23-31 (A) Stereoscopic ribbon drawing of the photosynthetic reaction center proteins of *Rhodopseudomonas viridis*. Bound chromophores are drawn as wire models. The H subunit is at the bottom; the L and M subunits are in the center. The upper globule is the cytochrome *c*. The view is toward the flat side of the L, M module with the L subunit toward the observer. (B) Stereo view of only the bound chromophores. The four heme groups He1–He4, the bacteriochlorophylls (Bchl) and bacteriopheophytins (BPh), the quinones Q_A and Q_B, and iron (Fe) are shown. The four hemes of the cytochrome are not shown in (B). From Deisenhofer and Michel.[320] (C) Stereoscopic view of the Bchl *b* molecules along the local twofold axis. The special pair (D_M, D_L) is in the center with its tetrapyrrole rings almost perpendicular to the plane of the paper; the monomeric chlorophylls are labeled B_B and B_A. The four histidine ligands to the magnesium ions of the bacteriochlorophylls as well as two tyrosines (M195 and M208) and three water molecules (large circles) are also shown. From Deisenhofer *et al.*[321] with permission.

the two molecules of bacteriopheophytin and below them the nonheme iron and the menaquinone, the first quinone acceptor Q_A. It corresponds to Q_A of PSII shown in Fig. 23-18.

Isolated reaction centers usually contain or will bind a second quinone, which may be ubiquinone-10 (Q_{10})[325a] and which is usually designated Q_B. Its binding site is to the left of the nonheme iron in Fig. 23-31 in a position symmetrically related to that of Q_A. The reaction centers also contain a carotenoid 1,2-dihydroneurosporene.[323] The reaction centers of purple bacterium, *Rhodospirillum rubrum*, each contain one molecule of spirilloxanthin; a variety of carotenoids are present in other species.[326] The reaction centers of a third purple bacterium, *Rhodobacter spheroides*, are closely similar in structure to that in Fig. 23-31 but lack the tetraheme.[327] Reaction centers of these bacteria accept electrons directly from a soluble cytochrome *c*.

Reaction center kinetics. After an 0.8-ps or shorter flash of light the decay of the singlet excited state of the bacteriochlorophyll dimer in isolated reaction centers can be followed by loss of its characteristic fluorescence.[328,329] The lifetime of this excited state in *R. sphaeroides* is only 4 ps indicating a rapid occurrence of the initial electron transfer of Eq. 23-31. A rise in absorbance at 1250 nm is interpreted as formation of the bacteriochlorophyll cation radical $BChl^+$ in the special pair. Other spectral changes support the formation of BPh^- as the first reduction product (A^- in Eq. 23-31). However, this is thought to occur in two steps[323] with the monomeric BChl (B_A in Fig. 23-31B) receiving the electron in ~2.8 ps and passing it to the pheophytin (Phe; ϕ_A in Fig. 23-31B) in ~0.7 ps (Eq. 23-32; corresponding steps for PSII of green plants are also shown).

The quinone Q_A (the secondary acceptor) is next reduced by the BPh^- radical in ~200 ps with development of a characteristic EPR signal[321,330] at $g = 1.82$. Over a much longer period of time (~320 ns) an electron passes from the tetraheme cytochrome subunit to the Chl^+ radical in the special pair.[323,323a] The relatively slow rate of this reaction may be related to the fact that the bacteriochlorophyll of the special pair is 2.1 nm (center-to-center) from the nearest heme in the cytochrome while BPh and Q_A are only 1.4 nm apart (see Fig. 23-31).[288,331] Over a period of ~24 μs after formation of the radical anion Q_A^- an electron from the Q_A^- radical is passed to Q_B, a weakly bound ubiquinone-9, to form the Q_B^- radical. Upon absorption of a second photon by the special BChl pair another electron is passed through the chain to form $Q_A^-Q_B^-$. Uptake of two protons with transfer of the second electron from Q_A^- to Q_B^- yields the ubiquinol QH_2, which dissociates from its binding site (Q_B) into the ubiquinone pool dissolved in the lipids of the membrane bilayer.[331a–g]

Why is this multistep sequence of electron transfers necessary? A variety of techniques such as femtosecond IR[332,333] and electronic[334–335a] spectroscopy, resonance Raman spectroscopy at low temperatures,[336] and study of many mutants[337–338c] have been directed toward an answer to this question. It has been generally accepted that light energy absorbed by any one of the "monomeric" Chl or pheophytins in the reaction centers is funneled "downhill" to the special pair within 0.1–0.2 ps to generate P*.[334,336] The ultrafast ~3 ps electron transfer from P* to the adjacent monomeric BChl or Chl is necessary to prevent loss of energy by fluorescence from P*. The subsequent energetically downhill transfer to a pheophytin and on to Q_A prevents reverse electron transfer, which could also lead to fluorescence. Both the efficiency and the quantum yield are very high.[339]

The rate of the ultrafast proton transfer becomes even higher at cryogenic temperature, suggesting quantum mechanical tunneling.[331,335,340–340b] The transfer is generally treated using Marcus theory (Chapter 16), which indicates a very small reorganization energy for the process. Another aspect of the process is a possible coupling of a vibrational mode of the protein matrix to the electron transfer. Femtosecond near-IR spectra show low-frequency vibrational modes of the excited-state reaction center chromophores which may facilitate electron transfer.[332,336] The transfer of an electron from the tetraheme cytochrome *c* of *R. viridis*[341] or from the small cytochrome c_2 of *Rhodobacter spheroides*[342] to Chl^+ of the special pair has similar characteristics but is slower than the initial electron transfer from P*. On the other hand, electron transfers from Q_A^- to Q_B^- involve two distinct steps and coupled uptake of two protons.[343] An unexplained fact is that photochemical electron transport through the reaction centers always occurs through the L-side to Q_A rather than the M-side.[343a–344] However, rapid electron transfer to the pheophytin on the B-side (M-side) has been observed following excitation with blue light. This may

PSII (Green plants)	*Rhodopseudomonas viridis*
H_2O	
↓	
$(Mn)_4$ Manganese cluster	Cytochrome *c*
↓	↓
Y_Z (Tyr)	Fe (Tetraheme cytochrome *c*)
↓	↓
Chl_2 (P680)	$BChl_2$ Special pair cation radical (P870)
↓	↓
Chl_2*	$BChl_2$* *hν* Electronic excitation of $BChl_2$
~10 ps ↓	2.8 ns ↓
$Chl_A(A_0)$	$BChl_A$
35-200 ps ↓	0.7 ns ↓
Phe (A_1)	$BPhe_A$
↓	200 ps ↓
Q_A (Plastoquinone)	Q_A (Menaquinone)
↓	24 μs ↓ 2 e^- in two steps
Q_B (Plastoquinone)	Q_B (Ubiquinone)

(23-32)

TABLE 23-4
Properties of Various Reaction Centers

	Rhodopseudomonas viridis (now *Blasatochloris*)	Green plants and Cyanobacteria PSII	PSI	Green sulfur bacteria, *heliobacteria*[a]
Subunits	L/M, H	D1/D2, CP43, CP47	PsaA/PsaB nine others	$(PscA)_2$
Masses, kDa	24/28, 33	38/39.4	83/83	65/65
Input	Cyt *c* (4 Fe)	H_2O, $(Mn)_4$, Y_Z, Y_D	Cyt *f*, plastocyanin or cyt *c*	Cyt c_1, c_2
Special pair	P870	P680 $(BChl\ b)_2$	P700 $(Chl\ a)_2$	P840 $(Chl\ a)_2$
Monomeric chlorophyll	$BChl_L$, $BChl_M$		A_0 Both Chl *a*	A_0 Both Chl *a*-like
Pheophytin	$BPhe_L$, $BPhe_M$		A_0 (Phe, Phe)	
Quinone	Q_B, Q_A (Ubiquinone, menaquinone-9)	Q_B, Q_A (Plastoquinone)	Q_B, Q_A (A_1) (Both phylloquinone in cyanobacteria, plastoquinone in chloroplasts)	Q_B, Q_A (Both menaquinone-7)[b]
Iron	Fe^{3+}		F_X (Fe_4S_4)	F_X
Output	Ubiquinone Cyt bc_1, Cyt c_2	Ubiquinone Cyt b_6f	F_A, F_B, Fd $NADP^+$	F_A, F_B, Fd NAD^+

[a] Nitschke, W., and Rutherford, A. W. (1991) *Trends Biochem. Sci.* **16**, 241–245
[b] Kjear, B., Frigaard, N.-U., Yang, F., Zybailov, B., Miller, M., Golbeck, J. H., and Scheller, H. V. (1998) *Biochemistry* **37**, 3237–3242

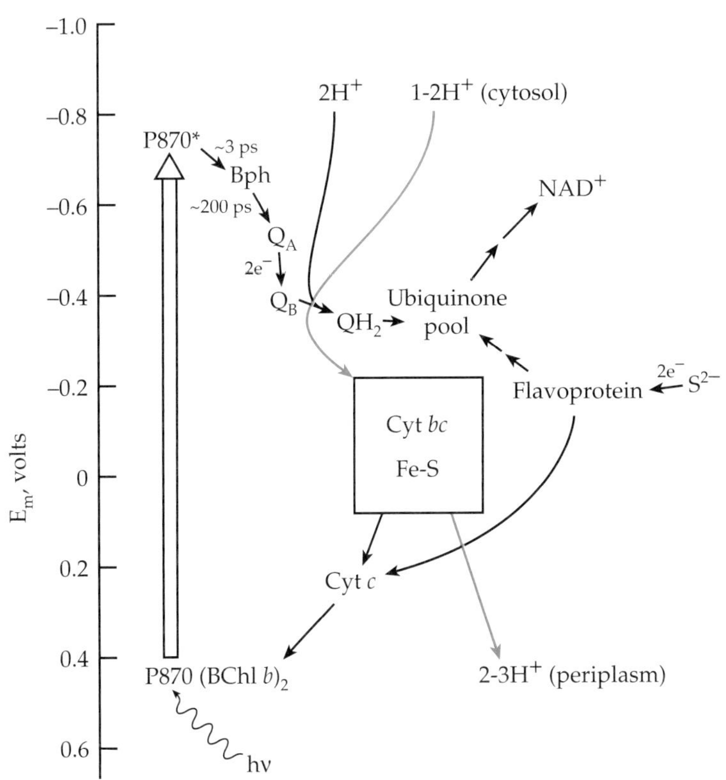

Figure 23-32 Simplified diagram of cyclic electron flow in purple bacteria. Two protons from the cytoplasm bind to Q_B^{2-} in the reaction center to form QH_2 (ubiquinol), which diffuses into the ubiquinone pool. From there it is dehydrogenated by the cytochrome bc_1 complex with expulsion of two protons into the periplasm. A third and possibly a fourth proton may be pumped (green arrows) across the membrane, e.g., via the Q cycle (Fig. 18-9). The protons are returned to the cytoplasm through ATP synthase with formation of ATP. Some electrons may flow to the reaction centers from such reduced substrates as S^{2-} and some electrons may be removed to generate NADPH using reverse electron transport.[345]

represent a photoprotective mechanism.[344a] In the PSI system, in which the two phylloquinones are tightly bound,[344b] both the A-side and the B-side seem to function in electron transfer.[344c]

Cyclic photophosphorylation in purple bacteria. QH_2 is eventually dehydrogenated in the cytochrome bc_1 complex, and the electrons can be returned to the reaction center by the small soluble cytochrome c_2, where it reduces the bound tetraheme cytochrome or reacts directly with the special pair in *Rhodobacter spheroides*. The overall reaction provides for a cyclic photophosphorylation (Fig. 23-32) that pumps 3–4 H^+ across the membrane into the periplasmic space utilizing the energy of the two photoexcited electrons. These protons can pass back into the cytoplasm via ATP synthases located in the same membrane with their catalytic centers in the cytosol (see Figs. 18-5 and 18-14).

Comparison with other reaction centers. Subunit L of the *R. viridis* reaction center was found to have a 25% sequence homology with a quinone-binding protein now known as D1, a component of the reaction center core of PSII of chloroplasts. This protein was identified as the specific target protein for inhibition by herbicides such as DCMU and atrazine (see Section 1 for structures). These compounds act as competitive inhibitors of quinone binding[346] and bind in the Q_B site in the *R. viridis* reaction center. This fact, together with the discovery that the core of PSII of green plants consists of a heterodimer of the related polypeptides D1 and D2, suggested that PSII is very similar to the bacterial center of Fig. 23-31. Both use a quinone as primary acceptor. However, P680 operates at a more positive potential (E_m ~1 V), consistent with the fact that it must provide an oxidizing agent able to oxidize H_2O to O_2 ($E°' = + 0.82$ V). Plastoquinone rather than menaquinone is the primary acceptor in PSII. A chlorophyll *a* dimer is apparently the initial electron donor. The methyl ester carbonyl groups on the edges of rings I (Fig. 23-20) of BChl may coordinate to other groups of the proteins.[288,346a] These ester groups are absent in Chl *a*.

The PSI reaction center (P700) of maize chloroplast was also found to contain a pair of homologous polypeptides with appropriately placed histidyl residues for chlorophyll binding.[347] FTIR spectra also indicate the presence of two Chl *a* molecules. Small shifts in IR frequencies upon electronic excitation suggest that one chlorophyll (P_1) is hydrogen-bonded through its 109-ester and 9-oxo groups (Fig. 23-20) while the chlorophyll (P_2) is free. This may account for the low value of E_m (Fig. 23-17). The charge on $P700^+$ appears to be carried in part on both chlorophylls. However, in the triplet state $^3P700^*$, which may be observed at low temperature, excitation is localized on P_1.[348] Although P700 operates at the low E_m of +0.49 V it produces a powerful reducing agent able to reduce ferredoxin. The first identified acceptors are other Fe–S centers present in integral membrane proteins.[345] The reaction centers of green bacteria[349] and PSI of cyanobacteria[245] have similar characteristics. As more genes have been sequenced and X-ray diffraction, electron crystallography, and electron microscopy with single-particle averaging have advanced, the fundamental similarity of all of the photosynthetic centers has been confirmed.[228,300,350–353] Gene sequences are often not highly conserved, but structures are more conserved. An evolutionary relationship of all of the reaction centers can be seen.[316,354]

PSI of cyanobacteria and green plants. The major reaction-center subunits **PsaA** and **PsaB** each have a C-terminal domain, resembling those of the L and M chains of purple bacteria, and a large N-terminal antenna-chlorophyll-binding domain. Cyanobacterial PSI contains ten other subunits, PsaC to PsaF, PsaI to PsaM, and PsaX. Thirty-one transmembrane helices have been assigned to the various subunits, several of which are in positions corresponding closely to those in the reaction centers of the purple bacteria (Fig. 23-33).[355–357] The PSI of higher plants is somewhat larger than that of cyanobacteria and contains somewhat different subunits.[356a,357,357a]

The electron donor to Chl^+ in PSI of chloroplasts is the copper protein plastocyanin (Fig. 2-16). However, in some algae either plastocyanin or a cytochrome *c* can serve, depending upon the availability of copper or iron.[345] Both Q_A and Q_B of PSI are phylloquinone in cyanobacteria but are plastoquinone-9 in chloroplasts. Mutant cyanobacteria, in which the pathway of phylloquinone synthesis is blocked, incorporate plastoquinone-9 into the A-site.[345a] Plastoquinone has the structure shown in Fig. 15-24 with nine isoprenoid units in the side chain. Spinach chloroplasts also contain at least six other plastoquinones. Plastoquinones C, which are hydroxylated in side-chain positions, are widely distributed. In plastoquinones B these hydroxyl groups are acylated. Many other modifications exist including variations in the number of isoprene units in the side chains.[358,359] There are about five molecules of plastoquinone for each reaction center, and plastoquinones may serve as a kind of electron buffer between the two photosynthetic systems.

Look at the Z scheme of Fig. 23-17. The lower end of each vertical arrow is located at an electrode midpoint potential E_m (or $E°'$) for the couple P^+/P, i.e., for a one-electron reduction of the Chl^+ or $BChl^+$ radical.[360] The top of the arrow is at the estimated value of E_m for the excited state P*. It is more negative than the ground-state value of E_m by the energy (in electron volts) of the light absorbed. This is a little misleading

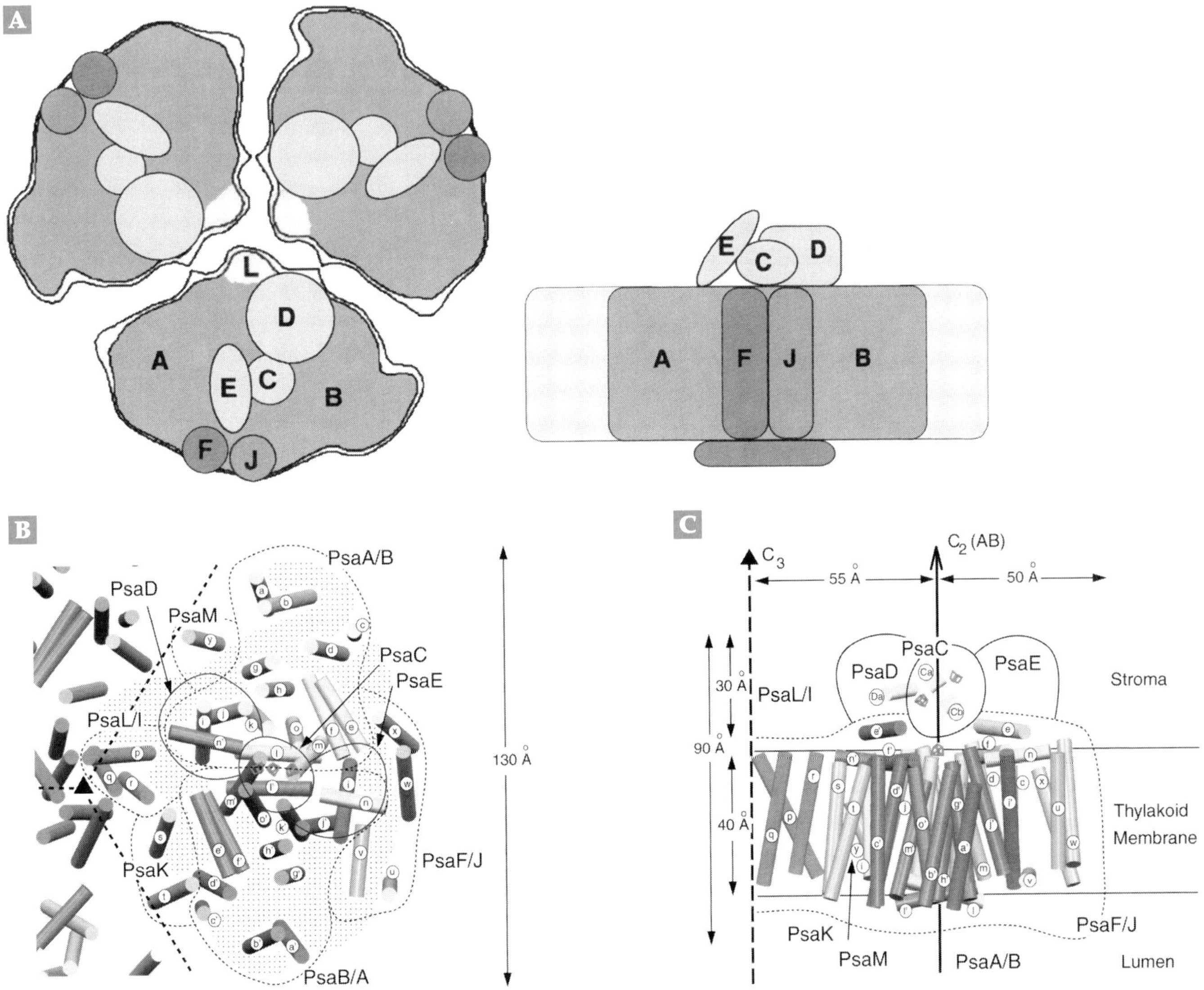

Figure 23-33 Models of the PSI reaction center of the cyanobacterium *Synechocystis* and of higher green plants. (A) Subunit arrangement in trimeric cyanobacterial PSI as gained from EM and crosslinking experiments with wild type and directed mutant forms. From Kruip *et al.*[355] Courtesy of Jochen Kruip: Left, top view; Right, side view. (B) Schematic representation (top view) of the structure of one monomer based on X-ray data at 0.6 nm resolution. The triangle at the left marks the threefold axis of a trimer and the heavy dashed lines the interfaces between monomers. Thirty-one helices, many of which are transmembrane, are shown as shaded cylinders. Fourteen of these are in the central core (light area) while the other 17 are in peripheral regions (light gray) that bind antenna chlorophyll molecules. (C) Side view, stroma at the top showing stromal subunits PsaC, PsaD, and PsaE. (D) Top view of the trimer showing locations of all α helices and core antenna chlorophylls as well as reaction center cofactors. From Schubert *et al.*[356] Courtesy of Nobert Krauss.

for it is not commonly appreciated that light carries entropy as well as energy. An important consequence of this fact is that not all of the energy of sunlight could be harnessed for chemical work. Knox[361] has calculated that at 700 nm at most 78% of the energy could be captured. See Parson[362] for further discussion. Nevertheless, the photoexcited P700* with E_m = – 1.26 V is able to reduce a series of membrane-bound Fe–S centers of $E°' \sim -0.5$ to -0.6 V. There are three of these designated F_X, F_A, and F_B. Center F_X is an Fe_4S_4 cluster located at almost the same position as the single Fe^{3+} of the *R. viridis* reaction center (Fig. 23-31B). F_A and F_B are also Fe_4S_4 clusters both of which are carried on the small 79-residue PsaC.[363,364] This protein binds to the reaction center on the stromal side as shown in Fig. 23-33. Close to it are two other subunits, PsaD and PsaE, which appear to assist the docking of ferredoxin or flavodoxin to PsaC and cluster F_X.[365,366]

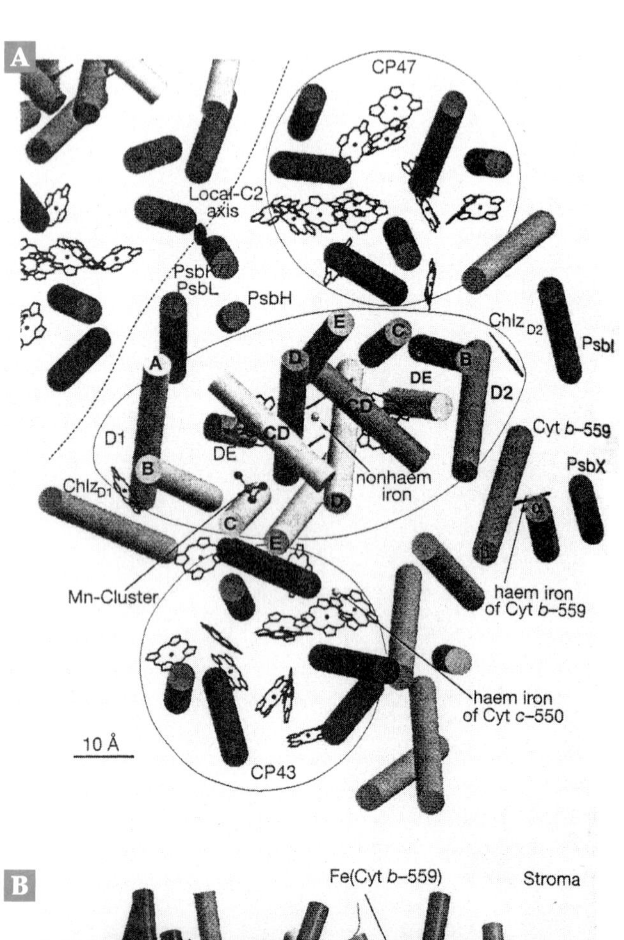

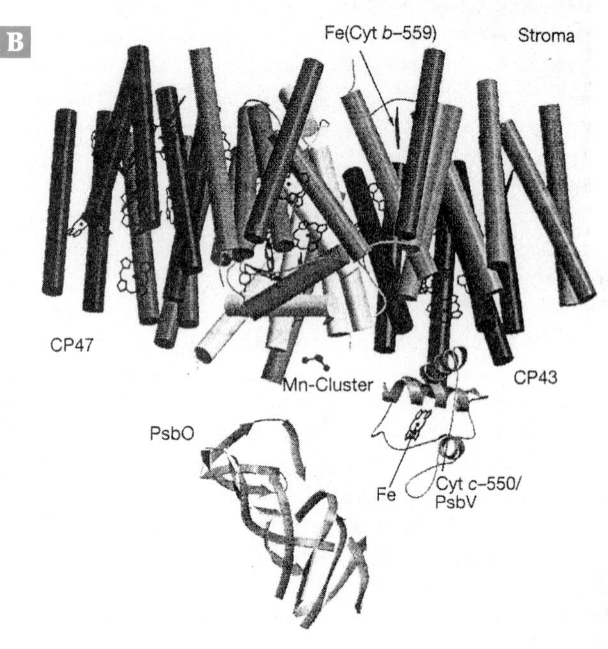

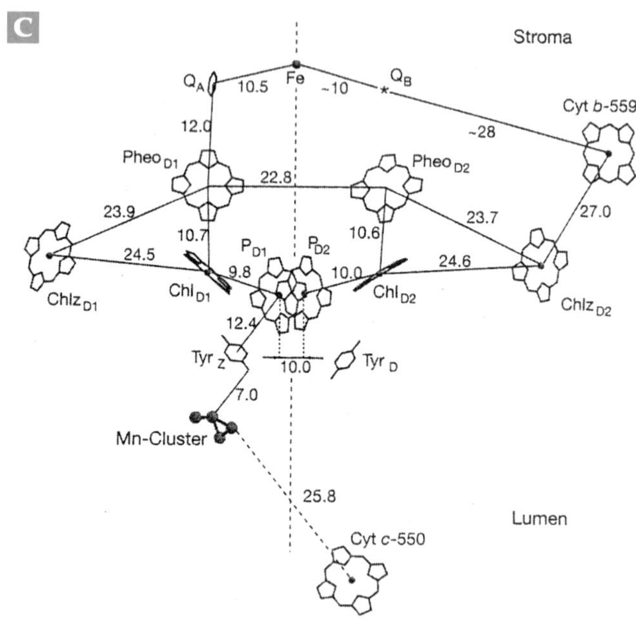

Figure 23-34 Structure of PSII with assignment of protein subunits and cofactors. (A) Arrangement of transmembrane α-helices and cofactors in PSII. One monomer of the dimer is shown completely, with part of the second monomer related by the local-C2 axis (filled ellipse on the dotted interface). Chlorophyll *a* head groups and hemes are indicated by black wire drawings. The view direction is from the luminal side, perpendicular to the membrane plane. The α-helices of D1, D2, and Cyt *b*-559 are labeled. D1/D2 are highlighted by an ellipse and antennae, and CP43 and CP47 by circles. Seven unassigned α-helices are shown in gray. The four prominent landmarks (three irons and the manganese (Mn) cluster) are indicated by arrows. (B) Side view of PSII monomer looking down the long axis of the D1/D2 subunits from the right side in (A), at slightly tilted membrane plane and rotated 180° so that the luminal side is bottom. PsbO (33K protein) is shown as a β-sheet structure, and Cyt *c*-550 as a helical model. (C) Arrangement of cofactors of the electron transfer chain located in subunits D1 and D2. View direction along the membrane plane. Full lines indicate center-to-center distances (nm) between the cofactors (uncertain to about ± 0.1 nm). The pseudo-C2 axis is shown by the vertical dotted line; it runs throught the nonheme iron Fe and is parallel to the local-C2 axis. The asterisk indicates the putative Q_B binding site. From Zouni *et al.*[371d] Courtesy of Athina Zouni.

The photosynthetic centers of green photosynthetic sulfur bacteria also have centers F_X, F_A, and F_B.

The soluble electron carriers released from the reaction centers into the cytoplasm of bacteria or into the stroma of chloroplasts are reduced single-electron carriers. Bacterial ferredoxin with two Fe_4S_4 clusters is formed by bacteria if enough iron is present. In its absence flavodoxin (Chapter 15), which may carry either one or two electrons, is used. In chloroplasts the carrier is the soluble **chloroplast ferredoxin** (Fig. 16-16,C), which contains one Fe_2S_2 center. Reduced ferredoxin transfers electrons to $NADP^+$ (Eq. 15-28) via **ferredoxin:NADP⁺ oxidoreductase**, a flavoprotein of known three-dimensional structure.[367–369]

PSII and formation of oxygen. The structure of PSII has been difficult to determine directly, but its core has been modeled in atomic detail using bacterial reaction centers as a guide.[370,371] More recently electron crystallography provided a three-dimensional image at 0.8 nm resolution.[314,371a] The resolution has been extended down to 4 nm by X-ray crystallography.[371b–e] The structure of the cyanobacterial PSII (Fig. 23-34) is very similar to that of green plants.[371e,f] PSII contains at least 17 protein subunits, all of which are encoded by chloroplast genes. The large structurally similar D1 and D2 form the core. They are encoded by genes *PsbA* and *PsbD* and have molecular masses of 38.0 and 39.4 kDa, respectively.[315] Both Q_A and Q_B are plastoquinone. It is of historical interest that these cofactors were first designated simply as Q, not for quinone but for *quencher*. This is because Q_A apparently quenched the fluorescence of P680, the reaction-center chlorophyll *a*. If chloroplasts are irradiated with 650-nm light, PSII is activated but PSI is not. Under these conditions Q_A becomes reduced, and the fluorescence of Chl^+ increases, presumably because the electron acceptor Q_A is absent. If PSI is activated by addition of far-red light, Q_A remains more oxidized, and fluorescence is quenched by a mechanism that appears to depend upon one of the additional bound chlorophylls, as well as the chlorophyll pair Chl_2 and the 9-kDa cytochrome b_{559}. The latter is an essential PSII subunit (*PsbE* gene),[372,373] which forms a tight complex with the D1/D2 pair.

Other subunits in PII include the 56- and 50-kDa antenna proteins CP43 (*PsbC* gene) and CP47 (*PsbB* gene). Three **extrinsic proteins**, which bind to the lumenal side of the thylakoid membrane, are the 33-kDa **manganese stabilizing protein** (*PsbO* gene),[374,375] cytochrome c_{550}, and a 12-kDa subunit (Fig. 23-34). The position of the larger two of these proteins are shown in the model in Fig. 23-34, which is based on 0.8-nm resolution data.[315] These extrinsic proteins seem to function together to facilitate binding not only of Mn ions but also of Ca^{2+} and Cl^-, both of which are essential for O_2 evolution.[376] Other smaller subunits are also present.[376a] Together with its antenna complexes PSII may form large supercomplexes with as many as 25 subunits.[376b] The distribution of complexes varies in the different regions of the thylakoid, e.g., the stroma lamellae and grana stacks (Fig. 23-19).[376c]

The four-electron dehydrogenation of two water

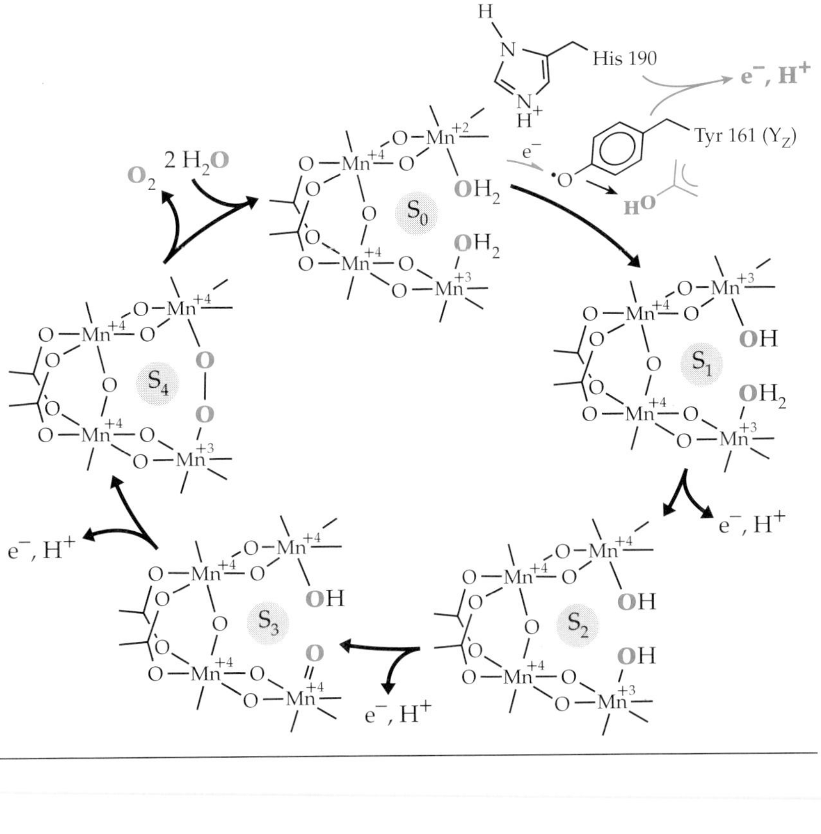

Figure 23-35 Proposed sequence of S-states of the manganese cluster of photosystem II. The successive states as two molecules of H_2O (green oxygen atoms) are converted to O_2 is shown with the successive states S_0–S_4 labeled. To save space and possible confusion tyrosine 161 (Y_Z) and the nearby His 190 are shown only by S_1. The Y_Z radical is thought to remove a hydrogen atom or H^+ from one bound H_2O and an electron from one Mn ion at each of the four S-states S_0–S_3 functioning in each case to eject a proton into the thylakoid lumen and to transfer an electron to P^+ of the reaction center. However, the exact sequence of e^- transfer and H^+ release may not be shown correctly. After Hoganson and Babcock.[392,392a]

molecules to give one molecule of O_2 by PSII is still not well understood. From experiments on oxygen evolution in the presence of repeated short flashes of light it was found that a four-quantum process is required.[376–380] There must be some way of storing oxidizing equivalents until enough are present to snap together an oxygen molecule. There is abundant evidence that manganese is required for this process and that the oxidation of H_2O occurs on a cluster of four atoms of manganese.[379]

The 33-kDa protein PsbO, which is present in all oxygen-forming photosynthetic organisms, is closely associated with the Mn_4 cluster. Removal of this protein leads to a gradual loss of two of the four Mn ions.[375] The structure of the Mn_4 cluster is not yet certain, but on the basis of EXAFS spectroscopy a pair of di-μ-oxo bridged Mn dimers linked by carboxylates and with a fifth μ-oxo bridge as in Fig. 23-35 has been proposed. The distance between Mn ions in the di-μ-oxo-Mn_2 groups is ~0.27 nm, and the planes of these groups are roughly parallel to the surface of the thylakoid membranes.[381,382] The protein groups that bind the Mn atoms include carboxylates, as shown in Fig. 23-35, and probably one or more histidine imidazole groups,[383] perhaps of His 332 and His 337 of the D1 chain.

The immediate donor of an electron to the reaction-center cation P^+ ($ChlZ^+$) of PSII was identified by EPR spectroscopy as a tyrosine radical.[384] On the basis of directed mutations this tyrosine, which is usually designated $\mathbf{Y_Z}$, is Tyr 161 of the D1 chain and is located ~1.2 nm from one of the chlorophylls of the $(Chl)_2$ pair.[371d,376,380,385–387] The two molecules in the pair are not in close contact, their central Mg atoms being ~1.0 nm apart. One of the two probably forms the $P680^{\bullet +}$ intermediate.[371d] Y_Z is also close to the Mn_4 cluster and to the imidazole group of His 190 of subunit D1 as is shown in Fig. 23-34C.[386,388] The $Y_Z^\bullet$ radical is able to accept an electron from the Mn_4 cluster within 30–1300 μs depending upon the oxidation state of the cluster (see Fig. 23-35).[388] If a proton is transferred synchronously from a bound H_2O, a neutral –OH group will be created on Y_Z. The proton may then be transferred to His 190, which can eject the proton on its other nitrogen atom into the lumen. Alternatively $Y_Z^\bullet$ may accept an electron to become tyrosinate $-O^-$, which then donates an electron to P^+, while His 190 accepts a proton directly from a bound H_2O.

In a mechanism proposed by Hoganson and Babcock (Fig. 23-35) four successive transfers, each of one H^+ + one e^-, leads to a three-electron oxidation of Mn ions, e.g., from the 2^+ and 3^+ oxidation states to all Mn^{4+}, and to joining of the two water oxygens to form a manganese peroxide linkage. Oxidation of the peroxide dianion to O_2 by the adjoining Mn^{4+} and Mn^{3+} ions completes the cycle. This mechanism is hypothetical, and various alternatives have been presented.[387–389b] Most assume a structure similar to that shown in Fig. 23-35. Some are based on nonenzymatic model reactions.[390,391] Chloride ions are essential to O_2 formation[393,394] especially in going from state S_2 to S_3 and S_3 to S_0. This suggests that Cl^- may function in passing electrons between Mn ions.[393,395] Calcium ions are also necessary, but it has been difficult to establish an exact function.[396–397a] A bicarbonate ion may also be an essential ligand in the Mn_4 cluster.[398,398a]

ATP synthesis in chloroplasts. The flow of electrons between PSII and PSI (Fig. 23-18) is of great importance for ATP formation. As previously mentioned, plastocyanin is usually the immediate donor to P700 and serves as a mobile carrier to bring electrons to this reaction center. In this function it is analogous to cytochrome *c* of mitochondrial membranes. The essentiality of plastocyanin was shown by study of copper-deficient *Scenedesmus* (Fig. 1-11). The photoreduction of CO_2 by H_2 is impaired in these cells, but the Hill reaction occurs at a normal rate.

Like mitochondria, chloroplasts (when illuminated) pump protons across their membranes (Fig. 23-18). However, while mitochondria pump protons to the outside, the protons accumulate on the inside of the thylakoids. The ATP synthase heads of coupling factor CF_1 are found on the outside of the thylakoids, facing the stromal matrix, while those of F_1 lie on the insides of mitochondrial membranes. However, the same mechanism of ATP formation is used in both chloroplasts and mitochondria (Chapter 18).

The cleavage of water at PSII also occurs on the inside of the thylakoids. The splitting of one water molecule leaves two protons (one per electron) inside the thylakoids, while the electrons are "photoejected" through the lipid bilayer to acceptor Q_A on the outside. The chlorophyll in PSI is likewise in contact with the inside of the bilayer with acceptor A_0 (Fig. 23-18) on the outside. Since the conversion of NAD^+ to NADH on the outside generates a proton, the overall reaction would be the pumping of one and a half protons per electron passing through the Z scheme.

The pathways involved in cyclic photophosphorylation in chloroplasts are not yet established. Electrons probably flow from the Fe–S centers Fd_x, Fd_a, or Fd_b back to cytochrome b_{563} or to the PQ pool as is indicated by the dashed line in Fig. 23-18. Cyclic flow around PSII is also possible. The photophosphorylation of inorganic phosphate to pyrophosphate (PP_i) occurs in the **chromatophores** (vesicles derived from fragments of infolded photosynthetic membranes) from *Rhodospirillum rubrum*. The PP_i formed in this way may be used in a variety of energy-requiring reactions in these bacteria.[399] An example is formation of NADH by reverse electron transport.

Protection of chloroplasts against radiation and oxygen. Carotenoids often act as accessory light-receiving pigments, but an additional function is protecting photosynthetic organisms against toxic effects of light.[400–403] Carotenoid photoprotection has been demonstrated in photosynthetic bacteria and in the reaction centers and light-harvesting complexes (LHC) of green plants. Excited chlorophyll molecules can pass from the singlet (^{1}Chl*) to the triplet (^{3}Chl*) state by intersystem crossing (Fig. 23-14). The triplet chlorophyll can then react with ordinary oxygen ($^3O_2$) to form singlet oxygen ($^1O_2$). The formation of ^{3}Chl* is favored when the intensity of sunlight is high and energy is absorbed in the LHCs faster than it can be utilized in the reaction centers. Carotenoids are able to quench excitation of both ^{3}Chl* and $^1O_2$. Strains of *Rhodobacter sphaeroides* that lack carotenoids rapidly form both ^{3}BChl* and BChl$^+$ cation radicals in their LH1 and LH2 complexes (Fig. 23-26). However, the presence of carotenoids quenches the triplet bacteriochlorophyll effectively and no formation of BChl$^+$ radicals is seen.[403] Another effect of excessive light energy is the reaction of O_2 with the highly reducing P700* of photoexcited PSI. This interaction can form triplet oxygen, which can react to generate superoxide anion radicals, H_2O_2, and hydroxyl radicals.[224,227,402]

In higher plants the quenching of both ^{3}Chl* and $^1O_2$ depends upon carotenoids[404] and also upon the large transthylakoid membrane potential that is generated by high light intensities.[405,406] High light intensity also induces a rapid reductive deoxygenation of epoxycarotenoids via the **xanthophyll** or **violaxanthin cycle** (Eq. 22-10).[401,407,407a,b] Epoxycarotenoids are found only in photosynthetic O_2 evolving organisms. Although occurring in response to light, the cyclic photodeoxygenation and reoxygenation is not a nonphotochemical process. Violaxanthin contains the epoxy structure at both ends of the molecule. Reduction of one end produces **antheroxanthin** and of both ends **zeaxanthin**. These three carotenoids are found in almost all higher green plants and algae. The de-epoxidation is mediated by ascorbic acid, occurs in the lumen of the thylakoids, and is favored by the low pH developed during illumination. Epoxidation (Eq. 22-10) is catalyzed by a monooxygenase located on the stromal side. The significance of the xanthophyll cycle is puzzling. There may be specific binding sites, perhaps in the inner antenna complex on CP29 (Fig. 23-34), that bind zeaxanthin or antheroxanthin. This could alter the antenna structure to form an "exciton trap" in which the fluorescence lifetime would be decreased and excitation energy would be dissipated rapidly as heat.[408] Zeaxanthin is also found in the macular area of the primate retina.[401]

Recent studies, using an *Arabidopsis* mutant defective in the xanthophyll cycle, point to a chlorophyll-binding protein PsbS, which participates in nonphotochemical quenching at high light intensity.[401a,b] Another *Arabidopsis* protein, which is probably a blue light receptor, participates in an avoidance response by which chloroplasts move to the side wall to avoid strong light.[401c]

Yet another carotenoid function in PSII has been proposed. Under some conditions, when electron flow from Tyr_z is blocked, the bound monomeric chlorophyll Z (Chlz) acts as a secondary electron donor to P680$^+$. The cytochrome b_{559} subunits may have a similar function. Both cyt b_{559} and the carotenoid are essential for assembly of PSII, and both may participate in a protective cycle.[409,410] Chloroplasts generate both O_2 and powerful reducing materials such as the membrane-bound FeS centers of PSI, which may form superoxide ions by single-electron donation to O_2. Probably for this reason, chloroplasts are rich in superoxide dismutase which converts superoxide to O_2 and H_2O_2. The latter can diffuse into peroxisomes and react with catalase and peroxidases. It can also be reduced to H_2O within the chloroplasts by ascorbic acid and ascorbate peroxidase.[224,227] The resulting dehydroascorbate (Box 18-D) can be reduced back to ascorbate by glutathione (Box 11-B) and dehydroascorbate reductase, in the following electron transfer sequence:

$$\text{NADPH} \rightarrow \text{G-S-S-G} \rightarrow \text{dehydroascorbate} \rightarrow H_2O_2 \quad (23\text{-}33)$$

Under extreme conditions of excess light energy **photoinhibition** is observed as a result of damage to the PSII structure.[411–415] The D1 polypeptide is cleaved, probably as a result of oxidation by $^1O_2$ and proteolysis. Damaged proteins are replaced and the PSII structure rebuilt, but the effect is a long-lasting decrease in photosynthetic efficiency. The cyanobacterium *Synechococcus* has three *PsbA* genes and resists UV-B radiation by exchanging a delicate D1 polypeptide with more resistant ones as necessary.[415] Other adaptations to varying light-intensity involve movement of light-harvesting complexes from the thylakoid stacks, which contain much PSII, to the stroma lamellae (Fig. 23-19), which contain more PSI. Some herbicides act by binding into the Q_B site in PSII. They may cause light-induced oxidative stress that kills the plant.[415a]

5. Control of Photosynthesis

The key reaction of the Calvin–Benson cycle of CO_2 reduction is the carboxylation of ribulose bisphosphate to form two molecules of 3-phosphoglycerate (Eq. 13-48). The properties of ribulose bisphosphate carboxylase (**rubisco**, Figs. 13-10 to 13-12), which catalyzes this reaction, are discussed in Chapter 13. It is controlled in part by CO_2 and by natural inhibitors,[416] but regulation of rubisco starts at the transcriptional level.

Light-induced transcription. Plants depend upon light both as a source of energy and also for control of development. Many genes are activated by light in response to at least three groups of photoreceptors. These are **phytochromes** (Section H) and the blue light responsive **cryptochromes** (Section I) and the ultraviolet light **UV-B photoreceptors**.[417] The synthesis of chlorophyll, of reaction center proteins, and of many enzymes are controlled by light-induced transcription.[418–420] Among these processes are synthesis of both the large and small subunits of rubisco (Fig. 13-10). The small subunits are synthesized in the cytoplasm in a precursor form. After illumination the concentration of the rubisco mRNA may be increased 100-fold.[421,422] On the other hand, the large subunit of the carboxylase is encoded in chloroplast DNA, and stimulation of its synthesis by light appears to be at the translational level.[423]

Light-induced control via the ferredoxin/thioredoxin system. Rubisco is activated by CO_2 (Chapter 13) and by fructose 6-*P* and is inhibited by fructose 1,6-P_2 (Fig. 23-36),[424] whose accumulation is a signal to turn off the carboxylase. Conversely, fructose 6-*P* in high concentrations turns on the Calvin–Benson cycle. Like the reactions of gluconeogenesis (Chapter 17), photosynthetic CO_2 incorporation is dependent on the highly regulated fructose-1,6-bisphosphatase. In chloroplasts it is activated by light through the mediation of reduced ferredoxin and thioredoxin.[424–427] The small mobile thioredoxin (Box 15-C) is reduced to its dithiol form by reduced ferredoxin[428–429a] and then reduces one or more disulfide linkages in the fructose 1,6-bisphosphatase to activate that enzyme (Fig. 23-36). Other light-activated enzymes of the Calvin–Benson cycle include sedoheptulose-1,7-bisphosphatase, the phosphoribulokinase that forms ribulose 1,5-bisphosphate and the $NADP^+$-dependent glyceraldehyde-3-phosphate dehydrogenase. $NADP^+$-dependent malate dehydrogenase, which has a major function in C_4 plants (see Fig. 23-38), is totally inactive in the dark but is activated by the ferredoxin–thioredoxin system in the light.[427] The activity of the LHCII complex is also affected.[429b]

Another aspect of chloroplast metabolism is synthesis of starch. Formation of ADP-glucose from glucose 1-phosphate is induced by 3-phosphoglycerate, a "feed-ahead" type of regulation (Fig. 23-36). Although fructose 2,6-bisphosphate is absent from chloroplasts, it has an important regulatory function in the cytoplasm of plants as it does in animals.[425,430] In the plant cytosol triose phosphates from the chloroplasts are converted to fructose 6-*P*, glucose 6-*P*, UDP-glucose, and sucrose. Inorganic phosphate P_i, which accumulates in plant vacuoles, also has a regulatory function.[431] It activates the kinase that converts fructose 6-*P* to fructose 2,6-P_2 and inhibits the phosphatase that converts the bisphosphate back to fructose 6-*P*. The accumulated fructose 2,6-P_2 inhibits fructose-1,6-bisphosphatase and slows the conversion of triose phosphates to sucrose (Eq. 23-34). Accumulation of fructose 6-*P* due to decreased utilization for sucrose formation will have a similar effect. However, both 3-phosphoglycerate and dihydroxyacetone phosphate have opposite effects and will act to remove the inhibition by lowering the fructose 2,6-P_2 level and to promote rapid sucrose formation (Eq. 23-34).[430]

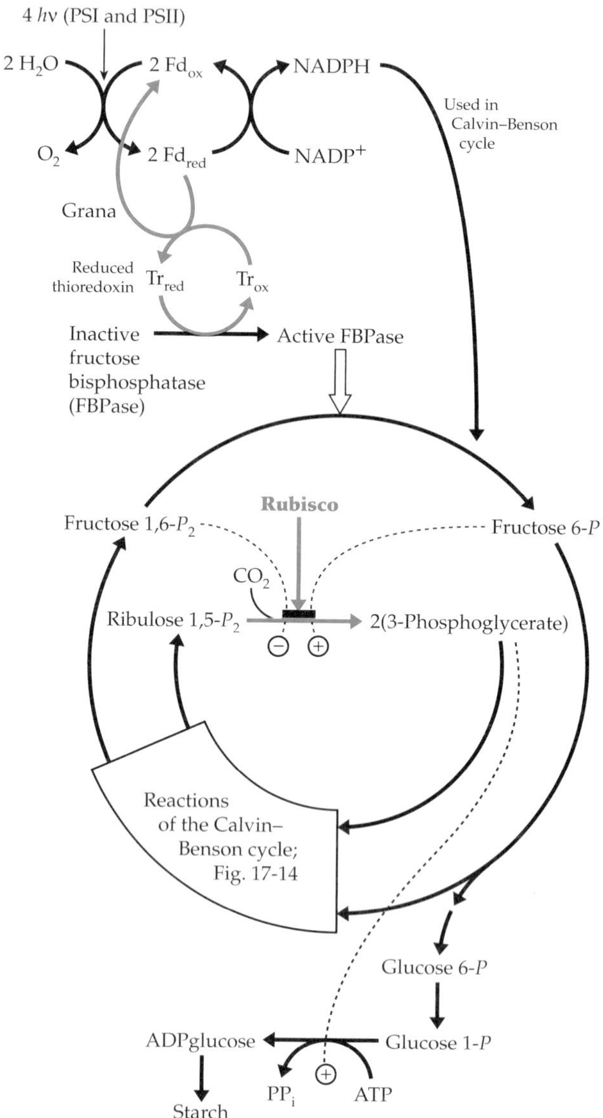

Figure 23-36 Some control mechanisms for photosynthetic assimilation of carbon dioxide. After Buchanan and Schurmann[424] with modifications.

$$\text{Triose-}P \longrightarrow \text{Fructose-1,6-}P_2 \xrightarrow{\ominus} \text{Fructose-6-}P \longrightarrow \longrightarrow \text{Sucrose}$$

ATP ⊕ ⊖ P_i ⊖ ⊕ H_2O Fructose-2,6-P_2 (23-34)

6. Photorespiration; C_3 and C_4 Plants

The first product of incorporation of CO_2 via the Calvin–Benson cycle is 3-phosphoglyceric acid (Box 17-E). It was the rapid appearance of radioactivity from $^{14}CO_2$ in phosphoglycerate and other three-carbon (C_3) compounds that permitted Calvin and associates, using green algae, to work out the complex cycle as it is shown in Fig. 17-14. Green algae, spinach, and many common crop plants are often known as **C_3 plants**. Another group of plants, mostly of tropical origin and capable of extremely fast growth, behave differently.[224,227,432] In these plants, which include sugar cane, maize, and crabgrass, radioactivity from ^{14}C-containing CO_2 is found first in the C_4 compounds oxaloacetate, malate, and aspartate. These **C_4 plants** are characterized by high efficiencies in photosynthesis, a fact that explains the rapid growth of crabgrass and the high yield of corn. Maximum rates of CO_2 incorporation may attain 40–60 mg of CO_2 per square decimeter of leaf surface per hour (~0.3 mmol CO_2 $m^{-1}s^{-1}$ or ~0.10 mol CO_2 per mol of total chlorophyll per second), more than twice that for common crop plants.

Like all other organisms plants respire in the dark, but illumination of C_3 plants markedly increases the rate of oxygen utilization. This light-enhanced respiration (**photorespiration**) may attain 50% of the net rate of photosynthesis. Photorespiration prevents plants from achieving a maximum yield in photosynthesis. For this reason, its understanding and control assume great importance in agriculture. It is difficult to measure the rate of photorespiration, and the literature on the subject often refers instead to the **CO_2 compensation point**. This is the CO_2 concentration (at a given constant light intensity) at which photosynthetic assimilation and respiration balance. (Similarly, the **light compensation point** is the light intensity at which the rate of photosynthetic CO_2 incorporation and that of respiration exactly balance.) Normal air has a CO_2 content of ~0.03% or 300 ppm. For common C_3 crop plants the CO_2 compensation point is ~40–60 ppm at 25°C. The C_4 plants are characterized by a much lower CO_2 compensation point, often less than 10 ppm. In strong sunlight the CO_2 level of air in a field of growing plants drops. Furthermore, as the temperature rises on a hot day, the CO_2 compensation point rises. The result is a serious decrease in efficiency of photosynthesis for the C_3 plants but not for the C_4 plants.

Metabolism of glycolic acid. The 2-carbon glycolic acid is formed in large quantities in the chloroplasts of C_3 plants and moves out into the cytosol.[433,434] The major source of this acid is phosphoglycolate whose formation is catalyzed by rubisco in the chloroplasts through competition of O_2 for the CO_2 binding site of the enzyme (Eq. 13-50). It is easy to understand why an increase in the O_2 pressure in air increases the CO_2 compensation point for a plant. Another less important source of glycolate is transketolase, which may yield glycolaldehyde as a side product (Eq. 17-15). Glycolaldehyde can be oxidized readily to glycolate. Glycolate is metabolized rapidly, some in the chloroplasts,[435] but most in the peroxisomes. There the flavoprotein glycolate oxidase converts it to glyoxylate with formation of H_2O_2 (Fig. 23-37).[436] Some of the hydrogen peroxide formed may react nonenzymatically, decarboxylating glyoxylate to formate and CO_2, but most is probably destroyed by peroxidases or catalase. The latter enzyme is lacking in chloroplasts, one reason why oxidation of glycolate must occur in the peroxisomes.

Glyoxylate undergoes transamination to glycine, which can be oxidatively decarboxylated (Fig. 15-20) in the mitochondria. It can also be converted to serine,[437] some of which returns to the peroxisomes to be oxidized to hydroxypyruvic acid and glyceric acid (Fig. 23-37). The latter can be synthesized into glucose. The net result is the stimulation of a large amount of metabolism that ultimately produces CO_2 and apparently accounts for the light-induced respirations of plants. Because much of the glycine formed in the peroxisomes is oxidatively decarboxylated in the mitochondria, photorespiration also generates large amounts of NH_3. This is recycled into amino acids within the photosynthetic cells (green lines in Fig. 23-37), an energy-requiring process.[438]

Although much metabolism occurs as a result of photorespiration, it appears to waste energy rather than to provide energy to the plant. Why then don't plants avoid this process? Wouldn't a small change in the structure of ribulose bisphosphate carboxylase allow plants to avoid photorespiration and to grow more efficiently? The answer is not clear. It has been difficult to create such modified plants, and there is a possibility that they would not grow well. One theory is that photorespiration protects plants when the CO_2 pressure is low and the absorbed light would damage the chloroplasts if there were not a way to utilize the accumulating reduced Fe–S proteins generated by PSI. Photorespiration provides a mechanism.[434] Most

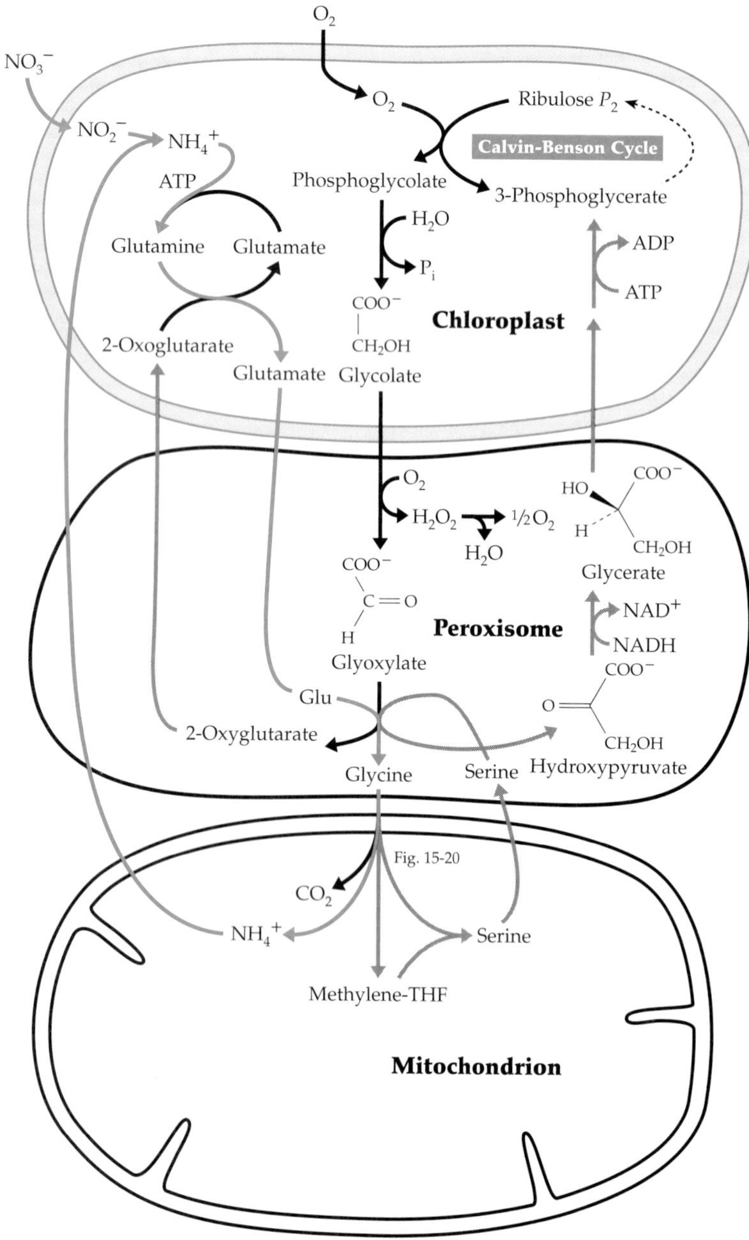

Figure 23-37 Production of glycolate by chloroplasts and some pathways of its metabolism in peroxisomes and in mitochondria. After Tolbert[436] and Givan *et al.*[438]

efforts to breed plants with lower photorespiration rates or to inhibit it chemically have failed.[433,439]

The C_4 cycle for concentration of carbon dioxide. The C_4 plants reduce their rate of photorespiration by using a CO_2 concentrating mechanism that enables them to avoid the competition from O_2. All species of C_4 plants have a characteristic internal leaf anatomy in which a single dense layer of dark green cells surrounds the vascular bundles in the leaves. This **bundle sheath** is surrounded by a loosely packed layer of cells, the **mesophyll**, an arrangement that is sometimes called the "Kranz anatomy." In C_4 plants there is a separation of the chemical reactions between the mesophyll and the bundle sheath cells. The incorporation of CO_2, as bicarbonate ion into oxaloacetate, occurs in the mesophyll cells, principally through the action of PEP carboxylase (Fig. 23-38). Oxaloacetate is reduced to malate by light-generated NADPH. Alternatively, it undergoes transamination to aspartate. Both malate and aspartate then diffuse out of the mesophyll cells and into bundle sheath cells where the malate undergoes oxidative decarboxylation via the malic enzyme (Eq. 14-42) to pyruvate (Fig. 23-38). Aspartate also can be converted to oxaloacetate, malate, and pyruvate in the same cells. The overall effect is to transport CO_2 from the mesophyll cells into the bundle sheath cells along with two reducing equivalents, which appear as NADPH following the action of the malic enzyme. The CO_2, the NADPH, and additional NADPH generated in the chloroplasts of the bundle sheath cells are then used in the Calvin–Benson cycle reactions to synthesize 3-phosphoglycerate and other materials. Of the CO_2 used in the bundle sheath cells, it is estimated that 85% comes via the C_4 cycle and only 15% enters by direct diffusion. The advantage to the cell is a higher CO_2 tension, less competition with O_2, and a marked reduction in photorespiration.

The pyruvate produced in the bundle sheath cell is largely returned to the mesophyll cells where it is acted upon by **pyruvate-phosphate dikinase**.[441] This unusual enzyme (Eq. 17-55) phosphorylates pyruvate to PEP while splitting ATP to AMP and PP_i. The latter is in turn degraded to P_i. In effect, two high-energy linkages are cleaved for each molecule of pyruvate phosphorylated. Because of this extra energy need, it is thought that cyclic photophosphorylation is probably more important in the chloroplasts of the mesophyll cells than in the bundle sheath cells. It also accounts for the fact that C_4 plants are less efficient than C_3 plants under cool or shaded conditions.[442] Other CO_2-concentrating mechanisms exist in plants.[442a] For example, cyanobacteria accumulate HCO_3^- ions in carboxysomes, polyhedral bodies to which rubisco adheres. An ABC type ATP-dependent transporter powers the bicarbonate accumulation.[442b]

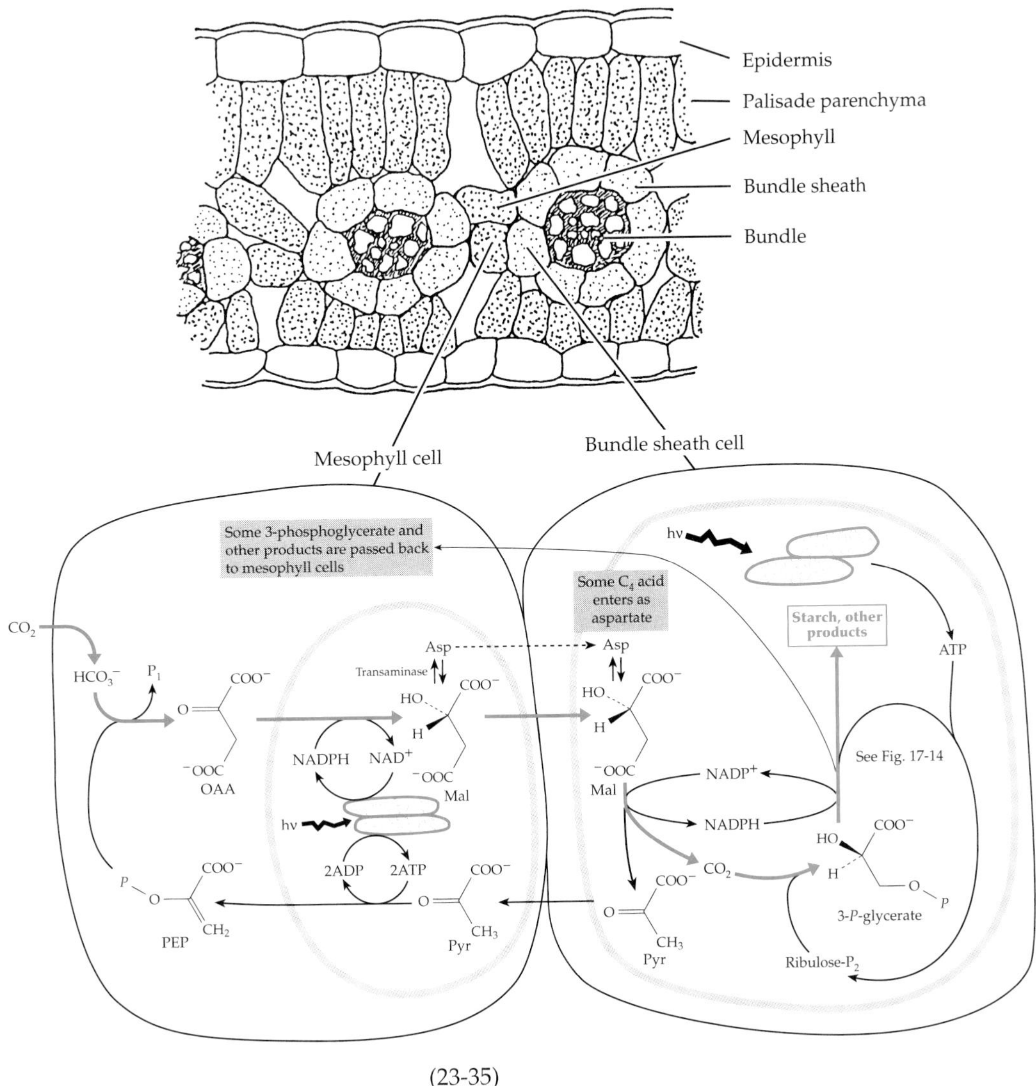

(23-35)

Figure 23-38 The C_4 cycle for concentrating CO_2 in the C_4 plants. From Haag and Renger with alterations.[440]

Metabolism in the family Crassulaceae. The crassulacean plants are a large group that includes many ornamental succulents such as *Sedum*. They have a remarkable metabolism by which large amounts of malic and isocitric acids are synthesized at night. During the day when photosynthesis occurs these acids disappear. The stomata in the leaves (Chapter 1) stay closed during the day and open only at night, an adaptation that permits the plants to live with little water. However, this requires that the plant accumulates carbon dioxide by night and incorporates it photosynthetically into organic compounds by day.[443] A possible mechanism is shown in Fig. 23-39. On the left side of the figure are reactions by which starch can be broken down at night to PEP. While it would also be possible to produce that compound by the glycolysis reactions, labeling studies have indicated that the pentose phosphate pathway is more important.[330] The PEP acts as the CO_2 acceptor to create oxaloacetate, which is then reduced to malic acid. A balanced fermentation reaction (Fig. 23-39; Eq. 23-35) can be written by using the NADPH formed in the conversion of glucose 6-*P* to ribulose 5-*P*. During the day when ATP and NADPH are available in abundance from photoreactions, the conversions on the right side of the figure can take place. The initial step, the release of CO_2 from malic acid by the malic enzyme, is the same as that employed by C_4 plants. In this case, it is used to release the CO_2 stored by night, making it available for incorporation via the Calvin–Benson cycle. The remaining pyruvate is reconverted to starch.

$$C_6H_{10}O_5 \text{ (starch)} + 2\,CO_2\text{ (g)} \rightarrow 2\,C_4H_4O_5^{2-} \text{ (malate)} + 4\,H^+ \quad \Delta G' \text{ (pH 7)} = -159 \text{ kJ mol}^{-1} \tag{23-35}$$

Many plants store substantial amounts of malate in their cytoplasm and in vacuoles. It apparently

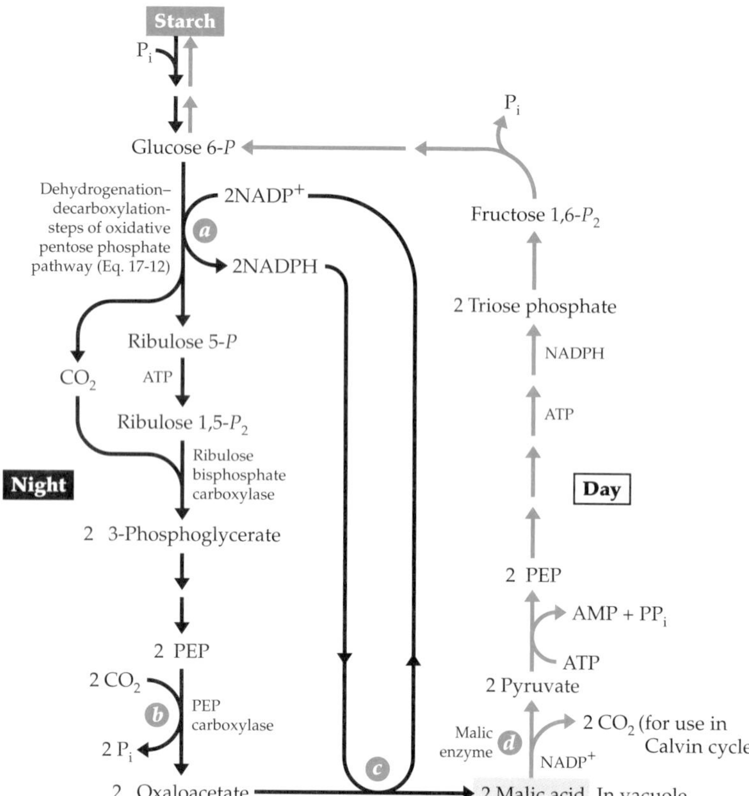

Figure 23-39 A proposed night–day cycle of crassulacean acid metabolism.

serves as a ready reserve for carbohydrate synthesis.

7. Photosynthetic Formation of Hydrogen

A system consisting of chloroplasts, ferredoxin, and hydrogenase has been used to generate H_2 photosynthetically.[444] This may be a prototype of a method of solar energy generation for human use. Another photochemical hydrogen-generating system makes use of both the nitrogen-fixing heterocysts and photosynthetic vegative cells of the cyanobacterium *Anabaena cylindrica*.[445] In this instance hydrogen production is accomplished by nitrogenase (Eq. 24-6). Photogeneration of H_2 by bacteria is just one of many kinds of photometabolism observed among photosynthetic microorganisms.

F. Vision

The light receptors of the eye perform a very different function from those of chloroplasts. Visual receptors initiate nerve impulses, and their primary requirement is a high sensitivity. By the use of stacked membranes containing a high concentration of an intensely absorbing molecule[446,447] the most sensitive visual receptors are able to trap nearly every photon that strikes them. The retina of the human eye contains more than 10^8 tightly packed receptor cells of two types. The **rod cells** are extremely sensitive. Used for night-time vision, they give a "black and white picture" and are concentrated around the periphery of the retina. The retina works as a coincidence detector. An ensemble of ~500 rods must register 5–7 isomerizations within a few tenths of a second in order to trigger a nerve impulse.[448] The less sensitive **cone cells**, which are most abundant in the center of the retina, are of three types with different spectral sensitivities. They provide color vision.

The retinal receptors have a very active metabolism. Human rod cells (Fig. 23-40) may live and function for a hundred years.[190] A self-renewal process leads to a casting off of the older membranous discs from the end of the rod[194] and replacement by new discs at the end nearest to the nucleus. The rod outer segment is surrounded by a plasma membrane. Within the membrane but apparently not attached to it are ~500 stacked discs of ~2 μm diameter and with a repeat distance between centers of ~32 nm. Each disc is enclosed by a pair of membranes each ~7 nm thick with a very narrow space between them. From electron micrographs it appears that this space within the discs is sealed off at the edges. Somewhat larger spaces separate adjacent discs.

The membranes of the rod discs are ~60% protein and 40% lipid (Table 8-3). About 80% of the protein is **rhodopsin** (visual purple), a lipoprotein that is insoluble in water but soluble in detergent solutions. Digitonin is widely used to disperse rhodopsin molecules because it causes no change in optical properties. In addition to rhodopsin, in the outer segment discs of frog retinal rods, there are ~65 molecules of phospholipid and smaller amounts of other materials for each molecule of rhodopsin (Table 8-3). The cone cells have a similar architecture but have a different shape and contain different light receptors. The receptors in the cones are present in deep indentations of the plasma membrane rather than in discs within the cytoplasm.

1. Visual Pigments

The rod pigment rhodopsin is readily available from cattle retinas and has been studied for many

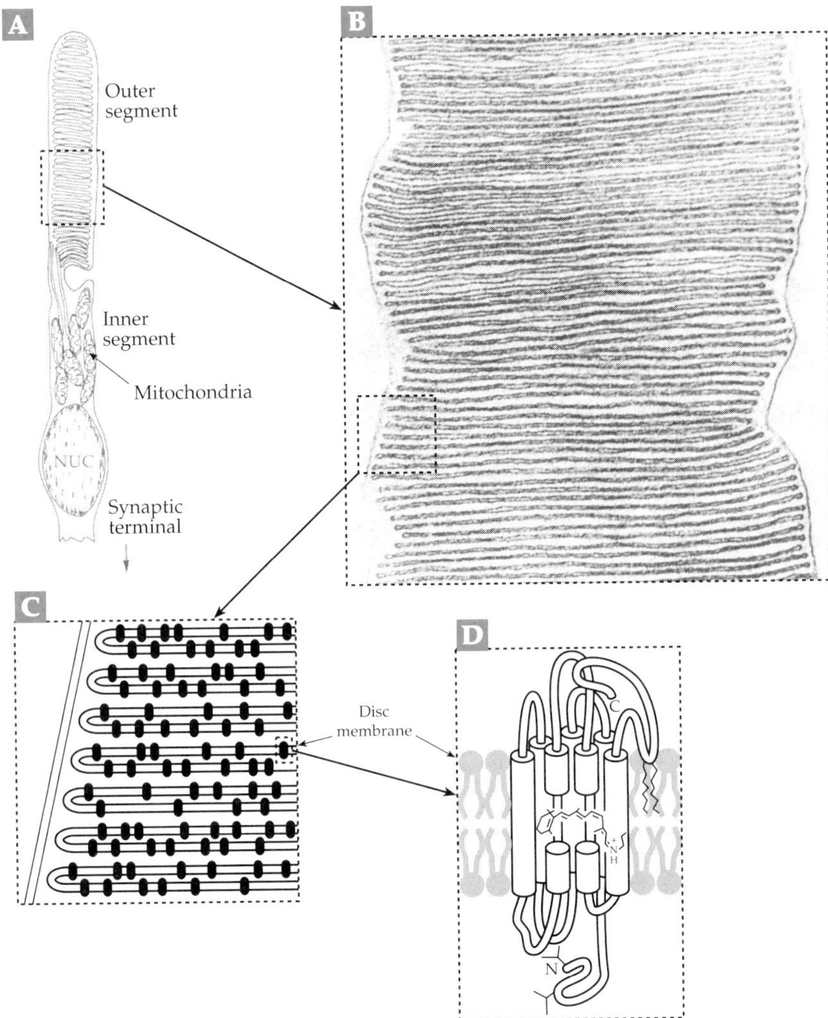

Figure 23-40 (A) Diagram of a vertebrate rod cell. From Abrahamson and Fager.[446] (B) Electron micrograph of a longitudinal section of the outer segment of a retinal rod of a rat. There are 600–2000 discs per rod and 2×10^4 and 8×10^5 rhodopsin molecules per disc. Courtesy of John E. Dowling. (C) Enlarged section; from Dratz and Hargrave.[451] (D) Schematic drawing of rhodopsin. The two α helices in the front have been partly cut away to reveal the 11-*cis* retinal in protonated Schiff base linkage to lysine 296. From Nathans.[448] Courtesy of Jeremy Nathans.

constructed for the three human cone pigments[452] and for the related bacterial protein **bacteriorhodopsin**.[453]

Transmembrane structure. From its circular dichroism rhodopsin was estimated to be 60% helical, and its amino acid sequence suggested that it contains seven parallel membrane-spanning helices (Fig. 23-41)[451] just as does bacteriorhodopsin (Section G; Fig. 23-45). Rhodopsin and other visual pigments are also members of the large family of **G-protein coupled receptors**, which includes the β_2 adrenergic receptor pictured in Fig. 11-6. It has been hard to determine the three-dimensional structure of rhodopsin or other receptors of this family. However, their relationship to bacteriorhodopsin, whose structure was obtained in 1975 by electron crystallography[454] and more recently by X-ray crystallography at 0.15 nm resolution,[455] permitted modeling based on similarities among the proteins.[456–458] New results of electron crystallography[459,460] and mass spectrometry[461] have been combined with studies of mutant forms of rhodopsin and other visual pigments to provide the picture given in Fig. 23-41A,B. Recently a higher resolution structure (Fig. 23-41D) has been obtained by X-ray diffraction.[461a–c] Mutations have been introduced in every part of the rhodopsin molecule, and the effects on photoreception, protein stability, and other properties have been observed.[448,450,462–465] Some of the essential residues identified are indicated in Fig. 23-41A, a schematic diagram showing the seven helices and connecting loops. Figure 23-41B shows an end view of the helix bundle with its retinal prosthetic group buried in the interior of the protein. Rhodopsin is roughly cylindrical with a length of 6–6.5 nm and a diameter of ~2.8 nm and is embedded in the phospholipid bilayer with its long axis perpendicular to the membrane surface (Fig. 23-40). The two oligosaccharide chains, which are attached near the N terminus, project into the intradiscal space on the side away from the cytoplasm. Palmitoyl groups on two cysteine side chains help to anchor the protein. Rhodopsin apparently exists in the mem-

years. It has a molecular mass of ~41 kDa of which ~2 kDa is contributed by two asparagine linked oligosaccharides.[195] Both bovine and human rhodopsins consist of a 348-residue protein known as **opsin** to which is bound a molecule of vitamin A aldehyde, **retinal**. Human and bovine opsins are 93% homologous. A totally synthetic 1057 base pair gene for bovine rhodopsin was made by Khorana and associates.[449,450] The gene was constructed with 28 unique sites for cleavage by specific restriction endonucleases. These have allowed easy specific mutation of the gene and production of a wide variety of mutant forms of opsin. Similar synthetic genes have been

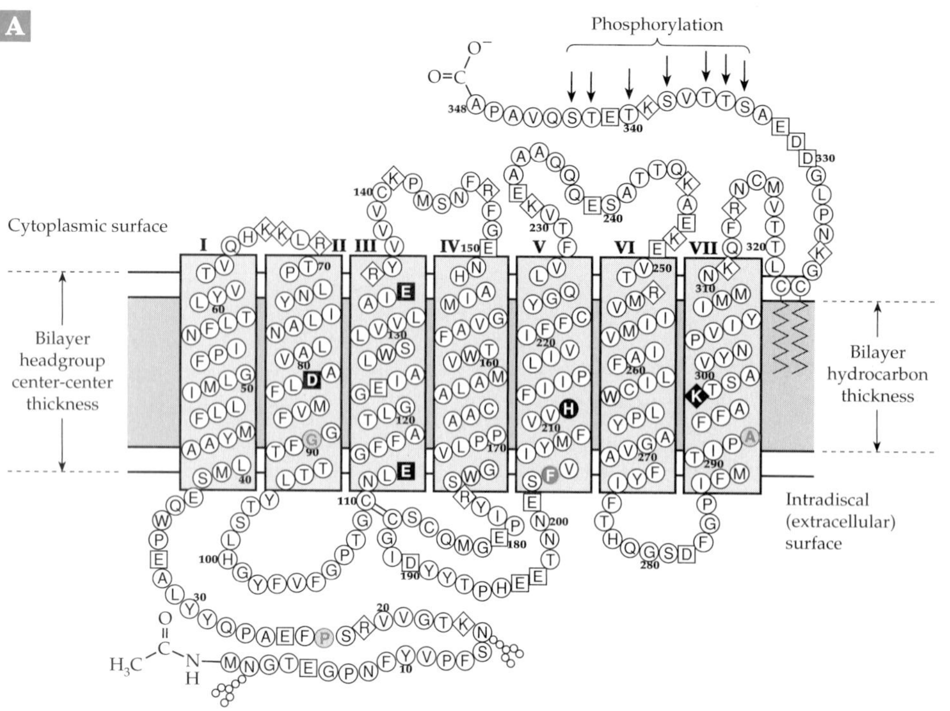

● A few residues essential to the mechanism of vision.

○ Point mutations causing serious diseases. P23H is a frequent cause of retinitis pigmentosa. Mutations G90D and A292E cause congenital night blindness.

● The mutation C203R in the green-sensitive pigment causes red-green color blindness.

Figure 23-41 (A) Model of the topological organization of bovine rhodopsin according to the consensus analysis of Baldwin.[456] Oligosaccharide chains are portrayed on the glycosylation sites Asn 2 and Asn 15. Glutamate 113 provides the counter ion for the *N*-protonated retinal Schiff base. Cysteines 110 and 187 form an essential disulfide linkage. Histidine 211 modulates the interconversion of the metarhodopsin forms I and II. Lysine 296 forms a Schiff base with 11-*cis*-retinal. Cysteines 322 and 323 are sites of palmitoylation, and as many as six serine and threonine hydroxyls (indicated by green arrows) may become phosphorylated during desensitization of rhodopsin. Aspartate 83, glutamate 134, and histidine 211 may be essential for proton movements. The point mutation C203R in the green-sensitive pigment causes red-green color blindness. The mutation

branes as monomers. From the composition it can be calculated[451] that the average center-to-center distance of the cylindrical molecules must be ~5.6 nm.

The visual chromophores. Rhodopsin has been an object of scientific interest for over 100 years.[462] Wald and associates[469,470] established that rhodopsin contains 11-*cis*-retinal bound to the opsin in Schiff base linkage (Eq. 23-36). When native rhodopsin is treated with sodium borohydride, little reduction is observed. However, after the protein is bleached by light, reduction of the Schiff base linkage becomes rapid, and the retinal is incorporated into a secondary amine, which was identified as arising from Lys 296.

In a crystal structure[471] 11-*cis*-retinal has the 12-*s*-*cis* conformation shown at the top in Eq. 23-36 rather than the 12-*s*-*trans* conformation at the center and in which there is severe steric hindrance between the 10–H and 13–CH_3. Nevertheless, ^1H– and ^{13}C–NMR spectroscopy suggest that the retinal in rhodopsin is in a twisted 12-*s*-*trans* conformation.[472,472a] The Schiff base of 11-*cis*-retinal with *N*-butylamine has an absorption maximum at ~360 nm but *N*-protonation, as in the structure in Eq. 23-36, shifts the maximum to 440 nm with ε_{max} = 40,600 M^{-1} cm^{-1} (Fig. 23-42). This large shift in the wavelength of the absorption maximum (the *opsin shift*) indicates that binding to opsin stabilizes the light-excited state by almost 33 kJ/mol compared to that in the free *N*-protonated Schiff base. This is evidently the result of a fixed negative charge, that of Glu 113, which is near the polyene chain of the retinal (Fig. 23-41B) and probably separated from it by a hydrogen-bonded water molecule.[473–477]

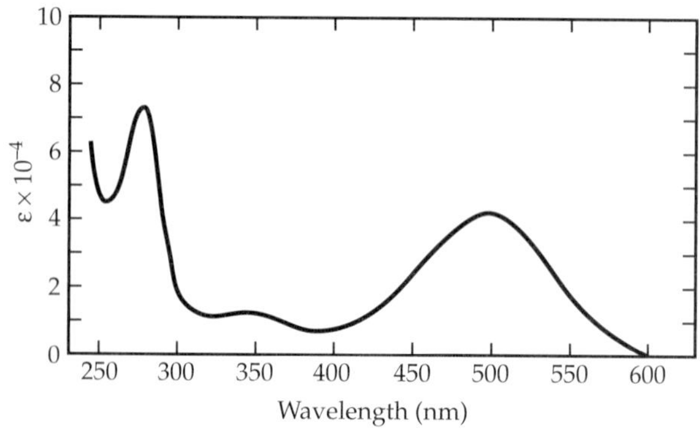

Figure 23-42 Absorption spectrum of cattle rhodopsin in aqueous dispersion with a nonionic detergent. From H. Shichi *et al.*[486,487]

P23H in rhodopsin is one of the most frequent causes of retinitis pigmentosa, and mutations G90D and A292E cause congenital night blindness. From Barnidge *et al.*[461] with modifications. (B) Structural model of rhodopsin based on the helix arrangement of Baldwin[456] and NMR constraints. The seven transmembrane helices, viewed from the cytoplasmic surface, are shown at three levels indicated by differences in shading. The α-carbons of residues of interest are shown in dots of various sizes, indicating the depth from the cytoplasmic domain. Gly 90, Glu 113, and Lys 296 are on the extracellular half of the transmembrane domain and close to each other in space. The 11-*cis*-retinal chromophore has been incorporated into the model using NMR constraints, which require a close interaction between Glu 113 and C_{12} of the chromophore. The relative position of the β-ionone ring and Trp 265 agrees well with crosslinking data of Zhang *et al.*[466] From Han and Smith.[467] (C) Stereoscopic view of the retinal-binding pocket of **bacteriorhodopsin** viewed from the cytoplasmic surface. The retinal, in Schiff base linkage to Lys 216, runs across the central cavity from top to bottom in this view. From Grigorieff *et al.*[468] (D) Ribbon drawing of bovine rhodopsin (stereoview). From Palczewski *et al.*[461a]

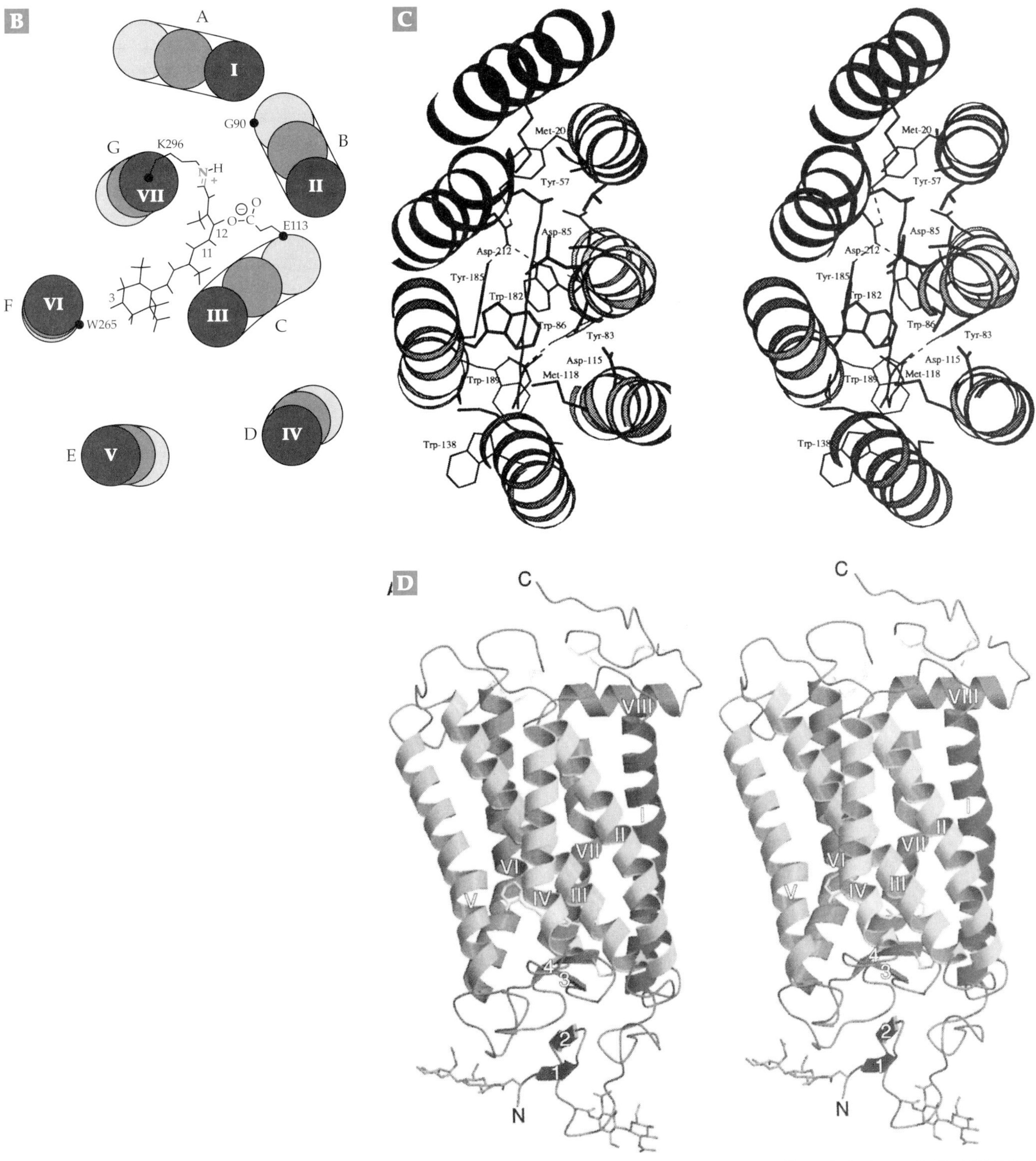

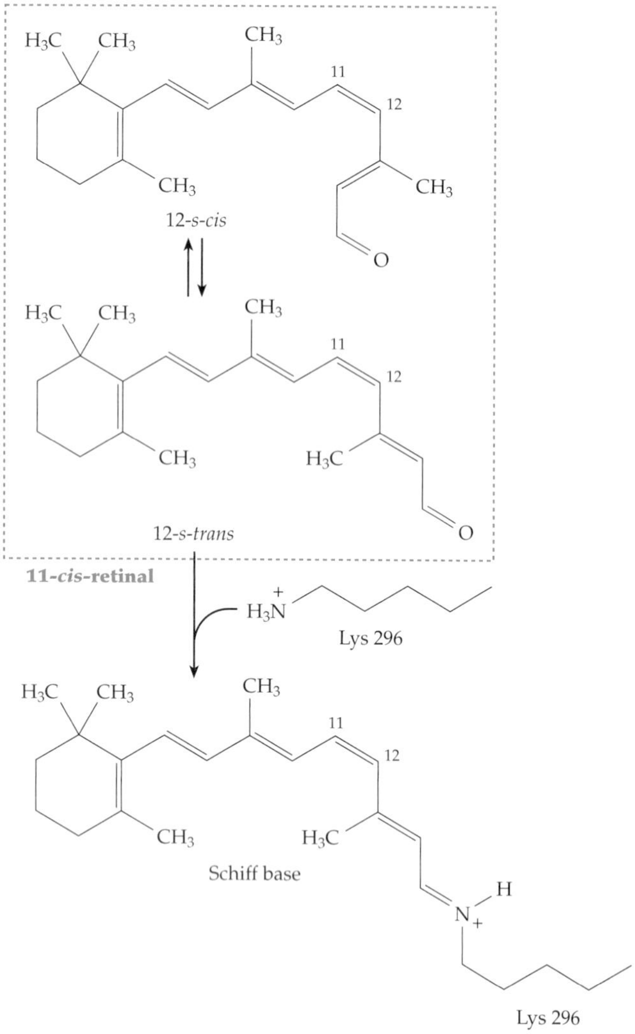

(23-36)

Three types of cone cells in the human retina are needed for color vision. Four genes specify the proteins for rhodopsin and for related cone photoreceptors absorbing blue (~425 nm), green (~530 nm), and red (~560 nm) light.[477–479b] All of the cone opsins also bind 11-*cis*-retinal. The rhodopsin gene is located on human chromosome 3, while that of the blue pigment is found on chromosome 7. However, the green and red sensitive pigment genes, whose sequences are 96% identical,[478] are close together on the q arm of the X-chromosome and near the gene for glucose-6-phosphate dehydrogenase. Examination of cloned DNA from persons with inherited red-green color-blindness shows that loss of a functional form of one of these genes is usually responsible for the problem.[480–482] Among Caucasians 8% of males and 1% of females differ from the normal in their color vision. About 30% of affected males are **dichromats** and lack either the red-sensitive pigment (they are protanopes) or the green-sensitive pigment (deuteranopes). They usually have a partial gene deletion. The other 70% often have hybrid genes created by errors in recombination during cell division.[483] A few deuteranopes have the point mutation C203R in the green pigment. John Dalton, of atomic theory fame, reasoned that his red-green confusion resulted from a blue tint in the vitreous humor of his eyes and ordered that they be dissected after his death (in 1884). There was no blue tint but DNA analysis performed more recently showed that Dalton was a deuteranope.[484] Defects in the blue-sensitive receptor are relatively rare affecting about 1 in 500 persons, while only one person in 100,000 has a total lack of color discrimination.[481,485]

All retinal-dependent visual pigments form Schiff bases with lysine side chains of the photoreceptor proteins. How can the same chromophore be "tuned" to absorb across the wavelength range of 360 to 635 nm? Modern techniques such as resonance Raman[477,479] and FTIR spectroscopies and study of mutant forms[488] have shown that interaction of the conjugated double bond system of the chromophores with immediately adjacent dipoles of side chain groups and peptide linkages is sufficient to account for the great variability in absorption maxima.

Visual pigments of many species have been investigated. Most vertebrate animals have a rhodopsinlike pigment plus a variable number of cone pigments. Mammals typically have only two, a short-wavelength pigment absorbing maximally in the ultraviolet, violet, or blue region[488a-c] and a long-wavelength pigment with maximum absorption in the green or red region.[489] The bottlenose dolphin has only a rod pigment of λ_{max} 524 nm.[489] In contrast, the chicken has four cone pigments with maximum absorption for violet, blue, green, and red.[490,490a] The red light receptor, called **iodopsin**, absorbs maximally at 571 nm. However, it binds chloride ions which induce an additional 40-nm red shift. The Cl^--binding site involves His 197 and Lys 200 which are present in an extracellular loop (Fig. 23-41A) and quite far from the bound retinal.[490] Human red and green color vision pigments, and also a green-sensitive pigment of the reptile *Gecko gecko*, undergo spectral shifts upon binding of Cl^- in the same site.[491] However, rhodopsin and most other visual pigments do not share this behavior.

Fishes live in a variety of environments and have a diversity of visual pigments. Goldfish have genes for five opsins, one of which gives rise to an ultraviolet light receptor. They are also sensitive to polarized light.[492] Related visual pigments occur throughout the animal kingdom. Even the eyespot of the alga *Chlamydomonas* (Fig. 1-11) contains rhodopsin with some sequence homology to invertebrate opsins.[493,494] The pineal glands of chickens and probably of reptiles[495–496a] as well as those of fish[497] also contain rhodopsinlike pigments. In a few freshwater marine species the visual pigments (**porphyropsins**) contain **3-dehydroretinal**. The peak positions of light absorption depend both upon the nature of the bound aldehyde and on the protein, the latter having the larger effect. Thus, retinal-based pigments absorb in the entire range 467–528 nm (18,900–21,400 cm^{-1}). The fruit fly, *Drosophila*, contains 3-hydroxy-11-*cis*-retinal in its rhodopsin and also contains other related photoreceptors.[498]

2. The Light-Induced Transformation

The retinal Schiff base chromophore is embedded in rhodopsin with its transition dipole moment parallel to the plane of the discs, i.e., perpendicular to the direction of travel of the incoming photons. Absorption of a photon leads to a sequence of readily detectable spectral changes.[37,461b,499,500] The relaxation times indicated in Eq. 23-37 are for 20°C.

Rhodopsin (498 nm, 20.1 x 10^3 cm^{-1})
$h\nu \downarrow \tau < 6 \times 10^{-12}$s
Bathorhodopsin (prelumirhodopsin; 543 nm, 18.4 x 10^3 cm^{-1})
$\downarrow \tau \sim 50$ ns
Lumirhodopsin (497 nm), 20.2 x 10^3 cm^{-1})
$\downarrow \tau \sim 50$ μs
Metarhodopsin I (478 nm, 20.9 x 10^3 cm^{-1})
$\downarrow\uparrow$ milliseconds
Metarhodopsin II (380 nm, 26.2 x 10^3 cm^{-1})
$\downarrow$ seconds
All-*trans*-retinal (387 nm, 25.8 x 10^3 cm^{-1}) + opsin (23-37)

The intermediate chemical species have been named bathorhodopsin, lumirhodopsin, and metarhodopsin I and II. At very low temperatures a transient form **photorhodopsin** with a wavelength maximum at ~580 nm may precede bathorhodopsin.[461b,501–502a] Furthermore, nanosecond photolysis of rhodopsin has revealed a **blue-shifted intermediate** that follows bathorhodopsin within ~40 ns and decays into lumirhodopsin.[500,503,504] The overall result is the light-induced isomerization of the bound 11-*cis*-retinal to all-*trans*-retinal (Eq. 23-38) and free opsin. The free opsin can then combine with a new molecule of 11-*cis*-retinal to complete the photochemical cycle.

What are the chemical structures of the intermediates in Eq. 23-37, and why are there so many of them? The answer to the last question is that the initial photochemical process is very fast. Subsequent conformational rearrangments and movement of protons are slower, occur in distinct steps, and give rise to the observed series of intermediates. To shed light on these processes many experiments have been done with analogs of retinal,[502,505–508] often using very rapid spectroscopic techniques.[37,508] These studies have shown that the isomerization of the Schiff base from 11-*cis* to all-*trans* occurs in the first very rapid step of

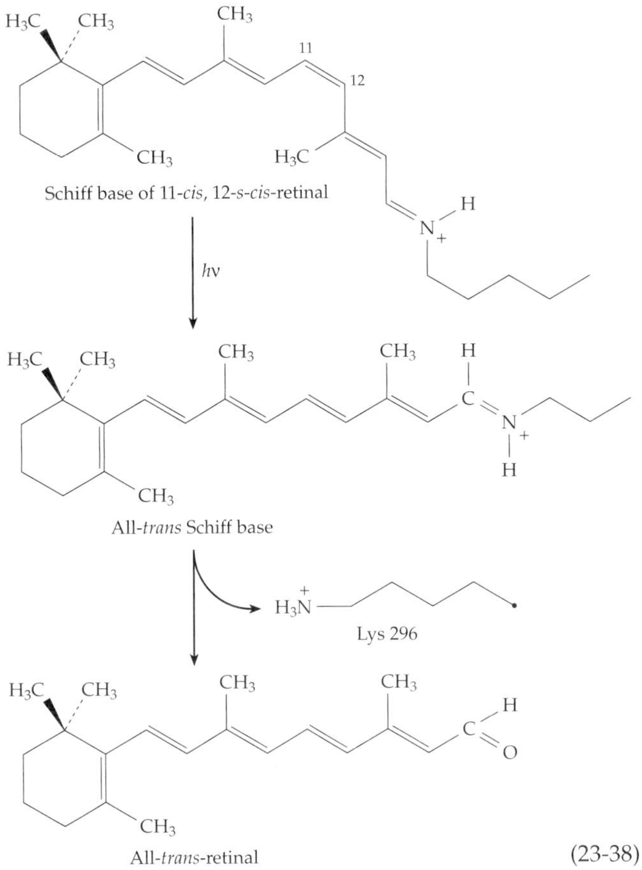

H_3C CH_3 CH_3 11 12 CH_3 H_3C H N^+

Schiff base of 11-*cis*, 12-*s*-*cis*-retinal

$h\nu$

H_3C CH_3 CH_3 CH_3 H C N + H CH_3

All-*trans* Schiff base

$H_3\overset{+}{N}$ Lys 296

H_3C CH_3 CH_3 CH_3 H C O CH_3

All-*trans*-retinal (23-38)

Eq. 23-37.[499,509–510a] Constrained analogs of 11-*cis* retinal also combine with opsin to form rhodopsinlike molecules with absorption maxima near 500 nm.

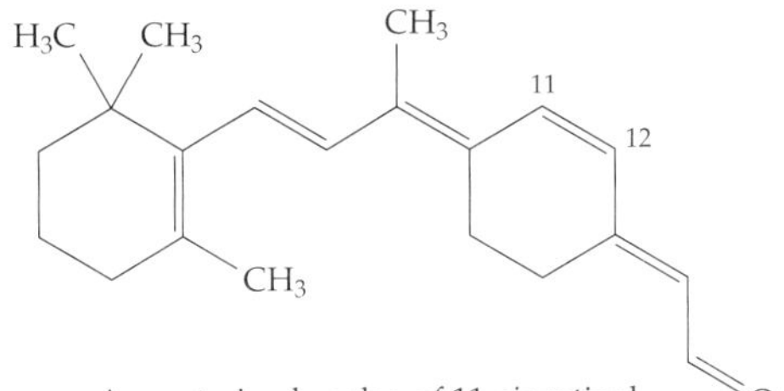

H_3C CH_3 CH_3 11 12 CH_3 O

A constrained analog of 11-*cis*-retinal

However, most of these analogs cannot isomerize and illumination does not produce bathorhodopsinlike molecules.[504,509,511]

In the photoexcited state the positive charge on the Schiff base is delocalized. For example, with some of the positive charge located on C5, rotation could occur around the more nearly single C11–C12 bond to give the all-*trans* isomer (Eq. 23-39). However, it seems more likely that simultaneous rotation occurs around two connected single and double bonds.[511a] The mechanism of this photoisomerization, which is among the fastest known chemical reactions, is still being investigated.[511b] In native rhodopsin the conversion to bathorhodopsin occurs with a high quantum

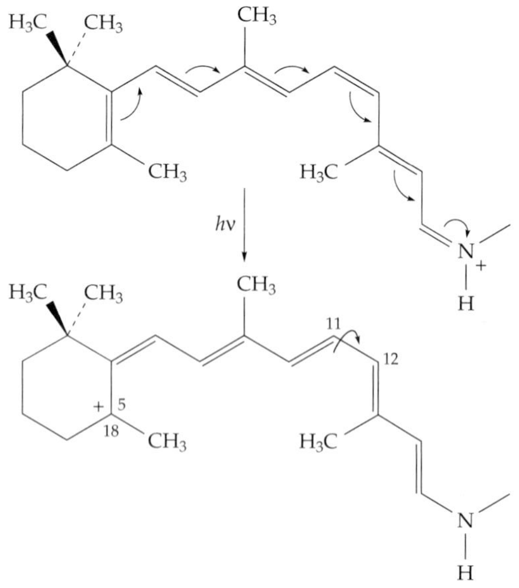

(23-39)

yield of 0.67 within 0.2 ps, a time comparable to the period of torsional vibrations of the retinal. This extreme speed suggests that the isomerization is a concerted process that is **vibrationally coherent**.[511c] Vibrational motion in the electronically excited state is utilized in the isomerization process.[506,512]

The reaction sequence of Eq. 23-37 can be slowed by lowering the temperature. Thus, at 70K illumination of rhodopsin leads to a **photostationary state** in which only rhodopsin, bathorhodopsin, and a third form, **isorhodopsin**, are present in a constant ratio.[510] Isorhodopsin (maximum absorption at 483 nm)[513] contains 9-*cis*-retinal and is not on the pathway of Eq. 23-37. Resonance Raman spectroscopy at low temperature supports a distorted all-*trans* structure for the retinal Schiff base in bathorhodopsin.[510] The same technique suggests the *trans* geometry of the C=N bond shown in Eqs. 23-38 and 23-39. Simple Schiff bases of 11-*cis*-retinal undergo isomerization just as rapidly as does rhodopsin.[514]

Some step in the sequence of Eq. 23-37 must initiate a chemical cascade that sends a nerve impulse out of the rod cell. This is accomplished through a GTP-dependent G protein cascade as outlined in Fig. 23-43. Light-activated rhodopsin initiates the cascade by interacting with the G protein **transducin**. Energy for the activation comes from the quantum of absorbed light. While the primary chemical reaction has long been recognized as the isomerization of the retinal Schiff base,[515] it is not obvious how this generates the signal for transducin to bind and become activated.

The seven helices of rhodopsin form a "box" around the bound retinal. The environment of the retinal is largely hydrophobic. However, there are also buried polar groups, some of which lie in conserved positions in more than 200 G-protein-coupled receptors[458] and internal water molecules whose vibrational spectra can be detected.[516] As in bacteriorhodopsin (Fig. 23-41, C) the buried polar groups and water molecules are doubtless hydrogen-bonded in an internal network. We can anticipate two effects of the isomerization reaction: (1) It will distort the shape of the box in which it occurs. (2) It will break some hydrogen bonds and allow new ones to form and may affect the balance of electrical charges within the protein. This in turn can lead to proton movements and alterations in the internal hydrogen-bonded network. Both of these anticipated effects have been observed.

Conformational changes induced in the rhodopsin protein by the isomerization of the retinal Schiff base include significant movement of the end of helix VI(F) at the cytosolic surface as well as smaller movements of other helices.[517,518] Spectroscopic measurements indicate that the Schiff base nitrogen remains protonated in both the bathorhodopsin and metarhodopsin I forms and in metarhodopsin II, the first long-lived form in the sequence.[518a,b] It seems likely that the proton has jumped via a bridging water molecule and the E113 carboxylate to the external (intradiscal) surface of the molecule. At the same time one or more protons are apparently taken up on the cytoplasmic side.[519] Study of mutant forms suggests that glutamate 134, near the cytosolic surface, and histidine 211 may be involved in proton transport.[520] By analogy with bacteriorhodopsin, aspartate 83 is probably also involved. The combination of conformational change plus altered charge distribution may be needed to create a binding surface with a suitable shape and charge constellation to bind tightly to transducin for the next step.

3. The Nerve Impulse

Which of the intermediates in Eq. 23-37 is responsible for initiation of a nerve impulse? Some evidence favored metarhodopsin I,[326] but its lifetime may be too short. On the other hand, the transformation of metarhodopsin I to metarhodopsin II is the slowest step that could trigger a nerve impulse, which must travel the length of the rod to the synapse in about one ms,[521] and metarhodopsin II is generally believed to be the activated signaling form of rhodopsin.[521a–c]

Transducin, cyclic GMP, and phosphodiesterase. The essential consequence of light absorption is an alteration in the membrane potential in the vicinity of the absorbed photon with the resulting propagation of a nerve impulse down the plasma membrane to the synapse by cable conduction (Chapter 30). The type of potential change that is transmitted differs among vertebrates and invertebrates.[522] In the case of mammalian photoreceptors the rod outer segment is permeable to sodium ions so that a large

dark current of sodium ions flows in through the plasma membrane and is pumped out by sodium pumps in the inner portion of the cell. Visual stimulation causes this permeability to Na^+ to be decreased with an increase in polarization of the membrane. Absorption of a single photon by rhodopsin blocks the outflow of ~10^6 sodium ions.

At one time calcium ions seemed to be the logical internal messenger between rhodopsin and the plasma membrane. If light absorption opened channels from the internal space of the rod discs, calcium ions could be released and diffuse quickly to the plasma membrane and block the entrance of sodium ions.[523] However, light *does not* increase the free [Ca^{2+}] in the cytoplasm but may decrease it from 500 nM to as low as 50 nM.[524,525] Stryer suggested that the essential messenger is **cyclic GMP** (Chapter 11)[526–528] and that a *decrease* in cGMP concentration initiates the nerve response. Cyclic GMP is apparently responsible for keeping the sodium ion channels open. Absorption of a photon in the rod disc (Fig. 23-43) produces activated rhodopsin R* (metarhodopsin II), which acts as an allosteric effector for the heterotrimeric G protein transducin whose structure and properties have been discussed in Chapter 11. Like proteins G_s and G_i of the adenylate cyclase system, transducin contains three subunits: α, 40 kDa, 350 residues; β, 36 kDa; and γ, ~8 kDa.[529] In the resting state they are associated as $T_{\alpha\beta\gamma}$ with a molecule of GDP bound to the α subunit. When activated rhodopsin R* binds to transducin (step *a*, Fig. 23-43) it catalyzes a rapid exchange of GTP for GDP (step *b*). This is followed by dissociation of $T_{\beta\gamma}$ from T_α GTP. The latter serves as an allosteric effector for a **cGMP phosphodiesterase** bound to the disc surface converting it to an active form (step *c*).[529a,b] The activated phosphodiesterase, an $\alpha\beta\gamma_2$ oligomer,[530] hydrolyzes the cGMP (step *d*, Fig. 23-43), reducing its concentration and thereby inhibiting the Na^+ outflow.

Because one molecule of activated cGMP phosphodiesterase can hydrolyze more than 10^5 molecules of cGMP per second the light response is highly amplified. There is also an earlier stage of amplification. Each molecule of light-activated rhodopsin (R*) is able to catalyze the exchange of GTP for GDP on hundreds of molecules of $T_{\alpha\beta\gamma}$ before R* passes on to other intermediates and releases all-*trans*-retinal from opsin (light green lines, Fig. 23-43).

Rhodopsin kinase, recoverin, and arrestin. Metarhodopsin II (R*) can become phosphorylated by rhodopsin kinase on as many as seven serine and threonine side chains on its cytoplasmic surface (Fig. 23-41).[531,532] The 45-kDa protein arrestin binds to such phosphorylated R*,[533–535a] which is rapidly deactivated and desensitized so that it is less likely to be immediately reactivated. This is important in the adaptation of the eye to bright light. At the same time T_α2GTP is hydrolyzed back to T_α2GDP and reforms $T_{\alpha\beta\gamma}$2GDP, and guanylate cyclase regenerates the cGMP.[536] At least four different arrestins are known. Some function in nonvisual tissues. In all cases they seem to serve as "uncouplers" of G protein-coupled receptors.[536a]

Recovery of the inhibited rhodopsin, which occurs most rapidly in dim light, depends upon calcium ions. In dim light both Ca^{2+} and Na^+ enter the visual cells through the cGMP-controlled channels. At the same time Ca^{2+} flows out through a Na^+/Ca^{2+} exchanger. When the channels are blocked by cGMP formed in

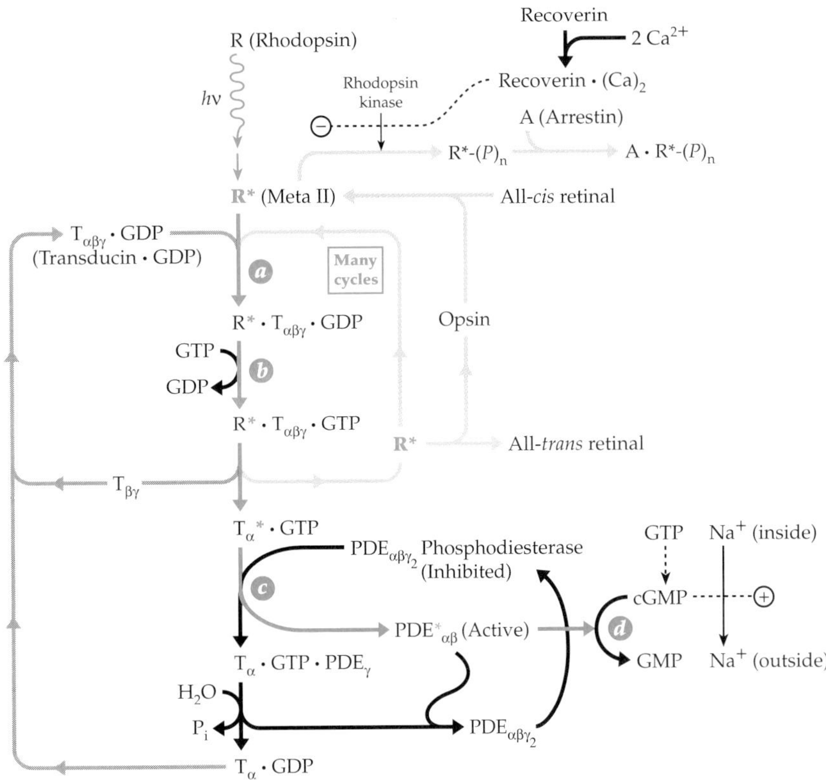

Figure 23-43 The light-activated transducin cycle. In step *a* photoexcited rhodopsin (R*) binds the GDP complex of the heterotrimeric transducin ($T_{\alpha\beta\gamma}$). After GDP–GTP exchange (step *b*) the activated transducin T*•GTP reacts with the inhibited phosphodiesterase ($PDE_{\alpha\beta\gamma_2}$) to release the activated phosphodiesterase ($PDE_{\alpha\beta}$). Based on scheme by Stryer[528] and other information.

response to light, [Ca^{2+}] falls as a result of continuing export by exchange with Na^+.[524,537] In the dark the [Ca^{2+}] concentration rises again and binds to a 23-kDa calcium sensor molecule called recoverin. Recoverin, like calmodulin (Fig. 6-8), has four EF-hand Ca^{2+}-binding motifs and also an unexpected feature. A **myristoyl group** attached at the N terminus is bound into a deep hydrophobic pocket in the protein. However, when two Ca^{2+} ions bind, recoverin undergoes a conformational change that allows the myristoyl group to be extruded and to bind to a membrane surface. This allows recoverin to move out of the cytosol to the surface of the disc membrane where it binds to and inhibits the activity of rhodopsin kinase,[538–541] increasing the sensitivity of photodetection.

Some details about cone cells and invertebrate vision. The biochemistry of retinal cones is less well known but is similar to that of rod cells. Cone pigments are present in the plasma membrane rather than in isolated discs (Fig. 23-40C). Different α, β, and γ subunits of transducin are formed in rods and cones.[522] Many differences are seen among various invertebrate visual systems. Inositol triphosphate (IP_3) and Ca^{2+} often serve as signals of photoexcitation. G proteins also play prominent roles.[522]

4. Regeneration of Visual Pigments; the Retinal Cycle

How is all-*trans*-retinal released from photobleached pigments and isomerized to 11-*cis*-retinal for the regeneration of the photopigments? Since new 11-*cis*-retinol is continuously brought in from the bloodstream and oxidized to retinal, isomerization can occur in other parts of the body. However, much of it takes place in the **pigment epithelium** of the retina, the layer of cells immediately behind the rod and cone cells. As indicated in Fig. 23-44, all-*trans*-retinal can leave the photoreceptor cells and, after reduction to retinal, be carried to the pigment epithelial cells by an **interphotoreceptor retinoid-binding** protein.[542,543] There it becomes esterified by the action of **lecithin:retinol acyltransferase**, an enzyme that transfers a fatty acyl group from lecithin to the retinol. The resulting retinyl esters are isomerized, and 11-*cis*-retinol is released.[543a] Some is stored as 11-*cis*-retinyl esters but enough is dehydrogenated to 11-*cis*-retinal to meet the needs of the photoreceptor cells and is transported back to them (Fig. 23-44). In the cephalopods the inner segment of the receptor cells contain a second pigment **retinochrome** that carries out a photochemical conversion of all-*trans*-retinal to 11-*cis*-retinal.[544,544a]

5. Diseases of the Retina

An important cause of blindness is **retinitis pigmentosa**, an inherited disease affecting about one in 3000 persons. Symptoms include progressive night blindness, degeneration of the rod cells, and gradual loss of cone cells and of nerve function in the retina. An autosomal dominant form of the disease arises from deletions or point mutations in the rhodopsin gene. In the United States 15% of cases arise from the mutation P23H.[545] By 1996 ~70 point mutations that cause retinitis pigmentosa had been discovered.[448,546–548] These mutations are found in all three of the rhodopsin domains: intradiscal, transmembrane, and cytosolic. Other rhodopsin point mutations such as G90D and A292E cause congenital night blindness.[549] Retinitis pigmentosa also arises from defects in **peripherin-RDS**, a structural component of rod cells identified originally from the gene *rds* (retinal degeneration slow) of the mouse.[545,550] Another form of congenital night blindness results from mutations in rhodopsin kinase.[551] A dominant **rod–cone dystrophy** is caused by a defect in the photoreceptor guanylate cyclase.[552] The most frequent cause of combined deafness and blindness in adults (**Usher syndrome**) is a defect in a cell adhesion molecule.[553] In **choroidemia**, another X-linked form of retinitis pigmentosa,

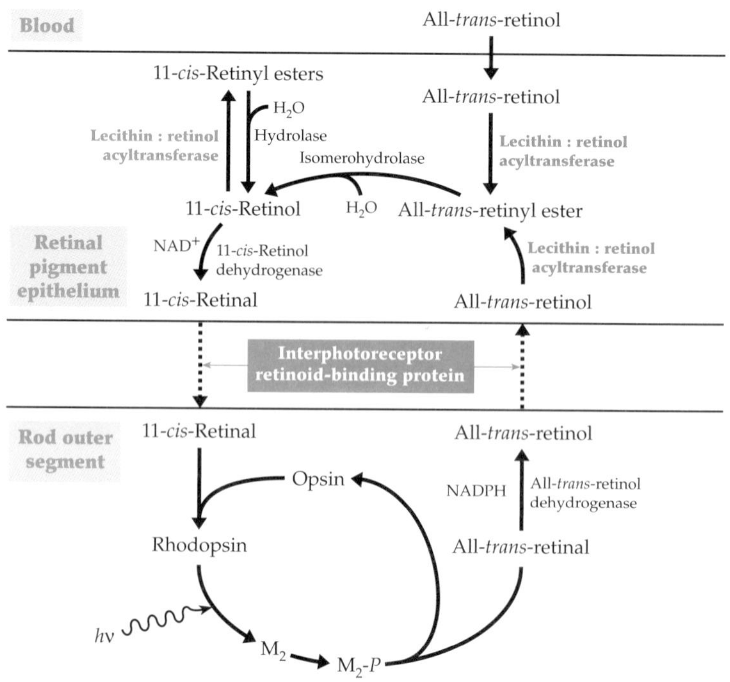

Figure 23-44 Reactions of retinol and the retinal cycle of mammalian rod cells. After Palczewski *et al.*[543]

the choroid layer behind the pigment epithelium also deteriorates. A geranylgeranyl transferase specific for the Rab family of G proteins is defective.[554]

The most frequent cause of vision loss in the elderly is **macular degeneration**. Mild forms of the disease occur in nearly 30% of those over 75 years of age and more serious forms in 7% of that age group. There are many causes, some hereditary.[555–556a] Excessive accumulation of fluorescent lipofuchsin, perhaps arising in part from Schiff base formation between retinal and phosphatidylethanolamine, is sometimes observed.[557]

6. Proteins of the Lens

The lens of the eye encloses cells that cannot be replaced and contains proteins that don't turn over and must last a lifetime.[558] The fiber cells, which make up the bulk of the lens, have no nuclei. They elongate and stretch to cover the central nucleus, the original fetal lens, like the layers of an onion, the edges of the cells interdigitated with the next cell like a piece in a child's construction set.[559] These cells are tightly packed with proteins in aggregates whose size is on the order of the wavelengths of light. The high concentration of proteins is needed to provide transparency and also a high refractive index.[560–562] The membranes of the lens cells acquire increasing amounts of a 28-kDa **major intrinsic protein** as they age.[563] Three classes of soluble lens proteins, called **crystallins**, are found in virtually all lenses. Alpha crystallins, which account for ~40% of the total soluble protein, are heterodimers of ~20-kDa subunits that associate into ~800-kDa complexes.[564] They have a chaperoninlike activity.[565] Beta crystallin, which may account for ~35% of the protein, as well as the γ crystallins are β-sheet proteins with "Greek key" folding motifs.[558]

In addition to the α, β, and γ crystallins many animals have recruited additional "taxon-specific" crystallins δ, ε, λ, etc., that have evolved from preexisting enzymes, chaperonins, or other proteins.[561,566–568] For example, avian and reptilian lenses contain a δ crystallin homologous to argininosuccinate lyase.[569] Many crystallins are derived from dehydrogenases, e.g., for lactate dehydrogenase (duck),[570] hydroxyacyl-CoA dehydrogenase (rabbit),[571] and aldehyde dehydrogenase (squid and octopus).[572] A high concentration of NADH may be present.[568] A crystallin of a diurnal gecko is a retinol-binding protein with bound 3-dehydroretinal (vitamin A_2), which probably acts as an ultraviolet filter that improves visual acuity and protects against ultraviolet damage.[567,572a] Human lenses contain small molecules that act as UV filters, e.g., glucosides of **3-hydroxykynurenine** (Fig. 25-11) and 4-(2-amino-3-hydroxyphenyl)-4-oxobutanoic acid. Lenses tend to discolor and become fluorescent with age, in part by irreversible reactions of crystallins with these compounds.[573–574a]

A common problem with lenses is **cataract**, a term that describes any loss of opacity or excessive coloration. There are many kinds of cataract, most of which develop in older persons.[560,562] Since lens proteins are so long-lived deamidation of some asparagine occurs. However, the reactions are slow. One of the asparagines in α crystalline has a half-life of 15–20 years, and some glutamines are undamaged after 60 years.[575]

G. Bacteriorhodopsin and Related Ion Pumps and Sensors

Under certain conditions, the salt-loving *Halobacterium salinarum* forms a rhodopsinlike protein, which it inserts into patches of **purple membrane** in the surface of the cell. These membranes, which may constitute up to 50% of the cell surface, contain light-operated proton pumps that translocate protons from the inside to the outside of the cells.[454,576–578] In this manner they may provide energy for a variety of cell functions including ion transport and ATP synthesis. The 248-residue retinal-containing **bacteriorhodopsin** makes up 75% of the mass of the membrane. Its molecules aggregate into a two-dimensional crystalline array in the purple patches of the membrane. This allowed determination of the three-dimensional structure to 0.7 nm resolution in 1975 by electron microscopy and neutron diffraction.[454] More recently the structure has been established at progressively higher resolution by electron crystallography[579–581] and X-ray diffraction.[455,582] The most recent studies have been focused on determination of the structural alterations in the protein that accompany the proton pumping.[582a–e] A step-by-step picture is emerging.[582f] Internal changes in the retinal chromophore, movements of protons, and alterations in the shapes of some of the protein helices are involved. The surface loops have been studied both by electron crystallography and by atomic force microscopy.[583] Each bacteriorhodopsin molecule is folded into seven closely packed α-helical segments which extend roughly perpendicular to the membrane. Although 100 residues shorter than rhodopsin, the folding pattern is very similar (Figs. 23-41C; 23-45). The protein molecules form an extremely regular array with phospholipid molecules (mostly of phosphatidylglycerol) filling the spaces between them. The retinal is buried in the interior of the protein and is bound as an *N*-protonated Schiff base with the side chain of lysine 216.

The retinal in bacteriorhodopsin (bR) exists in both all-*trans* and 13-*cis* configurations. The all-*trans* form has an absorption maximum at 568 nm and the 13-*cis* at 548 nm. The two exist in a slow nonphotochemically mediated equilibrium in the dark.[584,584a] However, in the light the all-*trans* bR_{568} undergoes a rapid photochemical cycle of reactions, which is presented in simplified form in Fig. 23-46. The subscript numbers designate the wavelength of maximum absorption. Many efforts have been made to determine the structures of the intermediates K, L, M, and O and to relate them to a mechanism of proton pumping.[585–585c] Both K and L contain 13-*cis* retinal. Therefore, as with rhodopsin (Eq. 23-39) the very first step is a photochemical isomerization. Intermediate M appears to contain a deprotonated Schiff base, but O is both *N*-protonated and again all-*trans*. It follows logically that the proton bound initially to the Schiff base is pumped out of the cell and is then replaced by a new proton in the O form. However, several questions must be answered if we are to understand this proton pump. Where in the sequence do proton transfers occur? How is the sequence driven by the absorbed light energy? Protons must enter the pump from the cytosol and exit on the exterior (periplasmic) side, flowing out against a concentration gradient. There must be a "gating" or "switch" mechanism that ensures that protons enter and leave the pump in the correct direction.[586]

Some aspects of a possible mechanism for pumping the single proton bound to the retinal Schiff base are included in Fig. 23-46. In bR_{568} the Schiff base bound proton is H-bonded, via a water bridge, to the carboxylate group of Asp 85 (Fig. 23-45B). The charge constellation in the interior of the protein, part of which is shown in this figure, is such that protonation of the Schiff base is stable and the pK_a of the protonated Schiff base is high, with estimates of 16 or above.[587] One cause of the high pK_a is the presence of the nearby negative charges on D85 and D212. Absorption of light and isomerization of the retinal causes a downward movement of the $=NH^+–$ group of the Schiff base[588] and facilitates movement of the Schiff base proton via the water molecule to the D85 carboxylate as indicated by the green arrows in Fig. 23-45B. Loss of the positive charge will instantly substantially raise the pK_a of D85 from a low value, while the loss of the negative charge will lower the pK_a of the Schiff base to closer to 7. The electrostatic interactions of the D212 and E194 carboxylates with the positive charge of R82 may also be altered. At some point in the sequence the interaction of R82 with the E194 carboxylate could cause E204, which is known by spectroscopic measurements to be protonated in the intermediate, to lose its proton to the outside. At some other point, perhaps between M_{412} and another intermediate, M_{408} (Fig. 23-46), a conformational switch must occur to limit flow of a proton back to E204 and to allow a new proton to enter from the cytosol. The D96 carboxylate is thought to accept this proton and to transfer it via a chain of

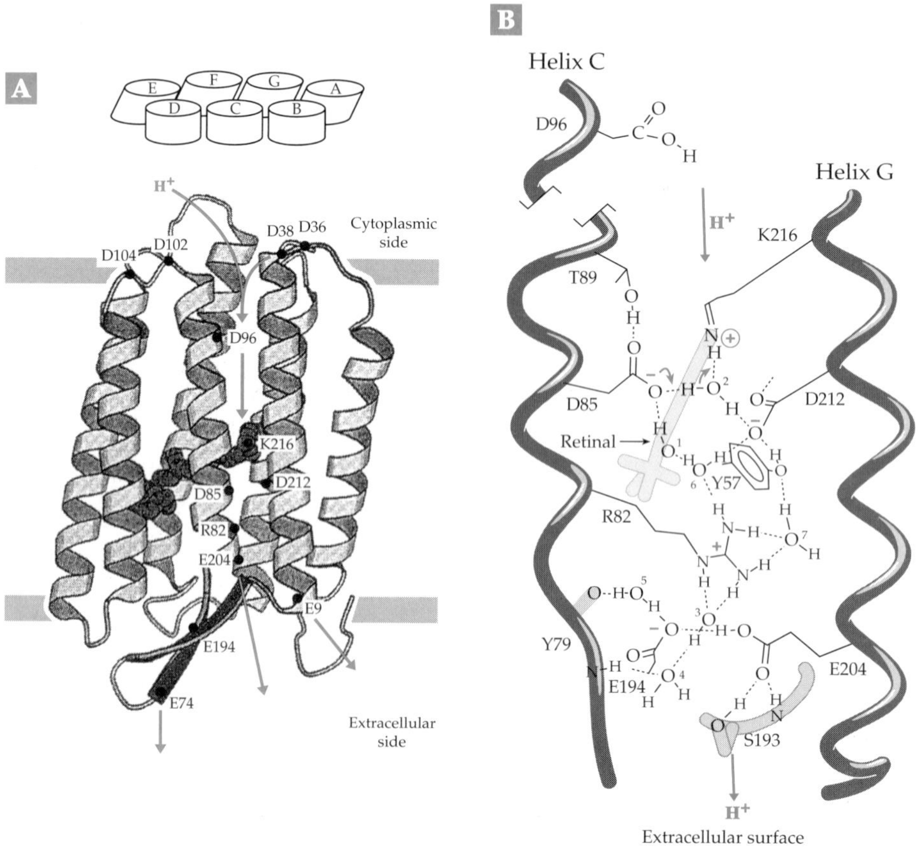

Figure 23-45 (A) Some aspects of the structure of bacteriorhodopsin. Ribbon diagram with the retinal Schiff base in ball-and-stick representation. At the top the helices are labeled as in Fig. 23-41. The locations of aspartate, glutamate, and arginine residues that might carry protons during the proton pumping action are indicated. Retinal is shown attached to lysine 216. From Kimura *et al.*[580] Courtesy of Yoshiaki Kimura. (B) Schematic drawing illustrating hydrogen-binding observed in the three-dimensional structure at 0.14 nm resolution. From Luecke *et al.*[455] The side chains shown are those thought to be involved in proton transport and in a hydrogen-bonded network with bound water molecules, principally between helices C and G. The positions of many of the hydrogen atoms in this network have not been established. They have been placed in reasonable positions in this figure but may be quite mobile. For another view of the hydrogen-bonded network see Fig. 23-41C.

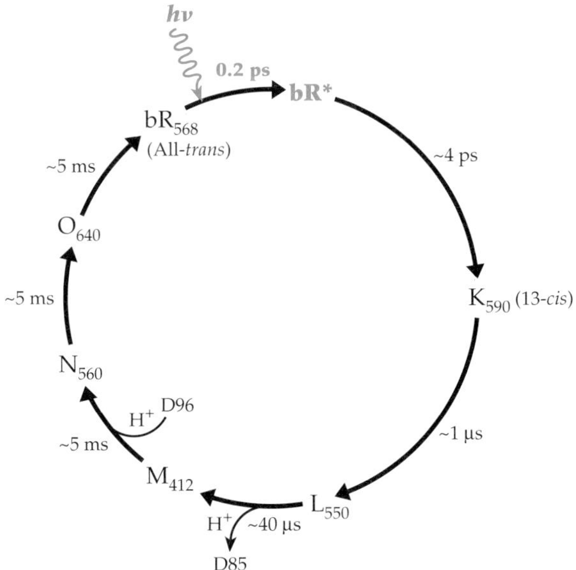

Figure 23-46 The photoreaction cycle of bacteriorhodopsin. After Bullough and Henderson.[585] The subscript numbers indicate the wavelengths of maximum absorption of each intermediate and the approximate lifetimes are given by the arrows. Resting bacteriorhodopsin as well as intermediates J and O have all-*trans* retinal but K through N are thought to all be 13-*cis*. A proton is transferred from L to aspartate 85 and then to the exterior surface of the membrane. A proton is taken up from the exterior surface via aspartate 96 to form N.

water molecules or water molecules plus protein groups to the Schiff base, which may isomerize back to all *trans* in going from intermediate N to O. Isomerization to all-*trans* O is a slow step. Both O_{640} and bR_{568} are all-*trans* and have a 15-*trans* Schiff base linkage. There may be a difference in protein conformation with the chromophore being more distorted in O than in bR_{568}. Many recent studies have provided additional information.[588a–j]

Halorhodopsin. In addition to bacteriorhodopsin there are three other retinal-containing proteins in membranes of halobacteria. From mutant strains lacking bacteriorhodopsin the second protein, **halorhodopsin**, has been isolated. It acts as a light-driven *chloride ion pump*, transporting Cl^- from outside to inside. Potassium ions follow, and the pump provides a means for these bacteria to accumulate KCl to balance the high external osmotic pressure of the environment in which they live.[578] The amino acid sequences of halorhodopsins from several species are very similar to those of bacteriorhodopsin as is the three-dimensional structure.[589] However, the important proton-carrying residues D85 and D96 of bacteriorhodopsin are replaced by threonine and alanine, respectively, in halorhodopsin.[590] Halorhodopsin (hR)

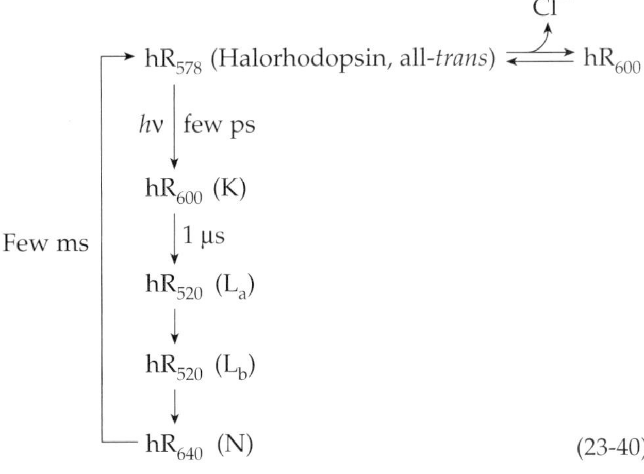

(23-40)

undergoes a light-dependent cycle (Eq. 23-40) that involves an all-*trans* to 13-*cis* photoisomerization with some intermediates resembling those of the bacteriorhodopsin cycle.[590a]

Sensory rhodopsins. The third and fourth light-sensitive proteins of halobacteria are **sensory rhodopsins** (SR)[578,591,591a,b] that are used by the bacteria to control **phototaxis**. These bacteria swim toward long-wavelength light, the maximum in the action spectrum being at ~580 nm. They are repelled by blue or ultraviolet light, the maximum in the action spectrum being at ~370 nm. Evidently the bacteria can detect either a decrease with time in red light intensity or an increase with time in blue light intensity. Either is interpreted as unfavorable and causes the bacteria to tumble and move in a new direction (see Chapter 19). Sensory rhodopsin I (SRI) appears to be able to provide both light responses. Absorption of orange light by SRI_{587}, which contains all-*trans*-retinal, yields SRI_{373}, in which the retinal Schiff bases have been isomerized to 13-*cis* as in bacteriorhodopsin. The red light response is proportional to the fraction of SRI_{373} present. However, this is converted spontaneously back to SR_{587} within seconds. Nevertheless, photoexcitation of SRI_{373} with blue light causes a faster reconversion and induces swimming reversals, the repellent response.[592,593] SRI exists in the bacterial membranes in a complex with a 57-kDa protein designated **halobacterial transducer I** (HtrI), which resembles bacterial chemotaxis receptors (Figs. 11-8 and 19-5) and is modulated by action of a methyltransferase.[591b,c] Interaction of SRI with HtrI depends upon a histidine residue, H166 of SRI. It may be part of a proton transfer pathway.[593]

Sensory rhodopsin II (SRII, also called phoborhodopsin) is specialized for repellant phototaxis.[591a] Blue light converts $SRII_{487}$ in < 1 ms to UV-absorbing $SRII_{360}$. It decays in ~100 ms to $SRII_{540}$ which reverts to the initial $SRII_{487}$ in ~0.5 s. The cycle is accompanied by swimming reversals that result in a repellent effect of light.[594–596] The three-dimensional structure is known.[593a,b]

Whereas retinal-based proton pumps all have the conserved residues D85 and D96 of bacteriorhodopsin, only the aspartate correspnding to D85 is conserved (as D73) in the sensory rhodopsins. D96 is replaced by tyrosine or phenylalanine.[586] In SRI D73 appears to be protonated and, therefore, does not form a counterion for the Schiff base iminium ion.[597] However, in SRII D73 is apparently unprotonated and available to -serve as a counterion and as a proton acceptor as in bacteriorhodopsin.[597a,b] There is also a corresponding aspartate (Asp 83) in rhodopsin (Fig. 23-41). This suggests a common signaling mechanism for rhodopsin and the sensory rhodopsins. Finally, there are retinal-containing proteins in fungi and in algae. They may serve as blue light receptors.[598,598a]

The photoactive yellow proteins (xanthopsins). A 124-residue, 14-kDa yellow protein isolated from a halophilic phototrophic purple bacterium, *Ectothiorhodospira halophila*, was at first thought to be a rhodopsin-type pigment. However, this photoactive yellow protein (PYP) carries a covalently bonded **coumaroyl (4-hydroxycinnamoyl) group** in thioester linkage,[599–601c] which is completely surrounded by the small protein.[601] The coumaroyl group, which was probably derived from coumaroyl-CoA (Fig. 25-8), is bound as a phenolate anion by hydrogen bonds to tyrosine and glutamate side chains (Eq. 23-41). After a laser flash at room temperature a readily observed intermediate I_1 (also called pR) absorbing maximally at 460–465 nm appears within ~3 ns and decays within a few milliseconds to a bleached intermediate I_2 (also called pB or pM)[602,603] with maximum absorption at ~355 nm. This returns to the original 446-nm form within a few seconds. Earlier intermediates I_0 and $I_0^‡$ have been identified by picosecond spectroscopy,[601c,604,605] and others have been identified at low temperatures.[602,606]

The structure of PYP is known to 0.1 nm resolution (Fig. 23-47).[601,607] Structures have also been determined for a very early intermediate by trapping at –100°C[607] and for I_1 (pR). The cofactor structures are shown in Eq. 23-41. The light-induced step is apparently the *cis–trans* isomerization,[608] and changes in hydrogen-bonding follow. The hydrogen bond between the phenolate ion of the coumaroyl group and glutamate 46 appears to break, and E46 may donate a proton to the phenolate group to form the 337-nm chromophore of I_2.[602,609,609a] The signaling mechanism may be similar to that in sensory rhodopsins.

Stentorin. A protein with a bound chromophore called stentorin mediates the light-avoidance response of the protozoan *Stentor*. Stentorin,[610] which is found in pigment granules in the cell surface, is a derivative of **hypericin**, a plant compound with antidepressant activity and the active ingredient in the herb St. John's wort.

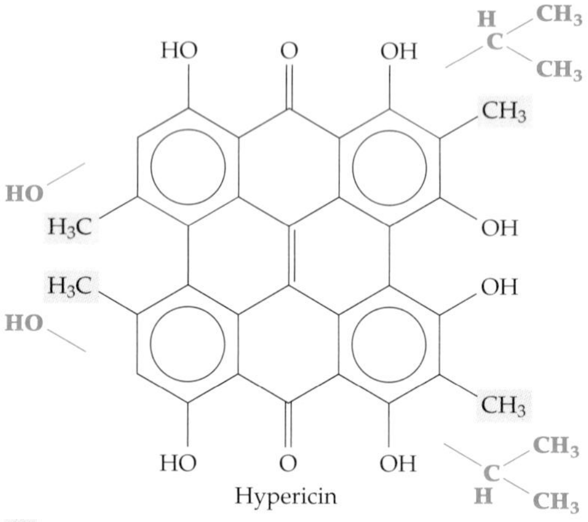

Hypericin

Replaced in stentorin by the green groups

Stentorin is covalently bonded to a 16-kDa protein in an acid-labile linkage. Its photocycle is not well investigated, but it is thought to initiate a response via

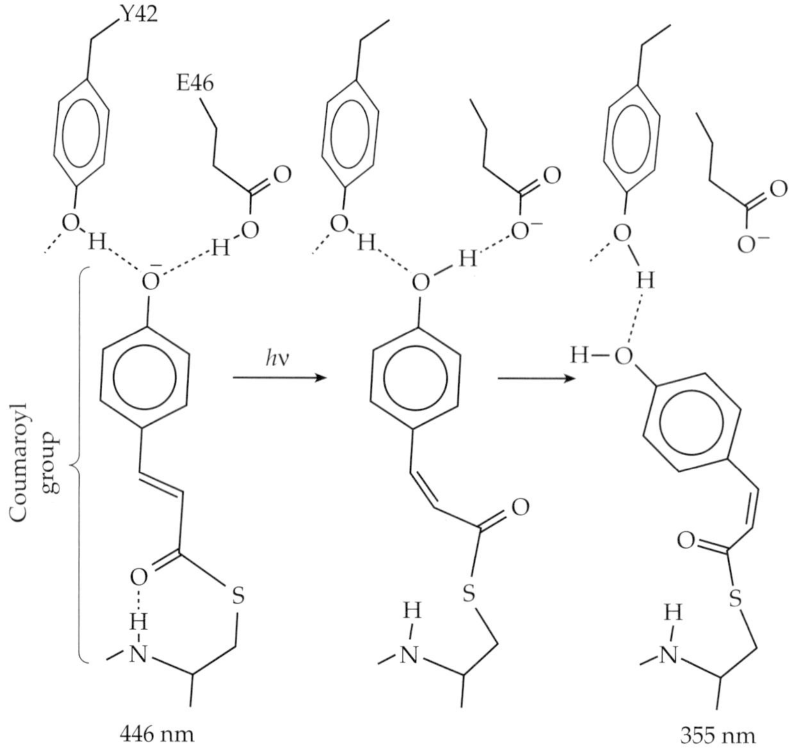

(23-41)

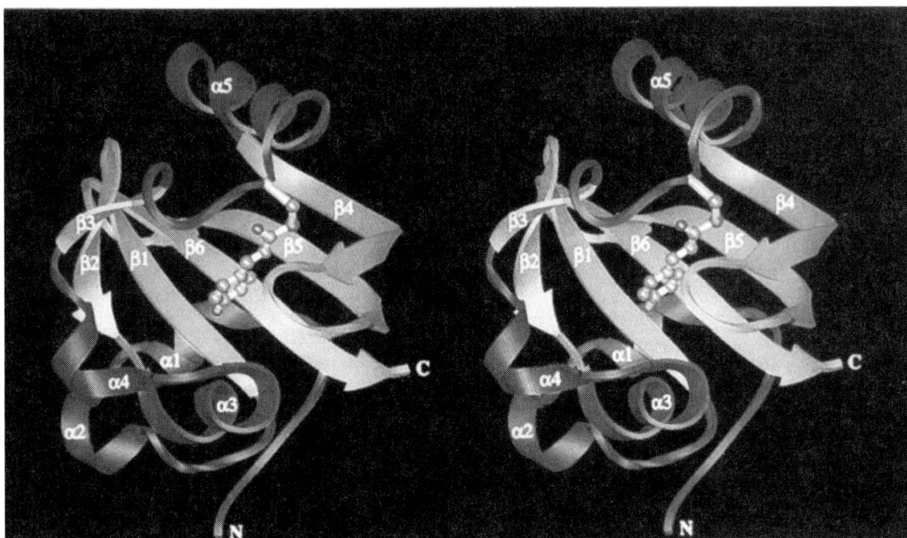

Figure 23-47 Ribbon drawing of the structure of the 125-residue yellow photoactive protein. The 4-hydroxycinnamoyl chromophore, which is attached to cysteine 69, is represented with balls and sticks. From Borgstahl *et al.*[601] Courtesy of Gloria Borgstahl.

proton transfer.[611] However, stentorin proteins apparently do not belong to the bacteriorhodopsin family.

Hypericin and related compounds have also aroused interest because of their antiviral and antitumor activities.[612,613] Hypericin is a strong photoactivator which produces singlet oxygen with a quantum efficiency of 0.73. However, antiviral activity may involve a radical mechanism.[613] Hypericin is attractive as a possible agent for photodynamic therapy (Section D,3). It can also receive energy from photoexcited firefly luciferin (Section J). A proposed application is to incorporate the gene for the enzyme luciferase of the firefly luminescence system into DNA from the virus HIV. This DNA could be used to promote synthesis of luciferase only in virus-infected cells. Addition of the nontoxic hypericin would lead to photoactivation of hypericin only in virus-infected cells, where the luciferin–luciferase complex would act as a "molecular flashlight" to activate the hypericin and destroy the cell.[613]

H. Phytochrome

In 1951, it was discovered that a flash of red light (maximum activity at 660 nm) during an otherwise dark period promoted a variety of responses in plants.[614] These included flowering, germination of seeds (e.g., those of lettuce), and the expansion of leaves in dark-grown pea seedlings. Interestingly, the effect of the short flash of red light could be *completely reversed* if followed by a flash of *far-red light* (730 nm). This discovery led to the isolation, in 1959, of the chromoprotein phytochrome, a kind of molecular switch that initiates a whole series of far-reaching effects in plants. The phototransformation[615] is completely reversible (Eq. 23-42; Fig. 23-48), and the switch can be thrown in one direction or the other many times in rapid succession by light flashes.

Green plants have a family of phytochromes. There are five genes for the ~125-kDa chains of about 1100 residues each in *Arabidopsis*,[618–619c] and the corres-

P_r
Phytochrome (red)

$h\nu$ (red) 667 nm ↑ ↓ $h\nu$ (far red) 730 nm

P_{fr}
Phytochrome (far red)

(23-42)

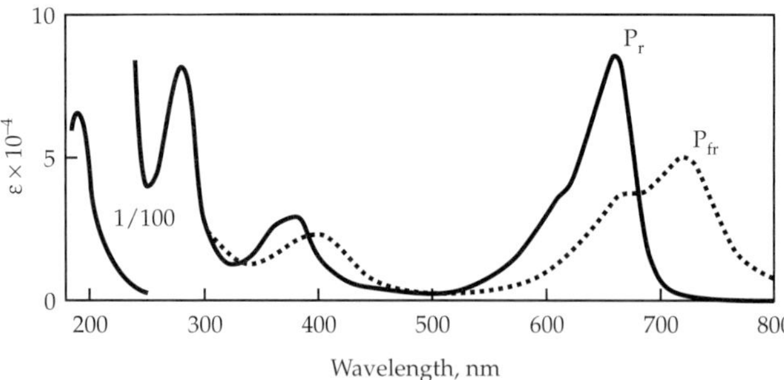

Figure 23-48 Absorption spectra of red (P_r) and far-red (P_{fr}) forms of purified oat phytochrome following saturating irradiations with red and far-red light. See Quail[616] and Anderson *et al.*[617]

sponding phytochromes A – E each appear to have distinct functions. The chromophore is an open tetrapyrrole closely related to phycocyanobilin and covalently attached to the peptide backbone near the N terminus through a cysteine side chain (top structure in Eq. 23-42). The initial photochemical reaction is thought to be the $Z \rightarrow E$ isomerizion around the C15 – C16 double bond, but there may also be rotation about the C14 – C15 single bond. The initial step occurs within a few microseconds and up to four intermediate species have been seen in the $P_r \rightarrow P_{fr}$ conversion and at least two other different ones in the $P_{fr} \rightarrow P_r$ conversion.[620–623c]

Phytochromes exist as two distinct domains, the N-terminal domain bearing the chromophore. However, the first 65 residues at the N terminus as well as the C-terminal domain probably interact with other proteins to transmit a signal. The slow responses to phytochrome are thought to involve regulation of transcription.[619b,623d] Thus, the synthesis of mRNA molecules specific for the small subunit of ribulose bisphosphate carboxylase and for the chlorophyll *a*/*b* binding protein increases in response to formation of P_{fr}. These responses are quite rapid occurring within 15–30 min.[614] Another response to P_{fr} is a decrease in the amount of the specific mRNA for phytochrome itself. That is, light induces a decrease in the concentration of this light-sensing molecule thereby decreasing the sensitivity of the system.[614]

Phytochrome is found not only in higher plants but also in algae, where it controls the movement of chloroplasts,[611] and also in cyanobacteria.[623e,f] Cyanobacterial phytochromes contain histidine kinase domains, which may function in a two-component system with a response regulator similar to protein CheY of the chemotaxis system in *E. coli* (Fig. 19-5).[624,625] Some nonphotosynthetic bacteria also use bacteriophytochromes for light sensing. In some cases biliverdin (Fig. 24-24) is the chromophore.[625a]

Phytochromes of higher plants also have histidine kinase-like N-terminal domains. Searches for associated signaling proteins have revealed a phytochrome-interacting factor in *Arabidopsis.* A possible partner for phytochrome B, it is a nuclear helix–loop–helix protein that may be a transcription factor.[626–627a] Phytochrome A may signal via a WD-repeat protein to control morphogenesis.[628]

One response under phytochrome control is the closing of leaflets of *Mimosa* at the onset of darkness. The response occurs within 5 min, too short a time to be the result of transcriptional control. This and the finding that some phytochrome is tightly bound to membranes have led to the proposal that one primary effect of phytochrome is to alter membrane properties. It is not certain whether it is P_r or P_{fr} that is active in causing a response, but P_{fr} seems to be the most likely candidate for the "active" form. According to one suggestion, phytochrome in plastid membranes may mediate the release of gibberelins stored within the plastids.[629]

I. Some Blue Light Responses

Numerous biological responses to light of wavelength 400–500 nm are known. These include phototropism in higher plants, the phototaxis of *Euglena,* and photorepair of DNA. On the basis of action spectra both carotenoids and flavins were long ago proposed as photoreceptors.[630–632] The action spectrum for opening of the stomates in coleoptiles matches the absorption spectrum of zeaxanthin.[631] On the other hand, genetic evidence[633,634] has strengthened the view that a flavin acts as the photoreceptor in the fungus *Phycomyces*. Recently compelling evidence for a flavoprotein receptor for phototropism in *Arabidopsis thaliana* has been obtained. Deficiency of a gene called *nph*1 (nonphototropic hypocotyl 1) is associated with loss of blue light-dependent phosphorylation of a 120-kDa protein. This protein was identified as the product of *nph*1 gene. The **nph1** protein is a soluble autophosphorylating Ser / Thr protein kinase with an N-terminal flavin-binding region. It apparently binds FMN and is a photoreceptor for phototropism in higher plants.[635–636b]

The complexity of the action spectra suggested the existence of more than one receptor.[634] In higher plants there are not only blue light receptors but also violet receptors and phytochrome. In addition to nph1 and a related protein **npl1**, *Arabidopsis* employs two cryptochromes (next section) and **phototropins**.[636c–e] These are also riboflavin 5'-phosphate (FMN)-dependent proteins. The action of light apparently causes addition of a highly conserved cysteinyl –SH group to the C4a position of the flavin.[636c] Phytochrome absorbs blue and ultraviolet light to some extent (Fig. 23-48) as well as red or far-red. This adds considerable complexity to the interpretation of light responses in plants.[637] The fern *Adiantum* contains a protein with a phytochrome photosensory domain fused to an NPH1 structure. It may mediate both real far-red and blue-light responses.[636] The protist *Euglena* (Fig. 1-9) makes use of a **photoactivated adenylate cyclase**, also a photoactivated enzyme, in a light avoidance response.[637a]

Fungi such as *Neurospora crassa* provide a simpler system for study of blue-light signaling than do green plants.[637] *Neurospora* contains no phytochromes. However, numerous genes including some involved in carotenoid biosynthesis and some that control the circadian cycle are regulated by blue light. Two mutants defective in riboflavin synthesis show a reduced sensitivity to blue light. A deficiency of either of two other genes *wc*-1 and *wc*-2 results in "blind" *Neurospora* unable to respond to light but able to grow. Proteins WC1 and WC2 are probably transcription factors, which act as a heterodimer. WC1, which contains bound FAD, is the photorecetor.[637,637b,c] Recently a rhodopsinlike protein NOP-1 of *Neurospora* has been identified.[637d]

1. Cryptochromes

The elusive nature of the principal blue-light receptor in plants gave rise to the name cryptochrome.[632] The gene for a cryptochrome in *Arabidopsis thaliana* was isolated by gene tagging and was cloned. It is surprisingly similar in sequence to the gene for the well-known **DNA photolyase** (Section 2).[638] It was soon recognized that cryptochromes, like photolyases, carry a bound flavin and also an antenna chromophore. The latter is probably a 5,10-methylenetetrahydrofolate, as in plant photolyases. It is a better light absorber than the flavin and passes electronic excitation to the flavin.

Two cryptochrome genes, *cry*1 and *cry*2, are present in *Arabidopsis*. The encoded proteins affect many aspects of plant growth. The cry-1 protein, together with NPH1, has a role in controlling phototropisin[639] while cryptochrome cry-2 affects flowering time, apparently via antagonistic signals from cry-2 and phytochrome B.[640–640b] cry-1 is also involved in controlling the daily rhythm of the plant, the **circadian cycle**. The circadian clock, which is discussed in Chapter 30, provides the organisms an oscillator with a period of about 24 hours. However, the oscillator must be **entrained** by the daylight cycle so that it remains in proper synchrony. The nature of the light signal and the mechanism of the entrainment are being investigated in many different organisms from fungi to human beings. In *Arabidopsis* the cycle is controlled by phytochromes A and B and by the cryptochrome cry-1.[641]

Cryptochrome genes have been found in many organisms. In the fly *Drosophila* cryptochrome appears to interact directly with the clock proteins that control the circadian cycle. Most important are products of two genes *per* (period) and *tim* (timeless). They are helix–loop–helix DNA binding proteins that form heterodimers, are translocated to the nucleus, and repress their own transcription. Morning light leads to a rapid disappearance of the TIM protein. The cryptochrome CRY appears to react directly with TIM to inactivate it. However, details remain to be learned.[642] The circadian clock mechanism appears to be universal and the cryptochrome-2 (*mcry*2 gene) appears to function in the mouse.[643,643a] A human cDNA clone was found to have a 48% identity with a relative of cryptochromes, the **(6–4) photolyase** of *Drosophila*. A second related human gene has also been found. The protein products of these two genes (*hcry*1, *hcry*2) lack photolyase activity. They too may encode cryptochromes.[644]

Where in the body is the light sensed for entraining the circadian cycle? Genes for CRY1 and CRY2 are specifically expressed in ganglion cells of the retina in mice. Severing of the optic nerve destroys both vision and light entrainment of mammals. However, in mice with the retinal degeneration (*rd*) syndrome all rod cells and virtually all cone cells are destroyed but the circadian rhythm is normal.[645] Furthermore, many blind persons with no conscious perception of light have normal light entrainment of their circadian cycle. For these reasons the ganglion cells of the retina, which are close to the location of the master circadian clock in the **suprachiasmatic nucleus** of the brain, are the most probable light sensory cells for the cycle[646,647] (see also Chapter 30). Recent evidence points to a retinal-based photoreceptor, **melanopsin**.[647a,b,c] However, vitamin A-deficient mice still display a normal circadian response.[647d]

2. Photolyases

A curious discovery was made many years ago. Bacteria given a lethal dose of ultraviolet radiation can often be saved by irradiating with visible or near ultraviolet light. This **photoreactivation**, which permits many bacteria to survive, results from the action of a **DNA photolyase**,[648,649] which often absorbs light maximally around 380 nm and carries out a photochemical reversal of Eq. 23-26, cutting the pyrimidine–pyrimidine covalent bonds of thymine dimers in DNA. The enzyme is present in cells in such small amounts, only 10–20 molecules per cell, that it was difficult to investigate until the gene had been cloned.[650,651] The significance cannot be doubted, for photoreactivation enzymes appear to be found in most organisms including some mammals. However, there is some doubt about the presence of a photolyase in the human body.

The *E. coli* DNA photolyase contains a blue flavin radical that arises from **FAD** and absorbs maximally at 580 nm (see also Chapter 15, Section B,6). The enzyme also contains a second chromophore in the form of bound 5,10-methenyltetrahydrofolylpolyglutamate with 3–6 γ-glutamyl residues.[652–653b] as shown in Fig. 23-49. The pterin coenzyme binds near the N terminus in a domain with an α/β folding pattern, while the FAD binds into a larger mostly helical domain. The pterin cofactor is not essential for repair activity, and it is generally agreed that because of its high molar extinction coefficient it acts as an effective **antenna**. It transfers energy in a nonradiative fashion to the $FADH^-$ anion located ~3 nm away.

The enzyme as isolated is in a stable blue radical form (Fig. 23-50; also Fig. 15-13) which must undergo a one-electron light-induced reduction to the anion $FADH^-$ before becoming active. A nearby indole ring

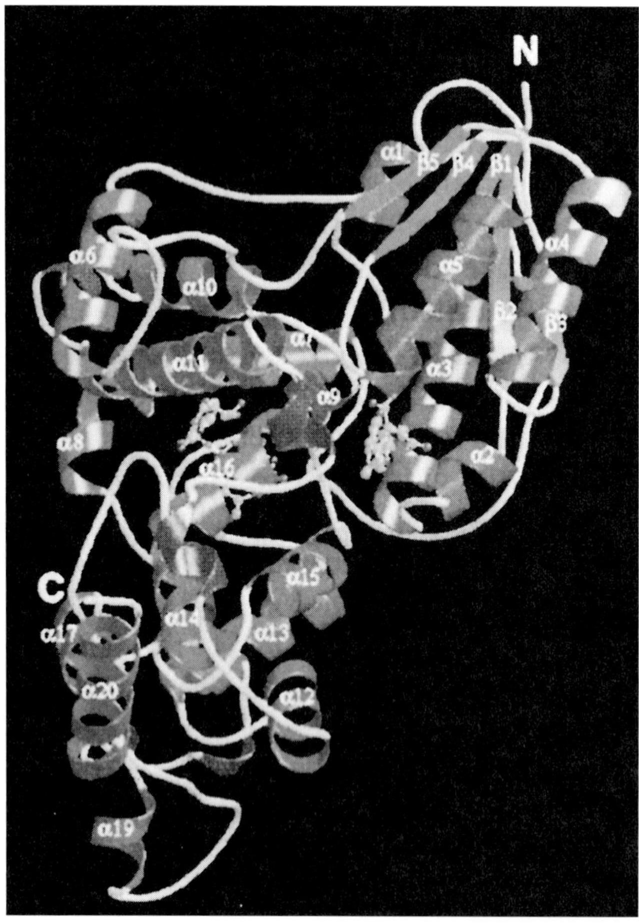

Figure 23-49 Overall view of the DNA photolyase structure from *E. coli*. The ribbon traces the 471-residue chain. The bound cofactors FAD (left) and 5,10-methenyltetrahydrofolate (right) are shown in ball-and-stick representation. From Park *et al.*[652] Courtesy of Johan Deisenhofer.

of Trp may donate the electron and be reoxidized by a tyrosyl ring.[654] The $FADH^-$ donates an electron to the pyrimidine dimer, initiating the sequence of radical reactions[654a–c] which cleaves both pyrimidine–pyrimidine bonds in the photodimers (Fig. 23-50).

The structures of all of the photolyases are thought to resemble that in Fig. 23-49. However, in one large group, which includes methanogenic bacteria, **8-hydroxy-5-deazariboflavin** acts as the antenna chromophore.[649] Another light-induced defect in DNA is the so-called 6–4 photoproduct, a different pyrimidine dimer. The 6–4 dimers are normally removed in most organisms by efficient **excision repair** (Chapter 27). However, a 6–4 photolyase was discovered in both *Arabidopsis* and *Drosophila* and has also been found in *Xenopus* and the rattlesnake.[655,655a] It has a structure similar to that of the *E. coli* photolyase and presumably acts by a related mechanism[191,656–657a] that uses the light-excited reduced flavin in an electron donation and return cycle as in Fig. 23-50. A homolog of the *Drosophila* 6–4 photolyase gene has been found in human cells, but there is uncertainty about its function.[658] Is it really a photolyase or is it a cryptochrome involved in the circadian cycle?

J. Bioluminescence

The emission of visible light by living beings is one of the most fascinating of natural phenomena. Luminescent bacteria, glowing toadstools, protozoa that can light up ocean waves, luminous clams, fantastically illuminated railroad worms,[659] and fireflies[660–661a] have all been the objects of the biochemists' curiosity.[662–664] The chemical problem is an interesting one. The firefly's light with a wavelength of 560 nm ($17,900\ cm^{-1}$) has an energy of 214 kJ/einstein. What kind of chemical reaction can lead to an energy yield that high? It is far too great to be provided by the splitting of ATP. Even the oxidation of NADH by oxygen would barely provide the necessary energy.

A clue comes from the fact that chemiluminescence is very common when O_2 is used as an oxidant in nonenzymatic processes. The slow oxidation of alcohols, aldehydes, and many nitrogen compounds (Eqs. 23-43, 23-44) is accompanied by emission of light

$$\text{Luminol} \xrightarrow[OH^-]{O_2} h\nu + C_6H_4(COO^-)_2 + N_2 \qquad (23\text{-}43)$$

Luminol, a synthetic luminescent compound

Figure 23-50 Proposed mechanism of DNA photoylase action. The antenna pigment (top) absorbs a photon, then transfers energy to the anion of bound $FADH_2$ (step *a*) which donates an electron to the thymine dimer in DNA (step *b*). After free-radical-driven steps an electron is returned to the blue FAD radical (step *c*).

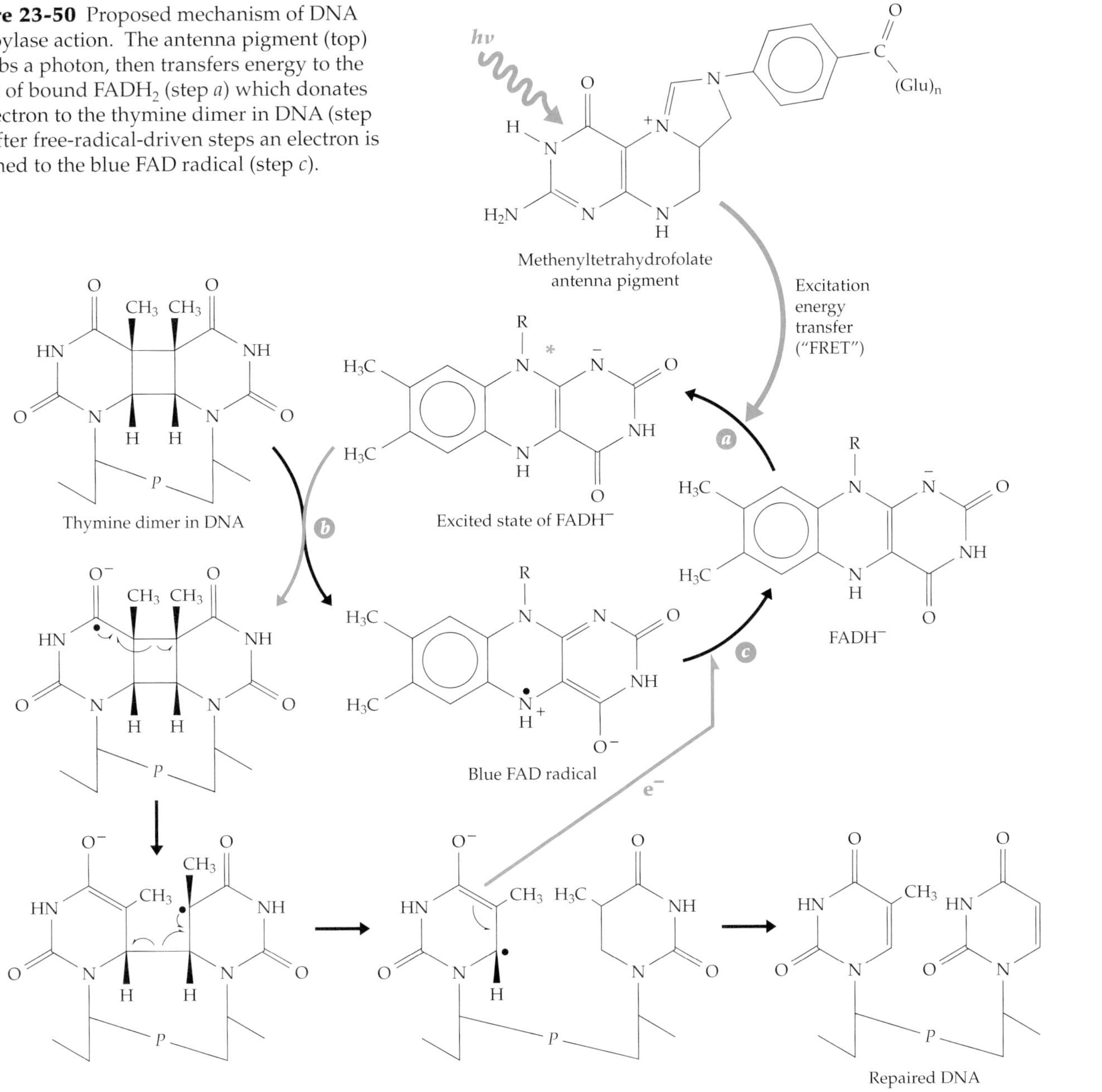

visible to the eye. Chemiluminescence is especially pronounced in those reactions that are thought to occur by radical mechanisms. The recombination of free radicals provides enough energy to permit the release of visible light. Cleavage of a peroxide linkage, e.g., in a dioxetane (Eq. 23-44),[665] is often involved.[666] For example, the reaction of Eq. 23-44 is used in a sensitive light-detected assay for alkaline phosphatase.

In view of these facts it is perhaps not so surprising that many organisms have mastered the ability to channel the energy released in an oxygenation reaction into light emission. Attempts to extract luminous materials from organisms date from the last century when the French physiologist, DuBois, in 1887 prepared both a cold-water extract and a hot-water extract of luminous clams.[662] He showed that the material in the cold-water extract, which he named **luciferase**, caused emission of light when a heat-stable material (which he called **luciferin**) present in hot-water extract was added. These names have been retained and are now used in a general way. Thus, the luciferins are a family of compounds whose structures have been determined for a number of species (Fig. 23-51). Firefly luciferin is a carboxylic acid, but it must be activated in an ATP-requiring reaction to give **luciferyl adenylate**, whose structure is shown in the figure. The latter emits light in the presence of O_2 and luciferase. It can be seen that the original carboxyl group becomes CO_2, while the ring becomes oxidized. In addition, the acyl adenylate linkage is broken. In

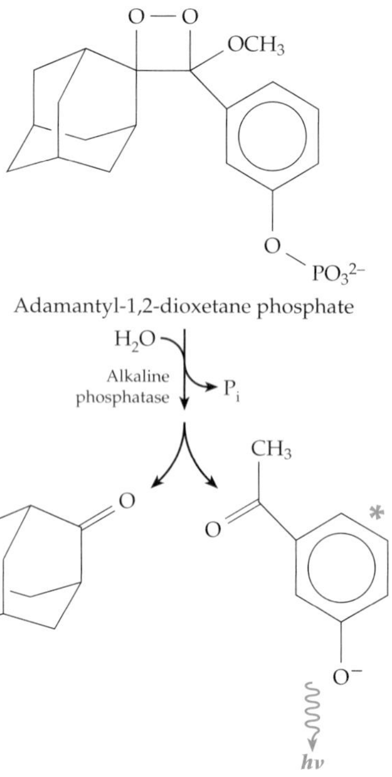

(23-44)

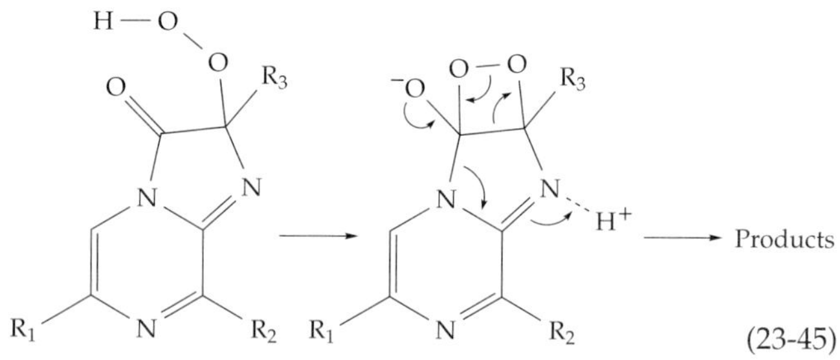

the "sea pansy" *Renilla reniformis* (a coelenterate) the luciferin has quite a different structure.[667,668] However, the reaction with O_2 to produce CO_2 and an oxidized product causes the light emission, just as in the firefly. The luciferin of *Renilla*, which is called **coelenterazine** (Fig. 23-51), is also found in the jellyfish *Aequorea*, the shrimp *Oplaphorus*,[669] the "firefly squid" *Watasenia scintillans*,[670] and other luminous organisms.

In *Renilla* the coelenterazine is stored as a coelenterazine sulfate, possibly having the structure shown. To convert this storage form to the active luciferin the sulfo group is transferred onto adenosine 3',5'-bisphosphate to form 3'-phosphoadenosine 5'-phosphosulfate, the reverse of step *d* of Eq. 17-38. The luciferin of the ostracod crustacean *Vargula hilgendorfii* has a structure (Fig. 23-51) close to that from *Renilla*. In *Vargula* (formerly *Cypridina*) the luciferin and luciferase are produced in separate glands and are secreted into the surrounding water where they mix and produce a bright cloud of light.[671]

In most mechanisms suggested for luciferase action O_2 reacts at the carbon atom that becomes the carbonyl group in the product to form an intermediate peroxide. In the case of *Renilla* luciferin this can easily be visualized as a result of flow of electrons (perhaps one at a time) from the pyrazine nitrogen (at the bottom of the structure in Fig. 23-51) into the O_2. According to one proposal, the peroxide group that is formed adds to the carbonyl to form a four-membered dioxetane ring as shown in Eq. 23-45 for coelenterazine peroxide. The latter opens, as indicated by the arrows, to give the products. This theory was tested using

(23-45)

$^{18}O_2$. In the case of *Vargula* luciferin the expected incorporation of one atom of ^{18}O into CO_2 was observed, but with firefly and *Renilla* luciferins no ^{18}O entered the CO_2. In these two cases, a somewhat different mechanism may hold.

The jellyfish *Aequorea* contains a **photoprotein**, which emits light only when calcium ions are present.[672,673] Since light emission can be measured with great sensitivity (modern photomultipliers can be used to count light quanta) the protein **aequorin** and related photoproteins[674a] are used as a sensitive indicator of calcium ion concentration.[674] (In a similar way the firefly luciferin–luciferase system, which requires ATP for activation, is widely used in an assay for ATP.)

To identify the chromophore in aequorin over 4000 kg of jellyfish were used to obtain 125 mg of electrophoretically pure photoprotein.[674] From this one mg of a chromophoric substance AF-350 (Fig. 23-51) was isolated and characterized as a product. The close relationship to the *Renilla* and *Vargula* luciferins is obvious, and it is thought that coelenterazine is present in aequorin and other photoproteins as a peroxide (as in Eq. 23-45). For this reason no additional oxygen is needed to complete the reaction when Ca^{2+} acts to alter the conformation of the protein.[675–676a] The structure of a photoactive intermediate from the coelenterazine-containing protein obelin (from *Obelia longissima*), however, shows only one oxygen atom attached to C2 (Fig. 23-51)[676a] Although coelenterazine is utilized by many cnidarians they apparently cannot synthesize the compound but must obtain it through their diet. The source of biosynthesis is unknown.[676b] Some dinoflagellates emit light from a 137-kDa luciferin that contains three homologous domains each of which binds a molecule of a tetrapyrrole.[676c] From its structure the latter appears to have arisen from chlorophyll (Fig. 23-20), whose ring has been opened to give a structure somewhat similar to that of phytochrome (Fig. 23-23).[676d]

The first step in the formation of light in the firefly is a reaction with ATP to form luciferyl adenylate (Eq. 23-46, step *a*).[676e] The proton on the carbon may then be removed making use of the electron accepting properties of the adjacent ring system and carbonyl group before addition of the O_2. The reactions should be compared to those catalyzed by oxygenases, e.g., Eq. 18-42. The large 62-kDa firefly luciferase has a

Firefly

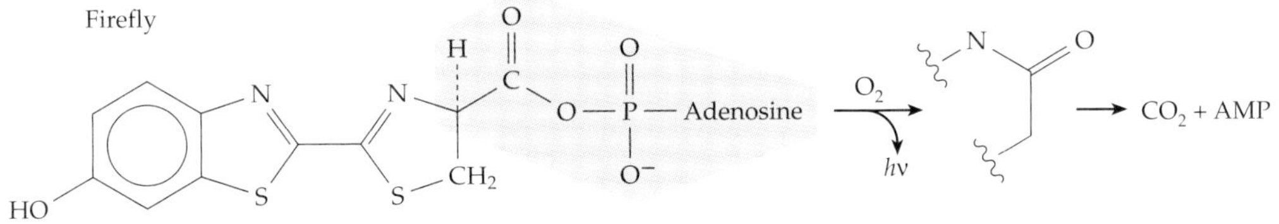

Luciferyl adenylate

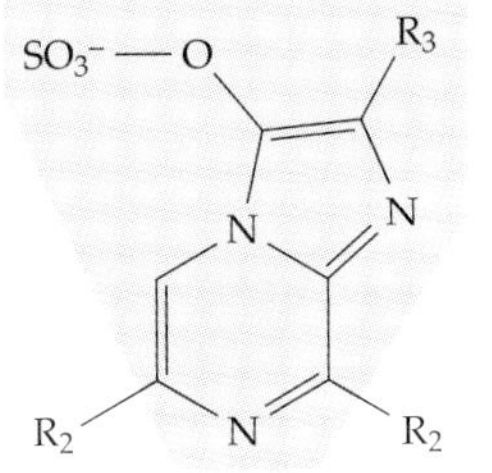

Luciferyl sulfate of *Renilla*

Renilla

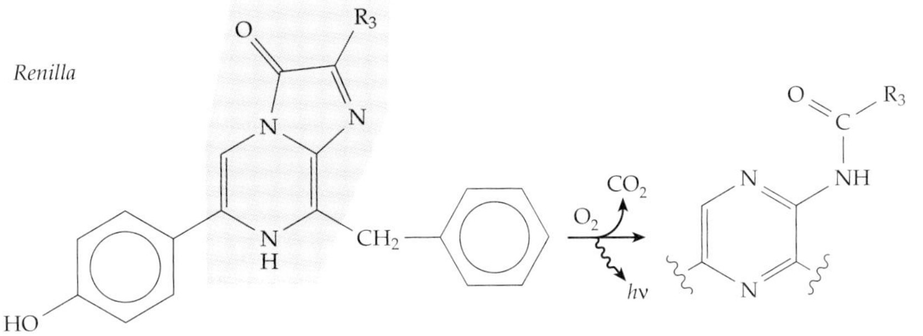

Vargula

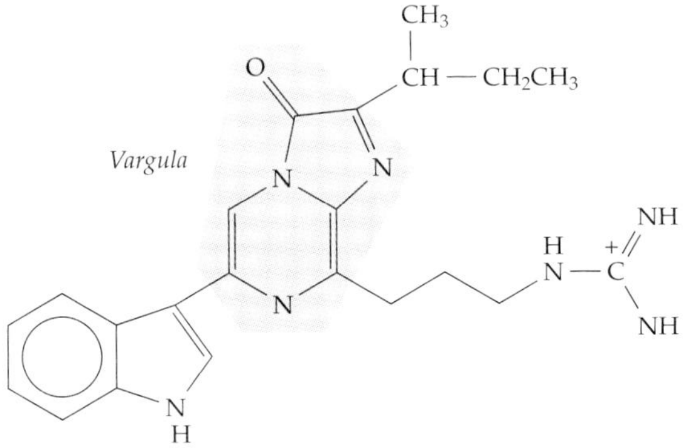

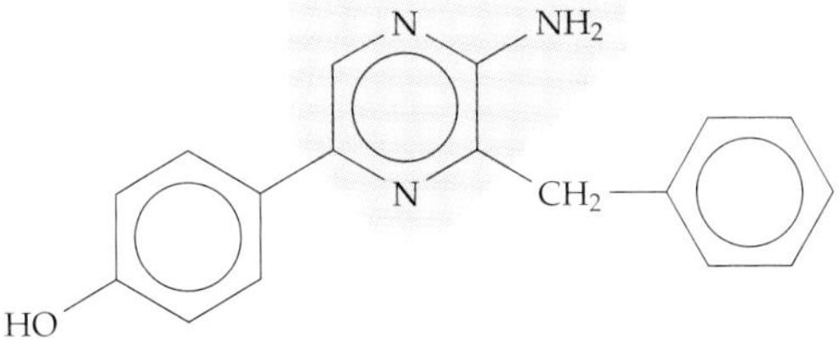

Compound AF-350 from *Aequorea*

Photoprotein **obelin** from *obelia* showing an oxy complex of bound coelenterazina. The oxygen atom is green. From Liu *et al.*[676a]

Latia

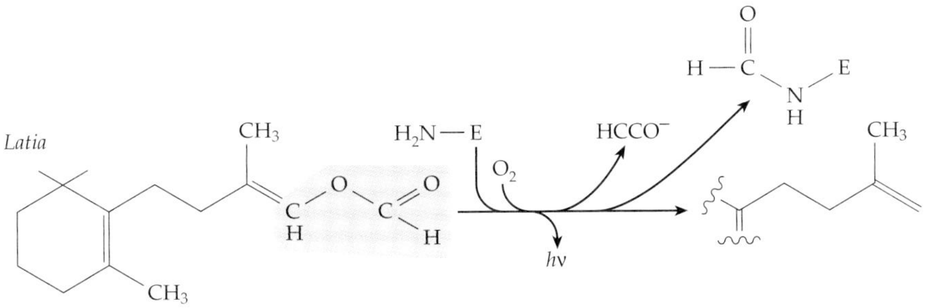

Figure 23-51 Structures of luciferins from several luminous organisms. The forms shown are the "activated" molecules ready to react with O_2. However, compound AF-350 is a breakdown product of the Ca^{2+}-activated luminous protein aequorin.

BOX 23-A THE GREEN FLUORESCENT PROTEIN AND OTHER LIGHT-EMITTING ANTENNAS

The Ca^{2+} -dependent luminescence of the jellyfish *Aquorea*, discussed in the main text, consists of blue light with a maximum intensity at 470 nm. However, the living organism has a more brilliant green luminescence. The excitation energy is transferred in a nonradiationless process to a green fluorescent protein with absorption maxima at 395 and 475 nm and an emission maximum for fluorescence at 508 nm.[a–c] A similar protein is used by *Renilla*.[d] The 238-residue green fluorescent protein (**GFP**) has a compact three-dimensional structure, an 11-stranded antiparallel β barrel with helices at one end and longer loops at the other, a "**β can**." The chromophore lies in the center of the cylinder. Numerous mutant forms have been made but only 15 residues in the terminal regions could be deleted without loss of fluorescence.[e]

The chromophore of GFP is formed spontaneously from -Ser-Tyr-Gly, residues 65–67 of the protein.[a,f,g] The entire protein has been synthesized chemically and forms the fluorescent chromophore just as in the protein produced biologically.[h] The reaction is autocatalytic, requiring only O_2 as the oxidant. Although in the living organism it accepts energy from the aqueorin chromophore a near ultraviolet lamp will elicit the fluorescence in the laboratory.

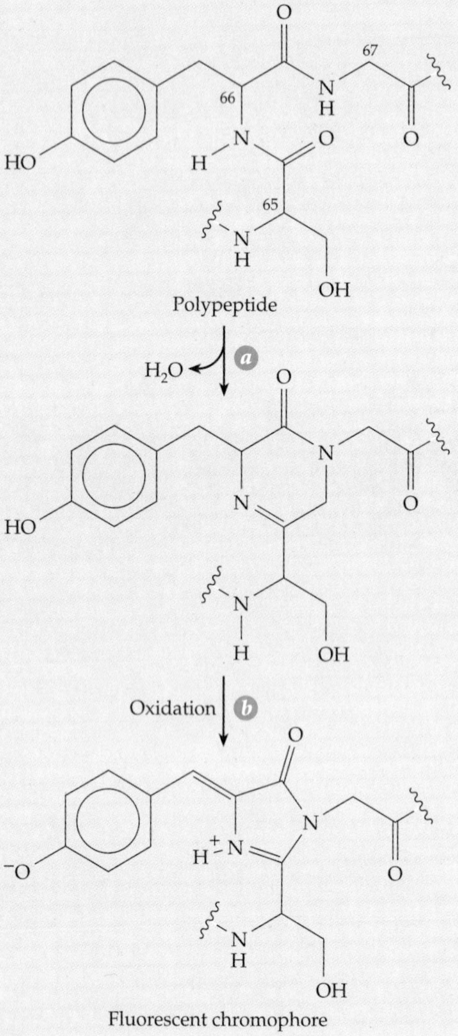

Polypeptide

H_2O ← *a*

Oxidation *b*

Fluorescent chromophore

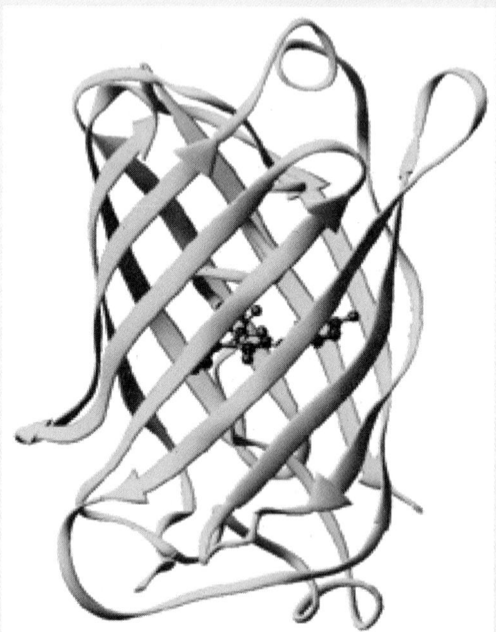

Ribbon drawing of the 238-residue green fluorescent protein showing the embedded chromophore as a ball-and-stick structure.[i] Courtesy of S. James Remington.

The green fluorescent protein is used widely in molecular biology as a fluorescent tag. Its rugged chemical nature, resistance to degradation by proteases and ability to form the chromophore autocatalytically from its own amino acids have permitted many applications. The entire GFP can be attached covalently to numerous cell components. Its gene can be spliced into the genome of an organism to form green-glowing worms, flies, and plants. Put behind a suitable promoter the fluorescent protein may be synthesized or not depending upon the control mechanism of a particular promoter (Chapter 28).[c,j–m]

The phenolic group of the GFP chromophore is apparently dissociated in the form absorbing at 395 nm and is in a tautomeric equilibrium with the other species. However, some histidine-containing replacement mutants have pH-dependent spectral changes in which the dipolar ionic form shown above, and absorbing at a longer wavelength, loses

BOX 23-A (continued)

a proton to form the anion. Observation of the excitation spectra for fluorescence of such mutant proteins within cells provides a new method for measuring the internal pH of cells and their organelles.[n–q] Some mutants emit blue or yellow light.[r,s] Two different color mutants have been fused with a molecule of calmodulin in such a way that the Ca^{2+}-induced conformational change in calmodulin allows fluorescence resonance energy transfer (FRET) between the two fluorophores. This provides a new type of calcium ion indicator.[s–u] A structurally similar red fluorescent protein, produced by a coral, extends the range of colors available as biological markers and may be useful in applications based on resonance energy transfer.[v,w,x]

Bioluminescent bacteria of the genus *Photobacterium* produce large amounts of a highly fluorescent 189-residue **lumazine protein** which contains bound 6,7-dimethyl-8-ribityllumazine (see Fig. 25-20).[y,z] Like the green fluorescent protein, it serves as a secondary light emitter receiving its energy by transfer from the flavin primary emitter. Its presence shifts the light-emission from the 495 nm of the luciferase to as low as 470 nm. *Vibrio fischeri* synthesizes a **yellow fluorescent protein** with either bound FMN or riboflavin. Its emission is at 542 nm, a longer wavelength than that of the luciferase emission. The value to the bacteria may be the higher quantum yield of fluorescence from the antenna emitter than from the luciferase. The luciferase fluorescence has a lifetime of 10 ns but on addition of the yellow fluorescent protein it is decreased to 0.25 ns with a greatly intensified emission.[u]

[a] Cubitt, A. B., Heim, R., Adams, S. R., Boyd, A. E., Gross, L. A., and Tsien, R. Y. (1995) *Trends Biochem. Sci.* **20**, 448–455

[b] Brejc, K., Sixma, T. K., Kitts, P. A., Kain, S. R., Tsien, R. Y., Ormö, M., and Remington, S. J. (1997) *Proc. Natl. Acad. Sci. U.S.A.* **94**, 2306–2311

[c] Chalfie, M., and Kain, S., eds. (1998) *Green Fluorescent Protein: Properties, Applications and Protocols*, Wiley-Liss, New York

[d] Hart, R. C., Matthews, J. C., Hori, K., and Cormier, M. J. (1979) *Biochemistry* **18**, 2204–2205

[e] Li, X., Zhang, G., Ngo, N., Zhao, X., Kain, S. R., and Huang, C.-C. (1997) *J. Biol. Chem.* **272**, 28545–28549

[f] Niwa, H., Inouye, S., Hirano, T., Matsuno, T., Kojima, S., Kubota, M., Ohashi, M., and Tsuji, F. I. (1996) *Proc. Natl. Acad. Sci. U.S.A.* **93**, 13617–13622

[g] Reid, B. G., and Flynn, G. C. (1997) *Biochemistry* **36**, 6786–6791

[h] Nishiuchi, Y., Inui, T., Nishio, H., Bódi, J., Kimura, T., Tsuji, F. I., and Sakakibara, S. (1998) *Proc. Natl. Acad. Sci. U.S.A.* **95**, 13549–13554

[i] Ormö, M., Cubitt, A. B., Kallio, K., Gross, L. A., Tsien, R. Y., and Remington, S. J. (1996) *Science* **273**, 1392–1395

[j] Chalfie, M., Tu, Y., Euskirchen, G., Ward, W. W., and Prasher, D. C. (1994) *Science* **263**, 802–805

[k] Ohashi, T., Kiehart, D. P., and Erickson, H. P. (1999) *Proc. Natl. Acad. Sci. U.S.A.* **96**, 2153–2158

[l] Yeh, E., Gustafson, K., and Boulianne, G. L. (1995) *Proc. Natl. Acad. Sci. U.S.A.* **92**, 7036–7040

[m] Hampton, R. Y., Koning, A., Wright, R., and Rine, J. (1996) *Proc. Natl. Acad. Sci. U.S.A.* **93**, 828–833

[n] Elsliger, M.-A., Wachter, R. M., Hanson, G. T., Kallio, K., and Remington, S. J. (1999) *Biochemistry* **38**, 5296–5301

[o] Robey, R. B., Ruiz, O., Santos, A. V. P., Ma, J., Kear, F., Wang, L.-J., Li, C.-J., Bernardo, A. A., and Arruda, J. A. L. (1998) *Biochemistry* **37**, 9894–9901

[p] Tsien, R. Y. (1998) *Ann. Rev. Biochem.* **67**, 509–544

[q] Wachter, R. M., Yarbrough, D., Kallio, K., and Remington, S. J. (2000) *J. Mol. Biol.* **301**, 157–171

[r] Yang, T.-T., Sinai, P., Green, G., Kitts, P. A., Chen, Y.-T., Lybarger, L., Chervenak, R., Patterson, G. H., Piston, D. W., and Kain, S. R. (1998) *J. Biol. Chem.* **273**, 8212–8216

[s] Pozzan, T. (1997) *Nature (London)* **388**, 834–835

[t] Romoser, V. A., Hinkle, P. M., and Persechini, A. (1997) *J. Biol. Chem.* **272**, 13270–13274

[u] Nagai, T., Sawano, A., Park, E. S., and Miyawaki, A. (2001) *Proc. Natl. Acad. Sci. U.S.A.* **98**, 3197–3202

[v] Yarbrough, D., Wachter, R. M., Kallio, K., Matz, M. V., and Remington, S. J. (2001) *Proc. Natl. Acad. Sci. U.S.A.* **98**, 462–467

[w] Garcia-Parajo, M. F., Koopman, M., van Dijk, E. M. H. P., Subramaniam, V., and van Hulst, N. F. (2001) *Proc. Natl. Acad. Sci. U.S.A.* **98**, 14392–14397

[x] Campbell, R. E., Tour, O., Palmer, A. E., Steinbach, P. A., Baird, G. S., Zacharias, D. A., and Tsien, R. Y. (2002) *Proc. Natl. Acad. Sci. U.S.A.* **99**, 7877–7882

[y] Hughes, R. E., Brzovic, P. S., Klevit, R. E., and Hurley, J. B. (1995) *Biochemistry* **34**, 11410–11416

[z] Petushkov, V. N., Gibson, B. G., and Lee, J. (1996) *Biochemistry* **35**, 8413–8418

two-domain structure that suggests that domain movement may be essential to bring reactants together.[660] The structure is homologous to those of acyl-CoA ligases and peptide synthetases which share a similarity in step *a*. Formation of the dioxetane intermediate is assisted by the loss of AMP (Eq. 23-46, step *b*). The electronically excited decarboxylation product interacts with groups in the protein. It apparently exists as an anion bound to acidic and basic groups of the protein. An equilibrium between oxo- and enolate forms is thought to regulate the color of the emitted light which can vary from red to yellow and green in various fireflies, other beetles, and larvae.[661,661a,677,677a] Oxyluciferin can be reconverted to luciferin for the next flash.[677b]

A very different light-producing reaction is used by the limpet *Latia*. The luciferin is an unusual terpene derivative (Fig. 23-51) that lacks any chromophore suitable for light emission.[678] Evidently oxidation of this luciferin causes electronic excitation of some other molecule, presumably a "purple protein" which is also needed for luminescence. A complex of luciferin plus the purple protein is believed to react with the luciferase (abbreviated E-NH_2 in Fig. 23-51). It is

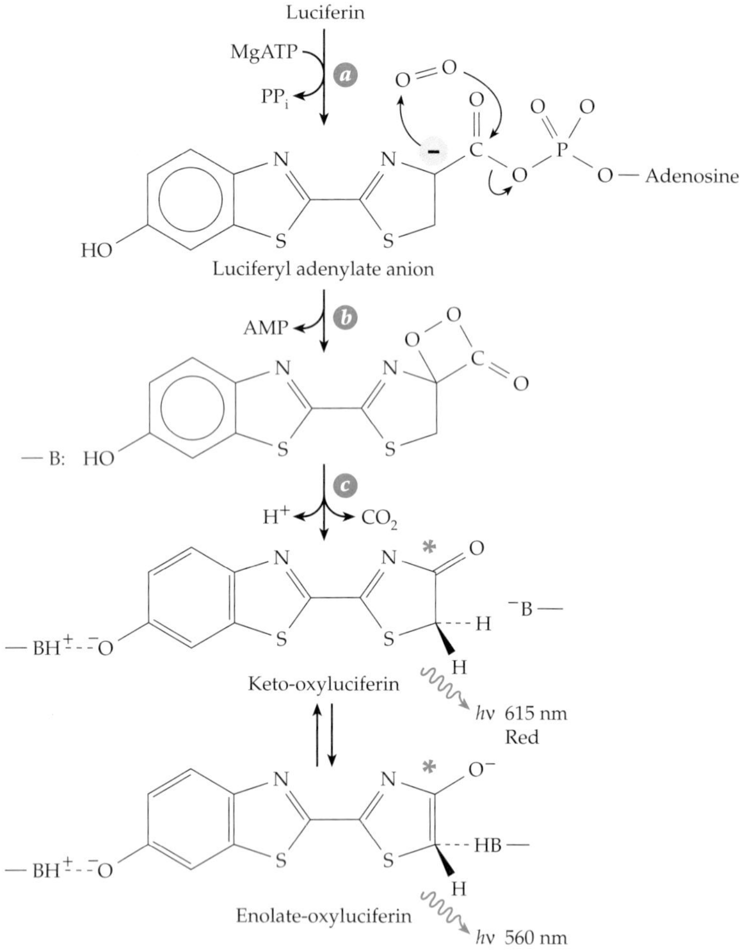

(23-46)

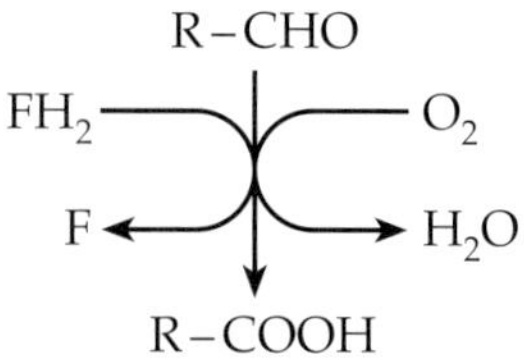

$$FH_2 + R\text{-}CHO + O_2 \rightarrow F + R\text{-}COOH + H_2O$$

(23-47)

thought that the formyl group is released from its enolic ester linkage in the luciferin. A Schiff base of the resulting aldehyde may form with the enzyme and react with oxygen. Yet another type of luciferin is found in dinoflagellates (Fig. 23-51).[679]

Luminescent bacteria all appear to obtain light from a riboflavin-5'-phosphate dependent oxygenase, which converts a long-chain aldehyde (usually *n*-tetradecanal) to a carboxylic acid (Eq. 23-47). Here FH_2 is the riboflavin 5'-*P*, which is thought to be supplied by a flavin reductase.[679a]

Bacterial luciferases are αβ heterodimers with subunit masses of ~40(α) and 35 (β) kDa.[664,680,681] In *Vibrio harveyi* these are encoded by the *lux A* and *lux B* genes. At least five other genes are essential for light production including two regulatory genes.[682,683] The tetradecanal and other long-chain aldehydes are supplied by reduction of the corresponding acyl-CoA (Eq. 23-48). A special thioesterase releases a myristoyl group from an acyl carrier protein, diverting it for luminescence in *V. harveyi*.[684] There is good evidence from ^{13}C NMR and electronic spectra for an enzyme-bound reduced flavin hydroperoxide as in Eq. 15-31. While this hydroperoxide can decompose slowly to flavin and H_2O_2 in the dark, it can also carry out the oxidation of the aldehyde with emission of light.[685,685a] The luminescent emission spectrum resembles the fluorescence spectrum of the 4a –OH adduct (Eq. 23-49), which is probably the light-emitting species.[686–688]

Cells of *Vibrio fischeri*, from the light organ of the fish *Monocentrus japonicus*, emit light only in dense cultures where a chemical inducer identified as N-(2-oxocaproyl)homoserine lactone[689,690] accumulates.

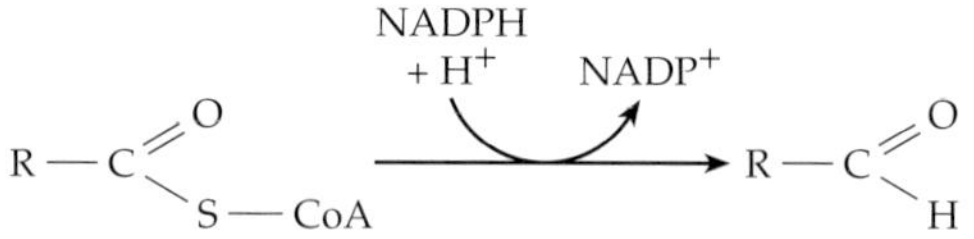

$$R\text{—}C(=O)\text{—}S\text{—}CoA \xrightarrow{NADPH + H^+ \;\to\; NADP^+} R\text{—}C(=O)\text{—}H$$

(23-48)

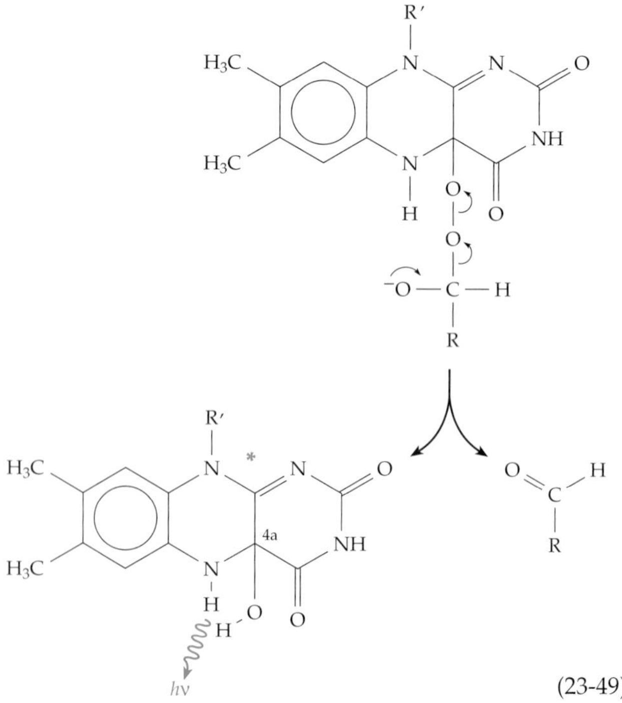

(23-49)

References

1. Wald, G. (1959) *Sci. Am.* **201**(Oct), 92–108
2. Calvert, J. G., and Pitts, J. N., Jr. (1966) *Photochemistry*, Wiley, New York
3. Suzuki, H. (1967) *Electronic Absorption Spectra and Geometry of Organic Molecules*, Academic Press, New York
4. Kraemer, K. H. (1997) *Proc. Natl. Acad. Sci. U.S.A.* **94**, 11–14
5. Kohen, E., Santus, R., and Hirschberg, J. G. (1995) *Photobiology*, Academic Press, San Diego
6. Murrell, J. N. (1967) *The Theory of the Electronic Spectra of Organic Molecules*, Academic Press, New York
7. Cantor, C. R., and Schimmel, P. R. (1980) *Biophysical Chemistry Part II*, Freeman, San Francisco, California (pp. 349–480)
8. Herzberg, G. (1950) *Molecular Spectra and Molecular Structure*, 2nd ed., Vol. I, Van Nostrand-Reinhold, Princeton, New Jersey
9. Brown, S. B., ed. (1980) *An Introduction to Spectroscopy for Biochemists*, Academic Press, New York
10. Bell, J. E., ed. (1980, 1981) *Spectroscopy in Biochemistry*, CRC Press, Boca Raton, Florida
11. Campbell, I. D., and Dwek, R. A. (1983) *Biological Spectroscopy*, Addison-Wesley, Reading, Massachusetts

11a. Steinmeyer, G., Sutter, D. H., Gallmann, L., Matuschek, N., and Keller, U. (1999) *Science* **286**, 1507–1512

11b. Hopkins, J.-M., and Sibbett, W. (2000) *Sci. Am.* **283**(Sep), 72–79

11c. Service, R. F. (2001) *Science* **292**, 1627–1628

11d. Ihee, H., Lobastov, V. A., Gomez, U. M., Goodson, B. M., Srinivasan, R., Ruan, C.-Y., and Zewail, A. H. (2001) *Science* **291**, 458–462

11e. Lattman, E. E. (2001) *Proc. Natl. Acad. Sci. U.S.A.* **98**, 6535–6536

11f. Drescher, M., Hentschel, M., Kienberger, R., Tempea, G., Spielmann, C., Reider, G. A., Corkum, P. B., and Krausz, F. (2001) *Science* **291**, 1923–1927

11g. Bhattacharjee, Y. (2001) *Nature (London)* **412**, 474–476

12. Kochakian, C. D. (1988) *Trends Biochem. Sci.* **13**, 359–362

12a. Burns, D. A., and Ciurczak, E. W., eds. (2001) *Handbook of Near-Infrared Analysis*, 2nd ed., Dekker, New York

13. Parker, F. S. (1983) *Applications of Infrared, Raman and Resonance Raman Spectroscopy in Biochemistry, Biology and Medicine*, Plenum, New York
14. Haris, P. I., and Chapman, D. (1992) *Trends Biochem. Sci.* **17**, 328–333
15. Surewicz, W. K., Mantsch, H. H., and Chapman, D. (1993) *Biochemistry* **32**, 389–394

15a. Gremlich, H.-U., and Yan, B., eds. (2000) *Infrared and Raman Spectroscopy of Biological Materials*, Dekker, New York

16. Reisdorf, W. C., Jr., and Krimm, S. (1996) *Biochemistry* **35**, 1383–1386
17. Wright, W. W., Laberge, M., and Vanderkooi, J. M. (1997) *Biochemistry* **36**, 14724–14732
18. Dagneaux, C., Liquier, J., and Taillandier, E. (1995) *Biochemistry* **34**, 16618–16623
19. Sanchez-Ruiz, J. M., and Martinez-Carrion, M. (1988) *Biochemistry* **27**, 3338–3342

19a. Zhao, W., and Wright, J. C. (1999) *J. Am. Chem. Soc.* **121**, 10994–10998

20. Zscherp, C., Schlesinger, R., Tittor, J., Oesterhelt, D., and Heberle, J. (1999) *Proc. Natl. Acad. Sci. U.S.A.* **96**, 5498–5503
21. Meskers, S., Ruysschaert, J.-M., and Goormaghtigh, E. (1999) *J. Am. Chem. Soc.* **121**, 5115–5122
22. Fraser, R. D. B., and MacRae, T. P. (1973) *Conformation in Fibrous Proteins and Related Synthetic Polypeptides*, Academic Press, New York (pp. 95–106)
23. Grasselli, J. G., Snavely, M. K., and Balkin, B. J. (1981) *Chemical Applications of Raman Spectroscopy*, Wiley, New York

23a. Baumruk, V., Pancoska, P., and Keiderling, T. A. (1996) *J. Mol. Biol.* **259**, 774–791

24. Burke, M. J., and Rougvie, M. A. (1972) *Biochemistry* **11**, 2435–2439
25. Anderson, T. S., Hellgeth, J., and Lansbury, P. T., Jr. (1996) *J. Am. Chem. Soc.* **118**, 6540–6546
26. Reilly, K. E., and Thomas, G. J., Jr. (1994) *J. Mol. Biol.* **241**, 68–82
27. de Jongh, H. H. J., Goormaghtigh, E., and Ruysschaert, J.-M. (1997) *Biochemistry* **36**, 13603–13610
28. Miles, H. T., Lewis, T. P., Becker, E. D., and Frazier, J. (1973) *J. Biol. Chem.* **248**, 1115–1117
29. Hendra, P., Jones, C., and Warnes, G. (1991) *Fourier Transform Raman Spectroscopy*, Ellis Horwood, New York
30. Overman, S. A., and Thomas, G. J., Jr. (1998) *Biochemistry* **37**, 5654–5665
31. Lippert, J. L., Tyminski, D., and Desmeules, P. J. (1976) *J. Am. Chem. Soc.* **98**, 7075–7080
32. Craig, W. S., and Gaber, B. P. (1977) *J. Am. Chem. Soc.* **99**, 4130–4134
33. Tsuboi, M., Overman, S. A., and Thomas, G. J., Jr. (1996) *Biochemistry* **35**, 10403–10410

33a. Wen, Z. Q., Overman, S. A., Bondre, P., and Thomas, G. J., Jr. (2001) *Biochemistry* **40**, 449–458

34. Emerson, R., Chalmers, R., and Cederstrand, C. (1957) *Proc. Natl. Acad. Sci. U.S.A.* **43**, 133–143
35. Emerson, R., and Arnold, W. (1932) *J. Gen. Physiol.* **16**, 191–205
36. Carey, P. R. (1982) *Biochemical Applications of Raman and Resonance Raman Spectroscopies*, Academic Press, New York
37. Lugtenburg, J., Mathies, R. A., Griffin, R. G., and Herzfeld, J. (1988) *Trends Biochem. Sci.* **13**, 388–393
38. Wright, J. C. (1982) in *Applications of Lasers to Chemical Problems* (Evans, T. R., ed), pp. 35–179, Wiley, New York
39. Chen, X. G., Li, P., Holtz, J. S. W., Chi, Z., Pajcini, V., Asher, S. A., and Kelly, L. A. (1996) *J. Am. Chem. Soc.* **118**, 9705–9715
40. Chi, Z., Chen, X. G., Holtz, J. S. W., and Asher, S. A. (1998) *Biochemistry* **37**, 2854–2864
41. Overman, S. A., and Thomas, G. J., Jr. (1999) *Biochemistry* **38**, 4018–4027
42. Hu, X., and Spiro, T. G. (1997) *Biochemistry* **36**, 15701–15712
43. Miura, T., and Thomas, G. J., Jr. (1994) *Biochemistry* **33**, 7848–7856
44. Miura, T., and Thomas, G. J., Jr. (1995) *Biochemistry* **34**, 9645–9654
45. Chan, S. S., Austin, R. H., Mukerji, I., and Spiro, T. G. (1997) *Biophys. J.* **72**, 1512–1520
46. Hu, X., Rodgers, K. R., Mukerji, I., and Spiro, T. G. (1999) *Biochemistry* **38**, 3462–3467
47. Chi, Z., and Asher, S. A. (1999) *Biochemistry* **38**, 8196–8203
48. Macdonald, I. D. G., Sligar, S. G., Christian, J. F., Unno, M., and Champion, P. M. (1999) *J. Am. Chem. Soc.* **121**, 376–380
49. Dong, S., and Spiro, T. G. (1998) *J. Am. Chem. Soc.* **120**, 10434–10440
50. Kitagawa, T., Nishina, Y., Kyogoku, Y., Yamano, T., Ohishi, N., Takai-Suzuki, A., and Yagi, K. (1979) *Biochemistry* **18**, 1804–1808
51. Kitagawa, T., Nishina, Y., Kyogoky, Y., Yamano, T., Ohishi, N., Takai-Suzuki, A., and Yagi, K. (1979) *Biochemistry* **18**, 1804–1808
52. Clarkson, J., Palfey, B. A., and Carey, P. R. (1997) *Biochemistry* **36**, 12560–12566
53. Benecky, M. J., Copeland, R. A., Hays, T. R., Lobenstine, E. W., Rava, R. P., Pascal, R. A., Jr., and Spiro, T. G. (1985) *J. Biol. Chem.* **260**, 11663–11670
54. Benecky, M. J., Copeland, R. A., Rava, R. P., Feldhaus, R., Scott, R. D., Metzler, C. M., Metzler, D. E., and Spiro, T. G. (1985) *J. Biol. Chem.* **260**, 11671–11678

54a. Altose, M. D., Zheng, Y., Dong, J., Palfey, B. A., and Carey, P. R. (2001) *Proc. Natl. Acad. Sci. U.S.A.* **98**, 3006–3011

55. Clark, R. J. H., D'Urso, N. R., and Zagalsky, P. F. (1980) *J. Am. Chem. Soc.* **102**, 6693–6698
56. Ozaki, Y., Pliura, D. H., Carey, P. R., and Storer, A. C. (1982) *Biochemistry* **21**, 3102–3108

56a. Vogel, K. M., Kozlowski, P. M., Zgierski, M. Z., and Spiro, T. G. (1999) *J. Am. Chem. Soc.* **121**, 9915–9921

56b. Franzen, S. (2001) *J. Am. Chem. Soc.* **123**, 12578–12589

57. Rospendowski, B. N., Kelly, K., Wolf, C. R., and Smith, W. E. (1991) *J. Am. Chem. Soc.* **113**, 1217–1225
58. Dähne, S. (1978) *Science* **199**, 1163–1167
59. Morley, J. O., Morley, R. M., and Fitton, A. L. (1998) *J. Am. Chem. Soc.* **120**, 11479–11488
60. Marder, S. R., Gorman, C. B., Meyers, F., Perry, J. W., Bourhill, G., Brédas, J.-L., and Pierce, B. M. (1994) *Science* **265**, 632–635
61. Horwitz, J., Strickland, E. H., and Billups, C. (1969) *J. Am. Chem. Soc.* **91**, 184–190
62. Siano, D. B., and Metzler, D. E. (1969) *J. Chem. Phys.* **51**, 1856–1861
63. Metzler, D. E., Harris, C. M., Johnson, R. J., Siano, D. B., and Thomson, J. A. (1973) *Biochemistry* **12**, 5377–5392
64. Metzler, C. M., Cahill, A. E., Petty, S., Metzler, D. E., and Lang, L. (1985) *Appl. Spectroscopy* **39**, 333–339
65. Horwitz, J., Strickland, E. H., and Billups, C. (1970) *J. Am. Chem. Soc.* **92**, 2119–2129
66. Horwitz, J., and Strickland, E. H. (1971) *J. Biol. Chem.* **246**, 3749–3752
67. Zucchelli, G., Dainese, P., Jennings, R. C., Breton, J., Garlaschi, F. M., and Bassi, R. (1994) *Biochemistry* **33**, 8982–8990
68. Tang, S. W., Coleman, J. E., and Myer, Y. P. (1968) *J. Biol. Chem.* **243**, 4286–4297
69. Hofrichter, J., and Eaton, W. A. (1976) *Annu Rev Biophys Bioeng.* **5**, 511–560
70. Arnone, A., Christen, P., Jansonias, J. N., and Metzler, D. E. (1985) in *Transaminases* (Christen, P., and Metzler, D. E., eds), pp. 349–357, Wiley, New York
71. Holmén, A., Broo, A., Albinsson, B., and Nordén, B. (1997) *J. Am. Chem. Soc.* **119**, 12240–12250
72. Peterson, J., Pearce, L. L., and Bominaar, E. L. (1999) *J. Am. Chem. Soc.* **121**, 5972–5980
73. Metzler, C. M., Mitra, J., Metzler, D. E., Makinen, M. W., Hyde, C. C., Rogers, P., and Arnone, A. (1988) *J. Mol. Biol.* **203**, 197–220
74. Morton, R. A. (1975) *Biochemical Spectroscopy*, Wiley, New York
75. Perkampus, H. H., Sandeman, I., and Timmons, C. J. (1966–1977) *UV Atlas of Organic Compounds*, Vol. 1–5, Plenum, New York
76. Lang, L., ed. (1961–1975) *Absorption Spectra in the Ultraviolet and Visible Region*, Academic Press, New York (A serial publication)
77. Petruska, J. (1961) *J. Chem. Phys.* **34**, 1120–1136
78. Stevenson, P. E. (1965) *J. Mol. Spectroscopy* **15**, 220–256
79. Strickland, E. H., Wilchek, M., Billups, C., and Horowitz, J. (1972) *J. Biol. Chem.* **247**, 572–580
80. Strickland, E. H., Billups, C., and Kay, E. (1972) *Biochemistry* **11**, 3657–3662

References

81. Pajcini, V., Chen, X. G., Bormett, R. W., Geib, S. J., Li, P., Asher, S. A., and Lidiak, E. G. (1996) *J. Am. Chem. Soc.* **118**, 9716–9726
82. Clark, L. B. (1995) *J. Am. Chem. Soc.* **117**, 7974–7986
83. Fülscher, M. P., and Roos, B. O. (1995) *J. Am. Chem. Soc.* **117**, 2089–2095
84. Fülscher, M. P., Serrano-Andrés, L., and Roos, B. O. (1997) *J. Am. Chem. Soc.* **119**, 6168–6176
85. Harders, H., Forster, S., Voelter, W., and Bacher, A. (1974) *Biochemistry* **13**, 3360–3364
86. Blackburn, G. M., and Gait, M. J., eds. (1996) *Nucleic Acids in Chemistry and Biology*, 2nd ed., Oxford Univ. Press, Oxford
87. Murrell, J. N. (1963) *The Theory of the Electronic Spectra of Organic Molecules*, Wiley, New York (Chapter 7)
88. Collins, K. D., and Stark, G. R. (1971) *J. Biol. Chem.* **246**, 6599–6605
89. Donovan, J. W. (1973) in *Methods in Enzymology*, Vol. 27, Part D (Hirs, C. H. W., and Timasheff, S. N., eds), pp. 497–525, Academic Press, New York
90. Metzler, D. E., Harris, C., Yang, I.-Y., Siano, D., and Thomson, J. A. (1972) *Biochem. Biophys. Res. Commun.* **46**, 1588–1597
91. Metzler, D. E., Metzler, C. M., and Mitra, J. (1986) *Trends Biochem. Sci.* **11**, 157–159
92. Kornblatt, J. A., Kornblatt, M. J., and Hoa, G. H. B. (1995) *Biochemistry* **34**, 1218–1223
93. Padrós, E., Dunach, M., Morros, A., Sobés, M., and Manosa, J. (1984) *Trends Biochem. Sci.* **9**, 508–510
94. Butler, W. L. (1979) *Methods Enzymol.* **56**, 501–515
95. Fasman, G. D., ed. (1996) *Circular Dichroism and the Conformational Analysis of Biomolecules*, Plenum, New York
96. Foss, J. G. (1963) *J. Chem. Educ.* **40**, 592–597
97. Johnson, W. C., Jr. (1988) *Ann. Rev. Biophys. Biophys. Chem.* **17**, 145–166
98. Strickland, E. H. (1974) *Crit. Revs. Biochem.* **2**, 113–175
99. Moffitt, W., Woodward, R. B., Moscowitz, A., Klyne, W., and Djerassi, C. (1961) *J. Am. Chem. Soc.* **83**, 4013–4018
100. Bayley, P. M. (1973) *Prog. Biophys. and Mol. Biol.* **27**, 1–76
101. Johnson, W. C., Jr., and Tinoco, I., Jr. (1972) *J. Am. Chem. Soc.* **94**, 4389–4390
102. Chen, Y.-H., Yang, J. T., and Chau, K. H. (1974) *Biochemistry* **13**, 3350–3359
103. Cantor, C. R., and Schimmel, P. R. (1980) *Biophysical Chemistry, Part II*, Freeman, San Francisco, California (pp. 409–433)
104. Tinoco, I., Jr., Maestre, M. F., and Bustamante, C. (1983) *Trends Biochem. Sci.* **8**, 41–44
105. Applequist, J. (1973) *J. Am. Chem. Soc.* **95**, 8255–8262
106. Applequist, J. (1987) *Am. Scientist* **75**, 58–68
107. Applequist, J., and Bode, K. A. (1999) *J. Phy. Chem. B* **103**, 1767–1773
108. Bode, K. A., and Applequist, J. (1998) *J. Am. Chem. Soc.* **120**, 10938–10946 , Erratum 13545
109. Bode, K. A., and Applequist, J. (1997) *Biopolymers* **42**, 855–860
110. Paterlini, M. G., Freedman, T. B., and Nafie, L. A. (1986) *J. Am. Chem. Soc.* **108**, 1389–1397
111. Bose, P. K., and Polavarapu, P. L. (1999) *J. Am. Chem. Soc.* **121**, 6094–6095
112. Baumruk, V., and Keiderling, T. A. (1993) *J. Am. Chem. Soc.* **115**, 6939–6942
113. Self, B. D., and Moore, D. S. (1997) *Biophys. J.* **73**, 339–347
114. Bell, A. F., Hecht, L., and Barron, L. D. (1997) *J. Am. Chem. Soc.* **119**, 6006–6013
115. Peters, K. S., Watson, T., and Marr, K. (1991) *Ann. Rev. Biophys. Biophys. Chem.* **20**, 343–362
116. Hung, R. R., and Grabowski, J. J. (1999) *J. Am. Chem. Soc.* **121**, 1359–1364
117. Meisel, A., Leonhardt, G., and Szargan, R. (1989) *X-Ray Spectra and Chemical Binding*, Springer-Verlag, New York
118. Binsted, N., Strange, R. W., and Hasnain, S. S. (1992) *Biochemistry* **31**, 12117–12125
119. Ito, E., Oji, H., Araki, T., Oichi, K., Ishii, H., Ouchi, Y., Ohta, T., Kosugi, N., Maruyama, Y., Naito, T., Inabe, T., and Seki, K. (1997) *J. Am. Chem. Soc.* **119**, 6336–6344
120. Vértes, A., Korecz, L., and Burger, K. (1979) *Mössbauer Spectroscopy*, Elsevier, Amsterdam
121. Popescu, C. V., Bates, D. M., Beinert, H., Münck, E., and Kiley, P. J. (1998) *Proc. Natl. Acad. Sci. U.S.A.* **95**, 13431–13435
122. Schünemann, V., Trautwein, A. X., Illerhaus, J., and Haehnel, W. (1999) *Biochemistry* **38**, 8981–8991
123. Clayton, R. K. (1970) *Light and Living Matter*, Vol. 1, McGraw-Hill, New York
124. Parker, C. A. (1968) *Photoluminescence of Solutions*, Elsevier, Amsterdam
125. Konev, S. V. (1967) *Fluorescence and Phosphorescence of Proteins and Nucleic Acids*, Plenum, New York
126. Lakowicz, J. R. (1985) *Principles of Fluorescence Spectroscopy*, Plenum, New York
127. Baeyens, W. R. G., De Keukeleire, D., and Korkidis, K., eds. (1991) *Luminescence Techniques in Chemical and Biochemical Analysis*, Dekker, New York

127a. Meyer-Almes, F.-J., and Auer, M. (2000) *Biochemistry* **39**, 13261–13268

128. Kotaki, A., and Yagi, K. (1970) *J. Biochem.* **68**, 509–516
129. Koziol, J., and Knobloch, E. (1965) *Biochim. Biophys. Acta.* **102**, 289–300
130. Chen, Y., and Barkley, M. D. (1998) *Biochemistry* **37**, 9976–9982
131. Beechem, J. M., and Brand, L. (1985) *Ann. Rev. Biochem.* **54**, 43–71
132. Anderson, S. R. (1991) *J. Biol. Chem.* **266**, 11405–11408
133. Hedstrom, J., Sedarous, S., and Prendergast, F. G. (1988) *Biochemistry* **27**, 6203–6208
134. Wahl, P., Auchet, J.-C., Visser, A. J. W. G., and Veeger, C. (1975) *Eur. J. Biochem.* **50**, 413–418
135. Vos, R., Engelborghs, Y., Izard, J., and Baty, D. (1995) *Biochemistry* **34**, 1734–1743
136. Broos, J., ter Veld, F., and Robillard, G. T. (1999) *Biochemistry* **38**, 9798–9803
137. Ross, J. B. A., Senear, D. F., Waxman, E., Kombo, B. B., Rusinova, E., Huang, Y. T., Laws, W. R., and Hasselbacher, C. A. (1992) *Proc. Natl. Acad. Sci. U.S.A.* **89**, 12023–12027
138. Farah, C. S., and Reinach, F. C. (1999) *Biochemistry* **38**, 10543–10551
139. Strambini, G. B., and Gonnelli, M. (1995) *J. Am. Chem. Soc.* **117**, 7646–7651
140. Zhou, J. S., and Hoffman, B. M. (1994) *Science* **265**, 1693–1696
141. Eftink, M. R., and Ghiron, C. A. (1976) *Biochemistry* **15**, 672–680
142. Calhoun, D. B., Vanderkooi, J. M., and Englander, S. W. (1983) *Biochemistry* **22**, 1533–1539
143. Wright, W. W., Owen, C. S., and Vanderkooi, J. M. (1992) *Biochemistry* **31**, 6538–6544
144. Vet, J. A. M., Majithia, A. R., Marras, S. A. E., Tyagi, S., Dube, S., Poiesz, B. J., and Kramer, F. R. (1999) *Proc. Natl. Acad. Sci. U.S.A.* **96**, 6394–6399
145. Hilinski, E. F., and Rentzepis, P. M. (1983) *Nature (London)* **302**, 481–487
146. Radda, G. K. (1971) *Curr. Top. Bioenerg.* **4**, 81–176
147. Lakowicz, J. R., Laczko, G., Gryczynski, I., and Cherek, H. (1986) *J. Biol. Chem.* **261**, 2240–2245
148. Ruggiero, A. J., Todd, D. C., and Fleming, G. R. (1990) *J. Am. Chem. Soc.* **112**, 1003–1014
149. Förster, T. (1948) *Ann Physik* **2**, 55–75
150. Stryer, L. (1968) *Science* **162**, 526–533
151. Wu, C.-W., and Stryer, L. (1972) *Proc. Natl. Acad. Sci. U.S.A.* **69**, 1104–1108
152. Ha, T., Enderle, T., Ogletree, D. F., Chemla, D. S., Selvin, P. R., and Weiss, S. (1996) *Proc. Natl. Acad. Sci. U.S.A.* **93**, 6264–6268
153. Weiss, S. (1999) *Science* **283**, 1676–1683
154. Ha, T., Zhuang, X., Kim, H. D., Orr, J. W., Williamson, J. R., and Chu, S. (1999) *Proc. Natl. Acad. Sci. U.S.A.* **96**, 9077–9082
155. Fu, P. K.-L., and Turro, C. (1999) *J. Am. Chem. Soc.* **121**, 1–7
156. Hogue, C. W. V., MacManus, J. P., Banville, D., and Szabo, A. G. (1992) *J. Biol. Chem.* **267**, 13340–13347
157. Xiao, M., Li, H., Snyder, G. E., Cooke, R., Yount, R. G., and Selvin, P. R. (1998) *Proc. Natl. Acad. Sci. U.S.A.* **95**, 15309–15314
158. Gordon, G. W., Berry, G., Liang, X. H., Levine, B., and Herman, B. (1998) *Biophys. J.* **74**, 2702–2713
159. Tsien, R. Y., and Miyawaki, A. (1998) *Science* **280**, 1954–1955
160. Wouters, F. S., Bastiaens, P. I. H., Wirtz, K. W. A., and Jovin, T. M. (1998) *EMBO J.* **17**, 7179–7189
161. Ward, L. D., Seckler, R., and Timasheff, S. N. (1994) *Biochemistry* **33**, 11900–11908
162. Farrar, S. J., Whiting, P. J., Bonnert, T. P., and McKernan, R. M. (1999) *J. Biol. Chem.* **274**, 10100–10104
163. Parkhurst, K. M., and Parkhurst, L. J. (1995) *Biochemistry* **34**, 293–300
164. Jares-Erijman, E. A., and Jovin, T. M. (1996) *J. Mol. Biol.* **257**, 597–617

164a. Norman, D. G., Grainger, R. J., Uhrín, D., and Lilley, D. M. J. (2000) *Biochemistry* **39**, 6317–6324

164b. Tóth, K., Brun, N., and Langowski, J. (2001) *Biochemistry* **40**, 6921–6928

165. Rhee, M.-J., Sudnick, D. R., Arkle, V. K., and Horrocks, W. DeW., Jr. (1981) *Biochemistry* **20**, 3328–3334
166. Szczesniak, M., Szczepaniak, K., Kwiatkowski, J. S., KuBulat, K., and Person, W. B. (1988) *J. Am. Chem. Soc.* **110**, 8319–8330
167. Vanderkooi, J. M., Kaposi, A., and Fidy, J. (1993) *Trends Biochem. Sci.* **18**, 71–76
168. Friedrich, J., Gafert, J., Zollfrank, J., Vanderkooi, J., and Fidy, J. (1994) *Proc. atl. Acad. Sci. U.S.A.* **91**, 1029–1033

168a. Bastiaens, P. I. H., and Pepperkok, R. (2000) *Trends Biochem. Sci.* **25**, 631–637

168b. Frohn, J. T., Knapp, H. F., and Stemmer, A. (2000) *Proc. Natl. Acad. Sci. U.S.A.* **97**, 7232–7236

168c. Klar, T. A., Jakobs, S., Dyba, M., Egner, A., and Hell, S. W. (2000) *Proc. Natl. Acad. Sci. U.S.A.* **97**, 8206–8210

168d. Weiss, S. (2000) *Proc. Natl. Acad. Sci. U.S.A.* **97**, 8747–8749

168e. Farber, S. A., Pack, M., Ho, S.-Y., Johnson, I. D., Wagner, D. S., Dosch, R., Mullins, M. C., Hendrickson, H. S., Hendrickson, E. K., and Halpern, M. E. (2001) *Science* **292**, 1385–1388

169. Ruth, J. L. (1984) *DNA* **3**, 123
170. Renz, M., and Kurz, C. (1984) *Nucleic Acids Res.* **12**, 3435–3444
171. Davenport, L., Dale, R. E., Bisby, R. H., and Cundall, R. B. (1985) *Biochemistry* **24**, 4097–4108
172. MacGregor, R. B., and Weber, G. (1986) *Nature (London)* **319**, 70–73

References

173. Tsien, R. Y., and Poenie, M. (1986) *Trends Biochem. Sci.* **11**, 450–455
174. Thomas, F., Serratrice, G., Béguin, C., Saint Aman, E., Pierre, J. L., Fontecave, M., and Laulhère, J. P. (1999) *J. Biol. Chem.* **274**, 13375–13383
175. Stewart, W. W. (1981) *Nature (London)* **292**, 17–21
176. Berland, K. M. (1997) *Biophys. J.* **72**, 1487–1488
177. Schwille, P., Meyer-Almes, F.-J., and Rigler, R. (1997) *Biophys. J.* **72**, 1878–1886
178. Korlach, J., Schwille, P., Webb, W. W., and Feigenson, G. W. (1999) *Proc. Natl. Acad. Sci. U.S.A.* **96**, 8461–8466
179. Marx, J. (1996) *Science* **273**, 430
180. Schröck, E., du Manoir, S., Veldman, T., Schoell, B., Wienberg, J., Ferguson-Smith, M. A., Ning, Y., Ledbetter, D. H., Bar-Am, I., Soenksen, D., Garini, Y., and Ried, T. (1996) *Science* **273**, 494–497
180a. Pathak, S., Cjoi, S.-K., Arnheim, N., and Thompson, M. E. (2001) *J. Am. Chem. Soc.* **123**, 4103–4104
181. Bruchez, M., Jr., Moronne, M., Gin, P., Weiss, S., and Alivisatos, A. P. (1998) *Science* **281**, 2013–2016
182. Chan, W. C. W., and Nie, S. (1998) *Science* **281**, 2016–2018
183. Turro, N. J., and Schuster, G. (1975) *Science* **187**, 303–312
184. Salem, L. (1976) *Science* **191**, 822–830
185. Bazhulina, N. P., Morozov, Y. V., Karpeisky, M. Y., Ivanov, V. I., and Kuklin, A. I. (1966) *Biofizika* **11**, 42–47
186. Bridges, J. W., Davies, D. S., and Williams, R. T. (1966) *Biochem. J.* **98**, 451–468
187. Johnson, G. F., Tu, J.-I., Bartlett, M. L. S., and Graves, D. J. (1970) *J. Biol. Chem.* **245**, 5560–5568
188. Shaltiel, S., and Cortijo, M. (1970) *Biochem. Biophys. Res. Commun.* **41**, 594–600
189. Loken, M. R., Hayes, J. W., Gohlke, J. R., and Brand, L. (1972) *Biochemistry* **11**, 4779–4786
190. Becker, M. M., and Wang, J. C. (1984) *Nature (London)* **309**, 682–687
191. Sancar, A. (1996) *Science* **272**, 48–49
191a. Lee, J.-H., Bae, S.-H., and Choi, B.-S. (2000) *Proc. Natl. Acad. Sci. U.S.A.* **97**, 4591–4596
192. Shaw, A. A., Falick, A. M., and Shetlar, M. D. (1992) *Biochemistry* **31**, 10976–10983
193. Pashev, I. G., Dimitrov, S. I., and Angelov, D. (1991) *Trends Biochem. Sci.* **16**, 323–326
194. Pearlman, D. A., Holbrook, S. R., Pirkle, D. H., and Kim, S.-H. (1985) *Science* **227**, 1304–1305
195. Cimino, G. D., Gamper, H. B., Isaacs, S. T., and Hearst, J. E. (1985) *Ann. Rev. Biochem.* **54**, 1151–1193
195a. Yoon, J.-H., Lee, C.-S., O'Connor, T. R., Yasui, A., and Pfeifer, G. P. (2000) *J. Mol. Biol.* **299**, 681–693
196. Hanson, K. M., Li, B., and Simon, J. D. (1997) *J. Am. Chem. Soc.* **119**, 2715–2721
197. Edelson, R. L. (1988) *Sci. Am.* **259**(Aug), 68–75
198. Berns, M. W. (1991) *Sci. Am.* **264**(Jun), 84–90
199. Amato, I. (1993) *Science* **262**, 32–33
200. Staros, J. V. (1980) *Trends Biochem. Sci.* **5**, 320–322
201. Jo, E., Blazyk, J., and Boggs, J. M. (1998) *Biochemistry* **37**, 13791–13799
202. Gronemeyer, H. (1985) *Trends Biochem. Sci.* **10**, 264–267
203. Dormán, G., and Prestwich, G. D. (1994) *Biochemistry* **33**, 5661–5673
204. Gribbon, P., and Hardingham, T. E. (1998) *Biophys. J.* **75**, 1032–1039
205. Peters, R. (1985) *Trends Biochem. Sci.* **10**, 223–227
206. Pagano, R. E., and Longmuir, K. J. (1983) *Trends Biochem. Sci.* **8**, 157–161
207. Anfinrud, P., de Vivie-Riedle, R., and Engel, V. (1999) *Proc. Natl. Acad. Sci. U.S.A.* **96**, 8328–8329
208. Goldbeck, R. A., Paquette, S. J., Björling, S. C., and Kliger, D. S. (1996) *Biochemistry* **35**, 8628–8639
209. Service, R. F. (1997) *Science* **276**, 1986–1987
210. Brunori, M., Cutruzzola, F., Savino, C., Travaglini-Allocatelli, C., Vallone, B., and Gibson, Q. H. (1999) *Trends Biochem. Sci.* **24**, 253–255
211. Wan, C., Fiebig, T., Kelley, S. O., Treadway, C. R., Barton, J. K., and Zewail, A. H. (1999) *Proc. Natl. Acad. Sci. U.S.A.* **96**, 6014–6019
212. Henderson, P. T., Jones, D., Hampikian, G., Kan, Y., and Schuster, G. B. (1999) *Proc. Natl. Acad. Sci. U.S.A.* **96**, 8353–8358
213. Block, S. M. (1992) *Nature (London)* **360**, 493–495
214. Svoboda, K., and Block, S. M. (1994) *Annu. Rev. Biophys. Biomol. Struct.* **23**, 247–285
215. Wang, M. D., Yin, H., Landick, R., Gelles, J., and Block, S. M. (1997) *Biophys. J.* **72**, 1335–1346
216. Arai, Y., Yasuda, R., Akashi, K.-i, Harada, Y., Miyata, H., Kinosita, K., Jr., and Itoh, H. (1999) *Nature (London)* **399**, 446–448
217. Fodor, S. P. A., Read, J. L., Pirrung, M. C., Stryer, L., Lu, A. T., and Solas, D. (1991) *Science* **251**, 767–773
218. McGall, G., Labadie, J., Brock, P., Wallraff, G., Nguyen, T., and Hinsberg, W. (1996) *Proc. Natl. Acad. Sci. U.S.A.* **93**, 13555–13560
219. Taubes, G. (1997) *Science* **276**, 1991–1993
220. Benaron, D. A., Cheong, W.-F., and Stevenson, D. K. (1997) *Science* **276**, 2002–2003
221. Bappart, S. A., Tearney, G. J., Bouma, B. E., Southern, J. F., Brezinski, M. E., and Fujimoto, J. G. (1997) *Proc. Natl. Acad. Sci. U.S.A.* **94**, 4256–4261
221a. Sharpe, J., Ahlgren, U., Perry, P., Hill, B., Ross, A., Hecksher-Sorensen, J., Baldock, R., and Davidson, D. (2002) *Science* **296**, 541–545
222. Jamin, N., Dumas, P., Moncuit, J., Fridman, W.-H., Teillaud, J.-L., Carr, G. L., and Williams, G. P. (1998) *Proc. Natl. Acad. Sci. U.S.A.* **95**, 4837–4840
223. Wetzel, D. L., and LeVine, S. M. (1999) *Science* **285**, 1224–1225
224. Foyer, C. H. (1984) *Photosynthesis*, Wiley, New York
225. Clayton, R. K. (1981) *Photosynthesis Physical Mechanisms and Chemical Patterns*, Cambridge Univ. Press, London
226. Gregory, R. P. F. (1989) *Biochemistry of Photosynthesis*, 3rd ed., Wiley, New York
227. Hall, D. O., and Rao, K. K. (1994) *Photosynthesis*, 5th ed., Cambridge Univ. Press, New York
228. Barber, J., and Andersson, B. (1994) *Nature (London)* **370**, 31–34
229. Kamminga, H. (1981) *Trends Biochem. Sci.* **6**, 164–165
230. van Niel, C. B. (1931) *Adv. Enzymol.* **1**, 263–328
231. Hill, R. (1937) *Nature (London)* **139**, 881–882
232. Gaffron, H. (1960) in *Plant Physiology*, Vol. 1B (Steward, F. C., ed), pp. 176–180, Academic Press, New York
233. Prince, R. C. (1996) *Trends Biochem. Sci.* **21**, 121–122
234. Arnon, D. I., Tsujimoto, H. Y., and McSwai, B. D. (1965) *Nature (London)* **207**, 1357–1372
235. Arnon, D. I. (1984) *Trends Biochem. Sci.* **9**, 258–262
236. Zito, F., Finazzi, G., Joliot, P., and Wollman, F.-A. (1998) *Biochemistry* **37**, 10395–10403
236a. Finazzi, G., Zito, F., Barbagallo, R. P., and Wollman, F.-A. (2001) *J. Biol. Chem.* **276**, 9770–9774
237. Haley, J., and Bogorad, L. (1989) *Proc. Natl. Acad. Sci. U.S.A.* **86**, 1534–1538
237a. Roberts, A. G., and Kramer, D. M. (2001) *Biochemistry* **40**, 13407–13412
237b. Deniau, C., and Rappaport, F. (2000) *Biochemistry* **39**, 3304–3310
237c. Sainz, G., Carrell, C. J., Ponamarev, M. V., Soriano, G. M., Cramer, W. A., and Smith, J. L. (2000) *Biochemistry* **39**, 9164–9173
237d. Fernández-Velasco, J. G., Jamshidi, A., Gong, X.-S., Zhou, J., and Ueng, R. Y. (2001) *J. Biol. Chem.* **276**, 30598–30607
237e. Bunney, T. D., van Walraven, H. S., and de Boer, A. H. (2001) *Proc. Natl. Acad. Sci. U.S.A.* **98**, 4249–4254
238. Peterman, E. J. G., Wenk, S.-O., Pullerits, T., Pålsson, L.-O., van Grondelle, R., Dekker, J. P., Rögner, M., and van Amerongen, H. (1998) *Biophys. J.* **75**, 389–398
239. Carrell, C. J., Schlarb, B. G., Bendall, D. S., Howe, C. J., Cramer, W. A., and Smith, J. L. (1999) *Biochemistry* **38**, 9590–9599
240. Xue, Y., Ökvist, M., Hansson, Ö., and Young, S. (1998) *Protein Sci.* **7**, 2099–2105
241. Schnackenberg, J., Than, M. E., Mann, K., Wiegand, G., Huber, R., and Reuter, W. (1999) *J. Mol. Biol.* **290**, 1019–1030
242. De la Cerda, B., Díaz-Quintana, A., Navarro, J. A., Hervás, M., and De la Rosa, M. A. (1999) *J. Biol. Chem.* **274**, 13292–13297
242a. Molina-Heredia, F. P., Hervás, M., Navarro, J. A., and De la Rosa, M. A. (2001) *J. Biol. Chem.* **276**, 601–605
242b. Baymann, F., Rappaport, F., Joliot, P., and Kallas, T. (2001) *Biochemistry* **40**, 10570–10577
243. Kuras, R., Guergova-Kuras, M., and Crofts, A. R. (1998) *Biochemistry* **37**, 16280–16288
244. Zhang, H., Huang, D., and Cramer, W. A. (1999) *J. Biol. Chem.* **274**, 1581–1587
245. Oh-oka, H., Iwaki, M., and Itoh, S. (1998) *Biochemistry* **37**, 12293–12300
246. Ohyama, K., Fukuzawa, H., Kohchi, T., Shirai, H., Sano, T., Sano, S., Umesono, K., Shiki, Y., Takeuchi, M., Chang, Z., Aota, S.-i, Inokuchi, H., and Ozeki, H. (1986) *Nature (London)* **322**, 572–574
247. Shinozaki, K., Ohme, M., Tanaka, M., Wakasugi, T., Hayashida, N., Matsubayashi, T., Zaita, N., Chunwongse, J., Obokata, J., Yamaguchi-Shinozaki, K., Ohto, C., Torazawa, K., Meng, B. Y., Sugita, M., Deno, H., Kamogashira, T., Yamada, K., Kusuda, J., Takaiwa, F., Kato, A., Tohdoh, N., Shimada, H., and Sugiura, M. (1986) *EMBO J.* **5**, 2043–2049
248. Maier, R. M., Neckermann, K., Igloi, G. L., and Kössel, H. (1995) *J. Mol. Biol.* **251**, 614–628
249. Wakasugi, T., Nagai, T., Kapoor, M., Sugita, M., Ito, M., Ito, S., Tsudzuki, J., Nakashima, K., Tsudzuki, T., Suzuki, Y., Hamada, A., Ohta, T., Inamura, A., Yoshinaga, K., and Sugiura, M. (1997) *Proc. Natl. Acad. Sci. U.S.A.* **94**, 5967–5972
249a. Cline, K. (2000) *Nature (London)* **403**, 148–149
250. Zhang, Z., Green, B. R., and Cavalier-Smith, T. (1999) *Nature (London)* **400**, 155–159
251. Kouranov, A., and Schnell, D. J. (1996) *J. Biol. Chem.* **271**, 31009–31012
252. Asai, T., Shinoda, Y., Nohara, T., Yoshihisa, T., and Endo, T. (1999) *J. Biol. Chem.* **274**, 20075–20078
253. Bölter, B., Soll, J., Schulz, A., Hinnah, S., and Wagner, R. (1998) *Proc. Natl. Acad. Sci. U.S.A.* **95**, 15831–15836

References

253a. Dabney-Smith, C., van den Wijngaard, P. W. J., Treece, Y., Vredenberg, W. J., and Bruce, B. D. (1999) *J. Biol. Chem.* **274**, 32351–32359
254. Douce, R., and Joyard, J. (1981) *Trends Biochem. Sci.* **6**, 237–240
255. Hoober, J. K. (1984) *Chloroplasts*, Plenum, New York
255a. von Wettstein, D. (2001) *Proc. Natl. Acad. Sci. U.S.A.* **98**, 3633–3635
256. Muhlethaler, K. (1971) in *Structure and Function of Chloroplasts* (Gibbs, M., ed), pp. 7–34, Springer-Verlag, Berlin and New York
257. Miller, K. R. (2000) *Sci. Am.*, 102–113
258. Seibert, M., DeWit, M., and Staehelin, L. A. (1987) *J. Cell Biol.* **105**, 2257–2265
259. Anderson, J. M., and Anderson, B. (1982) *Trends Biochem. Sci.* **7**, 288–292
260. Rebeiz, C. A., and Lascelles, J. (1982) in *Photosynthesis*, Vol. I (Govindjee, ed), pp. 699–780, Academic Press, New York
261. Miyashita, H., Ikemoto, H., Kurano, N., Adachi, K., Chihara, M., and Miyachi, S. (1996) *Nature (London)* **383**, 402
262. Caple, M. B., Chow, H., and Strouse, C. E. (1978) *J. Biol. Chem.* **253**, 6730–6737
263. Smith, K. M., Kehres, L. A., and Tabba, H. D. (1980) *J. Am. Chem. Soc.* **102**, 7149–7151
264. French, C. S., and Brown, J. S. (1972) in *Photosynthesis. Two Centuries after Its Discovery by Joseph Priestly*, 2nd ed. (Forti, G., Avron, M., and Melandri, A., eds), pp. 291–306, Junk, The Hague
265. Govindjee, R. (1974) *Sci. Am.* **231**(Dec), 68–82
266. Wolken, J. J. (1975) *Photoprocesses, Photoreceptors, and Evolution*, Academic Press, New York
267. Song, P.-S. (1978) *Trends Biochem. Sci.* **3**, 25–27
268. Hofmann, E., Wrench, P. M., Sharples, F. P., Hiller, R. G., Welte, W., and Diederichs, K. (1996) *Science* **272**, 1788–1791
269. Goodwin, T. W., ed. (1971) *Aspects of Terpenoid Chemistry and Biochemistry*, Academic Press, New York (pp. 346–348)
270. Schirmer, W., Bode, R., Sidler, W., and Zuber, H. (1985) *J. Mol. Biol.* **184**, 257–277
271. Szalontai, B., Gombos, Z., Csizmadia, V., Bagyinka, C., and Lutz, M. (1994) *Biochemistry* **33**, 11823–11832
272. Glazer, A. N. (1989) *J. Biol. Chem.* **264**, 1–4
273. Jung, L. J., Chan, F. C., and Glazer, A. N. (1995) *J. Biol. Chem.* **270**, 12877–12884
274. Killilea, S. D., O'Carra, P., and Murphy, R. F. (1980) *Biochem. J.* **187**, 311–320
275. Bishop, J. E., Nagy, J. O., O'Connell, J. F., and Rapoport, H. (1991) *Chemical Society Special Publication, No. 2* **113**, 8024–8035
276. Fairchild, C. D., and Glazer, A. N. (1994) *J. Biol. Chem.* **269**, 28988–28996
277. MacColl, R., Guard-Friar, D., and Ryan, T. J. (1990) *Biochemistry* **29**, 430–435
278. Wilk, K. E., Harrop, S. J., Jankova, L., Edler, D., Keenan, G., Sharples, F., Hiller, R. G., and Curmi, P. M. G. (1999) *Proc. Natl. Acad. Sci. U.S.A.* **96**, 8901–8906
279. Chang, W.-r, Jiang, T., Wan, Z.-l, Zhang, J.-p, Yang, Z.-x, and Liang, D.-c. (1996) *J. Mol. Biol.* **262**, 721–731
280. Liu, J.-Y., Jiang, T., Zhang, J.-P., and Liang, D.-C. (1999) *J. Biol. Chem.* **274**, 16945–16952
281. Reuter, W., Wiegand, G., Huber, R., and Than, M. E. (1999) *Proc. Natl. Acad. Sci. U.S.A.* **96**, 1363–1368
282. Rhie, G., and Beale, S. I. (1992) *J. Biol. Chem.* **267**, 16088–16093
283. Fairchild, C. D., and Glazer, A. N. (1994) *J. Biol. Chem.* **269**, 8686–8694
284. Terry, M. J., McDowell, M. T., and Lagarias, J. C. (1995) *J. Biol. Chem.* **270**, 11111–11118
285. Riley, C. T., Barbeau, B. K., Keim, P. S., Kézdy, F. J., Heinrikson, R. L., and Law, J. H. (1984) *J. Biol. Chem.* **259**, 13159–13165
286. Glazer, A. N. (1983) *Ann. Rev. Biochem.* **52**, 125–157
286a. Grossman, A. R., Bhaya, D., and He, Q. (2001) *J. Biol. Chem.* **276**, 11449–11452
287. Thornber, P. J., and Maxwell, J. P. (1981) *Trends Biochem. Sci.* **6**, 122–124
288. Deisenhofer, J., Michel, H., and Huber, R. (1985) *Trends Biochem. Sci.* **10**, 243–248
289. Brejc, K., Ficner, R., Huber, R., and Steinbacher, S. (1995) *J. Mol. Biol.* **249**, 424–440
290. Lao, K., and Glazer, A. N. (1996) *Proc. Natl. Acad. Sci. U.S.A.* **93**, 5258–5263
291. Hu, X., and Schulten, K. (1998) *Biophys. J.* **75**, 683–694
292. Karrasch, S., Bullough, P. A., and Ghosh, R. (1995) *EMBO J.* **14**, 631–638
293. Walz, T., Jamieson, S. J., Bowers, C. M., Bullough, P. A., and Hunter, C. N. (1998) *J. Mol. Biol.* **282**, 833–845
294. McDermott, G., Prince, S. M., Freer, A. A., Hawthornthwaite-Lawless, A. M., Papiz, M. Z., Cogdell, R. J., and Isaacs, N. W. (1995) *Nature (London)* **374**, 517–521
295. Prince, S. M., Papiz, M. Z., Freer, A. A., McDermott, G., Hawthornthwaite-Lawless, A. M., Cogdell, R. J., and Isaacs, N. W. (1997) *J. Mol. Biol.* **268**, 412–423
296. Hu, X., Damjanovic, A., Ritz, T., and Schulten, K. (1998) *Proc. Natl. Acad. Sci. U.S.A.* **95**, 5935–5941
297. Nagarajan, V., Alden, R. G., Williams, J. C., and Parson, W. W. (1996) *Proc. Natl. Acad. Sci. U.S.A.* **93**, 13774–13779
298. Fraser, N. J., Dominy, P. J., Ücker, B., Simonin, I., Scheer, H., and Cogdell, R. J. (1999) *Biochemistry* **38**, 9684–9692
298a. McLuskey, K., Prince, S. M., Cogdell, R. J., and Isaacs, N. W. (2001) *Biochemistry* **40**, 8783–8789
299. van Oijen, A. M., Ketelaars, M., Köhler, J., Aartsma, T. J., and Schmidt, J. (1999) *Science* **285**, 400–402
300. Kühlbrandt, W. (1995) *Nature (London)* **374**, 479–498
300a. van Grondelle, R., and Novoderezhkin, V. (2001) *Biochemistry* **40**, 15057–15068
300b. Alia, Matysik, J., Soede-Huijbregts, C., Baldus, M., Raap, J., Lugtenburg, J., Gast, P., van Gorkom, H. J., Hoff, A. J., and de Groot, H. J. M. (2001) *J. Am. Chem. Soc.* **123**, 4803–4809
300c. Simonetto, R., Crimi, M., Sandoná, D., Croce, R., Cinque, G., Breton, J., and Bassi, R. (1999) *Biochemistry* **38**, 12974–12983
301. Feick, R. G., and Fuller, R. C. (1984) *Biochemistry* **23**, 3693–3700
302. Li, Y.-F., Zhou, W., Blankenship, R. E., and Allen, J. P. (1997) *J. Mol. Biol.* **271**, 456–471
302a. Vassilieva, E. V., Antonkine, M. L., Zybailov, B. L., Yang, F., Jakobs, C. U., Golbeck, J. H., and Bryant, D. A. (2001) *Biochemistry* **40**, 464–473
303. Matthews, B. W., Fenna, R. E., Bolognesi, M. C., Schmid, M. F., and Olson, J. M. (1979) *J. Mol. Biol.* **131**, 259–285
304. Rémigy, H.-W., Stahlberg, H., Fotiadis, D., Müller, S. A., Wolpensinger, B., Engel, A., Hauska, G., and Tsiotis, G. (1999) *J. Mol. Biol.* **290**, 851–858
305. Fenna, R. E., and Matthews, B. W. (1975) *Nature (London)* **258**, 573–577
306. Douady, D., Rousseau, B., and Caron, L. (1994) *Biochemistry* **33**, 3165–3170
307. Pascal, A. A., Caron, L., Rousseau, B., Lapouge, K., Duval, J.-C., and Robert, B. (1998) *Biochemistry* **37**, 2450–2457
308. Green, B. R., Pichersky, E., and Kloppstech, K. (1991) *Trends Biochem. Sci.* **16**, 181–186
309. La Roche, J., van der Staay, G. W. M., Partensky, F., Ducret, A., Aebersold, R., Li, R., Golden, S. S., Hiller, R. G., Wrench, P. M., Larkum, A. W. D., and Green, B. R. (1996) *Proc. Natl. Acad. Sci. U.S.A.* **93**, 15244–15248
310. Kühlbrandt, W., Wang, D. N., and Fujiyoshi, Y. (1994) *Nature (London)* **367**, 614–621
311. Kleima, F. J., Gradinaru, C. C., Calkoen, F., van Stokkum, I. H. M., van Grendelle, R., and van Amerongen, H. (1997) *Biochemistry* **36**, 15262–15268
312. Simidjiev, I., Barzda, V., Mustárdy, L., and Garab, G. (1998) *Biochemistry* **37**, 4169–4173
313. Rhee, K.-H., Morris, E. P., Zheleva, D., Hankamer, B., Kühlbrandt, W., and Barber, J. (1997) *Nature (London)* **389**, 522–526
314. Rhee, K.-H., Morris, E. P., Barber, J., and Kühlbrandt, W. (1998) *Nature (London)* **396**, 283–286
315. Barber, J., Nield, J., Morris, E. P., and Hankamer, B. (1999) *Trends Biochem. Sci.* **24**, 43–45
315a. Voigt, B., Irrgang, K.-D., Ehlert, J., Beenken, W., Renger, G., Leupold, D., and Lokstein, H. (2002) *Biochemistry* **41**, 3049–3056
316. Schubert, W.-D., Klukas, O., Saenger, W., Witt, H. T., Fromme, P., and Krauss, N. (1998) *J. Mol. Biol.* **280**, 297–314
317. Croce, R., Zucchelli, G., Garlaschi, F. M., Bassi, R., and Jennings, R. C. (1996) *Biochemistry* **35**, 8572–8579
317a. Bibby, T. S., Nield, J., Partensky, F., and Barber, J. (2001) *Nature (London)* **413**, 590
317b. Rogl, H., Schödel, R., Lokstein, H., Kühlbrandt, W., and Schubert, A. (2002) *Biochemistry* **41**, 2281–2287
318. Thomas, B. A., McMahon, L. P., and Klotz, A. V. (1995) *Biochemistry* **34**, 3758–3770
318a. Heathcote, P., Fyfe, P. K., and Jones, M. R. (2002) *Trends Biochem. Sci.* **27**, 79–87
319. Deisenhofer, J., Epp, O., Miki, K., Huber, R., and Michel, H. (1985) *Nature (London)* **318**, 618–624
320. Deisenhofer, J., and Michel, H. (1991) *Ann. Rev. Biophys. Biophys. Chem.* **20**, 247–266
321. Deisenhofer, J., Epp, O., Sinning, I., and Michel, H. (1995) *J. Mol. Biol.* **246**, 429–457
322. Deisenhofer, J., and Michel, H. (1989) *Science* **245**, 1463–1473
323. Lancaster, C. R. D., and Michel, H. (1999) *J. Mol. Biol.* **286**, 883–898
323a. Chen, I.-P., Mathis, P., Koepke, J., and Michel, H. (2000) *Biochemistry* **39**, 3592–3602
324. Philipson, K. D., and Sauer, K. (1972) *Biochemistry* **11**, 1880–1885
325. Deisenhofer, J., Epp, O., Miki, K., Huber, R., and Michel, H. (1984) *J. Mol. Biol.* **180**, 385–398
325a. McAuley, K. E., Fyfe, P. K., Ridge, J. P., Cogdell, R. J., Isaacs, N. W., and Jones, M. R. (2000) *Biochemistry* **39**, 15032–15043
326. Siefermann-Harms, D. (1985) *Biochim. Biophys. Acta.* **811**, 325–355
327. El-Kabbani, O., Chang, C.-H., Tiede, D., Norris, J., and Schiffer, M. (1991) *Biochemistry* **30**, 5361–5369
328. Govindjee. (1978) *Photochemistry and Photobiology* **28**, 935–938
329. Blankenship, R. E., and Parson, W. W. (1978) *Ann. Rev. Biochem.* **47**, 635–653
330. Dutton, P. L., Leigh, J. S., Jr., and Reed, D. W. (1972) *Biochim. Biophys. Acta.* **292**, 654–664
331. Dutton, P. L., and Mosser, C. C. (1994) *Proc. Natl. Acad. Sci. U.S.A.* **91**, 10247–10250
331a. Li, J., Takahashi, E., and Gunner, M. R. (2000) *Biochemistry* **39**, 7445–7454
331b. Rabenstein, B., Ullmann, G. M., and Knapp, E.-W. (2000) *Biochemistry* **39**, 10487–10496
331c. Paddock, M. L., Ädelroth, P., Chang, C., Abresch, E. C., Feher, G., and Okamura, M. Y. (2001) *Biochemistry* **40**, 6893–6902

References

331d. Tandori, J., Baciou, L., Alexov, E., Maróti, P., Schiffer, M., Hanson, D. K., and Sebban, P. (2001) *J. Biol. Chem.* **276**, 45513–45515

331e. Gerencsér, L., and Maróti, P. (2001) *Biochemistry* **40**, 1850–1860

331f. Ädelroth, P., Paddock, M. L., Tehrani, A., Beatty, J. T., Feher, G., and Okamura, M. Y. (2001) *Biochemistry* **40**, 14538–14546

331g. Xu, Q., and Gunner, M. R. (2002) *Biochemistry* **41**, 2694–2701

332. Vos, M. H., Rappaport, F., Lambry, J.-C., Breton, J., and Martin, J.-L. (1993) *Nature (London)* **363**, 320–325

333. Hamm, P., Zurek, M., Mäntele, W., Meyer, M., Scheer, H., and Zinth, W. (1995) *Proc. Natl. Acad. Sci. U.S.A.* **92**, 1826–1830

334. Van Brederode, M. E., Jones, M. R., Van Mourik, F., Van Stokkum, I. H. M., and Van Grondelle, R. (1997) *Biochemistry* **36**, 6855–6861

335. Venturoli, G., Drepper, F., Williams, J. C., Allen, J. P., Lin, X., and Mathis, P. (1998) *Biophys. J.* **74**, 3226–3240

335a. van Brederode, M. E., van Stokkum, I. H. M., Katilius, E., van Mourik, F., Jones, M. R., and van Grondelle, R. (1999) *Biochemistry* **38**, 7545–7555

336. Cherepy, N. J., Shreve, A. P., Moore, L. J., Boxer, S. G., and Mathies, R. A. (1997) *Biochemistry* **36**, 8559–8566

337. Cua, A., Kirmaier, C., Holten, D., and Bocian, D. F. (1998) *Biochemistry* **37**, 6394–6401

338. Ivancich, A., Artz, K., Williams, J. C., Allen, J. P., and Mattioli, T. A. (1998) *Biochemistry* **37**, 11812–11820

338a. Takahashi, E., Wells, T. A., and Wraight, C. A. (2001) *Biochemistry* **40**, 1020–1028

338b. Kuglstatter, A., Ermler, U., Michel, H., Baciou, L., and Fritzsch, G. (2001) *Biochemistry* **40**, 4253–4260

338c. Eastman, J. E., Taguchi, A. K. W., Lin, S., Jackson, J. A., and Woodbury, N. W. (2000) *Biochemistry* **39**, 14787–14798

339. Tang, C.-K., Williams, J. C., Taguchi, A. K. W., Allen, J. P., and Woodbury, N. W. (1999) *Biochemistry* **38**, 8794–8799

340. Parson, W. W., Chu, Z. T., and Warshel, A. (1998) *Biophys. J.* **74**, 182–191

340a. Balabin, I. A., and Onuchic, J. N. (2000) *Science* **290**, 114–117

340b. Yakovlev, A. G., Shkuropatov, A. Y., and Shuvalov, V. A. (2002) *Biochemistry* **41**, 2667–2674

341. Dohse, B., Mathis, P., Wachtveitl, J., Laussermair, E., Iwata, S., Michel, H., and Oesterhelt, D. (1995) *Biochemistry* **34**, 11335–11343

342. Lin, X., Williams, J. C., Allen, J. P., and Mathis, P. (1994) *Biochemistry* **33**, 13517–13523

343. Stowell, M. H. B., McPhillips, T. M., Rees, D. C., Soltis, S. M., Abresch, E., and Feher, G. (1997) *Science* **276**, 812–816

343a. de Boer, A. L., Neerken, S., de Wijn, R., Permentier, H. P., Gast, P., Vijgenboom, E., and Hoff, A. J. (2002) *Biochemistry* **41**, 3081–3088

343b. Schulten, E. A. M., Matysik, J., Alia, Kiihne, S., Raap, J., Lugtenburg, J., Gast, P., Hoff, A. J., and de Groot, H. J. M. (2002) *Biochemistry* **41**, 8708–8717

344. Heller, B. A., Holten, D., and Kirmaier, C. (1995) *Science* **269**, 940–945

344a. Lin, S., Katilius, E., Haffa, A. L. M., Taguchi, A. K. W., and Woodbury, N. W. (2001) *Biochemistry* **40**, 13767–13773

344b. Purton, S., Stevens, D. R., Muhiuddin, I. P., Evans, M. C. W., Carter, S., Rigby, S. E. J., and Heathcote, P. (2001) *Biochemistry* **40**, 2167–2175

344c. Guergova-Kuras, M., Boudreaux, B., Joliot, A., Joliot, P., and Redding, K. (2001) *Proc. Natl. Acad. Sci. U.S.A.* **98**, 4437–4442

345. Nugent, J. H. A. (1984) *Trends Biochem. Sci.* **9**, 354–357

345a. Johnson, T. W., Zybailov, B., Jones, A. D., Bittl, R., Zech, S., Stehlik, D., Golbeck, J. H., and Chitnis, P. R. (2001) *J. Biol. Chem.* **276**, 39512–39521

346. Sinning, I. (1992) *Trends Biochem. Sci.* **17**, 150–154

346a. Witt, H., Schlodder, E., Teutloff, C., Niklas, J., Bordignon, E., Carbonera, D., Kohler, S., Labahn, A., and Lubitz, W. (2002) *Biochemistry* **41**, 8557–8569

347. Fish, L. E., Kuck, U., and Bogorad, L. (1985) *J. Biol. Chem.* **260**, 1413–1421

348. Breton, J., Nabedryk, E., and Leibl, W. (1999) *Biochemistry* **38**, 11585–11592

349. Francke, C., Permentier, H. P., Franken, E. M., Neerken, S., and Amesz, J. (1997) *Biochemistry* **36**, 14167–14172

350. Nitschke, W., and Rutherford, A. W. (1991) *Trends Biochem. Sci.* **16**, 241–245

351. Golbeck, J. H. (1993) *Proc. Natl. Acad. Sci. U.S.A.* **90**, 1642–1646

352. Karrasch, S., Typke, D., Walz, T., Miller, M., Tsiotis, G., and Engel, A. (1996) *J. Mol. Biol.* **262**, 336–348

353. Deisenhofer, J., and Norris, J. R., eds. (1993) *The Photosynthetic Reaction Center*, Vol. 1 and 2, Academic Press, San Diego, California

354. Blankenship, R. E., and Hartman, H. (1998) *Trends Biochem. Sci.* **23**, 94–97

355. Kruip, J., Chitnis, P. R., Lagoutte, B., Rögner, M., and Boekema, E. J. (1997) *J. Biol. Chem.* **272**, 17061–17069

356. Schubert, W.-D., Klukas, O., Krauss, N., Saenger, W., Fromme, P., and Witt, H. T. (1997) *J. Mol. Biol.* **272**, 741–769

356a. Jordan, P., Fromme, P., Witt, H. T., Klukas, O., Saenger, W., and Krauß, N. (2001) *Nature (London)* **411**, 909–916

356b. Kühlbrandt, W. (2001) *Nature (London)* **411**, 896–899

357. Kitmitto, A., Mustafa, A. O., Holzenburg, A., and Ford, R. C. (1998) *J. Biol. Chem.* **273**, 29592–29599

357a. Boekema, E. J., Jensen, P. E., Schlodder, E., van Breemen, J. F. L., van Room, H., Scheller, H. V., and Dekker, J. P. (2001) *Biochemistry* **40**, 1029–1036

358. Morton, R. A. (1971) *Biol. Rev. Cambridge Philos. Soc.* **46**, 47–96

359. Threlfall, D. R., and Whistance, G. R. (1971) in *Aspects of Terpenoid Chemistry and Biochemistry* (Goodwin, T. W., ed), pp. 372–374, Academic Press, New York

360. Blankenship, R. E., and Prince, R. C. (1985) *Trends Biochem. Sci.* **10**, 382–384

361. Knox, R. S. (1969) *Biophys. J.* **9**, 1351–1362

362. Parson, W. W. (1978) *Photochemistry and Photobiology* **28**, 389–393

363. Yu, L., Zhao, J., Lu, W., Bryant, D. A., and Golbeck, J. H. (1993) *Biochemistry* **32**, 8251–8258

364. Pandini, V., Aliverti, A., and Zanetti, G. (1999) *Biochemistry* **38**, 10707–10713

365. Barth, P., Lagoutte, B., and Sétif, P. (1998) *Biochemistry* **37**, 16233–16241

366. Meimberg, K., Lagoutte, B., Bottin, H., and Mühlenhoff, U. (1998) *Biochemistry* **37**, 9759–9767

367. Bruns, C. M., and Karplus, P. A. (1995) *J. Mol. Biol.* **247**, 125–145

368. Medina, M., Martínez-Júlvez, M., Hurley, J. K., Tollin, G., and Gómez-Moreno, C. (1998) *Biochemistry* **37**, 2715–2728

369. Martínez-Júlvez, M., Hermoso, J., Hurley, J. K., Mayoral, T., Sanz-Aparicio, J., Tollin, G., Gómez-Moreno, C., and Medina, M. (1998) *Biochemistry* **37**, 17680–17691

370. Xiong, J., Subramaniam, S., and Govindjee. (1996) *Protein Sci.* **5**, 2054–2073

371. Svensson, B., Etchebest, C., Tuffery, P., van Kan, P., Smith, J., and Styring, S. (1996) *Biochemistry* **35**, 14486–14502

371a. Büchel, C., Morris, E., Orlova, E., and Barber, J. (2001) *J. Mol. Biol.* **312**, 371–379

371b. Kuhl, H., Kruip, J., Seidler, A., Krieger-Liszkay, A., Bünker, M., Bald, D., Scheidig, A. J., and Rögner, M. (2000) *J. Biol. Chem.* **275**, 20652–20659

371c. Shen, J.-R., and Kamiya, N. (2000) *Biochemistry* **39**, 14739–14744

371d. Zouni, A., Witt, H.-T., Kern, J., Fromme, P., Krauß, N., Saenger, W., and Orth, P. (2001) *Nature (London)* **409**, 739–743

371e. Dismukes, G. C. (2001) *Science* **292**, 447–448

371f. Nield, J., Kruse, O., Ruprecht, J., da Fonseca, P., Büchel, C., and Barber, J. (2000) *J. Biol. Chem.* **275**, 27940–27946

372. Schweitzer, R. H., and Brudvig, G. W. (1997) *Biochemistry* **36**, 11351–11359

373. Merry, S. A. P., Nixon, P. J., Barter, L. M. C., Schilstra, M., Porter, G., Barber, J., Durrant, J. R., and Klug, D. R. (1998) *Biochemistry* **37**, 17439–17447

374. Betts, S. D., Ross, J. R., Pichersky, E., and Yocum, C. F. (1997) *Biochemistry* **36**, 4047–4053

375. Enami, I., Kamo, M., Ohta, H., Takahashi, S., Miura, T., Kusayanagi, M., Tanabe, S., Kamei, A., Motoki, A., Hirano, M., Tomo, T., and Satoh, K. (1998) *J. Biol. Chem.* **273**, 4629–4634

376. Babcock, G. T. (1993) *Proc. Natl. Acad. Sci. U.S.A.* **90**, 10893–10895

376a. Shi, L.-X., Lorkovic, Z. J., Oelmüller, R., and Schröder, W. P. (2000) *J. Biol. Chem.* **275**, 37945–37950

376b. Boekema, E. J., van Breemen, J. F. L., van Roon, H., and Dekker, J. P. (2000) *J. Mol. Biol.* **301**, 1123–1133

376c. Mamedov, F., Stefansson, H., Albertsson, P.-Å., and Styring, S. (2000) *Biochemistry* **39**, 10478–10486

377. Kok, B., Forbush, B., and McGloin, M. (1970) *Photochem. and Photobiol.* **11**, 457–475

378. Radmer, R., and Kok, B. (1975) *Ann. Rev. Biochem.* **44**, 409–433

379. Murata, N., and Miyao, M. (1985) *Trends Biochem. Sci.* **10**, 122–124

380. Lydakis-Simantiris, N., Dorlet, P., Ghanotakis, D. F., and Babcock, G. T. (1998) *Biochemistry* **37**, 6427–6435

381. Schiller, H., Dittmer, J., Iuzzolino, L., Dorner, W., Meyer-Klaucke, W., Solé, V. A., Nolting, H.-F., and Dau, H. (1998) *Biochemistry* **37**, 7340–7350

382. Chu, H.-A., Gardner, M. T., O'Brien, J. P., and Babcock, G. T. (1999) *Biochemistry* **38**, 4533–4541

383. Noguchi, T., Inoue, Y., and Tang, X.-S. (1999) *Biochemistry* **38**, 10187–10195

384. Barry, B. A., and Babcock, G. T. (1987) *Proc. Natl. Acad. Sci. U.S.A.* **84**, 7099–7103

385. Diner, B. A., Force, D. A., Randall, D. W., and Britt, R. D. (1998) *Biochemistry* **37**, 17931–17943

386. Hays, A.-M. A., Vassiliev, I. R., Golbeck, J. H., and Debus, R. J. (1998) *Biochemistry* **37**, 11352–11365

387. Ahlbrink, R., Haumann, M., Cherepanov, D., Bögershausen, O., Mulkidjanian, A., and Junge, W. (1998) *Biochemistry* **37**, 1131–1142

388. Mamedov, F., Sayre, R. T., and Styring, S. (1998) *Biochemistry* **37**, 14245–14256

References

388a. Bernát, G., Morvaridi, F., Feyziyev, Y., and Styring, S. (2002) *Biochemistry* **41**, 5830–5843
389. Siegbahn, P. E. M., and Crabtree, R. H. (1999) *J. Am. Chem. Soc.* **121**, 117–127
389a. Schlodder, E., and Witt, H. T. (1999) *J. Biol. Chem.* **274**, 30387–30392
389b. Geijer, P., Morvaridi, F., and Styring, S. (2001) *Biochemistry* **40**, 10881–10891
390. Baldwin, M. J., and Pecoraro, V. L. (1996) *J. Am. Chem. Soc.* **118**, 11325–11326
391. Limburg, J., Vrettos, J. S., Liable-Sands, L. M., Rheingold, A. L., Crabtree, R. H., and Brudvig, G. W. (1999) *Science* **283**, 1524–1527
392. Hoganson, C. W., and Babcock, G. T. (1997) *Science* **277**, 1953–1956
392a. Hillier, W., and Babcock, G. T. (2001) *Biochemistry* **40**, 1503–1509
393. Wincencjusz, H., Yocum, C. F., and van Gorkom, H. J. (1999) *Biochemistry* **38**, 3719–3725
394. Haddy, A., Hatchell, J. A., Kimel, R. A., and Thomas, R. (1999) *Biochemistry* **38**, 6104–6110
395. Kühne, H., Szalai, V. A., and Brudvig, G. W. (1999) *Biochemistry* **38**, 6604–6613
396. Seidler, A., and Rutherford, A. W. (1996) *Biochemistry* **35**, 12104–12110
397. Ädelroth, P., Lindberg, K., and Andréasson, L.-E. (1995) *Biochemistry* **34**, 9021–9027
397a. Vrettos, J. S., Stone, D. A., and Brudvig, G. W. (2001) *Biochemistry* **40**, 7937–7945
398. Klimov, V. V., Hulsebosch, R. J., Allakhverdiev, S. I., Wincencjusz, H., van Gorkom, H. J., and Hoff, A. J. (1997) *Biochemistry* **36**, 16277–16281
398a. Baranov, S. V., Ananyev, G. M., Klimov, V. V., and Dismukes, G. C. (2000) *Biochemistry* **39**, 6060–6065
399. Keister, D. L., and Raveed, N. J. (1974) *J. Biol. Chem.* **249**, 6454–6458
400. Schubert, H., Kroon, B. M. A., and Matthijs, H. C. P. (1994) *J. Biol. Chem.* **269**, 7267–7272
401. Demmig-Adams, B., Gilmore, A. M., and Adams, W. W., III. (1996) *FASEB J.* **10**, 403–412
401a. Demming-Adams, B., and Adams, W. W., III. (2000) *Nature (London)* **403**, 371–374
401b. Li, X.-P., Björkman, O., Shih, C., Grossman, A. R., Rosenquist, M., Jansson, S., and Niyogi, K. K. (2000) *Nature (London)* **403**, 391–395
401c. Kagawa, T., Sakai, T., Suetsugu, N., Oikawa, K., Ishiguro, S., Kato, T., Tabata, S., Okada, K., and Wada, M. (2001) *Science* **291**, 2138–2141
402. Havaux, M., and Niyogi, K. K. (1999) *Proc. Natl. Acad. Sci. U.S.A.* **96**, 8762–8767
403. Limantara, L., Fujii, R., Zhang, J.-P., Kakuno, T., Hara, H., Kawamori, A., Yagura, T., Cogdell, R. J., and Koyama, Y. (1998) *Biochemistry* **37**, 17469–17486
404. Telfer, A., Dhami, S., Bishop, S. M., Phillips, D., and Barber, J. (1994) *Biochemistry* **33**, 14469–14474
405. Bugos, R. C., and Yamamoto, H. Y. (1996) *Proc. Natl. Acad. Sci. U.S.A.* **93**, 6320–6325
406. Niyogi, K. K., Björkman, O., and Grossman, A. R. (1997) *Proc. Natl. Acad. Sci. U.S.A.* **94**, 14162–14167
407. Ruban, A. V., Lee, P. J., Wentworth, M., Young, A. J., and Horton, P. (1999) *J. Biol. Chem.* **274**, 10458–10465
407a. Frank, H. A., Bautista, J. A., Josue, J. S., and Young, A. J. (2000) *Biochemistry* **39**, 2831–2837
407b. Jahns, P., Wehner, A., Paulsen, H., and Hobe, S. (2001) *J. Biol. Chem.* **276**, 22154–22159
408. Gilmore, A. M., Shinkarev, V. P., Hazlett, T. L., and Govindjee. (1998) *Biochemistry* **37**, 13582–13593
409. Hanley, J., Deligiannakis, Y., Pascal, A., Faller, P., and Rutherford, A. W. (1999) *Biochemistry* **38**, 8189–8195
410. Buser, C. A., Diner, B. A., and Brudvig, G. W. (1992) *Biochemistry* **31**, 11449–11459
411. Barber, J., and Andersson, B. (1992) *Trends Biochem. Sci.* **17**, 61–66
412. Sharma, J., Panico, M., Shipton, C. A., Nilsson, F., Morris, H. R., and Barber, J. (1997) *J. Biol. Chem.* **272**, 33158–33166
413. Hagman, Å., Shi, L.-X., Rintamäki, E., Andersson, B., and Schröder, W. P. (1997) *Biochemistry* **36**, 12666–12671
414. Krieger, A., Rutherford, A. W., Vass, I., and Hideg, É. (1998) *Biochemistry* **37**, 16262–16269
415. Campbell, D., Eriksson, M.-J., Öquist, G., Gustafsson, P., and Clarke, A. K. (1998) *Proc. Natl. Acad. Sci. U.S.A.* **95**, 364–369
415a. Rutherford, A. W., and Krieger-Liszkay, A. (2001) *Trends Biochem. Sci.* **26**, 648–653
416. Kobza, J., and Seemann, J. R. (1988) *Proc. Natl. Acad. Sci. U.S.A.* **85**, 3815–3819
417. Puente, P., Wei, N., and Deng, X. W. (1996) *EMBO J.* **15**, 3732–3743
418. Fluhr, R., Kuhlemeier, C., Nagy, F., and Chua, N.-H. (1986) *Science* **232**, 1106–1112
419. Moses, P. B., and Chua, N.-H. (1988) *Sci. Am.* **258**(Apr), 88–93
420. Klein, R. R., and Mullet, J. E. (1990) *J. Biol. Chem.* **265**, 1895–1902
421. Morelli, G., Nagy, F., Fraley, R. T., Roger, S. G., and Chua, N.-H. (1985) *Nature (London)* **315**, 200–204
422. Giuliano, G., Pichersky, E., Malik, V. S., Timko, M. P., Scolnik, P. A., and Cashmore, A. R. (1988) *Proc. Natl. Acad. Sci. U.S.A.* **85**, 7089–7093
423. Inamine, G., Nash, B., Weissbach, H., and Brot, N. (1985) *Proc. Natl. Acad. Sci. U.S.A.* **82**, 5690–5694
424. Buchanan, B. B., and Schurmann, P. (1973) *Curr. Top. Cell. Regul.* **7**, 1–20
425. Wolosiuk, R. A., Ballicora, M. A., and Hagelin, K. (1993) *FASEB J.* **7**, 622–637
426. Kelly, G. (1982) *Trends Biochem. Sci.* **7**, 81–82
427. Johansson, K., Ramaswamy, S., Saarinen, M., Lemaire-Chamley, M., Issakidis-Bourguet, E., Miginiac-Maslow, M., and Eklund, H. (1999) *Biochemistry* **38**, 4319–4326
428. Hirasawa, M., Schürmann, P., Jacquot, J.-P., Manieri, W., Jacquot, P., Keryer, E., Hartman, F. C., and Knaff, D. B. (1999) *Biochemistry* **38**, 5200–5205
429. Staples, C. R., Gaymard, E., Stritt-Etter, A.-L., Telser, J., Hoffman, B. M., Schürmann, P., Knaff, D. B., and Johnson, M. K. (1998) *Biochemistry* **37**, 4612–4620
429a. Dai, S., Schwendtmayer, C., Schürmann, P., Ramaswamy, S., and Eklund, H. (2000) *Science* **287**, 655–658
429b. Rintamäki, E., Martinsuo, P., Pursiheimo, S., and Aro, E.-M. (2000) *Proc. Natl. Acad. Sci. U.S.A.* **97**, 11644–11649
430. Cseke, C., Balogh, A., Wong, J. H., Buchanan, B. B., Stitt, M., Herzog, B., and Held, H. W. (1984) *Trends Biochem. Sci.* **9**, 533–535
431. Bligny, R., Gardestrom, P., Roby, C., and Douce, R. (1990) *J. Biol. Chem.* **265**, 1319–1326
432. Zelitch, I. (1975) *Science* **188**, 626–633
433. Sommerville, C. R., and Ogren, W. L. (1982) *Trends Biochem. Sci.* **7**, 171–174
434. Heber, V., and Krause, G. H. (1980) *Trends Biochem. Sci.* **5**, 32–34
435. Goyal, A., and Tolbert, N. E. (1996) *Proc. Natl. Acad. Sci. U.S.A.* **93**, 3319–3324
436. Tolbert, N. E. (1973) *Curr. Top. Cell. Regul.* **7**, 21–50
437. Ho, C.-L., Noji, M., and Saito, K. (1999) *J. Biol. Chem.* **274**, 11007–11012
438. Givan, C. V., Joy, K. W., and Kleczkowski, L. A. (1988) *Trends Biochem. Sci.* **13**, 433–437
439. Kozaki, A., and Takeba, G. (1996) *Nature (London)* **384**, 557–560
440. Haag, E., and Renger, G. (1997) in *Bioenergetics* (Gräber, P., and Milazzo, G., eds), pp. 212–272, Birkhäuser Verlag, Basel
441. Burnell, J. N., and Hatch, M. D. (1985) *Trends Biochem. Sci.* **10**, 289–291
442. Moore, P. D. (1978) *Nature (London)* **272**, 400–401
442a. Voznesenskaya, E. V., Franceschi, V. R., Kiirats, O., Freitag, H., and Edwards, G. E. (2001) *Nature (London)* **414**, 543–546
442b. Maeda, S.-i, Price, G. D., Badger, M. R., Enomoto, C., and Omata, T. (2000) *J. Biol. Chem.* **275**, 20551–20555
443. Winter, K., and Smith, J. A. C., eds. (1996) *Crassulacean Acid Metabolism Biochemistry, Ecophysiology and Evolution*, Springer, Berlin
444. Benemann, J. R., Berenson, J. A., Kaplan, N. O., and Kamen, M. D. (1973) *Proc. Natl. Acad. Sci. U.S.A.* **70**, 2317–2320
445. Benemann, J. R., and Weare, N. M. (1974) *Science* **184**, 174–175
446. Abrahamson, E. W., and Fager, R. S. (1973) *Curr. Top. Bioenerg.* **5**, 125–200
447. Wald, G., and Brown, P. K. (1965) *Cold Spring Harb. Symp. on Quant. Biol.* **30**, 346
448. Nathans, J. (1992) *Biochemistry* **31**, 4923–4931
449. Ferretti, L., Karnik, S. S., Khorana, H. G., Nassal, M., and Oprian, D. D. (1986) *Proc. Natl. Acad. Sci. U.S.A.* **83**, 599–603
450. Khorana, H. G. (1992) *J. Biol. Chem.* **267**, 1–4
451. Dratz, E. A., and Hargrave, P. A. (1983) *Trends Biochem. Sci.* **8**, 128–131
452. Oprian, D. D., Asenjo.AB, Lee, N., and Pelletier, S. L. (1991) *Biochemistry* **30**, 11367–11372
453. Nassal, M., Mogi, T., Karnik, S. S., and Khorana, H. G. (1987) *J. Biol. Chem.* **262**, 9264–9270
454. Henderson, R. (1975) *J. Mol. Biol.* **93**, 123–138
455. Luecke, H., Schobert, B., Richter, H.-T., Cartailler, J.-P., and Lanyi, J. K. (1999) *J. Mol. Biol.* **291**, 899–911
456. Baldwin, J. (1993) *EMBO J.* **12**, 1693–1703
457. Baldwin, J. M., Schertler, G. F. X., and Unger, V. M. (1997) *J. Mol. Biol.* **272**, 144–64
458. Pogozheva, I. D., Lomize, A. L., and Mosberg, H. I. (1997) *Biophys. J.* **72**, 1963–1985
459. Schertler, G. F. X., and Hargrave, P. A. (1995) *Proc. Natl. Acad. Sci. U.S.A.* **92**, 11578–11582
460. Herzyk, P., and Hubbard, R. E. (1998) *J. Mol. Biol.* **281**, 741–754
461. Barnidge, D. R., Dratz, E. A., Sunner, J., and Jesaitis, A. J. (1997) *Protein Sci.* **6**, 816–824
461a. Palczewski, K., Kumasaka, T., Hori, T., Behnke, C. A., Motoshima, H., Fox, B. A., Trong, I. L., Teller, D. C., Okada, T., Stenkamp, R. E., Yamamoto, M., and Miyano, M. (2000) *Science* **289**, 739–745
461b. Okada, T., Ernst, O. P., Palczewski, K., and Hofmann, K. P. (2001) *Trends Biochem. Sci.* **26**, 318–331
461c. Teller, D. C., Okada, T., Behnke, C. A., Palczewski, K., and Stenkamp, R. E. (2001) *Biochemistry* **40**, 7761–7772
462. Khorana, H. G. (1993) *Proc. Natl. Acad. Sci. U.S.A.* **90**, 1166–1171
462a. Altenbach, C., Klein-Seetharaman, J., Cai, K., Khorana, H. G., and Hubbell, W. L. (2001) *Biochemistry* **40**, 15493–15500
463. Yeagle, P. L., Alderfer, J. L., and Albert, A. D. (1995) *Biochemistry* **34**, 14621–14625
464. Altenbach, C., Klein-Seetharaman, J., Hwa, J., Khorana, H. G., and Hubbell, W. L. (1999) *Biochemistry* **38**, 7945–7949
465. Struthers, M., Yu, H., Kono, M., and Oprian, D. D. (1999) *Biochemistry* **38**, 6597–6603
466. Zhang, H., Lerro, K. A., Yamamoto, T., Lien, T. H., Sastry, L., Gawinowicz, M. A., and Nakanishi, K. (1994) *J. Am. Chem. Soc.* **116**, 10165–10173

References

467. Han, M., and Smith, S. O. (1995) *Biochemistry* **34**, 1425–1432
468. Grigorieff, N., Ceska, T. A., Downing, K. H., Baldwin, J. M., and Henderson, R. (1996) *J. Mol. Biol.* **259**, 393–421
469. Wald, G. (1968) *Nature (London)* **219**, 800–807
470. Dowling, J. E. (1997) *Nature (London)* **387**, 356
471. Gilardi, R., Sperling, W., Karle, I. L., and Karle, J. (1971) *Nature (London)* **232**, 187–188
472. Shriver, J. W., Mateescu, G. D., and Abrahamson, E. W. (1979) *Biochemistry* **18**, 4785–4792
472a. Singh, D., Hudson, B. S., Middleton, C., and Birge, R. R. (2001) *Biochemistry* **40**, 4201–4204
473. Honig, B., Dinur, U., Nakanishi, K., Balogh-Nair, V., Gawinswicz, M. A., Arnaboldi, M., and Motto, M. G. (1979) *J. Am. Chem. Soc.* **101**, 7084–7086
474. Baasov, T., and Sheves, M. (1985) *J. Am. Chem. Soc.* **107**, 7524–7533
475. Sakmar, T. P., Franke, R. R., and Khorana, H. G. (1989) *Proc. Natl. Acad. Sci. U.S.A.* **86**, 8309–8313
475a. Lewis, J. W., Szundi, I., Fu, W.-Y., Sakmar, T. P., and Kliger, D. S. (2000) *Biochemistry* **39**, 599–606
476. Creemers, A. F. L., Klaassen, C. H. W., Bovee-Geurts, P. H. M., Kelle, R., Kragl, U., Raap, J., de Grip, W. J., Lugtenburg, J., and de Groot, H. J. M. (1999) *Biochemistry* **38**, 7195–7199
476a. Verhoeven, M. A., Creemers, A. F. L., Bovee-Geurts, P. H. M., De Grip, W. J., Lugtenburg, J., and de Groot, H. J. M. (2001) *Biochemistry* **40**, 3282–3288
477. Kochendoerfer, G. G., Lin, S. W., Sakmar, T. P., and Mathies, R. A. (1999) *Trends Biochem. Sci.* **24**, 300–305
478. Nathans, J., Thomas, D., and Hogness, D. S. (1986) *Science* **232**, 193–202
479. Kochendoerfer, G. G., Wang, Z., Oprian, D. D., and Mathies, R. A. (1997) *Biochemistry* **36**, 6577–6587
479a. Dukkipati, A., Vought, B. W., Singh, D., Birge, R. R., and Knox, B. E. (2001) *Biochemistry* **40**, 15098–15108
479b. Roorda, A., and Williams, D. R. (1999) *Nature (London)* **397**, 520–522
480. Nathans, J., Piantanida, T. D., Eddy, R. L., Shows, T. B., and Hogness, D. S. (1986) *Science* **232**, 203–210
481. Motulsky, A. G., and Deeb, S. S. (1995) in *The Metabolic and Molecular Bases of Inherited Disease*, 7th ed., Vol. 3 (Scriver, C. R., Beaudet, A. L., Sly, W. S., and Valle, D., eds), pp. 4275–4295, McGraw-Hill, New York
482. Nathans, J. (1989) *Sci. Am.* **260**(Feb), 42 – 49
483. Merbs, S. L., and Nathans, J. (1992) *Science* **258**, 464–466
484. Hunt, D. M., Dulai, K. S., Bowmaker, J. K., and Mollon, J. D. (1995) *Science* **267**, 984–988
485. Nathans, J., Davenport, C. M., Maumenee, I. H., Lewis, R. A., Hejtmancik, J. F., Litt, M., Lovrien, E., Weleber, R., Bachynski, B., Zwas, F., Klingaman, R., and Fishman, G. (1989) *Science* **245**, 831–838
486. Shichi, H., Lewis, M. S., Irreverre, F. and Stone, A. L. (1969) *J. Biol. Chem.* **244**, 529–536
487. Shichi, H., and Somers, R. L. (1974) *J. Biol. Chem.* **249**, 6570–6577
488. Fasick, J. I., Lee, N., and Oprian, D. D. (1999) *Biochemistry* **38**, 11593–11596
488a. Shi, Y., Radlwimmer, F. B., and Yokoyama, S. (2001) *Proc. Natl. Acad. Sci. U.S.A.* **98**, 11731–11736
488b. Janz, J. M., and Farrens, D. L. (2001) *Biochemistry* **40**, 7219–7227
488c. Dukkipati, A., Kusnetzow, A., Babu, K. R., Ramos, L., Singh, D., Knox, B. E., and Birge, R. R. (2002) *Biochemistry* **41**, 9842–9851
489. Fasick, J. I., and Robinson, P. R. (1998) *Biochemistry* **37**, 433–438
490. Imamoto, Y., Hirano, T., Imai, H., Kandori, H., Maeda, A., Yoshizawa, T., Groesbeek, M., Lugtenburg, J., and Shichida, Y. (1999) *Biochemistry* **38**, 11749–11754
490a. Imai, H., Hirano, T., Kandori, H., Terakita, A., and Shichida, Y. (2001) *Biochemistry* **40**, 2879–2886
491. Wang, Z., Asenjo, A. B., and Oprian, D. D. (1993) *Biochemistry* **32**, 2125–2130
492. Johnson, R. L., Grant, K. B., Zankel, T. C., Boehm, M. F., Merbs, S. L., Nathans, J., and Nakanishi, K. (1993) *Biochemistry* **32**, 208–214
493. Foster, K. W., Saranak, J., Derguini, F., Zarrilli, G. R., Johnson, R., Okabe, M., and Nakanishi, K. (1989) *Biochemistry* **28**, 819–824
494. Deininger, W., Kröger, P., Hegemann, U., Lottspeich, F., and Hegemann, P. (1995) *EMBO J.* **14**, 5849–5858
495. Max, M., Surya, A., Takahashi, J. S., Margolskee, R. F., and Knox, B. E. (1998) *J. Biol. Chem.* **273**, 26820–26826
496. Max, M., McKinnon, P. J., Seidenman, K. J., Barrett, R. K., Applebury, M. L., Takahashi, J. S., and Margolskee, R. F. (1995) *Science* **267**, 1502–1506
496a. Nakamura, A., Kojima, D., Imai, H., Terakita, A., Okano, T., Shichida, Y., and Fukada, Y. (1999) *Biochemistry* **38**, 14738–14745
497. Soni, B. G., Philp, A. R., Foster, R. G., and Knox, B. E. (1998) *Nature (London)* **394**, 27–28
498. Zuker, C. S. (1996) *Proc. Natl. Acad. Sci. U.S.A.* **93**, 571–576
499. Peters, K., Applebury, M. L., and Rentzepis, P. M. (1977) *Proc. Natl. Acad. Sci. U.S.A.* **74**, 3119–3123
500. Strassburger, J. M., Gärtner, W., and Braslavsky, S. E. (1997) *Biophys. J.* **72**, 2294–2303
501. Zhu, Y., and Liu, R. S. H. (1993) *Biochemistry* **32**, 10233–10238
502. Liu, R. S. H., and Shichida, Y. (1991) in *Photochemistry in Organized and Constrained Media* (Ramamurthy, V., ed), VCH Publishers, New York (Chapter 18)
502a. Ishiguro, M. (2000) *J. Am. Chem. Soc.* **122**, 444–451
503. Hug, S. J., Lewis, W. J., Einterz, C. M., Thorgeirsson, T. E., and Kliger, D. S. (1990) *Biochemistry* **29**, 1475–1485
504. Lewis, J. W., Pinkas, I., Sheves, M., Ottolenghi, M., and Kliger, D. S. (1995) *J. Am. Chem. Soc.* **117**, 918–923
505. Shichida, Y., Nakamura, K., Yoshizawa, T., Trehan, A., Denny, M., and Liu, R. S. H. (1988) *Biochemistry* **27**, 6495–6499
506. Wang, Q., Schoenlein, R. W., Peteanu, L. A., Mathies, R. A., and Shank, C. V. (1994) *Science* **266**, 422–424
507. Jäger, F., Fahmy, K., Sakmar, T. P., and Siebert, F. (1994) *Biochemistry* **33**, 10878–10882
508. DeLange, F., Bovee-Geurts, P. H. M., VanOostrum, J., Portier, M. D., Verdegem, P. J. E., Lugtenburg, J., and DeGrip, W. J. (1998) *Biochemistry* **37**, 1411–1420
509. Fukada, Y., Shichida, Y., Yoshizawa, T., Ito, M., Kodama, A., and Tsukida, K. (1984) *Biochemistry* **23**, 5826–5832
510. Bagley, K. A., Balogh-Nair, V., Croteau, A. A., Dollinger, G., Ebrey, T. G., Eisenstein, L., Hong, M. K., Nakanishi, K., and Vittitow, J. (1985) *Biochemistry* **24**, 6055–6071
510a. Borhan, B., Souto, M. L., Imai, H., Shichida, Y., and Nakanishi, K. (2000) *Science* **288**, 2209–2212
511. Ridge, K. D., Bhattacharya, S., Nakayama, T. A., and Khorana, H. G. (1992) *J. Biol. Chem.* **267**, 6770–6775
511a. Liu, R. S. H., and Hammond, G. S. (2000) *Proc. Natl. Acad. Sci. U.S.A.* **97**, 11153–11158
511b. González-Luque, R., Garavelli, M., Bernardi, F., Merchán, M., and Robb, M. A. (2000) *Proc. Natl. Acad. Sci. U.S.A.* **97**, 9379–9384
511c. Kim, J. E., Tauber, M. J., and Mathies, R. A. (2001) *Biochemistry* **40**, 13774–13778
512. Chosrowjan, H., Mataga, N., Shibata, Y., Tachibanaki, S., Kandori, H., Shichida, Y., Okada, T., and Kouyama, T. (1998) *J. Am. Chem. Soc.* **120**, 9706–9707
513. Loppnow, G. R., Miley, M. E., Mathies, R. A., Liu, R. S. H., Kandori, H., Shichida, Y., Fukada, Y., and Yoshizawa, T. (1990) *Biochemistry* **29**, 8985–8991
514. Freedman, K. A., and Becker, R. S. (1986) *J. Am. Chem. Soc.* **108**, 1245–1251
515. Yoshizawa, T., and Wald, G. (1963) *Nature (London)* **197**, 1279–1286
516. Nagata, T., Terakita, A., Kandori, H., Kojima, D., Shichida, Y., and Maeda, A. (1997) *Biochemistry* **36**, 6164–6170
517. Farrens, D. L., Altenbach, C., Yang, K., Hubbell, W. L., and Khorana, H. G. (1996) *Science* **274**, 768–770
518. Dunham, T. D., and Farrens, D. L. (1999) *J. Biol. Chem.* **274**, 1683–1690
518a. Bartl, F. J., Ritter, E., and Hofmann, K. P. (2001) *J. Biol. Chem.* **276**, 30161–30166
518b. Vogel, R., Fan, G.-B., Siebert, F., and Sheves, M. (2001) *Biochemistry* **40**, 13342–13352
519. Dickopf, S., Mielke, T., and Heyn, M. P. (1998) *Biochemistry* **37**, 16888–16897
520. Jäger, S., Szundi, I., Lewis, J. W., Mah, T. L., and Kliger, D. S. (1998) *Biochemistry* **37**, 6998–7005
521. Parkes, J. H., and Liebman, P. A. (1984) *Biochemistry* **23**, 5054–5061
521a. Yeagle, P. L., Choi, G., and Albert, A. D. (2001) *Biochemistry* **40**, 11932–11937
521b. Vought, B. W., Salcedo, E., Chadwell, L. V., Britt, S. G., Birge, R. R., and Knox, B. E. (2000) *Biochemistry* **39**, 14128–14137
521c. Pan, D., and Mathies, R. A. (2001) *Biochemistry* **40**, 7929–7936
522. Yarfitz, S., and Hurley, J. B. (1994) *J. Biol. Chem.* **269**, 14329–14332
523. Puckett, K. L., Aronson, E. T., and Goldin, S. M. (1985) *Biochemistry* **24**, 390–400
524. Dizhoor, A. M., Olshevskaya, E. V., Henzel, W. J., Wong, S. C., Stults, J. T., Ankoudinova, I., and Hurley, J. B. (1995) *J. Biol. Chem.* **270**, 25200–25206
525. Yau, K.-W., and Nakatani, K. (1985) *Nature (London)* **313**, 579–583
526. Stryer, L., Hurley, J. B., and Fung, B. K. K. (1981) *Trends Biochem. Sci.* **6**, 245–247
527. Stryer, L. (1987) *Sci. Am.* **257**(Jul), 42–50
528. Stryer, L. (1991) *J. Biol. Chem.* **266**, 10711–10714
529. Lambright, D. G., Sondek, J., Bohm, A., Skiba, N. P., Hamm, H. E., and Sigler, P. B. (1996) *Nature (London)* **379**, 311–319
529a. Granovsky, A. E., and Artemyev, N. O. (2001) *Biochemistry* **40**, 13209–13215
529b. Norton, A. W., D'Amours, M. R., Grazio, H. J., Hebert, T. L., and Cote, R. H. (2000) *J. Biol. Chem.* **275**, 38611–38619
530. Shimoda, Y., Hurley, J. B., and Miller, W. H. (1984) *Proc. Natl. Acad. Sci. U.S.A.* **81**, 616–619
531. Gibson, S. K., Parkes, J. H., and Liebman, P. A. (1998) *Biochemistry* **37**, 11393–11398
532. Palczewski, K., Buczylko, J., Lebioda, L., Crabb, J. W., and Polans, A. S. (1993) *J. Biol. Chem.* **268**, 6004–6013
533. Gurevich, V. V., Chen, C.-Y., Kim, C. M., and Benovic, J. L. (1994) *J. Biol. Chem.* **269**, 8721–8727

References

534. Smith, W. C., McDowell, J. H., Dugger, D. R., Miller, R., Arendt, A., Popp, M. P., and Hargrave, P. A. (1999) *Biochemistry* **38**, 2752–2761
535. Granzin, J., Wilden, U., Choe, H.-W., Labahn, J., Krafft, B., and Büldt, G. (1998) *Nature (London)* **391**, 918–921
535a. Oakley, R. H., Laporte, S. A., Holt, J. A., Caron, M. G., and Barak, L. S. (2000) *J. Biol. Chem.* **275**, 17201–17210
536. Baylor, D. (1996) *Proc. Natl. Acad. Sci. U.S.A.* **93**, 560–565
536a. Penn, R. B., Pascual, R. M., Kim, Y.-M., Mundell, S. J., Krymskaya, V. P., Panettieri, R. A., Jr., and Benovic, J. L. (2001) *J. Biol. Chem.* **276**, 32648–32656
537. Laitko, U., and Hofmann, K. P. (1998) *Biophys. J.* **74**, 803–815
538. Klenchin, V. A., Calvert, P. D., and Bownds, M. D. (1995) *J. Biol. Chem.* **270**, 16147–16152
539. Ames, J. B., Ishima, R., Tanaka, T., Gordon, J. I., Stryer, L., and Ikura, M. (1997) *Nature (London)* **389**, 198–202
540. Ames, J. B., Tanaka, T., Stryer, L., and Ikura, M. (1996) *Curr. Opin. Struct. Biol.* **6**, 432–438
541. Lange, C., and Koch, K.-W. (1997) *Biochemistry* **36**, 12019–12026
542. Palczewski, K., Jäger, S., Buczylko, J., Crouch, R. K., Bredberg, D. L., Hofmann, K. P., Asson-Batres, M. A., and Saari, J. C. (1994) *Biochemistry* **33**, 13741–13750
543. Palczewski, K., Van Hooser, J. P., Garwin, G. G., Chen, J., Liou, G. I., and Saari, J. C. (1999) *Biochemistry* **38**, 12012–12019
543a. McBee, J. K., Kuksa, V., Alvarez, R., de Lera, A. R., Prezhdo, O., Haeseleer, F., Sokal, I., and Palczewski, K. (2000) *Biochemistry* **39**, 11370–11380
544. Hara, R., Hara, T., Tokunaga, F., and Yoshizawa, T. (1981) *Photochem. Photobiol.* **33**, 883–891
544a. Terakita, A., Yamashita, T., and Shichida, Y. (2000) *Proc. Natl. Acad. Sci. U.S.A.* **97**, 14263–14267
545. Humphries, P., Kenna, P., and Farrar, G. J. (1992) *Science* **256**, 804–808
546. Dryja, T. P. (1995) in *The Metabolic and Molecular Bases of Inherited Disease*, 7th ed., Vol. 3 (Scriver, C. R., Beaudet, A. L., Sly, W. S., and Valle, D., eds), pp. 4297–4309, McGraw-Hill, New York
546a. Hwa, J., Klein-Seetharaman, J., and Khorana, H. G. (2001) *Proc. Natl. Acad. Sci. U.S.A.* **98**, 4872–4876
547. Maniloff, J. (1996) *Proc. Natl. Acad. Sci. U.S.A.* **93**, 10004–10006
548. Liu, X., Garriga, P., and Khorana, H. G. (1996) *Proc. Natl. Acad. Sci. U.S.A.* **93**, 4554–4559
549. Rao, V. R., Cohen, G. B., and Oprian, D. D. (1994) *Nature (London)* **367**, 639–642
550. Kajiwara, K., Berson, E. L., and Dryja, T. P. (1994) *Science* **264**, 1604–1608
551. Khani, S. C., Nielsen, L., and Vogt, T. M. (1998) *Proc. Natl. Acad. Sci. U.S.A.* **95**, 2824–2827
552. Tucker, C. L., Woodcock, S. C., Kelsell, R. E., Ramamurthy, V., Hunt, D. M., and Hurley, J. B. (1999) *Proc. Natl. Acad. Sci. U.S.A.* **96**, 9039–9044
553. Eudy, J. D., Weston, M. D., Yao, S., Hoover, D. M., Rehm, H. L., Ma-Edmonds, M., Yan, D., Ahmad, I., Cheng, J. J., Ayuso, C., Cremers, C., Davenport, S., Moller, C., Talmadge, C. B., Beisel, K. W., Tamayo, M., Morton, C. C., Swaroop, A., Kimberling, W. J., and Sumegi, J. (1998) *Science* **280**, 1753–1757
554. Seabra, M. C., Brown, M. S., and Goldstein, J. L. (1993) *Science* **259**, 377–381
555. Allikmets, R., Shroyer, N. F., Singh, N., Seddon, J. M., Lewis, R. A., Bernstein, P. S., Peiffer, A., Zabriskie, N. A., Li, Y., Hutchinson, A., Dean, M., Lupski, J. R., and Leppert, M. (1997) *Science* **277**, 1805–1807
556. Pennisi, E. (1998) *Science* **281**, 31
556a. Biswas, E. E. (2001) *Biochemistry* **40**, 8181–8187
557. Parish, C. A., Hashimoto, M., Nakanishi, K., Dillon, J., and Sparrow, J. (1998) *Proc. Natl. Acad. Sci. U.S.A.* **95**, 14609–14613
558. Norledge, B. V., Trinkl, S., Jaenicke, R., and Slingsby, C. (1997) *Protein Sci.* **6**, 1612–1620
559. Koretz, J. F., and Handelman, G. H. (1988) *Sci. Am.* **259**(Jul), 92–99
560. Tardieu, A., and Delaye, M. (1988) *Ann. Rev. Biophys. Biophys. Chem.* **17**, 47–70
561. Wistow, G. (1993) *Trends Biochem. Sci.* **18**, 301–306
562. Hejtmancik, J. F., Kaiser, M. I., and Piatigorsky, J. (1995) in *The Metabolic and Molecular Bases of Inherited Disease*, 7th ed., Vol. 3 (Scriver, C. R., Beaudet, A. L., Sly, W. S., and Valle, D., eds), pp. 4325–4349, McGraw-Hill, New York
563. König, N., Zampighi, G. A., and Butler, P. J. G. (1997) *J. Mol. Biol.* **265**, 590–602
564. Surewicz, W. K., and Olesen, P. R. (1995) *Biochemistry* **34**, 9655–9660
565. Clark, J. I., and Huang, Q.-L. (1996) *Proc. Natl. Acad. Sci. U.S.A.* **93**, 15185–15189
566. Piatigorsky, J., and Wistow, G. (1991) *Science* **252**, 1078–1079
567. Röll, B., Amons, R., and de Jong, W. W. (1996) *J. Biol. Chem.* **271**, 10437–10440
568. Zigler, J. S., Jr., and Rao, P. V. (1991) *FASEB J.* **5**, 223–225
569. Abu-Abed, M., Turner, M. A., Vallée, F., Simpson, A., Slingsby, C., and Howell, P. L. (1997) *Biochemistry* **36**, 14012–14022
570. Kraft, H. J., Hendriks, W., de Jong, W. W., Lubsen, N. H., and Schoenmakers, J. G. G. (1993) *J. Mol. Biol.* **229**, 849–859
571. Mulders, J. W. M., Hendriks, W., Blankesteijn, W. M., Bloemendal, H., and de Jong, W. W. (1988) *J. Biol. Chem.* **263**, 15462–15466
572. Zinovieva, R. D., Tomarev, S. I., and Piatigorsky, J. (1993) *J. Biol. Chem.* **268**, 11449–11455
572a. Werten, P. J. L., Röll, B., van Aalten, D. M. F., and de Jong, W. W. (2000) *Proc. Natl. Acad. Sci. U.S.A.* **97**, 3282–3287
573. Garner, B., Vazquez, S., Griffith, R., Lindner, R. A., Carver, J. A., and Truscott, R. J. W. (1999) *J. Biol. Chem.* **274**, 20847–20854
574. Tessier, F., Obrenovich, M., and Monnier, V. M. (1999) *J. Biol. Chem.* **274**, 20796–20804
574a. Hood, B. D., Garner, B., and Truscott, R. J. W. (1999) *J. Biol. Chem.* **274**, 32547–32550
575. Takemoto, L., and Boyle, D. (1998) *Biochemistry* **37**, 13681–13685
576. Stoeckenius, W. (1999) *Protein Sci.* **8**, 447–459
577. Racker, E., and Stoeckenius, W. (1974) *J. Biol. Chem.* **249**, 662–663
578. Stoeckenius, W. (1985) *Trends Biochem. Sci.* **10**, 483–486
579. Mitsuoka, K., Hirai, T., Murata, K., Miyazawa, A., Kidera, A., Kimura, Y., and Fujiyoshi, Y. (1999) *J. Mol. Biol.* **286**, 861–882
580. Kimura, Y., Vassylyev, D. G., Miyazawa, A., Kidera, A., Matsushima, M., Mitsuoka, K., Murata, K., Hirai, T., and Fujiyoshi, Y. (1997) *Nature (London)* **389**, 206–211
581. Subramaniam, S., Lindahl, M., Bullough, P., Faruqi, A. R., Tittor, J., Oesterhelt, D., Brown, L., Lanyi, J., and Henderson, R. (1999) *J. Mol. Biol.* **287**, 145–161
582. Pebay-Peyroula, E., Rummel, G., Rosenbusch, J. P., and Landau, E. M. (1997) *Science* **277**, 1676–1681
582a. Subramaniam, S., and Henderson, R. (2000) *Nature (London)* **406**, 653–657
582b. Edman, K., Nollert, P., Royant, A., Belrhali, H., Pebay-Peyroula, E., Hajdu, J., Neutze, R., and Landau, E. M. (1999) *Nature (London)* **401**, 822–826
582c. Royant, A., Edman, K., Ursby, T., Pebay-Peyroula, E., Landau, E. M., and Neutze, R. (2000) *Nature (London)* **406**, 645–648
582d. Luecke, H., Schobert, B., Cartailler, J.-P., Richter, H.-T., Rosengarth, A., Needleman, R., and Lanyi, J. K. (2000) *J. Mol. Biol.* **300**, 1237–1255
582e. Xiao, W., Brown, L. S., Needleman, R., Lanyi, J. K., and Shin, Y.-K. (2000) *J. Mol. Biol.* **304**, 715–721
582f. Kühlbrandt, W. (2000) *Nature (London)* **406**, 569–570
583. Müller, D. J., Sass, H.-J., Müller, S. A., Büldt, G., and Engel, A. (1999) *J. Mol. Biol.* **285**, 1903–1909
584. Smith, S. O., Myers, A. B., Pardoen, J. A., Winkel, C., Mulder, P. P. J., Lugtenburg, J., and Mathies, R. (1984) *Proc. Natl. Acad. Sci. U.S.A.* **81**, 2055–2059
584a. Patzelt, H., Simon, B., terLaak, A., Kessler, B., Kühne, R., Schmieder, P., Oesterhelt, D., and Oschkinat, H. (2002) *Proc. Natl. Acad. Sci. U.S.A.* **99**, 9765–9770
585. Bullough, P. A., and Henderson, R. (1999) *J. Mol. Biol.* **286**, 1663–1671
585a. Herbst, J., Heyne, K., and Diller, R. (2002) *Science* **297**, 822–825
585b. Kandori, H., Belenky, M., and Herzfeld, J. (2002) *Biochemistry* **41**, 6026–6031
585c. Maeda, A., Balashov, S. P., Lugtenburg, J., Verhoeven, M. A., Herzfeld, J., Belenky, M., Gennis, R. B., Tomson, F. L., and Ebrey, T. G. (2002) *Biochemistry* **41**, 3803–3809
586. Haupts, U., Tittor, J., Bamberg, E., and Oesterhelt, D. (1997) *Biochemistry* **36**, 2–7
587. Gat, Y., and Sheves, M. (1993) *J. Am. Chem. Soc.* **115**, 3772–3773
588. Moltke, S., Wallat, I., Sakai, N., Nakanishi, K., Brown, M. F., and Heyn, M. P. (1999) *Biochemistry* **38**, 11762–11772
588a. Sass, H. J., Büldt, G., Gessenich, R., Hehn, D., Neff, D., Schlesinger, R., Berendzen, J., and Ormos, P. (2000) *Nature (London)* **406**, 649–653
588b. Kandori, H., Yamazaki, Y., Shichida, Y., Raap, J., Lugtenburg, J., Belenky, M., and Herzfeld, J. (2001) *Proc. Natl. Acad. Sci. U.S.A.* **98**, 1571–1576
588c. Dioumaev, A. K., Brown, L. S., Needleman, R., and Lanyi, J. K. (2001) *Biochemistry* **40**, 11308–11317
588d. Oka, T., Yagi, N., Fujisawa, T., Kamikubo, H., Tokunaga, F., and Kataoka, M. (2000) *Proc. Natl. Acad. Sci. U.S.A.* **97**, 14278–14282
588e. Vonck, J. (2000) *EMBO J.* **19**, 2152–2160
588f. Rouhani, S., Cartailler, J.-P., Facciotti, M. T., Walian, P., Needleman, R., Lanyi, J. K., Glaeser, R. M., and Luecke, H. (2001) *J. Mol. Biol.* **313**, 615–628
588g. Allen, S. J., Kim, J.-M., Khorana, H. G., Lu, H., and Booth, P. J. (2001) *J. Mol. Biol.* **308**, 423–435
588h. Aharoni, A., Weiner, L., Lewis, A., Ottolenghi, M., and Sheves, M. (2001) *J. Am. Chem. Soc.* **123**, 6612–6616
588i. Spassov, V. Z., Luecke, H., Gerwert, K., and Bashford, D. (2001) *J. Mol. Biol.* **312**, 203–219
588j. Imasheva, E. S., Lu, M., Balashov, S. P., Ebrey, T. G., Chen, Y., Ablonczy, Z., Menick, D. R., and Crouch, R. K. (2001) *Biochemistry* **40**, 13320–13330
589. Havelka, W. A., Henderson, R., and Oesterhelt, D. (1995) *J. Mol. Biol.* **247**, 726–738

References

590. Chon, Y.-S., Kandori, H., Sasaki, J., Lanyi, J. K., Needleman, R., and Maeda, A. (1999) *Biochemistry* **38**, 9449–9455

590a. Sato, M., Kanamori, T., Kamo, N., Demura, M., and Nitta, K. (2002) *Biochemistry* **41**, 2452–2458

591. Seiff, F., Wallat, I., Ermann, P., and Heyn, M. P. (1985) *Proc. Natl. Acad. Sci. U.S.A.* **82**, 3227–3231

591a. Kandori, H., Furutani, Y., Shimono, K., Shichida, Y., and Kamo, N. (2001) *Biochemistry* **40**, 15693–15698

591b. Swartz, T. E., Szundi, I., Spudich, J. L., and Bogomolni, R. A. (2000) *Biochemistry* **39**, 15101–15109

591c. Wegener, A.-A., Klare, J. P., Engelhard, M., and Steinhoff, H.-J. (2001) *EMBO J.* **20**, 5312–5319

592. Jung, K.-H., and Spudich, J. L. (1996) *Proc. Natl. Acad. Sci. U.S.A.* **93**, 6557–6561

593. Zhang, X.-N., and Spudich, J. L. (1997) *Biophys. J.* **73**, 1516–1523

593a. Luecke, H., Schobert, B., Lanyi, J. K., Spudich, E. N., and Spudich, J. L. (2001) *Science* **293**, 1499–1503

593b. Royant, A., Nollert, P., Edman, K., Neutze, R., Landau, E. M., Pebay-Peyroula, E., and Navarro, J. (2001) *Proc. Natl. Acad. Sci. U.S.A.* **98**, 10131–10136

594. Hirayama, J., Imamoto, Y., Shichida, Y., Kamo, N., Tomioka, H., and Yoshizawa, T. (1992) *Biochemistry* **31**, 2093–2098

595. Spudich, E. N., Zhang, W., Alam, M., and Spudich, J. L. (1997) *Proc. Natl. Acad. Sci. U.S.A.* **94**, 4960–4965

596. Chizhov, I., Schmies, G., Seidel, R., Sydor, J. R., Lüttenberg, B., and Engelhard, M. (1998) *Biophys. J.* **75**, 999–1009

597. Rath, P., Olson, K. D., Spudich, J. L., and Rothschild, K. J. (1994) *Biochemistry* **33**, 5600–5606

597a. Bergo, V., Spudich, E. N., Scott, K. L., Spudich, J. L., and Rothschild, K. J. (2000) *Biochemistry* **39**, 2823–2830

597b. Ren, L., Martin, C. H., Wise, K. J., Gillespie, N. B., Luecke, H., Lanyi, J. K., Spudich, J. L., and Birge, R. R. (2001) *Biochemistry* **40**, 13906–13914

598. Bieszke, J. A., Braun, E. L., Bean, L. E., Kang, S., Natvig, D. O., and Borkovich, K. A. (1999) *Proc. Natl. Acad. Sci. U.S.A.* **96**, 8034–8039

598a. Sineshchekov, O. A., Jung, K.-H, and Spudich, J. L. (2002) *Proc. Natl. Acad. Sci. U.S.A.* **99**, 8689–8694

599. Hoff, W. D., Devreese, B., Fokkens, R., Nugteren-Roodzant, I. M., Van Beeumen, J., Nibbering, N., and Hellingwerf, K. J. (1996) *Biochemistry* **35**, 1274–1281

599a. Xie, A., Kelemen, L., Hendriks, J., White, B. J., Hellingwerf, K. J., and Hoff, W. D. (2001) *Biochemistry* **40**, 1510–1517

600. Kort, R., Hoff, W. D., Van West, M., Kroon, A. R., Hoffer, S. M., Vlieg, K. H., Crielaard, W., Van Beeumen, J. J., and Hellingwerf, K. J. (1996) *EMBO J.* **15**, 3209–3218

601. Borgstahl, G. E. O., Williams, D. R., and Getzoff, E. D. (1995) *Biochemistry* **34**, 6278–6287

601a. Brudler, R., Meyer, T. E., Genick, U. K., Devanathan, S., Woo, T. T., Millar, D. P., Gerwert, K., Cusanovich, M. A., Tollin, G., and Getzoff, E. D. (2000) *Biochemistry* **39**, 13478–13486

601b. Ren, Z., Perman, B., Srajer, V., Teng, T.-Y., Pradervand, C., Bourgeois, D., Schotte, F., Ursby, T., Kort, R., Wulff, M., and Moffat, K. (2001) *Biochemistry* **40**, 13788–13801

601c. Imamoto, Y., Kataoka, M., Tokunaga, F., Asahi, T., and Masuhara, H. (2001) *Biochemistry* **40**, 6047–6052

602. Imamoto, Y., Mihara, K., Hisatomi, O., Kataoka, M., Tokunaga, F., Bojkova, N., and Yoshihara, K. (1997) *J. Biol. Chem.* **272**, 12905–12908

603. Perman, B., Srajer, V., Ren, Z., Teng, T., Pradervand, C., Ursby, T., Bourgeois, D., Schotte, F., Wulff, M., Kort, R., Hellingwerf, K., and Moffat, K. (1998) *Science* **279**, 1946–1950

604. Genick, U. K., Devanathan, S., Meyer, T. E., Canestrelli, I. L., Williams, E., Cusanovich, M. A., Tollin, G., and Getzoff, E. D. (1997) *Biochemistry* **36**, 8–14

605. Ujj, L., Devanathan, S., Meyer, T. E., Cusanovich, M. A., Tollin, G., and Atkinson, G. H. (1998) *Biophys. J.* **75**, 406–412

606. Imamoto, Y., Kataoka, M., and Tokunaga, F. (1996) *Biochemistry* **35**, 14047–14053

607. Genick, U. K., Soltis, S. M., Kuhn, P., Canestrelli, I. L., and Getzoff, E. D. (1998) *Nature (London)* **392**, 206–209

608. Devanathan, S., Genick, U. K., Canestrelli, I. L., Meyer, T. E., Cusanovich, M. A., Getzoff, E. D., and Tollin, G. (1998) *Biochemistry* **37**, 11563–11568

609. Xie, A., Hoff, W. D., Kroon, A. R., and Hellingwerf, K. J. (1996) *Biochemistry* **35**, 14671–14678

609a. Unno, M., Kumauchi, M., Sasaki, J., Tokunaga, F., and Yamauchi, S. (2002) *Biochemistry* **41**, 5668–5674

610. Tao, N., Orlando, M., Hyon, J.-S., Gross, M., and Song, P.-S. (1993) *J. Am. Chem. Soc.* **115**, 2526–2528

611. Song, P.-S. (1983) *Annu Rev Biophys Bioeng.* **12**, 35–68

612. Smirnov, A., Fulton, D. B., Andreotti, A., and Petrich, J. W. (1999) *J. Am. Chem. Soc.* **121**, 7979–7988

613. Carpenter, S., Fehr, M. J., Kraus, G. A., and Petrich, J. W. (1994) *Proc. Natl. Acad. Sci. U.S.A.* **91**, 12273–12277

614. Quail, P. H. (1984) *Trends Biochem. Sci.* **9**, 450–453

615. Kelly, J. M., and Lagarias, J. C. (1985) *Biochemistry* **24**, 6003–6010

616. Quail, P. H. (1976) in *Plant Biochemistry*, 3rd ed. (Bonner, J., and Varner, J. E., eds), pp. 683–711, Academic Press, New York

617. Anderson, G. R., Jenner, E. L., and Mumford, F. E. (1970) *Biochim. Biophys. Acta.* **221**, 69–73

618. Quail, P. H., Boylan, M. T., Parks, B. M., Short, T. W., Xu, Y., and Wagner, D. (1995) *Science* **268**, 675–680

619. Ruddat, A., Schmidt, P., Gatz, C., Braslavsky, S. E., Gärtner, W., and Schaffner, K. (1997) *Biochemistry* **36**, 103–111

619a. Fankhauser, C. (2001) *J. Biol. Chem.* **276**, 11453–11456

619b. Nagy, F., and Schäfer, E. (2000) *EMBO J.* **19**, 157–163

619c. Smith, H. (2000) *Nature (London)* **407**, 585–591

620. Foerstendorf, H., Mummert, E., Schäfer, E., Scheer, H., and Siebert, F. (1996) *Biochemistry* **35**, 10793–10799

621. Andel, F., III, Lagarias, J. C., and Mathies, R. A. (1996) *Biochemistry* **35**, 15997–16008

622. Chen, E., Lapko, V. N., Lewis, J. W., Song, P.-S., and Kliger, D. S. (1996) *Biochemistry* **35**, 843–850

623. Kneip, C., Hildebrandt, P., Schlamann, W., Braslavsky, S. E., Mark, F., and Schaffner, K. (1999) *Biochemistry* **38**, 15185–15192

623a. Foerstendorf, H., Benda, C., Gärtner, W., Storf, M., Scheer, H., and Siebert, F. (2001) *Biochemistry* **40**, 14952–14959

623b. Andel, F., III, Murphy, J. T., Haas, J. A., McDowell, M. T., van der Hoef, I., Lugtenburg, J., Lagarias, J. C., and Mathies, R. A. (2000) *Biochemistry* **39**, 2667–2676

623c. Hennig, L., and Schäfer, E. (2001) *J. Biol. Chem.* **276**, 7913–7918

623d. Huq, E., Tepperman, J. M., and Quail, P. H. (2000) *Proc. Natl. Acad. Sci. U.S.A.* **97**, 9789–9794

623e. Park, C.-M., Kim, J.-I., Yang, S.-S., Kang, J.-G., Kang, J.-H., Shim, J.-Y., Chung, Y.-H., Park, Y.-M., and Song, P.-S. (2000) *Biochemistry* **39**, 10840–10847

623f. Schmitz, O., Katayama, M., Williams, S. B., Kondo, T., and Golden, S. S. (2000) *Science* **289**, 765–768

624. Yeh, K.-C., Wu, S.-H., Murphy, J. T., and Lagarias, J. C. (1997) *Science* **277**, 1505–1508

625. Jiang, Z., Swem, L. R., Rushing, B. G., Devanathan, S., Tollin, G., and Bauer, C. E. (1999) *Science* **285**, 406–409

625a. Bhoo, S.-H., Davis, S. J., Walker, J., Karniol, B., and Vierstra, R. D. (2001) *Nature (London)* **414**, 776–779

626. Smith, H. (1999) *Nature (London)* **400**, 710–712

627. Ni, M., Tepperman, J. M., and Quail, P. H. (1999) *Nature (London)* **400**, 781–784

627a. Sweere, U., Eichenberg, K., Lohrmann, J., Mira-Rodado, V., Bäurle, I., Kudla, J., Nagy, F., Schäfer, E., and Harter, K. (2001) *Science* **294**, 1108–1111

628. Hoecker, U., Tepperman, J. M., and Quail, P. H. (1999) *Science* **284**, 496–499

629. Evans, A., and Smith, H. (1976) *Proc. Natl. Acad. Sci. U.S.A.* **73**, 138–142

630. Song, P.-S., and Moore, T. A. (1974) *Photochem. Photobiol.* **19**, 435–441

631. Quiñones, M. A., Lu, Z., and Zeiger, E. (1996) *Proc. Natl. Acad. Sci. U.S.A.* **93**, 2224–2228

632. Cashmore, A. R., Jarillo, J. A., Wu, Y.-J., and Liu, D. (1999) *Science* **284**, 760–765

633. Presti, D., Hsu, W.-J., and Delbrück, M. (1976) *Photochem. Photobiol.* **26**, 403–405

634. Galland, P., and Lipson, E. D. (1987) *Proc. Natl. Acad. Sci. U.S.A.* **84**, 104–108

635. Christie, J. M., Reymond, P., Powell, G. K., Bernasconi, P., Raibekas, A. A., Liscum, E., and Briggs, W. R. (1998) *Science* **282**, 1698–1701

636. Nozue, K., Kanegae, T., Imaizumi, T., Fukuda, S., Okamoto, H., Yeh, K.-C., Lagarias, J. C., and Wada, M. (1998) *Proc. Natl. Acad. Sci. U.S.A.* **95**, 15816–15830

636a. Sakai, T., Kagawa, T., Kasahara, M., Swartz, T. E., Christie, J. M., Briggs, W. R., Wada, M., and Okada, K. (2001) *Proc. Natl. Acad. Sci. U.S.A.* **98**, 6969–6974

636b. Motchoulski, A., and Liscum, E. (1999) *Science* **286**, 961–964

636c. Christie, J. M., and Briggs, W. R. (2001) *J. Biol. Chem.* **276**, 11457–11460

636d. Kinoshita, T., Doi, M., Suetsugu, N., Kagawa, T., Wada, M., and Shimazaki, K.-i. (2001) *Nature (London)* **414**, 656–660

636e. Swartz, T. E., Wenzel, P. J., Corchnoy, S. B., Briggs, W. R., and Bogomolni, R. A. (2002) *Biochemistry* **41**, 7183–7189

637. Linden, H., and Macino, G. (1997) *EMBO J.* **16**, 98–109

637a. Iseki, M., Matsunaga, S., Murakami, A., Ohno, K., Shiga, K., Yoshida, K., Sugai, M., Takahashi, T., Hori, T., and Watanabe, M. (2002) *Nature (London)* **415**, 1047–1048

637b. Froehlich, A. C., Liu, Y., Loros, J. J., and Dunlap, J. C. (2002) *Science* **297**, 815–819

637c. Linden, H. (2002) *Science* **297**, 777–778

637d. Bieszke, J. A., Spudich, E. N., Scott, K. L., Borkovich, K. A., and Spudich, J. L. (1999) *Biochemistry* **38**, 14138–14145

References

638. Ahmad, M., and Cashmore, A. R. (1993) *Nature (London)* **366**, 162–166
639. Ahmad, M., Jarillo, J. A., Smirnova, O., and Cashmore, A. R. (1998) *Nature (London)* **392**, 720–723
640. Guo, H., Yang, H., Mockler, T. C., and Lin, C. (1998) *Science* **279**, 1360–1363
640a. Guo, H., Mockler, T., Duong, H., and Lin, C. (2001) *Science* **291**, 487–489
640b. Wang, H., Ma, L.-G., Li, J.-M., Zhao, H.-Y., and Deng, X. W. (2001) *Science* **294**, 151–158
641. Somers, D. E., Devlin, P. F., and Kay, S. A. (1998) *Science* **282**, 1488–1490
642. Ceriani, M. F., Darlington, T. K., Staknis, D., Más, P., Petti, A. A., Weitz, C. J., and Kay, S. A. (1999) *Science* **285**, 553–556
643. Thresher, R. J., Vitaterna, M. H., Miyamoto, Y., Kazantsev, A., Hsu, D. S., Petit, C., Selby, C. P., Dawut, L., Smithies, O., Takahashi, J. S., and Sancar, A. (1998) *Science* **282**, 1490–1494
643a. Hardin, P. E., and Glossop, N. R. J. (1999) *Science* **286**, 2460–2461
644. Hsu, D. S., Zhao, X., Zhao, S., Kazantsev, A., Wang, R.-P., Todo, T., Wei, Y.-F., and Sancar, A. (1996) *Biochemistry* **35**, 13871–13877
645. Lucas, R. J., Freedman, M. S., Munoz, M., Garcia-Fernández, J.-M., and Foster, R. G. (1999) *Science* **284**, 505–507
646. Miyamoto, Y., and Sancar, A. (1998) *Proc. Natl. Acad. Sci. U.S.A.* **95**, 6097–6102
647. Moore, R. Y. (1999) *Science* **284**, 2102–2103
647a. Hattar, S., Liao, H.-W., Takao, M., Berson, D. M., and Yau, K.-W. (2002) *Science* **295**, 1065–1070
647b. Berson, D. M., Dunn, F. A., and Takao, M. (2002) *Science* **295**, 1070–1073
647c. Barinaga, M. (2002) *Science* **295**, 955–957
647d. Thompson, C. L., Blaner, W. S., Van Gelder, R. N., Lai, K., Quadro, L., Colantunoni, V., Gottesman, M. E., and Sancar, A. (2001) *Proc. Natl. Acad. Sci. U.S.A.* **98**, 11708–11713
648. Hearst, J. E. (1995) *Science* **268**, 1858–1859
649. Sancar, A. (1994) *Biochemistry* **33**, 2–9
650. Snapka, R. M., and Sutherland, B. M. (1980) *Biochemistry* **19**, 4201–4208
651. Gindt, Y. M., Vollenbroek, E., Westphal, K., Sackett, H., Sancar, A., and Babcock, G. T. (1999) *Biochemistry* **38**, 3857–3866
652. Park, H.-W., Kim, S.-T., Sancar, A., and Deisenhofer, J. (1995) *Science* **268**, 1866–1872
653. Lipman, R. S. A., and Jorns, M. S. (1996) *Biochemistry* **35**, 7968–7973
653a. Durbeej, B., and Eriksson, L. A. (2000) *J. Am. Chem. Soc.* **122**, 10126–10132
653b. MacFarlane, A. W., IV, and Stanley, R. J. (2001) *Biochemistry* **40**, 15203–15214
654. Aubert, C., Mathis, P., Eker, A. P. M., and Brettel, K. (1999) *Proc. Natl. Acad. Sci. U.S.A.* **96**, 5423–5427
654a. Kay, C. W. M., Feicht, R., Schulz, K., Sadewater, P., Sancar, A., Bacher, A., Möbius, K., Richter, G., and Weber, S. (1999) *Biochemistry* **38**, 16740–16748
654b. Antony, J., Medvedev, D. M., and Stuchebrukhov, A. A. (2000) *J. Am. Chem. Soc.* **122**, 1057–1065
654c. Weber, S., Möbius, K., Richter, G., and Kay, C. W. M. (2001) *J. Am. Chem. Soc.* **123**, 3790–3798
655. Hitomi, K., Kim, S.-T., Iwai, S., Harima, N., Otoshi, E., Ikenaga, M., and Todo, T. (1997) *J. Biol. Chem.* **272**, 32591–32598
655a. Hitomi, K., Nakamura, H., Kim, S.-T., Mizukoshi, T., Ishikawa, T., Iwai, S., and Todo, T. (2001) *J. Biol. Chem.* **276**, 10103–10109
656. Kim, S.-T., Malhotra, K., Smith, C. A., Taylor, J.-S., and Sancar, A. (1994) *J. Biol. Chem.* **269**, 8535–8540
657. Zhao, X., Liu, J., Hsu, D. S., Zhao, S., Taylor, J.-S., and Sancar, A. (1997) *J. Biol. Chem.* **272**, 32580–32590
657a. Joseph, A., Prakash, G., and Falvey, D. E. (2000) *J. Am. Chem. Soc.* **122**, 11219–11225
658. Todo, T., Ryo, H., Yamamoto, K., Toh, H., Inui, T., Ayaki, H., Nomura, T., and Ikenaga, M. (1996) *Science* **272**, 109–112
659. Viviani, V. R., Bechara, E. J. H., and Ohmiya, Y. (1999) *Biochemistry* **38**, 8271–8279
660. Conti, E., Franks, N. P., and Brick, P. (1996) *Structure* **4**, 287–298
661. Branchini, B. R., Magyar, R. A., Murtiashaw, M. H., Anderson, S. M., and Zimmer, M. (1998) *Biochemistry* **37**, 15311–15319
661a. Branchini, B. R., Magyar, R. A., Murtiashaw, M. H., Anderson, S. M., Helgerson, L. C., and Zimmer, M. (1999) *Biochemistry* **38**, 13223–13230
662. McElroy, W. D., and Seliger, H. H. (1962) *Sci. Am.* **207**(Dec), 76–89
663. Campbell, A. K. (1986) *Trends Biochem. Sci.* **11**, 104–108
664. Fisher, A. J., Raushel, F. M., Baldwin, T. O., and Rayment, I. (1995) *Biochemistry* **34**, 6581–6586
665. Bronstein, I., and McGrath, P. (1989) *Nature (London)* **338**, 599–600
666. Adam, W., Bronstein, I., Edwards, B., Engel, T., Reinhardt, D., Schneider, F. W., Trofimov, A. V., and Vasil'ev, R. F. (1996) *J. Am. Chem. Soc.* **118**, 10400–10407
667. Hori, K., Wampler, J. E., Matthews, J. C., and Cormier, M. J. (1973) *Biochemistry* **12**, 4463–4468
668. Matthews, J. C., Hori, K., and Cormier, M. J. (1977) *Biochemistry* **16**, 85–91
669. Shimamura, O., and Johnson, F. H. (1978) *Proc. Natl. Acad. Sci. U.S.A.* **75**, 2611–2615
670. Tsuji, F. I. (1985) *Proc. Natl. Acad. Sci. U.S.A.* **82**, 4629–4632
671. Thompson, E. M., Nagata, S., and Tsuji, F. I. (1989) *Proc. Natl. Acad. Sci. U.S.A.* **86**, 6567–6571
672. Ray, B. D., Ho, S., Kemple, M. D., Prendergast, F. G., and Nageswara Rao, B. D. (1985) *Biochemistry* **24**, 4280–4287
673. Inouye, S., Noguchi, M., Sakaki, Y., Takagi, Y., Miyata, T., Iwanaga, S., Miyata, T., and Tsuji, F. I. (1985) *Proc. Natl. Acad. Sci. U.S.A.* **82**, 3154–3158
674. Shimomura, O., and Johnson, F. H. (1972) *Biochemistry* **11**, 1602–1608
674a. Markova, S. V., Vysotski, E. S., Blinks, J. R., Burakova, L. P., Wang, B.-C., and Lee, J. (2002) *Biochemistry* **41**, 2227–2236
675. Cormier, M. J., and Wampler, J. E. (1975) *Ann. Rev. Biochem.* **44**, 255–272
676. Hart, R. C., Matthews, J. C., Hori, K., and Cormier, M. J. (1979) *Biochemistry* **18**, 2204–2205
676a. Liu, Z.-J., Vysotski, E. S., Chen, C.-J., Rose, J. P., Lee, J., and Wang, B.-C. (2000) *Protein Sci.* **9**, 2085–2093
676b. Haddock, S. H. D., Rivers, T. J., and Robison, B. H. (2001) *Proc. Natl. Acad. Sci. U.S.A.* **98**, 11148–11151
676c. Li, L., Liu, L., Hong, R., Robertson, D., and Hastings, J. W. (2001) *Biochemistry* **40**, 1844–1849
676d. Nakamura, H., Kishi, Y., Shimomura, O., Morse, D., and Hastings, J. W. (1989) *J. Am. Chem. Soc.* **111**, 7607–7611
676e. Branchini, B. R., Murtiashaw, M. H., Magyar, R. A., and Anderson, S. M. (2000) *Biochemistry* **39**, 5433–5440
677. Thompson, J. F., Geoghegan, K. F., Lloyd, D. B., Lanzetti, A. J., Magyar, R. A., Anderson, S. M., and Branchini, B. R. (1997) *J. Biol. Chem.* **272**, 18766–18771
677a. Branchini, B. R., Murtiashaw, M. H., Magyar, R. A., Portier, N. C., Ruggiero, M. C., and Stroh, J. G. (2002) *J. Am. Chem. Soc.* **124**, 2112–2113
677b. Gomi, K., and Kajiyama, N. (2001) *J. Biol. Chem.* **276**, 36508–36513
678. Shimomura, O., and Johnson, F. H. (1968) *Biochemistry* **7**, 2574–2580
679. Nakamura, H., Kishi, Y., Shimomura, O., Morse, D., and Hastings, J. W. (1989) *J. Am. Chem. Soc.* **111**, 7607–7611
679a. Jeffers, C. E., and Tu, S.-C. (2001) *Biochemistry* **40**, 1749–1754
680. Xin, X., Xi, L., and Tu, S.-C. (1994) *Biochemistry* **33**, 12194–12201
681. Tanner, J. J., Miller, M. D., Wilson, K. S., Tu, S.-C., and Krause, K. L. (1997) *Biochemistry* **36**, 665–672
682. Baldwin, T. O., Berends, T., Bunch, T. A., Holzman, T. F., Rausch, S. K., Shamansky, L., Treat, M. L., and Ziegler, M. M. (1984) *Biochemistry* **23**, 3663–3667
683. Cohn, D. H., Mileham, A. J., Simon, M. I., Nealson, K. H., Rausch, S. K., Bonam, D., and Baldwin, T. O. (1985) *J. Biol. Chem.* **260**, 6139–6146
684. Lawson, D. M., Derewenda, U., Serre, L., Ferri, S., Szittner, R., Wei, Y., Meighen, E. A., and Derewenda, Z. S. (1994) *Biochemistry* **33**, 9382–9388
685. Kurfürst, M., Ghisla, S., and Hastings, J. W. (1983) *Biochemistry* **22**, 1521–152
685a. Lin, L. Y.-C., Sulea, T., Szittner, R., Kor, C., Purisima, E. O., and Meighen, E. A. (2002) *Biochemistry* **41**, 9938–9945
686. Eckstein, J. W., Hastings, J. W., and Ghisla, S. (1993) *Biochemistry* **32**, 404–411
687. Francisco, W. A., Abu-Soud, H. M., DelMonte, A. J., Singleton, D. A., Baldwin, T. O., and Raushel, F. M. (1998) *Biochemistry* **37**, 2596–2606
688. Li, H., Ortego, B. C., Maillard, K. I., Willson, R. C., and Tu, S.-C. (1999) *Biochemistry* **38**, 4409–4415
689. Engebrecht, J., and Silverman, M. (1984) *Proc. Natl. Acad. Sci. U.S.A.* **81**, 4154–4158
690. Eberhard, A., Burlingame, A. L., Eberhard, C., Kenyon, G. L., Nealson, K. H., and Oppenheimer, N. J. (1981) *Biochemistry* **20**, 2444–2449

Study Questions

1. Why is the Emerson enhancement effect (i.e., light at 650 nm plus 680 nm gives a higher rate of photosynthesis than either one alone) not observed with photosynthetic bacteria?

2. Agents that uncouple oxidative phosphorylation in mitochondria uncouple photoelectron transport and ATP formation in photosynthesis. Explain.

3. The action spectrum of photosynthesis, which describes the efficiency of photosynthesis as a function of the wavelength of incident light, has a valley around 550 nm. Why?

4. Plants exposed to $C^{18}O_2$ will have the ^{18}O first appear in a) carbohydrate; b) water; c) oxygen gas. (More than one answer may be true.)

5. Plants exposed to $H_2{}^{18}O$ will have the label first appear in a) oxygen gas; b) carbohydrate; c) CO_2. (More than one may be correct.)

6. The general equation describing the photosynthesis of glucose in higher plants is:

$$6\ CO_2 + 6\ H_2O \rightarrow C_6H_{12}O_6$$

We know that the oxygen gas comes from water, yet the equation shows only six atoms of oxygen in water on the left-hand side vs 12 in oxygen gas on the right. Explain.

7. The fructose bisphosphatase of green plants has an amino acid sequence which is very similar to those of the corresponding enzymes isolated from other sources such as yeast or mammals, except that the plant enzyme has an additional sequence of 20 or so amino acids that has no counterpart in the enzymes found in the other species. What function might this additional sequence have in the plant enzyme?

8. The following substances are either inhibitors or activators of rubisco, the enzyme that catalyzes the condensation of CO_2 with ribulose bisphosphate to yield 3-phosphoglycerate. State whether the substance should be an activator or an inhibitor of the enzyme and succinctly provide the logic supporting your conclusion.
   a) Fructose 1,6-bisphosphate
   b) Visible light
   c) NADPH

9. The reagent DCMU specifically inhibits electron transfer to plastoquinone in photosystem II. Discuss how the administration of this compound to a suspension of illuminated chloroplasts will affect the production of oxygen, ATP, and NADPH.

10. A chemical reagent is added to a solution of plant chloroplasts which immediately and specifically poisons photosystem II. What is the *short-term* effect of each of the following? Give a one-sentence defense for your conclusion.
    a) Cyclic photophosphorylation
    b) Noncyclic photophosphorylation
    c) Photorespiration
    d) NADPH production

11. If a C_3 and a C_4 plant are placed together in a sealed illuminated box, the C_3 plant withers and dies long before the C_4 plant. Explain.

12. What tricarboxylic acid cycle enzyme is analogous to the malate enzyme of bundle-sheath cells? What is the mechanism of the reaction?

13. There are two different forms of glyceraldehyde-3-phosphate dehydrogenase in higher plant cells.
    a) In which cell compartment is each one found?
    b) What are the reactions catalyzed by these two isozymes?
    c) Why are there two forms?

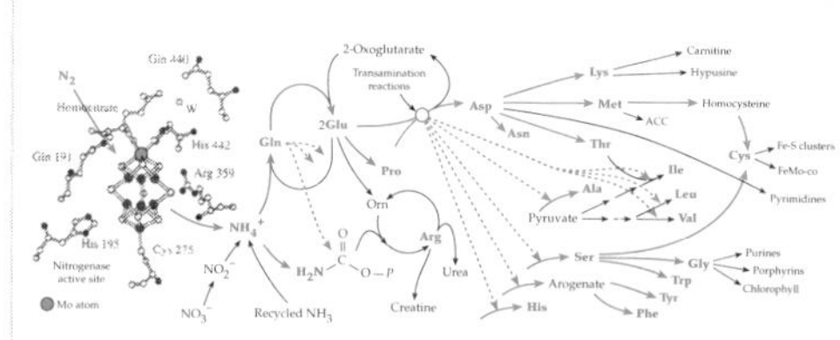

The air provides an abundant source of nitrogen for living organisms. Nitrogenase present in specialized bacteria utilizes the molybdenum- and iron-containing FeMo-co to reduce N_2 to two molecules of NH_3 (or NH_4^+). NH_3 is incorporated into the side chain of glutamine and much is transferred to the 5-carbon skeleton of 2-oxoglutarate to form glutamate. Nitrogen from glutamate and glutamine moves into the other amino acids via action of transaminases and glutamine amidotransferases. Thousands of compounds, a few of which are indicated here, are formed. (The 20 amino acid constituents of proteins are shown in green.) NH_3 from decaying materials is recycled, often after oxidation to nitrite or nitrate. Nitrates may also be formed by lightening and NH_3 industrially by catalytic reductions of N_2 by H_2 at high temperature and pressure (the Haber process).

Contents

The Metabolism of Nitrogen and Amino Acids

24

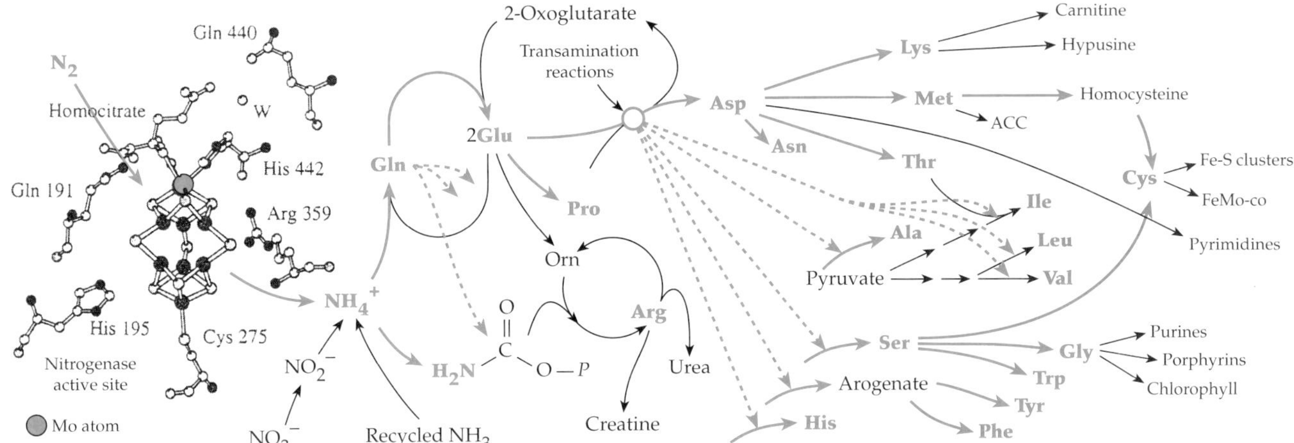

Because it is found in so many compounds and can exist in several oxidation states, nitrogen has a complex metabolism. The inorganic forms of nitrogen found in our surroundings range from the highly oxidized nitrate ion, in which N has an oxidation state of +5, to ammonia, in which the oxidation state is −3. Living cells both reduce and oxidize these inorganic forms. The organic forms of nitrogen are most often derived by incorporation of **ammonium ions** into amino groups or amide groups. Once it has been incorporated into an organic compound, nitrogen can be transferred into many other carbon compounds. Certain compounds including glutamic acid, aspartic acid, glutamine, asparagine, and carbamoyl phosphate are especially active in these transfer reactions. They constitute a **nitrogen pool** from which nitrogen can be withdrawn and to which it can be returned.

In addition to the pathways for synthesis and degradation of nitrogenous substances, many organisms have specialized metabolism for incorporation of excess nitrogen into relatively nontoxic excretion products. All of these aspects of nitrogen metabolism will be dealt with in this and the following chapter. We will look first at the reactions by which organic nitrogen compounds are formed from inorganic compounds, then at the reactions of the nitrogen pool. After that we will examine the specific reactions of synthesis and catabolism of individual nitrogenous compounds.

A. Fixation of N_2 and the Nitrogen Cycle

Most of the nitrogen of the biosphere exists as the unreactive N_2, which makes up 80% of the molecules of air. The "fixation" of N_2 occurs principally by the action of a group of bacteria known as **diazotrophs** and to a lesser extent by lightning, which forms oxides of nitrogen and eventually nitrate and nitrite. Human beings also contribute a smaller but significant share through production of chemical fertilizer by the Haber process. These reactions are an important part of the **nitrogen cycle**.[1,2] Quantitatively even more important are the biochemical processes of **nitrification**, by which ammonium ions from decaying organic materials are oxidized to NO_2^- and NO_3^- by soil bacteria (Fig. 24-1), and reactions of reduction and **assimilation** of nitrate and nitrite by bacteria, fungi, and green plants. Another reductive process catalyzed by **denitrifying bacteria** returns N_2 to the atmosphere (Fig. 24-1).

1. Reduction of Elemental Nitrogen

One of the most remarkable reactions of nitrogen metabolism is the conversion of dinitrogen (N_2) to ammonia. It was estimated that in 1974 this biological nitrogen fixation added 17 x 10^{10} kg of nitrogen to the earth (compared with 4 x 10^{10} kg fixed by chemical reactions).[3] The quantitative significance can be more easily appreciated by the realization that one square meter of land planted to nodulated legumes such as soybeans can fix 10–30 g of nitrogen per year.

Fixation of N_2 by *Clostridium pasteurianum* and a few other species was recognized by Winogradsky[4] in 1893. Subsequent nutritional studies indicated that both iron and molybdenum were required for the process. Inhibition by CO and N_2O was observed. While ammonia was the suggested product, the possibility remained that more oxidized compounds such as hydroxylamine were the ones first incorporated into organic substances. When cell-free preparations capable of fixing nitrogen were obtained in 1960 rapid progress became possible.[5] It was discovered that nitrogen-fixing bacteria are invariably able to reduce acetylene to ethylene, a catalytic ability that goes hand in hand with the ability to reduce N_2. A simple, sensitive **acetylene reduction test** permits easy measurement of the nitrogen-fixing potential of cells.

Application of this test revealed that nitrogen fixation is not restricted to a few species, but is a widespread ability of many prokaryotes. Most studied are *Azotobacter vinelandii*, Winogradsky's *C. pasteurianum*, *Klebsiella pneumoniae* (a close relative of *E. coli*), and several species of *Rhizobium*, the symbiotic bacterium of root nodules of legumes. The latter deserves special attention. Although some free-living rhizobia reduce N_2, the reaction usually takes place only in nodules developed by infected roots. Within these nodules the bacteria degenerate into **bacteroids**; and the special hemoglobin **leghemoglobin**,[6,7] whose sequence is specified by a plant gene,[8] is synthesized.

Legumes are not the only plants with nitrogen-fixing symbionts.[9] Some other angiosperms are hosts to nitrogen-fixing actinomycetes and some gymnosperms contain nitrogen-fixing blue-green algae. Leaf nodules of certain plants infected with *Klebsiella* fix nitrogen. While the nutritional significance is uncertain, nitrogen-fixing strains of *Klebsiella* have also been found in the intestinal tracts of humans in New Guinea. Of the free-living nitrogen-fixing organisms, cyanobacteria appear to be of most importance quantitatively. For example, in rice paddy fields cyanobacteria may fix from 2.4 to 10 g of nitrogen per square meter per year. Cyanobacteria in the oceans fix enormous amounts of nitrogen.[9a]

2. Nitrogenases

Cell-free nitrogenases have been isolated from a number of organisms. These enzymes all share the property of being inactivated by oxygen, a fact that impeded early work. Apparently nitrogen fixation occurs in anaerobic regions of cells. Leghemoglobin may protect the nitrogen-fixing enzymes in root nodules from oxygen. It probably also functions to deliver O_2 by facilitated diffusion to the aerobic mitochondria of the bacteroids at a stable, low partial pressure.[6,10] Some bacteria utilize protective proteins to shield the nitrogenase molecules when the O_2 pressure is too high.[10a]

Nitrogenases catalyze the six-electron reduction of N_2 to ammonia (Eq. 24-1) and are also able to reduce

$$N_2 + 6\,H^+ + 6\,e^- \rightarrow 2\,NH_3 \quad (24\text{-}1)$$

many other compounds. For example, the reduction of acetylene to ethylene (Eq. 24-2) is a two-electron process. Azide is reduced to N_2 and NH_4^+ in another two-electron reduction (Eq. 24-3). Cyanide ions yield methane and ammonia (Eq. 24-4).[11] Alkyl nitriles as well as N_2O and carbonyl sulfide (COS) are also reduced. Carbon dioxide is reduced slowly to CO[12], and nitrogenases invariably catalyze reduction of protons to H_2 (Eq. 24-5).

$$HC{\equiv}CH + 2\,H^+ + 2\,e^- \rightarrow H_2C{=}CH_2 \quad (24\text{-}2)$$

$$N{=}N{=}N^- + 4\,H^+ + 2\,e^- \rightarrow N_2 + NH_4^+ \quad (24\text{-}3)$$

$$^-C{\equiv}N + 8\,H^+ + 6\,e^- \rightarrow CH_4 + NH_4^+ \quad (24\text{-}4)$$

$$2\,H^+ + 2\,e^- \rightarrow H_2 \quad (24\text{-}5)$$

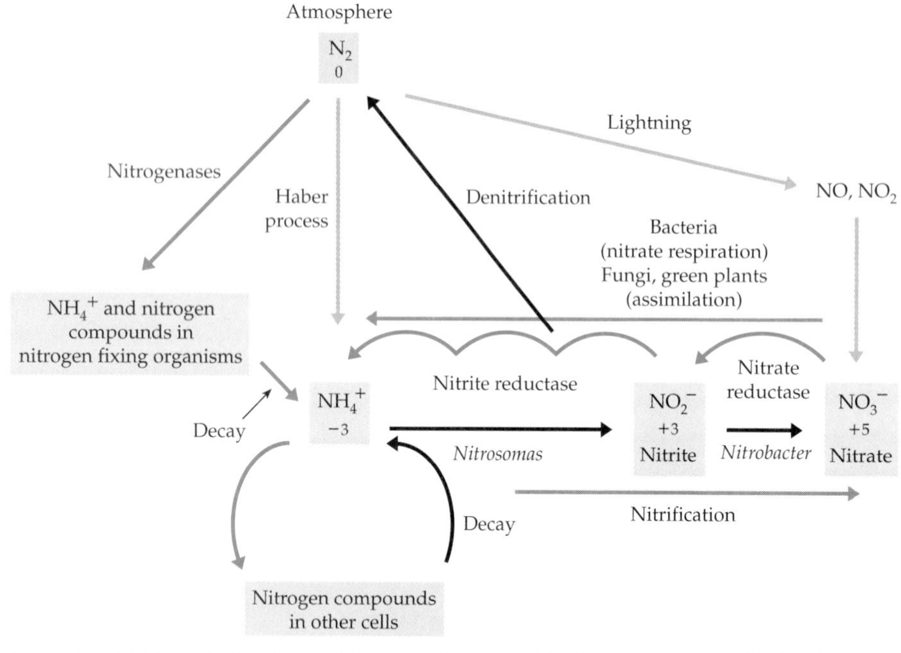

Figure 24-1 The nitrogen cycle. Conversion of N_2 (oxidation state 0) to NH_4^+ by nitrogen-fixing bacteria, assimilation of NH_4^+ by other organisms, decay of organic matter, oxidation of NH_4^+ by the nitrifying bacteria *Nitrosomas* and *Nitrobacter*, reduction of NO_3^- and NO_2^- back to NH_4^+, and release of nitrogen as N_2 by denitrifying bacteria are all part of this complex cycle.[1]

In early experiments it was found that sodium pyruvate was required for fixation of N_2 in cell-free extracts, and that large amounts of CO_2 and H_2 accumulated. Investigation showed that cleavage of pyruvate supplies cells with two important products: ATP and reduced ferredoxin. Pyruvate can be replaced by a mixture of ATP plus Mg^{2+} and reduced ferredoxin (Fd_{red}). Furthermore, the nonbiological reductant dithionite ($S_2O_4^{2-}$) can replace the reduced ferredoxin. Since ADP is inhibitory to the nitrogenase system, it is best in laboratory studies to supply ATP from an ATP-generating system such as a mixture of creatine phosphate, creatine kinase, and a small amount of ADP (Eqs. 6-65, 6-67).

The commonest type of nitrogenase can be separated easily into two components (Fig. 24-2). One of these, the **iron protein** (dinitrogenase reductase, azoferredoxin, or component II), is an extremely oxygen-sensitive iron-sulfur protein. It consists of two identical ~32-kDa peptide chains; those of *A. vinlandii* each contain 189 amino acid residues. The three-dimensional structure of the dimeric protein[13–16] shows that each subunit forms a nucleotide-binding domain with an ATP-binding site. About 2 nm away from this site is *a single Fe_4S_4 cluster which is shared symmetrically by the two subunits of the protein*. Each subunit contributes two thiolate groups from Cys 97 and Cys 132 as well as three N–H--S hydrogen bonds from NH groups at helix ends.[13]

The other component, the **molybdenum–iron protein** (dinitrogenase, molybdoferredoxin, or component I), contains both iron and molybdenum as well as labile sulfide. It is a mixed ($\alpha_2\beta_2$) tetramer of ~240-kDa mass and an analytical metal ion composition ~$Mo_2Fe_{30}S_{26}$. However, the X-ray structure[16–19a] suggests the composition $Mo_2Fe_{34}S_{36}$. The MoFe protein is a symmetric molecule in which each $\alpha\beta$ subunit contains two types of complex metal clusters. The active sites for N_2 reduction, which are embedded in the α subunits, contain the **FeMo-coenzyme** molecules, each with the metal composition $MoFe_7S_9$ and also containing a molecule of **homocitrate**.[20–22a] The other clusters, known as **P-clusters**, are shared between the α and β subunits, which for *A. vinlandii* contain 491 and 522

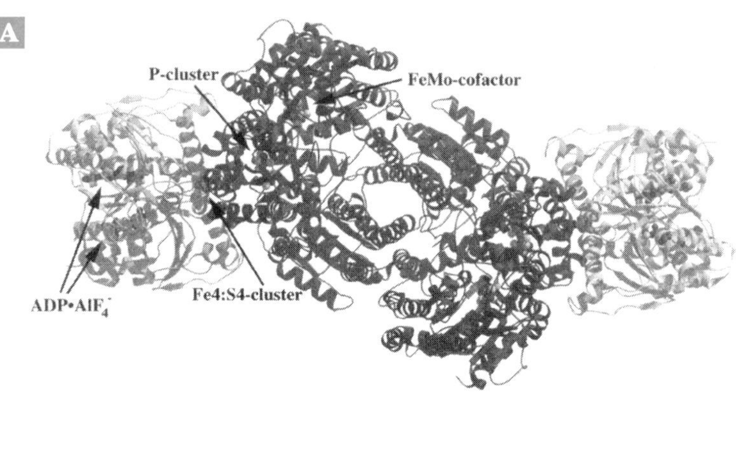

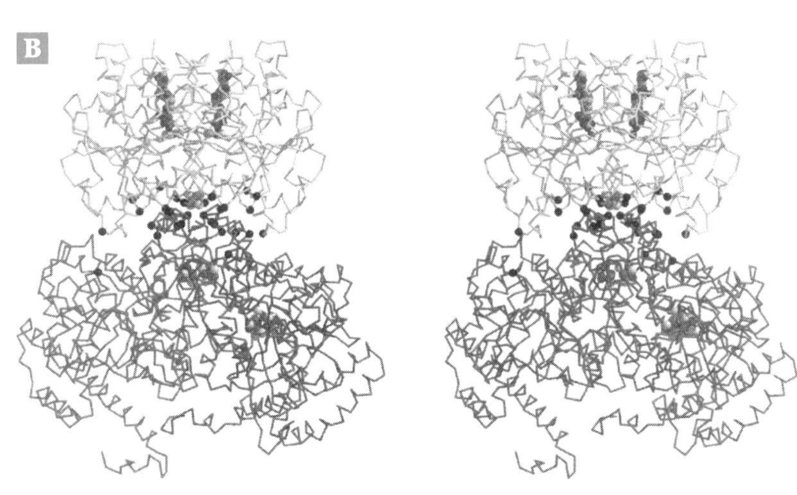

Figure 24-2 (A) Ribbon drawing of the three-dimensional structure of the nitrogenase from *Azotobacter vinlandii* viewed down the twofold axis of the molecule. The $\alpha\beta$ subunits of the MoFe-protein are in the center while the Fe-protein subunits are at the outer ends. The Fe_4S_4 cluster of the Fe-protein and the FeMo-co and P-cluster of the MoFe-protein are marked for the left-hand complex. The site of binding of ATP is identified by the bound ADP•AlF_4^- complex. (B) Stereoscopic view of one complete half of the nitrogenase complex at a 90° angle to the view in (A). The Fe-protein is at the top, and the MoFe-protein is below. The ADP•AlF_4^- complex is visible in the two symmetrically located binding sites of the Fe-protein. The shared Fe_4S_4 cluster is in the center above the P-cluster. The small black spheres mark α-carbons of residues that interact in forming the complex of Fe-protein and MoFe-proteins. When the Fe_4S_4 cluster accepts one electron from a molecule of ferredoxin or flavodoxin, the Fe-protein binds to the MoFe-protein and donates an electron to one of the two nearby P-clusters (one of which is in each of the $\alpha\beta$ subunits). At the same time both of the molecules of ATP bound to the Fe-protein are hydrolyzed. The oxidized Fe-protein then dissociates from the complex and is replaced by another reduced Fe-protein–ATP complex. The net result is that each electron is "pumped" from the Fe_4S_4 cluster of the Fe-protein into a P-cluster of one of the $\alpha\beta$ units of the MoFe-protein. Electrons then move from the P-clusters into the FeMo-coenzyme. From Schindelin *et al.*[19] Courtesy of Douglas C. Rees.

amino acid residues, respectively. Each P-cluster is actually a *joined pair* of cubane-type clusters, one Fe_4S_4 and one Fe_4S_3 with two bridging cysteine –SH groups and one iron atom bonded to three sulfide sulfur atoms (Fig. 24-3).[17,23] The FeMo-coenzyme can be released from the MoFe-protein by acid denaturation followed by extraction with dimethylformamide.[24] While homocitrate was identified as a component of the isolated coenzyme, the three-dimensional structure of FeMo-co was deduced from X-ray crystallography of the intact molybdenum–iron protein.[14,17,18]

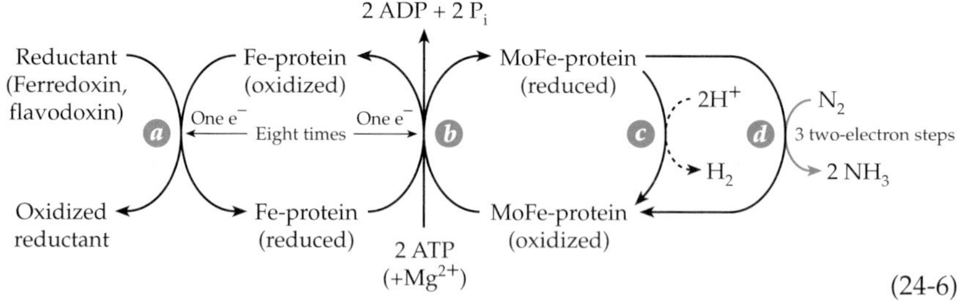

(24-6)

When the Fe-protein is reduced an EPR signal at g = 1.94, typical of iron–sulfur proteins (Fig. 16-17), is observed. This signal is altered by interaction with Mg and ATP, whereas ATP has no effect on the complex EPR signals produced upon reduction of MoFd. These are among the observations that led to the concept that the Fe-protein is an electron carrier responsible for reduction of the molybdenum in the MoFe-protein. The Mo(IV) or two atoms of Mo(III) formed in this way could then reduce N_2 in three two-electron steps with formation of Mo(VI) (Eq. 24-6). Three successive two-electron steps are required to completely reduce N_2 to two molecules of ammonia. An unexpected feature of nitrogenase action is that there is inevitably what was once regarded as a side reaction, the reduction of protons to H_2.[25,26] The amount of H_2 formed is variable and may be much greater than that of N_2 reduced. However, at high pressures of N_2 the ratio of H_2 formed to N_2 reduced is 1:1. This led to the suggestion[28] that H_2 formation is not a side reaction but an essential step in preparing the active site for the binding of N_2. Two protons that are bound somewhere on the reduced MoFe-protein could be reduced to H_2 in an obligatory step (Eq. 24-6, step *c*) that would, for example, cause a conformational change required for binding of N_2. If no reducible substrate (N_2, C_2H_2, etc.) is present, H_2 would still be formed slowly. Reducible substrates inhibit H_2 formation. However, addition of N_2 or any other reducible substrate causes an initial "burst" of H_2 to be released. This can be measured readily when a slow substrate such as CN^- is used as the inhibitor. The amount of H_2 released in the burst is stoichiometric with one H_2 per Mo being formed.[26] The overall stoichiometry for reduction of one N_2 becomes:

$$N_2 + 8\,e^- + 8\,H^+ + 16\,MgATP \rightarrow 2\,NH_3 + H_2 + 16\,MgADP + 16\,P_i \quad (24\text{-}7)$$

A second remarkable feature of nitrogenase is a requirement for hydrolysis of MgATP that is coupled to reduction of the MoFe-protein (Eq. 24-6). Two molecules of ATP are hydrolyzed to ADP and inorganic phosphate for each electron transferred. This large ATP requirement seems surprising in view of the fact that reduction of N_2 by reduced ferredoxin (Eq. 24-8) is thermodynamically spontaneous:

$$N_2 + 6\,Fd_{red} + 8\,H^+ \rightarrow 2\,NH_4^+ + 6\,Fd_{ox}^+$$
$$\Delta G'\ (pH\ 7) = -89.3\ kJ\ mol^{-1} \quad (24\text{-}8)$$

However, N_2 is exceedingly unreactive. In the commercial Haber process high pressure and temperature are needed to cause H_2 and N_2 to combine. Evidently cleavage of 16 molecules of ATP must be coupled to the nitrogenase reduction system to overcome the very high activation energy.

Not only are two molecules of ATP hydrolyzed to pump each electron, but the Fe-protein must receive electrons from a powerful (low $E^{\circ\prime}$) reductant such as reduced ferredoxin, reduced flavodoxin, or dithionite. *Klebsiella pneumoniae* contains a **pyruvate:flavodoxin oxidoreductase** (Eq. 15-35) that reduces either flavodoxin or ferredoxin to provide the low potential electron donor.[29,30] In some bacteria, e.g., the strictly aerobic *Azotobacter*, NADPH is the electron donor for reduction of N_2. The Fe-protein is thought to accept electrons from a chain that includes at least the ordinary bacterial ferredoxin (Fd) and a special one-electron-accepting **azotoflavin,** a flavoprotein that is somewhat larger than the flavodoxins (Chapter 15) and appears to play a specific role in N_2 fixation.[31] In *Clostridium* and *Rhizobium* reduced ferredoxins generated by cleavage of pyruvate reduce nitrogenase directly.[32]

The mechanism of nitrogenase action. The one-electron reduction of the Fe_4S_4 cluster of the Fe-protein (step *a* of Eq. 24-6) initiates the action. This reaction occurs before the Fe-protein forms a complex with the MoFe-protein. Following this initial reduction step the two molecules of ATP required for step *b* of Eq. 24-6 bind to the Fe-protein. One is bound to each subunit of this protein but neither is immediately adjacent to the shared Fe_4S_4 cluster, as can be seen from Fig. 14-3B. The binding to MgATP appears to induce a conformational change that permits the "docking" of the Fe-protein with the MoFe-protein to

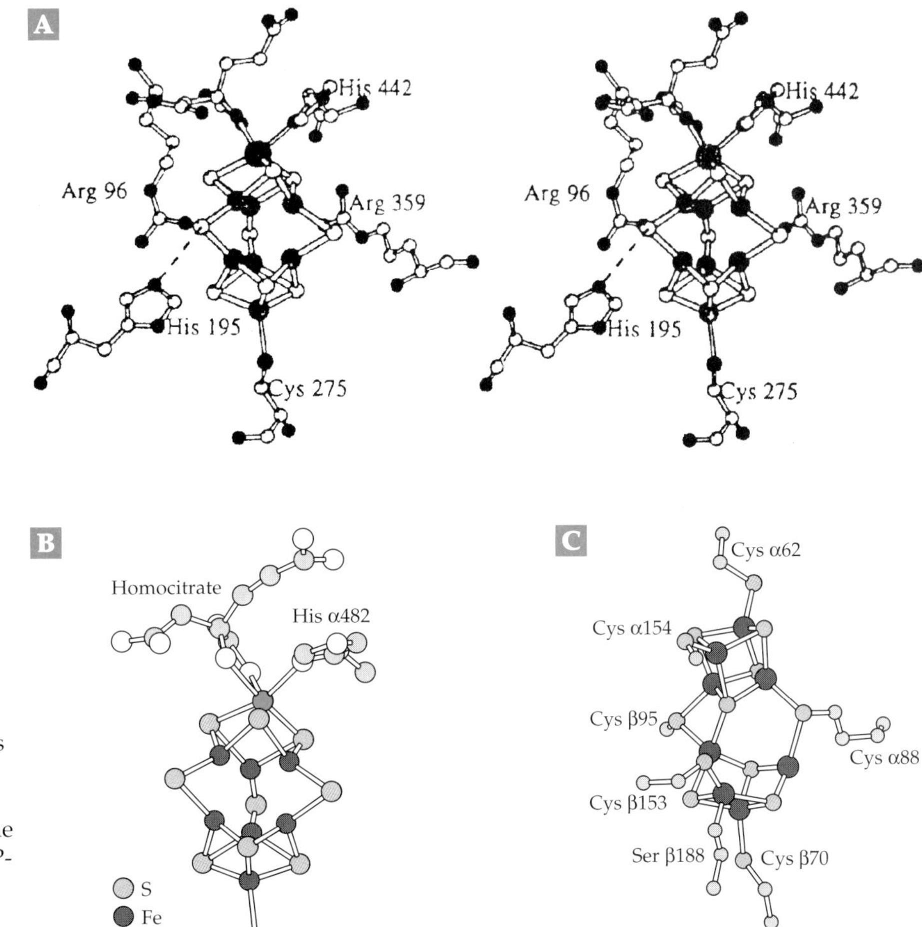

Figure 24-3 Structures of the metal-sulfide clusters of the MoFe-protein. (A) Stereoscopic view of the FeMo-co coenzyme with interacting side chains from the MoFe-protein of *A. vinlandii*. After Kim *et al.*[27] (B) FeMo-co with atom labels. From Kim *et al.*[18] (C) The structure of the oxidized form of the P-cluster. From Peters *et al.*[23] Recent studies by Einsle *et al.*[22a] indicate that the cluster probably also contains a nitrogen atom that is held within the cluster by coordination to six of the iron atoms.

form the complex in which the electron transfer of step *b* (Eq. 24-6) occurs. Abundant evidence indicates that electron transfer does not occur without the binding of MgATP.[33–35d] The electron transfer is coupled to the hydrolysis of the ATP, but the two reactions appear to be consecutive events. In a deletion mutant of the Fe-protein (lacking Leu 127) the hydrolysis of ATP does not occur, but the complex between Fe-protein and MoFe-protein is formed and electron transfer to the MoFe-protein takes place.[36–37a] The binding of the MgATP causes the midpoint redox potential to drop from −0.42 V to −0.62 V, assisting the transfer.[38] X-ray crystallographic studies reveal a distinct conformational change similar to those observed with G-proteins (Chapter 11) and involving movement of the Fe_4S_4 center into a better position for electron transfer.[19,38a]

After electron transfer the complex of the two proteins is thought to be tightly bonded when unhydrolyzed ATP is present.[35] This has allowed the direct observation and imaging of the complex at low resolution (~1.5 nm) using rapid synchrotron X-ray scattering measurements.[39] The ATP is hydrolyzed, and the Fe-proton is released from the complex.

Only one electron is transferred to the MoFe-protein in each catalytic cycle of the Fe-protein. Thus, the cycle must be repeated eight times to accomplish the reduction of $N_2 + 2\ H^+$. Where in the MoFe-protein does a transferred electron go? EPR spectroscopic and other experiments with incomplete and catalytically inactive molybdenum coenzyme[40] have provided a clear answer. The electron is transferred first to one of the two P-clusters, both of which are close to the Fe_4S_4 cluster of the Fe-protein. The transfer causes an observable change both in the spectroscopic properties and in the three-dimensional structure of the P-cluster.[23,40a] Since protons are needed at the active site for the reduction reactions (the FeMo-coenzyme), it is probable that hydrolysis of ATP in the Fe-protein is accompanied by transport of protons across the interface with the MoFe-protein. The electron transfer from the P-cluster on to the FeMo-co center would be assisted by a protic force resulting from ATP cleavage.

With defective FeMo-co (apparently lacking homocitrate) no reduction of N_2, acetylene, or protons is observed.[40] If intact FeMo-co is present, reduction of the cofactor can be observed. An $s = 3/2$ EPR signal arising from the Mo is seen,[40] and EXAFS measurements reveal decreased Mo – Fe distances as the coenzyme is reduced.[41] The molybdenum is probably present as Mo(VI) in the oxidized state of nitrogenase,[42] but after reduction it isn't clear whether it is Mo(III) or Mo(IV). Isolated FeMo-co exists in three identified oxidation states related by $E°'$ values of –0.17 and –0.465 V.[43] Only the middle state is EPR-active, but it is the most reduced state that is involved in N_2 reduction.[42,43] With its P-cluster and FeMo-co center each αβ unit of the MoFe-protein could store several electrons. Two or more might be stored in a P-cluster, and Mo(VI) could, in principle, accept three electrons to form Mo(III). However, it is a little hard to imagine storage of the eight electrons needed to reduce both N_2 and H_2 (Eq. 24-8). The reduction of N_2 may begin before all eight electrons have been transferred into the MoFe-protein.

Another uncertainty lies in the mode of binding of N_2 and other substrates. Does N_2 bind end-on to Mo, does it slide between Fe atoms within the coenzyme, or does it bind in some other way? While N_2 is unreactive, it forms nitrides with metals and complexes with some metal chelates. These complexes are generally of an end-on nature, e.g., N≡N–Fe. Stiefel suggested that N_2 first forms a complex of this type with an iron atom of the MoFe-protein.[44] Then an atom of Mo(IV) could donate two electrons to the N_2 (Eq. 24-9, step *a*) to form a complex of N_2 and Mo(VI). Addition of two protons (Eq. 24-9, step *b*) would yield a molecule of **diimide**, which would stay bound at the

$$\mathrm{N{\equiv}N} + \mathrm{Mo(IV)} \xrightarrow{a} \mathrm{Mo}{<}\overset{\mathrm{N}}{\underset{\mathrm{N}}{\|}} \xrightarrow[\ \ \to\ \mathrm{Mo(VI)}]{b,\ 2\,\mathrm{H}^+} \underset{\text{Diimide}}{\mathrm{H{-}N{=}N{-}H}} \qquad (24\text{-}9)$$

iron site while the molybdenum underwent another round of reduction. The diimide could be reduced to hydrazine and finally to ammonia (Eq. 24-10):

$$\mathrm{N_2} \xrightarrow{2\,\mathrm{H^+} + 2\,e^-} \mathrm{H{-}N{=}N{-}H} \xrightarrow{2\,\mathrm{H^+} + 2\,e^-} \mathrm{H_2N{-}NH_2} \xrightarrow{2\,\mathrm{H^+} + 2\,e^-} 2\,\mathrm{NH_3} \qquad (24\text{-}10)$$

Mo(VI) attracts electrons sufficiently strongly that protons bound to surrounding ligands, such as H_2O, tend to dissociate completely. Thus, the molybdate ion MoO_4^{2-} is not protonated. The same would be true of nitrogenous ligands of a protein that might be coordinated with the bound molybdenum. On the other hand, reduction to Mo(IV) would tend to favor protonation of ligands such as the His 442 imidazole seen in Fig. 24-3A. Concurrently with the electron transfer from molybdenum to N_2 these protons could be transferred to the N_2 molecule (Eq. 24-10). The fact that strictly *cis*-dideuteroethylene is formed from acetylene in the presence of ^2H_2O is in accord with this idea.

$$\begin{array}{ccc} \mathrm{H} & & \mathrm{H} \\ & \mathrm{C{=}C} & \\ {}^2\mathrm{H} & & {}^2\mathrm{H} \end{array}$$

However, looking at the FeMo-co molecule and the crowded surroundings of the Mo atom it may be more likely that reduction of N_2 occurs while it is bound to iron. Theoretical calculations as well as experimental data support this possibility.[44a] Recent crystallographic studies at a resolution of 0.12 nm revealed the presence of an atom, probably N, coordinated to six Fe atoms of FeMo-Co. This suggets, as previously proposed by Thorneley and Lowe,[44b,c] that a nitride ion (N_3^-) may be an intermediate in the formation of N_2.

Many mutant forms of nitrogenase have been investigated. Substitutions of His 195, Lys 191, and Gly 69 of the α chain affect reactions with various substrates.[45–45d] For example, the mutant obtained by substitution of His 195, whose imidazole forms an N–H- -S hydrogen bond to a central bridging sulfide atom of FeMo-co (Fig. 24-3A), with glutamine (H195Q mutant) reduces N_2 only very slowly.[45] However, it still reduces both acetylene and protons.[27,44a,45b] Thus, it may be that different modes of substrate binding are needed for the individual steps of Eq. 24-10.

Because of the practical significance to agriculture there is interest in devising better nonenzymatic processes for fixing nitrogen using nitrogenase models that mimic the natural biological reaction.[42,46–49a] One interesting catalyst is the following molybdenum complex Mo(III)(NRAr)$_3$ where R = C(C$_2$H$_3$)$_2$CH$_3$ and Ar = 3,5-dimethylphenyl.[47,50,51]

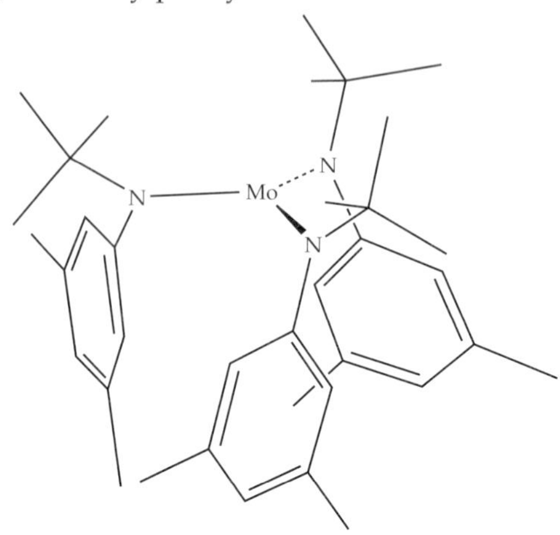

Many other synthetic complexes have been studied including cubic $MoFe_3S_4$ clusters.[46] However, no exact chemical model for the FeMo-co coenzyme has been developed, and the rates of reaction for all of the model reactions are much slower than those of nitrogenases.

Nitrogen fixation genes. At least 17 genes needed for nitrogen fixation are present in the 23-kb *nif* region of the *Klebsiella* chromosome[52,53] (Fig. 24-4). A similar gene cluster in *A. vinlandii*[54] contains five polygenic transcriptional units and one monogenic unit. The nitrogenase structural genes are *nifK, D,* and *H* as is indicated in Fig. 24-4. The *nifF* and *J* genes encode associated electron-transport proteins. *NifM* is needed to activate the Fe-protein in an unknown fashion. *NifS* encodes a cysteine desulfhydrase needed for assembly of Fe–S clusters in the nitrogenase and the nifU and nifY proteins assist the assembly.[54a,b] The chaperone GroEL is also required.[54c] *Nif Q, B, V, X, N, E_1,* and *H* are needed for synthesis of FeMo-co and for its incorporation into the MoFe-protein.[55,55a] *NifA* is an activator gene for the whole cluster including the *nifL* gene product, which is altered by the presence of O_2 or of glutamine. Accumulation of the latter in cells (see Section B,2) strongly represses transcription of the nitrogenase genes.

Legume nodules and cyanobacterial heterocysts. Nitrogen fixation requires an anaerobic environment. Free-living bacteria fix nitrogen only when anaerobic. However, Rhizobia produce their own anaerobic environment by symbiotic association with the roots of legumes.[10,57–59] Formation of root nodules is a genetically determined process, several nodulation (*nod*) genes of the bacterium being required along with an unknown number of plant genes.[57,58] Initiation of nodulation results from a two-way molecular conversation between root hairs of the plant and bacterial cells.[60,61] The roots secrete **flavonoid compounds** (Chapter 21) which are recognized by bacterial sensors and induce transcription of the *nod* genes. Several of these genes encode enzymes required for synthesis of **Nod factors**,[62,63] small β-linked *N*-acetyl-D-glucosamine oligosaccharides containing 3–5 sugar residues and an *N*-linked long-chain fatty-acyl substituent at the nonreducing terminus (**lipochitooligosaccharides**). See also Box 20-E. Genes *nodA, B, C* specify enzymes needed for synthesis of the oligosaccharide core present in all Nod factors.

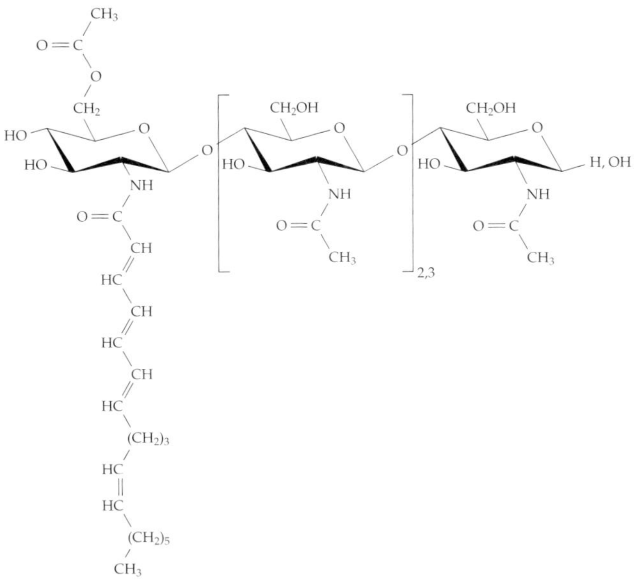

Structure of a Nod factor secreted by *Rhizobium leguminosarum*.[64]

Other Nod genes provide for modifications that restrict infection to specific species of legumes. For example, *nodS* encodes a methyltransferase and *nodU* a carbomoyltransferase.[65] NodH is a sulfotransferase.[66] NodD is a transcriptional activator that binds to DNA and induces the synthesis of the other Nod factors needed to initiate nodulation.[67] When an appropriate Nod factor is recognized, the root hairs on the legume

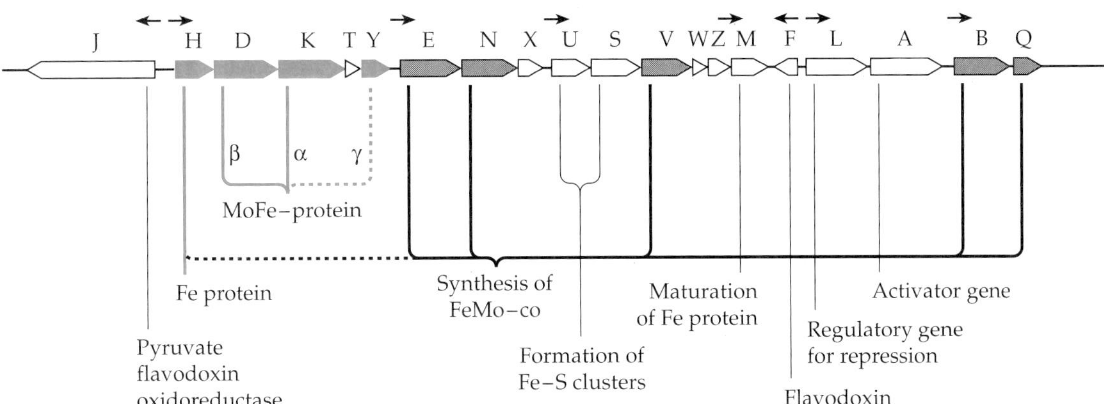

Figure 24-4 Sequence of *nif* genes of *Klebsiella pneumoniae*.[56] These precede the *his* operon directly at the right side. The nitrogenase structural genes are marked with green.

curl around the bacteria to initiate nodulation.[60,61] However, there are other factors. Infecting bacteria must reach a region of low oxygen in the plant. A hemoprotein **FixL** is a sensor kinase that regulates phosphorylation of transcription factor **FixJ**. This two-component system induces transcription of *nifA* (Fig. 24-4) and others.[68–69a] Nitrogen-fixing nodules, which are filled with the bacteroids derived from the infecting bacteria, synthesize leghemoglobin. The polypeptide chain of this protein is encoded by the plant, but its heme may be synthesized by bacteroid enzymes.[10,57] In at least one strain of *Rhizobium* the *nod* genes as well as the *fix* and *nif* genes are all carried on a 536-kb plasmid, which is almost as large as the whole 580-kb genome of *Mycoplasma genitalium* (Table 1-3).[70,71] This arrangement seems to have allowed these bacteria to form an unusually large number of Nod factors and to colonize a wider variety of hosts including a non-leguminous tree.

The H_2 that is produced in Eq. 24-6 (step *c*) may be used by bacteroids or by the plant cells. Some nodules evolve H_2, but in others it is utilized by hydrogenases as a source of energy.[57] From Eq. 24-6 it can be seen that up to 1/4 of the ATP utilized can, ideally, be recovered by use of the H_2 in this manner.

In cyanobacteria nitrogen fixation occurs in the **heterocysts**, specialized cells with thickened cell envelopes. They supply NH_4^+ to other cells in the filament of which they are a part. The cell envelopes prevent rapid diffusion of O_2 into the cells but do permit rapid enough entry of N_2 to maintain the observed rate of fixation of N_2.[72] In actinomycetes of the genus *Frankia*, which forms root nodules with woody plants, nitrogen fixation occurs in vesicles that are sheathed by multiple layers of **hopanoid lipids** (see Chapter 22).[73]

Genetic engineering. Because of the high cost of nitrogen fertilizers there is intense interest in improving biological nitrogen fixation. Ideas range from increasing the efficiency of nitrogenase by using fewer molecules of ATP, by limiting excessive evolution of H_2, or transferring the whole *nif* region of a bacterial genome into nonleguminous plants. The last proposal has generated much publicity, but it will probably be difficult because of the need to create an anaerobic environment suitable for nitrogen fixation. A crop plant engineered in this way might not resemble the hoped-for product. It would have an enormous energy requirement for nitrogen fixation, which would have to be met by photosynthesis. At present genetic engineering on *nif* genes to increase efficiency seems most likely to succeed.

Other nitrogenases. Although the well-characterized Mo-containing nitrogenase is responsible for most of their nitrogen fixation, bacteria often have alternative nitrogen fixation systems.[74] *Azotobacter vinlandii* produces three different nitrogenases in response to varying metal compositions in its surroundings.[75–76a] When the molybdenum level is adequate nitrogenase 1 is formed with its FeMo-coenzyme. In a low-molybdenum environment containing vanadium nitrogenase 2 is formed with an FeV-coenzyme.[75–77] If both molybdenum and vanadium are lacking, the bacteria form nitrogenase-3, which has an iron-only FeFe-coenzyme.

An unusual nitrogenase is formed by the chemolithotrophic *Streptomyces thermoautotrophicus*, which obtains energy from reduction of CO_2 or CO by H_2 (Eq. 17-50). These organisms form a MoFe nitrogenase that utilizes a manganese-containing superoxide dismutase to generate superoxide anion radicals. The latter transfer electrons to the MoFe protein in an ATP-dependent process. Electrons for generation of superoxide are formed using another molybdenum enzyme, a CO dehydrogenase containing molybdpterin cytosine dinucleotide (Fig. 16-31) and Fe–S centers.[78] The two systems function together as indicated by Eq. 24-11.

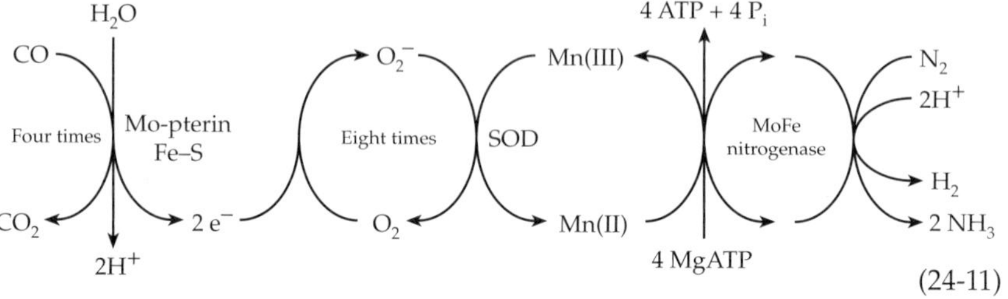

(24-11)

3. Interconversion of Nitrate, Nitrite, and Ammonium Ions

As is indicated in Fig. 24-1, the interconversions of nitrate and nitrite with ammonia and with organic nitrogen compounds are active biological processes. Two genera of nitrifying soil bacteria, which are discussed in Chapter 18, oxidize ammonium ions to nitrate. *Nitrosomas* carries out the six-electron oxidation to nitrite (Eq. 18-17) and *Nitrobacter* the two-electron oxidation of nitrite to nitrate (Eq. 18-18).[79]

The opposite sequence, reduction of nitrate and nitrite ions, provides a major route of acquisition of ammonia for incorporation into cells by bacteria, fungi, and green plants (Fig. 24-1). **Assimilatory** (biosynthetic) **nitrate reductases** catalyze the two-electron reduction of nitrate to nitrite (Eq. 16-61). This is thought to occur at the molybdenum atom of the large ~900-residue highly regulated[79a] molybdopterin-dependent enzyme. In green plants the reductant is usually NADH while in fungi it is more often NADPH.[80–82] In all cases the cofactors FAD, heme, and molybdopterin are bound to a single polypeptide chain with the molybdopterin domain near the N terminus, and the heme in the middle. The electron-accepting FAD domain is near the C terminus[83–85a] and is thought to transfer the two electrons through the following chain.

$$\text{NAD(P)H} \rightarrow \text{FAD} \rightarrow \text{Cyt } b \rightarrow \text{Mo-pterin} \rightarrow NO_3^- \quad (24\text{-}12)$$

Bacterial assimilatory nitrate reductases have similar properties.[86,86a] In addition, many bacteria, including *E. coli*, are able to use nitrate ions as an oxidant for **nitrate respiration** under anaerobic conditions (Chapter 18). The **dissimilatory nitrate reductases** involved also contain molybdenum as well as Fe–S centers.[85] The *E. coli* enzyme receives electrons from reduced quinones in the plasma membrane, passing them through cytochrome *b*, Fe–S centers, and molybdopterin to nitrate. The three-subunit $\alpha\beta\gamma$ enzyme contains cytochrome *b* in one subunit, an Fe_3S_4 center as well as three Fe_4S_4 clusters in another, and the molybdenum cofactor in the third.[87] Nitrate reduction to nitrite is also on the pathway of denitrification, which can lead to release of nitrogen as NO, N_2O, and N_2 by the action of **dissimilatory nitrite reductases**. These enzymes[87a] have been discussed in Chapters 16 and 18.

Assimilatory nitrite reductases of plants, fungi, and bacteria carry out the six-electron reduction of nitrite to ammonium ions (Eq. 24-13) using electron donors such as reduced ferredoxins or NADPH.

$$NO_2^- + 6\ e^- + 8\ H^+ \rightarrow NH_4^+ + 2\ H_2O \quad (24\text{-}13)$$

The enzymes from green plants and fungi are large multifunctional proteins,[80] which may resemble assimilatory sulfite reductases (Fig. 16-19). These contain **siroheme** (Fig. 16-6), which accepts electrons from either reduced ferredoxin (in photosynthetic organisms) or from NADH or NADPH. FAD acts as an intermediate carrier. It seems likely that the nitrite N binds to Fe of the siroheme and remains there during the entire six-electron reduction to NH_3. Nitroxyl (NOH) and hydroxylamine (NH_2OH) may be bound intermediates as is suggested in steps *a–c* of Eq. 24-14.

$$NO_2^- \xrightarrow[2H^+ \quad HO^-]{2e^- \ (a)} [HNO] \ (\text{Nitroxyl}) \xrightarrow[2H^+]{2e^- \ (b)} [NH_2OH] \xrightarrow[2H^+ \quad H_2O]{2e^- \ (c)} NH_3 \quad (24\text{-}14)$$

B. Incorporation of NH_3 into Amino Acids and Proteins

Prior to 1935, amino acids were generally regarded as relatively stable nutrient building blocks. That concept was abandoned as a result of studies of the metabolism of $^{15}NH_3$ and of ^{15}N-containing amino acids by Schoenheimer and Rittenberg[88] and more recent studies using ^{13}N by Cooper *et al.*[89,90] These investigations showed that nitrogen could often be shifted rapidly between one carbon skeleton and another. This confirmed proposals put forth earlier by Braunstein, Meister, and others who had pointed out that the C_4 and C_5 amino acids, aspartate and glutamate, which are closely related to the tricarboxylic acid cycle, are able to exchange their amino groups rapidly with those of other amino acids via transamination (Fig. 24-5, step *d*). Since ammonia can be incorporated readily into glutamate (Fig. 24-5, step *a*; see next section), a general means is available for the biosynthesis of amino acids. The citric acid cycle is able to provide any needed amount of 2-oxoglutarate for the synthesis of both glutamate and glutamine.[91–94]

Glutamine, and to a lesser extent asparagine, act as soluble, nontoxic carriers of additional ammonia in the form of their amide groups. An active synthase converts glutamate and ammonia to glutamine (Fig. 24-5, step *d*), and another enzyme transfers the amide nitrogen into aspartate, in an ATP-dependent reaction, to form asparagine (Fig. 24-5, step *e*). The amide nitrogen of glutamine is incorporated in a similar way into a great variety of other biochemical compounds, including carbamoyl phosphate (Fig. 24-5, step *f*; Section C,2), glucosamine (Eq. 20-5), NAD^+, *p*-aminobenzoate,

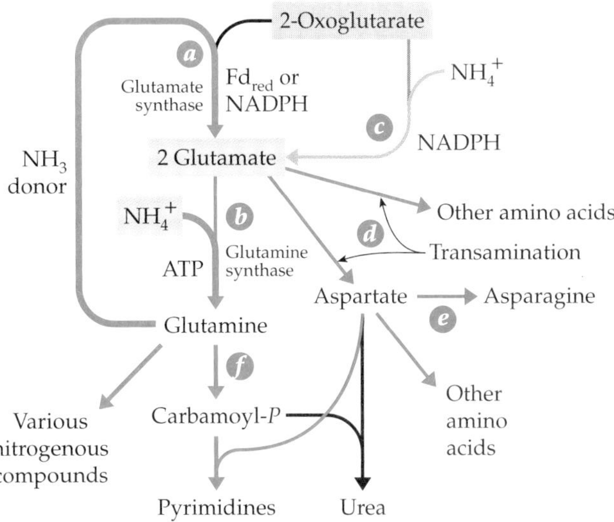

Figure 24-5 Major pathways of incorporation of nitrogen from ammonium ions into organic compounds, traced by green arrows.

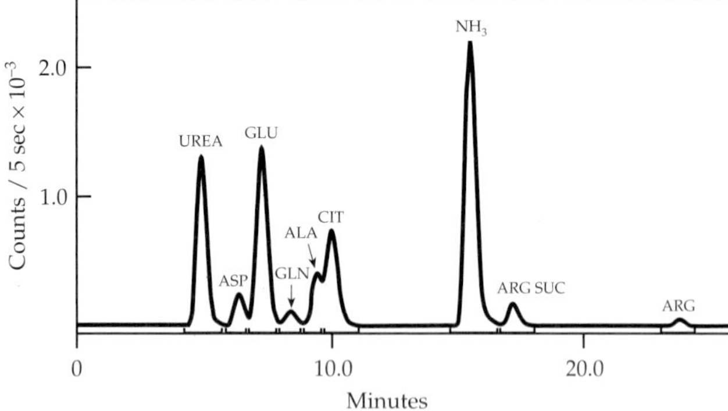

Figure 24-6 Elution profile of ^{13}N-containing metabolites extracted from liver 15 s after injection of $^{13}NH_3$ into the portal vein of an anesthetized adult male rat. CIT, citrulline; ARG SUC, argininosuccinate. From Cooper *et al.*[89]

histidine, CTP, and purines (Chapter 25). These reactions are catalyzed by a family of **amidotransferases**,[95–97d] which hydrolyze the glutamine to glutamate and NH_3. The last is entrapped until it reacts with the second substrate. **Asparagine synthetase** apparently first binds ATP and aspartate, which probably react to form β-aspartyl adenylate (β-aspartyl-AMP). Glutamine then binds and is hydrolyzed.[98,98a]

NH_3, ^-OOC, H_3N^+, H, C, O, O, P, O^-, O, Adenine, OH OH

β-Aspartyl adenylate

The liberated NH_3 can attack the β-aspartyl-AMP as indicated in the accompanying diagram to form asparagine. However, it is also possible that NH_3 is transferred via covalently bonded complexes[99] and is never free NH_3. An asparagine synthetase that utilizes free NH_3 as a nitrogen donor is also present in many organisms.

A third mechanism of synthesis, which was only recently recognized, appears to provide the sole source of asparagine for many bacteria.[98b] The asparagine-specific transfer RNA $tRNA^{Asn}$ is "mischarged" with aspartic acid to form Asp-$tRNA^{Asn}$. This compound is then converted to the properly aminoacylated Asn-$tRNA^{Asn}$ by a glutamine-dependent amidotransferase. (The entire ATP-dependent sequence is shown in Eq. 29-6.) The activated asparaginyl group is then transferred from Asn-$tRNA^{Asn}$ into proteins as they are synthesized.

Allantoin $\xrightarrow{^-OH}$ Allantoate (24-15)

Most green plants transport nitrogen from roots to growing shoots as asparagine. However, in peanuts **β-methyleneaspartate** is the major nitrogen carrier,[100] and in some legumes, including soybeans, **allantoin** and **allantoate** (Eq. 24-15) play this role. Allantoin arises from hydrolysis of purines (see Fig. 25-18), which are synthesized in root nodules of nitrogen-fixing plants.[101]

Glutamate, glutamine, and aspartate also play central roles in *removal* of nitrogen from organic compounds.[102] Transamination is reversible and is often the first step in catabolism of excess amino acids. 2-Oxoglutarate is the recipient of the nitrogen, and the glutamate that is formed can be deaminated to form ammonia which can then be incorporated into glutamine. Glutamate can also donate its nitrogen to form aspartate. In the brain glutamate is a major neurotransmitter but is toxic in excess. The astrocyte glial cells take up glutamate from the synaptic clefts between neurons, converting it to glutamine, which is then released into the extracellular space for reuptake by neurons.[103,104] In the animal body both aspartate and glutamine (via carbamoyl phosphate) are precursors of **urea**, the principal nitrogenous excretion product. These relationships are also summarized in Fig. 24-5, and details are provided in later sections.

While reductive amination of glutamate via glutamate synthase appears to be the major pathway for incorporation of nitrogen into amino groups, some direct amination of pyruvate and other 2-oxoacids in reactions analogous to that of glutamate dehydrogenase occurs in bacteria.[105,106] Another bacterial enzyme catalyzes reversible addition of ammonia to fumarate to form aspartate (p. 685).

An initially surprising conclusion drawn from the studies of Schoenheimer and Rittenberg was that proteins within cells are in a continuous steady state of synthesis and degradation. The initial biosynthesis, the processing, oxidative and hydrolytic degradative reactions of peptides, and further catabolism of amino acids all combine to form a series of metabolic loops as discussed in Chapter 17 and dealt with further in Chapters 12 and 29. Within cells some proteins are degraded much more rapidly than others, an important aspect of metabolic control. This is accomplished with the aid of the ubiquitin system (Box 10-C) and proteasomes (Box 7-A).[107] Proteins secreted into extracellular fluids often undergo more rapid turnover than do those that remain within cells.

1. Uptake of Amino Acids by Cells

While cells of autotrophic organisms can make all of their own amino acids, other organisms utilize many preformed amino acids. Human beings and other higher animals require several **essential amino acids** in their diets. Additional amounts of "nonessential" amino acids are also needed. It is true that amino groups can be transferred from one carbon skeleton to another among most of the amino acids. However, the body must take in enough amino groups to supply its need for all of the 20 amino acid components of proteins.[107a,b] Because of an unfavorable equilibrium constant, and the normally low concentration of NH_4^+, glutamate dehydrogenase (step *c* in Fig. 24-5) does not normally synthesize glutamate in the animal body. Its function is to deaminate excess glutamate. Furthermore, cells of some tissues take up amino acids that are made in other tissues. The active transport systems of bacteria have been described in Chapter 8. In mammals the absorption of amino acids takes place through epithelial cells of the intestinal tract, kidney tubules, and the brain (blood–brain barrier). Both Na^+-dependent transport (as for sugars; see Chapter 8) and Na^+-independent processes occur.[107c] Among the latter is the proposed **γ-glutamyl cycle,** which is described in Box 11-B. The cycle makes use of the γ-carboxyl group of glutamate, the same carboxyl that carries ammonia in the form of glutamine. Glutathione supplies the activated γ-glutamyl group. The amino acid to be transported reacts on the membrane surface by **transpeptidation**[108–109a] to form a **γ-glutamylamino acid** which enters the cytoplasm. It releases the free amino acid through an internal displacement by the free amino group of the glutamyl group. The natural tendency of the 5-carbon glutamate to undergo cyclization is used to provide the driving force for release of the bound amino acid. The cyclic product 5-oxoproline is then opened hydrolytically in an ATP-requiring reaction.[110] Cysteinylglycine formed in the initial transpeptidation is hydrolyzed by a peptidase, and glutathione is regenerated in two ATP-dependent steps as indicated in the scheme in Box 11-B.

The significance of the γ-glutamyl cycle is not fully understood. However, the finding of a mentally retarded individual who excretes 25–50 g / day of 5-oxoproline in the urine (possibly because of a defective 5-oxoprolinase) suggests that the pathway is a very active one.[111] A few persons deficient in γ-glutamyl transpeptidase have been found. They excrete glutathione and have a variety of medical problems.[109]

2. Glutamate Dehydrogenase and Glutamate Synthase

In animal tissues and in some bacteria the **glutamate dehydrogenase** reaction (Fig. 24-5, step *c*; see also Chapter 15)[112–115] provides a means of incorporating ammonia reversibly into glutamic acid. In eukaryotic cells the allosteric enzyme is found largely in the mitochondria.[115a] Glutamate dehydrogenase is also found in chloroplasts where it may function in glutamate synthesis when ammonia is present in excess.[116] The action of aminotransferases, both within and without mitochondria, distributes nitrogen from glutamate into most of the other amino acids. Especially active is aspartate aminotransferase (Eq. 14-24; Fig. 24-5, step *d*) which equilibrates aspartate and oxaloacetate with the 2-oxoglutarate–glutamate couple. However, the body obtains glutamate, as well as other amino acids, from foods, the initial source being largely green plants.

In plants as well as in *E. coli* and many other bacteria most glutamate is formed by **glutamate synthase**, which carries out reductive amination of 2-oxoglutarate (Fig. 24-5, reaction *a*). Glutamate synthase (also called **GOGAT**) utilizes the amide side chain of glutamine as the nitrogen donor. It is one of the previously mentioned amidotransferases in which glutamine is hydrolyzed to glutamate and NH_3 within the active site of the enzyme. Formation of a Schiff base and reduction probably occurs as in the reverse of reaction B of Table 15-1. However, one of the two glutamate molecules formed in reaction *a* of Fig. 24-5 must be reconverted by glutamine synthase to glutamine with the utilization of a molecule of ATP (Fig. 24-5, step *b*). Because of this coupling of ATP cleavage to the reaction the equilibrium in reaction *a* lies far toward the synthesis of glutamate. The low value of K_m for NH_4^+ that is characteristic of glutamine synthase favors glutamate synthesis even when little nitrogen is available.

Bacterial glutamate synthases are large oligomeric proteins containing flavin and Fe–S centers. That of *Azospirillum brasilense* consists of αβ units in which the 53-kDa β chains contain FAD and an NADPH binding site.[117–118a] The NADPH evidently transfers electrons to the FAD, which transfers them to an Fe_3S_4 center in the large 162-kDA α subunit. A molecule of bound FMN receives the electrons and reduces the iminoglutarate to glutamate. The site of binding and hydrolysis of glutamine is also present in subunit α. Chloroplasts of higher plants contain two glutamate synthases. One resembles the bacterial enzyme and utilizes NADPH as the reductant. The other requires reduced ferredoxin.[118–120]

Bacteria utilize both D-alanine and D-glutamate in the synthesis of their peptidoglycan layers (Fig. 8-29). Both D-amino acids are formed by racemases. That of alanine uses PLP (Chapter 15) but **glutamate racemase**[121–122a] does not. It may be able to remove the α-H of glutamate by utilizing the –COOH of the substrate, rather than the PLP ring, as an electron sink. Small amounts of D-amino acids occur also in animals.[123] Animal livers and kidneys contain **D-amino acid oxidase** and **D-aspartate oxidase,** which apparently function to metabolize D-amino acids from foods or those formed by brain activity (Chapter 30) or by aging.

3. Glutamine Synthetase

The formation of glutamine from glutamate (Eq. 24-16) also depends upon a coupled cleavage of ATP:

$$\text{Glu} \xrightarrow[\text{ATP} \quad \text{ADP} + \text{P}_i]{\text{NH}_4^+} \text{Gln} \qquad (24\text{-}16)$$

Glutamine synthase, as isolated from *E. coli*, contains 12 identical 51.6-kDa subunits arranged in the form of two rings of six subunits each with a center-to-center spacing of 4.5 nm. The units in one layer lie almost directly above those in the next,[104,124,125] the center-to-center spacing between the two layers is also 4.5 nm, and the array has 622 dihedral symmetry. The enzyme displays complex regulatory properties,[112,126–129] which are summarized in Fig. 24-7. The enzyme exists in two forms. **Active glutamine synthase** requires Mg^{2+} in addition to the three substrates glutamate, NH_4^+, and ATP. If the glutamate precursor, 2-oxoglutarate, is present in excess, the enzyme tends to remain in the active form because conversion to a modified form is inhibited; when the oxoglutarate concentration falls to a low value and glutamine accumulates, alteration is favored. The modifying enzyme **adenylyltransferase** (AT) in its active form AT_A transfers an adenylyl group from ATP to a tyrosine hydroxyl on glutamine synthase to give an adenylyl enzyme (GS-AMP). This **modified enzyme** requires Mn^{2+} instead of Mg^{2+} and is far more sensitive than the original enzyme to feedback inhibition by a series of end products of glutamine metabolism. All nine of the feedback inhibitors (serine, alanine, glycine, histidine, tryptophan, CTP, AMP, carbamoyl-*P*, and glucosamine 6-*P*) seem to bind to specific sites on the enzyme surface and to exert a cumulative inhibition. Serine, alanine, and glycine appear to be competitive inhibitors at the glutamate binding site.[130]

Relaxation of adenylylated glutamine synthase to the unmodified form is not catalyzed by a separate hydrolase but is promoted by a modified form of the adenylyltransferase AT_D. The active enzyme AT_A is

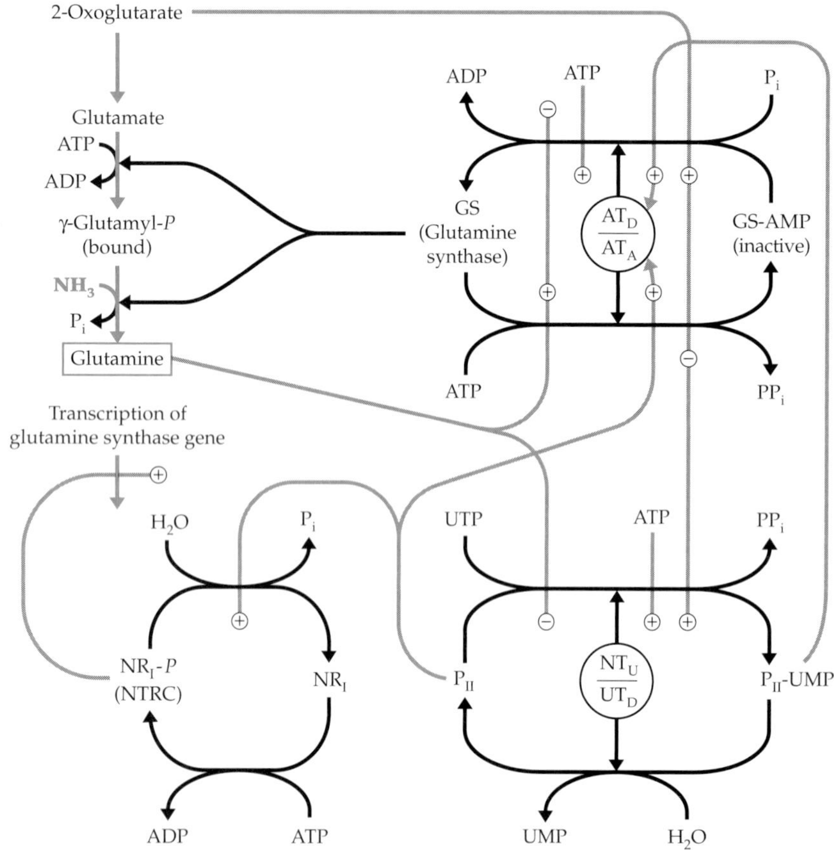

Figure 24-7 Regulation of glutamine synthase of *E. coli* using activation (+) and inhibition (–). Glutamine synthase (GS upper center) is converted by adenylylation of Tyr 397 into an inactive form GS-AMP by the action of an adenylyltransferase AT_A in complex with regulatory protein PII. PII is uridylylated at up to four sites by action of uridylyltransferase UT_U, which resides in the same polypeptide chain as a uridylyl removing enzyme UT_D (or UR). When PII carries a uridylyl group (PII-UMP), AT_A is transformed to AT_D, which reconverts the inactive GS-AMP to active GS by phosphorolytic removal of the adenylyl group. The ratios of AT_A/AT_D and UT_U/UT_D are controlled by the concentrations of the metabolites 2-oxoglutarate, a precursor, and glutamine, the immediate product. The amount of GS formed is controlled at the transcriptional level by an enhancer-binding transcription factor called NRI or NtrC (lower left). It is active when phosphorylated. Dephosphorylation of NRI-*P* is catalyzed by yet another protein and is stimulated by PII. Thus, PII both decreases synthesis of GS and promotes conversion of GS to its inactive form.

actually a complex AT • PII containing the regulatory protein PII. Subunit PII can be uridylylated on a tyrosine side chain by action of a 95-kDa **uridylyltransferase** (UT)[127,131] to form the modified glutamine synthetase AT • PII-UMP orAT_D. This form catalyzes phosphorolytic deadenylylation of glutamine synthetase, P_i displacing the adenylyl group to form ADP. Removal of the uridylyl group from PII-UMP is catalyzed by a fourth enzyme, UT_D (or UR), which is part of the same polypeptide chain as UT_U.[127] The cycle of interconversions of PII catalyzed by the UT_U and UT_D activities is shown at the lower right side of Fig. 24-7. From the allosteric modification reactions indicated by the gray lines, it is seen that glutamine not only promotes the adenylylation of glutamine synthetase but also inhibits the uridylylation of PII, thereby preventing AT_D from removing the adenylyl group from the synthetase. Furthermore, it allosterically inhibits the deadenylylation reaction itself. On the other hand, 2-oxoglutarate acts in the opposite way.

The glutamine synthase regulatory system has another important function. Protein PII stimulates the dephosphorylation of the enhancer-binding transcriptional regulator NRI-*P* (NtrC-*P*).[131,132] This slows transcription of the glutamine synthase gene (see Fig. 24-7) as well as a variety of other genes including those for the nitrogenase proteins in organisms that have them.[133] As a consequence, a deficiency of glutamine turns on a number of genes involved in nitrogen metabolism. Accumulation of glutamine promotes PII accumulation, modification of the synthase, and loss of gene activation.

Nitrogen can be transferred from glutamine into many other substrates.[102] Several antibiotic analogs of glutamine have been useful in studying these processes. Examples are the streptomyces antibiotics **L-azaserine** and 6-diazo-5-oxo-L-norleucine (DON).

L-Azaserine

These compounds act as alkylating agents; N_2 is released and a nucleophilic group from the enzyme becomes attached at the carbon atom indicated.[134] Other inhibitors bind noncovalently to form dead-end complexes.[134a]

4. Catabolism of Glutamine, Glutamate, and Other Amino Acids

Glutamine is hydrolyzed back to glutamate by glutaminases that are found both in eukaryotic tissues and in bacteria.[135,136] Liver contains an isozyme whose function appears to be to release NH_3 from glutamine for urea synthesis.[135,137] Glutamate dehydrogenase deaminates excess glutamate back to 2-oxoglutarate, which is degraded to succinyl-CoA and via β oxidation to malate, pyruvate, and acetyl-CoA. The last can reenter the citric acid cycle and be oxidized to CO_2 (Eq. 24-17). In fact, in mammalian tissues glutamate is essentially in equilibrium with 2-oxoglutarate and other citric acid cycle intermediates (see Box 17-C).

$$\text{Glu} \rightarrow \text{2-oxoglutarate} \rightarrow \text{succinyl-CoA} \rightarrow \rightarrow \rightarrow \text{malate} \rightarrow \text{pyruvate} \rightarrow \text{acetyl-CoA} \rightarrow CO_2 \quad (24\text{-}17)$$

Many other amino acids are degraded in similar ways. In most cases the sequence is initiated by transamination to the corresponding 2-oxoacid. Beta oxidation and breakdown to such compounds as pyruvate and acetyl-CoA follows.

Catabolism initiated by decarboxylation. An alternative pathway for glutamate degradation is through the γ-aminobutyrate shunt (Fig. 17-5). This pathway is initiated by a PLP-dependent decarboxylation rather than by a deamination or transamination. Since decarboxylases are known for most amino acids, there are usually alternative breakdown pathways initiated by decarboxylation. In many cases these pathways lead to important products. For example, γ-aminobutyrate functions in the brain as an important neurotransmitter. Dihydroxyphenylalanine is converted to noradrenaline and adrenaline, tryptophan to serotonin, and histidine to histamine. All of these are neurotransmitters (Chapter 30) and/or have other hormonal functions. A calmodulin-dependent glutamate decarboxylase occurs in higher plants, which accumulate γ-aminobutyrate in response to a variety of stresses.[138] However, the significance of this accumulation is unclear.

Fermentation of glutamate. Special problems face anaerobic bacteria subsisting on amino acids. Their energy needs must be met by balanced fermentations.[138a] For example, glutamate may be converted to CO_2, ammonia, acetate⁻, and butyrate⁻ according to the reactions of Fig. 24-8. The end result is described by Eq. 24-18.

$$2\ \text{Glutamate}^- + 2\ H_2O + H^+ \rightarrow 2\ CO_2 + 2\ NH_4^+ + 2\ \text{acetate}^- + \text{butyrate}^-$$
$$\Delta G'\ (\text{pH } 7) = -131\ \text{kJ} \quad (24\text{-}18)$$

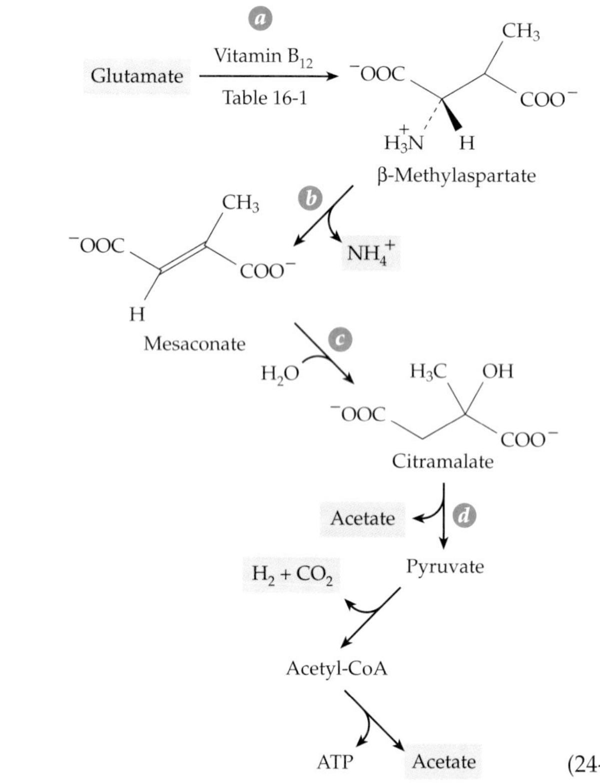

(24-19)

The sequence begins with the γ-aminobutyrate shunt reactions (Fig. 24-8, steps *a* and *b*), but succinic semialdehyde is reduced to γ-hydroxybutyric acid using the NADH generated in the trans-deamination process of step *c*. With the aid of a CoA-transferase (step *d*) two molecules of the CoA ester of this hydroxy acid are formed at the expense of two molecules of acetyl-CoA. Use is then made of a β,γ elimination of water (step *e*), analogous to that involved in the formation of vaccenic acid (Eq. 21-2). Isomerization (perhaps by the same enzyme that catalyzes elimination) forms crotonyl-CoA (step *f*). The latter undergoes dismutation, one-half being reduced to butyryl-CoA and one-half being hydrated and oxidized to acetoacetyl-CoA in the standard β-oxidation sequence. Acetoacetyl-CoA is cleaved to regenerate the two molecules of acetyl-CoA. The organism can gain one molecule of ATP through cleavage of the butyryl-CoA. Perhaps a second can be gained by oxidative phosphorylation between the NADH produced in the formation of acetoacetyl-CoA and the reduction of crotonyl-CoA to butyryl-CoA. The two processes take place at sufficiently different redox potentials to permit this kind of coupling.

Another fermentation of glutamate is initiated by the vitamin B_{12}-dependent isomerization of glutamate to β-methylaspartate (Eq. 24-19, step *a*).[138b] This rearrangement of structure permits α,β elimination of ammonia (step *b*), a process not possible in the original glutamate. Hydration to **citramalate** (step *c*) and aldol cleavage yields acetate and pyruvate. Acetate is one of the usual end products of the fermentation. The pyruvate can be cleaved to H_2, CO_2, and acetyl-CoA by the pyruvate-formate-lyase system (Fig. 15-16; Eq. 17-25), and cleavage of the acetyl-CoA can provide ATP. Alternatively, two molecules of acetyl-CoA can be coupled and reduced to butyryl-CoA. The reducing power generated in the cleavage of pyruvate is used to reduce crotonyl-CoA rather than being released as H_2. The stoichiometry is identical to that in Fig. 24-8. Still other fermentation mechanisms are used by some Clostridia to degrade glutamate.[138a] See study question number 16 at the end of this chapter.

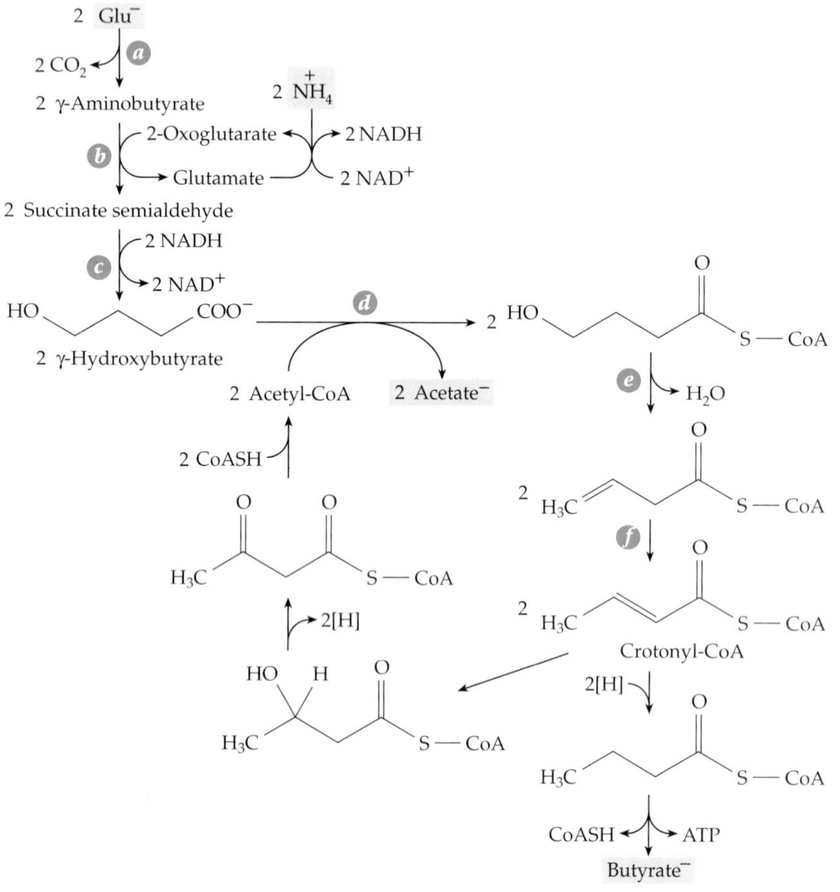

Figure 24-8 Fermentation of glutamate by *Clostridium aminobutylicum*.

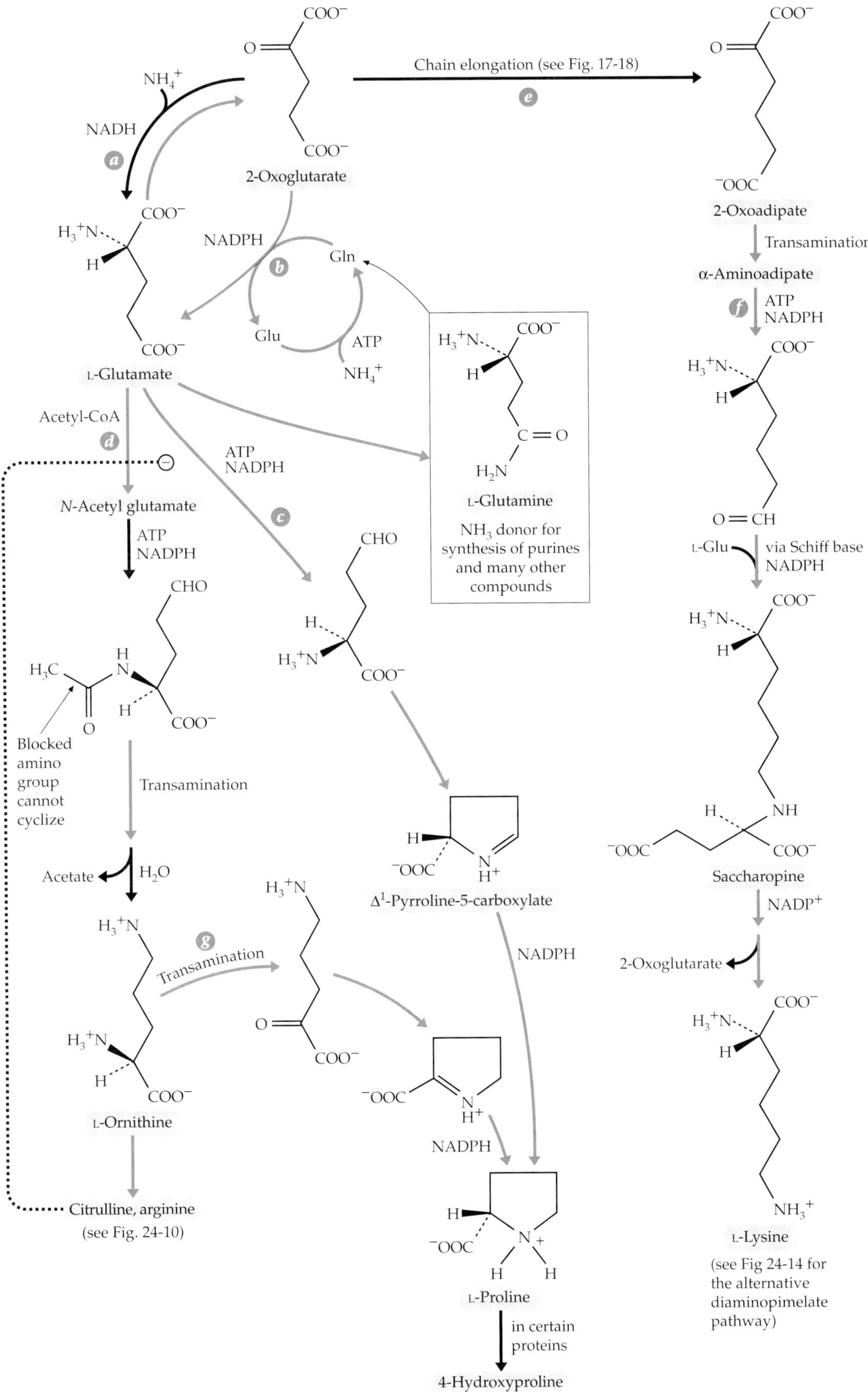

Figure 24-9 Biosynthesis of glutamate, glutamine, proline, and lysine from 2-oxoglutarate.

C. Synthesis and Catabolism of Proline, Ornithine, Arginine, and Polyamines

The 5-carbon skeleton of glutamic acid gives rise directly to those of proline, ornithine, and arginine. The reactions are outlined in Fig. 24-9. Arginine, in turn, is involved in the urea cycle, which is shown in detail in Fig. 24-10. Arginine is also a biosynthetic precursor of the polyamines. Another important biosynthetic product of glutamate metabolism is δ-aminolevulinate, a precursor to porphyrins (Eq. 24-44) in some organisms.[139]

1. Synthesis and Catabolism of Proline

The ATP-dependent reduction of the γ-carboxyl group of glutamate to an aldehyde by NADPH (reaction *c*, Fig. 24-9) is of a standard biosynthetic reaction type, the opposite of the oxidation reaction of Fig. 15-6. Like the latter it is thought to occur via an acyl phosphate intermediate.[140,141] The oxidation product, **glutamate semialdehyde**, cyclizes and can be converted to proline by further reduction (Fig. 24-9, step *c* and subsequent reactions in center of scheme). The pathway has been well established in bacteria and yeast by both biochemical and genetic experiments. In plants both the initial reduction and the cyclization are catalyzed by a bifunctional enzyme.[142,143] An alternative pathway important in animals is initiated by transamination of ornithine to the corresponding 2-oxoacid, spontaneous cyclization, and reduction to proline (Fig. 24-9, step *g*).[140,144] Selected prolines in collagen and in plant glycoproteins[145] are oxygenated to form 4-hydroxyproline (Eqs. 8-6, 18-51).

One route of catabolism of proline is essentially the reverse of its formation from glutamate. **Proline oxidase** yields Δ^1-pyrroline 5-carboxylate.[145a,b]

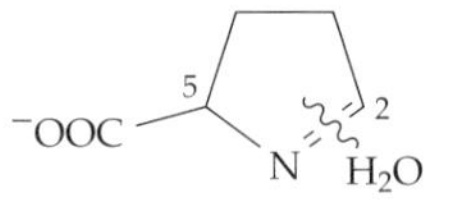

$^-$OOC, N, H_2O, 5, 2 — Δ^1-Pyrroline 5-carboxylate

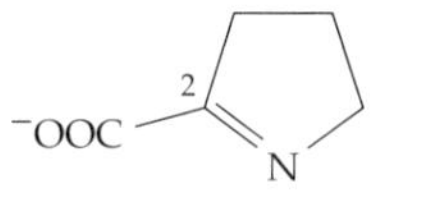

$^-$OOC, N, 2 — Δ^1-Pyrroline 2-carboxylate

The corresponding open-chain aldehyde, formed by hydrolysis, can be oxidized back to glutamate by pyrroline 5-carboxylate dehydrogenase.[145a–147] Lack of this enzyme is associated with the human genetic deficiency causing **hyperprolinemia**.[147–148a]

Alternatively, degradation can be initiated by oxidation on the other side of the ring nitrogen to form Δ^1-pyrroline 2-carboxylate. The metabolic fate of this compound is uncertain. A corresponding pathway for breakdown of 4-hydroxy-L-proline of collagen yields glyoxylate and pyruvate or malate and CO_2 (Eq. 24-20).[149] Oxidation on the other side of the ring nitrogen

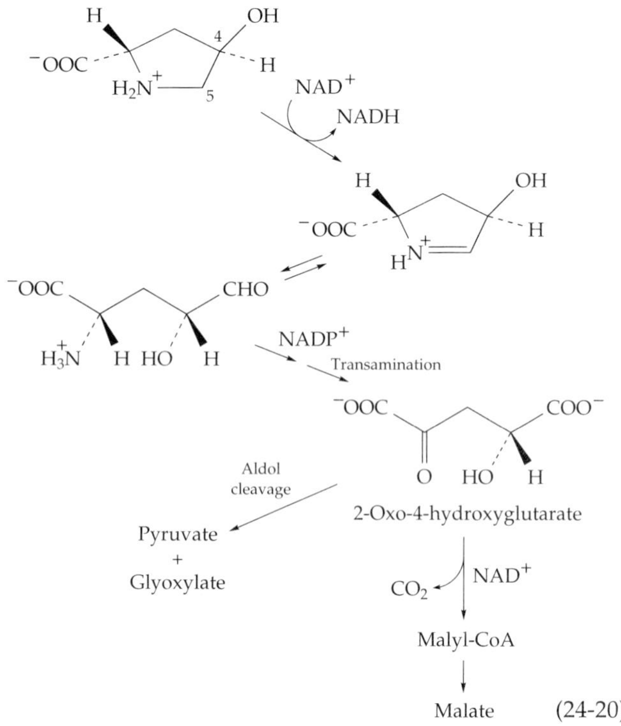

4-Hydroxyproline ($^-$OOC, H_2N^+, 4, 5, OH, H) → (NAD^+ → NADH) → ($^-$OOC, HN^+, OH, H) ⇌ ($^-$OOC, H_3N^+, CHO, H HO H) → ($NADP^+$, Transamination) → 2-Oxo-4-hydroxyglutarate ($^-$OOC, O, HO, H, COO$^-$)

2-Oxo-4-hydroxyglutarate → (Aldol cleavage) → Pyruvate + Glyoxylate

2-Oxo-4-hydroxyglutarate → (NAD^+; CO_2) → Malyl-CoA → Malate (24-20)

of hydroxyproline is utilized by some pseudomonads to convert the amino acid into 2-oxoglutarate. Anaerobic bacteria may reduce proline to 5-aminovalerate and couple this reaction to the oxidative degradation of another amino acid (Stickland reaction).

2. Synthesis of Arginine and Ornithine and the Urea Cycle

If the amino group of glutamate is blocked by acetylation prior to the reduction to the semialdehyde (Fig. 24-9, step *d*) cyclization is prevented. The γ-aldehyde group can be transaminated to an amino group and the acetyl blocking group removed to form **ornithine**. Ornithine is not usually a constituent of proteins, but it is sometimes formed by hydrolytic modification of arginine at specific sites in a protein. A 67-kDa urate-binding glycoprotein of plasma is reported to contain 43 residues of ornithine.[150,151] It is postulated that a special arginase is needed to form these residues, and that it may be lacking in some cases of gout in which the urate-binding capacity of blood is impaired. Ornithine appears to be present in specific sites in a few other proteins as well.[151] *Neurospora* grown in a minimal medium accumulates large amounts of both ornithine and arginine, over 98% of which is sequestered in vesicles within the cytoplasm.[152,153] This appears to be a way of accumulating a store of arginine that is protected from the active catabolism of that amino acid by the fungus. However, accumulation of ornithine in the human body, as a result of lack of **ornithine aminotransferase**

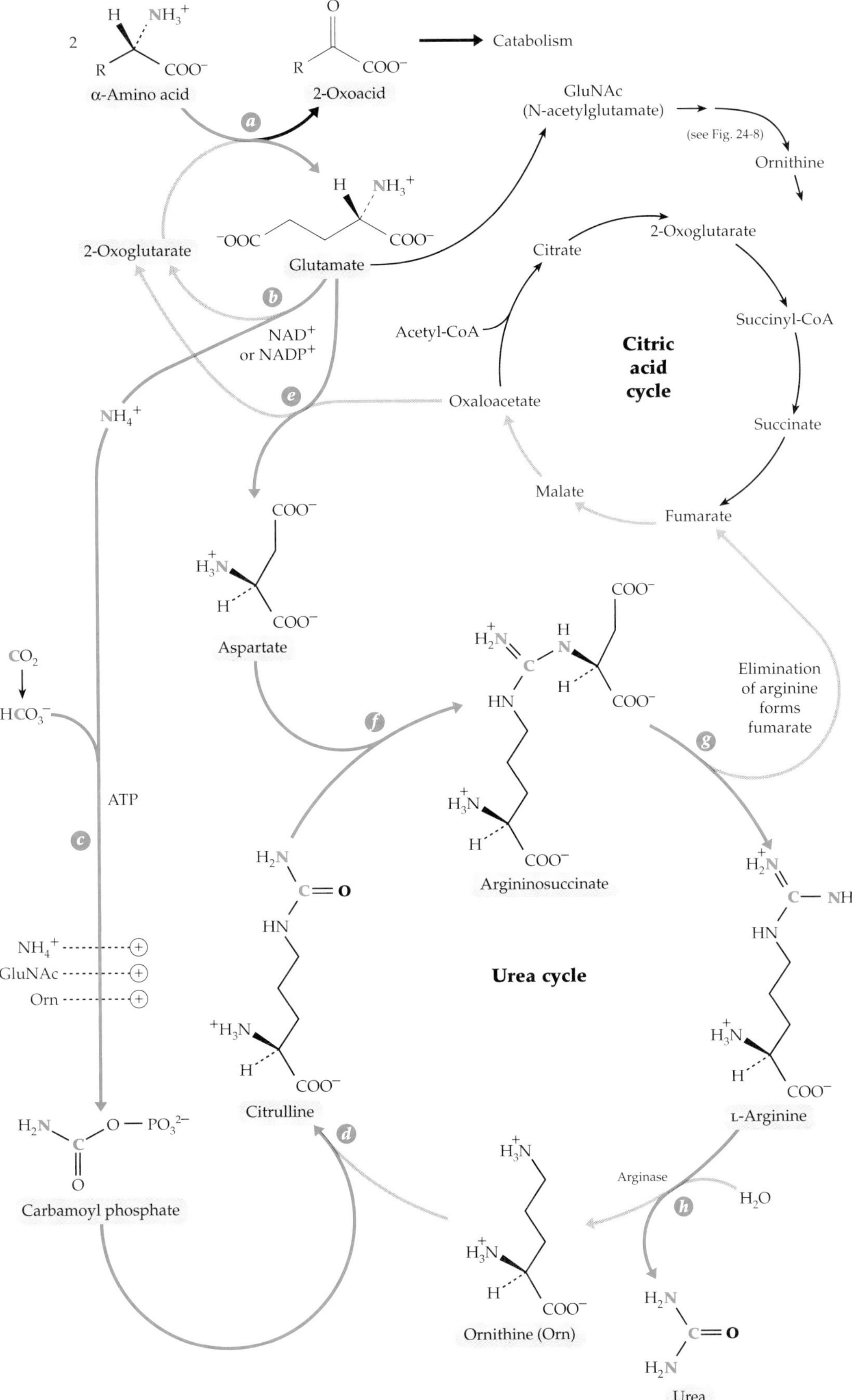

Figure 24-10 Biosynthesis of citrulline, arginine, and urea. The green arrows indicate reactions directly involved in deamination of amino acids and the synthesis of urea. N from amino acids and C from CO_2 are traced in green.

(Fig. 24-9, step *g*), causes gyrate atrophy of the choroid and retina, a disease that results in tunnel vision and blindness.[154,155] A major interest in arginine metabolism asises from its role in formation of urea in the human body. Study of arginine biosynthesis in bacteria has also been important in developing our understanding of regulation of gene expression.[156]

The urea cycle. In 1932, Krebs and Henseleit proposed that urea is formed in the liver by a cyclic process in which ornithine is converted first to **citrulline** and then to arginine.[157,158] The hydrolytic cleavage of arginine produces the urea and regenerates ornithine (Fig. 24-10, bottom). Subsequent experiments fully confirmed this proposal. Urea is the principal nitrogenous end product of metabolism in mammals and many other organisms, but the urea cycle reactions have other functions. As with the citric acid cycle, products other than urea can be withdrawn in any needed quantity. Most notably, the reactions of Fig. 24-10 provides for the biosynthesis of arginine in all organisms.[159,160] Also of physiological importance is the fact that the urea cycle involves both mitochondria and cytosolic enzymes.[161,162] This is illustrated in Fig. 24-11.

Let us trace the entire route of nitrogen removed by the liver from excess amino acids. Transaminases (step *a*, Fig. 24-10) transfer nitrogen to 2-oxoglutarate to form glutamate. Since urea contains two nitrogen atoms, two molecules of glutamate must donate their amino groups. One molecule is deaminated directly by glutamate dehydrogenase to form ammonia (step *b*). This ammonia is combined with bicarbonate (step *c*) to form carbamoyl phosphate, which transfers its carbamoyl group onto ornithine to form citrulline (step *d*). The second molecule of glutamate transfers its nitrogen by transamination to oxaloacetate (reaction *e*) to form aspartate. The aspartate molecule is incorporated intact into **argininosuccinate** by reaction with citrulline (reaction *f*). Undergoing a simple elimination reaction, the 4-carbon chain of argininosuccinate is converted to fumarate (step *g*) with arginine appearing as the elimination product. Finally, the hydrolysis of arginine (step *h*) yields urea and regenerates ornithine.

Carbamoyl phosphate synthetases. The first of the individual steps in the urea cycle is the formation of carbamoyl phosphate.[163] Carbon dioxide and ammonia equilibrate spontaneously with carbamic acid:

$$CO_2 + NH_4^+ \rightleftharpoons H_2N-C(=O)-OH + H^+ \quad \text{(24-21)}$$

Carbamic acid

Some bacteria have a kinase able to convert carbamate into carbamoyl phosphate starting with step *a* of Eq. 24-22. However, the equilibrium constant is low (0.04 at pH 9, 10°C), and it is now believed that carbamate kinase functions in the opposite direction, providing a means of synthesis of ATP for bacteria degrading arginine (Section C,5,d). The biosynthetic carbamoyl phosphate synthases harness the cleavage of *two* molecules of ATP to formation of one molecule of carbamoyl phosphate (reaction *c*, Fig. 24-10).[163] In bacteria such as *E. coli*, a single synthase provides carbamoyl phosphate for biosynthesis of both arginine and pyrimidines (Fig. 25-14). However, fungi and higher animals have at least two carbamoyl-*P* synthases. Synthase I provides substrate for formation of citrulline from ornithine (Fig. 24-10), while carbamoyl-*P* synthase II, which is part of a larger multifunctional protein,[164] functions in pyrimidine synthesis. Synthase I is found in mitochondria and synthase II in the cytoplasm. Mammalian carbamoyl phosphate sythase I consists of a single 160-kDa peptide.[163] A powerful allosteric effector for the liver synthase is **N-acetylglutamate** (Fig. 24-10), a precursor of ornithine.[165] The enzyme from certain marine elasmobranchs, such as the spiny dogfish *Squalus acanthias*, have carbamoyl-*P* synthase III, an enzyme with somewhat different molecular properties.[166] It probably functions in synthesis of urea, which is used by these animals to regulate osmotic pressure.[167,168] Synthase I utilizes only free NH_3. The others are amidotransferases and prefer glutamine as the ammonia donor. Carbamoyl-*P* synthase from *E. coli* consists of two subunits (~42 and 118 kDa, respectively) and can utilize *either* free ammonia or glutamine.[169,170] The light subunit has **glutaminase** activity; i.e., it is able to hydrolyze glutamine to ammonia. All of these synthatases presumably act by first phosphorylating bicarbonate to an enzyme-bound carboxyl phosphate,[163,171,172] which can then undergo a displacement of phosphate by NH_3 to give enzyme-

$$^-O-C(=O)-OH \xrightarrow[(a)]{ATP \to ADP} {}^{2-}O_3P-O-C(=O)-O^- \text{ (Carboxyl phosphate)}$$

Gln + H_2O → Glu + NH_3 (to carboxyl phosphate)

$$\xrightarrow[(b)]{\to P_i} H_2N-C(=O)-O^- \xrightarrow[(c)]{ATP \to ADP} H_2N-C(=O)-O-PO_3^{2-}$$

Carbamoyl phosphate

(24-22)

bound carbamate (Eq. 24-22, step *a*). Phosphorylation of the latter by ATP completes the reaction. In the single-chain enzymes the amidotransferase domain is at the N terminus.

Crystallographic study of a mutant form of the *E. coli* enzyme unable to act rapidly on glutamine showed that the latter released its ammonia to form a thioester with cysteine 269,[173,173a] suggesting a mechanism resembling that of serine proteases or papain (Chapter 12) for the glutaminase action. The X-ray crystallography also showed that the released NH_3 must travel 4.5 nm through the interior of the protein to the site of carbamate formation. The carbamate must travel another ~4.5 nm to the site from which carbamoyl phosphate is released.[170,172,173b] The C-terminal regions of the synthases undergo allosteric modification by a number of effectors.[163,173c] Both ornithine and IMP are activators for the *E. coli* enzyme, whereas UMP, a pyrimidine end product, exerts feedback inhibition. Phosphoribosyl pyrophosphate activates synthase II, and *N*-acetylglutamate activates the mammalian liver synthase I by binding near the C terminus.[165]

Citrulline and argininosuccinate. One NH_3 and one HCO_3^- for urea formation are provided by the carbamoyl group, which is transferred from carbamoyl-*P* to ornithine to form citrulline. The second nitrogen atom is transferred from glutamate to aspartate into argininosuccinate (steps *d* and *f*, Fig. 24-10). The equilibrium constant for ornithine transcarbamoylase (reaction *d*) is very high so that ornithine is completely converted to citrulline. The trimeric human enzyme is a trimer of 36-kDa subunits[174–175a] whose structural gene is on the X chromosome. Like many other mitochondrial matrix enzymes it is synthesized as a larger (40 kDa) pre-cursor, which enters the mitochondria in an energy-dependent process.[176] A genetic defect in this sex- linked gene is often lethal to boys, and even girls, heterozygous for the defect, sometimes have serious problems with accumulation of ammonia in the brain.[162,174,177]

The conversion of citrulline to argininosuccinate and the subsequent breakdown to fumarate and arginine take place in the cytosol (Fig. 24-11). The ureido

H_2N–$C(=^{18}O)$–NH–R $\xrightarrow[a]{ATP \rightarrow PP_i}$ HN=C(–NH–R)(–^{18}O–P(=O)(OH)–O–Adenosine), attacked by NH_2–Aspartate

$\xrightarrow{b}$ $^{-18}$O–P(=O)(OH)–O–Adenosine + HN=C(–NH–R)–NH–Asp (Argininosuccinate)

(24-23)

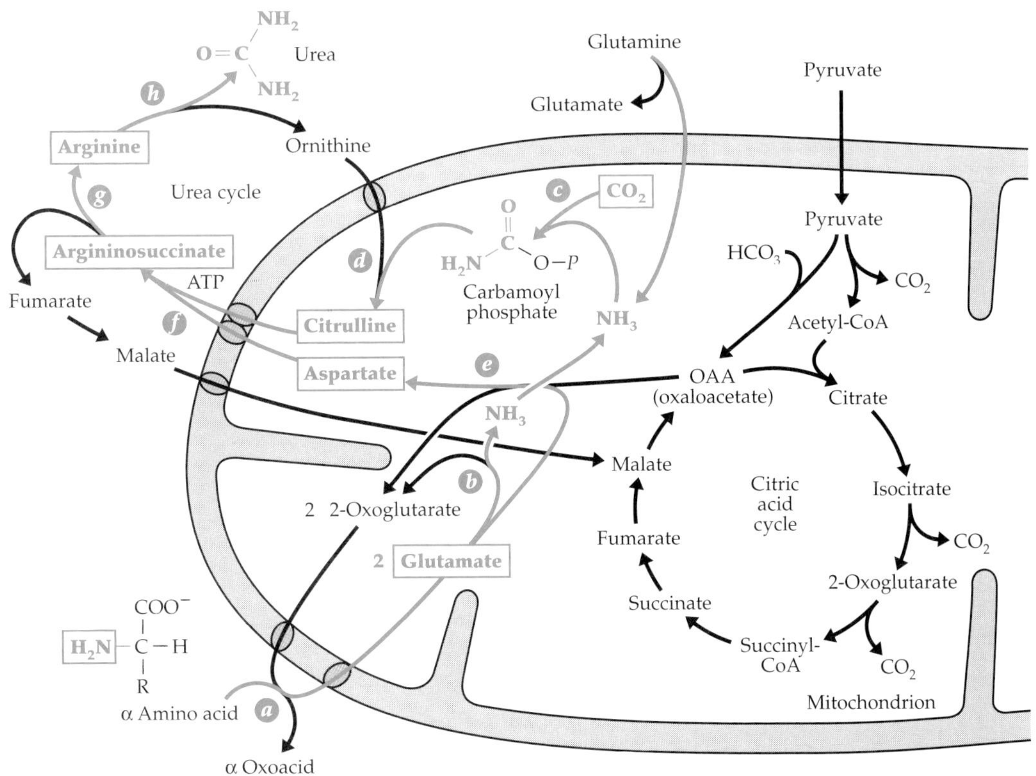

Figure 24-11 Integration of the urea cycle with mitochondrial metabolism. Green lines trace the flow of nitrogen into urea upon deamination of amino acids or upon removal of nitrogen from the side chain of glutamine.

group of citrulline is activated by ATP for the argininosuccinate synthase reaction (Eq. 24-23, step *a*). Thus, ^{18}O present in this group is transferred into AMP. A citrulline adenylate intermediate (center) is likely. **Argininosuccinate lyase** (reaction *g*, Fig. 24-10)[177a] catalyzes the elimination of arginine with formation of fumarate. It is entirely analogous to the bacterial aspartase that eliminates ammonia from aspartate to form fumarate.[178] Like the latter enzyme and fumarate hydratase (Chapter 13), argininosuccinase promotes a trans elimination.[179] The fumarate produced can be reconverted through reactions of the citric acid cycle to oxaloacetate, which can be reaminated to aspartate (Fig. 24-11). Aspartate is used to introduce amino groups in an entirely similar way in other metabolic sequences such as in the formation of adenylic acid from inosinic acid (Fig. 25-16).

The cleavage of arginine to ornithine and urea by the Mn^{2+}-containing **arginase** (Chapter 16)[180,181] converts the biosynthetic route to arginine into a cycle for the synthesis of urea. This cyclic pathway is unique to organisms that excrete nitrogenous wastes as urea, but the biosynthetic path is nearly ubiquitous.[182] Human adults excrete approximately 20 g of urea nitrogen per day. If this rate decreases, ammonia accumulates in the blood to toxic levels. Normally, plasma contains 0.03 mM ammonia, and only 2–3 times this level is required to produce toxic symptoms. Therefore, it is not surprising that five different well-documented hereditary enzyme deficiencies affect the urea cycle.[162,183] One of the most common, **argininosuccinic aciduria**, is a deficiency of the breakdown of argininosuccinic acid.[178] Both lethal and nonlethal variants of this disease are known. Human argininosuccinate lyase consists of two subunits. Defects may occur in either subunit but considerable genetic heterogeneity exists and intragenic complementation between the two subunits accounts for many of the nonlethal forms of the disorder.[177a,178a] A common feature of all of the hereditary defects of the urea cycle is an intolerance to high protein intake and mental symptoms. Toxic accumulation of ammonia in blood is often seen also in **alcoholic liver cirrhosis** as a result of a decreased capacity of the liver for synthesis of urea.

For some urea cycle defects a combination of a low-protein diet together with an arginine supplement prevents the ammonia intoxication while allowing normal growth. In other cases it is necessary to replace the natural dietary protein with a mixture of essential amino acids or with the corresponding 2-oxoacids, which can be converted to amino acids in the body with utilization of endogenous ammonia.[183] A specific treatment for lack of *N*-acetylglutamate synthetase, which forms the carbamoyl phosphate synthase activator *N*-acetylglutamate, is administration of the analog *N*-carbamoylglutamate. This also activates carbamoyl phosphate synthase and is not cleaved by acylases that would prevent the natural activator from being supplied artifically via the blood.[158]

Although the primary function of the urea cycle is usually regarded as the removal of NH_4^+ from the body, it also removes HCO_3^- in equal amounts (Eq. 24-24). This is essential for maintenance of neutral pH,

$$2\ HCO_3^- + 2\ NH_4^+ \longrightarrow H_2NCONH_2 + CO_2 + 3\ H_2O \qquad (24\text{-}24)$$

and Atkinson and Bourke suggested that removal of HCO_3^- is as important a function of the cycle as removal of NH_4^+.[184] However, there are strong arguments against this concept.[162]

Excretion of ammonia. While mammals excrete urea, many invertebrate organisms that live in water as well as some fishes simply excrete NH_3. Other organisms hydrolyze urea to NH_3. Even green plants recycle nitrogen via urea and the Ni^{2+}-dependent urease (Eq. 16-47). Two compounds that can be hydrolyzed by cells to urea and glyoxylate are allantoin and allantoic acid (Eq. 24-15). If cells of *Saccharomyces cerevisiae* are grown on either of these compounds as a sole source of nitrogen, they make a biotin-dependent **urea carboxylase** (Eq. 24-25). This enzyme facilitates the hydrolysis of urea by conversion to the more easily degraded allophanate (Eq. 24-25).

(24-25)

Catabolism of arginine. Arginine can also be converted back to glutamate and 2-oxoglutarate. The initial step is removal of the guanidino group to form ornithine. This occurs in the urea cycle and also in many bacteria[185] by the action of arginase (Fig. 24-5, step *h*). A parallel pathway involving conversion of arginine to N^1-succinylarginine, then on to succinylglutamate, and to free glutamate and succinate is used by some pseudomonads.[186] The alternative **arginine**

dihydrolase pathway, used by some bacteria and a few protozoa such as *Giardia*, is initiated by a different hydrolase that cleaves arginine to citrulline and ammonia.[187] Phosphorolysis of citrulline yields carbamoyl phosphate whose breakdown to CO_2 and ammonia (catalyzed by carbamate kinase, Eq. 24-22) can be utilized for generation of ATP by microorganisms that subsist on arginine.[188]

Degradation of L-arginine by *Streptomyces griseus* is initiated by a hydroxylase that causes decarboxylation and conversion of the amino acid into an amide (Eq. 24-26), a reaction analogous to that catalyzed by the flavin-dependent lysine oxygenase (Eq. 18-41). The

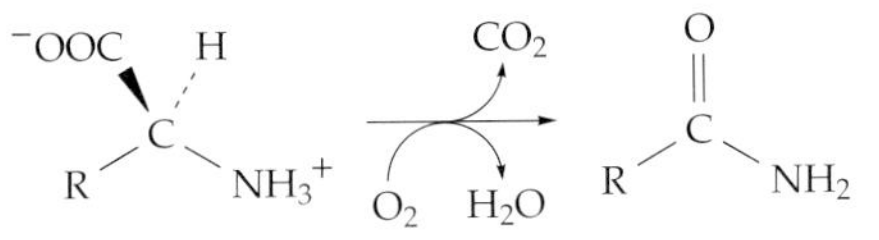

$$R{-}CH(COO^-)(NH_3^+) \xrightarrow[O_2 \;\; H_2O]{CO_2} R{-}C({=}O){-}NH_2 \quad (24\text{-}26)$$

product formed from arginine is γ-guanidinobutyramide, which is further degraded by the hydrolysis of the amide group and cleavage of the guanidino group to form urea and γ-aminobutyrate. *Pseudomonas putida* initiates degradation of arginine by decarboxylation to the corresponding 2-oxoacid and oxidative decarboxylation with a thiamin diphosphate-requiring enzyme to γ-guanidinobutyraldehyde. Dehydrogenation and hydrolysis lead, again, to γ-aminobutyrate.[189]

Specific arginine residues in proteins are methylated on their guanidino groups to give monomethylated and both symmetrically and asymmetrically dimethylated derivatives.[190,191] These methylated arginines also occur free in various mammalian tissues, where they may serve as endogenous regulators of nitric oxide synthases. A Zn^{2+}-containing dimethylarginase hydrolyzes the monomethyl and dimethyl arginines to citrulline and monomethyl or dimethyl amines.[191]

Insecticidal analogs of arginine. The toxic amino acid **L-canavanine** is synthesized by more than 1500 species of legumes including alfalfa and clover.[192–194] It is structurally similar to arginine, the 5-CH_2 group being replaced by O. However, the guanidino group is much less basic than in arginine. Canavanine is a natural insecticide, which in some plants accumulates to a level of 13% of the total dry matter.[193] Plants that store canavanine hydrolyze it to **canaline** and urea, which they use as a nitrogen source. Canaline is a toxic derivative of hydroxylamine and forms oximes with 2-oxoglutarate, other oxoacids, and PLP-containing enzymes. Although canavanine and canaline are effective insecticides, some beetles are adapted to these compounds to the extent that they feed exclusively on canavanine-containing seeds. The tobacco budworm is likewise resistant to these toxins and produces a **canavanine hydrolase** that converts canavanine to L-homoserine, a normal intermediate in

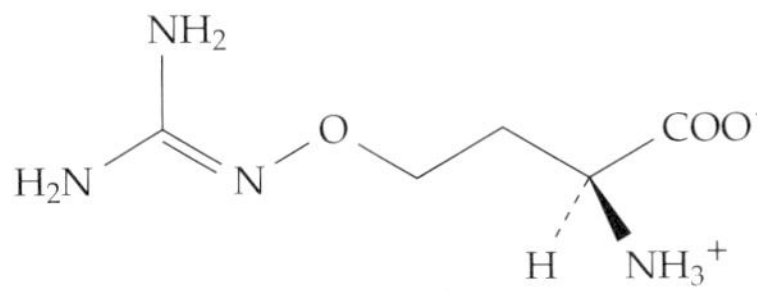

$$H_2N{-}C(NH_2){=}N{-}O{-}CH_2CH_2{-}CH(NH_3^+){-}COO^-$$

L-Canavanine

$$H_2N{-}O{-}CH_2CH_2{-}CH(NH_3^+){-}COO^-$$

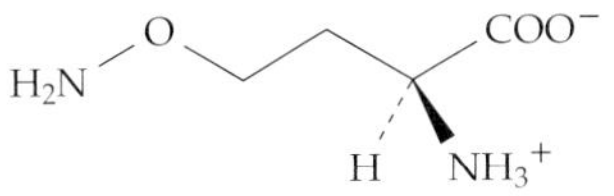

L-Canaline

the threonine, isoleucine, methionine biosynthetic pathway (Fig. 24-13), and **hydroxyguanidine**. The latter undergoes NADH-dependent reduction to guanidine[194] which can be catabolized.

3. Amidino Transfer and Creatine Synthesis

The terminal amidino group of arginine is transferred intact to a number of other substances in simple displacement reactions. An example is the formation of **guanidinoacetic acid** (Eq. 24-27, steps *a* and *b*). The amidino group appears to be transferred first to the SH group of cysteine 407 then to glycine in a double displacement mechanism.[195–197] Transmethylation from *S*-adenosylmethionine (Eq. 24-27, step *c*) converts guanidinoacetic acid to **creatine**, a compound of special importance in muscle. Creatine kinase reversibly transfers the phospho group of ATP to creatine to form the *N*-phosphate (Eq. 24-27, step *d*). **Creatine phosphate**, and in some invertebrates phosphoarginine,[198] serves as an important "energy buffer" for muscular contraction (Chapter 19). Through the reversible action of creatine kinase it is able rapidly to transfer its phospho group back onto ADP as fast as the latter is formed during the hydrolysis of ATP in the contraction process. An end product of creatine phosphate metabolism is the anhydride **creatinine** formed from creatine phosphate as is indicated in Eq. 24-27, step *e* as well as directly from creatine. The urinary creatinine excretion for a given individual is extremely constant from day to day, the amount excreted apparently being directly related to the muscle mass of the person. Another example of the transfer of amidino groups from arginine is found in the synthesis of streptomycin (Box 20-B).

A cyclic analog of creatine, **cyclocreatine**, when fed to animals, accumulates in large amounts in muscle, heart, and brain and is a long-acting phosphagen.[199]

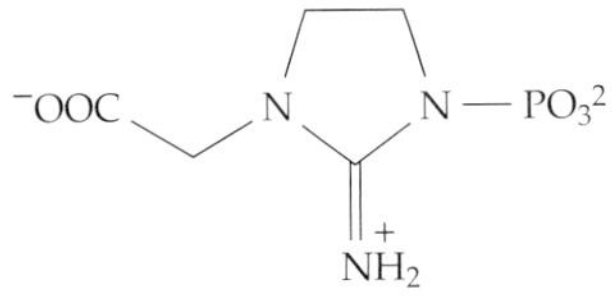

$$^-OOC{-}CH_2{-}N\langle\text{ring}\rangle N{-}PO_3^{2-}, \; C{=}NH_2^+$$

Cyclocreatine phosphate

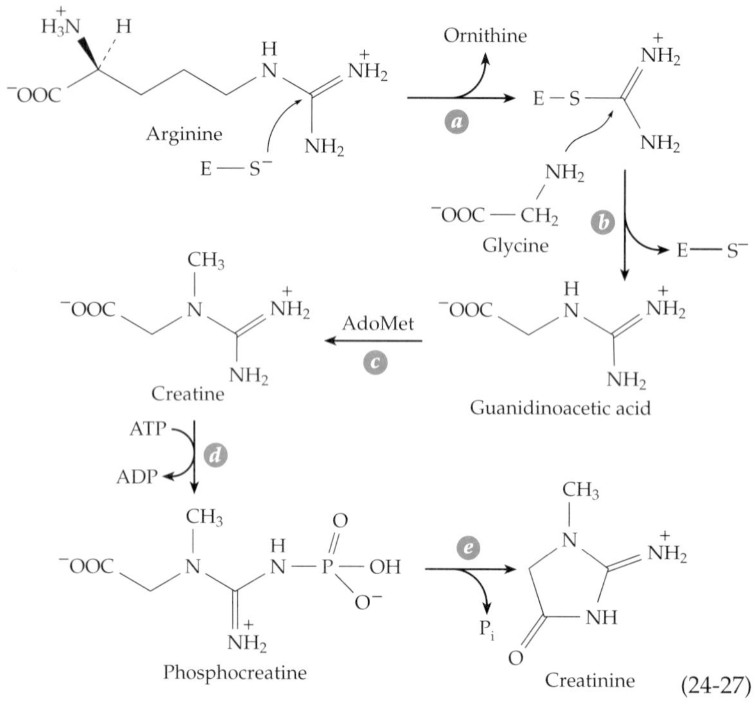

Arginine → (a, E—S⁻; Ornithine released) → E—S–amidino enzyme → (b, Glycine; E—S⁻ released) → Guanidinoacetic acid → (c, AdoMet) → Creatine → (d, ATP → ADP) → Phosphocreatine → (e, P_i) → Creatinine

(24-27)

several functions. They can substitute to some extent for cellular K^+ and Mg^{2+}, and they may play essential controlling roles in nucleic acid and protein synthesis. A specific role of spermidine in cell division seems likely.[207,207a] An absolute requirement for polyamines has been demonstrated for some bacteria such as *Hemophilus parainfluenzae*[208] and for mutants of *Aspergillus* and *Neurospora*. Polyamines are also essential for mammalian cells. Polyamines activate some enzymes including the serine/threonine protein kinase CK2.[209]

Mutants of *E. coli* have been constructed in which enzymes of all known biosynthetic pathways for polyamines are blocked by deletion of the genes for arginine decarboxylase (*SpeA*), agmatine ureahydrolase (*SpeB*), ornithine decarboxylase (*SpeC*), and adenosylmethionine decarboxylase (*SpeD*).[210] Even though polyamines cannot be detected in these cells they grow at one-third the normal rate. However, yeast cells require both putrescine and spermidine or spermine for growth.[211,211a] Another effect is seen in strains of yeast carrying the **"killer plasmid,"** a 1500-kDa double-stranded RNA plasmid that encodes a toxic protein, which is secreted and kills other susceptible strains of yeast. Yeast cells carrying the killer plasmid lose it when made deficient in polyamines.[212] The bacterial outer membrane porins OmF and OmC (Fig. 8-20) bind polyamines, especially spermine, and inhibit passage of ions. Polyamines may also modulate ion channels of heart, muscle, and neurons.[213] Both prokaryotic and eukaryotic cells have transporters that allow uptake of polyamines from their surroundings.[214,215]

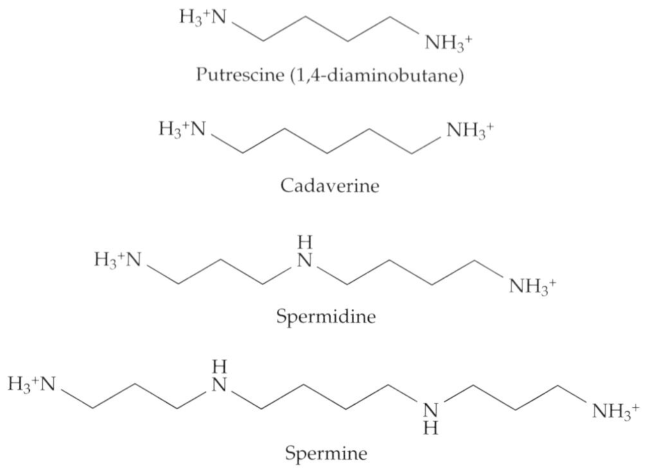

4. The Polyamines

A series of related polyamino compounds, which are derived in part from arginine, are present in all cells in relatively high, often millimolar, concentrations.[200–203] The content of polyamines in cells tends to be stoichiometric with that of RNA, and the polyamines are concentrated in the ribosomes and also in the nucleus. Two moles of polyamine are usually present per mole of any isolated tRNA.[202,204] The first satisfactory crystals of tRNA for X-ray structure determination were obtained in the presence of spermine. Spermidine is associated with RNA in the turnip yellow mosaic virus.[205] The T-even bacteriophage and most bacteria contain polyamines in association with DNA. Polyamines are able to interact with double helical nucleic acids by bridging between strands, the positively charged amino groups interacting with the phosphates of the nucleic acid backbones. Tsuboi suggested that the tetramethylene portion of the polyamine lies in the minor groove bridging three base pairs, and the trimethylene portions (one in spermidine, two in spermine) bridge adjacent phosphate groups in one strand.[206] Polyamines may also stabilize supercoiled or folded DNA.

The structures of polyamines are shown here as di- and tri-cations, but it should be realized that there are multiple positions for protonation and therefore various tautomers. Also, polyamines show extreme anticooperativity in proton binding, i.e., successive pK_a values range from very low to very high for the last proton to leave. Polyamines are thought to have

Putrescine (1,4-diaminobutane)

Cadaverine

Spermidine

Spermine

Major naturally occurring polyamines (as dications)

4-Aminobutylcadaverine

Silaffin (Lys) — n = 6-11

Polypeptide-bound polyamines of diatoms

A pentamine from *Thermus*, as tri-cation

Hydroxylamine-containing polyamine from snake venom

Biosynthesis. The 4-carbon putrescine arises most directly by decarboxylation of ornithine (Fig. 24-12, step *b*),[216] but it can also be formed by decarboxylation of arginine to agmatine followed by hydrolysis of the latter (Fig. 24-12, steps *c,d*). An alternative pathway utilizes an "agmatine cycle" in which agmatine is first hydrolyzed to ammonium ions and *N*-carbamoylputrescine. The latter transfers its ureido group to ornithine to form citrulline and releases free putrescine (Fig. 24-12, steps *f,g*). The citrulline is reconverted to arginine. This pathway appears to be important in plants.[217] Putrescine is normally present in all cells, and all cells are able to convert it on to spermidine. This is accomplished by decarboxylation of *S*-adenosylmethionine (Fig. 24-12, step *a*) and transfer of the propylamine group from the resulting decarboxylation product onto an amino group of putrescine (Fig. 24-12, step *h*).[218–221]:

The more complex spermine is found only in eukaryotes. It is formed by transfer of a second propylamine group onto spermidine (Fig. 24-12, step *i*). A historical note is that Anthony von Leeuwenhoek with one of his first microscopes observed crystals of the phosphate salt of spermine in human semen in 1678. The 5-carbon diamine cadaverine arises from decarboxylation of lysine (Fig. 24-12, step *d*). The extremely thermophilic bacterium *Thermus thermophilis* produces several additional polyamines including a pentamine (see above), a quaternary nitrogen compound.[222] Many other polyamines are known.[223–224c] Among these are a 4-aminobutylcadaverine isolated from root nodules of the adzuki bean[223] and very long partially aromatic hydroxylamine derivatives from venom of common funnel-web spiders (structures at top of page).[225] Cationic polypeptides called **silaffins**, with masses of ~3 kDa, apparently initiate the growth of the silica cell walls of diatoms (Box 4-B). These peptides contain polyamines consisting of 6 to 11 repeated *N*-methylpropylamine units covalently attached to lysine residues[224a,224b] and also many phosphoserines.[224d]

The synthesis of polyamines is tightly regulated. The PLP-dependent ornithine decarboxylase is present in very low concentrations[226] and apparently has the shortest half-life (~10 min) of any mammalian

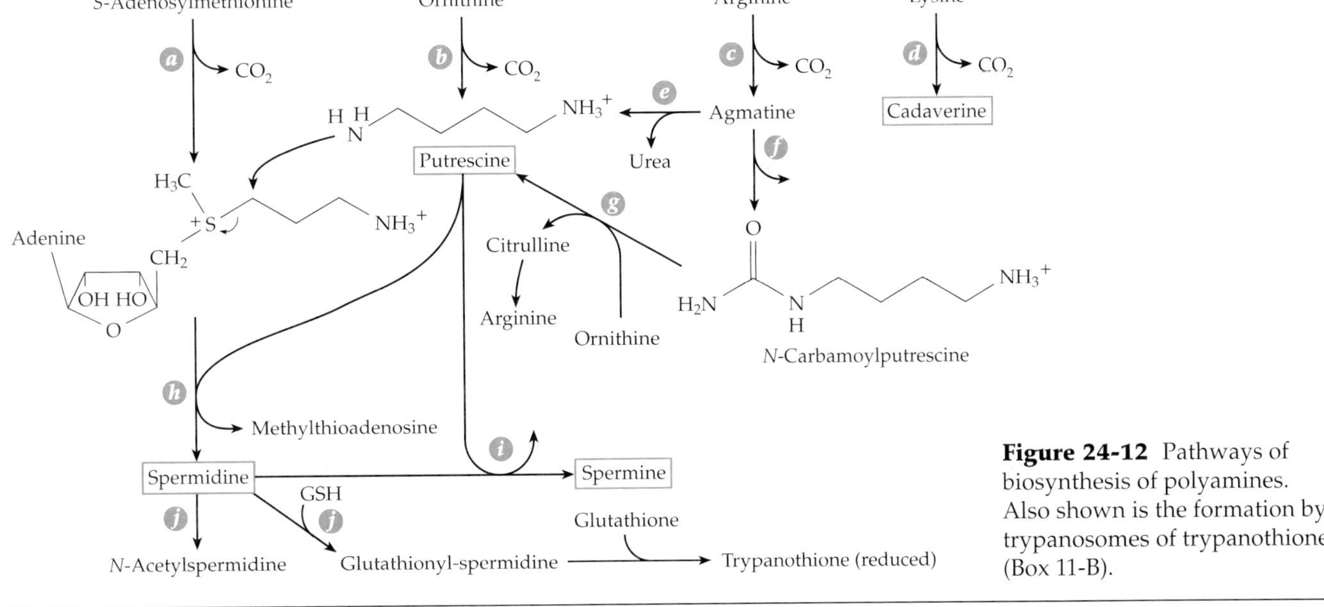

Figure 24-12 Pathways of biosynthesis of polyamines. Also shown is the formation by trypanosomes of trypanothione (Box 11-B).

enzyme.[227,228] Its concentration increases rapidly in most species with the onset of rapid growth, transformation to a neoplastic state, or initiation of cell differentiation. The rate of synthesis of the enzyme appears to be regulated by feedback repression by spermidine and by inactivation in response to a buildup of putrescine.[229] One mechanism of inactivation is the synthesis of a 26-kDa specific inhibitor called an **antizyme** in response to the presence of putrescine, spermidine, or spermine. The antizyme is ubiquitous in both prokaryotes and eukaryotes and keeps most of the ornithine decarboxylase bound and inactive and also promotes its degradation by 26S proteosomes.[230–230b] A polyamine-dependent protein kinase in *Physarum* phosphorylates the decarboxylase thereby inhibiting its activity.[231]

Breakdown. The catabolism of polyamines is less well understood than is their biosynthesis. Oxidative cleavages of spermine to spermidine and of the latter to 1,4-diaminobutane appear to occur in the animal body, and a substantial amount of this diamine is excreted in the urine.[201] Spermidine is acetylated on N^1 by acetyl-CoA and a spermidine *N*-acetyltransferase.[232,233] The resulting N^1-acetylspermidine is more readily cleaved by hepatic polyamine oxidase[233a] than is free spermine; again 1,4-diaminobutane is reformed together with an *N*-acetylaminopropionaldehyde (Eq. 24-28). This and other aldehydes formed from polyamines are very toxic but they may play essential roles in regulation of metabolism.[201] Transamination of 1,4-diaminobutane yields γ-aminobutyraldehyde which cyclizes (Eq. 24-29). Diamine oxidases of animal tissues oxidize 1,4-diaminobutane with formation of the same products.[234] Further metabolism of Δ^1-pyrroline yields γ-aminobutyrate, which can undergo transamination and oxidative metabolism as shown in Fig. 17-5. Other products[235] are also indicated in Eq. 24-29. Metabolism of other polyamines also begins by oxidation at the primary amino termini.[236] Formation of β-alanine, needed for synthesis of pantothenic acid, can also occur by oxidation of spermine.[236a]

N-Acetylspermidine

(a) −2 [H]

(b) → 1,4-Diaminobutane

N-Acetylaminopropionaldehyde (24-28)

(a) → Δ^1-Pyrroline

H_2O (b)

2[H] → 4-Aminobutanol

(c) −2[H]

2-Pyrrolidone → γ-Aminobutyrate; 5-Hydroxy-2-pyrrolidone (24-29)

When *E. coli* cells enter the stationary phase of the growth curve (Box 9-B), most of the spermidine is converted to **glutathionylspermidine** (γ-glutamylcysteinylglycylspermidine) in which glutathione and spermidine are joined by an amide linkage.[237–239] Trypanosomes join a second glutathione at the other end of the spermidine to form reduced **trypanathione**,[238] a compound also considered in Box 11-B. N^1-γ-Glutamylspermidine and related compounds have been found in proteolytic digests of certain proteins, suggesting that polyamines may be physiological substrates for transglutaminases.[240] Portions of polyamines are incorporated into a variety of products including **nicotine** (Fig. 30-22)[241] and the unusual amino acid **hypusine** (see p. 1386).[242]

Ornithine decarboxylase is specifically inhibited by the enzyme-activated inhibitor α-difluoromethylornithine, which can cure human infection with *Trypanosoma brucei* (African sleeping sickness) by interfering with polyamine synthesis.[243–244a] In combination with inhibitors of spermidine synthase or *S*-adenosylmethionine decarboxylase,[245] it can reduce polyamine levels and growth rates of cells. Another powerful inhibitor that acts on both ornithine and adenosylmethionine decarboxylases is the hydroxylamine derivative 1-aminooxy-3-aminopropane.[246]

$$H_2N{-}O{-}CH_2CH_2CH_2{-}NH_2$$

1-Aminooxy-3-aminopropane

Like difluoromethylornithine the compound at low concentrations is not toxic to cells but inhibits growth. It is hoped that adequate inhibition of growth of normal cells may allow more aggressive chemotherapeutic treatment of cancer.

D. Compounds Derived from Aspartate

The 4-carbon aspartate molecule is the starting point for synthesis of **pyrimidines** and of the amino acids **lysine**, **methionine**, **threonine**, **isoleucine**, and **asparagine**.[247,248] The pathways are summarized in Fig. 24-13. There are several branch points, and aspartate can be converted directly to asparagine, to carbamoylaspartate (the precursor of pyrimidines), or to β-aspartyl phosphate and aspartate semialdehyde. The latter can be converted in one pathway to lysine and in another to homoserine. Homoserine can yield either homocysteine and methionine or threonine. Although threonine is one of the end products and a constituent of proteins, it can also be converted further to 2-oxobutyrate, a precursor of isoleucine.

Most of the chemistry has been considered already. The reduction of aspartate via β-aspartyl phosphate[249,249a] and aspartate β-semialdehyde[250] is a standard one. Conversion to methionine can occur in two ways. In *E. coli* homoserine is succinylated with succinyl-CoA. The γ-succinyl group is then replaced by the cysteine molecule in a PLP-dependent γ-replacement reaction (Fig. 24-13). The product **cystathionine** (Eq. 14-33) undergoes elimination to form homocysteine. A similar pathway via *O*-phosphohomoserine occurs in chloroplasts of green plants.[251] A more direct γ replacement of the hydroxyl of homocysteine or *O*-phosphohomoserine by a sulfide ion has also been reported for both *Neurospora* and green plants.[252] Methylation of homocysteine to methionine (Fig. 24-13) has been considered previously, as has the conversion of homoserine to threonine by homoserine kinase[253] and the PLP-dependent **threonine synthase** (p. 746, Fig. 14-7).[254–255a] A standard PLP-requiring β elimination converts threonine to **2-oxobutyrate**, a precursor to isoleucine (Fig. 24-13).[256]

Formation of **asparagine** has been discussed in section B. Asparagine synthase of *E. coli*[98–99] cleaves ATP to AMP and PP_i rather than to ADP via an aspartyladenylate intermediate. In higher animals glutamine serves as the ammonia donor for synthesis of asparagine, but NH_4^+ can also function.[257] **L-Asparaginase**, a bacterial hydrolase, is an experimental antileukemic drug. It acts to deprive fast-growing tumor cells of the exogenous asparagine needed for rapid growth.[136,257a] Tissues with a low asparagine synthase activity are also damaged, limiting the clinical usefulness.

Aspartate can be decarboxylated either to α-alanine by a PLP-dependent enzyme[258] or to β-alanine by a pyruvoyl group-containing enzyme (Chapter 14). Beta-alanine is not only a component of the vitamin pantothenic acid but is found in the dipeptides carnosine (β-alanylhistidine) and anserine (β-alanyl-N^{δ}-methylhistidine) present in vertebrate muscles.[259] It is a crosslinking agent in insect cuticle.

Aspartate can be deaminated to fumarate by bacterial **L-aspartate oxidase**.[259a] This flavoprotein is structurally and mechanistically related to succinate dehydrogenase and can function as a soluble fumarate reductase (p. 1027). However, its main function appears to be to permit the intermediate iminoaspartate to react with dihydroxyacetone-*P* to form quinolinate, which can be converted to **NAD** (see Fig. 25-11).[259b]

1. Control of Biosynthetic Reactions of Aspartate

In *E. coli* there are three **aspartokinases** that catalyze the conversion of aspartate to β-aspartyl phosphate. All three catalyze the same reaction, but they have very different regulatory properties, as is indicated in Fig. 24-13. Each enzyme is responsive to a different set of end products.[247,260] The same is true for the two **aspartate semialdehyde reductases** which catalyze the third step. Both repression of transcription and feedback inhibition of the enzymes are involved. Two of the aspartokinases of *E. coli* are parts of bifunctional enzymes, which also contain the homoserine dehydrogenases that are needed to reduce aspartate semialdehyde in the third step. These aspartokinase–homoserine dehydrogenases I and II (Fig. 24-13) are encoded by *E. coli* genes *thrA* and *metL*, respectively, and have homologous sequences.[247,261–262a] The N-terminal portions are also homologous to the lysine-sensitive aspartokinase III which is encoded by the *lysC* gene.[263] In *Bacillus subtilis* the lysine-sensitive enzyme is known as aspartokinase II. It has an $\alpha_2\beta_2$ oligomeric structure and both α and β chains are encoded within a single gene.[264] There is no associated homoserine dehydrogenase. Both genetic organization and processing of the synthesized protein are thus different in these two bacteria.

2. Lysine Diaminopimelate, Dipicolinic Acid, and Carnitine

Lysine cannot be made at all by animals but is nutritionally essential. There are two distinct pathways for its formation in other organisms. The **α-aminoadipate pathway** (shown in Fig. 24-9) occurs in a few lower fungi, the higher fungi, and euglenids. The 5-carbon 2-oxoglutarate is the starting compound. Bacteria, other lower fungi, and green plants all use the **diaminopimelate** pathway (Fig. 24-14) which originates with the 4-carbon aspartate.

The α-aminoadipate pathway (Fig. 24-9) parallels

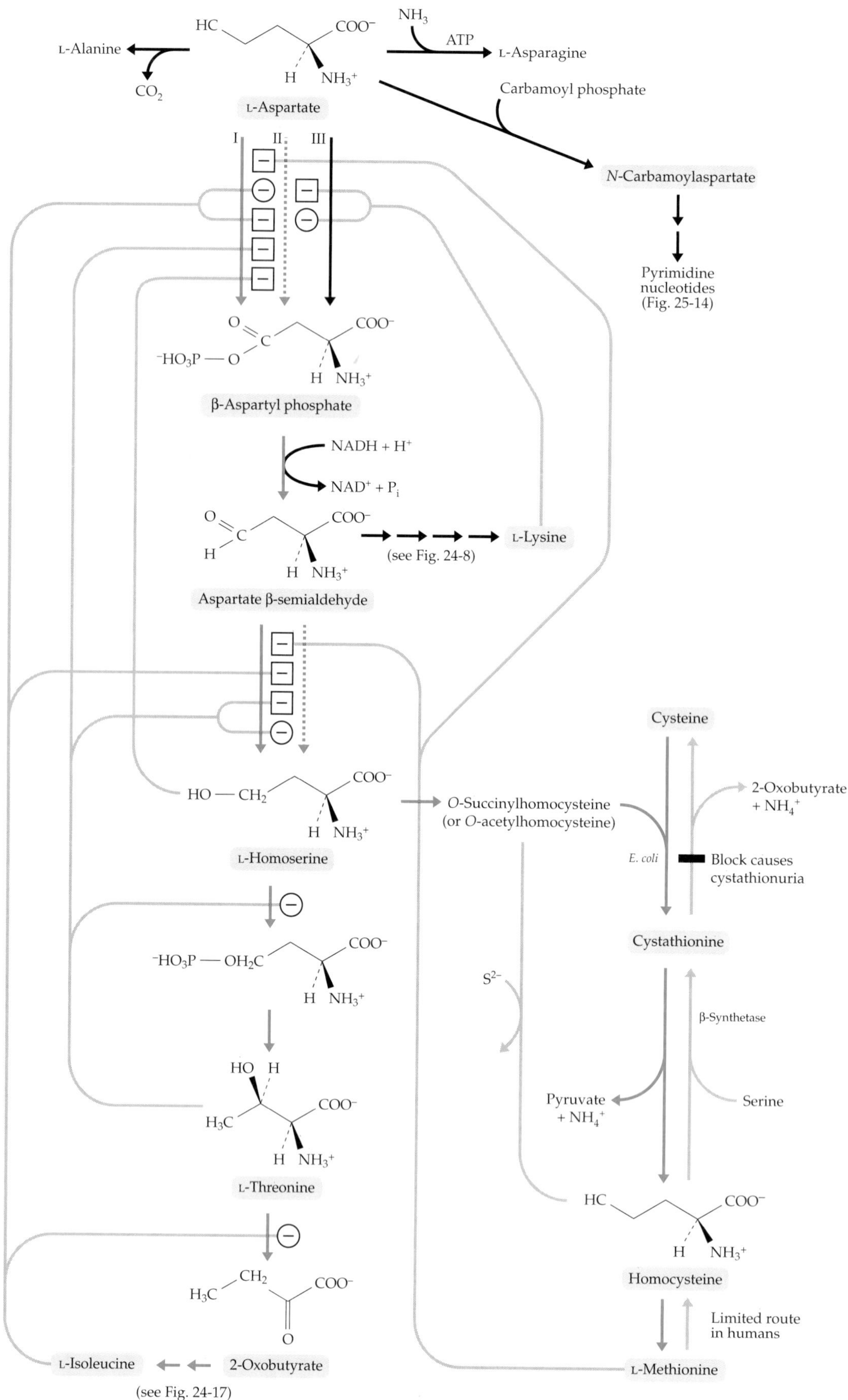

Figure 24-13 Some biosynthetic reactions of aspartate: ⊖, feedback inhibition and ⊟, feedback repression.

that of ornithine biosynthesis, 2-oxoglutarate undergoing chain elongation (Fig. 17-18) to 2-oxoadipate followed by transamination to α-aminoadipate. This is followed by ATP-dependent reduction to the aldehyde.[264a] The final step of transamination is not accomplished in the usual way (with a PLP-dependent enzyme), but through formation of a Schiff base with glutamate and reduction to **saccharopine**.[265] Oxidation now produces the Schiff base of lysine with 2-oxoglutarate.

In the diaminopimelate pathway of lysine synthesis (Fig. 24-14) aspartate is converted to aspartate semialdehyde, and a two-carbon unit is added via aldol condensation with pyruvate.[266–269] Decarboxylation at the end of the sequence yields lysine. A series of cyclic intermediates exist, but it is noteworthy that the initial product of the aldol condensation (bracketed in Fig. 24-14) is converted to diaminopimelic acid by a simple sequence involving α,β elimination of the hydroxy group, reduction with NADPH, and transamination.[267,268,270,271] The process is complicated by the natural tendency for ring closure. All gram-negative and many gram-positive bacteria use the succinylase pathway shown in Fig. 24-14. Succinylation serves to shift the equilibrium back in favor of open-chain compounds.[272–274] Some species of *Bacillus* use acetylation in the same way, while a few bacteria manage to use a dehydrogenase to reductively aminate tetrahydropimelate to diaminopimelate.[275,276] The diaminopimelate pathway is of special significance to prokaryotic organisms for the reason that **dipicolinic acid** is formed as an important side product and because of formation of **diaminopimelic acids**. The cyclic

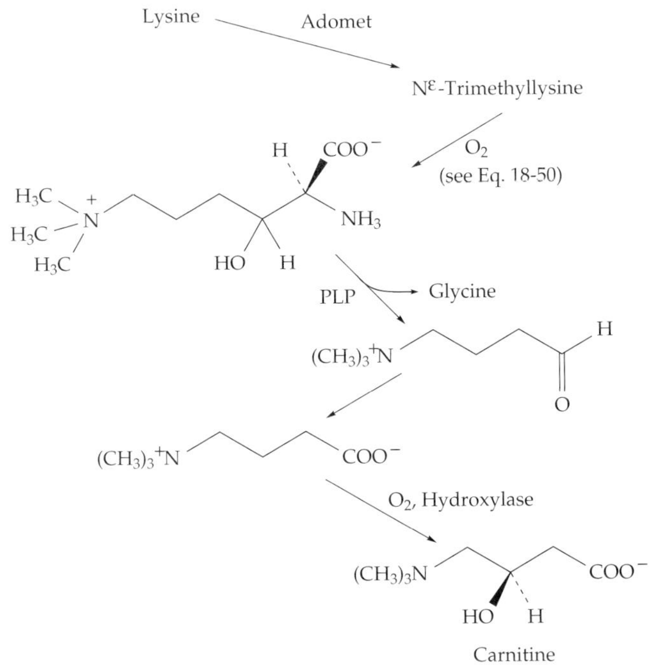

(24-30)

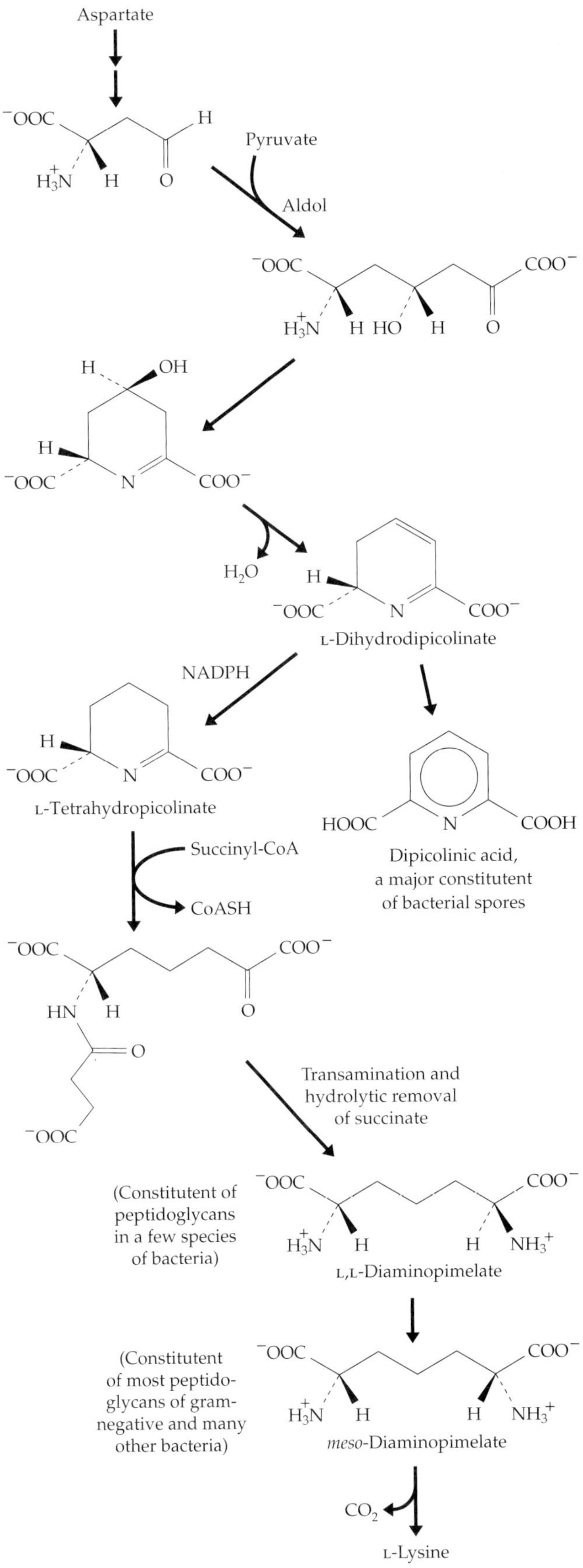

Figure 24-14 The biosynthesis of lysine by the diaminopimelate pathway.

dipicolinic acid is a major constituent of bacterial spores[277] but is rarely found elsewhere in nature. Both L, L- and meso-diaminopimelic acids[276,276a] are constituents of peptidoglycans of bacterial cell walls (Fig. 8-29).

Lysine is not only a constituent of proteins. It can also be trimethylated and converted to **carnitine** (p. 944). In mammals some specific lysyl side chains of proteins undergo *N*-trimethylation and proteolytic degradation with release of free trimethyllysine (Eq. 24-30).[278,279] The free trimethyllysine then undergoes hydroxylation by a 2-oxoglutarate–Fe^{2+}–ascorbate-dependent hydroxylase (Eq. 18-51) to form β-hydroxytrimethyllysine, which is cleaved by a PLP-dependent enzyme (Chapter 14). The resulting aldehyde is oxidized to the carboxylic acid and is converted by a second 2-oxoglutarate–Fe^{2+}–ascorbate-dependent hydroxylase to carnitine (Eq. 24-30; see also Eq. 18-50).

Hypusine (N^{ε}-(4-amino-2-hydroxybutyl)lysine)[242] occurs in mammalian initiation factor 4D, which is utilized in protein synthesis (Chapter 29) and is formed by transfer of the 4-carbon butylamine group from spermidine to a lysine side chain followed by hydroxylation.[280–282a] The lupine alkaloid lupinine[283] is formed from two C_5 units of cadaverine which arises by decarboxylation of lysine. Silaffins (pp. 178, 1381) also contain modified lysines.

Hypusine: N^{ε}-(4-amino-2-hydroxybutyl) lysine

3. The Catabolism of Lysine

An unusual feature of lysine metabolism is that the α-amino group does not equilibrate with the "nitrogen pool." Catabolism is initiated by deamination and proceeds by β oxidation.[284] At least six variations of the β-oxidation process have been proposed. The evolutionary differences concern the manner in which the two amino groups are moved from the carbon skeleton. In the seemingly simplest pathway (A in Fig. 24-15), which is used by *Flavobacterium fuscum*,[285] the ε-amino group is removed in a direct (but atypical) transamination. The resulting α-aminoadipate semialdehyde is oxidized to α-aminoadipate, which *is degraded in a sequence characteristic for the catabolism of amino acids.* Transamination is followed by oxidative decarboxylation of the resulting 2-oxoacid and β oxidation of the coenzyme A derivative. A decarboxylation step by which the terminal carboxyl group is removed is interposed in the β-oxidation sequence for lysine degradation.

Perhaps the initial transamination in pathway A is chemically difficult, for most organisms use more complex sequences to form 2-oxoadipate. In pathway B (which takes place in liver mitochondria and is believed to be the predominant pathway in mammals),[286,287] the ε-amino group is reductively coupled with 2-oxoglutarate to form saccharopine. The latter is in turn oxidized on the opposite side of the bridge nitrogen to form glutamic acid and α-aminoadipate semialdehyde. The overall process is the same as direct transamination and just the opposite of that occurring in the aminoadipate pathway of biosynthesis (Fig. 24-9). Absence of one or both of these dehydrogenases causes familial hyperlysinemia.[287,288]

Pathway C has been established for *Pseudomonus putida*[289] and is also followed to some extent in both plants and animals. In most animal tissues it may be used principally for degradation of D-lysine.[290] However, it is the major L-lysine oxidation pathway in brain.[291] In a fungal parasitic species of *Rhizoctonia* L-lysine is converted to saccharopine via pathway B; then using an $NADP^+$-dependent saccharopine oxidase the sequence is shunted to pathway C.[292] L-Pipicolic acid formed in this way also gives rise to various alkaloids including the α-mannosidase inhibitor swainsonine (Fig. 20-7).[290] Pathway C, like pathway B, makes use of transamination via a reduction-oxidation sequence. It is strictly internal, the oxidizing carbonyl group being formed by transamination of the α-amino group of lysine. Pathway D, apparently used by yeasts,[293] avoids cyclic intermediates by acetylation of the ε-amino group prior to transamination. The 2-oxo group is then effectively blocked by reduction to an alcohol, the blocking group is removed from the ε-amino group, and that end of the molecule is oxidized in a straightforward way to a carboxyl group. Now the hydroxyl introduced at position 2 is presumably oxidized back to the ketone, which again can be converted to give 2-oxoadipate.

Some bacteria, e.g., *Pseudomonas putida*,[294] degrade L-lysine with a flavin-dependent oxygenase (Eq. 18-41) to δ-aminovaleramide:

The product is hydrolyzed and oxidized to **glutaryl-CoA**, rejoining the pathways shown in Fig. 24-15. A remarkable and very different approach to lysine breakdown has been developed by clostridia which obtain energy from the fermentation of Eq. 24-31:

$$\text{L-Lysine} + 2\,H_2O \longrightarrow \text{butyrate}^- + \text{acetate}^- + 2\,NH_4^+ \quad (24\text{-}31)$$

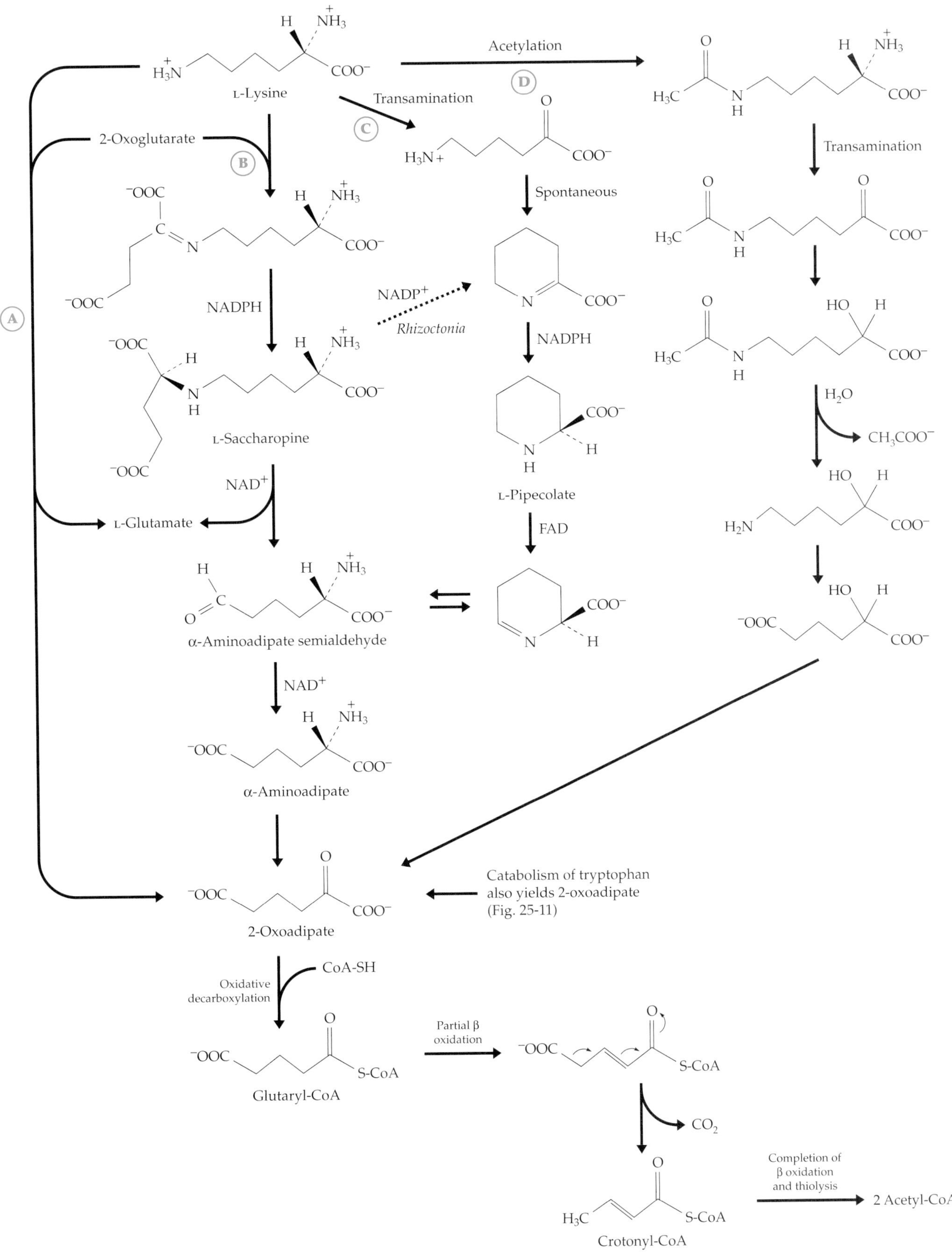

Figure 24-15 Catabolism of lysine.

The reaction is coupled to formation of one molecule of ATP from ADP and P_i. Two pathways have been worked out. In the first lysine is acted upon by a PLP-dependent **L-lysine 2,3-aminomutase** (Eq. 24-32, step *a*) to convert it to β-lysine (3,6-diaminohexanoate). The latter is further isomerized (Eq. 24-32, step *b*) by the vitamin B_{12} and PLP-dependent β-lysine mutase. Oxidative deamination to a 3-oxo compound (Eq. 24-32, step *c*) permits chain cleavage. The reader can easily propose the remaining reactions of chain cleavage, ATP synthesis, elimination of ammonia, and balancing of the redox steps. An alternative pathway begins with a racemase (Eq. 24-32, step *d*) and isomerization of the resulting D-lysine by another B_{12} and PLP-dependent enzyme (Eq. 24-32, step *e*).[294a] Oxidative deamination presumably occurs, but the mechanism for chain cleavage is not so obvious. It does occur between C-4 and C-5 as indicated by the dashed line in Eq. 24-32.

Another variation is used by *Pseudomonas* β4 (Eq. 24-31). Beta-lysine is acetylated on N-6, then undergoes transamination to a 2-oxo acid and removal of the first two carbons as acetyl-CoA. The resulting 4-aminobutyrate is then converted to succinate via succinate semialdehyde.[295]

Why are there so many pathways of lysine breakdown? The answer is probably related to the ease of spontaneous formation of cyclic intermediates as occurs in the pipecolate pathway (pathway C, Fig. 24-15). These intermediates may be too stable for efficient metabolism so the indirect pathways evolved. In the fermentation reactions additional constraints are imposed on the pathways by the need for balanced redox processes and a net Gibbs energy decrease.

L-Lysine
a PLP
L-β-Lysine
b
L-*erythro*-3,5-Diaminohexanoate
c Oxidative deamination
3-Oxo–5-aminohexanoate
d
D-Lysine
e PLP
5-Oxo-3-aminohexanoate
NH_4^+
Butyrate
Acetate
ATP

(24-32)

4. Metabolism of Homocysteine and Methionine

Autotrophic organisms synthesize methionine from aspartate as shown in the lower right side of Fig. 24-13. This involves transfer of a sulfur atom from cysteine into homocysteine, using the carbon skeleton of homoserine, the intermediate **cystathionine**, and two PLP-dependent enzymes, **cystathionine γ-synthase**[296,296a] and **cystathionine β-lyase**.[297] This **transsulfuration** sequence (Fig. 24-13, Eq. 14-33) is essentially irreversible because of the cleavage to pyruvate and NH_4^+ by the β-lyase. Nevertheless, this transsulfuration pathway operates in reverse in the animal body, which uses two different PLP enzymes, **cystathionine β-synthase** (which also contains a bound heme)[298–299c] and **cystathionine γ-lyase**[300] (Figs. 24-13, 24-16, steps *h* and *i*), in a pathway that metabolizes excess methionine.

For human beings methionine is nutritionally essential and comes entirely from the diet. However, the oxoacid analog of methionine can be used as a nutritional supplement. Dietary homocysteine can also be converted into methionine to a limited extent. Methionine is incorporated into proteins as such and as ***N*-formylmethionine** at the N-terminal ends of bacterial proteins (steps *a* and *b*, Fig. 24-16). In addition to its function in proteins methionine plays a major role in biological methylation reactions in all organisms. It is converted into **S-adenosylmethionine** (AdoMet or SAM; Fig. 24-16, step *e*; see also Eq. 17-37),[301–302b] which is the most widely used methyl group donor for numerous biological methylation reactions (Eq. 12-3). *S*-Adenosylmethionine is also the precursor of the special "wobble base" **queuine** (Fig. 5-33).[312]

The product of transmethylation, **S-adenosylhomocysteine**, is converted (step *g*) into homocysteine in an unusual NAD-dependent hydrolytic reaction (Eq. 15-14) by which adenosine is removed (step *g*).[302c] Homocysteine can be reconverted to methionine, as indicated by the dashed line in Fig. 24-16. This can be accomplished by the vitamin B_{12}- and tetrahydrofolate-dependent **methionine synthase**, (Eq. 16-43), which transfers a methyl group from methyl-tetrahydrofolate[303–303b]; by transfer of a methyl group from **betaine**, a trimethylated glycine (Eq. 24-33)[304], or by remethylation with AdoMet (Fig. 24-16).[304a]

$$\text{S-Adenosylhomocysteine} + \text{Betaine} \longrightarrow \text{Methionine} + \text{Dimethylglycine} \quad (24\text{-}33)$$

When present in excess methionine is toxic and must be removed. Transamination to the corresponding 2-oxoacid (Fig. 24-16, step *c*) occurs in both animals and plants. Oxidative decarboxylation of this oxoacid initiates a major catabolic pathway,[305] which probably involves β oxidation of the resulting acyl-CoA. In bacteria another catabolic reaction of methionine is γ-elimination of methanethiol and deamination to 2-oxobutyrate (reaction *d*, Fig. 24-16; Fig. 14-7).[306] Conversion to homocysteine, via the transmethylation pathway, is also a major catabolic route which is especially important because of the toxicity of excess homocysteine. A hereditary deficiency of cystathionine β-synthase is associated with greatly elevated homocysteine concentrations in blood and urine and often disastrous early cardiovascular disease.[299,307–309b] About 5–7% of the general population has an increased level of homocysteine and is also at increased risk of artery disease. An adequate intake of vitamin B_6 and especially of folic acid, which is needed for recycling of homocysteine to methionine, is helpful. However, if methionine is in excess it must be removed via the previously discussed transsulfuration pathway (Fig. 24-16, steps *h* and *i*).[310] The products are cysteine and 2-oxobutyrate. The latter can be oxidatively decarboxylated to propionyl-CoA and further metabolized, or it can be converted into leucine (Fig. 24-17) and cysteine may be converted to glutathione.[299a]

Methionine in plants can be converted to the sulfonium compound *S*-methyl-L-methionine, also called vitamin U. It has strong osmoprotectant activity and accumulates in many marine algae and some flowering plants.[311] Other organisms, including mammals, can use *S*-methylmethionine to methylate homocysteine, converting both reactants back to methionine[311a] enabling animals to meet some of their methionine need from this source.

A salvage pathway. Another product of *S*-adenosylmethionine is **5'-methylthioadenosine**, which can be formed by an internal displacement on the γ-methylene group by the carboxylate group (step *l*, Fig. 24-16). Methylthioadenosine also arises during formation of the compounds spermidine (Fig. 24-12) and ACC (Fig. 24-16). Mammalian tissues convert methylthioadenosine back to methionine by the sequence shown in Eq. 24-34. It undergoes phosphorolysis to 5'-methylthioribose whose ring is opened and

5'-Methylthioadenosine —(a; H_2O → Adenine)→ —(b; ATP → ADP)→ $H_3C-S-CH_2$-ribose-1-O-P (5'-Methylthioribose-1-*P*) —(c, Isomerase)→ H_3C-S-CH₂-CH(OH)-CH(OH)-CO-CH₂-O-P —(d)→ H_3C-S-CH₂CH₂-CO-C(OH)=CH-O⁻ —(e; O_2 → $HCOO^-$)→ H_3C-S-CH₂CH₂-CO-COO⁻ —(Transamination)→ Methionine; or —(f; → HCO_2^-, CO)→ H_3C-S-CH₂CH₂-COO⁻ (24-34)

converted to the 2-oxoacid analog of methionine.[313,314]

Step *c* of Eq. 24-34 may occur by ring opening to an enol phosphate which ketonizes to the observed product, but step *e* is a more complex multistep oxidative process.[314a,b] The last step is transamination to methionine with a glutamine-specific aminotransferase. Another enzyme from *Klebsiella* converts the same intermediate anion to methylthiopropionate, formate, and CO (Eq. 24-34, step *f*).[315]

The plant hormone ethylene. A major reaction of *S*-adenosylmethionine in plants is the formation of **ethylene**.[316,317] Ethylene has been recognized since 1858 as causing a thickening of stems of plants and a depression in the rate of elongation. In 1917, it was established that ethylene is formed in fruit and that addition of this gaseous compound hastened ripening. Ethylene is now an established plant hormone having a variety of effects including retardation of mitosis, inhibition of photosynthesis, and stimulation of respiration and of the enzyme phenyalanine ammonia-lyase (Eq. 14-45). These effects are indirectly a result of the action of ethylene on transcription of certain genes. In *Arabidopsis*, with which genetic studies are being made, ethylene binds to the N-terminal part of at least two receptor proteins, which have intracellular histidine kinase domains in the C-terminal parts.[318,319]

DNA-binding proteins specific for **ethylene-responsive elements** (EREs), having the conserved sequence AGCCGCC, are presumably phosphorylated by this kinase[320] and affect transcription of many genes. A protein homologous to one ethylene receptor of *Arabidopsis* has been identified in the tomato. A proline to leucine mutation at position 36, near the N terminus, destroys the sensitivity to ethylene and prevents ripening of this tomato. Another component in the ethylene signaling pathway in *Arabidopsis* is a protein serine / threonine kinase that resembles the mammalian raf kinase involved in the signaling cascade shown in Fig. 11-13.[318]

The formation of ethylene is often induced by the hormone **auxin** (Chapter 30), which stimulates activity of the synthase that forms **1-aminocyclopropane-1-carboxylate** (**ACC**) from *S*-adenosyl methionine (Eq. 14-27, step *j*: Fig. 24-16).[320a,b] Although ACC has been known as a minor plant product for over 25 years, it was much more recently identified as the immediate precursor of ethylene. ACC is often produced in response to stresses such as wounding, drought, or waterlogging of roots.[316,321] In the last of these cases the ACC is transferred through the xylem from the roots upward to shoots, which respond in characteristic ways to the ethylene that is released.

$$\text{ACC}\;(\text{cyclopropane with } NH_3^+ \text{ and } COO^-) + O_2 + \text{Ascorbate} \rightarrow H_2C = CH_2 + HCN + CO_2 + 2\,H_2O + \text{Dehydroascorbate} \quad (24\text{-}35)$$

The conversion of ACC to ethylene, HCN, and CO_2 is catalyzed by ACC oxidase, an Fe^{2+}-dependent enzyme of the isopenicillin-*N*-synthase (Eq. 18-52) subfamily of oxygenases. However, unlike most of

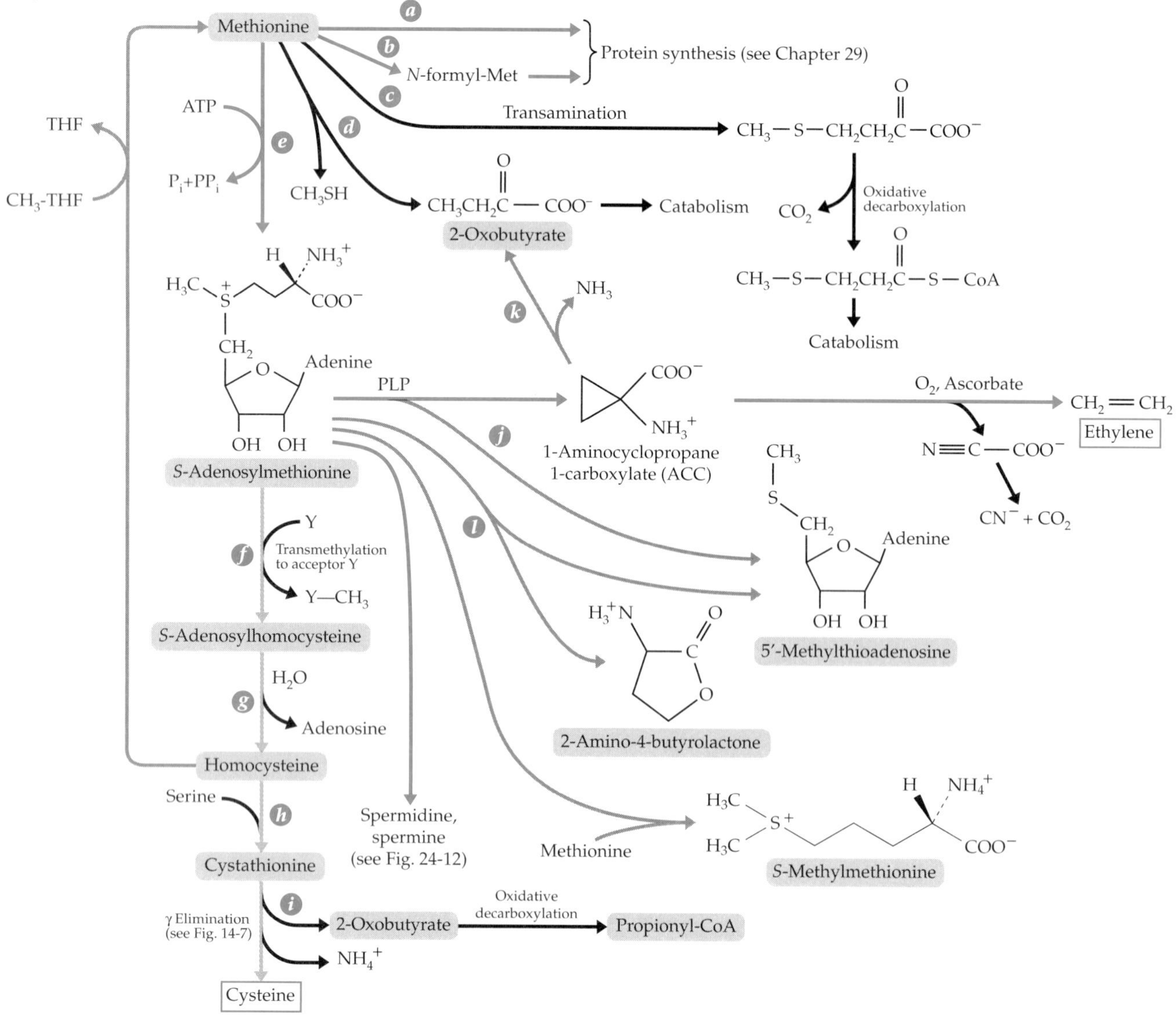

Figure 24-16 Some metabolic reactions of methionine. Biosynthetic reactions are indicated by green arrows.

COO^- NH_3^+ e^- H^+ COO^- $\overset{+}{N}H_2$ H^+ COO^- NH e^- COO^- NH $H_2C=CH_2$ $N{\equiv}C-COO^-$ $N{\equiv}C^- + CO_2$ (24-36)

these enzymes, which utilize 2-oxoglutarate as a cosubstrate (Eq. 18-52), ACC oxidase employs ascorbate and forms HCN or cyanide ions.[322–324b] It also requires CO_2 or bicarbonate as an activator.[324,325] A radical mechanism (Eq. 24-36) is probable,[317] with two electrons from ACC and two from ascorbate being utilized to reduce O_2 to 2 H_2O.

Ethylene is rather inert, but it is metabolized slowly, some of it to ethylene glycol.[326] Plants store *N*-malonyl-ACC as a metabolically inert pool. Excess ACC can be deaminated in a PLP-dependent reaction to 2-oxobutyrate (step *k*, Fig. 24-16), a process that also occurs in bacteria able to subsist on ACC.[327,327a] There may also be other mechanisms for ethylene formation, e.g., peroxidation of lipids during scenescence of leaves.[328] See also Chapter 31, Section G.

L-Threonine —(a, NH_4^+)→ 2-Oxobutyrate → Propionyl-CoA
L-Threonine —(b)→ Glycine + Acetaldehyde
L-Threonine —(c)→ 2-Amino-3-oxobutyrate
2-Amino-3-oxobutyrate —(d)→ Acetyl-CoA + Glycine
2-Amino-3-oxobutyrate —(e)→ CO_2 + Aminoacetone
Aminoacetone —(f)→ Methylglyoxal —(Eq. 13-33)→ D-Lactate
Aminoacetone —(g, NAD^+)→ *R*-1-Amino-2-propanol → → Vitamin B_{12}

(24-37)

5. Metabolism of Threonine

Excess threonine is degraded in several ways, one of which is a β elimination reaction catalyzed by L-threonine dehydratase (Eq. 24-37, step *a*). This PLP-requiring enzyme is produced in high amounts in *E. coli* grown on a medium devoid of glucose and oxygen. Under these circumstances the reaction provides a source of propionyl-CoA, which can be converted to propionate with generation of ATP. This **biodegradative threonine dehydratase** (threonine deaminase)[329,330] is allosterically activated by AMP, an appropriate behavior for a key enzyme in energy metabolism. A second **biosynthetic threonine dehydratase** is also produced by *E. coli*[331,332] and is specifically required for production of 2-oxobutyrate needed in the biosynthesis of isoleucine by bacteria, plants,[333] and other autotrophic organisms. In 1956, Umbarger[334] showed that this enzyme is inhibited by isoleucine, the end product of the synthetic pathway. This discovery was instrumental in establishing the concepts of feedback inhibition in metabolic regulation (Chapter 11) and of allostery.

A second catabolic reaction of L-threonine (Eq. 24-37, step *b*) is cleavage to glycine and acetaldehyde. The reaction is catalyzed by serine hydroxymethyltransferase (Eq. 14-30). Some bacteria have a very active D-threonine aldolase.[335] A quantitatively more important route of catabolism in most organisms is dehydrogenation (Eq. 24-37, step *c*)[336] to form 2-amino-3-oxobutyrate. This intermediate can be cleaved by another PLP-dependent enzyme to acetyl-CoA plus glycine (Eq. 24-38, step *d*). It can also be decarboxylated (Eq. 24-38, step *e*) to aminoacetone, a urinary excretion product, or oxidized by amine oxidases to **methylglyoxal** (Eq. 24-37, step *f*).[337] The latter can be converted to D-lactate through the action of glyoxalase (Eq. 13-33). Aminoacetone is also the source of 1-amino-2-propanol for the biosynthesis of vitamin B_{12} (Eq. 24-37, step *g*; Box 16-B).[338,338a]

E. Alanine and the Branched-Chain Amino Acids

As indicated in Fig. 24-17, pyruvate is the starting material for the formation of both L- and D-alanine and also the branched chain amino acids **valine**, **leucine**, and **isoleucine**.[339,340] The chemistry of the reactions has been discussed in the sections indicated in the figure. The first step is catalyzed by the thiamin diphosphate-dependent **acetohydroxyacid synthase** (acetolactate synthase), which joins two molecules of pyruvate or one of pyruvate and one of 2-oxobutyrate (Fig. 24-17; Fig. 14-3).[340a,b] In *E. coli* there are two isoenzymes encoded by genes *ilv B* and *ilv HI*. Both are regulated by feedback inhibition by valine, probably

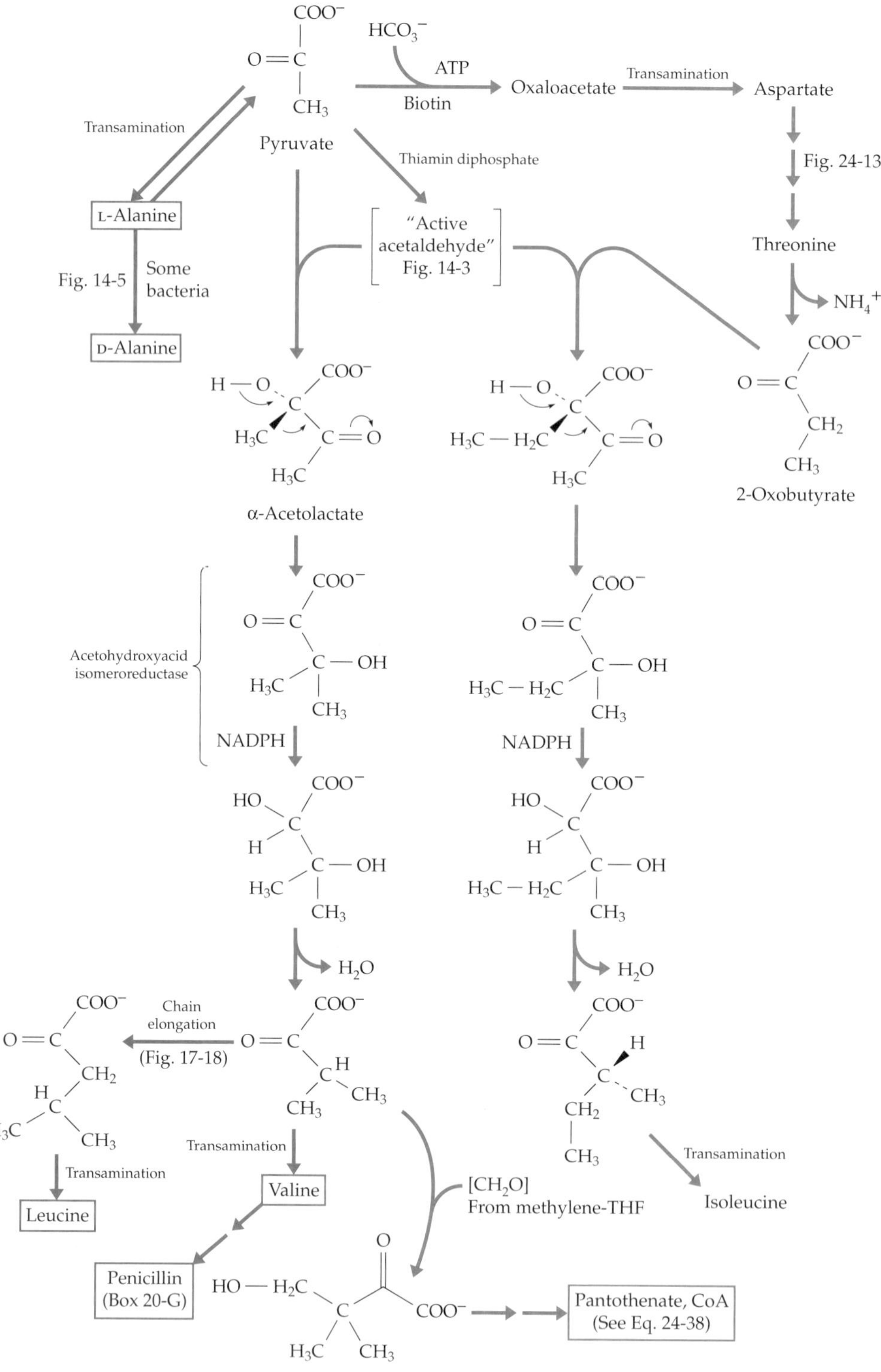

Figure 24-17 Biosynthesis of leucine, isoleucine, valine, and coenzyme A.

Chlorsulfuron

An imidazolinone herbicide

by an **attenuation** mechanism[341] (explained in Chapter 28). The enzymes are of some practical interest because they are specifically inhibited by two classes of herbicides, the **sulfonylureas**, of which chlorsulfuron is an example, and the **imidazolinones**.[342–345]

The second step in the synthesis, catalyzed by **acetohydroxyacid isomeroreductase**, involves shift of an alkyl group (Fig. 24-17). Neither this reaction nor the preceding one occurs in mammals. For this reason, the enzymes required are both attractive targets for herbicide design.[343,346] The third enzyme, **dihydroxy acid dehydratase**, catalyzes dehydration followed by tautomerization, resembling 6-phosphogluconate dehydratase (Eq. 13-32). The dihydroxyacid dehydratase from spinach contains an Fe_2S_2 cluster and may function by an aconitase type mechanism (Eq. 13-17).[347] In *Neurospora* isoleucine and valine are synthesized in the mitochondria.

Methylene-THF (Fig. 15-18) → [H₂C=O + 2-oxoisovalerate ($(H_3C)(CH_3)CH{-}CO{-}COO^-$)] ← Valine; Biosyntheses (Fig. 24-17)

↓ Aldol condensation (Eq. 13-36)

$HO{-}H_2C{-}C(H_3C)(CH_3){-}CO{-}COO^-$

↓ NADPH

$HO{-}H_2C{-}C(H_3C)(CH_3){-}CH(OH){-}COO^-$ D-Pantoate (See also Fig. 14-1)

$^-OOC{-}CH_2{-}CH_2{-}\overset{+}{N}H_3$ β-Alanine; ATP

↓

Pantothenate

↓ ATP

4′-Phosphopantethenate

↓ L-Cys, CTP → CMP + PP_i

4′-Phosphopantethenylcysteine

↓ → CO_2

4′-Phosphopantetheine

↓ ATP → PP_i

Diphospho-CoA

↓ ATP

Coenzyme A (Fig. 14-1) (24-38)

While the 2-oxobutyrate needed for isoleucine formation is shown as originating from threonine in Fig. 24-17, bacteria can often make it in other ways,[348] e.g., from glutamate via β-methylaspartate (Fig. 24-8) and transamination to the corresponding 2-oxoacid. It can also be made from pyruvate by chain elongation using acetyl-CoA (Fig. 17-18); citramalate and mesaconate (Fig. 24-8) are intermediates. This latter pathway is used by some methanogens as are other alternative routes.[348] The first step unique to the biosynthetic pathway to leucine is the reaction of the 2-oxo analog of valine with acetyl-CoA to form **α-isopropylmalate**, the first step in a chain elongation sequence leading to the oxoacid precursor of leucine (Figs. 17-18; 24-17). The third enzyme required in the chain elongation is a decarboxylating dehydrogenase similar to isocitrate dehydrogenase.[349]

An additional series of reactions,[350] which are shown in Eq. 24-38, leads to **pantoic acid**, **pantetheine**, **coenzyme A**, and related cofactors.[350a–j] The initial reactions of the sequence do not occur in the animal body, explaining our need for pantothenic acid as a vitamin.

Alanine also gives rise to a precursor of the vitamin **biotin** (Eq. 24-39) after a PLP-dependent decarboxylative condensation with the 7-carbon dicarboxylic acid unit of pimeloyl-CoA in a reaction analogous to that of Eq. 14-32.[351] The resulting alcohol is reduced to 7-oxo-8-aminopelargonic acid which is converted by transamination, with *S*-adenosylmethionine as the nitrogen donor,[351a] to 7,8-diaminopelargonic acid. This compound undergoes a two-step ATP-dependent cyclization[352–355] to form **dethiobiotin**. The final step, insertion of sulfur into dethiobiotin, is catalyzed by **biotin synthase**, a free-radical-dependent enzyme related to pyruvate formate lyase (Fig. 15-16). It transfers the sulfur from cysteine via an Fe–S cluster.[355a–c] Biosynthesis of **lipoic acid** involves a similar insertion of two sulfur atoms into octanoic acid.[356] See also p. 1410.

1. Catabolism

Degradation of amino acids most often begins with conversion, either by transamination[356a] or by NAD^+-dependent dehydrogenation,[357] to the corresponding 2-oxoacid and oxidative decarboxylation of the latter (Fig. 15-16). Alanine, valine, leucine, and isoleucine are all treated this way in the animal body. Alanine gives pyruvate and acetyl-CoA directly, but the others yield CoA derivatives that undergo

BOX 24-A MAPLE SYRUP URINE DISEASE AND JAMAICAN VOMITING SICKNESS

In a rare autosomal recessive condition (discovered in 1954) the urine and perspiration has a maple syrup odor.[a–c] High concentrations of the branched-chain 2-oxoacids formed by transamination of valine, leucine, and isoleucine are present, and the odor arises from decomposition products of these acids. The branched-chain amino acids as well as the related alcohols also accumulate in the blood and are found in the urine. The biochemical defect lies in the enzyme catalyzing oxidative decarboxylation of the oxoacids, as is indicated in Fig. 24-18. Insertions, deletions, and substitutions may be present in any of the subunits (Figs. 15-14, 15-15). The disease which may affect one person in ~200,000, is usually fatal in early childhood if untreated. Children suffer seizures, mental retardation, and coma. They may survive on a low-protein (gelatin) diet supplemented with essential amino acids, but treatment is difficult and a sudden relapse is apt to prove fatal. Some patients respond to administration of thiamin at 20 times the normal daily requirement. The branched-chain oxoacid dehydrogenase from some of these children shows a reduced affinity for the essential coenzyme thiamin diphosphate.[d]

Polled hereford calves in Australia develop maple syrup urine disease relatively often.[a,e] One cause was established as a mutation that introduces a stop codon that causes premature termination within the leader peptide during synthesis of the thiamin diphosphate-dependent E1 subunit. A similar biochemical defect in a mutant of *Bacillus subtilis*[f] causes difficulties for this bacterium, which requires branched-chain fatty acids in its membranes. Branched acyl-CoA derivatives are needed as starter pieces for their synthesis (Chapter 29). With the oxidative decarboxylation of the necessary oxoacids blocked, the mutant is unable to grow unless supplemented with branched-chain fatty acids.

Because persons may be born with defects in almost any gene, a variety of other problems leading to accumulation of organic acids are also known. **Methylmalonic aciduria** and propionic acidemia are discussed in Box 17-B. **Lactic acidemia** (Box 17-F) often results from a defect in pyruvate dehydrogenase. A rare defect of catabolism of leucine is **isovaleric acidemia**, a failure in oxidation of isovaleryl-CoA.[g] The symptoms of this disease are also present in the Jamaican vomiting sickness, caused by eating unripe ackee fruit. Although the ripe fruit is safe to eat, unripe fruit contains a toxin **hypoglycin A** with the following structure.[h–j] It is metabolized to an acyl-CoA derivative as shown.

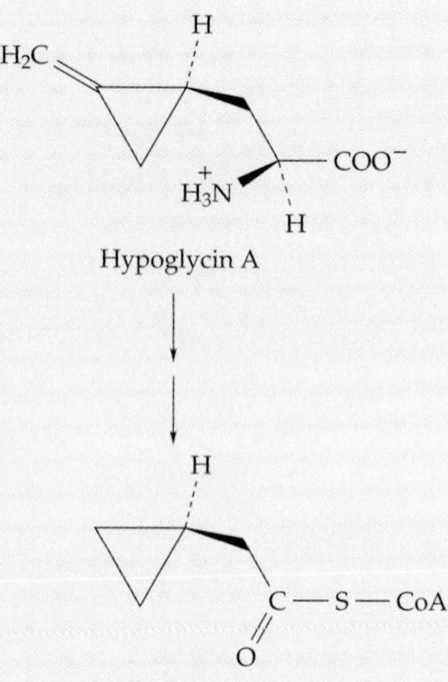

This is an enzyme-activated inhibitor of the medium-chain fatty acyl-CoA dehydrogenase required for β oxidation of fatty acids.[j,k] The compound also inhibits isovaleryl-CoA dehydrogenase, causing an accumulation of isovaleric acid in the blood. Depression of the central nervous system by isovaleric acid in the blood could be responsible for some symptoms.[h,i] However, death from the highly fatal Jamaican vomiting sickness comes from the hypoglycemic effect. Blood glucose levels may fall as low as 0.5 mM, one-tenth the normal concentration and patients must be treated by infusion of glucose.

---

[a] Patel, M. S., and Harris, R. A. (1995) *FASEB J.* **9**, 1164–1172

[b] Chuang, D. T., and Shih, V. E. (1995) in *The Metabolic and Molecular Bases of Inherited Disease*, 7th ed., Vol. 1 (Scriver, C. R., Beaudet, A. L., Sly, W. S., and Valle, D., eds), pp. 1239–1278, McGraw-Hill, New York

[c] Mamer, O. A., and Reimer, M. L. J. (1992) *J. Biol. Chem.* **267**, 22141–22147

[d] Chuang, D. T., Ku, L. S., and Cox, R. P. (1982) *Proc. Natl. Acad. Sci. U.S.A.* **79**, 3300–3304

[e] Zhang, B., Healy, P. J., Zhao, Y., Crabb, D. W., and Harris, R. A. (1990) *J. Biol. Chem.* **265**, 2425–2427

[f] Willecke, K., and Pardee, A. B. (1971) *J. Biol. Chem.* **246**, 5264–5272

[g] Mohsen, A.-W. A., and Vockley, J. (1995) *Biochemistry* **34**, 10146–10152

[h] Tanaka, K., Isselbacher, K. J., and Shih, V. (1972) *Science* **175**, 69–71

[i] Tanaka, K. (1972) *J. Biol. Chem.* **247**, 7465–7478

[j] Lai, M.-t, Liu, L.-d, and Liu, H.-w. (1991) *J. Am. Chem. Soc.* **113**, 7388–7397

[k] Lai, M.-t, Li, D., Oh, E., and Liu, H.-w. (1993) *J. Am. Chem. Soc.* **115**, 1619–1628

β and ω oxidation of fatty acyl-CoAs
2-Oxosuberate (see Eq. 21-1)
Oxidative decarboxylation
Methanogens
NAD^+
CO_2
Pimeloyl-CoA
PLP
L-Alanine
CoA–SH
CO_2
–2[H]
Transamination
AdoMet
2-Oxoacid
7,8-Diaminopelargonic acid
CO_2
ATP
$ADP + P_i$
Dethiobiotin
[S]
Biotin

(24-39)

β-oxidation within the mitochondria[358] via the schemes shown in Fig. 24-18. There are some variations from the standard β oxidation sequence for fatty acids shown in Fig. 17-1. In the case of valine the sequence proceeds only to the stage of addition of water to form the β-hydroxy derivative. The latter is converted to free 3-hydroxyisobutyrate, and β oxidation is then completed by oxidation to methylmalonate semialdehyde.[359] The latter is oxidatively decarboxylated to form *S*-methylmalonyl-CoA.[360] Further metabolism of the latter is indicated in Fig. 17-3. However, some methylmalonate semialdehyde may be decarboxylated to propionaldehyde, which could be oxidized to propionate. Either of these compounds could then be metabolized to propionyl-CoA.[361]

In the degradation of isoleucine, β oxidation proceeds to completion in the normal way with generation of acetyl-CoA and propionyl-CoA. However, in the catabolism of leucine after the initial dehydrogenation in the β-oxidation sequence, carbon dioxide is added using a biotin enzyme (Chapter 14). The double bond conjugated with the carbonyl of the thioester makes this carboxylation analogous to a standard β-carboxylation reaction. Why add the extra CO_2? The methyl group in the β position blocks complete β oxidation, but an aldol cleavage would be possible to give acetyl-CoA and acetone. However, acetone is not readily metabolized further. By addition of CO_2 the product becomes acetoacetate, which can readily be completely metabolized through conversion to acetyl-CoA.

An alternative pathway of leucine degradation in the liver is oxidative decarboxylation by a cytosolic oxygenase to form α-hydroxyisovalerate.[362]

α-Hydroxyisovalerate

This compound may be metabolized via the valine catabolic pathway of Fig. 24-18. A third pathway, present in some bacteria, begins with the vitamin B_{12}-dependent isomerization of leucine to β-leucine (Chapter 16), which can undergo transamination to 3-oxoisocaproate. This can be converted to its CoA ester by a CoA transferase and can undergo β cleavage by free CoA-SH to form acetyl-CoA and isobutyryl-CoA. The latter may enter the valine catabolic pathway (Fig. 24-18). Leucine has long been known as a regulator of protein degradation in muscle.[362a–e] Dietary protein deficiency leads to especially rapid degradation of the branched-chain amino acids. The daily turnover of proteins for a 70-kg adult ingesting 70 g of protein per day has been estimated as 280 g, most of which must be reused.[362d] This large turnover can lead to

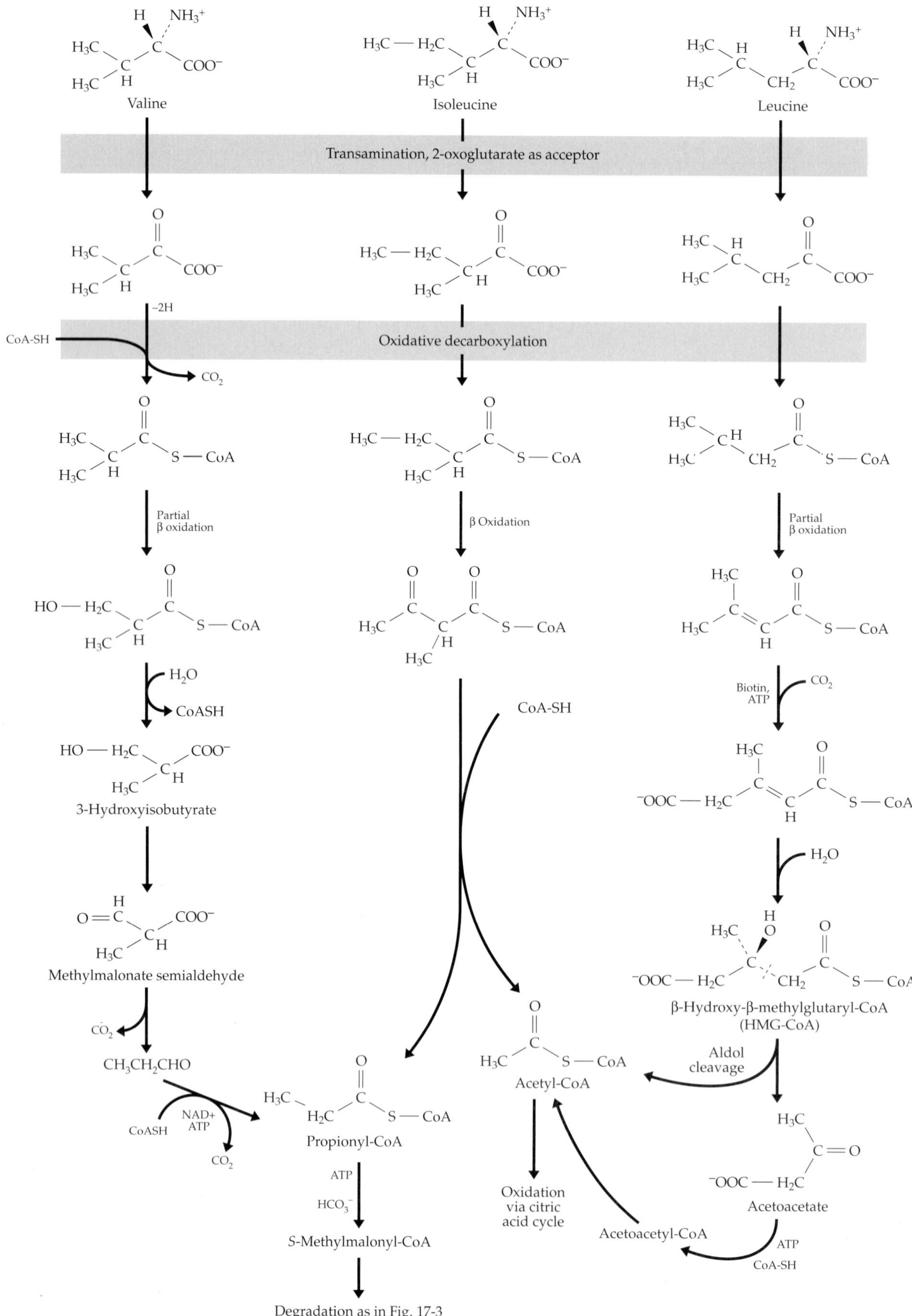

Figure 24-18 Catabolism of valine, leucine, and isoleucine.

excessive muscle wasting in disease states. A minor leucine metabolite found in muscle, β-hydroxy β-methylbutyrate has been proposed as a possible endogenous inhibitor of muscle breakdown.[362e,f] (See study question 17.)

Clostridium propionicum can use alanine as substrate for a balanced fermentation to form ammonium propionate, acetate, and CO_2 (Fig. 24-19).

2. Ketogenic and Glucogenic Amino Acids

According to a long-used classification amino acids are **ketogenic** if (like leucine) they are converted to acetyl-CoA (or acetyl-CoA and acetoacetate). When fed to a starved animal, ketogenic amino acids cause an increased concentration of acetoacetate and other ketone bodies in the blood and urine. On the other hand, **glucogenic** amino acids such as valine, when fed to a starved animal, promote the synthesis of glycogen (in the case of valine via methylmalonyl-CoA, succinate, and oxaloacetate). Examination of Fig. 24-18 shows that isoleucine is both ketogenic and glucogenic, a fact that was known long before the pathway of catabolism was worked out.

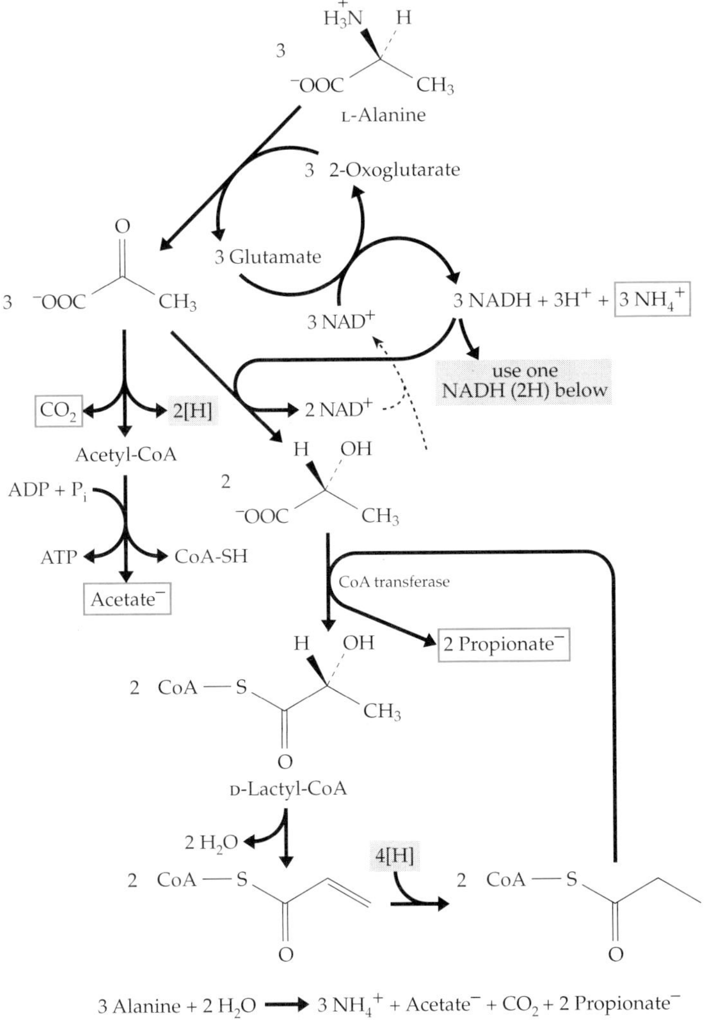

Figure 24-19 Fermentation of L-alanine by *Clostridium propionicum*. After Buckel.[363]

F. Serine and Glycine

Serine originates in a direct pathway from 3-phosphoglycerate (pathway *a*, Fig. 24-20) that involves dehydrogenation, transamination, and hydrolysis by a phosphatase. It can also be formed from glycine by the action of serine hydroxymethyltransferase (Eq. 14-30). This occurs in chloroplasts during photorespiration (Fig. 23-37)[364] and also with some methanogens and other autotropic bacteria and methylotrophs (Fig. 17-15). The glycine decarboxylase cycle shown in Fig. 15-20 provides another mechanism available in bacteria, plants, and animal mitochondria for reversible interconversion of glycine and serine. The principal route of catabolism of serine in many microorganisms is deamination to pyruvate (Fig. 24-20, step *b*),[364a] a reaction also discussed in Chapter 14 (Eq. 14-29). An alternative catabolic pathway is transamination to **hydroxypyruvate,** which as in plants (Fig. 23-37) can be reduced to D-glycerate and back to 3-phosphoglycerate.[365] That this pathway is important in human beings is suggested by the occurrence of a rare metabolic defect **L-glyceric aciduria** (or primary hyperoxaluria type II).[365–367] The biochemical defect may lie in the lack of reduction of hydroxypyruvate to D-glycerate. When hydroxypyruvate accumulates, lactate dehydrogenase effects its reduction to L-glycerate, which is excreted in large amounts (0.3–0.6 g / 24 h) in the urine. Surprisingly, the defect is accompanied by excessive production of oxalate from glyoxylate. This is apparently an indirect result of the primary defect in utilization of hydroxypyruvate. It has been suggested that oxidation of glyoxylate by NAD^+ is coupled to the reduction of hydroxypyruvate by NADH.[366] This and other hyperoxalurias are very serious diseases characterized by the formation of calcium oxalate crystals in tissues and often death from kidney failure before the age of 20.

1. Biosynthetic Pathways from Serine

L-Serine gives rise to many other substances (Fig. 24-20) including **sphingosine** and the **phosphatides**. In many bacteria conversion to

O-acetyl-L-serine (step *c*, Fig. 24-20) provides for the formation of **cysteine** by a β-replacement reaction.[368–369a] Serine is also the major source of glycine (step *d*) and of the single-carbon units needed for the synthesis of methyl and formyl groups. The enzyme **serine hydroxymethyltranferase** (step *d*) also provides the principal route of formation of glycine from serine,[370,371] but a lesser portion comes via phosphatidylserine, phosphatidylethanolamine, phosphatidylcholine, and free choline (step *e*). This pathway includes decarboxylation of phosphatidylserine by a pyruvoyl group-dependent enzyme (pp. 753–755). In contrast, in green plants the major source of ethanolamine is a direct PLP-dependent decarboxylation of serine (Fig. 24-20).[371a] Because the body's capacity to generate methyl groups is limited, **choline** under many circumstances is a dietary essential and has been classified as a vitamin. However, in the presence of adequate amounts of folic acid and vitamin B_{12}, it is not absolutely required. Choline can be used to re-form phosphatidylcholine (Fig. 21-5), while an excess can be dehydrogenated to **glycine betaine,** which is one of the osmoprotectant substances in plants (Eq. 24-40).[372,373] This quaternary nitrogen compound is one of a small number of substances that, like methionine, are able to donate methyl groups to other compounds and which are also capable of methylating homocysteine to form methionine. However, the product of transmethylation from betaine, dimethylglycine, is no longer a methylating agent. The two methyl groups are removed oxidatively as formic acid to produce glycine (Eq. 24-40). A third source of glycine is transamination of glyoxylate (step *f*, Fig. 24-20). The equilibrium constant for the reaction favors glycine strongly for almost any amino donor.

Figure 24-20 Metabolism of serine and glycine.

2. Metabolism of Glycine

While glycine may be formed from glyoxylate by transamination, the oxidation of glycine by an amino acid oxidase (Table 15-2) permits excess glycine to be converted to glyoxylate. That this pathway, too, is quantitatively important in humans is suggested by the existence of **type 1 hyperoxaluria.**[366] It is thought that a normal pathway for utilization of glyoxylate is blocked in this condition leading to its oxidation to oxalate. The biochemical defect is frequently the absence of a liver-specific alanine:glyoxylate aminotransferase that efficiently converts accumulating glyoxylate back to glycine. In some cases the disease arises because, as a result of

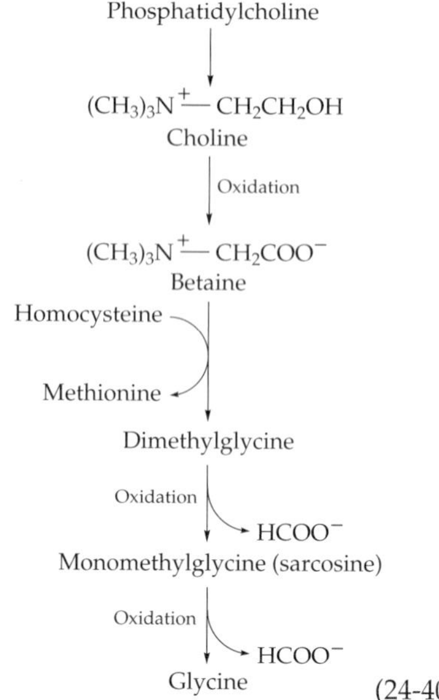

(24-40)

a mutation in its N-terminal targeting sequence the aminotransferase is targeted to mitochondria, where it functions less efficiently.[366a] Another possible defect lies in a thiamin-dependent enzyme that condenses glyoxylate with 2-oxoglutarate to form 2-hydroxy-3-oxoadipate (Eq. 24-41). The function of this reaction is uncertain, but the product could undergo decarboxylation and oxidation to regenerate 2-oxoglutarate. This would provide a cyclic pathway (closely paralleling the dicarboxylic acid cycle, Fig. 17-6) for oxidation of glyoxylate without formation of oxalate. Bear in mind that the demonstrated enzymatic condensation reactions of the rather toxic glyoxylate are numerous, and that its metabolism in most organisms is still not well understood.

An alternative route of catabolism is used by organisms such as *Diplococcus glycinophilus*, which are able to grow on glycine as a sole source of energy, of carbon, and of nitrogen,[374] and is also used in mitochondria of plants and animals.[374a] This glycine cleavage system, depicted in Fig. 15-20, involves decarboxylation, oxidation by NAD^+, release of ammonia, and transfer of the decarboxylated α-carbon of glycine to tetrahydrofolic acid (THF) to form methylene-THF. The C-1 methylene unit of the latter is used primarily for purine biosynthesis but can also

$^-$OOC—CHO (Glyoxylate) + 2-Oxoglutarate —(Thiamin diphosphate, − CO_2)→ 2-Hydroxy-3-oxoadipate

(24-41)

Glycine —(PLP, Lipoate, THF, NAD^+ (Fig. 15-20); − CO_2, − NH_4^+)→ Methylene-THF —(Fig. 15-18)→ CO_2

Methylene-THF + Glycine → Serine → Pyruvate + NH_4^+

(24-42)

be oxidized to CO_2 or can condense with another molecule of glycine (Fig. 15-18, step *c*, reverse) to form serine. This can in turn be converted to pyruvate and utilized for biosynthetic processes (Eq. 24-42).

Glycine can be reduced to acetate and ammonia by the selenium-dependent clostridial glycine reductase system (Eq. 15-61). A variety of additional products can be formed from glycine as is indicated in Fig. 24-20. **Hippuric acid** (Box 10-A), the usual urinary excretion product in the "detoxication" of benzoic acid, is formed via benzoyl-CoA (Eq. 24-43):

Benzoyl—S—CoA + $H_3\overset{+}{N}$—CH_2—COO^- → Benzoyl—NH—CH_2—COO^- (Hippuric acid) + CoA—SH

(24-43)

N-Methylation yields the monomethyl derivative **sarcosine**[375] and also dimethylglycine, compounds that may function as osmoprotectants (Box 20-C). Many bacteria produce **sarcosine oxidase**, a flavoprotein that oxidizes its substrate back to glycine and formaldehyde, which can react with tetrahydrofolate.[376–377a] The formation of porphobilinogen and the various pyrrole pigments derived from it and the synthesis of the purine ring (Chapter 25) represent two other major routes for glycine metabolism.

3. Porphobilinogen, Porphyrins, and Related Substances

In 1946, Shemin and Rittenberg[378] described one of the first successful uses of radiotracers in the study of metabolism. They reported that the atoms of the porphyrin ring in heme have their origins in the simple compounds acetate and glycine. As we now know, acetate is converted to succinyl-CoA in the citric acid cycle. Within the mitochondrial matrix of animal cells succinyl-CoA condenses with glycine to form **5(δ)-aminolevulinic acid** (Eq. 14-32),[379–381a] which is converted to **porphobilinogen** (Fig. 24-21), the immediate precursor to the porphyrins. The same pathways lead also to other tetrapyrroles including chlorophyll, the nickel-containing F_{430}, vitamin B_{12}, and other corrins.[382,383]

By degradation of ^{14}C-labeled porphyrins formed from labeled acetate and glycine molecules, Shemin and Rittenberg established the labeling pattern for the pyrrole ring that is indicated for porphobilinogen in Fig. 24-21. The solid circles mark those atoms that were found to be derived from methyl carbon atoms of acetate (bear in mind that acetyl groups of acetyl-CoA pass around the citric acid cycle more than once to introduce label from the methyl group of acetate into both the 2 and 3 positions of succinyl-CoA). Those atoms marked with open circles in Fig. 24-21 were found to be derived mainly from the methyl carbon of acetate and in small part from the carboxyl carbon. Atoms marked with asterisks came from glycine, while unmarked carbon atoms came from the carboxyl carbon of acetate.[384]

In cyanobacteria and in chloroplasts the intact 5-carbon skeleton of glutamate enters δ-aminolevulinate.[139,385,385a] A surprising finding was that the glutamate becomes coupled to one of the three known glutamate isoacceptor tRNAs that are utilized for protein synthesis. The aminoacyl-tRNA is formed in the usual manner with an ester linkage to the CCA end of the tRNA (Eq. 24-44; see also Eq. 17-36). This ester linkage can be reductively cleaved by NADPH to form glutamate 1-semialdehyde.[386] Isomerization of the glutamate semialdehyde to δ-aminolevulinate is accomplished by an aminomutase that is structurally and functionally related to aminotransferases.[139,387] The enzyme utilizes pyridoxamine phosphate (PMP) to transaminate the substrate carbonyl group to form 4,5-diaminovalerate plus bound PLP. A second transamination step yields the product and regenerates the PMP (Eq. 24-44, steps *c* and *d*).[387]

Glu —(a: $tRNA^{Glu}$)→ $t\text{-}RNA^{Glu}$–O–C(=O)–CH(NH_3^+)–$CH_2CH_2COO^-$

—(b: NADPH + H^+ → $NADP^+$; releases $tRNA^{Glu}$)→ Glutamate 1-Semialdehyde (H–C(=O)–CH(NH_3^+)–$CH_2CH_2COO^-$)

—(c: PMP → PLP)→ 4,5-Diaminovalerate (H_2C(5)(NH_3^+)–C4H(NH_3^+)–…–C1OO^-)

—(d: PLP → PMP)→ 5(δ)-Aminolevulinate (NH_3^+–CH_2(δ)–C(=O)–CH_2–CH_2(α)–COO^-)

(24-44)

Porphyrins. As indicated in Fig. 24-21, the conversion of two molecules of 5-aminolevulinate into porphobilinogen is a multistep reaction initiated by **5-aminolevulinate dehydratase** (porphobilinogen synthase).[381,388–390b] The enzyme binds two molecules of substrate in distinct sites known as the A site and the P site (Fig. 24-21). The substrate in the P site forms a Schiff base with a lysine side chain (K247 in the *E. coli* enzyme), while a bound Zn^{2+} is thought to polarize the carbonyl of the substrate in the A site. An aldol condensation (Fig. 24-21, step *a*) ensues and is followed by dehydration to form a carbon–carbon double bond and ring closure (step *b*). Tautomerization step (*c*) leads to porphobilinogen. The enzyme is a sensitive target for poisoning by lead ions.[381,390c] Condensation to form porphyrins requires two enzymes, **porphobilinogen deaminase** (hydroxymethylbilane synthase) and **uroporphyrinogen III cosynthase**. Porphobilinogen deaminase has a bound coenzyme (prosthetic group) consisting of two linked pyrromethane groups, also derived from porphobilinogen.[391] The first step in assembling the porphyrin ring is condensation of porphobilinogen with this coenzyme (Fig. 24-21, step *d*). To initiate this step ammonia is eliminated, probably not by the direct displacements, but by electron flow from the adjacent nitrogen in the same pyrrole ring as indicated in the figure to give an exocyclic double bond. The terminal ring of the coenzyme then adds to the double bond. The condensation process is repeated four times to produce **preuroporphyrinogen** (hydroxymethylbilane).[392–394a] This intermediate is a precursor of the symmetric uroporphyrin I (Fig. 16-5). In the presence of the cosynthase a different ring-closure reaction takes place. The five-membered ring in porphobilinogen has a symmetric arrangement of double bonds. Thus, a condensation reaction can occur at either of the positions α to the ring nitrogen. A sequence of condensation, tautomerization, cleavage, and reformation of the ring as shown in steps *e* to *h* of Fig. 24-21 leads to the unsymmetric uroporphyrinogen III with its characteristic pattern of the carboxymethyl and carboxyethyl side chains. A series of decarboxylation and oxidation reactions then leads directly to protoporphyrin IX.

The first of these decarboxylations is catalyzed by the cytoplasmic **uroporphyrinogen decarboxylase,** which removes the carboxylate groups of the four acetate side chains sequentially from the D, A, B, and C rings.[395–396a] A possible mechanism, utilizing a tautomerized ring, is illustrated in the accompanying structural formula.

The decarboxylated product, coproporphyrinogen (Fig. 16-5), enters the mitochondria and is acted upon by **coproporphyrinogen oxidase,** which oxidatively

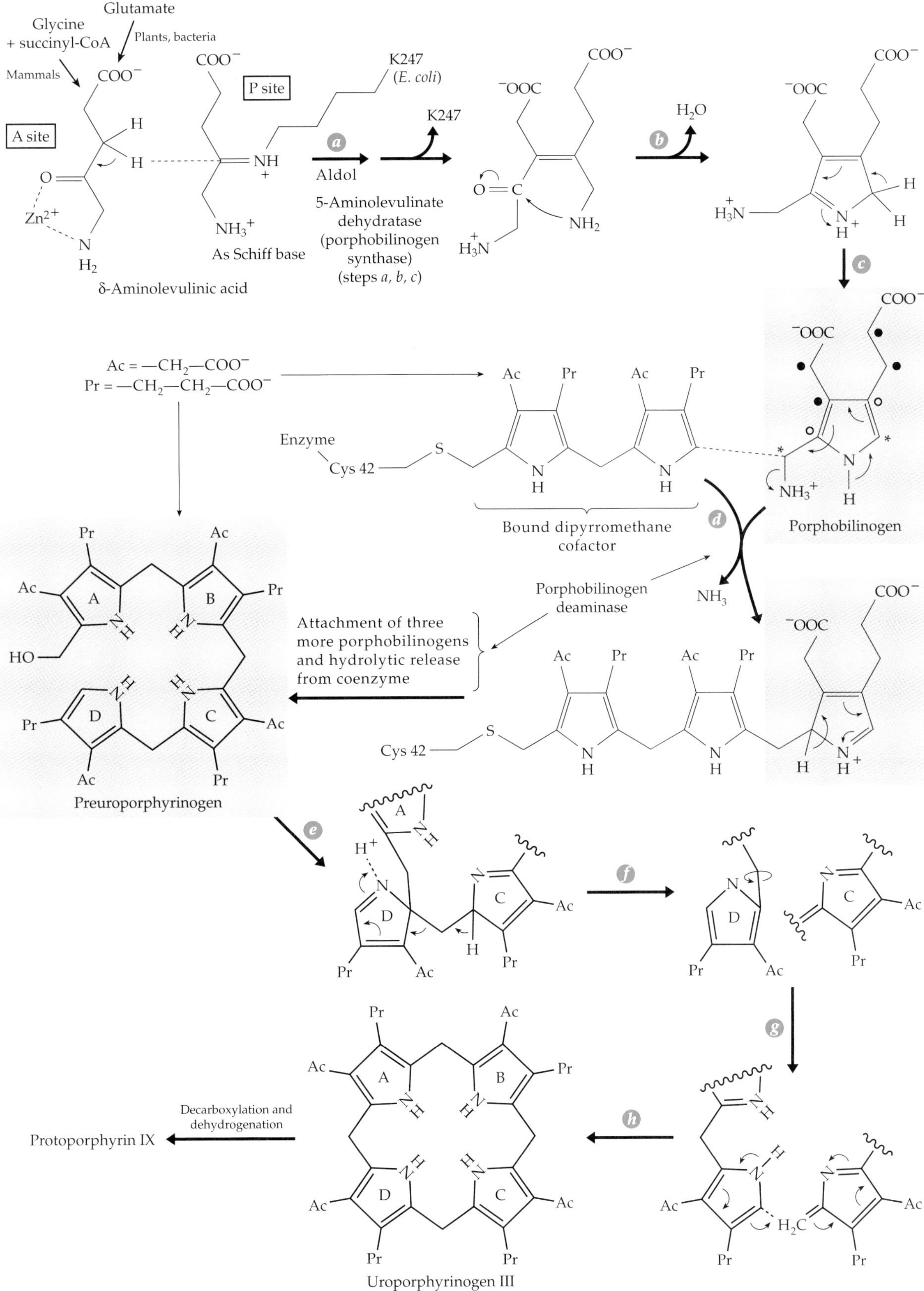

Figure 24-21 Biosynthesis of porphyrins, chlorins, and related compounds.

Tautomerized pyrrole ring of uroporphyrinogen undergoing decarboxylation

decarboxylates and oxidizes two of the propyl side chains to vinyl groups.[397] A flavoprotein, **protoporphyrinogen oxidase**, oxidizes the methylene bridges between the pyrrole rings[398,399] to form protoporphyrin IX. A somewhat different pathway from uroporphyrinogen is followed by sulfate-reducing bacteria.[400]

Ferrochelatase (protoheme ferro-lyase)[401–403] inserts Fe^{2+} into protoporphyrin IX to form heme. The enzyme is found firmly bound to the inner membrane of mitochondria of animal cells, chloroplasts of plants, and chromatophores of bacteria. While Fe^{2+} is apparently the only metallic ion ordinarily inserted into a porphyrin, the Zn^{2+} protoporphyrin chelate accumulates in substantial amounts in yeast, and Cu^{2+}–heme complexes are known (p. 843). Ferrochelatase, whose activity is stimulated by Ca^{2+}, appears to be inhibited by lead ions, a fact that may account for some of the acute toxicity of lead.[404]

Heme *b* is utilized for formation of hemoglobin, myoglobin, and many enzymes. It reacts with appropriate protein precursors to form the cytochromes *c*. Heme *b* is converted by prenylation to heme *o*[405] and by prenylation and oxidation to heme *a*.[405a] The porphyrin biosynthetic pathway also has a number of branches that lead to formation of corrins, chlorins, and chlorophylls as shown schematically in Fig. 24-22.

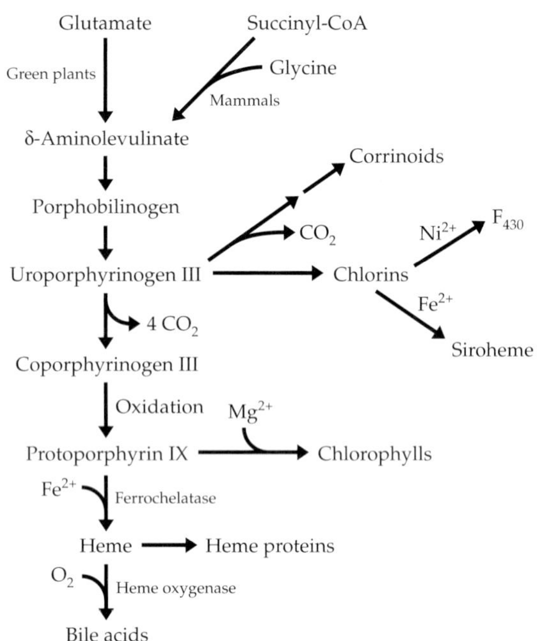

Figure 24-22 Abbreviated biosynthetic pathways from δ-aminolevulinate to heme proteins, corrins, chlorophylls, and related substances.

Corrins. The formation of vitamin B_{12}, other corrins, siroheme, and related chlorin chelates[406,407] requires a ring contraction with elimination of the methine bridge between rings A and D of the porphyrins (see Box 16-B). It is natural to assume that the methyl group at C-1 of the corrin ring might arise from the same precursor carbon atom as does the methine bridge in porphyrins, and it is easy to visualize a modified condensation reaction by which ring closure at step *e* in Fig. 24-21 occurs by nucleophilic addition to a C=N bond of ring A. However, ^{13}C-NMR data ruled out this possibility. When vitamin B_{12} was synthesized in the presence of ^{13}C-methyl-containing methionine, it was found that seven methyl groups contained ^{13}C. All of the "extra" methyl groups around the periphery of the molecule as well as the one at C-1 were labeled.[408] Other experiments established uroporphyrinogen III as a precursor of vitamin B_{12}. Therefore, it appeared that the ring first closed in a normal way and then reopened between rings A and D with removal of the carbon that forms the methylene bridge.[409] This turned out to be true, but with some surprises.

The complex pathways of corrin synthesis have been worked out in detail.[410–415a] This has been possible because of extensive use of ^{13}C and ^{1}H NMR and because the group of ~20 enzymes required has been produced in the laboratory from genes cloned from *Pseudomonas denitrificans*.[410] The first alterations of uroporphyrinogen are AdoMet-dependent methylations on carbon atoms. Surprisingly, one of these is on the bridge atom that is later removed. The details, including the insertion of Co^{2+} by a **cobaltochelatase**, are described by Battersby[410] and portrayed in Michal's *Biochemical Pathways*.[416]

Chlorophyll. The pathway of chlorophyll synthesis has been elucidated through biochemical genetic studies of *Rhodobacter spheroides*[417–418a] which produces bacteriochlorophyll, from studies of cyanobacteria,[419,420] and from investigations of green algae and higher plants,[421] which make chlorophyll *a*. The first step in the conversion of protoporphyrin IX into chlorophyll is the insertion of Mg^{2+} (Fig. 24-23, step *a*). This reaction does not occur readily spontaneously but is catalyzed by an ATP-dependent **magnesium protoporphyrin chelatase**.[419,422] Subsequently, the carboxyethyl side chain on ring C undergoes methylation (Fig. 24-23, step *b*) and β oxidation (step *c*).

Oxidative closure of ring E (step *d*) is followed by reduction of the vinyl group of ring B and of the double bond in ring D to form **chlorophyllide *a***. The latter is coupled with phytol, via phytyl diphosphate, to form chlorophyll *a*.[420] Chlorophyll *b* is derived from chlorophyll *a*, evidently by action of an as yet uncharacterized oxygenase, which converts the methyl group on ring B into a formyl group.[423,424] Bacteriochlorophylls also arise from chlorophyllide *a* and involve reduction of the double bond in ring B.[416,418,420] Most photosynthetic bacteria make bacteriochlorophylls esterified with the C_{20} phytol, but some substitute the unsaturated C_{20} geranylgeranyl group and a variety of other isoprenoid alcohols.

The porphyrias. The human body does not use all of the porphobilinogen produced, and a small amount is normally excreted in the urine, principally as coproporpyrins (Fig. 16-5). In a number of hereditary and acquired conditions blood porphyrin levels are elevated and enhanced urinary excretion (porphyria) is observed.[425–427] Porphyrias may be mild and almost without symptoms, but the intensely fluorescent free porphyrins are sometimes deposited under the skin and cause photosensitivity and ulceration. In extreme cases, in which the excreted porphyrins may color the urine a wine red, patients may have acute neurological attacks and a variety of other symptoms. Lucid accounts of such symptoms, experienced by King George III of England, have been written.[426,428] However, there are doubts about the conclusion that the king suffered from porphyria.[425]

Porphyria may result from several different enzyme deficiences in the porphyrin biosynthetic pathway. The condition is often hereditary but may be induced by drugs or other xenobiotic substances and may be continuous or intermittent.[425,426,429] In one type of congenital porphyria uroporphyrin I is excreted in large quantities. The biochemical defect appears to be a deficiency of the cosynthase that is required for formation of protoporphyrin IX. Another type of porphyria results from overproduction in the liver of

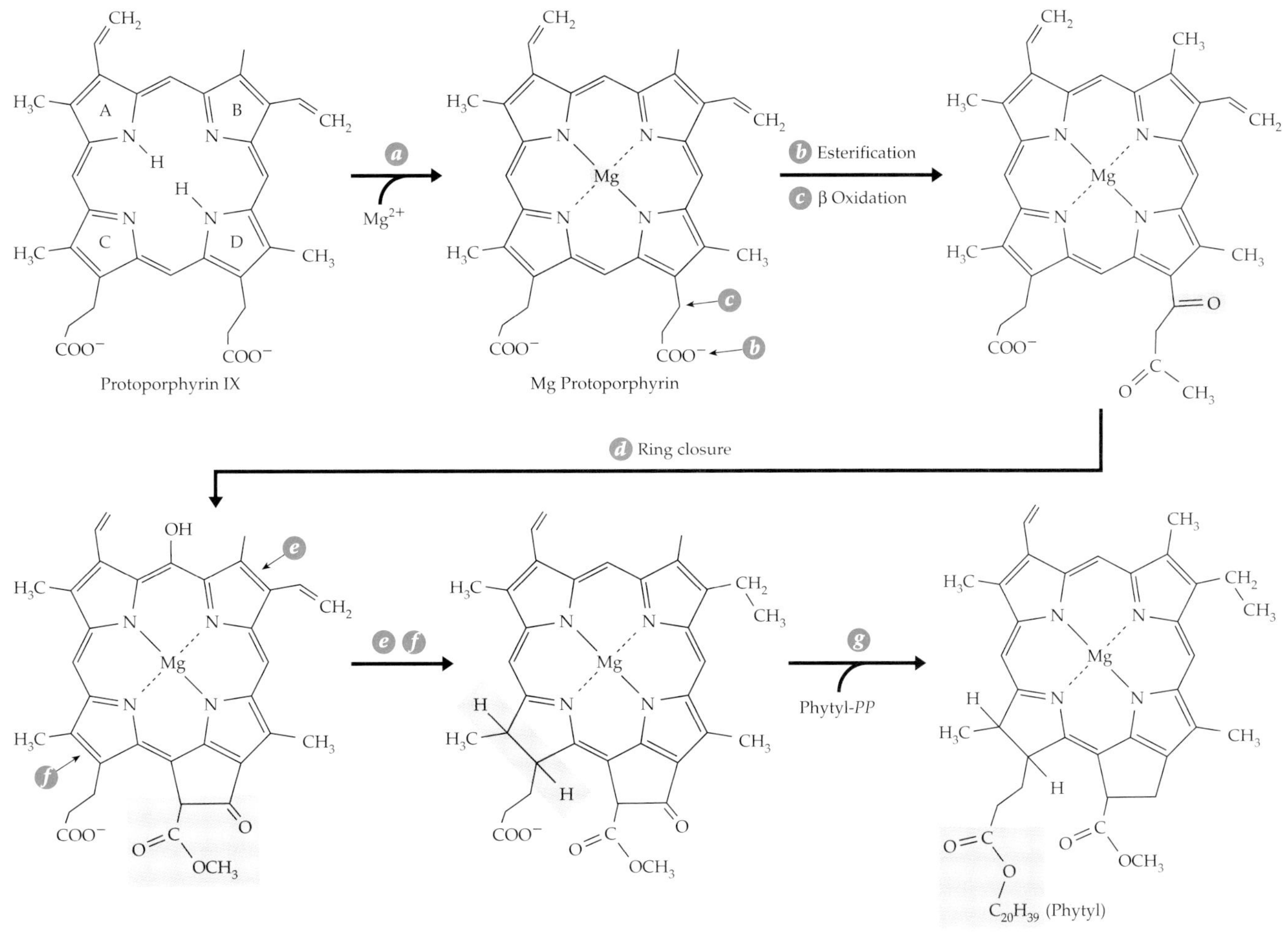

Figure 24-23 Outline of the biosynthetic pathways for conversion of protoporphyrin IX into the chlorophylls and bacteriochlorophylls. After Bollivar *et al.*[417]

δ-aminolevulinic acid, a compound with neurotoxic properties, possibly as a result of its similarity to the neurotransmitter γ-aminobutyrate.[426,427] This may account for some of the neurological symptoms of porphyria.

Some mild forms of intermittent porphyria may go unrecognized. However, ingestion of drugs can precipitate an acute attack, probably by inducing excessive synthesis of δ-aminolevulinate synthase. Among compounds having this effect are hexachlorobenzene and tetrachlorodibenzodioxin.

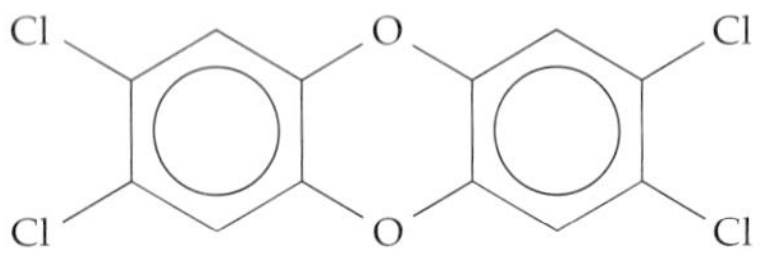

Cl O Cl

Cl O Cl

2,3,7,8-Tetrachlorodibenzo-*p*-dioxin

The latter is one of the most potent inducers of the synthase known.[430] The tendency for this dioxin to be present as an impurity in the herbicide 2,4,5-trichlorophenoxyacetic acid (2,4,5-T) has caused concern. For rodents this dioxin may be the most toxic small molecule known, the oral LD_{50} for guinea pigs being only 1 μg / kg body weight.[430] However, it is over 1000 times less toxic to humans.[430a] It is also a potent teratogenic agent. Synthesis of porphyrins in the liver is controlled by δ-aminolevulinate synthase. This key enzyme is sensitive to feedback inhibition by heme, but the increased synthesis of the enzyme induced by drugs can override the inhibition. Several times as much heme is synthesized in the erythroid cells of bone than in liver, but this is not subject to feedback regulation or to stimulation by drugs.[427] Heme is a potent and toxic regulator. Malaria mosquitos, which utilize blood for food but do not have a heme oxygenase (Fig. 24-24), detoxify heme by inducing its aggregation into an insoluble hydrogen-bonded solid known as β-hematin.[431]

The bile pigments. The enzymatic degradation of heme is an important metabolic process if only because it releases iron to be reutilized by the body. Some of the pathways are illustrated in Fig. 24-24. The initial oxidative attack is by the microsomal **heme oxygenases,**[432–434a] which catalyze the uptake of three molecules of O_2, formation of CO, and release of the chelated Fe. The electron transport protein NADPH-cytochrome P450 reductase brings electrons from NADPH to the oxygenase. An enzyme–substrate heme complex is formed with the oxygenase. Then the Fe^{3+} is reduced to Fe^{2+} which binds O_2 as in myoglobin or hemoglobin. The complex hydroxylates its own heme α-carbon (Fig. 24-24), the other oxygen atom being reduced to OH^- by the Fe^{2+} and an additional electron from NADPH. The same enzyme catalyzes the next steps in which the α carbon is split out as CO by reaction with two molecules of O_2 to form the open tetrapyrrole dicarbonyl compound **biliverdin**, one of the bile pigments (Fig. 24-24).[435–435a] When $^{18}O_2$ was used, it was found that the biliverdin contains two atoms of ^{18}O, and that the CO contains one. Heme from the cytochromes *c* appears to be degraded by the same enzymes after proteolytic release from the proteins to which it is bound.[436]

There are two human heme oxygenases. The first (HO-1) is synthesized principally in the liver and spleen. Its formation is strongly induced by heme. The second heme oxygenase (HO-2) is distributed widely among tissues, but it is most abundant in certain neurons in the brain.[437,437a] Its major function may be to generate CO, which is now recognized as a probable neurohormone (Chapter 30). Bacteria, such as *Corynebacterium diphtheriae,* employ their own heme oxygenase as a means of recovering iron that they need for growth.[438]

A large number of other open tetrapyrroles can be formed from biliverdin by reduction or oxidation reactions. Within our bodies biliverdin is reduced to **bilirubin**, which is transported to the liver as a complex with serum albumin. In the liver bilirubin is converted into glucuronides (Eq. 20-16), glycosylation occurring on the propionic acid side chains.[439] A variety of these bilirubin conjugates are excreted into the bile. In the intestine they are hydrolyzed back to free bilirubin, which is reduced by the action of intestinal bacteria to urobilinogen, stercobilinogen, and *meso*-bilirubinogen. These compounds are colorless but are readily oxidized by oxygen to **urobilin** and **stercobilin**. Some of the urobilin and other bile pigments is reabsorbed into the blood and excreted into the urine where it provides the familiar yellow color.

The yellowing of the skin known as **jaundice** can occur if the heme degradation system is overburdened (e.g., from excessive hemolysis), if the liver fails to conjugate bilirubin, or if there is obstruction of the flow of heme breakdown products into the intestinal tract. Bilirubin is toxic, and continued exposure to excessive bilirubin levels can cause brain damage.[434,439] Bilirubin has a low water solubility and tends to form complexes with various proteins, perhaps partly because it assumes folded conformations rather than the linear one shown in Fig. 24-24.[440] These properties make it difficult to excrete. Thousands of newborn babies are treated for jaundice every year by prolonged irradiation with blue or white light which isomerizes 4*Z*,15*Z* bilirubin to forms that are more readily transported, metabolized, and excreted.[441] A more difficult problem is posed by the fatal deficiency of the glucuronosyl transferase responsible for formation of bilirubin glucuronide. Efforts are being made to develop a genetic therapy.[439]

The open tetrapyrroles of algae and the chromophore of phytochrome (Chapter 23) are all derived from **phycoerythrobilin**, which is formed from biliverdin, as indicated in Fig. 24-24. The animal bile pigments have not been found in prokaryotes. However, *Clostridium tetanomorphum*, which accumulates uroporphyrinogen III, a precursor to vitamin B_{12}, and does not synthesize protoporphyrin IX, makes a blue bile pigment **bactobilin**. This is a derivative of uroporphyrin rather than of protoporphyrin.[442]

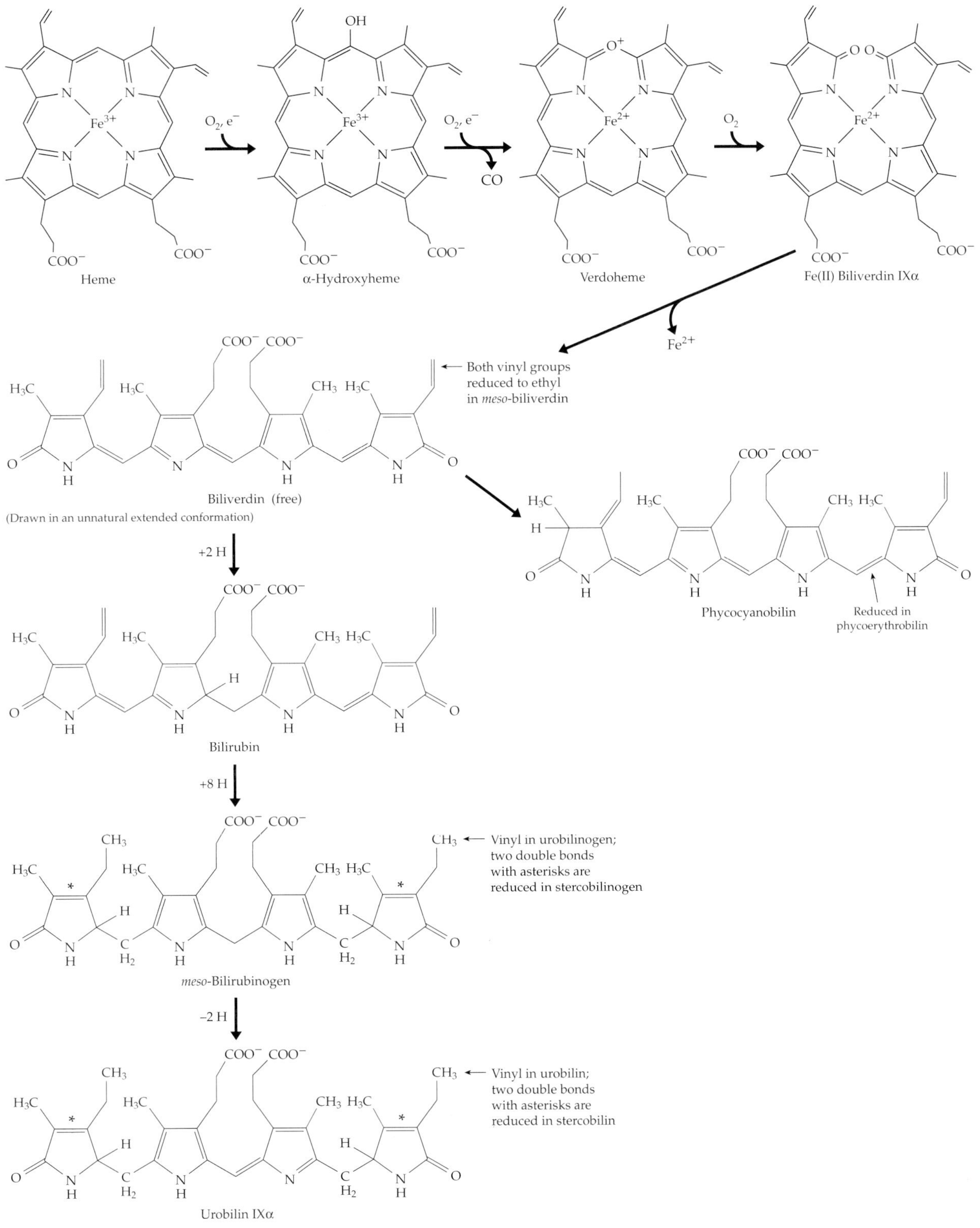

Figure 24-24 The degradation of heme and the formation of open tetrapyrrole pigments.

G. Cysteine and Sulfur Metabolism

Cysteine not only is an essential constituent of proteins but also lies on the major route of incorporation of inorganic sulfur into organic compounds.[443] Autotrophic organisms carry out the stepwise reduction of sulfate to sulfite and sulfide (H_2S). These reduced sulfur compounds are the ones that are incorporated into organic substances. Animals make use of the organic sulfur compounds formed by the autotrophs and have an active oxidative metabolism by which the compounds can be decomposed and the sulfur reoxidized to sulfate. Several aspects of cysteine metabolism are summarized in Fig. 24-25. Some of the chemistry of inorganic sulfur metabolism has been discussed in earlier chapters. Sulfate is reduced to H_2S by sulfate-reducing bacteria (Chapter 18). The initial step in *assimilative* sulfate reduction, used by autotrophs including green plants and *E. coli*, is the formation of adenosine 5'-phosphosulfate (APS) (step *a*, Fig. 24-25; see also Eq. 17-38).[444–444c] The sulfate-reducing bacteria reduce adenylyl sulfate directly to sulfite (Eq. 18-32, step *b*), but the assimilative pathway of reduction in *E. coli* proceeds through 3'-phospho-5'-adenylyl sulfate (PAPS), a compound whose function as "active sulfate" has been considered in Chapter 17. Reduction of PAPS to sulfite (Fig. 24-25, step *d*) is accomplished by an NADPH-dependent enzyme.

The same pathway is found in the alga *Chlorella*, but a second route of sulfate reduction occuring in green plants may be more important.[445] Adenylyl sulfate transfers its sulfo group to a thiol group of a carrier (Eq. 24-43, step *a*). The resulting thiosulfonate is reduced by a ferredoxin-dependent reductase. Finally, a sulfide group is transferred from the –S—S⁻ group of the reduced carrier directly into cysteine in a

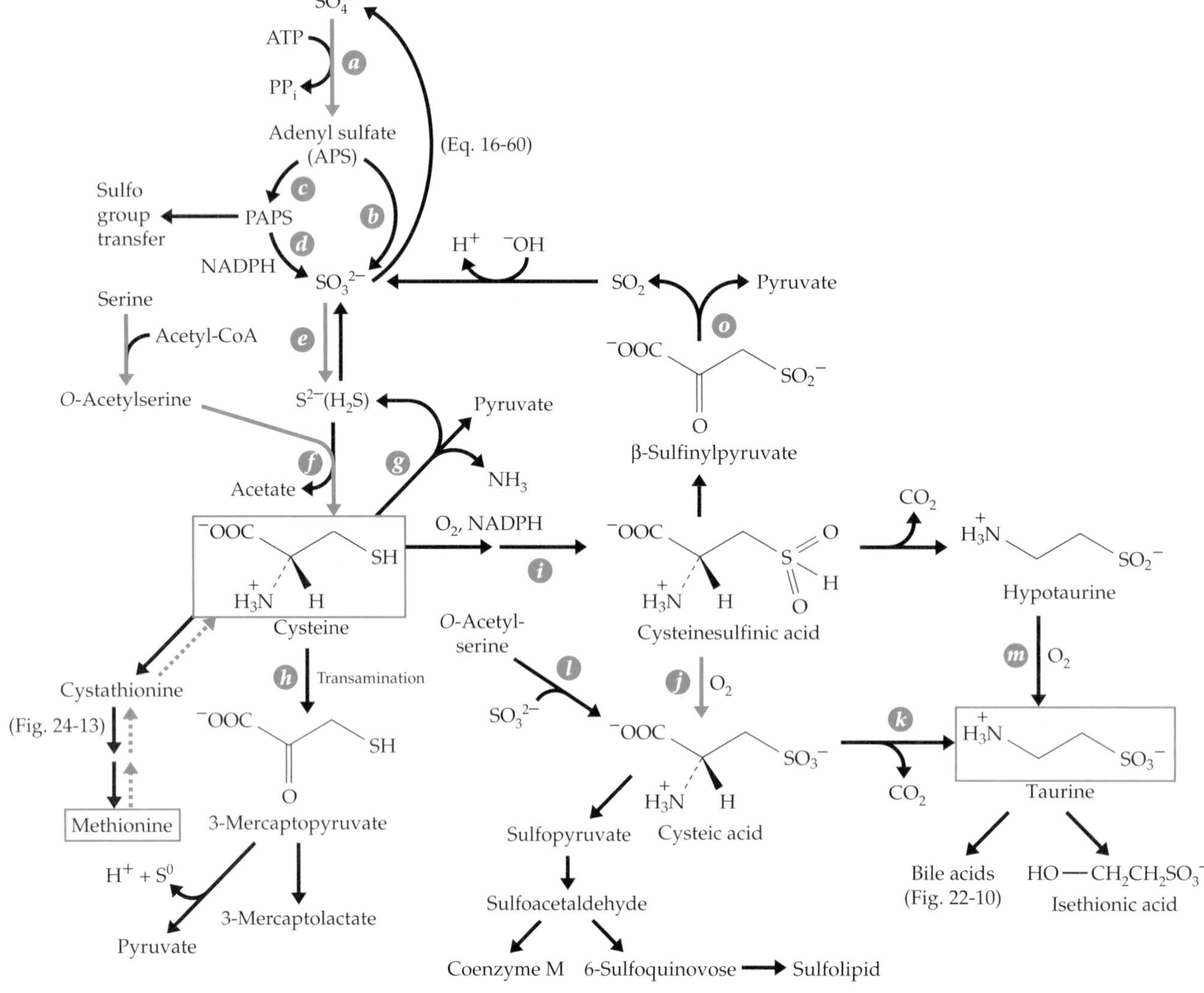

Figure 24-25 Pathways of biosynthesis (green arrows) and catabolism of cysteine as well as other aspects of sulfur metabolism. Solid arrows are major biosynthetic pathways. The dashed arrows represent more specialized pathways; they also show processes occurring in the animal body to convert methionine to cysteine and to degrade the latter.

Adenosine — O — P(=O)(O⁻) — O — S(=O)(=O) — O⁻

Adenylyl sulfate (APS)

a: AMP

Carrier — S — S(=O)(=O) — O⁻ Carrier — SH

$6\ Fd_{Red}$ b $6\ Fd_{ox}$

Carrier — S — S⁻

c

O-Acetylserine

Acetate Cysteine (24-43)

β-substitution reaction analogous to that described in the next paragraph.

1. Synthesis and Catabolism of Cysteine

Cysteine is formed in plants and in bacteria from sulfide and serine after the latter has been acetylated by transfer of an acetyl group from acetyl-CoA (Fig. 24-25, step *f*). This standard PLP-dependent β replacement (Chapter 14) is catalyzed by **cysteine synthase** (*O*-acetylserine sulfhydrase).[446,447] A similar enzyme is used by some cells to introduce sulfide ion directly into homocysteine, via either *O*-succinyl homoserine or *O*-acetyl homoserine (Fig. 24-13). In *E. coli* cysteine can be converted to methionine, as outlined in Eq. 16-22 and as indicated on the right side of Fig. 24-13 by the green arrows. In animals the converse process, the conversion of methionine to cysteine (gray arrows in Fig. 24-13, also Fig. 24-16), is important. Animals are unable to incorporate sulfide directly into cysteine, and this amino acid must be either provided in the diet or formed from dietary methionine. The latter process is limited, and cysteine is an essential dietary constituent for infants. The formation of cysteine from methionine occurs via the same transsulfuration pathway as in methionine synthesis in autotrophic organisms. However, the latter use cystathionine γ-synthase and β-lyase while cysteine synthesis in animals uses cystathionine β-synthase and γ-lyase.

Some bacteria degrade L-cysteine or D-cysteine[447a] via the PLP-dependent α,β elimination to form H_2S, pyruvate, and ammonia (reaction *g*, Fig. 24-25, Eq. 14-29). Another catabolic pathway is transamination (Fig. 24-25, rection *h*) to **3-mercaptopyruvate**.[448] The latter compound can be reductively cleaved to pyruvate and sulfide. Cysteine can also be oxidized by NAD^+ and lactate dehydrogenase to 3-mercaptopyruvate. An interesting PLP-dependent β-replacement reaction of cysteine leads to **β-cyanoalanine**, the lathyritic factor (Box 8-E) present in some plants.[449] This reaction also detoxifies the HCN produced during the biosynthesis of ethylene from ACC.

Cysteine and cystine are relatively insoluble and are toxic in excess.[450] Excretion is usually controlled carefully. However, in **cystinuria**, a disease recognized in the medical literature since 1810,[451] there is a greatly increased excretion of cystine and also of the dibasic amino acids.[451,452] As a consequence, stones of cystine develop in the kidneys and bladder. Patients may excrete more than 1 g of cystine in 24 h compared to a normal of 0.05 g, as well as excessive amounts of lysine, arginine, and ornithine. The defect can be fatal, but some persons with the condition remain healthy indefinitely. Cystinuria is one of several human diseases with altered membrane transport and faulty reabsorption of materials from kidney tubules or from the small intestine. Substances are taken up on one side of a cell (e.g., at the bottom of the cell in Fig. 1-6) and discharged into the bloodstream from the other side of the cell. In another rare hereditary condition, **cystinosis**, free cystine accumulates within lysosomes.[453]

2. Cysteine Sulfinate and Taurine

A quantitatively important pathway of cysteine catabolism in animals is oxidation to **cysteine sulfinate** (Fig. 24-25, reaction *i*),[450] a two-step hydroxylation requiring O_2, NADPH or NADH, and Fe^{2+}. Cysteine sulfinic acid can be further oxidized to **cysteic acid** (cysteine sulfonate),[454] which can be decarboxylated to **taurine**. The latter is a component of bile salts (Fig. 22-16) and is one of the most abundant free amino acids in human tissues.[455–457] Its concentration is high in excitable tissues, and it may be a neurotransmitter (Chapter 30). Taurine may have a special function in retinal photoreceptor cells. It is an essential dietary amino acid for cats, who may die of heart failure in its absence,[458] and under some conditions for humans.[459] In many marine invertebrates, teleosts, and amphibians taurine serves as a regulator of osmotic pressure, its concentration decreasing in fresh water and increasing in salt water. A similar role has been suggested for taurine in mammalian hearts. A chronically low concentration of Na^+ leads to increased taurine.[460] Taurine can be reduced to **isethionic acid**

(Fig. 24-25), another component of nervous tissue. Cysteic acid can arise in an alternative way from *O*-acetylserine and sulfite (reaction l, Fig. 24-25), and taurine can also be formed by decarboxylation of cysteine sulfinic acid to **hypotaurine** and oxidation of the latter (reaction *m*). Cysteic acid can be converted to the sulfolipid of chloroplasts (p. 387; Eq. 20-12).

Another route of metabolism for cysteine sulfinic acid is transamination to 3-sulfinylpyruvate, a compound that undergoes ready loss of SO_2 in a reaction analogous to the decarboxylation of oxaloacetate (reaction *o*, Fig. 24-25). This probably represents one of the major routes by which sulfur is removed from organic compounds in the animal body. However, before being excreted the sulfite must be oxidized to sulfate by the Mo-containing sulfite oxidase. The essentiality of sulfite oxidase is evidenced by the severe neurological defect observed in its absence (Chapter 16).

Most of the sulfate generated in the body is excreted unchanged in the urine, but a significant fraction is esterified with oligosaccharides and phenolic compounds. These sulfate esters are formed by sulfo transfer from PAPS (Eq. 17-38).

3. Mercaptopyruvate, Thiosulfate, and Assembly of Iron–Sulfur Centers

An important property of 3-mercaptopyruvate arises from electron withdrawal by the carbonyl group. This makes the SH group electrophilic and able to be transferred as SH^+, S^0, to a variety of nucleophiles (Eq. 24-44). Thus sulfite yields thiosulfate ($S_2O_3^{2-} + H^+$, Eq. 24-45, step *a*), cyanide yields thiocyanate (Eq. 24-45, step *b*), and cysteine sulfinate yields alanine thiosulfonate.[448,461] The reactions are catalyzed by **mercaptopyruvate sulfurtransferase**, an enzyme very similar to **thiosulfate sulfurtransferase**. The latter is a liver enzyme often called by the traditional

BOX 24-B SULFUR COMPOUNDS OF GARLIC, ONIONS, SKUNKS, ETC.

Many familiar odors and tastes come from sulfur-containing compounds. Crushing onions or garlic releases the pyridoxal phosphate-dependent enzyme **alliinase**. In garlic it acts upon the amino acid **alliin** (accompanying scheme) releasing, by β elimination, a sulfenic acid that dimerizes to form **allicin**, a chemically unstable molecule that accounts for the odor of garlic.[a,b] Among the breakdown products of allicin is the nonvolatile **ajoene**, a compound with anticoagulant activity and perhaps accounting for one aspect of the purported medical benefits of garlic. Another is an antibacterial activity.

Onions contain an amino acid that is a positional isomer of alliin. When acted upon by alliinase it produces 2-propene sulfenic acid, which isomerizes to the **lachrimatory factor** that brings tears to the eyes of onion cutters.[c] This, too, decomposes to form many other compounds.[a]

The defensive secretion of the striped skunk has intrigued chemists for over 100 years. The components were shown to contain

Alliin (garlic)

PLP → Pyruvate + NH_4^+

Sulfenic acid intermediate

Two molecules

Allyl-2- propenethiolsulfinate (allicin)

trans-Ojoene and other products

S-1-Propenyl-L-cysteine sulfoxide (onions)

2-Propene sulfenic acid

Lachrimatory factor

Decomposition

Various compounds

name **rhodanese**.[462–463b] It acts by a double displacement mechanism, a thiolate anion in the active site of the enzyme serving as a carrier for the S atom being transferred. Equation 24-46 illustrates the transfer of S^0 from thiosulfate to CN^-, converting that ion to the less toxic thiocyanate (also shown as step *c* in Eq. 24-45). Crystallographic studies show that the negative charge on the thiol and dithiol anions of rhodanese is balanced by the partial

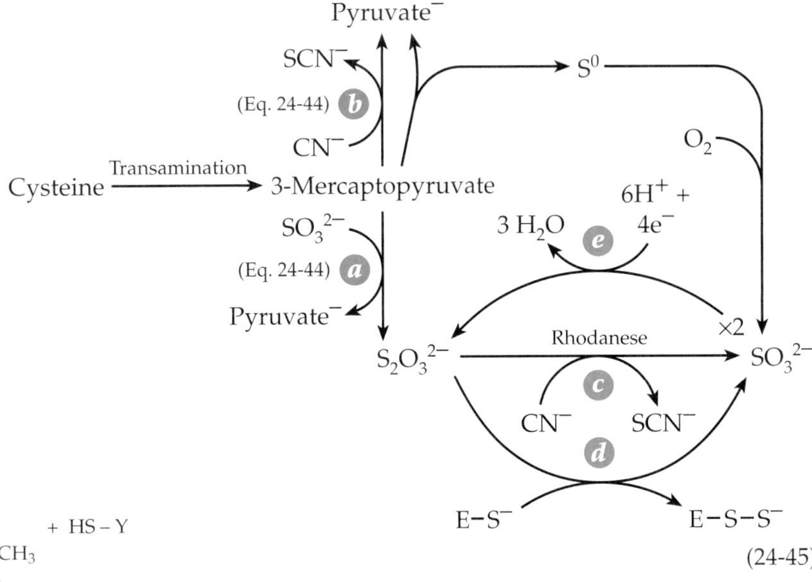

(24-45)

^-OOC–CH_2–S(H) + Y^- → Pyruvate + HS – Y

(24-44)

BOX 24-B (continued)

sulfur, and one was incorrectly identified and was long accepted as being butyl mercaptan. Modern capillary gas chromatography by Wood[d] has revealed the presence of seven major components, with the indicated structures. Two are simple volatile mercaptans, but three are thioacetates which hydrolyze in water only slowly, releasing smell for days or weeks from sprayed animals. Washing with mildly basic soap hastens the hydrolysis.

Many readers (~40%) may be aware that after eating asparagus a strong odor appears in their urine. These genetic "stinkers" secrete *S*-methyl thioacrylate, and related compounds, derived from a plant constituent.[e]

The sulfur compound sulfuraphane, extracted from fresh broccoli, has received attention in recent years because of its strong action in inducing synthesis of quinone reductase and glutathione *S*-transferases that help detoxify xenobiotics and may have significant anticancer activity.[f]

Sulforaphane;
1-isothiocyanate-(4*R*)-(methylsulfinyl)butane

[a] Block, E. (1985) *Sci. Am.* **252**(Mar), 114–119

[b] Jain, M. K., and Apitz-Castro, R. (1987) *Trends Biochem. Sci.* **12**, 252–254

[c] Imai, S., Tsuge, N., Tomotake, M., Nagatome, Y., Sawada, H., Nagata, T., and Kumagai, H. (2002) *Nature (London)* **419**, 685

[d] Wood, W. F. (1990) *J. Chem. Ecol.* **16**, 2057

[e] White, R. H. (1975) *Science* **189**, 810

[f] Zhang, Y., Talalay, P., Cho, C.-G., and Posner, G. H. (1992) *Proc. Natl. Acad. Sci. U.S.A.* **89**, 2399–2403

$$S-SO_3^{2-} + E-S^- \rightarrow SO_3^{2-} + E-S-S^- ; \quad E-S-S^- + CN^- \rightarrow E-S^- + SCN^- \quad (24\text{-}46)$$

positive charges at the N termini of two helices (see Fig. 2-20A) and by hydrogen bonds to protons of several side chains.[462] This evidently explains how the negative thiosulfate anion can react with another anion, $E-S^-$. Another interesting feature of this enzyme is that the monomer has a nearly perfect twofold axis of symmetry with respect to the protein folding pattern. However, the symmetry is lacking in the sequence and only one-half of the molecule contains an active site.[462]

Yet another enzyme able to release or transfer sulfur in the S^0 oxidation state is the PLP-dependent **cysteine desulfurase** that is encoded by the *nifS* gene of the nitrogenase gene cluster shown in Fig. 24-4. This enzyme releases S^0 from cysteine with formation of alanine[464–466] as is shown in Eq. 14-34 for release of Se^0 from selenocysteine. As with rhodanese an active site cysteine accepts the departing S^0 of cysteine to form an enzyme-bound persulfide. This protein may in turn transfer the sulfur into the forming Fe–S or Fe–S–Mo clusters.[464] Three PLP-dependent persulfide-forming sulfurtransferases related to the NifS protein have been found in *E. coli*. Similar enzymes are present in other organisms.[466a–d] A sulfur atom may be transferred from the bound persulfide anion to acceptor proteins involved in metal cluster formation. Some members of the nifS-like family act on cystine to release free thiocysteine (cysteine peroxide), which may also serve as a sulfur atom donor.[466e]

$^-OOC-CH(NH_3^+)-CH_2-S-SH$

Thiocysteine (cysteine persulfide)

Thiocysteine can also arise in a similar manner through action of cystathionine β lyase on cystine. Thiocysteine is eliminated with production of pyruvate and ammonia from the rest of the cystine molecule.[467] One of the nifS-like proteins of *E. coli* is thought to transfer a selenium atom from selenocysteine (pp. 823–827) into **selenophosphate**.[466a,f] The latter can be formed by transfer of a phospho group from ATP to selenide HSe^-. The other products of ATP cleavage are AMP and P_i. Reduction of Se^0 to HSe^- is presumably necessary.

Several additional proteins identified as necessary for metal-sulfide cluster formation are present in bacteria and in eukaryotes, both in the cytosol and in mitochondria. They may serve as intermediate sulfur carriers, as scaffolds or templates for cluster formation, or for insertion of intact Fe–S, Fe–S–Mo, or other types of clusters into proteins[468–473f] and into 2-selenouridine[473g] (see also p. 1617). Sulfurtransferases are also thought to be involved in insertion of sulfur atoms into organic molecules such as biotin, lipoic acid, or methanopterin.[474]

A reaction that is ordinarily of minor consequence in the animal body but which may be enhanced by a deficiency of sulfite oxidase is the reductive coupling of two molecules of sulfite to form thiosulfate (Eq. 24-45, step *e*). Several organic hydrodisulfide derivatives such as thiocysteine, thioglutathione, and thiotaurine occur in animals in small amounts. Another biosynthetic pathway, outlined in Eq. 24-47 converts sulfite and PEP into coenzyme M (Fig. 15-22).[475,475a] This cofactor is needed not only for methane formation (Fig. 15-2) but also for utilization of alkenes by soil bacteria.[475b]

$$H_2C{=}C(O-P)COO^- \text{ (PEP)} \xrightarrow{HSO_3^- \;\to\; P_i} {}^-O_3S-CH_2-C(H)(OH)-COO^- \text{ (Sulfolactate)}$$

$$\text{Sulfolactate} \rightarrow \text{Sulfopyruvate} \rightarrow \rightarrow \text{6-Sulfoquinovose (Fig. 4-4; Eq. 20-12)}$$

$$\text{Sulfopyruvate} \xrightarrow{-CO_2} {}^-O_3S-CH_2-CHO \text{ (Sulfoacetaldehyde)}$$

$$\text{Sulfoacetaldehyde} + \text{Cysteine} \rightarrow \text{Thiazolidine} \xrightarrow{2e^- + 2H^+} {}^-O_3S-CH_2-CH_2-S-CH_2-CH(NH_3^+)-COO^- \rightarrow {}^-O_3S-CH_2-CH_2-SH \text{ (Coenzyme M)} \quad (24\text{-}47)$$

References

1. Conn, E. E., Stumpf, P. K., Bruening, G., and Doi, R. H. (1987) *Outlines of Biochemistry*, 5th ed., Wiley, New York
2. Smil, V. (1997) *Sci. Am.* **277**(Jul), 76–81
3. Hardy, R. W. F., and Havelka, U. D. (1975) *Science* **188**, 633–643
4. Winogradsky, S. (1893) *C.R. Acad. Sci.* **116**, 1385–1388
5. Orme-Johnson, W. H. (1985) *Ann. Rev. Biophys. Biophys. Chem.* **14**, 419–459
6. Appleby, C. A., Nicola, N. A., Hurrell, J. G. R., and Leach, S. J. (1975) *Biochemistry* **14**, 4444–4450
7. Lee, H. C., Wittenberg, J. B., and Peisach, J. (1993) *Biochemistry* **32**, 11500–11506
8. Stouggard, J., Petersen, T. E., and Marcker, K. A. (1987) *Proc. Natl. Acad. Sci. U.S.A.* **84**, 5754–5757
9. Stacey, G., Burris, R. H., and Evans, H. J., eds. (1992) *Biological Nitrogen Fixation*, Chapman and Hall, London

9a. Capone, D. G., Zehr, J. P., Paerl, H. W., Bergman, B., and Carpenter, E. J. (1997) *Science* **276**, 1221–1229

10. Appleby, C. A. (1984) *Ann. Rev. Plant Physiol.* **35**, 443–478

10a. Lou, J., Moshiri, F., Johnson, M. K., Lafferty, M. E., Sorkin, D. L., Miller, A.-F., and Maier, R. J. (1999) *Biochemistry* **38**, 5563–5571

11. Lowe, D. J., Fisher, K., Thorneley, R. N. F., Vaughn, S. A., and Burgess, B. K. (1989) *Biochemistry* **28**, 8460–8466
12. Rasche, M. E., and Seefeldt, L. C. (1997) *Biochemistry* **36**, 8574–8585
13. Georgiadis, M. M., Komiya, H., Chakrabarti, P., Woo, D., Kornuc, J. J., and Rees, D. C. (1992) *Science* **257**, 1653–1659
14. Kim, J., and Rees, D. C. (1994) *Biochemistry* **33**, 389–397
15. Howard, J. B., and Rees, D. C. (1994) *Ann. Rev. Biochem.* **63**, 235–264
16. Peters, J. W., Fisher, K., and Dean, D. R. (1995) *Ann. Rev. Microbiol.* **49**, 335–366
17. Kim, J., and Rees, D. C. (1992) *Nature (London)* **360**, 553–560
18. Kim, J., Woo, D., and Rees, D. C. (1993) *Biochemistry* **32**, 7104–7115
19. Schindelin, H., Kisker, C., Schlessman, J. L., Howard, J. B., and Rees, D. C. (1997) *Nature (London)* **387**, 370–376

19a. Mayer, S. M., Lawson, D. M., Gormal, C. A., Roe, S. M., and Smith, B. E. (1999) *J. Mol. Biol.* **292**, 871–891

20. Hoover, T. R., Imperial, J., Ludden, P. W., and Shah, V. K. (1989) *Biochemistry* **28**, 2768–2771
21. Hoover, T. R., Imperial, J., Liang, J., Ludden, P. W., and Shah, V. K. (1988) *Biochemistry* **27**, 3647–3652
22. Imperial, J., Hoover, T. R., Madden, M. S., Ludden, P. W., and Shah, V. K. (1989) *Biochemistry* **28**, 7796–7799

22a. Einsle, O., Tezcan, F. A., Andrade, S. L. A., Schmid, B., Yoshida, M., Howard, J. B., and Rees, D. C. (2002) *Science* **297**, 1696 – 1700

23. Peters, J. W., Stowell, M. H. B., Soltis, S. M., Finnegan, M. G., Johnson, M. K., and Rees, D. C. (1997) *Biochemistry* **36**, 1181–1187
24. McLean, P. A., Wink, D. A., Chapman, S. K., Hickman, A. B., McKillop, D. M., and Orme-Johnson, W. H. (1989) *Biochemistry* **28**, 9402–9406
25. Jensen, B. B., and Burris, R. H. (1985) *Biochemistry* **24**, 1141–1147
26. Liang, J., and Burris, R. H. (1988) *Proc. Natl. Acad. Sci. U.S.A.* **85**, 9446–9450
27. Kim, C.-H., Newton, W. E., and Dean, D. R. (1995) *Biochemistry* **34**, 2798–2808
28. Cleland, W. W., Guth, J., and Burris, R. H. (1983) *Biochemistry* **22**, 5111–5122
29. Wahl, R. C., and Orme-Johnson, W. H. (1987) *J. Biol. Chem.* **262**, 10489–10496
30. Hoover, D. M., and Ludwig, M. L. (1997) *Protein Sci.* **6**, 2525–2537
31. Yoch, D. C. (1972) *Biochem. Biophys. Res. Commun.* **49**, 335–342
32. Carter, K. R., Rawlings, J., Orme-Johnson, W. H., Becker, R. R., and Evans, H. J. (1980) *J. Biol. Chem.* **255**, 4213–4223
33. Renner, K. A., and Howard, J. B. (1996) *Biochemistry* **35**, 5353–5358
34. Duyvis, M. G., Wassink, H., and Haaker, H. (1996) *J. Biol. Chem.* **271**, 29632–29636
35. Lanzilotta, W. N., Fisher, K., and Seefeldt, L. C. (1997) *J. Biol. Chem.* **272**, 4157–4165

35a. Chan, J. M., Wu, W., Dean, D. R., and Seefeldt, L. C. (2000) *Biochemistry* **39**, 7221–7228

35b. Jang, S. B., Seefeldt, L. C., and Peters, J. W. (2000) *Biochemistry* **39**, 14745–14752

35c. Clarke, T. A., Maritano, S., and Eady, R. R. (2000) *Biochemistry* **39**, 11434–11440

35d. Nyborg, A. C., Johnson, J. L., Gunn, A., and Watt, G. D. (2000) *J. Biol. Chem.* **275**, 39307–39312

36. Lanzilotta, W. N., Fisher, K., and Seefeldt, L. C. (1996) *Biochemistry* **35**, 7188–7196
37. Lanzilotta, W. N., and Seefeldt, L. C. (1996) *Biochemistry* **35**, 16770–16776

37a. Chiu, H.-J., Peters, J. W., Lanzilotta, W. N., Ryle, M. J., Seefeldt, L. C., Howard, J. B., and Rees, D. C. (2001) *Biochemistry* **40**, 641–650

38. Lanzilotta, W. N., and Seefeldt, L. C. (1997) *Biochemistry* **36**, 12976–12983

38a. Jang, S. B., Seefeldt, L. C., and Peters, J. W. (2000) *Biochemistry* **39**, 641–648

39. Grossman, J. G., Hasnain, S. S., Yousafzai, F. K., Smith, B. E., and Eady, R. R. (1997) *J. Mol. Biol.* **266**, 642–648
40. Ma, L., Brosius, M. A., and Burgess, B. K. (1996) *J. Biol. Chem.* **271**, 10528–10532

40a. Chan, J. M., Christiansen, J., Dean, D. R., and Seefeldt, L. C. (1999) *Biochemistry* **38**, 5779–5785

41. Christiansen, J., Tittsworth, R. C., Hales, B. J., and Cramer, S. P. (1995) *J. Am. Chem. Soc.* **117**, 10017–10024
42. Venters, R. A., Nelson, M. J., McLean, P. A., True, A. E., Levy, M. A., Hoffman, B. M., and Orme-Johnson, W. H. (1986) *J. Am. Chem. Soc.* **108**, 3487–3498
43. Shah, V. K., Ugalde, R. A., Imperial, J., and Brill, W. J. (1985) *J. Biol. Chem.* **260**, 3891–3894
44. Stiefel, E. I. (1973) *Proc. Natl. Acad. Sci. U.S.A.* **70**, 988–992

44a. Rod, T. H., and Noskov, J. K. (2000) *J. Am. Chem. Soc.* **122**, 12751–12763

44b. Thorneley, R. N., and Lowe, D. J. ((1984) *Biochem. J.* **224**, 887 – 894

44c. Durrant, M. C. (2002) *Biochemistry* **41**, 13934 – 13945; 13946 – 13955

45. Dilworth, M. J., Fisher, K., Kim, C.-H., and Newton, W. E. (1998) *Biochemistry* **37**, 17495–17505

45a. Fisher, K., Dilworth, M. J., Kim, C.-H., and Newton, W. E. (2000) *Biochemistry* **39**, 2970–2979

45b. Fisher, K., Dilworth, M. J., Kim, C.-H., and Newton, W. E. (2000) *Biochemistry* **39**, 10855–10865

45c. Christiansen, J., Seefeldt, L. C., and Dean, D. R. (2000) *J. Biol. Chem.* **275**, 36104–36107

45d. Sorlie, M., Christiansen, J., Lemon, B. J., Peters, J. W., Dean, D. R., and Hales, B. J. (2001) *Biochemistry* **40**, 1540–1549

46. Palermo, R. E., Singh, R., Bashkin, J. K., and Holm, R. H. (1984) *J. Am. Chem. Soc.* **106**, 2600–2612
47. Leigh, G. J. (1995) *Science* **268**, 827–828
48. Ferguson, R., Solari, E., Floriani, C., Osella, D., Ravera, M., Re, N., Chiesi-Villa, A., and Rizzoli, C. (1997) *J. Am. Chem. Soc.* **119**, 10104–10115
49. Nishibayashi, Y., Iwai, S., and Hidai, M. (1998) *J. Am. Chem. Soc.* **120**, 10559–10560

49a. Verma, A. K., and Lee, S. C. (1999) *J. Am. Chem. Soc.* **121**, 10838–10839

50. Laplaza, C. E., Johnson, A. R., and Cummins, C. C. (1996) *J. Am. Chem. Soc.* **118**, 709–710
51. Laplaza, C. E., and Cummins, C. C. (1995) *Science* **268**, 861–863
52. Ow, D. W., and Ausubel, F. M. (1983) *Nature (London)* **301**, 307–313
53. Thiel, T., Lyons, E. M., Erker, J. C., and Ernst, A. (1995) *Proc. Natl. Acad. Sci. U.S.A.* **92**, 9358–9362
54. Pulakat, L., Hausman, B. S., Lei, S., and Gavini, N. (1996) *J. Biol. Chem.* **271**, 1884–1889

54a. Yuvaniyama, P., Agar, J. N., Cash, V. L., Johnson, M. K., and Dean, D. R. (2000) *Proc. Natl. Acad. Sci. U.S.A.* **97**, 599–604

54b. Ribbe, M. W., Bursey, E. H., and Burgess, B. K. (2000) *J. Biol. Chem.* **275**, 17631–17638

54c. Ribbe, M. W., and Burgess, B. K. (2001) *Proc. Natl. Acad. Sci. U.S.A.* **98**, 5521–5525

55. Shah, V. K., Imperial, J., Ugalde, R. A., Ludden, P. W., and Brill, W. J. (1986) *Proc. Natl. Acad. Sci. U.S.A.* **83**, 1636–1640

55a. Rangaraj, P., Rüttimann-Johnson, C., Shah, V. K., and Ludden, P. W. (2001) *J. Biol. Chem.* **276**, 15968–15974

56. Jacobson, M. R., Brigle, K. E., Bennett, L. T., Setterquist, R. A., Wilson, M. S., Cash, V. L., Beynon, J., Newton, W. E., and Dean, D. R. (1989) *J. Bacteriol.* **171**, 1017–1027
57. Dilworth, M., and Glenn, A. (1984) *Trends Biochem. Sci.* **9**, 519–523
58. Marx, J. L. (1985) *Science* **230**, 157–158
59. Albrecht, C., Geurts, R., and Bisseling, T. (1999) *EMBO J.* **18**, 281–288
60. Fisher, R. F., and Long, S. R. (1992) *Nature (London)* **357**, 655–660
61. Vijn, I., das Neves, L., van Kammen, A., Franssen, H., and Bisseling, T. (1993) *Science* **260**, 1764–1765
62. Cedergren, R. A., Lee, J., Ross, K. L., and Hollingsworth, R. I. (1995) *Biochemistry* **34**, 4467–4477
63. Dunn, S. M., Moody, P. C. E., Downie, J. A., and Shaw, W. V. (1996) *Protein Sci.* **5**, 538–541
64. Spaink, H. P., Sheeley, D. M., van Brussel, A. A. N., Glushka, J., York, W. S., Tak, T., Geiger, O., Kennedy, E. P., Reinhold, V. N., and Lugtenberg, B. J. J. (1991) *Nature (London)* **354**, 125–130
65. Mergaert, P., D'Haeze, W., Geelen, D., Promé, D., Van Montagu, M., Geremia, R., Promé, J.-C., and Holsters, M. (1995) *J. Biol. Chem.* **270**, 29217–29223
66. Schultze, M., Staehelin, C., Röhrig, H., John, M., Schmidt, J., Kondorosi, E., Schell, J., and Kondorosi, A. (1995) *Proc. Natl. Acad. Sci. U.S.A.* **92**, 2706–2709
67. Fisher, R. F., and Long, S. R. (1993) *J. Mol. Biol.* **233**, 336–348
68. Monson, E. K., Ditta, G. S., and Helinski, D. R. (1995) *J. Biol. Chem.* **270**, 5243–5250
69. Soupène, E., Foussard, M., Boistard, P., Truchet, G., and Batut, J. (1995) *Proc. Natl. Acad. Sci. U.S.A.* **92**, 3759–3763

69a. Miyatake, H., Mukai, M., Park, S.-Y., Adachi, S., Tamura, K., Nakamura, H., Nakamura, K., Tsuchiya, T., Iizuka, T., and Shiro, Y. (2000) *J. Mol. Biol.* **301**, 415–431

70. Freiberg, C., Fellay, R., Bairoch, A., Broughton, W. J., Rosenthal, A., and Perre, X. (1997) *Nature (London)* **387**, 394–401

References

71. Downie, A. (1997) *Nature (London)* **387**, 352–354
72. Minchin, F. (1986) *Nature (London)* **320**, 483–484
73. Berry, A. M., Harriott, O. T., Moreau, R. A., Osman, S. F., Benson, D. R., and Jones, A. D. (1993) *Proc. Natl. Acad. Sci. U.S.A.* **90**, 6091–6094
74. Bishop, P. E. (1986) *Trends Biochem. Sci.* **11**, 225–227
75. Blanchard, C. Z., and Hales, B. J. (1996) *Biochemistry* **35**, 472–478
76. Chatterjee, R., Ludden, P. W., and Shah, V. K. (1997) *J. Biol. Chem.* **272**, 3758–3765
76a. Rüttimann-Johnson, C., Rangaraj, P., Shah, V. K., and Ludden, P. W. (2001) *J. Biol. Chem.* **276**, 4522–4526
77. Pau, R. N. (1989) *Trends Biochem. Sci.* **14**, 183–186
78. Ribbe, M., Gadkari, D., and Meyer, O. (1997) *J. Biol. Chem.* **272**, 26627–26633
79. DiSpirito, A. A., and Hooper, A. B. (1986) *J. Biol. Chem.* **261**, 10534–10537
79a. Kanamaru, K., Wang, R., Su, W., and Crawford, N. M. (1999) *J. Biol. Chem.* **274**, 4160 – 4165
80. Campbell, W. H., and Kinghorn, J. R. (1990) *Trends Biochem. Sci.* **15**, 315–319
81. Lu, G., Lindqvist, Y., Schneider, G., Dwivedi, U., and Campbell, W. (1995) *J. Mol. Biol.* **248**, 931–948
82. Garde, J., Kinghorn, J. R., and Tomsett, A. B. (1995) *J. Biol. Chem.* **270**, 6644–6650
83. Hyde, G. E., Crawford, N. M., and Campbell, W. H. (1991) *J. Biol. Chem.* **266**, 23542–23547
84. Ratnam, K., Shiraishi, N., Campbell, W. H., and Hille, R. (1995) *J. Biol. Chem.* **270**, 24067–24072
85. Cramer, S. P., Solomonson, L. P., Adams, M. W. W., and Mortenson, L. E. (1984) *J. Am. Chem. Soc.* **106**, 1467–1471
85a. Skipper, L., Campbell, W. H., Mertens, J. A., and Lowe, D. J. (2001) *J. Biol. Chem.* **276**, 26995 –27002
86. Lin, J. T., and Stewart, V. (1996) *J. Mol. Biol.* **256**, 423–435
86a. Butler, C. S., Charnock, J. M., Bennett, B., Sears, H. J., Reilly, A. J., Ferguson, S. J., Garner, C. D., Lowe, D. J., Thomson, A. J., Berks, B. C., and Richardson, D. J. (1999) *Biochemistry* **38**, 9000–9012
87. Augier, V., Guigliarelli, B., Asso, M., Bertrand, P., Frixon, C., Giordano, G., Chippaux, M., and Blasco, F. (1993) *Biochemistry* **32**, 2013–2023
87a. Zhao, Y., Lukoyanov, D. A., Toropov, Y. V., Wu, K., Shapleigh, J. P., and Scholes, C. P. (2002) *Biochemistry* **41**, 7464 – 7474
88. Bentley, R. (1985) *Trends Biochem. Sci.* **10**, 171–174
89. Cooper, A. J. L., Nieves, E., Coleman, A. E., Filc-DeRicco, S., and Gelbard, A. S. (1987) *J. Biol. Chem.* **262**, 1073–1080
90. Cooper, A. J. L., Nieves, E., Rosenspire, K. C., Filc-DeRicco, S., Gelbard, A. S., and Brusilow, S. W. (1988) *J. Biol. Chem.* **263**, 12268–12273
91. Chatham, J. C., Forder, J. R., Glickson, J. D., and Chance, E. M. (1995) *J. Biol. Chem.* **270**, 7999–8008
92. Martin, G., Chauvin, M.-F., and Baverel, G. (1997) *J. Biol. Chem.* **272**, 4717–4728
93. Chauvin, M.-F., Mégnin-Chanet, F., Martin, G., Lhoste, J.-M., and Baverel, G. (1994) *J. Biol. Chem.* **269**, 26025–26033
94. Ikeda, T. P., Shauger, A. E., and Kustu, S. (1996) *J. Mol. Biol.* **259**, 589–607
95. Zhalkin, H. (1993) *Adv. Enzymol.* **66**, 203–309
96. Boehlein, S. K., Richards, N. G. J., Walworth, E. S., and Schuster, S. M. (1994) *J. Biol. Chem.* **269**, 26789–26795
97. Muchmore, C. R. A., Krahn, J. M., Kim, J. H., Zalkin, H., and Smith, J. L. (1998) *Protein Sci.* **7**, 39–51
97a. Raushel, F. M., Thoden, J. B., and Holden, H. M. (1999) *Biochemistry* **38**, 7891–7899
97b. Bera, A. K., Smith, J. L., and Zalkin, H. (2000) *J. Biol. Chem.* **275**, 7975–7979
97c. Teplyakov, A., Obmolova, G., Badet, B., and Badet-Denisot, M.-A. (2001) *J. Mol. Biol.* **313**, 1093–1102
97d. Chittur, S. V., Klem, T. J., Shafer, C. M., and Davisson, V. J. (2001) *Biochemistry* **40**, 876–887
98. Boehlein, S. K., Stewart, J. D., Walworth, E. S., Thirumoorthy, R., Richards, N. G. J., and Schuster, S. M. (1998) *Biochemistry* **37**, 13230–13238
98a. Larsen, T. M., Boehlein, S. K., Schuster, S. M., Richards, N. G. J., Thoden, J. B., Holden, H. M., and Rayment, I. (1999) *Biochemistry* **38**, 16146–16157
98b. Min, B., Pelaschier, J. T., Graham, D. E., Tumbula-Hansen, D., and Söll, D. (2002) *Proc. Natl. Acad. Sci. U.S.A.* **99**, 2678–2683
99. Raghunathan, G., Miles, H. T., and Sasisekharan, V. (1993) *Biochemistry* **32**, 455–462
100. Winter, H. C., and Dekker, E. E. (1986) *J. Biol. Chem.* **261**, 11189–11193
101. Kohl, D. H., Schubert, K. R., Carter, M. B., Hagedorn, C. H., and Shearer, G. (1988) *Proc. Natl. Acad. Sci. U.S.A.* **85**, 2036–2040
102. Cooper, A. J. L., and Meister, A. (1985) in *Transaminases* (Christen, P., and Metzler, D. E., eds), pp. 533–563, Wiley, New York
103. Sibson, N. R., Dhankhar, A., Mason, G. F., Behar, K. L., Rothman, D. L., and Shulman, R. G. (1997) *Proc. Natl. Acad. Sci. U.S.A.* **94**, 2699–2704
104. Liaw, S.-H., Kuo, I., and Eisenberg, D. (1995) *Protein Sci.* **4**, 2358–2365
105. Kagan, Z. A., Kretovich, V. L., and Polyakov, V. A. (1966) *Biokhim.* **31**, 355–364
106. Baker, P. J., Waugh, M. L., Wang, X.-G., Stillman, T. J., Turnbull, A. P., Engel, P. C., and Rice, D. W. (1997) *Biochemistry* **36**, 16109–16115
107. Peters, J.-M., Harris, J. R., and Finley, D., eds. (1998) *Ubiquitin and the Biology of the Cell*, Plenum, New York
107a. Katagiri, M., and Nakamura, M. (1999) *Biochem. Edu.* **27**, 83–85
107b. Katagiri, M., and Nakamura, M. (2002) *IUBMB Life* **53**, 125 – 129
107c. Wipf, D., Ludewig, U., Tegeder, M., Rentsch, D., Koch, W., and Frommer, W. B. (2002) *Trends Biochem. Sci.* **27**, 139 – 147
108. Rajagopalan, S., Wan, D.-F., Habib, G. M., Sepulveda, A. R., McLeod, M. R., Lebovitz, R. M., and Lieberman, M. W. (1993) *Proc. Natl. Acad. Sci. U.S.A.* **90**, 6179–6183
109. Harding, C. O., Williams, P., Wagner, E., Chang, D. S., Wild, K., Colwell, R. E., and Wolff, J. A. (1997) *J. Biol. Chem.* **272**, 12560–12567
109a. Stein, R. L., DeCicco, C., Nelson, D., and Thomas, B. (2001) *Biochemistry* **40**, 5804–5811
110. Guo-jie, Breslow, E., and Meister, A. (1996) *J. Biol. Chem.* **271**, 32293–32300
111. Van Der Werf, P., Griffith, O. W., and Meister, A. (1975) *J. Biol. Chem.* **250**, 6686–6692
112. Reitzer, L. J., and Magasanik, B. (1987) in *Escherichia coli and Salmonella typhimurium* (Neidhardt, F. C., ed), pp. 302–318, Am. Soc. for Microbiology, Washington,DC
113. Kanamori, K., Weiss, R. L., and Roberts, J. D. (1988) *J. Biol. Chem.* **263**, 2817–2823
114. Knapp, S., de Vos, W. M., Rice, D., and Ladenstein, R. (1997) *J. Mol. Biol.* **267**, 916–932
115. Wang, X.-G., and Engel, P. C. (1995) *Biochemistry* **34**, 11417–11422
115a. Herrero-Yraola, A., Bakhit, S. M. A., Franke, P., Weise, C., Schweiger, M., Jorcke, D., and Ziegler, M. (2001) *EMBO J.* **20**, 2404–2412
116. Melo-Oliveira, R., Oliveira, I. C., and Coruzzi, G. M. (1996) *Proc. Natl. Acad. Sci. U.S.A.* **93**, 4718–4723
117. Vanoni, M. A., Edmondson, D. E., Zanetti, G., and Curti, B. (1992) *Biochemistry* **31**, 4613–4623
118. Vanoni, M. A., Fischer, F., Ravasio, S., Verzotti, E., Edmondson, D. E., Hagen, W. R., Zanetti, G., and Curti, B. (1998) *Biochemistry* **37**, 1828–1838
118a. Morandi, P., Valzasina, B., Colombo, C., Curti, B., and Vanoni, M. A. (2000) *Biochemistry* **39**, 727–735
119. Sakakibara, H., Watanabe, M., Hase, T., and Sugiyama, T. (1991) *J. Biol. Chem.* **266**, 2028–2035
120. Knaff, D. B., Hirasawa, M., Ameyibor, E., Fu, W., and Johnson, M. K. (1991) *J. Biol. Chem.* **266**, 15080–15084
121. Ho, H.-T., Falk, P. J., Ervin, K. M., Krishnan, B. S., Discotto, L. F., Dougherty, T. J., and Pucci, M. J. (1995) *Biochemistry* **34**, 2464–2470
122. Gallo, K. A., and Knowles, J. R. (1993) *Biochemistry* **32**, 3981–3990
122a. Glavas, S., and Tanner, M. E. (2001) *Biochemistry* **40**, 6199–6204
123. D'Aniello, A., D'Onofrio, G., Pischetola, M., D'Aniello, G., Vetere, A., Petrucelli, L., and Fisher, G. H. (1993) *J. Biol. Chem.* **268**, 26941–26949
124. Almassy, R. J., Janson, C. A., Hamlin, R., Xuong, N.-H., and Eisenberg, D. (1986) *Nature (London)* **323**, 304–309
125. Liaw, S.-H., and Eisenberg, D. (1994) *Biochemistry* **33**, 675–681
126. Mura, U., Chock, P. B., and Stadtman, E. R. (1981) *J. Biol. Chem.* **256**, 13022–13029
126a. Stadtman, E. R. (2001) *J. Biol. Chem.* **276**, 44357–44364
127. Garcia, E., and Rhee, S. G. (1983) *J. Biol. Chem.* **258**, 2246–2253
128. Son, H. S., and Rhee, S. G. (1987) *J. Biol. Chem.* **262**, 8690–8695
129. Jiang, P., Peliska, J. A., and Ninfa, A. J. (1998) *Biochemistry* **37**, 12802–12810
130. Liaw, S.-H., Pan, C., and Eisenberg, D. (1993) *Proc. Natl. Acad. Sci. U.S.A.* **90**, 4996–5000
131. Jiang, P., Peliska, J. A., and Ninfa, A. J. (1998) *Biochemistry* **37**, 12795–12801
132. Flashner, Y., Weiss, D. S., Keener, J., and Kustu, S. (1995) *J. Mol. Biol.* **249**, 700–713
133. Rosenfeld, S. A., and Brenchley, J. E. (1983) in *Amino Acids, Biosynthesis and Genetic Regulation* (Herrmann, K. M., and Somerville, R. L., eds), pp. 1–17, Addison-Wesley, Reading, Massachusetts
134. Buchanan, J. M. (1973) *Adv. Enzymol.* **38**, 91–183
134a. Gill, H. S., and Eisenberg, D. (2001) *Biochemistry* **40**, 1903–1912
135. Neu, J., Shenoy, V., and Chakrabarti, R. (1996) *FASEB J.* **10**, 829–837
136. Jakob, C. G., Lewinski, K., LaCount, M. W., Roberts, J., and Lebioda, L. (1997) *Biochemistry* **36**, 923–931
137. Watford, M. (1993) *FASEB J.* **7**, 1468–1474
138. Snedden, W. A., Koutsia, N., Baum, G., and Fromm, H. (1996) *J. Biol. Chem.* **271**, 4148–4153
138a. Locher, K. P., Hans, M., Yeh, A. P., Schmid, B., Buckel, W., and Rees, D. C. (2001) *J. Mol. Biol.* **307**, 297–308
138b. Chih, H.-W., and Marsh, E. N. G. (2001) *Biochemistry* **40**, 13060–13067
139. Smith, M. A., King, P. J., and Grimm, B. (1998) *Biochemistry* **37**, 319–329
140. Adams, E., and Frank, L. (1980) *Ann. Rev. Biochem.* **49**, 1005–1061

References

141. Leisinger, T. (1987) in *Escherichia coli and Salmonella typhimurium* (Neidhardt, F. C., ed), pp. 345–351, Am. Soc. for Microbiology, Washington,DC
142. Zhang, C.-s, Lu, Q., and Verma, D. P. S. (1995) *J. Biol. Chem.* **270**, 20491–20496
143. García-Ríos, M., Fujita, T., LaRosa, P. C., Locy, R. D., Clithero, J. M., Bressan, R. A., and Csonka, L. N. (1997) *Proc. Natl. Acad. Sci. U.S.A.* **94**, 8249–8254
144. Mestichelli, L. J. J., Gupta, R. N., and Spenser, I. D. (1979) *J. Biol. Chem.* **254**, 640–647
145. Ashford, D., and Neuberger, A. (1980) *Trends Biochem. Sci.* **5**, 245–248
145a. Becker, D. F., and Thomas, E. A. (2001) *Biochemistry* **40**, 4714–4721
145b. Farrant, R. D., Walker, V., Mills, G. A., Mellor, J. M., and Langley, G. J. (2001) *J. Biol. Chem.* **276**, 15107–15116
146. Small, W. C., and Jones, M. E. (1990) *J. Biol. Chem.* **265**, 18668–18672
147. Hu, C.-aA., Lin, W.-W., and Valle, D. (1996) *J. Biol. Chem.* **271**, 9795–9800
148. Phang, J. M., Yeh, G. C., and Scriver, C. R. (1995) in *The Metabolic and Molecular Bases of Inherited Disease*, 7th ed., Vol. 1 (Scriver, C. R., Beaudet, A. L., Sly, W. S., and Valle, D., eds), pp. 1125–1146, McGraw-Hill, New York
148a. Farrant, R. D., Walker, V., Mills, G. A., Mellor, J. M., and Langley, G. J. (2000) *J. Biol. Chem.* **276**, 15107 – 15116
149. Gupta, S. C., and Dekker, E. E. (1980) *J. Biol. Chem.* **255**, 1107–1112
150. Sletten, K., Aakesson, I., and Alvsaker, J. O. (1971) *Nature New Biol.* **231**, 118–119
151. Akers, H. A., and Dromgoole, E. V. (1982) *Trends Biochem. Sci.* **7**, 156–157
152. Umbarger, H. E. (1978) *Ann. Rev. Biochem.* **47**, 533–606
153. Davis, R. H., and Weiss, R. L. (1988) *Trends Biochem. Sci.* **13**, 101–104
154. Shah, S. A., Shen, B. W., and Brünger, A. T. (1997) *Structure* **5**, 1067–1075
155. Valle, D., and Simell, O. (1995) in *The Metabolic and Molecular Bases of Inherited Disease*, 7th ed., Vol. 1 (Scriver, C. R., Beaudet, A. L., Sly, W. S., and Valle, D., eds), pp. 1147–1185, McGraw-Hill, New York
156. Glansdorff, N. (1987) in *Escherichia coli and Salmonella typhimurium* (Neidhardt, F. C., ed), pp. 321–339, Am. Soc. for Microbiology, Washington, DC
157. Krebs, H. A., and Henseleit, K. (1932) *Hoppe-Seyler's Z. Physiol. Chem.* **210**, 33–66
158. Krebs, H. A. (1982) *Trends Biochem. Sci.* **7**, 76–78
159. Wakabayashi, Y., Yamada, E., Yoshida, T., and Takahashi, H. (1994) *J. Biol. Chem.* **269**, 32667–32671
160. Gessert, S. F., Kim, J. H., Nargang, F. E., and Weiss, R. L. (1994) *J. Biol. Chem.* **269**, 8189–8203
161. Watford, M. (1989) *Trends Biochem. Sci.* **14**, 313–314
162. Brusilow, S. W., and Horwich, A. L. (1995) in *The Metabolic and Molecular Bases of Inherited Disease*, 7th ed., Vol. 1 (Scriver, C. R., Beaudet, A. L., Sly, W. S., and Valle, D., eds), pp. 1187–1232, McGraw-Hill, New York
163. Lim, A. L., and Powers-Lee, S. G. (1996) *J. Biol. Chem.* **271**, 11400–11409
164. Guy, H. I., and Evans, D. R. (1996) *J. Biol. Chem.* **271**, 13762–13769
165. McCudden, C. R., and Powers-Lee, S. G. (1996) *J. Biol. Chem.* **271**, 18285–18294
166. Hong, J., Salo, W. L., Lusty, C. J., and Anderson, P. M. (1994) *J. Mol. Biol.* **243**, 131–140
167. Devaney, M. A., and Powers-Lee, S. G. (1984) *J. Biol. Chem.* **259**, 703–706
168. Anderson, P. M. (1981) *J. Biol. Chem.* **256**, 12228–12238
169. Stapleton, M. A., Javid-Majd, F., Harmon, M. F., Hanks, B. A., Grahmann, J. L., Mullins, L. S., and Raushel, F. M. (1996) *Biochemistry* **35**, 14352–14361
170. Miles, B. W., Banzon, J. A., and Raushel, F. M. (1998) *Biochemistry* **37**, 16773–16779
171. Powers, S. G., and Meister, A. (1978) *J. Biol. Chem.* **253**, 1258–1265
172. Thoden, J. B., Holden, H. M., Wesenberg, G., Raushel, F. M., and Rayment, I. (1997) *Biochemistry* **36**, 6305–6316
173. Thoden, J. B., Miran, S. G., Phillips, J. C., Howard, A. J., Raushel, F. M., and Holden, H. M. (1998) *Biochemistry* **37**, 8825–8831
173a. Thoden, J. B., Huang, X., Raushel, F. M., and Holden, H. M. (1999) *Biochemistry* **38**, 16158–16166
173b. Kim, J., Howell, S., Huang, X., and Raushel, F. M. (2002) *Biochemistry* **41**, 12575 – 12581
173c. Fresquet, V., Mora, P., Rochera, L., Ramón-Maiques, S., Rubio, V., and Cervera, J. (2000) *J. Mol. Biol.* **299**, 979–991
174. Shi, D., Morizono, H., Ha, Y., Aoyagi, M., Tuchman, M., and Allewell, N. M. (1998) *J. Biol. Chem.* **273**, 34247–34254
175. Tricot, C., Villeret, V., Sainz, G., Dideberg, O., and Stalon, V. (1998) *J. Mol. Biol.* **283**, 695–704
175a. Langley, D. B., Templeton, M. D., Fields, B. A., Mitchell, R. E., and Collyer, C. A. (2000) *J. Biol. Chem.* **275**, 20012–20019
176. Rosenberg, L. E., Kalousek, F., and Orsulak, M. D. (1983) *Science* **222**, 426–428
177. Maestri, N. E., Brusilow, S. W., Clissold, D. B., and Bassett, S. S. (1996) *N. Engl. J. Med.* **335**, 855–859
177a. Sampaleanu, L. M., Vallée, F., Thompson, G. D., and Howell, P. L. (2001) *Biochemistry* **40**, 15570–15580
178. Turner, M. A., Simpson, A., McInnes, R. R., and Howell, P. L. (1997) *Proc. Natl. Acad. Sci. U.S.A.* **94**, 9063–9068
178a. Yu, B., Thompson, G. D., Yip, P., Howell, P. L., and Davidson, A. R. (2001) *Biochemistry* **40**, 15581–15590
179. Garrard, L. J., Mathis, J. M., and Raushel, F. M. (1983) *Biochemistry* **22**, 3729–3735
180. Kanyo, Z. F., Scolnick, L. R., Ash, D. E., and Christianson, D. W. (1996) *Nature (London)* **383**, 554–557
181. Marathe, S., Yu, Y. G., Turner, G. E., Palmier, C., and Weiss, R. L. (1998) *J. Biol. Chem.* **273**, 29776–29785
182. Scolnick, L. R., Kanyo, Z. F., Cavalli, R. C., Ash, D. E., and Christianson, D. W. (1997) *Biochemistry* **36**, 10558–10565
183. Smith, I. (1981) *Nature (London)* **291**, 378–380
184. Atkinson, D. E., and Bourke, E. (1984) *Trends Biochem. Sci.* **9**, 297–300
185. Gargan, R., Rapoport, G., and Débarbouillé, M. (1995) *J. Mol. Biol.* **249**, 843–856
186. Jann, A., Stalon, V., Vander Wauven, C., Leisinger, T., and Haas, D. (1986) *Proc. Natl. Acad. Sci. U.S.A.* **83**, 4937–4941
187. Weickmann, J. L., Himmel, M. E., Squire, P. G., and Fahrney, D. E. (1978) *J. Biol. Chem.* **253**, 6010–6015
188. Abdelal, A. T. (1979) *Ann. Rev. Microbiol.* **33**, 139–168
189. Vanderbilt, A. S., Gaby, N. S., and Rodwell, V. W. (1975) *J. Biol. Chem.* **250**, 5322–5329
190. Gary, J. D., Lin, W.-J., Yang, M. C., Herschman, H. R., and Clarke, S. (1996) *J. Biol. Chem.* **271**, 12585–12594
191. Bogumil, R., Knipp, M., Fundel, S. M., and Vasák, M. (1998) *Biochemistry* **37**, 4791–4798
192. Rosenthal, G. A. (1984) *Sci. Am.* **250**(Jun), 164–171
193. Rosenthal, G. A. (1992) *Proc. Natl. Acad. Sci. U.S.A.* **89**, 1780–1784
194. Melangeli, C., Rosenthal, G. A., and Dalman, D. L. (1997) *Proc. Natl. Acad. Sci. U.S.A.* **94**, 2255–2260
195. Fritsche, E., Humm, A., and Huber, R. (1999) *J. Biol. Chem.* **274**, 3026–3032
196. Takata, Y., Konishi, K., Gomi, T., and Fujioka, M. (1994) *J. Biol. Chem.* **269**, 5537–5542
197. Humm, A., Fritsche, E., Steinbacher, S., and Huber, R. (1997) *EMBO J.* **16**, 3373–3385
198. Zhou, G., Somasundaram, T., Blanc, E., Parthasarathy, G., Ellington, W. R., and Chapman, M. S. (1998) *Proc. Natl. Acad. Sci. U.S.A.* **95**, 8449–8454
199. Turner, D. M., and Walker, J. B. (1987) *J. Biol. Chem.* **262**, 6605–6609
200. Cohen, S. S. (1971) *Introduction to the Polyamines*, Prentice-Hall, Englewood Cliffs, New Jersey
201. Tabor, C. W., and Tabor, H. (1976) *Ann. Rev. Biochem.* **45**, 285–306
202. Cohen, S. S. (1978) *Nature (London)* **274**, 209–210
203. Abraham, A. K., and Pihl, A. (1981) *Trends Biochem. Sci.* **6**, 106–107
204. Igarashi, K., Saisho, T., Yuguchi, M., and Kashiwagi, K. (1997) *J. Biol. Chem.* **272**, 4058–4064
205. Cohen, S. S., and Greenberg, M. L. (1981) *Proc. Natl. Acad. Sci. U.S.A.* **78**, 5470–5474
206. Tsuboi, M. (1964) *Bull. Chem. Soc. Jap.* **37**, 1514–1522
207. Balkema, G. W., Mangini, N. J., and Pinto, L. H. (1983) *Science* **219**, 1083–1087
207a. Chattopadhyay, M. K., Tabor, C. W., Tabor, H. (2002) *Proc. Natl. Acad. Sci. U.S.A.* **99**, 10330 — 10334
208. Herbst, E. J. et al. (1955) *J. Biol. Chem.* **214**, 175–
209. Leroy, D., Filhol, O., Delcros, J. G., Pares, S., Chambaz, E. M., and Cochet, C. (1997) *Biochemistry* **36**, 1242–1250
210. Hafner, E. W., Tabor, C. W., and Tabor, H. (1979) *J. Biol. Chem.* **254**, 12419–12426
211. Balasundaram, D., Tabor, C. W., and Tabor, H. (1991) *Proc. Natl. Acad. Sci. U.S.A.* **88**, 5872–5876
211a. Chattopadhyay, M. K., Tabor, C. W., and Tabor, H. (2002) *Proc. Natl. Acad. Sci. U.S.A.* **99**, 10330 – 10334
212. Cohn, M. S., Tabor, C. W., Tabor, H., and Wickner, R. B. (1978) *J. Biol. Chem.* **253**, 5225–5227
213. Iyer, R., and Delcour, A. H. (1997) *J. Biol. Chem.* **272**, 18595–18601
214. Kaouass, M., Gamache, I., Ramotar, D., Audette, M., and Poulin, R. (1998) *J. Biol. Chem.* **273**, 2109–2117
215. Antognoni, F., Del Duca, S., Kuraishi, A., Kawabe, E., Fukuchi-Shimogori, T., Kashiwagi, K., and Igarashi, K. (1999) *J. Biol. Chem.* **274**, 1942–1948
216. Momany, C., Ernst, S., Ghosh, R., Chang, N.-L., and Hackert, M. L. (1995) *J. Mol. Biol.* **252**, 643–655
217. Srivenugopal, K. S., and Adiga, P. R. (1981) *J. Biol. Chem.* **256**, 9532–9541
218. Bowman, W. H., Tabor, C. W., and Tabor, H. (1973) *J. Biol. Chem.* **248**, 2480–2486
219. Pegg, A. E., Wechter, R. S., Clark, R. S., Wiest, L., and Erwin, B. G. (1986) *Biochemistry* **25**, 379–384
219a. Tolbert, W. D., Ekstrom, J. L., Mathews, I. I., Secrist, J. A., III, Kapoor, P., Pegg, A. E., and Ealick, S. E. (2001) *Biochemistry* **40**, 9484–9494
220. Heby, O., and Persson, L. (1990) *Trends Biochem. Sci.* **15**, 153–158
221. Orr, G. R., Danz, D. W., Pontoni, G., Probhakaran, P. C., Gould, S. J., and Coward, J. K. (1988) *J. Am. Chem. Soc.* **110**, 5791–5799
222. Oshima, T., Hamasaki, N., Senshu, M., Kakinuma, K., and Kuwajima, I. (1987) *J. Biol. Chem.* **262**, 11979–11981

References

223. Fujihara, S., Abe, H., and Yoneyama, T. (1995) *J. Biol. Chem.* **270**, 9932–9938
224. Cacciapuoti, G., Porcelli, M., Bertoldo, C., De Rosa, M., and Zappia, V. (1994) *J. Biol. Chem.* **269**, 24762–24769
224a. Kröger, N., Deutzmann, R., and Sumper, M. (1999) *Science* **286**, 1129–1132
224b. Kröger, N., Deutzmann, R., Bergsdorf, C., and Sumper, M. (2000) *Proc. Natl. Acad. Sci. U.S.A.* **97**, 14133–14138
224c. Ober, D., and Hartmann, T. (1999) *Proc. Natl. Acad. Sci. U.S.A.* **96**, 14777–14782
224d. Kröger, N., Lorenz, S., Brunner, B., and Sumper, M. (2002) *Science* **298**, 584 – 585
225. Jasys, V. J., Kelbaugh, P. R., Nason, D. M., Phillips, D., Rosnack, K. J., Saccomano, N. A., Stroh, J. G., and Volkmann, R. A. (1990) *J. Am. Chem. Soc.* **112**, 6696–6704
226. McConlogue, L., Gupta, M., Wu, L., and Coffino, P. (1984) *Proc. Natl. Acad. Sci. U.S.A.* **81**, 540–544
227. Rom, E., and Kahana, C. (1994) *Proc. Natl. Acad. Sci. U.S.A.* **91**, 3959–3963
228. Li, R.-S., Abrahamsen, M. S., Johnson, R. R., and Morris, D. R. (1994) *J. Biol. Chem.* **269**, 7941–7949
229. Pena, A., Reddy, C. D., Wu, S., Hickok, N. J., Reddy, E. P., Yumet, G., Soprano, D. R., and Soprano, K. J. (1993) *J. Biol. Chem.* **268**, 27277–27285
230. Hayashi, S.-i, Murakami, Y., and Matsufuji, S. (1996) *Trends Biochem. Sci.* **21**, 27–30
230a. Zhu, C., Lang, D. W., and Coffino, P. (1999) *J. Biol. Chem.* **274**, 26425–26430
230b. Coffino, P. (2000) *Proc. Natl. Acad. Sci. U.S.A.* **97**, 4421–4423
231. Atmar, V. J., and Kuehn, G. D. (1981) *Proc. Natl. Acad. Sci. U.S.A.* **78**, 5518–5522
232. Coleman, C. S., Huang, H., and Pegg, A. E. (1995) *Biochemistry* **34**, 13423–13430
233. Coleman, C. S., and Pegg, A. E. (1997) *J. Biol. Chem.* **272**, 12164–12169
233a. Bina, C., Angelini, R., Federico, R., Ascenzi, P., and Mattevi, A. (2001) *Biochemistry* **40**, 2766–2776
234. Richards, J. C., and Spenser, I. D. (1978) *J. Am. Chem. Soc.* **100**, 7402–7404
235. Lundgren, D. W., and Fales, H. M. (1980) *J. Biol. Chem.* **255**, 4481–4486
236. Lee, Y., and Sayre, L. M. (1998) *J. Biol. Chem.* **273**, 19490–19494
236a. White, W. H., Gunyuzlu, P. L., Toyn, J. H. (2001) *J. Biol. Chem.* **276**, 10794 – 10800
237. Tabor, H., and Tabor, C. W. (1975) *J. Biol. Chem.* **250**, 2648–2654
238. Smith, K., Nadeau, K., Bradley, M., Walsh, C., and Fairlamb, A. H. (1992) *Protein Sci.* **1**, 874–883
239. Bollinger, J. M., Jr., Kwon, D. S., Huisman, G. W., Kolter, R., and Walsh, C. T. (1995) *J. Biol. Chem.* **270**, 14031–14041
240. Folk, J. E., Park, M. H., Chung, S. I., Schrode, J., Lester, E. P., and Cooper, H. L. (1980) *J. Biol. Chem.* **255**, 3695–3700
241. Leete, E., and McDonnell, J. A. (1981) *J. Am. Chem. Soc.* **103**, 658–662
242. Park, M. H., Cooper, H. L., and Folk, J. E. (1981) *Proc. Natl. Acad. Sci. U.S.A.* **78**, 2869–2873
243. Osterman, A. L., Brooks, H. B., Rizo, J., and Phillips, M. A. (1997) *Biochemistry* **36**, 4558–4567
244. Coleman, C. S., Stanley, B. A., and Pegg, A. E. (1993) *J. Biol. Chem.* **268**, 24572–24579
244a. Marali, S. (1999) *J. Biol. Chem.* **274**, 21017–21022
245. Pegg, A. E., Jones, D. B., and Secrist, J. A., III. (1988) *Biochemistry* **27**, 1408–1415
246. Hyvönen, T., Alakuijala, L., Andersson, L., Khomutov, A. R., Khomutov, R. M., and Eloranta, T. O. (1988) *J. Biol. Chem.* **263**, 11138–11144
247. Cohen, G. N., and Saint-Girons, I. (1987) in *Escherichia coli and Salmonella typhimurium* (Neidhardt, F. C., ed), pp. 429–442, Am. Soc. for Microbiology, Washington, DC
248. Shames, S. L., Ash, D. E., Wedler, F. C., and Villafranca, J. J. (1984) *J. Biol. Chem.* **259**, 15331–15339
249. Angeles, T. S., Hunsley, J. R., and Viola, R. E. (1992) *Biochemistry* **31**, 799–805
249a. Hadfield, A., Kryger, G., Ouyang, J., Petsko, G. A., Ringe, D., and Viola, R. (1999) *J. Mol. Biol.* **289**, 991–1002
250. Ouyang, J., and Viola, R. E. (1995) *Biochemistry* **34**, 6394–6399
251. Ravanel, S., Gakière, B., Job, D., and Douce, R. (1998) *Proc. Natl. Acad. Sci. U.S.A.* **95**, 7805–7812
252. Giovanelli, J., Mudd, S. H., and Datko, A. H. (1978) *J. Biol. Chem.* **253**, 5665–5677
253. Huo, X., and Viola, R. E. (1996) *Biochemistry* **35**, 16180–16185
254. Laber, B., Gerbling, K.-P., Harde, C., Neff, K.-H., Nordhoff, E., and Pohlenz, H.-D. (1994) *Biochemistry* **33**, 3413–3423
255. Curien, G., Job, D., Douce, R., and Dumas, R. (1998) *Biochemistry* **37**, 13212–13221
255a. Thomazeau, K., Curien, G., Dumas, R., and Biou, V. (2001) *Protein Sci.* **10**, 638–648
256. Chinchilla, D., Schwarz, F. P., and Eisenstein, E. (1998) *J. Biol. Chem.* **273**, 23219–23224
257. Horowitz, B., and Meister, A. (1972) *J. Biol. Chem.* **247**, 6708–6719
257a. Aghaiypour, K., Wlodawer, A., and Lubkowski, J. (2001) *Biochemistry* **40**, 5655–5664
258. Tate, S. S., and Meister, A. (1971) *Adv. Enzymol.* **35**, 503–543
259. Funkhouser, J. D., Abraham, A., Smith, V. A., and Smith, W. G. (1974) *J. Biol. Chem.* **249**, 5478–5484
259a. Tedeschi, G., Ronchi, S., Simonic, T., Treu, C., Mattevi, A., and Negri, A. (2001) *Biochemistry* **40**, 4738–4744
259b. Bossi, R. T., Negri, A., Tedeschi, G., and Mattevi, A. (2002) *Biochemistry* **41**, 3018–3024
260. Rafalski, J. A., and Falco, S. C. (1988) *J. Biol. Chem.* **263**, 2146–2151
261. Zakin, M. M., Duchange, N., Ferrara, P., and Cohen, G. N. (1983) *J. Biol. Chem.* **258**, 3028–3031
262. Wedler, F. C., and Ley, B. W. (1993) *J. Biol. Chem.* **268**, 4880–4888
262a. James, C. L., and Viola, R. E. (2002) *Biochemistry* **41**, 3726–3731
263. Cassan, M., Parsot, C., Cohen, G. N., and Patte, J.-C. (1986) *J. Biol. Chem.* **261**, 1052–1057
264. Chen, N.-Y., and Paulus, H. (1988) *J. Biol. Chem.* **263**, 9526–9532
264a. Ehmann, D. E., Gehring, A. M., and Walsh, C. T. (1999) *Biochemistry* **38**, 6171–6177
265. Fujioka, M., Takata, Y., Ogawa, H., and Okamoto, M. (1980) *J. Biol. Chem.* **255**, 937–942
266. Bhattacharjee, J. K. (1992) in *The Evolution of Metabolic Function* (Mortlock, R. P., ed), pp. 47–80, CRC Press, Boca Raton, Florida
267. Mirwaldt, C., Korndörfer, I., and Huber, R. (1995) *J. Mol. Biol.* **246**, 227–239
268. Blickling, S., Beisel, H.-G., Bozic, D., Knäblein, J., Laber, B., and Huber, R. (1997) *J. Mol. Biol.* **274**, 608–621
269. Blickling, S., Renner, C., Laber, B., Pohlenz, H.-D., Holak, T. A., and Huber, R. (1997) *Biochemistry* **36**, 24–33
270. Scapin, G., Blanchard, J. S., and Sacchettini, J. C. (1995) *Biochemistry* **34**, 3502–3512
271. Reddy, S. G., Scapin, G., and Blanchard, J. S. (1996) *Biochemistry* **35**, 13294–13302
272. Simms, S. A., Voige, W. H., and Gilvarg, C. (1984) *J. Biol. Chem.* **259**, 2734–2741
273. Beaman, T. W., Binder, D. A., Blanchard, J. S., and Roderick, S. L. (1997) *Biochemistry* **36**, 489–494
274. Beaman, T. W., Blanchard, J. S., and Roderick, S. L. (1998) *Biochemistry* **37**, 10363–10369
275. Scapin, G., Cirilli, M., Reddy, S. G., Gao, Y., Vederas, J. C., and Blanchard, J. S. (1998) *Biochemistry* **37**, 3278–3285
276. Cirilli, M., Zheng, R., Scapin, G., and Blanchard, J. S. (1998) *Biochemistry* **37**, 16452–16458
276a. Koo, C. W., Sutherland, A., Vederas, J. C., and Blanchard, J. S. (2000) *J. Am. Chem. Soc.* **122**, 6122–6123
277. Daniel, R. A., and Errington, J. (1993) *J. Mol. Biol.* **232**, 468–483
278. Henderson, L. M., Nelson, P. J., and Henderson, L. (1982) *Fed. Proc.* **41**, 2843–2847
279. Dunn, W. A., Rettura, G., Seifter, E., and Englard, S. (1984) *J. Biol. Chem.* **259**, 10764–10770
280. Golebiewski, W. M., and Spenser, I. D. (1984) *J. Am. Chem. Soc.* **106**, 1441–1442
281. Kang, K. R., Wolff, E. C., Park, M. H., Folk, J. E., and Chung, S. I. (1995) *J. Biol. Chem.* **270**, 18408–18412
282. Park, M. H., Joe, Y. A., and Kang, K. R. (1998) *J. Biol. Chem.* **273**, 1677–1683
282a. Wolff, E. C., Wolff, J., and Park, M. H. (2000) *J. Biol. Chem.* **275**, 9170–9177
283. Luckner, M. (1972) *Secondary Metabolism in Plants and Animals*, Academic Press, New York (p. 275)
284. Byron, C. M., Stankovich, M. T., and Husain, M. (1990) *Biochemistry* **29**, 3691–3700
285. Soda, K., Misono, H., and Yamamoto, T. (1968) *Biochemistry* **7**, 4102–4109
286. Grove, J. A., Linn, T. G., Willett, C. J., and Henderson, L. M. (1970) *Biochim. Biophys. Acta.* **215**, 191–194
287. Markovitz, P. J., and Chuang, D. T. (1987) *J. Biol. Chem.* **262**, 9353–9358
288. Markovitz, P. J., Chuang, D. T., and Cox, R. P. (1984) *J. Biol. Chem.* **259**, 11643–11646
289. Fangmeier, N., and Leistner, E. (1980) *J. Biol. Chem.* **255**, 10205–10209
290. Wickwire, B. M., Harris, C. M., Harris, T. M., and Broquist, H. P. (1990) *J. Biol. Chem.* **265**, 14742–14747
291. Mihalik, S. J., McGuinness, M., and Watkins, P. A. (1991) *J. Biol. Chem.* **266**, 4822–4830
292. Wickwire, B. M., Wagner, C., and Broquist, H. P. (1990) *J. Biol. Chem.* **265**, 14748–14753
293. Rothstein, M. (1965) *Arch. Biochem. Biophys.* **111**, 467–476
294. Chang, Y. F., and Adams, E. (1977) *J. Biol. Chem.* **252**, 7987–7991
294a. Tang, K.-H., Harms, A., and Frey, P. A. (2002) *Biochemistry* **41**, 8767 – 8776
295. Ohsugi, M., Kahn, J., Hensley, C., Chew, S., and Barker, H. A. (1981) *J. Biol. Chem.* **256**, 7642–7651
296. Brzovic, P., Holbrook, E. L., Greene, R. C., and Dunn, M. F. (1990) *Biochemistry* **29**, 442–451
296a. Steegborn, C., Laber, B., Messerschmidt, A., Huber, R., and Clausen, T. (2001) *J. Mol. Biol.* **311**, 789–801
297. Clausen, T., Huber, R., Laber, B., Pohlenz, H.-D., and Messerschmidt, A. (1996) *J. Mol. Biol.* **262**, 202–224
298. Kery, V., Bukovska, G., and Kraus, J. P. (1994) *J. Biol. Chem.* **269**, 25283–25288
299. Taoka, S., Ohja, S., Shan, X., Kruger, W. D., and Banerjee, R. (1998) *J. Biol. Chem.* **273**, 25179–25184

References

299a. Mosharov, E., Cranford, M. R., and Banerjee, R. (2000) *Biochemistry* **39**, 13005–13011
299b. Jhee, K.-H., Niks, D., McPhie, P., Dunn, M. F., and Miles, E. W. (2001) *Biochemistry* **40**, 10873–10880
299c. Taoka, S., Lepore, B. W., Kabil, Ö., Ojha, S., Ringe, D., and Banerjee, R. (2002) *Biochemistry* **41**, 10454 – 10461
300. Nishi, N., Tanabe, H., Oya, H., Urushihara, M., Miyanaka, H., and Wada, F. (1994) *J. Biol. Chem.* **269**, 1015–1019
301. Takusagawa, F., Kamitori, S., Misaki, S., and Markham, G. D. (1996) *J. Biol. Chem.* **271**, 136–147
302. Takusagawa, F., Kamitori, S., and Markham, G. D. (1996) *Biochemistry* **35**, 2586–2596
302a. McQueney, M. S., Anderson, K. S., and Markham, G. D. (2000) *Biochemistry* **39**, 4443–4454
302b. Sánchez del Pino, M. M., Corrales, F. J., and Mato, J. M. (2000) *J. Biol. Chem.* **275**, 23476–23482
302c. Elrod, P., Zhang, J., Yang, X., Yin, D., Hu, Y., Borchardt, R. T., and Schowen, R. L., (2002) *Biochemistry* **41**, 8134 – 8142
303. Chen, Z., Crippen, K., Gulati, S., and Banerjee, R. (1994) *J. Biol. Chem.* **269**, 27193–27197
303a. Bandarian, V., and Matthews, R. G. (2001) *Biochemistry* **40**, 5056–5064
303b. Olteanu, H., and Banerjee, R. (2001) *J. Biol. Chem.* **276**, 35558–35563
304. Garrow, T. A. (1996) *J. Biol. Chem.* **271**, 22831–22838
304a. Thomas, D., Becker, A., and Surdin-Kerjan, Y. (2000) *J. Biol. Chem.* **275**, 40718–40724
305. Livesey, G. (1984) *Trends Biochem. Sci.* **9**, 27– 29
306. Soda, K., Tanaka, H., and Esaki, N. (1983) *Trends Biochem. Sci.* **8**, 214–217
307. Welch, G. N., and Loscalzo, J. (1998) *N. Engl. J. Med.* **338**, 1042–1050
308. Durand, P., Lussier-Cacan, S., and Blache, D. (1997) *FASEB J.* **11**, 1157–1168
309. Watanabe, M., Osada, J., Aratani, Y., Kluckman, K., Reddick, R., Malinow, M. R., and Maeda, N. (1995) *Proc. Natl. Acad. Sci. U.S.A.* **92**, 1585–1589
309a. Jakubowski, H. (1999) *FASEB J.* **13**, 2277–2283
309b. Ragone, R. (2002) *FASEB J.* **16**, 401–404
310. Finkelstein, J. D., and Martin, J. J. (1986) *J. Biol. Chem.* **261**, 1582–1587
311. James, F., Nolte, K. D., and Hanson, A. D. (1995) *J. Biol. Chem.* **270**, 22344–22350
311a. Ranocha, P., Bourgis, F., Ziemak, M. J., Rhodes, D., Gage, D. A., and Hanson, A. D. (2000) *J. Biol. Chem.* **275**, 15962–15968
312. Slany, R. K., Bösl, M., Crain, P. F., and Kersten, H. (1993) *Biochemistry* **32**, 7811–7817
313. Furfine, E. S., and Abeles, R. H. (1988) *J. Biol. Chem.* **263**, 9598–9606
314. Wray, J. W., and Abeles, R. H. (1995) *J. Biol. Chem.* **270**, 3147–3153
314a. Dai, Y., Pochapsky, T. C., and Abeles, R. H. (2001) *Biochemistry* **40**, 6379–6387
314b. Al-Mjeni, F., Ju, T., Pochapsky, T. C., and Maroney, M. J. (2002) *Biochemistry* **41**, 6761 – 6769
315. Wray, J. W., and Abeles, R. H. (1993) *J. Biol. Chem.* **268**, 21466–21469
316. Adams, D. O., and Yang, S. F. (1981) *Trends Biochem. Sci.* **6**, 161–164
317. Pirrung, M. C. (1983) *J. Am. Chem. Soc.* **105**, 7207–7209
318. Chang, C. (1996) *Trends Biochem. Sci.* **21**, 129–133
319. Gamble, R. L., Coonfield, M. L., and Schaller, G. E. (1998) *Proc. Natl. Acad. Sci. U.S.A.* **95**, 7825–7829
320. Hao, D., Ohme-Takagi, M., and Sarai, A. (1998) *J. Biol. Chem.* **273**, 26857–26861
320a. Tatsuki, M., and Mori, H. (2001) *J. Biol. Chem.* **276**, 28051–28057
320b. McCarthy, D. L., Capitani, G., Feng, L., Gruetter, M. G., and Kirsch, J. F. (2001) *Biochemistry* **40**, 12276–12284
321. Boller, T., and Kende, H. (1980) *Nature (London)* **286**, 259–260
322. Peiser, G. D., Wang, T.-T., Hoffman, N. E., Yang, S. F., Liu, H.-w, and Walsh, C. T. (1984) *Proc. Natl. Acad. Sci. U.S.A.* **81**, 3059–3063
323. Barlow, J. N., Zhang, Z., John, P., Baldwin, J. E., and Schofield, C. J. (1997) *Biochemistry* **36**, 3563–3569
324. Zhang, Z., Barlow, J. N., Baldwin, J. E., and Schofield, C. J. (1997) *Biochemistry* **36**, 15999–16007
324a. Brunhuber, N. M. W., Mort, J. L., Christoffersen, R. E., and Reich, N. O. (2000) *Biochemistry* **39**, 10730–10738
324b. Thrower, J. S., Blalock, R., III, and Klinman, J. P. (2001) *Biochemistry* **40**, 9717–9724
325. Pirrung, M. C., Kaiser, L. M., and Chen, J. (1993) *Biochemistry* **32**, 7445–7450
326. Blomstrom, D. C., and Beyer, E. M., Jr. (1980) *Nature (London)* **283**, 66–68
327. Liu, H.-W., Auchus, R., and Walsh, C. T. (1984) *J. Am. Chem. Soc.* **106**, 5335–5348
327a. Yao, M., Ose, T., Sugimoto, H., Horiuchi, A., Nakagawa, A., Wakatsuki, S., Yokoi, D., Murakami, T., Honma, M., and Tanaka, I. (2000) *J. Biol. Chem.* **275**, 34557–34565
328. Bousquet, J.-F., and Thimann, K. V. (1984) *Proc. Natl. Acad. Sci. U.S.A.* **81**, 1724–1727
329. Rabinowitz, K. W., Niederman, R. A., and Wood, W. A. (1973) *J. Biol. Chem.* **248**, 8207–8215
330. Hirose, K., Kawata, Y., Yumoto, N., and Tokushige, M. (1991) *J. Biochem.* **110**, 971–975
331. Eisenstein, E. (1991) *J. Biol. Chem.* **266**, 5801–5807
332. Eisenstein, E., Yu, H. D., Fisher, K. E., Iacuzio, D. A., Ducote, K. R., and Schwarz, F. P. (1995) *Biochemistry* **34**, 9403–9412
333. Samach, A., Hareven, D., Gutfinger, T., Ken-Dror, S., and Lifschitz, E. (1991) *Proc. Natl. Acad. Sci. U.S.A.* **88**, 2678–2682
334. Umbarger, H. E. (1956) *Science* **123**, 848
335. Liu, J.-Q., Dairi, T., Itoh, N., Kataoka, M., Shimizu, S., and Yamada, H. (1998) *J. Biol. Chem.* **273**, 16678–16685
336. Epperly, B. R., and Dekker, E. E. (1991) *J. Biol. Chem.* **266**, 6086–6092
337. Ray, M., and Ray, S. (1985) *J. Biol. Chem.* **260**, 5913–5918
338. Kelley, J. J., and Dekker, E. E. (1984) *J. Biol. Chem.* **259**, 2124–2129
338a. Cheong, C.-G., Escalante-Semerena, J. C., and Rayment, I. (2002) *Biochemistry* **41**, 9079 – 9089
339. Barak, Z., Chipman, D. M., and Schloss, J. V., eds. (1990) *Biosynthesis of Branched Chain Amino Acids*, VCH Publ., Basel
340. Umbarger, H. E. (1987) in *Escherichia coli and Salmonella typhimurium* (Neidhardt, F. C., ed), pp. 352–366, Am. Soc. Microbiology, Washington, DC
340a. Lee, Y.-T., and Duggleby, R. G. (2001) *Biochemistry* **40**, 6836–6844
340b. Mendel, S., Elkayam, T., Sella, C., Vinogradov, V., Vyazmensky, M., Chipman, D. M., and Barak, Z. (2001) *J. Mol. Biol.* **307**, 465–477
341. Lawther, R. P., Calhoun, D. H., Adams, C. W., Hauser, C. A., Gray, J., and Hatfield, G. W. (1981) *Proc. Natl. Acad. Sci. U.S.A.* **78**, 922–925
342. Ray, T. B. (1986) *Trends Biochem. Sci.* **11**, 180–183
343. Halgand, F., Vives, F., Dumas, R., Biou, V., Andersen, J., Andrieu, J.-P., Cantegril, R., Gagnon, J., Douce, R., Forest, E., and Job, D. (1998) *Biochemistry* **37**, 4773–4781
344. Roy, T. B. (1986) *Trends Biochem. Sci.* **11**, 180–183
345. Schloss, J. V., van Dyk, D. E., Vasta, J. F., and Kutny, R. M. (1985) *Biochemistry* **24**, 4952–4959
346. Biou, V., Dumas, R., Cohen-Addad, C., Douce, R., Job, D., and Pebay-Peyroula, E. (1997) *EMBO J.* **16**, 3405–3415
347. Flint, D. H., and Emptage, M. H. (1988) *J. Biol. Chem.* **263**, 3558–3564
348. Ekiel, I., Smith, I. C. P., and Sprout, G. D. (1984) *Biochemistry* **23**, 1683–1687
349. Wallon, G., Kryger, G., Lovett, S. T., Oshima, T., Ringe, D., and Petsko, G. A. (1997) *J. Mol. Biol.* **266**, 1016–1031
350. Powers, S. G., and Snell, E. E. (1976) *J. Biol. Chem.* **251**, 3786–3793
350a. Stolz, J., and Sauer, N. (1999) *J. Biol. Chem.* **274**, 18747–18752
350b. Yun, M., Park, C.-G., Kim, J.-Y., Rock, C. O., Jackowski, S., and Park, H.-W. (2000) *J. Biol. Chem.* **275**, 28093–28099
350c. Matak-Vinkovic, D., Vinkovic, M., Saldanha, S. A., Ashurst, J. L., von Delft, F., Inoue, T., Miguel, R. N., Smith, A. G., Blundell, T. L., and Abell, C. (2001) *Biochemistry* **40**, 14493–14500
350d. Zheng, R., and Blanchard, J. S. (2000) *Biochemistry* **39**, 16244–16251
350e. Zheng, R., and Blanchard, J. S. (2001) *Biochemistry* **40**, 12904–12912
350f. Strauss, E., Kinsland, C., Ge, Y., McLafferty, F. W., and Begley, T. P. (2001) *J. Biol. Chem.* **276**, 13513–13516
350g. Strauss, E., and Begley, T. P. (2001) *J. Am. Chem. Soc.* **123**, 6449–6450
350h. Mootz, H. D., Finking, R., and Marahiel, M. A. (2001) *J. Biol. Chem.* **276**, 37289–37298
350i. Hoenke, S., Schmid, M., and Dimroth, P. (2000) *Biochemistry* **39**, 13233–13240
350j. Schneider, K., Dimroth, P., and Bott, M. (2000) *Biochemistry* **39**, 9438–9450
351. Alexeev, D., Alexeeva, M., Baxter, R. L., Campopiano, D. J., Webster, S. P., and Sawyer, L. (1998) *J. Mol. Biol.* **284**, 401–419
351a. Eliot, A. C., Sandmark, J., Schneider, G., and Kirsch, J. F. (2002) *Biochemistry* **41**, 12582 – 12589
352. Gibson, K. J., Lorimer, G. H., Rendina, A. R., Taylor, W. S., Cohen, G., Gatenby, A. A., Payne, W. G., Roe, D. C., Lockett, B. A., Nudelman, A., Marcovici, D., Nachum, A., Wexler, B. A., Marsilii, E. L., Turner, I. M., Sr., Howe, L. D., Kalbach, C. E., and Chi, H. (1995) *Biochemistry* **34**, 10976–10984
353. Käck, H., Sandmark, J., Gibson, K. J., Schneider, G., and Lindqvist, Y. (1998) *Protein Sci.* **7**, 2560–2566
354. Gibson, K. J. (1997) *Biochemistry* **36**, 8474–8478
355. Käck, H., Gibson, K. J., Lindqvist, Y., and Schneider, G. (1998) *Proc. Natl. Acad. Sci. U.S.A.* **95**, 5495–5500
355a. Birch, O. M., Hewitson, K. S., Fuhrmann, M., Burgdorf, K., Baldwin, J. E., Roach, P. L., and Shaw, N. M. (2000) *J. Biol. Chem.* **275**, 32277–32280
355b. Ugulava, N. B., Gibney, B. R., and Jarrett, J. T. (2001) *Biochemistry* **40**, 8343–8351
355c. Ollagnier-de-Choudens, S., Mulliez, E., Hewitson, K. S., and Fontecave, M. (2002) *Biochemistry* **41**, 9145 – 9152
356. Hayden, M. A., Huang, I., Bussiere, D. E., and Ashley, G. W. (1992) *J. Biol. Chem.* **267**, 9512–9515
356a. Yennawar, N. H., Conway, M. E., Yennawar, H. P., Farber, G. K., and Hutson, S. M. (2002) *Biochemistry* **41**, 11592 – 11601
357. Turnbull, A. P., Baker, P. J., and Rice, D. W. (1997) *J. Biol. Chem.* **272**, 25105–25111
358. Ikeda, Y., and Tanaka, K. (1983) *J. Biol. Chem.* **258**, 9477–9487

References

359. Hawes, J. W., Crabb, D. W., Chan, R. M., Rougraff, P. M., and Harris, R. A. (1995) *Biochemistry* **34**, 4231–4237
360. Kamoun, P. (1992) *Trends Biochem. Sci.* **17**, 175–176
361. Wolf, D. A., and Akers, H. A. (1986) *Trends Biochem. Sci.* **11**, 390–392
362. Sabourin, P. J., and Bieber, L. L. (1982) *J. Biol. Chem.* **257**, 7460–7467
362a. Hutson, S. M., Cree, T. C., and Harper, A. E. (1978) *J. Biol. Chem.* **253**, 8126–8133
362b. Tischler, M. E., Desautels, M., and Goldberg, A. L. (1982) *J. Biol. Chem.* **257**, 1613–1621
362c. Mordier, S., Deval, C., Béchet, D., Tassa, A., and Ferrara, M. (2000) *J. Biol. Chem.* **275**, 29900–29906
362d. Mitch, W. E., and Goldberg, A. L. (1996) *N. Engl. J. Med.* **335**, 1897–1905
362e. Nissen, S., Sharp, R., Ray, M., Rathmacher, J. A., Rice, D., Fuller, J. C., Jr., Connelly, A. S., and Abumrad, N. (1996) *J. Appl. Physiol.* **81**, 2095–2104
362f. Nissen, S. L., and Abumrad, N. N. (1997) *J. Nutr. Biochem.* **8**, 300–311
363. Buckel, W. (1992) *FEMS Microbiol. Rev.* **88**, 211–232
364. Ho, C.-L., Noji, M., Saito, M., and Saito, K. (1999) *J. Biol. Chem.* **274**, 397–402
364a. Xue, H.-H., Fujie, M., Sakaguchi, T., Oda, T., Ogawa, H., Kneer, N. M., Lardy, H. A., and Ichiyama, A. (1999) *J. Biol. Chem.* **274**, 16020–16027
365. Snell, K. (1986) *Trends Biochem. Sci.* **11**, 241–243
366. Williams, H. E., and Smith, L. H., Jr. (1971) *Science* **171**, 390–391
366a. Lumb, M. J., Drake, A. F., and Danpure, C. J. (1999) *J. Biol. Chem.* **274**, 20587–20596
367. Danpure, C. J., and Purdue, P. E. (1995) in *The Metabolic and Molecular Bases of Inherited Disease*, 7th ed., Vol. 2 (Scriver, C. R., Beaudet, A. L., Sly, W. S., and Valle, D., eds), pp. 2385–2424, McGraw-Hill, New York
368. Noji, M., Inoue, K., Kimura, N., Gouda, A., and Saito, K. (1998) *J. Biol. Chem.* **273**, 32739–32745
369. Tai, C.-H., Yoon, M.-Y., Kim, S.-K., Rege, V. D., Nalabolu, S. R., Kredich, N. M., Schnackerz, K. D., and Cook, P. F. (1998) *Biochemistry* **37**, 10597–10604
369a. Hindson, V. J., Moody, P. C. E., Rowe, A. J., and Shaw, W. V. (2000) *J. Biol. Chem.* **275**, 461–466
370. Renwick, S. B., Snell, K., and Baumann, U. (1998) *Structure* **6**, 1105–1116
371. Vatcher, G. P., Thacker, C. M., Kaletta, T., Schnabel, H., Schnabel, R., and Baillie, D. L. (1998) *J. Biol. Chem.* **273**, 6066–6073
371a. Rontein, D., Nishida, I., Tashiro, G., Yoshioka, K., Wu, W.-I., Voelker, D. R., Basset, G., and Hanson, A. D. (2001) *J. Biol. Chem.* **276**, 35523–35529
372. Rathinasabapathi, B., Burnet, M., Russell, B. L., Gage, D. A., Liao, P.-C., Nye, G. J., Scott, P., Golbeck, J. H., and Hanson, A. D. (1997) *Proc. Natl. Acad. Sci. U.S.A.* **94**, 3454–3458
373. Glaasker, E., Konings, W. N., and Poolman, B. (1996) *J. Biol. Chem.* **271**, 10060–10065
374. Klein, S. M., and Sagers, R. D. (1967) *J. Biol. Chem.* **242**, 297–300, 301–305
374a. Piper, M. D., Hong, S.-P., Ball, G. E., and Dawes, I. W. (2000) *J. Biol. Chem.* **275**, 30987–30995
375. Fujioka, M., Takata, Y., Konishi, K., and Ogawa, H. (1987) *Biochemistry* **26**, 5696–5702
376. Chlumsky, L. J., Sturgess, A. W., Nieves, E., and Jorns, M. S. (1998) *Biochemistry* **37**, 2089–2095
376a. Eschenbrenner, M., Chlumsky, L. J., Khanna, P., Strasser, F., and Jorns, M. S. (2001) *Biochemistry* **40**, 5352–5367
376b. Zhao, G., and Jorns, M. S. (2002) *Biochemistry* **41**, 9747–9750
377. Reuber, B. E., Karl, C., Reimann, S. A., Mihalik, S. J., and Dodt, G. (1997) *J. Biol. Chem.* **272**, 6766–6776
377a. Harris, R. J., Meskys, R., Sutcliffe, M. J., and Scrutton, N. S. (2000) *Biochemistry* **39**, 1189–1198
378. Shemin, D., and Rittenberg, D. (1946) *J. Biol. Chem.* **166**, 621–625
379. Gong, J., Hunter, G. A., and Ferreira, G. C. (1998) *Biochemistry* **37**, 3509–3517
379a. Tan, D., Barber, M. J., and Ferreira, G. C. (1998) *Protein Sci.* **7**, 1208–1213
380. Whiting, M. J., and Granick, S. (1976) *J. Biol. Chem.* **251**, 1340–1346
381. Warren, M. J., Cooper, J. B., Wood, S. P., and Shoolingin-Jordan, P. M. (1998) *Trends Biochem. Sci.* **23**, 217–221
381a. Ruiz de Mena, I., Fernández-Moreno, M. A., Bornstein, B., Kaguni, L. S., and Garesse, R. (1999) *J. Biol. Chem.* **274**, 37321–37328
382. Warren, M. J., and Scott, A. I. (1990) *Trends Biochem. Sci.* **15**, 486–491
383. Jordan, P. M., ed. (1991) *Biosynthesis of Tetrapyrroles*, Elsevier Science Publ., New York
384. Bevan, D. R., Bodlaender, P., and Shemin, D. (1980) *J. Biol. Chem.* **255**, 2030–2035
385. Jahn, D., Verkamp, E., and Söll, D. (1992) *Trends Biochem. Sci.* **17**, 215–218
385a. Moser, J., Schubert, W.-D., Beier, V., Bringemeier, I., Jahn, D., and Heinz, D. W. (2001) *EMBO J.* **20**, 6583–6590
386. Rieble, S., and Beale, S. I. (1991) *J. Biol. Chem.* **266**, 9740–9745
387. Hennig, M., Grimm, B., Contestabile, R., John, R. A., and Jansonius, J. N. (1997) *Proc. Natl. Acad. Sci. U.S.A.* **94**, 4866–4871
388. Dent, A. J., Beyersmann, D., Block, C., and Hasnain, S. S. (1990) *Biochemistry* **29**, 7822–7828
389. Jaffe, E. K., Volin, M., and Myers, C. B. (1994) *Biochemistry* **33**, 11554–11562
390. Cheung, K.-M., Spencer, P., Timko, M. P., and Shoolingin-Jordan, P. M. (1997) *Biochemistry* **36**, 1148–1156
390a. Erskine, P. T., Newbold, R., Brindley, A. A., Wood, S. P., Shoolingin-Jordan, P. M., Warren, M. J., and Cooper, J. B. (2001) *J. Mol. Biol.* **312**, 133–141
390b. Kervinen, J., Jaffe, E. K., Stauffer, F., Neier, R., Wlodawer, A., and Zdanov, A. (2001) *Biochemistry* **40**, 8227–8236
390c. Jaffe, E. K., Volin, M., Bronson-Mullins, C. R., Dunbrack, R. L. J., Kervinen, J., Martins, J., Quinlan, J. F., Jr., Sazinsky, M. H., Steinhouse, E. M., and Yeung, A. T. (2000) *J. Biol. Chem.* **275**, 2619–2626
391. Louie, G. V., Brownlie, P. D., Lambert, R., Cooper, J. B., Blundell, T. L., Wood, S. P., Warren, M. J., Woodcock, S. C., and Jordan, P. M. (1992) *Nature (London)* **359**, 33–39
392. Woodcock, S. C., and Jordan, P. M. (1994) *Biochemistry* **33**, 2688–2695
393. Porcher, C., Picat, C., Daegelen, D., Beaumont, C., and Grandchamp, B. (1995) *J. Biol. Chem.* **270**, 17368–17374
394. Warren, M. J., Gul, S., Aplin, R. T., Scott, A. I., Roessner, C. A., O'Grady, P., and Shoolingin-Jordan, P. M. (1995) *Biochemistry* **34**, 11288–11295
394a. Mathews, M. A. A., Schubert, H. L., Whitby, F. G., Alexander, K. J., Schadick, K., Bergonia, H. A., Phillips, J. D., and Hill, C. P. (2001) *EMBO J.* **20**, 5832–5839
395. Roméo, P.-H., Raich, N., Dubart, A., Beaupain, D., Pryor, M., Kushner, J., Cohen-Solal, M., and Goossens, M. (1986) *J. Biol. Chem.* **261**, 9825–9831
396. Whitby, F. G., Phillips, J. D., Kushner, J. P., and Hill, C. P. (1998) *EMBO J.* **17**, 2463–2471
396a. Martins, B. M., Grimm, B., Mock, H.-P., Huber, R., and Messerschmidt, A. (2001) *J. Biol. Chem.* **276**, 44108–44116
397. Proulx, K. L., Woodard, S. I., and Dailey, H. A. (1993) *Protein Sci.* **2**, 1092–1098
398. Camadro, J.-M., Thome, F., Brouillet, N., and Labbe, P. (1994) *J. Biol. Chem.* **269**, 32085–32091
399. Dailey, T. A., and Dailey, H. A. (1996) *Protein Sci.* **5**, 98–105
400. Ishida, T., Yu, L., Akutsu, H., Ozawa, K., Kawanishi, S., Seto, A., Inubushi, T., and Sano, S. (1998) *Proc. Natl. Acad. Sci. U.S.A.* **95**, 4853–4858
401. Dailey, H. A., Sellers, V. M., and Dailey, T. A. (1994) *J. Biol. Chem.* **269**, 390–395
402. Ferreira, G. C., Franco, R., Lloyd, S. G., Pereira, A. S., Moura, I., Moura, J. J. G., and Huynh, B. H. (1994) *J. Biol. Chem.* **269**, 7062–7065
402a. Ferreira, G. C., Franco, R., Mangravita, A., and George, G. N. (2002) *Biochemistry* **41**, 4809–4818
402b. Lu, Y., Sousa, A., Franco, R., Mangravita, A., Ferreira, G. C., Moura, I., and Shelnutt, J. A. (2002) *Biochemistry* **41**, 8253–8262
403. Smith, A. G., Santana, M. A., Wallace-Cook, A. D. M., Roper, J. M., and Labbe-Bois, R. (1994) *J. Biol. Chem.* **269**, 13405–13413
404. Daily, H. A., and Fleming, J. E. (1986) *J. Biol. Chem.* **261**, 7902–7905
405. Saiki, K., Mogi, T., Hori, H., Tsubaki, M., and Anraku, Y. (1993) *J. Biol. Chem.* **268**, 26927–26934
405a. Brown, K. R., Allan, B. M., Do, P., and Hegg, E. L. (2002) *Biochemistry* **41**, 10906–10913
406. Warren, M. J., Gonzalez, M. D., Williams, H. J., Stolowich, N. J., and Scott, A. I. (1990) *J. Am. Chem. Soc.* **112**, 5343–5345
407. Yap-Bondoc, F., Bondoc, L. L., Timkovich, R., Baker, D. C., and Hebbler, A. (1990) *J. Biol. Chem.* **265**, 13498–13500
408. Brown, C. E., Shemin, D., and Katz, J. J. (1973) *J. Biol. Chem.* **248**, 8015–8021
409. Scott, A. I., Townsend, C. A., Okada, K., and Kajiwara, M. (1974) *J. Am. Chem. Soc.* **96**, 8054–8069 and 8069–8080
410. Battersby, A. R. (1994) *Science* **264**, 1551–1557
411. Spencer, J. B., Stolowich, N. J., Santander, P. J., Pichon, C., Kajiwara, M., Tokiwa, S., Takatori, K., and Scott, A. I. (1994) *J. Am. Chem. Soc.* **116**, 4991–4992
412. Thibaut, D., Debussche, L., and Blanche, F. (1990) *Proc. Natl. Acad. Sci. U.S.A.* **87**, 8795–8799
413. Eisenreich, W., and Bacher, A. (1991) *J. Biol. Chem.* **266**, 23840–23849
414. Scott, A. I., Stolowich, N. J., Wang, J., Gawatz, O., Fridrich, E., and Müller, G. (1996) *Proc. Natl. Acad. Sci. U.S.A.* **93**, 14316–14319
415. Wang, J., Stolowich, N. J., Santander, P. J., Park, J. H., and Scott, A. I. (1996) *Proc. Natl. Acad. Sci. U.S.A.* **93**, 14320–14322
415a. Thompson, T. B., Thomas, M. G., Escalante-Semerena, J. C., and Rayment, I. (1999) *Biochemistry* **38**, 12995–13005
416. Michal, G., ed. (1999) *Biochemical Pathways*, Wiley–Spektrum Academischer Verlag, New York–Heidelberg, Germany
417. Bollivar, D. W., Suzuki, J. Y., Beatty, J. T., Dobrowolski, J. M., and Bauer, C. E. (1994) *J. Mol. Biol.* **237**, 622–640
418. Bollivar, D. W., Wang, S., Allen, J. P., and Bauer, C. E. (1994) *Biochemistry* **33**, 12763–12768

References

418a. Fujita, Y., and Bauer, C. E. (2000) *J. Biol. Chem.* **275**, 23583–23588

419. Jensen, P. E., Gibson, L. C. D., Henningsen, K. W., and Hunter, C. N. (1996) *J. Biol. Chem.* **271**, 16662–16667

420. Oster, U., Bauer, C. E., and Rüdiger, W. (1997) *J. Biol. Chem.* **272**, 9671–9676

421. Porra, R. J., and Meisch, H.-U. (1984) *Trends Biochem. Sci.* **9**, 99–104

422. Gibson, L. C. D., Willows, R. D., Kannangara, C. G., von Wettstein, D., and Hunter, C. N. (1995) *Proc. Natl. Acad. Sci. U.S.A.* **92**, 1941–1944

423. Tanaka, A., Ito, H., Tanaka, R., Tanaka, N. K., Yoshida, K., and Okada, K. (1998) *Proc. Natl. Acad. Sci. U.S.A.* **95**, 12719–12723

424. Scheumann, V., Schoch, S., and Rüdiger, W. (1998) *J. Biol. Chem.* **273**, 35102–35108

425. Warren, M. J., Jay, M., Hunt, D. M., Elder, G. H., and Röhl, J. C. G. (1996) *Trends Biochem. Sci.* **21**, 229–234

426. Brownlie, P. D., Lambert, R., Louie, G. V., Jordan, P. M., Blundell, T. L., Warren, M. J., Cooper, J. B., and Wood, S. P. (1994) *Protein Sci.* **3**, 1644–1650

427. Kappas, A., Sassa, S., Galbraith, R. A., and Nordmann, Y. (1995) in *The Metabolic and Molecular Bases of Inherited Disease*, 7th ed., Vol. 2 (Scriver, C. R., Beaudet, A. L., Sly, W. S., and Valle, D., eds), pp. 2103–2159, McGraw-Hill, New York

428. Macalpine, I., and Hunter, R. (1969) *Sci. Am.* **221**(Jul), 38–46

429. Lee, J.-S., and Anvret, M. (1991) *Proc. Natl. Acad. Sci. U.S.A.* **88**, 10912–10915

430. Poland, A., and Glover, E. (1973) *Science* **179**, 476–477

430a. Bate, R., ed. (1999) *What Risk*, Butterworth-Heinemann, Oxford; Boston

431. Bohle, D. S., Dinnebier, R. E., Madsen, S. K., and Stephens, P. W. (1997) *J. Biol. Chem.* **272**, 713–716

432. Maines, M. D. (1988) *FASEB J.* **2**, 2557–2568

433. Matera, K. M., Takahashi, S., Fujii, H., Zhou, H., Ishikawa, K., Yoshimura, T., Rousseau, D. L., Yoshida, T., and Ikeda-Saito, M. (1996) *J. Biol. Chem.* **271**, 6618–6624

434. Liu, Y., Moënne-Loccoz, P., Loehr, T. M., and Ortiz de Montellano, P. R. (1997) *J. Biol. Chem.* **272**, 6909–6917

434a. Sakamoto, H., Omata, Y., Palmer, G., and Noguchi, M. (1999) *J. Biol. Chem.* **274**, 18196–18200

435. Gray, C. H. (1983) *Trends Biochem. Sci.* **8**, 381–384

435a. Cunningham, O., Dunne, A., Sabido, P., Lightner, D., and Mantle, T. J. (2000) *J. Biol. Chem.* **275**, 19009–19017

436. Yoshinaga, T., Sassa, S., and Kappas, A. (1982) *J. Biol. Chem.* **257**, 7803–7807

437. Zakhary, R., Gaine, S. P., Dinerman, J. L., Ruat, M., Flavahan, N. A., and Snyder, S. H. (1996) *Proc. Natl. Acad. Sci. U.S.A.* **93**, 795–798

437a. Doré, S., Takahashi, M., Ferris, C. D., Hester, L. D., Guastella, D., and Snyder, S. H. (1999) *Proc. Natl. Acad. Sci. U.S.A.* **96**, 2445–2450

438. Wilks, A., and Schmitt, M. P. (1998) *J. Biol. Chem.* **273**, 837–841

439. Takahashi, M., Ilan, Y., Chowdhury, N. R., Guida, J., Horwitz, M., and Chowdhury, J. R. (1996) *J. Biol. Chem.* **271**, 26536–26542

440. Nogales, D., and Lightner, D. A. (1995) *J. Biol. Chem.* **270**, 73–77

441. McDonagh, A. F., Palma, L. A., and Lightner, D. A. (1980) *Science* **208**, 145–151

442. Brumm, P. J., Fried, J., and Friedmann, H. C. (1983) *Proc. Natl. Acad. Sci. U.S.A.* **80**, 3943–3947

443. Cooper, A. J. L. (1983) *Ann. Rev. Biochem.* **52**, 187–222

444. Klaassen, C. D., and Boles, J. W. (1997) *FASEB J.* **11**, 404–418

444a. Ullrich, T. C., and Huber, R. (2001) *J. Mol. Biol.* **313**, 1117–1125

444b. MacRae, I. J., Segel, I. H., and Fisher, A. J. (2001) *Biochemistry* **40**, 6795–6804

444c. Beynon, J. D., MacRae, I. J., Huston, S. L., Nelson, D. C., Segel, I. H., and Fisher, A. J. (2001) *Biochemistry* **40**, 14509–14517

445. Schmidt, A., Abrams, W. R., and Schiff, J. A. (1974) *Eur. J. Biochem.* **47**, 423–434

446. Benci, S., Vaccari, S., Mozzarelli, A., and Cook, P. F. (1997) *Biochemistry* **36**, 15419–15427

447. Cook, P. F., and Wedding, R. T. (1976) *J. Biol. Chem.* **251**, 2023–2029

447a. Soutourina, J., Blanquet, S., and Plateau, P. (2001) *J. Biol. Chem.* **276**, 40864–40872

448. Cooper, A. J. L., Haber, M. T., and Meister, A. (1982) *J. Biol. Chem.* **257**, 816–826

449. Akopyan, T. N., Braunstein, A. E., and Goryachenkova, E. V. (1975) *Proc. Natl. Acad. Sci. U.S.A.* **72**, 1617–1621

450. Weinstein, C. L., Haschemeyer, R. H., and Griffith, O. W. (1988) *J. Biol. Chem.* **263**, 16568–16579

451. Scriver, C. R. (1986) *N. Engl. J. Med.* **315**, 1155–1156

452. Colonge, M. J., Volpini, V., Bisceglia, L., Rousaud, F., de Sanctis, L., Beccia, E., Zelante, L., Testar, X., Zorzano, A., Estivill, X., Gasparini, P., Nunes, V., and Palacín, M. (1995) *Proc. Natl. Acad. Sci. U.S.A.* **92**, 9667–9671

453. Schneider, J. A. (1985) *N. Engl. J. Med.* **313**, 1473–1474

454. Weinstein, C. L., and Griffith, O. W. (1988) *J. Biol. Chem.* **263**, 3735–3743

455. Wright, C. E., Tallan, H. H., and Lin, Y. Y. (1986) *Ann. Rev. Biochem.* **55**, 427–453

456. Uchida, S., Kwon, H. M., Preston, A. S., and Handler, J. S. (1991) *J. Biol. Chem.* **266**, 9605–9609

457. Lombardini, J. B., Schaffer, S. W., and Azuma, J., eds. (1992) *Taurine Nutritional Value and Mechanisms of Action*, Plenum, New York

458. Pion, P. D., Kittleson, M. D., Rogers, Q. R., and Morris, J. G. (1987) *Science* **237**, 764–768

459. Geggel, H. S., Ament, M. E., Heckenlively, J. R., Martin, D. A., and Kopple, J. D. (1985) *N. Engl. J. Med.* **312**, 142–146

460. Thurston, H. H., Hauhart, R. E., and Naccarato, E. F. (1981) *Science* **214**, 1373–1374

461. Jarabak, R., and Westley, J. (1980) *Biochemistry* **19**, 900–904

462. Falany, C. N. (1997) *FASEB J.* **11**, 206–216

463. Miller, D. M., Delgado, R., Chirgwin, J. M., Hardies, S. C., and Horowitz, P. M. (1991) *J. Biol. Chem.* **266**, 4686–4691

463a. Gliubich, F., Gazerro, M., Zanotti, G., Delbono, S., Bombieri, G., and Berni, R. (1996) *J. Biol. Chem.* **271**, 21054–21061

463b. Bordo, D., Deriu, D., Colnaghi, R., Carpen, A., Pagani, S., and Bolognesi, M. (2000) *J. Mol. Biol.* **298**, 691–704

464. Zheng, L., Cash, V. L., Flint, D. H., and Dean, D. R. (1998) *J. Biol. Chem.* **273**, 13264–13272

465. Mihara, H., Kurihara, T., Yoshimura, T., Soda, K., and Esaki, N. (1997) *J. Biol. Chem.* **272**, 22417–22424

466. Nakai, Y., Yoshihara, Y., Hayashi, H., and Kagamiyama, H. (1998) *FEBS Lett.* **433**, 143–148

466a. Lacourciere, G. M., Mihara, H., Kurihara, T., Esaki, N., and Stadtman, T. C. (2000) *J. Biol. Chem.* **275**, 23769–23773

466b. Lange, H., Kaut, A., Kispal, G., and Lill, R. (2000) *Proc. Natl. Acad. Sci. U.S.A.* **97**, 1050–1055

466c. Clausen, T., Kaiser, J. T., Steegborn, C., Huber, R., and Kessler, D. (2000) *Proc. Natl. Acad. Sci. U.S.A.* **97**, 3856–3861

466d. Smith, A. D., Agar, J. N., Johnson, K. A., Frazzon, J., Amster, I. J., Dean, D. R., and Johnson, M. K. (2001) *J. Am. Chem. Soc.* **123**, 11103–11104

466e. Krupka, H. I., Huber, R., Holt, S. C., and Clausen, T. (2000) *EMBO J.* **19**, 3168–3178

466f. Ogasawara, Y., Lacourciere, G., and Stadtman, T. C. (2001) *Proc. Natl. Acad. Sci. U.S.A.* **98**, 9494–9498

467. Gentry-Weeks, C. R., Spokes, J., and Thompson, J. (1995) *J. Biol. Chem.* **270**, 7695–7702

468. Urbina, H. D., Silberg, J. J., Hoff, K. G., and Vickery, L. E. (2001) *J. Biol. Chem.* **276**, 44521–44526

469. Krebs, C., Agar, J. N., Smith, A. D., Frazzon, J., Dean, D. R., Huynh, B. H., and Johnson, M. K. (2001) *Biochemistry* **40**, 14069–14080

470. Garland, S. A., Hoff, K., Vickery, L. E., and Culotta, V. C. (1999) *J. Mol. Biol.* **294**, 8–

471. Tong, W.-H., and Rouault, T. (2000) *EMBO J.* **19**, 5692–5700

472. Kaut, A., Lange, H., Diekert, K., Kispal, G., and Lill, R. (2000) *J. Biol. Chem.* **275**, 15955–15961

473. Lutz, T., Westermann, B., Neupert, W., and Herrmann, J. M. (2001) *J. Mol. Biol.* **307**, 815–825

473a. Nakai, Y., Nakai, M., Hayashi, H., and Kagamiyama, H. (2001) *J. Biol. Chem.* **276**, 8314–8320

473b. Schwartz, C. J., Djaman, O., Imlay, J. A., and Kiley, P. J. (2000) *Proc. Natl. Acad. Sci. U.S.A.* **97**, 9009–9014

473c. Voisine, C., Cheng, Y. C., Ohlson, M., Schilke, B., Hoff, K., Beinert, H., Marszalek, J., and Craig, E. A. (2001) *Proc. Natl. Acad. Sci. U.S.A.* **98**, 1483–1488

473d. Wu, G., Mansy, S. S., Wu, S.-P., Surerus, K. K., Foster, M. W., and Cowan, J. A. (2002) *Biochemistry* **41**, 5024–5032

473e. Wu, G., Mansy, S. S., Wu, S-p., Surerus, K. K., Foster, M. W., and Cowan, J. A. (2002) *Biochemistry* **41**, 5024 – 5032

473f. Nuth, M., Yoon, T., and Cowan, J. A. (2002) *J. Am. Chem. Soc.* **124**, 8774 – 8775

473g. Mihara, H., Kato, S-i, Lacourciere, G. M., Stadtman, T. C., Kennedy, R. A. J. D., Kurihara, T., Tokumoto, U., Takahashi, Y., and Esaki, N. (2002) *Proc. Natl. Acad. Sci. U.S.A.* **99**, 6679 – 6683

474. Leimkühler, S., and Rajagopalan, K. V. (2001) *J. Biol. Chem.* **276**, 22024–22031

475. White, R. H. (1988) *Biochemistry* **27**, 7458–7462

475a. Tallant, T. C., Paul, L., and Krzycki, J. A. (2001) *J. Biol. Chem.* **276**, 4485–4493

475b. Krum, J. G., Ellsworth, H., Sargeant, R. R., Rich, G., and Ensign, S. A. (2002) *Biochemistry* **41**, 5005 – 5014

476. Barrios, A. M., and Lippard, S. J. (2000) *J. Am. Chem. Soc.* **122**, 9172–9177

477. Rutherfurd, K. J., Rutherfurd, S. M., Moughan, P. J., and Hendriks, W. H. (2002) *J. Biol. Chem.* **277**, 114–119

478. Hans, M., Bill, E., Cirpus, I., Pierik, A. J., Hetzel, M., Alber, D., and Buckel, W. (2002) *Biochemistry* **41**, 5873 – 5882

Study Questions

1. Bacterial glutamine synthetase is feedback inhibited by serine, glycine, and alanine. Explain specifically the connection between these amino acids and glutamine that would account for the logic of this inhibition.

2. Argininosuccinate synthetase catalyzes the following reaction:

$$H_2N{-}C({=}^{18}O){-}NH{-}CH_2{-}R_1 \ (I) + R_2{-}NH_2 \ (II) \underset{}{\overset{ATP \quad AMP + PP_i}{\rightleftharpoons}} H_2N{-}C({=}N{-}R_2){-}NH{-}CH_2{-}R_1$$

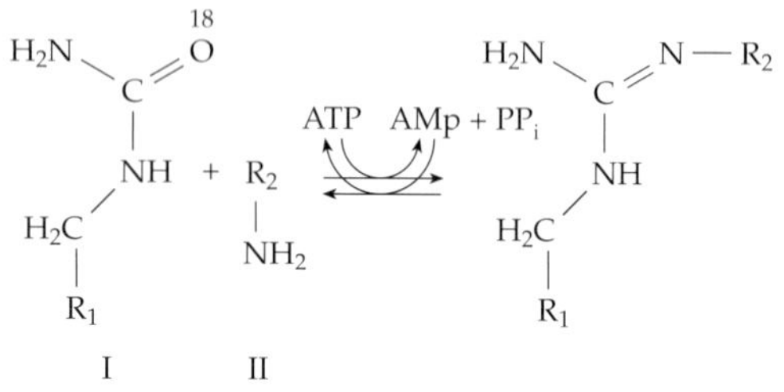

a) The reaction as shown is reversible. What metabolic stratagem is employed to drive the reaction from left to right?
b) It has been shown that the oxygen-18 from I (see structure) is transferred to the phosphate group of AMP. Propose a biochemical reaction mechanism to account for the transfer.

3. An organism has been discovered whose urea cycle does not include any reaction with aspartate. Both of the urea nitrogen atoms come directly from ammonia. All other components of the cycle are present and the net reaction is

$$CO_2 + 2\ NH_3 + 3\ ATP \rightarrow \text{urea} + 2\ ADP + 2\ P_i + AMP + PP_i$$

Explain how this is accomplished. Include plausible reaction mechanisms for any *new* steps proposed.

4. Which of the following compounds, if added to an active tissue preparation, might be expected to yield the greatest increase in urea production in terms of moles of urea produced per mole of added compound?

   a) Ammonia
   b) Bicarbonate
   c) Aspartate
   d) Ornithine

   More than one answer may be correct. Explain how you decided.

5. On a given diet yielding 2500 kcal per day, a 70-kg man excretes 30.0 g of urea daily. What percentage of his daily energy requirement is met by protein? Assume that 1.0 g of protein yields 4.0 kcal and 0.16 g of nitrogen as urea.

6. In many organisms the immediate biosynthetic precursor of L-lysine is α,ε-diaminopimelic acid (structure below). What type of enzyme would catalyze this reaction; what coenzyme would be required; and what type of enzyme–substrate complex would be formed?

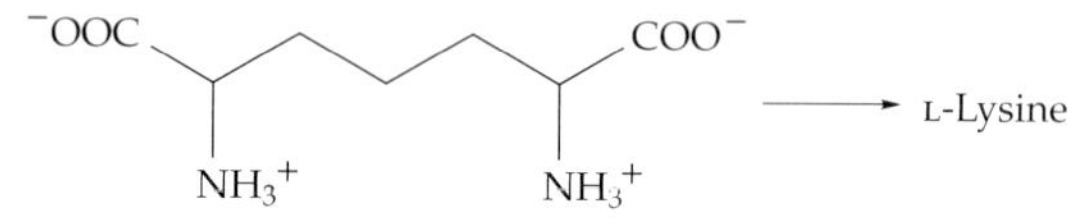

$$^{-}OOC{-}CH(NH_3^+){-}CH_2{-}CH_2{-}CH_2{-}CH(NH_3^+){-}COO^- \longrightarrow \text{L-Lysine}$$

α,ε-Diaminopimelic acid

7. A possible mechanism for the action of urease is pictured in Fig. 16-25 and Eq. 16-47. Carbamate is thought to be one intermediate. Can you suggest an alternative possibility for the initial nickel ion-dependent steps. See Barrios and Lippard.[476]

8. Leucine is known as a "ketogenic" amino acid. Explain what this means.

9. In some organisms leucine is not ketogenic. Why?

10. Here is a possible metabolic reaction for a fungus.

$$\text{L-Leucine} + \text{2-oxoglutarate}^{2-} + 2\tfrac{1}{2}\ O_2 \rightarrow \text{L-glutamate}^- + \text{citrate}^{3-} + H_2O + 2\ H^+$$

$\Delta G°'$ (pH 7) = –1026 kJ/mol

Suggest a metabolic pathway for this reaction. Is it thermodynamically feasible?

11. To be practical, the fungus should convert the L-glutamate back to 2-oxoglutarate using a glutamate dehydrogenase. Here are some values for Gibbs energies of formation from the elements under standard conditions (pH = 0).

2-Oxoglutarate^{2-}	–798.0 kJ/mol
L-Glutamate$^-$	–696.8 kJ/mol
NH_4^+	–79.5 kJ/mol

a) Calculate the apparent Gibbs energy change $\Delta G°'$ (pH 7) for the following reaction:

$$\text{L-Leucine} + 3\ O_2 \rightarrow \text{citrate}^{3-} + NH_4^+ + H_2O + 2\ H^+$$

b) Could this fungal reaction be used as a commercial process for making citric acid?

12. Two molecules of pyruvate can react to give a common precursor to valine, leucine, and pantoic acid. An isomerization step involving shift of a methyl group from one carbon to another is involved.

   a) Indicate as much as you can of the pathway for formation of valine showing coenzymes and mechanisms.
   b) Outline the reaction sequence by which the immediate oxoacid precursor of valine is converted to leucine.
   c) Outline a sequence by which the same oxoacid may be converted to pantoic acid using serine as an additional carbon atom source.

13. Explain how carboxylation can be coupled to cleavage of ATP and how this can be used to drive a metabolic sequence.

14. L-Serine is converted to pyruvate + NH_3 by serine dehydratase (deaminase) in a PLP-dependent reaction. However, using the same coenzyme selenocysteine is converted by selenocysteine lyase into L-alanine + elemental selenium Se^0. L-Cysteine may be converted by PLP-dependent enzymes into wither H_2S or into S^0 for transfer into metal clusters. Compare the chemical mechanisms.

15. Felinine is found in urine of cats, the highest amounts in males. The compound arises from a reaction of glutathione. Propose a route of synthesis. See Rutherfurd *et al.*[477]

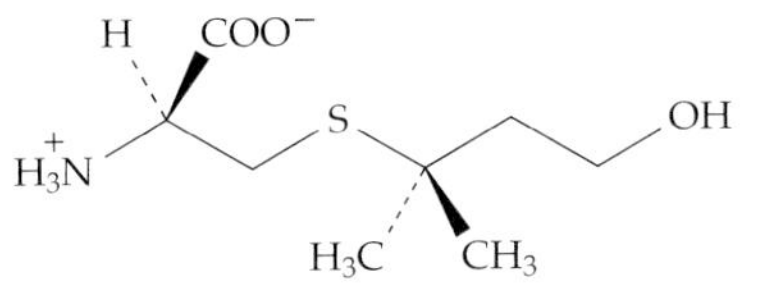

H COO⁻
$H_3\overset{+}{N}$
S
OH
H_3C CH_3

Felinine

16. Some clostridia ferment glutamate to ammonia, carbon dioxide, acetate, butyrate, and molecular hydrogen. Write a balanced equation and compare with Eq. 24-18 and Fig. 24-8. See Hans *et al.*[478]

17. How could β-hydroxy-β-methylbutyrate be formed in muscle? Could it be a physiologically important precursor to cholesterol?

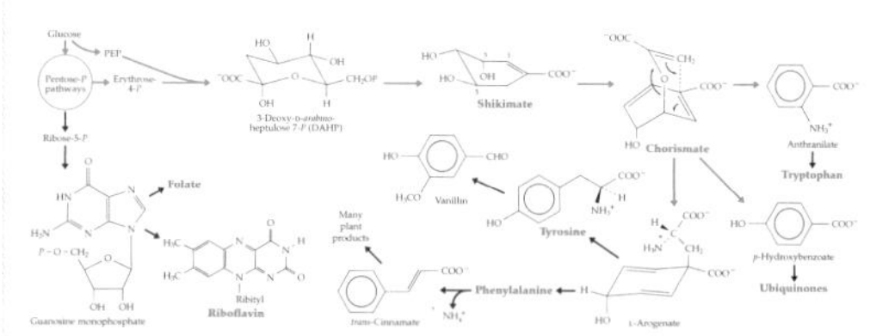

Aromatic compounds arise in several ways. The major route utilized by autotrophic organisms for synthesis of the aromatic amino acids, quinones, and tocopherols is the **shikimate pathway**. As outlined here, it starts with the glycolysis intermediate phosphoenolpyruvate (PEP) and erythrose 4-phosphate, a metabolite from the pentose phosphate pathway. Phenylalanine, tyrosine, and tryptophan are not only used for protein synthesis but are converted into a broad range of hormones, chromophores, alkaloids, and structural materials. In plants phenylalanine is deaminated to cinnamate which yields hundreds of secondary products. In another pathway ribose 5-phosphate is converted to pyrimidine and purine nucleotides and also to flavins, folates, molybdopterin, and many other pterin derivatives.

Contents

Boxes

25 Metabolism of Aromatic Compounds and Nucleic Acid Bases

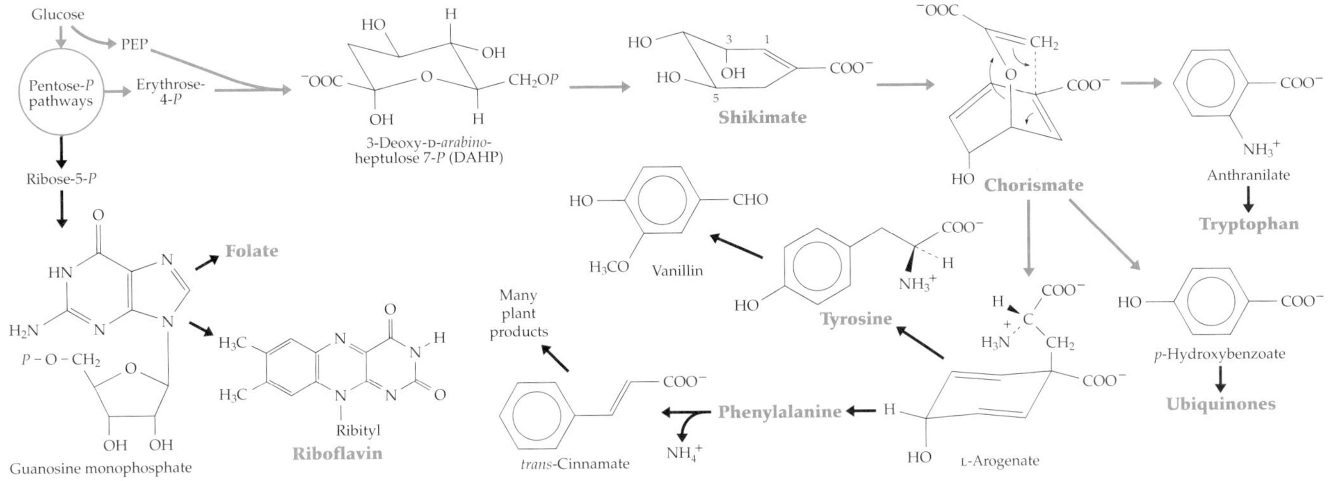

Aromatic rings are found in amino acids, purine and pyrimidine bases, vitamins, antibiotics, alkaloids, pigments of flowers and of the human skin, the lignin present in wood, and in many other substances. There are several biosynthetic pathways. One, which we met in Chapter 21, is the polyketide pathway. However, more important in most autotrophic organisms is the **shikimate pathway** which gives rise to phenylalanine, tyrosine, tryptophan, ubiquinone, plastoquinones, tocopherols, vitamin K, and other compounds.[1] The entire pathway, which is outlined in Fig. 25-1, is present in most bacteria and plants. However, animals are unable to synthesize the ring systems of the aromatic amino acids. Phenylalanine and tryptophan are dietary essentials. Tyrosine can also be formed in the animal body by the hydroxylation of phenylalanine. However, green plants lack phenylalanine hydroxylase and make tyrosine directly through the shikimate pathway (green arrows in Fig. 25-1).

There is interest in applying genetic engineering to increase the output of the shikimate pathway for production of industrially important aromatic compounds, e.g., the dye **indigo**, which is used in manufacture of blue denim (Box 25-C). Study of the enzymes involved has led to the development of potent inhibitors of the shikimate pathway which serve as widely used herbicides.[2,3]

A variety of pathways give rise to the nitrogen- and oxygen-containing heterocyclic rings of nature. All cells must be able to make pyrimidine and purine bases to be used in synthesis of nucleic acids and coenzymes. The pathway for synthesis of pyrimidine begins with aspartic acid and that for purines with glycine. In many organisms the pathway of purine formation is further enhanced because uric acid or a related substance is the major excretory product derived from excess nitrogen. This is true for both birds and reptiles, which excrete uric acid rather than urea, and for spiders which excrete guanine. In some plants, such as soy beans, the transport form of nitrogen is **allantoin** or **allantoic acid**, both of which are produced from uric acid.

A. The Shikimate Pathway of Biosynthesis

The shikimate pathway was identified through the study of ultraviolet light-induced mutants of *E. coli*, *Aerobacter aerogenes*, and *Neurospora*. In 1950, using the penicillin enrichment technique (Chapter 26), Davis obtained a series of mutants of *E. coli* that would not grow without the addition of aromatic substances.[4,5] A number of the mutants required five compounds: tyrosine, phenylalanine, tryptophan, *p*-aminobenzoic acid, and a trace of *p*-hydroxybenzoic acid. It was a surprise to find that the requirements for all five compounds could be met by the addition of shikimic acid, an aliphatic compound that was then regarded as a rare plant acid. Thus, shikimate was implicated as an intermediate in the biosynthesis of the three aromatic amino acids and of other essential aromatic substances.[6,7]

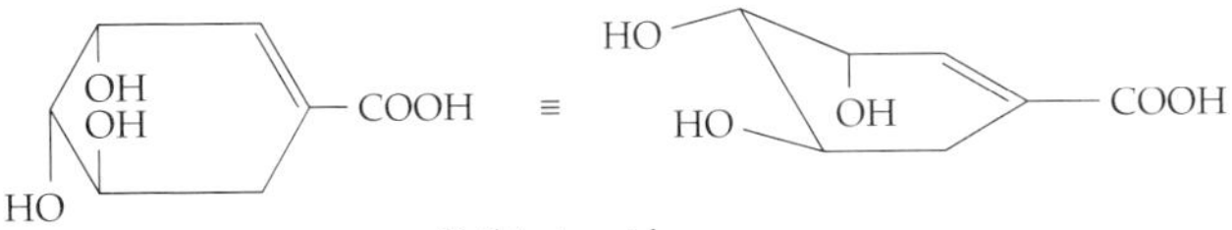

Shikimic acid

The mutants that grew in the presence of shikimic acid evidently had the biosynthetic pathway blocked

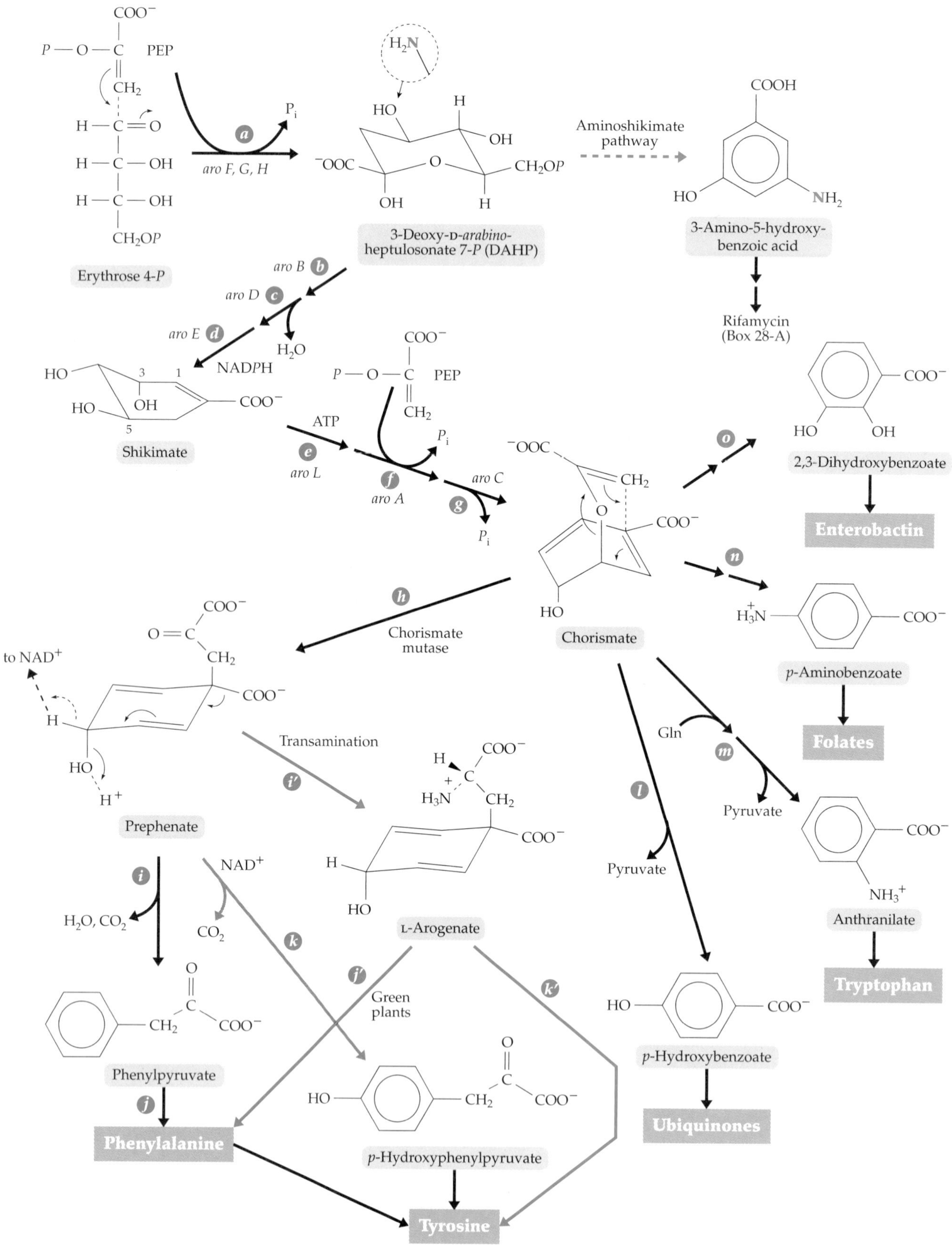

PEP
Erythrose 4-P
aro F, G, H
3-Deoxy-D-*arabino*-heptulosonate 7-P (DAHP)
Aminoshikimate pathway
3-Amino-5-hydroxy-benzoic acid
Rifamycin (Box 28-A)
aro B
aro D
aro E
H_2O
NADPH
Shikimate
PEP
ATP
aro L
aro A
aro C
Chorismate
2,3-Dihydroxybenzoate
Enterobactin
p-Aminobenzoate
Folates
Chorismate mutase
to NAD^+
Prephenate
Transamination
L-Arogenate
Gln
Pyruvate
Pyruvate
Anthranilate
Tryptophan
NAD^+
H_2O, CO_2
CO_2
Green plants
Phenylpyruvate
Phenylalanine
p-Hydroxyphenylpyruvate
p-Hydroxybenzoate
Ubiquinones
Tyrosine

Figure 25-1 Aromatic biosynthesis by the shikimate pathway. The symbols for several of the genes coding for the required enzymes are indicated. Their locations on the *E. coli* chromosome map are shown in Fig. 26-4. The aminoshikimate pathway which is initiated through 4-aminoDAHP leads to rifamycin and many other nitrogen-containing products.

at one or more earlier stages. Among these mutants, certain pairs were found that could not grow alone but that grew when plated together. The phenomenon is called **syntropism**. Mutant 83-2, which we now know to be blocked in the conversion of 5-dehydroshikimate to shikimate, accumulated dehydroshikimate and permitted mutant 83-1 or 83-3 to grow by providing it with a precursor that could be converted on to the end products (Eq. 25-1; the steps in this equation are lettered to correspond to those in Fig. 25-1). Eventually, the entire pathway was traced. The enzymes have all been isolated and studied[1] and the locations of the genes in the *E. coli* chromosome have been mapped[7–10] and are marked in Fig. 26-4.

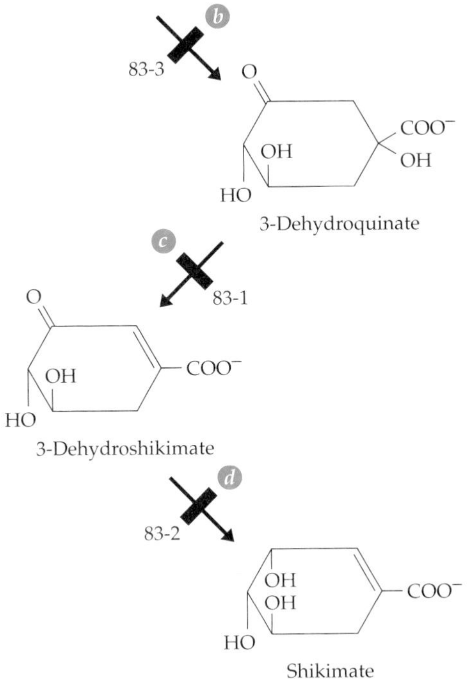

83-3 (b) → 3-Dehydroquinate; (c) 83-1 → 3-Dehydroshikimate; (d) 83-2 → Shikimate (25-1)

A variety of nitrogen-containing products, including rifamycin (Box 28-A), arise via the **aminoshikimate** pathway,[10a–d] which is also indicated in Fig. 25-1.

1. The Enzymes

The six carbons of the benzene ring of the aromatic amino acids are derived from the four carbons of erythrose 4-phosphate and two of the three carbons of phosphoenolpyruvate (PEP). The initial step in the pathway (Fig. 25-1, step *a*) is the condensation of erythrose 4-*P* with PEP and is catalyzed by 3-deoxy-D-*arabino*-heptulosonate-7-phosphate (DAHP) synthase. Closely analogous to an aldol condensation, the mechanism provides a surprise.[10] When PEP containing ^{18}O in the oxygen bridge to the phospho group reacts, the ^{18}O is retained in the eliminated phosphate; biochemical intuition would suggest that it should stay in the carbonyl group of the product. See Chapter 20, Section A,5.

Most bacteria and fungi have three isozymes of DAHP synthase, each controlled by feedback inhibition by one of the three products tyrosine, phenylalanine, or tryptophan. In *E. coli* these are encoded by genes *aro F*, *aro G*, and *aro H*, respectively.[11–12a] All of the enzymes contain one atom of iron per molecule and show spectral similarities to hemerythrin.[13]

The product of the DAHP synthase, 3-deoxy-D-*arabino*-heptulosonate 7-phosphate, is shown in its cyclic hemiacetal form at the beginning of Eq. 25-2. Its conversion to 3-dehydroquinate is a multistep process that is catalyzed by a single enzyme,[14,15] which is the product of *E. coli* gene *aro B*. The elimination of

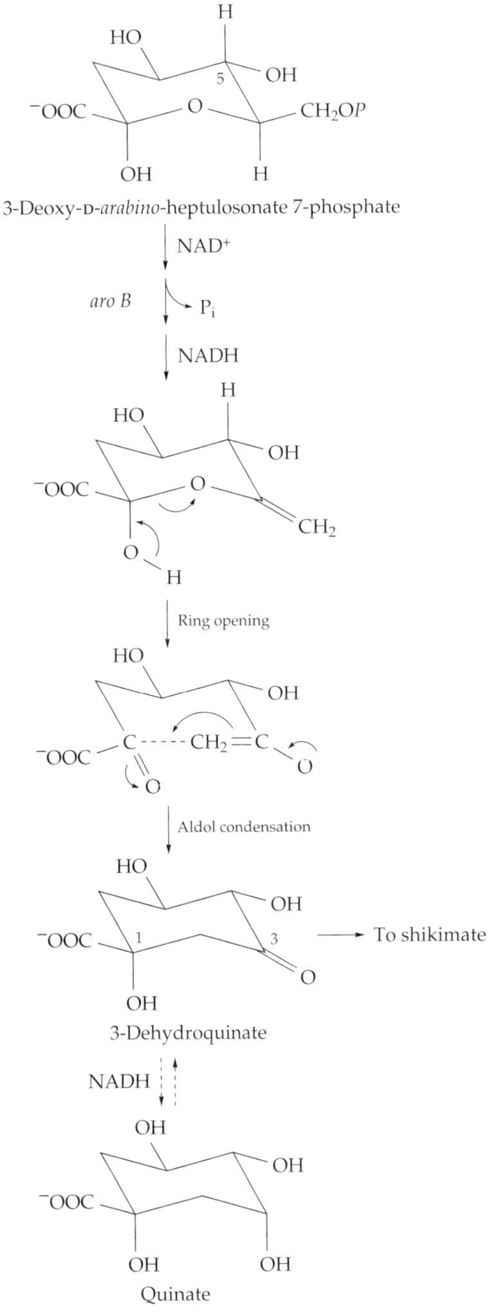

3-Deoxy-D-*arabino*-heptulosonate 7-phosphate → (NAD^+; *aro B*, P_i; NADH) → Ring opening → Aldol condensation → 3-Dehydroquinate → To shikimate; 3-Dehydroquinate ⇢ (NADH) Quinate (25-2)

inorganic phosphate in the second step of the sequence is assisted by a transient oxidation of the hydroxyl group at C-5 to a carbonyl group.[6,16,17] The enzyme contains bound NAD^+ and is activated by Co^{2+}. The last step in the dehydroquinate synthase sequence is the addition of an enolic intermediate to a carbonyl group, an aldol condensation, which forms the 6-carbon ring (Eq. 25-2). Also indicated in this equation, with dashed lines, is the reversible conversion of dehydroquinate to quinate. Although it is a "side product" quinic acid accumulates to high concentrations in many plants.[18]

Dehydration of 3-dehydroquinate (step *c*), the first step in Eq. 25-3, is the first of three elimination reactions needed to generate the benzene ring of the end products. This dehydration is facilitated by the presence of the carbonyl group. After reduction of the product to shikimate (step *d*)[19] a phosphorylation reaction (step *e*)[20,21] sets the stage for the future elimination of P_i. In step *f*, condensation with PEP adds three carbon atoms that will become the α, β, and

3-Dehydroquinate → (c, aro D, H_2O) → 3-Dehydroshikimate → (d, aro E, NADPH) → Shikimate → (e, aro L, ATP) → Shikimate-3-*P* → (f, aro A) → 5-Enoylpyruvyl-shikimate-3-*P* → (g, aro C, P_i) → Chorismate

(25-3)

carboxyl carbon atoms of phenylalanine and tyrosine. The reaction occurs by displacement of P_i from the α-carbon atom of PEP and resembles a reaction (Eq. 20-6, step *a*) in the synthesis of *N*-acetylmuramic acid.[22,23] When the reaction is carried out in ^3H-containing water, tritium enters the methylene group,[10,24] suggesting an addition–elimination mechanism (Eq. 25-4).[25] The enzyme 5-enoylpyruvylshikimate 3-phosphate synthase (**EPSP synthase**), which catalyzes this reaction, is strongly inhibited by the commercially important herbicide ***N*-(phosphonomethyl)glycine** (glyphosate).[3,26–27a]

N-(Phosphonomethyl) glycine (glyphosate)

A single mutation (Pro 101 → Ser) in the 427-residue protein from *E. coli* makes the enzyme more resistant to the herbicide.[28] Other mutations affect binding and catalysis.[29]

Elimination of P_i from 5-enolpyruvylshikimate 3-*P* (Eq. 25-3 and Fig. 25-1, step *g*) produces chorismate.[30] The 24-kDa chorismate synthase, which catalyzes this reaction, requires for activity a reduced flavin. Although there is no obvious need for an oxidation–reduction coenzyme, there is strong evidence that the flavin may play an essential role in catalysis, perhaps via a radical mechanism.[31–33b]

2. From Chorismate to Phenylalanine and Tyrosine

Chemical properties appropriate to a compound found at a branch point of metabolism are displayed by chorismic acid. Simply warming the compound in acidic aqueous solution yields a mixture of **prephenate** and ***para*-hydroxybenzoate** (corresponding to reactions *h* and *l* of Fig. 25-1). Note that the latter reaction is a simple elimination of the enolate anion of pyruvate. As indicated in Fig. 25-1, these reactions correspond to only two of several metabolic reactions of the chorismate ion. In *E. coli* the formation of **phenylpyruvate** (steps *h* and *i*, Fig. 25-1) is catalyzed by a single protein molecule with two distinctly different enzymatic activities: **chorismate mutase** and **prephenate dehydratase**.[34–36] However, in some organisms the enzymes are separate.[37] Both of the reactions catalyzed by these enzymes also occur spontaneously upon warming chorismic acid in acidic solution. The chorismate mutase reaction, which is unique in its mechanism,[37a] is discussed in Box 9-E. Stereochemical studies indicate that the formation of phenylpyruvate in Fig. 25-1, step *i*, occurs via a

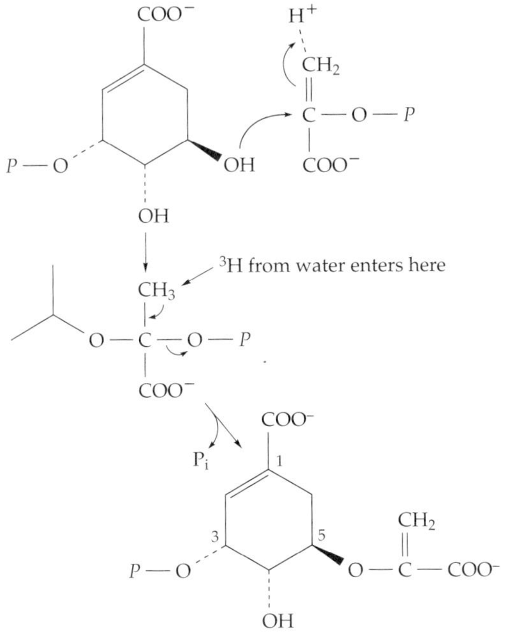

5-Enoylpyruvylshikimate 3-phosphate (25-4)

chair-like transition state.[38] Phenylpyruvate is transaminated to phenylalanine to complete the biosynthesis of that amino acid. Regulation in *E. coli* is accomplished in part by feedback inhibition of the bifunctional chorismate mutase-prephenate dehydratase.[38a]

In *E. coli* and many other bacteria a second bifunctional enzyme, **chorismate mutase-prephenate dehydrogenase** causes the isomerization of chorismate and the oxidative decarboxylation of prephenate to *p*-hydroxyphenylpyruvate (steps *h* and *k*, Fig. 25-1).[39] The latter can be converted by transamination to tyrosine.[40–42]

A slightly different pathway for tyrosine formation was found initially in cyanobacteria but has now been identified in a variety of organisms including higher green plants. In this pathway prephenate undergoes transamination to **L-arogenate** (pretyrosine), step *i*, Fig. 25-1.[43–45] In bacteria L-arogenate is oxidatively

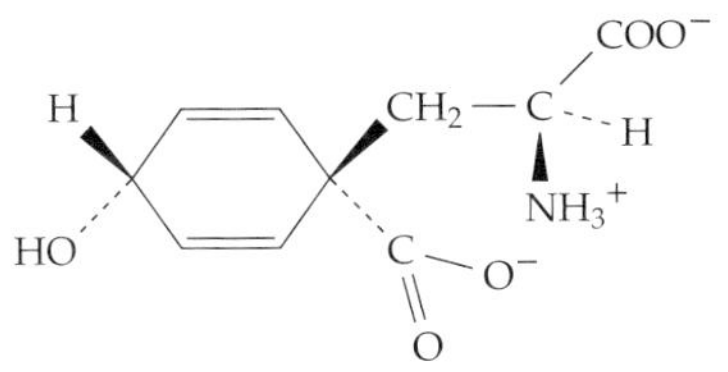

L-Arogenate (pretyrosine)

decarboxylated to tyrosine (step *k*, Fig. 25-1). However, in green plants L-arogenate undergoes decarboxylative elimination (step *j*) to give L-phenylalanine. This is a major reaction in green plants, which cannot form tyrosine by hydroxylation of phenylalanine[46,47] but which form a variety of additional products from phenylalanine by a pathway characteristic for plants (Fig. 25-8).

3. Anthranilate, Tryptophan, *para*-Aminobenzoate, and Dihydroxybenzoate

The three chemically similar reactions designated *m*, *n*, and *o* in Figs. 25-1 and 25-2 give rise to a variety of products. Step *m* leads to anthranilate and **L-tryptophan** and step *n* to the vitamin **folic acid.** Each of the three reactions *m*, *n*, and *o* involves addition of either NH_3 or HO^- at a position *ortho* or *para* to the carboxyl group of chorismate with elimination of the 5-OH group. The structures of the three intermediate products are shown in Fig. 25-2.[48–50] The three enzymes have significant similarities in amino acid sequence. Anthranilate synthase and *p*-aminobenzoate synthase are both two-subunit enzymes consisting of a 20-kDa subunit glutamine amidotransferase which is presumed to generate NH_3 (see Chapter 24,B).[49,51–53a] The second 50-kDa subunit of anthranilate synthase catalyzes the remaining steps in the reaction. However, *p*-aminobenzoate synthesis in *E. coli* requires an additional enzyme to catalyze the elimination of pyruvate in the final step[54] of synthesis.

The product of step *o* is known as **isochorismate.**[55,56] Isochorismate gives rise to a variety of products including **vitamin K**, salicylic acid,[56a] the iron chelator **enterobactin** (Fig. 16-1), and other siderophores. These are formed in *E. coli* via 2,3-dihydroxybenzoate as indicated in Fig. 25-2.[57–59a] The genes (*ent*) for the requisite enzymes are clustered at 14 min on the *E. coli* chromosome map (Fig. 26-4).

During the conversion of anthranilate to tryptophan, two additional carbon atoms must be incorporated to form the indole ring. These are derived from **phosphoribosyl pyrophosphate (PRPP)** which is formed from ribose 5-phosphate by transfer of a *pyrophospho* group from ATP.[60,61] The –OH group on the anomeric carbon of the ribose phosphate displaces AMP by attack on P_β of ATP (Eq. 25-5). In many organisms the enzyme that catalyzes this reaction is fused to subunit II of anthranilate synthase.[62] PRPP is also the donor of phosphoribosyl groups for biosynthesis of histidine (Fig. 25-13) and of nucleotides (Figs.

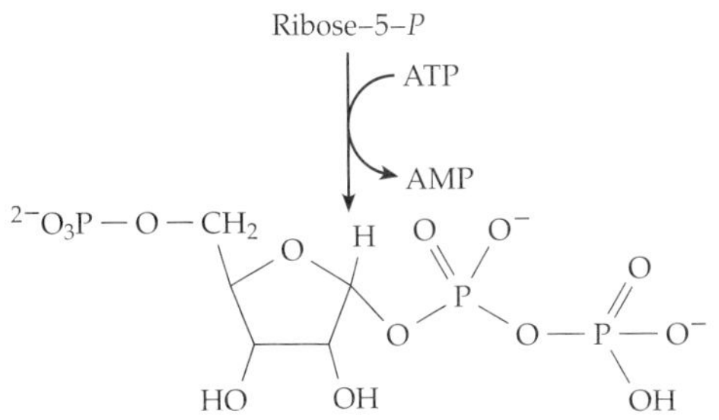

5-Phosphoribosyl 1-pyrophosphate (PRPP) (25-5)

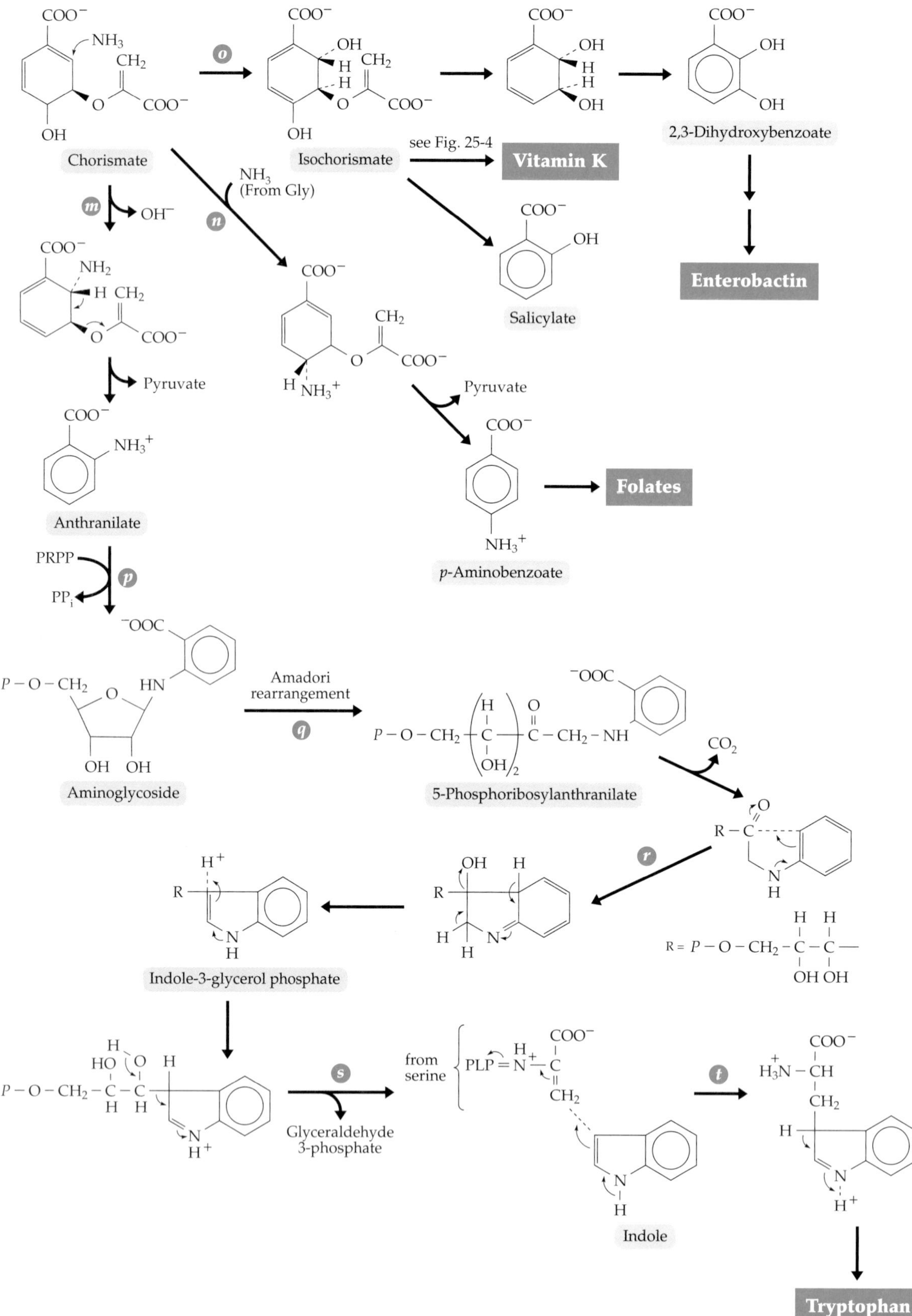

Figure 25-2 The biosynthesis of tryptophan from chorismate and related synthetic reactions.

25-14, 15). In tryptophan biosynthesis PRPP is converted into an aminoglycoside of anthranilic acid by displacement of its pyrophospho group by the amino group of anthranilate (Fig. 25-2, step *p*). The aminoglycoside then undergoes an **Amadori rearrangement** (Eq. 4-8; Fig. 25-2, step *q*). The product has an open chain. Decarboxylation and ring closure, as indicated in this figure, yields **indoleglycerol phosphate**.[63,64]

A β replacement reaction catalyzed by the PLP-dependent **tryptophan synthase** converts indoleglycerol phosphate and serine to tryptophan. Tryptophan synthase from *E. coli* consists of two subunits associated as an $\alpha_2\beta_2$ tetramer (Fig. 25-3). The α subunit catalyzes the cleavage (essentially a reverse aldol) of indoleglycerol phosphate to glyceraldehyde 3-phosphate and free indole (Fig. 25-2, step *s*).[67] The β subunit contains PLP. It presumably generates, from serine, the Schiff base of aminoacrylate, as indicated in Fig. 25-2 (step *t*). The enzyme catalyzes the addition of the free indole to the Schiff base to form tryptophan. The indole must diffuse for a distance of 2.5 nm through a tunnel to the active site where it condenses with the aminoacrylate Schiff base.[65–68c]

The genes encoding the seven enzymes of the tryptophan biosynthetic pathway are organized as a single operon in some bacteria.[69] Its regulation in enteric bacteria is discussed in Chapter 28, Section A,5. The α subunit of tryptophan synthase and the enzymes catalyzing the preceding two steps in tryptophan synthesis are all $(\beta\alpha)_8$-barrel proteins similar to the one shown in Fig. 2-28.[69,70] The biosynthetic pathway for tryptophan in the green plant *Arabidopsis* is the same as in bacteria. The enzymes appear to be present in the chloroplasts.[71]

While the tryptophan synthase of *E. coli* is made up of two different subunits, that of *Neurospora* is a single polypeptide chain. This is one of the first proteins for which it was proposed that there were originally two separate genes, as in *E. coli*, but that they became fused during the course of evolution. After this proposal was made, gene fusion was demonstrated experimentally in *Salmonella* by introduction of two consecutive "frame shift mutations" between two genes of histidine biosynthesis (Chapter 26, Section B,1). Because of the frame shift, the stop signal for protein synthesis is no longer read, with the result that the organism makes a single long protein corresponding to both genes. Gene fusion evidently occurs in nature frequently.[63] There are many instances known in which the two distinctly different catalytic activities are possessed by the same protein in some organisms but by separate enzymes in others. The gene for the α subunit of trytophan synthase in *Salmonella* was of historical importance as it was used to establish the colinearity of the genetic code and its amino acid sequence (Chapter 26, Section B,5).

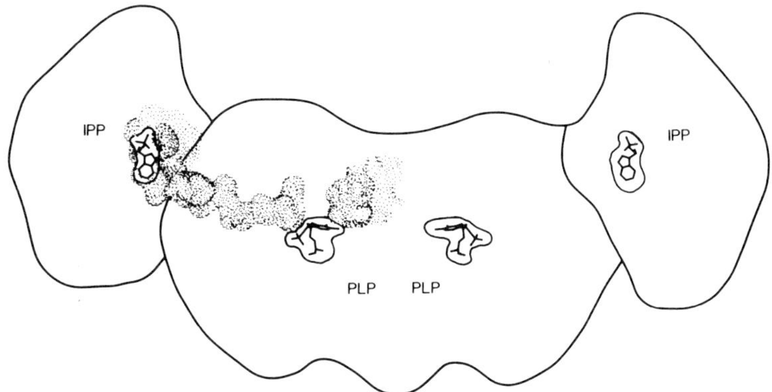

Figure 25-3 The structure of the two-enzyme $\alpha_2\beta_2$ complex tryptophan synthase.[65,66] The view is with the twofold axis of the $\alpha_2\beta_2$ complex vertical with the two α subunits at the ends and the β subunits in the center. The tunnel through which indole molecules released from indole propanol phosphate (IPP) in the α subunits to the pyridoxal phosphate (PLP) in the β subunits is shaded. Courtesy of C. Craig Hyde and Edith Wilson Miles.

4. Ubiquinones, Plastoquinones, Vitamin K, and Tocopherols

Radioactive carbon of [^{14}C]shikimate is efficiently incorporated into the quinones, vitamin K, and tocopherols. These chemically related redox agents (Fig. 15-24) also have a related biosynthetic origin, which has been elucidated in greatest detail for ubiquinone. In bacteria *p*-hydroxybenzoate is formed directly from chorismate (Fig. 25-1), but in plants it may originate from phenylalanine, *trans*-cinnamate, or *p*-coumarate as indicated in Fig. 25-4. The conversion of *p*-hydroxybenzoate on to the ubiquinones is also shown in this figure. A polyprenyl group is transferred onto a position ortho to the hydroxyl (see Chapter 22). Then a series of consecutive hydroxylation and *S*-adenosylmethionine-dependent transmethylation reactions lead directly to the ubiquinones.[72–73b] Several qui-nones that can serve as precursors to ubiquinones have been isolated from bacteria. Two of the corresponding quinols are shown as intermediates in Fig. 25-4. Chemical considerations suggest that both the methylations and hydroxylations occur on the reduced dihydroxy derivatives. A closely similar pathway is used for synthesis of ubiquinone[74–76] in mitochondria and in the membranes of the endoplasmic reticulum of fungal, plant, and animal cells.[77,78] In bacteria the decarboxylation step occurs early, as shown in Fig. 25-4, whereas in eukaryotes it occurs later.[75] Ubiquinone is poorly absorbed from the blood and it is apparently made in all aerobic tissues.[78] Ubiquinones are thought to serve as important antioxidant compounds in cell membranes. Dietary supplementation may be of value.[73a] Curiously, the nematode *C. elegans* slows its metabolism and lives longer if it has a defect in the hydroxylase catalyzing the next to last step in biosynthesis (Fig. 25-4).[73c] Mutants of *C. elegans* that cannot form their own ubiquinone-9 (containing nine prenyl units in the side chain) are unable to grow on bacteria that make ubiquinone-8. The worms appear to have both essential mitochondrial and nonmitochondrial requirements for ubiquinone.[73d] A ubiquinone deficiency with serious consequences can sometimes be caused in humans by inhibitors of 3-hydroxy-3-methylglutaryl-CoA reductase (HMG-CoA reductase) taken to lower blood cholesterol.[79]

Labeling experiments have shown that the plastoquinones of chloroplasts as well as the tocopherols each bear one methyl group (marked with an asterisk in Fig. 25-4) that originates from chorismate. The dihydroxy compound **homogentisate** is probably an intermediate.[80–83] It is a normal catabolite of tyrosine in the animal body (Fig. 25-5, Eq. 18-49). Both prenylation and methylation by AdoMet are required to complete the synthesis of the plastoquinones and tocopherols. Possible biosynthetic intermediates with one or more double bonds in the polyprenyl side chain have been found in plants and also in fish oils.[83a]

The vitamins K and other naphthoquinones arise from ***O*-succinylbenzoate**[84–86] whose synthesis from chorismate and 2-oxoglutarate depends upon a thiamine diphosphate-bound intermediate, as indicated in Fig. 25-4. Elimination of pyruvate yields *O*-succinylbenzoate. The remaining reactions of decarboxylation, methylation, and prenylation (Fig. 25-4) resemble those of ubiquinone synthesis.

B. Metabolism of Phenylalanine and Tyrosine

Figure 25-5 shows the principal catabolic pathways, as well as a few biosynthetic reactions, of phenylalanine and tyrosine in animals. Transamination to phenylpyruvate (reaction *a*) occurs readily, and the product may be oxidatively decarboxylated to phenylacetate. The latter may be excreted after conjugation with glycine (as in Knoop's experiments in which phenylacetate was excreted by dogs after conjugation with glycine, Box 10-A). Although it does exist, this degradative pathway for phenylalanine must be of limited importance in humans, for an excess of phenylalanine is toxic unless it can be oxidized to tyrosine (reaction *b*, Fig. 25-5). Formation of phenylpyruvate may have some function in animals. The enzyme **phenylpyruvate tautomerase**, which catalyzes interconversion of enol and oxo isomers of its substrate, is also an important immunoregulatory cytokine known as **macrophage migration inhibitory factor**.[86a]

The pterin-dependent hydroxylation of phenylalanine to tyrosine (Eq. 18-45)[87,87a] has received a great deal of attention because of the occurrence of the metabolic disease **phenylketonuria (PKU)**,[88–91b] in which this reaction is blocked. Infants born with phenylketonuria appear normal but mental retardation sets in rapidly. However, if these infants are identified promptly and are reared on a low-phenylalanine diet which supplies only enough of the amino acid for essential protein synthesis, most brain damage can be prevented. Throughout most of the world every infant born is now given a simple urine test to identify phenylketonuria. Tolerance to phenylalanine increases with age, and adults may return to a near normal diet. However, there may still be problems with increased phenylalanine levels during fever and infections. A high phenylalanine level during pregnancy may damage the unborn child. Temporary insertion of multitubular reactors containing phenylalanine-ammonia lyase (Eq. 14-45) can be of value[92] as is administration of the enzyme in encapsulated form.[93] The mechanism by which phenylalanine damages the brain is uncertain.

1. Catabolism of Tyrosine in Animals

The major route of degradation of tyrosine in animals begins with transamination (Fig. 25-5, reaction *c*) to ***p*-hydroxyphenylpyruvate**. The enzyme tyrosine aminotransferase[94] is induced in the liver in response to the action of glucorticoid hormones (Chapter 22). The synthesis of the enzyme is also controlled at the translational level, release of the newly formed protein from liver ribosomes being stimulated by cyclic AMP. The enzyme is subject to posttranscriptional

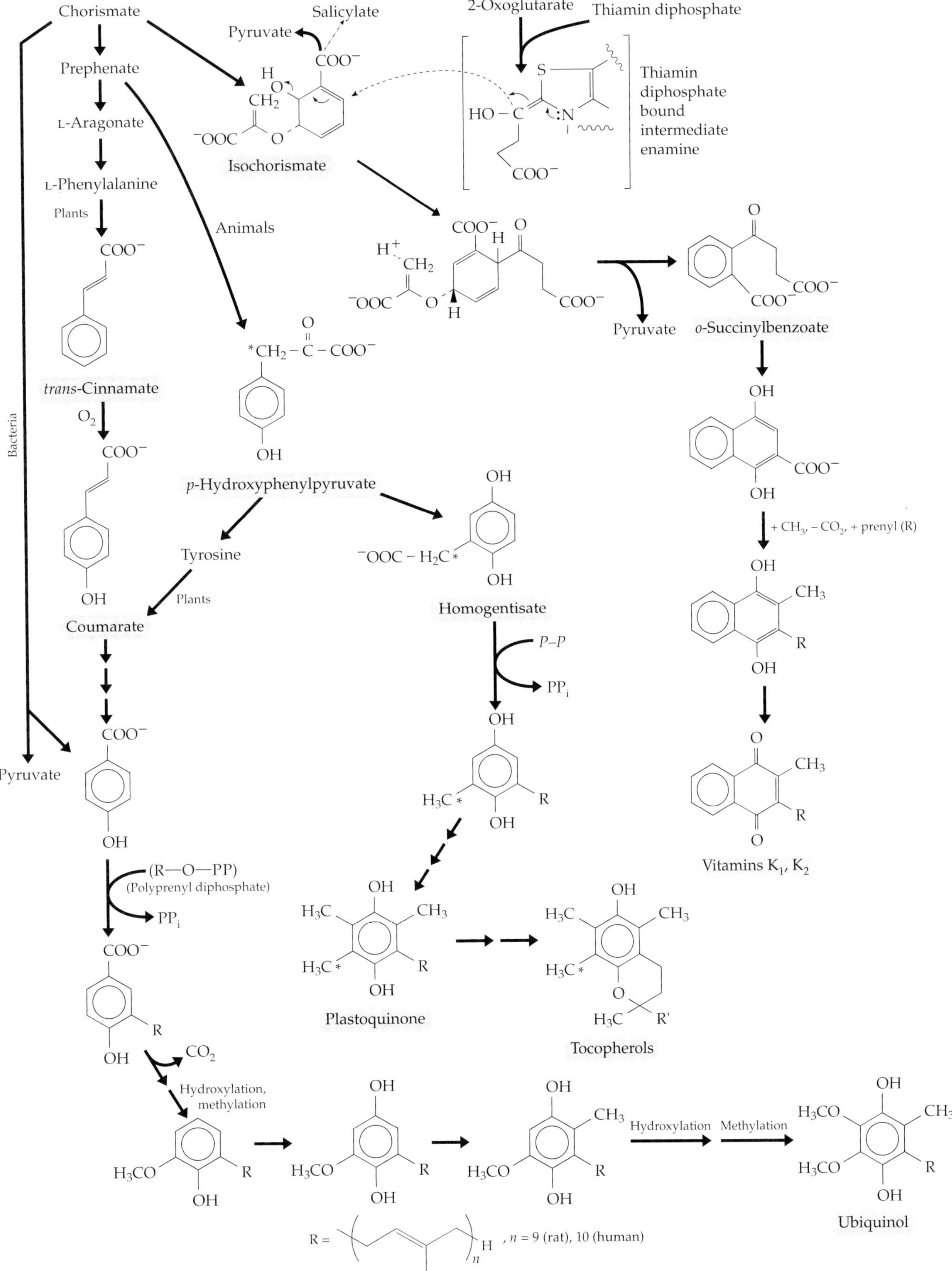

Figure 25-4 Pathways of biosynthesis of ubiquinones, plastoquinones, tocopherols, and vitamin K.

modification including phosphorylation and it undergoes unusually rapid turnover.[95,96]

The 2-oxoacid *p*-hydroxyphenylpyruvate is decarboxylated by the action of a dioxygenase (Eq. 18-49). The product **homogentisate** is acted on by a second dioxygenase, as indicated in Fig. 25-5, with eventual conversion to fumarate and acetoacetate. A rare metabolic defect in formation of homogentisate leads to tyrosinemia and excretion of **hawkinsin**[97] a compound postulated to arise from an epoxide (arene oxide) intermediate (see Eq. 18-47) which is detoxified by a glutathione transferase (Box 11-B).

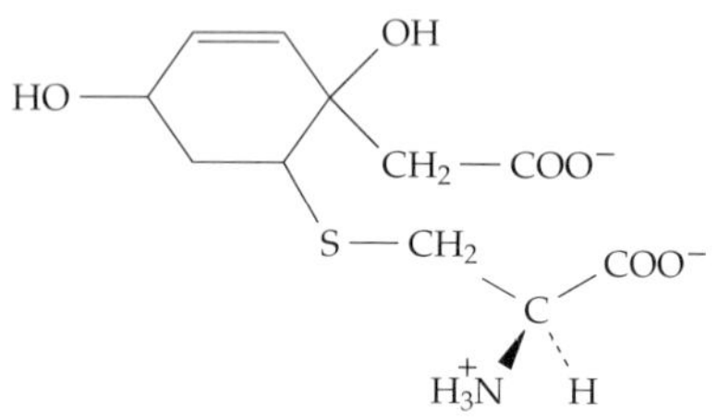

Hawkinsin

One of the first "inborn errors of metabolism" to be recognized was **alkaptonuria**, a lack of the oxygenase that cleaves the ring of homogentsic acid.[98] The condition is recognized by a darkening of the urine upon long standing (caused by oxidation of the homogentisate). Alkaptonuria was correctly characterized by Garrod (Box 1-D) in 1909 as a defect in the catabolism of tyrosine. Although relatively mild and not reducing the life span of individuals, it is nearly always accompanied by arthritis in later years and also by gray to bluish-black pigmentation of connective tissues, which may be visible through the shin or in the scleras (whites) of the eyes.[98] Absence of the next enzyme in the pathway, **maleylacetoacetate isomerase** causes one type of tyrosinemia.[99,99a] Absence of **fumarylacetoacetate hydrolase**, which acts on the product of the isomerase action causes the severe **type 1 hereditary tyrosinemia** which leads to accumulation of the toxic fumarylacetoacetate and its decarboxylation product succinylacetoacetate.[99,100]

2. The Thyroid Hormones

An important product of tyrosine metabolism in vetebrates is the thyroid hormone[101] of which the principal and most active forms are **thyroxine** (T_4) and **triiodothyronine** (T_3).[102] The thyroid gland is rich in iodide ion, which is actively concentrated from the plasma to ~1 μM free I^-.[103] This iodide reacts under the influence of a peroxidase (see Fig. 16-14 and accompanying discussion)[104] to iodinate tyrosyl residues of the very large ~660-kDa dimeric **thyroglobulin**, which is stored in large amounts in the lumen of the thyroid follicles.[105] Several of the tyrosine side chains (up to 15–25) are iodinated to form **mono-** and **diiodotyrosine** residues (Eq. 25-6), but only between four and eight of these, which are located at specific positions, are converted on to the hormones.[106,107]

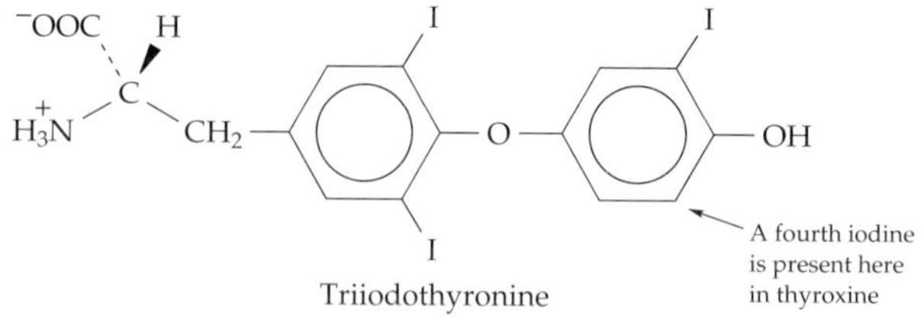

Triiodothyronine

A fourth iodine is present here in thyroxine

The coupling reaction by which the aromatic group from one residue of mono- or diiodotyrosine is joined in ether linkage with a second residue is also catalyzed readily by peroxidases. One dehydroalanine residue is formed for each molecule of hormone released.[108] A possible mechanism involves formation of an electron-deficient radical, which can undergo β elimination to produce a dehydroalanine residue and an aromatic radical. The latter could couple with a second radical to form triiodothyronine or thyroxine. However, as depicted in Eq. 25-6, the radical coupling may occur prior to chain cleavage. While β elimination (pathway A) has been favored, recent evidence suggests hydroxylation and cleavage to form a residue of aminomalonic semialdehyde in the thyroglobulin chain (pathway B).[108a] Alternatively, a PLP-dependent elimination of the radical could be used. Another possibility is oxidative attack on the 2-oxoacids derived from the iodotyrosines.

Thyroxine and triiodothyronine are released from thyroglobulin through the action of a series of proteases. Both the protease action and the release of the thyroid hormones into the bloodstream are stimulated by pituitary **thyrotropin (TSH)**.[109,110] Like glucagon thyrotropin is released from the pituitary in response to **thyrotropin-releasing hormone.**[111] Thyrotropin probably acts through cAMP-mediated mechanisms.[112] The hormones are carried throughout the body while bound to **thyroxine-binding globulin,** which serves as a carrier.[113] Some hormone is carried by other serum proteins such as **transthyretin** (thyroxine-binding prealbumin).[113,114] Both thyroxine and triiodothyronine have powerful hormonal effects on tissues, but the lag time for a response is shortest for triiodothyronine. Thus, it is thought that thyroxine undergoes loss of one iodine atom to form the more active triiodo form of the hormone within the target cells. Three **iodothyronine deiodinases**, all of which are selenoproteins, have been identified (Eq. 15-60).[115–116a]

Organically bound iodine is found in various invertebrates, but with one possible exception

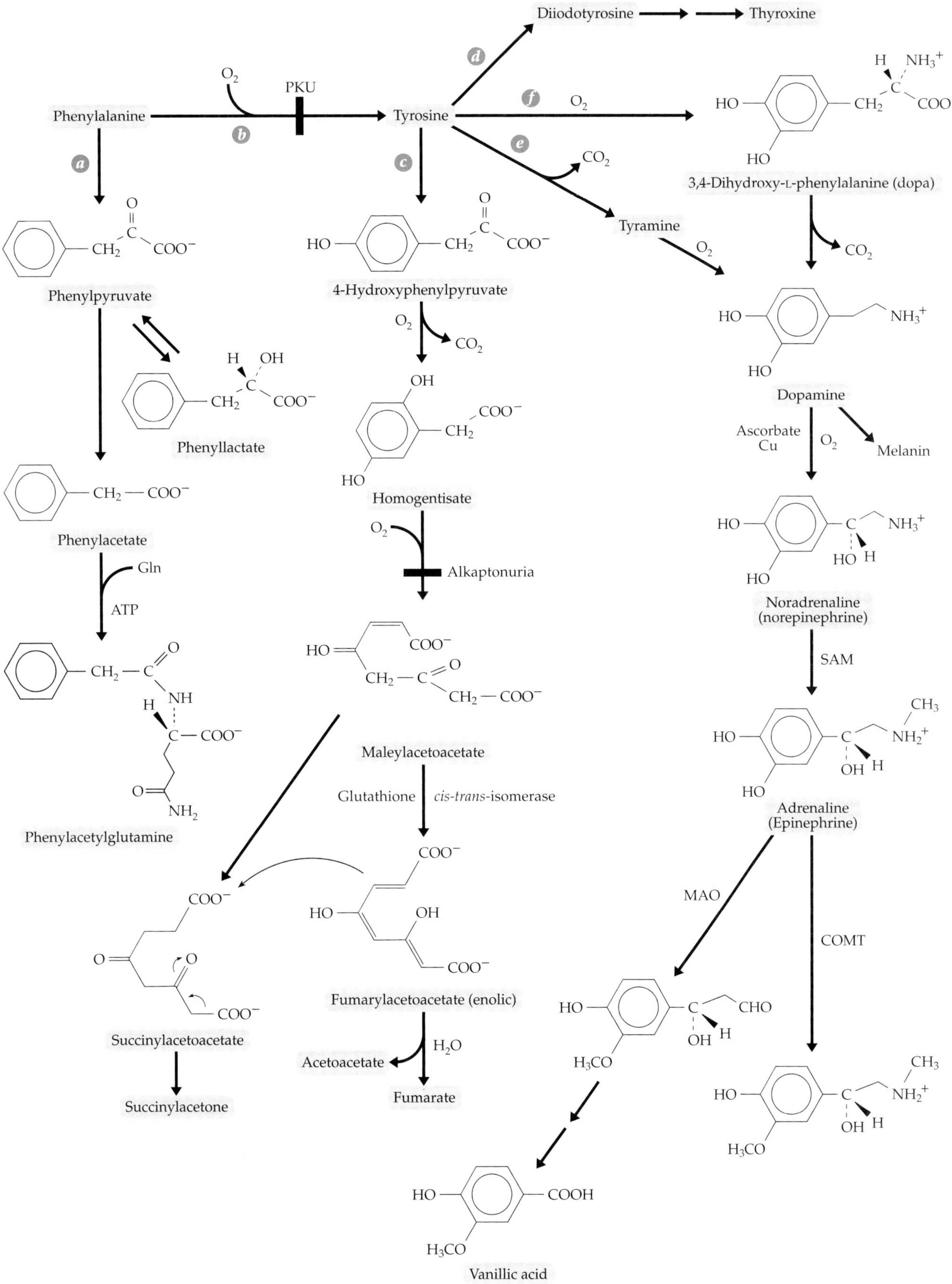

Figure 25-5 Some routes of metabolism of phenylalanine and tyrosine in animals.

$2\,I^- + 2\,H_2O_2$ → $2\,H_2O + 2\,HO^-$ (Two steps)

Dehydroalanine residue

Thyroxine (T_4) residue

Malonic semialdehyde residue

(25-6)

thyroxine is present only in vertebrates.[101] Why do we need this iodinated hormone? Halogen-free analogs in which the iodine atoms of thyroxine and triiodothyronine have been replaced by methyl or other alkyl groups are biologically active. Frieden[101] concluded that the role of iodine is related more to the evolution of biosynthetic and catabolic pathways than to chemical properties of the hormones themselves.

Thyroxine and triiodothyronine have many effects, a major one in mammals and birds being stimulation of energy metabolism in tissues. It has long been recognized that a deficiency of thyroid hormone is reflected in an overall lower basal metabolic rate (Chapter 6). Maley and Lardy observed that thyroxine uncouples oxidative phosphorylation (Chapter 18) in isolated mitochondria.[117] When mitochondria from animals receiving extra thyroxine were compared with those from control animals, an increased rate of electron transport was observed. However, there was little or no change in the P / O ratio. Thus, the hormone apparently increased the rate of electron transport without decreasing the overall efficiency of ATP synthesis.

Thyroid hormones also have a general effect on growth and development in all vertebrates. This is especially striking in amphibia in which thyroid hormones control the metamorphosis from the tadpole to adult stages.[101,118] Zebrafish, and presumably also other fishes, require thyroid hormone to complete their life cycles.[119] At present it is thought that most, if not all, effects of thyroid hormones are a result of their action on the transcription of genes.[120–123] Nuclear **thyroid hormone receptors** belong to a family of ligand-regulated transcription factors that respond to steroid, retinoid, and thyroid hormones (Table 22-1).[123–124] These proteins control many metabolic functions, often forming heterodimers with other receptors and also being activated by coactivators[125] or corepressors.[124,125a,b] Transcription of the genes for the thyroxine-synthesis proteins **thyroperoxidase**, **thyroglobulin**, and **iodide transporter**[125c] is regulated by a **thyroid transcription factor**.[126]

A number of thyroid-related diseases are known. Thyroid deficiency is often evident by enlargement of the thyroid gland (**goiter**). The deficiencies may involve inadequacy in dietary intake of iodine, transport of iodide into the thyroid, poor formation of iodinated thyroglobulin, inefficient coupling to form the iodinated thyronine residues,[127] or mutations in thyroid hormone receptors.[122,128] A major cause of goiter is a deficiency in the content of iodine in soil, a condition affecting about one billion (10^9) persons. A more severe effect of thyroid deficiency is the fetal brain damage called **cretinism**.[125c,129] Victims are mentally retarded, deaf-mute, and often with motor rigidity. In Grave disease, the commonest type of **hyperthyroidism**, the blood contains specific thyroid-stimulating autoantibodies.[130,131] These bind to the thyrotropin (TSH) receptors of the thyroid plasma membrane and stimulate excessive formation of thyroid hormone.

3. The Catecholamines

A combination of decarboxylation and hydroxylation of the ring of tyrosine produces derivatives of *o*-dihydroxybenzene (catechol), which play important roles as neurotransmitters and are also precursors to **melanin**, the black pigment of skin and hair. Catecholamines may be formed by decarboxylation of tyrosine into tyramine (step *e*, Fig. 25-5) and subsequent oxidation. However, the quantitatively more important route is hydroxylation by the reduced pterin-dependent tyrosine hydroxylase (Chapter 18) to 3,4-dihydroxyphenylalanine, better known as **dopa**. The latter is decarboxylated to **dopamine**.[131a] Hydroxylation of dopamine by an ascorbic acid and copper-requiring enzyme (Eq. 18-53) produces the important hormone **noradrenaline** (norepinephrine), which is methylated to form **adrenaline** (epinephrine).

There are two principal catabolic routes for destruction of these catecholamines as is illustrated for adrenaline in Fig. 25-5. **Monoamine oxidase** (MAO)

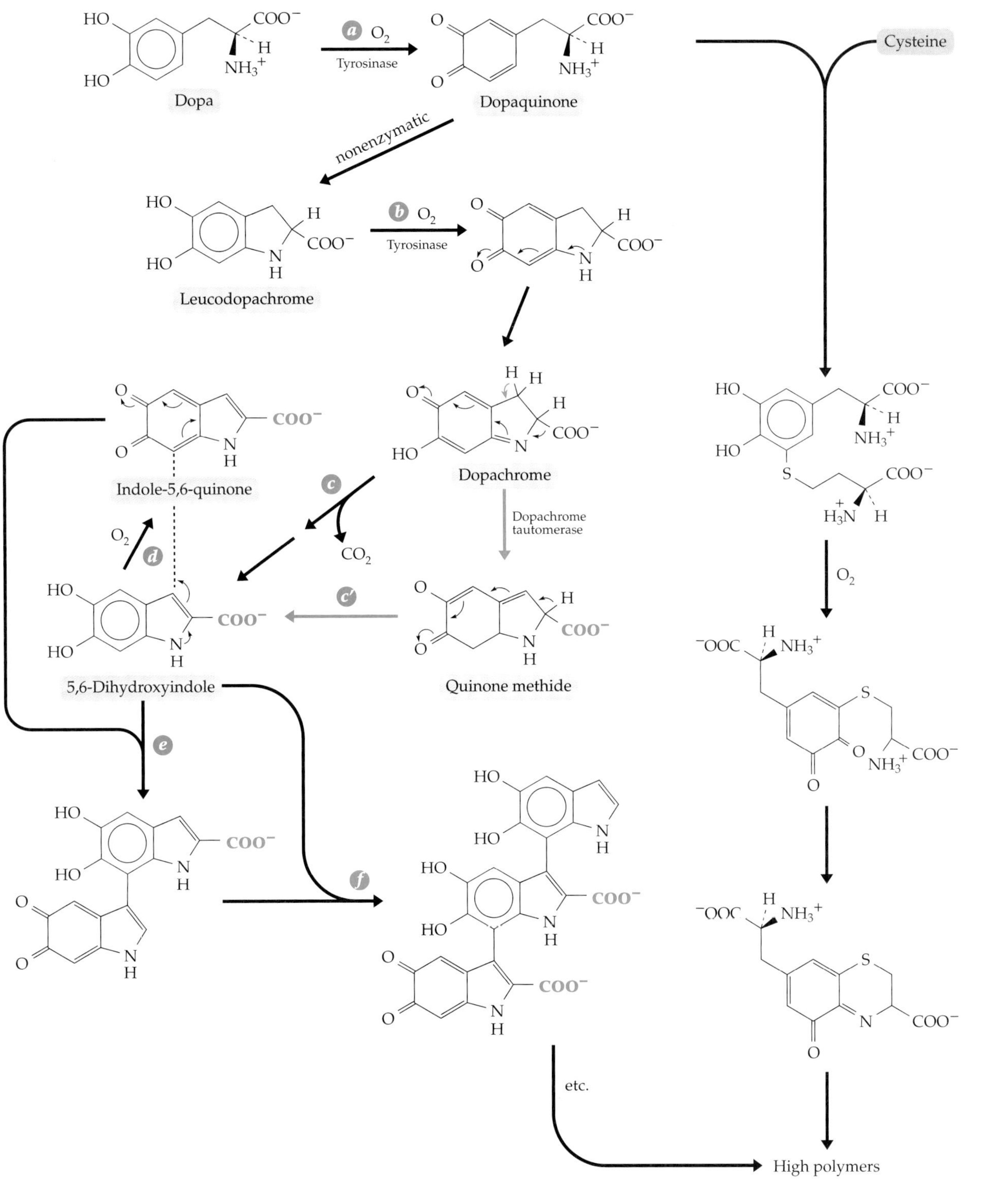

Figure 25-6 Postulated pathways for synthesis of the black pigment melanin and pigments (phaeomelanins) of reddish hair and feathers. Dopachrome reacts in two ways, with and without decarboxylation. The pathway without decarboxylation is indicated by green arrows. To the extent that this pathway is followed the green carboxylate groups will remain in the polymer. The black eumelanin is formed by reactions at the left and center while the reddish phaeomelanin is derived from polymers with cysteine incorporated by reactions at the right.

causes oxidative cleavage with deamination. Subsequent oxidative fission of the side chain together with methylation yields such end products as vanillic acid, which is excreted in the urine. The second catabolic route is immediate O-methylation by **catecholamine O-methyltransferase** (COMT), an active enzyme in neural tissues. The metabolites are relatively inactive physiologically and may be secreted as such or may undergo further oxidative degradation.

4. The Melanins

Dihydroxyphenylalanine (dopa) darkens rapidly when exposed to oxygen. The process is hastened greatly by tyrosinase (Chapter 16), which also catalyzes reaction *f* of Fig. 25-5, the oxidation of tyrosine to dopa. Tyrosinase is found in animals only in the organelles known as **melanosomes**, which are present in the melanin-producing **melanocytes** (Boxes 8-F; 25-A).[132] A series of enzymatic and nonenzymatic oxidation, decarboxylation, and coupling reactions forms the pigments. The initial steps for one pathway are indicated in Fig. 25-6. Oxidation of dopa to dopaquinone (step *a*)[133,134] is followed by an intramolecular addition reaction, together with tautomerization to the indole derivative, leucodopachrome. A second oxidation by tyrosinase (step *b*) is followed by decarboxylation and tautomerization to 5,6-dihydroxyindole (step *c*).[135] Alternatively, the tautomerization steps may take place without decarboxylation (green arrows, step *c'*).[133,136] In either case the dihydroxyindole that is formed can undergo a third oxidation step, also catalyzed by tyrosinase, to form indole-5,6-quinone (step *d*). Coupling of the last two products as indicated in step *e* yields a dimer, which is able to continue the addition of dihydroxyindole units (step *f*, etc.) with oxidation to form a high polymer of the black true melanin (**eumelanin**). However, the structure is not regular and is crosslinked. A related series of red polymers, the **phaeomelanins** found in red hair and feathers, are formed by addition of cysteine to dopaquinone.[137,138] Addition is possible at more than one position. The resulting adducts (only one is shown) can undergo oxidative ring closure in the manner indicated. Control of melanin formation is also complex. For example, more than 50 genetic loci affect the coat color of the house mouse.[139] Melanin in some fungi is formed by oxidative reactions of tetrahydroxynaphthalene formed via the polyketide pathways (Chapter 21).[140,140a] The melanin "inks" produced by cuttlefish and other cephalopods are formed in much the same way as melanins of skin.[140a]

Dopa is converted by at least some insects into *N*-β-alanyldopamine, which is a preferred substrate for the *o*-diphenol oxidase of the insect pupal cuticle. Oxidation of this substrate plays a crucial role in the hardening and darkening of the cuticle during pupal tanning.[141–142] There are many other oxidative reactions of tyrosine side chains within proteins. These include coupling of free radicals formed by peroxidases[143] or ultraviolet light[144,144a] to form dityrosines and other products. The walls of yeast ascospores,[145,146]

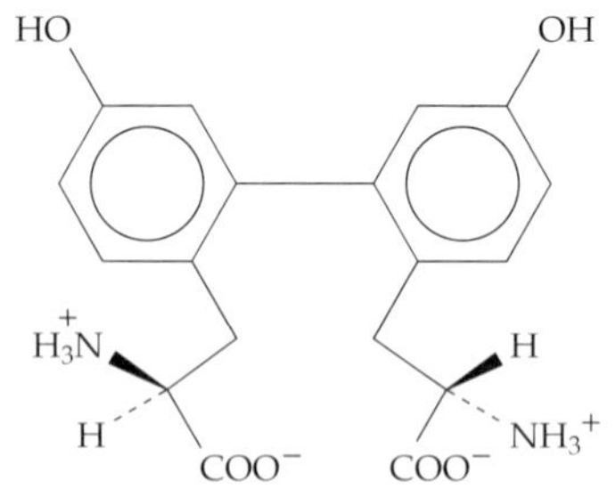

HO OH H_3N^+ H H NH_3^+ COO^- COO^-

Dityrosine

Other isomers are also formed

the cements formed by reef-building annelids,[147,148] and adhesive plaques of marine mussels[149,149a] all contain polyphenolic proteins. The 120-kDa "foot protein" of the mussel *Mytilus edulis* consists of tandemly repeated decapeptides, each containing 2 residues of lysine, 1–2 residues of dopa, 1–2 residues of *trans*-4-hydroxyproline, and 1 residue of *trans*-2,3, *cis*-3,4-dihydroxyproline.[149]

5. Microbial Catabolism of Phenylalanine, Tyrosine, and Other Aromatic Compounds

Bacteria and fungi play an essential role in the biosphere by breaking down the many aromatic products of plant metabolism.[150–153] These include vast amounts of lignin, alkaloids, flavonoid compounds, and other biochemically "inert" substances. Lignin is a major constituent of wood and a plant product second only to cellulose in abundance.

The chemical reactions used to degrade these aromatic compounds are numerous and complex. As was mentioned in Chapter 16, some fungi initiate the attack on lignin with peroxidases and produce soluble compounds that can be attacked by bacteria. In other cases elimination reactions may be used to initiate degradation. For example, some bacteria release phenol from tyrosine by β elimination (Fig. 14-5). However, more often hydroxylation and oxidative degradation of side chains lead to derivatives of benzoic acid or of the various hydroxybenzoic acids.[150,151,154–155a]

After the initial reactions many of the compounds are channeled into one of the major pathways illustrated in Fig. 25-7.[151,156,157] Dominant in aerobic bacteria is the conversion to **3-oxodipate** by one of the two convergent pathways shown. The products succinate and acetyl-CoA are readily oxidized by the citric

BOX 25-A SKIN COLOR

The principal pigment of human skin, hair, and eyes is **melanin**, which is synthesized in specialized cells, the **melanocytes**. They lie between the epidermis (outer layer) and the dermis (inner layer) as shown in Box 8-F. Melanocytes originate from embryonic nervous tissue and migrate into the skin by the third month of fetal life. They retain the highly branched morphology of neurons. Persons of different races all have the same numbers of melanocytes but the numbers and sizes of the pigmented melanosomes (Box 8-F) vary as does the content and chemical composition of the melanin.[a–d] Melanosomes not only are found in the dendrites of the melanocytes but are transferred from them into adjacent epithelial cells.[e,f]

Nevi (moles) are clusters of melanocytes that start to appear in the third year of life. They gradually increase in numbers but disappear in old age. **Freckles** appear beginning at about age six in genetically susceptible individuals. They are regions in which a higher concentration of melanin is formed.[a]

Both hair and the iris of the eye are also pigmented by melanin. Although dark eyes and dark hair are more prevalent among persons with dark skin there is no direct correlation. This is only one piece of evidence that the genetics of skin, eye, and hair coloration is complex. In mice over 150 different mutations occurring at more than 50 distinct genetic loci affect pigmentation.[f] Melanin formation begins with the action of tyrosinase. The human genome contains at least three genes for tyrosinase and related proteins.[b,g] The *Tyr* gene is absent in **oculocutaneous albinism**, the lack of pigment in eyes, hair, and skin. The tyrosinase-related protein 2 (TRP2), which has been identified as **dopachrome tautomerase** (see Fig. 25-6), is also a member of the tyrosinase family. Although a key enzyme in pigment synthesis, the amount of tyrosinase or of tyrosinase mRNA is the same in all skin types and colors.[h] Thus, differences in skin color must arise from differences in regulation.

Regulation of melanin formation is achieved in part by hormones, the **melanocyte-stimulating hormone** (MSH or melanotropin) being the most important.[a,f,g] The 13-residue pituitary hormone greatly increases pigmentation and stimulates differentiation of melanocytes. Other regulatory influences arise from interleukins, prostaglandins, interferons, tetrahydrobiopterins,[h] and protein kinase C.[i] Light also has a major effect, causing rapid tanning, especially in darker skin. Release of NO and cyclic GMP may be involved.[j]

Melanin and phaeomelanins have an important role in protecting skin from sunlight. This includes protection of light-sensitive vitamins, proteins, and DNA and RNA. The correlation of high pigmentation with the high intensity of light in tropical regions may reflect this property. Light-skinned persons of northern and southern latitudes, where light intensity is weaker, are less pigmented, allowing more adequate synthesis of vitamin D in the skin (Box 22-C).

A total lack of melanin as a result of a defective *Tyr* gene is seen in oculocutaneous albinism. Lacking protection from sunlight by melanin, albino individuals must shield their skin and eyes carefully. A second type of albinism results from mutations in the P gene, known in the mouse as the "pink-eyed dilution locus." In this condition synthesis of phaeomelanin is not impaired. Mutations in the *KIT* gene, which encodes a tyrosine kinase receptor lead to **piebaldism**, with white and dark splotched skin (or fur in animals).[b,k] While piebaldism is hereditary, **vitiligo** is an acquired autoimmune disease involving spotty loss of pigment and affecting 0.5 to 4% of the world's population. Melanocytes may be present in the affected areas but are unable to make melanin.[h,l]

---

[a] Lerner, A. B. (1961) *Sci. Am.* **205**(Jul), 98–108

[b] King, R. A., Hearing, V. J., Creel, D. J., and Oetting, W. S. (1995) in *The Metabolic and Molecular Bases of Inherited Disease*, 7th ed., Vol. 1 (Scriver, C. R., Beaudet, A. L., Sly, W. S., and Valle, D., eds), pp. 4353–4392, McGraw-Hill, New York

[c] Molnar, S. (1998) *Human Variation, Races, Types and Ethnic Groups*, 4th ed., Prentice Hall, Upper Saddle River, New Jersey (pp. 230–247)

[d] Robins, A. H. (1991) *Biological Perspectives on Human Pigmentation*, Cambridge Univ. Press, Cambridge

[e] Potterf, S. B., Muller, J., Bernardini, I., Tietze, F., Kobayashi, T., Hearing, V. J., and Gahl, W. A. (1996) *J. Biol. Chem.* **271**, 4002–4008

[f] Hearing, V. J., and Tsukamoto, K. (1991) *FASEB J.* **5**, 2902–2909

[g] Aroca, P., Urabe, K., Kobayashi, T., Tsukamoto, K., and Hearing, V. J. (1993) *J. Biol. Chem.* **268**, 25650–25655

[h] Schallreuter, K. U., Wood, J. M., Pittelkow, M. R., Gütlich, M., Lemke, K. R., Rödl, W., Swanson, N. N., Hitzemann, K., and Ziegler, I. (1994) *Science* **263**, 1444–1446

[i] Park, H.-Y., Russakovsky, V., Ohno, S., and Gilchrest, B. A. (1993) *J. Biol.Chem.* **268**, 11742–11749

[j] Roméro-Graillet, C., Aberdam, E., Biagoli, N., Massabni, W., Ortonne, J.-P., and Ballotti, R. (1996) *J. Biol. Chem.* **271**, 28052–28056

[k] Schmidt, A., and Beermann, F. (1994) *Proc. Natl. Acad. Sci. U.S.A.* **91**, 4756–4760

[l] Overwijk, W. W., Lee, D. S., Surman, D. R., Irvine, K. R., Touloukian, C. E., Chan, C.-C., Carroll, M. W., Moss, B., Rosenberg, S. A., and Restifo, N. P. (1999) *Proc. Natl. Acad. Sci. U.S.A.* **96**, 2982–2987

acid cycle and associated reactions. Many different compounds can be converted into the starting compounds shown at the top of the figure. Both D- and L-mandelate, toluene, benzyl alcohol, L-tryptophan, phenanthrene, naphthalene, and benzene can be converted to catechol and be metabolized via the catechol branch of the pathway. Benzoate, *p*-toluate, shikimate, and quinate can be metabolized via the protocatechuate branch. Halogenated compounds, e.g., 3-chlorocatechol, may sometimes be degraded via

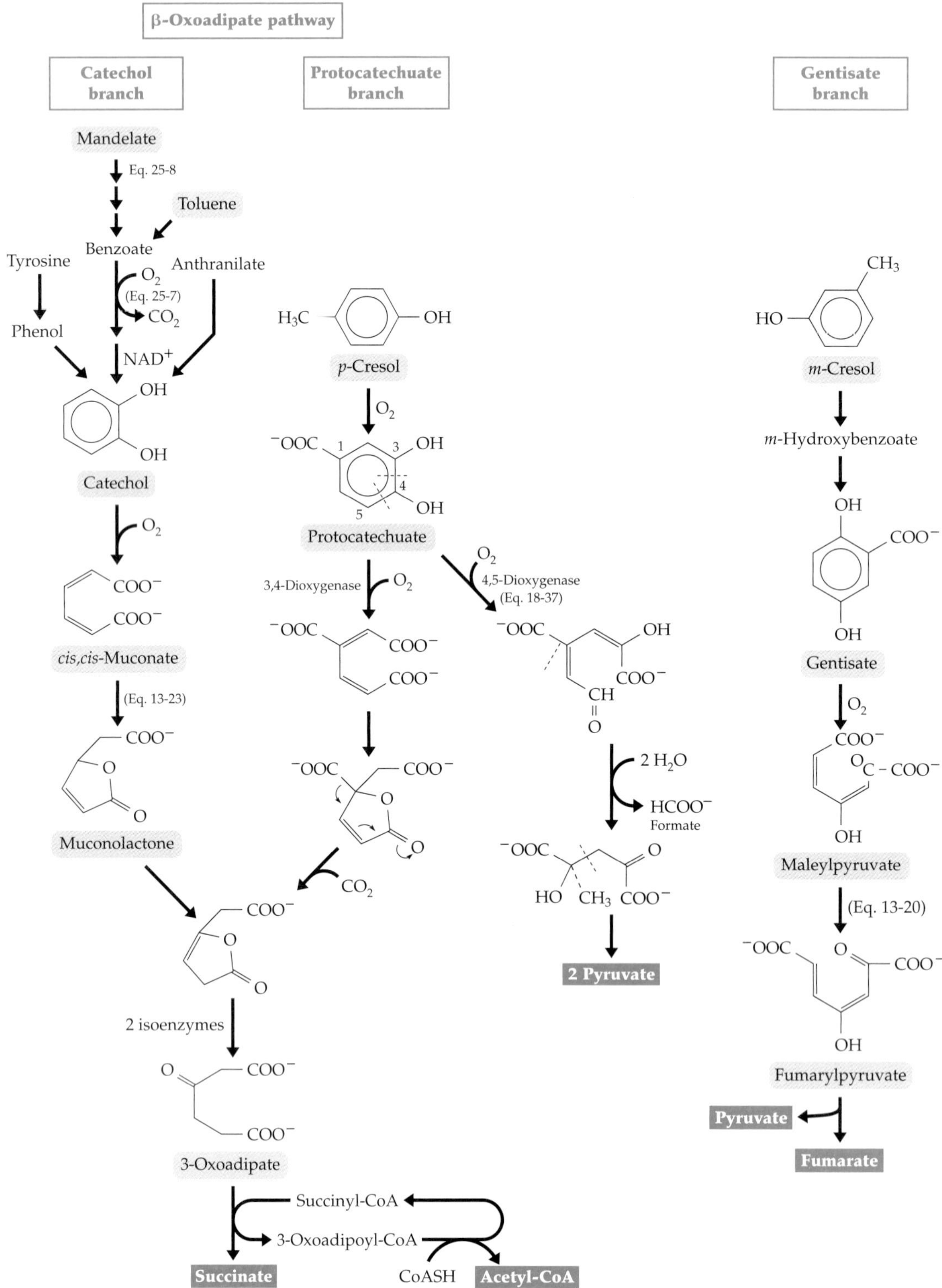

Figure 25-7 A few pathways of catabolism of aromatic substances by bacteria.

the same pathways[151,156] or in parallel or related pathways.[157–158c] However, chlorine atoms are sometimes eliminated as Cl^- at some point in the process.[159–160a] Other substrates, including *m*-hydroxybenzoate and sometimes anthranilate, are degraded via the gentisate pathway (Fig. 25-7).[157,161] Both benzoate and phenylacetate are sometimes degraded after conversion to coenzyme A thioesters.[161a,b]

Dioxygenases play a major role in all of these pathways of aromatic catabolism. In most cases a dioxygenase (Chapter 18) is required for the opening of the benzene ring. The pathways contain interesting isomerization steps, some of which have been discussed in Chapter 13, Section B. Microorganisms often have alternative choices in the chemistry of their attack. For example, tyrosine can be converted by one bacterium to homogentisate, as in animals (Fig. 25-5), or by other bacteria to protocatechuate, homoprotocatechuate, or gentisate (Fig. 25-7) before the ring is opened.[150] A single compound can be acted upon by more than one dioxygenase. Thus protocatechuate can be opened by a 3,4-dioxygenase or by a 4,5-dioxygenase (Fig. 18-22) leading to the branch point at protocatechuate in Fig. 25-7.

The initial hydroxylation of benzene, toluene, and other alkylbenzenes is accomplished by multicomponent aromatic ring dioxygenases that introduce two oxygen atoms to form diols.[158] Dioxygenation of benzoate yields a diol that can be **oxidatively decarboxylated** by reaction with NAD^+ (Eq. 25-7) to form catechol.[157,162] Toluene gives 3-methylcatechol

Benzoate → (O_2) → [^-OOC, OH, OH diol; H --- NAD^+] → Catechol (OH, OH)

(25-7)

whose ring is, however, opened by an extradiol 2,3-dioxygenase, a so called *meta*-cleavage.[163,164] The product, 2-hydroxy-6-oxohepta-2,4-dienoate, is cleaved hydrolytically as indicated on the structure

OH, COO^-, C=O, H_2O, CH_3

2-Hydroxy-6-oxohepta-2,4-dienoate

to give acetic acid and 2-hydroxypent-2,4-dienoate, which can be further metabolized. The hydrolytic cleavage is unusual[164] but is a β cleavage similar to the C–C bond cleavage by ribulose bisphosphate carboxylase (Eq. 13-48). Toluene can also be oxidized via benzoate through the β-oxoadipate pathway.

The plant acid *S*-mandelate must undergo conversion to *R*-mandelate by action of a racemase (Fig. 13-5) dehydrogenation, and side-chain cleavage as shown in Eq. 25-8 to form benzoate before it can be metabolized further.[165]

Although pseudomonads are well known for aerobic decomposition of aromatic compounds, some strains of *Pseudomonas*, as well as many other bacteria, are able to degrade aromatic compounds under completely anerobic conditions.[166,167] Benzoate can be converted to benzoyl-CoA and the ring can be partially reduced in two ATP- and NADH-dependent reactions (Eq. 25-9). The first of these reduction steps is unusual because ATP is apparently needed to drive the reaction.[166,166a,b] This is analogous to the need for ATP in nitrogen fixation (Eq. 24-6, step *b*).

Toluene, 3-chlorobenzoate, cinnamate, and 2-aminobenzoate can all be converted to benzoyl-CoA and be metabolized via the pathway of Eq. 25-9. Phenol, cresol, coumarate, protocatechuate, and vanillate can be converted to 4-hydroxybenzoyl-CoA and degraded in a similar fashion.[166] The breakdown of various forms of vitamin B_6 by bacteria is described in Section F (Eq. 25-24).

S-Mandelate (COO^-, H, OH)

↕ Mandelate dehydrogenase

Benzoylformate (COO^-, C=O)

↓ Thiamin diphosphate (CO_2 released)

Benzaldehyde

↓

Benzoate (O^-, C=O)

↓ O_2

[NAD^+, H, OH, OH, C(=O)–O^- diol]

(25-8)

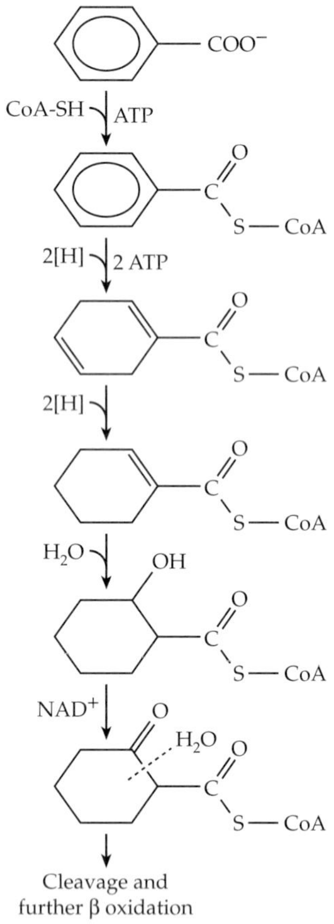

Cleavage and further β oxidation

(25-9)

6. Quinic and Gallic Acids

Quinic acid, a compound accumulated by many green plants, can be formed by reduction of 3-dehydroquinate (Eq. 25-2) in both plants and bacteria. Quinic acid can be converted into useful industrial products such as benzoquinone and hydroquinone, and its production by bacteria provides a convenient route to these compounds.[168] In the main shikimate pathway 3-dehydroquinate is dehydrated to 3-dehydroshikimate (Eq. 25-3). The latter can be dehydrated

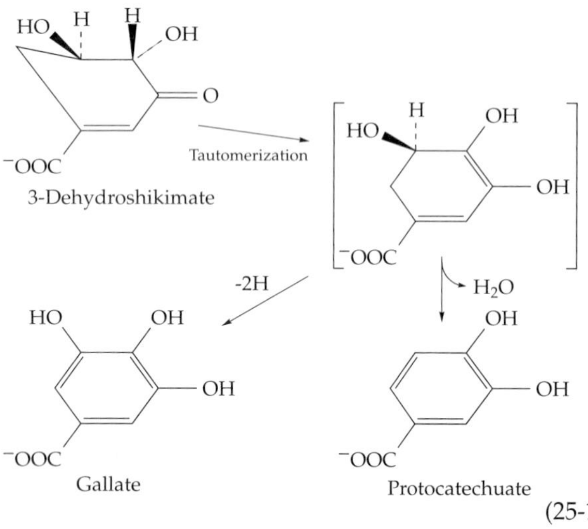

3-Dehydroshikimate → Tautomerization; -2H → Gallate; H_2O → Protocatechuate

(25-10)

further to **protocatechuate** (Eq. 25-10) either nonenzymatically[169] or by enzymatic action in bacteria.[170] Protocatechuate can be decarboxylated enzymatically to catechol, another compound of industrial value.[170] Nonenzymatic oxidation of 3-dehydroshikimate (Eq. 25-10) yields gallate.[169] Gallic acid derivatives are important plant constituents, but the biosynthetic origin has been obscure.[171] Gallate is probably formed from 3-dehydroshikimate as indicated in Eq. 25-10.[172] Esters and other derivatives of gallic acid constitute the **hydrolyzable tannins**. These materials accumulate in the vacuoles of the plants and are also deposited in the bark along with the **condensed tannins**, which are polymeric flavonoid compounds (Box 21-E).

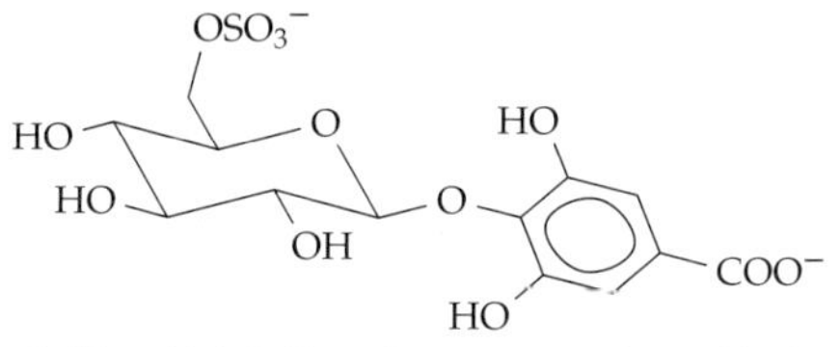

Gallic acid 4-*O*-(β-D-glucopyranosyl-6'-sulfate), the periodic leaf movement factor from *Mimosa*

7. The Metabolism of Phenylalanine and Tyrosine in Plants

Some of the pathways of animal and bacterial metabolism of aromatic amino acids also are used in plants. However, quantitatively more important are the reactions of the **phenylpropanoid pathway**,[173–174a] which is initiated by **phenylalanine ammonia-lyase** (Eq. 14-45).[175] As is shown at the top of Fig. 25-8, the initial product from phenylalanine is ***trans*-cinnamate**. After hydroxylation to 4-hydroxycinnamate (*p*-coumarate) and conversion to a coenzyme A ester,[175a] the resulting *p*-coumaryl-CoA is converted into mono-, di-, and trihydroxy derivatives including anthocyanins (Box 21-E) and other flavonoid compounds.[176] The dihydroxy and trihydroxy methylated products are the starting materials for formation of lignins and for a large series of other plant products, many of which impart characteristic fragrances. Some of these are illustrated in Fig. 25-8.

Benzoic and salicylic acids. Two of the simplest plant acids arising from *trans*-cinnamate are **benzoic acid**, accumulated in plums and cranberries, and **salicylic acid**, present in all green plants and accumulated as methyl esters or glycosides in some plants, e.g., those of the willow family. Salicylic acid is made by hydroxylation of benzoic acid,[177] which can be formed from *trans*-cinnamate by β oxidation as depicted in Fig. 25-8, but it may also arise from isochorismate as shown in Fig. 25-2.[178] Salicylic acid plays a central role in resistance of plants to a variety of

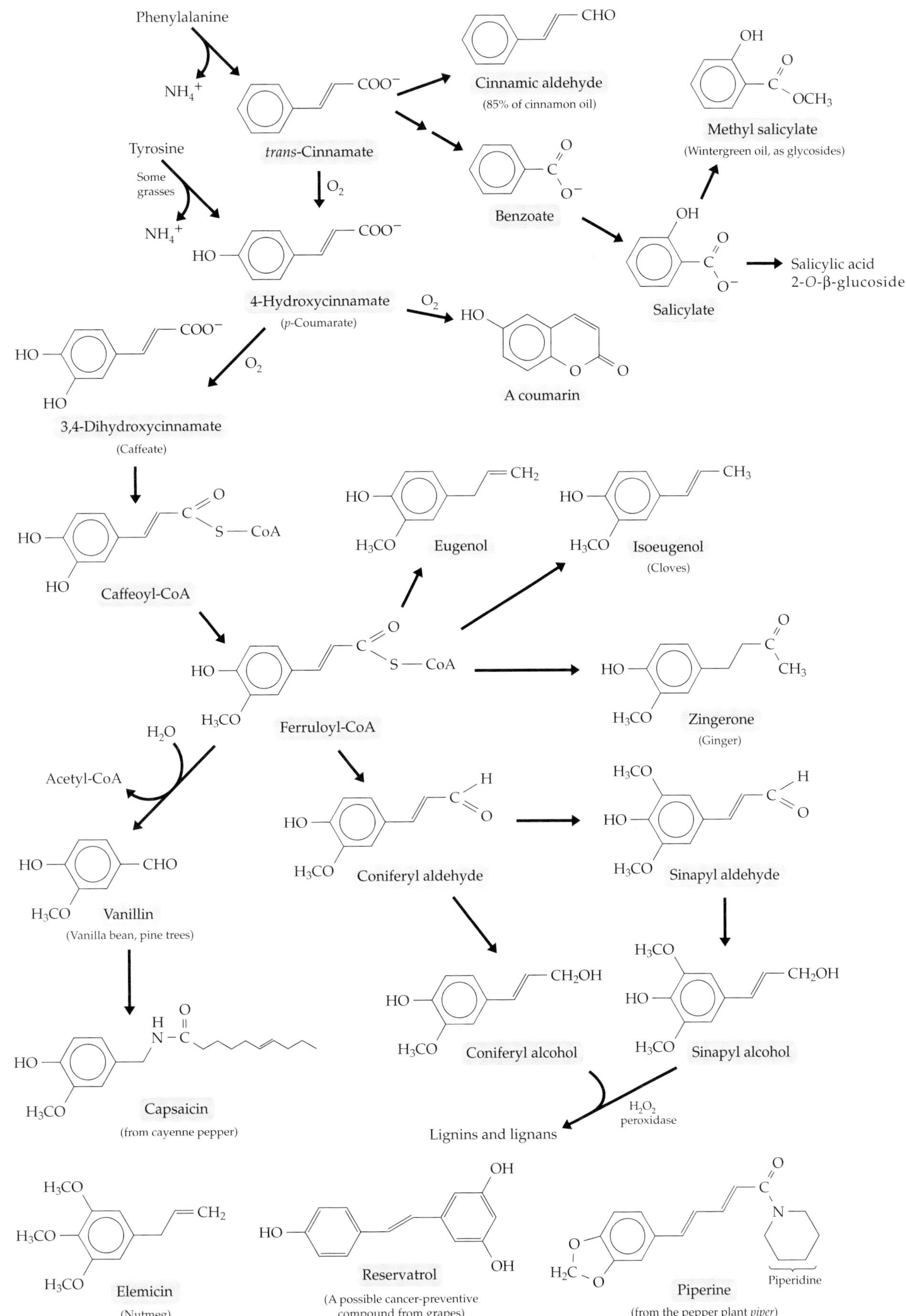

Figure 25-8 Formation of some plant metabolites from phenylalanine and tyrosine via the phenylpropanoid pathway.

diseases, a phenomenon called **aquired systemic resistance**.[179–181] A large number of defense-related genes are induced by salicylate leading to increased synthesis of phytoalexins, proteins, and lignins. The mechanism may be to inhibit catalases allowing the level of H_2O_2 to rise. Hydrogen peroxide not only is a precursor to potent antimicrobial compounds as HOCl (Eqs. 16-12, 16-13) and a participant in lignin synthesis, but it may directly activate transcription of disease resistance genes.[182] This mode of action appears to parallel an **acute-phase response** of the vertebrate immune system through which H_2O_2 activates the transcription factor NF-kB (Fig. 5-40; Chapter 28).[182] It is possible that salicylate also has an effect on transcription in the human body.[183]

Salicylic acid derivatives, such as aspirin (acetylsalicylic acid), have long been known as pain relievers for the human body. The major effect is thought to be inhibition of a cyclooxygenase (Eq. 21-16). Willow bark has been known since the 18th century to contain a pain reliever, which was identified as salicylate esters and salicyl alcohol derivatives such as salicin.

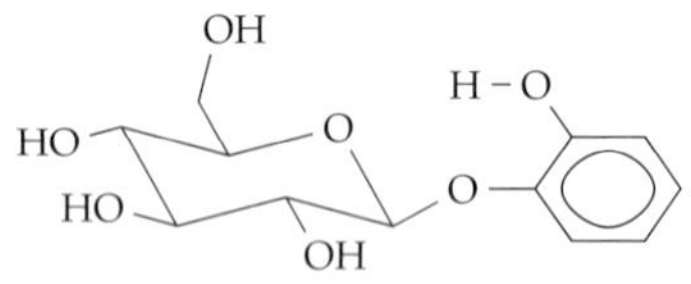

Salicin, a glucoside of salicyl alcohol

Vanilla and other plant products. One of the most widely used natural plant products is **vanilla extract**, which is obtained from cured, unripe fruit of the orchid *Vanilla planifolia*. The curing process releases **vanillin** and related compounds such as vanillic acid and 4-hydroxybenzaldehyde from glycosides. Because the flowers must be hand pollinated every day, natural vanilla extract is extremely expensive.[184,185] Most vanillin used in flavoring is obtained by hydrolysis of lignin, but production from glucose using bacterially produced enzyme reactors is possible.[185] Prince and Gunson, in an interesting article,[184] described the use of mass spectrometry and ^{13}C-enriched synthetic vanillin in the battles to distinguish natural vanilla extract from artificial mixtures of vanillin and other compounds and to camouflage the latter. Conversion of ferulic acid to vanillin in plants is apparently accomplished by β oxidation of acyl-CoA derivatives[186] as indicated in Fig. 25-8.

The ubiquitous plant compound **chlorogenic acid** (isolated from green coffee beans) is formed by transesterification with the glycoside cinnamoyl-glucose.[187] Coumaroyl-CoA is converted into monomeric and dimeric amides with **agmatine**, which provides barley plants with resistance to mildew.[188] Similar compounds with various polyamines and derived from *p*-coumaric, caffeic, ferulic, or sinapic acid appear to function in plant development. For example, caffeoylputrescine and caffeoyl-γ-aminobutyrate are accumulated specifically in sex organs of tobacco flowers.[189]

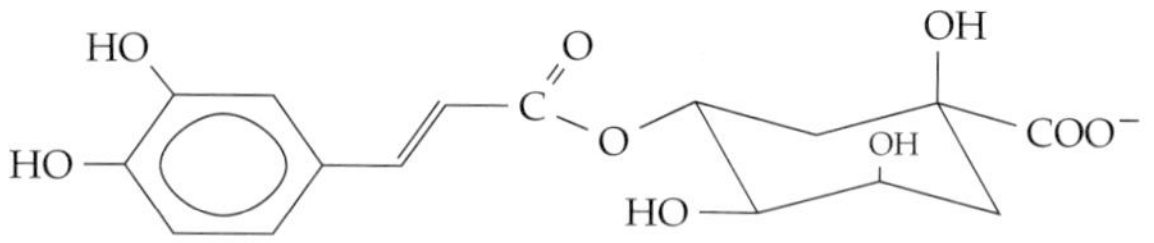

Chlorogenic acid (3-*O*-caffeoylquinic acid)

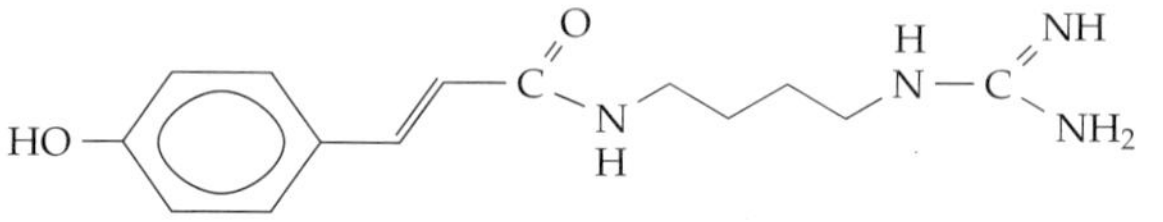

P-Coumaroylagmatine

Lignin, lignols, lignans, and phenolic coupling. Lignin is a complex material of relative molecular mass greater than 10,000. It is remarkably stable, being insoluble in hot 70% sulfuric acid. Lignin may be described as a "statistical polymer of oxyphenylpropane units." It arises from oxidative coupling of **coniferyl and sinapyl alcohols** (Figs. 25-8, 25-9) and related monomers known as **lignols**.[190–190c] The enzyme responsible for the polymerization may be a peroxidase, which catalyzes formation of lignin from the monomeric alcohols and H_2O_2. A radical generated by loss of an electron from a phenolate anion of coniferyl alcohol consists of a number of resonance forms in which the unpaired electron may be present not only on the oxygen but also at the positions marked by asterisks in the following structure:

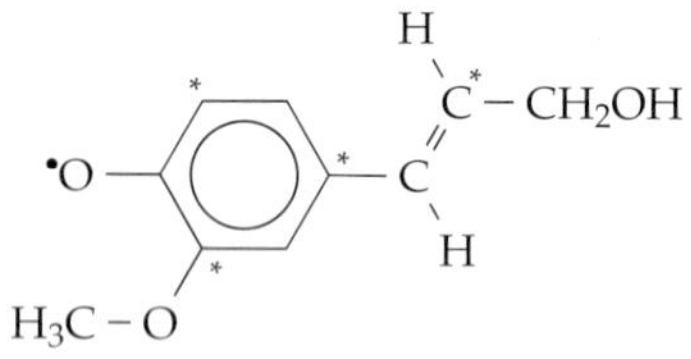

Coupling of such radicals yields a great variety of products. One type of dimerization gives the stable ether linked **pinoresorcinol** (Eq. 25-11). Through a complex sequence of reactions, it can be converted into other plant compounds including the phytoalexin **plicatic acid**, a major component of western

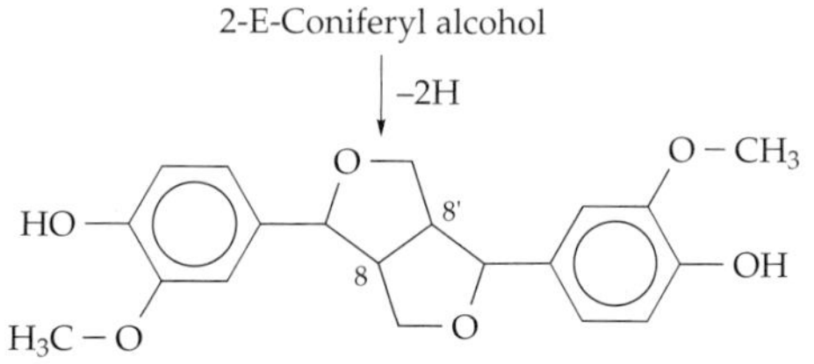

2-E-Coniferyl alcohol

↓ −2H

(25-11)

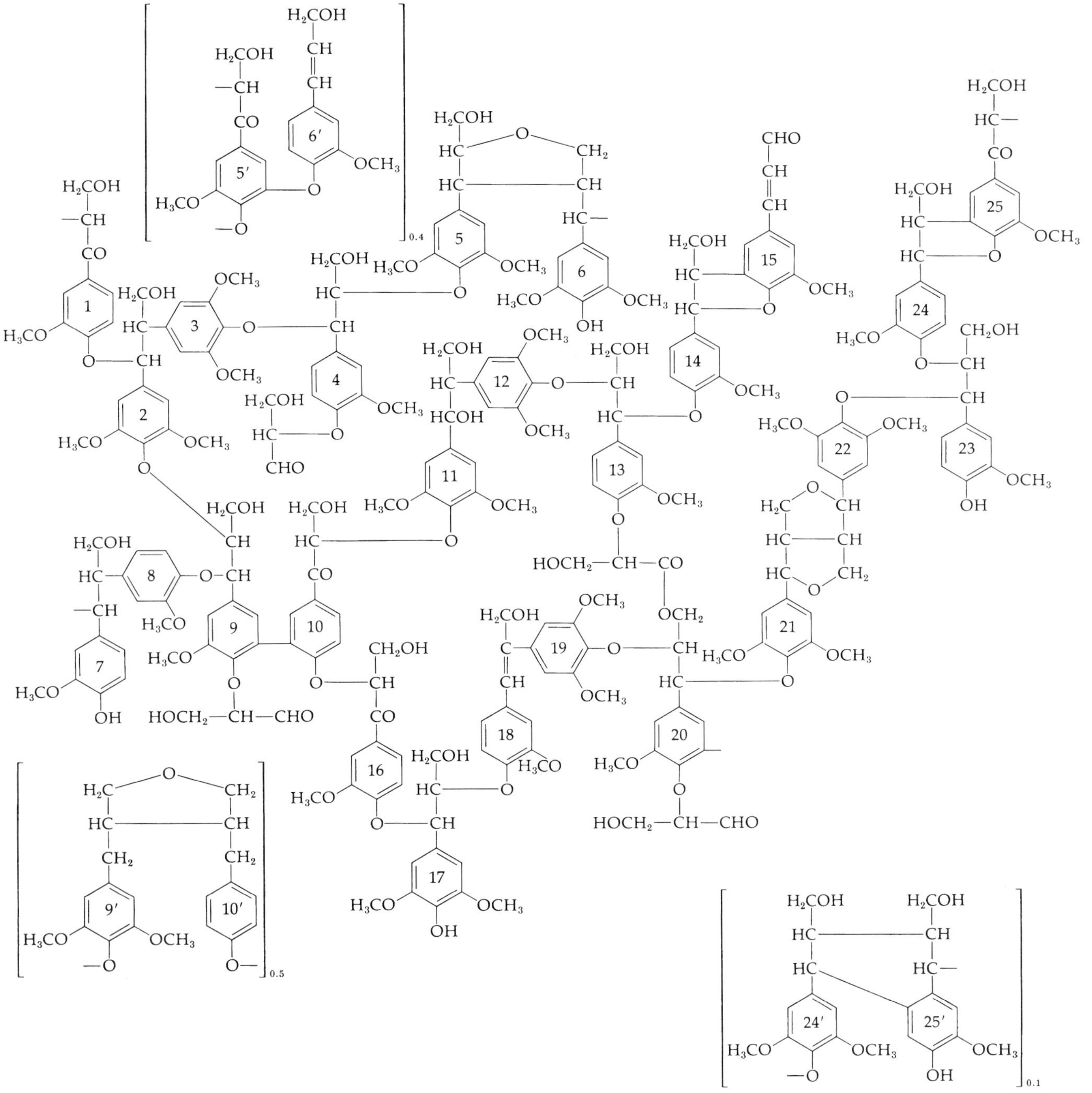

Figure 25-9 Proposed structure of beech lignin. There are 25 different C_9 units, of which several can, to some extent, be replaced by the three dimeric units in brackets. Redrawn from Nimz,[194] p. 317.

red cedar heartwood, and **podophyllotoxin**, found in the poisonous roots of the May apple (*Podophyllum peltatum*).[191] The little yellow "apple" is edible. Podophyllotoxin is used in cancer treatment. These dimeric 8,8'-carbon linked derivatives of lignols are called **lignans**[191a] while oligomers linked in other ways are **neolignans**.[192] The lignols are also incorporated covalently into **suberin**, a waxy layer of plant cell walls.[193]

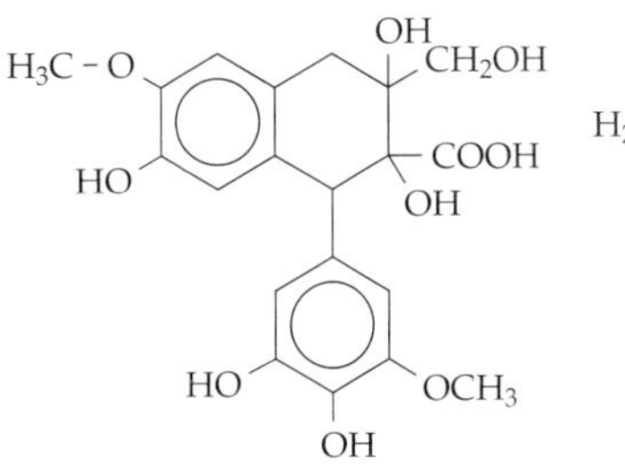

Plicatic acid

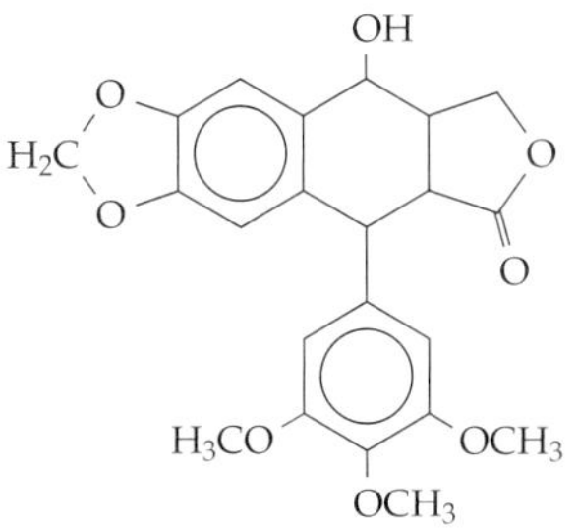

Podophyllotoxin

The lignols are synthesized within cells and are thought to move out into cell walls, possibly as phenolic glycosides.[190c,195] Cell wall peroxidases or laccases initiate polymerization.[173] The previously discussed lignans do not tend to polymerize, but other dimeric forms do. The dimers still contain hydroxyl groups capable of radical formation and addition to other units. At least ten types of intermonomer linkage other than that in the lignans are shown in Fig. 25-9. Lignin represents an enormous potentially valuable

BOX 25-B THE CYANOGENIC GLYCOSIDES

Cyanide-containing glycosides are synthesized by many higher plants including such crop plants as sorghum, cassava, and white clover.[a] The starting compounds are L-amino acids, most often phenylalanine, tyrosine, valine, or isoleucine.[b] The following sequence was proposed by Conn and others.[b–f] The conversion to an *N*-hydroxyamino acid in the first reaction (step *a*) is catalyzed by a cytochrome P450 hydroxylase system requiring NADPH and O_2. The same enzyme catalyzes a second hydroxylation that is followed by dehydration and decarboxylation (step *b*) to form an oxime.[f] The oxime is dehydrated to form a nitrile (step *c*) and a third hydroxylation (step *d*) produces an α-hydroxynitrile (cyanohydrin). Glycosylation by transfer from UDP-Glc or other sugar nucleotide (step *e*) forms the cyanogenic glycoside. If R = *p*-hydroxyphenyl in the foregoing equation and the sugar is glucose the product is **dhurrin**, present in **sorghum**. In **amygdalin**, present in bitter almonds and in pits of apricots, peaches, cherries, etc., two glucosyl units in β-1,6 linkage (gentiobiose) are attached to mandelonitrile.

Cyanogenic glycosides generate free cyanide by elimination when the glycosidic linkage is hydrolyzed.[a] This occurs with dhurrin at high pH and with others at pH=1, 70–100°C. Elimination of cyanide from the hydroxynitriles is catalyzed enzymatically.[g,h] Another cytochrome P450 dependent process utilizes oxidation to an oxime, as in the foregoing scheme, but converts the oxime to a **glucosinolate** in a two-step process.[i]

At one time amygdalin, sold as **Laetrile**, was promoted as a treatment for cancer, presumably based on the hope that the cancer cells would be poisoned by the released cyanide.[j] The tubers and leaves of the cassava plant provide a major source of food in many tropical countries. However, unless the cyanogenic glycosides are removed by boiling the tubers and pulping the leaves cassava is very toxic.[e,k]

---

a Vennesland, B., Castric, P. A., Conn, E. E., Solomonson, L. P., Volini, M., and Westley, J. (1982) *Fed. Proc.* **41**, 2639–

b Conn, E. E. (1979) *Nateuwissenschaften* **66**, 28–34

c Moller, B. L., and Conn, E. E. (1980) *J. Biol. Chem.* **255**, 3049–3056

d Moller, B. L., and Conn, E. E. (1979) *J. Biol. Chem.* **254**, 8575–8583

e Andersen, M. D., Busk, P. K., Svendsen, I., and Moller, B. L. (2000) *J. Biol. Chem.* **275**, 1966–1975

f Sibbesen, O., Koch, B., Halkier, B. A., and Moller, B. L. (1995) *J. Biol. Chem.* **270**, 3506–3511

g Lauble, H., Miehlich, B., Förster, S., Wajant, H., and Effenberger, F. (2001) *Protein Sci.* **10**, 1015–1022

h Lauble, H., Miehlich, B., Förster, S., Wajant, H., and Effenberger, F. (2002) *Biochemistry* **41**, 12043 – 12050

i Du, L., Lykkesfeld, J., Olsen, C. E., and Halkier, B. A. (1995) *Proc. Natl. Acad. Sci. U.S.A.* **92**, 12505–12509

j Newmark, J., Brady, R. O., Grimley, P. M., Gal, A. E., Waller, S. G., and Thistlethwaite, J. R. (1981) *Proc. Natl. Acad. Sci. U.S.A.* **78**, 6513–6516

k Ononogbu, I. C. (1980) *Trends Biochem. Sci.* **5**, X

industrial source of aromatic raw materials, whose utilization has proved difficult. Oxidative degradation of lignin produces **humic acid**, an important organic constituent of soils.[196]

Oxidative coupling of radicals derived from phenols has a much wider role in nature than in lignin formation. Many alkaloids and other plant and fungal metabolites are synthesized using this reaction.[197] Tyrosine radicals are thought to be involved in formation of thyroxine (Eq. 25-6), melanin (Fig. 25-6), crosslinkages of ferulic acid with polysaccharides of plant cell walls,[197a] dityrosine, and other protein crosslinkages (Section B,4). Autocatalytic reaction of tyrosine-derived aminoquinone radicals in the Cu^{2+}-containing active site of amine oxidases apparently generates the mature prosthetic group topaquinone (Eq. 15-53).[198–198b] Oxidative coupling of tryptophanyl or cysteinyl side chains generates tryptophan tryptophylquinone (p. 817) or **cysteine tryptophylquinone** (CTA).[198b,c] As discussed on pp. 885–886 the tyrosine-cysteine thioether-bridged prosthetic groups of galactose oxidase (Fig. 16-29) and several other enzymes are also self-processing.[198d,e]

8. Alkaloids

More than 12,000 miscellaneous nitrogen-containing compounds, known as alkaloids, are produced by plants.[174a,199] Alkaloids are often thought of simply as end products of nitrogen metabolism in plants. However, most plants do not make alkaloids, whereas certain families of plants make many. There are probably ecological reasons.[200] Alkaloids often have potent physiological effects on animals, and many have been used as medicines from ancient times. Some have been prized through centuries as hallucinogens and intoxicants.

There are several classes of alkaloids. Among these are purines such as xanthine and caffeine, terpenes (Chapter 22), polyketides (Chapter 21), and alkaloids derived from amino acids. The basic amino acids ornithine, arginine, histidine, and lysine as well as the aromatic amino acids, anthranilate, and nicotinate are some of the starting materials.[199,201] Robinson[202,203] in 1917 recognized that many alkaloids are derived directly from aromatic amino acids. He proposed that alkaloids arise from **Mannich reactions** (Eq. 25-12) in which an amine and an aldehyde (probably through a Schiff base) react with a nucleophilic carbon such as that of an enolate anion. Many of the

$$\geq C^- + H-C(=O)-R + H-C(R')=N-R'' \longrightarrow \geq C-CH(R)-N(R')(R'') \qquad (25\text{-}12)$$

amines are formed by decarboxylation of amino acids, and the aldehydes may arise by oxidative decarboxylation (transamination and decarboxylation) of amino acids. Thus, amino acids can provide both of the major reactants for alkaloid synthesis. Furthermore, nucleophilic centers in the aromatic rings, e.g., in positions para to hydroxyl substituents, are frequent participants in the proposed Mannich condensations. While Robinson's ideas on alkaloid biosynthesis were initially speculative, they have been confirmed by isotopic labeling experiments and more recently by isolation of the enzymes involved. Nevertheless, many questions remain. The postulated aldehydes are not proved intermediates. The condensations with 2-oxo acids may occur prior to decarboxylation.

An example is shown in Fig. 25-10. Dopa is decarboxylated to dopamine and is oxidized to 3,4-dihydroxybenzaldehyde. A Mannich reaction (via the Schiff base as shown) leads to ring closure. Oxidation of the ring produces an **isoquinoline** ring, a structural characteristic of a large group of alkaloids. Methylation produces **papaverine**, found in the opium poppy. A related alkaloid **morphine** (Fig. 25-10), at first glance, appears dissimilar. However, the biosynthetic route is similar. The initial Schiff base is formed from tyramine. Closure of the third ring together with hydroxylation and methylation yield ***R*-reticuline**, a precursor to many alkaloids. Its two rings are then oxidatively coupled through a C–C bond and an ether linkage.[204,204a] *S*-Reticuline is the precursor to another large family of alkaloids.[205] The six-membered ring of another alkaloid, **colchicine** (Box 7-D), originates from phenylalanine, while the seven-membered tropolone ring is formed from tyrosine by ring expansion.

C. Metabolism of Tryptophan and Histidine

The biosynthesis of tryptophan is outlined in Fig. 25-2. This amino acid not only assumes great importance in the structure and functioning of proteins but is converted into hormones, both in animals and plants, and into alkaloids in some plants. Some of the pathways are indicated in Figs. 25-11 and 25-12.

1. The Catabolism of Tryptophan

The primary catabolic pathway for tryptophan in animal cells is initiated (step *a*, Fig. 25-11) by **tryptophan 2,3-dioxygenase** (tryptophan pyrrolase; Eq. 18-38).[206] The enzyme is induced both by glucocorticoids and by tryptophan.[207] The related **indolamine 2,3-dioxygenase** catalyzes the same reaction of L-tryptophan but also acts on D-tryptophan and other substrates. It has different tissue distribution and reg- ulatory properties[208] and may play a role in

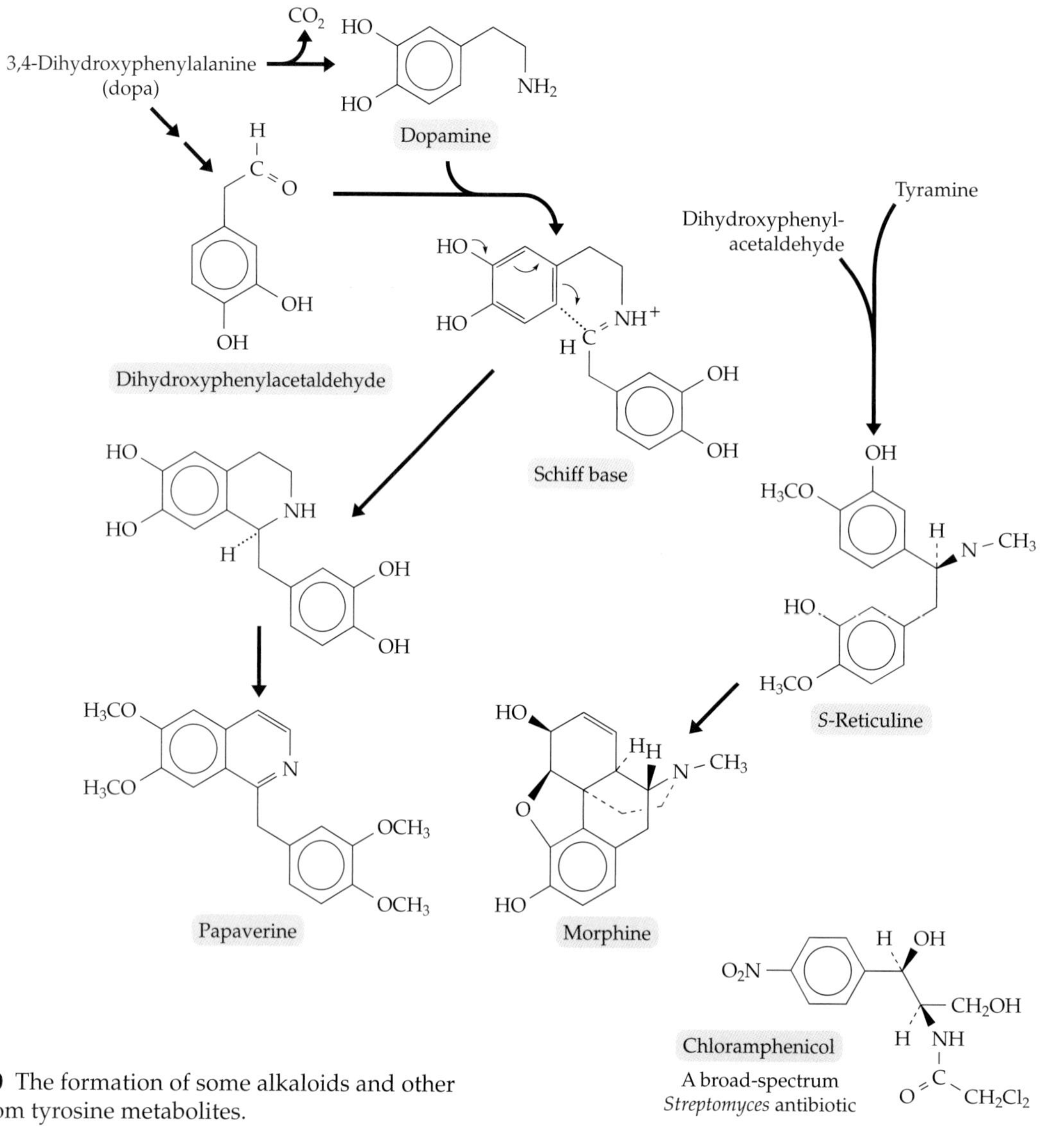

Figure 25-10 The formation of some alkaloids and other substances from tyrosine metabolites.

inflammatory responses.[206,209] An alternative pathway of tryptophan breakdown takes place in intestinal bacteria, which utilize tryptophan indolelyase (tryptophanase) to eliminate indole (step *b*, Fig. 25-11).[209a] The indole is hydroxylated to **indoxyl**, some of which is absorbed into the bloodstream and excreted in the urine as indoxyl sulfate.

Returning to the major tryptophan catabolic pathway, marked by green arrows in Fig. 25-11, formate is removed hydrolytically (step *c*) from the product of tryptophan dioxygenase action to form **kynurenine**, a compound that is acted upon by a number of enzymes. Kynureninase (Eq. 14-35) cleaves the compound to anthranilate and alanine (step *d*), while transamination leads to the cyclic **kynurenic acid** (step *e*). The latter is dehydroxylated in an unusual reaction to **quinaldic acid**, a prominent urinary excretion product.

Another major pathway of kynurenine metabolism (step *f*, Fig. 25-11) is hydroxylation to **3-hydroxykynurenine**, which in turn can undergo transamination to the cyclic **xanthurenic acid**. Xanthurenic acid is excreted from the human body, but in the malaria mosquito *Anopheles gambia* it acts as a mating factor for the malaria parasite *Plasmodium*.[210,211] In many insects, including *Anopheles*, 3-hydroxykynurenine is a precursor of insect eye pigments or "**omnochromes**."[210,212–213a] 3-Hydroxykynurenine also has neurotoxic properties.[213a] (See p. 1796.)

Cleavage of 3-hydroxykynurenine by kynureninase (step *g*, Fig. 25-11) forms 3-hydroxyanthranilate, which is opened under the action of another dioxygenase (step *h*) with eventual degradation to acetyl-CoA, as indicated. In insects the reactive 3-hydroxyanthranilate is utilized in "tanning" reactions, e.g., coupling to tyrosine residues to toughen insect cuticles and walls of cocoons.[214]

Tryptophan is hydroxylated to 5-hydroxytryptophan[213b] which is decarboxylated to **serotonin** (5-hydroxytryptamine), an important neurotransmitter

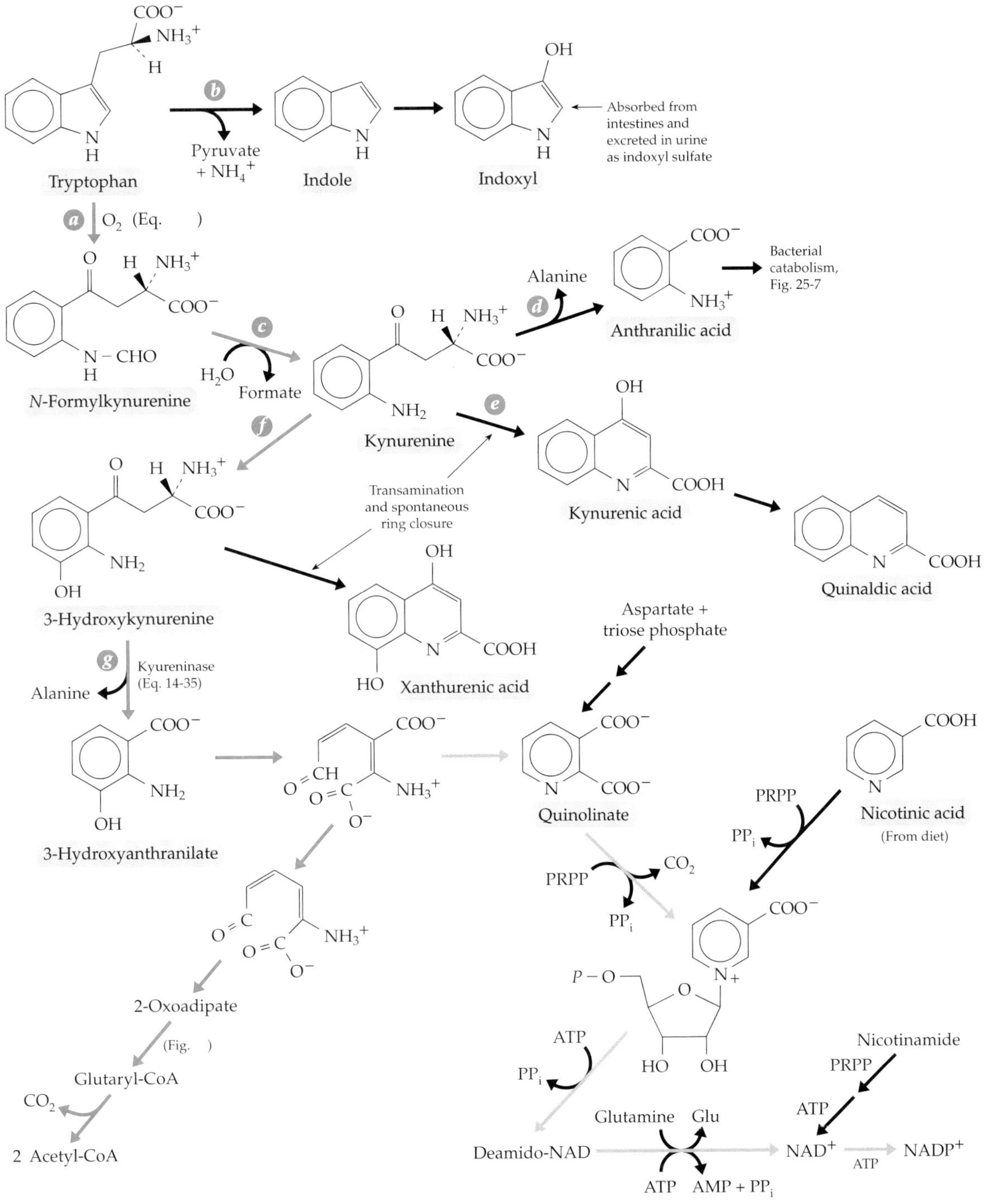

Figure 25-11 Some catabolic reactions of tryptophan and synthetic reactions leading to NAD and NADP.

substance[215,215a] and a regulatory component of plants and animals alike.[216] In the pineal gland serotonin is methylated and acetylated to **melatonin**, the pineal hormone[217–221] (Fig. 25-12).

The following dangerous tryptophan derivative was evidently formed in a fermentation used to produce tryptophan sold as a food supplement in 1990. More than 1,500 persons became ill and 27 died, perhaps as a direct result of toxicity of this compound.[222,223]

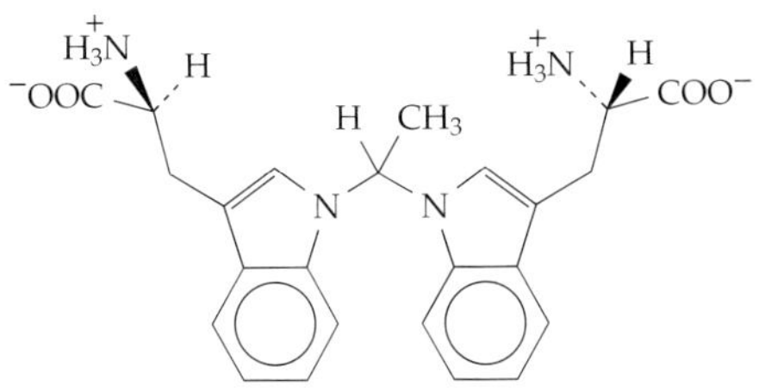

1,1'-Ethylidenebis[tryptophan]

Formation of NAD$^+$ and NADP$^+$. An alternative pathway, marked by shaded green arrows in Fig. 25-11, allows animals to form the nicotinamide ring of NAD$^+$ and NADP$^+$ from tryptophan.[224] The aldehyde produced by the ring opening reaction of step *h* can reclose (step *i*) to a pyridine ring in the form of **quinolinic acid**.[225] The latter, in a reaction that is also accompanied by decarboxylation, is coupled with a phosphoribosyl group of PRPP to form **nicotinate mononucleotide**.[225a] Adenylylation produces deamido NAD, which is converted to **NAD** by a glutamine- and ATP-dependent amination of the carboxyl group.[226]

As indicated in Fig. 25-11, free nicotinic acid can also be used to form NAD. Not surprisingly, nicotinic acid, an essential vitamin, is about 60 times more efficient than tryptophan as a source of NAD. Nevertheless, a high-tryptophan diet partially overcomes a deficiency in dietary intake of nicotinic acid. The effectiveness of a diet containing only maize as a source of protein in inducing the deficiency disease pellagra (Box 15-A) is in part a result of the low tryptophan content of maize protein. Nicotinic acid is rapidly converted in the liver to an amide with glycine, **nicotinuric acid**. Nicotinurate can be oxidatively cleaved by peptidylglycine monoxygenase to nicotinamide[227] in another alternative synthetic route to NAD.

An alternative pathway for synthesis of quinolinate from aspartate and a triose phosphate exists in bacteria and in plants and provides the major route of nicotinic acid synthesis in nature. In *E. coli* the reaction is catalyzed by two enzymes, one an FAD-containing L-aspartate oxidase which oxidizes aspartate to α-iminoaspartate.[228] The latter condenses with dihydroxyacetone-*P* to form quinolinate (Eq. 25-13).[229] There are at least two other pathways for synthesis of quinolinic acid as well as five or more salvage pathways for resynthesis of degraded pyridine nucleotide coenzymes.[224,230,231]

Although quinolinic acid provides an important source of nicotinamide coenzymes, in excess it is a neurotoxic **excitotoxin** (Chapter 30) that has been associated with epilepsy and with inflammatory neuropathalogical conditions resulting from encephalitis.[213,232,233]

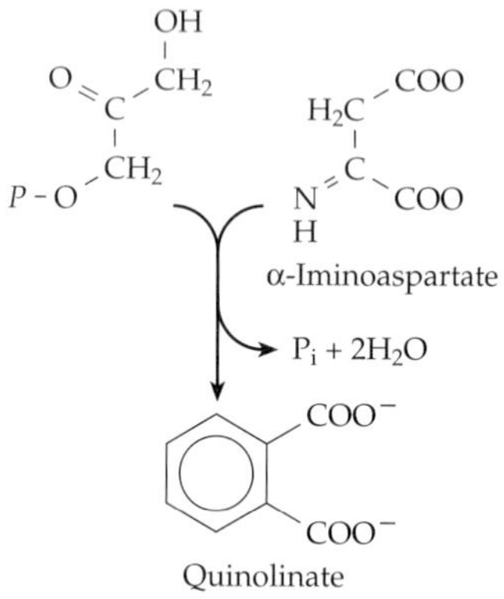

OH
CH_2
O=C
CH_2
P–O
H_2C COO
C
N COO
H
α-Iminoaspartate
$P_i + 2H_2O$
COO^-
COO^-
Quinolinate

(25-13)

Auxin. The important plant hormone **indole-3-acetic acid** (IAA; often called by the more general name **auxin**) is partially derived by oxidative decarboxylation (Fig. 25-12, top) catalyzed by tryptophan-2-monooxygenase,[234–234b] a flavoprotein similar to lysine monooxygenase (Eq. 18-41). The reduction product indole-3-ethanol also occurs in plants and is metabolically active.[235] However, most IAA in plants is not formed from tryptophan but from some precursor,[236,236a] perhaps indole-3-glycerol phosphate, which immediately precedes tryptophan in its biosynthesis (Fig. 25-2). Routes of synthesis from indole-3-acetonitrile, indoleacetaldoxime,[236b] indole-3-glucosinolate (glucobrassicin), indole-3-pyruvate, and tryptamine have also been reported.[237–238a] Nitrilases are found in

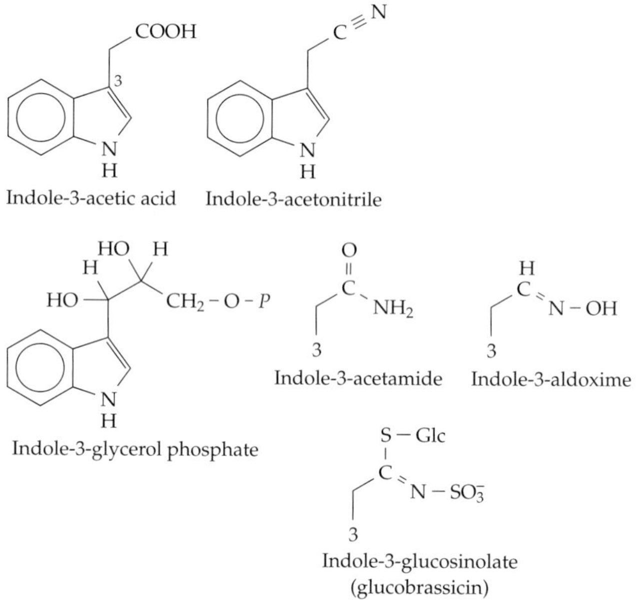

COOH
3
N
H
Indole-3-acetic acid

C≡N
N
H
Indole-3-acetonitrile

HO H
H
HO
CH_2-O-P
N
H
Indole-3-glycerol phosphate

O
C
NH_2
3
Indole-3-acetamide

H
C
N – OH
3
Indole-3-aldoxime

S – Glc
C
$N-SO_3^-$
3
Indole-3-glucosinolate
(glucobrassicin)

plants such as *Arabidopsis thaliana*[238] and also in symbiotic bacteria such as *Rhizobium*. Together with amidases they convert the acetonitrile, acetaldoximes, or acetamide derivatives to IAA,[237] which is transported throughout the plant.[238b]

About 95% of the IAA within plants is stored as conjugated forms that include amides with various amine acids and peptides[239] and glycosyl derivatives.[240] The gall-forming *Pseudomonas savastonoi* forms both IAA and conjugates such as $N^ε$-(indole-3-acetyl)-L-lysine, which aid these bacteria in colonizing olive and oleander plants.[241]

As a hormone IAA has a broad range of effects on plants, altering tissue differentiation, root growth, cell elongation, and cell division.[241a] The fastest observed response is an effect on cell elongation, which can be observed within 15–20 minutes.[242] In *A. thaliana* IAA causes very rapid transcription of at least five genes, one of which encodes 1-aminocyclopropane-carboxylase (ACC) synthase (Eq. 14-27).[243]

Alkaloids from tryptophan. The alkaloid **harmine**, which is found in several families of plants, can be formed from tryptophan and acetaldehyde (or pyruvate) in the same manner as is indicated for the formation of **papaverine** in Fig. 25-10. Some other characteristic plant metabolites such as **psilocybine**, an hallucinogenic material from the mushroom *Psilocybe aztecorum*, are formed directly from serotonin (Fig. 25-12). For many years **gramine** from barley was regarded as a curiosity because only one carbon atom separates the nitrogen atom from the indole ring. It is now believed that tryptophan is cleaved in a PLP-dependent reaction analogous to that of serine transhydroxymethylase (Eq. 14-30; Fig. 25-12). Other alkaloids arise in a more conventional fashion. Condensation of an isopentenyl group on the indole ring of tryptamine (Fig. 25-12) initiates the formation of **lysergic acid** and other ergot alkaloids.[244] The indole ring

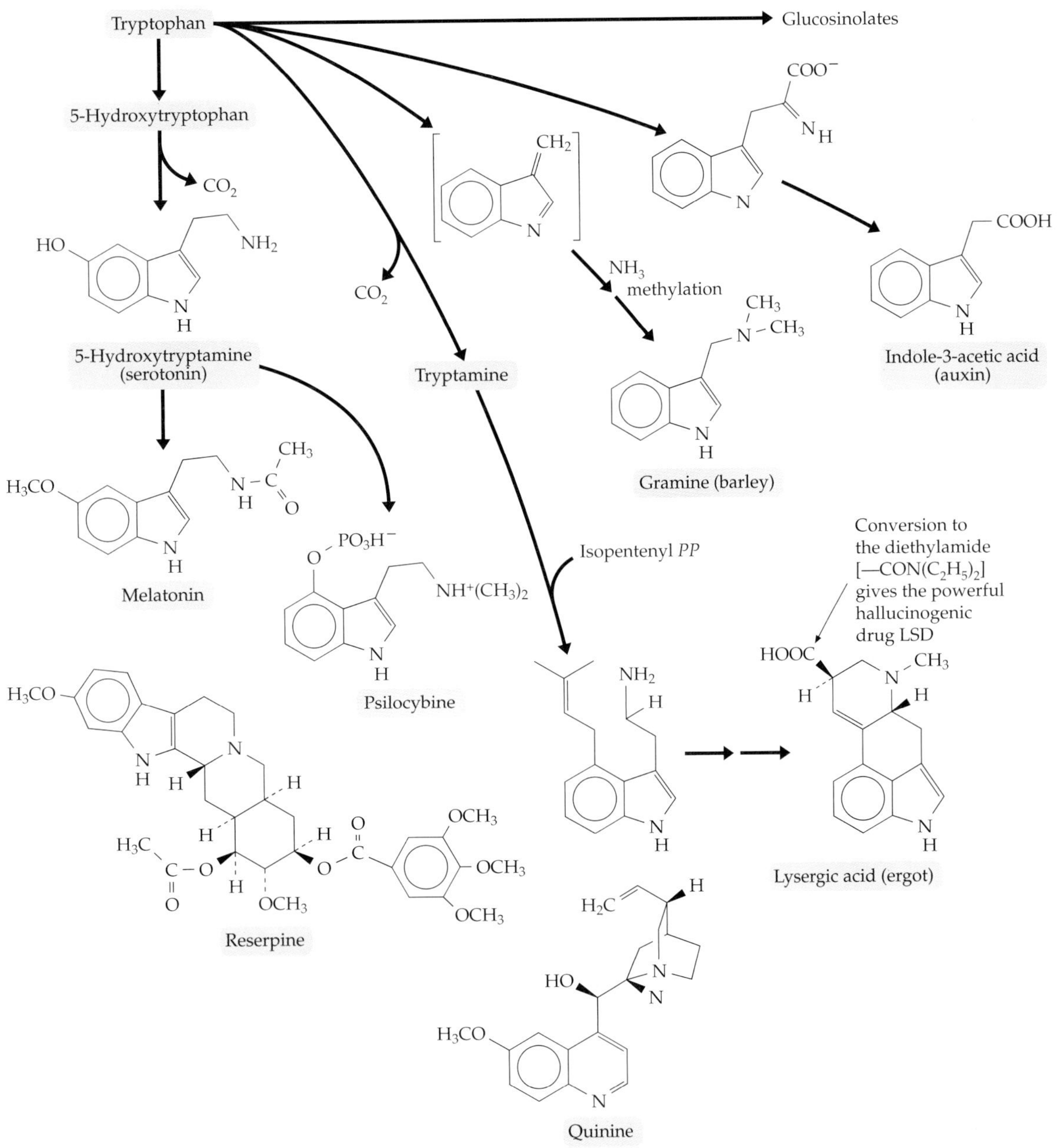

Figure 25-12 Structures and some biosynthetic pathways for some hormones, indole alkaloids, and other metabolites of tryptophan.

of tryptophan is clearly visible in the structure of **reserpine** (Fig. 25-12). This compound from *Rauwolfia* is of medical interest because of its effect in lowering blood pressure and in depleting nervous tissues of serotonin, dopamine, and noradrenaline. Reserpine also contains a benzene ring, which is derived from tryptophan by a ring expansion. The periwinkle alkaloids, including the antitumor drug **vincristine** (see Box 7-D), are formed by condensation of tryptamine with the complex glycosidic aldehyde secologamin. Additional reactions form vincristine and more than 100 other indole alkaloids.[245,245a]

Another group of plant metabolites derived from tryptophan are cyclic hydroxamic acids whose names

H
H_3C-O
in DIMBOA
O OH
N O
OH
DIBOA

BOX 25-C ROYAL PURPLE AND BLUE DENIM

Ancient people, including the Phoenicians, Chinese, and Peruvians, discovered a dyeing process that utilized secretions of certain marine molluscs, animals that were also eaten as food.[a] In processes that were perhaps closely guarded secrets, the molluscan secretions were heated for days in vats with water, salt, and additional additives including human urine, honey, etc. When the mixture was right, wool was dipped and allowed to air dry in sunlight to give the famous royal purple colors. In this ancient process as many as 10,000 molluscs were used to produce one gram of the dye.[a] In other parts of the world blue dyes were generated by fermentation of plants of the genus *Indigofera* and also of the European woad plant.[b,c]

By 1909 the chemical nature of the royal purple pigment dibromoindogotin (see scheme) had been established and by 1897 synthetic indigo production had already begun. Ancient indigo dyeing utilized the precursors, sulfate esters of **indoxyl** or of substituted indoxyl, metabolites of tryptophan (see Fig. 25-11). The sulfate esters were hydrolyzed by the sulfatase "purpurase" to give the tautomer of indoxyl that is shown in the accompanying scheme. Atmospheric oxygen converts these compounds to the corresponding oxidized dyes, the **indigotins**. However, they are very insoluble and unsuitable for dyeing. In modern indigo dyeing to form such fabrics as blue denim, the indigotin is reduced with sodium dithionite, about 2 kg of the latter being used to reduce 1 kg of the dye to the reduced leuco form.[c] Either wool or cotton can by dyed with this reduced form, air oxidation returning the dye to the blue oxidized form.

Ancient dyers also had to maintain the dye in the reduced form. In fermentations of the woad plant a species of thermophilic *Clostridium* apparently supplied the reducing agent. Padden *et al.* suggested that such bacterial reduction might be used today to avoid pollution by the by-products of dithionite reduction.[c] Use of engineered bacteria to form high yields of indole and indoxyl as a source of indigoid precursors has also been suggested.[b]

One more complexity needs to be considered. Some of the precursors are adducts (Y = $-SCH_3$, $-SO_2CH_3$ in the structures) and cannot be oxidized directly to the indigotins. Use of sunlight in a photochemical process was required in these cases.[a]

$O-SO_3^-$
Y
X
N
H
X = Br or H
Y = H–SCH_3 or –SO_2CH_3, etc.
X = Y = H in indoxyl sulfate
H_2O
"Purpurase"
HSO_3^-
O
H
Y
H, Br
N
H
O_2
If Y is not H
O
Y
N
O_2
If Y = H
$2H_2O$
hv
O
H
N
Br, H
H, Br
N
H
O
Royal purple (if X = Br)
Indigotin if X = H, insoluble
Reducing agent
O_2
OH
H
N
Br, H
H, Br
N
H
Reduced, soluble, and nearly colorless leuco form.

[a] McGovern, P. E., and Michel, R. H. (1990) *Acc. Chem. Res.* **23**, 152

[b] Ensley, B. D., Ratzkin, B. J., Osslund, T. D., Simon, M. J., Wackett, L. P., and Gibson, D. T. (1983) *Science* **222**, 167–169

[c] Padden, A. N., Dillon, V. M., John, P., Edmonds, J., Collins, M. D., and Alvarez, N. (1998) *Nature (London)* **396**, 225

are often abbreviated to DIBOA and DIMBOA. They are part of the defense system of grasses against insects and fungi. DIBOA is the main hydroxamic acid in rye while DIMBOA predominates in wheat and maize. The compounds arise from indole generated from indole-3-glycerol-*P* followed by action of four cytochrome P450 enzymes.[246]

2. The Metabolism of Histidine

The biosynthesis of histidine, which might be regarded as the "super catalyst" of enzyme active centers, begins with a remarkable reaction of ATP, the "super coenzyme" of cells. The reaction is a displacement by N-1 of the adenine ring on C-1 of PRPP (step *a*, Fig. 25-13). The resulting product undergoes a ring opening reaction, step *b*, followed by an Amadori rearrangement (step *c*). The rearrangement product is cleaved via reaction with ammonia released from glutamine with formation of **5-aminoimidazole-4-carboxamide**, whose ribotide is an established intermediate in the synthesis of ATP and other purines. Here it is recycled via ATP (Fig. 25-13). The other product of the cleavage contains the five carbons of the original ribosyl group of PRPP, together with one nitrogen and one carbon split out from the ATP molecule and the nitrogen donated by glutamine.[246a] Ring closure (step *e*) forms the imidazole group, which is attached to a glycerol phosphate molecule. The glycerol-*P* end of the molecule undergoes dehydration[247] and ketonization of the resulting enol to a product, which can be transaminated[247a,b] and dephosphorylated to histidinol. Dehydrogenation of this alcohol forms histidine.[248–249a]

Regulation of histidine synthesis. In all, ten different genes code for the enzymes of histidine biosynthesis in *Salmonella typhimurium*. They are clustered as the **histidine operon**, a consecutive series of genes which are transcribed into messenger RNA as a unit.[250,251] The gene symbols *HisA*, *HisB*, etc., are indicated in Fig. 25-13, and their positions on the *E. coli* gene map are indicated in Fig. 26-4. The gene *HisB* codes for a complex protein with two different enzymatic activities as shown in Fig. 25-13.

The presence of an excess of histidine in a bacterial cell brings about repression of synthesis of all of the

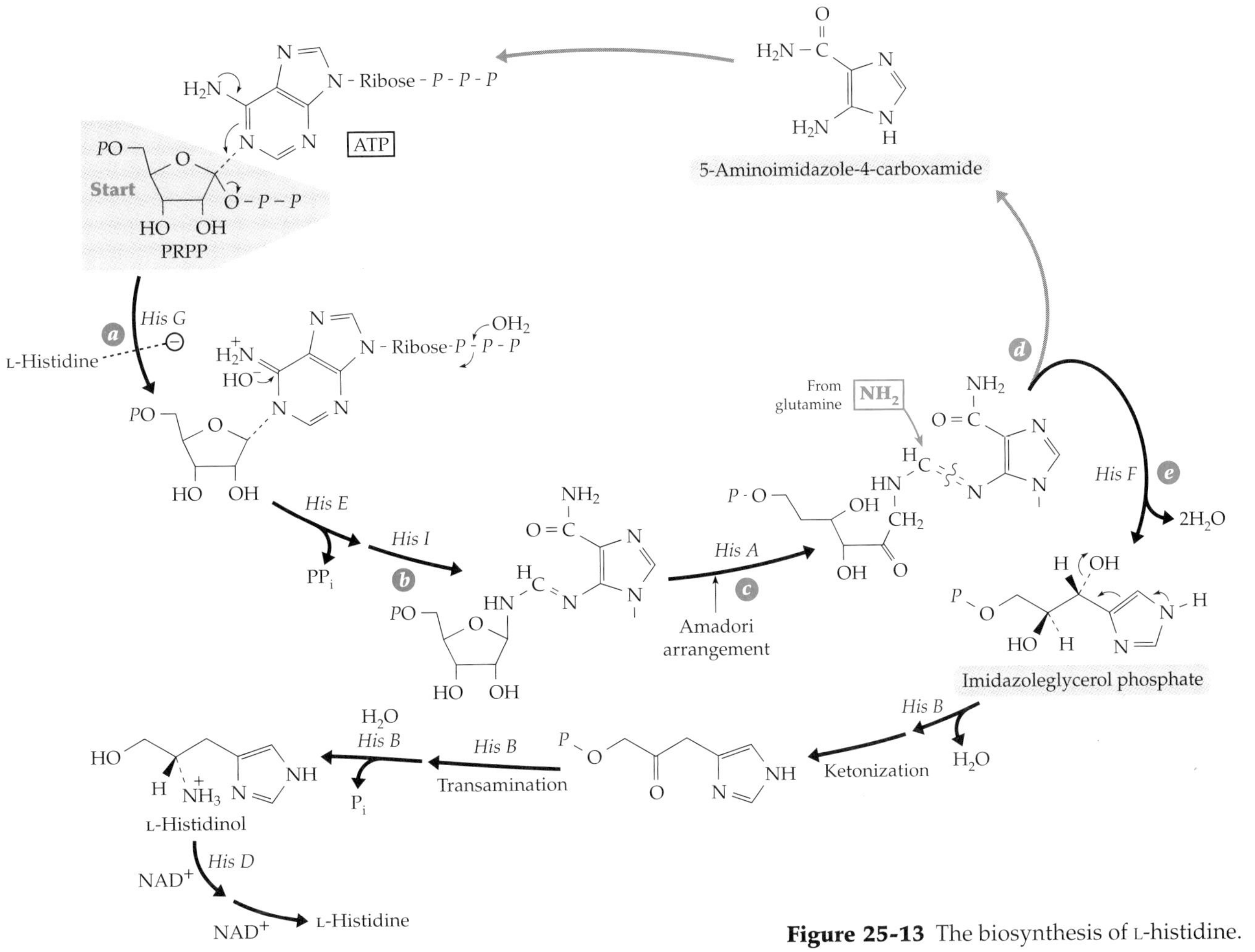

Figure 25-13 The biosynthesis of L-histidine.

enzymes encoded in the histidine operon. Details of the functioning of this and other operons are considered in Chapter 28. Histidine is also an allosteric inhibitor for the first enzyme of the biosynthetic sequence, i.e., step *a* of Fig. 25-13. Thus, instantaneous inhibition of the biosynthesis occurs if an excess of histidine accumulates. Similar patterns of both repression and feedback inhibition exist for many of the pathways of amino acid biosynthesis (Chapter 28).

Catabolism of histidine. The first steps of the major degradative pathway for histidine metabolism have already been discussed. Elimination of ammonia, followed by hydration and ring cleavage to **formiminoglutamate**, involves unusual reactions (Eq. 25-14)[252] which have been discussed earlier. Transfer of the formimino group to tetrahydrofolic acid and its further metabolism have also been considered (Chapter 15).

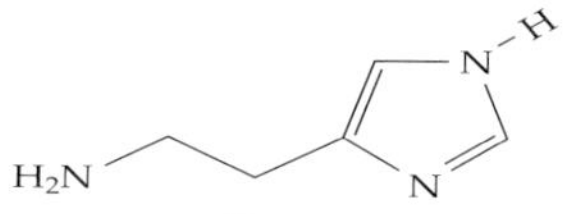

H_2N — N — N—H

Histamine

Other products from histidine include the hormonal substance **histamine** formed by decarboxylation, the oxidation product, imidazole acetic acid, and N^{δ}- and N^{ε}-methylhistidines. Histamine plays a role in release of gastric secretions and allergic responses (Chapter 5). Drugs (antihistamines) that inhibit its release are in widespread use. The unusual amino acid **diphthamide** has an unknown function in protein synthesis, occurring within the peptide chain of eukaryotic elongation factor 2 (Chapter 29).[253] Its biosynthesis from a specific histidine in EF-2 of all eukaryotes and archaebacteria requires four molecules of *S*-adenosylmethionine. The first transfers the four-carbon backbone of AdoMet to $C^{\varepsilon 1}$ of the histidyl group, a nucleophilic displacement resembling that of AdoMet-dependent C-methylation (Eq. 12-4). This is followed by transfer of three methyl groups, each from AdoMet, and finally an ATP-dependent amidation of the carboxyl group.[253] Diphthamide is the target for attack by diphtheria toxin (Box 29-A).

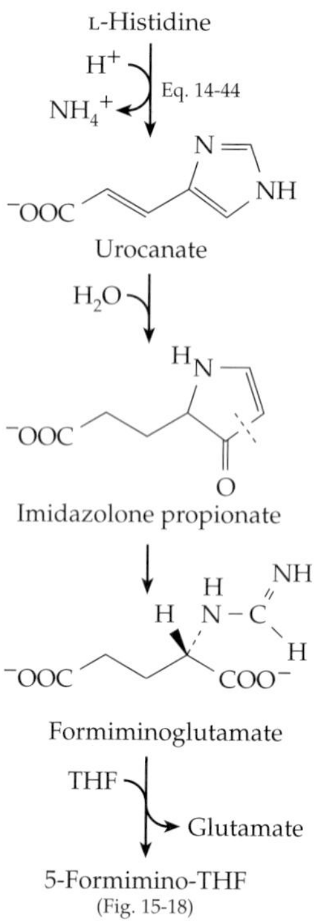

HN — CH_2 — N — N—H — H_3C, CH_3, H, N^+—CH_3 — C=O — NH_2; O=C, H

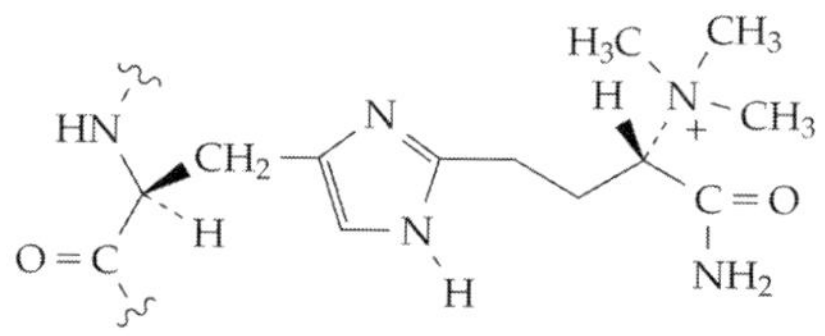

Dipthamide residue within a peptide chain.

L-Histidine

H^+; NH_4^+ — Eq. 14-44

^-OOC — N= — NH

Urocanate

H_2O

^-OOC — HN — O

Imidazolone propionate

H, H N–C(=NH)–H; ^-OOC — COO^-

Formiminoglutamate

THF → Glutamate

5-Formimino-THF
(Fig. 15-18)

(25-14)

D. Biosynthesis and Catabolism of Pyrimidines

L-Aspartate contributes four of the six ring atoms of pyrimidines including the nitrogen. The α-carboxylate group is eventually lost as CO_2, the decarboxylation helping to drive the synthetic sequence. Six enzymatic steps are required to form the product uridine 5'-phosphate (UMP) as shown in Fig. 25-14, steps *a–f*. UMP is then converted on to the cytidine, uridine, and thymidine nucleotides as shown. This pathway of pyrimidine synthesis has been conserved throughout evolution and is used by all but a few specialized organisms.[254–258] The first step is synthesis of carbamoyl phosphate by the glutamine-dependent carbamoyl phosphate synthetase, an allosteric enzyme discussed in Chapter 24 (Eq. 24-22).[259–260a] The next step is transfer of the carbamoyl group to aspartate (Fig. 25-14, step *b*). The product is able to cyclize immediately (step *c*) by elimination of water to form **dihydroorotate**.[260a] The highly controlled aspartate carbamoyltransferase has been discussed in Chapters 7 and 11. Although this is a monofunctional enzyme in bacteria, it is fused with two other proteins in mammalian cells. The resulting multifunctional enzyme (with 240-kDa subunits) catalyzes three consecutive steps: the synthesis of carbamoyl phosphate, the carbamoyltransferase reaction, and the Zn^{2+}-dependent ring-closing reaction that forms dihydroorotate.[261,262] This cyclic product is oxidized to **orotate** (Fig. 25-14, step *d*) by a flavoprotein oxidase, which in mammals utilizes ubiquinones as electron acceptors.[263–265a] A displacement reaction with PRPP (Fig. 25-14, step *e*; see also Fig. 25-13, step *a*) converts it into **orotidine 5'-phosphate**[266,267] with release of PP_i.

1. Synthesis of Pyrimidine Nucleotides

Orotidine 5'-phosphate undergoes an unusual decarboxylation (Fig. 25-14, step *f*), which apparently is not assisted by any coenzyme or metal ion but is enhanced over the spontaneous decarboxylation rate 10^{17}-fold. No covalent bond formation with the enzyme has been detected.[268] It has been suggested that the enzyme stabilizes a dipolar ionic tautomer of the substrate. Decarboxylation to form an intermediate ylid would be assisted by the adjacent positive charge.[269,270] Alternatively, a concerted mechanism may be assisted by a nearby lysine side chain.[270a–d] Hereditary absence of this decarboxylase is one cause of orotic aciduria. Treatment with uridine is of some value.[271]

We see that **uridine 5'-phosphate** (UMP) is formed from aspartate in a relatively direct and simple way. Phosphorylation with ATP in two steps produces UDP and UTP. The **cytosine nucleotides** are formed from UTP, the initial step being amination to CTP (step *h*, Fig. 25-14). This reaction resembles in many respects the conversion of citrulline to arginine, which depends upon ATP and involves transfer of the nitrogen of aspartate (Eq. 24-23). However, in the formation of CTP glutamine serves as the nitrogen donor (NH_4^+ can substitute). Observation of positional

O
HN
HO +N COO⁻
Ribose-*P*

Orotidine 5'-phosphate
(dipolar ionic tautomer)

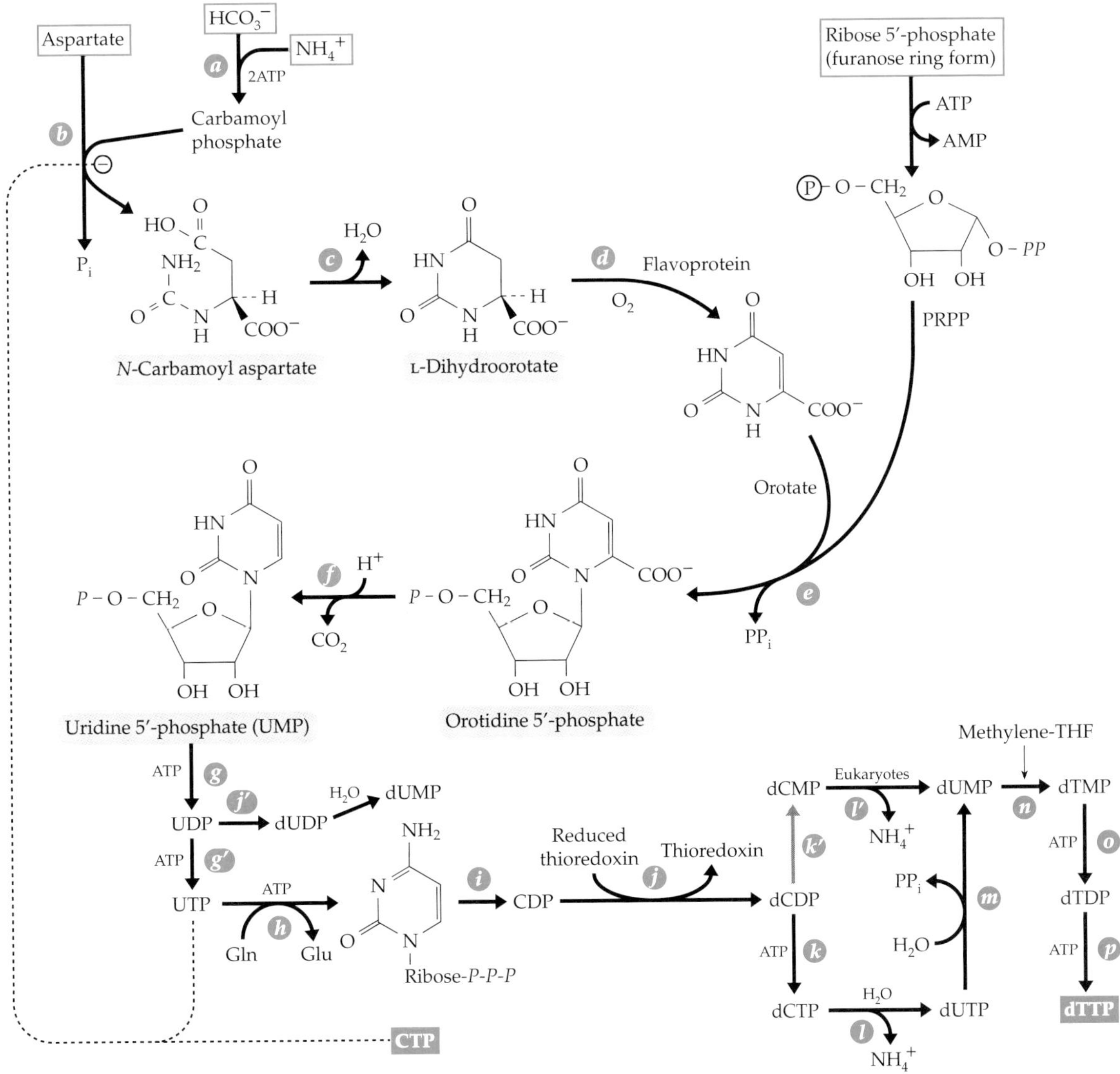

Figure 25-14 Assembly of the pyrimidine ring and biosynthesis of the pyrimidine nucleotide precursors of RNA and DNA.

isotope exchange with [γ-$^{18}O_4$]ATP suggested the occurrence of an enolic phosphate intermediate (Eq. 25-15)[272] as had also been suggested for adenylosuccinate synthase (Fig. 25-15, step *k*). CTP is incorporated directly into RNA and into such metabolic intermediates as CDP- choline, or it can be dephosphorylated to CDP (Fig. 25-14, step *i*). It is CDP that serves as the principal precursor for the deoxyribonucleotides **dCDP** and **thymidine diphosphate** (dTTP).

From Gln

H_3N PO_3^{2-} H^+ N O R H_2N NH_2 P_i

(25-15)

Deoxyribonucleotides. A chain involving NADPH, a flavoprotein, thioredoxin, and ribonucleotide reductase converts either the ribonucleoside diphosphates or triphosphates to the corresponding 2-deoxy forms (step *j*, Fig. 25-14) as indicated in Eq. 25-16.

NADPH
↓
FAD
(thioredoxin reductase)
↓
Thioredoxin
↓
Ribonucleotide reductase
↓
CDP → dCDP + H_2O (25-16)

Ribonucleotide reductases are discussed in Chapter 16. Some are iron-tyrosinate enzymes while others depend upon vitamin B_{12}, and reduction is at the nucleoside *triphosphate* level. Mammalian ribonucleotide reductase, which may be similar to that of *E. coli*, is regarded as an appropriate target for anticancer drugs. The enzyme is regulated by a complex set of feedback mechanisms, which apparently ensure that DNA precursors are synthesized only in amounts needed for DNA synthesis.[273] Because an excess of one deoxyribonucleotide can inhibit reduction of all ribonucleoside diphosphates, DNA synthesis can be inhibited by deoxyadenosine or by high levels of thymidine, despite the fact that both compounds are precursors of DNA.

Phosphorylation of dCDP to dCTP (step *k*, Fig. 25-14) completes the biosynthesis of the first of the pyrimidine precursors of DNA. The uridine nucleotides arise in two ways. Reduction of UDP yields dUDP (step *j'*, Fig. 25-14). However, the deoxycytidine nucleotides are more often hydrolytically deaminated (reactions *l* and *l'*).[274] Methylation of dUMP to form **thymidylate**, dTMP (step *n*, Fig. 25-14), is catalyzed by thymidylate synthase. The reaction involves transfer of a 1-carbon unit from methylene tetrahydrofolic acid with subsequent reduction using THF as the electron donor. A probable mechanism is shown in Fig. 15-21. See also Box 15-E. Some bacterial transfer RNAs contain 4-thiouridine (Fig. 5-33). The sulfur atom is introduced by a sulfurtransferase (the *ThiI* gene product in *E. coli*). The same protein is essential for thiamin biosynthesis (Fig. 25-21).[274a]

Formation of dUMP in eukaryotes may occur by hydrolytic removal of phosphate from dUDP or from the conversions dCDP → dCMP → dUMP (steps *k'* and *l'*, Fig. 25-14). A more roundabout pathway is employed by *E. coli*: dCDP → dCTP → dUTP → dUMP (steps *k*, *l*, and *m*, Fig. 25-14). One of the intermediates is dUTP. DNA polymerases are able to incorporate dUMP from this compound into polynucleotides to form uracil-containing DNA. The only reason that this does not happen extensively within cells is that dUTP is rapidly converted to dUMP by a pyrophosphatase (step *m*, Fig. 25-14). The uracil that is incorporated into DNA is later removed by a repair enzyme (Chapter 27). The presence of dUTP in DNA provides the basis for one of the most widely used methods of directed mutation of DNA (Chapter 26).

Bacteriophage-induced alterations in metabolism. Interesting alterations in nucleotide metabolism occur in cells of *E. coli* infected by T-even bacteriophage. Genes carried by the phage are transcribed, and the corresponding proteins are synthesized by the host cell.[273] A number of these viral gene products are enzymes affecting nucleotide metabolism. One enzyme catalyzes the hydrolytic conversion of dCTP to dCMP, and another promotes the synthesis of 4-hydroxymethyl-dCMP. Such virus-specified enzymes may be appropriate target sites for antiviral drugs.

2. Catabolism of Pyrimidine Nucleotides and Nucleosides

Nucleic acids within cells, as well as in the digestive tract, are continually under attack by many **nucleases**. Messenger RNA is degraded, often quite rapidly, as an essential part of the control of protein synthesis. Although DNA is very stable, nucleases are called upon to cut out damaged segments of single strands as part of essential repair processes (Chapter 27). Thus, there is an active breakdown of polynucleotides to mononucleotides, which are hydrolyzed to nucleosides by phosphatases. Nucleosides are converted to free bases by the action of **nucleoside phosphorylases** (Eq. 25-17). The further degradation of

Nucleoside
P_i
Free base
Ribose 1-*P*
or deoxyribose 1-*P* (25-17)

cytosine is initiated by deamination to uracil.[274b,c] Catabolism of uracil starts with reduction by NADPH according to Eq. 25-18 to form **β-alanine**.[275,275a] The latter can be oxidatively degraded to malonic semialdehyde and malonyl-CoA (see Fig. 17-3),[276] but it also serves as a biosynthetic precursor of pantothenic acid and coenzyme A (Eq. 24-38) and of the peptides carnosine and anserine.

Uracil —NADPH→ 5,6-Dihydrouracil
H_2O ↓
β-Ureidopropionate
↓ $NH_4^+ + CO_2$
H_3N^+–CH_2–CH_2–COO^-
β-Alanine (25-18)

Thymine undergoes degradation in a pathway analogous to that of Eq. 25-18, but with the formation of **3-aminoisobutyrate**. The latter can be oxidatively converted to methylmalonate (Eq. 25-19), which can enter the methylmalonyl pathways (Fig. 17-3).

H_4N^+–CH_2–$CH(CH_3)$–COO^-
↓
^-OOC–$CH(CH_3)$–COO^- (25-19)

3. Reuse or Salvage of Nucleic Acid Bases

Almost all organisms except for protozoa synthesize nucleotides via the *de novo* pathways of Figs. 25-14 and 25-15. However, they also receive nucleotides, nucleosides, and free nucleic acid bases from catabolism of RNA and DNA. Both the synthetic and degradative pathways are carefully controlled by cells to ensure that they can grow and repair their nucleic acids and not be inhibited by accumulation of an excess of any component. Furthermore, animals receive an additional supply of preformed bases from their foods. Protozoa receive all of their bases in this way. A large set of enzymes is needed to break down the nucleotides and nucleosides. Another group of enzymes resynthesize nucleotides through **salvage pathways**.[273,277] This keeps levels of inhibitory compounds low, and at the same time produces a small constant pool of nucleotide triphosphates ready for use in nucleic acid synthesis. These pathways are summarized in a general way in Eq. 25-20, which uses dark green arrows for biosynthetic pathways, light green for salvage, and black for degradative reactions (after Kornberg and Baker[273]). The very active salvage pathways of protozoa have provided a variety of targets for inhibitors aimed at parasitic protozoa such as trypanosomes and *Toxoplasma gondii*. The target enzymes often have structures sufficiently different from the corresponding human enzymes to allow for design of selective inhibitors that can serve as drugs.[278,279]

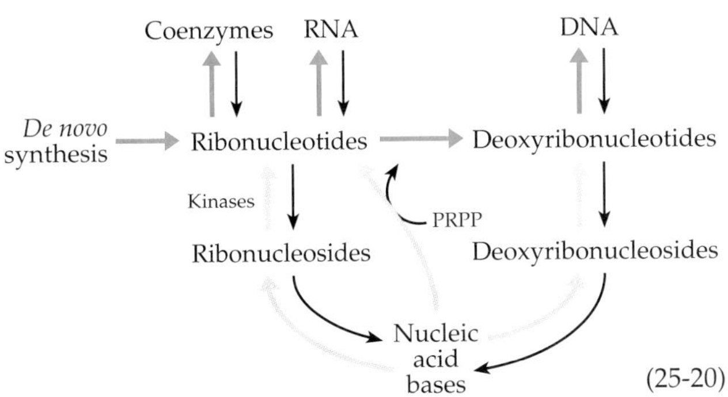

(25-20)

Just as orotic acid is converted to a ribonucleotide in step *e* of Fig. 25-14, other free pyrimidine and purine bases can react with PRPP to give monoribonucleotides plus PP_i. The reversible reactions, which are catalyzed by **phosphoribosyltransferases** (ribonucleotide pyrophosphorylases), are important components of the salvage pathways by which purine and pyrimidine bases freed by the degradation of nucleic acids are recycled.[273] However, thymine is usually *not* reused. Thymine will react with deoxribose 1-*P* to form thymidine plus inorganic phosphate (thymidine phosphorylase), and thymidine is rapidly phosphorylated by the action of successive kinases to dTTP; a substrate for DNA polymerases.[273] This has allowed biochemists to introduce radioactive thymine or thymidine into the DNA of an organism, an important experimental tool. Another important reaction of the salvage pathways for pyrimidines is the conversion of cytosine to uracil, the same kind of hydrolytic deamination represented by step *l* in Fig. 25-14.

E. Biosynthesis and Metabolism of Purines

The first decisive experiments shedding light on the biosynthetic origins of purines were done with pigeons, which form large amounts of uric acid. Labeling experiments established the complex pattern indicated in the box in the upper left-hand corner of Fig. 25-15. Two carbon atoms were found derived from glycine, one from CO_2, and two from formate. One nitrogen came from glycine, two from glutamine, and one from aspartate. In the case of adenine, the 6-NH_2 group was also found derived from aspartate.

1. The Enzymatic Reactions of Purine Synthesis

The detailed biosynthetic pathway, for which enzymes have now been isolated and studied, is indicated in Fig. 25-15. The first "committed step" in purine synthesis is the reaction of PRPP with glutamine to form **phosphoribosylamine** (step *a*).[280,281] This is another glutamine-dependent amination, pyrophosphate being displaced by ammonia generated *in situ* from glutamine. The amino group of the intermediate so-formed is coupled with glycine in a standard manner (step *b*),[282] and the resulting product is formylated by 10-formyltetrahydrofolate (step *c*).[283–285b] The latter can be generated from free formate,[286] accounting for the labeling pattern indicated in the box in Fig. 25-15. For many years it was accepted, incorrectly, that 5,10-methenyl-THF was the formyl donor for this reaction.

In step *d* of Fig. 25-15 a second glutamine-dependent amination takes place, possibly through aminolysis of an intermediate enol phosphate. An ATP-requiring ring closure and tautomerization (step *e*) serve to complete the formation of the imidazole ring.[287,287a] In many, perhaps all, eukaryotes a single multifunctional enzyme catalyzes steps *b*, *c*, and *e* of Fig. 25-15. The chicken enzyme has ~110-kDa subunits.[288] The product is a ribonucleotide of 5-amino-4-imidazole-4-carboxamide, (AIR), a compound that was isolated in 1945 from cultures of *E. coli* treated with sulfonamides. The latter are important drugs which are antagonists of *p*-aminobenzoate (Box 9-C) and interfere with completion of purine synthesis. This deprives the bacterial cells of essential folic acid derivatives. Its structure immediately suggested that 5-aminoimidazole-4-carboxamide might be a purine precursor. Later it was shown that it is actually the corresponding ribonucleotide (AIR or ZMP) that lies on the main route of purine synthesis. It also is an intermediate in the biosynthesis of thiamin (Fig. 25-21). Free 5-aminoimidazole-4-carboxamide participates in formation of histidine (Fig. 25-13).

The trinucleotide ZTP also accumulates, not only in bacteria but also in many eukaryotic cells. Bochner and Ames suggested it may be an **alarmone** signaling a deficit of folate coenzymes in the cell and causing a shutdown of protein synthesis. ZTP is synthesized by an unusual reaction, transfer of a pyrophosphate group from PRPP (phosphoribosyl pyrophosphate).

$$\text{ZMP} + \text{PRPP} \rightarrow \text{ZTP} + \text{Ribose 5'-}P \qquad (25\text{-}21)$$

This is similar to the reaction by which guanosine 5'-diphosphate 3'-diphosphate (ppGpp) is formed from GDP and ATP (Eq. 29-11).

In the next step of purine synthesis (Fig. 25-15, step *f*) a molecule of CO_2 is incorporated in an unusual type of carboxylation. It is shown in Fig. 25-15 as a single-step direct reaction of CO_2 with AIR. However, in many organisms it is a two-step ATP-dependent reaction to form a compound carboxylated on the 5-amino group. This rearranges[289–290a] to the product shown in Fig. 25-15.

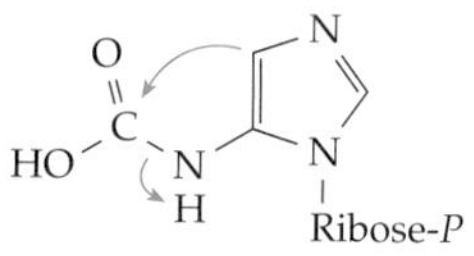

While the arrows on the foregoing structure suggest a possible mechanism of rearrangement, the implied four-membered ring transition state makes it unlikely. The reaction resembles biotin-dependent carboxylations, suggesting the possibility that the carboxylate releases CO_2, which moves and rebinds while trapped within a closed active site.[290] In a two-step amination reaction (steps *g* and *h*) nitrogen is transferred from aspartate in a manner strictly comparable to that in urea synthesis in which argininosuccinic acid is an intermediate (Fig. 24-10). As in urea formation, the carbon skeleton of the aspartate molecule is eliminated as fumarate (step *h*), leaving the nitrogen in the purine precursor. The final carbon atom is added from 10-formyltetrahydrofolic acid (step *i*).[291–292b] Spontaneous ring closure is followed by dehydration to **inosine 5'-phosphate** (IMP, inosinic acid), step *j*. Steps *i* and *j* are catalyzed by a single bifunctional enzyme.

IMP is converted via two different pathways to either AMP or GTP. Conversion to AMP (Fig. 25-15, steps *k* and *l*) occurs via another two-step aspartate-dependent amination.[292c,d] The intermediate is

adenylosuccinate.[293–294b] Positional isotope exchange studies using [γ-^{18}O]GTP suggested that the enolic 6-phospho-IMP is an intermediate comparable to that in Eq. 24-23.[295,296] X-ray studies have confirmed the prediction.[297] An NAD^+-dependent oxidation converts IMP to the corresponding **xanthine** ribonucleotide

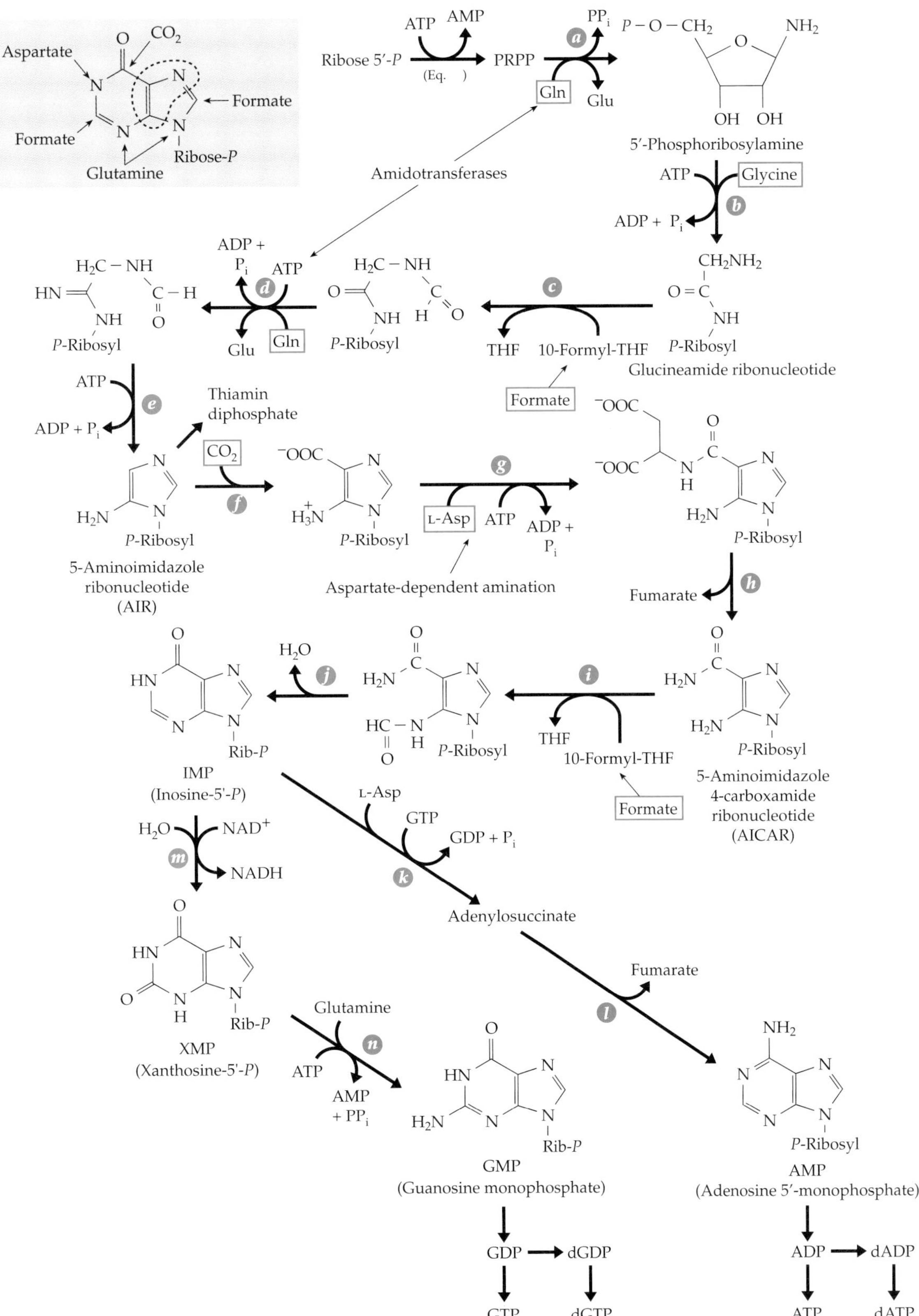

Figure 25-15 Biosynthesis of purine nucleotides from ribose 5'-phosphate.

(step *m*),[297a–d] which is aminated in a glutamine-dependent process,[298–300] as indicated (step *n*).

Synthesis of purines is under complex control.[273] Some of the mechanisms found in bacteria are outlined in Fig. 25-16. Both feedback inhibition and activation are involved. Very important is the fact that GTP is needed in the synthesis of ATP, and that ATP is needed for synthesis of GTP. This kind of control ensures that an excess of either nucleotide will not be formed for long. In bacteria all of the final end product nucleotides inhibit the initial reaction of step *a* in Fig. 25-15.

Modified purine nucleosides are important constituents of transfer RNAs (Fig. 5-33; Chapter 28, Section A,7). Among them are the 7-deazaguanosines **queuosine** (Fig. 5-33) and **archaeosine**, which contains a 7-formamidino group. These nucleosides are incorporated into tRNAs by an exchange mechanism catalyzed by **tRNA-guanine** transglycosylase. This exchanges a precursor of queuine (7-aminomethyl-7-deazaguanine) for guanine in selected residues in the tRNAs. In eukaryotic tRNAs free queuine is exchanged into the tRNAs whereas in Archaea the archaeosine precursor is incorporated.[301a,b] The conversion of the precursors to the final modified bases occurs in the tRNAs (Eq. 28-4). Many other purine derivatives are found in nature, e.g., **puromycin** (Box 29-B), which is formed from adenosine.[301c]

2. The Purine Nucleotide Cycle and Salvage Pathways for Purines

Muscular work is accompanied by the production of ammonia, the immediate source of which is adenosine 5'-phosphate (AMP).[301,302] This fact led to the recognition of another substrate cycle (Chapter 11) that functions by virtue of the presence of a biosynthetic pathway and of a degradative enzyme in the same cells (cycle A, Fig. 25-17). This **purine nucleotide cycle** operates in the brain[303,304] as well as in muscle. The key enzyme 5'-AMP aminohydrolase (AMP deaminase; step *a*, Fig. 25-17) also occurs in erythrocytes and many other tissues.[304,305] Persons having normal erythrocyte levels but an absence of this enzyme in muscles suffer from muscular weakness and cramping after exercise.[306]

Purine bases from ingested foods, or formed by catabolism of nucleic acids, are able to react with PRPP under the influence of phosphoribosyltransferases.[306a] Two such enzymes are known to act on purines. One converts adenine to AMP (Fig. 25-17, step *b*) and also acts upon 5-aminoimidazole-4-carboxamide. This enzyme may be especially important to parasitic protozoa such as *Leishmania*, which lack the *de novo* pathway of purine synthesis (Fig. 25-15).[278,306b]

AMP can be converted by the action of AMP 5'-nucleotidase to adenosine (step *c*, Fig. 25-17), which is thought to be an important local hormone or second

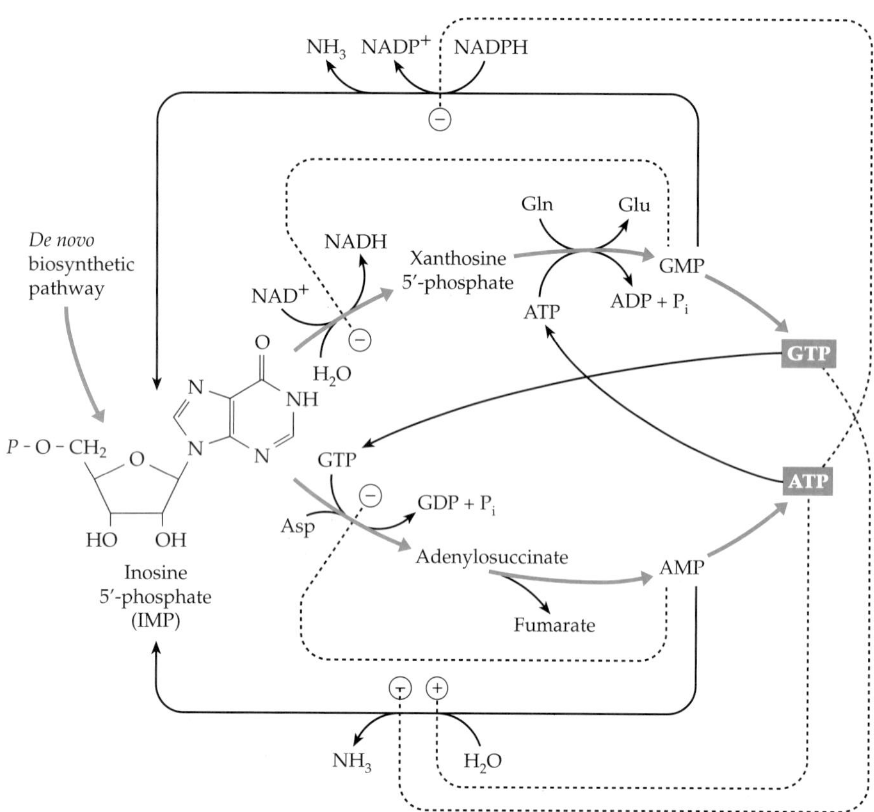

Figure 25-16 Control of the conversion of inosine 5'-phosphate to the adenine and guanine ribonucleotides and deoxyribonucleotides in bacteria by feedback inhibition and activation.

messenger.[307–309] Adenosine has a variety of effects on all organs of an animal. It affects heart rate, smooth muscle tone, and white blood cell function. It modulates the catabolic effects of hormones such as catecholamines and stimulation of the anabolic hormone insulin.[309] Adenosine receptors of at least three types are known.[310,311] Binding of adenosine to A_1 receptors inhibits adenylate cyclase, while binding to A_2 receptors stimulates this enzyme.[307] However, effects of adenosine on K^+ transport are probably more important.

Deamination of adenosine (step *d*) together with reconversion of the resulting inosine to IMP (steps *e* and *f*) completes a second cycle (cycle B, Fig. 25-17). Intense interest has been focused on adenosine deaminase because hereditary lack of this enzyme is linked to a severe immunodeficiency in which the numbers of B and T lymphocytes are inadequate to combat infection.[312] Until recently bone marrow transplantation in infancy was the only possible treatment for this otherwise fatal disease. Regular injections of adenosine deaminase covalently attached to polyethylene glycol (to delay removal from the bloodstream) have been used for some patients. Since 1990 **gene therapy**, transfer of an adenosine deaminase gene into white blood cells, has also been used with apparent success.[313–315] This topic is discussed in Chapter 31. Adenosine deaminase is a 40-kDa protein,[316,317] which exists as a complex with a large 200-kDa binding protein[315] which anchors the deaminase to cell membranes.

The basic cause of the severe immunodeficiency symptoms is uncertain. However, adenosine deaminase also catalyzes hydrolysis of 2'-deoxyadenosine, and in the absence of the enzyme both this compound and its trinucleotide precursor 2'-deoxy-ATP (dATP) accumulate in tissues.[312] Ribonucleotide reductase is allosterically inhibited by dATP, and this inhibition may interfere with DNA synthesis and with the rapid growth of lymphocytes needed in response to infections. Since T lymphocytes are more severely affected than B lymphocytes, it is necessary to postulate a difference in the extent to which these two cell types accumulate dATP.

The conversions of inosine to hypoxanthine (Fig. 25-17, step *e*), of guanosine to guanine (step *g*), and of other purine ribonucleosides and deoxyribonucleosides to free purine bases are catalyzed by **purine nucleoside phosphorylase**.[318–321b] Absence of this enzyme also causes a severe immune deficiency which involves the T cells. However, B cell function is not impaired.[312,315,322]

The last enzyme in cycle B of Fig. 25-17 (catalyzing step *f* and also step *b*) is the X-linked **hypoxanthine-guanine phosphoribosyltransferase** (HGPRT or HPRT).[322a,b] Its absence causes the **Lesch-Nyhan** syndrome characterized not only by overproduction of uric acid but by a serious disorder of the central nervous system. It causes both mental retardation and a compulsive form of self-mutilation of the gums and hands by biting.[323–325] The excessive production of uric acid is easy to understand because the accumulating hypoxanthine and guanine are both readily converted to uric acid by the reactions of Fig. 25-18. Patients with a partial deficiency in HGPRT escape the worst neurological symptoms but may have severe gouty arthritis (Box 25-D).[326] Efforts are being made to treat the disease by gene transfer.[327]

Trypanosomes and other parasitic protozoa are unable to synthesize purines and must obtain them from their hosts using salvage pathway. Selective inhibition of their HGPRT or of nucleoside hydrolases, which are absent from mammalian cells, are goals of drug development.[327a,b]

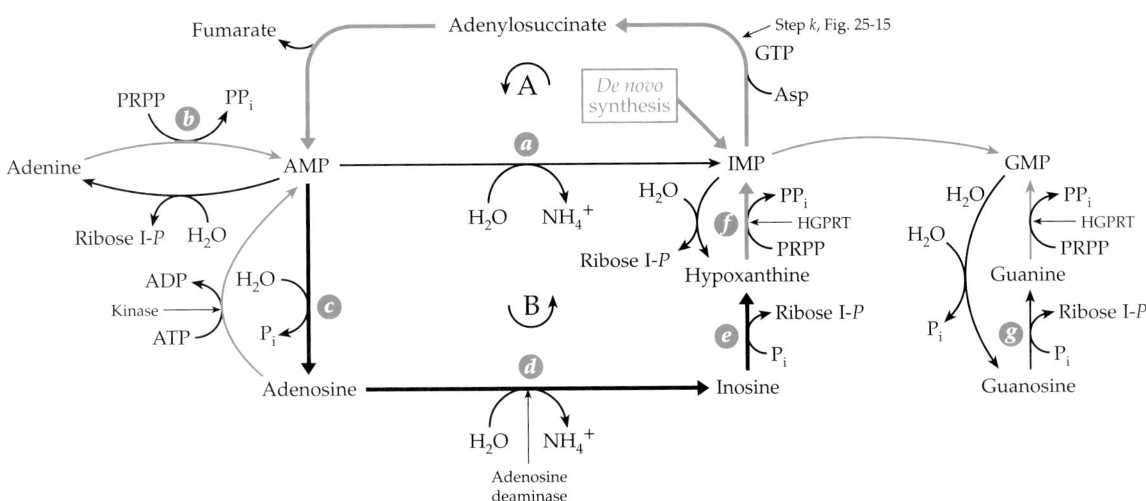

Figure 25-17 Some purine salvage pathways and related reactions. Green lines indicate biosynthetic pathways.

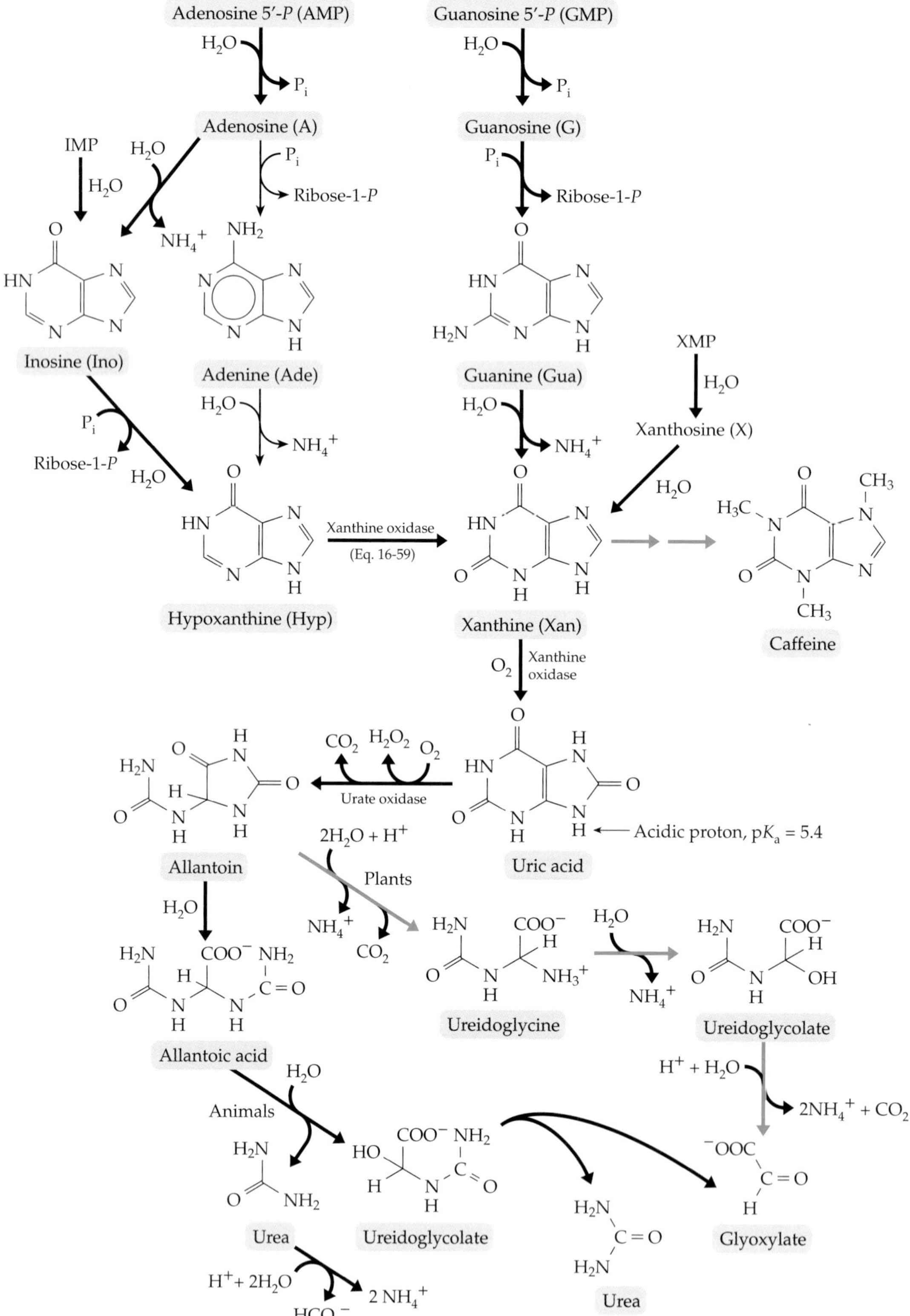

Figure 25-18 Pathways of catabolism of purine nucleotides, nucleosides, and free bases. Spiders excrete xanthine while mammals and birds excrete uric acid. Spiders and birds convert all of their excess nitrogen via the *de novo* pathway of Fig. 25-15 into purines. Many animals excrete allantoin, urea, or NH_4^+. Some legumes utilize the pathway marked by green arrows in their nitrogen transport via ureides.

3. Oxidative Metabolism of Purines

As indicated in Fig. 25-18, free adenine released from catabolism of nucleic acids can be deaminated hydrolytically to hypoxanthine, and guanine can be deaminated to xanthine.[328] The molybdenum-containing xanthine oxidase (Chapter 16) oxidizes hypoxanthine to xanthine and the latter on to uric acid. Some clostridia convert purine or hypoxanthine to xanthine by the action of a selenium-containing purine hydroxylase.[328a] Another reaction of xanthine occurring in some plants is conversion to the trimethylated derivative **caffeine**.[328b] One of the physiological effects of caffeine in animals is inhibition of pyrimidine synthesis.[329] However, the effect most sought by coffee drinkers may be an increase in blood pressure caused by occupancy of adenosine receptors by caffeine.[330]

Uric acid is the end product of purine metabolism in the human. Spiders excrete **xanthine**,[331,332] but in most animals **urate oxidase** converts uric acid to **allantoin**. Although urate oxidase contains no coenzymes or metal cofactors,[333] it catalyzes the reaction with O_2 to form a peroxide (Eq. 25-22),[334,335] a reaction resembling that of a reduced flavin (Eq. 15-31) or a reduced pterin. As in these other cases there may be an initial electron transfer between O_2 and urate to form a radical pair which couple. Elimination of H_2O_2, a feature also of flavoprotein oxidases, accomplishes oxidation of the urate ring. Hydration, ring opening, and decarboxylation complete the conversion to allantoin. The ease of formation of urate radicals permits uric acid to act as an effective oxidant. This may account for the fact that we long-lived human beings retain a high internal urate concentration.[335a,b]

BOX 25-D GOUT

A common metabolic derangement with an incidence of ~3 per 1000 persons is **hyperuricemia** or **gout**.[a,b] As with most metabolic defects, there is a family of diseases ranging from mild to severe. In acute gouty arthritis, a sudden attack occurs, usually in the night, when sodium urate crystals precipitate in one or more joints. In half the cases the victim is awakened by a terrible pain in the big toe. The disease most often strikes adult males. The heredity is apparently complex and not fully understood. The primary biochemical defect in gout is usually an overproduction of uric acid which, in some cases, may result from an overactive PRPP synthase.[c] In other cases a kidney defect interferes with excretion. The less severe saturnine gout, which occurs in relatively young persons of both sexes, is a result of chronic lead poisoning. It may involve deposition of guanine in the joints as a result of inhibition by Pb^{2+} of guanine aminohydrolase, the enzyme that hydrolyzes guanine to xanthine (Fig. 25-18).[d]

If properly controlled, simple gout may have few adverse effects. However, the severe neurological symptoms of Lesch–Nyhan syndrome (Section E,2 of text)[e] cannot be corrected by medication. Colchicine (Box 7-D), in a manner which is not understood, alleviates the painful symptoms of gout caused by the deposits of sodium urate in joints and tissues. It is also important to keep the dietary purine intake low and it is often necessary to inhibit xanthine oxidase. A widely used and effective inhibitor is the isomer of hypoxanthine known as **allopurinol**, which is taken daily in amounts of 100–600 mg or more.

O
HN
N
N
N
H

Allopurinol

Allopurinol and its oxidation product **oxypurinol**, a xanthine analog, both inhibit xanthine oxidase and patients receiving allopurinol excrete much of their purines as xanthine and hypoxanthine. Nucleotide derivatives of oxypurinol also inhibit the *de novo* purine biosynthetic pathway. The accumulating hypoxanthine is reused to a greater extent than normal, decreasing the total purine excretion. A number of other drugs stimulate increased excretion of uric acid.[e] Although many patients tolerate allopurinol for many years, some experience dangerous side effects.

a Becker, M. A., and Roessler, B. J. (1995) in *The Metabolic and Molecular Bases of Inherited Disease*, 7th ed., Vol. 2 (Scriver, C. R., Beaudet, A. L., Sly, W. S., and Valle, D., eds), pp. 1655–1677, McGraw-Hill, New York

b Kelley, W. N., and Wyngaarden, J. B. (1972) in *The Metabolic Basis of Inherited Disease*, 3rd ed. (Stanbury, J. B., Wyngaarden, J. B., and Fredrickson, D. S., eds), pp. 969–1002, McGraw-Hill, New York

c Becker, M. A., Kostel, P. J., and Meyer, L. J. (1975) *J. Biol. Chem.* **250**, 6822–6830

d Farkas, W. T., Stanawitz, T., and Scheider, M. (1978) *Science* **199**, 786–787

e Rossiter, B. J. F., and Caskey, C. T. (1995) in *The Metabolic and Molecular Bases of Inherited Disease*, 7th ed., Vol. 2 (Scriver, C. R., Beaudet, A. L., Sly, W. S., and Valle, D., eds), pp. 1679–1706, McGraw-Hill, New York

(25-22)

Allantoin is the excretory product in most mammals other than primates. Most fish hydrolyze allantoin to **allantoic acid,** and some excrete that compound as an end product. However, most continue the hydrolysis to form urea and glyoxylate using peroxisomal enzymes.[336] In some invertebrates the urea may be hydrolyzed further to ammonia. In organisms that hydrolyze uric acid to urea or ammonia, this pathway is used only for degradation of purines from nucleotides. Excess nitrogen from catabolism of amino acids either is excreted directly as ammonia or is converted to urea by the urea cycle (Fig. 24-10).

Plants also form the **ureides** allantoin and allantoic acid, and in some legumes, such as soy beans, these compounds account for 70–80% of the organic nitrogen in the xylem. They appear to function in nitrogen transport.[337] As indicated in Fig. 25-18, the hydrolysis to glyoxylate, NH_4^+, and CO_2 follows a different pathway than in animals. See also Chapter 24, Section C.

F. Pterins, Flavins, Dimethylbenzimidazole, Thiamin, and Vitamin B_6

Tracer studies have established that both folic acid and riboflavin originate from guanosine phosphates (Fig. 25-19).[338–341] All of the atoms of the purine ring are conserved in the products except for C-8 of the five-membered ring. The first step is opening of the 5-membered ring and the hydrolytic removal of formate (Fig. 25-19, step *a*). This is followed by an Amadori rearrangement (step *b*) and a simple ring closure between the resulting carbonyl and adjacent amino group (step c). The product is **7,8-dihydroneopterin triphosphate**. The single enzyme **GTP cyclohydrolase** catalyzes all three steps *a–c*.[342,342a–c] Dihydroneopterin is a central intermediate in pterin metabolism, being converted by one route into folate and methanopterin coenzymes and by another into biopterin, drosopterin, and others.[343]

An aldol cleavage (step *d*) followed by a series of reactions, shown at the left in in Fig. 25-19, leads to folate.[344–347] The reactions include the ATP-dependent conversion of 6-hydroxy-7,8-dihydropterin to its pyrophosphate ester.[347a] This is followed by coupling to *p*-aminobenzoate with elimination of PP_i to form dihydropteroic acid.[347b] ATP-dependent joining of glutamate and reduction yields **tetrahydrofolate**. Additional molecules of γ-linked glutamate are added to form the functional polyglutamate forms (p. 803).[347c,d] Gamma-glutamyl hydrolases provide essential turnover.[347e] Formation of biopterin is initiated by tautomerization of dihydroneopterin triphosphate (step *e*), a proton from the solvent binding to C-6 of the tetrahydropterin ring.[348] The single Zn^{2+}-dependent enzyme **6-pyruvoyltetrahydropterin synthase**[349–352] catalyzes the two consecutive tautomerization steps shown in Fig. 25-19 as well as elimination of the triphosphate group tripolyphosphate (step *f*), a reaction facilitated by the carbonyl group introduced in the preceding step.[349] The same enzyme promotes a final tautomerization to form 6-pyruvoyltetrahydropterin, a compound that is reduced by NADPH to tetrahydrobiopterin. The reaction is catalyzed by **sepiapterin reductase**.[353,354] Notice that biopterin, like the folates, is synthesized at the oxidation level of a tetrahydrobiopterin.

The *Drosophila* eye pigments **sepiapterin** and **drosopterins** (Figs. 15-17 and 25-19) arise from 6-pyruvoyltetrahydropterin.[355–357] Reduced glutathione appears to be the reducing agent needed to convert the 6-pyruvoyltetrahydropterin into the more reduced pyrimidinodiazepine (step *h*) with its 7-membered ring (Fig. 25-19).[357,358] Tetrahydrobiopterin can arise in mammalian cells, not only by the *de novo* pathway of Fig. 25-19 but also from salvage of sepiapterin.

Methanopterin (Fig. 15-17) is formed via a branch in the folate pathway. The 5-carbon chain that replaces the carboxyl group of *p*-aminobenzoate in the folates is derived by reaction of PRPP with *p*-aminobenzoate (Eq. 25-23).[359,360] White has proposed a

(25-23)

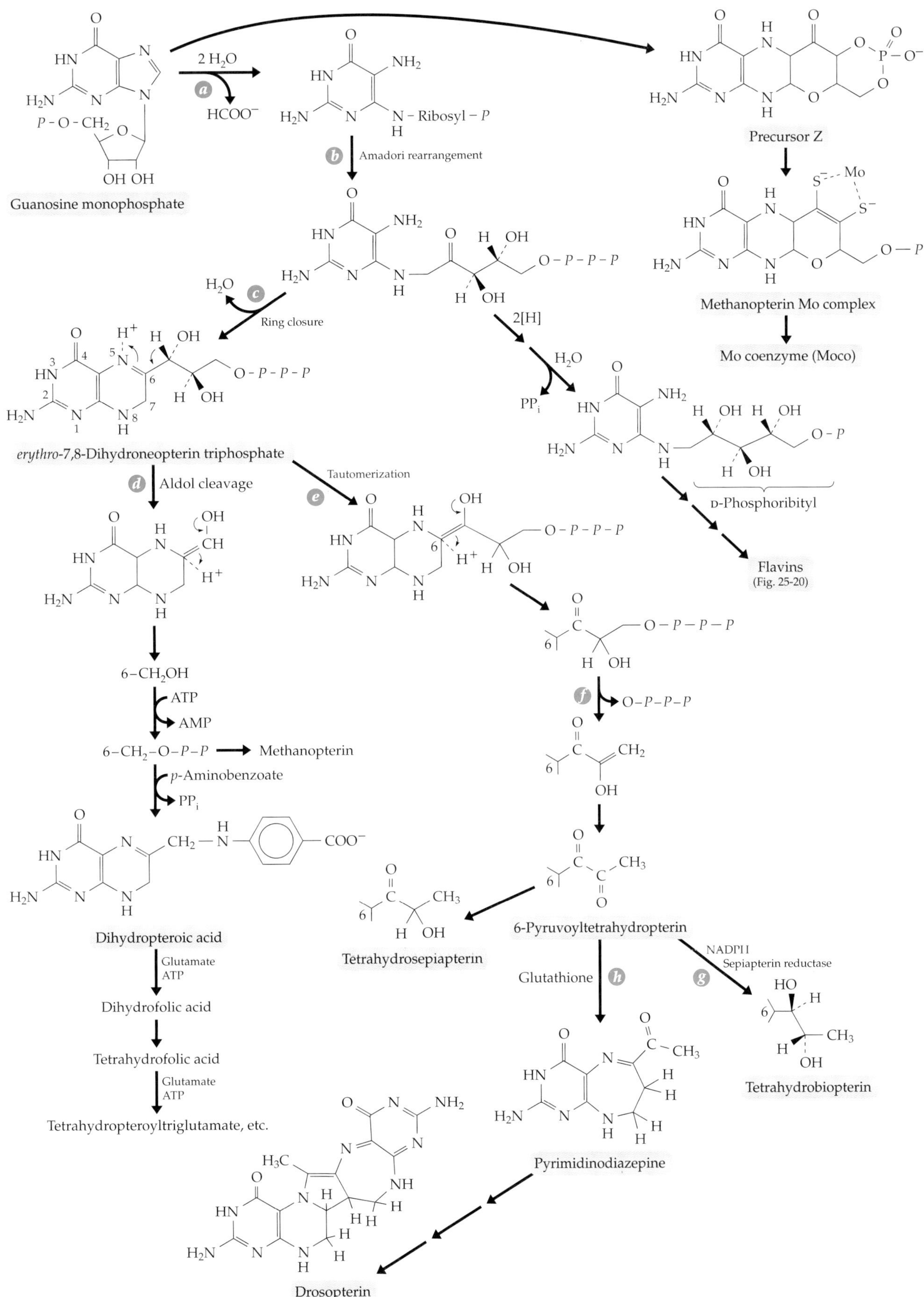

Figure 25-19 The biosynthesis of folic acid and other pterins.

detailed pathway for completion of the methanopterin synthesis.[359] After coupling with 6-hydroxymethyl-7,8-dihydropterin pyrophosphate (shown in abbreviated form in Fig. 25-19) with loss of PP_i, the ribose ring in the product of Eq. 25-23 is reductively opened to give the ribityl side chain of methanopterin (see Fig. 15-17). Hydrolytic removal of the phosphate is then followed by transfer of the α-linked ribose phosphate. The *S*-hydroxyglutarate (derived by reduction of 2-oxoglutarate) is added in an ATP-dependent reaction. The extra methyl groups at positions 7 and 9 are transferred from *S*-adenosylmethione and the compound is reduced to the tetrahydropterin state.

Biosynthesis of **molybdopterin** (Fig. 15-17) also begins with a guanosine derivative, quite possibly GMP as is shown in Fig. 25-19. However, the C-8 carbon atom is not lost as formate, as in step *a* of Fig. 25-19, but is incorporated into the side chain in the molybdopterin precursor Z. A pathway was proposed by Wuebbens and Rajagopalan.[361] The first stage in the reaction sequence,[361a–d] which is identical in most organisms, is formation of the metastable compound Z, a proposed structure of which is shown in Fig. 25-19. The conversion of this compound into molybdopterin requires opening of the cyclic phosphate ring and incorporation of two atoms of sulfur. These may both be released from cysteine as S^0 and carried by a sulfurtransferase as an enzyme-bound persulfide group (see Chapter 24, Section G,3).[361b] A thiocarboxylate group generated from a C-terminal carboxylate of a molybdopterin synthase subunit, as in the ThiS protein (p. 1463), may be the direct sulfur donor. It is probably formed in an ATP-dependent process.[361b] Incorporation of molybdenum, perhaps from MoO_4^{2-}, completes the synthesis of the molybdenum cofactor Moco.[361c,e] In some molybdoenzymes, e.g., xanthine oxidase, an additional sulfur atom is bound to the Mo atom (Fig. 16-32). This is also obtained from cysteine using a PLP-dependent sulfurtransferase similar to the NifS protein.[361f,g] In many bacteria molybdopterin is joined to GMP, AMP, IMP, or CMP to form a dinucleotide.[361h,i]

Both the fungus *Eremothecium* (Box 15-B) and mutants of *Saccharomyces* have been used to deduce the pathways of **riboflavin** synthesis outlined in Figure 25-20. The first reaction (step *a*) is identical to step *a* of Fig. 25-19 but is catalyzed by a different GTP cyclohydrolase.[362] Instead of an Amadori rearrangement it catalyzes the hydrolytic deamination and dephosphorylation (step *b*) to give the flavin precursor 4-ribitylamino-5-amino-2,6-dihydroxypyrimidine. Additional carbon atoms to build the benzene ring of riboflavin are supplied in two stages from ribulose 5-phosphate. Isotopic labeling showed that carbon atoms 1, 2, 3, and 5 of this compound are utilized as is marked in Fig. 25-20, while C-4 is eliminated as formic acid in a rearrangement (step *c*, Fig. 25-20). A plausible mechanism has been suggested.[363,363a,b] The product L-3,4-dihydroxy-2-butanone 4-phosphate condenses (step *d*) with the product formed from GMP, possibly via the Schiff base shown. Elimination of H_2O and inorganic phosphate followed by tautomerization gives **6,7-dimethyl-8-ribityllumazine**.[363a,364–364c] Completion of the flavin ring requires an additional four carbon atoms, which are supplied by a second molecule of 6,7-dimethyl-8-ribityllumazine, as indicated in Fig. 25-20.[364d] This disproportionation reaction appears remarkable but is less so when one considers that the bimolecular reaction to form riboflavin occurs spontaneously under mild conditions. The precursor 4-ribitylamino-5-amino-2,6-dihydroxypyrimidine is regenerated in this process (Fig. 25-20).

The enzyme complex that catalyses steps *d* to *f* of Fig. 25-20 has an unusual composition. An α_3 trimer of 23.5-kDa subunits is contained within an icosahedral shell of 60 16-kDa β subunits, similar to the protein coats of the icosahedral viruses (Chapter 7). The β subunits catalyze the formation of dimethylribityllumazine (steps *d*, *e*), while the α_3 trimer catalyzes the dismutation reaction of step *f*, the final step in riboflavin formation.[365] A separate bifunctional bacterial ATP-dependent synthetase phosphorylates riboflavin and adds the adenylyl group to form FAD.[366] Two separate mammalian enzymes are required.[367] Synthesis of **deazaflavins** of methanogens (Fig. 15-22) follows pathways similar to those of riboflavin. However, the phenolic ring of the deazaflavin originates from the shikimate pathway.[368]

Dimethylbenzimidazole, a constituent of vitamin B_{12} (Box 16-B), also arises from 6,7-dimethyl-8-ribityllumazine in a process resembling that of riboflavin synthesis, but in which the riboflavin formed is hydrolytically degraded to remove the pyrimidine ring and to form the imidazole ring.[369] Conversion to an α-ribazole and linkage to the aminopropyl group of the corrin ring is described by Thompson *et al.*[370] Various related cobamides are also formed by bacteria.[371] Synthesis of the corrin ring is described briefly in Chapter 24, and the chemistry of the ligands to cobalt at the "top" of the vitamin B_{12} molecules is considered in Chapter 16, Section B.

Thiamin. Investigation of the biosynthesis of thiamin has been difficult because only minute amounts are formed by microorganisms such as *E. coli* or yeast. Furthermore, significant differences in the routes of synthesis in different organisms have caused confusion.[372,372a] The pathways outlined in Fig. 25-21 are incomplete.

The pyrimidine portion of thiamin (Fig. 25-21) is distinct in structure from the pyrimidines of nucleic acids. In bacteria it originates from the purine precursor 5-aminoimidazole ribotide, which is converted into a hydroxymethylpyrimidine (Fig. 25-21)[373] which is

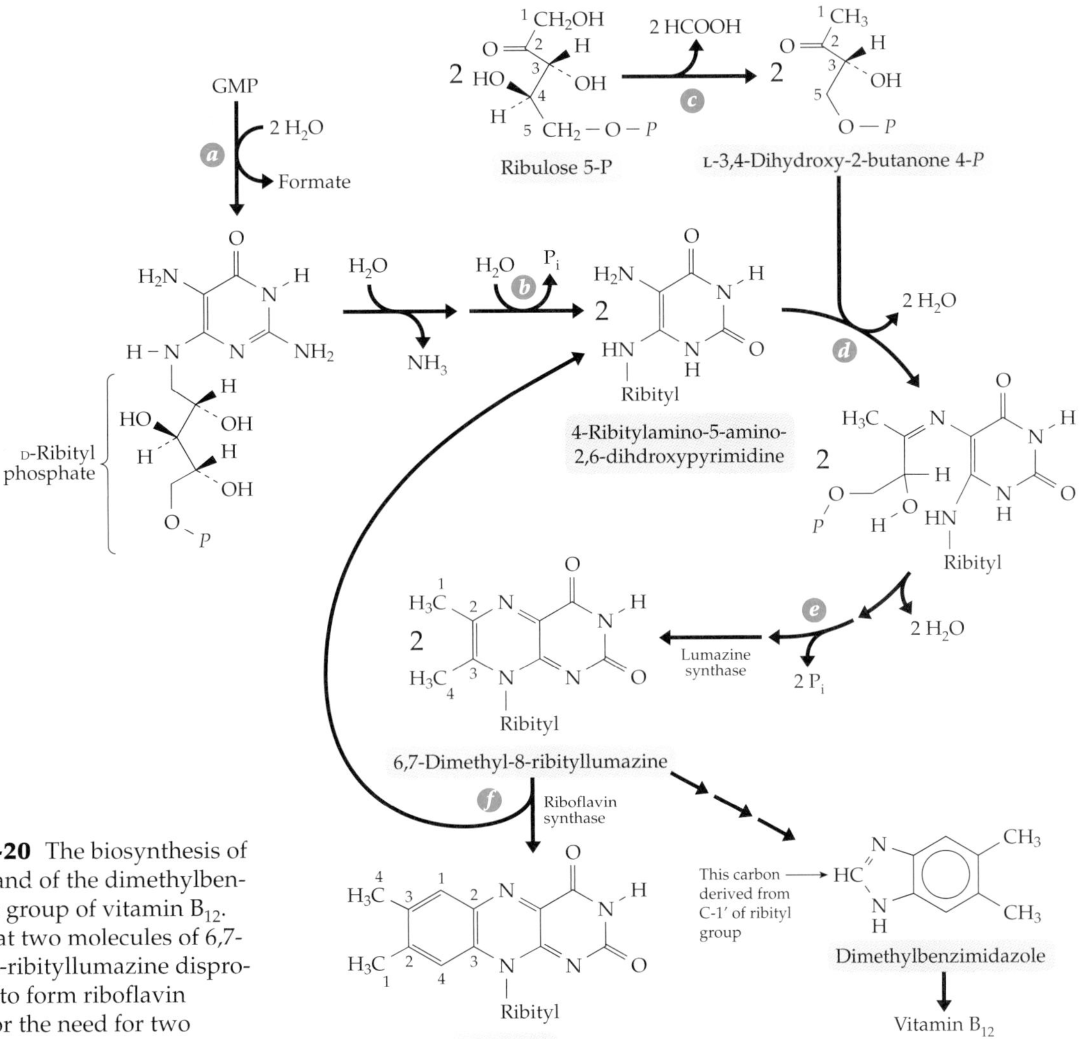

Figure 25-20 The biosynthesis of riboflavin and of the dimethylbenzimidazole group of vitamin B$_{12}$. The fact that two molecules of 6,7-dimethyl-8-ribityllumazine disproportionate to form riboflavin accounts for the need for two molecules of each reactant and product in many steps.

coupled with the thiazolium ring to form the vitamin. All of the carbon atoms of the substituted pyrimidine can be derived from the 5-aminoimidazole ribotide, but the pathway is uncertain. Both glycine and formate enter the pyrimidine, but labeling patterns are different in *E. coli* and in yeast.[372]

The thiazole ring is assembled on the 5-carbon backbone of **1-deoxyxylulose 5-phosphate**, which is also an intermediate in the alternative biosynthetic pathway for terpenes (Fig. 22-2) and in synthesis of vitamin B$_6$ (Fig. 25-21). In *E. coli* the sulfur atom of the thiazole comes from cysteine and the nitrogen from tyrosine.[374] The same is true for chloroplasts,[375] whereas in yeast glycine appears to donate the nitrogen.[372] The thiamin biosynthetic operon of *E. coli* contains six genes,[372a,376] one of which (***ThiS***) encodes a protein that serves as a sulfur carrier from cysteine into the thiazole.[374] The C-terminal glycine is converted into a thiocarboxylate:

$$\text{ThiS}-\overset{\text{H}}{\text{N}}-\text{CH}_2-\text{C}\!\begin{matrix}\text{O}\\ \text{S}-\text{H}\end{matrix}$$

Glycine

The *ThiI* gene, which encodes aother sulfurtransferase protein, is also needed.[374a] The enzymology of the insertion of this sulfur into the thiazole is uncertain but may resemble that involved in synthesis of biotin, lipoic acid, and molybdopterin.[374] Linkage of the two parts of the thiamin molecule (step *d*, Fig. 25-21) is catalyzed by thiamin phosphate synthase, evidently via an S$_N$2 type reaction.[377–377b]

Pyridoxol (vitamin B$_6$). Again 1-deoxyxylulose 5-*P* serves as a precursor.[378] In *E. coli* only two genes have been implicated in the condensation of this compound with 4-(phosphohydroxy)-L-threonine (Fig. 25-21, step *f*).[378a] One is an NAD$^+$-dependent dehydrogenase that acts on the second substrate prior to the condensation. Significant differences from the path-

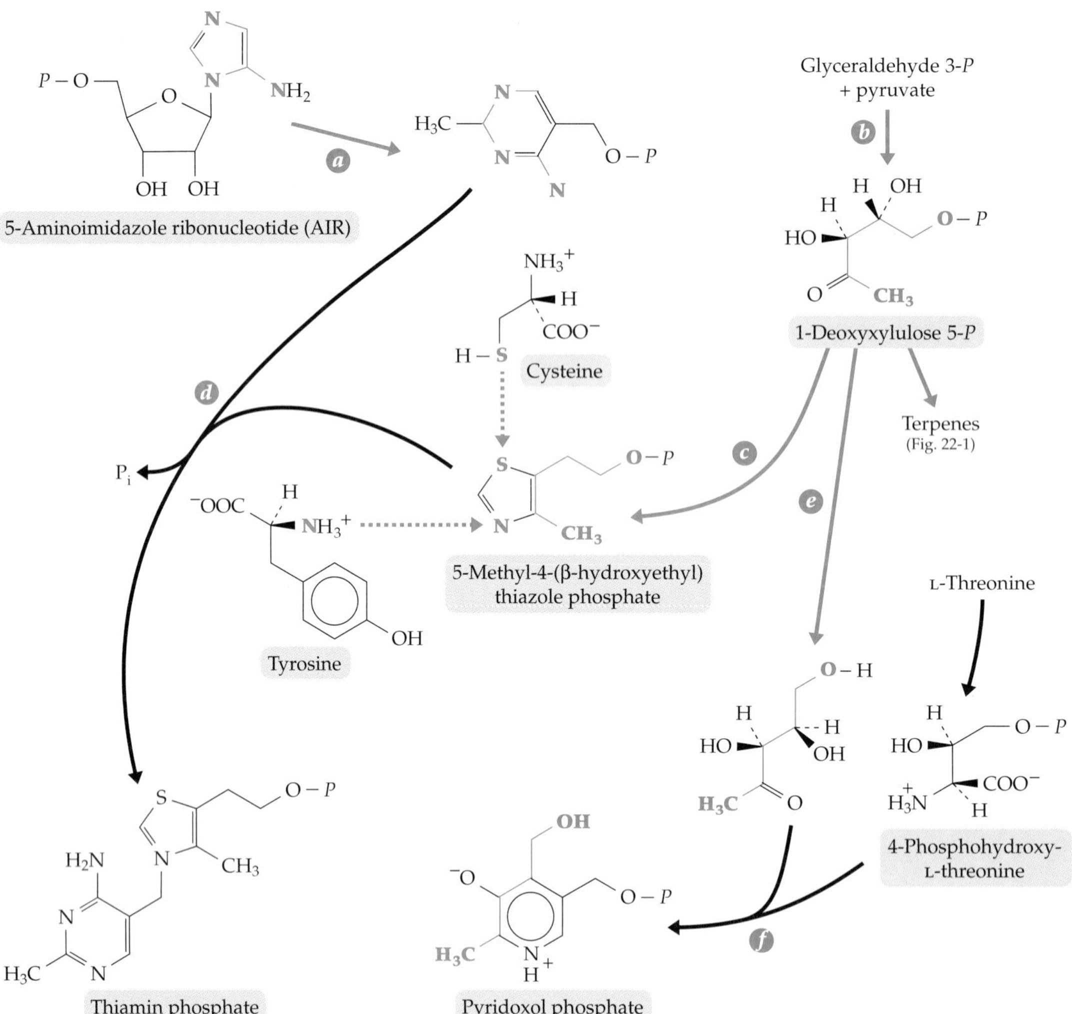

Figure 25-21 Proposed pathways for biosynthesis of thiamin phosphate and pyridoxol phosphate.

way shown in Fig. 25-21 exist in yeast and other fungi.[378b,c] Interconversion of pyridoxal phosphate and other forms of vitamin B_6 is discussed in Chapter 14.

The degradation of pyridoxol by bacteria has been investigated in detail.[379–381] In one pathway the hydroxymethyl group in the 5 position and the substituent in the 4 position are both oxidized in the early steps to carboxylate groups. Then, as indicated in Eq. 25-24, a decarboxylation is followed by the action of an unusual dioxygenase.

Isolated from a strain of *Pseudomonas*, this enzyme contains bound FAD, which must be reduced by external NADH. Like a typical dioxygenase the enzyme introduces two atoms of oxygen into the product. However, it also uses the reduced FAD to reduce the double bond system (either before or after the attack by oxygen).[381] Another enzyme of the same bacterium is remarkable in hydrolyzing the product of the oxygenation reaction to four different products without the accumulation of intermediates.

(25-24)

References

1. Floss, H. G. (1986) in *Recent Advances in Phytochemistry*, Vol. 20 (Conn, E. E., ed), pp. 13–55, Plenum, New York
2. Amrhein, N. (1986) in *Recent Advances in Phytochemistry*, Vol. 20 (Conn, E. E., ed), pp. 83–117, Plenum, New York
3. Marzabadi, M. R., Gruys, K. J., Pansegrau, P. D., Walker, M. C., Yuen, H. K., and Sikorski, J. A. (1996) *Biochemistry* **35**, 4199–4210
4. Davis, B. D. (1950) *Experientia* **6**, 4–50
5. Davis, B. D. (1950) *J. Biol. Chem.* **191**, 315–325
6. Weiss, U., and Edwards, J. M. (1980) *The Biosynthesis of Aromatic Compounds*, Wiley, New York
7. Pittard, A. J. (1987) in *Escherichia coli and Salmonella typhimurium*, Vol. I (Neidhardt, F. C., ed), pp. 368–394, Am. Soc. for Microbiology, Washington, DC
8. Gollub, E., Zalkin, H., and Sprinson, D. B. (1967) *J. Biol. Chem.* **242**, 5323–5328
9. Shultz, J., Hermodson, M. A., Garner, C. C., and Herrmann, K. M. (1984) *J. Biol. Chem.* **259**, 9655–9661
10. DeLeo, A. B., Dayan, J., and Sprinson, D. B. (1973) *J. Biol. Chem.* **248**, 2344–2353

10a. Yu, T.-W., Müller, R., Müller, M., Zhang, X., Draeger, G., Kim, C.-G., Leistner, E., and Floss, H. G. (2001) *J. Biol. Chem.* **276**, 12546–12555

10b. Guo, J., and Frost, J. W. (2002) *J. Am. Chem. Soc.* **124**, 10642–10643

10c. Eads, J. C., Beeby, M., Scapin, G., Yu, T.-W., and Floss, H. G. (1999) *Biochemistry* **38**, 9840–9849

10d. Arakawa, K., Müller, R., Mahmud, T., Yu, T-W., and Floss, H. G. (2002) *J. Am. Chem. Soc.* **124**, 10644 – 10645

11. Shultz, J., Hermodson, M. A., Garner, C. C., and Harrmann, K. M. (1984) *J. Biol. Chem.* **259**, 9655–9661
12. Dyer, W. E., Weaver, L. M., Zhao, J., Kuhn, D. N., Weller, S. C., and Harrmann, K. M. (1990) *J. Biol. Chem.* **265**, 1608–1614

12a. Wagner, T., Shumilin, I. A., Bauerle, R., and Kretsinger, R. H. (2000) *J. Mol. Biol.* **301**, 389–399

13. Herrmann, K. M., Schultz, J., and Hermodson, M. A. (1980) *J. Biol. Chem.* **255**, 7079–7081
14. Montchamp, J.-L., and Frost, J. W. (1997) *J. Am. Chem. Soc.* **119**, 7645–7653
15. Carpenter, E. P., Hawkins, A. R., Frost, J. W., and Brown, K. A. (1998) *Nature (London)* **394**, 299–302
16. Maitra, U. S., and Sprinson, D. B. (1978) *J. Biol. Chem.* **253**, 5426–5430
17. Widlanski, T., Bender, S. L., and Knowles, J. R. (1989) *Biochemistry* **28**, 7572–7582
18. Ranjeva, R., Refeno, G., Boudet, A. M., and Marmé, D. (1983) *Proc. Natl. Acad. Sci. U.S.A.* **80**, 5222–5224
19. Chaudhuri, S., and Coggins, J. R. (1985) *Biochem. J.* **226**, 217–223
20. Huang, L., Montoya, A. L., and Nester, E. W. (1975) *J. Biol. Chem.* **250**, 7675–7681
21. Krell, T., Coggins, J. R., and Lapthorn, A. J. (1998) *J. Mol. Biol.* **278**, 983–997
22. Ramilo, C., Appleyard, R. J., Wanke, C., Krekel, F., Amrhein, N., and Evans, J. N. S. (1994) *Biochemistry* **33**, 15071–15079
23. Kim, D. H., Tucker-Kellogg, G. W., Lees, W. J., and Walsh, C. T. (1996) *Biochemistry* **35**, 5435–5440
24. Lee, J. J., Asano, Y., Shiek, T.-L., Spreafico, F., Lee, K., and Floss, H. G. (1984) *J. Am. Chem. Soc.* **106**, 3367–3368
25. Jakeman, D. L., Mitchell, D. J., Shuttleworth, W. A., and Evans, J. N. S. (1998) *Biochemistry* **37**, 12012–12019
26. Stallings, W. C., Abdel-Meguid, S. S., Lim, L. W., Shieh, H.-S., Dayringer, H. E., Leimgruber, N. K., Stegeman, R. A., Anderson, K. S., Sikorski, J. A., Radgette, S. R., and Kishore, G. M. (1991) *Proc. Natl. Acad. Sci. U.S.A.* **88**, 5046–5050
27. Christensen, A. M., and Schaefer, J. (1993) *Biochemistry* **32**, 2868–2873

27a. Alibhai, M. F., and Stallings, W. C. (2001) *Proc. Natl. Acad. Sci. U.S.A.* **98**, 2944–2946

28. Stalker, D. M., Hiah, W. R., and Comai, L. (1985) *J. Biol. Chem.* **260**, 4724–4728
29. Shuttleworth, W. A., Pohl, M. E., Helms, G. L., Jakeman, D. L., and Evans, J. N. S. (1999) *Biochemistry* **38**, 296–302
30. Morrell, H., Clark, M. J., Knowles, P. F., and Sprinson, D. B. (1967) *J. Biol. Chem.* **242**, 82–90
31. Balasubramanian, S., Coggins, J. R., and Abell, C. (1995) *Biochemistry* **34**, 341–348
32. Macheroux, P., Petersen, J., Bornemann, S., Lowe, D. J., and Thorneley, R. N. F. (1996) *Biochemistry* **35**, 1643–1652
33. Lauhon, C. T., and Bartlett, P. A. (1994) *Biochemistry* **33**, 14100–14108

33a. Osborne, A., Thorneley, R. N. F., Abell, C., and Bornemann, S. (2000) *J. Biol. Chem.* **275**, 35825–35830

33b. Kitzing, K., Macheroux, P., and Amrhein, N. (2001) *J. Biol. Chem.* **276**, 42658–42666

34. Hudson, G. S., Wong, V., and Davidson, B. E. (1984) *Biochemistry* **23**, 6240–6249
35. Hermes, J. D., Tipton, P. A., Fisher, M. A., O'Leary, M. H., Morrison, J. F., and Cleland, W. W. (1984) *Biochemistry* **23**, 6263–6275
36. Christendat, D., Saridakis, V. C., and Turnbull, J. L. (1998) *Biochemistry* **37**, 15703–15712
37. MacBeath, G., Kast, P., and Hilvert, D. (1998) *Biochemistry* **37**, 10062–10073

37a. Guo, H., Cui, Q., Lipscomb, W. N., and Karplus, M. (2001) *Proc. Natl. Acad. Sci. U.S.A.* **98**, 9032–9037

38. Sogo, S. G., Widlanski, T. S., Hoare, J. H., Grimshaw, C. E., Berchtold, G. A., and Knowles, J. R. (1984) *J. Am. Chem. Soc.* **106**, 2701–2703

38a. Pohnert, G., Zhang, S., Husain, A., Wilson, D. B., and Ganem, B. (1999) *Biochemistry* **38**, 12212–12217

39. Christendat, D., and Turnbull, J. L. (1999) *Biochemistry* **38**, 4782–4793
40. Christendat, D., and Turnbull, J. (1996) *Biochemistry* **35**, 4468–4479
41. Turnbull, J., Cleland, W. W., and Morrison, J. F. (1990) *Biochemistry* **29**, 10245–10254
42. Turnbull, J., Morrison, J. F., and Cleland, W. W. (1991) *Biochemistry* **30**, 7783–7788
43. Zamir, L. O., Jensen, R. A., Arison, B. H., Douglas, A. W., Albers-Schönberg, G., and Bowen, J. R. (1980) *J. Am. Chem. Soc.* **102**, 4499–4504
44. Zamir, L. O., Tiberio, R., Fiske, M., Berry, A., and Jensen, R. A. (1985) *Biochemistry* **24**, 1607–1612
45. Xia, T., and Jensen, R. A. (1990) *J. Biol. Chem.* **265**, 20033–20036
46. Jung, E., Zamir, L. O., and Jensen, R. A. (1986) *Proc. Natl. Acad. Sci. U.S.A.* **83**, 7231–7235
47. Jensen, R. A. (1986) in *Recent Advances in Phytochemistry*, Vol. 20 (Conn, E. E., ed), pp. 57–81, Plenum, New York
48. Teng, C.-Y. P., and Ganem, B. (1984) *J. Am. Chem. Soc.* **106**, 2463–2464
49. Morollo, A. A., Finn, M. G., and Bauerle, R. (1993) *J. Am. Chem. Soc.* **115**, 816–817
50. Green, J. M., and Nichols, B. P. (1991) *J. Biol. Chem.* **266**, 12971–12975
51. Paluh, J. L., Zalkin, H., Betsch, D., and Weith, H. L. (1985) *J. Biol. Chem.* **260**, 1889–1894
52. Walsh, C. T., Erion, M. D., Walts, A. E., Delany III, J. J., and Berchtold, G. A. (1987) *Biochemistry* **26**, 4734–4745
53. Roux, B., and Walsh, C. T. (1992) *Biochemistry* **31**, 6904–6910

53a. Knöchel, T., Ivens, A., Hester, G., Gonzalez, A., Bauerle, R., Wilmanns, M., Kirschner, K., and Jansonius, J. N. (1999) *Proc. Natl. Acad. Sci. U.S.A.* **96**, 9479–9484

54. Liu, D. J., and Day, L. A. (1994) *Science* **265**, 671–674
55. Liu, J., Quinn, N., Berchtold, G. A., and Walsh, C. T. (1990) *Biochemistry* **29**, 1417–1425
56. Kozlowski, M. C., and Bartlett, P. A. (1991) *J. Am. Chem. Soc.* **113**, 5897–5898

56a. Wildermuth, M. C., Dewdney, J., Wu, G., and Ausubel, F. M. (2001) *Nature (London)* **414**, 562–565

57. Sakaitani, M., Rusnak, F., Quinn, N. R., Tu, C., Frigo, T. B., Berchtold, G. A., and Walsh, C. T. (1990) *Biochemistry* **29**, 6789–6798
58. Rusnak, F., Liu, J., Quinn, N., Berchtold, G. A., and Walsh, C. T. (1990) *Biochemistry* **29**, 1425–1435
59. Gehring, A. M., Mori, I., and Walsh, C. T. (1998) *Biochemistry* **37**, 2648–2659

59a. May, J. J., Wendrich, T. M., and Marahiel, M. A. (2001) *J. Biol. Chem.* **276**, 7209–7217

60. Schubert, K. R., Switzer, R. L., and Shelton, E. (1975) *J. Biol. Chem.* **250**, 7492–7500
61. Nosal, J. M., Switzer, R. L., and Becker, M. A. (1993) *J. Biol. Chem.* **268**, 10168–10175
62. Zalkin, H., Paluh, J. L., van Cleemput, M., Moye, W. S., and Yanofsky, C. (1984) *J. Biol. Chem.* **259**, 3985–3992
63. Eberhard, M., Tsai-Pflugfelder, M., Bolewska, K., Hommel, U., and Kirschner, K. (1995) *Biochemistry* **34**, 5419–5428
64. Knöchel, T. R., Hennig, M., Merz, A., Darimont, B., Kirschner, K., and Jansonius, J. N. (1996) *J. Mol. Biol.* **262**, 502–515
65. Hyde, C. C., Ahmed, S. A., Padlan, E. A., Miles, E. W., and Davies, D. R. (1988) *J. Biol. Chem.* **263**, 17857–17871
66. Schneider, T. R., Gerhardt, E., Lee, M., Liang, P.-H., Anderson, K. S., and Schlichting, I. (1998) *Biochemistry* **37**, 5394–5406
67. Rhee, S., Miles, E. W., and Davies, D. R. (1998) *J. Biol. Chem.* **273**, 8553–8555
68. Hyde, C. C., and Miles, E. W. (1990) *Bio/Technology* **8**, 27–32

68a. Miles, E. W. (2001) *Chem. Record* **1**, 140–151

68b. Weber-Ban, E., Hur, O., Bagwell, C., Banik, U., Yang, L.-H., Miles, E. W., and Dunn, M. F. (2001) *Biochemistry* **40**, 3497–3511

68c. Yamagata, Y., Ogasahara, K., Hioki, Y., Lee, S. J., Nakagawa, A., Nakamura, H., Ishida, M., Kuramitsu, S., and Yutani, K. (2001) *J. Biol. Chem.* **276**, 11062–11071

69. Sterner, R., Dahm, A., Darimont, B., Ivens, A., Liebl, W., and Kirschner, K. (1995) *EMBO J.* **14**, 4395–4402
70. Wilmanns, M., Hyde, C. C., Davies, D. R., Kirschner, K., and Jansonius, J. N. (1991) *Biochemistry* **30**, 9161–9169
71. Zhao, J., and Last, R. L. (1995) *J. Biol. Chem.* **270**, 6081–6087
72. Bentley, R., and Campbell, I. M. (1974) in *The Chemistry of the Quinonoid Compounds* (Patai, S., ed), pp. 683–736, Wiley, New York (Part 2)
73. Sippel, C. J., Goewert, R. R., Slachman, F. N., and Olson, R. E. (1983) *J. Biol. Chem.* **258**, 1057–1061

73a. Jonassen, T., and Clarke, C. F. (2000) *J. Biol. Chem.* **275**, 12381–12387

73b. Stenmark, P., Grünler, J., Mattsson, J., Sindelar, P. J., Nordlund, P., and Berthold, D. A. (2001) *J. Biol. Chem.* **276**, 33297–33300

References

73c. Miyadera, H., Amino, H., Hiraishi, A., Taka, H., Murayama, K., Miyoshi, H., Sakamoto, K., Ishii, N., Hekimi, S., and Kita, K. (2001) *J. Biol. Chem.* **276**, 7713–7716

73d. Hihi, A. K., Gao, Y., and Hekimi, S. (2002) *J. Biol. Chem.* **277**, 2202–2206

74. Marbois, B. N., and Clarke, C. F. (1996) *J. Biol. Chem.* **271**, 2995–3004

75. Hsu, A. Y., Poon, W. W., Shepherd, J. A., Myles, D. C., and Clarke, C. F. (1996) *Biochemistry* **35**, 9797–9806

76. Clarke, C. F., Williams, W., and Teruya, J. H. (1991) *J. Biol. Chem.* **266**, 16636 – 16644

77. Swiezewska, E., Dallner, G., Andersson, B., and Ernster, L. (1993) *J. Biol. Chem.* **268**, 1494–1499

78. Kalén, A., Appelkvist, E.-L., Chojnacki, T., and Dallner, G. (1990) *J. Biol. Chem.* **265**, 1158–1164

79. Folkers, K., Langsjoen, P., Willis, R., Richardson, P., Xia, L.-J., Ye, C.-Q., and Tamagawa, H. (1990) *Proc. Natl. Acad. Sci. U.S.A.* **87**, 8931–8934

80. Threfall, D. R. (1971) *Vitamins and Hormones* **29**, 153–200

81. Goodwin, T. W., and Merces, E. I. (1986) *Introduction to Plant Biochemistry*, 2nd ed., Pergamon, Oxford

82. Fernández-Cañón, J. M., and Peñalva, M. A. (1995) *J. Biol. Chem.* **270**, 21199–21205

83. Shintani, D., and DellaPenna, D. (1998) *Science* **282**, 2098–2100

83a. Yamamoto, Y., Fujisawa, A., Hara, A., and Dunlap, W. C. (2001) *Proc. Natl. Acad. Sci. U.S.A.* **98**, 13144–13148

84. Bentley, R., and Meganathan, R. (1987) in *Escherichia coli and Salmonella Typhimurium* (Nedihardt, F. C., ed), pp. 512–520, Am. Soc. for Microbiology, Washington, D.C.

85. Koike-Takeshita, A., Koyama, T., and Ogura, K. (1997) *J. Biol. Chem.* **272**, 12380–12383

86. Palmer, D. R. J., Garrett, J. B., Sharma, V., Meganathan, R., Babbitt, P. C., and Gerlt, J. A. (1999) *Biochemistry* **38**, 4252–4258

86a. Stamps, S. L., Taylor, A. B., Wang, S. C., Hackert, M. L., and Whitman, C. P. (2000) *Biochemistry* **39**, 9671–9678

87. Daubner, S. C., and Fitzpatrick, P. F. (1999) *Biochemistry* **38**, 4448–4454

87a. Andersen, O. A., Flatmark, T., and Hough, E. (2001) *J. Mol. Biol.* **314**, 279–291

88. Nyhan, W. L. (1984) *Trends Biochem. Sci.* **9**, 71–72

89. Knox, W. E. (1972) in *The Metabolic Basis of Inherited Disease*, 3rd ed. (Standbury, J. B., Wyngaarden, J. B., and Fredrickson, D. S., eds), pp. 266–295, McGraw-Hill, New York

90. Woo, S. L. C. (1989) *Biochemistry* **28**, 1–7

91. Kaufman, S. (1999) *Proc. Natl. Acad. Sci. U.S.A.* **96**, 3160–3164

91a. Levy, H. L. (1999) *Proc. Natl. Acad. Sci. U.S.A.* **96**, 1811–1813

91b. Gámez, A., Pérez, B., Ugarte, M., and Desviat, L. R. (2000) *J. Biol. Chem.* **275**, 29737–29742

92. Ambrus, C. M., Ambrus, J. L., Horvath, C., Pederson, H., Skarma, S., Kant, C., Mirand, E., Guthrie, R., and Paul, T. (1978) *Science* **201**, 837–839

93. Scriver, C. R., Kaufman, S., Eisensmith, R. C., and Woo, S. L. C. (1995) in *The Metabolic and Molecular Bases of Inherited Disease*, 7th ed., Vol. 1 (Scriver, C. R., Beaudet, A. L., Sly, W. S., and Valle, D., eds), pp. 1015–1075, McGraw-Hill, New York

94. Onuffer, J. J., Ton, B. T., Klement, I., and Kirsch, J. F. (1995) *Protein Sci.* **4**, 1743–1749

95. Natt, E., Kida, K., Odievre, M., Di Rocco, M., and Scherer, G. (1992) *Proc. Natl. Acad. Sci. U.S.A.* **89**, 9297–9301

96. Hargrove, J. L., and Grammer, D. K. (1985) in *Transaminases* (Christen, P., and Metzler, D. E., eds), pp. 511–525, Wiley, New York

97. Wilcken, B., Hammond, J. W., Howard, N., Bohane, T., Hocart, C., and Halpern, B. (1981) *N. Engl. J. Med.* **305**, 865–869

98. La Du, B. N. (1995) in *The Metabolic and Molecular Bases of Inherited Disease*, 7th ed., Vol. 1 (Scriver, C. R., Beaudet, A. L., Sly, W. S., and Valle, D., eds), pp. 1371–1386, McGraw-Hill, New York

99. Fernández-Cañón, J. M., and Peñalva, M. A. (1998) *J. Biol. Chem.* **273**, 329–337

99a. Polekhina, G., Board, P. G., Blackburn, A. C., and Parker, M. W. (2001) *Biochemistry* **40**, 1567–1576

100. Kubo, S., Sun, M., Miyahara, M., Umeyama, K., Urakami, K.-i, Yamamoto, T., Jakobs, C., Matsuda, I., and Endo, F. (1998) *Proc. Natl. Acad. Sci. U.S.A.* **95**, 9552–9557

101. Frieden, E. (1981) *Trends Biochem. Sci.* **6**, 50–53

102. Tata, J. R. (1990) *Trends Biochem. Sci.* **15**, 282–284

103. Levy, O., De la Vieja, A., Ginter, C. S., Riedel, C., Dai, G., and Carrasco, N. (1998) *J. Biol. Chem.* **273**, 22657–22663

104. Sun, W., and Dunford, H. B. (1993) *Biochemistry* **32**, 1324–1331

105. Surks, M. I., and Sievert, R. (1995) *N. Engl. J. Med.* **333**, 1688–1693

106. Rawtich, A. B., Chernoff, S. B., Litwer, M. R., Rouse, J. B., and Hamilton, J. W. (1983) *J. Biol. Chem.* **258**, 2079–2082

107. Gentile, F., Ferranti, P., Mamone, G., Malorni, A., and Salvatore, G. (1997) *J. Biol. Chem.* **272**, 639–646

108. Gavaret, J.-M., Cahnmann, H. J., and Nunez, J. (1981) *J. Biol. Chem.* **256**, 9167–9173

108a. Ma, Y.-A., Sih, C. J., and Harms, A. (1999) *J. Am. Chem. Soc.* **121**, 8967–8968

109. Shupnik, M. A., Chin, W. W., Ross, D. S., Downing, M. F., Habener, J. F., and Ridgway, E. C. (1983) *J. Biol. Chem.* **258**, 15120–15124

110. Van Heuverswyn, B., Streydio, C., Brocas, H., Refetoff, S., Dumont, J., and Vassart, G. (1984) *Proc. Natl. Acad. Sci. U.S.A.* **81**, 5941–5945

111. Perlman, J. H., Colson, A.-O., Wang, W., Bence, K., Osman, R., and Gershengorn, M. C. (1997) *J. Biol. Chem.* **272**, 11937–11942

112. Kosugi, S., Okajima, F., Ban, T., Hidaka, A., Shenker, A., and Kohn, L. D. (1992) *J. Biol. Chem.* **267**, 24153–24156

113. Nettleton, E. J., Sunde, M., Lai, Z., Kelly, J. W., Dobson, C. M., and Robinson, C. V. (1998) *J. Mol. Biol.* **281**, 553–564

114. Sebastiao, M. P., Saraiva, M. J., and Damas, A. M. (1998) *J. Biol. Chem.* **273**, 24715–24722

115. Silva, J. E., and Larsen, P. R. (1983) *Nature (London)* **305**, 712–713

116. Pallud, S., Lennon, A.-M., Ramauge, M., Gavaret, J.-M., Croteau, W., Pierre, M., Courtin, F., and St. Germain, D. L. (1997) *J. Biol. Chem.* **272**, 18104–18110

116a. Leonard, J. L., Visser, T. J., and Leonard, D. M. (2001) *J. Biol. Chem.* **276**, 2600–2607

117. Maley, G. F., and Lardy, H. A. (1955) *J. Biol. Chem.* **215**, 377–388

118. Shi, Y.-B., and Brown, D. D. (1993) *J. Biol. Chem.* **268**, 20312–20317

119. Brown, D. D. (1997) *Proc. Natl. Acad. Sci. U.S.A.* **94**, 13011–13016

120. Kumara-Siri, M. H., Shapiro, L. E., and Surks, M. I. (1986) *J. Biol. Chem.* **261**, 2844–2852

121. Apriletti, J. W., David-Inouye, Y., Eberhardt, N. L., and Baxter, J. D. (1984) *J. Biol. Chem.* **259**, 10941–10948

122. Brent, G. A. (1994) *N. Engl. J. Med.* **331**, 847–853

123. Wagner, R. L., Apriletti, J. W., McGrath, M. E., West, B. L., Baxter, J. D., and Fletterick, R. J. (1995) *Nature (London)* **378**, 690–697

123a. Wu, Y., Xu, B., and Koenig, R. J. (2001) *J. Biol. Chem.* **276**, 3929–3936

123b. Zhang, Y., Yin, L., and Hillgartner, F. B. (2001) *J. Biol. Chem.* **276**, 974–983

124. Hörlein, A. J., Näär, A. M., Heinzel, T., Torchia, J., Gloss, B., Kurokawa, R., Ryan, A., Kamei, Y., Söderström, M., Glass, C. K., and Rosenfeld, M. G. (1995) *Nature (London)* **377**, 397–404

125. Chen, Y., Chen, P.-L., Chen, C.-F., Sharp, Z. D., and Lee, W.-H. (1999) *Proc. Natl. Acad. Sci. U.S.A.* **96**, 4443–4448

125a. Zhu, X.-G., Hanover, J. A., Hager, G. L., and Cheng, S. (1998) *J. Biol. Chem.* **273**, 27058–27063

125b. Tagami, T., Park, Y., and Jameson, J. L. (1999) *J. Biol. Chem.* **274**, 22345–22353

125c. Riedel, C., Dohán, O., De la Vieja, A., Ginter, C. S., and Carrasco, N. (2001) *Trends Biochem. Sci.* **26**, 490–496

126. Shaw-White, J. R., Bruno, M. D., and Whitsett, J. A. (1999) *J. Biol. Chem.* **274**, 2658–2664

127. Vassart, G., Dumont, J. E., and Refetoff, S. (1995) in *The Metabolic and Molecular Bases of Inherited Disease*, 7th ed., Vol. 1 (Scriver, C. R., Beaudet, A. L., Sly, W. S., and Valle, D., eds), pp. 2883–2928, McGraw-Hill, New York

128. Collingwood, T. N., and and 19 other authors. (1998) *EMBO J.* **17**, 4760–4770

129. Hetzel, B. S. (1994) *N. Engl. J. Med.* **331**, 1770–1771

130. Hearn, M. T. W. (1980) *Trends Biochem. Sci.* **5**, 75–79

131. Utiger, R. D. (1991) *N. Engl. J. Med.* **325**, 278–279

131a. Bertoldi, M., Frigeri, P., Paci, M., and Voltattorni, C. B. (1999) *J. Biol. Chem.* **274**, 5514 – 5521

132. Potterf, S. B., Muller, J., Bernardini, I., Tietze, F., Kobayashi, T., Hearing, V. J., and Gahl, W. A. (1996) *J. Biol. Chem.* **271**, 4002–4008

133. Jiménez-Cervantes, C., Solano, F., Kobayashi, T., Urabe, K., Hearing, V. J., Lozano, J. A., and García-Borrón, J. C. (1994) *J. Biol. Chem.* **269**, 17993–18001

134. Leonard, L. J., Townsend, D., and King, R. A. (1988) *Biochemistry* **27**, 6156–6159

135. Sugimoto, H., Taniguchi, M., Nakagawa, A., Tanaka, I., Suzuki, M., and Nishihira, J. (1999) *Biochemistry* **38**, 3268–3279

136. Rodríguez-López, J. N., Tudela, J., Varón, R., García-Carmona, F., and García-Cánovas, F. (1992) *J. Biol. Chem.* **267**, 3801–3810

137. Thomson, R. H. (1974) *Angew. Chem. Int. Ed. Engl.* **13**, 305–312

138. Deibel, R. M. B., and Chedekel, M. R. (1984) *J. Am. Chem. Soc.* **106**, 5884–5888

139. Pawelek, J., Körner, A., Bergstrom, A., and Bologna, J. (1980) *Nature (London)* **286**, 617–619

140. Chen, J. M., Xu, S. L., Wawrzak, Z., Basarab, G. S., and Jordan, D. B. (1998) *Biochemistry* **37**, 17735–17744

140a. Liao, D.-I., Thompson, J. E., Fahnestock, S., Valent, B., and Jordan, D. B. (2001) *Biochemistry* **40**, 8696–8704

141. Saul, S. J., and Sugumaran, M. (1990) *J. Biol. Chem.* **265**, 16992–16999

141a. Hall, M., Scott, T., Sugumaran, M., Söderhäll, K., and Law, J. H. (1995) *Proc. Natl. Acad. Sci. U.S.A.* **92**, 7764–7768

141b. Asano, T., and Ashida, M. (2001) *J. Biol. Chem.* **276**, 11100–11112

142. Sugumaran, M., Semensi, V., Kalyanaraman, B., Bruce, J. M., and Land, E. J. (1992) *J. Biol. Chem.* **267**, 10355–10361

References

143. McCormick, M. L., Gaut, J. P., Lin, T.-S., Britigan, B. E., Buettner, G. R., and Heinecke, J. W. (1998) *J. Biol. Chem.* **273**, 32030–32037
144. Malencik, D. A., and Anderson, S. R. (1994) *Biochemistry* **33**, 13363–13372
144a. Kanwar, R., and Balasubramanian, D. (2000) *Biochemistry* **39**, 14976–14983
145. Briza, P., Winkler, G., Kalchhauser, H., and Breitenbach, M. (1986) *J. Biol. Chem.* **261**, 4288–4294
146. Briza, P., Ellinger, A., Winkler, G., and Breitenbach, M. (1990) *J. Biol. Chem.* **265**, 15118–15123
147. Taylor, S. W., Waite, J. H., Ross, M. M., Shabanowitz, J., and Hunt, D. F. (1994) *J. Am. Chem. Soc.* **116**, 10803–10804
148. Waite, J. H., Jensen, R. A., and Morse, D. E. (1992) *Biochemistry* **31**, 5733–5738
149. Papov, V. V., Diamond, T. V., Biemann, K., and Waite, J. H. (1995) *J. Biol. Chem.* **270**, 20183–20192
149a. Yu, M., Hwang, J., and Deming, T. (1999) *J. Am. Chem. Soc.* **121**, 5825–5826
150. Dagley, S. (1978) *Naturwissenschaften* **65**, 85–95
151. Allewell, N. (1989) *Trends Biochem. Sci.* **14**, 473–474
152. Dagley, S., and Nicholson, D. E. (1970) *An Introduction to Metabolic Pathways*, Wiley, New York
153. Gibson, D. T., ed. (1984) *Microbial Degradation of Organic Compounds*, Dekker, New York
154. Nakai, C., Kagamiyama, H., Nozaki, M., Nakazawa, T., Inouye, S., Ebina, Y., and Nakazawa, A. (1983) *J. Biol. Chem.* **258**, 2923–2928
155. Correll, C. C., Batie, C. J., Ballou, D. P., and Ludwig, M. L. (1985) *J. Biol. Chem.* **260**, 14633–14635
155a. Palumbo, A., Poli, A., Cosmo, A. D., and d'Ischia, M. (2000) *J. Biol. Chem.* **275**, 16885–16890
156. Blasco, R., Wittich, R.-M., Mallavarapu, M., Timmis, K. N., and Pieper, D. H. (1995) *J. Biol. Chem.* **270**, 29229–29235
157. Doelle, H. W. (1975) *Bacterial Metabolism*, Academic Press, New York
158. Werlen, C., Kohler, H.-P. E., and van der Meer, J. R. (1996) *J. Biol. Chem.* **271**, 4009–4016
158a. Copley, S. D. (2000) *Trends Biochem. Sci.* **25**, 261–265
158b. Anandarajah, K., Kiefer, P. M., Donohoe, B. S., and Copley, S. D. (2000) *Biochemistry* **39**, 5303–5311
158c. Tame, J. R. H., Namba, K., Dodson, E. J., and Roper, D. I. (2002) *Biochemistry* **41**, 2982–2989
159. Seah, S. Y. K., Terracina, G., Bolin, J. T., Riebel, P., Snieckus, V., and Eltis, L. D. (1998) *J. Biol. Chem.* **273**, 22943–22949
160. Scholten, J. D., Chang, K.-H., Babbitt, P. C., Charest, H., Sylvestre, M., and Dunaway-Mariano, D. (1991) *Science* **253**, 182–185
160a. Zhang, W., Wei, Y., Luo, L., Taylor, K. L., Yang, G., and Dunaway-Mariano, D. (2001) *Biochemistry* **40**, 13474–13482
161. Harpel, M. R., and Lipscomb, J. D. (1990) *J. Biol. Chem.* **265**, 22187–22196
161a. Olivera, E. R., Minambres, B., Garcia, B., Muniz, C., Moreno, M. A., Ferrández, A., Díaz, E., García, J. L., and Luengo, J. M. (1998) *Proc. Natl. Acad. Sci. U.S.A.* **95**, 6419–6424
161b. Zaar, A., Eisenreich, W., Bacher, A., and Fuchs, G. (2001) *J. Biol. Chem.* **276**, 24997–25004
162. Yamaguchi, M., and Fujisawa, H. (1982) *J. Biol. Chem.* **257**, 12497–12502
163. Zylstra, G. J., and Gibson, D. T. (1989) *J. Biol. Chem.* **264**, 14940–14946
164. Díaz, E., and Timmis, K. N. (1995) *J. Biol. Chem.* **270**, 6403–6411
165. Petsko, G. A., Kenyon, G. L., Gerlt, J. A., Ringe, D., and Kozarich, J. W. (1993) *Trends Biochem. Sci.* **18**, 372–376
166. Harwood, C. S., and Gibson, J. (1997) *J. Bacteriol.* **179**, 301–309
166a. Boll, M., Laempe, D., Eisenreich, W., Bacher, A., Mittelberger, T., Heinze, J., and Fuchs, G. (2000) *J. Biol. Chem.* **275**, 21889–21895
166b. Boll, M., Fuchs, G., and Lowe, D. J. (2001) *Biochemistry* **40**, 7612–7620
167. Härtel, U., Eckel, E., Koch, J., Fuchs, G., Linder, D., and Buckel, W. (1993) *Arch Microbiol* **159**, 174–181
168. Draths, K. M., Ward, T. L., and Frost, J. W. (1992) *J. Am. Chem. Soc.* **114**, 9725–9726
169. Richman, J. E., Chang, Y.-C., Kambourakis, S., Draths, K. M., Almy, E., Snell, K. D., Strasburg, G. M., and Frost, J. W. (1996) *J. Am. Chem. Soc.* **118**, 11587–11591
170. Draths, K. M., and Frost, J. W. (1991) *J. Am. Chem. Soc.* **113**, 9361–9363
171. Haslam, E. (1986) in *Recent Advances in Phytochemistry*, Vol. 20 (Conn, E. E., ed), pp. 163–200, Plenum, New York
172. Werner, I., Bacher, A., and Eisenreich, W. (1997) *J. Biol. Chem.* **272**, 25474–25482
173. Lauvergeat, V., Kennedy, K., Feuillet, C., McKie, J. H., Gorrichon, L., Baltas, M., Boudet, A. M., Grima-Pettenati, J., and Douglas, K. T. (1995) *Biochemistry* **34**, 12426–12434
174. Ellis, B. E., Kuroki, G. W., and Stafford, H. A., eds. (1994) *Genetic Engineering of Plant Secondary Metabolism*, Plenum, New York
174a. Crouteau, R., Kutchan, T. M., and Lewis, N. G. (2000) in *Biochemistry and Molecular Biology of Plants* (Buchanan, B., Gruissem, W., and Jones, R., eds), pp. 1250–1318, Am. Soc. Plant Physiologists, Rockville, Maryland
175. Schwede, T. F., Rétey, J., and Schulz, G. E. (1999) *Biochemistry* **38**, 5355–5361
175a. Stuible, H.-P., and Kombrink, E. (2001) *J. Biol. Chem.* **276**, 26893–26897
176. Becker-André, M., Schulze-Lefert, P., and Hahlbrock, K. (1991) *J. Biol. Chem.* **266**, 8551–8559
177. León, J., Shulaev, V., Yalpani, N., Lawton, M. A., and Raskin, I. (1995) *Proc. Natl. Acad. Sci. U.S.A.* **92**, 10413–10417
178. Lee, H.-i, León, J., and Raskin, I. (1995) *Proc. Natl. Acad. Sci. U.S.A.* **92**, 4076–4079
179. Delaney, T. P., Uknes, S., Vernooij, B., Friedrich, L., Weymann, K., Negrotto, D., Gaffney, T., Gut-Rella, M., Kessmann, H., Ward, E., and Ryals, J. (1994) *Science* **266**, 1247–1250
180. Chen, Z., Malamy, J., Henning, J., Conrath, U., Sánchez-Casas, P., Silva, H., Ricigliano, J., and Klessig, D. F. (1995) *Proc. Natl. Acad. Sci. U.S.A.* **92**, 4134–4137
181. Durner, J., and Klessig, D. F. (1996) *J. Biol. Chem.* **271**, 28492–28501
182. Durner, J., and Klessig, D. F. (1995) *Proc. Natl. Acad. Sci. U.S.A.* **92**, 11312–11316
183. Jurivich, D. A., Pachetti, C., Qiu, L., and Welk, J. F. (1995) *J. Biol. Chem.* **270**, 24489–24495
184. Prince, R. C., and Gunson, D. E. (1994) *Trends Biochem. Sci.* **19**, 521
185. Li, K., and Frost, J. W. (1998) *J. Am. Chem. Soc.* **120**, 10545–10546
186. Rosen, M. A., Shapiro, L., and Patel, D. J. (1992) *Biochemistry* **31**, 4015–4026
187. Villegas, R. J. A., and Kojima, M. (1986) *J. Biol. Chem.* **261**, 8729–8733
188. von Röpenack, E., Parr, A., and Schulze-Lefert, P. (1998) *J. Biol. Chem.* **273**, 9013–9022
189. Balint, R., Cooper, G., Staebell, M., and Filner, P. (1987) *J. Biol. Chem.* **262**, 11026–11031
190. Crawford, R. L. (1981) *Lignin Biodegradation and Transformation*, Wiley, New York
190a. Humphreys, J. M., Hemm, M. R., and Chapple, C. (1999) *Proc. Natl. Acad. Sci. U.S.A.* **96**, 10045–10050
190b. Li, L., Popko, J. L., Umezawa, T., and Chiang, V. L. (2000) *J. Biol. Chem.* **275**, 6537–6545
190c. Lim, E.-K., Li, Y., Parr, A., Jackson, R., Ashford, D. A., and Bowles, D. J. (2001) *J. Biol. Chem.* **276**, 4344–4349
191. Fujita, M., Gang, D. R., Davin, L. B., and Lewis, N. G. (1999) *J. Biol. Chem.* **274**, 618–627
191a. Xia, Z.-Q., Costa, M. A., Pélissier, H. C., Davin, L. B., and Lewis, N. G. (2001) *J. Biol. Chem.* **276**, 12614–12623
192. Lewis, N. G., and Davin, L. B. (1994) in *Isopentenoids and Other Natural Products* (Nes, W. D., ed), pp. 202–246, American Chemical Society, Washington D.C.
193. von Wettstein-Knowles, P. M. (1993) in *Lipid Metabolism in Plants* (Moore, T. S., Jr., ed), pp. 127–166, CRC Press, Boca Raton, Florida
194. Nimz, H. (1974) *Angew. Chem. Int. Ed. Engl.* **13**, 313–321
195. Ralph, J., Hatfield, R. D., Quideau, S., Helm, R. F., Grabber, J. H., and Jung, H.-J. G. (1994) *J. Am. Chem. Soc.* **116**, 9448–9456
196. Steelink, C. (1972) *Rec. Adv. Phytochem.* **4**, 239–271
197. Huang, K.-x, Fujii, I., Ebizuka, Y., Gomi, K., and Sankawa, U. (1995) *J. Biol. Chem.* **270**, 21495–21502
197a. Henriksen, A., Smith, A. T., and Gajhede, M. (1999) *J. Biol. Chem.* **274**, 35005–35011
198. Nakamura, N., Matsuzaki, R., Choi, Y.-H., Tanizawa, K., and Sanders-Loehr, J. (1996) *J. Biol. Chem.* **271**, 4718–4724
198a. Schwartz, B., Olgin, A. K., and Klinman, J. P. (2001) *Biochemistry* **40**, 2954–2963
198b. Klinman, J. P. (2001) *Proc. Natl. Acad. Sci. U.S.A.* **98**, 14766–14768
198c. Datta, S., Mori, Y., Takagi, K., Kawaguchi, K., Chen, Z.-W., Okajima, T., Kuroda, S., Ikeda, T., Kano, K., Tanizawa, K., and Mathews, F. S. (2001) *Proc. Natl. Acad. Sci. U.S.A.* **98**, 14268–14273
198d. Firbank, S. J., Rogers, M. S., Wilmot, C. M., Dooley, D. M., Halcrow, M. A., Knowles, P. F., McPherson, M. J., and Phillips, S. E. V. (2001) *Proc. Natl. Acad. Sci. U.S.A.* **98**, 12932–12937
198e. Xie, L., and van der Donk, W. A. (2001) *Proc. Natl. Acad. Sci. U.S.A.* **98**, 12863–12865
199. Roberts, M. F., and Wink, M., eds. (1998) *Alkaloids Biochemistry, Ecology, and Medicinal Applications*, Plenum, New York
200. Harborne, J. B. (1993) *Introduction to Ecological Biochemistry*, 4th ed., Academic Press, San Diego, California
201. Nakajima, K., Yamashita, A., Akama, H., Nakatsu, T., Kato, H., Hashimoto, T., Oda, J., and Yamada, Y. (1998) *Proc. Natl. Acad. Sci. U.S.A.* **95**, 4876–4881
202. Robinson, R. (1917) *J. Chem. Soc.* **111**, 876–899
203. Robinson, R. (1955) *The Structural Relations of Natural Products*, Oxford Univ. Press, London
204. Lenz, R., and Zenk, M. H. (1995) *J. Biol. Chem.* **270**, 31091–31096
204a. Grothe, T., Lenz, R., and Kutchan, T. M. (2001) *J. Biol. Chem.* **276**, 30717–30723
205. Kutchan, T. M., and Dittrich, H. (1995) *J. Biol. Chem.* **270**, 24475–24481
206. Hayaishi, O. (1993) *Protein Sci.* **2**, 472–475
207. Young, S. N., Oravec, M., and Sourkes, T. L. (1974) *J. Biol. Chem.* **249**, 3932–3936
208. Sono, M. (1989) *Biochemistry* **28**, 5400–5407
209. Taylor, M. W., and Feng, G. (1991) *FASEB J.* **5**, 2516–2522

References

209a. Phillips, R. S., Johnson, N., and Kamath, A. V. (2002) *Biochemistry* **41**, 4012–4019
210. Billker, O., Lindo, V., Panico, M., Etienne, A. E., Paxton, T., Dell, A., Rogers, M., Sinden, R. E., and Morris, H. R. (1998) *Nature (London)* **392**, 289–292
211. Garcia, G. E., Wirtz, R. A., Barr, J. R., Woolfitt, A., and Rosenberg, R. (1998) *J. Biol. Chem.* **273**, 12003–12005
212. Kayser, H. (1985) in *Comprehensive Insect Physiology, Biochemistry and Pharmacology*, Vol. 10 (Kerkut, G. A., and Gilbert, L. I., eds), pp. 368–416, Pergamon, Oxford
213. Stone, T. W., ed. (1988) *Quinolinic Acid and the Kynurenines*, CRC Press, Boca Raton, Florida
213a. Aquilina, J. A., Carver, J. A., and Truscott, R. J. W. (1999) *Biochemistry* **38**, 11455–11464
213b. Wang, L., Erlandsen, H., Haavik, J., Knappskog, P. M., and Stevens, R. C. (2002) *Biochemistry* **41**, 12569 – 12574
214. Manthey, M. K., Pyne, S. G., and Truscott, R. J. W. (1992) *Proc. Natl. Acad. Sci. U.S.A.* **89**, 1954–1957
215. Schmidt, A. W., and Peroutka, S. J. (1989) *FASEB J.* **3**, 2242–2249
215a. McKinney, J., Teigen, K., Froystein, N. Å., Salaün, C., Knappskog, P. M., Haavik, J., and Martínez, A. (2001) *Biochemistry* **40**, 15591–15601
216. Hajduch, E., Rencurel, F., Balendran, A., Batty, I. H., Downes, C. P., and Hundal, H. S. (1999) *J. Biol. Chem.* **274**, 13563–13568
217. Klein, D. C., and Namboodiri, M. A. A. (1982) *Trends Biochem. Sci.* **7**, 98–102
218. Utiger, R. D. (1992) *N. Engl. J. Med.* **327**, 1377–1379
219. Ishida, I., Obinata, M., and Deguchi, T. (1987) *J. Biol. Chem.* **262**, 2895–2899
220. Rodriguez, I. R., Mazuruk, K., Schoen, T. J., and Chader, G. J. (1994) *J. Biol. Chem.* **269**, 31969–31977
221. Slominski, A., Baker, J., Rosano, T. G., Guisti, L. W., Ermak, G., Grande, M., and Gaudet, S. J. (1996) *J. Biol. Chem.* **271**, 12281–12286
222. Mayeno, A. N., Lin, F., Foote, C. S., Loegering, D. A., Ames, M. M., Hedberg, C. W., and Gleich, G. J. (1990) *Science* **250**, 1707–1708
223. Aldhous, P. (1991) *Nature (London)* **353**, 490
224. Moat, A. G., and Foster, J. W. (1987) in *Pyridine Nucleotide Coenzymes: Chemical, Biochemical and Medical Aspects*, Vol. B (Dolphin, D., Avramovic, O., and Poulson, R., eds), pp. 1–24, Wiley (Interscience), New York
225. Keys, L. D., III, and Hamilton, G. A. (1987) *J. Am. Chem. Soc.* **109**, 2156–2163
225a. Saridakis, V., Christendat, D., Kimber, M. S., Dharamsi, A., Edwards, A. M., and Pai, E. F. (2001) *J. Biol. Chem.* **276**, 7225–7232
226. Rizzi, M., Nessi, C., Mattevi, A., Coda, A., Bolognesi, M., and Galizzi, A. (1996) *EMBO J.* **15**, 5125–5134
227. Merkler, D. J., Glufke, U., Ritenour-Rodgers, K. J., Baumgart, L. E., DeBlassio, J. L., Merkler, K. A., and Vederas, J. C. (1999) *J. Am. Chem. Soc.* **121**, 4904–4905
228. Tedeschi, G., Zetta, L., Negri, A., Mortarino, M., Ceciliani, F., and Ronchi, S. (1997) *Biochemistry* **36**, 16221–16230
229. Nasu, S., Wicks, F. D., and Gholson, R. K. (1982) *J. Biol. Chem.* **257**, 626–632
230. White, H. B., III. (1982) in *The Pyridine Nucleotide Coenzymes* (Everse, J., Anderson, B., and You, K.-S., eds), pp. 225–242, Academic Press, New York
231. Tritz, G. J. (1987) in *Escherichia coli and Salmonella typhimurium* (Neidhardt, F. C., ed), pp. 557–563, Am. Soc. for Microbiology, Washington,DC
232. Malherbe, P., Köhler, C., Da Prada, M., Lang, G., Kiefer, V., Schwarcz, R., Lahm, H.-W., and Cesura, A. M. (1994) *J. Biol. Chem.* **269**, 13792–13797
233. Heyes, M. P., Saito, K., Lackner, A., Wiley, C. A., Achim, C. L., and Markey, S. P. (1998) *FASEB J.* **12**, 881–896
234. Emanuele, J. J., and Pitzpatrick, P. F. (1995) *Biochemistry* **34**, 3716–3723
234a. Gadda, G., Dangott, L. J., Johnson, W. H., Jr., Whitman, C. P., and Fitzpatrick, P. F. (1999) *Biochemistry* **38**, 5822–5828
234b. Zhao, Y., Christensen, S. K., Fankhauser, C., Cashman, J. R., Cohen, J. D., Weigel, D., and Chory, J. (2001) *Science* **291**, 306–309
235. Brown, H. M., and Purves, W. K. (1976) *J. Biol. Chem.* **251**, 907–913
236. Normanly, J., Cohen, J. D., and Fink, G. R. (1993) *Proc. Natl. Acad. Sci. U.S.A.* **90**, 10355–10359
236a. Mikkelsen, M. D., Hansen, C. H., Wittstock, U., and Halkier, B. A. (2000) *J. Biol. Chem.* **275**, 33712–33717
236b. Kato, Y., Nakamura, K., Sakiyama, H., Mayhew, S. G., and Asano, Y. (2000) *Biochemistry* **39**, 800–809
237. Kobayashi, M., Suzuki, T., Fujita, T., Masuda, M., and Shimizu, S. (1995) *Proc. Natl. Acad. Sci. U.S.A.* **92**, 714–718
238. Bartling, D., Seedorf, M., Schmidt, R. C., and Weiler, E. W. (1994) *Proc. Natl. Acad. Sci. U.S.A.* **91**, 6021–6025
238a. Kendrew, S. G. (2001) *Trends Biochem. Sci.* **26**, 218
238b. Jones, A. M. (1998) *Science* **282**, 2201 – 2202
239. Bartel, B., and Fink, G. R. (1995) *Science* **268**, 1745–1748
240. Szerszen, J. B., Szczyglowski, K., and Bandurski, R. S. (1994) *Science* **265**, 1699–1701
241. Roberto, F. F., Klee, H., White, F., Nordeen, R., and Kosuge, T. (1990) *Proc. Natl. Acad. Sci. U.S.A.* **87**, 5797–5801
241a. Crozier, A., Kamiya, Y., Bishop, G., and Yokota, T. (2000) in *Biochemistry and Molecular Biology of Plants* (Buchanan, B., Gruissem, W., and Jones, R., eds), pp. 850–929, Am. Soc. Plant Physiologists, Rockville, Maryland
242. Abel, S., Nguyen, M. D., and Theologis, A. (1995) *J. Mol. Biol.* **251**, 533–549
243. Abel, S., Nguyen, M. D., Chow, W., and Theologis, A. (1995) *J. Biol. Chem.* **270**, 19093–19099
244. Shibuya, M., Chou, H.-M., Fountoulakis, M., Hassam, S., Kim, S.-U., Kobayashi, K., Otsuka, H., Rogalska, E., Cassady, J. M., and Floss, H. G. (1990) *J. Am. Chem. Soc.* **112**, 297–304
245. Mizukami, H., Nordlöv, H., Lee, S.-L., and Scott, A. I. (1979) *Biochemistry* **18**, 3760–3763
245a. Stocking, E. M., Williams, R. M., and Sanz-Cervera, J. F. (2000) *J. Am. Chem. Soc.* **122**, 9089–9098
246. Frey, M., Chomet, P., Glawischnig, E., Stettner, C., Grün, S., Winklmair, A., Eisenreich, W., Bacher, A., Meeley, R. B., Briggs, S. P., Simcox, K., and Gierl, A. (1997) *Science* **277**, 696–699
246a. Beismann-Driemeyer, S., and Sterner, R. (2001) *J. Biol. Chem.* **276**, 20387–20396
247. Parker, A. R., Moore, J. A., Schwab, J. M., and Davisson, V. J. (1995) *J. Am. Chem. Soc.* **117**, 10605–10613
247a. Sivaraman, J., Li, Y., Larocque, R., Schrag, J. D., Cygler, M., and Matte, A. (2001) *J. Mol. Biol.* **311**, 761–776
247b. Haruyama, K., Nakai, T., Miyahara, I., Hirotsu, K., Mizuguchi, H., Hayashi, H., and Kagamiyama, H. (2001) *Biochemistry* **40**, 4633–4644
248. Reizer, J., Michotey, V., Reizer, A., and Saier, M. H., Jr. (1994) *Protein Sci.* **3**, 440–450
249. Nagai, A., Ward, E., Beck, J., Tada, S., Chang, J.-Y., Scheidegger, A., and Ryals, J. (1991) *Proc. Natl. Acad. Sci. U.S.A.* **88**, 4133–4137
249a. Grubmeyer, C., and Teng, H. (1999) *Biochemistry* **38**, 7355–7362
250. Brenner, M., and Ames, B. N. (1971) in *Metabolic Pathways*, 3rd ed. (Vogel, H. J., ed), Academic Press, New York
251. Winkler, M. E. (1987) in *Escherichia coli and Salmonella typhimurium* (Neidhardt, F. C., ed), pp. 395–411, Am. Soc. for Microbiology, Washington,DC
252. Levy, H. L., Taylor, R. G., and McInnes, R. R. (1995) in *The Metabolic and Molecular Bases of Inherited Disease*, 7th ed., Vol. 1 (Scriver, C. R., Beaudet, A. L., Sly, W. S., and Valle, D., eds), pp. 1107–1123, McGraw-Hill, New York
253. Chen, J.-Y. C., and Bodley, J. W. (1988) *J. Biol. Chem.* **263**, 11692–11696
254. Jones, M. E. (1980) *Ann. Rev. Biochem.* **49**, 253–279
255. Gao, G., Nara, T., Nakajima-Shimada, J., and Aoki, T. (1999) *J. Mol. Biol.* **285**, 149–161
256. Serre, V., Guy, H., Liu, X., Penverne, B., Hervé, G., and Evans, D. (1998) *J. Mol. Biol.* **281**, 363–377
257. Neuhard, J., and Nygaard, P. (1987) in *Escherichia coli and Salmonella typhimurium* (Neidhardt, F. C., ed), pp. 446–473, Am. Soc. for Microbiology, Washington,DC
258. Switzer, R. L., and Quinn, C. L. (1993) in *Bacillus subtilis and Other Gram-Positive Bacteria* (Sonenshein, A. L., Hoch, J. A., and Losick, R., eds), pp. 343–358, American Society for Microbiology, Washington, D. C.
259. Guy, H. I., and Evans, D. R. (1996) *J. Biol. Chem.* **271**, 13762–13769
259a. Thoden, J. B., Raushel, F. M., Wesenberg, G., and Holden, H. M. (1999) *J. Biol. Chem.* **274**, 22502–22507
260. Braxton, B. L., Mullins, L. S., Rauschel, F. M., and Reinhart, G. D. (1999) *Biochemistry* **38**, 1394–1401
260a. Thoden, J. B., Phillips, G. N., Jr., Neal, T. M., Raushel, F. M., and Holden, H. M. (2001) *Biochemistry* **40**, 6989–6997
261. Williams, N. K., Manthey, M. K., Hambley, T. W., O'Donoghue, S. I., Keegan, M., Chapman, B. E., and Christopherson, R. I. (1995) *Biochemistry* **34**, 11344–11352
262. Zimmermann, B. H., Kemling, N. M., and Evans, D. R. (1995) *Biochemistry* **34**, 7038–7046
263. Hines, V., and Johnston, M. (1989) *Biochemistry* **28**, 1227–1234
264. Björnberg, O., Rowland, P., Larsen, S., and Jensen, K. F. (1997) *Biochemistry* **36**, 16197–16205
265. Rowland, P., Björnberg, O., Nielsen, F. S., Jensen, K. F., and Larsen, S. (1998) *Protein Sci.* **7**, 1269–1279
265a. Argyrou, A., Washabaugh, M. W., and Pickart, C. M. (2000) *Biochemistry* **39**, 10373–10384
266. Scapin, G., Ozturk, D. H., Grubmeyer, C., and Sacchettini, J. C. (1995) *Biochemistry* **34**, 10744–10754
267. Tao, W., Grubmeyer, C., and Blanchard, J. S. (1996) *Biochemistry* **35**, 14–21
268. Radzicka, A., and Wolfenden, R. (1995) *Science* **267**, 90–93
269. Beak, P., and Siegel, B. (1976) *J. Am. Chem. Soc.* **98**, 3601–3606
270. Smiley, J. A., and Jones, M. E. (1992) *Biochemistry* **31**, 12162–12168
270a. Appleby, T. C., Kinsland, C., Begley, T. P., and Ealick, S. E. (2000) *Proc. Natl. Acad. Sci. U.S.A.* **97**, 2005–2010

References

270b. Harris, P., Poulsen, J.-C. N., Jensen, K. F., and Larsen, S. (2000) *Biochemistry* **39**, 4217–4224

270c. Miller, B. G., Snider, M. J., Short, S. A., and Wolfenden, R. (2000) *Biochemistry* **39**, 8113–8118

270d. Phillips, L. M., and Lee, J. K. (2001) *J. Am. Chem. Soc.* **123**, 12067–12073

271. Girot, R., Hamet, M., Perignon, J.-L., Guesnu, M., Fox, R. M., Cartier, P., Durandy, A., and Griscelli, C. (1983) *N. Engl. J. Med.* **308**, 700–704

272. Lewis, D. A., and Villafrance, J. J. (1989) *Biochemistry* **28**, 8454–8459

273. Kornberg, A., and Baker, T. A. (1992) *DNA Replication*, 2nd ed., Freeman, New York

274. Weiner, K. X. B., Ciesla, J., Jaffe, A. B., Ketring, R., Maley, F., and Maley, G. F. (1995) *J. Biol. Chem.* **270**, 18727–18729

274a. Kambampati, R., and Lauhon, C. T. (2000) *J. Biol. Chem.* **275**, 10727–10730

274b. Johansson, E., Mejlhede, N., Neuhard, J., and Larsen, S. (2002) *Biochemistry* **41**, 2563–2570

274c. Snider, M. J., Reinhardt, L., Wolfenden, R., and Cleland, W. W. (2002) *Biochemistry* **41**, 415–421

275. Kvalnes-Krick, K. L., and Traut, T. W. (1993) *J. Biol. Chem.* **268**, 5686–5693

275a. Dobritzsch, D., Schneider, G., Schnackerz, K. D., and Lindqvist, Y. (2001) *EMBO J.* **20**, 650–660

276. Kim, Y. S., and Bang, S. K. (1985) *J. Biol. Chem.* **260**, 5098–5104

277. Nygaard, P. (1993) in *Bacillus subtilis and Other Gram-Positive Bacteria* (Sonenshein, A. L., Hoch, J. A., and Losick, R., eds), pp. 359–378, American Society for Microbiology, Washington, D. C.

278. Iovannisci, D. M., Goebel, D., Allen, K., Kaur, K., and Ullman, B. (1984) *J. Biol. Chem.* **259**, 14617–14623

279. Schumacher, M. A., Carter, D., Scott, D. M., Roos, D. S., Ullman, B., and Brennan, R. G. (1998) *EMBO J.* **17**, 3219–3232

280. Muchmore, C. R. A., Krahn, J. M., Kim, J. H., Zalkin, H., and Smith, J. L. (1998) *Protein Sci.* **7**, 39–51

281. Smith, J. L., Zaluzec, E. J., Wery, J.-P., Niu, L., Switzer, R. L., Zalkin, H., and Satow, Y. (1994) *Science* **264**, 1427–1433

282. Wang, W., Kappock, T. J., Stubbe, J., and Ealick, S. E. (1998) *Biochemistry* **37**, 15647–15662

283. Benkovic, S. J. (1984) *Trends Biochem. Sci.* **9**, 320–322

284. Ononogbu, I. C. (1980) *Trends Biochem. Sci.* **5**, X

285. Shim, J. H., and Benkovic, S. J. (1998) *Biochemistry* **37**, 8776–8782

285a. Shim, J. H., and Benkovic, S. J. (1999) *Biochemistry* **38**, 10024–10031

285b. Gooljarsingh, L. T., Ramcharan, J., Gilroy, S., and Benkovic, S. J. (2001) *Proc. Natl. Acad. Sci. U.S.A.* **98**, 6565–6570

286. Kastanos, E. K., Woldman, Y. Y., and Appling, D. R. (1997) *Biochemistry* **36**, 14956–14964

287. Schrimsher, J. L., Schendel, F. J., Stubbe, J., and Smith, J. M. (1986) *Biochemistry* **25**, 4366–4371

287a. Mueller, E. J., Oh, S., Kavalerchik, E., Kappock, T. J., Meyer, E., Li, C., Ealick, S. E., and Stubbe, J. (1999) *Biochemistry* **38**, 9831–9839

288. Schrimsher, J. L., Schendel, F. J., and Stubbe, J. (1986) *Biochemistry* **25**, 4356–4365

289. Firestine, S. M., and Davisson, V. J. (1994) *Biochemistry* **33**, 11917–11926

290. Meyer, E., Kappock, T. J., Osuji, C., and Stubbe, J. (1999) *Biochemistry* **38**, 3012–3018

290a. Thoden, J. B., Kappock, T. J., Stubbe, J., and Holden, H. M. (1999) *Biochemistry* **38**, 15480–15492

291. Szabados, E., Hindmarsh, E. J., Phillips, L., Duggleby, R. G., and Christopherson, R. I. (1994) *Biochemistry* **33**, 14237–14245

292. Rayl, E. A., Moroson, B. A., and Beardsley, G. P. (1996) *J. Biol. Chem.* **271**, 2225–2233

292a. Wall, M., Shim, J. H., and Benkovic, S. J. (2000) *Biochemistry* **39**, 11303–11311

292b. Vergis, J. M., Bulock, K. G., Fleming, K. G., and Beardsley, G. P. (2001) *J. Biol. Chem.* **276**, 7727–7733

292c. Brosius, J. L., and Colman, R. F. (2000) *Biochemistry* **39**, 13336–13343

292d. Toth, E. A., Worby, C., Dixon, J. E., Goedken, E. R., Marqusee, S., and Yeates, T. O. (2000) *J. Mol. Biol.* **301**, 433–450

293. Silva, M. M., Poland, B. W., Hoffman, C. R., Fromm, H. J., and Honzatko, R. B. (1995) *J. Mol. Biol.* **254**, 431–446

293a. Poland, B. W., Lee, S.-F., Subramanian, M. V., Siehl, D. L., Anderson, R. J., Fromm, H. J., and Honzatko, R. B. (1996) *Biochemistry* **35**, 15753–15759

294. Lee, T. T., Worby, C., Bao, Z.-Q., Dixon, J. E., and Colman, R. F. (1999) *Biochemistry* **38**, 22–32

294a. Guicherit, O. M., Cooper, B. F., Rudolph, F. B., and Kellems, R. E. (1994) *J. Biol. Chem.* **269**, 4488–4496

294b. Iancu, C. V., Borza, T., Choe, J. Y., Fromm, H. J., and Honzatko, R. B. (2001) *J. Biol. Chem.* **276**, 42146–42152

295. Bass, M. B., Fromm, H. J., and Rudolph, F. B. (1984) *J. Biol. Chem.* **259**, 12330–12333

296. Webb, M. R., Reed, G. H., Cooper, B. F., and Rudolph, F. B. (1984) *J. Biol. Chem.* **259**, 3044–3046

297. Choe, J.-Y., Poland, B. W., Fromm, H. J., and Honzatko, R. B. (1999) *Biochemistry* **38**, 6953–6961

297a. Fleming, M. A., Chambers, S. P., Connelly, P. R., Nimmesgern, E., Fox, T., Bruzzese, F. J., Hoe, S. T., Fulghum, J. R., Livingston, D. J., Stuver, C. M., Sintchak, M. D., Wilson, K. P., and Thomson, J. A. (1996) *Biochemistry* **35**, 6990–6997

297b. Markham, G. D., Bock, C. L., and Schalk-Hihi, C. (1999) *Biochemistry* **38**, 4433–4440

297c. Barnes, B. J., Eakin, A. E., Izydore, R. A., and Hall, I. H. (2000) *Biochemistry* **39**, 13641–13650

297d. Kerr, K. M., Digits, J. A., Kuperwasser, N., and Hedstrom, L. (2000) *Biochemistry* **39**, 9804–9810

298. Zhang, R.-g, Evans, G., Rotella, F. J., Westbrook, E. M., Beno, D., Huberman, E., Joachimiak, A., and Collart, F. R. (1999) *Biochemistry* **38**, 4691–4700

299. Whitby, F. G., Luecke, H., Kuhn, P., Somoza, J. R., Huete-Perez, J. A., Phillips, J. D., Hill, C. P., Fletterick, R. J., and Wang, C. C. (1997) *Biochemistry* **36**, 10666–10674

300. Farazi, T., Leichman, J., Harris, T., Cahoon, M., and Hedstrom, L. (1997) *J. Biol. Chem.* **272**, 961–965

301. Lowenstein, J., and Tornheim, K. (1971) *Science* **171**, 397–400

301a. Bai, Y., Fox, D. T., Lacy, J. A., Van Lanen, S. G., and Iwata-Reuyl, D. (2000) *J. Biol. Chem.* **275**, 28731–28738

301b. Watanabe, M., Nameki, N., Matsuo-Takasaki, M., Nishimura, S., and Okada, N. (2001) *J. Biol. Chem.* **276**, 2387–2394

301c. Tercero, J. A., Espinosa, J. C., Lacalle, R. A., and Jiménez, A. (1996) *J. Biol. Chem.* **271**, 1579–1590

302. Coffee, C. J., and Kofke, W. A. (1975) *J. Biol. Chem.* **250**, 6653–6658

303. Schultz, V., and Lowenstein, J. M. (1978) *J. Biol. Chem.* **253**, 1938–1943

304. Merkler, D. J., and Schramm, V. L. (1993) *Biochemistry* **32**, 5792–5799

305. Sollitti, P., Merkler, D. J., Estupiñán, B., and Schramm, V. L. (1993) *J. Biol. Chem.* **268**, 4549–4555

306. Fishbein, W. N., Armbrustmacher, V. W., and Griffin, J. L. (1978) *Science* **200**, 545–548

306a. Craig, S. P., III, and Eakin, A. E. (2000) *J. Biol. Chem.* **275**, 20231–20234

306b. Phillips, C. L., Ullman, B., Brennan, R. G., and Hill, C. P. (1999) *EMBO J.* **18**, 3533–3545

307. Stiles, G. L. (1992) *J. Biol. Chem.* **267**, 6451–6454

308. Newby, A. C. (1984) *Trends Biochem. Sci.* **9**, 42–44

309. Linden, J. (1991) *FASEB J.* **5**, 2668–2676

310. Palmer, T. M., Benovic, J. L., and Stiles, G. L. (1995) *J. Biol. Chem.* **270**, 29607–29613

311. Erb, L., Garrad, R., Wang, Y., Quinn, T., Turner, J. T., and Weisman, G. A. (1995) *J. Biol. Chem.* **270**, 4185–4188

312. Osborne, W. R. A. (1981) *Trends Biochem. Sci.* **6**, 80–83

313. Anderson, W. F. (1995) *Sci. Am.* **273**(Sep), 124–128

314. Marshall, E. (1995) *Science* **269**, 1050–1055

315. Hershfield, M. S., and Mitchell, B. S. (1995) in *The Metabolic and Molecular Bases of Inherited Disease*, 7th ed., Vol. 2 (Scriver, C. R., Beaudet, A. L., Sly, W. S., and Valle, D., eds), pp. 1725–1768, McGraw-Hill, New York

316. Sideraki, V., Mohamedali, K. A., Wilson, D. K., Chang, Z., Kellems, R. E., Quiocho, F. A., and Rudolph, F. B. (1996) *Biochemistry* **35**, 7862–7872

317. Wang, Z., and Quiocho, F. A. (1998) *Biochemistry* **37**, 8314–8324

318. Mao.C, Cook, W. J., Zhou, M., Federov, A. A., Almo.SC, and Ealick, S. E. (1998) *Biochemistry* **37**, 7135–7146

319. Cook, W. J., Ealick, S. E., Krenitsky, T. A., Stoeckler, J. D., Helliwell, J. R., and Bugg, C. E. (1985) *J. Biol. Chem.* **260**, 12968–12969

320. Benveniste, P., and Cohen, A. (1995) *Proc. Natl. Acad. Sci. U.S.A.* **92**, 8373–8377

321. Deng, H., Kurz, L. C., Rudolph, F. B., and Callender, R. (1998) *Biochemistry* **37**, 4968–4976

321a. Fedorov, A., Shi, W., Kicska, G., Fedorov, E., Tyler, P. C., Furneaux, R. H., Hanson, J. C., Gainsford, G. J., Larese, J. Z., Schramm, V. L., and Almo, S. C. (2001) *Biochemistry* **40**, 853–860

321b. Tebbe, J., Bzowska, A., Wielgus-Kutrowska, B., Schröder, W., Kazimierczuk, Z., Shugar, D., Saenger, W., and Koellner, G. (1999) *J. Mol. Biol.* **294**, 1239–1255

322. McRoberts, J. A., and Martin, D. W., Jr. (1980) *J. Biol. Chem.* **255**, 5605–5615

322a. Xu, Y., and Grubmeyer, C. (1998) *Biochemistry* **37**, 4114–4124

322b. Canyuk, B., Focia, P. J., and Eakin, A. E. (2001) *Biochemistry* **40**, 2754–2765

323. Focia, P. J., Craig, S. P., III, Nieves-Alicea, R., Fletterick, R. J., and Eakin, A. E. (1998) *Biochemistry* **37**, 15066–15075

324. Wilson, J. M., and Kelley, W. N. (1984) *J. Biol. Chem.* **259**, 27–30

325. Rossiter, B. J. F., and Caskey, C. T. (1995) in *The Metabolic and Molecular Bases of Inherited Disease*, 7th ed., Vol. 2 (Scriver, C. R., Beaudet, A. L., Sly, W. S., and Valle, D., eds), pp. 1679–1706, McGraw-Hill, New York

326. Davidson, B. L., Pashmforoush, M., Kelley, W. N., and Palella, T. D. (1989) *J. Biol. Chem.* **264**, 520–525

327. Willis, R. C., Jolly, D. J., Miller, A. D., Plent, M. M., Esty, A. C., Anderson, P. J., Chang, H.-C., Jones, O. W., Seegmiller, J. E., and Friedmann, T. (1984) *J. Biol. Chem.* **259**, 7842–7849

References

327a. Héroux, A., White, E. L., Ross, L. J., and Borhani, D. W. (1999) *Biochemistry* **38**, 14485–14494
327b. Versées, W., Decanniere, K., Pellé, R., Depoorter, J., Brosens, E., Parkin, D. W., and Steyaert, J. (2001) *J. Mol. Biol.* **307**, 1363–1379
328. Yuan, G., Bin, J. C., McKay, D. J., and Snyder, F. F. (1999) *J. Biol. Chem.* **274**, 8175–8180
328a. Self, W. T., and Stadtman, T. C. (2000) *Proc. Natl. Acad. Sci. U.S.A.* **97**, 7208–7213
328b. Kato, M., Mizuno, K., Crozier, A., Fujimura, T., and Ashihara, H. (2000) *Nature (London)* **406**, 956–957
329. Rumsby, P. C., Kato, H., Waldren, C. A., and Patterson, D. (1982) *J. Biol. Chem.* **257**, 11364–11367
330. Onrot, J., Goldberg, M. R., Biaggioni, I., Hollister, A. S., Kincaid, D., and Robertson, D. (1985) *N. Engl. J. Med.* **313**, 549–554
331. Florkin, M. (1949) *Biochemical Evolution*, Academic Press, New York
332. Henderson, J. F., and Paterson, A. R. P. (1973) *Nucleotide Metabolism*, Academic Press, New York
333. Kahn, K., and Tipton, P. A. (1997) *Biochemistry* **36**, 4731–4738
334. Kahn, K., Serfozo, P., and Tipton, P. A. (1997) *J. Am. Chem. Soc.* **119**, 5435–5442
335. Kahn, K., and Tipton, P. A. (1998) *Biochemistry* **37**, 11651–11659
335a. Enomoto, A., and 18 other authors (2002) *Nature (London)* **417**, 447–452
335b. Hediger, M. A., (2002) *Nature (London)* **417**, 393, 395
336. Hayashi, S., Fujiwara, S., and Noguchi, T. (1989) *J. Biol. Chem.* **264**, 3211–3215
337. Reynolds, P. H. S., Boland, M. J., Blevins, D. G., Randall, D. D., and Schubert, K. R. (1982) *Trends Biochem. Sci.* **7**, 366–368
338. Burg, A. W., and Brown, G. M. (1968) *J. Biol. Chem.* **243**, 2349–2358
339. Shiota, T., Palumbo, M. P., and Tsai, L. (1967) *J. Biol. Chem.* **242**, 1961–1969
340. Brown, G. M., and Williamson, J. M. (1987) in *Escherichia coli and Salmonella typhimurium* (Neidhardt, F. C., ed), pp. 521–538, Am. Soc. for Microbiology, Washington,DC
341. Bracher, A., Eisenreich, W., Schramek, N., Ritz, H., Götze, E., Herrmann, A., Gütlich, M., and Bacher, A. (1998) *J. Biol. Chem.* **273**, 28132–28141
342. Nar, H., Huber, R., Auerbach, G., Fischer, M., Hösl, C., Ritz, H., Bracher, A., Meining, W., Eberhardt, S., and Bacher, A. (1995) *Proc. Natl. Acad. Sci. U.S.A.* **92**, 12120–12125
342a. Bracher, A., Fischer, M., Eisenreich, W., Ritz, H., Schramek, N., Boyle, P., Gentili, P., Huber, R., Nar, H., Auerbach, G., and Bacher, A. (1999) *J. Biol. Chem.* **274**, 16727–16735
342b. Bracher, A., Schramek, N., and Bacher, A. (2001) *Biochemistry* **40**, 7896–7902
342c. Maita, N., Okada, K., Hatakeyama, K., and Hakoshima, T. (2002) *Proc. Natl. Acad. Sci. U.S.A.* **99**, 1212–1217
343. Ahn, C., Byun, J., and Yim, J. (1997) *J. Biol. Chem.* **272**, 15323–15328
344. Haussmann, C., Rohdich, F., Schmidt, E., Bacher, A., and Richter, G. (1998) *J. Biol. Chem.* **273**, 17418–17424
345. Hampele, I. C., D'Arcy, A., Dale, G. E., Kostrewa, D., Nielsen, J., Oefner, C., Page, M. G. P., Schönfeld, H.-J., Stüber, D., and Then, R. L. (1997) *J. Mol. Biol.* **268**, 21–30
346. Sun, X., Bognar, A. L., Baker, E. N., and Smith, C. A. (1998) *Proc. Natl. Acad. Sci. U.S.A.* **95**, 6647–6652
347. Roy, K., Mitsugi, K., and Sirotnak, F. M. (1996) *J. Biol. Chem.* **271**, 23820–23827
347a. Bermingham, A., Bottomley, J. R., Primrose, W. U., and Derrick, J. P. (2000) *J. Biol. Chem.* **275**, 17962–17967
347b. Baca, A. M., Sirawaraporn, R., Turley, S., Sirawaraporn, W., and Hol, W. G. J. (2000) *J. Mol. Biol.* **302**, 1193–1212
347c. Sun, X., Cross, J. A., Bognar, A. L., Baker, E. N., and Smith, C. A. (2001) *J. Mol. Biol.* **310**, 1067–1078
347d. Ravanel, S., Cherest, H., Jabrin, S., Grunwald, D., Surdin-Kerjan, Y., Douce, R., and Rébeillé, F. (2001) *Proc. Natl. Acad. Sci. U.S.A.* **98**, 15360–15365
347e. Chave, K. J., Auger, I. E., Galivan, J., and Ryan, T. J. (2000) *J. Biol. Chem.* **275**, 40365–40370
348. Smith, G. K., Cichetti, J. A., Chandrasurin, P., and Nichol, C. A. (1985) *J. Biol. Chem.* **260**, 5221–5224
349. Switchenko, A. C., and Brown, G. M. (1985) *J. Biol. Chem.* **260**, 2945–2951
350. Oppliger, T., Thöny, B., Nar, H., Bürgisser, D., Huber, R., Heizmann, C. W., and Blau, N. (1995) *J. Biol. Chem.* **270**, 29498–29506
351. Bürgisser, D. M., Thöny, B., Redweik, U., Hess, D., Heizmann, C. W., Huber, R., and Nar, H. (1995) *J. Mol. Biol.* **253**, 358–369
352. Ploom, T., Thöny, B., Yim, J., Lee, S., Nar, H., Leimbacher, W., Richardson, J., Huber, R., and Auerbach, G. (1999) *J. Mol. Biol.* **286**, 851–860
353. Citron, B. A., Milstien, S., Gutierrez, J. C., Levine, R. A., Yanak, B. L., and Kaufman, S. (1990) *Proc. Natl. Acad. Sci. U.S.A.* **87**, 6436–6440
354. Auerbach, G., Herrmann, A., Gütlich, M., Fischer, M., Jacob, U., Bacher, A., and Huber, R. (1997) *EMBO J.* **16**, 7219–7230
355. Nichol, C. A., Smith, G. K., and Duch, D. S. (1985) *Ann. Rev. Biochem.* **54**, 729–764
356. Tanaka, K., Akino, M., Hagi, Y., Doi, M., and Shiota, T. (1981) *J. Biol. Chem.* **256**, 2963–2972
357. Wiederrecht, G. J., and Brown, G. M. (1984) *J. Biol. Chem.* **259**, 14121–14127
358. Jacobson, K. B., Dorsett, D., Pfleiderer, W., McCloskey, J. A., Sethi, S. K., Buchanan, M. V., and Rubin, I. B. (1982) *Biochemistry* **21**, 1238–1243
359. White, R. H. (1996) *Biochemistry* **35**, 3447–3456
360. Rasche, M. E., and White, R. H. (1998) *Biochemistry* **37**, 11343–11351
361. Wuebbens, M. M., and Rajagopalan, K. V. (1995) *J. Biol. Chem.* **270**, 1082–1087
361a. Unkles, S. E., Heck, I. S., Appleyard, M. V. C. L., and Kinghorn, J. R. (1999) *J. Biol. Chem.* **274**, 19286–19293
361b. Leimkühler, S., Wuebbens, M. M., and Rajagopalan, K. V. (2001) *J. Biol. Chem.* **276**, 34695–34701
361c. Leimkühler, S., and Rajagopalan, K. V. (2001) *J. Biol. Chem.* **276**, 1837–1844
361d. Schrag, J. D., Huang, W., Sivaraman, J., Smith, C., Plamondon, J., Larocque, R., Matte, A., and Cygler, M. (2001) *J. Mol. Biol.* **310**, 419–431
361e. Kuper, J., Palmer, T., Mendel, R. R., and Schwarz, G. (2000) *Proc. Natl. Acad. Sci. U.S.A.* **97**, 6475–6480
361f. Watanabe, T., Ihara, N., Itoh, T., Fujita, T., and Sugimoto, Y. (2000) *J. Biol. Chem.* **275**, 21789–21792
361g. Bittner, F., Oreb, M., and Mendel, R. R. (2001) *J. Biol. Chem.* **276**, 40381–40384
361h. Temple, C. A., and Rajagopalan, K. V. (2000) *J. Biol. Chem.* **275**, 40202–40210
361i. Lake, M. W., Temple, C. A., Rajagopalan, K. V., and Schindelin, H. (2000) *J. Biol. Chem.* **275**, 40211–40217
362. Ritz, H., Schramek, N., Bracher, A., Herz, S., Eisenreich, W., Richter, G., and Bacher, A. (2001) *J. Biol. Chem.* **276**, 22273–22277
363. Kis, K., Volk, R., and Bacher, A. (1995) *Biochemistry* **34**, 2883–2892
363a. Kelly, M. J. S., Ball, L. J., Krieger, C., Yu, Y., Fischer, M., Schiffmann, S., Schmieder, P., Kühne, R., Bermel, W., Bacher, A., Richter, G., and Oschkinat, H. (2001) *Proc. Natl. Acad. Sci. U.S.A.* **98**, 13025–13030
363b. Illarionov, B., Eisenreich, W., and Bacher, A. (2001) *Proc. Natl. Acad. Sci. U.S.A.* **98** 7224–7229
364. García-Ramírez, J. J., Santos, M. A., and Revuelta, J. L. (1995) *J. Biol. Chem.* **270**, 23801–23807
364a. Jordan, D. B., Bacot, K. O., Carlson, T. J., Kessel, M., and Viitanen, P. V. (1999) *J. Biol. Chem.* **274**, 22114–22121
364b. Braden, B. C., Velikovsky, C. A., Cauerhff, A. A., Polikarpov, I., and Goldbaum, F. A. (2000) *J. Mol. Biol.* **297**, 1031–1036
364c. Meining, W., Mörtl, S., Fischer, M., Cushman, M., Bacher, A., and Ladenstein, R. (2000) *J. Mol. Biol.* **299**, 181–197
364d. Illarionov, B., Kemter, K., Eberhardt, S., Richter, G., Cushman, M., and Bacher, A. (2001) *J. Biol. Chem.* **276**, 11524–11530
365. Ritsert, K., Huber, R., Turk, D., Ladenstein, R., Schmidt-Bäse, K., and Bacher, A. (1995) *J. Mol. Biol.* **253**, 151–167
366. Efimov, I., Kuusk, V., Zhang, X., and McIntire, W. S. (1998) *Biochemistry* **37**, 9716–9723
367. Bowers-Komro, D. M., Yamada, Y., and McCormick, D. B. (1989) *Biochemistry* **28**, 8439–8446
368. Eisenreich, W., Schwarzkopf, B., and Bacher, A. (1991) *J. Biol. Chem.* **266**, 9622–9631
369. Hörig, J. A., Renz, P., and Heckmann, G. (1978) *J. Biol. Chem.* **253**, 7410–7414
370. Thompson, T. B., Thomas, M. G., Escalante-Semerena, J. C., and Rayment, I. (1998) *Biochemistry* **37**, 7686–7695
371. Trzebiatowski, J. R., and Escalante-Semerena, J. C. (1997) *J. Biol. Chem.* **272**, 17662–17667
372. Himmeldirk, K., Sayer, B. G., and Spenser, I. D. (1998) *J. Am. Chem. Soc.* **120**, 3581–3589
372a. Miranda-Rios, J., Navarro, M., and Soberón, M. (2001) *Proc. Natl. Acad. Sci. U.S.A.* **98**, 9736–9741
373. Estramareix, B., and Thérisod, M. (1984) *J. Am. Chem. Soc.* **106**, 3857–3860
374. Taylor, S. V., Kelleher, N. L., Kinsland, C., Chiu, H.-J., Costello, C. A., Backstrom, A. D., McLafferty, F. W., and Begley, T. P. (1998) *J. Biol. Chem.* **273**, 16555–16560
374a. Mueller, E. G., Palenchar, P. M., and Buck, C. J. (2001) *J. Biol. Chem.* **276**, 33588–33595
375. Julliard, J.-H., and Douce, R. (1991) *Proc. Natl. Acad. Sci. U.S.A.* **88**, 2042–2045
376. Kelleher, N. L., Taylor, S. V., Grannis, D., Kinsland, C., Chiu, H.-J., Begley, T. P., and McLafferty, F. W. (1998) *Protein Sci.* **7**, 1796–1801
377. Chiu, H.-J., Reddick, J. J., Begley, T. P., and Ealick, S. E. (1999) *Biochemistry* **38**, 6460–6470
377a. Peapus, D. H., Chiu, H.-J., Campobasso, N., Reddick, J. J., Begley, T. P., and Ealick, S. E. (2001) *Biochemistry* **40**, 10103–10114
377b. Reddick, J. J., Nicewonger, R., and Begley, T. P. (2001) *Biochemistry* **40**, 10095–10102
378. Cane, D. E., Hsiung, Y., Cornish, J. A., Robinson, J. K., and Spenser, I. D. (1998) *J. Am. Chem. Soc.* **120**, 1936–1937
378a. Cane, D. E., Du, S., and Spenser, I. D. (2000) *J. Am. Chem. Soc.* **122**, 4213–4214
378b. Osmani, A. H., May, G. S., and Osmani, S. A. (1999) *J. Biol. Chem.* **274**, 23565–23569
378c. Gupta, R. N., Hemscheidt, T., Sayer, B. G., and Spenser, I. D. (2001) *J. Am. Chem. Soc.* **123**, 11353–11359

References

379. Huynh, M. S., and Snell, E. E. (1985) *J. Biol. Chem.* **260**, 2379–2383
380. Nelson, M. J. K., and Snell, E. E. (1986) *J. Biol. Chem.* **261**, 15115–15120
381. Chaiyen, P., Brissette, P., Ballou, D. P., and Massey, V. (1997) *Biochemistry* **36**, 13856–13864
382. Fleming, S. M., Robertson, T. A., Langley, G. J., and Bugg, T. D. H. (2000) *Biochemistry* **39**, 1522–1531
383. Krieger, C. J., Roseboom, W., Albracht, S. P. J., and Spormann, A. M. (2001) *J. Biol. Chem.* **276**, 12924–12927

Study Questions

1. The vitamin niacin (nicotinic acid) is converted to NAD through the intermediate, desamido NAD.

   a) Propose an enzymatic synthesis of desamido NAD beginning from niacin and ribose 5-phosphate
   b) Propose an enzymatic synthesis of NAD beginning from desamido NAD

2. AMP is formed from IMP in an ATP requiring reaction. The introduced nitrogen atom is derived directly from aspartate.

   a) Propose a reaction mechanism based on an analogous reaction sequence occurring during amino acid metabolism.
   b) Identify the analogous enzymes upon which you have based your answer by names of the enzymes or by the names of the substrates and products.

3. One of the mechanisms proposed for the decarboxylation of orotidine 5'-phosphate to UMP involves initial addition of an enzyme nucleophile to the pyrimidine ring. Describe and criticize this mechanism. Hint: The proposal has some similarity to the thymidylate synthase mechanism.

4. Name the enzymes and describe the chemical reaction that occurs for each of the four steps in the following pathway:

$$\text{UDP} \xrightarrow{1} \text{dUDP} \xrightarrow{2} \text{dUTP} \xrightarrow{3} \text{dUMP} \xrightarrow{4} \text{dTMP}$$

5. The following reaction is catalyzed by a dioxygenase from *E. coli*. The dioxygenase reaction opens the ring and the intermediate is cleaved hydrolytically. Propose a structure for the intermediate and a mechanism for the hydrolytic cleavage. See Fleming *et al.*[382]

COO⁻ OH OH O_2, H_2O ^-OOC COO⁻ OH H_2C COO⁻

6. Anaerobic breakdown of toluene by denitrifying bacteria begins by the addition of toluene to fumarate to form benzylsuccinate. Benzylsuccinate synthase has an amino acid sequence homologous to that of pyruvate formate lyase (p. 799–801), contains a glycyl radical, and is activated in a manner similar to activation of pyruvate formate lyase. Propose a mechanism for formation of benzylsuccinate. See Krieger *et al.*[383]

COO⁻ COO⁻

Benzylsuccinate

7. Write out, using structural formulas, a step-by-step reaction sequence for the conversion of *O*-succinylbenzoate into isochorismate as indicated in Fig. 25-4.

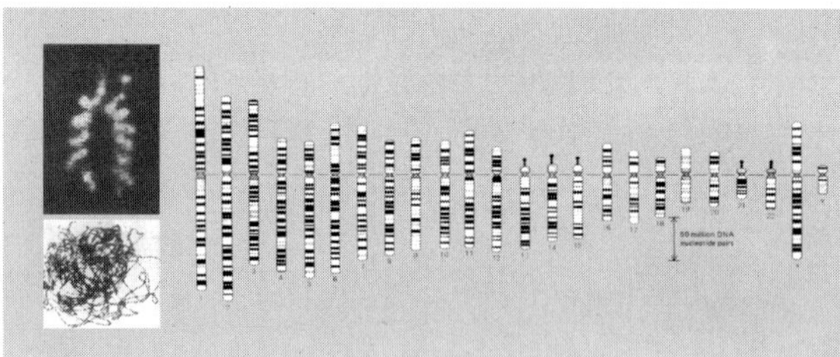

Right: The human **karyotype**, a full set of chromosomes numbered according to size and content and showing characteristic banding patterns. The 22 autosomes plus either an X or Y chromosome make up the haploid set which, for the female, contains ~ 3,500 Mbp of DNA. Each diploid cell contains 46 chromosomes. From Alberts *et al.* "Molecular Biology of the Cell", Third edition, Garland, New York, 1994, p 356.

Upper Left: A pair of mitotic sister chromatids in a section stained with an antibody to topoisomerase II. Notice that the two chromatids are coiled with opposite helical handedness. Lower Left: Meiotic chromosomes of a lily at the pachytene stage in which sister chromatids are connected along their length in a **synaptonemal complex.** From Kleckner, N. (1996) *Proc. Natl. Acad. Sci. U.S.A.* **93**, 8167–8174

Contents

Biochemical Genetics 26

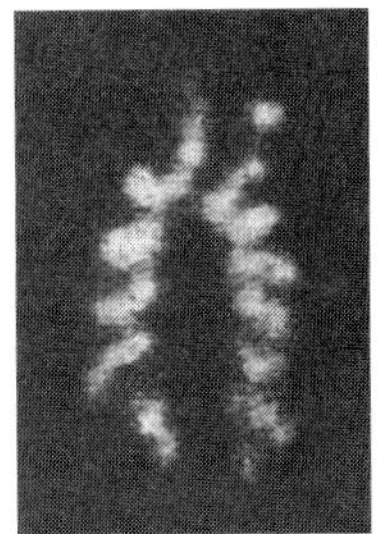

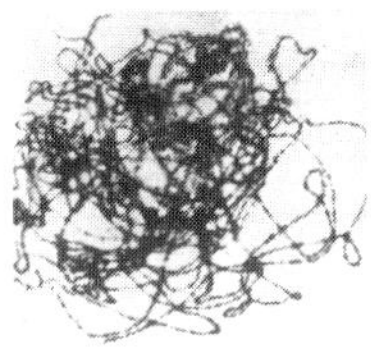

The most exciting biological discoveries of the 20th century include the unraveling of the genetic code (Chapter 5) and the understanding of the ways in which nucleic acids and proteins are synthesized. The biosynthesis of both nucleotides and amino acids has been considered in Chapters 24 and 25 and the basic chemistry of the polymerization processes in Chapters 12 and 17. Chapters 27–29 deal with the ways in which these polymerization reactions are controlled and by which the correct sequences of nucleotides or amino acids are obtained. The understanding of these matters is a development of genetics[1–2b] as well as of biochemistry; hence this introductory and historical chapter.

A. Historical Sketch

The discovery of deoxyribonucleic acid dates to 1869, when Miescher isolated a new chemical substance from white blood cells that he obtained from pus and later from sperm cells.[3] The material, which became known as nucleic acid, occurred in both plants and animals, thymus glands and yeast cells being among the best sources. Chemical studies indicated that the nucleic acids isolated from thymus glands and from yeast cells were different. As we now know, thymus nucleic acid was primarily DNA and yeast nucleic acid primarily RNA. For a while it was suspected that animals contained only DNA and plants only RNA, and it was not until the early 1940s that it was established that both substances were present in all organisms.[3–6]

1. DNA as the Genetic Material

In 1928, Griffith, using cells of *Diplococcus pneumoniae*, showed that genetic information that controls properties of the capsular polysaccharides could be transferred from one strain of bacteria to another. A material present in killed cells and in cell-free extracts permanently altered the capsular properties of cells exposed to the material.[7] This **transformation** of bacteria remained a mystery for many years. At the time of the experiments there was no hint of the genetic role of nucleic acids, which were generally regarded as strange materials. Furthermore, the covalent structure of nucleic acids was uncertain. A popular idea was that a tetranucleotide served as a repeating unit for some kind of regular polymer. Genes were most often thought to be protein in nature.

However, in 1944, Avery and associates showed that purified DNA extracted from pneumococci could carry out transformation.[8–12] The transforming principal appeared to contain little protein. It was not inactivated by proteolytic enzymes but was inactivated by deoxyribonuclease. Avery was 67 at the time of this discovery, refuting the popular contention that all important scientific discoveries are made by young people.

Other experiments also pointed to the conclusion that DNA was the genetic material. DNA was found localized in the nuclei of eukaryotic cells. The absolute amount per cell was constant for a given species. Studies of bacteriophage replication pointed strongly to DNA as the genetic material.[11] In 1952 Hershey and Chase showed that when a phage particle infects a cell the viral DNA enters the bacterium, but the protein "coat" remains outside.[13] This was demonstrated by preparing two types of labeled bacteriophage T2. One contained ^{32}P, which had been incorporated into the DNA, and the other ^{35}S incorporated into the proteins. Cells of *E. coli* were infected with the labeled phage preparations and were then agitated violently in a blender to shear the phage particles off of the bacteria. Over 80% of the ^{35}S was removed from the bacteria by this treatment, but most of the ^{32}P entered the bacteria and could be recovered in the next generation of progeny bacteriophage.[14]

2. The Double Helix

Development of newer methods of investigation of the chemical composition of nucleic acids led Chargaff to an important discovery. Although the base composition of DNA was extremely variable from organism to organism, the molar ratio of adenine to thymine was nearly always 1:1, as was that of cytosine to guanine![15] This observation provided the basis for the concept of base pairing in the structure of DNA. The final important information was supplied by X-ray diffraction studies of stretched fibers of DNA, which showed that molecules of DNA were almost certainly helical structures containing more than one chain. The crucial experiments were done by Rosalind Franklin[16–17b] and M. H. F. Wilkins, whose data were used by Watson and Crick in 1952 in constructing their model of the double helical structure[12,18–20] (Fig. 5-6). Once established, the structure of DNA itself suggested both the nature of the genetic code and a replication mechanism. *The genetic code had to lie in the sequence of nucleotides. Base pairing provided a mechanism by which the two mutually complementary strands could be separated. Biosynthesis of a new complementary strand alongside each one would result in a precise replication of each gene and of the entire genome.* In a similar fashion, RNA could be synthesized alongside a DNA "template" and could then be used to direct protein synthesis.

The presence of RNA in the cytoplasm had been linked to protein synthesis by experiments done in the early 1940s. After the discovery of the double helix, the concept followed quickly that DNA was the master "blueprint" from which secondary blueprints or **transcripts** of RNA could be copied. The RNA copies, later identified as **messenger RNA** (mRNA), provided the genetic information for specifying protein sequence. The flow of information from DNA to RNA to proteins could be symbolized as in Eq. 26-1.

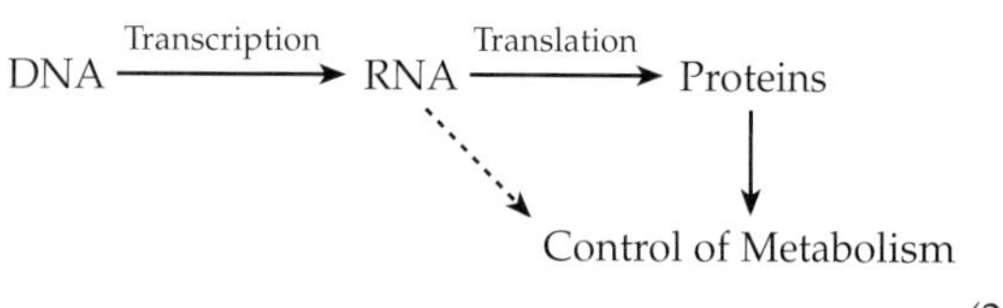

(26-1)

Proteins, in one way or another, control nearly all of metabolism. This includes the reactions that form the nucleotide precursors of the nucleic acids and that lead to polymerization of the amino acids and nucleotides, reactions catalyzed by **protein enzymes** and **ribozymes**. Thus, the flow of information from DNA to proteins is only part of a larger loop of metabolic processes (p. 973). Genetic information flows from DNA out into the cell, and copies of the master blueprint are passed nearly unchanged from generation to generation. The simple concepts implied by Eq. 26-1 quickly caught the imagination of the entire community of scientists and led to a rapid blossoming of the field of biochemical genetics.[20a,b]

3. Ribonucleic Acids and Proteins

By 1942 it was clear from ultraviolet cytophotometry developed by Caspersson[21] and from cytochemical work of Brachet[22,23] that RNA had something to do with protein synthesis. Use of radioautography with ^{3}H-containing uridine showed that RNA was synthesized in the nucleus of eukaryotic cells and was transported out into the cytoplasm.[24,25]

Ribosomes were discovered by electron microscopists examining the structure of the endoplasmic reticulum using ultrathin sectioning techniques. Their presence in cells was established by 1956, and the name ribosome was proposed in 1957. At first it was difficult to study protein synthesis *in vitro* using isolated ribosomes. No *net* synthesis could be detected until Hoagland and associates measured the rate of incorporation of ^{14}C-labeled amino acids into protein.[26] This sensitive method permitted measurement of very small amounts of protein synthesis in cell-free preparations from rat liver and paved the way toward studies with ribosomes themselves.

Immediately after the Watson–Crick proposals were made in 1953, it was generally thought that ribosomal RNA (rRNA), which constitutes up to 90% of the total RNA of some cells, carried the genetic message from the nucleus to the cytoplasm. By 1960 it seemed unlikely. For one thing the size and composition of rRNA was similar for different bacteria, despite differences in base composition of the DNA (Chapter 5).[27] It had been concluded that a relatively unstable, short-lived form of RNA must carry the message. Ribosomal RNA, however, was quite stable.[28]

Messenger RNA. In 1956, Volkin and Astrachan[29,30] detected a rapidly labeled and labile RNA in phage-infected bacterial cells. Studies of enzyme induction also suggested the existence of mRNA. Many bacteria, including *E. coli*, when grown on glucose as the sole source of energy and then suddenly switched to lactose, are unable to utilize the new sugar immediately. However, transfer to lactose induces, within a period of two minutes, the synthesis of new proteins needed for the metabolism of this sugar. Among the new proteins is a **permease** for lactose and a **β-galactosidase** (Chapter 12)[31] that cleaves the disaccharide to glucose and galactose. When the lactose is exhausted, the level of the induced enzymes drops almost as quickly. These results suggested that the RNA that carries the genetic message for synthesis of the new enzymes must be unstable. It must be produced rapidly in response to the presence of the inducing sugar and must disappear rapidly in its absence.

In 1961, Jacob and Monod postulated messenger RNA (mRNA) as a short-lived polynucleotide.[30,32,33] An abundance of additional evidence supported the proposal. For example, RNA molecules produced after infection of *E. coli* by bacteriophage T4 underwent hybridization (Chapter 5) with denatured DNA of the bacteriophage. Furthermore, this virus-specific mRNA became associated with preexisting bacterial ribosomes and provided the template for synthesis of phage proteins.[34] The experiment provided direct evidence for transcription of mRNA from genes of the viral DNA.

Transfer RNA. Crick[35] suggested, in 1957, that special "adapter" molecules might be needed to align amino acids with their codons in the RNA transcript. He thought that the adapters might be polynucleotides. At the same time chemical studies of the RNA of cells revealed that a low-molecular-weight RNA made up 15% of the total RNA of *E. coli*. This RNA was recognized in the same year as constituting the needed adapters, when Hoagland demonstrated enzymatic "activation" of amino acids and their subsequent incorporation into protein. The name transfer RNA (tRNA; Figs. 5-30, 5-31) was proposed.[36]

4. Deciphering the Genetic Code

The genetic code consists of triplets of DNA base pairs (codons), each corresponding to a single amino acid. The triplets are consecutive, do not overlap, and are not separated by any "punctuation." Although this represented one of the simplest possibilities for a code, it required much effort more than a period of more than 10 years to prove it. Even after the triplet nature of the genetic code became evident, many questions remained. Were all of the 64 possible codons used by the living cell? If so, were they all used to code for amino acids or were some set aside for other purposes? How many codons were used for a single amino acid? Was the code universal or did different organisms use different codes? How could one decipher the code? Despite the complexity of these questions, they seem to have been almost completely answered.

In an important experiment[37] Nirenberg and Matthaei, in 1961, isolated ribosomes from *E. coli* and mixed them with crude extracts of soluble materials, also from *E. coli* cells. The extracts included tRNA molecules and aminocyl-tRNA synthases. The 20 amino acids, ATP, and an ATP-generating system (PEP + pyruvate kinase) were added. Nirenberg showed that under such conditions protein was synthesized by ribosomes in response to the presence of added RNA. For example, RNA from tobacco mosaic virus (Chapter 7) was very effective in stimulating protein synthesis. The crucial experiment, which was done originally simply as a "control," was one in which a synthetic polynucleotide consisting solely of uridylic acid units was substituted for mRNA. In effect, this was a synthetic mRNA containing only the codon UUU repeated over and over. The ribosomes read this code and synthesized a peptide containing only phenylalanine. Thus, poly(U) gave polyphenylalanine, and *UUU was identified as a codon specifying phenylalanine.* The first nucleotide triplet had been identified! In the same manner CCC was identified as a proline codon and AAA as a lysine codon. Study of mixed copolymers containing two different nucleotides in a random sequence suggested other codon assignments. A few years later, after Khorana had supplied the methods for synthesis of oligonucleotides and of regular alternating polymers of known sequence, the remaining codons were identified.

Another important technique was based on the observation that synthetic trinucleotides induced the binding to ribosomes of tRNA molecules that were "charged" with their specific amino acids.[38,39] For example, the trinucleotides UpUpU and ApApA stimulated the binding to ribosomes of ^{14}C-labeled phenylalanyl-tRNA and lysyl-tRNA, respectively. The corresponding dinucleotides had no effect, an observation that not only verified the two codons but also provided direct evidence for the triplet nature of the genetic code. Another powerful approach was the use of artificial RNA polymers, synthesized by combined chemical and enzymatic approaches.[40] For example, the polynucleotide CUCUCUCUCU · · · led to the synthesis by ribosomes of a regular alternating polypeptide of leucine and serine.

Table 5-5 shows the codon assignments, as we now know them, for each of the 20 amino acids. Table 5-6 shows the same 64 codons in a rectangular array. In addition to those codons assigned to specific amino acids, three are designated as **chain termination codons**: UAA, UAG, and UGA. These are frequently referred to as "nonsense" codons. The termination codons UAA and UAG are also known as *ochre* and *amber*, respectively, although these names have no scientific significance.[41] The codons AUG (methionine) and much less often GUG (valine) serve as the **initiation codons** in protein synthesis. Consequently, the N-terminal amino acid in most newly synthesized eukaryotic proteins is methionine, and in bacterial proteins it is *N*-formylmethionine. As explained in Chapter 29, *N*-formylmethionyl-tRNA is specifically bound to initiation sites containing the AUG codon in bacterial mRNA-ribosome complexes.

A number of studies suggested that the genetic code as worked out for *E. coli* might be universal. For example, in the laboratories of Wittman and of Fraenkel-Conrat, RNA extracted from tobacco mosaic virus was treated with nitrous acid, a procedure known to deaminate many cytosine residues to uracil (Eq. 5-12). Such treatment could change the codon UCU (serine) to UUU (phenylalanine) and the codon CCC (proline) to CUC (leucine). When the nitrous acid-treated RNA was used to infect tobacco plants and virus particles were prepared in quantity from the resultant mutant strains, it was found that the amino acid sequence of the virus coat protein had been altered,[42] and that many of the alterations were exactly those that would be predicted from Table 5-6. Likewise, the amino acid substitutions in known defects of hemoglobin (Fig. 7-23) could be accounted for, in most cases, by single base alterations. Thus, hemoglobin S arose as a result of the following change in the sixth codon of the globin β chain gene: GAG (Glu) $\rightarrow$ GTG (Val).[43] Another argument favoring a universal code was based on the observation that mRNA coding for a globin chain could be translated by ribosomes and tRNA molecules from *E. coli*. The resulting protein was authentic mammalian globin.[44]

As often happens, a well-established conclusion may have to be modified. There are exceptions to the universal genetic code in mitochondrial DNA and in some protozoa (Chapter 5).[45]

B. Genetic Methods

Our present knowledge of molecular biology has depended greatly on the methods of genetics. The following introduction begins with a consideration of mutations.

1. Mutations

Changes in the structure of DNA occur only rarely. The average gene may be duplicated 10^6 times before some mistake results in a single detectable mutation.[46] Nevertheless, by using bacteria or bacterial viruses, it is possible to screen enormous numbers of individuals for the occurrence of mutations. If one million virus particles are spread on an agar plate under conditions where mutation in a certain gene can be recognized, on the average one mutant is found. The most common mutations are **base-pair switches** or **point mutations** that result from incorporation of the wrong base during replication or repair. In these mutations one base of a triplet codon is replaced by another to form a different codon, causing the substitution of one amino acid by another in the corresponding protein as was seen for hemoglobin S. Changes involving replacement of one pyrimidine by another (C $\rightarrow$ T or T $\rightarrow$ C) or of one purine by another are sometimes called **transition mutations**, whereas if a pyrimidine is replaced by a purine, or vice versa, the mutation is known as a **transversion**. An example is the previously mentioned A $\rightarrow$ T in hemoglobin S. Transition mutations are by far the most common, one possible cause being pairing with a minor tautomer of one of the bases (Chapter 5). For example, A could pair with a minor tautomer of C, causing a mutation from T to C. Note that substitution of an incorrect base in one strand will lead, in the next round of replication, to correct pairing again but with an AT pair replaced by GC, or vice versa, in one of the daughter DNA duplex strands. A base substitution does not always cause an amino acid replacement because of the "degeneracy" of the code, i.e., the fact that more than one codon specifies a given amino acid.

From the observed rate of appearance of point mutations (one mutation per 10^6 gene duplications), we can estimate that one mutation occurs per 10^9 replications at a single nucleotide site. Point mutants tend to "back mutate," often at almost the same rate as is observed for the forward mutation. That is, one in 10^9 times a mutation of the same nucleotide will take place to return the code to its original form. The phenomenon is easy to understand. For example, if T should be replaced by C because the latter formed a minor tautomer and paired with A, the mutation would appear in progeny duplexes as a GC pair. When this pair was replicated, there would be a finite probability that the C of the parental DNA strand would again assume the minor tautomeric structure and pair with A instead of G, leading to a back mutation.

Although the rates of spontaneous mutation are low, they can be greatly increased by mutagenic chemicals (Chapter 27) or by irradiation. It is perfectly practical to measure the rates of both forward and back mutation. When this was done, it was found that certain chemicals, e.g., acridine dyes, induce mutations that undergo reverse mutation at a very much lower frequency than normal. It was eventually shown that these mutations resulted either from **deletions** of one or more nucleotides from the chain or from **insertions** of extra nucleotides. Deletion and insertion mutations often result from errors during genetic recombination and repair at times when the DNA chain is broken.

Mutations involving deletion or insertion of one or a few nucleotides are called **frame-shift mutations**.

Messenger RNA is read by the protein synthesizing machinery from some starting point. As is illustrated in Chapter 5, Section E,1 the codons are read three bases at a time, and the proper amino acid corresponding to each codon is inserted. *When a deletion or insertion in the mRNA is met, all subsequent codons may be misread because the reading frame has shifted forward or backward by one or two nucleotides.* The protein synthesized bears little resemblance to that formed by the nonmutant organism and is usually completely nonfunctional. Mutations are considered further in Chapter 27.

2. Mapping the Chromosome of a Bacteriophage

Intensive work on the "T-even" phage T2, T4, and T6 (Box 7-C) was begun in 1938 by Delbrück and associates. The genetic information for these viruses is carried in a single linear DNA molecule, which in the case of T4 contains ~1.7×10^5 base pairs (170 kb), enough for about 200 genes. Before the sequences of the viral DNA were known, the positions of more than 60 of these genes were mapped in the following way. When a bacteriophage infects a cell of *E. coli*, it injects its DNA through the cell wall and into the cytoplasm. About 20 minutes later the cell bursts, and ~100 fully formed replicas of the original virus particle are released. This rapid rate of production of progeny is so fast that it is possible to carry out in a test tube in 20 min a genetic experiment that would require the entire population of the earth if done with humans. The approach is explained nicely by Seymour Benzer, the man who first mapped the fine structure of a gene.[47] Bacteriophage particles, like bacteria, can be "plated out" on agar plates, which must contain a uniform suspension of bacteria susceptible to the virus. Wherever a virus particle lies, a bacterium is infected. Soon the infection spreads to neighboring bacteria with production of a transparent "plaque" (Fig. 26-1). The number of active virus particles present in a suspension can be determined easily by plating and counting of the plaques.

Mutant bacteriophages can be identified in various ways. Some biochemical traits affect the appearance of the plaque. Other easily detected traits include alteration in the specificity toward strains of the host bacterium. A key discovery that made genetic mapping possible for bacteriophage was that *genetic recombination between two phage particles can take place within a host bacterium.* When large numbers of bacteriophages of two different mutant strains were grown and were mixed together in excess with many bacteria, a few of the progeny phage were found to contain both mutant traits in the same virus and an equal number were "wild type." Although recombinations between mutations that are located close together in the DNA are rare, their frequency still greatly exceeds that of new mutations. While this type of experiment gave no hint about the nature of the events involved, it showed that recombination had occurred.

Study of recombination frequencies between different strains of phage soon revealed that some sites of mutation are **closely linked**. Recombination between these sites occurs only rarely. Other sites are weakly linked, and recombination occurs often. This behavior was reminiscent of that established many years earlier for genes of the fruit fly *Drosophila*, maize, and other higher organisms. Recombination by "crossing over" in the chromosomes of *Drosophila* was established by Morgan and associates in 1911.[1,2] The basic idea behind chromosome mapping in any organism is the assumption that *recombination frequencies between two mutations are directly proportional to the distance between them on the genetic map.* For the T4 phage a recombination frequency of 1% is taken as one unit. The total T4 map is 700 units long. The fact that this is greater than 100% means that if genes are located at opposite ends of the chromosome multiple recombination events can occur between them. However, a maximum of 50% crossing over is observed for distant gene pairs, and the approximate linearity of map distance and recombination frequency holds only for distances of 10 units or less.[48]

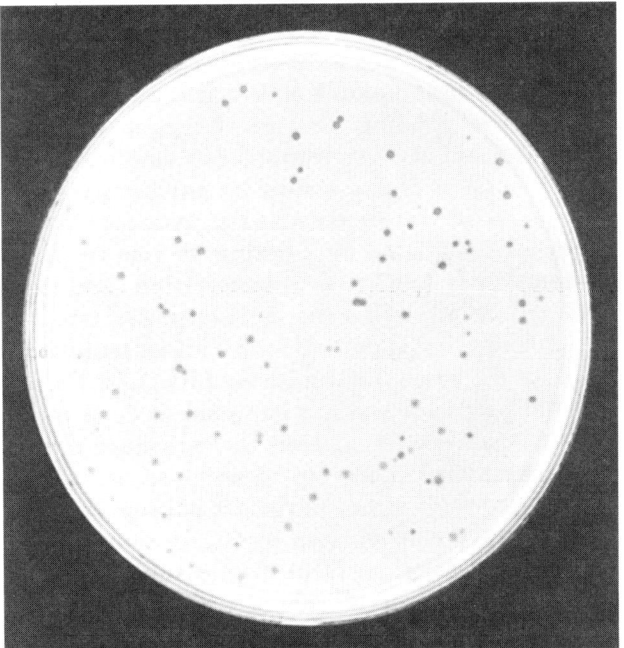

Figure 26-1 Plaques formed by bacteriophage ϕ11 growing on *Staphylococcus aureus*. Each transparent (dark) plaque is the result of lysis of bacteria by the progeny of a single bacteriophage particle. Courtesy of Peter Pattee

How can recombinant bacteriophage be identified rapidly? Benzer used two strains of *E. coli*, the B strain and the K strain, as hosts. Mutants in gene *rII* form characteristic plaques on strain B but do not grow on strain K. To determine the recombination frequency between two different *rII* mutants, the viruses were added to liquid cultures of B cells (in which they replicate), and recombination was allowed to occur. Recombination permitted the emergence not only of a phage containing *both* mutations but also of a wild-type phage in which both mutations had been eliminated by the recombination process. Since only recombinants of the latter type grow in strain K, it was possible to detect a single recombinant among one billion progeny. Since the total DNA length in phage T4 is 166 kb, there are 237 base pairs for each of the 700 units of map length. Thus, a recombination frequency at 0.01% between two mutations meant that the two mutations were no more than three base pairs apart in the DNA. Benzer concluded that he had observed the expected recombination frequencies for mutations even of immediately adjacent bases in the DNA.

To make fine genetic mapping practical, a series of bacteriophage containing deletion mutations involving large segments of the *rII* gene were isolated. Using these, it was possible to establish in which segments of the gene a particular mutation lay. Then, recombination experiments with previously identified mutations in that same general region allowed the mutations to be pinpointed. In this way Benzer identified over 300 sites of mutations within the *rII* region. He concluded that the minimum distance between two mutable sites was compatible with the Watson–Crick structure of the gene.[47]

3. The Cistron

How can one tell whether two mutations are in the same gene or in nearby or adjacent genes? The answer can be supplied by a test of **complementation**. If two mutant bacteriophage are altered in different genes, they can often reproduce within a host if the bacterium is infected with both of the mutant phage. Since each one has a good gene for one of the two proteins involved, recombinant phage, in which all of the gene functions are fulfilled, will be formed. On the other hand, if both mutant phages are defective in the same gene (although at different locations), they usually cannot complement each other at high frequencies in a coinfection. The experiment is referred to as a ***cis–trans* comparison**. The coinfection with the two different mutants is the trans test. A control, the *cis* test, uses a recombinant containing both of the mutations in the same DNA and coinfection with a standard phage. Normal replication is expected in this instance.

When the complementation (or *cis–trans*) test was applied to various mutants in the *rII* region, it became clear that there are two genes, *rIIA* and *rIIB*. The name **cistron** was proposed by Benzer to represent that length of DNA identifiable in this fashion as a genetic unit. *For most purposes the terms gene and cistron are nearly synonymous.* When mapping of the *rII* region was done, there was no information about the functions of the proteins specified by these two cistrons. However, both the *rIIA* and *rIIB* proteins have since been shown to become incorporated in the membranes of phage-infected bacterial cells.[49,50] There they affect the ease of lysis of the infected cells and, in that manner, cause *rII* plaques to be larger and to have sharper edges than standard plaques.

4. Nutritional Auxotrophs

The beginning of biochemical genetics is often attributed to Beadle and Tatum, who, in 1940, discovered mutants of *Neurospora* with specific growth requirements. They X-rayed one parent strain to form mutants, then tested individual spores for their possible need of a specific nutrient for growth. The 299th spore tested required pyridoxine! Many other mutants requiring vitamins, amino acids, and nucleic acid bases were then discovered.[1,51] A few years later similar **nutritional auxotrophs**, as the mutants are called, were discovered for *E. coli*. Ordinary wild-type cells of *E. coli* can grow on a minimal medium containing a carbon compound as a source for energy together with inorganic nutrients. Irradiation with ultraviolet light or treatment with mutagenic chemicals produces many mutant cells that fail to grow on such a minimal medium. However, addition of one or more specific compounds, such as an amino acid or vitamin, usually permits growth. Selection of such nutritional auxotrophs can be accomplished by plating out large numbers of the irradiated or chemically treated cells on a solid, rich nutrient medium. Colonies (**clones**) are allowed to develop by multiplication of the individual bacteria. The auxotrophs are selected by **replica plating**.[2] In this procedure a sterile velveteen pad is pressed against a nutrient agar plate containing small colonies of bacteria and is used to "print" replica plates containing a minimal medium. The colonies on the initial and replica plates are compared and the colonies of auxotrophs (which do not grow on the minimal medium) are selected. In a second stage, the auxotrophs may be replica plated to minimal medium supplemented with various nutrients (amino acids, purines, pyrimidines, vitamins, etc.). Selection is made easier by pretreatment of the irradiated cells suspended in minimal medium with penicillin. Penicillin kills the growing cells, but the auxotrophs, which do not grow on the minimal medium, survive. The penicillin is then destroyed by adding penicillinase (a β-lactamase; Box 20-G) leaving a suspension much enriched in the percentage of auxotrophic mutants.[2]

A nutritional auxotroph of a bacterium often has a defect in a gene specifying a protein needed for the biosynthesis of the required nutrient. Individual genes recognized in this way are named with a genetic symbol. For example, gene *trpA* specifies one of the two protein subunits of tryptophan synthase. Other kinds of mutations, e.g., those affecting motility or other properties of the cells, can also be detected and are given appropriate symbols. A few of these genetic symbols are indicated on the *E. coli* chromosome map in Fig. 26-4, and many others are used throughout this book. On the basis of such nutritional experiments Beadle, by 1945, had proposed his famous one-gene-one-enzyme hypothesis.[1]

5. Establishing the Correspondence of a Genetic Map and an Amino Acid Sequence

Although the studies of the *rII* region of the T4 chromosome established that genetic mapping could be carried to the level of individual nucleotides in the DNA, it was still necessary to prove a linear correspondence between the nucleotide sequence in the DNA and the amino acid sequence in proteins. This was accomplished by Yanofsky[52,53] and associates through study of the enzyme tryptophan synthase of *E. coli*. Tryptophan synthase (Fig. 25-3) consists of two subunits, α and β, the former containing only 268 amino acids and encoded by the *trpA* gene. To obtain a fine structure map of the A gene, a large series of tryptophan auxotrophs unable to grow in the absence of added tryptophan were isolated. Genetic crosses were carried out with the aid of a **transducing bacteriophage** Plkc. Transducing bacteriophage (Section E,3), while multiplying in susceptible bacteria, sometimes incorporate a portion of the bacterial chromosome into their own DNA. Then, when the virus infects other bacteria, some of the genetic information can be transferred through recombination into the chromosome of bacteria that survive infection. Use of a series of deletion mutants, as in the *rII* mapping, permitted division of the A gene into a series of segments, and observation of recombination frequencies permitted fine structure mapping.

The second part of the proof of colinearity of DNA and protein sequences was the determination of the complete amino acid sequence of tryptophan synthase and peptide mapping (Chapter 3) of fragments of the mutant enzymes. From the peptide maps it was possible to identify altered peptides and to establish the exact nature of the amino acid substitutions present in a variety of different tryptophan auxotrophs. When this was done, it was found that those mutations that mapped very close together had amino acid substitutions at adjacent or nearly adjacent sites in the peptide chain.

The same problem was approached by Sarabhai and associates[54] through the nonsense mutations (Section 6), which lead to premature chain termination during protein synthesis. During late stages of the infection of *E. coli* by phage T4, most of the protein synthesis is that of a single protein of the virus head. Synthesis of protein by infected cells was allowed to proceed in the presence of specific ^{14}C-labeled amino acids. Then cell extracts were digested with trypsin or chymotrypsin, the head-protein peptides were separated by electrophoresis, and autoradiograms were prepared. A series of T4 nonsense mutants that mapped within the head-protein gene were shown to give rise to incomplete head-protein chains. The peptide fragments were of varying lengths. By examining the radioautograms prepared from the enzymatically fragmented peptides, it was possible to arrange the mutants in a sequence based on the length of peptide formed and to show that this was the same as that deduced by genetic mapping. More recently the colinearity of codon and amino acid sequences has been verified repeatedly by comparison of experimentally determined nucleotide sequences in RNA and DNA molecules with the corresponding amino acid sequences for thousands of proteins.

Before the triplet nature of codons had been established, Crick and associates used frame-shift mutations in a clever way to demonstrate that the genetic code did consist of triplets of nucleotides.[7,55,55a] Consider what will happen if two strains of bacteria, each containing a frame-shift mutation (e.g., a −1 deletion), are mated. Genetic recombination can occur to yield mutants containing *both* of the frame-shift mutations. It would be difficult to recognize such recombinants because, according to almost any theory of coding, they would still produce completely defective proteins. However, Crick *et al.* introduced a third frame-shift mutation of the same type into the same gene and observed that the recombinants containing all three deletions (or insertions) were able to synthesize at least partially active proteins. Thus, while introduction of one or two single nucleotide deletions completely inactivates a gene, deletion of three nucleotides close together within a gene shortens the total message by just three nucleotides. The gene will contain only a short region in which the codons are scrambled. The reading frame for the remainder of the protein will not be changed. The protein specified will often be functional because it has a normal sequence except for a small region where some amino acid substitutions will be found and where one amino acid will be completely missing.

6. Conditionally Lethal Mutations

Studies of plaque morphology and of nutritional auxotrophs are directed narrowly at one gene or group of genes. It is desirable to have a general means of detecting mutations in the many other genes present within cells. However, most mutations are **lethal,** and this effect cannot be overcome by adding any nutrient. Lethal mutations are very common in higher organisms, but since eukaryotic cells have pairs of homologous chromosomes, they can be carried in one chromosome and the individual survives. With bacteria and viruses there is only one chromosome, and lethal mutants cannot survive.

Nutritional auxotrophs can be described as **conditionally lethal mutants**; they survive only if the medium is supplemented with the nutrient, whose synthesis depends upon the missing enzyme. Other kinds of conditional lethal mutations permit study of almost every gene in an organism. For example, **temperature-sensitive** (*ts*) mutants grow perfectly well at a low temperature, e.g., 25°C, but do not grow at a higher temperature, e.g., 42°C.[41,56] Many temperature-sensitive mutations involve an amino acid replacement that causes the affected protein to be less stable to heat than is the wild type protein. Others involve a loss in protein-synthesizing ability for reasons that may be obscure. Temperature-sensitive mutations occur spontaneously in nature, an example being the gene that controls hair pigment in Siamese cats.[41] The gene (or gene product) is inactivated at body temperatures but is active in the cooler parts of the body, such as the paws, tail, and nose, with the result that the cat's hair is highly pigmented only in those regions.

Screening for conditionally lethal temperature-sensitive mutants of bacteriophage T4 permitted isolation of hundreds of mutants involving sites at random over the entire viral chromosome. Complementation studies permitted assignment of these to individual genes, which at first were identified only by number (Fig. 26-2). Later specific functions were associated with the genes.[57,58] For example, the product of gene 42 was identified as an enzyme required in the synthesis of hydroxymethyl-dCMP (Chapter 5). Genes 20–24, among others, must code for head proteins because mutants produce normal tails but no heads. Gene 23 codes for the major head subunits, while gene 20 has something to do with "capping" the end of the head. These mutants produce cylindrical "polyheads" in place of the normal heads. Mutants of genes 25–29 have defective base plates and do not form tails, while mutants 34–38 lack tail fibers. The specific ways in which some of these gene products are assembled to form the base plates and tails of the phage are indicated in Fig. 7-29. The positions of the *rIIA* and *rIIB* genes are also shown in Fig. 26-2.

A second type of conditionally lethal mutation leads to alteration of an amino acid codon to one of the three **chain termination codons** UAG, UAA, and UGA (Table 5-3).[60,61] These are often called **nonsense mutations** in contrast to **missense mutations** in which one amino acid is replaced by another. A chain termination mutant synthesizes only part of the product of the defective gene because of the presence of the termination codon. A remarkable aspect of chain termination mutations is that they can be **suppressed** by other mutations in distant parts of the virus or bacterial chromosome. Many otherwise lethal mutations of bacteriophage T4 were discovered by their ability to grow in mutant strains of *E. coli*, which contained **suppressor genes**,[62] and their inability to grow in the normal B strain. Three different suppressor genes *supD*, *supE*, and *supF* were found to suppress mutations that formed UAG. These are commonly known as *amber* suppressor genes. A second group of *ochre* suppressor genes, including *supB* and *supC*, suppressed mutations that formed codon UAA. Suppressors for mutations that form codon UGA have also been found.[63] Like the temperature-sensitive mutants, *amber* and *ochre* mutants can be obtained in almost any genes of a bacterial virus. Chain termination mutants of unessential genes in bacteria can be recognized by transferring the genes by conjugation or by viral transduction into a strain (sup^+) that contains a desired suppressor gene.

Conditionally lethal mutants have been of great value in developing our understanding of the genetics of bacterial viruses. They have also provided a powerful technique for approaching complex problems of bacterial physiology. For example, we may ask how many genes are required for a bacterium to sense the presence of a food and to swim toward it (Chapter 19). Even though few clues as to the basic chemistry underlying these phenomena can be obtained in this way, the use of temperature-sensitive mutants and complementation tests permits us to establish the total number of genes involved in these complex processes and to map their positions on a bacterial or viral chromosome. This is often an important step toward a more complete understanding of a biological phenomenon.

7. The Nature of Suppressor Genes

How can one mutation be suppressed by a second mutation at a different point in the chromosome? Rarely, a mutation is suppressed by a second mutation within the *same* gene. Such **intragenic complementation** sometimes occurs when a mutation leads to an amino acid replacement that disrupts the structural stability or function of a protein. Sometimes a mutation at another site involving a residue, which interacts with the first amino acid replaced, will alter the inter-

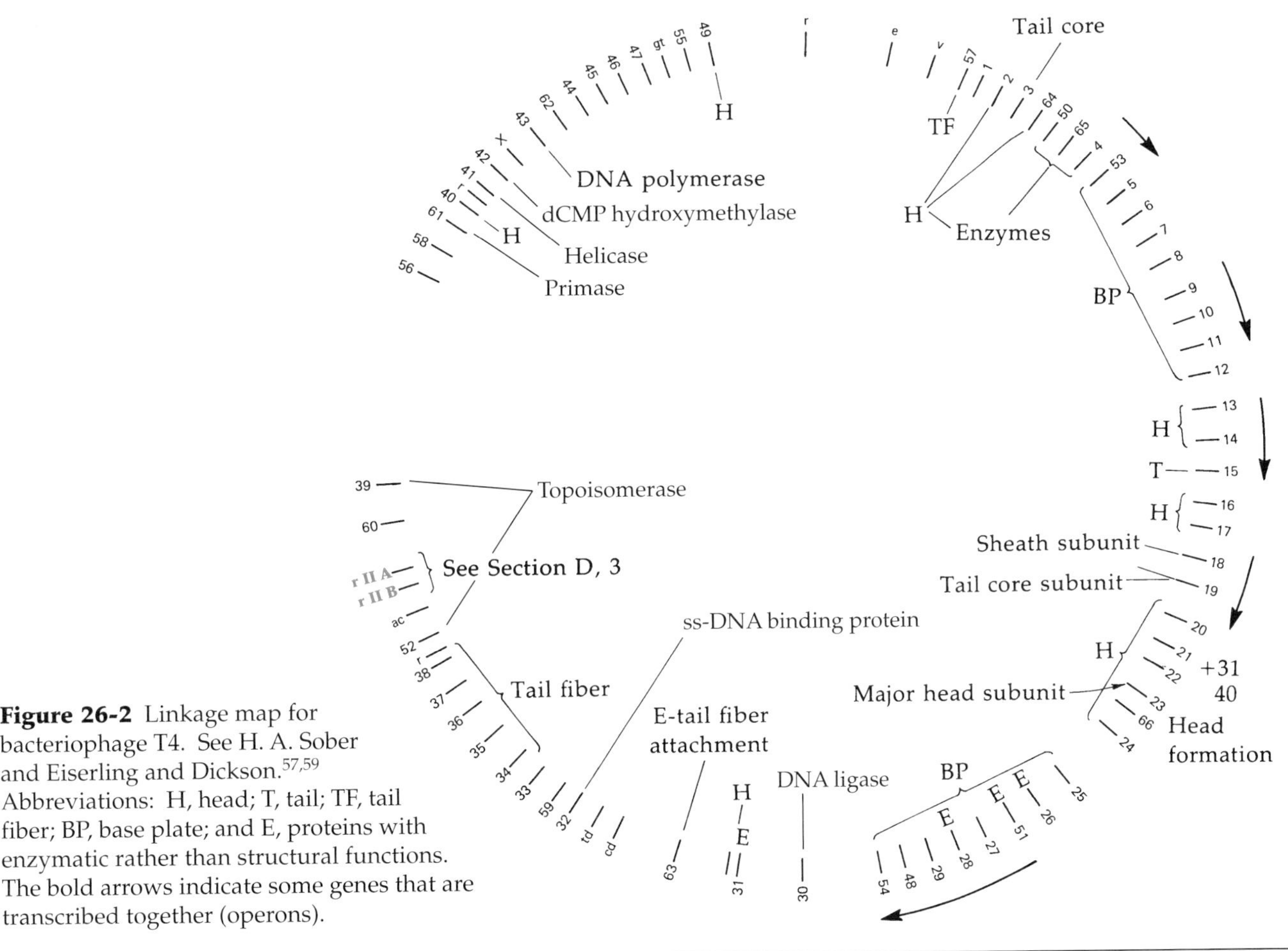

Figure 26-2 Linkage map for bacteriophage T4. See H. A. Sober and Eiserling and Dickson.[57,59] Abbreviations: H, head; T, tail; TF, tail fiber; BP, base plate; and E, proteins with enzymatic rather than structural functions. The bold arrows indicate some genes that are transcribed together (operons).

action between the two residues in a way that restores function to the protein. For example, if the first amino acid side chain is small and is replaced by mutation with a larger side chain, a second mutation leading to a decrease in the size of another side chain may permit the protein to fold and function properly. An example was found among mutants of tryptophan synthase.[64] Mutants in which Gly 211 of the α chain was replaced by Glu or Tyr 175 of the same chain by Cys both produced inactive enzymes. However, the double mutant with both replacements synthesized active tryptophan synthase. It is known now that these residues are adjacent to one another and form part of the binding site for the substrate indole-3-glycerol phosphate (Figs. 25-2; 25-3). Only the double mutant permits the substituted side chains to pack properly. In other cases, intragenic suppression involves changes in the subunit interactions in oligomeric proteins. These changes may affect the formation of correct quaternary structures of the proteins.

As discussed in the preceding section, the best known suppressor genes are those that suppress chain termination mutations (Section 6). These genes often encode mutant forms of tRNA molecules, which allow incorporation of an amino acid rather than chain termination to occur. They are discussed further in Chapter 29. Suppressor genes are not limited to bacteria. For example, the vermilion eye color mutation of *Drosophila* leads to a loss of brown eye pigments because of the inactivity of tryptophan 2,3-dioxygenase (Eq. 18-38). However, synthesis of the tryptophan dioxygenase from the *vermilion* mutant is inhibited by $tRNA_2^{Trp}$, one of the two tryptophanyl tRNAs. The suppressor mutation alters the tRNA in such a way that the inhibition is relieved.[61]

C. Plasmids, Episomes, and Viruses

The small pieces of DNA known as plasmids, which replicate independently of the chromosomes, have been discussed briefly in Chapter 5. Plasmids share a number of properties with viruses, and both are important to the techniques of contemporary molecular biology and genetic engineering. Bacterial plasmids may be present as one or several copies for each chromosome. Episomes are plasmids that are able to become integrated into the bacterial chromosome. Some extrachromosomal elements are episomes in one host and plasmids in another. Bacterial plasmids may be infectious (transferable) or noninfectious. In the former case, they are able to transfer their DNA into another cell and are known as **sex factors** (F agents or fertility factors). A sex factor is able to integrate into a chromosome and later to come out and transfer other genes with it. In this property it resembles a transducing phage.

Plasmids and episomes vary in size. The F sex factor is a 100-kb circular molecule of supercoiled DNA. Colicinogenic factors,[65] which may also be present in *E. coli* in as many as 10–15 copies per bacterial chromosome, are often much smaller (6–7.5 kb). Some larger colicinogenic plasmids are also sex factors. They carry genes for toxic protein antibiotics known as **colicins** (Box 8-D) which attack other strains of *E. coli*, providing a selective advantage for the strain producing the colicin. They also carry a gene or genes conferring on the host bacterium resistance to antibiotics such as penicillin and chloramphenicol. Penicillin is inactivated because the plasmids carry a gene encoding a penicillinase that hydrolytically cleaves the β-lactam ring (Box 20-G). Chloramphenicol (Fig. 25-10) is inactivated by the action of chloramphenicol *O*-acyltransferase.

1. Bacterial Sex Factors

Bacteria usually reproduce by a simple cell division. The DNA in the chromosome is doubled in quantity and the cell divides, each daughter cell receiving an identical chromosome. However, in 1946 Lederberg and Tatum showed that sexual reproduction is also possible.[66,67] They studied nutritional auxotrophs of *E. coli* strain K-12, which lacked the ability to synthesize amino acids or vitamins. When cells of two different mutants were mixed together and allowed to grow for a few generations, a few individual bacteria regained the ability to grow on a minimal medium. Since each of the two strains had one defective gene, the creation of an individual with neither of the two defects required combining of genetic traits from both strains. The existence of bacterial conjugation was recognized. Later it was established that true **genetic recombination** had occurred, i.e., genes from the two mating cells had been integrated into a single molecule of bacterial DNA.

This transfer of DNA between bacterial cells requires the presence of a plasmid sex factor (F agent), whose presence confers a male character to the individual cell. The F agent is large enough to contain about 90 genes and has a length of ~30 μm, ~2.5% that of the *E. coli* chromosome. Among other things, the *E. coli* F agent contains the genes needed to direct the synthesis of the **F pili** (sex pili). These tiny appendages, 8.5 nm in diameter (see Fig. 7-9), grow out quickly during a period of 4–5 min to a length of about 1.1 μm. The end of an F pilus becomes attached to a female cell (a bacterium lacking the F agent) and may induce the transfer of DNA into the female cell. The mechanism of transfer has been uncertain. It may involve formation of a cytoplasmic bridge between cells in close contact. The pilus may be retracted into the membrane of the male cell, pulling the two cells close together. The DNA probably flows through the pilus into the female.[68,69]

On rare occasions an F agent becomes integrated into the chromosome of a bacterium. Both the F agent and the chromosome have been shown by electron microscopy to be circular. The integration process requires the enzymatic cleavage of the DNA of both the chromosome and the F agent and the rejoining of the ends in such a way that a continuous circle is formed (Fig. 26-3). The enzymes that catalyze these reactions are considered in Chapter 27. Different F agents can be incorporated into the chromosome at different points around the circle. A strain of bacteria containing an integrated F agent is known as an *Hfr* (high frequency of recombination) strain.

When an Hfr strain conjugates with an F^- (female), replication of the entire male chromosome commences at some point near the end of the integrated F agent, and genes of the bacterial chromosome followed by those of the F factor are transferred into the female. Only a single strand of DNA (customarily referred to as the **plus strand**) is transferred from the donor cell and into the recipient cell (Fig. 26-3). There the complementary **minus strand** is synthesized to form a complete double-stranded DNA molecule bearing the genes from the Hfr cell. Only rarely does a copy of the entire chromosome of the donor cell enter the female cell. More often the DNA strand, or perhaps the pilus itself, breaks and only part of the chromosome is transferred.

Partial chromosome transfer from a male cell transforms the F^- cell into a partial diploid (**merozygote**) containing double the usual number of some of the genes. Within this partial diploid genetic recombination between the two chromosomes takes place (Fig. 26-3) by the mechanisms discussed in Chapter 27. The end result of the recombination process is that the daughter cells formed by subsequent division contain only single chromosomes with the usual number of genes. However, some genes come from each of the two parental strains. Thus, an F^- mutant unable to grow on a medium deficient in a certain nutrient may receive a gene from the male and now be able to grow on a minimal medium. Even though the number of such recombinants is small, they are easily selected from the very large number of mutant bacteria that are mixed together initially.

One result of DNA transfer from Hfr into F^- bacteria is sometimes the introduction of a complete copy of the F agent into the female bacterium. Since this con-

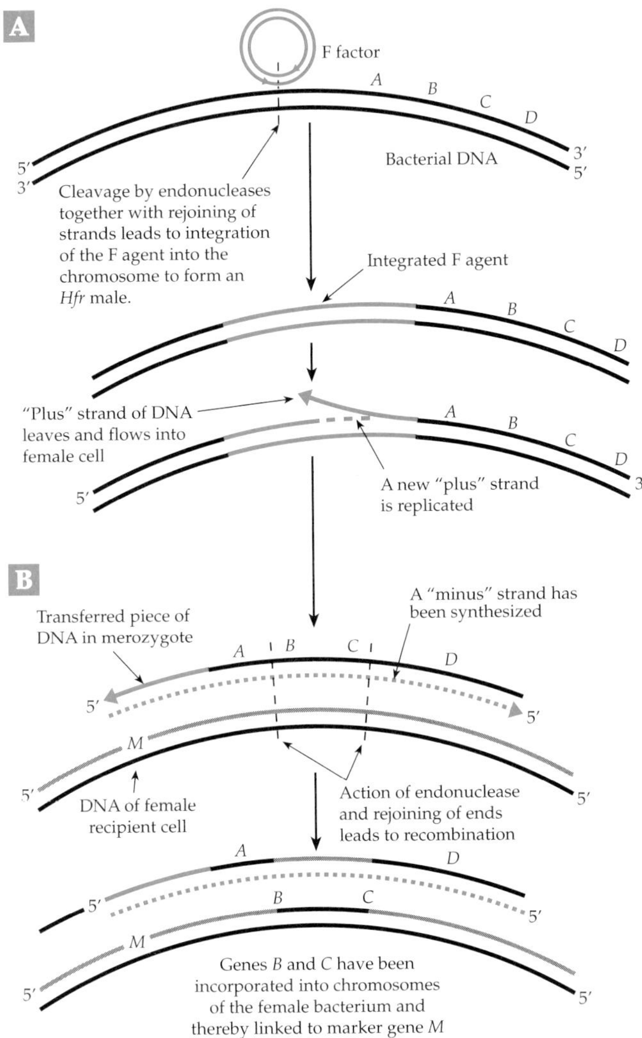

Figure 26-3 Integration of an F agent into a bacterial chromosome and transfer of some bacterial genes into another cell. (A) Incorporation of the F agent into *E. coli* genome and transfer of the "plus" strand of DNA out to a female recipient cell. (B) Genetic recombination between a piece of transferred DNA and the genome of the recipient cell.

verts the recipient into a male, Brinton referred to "bacterial sex as a virus disease." In fact, infectious plasmids and viruses display many similarities. For example, filamentous bacteriophages (Chapter 5; Fig. 7-7) adsorb to the F pili of male bacteria and the DNA, flowing in a direction opposite to that in bacterial conjugation, enters the cell.[70] The bacteriophage carry genes for the protein subunits of their coats (Figs. 26-2, 7-29), while F factors carry genes for synthesis of pilins. Pilins accumulate within the cell membrane and are extruded to generate F pili, just as viral subunits are extruded to form the virus coats. There is also a close similarity between episomes that can be integrated into bacterial chromosomes and the **temperate bacteriophages** considered in the next section.

2. Temperate Bacteriophage; Phage Lambda

When DNA from a typical bacteriophage enters a bacterial cell, it seizes control of the metabolic machinery of the cell almost immediately and directs it entirely toward the production of new virus particles. This leads within a period of about 20 min to the production of one or two hundred progeny viruses and to the lysis and death of the cell. However, the DNA from a temperate phage may become repressed and, like an F factor, be integrated with the bacterial genome (Fig. 26-3). In the resulting **prophage** or **lysogenic** state, the repressed phage DNA is replicated as part of the bacterial genome but does no harm to the host cells unless some factor "activates" the incorporated genetic material by release of the repression. Replication of the phage and lysis of the bacterium then ensues. Temperate phage may also exist as plasmids (e.g., plasmid P1).

The decision for lysis or lysogeny, which is very important for the survival of the bacteriophage, is governed primarily by the nutritional status of the host. For a bacterium growing in a relatively rich environment such as the colon, lysis will increase the chances of daughter phage encountering host bacteria. However, in soils, *E. coli* grows very slowly, and a bacteriophage capable of entering a lysogenic state has an increased chance of survival until the host bacterium finds a richer growth medium.

The best known temperate phage is **phage lambda** of *E. coli*.[70–72] A tailed virus resembling the T-even phages (Box 7-C), phage λ has a smaller (~48.5 kb) DNA genome.[73] Within the bacterial cell the ends of the λ DNA may be joined to form a circular replicative form of the virus. In ~30% of the infected cells the λ DNA becomes integrated into the *E. coli* chromosome at the special site, *att* λ, which is located at 17 min on the *E. coli* chromosome map (Fig. 26-4). The incorporated phage DNA now occupies a linear segment amounting to about 1.2% of the total length of the *E. coli* chromosome. It is replicated along with the rest of the chromosome and for, the most part, goes unnoticed.

The host, *E. coli* K12, contains useful *amber* suppressors that make it easy to detect mutations in the bacteriophage. The integrated prophage can undergo mutations of almost any type, including large deletion mutations, and can still be investigated through complementation studies with other strains of virus. Thus, a family of modified **defective λ phage** was developed. When the λ prophage is excised from the bacterial chromosome, adjacent bacterial genes are occasionally carried with it. This allowed development of **λ transducing phage,** which can carry genes and transfer them into bacteria lacking these genes. More recently an important series of cloning vehicles have been derived from phage λ.

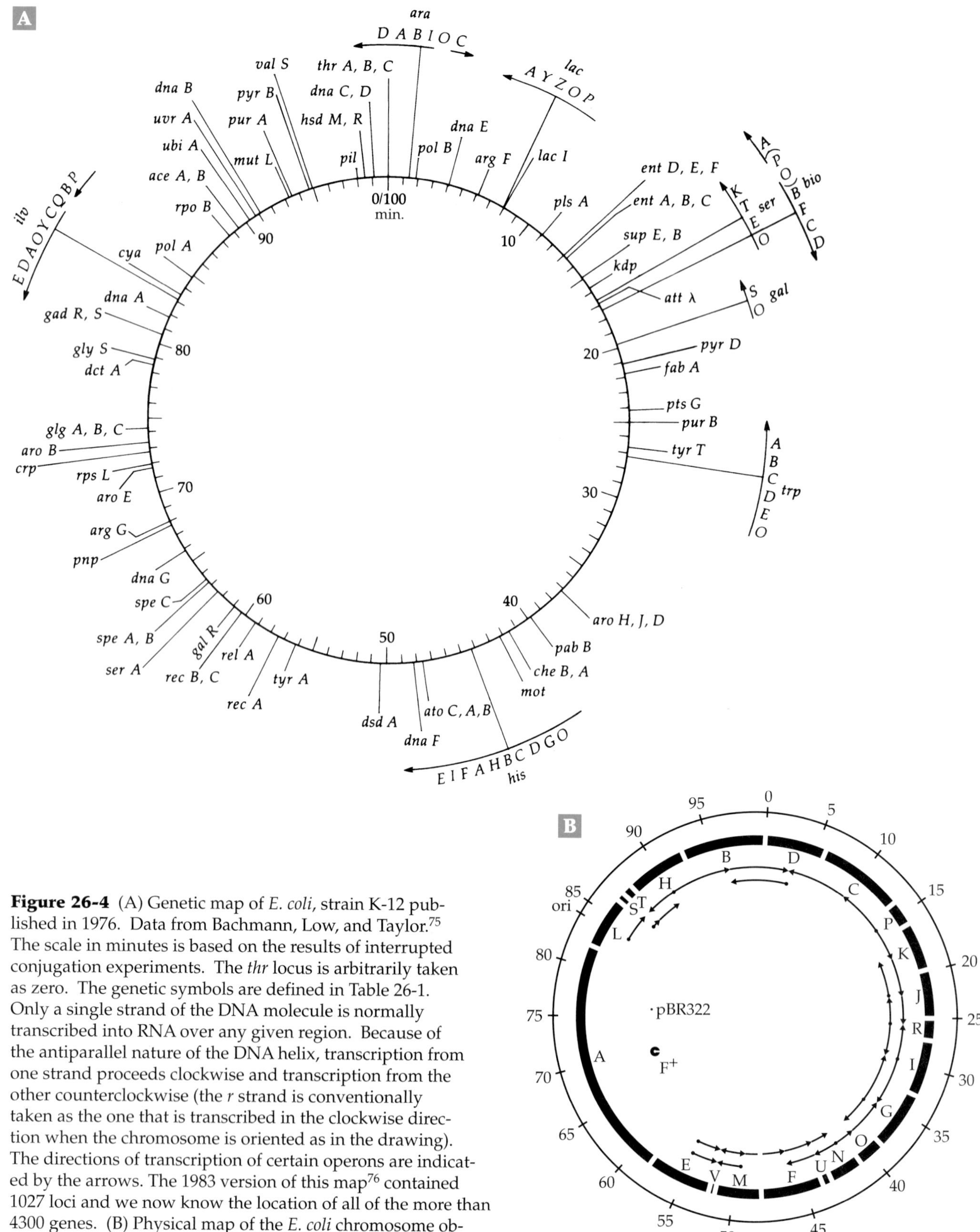

Figure 26-4 (A) Genetic map of *E. coli*, strain K-12 published in 1976. Data from Bachmann, Low, and Taylor.[75] The scale in minutes is based on the results of interrupted conjugation experiments. The *thr* locus is arbitrarily taken as zero. The genetic symbols are defined in Table 26-1. Only a single strand of the DNA molecule is normally transcribed into RNA over any given region. Because of the antiparallel nature of the DNA helix, transcription from one strand proceeds clockwise and transcription from the other counterclockwise (the *r* strand is conventionally taken as the one that is transcribed in the clockwise direction when the chromosome is oriented as in the drawing). The directions of transcription of certain operons are indicated by the arrows. The 1983 version of this map[76] contained 1027 loci and we now know the location of all of the more than 4300 genes. (B) Physical map of the *E. coli* chromosome obtained by study of large 20–100 kb fragments obtained by digestion with restriction endonuclease *Not*I and separated by pulsed field gel electrophoresis. The fragments A–V were ordered using genetic information and overlapping fragments from partial *Not*I digests (arrows). On the outside the genetic map has been superimposed after distortion of the scale in minutes to make the two maps coincident. The 100-kb F⁺ plasmid and the 4.5-kb plasmid pBR322 are also drawn to show relative sizes. From Smith *et al.*[77] A finer restriction map, based on 3400 cloned fragments, was prepared by Kohara, Akiyama, and Isono in 1987.[78]

TABLE 26-1
Some Genes of *E. coli*[a]

Gene symbol	Mnemonic	Map position (min)[a]	Phenotypic trait affected
aceA	Acetate	89	Isocitrate lyase
aceB	Acetate	89	Malate synthetase A
araA	Arabinose	1	L-Arabinose isomerase
araB	Arabinose	1	L-Ribulokinase
araC	Arabinose	1	Regulatory gene
araD	Arabinose	1	L-Ribulose-5-phosphate-4-epimerase
araI	Arabinose	1	Initiator locus
araO	Arabinose	1	Operator locus
argF	Arginine	6	Ornithine carbamoyltransferase
argG	Arginine	68	Argininosuccinic acid synthetase
aroB	Aromatic	73	Dehydroquinate synthetase
aroD	Aromatic	37	Dehydroquinate dehydratase
aroE	Aromatic	71	Dehydroshikimate reductase
aroH	Aromatic	37	DAHP synthetase (tryptophan-repressible isoenzyme)
aroJ	Aromatic	37	Probable operator locus for *aro*H
atoA	Acetoacetate	48	Coenzyme A transferase
atoB	Acetoacetate	48	Thiolase II
atoC	Acetoacetate	48	Regulatory gene
*att*λ	Attachment	17	Integration site for prophage λ
bioA	Biotin	17	Group II; 7-oxo-8-aminopelargonic acid (7 KAP) → 7,8-diaminopelargonic acid (DAPA)
bioB	Biotin	17	Conversion of dethiobiotin to biotin
bioC	Biotin	17	Block prior to pimeloyl-CoA
bioD	Biotin	17	Dethiobiotin synthetase
bioF	Biotin	17	Pimeloyl-CoA → 7 KAP
bioO	Biotin	17	Operator for genes *bio*B through *bio*D
bioP	Biotin	17	Promoter site for genes *bio*B through *bio*D
cheA	Chemotaxis	42	Chemotactic motility
cheB	Chemotaxis	42	Chemotactic motility
crp		73	Cyclic adenosine monophosphate receptor protein
cya		83	Adenylate cyclase
dctA		79	Uptake of C_4-dicarboxylic acids
dnaA	DNA	82	DNA synthesis; initiation defective
dnaB	DNA	91	DNA synthesis
dnaC	DNA	99	*dnaD*; DNA synthesis; initiation defective
dnaE	DNA	4	*polC*, DNA polymerase III and mutator activity
dnaF	DNA	48	*nrdA*; ribonucleoside diphosphate reductase
dnaG	DNA	66	DNA synthesis
dsdA	D-Serine	50	D-Serine deaminase
entA	Enterochelin	13	2,3-Dihydro-2,3-dihydroxybenzoate dehydrogenase
entB	Enterochelin	13	2,3-Dihydro-2,3-dihydroxybenzoate synthetase
entC	Enterochelin	13	Isochorismate synthetase
entD,E,F	Enterochelin	13	Unknown steps in conversion of 2,3-dihydroxybenzoate to enterochelin
fabA		22	β-Hydroxydecanoylthioester dehydratase
gadR		81	Regulatory gene for *gadS*
gadS		81	Glutamic acid decarboxylase
galE	Galactose	17	Uridine diphosphogalactose 4-epimerase
galK	Galactose	17	Galactokinase
galO	Galactose	17	Operator locus
galT	Galactose	17	Galactose 1-phosphate uridyltransferase
galR	Galactose	61	Regulatory gene
glgA	Glycogen	74	Glycogen synthetase
glgB	Glycogen	74	α-1,4-Glucan: α-1,4-glucan 6-glucosyltransferase
glgC	Glycogen	74	Adenosine diphosphate glucose pyrophosphorylase
glyS	Glycine	79	Glycyl-transfer RNA synthetase
hisA	Histidine	44	Isomerase
hisB	Histidine	44	Imidazole glycerol phosphate dehydrase: histidinol phosphatase
hisC	Histidine	44	Imidazole acetol phosphate aminotransferase
hisD	Histidine	44	Histidinol dehydrogenase
hisE	Histidine	44	Phosphoribosyl-adenosine triphosphate-pyrophospho-hydrolase
hisF	Histidine	44	Cyclase
hisG	Histidine	44	Phosphoribosyl-adenosine triphosphate-pyro-phosphorylase
hisH	Histidine	44	Amidotransferase
hisI	Histidine	44	Phosphoribosyl-adenosine monophosphate-hydrolase
hisO	Histidine	44	Operator locus
dsdM	Host specificity	98	Host modification activity: DNA methylase M
hsdR	Host specificity	98	Host restriction activity: endonuclease R
ilvA	Isoleucine–valine	83	Threonine deaminase (dehydratase)
ilvB	Isoleucine–valine	83	Acetohydroxy acid synthetase I
ilvC	Isoleucine–valine	83	α-Hydroxy-β-oxo acid reductoisomerase
ilvD	Isoleucine–valine	83	Dehydrase
ilvE	Isoleucine–valine	83	Aminotransferase B
ilvO	Isoleucine–valine	83	Operator locus for genes *ilv*A,D,E
ilvP	Isoleucine–valine	83	Operator locus for gene *ilv*B
ilvQ	Isoleucine–valine	83	Induction recognition site for *ilv*C
ilvY	Isoleucine–valine	83	Positive control element for *ilv*C induction
kdp	K accumulation	16	Defect in potassium ion uptake
lacA	Lactose	8	Thiogalactoside transacetylase
lacI	Lactose	8	Regulator gene
lacO	Lactose	8	Operator locus
lacP	Lactose	8	Promoter locus
lacY	Lactose	8	Galactoside permease (M protein)
lacZ	Lactose	8	β-Galactosidase
mot	Motility	42	Flagellar paralysis
mutL	Mutator	93	Generalized high mutability (AT → GC)
pabB	*p*-Aminobenzoate	40	Requirement
pil	Pili	98	Presence or absence of pili (fimbriae)
plsA	Phospholipid	11	Glycerol-3-phosphate acyltransferase
pnp		68	Polynucleotide phosphorylase
polA	Polymerase	85	DNA polymerase I
polB	Polymerase	2	DNA polymerase II
ptsG	Phosphotranserase	24	Catabolite repression system
purA	Purine	93	Adenylosuccinic acid synthetase
purB	Purine	25	Adenylosuccinase
pyrB	Pyrimidine	95	Aspartate carbamoyltransferase
pyrD	Pyrimidine	21	Dihydroorotic acid dehydrogenase
recA	Recombination	58	Ultraviolet sensitivity and competence for genetic recombination
recB	Recombination	60	Ultraviolet sensitivity, genetic recombination; exonuclease V subunit
recC	Recombination	60	Ultraviolet sensitivity, genetic recombination; exonuclease V subunit
relA	Relaxed	59	Regulation of RNA synthesis
rpoB	RNA polymerase	89	RNA polymerase: β subunit (*rif* gene)
rpsL	Ribosomal protein, small	72	Ribosomal protein S12 (*strA* gene, streptomycin resistance)
serA	Serine	62	3-Phosphoglyceric acid dehydrogenase
serO	Serine	20	Operator locus
serS	Serine	20	Seryl transfer RNA synthetase
speA	Spermidine	63	Arginine decarboxylase
speB	Spermidine	63	Agmatine ureohydrolase
speC	Spermidine	63	Ornithine decarboxylase
supB	Suppressor	15	Suppressor of *ochre* mutations
supE	Suppressor	15	Suppressor of *amber* mutations (*su-2*)
thrA	Threonine	0	Aspartokinase I-homoserine dehydrogenase I complex
thrB	Threonine	0	Homoserine kinase
thrC	Threonine	0	Threonine synthetase
trpA	Tryptophan	27	Tryptophan synthetase, A protein
trpB	Tryptophan	27	Tryptophan synthetase, B protein
trpC	Tryptophan	27	*N*-(5-Phosphoribosyl) anthranilate
trpD	Tryptophan	27	Phosphoribosyl anthranilatetransferase
trpE	Tryptophan	27	Anthranilate synthetase
trpO	Tryptophan	27	Operator locus
tyrA	Tyrosine	56	Chorismate mutase T-prephenate dehydrogenase
tyrT	Tyrosine	27	Tyrosine transfer RNA_1 (*su-3* gene; amber suppressor)
ubiA	Ubiquinone	90	4-Hydroxybenzoate → 3-octaprenyl 4-hydroxybenzoate
uvrA	Ultraviolet	91	Repair of ultraviolet radiation damage to DNA, UV endo-nuclease
valS	Valine	95	Valyl-transfer RNA synthetase

[a] This list contains 126 of more than 1027 genes that had been mapped by 1983. (Bachmann, B. J. (1983) *Bacteriol. Rev.* **47**, 180–230). Their positions are shown diagrammatically in Fig. 26-4.

D. Mapping of Chromosomes

Let us now consider how knowledge of bacterial sex factors and of phage λ permitted the mapping of bacterial chromosomes. Together with the use of restriction endonucleases these techniques gave us the first precise physical maps of bacterial chromosomes and pointed the way toward the determination of complete genome sequences.

1. The Chromosome Map of *E. coli*

There are about 4,639,221 nucleotide pairs in the circular DNA molecule that is the chromosome of *E. coli* strain K-12. We now know the complete sequence, which includes all of the individual genes that are present.[74] However, our first knowledge of the location of these genes in the chromosome depended upon construction of a **linkage map** (Fig. 26-4A). Construction of this map, with 126 genes, began with the study of nutritional auxotrophs whose defective genes are located at many points on the chromosome. By 1983, 1027 genes had been mapped. In 1997, when the complete nucleotide sequence became known, 4288 protein coding genes could be recognized.[74] The map in Fig. 26-4 was established 30 years earlier, largely by use of interrupted bacterial mating.[79,80] In this procedure *Hfr* cells carrying specific mutations are mixed with wild-type F cells, and conjugation is allowed to proceed for a certain length of time. Then the cells are agitated violently, e.g., in a Waring blender. This breaks all of the conjugation bridges and interrupts the mating process. Mating is interrupted at different times, and the recipient bacteria are tested for the presence of genes transferred from the donor strain. Using this technique it was found that complete transfer of the chromosome takes ~100 min at 37°C, and that the approximate location of any gene on the chromosome can be determined by the length of time required for transfer of that gene into the recipient cell. It is a little more complex than this. Because complete chromosome transfer is rare, substrains of *E. coli* K-12 with an F agent integrated at different points were used. With certain F factors those genes lying clockwise around the circle in Fig. 26-4 immediately beyond the point of integration are transferred quickly and with high frequency.

The **time-of-entry map** in Fig. 26-4A is based not only on interrupted matings but also on the use of **transduction** by bacteriophage P1.[76,79] Transduction by phage permits the tranfer of a short fragment of DNA, about 2 min in length, on the *E. coli* map. Joint transduction, i.e., joint incorporation of two genes into the chromosome of the receptor, occurs with a frequency related to the map distance between these two genes. Thus, finer mapping was done within many segments of the *E. coli* chromosome. Meanings of the gene symbols used in the figure are given in Table 26-1. Similar maps were prepared for *Salmonella typhimurium* and *Bacillus subtilis*.

2. Restriction Endonucleases

Many of the procedures for cloning genes, synthesizing more copies of a DNA ("amplifying" the DNA), making genetic maps, and generating mutants, are dependent upon **restriction endonucleases**. The name comes from a property of bacteria, which often can digest and destroy DNA of invading viruses or DNA that has been injected during mating with a bacterium of an incompatible strain. Investigation of this phenomenon, known as **restriction**, revealed that the DNA of viruses that are able to replicate within a particular host is *marked* in some fashion at specific sites in the molecule. The marking often consists of the presence of methyl groups. Properly methylated DNA is not degraded, but unmethylated DNA is cleaved by a highly specific endonuclease at the same sites that are normally methylated. Each species of bacteria (and often an individual strain within a species) has its own restriction enzymes. Restriction enzymes are very specific and cut DNA chains at unique base sequences. Three types are recognized.[80a]

Type I restriction enzymes, such as those encoded in the chromosome of *E. coli*, are large 300- to 400-kDa proteins composed of at least three kinds of polypeptide chain. They bind at specific sites of a foreign DNA and apparently cleave the chain randomly nearby. They require ATP, Mg^{2+}, and *S*-adenosylmethionine and have the unusual property of promoting the hydrolysis of large amounts of ATP.[81,82] The significance of these properties is still unknown.

The type II restriction endonucleases, which are the ones most widely used in molecular biology, are relatively small 50- to 100-kDa monomeric or dimeric proteins. About 2400 different enzymes with 188 different specificities had been isolated by 1995.[83,84] The sites of attack, in most instances, are nucleotide sequences with a twofold axis of local symmetry.[85] For example, the following sites of cleavage have been identified for two restriction endonucleases encoded by the DNA of R-factor plasmids of *E. coli* and for a restriction enzyme from *Hemophilus influenzae*. In the diagrams ↓ are sites of cleavage, * are sites of methylation, and • are local twofold axes (centers of palindromes); N can be any nucleotide with a proper base pairing partner.

Restriction enzymes often create breaks in each of the two strands in positions symmetrically arranged around the local twofold axis. This is what we might expect of a dimeric enzyme that binds in the major or minor groove of the double helix, each active site

E. coli R factor (*Eco*RI)

```
         ↓  *
5'-NNNGAATTCNNN
3'-NNNCTTAAGNNN
          *  ↑
```

E. coli R factor (*Eco*RII)

```
          ↓  *
5'-NNNNNCCAGGNNNN-3'
3'-NNNNNGGTCCNNNN-5'
            *  ↑
```

H. influenzae (*Hin*dIII)

```
        *↓
5'-NNNAAGCTTNNN-3'
3'-NNNTTCGAANNN-5'
             ↑*
```

attacking one of the polynucleotide chains. In fact, the two 277-residue subunits of the *Eco*RI enzyme[86] bind primarily in the major grooves of the DNA, one active site on each strand. Each recognition unit makes 12 hydrogen bonds to the DNA. Each base pair forms two of these hydrogen bonds. Four arginines and two glutamates participate.[87] This provides a net charge of +2, which may be important for electrostatic interaction of the protein with backbone phosphate groups of the DNA. The binding of the protein affects the conformation of the DNA, widening the major groove from that in B DNA, and causing a torsional kink with some unwinding of the double helix.[88] It appears that the specificity for the GAATTC hexanucleotide is in part a result of direct complementary interactions between functional groups in the major groove (see Fig. 5-3), bound water molecules, and amino acid side chains from the enzyme (Fig. 26-5). Methylation of the 6-amino groups of the adenines in the center of the recognition sequence prevents cleavage by the *Eco*RI endonuclease, but appears to alter the interaction with the protein only slightly.[89]

TABLE 26-2
Some Commonly Used Restriction Endonucleases, Their Sources, and Cleavage Sites

Enzyme	Source	Cleavage Site
Alu I	*Arthrobacter luteus*	5' - - AG↓CT - - 3'
*Bam*HI	*Bacillus amyloliquefaciens H*	G↓GATCC
Bcl I	*Bacillus caldolyticus*	T↓GATCA
Bgl II	*Bacillus globigii*	A↓GATCT
*Cfr*10I	*Citrobacter freundii*	Pu↓CCGGPy
*Eco*RI	*Escherichia coli*	G↓AATTC
*Eco*RV	*Escherichia coli*	GAT↓ATC
Hae III	*Haemophilus aegypticus*	GG↓CC
Hind III	*Haemophilus influenzae*	A↓AGCTT
Kpn I	*Klebsiella pneumoniae*	GGTAC↓C
Mbo I	*Moraxella bovis*	↓GATC
Pst I	*Providencia stuartii*	CTGCA↓G
Sal I	*Streptomyces albus*	G↓TCGAC
*Sau*3AI	*Streptococcus aureus*	↓GATC
Sfi I	*Streptomyces fimbriatus*	GGCCNNNN↓NGGC

Although they often share little sequence similarity and have quite different specificiities, many restriction enzymes have similar three-dimensional structures as well as mechanisms of action. This is true for the *Eco*RI, *Bam*HI (Fig. 26-5),[83,90] *Eco*RV,[91,91a] and *Cfr*10I enzymes,[84] and presumably many others. The specifically shaped and tightly packed active sites in the enzyme–substrate complexes ensure specificity. For example, the *Eco*RV endonuclease cleaves DNA at its recognition site at least a million times faster than at any other DNA sequence.[91] As mentioned in Chapter 12, restriction endonucleases require a metal ion, preferably Mg^{2+}, and probably act via a hydroxyl ion generated from Mg^{2+}–OH_2 at the active site. Three conserved active site residues, Asp 91, Glu 111, and Lys 113, in the *Eco*RI endonuclease interact with the DNA near the cleavage site. Lys 113 is replaced by Glu 113 in the *Bam*HI enzyme.[83,90]

The corresponding conserved residues in the smaller *Eco*RV enzyme are Asp 74, Asp 90, and Lys 92. They are shown in the following diagram that represents one of several possible metal-ion dependent mechanisms.[91–93] The metal-coordinated hydroxyl ion is generated by proton transfer to the $-NH_2$ group of Lys 92 and carries out an in-line attack on the backbone phospho group of thymidine at the cleavage point. At least two metal ions are needed, and three may be present, as shown in this diagram from Sam and Perona.[93]

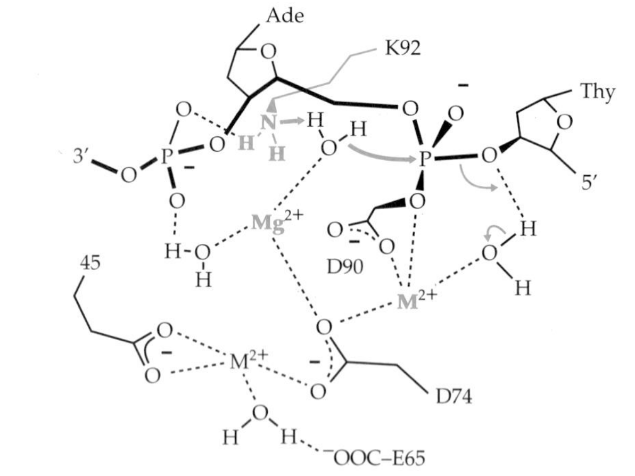

Restriction enzymes that cleave DNA at a large number of specific sequences are available commercially. A few are listed in Table 26-2. Another group of restriction enzymes have similar recognition sequences but cut the dsDNA that they recognize at a specific neighboring site rather than within the recognition sequence. An example is *Fok*I, which recognizes the nonpalindromic $\frac{\text{GGATG}}{\text{CCTAC}}$, but cuts the chains 9 and 13 base pairs to the right. This enzyme has been used by Szybalski and associates to devise a system for cutting ssDNA precisely at a desired point and converting it to ds fragments.[95]

3. Restriction Mapping

The calibration of the *E. coli* genetic map in minutes was a temporary expedient. It was followed by **physical maps** expressed directly as micrometers of DNA length (total length ~1.6 mm) or thousands of nucleotide units (kb). A physical map obtained by **restriction enzyme mapping** is shown in Fig. 26-4B. To obtain this map DNA fragments were prepared using specific restriction endonucleases (Section E, 1).

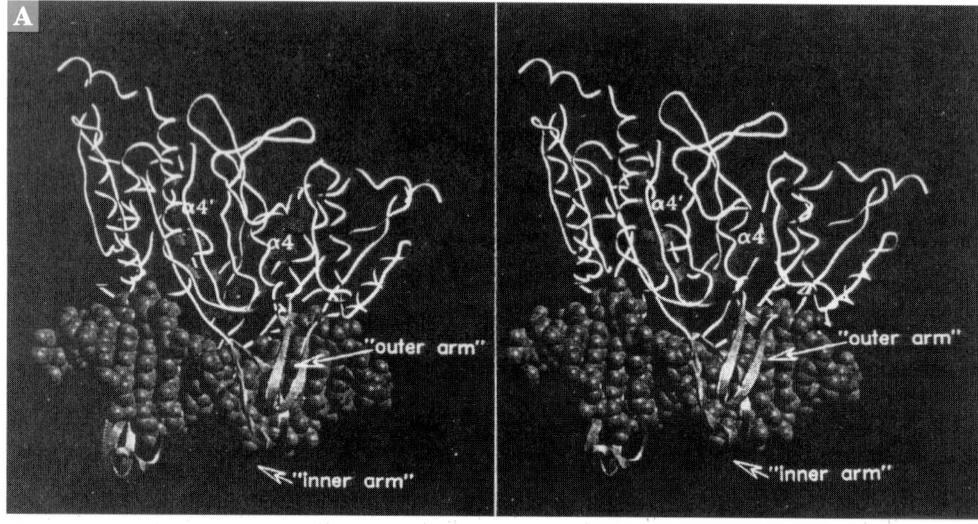

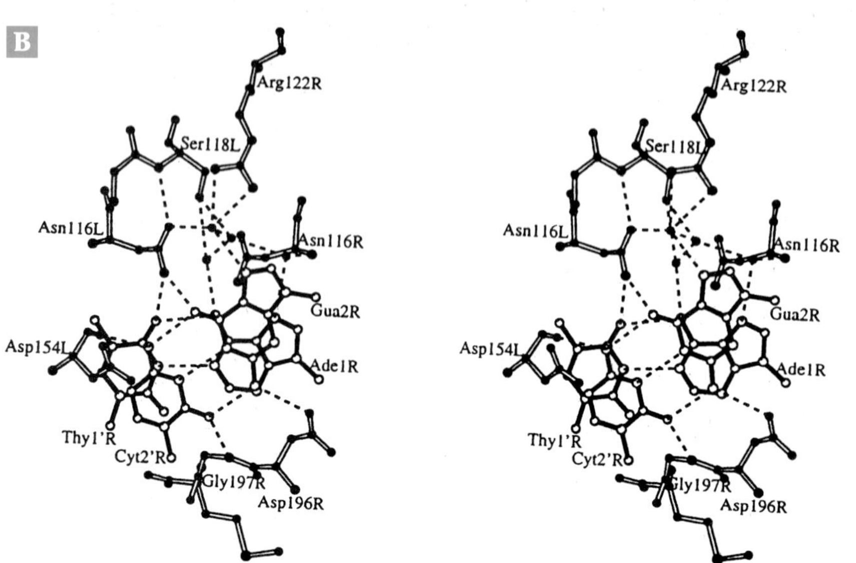

Figure 26-5 (A) Stereoscopic ribbon drawing of the dimeric *Eco*RI restriction endonuclease in a complex with DNA. The equivalent helices marked α4 and α4' point into the major groove of the DNA double helix while the inner and outer "arms" wrap around the DNA. From Bozic *et al.*[84] based on coordinates of Kim *et al.*[94] (B) Stereoscopic view of the two base pairs T•A and C•G of the right end of the recognition motif 5'-CGATCC-3' (Table 26-2) bound to the *Bam*HI restriction endonuclease. The third base pair C•G lies below the two that are shown. Notice the numerous hydrogen bonds, some of which bind atoms of the DNA directly to atoms of the protein and also hydrogen bonds to water molecules (filled circles). The tight packing of complementary charged and dipolar groups of protein, nucleic acid, water, and Mg^{2+} ions (not seen in this drawing) thoughout the complex accounts for the high specificity of these enzymes. From Newman *et al.*[83] Courtesy of Aneel Aggarwal.

The fragments were aligned using genetic markers, and their lengths were estimated by their electrophoretic mobilities.[77] The time-of-entry map has been added on a distorted scale in Fig. 26-4B. There are, on the average, 46.4 kb of DNA per minute, but this amount varies around the chromosome between 38 and 61 kb / min.

Mapping with restriction endonucleases was for many years an essential step in determination of the complete sequence of a piece of DNA. To make a restriction map the DNA, which may have been cut from a chromosome by a restriction endonuclease, is cloned. This permits isolation of a large amount of the DNA, which is then cut by other restriction endonucleases with differing specificities. Overlapping fragments resulting from the cleavages by single restriction enzymes are ordered to provide a map such as that of yeast mitochondrial DNA shown in Fig. 5-48. Any piece from the mapped DNA can now be cloned, and the exact sequence determined. The development of pulsed-field electrophoresis (Chapter 5), with its ability to separate DNA fragments 2000 kb or greater in length, allows restriction mapping with enzymes that cut at rare intervals to give very large fragments. For example, the *Not*I restriction endonuclease cuts the 4.7 Mb *E. coli* K12 genome into 22 fragments that were used to construct the complete restriction map of Fig. 26-4B.[77] Sequences of many viruses and plasmids, mitochondrial and plastid DNAs, and several bacterial genomes have been determined by use of restriction mapping and sequencing of the restriction fragments. Restriction fragment patterns have also been important to determine eukaryotic genome maps including the first genetic linkage map,[96] the first physical map of the human genome,[97] and the complete human genome sequences (Section G).

4. Electron Microscopy

Physical mapping by electron microscopy has been applied to bacteriophage, which can be obtained with large deletions in various parts of the genome.[98,99] The method can also be applied to cloned pieces of DNA. DNA might be isolated from two different phage strains, for example, from wild-type λ and from a mutant phage with a particular gene or genes deleted. The λ DNA can be denatured readily and separated into *r* strands and *l* strands by isopycnic centrifugation. If the isolated *l* strand of one strain is mixed with the *r* strand of another strain and annealed, a double-stranded DNA will be formed and, if there is a deletion in one strain, the homologous region in the normal λ DNA will form a single-stranded loop that can be visualized in the electron microscope. Figure 26-6 shows an example of a micrograph of such a **heteroduplex** molecule with a deletion loop and also a "bubble," where a segment of nonhomologous

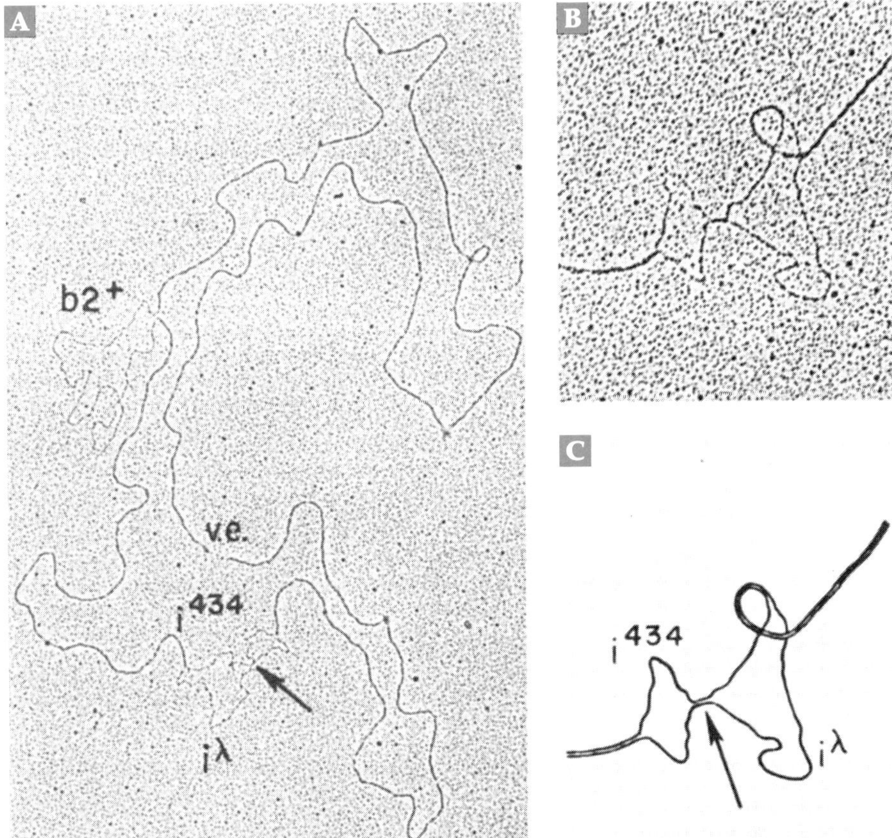

Figure 26-6 (A) Electron micrograph of a heteroduplex DNA molecule constructed from complementary strands of phages λb2 and λ*imm*434. In λb2 a segment of λ DNA has been deleted producing a deletion loop (labeled b2) and in λ*imm*434 a piece of DNA from phage 434 has been substituted for λ DNA resulting in a "nonhomology bubble" (labeled i^{434} / i^{λ}). The vegetative (cohesive) ends of the DNA are labeled v.e. (B) Enlargement of the nonhomology bubble. (C) Interpretative drawing of view in (B). Arrow marks a short (20–150 nucleotide) region of apparent homology. From Westmoreland, Szybalski, and Ris.[98]

DNA has been substituted in one strand.[98] Since distances can be measured accurately on the electron micrographs, rather precise (±50–100 bp) physical maps can be obtained. The chromosome map of phage λ was mapped in this way initially; now its complete nucleotide sequence is known. Another electron microscopic method is useful for location of AT-rich regions that denature readily. In a suitable concentration of formamide these regions melt to form visible single-stranded **denaturation loops** similar to the bubbles in Figure 26-6.

An important technique is to hybridize pieces of mRNA with DNA. If this is done with denatured (single-stranded) DNA and processed mRNA, which has been transcribed from genes with intervening sequences, the intervening sequences will form single-stranded loops in the DNA–RNA hybrid. A related technique depends upon the increased stability of DNA–RNA hybrids in high concentrations of formamide. Under these conditions an RNA segment will hybridize with its complementary strand of the DNA duplex displacing the other strand of DNA, which then appears as a visible **R-loop**. Intervening sequences appear as undisturbed DNA duplexes.

5. Optical Mapping

The long DNA molecules of bacterial or eukaryotic chromosomes are easily broken by vigorous stirring. However, if handled carefully very large fluorescently stained DNA molecules of 0.4–1.4 Mb lengths can be stretched out on a glass surface, and their lengths measured by optical microscopy. The technique depends upon binding of one of the ends of the DNA to the glass surface. If biotin is covalently attached to one end of the DNA, it will bind to a steptavidin-coated plate. However, unaltered DNA also binds to a vinyl-silane or a trichlorosilane coating under suitable conditions.[100,101] If DNA is incubated with a submerged silanized coverslip, which is then pulled out of the liquid mechanically at a constant speed, the DNA molecules are "dyamically combed" so that they are aligned for easy observation and measurement.[101] DNA can be transferred directly from agarose gels used for electrophoretic separation to the plates.[102] Used in combination with restriction enzymes these procedures allow rapid automated construction of physical maps. An example of whole-genome optical mapping is provided by the restriction map of the genome of the radiation-resistant bacterium *Deinococcus radiodurans*. The genome consists of two circular DNA molecules of 2.6 and 0.415 megabases and a smaller 176-kilobase DNA.[103] These were mapped without the laborious subcloning required by conventional restriction mapping. For example, the *E. coli* restriction map required analysis of 3400 phage clones.

Mapping of eukaryotic chromosomes has involved additional methods which are discussed in Sections E and G,1. These incluse **radiation hybrid mapping**,[104] use of **meiotic recombination**, identification of **restriction fragment length polymorphisms** (**RFLPs**; described in Section E,7), and use of **expressed sequence tags** (**ESTs**), short DNA sequences deduced from mRNA molecules transcribed from the DNA.[105,106]

E. Cloning, Modifying, and Transferring Genes

A true revolution in biology and in medicine is in progress as a result of our ability to clone, sequence, mutate, and manipulate genes at will. Methods of sequence determination are discussed in Chapter 5 as is the laboratory synthesis of oligonucleotides and of complete genes. Both of these techniques are essential to present-day genetic engineering as are the techniques of cloning,[99,107–113] which are considered in this section.

A diploid cell contains only two copies of many genes, and these two copies are often not identical. In a gram of any tissue, which may contain 10^9 cells, we will have only ~1 ng of a 1 kb (1 kilobase) gene. To isolate this gene we would have to fish it out from a huge excess of other genes. Present-day cloning techniques offer a way to locate the gene, increase its quantity by many orders of magnitude, learn its sequence, induce any desired mutations at any points, and transfer the gene into other organisms in such a form that it can be expressed, i.e., be transcribed and induce synthesis of proteins. The methods have taken years to develop and continue to be improved. Only some basic procedures and concepts are described here. Numerous manuals[99,108,109,111,114–118] as well as commercial "kits" are available.

1. Joining DNA Fragments

The cloning and manipulation of genes usually depends upon the precise cutting of DNA into discrete fragments by restriction endonucleases. Many restriction enzymes generate **cohesive ends** (sticky ends). Thus, *Eco*RI produces DNA fragments with the single-stranded "tails" shown here at the 5'-ends of the cut duplexes:

```
5′ ----- G 3′                      5′pA A T T C ----- 3′
3′ ----- C T T A Ap 5′                       3′G ----- 5′
```

These cohesive ends can be used to join together different restriction fragments. It is easy to see that the complementary single-stranded tails can form base pairs to regenerate the original hexanucleotide sequence cleaved by *Eco*RI. There will still be nicks between G and A at the specific cleavage points, but these nicks can be closed enzymatically using DNA ligase. Thus, the original DNA cleaved by *Eco*RI can be reformed, or another piece of DNA that also has tails generated by *Eco*RI can be grafted onto an end.

Many of the other enzymes in Table 26-2 also form cohesive ends. Five of them (*Bam*HI, *Bcl*I, *Bgl*II, *Mbo*I, and *Sau*3A) have at the center of their recognition sites the same tetranucleotide: GATC. Enzymes *Sau*3A and *Mbo* I are called **isoschizomers** because they have just the same 4-base recognition sequence and also yield the same restriction patterns. Notice that they will both cut all of the *Bam*HI, *Bcl*I, or *Bgl*II sites, but *Bam*HI and *Bgl*II will not cut all *Mbo*I or *Sau*3A sites. However, cohesive ends made by any of these enzymes can be joined. The gaps left during the joining of certain of the fragments can be ligated enzymatically. *Sau*3A will cut at either methylated or unmethylated sites but *Mbo*I will not cut at methylated sites.

Two enzymes (*Kpn*I and *Pst*I in the list in Table 26-2) form 3'-cohesive ends rather than 5'-cohesive ends. In addition, there are three (*Alu*I, *Eco*RV, and *Hae*III) that cut at the local twofold axis; they form no cohesive ends but leave **blunt ends** (flush ends). Blunt end fragments are also much used in genetic engineering. "Linkers" that provide cohesive ends can be added.[119] The *Sfi*I endonuclease cuts between two 4-bp palindromes in a 13-bp recognition sequence (Table 26-2).[120]

Some useful enzymes. Several enzymes of use in cloning[121] are listed in Table 26-3. The detailed chemistry of most of these is discussed in Chapter 27. Among these are the **ligases** that allow DNA fragments to be joined. They act on DNA strands with adjacent 3'-OH and 5'-phosphate termini. The *E. coli* ligase seals single stranded nicks using NAD^+ as an energy source (Eq. 27-5). It is therefore able to ligate DNA fragments with cohesive ends. The T4 DNA ligase, which is obtained from *E. coli* infected with phage T4, not only can seal nicks but can ligate pieces of DNA with blunt ends. Its activity is linked to cleavage of ATP. If two DNA strands are joined, but with gaps in one or both strands, the gaps can be filled efficiently by the 109-kDa **DNA polymerase I** from *E. coli*. Most often the 76-kDa **Klenow fragment**, which is lacking the 5' $\rightarrow$ 3' exonuclease activity, is used. **T4 DNA polymerase** has similar properties.

A problem with DNA fragments with cohesive ends is that they spontaneously form closed circles, a process that may compete with a desired joining to another piece of DNA. One solution to this problem is to hydrolyze off the 5'-phosphate groups with an **alkaline phosphatase** (Chapter 12). This prevents formation of covalently closed circles. However, pieces of DNA that retain their 5'-phosphate groups can be ligated to these dephosphorylated pieces. **T4 polynucleotide kinase**[122] can be used to put a phospho group back onto the 5' end of a chain. A ^{32}P end label can be added to such a polynucleotide using ^{32}P-labeled ATP.

Forming homopolymeric tails. Chromosomal DNA may be cleaved with restriction enzymes that leave blunt ends or it may be cleaved randomly by shearing. In either case the blunt ends can be treated first with **λ-exonuclease**, then with **terminal deoxynucleotidyl transferase**[123] isolated from calf thymus. The exonuclease treatment cuts off a few nucleotides from the 5' termini leaving short single-stranded 3'–OH termini. The terminal transferase, a polynucleotide polymerase that acts on ssDNA, is nonspecific and requires no template. Using an appropriate nucleotide triphosphate, it will add a single-stranded tail of either deoxyribonucleotides or ribonucleotides to the exposed 3' termini of a polynucleotide of three or more residues. If deoxyATP is used, a 3' poly(dA) tail will be added to each 5' terminus. Such a poly(dA)-tailed DNA fragment can be annealed and ligated to DNA carrying poly(dT) tails. This approach has been used widely to insert a piece of DNA into a cloning vehicle. For example, a circular plasmid (Fig. 26-7) can be opened by a single cleavage with *Eco*RI or other suitable restriction enzyme. The opened plasmid is treated with exonuclease, and poly(dA) tails are added. The piece of DNA to be cloned is tailed with poly(dT). After annealing and ligation recombinant plasmids carrying the **passenger DNA** will be formed.

If DNA is cleaved with *Pst*I or *Kpn*I (Table 26-2), the resulting 3' cohesive ends can be extended with a poly(dC) tail. If the cloning vehicle also has a site for *Pst*I or *Kpn*I, it can be opened and poly(dG) tails can be applied. A useful feature is that after annealing, filling in the gaps, and ligation the original *Pst*I or *Kpn*I sites are restored. This provides for easy recovery of the cloned fragments (Fig. 26-8).

Preparing material for cloning. DNA may be prepared for cloning by (1) random cleavage by shearing or by enzymatic attack, (2) cleavage by one or more restriction endonucleases, (3) preparation of cDNA from mRNA, or (4) nonenzymatic chemical synthesis of DNA segments. The use of random cleavage has largely been replaced by cleavage with restriction enzymes. A major problem is the separation of the very large number of different restriction fragments formed from a large piece of DNA or from an entire genome. The creation of "libraries" of such fragments is described in Section 5. Considerable simplification comes from separation of individual eukaryotic chromosomes before the library is prepared. Careful purification of DNA to be used in cloning is helpful. This may be done by electrophoresis in agarose or polyacrylamide gels or using HPLC. One technique is to embed cells directly in a gel, to diffuse in proteases and restriction enzymes that lyse the cells and release the DNA, and cleave it, and then to conduct the electrophoresis (see Chapter 5, Section H,1). DNA fragments of very large size can be separated. In addition to isolation of a fragment to be cloned the cloning vehicle must be prepared. This often involves release of a plasmid by lysis of bacteria that carry it and isolation using a suitable column. Likewise, after the DNA has been cloned and the content of DNA has been increased by growing a large bacterial culture, the plasmids must be released and purified, and the cloned DNA excised with a restriction enzyme. Alternatively, the cloning vehicle may be a virus, which must be isolated and disrupted to release the DNA. Use of the polymerase chain reaction (PCR; Fig. 5-47) allows cloning and amplification of DNA fragments with a minimum of purification.

2. Cloning Vehicles (Vectors)

Many cloning vehicles, more commonly referred to as **vectors**, originated with naturally occurring, independently replicating plasmids or viruses (replicons). More recently artificial chromosomes have been developed as cloning vehicles. Plasmids and viruses have been extensively engineered to provide convenience and safety. A large number of specialized

TABLE 26-3
Some Enzymes Used in Molecular Cloning

Name	Source	Reaction
T4 DNA polymerase	*E. coli* infected with bacteriophage T4	5'→ 3' chain growth 3'→ 5' exonuclease
E. coli DNA polymerase I and Klenow fragment	*E. coli*	5'→ 3' chain growth 3'→ 5' exonuclease 5'→ 3' exonuclease (lacking in Klenow fragment)
Reverse transcriptase	RNA tumor viruses, e.g., avian myoblastosis virus	5'→ 3' DNA chain growth
Ribonuclease H	*E. coli*	Cuts RNA in DNA-RNA hybrid
Lambda and T7 exonucleases	Bacteriophages	
Bal 31 nuclease	*Alteromonas espejiano*, a marine bacterium	Degrades both 3' and 5' termini of dsDNA
T7 RNA polymerase	Bacteriophage T7	DNA-dependent RNA polymerase
Terminal deoxyribonucleotide transferase	Thymus gland, plants	Limited 5'→ 3' chain growth; template independent addition of tails to DNA fragments
T4 DNA ligase	*E. coli* carrying an engineered λ phage	Ligation of DNA, either blunt or cohesive ends; uses ATP
E. coli DNA ligase	*E. coli*	Ligation of DNA with cohesive ends; uses NAD^+ as energy source
RNA ligase	Bacteriophage T4	Ligation of RNA and DNA
T4 polynucleotide kinase	Bacteriophage T4	Phosphorylation of 5′-OH terminus of a polynucleotide (DNA or RNA)
*Eco*RI methylase	*E. coli*	Transfer CH_3 from *S*-adenosylmethionine to adenines in *Eco*RI sites

See Chapter 12 for general discussion of nucleases.

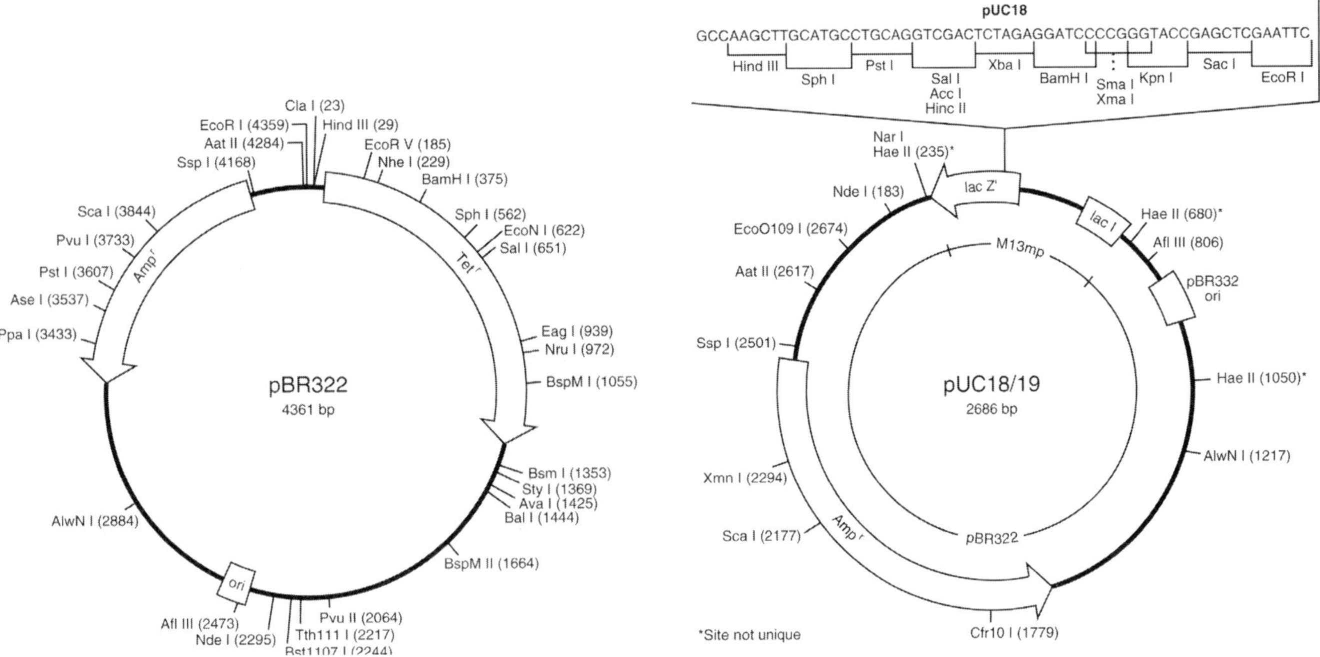

Figure 26-7 Genetic map of cloning plasmids pBR322 and pUC18. Abbreviations: ori, origin of replication; Amp^r, ampicillin resistance gene; Tet^r, tetracycline resistance gene. Other abbreviations are for sites cleaved by specific restriction endonucleases, a few of which are defined in Table 26-2. The nucleotide sequence numbers and directions of transcription are also indicated. Reproduced by permission of Amersham Pharmacia Biotech Inc.

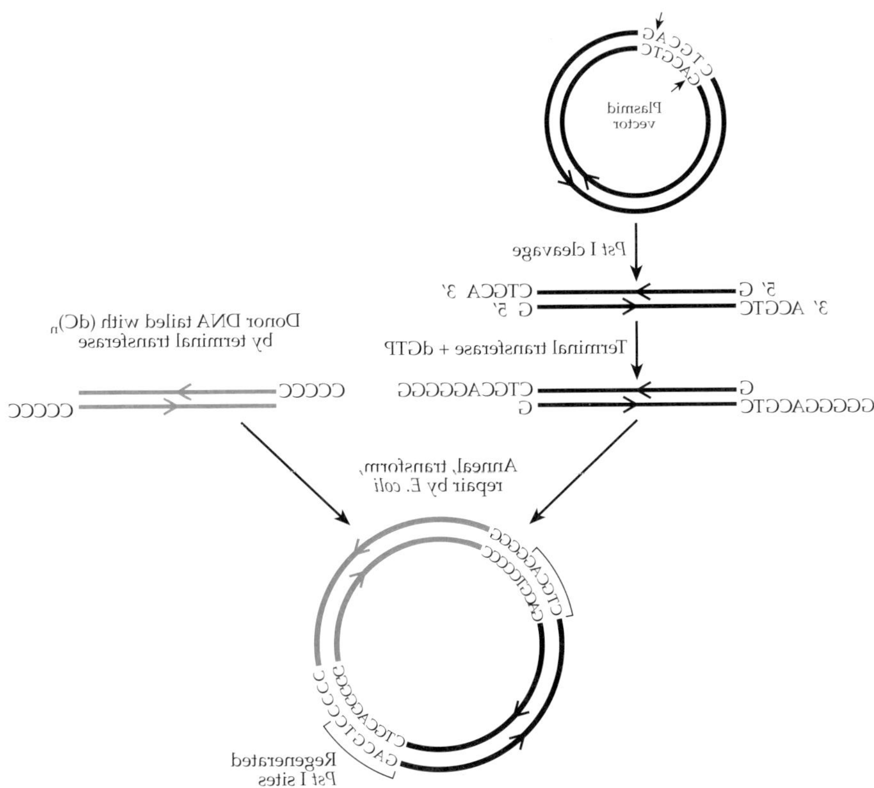

Figure 26-8 Regeneration of a *Pst* I site by dG:dC tailing with terminal transferase. See Glover.[99] Arrowheads indicate 5′→3′ directions.

vehicles have been devised. Only a few will be described briefly here. Suitable books and manuals must be consulted for details.[5,99,108–110,114,115,121,123a]

Plasmids related to ColE1. The small colicinogenic plasmid, ColE1 of *E. coli*, is attractive for cloning because there are 20 copies per bacterial cell. If chloramphenicol is added during the logarithmic phase of growth, the *E. coli* cells will make 1000–3000 copies. Thus, a relatively small culture of the bacteria containing the cloned DNA in a ColE1 plasmid will yield a large amount of the desired DNA. In the past one of the most widely used cloning vehicles[124] has been plasmid pBR322, which was derived from a close relative of ColEl. The genome size of the original plasmid was reduced by deletion of genes unnecessary for its successful replication. Transposons that might permit accidental transfer of DNA to other organisms were inactivated. Unneeded or undesirable restriction enzyme sites were eliminated, and useful restriction sites were introduced by point mutations. The resulting pBR322 has only one site of cleavage each for *Bam*HI, *Sal*I, *Pst*I, *Pvu*I, and *Eco*RI. These are at known positions in the 4363-nucleotide plasmid.[125]

Plasmid pBR322 contains two different antibiotic-resistance genes that were brought in from bacterial R-factors. These are used in selecting bacterial colonies that carry the desired recombinant plasmids. One of these is the β-lactamase gene, which confers resistance to ampicillin (*Amp*^{r}); the other provides resistance to tetracycline (*Tet*^{r}). Their positions are indicated in the pBR322 gene map (Fig. 26-7) as is the essential origin of replication (*ori*). The drug resistance genes are used as follows. If a unique restriction site such as that for *Bam*HI or *Sal*I that lies within the *Tet*^{r} gene is used to introduce the passenger DNA, the resistance to tetracycline is lost but that to ampicillin is retained. Thus, after incubation with the recombinant plasmids under conditions that favor their uptake by the host bacteria, the bacteria are plated onto an ampicillin-containing medium. Only those harboring the pBR322 plasmid with its *Amp*^{r} gene can grow. After these have produced small colonies, a replica plate is made on a tetracycline-containing medium. On this medium the desired recombinants do *not* grow because the *Tet*^{r} gene has been inactivated. This allows selection of colonies containing passenger DNA (Fig. 26-9). A further selection procedure is required to establish that the piece of DNA inserted into the recombinant plasmid is one that is desired.

Typical cloning procedure with pBR322. In much simplifed form the procedure might go as follows: (1) Purchase or isolate plasmid. (2) Cleave plasmid with *Bam*HI; heat at 70°C to inactivate the enzyme. (3) Treat with alkaline phosphatase to remove the 5'-phospho groups. (4) Mix with passenger DNA with cohesive ends generated by *Bam*HI, anneal, and join with DNA ligase. Although the resulting circular recombinant DNA contains a nick as a result of the missing 5'-phospho group, it will be taken up by bacteria and repaired. (5) Incubate joined DNA with cells of host *E. coli* that have been made permeable to DNA by treatment with Ca^{2+} ions. This type of transformation is called **transfection** and is widely used in cloning. (6) Plate transfected cells onto agar containing the first antibiotic, in this case ampicillin. (7) Make replica plate on medium containing second antibiotic, in this case tetracycline. (8) Screen selected colonies for desired DNA fragment. In one procedure a small sample from each of the selected colonies is placed onto spots on a nitrocellulose filter. Several colonies can be placed on one filter and the bacteria lysed, hybridized with a radioactive probe, and then viewed by autoradiography.

Selecting clones using β-galactosidase in pUC cloning vehicles. The newer **pUC** vehicles[126] contain the origin of replication and the ampicillin-resistance gene from pBR322. In addition, a segment of DNA from the *E. coli lac* operon (Fig. 28-2) has been grafted into an intergenic region (Fig. 26-7). It contains the lac control region as well as the coding sequence for the first 145 residues of β-galactosidase. Within bacteria containing the pUC DNA an N-terminal portion of β-galactosidase is synthesized. The specially designed host cell contains (in an episome) the gene for another defective β-galactosidase, one lacking the N-terminal portion. This defective enzyme, together with the N-terminal portions encoded in the pUC DNA, forms an active galactosidase. When the chromogenic substrate 5-chloro-4-bromo-3-indolyl-β-D-galactoside ("X-gal") and a suitable inducer, such as isopropylthio-β-galactoside (IPTG), are present, the unoccupied plasmid vehicles generate blue colonies. However, if passenger DNA is inserted within the galactosidase gene segment, formation of the enzyme will be disrupted and white colonies will appear. The restriction sites for cloning are placed in a **polylinker** near the 5' terminus of the galactosidase gene. The polylinker has been carefully engineered to maintain the correct reading frame and to avoid disruption of the galactosidase activity. It contains several different restriction sites for insertion of passenger DNA (Fig. 26-7). Such insertion does destroy the galactosidase activity allowing the user to detect the recombinant DNA from the white plaques.

Filamentous bacteriophages. An important series of cloning vehicles have been derived from the circular replicating forms of the filamentous bacteriophage M13 (Chapter 5; Fig. 7-7; Chapter 27).[127] Although the genome contains only a short intergenic region that can be deleted, up to 50 kb of passenger DNA can be inserted into these vehicles. Since long inserted sequences may be deleted spontaneously, M13 is most useful for cloning about 300- to 400-nucleotide chains. Many of the M13 vehicles also use the β-galactosidase blue-white screening technique. These modified viruses are highly infective, but the infected *E. coli* cells are not killed. Rather they produce large numbers of virus particles with single-stranded DNA representing one of the two DNA chains of the parental phage. These are widely used for sequencing by the chain-termination procedure of Sanger *et al.* (Chapter 5). The procedure requires a primer sequence. If M13 recombinants are sequenced, the primer consists of a synthetic oligonucleotide that can be annealed to the galactosidase gene fragment at its 3' end just in front of the DNA segment that is to be sequenced.

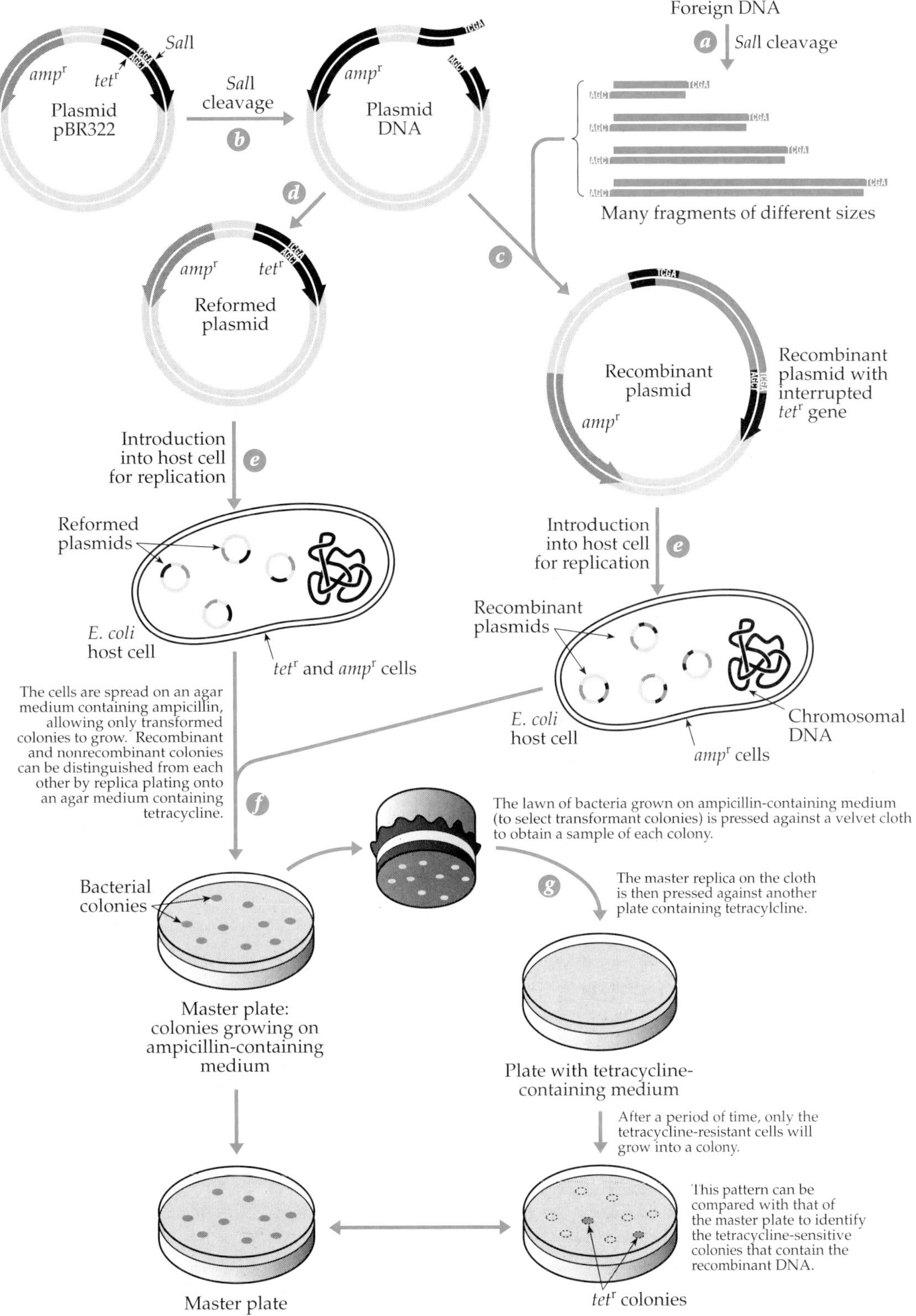

Figure 26-9 A classical scheme for cloning DNA in a pBR322 plasmid vehicle. A DNA sample is digested (step *a*) with one of the restriction endonucleases (e.g., *Sal*I), that cuts the cloning vehicle within the *Tet*^{r} gene (see Fig. 26-7). The plasmid is also cut with the *Sal*I restriction enzyme (step *b*). After mixing with the digested DNA sample and annealing, recombinant molecules are formed (step *c*). Some plasmids are reformed (step *d*). Both recombinant and reformed plasmids transform *E. coli* cells (step *e*). The transformed cells are plated on an agar medium containing ampicillin (step *f*). Only cells containing the ampicillin-resistance gene grow (step *g*). A replica plate is made and is pressed onto another plate containing tetracycline. The recombinant colonies do not grow on the medium because the tetracycline-resistance gene has been interrupted. By comparing the two plates recombinant colonies can be selected. These can be tested, using a suitable probe, to determine whether they carry a desired gene. After Atherly, Girton, and McDonald.[2]

Lambda cloning vehicles. Many cloning vehicles have been derived from the 48,502-bp[128] DNA from the temperate *E. coli* bacteriophage λ. The DNA from phage particles are taken up efficiently by *E. coli* cells, much more so than by transfection. The virus has a complex life cycle, which is discussed in Chapter 28. Within the phage head the λ DNA exists in a folded linear form with 12-base 5' cohesive ends (Fig. 28-11). After entrance into the bacterial cell the DNA cyclizes through its cohesive ends and is ligated by the *E. coli* ligase. Replication of the circular forms ensues. Later in the cycle, rolling circle replication (Eq. 27-7 and associated discussion) produces long concatamers with several phage genomes joined as a single chain. This DNA is "packaged" into new phage heads. As this is done, a nuclease cuts the concatamers at *cos* sites forming monomeric genomes with cohesive ends. This is the **lytic cycle** of the phage. In the alternative **lysogenic cycle** the DNA becomes integrated into the *E. coli* genome. Maintenance of the lysogenic state depends in part on gene *cI*, which encodes a repressor (see Chapter 28) that prevents expression of the genes required in the lytic pathway. Since only the lytic cycle is needed for cloning, it is convenient to place a cloning site within the *cI* gene. The screening of recombinant phage particles is done by examination of plaques. A phage without inserted DNA will be able to undergo both the lytic and lysogenic cycles and will form turbid plaques. However, if passenger DNA is inserted in the *cI* gene, the lysogenic cycle is prevented, and clear plaques are formed.

Of the ~50 genes present in native λ, only about half are necessary for replication in the lytic cycle. Thus, it is possible to delete about 1/3 of the genome to make room for more passenger DNA. However, to form mature phage particles the length of the DNA must be at least 75% of the native length. No more than 110% of the native amount may be present. The total DNA must fall between 38 and 53 kb in length. To accommodate these packaging requirements **replacement vectors** containing unnecessary "stuffer DNA" between two lambda "arms" are used. The unneeded stuffer piece has the same kind of restriction site or sites at each end so that it can easily be cut out and replaced by the passenger DNA. This permits cloning of DNA segments up to about 22 kb in length. Lambda vehicles have all been engineered to eliminate undesired restriction sites and to reduce the number of sites for *Eco*RI and other restriction enzymes commonly used for cloning. The widely used **Charon series**[129] have been further engineered so that they will grow only in strains of bacteria that cannot survive in the human intestinal tract. For example, amber mutations (Section B,6) are incorporated into genes needed for phage assembly, and the bacterial hosts must contain an amber suppressor gene. The bacteria are also nutritional auxotrophs with absolute requirements for thymidine and diaminopimelic acid in the medium. The latter compound is not found in the intestinal tract. The purpose of these alterations is to prevent the spread of recombinant DNA into the environment.

To sequence DNA carried in a lambda vehicle or to study it in other ways, it is often necessary to cut it with restriction enzymes, to prepare a restriction fragment map, and to subclone the fragments into a plasmid vehicle. Lambda vehicles, which will automatically transfer the passenger DNA into an M13 vehicle when propagated in a host carrying a special helper virus, have been devised.[130] The helper virus encodes proteins that recognize and cleave sequences that mark the initiation and termination of M13 DNA synthesis. These are used to mark the ends of the passenger DNA. As DNA synthesis occurs the displaced passenger DNA and M13 genes are excised, circularized, and converted into a replicating form of an M13 cloning vehicle.

Cosmid vehicles and in vitro packaging. Cosmids[131] are hybrids of a plasmid vehicle and phage λ. They contain the *cos* sites that are cleaved during packaging of λ DNA. A cosmid cleaved at a restriction site will form upon ligation a range of different sized DNA molecules that contain *cos* sites on both sides of a piece of passenger DNA, which may be up to 45 kb in length (Fig. 26-10). This can be cut at the *cos* sites and packed into heads using an *in vitro* packaging system. In this system the unassembled subunits of the phage particle are produced in special strains of bacteria and

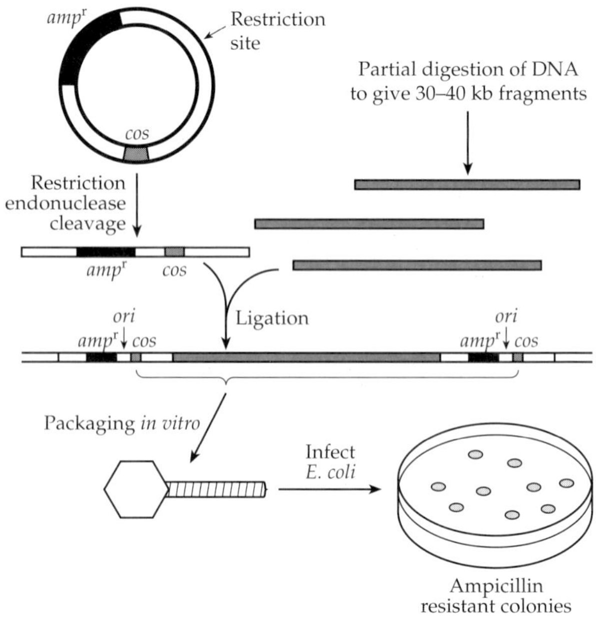

Figure 26-10 Cloning DNA in cosmids. See Glover.[99]

are allowed to assemble and package the cosmid DNA. Since cosmids contain relatively large pieces of DNA, they are useful for preparing sequentially overlapping clones that allow the investigator to "walk" along the DNA hunting for a target gene. Cosmid vehicles have been designed to allow both efficient genomic walking and restriction mapping.[132]

3. Expression of Cloned Genes in Bacteria

A major goal in recombinant DNA technology is the production of useful foreign proteins by bacteria, yeast, or other cultured cells. Protein synthesis depends upon both transcription and translation of the cloned genes and may also involve secretion of proteins from the host cells. The first step, transcription, is controlled to a major extent by the structures of promoters and other control elements in the DNA (Chapter 28). Since eukaryotic promoters often function poorly in bacteria, it is customary to put the cloned gene under the control of a strong bacterial or viral λ promoter. The latter include the λ promoter P_L (Fig. 28-8) and the *lac* (Fig. 28-2) and *trp* promoters of *E. coli*. These are all available in cloning vehicles.

It is often useful to create hybrid proteins fused to the *E. coli* β-galactosidase gene. If another gene is spliced in at either the N terminus or the C terminus of the galactosidase (*lac*Z) gene but is kept under control of the *lac* promoter, the resulting hybrid protein will have galactosidase activity, which can be used for screening. In addition, the hybrid protein will often react with antibodies directed against the protein whose gene is being cloned. Another kind of hybrid fuses the cloned gene to that of β-lactamase, for example at the *Pst I* site of plasmid pBR322 (Fig. 26-6). The β-lactamase activity will be gone, but the hybrid protein will be secreted because β-lactamase normally is secreted into the periplasmic space and its N-terminal signal sequence is now fused to the cloned protein. Engineering of a suitable site for cleavage by a protease can release the foreign protein in an active form. A variety of other **expression systems**, often using **reporter gene** products, have been developed.[133]

Bacteria often degrade foreign proteins by hydrolytic attack. One way in which such damage has been minimized is to clone multiple fused copies of the gene for the desired product. The resulting polyprotein may be resistant to degradation, and if the gene has been correctly engineered, may be cut apart by cyanogen bromide (Eq. 3-17) or by a specific protease.

For many years most cloning was done in *E. coli*, but cloning systems have now been developed for many other bacteria including *Bacillus* and other gram-positive bacteria, and also for yeast, insect cells, animals, and plants.[133–135]

4. Cloning and Transferring Eukaryotic Genes

Eukaryotic genes cloned into bacterial plasmids are often poorly expressed. It is advantageous to clone such genes in eukaryotic cells, where the cutting and splicing of hnRNA to remove intervening sequences during formation of mRNA (Chapter 28) does occur. This permits expression of the cloned gene, something that is possible in bacteria only if cDNA that lacks the introns is cloned. The need for posttranslational modification of many proteins also interferes with expression in bacteria. Many methods of gene transfer and cloning have been developed.[136] The yeast *Saccharomyces cerevisiae* is often an ideal host for cloning. It grows rapidly in either its haploid or diploid stage (Chapter 1). Some strains carry a 2-μm 6.3-kb circular plasmid with 50–100 copies per cell.[109] This has been developed as a cloning vehicle. Recombinant plasmids can be used as **shuttle vehicles** for transferring genes cloned in *E. coli* into the yeast plasmid. Genes may also be cloned as minichromosomes, such as **yeast artificial chromosomes (YACs)**. Artificial chromosomes contain origins of replication from yeast, human, or bacterial chromosomes as well as telomeres and centromeres[137] (see also Chapter 27). YACs have been widely used and became popular because they can accomodate 600 kbp or more of DNA.[137–139] However, their use in the human genome project resulted in serious problems of instability.[140] Bacterial artificial chromosomes (BACs), which accommodate only 200–300 kbp, **P1 artificial chromosomes (PACs)**,[141,142] and human minochromosomes are more stable.[143] Another problem, which affects the use of yeast for production of eukaryotic proteins, is the tendency for poor removal of introns.

Plant genes. Much of the cloning in higher plant cells has made use of the **Ti plasmid** of *Agrobacterium tumefaciens*, a soil bacterium that enters wounds in dicotyledenous plants causing tumors known as **crown galls**.[144–146c] A related species *A. rhizogenes* harbors a similar plasmid that causes "hairy root" disease.[146c,147] The infecting bacteria respond to the synthesis of certain phenolic compounds such as **acetosyringone**, which are produced in plant wounds, by entering the plant cells.[148,149] Only bacteria carrying the Ti (tumor-inducing) plasmid cause tumors. The plasmid carries a 13-kb region

O CH3
H3CO OCH3
OH

Acetosyringone, a compound that induces infection by *Agrobacterium* (transferred region) known as T-DNA that encodes enzymes for production of the auxin indoleacetic acid (Fig. 25-12), the cytokinin isopentenyl-AMP, and the compounds known as **opines**. These are reduction products of Schiff bases of amino acids and 2-oxoacids or sugars.

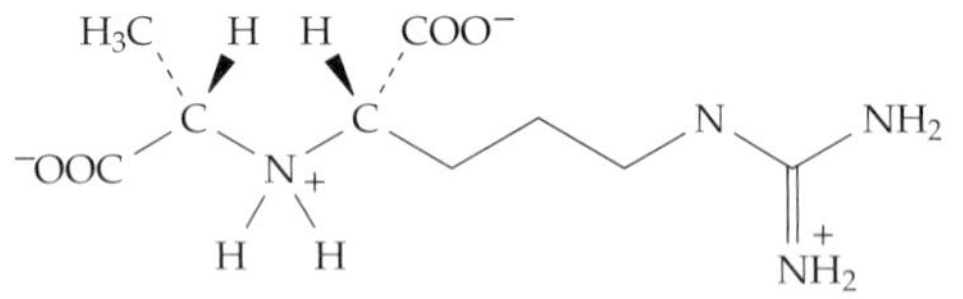

Octopine: derived from pyruvate and arginine
Histopine: derived from pyruvate and histidine[150]
Lysopine: derived from pyruvate and lysine
Nopaline: derived from 2-oxoglutarate and arginine
Leucopine: derived from 2-oxoglutarate and leucine

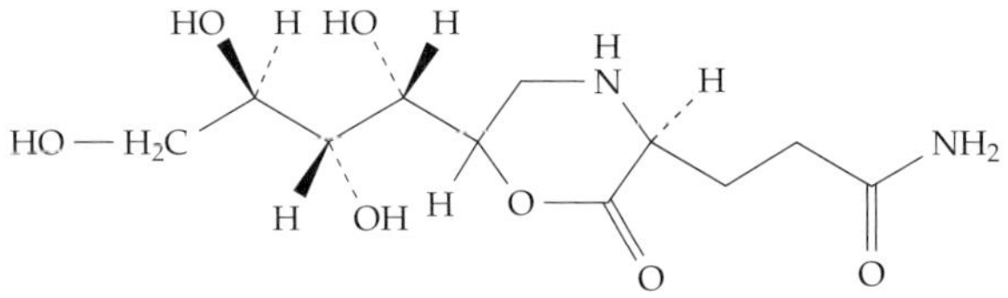

Agropine: derived from D-mannose and L-glutamine

The auxin and cytokinin, whose production is normally controlled, are now formed in large amounts and cause uncontrolled tumor growth. The opines are used by *Agrobacterium* as a unique source of energy and of metabolites for biosynthesis. The host plant cells, however, cannot catabolize the opines.

Upon entrance into a plant cell the T-region is excised from the Ti plasmid and can become integrated into the DNA of the host plant much as occurs during bacterial conjugation (Fig. 26-3).[151–154a] By deleting the genes for synthesis of auxin, cytokinin, and opines, the Ti plasmid loses its tumorigenic property but is still able to transfer genes into the plant genome. Fusion of an *E. coli* plasmid vehicle into the modified Ti plasmid creates a useful plant shuttle vehicle. Also see below under "physical methods."

Transferring genes with engineered animal viruses. There is special interest in transferring cloned DNA into human cells to correct genetic defects.[155] Transfer of human genes into other animals is also important for a variety of reasons. For example, human proteins of therapeutic value could be produced in animal milk. Animals can be engineered to have defects that mimic those in humans and which can then be studied in animals.[156]

A number of different viruses have been used to transfer foreign genes into eukaryotic cells and also to create stable plasmids for cloning. One of the first was the 5380 bp SV40. The relatively small size of SV40 limits the amount of DNA that can be incorporated. However, many of the functions of this virus can be performed by simultaneous infection with a **helper virus**, often an adenovirus which is itself defective. With the helper virus making proteins essential for SV40 replication, all but 85 base pairs at the origin of replication of SV40 can be deleted and replaced by other DNA. More recently engineered **adenoviruses** and **retroviruses** have been widely used. In both cases genes required for replication of the viruses have been deleted.[155] Adenovirus vehicles can carry 7 kbp or more of passenger DNA. They are efficiently taken up into endosomes and into the nucleus by both replicating and nonreplicating cells. The DNA is not incorporated into the genome of the host cell.[155,157,158] Replication-deficient avian **retroviruses**, or **retrotransposons** (Chapter 28), can carry up to 9 kbp of passenger DNA and are incorporated into the host genome by the process of homologous recombination (Chapter 27).[155,159,160] Both adenovirus and retroviral vehicles suffer from serious problems. The adenovirus proteins can induce fatal inflammatory reactions,[160a] and DNA transfer from the retroviral vehicles is inefficient. A nonpathogenic human parvovirus, the **adeno-associated virus** (AAV), is also being developed as a gene therapy vector.[161,161a,b] Another possibility is to engineer a **lentivirus**, even HIV, to provide efficient integration (Chapter 28).[162,162a] Safety concerns have delayed human testing of this possibility.

A group of insect viruses, the **baculoviruses**, are being used in cultured insect cells for large-scale production of proteins. The cloned genes are placed under the control of the promoter region of the gene that encodes the major viral coat protein.[135,163] The **baculovirus** vehicles can carry over 15 kbp of passenger DNA and may also be useful in human gene therapy.[164,165] Filamentous bacteriophages have been reengineered for the same purpose.[166]

Nonviral methods for DNA incorporation into the human genome may utilize **transposons** (Chapter 27)[166a] or mobile **group II introns** (Chapter 28).[166b,c]

Physical methods of gene transfer. Genes can often be transferred without the use of a cloning vehicle. This is especially important for certain plant cells, such as those of cereal grains, for which transfer of genes via the Ti plasmid has been difficult.[167] If DNA, which may be in a plasmid, is coprecipitated with calcium phosphate, it can often be taken up directly either by animal cells or by plant protoplasts.[168,169] Polycations also facilitate DNA uptake; cationic **liposomes** seem to be especially effective.[170] In the widely used **electroporation** technique a short electrical pulse of a few hundred volts / cm is applied to create transient pores in the plasma membrane through which the DNA can enter a cell.[108,171–175] Chromosomes can be transferred by cell fusion and either entire chromosomes, isolated DNA, or cloned DNA can be transferred into egg cells by microinjection.[176] Following transfer of DNA or of intact chromosomes, recombination sometimes leads to stable incorporation of some of the transferred genes into the host cell's genome.

A very important technique is the use of high-velocity microprojectiles shot from a particle gun. Spherical tungsten particles of ~0.5–1 μm diameter are coated with recombinant plasmids containing the genes to be transferred. The particles are then shot, using a gunpowder cartridge, into intact recipient cells. The particles penetrate cell membranes, mitochondria, and chloroplasts without serious damage if the number of particles is not too great.[177,178] The technique is very useful for genetic engineering of plants.[179]

5. Genomic Libraries

The human genome contains about 3×10^6 kb of DNA, about the average amount for a eukaryotic organism. If the entire genome is digested to completion with a restriction endonuclease, whose cleavage sites are distributed more or less randomly, the resulting restriction fragments constitute a "library" for that genome. If the average length of a fragment is 17 kb, about 1.8×10^5 unique fragments will be produced. To make a practical library these must be cloned into a suitable vehicle. Derivatives of phage λ or cosmids are most often used. The cloned fragments can then be packaged using the *in vitro* packaging system to form infectious phage particles, which can be propagated as plasmids in *E. coli* cells.

What is the probability that a given fragment among the total produced will be found in one of the recombinant phage in the library? From simple probability theory the number of clones that must be isolated and screened is given by Eq. 26-2.[180] Here N is the number of clones needed, p is the probability of

$$N - \ln(1-p)\ \ln(1-f) \qquad (26\text{-}2)$$

having the desired fragment in the library, and f is the fractional proportion of the genome represented by the fragment sought. For our example of a 17-kb fragment of the human genome this is 17 / 3000. From this equation one can calculate that to have a 99% chance of finding our fragment ($p = 0.99$) we need 8×10^5 clones.[99] What the equation does not show is that there will probably be some long fragments that cannot be cloned in the selected vehicle. These will be missing from the library.

Large numbers of clones obtained can be screened rapidly by colony hybridization using a labeled DNA probe. Thus, if it is desired to isolate a gene for a particular protein and some part of that protein has been sequenced, a synthetic DNA probe can be made. The phage containing the recombinant fragments can be plated, and after plaques form a nitrocellulose filter can be laid on the plate to form a replica. After release of DNA from the phage, denaturation by NaOH, and neutralization the single-stranded DNA fragments can be hybridized with the probe. Another screening method uses a probe carried in a plasmid that promotes homologous recombination between the probe sequence and restriction fragments with a similar sequence.[181] A problem that arises in preparing genomic libraries is that certain sequences, e.g., those involving highly polymorphic regions and inverted repetions, often cannot be propagated in most lambda cloning vehicles.[182]

In addition to genomic libraries, **cDNA libraries** can be prepared from mixed mRNAs. The total RNA of cells is isolated and passed through an affinity column containing oligo(dT) chains. These bind to the 3'-poly(A) tails of the mRNAs, allowing them to be isolated. The mixed mRNAs can then be cloned using a poly(AT)-tailed plasmid vehicle and a reverse transcriptase.[183,184]

6. Probes

The first step in screening the recombinant DNA in a library is use of some probe for detecting the desired DNA fragment. The most direct way is to synthesize a radioactive or fluorescent labeled oligonucleotide[185] complementary to a short known sequence in the protein. The number of codons for a single amino acid varies from 1 to 6 (Table 5-6). It is therefore desirable to prepare a probe complementary to segments of DNA containing a high proportion of codons for Trp and Met (1 codon each) and Asn, Asp, Cys, Glu, Gln, His, Lys, Phe, and Tyr (2 codons each). A popular procedure is to synthesize a mixture of probes containing all of the possible nucleotide sequences coding for the selected sequence of amino acids. The probe may be a mixture of more than 1024 different nucleotide sequences.[186] See also Chapter 5, Section H,4.

Antibodies are another popular type of probe. Antibodies to a specific protein may be utilized in isolating mRNA from ribosomes that are making that protein (Chapter 29). Thus one or more strongly binding antibodies may have already been obtained before the library clones are to be screened. To use this technique for screening recombinant DNA, the cloning must be done in a vehicle that causes expression of the gene, e.g., as proteins fused to *E. coli* galactosidase. One type of expression library is created by insertion of the cDNAs into copies of a bacteriophage gene that permits the expressed proteins to be displayed on surfaces of the phages[186a] (see Fig. 3-16).

7. Studies of Restriction Fragments

By cleavage with the correct restriction enzyme, cloned DNA fragments can be released from the vehicle in which they are carried in the library. What can be done with these fragments? The first obvious use is to sequence a gene that has been located with a probe. In many instances the gene will be longer than the cloned piece. However, the isolated restriction fragment, if labeled and denatured, becomes a highly specific probe for locating other restriction fragments that overlap it. For example, a fragment from an *Eco*RI library may bind to two or more fragments from a *Hin*dIII library. This permits "walking" along the chromosome to locate adjacent fragments. Cosmid vehicles that facilitate ^{32}P-labeling at the ends of the passenger DNA are useful.[187] A related approach called "jumping" depends upon converting very large DNA restriction fragments into circular molecules, digesting with restriction enzymes, and cloning the junction fragments of the circles. These fragments contain segments that may have been separated by as much as 100 kb in the genomic DNA, and enable the investigator to walk or jump again from a new location.[188]

Locating mutations. The study of restriction fragments provides a way of locating many mutations. For point mutations in genes for known proteins, sequencing reveals the exact defect. Some mutations, especially deletions, lead to changes in the lengths of restriction fragments. If the mutation causes loss of a restriction site, a longer piece of DNA will be present than in a digest of normal DNA. Such differences in length of restriction fragments are usually referred to as **restriction fragment length polymorphism** (RFLP, usually used in the plural as **RFLPs**). These polymorphisms are readily detectable by differences in mobility on gel electrophoresis.[96,189] They can often be mapped to particular chromosomes by hybridization *in situ* (Fig. 26-14),[190,191] by study of naturally occurring translocations of chromosome fragments, or other techniques.[192] An example is provided by the human hemoglobin abnormalities known as **thalassemias** (Chapter 32). Here, deletions remove certain restriction sites leading to observation of RFLP. Occasionally a point mutation is linked to RFLP at a nearby site. For example, in the United States most carriers of the sickle cell trait have a 13-kb *Hpa*I fragment that carries the globin S gene. However, noncarriers have their globin gene on a 7.6-kb *Hpa*I fragment.[99] Although the association is fortuitous, the linkage between the hemoglobin S gene and the mutated restriction site is broken only rarely by crossing-over during meiosis. RFLPs have been linked to many other human genetic defects and have also provided the basis for the first linkage maps of human chromosomes.[96,193]

Positional cloning. By using enzymes that cut at relatively rare sites, genomic DNA can be cut into very large restriction fragments. From studies of inheritance within families carrying specific genetic traits it is sometimes possible to find linkages between those traits and polymorphic restriction fragments.[194–196] This has been accomplished for a number of defective human genes including those responsible for sickle cell anemia,[197] cystic fibrosis (Box 26-A), Duchenne muscular dystrophy (Box 19-A), Huntington's disease,[198,199] X-linked chronic granulomatous disease[200] (p. 1072), neurofibromatosis (elephant man's disease),[201] the hereditary cancer retinoblastoma,[202] and others. By 1997 nearly 100 hereditary disease loci had been located by positional cloning.[203] These astonishing successes provided a major impetus for what became the Human Genome Project (Section G).[193]

Serious problems were met in actually locating these disease genes. Crossing-over is infrequent, occurring only at about 50 locations during each meiosis.[192] Therefore, linkage analysis does not tell us with any precision how close the linked gene is to a known DNA probe within a restriction fragment that may be up to 2000 kb in length. Finer restriction mapping or chromosome walking can be used to locate the precise piece of DNA that is defective.[204] This can still be a formidable problem. However, if the defective protein can be identified it can be sequenced. A specific oligonucleotide probe can be made for its gene and can be used to establish the exact chromosome location.

8. Directed Mutation

In addition to the developments of cloning and sequencing of DNA, a third technique is essential to the present revolution in molecular genetics. That is the ability to mutate any gene at any point in a specific way. Because of its precise nature the technique is called **directed mutation** in this book. However, the term **site-directed mutagenesis** is often used. Mutations can be introduced randomly in DNA in many ways including treatment with nitrous acid, bisulfite, formic acid, or hydrazine or by incorporation of nucleotide analogs.[205] Efficient procedures have been devised for isolating the mutants.[206] For many purposes **oligonucleotide-directed mutation** is the preferred technique.[207–210] An oligonucleotide of ~16–20 nucleotide length is synthesized with a sequence complementary to the coding strand containing the desired site of mutation. At that site the codon for the new amino acid is present. Despite this mismatch the oligonucleotide can be successfully hybridized with a single-stranded DNA such as that cloned in an M13 vehicle. Now the Klenow fragment of DNA polymerase I or a viral DNA polymerase (Chapter 27) is used to convert the single-stranded circular DNA into a double-stranded replicating form. Many of the single-stranded progeny will contain the mutated DNA. They can be screened with a labeled probe made from the oligonucleotide used to induce the mutation. It will hybridize most tightly to the correctly mutated gene.[207] Use of the PCR reaction simplifies the procedure.[210–212] In another screening procedure the template strand is synthesized in an M13 phage vehicle using uracil rather than thymine. The circular heteroduplex obtained after synthesis with T4 DNA polymerase and T4 DNA ligase is taken up by *E. coli* cells, which select against the uracil-containing strand and, therefore, in favor of the mutated strand.[208]

A third approach is to completely synthesize a gene for the protein under study. The sequence does not have to be exactly the natural one but can be made with restriction enzyme cleavage sites that permit easy excision and readdition of particular fragments. Synthetic fragments containing various mutations can then be grafted in at will. Genes of this type have been made for rhodopsin and related proteins (Chapter 23) and for numerous other proteins.

What are the uses of directed mutation? As we have seen in previous chapters, the technique is being used in every area of biochemistry to bring new understanding of protein functions and of the chemical basis of disease. Together with complete synthesis of genes it provides the basis for genetic engineering of specific proteins of plants, animals, and microorganisms. Many protein products can probably be improved. For example, enzymes can be made more stable.[213] Specificities can be changed, but this is difficult.

Targeting and replacing genes. One goal of human genetic therapy is to replace a defective gene in body cells with a good gene. Is this really possible? It is essential to engineer the DNA that is to be transferred, so that it contains all of the components needed for efficient expression in the host following its incorporation as a **transgene**. A transcription-initiation region with suitable promoter, both 5' and 3' untranslated regions, start and stop codons, and polyadenylation site (Chapter 28) must all be present. It is hoped that a correctly constructed promoter region will allow the transgene to be picked up by the machinery of homologous recombination (discussed in Chapter 27) and be incorporated into the host's DNA and expressed in the appropriate tissues.[214,215] Only some cells will take up the new gene and discard the old. However, it may happen enough to benefit a patient. Targeted gene replacement has been very successful in mice.[216]

Knockout mice. If targeted DNA is injected into a fertilized mouse egg, there is a chance that the mouse will have the targeted gene replaced in one chromosome, and that it will be stably transmitted to some of its progeny. If the transgene is totally nonfunctional, the mouse will be a "knockout mouse," suffering from a hereditary defect that can be transmitted through carriers such as its mother. The standard knockout technique is to inactivate the gene of choice in cultured embryonic stem cells and to inject these into mouse embryos. Some progeny will carry the inactivated gene in their germ cells. A refinement of the technique utilizes the **Cre recombinase** or related enzymes discussed in Chapter 27 to selectively remove pieces of DNA from genes in specialized tissues of mice.[217,218] By 1996 several hundred different knockout mice had been created.[215] Nevertheless, interpretations of results of gene knockout are sometimes complex.[219]

F. The Genetics of Eukaryotic Organisms

Whereas DNA synthesis takes place almost continuously in a rapidly growing bacterium, replication of DNA occupies a more limited part of the **cell cycle** of eukaryotes (Fig. 11-15). In a mammalian cell mitosis proper (Fig. 26-11) may require about one hour. It is followed by the "gap" period, G_1, whose length is variable and depends upon the cell type, the nutritional state of the cell, and other factors. About 10 h is typical. During the S phase (~9 h) active DNA replication takes place. This is followed by a second gap (G_2) that occupies 4 h in the 24 h cell cycle shown in Fig. 11-15. The length of the different segments of the cell cycle varies widely among different organisms. Indeed, the concept of a cell cycle can be criticized.[219a] It is only in a rapidly growing culture that all, or most, cells can follow the same cycle. In the adult body most cells are inhibited from division (or are not stimulated to divide) most of the time.

1. Mitosis

The distribution of chromosomes to daughter cells of somatic cells undergoing division is accomplished by mitosis whose successive phases are referred to as **prophase**, **metaphase**, **anaphase**, and **telophase** (Fig. 26-11). As the chromosomes condense during prophase, it is seen that each one actually consists of two separate entities coiled together. These are the identical **chromatids,** which are formed from the two identical double-stranded DNA molecules formed by replication of the DNA of the chromosome during the S phase of the cell cycle. As the folding of the chromosomes occurs (during prophase), the nuclear envelope completely fragments or dissolves in many species.

An important event that *precedes* the main stages of mitosis is the formation of **poles** in the cell. In animal cells, the poles are formed by the **centrioles,** which move apart and take up positions at opposite sides of the cell. Each of the centrioles is accompanied by a smaller "daughter" centriole lying at right angles to the larger parent. In plant cells, which lack centrioles, a more diffuse pole is formed. As the cell prepares for mitosis, fine microtubules (15 nm diameter) can be seen radiating from the poles. At the end of prophase the microtubules run from one pole to the other to form the **spindle**. Microtubules also become attached to the chromosomes at the **centromeres**.

At metaphase the chromosomes are precisely lined up in the center of the cell to form the **metaphase plate**. Now each centromere divides, permitting the sister chromatids to be completely separated. A protein complex **cohesin**, which holds the sister chromatids together, undergoes proteolysis by a **separase** at this stage.[220,220a,b] During anaphase the separated chromatids, now referred to as **daughter chromosomes**, move to opposite poles as if pulled by contraction of the spindle fibers. Telophase is the final stage in which new nuclear envelopes are formed around each set of daughter chromosomes. In humans and many other species the cell pinches in two. In plants and fungi new plasma membranes and cell walls are constructed through the center of the cell. A partitioning of mitochondria and of Golgi components between cells must also occur.[220c]

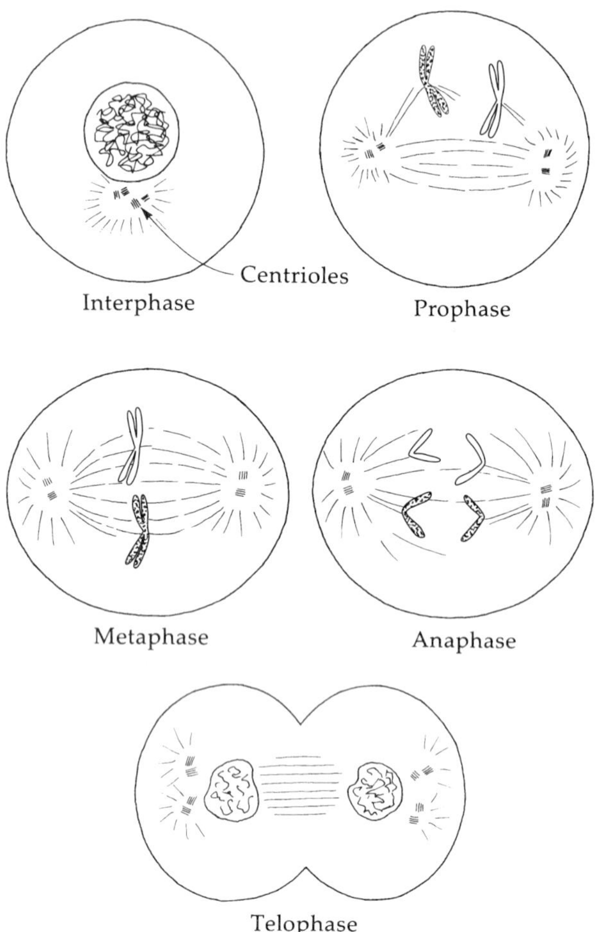

Figure 26-11 Mitosis. Illustrated for a cell with one homologous pair of chromosomes. After Mazia.[221]

The foregoing description overlooks the extreme complexity of mitosis, each stage of which must occur with precision and in the correct sequence.[222–225a] The replication of DNA, which takes place in the S phase of the cell cycle (and is discussed in Chapter 27) must be completed before mitosis begins. This is followed by condensation of the DNA into chromosomes (Chapter 27), breakdown of the nuclear membrane,[226,226a] assembly of the **kinetochores** by which the chromosomes attach to the spindle,[222] assembly of the spindle, attachment of chromosomes to the spindle, segregation of the chromosomes to opposite poles in anaphase, and finally the cleavage of the cell.

Cyclin-dependent kinases. As is shown in Fig. 11-15, the cell cycle is controlled by a series of complexes of the 30- to 45-kDa proteins called **cyclins** with **cyclin-dependent protein kinases** (**CDKs**). These kinases contain ~300-residue catalytic cores that resemble protein kinase A (Fig. 12-32). Like that kinase they transfer phospho groups from ATP to serine and threonine side chains of target proteins.[227–228] The kinases are inactive until a complex with the appropriate cyclin is formed and is activated by phosphorylation.[229] One or more additional kinases are required for this activation. Each stage in the cell cycle is controlled by one or more different cyclin–CDK complexes (Fig. 11-15). One of the best known CDKs is human **CDK2**, which functions in a complex with cyclin A during the S phase of the cycle.[228,230–232] Binding of the cyclin greatly alters the conformation of CDK2, opening the catalytic cleft and exposing threonine 160. Its hydroxyl group can be phosphorylated by the action of the **CDK-activating kinase** (**CAK**) with a 100-fold increase in catalytic activity.[228] Control of the CDKs also depends upon proteins that act as specific inhibitors and upon precise elimination of both cyclins and inhibitors via the ubiquitin system.[228a,b,c]

Checking for completion of replication and for DNA damage. The first checkpoint in the cell cycle is the **start** or **G_1 DNA damage checkpoint** (Fig. 11-15). Replication of DNA does not begin until the cell has had time to repair as much damage to DNA as is possible.[228d] As mentioned in Box 11-D, the cancer suppressor protein p53 is an essential component of the checking process.[224] A second checkpoint, the **G_2 checkpoint**, at the end of the G_2-phase (Fig. 11-15) requires verification that all DNA of all chromosomes has been replicated, checked for damage, and repaired if necessary. Control of these processes is accomplished by a mechanism first identified in

fission yeasts, such as *Schizosaccharomyces pombe*, and which has been conserved in metazoa.[223,233–236d] The CDK known as **cdc2** (fission yeasts), CDC28 in budding yeast, or CDC2 or CDK1 in mammals, functions from G_2 through the DNA damage checkpoint to mitosis (Fig. 11-15) and is sometimes described as the master cell-cycle switch.[236a] The catalytic subunit of cdc2, a serine / threonine protein kinase, is often called $p34^{cdc2}$. It is inactive unless complexed with a suitably phosphorylated cyclin (Eq. 26-3) and phosphorylated on Thr 161 by the action of kinase CAK (Eq. 26-3). However, when cdc2 becomes phosphorylated on Tyr 15 (in *Schizosaccharomyces pombe*) by the action of protein kinase wee 1 (Eq. 26-3) and on Thr 14 by another kinase, cdc2 is inhibited and mitosis is arrested at the G_2 checkpoint. This allows time to verify completion of replication as well as for repair before the replicated DNA strands are separated in mitosis. Hydrolytic removal of the phospho groups from Tyr 15 and Thr 14 of cdc2 by the action of phosphatase cdc25 then allows mitosis to begin.[223,237,237a] This phosphatase is also controlled by a phosphorylation-dephosphorylation cycle. The checkpoint kinase **Chk1** phosphorylates and inhibits cdc25.[238,238a]

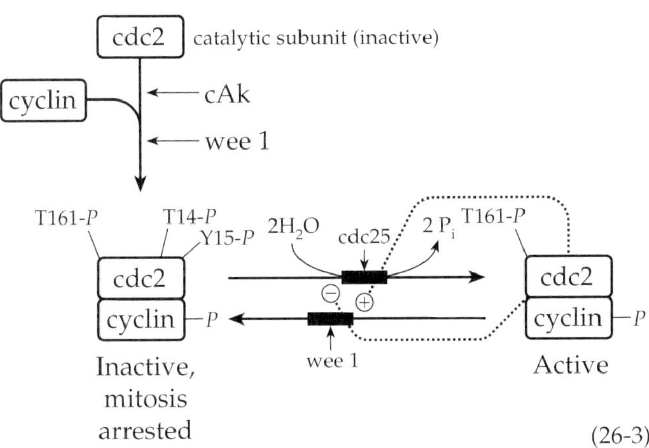

(26-3)

When damaged DNA is present, protein p53 accumulates, just as at the G_1 checkpoint, and activates transcription of the seven phosphoserine-binding proteins of the 14-3-3 family.[224] These bind to the phosphorylated cdc25 phosphatase preventing activation of cdc2 by dephosphorylation.[224,238–240] In fission yeast one of the 14-3-3 proteins, **Rad 24**, apparently binds to the phosphorylated cdc25 and induces its export from the nucleus preventing mitosis.[241] Dephosphorylation of cdc25 and return to the nucleus allows mitosis to occur. A somewhat different regulatory mechanism is used by the budding yeast *Saccharomyces cerevisiae*.[242] In animal cells there are three cdc25 isoforms.[240,240a,b] In each case we probably have only a glimpse of a very complex process of checking for DNA damage and repair (see also Chapter 27).

Mitotic spindle formation and the spindle assembly checkpoint. Separation of the two copies of each replicated chromosome depends upon the spindle fibers. Their formation is preceded by replication of centrioles, if present, and formation of the two poles of the cell.[243–244b] Both γ-tubulin (Fig. 7-34) and acidic Ca^{2+}-binding proteins called **centrins** are involved.[244c–e] The microtubules appear to grow outward from the poles with their minus ends at the poles and their plus ends (Fig. 7-33) available for binding to the kinetochores, specialized protein complexes that assemble around the centromeric DNA (Chapter 27).[245–247] Each chromosome has two kinetochores, one for each daughter chromatid. These must be attached to spindle fibers coming from opposite poles. It has usually been assumed that random encounters of microtubule plus ends with kinetochores leads to correct linkage.[245,246] However, neither centrosomes or kinetochores are always essential to spindle assembly, and self-assembly occurs by motor-driven sorting according to polarity of the microtubules.[246,247] In addition to microtubules formation of spindles may require specialized matrix proteins.[247a]

The spindle isn't formed until replication is complete. A small G protein called **Ran**, a relative of Ras (Fig. 11-7), regulates spindle formation. Ran, in turn, depends upon a nucleotide exchange factor called **RCC1** (**regulator of chromatin condensation**).[248–250] RCC1 may signal that the replicated DNA is folding into chromosomes indicating that replication is complete. At the **spindle assembly checkpoint** (Fig. 11-15) the cell verifies that the metaphase spindle has been assembled correctly.[250a] All of the microtubules that pull the sister chromatids toward one pole or the other must be correctly attached to a kinetochore.[220,250b] In addition, there are interdigitated microtubules coming from both poles. Specific motor molecules (Chapter 19) then push the two poles apart. During the assembly and complex movements of the spindle both cytosolic dynein and four different types of kinesin-like motor molecules are required.[251,252] One of these is Kar3 (see Fig. 19-17). After assembly of the spindle is complete, a signal must be sent to the mitotic apparatus to move into anaphase. A clue to the nature of the signal has come from the observation that a single kinetochore lacking a spindle fiber connection causes arrest of mitosis.[253] Apparently unattached kinetochores send a "wait" signal, perhaps via the cytoskeleton.[253–255]

Anaphase. After the spindle has been checked, a sudden loss of cohesion between the sister chromatid pairs allows them to move toward the opposite poles. This process is catalyzed by the **anaphase-promoting complex** (**APC**, or cyclosome) and its activator protein **Cdc20**, a large multiprotein complex.[225] The APC also promotes proteolytic breakdown of cyclins and other proteins by a ubiquitin- and proteosome-dependent mechanism.[225,256–259c] Its E_3-ubiquitin ligase (Box 10-C) targets the mitotic cyclins and other proteins for destruction.[260] A specific E_2 ubiquitin-conjugating enzyme (Box 10-C) is required for degradation of cyclin B and exit from mitosis.[261] The centrosome also plays an active role in cytokinesis, the final step in cell division.[261a]

The cell cycle encompasses so many different processes that it is clearly impossible to describe it by the single diagram of Fig. 11-15 or by the text written here. The cycle is influenced by a host of growth factors and external stimuli, many of which act on transcription of cyclins and other essential proteins. Transcription factors such as those of the Fos / Jun (AP-1) family in response to the MAP cascade (Fig. 11-13) are among those that control the transcription of cyclins.[262–263c] However, during mitosis most transcription of any genes is repressed.[264]

Among other factors influencing the cell cycle is the size of the cell and the availability of nutrients including purine and pyrimidine nucleotides.[263b,c,264a] Lack of cholesterol decreases the cdc2 kinase activity and causes apoptosis.[265] A cell cycle regulator in *S. pombe* known as **suc1** is essential for cell cycle progression. Although its three-dimensional structure is known, its function (like that of its human homolog CksHs2) is uncertain.[266,267]

2. Meiosis

The mechanism by which chromosomes are distributed during the formation of **gametes** (egg and sperm cells) is known as meiosis (Chapter 1; Fig. 26-12). Formation of gametes involves a halving of the chromosome content of a cell, each gamete receiving only one chromosome of each homologous pair. Genes found in the same chromosome are said to be **linked** because of their tendency to be passed together to the offspring. Genes present in different chromosomes are not linked, and their inheritance follows the pattern of **random segregation** established in Mendel's famous studies.

The simple fact that the genetic material is put up in several different packages (chromosomes) is sufficient to provide for considerable mixing of genetic information between different individuals in sexual reproduction. However, it doesn't provide a means for exchanging genes on the same chromosomes. Mixing of genetic information within chromosomes occurs by genetic recombination occurring during **crossing-over**, an aspect of meiosis with an essential biological role. In the S phase preceding meiosis, DNA is duplicated just as it is prior to mitosis. This provides sufficient genetic material to produce *four haploid cells*. These are formed during meiosis by two consecutive cell divisions (Fig. 26-12). Crosssing-over occurs prior to the first of these divisions, at the four-strand stage. The two homologous chromosomes of a pair come together to form what is called a **bivalent** or **tetrad** made up of four chromatids. For each chromosome, at least one chromatid is seen to come into intimate contact with a chromatid in the other homologous chromosome at points known as **chiasmata** (Fig. 26-13). During metaphase of the first meiotic cell division the homologous chromosomes, each still containing two chromatids, separate. Each chromatid now carries with it some genetic information that was previously found in the other member of the homologous pair and vice versa (Fig. 26-12). Now, without further replication of DNA in the second meiotic cell division, the chromatids separate to form haploid cells.

The process of crossing-over provides a means by which genes that are linked on the same chromosome can be separated, providing offspring with mixtures of genetic traits other than those predicted by simple Mendelian theory. The effects of crossing-over were first studied extensively by T. H. Morgan with the fruit fly *Drosophila melanogaster*. Morgan discovered his first mutant, a white-eyed fly,[267a] in 1910. The first genetic maps were made by assuming a direct relationship between the frequency of crossing-over and the linear distance between genes in a chromosome. Thus, the same approach to genetic mapping that was used later with *E. coli*, i.e., the measurement of recombination frequencies, was applied much earlier to crossing-over in the chromosomes of *Drosophila*. Extensive genetic maps involving many mutations were obtained for the four chromosomes of this organism, and similar techniques have been applied to many other organisms. The unit of distance in these chromosome maps is the **morgan** (named for T. H. Morgan[2,267a]). One centimorgan is the distance that allows recovery of 1% of recombinant progeny.[2] In the human chromosomes, this is ~1000 kb (1 Mb).

Biochemical and genetic studies of meiosis have been conducted in many organisms including fission[268,269] and budding[270–271a] yeasts, *Drosophila*,[272,273] starfish,[274] *Xenopus*,[275] and the mouse.[276] Meiosis can be viewed as a modification of mitosis but with the added initial step of crossing over and recombination. In addition, the S-phase of the cell cycle is absent in the second meiotic division. As was mentioned in Chapter 1, meiosis may occur at different stages of the life cycle of organisms. An important advantage in using fungi for genetic studies is that, like prokaryotes, they are haploid during much of their life cycle. Biochemical defects such as the inability to synthesize a particular nutrient can be recognized readily at this stage. At the same time genetic crosses can be made, and crossing-over frequencies can be measured and used for genetic mapping. The onset of meiosis may

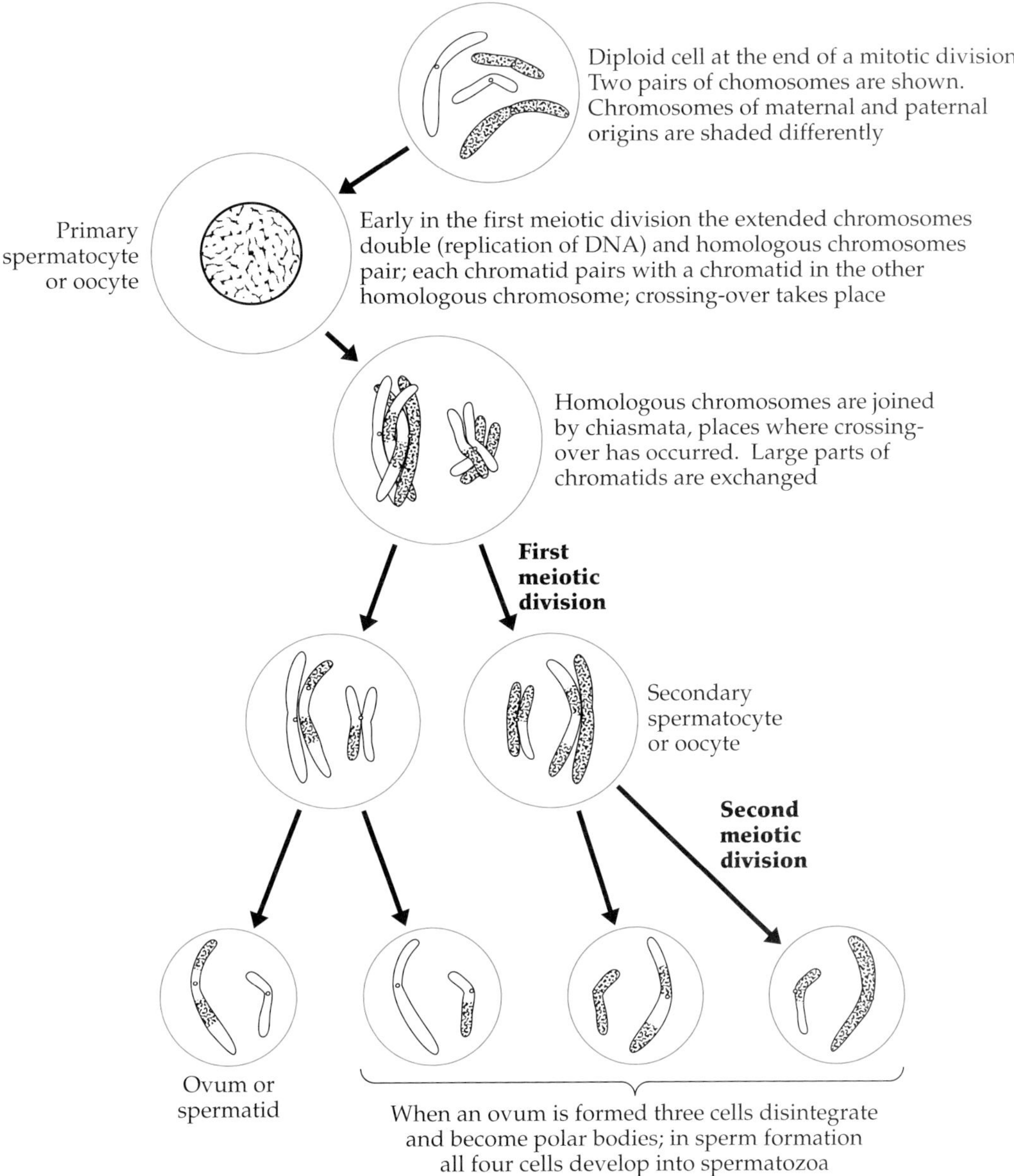

Figure 26-12 Meiosis. Cell division leading to formation of haploid gametes.

vary in time with diploid organisms. Meiosis in males may occur quickly. However, after chromatid pairing and recombination meiotic divisions of oocytes are arrested at the G_2 stage, in some species for many years.[274,275] The arrest is ended by hormonal stimulation, e.g., via progesterone acting through a cyclin B–cdc2 complex.[276] **Meiosis-activating sterols**, intermediates in lanosterol metabolism (Fig. 22-8) also accumulate and are thought to participate in control of meiosis.[276a] In haploid strains of fission yeasts sexual development is induced by starvation, especially of nitrogen. Cells of opposite mating types then fuse to form zygotes, which usually undergo meiosis immediately. Starvation is apparently signaled by the **cyclic AMP–protein kinase A cascade** (Fig. 11-13).[268]

During the prophase of the first meiotic division (meiosis I) two homologous pairs of partially "condensed" chromosomes must find each other and pair with appropriate orientation. A protein in the telomeres of the chromosomes seems to be involved.[269,277] The key structure in meiotic crossing-over is the ribbonlike **synaptonemal complex** formed by the pairs of homologous chromatids.[271,278–279b] This complex, in which a proteinaceous core or **axial element** separates the greatly extended chromatid pairs (Fig. 26-13), is fully formed in the **pachytene stage** of meiosis. Formation of the synaptonemal complex is preceded by development of a few double-stranded breaks in

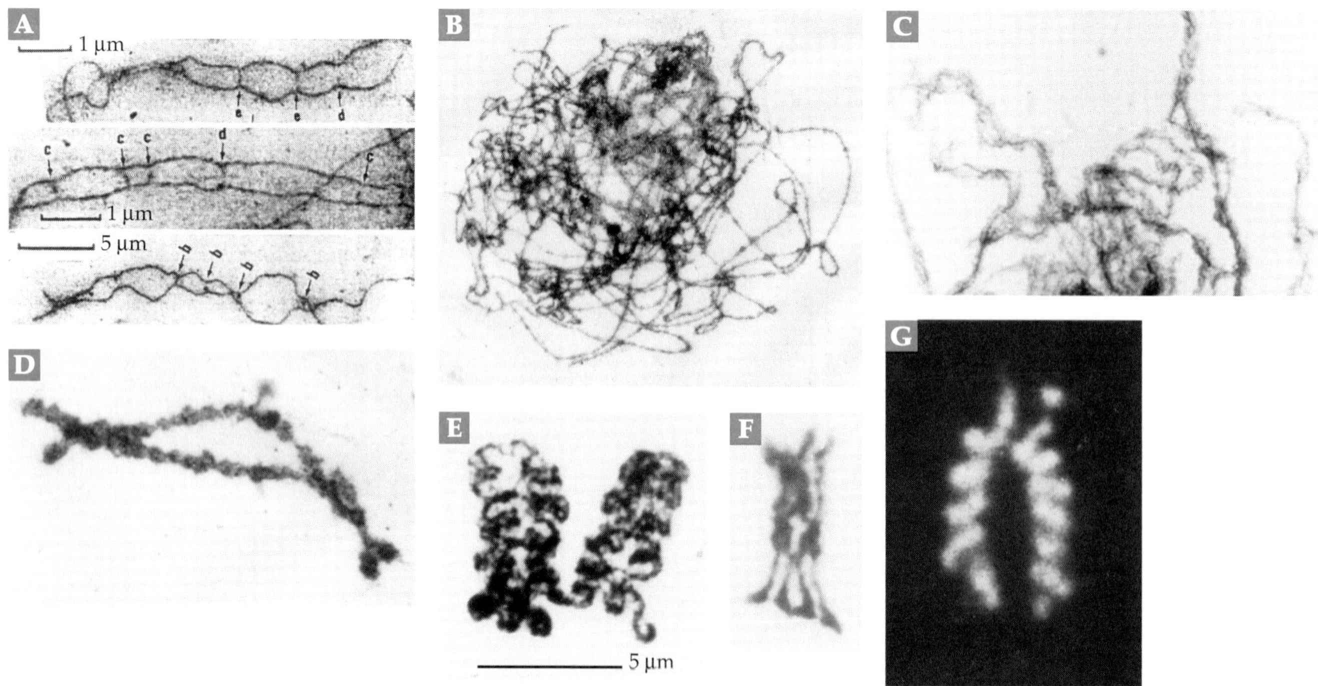

Figure 26-13 Synaptonemal complexes. (A) Aligned pairs of homologous chromatids lying ~0.4 μm apart in *Allium cepa*. Arrows indicate "recombination nodules" which may be involved in initiating formation of crossovers. Portions of meiotic chromosomes of lily are shown at successive stages: (B) Pachytene. (C) Portion of diplotene nucleus. (D) A bivalent at diplotene. (E) Two bivalents at diakinesis. Pairs of sister chromatids are coiled with appropriate handedness. (F) Sister chromatid cores are far apart in preparation for separation. A chiasma is present between the two central strands. (B) through (F) courtesy of Stephen Stack.[279,279d] (G) Pair of sister chromatids coiled with opposite handedness at metaphase. These are immunostained with anti-topoisomerase II antibodies. From Boy de la Tour and Laemmli.[280] Courtesy of U. K. Laemmli.

the DNA.[278,279c] The 3' ends of the DNA chains then invade the homologous chromatids to initiate the exchange via **Holliday junctions** (Fig. 5-28 and Fig. 27-11). The details of the recombination process are considered further in Chapter 27. The first meiotic division is a long process, for example, lasting ten days in mouse spermatocytes.[281] As cells pass to the metaphase stage of meiosis I, the chromosomes become much more compact, but the attachments between the homologous chromatids are still visible as chiasmata (Fig. 26-13). The chromatids then separate, the two homologs appearing as coils of opposite handedness. Within these coils the pairs of sister chromatids continue to be held together at their centromeres through metaphase and anaphase of meiosis I. One or more specific proteins, which must release their hold in meiosis II, are required.[270,277] A leading cause of embryonic and fetal death as well as severe mental and physical problems after birth is incorrect segregation of chromosomes during meiosis. If a gamete contains two copies of any chromosome the embryo will have three. Down syndrome, trisomy of chromosome 21, is the most frequent example. The spindle assembly checkpoint during meiosis I may sometimes be at fault.[281a,b]

3. Polytene Chromosomes

While most cells of higher organisms are normally diploid, the chromosome number may sometimes be doubled or increased even more. A cell with twice the diploid number of chromosomes is **tetraploid,** and with higher multiples of the haploid number it is **polyploid**. Plant breeders have succeeded in producing many tetraploid varieties of flowering plants often with increased size. One tetraploid mammal, the red viscacha rat, is known.[281] While most of our body cells are diploid, we, too, have polyploid cells. For example, some are always found in the liver. The most spectacular example of an increase in the normal DNA content of cells is provided by the giant **polytene chromosomes** of dipteran (fly) larvae. The DNA of cells in the salivary glands and some other parts of these organisms doubles as many as 11–14 times without cell division to give a several thousandfold (i.e.,

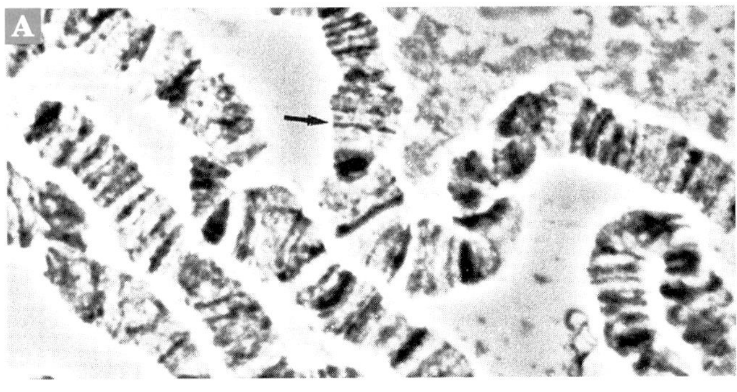

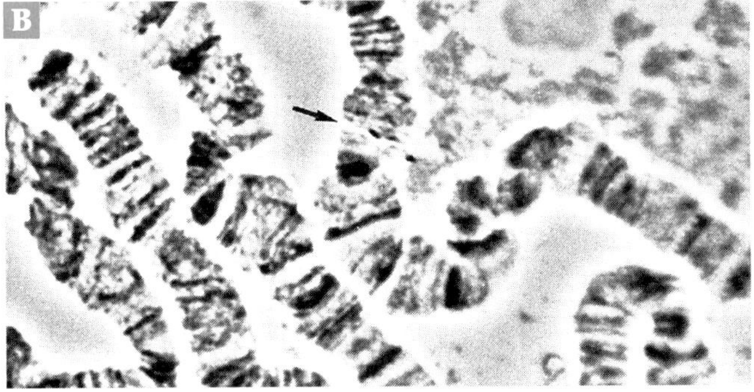

Figure 26-14 Microdissection of a *Drosophila* salivary gland chromosome. (A) Before the cut. (B) After cutting a medium size band. The arrow indicates the site of the cut. From Pirrotta.[282]

2^{14}-fold) increase. The supercoiled, duplicated DNA molecules all line up side by side in a much more extended form than in ordinary chromosomes. They can be seen readily by light microscopy. The total length of the four giant chromosomes of *Drosophila* is ~2 μm, compared to 7.5 cm in diploid cells. The giant chromosomes have a banded structure, ~5000 bands being visible along the length of the chromosome (Fig. 26-14). An average band contains ~36 kb of DNA in each of the strands. Since it has been possible to correlate visible changes in the appearance of these bands with particular mutations in the DNA, study of polytene chromosomes provided a second important method of mapping genes of the fruit fly. The maps produced by the two methods agree well.

Another use of polytene chromosomes is microdissection of DNA for cloning (Fig. 26-14). A piece of DNA containing 100–400 kb can be cut out of any desired spot, cleaved with restriction enzymes, and cloned.[282] Since it has been estimated that *Drosophila* may contain only ~9000 –17,000 genes there may be 2–3 genes per band in these chromosomes.[283] The technique has been extended to human and other mammalian chromosomes.[284]

4. Cytoplasmic Inheritance

Not all hereditary traits follow the Mendelian patterns expected for chromosomal genes. Some are inherited directly from the maternal cell because their genes are carried in the cytoplasm rather than the nucleus. There are three known locations for cytoplasmic genes: the mitochondria, the chloroplasts, and certain other membrane-associated sites.[285,286] An example of the last is found in "killer" strains of yeast. Cells with the killer trait release a toxin that kills sensitive cells but are themselves immune. The genes are carried in **double-stranded RNA** rather than DNA, but are otherwise somewhat analogous to the colicin factors of enteric bacteria (Box 8-D). Similar particles (κ factors) are found in *Paramecium*.[287]

Mitochondrial and chloroplast genes are discussed in Chapters 18 and 23, respectively.

G. The Human Genome Project

The discovery of site-specific restriction endonucleases in 1970 and the development of efficient DNA sequencing methods in 1977 sparked a revolution in biology. On Oct 1, 1990, the 15-year project to sequence the complete human genome was officially begun.[288–290] The project is far ahead of schedule. "First drafts" of the genome were published in 2001.[290a–e] A more complete version will include **annotation**, a listing of predicted exons, mRNA transcripts,[290f] functions of recognizable genes, etc.[290e] Such documentation was provided first for chromosome 22[290e] and is now available for other human genes.[290g] It was a surprise to discover, at least initially, that only 30,000–40,000 protein-coding genes could be recognized.[290a,290h] Earlier 50,000–150,000 genes had been predicted (Table 3-1). A nearly complete version of the human genome is anticipated for spring of 2003.[290i]

Included in the genome project are completion of the sequences for *Caenorhabditis elegans*,[290j] *Drosophila melanogaster*, and the mouse (*Mus musculus*).[291,291a–e] However, substantial difficulties remain in filling hard-to-sequence gaps, and some new approaches may need to be developed.[292] More recently the project has been extended to include many additional species. Among them are the rat,[292a] the zebrafish,[292b] the pufferfish *Fugu rubripes*,[292c] the human malaria parasite *Plasmodium falciparum*,[292d,e] the rodent malaria *Plasmodium yoelii, yoelli*,[292f] and the mosquito *Anopheles gambia*, which carries the parasite.[292g,h] Plant genome sequences are also being determined (p. 1511).

1. The Mammalian Genome and Human Health

Human beings and other mammals all have about

3500 Mbp of DNA containing perhaps 40,000 protein-coding genes. However, ~ 97% of the genome is repetitive DNA and other DNA that, at present, has no recognizable function. The genome project will reveal all of these sequences and will doubtless provide us with many surprises.

A primary goal of the genome project is to understand the relationships between gene sequences and human diseases and health. Until recently little was known about the locations or structures of genes in human chromosomes. Human genetic diseases provided the first clues. Although systematic genetic experiments cannot be done with human beings, almost the entire population is under some kind of medical care. Genetic defects are being detected more and more often, and the inheritance within families is traced with increasing frequency. Consequently, a huge body of knowledge of mutations to the human genome is available.[293] By 1995 ~4500 human genetic disorders had been discovered.[294] Many more have been found since then.

The first mapping of human genetic defects came with the recognition that sex-linked traits are encoded on the X-chromosome. Some linkage analysis was also possible from studies of inheritance within families. For example, among individuals who have two X-linked traits, e.g., color blindness plus one other, naturally occurring crossing-over occasionally breaks the linkage in some individuals within the family. **Somatic cell fusion**[191] provided an additional approach. Human lymphocytes can be fused with rodent cells under the influence of inactivated Sendai virus, which causes the cells to adhere and then to fuse. From such human–mouse or human–hamster hybrid cells, strains in which the nuclei have also fused can be selected. Although such cell lines can be propagated for many generations, they tend to lose chromosomes, especially those of human origin. By observing loss of particular human enzymes or other proteins (separable from the hamster enzymes by electrophoresis), it was possible to assign genes to specific chromosomes. This also required identification of the chromosomes lost at each stage in the experiment. New staining techniques made it possible to identify each of the 22 pairs of human autosomes as well as the X and Y chromosomes (Fig. 26-15).[295] Using a variety of techniques, 500 loci in human chromosomes had been mapped by 1983.[296]

Separation of individual human chromosomes with a fluorescence-activated cell sorter (Box 3-B)[299] permitted the preparation of libraries of cloned DNA from individual chromosomes. Fragments, of an average length of 4 kbp, from digestion by *Eco*R I or *Hin*dIII were packaged into the Charon 21A cloning vehicles and "amplified" by culturing infectious phage particles in *E. coli*.[282] However, this approach was inadequate for sequencing the entire genome. By 1987 the use of yeast artificial chromosomes (YACs)[300] and later bacterial artificial chromosomes (BACs)[301,302] and radiation hybrid mapping[104,303] provided the means for sequencing the 34- to 260-Mbp human chromosomes.[288,304] Also essential was the development of high-speed automatic sequencing machines[305–305b] and of computers adequate to assist in compiling the sequences.[302,306,306a] Sets of overlapping and redundant clones define a continuous sequenced segment that is called a **contig**. As the contigs are enlarged the gaps are filled.

In 1987, the first global genetic linkage map of the human genome was published.[288,307] It was based on 403 polymorphic loci, of which 393 were RFLPs studied in 21 three-generation families. By 1996 a linkage map containing 5264 loci, in the form of polymorphism in **short tandem** repeats $(AC/TG)_n$ represented by **microsatellite DNA**, was available.[304] Also mapped by 1998 were 2227 **single nucleotide polymorphisms** (SNPs), where the two alternative bases occur with a frequency >1%.[308] This was increased to 1.4 million by 2001.[308a] As many as 5.3 million common SNPs may occur, each with a frequency of 10–50%. On the average there may be one SNP for every 600 base pairs.[308b] These SNPs account for a large fraction of the diversity in human DNA. By 1995 a physical map with >15,000 **sequence tagged sites** (STSs) with an average spacing of 199 kb had been constructed.[303,309] An STS is defined as a cloned fragment in a YAC (or BAC) library that has been amplified by PCR and tested to establish that it contains a known locus. DNA sequences known as **expressed sequence tags** have also been mapped (see p. 1490). Using STSs, RFLPs, ESTs, and the growing contigs, more than 16,000 human genes had been mapped by 1996.[310] By 1998 more than 1,060,000 ESTs had been reported.[311] In 1999, the nucleotide sequence of the smallest human chromosome, the 33.4-Mbp chromosome 22, was completed. It contains at least 545 genes and 134 pseudogenes.[312] In the entire genome >1000 seed contigs had been assembled, and completion of the first phase of the Human Genome Project was in sight.[313] In 2001 a whole-genome clone-based physical map was published.[313a]

2. Understanding Gene Sequences

The vast array of sequence data coming from the human genome and from genes of other species are deposited, as they are reported in the scientific literature, in the Human Genome Central. It can be found on the Web at the National Center for Biotechnology Information (NCBI) at http://www.ncbi.nlm.nih.gov/genome/guide and European Bioinformatics Institute (EBI) at http://www.ensembl.org/genome/central/. The data have been doubling every 15 months. By

2000 almost 1.3 million human ESTs were included as were sequence data on more than 25,000 other species.[311,314] A problem is that there are many errors in the data with some seriously incorrect sequences.[315] It is hard to manage the mass of new data, but in time most of the errors will probably be corrected. Books[316–319] and computer programs[320,321–321c] are available to help understand genomes. The widely used programs **BLAST** and **FASTA**, available on the Web or in computational packages, routinely compare protein-coding genes with known genes in order to predict function.[322–325] An inexpensive high-powered desktop computer and an internet connection will enable a person to do complex biocomputing.[325a] See also Chapter 5, Section H,7.

Genes must be transcribed, and most transcripts must be spliced, modified, and translated by the ribosomal machinery. Genes cannot be fully understood without considering these processes, which are dealt with in Chapters 28 and 29.

Human variation. Between any two unrelated human beings there are on the average one base pair that is different out of every 500–1000 nucleotides. This amounts to $\sim 4 \times 10^6$ differences in the whole genome.[305,326] In addition to these single nucleotide polymorphisms (SNPs) there are many differences in the >100,000 nearly randomly dispersed short tandem repeats of microsatellite DNA.[189,327] The latter form the basis of DNA fingerprinting (Box 5-D), and together with the SNPs are helpful in tracking genetic diseases. One of the more difficult goals is to identify genes that confer increased susceptibility to cancer and to the complex syndromes of diabetes and mental problems. Searching for correlations with SNPs has proved difficult.[327a,b] A new approach termed **haplotype mapping** may permit correlations of disease susceptibility with larger blocks of conserved DNA sequence known as haplotypes.[327c,d] Some controversy surrounds possible uses of these maps, as also is the case for proposals to conduct extensive genetic testing of populations including newborn infants.[327e,f,g] Another planned project is the sequencing of 1000 individual human genomes.[327h]

Studies of DNA have also shed light on human evolution.[328] Mitochondrial DNA,[329] as well as a variety of nuclear genes,[330] is being studied in attempts to establish approximate dates of evolutionary divergence. If we assume a constant rate of incorporation of nucleotide substitutions during evolution, we can use sequence data as a **molecular clock** to construct phylogenetic trees such as that of Fig. 1-5, which is based on the gene for 16S ribosomal RNA. A problem is that mutation rates of genes are not all the same, and there are not yet enough data to draw firm conclusions. Estimates based upon the maternally transmitted mitochondrial DNA (see Chapter 18) suggest that we are all descended from Africans who lived 100,000 to 200,000 years ago. Limited data from the Y chromosome, which is transmitted through males, agree with the mitochondrial DNA data.[328,328a] However, study of the pyruvate dehydrogenase gene has been interpreted to indicate a more ancient divergence of African and other peoples.[331] Analysis of DNA from a fossil bone shows that the mtDNA of Neandertal people differed from modern mtDNAs in an average of 26 positions. From this figure the mitochondrial DNA molecular clock predicts that the extinct Neandertal line diverged from ours ~ 550,000 to 690,000 years ago. The most ancient mtDNA found is Australian.[331a,b] Unfortunately, DNA in fossils is very unstable and has not been recovered from older fossils, e.g., from dinosaurs. However, the search for ancient DNA goes on.[331c,d]

We diverged from our closest ape relative, the chimpanzee, about 4–6 million years ago.[196,332] Chimpanzee and bonobo DNA sequences are ~98.8% identical to those of humans.[332a–e] If differences in inserted and deleted segments of genes are included, however, the identity drops to 95%.[332f] One of the longest DNA sequences to be compared among humans and apes is a 7-kb length around the pseudogene in the β globin cluster (Fig. 27-10). In this sequence chimpanzees are closest to humans with gorillas being the next closest.[195] We may well ask in what way we differ from these apes? Some specific differences have been found. Notably, human beings do not hydroxylate the glycolyl groups of sialic acids (Chapter 4) to form *N*-glycolylneuraminic acid residues on glycoproteins.[328,333] Could this really be the most important difference between us and the apes? More genomic analysis may tell.

What does DNA analysis tell us about race? Most investigators conclude that there is only one human race with no detectable boundaries between the group commonly referred to as races.[334–336] As Pääbo put it, "in terms of the variation at most loci, we all seem to be Africans, either living on that continent or in recent exile."[328] The differences in skin color seem to reflect adaptation to the environment in which people live.[336a] Variations in the level of melanocyte-stimulating hormone receptor, one regulator of skin color (Box 25-A), are especially high in Africans.[328]

DNA analysis has also been useful in tracing human migration.[337–338e] For example, a genetic marker in the Y chromosome is carried by 85% of native Americans, suggesting that they are all descended from a man who lived ~ 20,000 years ago, probably an immigrant from Siberia.[339] Contrary to usual assumptions women, more often than men, seem to have spread their DNA to new locations in the world.[327b] Studies of cattle and of the wild ox reveal information about domestification of these animals about 10,000 years ago in Europe, Asia, and Africa.[340]

Other evolutionary relationships. Studies of chromosome banding, chromosome maps, restriction fragments, and detailed sequences provide many insights into relationships among species. For example, the chromosome banding pattern and also DNA sequences show close similarities between human beings and the mouse.[194,312] The latter is often regarded as the premier organism for the study of mammalian genetics and development.[291a–c,341] Dense genetic maps are available for both the mouse and rat as are moderate-resolution maps for livestock, companion animals, and other mammals.[327,342,343] Comparative gene maps are being constructed for more than 40 mammals[344] and other species of animals, plants, and fungi. Comparisons of these genomes reveal much of interest. For example, the pufferfish *Fugu rupripes* has a genome only 1/9 the size of the human genome. However, both species seem to contain about the same number of genes. Many of them can be directly correlated and some human-disease-causing mutations have been identified first in the Fugu.[344a]

Evolutionary history is being rewritten in molecular terms. Comparison of sequences of individual proteins allows evolutionary relationships of their genes to be traced. Many families of **homologous** genes can be identified.[344b] These include both **orthologs**, genes in different species that have evolved from a common ancestor, and **paralogs**, genes related by gene duplication within a genome. Orthologs have the same function in different organisms, but paralogs have different functions within a single species.[345]

Sophisticated molecular clock studies suggest that the evolution of metazoan organisms began earlier than had been supposed. Ancestral primates appeared at least 65 million years ago.[344] Gene sequence data for many species suggest that a great variety of mammals lived 100 million years ago in the age of dinosaurs,[346] a view also supported by new fossil evidence.[344,347] Metazoans appeared earlier than the "Cambrian explosion" generally thought to have occurred ~550 million years ago.[348] New geochemical data suggest that cyanobacteria diverged from other bacteria as early as 2.1×10^9 years ago.[349] Gram-negative bacteria diverged from gram-positive microbes ~3.2×10^9 years ago,[350] *Salmonella* from *Escherichia* only 0.1×10^9 years ago. However, DNA analysis shows that within this latter time period many genes from other microorganisms have been inserted into the *E. coli* chromosome and into other bacterial chromosomes.[351,352] Some of these transfers have occurred with the help of bacteriophages.[353] A puzzle is the fact that among eukaryotic cells the enzymes catalyzing the genetic **information transfer** via transcription and protein synthesis resemble those of archaea. However, **operational enzymes** that catalyze other basic metabolic processes tend to be more similar to those of eubacteria such as *E. coli*.[353,354] So much gene transfer between organisms has occurred that it is difficult to establish the earliest parts of a phylogenetic tree of the type shown in Fig. 1-5. Another factor that confuses our study of bacteria is that less than 1% of all living microorganisms have been grown in pure cultures.[355,356]

We are still dangerously ignorant of the complexity of the microbial world, which both threatens us with diseases and sustains our environment. We do have complete genome sequences for more than 60 different bacteria with hundreds more expected within a few years.[356,356a] They contain from 0.58 Mbp (*Mycoplasma genitalium*) to 8.7 Mbp for the antibiotic-producing *Streptomyces coelicolor*,[356b] a size shared by the legume symbiont *Bradyzhizobium japonicum*.[356c] The latter is one of many bacteria that have genomes split into two or more parts, often a major chromosome plus one or more plasmids. For example, the 6.7-Mbp genome of *Sinorhizobium melitoti*, an alfalfa symbiont, comprises a 3.65-Mbp chromosome and 1.35- and 1.68-Mbp megaplasmids.[356d] The 5.67-Mbp genome of *Agrobacterium tumefaciens*, much used in genetic engineering (Section E,4), consists of a circular chromosome, a linear chromosome, and two plasmids.[146a–c] The genome sequence revealed a close similarity to those of the above-mentioned legume symbionts.

Sequences of many pathogenic bacteria are known.[356e–o] These include the causative agents of cholera,[356c] typhoid fever,[356f] plague,[356g] brucellosis,[356h] leprosy,[356i] tuberculosis,[356i] and anthrax.[356j] Also included is a virulent strain of *Streptococcus pneumonia* (respiratory infections, ear aches, meningitis),[356k] *Pseudomonas aeruginosa* (a common "opportunistic" pathogen),[356l] *Listeria monocytogenes*, which causes a severe food-borne disease (Box 19-C),[356m] the pathogenic *E. coli* 0157:H7 (see Fig. 1-2),[356n] and a tiny mucosal pathogen *Ureaplasma urealyticum*.[256o] Many surprises were found in the genome sequences. For example, *E. coli* 0157:H7 is related to the laboratory strain *E. coli* K-12, and the two strains share a ~4.1 Mbp common "backbone" sequence. However, the pathogenic strain has hundreds of "islands" of additional DNA spread throughout the genome and amounting to ~1.54 Mbp. Many of these carry genes associated with virulence.[356n]

Specialized features appear in virtually every genome.[356a] For example, the mycoplasma *U. urealyticum* contains genes for enzymes that allow the bacterium to obtain almost all of its ATP from hydrolysis of urea.[356o] Both *Caulobacter crescentus*[356p] and *E. coli* develop specialized structures involved in motility (Figs. 19-3, 32-1). *C. crescentus* and also spore-forming bacteria have alternative developmental plans (Fig. 32-1). At the low end of the genome size range are species of *Buchnera*, which are endocellular symbionts of aphids.[356q] Their genome is only slightly larger than that of *M. genitalium*. The extremely salt-tolerant *Halobacterium*[356r] and the heat-tolerant *Thermoplasma volcanium*[356s] are among the archaea for which complete genome sequences are known. Species of *Xanthomonas*, whose genomes have been sequenced, are economically important plant pathogens. *Xanthomonas campestris* is also grown commercially to produce xanthan gum (p. 179).[358a] Determination of the genomes of mycobacteria has been challenging.[356i,t] Among these slow-growing organisms is *Mycobacterium tuberculosis*, the causative agent of human tuberculosis. A large fraction of its ~4000 coding genes is devoted to metabolism of lipids and to the synthesis of unusual proteins and lipids of its cell wall.[356t]

The larger genomes code for many proteins of unknown function, but over 80% of the "ORFs" (presumed genes) of *Haemophilus influenzae* have been identified, as have their presumed functions.[357,358] The encoded proteins appear to catalyze 488 metabolic reactions on 343 different metabolites. Together these systems provide a **metabolic genotype**.[357] Results of such analyses are accumulating in **metabolic databases**.[359,359a] What is the minimum number of metabolic reactions needed for support of life? Transposon insertions can inactivate all but from 265 to 350 of the protein-encoding genes of *M. genitalium* without killing the bacteria.[360] From comparisons of a variety of bacterial chromosomes it seems likely that ~256 of these genes are truly essential.[350,361] This conclusion leads to an interesting question. Is it ethical to now try to generate such a minimal bacterium?[361a] Are there hazards, e.g., that its escape might endanger our health or the environment? On the other hand, genetic engineering of bacteria, which is already practiced, can provide useful improvements in bacteria used in foods and in industry.[362,362a]

Modeling with the aid of data available on the World Wide Web is leading to development of new mathematical descriptions of metabolic networks.[359a] An ambitious new project is to model the entire *E. coli* cell. Many experimental data will be required and it has been estimated that ten years will be needed. The effort involves investigators in many laboratories and will be at least ten times as complex as the determination of the human genome.[362b]

Metabolic studies of eukaryotic cells. The yeast, *Saccharomyces cerevisiae*, contains ~6200 genes of which, until recently, only 40% had been assigned a function. Now a variety of methods are being employed to understand this little fungus.[363–365] A useful approach is to systematically inactivate or "knock out" genes. Davis and associates[363] used a PCR-based strategy to delete one gene at a time of 2026 yeast genes. Of these genes 1620 were found not essential for growth in a rich medium. Ross-McDonald and coworkers engineered a transposon, a 274-bp derivative of the *E. coli* Tn3, and allowed it to be inserted into the genes of yeast cells by homologous recombination (Chapter 27, Section D). The transposon carried DNA for a short peptide tag in the form of a specific immunological epitope that could identify the transformed cells. More than 11,000 strains with disruptions in nearly 2000 genes were obtained. These and other deletion mutants are now available for study.[364a,365] A second yeast, *Schizosaccharomyces pombe*, has assumed major scientific importance in studies of the cell cycle and of metabolism.[365a] Its 13.8-Mbp genome is only a little smaller than that of *S. cerevisiae*, but it has ~1400 fewer recognized genes, a total of 4824.[365b] There are smaller eukaryotic genomes. That of the tiny marine chordate *Oikopleura dioica* may be only 51 Mbp. However, there may be a total of ~15,000 genes.[365c]

Transposon-induced mutations have also been created in nearly one-fourth of the ~12,000 genes of *Drosophila melanogaster*.[283] Studies of the expression of genes in both *Drosophila* and in the nematode *Caenorhabditis elegans* are directed toward understanding of development and differentiation. Of the nematode's predicted 19,293 genes, only 7% have been studied at the biochemical level. To understand what happens during development we need methods for studying simultaneously the expression of all of these genes. One approach is to look at messenger RNAs that are formed at different times during development. More than 9000 mRNAs have been identified in cells of *C. elegans*, and their patterns of expression have been observed using DNA microarrays.[366,367] A similar technique has been applied to yeast. DNA sequences of fragments of ~6400 yeast genes were amplified by PCR and printed onto a glass plate to form a "DNA chip." From mRNAs formed at different times during growth, fluorescent cDNA copies were made, and their amounts were checked by use of the DNA chip.[368] Another approach is to look directly at the proteins formed. Walhout and coworkers have devised a large-scale automated system for cloning all of the genes of *C. elegans*, expressing the protein products, and testing them in the yeast two-hybrid system (Box 29-F) for protein–protein interactions.[369] Another project is to use large-scale sequence comparisons between proteins of *C. elegans* and of other organisms to identify nematode genes that encode extracellular matrix proteins involved in cell adhesion and to trace their evolution.[370]

The genomes and the metabolism of the two insects *Drosophila melanogaster* and *Anopheles gambia*[370a] can now be compared. Many differences can be seen but almost half of the genes are orthologs. Many of these can be related also to those of pufferfish, mice,[370b] humans, and other species.

Among plant genomes that of *Arabidopsis thaliana* has been studied most. The sequences of its five chromosomes have been determined and analyzed.[371–375c]

About 25,498 genes encoding proteins from 11 different families have been found.[375a] More than 14,000 ESTs were established from cDNAs.[375b] Many of the genes represent new families, some of which may be peculiar to plants. However, many others are homologous to those of *C. elegans* and *H. sapiens*. For example, developmentally important homeotic genes, marked by a **homeo sequence** as in animal genes, are present as are thousands of cell surface receptors. However, only Ser / Thr and histidine kinases are present in *Arabidopsis*. No tyrosine kinases have been identified.[375]

The ~125-Mbp *Arabidopsis* genome is tiny compared to the 3000 Mbp of DNA present in the genomes of maize and of many other plants.[374] However, the rice genome, only ~15% as large as that of maize,[376] has been chosen for complete sequencing.[375d,e] Draft sequences for the genomes of two subspecies of rice (*Oryza sativa L.*), ssp. *indica*[375f] and ssp. *japonica*,[375g] were published in 2002. The much larger genomes of other cereal grains (2500, 4900, and 16,000 for maize, barley, and wheat, respectively) will probably all be sequenced within a few years. The genes of maize are being mapped and studied using transposon-tagging

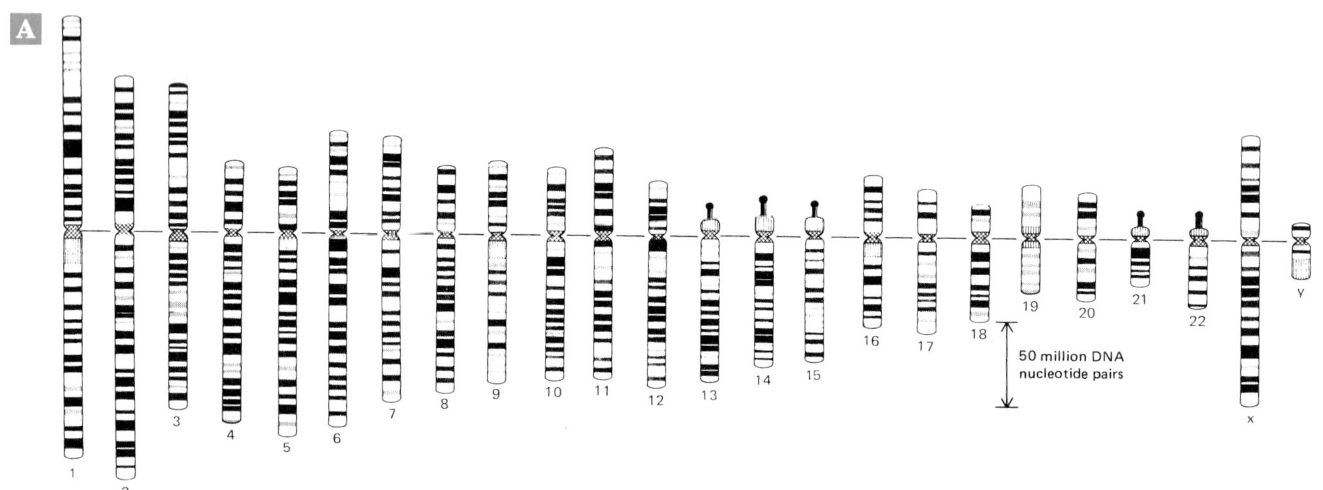

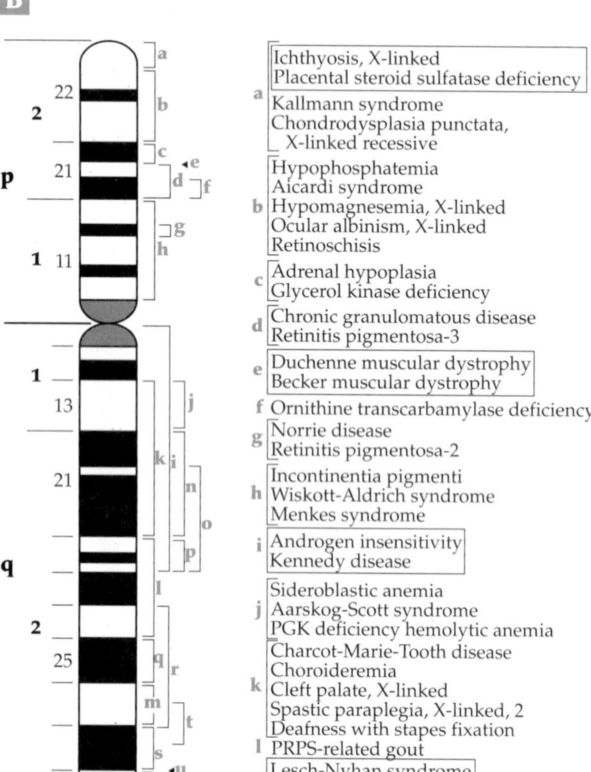

Figure 26-15 (A) Numbering and staining patterns (G-banding) of human chromosomes. The horizontal line marks the centromeres. Chromosomes 13, 14, 15, 21, and 22 have nucleolar organizers located at the constrictions in their short arms. From Alberts *et al.*[297] Adapted from U. Franke.[298] (B) Map of genetic defects identified in X chromosomes by 1995. From McKusick and Amberger.[294]

methods.[377] The smaller, more compact genomes may have just as many genes as the longer ones but have less repetitive DNA.[378] Some ferns have 307,000 Mbp of DNA, nearly 100 times that of a human; bony fishes have ~307,000 Mbp and the ameba >200,000 Mbp. These organisms appear to lose unneeded repetitive DNA faster than those with smaller genomes.[378]

3. Understanding Human Genetic Diseases

Genetic diseases have always been with us, but it was not until 1949 that the first disease, sickle cell anemia (Box 7-B), was understood at the molecular level. A single base substitution in DNA and the resultant single amino acid substitution in hemoglobin causes this disastrous disease. It was soon recognized that defects in single proteins are causes of other inherited diseases. Many of the hundreds of other known genetic disorders[201,203,379] are discussed elsewhere in this book. Among them are muscular dystrophies and cardiomyopathies (Box 19-A),[380] lysosomal deficiency diseases (Chapter 22), problems with ion transporters[381–383] and channels (cystic fibrosis, Box 26-A), defective collagens (Box 8-E),[384] neurological disorders (Chapter 30), and defects in defense systems (X-linked granulomatous disease, Chapter 18, Section G). Many of these were first recognized by their frequent occurrence in boys. Some of these X-linked deficiencies are mapped in Fig. 26-15B.

One insight into molecular disease was the recognition that mutations that cause many diseases, e.g.,

BOX 26-A CYSTIC FIBROSIS

One of the commonest of genetic diseases, cystic fibrosis affects persons all over the world. The incidence is unusually high in persons of northern European descent, about one in 2500 children being born with the defect. The inheritance pattern showed that cystic fibrosis is recessive and is caused by a single-gene defect that is carried by almost 5% of white Americans. In the United States there are ~30,000 persons with the disease. Many die in early childhood and even with careful treatment only 50% live into their late twenties or beyond.[a,b] Through extensive linkage analysis the cystic fibrosis gene was mapped to chromosome 7, and its location was narrowed further to a 1600-kbp region between the oncogene *met* and another marker designated J3.11.[c] Random searching located closer markers, and "chromosome walking and jumping" led to identification and characterization of the gene in 1989.[c,d] The large 250-kbp gene contains 27 exons. The transcribed mRNA is 6129 bp in length, and the gene product is a 1480-residue amino acid protein,[e] which is known as the **cystic fibrosis transmembrane conductance regulator** (**CFTR**).[f]

Children with cystic fibrosis lose excessive amounts of salt in perspiration and become dehydrated readily. A salty taste of the skin and an elevated chloride concentration of sweat are traditional diagnostic symptoms.[a] More serious problems arise from progressive respiratory failure and inadequate pancreatic secretion. Lung infections with *Pseudomonas aeruginosa* are the major cause of death.[g] The CFTR gene is expressed in many tissues, especially those of the mucous membranes. An alternatively spliced isoform may form chloride channels in heart.[h,i] As mentioned in Chapter 8, Section C,5 the CFTR protein is a member of the ATP-dependent ABC transporter family. However, it is atypical because it also contains a regulated chloride channel.[j] In secretory epithelia of intestines, pancreas, lungs, sweat glands, and kidneys Cl^- enters epithelial cells through their basolateral surfaces using an $Na^+ + K^+ + 2\ Cl^-$ cotransporter and exits the cells through their apical surfaces using the CFTR channel. Absorptive epithelia also contain both the cotransporter and the CFTR channel, but Cl^- flows into the cells from the exterior surface, and the distribution of the cotransporter and CFTR between basolateral and apical surfaces is opposite to that in secretory cells.[f]

From the amino acid sequence the CFTR protein is predicted to form two 6-helix transmembrane domains, two ~240-residue cytosolic ATP-binding domains, and a cytosolic regulatory (R) domain that contains at least five serine residues that may be phosphorylated by the cAMP-dependent **protein kinase A** (Chapter 11, Section C; Fig. 11-4).[k,l,m] See adjacent scheme. The two ATP-binding domains resemble those of myosin and other ATP-hydrolyzing proteins. The chloride channel, which is probably formed by helices M1, M3, and M6,[f] remains closed unless serine residues of the regulatory domain are phosphorylated. However, opening of the channels also requires binding of ATP in the nucleotide-binding domains. Binding of either vanadate or BeF_x stabilizes the open state of the channels.[k] There are four small 55- to 65-residue cytosolic loops, labeled 1–4 in the diagram. Transmembrane loop 3 also appears to function in regulation of the channel,[n] which involves a complex regulatory mechanism. Two sites of *N*-linked glycosylation are present in the short extracellular loop between helices M7 and M8.

cystic fibrosis, while affecting a single protein, occur at many places in the gene that encodes the protein. Not all mutations are caused by base substitutions; they are often a result of deletion or insertion of DNA. A whole group of diseases are caused by accumulation of repetitive DNA, often of nucleotide triplets, within genes or in control regions of genes. Another important insight is that understanding a newly discovered and very rare disease may help us to understand other related disorders. For example, after the Duchenne muscular dystrophy gene was located, the encoded protein dystrophin was found to have mutations resulting in other milder dystrophies (Box 19-A).

Mutations are only rarely beneficial, but we know that many mutations alter the properties of proteins very little. We can anticipate that most genes may undergo mutations that cause some loss of good health and vitality without being diagnosed as causing disease.

We have also come to understand that many complex diseases such as diabetes, **polycystic ovary syndrome**,[385] **Crohn's disease** (inflammatory bowel disease),[386] and schizophrenia are in fact multiple diseases. Diabetes is a syndrome that can arise from causes such as defective insulin receptors or defective glucose transporters or from as yet unknown metabolic problems (Chapter 17).[387] Many cancers have a

BOX 26-A CYSTIC FIBROSIS (continued)

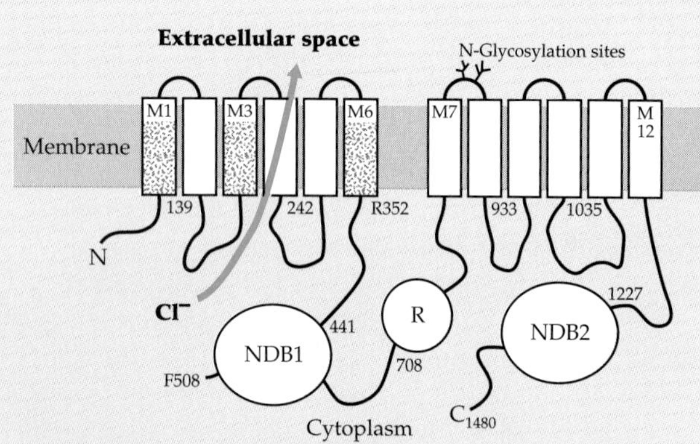

The CFTR protein can undergo endocytosis into clathrin-coated vesicles as part of its regulatory mechanism.[o] Since HCO_3^- is usually exchanged for Cl^- in epithelial ion transport, regulation of HCO_3^- uptake is also a significant aspect of CFTR function.[p]

Mutations that cause cystic fibrosis are found at many locations in the gene. However, ~70% of the mutations are caused by the absence of phenylalanine 508, as a result of a three-nucleotide deletion, in the first nucleotide-binding domain.[e,k,p] This deletion causes misfolding of the CFTR to give an inactive protein.[p] Hundreds of other mutations, some in the regulatory domain and some in the cytosolic loops,[q,r] also cause the disease. Cystic fibrosis is one of the diseases for which targeted gene transfer may become an effective treatment.[a,s] Cystic fibrosis induced in mice by targeted disruption of the CFTR gene has been successfully treated by gene therapy.[t] However, the gene is large and efficiency of transfer into animals is low. Nevertheless, human gene therapy for cystic fibrosis is being pursued cautiously.[s]

The CFTR protein has additional medical significance. Its stimulation by bacterial toxins is responsible for **secretory diarrhea**, which kills 3 million young children annually.[f]

a Welsh, M. J., and Smith, A. E. (1995) *Sci. Am.* **273**(Dec), 52–59
b Koh, J., Sferra, T. J., and Collins, F. S. (1993) *J. Biol. Chem.* **268**, 15912–15921
c Marx, J. L. (1989) *Science* **245**, 923–925
d Riordan, J. R., Rommens, J. M., Kerem, B.-s, Alon, N., Rozmahel, R., Grzelczak, Z., Zielenski, J., Lok, S., Plavsic, N., Chou, J.-L., Drumm, M. L., Iannuzzi, M. C., Collins, F. S., and Tsui, L.-C. (1989) *Science* **245**, 1066–1073
e Collins, F. S. (1992) *Science* **256**, 774–779
f Akabas, M. H. (2000) *J. Biol. Chem.* **275**, 3729–3732
g Ernst, R. K., Yi, E. C., Guo, L., Lim, K. B., Burns, J. L., Hackett, M., and Miller, S. I. (1999) *Science* **286**, 1561–1565
h Hart, P., Warth, J. D., Levesque, P. C., Collier, M. L., Geary, Y., Horowitz, B., and Hume, J. R. (1996) *Proc. Natl. Acad. Sci. U.S.A.* **93**, 6343–6348
i Howell, J. M., and Gawthorne, J. M., eds. (1987) *Copper in Animals and Man*, Vol. I and II, CRC Press, Boca, Raton,Florida
j Li, C., Ramjeesingh, M., Wang, W., Garami, E., Hewryk, M., Lee, D., Rommens, J. M., Galley, K., and Bear, C. E. (1996) *J. Biol. Chem.* **271**, 28463–28468
k Gadsby, D. C., and Nairn, A. C. (1994) *Trends Biochem. Sci.* **19**, 513–518
l Ikuma, M., and Welsh, M. J. (2000) *Proc. Natl. Acad. Sci. U.S.A.* **97**, 8675–8680
m Ostedgaard, L. S., Baldursson, O., and Welsh, M. J. (2001) *J. Biol. Chem.* **276**, 7689–7692
n Aleksandrov, L., Mengos, A., Chang, X.-b, Aleksandrov, A., and Riordan, J. R. (2001) *J. Biol. Chem.* **276**, 12918–12923
o Weixel, K. M., and Bradbury, N. A. (2000) *J. Biol. Chem.* **275**, 3655–3660
p Luo, X., Choi, J. Y., Ko, S. B. H., Pushkin, A., Kurtz, I., Ahn, W., Lee, M. G., and Muallem, S. (2001) *J. Biol. Chem.* **276**, 9808–9816
q Seibert, F. S., Jia, Y., Mathews, C. J., Hanrahan, J. W., Riordan, J. R., Loo, T. W., and Clarke, D. M. (1997) *Biochemistry* **36**, 11966–11974
r Cotten, J. F., Ostedgaard, L. S., Carson, M. R., and Welsh, M. J. (1996) *J. Biol. Chem.* **271**, 21279–21284
s Verma, I. M., and Somia, N. (1997) *Nature (London)* **389**, 239–242
t Hyde, S. C., Gill, D. R., Higgins, C. F., Trezise, A. E. O., MacVinish, L. J., Cuthbert, A. W., Ratcliff, R., Evans, M. J., and Colledge, W. H. (1993) *Nature (London)* **362**, 250–255

strong hereditary component, and many of these are being mapped to specific DNA locations.[388] Some specific proteins, such as **Ras** (Chapter 11) that are mutated in many cancers were first recognized as avian oncogenes. The tumor suppressors **Rb** (retinoblastoma protein, Box 11-D)[388,389] and **p53** (Fig. 11-15)[390,391] are major sites of mutation in cancer. See also Chapters 11 and 31. Since cancers contain multiple mutations, they are complex diseases. However, many specific susceptibility loci are being located, including some for breast cancer (Box 11-D),[392] prostate cancer,[393] and familial adenomatous polyposis, a hereditary disease leading to colon cancer.[388,394] Cancer has long been known to be associated with chromosome instability including deletion and insertion mutations at simple repeat sequences, frame-shift mutations,[395] DNA breakage, translocation,[396] and losses or gains of whole chromosomes.[397]

From a practical viewpoint the understanding that we are gaining will help us to provide better treatments of genetic diseases. At present almost every human genetic disease can be mimicked in a knockout mouse.[291a,398] A turnabout is that rare human hair-

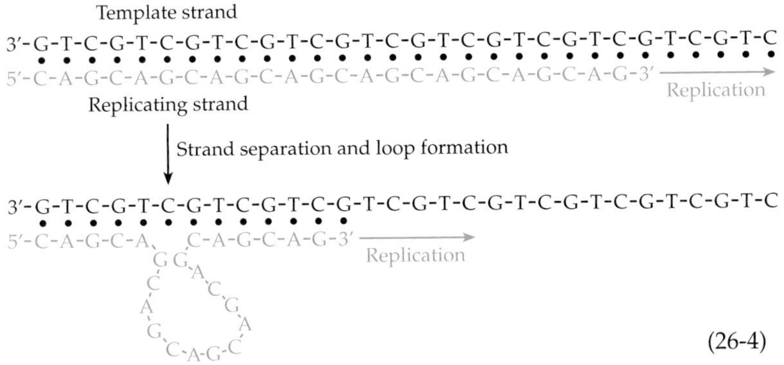

Template strand
3′-G-T-C-G-T-C-G-T-C-G-T-C-G-T-C-G-T-C-G-T-C-G-T-C-G-T-C-G-T-C
5′-C-A-G-C-A-G-C-A-G-C-A-G-C-A-G-C-A-G-C-A-G-C-A-G-3′ → Replication
Replicating strand

Strand separation and loop formation

3′-G-T-C-G-T-C-G-T-C-G-T-C-G-T-C-G-T-C-G-T-C-G-T-C-G-T-C-G-T-C
5′-C-A-G-C-A C-A-G-C-A-G-3′ → Replication
(loop: G G A C G A C G A C A C)

(26-4)

TABLE 26-4
Some Human Triplet Repeat and Related Diseases[a]

Name	Repeat sequence (5' to 3')	Transmission[b]	Location
Fragile X syndrome (FRAXA)	$(CGG)_n$	XD	by 5' side of gene *FMR*-1 + C_pG island methylation
Type E (FRAXE)	$(CGG)_n$	XR	5' side of gene *FMR*-2
Synpolydactyly	$(GCG)_n$		HOXD13 gene (polyalanine)
Myotonic dystrophy (DM1)	$(CTG)_n$	AD	3' untranslated region of gene for cAMP-dependent protein kinase
DM2	$(CCTG)_n$	AD	Intron in a zinc finger protein gene
Friedreich ataxia	$(GAA)_n$	AR	Intron in frataxin, a mitochondrial protein
Huntington disease (HD)	$(CAG)_n$	AD	Huntingtin (polyglutamine)
Spinocerebellar ataxia, SCA-1	$(CAG)_n$	AD	Ataxin-1 (polyglutamine)
SCA-2	$(CAG)_n$	AD	Ataxin-2 (polyglutamine)
Machado–Joseph disease (SCA-3)	$(CAG)_n$	AD	Ataxin-3
SCA-6	$(CAG)_n$	AD	Calcium channel
Spinobulbar muscular atrophy (Kennedy disease; SBMA)	$(CAG)_n$	XR	Atrophin (androgen receptor)
Dentato-rubro-pallido-luysian atrophy (DRPLA)	$(CAG)_n$	AD	
Progressive myoclonus epilepsy	(G+C)-rich oligonucleotide repeat		Cystatin B

[a] See Mandel, J.-L. (1997) *Nature (London)* **386**, 767–769 and Richards, R. I., and Sutherland, G. R. (1997) *Trends Biochem. Sci.* **22**, 432–436
[b] XD, X dominant; XR, X recessive; AD, autosomal dominant; AR, autosomal recessive

less people have a mutation in a gene homologous to the well-known *hairless* gene of mice.[399] Genes homologous to those of many human genetic defects have also been identified in yeast.[400] Studies of both mutant mice and of mutant yeast cells can also help in understanding diseases and in devising therapies.

Triplet repeat diseases. With the exception of Down disease (extra chromosome 21, affecting 1 in 600 children) the most prevalent cause of mental retardation is the **fragile X syndrome,** which affects ~1 of 2000 newborn males. A fragile site,[401,402] where the X chromosome breaks easily, marks the location of the defect. Identification of nearby RFLPs led to cloning of the gene in 1991. The defect was found to be a repeated trinucleotide sequence 5'-$(CGG)_n$ in the DNA. The value of *n* varies, in most healthy individuals averaging ~ 30. However, for some normal individuals *n* may be 200–300. These persons may transmit the fragile X disease to their offspring in whom *n* increases from one generation to the next with increasing severity of the disease. There may be 2000 or more CGG triplets.[403] The defect lies at the 5' end in an untranslated part of the gene for the fragile X mental retardation protein (FMRP). A cytoplasmic RNA-binding protein, FMRP, may enter the nucleus and have an as yet uncertain function. In rare cases the fragile X syndrome arises from deletions or missense mutations in the *FMRP* gene.[404] There are actually two similar genes *FMRP*1 and *FMRP*2. Mutations in the latter are associated with a milder form of fragile X disease. FMRP genes with expanded $(CGG)_n$ tracts are not expressed, evidently because the mutation induces methylation of an adjacent "CpG island" (see Chapter 27) as well as of sites within the FMRP gene. Both FMRP proteins are apparently needed for normal brain function. The FMRP1 defect is genetically dominant, and female heterozygotes also suffer from the fragile X syndrome. However, FMRP2, also encoded on the X chromosome, is recessive.[405] FMRP appears to function in neuronal dendrites (Chapter 30) where it binds to polysomal aggregates and participates in regulation of translation of mRNA.[405a–c] Like DNA, mRNA may contain **G quartets** (Fig. 5-8, p. 227).[405d]

Twelve or more additional triplet repeat diseases, many with neurological symptoms, have been identified (Table 26-4).[405–407a] These involve other trinucleotide repeats 5'-$(GCG)_n$, 5'-$(CTG)_n$, 5'-$(GAA)_n$, and 5'-$(CAG)_n$. In **synpolydactyly**, an inherited developmental defect causing malformation of hands and feet, an expansion of a GCG trinucleotide occurs within the gene *HoxD*13. This results in incorporation of a polyalanine tract near the N terminus of the protein.[408] **Myotonic dystrophy** DM1 (Box 19-A) results from expansion of CTG to 6 kbp or more within the untranslated 3' region of a gene for cAMP-dependent protein kinase.[405] The mRNA transcripts accumulate in the nucleus and may bind to a CUG-binding protein that is involved in splicing other mRNAs, thereby poisoning the cell.[409] DM2 is caused by expansion of a CCTG quartet in an intron of a zinc-finger protein.[408a] Expansion of the GAA triplet is associated with the neurological disease **Friedreich ataxia**, which has a prevalence of ~1 in 50,000. The defect lies within an intron found in the gene for the 210-residue mitochondrial protein **frataxin**. The function of the protein is unknown.[409a] However, the defect in Freidrich ataxia leads to a deficit in mitochondrial ATP synthesis.[409b] Studies of a corresponding protein in yeast suggests that frataxin is an iron storage protein.[409c–e] Apparently the DNA defect in the intron interferes with splicing of the mRNA transcript.[410–413] The polyGAA strand in the triplet repeat region of the DNA is able to form various alternative structures including a parallel (GAA)•(TTC) duplex.[410a] Such structures in mRNA may interfere with proper splicing.

A defect in an α-tocopherol-transfer protein causes a similar set of neurological symptoms. Oxidative damage may therefore be a component of this disease.[410b]

Expansion of $(CAG)_n$ sequences causes a series of neurodegenerative diseases (Table 26-4),[413–416c] the commonest of which is **Huntington disease** (HD).[414–414c] In Huntington's disease the $(CAG)_n$ tract is found in the **huntington** gene. The protein is enormous, with over 3140 residues. It is essential for nerve development, but the function remains uncertain. The first exon at the HD gene encodes a polyglutamine tract of 6–35 residues, which when expanded to 36–100 or more causes the disease. A similar situation holds for the **spinocerebellar ataxias** (SCA) and other $(CAG)_n$ diseases (Table 26-4).[414,417–417b] The encoded proteins have a variety of functions. The SCA-1 protein ataxin-1 functions in the nuclear matrix.[418] The X-linked **spinobulbar muscular atrophy** (SBMA) gene encodes an androgen receptor protein.[414,417] The polyglutamine sequences inserted in those proteins seem to be toxic, but the mechanism of toxicity is uncertain. It may result from aggregation of the proteins within cells.[414] It has also been suggested that cleavage of these chains by **caspases** (cysteine proteases; Chapter 12) may produce truncated proteins, which induce apoptosis.[415]

A form of epilepsy (Table 27-4) appears to be a result of repeats of a (G + C)-rich sequence that may be a dodecamer.[405] Dinucleotide repeats and other "minisatellite" DNA sequences are also associated with instability of DNA and may undergo expansion.[419–421] A pentanucleotide repeat $(CCTTT)_n$ is associated with increased expression of the nitric oxide synthase gene NOS2A. Persons with $n = 14$ were found to have enhanced resistance to development of diabetic retinopathy. This seems to be a case of a beneficial "gain of function" mutation.[422]

How do repeat sequences expand from one generation to the next? There is probably more than one mechanism. One is **strand slippage** during DNA replication. If single-strand loops are present at any time one strand could be displaced (slipped) relative to the other. Replication could then either expand the repetitive sequence or cause a deletion (Eq. 26-4).[405,423–425] Expansion could also occur by gene conversion during homologous recombination (Chapter 27, Section D)[401,426] or during DNA repair.[421,425] Repeat sequences may also prevent proper formation of nucleosomes.[402]

Cloned genes and diagnosis. The first areas in which cloned DNA has affected medicine are in the diagnosis of genetic diseases and in the production of medicinally useful proteins in bacteria, yeast, or cultured cells. One of the first applications was for the diagnosis of the sickle-cell trait by use of PCR on DNA isolated from blood.[43] This was followed quickly by methods for recognition of other inherited diseases and by automated procedures. Intense efforts are now being made to develop DNA "chips" (usually small glass plates with an array of DNA fragments bound to the surface) that can recognize a great variety of defects. For example, DNA polymorphisms in the 16.6-kbp human mitochondrial genome can be recognized by a plate containing 135,000 oligonucleotide probes assembled in a regular grid by photolithography and solid-state synthesis (Chapter 3).[427] Cancers may be classified quickly.[428] Many systems are under development for binding and recognizing genomic DNA, cDNA, or mRNA using microchip arrays of up to 400,000 or more oligonucleotides on a 2 cm x 2 cm plate.[429–430b] One chip combines PCR with use of "zip-code primers" that direct the PCR products to specific zip-code addresses on the chip to give a universal array able to detect low abundance of mutations in any gene of interest.[431] Mass spectrometry is also being harnessed to identify oligonucleotides bound at any address on a chip.[432] Commercial chips are expensive, $100 – 2000 apiece, and good for only one use. The price will fall. To build a machine to make your own chips go to http//cmgm.stanford.edu/pbrown/mguide/.[433,434]

An alternative to DNA chips is to miniaturize DNA sequencing and analysis machinery. Using nanoliter droplets of fluid passing through microchannels built by photolithographic techniques of computer chip construction, these nanolaboratories may be the size of a credit card but able to cleave DNA and conduct PCR reactions, gel electrophoresis, and sequence determinations.[435,436]

Vaccines, hormones, and other medicines. A myriad of products of recombinant DNA technology are already in use in medical diagnostics.[437] In the past problems have arisen because vaccines and hormones can cause allergic reactions and may harbor viruses. A small percentage of diabetics are allergic to animal insulins, but human insulin produced in bacteria is now available. A number of children receiving human growth hormone isolated from cadavers contracted the fatal Creutzfeld–Jakob disease,[438] a neurological disorder caused by a prion (Chapter 29). One way in which AIDS has been spread is through contamination of the blood-clotting factors VIII or IX needed by hemophiliacs. These sources of contamination are being eliminated by the use of bacterially produced proteins.[439]

Recombinant DNA techniques can be used in two ways in the production of vaccines. The first is to find a protein in the virus or other infective agent that is a good inducer of antibody formation, i.e., a good antigen. This protein, or even a fragment of it, can then be produced from its cloned gene or can be made synthetically. Since the cloned protein can be purified highly, it may make a superior vaccine to those made from killed cells, inactivated virus particles, or mixtures of proteins.[440] The first commercial vaccine of this type was against viral hepatitis B, a major cause of liver cancer.[441] Particles consisting of viral envelope proteins can be produced in yeast and be used for vaccination. A DNA encoding these proteins has been transferred, using the Ti plasmid (Section E, 4), into lettuce. Human volunteers produced anti-hepatitis antibodies after eating the lettuce.[442] This suggests that vaccination through ingestion of antigenic proteins in food crops may be feasible. Injection of a small piece of DNA carrying the gene may also lead to antibody formation.[443,444] Using either purified antigens, proteins, or DNA, it may be possible to develop effective vaccines against Rocky Mountain spotted fever,[445] a rickettsial disease, and against malaria, for which there are no satisfactory vaccines.

The vaccinia virus, formerly used to vaccinate against smallpox, has been engineered for use against other diseases.[446] Much of its 187-kb DNA can be excised and replaced with passenger DNA. The virus particles are stable and highly infectious. The vaccinia virus is unusual in carrying genes for both RNA and DNA polymerases and other proteins that permit it to undergo replication and transcription of its own genes in the cytoplasm of the infected cells. One application of a recombinant vaccinia virus is oral vaccination of wild foxes and racoons against rabies.[447]

A major advance based on cloned genes is the production of new medicines previously unavailable or available in only small amounts. Among these are the **interferons**[287] (Chapter 31) and many hormones such as the **interleukins** produced by lymphocytes[448] and the **atrial natriuretic hormones** (Chapter 23). Another candidate is the **α1-protease inhibitor** (Chapter 12). Perhaps better inhibitors than the natural one can be devised and produced in bacteria.[449]

4. Gene Therapy

A few years ago it seemed like fantasy, but there is little doubt that we will soon be able to routinely treat some genetic illnesses by introducing new copies of genes into the body. A current goal is to insert cloned genes into body cells (somatic cells) to correct specific hereditary defects. For example, juvenile diabetics would benefit from introduction of genes for insulin production into cells that could replace their atrophied pancreatic beta cells. At present we don't know how to do this. However, genes have been transferred into human beings lacking adenosine deaminase and showing severe **combined immunodeficiency** (Chapter 31) and those lacking hypoxanthine guanine phosphoribosyltransferase and displaying the **Lesch–Nyhan syndrome** (Chapter 25, Section E,2). Corrections of **Gaucher disease** and other deficiencies of lysosomal enzymes is also an early goal (Chapter 20, Section G,2).

By 1986 more than 5000 children had received bone marrow transplants from close relatives to correct severe combined immunodeficiency caused by a defective adenosine deaminase gene.[450] However, the patients must receive chemotherapy or irradiation to suppress their immune system before the transplantation. Hospitalization may last for 30–60 days. Using genetic therapy some bone marrow cells can be removed from the patient with a needle. The cells can be treated to introduce the corrected genes using a suitable retrovirus. Clones of corrected bone marrow stem cells, which will give rise to lymphocytes in the body, can be selected, cutured and reintroduced into the patient. As discussed on p. 1498, retrovirus vehicles that locate and become integrated at appropriate sites in the genome have been developed.[216,451] Use of homologous recombination (Chapter 27) introduces the cloned gene into its normal chromosomal location.[452] The first children were treated by transfer of the adenosine deaminase gene in 1991. The procedure has been partially successful.[453,454]

By 1999 more than 400 clinical gene therapy trials were planned or in progress.[455] Nevertheless, development of suitable vehicles for gene delivery has been slow.[456,456a] Uncertainty about the safety of adenovirus vectors is one problem.[160a,454,457,457a] Poor efficiency of gene transfer is another. A glycogen storage disease of knockout mice has been cured by transfer of human α-glucosidase (Box 20-D) using an adenovirus vector.[458] Mice have also been used in developing gene therapy for hemophilia,[456a,457a] sickle-cell disease,[458a] and aspartylglycosaminuria.[458b] However, gene targeting in animals other than the mouse has been difficult.[459] Genetic therapy may be most effective when the gene is transferred into stem cells, which can then take up residence within the body (Chapter 32). Young stem cells, which can be obtained from unbilical cord blood at birth, can be used.[460,461] For many diseases therapy will probably be best soon after birth, or even prior to birth during the second trimester of pregnancy.[455]

An alternative approach is to synthesize highly specific hydrolysis-resistant DNA analogs that can form triple helical structures with DNA (Chapter 5). If these can be made specific enough they might bind to a targeted DNA site, such as the sickle cell anemia locus and induce a "back mutation" from the faulty A•T base pair (see Fig. 7-23) to a T•A pair in at least some of the hematopoietic cells that give rise to hemoglobin.[462] Another possibility is to make oligonucleotide analogs that serve as mimics of **antisense RNA** (discussed in Chapter 28). This could impede translation of bad mRNAs, such as that giving rise to polyglutamine chains in the triple repeat diseases.

At this time no effort is being made to alter the DNA in human germ cells. It seems undesirable to experiment with such changes.[160a,462a] However, as methods are developed for genetic therapy of somatic cells, we will rear more and more healthy carriers of serious genetic defects. Eventually we may need to develop therapy for germ cells.

5. Genetic Engineering of Bacteria, Plants, and Animals

Many improvements in bacteria used in industrial fermentations have been made.[463] The number of copies of a useful gene may be increased, and repressed genes may be made more active by deletion mutations in regulatory genes (Chapter 28). New genes are being transferred between bacteria and into plants and animals. For example, *Bacillus thuringensis* produces crystalline protein toxins, which are harmless to mammals but active against many insects. The *Bt* gene for this toxin has already been transferred into many different plants and is having a major effect on agriculture in allowing for decreased use of insecticides.[464,465] Genetic engineering may allow the toxin to be made more specific for particular insect species.

Genetic engineering of plant genes[466] may improve the quality of storage proteins in cereal grains[467] and the flavors of fruits,[468] provide more drought-resistant plants,[469] and offer increased resistance of crop plants to particular herbicides[464] (Chapter 25, Section A,1) and to viruses.[470] Many things can be done to enhance the nutritional qualities of foods.[464] For example, rice has been engineered to produce and store β-carotene in the grain, a development that may benefit 400 million people in the world deficient in vitamin A and often suffering from infections and blindness.[471] (See also p. 1240.) Specialty products such as poly-β-hydroxybutyrate and many others can also be produced in plants.[472] Genetic engineering of animals has produced not only knockout mice but the possibility for nutritional improvement[156] and for production of useful products such as vaccines in milk.

6. Ethical Problems

With a host of new medicines and agricultural products coming and with the ability to alter genomes at will, we face new ethical problems. For example, who owns the human genome?[473–475] Should patenting of human genes be allowed?[476,477] How private are genetic data?[478] Who has access to human DNA data?[479] Insurance companies?[480,481] Should routine screening with new genetic tests be used (as is now the case for phenylketonuria, diabetes, and others), even if no effective treatment is available?[482,483]

Ethical questions arise in application of experimental forms of genetic therapy to patients dying of inherited disorders. What if the therapy simply provides a longer life of suffering to patients? Some biologists foresee a future in which human beings learn to control their own genes and prevent genetic deterioration resulting from the accumulation of harmful mutations. They find it exciting to think that we can, in an intelligent way, elect to continue our evolution in a desired direction. Others caution that our present knowledge is such that attempts to eliminate all "bad" genes from the population might be disastrous.[484] They point to the hemoglobin S gene (Box 7-B) and the role it once had in preserving life in a malaria-infested environment and urge that at this stage we allow maximum heterogeneity of genetic types. There are other dangers in allowing genetics to control human lives. In the past "eugenic" doctrines have been used to justify racist laws and (in Nazi Germany) genocide.

Problems have resulted from the availability of hormones and other products of genetic engineering. For example, should parents be allowed to request human growth hormone for short children who do not have a pituitary deficiency? Should it be available to athletes and to aging people? Use of stem cells in genetic therapy raises another set of questions. Present guidelines of the National Institutes of Health (NIH), which supports much of the gene therapy research, allows use of stem cells in therapy but doesn't allow their use in reproductive cloning or the combining of human stem cells with animal embryos.[485] Dangers from retroviruses may lurk in attempts to transplant animal organs (e.g., pig hearts) into people.[486–488]

In attempting to improve crop plants, proteins causing allergic reactions in some individuals may be tranferrred.[489] This is a compelling reason that mandatory monitoring and labeling of genetically modified foods is recommended.[490] The danger is emphasized by the highly popularized contamination of foods with "starlink" maize.

Should we rush commercial developments of genetic engineering to increase food supply at a time when there is an excess worldwide? Should we engineer increased herbicide-resistance into plants when drinking water supplies are being contaminated by herbicides already in use? On the other hand, less toxic herbicides effective at low levels can be used on plants engineered to resist them.[465] Will we generate superweeds resistant to herbicides by gene transfer? With so many genes being engineered, are there other dangers in their release into the environment?[465] A **terminator** technology that produces plants with sterile seeds has been rejected because it would force poor farmers to purchase seeds year after year from multinational corporations. Yet, it could eventually have beneficial consequences in preventing cross-fertilization with genetically modified plants.[491] Since plants with the *Bt* gene kill insects, will their widespread use decimate populations of butterflies or of various beneficial insects or adversely affect soil organisms in the **rhizosphere**?[492,493]

The darkest prospect may be the possibility that military organizations will develop new biological warfare weapons using recombinant DNA methods.[494] An international treaty bans such weapons. However, governments often respond with the argument "We must do it because we know that 'they' are doing it." Couldn't such an attitude bring on unparalleled disaster? Could new types of viruses spread throughout the world and literally tear the human genome to bits? Does any kind of imagined danger justify preparation to attack populations with new viruses or new toxins? What about the smallpox virus? Should remaining stocks in the United States and Russia be destroyed?[495,496]

References

1. Horowitz, N. H. (1995) *Protein Sci.* **4**, 1017–1019
2. Atherly, A. G., Girton, J. R., and McDonald, J. F. (1999) *The Science of Genetics*, Saunders, Philadelphia, Pennsylvania

2a. Alberts, B., Johnson, A., Lewis, J., Raff, M., Roberts, K., and Walter, P. (2002) *Molecular Biology of the Cell*, 4th ed., Garland Science, New York

2b. Lander, E. S., and Weinberg, R. A. (2000) *Science* **287**, 1777 – 1782

3. Mirsky, A. E. (1968) *Sci. Am.* **218**(Jun), 78–88
4. Portugal, F. H., and Cohen, J. S. (1977) *A Century of DNA: A History of the Discovery of the Structure and Function of the Genetic Substance*, MIT Press, Cambridge, Massachusetts
5. Adams, R. L. P., Knowler, J. T., and Leader, D. P. (1992) *The Biochemistry of the Nucleic Acids*, 11th ed., Chapman & Hall, London
6. Olby, R. (1975) *The Path to the Double Helix*, Univ. of Washington Press, Seattle, Washington
7. Stent, G. S., and Calendar, R. (1978) *Molecular Genetics*, 2nd ed., Freeman, San Francisco, California
8. Chargaff, E. (1971) *Science* **172**, 637–642
9. Olby, R. (1974) *Nature (London)* **248**, 782–785
10. McCarty, M. (1985) *The Transforming Principle*, Norton, New York
11. Cohen, S. S. (1984) *Trends Biochem. Sci.* **9**, 334–336
12. Chambers, D. A., Reid, K. B. M., and Cohen, R. L. (1994) *FASEB J.* **8**, 1219–1226
13. Hershey, A. D., and Chase, M. (1952) *J. Gen. Physiol.* **36**, 39–56
14. Hayes, W. (1968) *The Genetics of Bacteria and Their Viruses; Studies in Basic Genetics and Molecular Biology*, 2nd ed., Wiley, New York
15. Chargaff, E. (1950) *Experientia* **6**, 201–209
16. Brinton, C. C. (1971) *Crit. Rev. Microbiol* **1**, 105–160
17. Sayre, A. (1975) *Rosalind Franklin and DNA*, Norton, New York

17a. Maddox, B. (2002) *Rosalind Franklin: The Dark Lady of DNA*, HarperCollins, New York

17b. Blow, D. (2002) *Nature (London)* **418**, 725 – 726

18. Watson, J. D., and Crick, F. H. C. (1953) *Nature (London)* **171**, 737–738
19. Crick, F. (1974) *Nature (London)* **248**, 766–769
20. Watson, J. D. (1968) *The Double Helix*, Atheneum, New York

20a. Morange, M. (2000) *A History of Molecular Biology*, Harvard Univ. Press, Cambridge,MA (translated by Matthew Cobb)

20b. Echols, H. (2001) *Operators and Promoters: The Story of Molecular Biology and Its Creators*, Univ. of California Press, Berkeley, California (edited by Carol A. Gross)

21. Caspersson, T. (1941) *Naturwissenschaften* **29**, 33–43
22. Brachet, J. (1942) *Arch. Biol.* **53**, 207–257
23. Brachet, J. (1987) *Trends Biochem. Sci.* **12**, 244–246
24. Prescott, D. M. (1964) *Prog. Nucleic Acid Res. Mol. Biol.* **3**, 33–57
25. Sirlin, J. L. (1972) *Biology of RNA*, Academic Press, New York
26. Hoagland, M. B., Keller, E. B., and Zamecnik, P. (1956) *J. Biol. Chem.* **218**, 345–358
27. Belozersky, A. N., and Spirin, A. S. (1960) in *The Nucleic Acids*, Vol. 3 (Chargaff, E., and Davidson, J. N., eds), pp. 147–185, Academic Press, New York
28. Davern, C. I., and Meselson, M. (1960) *J. Mol. Biol.* **2**, 153–160
29. Volkin, E., and Astrachan, L. (1956) *Virology* **2**, 149–161
30. Volkin, E. (1995) *Trends Biochem. Sci.* **20**, 206–209
31. Rouwenhorst, R. J., Pronk, J. T., and van Dijken, J. P. (1989) *Trends Biochem. Sci.* **14**, 416–418
32. Jacob, F., and Monod, J. (1961) *J. Mol. Biol.* **3**, 318–356
33. Cohen, G. N. (1995) *FASEB J.* **9**, 981–982
34. Brenner, S., Jacob, F., and Meselson, M. (1961) *Nature (London)* **190**, 576–581
35. Crick, F. H. C. (1957) *Biochem. Soc. Symp.* **14**, 25–26
36. Hoagland, M. (1996) *Trends Biochem. Sci.* **21**, 77–80
37. Nirenberg, M. W., and Matthaei, J. H. (1961) *Proc. Natl. Acad. Sci. U.S.A.* **47**, 1588–1602
38. Nirenberg, M., and Leder, P. (1964) *Science* **145**, 1399–1407
39. Matthaei, J. H., Voigt, H. P., Heller, G., Neth, R., Schöch, G., Kübler, H., Amelunxen, F., Sander, G., and Parmeggiani, A. (1966) *Cold Spring Harbor Symp. Quant. Biol.* **31**, 25–38
40. Khorana, H. G., Büchi, H., Ghosh, H., Gupta, N., Jacob, T. M., Kössel, H., Morgan, R., Narang, S. A., Ohtsuka, E., and Wells, R. D. (1966) *Cold Spring Harbor Symp. Quant. Biol.* **31**, 39–49
41. Edgar, R. S., and Epstein, R. H. (1965) *Sci. Am.* **212**(Feb), 70–78
42. Fraenkel-Conrat, H. (1964) *Sci. Am.* **211**(Oct), 47–54
43. Wu, D. Y., Ugozzoli, L., Pal, B. K., and Wallace, R. B. (1989) *Proc. Natl. Acad. Sci. U.S.A.* **86**, 2757–2760
44. Lewin, B. (1974) *Gene Expression*, Vol. 2, Wiley, New York (p. 258)
45. Knight, R. D., Freeland, S. J., and Landweber, L. F. (1999) *Trends Biochem. Sci.* **24**, 241–247
46. Seid-Akhavan, M., Winter, W. P., Abramson, R. K., and Rucknagel, D. L. (1976) *Proc. Natl. Acad. Sci. U.S.A.* **73**, 882–886
47. Benzer, S. (1962) *Sci. Am.* **206**(Jan), 70–84
48. Hayes, W. (1968) *The Genetics of Bacteria and Their Viruses*, 2nd ed., Wiley, New York (pp. 52–54)
49. Weintraub, S. B., and Frankel, F. R. (1972) *J. Mol. Biol.* **70**, 589–615
50. Ennis, H. L., and Kievitt, K. D. (1973) *Proc. Natl. Acad. Sci. U.S.A.* **70**, 1468–1472
51. Beadle, G. W., and Tatum, E. L. (1941) *Proc. Natl. Acad. Sci. U.S.A.* **27**, 499–
52. Yanofsky, C. (1967) *Sci. Am.* **216**(May), 80–94
53. Yanofsky, C. (1976) in *Reflections on Biochemistry* (Kornberg, A., ed), pp. 263–269, Pergamon, New York
54. Sarabhai, A. S., Stretton, A. O. W., Brenner, S., and Bolle, A. (1964) *Nature (London)* **210**, 13–17
55. Crick, F. H. C., Barnett, L., Brenner, S., and Watts-Tobin, R. J. (1961) *Nature (London)* **192**, 1227–1232

55a. Atkins, J. F., and Gesteland, R. F. (2001) *Nature (London)* **414**, 693

56. Watson, J. O. (1987) *Molecular Biology of the Gene*, 4th ed., Benjamin, New York
57. Eiserling, F. A., and Dickson, R. C. (1972) *Ann. Rev. Biochem.* **41**, 467–502
58. Wood, W. B., and Edgar, R. S. (1967) *Sci. Am.* **217**(Jul), 60–74
59. Sober, H. A., ed. (1968) *CRC Handbook of Biochemistry*, 1st ed., Chem. Rubber Publ. Co., Cleveland, Ohio (p. I25)
60. Garen, A. (1968) *Science* **160**, 149–159
61. Littauer, U. Z., and Inouye, H. (1973) *Ann. Rev. Biochem.* **42**, 439–470
62. Hartman, P. E., and Roth, J. R. (1973) *Adv. Genet.* **17**, 1–105
63. Lewin, B. (1974) *Gene Expression*, Vol. 1, Wiley, New York (p. 213)
64. Yanofsky, C., and Crawford, I. P. (1972) in *The Enzymes*, 3rd ed., Vol. 7 (Boyer, P. D., ed), Academic Press, New York (pp. 1–31)
65. Sherratt, D. (1975) *Nature (London)* **254**, 559–560
66. Lederberg, J., and Tatum, E. L. (1946) *Nature (London)* **158**, 558
67. Kornberg, A. (1974) *DNA Synthesis*, Freeman, San Francisco, California (p. 242–248)
68. Harrington, L. C., and Rogerson, A. C. (1990) *J. Bacteriol.* **172**, 7263–7264
69. Eisenbrandt, R., Kalkum, M., Lai, E.-M., Lurz, R., Kado, C. I., and Lanka, E. (1999) *J. Biol. Chem.* **274**, 22548–22555
70. Hendrix, R. W., Roberts, J. W., Stahl, F. W., and Weisberg, R., eds. (1983) *Lambda II*, Cold spring Harbor Lab, Cold Spring Harbor, New York
71. Hershey, A. D., ed. (1971) *The Bacteriophage Lambda*, Cold Spring Harbor Lab, Cold Spring Harbor, New York
72. Echols, H. (1971) *Ann. Rev. Biochem.* **40**, 827–854
73. O'Brien, S. J., ed. (1993) *Genetic Maps*, 6th ed., Cold Spring Harbor Lab. Press, Cold Spring Harbor, New York ((Books 1–6))
74. Blattner, F. R., Plunkett, G., III, Bloch, C. A., Perna, N. T., Burland, V., Riley, M., Collado-Vides, J., Glasner, J. D., Rode, C. K., Mayhew, G. F., Gregor, J., Davis, N. W., Kirkpatrick, H. A., Goeden, M. A., Rose, D. J., Mau, B., and Shao, Y. (1997) *Science* **277**, 1453–1462
75. Bachmann, B. J., Low, K. B., and Taylor, A. L. (1976) *Bacteriol. Rev.* **40**, 116–167
76. Bachmann, B. J. (1983) *Bacteriol. Rev.* **47**, 180–230
77. Smith, C. L., Econome, J. G., Schutt, A., Klco, S., and Cantor, C. R. (1987) *Science* **236**, 1448–1453
78. Kohara, Y., Akiyama, K., and Isono, K. (1987) *Cell* **50**, 495–508
79. Low, K. B. (1987) in *Escherichia coli and Salmonella typhimurium* (Neidhardt, F. C., ed), Am. Soc. for Microbiology, Washington, D. C.
80. Willetts, N., and Skurray, R. (1987) in *Escherichia coli and Salmonella typhimurium*, Vol. 2 (Neidhardt, F. C., ed), pp. 1110–1133, Am. Soc. for Microbiology, Washington, D. C.

80a. Bist, P., Sistla, S., Krishnamurthy, V., Acharya, A., Chandrakala, B., and Rao, D. N. (2001) *J. Mol. Biol.* **310**, 93 – 109

81. Davies, G. P., Martin, I., Sturrock, S. S., Cronshaw, A., Murray, N. E., and Dryden, D. T. F. (1999) *J. Mol. Biol.* **290**, 565–579
82. Janscak, P., MacWilliams, M. P., Sandmeier, U., Nagaraja, V., and Bickle, T. A. (1999) *EMBO J.* **18**, 2638–2647
83. Newman, M., Strzelecka, T., Dorner, L. F., Schildkraut, I., and Aggarwal, A. K. (1995) *Science* **269**, 656–663
84. Bozic, D., Grazulis, S., Siksnys, V., and Huber, R. (1996) *J. Mol. Biol.* **255**, 176–186
85. Luria, S. E. (1970) *Sci. Am.* **222**(Jan), 88–102
86. McClarin, J. A., Frederick, C. A., Wang, B.-C., Greene, P., Boyer, H. W., Grable, J., and Rosenberg, J. M. (1986) *Science* **234**, 1526–1541
87. Hager, P. W., Reich, N. O., Day, J. P., Coche, T. G., Boyer, H. W., Rosenberg, J. M., and Greene, P. J. (1990) *J. Biol. Chem.* **265**, 21520–21526
88. Frederick, C. A., Grable, J., Melia, M., Samudzi, C., Jen-Jacobson, L., Wang, B.-C., Greene, P., Boyer, H. W., and Rosenberg, J. M. (1984) *Nature (London)* **309**, 327–331
89. Frederick, C. A., Quigley, G. J., van der Marel, G. A., van Boom, J. H., Wang, A. H.-J., and Rich, A. (1988) *J. Biol. Chem.* **263**, 17872–17879

References

90. Newman, M., Strzelecka, T., Dorner, L. F., Schildkraut, I., and Aggarwal, A. K. (1994) *Nature (London)* **368**, 660–664
91. Kostrewa, D., and Winkler, F. K. (1995) *Biochemistry* **34**, 683–696
91a. Horton, N. C., Otey, C., Lusetti, S., Sam, M. D., Kohn, J., Martin, A. M., Ananthnarayan, V., and Perona, J. J. (2002) *Biochemistry* **41**, 10754–10763
92. Stanford, N. P., Halford, S. E., and Baldwin, G. S. (1999) *J. Mol. Biol.* **288**, 105–116
93. Sam, M. D., and Perona, J. J. (1999) *Biochemistry* **38**, 6576–6586
94. Kim, Y., Grable, J. C., Love, R., Greene, P. J., and Rosenberg, J. M. (1990) *Science* **249**, 1307–1309
95. Kim, S. C., Podhajska, A. J., and Szybalski, W. (1988) *Science* **240**, 504–506
96. Botstein, D., White, R. L., Skolnick, M., and Davis, R. W. (1980) *Am. J. Hum. Genet.* **32**, 314–331
97. Cohen, D., Chumakov, I., and Weissenbach, J. (1993) *Nature (London)* **366**, 698–701
98. Westmoreland, B. C., Szybalski, W., and Ris, H. (1969) *Science* **163**, 1343–1348
99. Glover, D. M., ed. (1987) *DNA Cloning: A Practical Approach*, Vol. III, IRL Press, Oxford
100. Allemand, J.-F., Bensimon, D., Jullien, L., Bensimon, A., and Croquette, V. (1997) *Biophys. J.* **73**, 2064–2070
101. Michalet, X., Ekong, R., Fougerousse, F., Rousseaux, S., Schurra, C., Hornigold, N., van Slegtenhorst, M., Wolfe, J., Povey, S., Beckmann, J. S., and Bensimon, A. (1997) *Science* **277**, 1518–1523
102. Samad, A. H., Cai, W. W., Hu, X., Irvin, B., Jing, J., Reed, J., Meng, X., Huang, J., Huff, E., Porter, B., Shenkar, A., Anantharaman, T., Mishra, B., Clarke, V., Dimalanta, E., Edington, J., Hiort, C., Rabbah, R., Skiada, J., and Schwartz, D. C. (1995) *Nature (London)* **378**, 516–517
103. Lin, J., Qi, R., Aston, C., Jing, J., Anantharaman, T. S., Mishra, B., White, O., Daly, M. J., Minton, K. W., Venter, J. C., and Schwartz, D. C. (1999) *Science* **285**, 1558–1561
104. Cox, D. R., Burmeister, M., Price, E. R., Kim, S., and Myers, R. M. (1990) *Science* **250**, 245–250
105. Adams, M. D., Kelley, J. M., Gocayne, J. D., Dubnick, M., Polymeropoulos, M. H., Xiao, H., Merril, C. R., Wu, A., Olde, B., Moreno, R. F., Kerlavage, A. R., McCombie, W. R., and Venter, J. C. (1991) *Science* **252**, 1651–1656
106. Mann, M. (1996) *Trends Biochem. Sci.* **21**, 494–495
107. Watson, J. D., Gilman, M., Witkowski, J., and Zoller, M., eds. (1992) *Recombinant DNA*, 2nd ed., Scientific American Publ. Co., New York
108. Sambrook, J., and Russell, D. (2000) *Molecular Cloning: A Laboratory Manual*, 3rd ed., Cold Spring Harbor Lab. Press, Plainview, New York
109. Berger, S. L., and Kimmel, A. R., eds. (1987) *Guide to Molecular Cloning Techniques*, Academic Press, San Diego
110. Perbal, B. (1988) *A Practical Guide to Molecular Cloning*, Wiley, New York
111. Cantor, C. R., and Schimmel, P. R. (1980) *Biophysical Chemistry, Part II*, Freeman, San Francisco, California (pp. 409–433)
112. Greene, J. J., and Rao, V. B., eds. (1998) *Recombinant DNA Principles and Methodologies*, Dekker, New York
113. Miller, J. H. (1992) *A Short Course in Bacterial Genetics*, Cold Spring Harbor Lab. Press, Cold Spring Harbor, New York
114. Wu, R., ed. (1987) *Methods in Enzymology, "Recombinant DNA,"* Vols. 153–155, Academic Press, San Diego
115. Pouwels, P. H., Enger-Valk, B. E., and Brammar, W. J., eds. (1985) *Cloning Vectors: A Laboratory Manual*, Elsevier, Amsterdam
116. Adolph, K. W., ed. (1993) *Methods in Molecular Genetics: Gene and Chromosome Analysis*, Vols. 1 and 2, Academic Press, San Diego, California
117. Birren, B., Green, E. D., Klapholz, S., Myers, R. M., and Roskams, J., eds. (1997) *Analyzing DNA: A Laboratory Manual*, Cold Spring Harbor Lab. Press, Cold Spring Harbor, New York
118. Birren, B., Green, E. D., Klapholz, S., Myers, R. M., Riethman, H., and Roskams, J., eds. (1999) *Cloning Systems: A Laboratory Manual*, Cold Spring Harbor Lab. Press, Cold Spring Harbor, New York
119. Lathe, R., Kieny, M. P., Skory, S., and Lecocq, J. P. (1984) *DNA* **3**, 173–182
120. Nobbs, T. J., and Halford, S. E. (1995) *J. Mol. Biol.* **252**, 399–411
121. Winnacker, E. L. (1987) *From Genes to Clones*, VCH, New York
122. Midgley, C. A., and Murray, N. E. (1985) *EMBO J.* **4**, 2695–2703
123. Bollum, F. J. (1981) *Trends Biochem. Sci.* **6**, 41–43
123a. Shanks, O. C., Bissonette, L., and Ream, W. (2000) *Science* **289**, 413
124. Bolivar, F., Rodriguez, R. L., Coreene, P. J., Betlach, M. C., Heyneker, H. L., Boyer, H. W., Crosa, J. H., and Falkow, S. (1977) *Gene* **2**, 95–113
125. Peden, K. W. C. (1983) *Gene* **22**, 277–280
126. Vieira, J., and Messing, J. (1982) *Gene* **19**, 259–268
127. Messing, J. (1981) *Nucleic Acids Res.* **9**, 309–321
128. Sanger, F., Coulson, A. R., Friedmann, T., Air, G. M., Barrell, B. G., Brown, N. L., Fiddes, J. C., Hutchison, C. A., III, Slocombe, P. M., and Smith, M. (1978) *J. Mol. Biol.* **125**, 225–246
129. Blattner, F. R., Williams, B. G., Blechi, A. E., Denniston-Thompson, K., Faber, H. E., Furlong, L.-A., Grunwald, D. J., Kiefer, D. O., Moore, D. D., Schumm, J. W., Sheldon, E. L., and Smithies, O. (1977) *Science* **196**, 161–169
130. Short, J. M. (1988) *Nucleic Acids Res.* **15**
131. Collins, J., and Hohn, B. (1978) *Proc. Natl. Acad. Sci. U.S.A.* **75**, 4242–4246
132. Wahl, G. M., Lewis, K. A., Ruiz, J. C., Rothenberg, B., Zhao, J., and Evans, G. A. (1987) *Proc. Natl. Acad. Sci. U.S.A.* **84**, 2160–2164
133. Fernandez, J. M., and Hoeffler, J. P., eds. (1998) *Gene Expression Systems*, Academic Press, San Diego, California
134. Moss, B., Elroy-Stein, O., Mizukami, T., Alexander, W. A., and Fuerst, T. R. (1990) *Nature (London)* **348**, 91–92
135. O'Reilly, D. R., Miller, L. K., and Luckow, V. A. (1992) *Baculovirus Expression Vectors*, Freeman, New York
136. Kriegler, M. (1990) *Gene Transfer and Expression*, Stockton Press, New York
137. Burke, D. T., Carle, G. F., and Olson, M. V. (1987) *Science* **236**, 806–812
138. Murray, A. W., and Szostak, J. W. (1987) *Sci. Am.* **257**(Nov), 62–68
139. Forget, B. G. (1993) *Proc. Natl. Acad. Sci. U.S.A.* **90**, 7909–7911
140. Anderson, C. (1993) *Science* **259**, 1684–1687
141. Sternberg, N. (1990) *Proc. Natl. Acad. Sci. U.S.A.* **87**, 103–107
142. Pierie, J. C., and Sternberg, N. L. (1992) *Methods Enzymol.* **216**, 549–
143. Heller, R., Brown, K. E., Burgtorf, C., and Brown, W. R. A. (1996) *Proc. Natl. Acad. Sci. U.S.A.* **93**, 7125–7130
144. Chilton, M.-D. (1983) *Sci. Am.* **248**(Jun), 51–59
145. Weiler, E. W., and Schröder, J. (1987) *Trends Biochem. Sci.* **12**, 271–275
146. Zambryski, P. C. (1992) *Annu. Rev. Plant Physiol. Plant Mol. Biol.* **43**, 465–490
146a. Wood, D. W., and 50 other authors. (2001) *Science* **294**, 2317–2323
146b. Goodner, B., and 30 other authors. (2001) *Science* **294**, 2323–2328
146c. Moriguchi, K., Maeda, Y., Satou, M., Hardayani, N. S. N., Kataoka, M., Tanaka, N., and Yoshida, K. (2001) *J. Mol. Biol.* **307**, 771–784
147. Slightom, J. L., Durand-Tardif, M., Jouanin, L., and Tepfer, D. (1986) *J. Biol. Chem.* **261**, 108–121
148. Hess, K. M., Dudley, M. W., Lynn, D. G., Joerger, R. D., and Binns, A. N. (1991) *Proc. Natl. Acad. Sci. U.S.A.* **88**, 7854–7858
149. Stachel, S. E., Messens, E., Van Montagu, M., and Zambryski, P. (1985) *Nature (London)* **318**, 624–629
150. Bates, H. A., Kaushal, A., Deng, P.-N., and Sciaky, D. (1984) *Biochemistry* **23**, 3287–3290
151. Tinland, B., Schoumacher, F., Gloeckler, V., Bravo-Angel, A. M., and Hohn, B. (1995) *EMBO J.* **14**, 3585–3595
152. Scheiffele, P., Pansegrau, W., and Lanka, E. (1995) *J. Biol. Chem.* **270**, 1269–1276
153. Bundock, P., den Dulk-Ras, A., Beijersbergen, A., and Hooykaas, P. J. J. (1995) *EMBO J.* **14**, 3206–3214
154. Fullner, K. J., Lara, J. C., and Nester, E. W. (1996) *Science* **273**, 1107–1109
154a. Ward, D. V., Draper, O., Zupan, J. R., and Zambryski, P. C. (2002) *Proc. Natl. Acad. Sci. U.S.A.* **99**, 11493–11500
155. Crystal, R. G. (1995) *Science* **270**, 404–410
156. Westphal, H. (1989) *FASEB J.* **3**, 117–120
157. Cotton, M., Wagner, E., Zatloukal, K., Phillips, S., Curiel, D. T., and Birnstiel, M. L. (1992) *Proc. Natl. Acad. Sci. U.S.A.* **89**, 6094–6098
158. Kojima, H., Ohishi, N., and Yagi, K. (1998) *Biochem. Biophys. Res. Commun.* **246**, 868–872
159. Yu, S.-F., von Rüden, T., Kantoff, P. W., Garber, C., Seiberg, M., Rüther, U., Anderson, W. F., Wagner, E. F., and Gilboa, E. (1986) *Proc. Natl. Acad. Sci. U.S.A.* **83**, 3194–3198
160. Chakraborty, A. K., Zink, M. A., Boman, B. M., and Hodgson, C. P. (1993) *FASEB J.* **7**, 971–977
160a. Ferber, D. (2001) *Science* **294**, 1638–1642
161. Daly, T. M., Vogler, C., Levy, B., Haskins, M. E., and Sands, M. S. (1999) *Proc. Natl. Acad. Sci. U.S.A.* **96**, 2296–2300
161a. Yan, Z., Zhang, Y., Duan, D., and Engelhardt, J. F. (2000) *Proc. Natl. Acad. Sci. U.S.A.* **97**, 6716–6721
161b. Xie, Q., Bu, W., Bhatia, S., Hare, J., Somasundaram, T., Azzi, A., and Chapman, M. S. (2002) *Proc. Natl. Acad. Sci. U.S.A.* **99**, 10405–10410
162. Amado, R. G., and Chen, I. S. Y. (1999) *Science* **285**, 674–676
162a. Lois, C., Hong, E. J., Pease, S., Brown, E. J., and Baltimore, D. (2002) *Science* **295**, 868–872
163. Jasny, B. R. (1987) *Science* **238**, 1653
164. Hofmann, C., Sandig, V., Jennings, G., Rudolph, M., Schlag, P., and Strauss, M. (1995) *Proc. Natl. Acad. Sci. U.S.A.* **92**, 10099–10103
165. Yin, H. L., and Stull, J. T. (1999) *J. Biol. Chem.* **274**, 32529–32530
166. Poul, M.-A., and Marks, J. D. (1999) *J. Mol. Biol.* **288**, 203–211
166a. Kaminski, J. M., Huber, M. R., Summers, J. B., and Ward, M. B. (2002) *FASEB J.* **16**, 1242–1247
166b. Saldanha, R., Chen, B., Wank, H., Matsuura, M., Edwards, J., and Lambowitz, A. M. (1999) *Biochemistry* **38**, 9069–9083
166c. Guo, H., Karberg, M., Long, M., Jones, J. P., III, Sullenger, B., and Lambowitz, A. M. (2000) *Science* **289**, 452–457
167. Cocking, E. C., and Davey, M. R. (1987) *Science* **236**, 1259–1262

References

168. Smith, H. O., Danner, D. B., and Deich, R. A. (1981) *Ann. Rev. Biochem.* **50**, 41–68
169. Wahl, G. M., de Saint Vincent, B. R., and DeRose, M. L. (1984) *Nature (London)* **307**, 516–520
170. Felgner, P. L., and Ringold, G. M. (1989) *Nature (London)* **337**, 387–388
171. Miller, J. F., Dower, W. J., and Tompkins, L. S. (1988) *Proc. Natl. Acad. Sci. U.S.A.* **85**, 856–860
172. Solioz, M., and Bienz, D. (1990) *Trends Biochem. Sci.* **15**, 175–177
173. Hengen, P. N. (1995) *Trends Biochem. Sci.* **20**, 248–249
174. Collombet, J.-M., Wheeler, V. C., Vogel, F., and Coutelle, C. (1997) *J. Biol. Chem.* **272**, 5342–5347
175. Chang, D. C., Hunt, J. R., Zheng, Q., and Gao, P. Q. (1992) in *Guide to Electroporation and Electrofusion* (Chang, D. C., ed), pp. 303–326, Academic Press, San Diego, California
176. Klobutcher, L. A., and Ruddle, F. H. (1981) *Ann. Rev. Biochem.* **50**, 533–554
177. Klein, T. M., Wolf, E. D., Wu, R., and Sanford, J. C. (1987) *Nature (London)* **327**, 70–73
178. Butow, R. A., and Fox, T. D. (1990) *Trends Biochem. Sci.* **15**, 465–468
179. Kohli, A., Leech, M., Vain, P., Laurie, D. A., and Christou, P. (1998) *Proc. Natl. Acad. Sci. U.S.A.* **95**, 7203–7208
180. Hohn, B., and Murray, K. (1977) *Proc. Natl. Acad. Sci. U.S.A.* **74**, 3259–3263
181. Lutz, C. T., Hollifield, W. C., Seed, B., Davie, J. M., and Huang, H. V. (1987) *Proc. Natl. Acad. Sci. U.S.A.* **84**, 4379–4383
182. Wyman, A. R., Wolfe, L. B., and Botstein, D. (1985) *Proc. Natl. Acad. Sci. U.S.A.* **82**, 2880–2884
183. Okayama, H., and Berg, P. (1982) *Mol. Cell. Biochem.* **2**, 161–170
184. Gubler, U., and Hoffman, B. J. (1983) *Gene* **25**, 263–269
185. Mason, W. T., ed. (1999) *Fluorescent and Luminescent Probes*, 2nd ed., Academic Press, San Diego, California
186. Keller, G. H., and Manak, M. M. (1989) *DNA Probes*, Stockton Press, New York
186a. Santi, E., Capone, S. Mennuni, C., Lahm, A., Tramontano, A, Luzzago, A., and Nicosia, A. (2000) *J. Mol. Biol.* **296**, 497 – 508
187. Little, P. F. R., and Cross, S. H. (1985) *Proc. Natl. Acad. Sci. U.S.A.* **82**, 3159–3163
188. Richards, J. E., Gilliam, T. C., Cole, J. L., Drumm, M. L., Wasmuth, J. J., Gusella, J. F., and Collins, F. S. (1988) *Proc. Natl. Acad. Sci. U.S.A.* **85**, 6437–6441
189. Housman, D. (1995) *N. Engl. J. Med.* **332**, 318–320
190. Viegas-Pequignot, E., Dutrilleax, B., Magdelenat, H., and Coppey-Moisam, M. (1989) *Proc. Natl. Acad. Sci. U.S.A.* **86**, 582–586
191. Abbott, C., and Povey, S. (1995) *Somatic Cell Hybrids*, IRL Press, Oxford
192. Williamson, B. (1981) *Nature (London)* **293**, 10–11
193. White, R., and Lalouel, J.-M. (1988) *Sci. Am.* **258**(Feb), 40–48
194. Gusella, J. F. (1986) *Ann. Rev. Biochem.* **55**, 831–854
195. Davies, K. E., ed. (1986) *Human Genetic Diseases*, IRL Press, Washington, D.C.
196. Martin, J. B. (1987) *Science* **238**, 765–772
197. Saiki, R. K., Scharf, S., Faloona, F., Mullis, K. B., Horn, G. T., Erlich, H. A., and Arnheim, N. (1985) *Science* **230**, 1350–1354
198. Gusella, J. F., Tanzi, R. E., Bader, P. I., Phelan, M. C., Stevenson, R., Hayden, M. R., Hofman, K. J., Faryniarz, A. G., and Gibbons, K. (1985) *Nature (London)* **318**, 75–78
199. Gilliam, T. C., Bucan, M., MacDonald, M. E., Zimmer, M., Haines, J. L., Cheng, S. V., Pohl, T. M., Meyers, R. H., Whaley, W. L., Allitto, B. A., Faryniarz, A., Wasmuth, J. J., Frischauf, A.-M., Conneally, P. M., Lehrach, H., and Gusella, J. F. (1987) *Science* **238**, 950–952
200. Royer-Pokora, B., Kunkel, L. M., Monaco, A. P., Goff, S. C., Newburger, P. E., Baehner, R. L., Cole, F. S., Curnutte, J. T., and Orkin, S. H. (1986) *Nature (London)* **322**, 32–38
201. Reich, N. O., and Mashboon, N. (1993) *J. Biol. Chem.* **268**, 9191–9193
202. Lee, W.-H., Shew, J.-Y., Hong, F. D., Sery, T. W., Donoso, L. A., Young, L.-J., Bookstein, R., and Lee, E. Y.-H. P. (1987) *Nature (London)* **329**, 642–645
203. Thomson, G., and Esposito, M. S. (1999) *Trends Biochem. Sci.* **24**, M17–M20
204. Little, P. (1986) *Nature (London)* **321**, 558–559
205. Zaccolo, M., Williams, D. M., Brown, D. M., and Gherardi, E. (1996) *J. Mol. Biol.* **255**, 589–603
206. Myers, R. M., Lerman, L. S., and Maniatis, T. (1985) *Science* **229**, 242–247
207. Botstein, D., and Shortle, D. (1985) *Science* **229**, 1193–1201
208. Kunkel, T. A., Roberts, J. D., and Zakour, R. A. (1987) *Methods Enzymol.* **154**, 367–382
209. Zoller, M. J., and Smith, M. (1983) *Methods Enzymol.* **100**, 468–500
210. Higuchi, R., Krummel, B., and Saiki, R. (1988) *Nucleic Acids Res.* **16**, 7351–7367
211. Jones, D. H., and Howard, B. H. (1990) *BioTechniques* **8**, 178–183
212. Jones, D. H., Sakamoto, K., Vorce, R. L., and Howard, B. H. (1990) *Nature (London)* **344**, 793–794
213. Wells, J. A., and Powers, D. B. (1986) *J. Biol. Chem.* **261**, 6564–6570
214. Shuldiner, A. R. (1996) *N. Engl. J. Med.* **334**, 653–655
215. Jasin, M., Moynahan, M. E., and Richardson, C. (1996) *Proc. Natl. Acad. Sci. U.S.A.* **93**, 8804–8808
216. Capecchi, M. R. (1994) *Sci. Am.* **270**(Mar), 52–59
217. Travis, J. (1992) *Science* **256**, 1392–1394
218. Barinaga, M. (1994) *Science* **265**, 26–28
219. Weissmann, C., and Aguzzi, A. (1999) *Science* **286**, 914–915
219a. Smith, J. A., and Martin, L. (1973) *Proc. Natl. Acad. Sci. U.S.A.* **70**, 1263–1267
220. Nasmyth, K. (2002) *Science* **297**, 559–565
220a. Hauf, S., Waizenegger, I. C., and Peters, J.-M. (2001) *Science* **293**, 1320–1323
220b. Campbell, J. L., and Cohen-Fix, O. (2002) *Trends Biochem. Sci.* **27**, 492–495
220c. Seemann, J., Pypaert, M., Taguchi, T., Malsam, J., and Warren, G. (2002) *Science* **295**, 848–851
221. Mazia, D. (1961) *Sci. Am.* **205**(Sep), 101–120
222. Earnshaw, W. C., and Mackay, A. M. (1994) *FASEB J.* **8**, 947–956
223. Russell, P. (1998) *Trends Biochem. Sci.* **23**, 399–402
224. Piwnica-Worms, H. (1999) *Nature (London)* **401**, 535–537
225. Nasmyth, K. (1999) *Trends Biochem. Sci.* **24**, 98–104
225a. Cortez, D., and Elledge, S. J. (2000) *Nature (London)* **406**, 354–356
226. Warren, G. (1985) *Trends Biochem. Sci.* **10**, 439–443
226a. Lippincott-Schwartz, J. (2002) *Nature (London)* **416**, 31–32
227. Morgan, D. O. (1995) *Nature (London)* **374**, 131–133
227a. Kong, M., Barnes, E. A., Ollendorff, V., and Donoghue, D. J. (2000) *EMBO J.* **19**, 1378–1388
227b. Healy, J. M. S., Menges, M., Doonan, J. H., and Murray, J. A. H. (2001) *J. Biol. Chem.* **276**, 7041–7047
228. Pavletich, N. P. (1999) *J. Mol. Biol.* **287**, 821–828
228a. Nash, P., Tang, X., Orlicky, S., Chen, Q., Gertler, F. B., Mendenhall, M. D., Sicheri, F., Pawson, T., and Tyers, M. (2001) *Nature (London)* **414**, 514–521
228b. Bartek, J., and Lukas, J. (2001) *Science* **294**, 66–67
228c. Schwab, M., and Tyers, M. (2001) *Nature (London)* **413**, 268–269
228d. Zhou, B.-B. S., and Elledge, S. J. (2000) *Nature (London)* **408**, 433–439
229. Fotedar, A., Cannella, D., Fitzgerald, P., Rousselle, T., Gupta, S., Dorée, M., and Fotedar, R. (1996) *J. Biol. Chem.* **271**, 31627–31637
230. De Bondt, H. L., Rosenblatt, J., Jancarik, J., Jones, H. D., Morgan, D. O., and Kim, S.-H. (1993) *Nature (London)* **363**, 595–602
231. Jeffrey, P. D., Russo, A. A., Polyak, K., Gibbs, E., Hurwitz, J., Massagué, J., and Pavletich, N. P. (1995) *Nature (London)* **376**, 313–320
232. Shiffman, D., Brooks, E. E., Brooks, A. R., Chan, S. C., and Milner, P. G. (1996) *J. Biol. Chem.* **271**, 12199–12204
233. Nurse, P. (1990) *Nature (London)* **344**, 503–507
234. Pines, J. (1993) *Trends Biochem. Sci.* **18**, 195–197
235. Solomon, M. J. (1994) *Trends Biochem. Sci.* **19**, 496–500
236. Lew, J., and Wang, J. H. (1995) *Trends Biochem. Sci.* **20**, 33–37
236a. Crawford, D. F., and Piwnica-Worms, H. (2001) *J. Biol. Chem.* **276**, 37166–37177
236b. Karsenti, E., and Vernos, I. (2001) *Science* **294**, 543–547
236c. Kastan, M. B. (2001) *Nature (London)* **410**, 766–767
236d. Davenport, R. J. (2001) *Science* **292**, 2415–2417
237. Morris, M. C., and Divita, G. (1999) *J. Mol. Biol.* **286**, 475–487
237a. Jacobs, H. W., Keidel, E., and Lehner, C. F. (2001) *EMBO J.* **20**, 2376–2386
238. Lopez-Girona, A., Furnari, B., Mondesert, O., and Russell, P. (1999) *Nature (London)* **397**, 172–175
238a. Yang, Q., Manicone, A., Coursen, J. D., Linke, S. P., Nagashima, M., Forgues, M., and Wang, X. W. (2000) *J. Biol. Chem.* **275**, 36892–36898
239. Reynolds, R. A., Yem, A. W., Wolfe, C. L., Deibel, M. R., Jr., Chidester, C. G., and Watenpaugh, K. D. (1999) *J. Mol. Biol.* **293**, 559–568
240. Hagting, A., Karlsson, C., Clute, P., Jackman, M., and Pines, J. (1998) *EMBO J.* **17**, 4127–4138
240a. Qian, Y.-W., Erikson, E., and Maller, J. L. (1998) *Science* **282**, 1701–1704
240b. Körner, K., Jérôme, V., Schmidt, T., and Müller, R. (2001) *J. Biol. Chem.* **276**, 9662–9669
241. Pines, J. (1999) *Nature (London)* **397**, 104 – 105
242. Sanchez, Y., Bachant, J., Wang, H., Hu, F., Liu, D., Tetzlaff, M., and Elledge, S. J. (1999) *Science* **286**, 1166–171
243. Kellogg, D. R., Moritz, M., and Alberts, B. M. (1994) *Ann. Rev. Biochem.* **63**, 639–674
244. Geier, B. M., Wiech, H., and Schiebel, E. (1996) *J. Biol. Chem.* **271**, 28366–28374
244a. Murray, A. W. (2001) *Science* **291**, 1499–1502
244b. Hinchcliffe, E. H., Miller, F. J., Cham, M., Khodjakov, A., and Sluder, G. (2001) *Science* **291**, 1547–1550
244c. Middendorp, S., Paoletti, A., Schiebel, E., and Bornens, M. (1997) *Proc. Natl. Acad. Sci. U.S.A.* **94**, 9141–9146
244d. do Carmo Avides, M., and Glover, D. M. (1999) *Science* **283**, 1733–1735

References

244e. Flory, M. R., Moser, M. J., Monnat, R. J., Jr., and Davis, T. N. (2000) *Proc. Natl. Acad. Sci. U.S.A.* **97**, 5919–5923
245. Nicklas, R. B. (1997) *Science* **275**, 632–637
246. Heald, R., Tournebize, R., Blank, T., Sandaltzopoulos, R., Becker, P., Hyman, A., and Karsenti, E. (1996) *Nature (London)* **382**, 420–425
246a. Fraschini, R., Beretta, A., Sironi, L., Musacchio, A., Lucchini, G., and Piatti, S. (2001) *EMBO J.* **20**, 6648–6659
247. Hyams, J. (1996) *Nature (London)* **382**, 397–398
247a. Bloom, K. (2002) *Proc. Natl. Acad. Sci. U.S.A.* **99**, 4757–4759
248. Dasso, M. (1993) *Trends Biochem. Sci.* **18**, 96–101
249. Ohba, T., Nakamura, M., Nishitani, H., and Nishimoto, T. (1999) *Science* **284**, 1356–1358
250. Wilde, A., and Zheng, Y. (1999) *Science* **284**, 1359–1362
250a. Gruneberg, U., Campbell, K., Simpson, C., Grindlay, J., and Schiebel, E. (2000) *EMBO J.* **19**, 6475–6488
250b. Martin-Lluesma, S., Stucke, V. M., and Nigg, E. A. (2002) *Science* **297**, 2267–2270
251. Barton, N. R., and Goldstein, L. S. B. (1996) *Proc. Natl. Acad. Sci. U.S.A.* **93**, 1735–1742
252. Lockhart, A., and Cross, R. A. (1996) *Biochemistry* **35**, 2365–2373
253. Pennisi, E. (1998) *Science* **279**, 477–478
254. Elledge, S. J. (1998) *Science* **279**, 999–1000
255. Hwang, L. H., Lau, L. F., Smith, D. L., Mistrot, C. A., Hardwick, K. G., Hwang, E. S., Amon, A., and Murray, A. W. (1998) *Science* **279**, 1041–1044
256. Yamano, H., Gannon, J., and Hunt, T. (1996) *EMBO J.* **15**, 5268–5279
257. Bachant, J. B., and Elledge, S. J. (1999) *Nature (London)* **398**, 757–758
258. Funabiki, H., Kumada, K., and Yanagida, M. (1996) *EMBO J.* **15**, 6617–6628
259. Page, A. M., and Hieter, P. (1999) *Ann. Rev. Biochem.* **68**, 583–609
259a. Prinz, S., and Amon, A. (1999) *Nature (London)* **402**, 133–135
259b. Nakaseko, Y., and Yanagida, M. (2001) *Nature (London)* **412**, 291–292
259c. Morgan, D. O., and Roberts, J. M. (2002) *Nature (London)* **418**, 495–496
260. Grossberger, R., Gieffers, C., Zachariae, W., Podtelejnikov, A. V., Schleiffer, A., Nasmyth, K., Mann, M., and Peters, J.-M. (1999) *J. Biol. Chem.* **274**, 14500–14507
261. Jiang, F., and Basavappa, R. (1999) *Biochemistry* **38**, 6471–6478
261a. Piel, M., Nordberg, J., Euteneuer, U., and Bornens, M. (2001) *Science* **291**, 1550–1553
262. Wisdom, R., Johnson, R. S., and Moore, C. (1999) *EMBO J.* **18**, 188–197
263. Takenaka, K., Moriguchi, T., and Nishida, E. (1998) *Science* **280**, 599–602
263a. Wilkinson, M. G., and Millar, J. B. A. (2000) *FASEB J.* **14**, 2147–2157
263b. Graves, L. M. (2000) *Nature (London)* **403**, 328–332
263c. Whitmarsh, A. J., and Davis, R. J. (2000) *Nature (London)* **403**, 255–256
264. Gottesfeld, J. M., and Forbes, D. J. (1997) *Trends Biochem. Sci.* **22**, 197–202
264a. Goldstein, L. S. B. (2001) *Science* **291**, 2102–2103
265. Martínez-Botas, J., Suárez, Y., Ferruelo, A. J., Gómez-Coronado, D., and Lasunción, M. A. (1999) *FASEB J.* **13**, 1359–1370
266. Bourne, Y., Arvai, A. S., Bernstein, S. L., Watson, M. H., Reed, S. I., Endicott, J. E., Noble, M. E., Johnson, L. N., and Tainer, J. A. (1995) *Proc. Natl. Acad. Sci. U.S.A.* **92**, 10232–10236
267. Parge, H. E., Arvai, A. S., Murtari, D. J., Reed, S. I., and Tainer, J. A. (1993) *Science* **262**, 387–395
267a. Rubin, G. M., and Lewis, E. B. (2000) *Science* **287**, 2216–2218
268. Yamamoto, M. (1996) *Trends Biochem. Sci.* **21**, 18–22
269. de Lange, T. (1998) *Nature (London)* **392**, 753–754
270. Conrad, M. N., Dominguez, A. M., and Dresser, M. E. (1997) *Science* **276**, 1252–1255
271. McKim, K. S., Green-Marroquin, B. L., Sekelsky, J. J., Chin, G., Steinberg, C., Khodosh, R., and Hawley, R. S. (1998) *Science* **279**, 876–878
271a. Lindgren, A., Bungard, D., Pierce, M., Xie, J., Vershon, A., and Winter, E. (2000) *EMBO J.* **19**, 6489–6497
272. Orr-Weaver, T. L. (1995) *Proc. Natl. Acad. Sci. U.S.A.* **92**, 10443–10449
273. McKim, K. S., and Hawley, R. S. (1995) *Science* **270**, 1595–1601
274. Picard, A., Galas, S., Peaucellier, G., and Dorée, M. (1996) *EMBO J.* **15**, 3590–3598
275. Lenormand, J.-L., Dellinger, R. W., Knudsen, K. E., Subramani, S., and Donoghue, D. J. (1999) *EMBO J.* **18**, 1869–1877
276. Sette, C., Barchi, M., Bianchini, A., Conti, M., Rossi, P., and Geremia, R. (1999) *J. Biol. Chem.* **274**, 33571–33579
276a. Xu, R., Wilson, W. K., and Matsuda, S. P. T. (2002) *J. Am. Chem. Soc.* **124**, 918–919
277. Watanabe, Y., and Nurse, P. (1999) *Nature (London)* **400**, 461–464
278. Roeder, G. S. (1995) *Proc. Natl. Acad. Sci. U.S.A.* **92**, 10450–10456
278a. Albini, S. M., and Jones, G. H. (1987) *Chromosoma* **95**, 324–328
279. Kleckner, N. (1996) *Proc. Natl. Acad. Sci. U.S.A.* **93**, 8167–8174
279a. Nabeshima, K., Kakihara, Y., Hiraoka, Y., and Nojima, H. (2001) *EMBO J.* **20**, 3871–3881
279b. Yuan, L., Liu, J.-G., Hoja, M.-R., Wilbertz, J., Nordqvist, K., and Höög, C. (2002) *Science* **296**, 1115–1118
279c. Gerton, J. L., DeRisi, J., Shroff, R., Lichten, M., Brown, P. O., and Petes, T. D. (2000) *Proc. Natl. Acad. Sci. U.S.A.* **97**, 11383–11390
279d. Stack, S. M. (1991) *Genome* **34**, 900–908
280. Boy de la Tour, E., and Laemmli, U. K. (1988) *Cell* **55**, 937–944
281. Gallardo, M. H., Bickham, J. W., Honeycutt, R. L., Ojeda, R. A., and Köhler, N. (1999) *Nature (London)* **401**, 341
281a. Shonn, M. A., McCarroll, R., and Murray, A. W. (2000) *Science* **289**, 300–303
281b. Sluder, G., and McCollum, D. (2000) *Science* **289**, 254–255
282. Pirrotta, V. (1984) *Trends Biochem. Sci.* **9**, 220–221
283. Burtis, K. C., and Hawley, R. S. (1999) *Nature (London)* **401**, 125–127
284. Kao, F.-t, and Yu, J.-w. (1991) *Proc. Natl. Acad. Sci. U.S.A.* **88**, 1844–1848
285. Sager, R. (1965) *Sci. Am.* **212**(Jan), 71–79
286. Goodenough, U. W., and Levine, R. P. (1970) *Sci. Am.* **223**(Nov), 22–29
287. Nagata, S., Taira, H., Hall, A., Johnsrud, L., Streuli, M., Ecsodi, J., Boll, W., Cantell, K., and Weissmann, C. (1980) *Nature (London)* **284**, 316–320
288. Olson, M. V. (1993) *Proc. Natl. Acad. Sci. U.S.A.* **90**, 4338–4344
289. Collins, F. S. (1995) *Proc. Natl. Acad. Sci. U.S.A.* **92**, 10821–10823
290. Guyer, M. S., and Collins, F. S. (1995) *Proc. Natl. Acad. Sci. U.S.A.* **92**, 10841–10848
290a. International Human Genome Sequencing Consortium. (2001) *Nature (London)* **409**, 860–921
290b. Pennisi, E. (2001) *Science* **291**, 1177–1180
290c. Venter, J. C., and many others. (2001) *Science* **291**, 1304–1351
290d. Olivier, M., and 54 other authors. (2001) *Science* **291**, 1298–1302
290e. Shoemaker, D. D., and 34 other authors. (2001) *Nature (London)* **409**, 922–927
290f. Caron, H., and 12 other authors. (2001) *Science* **291**, 1289–1292
290g. Deloukas, P., and 126 other authors. (2001) *Nature (London)* **414**, 865–871
290h. Claverie, J.-M. (2001) *Science* **291**, 1255–1257
290i. Pennisi, E. (2002) *Science* **296**, 1600–1601
290j. Kim, S. K., Lund, J., Kiraly, M., Duke, K., Jiang, M., Stuart, J. M., Eizinger, A., Wylie, B. N., and Davidson, G. S. (2001) *Science* **293**, 2087–2092
291. Collins, F. S., Patrinos, A., Jordan, E., Chakravarti, A., Gesteland, R., and Walters, L. (1998) *Science* **282**, 682–689
291a. Malakoff, D. (2000) *Science* **288**, 248–253
291b. The RIKEN Genome Exploration Research Group Phase II Team and the FANTOM Consortium. (2001) *Nature (London)* **409**, 685–690
291c. The International Mouse Mutagenesis Consortium. (2001) *Science* **291**, 1251–1255
291d. Gregory, S. G., and 85 other authors (2002) *Nature (London)* **418**, 743–750
291e. Mural, R. J., and 176 other authors (2002) *Science* **296**, 1661–1671
292. Roach, J. C., Siegel, A. F., van den Engh, G., Trask, B., and Hood, L. (1999) *Nature (London)* **401**, 843–845
292a. Marshall, E. (2001) *Science* **291**, 1872
292b. Vogel, G. (2000) *Science* **290**, 1671
292c. Aparicio, S., and 40 other authors (2002) *Science* **297**, 1301–1310
292d. Gardner, M. J., and 44 other authors (2002) *Nature (London)* **419**, 498–511
292e. Doolittle, R. F. (2002) *Nature (London)* **419**, 493–494
292f. Carlton, J. M., and 43 other authors (2002) *Nature (London)* **419**, 512–519
292g. Holt, R. A., and 121 other authors (2002) *Science* **298**, 129–149
292h. Kaufman, T. C., Severson, D. W., and Robinson, G. E. (2002) *Science* **298**, 97–115
293. White, R., and Caskey, C. T. (1988) *Science* **240**, 1483–1488
294. McKusick, V. A., and Amberger, J. S. (1995) in *The Metabolic and Molecular Bases of Inherited Disease*, 7th ed., Vol. 1 (Scriver, C. R., Beaudet, A. L., Sly, W. S., and Valle, D., eds), pp. 119–125, McGraw-Hill, New York
295. Sawyer, J. R., and Hozier, J. C. (1986) *Science* **232**, 1632–1635
296. Tunnacliffe, A., Benham, F., and Goodfellow, P. (1984) *Trends Biochem. Sci.* **9**, 5–7
297. Alberts, B., Bray, D., Lewis, J., Raff, M., Roberts, K., and Watson, J. D. (1994) *Molecular Biology of the Cell*, 3rd ed., Garland, New York
298. Franke, U. (1981) *Cytogenet. Cell Genet.* **31**, 24–32
299. Gray, J. W., and Langlois, R. G. (1986) *Ann. Rev. Biophys. Biophys. Chem.* **15**, 195–235
300. Chumakov, I., and plus 35 other authors (1992) *Nature (London)* **359**, 380–387
301. Shizuya, H., Birren, B., Kim, U.-J., Mancino, V., Slepak, T., Tachiiri, Y., and Simon, M. (1992) *Proc. Natl. Acad. Sci. U.S.A.* **89**, 8794–8797
302. Venter, J. C., Adams, M. D., Sutton, G. G., Kerlavage, A. R., Smith, H. O., and Hunkapiller, M. (1998) *Science* **280**, 1540–1542
303. Marx, J. (1995) *Science* **270**, 1919–1920

References

304. Dib, C., Fauré, S., Fizames, C., Samson, D., Drouot, N., Vignal, A., Millasseau, P., Marc, S., Hazan, J., Seboun, E., Lathrop, M., Gyapay, G., Morissette, J., and Weissenbach, J. (1996) *Nature (London)* **380**, 152–154, iii–v + supplement
305. Rogers, J. (1999) *Science* **286**, 429–432
305a. Liu, S., Ren, H., Gao, Q., Roach, D. J., Loder, R. T., Jr., Armstrong, T. M., Mao Q, Blaga, I., Barker, D. L., and Jovanovich, S. B. (2000) *Proc. Natl. Acad. Sci., U.S.A.* **97**, 5369 – 5374
305b. Paegel, B. M., Emrich, C. A., Wedemayer, G. J., Scherer, J. R., and Mathies, R. A. (2002) *Proc. Natl. Acad. Sci. U.S.A.* **99**, 574 – 579
306. Marshall, E. (1999) *Science* **284**, 1906–1909
306a. Green, P. (2002) *Proc. Natl. Acad. Sci. U.S.A.* **99**, 4143 – 4144
307. Donis-Keller, H., and 32 other authors (1987) *Cell* **51**, 319–337
308. Wang, D. G., and 26 other authors (1998) *Science* **280**, 1077–1082
308a. The International SNP Map Working Group. (2001) *Nature (London)* **409**, 928–933
308b. Patil, N., and 21 other authors (2001) *Science* **294**, 1719–1723
309. Hudson, T. J., and 50 other authors (1995) *Science* **270**, 1945–1954
310. Schuler, G. D., and 103 other authors (1996) *Science* **274**, 540–546
311. Wang, S. M., and Rowley, J. D. (1998) *Proc. Natl. Acad. Sci. U.S.A.* **95**, 11909–11914
312. Dunham, I., Shimizu, N., Roe, B. A., Chossoe, S., and 117 other authors (1999) *Nature (London)* **402**, 489–495
313. Butler, D. (1999) *Nature (London)* **402**, 447–448
313a. The International Human Genome Mapping Consortium. (2001) *Nature (London)* **409**, 934–941
314. Pandey, A., and Lewitter, F. (1999) *Trends Biochem. Sci.* **24**, 276–280
315. Pennisi, E. (1999) *Science* **286**, 447–450
316. Baxevanis, A. D., and Ouellette, B. F. F., eds. (1998) *Bioinformatics: A Practical Guide to the Analysis of Genes and Proteins*, Wiley, New York
317. Bishop, M. J., ed. (1999) *Genetics Databases*, Academic Press, San Diego, California
318. Durbin, R., Eddy, S., Krogh, A., and Mitchinson, G., eds. (1998) *Biological Sequence Analysis*, Cambridge Univ. Press., Cambridge
319. Bishop, M. J., ed. (1998) *Human Genome Computing*, 2nd ed., Academic Press, San Diego, California
320. Calvet, J. P. (1998) *Science* **282**, 1057–1058
321. Kereil, D. P., and Etzold, T. (1999) *Trends Biochem. Sci.* **24**, 155–157
321a. Tamames, J., and Tramontano, A. (2000) *Trends Biochem. Sci.* **25**, 402–403
321b. Jones, D. T., and Swindells, M. B. (2002) *Trends Biochem. Sci.* **27**, 161–164
321c. Zafar, N., Mazumder, R., and Seto, D. (2001) *Trends Biochem. Sci.* **26**, 514–516
322. Boguski, M. S. (1995) *N. Engl. J. Med.* **333**, 645–647
323. Borodovsky, M., Koonin, E. V., and Rudd, K. E. (1994) *Trends Biochem. Sci.* **19**, 309–313
324. Brown, N. P., Whittaker, A. J., Newell, W. R., Rawlings, C. J., and Beck, S. (1995) *J. Mol. Biol.* **249**, 342–359
325. Pennisi, E. (1998) *Science* **280**, 1692–1693
325a. Matthiessen, M. W. (2002) *Trends Biochem. Sci.* **27**, 586 – 588
326. Little, P. (1999) *Nature (London)* **402**, 467–468
327. O'Brien, S. J., Menotti-Raymond, M., Murphy, W. J., Nash, W. G., Wienberg, J., Stanyon, R., Copeland, N. G., Jenkins, N. A., Womack, J. E., and Graves, J. A. M. (1999) *Science* **286**, 458–481
327a. Gura, T. (2001) *Science* **293**, 593–595
327b. Buetow, K. H., Edmonson, M., MacDonald, R., Clifford, R., Yip, P., Kelley, J., Little, D. P., Strausberg, R., Koester, H., Cantor, C. R., and Braun, A. (2001) *Proc. Natl. Acad. Sci. U.S.A.* **98**, 581–584
327c. Couzin, J. (2002) *Science* **296**, 1391,1393
327d. Adam, D. (2001) *Nature (London)* **412**, 105
327e. Yan, H., Kinzler, K. W., and Vogelstein, B. (2000) *Science* **289**, 1890–1892
327f. Wright, A. F., and Van Heyningen, V. (2001) *Nature (London)* **414**, 705–706
327g. Marshall, E. (2001) *Science* **294**, 2272–2274
327h. Trager, R. S. (2002) *Science* **298**, 947
328. Pääbo, S. (1999) *Trends Biochem. Sci.* **24**, M13–M16
328a. Templeton, A. R. (2002) *Nature (London)* **416**, 45–51
329. Olsen, G. J., and Woese, C. R. (1993) *FASEB J.* **7**, 113–123
330. Shenk, M. A., and Steele, R. E. (1993) *Trends Biochem. Sci.* **18**, 459–463
331. Harris, E. E., and Hey, J. (1999) *Proc. Natl. Acad. Sci. U.S.A.* **96**, 3320–3324
331a. Adcock, G. J., Dennis, E. S., Easteal, S., Huttley, G. A., Jermiin, L. S., Peacock, W. J., and Thorne, A. (2001) *Proc. Natl. Acad. Sci. U.S.A.* **98**, 537–542
331b. Relethford, J. H. (2001) *Proc. Natl. Acad. Sci. U.S.A.* **98**, 390–391
331c. Jones, M. (2001) *The Molecule Hunt: Archaeology and the Search for Ancient DNA*, Allen Lane,
331d. Fish, S. A., Shepherd, T. J., McGenity, T. J., and Grant, W. D. (2002) *Nature (London)* **417**, 432–436
332. Kahn, P., and Gibbons, A. (1997) *Science* **277**, 176–178
332a. Gibbons, A. (2001) *Science* **292**, 627–629
332b. Fujiyama, A., Watanabe, H., Toyoda, A., Taylor, T. D., Itoh, T., Tsai, S.-F., and 11 other authors. (2002) *Science* **295**, 131–134
332c. Enard, W., Khaitovich, P., Klose, J., Zöllner, S., and 9 other authors. (2002) *Science* **296**, 340–343
332d. Cyranoski, D. (2002) *Nature (London)* **418**, 910–912
332e. Stone, A. C., Griffiths, R. C., Zegura, S. L., and Hammer, M. F. (2002) *Proc. Natl. Acad. Sci. U.S.A.* **99**, 43–48
332f. Britten, R. J. (2002) *Proc. Natl. Acad. Sci. U.S.A.* **99**, 13633–13635
333. Gibbons, A. (1998) *Science* **281**, 1432–1434
334. Kennedy, E. P. (1992) *Ann. Rev. Biochem.* **61**, 1–28
335. Barbujani, G., Magagni, A., Minch, E., and Cavalli-Sforza, L. L. (1997) *Proc. Natl. Acad. Sci. U.S.A.* **94**, 4516–4519
336. Molnar, S. (1998) *Human Variation, Races, Types and Ethnic Groups*, 4th ed., Prentice Hall, Upper Saddle River, New Jersey (pp. 230–247)
336a. Jablonski, N. G., and Chaplin, G. (2002) *Sci. Am.* **287**(Oct), 74–81
337. Rannala, B., and Mountain, J. L. (1997) *Proc. Natl. Acad. Sci. U.S.A.* **94**, 9197–9201
338. Sugimoto, C., Kitamura, T., Guo, J., Al-Ahdal, M. N., Shchelkunov, S. N., Otova, B., Ondrejka, P., Chollet, J.-Y., El-Safi, S., Ettayebi, M., Grésenguet, G., Kocagöz, T., Chaiyarasamee, S., Thant, K. Z., Thein, S., Moe, K., Kobayashi, N., Taguchi, F., and Yogo, Y. (1997) *Proc. Natl. Acad. Sci. U.S.A.* **94**, 9191–9196
338a. Pääbo, S. (2001) *Science* **291**, 1219–1220
338b. Pennisi, E. (2001) *Science* **291**, 1733–1734
338c. Thomson, R., Pritchard, J. K., Shen, P., Oefner, P. J., and Feldman, M. W. (2000) *Proc. Natl. Acad. Sci. U.S.A.* **97**, 7360–7365
338d. Gibbons, A. (2000) *Science* **290**, 1080–1081
338e. Cann, R. L. (2001) *Science* **291**, 1742–1747
339. De Mendoza, D. H., and Braginski, R. (1999) *Science* **283**, 1439–1440
340. Bradley, D. G., MacHugh, D. E., Cunningham, P., and Loftus, R. T. (1996) *Proc. Natl. Acad. Sci. U.S.A.* **93**, 5131–5135
341. Makalowski, W., and Boguski, M. S. (1998) *Proc. Natl. Acad. Sci. U.S.A.* **95**, 9407–9412
342. Nadeau, J. H., Grant, P. L., Mankala, S., Reiner, A. H., Richardson, J. E., and Eppig, J. T. (1995) *Nature (London)* **373**, 363–365
343. Collins, F. S., and Jegalian, K. G. (1999) *Sci. Am.* **281**(Dec), 86–91
344. O'Brien, S. J., and Stanyon, R. (1999) *Nature (London)* **402**, 365–366
344a. Hedges, S. B., and Kumar, S. (2002) *Science* **297**, 1283 – 1285
344b. Knight, J. (2002) *Nature (London)* **417**, 374–376
345. Tatusov, R. L., Koonin, E. V., and Lipman, D. J. (1997) *Science* **278**, 631–637
346. Gibbons, A. (1998) *Science* **280**, 675–676
347. Rowe, T. (1999) *Nature (London)* **398**, 283–284
348. Bromham, L., Rambaut, A., Fortey, R., Cooper, A., and Penny, D. (1998) *Proc. Natl. Acad. Sci. U.S.A.* **95**, 12386–12389
349. Knoll, A. H. (1999) *Science* **285**, 1025–1026
350. Maniloff, J. (1996) *Proc. Natl. Acad. Sci. U.S.A.* **93**, 10004–10006
351. Lawrence, J. G., and Ochman, H. (1998) *Proc. Natl. Acad. Sci. U.S.A.* **95**, 9413–9417
352. Doolittle, W. F. (1999) *Trends Biochem. Sci.* **24**, M5–M8
353. Lake, J. A., Jain, R., and Rivera, M. C. (1999) *Science* **283**, 2027–2028
354. Pennisi, E. (1998) *Science* **280**, 672–674
355. Stein, J. L., and Simon, M. I. (1996) *Proc. Natl. Acad. Sci. U.S.A.* **93**, 6228–6230
356. Davies, J. (1999) *Trends Biochem. Sci.* **24**, M2–M5
356a. Doolittle, R. F. (2002) *Nature (London)* **416**, 697–700
356b. Bentley, S. D. and 42 other authors (2002) *Nature (London)* **417**, 141–147
356c. Barnett, M. J., and 25 other authors (2001) *Proc. Natl. Acad. Sci. U.S.A.* **98**, 9883–9888
356d. Galibert, F., and 55 other authors (2001) *Science* **293**, 668–672
356e. Heidelberg, J. F., and 31 other authors (2000) *Nature (London)* **406**, 477–483
356f. Parkhill, J., and 40 other authors (2001) *Nature (London)* **413**, 848–852
356g. Parkhill, J., and 34 other authors (2001) *Nature (London)* **413**, 523–527
356h. Moreno, E., and Moriyón, I. (2002) *Proc. Natl. Acad. Sci. U.S.A.* **99**, 1–3
356i. Cole, S. T., and 43 other authors (2001) *Nature (London)* **409**, 1007–1011
356j. Enserink, M. (2002) *Science* **295**, 1442–1443
356k. Tettelin, H., and 38 other authors (2001) *Science* **293**, 498–506
356l. Stover, C. K., and 30 other authors (2000) *Nature (London)* **406**, 959–964
356m. Glaser, P., and 54 other authors (2001) *Science* **294**, 849–853
356n. Perna, N. T., and 27 other authors (2001) *Nature (London)* **409**, 529–533
356o. Glass, J. I., Lefkowitz, E. J., Glass, J. S., Heiner, C. R., Chen, E. Y., and Cassell, G. H. (2000) *Nature (London)* **407**, 757–761
356p. Nierman, W. C., and 36 other authors (2001) *Proc. Natl. Acad. Sci. U.S.A.* **98**, 4136–4141
356q. Shigenobu, S., Watanabe, H., Hattori, M., Sakaki, Y., and Ishikawa, H. (2000) *Nature (London)* **407**, 81–86
356r. Ng, W. V., Kennedy, S. P., Mahairas, G. G., and 40 other authors. (2000) *Proc. Natl. Acad. Sci. U.S.A.* **97**, 12176 – 12181

References

356s. Kawashima, T., Amano, N., Koike, H., Makino, S.-i, Higuchi, S., and 10 other authors. (2000) *Proc. Natl. Acad. Sci. U.S.A.* **97**, 14257–14262

356t. Cole, S. T., and 41 other authors. (1998) *Nature (London)* **393**, 537 – 544

357. Edwards, J. S., and Palsson, B. O. (1999) *J. Biol. Chem.* **274**, 17410–17416

358. Strauss, E. J., and Falkow, S. (1997) *Science* **276**, 707–711

358a. da Silva, A. C. R., and 64 other authors. (2002) *Nature (London)* **417**, 459 – 463

359. Karp, P. D. (1998) *Trends Biochem. Sci.* **23**, 114–116

359a. Covert, M. W., Schilling, C. H., Famili, I., Edwards, J. S., Goryanin, I. I., Selkov, E., and Palsson, B. O. (2001) *Trends Biochem. Sci* **26**, 179 – 186

360. Hutchison, C. A., III, Peterson, S. N., Gill, S. R., Cline, R. T., White, O., Fraser, C. M., Smith, H. O., and Venter, J. C. (1999) *Science* **286**, 2165–2169

361. Mushegian, A. R., and Koonin, E. V. (1996) *Proc. Natl. Acad. Sci. U.S.A.* **93**, 10268–10273

361a. Szostak, J. W., Bartel, D. P., and Luisi, P. L., (2001) *Nature (London)* **409**, 387 – 390

362. Cho, M. K., Magnus, D., Caplan, A. L., McGee, D., and Ethics of Genomics Group. (1999) *Science* **286**, 2087–2090

362a. Stephanopoulos, G., and Kelleher, J. (2001) *Science* **292**, 2024 – 2025

362b. Holden, C. (2002) *Science* **297**, 1459 – 1460

363. Winzeler, E. A., and 51 other authors (1999) *Science* **285**, 901–906

364. Hieter, P. (1999) *Nature (London)* **402**, 362–363

364a. Ross-Macdonald, P., and 17 other authors (1999) *Nature (London)* **402**, 413–418

365. Glaever, G., and 72 other authors. (2002) *Nature (London)* **418**, 387 – 391

365a. Eisen, J. A. (2002) *Nature (London)* **415**, 845 – 848

365b. Wood, V., and 132 other authors (2002) *Nature (London)* **415**, 871– 880

365c. Seo, H.-C., Kube, M., Edvardsen, R. B., Jensen, M. F., Beck, A., Spriet, E., Gorsky, G., Thompson, E. M., Leharch, H., Reinhardt, R., and Chourrout, D. (2001) *Science* **294**, 2506

366. Fields, S., Kohara, Y., and Lockhart, D. J. (1999) *Proc. Natl. Acad. Sci. U.S.A.* **96**, 8825–8826

367. Tabara, H., Motohashi, T., and Kohara, Y. (1996) *Nucleic Acids Res.* **24**, 2119–2124

368. DeRisi, J. L., Iyer, V. R., and Brown, P. O. (1997) *Science* **278**, 680–686

369. Walhout, A. J. M., Sordella, R., Lu, X., Hartley, J. L., Temple, G. F., Brasch, M. A., Thierry-Mieg, N., and Vidal, M. (2000) *Science* **287**, 116–122

370. Hutter, H., Vogel, B. E., Plenefisch, J. D., Norris, C. R., Proenca, R. B., Spieth, J., Guo, C., Mastwal, S., Zhu, X., Scheel, J., and Hedgecock, E. M. (2000) *Science* **287**, 989–994

370a. Zdobnov, E. M., and 35 other authors. (2002) *Science* **298**, 149 – 159

370b. Miki, R., and 29 other authors. (2001) *Proc. Natl. Acad. Sci. U.S.A.* **98**, 2199 – 2204

371. Lin, X., and 36 other authors (1999) *Nature (London)* **402**, 761–768

372. European Union Arabidopsis Genome Sequencing Consortium, and Cold Spring Harbor, W. U. in St. L. and P. E. B. A. S. C. (1999) *Nature (London)* **402**, 769–777

373. Mathé, C., Peresetsky, A., Déhais, P., Van Montagu, M., and Rouzé, P. (1999) *J. Mol. Biol.* **285**, 1977–1991

374. Gura, T. (2000) *Science* **287**, 412–414

375. Meyerowitz, E. M. (1999) *Nature (London)* **402**, 731–732

375a. The Arabidopsis Genome Initiative. (2000) *Nature (London)* **408**, 796 – 815

375b. Seki, M., and 19 other authors (2002) *Science* **296**, 141– 145

375c. Pennisi, E. (2000) *Science* **290**, 32 – 35

375d. Normile, D., and Pennisi, E. (2002) *Science* **296**, 32 – 39

375e. Bennetzen, J. (2002) *Science* **296**, 60 – 63

375f. Yu, J., and 99 other authors (2002) *Science* **296**, 79 – 92

375g. Goff, S. A., and 54 other authors (2002) *Science* **296**, 92 – 100

376. Somerville, C., and Somerville, S. (1999) *Science* **285**, 380–383

377. Pennisi, E. (1998) *Science* **282**, 652–654

378. Capy, P. (2000) *Science* **287**, 985–986

379. Scriver, C. R., Kaufman, S., Eisensmith, R. C., and Woo, S. L. C. (1995) in *The Metabolic and Molecular Bases of Inherited Disease*, 7th ed., Vol. 1 (Scriver, C. R., Beaudet, A. L., Sly, W. S., and Valle, D., eds), pp. 1015–1075, McGraw-Hill, New York

380. Barinaga, M. (1998) *Science* **281**, 32–34

381. Schultheis, P. J., Lorenz, J. N., Meneton, P., Nieman, M. L., Riddle, T. M., Flagella, M., Duffy, J. J., Doetschman, T., Miller, M. L., and Shull, G. E. (1998) *J. Biol. Chem.* **273**, 29150–29155

382. Chairoungdua, A., Segawa, H., Kim, J. Y., Miyamoto, K.-i, Haga, H., Fukui, Y., Mizoguchi, K., Ito, H., Takeda, E., Endou, H., and Kanai, Y. (1999) *J. Biol. Chem.* **274**, 28845–28848

383. Chambers, E. J., Bloomberg, G. B., Ring, S. M., and Tanner, M. J. A. (1999) *J. Mol. Biol.* **285**, 1289–1307

384. Annunen, S., Paassilta, P., Lohiniva, J., Perälä, M., Pihlajamaa, T., Karppinen, J., Tervonen, O., Kröger, H., Lähde, S., Vanharanta, H., Ryhänen, L., Göring, H. H. H., Ott, J., Prockop, D. J., and Ala-Kokko, L. (1999) *Science* **285**, 409–412

385. Ward, C. J., Turley, H., Ong, A. C. M., Comley, M., Biddolph, S., Chetty, R., Ratcliffe, P. J., Gatter, K., and Harris, P. C. (1996) *Proc. Natl. Acad. Sci. U.S.A.* **93**, 1524–1528

386. Hugot, J.-P., 16 other authors, and Groupe d'Etude Thérapeutique des Affections Inflammatoires Digestives. (1996) *Nature (London)* **379**, 821–823

387. Ghosh, S., and 39 other authors (1999) *Proc. Natl. Acad. Sci. U.S.A.* **96**, 2198–2203

388. Kinzler, K. W., and Vogelstein, B. (1993) *Nature (London)* **363**, 495

389. Brehm, A., and Kouzarides, T. (1999) *Trends Biochem. Sci.* **24**, 142–145

390. Cho, Y., Gorina, S., Jeffrey, P. D., and Pavletich, N. P. (1994) *Science* **265**, 346–355

391. Oren, M. (1999) *J. Biol. Chem.* **274**, 36031–36034

392. Venkitaraman, A. R. (1999) *Science* **286**, 1100–1102

393. Smith, J. R., and 22 other authors (1996) *Science* **274**, 1371–1374

394. Novelli, M. R., Williamson, J. A., Tomlinson, I. P. M., Elia, G., Hodgson, S. V., Talbot, I. C., Bodmer, W. F., and Wright, N. A. (1996) *Science* **272**, 1187–1190

395. Rampino, N., Yamamoto, H., Ionov, Y., Li, Y., Sawai, H., Reed, J. C., and Perucho, M. (1997) *Science* **275**, 967–969

396. Rabbitts, T. H. (1994) *Nature (London)* **372**, 143–149

397. Lengauer, C., Kinzler, K. W., and Vogelstein, B. (1998) *Nature (London)* **396**, 643–649

398. Kappel, C. A., Bieberich, C. J., and Jay, G. (1994) *FASEB J.* **8**, 583–592

399. Ahmad, W., and 17 other authors (1998) *Science* **279**, 720–724

400. Bassett, D. E., Jr., Boguski, M. S., and Hieter, P. (1996) *Nature (London)* **379**, 589–590

401. Richards, R. I., and Sutherland, G. R. (1997) *Trends Biochem. Sci.* **22**, 432–436

402. Wang, Y.-H., Gellibolian, R., Shimizu, M., Wells, R. D., and Griffith, J. (1996) *J. Mol. Biol.* **263**, 511–516

403. Weisman-Shomer, P., Naot, Y., and Fry, M. (2000) *J. Biol. Chem.* **275**, 2231–2238

404. Fridell, R. A., Benson, R. E., Hua, J., Bogerd, H. P., and Cullen, B. R. (1996) *EMBO J.* **15**, 5408–5414

405. Mandel, J.-L. (1997) *Nature (London)* **386**, 767–769

405a. Moine, H., and Mandel, J.-L. (2001) *Science* **294**, 2487– 2488

405b. Greenough, W. T., Klintsova, A. Y., Irwin, S. A., Galvez, R., Bates, K. E., and Weiler, I. J. (2001) *Proc. Natl. Acad. Sci. U.S.A.* **98**, 7101–7106

405c. Sohn, E. (2001) *Science* **294**, 1809

405d. Deng, J., Xiong, Y., and Sundaralingam, M. (2001) *Proc. Natl. Acad. Sci. U.S.A.* **98**, 13665–13670

406. Wells, R. D. (1996) *J. Biol. Chem.* **271**, 2875–2878

407. Timchenko, L. T., and Caskey, C. T. (1996) *FASEB J.* **10**, 1589–1597

407a. Sinden, R. R. (2001) *Nature (London)* **411**, 757– 758

408. Muragaki, Y., Mundlos, S., Upton, J., and Olsen, B. R. (1996) *Science* **272**, 548–551

408a. Liquori, C. L., Ricker, K., Moseley, M. L., Jacobsen, J. F., Kress, W., Naylor, S. L., Day, J. W., and Ranum, L. P. W. (2001) *Science* **293**, 864 – 867

409. Singer, R. H. (1998) *Science* **280**, 696–697

409a. Dhe-Paganon, S., Shigeta, R., Chi, Y.-I., Ristow, M., and Shoelson, S. E. (2000) *J. Biol. Chem.* **275**, 30753 – 30756

409b. Lodi, R., Cooper, J. M., Bradley, J. L., Manners, D., Styles, P., Taylor, D. J., and Schapira, A. H. V. (1999) *Proc. Natl. Acad. Sci. U.S.A.* **96**, 11492 – 11495

409c. Chen, O. S., and Kaplan, J. (2000) *J. Biol. Chem.* **275**, 7626 – 7632

409d. Foury, F., and Talibi, D. (2001) *J. Biol. Chem.* **276**, 7762 – 7768

409e. Gakh, O., Adamec, J., Gacy, A. M., Twesten, R. D., Owen, W. G., and Isaya, G. (2002) *Biochemistry* **41**, 6798 – 6804

410. Campuzano, V., and 26 other authors (1996) *Science* **271**, 1423–1427

410a. LeProust, E. M., Pearso, C. E., Sinden, R. R., and Gao, X. (2000) *J. Mol. Biol.* **302**, 1063 – 1080

410b. Rosenberg, R. N. (1995) *N. Engl. J. Med.* **333**, 1351– 1353

411. Mariappan, S. V. S., Catasti, P., Silds, L. A., III, Bradbury, E. M., and Gupta, G. (1999) *J. Mol. Biol.* **285**, 2035–2052

412. Dürr, A., Cossee, M., Agid, Y., Campuzano, V., Mignard, C., Penet, C., Mandel, J.-L., Brice, A., and Koenig, M. (1996) *N. Engl. J. Med.* **335**, 1169–1175

413. Rosenberg, R. N. (1996) *N. Engl. J. Med.* **335**, 1222–1224

414. Perutz, M. F. (1999) *Trends Biochem. Sci.* **24**, 58–63

414a. Holbert, S., Denghien, I., Kiechle, T., Rosenblatt, A., Wellington, C., Hayden, M. R., Margolis, R. L., Ross, C. A., Dausset, J., Ferrante, R. J., and Néri, C. (2001) *Proc. Natl. Acad. Sci. U.S.A.* **98**, 1811– 1816

414b. Peters, M. F., and Ross, C. A. (2001) *J. Biol. Chem.* **276**, 3188 – 3194

414c. Bates, G. P. (2001) *Nature (London)* **413**, 691– 694

415. Wellington, C. L., and 20 other authors (1998) *J. Biol. Chem.* **273**, 9158–9167

References

416. Karlin, S., and Burge, C. (1996) *Proc. Natl. Acad. Sci. U.S.A.* **93**, 1560–1565
416a. Kaytor, M. D., and Warren, S. T. (1999) *J. Biol. Chem.* **274**, 37507–37510
416b. Perutz, M. F., and Windle, A. H. (2001) *Nature (London)* **412**, 143–144
416c. Chai, Y., Wu, L., Griffin, J. D., and Paulson, H. L. (2001) *J. Biol. Chem.* **276**, 44889–44897
417. Diamond, M. L., Robinson, M. R., and Yamamoto, K. R. (2000) *Proc. Natl. Acad. Sci. U.S.A.* **97**, 657–661
417a. Fernandez-Funez, P., and 14 other authors (2000) *Nature (London)* **408**, 101–106
417b. Mushegian, A. R., Vishnivetskiy, S. A., and Gurevich, V. V. (2000) *Biochemistry* **39**, 6809–6813
418. Skinner, P. J., Koshy, B. T., Cummings, C. J., Klement, I. A., Helin, K., Servadio, A., Zoghbi, H. Y., and Orr, H. T. (1997) *Nature (London)* **389**, 971–974
419. Collick, A., Norris, M. L., Allen, M. J., Bois, P., Barton, S. C., Surani, M. A., and Jeffreys, A. J. (1994) *EMBO J.* **13**, 5745–5753
420. Sutherland, G. R., and Richards, R. I. (1994) *N. Engl. J. Med.* **331**, 191–193
421. McMurray, C. T. (1999) *Proc. Natl. Acad. Sci. U.S.A.* **96**, 1823–1825
422. Warpeha, K. M., Xu, W., Liu, L., Charles, I. G., Patterson, C. C., Ah-Fat, F., Harding, S., Hart, P. M., Chakravarthy, U., and Hughes, A. E. (1999) *FASEB J.* **13**, 1825–1832
423. Petruska, J., Hartenstine, M. J., and Goodman, M. F. (1998) *J. Biol. Chem.* **273**, 5204–5210
424. Pearson, C. E., Eichler, E. E., Lorenzetti, D., Kramer, S. F., Zoghbi, H. Y., Nelson, D. L., and Sinden, R. R. (1998) *Biochemistry* **37**, 2701–2708
425. Iyer, R. R., Pluciennik, A., Rosche, W. A., Sinden, R. R., and Wells, R. D. (2000) *J. Biol. Chem.* **275**, 2174–2184
426. Jakupciak, J. P., and Wells, R. D. (1999) *J. Biol. Chem.* **274**, 23468–23479
427. Chee, M., Yang, R., Hubbell, E., Berno, A., Huang, X. C., Stern, D., Winkler, J., Lockhart, D. J., Morris, M. S., and Fodor, S. P. A. (1996) *Science* **274**, 610–614
428. Golub, T. R., Slonim, D. K., Tamayo, P., Huard, C., Gaasenbeek, M., Mesirov, J. P., Coller, H., Loh, M. L., Downing, J. R., Caligiuri, M. A., Bloomfield, C. D., and Lander, E. S. (1999) *Science* **286**, 531–537
429. Service, R. F. (1998) *Science* **282**, 396–399
430. Gerhold, D., Rushmore, T., and Caskey, C. T. (1999) *Trends Biochem. Sci.* **24**, 168–173
430a. Pennisi, E. (2002) *Science* **297**, 1985, 1987
430b. Schena, M. (2002) *Microarray Analysis*, Wiley, New York
431. Gerry, N. P., Witowski, N. E., Day, J., Hammer, R. P., Barany, G., and Barany, F. (1999) *J. Mol. Biol.* **292**, 251–262
432. Tang, K., Fu, D.-J., Julien, D., Braun, A., Cantor, C. R., and Köster, H. (1999) *Proc. Natl. Acad. Sci. U.S.A.* **96**, 10016–10020
433. Marshall, E. (1999) *Science* **286**, 444–447
434. Dalton, R. (2000) *Nature (London)* **403**, 234
435. Burns, M. A., Johnson, B. N., Brahmasandra, S. N., Handique, K., Webster, J. R., Krishnan, M., Sammarco, T. S., Man, P. M., Jones, D., Heldsinger, D., Mastrangelo, C. H., and Burke, D. T. (1998) *Science* **282**, 484–487
436. Service, R. F. (1998) *Science* **282**, 399–401
437. Landegren, U., Kaiser, R., Caskey, C. T., and Hood, L. (1988) *Science* **242**, 229–237
438. Kolata, G. (1986) *Science* **234**, 22–24
439. Schnieke, A. E., Kind, A. J., Ritchie, W. A., Mycock, K., Scott, A. R., Ritchie, M., Wilmut, I., Colman, A., and Campbell, K. H. S. (1997) *Science* **278**, 2130–2133
440. Arnon, R. (1986) *Trends Biochem. Sci.* **11**, 521–524
441. McAleer, W. J., Buynak, E. B., Maigetter, R. Z., Wampler, D. E., Miller, W. J., and Hilleman, M. R. (1984) *Nature (London)* **307**, 178–180
442. Kapusta, J., Modelska, A., Figlerowicz, M., Pniewski, T., Letellier, M., Lisowa, O., Yusibov, V., Koprowski, H., Plucienniczak, A., and Legocki, A. B. (1999) *FASEB J.* **13**, 1796–1799
443. Weiner, D. B., and Kennedy, R. C. (1999) *Sci. Am.* **281**(Jul), 50–57
444. Taubes, G. (1997) *Science* **278**, 1711–1714
445. McDonald, G. A., Anacker, R. L., and Garjian, K. (1987) *Science* **235**, 83–85
446. Brown, F., Schild, G. C., and Ada, G. L. (1986) *Nature (London)* **319**, 549–550
447. Rupprecht, C. E., Wiktor, T. J., Johnston, D. H., Hamir, A. N., Dietzschold, B., Wunner, W. H., Glickman, L. T., and Koprowski, H. (1986) *Proc. Natl. Acad. Sci. U.S.A.* **83**, 7947–7950
448. Waldmann, T. A. (1986) *Science* **232**, 727–732
449. Travis, J., Owen, M., George, P., Carrell, R., Rosenberg, S., Hallewell, R. A., and Barr, P. J. (1985) *J. Biol. Chem.* **260**, 4384–4389
450. Parkman, R. (1986) *Science* **232**, 1373–1378
451. Sandig, V., Youil, R., Bett, A. J., Franlin, L. L., Oshima, M., Maione, D., Wang, F., Metzker, M. L., Savino, R., and Caskey, C. T. (2000) *Proc. Natl. Acad. Sci. U.S.A.* **97**, 1002–1007
452. Verma, I. M., and Somia, N. (1997) *Nature (London)* **389**, 239–242
453. Anderson, W. F. (1995) *Sci. Am.* **273**(Sep), 124–128
454. Marshall, E. (1995) *Science* **269**, 1050–1055
455. Zanjani, E. D., and Anderson, W. F. (1999) *Science* **285**, 2084–2088
456. Finkel, T., and Epstein, S. E. (1995) *FASEB J.* **9**, 843–851
456a. Miller, D. G., and Stamatoyannopoulos, G. (2001) *N. Engl. J. Med.* **344**, 1782–1783
457. Miller, H. I. (2000) *Science* **287**, 591–592
457a. Gura, T. (2001) *Science* **291**, 1692–1697
458. Amalfitano, A., McVie-Wylie, A. J., Hu, H., Dawson, T. L., Raben, N., Plotz, P., and Chen, Y. T. (1999) *Proc. Natl. Acad. Sci. U.S.A.* **96**, 8861–8866
458a. Pawliuk, R., Westerman, K. A., Fabry, M. E., Payen, E., Tighe, R., Bouhassira, E. E., Acharya, S. A., Ellis, J., London, I. M., Eaves, C. J., Humphries, R. K., Beuzard, Y., Nagel, R. L., and Leboulch, P. (2001) *Science* **294**, 2368–2371
458b. Dunder, U., Kaartinen, V., Valtonen, P., Väänänen, E., Kosma, V.-M., Heisterkamp, N., Groffen, J., and Mononen, I. (2000) *FASEB J.* **14**, 361–367
459. Capecchi, M. R. (2000) *Proc. Natl. Acad. Sci. U.S.A.* **97**, 956–957
460. Marshall, E. (1996) *Science* **271**, 586–588
461. Kohn, D. B., and Parkman, R. (1997) *FASEB J.* **11**, 635–639
462. Broitman, S., Amosova, O., Dolinnaya, N. G., and Fresco, J. R. (1999) *J. Biol. Chem.* **274**, 21763–21768
462a. Knight, J. (2001) *Nature (London)* **413**, 12–15
463. Alper, J. (1999) *Science* **283**, 1625–1626
464. Kishore, G. M., and Shewmaker, C. (1999) *Proc. Natl. Acad. Sci. U.S.A.* **96**, 5968–5972
465. Trewavas, A. (1999) *Nature (London)* **402**, 231–232
466. Goodman, R. M., Hauptli, H., Crossway, A., and Knauf, V. C. (1987) *Science* **236**, 48–54
467. Jones, M. G. K. (1985) *Nature (London)* **317**, 579–580
468. Roberts, L. (1988) *Science* **241**, 1290
469. Frommer, W. B., Ludewig, U., and Rentsch, D. (1999) *Science* **285**, 1222–1223
470. Kling, J. (1996) *Science* **274**, 180–181
471. Gura, T. (1999) *Science* **285**, 994–995
472. Simon Moffat, A. (1992) *Science* **256**, 770–771
473. Thomas, S. M., Davies, A. R. W., Birtwistle, N. J., Crowther, S. M., and Burke, J. F. (1996) *Nature (London)* **380**, 387–388
474. Marshall, E. (1996) *Science* **273**, 1788–1789
475. Enserink, M. (1998) *Science* **281**, 890–891
476. Dickson, D. (1993) *Nature (London)* **366**, 391
477. Poste, G. (1995) *Nature (London)* **378**, 534–536
478. Fuller, B. P., Ellis Kahn, M. J., Barr, P. A., Biesecker, L., Crowley, E., Garber, J., Mansoura, M. K., Murphy, P., Murray, J., Phillips, J., Rothenberg, K., Rothstein, M., Stopfer, J., Swergold, G., Weber, B., Collins, F. S., and Hudson, K. L. (1999) *Science* **285**, 1359–1361
479. Butler, D. (1998) *Nature (London)* **391**, 727
480. Hudson, K. L., Rothenberg, K. H., Andrews, L. B., Kahn, M. J. E., and Collins, F. S. (1995) *Science* **270**, 391–393
481. Lapham, E. V., Kozma, C., and Weiss, J. O. (1996) *Science* **274**, 621–624
482. Fost, N. (1992) *FASEB J.* **6**, 2813–2817
483. Holtzman, N. A. (1999) *Science* **286**, 409
484. Wills, C. (1970) *Sci. Am.* **222**(Mar), 98–107
485. Vogel, G. (1999) *Science* **286**, 2050–2051
486. Stoye, J. P. (1997) *Nature (London)* **386**, 126–127
487. Weiss, R. A. (1999) *Science* **285**, 1221–1222
488. Paradis, K., Langford, G., Long, Z., Heneine, W., Sandstrom, P., Switzer, W. M., Chapman, L. E., Lockey, C., Onions, D., XEN 111 Study Group, and Otto, E. (1999) *Science* **285**, 1236–1241
489. Enserink, M. (1999) *Science* **286**, 1662–1668
490. Haslberger, A. G. (2000) *Science* **287**, 431–432
491. Jackson, A., and Inglehearn, C. (1999) *Nature (London)* **402**, 457
492. Crawley, M. J. (1999) *Nature (London)* **400**, 501–502
493. Saxena, D., Flores, S., and Stotzky, G. (1999) *Nature (London)* **402**, 480
494. Smith, R. J. (1984) *Science* **224**, 1215–1216
495. Many, B. W. J., Almond, J. W., Berns, K. I., Chanock, R. M., Lvov, D. K., Pettersson, R. F., Schatzmayr, H. G., and Fenner, F. (1993) *Science* **262**, 1223–1224
496. Joklik, W. K., Moss, B., Fields, B. N., Bishop, D. H. L., and Sandakhchiev, L. S. (1993) *Science* **262**, 1225–1226
497. Jesson, L. K., and Barrett, S. C. H. (2002) *Nature (London)* **417**, 707
498. Human Genome Organization Ethics Committee. (2000) *Science* **290**, 49
499. Williamson, R., and Duncan, R. (2002) *Nature (London)* **418**, 585–586
500. Adam, D. (2002) *Nature (London)* **417**, 370
501. Aldhous, P. (2002) *Nature (London)* **418**, 355–356
502. Rothenberg, K. H., and Terry, S. F. (2002) *Science* **297**, 196–197
503. Nowlan, W. (2002) *Science* **297**, 195–196
504. McDowell, N. (2002) *Nature (London)* **416**, 571
505. Stokstad, E. (2002) *Science* **297**, 1797, 1799
506. Kaiser, J. (2001) *Science* **292**, 34–36
507. Enserink, M. (2002) *Science* **297**, 30–31
508. Clarke, T. (2002) *Nature (London)* **419**, 429–430
509. Rieger, M. A., Lamond, M., Preston, C., Powles, S. B., and Roush, R. T. (2002) *Science* **296**, 2386–2388
510. Spurgeon, D. (2001) *Nature (London)* **409**, 749

Study Questions

1. The two chromatids in Fig. 26-13G are said to be coiled with opposite handedness. Can you draw this conclusion from Fig. 26 alone? What are the biological implications for mitosis?[280] Does the DNA have a differing chirality at the molecular level? Compare this observation with the existence of snail shells or flowers[497] with both right and left handedness within the same species.

2. Will the sequencing of the human genome ever be complete?

3. Should patenting of human genes be allowed? Under what circumstances should patents be allowed on genetically engineered genes?

4. Should the human genome be regarded as a **common heritage** such that there is a guarantee that the medical and other benefits arising from genetic research are available to all persons on earth? See Human Genome Organization Ethics Committee.[498]

5. DNA testing (Box 5-D) is widely used by police throughout the world. It has been estimated that if ten loci in the DNA are tested the chance of a random match between two people is one in a billion. In the United Kingdom it is planned to hold DNA profiles on record for one of every 15 people. Is this wise? Or should DNA profiles be recorded for all people?[499] If DNA profiles are on record how can we be sure that they are not used dishonestly? Should police have access to DNA data bases? See Adam.[500]

6. Should "race" be used as a variable in biomedical studies? See Aldhous.[501]

7. Should insurance companies be allowed to have access to genetic information about insured people? Companies usually obtain other medical information. See Rothenberg and Terry,[502] Adam,[500] and Nowlan.[503]

8. Should genetically engineered fish be allowed in "farms" that are set up in ocean waters? See McDowell[504] and Stokstad.[505]

9. Do transgenic trees pose a threat to natural ecosystems? See Kaiser.[506]

10. Should we attempt to replace wild populations of mosquitoes with genetically engineered mosquitoes that can not transmit malaria? See Enserick[507] and Clarke.[508]

11. Could dispersal of pollen from genetically modified plants lead to undesirable "genetic pollution" of the environment? See Rieger *et al.*[509]

12. To what extent is genetic modification of plants and animals equivalent to changes made by conventional breeding? See Spurgeon.[510]

13. Choose one of the ethical questions that can be raised about application of our new knowledge about the genome (e.g., see pp. 1518, 1519, Chapter 32, and study questions on this page). Study literature available to you and prepare a recommendation to the public, Congress, or to local regulatory agencies. Follow a scientific approach. Try to find true facts that can be verified. Consider all viewpoints. State some of the uncertainties in your recommendation. Present your proposal to a class or to a friend for criticism. Then publish your view in a newspaper if you wish to.

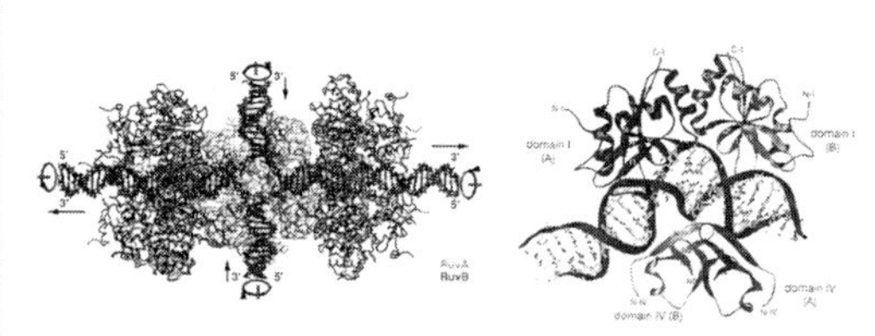

Many proteins interact with DNA. These include polymerases that replicate DNA, helicases that unwind double helices, topoisomerases that cut and reseal DNA strands to avoid entangling, and enzymes that repair damage to DNA. At left is a motor complex consisting of the tetrameric binding protein RuvA (light shading) and the hexameric helicase RuvB (two copies in darker shading). See also Fig. 27-26B. From Putnam *et al.* (2001) *J. Mol. Biol.*, **311**, 297–310. Right bacterial protein (from *Thermus aquaticus*) MutS recognizes mispaired bases in DNA and initiates their removal. The DNA is bound in a bent conformation. In this duplex one strand contains an unpaired thymine (at top of bend), which would cause a mutation if not removed. From Obmolova *et al.* (2000) *Nature*, **407**, 703–7.

Contents

27 Organization, Replication, Transposition, and Repair of DNA

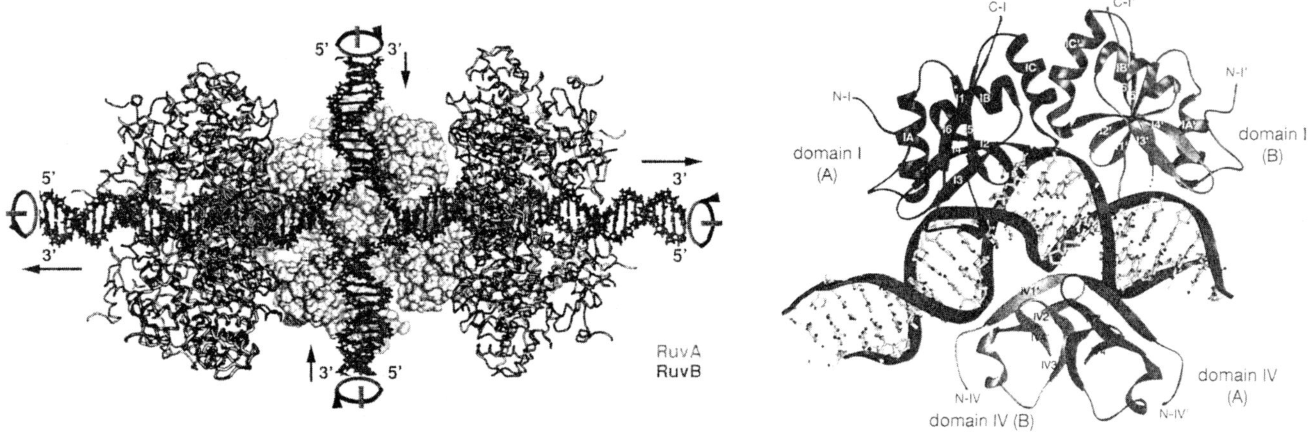

Both the replication and transcription of DNA are complex processes. Although the basic chemistry is relatively simple many enzymes and other proteins are required. In part this reflects organizational and topological problems[1] associated with the huge amount of DNA present as a single molecule within a chromosome.

A. The Topology and Environment of DNA

Although we can isolate DNA in the form of simple double helical fragments, the topology of natural DNA is always more complex. Covalently closed circular DNA such as that in plasmids, mitochondria, and bacterial chromosomes is supercoiled (Chapter 5) and bound to proteins. The DNA of chromosomes and bacteriophage particles is folded further into more compact forms. For example, the chromosome of *E. coli* contains DNA about 1.5 mm in length folded within a cell that is only 2 µm long. The diploid length of DNA in a 20 µm cell of a human is about 1.5 meters. At the time of cell division human DNA must all be replicated and packaged into chromosomes, 23 pairs in each cell. The density of the compacted DNA varies. A bacterial nucleoid may contain 10–30 mg / ml of DNA.[2] In chromatin of a eukaryotic nucleus there may be 200 mg / ml of DNA and in nucleosomes 330–400 mg / ml.[2a] The tightly compacted head of the T4 bacteriophage (Box 7-C) contains 520 mg / ml.[3]

1. DNA in Viruses

In the simplest filamentous DNA viruses such as M13 the DNA is coated by a helical protein sheath (Fig. 7-7), as it is extruded from a cell. The sheath is peeled off as the virus enters another cell. However, in the large tailed phage (Box 7-C), which contains ~160 kb of polynucleotide chains, the DNA is closely packed within the heads. In a model that seems to accommodate most experimental results, the DNA rod bends sharply into a series of folds, which are laid down around the long axis of the head in spirally arranged shells (Fig. 27-1).[4,4a] The end of the DNA that enters the phage head first appears to be located in the center with successive shells of DNA around it.[4] In the large bacteriophage G the DNA appears to be folded to form 12 icosahedrally arranged pear-shaped rings in the corners of the capsid.[5] The 2.0 nm diameter double helical segments of DNA lie roughly parallel and are separated by only 0.5–1.0 nm of solvent,[6] which contains cations such as the polycation of spermidine. Another possibility is that the DNA may be wound as on a spool of thread.[7] The DNA chains have an external diameter of ~2 nm with 0.6 nm additional for a hydration layer. In a phage head the adjacent parallel chains are 2.6–2.7 nm apart. Thus, the packing is very tight. Even so, capsids tend to be only about half filled with DNA. An exception is provided by the tobacco mosaic virus (Fig. 7-8) in which the RNA genome is held precisely by protein subunits, which dissociate to release the RNA during infection.[3]

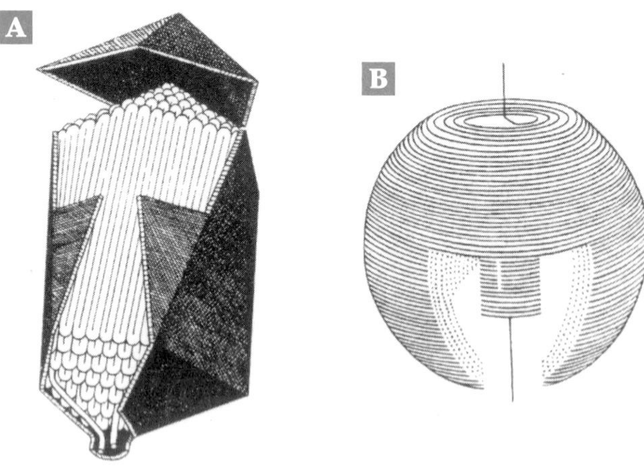

Figure 27-1 Possible ways of packing DNA into the heads of bacteriophage particles. (A) Spiral-fold. (B) Concentric shell model From Black *et al.*[4]

2. Bacterial Chromosomes and Plasmids

Most DNA in living organisms, whether bacteria or eukaryotes, is underwound. That is, the superhelix density (Chapter 5) is about –0.05 or one supercoil per 200 base pairs. In eukaryotes this negative supercoiling can be accounted for by the winding of DNA around the histones within the nucleosomes (Figs. 5-21, 27-4). The situation in bacteria is not as clear. There are many bacterial DNA-binding proteins,[8,9] but one of them known as HU is particularly abundant.[8–12] In *E. coli* it exists as mixed dimers and tetramers of 9.5 (α) and 9.2 (β) kDa subunits. Each HU tetramer can bind ~60 bp of DNA. There are about 60,000 HU monomers per cell, enough to coat ~20% of the genome. Possible modes of interaction with DNA have been proposed on the basis of the X-ray structure of HU.[8,13–14a] Binding to HU causes the DNA to be more tightly wound, introducing additional negative supercoils. The resulting unreleased torsional stress may be important in the functioning of the DNA.[15–17a] Binding is strongest to four-way junctions,[17a] and to DNA with nicks and gaps or to structures induced by supercoiling.[17b] Other basic histonelike proteins may also bind to the DNA. However, there are no structures that resemble eukaryotic nucleosomes.

If bacterial cells are lysed under certain conditions, e.g., in 1 M NaCl or in the presence of a "physiological" 5 mM spermidine, the entire bacterial chromosome can be isolated.[10] The DNA in these isolated chromosomes retains some torsional tension that, however, can be relaxed by nicking with nucleases or by γ-irradiation. However, a single nick relaxes the DNA very little. The explanation appears to be that the DNA is held by proteins of the nucleoid matrix in a series of loops (Fig. 27-2). A single nick relaxes just one loop. On this basis there are 43 ± 10 loops per genome with ~100 kb of DNA per loop.[18] A 136-residue protein **H-NS** is involved in condensation of bacterial DNA.[18a] It may act as a scaffolding protein, but it also functions in controlling transcription.[18b]

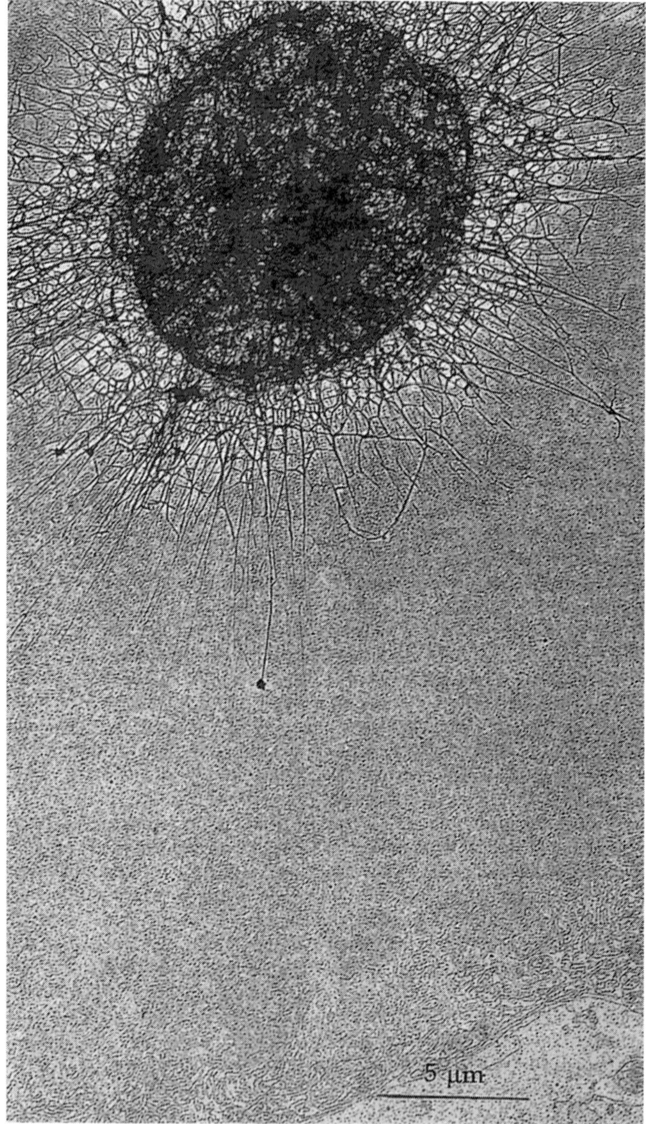

Figure 27-2 Electron micrograph of a bacterial nucleoid. The DNA is usually contained within the 'cage', but has been spread, using Kleinschmidt's procedure, to yield a surrounding skirt. The cage contains a protein network which includes elements of the cytoskeleton, enclosing the residual nuclear substructures. The denser fibrils radiating from this cage disappear to nuclease digestion and are probably aggregates of DNA fibers which merge with individual DNA strands at the extremities of the skirt. These strands are highly supercoiled, indicative of intact DNA. From Jackson and Patel, provided by Dr. S. J. McCready.[19]

3. Protamines, Histones, and Nucleosomes

Within bacterial cells the negatively charged phosphate groups of the DNA are neutralized to a large extent by the positively charged polyamines,[19a] by cations such as K^+ and Mg^{2+}, and by basic proteins such as HU. Within the mature heads of sperm cells of fish, the tightly packaged DNA is neutralized by the **protamines**, small ~5-kDa proteins rich in arginine.[4a,20–21a] Similar basic proteins are found in mammalian sperm.[22,23] However, within most eukaryotic cells, the charges on DNA are balanced principally by a group of basic proteins, first isolated and named **histone** by Kossel in 1884.[24]

There are five classes of histones, which range in molecular mass from ~11 to 21.5 kDa:[25,26]

H1 (including $H1^0$, H5; lysine-rich "linker" histone)
H2A, H2B (moderately lysine-rich) } "core" histones
H3, H4 (arginine-rich) }

All of the core histones share a conserved 65-residue **histone fold.**[27,28] The arginine-rich histones have a strongly conserved amino acid sequence, histone H4 from pea seedlings differing from that of the bovine thymus by only two amino acids. On the other hand, the lysine-rich H1 is almost species-specific in its sequence. Differentiated tissues contain at least seven variant forms of histone H1 including proteins designated $H1^0$, H1t, and H5.[29–31]

The N-terminal 25–40 amino acids of the core histones are positively charged and highly conserved.[32] The 135-residue histone H3 of calf thymus carries a net charge of +18 within the first 53 residues. This is probably the portion that binds to DNA. On the other hand, the carboxyl terminal end is hydrophobic and only slightly basic.[33] Histones undergo substantial amounts of **micromodification** including phosphorylation, acetylation, and methylation.[33a,b] Mono-, di-, and tri-methyllysine residues may be present.[33c,d] The core histones all undergo acetylation on specific lysyl side chains. Nuclear histone acetyltransferases[34–38b] transfer acetyl groups from acetylCoA and hydrolytic deacetylases may remove them.[39–40b] The amount of acetylation varies during different stages of the cell cycle, suggesting a regulatory role.[41] Acetylation sites in H3 and H4 are highly conserved in all eukaryotes.[42]

A small fraction of histone H2A undergoes phosphorylation and dephosphorylation continuously,[43] but H1 and H3 are phosphorylated and dephosphorylated at specific stages of the cell cycle. Phosphorylation of H1 has been thought essential for "condensation" of chromatin,[44,45] the folding into the tightly packed chromosome structures. However, more recent experiments point to the N terminus of histone H2B as the required site of phosphorylation for chromosome condensation.[45a] In addition, histone H1 may interact with membrane lipids.[46] Histone H2A exhibits the greatest heterogeneity and appears to function in regulation of transcription, of gene silencing, and of repair of double-strand breaks in DNA.[46a] Each of the histones appears to be regulated separately. In animal chromatin ~10% of histone H2A and small fractions of H2B and of tissue-specific histones are covalently linked to ubiquitin (Box 10-C). However, this monoubiquitination may not be related to proteolytic degradation.[47] Archea contain histones that dimerize and bind DNA to form nucleosomes.[47a,b]

Nucleosomes. An early idea of the function of histones was that they serve as gene repressors. To some extent this view is still valid. However, the large quantity of histone and uniform distribution over the DNA suggested some other role. This was clarified when electron micrographs showed that chromatin fibers form **nucleosomes,**[48–50] regular repeating structures resembling beads on a string. The same structure is seen in the "minichromosomes" formed from virus SV40 (Fig. 27-3).[51–53] Two molecules each of histones H2A, H2B, H3, and H4 form the core of the nucleosome around which ~146 bp of dsDNA is coiled into approximately two negative, left-handed toroidal superhelical turns (Figs. 5-21, 27-4).

Digestion of chromatin by nucleases causes rapid cleavage into ~200-bp fragments and slower cleavage to 146 ± 20-bp fragments. This suggested that ~200-bp segments of DNA are folded around a histone octomer, contracting the 68 nm extended length of relaxed B-DNA into a 10-nm nucleosome. A short linker region of variable length, up to 80 bp, lies between the nucleosomes.[54] The fifth histone, H1 (or H5 in some species), may bind to this linker DNA. A nucleosome with bound H1 is sometimes called a **chromatosome**.[55]

The superhelix density of ~0.05 observed for DNA extracted from eukaryotic cells is just equal to one negative superhelix turn per nucleosome. For example, the number of nucleosomes seen in the minichromosome of Fig. 27-3 matches the numbers of supercoils in the SV40 DNA (Fig. 5-20). If there are two negative supercoils per nucleosome, as shown in Fig. 27-4, the DNA in the nucleosome must be wound more tightly than in relaxed DNA (10.0 bp per turn instead of the 10.6 of relaxed DNA).[56] NMR data suggest that within the nucleosome the regular base pairing in the DNA may be partially disrupted and that some parts of the histone have a high degree of mobility.[57]

Although nucleosomes are distributed rather evenly along the DNA of a cell, there are some DNA sequences that favor nucleosome formation. The resulting **positioned nucleosomes** are often found in the vicinity of gene promoters, enhancers, and other

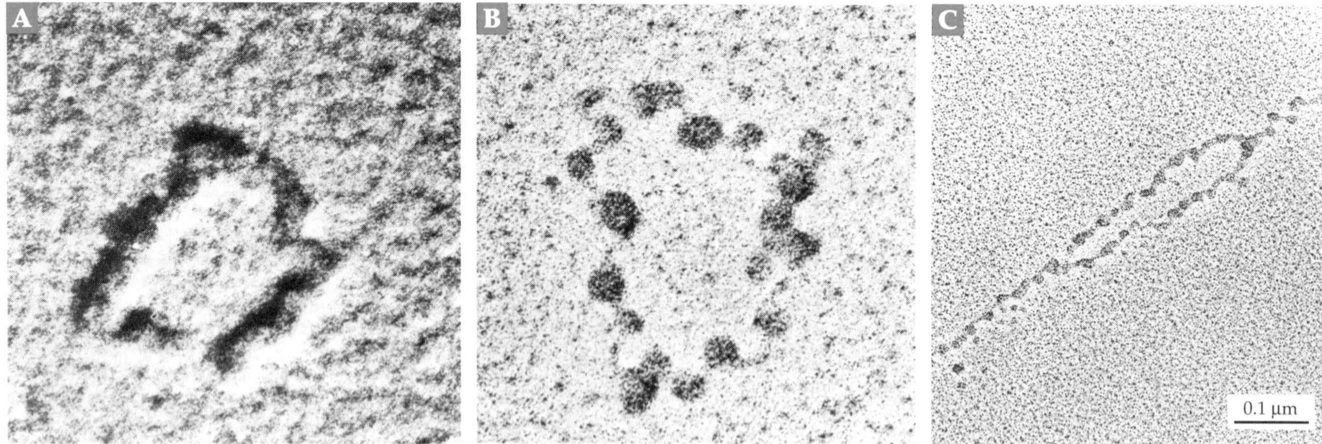

Figure 27-3 (A) Electron micrograph of "minichromosome" formed from virus SV40 growing in monkey cells in culture.[51] In this native form the nucleoprotein fiber is ~11 nm in diameter and ~210 nm in length. (B) Beaded form of minichromosome observed when the ionic strength was lowered. The 20 beads have diameters of ~11 nm and are joined by bridges roughly 2 nm in diameter and 13 nm long. Deproteinization and relaxation of the DNA revealed that the overall length of the DNA present is seven times the length of the native minichromosome. (C) Electron micrograph of chromatid of a blastoderm-stage embryo of *Drosophila melanogaster* in the process of replication. Nucleosomal particles are visible immediately adjacent to the replication forks. Courtesy of Steven L. McKnight and Oscar L. Miller, Jr.

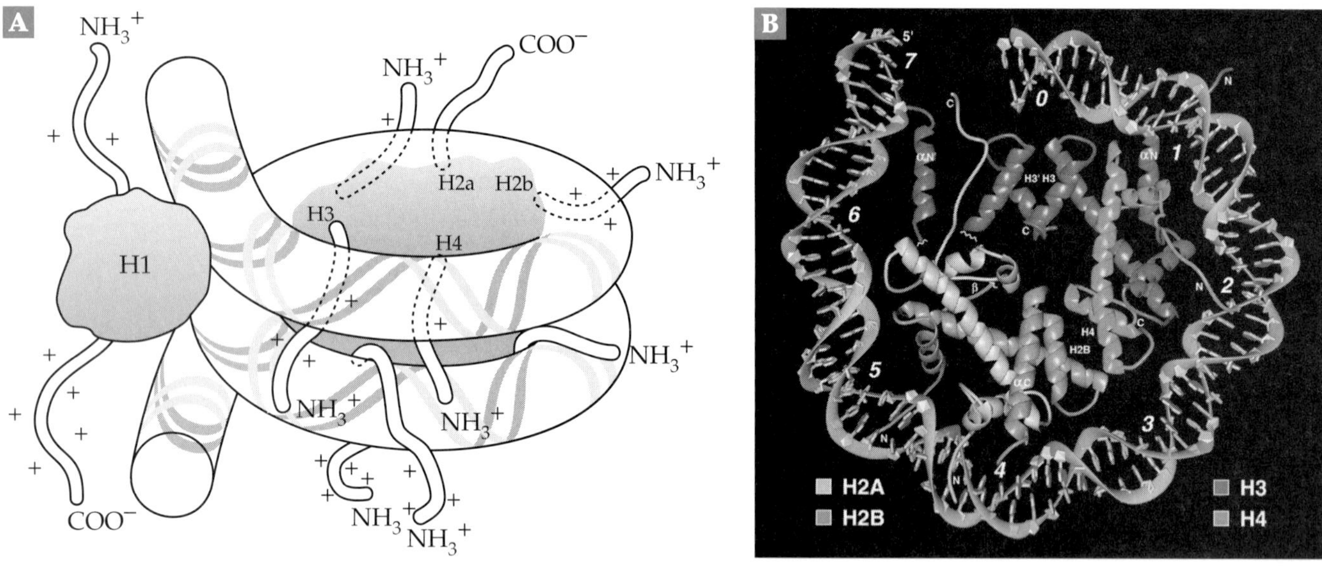

Figure 27-4 (A) A nucleosome is formed when dsDNA, shown schematically as a tube, wraps roughly twice around a histone octomer. This complex, or nucleosomal core particle, includes at its center two copies of histone H3, two of histone H4, and two H2A–H2B dimeric pairs, one of which is not visible. Ends of each histone molecule are thought to protrude like tails from the core, ready to interact with other molecules. In many organisms, histone H1, portrayed at left in one possible position, helps to anchor DNA to the core and promotes further compaction of the DNA into a 30-nanometer fiber. See Grunstein.[49] (B) Nucleosome core particle: 73-bp half. The view is down the superhelix axis with the pseudo dyad axis aligned vertically. The central base pair through which the dyad passes is above the superhelix axis location labeled 0. Each additional numerical label 1–7 represents one further DNA double helix turn. The complete histone proteins (except for the tail regions) that are primarily associated with the 73-bp superhelix half are shown. The two copies of each histone pair are distinguished as unprimed and primed, where the histone fold of the unprimed copy is primarily associated with the 73-bp DNA half and the primed copy with the 72-bp half. Four-helix bundles are labeled as H3', H3, and H2B, H4; histone-fold extensions of H3 and H2B are labeled as αN', αN, and αC, respectively; the interface between the H2A docking domain and the H4 C terminus as β; and N- and C-terminal tail regions as N or C. From Luger *et al*.[50]

control sequences.[58–61a] It seems likely that positioning of nucleosomes is related to control of transcription or of other activities. The same basic structure for chromatin has been found in animals, fungi, and green plants.

The linker histones. A nucleosome is pictured in Fig. 27-4A as if held in a compact configuration by the binding of histone H1 at a position that marks a pseudo twofold axis that lies in the plane of the nucleosome. However, this is only one of several possible locations for linker histones of the H1, H5 family.[62–67] The structure of the linker histones is somewhat different from that of core histones. They have an 80-residue globular domain with long N-terminal *and* C-terminal chains, both of which are rich in basic residues and evidently available for binding to DNA.[68] Perhaps they bind both to the DNA entering the nucleosome and to that leaving the nucleosome, reducing electrostatic repulsion of those two parts of the DNA superhelix.[69]

Another possible location for histone H1 or H5 is above the histone surface as shown in Fig. 27-4 and *inside* the DNA loop.[65] A third suggested location for the globular linker core is *between* the two turns of the DNA strand.[66] While one function of the linker histones may be to stabilize mononucleosomes, they may also play a role in compaction of the DNA into the 30-nm fibers universally seen in nuclei of cells.[62,64–64b]

Histone H1 can also be regarded as a general repressor, holding chromatin tightly folded and preventing transcription. The possible roles of acetylation, phosphorylation, methylation, ubiquitination, and other modifications of histones in controlling transcription, replication, and DNA repair are receiving increasing attention.[70–73] **Active chromatin**, where transcription is occurring, has an altered nucleosome structure and increased susceptibility to nuclease action. It appears to be less tightly packed than inactive chromatin and to contain regions called **hypersensitive sites** that are accessible to nucleases or chemical modification reagents.[74–76] There seems to be a direct link between increased acetylation of histones and enhanced initiation of transcription by RNA polymerase II.[35,38,77–82] Conversely, deacetylation is associated with repression of transcription. Both histone acetylase and deacetylase activities have been found in transcriptional regulators.[80–82]

The observation that H1 becomes phosphorylated during the initiation step of mitosis suggests another control mechanism for its repressor functions.[83] Several multiprotein complexes that "remodel" chromatin have been identified.[73,84] These complexes contain

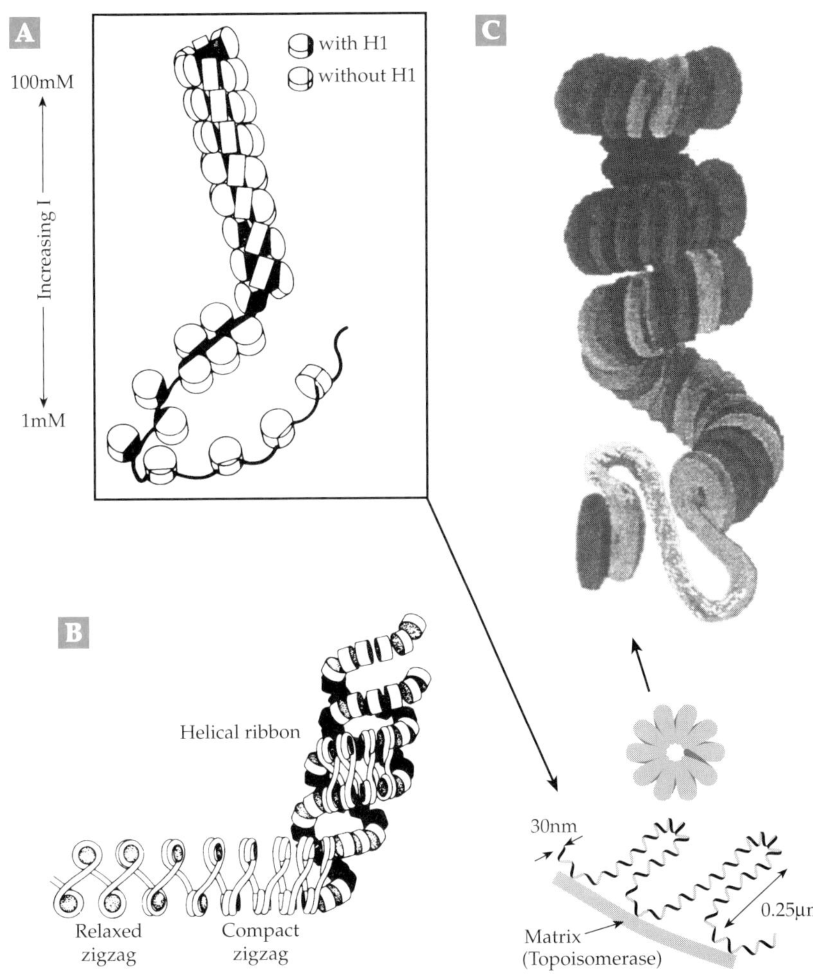

Figure 27-5 (A, B) Two possible models of the 30-nm chromatin fiber.[55] (A) Thoma *et al.*[85] (B) Woodcock *et al.*[64,87] The fully compacted structure is seen at the top of each figure. The bottom parts of the figures illustrate proposed intermediate steps in the ionic strength-induced compaction. (C) Possible organization of the DNA within a metaphase chromosome. Six nucleosomes form each turn of a solenoid in the 30-nm filament as in (A). The 30-nm filament forms ~30 kb-loop domains of DNA and some of these attach at the base to the nuclear matrix that contains topoisomerase II. About ten of the loops form a helical radial array of 250-nm diameter around the core of the chromosome. Further winding of this helix into a tight coil ~700 nm in diameter, as at the top in (C), forms a metaphase chromatid. From Manuelidis[91].

ATP-dependent **helicase** activities that open up DNA for transcription (see Chapter 28, Section B).

Folding of nucleosome chains; chromosomes. Electron microscopy shows that chromatin is packed into the nucleus largely as fibers of ~30- to 36-nm diameter.[55,84a] Thoma *et al.*[85] proposed that the fibers could be formed by winding the string of nucleosomes into a simple one-start helix with six nucleosomes per turn and a pitch of ~11 nm (Fig. 27-5A). The H1 molecules would be close together in the center. An alternative model, one of several suggested,[55,86] is shown in Fig. 27-5B. Its zigzag pattern of adjacent nucleosomes would generate a two-start helix.[64,87,88] Another alternative model envisions larger solenoids with interdigitated nucleosomes.[89]

A somewhat similar structure appears to be present in the specialized eukaryotic **lampbrush chromosomes** (Fig. 27-6), which are observed during the meiotic prophase of oocytes. They have been studied intensively in amphibians such as *Xenopus*. A lampbrush chromosome is actually a homologous pair of chromosomes, each one in turn consisting of two closely associated chromatids. The chromosomes are highly expanded, and about 5% of the DNA is extended in the form of ~4000 perfectly paired loops visible with an electron microscope. Each loop consists of ~50 μm or ~150 kb of extended DNA. No evidence of any breaks in the DNA is seen, a fact that supports the belief that a single DNA molecule extends from one end of the chromosome to the other through all of the loops.

Like the puffs of polytene chromosomes (Chapter 28), which may have a similar structure, lampbrush chromosomes appear to be actively engaged in transcription. Approximately 3% of the DNA may be functional in producing mRNA that is accumulated within the oocyte and is used as a template for protein synthesis during early embryonic development.[90]

A different arrangement is present in metaphase chromosomes, which appear as two dense parallel sister chromatids of ~700-nm diameter.[91] The DNA must be highly folded. In the model shown in Fig. 27-5C the 30-nm fiber is folded into ~30-kb loops, each one formed from ~25 turns of the 30-nm helix. The loops then form a helical array 250 nm in diameter with ~ ten loops (300 kb) per turn. This helix is further wound into a tight helix of ~700-nm diameter. A single turn of this helix may contain as much as 9 Mb of DNA.[91] About four hundred coils (an average of 18 coils per human haploid chromosome) could accommodate the entire genome. A group of five proteins, some of which are designated **SMC** (structural maintenance of chromosomes) proteins, form a complex called **cohesin**. SMC proteins, large multidomain proteins found in all eukaryotes, are also present in bacteria.[91a,b]

Interphase chromatin must be much less tightly

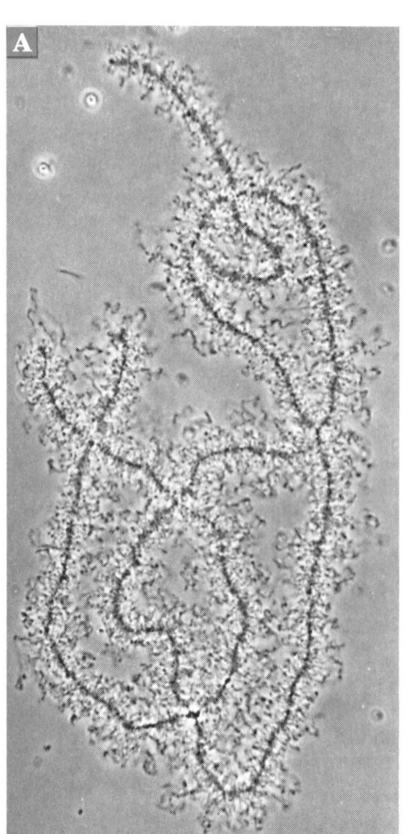

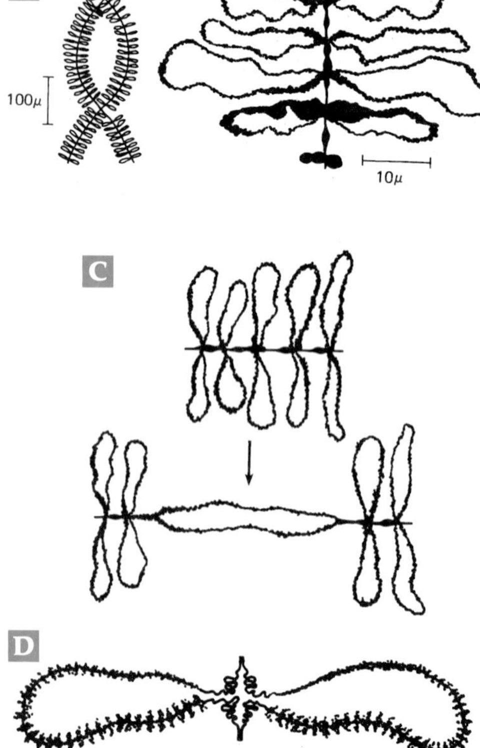

Figure 27-6 (A) Photomicrograph of a lampbrush chromosome from the nucleus of an oocyte of the newt *Triturus*. From L. M. Mays, *Genetics, A Molecular Approach*, Macmillan, New York, 1981, p. 227. (B–D) Diagrammatic views of lampbrush chromosomes. (B) The two homologous chromosomes (left) are held together by two chiasmata. A portion of the central chromosome axis (right) shows that two loops with identical morphology emerge at a given point, evidence that each chromosome has already split into two chromatids. (C) Accidental stretching of a chromosome reveals the continuity of the loop axis with the central axis. (D) A single loop pair, showing the single DNA molecules on which RNA chains (indicated by fuzzy shading) are being transcribed. From J. Gall, *Brookhaven Symp. Biol.* **8**, 17 (1955).

packed and contains regions in which large loops, e.g., of 20–120-kb size, are uncoiled enough to allow transcription factors and other proteins to locate their target sequences.[91c] Many models of interphase chromatin have been proposed.[92,93]

4. The Cell Nucleus

It has been clear for many years that the nucleus has a well-organized structure. However, techniques such as fluorescent labeling have only recently been used to provide important details. These studies show that individual interphase chromosomes occupy discrete territories within the nucleus.[94–96] Parts of the chromosomes may be unfolded and active in transcription, while others are more tightly coiled. Some parts, known as **heterochromatin**,[97–98a] are very tightly coiled and metabolically almost inert. These regions include the highly repetitive DNA of telomeres and centromeres as well as other regions and complete inactivated female X chromosomes. Replication, transcription, and RNA splicing complexes are found at distinct locations within cells.[99–102a] They are apparently fixed, perhaps attached to the inner nuclear membrane, while the DNA passes through the complexes.

The nuclear matrix. The lipid bilayers, the histones and other soluble proteins, and the DNA can all be removed from nuclei by extraction and enzymatic digestion. An insoluble residue, the **nuclear matrix,** is left.[103–107] Largely protein in nature, this matrix is spread throughout the nucleus. Remnants of the membranes remain in the form of proteins that were in or along the bilayer. The nucleolus is clearly defined. The DNA appears to be bound to the nuclear matrix proteins. A specific 320-kb piece of a *Drosophila* chromosome has been mapped and used to locate nontranscribed scaffold (or matrix) attachment regions of DNA bound to matrix proteins. These were found at intervals of 26–112 kb, the intervening loops containing up to five or more genes.[105,106,108,109] A 120-kDa protein together with topoisomerase II (Section C,2)[102,104] may be components of a **nuclear scaffold** that constrains the loops of DNA. The scaffold may also provide locations for the complexes of proteins involved in replication and other processes.[103] The **matrix attachment regions** (MARs) may also act as **insulators** that shield promoters for transcription within certain loops from control elements such as **enhancers** that may be present in adjacent loops.[91b,110] At least one nuclear matrix component becomes phosphorylated and moves to the nuclear poles during mitosis.[111]

Other nonhistone nuclear proteins. Polyacrylamide gel electrophoresis revealed more than 450 components in HeLa cell nuclei. Most are present in small amounts of <10,000 molecules per cell and are not detectable in cytoplasm.[112] Among the more acidic proteins are many enzymes including RNA polymerases. There are also gene repressors, hormone-binding proteins, protein kinases, and topoisomerases.[113] Among the six most abundant nonhistone nuclear proteins in the rat are the cytoskeletal proteins myosin, actin, tubulin, and tropomyosin.[114]

A group of small (<30 kDa) proteins, the **high mobility group (HMG)** proteins,[112,115] can be extracted from chromatin with 0.35 M NaCl. Two pairs, HMG-1 + HMG-2 R (renamed **HMGB**)[115a,b] and HMG-14 + HMG-17 (renamed **HMGN**),[115b,c] are present in nuclei of all mammals and birds. HMG-14

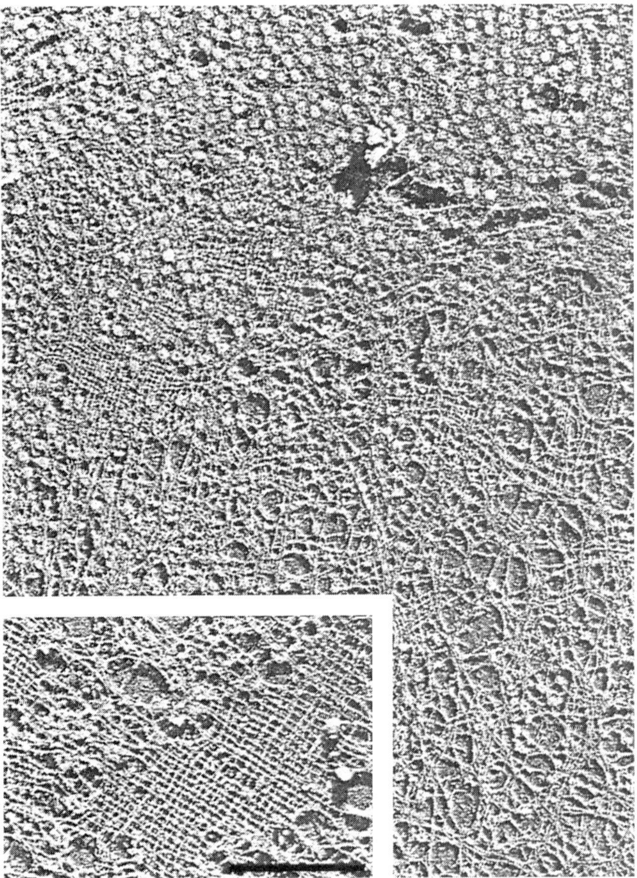

Figure 27-7 Native nuclear lamina of *Xenopus* oocytes. Freeze-dried metal-shadowed nuclear envelope extracted with Triton X-100, revealing the nuclear lamina meshwork partially covered with arrays of nuclear pore complexes. Inset, relatively well-preserved area of the meshwork of nearly orthogonal filaments from which pore complexes have been mechanically removed. Bar, 1 µm. From Aebi *et al.*[121]

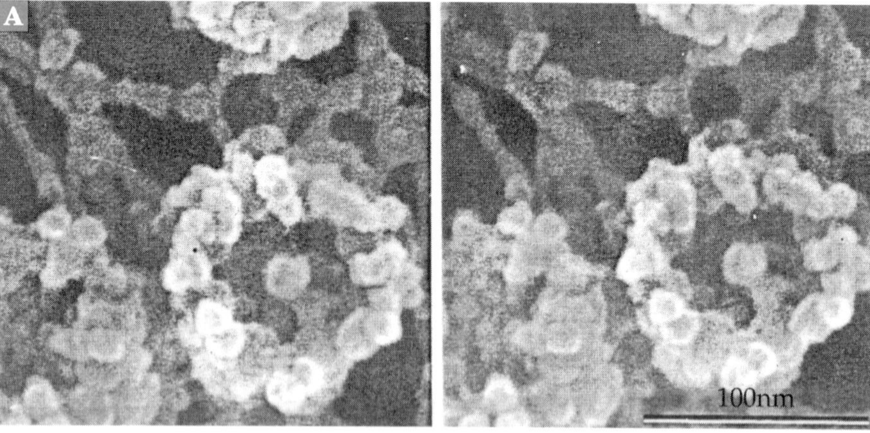

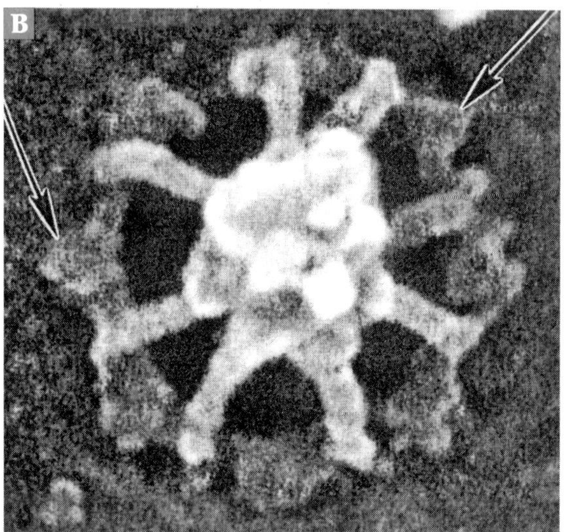

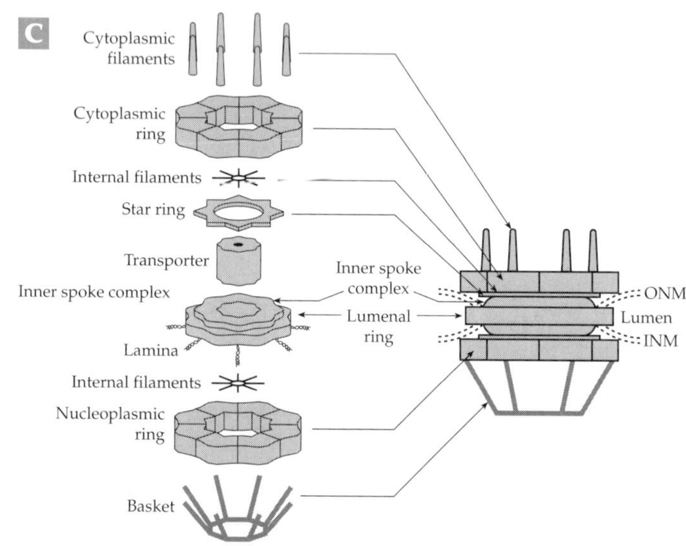

Figure 27-8 Electron micrographs of nuclear pores from a *Xenopus* oocyte. (A) Cytoplasmic face of a detergent-extracted membrane. Stereo view showing structure of a single pore. Lamin fibers are seen in the background. (B) Nucleoplasmic face of a trypsin-treated sample. The trypsin-resistant basket is clearly visible. Arrows point to the triangular pieces of the nucleoplasmic ring. (C) Diagram illustrating various structural features of the pores and the way in which they may fit together. From Goldberg and Allen.[120] Courtesy of Martin W. Goldberg.

and HMG-17 appear to be concentrated where active transcription is occurring. HMG-17 undergoes a variety of modifications including acetylation, methylation, phosphorylation, ubiquitination, glycosylation, and ADP-ribosylation, suggesting that it may assist in regulation of transcription. Other members of the HMG family bind more specifically to certain DNA sequences or structures.[115d,e]

Nuclear membrane, pores, and lamina. The nuclear membrane consists of two bilayer membranes separated by a 40- to 60-nm perinuclear space.[116,117] Both of the membranes and the intervening space are penetrated by large proteinaceous **nuclear pores**.[118–120b] The two membranes are fused together around the pores. On the inside of the innermost nuclear membrane is a relatively insoluble meshwork of intermediate filaments, the **nuclear lamina** (Fig. 27-7). It acts as a scaffold for the pores and may also interact with chromatin.[113,121–123a] Three proteins, the **lamins** A, B, and C, are the major components of the nuclear lamina. The A and C lamins, which differ only at their C termini, are homologous to cytoplasmic intermediate filament proteins and may exist as a network of coiled-coil polymers.[124] Lamins are phosphorylated, and the lamina is disassembled during the prophase of mitosis, apparently under the influence of the cdc2–cyclin B complex (Eq. 26-3),[125] and reappears in telophase at the end of mitosis.

Nuclear pores consist of octameric rings of protein subunits with a complex structure and an outside diameter of 120 nm and an inside diameter of ~80 nm (Fig. 27-8). A pore may consist of as many as 80 to 100 different proteins and have a mass of ~125 MDa.[119,126,126a,b] The pores are sometimes seen to be blocked by ~35 nm granules, perhaps pre-ribosomes. Transport through nuclear pores occurs in both directions. Numerous proteins enter the nucleus where many bind to RNA and are then exported as ribosomal subunits. Transfer RNAs and messenger RNAs must also be exported. A family of nuclear transporters known as **importins** and **exportins** mediate the movement of protein and RNA through the pores.[119,126b,127–130] They depend upon the G protein **Ran**, which also functions in spindle formation (p. 1503), and the hydrolysis of GTP.[131–132b] Large conformational changes may be linked to the transport cycle[133,134] and also have been observed in response to changes in the Ca^{2+} level.[135,136]

The nucleolus. This organelle is a dynamic structure, which breaks down and reforms in each mitotic cycle. It is organized around clusters of genes for the 28S, 18S, and 5.8S ribosomal RNA subunits. Both chromatin and ribonucleoproteins that transport the rRNA out to the cytoplasm are present.[95,137] The major nucleolar protein **nucleolin** is characterized by its glycine-rich C terminus, which contains seven repeats of GGRGG and also contains N^{γ},N^{γ}-dimethylarginine.[138,139] Nucleolin may control transcription of the DNA that carries the rRNA genes. It appears to be required for ribosome synthesis (Chapter 29) and for attachment to the nuclear matrix. A variety of other proteins in smaller amounts are also needed in nucleoli. In addition to the true nucleoli other regions described as **speckles** and **coiled bodies** may develop around RNA splicing centers.[95,139a]

B. Organization of DNA

A vast amount of new information on DNA sequences and structures is available for yeast, *Arabidopsis*,[140] *Drosophila*, the nematode *Caenorhabditis elegans*,[141] human beings, and other species (see Chapter 26). Nevertheless, it may be worthwhile to consider older discoveries, some dating back 20 years or more.

When DNA is cut into ~10-kb fragments by shearing and the fragments are denatured by heat, the renaturation of the resulting single-stranded fragments upon cooling takes place in two or more steps. Some material reforms double helices rapidly, whereas other material is slow to renature (Fig. 5-46). At least four kinetically distinct fractions have been recognized: (1) About 70% of mammalian DNA appears to exist largely as single copies, i.e., with unique sequences. (2) About 20% of the total DNA is moderately repetitious, containing sequences that may be present 10^3–10^5 times. (3) About 5% reassociates very rapidly and is identified as highly repetitive **satellite DNA** of which there may be ~10^6 copies. (4) A smaller fraction contains long palindromic sequences. As a consequence, the single-stranded pieces can fold back almost instantaneously to form hairpin structures.[142,143] These differences also show up in density gradient centrifugation in CsCl, whether on large pieces of DNA[144] in the presence of ligands such as Ag^+ or on smaller restriction fragments which have often been studied by polyacrylamide gel electrophoresis.

If there are ~35,000 human genes, protein or RNA coding sequences must occupy only 2–3% of the genome.[145] The pufferfish *Fuga rubripes* probably has almost as many genes as we but only ~13% as much DNA. In further contrast the newt *Triturus cristatus* has six times as much DNA as a human.[146] The compact genome of the green plant *Arabidopsis thaliana* occupies only 120 Mb. In contrast are the 415-, 2500-, and 5300-Mb genomes of rice, maize, and barley, respectively.[147] What is the function of all of the apparently noncoding DNA present in some organisms? It is often viewed as "junk," whose only function may be to facilitate evolutionary changes in the genome. However, there is doubtless important undiscovered information in these regions.[145] It has been very hard to determine sequences, in part because of the large amount of highly repetitive DNA. Some regions have been "unclonable" in prokaryotic systems because of the presence of transposon-like sequences or "kinkable" elements (TG•CA steps), palindromes, etc.[145a,b]

1. Repetitive DNA

Rapidly renaturing DNA fragments often have a different base composition than the bulk of the DNA and, consequently, often separate as small satellite bands upon centrifugation in a CsCl gradient. Satellite DNA is usually associated with regions of the chromosome that do not unravel in telophase as does the bulk of the DNA. Satellite DNA usually consists of short highly repetitive sequences,[148] which occur in large clusters of up to 100 Mb of DNA, often near centromeres or telomeres or on a Y chromosome.[146] The DNA of a satellite band from the kangaroo rat contains the sequence 5'-GGACACAGCG-3' repeated so often that it accounts for 11% of the entire DNA of the cell. Longer repeating sequences ~170 bp are also often present as are **microsatellites** of 2- to 5-base-pair repeats. At least 30,000 microsatellite loci are present in the human genome.[149]

Centromeres. The attachment of spindle fibers to chromosomes depends upon the segments of DNA known as centromeres to organize the attached kinetochores. In the yeast *Saccharomyces cerevisiae* a 120-bp region containing three short conserved sequences is present in the centrosomes of all ten chromosomes.[150,151] This may fold into a distinctive looped structure.[152] Human centrosomes are large and complex,[153] but the DNA is highly repetitive, giving rise to **α-satellite DNA**.[154–157b] Sequences such as $(TGGAA)_n$ are repeated many times. Such sequences can form self-complementary looped structures containing some unpaired guanines that intercalate and stack between sheared G•A pairs.[158,158a] Complex regional centromeres involving kilobases or megabases of DNA have also been identified in fission yeasts, *Drosophila*, and green plants.[156,157] The great variability indicates that centromeric sequences undergo rapid evolution. This may be related to the fact that of the four cells formed by female meiosis only one becomes an egg.[158b] A series of unique **centromeric proteins** (CENP-A to CENP-G) bind to the DNA sequences of centromeres[159–162a] and direct the formation of the kinetochores. Even for the simpler centromere of budding yeasts, kinetochores have a complex structure.[150] The CENP proteins were first identified as autoantigens in sera of patients with the autoimmune disorder **scleroderma** (Chapter 31).[160,162]

Telomeres. The DNA sequences at the chromosome ends have a TG-rich strand, such as the $(TTGGGG)_{50-70}$ (*Tetrahymena*)[163] and the $(TTAGGG)_n$ of both human and trypanosome chromosomes.[164–166] The complementary DNA strand is CA-rich. The *S. cerevisiae* telomers have ~350 base pairs containing the sequences (TG_{1-3}/$C_{1-3}A$) as well as one or more copies of a 6.7-kb nonrepetitive sequence and other elements.[166,167] In many species the repetitive telomeric sequences have 3' poly(G) tails at the ends of the DNA molecules. These tails are able to form G quartet structures (Fig. 5-26 and Chapter 5, Section C,4). A variety of telomere-binding proteins have been isolated.[167a–d] Some of these bind to G quartet structures[168–169a] and some, such as **RAP1** of yeast,[170–171a] to double-stranded telomere repeat regions. Special problems associated with replication of telomeres are discussed in Section C,8.

Short interspersed sequences (SINES). Much of the reiterated DNA is present in repeated segments 100–500 bp in length that lie between 1- to 2-kb segments of unique DNA.[172–174] The best known example is the human ***Alu*** family,[175–177b] so-called because it contains a site for cleavage by restriction enzyme *Alu*I. Sequences of the *Alu* family also exists in other primates and in rodents. The ~300-bp *Alu* sequence is reiterated over 500,000 times in the human genome with various sequence alterations, but an 80–90% homology. This sequence (Fig. 27-9) consists of two similar ~130-bp segments called the "left monomer" and "right monomer." The right monomer contains a 31-bp insertion and the left end carries a poly(dA) sequence. In addition there is a short 7- to 20-bp sequence, which is variable between different *Alu* sequences but is directly repeated at each end of a given *Alu* sequence.

The *Alu* sequence has strong homology with the 7S RNA (Table 5-4) that is part of the signal recognition particle involved in transport of newly synthesized peptide chains across the membranes of the ER (Chapter 10). *Alu* sequences are transcribed into hnRNA, the precursors to mRNA. Some *Alu* sequences are present in intervening sequences (introns) within genes and others are in noncoding sequences between genes. Sharp[176] suggested that specific proteins in the nucleus may bind to the *Alu* sequences, preventing hnRNA from leaving the nucleus before it has been processed to remove introns and other sequences absent in mRNA. However, the presence of the poly(dA) regions and the direct terminal repeats suggests that the whole *Alu* sequence is **pseudogene** derived from a **retroposon**, a type of transposon that originated from an RNA molecule. *Alu* and other SINES contain an RNA polymerase III promoter and are transcribed. Active retroposons form reverse transcripts (cDNAs) that can be integrated at various points in the genome (see Section D,4 and Chapter 28). One theory is that retroposons have no biological function but have invaded the genome at random locations.

Many SINES and other families of repetitive sequences have been characterized by the presence of a restriction enzyme cleavage site in each copy of the sequence. *Eco*RI cuts the previously mentioned α-satellite DNA.[177] A 319-bp reiterated sequence in the human genome surrounds a *Hinf* site,[178] and *Sau*3A cleaves a 849-bp sequence with ~1000 copies per haploid genome.[179] In the hermit crab 30% of the genome consists of repeated sequences, one 156-bp unit occuring ~7 million times. Many identical 14-bp GC-rich inverted repeats are present.[180] *Neurospora* and yeast both contain many copies of the GC-rich palindromic sequence 5'-CC<u>CTGCAG</u>TA<u>CTGCAG</u>GG-3', which contains the two underlined *Pst*I sites.[181] Some repetitive sequences such as the **CAT** family and the **homeodomain sequence** (Chapter 28, Section C,6) are found in the control regions preceding the 5' ends of groups of genes that are regulated coordinately.[182] Some repetitive sequences seem to be unstable in the genome and may be excised and lost or may increase in number during aging.[183] Repeated DNA

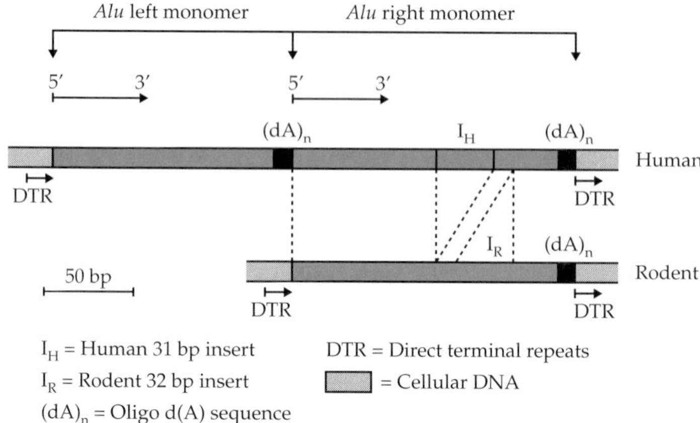

Figure 27-9 Structure of human and rodent *Alu* sequences in DNA. From Ullu.[175]

sequences have apparently originated not only with 7S RNA but also from mRNA or tRNA molecules.[172] A characteristic SINE in rodent DNA is known as the ID (identifier) sequence because it was once thought to be a marker for genes transcribed in neural tissues. Rat DNA contains ~130,000 copies of the ID sequence.[172a] Other **clustered repeats** are frequently found between genes that are present in large numbers. These include the genes for ribosomal RNA, tRNA, small nuclear RNAs, and histones. Triplet repeats are considered in Chapter 26, Section G,3.

Long interspersed repeat sequences (LINES). These moderately repetitive sequences may be several kb in length. Just one type seems to be abundant in each mammalian species.[122] The human L1 sequence is over 6 kb in length. Like the *Alu* sequence it has a poly(dA) sequence at the 3'-end, but it does not contain short terminal direct repeats. These interspersed repeat sequences usually contain genes which may be functional, but many of the copies contain pseudogenes or genes that are randomly truncated, having lost a segment from the 5' end. These sequences, too, seem to have been dispersed throughout the genome by retroposons.[122,184,185] Human chromosome 22, sequenced in 1999, contains within its 33.4 Mb of euchromatin at least 545 genes, 134 pseudogenes, and 8043 L1 sequences (9.7% of the DNA), as well as 20,188 *Alu* sequences (16.8% of the DNA) and many other interspersed repeats.[186] Repetitive DNA sequences are less common in prokaryotes, but they do exist. For example, many dispersed and clustered repeating units are present in *Halobacterium* DNA.[187]

2. Genes for Ribosomal RNA and Small RNA Molecules

Most genes are present as one copy each per haploid genome. However, there are many copies of the genes for ribosomal RNA and tRNA. In *Xenopus* DNA there are ~450 repeats of the 28 S and 18 S rRNA genes on one chromosome and ~24,000 copies of the 5 S RNA genes at the ends of the long arms of most of the chromosomes.[188] Nontranscribed spacer regions lie between the repeats of the 28S and 18S gene pairs, as can be seen with the electron microscope in Fig. 28-17. The gene for 5 S RNA has a high GC content. A denaturation map of the DNA shows easily melted regions separated by shorter 120-bp sequences, apparently of high GC content and presumably coding of the 5 S RNA. The easily denatured AT-rich spacers are ~630 bases long. Using restriction enzymes much of this DNA can be cut into segments that contain repeats within repeats. One 15-unit polynucleotide contains the sequence $A_4CUCA_3CU_3G$ repeated about 30 times.[189]

About 200 copies of rRNA genes per haploid genome are located at the constrictions in the short arms of human chromosomes 13, 14, 15, 21, and 22 (Chapter 26, banner). As in *Xenopus*, clusters of tandemly repeated 5S RNA genes are found at the ends of the long arms of most chromosomes. A similar organization of rRNA genes is found in the rat.[190] The ten different chromosome ends carrying rRNA genes in the human diploid nucleus come together to form the nucleolus, the site of synthesis of ribosomal subunits.[191] At first there are ten small nucleoli, but these fuse to form the single highly structured but membraneless nucleolus. The rRNA genes of the macronucleus of *Tetrahymena* have been "amplified" and are found on linear 21-kb palindromic molecules, about 10^4 copies being present per cell. Initiation of transcription begins near the center and proceeds outward in both directions.[192] These short chromosomes contain typical telomeric ends.

Genes for the tRNAs are spread throughout the genome of bacteria, mitochondria (Fig. 18-3), chloroplasts,[193] and eukaryotic nuclei. They sometimes occur in clusters but more often are far apart. In *Drosophila* there are probably at least 600 tRNA genes, many occurring in pairs of opposite polarity, i.e., as inverted repeat sequences. The genes for small nuclear RNAs U1–U6 (Chapter 28) are organized in a variety of ways. The ~100 human U1 genes occur on a single chromosome, perhaps organized in a tandem array and interspersed with as many as 10,000 defective pseudogenes.[194]

3. Other Gene Clusters and Pseudogenes

Closely related structural genes often occur in clusters.[195] Among these are clusters of **immunoglobulin** genes (Fig. 31-17) and clustered genes for the α and β **globins** (Fig. 27-10), which encode the protein sequences for the hemoglobins. The human α globin gene cluster occupies about 30 kb on chromosome 16 and the β globin genes 60 kb on chromosome 11.[196–197d] The α1 and α2 genes (Fig. 27-10) encode identical peptides, while the related ζ gene encodes the corresponding subunit of embryonic hemoglobin. The β cluster includes, in addition to the adult gene, a pair of fetal globin genes (γ^G and γ^A) differing by only one amino acid (Gly vs Ala) at position 136, the embryonic ε chain, and the minor adult δ chain.

Besides the functional genes the globin cluster contains **pseudogenes**, which are given the prefix ψ in Fig. 27-10. These are nonfunctional genes, which appear to encode peptides homologous to the known globins. However, they contain mutations that prevent expression. For example, deletion of a single nucleotide near the beginning of the pseudogene will scramble the genetic message by changing the reading

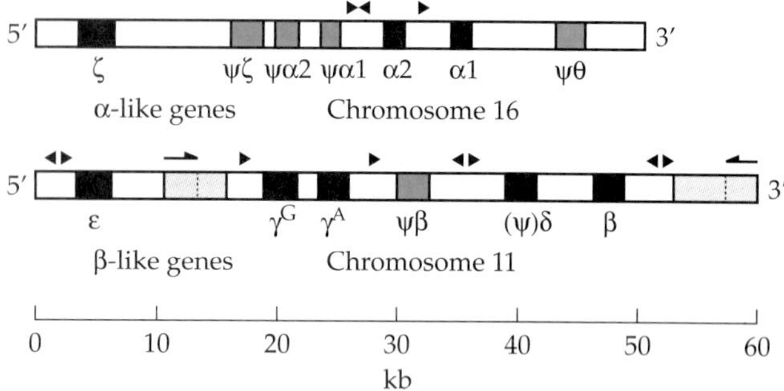

Figure 27-10 Organization of the globin genes on human chromosomes 11 and 16. The composition of the various embryonic, fetal, and adult hemoglobins is also indicated. Closed boxes indicate active genes and open boxes pseudogenes. The triangles (▸) indicate *Alu* repetitive sequences and their orientation. The shaded boxes indicate Kpn repeat sequences and the half-arrows their respective orientation. The Kpn sequence between the ε and γ^G genes in fact consists of two tandemly linked Kpn repeats. From Karlsson and Nienhuis[196] and Proudfoot.[197]

frame. Globin pseudogene ζψ has a nonsense mutation at codon 6, ψα1 contains a whole set of mutations that prevent transcription and translation, while ψα2 is mutated almost to the point of being unrecognizable as a globin relative.[198]

Do pseudogenes have a physiological function? The presence of pseudogenes in close proximity to functional genes suggests that there may be an as yet unrecognized controlling function.[198] On the other hand, they may be relics of evolution. If a gene family arises by duplication and mutation of duplicated copies, we may expect to find gene copies that have not been selected as useful but are still present in the genome.

There are regions of the genome that seem to be largely single copy DNA with few reiterated sequences or gene clusters.[144] While some gene families exist as clusters others are dispersed throughout the genome.

4. Introns, Exons, and Overlapping Genes

A major difference between prokaryotes and eukaryotes is the presence of intervening sequences (introns) between the coding sequences (exons) in eukaryotes. Introns are especially numerous in higher organisms. For example, the gene for the myosin heavy chain present in rat embryonic skeletal muscle[199] encodes a 1939-residue peptide but occupies a length of 22 kb of DNA. The gene is split into 41 exons, whose transcribed RNA must be cut and spliced at 40 places to form the mRNA. By comparison the corresponding gene from the nematode *Caenorhabdatis elegans* is far less fragmented. It is not clear why some genes have so many introns and others so few. The very compact chromosomes of viruses and most bacteria do not contain introns. However, they are sometimes present in mitochondrial and chloroplast genes.

A surprise, which was first recognized in viral RNA and DNA, is that genes sometimes overlap. For example, two proteins, one long and one shorter, are synthesized starting at the same point in the RNA genome of phage Qβ.[200] In the DNA of phage φX174 the third nucleotide of the stop signals for some genes are also the first nucleotides for the start signal for translation of the next gene. Pairs of genes have been found in which one of the genes of the pair is found completely inside the other gene but is translated in a different reading frame.

A four-base overlap between dihydrofolate reductase and thymidylate synthase has been found[201] in the DNA of phage T4. A transposable DNA insertion sequence (see Section D,5) in *E. coli* encodes two genes, one of which is contained within the other and which is transcribed from the opposite strand of DNA.[202] The double-stranded RNA of a reovirus produces two peptides from the same sequence using two different AUG initiation codons in different reading frames.[203]

Overlapping obviously limits severely the mutational alterations that are allowable. Perhaps it is for this reason that eukaryotic genes seem to be dispersed in widely separated locations and overlap is rare. However, the human type IV collagen genes for the α1 and α2 chains are encoded on opposite strands with their 5' flanking regions overlapping.[204] Some introns contain genes.[205,206] Chlorarachniophyte algae contain a multimembraned chloroplast thought to be a vestigial remnant of an endosymbiont. Its 380-kb genome consists of three short chromosomes that encode overlapping genes and contain the shortest introns (18–20 base pairs) known.[207]

5. DNA of Organelles

Mitochondrial DNA (mtDNA), discussed in Chapter 18, varies in size from <6 kb to 367 kb.[208] The small circular mtDNA of animals is extremely efficiently packed with genes for tRNAs, rRNA, and a small number of protein subunits (Fig. 18-3). However, the 78-kb yeast mtDNA contains many long AT-rich spacers as well as long introns, some of which contain genes for splicing enzymes. Otherwise, it is similar to animal mitochondrial DNA. Mitochondrial genomes of higher plants are much larger; that of maize is a 570-kb circle, which contains both direct and inverted repeat sequences. Recombination between these sequences is apparently the origin of smaller incomplete circular genomes that are also present. Plant mtDNA appears to contain the genes present in animal mitochondria plus additional genes. For example, one encodes a 5S RNA. The most unusual mtDNA is found in hemoflagellates such as *Trypanosoma* and *Leishmania*. The single large mitochondrion or **kinetoplast** located near the base of the flagellum contains a network of catenated circular DNA molecules (Fig. 5-16).[209,210] The genetic makeup is similar to that of other mtDNA, but the rRNA genes are unusually short. Chloroplast DNA, which varies in size from 120 to 160 kb, is discussed in Chapter 23.

6. Methylation of DNA

As mentioned in Chapter 5, a significant fraction of the pyrimidine and purine bases in DNA is methylated. One function of such methylation in bacteria, discussed in Chapter 26, is to protect against the action of restriction endonucleases.[211] For example, the gene for the well-known *Eco*RI endonuclease is carried in *E. coli* cells by an R factor. This plasmid also carries (just 29 base pairs away) the gene for a 326-residue **N⁶-adenine methylase**.[212–213a] This enzyme uses *S*-adenosylmethionine (AdoMet) to methylate the two adenines (marked by asterisks) in the six-base-pair recognition sequence 5'-G A A* T T C / 3'-C T T A* AG converting them to N^6-methyladenines (m^6A). Other DNA methyltransferases place methyl groups on N-4 of cytosine or on C-5 of cytosine.[214–215a] The latter utilizes a mechanism illustrated in Eq. 12-4.[216,217] Such enzymes are components of both type II and type I restriction-methylation systems.[218] However, most of the m^6A in the *E. coli* chromosome arises from action of a different methylase, one that recognizes the palindromic sequence 5'-GATC and methylates adenines in both chains.[219] This **DNA adenine methylase**, a product of the *dam* gene, plays an important role in mismatch repair, transposition, regulation of transcription, and initiation of DNA replication.[220] The same methylase regulates at least 20 genes induced during infection by *Salmonella typhimurium*. Some of these genes are essential to virulence.[221] A similar methyltransferase appears to control differentiation of the stalked *Caulobacteria* cells.[220]

CG doublets. The only modified base commonly found in eukaryotes is 5-methylcytosine,[222,223] which upon deamination becomes thymine (Eq. 27-1). Most methylation occurs when C is followed by G. Usually 60–90% of all 5'-CG sequences (CpG sequences) in eukaryotic DNA are methylated. However, the fraction of methylated cytosine varies from almost zero for *Drosophilia, Caenorhabditis,* and *Saccharomyces* to as much as 30% in higher plants.[224]

When CG pairs are methylated on both cytosines,

5-Methylcytosine (m^5C) $\xrightarrow{H_2O \quad NH_3}$ Thymine (T) (27-1)

an interesting result arises upon replication of the DNA. The methyl groups don't prevent replication, but only one of the DNA strands in each daughter duplex is methylated (Eq. 27-2). However, additional methylation of DNA occurs within three hours of DNA replication by a **maintenance methyltransferase** that recognizes methylated CpG sequences in the old DNA strand and methylates the cytosine in the 3'-GpC of the newly synthesized strand.[220,225,226] Mammalian cells contain at least three different DNA (cytosine-5) methyltransferases.[227–229b] Enzymatic **demethylation**, which converts m^5C residues back to cytosine, may also occur.[230–233] It has been difficult to demonstrate[233a] but enzymatic demethylation of 5-methylcytosine has been reported.[233b,c] The observed loss of methylation during embryonic development (Chapter 32) may be a result of loss of the maintenance methylase. Most CpG doublets are found in large "islands" of several hundred bases to about 2 kb in length, which are unusually rich in G + C.[234,235] These islands lie near the 5' ends of ~60% of all human genes[236] and near origins of replication.[237] They are also found in regions of the DNA that are compacted into heterochromatin. The genes in methylated regions tend to be "silent," i.e., they are not actively transcribed.[236] Demethylation may permit transcription.

The degree of methylation of the CG doublets is variable both in position within a chromosome and

5′ 3′
H_3C—C⋮G
G⋮C—CH_3
DNA replication
H_3C—C⋮G / G⋮C   C⋮G / G⋮C—CH_3
Hemimethylated DNA
Maintenance methylation
H_3C—C⋮G / G⋮C—CH_3 H_3C—C⋮G / G⋮C—CH_3
Normal methylation pattern retained (27-2)

with stage of development. Cytosine methylation is essential to embryonic development,[238] and mice lacking the maintenance DNA methytransferase are developmentally retarded and die at mid-gestation.[239] Not all CG sequences are methylated and various patterns of DNA methylation are generated at different stages of development by rounds of methylation and by the action of demethylases.[232] The maintenance methylase (Eq. 27-2) ensures that a stable methylation pattern persists until altered by new rounds of methylase or demethylase action. A parallel is found in the prokaryotic *Caulobacter* in which three chromosomal sites are fully methylated in swarmer cells, become hemimethylated in stalked cells, and are fully methylated again just prior to cell division.[220] See Fig. 32-1.

DNA methylation affects transcription, either directly by preventing the binding of transcription factors or indirectly via a series of binding proteins specific for methylated CG doublets. In early stages of embryonic development there is very little methylation, but some genes are quickly silenced as methylation takes place. Heterochromatic regions including inactivated X chromosomes are heavily methylated. However, additional alteration in chromatin is required for complete silencing of genes.[236,240] Recent studies indicate that an abundant mammalian protein binds to the methylated DNA along with a **histone deacetylase**.[239,241–244] The latter acts on acetylated histones to free lysine side chains, which may interact in an inhibitory manner with the DNA.

In female mammalian cells most of the genes on one of the two X-chromosomes are completely inactivated. DNA methylation plays a major role in this process.[244,245] A perfect correlation has been observed between 5'-methylation of cytosines in CpG islands and inactivation of X-chromosome genes.[246] Methylation may also play a role in recombination and repair.[247] Methylation of DNA decreases with increasing age.[248] It increases as a result of oncogenic transformation of cells.[249] Some other modifications of DNA largely limited to bacteriophages are discussed on p. 234.[247,250]

Imprinting. With the exception of X-linked genes each person has two copies of each gene, one of maternal and one of paternal origin. Both copies of most of these genes are expressed. However, a few of the genes receive from one parent or the other an **imprint**, a mark that distinguishes the parental origin.[251–253] Such imprints are maintained in cells through embryonic development but are erased in embryonic gonads to allow for a new imprint in the germ cells. Imprinting depends upon DNA methylation, and all imprinted genes show the presence of differentially methylated regions.[253] See also Chapter 32, Section A,1.

C. Replication

Following the discovery of the double helix and the enthusiasm that it engendered many people thought that the synthesis of DNA was simple. The nucleotide precursors would align themselves along separated DNA template strands and perhaps spontaneously snap onto the growing chains. In fact, replication is a complex process that requires the cooperative action of many different gene products and perhaps an association with membrane sites. The matter is made more complex by the fact that some of the enzymes involved in replication are also required in the processes of genetic recombination, in repair of damaged DNA molecules, and in defensive systems of cells.

1. Early Studies

That the DNA content doubles prior to cell division was established by microspectrophotometry. It was clear that both daughter cells must receive one or more identical molecules of DNA. However, it was not known whether the original double-stranded DNA molecule was copied in such a way that an entirely new double-stranded DNA was formed or whether, as we now know to be the case, the two chains of the original molecule separated. The latter is called **semiconservative** replication, each of the separated strands having a new complementary strand synthesized along it to form the two identical double-stranded molecules.

The first definitive evidence for semiconservative replication was reported by Meselson and Stahl[254] in 1958. Cells of *E. coli* were grown on a medium containing isotopically pure $^{15}NH_4^+$ ions as the sole source of nitrogen. After a few generations of growth in this medium the DNA contained exclusively ^{15}N. Then the cells were transferred abruptly to a medium containing $^{14}NH_4^+$ and were allowed to grow and to double and quadruple in number. At various stages DNA was isolated and subjected to ultracentrifugation in a CsCl gradient. Small but easily detectable differences in density led to separation of dsDNA molecules containing only ^{15}N from those containing partly ^{15}N and from those containing only ^{14}N. At the beginning of the experiment only DNA containing entirely ^{15}N was present. However, after one generation of growth in the ^{14}N-containing medium, the density of *all* the DNA was such as to indicate a content of one-half ^{14}N and one-half ^{14}N. After a second generation of growth half of the DNA still contained both nitrogen isotopes in equal quantity, whereas half contained only ^{14}N, exactly the result expected for semiconservative replication. A similar experiment using 5-bromodeoxyuridine, a thymidine analog, is shown in Fig. 27-11.

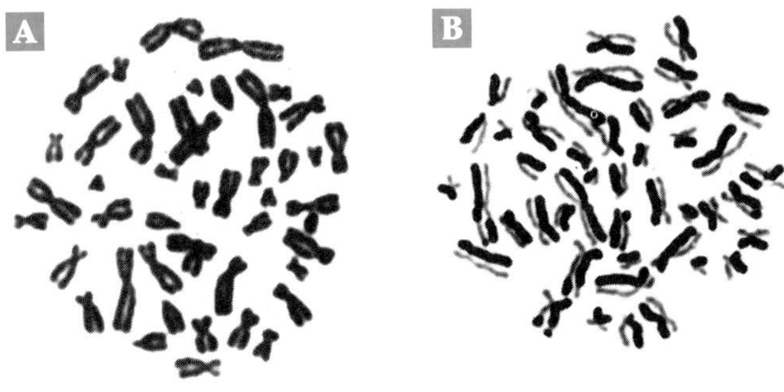

Figure 27-11 (A) Human chromosomes after one replication in the presence of 5-bromodeoxyuridine (BrdU). Both chromatids of each chromosome contain BrdU in one strand of the DNA duplex and normal thymidine in the other. (B) After two replications in the presence of BrdU one chromatid of each chromosome contains BrdU in both strands of the duplex and stains strongly with a special differential staining procedure. The other chromatid contains only normal thymidine in one strand of the duplex and is not stained. Courtesy of Carolina Biologicals.

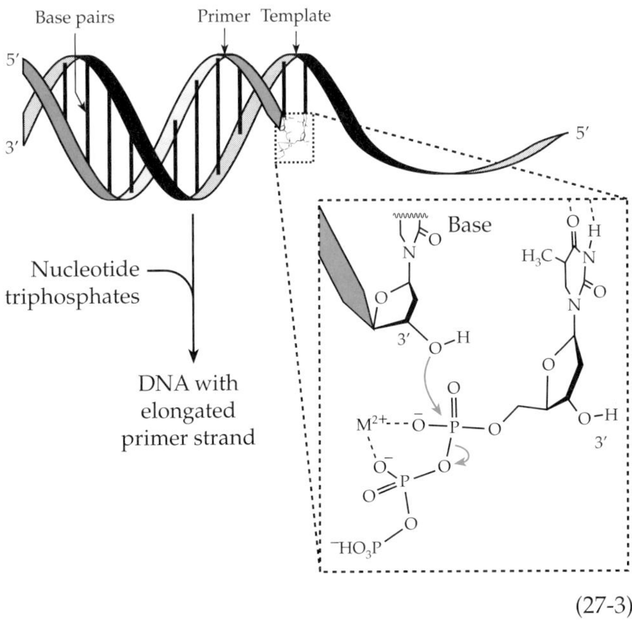

(27-3)

Autoradiography. Later a technique of direct autoradiography of DNA using ^{3}H-labeled thymidine[255] was applied by Cairns.[256,257] Cells of *E. coli* were grown on a medium containing the radioactive thymidine for various times but typically for 1 h (~2 generations). The cells were then ruptured, the DNA was spread on a thin membrane filter, and autoradiograms were prepared. When the DNA contained [^{3}H]thymidine the exposed trace in the autoradiogram could be followed around the entire 1.1–1.4 mm circumference of the spread DNA molecule. Molecules partially labeled and in the process of replication could also be identified. After two hours of growth in the presence of ^{3}H-labeled thymidine about half of the bacterial DNA was fully labeled, but half contained regions that were labeled only half as heavily. They presumably contained ^{3}H label in a single strand and, therefore, represented unreplicated regions. All of the molecules had undergone one round of replication with tritiated thymidine to yield lightly labeled molecules; parts of the molecules had not completed the second round. The more heavily labeled regions were interpreted as fully replicated. The shapes of the "replication figures" suggested that DNA is synthesized in a continuous manner starting from one point and continuing around the circular molecule at a constant rate. Although subsequent experiments (considered in Section 4) show that replication is usually **bidirectional**, the experiments of Cairns were important because they introduced a technique for direct visualization of replication *in vivo*.

The chemistry of DNA polymerization. What are precursors of DNA? Early experiments showed that the nucleoside [^{3}H]thymidine was efficiently incorporated into DNA, but for energetic reasons it seemed unlikely that thymidine was an immediate precursor. Evidence favoring the nucleoside triphosphates was provided in 1958 when Arthur Kornberg identified a DNA polymerase from *E. coli*. Kornberg's enzyme, now known as **DNA polymerase I**, was isolated in the amount of 600 mg from 90 kg of bacterial cells[258,259] (over 400 molecules of enzyme per cell). The 928-residue enzyme displayed many of the properties expected of a DNA-synthesizing enzyme. It requires a **template strand** of DNA as well as a shorter **primer strand**. As indicated in Eq. 27-3, the enzyme recognizes the 3' end of the primer strand and binds the proper nucleoside triphosphate to pair with the next base in the template strand. Then it catalyzes the displacement of pyrophosphate, at the same time linking the new nucleotide unit onto the 3' end of the primer strand. Continuing in this way, the enzyme is able to turn a single-stranded template DNA into a double-stranded DNA in which the newly synthesized strand contains, at each point, the base complementary to the one in the template strand.[259a]

Although the action of the DNA polymerase I, according to Eq. 27-3, provided a straightforward way to form a complementary strand of DNA, it did not explain how double-stranded DNA could be copied. One problem is that the two strands must be separated and unwound. If unwinding and replication occured at a single **replication fork** in the DNA, as indicated by Cairns' experiment, the entire molecule would have to spin at a speed of 300 revolutions per second to permit replication of the *E. coli* chromosome in 20 min. It also required that some kind of a **swivel**, or at least a nick in one chain, be present in the chromosome as indicated in the following drawing.

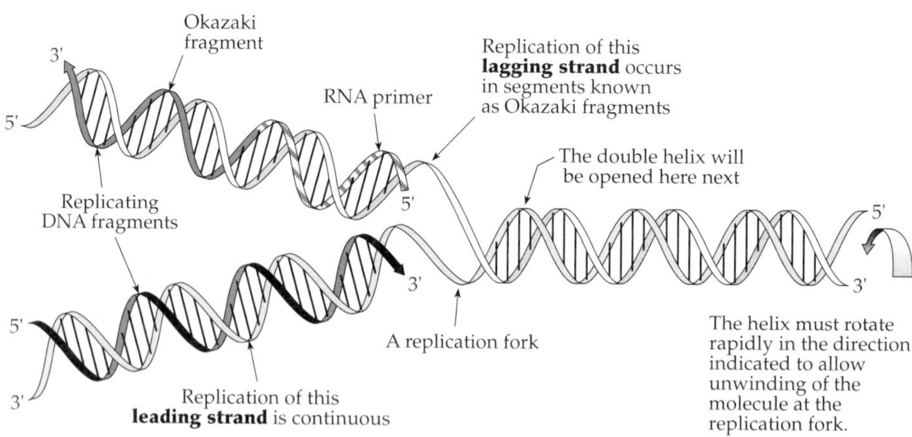

Another problem was posed by the fact that the two chains in DNA have opposite orientations. Thus, at the replication fork one of the new chains might be expected to grow by addition of a new nucleotide at the 3' end, while the other chain would grow at the 5' end. *If so, there should be two DNA polymerases*, one specific for polymerization at each of the two kinds of chain end. Nevertheless, despite intensive search the only DNA polymerases found added new units only at the 3' end.

Discontinuous replication and RNA primers. In 1968, Okazaki reported that during the time that replication of DNA is taking place bacterial cells contain short fragments of DNA. These are now called **Okazaki fragments** or **replication fragments**.[260] A second development was the discovery of the enzyme **DNA ligase**,[261,262] which is able to join two pieces of DNA to form a continuous chain. These two discoveries provided an explanation for the lack of a second kind of DNA polymerase. One strand, the **leading strand**, of the replicating DNA could be synthesized continuously in the 5' to 3' direction while the other strand, the **lagging strand**, would have to be synthesized in segments (the Okazaki fragments), which could then be joined by the DNA ligase. In 1971 Brutlag *et al.* reported that initiation of synthesis of the DNA of phage M13 in *E. coli* required formation of a short segment of RNA as a primer.[263] It was subsequently shown that *with few exceptions priming by RNA synthesis is always required for replication.*

Newer studies have revealed great complexity in the mechanisms of replication. As for synthesis of any polymer there are distinct steps of *initiation*, *elongation*, and *termination*. Topological problems associated with the unwinding and rewinding of the double helices and with disconnection of catenated circles and untying of knots are solved with the aid of special enzymes, the **helicases** and **topoisomerases**. Replication requires both DNA and RNA polymerases, a ligase, and ancillary proteins, some of whose functions aren't yet clear. Many of these proteins associate to form large multiprotein complexes, which are given names such as **primosome** (for priming) and **replisome**. Many bacteriophages and plasmids also replicate within cells of *E. coli* utilizing bacterial proteins as well as proteins encoded in the viral or plasmid genome.

2. DNA Polymerases

Immediately after its identification DNA polymerase I was generally assumed to be the principal chain elongation enzyme. However, an *amber* mutant of *E. coli* deficient in DNA polymerase I (gene *pol*A; Fig. 26-4A) synthesized DNA normally. This finding stimulated an intensive search for new polymerases. Two were found: DNA polymerases II (gene *pol B*)[264] and III. Both are present in amounts less than 25% of that of DNA polymerase I.[265,266] Both have properties similar to those of polymerase I, but there are important differences. By now DNA polymerases have been isolated from many organisms, many genes have been cloned and many sequences, both of bacterial and eukaryotic polymerases are known. Comparisons of both sequences and three-dimensional structures,[266a,b] a few of which are shown in Fig. 27-12, suggest that the polymerases belong to at least six families (Table 27-1). These include the RNA-dependent DNA polymerases known as **reverse transcriptases** as well as some **RNA polymerases**.[267–268b]

Some of the polymerases exist as single polypeptide chains, while others function only as large complexes. In every case a two-metal ion catalytic mechanism with in-line nucleotidyl transfer,[269] illustrated in Fig. 27-13, appears to be used by the enzymes.[267,270] Two-metal ion catalysis is also observed for phosphatases and ribozymes (Chapter 12).

Exonuclease activities, proofreading, and editing. DNA polymerase I not only catalyzes the growth of DNA chains at the 3' end of a primer strand but also, at about a 10-fold slower rate, the hydrolytic removal of nucleotides from the 3' end (**3'- 5' exonuclease activity**). The same enzyme also catalyzes hydrolytic removal of nucleotides from the 5' end of DNA chains. This latter **5'- 3' exonuclease activity**, the DNA polymerase activity, and the 3'-5' exonuclease activity all arise from separate active sites in the protein. DNA polymerases II and III do not catalyze

TABLE 27-1
Families of DNA Polymerases[a,]

Class	Name	Function	Molecular mass (kDa)
A	*E. coli* polymerase I (Pol I)	DNA excision repair	103
	Klenow fragment		68
	Bacillus subtilis Pol I[b]	DNA excision repair	
	Thermus aquaticus DNA polymerase (Taq)[c,d]	DNA excision repair	
	T7 DNA polymerase[e]	Virus replication	80
	T7 RNA polymerase[f,g]		99
	Eukaryotic Pol γ (gamma)	Mitochondrial replication	
	Eukaryotic Pol θ (theta)	DNA repair	
B	Eukaryotic Pol α (alpha)	DNA replication	180 (core)
	Eukaryotic Pol δ (delta)	DNA replication	
	Eukaryotic Pol ε (epsilon)	DNA replication	
	Eukaryotic Pol ζ (zeta)	Bypass synthesis	
	Bacteriophage T4 DNA Pol + accessory proteins[h,i]	DNA replication	43
	E. coli Pol II[j]		90
C	Bacterial DNA Pol III[p] + accessory proteins	DNA replication	~900 (holoenzyme)
D	Euryarchaeotic Pol II		
X	Eukaryotic DNA Pol β[n,o]	DNA repair	39 x 2
	Eukaryotic Pol λ (lambda)	Base excision repair	
	Eukaryotic Pol μ (mu)	Non-homologous end-joining	
	Eukaryotic Pol σ (sigma)	Sister chromatid cohesion	
Y	*E. coli* UmuC protein	SOS response	
	Eukaryotic Pol η (eta, XP-V, RAD 30)	Bypass synthesis	
	Eukaryotic Pol ι (iota)	Bypass synthesis	
	Eukarytoic Pol κ (kappa)	Bypass synthesis	
Reverse transcriptase family			
	HIV reverse transcriptase[k,l]		
	Telomerase[m]		
	RNA-dependent RNA polymerases		

[a] Based on reviews by Burgers, P. M. J. *et al.* (2001) *J. Biol. Chem.* **276**, 43487–43490 and Steitz, T. A. (1999) *J. Biol. Chem.* **274**, 17395–17398

[b] Kiefer, J. R., Mao, C., Braman, J. C., and Beese, L. S. (1998) *Nature (London)* **391**, 304–307

[c] Eom, S. H., Wang, J., and Steitz, T. A. (1996) *Nature (London)* **382**, 278–281

[d] Li, Y., Korolev, S., and Waksman, G. (1998) *EMBO J.* **17**, 7514–7525

[e] Doublié, S., Tabor, S., Long, A. M., Richardson, C. C., and Ellenberger, T. (1998) *Nature (London)* **391**, 251–258

[f] Sousa, R., Chung, Y. J., Rose, J. P., and Wang, B.-C. (1993) *Nature (London)* **364**, 593–599

[g] Sastry, S., and Ross, B. M. (1999) *Biochemistry* **38**, 4972–4981

[h] Wang, J., Yu, P., Lin, T. C., Konigsberg, W. H., and Steitz, T. A. (1996) *Biochemistry* **35**, 8110–8119

[i] Jing, D. H., Dong, F., Latham, G. J., and von Hippel, P. H. (1999) *J. Biol. Chem.* **274**, 27287–27298

[j] Cai, H., Yu, H., McEntee, K., Kunkel, T. A., and Goodman, M. F. (1995) *J. Biol. Chem.* **270**, 15327–15335

[k] Morris, M. C., Berducou, C., Mery, J., Heitz, F., and Divita, G. (1999) *Biochemistry* **38**, 15097–15103

[l] Rodgers, D. W., Gamblin, S. J., Harris, B. A., Ray, S., Culp, J. S., Hellmig, B., Woolf, D. J., Debouck, C., and Harrison, S. C. (1995) *Proc. Natl. Acad. Sci. U.S.A.* **92**, 1222–1226

[m] Lundblad, V. (1998) *Proc. Natl. Acad. Sci. U.S.A.* **95**, 8415–8416

[n] Sawaya, M. R., Pelletier, H., Kumar, A., Wilson, S. H., and Kraut, J. (1994) *Science* **264**, 1930–1935

[o] Sawaya, M. R., Prasad, R., Wilson, S. H., Kraut, J., and Pelletier, H. (1997) *Biochemistry* **36**, 11205–11215

[p] Kornberg, A., and Baker, T. A. (1992) *DNA Replication*, 2nd ed., Freeman, New York

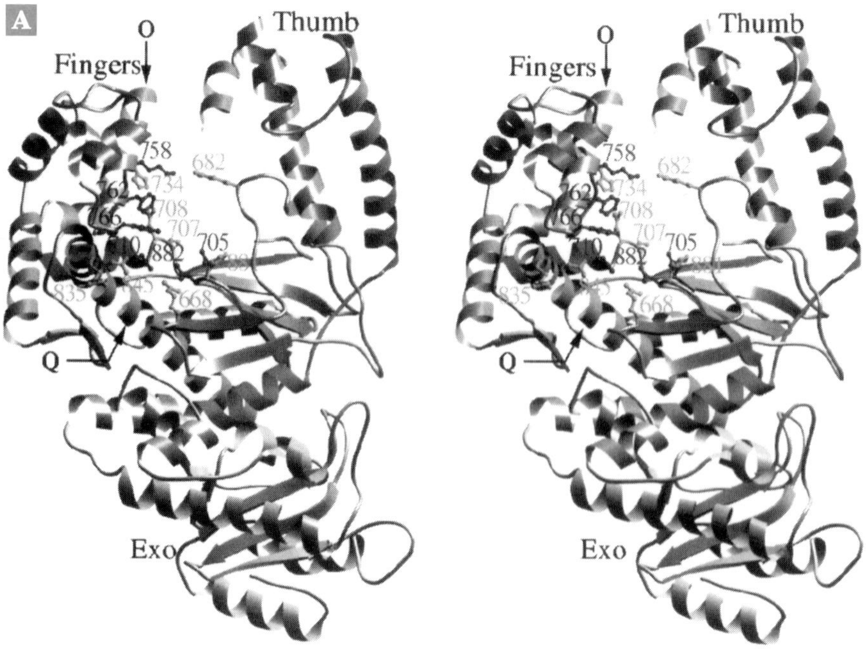

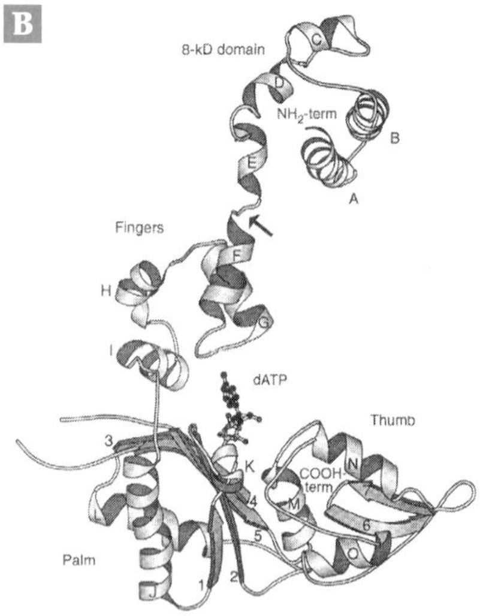

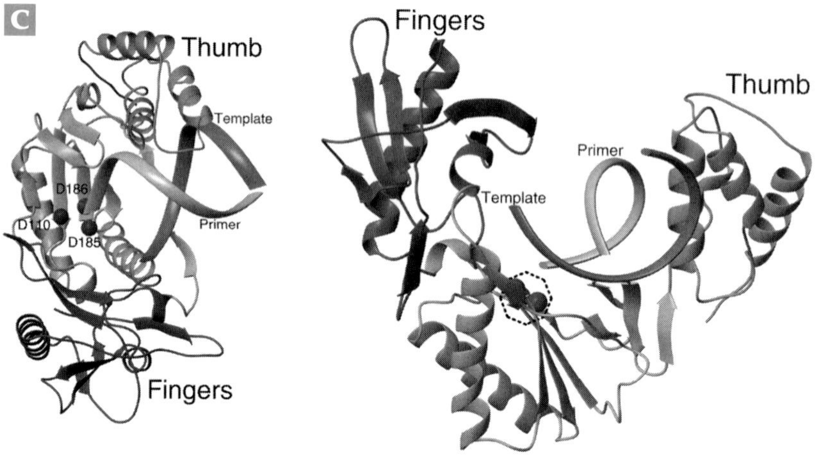

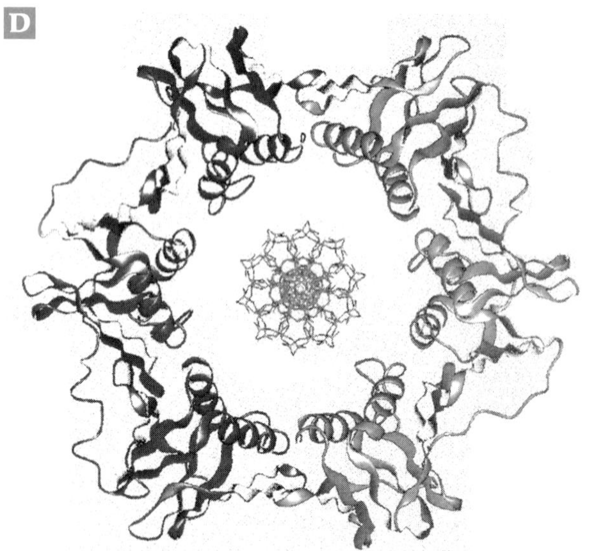

Figure 27-12 (A) Stereoscopic MolScript ribbon drawing showing three-dimensional structure of the 605-residue Klenow fragment of *E. coli* DNA polymerase I and illustrating the characteristic "right hand" structure of DNA polymerases. Fingers and thumb domains are marked. The polymerase active site is between these in the palm region and involves essential catalytic residues Asp 705, Asp 882, and Lys 758. The 3'-5' exonuclease domain is in the foreground. From Astatke *et al.*[271] Courtesy of Catherine M. Joyce. (B) Ribbon drawing of rat DNA polymerase β with dATP in the active site. From Sawaya *et al.*[273] Courtesy of Michael R. Sawaya. (C) Ribbon drawings of two views of the HIV reverse transcriptase with a DNA fragment in the active site. The 3' end of the primer is adjacent to the catalytic triad Asp 110, Asp 185, and Asp 186 (dark spheres). From Singh and Modak.[272] Courtesy of Kamlendra Singh. (D) Ribbon drawing of PCNA, the eukaryotic processivity factor or "sliding clamp" (see pp. 1549, 1562). Courtesy of John Kuriyan.[272a]

hydrolysis from the 5' end.

Treatment with proteolytic enzymes cuts a 323-residue piece containing the 5'-3' exonuclease from DNA polymerase I leaving a larger C-terminal piece known as the **Klenow fragment**. This fragment retains the polymerase activity as well as the 3'-5' exonuclease activity and is widely used in genetic engineering. Its three-dimensional structure[271,275] is shown in Fig. 27-12A. In the Klenow fragment the large C-terminal domain contains the polymerase. The N-terminal domain contains the 3', 5'-exonuclease activity, which is thought to fulfill a **proofreading** and **editing** function.[275a] The polymerase acts at the 3' end of the growing DNA chain. Before moving on to the next position, the enzyme verifies that the correct base pair has been formed in the preceding polymerization event. If it has not, the exonuclease action removes the incorrect nucleotide and allows the polymerase to add the correct one. Thus, each base pair is checked twice, first before polymerization and then after polymerization.

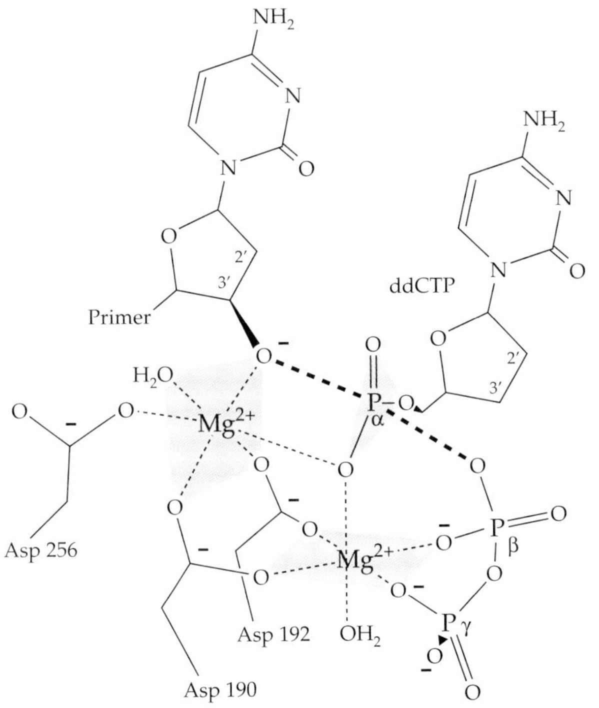

A puzzle was the fact that structural studies indicate that the editing center is over 3 nm away from the catalytic center in Pol I.[267] The proposed solution to the puzzle is illustrated in Fig. 27-14. When the catalytic center "identifies" a nucleotide triphosphate as able to form a proper Watson–Crick nucleotide pair, it catalyzes the formation of the new nucleotide linkage. Then it releases the 3'-chain end, which sometimes "melts" and over a 10- to 100-ms time interval is able to reach over into the exonuclease site where the newly added nucleotide may be hydrolytically removed. However, if the newly formed nucleotide is properly paired, it will be less likely to melt, and the new nucleotide will be more likely to be retained.[259a,265,267] DNA polymerase I and other related polymerases utilize **processive mechanisms**, moving from one site to the next without diffusing away from the DNA. The schematic picture shown in Fig. 27-14 also indicates how the 5'-3' exonuclease activity can come into play, when the polymerase reaches the end of a single-stranded gap.

Figure 27-13 Proposed mechanism and transition state structure for the synthetic nucleotidyltransfer activity of DNA polymerase β (and other DNA polymerases). The chain-terminating inhibitor dideoxy CTP is reacting with the 3'-OH group of a growing polynucleotide primer chain. This –OH group (as $-O^-$) makes an in-line nucleophilic attack on P_α of the dideoxy-CTP. Notice the two metal ions, which interact with the phospho groups and which are held by three aspartate side chains. Two of the latter, Asp 190 and Asp 256, are present in similar positions in all of the polymerases. The active centers for the hydrolytic 3'-5' and 5'-3' exonuclease activities of some of the polymerases also appear to involve two-metal catalysis and in-line displacement. See Sawaya *et al.*[274]

Other Class A polymerases. The *Thermus aquaticus* (***Taq***) polymerase is best known for its widespread use in the polymerase chain reaction (PCR; Fig. 5-47). Like *E. coli* I the enzyme is a large multidomain protein. The structure of the catalytic domains of the two enzymes are nearly identical, but the *Taq* polymerase has poor 3'-5' editing activity.[276] The enzyme has been carefully engineered to improve its characteristics for use in the PCR reaction.[277]

Some bacteriophage encode their own DNA polymerases. However, they usually rely on the host cell to provide accessory proteins. The sequence of the DNA polymerase from phage T7 is closely homologous to that of the Klenow fragment and the 3D structures are similar. The 80-kDa T7 polymerase requires the 12-kDa thioredoxin from the host cell as an additional subunit. It has been genetically engineered to improve its usefulness in DNA sequencing.[278]

About 45% of the sequence of the **RNA polymerase** encoded by phage T7, which transcribes RNA from the phage DNA, is also similar to that of the Klenow fragment. Sequences of these DNA polymerases are distantly related to those of reverse transcriptases.[279,280] The 136-kDa polymerase γ functions in mitochondria but is encoded in a nuclear gene. It is the only DNA polymerase that is inhibited by antiviral nucleotide analogs such as AZT (Box 28-C).[280a,b]

Polymerases of Class B. Although *E. coli* polymerase II is a member of this family, relatively little is known about its function. It may participate in DNA repair in the "SOS" response (Section E).[265]

The catalytic subunits of the major *eukaryotic* DNA

polymerases as well as of archaeal DNA polymerases are members of the B family.[267,281–282a] Eukaryotic cells contain at least 13 DNA polymerases which are designated by Greek letters (Table 27-1). Polymerases α, δ, and ε are essential for nuclear DNA synthesis.[283–286b] They function together with accessory proteins in primase and replisome complexes considered in Section 8. Others participate in DNA repair (pp. 1583, 1584).

Phage T4 encodes a DNA polymerase much used in the laboratory because of its ability to polymerize using a long single-stranded template. It also depends upon accessory factors provided by the bacterial host (Section 5).[287,287a]

Other DNA polymerases. **Reverse transcriptases** synthesize DNA using an RNA template strand. They are best known for their function in retroviruses (Chapter 28). The HIV reverse transcriptase is a heterodimer of 51- and 66-kDa subunits. The larger subunit contains a **ribonuclease H** domain.[288–289a] The enzyme is a prime target for drugs such as AZT and others.[290,291] A different reverse transcriptase is found in all eukaryotic cells in **telomerase**, an enzyme essential for replication of chromosome ends. Reverse transcriptases have also been found in rare L1 sequences that are functioning retrotransposons (Section D).[292]

The ~335-residue catalytic subunit of **eukaryotic polymerase β**, which has a DNA repair function, is the simplest known DNA polymerase. The active enzyme (Fig. 27-12B) is as small as 38 kDa. It lacks proofreading and is less accurate than the eukaryotic replicative polymerases listed in Table 27-1.[293–293c] Structural analysis revealed a folding pattern (Class X) related to that of a **nucleotidyltransferase superfamily** that includes enzymes such as terminal deoxynucleotidyltransferase (Chapter 12) and the glutamine synthase adenylytransferase (Fig. 24-7).[294] However, its active site (Fig. 27-13) is similar to that of other DNA polymerases.

DNA polymerase III. This Class C enzyme is the major bacterial polymerase for DNA replication. In its complete holoenzyme form it can synthesize new DNA strands at rates as high as ~1 kilobase s^{-1} without dissociation from a template.[294a] A genetic approach has provided important information about DNA replication.[265,266] A series of temperature-sensitive mutants of *E. coli* unable to carry out DNA synthesis were obtained. From these mutations, genes *dnaA, F, G, I, J, K, L, P, I, X,* and *Z* were identified and located at various points on the chromosome map. Genes *C* and *D* map close together at 89 min, and it now appears that they are one gene. Gene *F* encodes a ribonucleotide reductase (Eq. 16-22). The functions of genes *A*–*E, G, X,* and *Z* are discussed in the following sections. None of these genes code for DNA polymerase I, but gene *dnaE* was identified as that of DNA polymerase III, which is now known to be the major polymerase in bacterial DNA replication.[265,295] To obtain rapid error-

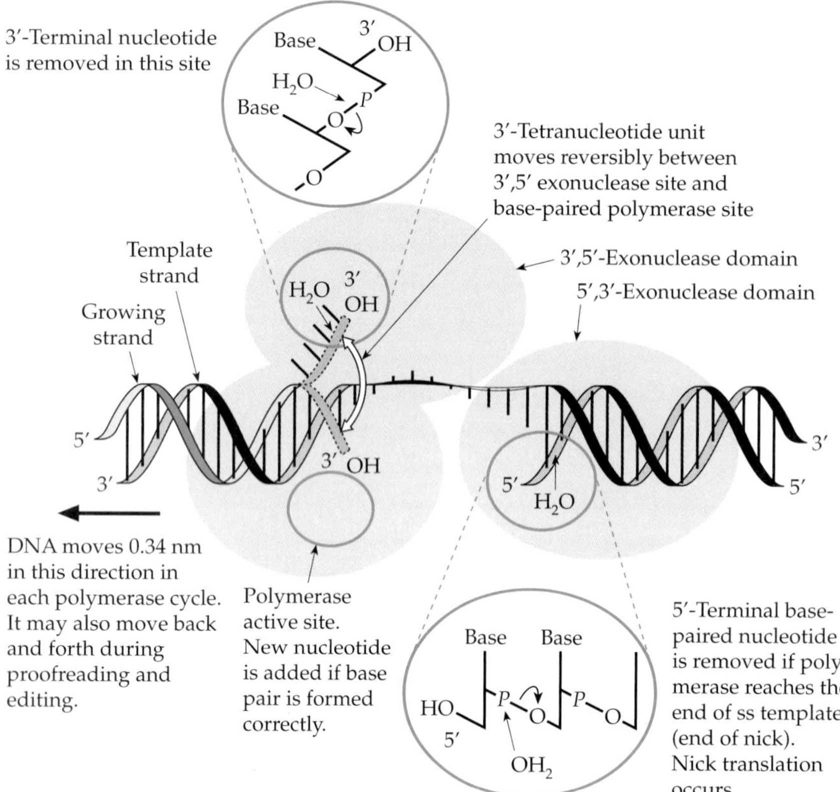

Figure 27-14 Schematic representation of DNA polymerase action on a nicked strand of DNA in which the nick has been enlarged. At the catalytic center new nucleotide units are added at the 3' end of a growing strand. At the 3'-5' exonuclease site the 3' terminal nucleotide may be removed hydrolytically. This will happen to the greatest extent if the nucleotide is poorly paired in the duplex. At the 5'-3' exonuclease site nucleotides are hydrolytically removed from the 5' end of a strand in the chain.[265,267]

free synthesis it must be combined with a number of other subunits to form **polymerase III holoenzyme** (or replisome). Ten different subunits, some as two or more copies, form the holoenzyme.[294a,296–299] Subunits are listed in Table 27-2. The polymerase catalytic center is in the 132-kDa α subunit. The 27.5-kDa ε subunit contains the 3'-5' exonuclease editing activity.[298,300,300a] Mutation in its gene (*dnaQ*) leads to a high spontaneous mutation rate in bacteria. Together with the θ subunit, α and ε form the polymerase III core. This complex has polymerase activity and improved proofreading ability but is still unable to act rapidly, accurately, and processively. Full catalytic activity requires at least the additional β, γ, δ, and τ subunits (Table 27-2). The presence of the τ subunit causes the core complex to dimerize to form Pol III′ (Eq. 27-4, step *a*). Pol III′ can add only about ten nucleotides to a growing DNA chain before it dissociates.[301] The presence of the β_2 dimer, known as the **processivity factor** or **sliding clamp**,[301a,b] is essential for highly processive polymerization. The β protein forms a ring around the duplex DNA and interacts with the polymerase clamping it tightly to the DNA. Putting the β_2 clamp onto the DNA is an ATP-dependent process that first involves binding of the **γ-complex** or **clamp loader** (Eq. 27-4, step *b*), ATP-dependent opening of the β_2 ring, and insertion of the DNA duplex (step *c*).[301a,c] This complex may form a replisome structure that acts simultaneously on the leading and lagging strands (see Fig. 27-19).[302,303] The *Bacillus subtilis* replisome appears to contain two different catalytic (α) subunits, perhaps one for each strand.[303a]

3. Other Replication Proteins

DNA ligases. These enzymes, which are essential to replication, have a specific function of repairing "nicked" DNA.[261,304–307a] Such DNA, as indicated in Eq. 27-5, has a break in one strand and contains a 3'-hydroxyl group and a 5'-phosphate group, which must be rejoined. The ligase from *E. coli* activates the phosphate group in an unusual way by transfer of an adenylyl group from NAD$^+$, with displacement of nicotinamide mononucleotide (Eq. 27-5, step *a*). The reaction is completed by displacement of AMP as indicated in Eq. 27-5, step *b*. Cells infected by bacteriophage T4 synthesize a virally encoded ligase, which utilizes ATP rather than NAD$^+$ as the activating reagent. The ~190-kDa mammalian DNA ligase I has been found deficient in some patients with the Bloom syndrome, a condition associated with poor DNA repair and a high incidence of cancer (see also p. 1550).[304]

Single-strand binding proteins. Genetic analysis of replication of the DNA of phage T4 within cells of *E. coli* revealed that at least five genes of the virus are required. One of these, gene 43 specifies the T4 DNA polymerase, while gene 32 codes for a single-strand binding protein, also known as the DNA unwinding, melting, or helix-destabilizing protein. It has a greater affinity for ssDNA than for dsDNA and binds to a length of ssDNA causing unwinding of the double helix and exposure of the purine and pyrimidine bases of the template strand.[308] The protein is required for replication, genetic recombination, and repair of DNA. Similar proteins are encoded in the genomes of many viruses.[309] The 87-residue single-strand binding protein encoded by gene 5 of phage M13 forms a dimer, which completely coats newly synthesized ssDNA preventing the DNA polymerase system of the host

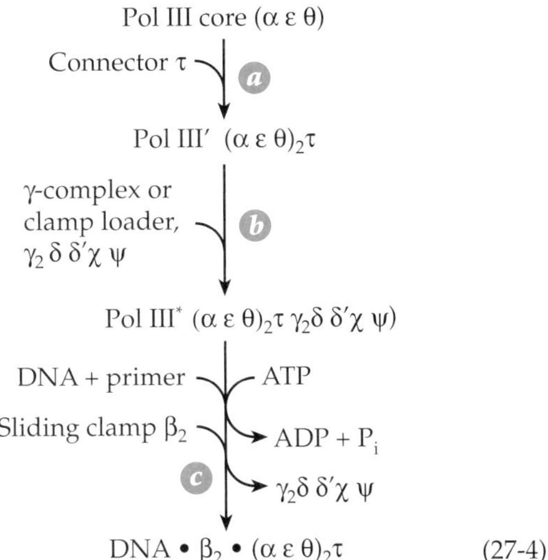

(27-4)

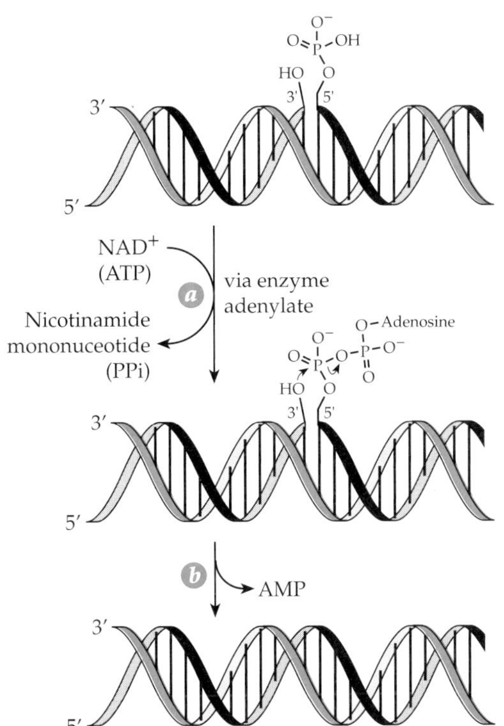

(27-5)

bacteria from converting it into dsDNA. The polynucleotide chain binds into a groove in the protein with one tyrosine intercalated between the DNA bases.[310,311]

The *E. coli* single-strand binding protein, another helix-destabilizing protein that is usually called simply **SSB**, is a tetramer of 18.5-kDa subunits.[265,312,313] It is essential to DNA replication. About 35 nucleotides may bind to each tetramer.[314] The situation is not as clear in eukaryotes where DNA is largely coiled around histones in the nucleosomes. Several single-strand binding proteins have been identified,[315] but the need for SSB proteins in eukaryotic nuclear replication is uncertain.[316] A human mitochondrial SSB resembles that of *E. coli*.[317]

Helicases and topoisomerases. Cells of *E. coli* contain at least 12 DNA-dependent ATPases that cause unwinding of DNA at the expense of hydrolysis of ATP.[318] The activity of these **helicases** is essential to replication, repair, and recombination of DNA in all organisms.[319–320b] The primary replicative helicase of *E. coli*, which unwinds DNA ahead of the replication fork, is encoded by gene *dnaB*. The active form is a hexamer, which exists in at least two conformational states.[321,321b] A segment of ~20 nucleotide units of DNA binds to one hexamer.[322] Helicases have ATPase activity, and the dnaB hexamer contains six ATP-binding sites. However, only three of them may be occupied.[323,324] Many helicases have a hexameric ringlike structure; that of *Bacillus subtilis* is seen clearly in electron micrographs.[325] Three-dimensional structures are known for some.[321a,325a] Although these enzymes may bind to duplex DNA, they also bind to and move along single-stranded DNA in either the 5'→3' or 3'→5' direction. The directionality of a helicase can be determined by annealing two small pieces of ssDNA to the 5' and 3' ends of a longer strand of ssDNA. A 5'→3' helicase will translocate along the long ssDNA and displace the oligonucleotide annealed at the 3' end of the strand, while a 3'→5' helicase will displace the oligonucleotide annealed at the 5' end.[265]

The dnaB protein is a 5'→3' helicase. However, the first helicase identified, the product of the *E. coli rep* gene, is a 3'→5' helicase. It is one of the host proteins needed for the propagation of phages such as ϕX174 and M13. It catalyzes the unwinding of the double-stranded replication forms of these viruses. It binds to a stretch of ~20 nucleotides in a single-stranded region of nicked DNA. The hydrolysis of ATP moves the enzyme along the bound strand in a 3' to 5' direction opening up the DNA at a replication fork. Another *E. coli* 3'→5' helicase is protein *priA* (also called n'), a component of some primasome structures involved in replication of viruses.[318]

The bacteriophage T4 gene 41 protein, a 5'→3' helicase, functions together with the gene 61 primase in replication of that virus.[326–326b] The phage T7 gene 4 protein and virus SV40 large T antigen are also hexameric ringlike helicases. The *E. coli* protein **RecQ** is required for various aspects of recombination and is the prototype of a large group of helicases present in both prokaryotes and eukaryotes.[326c] The bacterial ruvB protein is a hexameric helicase that propels branch migration in Holliday junctions (Fig. 5-28) during genetic recombination,[327–327b] while helicase rho is required for termination of RNA synthesis.[328] Numerous eukaryotic helicases have been identified and purified.[329–331] Helicase DNA2 is needed for DNA replication in nuclear extracts from yeast.[332] Human 3'→5' DNA helicases, members of the RecQ family, are defective in some patients with **Bloom syndrome** and **Werner Syndrome**[332a] (Box 27-A). Bloom syndrome causes growth retardation, immunodeficiency, sensitivity to sunlight, and a predisposition to skin cancers and leukemias.[331a,b] The yeast (*S. cerevisiae*) genome contains genes for 134 different proteins that are probably helicases.[332b] RNA helicases are also known.[333]

A characteristic of all helicases is their ATPase activity, which apparently provides energy for "melting" the DNA. The mechanisms are not clear, but rapid separation of the stacked and hydrogen-bonded base pairs may be impossible without some assistance from an ATP-dependent process. In the case of the rep protein hydrolysis of two molecules of ATP seems to be required to melt one base pair.[323] It isn't clear whether one strand of DNA passes through the center of the oligomeric ring, as shown in Fig. 27-15, or whether both strands pass through. Helicases vary in their amino acid sequences, but they all possess several characteristic **signature sequences** including the Walker A and B motifs found in other ATPases, in

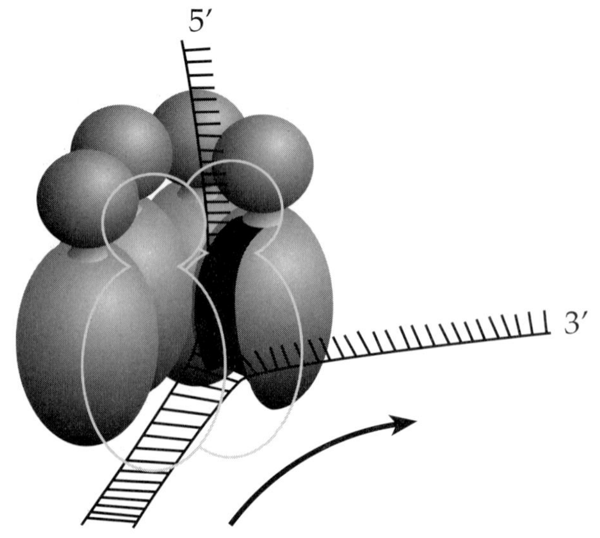

Figure 27-15 Model of DnaB hexamer proposed by Jezewska *et al.*[322] The arrow indicates the direction of movement of the DNA relative to the position of the helicase.

sequences related to the *E. coli* **recA** proteins (Section D) and in synthases (Chapter 18).[328,334]

While helicases can cause the unwinding of linear DNA duplexes, they cannot alter the linking number of covalently closed circular double-stranded DNA. The latter is the function of **topoisomerases**, which have been found in organisms from bacteria to higher plants and animals and may also be encoded by viruses.[265,335–337] There are two basic types of DNA topoisomerase. Those of type I change the linking number in steps of 1. One way that this might be accomplished is for an enzyme to nick one strand in the DNA allowing one of the cut ends to swivel around the unbroken strand, then to reseal the chain. However, it was found that topoisomerases can also cause catenation or decatenation of circular duplex DNA as long as at least one of the reacting DNA molecules is nicked. This observation suggested that a topoisomerase binds to a single-stranded region at a nick and cuts the chain but does not release the ends. This permits either a single strand or a duplex to pass through the broken strand, which is then resealed.

Topoisomerases of type I usually act most rapidly on negatively supercoiled DNA. They relax it by decreasing the number of negatively supercoiled turns one at a time. Negative supercoiling presumably facilitates binding of the enzyme to a single-stranded region by unwinding of the duplex. No ATP or other obvious source of energy is needed by type I topoisomerases. The chain cleavage involves a simple nucleophilic displacement by an –OH group of a tyrosine side chain (Tyr 319 of *E. coli* topoisomerase I),[338] which attacks a phosphorus atom in the DNA chain (Eq. 27-6). The result is covalent attachment of the enzyme to the 5'-end of the cut strand in type IA topoisomerases and at the 3'-end in type IB topoisomerases.[336,337,339] After the passage of the other

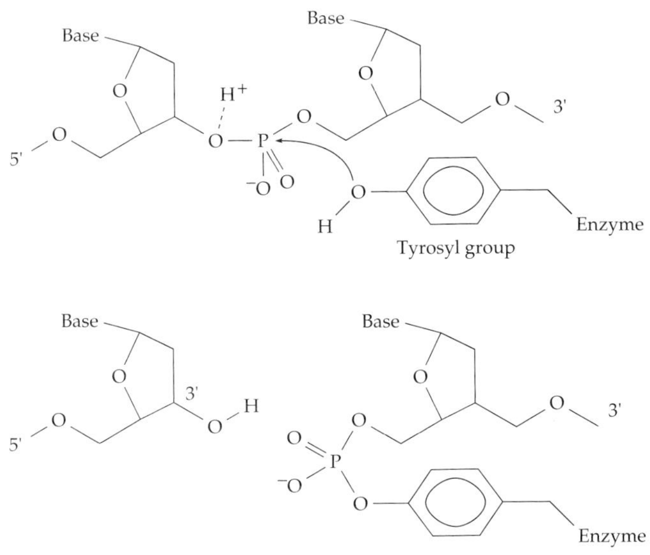

(27-6)

TABLE 27-2
Some Proteins of DNA Replication in *E. coli*[a]

Name	Gene	Mass of monomer (kDa)	Map location (min) Fig. 26-4
Polymerase I (also has 3'-5' and 5'-3' exonuclease, and RNase activities)	*polA*	103	87
Polymerase III			
Core			
α subunit (polymerase)	*dnaE*	130	4
ε subunit (3'-5' exonuclease)	*dnaQ*	27.5	5
θ subunit	*hol E*	10	
Sliding clamp, β_2	*dnaN*	37 × 2	83
Connector τ (ATPase)	*dnaX*	71	
Clamp loader (γ complex)			
γ subunit	*dnaX*	47.5 × 2	11
σ subunit		35	
σ' subunit		33	
χ subunit		15	
ψ subunit		12	
DNA binding proteins			
Single-strand, SSB	*ssb*	18.5 × 4	92
Double-strand, HU			
α subunit		9.5 × 2	
β subunit		9.5 × 2	
Helicases (ATP-dependent)			
Primary replicative	*dnaB*	52 × 6	92
Dna C protein	*dnaC*		99
PriA (n'), primosome	*priA*	76	88
Rep	*rep*	76.4	85
Initiation and priming proteins			
Dna A protein	*dnaA*	52	83
Primase (an RNA polymerase)	*dnaG*	60	67
PriB (n) primosome	*priB*		96
PriC (n'') primosome	*priC*		
DnaT (primosome assembly)	*dnaT*		99
Ribonuclease HI	*rnhA*		
DNA ligase	*lig*	75	52
Topoisomerases			
Type I	*topA*		28
Type II, DNA gyrase ($\alpha_2\beta_2$)			
Subunit α	*gyrA*		97
Subunit β	*gyrB*		90

[a] In large part from Kornberg, A., and Baker, T. A. (1992) *DNA Replication*, 2nd ed., Freeman, New York

strand through the gap in the cut strand of a type IB topoisomerase (Fig. 27-16) the free 3'-OH oxygen atom (or 5'-OH) attacks the phosphorus atom in the phosphotyrosine diester to reform the chain and release the enzyme. X-ray diffraction studies show that both the Type IA *E. coli* topoisomerase I and the human type IB topoisomerase[339–341a] are large proteins with holes of appropriate diameter for a DNA double helix. As illustrated in Fig. 27-16B, the protein may open to allow a double helix to enter and occupy a suitable position for cleavage of one chain by the active-site Tyr 319. Topoisomerases are metalloenzymes, usually functioning best with Mg^{2+}. *E. coli* topoisomerase I also contains 3–4 tightly bound Zn^{2+} ions.

Topoisomerases of type II change the linking number by 2 in either the positive or negative direction and hydrolyze ATP in the process. The best known of these is the *E. coli* **DNA gyrase**, an $\alpha_2\beta_2$ dimer of 97-kDa (α) and 90-kDa (β) subunits.[343,344] The enzyme catalyzes the ATP-dependent introduction of negative supercoils into DNA. It also relaxes negatively supercoiled DNA slowly in the absence of ATP. Type II topoisomerases are found in all organisms.[335,345] They are encoded by some bacteriophage such as T4[346] and by plasmids.[347] However, most differ from bacterial DNA gyrase in not coupling DNA supercoiling to ATP hydrolysis. They require ATP but like topoisomerase I cause a relaxation of the supercoiling. Strands of one segment of DNA (called the "gate" or **G-segment**) are cleaved by the enzyme with staggered cuts four base pairs apart. Another segment of DNA (the "transport" or **T-segment**) is then passed through the gate and is thought to be released from a second gate in the complex (Fig. 27-16C).[342,348] The enzyme subunits bind through phosphotyrosine linkages as in Eq. 27-5 to the 5' phospho groups of the two cleaved chains, while the subunits bind ATP and may like tiny muscles twist the DNA.[349] Topoisomerases II are large dimeric proteins. The subunits of yeast topoisomerase II (Fig. 27-16C) are 1200-residue multidomain proteins.[342,350] Mechanisms of DNA cleavage by types I and II topoisomerases appear to be related.[351] However, the ATP-dependent conformational changes involved in a two-gate mechanism are unique to topoisomerases II.[348,348a] The bacterium *Sulfolobus* contains a type I topoisomerase that is called **reverse gyrase** because it utilizes ATP to introduce *positive* supercoils into DNA.[352] This is in contrast to gyrase, a type II topoisomerase that introduces negative supercoils.

Type II topoisomerases are essential and function in replication, DNA repair, transcription, and chromosome segregation at mitosis.[345,349] Yeast with a *top2* mutation dies during mitosis with hopelessly entangled daughter chromosomes.[353] A fluorescent antibody to eukaryotic topoisomerase II binds to chromosomes, probably at the bases of the radial loops present during mitosis,[353] and also to centrosomes[353a] Higher organisms contain more than one topoisomerase II.[354] Their specific functions are uncertain, but one appears to be to unknot entangled chromosomal DNA. In the crowded conditions of a cell nucleus topoisomerase can also cause inadvertent *formation* of knots.[355]

The functional role of topoisomerases of type I is less clear. Staining with fluorescent antibodies to the enzyme has revealed its presence in the transcriptionally active "puffs" of polytene chromosomes (p. 1635)[356] and in centromeres of mitotic cells.[357] A current hypothesis is that in *E. coli* class I topoisomerases act to relax negatively supercoiled strands of DNA behind transcription complexes, while gyrase acts to generate superhelical twists, which favor opening of the duplex ahead of transcription complexes.[354,358] Transcription of a supercoiled rRNA gene *in vitro* is diminished by the selective topoisomerase I inhibitor **camptothecin**, one of a group of antitumor drugs directed against topoisomerase of both types I and II.[349,354,359]

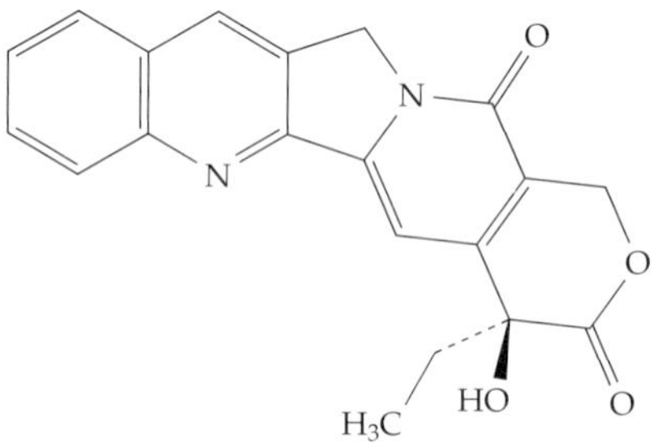

Camptothecin, a topoisomerase I inhibitor

In the autoimmune diseases scleroderma and systematic lupus erythematosus antigens to nuclear proteins or nucleic acids are present in the blood. Many patients with severe scleroderma have an antibody against topoisomerase I.[360]

Primases, initiator proteins, and ribonucleases. The priming segment needed for initiation of DNA replication is either a short segment of RNA or an oligonucleotide containing a mixture of ribonucleotide and deoxyribonucleotide units. The enzyme forming the primer is an RNA polymerase called **primase**. In *E. coli* it is encoded by gene *dna*G.[361–362a] Under some circumstances other RNA polymerases can act as the primase. Bacteriophages and plasmids may also encode primases (Table 27-1). For example, gene 61 of phage T4 (Fig. 26-2) encodes a primase, which together with the T4 helicase forms the priming particle or **primosome**.[363] The phage T7 gene 4 encodes a 63-kDa mutlifunctional protein that is both a primase and a helicase.[364–366b] The primase active site is on the outside of the hexameric ring. Additional proteins representing products of genes *dna*A and C are also required for initiation of replication in *E. coli*. Several molecules of the dnaA **initiator protein** bind to a specific DNA **origin** sequence and participate in assembling a primosome that also contains the hexameric dnaB helicase and, transiently, protein dnaC.[367–370] Replication of some single-stranded phages, such as ϕX174, in *E. coli* also require the *E. coli* priA, priB, and priC proteins (Table 27-2).[265,371,372] For successful completion of replication chaperonins the products of genes *dna*J and *grp* E are needed[373] as is a ribonuclease that digests the primer segments after they have been used as replication initiators.[374–376]

4. Replication of Bacterial DNA

The basic mechanisms of replication implied in Eq. 27-3 seems to be universal, but several questions had to be asked. "Is replication initiated at a fixed point or points in a chromosome?" and "Does replication occur in one direction only or do two forks form at the point

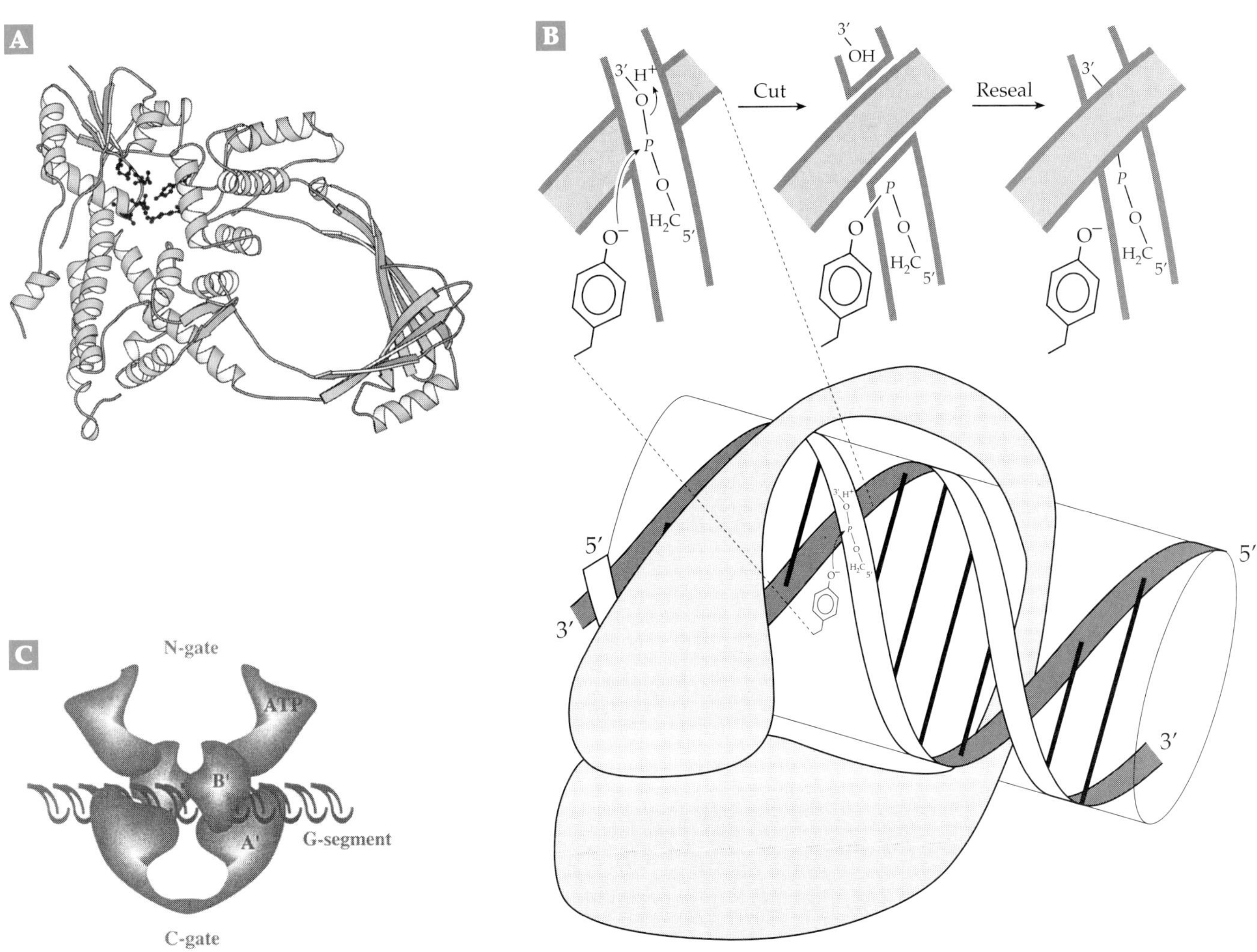

Figure 27-16 (A) Ribbon drawing of a large 67-kDa fragment of the 97-kDa (864-residue) *E. coli* topoisomerase I showing the position of the active-site tyrosine 319 and an adjacent arginine. From Lima *et al.*[341] Courtesy of Alfonso Mondragon. (B) Schematic diagram indicating a way in which topoisomerases of type 1 may pass one strand of DNA through another. The protein is shown binding to a single strand of a DNA duplex. This binding is facilitated by negative supercoiling. The enzyme then cuts the same strand by means of a nucleophilic displacement on a phosphorus atom using a tyrosinate side chain. The other cut end is held noncovalently by the enzyme, while the second strand passes through the gap. Then the gap is resealed by a reversal of the cleavage reaction. From Lima *et al.*[341] (C) Schematic model of a type II topoisomerase bound to a G-segment of DNA. This double helix is cut and another double helical strand, the T-strand, enters the N-gate. The gate then closes, and the central gate opens to allow the T-strand to pass through and exit through the C-gate. The shapes of the three domains are based on crystallographic data for the yeast enzyme. The ATPase, B′, and A′ domains consist of residues 1 to 409, 410 to 660, and 660 to 1200, approximately. From Olland and Wang.[342] Courtesy of James C. Wang. See also Champoux.[340c]

of origin and travel in opposite directions?" To answer these questions both genetic methods and electron microscopy have been employed.

Directions of replication. One technique for establishing the direction of replication in *E. coli* was to insert a λ prophage at the *att* site (Fig. 26-4, 17 min) and DNA from phage Mu-1 at a variety of other sites around the chromosome.[377] Phage Mu-1 was especially useful because it can be integrated at many different locations within the well-mapped genome. Integration within a gene inactivates that gene (an addition mutation) and allows the localization of the Mu prophage. Bacteria were prepared containing both λ and Mu-1 prophage, the latter at various sites. The bacteria were also auxotrophic for certain amino acids. Because of this, replication could be stopped by amino acid starvation. The bacteria usually completed any replication cycle in progress and then stopped. When the missing amino acids were added, replication began, again starting from the replication origin. Bromouracil, which enters DNA in place of thymine, was added at the same time. Consequently, the newly synthesized DNA strands were denser than the parent strands (see Fig. 27-11). After various times of replication the newly formed strands were separated by centrifugation in a CsCl gradient and were tested for hybridization with both λ and Mu-1 DNA. Since the cells did not all begin replication at the same time after addition of amino acids, a variety of lengths of newly replicated DNA were present. Nevertheless, from the observed ratios of Mu-1 DNA to λ DNA for the various strains it was possible to map the progress of replication beginning at an **origin** *ori*C near gene *ilv* at 74 min (Fig. 26-4). Replication was found to progress bidirectionally around the chromosome and to terminate between genes *trp* and *his* at ~25 min.

The use of autoradiographic methods confirmed bidirectional replication. Strains of amino acid auxotrophs with small nucleoside triphosphate pools were used. The addition of amino acids after starvation led to initiation of replication with only a 6-min lag. The cells were labeled with [^{3}H]thymidine, and after the replication forks had moved a short distance from the origin of replication the cells were given a pulse of "super-hot" [^{3}H]thymidine. Using autoradiography it was possible to observe the clearly bidirectional replication forks[378] (Fig. 27-17). Replication in other bacteria is also bidirectional.

Origins of replication. Replication of the *E. coli* chromosome begins and proceeds bidirectionally from its defined origin *ori*C. Replication of linear phage T7 is also bidirectional and begins at a point 17% of the way from one end.[265] In mammalian mitochondrial DNA the origin of replication for the H-strand is in the D-loop but that for the L-strand is 2 / 3 of the way around the circular chromosome within a cluster of tRNA genes (Fig. 18-3).[381,382] The single-stranded circular DNA genomes of Ff, ϕX174, and G4 phage also have distinct origins for initiation of replication to give RF circles.[265]

Most origins have quasi-palindromic nucleotide sequences, perhaps so that DNA can be looped out from the main duplex as is shown in Fig. 27-18A and B. The lengths of *ori* sequences vary, as does the complexity of their possible folding patterns. Plasmids have been constructed which not only contain the *E. coli* origin, but are dependent upon that origin for their own replication.[382a] Study of those plasmids indicate

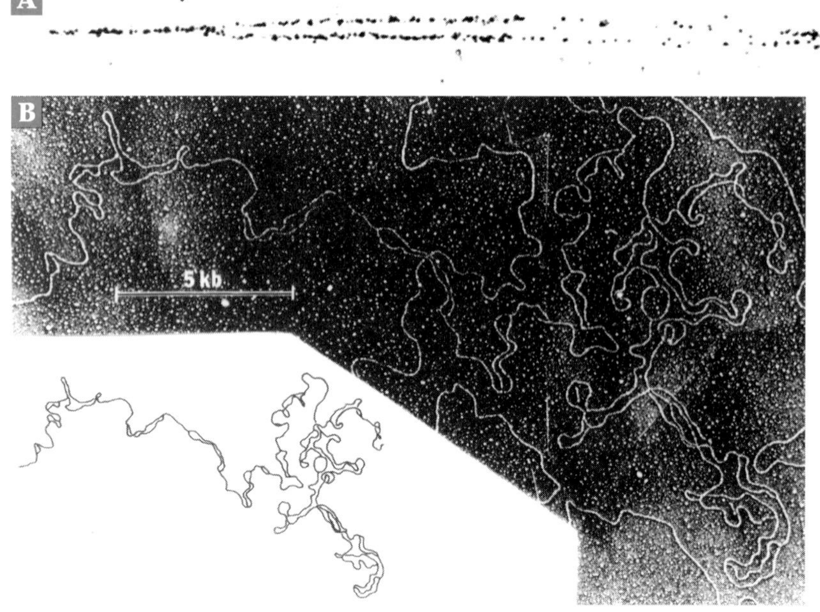

Figure 27-17 Bidirectional replication. (A) Replication forks in the *E. coli* chromosome. The autoradiographic pattern was produced by a chromosome that initiated replication with [^{3}H]thymine (5 Ci / mmol) and was subsequently labeled with [^{3}H]thymidine (52 ci / mmol) for 6 min. The total length of the grain track is 370 μm. From Kuempel *et al.*[379] (B) Fragment of replicating chromosomal DNA from cleavage nuclei of *Drosophila melanogaster*. The DNA, which was spread in the presence of formamide, contains several "eyes" formed where the DNA has been replicated. From Kriegstein and Hogness.[380]

that a 245-bp *ori*C sequence is essential.[265,383,383a] This sequence, which is shown in Fig. 27-18, contains several repeated oligonucleotides including 11 GATC sequences, which are sites of adenine N^6-methylation (see Section B, 6), and four "9-mers" (commonly known as dnaA "boxes") with the consensus sequence

$$\mathrm{TTAT}{}^{A}_{C}\mathrm{CA}{}^{A}_{C}\mathrm{C}$$

These appear in both 5' to 3' and 3' to 5' orientations, allowing them to form two base paired "stems." In addition, there are three direct repeats of a 13-residue consensus sequence 5'-GATCTNTTNTTTT (shaded in green in Fig. 27-18), which form an AT rich duplex. Other bacterial replication origins often follow a pattern similar to that in Fig. 27-18.[384] However, the origin for *Mycoplasma genitalium* has been hard to detect.[385] Replication origins of archaea have characteristics similar to those of bacteria and of organelles.[385a]

Priming and initiation of DNA synthesis. The first step in initiating a new round of replication and a new cell cycle in *E. coli* appears to be the binding of the initiator protein, dnaA, to the 9-mers in the negatively supercoiled origin.[295,383,386–388] The resulting complex is visible in an electron microscope and may consist of a core of up to 20–40 molecules of dnaA protein with the DNA wrapped around them (Fig. 27-19). ATP is also required and is bound to the protein where it hydrolyzes slowly. Similar initiator proteins are found

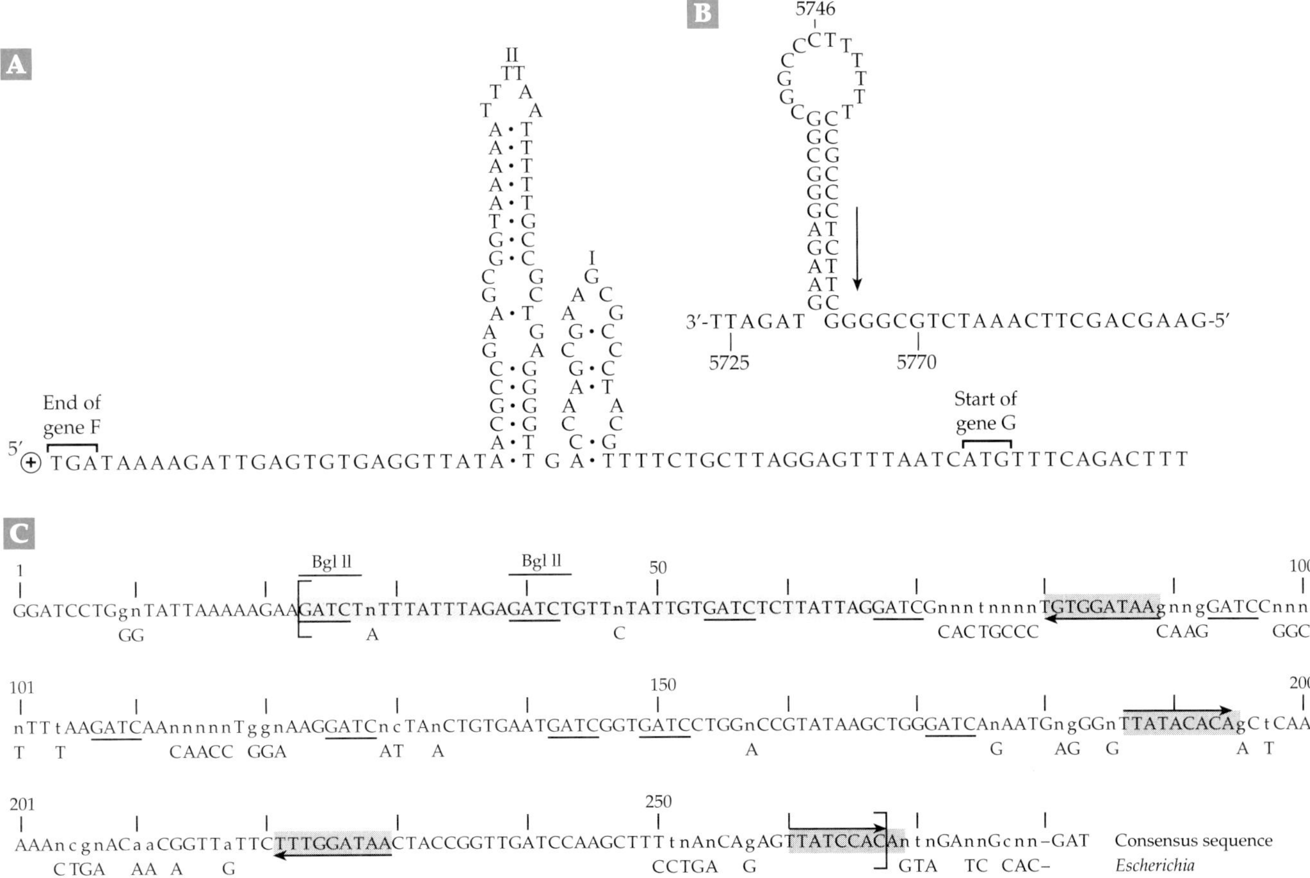

Figure 27-18 Nucleotide sequences and proposed secondary structures for some origins of replication. (A) Bacteriophage ϕX174. The ends of the neighboring genes F and G are shown. From Kornberg and Baker.[265] (B) Human mitochondrial DNA; origin of replication of the L strand. The sequence shown is for the H strand. The arrow indicates the direction of replication. From Hixson *et al.*[381] (C) Bacterial *ori* region. The consensus sequence of origins of replication that will function in *E. coli*. Derived from sequences from *E. coli, Salmonella typhimurium, Enterobacter aerogenes, Klebsiella pneumoniae,* and *Erwinia carotovora.* In the consensus sequence a large capital letter means that the same nucleotide is present in all six origins; a lowercase letter means that the same nucleotide is present in three of the five origins; n means that any of the four nucleotides may be present; – represents a deletion, G being present here in the *Enterobacter* origin. Underlined are *dam* methylation sites. Green shaded sequences, three 13-mers in a region that undergoes easy opening of the duplex for insertion of replication proteins; gray boxes with arrows, four 9-mers that are specific sites for binding of replication protein A. Brackets enclose a 245-bp minimum origin. From Kornberg and Baker.[265] Data from Zyskind *et al.*[390]

in many bacteria and are encoded by some viruses and plasmids. In archaea, as well as in humans, the initiator proteins seem to combine functions of the *E. coli* dnaA protein with those of SSB.[391,391a]

Following the binding of the dnaA protein the hexameric helicase dnaB (Fig. 27-15) is loaded onto the adjacent DNA in the region of the 13-residue repeated sequences, which are labeled 1, 2, and 3 in Fig. 27-19. As shown in this figure introduction of the helicase is assisted by protein dnaC, which forms a complex with the helicase. Additional binding of protein HU and a temperature of >30°C are essential for tight binding of the proteins. Although the dnaC protein is needed for formation of the prepriming complex, it dissociates after the dnaB protein becomes firmly bound. One dnaB hexamer binds to each single strand of the DNA duplex with opposite orientations (Fig. 27-19).[383a] With the dnaB helicase in place on each strand, this ATP-driven enzyme processes along the DNA, unwinding the duplex in both di-rections. (Perhaps it may be more accurate to say that the *DNA moves through the helicase*.) The resulting "bubble" is held open by binding of SSB tetramers. The primase (dnaG protein) then binds adjacent to the helicase and synthesizes the RNA primer along each strand of DNA. As the helicase processes to the right in the fork shown in Fig. 27-19, the complex with the dnaA protein dissociates, permitting primer synthesis into the origin region. Alternatively, RNA polymerase (Chapter 28) can prime replication by initiating transcription on both strands of the DNA. Suitable promoters are present and oriented in opposite directions on the two strands.[392]

It has long been postulated that the bacterial chromosome is attached to the plasma membrane. At least one such attachment site may be at or near the origin of replication.[393] Furthermore, the exchange of ADP for ATP in the dnaA protein is catalyzed specifically by cardiolipin and phosphatidylglycerol containing the unsaturated oleic acid.[386,393] Inositol polyphosphates may also play a role.[394]

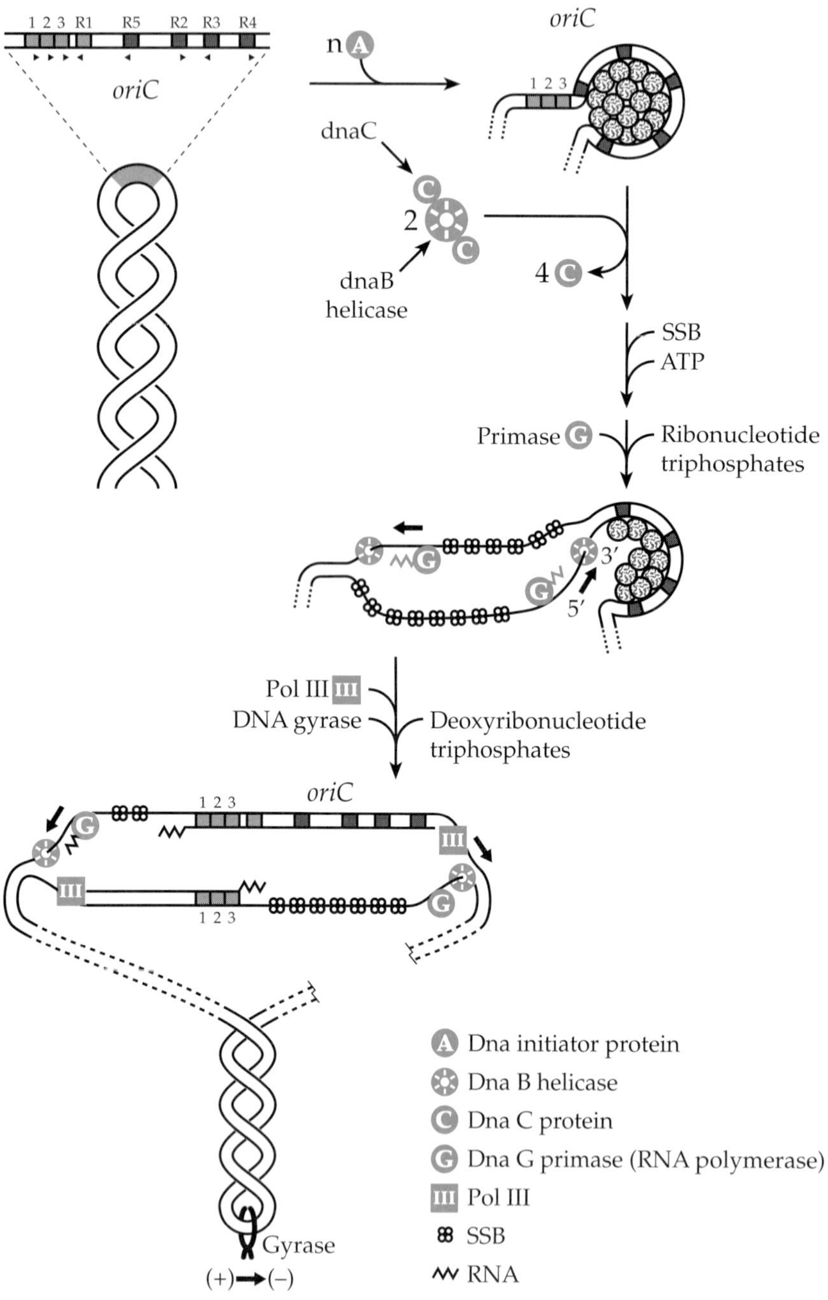

Figure 27-19 Hypothetical scheme for initiation of bidirectional replication in *E. coli*. The closed boxes R1 through R5 represent the 9-residue recognition sequences for the *E. coli* dnaA protein. The open boxes 1, 2, and 3 represent the three 13-residue repeats, possible sites for binding of the dnaB-dnaC protein complex. From McMacken *et al*.[383] Redrawn in simplified form.

Elongation of DNA chains. DNA polymerase III in its holoenzyme form is the major polymerase for DNA replication. It elongates the primer chains rapidly and processively leaving only very small gaps at the ends of single-stranded regions. The rate of elongation, which is ~3 nucleotides / s for 8 kb *oriC* plasmids, may be determined by the rate of action of the dnaB helicase. A completely unwound *ori*C plasmid, bound to SSB, undergoes primer elongation ~ 10 times faster.[383] However, the rate for the intact *E. coli* replisome is nearly 1000 nucleotide s^{-1} with a rate of misincorporation of only one in 10^9 nucleotides.[394a]

Small *ori*C plasmids need to be primed at only one location, but the large bacterial chromosome must undergo priming at many sites on the lagging strand to permit DNA polymerase III to act on that strand with formation of the Okazaki fragments. The DNA polymerase complex may be a dimer that works on both strands at once. The lagging strand may be looped out to allow it to lie parallel with the leading strand (Fig. 27-20). The appearance of the electron micrograph in Fig. 27-20B supports this suggestion. However, the manner in which the lagging strand can be shifted to bring the next primed initiation site to the replication complex is not clear. DNA polymerase III holoenzyme itself may be organized as an asymmetric oligomer[395] that operates on both strands in a complex such as is shown in Fig. 27-20. The primase, either the dnaG protein[265,383] or the primosome used by phage ϕX174 (Section 5),[396] may synthesize the RNA primers on the lagging strand.

Termination of replication. As each Okazaki fragment is completed along the template for the lagging strand, the RNA primer piece is digested out, replaced by DNA, and the nick sealed by action of DNA ligase. Ribonuclease H, which is found in both bacteria and eukaryotic cells, specifically degrades the RNA component in these RNA-DNA hybrid regions.[396a] In bacteria another mechanism for primer removal is available. The 5'- 3' exonuclease activity of DNA polymerase I will cut out the RNA segment, while the 5'- 3' polymerase activity of the same enzyme will fill the gap.

Replication of *oriC* plasmids may occur by simply allowing replication of the leading strands at both replication forks to continue all the way around the circle.[397] However, in *E. coli* bidirectional replication continues only until the two replication forks converge. This can occur anywhere between two **terminators**, T1 and T2, located at 28.1 min and 35.6 min. The terminators slow replication in the counterclockwise and clockwise directions, respectively. A gene (*tus*) near T1 encodes a **terminator utilization substance**, a DNA-binding protein that associates with T1 and T2 and causes termination.[398–401] Another problem may be the separating of catenated DNA circles by action of a topoisomerase. Finally, it is essential to **partition** the original chromosome and its replica, one to each daughter cell. This requires at least three other gene products including one large 170-kDa protein.[402]

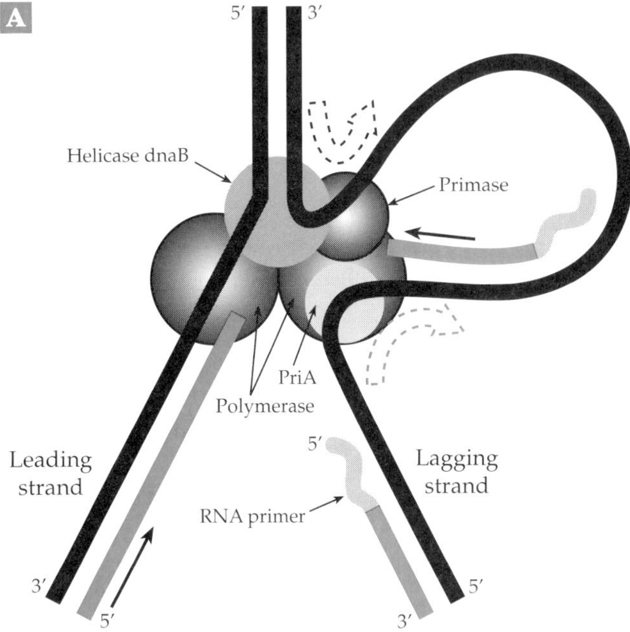

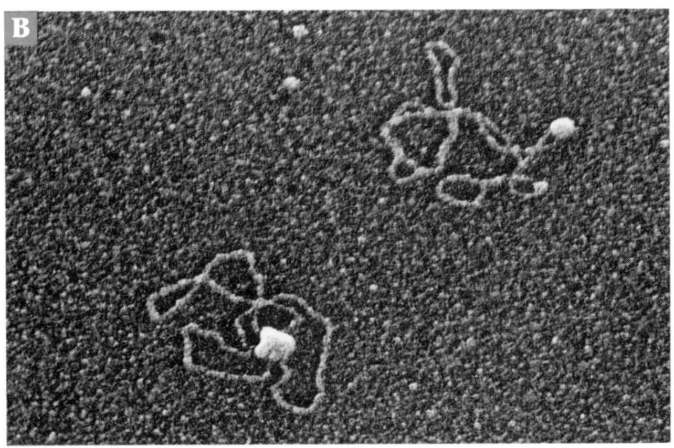

Figure 27-20 (A) Hypothetical replisome for concurrent replication of leading and lagging strands by a dimeric polymerase associated with helicase dnaB and a primosome. Open arrows indicate directions of movement of DNA, which is forming a loop as the polymerase fills a gap to complete an Okazaki fragment. The primase will then form a new primer and a new loop. From Kornberg and Baker.[265] (B) Electron micrograph of the primosome bound to covalently closed ϕX174 duplex replicative form. These enzymatically synthesized duplexes invariably contain a single primosome with one or two associated small DNA loops. From A. Kornberg in Hubscher and Spadari,[266] pp. 9,10.

5. The Replication of Viral DNA

The replication of viral DNA usually depends upon the genes of both the host organism and the virus. For example, *ts* mutations in the *E. coli* genes *dnaB*, *D*, *E*, *F*, and *G* lead to a loss in ability to support growth of phage λ as well as loss of ability to reproduce under conditions where the *ts* gene products are inactivated. However, the phage can replicate in *E. coli* with mutated genes *dnaA* and *C* because phage λ encodes its own initiator proteins by genes *O* and *P* (marked on the gene map in Fig. 26-4). In addition to these two proteins, seven *E. coli* proteins are required to initiate replication at the lambda origin *ori* λ and to complete replication.[403–407] The *E. coli* dnaB helicase and the dnaC protein are needed, as in Fig. 27-19. The chaperonins dnaJ, dnaK, and GrpE are also necessary for replication of phage λ and other viruses.[408,409] As we have seen (Section 2) many viruses contain genes specifying their own DNA polymerases and primases, which function in cooperation with host proteins. For many dsDNA viruses the origins of replication, priming reactions, and chain elongation processes closely resemble those of *E. coli*.[363,410]

In contrast, the first step in replication of the filamentous Ff viruses (f1, fd, and M13) or the small icosahedral ϕX174 or G4 is conversion of the single-stranded closed circular DNA molecules of the infecting virus particles into circular double-stranded **replicative forms** (RF).[265,411,412] This occurs as the DNA enters the bacterial cell and is accomplished entirely by the enzymes of the host cell. Phage G4 DNA, whose replication has the simplest known requirements, contains a tight hairpin region at its origin. The rest of the DNA must be coated by SSB for replication to occur. The hairpin resists melting and serves as a binding site for *E. coli* primase. This is the only known case in which no other priming proteins are needed.[265] Primase synthesizes up to a 29-ribonucleotide primer after which DNA polymerase III holoenzyme copies the rest of the chain.

Replication of the closely related ϕX174 is more complex. It requires assembly of a **primosome** made up of at least seven host proteins[265,371,413]: dnaB, dnaC, primase (dnaG), and proteins priB, A, C (n, n′, n″), and dnaT (i), Table 27-2. The 76-kDa helicase priA (n′) may locate the **primosome assembly sequences**, which are ~70 nucleotides in length,[414] and displace SSB from them. These sequences can adopt secondary structures with a pair of hairpin loops.[265] Kornberg and associates suggested that the same kind of primosome formed at these sites in the ϕX DNA may participate in replication of the lagging strand of the chromosomal DNA. If so, helicase priA presumably functions in the replisome on one strand and dnaB helicase on the other as depicted in Fig. 27-20.

In the second and third stages of replication ϕX174 RF molecules are themselves replicated and are then used for synthesis of new viral (+) strands. At both stages a virally encoded **gene A protein**, which has endonuclease activity, nicks the duplex. Cutting the (+) strand it leaves a free 3'-OH on DNA residue 4305, while the 5'-phospho group of residue 4306 becomes covalently attached to a tyrosyl residue in the A protein.[411,415] The free 3'-OH serves as the primer for a **rolling-circle synthesis** (Eq. 27-7).[412,416,416a] As a new viral strand is synthesized along the complementary (-) strand as a template, the original viral DNA (+) strand is displaced (Eq. 27-7) as a single-stranded tail.

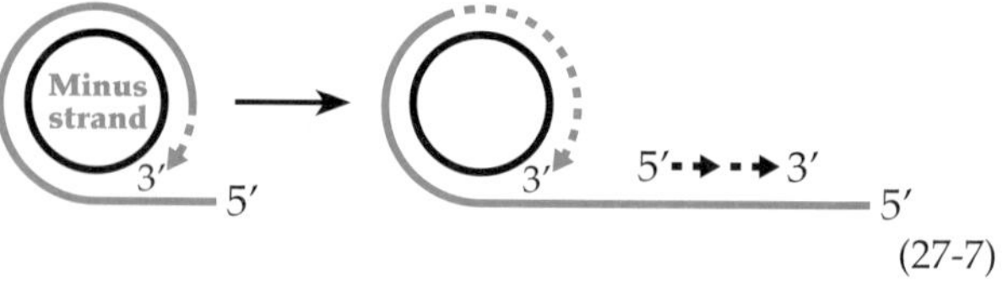

(27-7)

A strand complementary to the single-stranded tail is then formed in segments. A complete turn of the circle produces a viral strand twice the normal length. Cleavage by the endonuclease activity of the gene A protein and closure of the circle completes the replication. The displaced (+) strand can be cut off and either incorporated into a progeny phage or converted into another RF circle. The A protein, attached to the 5'-terminus of the (+) strand, is involved in either case. It can participate in repeated sequences of initiation and termination of viral (+) strand synthesis.[265,411] Once double-stranded circles are formed, they undergo several replications to give additional RF circles, which serve as templates for the synthesis of many single strands of viral (+) DNA, which are incorporated into the mature viruses. This synthesis of additional RF circles requires transcription of some viral RF genes.

In the final stage of replication the single-stranded (+) chains formed by the rolling-circle mechanism are packaged into phage particles. The gene 5 single-strand binding protein of M13 coats the DNA chains as they are formed, evidently preventing their conversion to RF circles. In the case of ϕX174 the new single-stranded DNA circles are packaged as they are synthesized (Fig. 7-28) to form complete icosahedral virus particles. See Kornberg and Baker[265] for details about these and many other virus replication systems.

Replication of the larger tailed viruses, which have many genes, is complex and varied. The lytic phage λ resembles the smaller viruses in using the host replication enzymes.[265,417,418] In the final stages a rolling-circle mechanism is utilized to form **concatemers** consisting of linear DNA duplexes with numerous successive copies of the viral DNA. The ssDNA that is formed in the rolling circle is converted to dsDNA as it is formed. Finally a **terminase** cleaves the DNA at specific *cos* sites, using staggered cuts, to form cohesive ends.[417,419–420a] However, there are uncertainties.[421] The linear dsDNA enters an empty preformed procapsid, apparently pumped in an ATP-dependent fashion, perhaps by a rotating portal ring.[421a]

6. Packaging of Viral Genomes

The construction of intact virus particles from the genomic DNA and protein subunits is often a complex process. It is simplest for the small filamentous ssDNA viruses (Fig. 7-7). The subunits are synthesized as soluble proteins, which enter the cell membrane, then lose their leader sequences. As the viral DNA coated by the viral gene 5 ssDNA-binding protein enters the membrane, the binding protein is replaced by the coat subunits.[422–424]

The process is somewhat more complex for the icosahedral viruses. In the φX, G4, α3 family the icosahedral procapsid is constructed with the aid of both internal and external **scaffolding proteins**[425,426] as is illustrated in Fig. 7-28. In phage that replicate via concatameric dsDNA the terminase that cleaves the DNA also interacts in a precise way with the packaging apparatus of the prohead.[427–429] For the tailed phage the ring-shaped oligomeric head-tail connector (Fig. 7-29), together with an ATPase, may function as a rotatory pump to feed the DNA into the prohead.[430–430c] This has been demonstrated for phage φ29.[430a–c] In some cases the terminase produces new phage DNA of unit size, but in other cases, e.g., with phage T4, the DNA may be cut more randomly when the head is full or when another piece is needed to fill it.[431] After the DNA is packaged the virus capsids usually expand and become stronger.[432,432a]

7. Plasmids

Most bacteria contain plasmids which are self-replicating but stably maintained at well-defined numbers of copies per cell.[265,433] They are usually not essential to the cell but may carry traits such as antibiotic resistance or toxin formation that benefit the bacterial host. A plasmid always carries in its DNA an origin (*ori*) of replication and a gene, usually designated *rep*, for an initiator protein. It usually encodes other proteins as well but may depend largely on host proteins for replication. Plasmids may use the *oriC* copied from the bacterial host's DNA, the origin from phage λ, yeast autonomously replicating sequences (ARSs, Section 10), or other origins. Replication of the small 6.6-kb plasmid ColE1, which is present at ~20 copies per cell, depends entirely on the host-cell replication machinery.[265] However, the control of copy number depends upon synthesis of **antisense RNA** and its reaction with the plasmid DNA (see also Chapter 28, p. 1615).[434,435] Similar copy number control is used by the larger ~100-kb resistance factor **R plasmids**.[436] Some plasmids use replication systems very similar to those of viruses such as φX174, often using rolling-circle replication.[437] However, the plasmids lack the proteins for virus coat formation and maturation.[438]

The F factor plasmids, discussed in Chapter 26, are large 100-kb circular DNA molecules containing ~60 genes, about 20 of which encode proteins involving transfer of DNA into another bacterial cell (Fig. 26-3).[265,439,440] F plasmids display strict copy number control with only 1–2 copies per host chromosome. The controls lie in a region known as the partition locus, which resembles regions of the host chromosomes that are involved in partition of the bacterial genome. They have repetitive sequences suggesting a similarity to centromeres of eukaryotic chromosomes.[441]

8. Chromosome Ends

The T-odd bacteriophages T1, T3, T5, and T7 are medium-sized phage with linear duplex DNA genomes. Replication of linear DNA in these and in many other genomes presents a problem. Even if the RNA primer segment is made at the very 3' end of the template strand, there will be a gap in the final replicated strand when the primer is digested out. Since there is no known enzyme that will add to the 3' end of a chain, this gap will remain unfilled. The problem is solved by **terminal redundancy**, the presence of a common 260-nucleotide

sequence at both ends. Several daughter DNA molecules with gaps at the 5' ends can be joined by their cohesive ends to form a long **concatamer**. DNA polymerase I fills any gaps, and the chains can then be ligated and cleaved at different points to generate complete 5' ends.[442]

Another mechanism, which is utilized by some single-stranded parvoviruses,[416,443] obviates the need for an RNA primer by use of a palindromic sequence to form a hairpin loop (Eq. 27-8).

Yet another solution to this problem is used by some viruses. **Phage φ29** of *Bacillus subtilis* primes the replication of its 19,285 bp dsDNA at both ends by a **terminal protein,** which is linked covalently through its Ser 232 –OH group to dAMP.[443a] The 3'–OH of the deoxyadenosyl group primes the DNA replication. In a similar fashion replication of the eukaryotic **adenoviruses**, whose genome is a 35- to 36-kb linear DNA duplex, starts at the ends and is primed by one residue of dCMP covalently attached through a 5'-phosphodiester linkage to a serine side chain in a 80-kDa **preterminal protein**. It substitutes for the RNA oligonucleotides that prime most DNA synthesis.[444–447] The dCMP pairs with guanine at the 3' terminus of the template strand and provides the initiating 3'-OH group. During the replication the

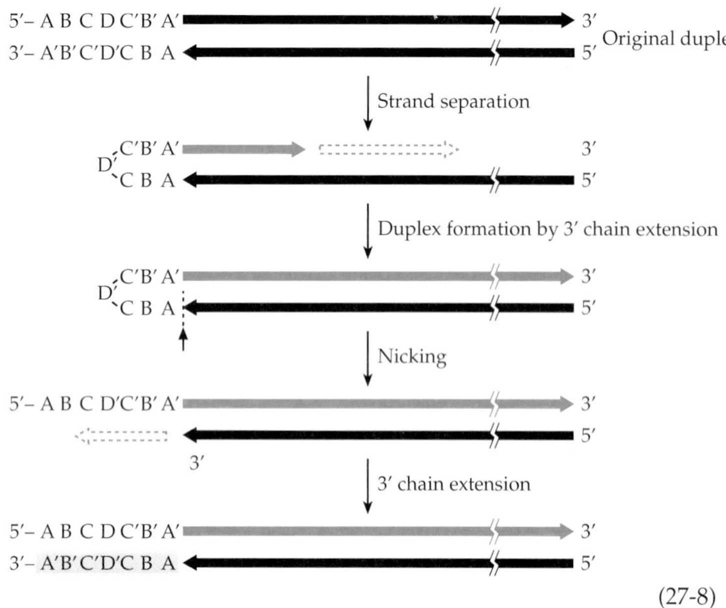

(27-8)

preterminal protein is cleaved to the 55-kDa **terminal protein,** which remains covalently attached, one molecule at the 5' end of each strand. The genome can be replicated *in vitro* by five proteins: the virally encoded preterminal protein, DNA polymerase, DNA-binding protein, and two cellular transcription factors that bind in the adenovirus origin region.

The chromosome end problem is solved in another way in eukaryotes. As discussed in Section B,1, **telomeres** (chromosome ends) contain repeated sequences of variable length. One DNA strand is always G-rich. For example, in human cells the sequence 5'-(TTAGGG)$_n$ –3', where n may be ~20, oc- curs at the 3' ends of the G-rich strands. The other strand, whose 5' end is at the telomere end, is C-rich and has the complementary sequence 3'-(AACTCCC)$_n$ –5'. The 3' end of the G-rich strand is always longer by 12–16 nucleotides than the end of the C-rich strand. This 3' extension may fold back to form non-Watson–Crick structures that apparently involve G-quartets (Chapter 5).[447a,b] The shorter 5' end is thought to result from the need for a short RNA primer during replication. As shown in Fig. 27-21, when replicated in the normal fashion the full G-rich leading strand will be formed, but the C-rich lagging strand will be 8–12 nucleotides short, when the RNA primer is digested away. A result of this is that human somatic cells gradually lose telomeric repeats. However, in tumor cells, germline cells, and unicellular organisms the enzyme **telomerase** prevents this telomere loss.[448–449a]

Telomerase is a reverse transcriptase that copies the DNA sequence of the telomeric repeats from a small **guide RNA** that is part of the enzyme. The first telomerase studied was the relatively abundant enzyme from *Tetrahymena*. It contains a 159-nucleotide RNA with the sequence 5'-CAACCCCAA-3' at positions 43–51. This sequence is complementary to the 5'-TTGGGG-3' repeat sequence of the *Tetrahymena* telomeres.[450,451] A 127-kDa human protein contains a similar guide RNA with the sequence 5'-CUAACCCUACC-3', which is complementary to the human telomere repeat sequence as is illustrated in Fig. 27-21.[452–454] Telomerases[455,456] evidently allow the cell to elongate the telomere 5'-ends using the

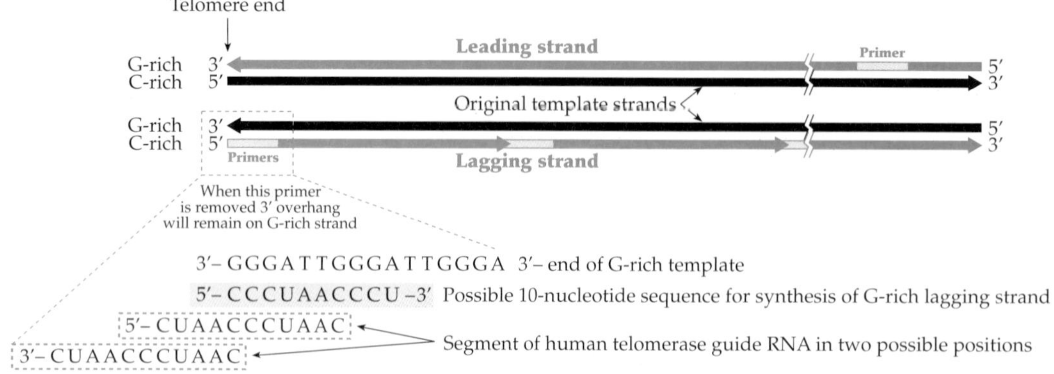

Figure 27-21 Aspects of telomere synthesis. The end of the chromosome and the 5' end of the C-rich strand is at the left. This is the template for replication of the leading G-rich strand (green). The primer lies far back in the chromosome. The C-rich strand is replicated in segments from several RNA oligonucleotide primers, one of which lies at the 5'-terminus. This first primer is removed by RNase activity leaving a 12–16 nucleotide 3'-overhang. The telomerase guide RNA can hybridize with the 3'-end of the G-rich strands providing a template that allows additional growth of the G-rich strand and extension of the C-rich strand also in the next replication.

guide sequence and the reverse transcriptase activity of the telomerase. Any number of additional repeats may be added to the 5' ends. The shortened 3' ends can also be lengthened in the next round of replication.

The control of telomerase must be important. The enzyme is active in early embryonic cells and some stem cells. However, most normal cells have little or no telomerase activity and lose telomere length throughout their lifespan with eventual growth arrest.[448,454,457] On the other hand, excessive telomerase activity may induce cancer.[451,458,459] Certain mutations in the telomerase guide RNA can cause greatly increased telomerase activity.[459] The control of telomerase is still poorly understood but involves specific telomere-binding proteins.[448,454,460–460b]

9. Mitochondrial and Chloroplast DNA

Replication of the ~16-kb mammalian mtDNA begins with RNA priming within a small **displacement loop** or D-loop. One daughter strand, the heavier or H strand, starts to grow on the primer. As it does, the parental H strand is displaced and the D-loop is enlarged. The H-strand grows until ~70% of the parental H strand has been displaced and the L-strand origin (Fig. 18-3) is uncovered. Then a new light L strand is laid down to form the second daughter duplex. The rate of formation of the new L strand is only 10 nucleotides / s, an hour being required to complete the process. The DNA formed is initially relaxed, another 40 min being needed to introduce the 100 superhelical turns present in the finished chromosome.[461]

The kinetoplast DNA of trypanosomes (Fig. 5-16) consists of thousands of catenated circular DNA molecules. Among these the smaller minicircles always contain the sequence GGGGTTGGTGTA at their origins of replication. The minicircles are individually removed from the mass prior to replication. The two progeny circles are then both recatenated into the mass.[462]

Chloroplasts contain large 120- to 169-kb circular genomes encoding about 100 proteins (Chapter 23). A characteristic feature of most chloroplast DNA is the presence of long inverted repeat sequences (10,058 bp in the liverwort, 25,339 bp in tobacco).[463,464] These are separated by 19,813 and 81,095 bp single copy regions in the liverwort and by similar sized regions in tobacco. Plastid DNA exists as a mixture of monomeric molecules with smaller amounts of dimers, trimers, and tetramers.[464]

Ethidium bromide inhibits the replication of chloroplast DNA and causes partial degradation of existing DNA in chloroplasts without interfering with replication of DNA in the nucleus. The effect is similar to that of the same drug on mitochondrial DNA. However, cells of *Chlamydomonas* treated with ethidium bromide are able later to regenerate their chloroplast DNA. This result has been interpreted to mean that there may be one or a few "master copies" of chloroplast DNA in specially protected locations. The result should also be considered in relationship to the following observation. Although nuclear and organelle DNA molecules replicate at different times in the cell cycle, constant proportions of the organelle and nuclear DNA tend to be maintained. Thus, there must be some kind of control mechanism leading to a coupling of DNA replication in nuclei, mitochondria, and chloroplasts.[465]

10. Replication of Eukaryotic Nuclear and Viral DNA

Replication in eukaryotes is similar in many ways to that in bacteria.[284,466–467a] However, the ~10^6 kb of DNA in a typical eukaryotic genome is divided into many **replicons**, segments of DNA 30–150 kb in length, each having its own origin of replication. In the relatively small ~14-Mb yeast genome there are ~400 replicons,[468,468a] but in mammalian DNA and also in plant DNA[464] there are probably thousands. DNA synthesis is initiated at different times during the S-phase of the cell cycle at the various origins in an ordered pattern.[469,469a,b] An important unanswered question is how the cell is able to replicate all segments of all of the chromosomes just once before entering mitosis.

Many of the proteins of eukaryotic replication are closely related in sequences and functions to those of bacteria. There are initiator proteins analogous to *E. coli* dnaA (Fig. 27-19). The DNA polymerases have been discussed in Section C, 2 (see Table 27-1). Eukaryotic polymerases α and δ and possibly ε are essential for replication.[470,471] Polymerase α, which is inhibited specifically by the fungal metabolite **aphidocolin**,[472] is a complex of a ~170-kDa DNA polymerase core, an RNA-synthesizing primase consisting

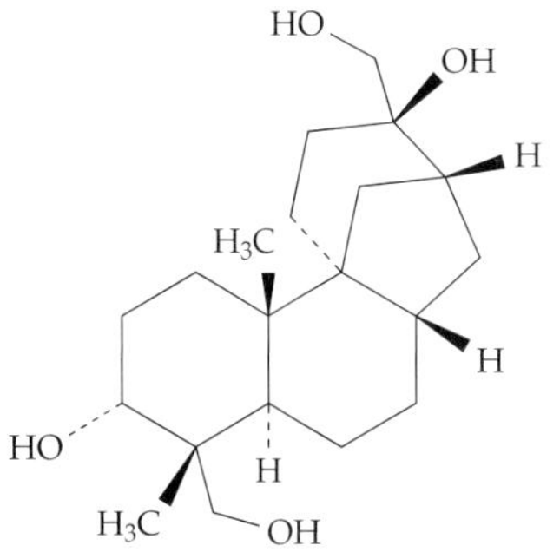

HO OH H H₃C H HO H H₃C OH

Aphidocolin

of 58- and 49-kDa subunits, and a 70-kDa subunit of uncertain function.[473–474a] The complex makes an RNA-DNA primer consisting of ~10 nucleotides of RNA and ~30 of DNA.[475] This pol α/primer is replaced early in replication by the highly processive polymerase δ and, perhaps, under some conditions by polymerase ε. The ringlike processivity factor or "clamp" that is provided in *E. coli* by protein dnaB is called the **proliferating cell nuclease antigen (PCNA)** in eukaryotes.[476] It is loaded onto the DNA by a **clamp loader**, the 5-subunit replication factor C (RFC).[476a] As in *E. coli* (Fig. 27-19) an SSB type protein known as **replication protein A (RPA)** is also essential.[467,477] PCNA is not only essential to eukaryotic replication but is also required for recombination and repair.[476] The polα/primer primes leading strand synthesis initially but then switches to replication origins on the lagging strand where, together with other proteins, it primes the formation of the Okazaki fragments. Polα/primer is also a logical participant in the control of the initiation of the S-phase of the cell cycle.[478]

Eukaryotic viruses. Investigation of viruses provided the first insights into eukaryotic DNA replication. Most of the factors needed for replication of the DNA of adenoviruses, simian virus 40 (SV40), and polyomavirus[447] within animal cells are supplied by the host. Replication of the 5-kb SV40 DNA, whose DNA forms typical nucleosomes (Fig. 27-3), appears to be an excellent model for eukaryotic replication in general.[479–482] The single SV40 origin of replication is a 64-bp sequence containing the 5-bp sequence GAGG C four times as pairs of inverted repeats. These are recognized by the 95-kDa virally encoded initiation protein which also has helicase activity and is known as the **T antigen**.[183,183a] A nearby 17-bp sequence containing only AT pairs is presumably the region of entry of the host cell's polymerase α/primer. Single-stranded DNA regions are coated with the replication protein A. After the primer is formed, the RFC complex loads the sliding clamp PCNA, and polymerase α is replaced by polymerase δ on both leading and lagging DNA strands permitting highly processive bidirectional chain elongation. Topoisomerase activity is required to decatenate the replicated chromosomes. Since SV40 DNA forms typical nucleosome (Fig. 27-3), its replication is thought to mimic chromosomal replication quite closely. The more complex herpes simplex virus HSV-1 has a 153-kb genome, a linear DNA duplex. It has ~75 genes and encodes its own DNA polymerase, origin-binding protein, SSB, and other proteins needed for replication within eukaryotic cells.[484,485]

Artificial chromosomes. Another approach to understanding eukaryotic replication, similar to the use of *ori*C plasmids in *E. coli*, is to study **autonomously replicating sequences (ARSs)**[469,486] and plasmids[468] and **artificial chromosomes**[487] made from them. ARS sequences were first found in the budding yeast *S. cerevisiae*. Plasmids containing an ARS, whose core consensus sequence is 5'-(A/T)TT TAT(A/G)TTT(A/T), replicate autonomously during S-phase. Such plasmids have been genetically engineered, providing them with telomeres and some kind of functional centromere, to form artificial chromosomes. **Yeast artificial chromosomes (YACs)** have become extremely important as cloning vehicles (Chapter 26), and they also serve as important tools for studying eukaryotic replication and its control. They can be cultured in yeast cells or can be transferred into animal cells, etc.

As mentioned in Section B, 1, human centromeres are rich in the repetitive α-satellite DNA. By joining α-satellite DNA-containing fragments of the X-chromosome to cloned telomeric DNA, human **minichromosomes** have been created.[488] These have been developed into **human artificial chromosomes**,[489] which may be practical vehicles for gene transfer in human therapy.

Replication of nuclear DNA. The budding yeast *Saccharomyces cerevisiae* has permitted the most detailed picture of DNA replication in a eukaryote. The complete genome sequences are known and the ARSs have been physically mapped.[468a] For example, in chromosome VI there are nine origins that differ in frequency of initiation and which replicate sequentially during the S-phase of the cell cycle.[490] The initiation (**replicator**) regions surround the 11-bp consensus sequence of the ARSs, each occupying at most ~150 bp. However, in metazoa and even in the fission yeast *Schizosaccharomyces pombe* the ARSs range from 500 to 1500 bp in length. These origin regions frequently overlap the **promoter** sequences, which control initiation of transcription (Chapter 28).[491] This association with transcription origins has also been observed in metazoan cells, where replication origins are often clustered.[492] However, there is no sequence homology between the ARSs of *S. cerevisiae* and replication origins in other species, even those of *S. pombe*.[490]

The study of replication in yeast ARSs and artificial chromosomes has revealed that initiation of replication requires not only an initiator protein but a complex of six proteins that form an **origin recognition complex (ORC)**.[493–495a] This complex, which is essential to initiation of replication, may be joined by additional proteins in a **prereplication complex**. At least some of the ORC proteins have their homologous counterparts in metazoa, suggesting a highly conserved initiation machinery.[495a–c]

One or more of the proteins that bind to the ORC may constitute a **license** to replicate. The licensing concept states that when replication occurs the license is destroyed and the origin involved cannot initiate replication again without a new license. The **replication licensing factor** (**LRF**) is postulated to be unable to pass through the nuclear membrane.[494,496] It can only reach the replication origin after the S-phase has concluded and mitosis has taken place. At this time the membrane has been disrupted. A second signal, the **S-phase promoting factor** (**SPF**), cannot act without an intact nucleus and a license in place.[469,495,497–500] This system ensures that DNA is replicated only once per cell cycle. Among the proteins involved in the licensing is a group of **minichromosome maintenance** (**MCM**) **proteins**, so-named because of their importance to replication of ARSs and artificial chromosomes.[495] Six of these proteins (MCM2–MCM7) can form a hexameric complex with one subunit of each type as well as other complexes, e.g., $(MCM4,6,7)_2$.[500a–c] The latter acts as an ATP-dependent helicase. A somewhat simpler MCM complex is found in archaea.[500d] Some of the cell cycle proteins (Chapter 26, Section F,1), including Cdc6 and the protein kinases Cdc7 and Cdc28 as well as other proteins, are also required for regulation of replication origins. Proteins homologous to those of yeast have been identified in humans and other eukaryotes. Licensing of replication involves association of the MCM helicases with each ORC during the G1 phase of the cell cycle. Binding of the initiation factor Cdc6/18 and of a recently discovered loading factor **Cdt-1** apparently completes the licensing. Once licensing has occurred both cdc6/18 and Cdt-1 can dissociate from the DNA.[496] Removal of Cdt is facilitated by its binding to another protein **Geminen**, found first in the frog *X. laevis*.[500e–g] The ORC complexes may remain at the origins. It has been estimated that a yeast cell contains ~400–600 molecules of the very stable ORC, about one ORC per replication origin. However, a large excess of MCM proteins may be present. Their concentrations may regulate the number of ORC molecules that associate with DNA. Replication of DNA is not the only aspect of cell growth. For example, as DNA is replicated histones must be synthesized and assembled. This synthesis occurs during the S-phase and is tightly coupled to replication.[500h]

Initiation of replication in metazoans is still confusing. Almost any piece of DNA will be replicated if introduced into a *Xenopus* egg, where initiation appears to occur just once at a random position.[501,501a,b] However, in differentiated tissue the origins of replication seem to be fewer in number and more specifically located. A possible explanation is that high concentrations of ORC and MCM proteins in the embryo may lead to many relatively nonspecific origins and a replicon size of ~7 kb. The lower concentrations of these auxiliary factors in somatic cells may lead to fewer but more specific origins with a replicon size of ~170 kb.[495] Three distinct mammalian origins have been studied in detail. One is in the β globin locus (Fig. 27-10).[500] A second is near the dihydrofolate reductase gene.[502] A third, which is activated early in S-phase, is at the 3' end of the lamin B2 gene.[503] The latter has been localized to a 500-bp region. These findings suggest that the replicon concept, as developed for yeast, may be generally applicable.

Replication of the intact genome of *Drosophila* has been studied in rapidly dividing nuclei by electron microscopy.[504] The replication rate in these nuclei is ~300,000 bases / s, but it has been estimated that replication forks in animal chromosomes move no faster than ~50 bases / s. Thus, we would anticipate at least 6000 forks, or one fork per 10 kK bases. Indeed, this number of forks has been observed.[505] They occur in pairs with many short regions containing single-stranded DNA as if one strand at the fork is replicated more rapidly than the other as in mitochondrial DNA. The arrangement of the ssDNA regions at the two forks in a pair suggests bidirectional replication. However, replication forks are rarely seen in higher eukaryotes, but extensive regions of single-stranded DNA are often visible. Benbow and associates suggested that in higher eukaryotes the strands of duplex DNA may be separated throughout a whole looped domain of DNA. Replication could then occur with initiation at many points along each strand.[505]

Replication reactions are similar in bacteria and eukaryotes, but some details differ. In eukaryotes at least two DNA polymerases, α and δ, are required. In budding yeast polymerase ε is also essential.[506] Both polymerases δ and ε may replicate separate strands at the replication fork.[506a] Processing of Okazaki fragments also differs from that in bacteria, where either RNase H or the 5' to 3' exonuclease activity of DNA pol I removes the RNA primer (Fig. 27-14). This exonuclease activity is lacking in eukaryotic polymerases. Replication primers are removed in a two-step process by **RNase HI**, which makes an endonucleolytic attack that removes all but one nucleotide residue of the primer in a single piece, leaving a 5'-phospho group on the remaining ribonucleotides. That residue is removed by a 5' to 3' exonuclease designated RTH-1 nuclease (Eq. 27-9).[467,507,507a,b] This is a homolog of the yeast RAD27 protein. A polynucleotide kinase may then phosphorylate the 5' end of the DNA fragment.[508]

Another difference between bacterial and eukaryotic replication is the presence of nucleosomes in eukaryotes. Some evidence suggests that nucleosomes may open and close to allow replication forks to pass through.[509] Studies of SV40 minichromosomes indicate that passage of the replication machinery does destabilize nucleosomes, which must be partially reconstructed about 260 nucleotides past the elongation point.[510] Another factor is the variable extent and location of modifications to histones, in particular to

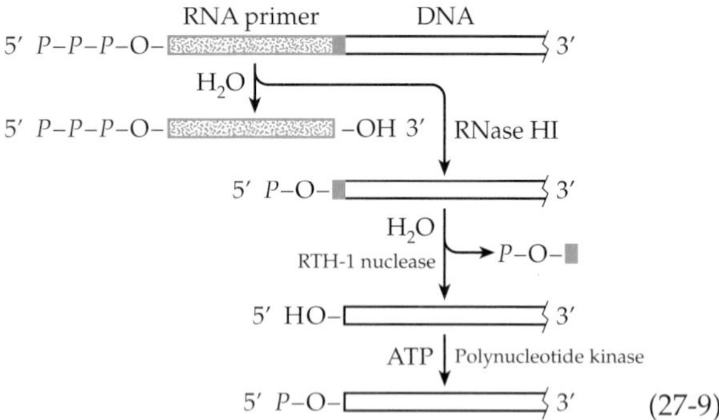

(27-9)

the H3 and H4 histone tails (Section A,3). A code has been proposed according to which certain modifications would favor transcription or mitosis, while lack of modification would silence the genes.[72]

Much of the control of replication is at the initiation stage. Growth factors and other mitogenic stimuli acting at the plasma membrane can stimulate expression of such nuclear proteins as those encoded by the proto-oncogenes *c-myc*, *c-myb*, and *c-fos*. These may initiate a regulatory cascade (Fig. 11-13) and trigger mitosis.[511] As indicated in Chapter 26, the 34-kDa protein kinase encoded by fission yeast gene *cdc*2 (budding yeast CDC28) is essential for progression of the cell cycle through the G1 phase into mitosis (Eq. 26-3). A single oscillation in this kinase activity induced by a B-type cyclin can promote both replication and mitosis. However, in *S.cerevisiae* there are 14 different cyclinlike proteins, and their individual functions are not clear.[512] The signal that is sent to the ORCs is likewise unclear.[513] However, theoretical models involving Eq. 26-3 and many additional components have been proposed.[514] Multiple phosphorylations may occur, some on the RPA initiator protein.[515] Many proteins required for replication, including DNA polymerase and primase, are associated with the nuclear matrix.[516] The nuclear membrane may also be important in controlling replication.

D. Integration, Excision, and Recombination of DNA

The exchange of genetic information between chromosomes, plasmids, and viruses occurs in many ways, which are described collectively as recombination.[517–520] Mutants of *E. coli* deficient in recombinational ability often have defects in genes designated *recA, B, C*, etc. (for recombination), or *ruvA, B, C . . .* (for resistance to ultraviolet light). Some of these mutants are unusually sensitive to ultraviolet light because of their inability to repair damage to DNA. Several of the recombination enzymes are used for repair of ultraviolet damage and of double-strand breaks in DNA arising from other causes.[521,522] In eukaryotes recombination occurs during meiosis. Many viruses, including phage, also carry genes for their own general recombination systems. In addition, the DNA of some viruses, such as the temperate phage λ, undergoes recombination with the host DNA. This can happen during the processes of integration of the viral DNA into the host genome or excision of the viral DNA during the lysogenic cycle of replication (Chapter 28, Section B). Recombination occurs around specific sites in the chromosomes of both the virus and its host and is called **site-specific recombination**. Genetic recombination is essential to the development of genomic diversity, to the survival of a species, and to evolution.

1. Recombination Mechanisms

How can the homologous regions of two different DNA duplexes be brought together? As illustrated schematically in Eq. 27-10, the strand exchange must occur at exactly the same point in each duplex. An early attempt to explain this postulated a "copy choice" mechanism of replication. It was assumed that replication occurred along one DNA strand up to some random point at which the polymerase jumped and

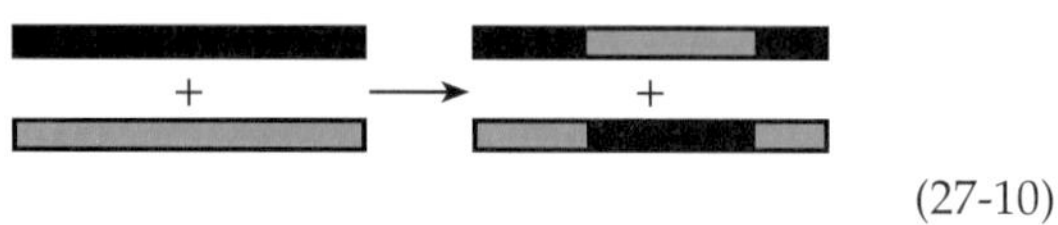

(27-10)

began to copy from the second of a pair of homologous chromosomes. The newly formed DNA molecule would be complementary to different parts of both parental DNA duplexes. To test the idea, Meselson and Weigle infected *E. coli* with two strains of phage λ containing ^{13}C- and ^{15}N-labeled DNA, respectively.[523] Recombinant DNA was found to contain some ^{13}C and some ^{15}N, as judged by density gradient centrifugation. It was clear that DNA from both parents was incorporated into DNA of recombinant progeny, a finding that ruled out the copy choice hypothesis.

If recombination occurs instead by enzymatic cutting of two homologous duplex DNA molecules followed by rejoining, how is it possible to avoid inactivation of genes by addition or deletion of genetic material? Recombination cannot depend upon the random action of a nonspecific enzyme with random rejoining. Yet, general recombination can occur at any point and with a roughly constant frequency throughout the DNA chain. The explanation of these facts lies in the occurrence of base pairing between at least some short homologous regions of strands of the two different DNA duplexes.

The Holliday recombination intermediate. In 1964, Holliday suggested a recombination process that would give rise to characteristic H-shaped intermediates.[524] Recombination could be initiated at special points on the duplexes, recognizable by a recombination enzyme (Fig. 27-22). A short amount of unraveling would be followed by strand exchange with the two broken strands being rejoined by a ligase as indicated in Fig. 27-22. The crossover points would then migrate up or down the chains as the two helices turned about their own axes. Long regions of heteroduplex DNA could be generated in this way, and the process could be terminated at a random distance from the starting point, accounting for the observed uniformity of genetic recombination events. Chain cleavage and rejoining of two of the strands would terminate the process. If these were the same strands broken in the initiation event (cleavage at points *aa'* in Fig. 27-22), genes lying outside the heteroduplex region would not be recombined, but cleavage of the other chains (at points *bb'*) would lead to their recombination. Intermediates of the type predicted by the Holliday model were soon observed by electron microscopy (Fig. 27-23).[525] Three-dimensional structures have been determined by X-ray crystallography[525a] and have been studied by atomic force microscopy.[525b]

The cross-stranded structure shown in Fig. 27-22 can be formed with all base pairs in both duplexes intact.[526,527] All that is required is formation of a nick in each of the two polynucleotide chains and a rejoining of the backbones across the close gap between the duplexes. This model also accounts for the cutting of the two crossed strands at exactly equivalent points to terminate the process. Various mechanisms of recombination exist, and most make use of the key Holliday recombination intermediate (Fig. 27-22, Fig. 5-28, Eq. 27-11).[521,528–530a] Such four-way junctions can arise in several ways.[530a–d] For example, a 3' or 5' single-stranded tail in a piece of dsDNA can "invade"

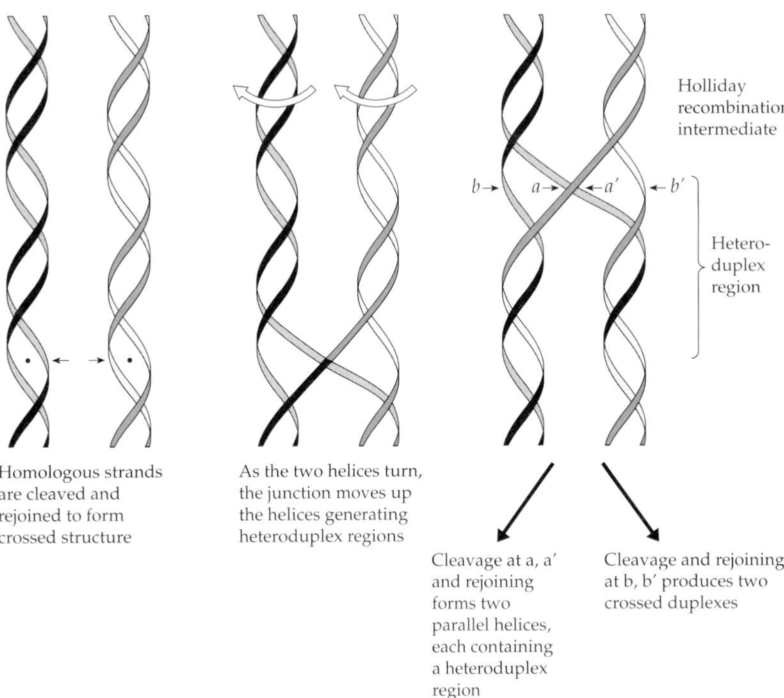

Figure 27-22 A recombination mechanism involving single-stranded exchanges. After Holliday.[524]

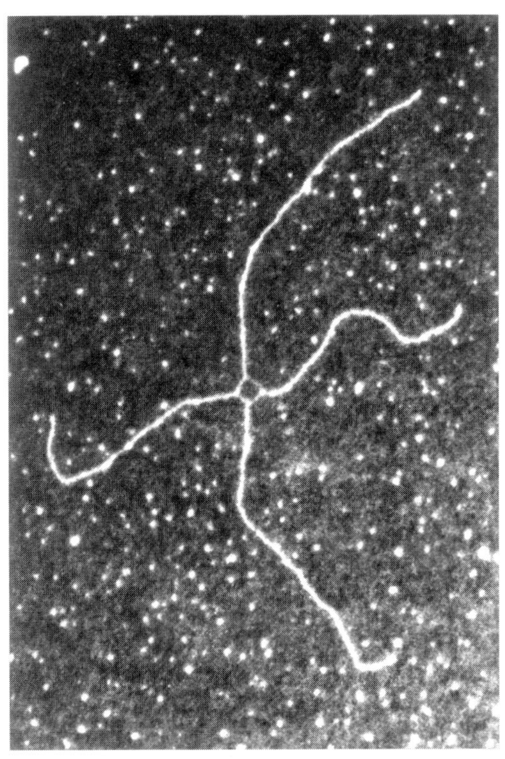

Figure 27-23 A chi form of DNA from the colicin E1 plasmid. These forms are thought to be derived from recombination intermediates of the Holliday type, which appear as "figure eight"-shaped molecules twice the length of the colicin genome. This figure eight form was cut at a specific site that occurs only once in the genome (twice in the figure eight) by restriction enzyme *Eco*R1 to give the chi form. The pairs of long and short arms are believed to represent homologous duplexes. The single strands in the crossover have pulled apart revealing the strand connections clearly. Such a structure would be expected from the Holliday intermediate (upper right corner of Fig. 27-22), e.g., if one of the two vertical duplexes were rotated end over end. From Potter and Dressler.[525]

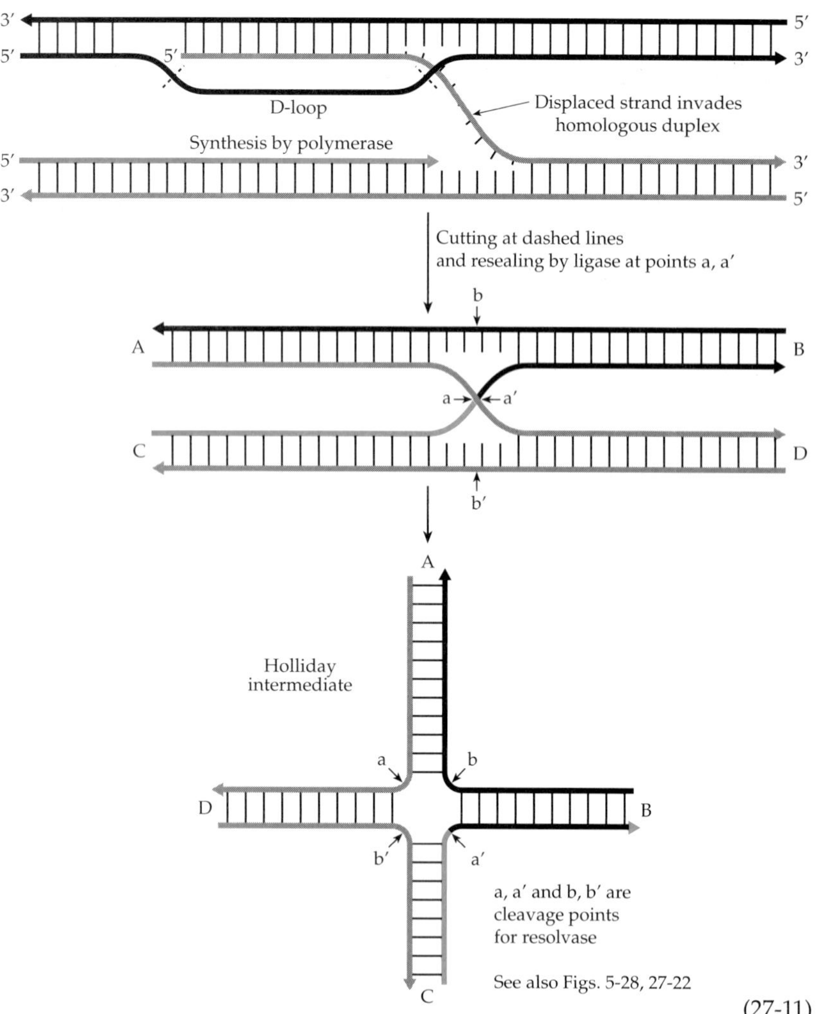

(27-11)

E. coli depends upon a dsDNA nuclease and an unwinding complex consisting of proteins RecB, C, and D.[531–533] The complex is a powerful *exonuclease*, which can digest the ends of a DNA duplex. It degrades the 3' ends most rapidly, leaving 5'-tails that can invade other homologous duplexes as in Eq. 27-11. The RecBCD complex is also an ATP-dependent helicase, which unwinds the DNA, preparing ssDNA for reaction with the strand-exchange protein **RecA**. The RecBCD complex also functions to completely degrade foreign dsDNA such as that from invading bacteriophages.[532] Why doesn't it also degrade the genomic DNA of the *E. coli* cell in which it functions? The answer lies in an eight-base DNA sequence, a recom-bination "hot spot" called **chi (χ)**: 5'–GCTGGTG G–3'. This χ sequence occurs 761 times in the leading strands for DNA replication in the *E. coli* genome.[533a–534a] When the RecBCD complex reaches a χ sequence, when approaching it from the 3' end, the enzyme stops its exonuclease action by inactivating the nucleolytic activity of the D subunit and promotes recombination about five- to ten-fold as fast as at other sites.[533,535]

another dsDNA that has a homologous sequence as indicated in Eq. 27-11. In this drawing a 5' tail has been displaced during repair of a gap in one strand. The resulting D loop may be trimmed out and a new connection made to give the Holliday intermediate. The cleavage points a, a', b, b' marked in Eq. 27-11 correspond to those in Fig. 27-22. Holliday junctions may also be formed in stalled replication forks and must be removed to allow replication and transcription to continue.[530e] The existence of the Holliday intermediate has been supported not only by electron micrographs such as that of Fig. 27-23 but also by the identification of endonucleases that carry out the necessary cleavages of synthetic Holliday intermediates that have been made artificially (see Chapter 5.) Endonucleases with a high specificity for Holliday junctions have been found in bacteria, among proteins encoded by viruses, and in a wide variety of eukaryotic cells. Additional proteins including helicases, DNA-binding proteins, and specialized **strand exchange proteins** are also required to catalyze the individual steps in the recombination process.

The main **RecBCD pathway** of recombination in

RecA and other strand-exchange proteins. The 352-residue product of the *E. coli RecA* gene is a multifunctional **recombinase,** which is required both for recombination and also for DNA repair.[536–540a] In its repair function the RecA protein acts as a DNA-dependent protease that cleaves a number of repressors in response to damage to DNA. It has a quite different role in recombination where it (1) brings a piece of single-stranded DNA (an end or a gap) together with a duplex; (2) locates homologous sequences; and (3) forms a synaptic complex in which strand exchange can occur. Electron microscopic observations[536] show that the RecA protein binds to either single-stranded or duplex DNA in a cooperative manner to form long rodlike spiral filaments (Fig. 27-24). Measurement of the lengths of RecA protein-covered duplexes shows that the DNA is underwound and stretched by about 50%. It contains ~18 nucleotides per turn.[536,539,539a] Similar filaments are formed with single-stranded DNA, 3–4 nucleotides being bound per RecA protein monomer. Formation of this ssDNA complex, which may be regarded as an initiation

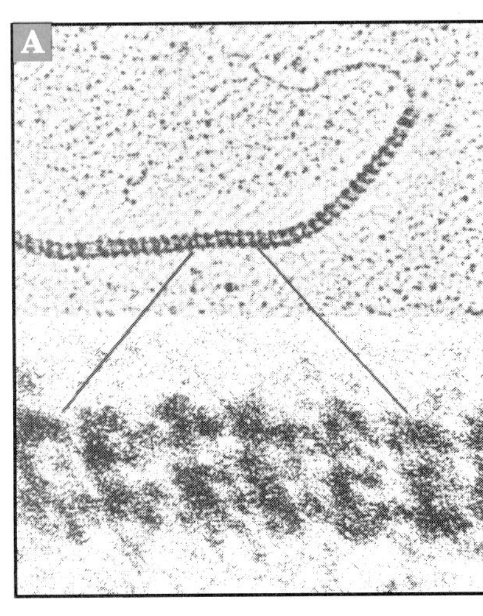

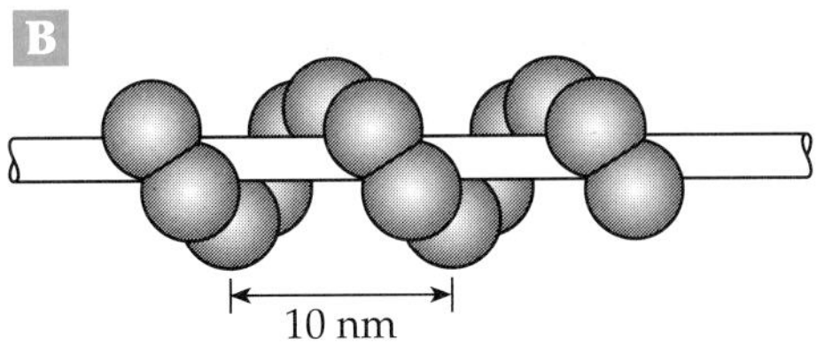

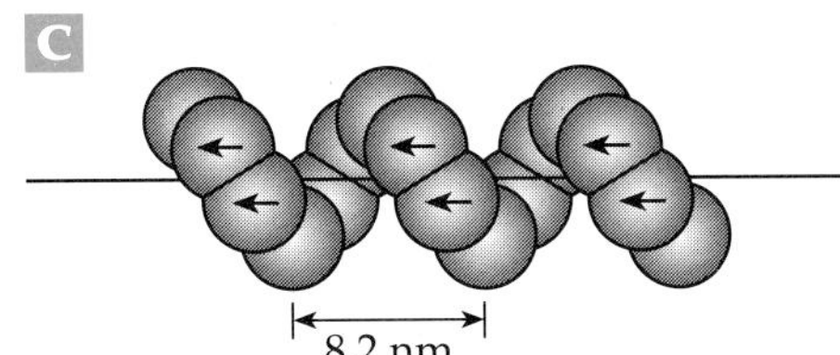

Figure 27-24 Structures of RecA protein spiral filaments. (A) RecA protein filament formed on circular duplex DNA in the presence of ATP(γ-S), shadowed with Pt and seen by electron microscopy. (B) Diagram of RecA bound to duplex DNA in the presence of ATP(γ-S), as determined by electron microscopy. RecA monomers are shown as spheres, but their exact shape is unknown. (C) Diagram of RecA spiral filament in crystals of RecA protein free of DNA, based on X-ray crystallographic data. Arrows indicated alignment of monomers. From Howard-Flanders, West, and Stasiak.[536]

complex for recombination, requires MgATP and is faciliated by prior coating of the DNA with SSB protein. The RecA protein subunits are added in the 5'- 3' direction of the DNA, and SSB is displaced in the process.[541]

The initiation complex binds to duplex DNA rapidly and more slowly promotes strand exchange. In related reactions a single-stranded SSB-coated circular DNA will bind to RecA protein, then exchange strands with a linear duplex (Eq. 27-12). The strand exchange requires ATP and advances in the 5' to 3'

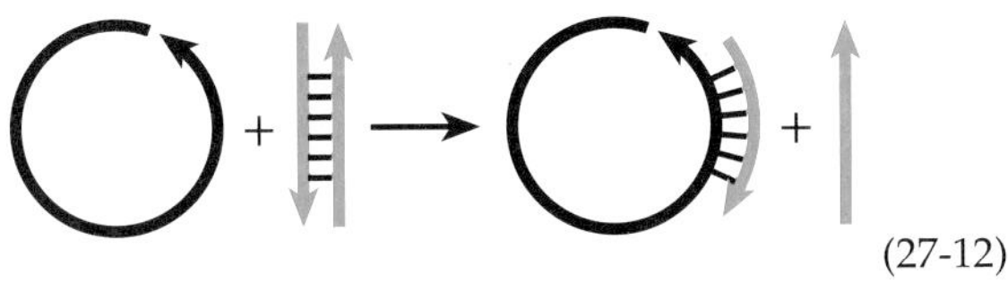

(27-12)

direction along the original single strand at the rate of a few bases / s. Strand exchange can also occur between two duplexes if there is a suitable gap in one strand, e.g., as is illustrated in Eq. 27-13.

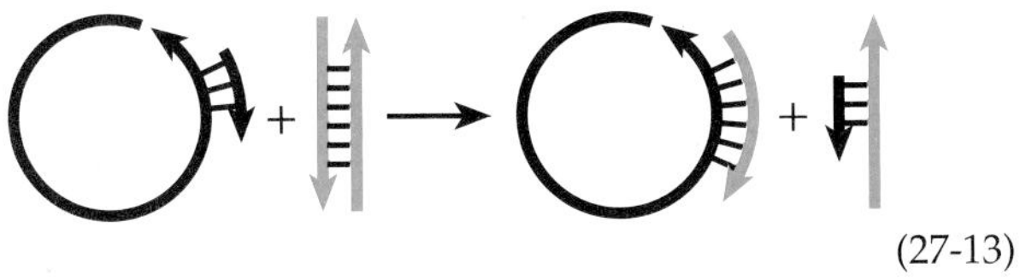

(27-13)

A possible mechanism of strand exchange is illustrated in Figs. 27-25 and 27-26. The RecA protein has binding sites that can accommodate nucleotides from two DNA molecules, one single-stranded and the other a duplex. It may also accommodate two DNA duplexes.[542] As shown in Fig. 27-25, the RecA protein could test the hydrogen-bonding between many base pairs at once in a search of homologous regions. The two DNA chains would have to either slide past each other or repeatedly dissociate and reassociate[543] until a homologous region was found. Then strand exchange could occur. As is shown in Fig. 27-26, the single strand may be wound into the major groove of the duplex to form an interwound triplex. The matching of hydrogen-bonding atoms may be an attractive way of searching for homology, but the actual search seems to substitute speed for precision. Base substitutions are quite permissive. The need for precise hydrogen-bonding has not been demonstrated, and the exact recognition mechanisms in homologous recombination remain uncertain.[544–544b] Whole chromosomes must be aligned and checked rapidly in the homology search.[545]

Many proteins similar to the RecA protein and with similar functions have been found.[546,547] These include the products of gene *uvsX* of phage T4,[548,549] the β protein of phage lambda,[550] the yeast RAD51[551,551a] and human RAD51 proteins,[552,553] a meiosis-specific human homolog of the RecA protein,[554] and corresponding proteins from plastids of higher plants.[555] Both the UvsX protein of phage T4 and human RAD51 protein yield strands of coated DNA similar to those in Fig. 27-24.

Processing the Holliday junction. Completion of the recombination process requires "resolution" of the Holliday intermediate by endonuclease action followed by ligation and perhaps by gap repair. The major recombination pathway in *E. coli* employs a binding protein, a nuclease, and a helicase encoded by genes *RuvA*, *B*, and *C*.[528] **RuvA** is a DNA binding protein specific for symmetric Holliday junctions.[529] **RuvB** is a closely associated ATP-dependent helicase.[556–558] On the basis of genetic and X-ray crystallographic evidence it is now evident that some of the functions previously attributed to RecA are carried out by the RuvABC complex. As indicated in Fig. 27-26B, RuvA binds to the Holliday junctions, holding it in the symmetric square configuration in which branch

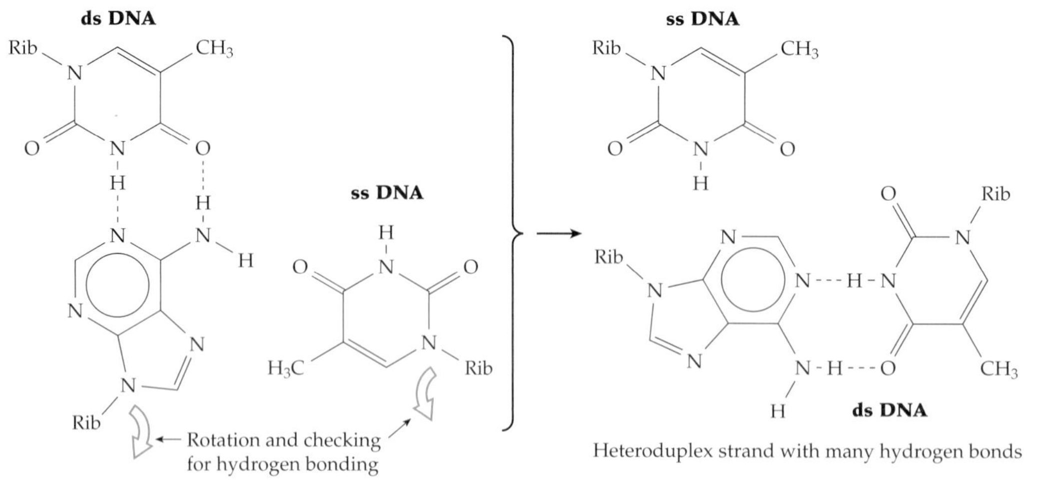

Figure 27-25 A possible mechanism for homologous pairing of an ssDNA with a duplex DNA and strand exchange. The ssDNA (right) binds together with a hydrogen-bonded duplex (left). The RecA protein rotates the bases into the heteroduplex configuration, where hydrogen bonds may be formed in many of the base pairs. After Howard-Flanders, West, and Stassiak.[536]

migration is possible. Two molecules of the oligomeric RuvB helicase apparently rotate the DNA, causing branch migration and movement of the DNA through the RuvAB complex.[559,559a,b] Under some circumstances a different helicase, encoded by *E. coli* gene **RecG,** moves Holliday junctions in the opposite direction.[560] In addition to RuvA a variety of other Holliday junction-binding proteins are known.[561] These include **p53** and the nuclear HMG proteins.[560,561]

RuvC is an endonuclease that is highly specific for Holliday junctions. It is a **resolvase** that cuts at either points a,a′ or b,b′ of Eq. 27-11 to form either "patched" or "spliced" recombinant DNA (Fig. 27-26C). Similar resolvases process bacteriophage DNA[562–564] and have also been found in yeasts and in mammals.[565,566] All are dimeric metal ion-dependent proteins.[567]

2. Nonreciprocal Recombination and Unequal Crossing-Over

The phenomenon of **gene conversion** or **nonreciprocal recombination**[568,569] was first recognized in genetic studies of fungi for which the four haploid miotic products can be examined individually (tetrad analysis; p. 20). Instead of the normal Mendelian ratio of 2:2 for the gene distribution in the progeny at a heterozygous locus a ratio of 3:1 is sometimes observed. One of the recombinant chromosomes appears to have been altered to a parental type. A reasonable mechanism by which this can occur arises from the fact that heteroduplex regions, which are present in recombination intermediates, contain defects in base pairing. One strand of the heteroduplex will have a base that does not properly pair with the base in the other strand, or will have an extra base that loops out from the heteroduplex. Since cells contain repair mechanisms that search for defects and carry out a repair process, there is a likelihood that one strand in the heteroduplex region will be altered to restore perfect base pairing, thus causing the observed gene conversion. "Flanking" genetic markers outside of the heteroduplex region are unaffected by gene conversion, and during meiosis crossing-over between these markers occurs in about 50% of gene conversion events as would be predicted from the model of Fig. 27-21. Data from yeast show that nonreciprocal recombination during meiosis may also result from double-strand breaks and gap formation followed by repair synthesis using both strands of the homologous chromosome as templates.[569]

Recombination is not limited to meiosis but can occur between homologous chromosomes during mitosis, during the G_1 period preceding mitosis, or even during the G_2 period.[570,571] Certain mutations in yeast abolish meiotic recombination but have much less effect on mitotic recombination.[572] Thus, the two processes are not identical. It has been suggested that mitotic recombination is utilized to maintain sequence homogeneity between repeated eukaryotic genes.[572,573]

Since DNA contains many repeated sequences, crossing-over sometimes occurs between locations that are not the same in the two duplexes. Such **unequal crossing-over** has the effect of lengthening one duplex and shortening the other. This may be very important in evolution. It may also, surprisingly, function to preserve homogeneity of chromosomes within a species.[574,575] For example, tandem arrays of ribosomal RNA genes (Section B,3) in yeast have 140 identical copies of their 9-kb repeat unit.[575] Unequal crossing-over between either sister chromatids or homologous chromosomes, when repeated often enough, can lead statistically to a highly homogeneous population.

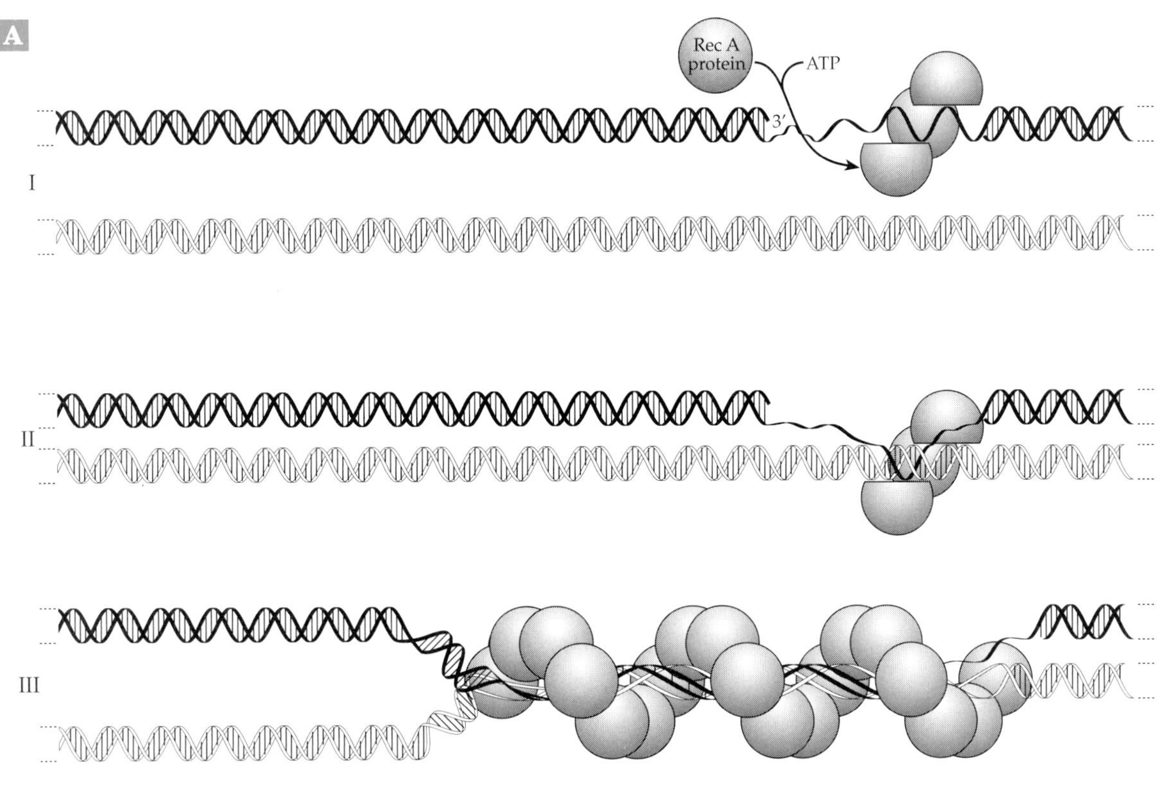

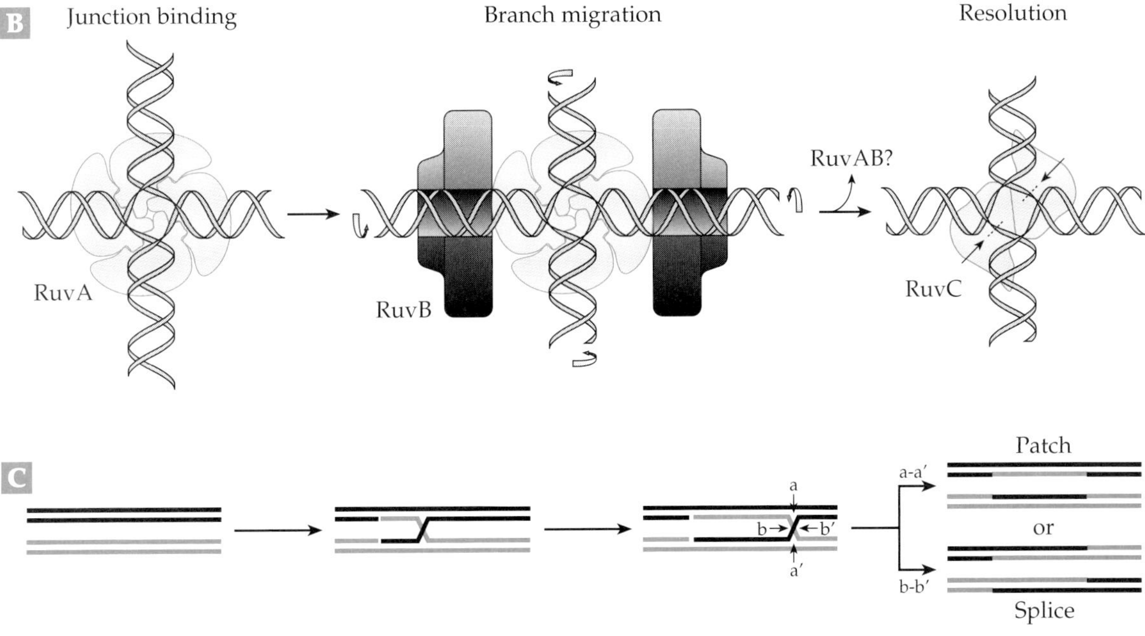

Figure 27-26 (A) Model for genetic recombination proposed by Howard-Flanders *et al.*[536] (I) RecA protein binds cooperatively to the single strand in a gapped duplex to form an initiation complex in preparation for pairing. (II) The initiation complex binds to the intact duplex, making transient contacts until a homologous site is reached. For clarity, the initiation complex is drawn with only a few protein monomers, but in reality it is likely to extend over hundreds or thousands of nucleotides. (III) When homologous contacts are made and the strands become paired locally, the initiation complex acts as nucleus for further cooperative binding, which extends the RecA spiral filament around all three or perhaps all four interacting strands. (B) Arrangement of the proteins and DNA during three stages of recombination catalyzed by the RuvABC system. The two RuvB hexameric rings are shown in cross section with the DNA passing through their centers. After Rafferty *et al.*[529] See also chapter banner, p. 1527. (C) Scheme of DNA rearrangement during homologous recombination in *E. coli*.

3. Site-Specific Recombination and the Integration and Excision of DNA

Recombination at specific sites in DNA is responsible for integration of DNA from viruses into the genome and for the cutting out of viral DNA and other pieces of DNA from the genome. The temperate bacteriophage λ and the F factors and R factors of bacteria can all be integrated into the genomic DNA of the host in this way. Genes encoded by the phage or plasmid are required. In the case of phage λ the viral genes *int* and *xis* are required for integration and excision, respectively.[576] These are not the same as the enzymes of the *rec* loci of the bacterium or the general recombination genes *exo* and *bet* of the phage. In addition, both integration and excision require an *E. coli* protein called integration host factor (**IHF**), a DNA bending protein resembling the DNA-binding HU.[18b,265,577,577a]

Integration of λ DNA (Fig. 27-27) occurs at the ~25-bp site *att B* in *E. coli* (Fig. 26-4) and the ~240-bp site *att P* in the λ chromosome (Fig. 28-11). These two sites contain identical 15-bp core sequences within which the recombination occurs. In a manner similar to that of the recA protein a homologous region is located by the complex of the Int (**integrase**) protein and IHF. Several molecules of Int protein bind and, together with the IHF protein, hold the phage DNA in a nucleosomelike structure (an **intasome**) in which the recombination occurs.[406,578–580b] Strand cleavage and rejoining occur within the short core sequence (Fig. 27-27). Parts of both *att P* and *att B* are recombined to give sites *att L* (left) and *att R* (right) in the DNA of the integrated prophage. In the integration complex the two core *att* sequences are aligned, and single-strand cuts are made at one of the points *a* or *b* that are indicated by the small arrows located on opposite strands and seven bp apart in the core sequence shown in Fig. 27-27. Rejoining of strands from the opposite duplex yields a Holliday intermediate. That this really occurs is shown by the fact that the Int protein cleaves synthetic Holliday intermediates derived from the *att* core and reseals the strands to give the expected products.[581] Cleavage of the Holliday intermediate at points a (Fig. 27-27) will lead to excision of the viral circle, but cutting at points b followed by resealing with opposite strands, as is observed, will yield integrated prophage. Although Int and IHF proteins are sufficient to promote integration, the **excisionase** encoded by phage λ gene *xis* is needed together with the Int protein for excision of the λ prophage.[576,582]

The integrase (tyrosine recombinase) family. The lambda integrase is the first recognized member of a family of a hundred or more closely related enzymes that are involved not only in integration and excision of phage DNA but also in converting multimeric forms of bacterial and plasmid chromosomes into monomers. One of the best known integrases is the 38.5-kDa **Cre recombinase**, which functions to keep the lysogenic phage P1 in a monomeric form by recombination between pairs of 34-bp core sequences designated *loxP*.

```
17                                  1 1'                                17'
5'-A-T-A-A-C-T-T-C-G-T-A-T-A-G-C-A-T-A-C-A-T-T-A-T-A-C-G-A-A-G-T-T-A-T-3'
3'-T-A-T-T-G-A-A-G-C-A-T-A-T-C-G-T-A-T-G-T-A-A-T-A-T-G-C-T-T-C-A-A-T-A-5'
                                     1'
```

A *loxP* site

Since the reaction doesn't require accessory protein factors and can be performed *in vitro* with a variety of DNA substrates, the *Cre–loxP* system is much used in genetic engineering.[583,584] A pair of related integrase subunits known as **XerC** and **XerD** perform a similar function for the *E. coli* chromosome as well as for multicopy plasmids.[585,586] The XerC / XerD system is

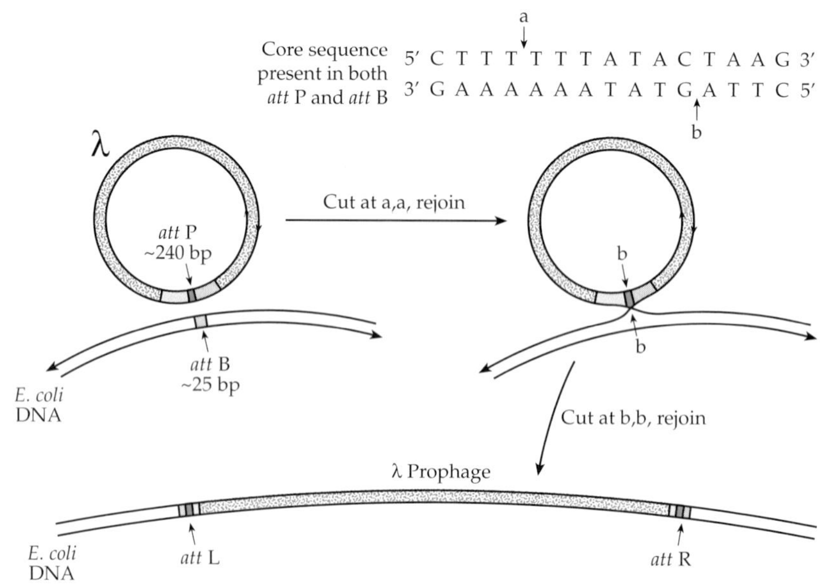

Figure 27-27 Integration of the phage λ genome into the *E. coli* chromosome at site *att*B. The same recognition sequences are present at *attB* and *attP*. These are cut at points *a* with rejoining to give the structure at center right. This is cleaved at points *b* with rejoining to give the integrated prophage.

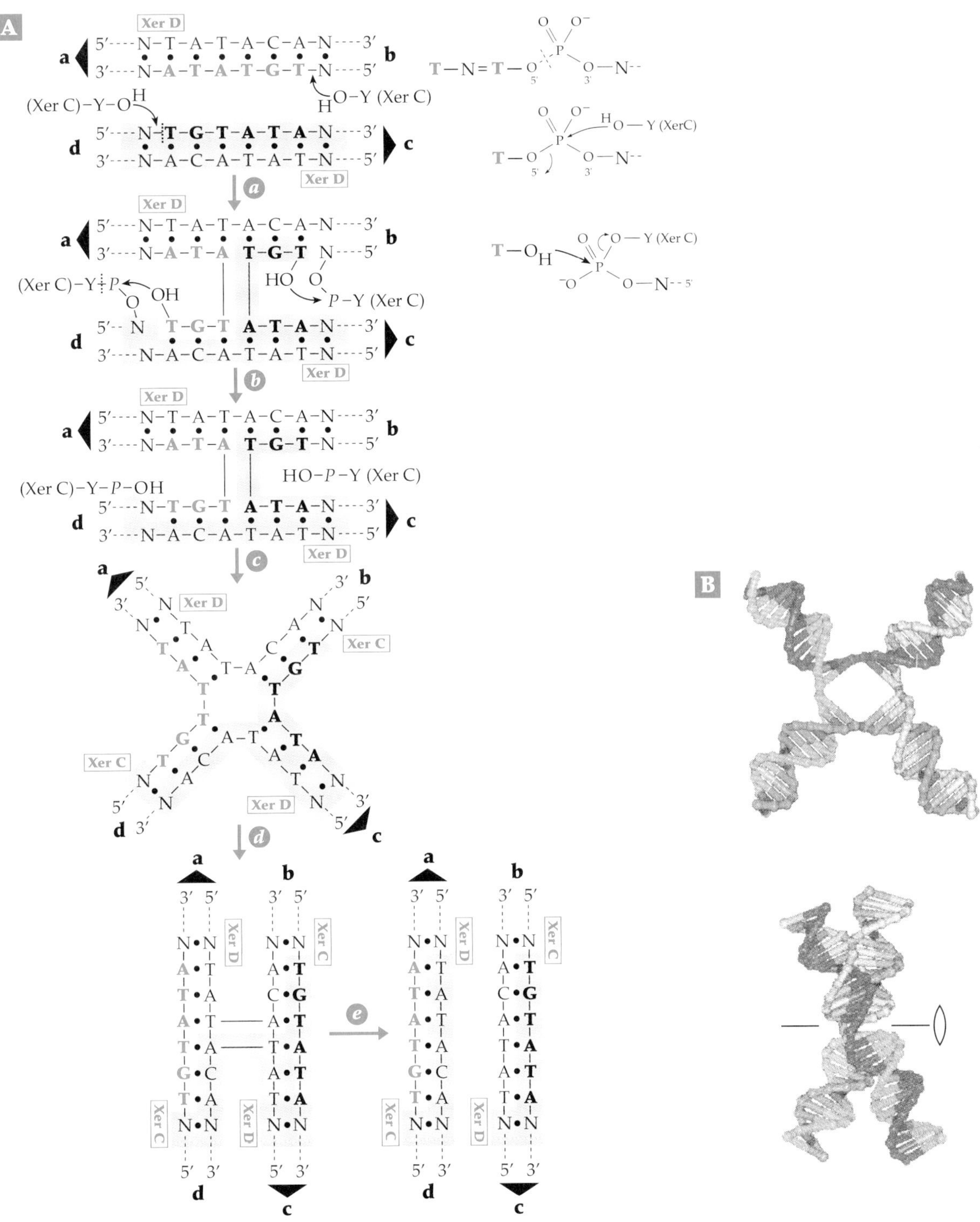

Figure 27-28 (A) Action of the integrase XerC / XerD on a pair of *E. coli dif* sites, containing the central six bp sequences TATACA / ATATGT, which are shown in an antiparallel orientation. In step *a* the active site tyrosine hydroxyl groups (Y-OH) of a pair of XerC subunits carry out transesterification reactions on the 5'-terminal thymidylate residues of the central hexanucleotide sequences to yield 3'-phosphodiester linkages to the XerC tyrosines. In step *b* the cut 5'-ends, containing free thymidylate 5'–OH groups, fold back to form new base pairs, and the strands are resealed in a second transesterification. This generates a Holliday junction, which *isomerizes* (steps *c* + *d*) to an isomeric species that is acted on by a second pair of transesterification steps (*e*) that are catalyzed by protein XerD, again with folding back of the central trinucleotides. After Arciszewska *et al.*[586] (B) The two isoforms of an antiparallel stacked X Holliday junction are shown. These can be reached from the symmetric square form shown in Fig. 5-28 and schematically in (A) by folding into X-conformations in which all base pairs are stacked either in pairs a,b and c,d or a,c and b,d. From Eichman *et al.*[525a]

atypical because it utilizes a pair of integrase subunits rather than just one. However, the basic chemistry (Fig. 27-28) is the same for the entire family.[587,587a] As with the λ integrase (Fig. 27-27) the XerC / XerD complex acts on a pair of identical core sequences that are aligned in an antiparallel fashion. Active sites in all of the integrases contain the conserved amino acid sequence Arg-His-Arg-Tyr. All of these residues are essential for catalysis.[585] Staggered cuts are made sequentially in the core sequences, e.g., at points a and b in Fig. 27-27 and adjacent to the 5'-terminal thymidylate residues in Fig. 27-28A. A transesterification reaction forms 3'-phosphotyrosyl linkages from the cut DNA to the integrase protein. The freed 5'-OH groups on the other cut end fold back and recombine with the bound 3' ends of the second duplex to generate a Holliday junction. In the mechanism proposed in this figure, three nucleotide units are involved in the folding back. The hydrogen bonds of their initial base pairs are broken, and new bonds are formed. This "base swapping" process is the equivalent of a short branch migration and verifies homology of the two recombination sites.

The Holliday junction can isomerize readily between two forms. In one a base-stacked double helix runs between ends a and b in Fig. 27-28B and another runs between ends c and d. The core hexanucleotide sequences lie between the marked XerC and XerD cleavage sites. In the other isomer (lower drawing) one helix has ends a and d and the other b and c. Following the isomerization (steps *c* and *d* in Fig. 27-28A) the XerD active sites act in two transesterification reactions (not shown but analogous to those in steps *a* and *b*) with base swapping of the trinucleotides at the cut ends. This generates the two separate recombinant duplexes. These might be two circular chromosomes or plasmids formed by recombination from a double length chromosome or plasmid. The previously mentioned λ amd Cre recombinases appear to act by closely similar mechanisms.

Tyrosine recombinases of the lambda family also function in eukaryotes. Best known is the **FLP (Flip) recombinase,** which is encoded by the 2-μm plasmid of *Saccharomyces cerevisiae* and is thought to function in amplifying the number of plasmid copies.[265] The 6.3-kbp plasmid contains a unique DNA sequence that lies between two 599-bp repeats in inverted orientation. Embedded in each repeat is an FLP recombination target (**FRT**) sequence, which is recognized by the plasmid recombinase. Each FRT segment includes inverted repeats 13 bp in length with an 8-bp spacer between them. As with other integrase systems the 8-bp spacer or **strand exchange region** is **asymmetric** and establishes the orientation of the recombination sites.[587–590] The role of the recombinase is to invert one of the 599-bp repeats with respect to the other (see Eq. 27-15). This switches replication of the plasmid to a rolling circle pattern.[265]

The resolvase / invertase family and invertible DNA sequences. A second large family of recombinases act by cleaving a target DNA sequence hydrolytically leaving a free 3'-OH end (Eq. 27-14, step *a*). This free end then attacks a phosphodiester linkage in a second strand of DNA, cleaving that strand with an in-line nucleophilic displacement (step *b*). Active sites usually contain a characteristic cluster of aspartate and

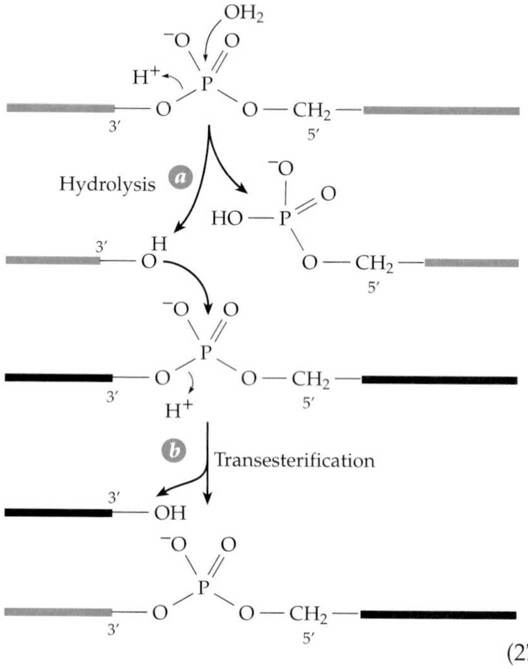

(27-14)

glutamate (DDE) side chains, which probably act together with a metal ion, perhaps as in Fig. 27-13.[591] Enzymes in this resolvase/invertase family act either to resolve cointegrates in transposon action (next section) or to invert DNA sequences.

If recombination occurs within a piece of DNA at two homologous sites such as the *attL* and *attR* sites at the boundaries of the λ prophage, the intervening DNA will be excised as a circular particle (Eq. 27-15). In this instance the two homologous regions must be repeated in the same direction, as is indicated by the arrow structures in Eq. 27-15. If the homologous sequences are oriented in opposite directions, i.e., they are inverted repeats, excision will not occur but the piece of DNA between the repeats will be inverted (Eq. 27-16).

A number of such invertible

FLP recombination target (FRT) sequence

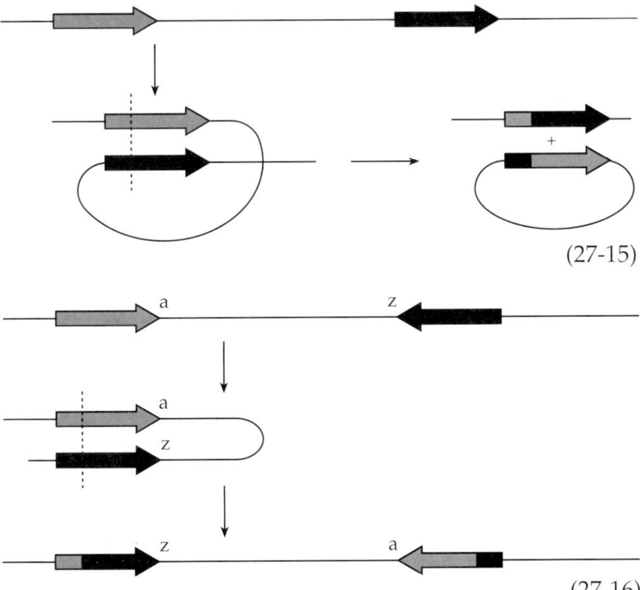

DNA recombination systems are known and are sometimes used to control specific genes.[592] For example, many strains of *Salmonella* have two types of flagella, which are composed of flagellin subunits encoded by genes H1 and H2 respectively.[592–594] On rare occasions an individual bacterium switches from one flagellar "phase" to the other. This occurs by recombination, as in Eq. 27-16, between two 26-recombination sites *hixL* and *hixR*, each of which contains a 14-bp inverted repeat. The 993-bp invertible segment encodes a recombinase gene called *hin* and a promoter, i.e., an mRNA initiation site, which has a specific orientation. In one orientation mRNA is transcribed from a short operon that includes the right inverted repeat IRR, the *H2* flagellin gene and gene *rh*1, which encodes a repressor for flagellin gene *H1*. Consequently, only gene *H2* is expressed. In the other orientation the RNA transcription is in the opposite direction so that neither *H2* nor *rh*1 is expressed. However, *H1*, which is located elsewhere, is expressed freely.

Two other proteins are required for efficient inversion by the Hin recombinase. A dimer of a 98-residue helix–turn–helix DNA binding protein called **Fis** (factor for inversion stimulation), a relative of protein HU,[18b] must bind to an enhancer, a 65-bp DNA segment. Binding of Fis to the enhancer helps to hold the supercoiled DNA and the recombinase in a correct orientation for reaction.[576,595,596] Protein HU is also needed. The same Fis protein binds to an enhancer for a 3-kbp invertible DNA sequence, which controls alternative host preferences for bacteriophage Mu (Fig. 27-29).[576,596] The chemistry of the inversion reaction is related to that of the replicative transposons discussed in Section 4.

Microorganisms sometimes control the synthesis of surface proteins using segments of invertible DNA. The pathogenic bacterium *Campylobacter fetus* utilizes DNA rearrangements to allow one of a large family of surface layer (S-layer) proteins to be formed.[597] The yeast FLP recombinase, mentioned in the preceding section, also inverts the sequence flanked by the 599-bp repeats.[589]

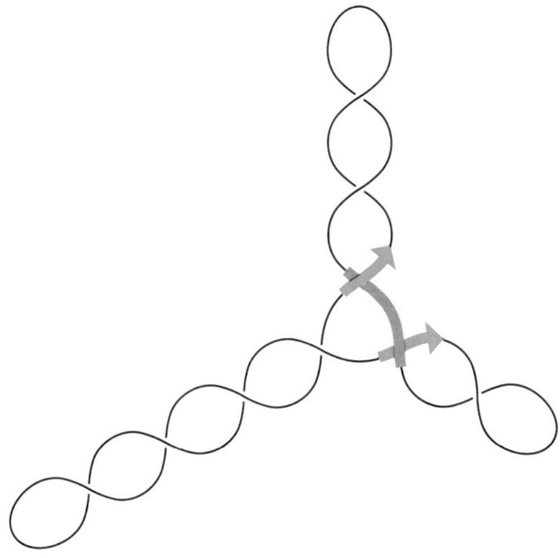

Figure 27-29 Formation of a synaptic complex of a supercoiled circular DNA containing the sites *gix* (green), which pass over and under the enhancer (gray). The recombinase Gin and the enhancer-binding Fis form a synaptic complex with DNA in this form as seen directly by electron microscopy. From Sadowski.[576]

4. Transposons and Insertion Sequences

The first evidence that some genes can move from one location to another within the genome came from studies of *Zea mays* by Barbara McClintock in the late 1940s.[598–602] She concluded that the variegated kernels found in some colored maize were a result of **controlling elements**, which could move from place to place turning on or inhibiting expression of various genes including some of those determining anthocyanin pigment formation. Two of these systems have been

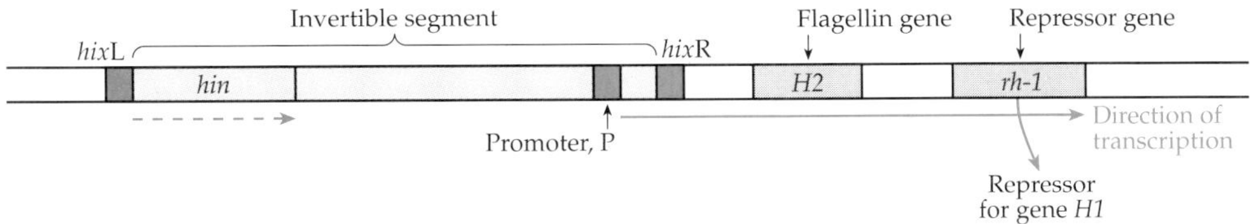

studied especially intensively: the "**activator (Ac)–dissociation (Ds)**" system, discovered by McClintock and the "enhancer (En)–inhibitor (I)" system, discovered by Peterson[599,600] and independently by McClintock.[603] Each contains two genetic elements (segments of DNA) of which Ac and En are autonomous, i.e., they can move by themselves. Ds and I, however, cannot move unless the other element of the pair is also present. Both Ds and En have now been cloned and sequenced.

The general importance of transposable genetic elements was not appreciated by most molecular biologists until about 20 years after McClintock's discoveries. Then several moveable **insertion sequences** (Is elements) were found in enteric bacteria. Like the controlling elements of maize these small (0.8–1.4 kb) sequences can move and insert themselves at many points in the genome, often inactivating genes which they enter.[602–608] The *E. coli* genome contains eight copies of IS1, and five of IS2, as well as several others. Most species of *Shigella* contain more than 40 copies of IS1.[604] Mobile elements similar to those present in maize also exist in archaea.[604a]

Following the discovery of the IS elements it was found that transposable elements named **transposons** could transfer resistance to antibiotics between bacteria. All of these transposable elements have inverted repeat sequences at the ends. For example, IS1 contains the following sequence at both ends but with opposite orientation as if in a

$$\text{CAACTTA}^{\text{C}}_{\text{T}}\text{TGAT}$$

palindrome.[605] Some complex transposons have an IS sequence at each end. For example, Tn10 contains IS10, a relative of IS1, at both ends.[608] These provide the characteristic inverted repeat termini. Figure 27-30 shows a schematic drawing of a bacterial drug resistance plasmid containing IS1, IS2, and IS10 as well as transposons Tn3, Tn10, and Tn55. The two IS1 elements surround the large resistance determinant, which can be transferred as a block.[602] To be autonomously mobile a transposon must contain a **transposase** that enables it to be transferred. It usually carries other genes as well and may also contain one or more signals for transcriptional regulation such as promoters.

The chemistry of transposition is more complex than that of simple site-specific recombination. Transposition can occur at many sites in a genome, and no homology with the transposon termini is required. Transposition is accompanied by *duplication of a short sequence of the recipient DNA* exactly at the ends of the transposon. Usually 5, 9, or 11 base pairs are duplicated as in the hypothetical example of Eq. 27-17. This happens because staggered cuts 5, 9, or 11 bp apart in the recipient DNA are made during recombination. These are indicated by the small arrows in Eq. 27-17. Transposons causing 5-bp duplications are Tn3, Tn7,[608a] γδ, phage Mu of *E. coli*, and Ty1 of yeast[609]; IS1, Tn10,[609a] and Tn5 of *E. coli* cause 9-bp duplications.[610–611b] When a transposon of one major group moves to a new location, the original copy remains. In this case transposition involves a combination of replication and site-specific (for the transposon) recombination. As a consequence, a circular DNA molecule containing the transposon will often react with a second circular DNA to form a large circle, a **cointegrate**, which contains two copies of the transposon. However, another group of transposons utilize a "cut-and-paste" mechanism that doesn't require extensive DNA duplication.

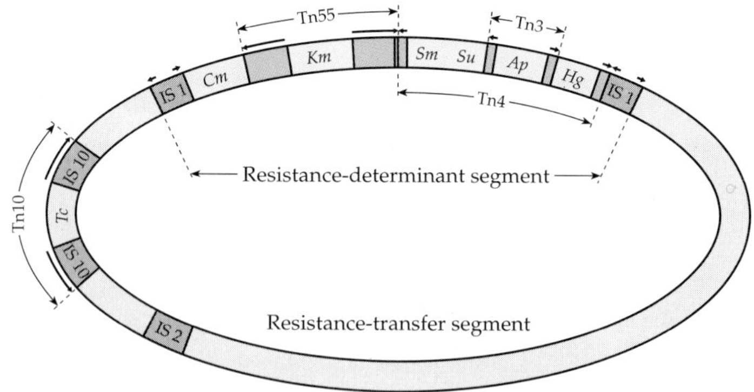

Figure 27-30 Transposons in an antibiotic-resistance plasmid. The plasmid appears to have been formed by the joining of a resistance-determinant segment and a resistance-transfer segment; there are insertion elements (*IS1*) at the junctions, where the two segments sometimes dissociate reversibly. Genes encoding resistance to the antibiotics chloramphenicol (*Cm*), kanamycin (*Km*), streptomycin (*Sm*), sulfonamide (*Su*), and ampicillin (*Ap*) and to mercury (*Hg*) are clustered on the resistance-determinant segment, which consists of multiple transposable elements; inverted-repeat termini are designated by arrows pointing outward from the element. A transposon encoding resistance to tetracycline (*Tc*) is on the resistance-transfer segment. Transposon Tn3 lies within Tn4. Each transposon can be transferred independently. From Cohen and Shapiro.[602]

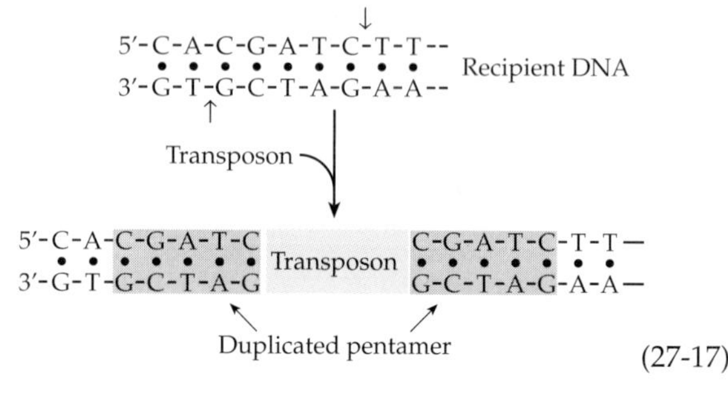

(27-17)

Cut-and-paste (nonreplicative) transposons. The transposases of *E. coli* Tn5, Tn7, and Tn10 act by hydrolytically cutting both strands of duplex DNA at the transposon ends leaving the phospho groups attached to the 5' cut ends, as is depicted in detail in Eq. 27-14, step *a*. The two 3' ends then carry out transesterification reactions, as in Eq. 27-14, step *b*. These two steps are used to nick both strands of the DNA carrying the transposon and to join them to a target DNA sequence to give a branched intermediate (Eq. 27-18, step *a*). Nonreplicative transposons apparently cut off two arms, e.g., A and B, and heal up the small gaps by repair synthesis, leaving the transposon in a new location between C and D. The gap repair accounts for the duplication of the end sequences of the cut target DNA.[610–612a]

Replicative transposons. In 1979 Shapiro proposed the mechanism illustrated in Fig. 27-31 for replicative transposons. The two inversely repeated segments (green) at the ends of the transposon are aligned with the recipient DNA whose ends are labeled C and D. In fact, the recombining DNA molecules must be supercoiled.[1,613] Staggered cuts are made in the recipient DNA at points *a* and *b*, which are 5, 9, or 11 bp apart, depending upon the specific recombinase. Nicks are also made in the transposon ends. The 3' ends from the transposon are resealed

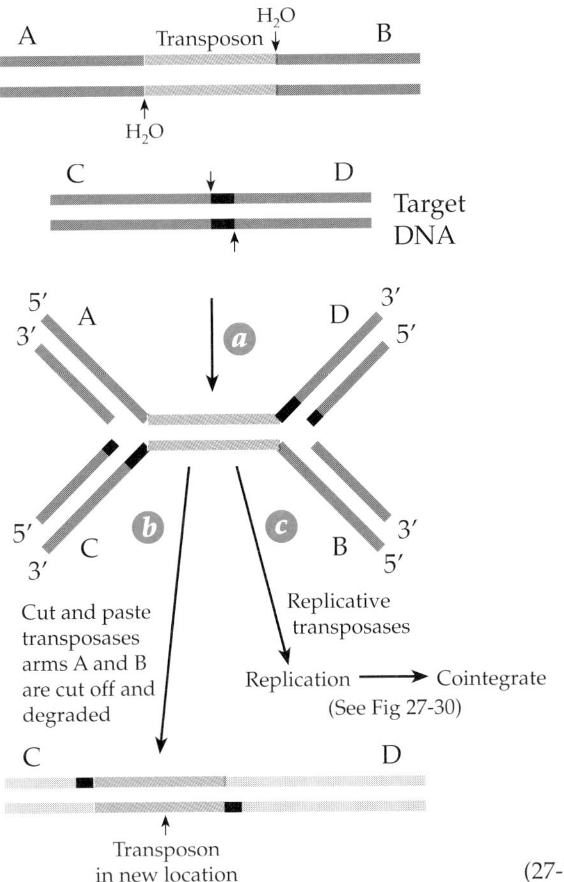

(27-18)

with the 5' ends from the recipient DNA (step *a*) to give a structure that in effect has two replication forks. Replication (step *b*) yields the cointegrate, which contains two copies of the transposon as indicated. In a third step (step *c*) recombination between the two integrated transposons yields a copy of the original transposon-containing donor and the recipient DNA, which now also contains a copy of the transposon.

When a transposon reacts with another part of the same DNA circle there are two possibilities. The piece of DNA lying between the transposon and the recipient site may be excised as a circle containing a copy of the transposon. Alternatively, there will be inversion of that sequence as well as replication of the transposon (compare with Eqs. 27-15, 27-16).

The closely related transposons Tn3 and γδ are understood best. They contain not only a transposase gene but also a **resolvase** gene.[613–616] The transposase carries out the recombination reactions that yield the cointegrates, while the resolvase catalyzes site-specific recombination between the two transposons in the cointegrate to complete the transposition. Several subunits of the resolvases bind the two **resolution (res) sites** of the supercoiled DNA in a parallel orientation with the DNA supercoiled as in Fig. 27-29. However, no enhancer-binding protein is needed, and the two *res* sites must be supercoiled. Purified γδ resolvase uses hydroxyl groups of the serine-10 side chains as the nucleophiles to cleave the DNA by displacement at specific *res* sites to give transient enzyme-bound phosphodiester linkages.[615]

Cleavage points
5′-T-A-T-A-3′
3′-A-T-A-T-5′
Center of a *res* site

H–O–Ser^{10}–
O, O^-, P
–T–O (3′) … O–CH_2–A-3′ (5′)

The synaptic complex contains 240 bp of DNA and at least two resolvase dimers. All four DNA chains are cut to give eight ends. Four of these are bound to the serine side chains in phosphodiester linkage. In the second step the freed 3'-OH groups react with the bound ends of the other duplex via a transesterification reaction to form the recombinant chains.

The resolvases act on supercoiled cointegrated DNA molecules that contain two directly repeated res sites to produce two singly linked circles (which are still supercoiled) each containing one res site as shown in Fig. 27-32. The two *res* (resolution) sites within the transposons are aligned, the open circle of DNA shown at the upper left being folded as shown in the lower part of the drawing. The DNA substrate is not knotted. However, after recombination it is catentated and will require action of a topoisomerase to separate

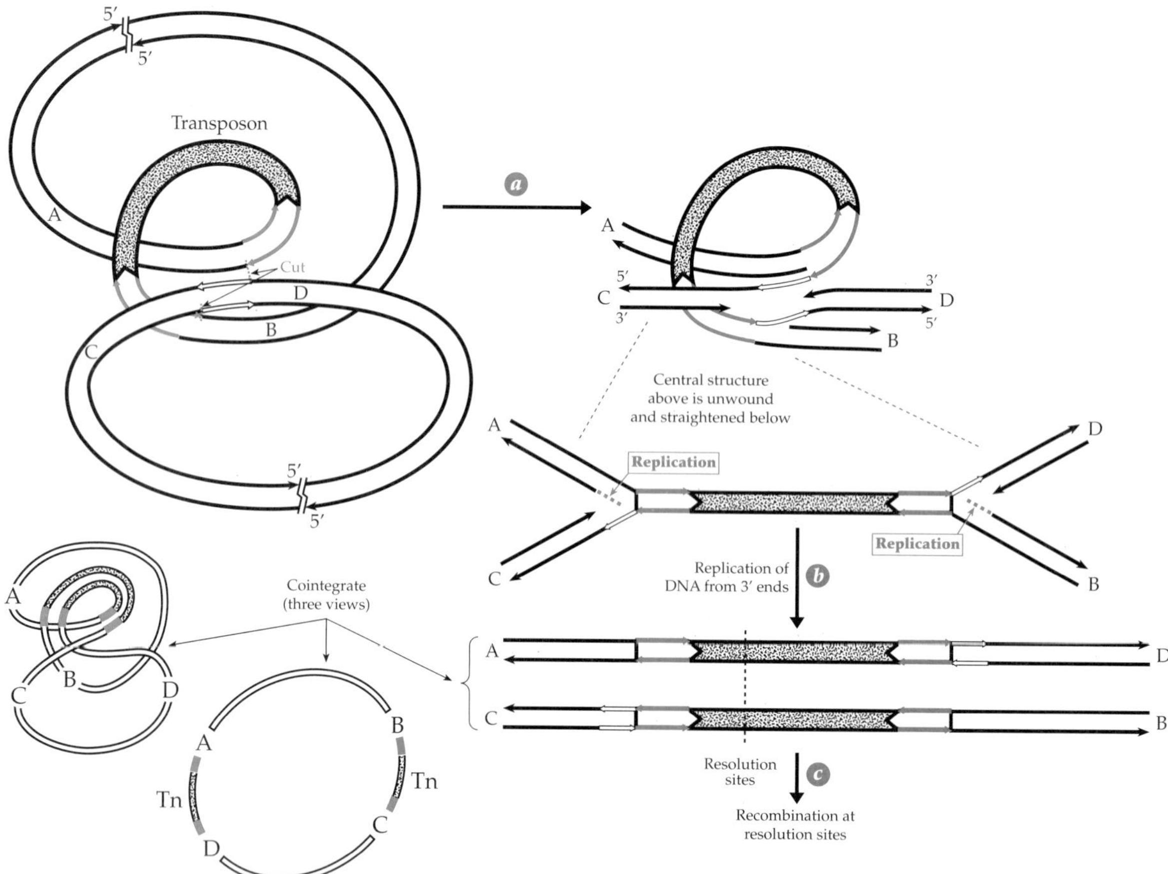

Figure 27-31 Scheme for integration of a transposon (stippled duplex) present in a piece of DNA with ends A and B into another piece of DNA with ends C and D and containing a suitable recognition sequence (open bars). Inversely repeated sequences in the transposon are shown as solid bars with a direction arrowhead. Arrowheads point toward 5′ strand ends. Cleavage and rejoining at points a,a and b,b yield an intermediate with two replication forks. Replication through the transposon yields one unchanged DNA segment with ends A and B and a transposon inserted into the other DNA segment. If A is continuous with B, a cointegrate structure is formed. See Cohen and Shapiro.[602]

the two products. Occasionally additional recombination events occur, perhaps processively along interwound double helices. This produces various knotted products.[617] An electron micrograph of one of these is shown in Fig. 5-17.

The temperate bacteriophage Mu. The most efficient transposon known is the 37-kb genome of the **mutator bacteriophage Mu**.[265,618–621] Once it becomes integrated into a bacterial chromosome, it replicates by repeated rounds of transposition within the host bacterium. During the lytic cycle some of the replicating DNA is excised as extrachromosomal circles of various sizes, which are packaged into virus particles by a headfull mechanism. The circles contain copies of some host DNA, but this is left behind when a virus particle infects a new cell.[265] The phage Mu DNA is integrated into host DNA by a cut-and-paste mechanism in a transpososome that contains a tetramer of the virally encoded transposase (MuA protein) bound to a supercoiled DNA. The transpososome resembles that in Fig. 27-29 and contains cleavage sites at the ends of the transposon and also an enhancer sequence.[620,621] Several steps involving conformational alterations occur in the transpososome. One is an ATP-dependent action of a second Mu-encoded protein MuB.[622] During the lytic cycle replicative transposition predominates.

Some other transposons. Transposons have a variety of biological functions. For example, haploid cells of the yeast *S. cerevisiae* exist as one of two mating types *a* or α. The mating type is established by transposition of one of two "cassettes" of genes from two different "silent" locations to a location from which they can be expressed.[623,624] See Chapter 28.

One of the best known eukaryotic transposons is the P element of *Drosophila*, which transposes only within the germ line cells of developing embryos, somatic cells being unaffected.[265,625,626] It belongs to

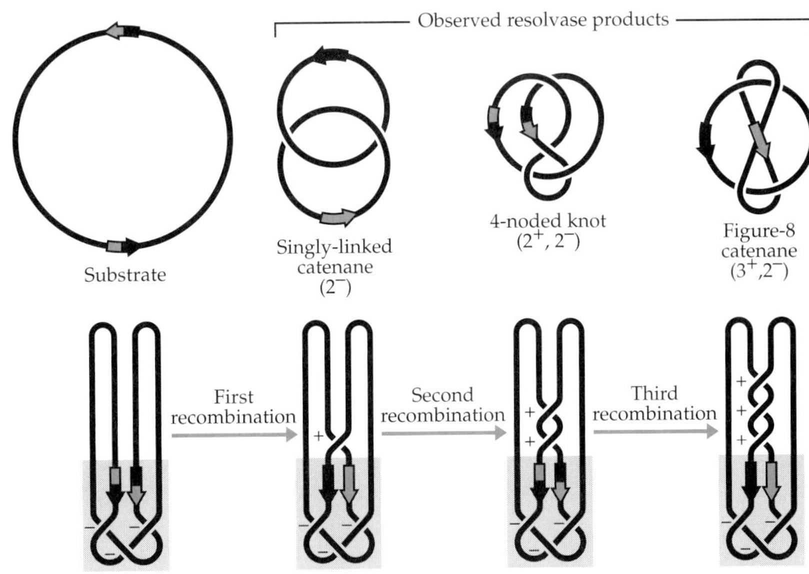

Figure 27-32 Scheme for resolution of an unknotted cointegrate molecule by a resolvase that cuts the transposons at resolution (*res*) sites and recombines them. The resolvase may act only once or repeatedly as shown. In the upper row, the duplex DNA substrate and products are represented in standard topological form as they might appear after nicking. In the lower row the DNAs are depicted as folded forms bound to the resolvase with the two directly repeated *res* sites (thick arrows) dividing the substrate into two domains (thick and thin regions). The substrate at synapsis has three (–) supercoils that entail crossing of the two domains. Successive rounds of recombination, each introducing a single (+) interdomainal node (see Fig. 5-17), are drawn in the lower row. Bound resolvase maintains the three synaptic supercoils. After dissociation from the resolvase at any stage, the product supercoil nodes either cancel with ones of opposite sign or are removed by subsequent nicking. The node composition is indicated in parentheses. From Wasserman *et al.*[617]

the same family as Barbara McClintock's Ac element of maize and the *Tc*1 family of nematodes.[627] P elements use a nonreplicative cut-and-paste method of transposition. The 87-kDa transposase protein requires GTP and Mg^{2+} for activity.[628] It was only in the past few decades that P elements have been found in *D. melanogaster*. They may have entered this fruit fly from another species, possibly transferred by a mite.[627,629]

A second *Drosophila* transposon called **mariner**[630] typifies the *mariner* / *Tc*1 transposon superfamily, which also contains members from nematodes,[631] other invertebrates, fishes,[632] amphibia,[633] and possibly human beings.[634] These transposons encode a transposase containing a D, D, D or D, D, E motif[630] but no other proteins. They contain short ~30-bp terminal inverted repeats and become inserted into host TA sequences.[631] Movement of some repetitive sequences of the LINE[635] and SINE[636] families within the human genome may be assisted by *mariner* transposons.[637]

The maize transposon *ac* is widely used as a means of inactivating genes and placing a "tag" that can be used to map the gene and to permit it to be cloned and sequenced.[598,638] Although initially of use only in maize the method has been extended to other plants,[639–641] and genetically engineered transposons have allowed it to be utilized in animals.[642]

A different kind of transposition controls self-sterility in maize. The cause of self-sterility in one strain has been traced to the presence of two linear episomes called S-1 (614 kb) and S-2 (514 kb) within mitochondria.[643] These have inverted terminal 208 bp repeats. On rare occasions they recombine with the circular mtDNA converting it to a linear form with the episomes covalently linked to one end. The change is accompanied by reversion to fertility.

As described in Section B,1, mammalian DNA contains many **retrotransposons** (retroposons) that lie within short direct repeats characteristic of transposons. However, they contain a poly(A) tail at the 3' end, an indication of their relationship to RNA transcripts, and are discussed in Chapter 28.

5. Other Causes of Genetic Recombination

Because their integration sites in host DNA do not depend upon homology with the transposon ends, transposition is sometimes called "**illegitimate recombination.**" Although no homology is required, there are prefered sites for integration. For example, Tn10 is transposed most readily into certain "hot spots" in the *Salmonella* DNA among which is the sequence 5'-GC^{m5}CAGGC.[608] Illegitimate recombination can also be induced by other processes that involve DNA chain cleavage, e.g., by topoisomerases.[644] Whenever a DNA chain breaks, it must be repaired, a process that often also involves recombination. Recombination is often observed to occur between direct repeat sequences, which are a major cause of instability in the genome.[645]

Under some circumstances selected segments of the genome are **amplified** by repeated replication of a gene or genes.[646,647] Amplication of specific genes occurs in viruses,[648,649] in bacteria where it may provide for adaptation to conditions of stress,[650] and in eukaryotes.[651] In oocytes of amphibia, such as *Xenopus*, excess DNA accumulates around the nucleoli and later breaks up to form 1000 or more separate nucleoli. As many as 3000 copies of the rDNA (which forms a distinct satellite band upon centrifugation) may be present. Much of this DNA exists as extrachromosomal rings containing 1–20 rDNA units. Using these genes as many as 10^{12} ribosomes per oocyte are synthesized.

Tetrahymena contains only one set of rRNA genes per haploid genome in its diploid **micronucleus**. Following sexual conjugation the chromosomes of the micronucleus undergo multiple replications to form polytene chromosomes containing thousands of copies. However, about 5% of the resulting DNA is excised as linear pieces with characteristic inverted repeat sequences and 3' single-stranded tails:

$$5'\text{--}\overrightarrow{C_4A_4C_4A_4C_4}\text{--//--}G_4T_4G_4T_4G_4T_4G_4T_4G_4\text{--}3'$$
$$3'\text{--}G_4T_4G_4T_4G_4T_4G_4T_4G_4\text{--//--}\overleftarrow{C_4A_4C_4A_4C_4}\text{--}5'$$

These tails are joined by a protein to form the circles, which are segregated in the **macronucleus**. The other 95% of the amplified DNA is degraded. At the next meiosis the macronucleus is discarded entirely, and a new one is formed at the next diploid stage.

Toxic drugs often cause cells to amplify genes that help resist the drug.[647] This can be a major problem in the chemotherapy of cancer. For example, a culture of human leukemia cells grown in the presence of increasing concentrations of methotrexate increased its level of dihydrofolate reductase 240-fold.[652] The cause is an increase in the number of copies of a chromosomal region containing the gene.[653] Cancer cells tend to amplify oncogenes such as c-*myc*.[654]

There are several mechanisms for gene amplification. The formation and breakup of polytene chromosomes in *Tetrahymena* is one. The circular copies of rDNA in *Xenopus* and many other species are generated by a rolling circle mechanism similar to that in Eq. 27-7. If each circle excised from the original chromosomes contains an origin of replication, many copies can be formed. Another possible mechanism is replication of a local region of DNA several times followed by excision of pieces of the DNA. A mechanism that may give rise to homogeneously staining regions is unequal crossing-over (Section 2) repeated several times within a gene cluster. Transposition can cause excision of DNA in a circular form, which can be amplified by a rolling-circle mechanism.

E. Damage and Repair of DNA

A characteristic of living things is their high degree of mutability. Harmful mutations take a toll of human life at an early age, and the very high incidence of cancer in older persons is largely a result of the accumulation of somatic mutations. Mutations are a major factor in aging and are continuously introducing new genetic defects into the population. Mutations can be described as **base substitutions**, **deletions**, or **additions**. Base substitution mutations are classified as **transitions**, in which a pyrimidine in one strand is replaced by a different pyrimidine. In the complementary strand a purine is replaced by the other purine, e.g.,

$$C\cdot G \rightarrow T*G \xrightarrow{\text{Replication}} T\cdot A + C\cdot G \qquad (27\text{-}19)$$

$$A\cdot T \rightarrow \underset{\text{Mispair}}{G*T} \longrightarrow \underset{\text{Mutation}}{G\cdot C} + A\cdot T \qquad (27\text{-}20)$$

In a *transversion* a purine in one chain is replaced by a pyrimidine, while the pyrimidine in the complementary chain is replaced by a purine:

$$C\cdot G \rightarrow A*G \longrightarrow A\cdot T + C\cdot G \qquad (27\text{-}21)$$

Here the central asterisks designate mispairs and the green shade marks mismatched bases and also the resulting mutant base pair formed in the next replication cycle.

1. Causes of Mutations

DNA can be damaged in many ways.[655] Spontaneous hydrolysis of the glycosidic bonds between nucleic acid bases and the deoxyribose to which it is connected cause the loss of ~10^5 purines and pyrimidine rings per day from the DNA in a mammalian cell.[656] About 100 residues per day of cytosine are deaminated by such agents as nitrite or bisulfite (Chapter 5, Section H,3) to form uracil. Like thymine, uracil will pair with adenine causing a C • G → T • G transition mutation as in Eq. 27-19. A few adenines per day are also deaminated to form hypoxanthine. Oxidized bases are formed by attack of HO$^\bullet$ radicals and other species of reduced oxygen.[657–659] Alkylating agents from the external environment as well as *S*-adenosylmethionine carry out slow, nonspecific alkylation of purine and pyrimidine bases. Polycyclic aromatic hydrocarbons and other carcinogens are converted to metabolites that alkylate DNA, and alkylated bases often mispair during replication.[660,661] Ultraviolet light induces formation of photohydrates (Eq. 23-25),[662] pyrimidine dimers (Eq. 23-26), and other photochemical products.[663] X-rays and gamma rays cleave nucleic acid bases and break chromosomes.[663a]

Natural radioactivity has a measurable effect.[663b] Mistakes caused by mispairing[664] or misalignment[665] are made during replication and recombination and even during repair of DNA. Errors in replication are especially numerous in highly repetitive DNA sequences. Some errors probably arise as a result of tautomerization and others from incorporation of uracil in place of thymine. Thus, to keep its DNA in repair a cell must continuously deal with missing bases, wrong bases, altered bases or sugars, pyrimidine dimers and other crosslinkages, deletions, and insertions.

2. Fidelity of Replication

During DNA replication in *E. coli* only one mistake is made on the average during polymerization of 10^9–10^{10} nucleotides.[666–668] The rate varies among different sites in the genome.[668a] In eukaryotes the error rate may be only 1 in 10^{-10} base pair or less per generation.[669,669a] Early workers often attributed the specificity in base pairing and the resultant high precision of replication entirely to the strength of the two or three hydrogen bonds formed together with the stabilization provided by the adjacent helix. However, the Gibbs energy of formation of the base pairs is small (Chapter 5), and the additional energy of binding to the end of an existing helix is insufficient to account for the specificity of pairing.[670,671] Thus, according to Eq. 6-30 a difference in ΔG of binding between the correct nucleotide and an incorrect one of 11 kJ / mol would give an error rate of 1 in 100.

The role of polymerases. The polymerase enzymes play a major role in ensuring correct pairing during replication, transcription, and protein synthesis. RNA and DNA polymerases are large molecules. The binding site on the enzyme can completely surround the double helix. Water will be excluded as the enzyme folds around the base pairs. This may have the effect of greatly increasing differences in the Gibbs energies of binding and thereby enhancing selectivity.[671] It has traditionally been assumed that formation of proper hydrogen bonds is essential for the base selection. A hypothetical active site is portrayed in Fig. 27-33. A guanine ring of the template strand of DNA is portrayed at the point where the complementary DNA strand is growing from the 3' end. The proper nucleoside triphosphate must be fitted in to form the correct GC base pair before the displacement reaction takes place to link the new nucleotide unit to the growing chain. Let us suppose that the enzyme possesses binding sites for the deoxyribose unit of the template nucleotide and for the sugar unit of the incoming nucleoside triphosphate and that the two binding sites are held at a fixed distance one from the other. As indicated in Fig. 27-33, some group H-Y might also be present at each binding site to hydrogen bond to the nitrogen or oxygen indicated by the heavy green arrows. All four of the bases could form hydrogen bonds of this type in the same position. Hydrophobic interactions could provide additional stabilization. With such an arrangement the correct nucleoside triphosphate could be selected no matter which one of the four bases occupied the binding site on the left side of the figure. (The outlines of the thymine and adenine rings have been drawn in with dotted and dashed lines, respectively.) If a purine is present on the left side, as is shown in the drawing, there is room on the right side only for a pyrimidine ring. Thus, A and G are excluded, and the choice is only between C and U (or T). However, U will be excluded because the dipoles needed to form the hydrogen bonds point in the wrong direction. These dipolar groups are hydrated in solution. They are unlikely to give up their associated water molecules unless hydrogen bonds can be formed within the base pair. Not only would a molecule of U (or T) be unable to form the stabilizing hydrogen bonds within the vacant site but also the electrostatic repulsion of the like ends of the dipoles would tend to prevent association. This would lower the affinity of the polymerase for mispaired bases. Verification that the proper base pair has been formed could be accomplished by using its tautomeric properties to sense an electronic

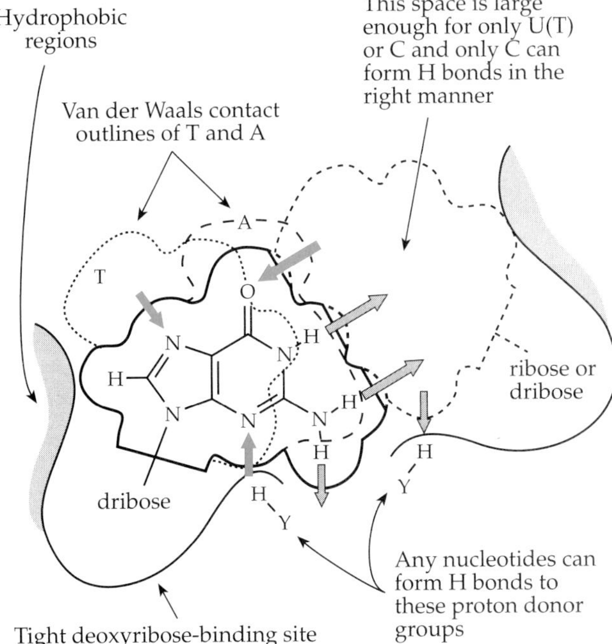

Figure 27-33 Selecting the right nucleotide for the next unit in a growing RNA or DNA chain. A deoxyguanosine unit of the template chain is shown bound to a hypothetical site of a DNA polymerase.

displacement through the hydrogen-bonded network.[672,673] Considerable experimental evidence suggests, however, that hydrogen bonding may not be important to the selectivity of DNA polymerases.[666,674–676] For example, the triphosphate of the following analog of thymidine is efficiently incorporated by *E. coli* DNA polymerase I into DNA opposite the thymine base-pair partner adenine.[675]

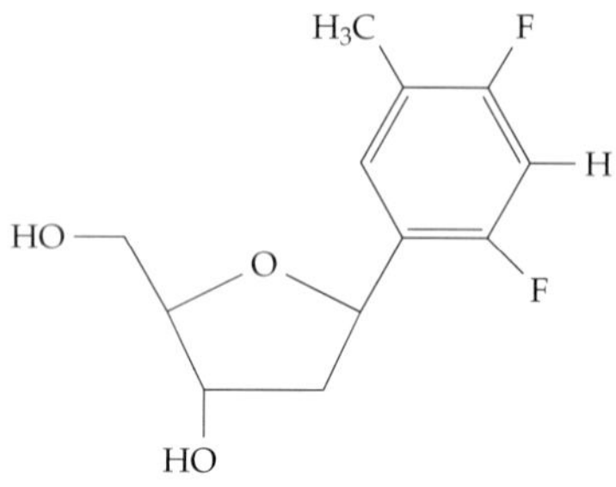

H_3C F H F HO O HO

Structural analog of thymidine

Similar results have been obtained for a variety of other analogs with poor hydrogen bonding characteristics. Apparently *shape* is most important.[666,674,676,676a] Binding strengths in transition state structures must also be considered.[676b]

Editing (proofreading). As is discussed in Section C,2, base-pairing is checked after a new monomer is added at the 3' end of a growing polynucleotide chain, as well as before polymerization occurs. If the wrong base pair has been formed, the newly created linkage is hydrolyzed, and the incorrect nucleotide is released. It has been estimated[677] that in *E. coli* the error rate for DNA polymerase III holoenzyme is $\sim 1 \times 10^{-7}$ per base pair of which proofreading by the ε subunit may provide $\sim 10^{-2}$. Additional mismatch repair reduces the error rate 200- to 300-fold to give the overall error rate of $<10^{-9}$. Fidelity of replication, which seems to be higher on the lagging strand than on the leading strand,[678] may also be improved by operation of other proteins, such as the sliding clamp, that are part of the replication apparatus.[679] Replicative proteins also appear to be designed to minimize **frameshift mutations** that could arise by slippage of the template versus the replicated strand in long runs of dA•dT pairs.[680]

3. Repair of Damaged DNA

A final check of the fidelity of replication is made after a new strand has been formed. Mismatched base pairs are identified, and the incorrect nucleotides are cut out and replaced by correct ones.[655,670,681–683] Some of the thymine dimers created by the action of light are also repaired photochemically by photolyases (see Chapter 23). **Photoreactivation** was the first DNA repair process recognized.[684] However, most thymine dimers in human DNA must be excised and replaced. Loops, gaps, and double-stranded breaks are also sensed, and appropriate corrections are made. If necessary the sequence of a badly damaged segment of DNA may be copied from the DNA in a sister chromatid before mitosis is completed or from the homologous chromosome.[685]

Much of our understanding of DNA repair comes from investigations of mutant strains of *E. coli*, which display an elevated incidence of mutations. On this basis, the "mutator" genes *dam, mutD, mutH, mutL, mutS, mutU* (also called *uvrD*), and *mutY* were implicated in DNA repair.[677,686] An additional series of genes, *uvrA, B, C*, and *D*, were identified by their participation in resistance to ultraviolet radiation damage and were also recognized as being involved in DNA repair. Many corresponding genes were found later in the yeast, *Drosophila*, and human genomes. The study of human DNA repair has also been facilitated by the recognition of a group of human inherited defects (Box 27-A). There are 130 known human DNA repair genes.[686a]

Methyl-directed and other mismatch repair. In *E. coli* the methylase encoded by gene *dam* (Fig. 29-4), within ~7 min of replication, methylates adenine at N-6 in the sequence GATC, which occurs at many locations.[670,686,687] This methylation provides a label for the template strand in a newly replicated duplex. The GATC sites in the newly synthesized strand will not yet be methylated, and repair enzymes recognize it as the strand on which to carry out excision repair. Another system, which may function during recombination, acts on fully *dam* methylated DNA.[655] Since eukaryotes do not use methylation of GATC sites to distinguish the old template and newly replicated DNA strands, other mechanisms must operate.[669]

Ten proteins are required for methyl-directed mismatch correction. Of the required *mut* genes, *mutD* (also called *dnaQ*) was found to be the structural gene for the proofreading subunit ε of DNA polymerase III (Table 27-2).[688] *MutH, L*, and *S* as well as ATP are also needed, as is a helicase, an exonuclease, DNA pol III, and a ligase. Homodimers of proteins MutS and MutL preferentially bind to DNA with mismatched base pairs such as Pur-Pur (e.g., A•G) or Pyr•Pyr (less well recognized). Some of these mismatched pairs may contain modified bases. After **recognition** by the MutS•MutL complex (see banner, p. 1527)[688a–c] the endonuclease MutH is activated and cuts the DNA chain at the nearest unmethylated GATC sequence.[689] An exonuclease then degrades the chain past the mismatched base leaving a gap that must be filled by the action of DNA polymerase and DNA ligase. This **long-patch mismatch repair** (**MMR**) is utilized also by eukaryotes,[688a,b,690–692a] using proteins homologous to *E. coli* proteins MutS and MutL. However, methylation is not involved. In both yeast and human beings there are at least six MutS homologs, some of which have functions other than repair. For example, in yeast MutS- and MutL-related proteins are essential for normal levels of meiotic crossing-over.[692] Defects in human mismatch proteins may cause hereditary non-polyposis colorectal cancer (Box 27-A).[692b]

Excision repair. The *E. coli* mismatch repair is a type of excision repair. However, a different **nucleotide excision repair** system (**NER**) is utilized by all organisms from bacteria to human to remove a variety of defects. These include thymine dimers, photohydrates, oxidized bases, adducts of cisplatin (Box 5-B), mutagens derived from polycyclic aromatic compounds,[683] and poorly recognized C•C mismatched pairs.[692] In *E. coli* this excision repair process depends upon proteins encoded by genes ***UvrA, B, C,*** and ***D*** and also DNA polymerase I and DNA ligase.[693–695a] A dimer of protein UvrA forms a complex with helicase UvrB (Eq. 27-22).[696,696a]

$$
\begin{array}{l}
(\mathrm{UvrA})_2 \xrightarrow{\mathrm{UvrB}} (\mathrm{UvrA})_2 \cdot \mathrm{UvrB} \\
\quad \text{DNA} \downarrow \quad (\mathrm{UvrA})_2 \text{ released and recycled} \\
\mathrm{UvrB} \cdot \mathrm{DNA} \\
\quad \mathrm{UvrC} \downarrow \\
\mathrm{UvrB} \cdot \mathrm{C} \cdot \mathrm{DNA} \\
\quad \downarrow \; \mathrm{UvrB}, \mathrm{UvrC} \\
\mathrm{UvrB} \cdot \mathrm{C} \cdot \mathrm{DNA}
\end{array}
\qquad (27\text{-}22)
$$

The helicase, driven by ATP hydrolysis, may move along the DNA chain with its associated UvrA protein in search of defects.[696] When one is located UvrB binds tightly.[697] UvrA then dissociates, and the nuclease UvrC binds and cleaves the DNA chain in two places as in the following scheme. One is at the fourth or fifth phosphodiester linkage in the 3' direction from the defect. The other is at the eighth phosphodiester on the 5' side.[698] The resulting gap is filled by a DNA polymerase and DNA ligase. Another helicase, encoded by *UvrD*, is also required.[699]

Cut ↓ (5' side), Cut ↓ (at Defect), Cut ↓ ↓ (3' side)

5′ T*P*T*P*T*P*T*P*T*P*T*P*T*P*T*P*T*P*T*P*T*P*T*P*T*P*T*P*T*P*T*P*T 3′

Defect

3′ ⊥*P*⊥*P*⊥*P*⊥*P*⊥*P*⊥*P*⊥*P*⊥*P*⊥*P*⊥*P*⊥*P*⊥*P*⊥*P*⊥*P*⊥*P*⊥*P*⊥ 5′

UvrBC preincision complex

Both yeast and human cells have similar but more complex systems of nucleotide excision repair.[700–703] The dual incision steps require at least six components. Several of these (XPA, C, F, G) have been found defective in various forms of the inherited disease **xeroderma pigmentosum** (Box 27-A), in which there is a high incidence of UV-induced skin cancer.[700,703a] Replication protein A (Section C,10), a three-subunit ssDNA binding protein, is also essential.[704] In eukaryotic excision repair the DNA is cleaved at the same position as in the bacterial preincision complex (preceding scheme), on the 3' side of the defect but further out on the 5' side, the excised polynucleotide being ~24–32 nucleotides in length. Repair of the resulting gap in the DNA duplex is accomplished by synthesis mediated by DNA polymerase δ or ε and action of a DNA ligase.

One of the lesions removed by nucleotide excision is the thymine photodimer.[705] In the fission yeast *S. pombe* an **alternative excision repair** system, specialized for removal of thymine dimers and 6–4 photoproducts (Chapter 23), produces two-nucleotide gaps with 3'-OH and 5'-phospho-group ends.[702,705a] Alternative NER pathways are also employed by bacteria.[705b]

Base excision repair (BER). The *N*-glycosyl linkages of the purine and pyrimidine bases to the deoxribose residues of the sugar–phosphate backbone of DNA are subject to spontaneous hydrolysis, one important source of damage to DNA. Similar hydrolytic reactions are catalyzed by **DNA glycosylases**, which remove many mismatched or damaged bases.[706–708a] At least seven enzymes of this type are present in cells of *E. coli*. One of them, a **uracil-DNA glycosylase**, hydrolyzes the glycosyl linkage wherever uracil has been incorporated accidentally in place of thymine or has been produced by deamination of cytosine.[706,709–711d] Acting via a nucleophilic displacement mechanism (Eq. 27-23, black arrows), it removes the uracil, leaving the DNA backbone intact but with an **apyrimidinic site** in one chain. The resulting apyrimidinic (apurinic sites) are recognized and cleaved by **apurinic / apyrimidinic DNA endonucleases** (AP nucleases).[708,712–714] These enzymes cut the DNA backbone, some leaving a 3'-phosphate group and a 5'-OH terminus and others a 3'-OH and 5'-phosphate. The resulting gap can be filled, for example by the action of DNA polymerase I and a ligase. To prevent the eliminated uracil from being converted to dUTP and reincorporated into DNA a **deoxyUTPase**, essential to both *E. coli* and yeast, hydrolyzes dUTP to dUMP and inorganic phosphate, decreasing the concentration of dUTP.[715] Other enzymes, known as **DNA glycosylase/AP lyases,**[706,716–718] use an amino group of an enzyme side chain as a nucleophile in a ring-opening reaction that is followed by β elimination of the nucleotide base with formation of a Schiff base intermediate (green arrow, Eq. 27-23). An example is the bacteriophage T4 enzyme **T4 endonuclease V**.[706,719] Others include the *E. coli* **endonuclease III**, encoded by the **Nth** gene, its yeast homolog **Nth-Spo**,[717–718] and human DNA pol ι (iota).[718a] After the Schiff base is formed, the DNA backbone is cut on the 3' side of the abasic site by a second β elimination reaction. A δ elimination (green dashed arrows in Eq. 27-23) may also occur, cleaving the DNA chain on the 5' side of the lesion. Alternatively, a hydrolytic cleavage is possible as is also indicated in Eq. 27-23. Either type of DNA glycosylase forms a single-nucleotide gap or a gap missing just a few nucleotides. This gap can also be filled by polymerase and ligase action.

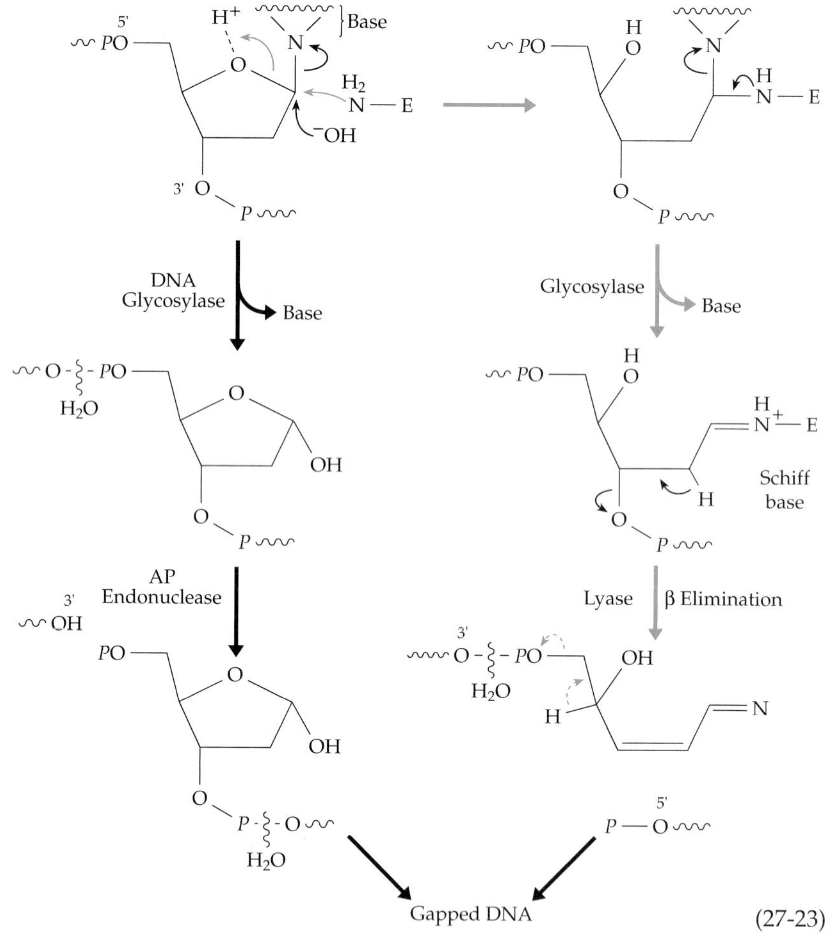

Both NER and BER forms of excision repair remove a great variety of defects, many of which are a result of oxidative damage.[657,720] Most prominent among these is **7,8-dihydro-8-oxoguanine** (8-OG), which is able to base pair with either cytosine (with normal Watson–Crick hydrogen bonding) or with adenine, which will yield a purine–purine mismatch and a C•G → A•T transversion mutation (Eq. 27-24), a frequent mutation in human cancers.[721,722]

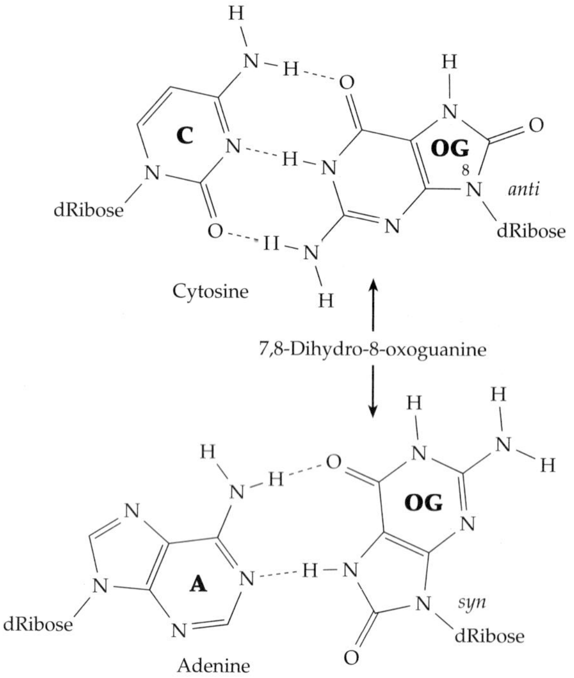

(27-23)

In *E. coli* three enzymes protect against 8-OG DNA mismatches. A glycosyltransferase encoded by the *mutM* (or *fpg*) gene removes 8-OG from DNA.[723,724] Because some A*OG base pairs may remain, a second glycosyltransferase (**MutY**) removes adenine from them.[723,725–725b] Interestingly, MutY contains an Fe_4S_4 cluster, essential to its function.[725,726] Both yeast and human cells have a corresponding enzyme.[727,728] The third *E. coli* enzyme **(MutT)** is a nucleoside triphosphate pyrophosphohydrolase, similar to the previously mentioned dUTPase. It preferentially hydrolyzes free 8-oxo-dGTP, preventing its incorporation into DNA.[729]

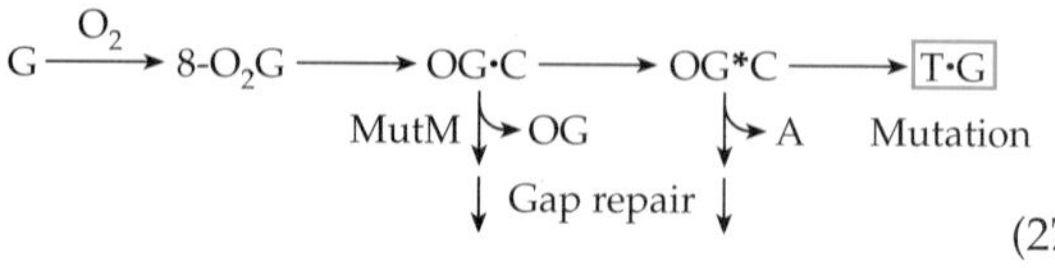

$$G \xrightarrow{O_2} 8\text{-}O_2G \longrightarrow OG{\cdot}C \longrightarrow OG{*}C \longrightarrow \boxed{T{\cdot}G}\ \text{Mutation}$$

MutM → OG; Gap repair; → A

(27-24)

Other DNA glycosylases remove thymine rings that have been converted to saturated or fragmented forms by oxidizing agents or ionizing radiation.[720,730,731] Among these are 5,6-hydrated thymine and urea, which are still attached in ribosyl linkage. Purines such as hypoxanthine and 3-methyladenine can all be removed by glycosylases.

Reactions of alkylated bases. The O^6 alkylation of guanine rings in DNA is highly mutagenic, presumably because mispairing with thymine will cause C•G → T•A mutations.

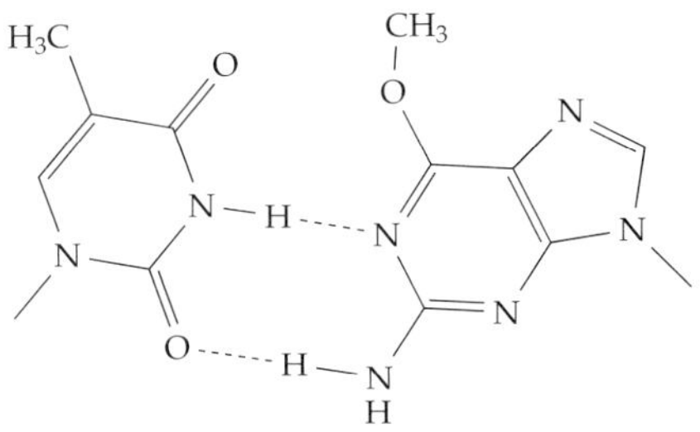

Thymine–O^6-methylguanine pair

Both bacteria and higher eukaryotes have an enzyme-like **O^6-methylguanine–DNA methyltransferase,** whose synthesis may be induced by culture of cells in the presence of alkylating agents. This 354-residue Zn^{2+}-containing **Ada protein** acts as both acceptor and catalyst for transfer of the methyl group off from the O^6-methylguanine onto the sulfur atom of a cysteine side chain in the protein.[732–734b] The presence of alkylating agents also induces synthesis of DNA glycosylases that remove 3-methyladenine and other alkylated bases.[735,736] 5-Methylcytosine, present in CpG islands in human DNA, will occasionally be hydrolyzed to thymine, giving T*G mismatches. The mismatched thymines are removed in cells of *E. coli* by a specialized endonuclease (**vsr** gene),[737] which hydrolyzes the phosphodiester linkage preceding the mismatch. In eukaryotic cells a **thymine–DNA glycosylase**[738,739] accomplishes the removal.

Repair of double-strand breaks. Following replication both in bacteria and in eukaryotes, recombinational exchanges may occur between sister chromatid duplexes or between homologous pairs of chromosomes.[740] A newly replicated chain may have gaps because of defects in the template strand from which it was copied. Copying from a sister duplex may permit assembly of a correct chromosome sequence and survival for the cell.[551a] Recombinational repair is also a major mechanism for preventing loss of genetic information from **double-strand breaks**. Such breaks are induced by ionizing radiation or chemical damage. They are often created at replication forks, stalling DNA synthesis with potentially disastrous consequences to the cell. Repair of these breaks, which is required for completion of replication, depends upon a large complex of replication and recombination proteins.[741–741b] In *E. coli* this includes protein RecF.[741,742] Repair of double-strand breaks by homologous recombination may be the most frequently employed type of DNA repair in bacteria, yeasts such as *S. cerevisiae*[530d] and also in mammals and other organisms.[742a–d]

A second repair pathway, called **nonhomologous end-joining** (**NHEJ**), is also utilized and may be relatively more important in higher organisms.[743–746b] The NHEJ mechanism repairs double-strand breaks caused by ionizing radiation and is also employed in specialized genomic rearrangements of developing B and T cells of the immune system. This V(D)J recombination (Chapter 31) is used to create the huge array of antibodies and antigen receptors required for immunological recognition and protection.[745,746a,747,748] NHEJ doesn't depend upon templates but simply rejoins duplex ends. Errors can be made if the ends have been damaged or if the wrong ends are joined. Several special proteins are required. One is a DNA-activated protein kinase (**DNA-PK**)[749] and a heterodimeric **end-binding protein** with subunits **KU70** and **KU80**.[745,750,750a] The serine / threonine protein kinase ATM (Box 27-A) is also activated and phosphorylates protein p53 in response to γ-irradiation of cells.[750] This is thought to be part of a signaling pathway for control of the cell cycle following DNA damage. Mitosis may be delayed while repair is completed. Alternatively, the damaged cell may be killed by apoptosis.

Subjects of current interest are the mechanisms by which completion of DNA synthesis is signaled at the various checkpoints in the cell cycle.[750b,c] These include points in addition to those marked in Fig. 11-15. At every point signals must accumulate to indicate that replication must be delayed to allow repair or that the cell must be allowed to die by apoptosis. Failure to accomplish repair may lead to cancer.[750d] Even one double-strand break will prevent the completion of mitosis.[750b] Among other problems faced by a dividing cell are slow replication and stalling at replication forks.[750c] A range of chemical alterations in chromatin and other proteins are associated with DNA damage and repair.[750b,e–h] Among these are conjugation of PCNA with ubiquitin and SUMO (Fig. 27-12D).[750i]

The SOS response and translesion repair. If cells of bacteria or eukaryotes are heavily damaged by UV, X-ray irradiation, or mutagenic chemicals, an emergency or **SOS response** is initiated.[711a,741,751–753a] In *E. coli* the two proteins specified by genes *lexA* and *recA* initiate the response. The *lexA* protein is a repressor that prevents transcription of a group of SOS genes (see Chapter 28). It is thought that some product from damaged DNA activates the RecA protease activity. The activated RecA protein then cleaves the lexA protein allowing transcription of the SOS genes. The SOS response is transient but complex. It includes increased recombinational activity, alterations in replication initiation, inhibition of nucleases, and induction of an **error-prone DNA synthesis**. The cell now replicates DNA more rapidly than normal but with an increased frequency of errors.[754,754a] For example, it will bypass thymine dimers and other defects, and it will continue the DNA synthesis even though incorrect bases have been put into the chain opposite the thymine dimer. Later the errors may be corrected by recombinational repair or by photoreactivation. In *E. coli* this translesional repair depends upon DNA polymerase III, the RecA protein, and two proteins encoded by genes *umuC* and *umuD*. These genes were recognized as providing resistance to mutations induced by UV radiation.[753–754b] Study of similar genes in yeast led to the discovery of two new and unusual DNA polymerases essential for translesional repair. DNA polymerase ζ (zeta), encoded by gene *REV3* and *REV7* subunits, bypasses abasic sites by inserting a dCMP residue into the growing DNA chain. Because C may be the wrong base the process is "error-prone." Polymerase η (eta) bypasses thymine photodimers by placing two consecutive dAMP residues in the growing strand. This is an error-free process.[755,756] Genes for human polymerases ζ and η have also been identified,[757–759b] and additional polymerases have been found in yeast and other organisms.[760–760e] The human XP-V gene for polymerase η is defective in xeroderma pigmentosa variant type (Box 27-A). Polymerase ι may act sequentially with pol η to bypass highly distorting lesions.[760b] The very imprecise pol θ may also be used to bypass hard-to-remove lesions.[760c] Polymerase λ is thought to play a role in DNA repair during meiosis,[760d] while pol κ is needed in some way to provide cohesion between sister chromatids.[760e] Some of these DNA polymerases are very inaccurate.[760a] It has been suggested that human pol ι may participate in hypermutation of immunoglobin variable genes.[760f] Polymerase ϕ of yeast may be required for synthesis of ribosomal RNA.[760g]

Poly (ADP-ribose). A eukaryotic peculiarity, which is not well understood, is the synthesis of poly(ADP-ribose) chains attached to many sites in nuclear proteins (see also Eq. 15-16). Increased synthesis is observed following damage to DNA.[761–763a] The poly (ADP-ribose) polymerase binds to DNA near strand breaks or nicks and, using NAD^+ as a substrate, synthesizes the highly branched polymer attached to a small number of nuclear target proteins.

F. Mutagens in the Environment

More than 500 new chemicals are introduced into the environment industrially each year. Some widely used drugs, e.g., **hycanthone** (Fig. 5-22), are mutagenic. Powerful mutagens are present naturally in some foods.[764–766] Others have been added through ignorance. Although many of these have now been removed, the problem cannot be ignored.

One way in which chemical compounds can induce base substitution mutation is through their incorporation into the structure of DNA itself. Thus, 5-bromodeoxyuridine (or bromouracil) can replace thymidine in DNA, where it serves as an efficient mutagenic agent.[767] 2-Aminopurine, an analog of adenine, pairs with thymine, just as does adenine when incorporated into DNA.

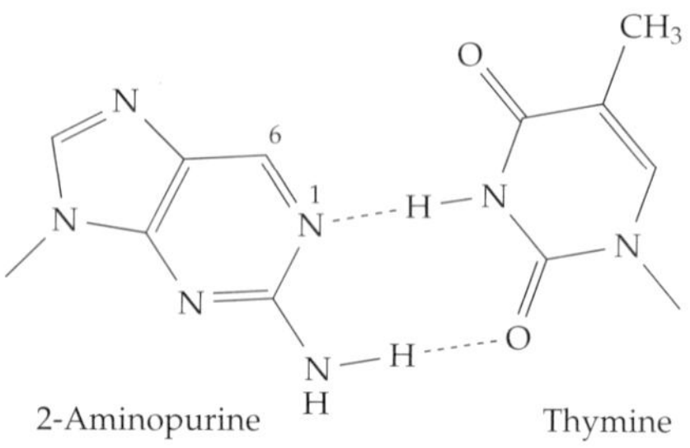

2-Aminopurine Thymine

However, if it protonates on N-1, it can pair with cytosine, causing mutation.[768,769]

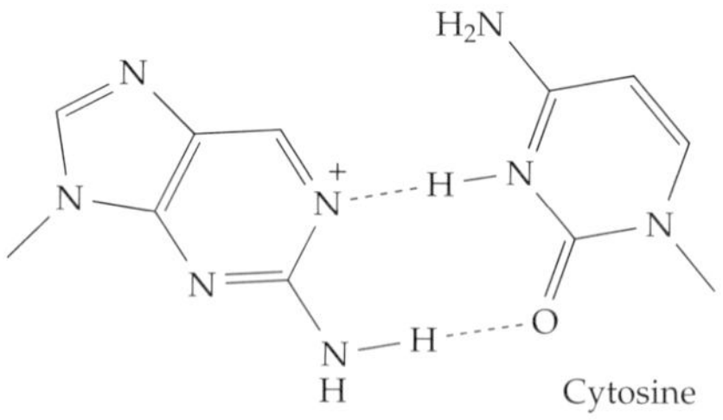

Cytosine

Another mutagenic base is N^6-hydroxylaminopurine.[770]

Many alkylating agents are powerful mutagens. Alkylation can occur at many places in DNA, but the N-7 position of guanine is especially susceptible (Eq. 5-18). The resultant positive charge on the imidazole ring portion of the alkylated guanine causes hydrolysis of the *N*-glycosyl linkage and depurination. However, this may be a lethal rather than a mutagenic event. The previously mentioned methylation on O-6 of guanine is probably more important in inducing mutations.[771]

Among the most biologically reactive alkylating agents are the nitrogen and sulfur "mustards" such as bis-(2-chloroethyl)sulfide. These toxic bifunctional compounds cause lethal crosslinking of DNA chains

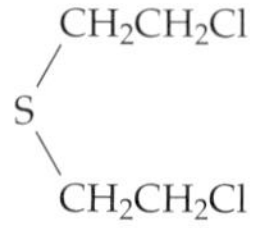

$S(CH_2CH_2Cl)_2$

bis-(2-chloroethyl)sulfide

(see Eq. 5-20). The monofunctional half-mustards are mutagenic but are less acutely toxic. Another group of alkylating agents, the **nitrosamines** (Chapter 5), are

BOX 27-A DEFICIENCIES IN HUMAN DNA REPAIR

Problems with human repair systems account for several serious diseases. One is the rare skin condition **xeroderma pigmentosum** (**XP**), an autosomal recessive hereditary defect. Homozygous individuals are extremely sensitive to ultraviolet radiation and have a high incidence of multiple carcinomas.[a–e] Defects in at least seven human genes, listed in the accompanying table, can cause XP. Most of the genes have been cloned and found to be homologs of genes for nucleotide excision repair (NER). In rodents the *ERCC* genes have the same functions as do genes for resistance to ultraviolet light in yeast (*RAD* genes). This information has elucidated the functions of the XP genes and the basis for this group of human diseases.[a,b,f]

The XP-A protein (see table) is a zinc-finger-containing DNA-binding protein, which interacts with human replication protein A (Section C,10). It may also interact with proteins corresponding to the rodent ERCC1 and ERCC4 proteins to form a complex that recognizes DNA defects.[g–i] Proteins XPC and XPE may be specialized DNA binding proteins for cyclobutane dimers and other photoproducts.[j,k] The XPG protein is an endonuclease that probably cuts on the 3' side of the DNA defect.[a,l] In human cell lines with a variety of XP mutations the DNA problems have been corrected by transfer of the corresponding normal genes into the cells.[m,n]

While defects in protein XPD often cause typical XP symptoms, some defects in the same protein lead to **trichothiodystrophy** (TTD, brittle hair disease). The hair is sulfur deficient, and scaly skin (ichthyosis, Box 8-F), mental retardation, and other symptoms are observed.[o] Like their yeast counterparts (proteins RAD3 and RAD25), XPB and XPD are both DNA helicases.[o] They also constitute distinct subunits of the human transcription factor TFIIH[p], which is discussed in Chapter 28. It seems likely that XPD is involved in **transcription-coupled repair** (**TCR**) of DNA.[o,q–s] This is a subpathway of the nucleotide excision repair (NER) pathway, which allows for rapid repair of the transcribed strand of DNA. This is important in tissues such as skin, where the global NER process may be too slow to keep up with the need for rapid protein synthesis. Transcription-coupled repair also appears to depend upon proteins CSA and CSB, defects which may result in the rare **cockayne syndrome**.[b,o,t,u] Patients are not only photosensitive but have severe mental and physical retardation including skeletal defects and a wizened appearance.

In a variant form of XP, designated XP-V, nucleotide excision repair is normal, but DNA replication is very slow. Postreplicational translesional repair (bypass repair) is also slow, and patients are cancer-prone.[v,w] The recently discovered DNA polymerase η may be defective.[x]

Children with **ataxia telangiectasia** (**AT**) have progressive neurological problems, a weak immune system, premature aging, and a high incidence of cancer.[y] Their skin fibroblasts are deficient in the ability to repair X-ray damage, which causes many double-strand breaks. Apparently in this disease cells do not wait until DNA repair has been carried out after exposure to ionizing radiation but attempt to replicate the damaged DNA.[y] The defective protein **ATM** (ataxia telangiectasia, mutated) has been identified as a large 370-kDa Ser / Thr protein kinase with a carboxyl terminal domain similar to phosphatidylinositol 3-kinase.[z,aa] It appears to play a crucial role in the cell cycle DNA damage checkpoints (Fig. 11-15) by participating in the detection of double-strand breaks and in delay of replication, while they are repaired by homologous recombination.[y,bb,cc] Although AT is an autosomal recessive disease, women heterozygous for the ATM gene have an increased susceptibility to breast cancer. This observation led to the discovery that the proteins encoded by the well known breast cancer genes *Brca1* and *Brca2* form a complex with ATM. Phosphorylation of BRCA1 by ATM may initiate a signaling pathway through the p53, c-Abl, and Chk2 proteins that cause cell cycle arrest (Fig. 11-15). BRCA1 and BRCA2 also form a complex with protein RAD51, a RecA homolog necessary for homologous recombination. BRCA1 also may be essential to transcription-coupled repair.[dd,de]

There are two major forms of hereditary susceptibility to colon cancer.[ee] Familial adenomatous polyposis is caused by defects in the *APC* gene (see Chapter 32). The more common **hereditary non-polyposis colorectal cancer** (**HNPCC**), which includes many endometrial, stomach, and urinary tract tumors, results from defects in DNA mismatch repair.[ff–jj] The proteins hMSH2 and hMSL1 are homologs of the *E.coli* MutS and MutL (main text).

Cells of patients with **Bloom syndrome** (**BS**) have many chromosome breaks and a high frequency of sister chromatid exchanges, perhaps in an effort to correct these breaks. The body is small but well-proportioned.[kk] A somewhat similar disease, the **Werner syndrome** (**WS**), is associated with premature aging.[ll] The Bloom's protein **BLM** and the WS gene product **WRN** are both helicases related to *E.coli* RecQ. Protein BLM colocalizes with replication protein A as discrete foci in the meiotic synaptonemal complex.[mm] Protein WRN also seems to be associated with DNA replication. Defects

BOX 27-A DEFICIENCIES IN HUMAN DNA REPAIR (continued)

appear to increase homologous and illegitimate recombination.[nn] Both proteins may also function in transcription.[oo]

Many other diseases leading to a high incidence of cancer are known. Among them is the Nijmegan breakage syndrome, in which chromosomes are hypersensitive to breakage by ionizing radiation. The gene has been identified by positional cloning, and its protein is apparently involved in repair of double strand breaks.[pp] **Fanconi anemia**, **Gardner syndrome**, and hereditary **retinoblastoma** (Box 11-D) may also involve defects in DNA repair.

Some Human Hereditary Defects of DNA Repair

Human disease	Human gene involved		Yeast gene	Function
Xeroderma pigmentosum (XP)				
XP-A	*XPA*		*RAD*14	DNA-binding, damage recognition
XP-B	*XPB*	*ERCC3*	*RAD*25 (*SSL2*)	DNA helicase
XP-C	*XPC*		*RAD4*	DNA-binding, thymine dimers
XP-D	*XPD*	*ERCC2*	*RAD3*	DNA helicase
XP-E	*XPE*		*RAD16*	
XP-F	*XPF*		*RAD*10	DNA nuclease complex
XP-G	*XPG*		*RAD*2	DNA nuclease
XP-V	*XPV*		*RAD*30	DNA polymerase η
Cockayne syndrome (CS)				
CSA	*CSA*			Interaction with helicase CSB and with TFIIH
CSB	*CSB*			DNA helicase
Trichothiodystrophy (TID, brittle hair disease)				
	XPDE			
Ataxia telangiectasia (AT)				
	ATM			Cell cycle delay for repair of ds breaks
Human nonpolyposis colorectal cancer (HNPCC)				
	*hMSH*2		*MSH*2	Mismatch repair
	*hMSL*1		*MSL*1	
	GTBP			

[a] Hoeijmakers, J. H. J., and Bootsma, D. (1994) *Nature (London)* **371**, 654–655
[b] Tanaka, K., and Wood, R. D. (1994) *Trends Biochem. Sci.* **19**, 83–86
[c] de Vries, A., van Oostrom, C. T. M., Hofhuis, F. M. A., Dortant, P. M., Berg, R. J. W., de Gruijl, F. R., Wester, P. W., van Kreijl, C. F., Capel, P. J. A., van Steeg, H., and Verbeek, S. J. (1995) *Nature (London)* **377**, 169–173
[d] Nakane, H., and 19 other authors. (1995) *Nature (London)* **377**, 165–168
[e] Lehmann, A. R. (1995) *Trends Biochem. Sci.* **20**, 402–405
[f] Bootsma, D., and Hoeijmakers, J. H. J. (1993) *Nature (London)* **363**, 114–115
[g] Bankmann, M., Prakash, L., and Prakash, S. (1992) *Nature (London)* **355**, 555–558
[h] Li, L., Elledge, S. J., Peterson, C. A., Bales, E. S., and Legerski, R. J. (1994) *Proc. Natl. Acad. Sci. U.S.A.* **91**, 5012–5016
[i] Lee, S.-H., Kim, D.-K., and Drissi, R. (1995) *J. Biol. Chem.* **270**, 21800–21805
[j] Emmert, S., Kobayashi, N., Khan, S. G., and Kraemer, K. H. (2000) *Proc. Natl. Acad. Sci. U.S.A.* **97**, 2151–2156
[k] Hwang, B. J., and Chu, G. (1993) *Biochemistry* **32**, 1657–1666
[l] O'Donovan, A., Scherly, D., Clarkson, S. G., and Wood, R. D. (1994) *J. Biol. Chem.* **269**, 15965–15968
[m] Evans, M. K., Robbins, J. H., Ganges, M. B., Tarone, R. E., Nairn, R. S., and Bohr, V. A. (1993) *J. Biol. Chem.* **268**, 4839–4847

BOX 27-A (continued)

[n] Tebbs, R. S., Zhao, Y., Tucker, J. D., Scheerer, J. B., Siciliano, M. J., Hwang, M., Liu, N., Legerski, R. J., and Thompson, L. H. (1995) *Proc. Natl. Acad. Sci. U.S.A.* **92**, 6354–6358

[o] van Gool, A. J., van der Horst, G. T. J., Citterio, E., and Hoeijmakers, J. H. J. (1997) *EMBO J.* **16**, 4155–4162

[p] Schaeffer, L., Roy, R., Humbert, S., Moncollin, V., Vermeulen, W., Hoeijmakers, J. H. J., Chambon, P., and Egly, J.-M. (1993) *Science* **260**, 58–63

[q] Guzder, S. N., Sung, P., Prakash, S., and Prakash, L. (1995) *J. Biol. Chem.* **270**, 17660–17663

[r] Winkler, G. S., Araújo, S. J., Fiedler, U., Vermeulen, W., Coin, F., Egly, J.-M., Hoeijmakers, J. H. J., Wood, R. D., Timmers, H. T. M., and Weeda, G. (2000) *J. Biol. Chem.* **275**, 4258–4266

[s] Reardon, J. T., Bessho, T., Kung, H. C., Bolton, P. H., and Sancar, A. (1997) *Proc. Natl. Acad. Sci. U.S.A.* **94**, 9463–9468

[t] Mu, D., and Sancar, A. (1997) *J. Biol. Chem.* **272**, 7570–7573

[u] Cooper, P. K., Nouspikel, T., Clarkson, S. G., and Leadon, S. A. (1997) *Science* **275**, 990–993

[v] Masutani, C., Araki, M., Yamada, A., Kusumoto, R., Nogimori, T., Maekawa, T., Iwai, S., and Hanaoka, F. (1999) *EMBO J.* **18**, 3491–3501

[w] Johnson, R. E., Kondratick, C. M., Prakash, S., and Prakash, L. (1999) *Science* **285**, 263–265

[x] Masutani, C., Kusumoto, R., Yamada, A., Dohmae, N., Yokoi, M., Yuasa, M., Araki, M., Iwai, S., Takio, K., and Hanaoka, F. (1999) *Nature (London)* **399**, 700–704

[y] Beamish, H., Williams, R., Chen, P., and Lavin, M. F. (1996) *J. Biol. Chem.* **271**, 20486–20493

[z] Lavin, M. F., Khanna, K. K., Beamish, H., Spring, K., and Watters, D. (1995) *Trends Biochem. Sci.* **20**, 382–383

[aa] Chong, M. J., Murray, M. R., Gosink, E. C., Russell, H. R. C., Srinivasan, A., Kapsetaki, M., Korsmeyer, S. J., and McKinnon, P. J. (2000) *Proc. Natl. Acad. Sci. U.S.A.* **97**, 889–894

[bb] Morrison, C., Sonoda, E., Takao, N., Shinohara, A., Yamamoto, K.-i, and Takeda, S. (2000) *EMBO J.* **19**, 463–471

[cc] Cortez, D., Wang, Y., Qin, J., and Elledge, S. J. (1999) *Science* **286**, 1162–1166

[dd] Gowen, L. C., Avrutskaya, A. V., Latour, A. M., Koller, B. H., and Leadon, S. A. (1998) *Science* **281**, 1009–1012

[de] Davies, A. A., Masson, J.-Y., McIlwraith, M. J., Stasiak, A. Z., Stasiak, A., Venkitaraman, A. R., and West, S. C. (2001) *Molecular Cell* **7**, 273–282

[ee] Peltomäki, P., Aaltonen, L. A., Sistonen, P., Pylkkänen, L., Mecklin, J.-P., Järvinen, H., Green, J. S., Jass, J. R., Weber, J. L., Leach, F. S., Petersen, G. M., Hamilton, S. R., de la Chapelle, A., and Vogelstein, B. (1993) *Science* **260**, 810–819

[ff] Papadopoulos, N., and 19 other authors. (1994) *Science* **18**, 1625–1629

[gg] Prolla, T. A., Pang, Q., Alani, E., Kolodner, R. D., and Liskay, R. M. (1994) *Science* **265**, 1091–1092

[hh] Kolodner, R. D. (1995) *Trends Biochem. Sci.* **20**, 397–401

[ii] Palombo, F., Gallinari, P., Iaccarino, I., Lettieri, T., Hughes, M., D'Arrigo, A., Truong, O., Hsuan, J. J., and Jiricny, J. (1995) *Science* **268**, 1912–1914

[jj] Mellon, I., Rajpal, D. K., Koi, M., Boland, C. R., and Champe, G. N. (1996) *Science* **272**, 557–560

[kk] German, J., Roe, A. M., Leppert, M. F., and Ellis, N. A. (1994) *Proc. Natl. Acad. Sci. U.S.A.* **91**, 6669–6673

[ll] Lebel, M., Spillare, E. A., Harris, C. C., and Leder, P. (1999) *J. Biol. Chem.* **274**, 37795–37799

[mm] Walpita, D., Plug, A. W., Neff, N. F., German, J., and Ashley, T. (1999) *Proc. Natl. Acad. Sci. U.S.A.* **96**, 5622–5627

[nn] Yamagata, K., Kato, J.-I., Shimamoto, A., Goto, M., Furuichi, Y., and Ikeda, H. (1998) *Proc. Natl. Acad. Sci. U.S.A.* **95**, 8733–8738

[oo] Lee, S.-K., Johnson, R. E., Yu, S.-L., Prakash, L., and Prakash, S. (1999) *Science* **286**, 2339–2342

[pp] Dong, Z., Zhong, Q., and Chen, P.-L. (1999) *J. Biol. Chem.* **274**, 19513–19516

highly mutagenic.[772,773] Much used in the laboratory as a mutagen is ***N*-methyl-*N'*-nitro-*N*-nitroso-guanidine**:

$$H_2\overset{+}{N}{=}C(-NH-NO_2)(-N(CH_3)-NO)$$

N-Ethyl-*N*-nitrosourea is one of the most potent carcinogens known.[773,774] These compounds, as well as diazomethane and other related substances, probably act via a common intermediate: $CH_3N_2^+$ or $C_2H_5N_2^+$.[775]

Any secondary amine will react with nitrous acid to form a nitrosamine (Eq. 27-25). Tertiary amines can also react with loss of one alkyl group. This can occur in the stomach, and the nitrosamines may be absorbed into the system. All plants contain some nitrate and some, such as spinach and beets, have large amounts. Bacon and other cured R′ R meats contain both nitrites and nitrates, and many drugs and natural food

$$RR'NH \xrightarrow{HNO_2 \;\; -H_2O} RR'N{-}N{=}O \quad \text{(Nitrosamine)} \qquad (27\text{-}25)$$

constituents are secondary amines. Cigarette smoke also contains a nitrosamine.[776] There is a possibility that these substances may induce human cancer.[772,777]

About 3% of residents of North America excrete feces that contain mutagenic unsaturated glyceryl ethers, which have been named **fecapentaenes**.[778,779]

HO–CH₂–CH(OH)–CH₂–O–CH=CH–CH=CH–CH=CH–CH=CH–CH=CH–CH₂–CH₃

A fecapentaene

They could be among the causative agents of colorectal cancer. Protonation of a double bond will produce a reactive carbocation, which may be an active alkylating agent.[779]

Many halogenated compounds are carcinogenic. Among these is **1,2-dibromoethane,** which has been produced in the United States in quantities as great as 10^8 kg / year. It has been widely used as a fumigant for foods, as an industrial solvent, and as an additive to gasoline. It is a **procarcinogen**, whose carcinogenic properties are expressed only after **metabolic activation**. One pathway of activation is reaction with glutathione to form a thioether, which can cyclize to a sulfonium compound.[780] The latter alkylates nucleophilic groups in DNA as in Eq. 27-26. Other bifunctional electrophiles such as acrolein, malondialdehyde, vinyl chloride, and urethane are procarcinogens.[781–784] Styrene, another procarcinogen, is converted to the carcinogenic styrene oxide by action of a cytochrome P450.[785] Metabolites of glucose such as **pyruvaldehyde**, **methylglyoxal**, and other reactive aldehydes can also attack DNA.[786,787]

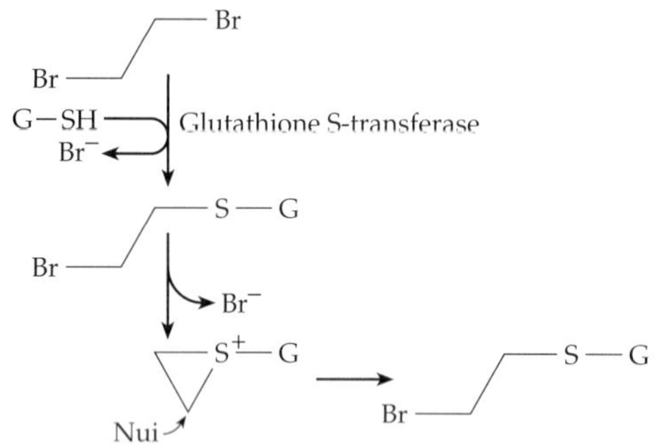

Br
Br
G−SH
Br⁻
Glutathione S-transferase
S—G
Br
Br⁻
S⁺—G
Nui
S—G
Br

(27-26)

Less common than base pair exchanges are frame-shift mutations. A characteristic of such mutations is that they do not revert (back mutate) as readily as do base substitution mutations, and reversion is not induced by chemicals known to cause base substitutions. However, reversion of frame-shift mutations is induced by acridines and other flat molecules that are known to act as intercalating agents in DNA helices and which promote frame-shift mutations. They are especially effective in causing mutations of regions in which long repeated sequences of a single base such as $(A)_n$ or $(G)_n$ occur.[770,788] For example, deletion of two base pairs from a "hot spot" (site of frequent mutation) in the *Salmonella* histidine[789] operon with the following sequence is induced by 2-nitrosofluorene and causes reversion of a (–1) histidine-requiring mutant: 5'-CGCGCGCG. Whereas simple intercalating agents are often not very mutagenic, compounds that are both an intercalating agent and an alkylating agent are especially potent. An example is the following compound, which contains a half-mustard side chain:

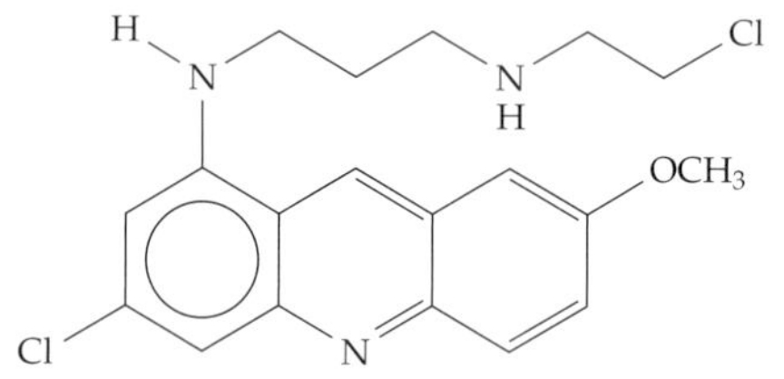

H
N
N
H
Cl
OCH_3
Cl
N

When the CH_2Cl group of the side chain is replaced by CH_2OH, the compound is 100 times less mutagenic.

The carcinogenic **aflatoxins**, which are produced by *Aspergillus flavus*, may be present in infected peanuts and other foodstuffs.[790] Like many other compounds that are carcinogenic or mutagenic, the aflatoxins are not unusually reactive chemically.

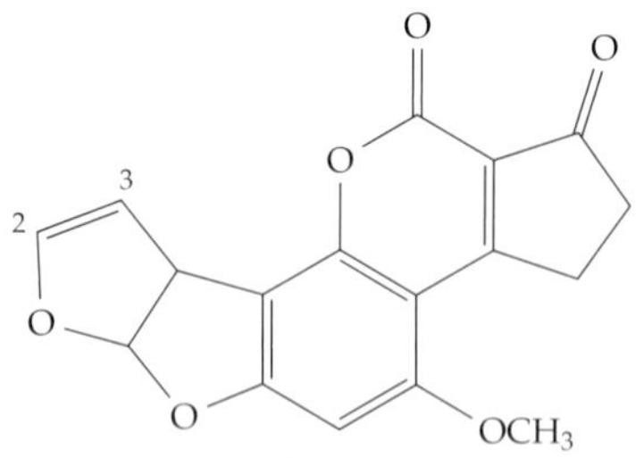

O
O
O
3
2
O
O
OCH_3
Aflatoxin B_1

However, they are activated in the animal body by hydroxylation and formation of 2,3-epoxides. The latter may react with N-7 of guanyl residues in DNA (Eq. 27-27).[791,792]

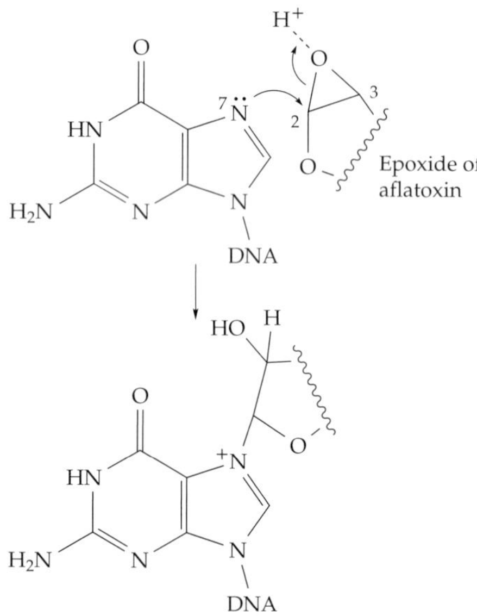

H^+
O
O
7
N
3
HN
2
O
Epoxide of aflatoxin
H_2N
N
N
DNA
HO
H
O
O
HN
⁺N
H_2N
N
N
DNA

(27-27)

Benz(*a*)anthracene is also converted in the body to a carcinogenic epoxide (Eq. 27-28).[793] Benzo(*a*)pyrene was isolated from coal tar in 1929 and in 1930 provided the first demonstration of the carcinogenicity of a pure chemical compound.[794] It can also be activated by conversion in the ER to the 7,8-dihydrodiol 9,10-

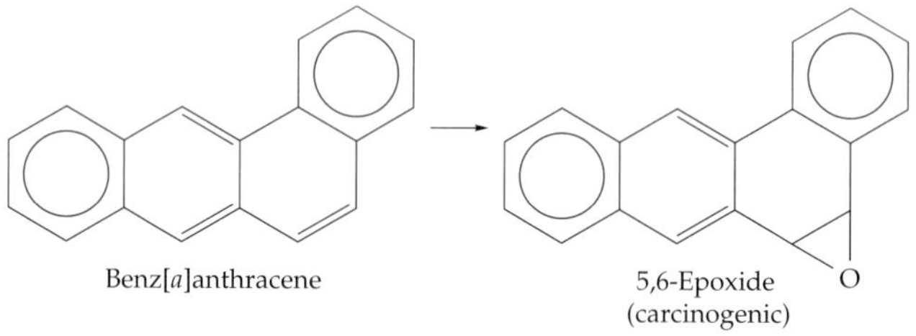

Benz[*a*]anthracene
5,6-Epoxide (carcinogenic)
O

(27-28)

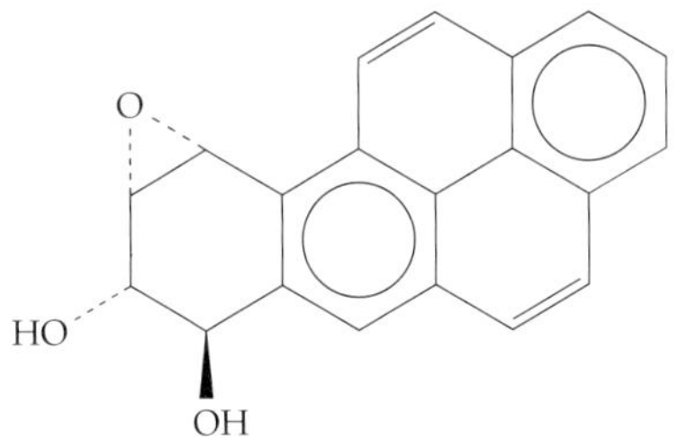

Benzo(*a*)pyrene diol epoxide

epoxide.[795–797] Hydroxylation activates 2-acetylaminofluorene to a carcinogenic *N*-sulfate (Eq. 27-29),[788,798] which reacts with guanine rings.[788,799]

Although it seems incongruous, many of the most potent antitumor drugs are powerful mutagens.

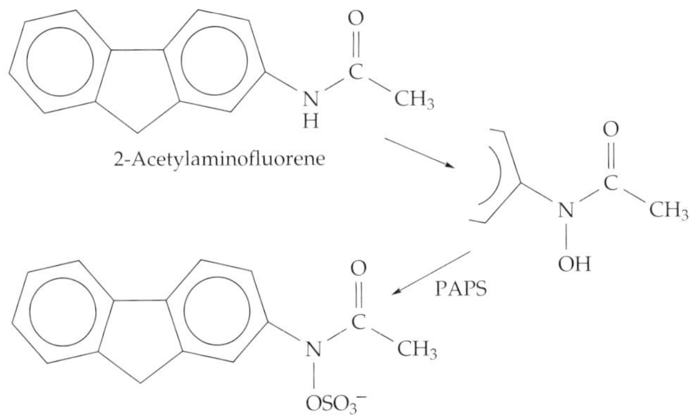

2-Acetylaminofluorene

PAPS

Carcinogen

(27-29)

Among these are intercalating agents such as daunomycin (Figs. 5-22, 5-23), neocarzinostatin, and bleomycin (Box 5-B). These are alkylating reagents[800] or attack DNA in other ways. The fact that such compounds are in use for chemotherapy emphasizes the need for new approaches to cancer treatment.

How can compounds be recognized as mutagenic? This is an important question that is difficult to answer and which has led to many controversies. For example, how dangerous is formaldehyde in the environment?[801] Butadiene?[802] Dioxin?[790,803] Is fluoride a carcinogen?[804,805]

A quick test for mutagenic activity makes use of **tester strains** of bacteria developed by Ames and associates. These are *Salmonella* mutants that are unable to synthesize their own histidine, but which can grow when a mutagenic agent produces a back mutation.[806,807] One of the strains can be mutated by agents causing base exchanges, while the other three, which contain different types of frame-shift mutations, are affected differently by various mutagens. About 10^9 bacteria are spread on a Petri plate, and a small amount of the mutagenic chemical is introduced in the center of the plate. Where back mutation has occurred, a colony of the bacteria appears. The strains all carry a mutation in the main DNA excision-repair system so that most mutations are not repaired, and the test is very sensitive. Addition of a liver homogenate plus an NADPH-generating system to the sample tested allows activation of many aromatic chemicals by hydroxylation.[808] Feeding of mutagens to *Drosophila* eye color mutants permits testing for back mutation in a eukaryote.[809]

The bacterial tests have been widely used and have been of great value. For example, they revealed that certain chemicals that were being used as flame retardants in children's clothing are mutagens,[810] and that mutagens can be generated during cooking of meat and other foods.[811] However, there are good reasons for using other methods of identifying mutagens together with the bacterial tests.[812] Government regulatory agencies in the United States have relied largely on tests with rodents. Compounds are tested at very high doses with these short-lived animals. Extrapolation to human exposures at very low levels has been criticized.[764,765,813,814] However, rodent tests have identified many true carcinogens.[815] A broader range of tests are now in use.[816] Direct monitoring of the accumulation of defects in animal and human cells is now possible. For example, in the **^{32}P-postlabeling technique** DNA is enzymatically digested to the nucleotide level, and the adducts with mutagens are labeled with ^{32}P, separated, and their quantities measured.[817–819] Laser-excited fluorescence from such adducts, GC / mass spectrometry, and immunological methods can also be used to identify DNA adducts.[819] Careful vigilance is needed to keep our mutation rate at as low a level as possible.

References

1. Wasserman, S. A., and Cozzarelli, N. R. (1986) *Science* **232**, 951–960
2. Reich, Z., Wachtel, E. J., and Minsky, A. (1994) *Science* **264**, 1460–1463

2a. Daban, J.-R. (2000) *Biochemistry* **39**, 3862–3866

3. Moody, M. F. (1999) *J. Mol. Biol.* **293**, 401–433
4. Black, L. W., Newcomb, W. W., Boring, J. W., and Brown, J. C. (1985) *Proc. Natl. Acad. Sci. U.S.A.* **82**, 7960–7964

4a. Hud, N. V., and Downing, K. H. (2001) *Proc. Natl. Acad. Sci. U.S.A.* **98**, 14925–14930

5. Sun, M., and Serwer, P. (1997) *Biophys. J.* **72**, 958–963
6. Marx, K. A., and Reynolds, T. C. (1982) *Proc. Natl. Acad. Sci. U.S.A.* **79**, 6484–6488
7. Widom, J., and Baldwin, R. L. (1983) *J. Mol. Biol.* **171**, 419–437
8. Kamashev, D., Balandina, A., and Rouviere-Yaniv, J. (1999) *EMBO J.* **18**, 5434–5444
9. Kobryn, K., Lavoie, B. D., and Chaconas, G. (1999) *J. Mol. Biol.* **289**, 777–784
10. Schmid, M. B. (1988) *Trends Biochem. Sci.* **13**, 131–135
11. Claret, L., and Rouviere-Yaniv, J. (1996) *J. Mol. Biol.* **263**, 126–139
12. Esser, D., Rudolph, R., Jaenicke, R., and Böhm, G. (1999) *J. Mol. Biol.* **291**, 1135–1146
13. Tanaka, I., Appelt, K., Dijk, J., White, S. W., and Wilson, K. S. (1984) *Nature (London)* **310**, 376–381
14. White, S. W., Wilson, K. S., Appelt, K., and Tanaka, I. (1999) *Acta Crystallogr. D* **55**, 801–809

14a. Grove, A., and Saavedra, T. C. (2002) *Biochemistry* **41**, 7597–7603

15. Ueguchi, C., Suzuki, T., Yoshida, T., Tanaka, K.-i, and Mizuno, T. (1996) *J. Mol. Biol.* **263**, 149–162
16. Starich, M. R., Sandman, K., Reeve, J. N., and Summers, M. F. (1996) *J. Mol. Biol.* **255**, 187–203
17. Krueger, J. K., McCrary, B. S., Wang, A. H.-J., Shriver, J. W., Trewhella, J., and Edmondson, S. P. (1999) *Biochemistry* **38**, 10247–10255

17a. Kamashev, D., and Rouviere-Yaniv, J. (2000) *EMBO J.* **19**, 6527–6535

17b. Grove, A., and Lim, L. (2001) *J. Mol. Biol.* **311**, 491–502

18. Pettijohn, D. E. (1988) *J. Biol. Chem.* **263**, 12793–12796

18a. Renzoni, D., Esposito, D., Pfuhl, M., Hinton, J. C. D., Higgins, C. F., Driscoll, P. C., and Ladbury, J. E. (2001) *J. Mol. Biol.* **306**, 1127–1137

18b. Schröder, O., and Wagner, R. (2000) *J. Mol. Biol.* **298**, 737–748

19. Jackson, D. A., and Patel, S. B. (1982) *Trends Biochem. Sci.* **7**, 272–274

19a. Korolev, N., Lyubartsev, A. P., Nordenskiöld, L., and Laaksonen, A. (2001) *J. Mol. Biol.* **308**, 907–917

20. Hunt, J. G., Kasinsky, H. E., Elsey, R. M., Wright, C. L., Rice, P., Bell, J. E., Sharp, D. J., Kiss, A. J., Hunt, D. F., Arnott, D. P., Russ, M. M., Shabanowitz, J., and Ausió, J. (1996) *J. Biol. Chem.* **271**, 23547–23557
21. Harborne, J. B. (1993) *Introduction to Ecological Biochemistry*, 4th ed., Academic Press, San Diego, California

21a. Prieto, C., Saperas, N., Arnan, C., Hills, M. H., Wang, X., Chiva, M., Aligué, R., Subirana, J. A., and Ausió, J. (2002) *Biochemistry* **41**, 7802–7810

22. Bellvé, A. R., McKay, D. J., Renaux, B. S., and Dixon, G. H. (1988) *Biochemistry* **27**, 2890–2897
23. de Yebra, L., Ballescà, J. L., Vanrell, J. A., Bassas, L., and Oliva, R. (1993) *J. Biol. Chem.* **268**, 10553–10557
24. Doenecke, D., and Karlson, P. (1984) *Trends Biochem. Sci.* **9**, 404–405
25. Kornberg, R. D., and Thomas, J. O. (1974) *Science* **184**, 865–868
26. Isenberg, I. (1979) *Ann. Rev. Biochem.* **48**, 159–191
27. Arents, G., and Moudrianakis, E. N. (1995) *Proc. Natl. Acad. Sci. U.S.A.* **92**, 11170–11174
28. Zemzoumi, K., Frontini, M., Bellorini, M., and Mantovani, R. (1999) *J. Mol. Biol.* **286**, 327–337
29. Wellman, S. E., Song, Y., and Mamoon, N. M. (1999) *Biochemistry* **38**, 13112–13118
30. Lindner, H., Sarg, B., Hoertnagl, B., and Helliger, W. (1998) *J. Biol. Chem.* **273**, 13324–13330
31. Zlatanova, J., and Doenecke, D. (1994) *FASEB J.* **8**, 1260–1268
32. Hansen, J. C., Tse, C., and Wolffe, A. P. (1998) *Biochemistry* **37**, 17637–17641
33. DeLange, R. J., Hooper, J. A., and Smith, E. L. (1973) *J. Biol. Chem.* **248**, 3261–3274

33a. Wu, J., and Grunstein, M. (2000) *Trends Biochem. Sci.* **25**, 619–623

33b. Jenuwein, T., and Allis, C. D. (2001) *Science* **293**, 1074–1080

33c. Rea, S., Eisenhaber, F., O'Carroll, D., Strahl, B. D., Sun, Z.-W., Schmid, M., Opravil, S., Mechtler, K., Ponting, C. P., Allis, C. D., and Jenuwein, T. (2000) *Nature (London)* **406**, 593–599

33d. Jacobs, S. A., and Khorasanizadeh, S. (2002) *Science* **295**, 2080–2083

34. Kelner, D. N., and McCarty, K. S., Sr. (1984) *J. Biol. Chem.* **259**, 3413–3418
35. Sternglanz, R. (1996) *Trends Biochem. Sci.* **21**, 357–358
36. Strenglanz, R., and Schindelin, H. (1999) *Proc. Natl. Acad. Sci. U.S.A.* **96**, 8807–8808
37. Champagne, N., Bertos, N. R., Pelletier, N., Wang, A. H., Vezmar, M., Yang, Y., Heng, H. H., and Yang, X.-J. (1999) *J. Biol. Chem.* **274**, 28528–28536
38. Wade, P. A., Pruss, D., and Wolffe, A. P. (1997) *Trends Biochem. Sci.* **22**, 128–132

38a. Marmorstein, R. (2001) *J. Mol. Biol.* **311**, 433–444

38b. Kruhlak, M. J., Hendzel, M. J., Fischle, W., Bertos, N. R., Hameed, S., Yang, X.-J., Verdin, E., and Bazett-Jones, D. P. (2001) *J. Biol. Chem.* **276**, 38307–38319

39. Hay, C. W., and Candido, E. P. M. (1983) *J. Biol. Chem.* **258**, 3726–3734
40. Kölle, D., Brosch, G., Lechner, T., Pipal, A., Helliger, W., Taplick, J., and Loidl, P. (1999) *Biochemistry* **38**, 6769–6773

40a. Huang, X., and Kadonaga, J. T. (2001) *J. Biol. Chem.* **276**, 12497–12500

40b. Furumai, R., Komatsu, Y., Nishino, N., Khochbin, S., Yoshida, M., and Horinouchi, S. (2001) *Proc. Natl. Acad. Sci. U.S.A.* **98**, 87–92

41. Osley, M. A. (1991) *Ann. Rev. Biochem.* **60**, 827–861
42. Gorovsky, M. A. (1986) in *Molecular Biology of Ciliated Protozoas* (Gall, J. G., ed), Academic Press, Orlando, Florida
43. Prentice, D. A., Loechel, S. C., and Kitos, P. A. (1982) *Biochemistry* **21**, 2412–2420
44. Swank, R. A., Th'ng, J. P. H., Guo, X.-W., Valdez, J., Bradbury, E. M., and Gurley, L. R. (1997) *Biochemistry* **36**, 13761–13768
45. Allera, C., Lazzarini, G., and Patrone, E. (1995) *Biochemistry* **34**, 301–311

45a. de la Barre, A.-E., Angelov, D., Molla, A., and Dimitrov, S. (2001) *EMBO J.* **20**, 6383–6393

46. Koiv, A., Palvimo, J., and Kinnunen, P. K. J. (1995) *Biochemistry* **34**, 8018–8027

46a. Ausió, J., and Abbott, D. W. (2002) *Biochemistry* **41**, 5945–5949

47. Chen, H. Y., Sun, J.-M., Zhang, Y., Davie, J. R., and Meistrich, M. L. (1998) *J. Biol. Chem.* **273**, 13165–13169

47a. Soares, D. J., Sandman, K., and Reeve, J. N. (2000) *J. Mol. Biol.* **297**, 39–47

47b. Bell, S. D., Botting, C. H., Wardleworth, B. N., Jackson, S. P., and White, M. F. (2002) *Science* **296**, 148–151

48. Kornberg, R. D., and Klug, A. (1981) *Sci. Am.* **244**(Feb), 52–64
49. Grunstein, M. (1992) *Sci. Am.* **267**(Oct), 68–74B
50. Luger, K., Mäder, A. W., Richmond, R. K., Sargent, D. F., and Richmond, T. J. (1997) *Nature (London)* **389**, 251–260
51. Griffith, J. D. (1975) *Science* **187**, 1202–1203
52. Krude, T., and Knippers, R. (1994) *J. Biol. Chem.* **269**, 21021–21029
53. Givens, R. M., Saavedra, R. A., and Huberman, J. A. (1996) *J. Mol. Biol.* **257**, 53–65
54. Sen, D., Mitra, S., and Crothers, D. M. (1986) *Biochemistry* **25**, 3441–3447
55. Felsenfeld, G., and McGhee, J. D. (1986) *Cell* **44**, 375–377
56. Wasserman, S. A., White, J. H., and Cozzarelli, N. R. (1988) *Nature (London)* **334**, 448–450
57. McMurray, C. T., van Holde, K. E., Jones, R. L., and Wilson, W. D. (1985) *Biochemistry* **24**, 7037–7044
58. Fitzgerald, D. J., and Anderson, J. N. (1999) *J. Mol. Biol.* **293**, 477–491
59. Lowary, P. T., and Widom, J. (1998) *J. Mol. Biol.* **276**, 19–42

59a. Thåström, A., Lowary, P. T., Widlund, H. R., Cao, H., Kubista, M., and Widom, J. (1999) *J. Mol. Biol.* **288**, 213–229

60. Flaus, A., and Richmond, T. J. (1998) *J. Mol. Biol.* **275**, 427–441
61. Widlund, H. R., Cao, H., Simonsson, S., Magnusson, E., Simonsson, T., Nielsen, P. E., Kahn, J. D., Crothers, D. M., and Kubista, M. (1997) *J. Mol. Biol.* **267**, 807–817

61a. Negri, R., Buttinelli, M., Panetta, G., De Arcangelis, V., Di Mauro, E., and Travers, A. (2001) *J. Mol. Biol.* **307**, 987–999

62. Leuba, S. H., Zlatanova, J., and van Holde, K. (1993) *J. Mol. Biol.* **229**, 917–929
63. Crane-Robinson, C. (1997) *Trends Biochem. Sci.* **22**, 75–77
64. Bednar, J., Horowitz, R. A., Grigoryev, S. A., Carruthers, L. M., Hansen, J. C., Koster, A. J., and Woodcock, C. L. (1998) *Proc. Natl. Acad. Sci. U.S.A.* **95**, 14173–14178

64a. Vila, R., Ponte, I., Jiménez, M. A., Rico, M., and Suau, P. (2000) *Protein Sci.* **9**, 627–636

64b. Bharath, M. M. S., Ramesh, S., Chandra, N. R., and Rao, M. R. S. (2002) *Biochemistry* **41**, 7617–7627

65. Pruss, D., Bartholomew, B., Persinger, J., Hayes, J., Arents, G., Moudrianakis, E. N., and Wolffe, A. P. (1996) *Science* **274**, 614–617
66. Zhou, Y.-B., Gerchman, S. E., Ramakrishnan, V., Travers, A., and Muyldermans, S. (1998) *Nature (London)* **395**, 402–405
67. Hayes, J. J. (1996) *Biochemistry* **35**, 11931–11937
68. Ramakrishnan, V., Finch, J. T., Graziano, V., Lee, P. L., and Sweet, R. M. (1993) *Nature (London)* **362**, 219–223
69. DeLucia, F., Alilat, M., Sivolob, A., and Prunell, A. (1999) *J. Mol. Biol.* **285**, 1101–1119
70. Pennisi, E. (1997) *Science* **275**, 155–157
71. Polach, K. J., and Widom, J. (1995) *J. Mol. Biol.* **254**, 130–149
72. Strahl, B. D., and Allis, C. D. (2000) *Nature (London)* **403**, 41–45
73. Cairns, B. R. (1998) *Trends Biochem. Sci.* **23**, 20–24
74. Kellenberger, E. (1987) *Trends Biochem. Sci.* **12**, 105–107

References

75. Gross, D. S., and Garrard, W. T. (1988) *Ann. Rev. Biochem.* **57**, 159–197
76. Felsenfeld, G., Boyes, J., Chung, J., Clark, D., and Studitsky, V. (1996) *Proc. Natl. Acad. Sci. U.S.A.* **93**, 9384–9388
77. Wu, C. (1997) *J. Biol. Chem.* **272**, 28171–28174
78. Grunstein, M. (1997) *Nature (London)* **389**, 349–352
79. Edmondson, D. G., and Roth, S. Y. (1996) *FASEB J.* **10**, 1173–1182
80. Nightingale, K. P., Wellinger, R. E., Sogo, J. M., and Becker, P. B. (1998) *EMBO J.* **17**, 2865–2876
81. Zhang, W., Bone, J. R., Edmondson, D. G., Turner, B. M., and Roth, S. Y. (1998) *EMBO J.* **17**, 3155–3167
82. Utley, R. T., Ikeda, K., Grant, P. A., Côté, J., Steger, D. J., Eberharter, A., John, S., and Workman, J. L. (1998) *Nature (London)* **394**, 498–502
83. Mueller, R. D., Yasuda, H., and Bradbury, E. M. (1985) *J. Biol. Chem.* **260**, 5081–5086
84. Muchardt, C., and Yaniv, M. (1999) *J. Mol. Biol.* **293**, 187–198
84a. Horn, P. J., and Peterson, C. L. (2002) *Science* **297**, 1824 – 1827
85. Thoma, F., Koller, T. H., and Klug, A. (1979) *J. Cell Biol.* **83**, 403–427
86. Walker, P. R., and Sikorska, M. (1987) *J. Biol. Chem.* **262**, 12218–12222
87. Woodcock, C. L. F., Frado, L.-L. Y., and Rattner, J. B. (1984) *J. Cell Biol.* **99**, 42–52
88. Rydberg, B., Holley, W. R., Mian, I. S., and Chatterjee, A. (1998) *J. Mol. Biol.* **284**, 71–84
89. Daban, J.-R., and Bermúdez, A. (1998) *Biochemistry* **37**, 4299–4304
90. Sirlin, J. L. (1972) *Biology of RNA*, Academic Press, New York (pp. 162–164)
91. Manuelidis, L. (1990) *Science* **250**, 1533–1540
91a. Uhlmann, F. (2002) *Nature (London)* **417**, 135–136
91b. Antes, T. J., Namciu, S. J., Fournier, R. E. K., and Levy-Wilson, B. (2001) *Biochemistry* **40**, 6731–6742
91c. Löwe, J., Cordell, S. C., and van den Ent, F. (2001) *J. Mol. Biol.* **306**, 25–35
92. Münkel, C., Eils, R., Dietzel, S., Zink, D., Mehring, C., Wedemann, G., Cremer, T., and Langowski, J. (1999) *J. Mol. Biol.* **285**, 1053–1065
93. van Holde, K., and Zlatanova, J. (1995) *J. Biol. Chem.* **270**, 8373–8376
94. Nagele, R., Freeman, T., McMorrow, L., and Lee, H.-y. (1995) *Science* **270**, 1831–1835
95. Lamond, A. I., and Earnshaw, W. C. (1998) *Science* **280**, 547–553
96. Wagner, R. P., Maguire, M. M., and Stallings, R. L. (1993) *Chromosomes: A Synthesis*, Wiley-Liss, New York
97. Cook, K. R., and Karpen, G. H. (1994) *Proc. Natl. Acad. Sci. U.S.A.* **91**, 5219–5221
98. Motta, M. C., Landsberger, N., Merli, C., and Badaracco, G. (1998) *J. Biol. Chem.* **273**, 18028–18039
98a. Ringrose, L., and Paro, R. (2001) *Nature (London)* **412**, 493–494
99. Wei, X., Samarabandu, J., Devdhar, R. S., Siegel, A. J., Acharya, R., and Berezney, R. (1998) *Science* **281**, 1502–1505
100. Baskin, Y. (1995) *Science* **268**, 1564–1565
101. Lemon, K. P., and Grossman, A. D. (1998) *Science* **282**, 1516–1519
102. Hirano, T. (1995) *Trends Biochem. Sci.* **20**, 357–361
102a. Misteli, T. (2001) *Science* **291**, 843 – 847
103. Tsutsui, K., Tsutsui, K., and Muller, M. T. (1988) *J. Biol. Chem.* **263**, 7235–7241
104. Nelson, W. G., Pienta, K. J., Barrack, E. R., and Coffey, D. S. (1986) *Ann. Rev. Biophys. Biophys. Chem.* **15**, 457–475
105. Pemov, A., Bavykin, S., and Hamlin, J. L. (1998) *Proc. Natl. Acad. Sci. U.S.A.* **95**, 14757–14762
106. Benham, C., Kohwi-Shigematsu, T., and Bode, J. (1997) *J. Mol. Biol.* **274**, 181–196
107. Pederson, T. (1998) *J. Mol. Biol.* **277**, 147–159
108. Mirkovitch, J., Spierer, P., and Laemmli, U. K. (1986) *J. Mol. Biol.* **190**, 255–258
109. Kay, V., and Bode, J. (1994) *Biochemistry* **33**, 367–374
110. Udvardy, A. (1999) *EMBO J.* **18**, 1–8
111. Harborth, J., Wang, J., Gueth-Hallonet, C., Weber, K., and Osborn, M. (1999) *EMBO J.* **18**, 1689–1700
112. Peterson, J. L., and McConkey, E. H. (1976) *J. Biol. Chem.* **251**, 548–554
113. Hnilica, L. S., ed. (1983) *Chromosomal Nonhistone Proteins*, Vol. III, CRC Press, Boca Raton, Florida
114. Quick, D. P., Orchard, P. J., and Duerre, J. A. (1981) *Biochemistry* **20**, 4724–4729
115. Landsman, D., McBride, O. W., Soares, N., Crippa, M. P., Srikantha, T., and Bustin, M. (1989) *J. Biol. Chem.* **264**, 3421–3427
115a. Müller, S., Scaffidi, P., Degryse, B., Bonaldi, T., Ronfani, L., Agresti, A., Beltrame, M., and Bianchi, M. E. (2001) *EMBO J.* **20**, 4337–4340
115b. Thomas, J. O., and Travers, A. A. (2001) *Trends Biochem. Sci.* **26**, 167–174
115c. Bustin, M. (2001) *Trends Biochem. Sci.* **26**, 431–437
115d. Banks, G. C., Li, Y., and Reeves, R. (2000) *Biochemistry* **39**, 8333–8346
115e. Webb, M., Payet, D., Lee, K.-B., Travers, A. A., and Thomas, J. O. (2001) *J. Mol. Biol.* **309**, 79–88
116. Dingwall, C., and Laskey, R. (1992) *Science* **258**, 942–947
117. Ashery-Padan, R., Weiss, A. M., Feinstein, N., and Gruenbaum, Y. (1997) *J. Biol. Chem.* **272**, 2493–2499
118. Davis, L. I. (1995) *Ann. Rev. Biochem.* **64**, 865–896
119. Görlich, D. (1998) *EMBO J.* **17**, 2721–2727
120. Goldberg, M. W., and Allen, T. D. (1996) *J. Mol. Biol.* **257**, 848–865
120a. Raska, I., Aebi, U., and Earnshaw, W. C. (2000) *EMBO J.* **19**, 3843 – 3848
120b. Shahin, V., Danker, T., Enss, K., Ossig, R., and Oberleithner, H. (2001) *FASEB J.* **15**, 1895 – 1901
121. Aebi, U., Cohn, J., Buhle, L., and Gerace, L. (1986) *Nature (London)* **323**, 560–564
122. Worman, H. J., Lazaridis, I., and Georgatos, S. D. (1988) *J. Biol. Chem.* **263**, 12135–12141
123. Gerace, L. (1986) *Trends Biochem. Sci.* **11**, 443–446
123a. Georgatos, S. D. (2001) *EMBO J.* **20**, 2989 – 2994
124. McKeon, F. D., Kirschner, M. W., and Caput, D. (1986) *Nature (London)* **319**, 463–468
125. Schneider, U., Mini, T., Jenö, P., Fisher, P. A., and Stuurman, N. (1999) *Biochemistry* **38**, 4620–4632
126. Favreau, C., Worman, H. J., Wozniak, R. W., Frappier, T., and Courvalin, J.-C. (1996) *Biochemistry* **35**, 8035–8044
126a. Rout, M. P., and Aitchison, J. D. (2001) *J. Biol. Chem.* **276**, 16593–16596
126b. Allen, N. P. C., Huang, L., Burlingame, A., and Rexach, M. (2001) *J. Biol. Chem.* **276**, 29268–29274
127. Weis, K. (1998) *Trends Biochem. Sci.* **23**, 185–189
128. Cingolani, G., Petosa, C., Weis, K., and Müller, C. W. (1999) *Nature (London)* **399**, 221–228
129. Chook, Y. M., and Blobel, G. (1999) *Nature (London)* **399**, 230–237
130. Panté, N., and Aebi, U. (1996) *Science* **273**, 1729–1732
131. Moore, M. S. (1998) *J. Biol. Chem.* **273**, 22857–22860
132. Vetter, I. R., Nowak, C., Nishimoto, T., Kuhlmann, J., and Wittinghofer, A. (1999) *Nature (London)* **398**, 39–46
132a. Geyer, M., Assheuer, R., Klebe, C., Kuhlmann, J., Becker, J., Wittinghofer, A., and Kalbitzer, H. R. (1999) *Biochemistry* **38**, 11250–11260
132b. Goldberg, M. W., Rutherford, S. A., Hughes, M., Cotter, L. A., Bagley, S., Kiseleva, E., Allen, T. D., and Clarke, P. R. (2000) *J. Mol. Biol.* **300**, 519–529
133. Akey, C. W. (1995) *J. Mol. Biol.* **248**, 273–293
134. Kiseleva, E., Goldberg, M. W., Daneholt, B., and Allen, T. D. (1996) *J. Mol. Biol.* **260**, 304–311
135. Perez-Terzic, C., Pyle, J., Jaconi, M., Stehno-Bittel, L., and Clapham, D. E. (1996) *Science* **273**, 1875–1877
136. Stoffler, D., Goldie, K. N., Feja, B., and Aebi, U. (1999) *J. Mol. Biol.* **287**, 741–752
137. Cook, P. R. (1999) *Science* **284**, 1790–1795
138. Chang, J.-H., Dumbar, T. S., and Olson, M. O. J. (1988) *J. Biol. Chem.* **263**, 12824–12827
139. Serin, G., Joseph, G., Ghisolfi, L., Bauzan, M., Erard, M., Amalric, F., and Bouvet, P. (1997) *J. Biol. Chem.* **272**, 13109–13116
139a. Wang, I.-F., Reddy, N. M., and Shen, C.-K. J. (2002) *Proc. Natl. Acad. Sci. U.S.A.* **99**, 13583 – 13588
140. Somerville, C., and Somerville, S. (1999) *Science* **285**, 380–383
141. C. elegans Sequencing Consortium. (1998) *Science* **282**, 2012–2018
142. Long, E. O., and Dawid, I. B. (1980) *Ann. Rev. Biochem.* **49**, 727–764
143. Jelinek, W. R., and Schmid, C. W. (1982) *Ann. Rev. Biochem.* **51**, 813–844
144. Bernardi, G., Olofsson, B., Filipski, J., Zerial, M., Salinas, J., Cuny, G., Meunier-Rotival, M., and Rodier, F. (1985) *Science* **228**, 953–956
145. Nowak, R. (1994) *Science* **263**, 608–610
145a. Razin, S. V., Ioudinkova, E. S., Trifonov, E. N., and Scherrer, K. (2001) *J. Mol. Biol.* **307**, 481–486
145b. Mashkova, T. D., Oparina, N. Y., Lacroix, M.-H., Fedorova, L. I., Tumeneva, I. G., Zinovieva, O. L., and Kisselev, L. L. (2001) *J. Mol. Biol.* **305**, 33–48
146. Charlesworth, B., Sniegowski, P., and Stephan, W. (1994) *Nature (London)* **371**, 215–220
147. Barakat, A., Matassi, G., and Bernardi, G. (1998) *Proc. Natl. Acad. Sci. U.S.A.* **95**, 10044–10049
148. Rosenberg, H., Singer, M., and Rosenberg, M. (1978) *Science* **200**, 394–402
149. Inoue, S., Kitajima, K., and Inoue, Y. (1996) *J. Biol. Chem.* **271**, 24341–24344
150. Pluta, A. F., Mackay, A. M., Ainsztein, A. M., Goldberg, I. G., and Earnshaw, W. C. (1995) *Science* **270**, 1591–1594
151. Xiao, Z.-X., and Fitzgerald-Hayes, M. (1995) *J. Mol. Biol.* **248**, 255–263
152. Tal, M., Shimron, F., and Yagil, G. (1994) *J. Mol. Biol.* **243**, 179–189
153. Ferrer, N., Azorín, F., Villasante, A., Gutiérrez, C., and Abad, J. P. (1995) *J. Mol. Biol.* **245**, 8–21
154. Catasti, P., Gupta, G., Garcia, A. E., Ratliff, R., Hong, L., Yau, P., Moyzis, R. K., and Bradbury, E. M. (1994) *Biochemistry* **33**, 3819–3830
155. Chou, S.-H., Zhu, L., and Reid, B. R. (1996) *J. Mol. Biol.* **259**, 445–457
156. Copenhaver, G. P., Nickel, K., Kuromori, T., Benito, M.-I., Kaul, S., Lin, X., Beven, M., Murphy, G., Harris, B., Parnell, L. D., McCombie, W. R., Martienssen, R. A., Marra, M., and Preuss, D. (1999) *Science* **286**, 2468–2474

References

157. Jiang, J., Nasuda, S., Dong, F., Scherrer, C. W., Woo, S.-S., Wing, R. A., Gill, B. S., and Ward, D. C. (1996) *Proc. Natl. Acad. Sci. U.S.A.* **93**, 14210–14213
157a. Schueler, M. G., Higgins, A. W., Rudd, M. K., Gustashaw, K., and Willard, H. F. (2001) *Science* **294**, 109–115
157b. Pennisi, E. (2001) *Science* **294**, 30–31
158. Zhu, L., Chou, S.-H., and Reid, B. R. (1995) *J. Mol. Biol.* **254**, 623–637
158a. Gao, Y.-G., Robinson, H., Sanishvili, R., Joachimiak, A., and Wang, A. H.-J. (1999) *Biochemistry* **38**, 16452–16460
158b. Henikoff, S., and Malik, H. S. (2002) *Nature (London)* **417**, 227
159. Pluta, A. F., Cooke, C. A., and Earnshaw, W. C. (1990) *Trends Biochem. Sci.* **15**, 181–185
160. Sugimoto, K., Hagishita, Y., and Himeno, M. (1994) *J. Biol. Chem.* **269**, 24271–24276
161. Pluta, A. F., and Earnshaw, W. C. (1996) *J. Biol. Chem.* **271**, 18767–18774
162. Fukagawa, T., Pendon, C., Morris, J., and Brown, W. (1999) *EMBO J.* **18**, 4196–4209
162a. Takahashi, K., Chen, E. S., and Yanagida, M. (2000) *Science* **288**, 2215–2219
163. Blackburn, E. H., and Szostak, J. W. (1984) *Ann. Rev. Biochem.* **53**, 163–194
164. Moyzis, R. K. (1991) *Sci. Am.* **265**(Aug), 48–55
165. Blackburn, E. H. (1991) *Trends Biochem. Sci.* **16**, 378–381
166. Zakian, V. A. (1995) *Science* **270**, 1601–1607
167. Vega-Palas, M. A., Venditti, S., and Di Mauro, E. (1998) *J. Biol. Chem.* **273**, 9388–9392
167a. de Lange, T. (2001) *Science* **292**, 1075–1076
167b. Stansel, R. M., de Lange, T., and Griffith, J. D. (2001) *EMBO J.* **20**, 5532–5540
167c. Tomaska, L., Makhov, A. M., Nosek, J., Kucejova, B., and Griffith, J. D. (2001) *J. Mol. Biol.* **305**, 61–69
167d. Mitton-Fry, R. M., Anderson, E. M., Hughes, T. R., Lundblad, V., and Wuttke, D. S. (2002) *Science* **296**, 145–147
168. Schierer, T., and Henderson, E. (1994) *Biochemistry* **33**, 2240–2246
169. Frantz, J. D., and Gilbert, W. (1995) *J. Biol. Chem.* **270**, 20692–20697
169a. Baumann, P., and Cech, T. R. (2001) *Science* **292**, 1171–1175
170. Shore, D. (1998) *Science* **281**, 1818–1819
171. König, P., and Rhodes, D. (1997) *Trends Biochem. Sci.* **22**, 43–47
171a. Rossetti, L., Cacchione, S., De Menna, A., Chapman, L., Rhodes, D., and Savino, M. (2001) *J. Mol. Biol.* **306**, 903–913
172. Ohshima, K., and Okada, N. (1994) *J. Mol. Biol.* **243**, 25–37
172a. Kim, J., Martignetti, J. A., Shen, M. R., Brosius, J., and Deininger, P. (1994) *Proc. Natl. Acad. Sci. U.S.A.* **91**, 3607–3611
173. Ünsal, K., and Morgan, G. T. (1995) *J. Mol. Biol.* **248**, 812–823
174. Izsvák, Z., Ivics, Z., Garcia-Estefania, D., Fahrenkrug, S. C., and Hackett, P. B. (1996) *Proc. Natl. Acad. Sci. U.S.A.* **93**, 1077–1081
175. Ullu, E. (1982) *Trends Biochem. Sci.* **7**, 216–219
176. Sharp, P. A. (1983) *Nature (London)* **301**, 471–472
177. Adams, D. S., Eickbash, T. H., Herrea, R. S., and Lizandi, P. M. (1986) *J. Mol. Biol.* **187**, 465–478
177a. Carroll, M. L., Roy-Engel, A. M., Nguyen, S. V., Salem, A.-H., Vogel, E., Vincent, B., Myers, J., Ahmad, Z., Nuguyen, L., Sammarco, M., Watkins, W. S., Henke, J., Makalowski, W., Jorde, L. B., Dininger, P. L., and Batzer, M. A. (2001) *J. Mol. Biol.* **311**, 17–40
177b. Martinez, J., Dugaiczyk, L. J., Zielinski, R., and Dugaiczyk, A. (2001) *J. Mol. Biol.* **308**, 587–596
178. Shimizu, Y., Yoshida, K., Ren, C.-S., Fujinaga, K., Rejagopalan, S., and Chinnadurai, G. (1983) *Nature (London)* **302**, 587–590
179. Kiyama, R., Matsui, H., and Oishi, M. (1986) *Proc. Natl. Acad. Sci. U.S.A.* **83**, 4665–4669
180. Fowler, R. F., and Skinner, D. M. (1985) *J. Biol. Chem.* **260**, 1296–1303
181. Nargang, F. E., Bell, J. B., Stohl, L. L., and Lambowitz, A. M. (1983) *J. Biol. Chem.* **258**, 4257–4260
182. Wildeman, A. G., Rasquinha, I., and Nazar, R. N. (1986) *J. Biol. Chem.* **261**, 13401–13403
183. Reis, R. J. S., Lumpkin, C. K., Jr., McGill, J. R., Riabowol, K. T., and Goldstein, S. (1983) *Nature (London)* **301**, 394–398
184. Casavant, N. C., and Hardies, S. C. (1994) *J. Mol. Biol.* **241**, 390–397
185. Dhellin, O., Maestre, J., and Heidmann, T. (1997) *EMBO J.* **16**, 6590–6602
186. Dunham, I., Shimizu, N., Roe, B. A., Chossoe, S., and 117 other authors. (1999) *Nature (London)* **402**, 489–495
187. Sapienza, C., and Doolittle, W. F. (1982) *Nature (London)* **295**, 384–389
188. Wellauer, P. K., Reeder, R. H., Carroll, D., Brown, D. D., Deutch, A., Higashinakagawa, T., and Dawid, I. B. (1974) *Proc. Natl. Acad. Sci. U.S.A.* **71**, 2823–2827
189. Carroll, D., and Brown, D. D. (1976) *Cell* **7**, 467–475
190. Mroczka, D. L., Cassidy, B., Busch, H., and Rothblum, L. E. (1984) *J. Mol. Biol.* **174**, 141–162
191. Sommerville, J. (1985) *Nature (London)* **318**, 410–411
192. Budarf, M. L., and Blackburn, E. H. (1986) *J. Biol. Chem.* **261**, 363–369
193. Karabin, G. D., and Hallick, R. B. (1983) *J. Biol. Chem.* **258**, 5512–5518
194. Mattaj, I. W. (1984) *Trends Biochem. Sci.* **9**, 435–437
195. Dandekar, T., Snel, B., Huynen, M., and Bork, P. (1998) *Trends Biochem. Sci.* **23**, 324–328
196. Karlsson, S., and Nienhuis, A. W. (1985) *Ann. Rev. Biochem.* **54**, 1071–1108
197. Proudfoot, N. (1986) *Nature (London)* **321**, 730–731
197a. Wang, Z., and Liebhaber, S. A. (1999) *EMBO J.* **18**, 2218–2228
197b. Li, J., Noguchi, C. T., Miller, W., Hardison, R., and Schechter, A. N. (1998) *J. Biol. Chem.* **273**, 10202–10209
197c. Filipe, A., Li, Q., Deveaux, S., Godin, I., Roméo, P.-H., Stamatoyannopoulos, G., and Mignotte, V. (1999) *EMBO J.* **18**, 687–697
197d. Ristaldi, M. S., Drabek, D., Gribnau, J., Poddie, D., Yannoutsous, N., Cao, A., Grosveld, F., and Imam, A. M. A. (2001) *EMBO J.* **20**, 5242–5249
198. Vanin, E. F., Goldberg, G. I., Tucker, P. W., and Smithies, O. (1980) *Nature (London)* **286**, 222–226
199. Strehler, E. E., Strehler-Page, M.-A., Prriard, J.-C., Periasamy, M., and Nadal-Ginard, B. (1986) *J. Mol. Biol.* **190**, 291–317
200. Kolata, G. B. (1977) *Science* **196**, 1187–1188
201. Purohie, S., and Mathews, C. K. (1984) *J. Biol. Chem.* **259**, 6261–6266
202. Rak, B., Lusky, M., and Hable, M. (1982) *Nature (London)* **297**, 124–128
203. Ernst, H., and Shatkin, A. J. (1985) *Proc. Natl. Acad. Sci. U.S.A.* **82**, 48–52
204. Soininen, R., Huotari, M., Hostikka, S. L., Prockop, D. J., and Tryggvason, K. (1988) *J. Biol. Chem.* **263**, 17217–17220
205. Henikoff, S., Keene, M. A., Fechtel, K., and Fristrom, J. W. (1986) *Cell* **44**, 33–42
206. Borst, P., and Grivell, L. A. (1981) *Nature (London)* **289**, 439–440
207. Gilson, P. R., and McFadden, G. I. (1996) *Proc. Natl. Acad. Sci. U.S.A.* **93**, 7737–7742
208. Gray, M. W., Burger, G., and Lang, B. F. (1999) *Science* **283**, 1476–1481
209. Chen, K. K., and Donelson, J. E. (1980) *Proc. Natl. Acad. Sci. U.S.A.* **77**, 2445–2449
210. de la Cruz, V. F., Lake, J. A., Simpson, A. M., and Simpson, L. (1985) *Proc. Natl. Acad. Sci. U.S.A.* **82**, 1401–1405
211. Jeltsch, A., Friedrich, T., and Roth, M. (1998) *J. Mol. Biol.* **275**, 747–758
212. Newman, A. K., Rubin, R. A., Kim, S.-H., and Modrich, P. (1981) *J. Biol. Chem.* **256**, 2131–2139
213. Surby, M. A., and Reich, N. O. (1996) *Biochemistry* **35**, 2201–2208
213a. Gowher, H., and Jeltsch, A. (2000) *J. Mol. Biol.* **303**, 93–110
214. Schumann, J., Walter, J., Willert, J., Wild, C., Koch, D., and Trautner, T. A. (1996) *J. Mol. Biol.* **257**, 949–959
215. Ahmad, I., and Rao, D. N. (1996) *J. Mol. Biol.* **259**, 229–240
215a. Pues, H., Bleimling, N., Holz, B., Wölcke, J., and Weinhold, E. (1999) *Biochemistry* **38**, 1426–1434
216. Erlanson, D. A., Chen, L., and Verdine, G. L. (1993) *J. Am. Chem. Soc.* **115**, 12583–12584
217. Lau, E. Y., and Bruice, T. C. (1999) *J. Mol. Biol.* **293**, 9–18
218. Powell, L. M., Connolly, B. A., and Dryden, D. T. F. (1998) *J. Mol. Biol.* **283**, 947–961
219. Herman, G. E., and Modrich, P. (1982) *J. Biol. Chem.* **257**, 2605–2612
220. Zweiger, G., Marczynski, G., and Shapiro, L. (1994) *J. Mol. Biol.* **235**, 472–485
221. Heithoff, D. M., Sinsheimer, R. L., Low, D. A., and Mahan, M. J. (1999) *Science* **284**, 967–970
222. Burdon, R. H., and Adams, R. L. P. (1980) *Trends Biochem. Sci.* **5**, 294–297
223. Doerfler, W. (1983) *Ann. Rev. Biochem.* **52**, 93–124
224. Antequera, F., Tamame, M., Villaneuva, J. R., and Santos, T. (1984) *J. Biol. Chem.* **259**, 8033–8036
225. Selker, E. U., Jensen, B. C., and Richardson, G. A. (1987) *Science* **238**, 48–53
226. Holliday, R. (1989) *Sci. Am.* **260**(Jun), 60–73
227. Hsu, D.-W., Lin, M.-J., Lee, T.-L., Wen, S.-C., Chen, X., and Shen, C.-K., J. (1999) *Proc. Natl. Acad. Sci. U.S.A.* **96**, 9751–9756
228. Yoder, J. A., Soman, N. S., Verdine, G. L., and Bestor, T. H. (1997) *J. Mol. Biol.* **270**, 385–395
229. Pradhan, S., Bacolla, A., Wells, R. D., and Roberts, R. J. (1999) *J. Biol. Chem.* **274**, 33002–33010
229a. Gowher, H., and Jeltsch, A. (2001) *J. Mol. Biol.* **309**, 1201–1208
229b. Saito, Y., Kanai, Y., Sakamoto, M., Saito, H., Ishii, H., and Hirohashi, S. (2002) *Proc. Natl. Acad. Sci. U.S.A.* **99**, 10060–10065
230. Gjerset, R. A., and Martin, D. W., Jr. (1982) *J. Biol. Chem.* **257**, 8581–8583
231. Ramchandani, S., Bhattacharya, S. K., Cervoni, N., and Szyf, M. (1999) *Proc. Natl. Acad. Sci. U.S.A.* **96**, 6107–6112
232. Cervoni, N., Bhattacharya, S., and Szyf, M. (1999) *J. Biol. Chem.* **274**, 8363–8366
233. Matsuo, K., Silke, J., Georgiev, O., Marti, P., Giovannini, N., and Rungger, D. (1998) *EMBO J.* **17**, 1446–1453
233a. Smith, S. S. (2000) *J. Mol. Biol.* **302**, 1–7
233b. Bhattacharya, S. K., Ramchandani, S., Cervoni, N., and Szyf, M. (1999) *Nature (London)* **397**, 579–583
233c. Cedar, H., and Verdine, G. L. (1999) *Nature (London)* **397**, 568–569
234. Bird, A. P. (1986) *Nature (London)* **321**, 209–213

References

235. Shiraishi, M., Lerman, L. S., and Sekiya, T. (1995) *Proc. Natl. Acad. Sci. U.S.A.* **92**, 4229–4233
236. Pieper, R. O., Patel, S., Ting, S. A., Futscher, B. W., and Costello, J. F. (1996) *J. Biol. Chem.* **271**, 13916–13924
237. Rein, T., Kobayashi, T., Malott, M., Leffak, M., and DePamphilis, M. L. (1999) *J. Biol. Chem.* **274**, 25792–25800
238. Mayer-Jung, C., Moras, D., and Timsit, Y. (1998) *EMBO J.* **17**, 2709–2718
239. Wakefield, R. I. D., Smith, B. O., Nan, X., Free, A., Soteriou, A., Uhrin, D., Bird, A. P., and Barlow, P. N. (1999) *J. Mol. Biol.* **291**, 1055–1065
240. Davey, C., Pennings, S., and Allan, J. (1997) *J. Mol. Biol.* **267**, 276–288
241. Bester, T. H. (1998) *Nature (London)* **393**, 311–312
242. Razin, A. (1998) *EMBO J.* **17**, 4905–4908
243. Naruse, Y., Aoki, T., Kojima, T., and Mori, N. (1999) *Proc. Natl. Acad. Sci. U.S.A.* **96**, 13691–13696
244. O'Neill, L. P., Keohane, A. M., Lavender, J. S., McCabe, V., Heard, E., Avner, P., Brockdorff, N., and Turner, B. M. (1999) *EMBO J.* **18**, 2897–2907
245. Herzing, L. B. K., Romer, J. T., Horn, J. M., and Ashworth, A. (1997) *Nature (London)* **386**, 272–275
246. Jegalian, K., and Page, D. C. (1998) *Nature (London)* **394**, 776–780
247. Gommers-Ampt, J. H., and Borst, P. (1995) *FASEB J.* **9**, 1034–1042
248. Wilson, V. L., Smith, R. A., Ma, S., and Cutler, R. G. (1987) *J. Biol. Chem.* **262**, 9948–9951
249. Bakin, A. V., and Curran, T. (1999) *Science* **283**, 387–390
250. Song, H. K., Sohn, S. H., and Suh, S. W. (1999) *EMBO J.* **18**, 1104–1113
251. Sapienza, C. (1990) *Sci. Am.* **263**(Oct), 52–60
252. Barlow, D. P. (1993) *Science* **260**, 309–310
253. Birger, Y., Shemer, R., Perk, J., and Razin, A. (1999) *Nature (London)* **397**, 84–88
254. Meselson, M., and Stahl, F. W. (1958) *Proc. Natl. Acad. Sci. U.S.A.* **44**, 671–682
255. Taylor, J. H. (1997) *Trends Biochem. Sci.* **22**, 447–450
256. Cairns, J. (1963) *J. Mol. Biol.* **6**, 208–213
257. Zimm, B. H. (1999) *Trends Biochem. Sci.* **24**, 121–123
258. Kornberg, A. (1969) *Science* **163**, 1410–1418
259. Lehman, I. R., Bessman, M. J., Simms, E. S., and Kornberg, A. (1958) *J. Biol. Chem.* **233**, 163–170
259a. Patel, P. H., Suzuki, M., Adman, E., Shinkai, A., and Loeb, L. A. (2001) *J. Mol. Biol.* **308**, 823–837
260. Sugimoto, K., Okazaki, T., and Okazaki, R. (1968) *Proc. Natl. Acad. Sci. U.S.A.* **60**, 1356–1362
261. Modrich, P., Anraku, Y., and Lehman, I. R. (1973) *J. Biol. Chem.* **248**, 7495–7501
262. Lehman, I. R. (1974) *Science* **186**, 790–797
263. Brutlag, D., Schekman, R., and Kornberg, A. (1971) *Proc. Natl. Acad. Sci. U.S.A.* **68**, 2826–2829
264. Cai, H., Yu, H., McEntee, K., Kunkel, T. A., and Goodman, M. F. (1995) *J. Biol. Chem.* **270**, 15327–15335
265. Kornberg, A., and Baker, T. A. (1992) *DNA Replication*, 2nd ed., Freeman, New York
266. Hubscher, U., and Spadari, S., eds. (1984) *Proteins Involved in DNA Replication*, Plenum, New York
266a. Friedberg, E. C., Feaver, W. J., and Gerlach, V. L. (2000) *Proc. Natl. Acad. Sci. U.S.A.* **97**, 5681–5683
266b. Burgers, P. M. J., Koonin, E. V., Bruford, E., Blanco, L., Burtis, K. C., Christman, M. F., Copeland, W. C., Friedberg, E. C., and 13 other authors. (2001) *J. Biol. Chem.* **276**, 43487–43490
267. Steitz, T. A. (1999) *J. Biol. Chem.* **274**, 17395–17398
268. Sousa, R. (1996) *Trends Biochem. Sci.* **21**, 186–190
268a. Yang, G., Franklin, M., Li, J., Lin, T.-C., and Konigsberg, W. (2002) *Biochemistry* **41**, 10256–10261
268b. Hübscher, U., Maga, G., and Spadari, S. (2002) *Ann. Rev. Biochem.* **71**, 133–163
269. Brody, R. S., and Frey, P. A. (1981) *Biochemistry* **20**, 1245–1252
270. Steitz, T. A. (1998) *Nature (London)* **391**, 231–232
271. Astatke, M., Ng, K., Grindley, N. D. F., and Joyce, C. M. (1998) *Proc. Natl. Acad. Sci. U.S.A.* **95**, 3402–3407
272. Singh, K., and Modak, M. J. (1998) *Trends Biochem. Sci.* **23**, 277–281
272a. Krishna, T. S. R., Kong, X-P, Gary, S., Burgers, P. M., and Kuriyan, J. (1994) *Cell* **79**, 1233–1243
273. Sawaya, M. R., Pelletier, H., Kumar, A., Wilson, S. H., and Kraut, J. (1994) *Science* **264**, 1930–1935
274. Sawaya, M. R., Prasad, R., Wilson, S. H., Kraut, J., and Pelletier, H. (1997) *Biochemistry* **36**, 11205–11215
275. Beese, L. S., Derbyshire, V., and Steitz, T. A. (1993) *Science* **260**, 352–355
275a. Spratt, T. E. (2001) *Biochemistry* **40**, 2647–2652
276. Eom, S. H., Wang, J., and Steitz, T. A. (1996) *Nature (London)* **382**, 278–281
277. Li, Y., Mitaxov, V., and Waksman, G. (1999) *Proc. Natl. Acad. Sci. U.S.A.* **96**, 9491–9496
278. Tabor, S., and Richardson, C. C. (1990) *J. Biol. Chem.* **265**, 8322–8328
279. Sousa, R., Chung, Y. J., Rose, J. P., and Wang, B.-C. (1993) *Nature (London)* **364**, 593–599
280. Osumi-Davis, P. A., Sreerama, N., Volkin, D. B., Middaugh, C. R., Woody, R. W., and Woody, A.-Y. M. (1994) *J. Mol. Biol.* **237**, 5–19
280a. Longley, M. J., Ropp, P. A., Lim, S. E., and Copeland, W. C. (1998) *Biochemistry* **37**, 10529–10539
280b. Fan, L., and Kaguni, L. S. (2001) *Biochemistry* **40**, 4780–4791
281. Kelman, Z., Pietrokovski, S., and Hurwitz, J. (1999) *J. Biol. Chem.* **274**, 28751–28761
282. Hopfner, K.-P., Eichinger, A., Engh, R. A., Laue, F., Ankenbauer, W., Huber, R., and Angerer, B. (1999) *Proc. Natl. Acad. Sci. U.S.A.* **96**, 3600–3605
282a. Rodriguez, A. C., Park, H.-W., Mao, C., and Beese, L. S. (2000) *J. Mol. Biol.* **299**, 447–462
283. Maga, G., and Hübscher, U. (1996) *Biochemistry* **35**, 5764–5777
284. Bambara, R. A., Murante, R. S., and Henricksen, L. A. (1997) *J. Biol. Chem.* **272**, 4647–4650
285. Mozzherin, D. J., and Fisher, P. A. (1996) *Biochemistry* **35**, 3572–3577
286. Dua, R., Levy, D. L., and Campbell, J. L. (1998) *J. Biol. Chem.* **273**, 30046–30055
286a. MacNeill, S. A., Baldacci, G., Burgers, P. M., and Hübscher, U. (2001) *Trends Biochem. Sci.* **26**, 16–17
286b. Xie, B., Mazloum, N., Liu, L., Rahmeh, A., Li, H., and Lee, M. Y. W. T. (2002) *Biochemistry* **41**, 13133–13142
287. Alley, S. C., Trakselis, M. A., Mayer, M. U., Ishmael, F. T., Jones, A. D., and Benkovic, S. J. (2001) *J. Biol. Chem.* **276**, 39340–39349
287a. Pietroni, P., Young, M. C., Latham, G. J., and von Hippel, P. H. (2001) *J. Mol. Biol.* **309**, 869–891
288. Davies, J. F., II, Hostomska, Z., Hostomsky, Z., Jordan, S. R., and Matthews, D. A. (1991) *Science* **252**, 88–95
289. Wang, J., Smerdon, S. J., Jäger, J., Kohlstaedt, L. A., Rice, P. A., Friedman, J. M., and Steitz, T. A. (1994) *Proc. Natl. Acad. Sci. U.S.A.* **91**, 7242–7246
289a. Sarafianos, S. G., Das, K., Tantillo, C., Clark, A. D., Jr., Ding, J., Whitcomb, J. M., Boyer, P. L., Hughes, S. H., and Arnold, E. (2001) *EMBO J.* **20**, 1449–1461
290. Tantillo, C., Ding, J., Jacobo-Molina, A., Nanni, R. G., Boyer, P. L., Hughes, S. H., Pauwels, R., Andries, K., Janssen, P. A. J., and Arnold, E. (1994) *J. Mol. Biol.* **243**, 369–387
291. Carroll, S. S., Olsen, D. B., Bennett, C. D., Gotlib, L., Graham, D. J., Condra, J. H., Stern, A. M., Shafer, J. A., and Kuo, L. C. (1993) *J. Biol. Chem.* **268**, 276–281
292. Mathias, S. L., Scott, A. F., Kazazian, H. H., Jr., Boeke, J. D., and Gabriel, A. (1991) *Science* **254**, 1808–1810
293. Pelletier, H., Sawaya, M. R., Wolfle, W., Wilson, S. H., and Kraut, J. (1996) *Biochemistry* **35**, 12742–12761
293a. Shah, A. M., Conn, D. A., Li, S.-X., Capaldi, A., Jäger, J., and Sweasy, J. B. (2001) *Biochemistry* **40**, 11372–11381
293b. Liu, J., and Tsai, M.-D. (2001) *Biochemistry* **40**, 9014–9022
293c. Eckert, K. A., Mowery, A., and Hile, S. E. (2002) *Biochemistry* **41**, 10490–10498
294. Holm, L., and Sander, C. (1995) *Trends Biochem. Sci.* **20**, 345–347
294a. Song, M.-S., Pham, P. T., Olson, M., Carter, J. R., Franden, M. A., Schaaper, R. M., and McHenry, C. S. (2001) *J. Biol. Chem.* **276**, 35165–35175
295. Kornberg, A. (1988) *J. Biol. Chem.* **263**, 1–4
296. Kelman, Z., and O'Donnell, M. (1995) *Ann. Rev. Biochem.* **64**, 171–200
297. Kim, D. R., and McHenry, C. S. (1996) *J. Biol. Chem.* **271**, 20699–20704
298. Perrino, F. W., Harvey, S., and McNeill, S. M. (1999) *Biochemistry* **38**, 16001–16009
299. Turner, J., Hingorani, M. M., Kelman, Z., and O'Donnell, M. (1999) *EMBO J.* **18**, 771–783
300. Brenowitz, S., Kwack, S., Goodman, M. F., O'Donnell, M., and Echols, H. (1991) *J. Biol. Chem.* **266**, 7888–7892
300a. DeRose, E. F., Li, D., Darden, T., Harvey, S., Perrino, F. W., Schaaper, R. M., and London, R. E. (2002) *Biochemistry* **41**, 94–110
301. LaDuca, R. J., Crute, J. J., McHenry, C. S., and Bambara, R. A. (1986) *J. Biol. Chem.* **261**, 7550–7557
301a. Leu, F. P., and O'Donnell, M. (2001) *J. Biol. Chem.* **276**, 47185–47194
301b. Matsumiya, S., Ishino, Y., and Morikawa, K. (2001) *Protein Sci.* **10**, 17–23
301c. Bertram, J. G., Bloom, L. B., Hingorani, M. M., Beechem, J. M., O'Donnell, M., and Goodman, M. F. (2000) *J. Biol. Chem.* **275**, 28413–28420
302. Hingorani, M. M., Bloom, L. B., Goodman, M. F., and O'Donnell, M. (1999) *EMBO J.* **18**, 5131–5144
303. Kuriyan, J., and O'Donnell, M. (1993) *J. Mol. Biol.* **234**, 915–925
303a. Dervyn, E., Suski, C., Daniel, R., Bruand, C., Chapuis, J., Errington, J., Jannière, L., and Ehrlich, S. D. (2001) *Science* **294**, 1716–1718
304. Lindahl, T., and Barnes, D. E. (1992) *Ann. Rev. Biochem.* **61**, 251–281

References

305. Husain, I., Tomkinson, A. E., Burkhart, W. A., Moyer, M. B., Ramos, W., Mackey, Z. B., Besterman, J. M., and Chen, J. (1995) *J. Biol. Chem.* **270**, 9683–9690
306. Arabshahi, A., and Frey, P. A. (1999) *J. Biol. Chem.* **274**, 8586–8588
307. Doherty, A. J., and Wigley, D. B. (1999) *J. Mol. Biol.* **285**, 63–71
307a. Doherty, A. J., and Dafforn, T. R. (2000) *J. Mol. Biol.* **296**, 43–56
308. Shamoo, Y., Friedman, A. M., Parsons, M. R., Konigsberg, W. H., and Steitz, T. A. (1995) *Nature (London)* **376**, 362–366
309. Folmer, R. H. A., Nilges, M., Konings, R. N. H., and Hilbers, C. W. (1995) *EMBO J.* **14**, 4132–4142
310. Guan, Y., Zhang, H., and Wang, A. H.-J. (1995) *Protein Sci.* **4**, 187–197
311. Olah, G. A., Gray, D. M., Gray, C. W., Kergil, D. L., Sosnick, T. R., Mark, B. L., Vaughan, M. R., and Trewhella, J. (1995) *J. Mol. Biol.* **249**, 576–594
312. Lohman, T. M., and Ferrari, M. E. (1994) *Ann. Rev. Biochem.* **63**, 527–570
313. Raghunathan, S., Ricard, C. S., Lohman, T. M., and Waksman, G. (1997) *Proc. Natl. Acad. Sci. U.S.A.* **94**, 6652–6657
314. Lohman, T. M., Bujalowski, W., and Overman, L. B. (1988) *Trends Biochem. Sci.* **13**, 250–255
315. Cobianchi, F., SenGupta, D. N., Zmudzka, B. Z., and Wilson, S. H. (1986) *J. Biol. Chem.* **261**, 3536–3543
316. Richter, A., Sapp, M., and Knippers, R. (1986) *Trends Biochem. Sci.* **11**, 283
317. Kelly, T. J., Simancek, P., and Brush, G. S. (1998) *Proc. Natl. Acad. Sci. U.S.A.* **95**, 14634–14639
318. Subramanya, H. S., Bird, L. E., Brannigan, J. A., and Wigley, D. B. (1996) *Nature (London)* **384**, 379–383
319. Lohman, T. M., and Bjornson, K. P. (1996) *Ann. Rev. Biochem.* **65**, 169–214
320. Marians, K. J. (1997) *Structure* **5**, 1129–1134
320a. Soultanas, P., and Wigley, D. B. (2001) *Trends Biochem. Sci.* **26**, 47–54
320b. Soultanas, P., Dillingham, M. S., Wiley, P., Webb, M. R., and Wigley, D. B. (2000) *EMBO J.* **19**, 3799–3810
321. Yu, X., Jezewska, M. J., Bujalowski, W., and Egelman, E. H. (1996) *J. Mol. Biol.* **259**, 7–14
321a. Donate, L.-E., Llorca, O., Bárcena, M., Brown, S. E., Dixon, N. E., and Carazo, J.-M. (2000) *J. Mol. Biol.* **303**, 383–393
321b. Kaplan, D. L. (2000) *J. Mol. Biol.* **301**, 285–299
322. Jezewska, M. J., Rajendran, S., Bujalowska, D., and Bujalowski, W. (1998) *J. Biol. Chem.* **273**, 10515–10529
323. Hsieh, J., Moore, K. J. M., and Lohman, T. M. (1999) *J. Mol. Biol.* **288**, 255–274
324. Biswas, E. E., and Biswas, S. B. (1999) *Biochemistry* **38**, 10929–10939
325. Bárcena, M., San Martín, C., Weise, F., Ayora, S., Alonso, J. C., and Carazo, J. M. (1998) *J. Mol. Biol.* **283**, 809–819
325a. Niedenzu, T., Röleke, D., Bains, G., Scherzinger, E., and Saenger, W. (2001) *J. Mol. Biol.* **306**, 479–487
326. Richardson, R. W., and Nossal, N. G. (1989) *J. Biol. Chem.* **264**, 4725–4731
326a. Jones, C. E., Mueser, T. C., and Nossal, N. G. (2000) *J. Biol. Chem.* **275**, 27145–27154
326b. Delagoutte, E., and von Hippel, P. H. (2001) *Biochemistry* **40**, 4459–4477
326c. Harmon, F. G., and Kowalczykowski, S. C. (2001) *J. Biol. Chem.* **276**, 232–243
327. Stasiak, A., Tsaneva, I. R., West, S. C., Yu, X., and Engelman, E. H. (1994) *Proc. Natl. Acad. Sci. U.S.A.* **91**, 7618–7622
327a. Parsons, C. A., Stasiak, A., Bennet, R. J., and West, S. C. (1995) *Nature (London)* **374**, 375–378
327b. Yu, X., West, S. C., and Egelman, E. H. (1997) *J. Mol. Biol.* **266**, 217–222
328. West, S. C. (1996) *Nature (London)* **384**, 316–317
329. Seo, Y.-S., and Hurwitz, J. (1993) *J. Biol. Chem.* **268**, 10282–10295
330. Poll, E. H. A., Harrison, J., Umthun, A., Dobbs, D. L., and Benbow, R. M. (1994) *Biochemistry* **33**, 3841–3847
331. Karow, J. K., Chakraverty, R. K., and Hickson, I. D. (1997) *J. Biol. Chem.* **272**, 30611–30614
331a. Wang, W., Seki, M., Narita, Y., Sonoda, E., Takeda, S., Yamada, K., Masuko, T., Katada, T., and Enomoto, T. (2000) *EMBO J.* **19**, 3428–3435
331b. Wu, L., and Hickson, I. D. (2001) *Science* **292**, 229–230
332. Braguglia, D., Heun, P., Pasero, P., Duncker, B. P., and Gasser, S. M. (1998) *J. Mol. Biol.* **281**, 631–649
332a. Orren, D. K., Theodore, S., and Machwe, A. (2002) *Biochemistry* **41**, 13483 – 13488
332b. Egelman, E. H. (2001) *Nature (London)* **409**, 573,575
333. de la Cruz, J., Kressler, D., Tollervey, D., and Linder, P. (1998) *EMBO J.* **17**, 1128–1140
334. Gorbalenya, A. E., and Koonin, E. V. (1993) *Current Opinion in Structural Biology* **3**, 419–429
335. Wang, J. C. (1996) *Ann. Rev. Biochem.* **65**, 635–692
336. Roca, J. (1995) *Trends Biochem. Sci.* **20**, 156–160
337. Cheng, C., and Shuman, S. (1999) *Biochemistry* **38**, 16599–16612
338. Chen, S.-J., and Wang, J. C. (1998) *J. Biol. Chem.* **273**, 6050–6056
339. Stewart, L., Redinbo, M. R., Qiu, X., Hol, W. G. J., and Champoux, J. J. (1998) *Science* **279**, 1534–1541
340. Redinbo, M. R., Stewart, L., Champoux, J. J., and Hol, W. G. J. (1999) *J. Mol. Biol.* **292**, 685–696
340a. Redinbo, M. R., Champoux, J. J., and Hol, W. G. J. (2000) *Biochemistry* **39**, 6832–6840
340b. Dekker, N. H., Rybenkov, V. V., Duguet, M., Crisona, N. J., Cozzarelli, N. R., Bensimon, D., and Croquette, V. (2002) *Proc. Natl. Acad. Sci. U.S.A.* **99**, 12126 – 12131
340c. Champoux, J. J. (2002) *Proc. Natl. Acad. Sci. U.S.A.* **99**, 11998 – 12000
341. Lima, C. D., Wang, J. C., and Mondragón, A. (1994) *Nature (London)* **367**, 138–146
341a. Changela, A., DiGate, R. J., and Mondragón, A. (2001) *Nature (London)* **411**, 1077–1081
342. Olland, S., and Wang, J. C. (1999) *J. Biol. Chem.* **274**, 21688–21694
343. Smith, C. V., and Maxwell, A. (1998) *Biochemistry* **37**, 9658–9667
344. Williams, N. L., and Maxwell, A. (1999) *Biochemistry* **38**, 13502–13511
345. Nichols, M. D., DeAngelis, K., Keck, J. L., and Berger, J. M. (1999) *EMBO J.* **18**, 6177–6188
346. Huang, W. M., Wei, L. S., and Casjens, S. (1985) *J. Biol. Chem.* **260**, 8973–8977
347. Li, Z., Hiasa, H., Kumar, U., and DiGate, R. J. (1997) *J. Biol. Chem.* **272**, 19582–19587
348. Baird, C. L., Harkins, T. T., Morris, S. K., and Lindsley, J. E. (1999) *Proc. Natl. Acad. Sci. U.S.A.* **96**, 13685–13690
348a. Williams, N. L., Howells, A. J., and Maxwell, A. (2001) *J. Mol. Biol.* **306**, 969–984
349. Morris, S. K., and Lindsley, J. E. (1999) *J. Biol. Chem.* **274**, 30690–30696
350. Berger, J. M., Gamblin, S. J., Harrison, S. C., and Wang, J. C. (1996) *Nature (London)* **379**, 225–232
351. Berger, J. M., Fass, D., Wang, J. C., and Harrison, S. C. (1998) *Proc. Natl. Acad. Sci. U.S.A.* **95**, 7876–7881
352. Kikuchi, A., and Asai, K. (1984) *Nature (London)* **309**, 677–681
353. North, G. (1985) *Nature (London)* **316**, 394–395
353a. Barthelmes, H. U., Grue, P., Feineis, S., Straub, T., and Boege, F. (2000) *J. Biol. Chem.* **275**, 38823–38830
354. Froelich-Ammon, S. J., and Osheroff, N. (1995) *J. Biol. Chem.* **270**, 21429–21432
355. Sogo, J. M., Stasiak, A., Martínez-Robles, M. L., Krimer, D. B., Hernández, P., and Schvartzman, J. B. (1999) *J. Mol. Biol.* **286**, 637–643
356. Fleischmann, G., Pflugfelder, G., Steiner, E. K., Javaherian, K., Howard, G. C., Wang, J. C., and Elgin, S. C. R. (1984) *Proc. Natl. Acad. Sci. U.S.A.* **81**, 6958–6962
357. Brooks, M., and Dumas, L. B. (1989) *J. Biol. Chem.* **264**, 3602–3610
358. Massé, E., and Drolet, M. (1999) *J. Mol. Biol.* **294**, 321–332
359. Bailly, C., Carrasco, C., Hamy, F., Vezin, H., Prudhomme, M., Saleem, A., and Rubin, E. (1999) *Biochemistry* **38**, 8605–8611
360. Shero, J. H., Bordwell, B., Rothfield, N. F., and Earnshaw, W. C. (1986) *Science* **231**, 737–740
361. Hiasa, H., and Marians, K. J. (1999) *J. Biol. Chem.* **274**, 27244–27248
362. Sun, W., and Godson, G. N. (1998) *J. Mol. Biol.* **276**, 689–703
362a. Keck, J. L., Roche, D. D., Lynch, A. S., and Berger, J. M. (2000) *Science* **287**, 2482–2486
363. Jing, D. H., Dong, F., Latham, G. J., and von Hippel, P. H. (1999) *J. Biol. Chem.* **274**, 27287–27298
364. Frick, D. N., and Richardson, C. C. (1999) *J. Biol. Chem.* **274**, 35889–35898
365. Doublié, S., Tabor, S., Long, A. M., Richardson, C. C., and Ellenberger, T. (1998) *Nature (London)* **391**, 251–258
366. Hacker, K. J., and Johnson, K. A. (1997) *Biochemistry* **36**, 14080–14087
366a. Picha, K. M., Ahnert, P., and Patel, S. S. (2000) *Biochemistry* **39**, 6401–6409
366b. VanLoock, M. S., Chen, Y.-J., Yu, X., Patel, S. S., and Egelman, E. H. (2001) *J. Mol. Biol.* **311**, 951–956
367. Skarstad, K., Thöny, B., Hwang, D. S., and Kornberg, A. (1993) *J. Biol. Chem.* **268**, 5365–5370
368. Tougu, K., Peng, H., and Marians, K. J. (1994) *J. Biol. Chem.* **269**, 4675–4682
369. Tougu, K., and Marians, K. J. (1996) *J. Biol. Chem.* **271**, 21398–21405
370. Komori, H., Matsunaga, F., Higuchi, Y., Ishiai, M., Wada, C., and Miki, K. (1999) *EMBO J.* **18**, 4597–4607
371. Ng, J. Y., and Marians, K. J. (1996) *J. Biol. Chem.* **271**, 15642–15648
372. Allen, G. C., Jr., and Kornberg, A. (1993) *J. Biol. Chem.* **268**, 19204–19209
373. Hupp, T. R., and Kaguni, J. M. (1993) *J. Biol. Chem.* **268**, 13143–13150
374. Katayanagi, K., Ishikawa, M., Okumura, M., Ariyoshi, M., Kanaya, S., Kawano, Y., Suzuki, M., Tanaka, I., and Morikawa, K. (1993) *J. Biol. Chem.* **268**, 22092–22099
375. Keck, J. L., Goedken, E. R., and Marqusee, S. (1998) *J. Biol. Chem.* **273**, 34128–34133
376. Ohtani, N., Haruki, M., Morikawa, M., Crouch, R. J., Itaya, M., and Kanaya, S. (1999) *Biochemistry* **38**, 605–618
377. Bird, R. E., Louarn, J., Martuscelli, J., and Caro, L. (1972) *J. Mol. Biol.* **70**, 549–566
378. Prescott, D. M., and Kuempel, P. L. (1972) *Proc. Natl. Acad. Sci. U.S.A.* **69**, 2842–2845

References

379. Kuempel, P. L. et al. (1972) in *DNA Synthesis in Vitro* (Wells, R., and Inman, R., eds), pp. 463–472, University Park Press, Baltimore, Maryland
380. Kriegstein, H. J., and Hogness, D. S. (1974) *Proc. Natl. Acad. Sci. U.S.A.* **71**, 135–139
381. Hixson, J. E., Wong, T. W., and Clayton, D. A. (1986) *J. Biol. Chem.* **261**, 2384–2390
382. Lee, D. Y., and Clayton, D. A. (1998) *J. Biol. Chem.* **273**, 30614–30621
382a. Krüger, R., Konieczny, I., and Filutowicz, M. (2001) *J. Mol. Biol.* **306**, 945 – 955
383. McMacken, R., Silver, L., and Georgopoulos, C. (1987) in *Escherichia coli and Salmonella typhimurium* (Neidhardt, F. C., ed), pp. 565–612, Am. Soc. Microbiol., Washington,DC
383a. Smelkova, N., and Marians, K. J. (2001) *J. Biol. Chem.* **276**, 39186–39191
384. Krause, M., Rückert, B., Lurz, R., and Messer, W. (1997) *J. Mol. Biol.* **274**, 365–380
385. Lobry, J. R. (1996) *Science* **272**, 745–746
385a. Kelman, Z. (2000) *Trends Biochem. Sci.* **25**, 521–523
386. Yung, B. Y.-M., and Kornberg, A. (1988) *Proc. Natl. Acad. Sci. U.S.A.* **85**, 7202–7205
387. Sutton, M. D., Carr, K. M., Vicente, M., and Kaguni, J. M. (1998) *J. Biol. Chem.* **273**, 34255–34262
388. Sutton, M. D., and Kaguni, J. M. (1997) *J. Mol. Biol.* **274**, 546–561
389. Radzicka, A., and Wolfenden, R. (1988) *Biochemistry* **27**, 1664–1670
390. Zyskind, J. W., Cleary, J. M., Brusilow, W. S. A., Harding, N. E., and Smith, D. W. (1983) *Proc. Natl. Acad. Sci. U.S.A.* **80**, 1164–1168
391. Chédin, F., Seitz, E. M., and Kowalczykowski, S. C. (1998) *Trends Biochem. Sci.* **23**, 273–277
391a. Bochkareva, E., Belegu, V., Korolev, S., and Bochkarev, A. (2001) *EMBO J.* **20**, 612–618
392. Lother, H., and Messer, W. (1981) *Nature (London)* **294**, 376–378
393. Xia, W., and Dowhan, W. (1995) *Proc. Natl. Acad. Sci. U.S.A.* **92**, 783–787
394. York, J. D., Saffitz, J. E., and Majerus, P. W. (1994) *J. Biol. Chem.* **269**, 19992–19999
394a. Li, X., and Marians, K. J. (2000) *J. Biol. Chem.* **275**, 34757–34765
395. Maki, H., Maki, S., and Kornberg, A. (1988) *J. Biol. Chem.* **263**, 6570–6578
396. Lasken, R. S., and Kornberg, A. (1988) *J. Biol. Chem.* **263**, 5512–5518
396a. Chapados, B. R., Chai, Q., Hosfield, D. J., Qiu, J., Shen, B., and Tainer, J. A. (2001) *J. Mol. Biol.* **307**, 541 – 556
397. O'Donnell, M. E., and Kornberg, A. (1985) *J. Biol. Chem.* **260**, 12884–12889
398. Lee, E. H., and Kornberg, A. (1992) *J. Biol. Chem.* **267**, 8778–8784
399. Kamada, K., Horiuchi, T., Ohsumi, K., Shimamoto, N., and Morikawa, K. (1996) *Nature (London)* **383**, 598–603
400. Duggin, I. G., Andersen, P. A., Smith, M. T., Wilce, J. A., King, G. F., and Wake, R. G. (1999) *J. Mol. Biol.* **286**, 1325–1335
401. Manna, A. C., Pai, K. S., Bussiere, D. E., White, S. W., and Bastia, D. (1996) *Proc. Natl. Acad. Sci. U.S.A.* **93**, 3253–3258
402. Yamazoe, M., Onogi, T., Sunako, Y., Niki, H., Yamanaka, K., Ichimura, T., and Hiraga, S. (1999) *EMBO J.* **18**, 5873–5884
403. Dodson, M., McMacken, R., and Echols, H. (1989) *J. Biol. Chem.* **264**, 10719–10725
404. Learn, B. A., Um, S.-J., Huang, L., and McMacken, R. (1997) *Proc. Natl. Acad. Sci. U.S.A.* **94**, 1154–1159
405. Wickner, S., and McKenney, K. (1987) *J. Biol. Chem.* **262**, 13163–13167
406. Echols, H. (1990) *J. Biol. Chem.* **265**, 14697–14700
407. Murialdo, H. (1991) *Ann. Rev. Biochem.* **60**, 125–153
408. Wickner, S. H. (1990) *Proc. Natl. Acad. Sci. U.S.A.* **87**, 2690–2694
409. Wickner, S., Skowyra, D., Hoskins, J., and McKenney, K. (1992) *Proc. Natl. Acad. Sci. U.S.A.* **89**, 10345–10349
410. Nossal, N. G. (1992) *FASEB J.* **6**, 871–878
411. Brown, D. R., Roth, M. J., Reinberg, D., and Hurwitz, J. (1984) *J. Biol. Chem.* **259**, 10545–10555
412. Higashitani, A., Greenstein, D., Hirokawa, H., Asano, S., and Horiuchi, K. (1994) *J. Mol. Biol.* **237**, 388–400
413. Masai, H., Nomura, N., Kubota, Y., and Arai, K.-i. (1990) *J. Biol. Chem.* **265**, 15124–15133
414. Abarzua, P., Soeller, W., and Marians, K. J. (1984) *J. Biol. Chem.* **259**, 14286–14292
415. Roth, M. J., Brown, D. R., and Hurwitz, J. (1984) *J. Biol. Chem.* **259**, 10556–10568
416. Mok, M., and Marians, K. J. (1987) *J. Biol. Chem.* **262**, 2304–2309
416a. Campos-Olivas, R., Louis, J. M., CLérot, D., Gronenborn, B., and Gronenborn, A. M. (2002) *Proc. Natl. Acad. Sci. U.S.A.* **99**, 10310–10315
417. Cue, D., and Feiss, M. (1998) *J. Mol. Biol.* **280**, 11–29
418. Yang, Q., Berton, N., Manning, M. C., and Catalano, C. E. (1999) *Biochemistry* **38**, 14238–14247
419. Higgins, R. R., and Becker, A. (1995) *J. Mol. Biol.* **252**, 31–46
420. Yeo, A., and Feiss, M. (1995) *J. Mol. Biol.* **245**, 126–140
420a. Cue, D., and Feiss, M. (2001) *J. Mol. Biol.* **311**, 233–240
421. Franklin, J. L., Haseltine, D., Davenport, L., and Mosig, G. (1998) *J. Mol. Biol.* **277**, 541–557
421a. Rao, V. B., and Mitchell, M. S. (2001) *J. Mol. Biol.* **314**, 401 – 411
422. Marvin, D. A. (1998) *Current Opinion in Structural Biology* **8**, 150–158
423. Papavoine, C. H. M., Christiaans, B. E. C., Folmer, R. H. A., Konings, R. N. H., and Hilbers, C. W. (1998) *J. Mol. Biol.* **282**, 401–419
424. Rakonjac, J., Feng, J.-n, and Model, P. (1999) *J. Mol. Biol.* **289**, 1253–1265
425. Burch, A. D., Ta, J., and Fane, B. A. (1999) *J. Mol. Biol.* **286**, 95–104
426. Dokland, T., Bernal, R. A., Burch, A., Pletnev, S., Fane, B. A., and Rossmann, M. G. (1999) *J. Mol. Biol.* **288**, 595–608
427. Chai, S., Lurz, R., and Alonso, J. C. (1995) *J. Mol. Biol.* **252**, 386–398
428. Valle, M., Kremer, L., Martínez-A, C., Roncal, F., Valpuesta, J. M., Albar, J. P., and Carrascosa, J. L. (1999) *J. Mol. Biol.* **288**, 899–909
429. Caetano de Sousa, P., Jr., Tuma, R., Prevelige, P. E., Jr., Silva, J. L., and Foguel, D. (1999) *J. Mol. Biol.* **287**, 527–538
430. Grimes, S., and Anderson, D. (1997) *J. Mol. Biol.* **266**, 901–914
430a. Rodríguez-Casado, A., Moore, S. D., Prevelige, P. E., Jr., and Thomas, G. J., Jr. (2001) *Biochemistry* **40**, 13583–13591
430b. Simpson, A. A., Tao, Y., Leiman, P. G., Badasso, M. O., He, Y., Jardine, P. J., Olson, N. H., Morais, M. C., Grimes, S., Anderson, D. L., Baker, T. S., and Rossmann, M. G. (2000) *Nature (London)* **408**, 745–750
430c. Smith, D. E., Tans, S. J., Smith, S. B., Grimes, S., Anderson, D. L., and Bustamante, C. (2001) *Nature (London)* **413**, 748–752
431. Leffers, G., and Rao, V. B. (1996) *J. Mol. Biol.* **258**, 839–850
432. Jardine, P. J., and Coombs, D. H. (1998) *J. Mol. Biol.* **284**, 661–672
432a. Conway, J. F., Wikoff, W. R., Cheng, N., Duda, R. L., Hendrix, R. W., Johnson, J. E., and Steven, A. C. (2001) *Science* **292**, 744–748
433. Burian, J., Stuchlík, S., and Kay, W. W. (1999) *J. Mol. Biol.* **294**, 49–65
434. Merlin, S., and Polisky, B. (1995) *J. Mol. Biol.* **248**, 211–219
435. Paulsson, J., and Ehrenberg, M. (1998) *J. Mol. Biol.* **279**, 73–88
436. Nordström, K., and Wagner, E. G. H. (1994) *Trends Biochem. Sci.* **19**, 294–300
437. Kramer, M. G., Khan, S. A., and Espinosa, M. (1997) *EMBO J.* **16**, 5784–5795
438. Novick, R. P. (1998) *Trends Biochem. Sci.* **23**, 434–438
439. Willetts, N., and Skurray, R. (1987) in *Escherichia coli and Salmonella typhimurium*, Vol. 2 (Neidhardt, F. C., ed), pp. 1110–1133, Am. Soc. for Microbiology, Washington, D. C.
440. Lanka, E., and Wilkins, B. M. (1995) *Ann. Rev. Biochem.* **64**, 141–169
441. Yates, P., Lane, D., and Biek, D. P. (1999) *J. Mol. Biol.* **290**, 627–638
442. White, J. H., and Richardson, C. C. (1987) *J. Biol. Chem.* **262**, 8851–8860
443. Willwand, K., Mumtsidu, E., Kuntz-Simon, G., and Rommelaere, J. (1998) *J. Biol. Chem.* **273**, 1165–1174
443a. Dufour, E., Méndez, J., Lázaro, J. M., de Vega, M., Blanco, L., and Salas, M. (2000) *J. Mol. Biol.* **304**, 289–300
444. King, A. J., and van der Vliet, P. C. (1994) *EMBO J.* **13**, 5786–5792
445. Pombo, A., Ferreira, J., Bridge, E., and Carmo-Fonseca, M. (1994) *EMBO J.* **13**, 5075–5085
446. Monaghan, A., and Hay, R. T. (1996) *J. Biol. Chem.* **271**, 24242–24248
447. Voyles, B. A. (1993) *The Biology of Viruses*, Mosby, St. Louis, Missouri
447a. Parkinson, G. N., Lee, M. P. H., and Neidle, S. (2002) *Nature (London)* **417**, 876 – 880
447b. Patel, D. J. (2002) *Nature (London)* **417**, 807 – 808
448. Greider, C. W. (1996) *Ann. Rev. Biochem.* **65**, 337–365
449. Kowald, A. (1997) *J. Mol. Biol.* **273**, 814 – 825
449a. Aigner, S., Lingner, J., Goodrich, K. J., Grosshans, C. A., Shevchenko, A., Mann, M., and Cech, T. R. (2000) *EMBO J.* **19**, 6230–6239
450. Shippen-Lentz, D., and Blackburn, E. H. (1990) *Science* **247**, 546–552
451. Greider, C. W., and Blackburn, E. H. (1996) *Sci. Am.* **274**(Feb), 92–97
452. Feng, J., Funk, W. D., Wang, S.-S., Weinrich, S. L., Avilion, A. A., Chiu, C.-P., Adams, R. R., Chang, E., Allsopp, R. C., Yu, J., Le, S., West, M. D., Harley, C. B., Andrews, W. H., Greider, C. W., and Villeponteau, B. (1995) *Science* **269**, 1236–1241
453. Bachand, F., and Autexier, C. (1999) *J. Biol. Chem.* **274**, 38027–38031
454. Shay, J. W., and Wright, W. E. (1999) *Science* **286**, 2284–2285
455. Singer, M. S., and Gottschling, D. E. (1994) *Science* **266**, 404–409
456. Lingner, J., Hughes, T. R., Shevchenko, A., Mann, M., Lundblad, V., and Cech, T. R. (1997) *Science* **276**, 561–567
457. Lee, H.-W., Blasco, M. A., Gottlieb, G. J., Horner, J. W., II, Greider, C. W., and DePinho, R. A. (1998) *Nature (London)* **392**, 569–574
458. Lustig, A. J. (1999) *Proc. Natl. Acad. Sci. U.S.A.* **96**, 3339–3341
459. McEachern, M. J., and Blackburn, E. H. (1995) *Nature (London)* **376**, 403–409
460. Smith, S., Giriat, I., Schmitt, A., and de Lange, T. (1998) *Science* **282**, 1484–1487
460a. Blackburn, E. H. (2000) *Nature (London)* **408**, 53–56

References

460b. Lundblad, V. (2000) *Science* **288**, 2141–2142
461. King, T. C., and Low, R. L. (1987) *J. Biol. Chem.* **262**, 6204–6213
462. Ntambi, J. M., Shapiro, T. A., Ryan, K. A., and Englund, P. T. (1986) *J. Biol. Chem.* **261**, 11890–11895
463. Palmer, J. D. (1983) *Nature (London)* **301**, 92–93
464. Dey, P. M., and Harborne, J. B., eds. (1997) *Plant Biochemistry*, Academic Press, San Diego, California
465. Klein, A., and Bonhoeffer, F. (1972) *Ann. Rev. Biochem.* **41**, 301–332
466. DePamphilis, M. L., ed. (1996) *DNA Replication in Eukaryotic Cells*, Cold Spring Harbor Lab. Press, Cold Spring Harbor, New York
466a. DePamphilis, M. L., ed. (1999) *Concepts of Eukaryotic DNA Replication*, Cold Spring Harbor Lab. Press, Cold Spring Harbor, New York
467. Waga, S., and Stillman, B. (1998) *Ann. Rev. Biochem.* **67**, 721–751
467a. Bell, S. P., and Dutta, A. (2002) *Ann. Rev. Biochem.* **71**, 333–374
468. Campbell, J. L. (1988) *Trends Biochem. Sci.* **13**, 212–217
468a. Wyrick, J. J., Aparicio, J. G., Chen, T., Barnett, J. D., Jennings, E. G., Young, R. A., Bell, S. P., and Aparicio, O. M. (2001) *Science* **294**, 2357–2360
469. Diller, J. D., and Raghuraman, M. K. (1994) *Trends Biochem. Sci.* **19**, 320–325
469a. Raghuraman, M. K., Winzeler, E. A., Collingwood, D., Hunt, S., Wodicka, L., Conway, A., Lockhart, D. J., Davis, R. W., Brewer, B. J., and Fangman, W. L. (2001) *Science* **294**, 115–121
469b. Stillman, B. (2001) *Science* **294**, 2301–2304
470. Weiser, T., Gassmann, M., Thömmes, P., Ferrari, E., and Hafkemeyer, P. (1991) *J. Biol. Chem.* **266**, 10420–10428
471. Sugino, A. (1995) *Trends Biochem. Sci.* **20**, 319–323
472. Spadari, S., Sala, F., and Pedrali-Noy, G. (1982) *Trends Biochem. Sci.* **7**, 29–32
473. Copeland, W. C., and Tan, X. (1995) *J. Biol. Chem.* **270**, 3905–3913
474. Arezi, B., Kirk, B. W., Copeland, W. C., and Kuchta, R. D. (1999) *Biochemistry* **38**, 12899–12907
474a. Arezi, B., and Kuchta, R. D. (2000) *Trends Biochem. Sci.* **25**, 572–576
475. Copeland, W. C., and Wang, T. S.-F. (1993) *J. Biol. Chem.* **268**, 11028–11040
476. Kelman, Z., and Hurwitz, J. (1998) *Trends Biochem. Sci.* **23**, 236–238
476a. Shiomi, Y., Usukura, J., Masamura, Y., Takeyasu, K., Nakayama, Y., Obuse, C., Yoshikawa, H., and Tsurimoto, T. (2000) *Proc. Natl. Acad. Sci. U.S.A.* **97**, 14127–14132
477. Bochkarev, A., Pfuetzner, R. A., Edwards, A. M., and Frappier, L. (1997) *Nature (London)* **385**, 176–181
478. Foiani, M., Lucchini, G., and Plevani, P. (1997) *Trends Biochem. Sci.* **22**, 424–427
479. Kelly, T. J. (1988) *J. Biol. Chem.* **263**, 17889–17892
480. Hurwitz, J., Dean, F. B., Kwong, A. D., and Lee, S.-H. (1990) *J. Biol. Chem.* **265**, 18043–18046
481. Waga, S., and Stillman, B. (1994) *Nature (London)* **369**, 207–212
482. Challberg, M. D., and Kelly, T. J. (1989) *Ann. Rev. Biochem.* **58**, 671–717
483. SenGupta, D. J., and Borowiec, J. A. (1992) *Science* **256**, 1656–1661
483a. Weisshart, K., Förster, H., Kremmer, E., Schlott, B., Grosse, F., and Nasheuer, H.-P. (2000) *J. Biol. Chem.* **275**, 17328–17337
484. Lehman, I. R., and Boehmer, P. E. (1999) *J. Biol. Chem.* **274**, 28059–28062
485. Murata, L. B., and Dodson, M. S. (1999) *J. Biol. Chem.* **274**, 37079–37086
486. Toyn, J. H., Toone, W. M., Morgan, B. A., and Johnston, L. H. (1995) *Trends Biochem. Sci.* **20**, 70–73
487. Murray, A. W., and Szostak, J. W. (1987) *Sci. Am.* **257**(Nov), 62–68
488. Farr, C. J., Bayne, R. A. L., Kipling, D., Mills, W., Critcher, R., and Cooke, H. J. (1995) *EMBO J.* **14**, 5444–5454
489. Warburton, P. E., and Kipling, D. (1997) *Nature (London)* **386**, 553–555
490. Shirahige, K., Hori, Y., Shiraishi, K., Yamashita, M., Takahashi, K., Obuse, C., Tsurimoto, T., and Yoshikawa, H. (1998) *Nature (London)* **395**, 618–621
491. Gómez, M., and Antequera, F. (1999) *EMBO J.* **18**, 5683–5690
492. Kim, S.-M., and Huberman, J. A. (1999) *J. Mol. Biol.* **288**, 867–882
493. Gavin, K. A., Hidaka, M., and Stillman, B. (1995) *Science* **270**, 1667–1671
494. Marx, J. (1995) *Science* **270**, 1585–1587
495. Tye, B. K. (1999) *Ann. Rev. Biochem.* **68**, 649–686
495a. Dhar, S. K., Delmolino, L., and Dutta, A. (2001) *J. Biol. Chem.* **276**, 29067–29071
495b. Blow, J. J. (2001) *EMBO J.* **20**, 3293–3297
495c. Krude, T. (2000) *J. Biol. Chem.* **275**, 13699–13707
496. Prokhorova, T. A., and Blow, J. J. (2000) *J. Biol. Chem.* **275**, 2491–2498
497. Coverley, D., and Laskey, R. A. (1994) *Ann. Rev. Biochem.* **63**, 745–776
498. Chong, J. P. J., Thömmes, P., and Blow, J. J. (1996) *Trends Biochem. Sci.* **21**, 102–106
499. Stillman, B. (1994) *J. Biol. Chem.* **269**, 7047–7050
500. Aladjem, M. I., Rodewald, L. W., Kolman, J. L., and Wahl, G. M. (1998) *Science* **281**, 1005–1009
500a. Tye, B. K., and Sawyer, S. (2000) *J. Biol. Chem.* **275**, 34833–34836
500b. Sato, M., Gotow, T., You, Z., Komamura-Kohno, Y., Uchiyama, Y., Yabuta, N., Nojima, H., and Ishimi, Y. (2000) *J. Mol. Biol.* **300**, 421–431
500c. Lee, J.-K., and Hurwitz, J. (2001) *Proc. Natl. Acad. Sci. U.S.A.* **98**, 54–59
500d. Chong, J. P. J., Hayashi, M. K., Simon, M. N., Xu, R.-M., and Stillman, B. (2000) *Proc. Natl. Acad. Sci. U.S.A.* **97**, 1530–1535
500e. Wohlschlegel, J. A., Dwyer, B. T., Dhar, S. K., Cvetic, C., Walter, J. C., and Dutta, A. (2000) *Science* **290**, 2309–2312
500f. Lygerou, Z., and Nurse, P. (2000) *Science* **290**, 2271–2273
500g. Blow, J. J., and Tada, S. (2000) *Nature (London)* **404**, 560–561
500h. Keller, C., and Krude, T. (2000) *J. Biol. Chem.* **275**, 35512–35521
501. DePamphilis, M. L. (1993) *J. Biol. Chem.* **268**, 1–4
501a. Lucas, I., Chevrier-Miller, M., Sogo, J. M., and Hyrien, O. (2000) *J. Mol. Biol.* **296**, 769–786
501b. Gilbert, D. M. (2001) *Science* **294**, 96–100
502. Wu, J.-R., and Gilbert, D. M. (1996) *Science* **271**, 1270–1272
503. Dimitrova, D. S., Giacca, M., Demarchi, F., Biamonti, G., Riva, S., and Falaschi, A. (1996) *Proc. Natl. Acad. Sci. U.S.A.* **93**, 1498–1503
504. Kasamatson, H., and Vinograd, J. (1973) *Nature (London)* **241**, 103–105
505. Gaudette, M. F., and Benbow, R. M. (1986) *Proc. Natl. Acad. Sci. U.S.A.* **83**, 5953–5957
506. Turchi, J. J., Huang, L., Murante, R. S., Kim, Y., and Bambara, R. A. (1994) *Proc. Natl. Acad. Sci. U.S.A.* **91**, 9803–9807
506a. Karthikeyan, R., Vonarx, E. J., Straffon, A. F. L., Simon, M., Faye, G., and Kunz, B. A. (2000) *J. Mol. Biol.* **299**, 405–419
507. Huang, L., Rumbaugh, J. A., Murante, R. S., Lin, R. J. R., Rust, L., and Bambara, R. A. (1996) *Biochemistry* **35**, 9266–9277
507a. Gomes, X. V., and Burgers, P. M. J. (2000) *EMBO J.* **19**, 3811–3821
507b. Bae, S.-H., and Seo, Y.-S. (2000) *J. Biol. Chem.* **275**, 38022–38031
508. Pohjanpelto, P., and Hölttä, E. (1996) *EMBO J.* **15**, 1193–1200
509. Bonne-Andrea, C., Wong, M. L., and Alberts, B. M. (1990) *Nature (London)* **343**, 719–726
510. Gasser, R., Koller, T., and Sogo, J. M. (1996) *J. Mol. Biol.* **258**, 224–239
511. Murray, A. W. (1987) *Nature (London)* **327**, 14–15
512. Fisher, D. L., and Nurse, P. (1996) *EMBO J.* **15**, 850–860
513. Santocanale, C., Neecke, H., Longhese, M. P., Lucchini, G., and Plevani, P. (1995) *J. Mol. Biol.* **254**, 595–607
514. Novak, B., and Tyson, J. J. (1997) *Proc. Natl. Acad. Sci. U.S.A.* **94**, 9147–9152
515. Brush, G. S., Morrow, D. M., Hieter, P., and Kelly, T. J. (1996) *Proc. Natl. Acad. Sci. U.S.A.* **93**, 15075–15080
516. Collins, J. M., and Chu, A. K. (1987) *Biochemistry* **26**, 5600–5607
517. Stahl, F. W. (1987) *Sci. Am.* **256**(Feb), 91–101
518. Kucherlapati, R., and Smith, G. R., eds. (1988) *Genetic Recombination*, Am. Soc. for Microbiology, Washington, DC
519. Kowalczykowski, S. C. (1991) *Ann. Rev. Biophys. Biophys. Chem.* **20**, 539–575
520. Edelmann, W., and Kucherlapati, R. (1996) *Proc. Natl. Acad. Sci. U.S.A.* **93**, 6225–6227
521. Shinohara, A., and Ogawa, T. (1995) *Trends Biochem. Sci.* **20**, 387–391
522. Haber, J. E. (1999) *Nature (London)* **398**, 665–667
523. Meselson, M., and Weigle, J. J. (1961) *Proc. Natl. Acad. Sci. U.S.A.* **47**, 857–868
524. Holliday, R. (1964) *Genet. Res. Camb.* **5**, 282–304
525. Potter, H., and Dressler, D. (1976) *Proc. Natl. Acad. Sci. U.S.A.* **73**, 3000–3004
525a. Eichman, B. F., Vargason, J. M., Mooers, B. H. M., and Ho, P. S. (2000) *Proc. Natl. Acad. Sci. U.S.A.* **97**, 3971–3976
525b. Sha, R., Liu, F., and Seeman, N. C. (2002) *Biochemistry* **41**, 5950–5955
526. Sigal, N., and Alberts, B. (1972) *J. Mol. Biol.* **71**, 789–793
527. Meselson, M. (1972) *J. Mol. Biol.* **71**, 795–798
528. Shinagawa, H., and Iwasaki, H. (1996) *Trends Biochem. Sci.* **21**, 107–111
529. Rafferty, J. B., Sedelnikova, S. E., Hargreaves, D., Artymiuk, P. J., Baker, P. J., Sharples, G. J., Mahdi, A. A., Lloyd, R. G., and Rice, D. W. (1996) *Science* **274**, 415–421
530. Ishimori, K., Sommer, S., Bailone, A., Takahashi, M., Cox, M. M., and Devoret, R. (1996) *J. Mol. Biol.* **264**, 696–712
530a. McIlwraith, M. J., and West, S. C. (2001) *J. Mol. Biol.* **305**, 23–31
530b. Postow, L., Ullsperger, C., Keller, R. W., Bustamante, C., Vologodskii, A. V., and Cozzarelli, N. R. (2001) *J. Biol. Chem.* **276**, 2790–2796
530c. Neschastnova, A. A., Markina, V. K., Popenko, V. I., Danilova, O. A., Sidorov, R. A., Belitsky, G. A., and Yakubovskaya, M. G. (2002) *Biochemistry* **41**, 7795–7801
530d. McIlwraith, M. J., Van Dyck, E., Masson, J.-Y., Stasiak, A. Z., Stasiak, A., and West, S. C. (2000) *J. Mol. Biol.* **304**, 151–164
530e. Bolt, E. L., Sharples, G. J., and Lloyd, R. G. (2000) *J. Mol. Biol.* **304**, 165–176

References

531. Farah, J. A., and Smith, G. R. (1997) *J. Mol. Biol.* **272**, 699–715
532. Wang, J., Chen, R., and Julin, D. A. (2000) *J. Biol. Chem.* **275**, 507–513
533. Anderson, D. G., and Kowalczykowski, S. C. (1998) *J. Mol. Biol.* **282**, 275–285
533a. Arnold, D. A., Handa, N., Kobayashi, I., and Kowalczykowski, S. C. (2000) *J. Mol. Biol.* **300**, 469–479
534. Blattner, F. R., Plunkett, G., III, Bloch, C. A., Perna, N. T., Burland, V., Riley, M., Collado-Vides, J., Glasner, J. D., Rode, C. K., Mayhew, G. F., Gregor, J., Davis, N. W., Kirkpatrick, H. A., Goeden, M. A., Rose, D. J., Mau, B., and Shao, Y. (1997) *Science* **277**, 1453–1462
534a. Flores, M.-J., Bierne, H., Ehrlich, S. D., and Michel, B. (2001) *EMBO J.* **20**, 619–629
535. Köppen, A., Krobitsch, S., Thoms, B., and Wackernagel, W. (1995) *Proc. Natl. Acad. Sci. U.S.A.* **92**, 6249–6253
536. Howard-Flanders, P., West, S. C., and Stasiak, A. (1984) *Nature (London)* **309**, 215–220
537. Story, R. M., Weber, I. T., and Steitz, T. A. (1992) *Nature (London)* **355**, 318–325
538. Cui, X., Wise, R. P., and Schnable, P. S. (1996) *Science* **272**, 1334–1336
539. Arenson, T. A., Tsodikov, O. V., and Cox, M. M. (1999) *J. Mol. Biol.* **288**, 391–401
539a. Egelman, E. H. (2001) *J. Mol. Biol.* **309**, 539–542
540. Hegner, M., Smith, S. B., and Bustamante, C. (1999) *Proc. Natl. Acad. Sci. U.S.A.* **96**, 10109–10114
540a. Lusetti, S. L., and Cox, M. M. (2002) *Ann. Rev. Biochem.* **71**, 71 – 100
541. Register, J. C., III, and Griffith, J. (1985) *J. Biol. Chem.* **260**, 12308–12312
542. Zaitsev, E. N., and Kowalczykowski, S. C. (1999) *J. Mol. Biol.* **287**, 21–31
543. Adzuma, K. (1998) *J. Biol. Chem.* **273**, 31565–31573
544. Rao, B. J., Chiu, S., Bazemore, L. R., Reddy, G., and Radding, C. M. (1995) *Trends Biochem. Sci.* **20**, 109–113
544a. Voloshin, O. N., Wang, L., and Camerini-Otero, R. D. (2000) *J. Mol. Biol.* **303**, 709–720
544b. Rice, K. P., Eggler, A. L., Sung, P., and Cox, M. M. (2001) *J. Biol. Chem.* **276**, 38570–38581
545. Engels, W. R., Preston, C. R., and Johnson-Schlitz, D. M. (1994) *Science* **263**, 1623–1625
546. Kowalczykowski, S. C., and Eggleston, A. K. (1994) *Ann. Rev. Biochem.* **63**, 991–1043
547. Beernink, H. T. H., and Morrical, S. W. (1999) *Trends Biochem. Sci.* **24**, 385–389
548. Birkenkamp, K., and Kemper, B. (1996) *J. Mol. Biol.* **259**, 622–631
549. Ando, R. A., and Morrical, S. W. (1999) *Biochemistry* **38**, 16589–16598
550. Passy, S. I., Yu, X., Li, Z., Radding, C. M., and Egelman, E. H. (1999) *Proc. Natl. Acad. Sci. U.S.A.* **96**, 4279–4284
551. Sung, P. (1994) *Science* **265**, 1241–1243
551a. Masson, J.-Y., and West, S. C. (2001) *Trends Biochem. Sci.* **26**, 131 – 136
552. Benson, F. E., Stasiak, A., and West, S. C. (1994) *EMBO J.* **13**, 5764–5771
553. Baumann, P., and West, S. C. (1998) *Trends Biochem. Sci.* **23**, 247–251
554. Li, Z., Golub, E. I., Gupta, R., and Radding, C. M. (1997) *Proc. Natl. Acad. Sci. U.S.A.* **94**, 11221–11226
555. Cerutti, H., Osman, M., Grandoni, P., and Jagendorf, A. T. (1992) *Proc. Natl. Acad. Sci. U.S.A.* **89**, 8068–8072
556. Rafferty, J. B., Ingleston, S. M., Hargreaves, D., Artymiuk, P. J., Sharples, G. J., Lloyd, R. G., and Rice, D. W. (1998) *J. Mol. Biol.* **278**, 105–116
557. George, H., Mézard, C., Stasiak, A., and West, S. C. (1999) *J. Mol. Biol.* **293**, 505–519
558. Qiu, X.-B., Lin, Y.-L., Thome, K. C., Pian, P., Schlegel, B. P., Weremowicz, S., Parvin, J. D., and Dutta, A. (1998) *J. Biol. Chem.* **273**, 27786–27793
559. Rafferty, J. B., Sedelnikova, S. E., Hargreaves, D., Artymiuk, P. J., Baker, P. J., Sharples, G. J., Hahdi, A. A., Lloyd, R. G., and Rice, D. W. (1996) *Science* **274**, 415–421
559a. Putnam, C. D., Clancy, S. B., Tsuruta, H., Gonzalez, S., Wetmur, J. G., and Tainer, J. A. (2001) *J. Mol. Biol.* **311**, 297–310
559b. Yamada, K., Kunishima, N., Mayanagi, K., Ohnishi, T., Nishino, T., Iwasaki, H., Shinagawa, H., and Morikawa, K. (2001) *Proc. Natl. Acad. Sci. U.S.A.* **98**, 1442–1447
560. Lee, S., Cavallo, L., and Griffith, J. (1997) *J. Biol. Chem.* **272**, 7532–7539
561. Zlatanova, J., and van Holde, K. (1998) *FASEB J.* **12**, 421–431
562. Golz, S., and Kemper, B. (1999) *J. Mol. Biol.* **285**, 1131–1144
563. Raaijmakers, H., Vix, O., Töro, I., Golz, S., Kemper, B., and Suck, D. (1999) *EMBO J.* **18**, 1447–1458
564. Bolt, E. L., Sharples, G. J., and Lloyd, R. G. (1999) *J. Mol. Biol.* **286**, 403–415
565. Whitby, M. C., and Dixon, J. (1998) *J. Biol. Chem.* **273**, 35063–35073
566. Komori, K., Sakae, S., Shinagawa, H., Morikawa, K., and Ishino, Y. (1999) *Proc. Natl. Acad. Sci. U.S.A.* **96**, 8873–8878
567. Kvaratskhelia, M., George, S. J., Cooper, A., and White, M. F. (1999) *Biochemistry* **38**, 16613–16619
568. Fincham, J. R. S., and Oliver, P. (1989) *Nature (London)* **338**, 14–15
569. Nicolas, A., Treco, D., Schultes, N. P., and Szostak, J. W. (1989) *Nature (London)* **338**, 35–39
570. Fabre, F. (1978) *Nature (London)* **272**, 795–798
571. Groden, J., Nakamura, Y., and German, J. (1990) *Proc. Natl. Acad. Sci. U.S.A.* **87**, 4315–4319
572. Jackson, J. A., and Fink, G. R. (1981) *Nature (London)* **292**, 306–311
573. Klein, H. L. (1984) *Nature (London)* **310**, 748–753
574. Robertson, M. (1981) *Nature (London)* **293**, 333–334
575. Szostak, J. W., and Wu, R. (1980) *Nature (London)* **284**, 426–430
576. Sadowski, P. D. (1993) *FASEB J.* **7**, 760–767
577. Wang, S., Cosstick, R., Gardner, J. F., and Gumport, R. I. (1995) *Biochemistry* **34**, 13082–13090
577a. Read, E. K., Gumport, R. I., and Gardner, J. F. (2000) *J. Biol. Chem.* **275**, 33759–33764
578. Nash, H. A. (1990) *Trends Biochem. Sci.* **15**, 222–227
579. Tirumalai, R. S., Kwon, H. J., Cardente, E. H., Ellenberger, T., and Landy, A. (1998) *J. Mol. Biol.* **279**, 513–527
580. Crisona, N. J., Weinberg, R. L., Peter, B. J., Sumners, D. W., and Cozzarelli, N. R. (1999) *J. Mol. Biol.* **289**, 747–775
580a. Bankhead, T., and Segall, A. M. (2000) *J. Biol. Chem.* **275**, 36949–36956
580b. Lorbach, E., Christ, N., Schwikardi, M., and Dröge, P. (2000) *J. Mol. Biol.* **296**, 1175–1181
581. Hsu, P. L., and Landy, A. (1984) *Nature (London)* **311**, 721–726
582. Dorgai, L., Sloan, S., and Weisberg, R. A. (1998) *J. Mol. Biol.* **277**, 1059–1070
583. Guo, F., Gopaul, D. N., and Van Duyne, G. D. (1997) *Nature (London)* **389**, 40–46
584. Gopaul, D. N., Guo, F., and Van Duyne, G. D. (1998) *EMBO J.* **17**, 4175–4187
585. Subramanya, H. S., Arciszewska, L. K., Baker, R. A., Bird, L. E., Sherratt, D. J., and Wigley, D. B. (1997) *EMBO J.* **16**, 5178–5187
586. Arciszewska, L. K., Grainge, I., and Sherratt, D. J. (1997) *EMBO J.* **16**, 3731–3743
587. Jayaram, M. (1994) *Trends Biochem. Sci.* **19**, 78–82
587a. Ferreira, H., Sherratt, D., and Arciszewska, L. (2001) *J. Mol. Biol.* **312**, 45–57
588. Zhu, X.-D., Pan, G., Luetke, K., and Sadowski, P. D. (1995) *J. Biol. Chem.* **270**, 11646–11653
589. Huffman, K. E., and Levene, S. D. (1999) *J. Mol. Biol.* **286**, 1–13
590. Lee, J., Jayaram, M., and Grainge, I. (1999) *EMBO J.* **18**, 784–791
591. Allingham, J. S., Pribil, P. A., and Haniford, D. B. (1999) *J. Mol. Biol.* **289**, 1195–1206
592. Watson, M. (1984) *Trends Biochem. Sci.* **9**, 82–83
593. Lim, H. M., Hughes, K. T., and Simon, M. I. (1992) *J. Biol. Chem.* **267**, 11183–11190
594. Feng, J.-A., Johnson, R. C., and Dickerson, R. E. (1994) *Science* **263**, 348–355
595. Kostrewa, D., Granzin, J., Koch, C., Choe, H.-W., Raghunathan, S., Wolf, W., Labahn, J., Kahmann, R., and Saenger, W. (1991) *Nature (London)* **349**, 178–180
596. Safo, M. K., Yang, W.-Z., Corselli, L., Cramton, S. E., Yuan, H. S., and Johnson, R. C. (1997) *EMBO J.* **16**, 6860–6873
597. Dworkin, J., and Blaser, M. J. (1997) *Proc. Natl. Acad. Sci. U.S.A.* **94**, 985–990
598. Fedoroff, N., ed. (1992) *The Dynamic Genome Barbara McClintock's Ideas in the Century of Genetics*, Cold Spring Harbor Lab. Press, Cold Spring Harbor, New York
599. Fedoroff, N. V. (1984) *Sci. Am.* **250**(Jun), 85–98
600. Burr, B., and Burr, F. A. (1983) *Trends Biochem. Sci.* **8**, 429–431
601. Kingsman, A. J., Chater, K. F., and Kingsman, S. M., eds. (1988) *Transposition*, Cambridge Univ. Press, London and New York
602. Cohen, S. N., and Shapiro, J. A. (1980) *Sci. Am.* **243**(Feb), 40–49
603. Watson, M. (1985) *Trends Biochem. Sci.* **10**, 178
604. Ohtsubo, H., Nyman, K., Doroszkiewicz, W., and Ohtsubo, E. (1981) *Nature (London)* **292**, 640–643
604a. Redder, P., She, Q., and Garrett, R. A. (2001) *J. Mol. Biol.* **306**, 1–6
605. Gamas, P., Galas, D., and Chandler, M. (1985) *Nature (London)* **317**, 458–460
606. Kröger, M., and Hobom, G. (1982) *Nature (London)* **297**, 159–162
607. Phadnis, S. H., and Berg, D. E. (1987) *Proc. Natl. Acad. Sci. U.S.A.* **84**, 9118–9122
608. Lee, S. Y., Butler, D., and Kleckner, N. (1987) *Proc. Natl. Acad. Sci. U.S.A.* **84**, 7876–7880
608a. Rao, J. E., and Craig, N. L. (2001) *J. Mol. Biol.* **307**, 1161–1170
609. Farabaugh, P. J., and Fink, G. R. (1980) *Nature (London)* **286**, 352–356
609a. Pribil, P. A., and Haniford, D. B. (2000) *J. Mol. Biol.* **303**, 145–159
610. Sakai, J., Chalmers, R. M., and Kleckner, N. (1995) *EMBO J.* **14**, 4374–4383
611. Bhasin, A., Goryshin, I. Y., and Reznikoff, W. S. (1999) *J. Biol. Chem.* **274**, 37021–37029
611a. Davies, D. R., Goryshin, I. Y., Reznikoff, W. S., and Rayment, I. (2000) *Science* **289**, 77–85
611b. Bhasin, A., Goryshin, I. Y., Steiniger-White, M., York, D., and Reznikoff, W. S. (2000) *J. Mol. Biol.* **302**, 49–63
612. Gary, P. A., Biery, M. C., Bainton, R. J., and Craig, N. L. (1996) *J. Mol. Biol.* **257**, 301–316
612a. Williams, T. L., and Baker, T. A. (2000) *Science* **289**, 73 – 74
613. Benjamin, K. R., Abola, A. P., Kanaar, R., and Cozzarelli, N. R. (1996) *J. Mol. Biol.* **256**, 50–65
614. Arnold, P. H., Blake, D. G., Grindley, N. D. F., Boocock, M. R., and Stark, W. M. (1999) *EMBO J.* **18**, 1407–1414

References

615. Boocock, M. R., Zhu, X., and Grindley, N. D. F. (1995) *EMBO J.* **14**, 5129–5140
616. Watson, M. A., Boocock, M. R., and Stark, W. M. (1996) *J. Mol. Biol.* **257**, 317–329
617. Wasserman, S. A., Dungan, J. M., and Cozzarelli, N. R. (1985) *Science* **229**, 171–174
618. Savilahti, H., Rice, P. A., and Mizuuchi, K. (1995) *EMBO J.* **14**, 4893–4903
619. Krementsova, E., Giffin, M. J., Pincus, D., and Baker, T. A. (1998) *J. Biol. Chem.* **273**, 31358–31365
620. Naigamwalla, D. Z., Coros, C. J., Wu, Z., and Chaconas, G. (1998) *J. Mol. Biol.* **282**, 265–274
621. Jiang, H., Yang, J.-Y., and Harshey, R. M. (1999) *EMBO J.* **18**, 3845–3855
622. Stellwagen, A. E., and Craig, N. L. (1998) *Trends Biochem. Sci.* **23**, 486–490
623. Dranginis, A. M. (1986) *Trends Biochem. Sci.* **11**, 328–331
624. Sugawara, N., Ivanov, E. L., Fishman-Lobell, J., Ray, B. L., Wu, X., and Haber, J. E. (1995) *Nature (London)* **373**, 84–86
625. Flavell, A. (1986) *Nature (London)* **320**, 397
626. Sherratt, D. J., ed. (1995) *Mobile Genetic Elements*, Oxford Univ. Press, New York
627. Gloor, G. B., Nassif, N. A., Johnson-Schlitz, D. M., Preston, C. R., and Engels, W. R. (1991) *Science* **253**, 1110–1117
628. Beall, E. L., and Rio, D. C. (1998) *EMBO J.* **17**, 2122–2136
629. Marx, J. (1991) *Science* **253**, 1092–1093
630. Lohe, A. R., De Aguiar, D., and Hartl, D. L. (1997) *Proc. Natl. Acad. Sci. U.S.A.* **94**, 1293–1297
631. van Pouderoyen, G., Ketting, R. F., Perrakis, A., Plasterk, R. H. A., and Sixma, T. K. (1997) *EMBO J.* **16**, 6044–6054
632. Goodier, J. L., and Davidson, W. S. (1994) *J. Mol. Biol.* **241**, 26–34
633. Lam, W. L., Seo, P., Robison, K., Virk, S., and Gilbert, W. (1996) *J. Mol. Biol.* **257**, 359–366
634. Oosumi, T., and Belknap, W. R. (1995) *Nature (London)* **378**, 672
635. Holzman, D. (1991) *Science* **254**, 1728–1729
636. Lykke-Andersen, J., Garrett, R. A., and Kjems, J. (1997) *EMBO J.* **16**, 3272–3281
637. Morgan, G. T. (1995) *J. Mol. Biol.* **254**, 1–5
638. Jones, J. D. G., Carland, F. M., Maliga, P., and Dooner, H. K. (1989) *Science* **244**, 204–207
639. Spaink, H. P. (1999) *Nature (London)* **402**, 135–136
640. Chopra, S., Brendel, V., Zhang, J., Axtell, J. D., and Peterson, T. (1999) *Proc. Natl. Acad. Sci. U.S.A.* **96**, 15330–15335
641. Koes, R., Souer, E., van Houwelingen, A., Mur, L., Spelt, C., Quattrocchio, F., Wing, J., Oppedijk, B., Ahmed, S., Maes, T., Gerats, T., Hoogeveen, P., Meesters, M., Kloos, D., and Mol, J. N. M. (1995) *Proc. Natl. Acad. Sci. U.S.A.* **92**, 8149–8153
642. Morgan, B. A., Conlon, F. L., Manzanares, M., Millar, J. B. A., Kanuga, N., Sharpe, J., Krumlauf, R., Smith, J. C., and Sedgwick, S. G. (1996) *Proc. Natl. Acad. Sci. U.S.A.* **93**, 2801–2806
643. Schardl, C. L., Lonsdale, D. M., Pring, D. R., and Rose, K. R. (1984) *Nature (London)* **310**, 292–296
644. Ikeda, H. (1986) *Proc. Natl. Acad. Sci. U.S.A.* **83**, 922–926
645. Bi, X., and Liu, L. F. (1996) *J. Mol. Biol.* **256**, 849–858
646. Stark, G. R., and Wahl, G. M. (1984) *Ann. Rev. Biochem.* **53**, 447–491
647. Schimke, R. T. (1988) *J. Biol. Chem.* **263**, 5989–5992
648. Syu, L.-J., and Fluck, M. M. (1997) *J. Mol. Biol.* **271**, 76–99
649. Wu, C. H. H., and Black, L. W. (1995) *J. Mol. Biol.* **247**, 604–617
650. Andersson, D. I., Slechta, E. S., and Roth, J. R. (1998) *Science* **282**, 1133–1135
651. Huang, T., and Campbell, J. L. (1995) *J. Biol. Chem.* **270**, 9607–9614
652. Sheriff, S., and Herriott, J. R. (1981) *J. Mol. Biol.* **145**, 441–451
653. Subramanian, V., Liu, T.-N., Yeh, W.-K., Narro, M., and Gibson, D. T. (1983) *J. Biol. Chem.* **256**, 2723–2730
654. Alitalo, K. (1985) *Trends Biochem. Sci.* **10**, 194–197
655. Friedberg, E. C., Walker, G. C., and Siede, W. (1995) *DNA Repair and Mutagenesis*, ASM Press, Materials Park, Ohio
656. Kuhnlein, U. (1985) *J. Biol. Chem.* **260**, 14918–14924
657. Beckman, K. B., and Ames, B. N. (1997) *J. Biol. Chem.* **272**, 19633–19636
658. Purmal, A. A., Lampman, G. W., Bond, J. P., Hatahet, Z., and Wallace, S. S. (1998) *J. Biol. Chem.* **273**, 10026–10035
659. Jourdan, M., Garcia, J., Defrancq, E., Kotera, M., and Lhomme, J. (1999) *Biochemistry* **38**, 3985–3995
660. Huff, A. C., and Topal, M. D. (1987) *J. Biol. Chem.* **262**, 12843–12850
661. Rosenblatt, J., and Mitchison, T. J. (1998) *Nature (London)* **393**, 739–740
662. Boorstein, R. J., Hilbert, T. P., Cunningham, R. P., and Teebor, G. W. (1990) *Biochemistry* **29**, 10455–10460
663. Doetsch, P. W., Zastawny, T. H., Martin, A. M., and Dizdaroglu, M. (1995) *Biochemistry* **34**, 737–742
663a. Zhou, H., Suzuki, M., Randers-Pehrson, G., Vannais, D., Chen, G., Trosko, J. E., Waldren, C. A., and Hei, T. K. (2001) *Proc. Natl. Acad. Sci. U.S.A.* **98**, 14410–14415
663b. Forster, L., Forster, P., Lutz-Bonengel, S., Willkomm, H., and Brinkmann, B. (2002) *Proc. Natl. Acad. Sci. U.S.A.* **99**, 13950 – 13954
664. Lai, M.-D., and Beattie, K. L. (1988) *Biochemistry* **27**, 1722–1728
665. Kunkel, T. A., and Soni, A. (1988) *J. Biol. Chem.* **263**, 14784–14789
666. Echols, H., and Goodman, M. F. (1991) *Ann. Rev. Biochem.* **60**, 477–511
667. Kunkel, T. A. (1992) *J. Biol. Chem.* **267**, 18251–18254
668. Timsit, Y. (1999) *J. Mol. Biol.* **293**, 835–853
668a. Viswanathan, M., Lacirignola, J. J., Hurley, R. L., and Lovett, S. T. (2000) *J. Mol. Biol.* **302**, 553–564
669. Radman, M., and Wagner, R. (1988) *Sci. Am.* **259**(Aug), 40–46
669a. Patel, P. H., Kawate, H., Adman, E., Ashbach, M., and Loeb, L. A. (2001) *J. Biol. Chem.* **276**, 5044–5051
670. Fersht, A. R. (1980) *Trends Biochem. Sci.* **5**, 262–265
671. Petruska, J., Goodman, M. F., Boosalis, M. S., Sowers, L. C., Cheong, C., and Tinoco, I., Jr. (1988) *Proc. Natl. Acad. Sci. U.S.A.* **85**, 6252–6256
672. Metzler, D. E. (1977) *Biochemistry; The Chemical Reactions of Living Cells*, 1st ed., Academic Press, New York (pp. 912–915)
673. Metzler, D. E. (1979) *Adv. Enzymol.* **50**, 1–40
674. Goodman, M. F. (1997) *Proc. Natl. Acad. Sci. U.S.A.* **94**, 10493–10495
675. Moran, S., Ren, R. X.-F., and Kool, E. T. (1997) *Proc. Natl. Acad. Sci. U.S.A.* **94**, 10506–10511
676. Matray, T. J., and Kool, E. T. (1998) *J. Am. Chem. Soc.* **120**, 6191–6192
676a. Kool, E. T. (2002) *Ann. Rev. Biochem.* **71**, 191 – 219
676b. Showalter, A. K., and Tsai, M.-D. (2002) *Biochemistry* **41**, 10571 – 10576
677. Mo, J.-Y., and Schaaper, R. M. (1996) *J. Biol. Chem.* **271**, 18947–18953
678. Fijalkowska, I. J., Jonczyk, P., Tkaczyk, M. M., and Bialoskorska, M. (1998) *Proc. Natl. Acad. Sci. U.S.A.* **95**, 10020–10025
679. Kroutil, L. C., Frey, M. W., Kaboord, B. F., Kunkel, T. A., and Benkovic, S. J. (1998) *J. Mol. Biol.* **278**, 135–146
680. Seki, M., Akiyama, M., Sugaya, Y., Ohtsubo, E., and Maki, H. (1999) *J. Biol. Chem.* **274**, 33313–33319
681. Modrich, P. (1987) *Ann. Rev. Biochem.* **56**, 435–466
682. Bohr, V. A., and Wassermann, K. (1988) *Trends Biochem. Sci.* **13**, 429–433
683. Thoma, F. (1999) *EMBO J.* **18**, 6585–6598
684. Friedberg, E. C. (1995) *Trends Biochem. Sci.* **20**, 381
685. Arbel, A., Zenvirth, D., and Simchen, G. (1999) *EMBO J.* **18**, 2648–2658
686. Modrich, P. (1989) *J. Biol. Chem.* **264**, 6597–6600
686a. Wood, R. D., Mitchell, M., Sgouros, J., and Lindahl, T. (2001) *Science* **291**, 1284 – 1289
687. Modrich, P., and Lahue, R. (1996) *Ann. Rev. Biochem.* **65**, 101–133
688. Schaaper, R. M. (1993) *J. Biol. Chem.* **268**, 23762–23765
688a. Obmolova, G., Ban, C., Hsieh, P., and Yang, W. (2000) *Nature (London)* **407**, 703–711
688b. Lamers, M. H., Perrakis, A., Enzlin, J. H., Winterwerp, H. H. K., de Wind, N., and Sixma, T. K. (2000) *Nature (London)* **407**, 711–717
688c. Bowers, J., Tran, P. T., Joshi, A., Liskay, R. M., and Alani, E. (2001) *J. Mol. Biol.* **306**, 957 – 968
689. Ban, C., and Yang, W. (1998) *EMBO J.* **17**, 1526–1534
690. Modich, P. (1997) *J. Biol. Chem.* **272**, 24727–24730
691. Wang, H., Lawrence, C. W., Li, G.-M., and Hays, J. B. (1999) *J. Biol. Chem.* **274**, 16894–16900
692. Nakagawa, T., Datta, A., and Kolodner, R. D. (1999) *Proc. Natl. Acad. Sci. U.S.A.* **96**, 14186–14188
692a. Coïc, E., Gluck, L., and Fabre, F. (2000) *EMBO J.* **19**, 3408–3417
692b. Rich, T., Allen, R. L., and Wyllie, A. H. (2000) *Nature (London)* **407**, 777–783
693. Ahn, B., and Grossman, L. (1996) *J. Biol. Chem.* **271**, 21462–21470
694. Sancar, A. (1996) *Ann. Rev. Biochem.* **65**, 43–81
695. Zou, Y., and Van Houten, B. (1999) *EMBO J.* **18**, 4889–4901
695a. Verhoeven, E. E. A., Wyman, C., Moolenaar, G. F., Hoeijmakers, J. H. J., and Goosen, N. (2001) *EMBO J.* **20**, 601–611
696. Theis, K., Chen, P. J., Skorvaga, M., Van Houten, B., and Kisker, C. (1999) *EMBO J.* **18**, 6899–6907
696a. Villani, G., and Le Gac, N. T. (2000) *J. Biol. Chem.* **275**, 33185–33188
697. Shi, Q., Thresher, R., Sancar, A., and Griffith, J. (1992) *J. Mol. Biol.* **226**, 425–432
698. Verhoeven, E. E. A., van Kesteren, M., Moolenaar, G. F., Visse, R., and Goosen, N. (2000) *J. Mol. Biol.* **275**, 5120–5123
699. Ali, J. A., Maluf, N. K., and Lohman, T. M. (1999) *J. Mol. Biol.* **293**, 815–834
700. Wood, R. D. (1997) *J. Biol. Chem.* **272**, 23465–23468
701. Sancar, A. (1995) *J. Biol. Chem.* **270**, 15915–15918
702. Doetsch, P. W. (1995) *Trends Biochem. Sci.* **20**, 384–386
703. Guzder, S. N., Bailly, V., Sung, P., Prakash, L., and Prakash, S. (1995) *J. Biol. Chem.* **270**, 8385–8388
703a. Cappelli, E., Degan, P., Thompson, L. H., and Frosina, G. (2000) *Biochemistry* **39**, 10408–10412

References

704. Lao, Y., Gomes, X. V., Ren, Y., Taylor, J.-S., and Wold, M. S. (2000) *Biochemistry* **39**, 850–859
705. Zamble, D. B., Mu, D., Reardon, J. T., Sancar, A., and Lippard, S. J. (1996) *Biochemistry* **35**, 10004–10013
705a. Kunz, C., and Fleck, O. (2001) *J. Mol. Biol.* **313**, 241–253
705b. Van Houten, B., Eisen, J. A., and Hanawalt, P. C. (2002) *Proc. Natl. Acad. Sci. U.S.A.* **99**, 2581–2583
706. Seeberg, E., Eide, L., and Bjorås, M. (1995) *Trends Biochem. Sci.* **20**, 391–397
707. McCullough, A. K., Dodson, M. L., and Lloyd, R. S. (1999) *Ann. Rev. Biochem.* **68**, 255–285
708. Wilson, D. M., III, and Thompson, L. H. (1997) *Proc. Natl. Acad. Sci. U.S.A.* **94**, 12754–12757
708a. Parker, A., Gu, Y., Mahoney, W., Lee, S.-H., Singh, K. K., and Lu, A.-L. (2001) *J. Biol. Chem.* **276**, 5547–5555
709. Savva, R., McAuley-Hecht, K., Brown, T., and Pearl, L. (1995) *Nature (London)* **373**, 487–493
710. Slupphaug, G., Mol, C. D., Kavli, B., Arvai, A. S., Krokan, H. E., and Tainer, J. A. (1996) *Nature (London)* **384**, 87–92
711. Drohat, A. C., Xiao, G., Tordova, M., Jagadeesh, J., Pankiewicz, K. W., Watanabe, K. A., Gilliland, G. L., and Stivers, J. T. (1999) *Biochemistry* **38**, 11876–11886
711a. Otterlei, M., Kavli, B., Standal, R., Skjelbred, C., Bharati, S., and Krokan, H. E. (2000) *EMBO J.* **19**, 5542–5551
711b. Garvish, J. F., and Lloyd, R. S. (2000) *J. Mol. Biol.* **295**, 479–488
711c. Werner, R. M., and Stivers, J. T. (2000) *Biochemistry* **39**, 14054–14064
711d. Dinner, A. R., Blackburn, G. M., and Karplus, M. (2001) *Nature (London)* **413**, 752–755
712. Mol, C. D., Kuo, C.-F., Thayer, M. M., Cunningham, R. P., and Tainer, J. A. (1995) *Nature (London)* **374**, 381–386
713. Otterlei, M., Warbrick, E., Nagelhus, T. A., Haug, T., Slupphaug, G., Akbari, M., Aas, P. A., Steinsbekk, K., Bakke, O., and Krokan, H. E. (1999) *EMBO J.* **18**, 3834–3844
714. Carey, D. C., and Strauss, P. R. (1999) *Biochemistry* **38**, 16553–16560
715. Harris, J. M., McIntosh, E. M., and Muscat, G. E. O. (1999) *J. Mol. Biol.* **288**, 275–287
716. Williams, S. D., and David, S. S. (1999) *Biochemistry* **38**, 15417–15424
717. Thayer, M. M., Ahern, H., Xing, D., Cunningham, R. P., and Tainer, J. A. (1995) *EMBO J.* **14**, 4108–4120
717a. Dizdaroglu, M., Bauche, C., Rodriguez, H., and Laval, J. (2000) *Biochemistry* **39**, 5586–5592
718. Karahalil, B., Roldán-Arjona, T., and Dizdaroglu, M. (1998) *Biochemistry* **37**, 590–595
718a. Bebenek, K., Tissier, A., Frank, E. G., McDonald, J. P., Prasad, R., Wilson, S. H., Woodgate, R., and Kunkel, T. A. (2001) *Science* **291**, 2156–2159
719. Fuxreiter, M., Warshel, A., and Osman, R. (1999) *Biochemistry* **38**, 9577–9589
720. Croteau, D. L., and Bohr, V. A. (1997) *J. Biol. Chem.* **272**, 25409–25412
721. Lipscomb, L. A., Peek, M. E., Morningstar, M. L., Verghis, S. M., Miller, E. M., Rich, A., Essigmann, J. M., and Williams, L. D. (1995) *Proc. Natl. Acad. Sci. U.S.A.* **92**, 719–723
722. Bruner, S. D., Norman, D. P. G., and Verdine, G. L. (2000) *Nature (London)* **403**, 859–866
723. Hatahet, Z., Zhou, M., Reha-Krantz, L. J., Morrical, S. W., and Wallace, S. S. (1998) *Proc. Natl. Acad. Sci. U.S.A.* **95**, 8556–8561
724. Bhagwat, M., and Gerlt, J. A. (1996) *Biochemistry* **35**, 659–665
725. Porello, S. L., Cannon, M. J., and David, S. S. (1998) *Biochemistry* **37**, 6465–6475
725a. Miyako, K., Takamatsu, C., Umeda, S., Tajiri, T., Furuichi, M., Nakabeppu, Y., Sekiguchi, M., Hamasaki, N., Takeshige, K., and Kang, D. (2000) *J. Biol. Chem.* **275**, 12326–12330
725b. Zharkov, D. O., Gilboa, R., Yagil, I., Kycia, J. H., Gerchman, S. E., Shoham, G., and Grollman, A. P. (2000) *Biochemistry* **39**, 14768–14778
726. Manuel, R. C., Czerwinski, E. W., and Lloyd, R. S. (1996) *J. Biol. Chem.* **271**, 16218–16226
727. Guibourt, N., Castaing, B., Auffret Van Der Kemp, P., and Boiteux, S. (2000) *Biochemistry* **39**, 1716–1724
728. Kang, D., Nishida, J.-i, Iyama, A., Nakabeppu, Y., Furuichi, M., Fujiwara, T., Sekiguchi, M., and Takeshige, K. (1995) *J. Biol. Chem.* **270**, 14659–14665
729. Bessman, M. J., Frick, D. N., and O'Handley, S. F. (1996) *J. Biol. Chem.* **271**, 25059–25062
730. Masaoka, A., Terato, H., Kobayashi, M., Honsho, A., Ohyama, Y., and Ide, H. (1999) *J. Biol. Chem.* **274**, 25136–25143
731. Kung, H. C., and Bolton, P. H. (1997) *J. Biol. Chem.* **272**, 9227–9236
732. Demple, B., and Karran, P. (1983) *Trends Biochem. Sci.* **8**, 137–139
733. Spratt, T. E., Wu, J. D., Levy, D. E., Kanugula, S., and Pegg, A. E. (1999) *Biochemistry* **38**, 6801–6806
734. Myers, L. C., Verdine, G. L., and Wagner, G. (1993) *Biochemistry* **32**, 14089–14094
734a. Hashimoto, H., Inoue, T., Nishioka, M., Fujiwara, S., Takagi, M., Imanaka, T., and Kai, Y. (1999) *J. Mol. Biol.* **292**, 707–716
734b. Lin, Y., Dötsch, V., Wintner, T., Peariso, K., Myers, L. C., Penner-Hahn, J. E., Verdine, G. L., and Wagner, G. (2001) *Biochemistry* **40**, 4261–4271
735. Asaeda, A., Ide, H., Asagoshi, K., Matsuyama, S., Tano, K., Murakami, A., Takamori, Y., and Kubo, K. (2000) *Biochemistry* **39**, 1959–1965
736. Hollis, T., Ichikawa, Y., and Ellenberger, T. (2000) *EMBO J.* **19**, 758–766
737. Lutsenko, E., and Bhagwat, A. S. (1999) *J. Biol. Chem.* **274**, 31034–31038
738. Neddermann, P., Gallinari, P., Lettieri, T., Schmid, D., Truong, O., Hsuan, J. J., Wiebauer, K., and Jiricny, J. (1996) *J. Biol. Chem.* **271**, 12767–12774
739. Um, S., Harbers, M., Benecke, A., Pierrat, B., Losson, R., and Chambon, P. (1998) *J. Biol. Chem.* **273**, 20728–20736
740. Haber, J. E. (1999) *Trends Biochem. Sci.* **24**, 271–275
741. Cox, M. M., Goodman, M. F., Kreuzer, K. N., Sherratt, D. J., Sandler, S. J., and Marians, K. J. (2000) *Nature (London)* **404**, 37–41
741a. Connelly, J. C., and Leach, D. R. F. (2002) *Trends Biochem. Sci.*, **27**, 410 – 418
741b. Hopfner, K.-P., Craig, L., Moncalian, G., Zinkel, R. A., Usui, T., Owen, B. A. L., Karcher, A., Henderson, B., Bodmer, J.-L., McMurray, C. T., Carney, J. P., Petrini, J. H. J., and Tainer, J. A. (2002) *Nature (London)* **418**, 562 – 566
742. Webb, B. L., Cox, M. M., and Inman, R. B. (1999) *J. Biol. Chem.* **274**, 15367–15374
742a. Johnson, R. D., and Jasin, M. (2000) *EMBO J.* **19**, 3398–3407
742b. Arnaudeau, C., Lundin, C., and Helleday, T. (2001) *J. Mol. Biol.* **307**, 1235–1245
742c. Kondo, T., Wakayama, T., Naiki, T., Matsumoto, K., and Sugimoto, K. (2001) *Science* **294**, 867–870
742d. Hoege, C., Pfander, B., Moldovan, G.-L, Pyrowolakis, G., and Jentsch, S. (2002) *Nature (London)* **419**, 135 – 141
743. Yamaguchi-Iwai, Y., Sonoda, E., Sasaki, M. S., Morrison, C., Haraguchi, T., Hiraoka, Y., Yamashita, Y. M., Yagi, T., Takata, M., Price, C., Kakazu, N., and Takeda, S. (1999) *EMBO J.* **18**, 6619–6629
743a. Weller, G. R., Kysela, B., Roy, R., Tonkin, L. M., Scanlan, E., Della, M., Devine, S. K., Day J. P., Wilkinson, A., d'Adda di Fagagna, F., Devine, K. M., Bowater, R. P., Jeggo, P. A., Jackson, S. P., and Doherty, A. J. (2002) *Science* **297**, 1686 – 1689
744. Hays, S. L., Firmenich, A. A., and Berg, P. (1995) *Proc. Natl. Acad. Sci. U.S.A.* **92**, 6925–6929
745. Critchlow, S. E., and Jackson, S. P. (1998) *Trends Biochem. Sci.* **23**, 394–398
746. Wilson, T. E., and Lieber, M. R. (1999) *J. Biol. Chem.* **274**, 23599–23609
746a. Fugmann, S. D. (2002) *Nature (London)* **416**, 691–694
746b. Ooi, S. L., Shoemaker, D. D., and Boeke, J. D. (2001) *Science* **294**, 2552–2556
747. Jackson, S. P., and Jeggo, P. A. (1995) *Trends Biochem. Sci.* **20**, 412–415
748. Chu, G. (1997) *J. Biol. Chem.* **272**, 24097– 24100
749. Hammarsten, O., DeFazio, L. G., and Chu, G. (2000) *J. Biol. Chem.* **275**, 1541–1550
750. Brown, K. D., Lataxes, T. A., Shangary, S., Mannino, J. L., Giardina, J. F., Chen, J., and Baskaran, R. (2000) *J. Biol. Chem.* **275**, 6651–6656
750a. Walker, J. R., Corpina, R. A., and Goldberg, J. (2001) *Nature (London)* **412**, 607–614
750b. Rouse, J., and Jackson, S. P. (2002) *Science* **297**, 547 – 551
750c. Kolodner, R. D., Putnam, C. D., and Myung, K. (2002) *Science* **297**, 552 –557
750d. Carr, A. M. (2002) *Science* **297**, 557 – 558
750e. Cha, R. S., and Kleckner, N. (2002) *Science* **297**, 602 – 606
750f. Sogo, J. M. Lopes, M., and Foiani, M. (2002) *Science* **297**, 599 – 602
750g. Celeste, A., and 20 other authors (2002) *Science* **296**, 922 – 927
750h. Bird, A. W., Yu, D. Y., Pray-Grant, M. G., Qiu, Q., Harmon, K. E., Megee, P. C., Grant, P. A., Smith, M. M., and Christman, M. F. (2002) *Nature (London)* **419**, 411 – 415
750i. Pickart, C. M. (2002) *Nature (London)* **419**, 120 – 121
751. Flanders, P. H. (1981) *Sci. Am.* **245**(Nov), 72–80
752. Kenyon, C. J. (1983) *Trends Biochem. Sci.* **8**, 84–87
753. Walker, G. C. (1995) *Trends Biochem. Sci.* **20**, 416–420
753a. Yasuda, T., Morimatsu, K., Kato, R., Usukura, J., Takahashi, M., and Ohmori, H. (2001) *EMBO J.* **20**, 1192–1202
754. McDonald, J. P., Peat, T. S., Levine, A. S., and Woodgate, R. (1999) *J. Mol. Biol.* **285**, 2199–2209
754a. Goodman, M. F. (2002) *Ann. Rev. Biochem.* **71**, 17 – 50
754b. Livneh, Z. (2001) *J. Biol. Chem.* **276**, 25639–25642
755. Johnson, R. E., Washington, M. T., Prakash, S., and Prakash, L. (1999) *Proc. Natl. Acad. Sci. U.S.A.* **96**, 12224–12226
756. Baynton, K., and Fuchs, R. P. P. (2000) *Trends Biochem. Sci.* **25**, 74–79
757. Murakumo, Y., Roth, T., Ishii, H., Rasio, D., Numata, S.-i, Croce, C. M., and Fishel, R. (2000) *J. Biol. Chem.* **275**, 4391–4397
758. Wood, R. D. (1999) *Nature (London)* **399**, 639–640
759. Daube, S. S., Tomer, G., and Livneh, Z. (2000) *Biochemistry* **39**, 348–355
759a. Matsuda, T., Bebenek, K., Masutani, C., Rogozin, I. B., Hanaoka, F., and Kunkel, T. A. (2001) *J. Mol. Biol.* **312**, 335–346

References

759b. Kusumoto, R., Masutani, C., Iwai, S., and Hanaoka, F. (2002) *Biochemistry* **41**, 6090–6099
760. Hübscher, U., Nasheuer, H.-P., and Syväoja, J. E. (2000) *Trends Biochem. Sci.* **25**, 143–147
760a. Friedberg, E. C., Wagner, R., and Radman, M. (2002) *Science* **296**, 1627–1630
760b. Haracska, L., Johnson, R. E., Unk, I., Phillips, B. B., Hurwitz, J., Prakash, L., and Prakash, S. (2001) *Proc. Natl. Acad. Sci. U.S.A.* **98**, 14256–14261
760c. Johnson, R. E., Prakash, S., and Prakash, L. (2000) *Proc. Natl. Acad. Sci. U.S.A.* **97**, 3838–3843
760d. García-Díaz, M., Domínguez, O., López-Fernández, L. A., de Lera, L. T., Saníger, M. L., Ruiz, J. F., Párraga, M., García-Ortiz, M. J., Kirchhoff, T., del Mazo, J., Bernad, A., and Blanco, L. (2000) *J. Mol. Biol.* **301**, 851–867
760e. Wang, Z., Castano, I. B., De Las Penas, A., Adams, C., and Christman, M. F. (2000) *Science* **289**, 774–779
760f. Frank, E. G., Tissier, A., McDonald, J. P., Rapic-Otrin, V., Zeng, X., Gearhart, P. J., and Woodgate, R. (2001) *EMBO J.* **20**, 2914–2922
760g. Shimizu, K., Kawasaki, Y., Hiraga, S.-I., Tawaramoto, M., Nakashima, N., and Sugino, A. (2002) *Proc. Natl. Acad. Sci. U.S.A.* **99**, 9133–9138
761. de Murcia, G., and de Murcia, J. M. (1994) *Trends Biochem. Sci.* **19**, 172–176
761a. Jacobson, M. K., and Jacobson, E. L. (1999) *Trends Biochem. Sci.* **24**, 415–417
762. Lindahl, T., Satoh, M. S., Poirier, G. G., and Klungland, A. (1995) *Trends Biochem. Sci.* **20**, 405–411
763. Aoufouchi, S., Yélamos, J., and Milstein, C. (1999) *J. Mol. Biol.* **290**, 943–949
763a. Smith, S. (2001) *Trends Biochem. Sci.* **26**, 174–179
764. Ames, B. N., Magaw, R., and Gold, L. S. (1987) *Science* **236**, 271–280
765. Gold, L. S., Slone, T. H., Stern, B. R., Manley, N. B., and Ames, B. N. (1992) *Science* **258**, 261–265
766. Miller, J. A., and Miller, E. C. (1976) *Fed. Proc.* **35**, 1316–1321
767. Yu, H., Eritja, R., Bloom, L. B., and Goodman, M. F. (1993) *J. Biol. Chem.* **268**, 15935–15943
768. Goodman, M. F., and Ratliff, R. L. (1983) *J. Biol. Chem.* **258**, 12842–12846
769. Sowers, L. C., Fazakerley, G. V., Eritja, R., Kaplan, B. E., and Goodman, M. F. (1986) *Proc. Natl. Acad. Sci. U.S.A.* **83**, 5434–5438
770. Ripley, L. S., and Clark, A. (1986) *Proc. Natl Acad. Sci. U.S.A.* **83**, 6954–6958
771. Ginell, S. L., Kuzmich, S., Jones, R. A., and Berman, H. M. (1990) *Biochemistry* **29**, 10461–10465
772. Wolff, I. A., and Wasserman, A. E. (1972) *Science* **177**, 15–18
773. Russell, W. L., Kelly, E. M., Hunsicker, P. R., Bangham, J. W., Maddux, S. C., and Phipps, E. L. (1979) *Proc. Natl. Acad. Sci. U.S.A.* **76**, 5818–5819
774. Shaver-Walker, P. M., Urlando, C., Tao, K. S., Zhang, X. B., and Heddle, J. A. (1995) *Proc. Natl. Acad. Sci. U.S.A.* **92**, 11470–11474
775. Wurdeman, R. L., Church, K. M., and Gold, B. (1989) *J. Am. Chem. Soc.* **111**, 6408–6412
776. Klein-Szanto, A. J. P., Iizasa, T., Momiki, S., Garcia-Palazzo, I., Caamano, J., Metcalf, R., Welsh, J., and Harris, C. C. (1992) *Proc. Natl. Acad. Sci. U.S.A.* **89**, 6693–6697
777. Craddock, V. M. (1983) *Nature (London)* **306**, 638
778. Hirai, N., Kingston, D. G. I., Van Tassell, R. L., and Wilkins, T. D. (1982) *J. Am. Chem. Soc.* **104**, 6149–6150
779. Gupta, I., Suzuki, K., Bruce, W. R., Krepinsky, J. J., and Yates, P. (1984) *Science* **225**, 521–522
780. Ozawa, N., and Guengerich, F. P. (1983) *Proc. Natl. Acad. Sci. U.S.A.* **80**, 5266–5270
781. Langouët, S., Mican, A. N., Müller, M., Fink, S. P., Marnett, L. J., Muhle, S. A., and Guengerich, F. P. (1998) *Biochemistry* **37**, 5184–5193
782. Leonard, G. A., McAuley-Hecht, K. E., Gibson, N. J., Brown, T., Watson, W. P., and Hunter, W. N. (1994) *Biochemistry* **33**, 4755–4761
783. Weisenseel, J. P., Moe, J. G., Reddy, G. R., Marnett, L. J., and Stone, M. P. (1995) *Biochemistry* **34**, 50–64
784. Palejwala, V. A., Simha, D., and Humayun, M. Z. (1991) *Biochemistry* **30**, 8736–8743
785. Zegar, I. S., Setayesh, F. R., DeCorte, B. L., Harris, C. M., Harris, T. M., and Stone, M. P. (1996) *Biochemistry* **35**, 4334–4348
786. Bucala, R., Model, P., and Cerami, A. (1984) *Proc. Natl. Acad. Sci. U.S.A.* **81**, 105–109
787. Shires, T. K., Tresnak, J., Kaminsky, M., Herzog, S. L., and Truc-pham, B. (1990) *FASEB J.* **4**, 3340–3346
788. Napolitano, R. L., Lambert, I. B., and Fuchs, R. P. P. (1994) *Biochemistry* **33**, 1311–1315
789. Isono, K., and Yourno, J. (1974) *Proc. Natl. Acad. Sci. U.S.A.* **71**, 1612–1617
790. Stoloff, L. (1987) *Science* **237**, 1283
791. Baertschi, S. W., Raney, K. D., Stone, M. P., and Harris, T. M. (1988) *J. Am. Chem. Soc.* **110**, 7929–7931
792. Iyer, R. S., Coles, B. F., Raney, K. D., Thier, R., Guengerich, F. P., and Harris, T. M. (1994) *J. Am. Chem. Soc.* **116**, 1603–1609
793. Heidelberger, C. (1975) *Ann. Rev. Biochem.* **44**, 79–121
794. Phillips, D. H. (1983) *Nature (London)* **303**, 468–472
795. Flowers, L., Bleczinski, W. F., Burczynski, M. E., Harvey, R. G., and Penning, T. M. (1996) *Biochemistry* **35**, 13664–13672
796. Schurter, E. J., Sayer, J. M., Oh-hara, T., Yeh, H. J. C., Yagi, H., Luxon, B. A., Jerina, D. M., and Gorenstein, D. G. (1995) *Biochemistry* **34**, 9009–9020
797. Mao, B., Gu, Z., Gorin, A., Chen, J., Hingerty, B. E., Amin, S., Broyde, S., Geacintov, N. E., and Patel, D. J. (1999) *Biochemistry* **38**, 10831–10842
798. Tang, M.-s, Bohr, V. A., Zhang, X.-s, Pierce, J., and Hanawalt, P. C. (1989) *J. Biol. Chem.* **264**, 14455–14462
799. Cho, B. P., and Zhou, L. (1999) *Biochemistry* **38**, 7572–7583
800. Eis, P. S., Smith, J. A., Rydzewski, J. M., Case, D. A., Boger, D. L., and Chazin, W. J. (1997) *J. Mol. Biol.* **272**, 237–252
801. Perera, F., and Petito, C. (1982) *Science* **216**, 1285–1291
802. Rall, D. P. (1992) *Science* **257**, 1330
803. Hanson, D. J. (1991) *Chem. Eng. News* **Aug 12**, 7–14
804. Marshall, E. (1990) *Science* **247**, 276–277
805. Hileman, B. (1990) *Chem. Eng. News* **May 7**, 4
806. Ames, B. N. (1979) *Science* **204**, 587–593 (see also discussion in Science 224, 659–670, 757–760 (1984))
807. Devoret, R. (1979) *Sci. Am.* **241**(Aug), 40–49
808. Muench, K. F., Misra, R. P., and Humayun, M. Z. (1983) *Proc. Natl. Acad. Sci. U.S.A.* **80**, 6–10
809. Green, M. M., Todo, T., Ryo, H., and Fujikawa, K. (1986) *Proc. Natl. Acad. Sci. U.S.A.* **83**, 6667–6671
810. Gold, M. O., Blum, A., and Ames, B. N. (1978) *Science* **200**, 785–787
811. Commoner, B., Vithayathil, A. J., Dolara, P., Nair, S., Madyastha, P., and Cuca, G. C. (1978) *Science* **201**, 913–916
812. Tennant, R. W., Margolin, B. H., Shelby, M. D., Zeiger, E., Haseman, J. K., Spalding, J., Caspary, W., Resnick, M., Stasiewicz, S., Anderson, B., and Minor, R. (1987) *Science* **236**, 933–941
813. Epstein, S. S., Ames, B. N., and Gold, L. S. (1988) *Science* **240**, 1043–1047
814. Ashby, J., and Morrod, R. S. (1991) *Nature (London)* **352**, 185–186
815. Weinstein, I. B. (1991) *Science* **251**, 387–388
816. Stone, R. (1995) *Science* **268**, 356–357
817. Randerath, K., Reddy, M. V., and Gupta, R. C. (1981) *Proc. Natl. Acad. Sci. U.S.A.* **78**, 6126–6129
818. Kadlubar, F. F. (1992) *Nature (London)* **360**, 189
819. Schut, H. A. J., and Shiverick, K. T. (1992) *FASEB J.* **6**, 2942–2951

Study Questions

1. Demethylation of 5-methylcytosine in DNA during early embryonic development has been proposed (see Chapter 32). Can you suggest one or more mechanisms by which such demethylation could occur?

2. Describe the structures and functions of histones and of nucleosomes in eukaryotic cells. Are there comparable proteins and structures in bacteria? Do you think that our knowledge of histones and nucleosomes is nearly complete?

3. List the major molecular components required for replication of DNA in *E. coli*. Describe briefly the functions of each protein or other component.

4. Compare replication in bacteria and in eukaryotes.

5. Compare systhesis of the leading and lagging strands in the elongation phase of DNA replication. Explain why DNA polymerases may have difficulty in replicating the 3'-end of the *lagging* strand of **linear** DNA. How has this problem been solved in many bacterial and viral systems? In eukaryotic cells?

6. The circular chromosome of an *E. coli* cell contains 4.6×10^6 base pairs. If a replication fork moves at a rate of ~1000 nucleotides per second, how much time will be required for replication of the DNA? Cells of *E. coli* can divide every 20 minutes under favorable conditions. How can you explain this rapid rate of growth?

7. DNA ligase, whose reaction is reversible, is able to relax supercoiled circular DNA in the presence of AMP but not in its absence. Outline the chemical mechanism of the ligase reaction. Why is it dependent on AMP? What other DNA ligase mechanism is known?

8. DNA polymerases involved in replication require a primer. Why? What is the nature of the primer?

9. Why is it essential for a cell to have several different mechanisms of DNA repair? Describe some of these mechanisms.

10. Why do cells use error-prone DNA polymerases under some circumstances?

11. Do you see a relationship of some types of DNA repair to the chemical events during meiosis?

12. Is O^6-methyltransferase an enzyme?

13. Why are high rates of mutation observed in regions of DNA that contain 5-methylcytosine?

14. Why is uracil-DNA glycosylase important in DNA repair? Is it important for DNA replication?

15. Why do cells exposed to visible light following irradiation by ultraviolet light have a greater survival rate than cells kept in the dark after UV irradiation?

16. Can exposure of *E. coli* to nitrous acid (HNO_2) lead to mutation of a $tRNA^{GLY}$ to an amber suppressor? The Gly codons are GGX (where X = any nucleotide) and the amber codon is UAG.

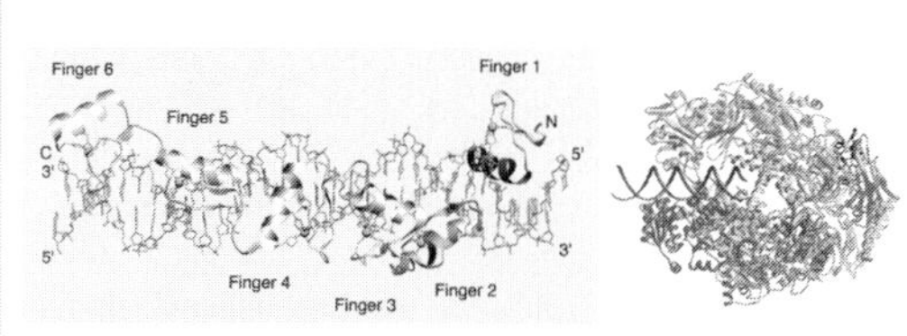

Left. The N-terminal 190-residue fragment of transcription factor TFIIIA of *Xenopus laevis* bound to a 31 bp DNA segment of the promoter region for 5S ribosomal RNA. Six zinc finger motifs (zinc atoms are green) bind in several ways into the major groove of DNA and across the minor groove. From Nolte *et al*, *Proc. Natl. Acad. Sci. USA* **95**, 2938–2943, 1998. Courtesy of Raymond S. Brown. Right. Ribbon drawing of the three-dimensional structure of a 10-subunit form of yeast RNA polymerase II, which transcribes genes to form messenger RNA. A 20 base pair segment of B-DNA has been modeled but the transcription bubble is not shown. The active site Mg^{2+} is green. From Cramer *et al*, *Science* **288**, 640–649, 2000. Courtesy of Roger D. Kornberg.

Contents

28 The Transcription of Genes

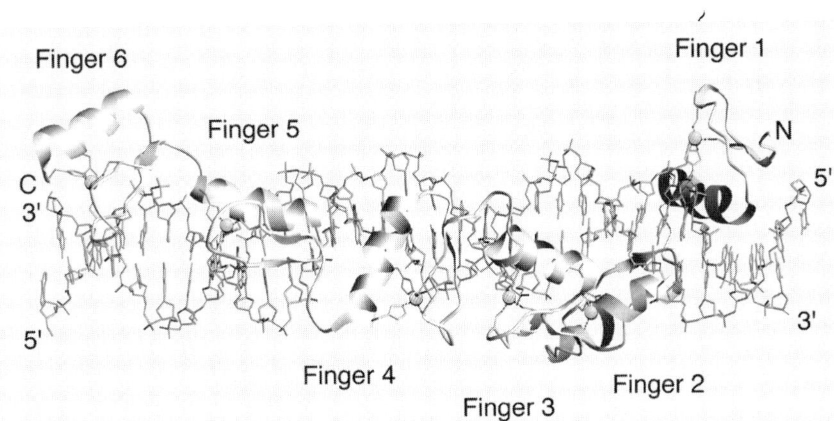

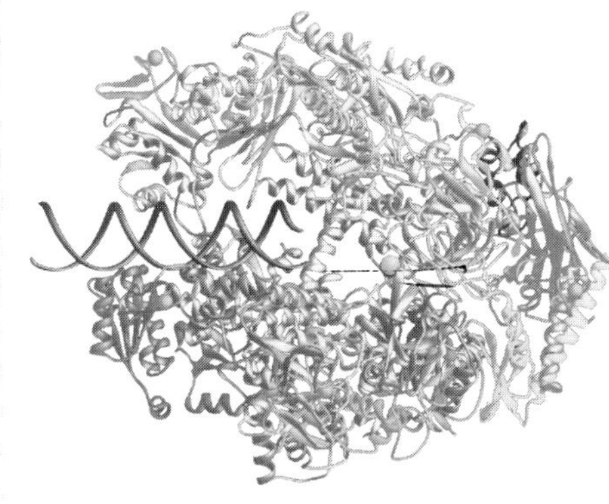

The copying of genetic information from DNA into messenger RNA is the initial step in the chain of reactions leading to synthesis of the multitude of proteins and specialized RNA molecules needed by cells. The requirement for these macromolecules varies with conditions, and in eukaryotic cells, with the stage of differentiation. Therefore, it is not surprising that transcription is highly controlled.

Cells make four principal kinds of RNA: ribosomal (rRNA), transfer (tRNA), messenger (mRNA), and a variety of small RNAs. The last, which range in length from a few up to several hundred nucleotide units,[1–1b] are designated variously as sRNAs, ncRNAs, miRNAs, siRNAs, snRNAs, and snoRNAs. The abbreviation *s*, *nc*, *mi*, *si*, *sn*, and *sno* stand for small, noncoding, micro, silencing, small nuclear, and small nucleolar, respectively. All of these RNAs are synthesized as larger transcripts, which undergo cleavage and other modifications within the cell. Therefore, a second major topic in this chapter is the processing of RNA precursors. We will also consider the fact that cells may be hosts for RNA viruses, may occasionally harbor RNA plasmids, and must sometimes transcribe viral DNA.

The absence of a nuclear membrane is a characteristic of bacteria that has a profound effect on transcription. Bacterial transcripts are processed rapidly, and their 5' ends often enter ribosomes and are directing protein synthesis, while the 3' ends of the genes are still being transcribed. In contrast, most eukaryotic RNA transcripts must be processed and transported out of the nucleus before they can function. As consequence, many aspects of the control of transcription differ between prokaryotes and eukaryotes.

A. Transcription and Processing of RNA in Bacteria

Even after the existence of mRNA had been recognized, it was not obvious how formation of single-stranded (ss) RNA would be accomplished, using a double-stranded (ds) template. The fact that purified RNA polymerases can synthesize RNA from the four ribonucleoside triphosphates using ssDNA as the template suggested that transcription, like DNA replication, involves base pairing. In line with this conclusion was the fact that the ssDNA obtained from bacteriophage ϕX174 was converted by RNA polymerases into a dsRNA–DNA hybrid. However, when dsDNA served as the template, free ssRNA was formed. Thus, it appeared likely that at the site of the polymerase action the dsDNA was momentarily pulled apart into single strands and that one of these was copied by the polymerase. More recent experiments have confirmed this view.

1. The *lac* Operon

Much of the terminology used to describe the control of transcription originated with Jacob and Monod. Based on studies of the induction of enzymes in bacteria they proposed the **operon model**.[1c–3] An operon is a regulated cluster of genes, one of which is shown diagrammatically in Fig. 28-1. This is the *lac* operon of *E. coli*. Found at position 8 min on the genetic map of Fig. 26-4, it is probably the most intensively studied group of *E. coli* genes. Three structural genes encode the amino acid sequences of β-galactosidase (*lacZ*),[3a] permease (*lacY*),[4] and a transacetylase (acetyltransferase, *lacA*), which transfers acetyl groups from acetyl-CoA to β galactosides. To account for the apparently synchronous control of these three genes, Jacob and Monod proposed that they function as a **transcriptional unit** the operon, which encodes a single molecule of mRNA. They proposed that each operon is controlled by a segment of the DNA molecule located at the beginning of the operon, i.e., at the 5' end of the coding chain or 3' end of the template chain. The first part of this **control region** they called the **promoter** (*P*). The promoter is the site of the initial binding of the RNA polymerase to the DNA, the binding constants for the association being very high. The rates of association and of initiation may be in fluenced strongly by various other proteins. One of these, the **catabolite gene activator protein** (**CAP**; also called cAMP receptor protein, CRP), is important to the *lac* operon. It also binds in the promoter region (Fig. 28-1) and stimulates transcription.

Repression and induction. Immediately adjacent to the promoter is the **operator** (*O*), which is a binding site for a **repressor** (R). When the operator is free, transcription is initiated and proceeds through the operator region and on to the genes coding for the three proteins. On the other hand, if the repressor is bound to the operator, transcription is blocked. When the operon model was first proposed, the chemical nature of the repressor was unknown. However, many repressors have been identified as oligomeric proteins able to undergo allosteric alteration. The *lac* repressor is made up of four identical 360-residue subunits. Each subunit has a helix–turn–helix binding domain that is specific for the DNA sequence of the operator and an allosteric binding site for an effector.[5,6] The drawing in Fig. 28-1 is simplified to show only two of the four subunits (see also Chapter 5, Section F,1).

The *lac* operon is ordinarily subject to repression and is activated by the presence of an **inducer**, now known to be **allolactose**, D-Gal*p*-β1→6-D-Glc. However, in experimental work artificial inducers such as **isopropyl-β-D-thiogalactoside** (**IPTG**) are most often used. Jacob and Monod postulated that the free repressor protein binds to the operator. In the presence of the inducer a conformational change takes place, destroying the affinity of the repressor protein for the operator site. Thus, in the presence of inducer the operator is not blocked, and the transcription takes place. Such an operon is said to be **negatively controlled** and **inducible**.

Important to the control of the operon is the **regulatory gene,** which codes for the synthesis of the repressor protein. In the case of the *lac* operon, the regulatory gene (the *I* gene) is located immediately preceding the operon itself (Fig. 28-1). However, for some operons the regulatory gene is located a considerable distance away. For example, the *gal* operon of *E. coli*, which codes for enzymes of galactose metabolism, is found at map position 17 min, while the regulatory gene is at 61 min.[7]

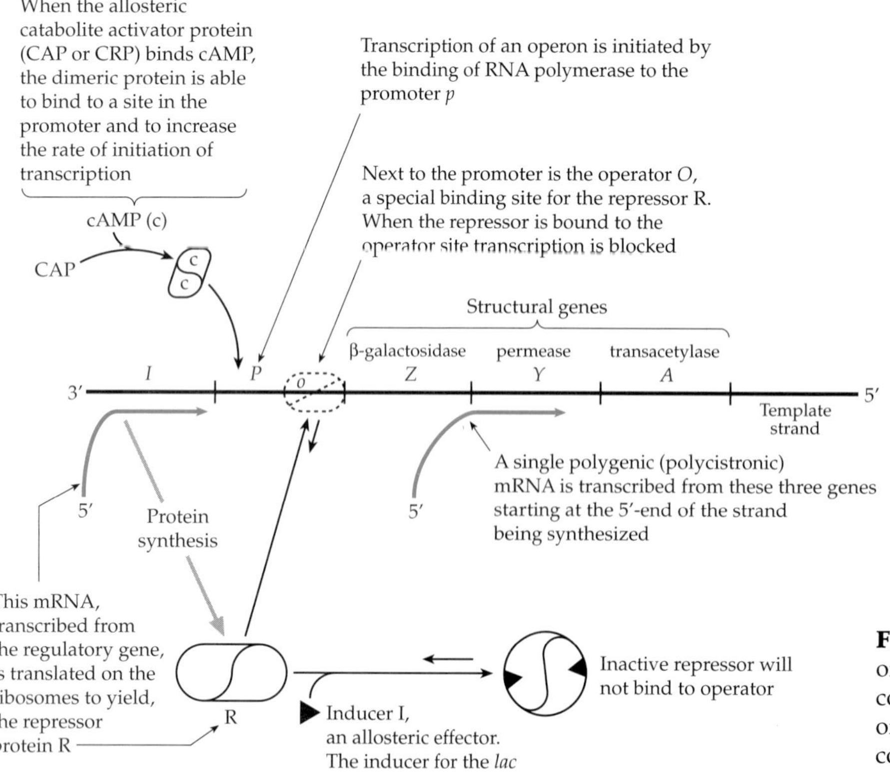

Figure 28-1 Schematic representation of the *lac* operon of *E. coli* and of its control. Here only the template strand of the DNA is shown. However, the coding (nontranscribed) strand is usually the one labeled, as in Fig. 28-2.

Regulatory genes are normally transcribed at a slow but steady rate, presumably because RNA polymerase initiates RNA chains infrequently at the promoter sites of regulatory genes. Thus, each cell of *E. coli* normally contains only about ten molecules of the *lac* repressor protein. A mutation in a regulatory gene may lead to a defective repressor, which no longer binds at the operator. Then, transcription of the operon is uncontrolled, and mRNA is produced in greater amounts. In such a mutant strain (designated I^- in contrast to the normal I^+ strain) production of the enzyme representing the gene product becomes constitutive, just as is the formation of the enzymes of the central pathways of metabolism. The latter enzymes also appear to be produced regularly in large amounts without control by a repressor, establishing that *transcription* rather than translation regulates the expression of these genes. The operon can also become unregulated, i.e., **constitutively expressed**, if a mutation occurs in the operator site and the repressor can no longer bind because of the altered DNA sequence.

Nucleotide sequence of the lac control region. The sequence of the *E. coli* DNA representing the promoter-operator region of the *lac* operon is shown in Fig. 28-2. It includes the end of the *I* gene, at the left, and the beginning of the Z gene, at the right.[2] The series of codons representing the peptide sequence Glu-Ser-Gly-Gln-Stop at the left-hand end corresponds to the known C-terminal sequence of the repressor, while the three codons at the right are those of formyl-Met-Thr-Met, the known N-terminal sequence of the Z gene product β-galactosidase. Detailed genetic mapping of the region in 1973 made it possible to assign operator and promoter regions with confidence as indicated. The mRNA transcript begins in the operator region as shown in the figure. The initiation codon for the Z gene is 39 bases from the end of the transcript. In this figure the original numbering of the nucleotides is printed in black. Now it is customary to number from the transcription initiation position (+1). Green numbers are used in this way in Fig. 28-2. Positions to the left and toward the 5' end of the nontranscribed coding strand precede the initiation position. They are referred to as **upstream** and are numbered with negative integers. Positions to the right are **downstream** and numbered with positive integers.

The operator region was located by digesting the DNA with deoxyribonuclease in the presence of the repressor protein.[8] The bound repressor protected a region of 27 base pairs as indicated in the figure. The operator is centered on a region of local twofold rotational symmetry (Chapter 5; Fig. 5-34). The symmetry is not perfect, the sequence being **quasipalindromic**. The following precisely symmetric synthetic sequence, which contains an 11-bp inverted repeat of the left half of the *lac* operator sequence, binds *lac* repressor 8-fold more tightly than does the natural *E. coli* operator.[9]

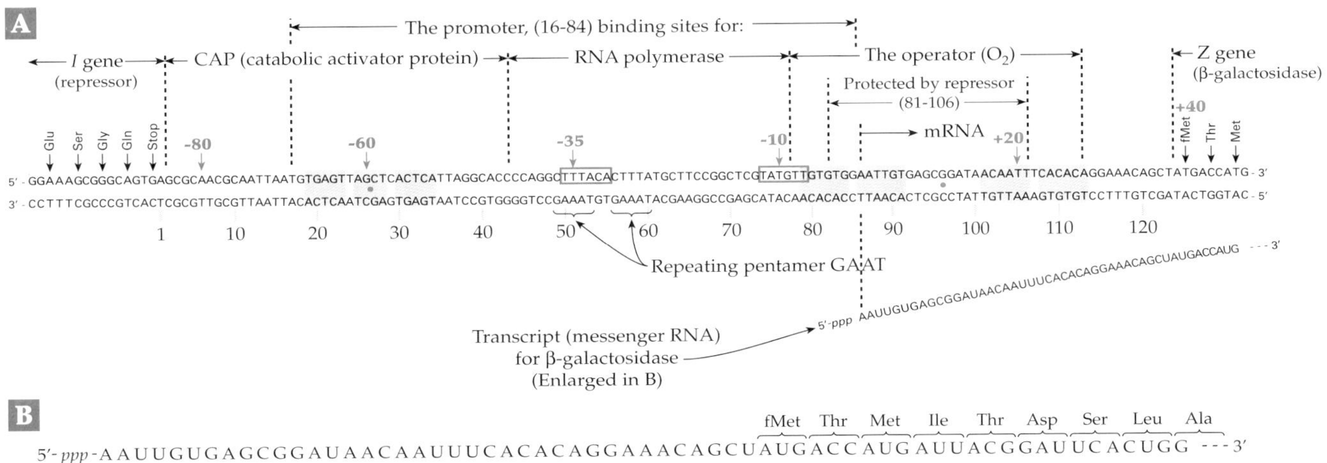

Figure 28-2 (A) Nucleotide sequence of the *lac* promoter-operator region of the *E. coli* chromosome.[7] The proposed locations of the *I* gene, the promoter (which contains CAP and RNA polymerase binding sites), the operator, and the beginning of the Z gene (β-galactosidase) are shown. Note the two regions of local 2-fold rotational symmetry, which are marked by bars and central dots and the repeating pentamer. Positions upstream (–) or downstream (+) from the +1 start position for transcription are marked in green. The –10 (Pribnow) and –35 promoter elements are boxed on the coding strand (top). Labels are usually applied to the coding strand. The mRNA is copied from the complementary template strand (bottom). (B) The sequence of an mRNA molecule initiated in the *lac* promoter-operator region of a mutant strain of *E. coli* with an altered promoter.[10] The peptide initiation amino acid is identified by the symbol fMet, and the successive amino acids from the known N-terminal sequence of β-galactosidase have been matched with the codons.

5′-G-A-A-T-T-G-T-G-A-G-C-G-C-T-C-A-C-A-A-T-T-C
3′-C-T-T-A-A-C-A-C-T-C-G-C-G-A-G-T-G-T-T-A-A-G

The dyad symmetry of the operator sequence is probably important in providing tight binding to two subunits of the symmetric tetrameric protein.[11–13] It is also possible that repressor molecules move along DNA chains in a one-dimensional diffusion process, and that the symmetry of the operator site facilitates recognition by a protein moving from either direction.[14,15]

The repressor structure. The *lac* repressor protein is a member of the **helix–turn–helix** family of DNA binding proteins (Fig. 28-3; see also Fig. 5-35). The first helix and the turn of this motif fit into the major groove of the DNA, the side chains from the helix interacting with specificity-determining groups in the major groove. The protein consists of three domains; the N-terminal DNA binding "head" (residues ~1– ~50), a core domain (residues 62–340), and a leucine heptad repeat domain (residues 340–360) that forms the dimer–dimer interface. The dimeric form of the repressor binds to the palindromic operator sequence. The tetramer can bind to two operator sequences.[16–20] The inducer IPTG binds to the core domain near the ONPF site shown in Fig. 28-3. The sequence, structure, and binding site of the core domain resemble those of sugar transport proteins such as the galactose-binding chemoreceptor protein (Fig. 4-18A). Binding of the inducer causes the conformational change that appears to disrupt the interactions of the "hinge helices," seen in the center of Fig. 28-3A, with the DNA. This causes the repressor to dissociate from the operator (Fig. 28-1) and allows RNA polymerase to bind and to initiate transcription.

The *E. coli lac* repressor is one of the most investigated of all proteins. For example, 4000 single-amino-acid mutants have been prepared and studied.[6,17,21] Suppressor mutations were used to determine the function of various portions of the protein.[22,23] Many of the mutant proteins were created using *amber* mutations that were induced in the gene at many positions. The mutated genes were transferred into plasmids for cloning. Each plasmid was used to infect five different strains of bacteria, each carrying a suppressor mutation that would introduce a different amino acid when the (termination) codon UAG was encountered (see Chapter 29, Section C,3). From these infected bacteria large quantities of the mutant forms of the *lac* repressor were isolated. It was found that many mutations near the N-terminal end interfered with binding of the repressor to DNA, whereas mutations near the center interfere with binding to the inducer.

In addition to the main *lac* operator O_1, which is marked in Fig. 28-2, there are two weaker auxiliary operator sequences designated O_2 and O_3 located 401

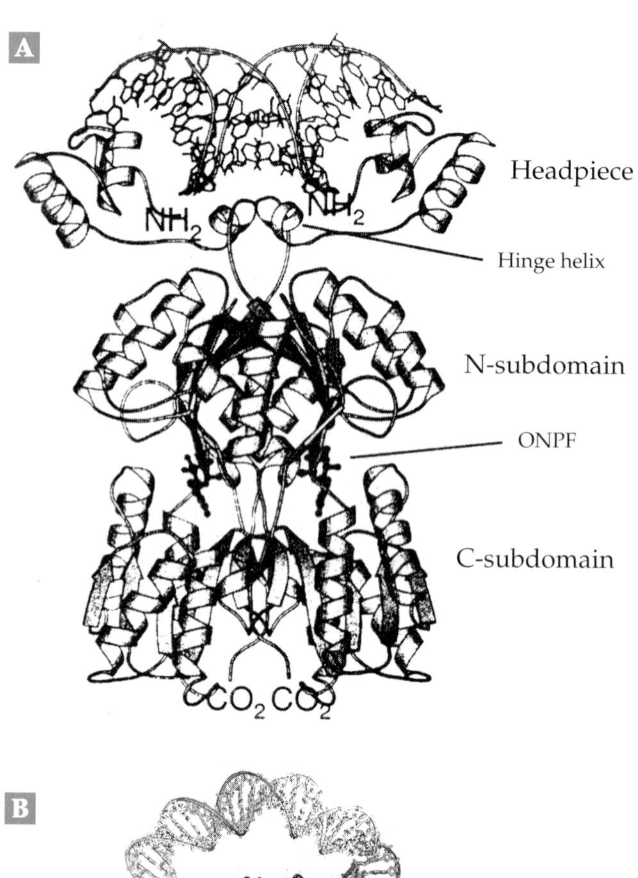

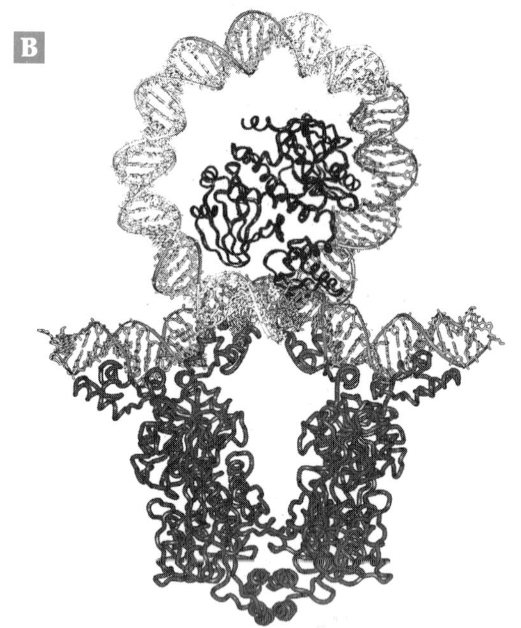

Figure 28-3 (A) Ribbon view of the dimeric *lac* repressor bound to a natural operator and to the anti-inducer *o*-nitrophenylfucoside (ONPF). The headpiece (residues 2–46) and the hinge helix (residues 50–58) form the DNA-binding domains. The core (residues 62–330), which is divided into N- and C-terminal subdomains, forms the binding site for ONPF. The C-terminal residues 334–360, which form a tetramerization domain, are absent from this MolScript drawing. Notice that the hinge helices bind to and widen the minor groove at the center of the operator. From Lewis *et al.*[5a] (B) Model of a 93-bp DNA loop corresponding to residues –82 to +11 of the *lac* operon (Fig. 28-2) bound to the tetrameric *lac* repressor. The active sites of the repressor are bound to the major operator O_1 and to the secondary operator O_3. From Lewis *et al.*[5]

bp downstream of O_1 in the lacZ coding region, and 92 bp upstream of O_1, respectively.[24] The DNA can bind to both O_1 and either O_2 or O_3 with a loop between them as in Fig. 28-3B. Binding of the 10–20 copies of the lac repressor present in a cell of *E. coli* to the operator sequence is so tight that expression of the genes controlled is reduced 1000-fold. However, when placed in front of other operators or in different positions relative to the Pribnow sequence (Fig. 28-2), its effectiveness varied greatly. The extent of repression appears to be affected both by thermodynamic factors[12,13] and by relative rates of repressor binding and of RNA polymerase movement (see also Eq. 28-1).[25]

2. Initiation of Transcription

The rate of RNA synthesis varies from one operon to another. Sequences of promoters, operators, and other control sequences as well as the state of repressors and activator proteins all affect these rates.[10,26,27] However, in every instance the first steps in transcription involve the binding of RNA polymerase to DNA.

Bacterial RNA polymerase. Most RNA polymerases (**RNAPs**) are large multisubunit proteins. However, bacterial viruses sometimes induce their own RNA polymerases, and these may be monomeric. For example, the 99-kDa (883-residue) phage T7-encoded polymerase is a single peptide chain with a structure and two-metal-ion active site resembling those of *E. coli* DNA polymerase I.[28–30a] It is able to carry out all of the steps of the transcription cycle of the virus. In contrast, the most studied bacterial RNAP, that from *E. coli*, consists of five kinds of subunits[31–33d] with the composition $\alpha_2\beta\beta'\sigma\omega$. A similar composition has been found for RNAPs of other bacteria.[33e] Functions of the five subunits can be correlated directly with components of archaeal and eukaryotic RNAPs.[33f] However, the latter contain additional subunits. The two α subunits in the *E. coli* enzyme have identical sequences, but their locations and interactions are different.[33a–c]

Gene symbol	Subunit	Molecular mass (kDa)	Number of amino acid residues
α	*rpo*A	36.5	329
β	*rpo*B	150.6	1342
β′	*rpo*C	155.2	1407
σ	*rpo*D	70.2	
ω	*rpo*Z	6	

The three-dimensional structure of the *E. coli* RNAP bound to DNA in an initiation complex shows that the enzyme forms a groove into which the DNA can fit. It can then close to form a tunnel in which the template and nontemplate strands are separated (Fig. 28-4). The polymerase is present in large amounts, ~3000 molecules per cell in E. coli.[26] The β and β′ subunits of the *E. coli* enzyme each contain an essential zinc ion. The Zn^{2+} in the β′ subunit is present in a zinc finger motif near the N terminus.[34] It is thought to interact with the template strand of the DNA. The Zn^{2+} in the β strand is more loosely bound.[33,35] The active site is largely in the β subunit. In the assembly of the RNAP complex a dimer of the small α subunit binds to β, after which β′ is added.[36] The sequences of the β and β′ subunits have several highly conserved regions with homologous sequences from bacteria and from functionally equivalent regions in eukaryotic RNAPs. Three aspartate residues in the sequence NADFDGD may chelate two Mg^{2+} ions as in the active site of DNA polymerases (Fig. 27-13).[37] The basic chemistry of all of the polymerases may be similar, but the modular structure of the bacterial RNAP differs markedly from that of the DNA Pol I family.[38]

Of the RNA polymerase subunits σ (sigma) plays a unique role in initiation of transcription. It is required for the recognition of promoter sites.[39] However, it is not needed for elongation of an RNA chain and dissociates from the $\alpha_2\beta\beta'$ core complex soon after transcription is initiated. In a given bacterial species there is one predominant σ factor, but there are often smaller amounts of other σ factors with homologous sequences.[26,40,40a] In *E. coli* σ^{70} (where the superscript number is the molecular mass in kDa) is predominant, but other specialized σ subunits recognize different groups of promoters. For example, $\sigma^N(\sigma^{54})$ binds to promoters that allow transcription of genes involved in assimilation of nitrogen[41,41a] as well as in aromatic catabolism.[42] Synthesis of protein σ^S of *E. coli* is induced by stress such as carbon starvation.[43] At high temperatures (e.g., 40–49°C) σ^{32} is synthesized and permits transcription of genes for "heat shock proteins."[44] Actively growing cells of *Bacillus subtilis* contain at least five different sigma factors. An additional four control gene expression during spore formation.[45]

Promoter sequences. In 1975, Pribnow pointed out[46] that a series of six known promoters had a conserved 7-base sequence beginning six nucleotides upstream from the initiation site for transcription. Although this sequence varies somewhat from one promoter to another, it has been found in hundreds of *E. coli* promoters. This is called the **–10 region**, the **Pribnow sequence,** or Pribnow box (the last in recognition of the fact that people like to draw boxes around these special sequences). A typical 6-base consensus Pribnow sequence is 5'-TATAAT as written for the coding strand, whose sequence corresponds to that of the mRNA. Only three of these bases are highly conserved: 5'-TA_ _ _T. For example, in the *lac* promoter (Fig. 28-2) the sequence is TATGTT. The nucleotides of the –10 consensus sequence are present with the following frequencies (as percentages): T(80)A(95)T(45)A(60)A(50)T(96).[47] The position of the –10 sequence is not exactly the same in all promoters but usually starts 5–9 bp upstream of the start position for transcription. About 16–19 bp upstream from the Pribnow sequence is another conserved sequence, the **–35 region**. A consensus sequence is TTGACA, the TTG sequence to the left being the most highly conserved. Both –10 and –35 regions are needed for efficient promoter activity. Lewin suggested the following "optimal" promoter sequence.[47] The location of these two regions in transcription initiation complex is shown in Fig. 28-4B.

```
        –35                        –10           Start
                                                   ↓
5'---|TTGAC⟨    16-19 bp     ⟩TATAAT⟨  5-9 bp  ⟩AT---3'
```

Naturally occurring promoters usually do not have the exact –10 and –35 consensus sequences, but artificially constructed promoters containing them are highly effective *in vivo*.[26] The fact that most promoters depart from these "ideal" sequences is not surprising because cells need varying amounts of different proteins. Promoter strengths vary over a range of at least 10^4. Much of this variation comes from variations in the specific –10 and –35 sequences, which appear to be specifically recognized by RNA polymerase. This variation includes an upward extension of the –10 region[48] for some promoters. Activator-binding sites are also often present in an **upstream activating region** (**UAR**; see following diagram). This may extend from the –35 sequence through the remainder of the promoter region and as far upstream as –200 to –500 bp. For example, the CAP-binding site in the *lac* operon DNA is centered at –60 (Fig. 28-2). In spite of the variation it was possible to locate 2584 operators (of which only 392 were previously known) and to predict the location of 2405 promoters, when the complete *E. coli* genome sequence became known.[49]

The binding of the various RNAP subunits, repressors, and activators has been studied using mutant promoter sequences,[31,37] antibiotic-binding sites,[50,51] chemical crosslinking reagents,[34,37] and a cysteine-tethered Fe-dependent DNA-cutting reagent.[32,52] Other "footprinting" techniques, e.g., observing cleavage of the DNA by hydroxyl radicals generated by reduction of H_2O_2 by Fe(II) (Fig. 5-50), have also been employed. It was shown that RNA polymerase binds to both the –10 and –35 sequences and also to sequences further upstream. The σ^{70} subunit associates with the DNA, principally the transcribed strand, along a region from about the –25 to the +12 position relative to the transcription start site.[32–32b] The α subunits bind to an UP element from ~–40 to –60 via their C-terminal domains (CTDs). See Fig. 28-4B.[53–54d]

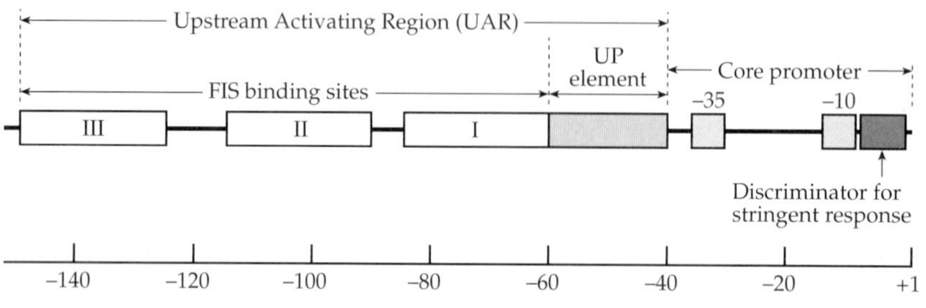

Control of stable RNA synthesis. Whereas most mRNA has a relatively short lifetime, the stable ribosomal RNAs and transfer RNAs have much longer lives. Furthermore, in *E. coli* their transcription is coordinately controlled by seven *rrnB* P1 promoters.[55] The genes for stable RNA have promoters with the usual –10 and –35 sequences, but they contain a complex upstream activating sequence that includes the UP element and three binding sites for protein FIS (**the factor for inversion stimulation**).[56] This name reflects a second function, that of promoting inversion of a DNA segment in the Hin recombinase system (Chapter 27, Section D,3). FIS is a dsDNA-bending protein.[55,57] A dimer of 11.2-kDa subunits, it is an abundant protein. Like HU, IHF, H–NS, and Dps, it coats a significant fraction of the DNA in the *E. coli* chromosome.[58] It binds to the FIS sites using a helix–turn–helix motif. In addition, there is a GC-rich region at positions –7 to –1 with the consensus sequence 5'-GCGCC_C. It has been suggested that this **discriminator** is involved in the **stringent response**, the diminished rate of stable RNA synthesis observed during amino acid starvation.[59,59a]

The very complex stringent response, which involves ribosomes in the synthesis of guanosine 5'-triphosphate-3'-diphosphate (pppGpp) under some conditions,[60] is dealt with further in Chapter 29 (Section C,8). The regulator pppGpp, whose concentration may rise from ~50 μM to ~1 mM within minutes after deprivation of amino acids, may react directly with RNA polymerase at an allosteric site to inhibit transcription.[61–62a] Another possibility is that pppGpp acts on another protein that binds to the discriminator sequence. However, the fact that a phage T7 gene, which is also under stringent control, lacks the discriminator sequence argues against this.[63] pppGpp, whose concentration is usually high under conditions of slow growth, may be a major growth-controlling effector, which acts by inhibiting the replication of the rRNA needed for ribosome formation and protein synthesis when nutrients are scarce. However, some operons are activated by the same effector.[62]

The initiation reaction. A promoter not only locates the site of initiation but also determines the direction of transcription and, therefore, the strand of the DNA duplex that is to serve as the template. The requirement for two specific recognition sequences ensures this directionality. The RNA polymerase may bind randomly to DNA, then move rapidly along the double helix until it locates a strong binding site[64–66] where it binds to the recognition sequences of the promoter through specific interactions in the major groove of the DNA helix (see Fig. 5-3).

A satisfactory mathematical model for initiation of transcription supposes that the polymerase and DNA bind reversibly to form a complex with formation constant K_f. This initial specific polymerase–promoter complex is referred to as a **closed complex** because it is thought that the bases in the DNA chain are all still paired. It is postulated that in a rate-determining step the closed complex is converted into an **open complex,** which is ready to initiate mRNA synthesis (Eq. 28-1).[26,67] In the open complex the hydrogen bonds holding together the base pairs have been broken, and the bases of the template chain are available for pairing with incoming ribonucleotide triphosphates.

$$\begin{array}{l} \text{E(polymerase)} + \text{P(promoter)} \overset{K_f}{\rightleftharpoons} \text{EP}_c \text{ (closed)} \\ \qquad\qquad\qquad\qquad\qquad\qquad\qquad \downarrow k \\ \qquad\qquad\qquad\qquad\qquad\qquad\qquad \text{EP}_o \text{ (open)} \\ \qquad\qquad\qquad\qquad\qquad\qquad\qquad \downarrow \\ \qquad\qquad\qquad\qquad\qquad\qquad\qquad \downarrow \\ \qquad\qquad\qquad\qquad\qquad\qquad \text{Transcription} \end{array} \tag{28-1}$$

It is clear from Eq. 28-1 that the efficiency of initiation depends upon both the affinity K_f and the rate constant k for opening of the double helix. Notice that the Pribnow sequence is AT-rich; therefore, opening of the helix at this point would be easier than in a GC-rich region. Thus, the Pribnow sequence may represent a point of entry of RNA polymerase to form the open complex.[67] Other upstream A•T tracts are often present frequently at about the –43 position in the UP element. They also seem to strengthen promoter activity.[68] The open complex is thought to undergo some kind of isomerization to form an **initial transcribing**

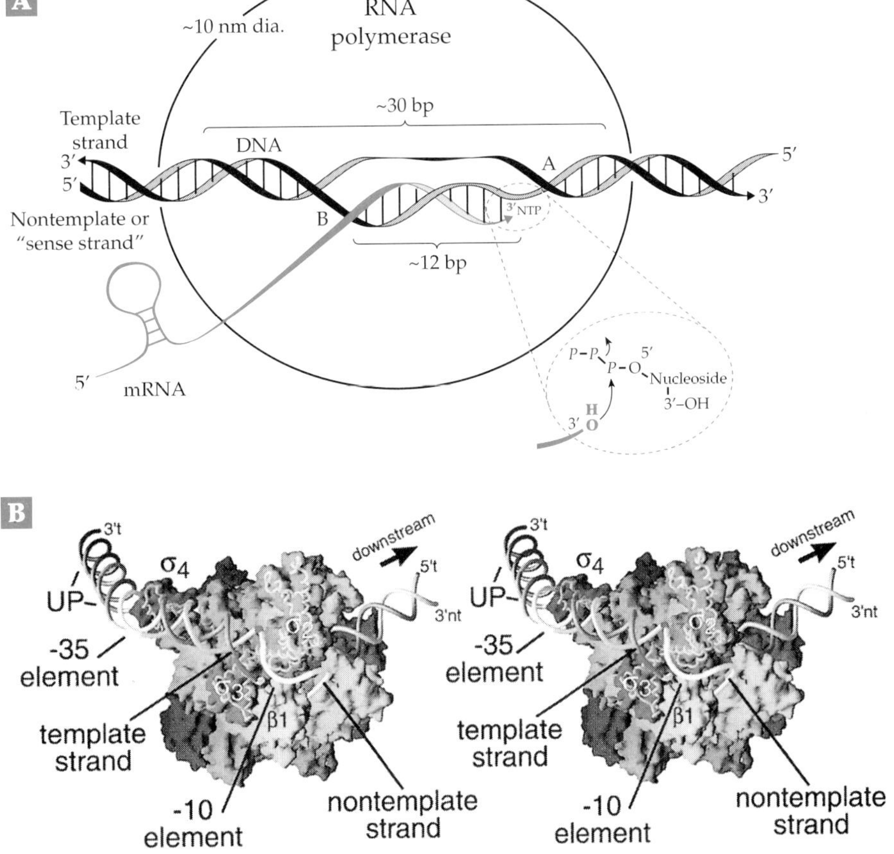

Figure 28-4 (A) Hypothetical structure of a "transcription bubble" formed by an RNA polymerase. Shown is a double-stranded length of DNA with the unwound bubble in the center. This contains a short DNA–RNA hybrid helix formed by the growing mRNA. The DNA double helix is undergoing separation at point A as is the hybrid helix at point B. NTP is the ribonucleotide triphosphate substrate. See Yager and von Hippel.[71] (B) Stereoscopic view of the structure of RNA polymerase from *Thermus aquaticus* in a complex with a promoter DNA. Included are the αI, αII, ω, β, β′, and σ subunits. However, the α C-terminal domains have been omitted. The template (t) strand passes through a tunnel, which is formed by the β and β′ subunits and two of the structural domains of the σ subunit. The nontemplate (nt) strand follows a different path. The position of the –10, –35, and UP elements of the DNA are marked. From Murakami *et al.*[33d] Courtesy of Seth A. Darst.

complex in which the first chemical steps in RNA formation occur.[69,70] These initial steps may involve rearrangements of subunit interactions and untwisting of DNA by torsional movements between subunits.[57]

Initiation of an RNA chain begins by reaction within the transcription bubble of either ATP or GTP with a second ribonucleotide triphosphate (Eq. 28-2) to form a dinucleotide still bearing a triphosphate at the 5' end. Further addition of nucleotide units at the 3' end by the same type

$$\text{ATP(GTP)} + \text{NTP} \rightarrow \text{PP}_i + ppp\text{Pu}p\text{N} \quad (28\text{-}2)$$

of reaction leads to rapid transcription at a rate of ~50 nucleotides s^{-1} at 25°C. This is about one-thirtieth the rate of replication. The action of the RNA polymerase is apparently processive, a single molecule of the enzyme synthesizing the entire mRNA transcript.

3. Elongation of RNA Transcripts

After the newly initiated transcript has grown to 8–9 nucleotides, the sigma factor is lost from the RNA polymerase complex, the complex becomes very stable toward increased salt concentrations, and transcription proceeds processively in a **stably elongating mode**. As it does, the strands of the DNA duplex are pulled apart ahead of the polymerase and close up again behind the polymerase, the polymerase itself moving in a "transcription bubble" (Fig. 28-4). This is thought to lie within the ~10 nm diameter RNA polymerase complex and to encompass ~30 bp of DNA.[71] At the leading edge (point A in Fig. 28-4) a "separator" opens the DNA, which then closes behind the bubble. Behind the polymerase active site (next to substrate NTP in Fig. 28-4) the transcribed RNA forms a short hybrid helix with the DNA, presumably with an A type structure (Chapter 5). A severe topological problem is avoided if the transcribed RNA is separated at point B (Fig. 28-4) as the polymerase moves along the double helix. The driving force for the polymerization lies largely in the hydrolysis of the inorganic PP_i formed in the polymerization (Eq. 17-57). Stabilization of the RNA transcript by formation of loops and other secondary structure may also be a factor.

Although this picture seems clear and simple, many uncertainties remain. Transcription does not proceed evenly but by pauses and spurts. This has suggested the possibility of an "inchworm" type of movement of RNA polymerase.[72–75] However, the observations may also be explained by variations in the sequence. There are both pausing or stalling sites[76] and terminator sequences. The concentrations of the needed ribonucleotide triphosphate precursors will also affect the kinetics. In addition, defects in the DNA will be met. Transcriptionally linked repair (Box 27-A)[76a] may have to be called into play before transcription can continue.[77,78] Mispairing of bases can also occur in the growing RNA chain necessitating a pause during which the mismatch is recognized. Specialized proteins **GreA** and **GreB** participate in an editing step during which the RNAP backtracks for a few nucleotides, while a piece is hydrolyzed off from the 3' end before transcription can continue.[69,79,80]

If the transcription complex moves straight along a DNA double helix, separation of the strands will create positive supercoils (overwound DNA) in front of it and negative supercoils behind the bubble.[81] Experimental data support this prediction.[82] In *E. coli* the transcription of a plasmid generates positively supercoiled plasmid DNA when DNA gyrase (Chapter 27) is inhibited selectively.[83] A similar result was observed in yeast.[84] These and other data suggest that DNA gyrase may act to remove these positive supercoils, and that topoisomerase I may function in removing the negative supercoils generated behind the transcription bubble.

The electron micrograph in Fig. 28-5 shows RNA polymerase complexes apparently moving along a DNA strand with ribosomes assembled on the RNA and

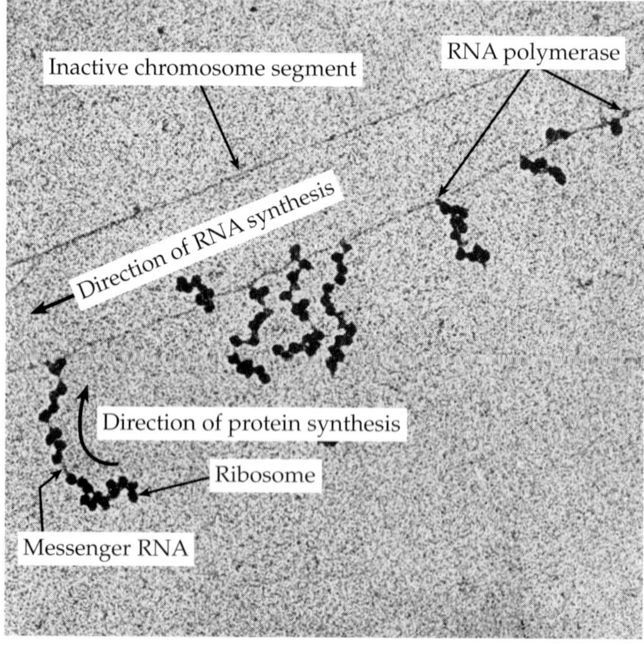

Figure 28-5 Electron micrograph showing transcription from an unidentified operon in *E. coli*. Note the DNA duplexes (horizontal) and the mRNA chains with ribosomes attached. The mRNA chains are shorter at the right side where transcription begins and larger to the left where transcription has proceeded for a longer time. From O. L. Miller, Jr.[85]

presumably synthesizing proteins. Actually, the DNA may be moving through the polymerase complexes.

4. Control of Various Bacterial Operons

With more than 2400 promoters an *E. coli* cell can be expected to utilize a great variety of control mechanisms. The best known of these involve the basic biosynthetic pathways and energy-yielding reactions. Here is a small sample.

Many repressors. Bacteria tend to keep most operons relatively inactive by use of repressors, most of which are proteins. Repressor proteins come in a variety of sizes and three-dimensional structures. Most are oligomers, and all have a DNA-binding motif, often at the N terminus but sometimes at the C terminus or elsewhere. The most studied and perhaps most abundant family[86] have the helix–turn–helix (HTH) motif. Within this family three-dimensional structures have been established for the *lac* repressors, the *trp* repressor (Fig. 5-35), the 66-residue *cro* repressor,[87–89] and a 92-residue DNA-binding "head piece" of the 236-residue lambda (cI) repressor.[90–91] The latter two proteins, both of which occur as symmetric dimers, are involved in preserving the lysogenic state of the λ prophage (Section B,1). The related 71-residue *cro* repressor from phage 434[92] and the head piece from the phage 434 repressor have similar structures.[93,94] All of these proteins have the N-terminal HTH structural motif, which binds to DNA and recognizes the appropriate sequence by interactions in the major groove of the DNA. Experiments in which the recognition helix of the 434 repressor was replaced with the corresponding helix from a phage P22 repressor supported the concept of an N-terminal HTH DNA-binding domain.[95,96] The hybrid repressor bound to the P22 operator rather than to that of phage 434. A similar HTH domain is present near the C terminus in the catabolite-activator protein (Fig. 28-6). The fit of the helix–turn–helix recognition corner varies from one protein to another. The small 53-residue **Arc** repressor of bacteriophage P22 of *Salmonella* has a very different interaction with DNA: the β sheet of an arc dimer fits into the major groove of the operator DNA making specific contacts.[97,98] A similar interaction characterizes the dimeric methionine repressor.[99]

Basic to the functions both of repressors and of activator proteins are allosterically induced conformational changes caused by the binding of inducers or corepressors. The changes begin at the binding sites of these small effector molecules but are transmitted to the DNA-binding heads. In several cases the conformational changes have been observed by X-ray crystallography and are seen to involve a movement of the recognition helices. This is also true for the *lacI* repressor (Fig. 28-3) and for the *TrpR* repressor (Fig. 5-35B) Notice that the tryptophan binds to the aporepressor immediately adjacent to the DNA-binding site, where it may control not only the shape but perhaps also the charge distribution within the recognition motif.

Another important factor in determining the strength of a repressor–operator interaction is the twist of the DNA or any other distortion of its regular helical structure. For example, in the center of the *cro* 434-operator complex the DNA is wound, while at the ends it is underwound.[92] Conformational changes in either the repressor or in the DNA or in both may be needed to provide optimal binding. As is seen in Fig. 28-3 the *lac* repressor causes a distinct bend in the DNA.

A theoretical possibility would be for the DNA in an operator site to be extruded as a cruciform structure like that in Fig. 5-34A. Such structures bind well to certain oligomeric proteins (e.g., see Fig. 27-26B), and they do appear to form in some promoters.[100] However, crystallographic structure determinations have ruled out cruciform structures for many repressors. A change from a linear helical duplex to cruciform structure would require a substantial unwinding of the helix and would mean that negatively superhelical DNA molecules would bind repressor much more tightly than does DNA without superhelical turns. Negative supercoiling does facilitate the *lac* operator–repressor interaction. However, there is only a 50–90° unwinding of the DNA.[101] Some AT-rich palindromes are readily converted to cruciform structures when negative supercoiling is increased. Placement of a promoter sequence within such a structure represses transcription unless the supercoiling is relaxed.[102]

Binding of repressors and activator proteins to DNA control sequences is being studied in many ways. Among them are ^1H and ^{19}F NMR measurements on *lac* and *cro* repressors with specific tyrosine side chains replaced by deuterated tyrosine or 3-fluorotyrosine[103] or with 5-F-uracil replacing specific thymines in an operator sequence.[104] Footprinting shows that λ and *cro* repressors bind to only one side of the double helix, as is depicted in Fig. 5-50. Addition of dimethyl sulfate to growing *E. coli* causes methylation at many sites in the DNA and has been used to obtain footprints *in vivo*.[105] Not only are certain guanosines in a promoter protected from methylation when active transcription is occurring but also guanosines on the opposite side of the double helix become unusually reactive. This suggested that the DNA helix is bent in the transcriptional initiation complex and that the resulting distortion makes the bases on the outside of the bend more reactive.[105] There is also evidence that the repressor binds to the opposite face of the DNA that binds RNA polymerase in the –35 region of the promoter.[106]

Repressors may have similar recognition domains but may vary greatly both in size and in the functioning of their other domains, which may react both with small allosteric effectors and with other proteins. The repressor **BirA** of the *E. coli* biotin synthesis operon is an enzyme. The 321-residue protein activates biotin to form biotinyl 5'-adenylate and transfers the biotinyl group to proteins such as acetyl-CoA carboxylase[107–109a] and also represses transcription.

Inducible operons. The operon model as presented in Fig. 28-1 describes the negatively controlled inducible *lac* operon. There are many other examples of this type of control. Among them are control of the utilization of *N*-acetylglucosamine and xylose.[110] Ten genes for catabolism of inositol are encoded in two negatively controlled operons in *Bacillus subtilis*.[111] Similar controls are used by bacteria to protect against the antibiotic tetracycline. The **TetR** repressor controls a membrane associated protein TetA, which acts as an energy-dependent pump to remove tetracycline from the bacterial cell. Synthesis of TetA is normally repressed, but tetracycline, if present, acts as an inducer to cause TetR to dissociate from its operator site, allowing transcription of the transporter gene.[112–113a] The gene order, gene content, and regulatory mechanisms in an operon are often poorly conserved among related species of bacteria. However, functional and regulatory relationships may be maintained.[113b]

Feedback repression. A simple modification of the operon model accounts for **feedback repression** by end products of biosynthetic sequences (Fig. 11-1). The product, e.g., an amino acid, vitamin, purine, or pyrimidine base, acts as a corepressor binding to the aporepressor and causing an allosteric modification that inhibits transcription. It is not the aporepressor but the effector–repressor complex that binds to the operon and blocks transcription. An example is the tryptophan–repressor complex, whose 3-dimensional structure is shown in Fig. 5-35. If free tryptophan accumulates within a bacterial cell, it binds to the allosteric site in the *TrpR* repressor, inducing a conformational change that permits the tryptophan–repressor complex to bind to at least three 21-bp operator sites. Binding of the repressor–corepressor complex at the *aroH* operator locus represses the genes for the first three steps in the aromatic biosynthetic pathway (Fig. 25-1, step *a*). Binding at the *trpEDBCA* operator represses the genes for conversion of chorismate to tryptophan (Fig. 25-2), and binding at the *trpR* locus lowers the rate of synthesis of the trp repressor, providing a counteracting effect that may be important in controlling the growth rate. The *trp* operon is also well known for another type of control called attenuation (see Fig. 28-9 and associated text).

Feedback repression controls methionine synthesis in *E. coli*. The corepressor *S*-adenosylmethionine binds to an aporepressor that recognizes the following palindromic sequence in the DNA. This sequence occurs,

```
 ————————→
A G A C G T C T
       •
T C T G C A G A
     ←————————
```

with minor variation, in front of four different operons encoding enzymes of methionine biosynthesis.[99,114] Together they form the methionine **regulon**. Other *E. coli* operons that are negatively regulated by operator–repressor interactions include those involved in biosynthesis of phenylalanine, tyrosine, arginine, threonine, and isoleucine. The *tyr* repressor modulates gene expression in at least eight operons, largely by repression.[115] The *ile* repressor acts on both the *ile* and *thr* operons.[116] The purine repressor, 341-residue PurR, belongs to the LacI family of repressors. Its C-terminal domain also has some sequence similarity to that of periplasmic sugar-binding proteins.[117] PurR represses several steps in the biosynthesis of IMP and in its conversion to GMP and AMP (Fig. 25-15).[118,119] In *E. coli* twelve biosynthetic genes, organized as nine transcriptional units, provide for the synthesis of arginine in eight enzymatic steps (Fig. 24-10).[120,121] As with the *trp* operons, synthesis of the repressor ArgR is autoregulated. The hexameric repressor has an N-terminal winged HTH DNA-binding domain.[121–123]

As is pointed out in Chapter 16, the acquisition of iron and control of its concentration is of crucial importance to bacteria. In *E. coli* the Fe^{2+}-binding protein **Fur** (ferric uptake regulator) represses promoters controlling siderophore biosynthesis as well as other responses. It is a **global regulator** that controls ~40 transcriptional units.[124] Similar proteins repress synthesis of the diphtheria toxin by *Corynebacterium diphtheriae*,[125,126] uptake of iron in these bacteria and in *Mycobacterium tuberculosis*,[127] and uptake of molybdate.[127a]

Positive control by activator proteins. Cyclic AMP in bacterial cells mediates the phenomenon of **catabolite repression**. This is the inhibition of the transcription of genes for enzymes needed in catabolism of lactose and other energy-yielding substrates, when the more efficient energy source glucose is present. Glucose causes a decrease in the concentration of cAMP by a complex mechanism[59,128] that may also cause a decrease in the concentration of inducer. When the glucose concentration decreases, the concentration of cAMP rises and stimulates the initiation of transcription in many operons. This is accomplished through the mediation of the 209-residue catabolite gene activator protein, CAP (also known as cyclic AMP receptor protein, CRP). This protein is a "global" regulator of gene expression that activates transcription at over 100 promoters in *E. coli*.[129] The CAP–cAMP complex binds to the *lac* promoter adjacent to the RNA polymerase site at the palindromic sequence in the DNA as is indicated in Fig. 28-2. The CAP molecule is a 45-kDa dimer of identical subunits, which resembles the repressors in having the HTH reading head that binds to the DNA.[130–132] However, the HTH motif is at the C terminus. The cAMP binds to two sites in each monomer (Fig. 28-6). Tightly bound cAMP molecules in an *anti* conformation bind in the center of the large N-terminal domains in the major allosteric sites. At higher cAMP concentration the second sites in the C-terminal DNA-binding domains are occupied.[131]

The CAP binds to DNA with the consensus sequence 5′-AAATGTGATCT/5′-AGATCACATTT, which may be located at variable distances from the promoter.[133] How does binding of the CAP–cAMP complex increase the rate of initiation of mRNA transcription? The anwer evidently lies in direct interaction between CAP and the N-terminal domain of the RNAP α subunit.[54d,129] Binding of CAP induces a 90° bend in the DNA, which may facilitate the protein–protein interaction and may lead to looping.[130,134]

The galactose (*gal*) operon of *E. coli* is negatively controlled and inducible by D-galactose or D-fucose, which bind to the *gal* repressor. There are two overlapping promoter sites, one of which is stimulated

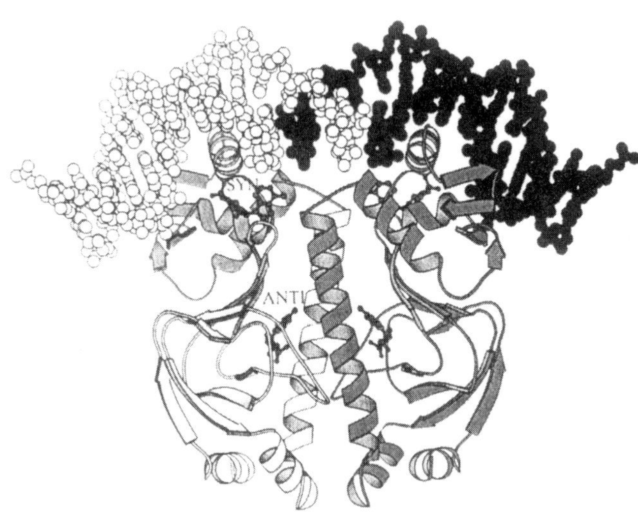

Figure 28-6 MolScript ribbon drawing of the CAP dimer bound to DNA with two molecules of the coactivator cAMP bound per monomer. A *syn*-cAMP molecule is bound to the HTH domain and a loop from the N-terminal domain, while the second *anti*-cAMP is bound more tightly in the center of the larger N-terminal domain. The DNA sequence for each half site is 5'-ATGTCACATTAATTGCGTTGCGC-3'. From Passner and Steitz.[131] Courtesy of Thomas A. Steitz.

by adjacent binding of a CAP molecule.[135,136] A surprise came from the discovery that the operator was *upstream* from the promoter, that is, it comes before both the promoter and the structural genes to be transcribed. Later, a second operator sequence was found 90 bp away from the first and within the first structural gene. This suggested that the dimeric gal repressor binds the two operators to form a loop that blocks transcription.

Another example of positive regulation by CAP is provided by the seven proteins required for uptake of maltose and its catabolism by *E. coli*. These are encoded in two operons that are controlled as a single regulon. An apo-activator protein becomes an activator when it binds maltose.[137,138]

An *E. coli* protein known as **FRN** (for fumarate nitrate reduction) is a global transcription regulator homologous to CAP. It is active only under anaerobic conditions in which it controls more than 100 genes.[139–141] FRN contains an $[Fe_4S_4]^{2+}$ cluster, which is required for dimerization and binding to DNA. Exposure to O_2 converts the cluster into an $[Fe_2S_2]^{2+}$ cluster with loss of activator activity. The photosynthetic *Rhodospirillum rubrum* is able to adapt to growth on carbon monoxide as a carbon source. A CAP-like transcriptional activator **CooA** contains heme. It acts as a sensor for CO, which activates transcription, as does cyclic AMP with CAP.[142,142a] Sequence homologies suggest that several other bacterial activator proteins also have the HTH DNA-binding motif near their C termini.

There are other types of transcriptional activators in bacteria. One is transcription factor 1 (TF1) encoded by a *Bacillus subtilis* phage. It is a member of the protein HU family (Chapter 27). However, unlike the nonspecific HU it binds to some sites specifically and activates transcription.[143] The *E. coli* Ada protein is the acceptor protein in removal of methyl groups from DNA (Chapter 27). The same protein is an inducer of transcription of DNA repair enzymes in the large ada regulon. Methylation of Cys 69 of the Ada protein itself converts it into a gene activator.[144]

Control by looping. The arabinose utilization operon of *E. coli*, *araBAD*, encodes proteins needed for uptake of arabinose and conversion to D-xylulose 5-*P*. The repressor AraC in the absence of arabinose binds at operator 1 (O_1) to prevent further synthesis of repressor (autorepression) and also at the *araI* region to repress transcription of operon *araBAD*. The operator 2 (O_2) site, which is 211 bp upstream from *ara*I, is also needed for this repression.[145–147b] A loop is apparently formed by repressor binding (Fig. 28-7). Binding of arabinose to the repressor converts it into an activator, which stimulates initiation of transcription at the *BAD* promoter. Further stimulation is provided by the CAP-cAMP complex, which binds next at *araI*.

Looping is a recognized control mechanism for a number of other operons as well.[148]

Bacterial enhancers. Positive regulatory DNA sequences that are distant from the genes controlled are often called enhancers.[149,150] Their function is usually independent of position over a range of hundreds or more base pairs either upstream or downstream of the transcription initiation site. Quite common in eukaryotes (Section C,4), enhancers are less often found in bacteria. However, the binding sites for the **nitrogen regulatory protein C** (**NtrC** or Nr_I) of *E. coli* has the characteristics of an enhancer.[151,152] It functions with the rather complex glutamine synthetase (*glnALG*) operon in a major control point for nitrogen metabolism. The enzymology is illustrated in Fig. 24-7. When the supply of nitrogen from NH_3 is low the NtrC protein, a product of gene *glnG*, binds to the enhancer, which is located over 100 bp upstream and is thought to contact the σ^N (σ^{54}) subunit of the RNAP by formation of a loop.[153,153a] The process has been visualized by scanning force microscopy.[152] Another interesting aspect of this control system is the activation of Ntrc by phosphorylation of a specific aspartate side chain (Asp 54). The NtrC-*P* form is the active enhancer-binding protein. NtrC is a member of the family of two-component sensor-response regulator pairs, which frequently control bacterial metabolism and behavior (Chapter 11, Section C, 2; Fig. 19-5). The sensor protein is **NtrB** (NR_{II}), which is an autophosphorylating histidine kinase similar to the CheY protein of bacterial chemotaxis (Fig. 19-5).

NtrC-*P* dimerizes and binds to the enhancer sequence, where it appears to catalyze an ATP-dependent isomerization of the closed to open forms of the transcription initiation complex (Eq. 28-1).[153,154] The isomerization may depend upon looping.[152] Other operons that utilize the σ^N subunit of RNAP often also have upstream or downstream enhancers.[155,156]

Other two-component control systems. More than 60 different sensor–response regulator pairs have been discovered in bacteria. Many are associated with nutrition. For example, more than 30 genes of the phosphate (*pho*) regular are controlled by the sensor kinase **phoR**, which detects low phosphate ion concentrations and phosphorylates the response regulator **phoB**.[157–159] Protein **ArB** senses changes in environmental O_2 levels, and response regulator **ArcA** regulates ~30 operons in response.[159a] One effect is to activate ~30 genes needed for the conjugative transfer of DNA (Fig. 26-3).[160] Transcription of rhizobial *nif* and *fix* genes (Fig. 24-4) is controlled by the O_2 sensor **FixL** and its response regulator **FixJ**.[161] FixL is a heme protein whose kinase is active only when the heme is deoxygenated. The *E. coli* proteins discussed on p. 1075 mediate transcriptional responses to accumulation of superoxide anions or hydrogen peroxide.[161a]

In the absence of O_2 the *E. coli* FNR protein induces proteins of the anaerobic respiration pathways. Nitrate also has its own two-component system that senses nitrate availability and activates transcription of enzymes catalyzing nitrate respiration.[162] An expanded two-component system induces sporulation in *Bacillus subtilis* in response to poor growth conditions.[163,164] The crystal structure of one of two response regulators (**Spo0F**) has a structure closely related to that of CheY and the nitrate response regulator NarL.

Antisense RNA. Another mechanism of control of either transcription or of plasmid replication involves small molecules of RNA that are transcribed from the opposite strand than the template strand used for mRNA synthesis.[1,1a,165–166b] These **antisense** RNA molecules have at least some part of their sequence complementary to that of the mRNA and to the corresponding sequence in DNA. A well-studied example is control of the copy number of the colicin E1 and other plasmids of *E. coli*.[167–169] Two transcripts, RNAI and RNAII, are initiated upstream from the origin of replication (Fig. 28-8). RNA II is a 555-nucleotide primer of replication. It is synthesized as a longer transcript that is cut by RNase H at *ori*. This

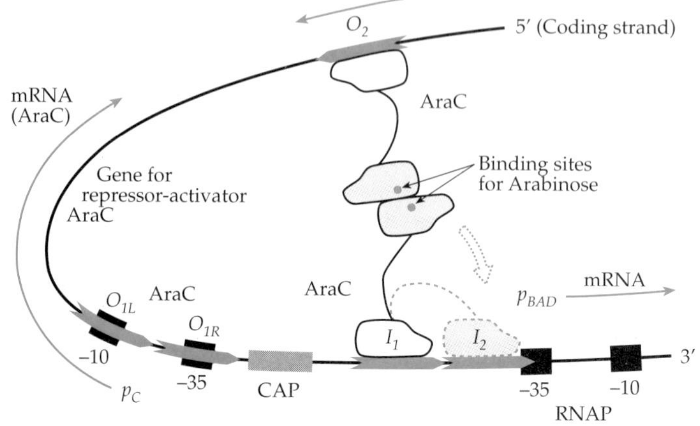

Figure 28-7 Scheme showing regulatory region of the *araCBAD* operon of *E. coli*. In the absence of arabinose the protein AraC acts as a repressor, which binds to upstream region I of the promoter and also to operator O_2, forming a loop of ~210 bp of DNA. Binding of arabinose to the N-terminal domains of the dimeric repressor allows the dimer to dissociate and the N-terminal domain of one subunit to bind to region I_2 of the promoter for genes *B,A,D* (dashed outline) activating their transcription. The AraC protein also binds to promoter P_c, repressing its own synthesis. After Zhang *et al.*[146]

cleavage is inhibited if the 108 nucleotide RNA I forms a duplex with the 5' end of the RNA II, which has a complementary base sequence. The process is more complex than this because both RNA II and RNA I have complex secondary structures and are brought together with the help of the small protein product of gene *gro*, which permits them to recognize each other prior to duplex formation.

The major outer membrane porins of *E. coli* (Fig. 8-20) are encoded by genes *ompC* and *ompF*. A small 174-nucleotide RNA called **mRNA-interfering complementary RNA** (mic RNA), whose gene is upstream of the *ompC* promoter, is transcribed in the opposite direction from that of the porin gene.[167] It is homologous with the 5' end of *ompF* mRNA, and its function is evidently to inhibit translation of the *ompF* mRNA. Since *ompC* and the *micF*RNA gene are apparently regulated coordinately, synthesis of the *ompF* product is inhibited if the *omp*C product is being synthesized, as happens when the bacteria are growing in a medium of high osmotic strength.[166b,170,171]

Many other examples of regulation by antisense RNA are being discovered. Small noncoding RNA molecules (**ncRNAs**) often serve as templates or **guides**[1,171a] in processes ranging from synthesis of telomere ends (Fig. 27-21) to editing,[172] modifying,[173] and splicing mRNA[174] (Section D). Both in bacteria[174a] and in eukaryotes (p. 1640) dsRNAs are often formed and subsequently cleaved to single-stranded antisense RNAs that act as guides to initiate hydrolytic destruction of defective, toxic, or unwanted mRNAs.[171a] The recognition of these natural regulatory mechanisms has led to keen interest in artifical regulation by antisense RNA. Synthetic antisense RNA injected into cells will inhibit expression of selected genes. There is the possibility of effective therapy against viruses, cancer, and inflammation if suitable antisense RNA could be generated within eukaryotic cells or introduced as drugs.[175–176b] Such drugs, which are typically ~15 nucleotides in length, are most satisfactory if they are stable, enter cells, and interact specifically with complementary sequences of DNA or RNA.[177] Stability can be improved by use of linkages other than the natural phosphodiester.[178–181] Phosphorothioates, in which sulfur replaces one nonbridging oxygen atom on phosphorus, is a favorite. Synthetic antisense oligonucleotide mimics may cause adverse reactions with proteins within cells.[178,182] Nevertheless, future successes seem likely.[182a] See also RNA interference (p. 1640).

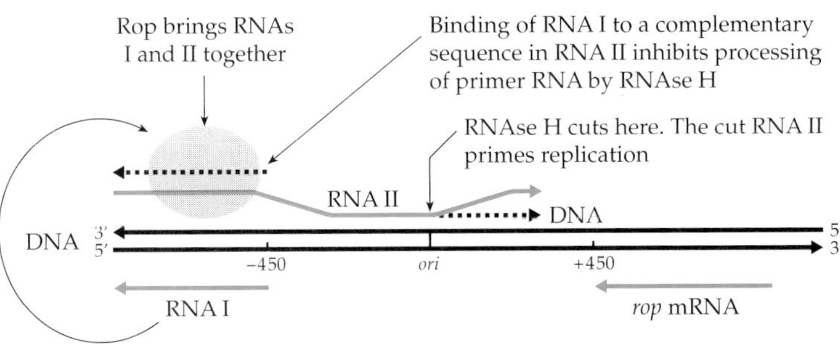

Figure 28-8 Simplified scheme for control of replication of the ColE1 type plasmid by antisense RNA. The primer for DNA synthesis is RNA II whose initial transcript extends past the replication *ori*. It is cut by RNase H at *ori* and then primes replication of the upper strand as shown in the figure. The antisense RNA is RNA I. It binds to protein Rop whose gene location is also indicated in the figure. Rop assists RNA I and RNA II in undergoing a complementary interaction. However, both RNAs apparently maintain a folded tertiary structure, and only some segments interact. The interaction with the Rop protein evidently in some way prevents initiation of replication until the Rop concentration falls because of replication of the host cell.[167,168]

5. Termination of Transcription in Bacteria

Encoded in DNA are not only the initiation signals for transcription but also termination signals or **terminators**.[183] Some of these are sensed by the bacterial RNA polymerase itself while the "reading" of others requires auxiliary proteins. Terminators can be either constitutive or regulatable. The simplest terminators result from GC-rich regions of dyad symmetry in the DNA. The RNA transcript is able to form a stable hairpin loop, possibly within the transcription bubble. If such a loop is followed closely by a series of uracils, the RNA and the polymerase will dissociate from the DNA template terminating transcription. The low stability of AU base pairs may facilitate dissociation, but RNA polymerase may also recognize the terminator loop. Sometimes a terminator will have a series of adenines preceding the loop. This is often a bidirectional terminator; the transcript from the other strand of DNA will have a loop followed by a series of U's. There are many more complex termination mechanisms.

Attenuation. A major mechanism of feedback repression, known as attenuation, depends not upon a repressor protein but upon control of premature termination. It was first worked out in detail by Yanofsky *et al.* for the *trp* operon of *E. coli* and related bacteria.[184–186] Accumulation of tryptophan in the cell represses the *trp* biosynthetic operon by the action of accumulating tryptophanyl-tRNA^{Trp}, which specifically induces termination in the *trp* operon. Other specific "charged" aminoacyl-tRNA molecules induce termination at other amino acid synthesis operons. The first structural gene in the *trp* operon, *trp*E, is preceded at the 5' end by a 162-bp **leader sequence,** which is transcribed into mRNA. Within this RNA are two adjacent hairpin loops. The second of these loops (labeled **3:4** in Fig. 28-9) has a GC-rich stem and is followed by eight consecutive U's. It is a typical efficient terminator. An RNA polymerase, having just passed this sequence, will interact with the looped RNA formed behind it and will dissociate from the DNA to terminate transcription. However, if the terminator loop is prevented from forming, transcription will continue, and the structural genes of the operon will be expressed.

The *trp* operon contains a short gene for a **leader peptide** preceding the terminator. Its RNA transcript is shaded, and its initiation codon AUG and termination codon AAU are boxed in Fig. 28-9. In bacteria translation of mRNA begins while transcription is still in progress (Fig. 28-5). The 5' end of the mRNA may enter a ribosome before the RNA polymerase reaches the terminator loop, after which the leader peptide will be synthesized. The ninth and tenth codons of the leader peptide gene are of special importance. They lie at the beginning of the **1:2** or "protector" loop (Fig. 28-9) and code for tryptophan. If the level of tryptophan is high, tryptophanyl-tRNA^{Trp} will be formed and the mRNA will move rapidly through the ribosome, these tryptophan codons will pass through, and tryptophan will be incorporated into the peptide. The 1:2 loop will be opened, but the terminator loop will remain intact. The result will be termination of transcription. However, if the tryptophan concentration is low, there will be a shortage of charged tRNA^{Trp}, and peptide synthesis will be "stalled" with these tryptophan codons in the active sites of the ribosome. This will allow time for the attentuator region to assume the alternative secondary structure shown in Fig. 28-9B. Here the **1** limb of the protector loop is stalled in the ribosome allowing the **2** limb to form the altenative **2:3** loop. The terminator has been destroyed, and transcription continues through the rest of the operon. Thus a low level of tryptophan (and of tryptophanyl-tRNA) favors transcription of the *trp* synthetic operon.

Cells of *Bacillus subtilis* also synthesize a *trp* operon transcript that can form either an antiterminator or a terminator loop (Fig. 28-9C).[187,187a–c] Tryptophan, when present in a high enough concentration, binds to a *trp* RNA-binding attenuation protein (*TRAP*). This is an 11-subunit protein, which has 11 tryptophan-binding pockets and also 11 binding sites for GAG or UAG RNA triplets. When tryptophan accumulates within the cell it binds to TRAP, which then wraps ~53 residues of RNA transcript containing 11 GAG or UAG triplets around its perimeter (Fig. 28-9D). This prevents formation of the antiterminator loop but allows the terminator loop to form. At low tryptophan concentrations the antiterminator loop is formed, preventing formation of the terminator loop.[188]

Attenuation is also an important mechanism of control of transcription of biosynthetic operons for histidine, phenylalanine, leucine, isoleucine, and threonine.[189] Like the *trp* attenuator region, attenuators for these operons contain codons for the amino acid whose synthesis is being regulated: seven Phe codons in the *phe* attenuator, seven His codons in the *his* attenuator, four Leu codons in the *leu* attenuator. The *thr* operon, which is sensitive to both threonine and isoleucine, has eight Thr and four Ile codons, while the *ilv* attenuator has four Leu, five Ile, and six Val codons permitting feedback repressor by three kinds of charged tRNA.[190] The pyrimidine synthesis operon *pyr* has three attenuator sequences, one right after the promoter and two others, one just before each of the two genes in the operon. This permits partially independent control of the two genes.[191]

Rho and other termination factors. Termination proteins can also react with specific regions of DNA or of an RNA transcript to terminate transcription.[183] The best known termination factor is the rho protein; a hexamer of 45-kDa subunits. It interacts with transcripts at specific termination sequences, which are often C-rich, and in a process accompanied by hydrolysis of ATP causes release of both RNA and the polymerase from the DNA.[192,193] Additional *E. coli* proteins, products of genes ***nus A*** and *nus G*, cooperate with the rho factor at some termination sequences.[194–196c] The rho hexamer is a helicase that moves along the RNA transcript in the 5' → 3' direction driven by ATP hydrolysis. If it locates an appropriate termination signal, it may utilize its helicase activity to uncoil the DNA–RNA hybrid segment within the transcription bubble (Fig. 28-4).[197–198b]

Cells also contain **antitermination proteins,** which prevent termination of transcription of rRNA or tRNA genes at the many loops of secondary structure that are possible with these transcripts.[59,199–200b] These antitermination factors are also important in regulating transcription during the lytic phase of growth of phage λ (see Section B,1). Also important are rates of hydrolytic degradation of mRNA molecules.[201]

6. Effects of Antibiotics

The antibiotic **rifamycin** (Box 28-A) appears to interfere with initiation by competing for the binding of the initial purine nucleoside 5'-triphosphate. The same bacterial RNAP that synthesizes mRNA also transcribes both rRNA and the tRNAs. Thus, the synthesis of all forms of RNA is inhibited by rifamycin. When a population of bacteria is subjected to this antibiotic, a few individuals survive. These rifamycin-resistant mutants are no longer sensitive to the antibiotic. Among them are some mutants that produce an

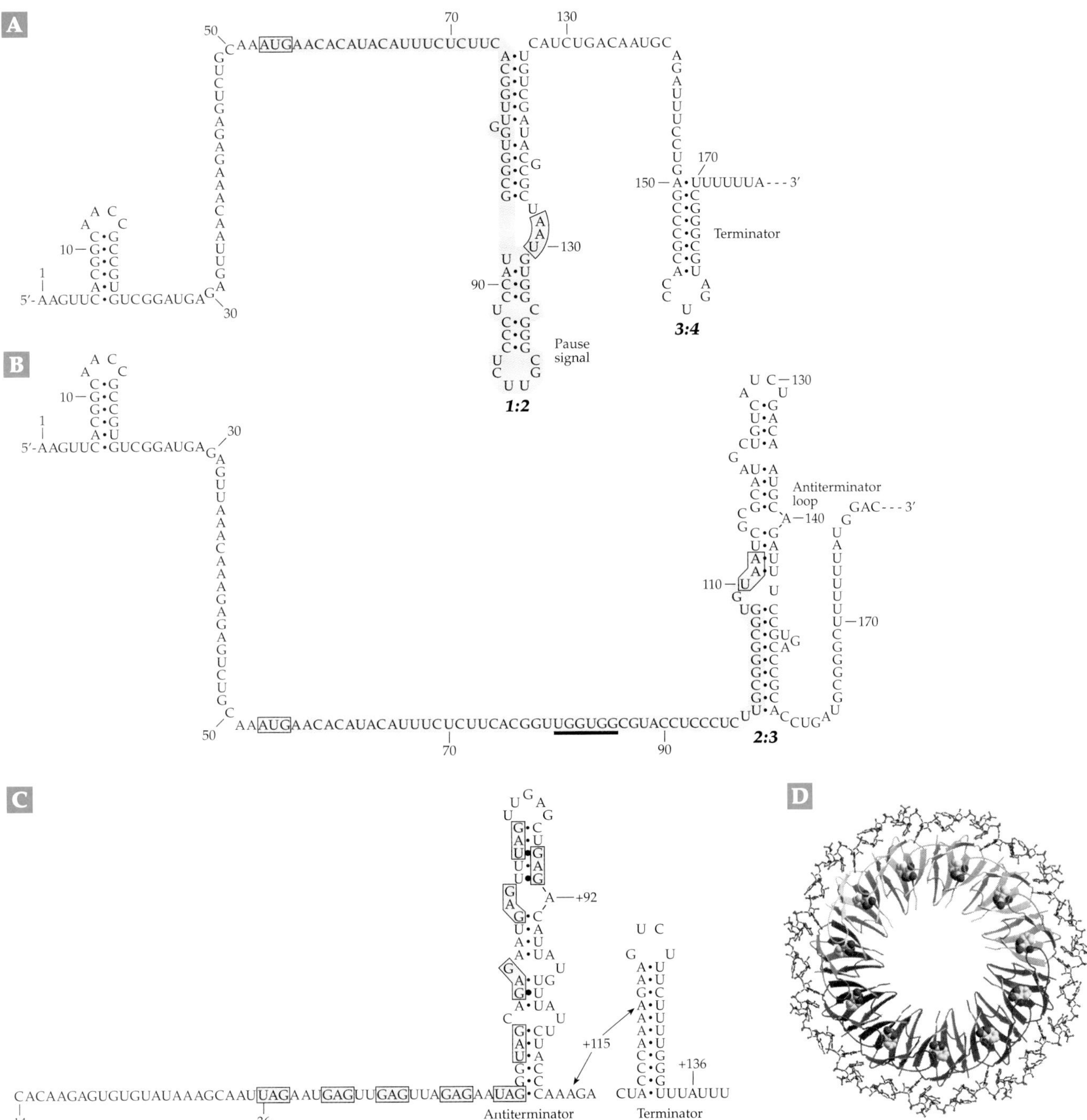

Figure 28-9 (A,B) Alternative leader RNA structures that mediate control of attenuation in the *trp* operon of *Serratia marcescens*. The leader peptide initiation and termination codons are *boxed*. Tryptophan codons are indicated by *underlining* in *B*. Formation of the 5' hairpin structure is predicted by computer analysis but is not implicated in control by attenuation. A structure **1:2** is thought to serve as a transcription pause signal. Structure **3:4** is predicted to form when there is an adequate supply of charged $tRNA^{Trp}$ and is thought to function as the transcription termination signal recognized by RNA polymerase. The **2:3** structure, or **antiterminator loop,** is predicted to form when charged $tRNA^{Trp}$ is unavailable. Its formation presumably precludes formation of structure **3:4**, thereby allowing RNA polymerase to continue transcription into the structural genes of the operon. From Kuroda and Yanofsky.[186] (C) Antiterminator and terminator loops, one of which may form from the leader sequence of the *Bacillus subtilis trp* operon mRNA. Numbering refers to the start point of transcription. The triplet repeats involved in attenuation are shaded. From Baumann *et al.*[187] (D) Structure of the 11-subunit tryptophan RNA-binding attenuation protein (TRAP) as a ribbon diagram with 11 molecules of L-tryptophan shown as van der Waals spheres. The apparently circular RNA structure reflects the fact that the gap between the beginning and end of the bound RNA segment is averaged (randomized) over eleven binding sites in the crystal structure. The 53-residue RNA containing 11 triplet repeats of GAG or UAG is bound around the perimeter and is shown as a ball-and stick model. From Antson *et al.*[188]

BOX 28-A THE ANTIBIOTICS RIFAMYCIN, RIFAMPICIN, AND ACTINOMYCIN D

Rifamycin, produced by *Streptomyces mediterranei*, is of medical value because it affects **acid-fast** as well as gram-positive bacteria. The semisynthetic rifampicin has been especially useful in the treatment of tuberculosis. The ether linkage at the bottom of the ring at the right in the structural formula is cleaved, and the resulting hydroquinone is oxidized to a quinone within the bacteria.[a] At a concentration of 2×10^{-8} M rifampicin inhibits bacterial RNA polymerase 50%. It does not prevent the binding of polymerase to DNA but inhibits initiation of transcription. Mutants of *E. coli* resistant to rifampicin produce RNA polymerase whose β subunit has been altered, sometimes with a change in electrophoretic mobility. The related antibiotic **streptolydigin** also binds to the subunit of RNA polymerase and blocks elongation, resistant mutants mapping close to *rif* mutants.

The actinomycins, which are also produced by *Streptomyces*, not only kill bacteria but also are among the most potent antitumor agents known.[b] However, because of their extreme toxicity they are of little medicinal value. Actinomycin D, which is a specific inhibitor of RNA polymerase, contains a planar phenoxazone chromophore bearing two carboxyl groups, each one linked to an identical cyclic peptide made up of L-threonine, D-valine, L-proline, sarcosine (*N*-methylglycine), and L-methylvaline.

An ester linkage joins the methylvaline residue of the peptide to the side-chain hydroxyl of threonine. Two *cis* peptide linkages are present. Ignoring the obvious asymmetry of the phenoxazine ring, actinomycin possesses approximate twofold symmetry.

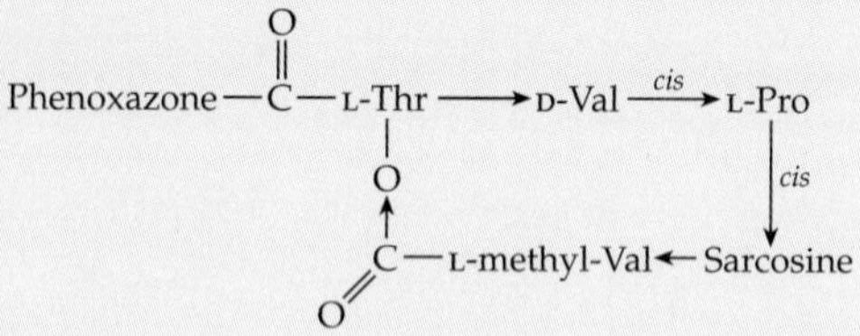

Phenoxazone—C(=O)—L-Thr ⟶ D-Val —cis⟶ L-Pro
L-Thr—O ← C(=O)—L-methyl-Val ← Sarcosine (L-Pro —cis⟶ Sarcosine)

The antibiotic binds tightly to double-stranded DNA in regions containing guanine. A 2:1 deoxyguanosine–actinomycin complex has been crystallized, and the structure has been determined by X-ray diffraction.[c–e] The phenoxazine ring lies at the center of the complex, one peptide loop extending above it. The twofold symmetry is present in the dideoxyguanosine complex as well as in actinomycin itself. The phenoxazine ring lies between the two flat guanosine rings in van der Waals contact. The two amino groups of the guanine rings form strong hydrogen bonds with the carbonyl groups of the threonine residues. There are also weaker nonlinear hydrogen bonds from the N-3 atoms of the guanines to the NH groups of the same threonines.

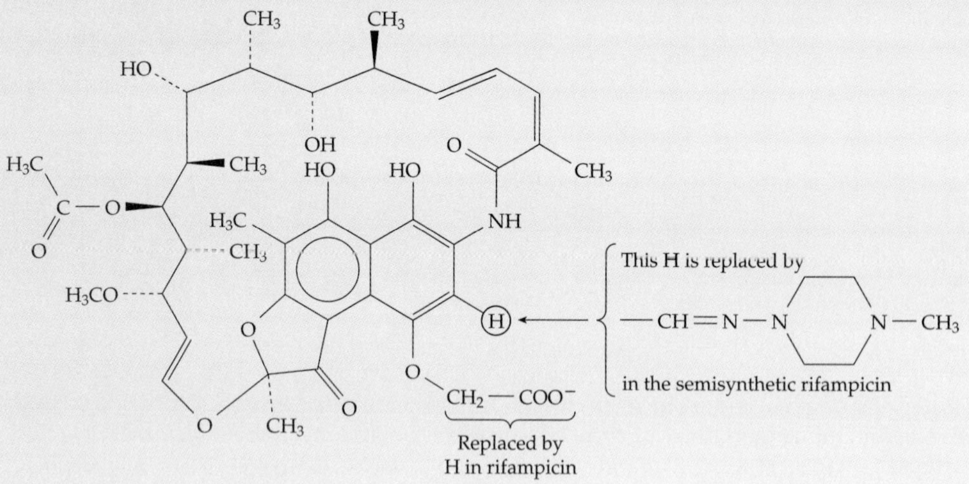

CH_3 CH_3 HO H_3C CH_3 OH HO HO O CH_3 C—O O H_3C NH CH_3 H_3CO O (H) O O O CH_3 O $CH_2—COO^-$

Replaced by H in rifampicin

This H is replaced by —CH=N—N(piperazine)N—CH_3 in the semisynthetic rifampicin

RNAP with an altered β subunit. Since the mutant polymerases do not bind rifamycin, it was concluded that rifamycin binds to the β subunit and that the rifamycin-resistance gene *rpo*B or *rif* (which maps at 89 min) is the gene for the β subunits of RNA polymerase.

Streptolydigin inhibits both initiation and elongation. **Actinomycin D**, by binding to DNA, inhibits DNA polymerases as well as RNA polymerases, the latter at a concentration of only 10^{-6} M (Box 28-A). The eukaryotic RNA polymerases are not inhibited by rifamycin, but RNA polymerases II and III are completely inhibited by the mushroom poison **α-amanitin** (see Box 28-B). Inhibitors of DNA gyrase (Chapter 27) also interfere with transcription as do chain terminators such as cordycepin (3'-deoxyadenosine) and related nucleosides.

BOX 28-A (continued)

A symmetric pair of hydrogen bonds join the two carbonyl and NH groups of the D-valine residues in the peptide loops.

Model building studies show that a similar complex can be formed with double-stranded DNA.[f] While the amino groups of the guanine rings (see drawing) are hydrogen-bonded to the actinomycin, the other hydrogen atoms of the same amino groups as well as the N-1 hydrogen atoms and the carbonyl groups of the guanine ring are available for hydrogen bonding to form a GC base pair. Thus, the structure above can be modified readily into part of the double-stranded DNA molecule in which the phenoxazone ring of actinomycin is intercalated between two CG pairs. To do this the normal DNA structure has to be unwound by 18° at the point of insertion of the extra ring. Binding also occurs at other sites.[f] Sobell suggested that actinomycin binds to a premelted region of the DNA helix within the transcription bubble and immobilizes it. This interferes with the elongation of growing RNA chains.[g]

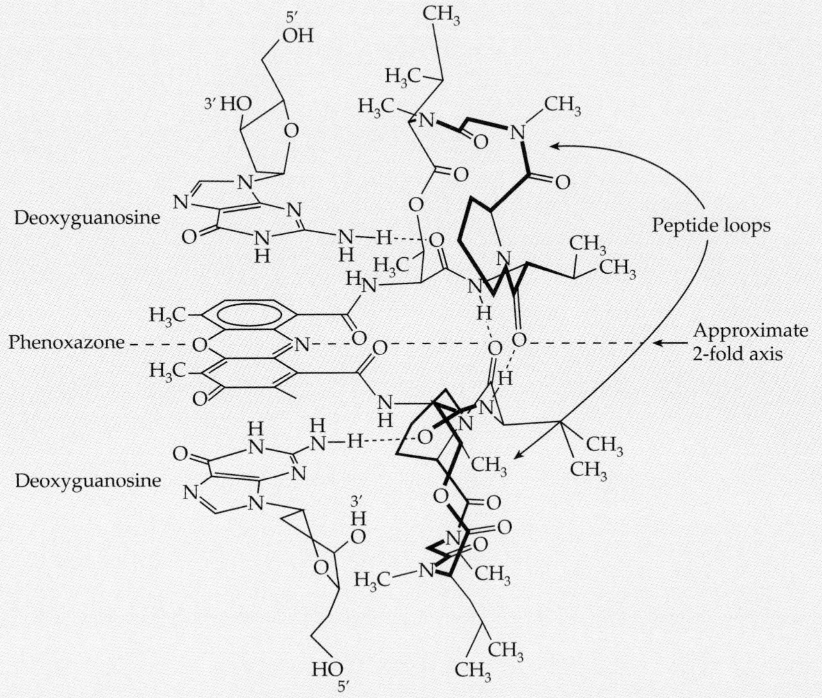

[a] Goldberg, I. H., and Friedman, P. A. (1971) *Ann. Rev. Biochem.* **40**, 775–810

[b] Perlman, D. (1970) in *Medicinal Chemistry*, 3rd ed. (Burger, A., ed), pp. 309–316, Wiley (Interscience), New York (Part 1)

[c] Sobell, H. M. (1973) *Prog. Nucleic Acid Res. Mol. Biol.* **13**, 153–190

[d] Sobell, H. M., Jain, S. C., Sakore, T. D., and Ponticello, G. (1971) *Cold Spring Harbor Symp. Quant. Biol.* **36**, 263–270

[e] Sobell, H. M. (1974) *Sci. Am.* **231**(Aug), 82–91

[f] Robinson, H., Gao, Y.-G., Yang, X.-l, Sanishvili, R., Joachimiak, A., and Wang, A. H.-J. (2001) *Biochemistry* **40**, 5587–5592

[g] Sobell, H. M. (1985) *Proc. Natl. Acad. Sci. U.S.A.* **82**, 5328–5331

7. Processing of Bacterial RNA

Newly formed rRNA and tRNA molecules are usually not functional but must undergo chain cleavage, methylation, and other alterations before they are "mature." However, most bacterial mRNA does not require processing. Bacteria produce a series of mRNA molecules of variable length, some corresponding to polycistronic (polygenic) and some to monocistronic operons. Most of the mRNA molecules produced are unstable with an average lifetime of about two min; however, some, such as those produced in bacteria about to undergo sporulation, survive much longer. Bacterial mRNA sometimes does undergo processing before it reaches the ribosomes. For example, following infection of *E. coli* cells by phage T7, ribonuclease RNase III cleaves the large 7000 nucleotide "early RNA" transcript from the virus DNA into five defined fragments.[202] Each fragment presumably carries the message for a single viral gene. Genes containing introns have been identified in several archaebacteria[203–205] and in certain phage.[206] The corresponding transcripts must be spliced, as are most eukaryotic transcripts.

Ribosomal RNA. Quantitatively the most im portant RNA, making up 90% of that present in cells, is ribosomal RNA. Synthesis of rRNA must be rapid, for an *E. coli* cell produces 5–10 new ribosomes per second, or 2 x 10^4 molecules of RNA per generation. Bacterial ribosomes contain three pieces of RNA. These are designated, according to their sedimentation constants, as 5S, 16S, and 23S and contain about 120, 1700, and 3300 nucleotides, respectively. All three pieces appear in cells as parts of larger **pre-rRNA** precursor molecules with extra nucleotide sequences at both the 3' and 5' ends.[207,208]

There are seven rRNA regions in the *E. coli* chromosome.[208a] Each region consists of a single transcriptional unit containing a gene each for 16S, 23S, and 5S rRNA with interspersed tRNA genes as follows: 16S, tRNA, 23S, tRNA, 5S, tRNA. A single transcript (which can, in certain mutant strains, appear as a 30S molecule) is cut by the endonuclease RNAse III into the smaller pre-rRNA molecules.[209,210] Other nucleases trim these to their final sizes, and methylases act to modify 24 residues in *E. coli* rRNAs.[208,211–213] Most RNAs of all organisms contain **pseudouridine** (ψ), which is formed by isomerization of specific uridines present in the RNAs (Eq. 28-3). In *E. coli* there is one pseudouridine in the 16S ribosome RNA and nine in the 23S RNA as well as one or more in most tRNAs.[214–217] The isomerization depends upon a carboxylate group of the enzyme, which evidently adds to the 6-position of the uracil ring to form a pivot around which the ring can rotate after it is eliminated from its attachment to the RNA and before it is reattached with a C–N linkage (Eq. 28-3).[217] Both bacteria and eukaryotes contain several pseudouridine synthases, which act to isomerize specific uridine residues in the precursor RNAs.[218,218a,b] In eukaryotes special **guide RNAs** direct the pseudouridine synthases to specific locations in their substrates.[174,219] The same thing is true for 2'-*O*-methylases that modify selected ribose rings in precursor RNAs.[219,220]

Transfer RNA. The genes for tRNA molecules in both bacteria and mammalian cells are grouped in clusters, which are transcribed as large precursors sometimes containing more than one kind of tRNA or containing tRNA fused to rRNA or mRNA sequences. At least three different nucleases are needed for cutting and trimming to form the mature tRNA. These enzymes may not always act in the same sequence. Thus, for some but not all tRNAs cleavage near the 3' end is needed before cleavage can take place at the 5' end.

The best known processing nuclease is **RNase P,** which cleaves bacterial tRNA precursors to create the 5' ends of the mature tRNAs. All of the 64 tRNA precursors present in *E. coli* are cleaved by this unusual enzyme,[221–222c] which contains an essential piece of RNA (Chapter 12, Section D,6). Cleavage of polycistronic tRNA precursors by RNase P or of the previously mentioned rRNA and tRNA pecursor by RNAses P and III releases mature tRNAs carrying extra nucleotides at the 3' ends are removed by exoribonucleases, a surprising number of which are present in cells. In *E. coli* they include RNases II, D, BN, and T,[223,224] as well as polynucleotide phosphorylase.[225]

The structure of the precursor to the minor *E. coli* tyrosine $tRNA_1$ is shown in Fig. 28-10. This is encoded by the *amber* suppressor gene *Sup*F (see Chapter 29). Transcription of its gene is initiated by GTP 43 bp upstream of the 5' end of the mature tRNA and usually terminates at a ρ-dependent signal 225 bp beyond the CCA terminus of the tRNA. An endonuclease cuts the transcript a few nucleotides beyond the CCA end. It is then trimmed to an ~130-nucleotide piece still containing 2–3 extra nucleotides at the 3' end. This intermediate is cut by RNase P at the 5' end after which final trimming is done at the 3' end (Fig. 28-10).[226,227]

An important chemical achievement was the synthesis by H. G. Khorana and associates of the double-stranded DNA segment coding for this *E. coli* tyrosine tRNA[228] and its precursor.[229] This was one of the first synthetic genes (Chapter 5). The synthesis was extended to include the gene termination region, which lies beyond the CCA end of the tRNA. Two noteworthy features appeared. There is a local center of dyad symmetry (indicated by vertical bars and a central dot in Fig. 28-10), which may serve as a termination signal. The operator is located in the 29-nucleotide sequence *preceding* the tyrosine tRNA gene.[230–232]

The 3'-terminal group of three nucleotides, CCA, is invariant among all tRNA molecules and is labile, undergoing active removal and resynthesis. The rate of this turnover is sufficient to involve about 20% of the tRNA molecules of a cell per generation, but it is very much slower than the rate of participation of the tRNA molecules in protein synthesis. The physiological significance of end turnover is unknown.[233] While this CCA sequence is encoded in bacterial tRNA genes, it is added in a separate reaction in eukaryotes.[234]

In addition to the cutting and trimming of precursors by nucleases, extensive modification of purine and pyrimidine bases is required to generate mature tRNAs.[235] Some of these modification reactions are

Uridine in an RNA chain — Rotate — Two steps — Tautomerization — Pseudouridine in RNA (28-3)

discussed in Chapter 5 (see Fig. 5-33). In Fig. 28-10 the modifications in the mature tyrosine suppressor tRNA are shown, and in Fig. 5-30 those in *E. coli* phenylalanine tRNA are indicated. Modification usually begins with tRNA precursors. For example, the precursor in Fig. 28-10 is methylated to form ribothymidine at position 63.[236] Pseudouridine is then introduced at positions 40 and 64 by isomerization of the uridines present in the initial transcript (Eq. 28-3).[237] The Tψ pair at position 63 and 64 of Fig. 28-10 is almost universally found in tRNA,[238] but the positions are

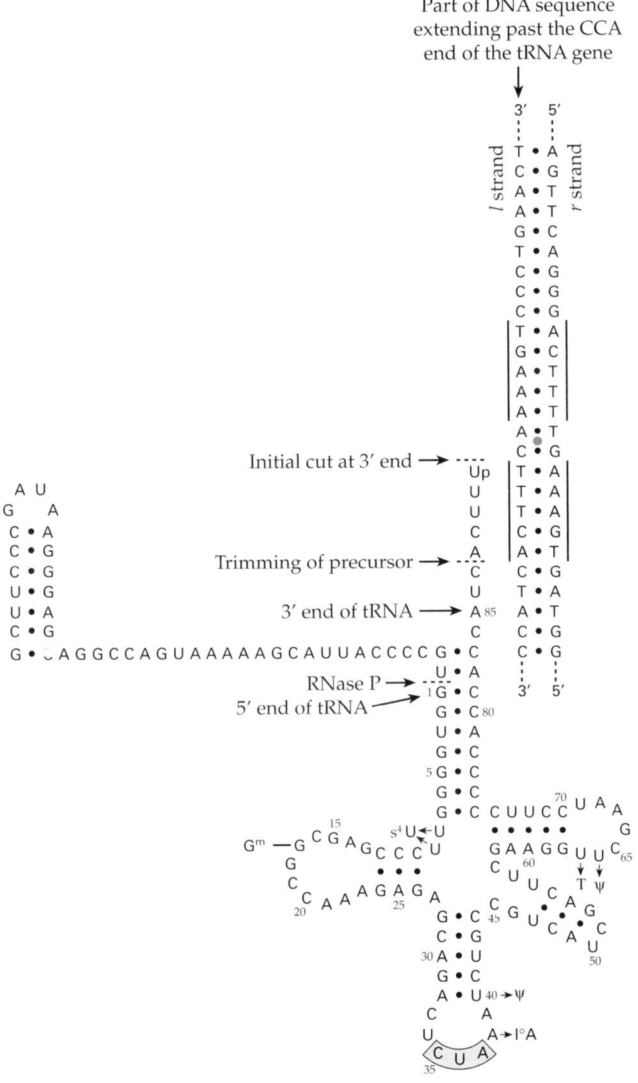

Figure 28-10 Sequence of an *E. coli* tyrosine tRNA precursor drawn in a hypothetical secondary structure. Nucleotides found modified in the mature tRNA are indicated with their modifications (S^4, 4-thiouridine; G^m, 2′-*O*-methylguanosine; I^0, N^6-isopentenyladenosine; ψ, pseudouridine; T, ribothymidine; see also Fig. 5-33).[241] A partial sequence of the tRNA gene past the CCA end is also shown. Note the region of local 2-fold rotational symmetry (indicated by the bars and the dot). The anticodon 3′-CUA (shaded) of this suppressor tRNA pairs with termination codon 5′-UAG.

usually designated as 54 and 55 as in Fig. 5-30. Position 8 in most tRNAs is occupied by 4-thiouridine. The sulfur atom is transferred from cysteine as S^0 using a PLP-dependent mechanism similar to that in Eq. 14-34 and involving an **S-sulfonylcysteine (persulfide) intermediate**.[239,240] Some modifications are completed on the mature tRNA.[241] Some tRNAs require RNA splicing for maturation (see Section D,5).[205]

The hypermodified nucleoside **queuosine** is found in the first (wobble) position of anticodons of most eukaryotic tRNAs for Asn, Asp, His, and Tyr and also in most bacteria.[242–244] Bacteria apparently make aminomethyl-7-deazaguanine from guanine (Eq. 28-4, step *a*) and transfer this compound into the appropriate position in tRNA (step *b*) by a tRNA-guanine transglycosylase.[245–246a] The incorporated nucleoside is then converted to queuine by incorporation of the 5-carbons of the ribosyl group in *S*-adenosylmethionine (Eq. 28-4, step *c*) to form an intermediate epoxide.[247] This is converted to queuine in the vitamin B_{12}-dependent step *d*. Eukaryotes are unable to form queuine and must obtain it as a nutrient factor or from intestinal flora.[244] It is exchanged into tRNA by action of the transglycosylase.[245] Queuine might be considered a

Guanine
Guanine–tRNA
Guanine
Adenine
S-Adenosylmethionine
Methionine
Adenine
Vitamin B_{12}
Nucleotide of queuine in tRNA

(28-4)

vitamin. However, germ-free mice reared on a queuine-deficient diet seemed normal. Therefore, the essentiality of queuine in the human diet is in doubt.[242,243]

B. Viral RNA in Prokaryotes

Bacteria not only transcribe their own genes but sometimes transcribe, or assist in transcribing, genes of invading DNA viruses or of integrated proviruses. In addition, they assist in replication of RNA viruses, another process that requires RNA synthesis. Viruses sometimes make use of host RNA polymerases but often synthesize their own catalytic subunits. Bacteriophage T4 uses the *E. coli* RNA polymerase and σ factors but modifies their action through the binding of several phage-encoded proteins.[248] In contrast, phage T7 encodes its own relatively simple RNAP whose initiation complex (Section A,2)[29] and elongation complexes have been studied.[249–249b]

1. The Lysogenic State of Phage λ

The study of bacteriophage lambda has provided many insights into biological process.[250] As we have seen (Chapter 27), the DNA of phage λ can become incorporated into the genomic DNA of *E. coli*. The resulting prophage contains many genes (Fig. 28-11), but they remain largely unexpressed until the SOS signal (Chapter 27) is generated. Certain prophage genes are then expressed with the result that the λ DNA is excised as a replicating virus.[251] How can the λ genes remain unexpressed in the prophage but be expressed rapidly at the time of excision? Part of the answer has been found[252–254] in the *cI* and *Cro* repressors. The short L1 operon (Fig. 28-11) of the λ prophage is transcribed continuously by the *E. coli* RNA polymerase. This operon contains genes *cI* and *rex*, which are transcribed from the *l* strands of the prophage DNA as indicated in Fig. 28-11. The protein C_I (or CI) specified by gene *cI* is the **lambda repressor,** which binds to two operator sites in the prophage DNA. One operator (o_L) is to the left and the other (o_R) to the right of the *cI* gene. From a study of fragments of DNA protected by the repressor, it was concluded that each operator has three subsites, which are filled from left to right at o_L and from right to left at o_R successively by up to six repressor monomers. Each presumed subsite has a similar 17-bp quasipalindromic sequence to which a dimeric repressor can bind. The binding is cooperative, probably because the repressor molecules contact each other, apparently binding the DNA into a loop.

The right operator o_R controls not only the R1 operon but also the L1 operon, which encodes the λ repressor (*cI* gene). The first of the three subsites in the operator is adjacent to the L1 promoter P_{RM}, and binding of λ repressor activates that promoter at the same time that binding to the adjacent subsites blocks transcription of the R1 operon. Thus, the λ repressor positively controls its own synthesis. At the same time, blocking of promoters P_R and P_L prevents synthesis of virally encoded enzymes that catalyze excision of the λ DNA and replication and transcription of the rest of the genes.

The SOS signal causes rapid hydrolytic cleavage of the λ repressor and transcription of the other λ operons. The matter is more complex than this. Gene products *cIII* and *cII*, from the **early left** and **early right operons**, respectively, stimulate the transcription of *cI* and are needed for establishing the lysogenic state initially.[256] Once established these genes do not function since they are never transcribed. There are only a few molecules of the λ repressor present in a cell, but this is ordinarily sufficient to maintain the prophage state. On the other hand, irradiation of the bacterium with ultraviolet light activates the SOS response and results in rapid hydrolytic cleavage of the λ repressor and transcription of other phage operons.

Of special significance to the lytic cycle is the Cro repressor gene *cro*, found at the beginning of operon R1. Although it binds to the same operator sequences as does the λ repressor, the Cro repressor has opposite effects.[254] It represses transcription of operon L2 and hence synthesis of λ repressor, but it positively activates P_R and P_L. The earliest proteins synthesized during lytic development are the Cro protein and the product of the first gene *N* in the left operon L2. The N protein, an antiterminator that permits transcription to continue on past points t_L and t_R, is an unstable, short-lived molecule of $t_{1/2}$ ~2 min.[257] Leftward transcription proceeds through genes *exo* and β, which are involved with recombination, and *xis*, which is required for excision. When the DNA is integrated into the *E. coli* chromosome, it is cut at points *aa'* (Fig. 28-11) and is inserted just to the right of the *gal* operon (Fig. 26-4). Prophage transcription can now continue past point a' and into the genes of the bacterium. Translation of the mRNA formed from this early left operon generates the enzymes needed to free the prophage and to permit reformation of the circular replicative form of the phage DNA. The excision is also made near point a', and it is easy to see how the nearby *gal* genes can sometimes be included in the excised λ genome.

The product of gene *N* also permits rightward transcription through genes *O*, *P*, and *Q* and at a slower rate on along the rest of the chromosome to point a. Genes *O* and *P* code for proteins that permit the host replication system to initiate formation of new λ DNA molecules. Replication begins at the point *ori* and occurs in both directions. Gene *Q* codes for a protein that activates transcription of the **late genes** beginning at promoter P_R.

As indicated in Fig. 28-11, the chromosome can be divided into four major operons, the short one that produces repressor, and the early left, early right, and late operons. The early operons code largely for replication and recombination enzymes and control proteins. The late operon is concerned with production of proteins needed for assembly of the virus particles and must be transcribed at an even higher rate; hence the need for the product of gene *Q*. Within the late operon, genes *A* to *F* are involved in packaging of λ DNA and in formation of heads, while genes *S* to *J* are concerned with the production and assembly of tails. Genes *S* and *R* produce proteins that lead to destruction of the host membrane and to lysis of the cell. During the late stages of lytic growth the early genes are largely shut off by the Cro repressor. We can see that even in a virus the control of transcription can be a complex process.

2. Replication of RNA Bacteriophages

The small icosahedral RNA-containing bacteriophage are of interest because of the small number of

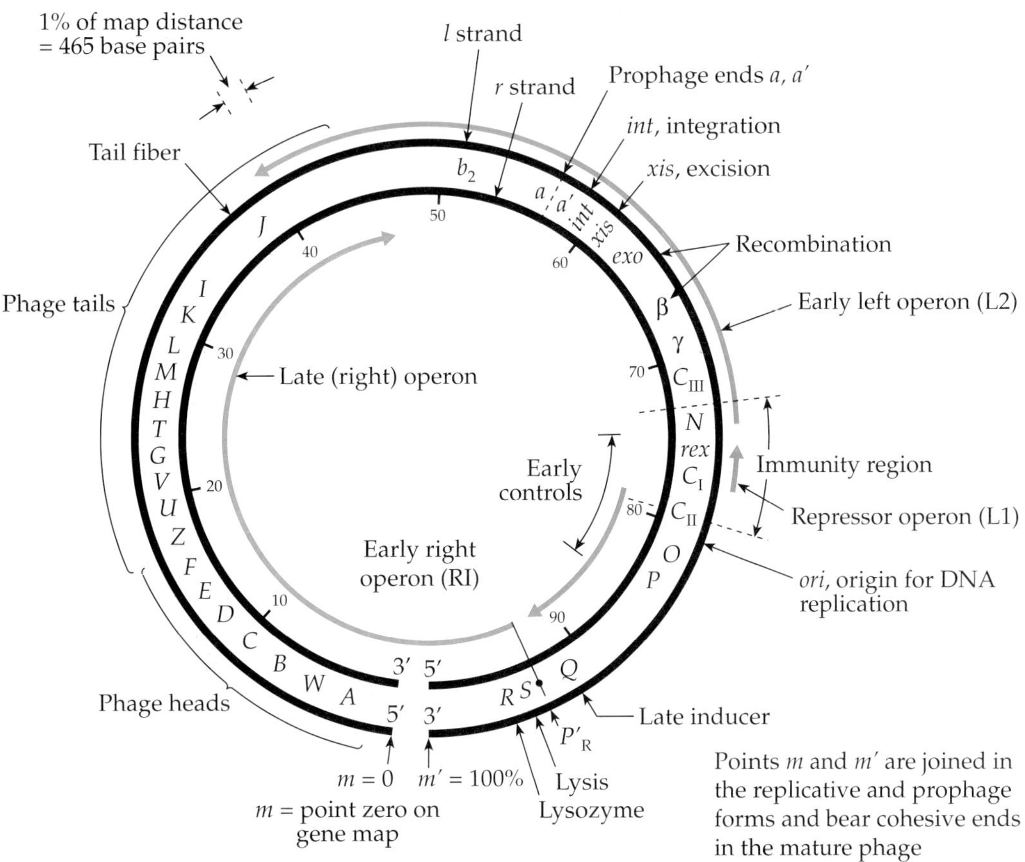

Expanded diagram of immunity region

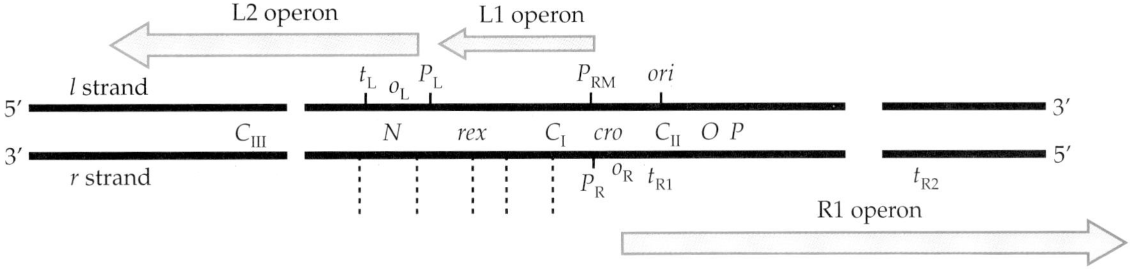

Figure 28-11 Genetic and physical map of the λ phage genome. After Szybalski. See Honigman *et al.*[255] for a more detailed diagram of the immunity region. The gene for the lambda repressor is labeled C_I.

genes and the possibility of obtaining a detailed understanding of their replication. The four genes in the 3569-nucleotide MS2 RNA (Fig. 29-17) code for the A protein (maturation protein), the coat protein, a replicase (RNA polymerase) subunit, and a protein needed for lysis of the host cell. The last gene overlaps both the replicase gene and that of the coat protein.[258,259] The somewhat larger phage Qβ contains a 4.2-kb RNA genome. One subunit of the Qβ replicase is encoded by the virus, but three bacterial proteins are needed to form the complete replicase.[260] They are ribosomal protein S1 and elongation factors EF-Tu and EF-Ts, proteins that normally function within *E. coli* in translation of mRNA (Chapter 29). Their ability to associate with RNA has been exploited by the phage for a quite different purpose.

Replication of a single-stranded virus must take place in two steps. From the (+) strand present in the virus a complementary (–) strand is first formed. Initiation of this step requires another bacterial host factor Hfq[260,261] and GTP. The (–) strands formed do not associate with the (+) strands. They are apparently released from the replicase in a single-stranded form and presumably fold into highly structured molecules with many hairpin loops (as for the (+) strand of MS2 RNA shown in Fig. 29-17). The (–) strands are then copied (Hfq is not needed for this) to make a large number of new (+) strands for incorporation into the finished virus particles. The Qβ replicase is able to synthesize *in vitro* complete complementary strands to either (+) or (–) viral RNA molecules. However, the system is specific for the viral RNA and will not copy any arbitrary nucleotide sequence, certain sequences at the 3' end being essential for initiation of replication. During replication in the test tube mistakes are made including premature termination and mispairing of bases.[262,263] Thus mutation takes place, and it is possible to select RNA molecules much smaller than the original viral RNA that will be replicated readily by the Qβ replicase system. One such fragment contains only 114 nucleotides in a known sequence.[264]

C. Transcription in Eukaryotic Cells and in Archaea

There are three primary domains of life, represented by the bacteria, archaea, and eukaryotes. Some of the clearest evidence for the independent evolution of these three groups of organisms is found in the transcriptional apparatus. While the basic chemistry is the same, the details of initiation and control of transcription in bacteria and in eukaryotes are very different.[264a] The archaea share characteristics of both bacteria and eukaryotes. Archaeal RNA polymerases have a complexity similar to that of eukaryotes and also share a similar mechanism of initiation of transcription.[265–266b] Several of the protein transcription factors of archaea also resemble those of eukaryotes.[267,268] However, in a comparison of DNA sequences from the complete genomes of four archaeal species, it was found that of 280 predicted transcription factors or transcription-associated proteins 168 were homologous to bacterial proteins and only 51 to eukaryotic proteins.[268] This tends to confirm the ancient divergence of the three primary domains of life.

In bacteria transcription and translation are closely linked. Polyribosomes may assemble on single DNA strands as shown in Fig. 28-5. It has often been assumed that RNA synthesis occurs on loops of DNA that extend out into the cytosol. However, recent studies indicate that most transcription occurs in the dense nucleoid and that assembly of ribosomes takes place in the cytosol.[268a] In a similar way eukaryotic transcription occurs in the nucleus and protein synthesis in the cytosol. Nevertheless, some active ribosomes are present in the nucleus.[268b]

1. Eukaryotic Nuclei and Transcription

In cells with true membrane-enclosed nuclei the messenger RNA molecules are relatively long lived. They must move out from the nucleus to the sites of protein synthesis in the cytoplasm. In addition to the need for eukaryotic mRNA to travel further and to last longer than that of bacteria, a number of other differences are evident. Eukaryotic mRNAs are usually transcribed from single genes. Polygenic operons are uncommon in most animals but are numerous in *C. elegans*.[268c] Eukaryotic cells appear to rely less on negative control through specialized protein repressors than do bacteria but use a greater variety of positive control mechanisms. However, most genes are repressed by being held in a **silent state**.

Another characteristic of eukaryotes is the extensive processing of transcripts. Most primary transcripts that give rise to mRNA appear first in the nucleus as **heterogeneous nuclear RNA (hnRNA)**. Like mRNA it has a base composition resembling that of DNA. The molecular size varies from 1.5 to 30 kb or more. It turns over rapidly, most of it having a half-life of ~10 min. However, some may last as long as 20 h. Only about 5% of the hnRNA ever leaves the nucleus as mRNA, most being degraded without export to the cytoplasm.[269] The processing consists of **capping** at the 5' ends, removal of introns (splicing), **cleavage** by nucleases, **polyadenylation** at the 3' ends, **methylation**, formation of pseudouridines, other covalent base modifications, and sometimes **editing**. Because of the complexity of eukaryotic transcription there are many points at which control can be exerted during initiation of transcription, termination of transcription, splicing and methylation, transport of mRNA out of the nucleus, and the degradation of mRNA. However, the first points of control are found in the nucleosomes and in the structure of the chromatin.

It has been recognized for many years that genes in heterochromatin (Chapter 27) are usually not transcribed. In most chromatin the great majority of the genes are silent most of the time. Genes are repressed by the folding of the DNA into nucleosomes and by the further folding into higher order folds or coils (Fig. 27-5).[270–272] Various activating transcription factors as well as RNA polymerases must bind to the DNA, displacing it from the histones around which it is wrapped in the nucleosomes. The processes by which inactive nucleosomal DNA becomes active in response to external signals are beginning to be understood. Chemical alterations in the histones in the nuclear matrix and in other nuclear proteins and also in the DNA itself may be involved.[273–275] As pointed out in Chapter 27, the CpG "islands" that lie upstream of many genes are heavily methylated in the silent heterochromatin. Repressor proteins may bind to methyl-CpG groups.[276,276a]

In recent years attention has been focused on the N-terminal "tails" of histones H3 and H4 (Fig. 27-4) in which lysine side chains undergo reversible acetylation and which may also be phosphorylated and

BOX 28-B POWERFUL POISONS FROM MUSHROOMS

Several deadly species of the genus *Amanita* produce colorless toxic octapeptides, the **amanitins**.[a,b] Two residues of glycine, one of L-isoleucine, one of the unusual L-dihydroxyisoleucine, one of L-asparagine, and one of L-hydroxyproline are present in α-amanitin. In the center a modified tryptophan residue has been combined oxidatively with an SH group of a cysteine residue. If the dihydroxyisoleucine residue of α-amanitin is replaced with unhydroxylated leucine, the resulting compound, known as amanullin, is nontoxic. The LD_{50} for mice is 0.3 mg kg^{-1} and 50 g of fresh *Amanita phalloides* may be sufficient to kill a person. Amanitins act slowly, and it is impossible to kill mice in less than 15 h, no matter how high the dose.

α-Amanitin completely blocks transcription by eukaryotic RNA polymerases II and III. Polymerase II is the major nuclear RNA polymerase, and its inhibition prevents almost all protein synthesis by the cell. Note that the amanitin molecule is semisymmetric overall, much as is a actinomycin (Box 28-A), with an aromatic group protruding from behind in the center.

The same mushrooms contain several fast-acting toxic heptapeptides, the **phalloidins**, whose structures are similar to those of the amanitins. However, they contain a reduced sulfur atom (—S—) in the cross-bridge. They are specifically toxic to the liver.[c] The same mushrooms also contain an antidote to the phalloidins, **antamanide**. This cyclic decapeptide, like the toxins, is made up entirely of L-amino acids, and it apparently competes for the

Pro → Ala → Phe → Phe → Pro
↑ ↓ Antamanide
Pro ← Val ← Phe ← Phe ← Pro

binding site of the phalloidins. Unfortunately, it is of little value in treating cases of mushroom poisoning. Antamanide is a specific sodium-binding ionophore.

[a] Wieland, T., and Wieland, O. (1972) *Microb. Toxins* **8**, 249–280
[b] Wieland, T., and Faulstich, H. (1983) *Handbook of Natural Toxins*, Marcel Dekker, New York
[c] Wieland, T., Nassal, M., Kramer, W., Fricker, G., Bickel, U., and Kurz, G. (1984) *Proc. Natl. Acad. Sci. U.S.A.* **81**, 5232–5236

methylated.[33b,271,275a–c] The resulting modifications in shape and electrical charge can affect the ways in which the histones interact with the DNA, with each other, and with other proteins in transcription complexes.[271a,275a] The histone tails may interact with adjacent molecules either to pack the chromatin more tightly or to loosen it and allow transcription to take place. Active chromatin has long been known to be highly acetylated, while silent chromatin has a low degree of acetylation, but a high degree of methylation of both histones and DNA (pp. 1541–1542). Methylation of histone H3 on lysine 9 (H3-Lys^9 or H3-K9) is especially significant.[275d,e] See Chapter 32, Section A,1 and C,1 for further discussion.

Several transcriptional activators form complexes with acetylating enzymes, the **histone acetyltransferases** (**HATs**), while transcriptional repressor proteins often associate with **histone deacetylases**.[277–279c] The deacetylases are often found in very large complexes. For example, the mammalian complex **Sin3** contains two histone deacetylases plus at least five other subunits, some of which evidently bind to histones.[277–278a] Sodium butyrate in millimolar concentrations is a powerful inhibitor of these deacetylases.[280] Special **silencer** sequences in the DNA are sometimes present and provide sites for binding of transcriptional repressors. Among these are **silent information repressors** (**Sir** proteins). They regulate regions of DNA that can be converted to a heterochromatin-like state. They have been located in silenced mating type loci in yeast (see p. 1880), in telomeres, and in DNA containing ribosomal RNA genes.[280a,b] Sir proteins have an unusual histone deacetylase activity. The acetyl groups removed from histones are transferred by reaction with NAD^+ to ADP ribose (see Eq. 15-16).[280c,d] Regions of silenced DNA are often set apart by **insulator** or bounding regions.[280e–g]

An important mechanism of silencing some genes is the use of an antisense RNA strand, as is illustrated for a bacterial system in Fig. 28-8. This **RNA interference** is also used in animals and plants, often as a way of blocking replication of viruses.[1,1a,280h–j] The small 20–25 nucleotide siRNAs that function in this way are abundant in *C. elegans* and in *Drosophila*.

Histone acetyltransferases also form large complexes that acetylate not only histones but also other nuclear proteins.[281–283] The **SAGA HAT** complex of *S. cerevisiae* has a molecular mass of ~2.0 MDa and contains at least 14 subunits.[283a] Large acetylating complexes have also been identified in *Tetrahymena*, *Drosophila*, *Arabidopsis*, and mammalian species.[283] Some subunits of these complexes have been identified as previously known transcription factors. Such multiprotein complexes are sometimes described as **cis-regulatory elements** (CREs).[283b,c]

Changes in the structural properties of chromatin observed during silencing of genes or during their activation are often described as **chromatin remodeling**. Large multisubunit complexes are involved. Their action is characteristically dependent upon ATP hydrolysis.[284–284b] Complexes **SWI/SNF** and **RSC**, first found in yeast but also present in human cells, appear to participate in the disruption of nucleosomes needed for initiation of transcription.[284–287c] Among other distinctly different remodeling complexes are **ISWI** of *Drosophila*, its human homolog, and the human Williams syndrome transcription factor **WCRF**.[284] The ATP-dependent component in these complexes has a conserved sequence that is shared with DNA helicases. RNA helicases are also required for all processes that form, modify, or utilize RNA.[288,288a–c] However, the chromatin remodeling complexes appear not to unwind DNA but to open the nucleosomal DNA for initiation of transcription.[272,284] They may act in a processive fashion and be coupled to transcription. **Peptidyl-prolyl isomerases**, such as the cyclophilins (Box 9-F) may also be essential components of chromatin-remodeling complexes.[289]

Just as they participate in driving the cell cycle, ubiquitin and proteasomes also function in the control of transcription. In many cases specific transcription factors are targeted for destruction after they are used to activate or repress a gene.[289a–c] However, a 19S regulatory complex, which consists of a base and a lid of the proteasome (Box 7-A), may participate directly in control of transcription rather than in mediating proteolysis.[289a–e] In addition to ubiquitin a 97-residue relative designated SUMO-1 is linked to proteins by enzymes resembling E_1 and E_2 of the ubiquitin system (Box 10-C). Conjugation with **SUOMO-1** regulates some transcription factors and has other functions,[289f,g] e.g., participation in control of nuclear pores. The 81-residue protein **NEDD8**, which is 80% homologous with ubiquitin, controls some transcriptional processes in heart and skeletal muscle.[289h,i] Furthermore, ubiquitin-like sequences (UBX domain) are present in the C-terminal ends of a variety of specific proteins.[289j]

2. RNA Polymerases

Eukaryotic nuclei contain at least three RNA polymerases[269,290–292] which have the following functions:

Polymerase I	Formation of large pre-rRNAs
Polymerase II	Transcription of most genes to give precursors to mRNA and most small nuclear RNAs (snRNAs and small nucleolar RNAs (snoRNAs)
Polymerase III	Formation of 5S rRNA, tRNAs, and small RNA U6

Polymerase I is localized in the nucleolus.[293,294] Mitochondria contain a fourth RNA polymerase[295,295a] and chloroplasts a fifth.[296]

Like the bacterial polymerase, eukaryotic RNA polymerases are large 500–600 kDa aggregates of 9–14 subunits each. Yeast and human RNA Pol IIs each contain twelve subunits (Fig. 28-12).[290] There are two large nonidentical subunits, which in mammalian cells have masses of 214 kDa and 140 kDa and are homologous to the β′ and β core subunits of the *E. coli* polymerase, respectively.[291,297–299c] The active site contains one or two catalytic Mg^{2+} ions.[290] The largest subunit has an unusual singly glycosylated C-terminal domain (known as the **CTD**). It contains the repeating sequence $(YSPTSPS)_n$. The number of repeats varies: $n = 18$ in plasmodia, 27 in yeast, 45 in *Drosophila*, and 52 in mammals.[300] The numerous serine side chains in this tail domain undergo phosphorylation and dephosphorylation to varying extents during each catalytic cycle.[301] This may be a way of easing the transcriptional complex through nucleosomes,[302] but its most important function appears to be the linking of transcription to pre-mRNA processing.[303,304]

RNA polymerases I and III have properties similar to those of Pol II.[304a,b] Polymerases II and III are very sensitive to inhibition by the lethal mushroom poison **amanitin** (Box 28-B). However, both RNA polymerase I and RNA polymerases of mitochondria resemble the bacterial enzyme in being resistant. Genes transcribed by polymerase I, II, and III are often referred to as genes of classes I, II, and III, respectively. While mRNA is transcribed from class II genes, rRNAs, tRNAs, and some small RNAs, which must undergo processing but are not polyadenylated, are transcribed from genes of classes I and III. Each type of nuclear RNA functions in its own sites as independent "factories."[305,306]

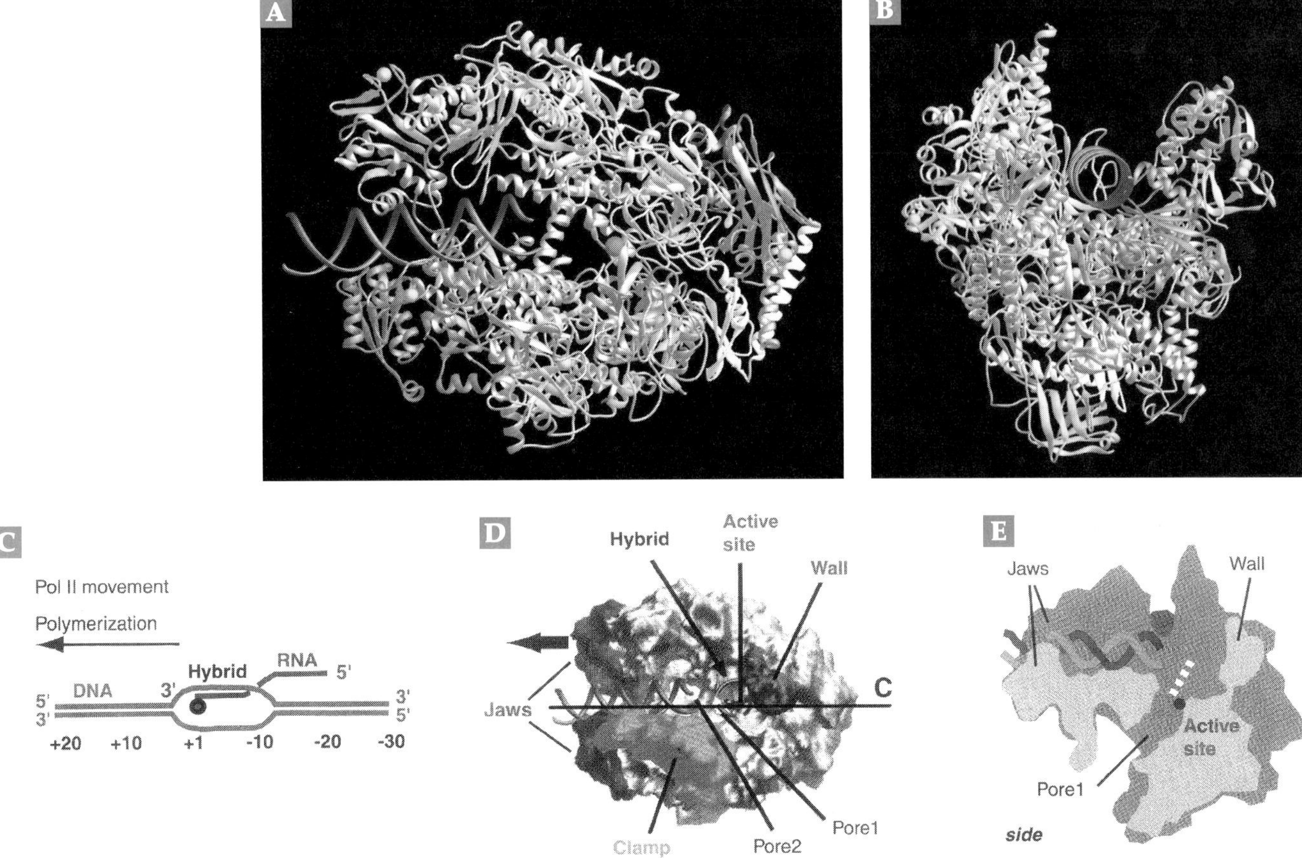

Figure 28-12 Three-dimensional structure of yeast RNA polymerase II. (A) View from the "top," with backbones of ten subunits (of 12) in the 514-kDa protein shown as ribbon drawings. A 20-base-pair segment of B-DNA has been modeled in a location indicated by electron crystallography. (B) Side view looking toward the end of the DNA. Eight zinc atoms as well as an active-site magnesium (green) are visible. (C) Schematic drawing showing the transcription bubble as proposed for a transcribing polymerase. (D) Surface representation of the polymerase viewed as in (A). (E) Side view of a section cut along the line marked C in D. The dashed white line represents the axis of the DNA-RNA hybrid segment. The hybrid axis must lie at an anlge with respect to the axis of the incoming DNA. Pore 1 may be a route for exit of RNA during "backtracking." The nucleotide triphosphate substrates may also enter via pore 1. From Cramer *et al.*[290] Courtesy of Roger D. Kornberg.

3. Transcriptional Units and Initiation of RNA Synthesis

Typical simple **transcriptional units** for class II eukaryotic genes contain the following elements: (a) site of initiation, (b) **TATA sequence** (Goldberg–Hogness sequence) at position ~–30 bp, (c) **upstream regulatory elements**, (d) **enhancers**, (e) a series of **coding sequences** or exons separated by introns, an **AATAAA sequence** that in the RNA transcript may establish the 3' polyadenylation site, (f) a termination region.[307]

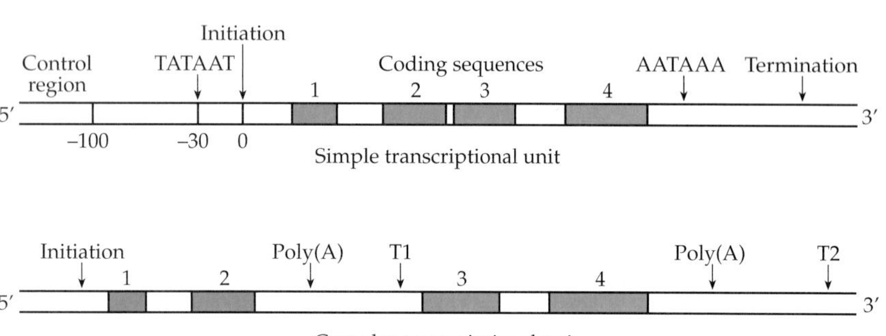

The preceding paragraph describes a **simple transcriptional unit**. There are also **complex transcriptional units**.[308] For example, the terminator sequence T1 may be followed by additional exons and a second polyadenylation signal and second termination sequence T2. Termination may sometimes occur at T1 and sometimes at T2 resulting, after splicing, in two mRNAs, one containing only exons 1 and 2 and the other all four exons. In some cases two or more different modes of splicing may occur with one or more exons omitted from the final processed mRNA. Thus, a single transcriptional unit can give rise to two or more different proteins which share some common sequences.[309–312] Multiple initiation sites sometimes exist as well.

Initiation of RNA synthesis is a complex process, which is summarized diagrammatically in Eq. 28-5. Some details are given in Sections 4–7. Elongation of the RNA being formed is also complex, often requiring splicing and other processing steps before synthesis can be terminated. These steps, which are discussed in Section 8 and in Section D, also depend upon large complexes of proteins, which are directly coupled to the RNA polymerase.[312a–c]

4. Promoters, Transcription Factors, Enhancers, and Activators

Eukaryotic promoter sequences are less well defined than are those of bacteria, and the initiation points for transcription are more variable. Most promoters contain the TATA sequence, which is usually required for binding of polymerase II. In addition, there are upstream regulatory elements (Table 28-1). Many promoters contain the sequence 5'- **CCAAT** at about –75, and others have 5'- **GGGCGG** or similar sequences at close to –100.[273, 273a,b] A sequence found upstream from many yeast genes is 5'- **TGACTC** or the longer semipalindromic 5'-**ATGACTCAT**.[313,314] Some sequences are unique to small sets of genes such as the "heat-shock" genes[315] considered in Section 6. Most upstream sequences are not polymerase-binding sites but attachment sites for additional protein transcription factors. Computer programs that help locate pro- moter sequences have been devised.[316,317]

The SV40 early promoter and transcription factor Sp1. The study of transcription in eukaryotes has been difficult because purified polymerases do not initiate transcription at most promoter sites. As a consequence, much of the early work was done with viruses such as adenovirus and SV40. Their genes are transcribed by RNA polymerase II and have unusually effective promoters. A protein known as **Sp1**, isolated from human cultured cells, protects an SV40 promoter from digestion by DNase. Sp1, which is now known as an **acessory factor** or **coactivator** for transcription,[318,318a] protects a region that extends from about –45 to –104 bp and contains the hexanucleotide GGGCGG sequence repeated six times. The Sp1 protein was isolated as a mixture of related 95- and 105-kDa peptides[319] and was found to contain two DNA-binding zinc-finger domains (Figs. 5-37, 5-38).[320] The Sp1 protein is synthesized in most cells and binds not only to SV40 promoters but also to many promoters of host cells[321] (an example is provided by the mouse mitochondrial aspartate aminotransferase gene whose sequence is shown in Fig. 5-4). In genes that lack a TATA sequence the binding of Sp1 or the related Sp2, 3, and 4 to GGGCGG or similar GC-rich sequences is essential to initiation of transcription.[322,323] A possible role is to assist in nucleosome remodeling.[322] Sp1 also binds, together with other transcription factors, to certain enhancer sequences.[324,324a,b] Its effects are modulated by posttranscriptional phosphorylation and glycosylation.[321]

The TATA binding protein and general transcription initiation factors. A slow basal level of transcription can be observed when all but a small part of the control region at the 5' end of a gene is deleted.[325] This minimum promoter, which includes the TATA sequence, is the binding site of both the RNA

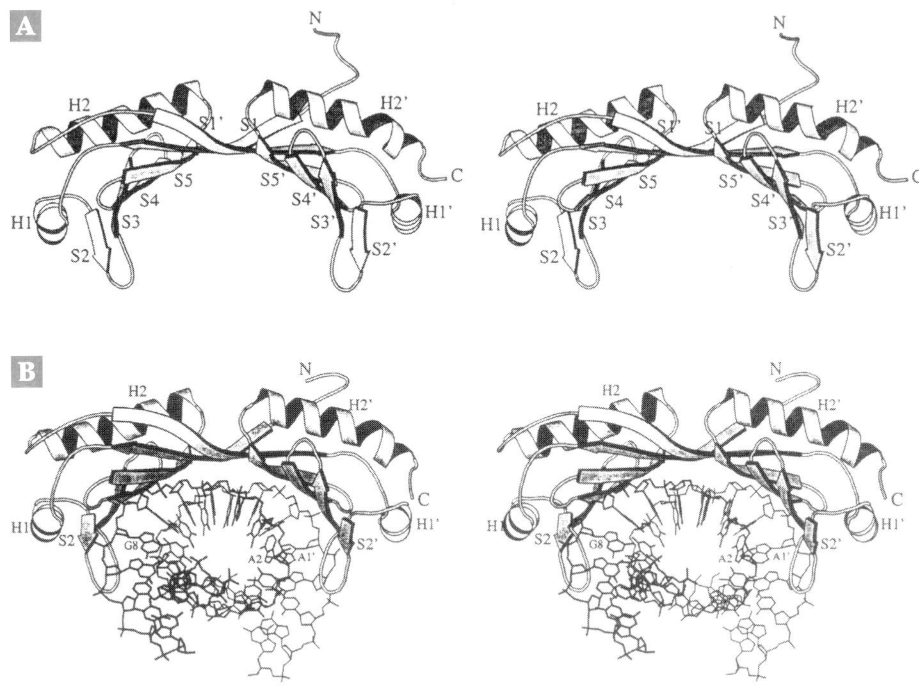

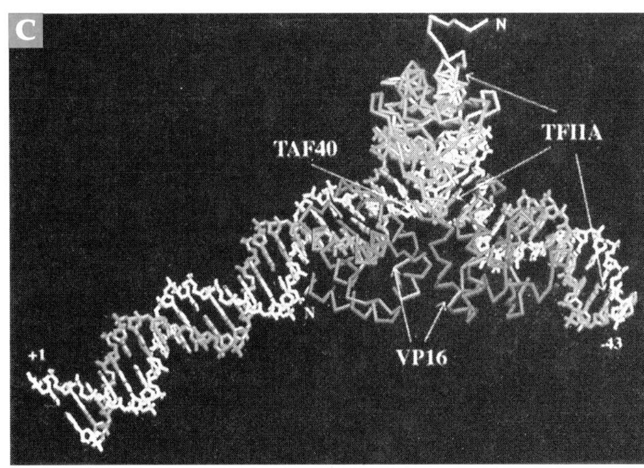

Figure 28-13 (A) Stereoscopic ribbon drawing of the phylogenetically conserved 180-residue C-terminal portion of the TATA-binding protein (TBP) from *Arabidopsis thaliana*. The sequence consists of two direct repeats, giving the protein an approximate twofold symmetry. From Nikolov *et al.*[337] (B) Structure of the corresponding C-terminal core (residues 155–335) of the human TATA-binding protein (TBP) bound to the TATA sequence of a promoter in adenovirus DNA. From Nikolov *et al.*[327] (C) Structure of human transcription factor IIB bound to a TBP from *Arabidopsis thaliana*, which, in turn, is bound to an adenovirus TATA sequence. Hypothetical B DNA extensions have been modeled at both ends of the DNA segment. The +1 at the left end is the transcription start site and the –43 upstream end is to the right. From Nikolov *et al.*[338] Courtesy of Stephen K. Burley.

polymerase and of transcription initiation factors that are designated TFII-A, -B, -D, -E, -F, and H. Because they affect many genes these are called **general transcription factors**.[326–329] TFIID is a large complex of a DNA-binding subunit known as the **TATA-binding protein (TBP)** together with 8–12 additional tightly bound subunits known as **TBP-associated factors** (**TAF_{II}s**). Many of the TAF_{II}s have histone-fold structures. Some possess histone acetyltransferase and other enzymatic activities.[329a] They may stabilize initiation complexes on specific gene promoters.[330–332a] TBP binds specifically to the TATA sequence (Table 28-1),[333] which is found in most promoters for RNA polymerases I, II, and III. Its three-dimensional structure resembles a saddle, which sits astride the TATA sequence (Fig. 28-13). The DNA is bent, untwisted by ~117°, and the minor groove broadened to allow a good fit.[327,328,334] TFIIB is thought to bind first to the DNA–TBP complex, after which the RNA polymerase II complex binds and becomes positioned on a promoter site. Other factors, including the ATP-dependent bidirectional helicase TFIIH,[335,335a] also add (Eq. 28-5). ATP may be needed for more than one step in initiation.[336] TFIID contains TAF_{II} subunits. They may bind along with TBP[336a] as indicated in Eq. 28-5, or they may bind at a later point in an assembly pathway. It is often stated in current literature that the growing initiation complex "recruits" the next subunit. It is important to realize that this simply means that the next subunit that strikes the complex by diffusion sticks, perhaps with cooperativity in binding. The word recruit doesn't mean that the complex advertises a vacancy for the next subunit. The very large complexes (**transcriptosomes**) that are formed vary in composition and in assembly pathways. Proteins related to TBP bind to different promoters.[310] It is also

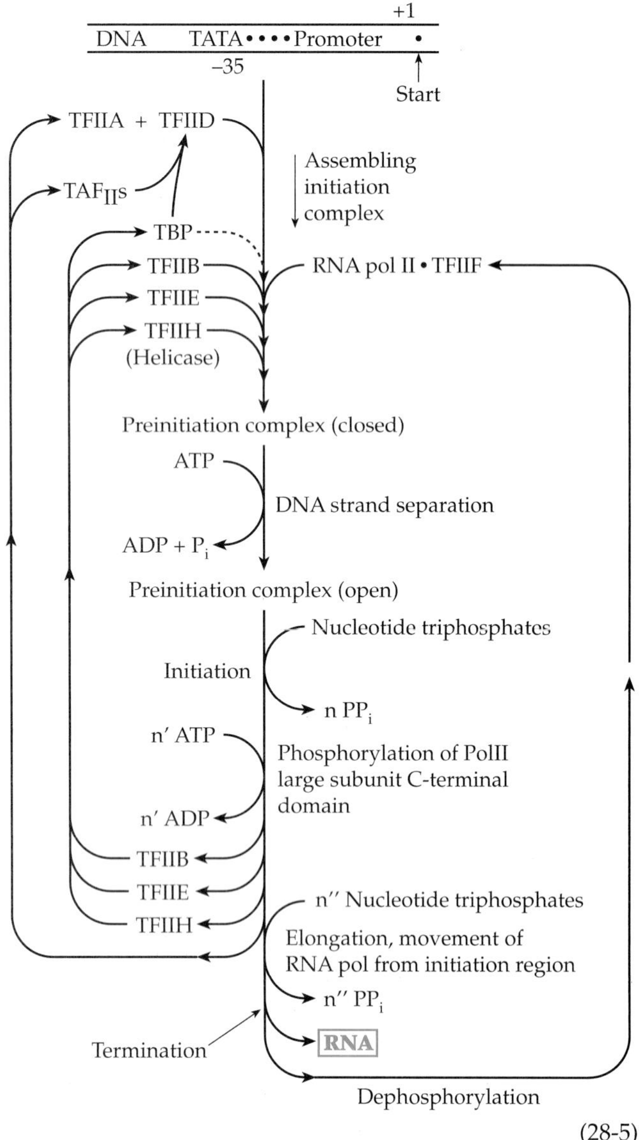

(28-5)

important to recognize that some promoters bind transcription factors tightly while others do so only weakly. Some are constitutive, always functioning, but most are inducible, acting only upon appropriate stimulation. The many operators present in a genome and the many species of eukaryotes present on earth ensure a vast variety of detailed pathways and control mechanisms.

Transcriptional activators. Many proteins serve as activators of transcription, causing larger increases in rate over those observed with TBP alone. Some of these are listed in Table 28-2.[338a] The table also lists two proteins (Sp1 and NF1), and the DNA sequence CCAAT, which control constitutive or continuously active genes. A large group of transcription factors are active in development. Receptors may be resident in cytoplasm, cell membrane or nucleus, as indicated in Table 28-2. Some cytoplasmic factors are *latent*, becoming active only following stimulation by an external signal. All of the factors in Table 28-2 are positive-acting.[338a] However, some negative-acting factors are known. One, designated **Ssn6-Tup1** in *S. cerevisiae* is a global repressor, affecting many genes.[338b] For example, it opposes the activator GAL4. Proteins related to Ssn6-Tup1 are found in flies, worms, and mammals.

Genetic studies indicate that gene activator proteins often bind to TFIIB, TFIID, and TFIIH.[272] The coactivator **TafII130** (which binds to TFIID), and the bound transcriptional activator Sp1 apparently interact with the protein **huntingtin** in regulation of transcription in the brain.[338c] A defect in huntingtin leads to the fatal neurodegenerative Huntington disease (pp. 1516, Chapter 30). Other activator proteins bind to upstream activator sequences, as in prokaryotes. Among the most studied of these is GAL4, an 881-residue yeast protein that binds to a specific 17-bp palindromic upstream site near the TATA sequence. It activates transcription of genes needed for galactose metabolism.[339–343] The GAL4 protein contains a binuclear metal cluster composed of two Zn^{2+} ions and six cysteine side chains, two of which bridge the pair of metal ions.[341,343] GAL4 is able to activate genes of *Drosophila* and of human cells. The specific GAL4-binding sequence has been introduced into 5' control regions of genes in various positions.[344] It was found that neither an exact distance nor alignment between the GAL4 binding site and the TATA sequence is required, but activation is best when this distance is not too large. The explanation for the lack of a requirement for alignment seems to lie in the flexibility of the C-terminal segment of GAL4, which carries a large negative charge and may bind to the repeated C-terminal sequence of RNA polymerase II to activate it.[345] Many other promoter-specific activators are known.

Mediators and coactivators. Transcriptional activators that act in a crude cell-free system often do not function with purified DNA, RNA polymerase, and the basal transcription factors as indicated in Eq. 28-5. Studies with yeast, *Drosophila*, and human cells revealed that additional large multisubunit complexes known as mediators are needed.[272,346–348] A yeast mediator complex consists of 20 subunits.[349–350b] Many activator proteins bind to the DNA sequences known as enchancers, discussed in the next section. Mediator complexes may also interact with enhancer-bound activators. Individual proteins, such as the TAF subunits, that bind to and cooperate with activator proteins are often called coactivators.[351]

Enhancers. Complex DNA sequences called enhancers help to regulate transcription of many eukaryotic genes. The first of these was discovered in an upstream control region of the virus SV40 DNA and consists of two repeats of a 72-bp sequence.[47,352,353] The presence of an enhancer sequence may cause as much as 100- to 1000-fold increase in the rate of transcription as compared with the same transcriptional unit from which the enhancer has been deleted. A surprising fact is that enhancers as far as 1–2 kbp upstream or even far downstream of the promoter and in either of the two possible orientations are effective. This finding suggested that enhancers induce long-range conformational alterations in DNA. Alternatively, they might contain points of entry for RNA polymerase or for an initiation factor that could move along the DNA to the promoter region. However, the synthetic DNA molecule shown in Fig. 28-14 contains two copies of an enhancer in opposite orientations in one strand but none in the other strand.[354] The

TABLE 28-1
Nucleotide Sequences of Some Commonly Found Regulatory Elements in Promoters and Upstream Activator Sites

Sequence[a] transcription	*Description. Positions are relative to the start site*
RNA polymerase binding sites:	
5'-**TAT**AA**T**[b]	Bacterial –10 or Pribnow promoter sequence
5'-**TT**GACA	Bacterial –35 region promoter sequence
5'-(C/T)TTA(A/T)Ann	Archaeal –30 region, TBP binding site[c]
5'-TATA(A/T)A(A/T) or 5'-TATA@A@n[d]	Eukaryotic –30 region; yeast –60 to –120 region
Upstream eukaryotic promoter sites:	
5'-GGGCGG	–100 region, Sp1
5'-CCAAT	–75 region, CTF
Small GC clusters ~5 bp apart	Binding site for TFIIIA
5'-GCGCC-C	~ –5; "discriminator" sequence: inhibition of gene expression by ppGpp
Enhancer elements and transcription factors:	
5'-ATGA(C/G)TCAT	AP-1, cJun, GCN4 (yeast)
5'-CCCCAGGC	AP-2
5'-CAC(G/T)	Myc / Max heterodimer
5'-ATGACGTCAT	CRE (cAMP responsive element)
5'-GGTCAnnnTGACC	Estrogen-responsive element
5'-GGGTGAnnnGGGTGA	Vitamin D-responsive element; direct repeats
5'-CC(A/T)$_6$GC	SRE (serum response element)
5'-ATGCAAAT	Homeotic genes; Oct-2
5'-GGTCAnnnTGACC	ERE (erythroid responsive element)

[a] The sequences are all for the sense strand of the DNA, n = any base. See Keller, W., König, P., and Richmond, T. J. (1995) *J. Mol. Biol.* **254**, 657–667; Adams, R. L. P., Knowler, J. T., and Leader, D. P. (1992) *The Biochemistry of the Nucleic Acids*, 11th ed., Chapman & Hall, London; Lewin, B. (2000) *Genes VII*, Oxford Univ. Press, New York.

[b] Consensus sequence. The bases in boldface are the most highly conserved.

[c] From DeDecker, B. S., O'Brien, R., Fleming, P. J., Geiger, J. H., Jackson, S. P., and Sigler, P. B. (1996) *J. Mol. Biol.* **264**, 1072–1084.

[d] The symbol @ refers to either A/T or T/A. See Juo, Z. S., Chiu, T. K., Leiberman, P. M., Baikalov, I., Berk, A. J., and Dickerson, R. E. (1996) *J. Mol. Biol.* **261**, 239–254.

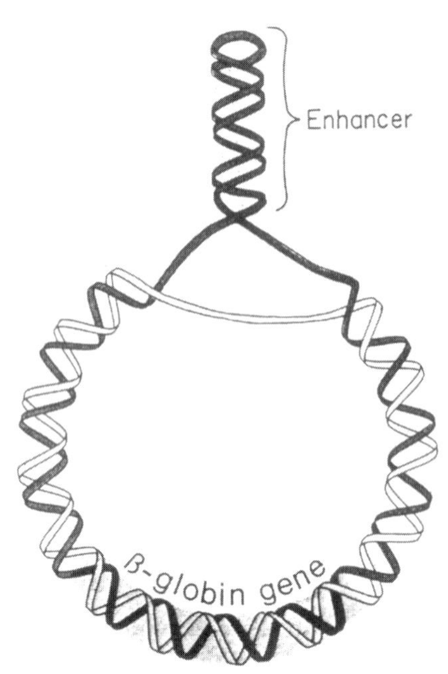

Figure 28-14 A "tailed circle" consisting of an enhancer linked to, but topologically separated from, a gene. One of the DNA strands of this plasmid bears two copies of an SV40 enhancer sequence, one copy inverted with respect to the other. This extra region protrudes from the circle and self-pairs to form a functional enhancer. The main body of the circle contains the β-globin gene, transcription of which is increased by the enhancer. Twisting of the enhancer has no effect on the winding of the strands on the main body of the circle; nevertheless, the enhancer efficiently increases β-globin transcription. From Ptashne.[355]

TABLE 28-2
Functional Classification of Positive-Acting Eukaryotic Transcription Factors[a]

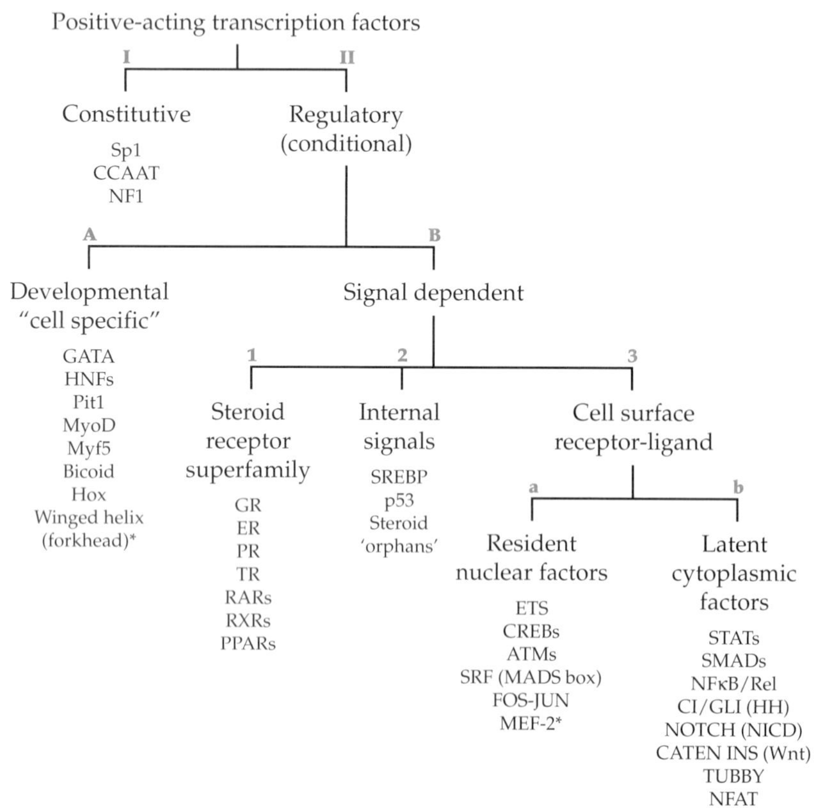

[a] Major groups are labeled in large type and specific examples are listed below (green). Some bacterial proteins are included. Proteins designated by asterisks can be trapped in the cytoplasm by phosphorylation of serine site chains. From Brivanlou and Darnell.[338a]

enhancer is thus topologically separated from the globin gene present in the DNA. Nevertheless, the enhancer functioned efficiently. This suggested that enhancers are protein-binding sequences and that the bound proteins may, perhaps, close a loop to affect the transcription initiation complex.[149,355,356] An enhancer may control a whole loop of DNA in a chromosome affecting many genes.[357] Enhancer DNA sequences are often complex or *modular*, consisting of several shorter elements. For example, the SV40 enhancer contains 8- to 10-bp sequences that are repeated with minor variations and are also present in other enhancers. These are sites for the binding of activator proteins, some of which carry messages from signaling cascades such as the MAP kinase pathway of Fig. 11-13. These pathways are central to the control of cell growth. Their activation is recognized as a mitogenic response. Some enhancers (*nonmodular*) are more compact and directly tied to the RNA pol complex. An example is the human interferon-β (IFN-β) enhancer.[357a,b] It responds cooperatively to three separate activator proteins: NF-κB, interferon regulatory factors, and ATF-2/c-Jun together with an architectural HMG protein. These form an **enhanceosome**, which interacts directly with the transcriptosome complex. The whole complex is sometimes referred to as a cis-regulatory module (CRM). The DNA domains affected by enhancers may be separated by insulator or boundary regions (p. 1626).

5. The Variety of DNA-Binding Proteins

A large number of proteins, often present in very small amounts, bind to DNA and may affect transcription. Some, such as the histones, are relatively nonselective. The more specific transcription factors are often capable of binding tightly, but may do so only upon allosteric activation. Architectures of transcription factors vary. Some place an α helix in the major groove of B-DNA while for others β strands or peptide loops may interact with the DNA in the wide groove. Many transcription factors bind only to a bent or distorted DNA helix, often with a broadened groove. Intercalation of groups from other domains of a transcription factor within the minor groove may help to bring about necessary distortion.[358] Transcription factors sometimes, perhaps often, have other roles, e.g., as enzymes. For example, a pterin dehydratase acts as a transcriptional cofactor in liver cells.[359]

While architectures vary greatly, the DNA-binding domain or domains are usually rich in positively charged side chains, which may interact directly by hydrogen bonding with the charged phosphate groups of the DNA backbone or indirectly via bound water molecules (see Fig. 5-36).

Leucine zipper transcription factors. A large family of DNA enhancer-binding proteins, which are involved in regulating cell growth in specialized cells of differentiated tissues and also in yeast cells, have related structure domains with a characteristic coiled coil that holds two subunits together.[360,361] An example is the Max structure shown in Fig. 2-21. One of the first proteins of this type to be recognized was the mammalian phorbol ester-induced **cell activator protein AP-1**, which is a heterodimer of the protein **cJun**, encoded by cellular protooncogene *c-jun*, and **cFos**, encoded by protooncogene *c-fos*.[362] The heterodimer binds to the palindromic sequence 5'-TGACTCA. Each of the monomeric proteins c-jun and c-fos, as well as other members of the leucine zipper family, has an N-terminal DNA-binding domain rich in positively charged basic amino acid side chains, an **activation domain** that can interact with other proteins in the initiation complex, and the leucine-rich dimerization domain.[363] The parallel coiled-coil structure (Fig. 2-21) allows for formation of either homodimers or heterodimers. However, cFos alone does not bind to DNA significantly and the cJun/cFos heterodimer binds much more tightly than does cJun alone.[364] The yeast transcriptional activator protein GCN4 binds to the same 5'-TGACTCA sequence as does the mammalian AP-1 and also has a leucine zipper structure.[360,364,365]

Several mammalian leucine zipper proteins bind to the CCAAT sequence (Table 28-1) and are, therefore, as a family designated C/EBP.[361,366,367] A 30-residue segment of C/EBP contains four leucine residues at 7-residue intervals. When plotted as a helical wheel (Fig. 2-20) the four leucines are aligned on one side.[368] Similar sequences are present in the proteins cMyc, cJun, and cFos and in GCN4. These observations suggested that if the peptide sequence forms an α helix, the leucine side chain from two identical subunits or closely related proteins might interdigitate in a knobs-in-holes fashion to form the leucine zipper (Fig. 2-21).[368] The structures of these leucine zipper proteins have now been thoroughly investigated and verified by X-ray and NMR methods.[360,369–371] Mutational alterations in the zipper regions of these proteins decreases both activation[372] and dimerization.[373] The CCAAT sequence is found in many enhancers[366,374,375] and is present in ~25% of all eukaryotic promoters that function in differentiated tissues. A trimeric protein known as NF-Y binds specifically to this sequence.[366] Its subunits contain glutamine-rich domains and histone-fold domains that suggest formation of nucleosome-like structures.[366,376]

Another transcriptional activator of the leucine zipper class is **Myc**, a product of the *c-myc* oncogene (Chapter 11) and a key regulator of both cell growth and programmed cell death (apoptosis).[377–379a] It binds to the sequence 5'-CACGTG, and its binding is greatly enhanced by formation of the heterodimer **Myc/Max** (Fig. 2-21). Max is not an activator and may dimerize with certain other proteins to become an inhibitor of transcription. However, it is ubiquitously present and ready to join with Myc to activate an appropriate series of genes. A related enhancer sequence 5'-TGACGTCA is a **cyclic AMP response element** (CRE) that functions within hormone-responsive tissues that use cAMP as a second messenger.[364,380] Cyclic AMP activates protein kinase A (Fig. 11-4), whose catalytic subunit diffuses into the nucleus and phosphorylates the cyclic AMP-response element-binding protein **CREB**, a coactivator that binds to the CRE and which also contacts the general transcriptional factor complex.[365,381–382b] Transcriptional responses to cAMP are quite complex. For example, activation of the phosphoenolpyruvate carboxykinase (PEPCK) gene is maximal only when CREB, C/EBP, and AP-1 are all bound at adjacent DNA sites.[382c] CREB (also known as ATF) is a family of proteins that control the activities of hundreds of genes. Participating in this control is a coactivator, the CREB-binding protein (**CBP**),[382d] which is, in turn, subject to control by methylation.[383]

Control of growth. A large variety of transcription factors control cell growth. Some of these are indicated in Figs. 11-14 and 11-15 and Table 28-2. Since growth in cell numbers requires completion of the cell cycle, the specialized transcription factors involved are necessary. As is indicated in Fig. 11-5, factor E2F is of central importance. In fact, there are at least six mammalian E2F proteins, five of which have both a conserved DNA-binding domain and conserved activation domains.[383a,b] E2F1, 2, and 3 can all induce the S phase of the cycle. E2F6, in contrast, appears to bind to EF2-binding sites in DNA and also to Myc-binding sites to silence these genes and to help keep the cell in the G_0 state.[383c,d] As shown in Fig. 11-5, the retinoblastoma protein Rb also binds to EF2 and represses transcription. However, it allows transcription when phosphorylated.[383e]

Among the many other proteins that influence growth are cFos and cJun, which may be activated by the MAP kinase pathway (Fig. 11-13).[383f,g] Binding of cytokines (pp. 1571, 1845) activates signaling pathways from cell membrane receptors to two other families of transcription factors. These are the **STATS** (p. 1845)[383h–j] and **SMADS**.[383h,k,l] Upon activation STATS and SMADS move from the cytosol into the nucleus where they find their binding sites on DNA.

Response elements. DNA binding sites for activator proteins are often described as "response elements."[47] Thus, the site for the cAMP-responsive protein CREB is the response element CRE. The binding site for AP-1 is **TRE**, named after the phorbol ester TPA. HSE is the heat shock response element, GRE the glucocorticoid response element (or the glucose response element), and **SRE** the sterol regulatory element.[384,384a] Response elements tend to be present in many enhancers and cooperate with other enhancer-binding proteins to activate groups of genes. An especially large group of genes respond to the **serum-response element** (also SRE), which is found within the *c-fos* promoter region.[384b,c] It is the DNA binding site for the ubiquitous serum protein SRF (serum response factor),[385–387] which is involved in growth control, cell cycle progression, and wound repair.

Zinc-containing transcription factors. The zinc finger domain (Fig. 5-37), which is also designated Krüppel-like finger in reference to a *Drosophila* protein,[388,388a] is a repeated motif present in many transcription factors (see Fig. 5-38).[389–391] In each finger a Zn^{2+} ion is coordinated by two –SH groups and two imidazole groups to form the Cys_2His_2Zn domain. One of the first proteins in which zinc fingers were recognized is TFIIIA, one of the factors that controls transcription of 5S RNA genes. A large 30-kDa N-terminal domain contains nine ~30-residue repeats of the sequence $XF/YXCX_{2-4}CX_3FX_{4-5}LX_{2-3}HX_{3-4}HX_{2-6}$ where X can be any amino acid. The fact that TFIIIA binds 7–11 Zn^{2+} ions per polypeptide chain suggested that the repeated sequences might be Zn^{2+}-binding domains in which each Zn^{2+} is held by two cysteine and two histidine side chains. Each of the nine Zn^{2+}-binding domains might constitute a metal-binding "finger" able to interact with about five bases in the DNA.[391–394] In agreement with this idea was the observation of an ~5-bp repeat of guanine clusters in the DNA.[395] The three-dimensional structures of a large fragment of the TFIIIA N-terminal domain (see title page banner for this chapter)[396] and of numerous other zinc finger proteins are known.

Yeast proteins often contain a pair of zinc fingers, but in the nematode *Caenorhabditis elegans* and in *Drosophila* there are more proteins with three or more zinc fingers.[391] The previously discussed Sp1 has three.[390] In *C. elegans* there are more than 100 genes that encode proteins with the Cys_2His_2 zinc-binding motif.[391] In addition, there are many proteins with four-cysteine zinc-binding motifs. These include Cys_4 Zn proteins of the **GATA family** of transcription factors, which are found in fungi, plants, and animals.[397,397a] GATA-1 is a specific transcription factor for regulation of erythroid genes. It binds to the consensus sequence 5'(T/A)GATA(A/G) found in globin genes.[398] In fungi members of the GATA family regulate nitrogen metabolism, biosynthesis of siderophores, and uptake of iron.[397,399] Another family (LIM) has $Cys_2HisCysZn$ domains.[391] A widely distributed motif in transcriptional repressors is a $Cys_3HisCys_4Zn_2$ or RING finger domain.[400]

If zinc-containing domains lose their Zn^{2+} they do not bind tightly to DNA. Regulation of the flow of zinc ions from storage sites in metallothioneins (Box 6-E) into transcription factors as well as into more than 300 enzymes poses interesting mechanistic questions.[401]

Winged helix transcription factors. Liver-specific expression of certain genes in rats depends upon **hepatocyte nuclear factor-3** (HNF-3). Related proteins are encoded by the **forkhead family** of genes in *Drosophila*. These proteins have characteristic C-terminal DNA-binding domains, each consisting of three helices, one of which fits into the major groove of DNA. Also present is a twisted three-strand β structure and two flexible loops or "wings."[402,403] The structure of the DNA-binding domain is similar to that of histone H5 and also resembles the HTH domains of prokaryotic repressors (Figs. 5-35, 28-3) and of CAP (Fig. 28-6).[404] There are many members of the HNF-3/Forkhead family of proteins and of the related **Ets-domain** transcription factors.[405–406a]

The NF-κB/Rel proteins. Nuclear factor NF-κB plays a crucial role in cellular immune responses and in inflammatory disease.[407–408c] This transcription factor was first recognized for its function in regulating transcription of the κ light chains of immunoglobulins. It is a member of the larger NF-κB/Rel family, which act in concert with a group of DNA-binding inhibitors of the IκB family.[409] The structure of an NF-κB dimer bound to its DNA target, whose consensus sequence is 5'-GGGRNYYYCC, is shown in Fig. 5-40. Its architecture[410] is quite unlike that of other transcription factors discussed in this book.

HMG proteins as transcription factors. The abundant high mobility group (HMG) nuclear proteins (Chapter 27) bind to DNA, some of them to four-way junctions. The latter may be present in cruciform structures thought to play a role in regulation of transcription.[411] They are often regarded as modulators of chromatin structure. The ~80-residue HMG domain contains three helices and binds into a flattened, underwound, and bent DNA minor groove.[412,413] HMG proteins also act as transcription factors, which may interact directly with TBP, p53, steroid hormone receptors, and enhancers.[414] Cooperative binding with other DNA-binding proteins is characteristic of the effects of HMG proteins.[414a] Such interactions may be affected by acetylation. Members of the enhancer-binding HMG-14 / -17 family undergo acetylation at seven specific sites.[415]

6. The Variety of Transcriptional Responses

Every protein has specialized functions, and specific regulatory mechanisms often control transcription of its genes. A cell must respond to a large number of stimuli, and responses often include activation or repression of transcription.[415a] In some cases an internal signal, such as a change in concentration of a nutrient or a key metabolite, provides the stimulus. In other cases an external stimulus such as heat or light is the inducer. A few examples follow. Others are mentioned throughout the book.

Many types of hormonal response, including those of insulin[415b,416] (Chapter 11), are transcriptionally mediated. This is also true for plants (Chapter 30). Defensive responses of both animals and plants (Chapter 31) are mediated in part by transcriptional responses.

Nuclear hormone receptors. Among the best known transcription factors are a large family of hormone receptors, which not only bind specific hormones but also contain in a central domain a pair of Cys_4Zn fingers that interact with response elements (discussed in Chapter 22).[416a–c] A subfamily of these binding proteins includes receptors for glucocorticoids, progesterone,[417] androgens, and mineralocorticoids. Another member binds the insect hormone 20-hydroxyecdysone (Fig. 22-12) and regulates the puffing seen in giant salivary gland chromosomes of *Drosophila*.[418,418a] A larger subfamily binds estrogens, vitamin D_3, thyroid hormone,[419] and retinoic acid.[420] The vitamin D receptor appears to serve also as a **bile acid sensor**.[420a] The same subfamily also includes many "orphan receptors."[421–424] The latter have been discovered by DNA sequence comparisons and have led through "endocrinology in reverse" to discovery of new hormonal signaling pathways.[425] The binding of a nuclear receptor to its response element in DNA is well illustrated by the estrogen receptor,[426–427b] which binds to a palindromic **estrogen response element** with the consensus sequence 5'-GGTCAnnnTCACC. It is regulated both by hormone-binding and by phosphorylation. The latter is catalyzed by a cyclinA-CDK2 complex in response to cell cycle alterations.[428] The glucocorticoid receptor protein forms a complex with a second protein, which has been identified as the 90-kDa chaperonin hsp90.[429] A **sterol regulatory element-binding protein** functions in a more general way to activate over 20 different genes that encode enzymes needed for synthesis of cholesterol and unsaturated fatty acids by animal cells.[429a,b] The steroid hormones often bind to their receptors in the cytosol, and the resulting complex is translocated into the nucleus (p. 1264).

Nutrient control. In addition to the sterols and fat-soluble vitamins, other dietary constituents are also recognized by transcriptional activators or repressors.[430] These include glucose,[431,432] amino acids,[433] phosphate ions,[434,435] and various metal ions. Best known among the latter is iron. The mammalian **iron response element** (IRE) is a hairpin loop RNA structure, which like the bacterial attenuator system (Fig. 28-9) functions posttranscriptionally.[436–438a] Iron regulatory proteins (IRP1 and IRP2), which contain Fe_4S_4 clusters, bind to the IRE sequences and inhibit translation. IRP1 is identical to cytosolic aconitase. A high intracellular iron concentration promotes assembly of the Fe_4S_4 cluster and binding to the IRE (see also Chapter 16). In the green algae *Chlamydomonas* a **copper response element** (GTAC) in DNA induces expression of genes important to copper uptake.[439] Many nutritional response systems have been recognized first in bacteria. For example, *E. coli* controls uptake of molybdate[127a,438,440] and of phosphate[437,441] as well as of sugars and ammonia (discussed in Section A,4).

Stress proteins. Ritossa reported in 1962 that when *Drosophila* were suddenly warmed from 25°C to 36–37°C, a series of new puffs could be observed within the polytene chromosomes of the salivary glands.[442] These same puffs, which were also seen after other stresses, appeared within a few minutes and were associated with formation of new mRNAs. This **heat-shock** phenomenon was subsequently found to be universal. It is observed in all species of animals as well as in plants and bacteria.[443–448]

The principal **heat-shock proteins** (hsp), encoded by the new mRNAs, belong to five conserved classes: hsp100, hsp90, hsp70, hsp60, and small heat shock proteins. The function of some proteins as chaperonins has been discussed in Chapter 10. The *E. coli* chaperonin GroEL, a member of the hsp60 group, forms cylindrical aggregates with chambers in which proteins may fold (Box 7-A).[449] In a similar way a 16.5-kDa small heat-shock protein of *Methanococcus* *jannaschii* forms 24-subunit hollow spheres with octahedral geometry.[450] The structure of the 68- to 70-kDa hsp 70 (dnaK) in *E. coli* has been conserved with high homology throughout evolution. Using DNA chip technology 77 heat-shock genes have been identified recently in *E. coli*.[451]

Induction of heat-shock proteins depends upon a heat-shock promoter element (**HSE**) that binds an activating transcription factor **HSF**.[452–455] An increase in temperature not only induces synthesis of heat-shock proteins but represses synthesis of most other proteins. Thus, in *E. coli* or *Salmonella* a shift from 30°C to 42°C causes the appearance of 13 heat-shock proteins. At 50°C synthesis of almost all other proteins stops. In *E. coli* transcription of heat-shock genes is controlled by alternative factors, σ^{32} and σ^{E}.[456,456a]

In *S. typhimurium* a series of unusual nucleotides such as P^1, P^4-diadenosine 5'-tetraphosphate (ApppA or Ap_4A) accumulate. The related compounds ApppGpp (with a 3' pyrophosphate), ApppG, AppG, and ApppA also accumulate and appear not only in bacteria but in eukaryotes as well.[457] They are formed as a side product in aminoacylation of transfer RNAs (Chapter 29).[458,459] Lee *et al.*[460] proposed that Ap_4A and related nucleotides are formed in response to oxidative stress and serve as **alarmones** that signal the need to reduce transcription of most genes and to increase transcription of genes for protective proteins. However, there is little correlation between the Ap_4A concentration and the heat shock response.[461] See also p. 1715.

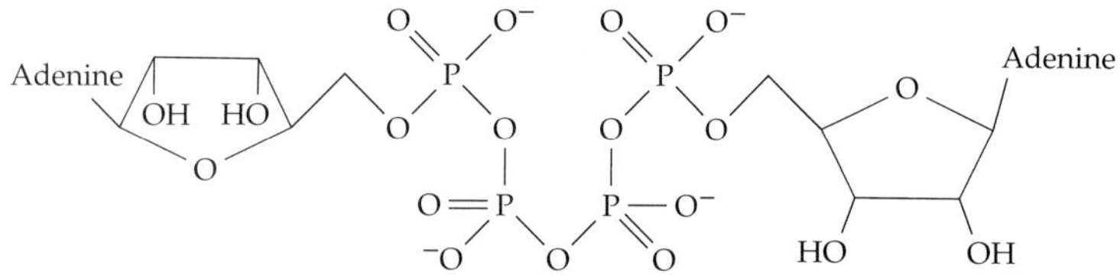

P^1,P^4-Diadenosine 5'-tetraphosphate (ApppA or AP_4A)

Another type of **stress response protein** is related to glutathione *S*-transferase.[462] Cells of *E. coli* also accumulate a 15.8-kDa **universal stress protein** in response to most types of stress. It is an autophosphorylating phosphoprotein, a member of a little-studied group of phosphoproteins normally present in very small amounts.[463] Cells also respond to various other types of stress such as deprivation of glucose, hypoxia,[464] ultraviolet irradiation, presence of hydrogen peroxide,[465,465a] or change in osmotic pressure.[466] Salt-tolerant plants synthesize new proteins in response to increased salinity of water.[467]

Responses to low oxygen tension in tissues (**hypoxia**) are important to all aerobic organisms.[464,467a–d] In mammals transcription of hypoxia-responsive genes is regulated by **hypoxia inducible factors** HIF-1 and HIF-2. A subunit of HIF-1 undergoes 2-oxoglutarate-dependent hydroxylation on proline and asparagine residues. This may be a step in induction of ubiquitination and destruction of this component of the transcription factor complex. The human **von Hippel-Lindau** (VHL) tumor suppressor is a ubiquitin E3 ligase, which is also present in this transcription factor complex.[467b,d–h] One subunit of HIF-1 also interacts with the tumor suppressor p53.[467i] Together with the VHL protein and the elongation factor elongin (p. 1637)[467j] HIF participates in controlling both production of red blood cells and growth of new blood vessels (angiogenesis, Chapter 32, Section D).[467k]

Light-induced transcription. Light has a strong effect on transcription, especially in plants and photosynthetic bacteria. The photosystem II subunits D1, D2, CP47, and CP43 (see Fig. 23-34) are encoded in the chloroplast genome. D1 and D2 are unstable in light, and their rate of synthesis is increased as a result of elevated levels of transcription that are induced by a blue light response.[468] The light-induced conversion of phytochrome to its far-red absorbing form **Pfr** (Eq. 23-42) causes increased transcription of a variety of plant genes.[469] See also Chapter 23.

Homeotic genes and homeodomain proteins. Geneticists discovered in *D. melanogaster* and other species genes that establish the placement of antennae and legs on particular segments and in general to specify the body plan.[470,471] These homeotic genes encode a series of proteins containing a 60-residue **homeodomain**, a DNA-binding domain of the helix–turn–helix class.[470,472] Some homeodomain-containing proteins bind to DNA containing the octameric sequence shown in Table 28-1 and are known as **octamer-binding transcription factors** (Oct). One of these (Oct-2) is specifically needed for activation of immunoglobulin genes,[173,174] while Oct-1 binds to promoters of various other genes including that of histone H2B, U1, and U2 snRNAs.[475,475a] Another transcription factor, **Pit-1**, which activates genes for growth hormone and for prolactin in the pituitary, binds to the same octamer.[476,476a] Homeotic genes are considered further in Chapter 32 and immunoglobulin genes in Chapter 31.

7. Transcription by RNA Polymerases I and III

Promoters for RNA Pol I, like those of Pol II, lie upstream of the initiation site for transcription. At least two transcription factors have been identified[47,477–478a] and vary among species. The human factors bind to a G•C-rich DNA sequence in the –45 to

+20 region and to a related upstream control element **UCE** at –180 to –107.

In vertebrate cells transcription by RNA pol III is controlled by three types of promoter: class 1 (5S RNA), class 2 (tRNAs), and class 3 (U6 snRNA).[47,479,480] Initiation of transcription of class 1 genes requires the 40-kDa **transcription factor TF IIIA**.[479] This was the first eukaryotic transcription factor to be purified and characterized,[481,482] and one of the first of the zinc finger proteins to be recognized. It was a surprise to find that TFIIIA does not bind to the promoter region but to a DNA sequence in the center of the 5S RNA gene between positions +55 and +80.[47] The presence of the TFIIIA binding site in the middle of the gene it controls suggests that TFIIIA interacts with other proteins that bind in the promoter region to form a loop. TFIIIA is involved in initiation of transcription of the ~24,000 oocyte type 5S RNA genes of *Xenopus*, but the ~400 somatic type 5S RNA genes are not activated in the same way. TFIIIA is also unusual in that it binds not only to DNA but also to a specific CCUGG sequence in the transcribed 5S rRNA. This RNA is stored as a 7S ribonucleoprotein particle until it is needed after the oocyte is fertilized and begins rapid protein synthesis.[483]

After TFIIIA binds, proteins TFIIIC and then TFIIIB also bind. Although promoters of classes 2 and 3 do not require TFIIIA, all three classes depend upon TFIIIB and TFIIIC.[484] The TATA-binding protein TBF is one of three components present in TFIIIB, which may be regarded as the true initiation factor.[47,485] Both TFIIIA and TFIIIC can be described as **assembly factors**.[47] A silkworm RNA pol III has been reported to require a transcription factor consisting of RNA.[486]

8. Elongation, Modification, and Termination of Transcription

As in prokaryotic transcription (Fig. 28-4) elongation by RNA polymerase II occurs within a transcription bubble of ~20–30 nucleotides in length.[487] Most transcriptionally active DNA is still in the form of nucleosomes, which must be unwound as the transcription bubble moves. Details are still uncertain.[269,488] All of the major steps in processing of the pre-mRNA transcripts, which include capping, splicing, 3'-end cleavage, and polyadenylation (Eq. 28-6), are coupled to transcription. This is apparently accomplished, in part, by physical connections of the necessary proteins to the CTD domain of RNA polymerase II.[304,312a,b] While pre-mRNA usually undergoes all of the steps of Eq. 28-6, rRNA and tRNAs are not capped or polyadenylated and often are not spliced.

Elongation of the RNA chain depends upon five **general elongation factors**, designated **P-TEFb**, **SII** (TFIIS), **TFIIF**, **Elongin** (SIII), and **ELL**.[489–492] Many of

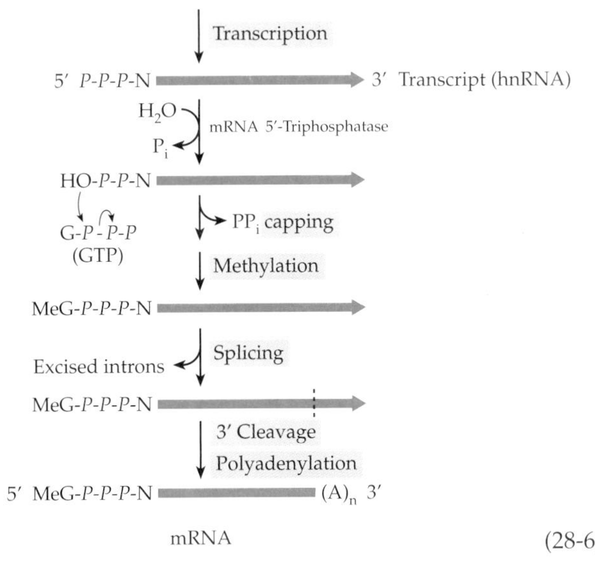

(28-6)

these subunits of the transcriptosome function in the suppression of pausing in the uneven movement of the template DNA through the complex. Some may have to deal with torsional strain induced in the transcribed DNA.[493] Elongin is a heterotrimer of subunits A, B, and C.[494] Most of the other elongation factors are heterodimers,[489] but SII is a monomer that contains two conserved Zn^{2+}-binding **zinc ribbon** motifs.[492] It promotes cleavage of the growing transcript within the transcription bubble at stalled sites, allowing transcription to be restarted from a fresh 3' end. Defects in elongation factors P-TEFb, ELL, and **CSB** (Cockayne syndrome complementation group) have been correlated with human diseases including cancer.[495]

Despite the complexity of the processes represented by Eq. 28-6, a yeast cell is able to transcribe genes at rates of about one in every 6–8 s.[496] This can be compared with a rate of about once in 2–3 s for RNA polymerase of *E. coli*.

The least well defined step in eukaryotic transcription is termination, which follows the various steps of processing, discussed in Section D. The final 3' end of processed transcripts of RNA pol II action in mammals is marked by the sequence AAUAAA, which is found about 10–30 nucleotides upstream from the end.[497] This is usually followed by a polyU or GU-rich sequence.[498] In yeast the termination and polyadenylation signals are less clear.[499,500] The initial transcript almost always continues beyond the AAUAAA signal, sometimes for hundreds of nucleotides. However, the excess RNA is rapidly degraded by a large complex of proteins.[499–502] The precise 3' end cleavage is energy-dependent, requiring creatine phosphate rather than ATP or GTP.[500] Transcription termination by RNA polymerases I and III is more like that of bacteria. Terminator sequences are present in the DNA, and terminator proteins interact with them.[480,503–504b]

9. Conformational Properties of RNA

Newly formed RNA transcripts fold quickly into structures of complex shapes,[505–508b] the folding being influenced by interactions with proteins and with other RNA molecules. RNA chains are flexible, with many sterically allowed conformations.[509] As with proteins folding probably begins with a nucleation event, perhaps involving monovalent or divalent metal ions[510–512] and continues rapidly.[513] Folding is affected by hydrogen bonding,[514] base stacking, and binding of ions, and by formation of pseudoknots (Fig. 5-29).[515]

As is apparent from the structures of tRNAs (Figs. 5-30, 5-31, 28-20), the *Tetrahymena* self-splicing ribozyme (Fig. 12-26), and ribosomal RNA structures (Fig. 29-2), a large fraction of a folded RNA exists as hairpin or **stem –loop** structures. These are A-type structures with largely Watson–Crick base pairs. However, mismatched pairs, triples, and quadruples of bases are also formed. Recently discovered RNA structural elements include **base platforms**, formed by pairing of adjacent bases,[516] interdigitation of unpaired bases (also seen in DNA; Fig. 5-27), and wobble pairs (Chapter 5).[517] Hydrogen bonding between riboses of consecutive nucleotides in two strands may help to form a **ribose zipper**.[437,507] Guanine-rich tetraplexes (pp. 208, 227), cytosine-rich i-motif structures (p. 228), and water-mediated U•C base pairs also arise in RNAs.[517a–c]

The terminal loops, which usually contain the consensus sequence **GNRA**, may constitute up to one-third of the entire molecule.[518] These loops interact with many binding proteins, such as those in the snRNA-protein particles.[519] GNRA loops may also dock into the shallow groove of RNA helices.[518] Adenosines that are not paired in double helices, e.g., those in GNRA loops, are able to interact in a variety of ways with other parts of an RNA molecule or with other molecules. They are involved in helix packing interactions in virtually every RNA studied.[517d] Although examples are still rare, specific mRNA molecules may provide binding pockets for small regulatory molecules, e.g., amino acids and thiamin.[517e]

D. Processing of Eukaryotic RNAs

All RNA found in eukaryotes undergoes major alterations prior to functioning. The cutting out of rRNA and tRNA molecules from larger precursors resembles that in bacteria, but subsequent processing is much more complex, as is that of mRNA.

1. Ribosomal RNA

Eukaryotic ribosomes contain four pieces of RNA (Tables 5-4 and 29-1), which are usually designated by their sedimentation coefficients. The 18S, 5.8S, and 28S RNAs are encoded as single transcriptional units with spacers separating the sequences that encode the mature RNAs. A typical animal cell contains several hundred copies of this transcriptional unit, all located in the DNA in the nucleolus (Fig. 28-15), and each having its own set of promoter sequences, enhancers, and transcription factors.[47,520–522] The promoter sequences vary substantially among different species.[523] The primary transcripts from these units are the sole product of RNA polymerase I.

Electron micrographs of portions of unwound cores of nucleoli have revealed fibrils of RNA coated with protein growing from the DNA strands of the pre-rRNA genes (Fig. 28-16), ~80–100 RNA chains of different length being transcribed concurrently from a single gene. The overall gene length in the electron microscope is 2.3 µm, only a little less than the calculated length for a fully extended DNA molecule in

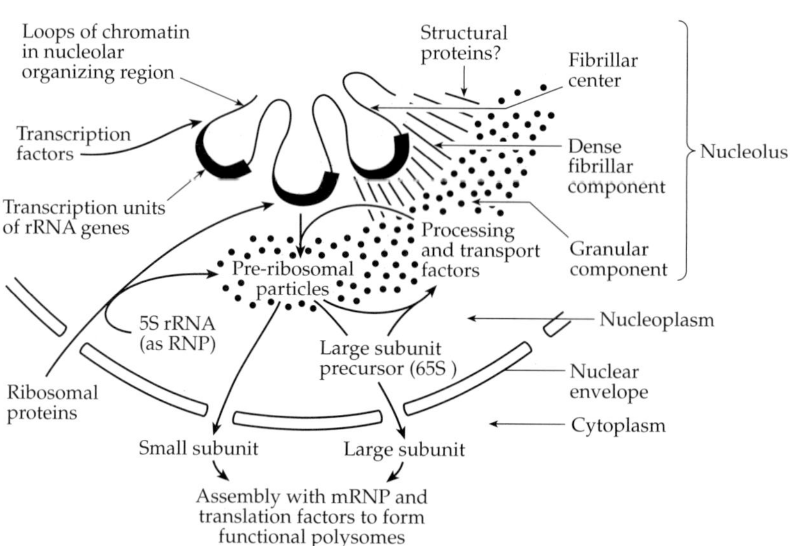

Figure 28-15 Steps in ribosome formation in the nucleolus. From Sommerville.[524] Abbreviations are NOR, nucleolar organizing regions; RNP, ribonucleoproteins.

the B form. However, judging by the lengths of the transcripts formed, the pre-rRNA chains are folded extensively.

The primary eukaryotic rRNA transcripts extend several hundred nucleotides past the 3' termini of the mature 28S RNA molecules. As the primary transcripts, which are formed in the core region of the nucleolus, move away into the outer cortex of the nucleolus, cleavage occurs in a number of steps (Eq. 28-7).[525,526] Electron microscopy provided the first direct confirmation of the relationship of one pre-rRNA molecule to another (Fig. 28-17).[527,528] The 18S portion of the 45S RNA seen in Fig. 28-17 lies nearest to the 5'-end just as does the 16S rRNA in the large transcript of the prokaryotic rRNA genes (Section A,7).

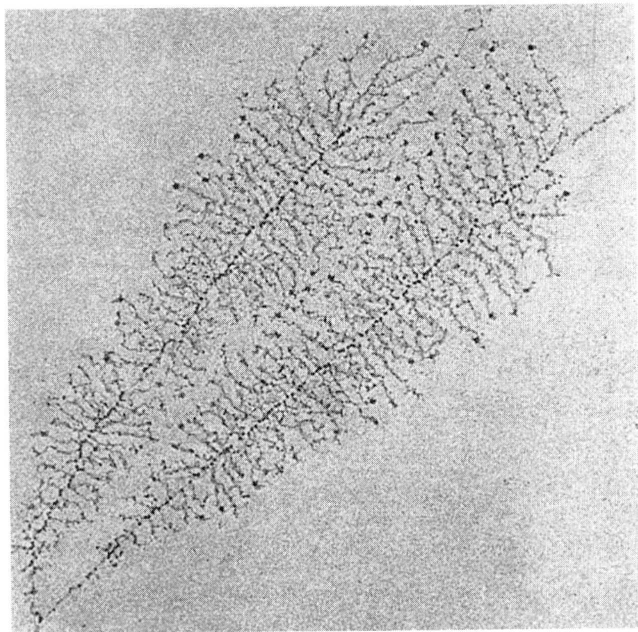

Figure 28-16 Ribosomal RNA genes from an embryo of *Drosophila melanogaster* in the process of transcription. The densely packed ribonucleoprotein strands are shorter where transcription begins and contain increasing lengths of rRNA where transcription has proceeded for a longer time. Also note the characteristic granular knobs at the tips of the strands. From Miller.[85]

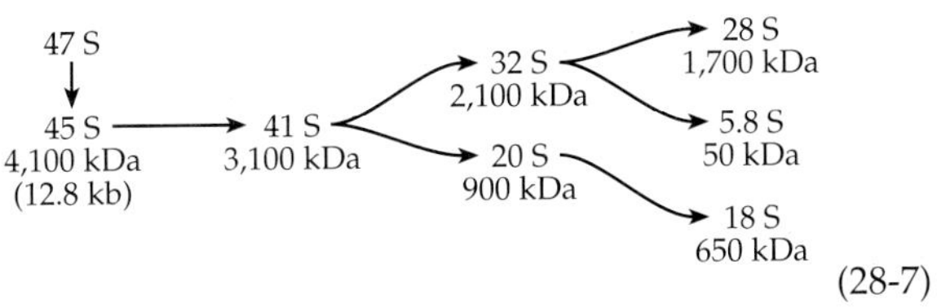

(28-7)

As is indicated in Fig. 28-15, transcription is thought to occur from the loops of DNA that form the nucleolar organizing region. The 100-kDa **nucleolin**, the major protein of the nucleolus, binds to the nontranscribed spacer sequences in the DNA.[529–530] It also binds to the newly formed transcripts, as do various proteins that enter the nucleus from the cytoplasm.[524,531] More than 270 proteins, many of which participate in synthesis of ribosomes, have been detected in the nucleolus.[531a] Some of these proteins, acting together with the snoRNAs discussed in the next section, catalyze hydrolytic cleavage of the pre-rRNA molecules. For completion of pre-ribosomal particles additional protein molecules enter the nucleolus and associate with the pre-rRNA particles, then diffuse out of the nucleus.

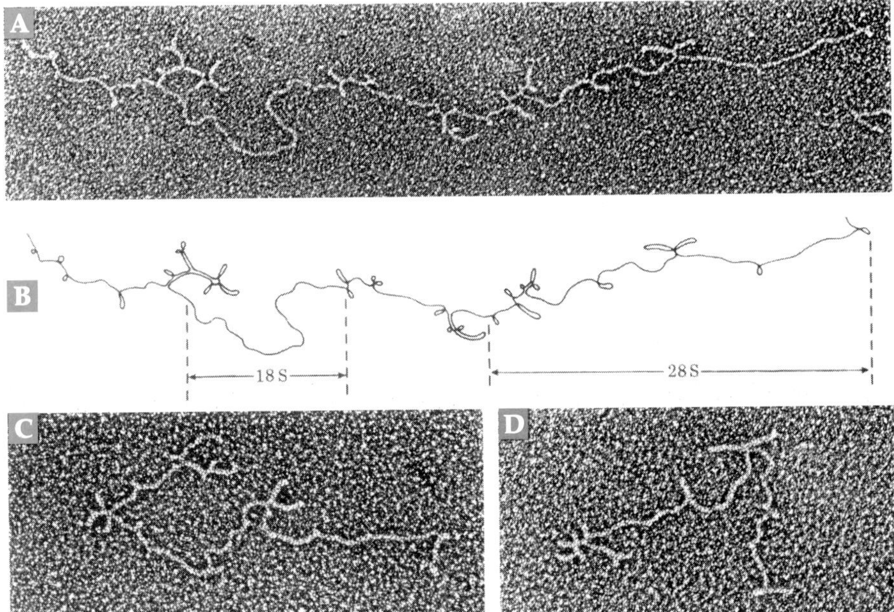

Figure 28-17 (A) Electron micrograph of the 45S precursor of rRNA from HeLa cells after spreading from 80% formamide and 4 M urea. The molecule is shown in reverse contrast. (B) Tracing of molecule in (A) showing several regions of secondary structure as hairpin loops. The 28S and 18S rRNA regions are indicated. (C) 32S rRNA. (D) 28S rRNA. Notice that the same secondary structure can be seen in the 28S RNA as in its 32S and 45S precursors. From Wellauer and Dawid.[527]

2. Small Nuclear, and Nucleolar, and Cytoplasmic RNAs

The nuclei of all eukaryotic cells contain a group of small nuclear RNAs (**snRNAs**), some of which (**snoRNAs**) are localized to the nucleolus and function there.[532–535e] At least ten of these are always present (Table 28-3), and yeast appears to contain more than 100. Some are present in small amounts, and it may be that a large number of snRNAs will also be found in other eukaryotes. The uridine-rich or U series of snRNA are especially abundant (10^4–10^6 molecules per nucleus). Several of them (U1,2,4–6,11,12,16,18) function in RNA splicing. Species U1–U10 contain from ~60 to ~215 nucleotides but some snRNAs from yeast have over 300 nucleotides and one over 1000.[532] The most abundant snRNAs, the metabolically stable U1, U2, U4, U5, and U6, are exported from the nucleus. In the cytoplasm each snRNA associates with a complex of several proteins to form ribonucleoprotein particles (**snRNP particles**). The proteins, known as **Sm proteins**, range in mass from 11 to 70 kDa[536] and are designated B, B′, D1, D2, D3, E, F, and G.[535,537,537a] The proteins may associate to a complex B_2D_2EFG, which then binds the snRNAs. Whereas many proteins are found in most of the snRNP particles, some are associated with specific RNAs. Thus, mature U1 snRNP may have a partial stoichiometry $U1A_2B_2C_2D_4E$.[536,538] The snRNA particle, known as U4 / U6snRNP, contains two snRNAs. After returning to the nucleus the four snRNP complexes U1, U2, U5, and U4 / U6, together with the pre-mRNA molecules, associate via an ordered assembly pathway that gives rise to the large **spliceosomes** in which the removal of introns takes place.[539] Some of the snRNP complexes, e.g. the spliceosomal U6 snRNP and snRNPs involved in mRNA degradation, contain Sm-like (Lsm) proteins.[537a]

More recently small 20–25 nucleotide antisense micro RNAs (**miRNAs**) have become recognized as important in control of the breakdown of mRNA, in formation of heterochromatin, and in control of specific stages in development (Chapter 32), and in cellular defense mechanisms (Chapter 31, Section G).[1a,537b] In animals the 21–23 nt miRNAs are formed by enzymatic cleavage of double-stranded dsRNA by a ribonuclease III-type enzyme called **Dicer**.[171a,537c] In the resulting fragments, which have 5′-phosphate termini, one strand is antisense to a "target" sequence in mRNA. **Small temporal RNAs** (stRNAs) that guide development are cut by Dicer from RNA stem-loop structures. In other cases, e.g., in formation of the siRNAs that silence individual genes, an RNA-dependent RNA polymerase acts on an RNA transcript to form a long dsRNA that can be cleaved by Dicer. This often happens with foreign RNAs, e.g., from viruses or trans-genes. Unwinding of the small fragments formed by Dicer, in a process that may require ATP, provides the single-stranded antisense siRNA molecules. Such antisense molecules bind to their target RNAs together with a group of proteins to form **RNA-induced silencing complexes** (RISCs). These complexes contain an RNAse (different from Dicer) that cuts the dsRNA of an siRNA-target complex, initiating destruction of the mRNA and silencing of its gene.[1a,171a,539a] This **RNA-interference** is widely utilized by plants and animals. It has also provided the basis for development of many practical tools for understanding gene sequences, for genetic engineering, and for design of new drugs.

Processing of ribosomal RNA. Transcripts of rRNA genes vary in size from ~35 to 47S (6–15 kb) and often contain spacer regions at both ends as well as between the 18S, 5.8S, and 28S sequences. For example, human 47S transcripts have 414 extra nucleotides at the 5' ends.[540] One group of snoRNAs participates in the hydrolytic cleavage of pre-rRNA. RNAs U3, U14, snR10, snR30, as well as MRP RNA (Table 28-3) are always required.[541–545] Also needed is U22, an intron-encoded RNA.[531] The reactions represented by Eq. 28-7 are best known for *Saccharomyces cerevisiae*.[525,541–541c] Four spacers, the **5'-external transcribed** spacer (5'-ETS), the first and second **internal transcribed spacers** (ITS1 and ITS2), and the 3'-external transcribed sequence (3'-ETS), must be removed.[525,541,546] Removal of the 5'-ETS depends upon the snoRNA U3, which contains two highly conserved sequences able to form base-paired structures with the 5' end of the 5.8S rRNA region of the pre-rRNA gene.[542–544,546a] Both U8 and U22 are also needed for cleavage of the pre-rRNA.[546b] Although the exact functions of snoRNAs and their associated proteins in the cleavage of pre-rRNA are still uncertain, these RNAs probably act as guide molecules for the cleavage reactions. They may have ribozyme activity and may perhaps be chaperones.

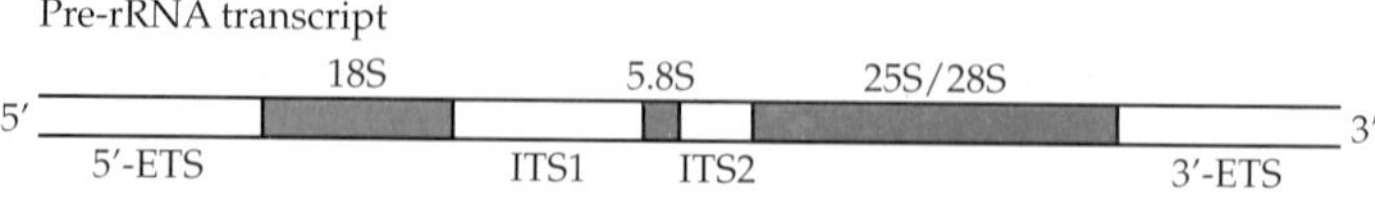

Cleavage at the 5' end of the 5.8S region requires RNase MRP, a relative of the RNase P that cleaves at the 5' ends of tRNAs (Fig. 28-10).[525,547] MRP (**mitochondrial processing protein**) also cleaves primers for mitochondrial DNA replication. The importance of the enzyme is emphasized by the existence of a hereditary defect in the MRP RNA (Table 28-3) that causes abnormalities in bone, cartilage, hair, and the immune system.[547a] Most bacterial rRNA genes have a tRNA gene in the position corresponding to that of 5.8S RNA

TABLE 28-3
Some Eukaryotic Small Nuclear and Nucleolar RNA Molecules

Designation	Number of nucleotides Vertebrate	Yeast
U1[a]	164	568
U2[a]	188	1175
U3[b]	206 – 228	333
U4[a]	142 – 146	160
$U4_{atac}$[c]	131	
U5[a]	116 – 118	183 or 196
U6[a,d]	116	214
$U6_{atac}$[c]	125	
U7[e]	57 – 58	
U8[b]	136 – 140	
U9	130	
U10	60	
U11[c]	131 – 135	
U12[c]	150	
U13[b,f]	105	
U14[b,f]	87– 96	125 – 128
U18[b,f]	67– 70	102
X[b]	150	
U20[b,f]	80	
U21[b,f]	93	
U22(Y)[b]	125	
U24[b,f,g]	77	
U32 – U40[b,f,h]		
SnR10[b]		245
SnR30[b]		605
SnR38[f]		93
SnR39[f]		85
SnR40[f]		96
SnR41[f]		
MRP RNA (RNA 7–2)	260 – 280	339

[a] Major spliceosomal RNAs.
[b] Fibrillarin-associated SnoRNAs that function in pre-ribosomal RNA processing. See Morrissey, J. P., and Tollervey, D. (1995) *Trends Biochem. Sci.* **20**, 78–82.
[c] Function in AT–AC spliceosomes. See Tarn, W.-Y., and Steitz, J. A. (1997) *Trends Biochem. Sci.* **22**, 132–137 and Fournier, M. J., and Maxwell, E. S. (1993) *Trends Biochem. Sci.* **18**, 131–135.
[d] γ-Monomethyl cap.
[e] Required for 3'-end formation in histone mRNAs.
[f] SnoRNAs with long complementarities to rRNA. C and D sequences are present. See Bachellerie, J.-P., Michot, B., Nicoloso, M., Balakin, A., Ni, J., and Fournier, M. J. (1995) *Trends Biochem. Sci.* **20**, 261–264 and Nicoloso, M., Qu, L.-H., Michot, B., and Bachellerie, J.-P. (1996) *J. Mol. Biol.* **260**, 178–195.
[g] Polyadenylated, noncoding.
[h] Participate in 2'-*O*-ribose methylation.

in eukaryotes. This provides an RNase P-dependent cleavage mechanism, which is alternative to action of other nucleases.[525]

The most abundant protein in the fibrillar regions of the nucleus, where the early stages of pre-rRNA processing occur, is **fibrillarin**.[541,548,549] Many of the snoRNAs are closely associated with this protein. Fibrillarin is also well known as an autoantigen, which can induce formation of destructive antibodies that cause **scleroderma** (Chapter 31, Section F).

Modification guide RNAs. A second group of snoRNAs function in methylation, pseudouridine formation, and other RNA modifications (Section 6).[174,525,541,548] These snoRNAs have long sequences complementary to highly conserved regions of pre-rRNA, enabling them to form helical regions that may guide the docking with modification enzymes. Many of them also contain characteristic conserved sequences: **C**, 5'-UGAUGA; **D**, 5'-CUGA; **H**, 5'-AnAnnA; and 5'-ACA. Sequences C and D are present in snoRNAs that act as methylation guides, while the H and ACA sequences characterize guide RNAs for pseudouridine formation.[174]

Transcription and processing of snRNAs and snoRNAs. In higher organisms each of the snRNAs has several genes,[538] e.g., there are 50 –60 U1 genes in the human haploid genome. However, in yeast there are often single copies.[541] All of the snRNA genes, except for that of U6,[550] are transcribed by RNA polymerase II. The transcripts, which are capped at the 5'-end but are not polyadenylated, pass into the cytoplasm, where they undergo further processing and become associated with proteins. The 3' ends are trimmed, the 7-methylguanosine of the cap is methylated further, and methylation may occur on other bases as well.[551] SnoRNAs are not capped. It was a great surprise to discover that many of the snoRNA genes lie within introns that occur in abundantly expressed genes for functionally unrelated proteins.[173,541,548]

Patients with the autoimmune disease systemic **lupus erythematosus** make autoantibodies directed against the Sm proteins of snRNP particles.[552,553] Antibodies from different patients vary in their specificities, making these antibodies a useful tool in the isolation and study of snRNAs and their protein complexes.[552,554]

3. Processing of 5S RNA and tRNAs

The genes for 5S ribosomal RNA and all of the tRNAs are transcribed by RNA polymerase III. In the yeast genome the 5S RNA genes are located in the spacers between the transcriptional units containing the other rRNAs. However, in animals the 5S RNA genes are separate from the other rRNA genes and are not located in the nucleolus. In *Drosophila* ~500 copies of the 5S RNA gene are located in the right arm of chromosome 2. In *Xenopus* ~400 genes are active in somatic cells and another 24,000 only in oocytes. These are arranged as large transcriptional units, each containing several thousand copies of the 120-bp 5S RNA gene separated by 720-bp spacers. Cleavage and trimming are required to form the final product, but there are usually no introns to be removed.

Most eukaryotic tRNAs are formed from monomeric precursors, each gene acting as a transcriptional unit. Processing is similar to that in bacteria (Section A, 7). Eukaryotic RNase P usually cleaves the 5' end,[555–558c] and another enzyme cuts at the 3' end.[542,556] The 3' CCA sequence of the mature tRNA is usually not present in the primary transcript but is added.[559,559a,b] As in bacteria (p. 1620) extensive modification of bases also occurs in the tRNA precursors of eukaryotes.[235,560–562] Many tRNA genes contain introns, which must be removed by splicing (Section 5).

4. Messenger RNA, Caps, and Polyadenylate Tails

The first processing event (Eq. 28-6) for most of the pre-mRNA and snRNA transcripts made by RNA polymerase II is addition to the 5' end of a "cap," a terminal structure containing 7-methylguanosine from which a proton has dissociated to form a dipolar ion.[563–565] The cap structure may be abbreviated 5′-m^{7}G(5')pppNm —. The 5' terminal ribose is often methylated on O2', as shown below. More complex caps are methylated at additional sites, e.g., the guanine may be dimethylated on the 2-NH_2 group.[551] Most snRNAs, including the U1–U5 and U7–U13 snRNAs, have such 2,2,7-trimethylguanosine caps.[551,566] Many viral transcripts, including those of the much-studied vaccinia virus, have similar caps.[564] However, U6 and some other snRNAs, which are transcribed by RNA polymerase III, have γ-monomethyl (me-ppp) caps[567] or undergo a series of additions and deletions of uridylate residues at the 3' ends.[566,568]

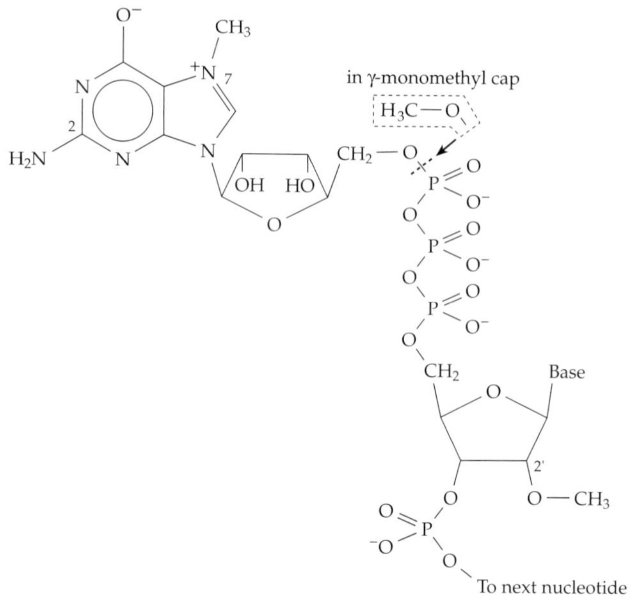

Cap synthesis occurs as follows. The 5' end of an RNA transcript initially contains a triphosphate group arising from the fact that a nucleotide triphosphate serves as the primer in initiating transcription. The terminal phospho group is removed by a triphosphatase leaving a diphosphate, which is then guanylated by GTP (Eq. 28-6).[569] The capped transcripts are exported from the nucleus, after which additional methylation may follow.[570]

The cap structure affects several processes.[565,571] A family of cap-binding proteins recognize the structure and may facilitate splicing as well as export from the nucleus.[571a] The cap is very important for ribosome binding and initiation of translation (Chapter 29). The trimethylated caps of snRNAs, on the other hand, may be signals for retention in the nucleus where they function.[563] Following capping is the often elaborate process of splicing to remove introns (Section 5).

A poly(A) "tail" consisting of ~250 residues of adenylic acid is added next by poly(A) polymerase, a component of an enzyme complex that also cleaves the RNA chains.[545,571b] Most eukaryotic mRNA is polyadenylated with the exception of that encoding histones. The function of the poly(A) is unclear. It is needed for transport of mRNA out of the nucleus, but it does confer a greatly increased stability to the mRNA in the cytoplasm where the adenylate units are gradually removed.[307,308] In contrast, in chloroplasts and plant mitochondria polyadenylation is required for rapid degradation of mRNA.[571c,d] Polyadenylation may also increase the efficiency of translation.[572] Polyadenylation occurs rapidly within ~1 min after transcription is completed.

Mature mRNA molecules vary in lifetime.[573,574] Some last for hours or days. Among the latter are mRNAs of maternal origin that accumulate in oocytes and are utilized during the early stages of embryonic development.[575,575a,b] Other mRNAs, e.g., transcripts of the *c-fos* and *c-myc* proto-oncogene products, have half-lives of 30 min.[573,576] Some mRNA molecules are degraded while attached to ribosomes in response to recognition of the synthesized peptide (Chapter 29). Longer lived mRNA molecules may be protected by RNase inhibitors.[577] Hydrolytic removal of caps often initiates degradation, and Sm-like protein complexes participate.[578–579a]

5. Splicing

An essential modification of the precursor forms

of large rRNA, tRNA, and mRNA molecules is the splicing out of intervening sequences. This occurs prior to polyadenylation of mRNA and is usually slow, the half-life of introns varying from a few seconds to 10–20 min.[551] Splicing occurs by at least four distinctly different pathways.[47,580]

Self-splicing RNA. The precursor to the 26S rRNA of *Tetrahymena* contains a 413-nucleotide intron, which was shown by Cech and coworkers to be self-splicing, i.e., not to require a protein catalyst for maturation.[581,582] This pre-rRNA is a ribozyme with true catalytic properties (Chapter 12). It folds into a complex three-dimensional structure which provides a binding site for free guanosine whose 3'–OH attacks the phosphorus at the 5' end of the intron as shown in Fig. 28-18A, step *a*. The reaction is a simple displacement on phosphorus, a transesterification similar to that in the first step of pancreatic ribonuclease action (Eq. 12-25). The resulting free 3'–OH then attacks the phosphorus atom at the other end of the intron (step *b*) to accomplish the splicing and to release the intron as a linear polynucleotide. The excised intron undergoes a third transesterification reaction, of uncertain significance (step *c*), to form a circular polynucleotide and a short displaced 15-residue oligonucleotide. The *Tetrahymena* pre-rRNA intron is a member of a group of similar **Group I introns**, many of which are found in fungal mitochondrial pre-mRNA and pre-rRNA.[583] All are excised by a similar mechanism. Many are self-splicing, but others require a protein catalyst.[584–585a] A similar splicing sequence is involved in removal of a 1017-nucleotide intron from the thymidylate kinase gene of phage T4 and other introns in T-even phage. The later are among the relatively rare introns in prokaryotic systems.[583]

All Group I introns have several small conserved sequences, which suggest a common folded tertiary structure as is indicated in Fig. 28-19A. The conserved sequences are labeled **A**, **B**, **9L**, **2**, **9R**, and **9R′**. **A** is paired with **B**, **9R** with **9R′,** and **9L** with **2**. The sites of chain cleavage at the 3' and 5' ends of the intron are indicated by the heavy arrows. They are evidently selected by formation of the double-stranded regions.[586–588]

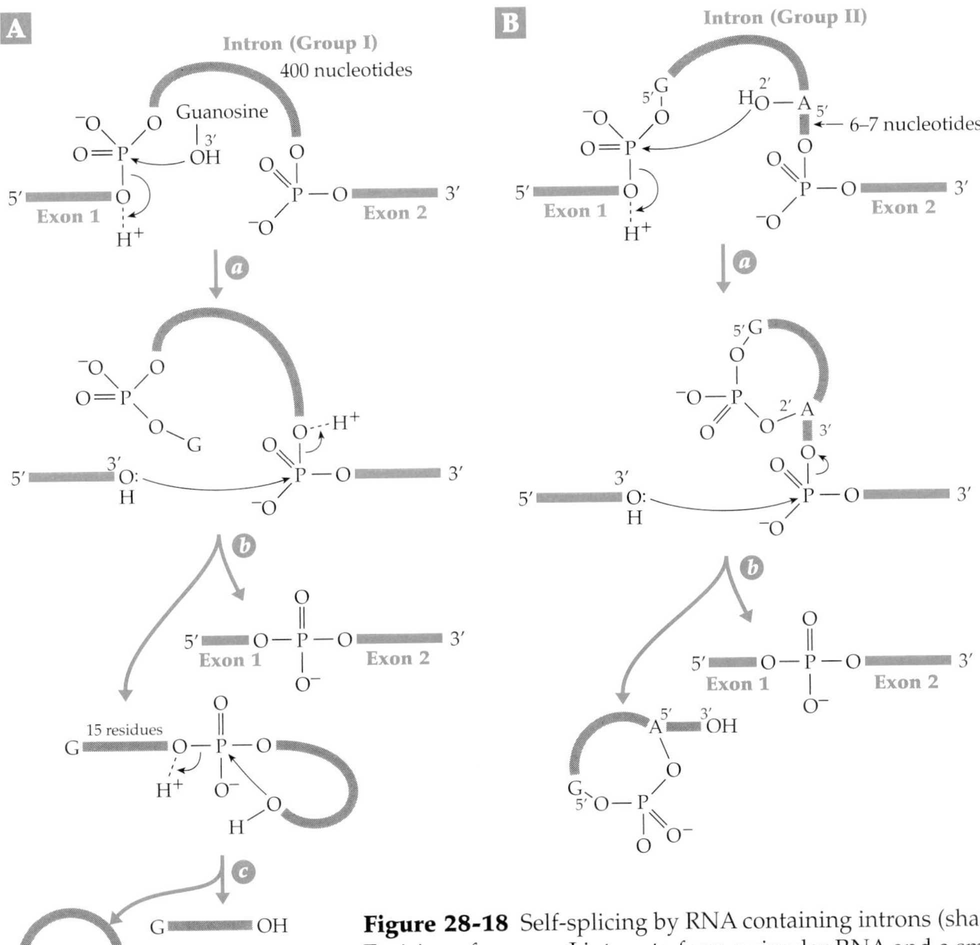

Figure 28-18 Self-splicing by RNA containing introns (shaded) of groups I or II. (A) Excision of a group I intron to form a circular RNA and a small oligonucleotide. (B) Excision of a group II intron from yeast mitochondrial pre-mRNA as a circular RNA. See Cech and Bass.[581]

Another type of intron (Group II) also undergoes self-splicing.[589–590d] The best known example is the last intron in the yeast mitochondrial pre-mRNA. The splicing pathway shown in Fig. 28-18B is similar chemically to that of the group I introns. However, the initial attack is not by free guanosine, but by the 2' OH of an internal adenosine, the intermediate product having a **lariat structure** with a loop at the end. Otherwise, processing is similar to that of group I introns. The same pathway with lariat formation is followed by the more widely used removal of introns from pre-mRNA in spliceosomes (see Fig. 28-22).

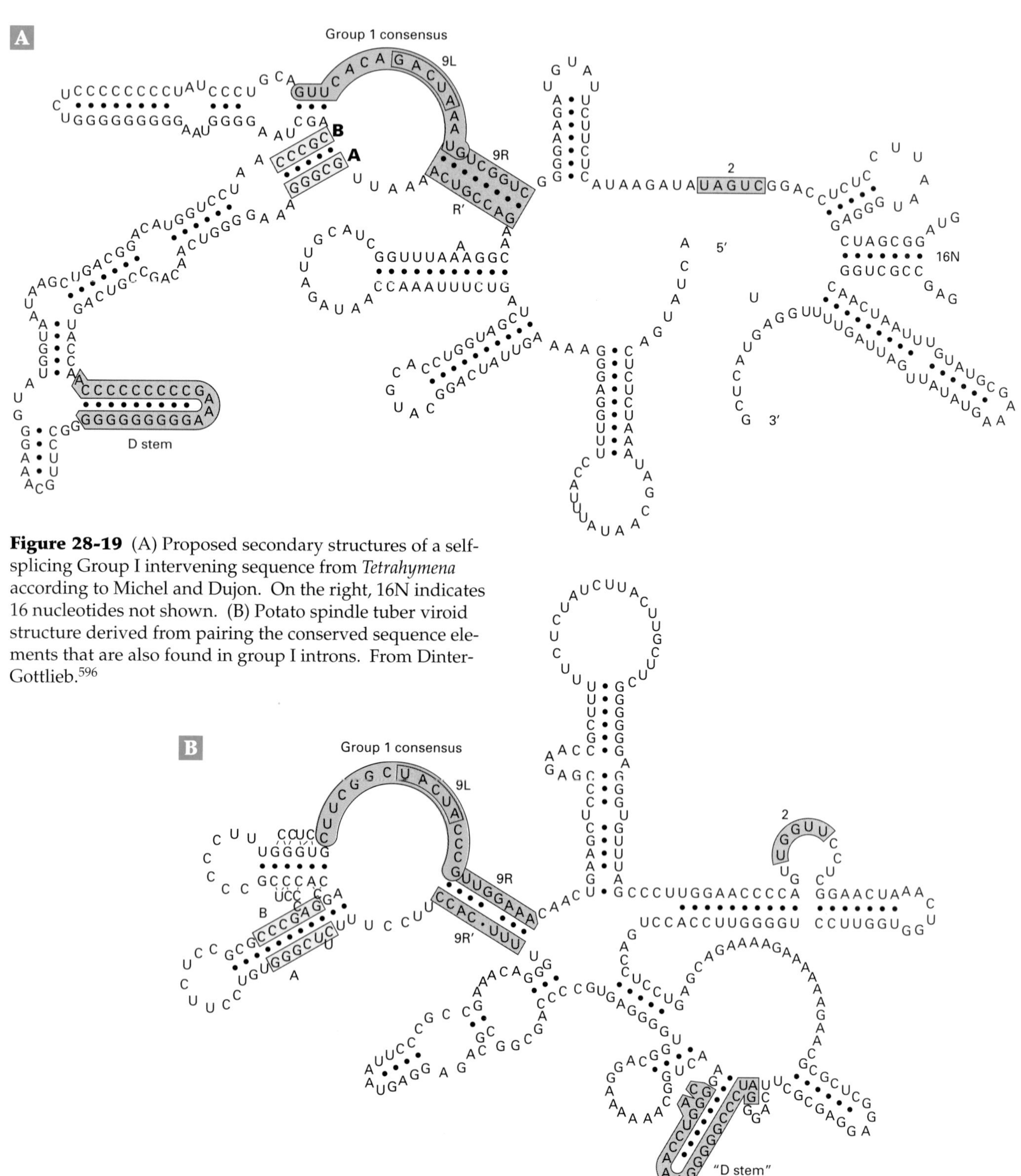

Figure 28-19 (A) Proposed secondary structures of a self-splicing Group I intervening sequence from *Tetrahymena* according to Michel and Dujon. On the right, 16N indicates 16 nucleotides not shown. (B) Potato spindle tuber viroid structure derived from pairing the conserved sequence elements that are also found in group I introns. From Dinter-Gottlieb.[596]

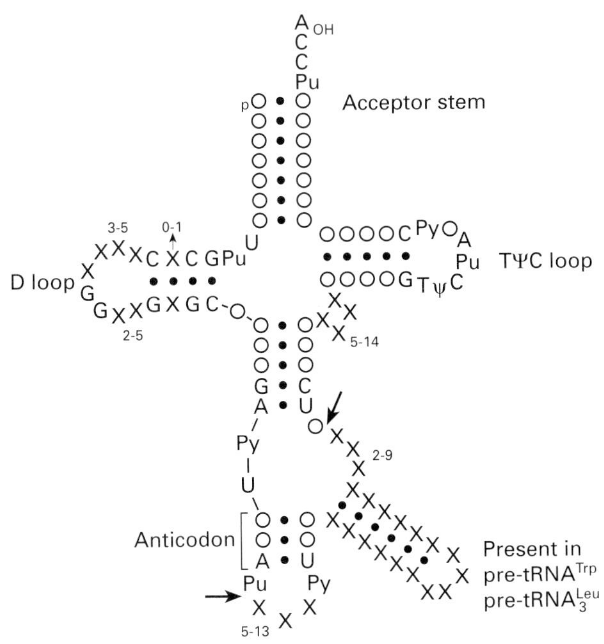

Figure 28-20 Composite structure representing several tRNA precursors arranged in a similar secondary structure (see also Fig. 5-30). The arrows indicate splice points. Variable positions are designated (O) for the mature tRNA and (X) for the intervening sequence and also in loops where insertions or deletions occur. From Ogden *et al.*[603]

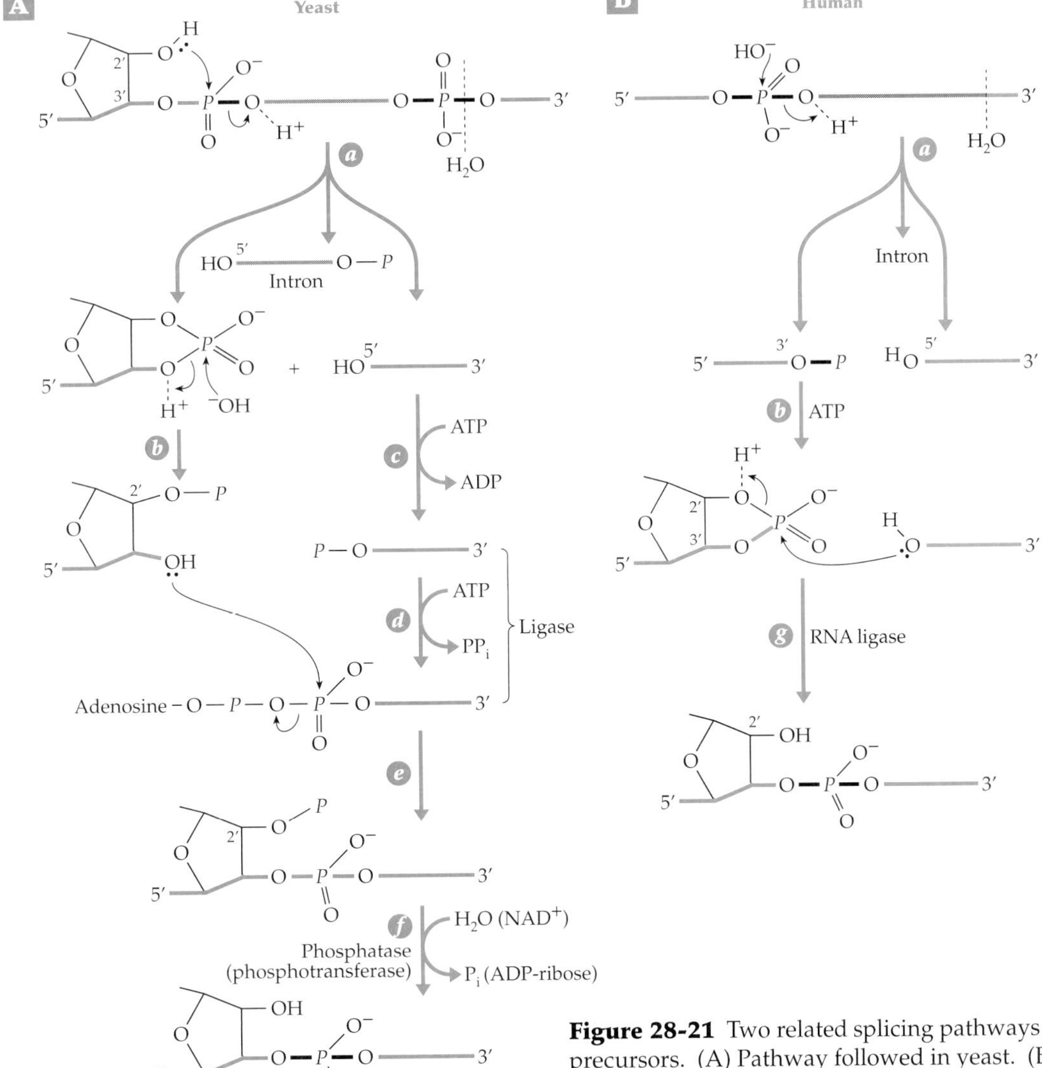

Figure 28-21 Two related splicing pathways for tRNA precursors. (A) Pathway followed in yeast. (B) Related pathway used by HeLa cells.[598,602]

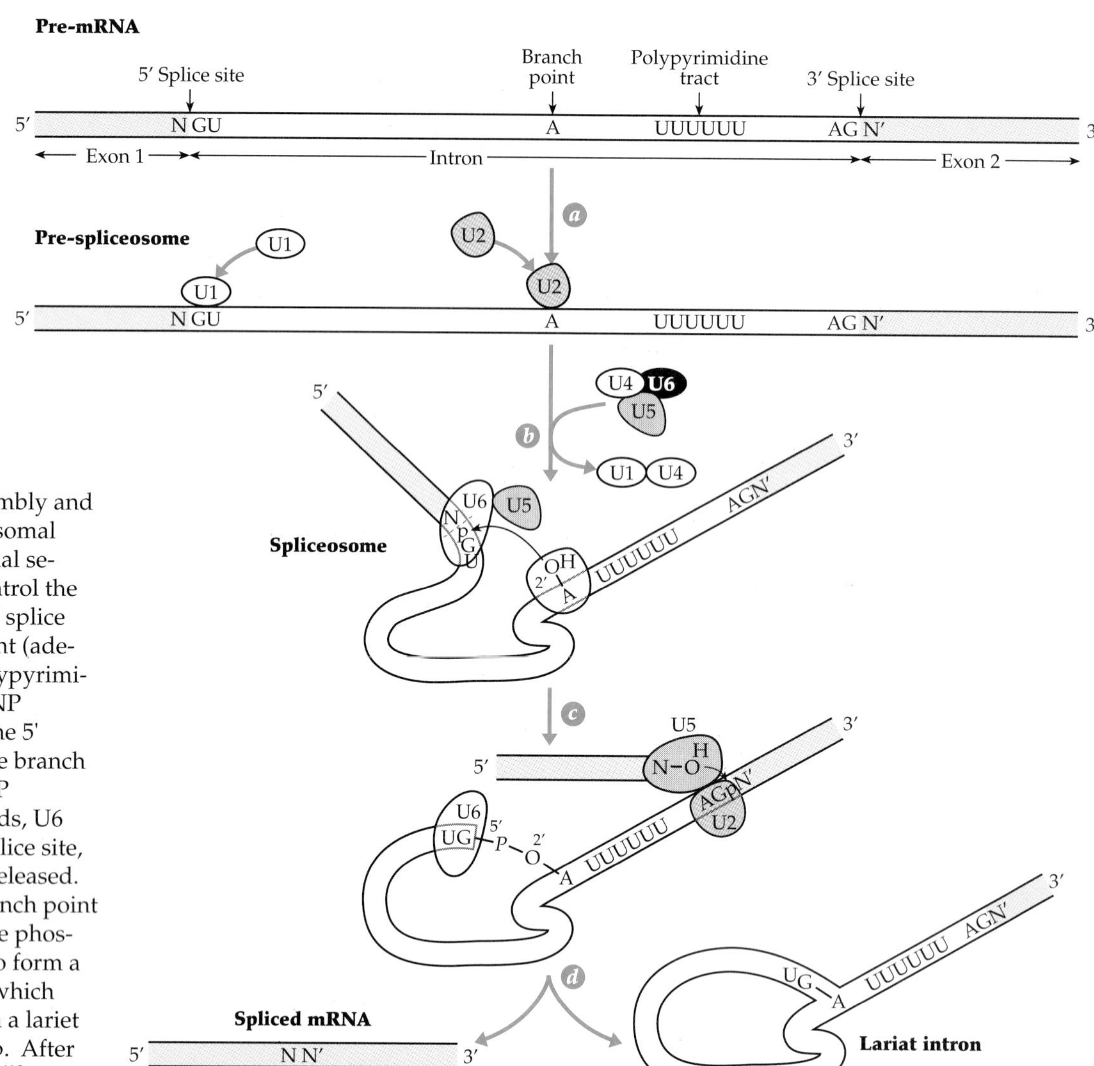

Figure 28-22 Assembly and action of the spliceosomal complex. Four special sequence elements control the process: the 5' and 3' splice sites, the branch point (adenosine A), and a polypyrimidine tract. The snRNP particle U1 locates the 5' splice site and U2 the branch point. The tri-snRNP U4•U6•U5 then binds, U6 recognizing the 5' splice site, and U1 and U4 are released. The 2'-OH of the branch point adenosine attacks the phosphodiester linkage to form a lariet intermediate, which releases the intron in a lariet form in the final step. After Valcárcel and Green.[612]

Relationship to viroids and virusoids. The smallest of viruses are the naked ~250–460 nucleotide single-stranded circular RNA molecules called viroids (Chapter 5).[591–593] Viroids are closely related to group I introns. The conserved paired sequence characteristic of the group I introns are also present in viroids (Fig. 28-19B), and it looks as if viroids may be "escaped introns." Another group of "plant satellite RNAs" or virusoids replicate only with the help of larger RNA viruses. These satellite RNAs are replicated by a rolling-circle mechanism.[593a] The resulting long RNA molecules are self-cleaving, cutting themselves to form the unit length satellite RNAs.[594,595] These are the simplest known self-cleaving RNA molecules. They have been discussed in Chapter 12.

Pre-tRNAs. In the removal of type I introns the formation of specific stem and loop structures directs the splicing reactions (Fig. 28-18).[47,597] Stems and loop structures already exist in tRNA precursors. Cleavage sites are usually located just to the 3' side of the anticodon as is seen in the pre-tRNA molecule and in the composite structure shown in Fig. 28-20. The chemistry of the splicing process[205,598–601] is shown in Fig. 28-21. Cleavage at the two splice sites excises the intron. Then the two ends are rejoined. In yeast cleavage at the 5' end of the intron is by a multimeric enzyme with a pancreatic ribonuclease-type of action that leaves a 2',3'-cyclic phosphodiester (Fig. 28-21A, step *a*) that is opened hydrolytically (step *b*). Another hydrolytic cleavage, catalyzed by a different subunit of the enzyme complex, occurs at the 3' end of the intron. The resulting 5' OH is phosphorylated by GTP (step *c*),[601] and the resulting 5'-phospho group reacts with ATP to displace PP_i and to form a transient adenosine-5'-diphosphate terminus (step *d*). This reacts with the 3' OH formed at the 5' splice site (step *e*) to create a phosphodiester linkage between the two pieces of RNA. A phosphotransferase then removes the superfluous 2' phosphate (step *f*).[601a] Steps *d* and *e* are catalyzed by an **RNA ligase**.[598,599]

In HeLa cells and presumably in normal mammalian nuclei, the initial cleavage at the 5' end of the intron leaves a 3' phospho group (Fig. 28-18B, step *a*), which is cyclized, probably in an ATP-dependent process, to the 2',3'-cyclic phosphate (step *b*). This is ligated to the other piece of RNA by a direct displacement on the cyclic phospho group (step *c*).[602]

The spliceosome. The hnRNA of nuclei, which includes all of the pre-mRNA, is associated with proteins, which sometimes form very large 200S particles.[604] After limited cleavage with nucleases they tend to sediment in the 30S–40S range and to contain a variety of proteins.[605,606] Some of the proteins may have been involved in control of transcription.[606] Others participate in splicing. The smaller snRNP particles then appear to come into the nucleus and displace much, but not all, of the protein present in the pre-mRNA ribonucleoprotein particles.[605]

The 50–60S spliceosome complexes, with their protein and RNA components, are reminiscent of ribosomes. Although smaller in size, they can be vizualized by electron microscopy.[607] Each spliceosome is constructed from the four snRNP particles and additional proteins (Fig. 28-22).[539,608–612] Each spliceosome can accommodate ~500 nucleotides of pre-mRNA. As the pre-mRNA moves through the spliceosome the splice sites, which have only weakly conserved consensus sequences, must be located. The two ends of the introns, which may be much longer than the average 137-nucleotide exon, must be brought close together in the spliceosome.[613] The exact splice sites are usually located by an invariant **GU** at the 5' end and **AG** at the 3' end. The sequence of the first 18 nucleotides of U1 snRNA is largely complementary to that of the 5' splice site, which has the consensus sequence AG:**GU**RAGU, where the colon marks the junction. The **GU** is invariant.

After the U1 snRNP binds to the pre-mRNA (step *a*, Fig. 28-22)[614] the U2 snRNP binds to another almost invariant sequence CUR**A**CU found 20 to 55 nucleotides upstream of the 3' junction.[608,615–617] The A in this sequence becomes a branch point. It is brought close to the 5' splice site with the aid of a preassembled complex of snRNPs U4, U6, and U5. In this complex U4 and U6 are tightly paired, additional proteins are also present,[618–621] and enhancers may be located in adjacent exons.[617] Upon binding of U6 to the 5' splice site, the U1 and U4 snRNPs are released (step *b*, Fig. 28-22) and the 2'-OH of the branch point adenosine attacks the backbone phosphorus atom (step *c*) at the 5' splice junction forming a lariat intermediate. The 3' end created at the 5' junction must now be held and brought close to the 3' splice junction, which is located with the aid of U5 snRNP.[622] The 3' splice junction, utilized in the second splicing step (step *d*, Fig. 28-22) has the consensus sequence (T/C)N(C/T)AG:G.

The first splicing step is dependent upon a divalent metal ion, but the second is not.[623] Both steps appear to be in-line nucleophilic displacement reactions.[624] Additional **splicing factors** are needed for formation of the U4, U6, U5 complex and its function in the second splicing step, which also appears to require ATP.[621,622]

A small fraction of eukaryotic mRNA introns are characterized by **AU** and **AC** (rather than GU and AG) ends. The spliceosomes that act on these introns contain modified snRNAs U4 and U6, which are designated $U4_{atac}$ and $U6_{atac}$. They also require U11 and U12 snRNPs.[534,625,626]

Since pre-mRNAs usually contain many introns, a series of splicing events must occur. These apparently take place consecutively beginning at the 5' end. Similar splicing pathways are followed in yeast, higher plants, insects,[627] and mammals.

Alternative splicing pathways. RNA that contains many introns can undergo splicing in more than one way. Many examples of alternative splicing have been discovered.[612] The mammalian isoenzyme forms of pyruvate kinase called M_1, M_2, L, and R are all tetramers of 60-kDa subunits. The M_1 and M_2 forms are encoded by a single gene. The two mRNAs contain 1593-nucleotide coding regions, which are identical except for a 160-nucleotide sequence that determines the amino acid sequence in a region responsible for intersubunit contact. The difference between the M_1 and M_2 forms involves a choice of two alternative exon regions, one or the other of which is omitted during splicing.[628] Other examples have been found in human collagen,[629] in fibronectin,[630] in neuropeptide formation (Chapter 30),[631] and among human proline-rich salivary proteins,[632] cytoskeletal tropomyosin,[633] platelet-derived growth factor,[634] coagulation factor X,[635] and porphobilinogen deaminase.[636] Alternative splicing is very common in transcripts of viral DNA (Section E).

Alternative splicing could have arisen accidentally, but it is controlled by proteins. Best known is the **alternative splicing factor** (**ASF** or SF2).[612,637] It was first recognized by its function in *Drosophila melanogaster*, where the sex of individuals is determined by alternative splicing of an mRNA.[638] In addition to ASF other serine- and arginine-rich **SR proteins** participate in alternative splice site selection.[612]

Trans splicing. Every mRNA in trypanosomes has, at its 5' end, a short 35-nucleotide sequence that is not encoded in the transcribed gene. It was found that for each mRNA molecule two transcripts are formed and are spliced together, always with the 5' piece from a short SL (spliced leader) gene being joined to each of the others.[639] This trans splicing has since been observed in many plants, animals, and protists.[640]

Among these are the nematode *Caenorhabditis*,[639,641] flatworms,[642] yeast,[643] plant chloroplasts and mitochondria,[640,644] and mammalian cells.[645]

6. Modification and Editing of RNAs

Both mRNA and rRNA undergo rapid methylation of selected residues. About 1–3 internal adenylate residues per kb are methylated at their N^6 positions.[646] These are usually the central adenylates in the sequences GAC or AAC. The methylated sites are not uniformly distributed but are clustered, sometimes in the 3' untranslated ends of the RNA. Many more residues (55 in yeast and ~100 in vertebrates) in rRNA are methylated on selected ribose 2'-OH groups.[173,535c–e,647–648a] About an equal number of uridine residues are converted to pseudouridines (Eq. 28-3). Methylation sites are apparently selected by the fibrillarin-associated snoRNAs U32–U40 (Table 28-3). These **methylation-guide** snoRNAs contain 10- to 14-nucleotide sequences that are complementary to segments of the rRNA that contain the methylation sites, and evidently provide rigid helical regions that are targets for the methylase action. The snoRNAs contain the previously mentioned C and D sequences. Methylation occurs on the ribose of the nucleotide that is base-paired with the fifth nucleotide that is upstream of the D sequence in the snoRNA. Each snoRNA directs methylation of a different ribose.[173] In a similar manner the snoRNAs containing the ACA motif located three nucleotides upstream of their 3' ends appear to direct the conversion of uridines to **pseudouridines**, as in Eq. 28-3.

Editing of RNA transcripts. Interpretation of the genetic code utilized by the mitochondria of trypanosomes and other kinetoplastid protozoa was confounded by the discovery that the DNA sequences of many genes, including the COIII cytochrome oxidase gene present in the kinetoplast maxi circle DNA, do not appear to encode the correct amino acid sequence. In fact, the RNA transcripts are extensively edited, mostly by insertion of multiple uridine residues at many positions[649,650] and by occasional deletions at others. Editing of some mRNAs causes 45% of the message to be rewritten.[651] Several additional kinds of editing were soon discovered, not only in protozoa but also in mammals, plants, and archaea.[652,652a]

A frequent editing change is the hydrolytic deamination of a C to a U residue.[652b] For example, human apolipoprotein B is synthesized in two forms: apoB100, a full-length 512-kDa protein made in the liver and used for transport of cholesterol and triglycerides, and a shorter 241-kDa form, apoB48, used in absorption of dietary lipids (Chapter 21, Section A1). ApoB100 is synthesized from a full-length mRNA, but apoB48 is made according to a shortened mRNA in which a glutamine codon (CAA) has been converted by editing to the translation stop codon UAA.[652,653] A special enzyme deaminates only cytidine 6666 of the mRNA.[654] C to U editing occurs in chloroplasts and mitochondria of plants.[655–657] In *Arabidopsis* mitochondria 456 different C to U conversions have been identified in mRNAs.[656]

Deamination of adenosine residues produces inosine, which occurs in brain mRNA once in ~17,000 ribonucleotides.[658] Some ionotropic glutamate receptors in the brain have subunits translated from inosine-containing mRNAs. A glutamine codon CAG is edited to CIG, an arginine codon. The arginine codon AGA is converted to the glycine codon IGA at another site, ATT is converted to ITT and TAC to TIC. All of these changes affect the properties of the glutamate-activated ion channel.[659] The adenosine deaminases involved in these editing events are usually specific for double-stranded helical segments of RNA, e.g., for the stems in stem–loop structures.[660–662] At least one human adenosine deaminase not only binds to RNA but also contains a DNA-binding domain specific for Z-DNA.[663]

Returning to the trypanosomes and their relatives, mitochondrial RNAs undergo extensive insertion and deletion of U's. The editing site is located by a **guide RNA** (gRNA), which directs the hydrolytic cleavage of the chain and either the addition of U's to the 3' cut end by transfer from UTP or hydrolytic deletion of U's from the 3' cut end. The chain ends are then rejoined by an RNA ligase.[172,664–667c] The functional significance of the editing of kinetoplast mRNA is uncertain. However, at least some of the edited mRNA is translated to give proteins that are presumably used.[668] While trypanosomes usually insert only U's, the slime mold *Physarum polycephalum* may insert dinucleotides such as AA, AU, CU, or GU[669] and may also add nucleotides at the 3'-ends of RNAs.[669a]

A to G editing occurs in RNA of *Drosophila*.[670] Yet another type of editing has been observed in viral RNA from paramyxovirus. The virally encoded RNA polymerase sometimes "stutters" reading the same template base two or more times, with a resulting insertion of a base.[671] Editing of transcripts usually serves an essential biological function, creating the correct sequence for translation of the mRNA and often generating multiple isoforms of proteins.

Finishing the transcripts. Additional modifications must be made to some mRNAs, and there will doubtless be many surprises as the details are worked out. One detail, which was discovered in the 1980s, is the specific function of snRNA U7 in recognition of the 3' end of pre-mRNAs for histones. The U7 RNA apparently base-pairs with a sequence near the 3' end cleavage site, acting as a cutting guide.[47,672,673]

Newly synthesized mRNA emerges from the nuclear pores as nucleoprotein complexes containing as many as ten different proteins. A major component is a 78-kDa polypeptide thought to be associated with the poly(A) tail. These ribonucleoproteins are sometimes stored for long periods of time, for example, in mature seed embryos and in amphibian oocytes.[575a,674] They may also travel rapidly for long distances, e.g., down nerve axons[675] or from cell to cell in plants via transport in the phloem.[676]

E. Transcription of Mitochondrial, Chloroplast, and Viral Genes

In the compact 16-kb chromosomes of mammalian mitochondria, the genes are tightly packed against one another (Fig. 18-3).[677,678] Most genes are transcribed using heavy H strands as templates and specialized bacteriophage-type RNA polymerases encoded in nuclear DNA.[679] A single promoter in the D loop region (Fig. 18-3) is used to make a long transcript from the entire H strand.[680] The transcript is then cut precisely by mitochondrial RNase P at the 5' ends of the tRNAs.[681] Similarly precise cleavage must occur at the 3' ends because there are often no nucleotides or only one to a few nucleotides separating adjacent genes. The nucleases involved differ from those used in nuclear tRNA 3' processing.[682] In animal mitochondria tRNA genes sometimes overlap by one nucleotide. Polyadenylation of the transcripts provides a form of editing that is required to create a UAA translational termination signal to which termination proteins bind.[683] The tRNA genes lack the 3' terminal CCA, which must be added. The lighter L strand has its own promoter, also located in the D loop. Both promoters contain the conserved nonanucleotide sequence 5'-ATATAAGTA. The pre-mRNAs created by these cleavages are not capped but are polyadenylated. Since the promoters are simple and the transcription factors few, mitochondrial transcription is controlled largely by mRNA stability, translation, and posttranslational events.[684]

The 70- to 100-kb circular mitochondrial DNA molecules of yeast and of higher plants contain more genes than do animal mitochondria, but most of the increased size is accounted for by intergenic spacers and by a few long introns. All yeast mitochondrial genes except for that of $tRNA^{Thr}$ are transcribed from a single strand. About 20 different primary transcripts have been identified in *Saccharomyces cerevisiae*. These originate at several points in the genome but always at the sequence 5-ATATAAGTA, the 3'A corresponding to the 5' nucleotide of the transcript.[678] One of these sequences is located at the origin of replication, suggesting the possibility that a normal RNA transcript provides the primer for DNA replication in yeast mitochondria.[677]

The 100- to 160-kb chloroplast genomes (Chapter 23, Section E,2) also have many prokaryotic features. They encode ~50 proteins as well as the tRNAs and rRNAs. Promoter and terminator sequences resemble those of bacteria and protein sequences are often homologous to those in bacteria. This applies, for example, to the α, β, and β' subunits of RNA polymerase.[685]

1. Viral Transcription and Replication

Because viruses contain small genomes, study of transcription of viral DNA and of replication of RNA viruses has played an important role in helping us to understand transcription in eukaryotes.[47,686–688] An example is the discovery of the virus SV40 enhancer, which has been discussed in Section C,4. Study of viral life cycles is also essential to future progress in fighting viral diseases. Each of the many different viruses has its own often very complex life cycle. Only a few details can be given here. For lucid summaries see Voyles.[259]

Eukaryotic DNA viral genomes, like that of phage λ, usually contain early and late transcriptional units. The small papovaviruses, such as SV40 and polyoma virus, have 5.2 kb genomes. Like the small RNA viruses, they make use of overlapping genes and alternative RNA processing. In SV40 DNA there are two overlapping promoters called **early-early** and **late-early**. The first of these contains a TATA sequence, and both promoters also depend upon a 21-bp repeat segment as well as the SV40 enhancer. At least two proteins, one that binds to the enhancer and one that binds to the 21-bp repeat, are needed for initiation of early transcription. The early-late promoter lacks the TATA sequence but requires the 21-bp repeat and enhancer. A 94-kDa encoded protein called the **large T-antigen** (Chapter 27, Section C,10) is one of the regulators of transcription as well as of DNA replication.[689,690] This protein is also sufficient to transform rodent cells in culture. Although predominantly nuclear it is also inserted into the membrane where it acts as an antigen.

The large icosahedral adenoviruses cause respiratory infections in humans and attack and may cause cancer in many other vertebrate species including birds and amphibians. The 35.9-kb genome of human adenovirus-2 encodes at least 30 proteins, 10 of which appear in the virion. One of these is covalently linked to the 5' end of the DNA. As with smaller DNA viruses extensive use is made of alternative splicing of the transcribed RNA. There are at least six early transcriptional units, each with its own promoter. A variety of mRNAs are created using the various coding segments and a complex array of regulatory sequences control transcription. A 32-kDa phosphoprotein transcription factor designated E1A is encoded by a "pre-early" gene. It is required along with host-encoded transcription factors for viral transcription.[691,692]

The late region of the adenovirus genome encodes structural proteins for the virus coat. Most of its transcripts begin about 16.5% of the way along the 36.5-kb dsDNA. However, the initial transcripts are cleaved at several different positions to yield a series of different 3'-poly(A)-terminated transcripts. In an exceedingly complex process the transcripts undergo splicing out of genes at their 5'-ends so that the final mRNAs typically code for single proteins. At the 5' ends the cap is joined to short segments from the original mRNA 5' end.[687,688,693]

Viruses SV40, polyoma, and some strains of adenoviruses are oncogenic in some species and cause transformations of cells in culture. Transformed cells always contain integrated viral DNA. That of SV40 can be incorporated at many different sites in the host genome. The integrated DNA does not always include the complete SV40 genome, and parts of the DNA may be inverted, deleted, or scrambled. Integration is not an essential part of the viral life cycle and has no effect on the infective properties of the viruses.[688] Cells transformed by adenoviruses usually also contain only a fragment of the viral genome in their DNA. However, one small set of genes from early region 1A is present in all transformed cells. The encoded proteins appear to be modulators of transcription and may cause cancer by promoting uncontrolled transcription of certain genes.[694]

2. Replication of RNA Viruses

The RNA (+) strands present in many RNA viruses often serve immediately after infection as a messenger RNA. However, replication requires formation of (–) strands of viral RNA from which new (+) strands can be transcribed for assembly into new virus particles. Other RNA viruses contain (–) strands of RNA or double-stranded RNA and, therefore, have significantly different life cycles.[259]

Small RNA viruses. The human polio virus, the common cold virus (rhinoviruses, Fig. 7-15), and other picorna viruses have 7.2- to 7.5-kb genomes with considerable homology (50% between the polio and rhinoviruses) and similar overall structures.[695] The polio genome encodes eight different proteins, one being a small 22-residue peptide that becomes covalently linked to the 5'-end of the RNA through a phosphodiester linkage to the side chain of a tyrosine. Cleavage of this linkage by a host enzyme allows the viral RNA (+) strand, which is polyadenylated at the 3' end, to serve as an mRNA for synthesis of a single large 220-kDa polyprotein. This is cleaved by a host protease at several Gln-Gly bonds to form several proteins. These include the 22-residue RNA-linked peptide, two capsid proteins, and a capsid precursor protein. The latter is cleaved during capsid assembly by a viral protease at an Asn-Ser bond to give two more capsid proteins. These four proteins have masses of 7, 26, 29, and 32 kDa. A viral protease, a large 58-kDa replicase, and a 37-kDa protein of unknown function are also cut from the polyprotein. A host protein initiates cleavage of the polyprotein, but the virally encoded protease later takes over this function.

One of the best understood of the many viral pathogens of plants is the tobacco mosaic virus (Fig. 7-8). Its 6.7-kb positive strand RNA encodes a replicase, coat protein, and at least one other protein.[696]

Influenza viruses. These negative-stranded viruses are classified into types A, B, and C, but it is only type A that infects nonhuman species including birds, horses, pigs, seals, mink, and whales.[697–699] Type A influenza viruses have also caused the great pandemics such as those in 1918–1919 and in 1968. Influenza viruses are surrounded by a lipid bilayer in which the virally encoded **hemagglutinin** and a **neuraminidase** (p. 186) are embedded. The inside of the bilayer is coated with a matrix protein and within this coat eight pieces of RNA of total length 13.6 kb are coiled together with a basic nucleoprotein. Also present are ten molecules each of three other proteins. The eight pieces of RNA vary in length from 900 to 2500 nucleotides. Seven of them encode one each of the seven virion proteins. One encodes an additional nonstructural protein while the smallest piece, using overlapping nucleotide sequences, encodes two nonstructural proteins.[697]

The existence of a fragmented genome evidently underlies the ability of influenza A viruses to undergo rapid changes in antigenic behavior. If a cell is coinfected with two strains of virus, the eight fragments act as independent chromosomes, which can be reassorted into new combinations in the progeny viruses. As a consequence, it is difficult to develop safe, live virus vaccines. A large reservoir of infection among migratory water birds and other animals facilitates the appearance of new strains and their rapid spread throughout the world.[698]

The first step in the replication of influenza viruses, which takes place in the cytoplasm, is the synthesis of (+) strands that can serve both as mRNA for synthesis of proteins and as templates for synthesis of new (–) strands. Three of the capsid proteins form the required RNA polymerase. This "transcriptase" is primed preferentially by 5'-capped 10- to 13-nucleotide segments of RNA that have been cut by a viral nuclease from host mRNAs.[700] The mRNAs made from viral RNA are polyadenylated and are translated by the host cell's ribosomes. However, some transcripts are used as templates to form viral (–) strands, which are not polyadenylated and which contain uncapped pppA at the 5'-ends.

HIV-1 and other retroviruses. Because of their association with viral oncogenes (Chapter 11) and because of the **human immunodeficiency virus** (HIV-1) and the AIDS epidemic a great deal of attention is focused on retroviruses.[701–701b] Each retrovirus particle contains *two* identical single-stranded (+) RNA molecules, which may be as long as 10 kb. Their unique characteristic is that they induce synthesis of DNA, which must be integrated into the host genome before new viral (+) strands are transcribed. Retroviruses may sometimes cause cancer and may carry oncogenes (Chapter 11). Study of the **Rous sarcoma virus** (RSV), which infects chickens, and of the related **avian myeloblastosis virus** (AMV) and of HIV has revealed a common structure and a complex life style that are largely shared by all known retroviruses.

The organization of retroviruses[687,688,702] always includes a sequence of genes designated *gag* (glycoprotein antigen core proteins), *pol* (polymerase), and *env* (envelope) (Fig. 28-23). These are often followed by an oncogene.[703] In RSV this is the *src* gene (Chapter 11). At each end of the retrovirus gene sequence is a short direct repeat labeled R in Fig. 28-23. In RSV the R sequence is 21 nucleotides in length[688] and in HIV (see Fig. 28-23)[704] it is 98 nucleotides long.[705,706] The 5' end of the viral RNA is capped, and the 3' end is polyadenylated. The dsDNA of the integrated form of the virus (Fig. 28-23) is longer and at each end is bounded by **long terminal repeats** (LTRs). Each LTR consists of a sequence, designated U3, that is present next to R at the 3' end of the viral RNA. In the LTR this is followed by sequence R and then by U5, a unique sequence that came from the 5′ end of the viral RNA. Each ds-LTR begins and ends with a short inversely repeated segment:

5'-TGT —— ACA in RSV
5'-CTG —— CAG in HIV

The integrated provirus is always bounded by a sequence of host DNA that is repeated without inversion at the opposite end. For RSV this is a 5-bp sequence.

The LTRs in RSV are 569 bp in length, and those in HIV are 634 bp in length, 83 bp coming from U5, 98 from R, and 453 from U3.[706] The LTRs themselves often contain promoter and other control elements and even entire genes. The organization of a retrovirus (Fig. 28-23) reflects the complex mode of replication, which is presented in simplified form in Fig. 28-24. The key enzyme is the RNA-directed **reverse transcriptase** (Fig. 27-12).[707–709] The initial synthesis of DNA by this enzyme is primed by a tRNA. RSV uses $tRNA^{Trp}$ and HIV $tRNA^{Lys}$ for this purpose.[710–712] The 3' end of the tRNA, including the nucleotides forming both the acceptor stem and the stem of the TΨC loop (Fig. 5-30), must relinguish its normal base-pairing to form ~18 Watson–Crick base pairs with a **primer binding site** (labeled PB in Fig. 28-24) near the 5' end of the retroviral DNA. Because synthesis of the (–) strand of the retroviral DNA begins so close to the 5' end of the template (Fig. 28-24), only a short piece of DNA, including sequences U5' and R′ complementary to U5 and R, can be formed. For replication to continue RNA must be removed from hybrid regions. This is accomplished by the **RNaseH** activity of the reverse transcriptase.[708,709,713,713a] After removal of the RNA the primer tRNA must undergo a **strand transfer**, in which it shifts from the 5' end of the viral RNA template to the 3' end (Fig. 28-24), utilizing pairing between the right-hand R sequence of the template and the R′ sequence of the growing cDNA copy.[703,711,714] This transfer is sometimes to the second of the pair of identical RNA molecules in the virus, providing a way of increasing diversity by recombination. A second strand transfer of the growing (+) strand is needed to complete the dsDNA, which now contains the two identical LTRs. The 5' and 3' ends of the template RNA are doubtless held close together to facilitate strand transfer.

Integration of the dsDNA into the host DNA can occur at many places. The mechanism of integration probably resembles that used by phage λ (Fig. 27-27) and accounts for the duplication of host sequences at the two ends of the integrated virus. A virally encoded integrase catalyzes the process (see also Chapter 27, Section D,3).[715–717] It is the integrated virus that is transcribed to form new (+) viral RNA strands.

Integrated retroviruses are usually transcribed as full-length RNA copies, which may or may not have introns spliced out. The smaller spliced pieces encode the *env* and other genes such as *src* (Fig. 28-25). The *gag-pol* region is translated as a polyprotein that is cleaved into a number of pieces. These include four proteins of the virus core (encoded by *gag*), the reverse transcriptase with its associated RNaseH, and an **integrase**[718,718a] (all encoded by *pol*).[259] There is also an **aspartic protease** only 99 residues in length within *pol* (Fig. 28-25; Chapter 12, Box 12-C). The promoter and control region for transcription is located in the U3 region and is placed into a position where it can function only upon synthesis of the first LTR. The gene *env* encodes the major viral envelope protein and is translated from a spliced mRNA (Fig. 28-25).

Accessory regulatory genes. HIV and some related retroviruses such as HTLV-1 (which causes rare T-cell leukemias)[719] are distinguished from other retroviruses by a marked increase in the rate of DNA transcription within infected cells as compared with uninfected cells. This is thought to be a result of synthesis of virally encoded proteins that are trans-acting

A RNA form
5′–m7G
Cap
R
U5
gag
pol
env
U3
R
AAAA - - -
Poly A tail
B DNA form
U3
R
U5
gag
pol
env
U3
R
U5
LTR
250-1400 bp
LTR
C Integrated DNA form
Host DNA
U3
R
U5
gag
pol
env
U3
R
U5
Host DNA
LTR
LTR
4-6 bp repeat
of target DNA

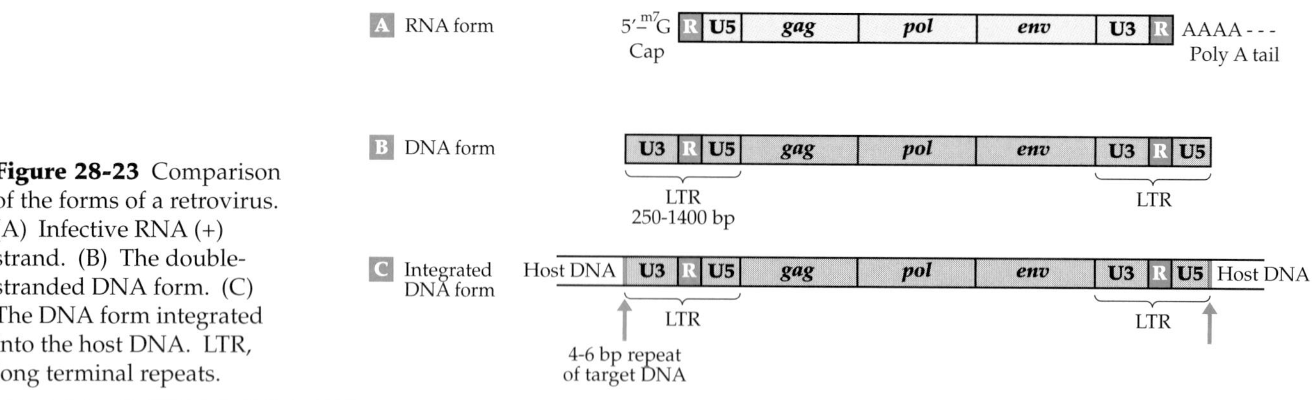

Figure 28-23 Comparison of the forms of a retrovirus. (A) Infective RNA (+) strand. (B) The double-stranded DNA form. (C) The DNA form integrated into the host DNA. LTR, long terminal repeats.

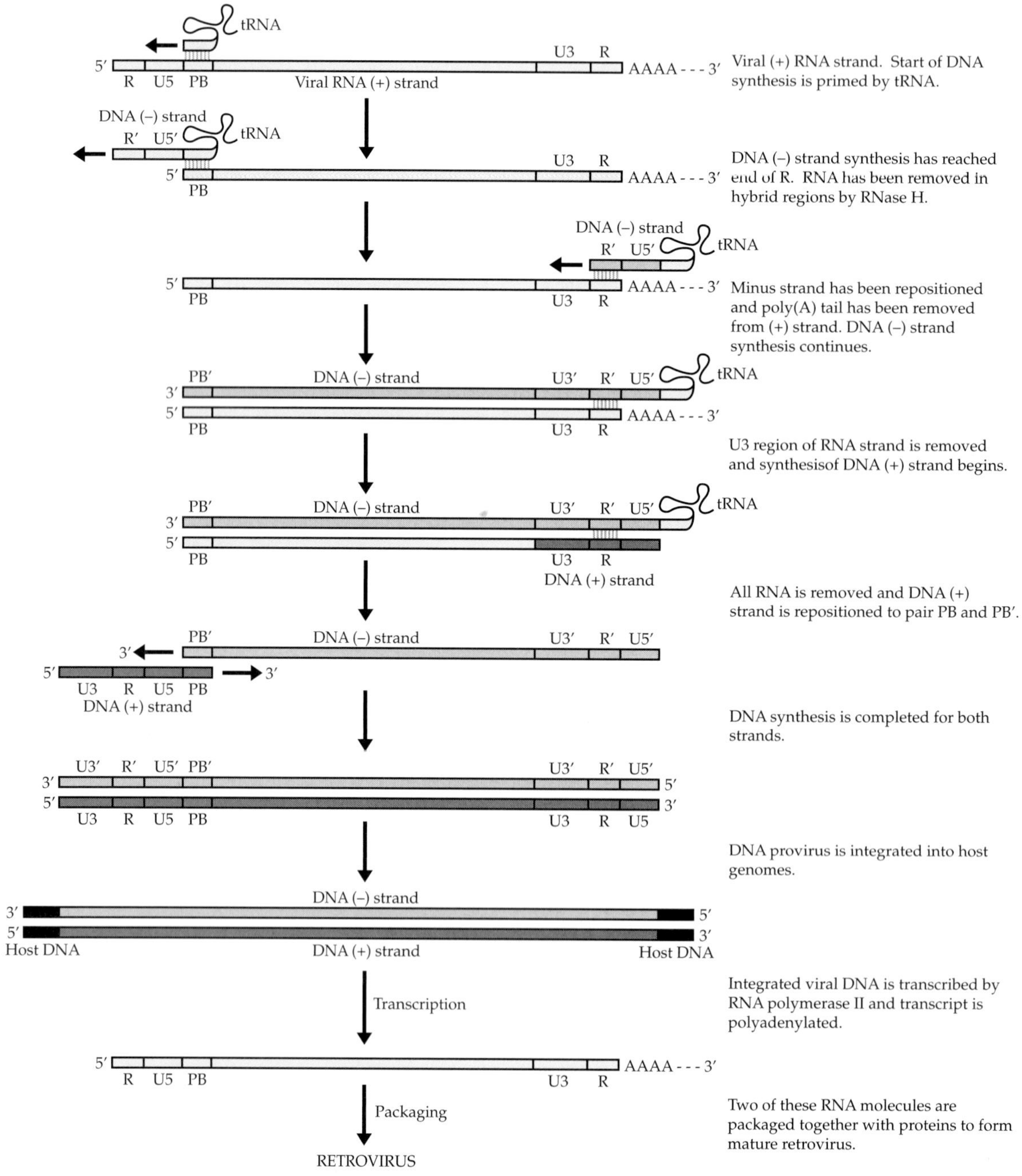

tRNA
5′
R U5 PB
Viral RNA (+) strand
U3 R
AAAA - - - 3′
Viral (+) RNA strand. Start of DNA synthesis is primed by tRNA.
DNA (–) strand
R′ U5′
tRNA
DNA (–) strand synthesis has reached end of R. RNA has been removed in hybrid regions by RNase H.
Minus strand has been repositioned and poly(A) tail has been removed from (+) strand. DNA (–) strand synthesis continues.
PB′
DNA (–) strand
U3′ R′ U5′
U3 region of RNA strand is removed and synthesisof DNA (+) strand begins.
DNA (+) strand
All RNA is removed and DNA (+) strand is repositioned to pair PB and PB′.
U3 R U5 PB
DNA synthesis is completed for both strands.
U3′ R′ U5′ PB′
DNA provirus is integrated into host genomes.
Host DNA
DNA (+) strand
Host DNA
Integrated viral DNA is transcribed by RNA polymerase II and transcript is polyadenylated.
Transcription
Packaging
Two of these RNA molecules are packaged together with proteins to form mature retrovirus.
RETROVIRUS

Figure 28-24 Simplified scheme for replication of the RNA genome of a retrovirus. See Sugden.[703] PB, Primer-binding site.

regulators of transcription. The HIV genome map (Fig. 28-26) shows the positions of the nine recognized genes marked in the three reading frames. In addition to the *gag*, *pol*, and *env* genes there are genes for six accessary proteins: Tat, Rev, Vif, Vpr, Vpu, and Nef.[720–722]

Transcription is initiated at the promoter in the 3' LTR. This contains a TATAA sequence, an SP1 binding site, and an enhancer that binds transcription factor NF-κB (Fig. 5-40). The full-length 9-kb transcript contains, according to Frankel and Young,[722] the following

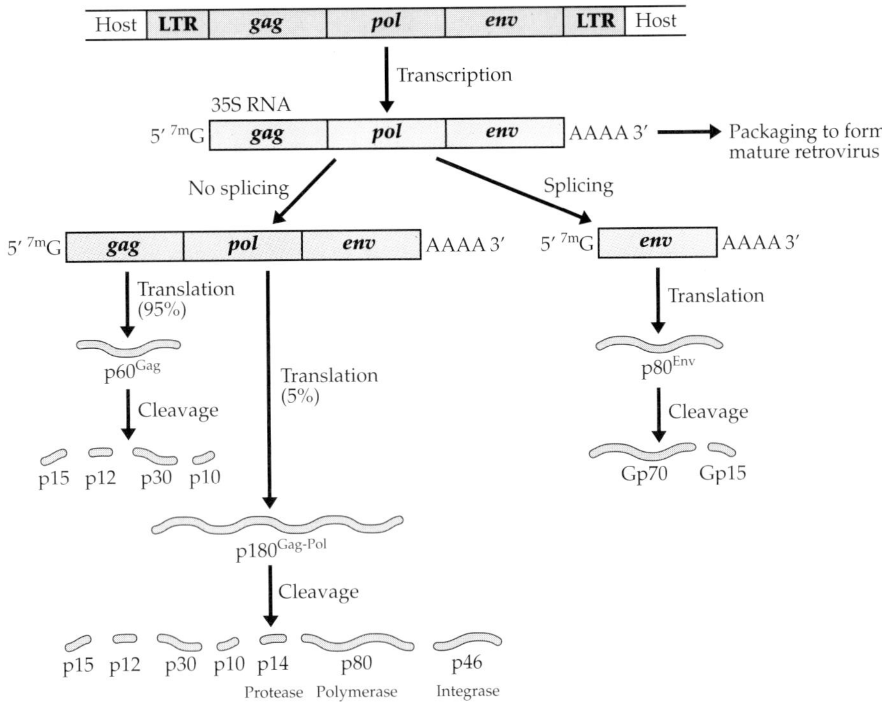

Figure 28-25 Gene expression from a typical retrovirus that has been integrated into a host's genome. This figure illustrates how a variety of proteins are encoded by a single rather short piece of DNA. After Voyles.[259]

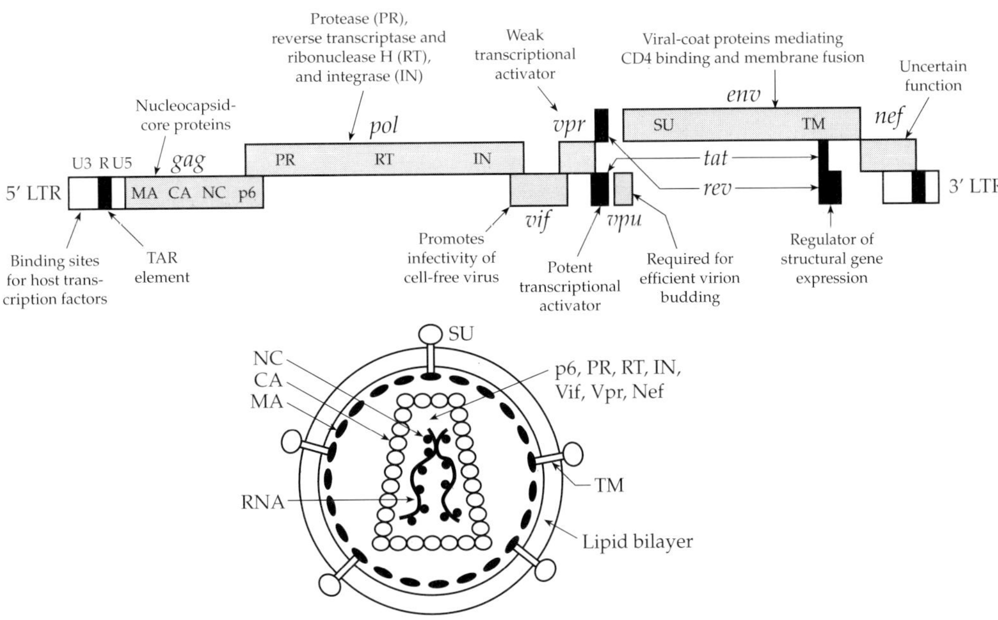

Figure 28-26 Simplified genetic map of the AIDS virus HIV-1. All three reading frames are utilized to encode nine genes, which give rise to 15 proteins. After Frankel and Young.[722]

BOX 28-C SYNTHETIC ANTIVIRAL COMPOUNDS

Most bacterial infections can be treated successfully with antibiotics, but the development of satisfactory antiviral agents has been slow. Yet we are susceptible to many dangerous virus diseases, and viruses also take a huge toll among domestic animals and plants.[a] The first antiviral drug, 5-iodo-2'-deoxyuridine (idoxuridine), was introduced in 1962 and was used for 20 years by ophthalmologists to treat serious eye infections by the herpes simplex virus (HSV).

More recently 9-β-D-arabinofuranosyladenine (Ara-A) has become a preferred drug in treatment of ocular herpes infections. This compound, which is a naturally occurring antibiotic, can also be administered intravenously for life-threatening infections such as herpes encephalitis.[a] Ara-A is quite toxic but the guanine derivative 9-(2-hydroxyethoxymethyl) guanine (acyclovir) is less so. Another acyclic 2'-deoxyguanine analog, 9-(1,3-dihydroxypropoxymethyl)-guanine (DHPG), is more soluble, more potent, and has a broader range of effectiveness.[a–c]

One of the first effective drugs against RNA viruses was ribavirin (1-β-D-ribofuranosyl-1,2,4-triazole-3-carboxamide), an analog of guanosine. It has a broad range of action and is used to treat severe viral pneumonia and bronchitis caused by respiratory syncytial virus (RSV). This common virus usually is mild but may cause death in infants and children. Ribavirin may also be of some value in the treatment of AIDS.

How do these compounds exert their antiviral effects? The nucleoside analogs are usually phosphorylated to the corresponding mono-, di-, and triphosphate derivatives by cellular enzymes. Thus, Ara-A yields Ara-ATP, which inhibits the herpes virus-encoded DNA polymerase. Ara-A may also enter the viral DNA. In addition Ara-A may inhibit

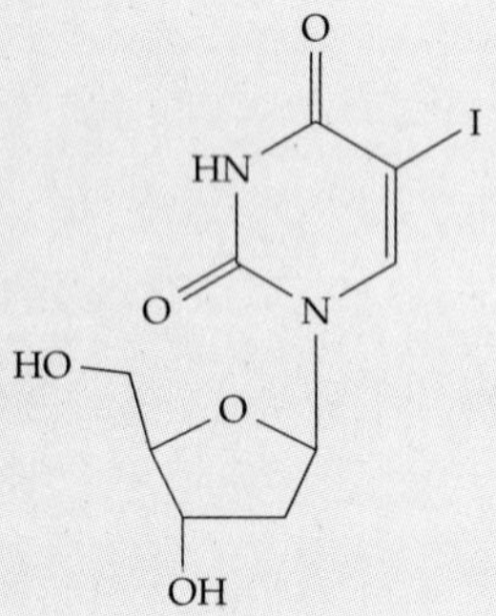

Idoxuridine

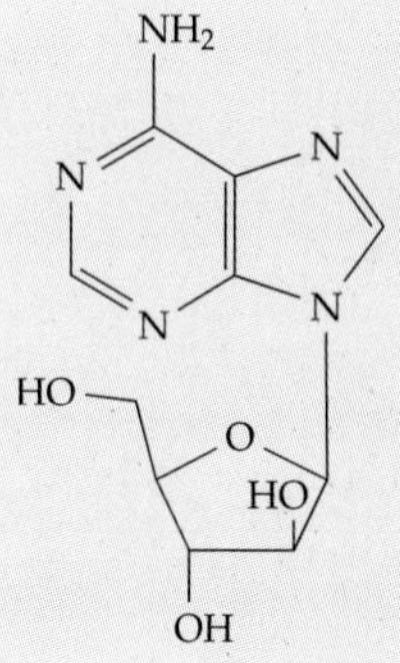

Ara-A

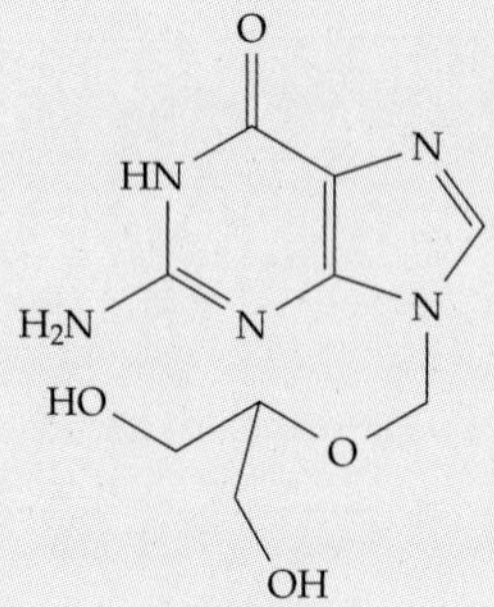

DHPG

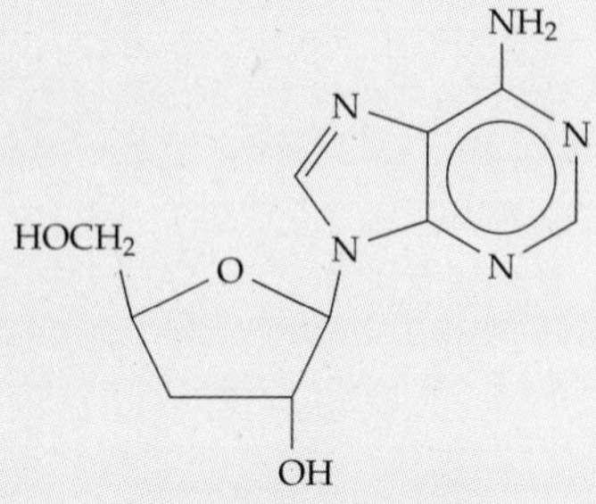

Cordycepin (3'-deoxyadenosine) preferentially inhibits synthesis of ribosomal and tRNA

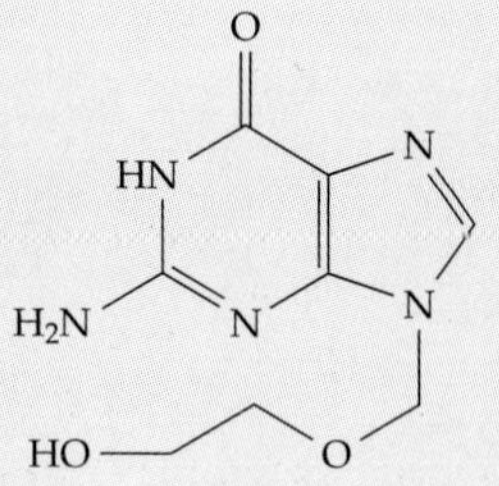

Acyclovir

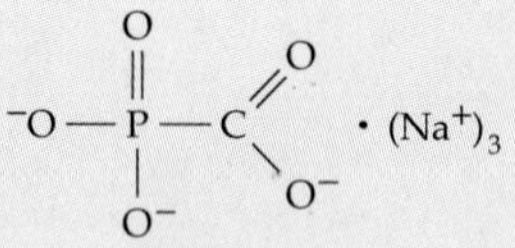

Trisodium phosphonoformate (Foscarnet)

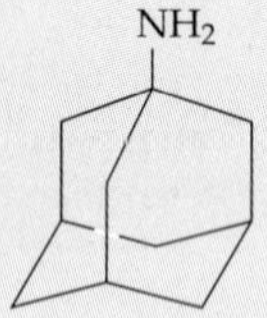

Amantadine

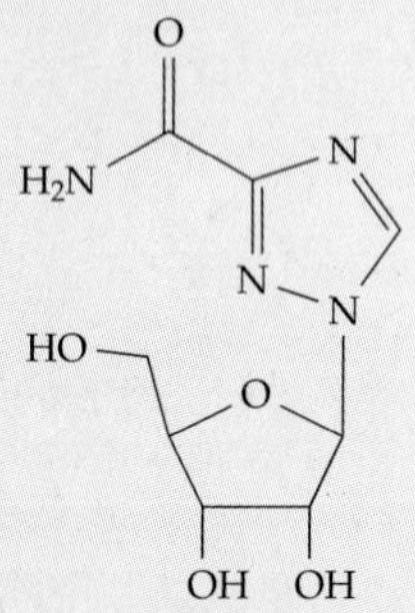

Ribavirin

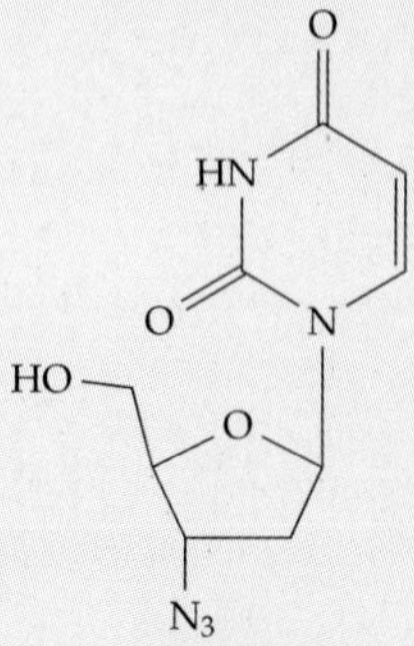

3'-Deoxy-3'-azidothymidine (AZT)

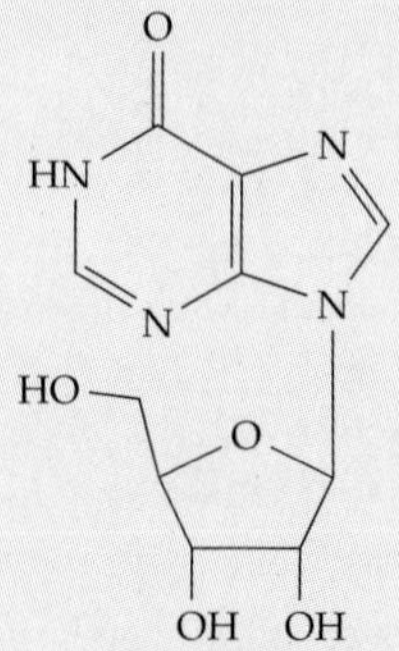

Dideoxyadenosine

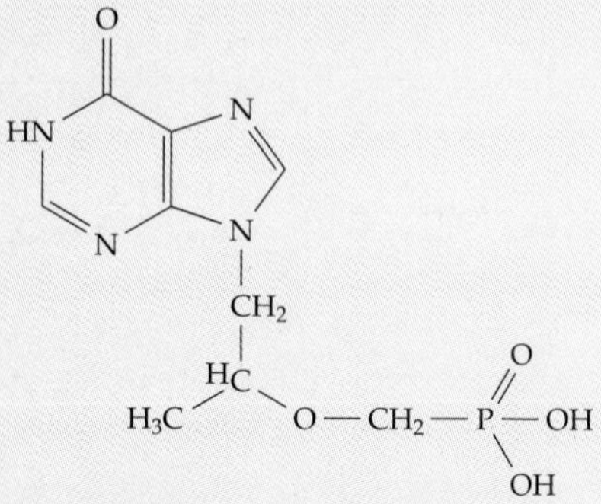

(*R*)-9-(2-Phosphonylmethoxypropyl) adenine

BOX 28-C (continued)

polyadenylation of virally induced mRNA. Some analogs, such as acyclovir, undergo little conversion to the monophosphate by cellular kinases but are efficiently phosphorylated by herpes virus-encoded thymidine kinase. Thus, acyclovir does little damage to uninfected cells.[a] Ribavirin 3'-monophosphate may inhibit IMP dehydrogenase (Fig. 25-16, left) thereby interfering with GTP production. At the same time ribavirin triphosphate competes with GTP to inhibit virally encoded RNA polymerase.[a]

Phosphonoformate is a pyrophosphate analog and inhibits both DNA polymerases and reverse transcriptase. However, toxicity may prevent long-term treatment of AIDS patients. Amantadine has a narrow antiviral specificity. It specifically inhibits initiation of the replication of influenza virus RNA of type A (but not of type B). Active only against retroviruses, 3'-azidothymidine is a reverse transcriptase inhibitor, which acts by a chain termination mechanism. It was synthesized in the early 1960s but only recently has been used in treatment of AIDS victims. More recently a series of 2',3'-dideoxynucleosides, such as dideoxyinosine, have also been used.[d] Acyclic phosphonates, such as phosphonylmethoxypropyladenine, avoid the need for metabolic phosphorylation of the drug.[e]

Development of synthetic antiviral compounds is hardly beyond its infancy.[f] Serious problems must be overcome with most of these compounds. Toxicity (sometimes carcinogenicity), development of resistance by viruses, and enzymatic destruction limit the utility of most drugs. For example, adenosine deaminase destroys Ara-A quite rapidly. With our rapidly advancing knowledge of viral life cycles and protein and nucleic acid structures many new drug targets have been identified.[g] Among the targets for HIV are the reverse transcriptase,[h] protease,[i] and integrase.[j,k] Computer-assisted design, as well as new techniques of synthesis and screening, have allowed development of many non-nucleoside inhibitors.

Oligonucleotide phosphoramidates and other triplex-forming compounds may be designed to bind to specific DNA targets.[l]

[a] Robins, R. K. (1986) *Chem. Eng. News* **64**, 28–40

[b] Cheng, Y.-C., Grill, S. P., Dutschman, G. E., Nakayama, K., and Bastow, K. F. (1983) *J. Biol. Chem.* **258**, 12460–12464

[c] Biron, K. K., Fyfe, J. A., Stanat, S. C., Leslie, L. K., Sorrell, J. B., Lambe, C. U., and Coen, D. M. (1986) *Proc. Natl. Acad. Sci. U.S.A.* **83**, 8769–8773

[d] Sandberg, J. A., and Slikker, W., Jr. (1995) *FASEB J.* **9**, 1157–1163

[e] Tsai, C.-C., Follis, K. E., Sabo, A., Beck, T. W., Grant, R. F., Bischofberger, N., Benveniste, R. E., and Black, R. (1995) *Science* **270**, 1197–1199

[f] Mitsuya, H., Yarchoan, R., and Broder, S. (1990) *Science* **249**, 1533–1544

[g] Richman, D. D. (1996) *Science* **272**, 1886–1888

[h] Althaus, I. W., Chou, J. J., Gonzales, A. J., Deibel, M. R., Chou, K.-C., Kezdy, F. J., Romero, D. L., Palmer, J. R., Thomas, R. C., Aristoff, P. A., Tarpley, W. G., and Reusser, F. (1993) *Biochemistry* **32**, 6548–6554

[i] Rosin, C. D., Belew, R. K., Walker, W. L., Morris, G. M., Olson, A. J., and Goodsell, D. S. (1999) *J. Mol. Biol.* **287**, 77–92

[j] Robinson, W. E., Jr., Reinecke, M. G., Abdel-Malek, S., Jia, Q., and Chow, S. A. (1996) *Proc. Natl. Acad. Sci. U.S.A.* **93**, 6326–6331

[k] Goldgur, Y., Craigie, R., Cohen, G. H., Fujiwara, T., Yoshinaga, T., Fujishita, T., Sugimoto, H., Endo, T., Murai, H., and Davies, D. R. (1999) *Proc. Natl. Acad. Sci. U.S.A.* **96**, 13040–13043

[l] Giovannangeli, C., Perrouault, L., Escudé, C., Gryaznov, S., and Hélène, C. (1996) *J. Mol. Biol.* **261**, 386–398

essential regions, whose positions may vary somewhat from one isolate of the virus to another:

A complex pattern of splicing produces more than 30 mRNAs.[723] When viral RNAs are first produced most are doubly spliced, allowing the split genes *tat* and *rev* to be expressed by synthesis of Tat and Rev. The *trans*-activator protein Tat is a small 86-residue cysteine-rich protein, which binds the Tar stem–loop structure and greatly stimulates transcription.[724–725a]

The 116-residue *rev* (regulator of expression of virion genes) gene product is also a transactivator, which is needed for rapid production of singly spliced (4 kb) or unspliced (9kb) *gag-pol* mRNA required for formation of virus structural proteins.[726,727] The effect of Rev is probably on transport from the nucleus rather than on splicing.

Transcription is repressed by the 206-residue N-terminal myristoylated protein, Nef, a phosphoprotein that associates with cytoplasmic membranes.[728–729a] It has been difficult to learn its exact function, but it seems to be required for maintenance of the integrated provirus for long periods of time without extensive replication. Mutations in gene *nef* do not eliminate the ability of the virus to replicate in T lymphocytes and to kill them.

The 23-kDa protein Vif (**viral infectivity factor**) is not needed for growth but is essential for infectivity.[729b] Other genes in HIV are *vpu*, which encodes an 81-residue integral membrane protein (**virion protein U**), and *vpr*, which encodes the 96-kDa **virion protein R**. Several possible functions have been proposed for these small proteins.[722,729b,c]

HIV-1 is a member of the group of slow viruses or **lentiviruses**.[730] Other lentiviruses include the human HIV-2,[731] an immunodeficiency virus that attacks cats causing leukemia,[732] and the human leukemia virus

Position from 5' end	Description
1–55	TAR, a 59-residue stem-loop structure; binding site for Tat
182–199	PB, primer binding site
240–350	Packaging signal. Binds envelope protein NC
248–271	Dimerization site with "kissing loop"
290	Major splice donor site, used to form all spliced mRNAs
1631–1673	Gag-Pol frameshifting region where –1 ribosomal frameshifting occurs to allow 5–10% synthesis of Gag-Pol polyprotein
7362–7596	Rev response element. Binding site for Rev
5358 and 7971	Two major splice acceptor sites; other minor sites are also used
9205–9210	Polyadenylation signal

HTLV-1.[733] Another lentivirus causes two diseases of sheep, **maedi**, a pulmonary disease, and **visna**, a paralytic condition somewhat similar to multiple sclerosis.[734] Because of its slow development there has been doubt as to the cause of AIDS, but there is now little doubt that HIV-1 is the true culprit.

Success in treating AIDS may depend upon better understanding of the complex life cycle of HIV-1,[722,730,735] which is summarized in Fig. 28-27. The cycle begins with the binding of the virion envelope protein to the immunoglobulin-like surface protein **CD4**, which is found principally on the type T4 helper T cells (Chapter 31). Binding of CD4 to the HIV envelope proteins appears to activate the T cells to enter the cell cycle and to take up and integrate the virus. The virus infection destroys these CD4+ lymphocytes with a half-life of less than two days.[735]

A major effort is being made to devise a vaccine against HIV. However, rapid mutation of the *env* gene makes it difficult to accomplish.[736,736a] This high rate of mutation appears to be a result of a high frequency of errors by the HIV reverse transcriptase.[737] There is

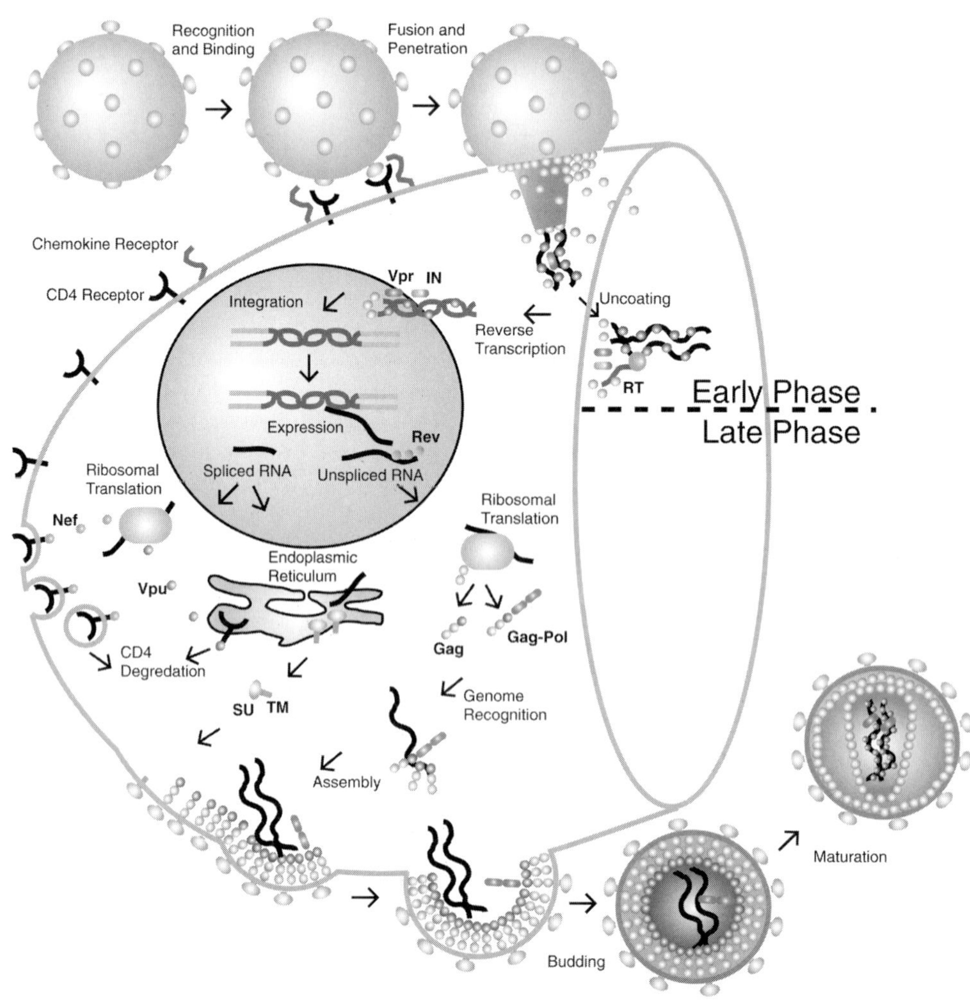

Figure 28-27 General features of the HIV-1 replication cycle. The early phase (upper portion of the diagram) begins with CD4 recognition and involves events up to and including integration of the proviral DNA, and the late phase includes all events from transcription of the integrated DNA to virus budding and maturation. From Turner and Summers.[735]

hope from the fact that some individuals are naturally resistant to HIV infection.[738,739]

3. Retrotransposons

Transposition of DNA, which is discussed in Chapter 27, Section D,4, may seem to be a rare and relatively unimportant event in our body cells. However, transposon DNA accounts for 35% or more of the human genome[740] and apparently plays a major role in evolution. Like other transposons, the DNA sequences known as retrotransposons also move about within DNA. However, they use an indirect mechanism that involves synthesis of mRNA and reverse transcription.[740,741] The reverse transcribed complementary DNA may be inserted back into the genome at new locations. The necessary chemical reactions parallel those involved in the replication of retroviruses (Fig. 28-23, 28-24). Retrotransposens, truncated retrotransposons, and related sequences constitute as much as 16% of the human genome.[741]

There are two classes of retrotrotransposons: those with long terminal repeats (LTRs) and those without (LTRs). The first group is closely related to retroviruses, but its members lack genes for envelope proteins. They do carry *gag* and *pol* genes similar to those of retroviruses (Fig. 28-3). Most retrotransposons are defective and do not move. Over evolutionary time they accumulate in the genome, sometimes to the extent that the genome size grows enormously. This has happened often during the evolution of plants, some of which (e.g., certain lilies) have 40 times more DNA per cell than do humans (Table 1-3).[742] Although most retrotransposons are inactive, some of them occasionally jump to new locations where they may mutate a gene and may sometimes cause disease. However, their major significance is probably in facilitating evolution, perhaps including the formation of new species.[743]

The non-LTR transposons are exemplified by the 6–7 kbp LINES (p. 1539)[741,743] and the short 90–400 bp SINES (p. 1538, Fig. 27-9).[744] Mammalian genomes contain ~50,000 truncated members of the LINE-1 (L1) family and 3000–5000 full-length L1s. Only a few of these are active in our present population. The RNA intermediates that participate in retroposition of LINES are generated by RNA polymerase II, while RNA polymerase III forms the RNA intermediates for propagation of SINES. Participation of these RNAs in trans splicing processes can modify existing genes, contributing to the remodeling of the genome.[744]

Yet another group of mobile elements in the genome are **intein genes**, which encode protein-splicing polypeptides (see Box 29-D). Many inteins also have **homing endonuclease** activity and cleave DNA at specific insertion sequences, initiating incorporation of intein DNA into new locations in the genome.[745] Group II introns, which are found in bacteria and in organelles of fungi and plants, may also act as mobile DNA elements.[746]

References

1. Storz, G. (2002) *Science* **296**, 1260–1262
1a. Zamore, P. D. (2002) *Science* **296**, 1265–1269
1b. Ruvkun, G. (2001) *Science* **294**, 797–799
1c. Cohen, G. N. (1995) *FASEB J.* **9**, 981–982
2. Dickson, R. C., Abelson, J., Barnes, W. M., and Reznikoff, W. S. (1975) *Science* **187**, 27–35
3. Müller-Hill, B. (1996) *The lac Operon. A Short History of a Genetic Paradigm*, de Gruyter, Berlin
3a. Juers, D. H., Heightman, T. D., Vasella, A., McCarter, J. D., Mackenzie, L., Withers, S. G., and Matthews, B. W. (2001) *Biochemistry* **40**, 14781–14794
4. Page, M. G. P., and Rosenbusch, J. P. (1988) *J. Biol. Chem.* **263**, 15906–15914
5. Lewis, M., Chang, G., Horton, N. C., Kercher, M. A., Pace, H. C., Schumacher, M. A., Brennan, R. G., and Lu, P. (1996) *Science* **271**, 1247–1254
5a. Bell, C. E., and Lewis, M. (2001) *J. Mol. Biol.* **312**, 921–926
6. Markiewicz, P., Kleina, L. G., Cruz, C., Ehret, S., and Miller, J. H. (1994) *J. Mol. Biol.* **240**, 421–433
7. Nakanishi, S., Adhya, S., Gottesman, M., and Pastan, I. (1973) *J. Biol. Chem.* **248**, 5937–5942
8. Gilbert, W., and Maxam, A. (1973) *Proc. Natl. Acad. Sci. U.S.A.* **70**, 3581–3584
9. Simons, A., Tils, D., van Wilcken-Bergmann, B., and Muller-Hill, B. (1984) *Proc. Natl. Acad. Sci. U.S.A.* **81**, 1624–1628
10. von Hippel, P. H., Bear, D. G., Morgan, W. D., and McSwiggen, J. A. (1984) *Rev. Biochem.* **53**, 389–446
11. Rastinejad, F., Artz, P., and Lu, P. (1993) *J. Mol. Biol.* **233**, 389–399
12. Levandoski, M. M., Tsodikov, O. V., Frank, D. E., Melcher, S. E., Saecker, R. M., and Record, M. T., Jr. (1996) *J. Mol. Biol.* **260**, 697–717
13. Frank, D. E., Saecker, R. M., Bond, J. P., Capp, M. W., Tsodikov, O. V., Melcher, S. E., Levandoski, M. M., and Record, M. T., Jr. (1997) *J. Mol. Biol.* **267**, 1186–1206
14. von Hippel, P. H., and Berg, O. G. (1989) *J. Biol. Chem.* **264**, 675–678
15. Shimamoto, N. (1999) *J. Biol. Chem.* **274**, 15293–15296
16. Harada, K., and Frankel, A. D. (1995) *EMBO J.* **14**, 5798–5811
17. Pace, H. C., Kercher, M. A., Lu, P., Markiewicz, P., Miller, J. H., Chang, G., and Lewis, M. (1997) *Trends Biochem. Sci.* **22**, 334–339
18. Horton, N., Lewis, M., and Lu, P. (1997) *J. Mol. Biol.* **265**, 1–7
19. Slijper, M., Bonvin, A. M. J. J., Boelens, R., and Kaptein, R. (1996) *J. Mol. Biol.* **259**, 761–773
20. Barry, J. K., and Matthews, K. S. (1999) *Biochemistry* **38**, 3579–3590
21. Suckow, J., Markiewicz, P., Kleina, L. G., Miller, J., Kisters-Woike, B., and Müller-Hill, B. (1996) *J. Mol. Biol.* **261**, 509–523
22. Kolata, G. B. (1976) *Science* **191**, 373
23. Sommer, H., Lu, P., and Miller, J. H. (1976) *J. Biol. Chem.* **251**, 3774–3779
24. Müller, J., Oehler, S., and Müller-Hill, B. (1996) *J. Mol. Biol.* **257**, 21–29
25. Lanzer, M., and Bujard, H. (1988) *Proc. Natl. Acad. Sci. U.S.A.* **85**, 8973–8977
26. McClure, W. R. (1985) *Ann. Rev. Biochem.* **54**, 171–204
27. Auble, D. T., Allen, T. L., and deHaseth, P. L. (1986) *J. Biol. Chem.* **261**, 11202–11206
28. Place, C., Oddos, J., Buc, H., McAllister, W. T., and Buckle, M. (1999) *Biochemistry* **38**, 4948–4957
29. Cheetham, G. M. T., and Steitz, T. A. (1999) *Science* **286**, 2305–2309
30. Brieba, L. G., and Sousa, R. (2000) *Biochemistry* **39**, 919–923
30a. Brieba, L. G., Gopal, V., and Sousa, R. (2001) *J. Biol. Chem.* **276**, 10306–10313
31. Opalka, N., Mooney, R. A., Richter, C., Severinov, K., Landick, R., and Darst, S. A. (2000) *Proc. Natl. Acad. Sci. U.S.A.* **97**, 617–622

References

32. Owens, J. T., Chmura, A. J., Murakami, K., Fujita, N., Ishihama, A., and Meares, C. F. (1998) *Biochemistry* **37**, 7670–7675
32a. Vuthoori, S., Bowers, C. W., McCracken, A., Dombroski, A. J., and Hinton, D. M. (2001) *J. Mol. Biol.* **309**, 561–572
32b. Kuznedelov, K., Minakhin, L., Niedziela-Majka, A., Dove, S. L., Rogulja, D., Nickels, B. E., Hochschild, A., Heyduk, T., and Severinov, K. (2002) *Science* **295**, 855–857
33. Brodolin, K., Mustaev, A., Severinov, K., and Nikiforov, V. (2000) *J. Biol. Chem.* **275**, 3661–3666
33a. Ebright, R. H. (2000) *J. Mol. Biol.* **304**, 687–698
33b. Minakhin, L., Bhagat, S., Brunning, A., Campbell, E. A., Darst, S. A., Ebright, R. H., and Severinov, K. (2001) *Proc. Natl. Acad. Sci. U.S.A.* **98**, 892–897
33c. Murakami, K. S., Masuda, S., and Darst, S. A. (2002) *Science* **296**, 1280–1284
33d. Murakami, K. S., Masuda, S., Campbell, E. A., Muzzin, O., and Darst, S. A. (2002) *Science* **296**, 1285–1290
33e. Vassylyev, D. G., Sekine, S.-i, Laptenko, O., Lee, J., Vassylyeva, M. N., and Borukhov, S. (2002) *Nature (London)* **417**, 712–718
33f. Lin, S., Katilius, E., Haffa, A. L. M., Taguchi, A. K. W., and Woodbury, N. W. (2001) *Biochemistry* **40**, 13767–13773
34. Nudler, E., Avetissova, E., Markovtsov, V., and Goldfarb, A. (1996) *Science* **273**, 211–217
35. Wu, F. Y. H., Huang, W.-J., Sinclair, R. B., and Powers, L. (1992) *J. Biol. Chem.* **267**, 25560–25567
36. Katayama, A., Fujita, N., and Ishihama, A. (2000) *J. Biol. Chem.* **275**, 3583–3592
37. Zaychikov, E., Martin, E., Denissova, L., Kozlov, M., Markovtsov, V., Kashlev, M., Heumann, H., Nikiforov, V., Goldfarb, A., and Mustaev, A. (1996) *Science* **273**, 107–109
38. Mustaev, A., Kozlov, M., Markovtsov, V., Zaychikov, E., Denissova, L., and Goldfarb, A. (1997) *Proc. Natl. Acad. Sci. U.S.A.* **94**, 6641–6645
39. Buckle, M., Pemberton, I. K., Jacquet, M.-A., and Buc, H. (1999) *J. Mol. Biol.* **285**, 955–964
40. Helmann, J. D., and Chamberlin, M. J. (1988) *Ann. Rev. Biochem.* **57**, 839–872
40a. Chadsey, M. S., and Hughes, K. T. (2001) *J. Mol. Biol.* **306**, 915–929
41. Carmona, M., Claverie-Martin, F., and Magasanik, B. (1997) *Proc. Natl. Acad. Sci. U.S.A.* **94**, 9568–9572
41a. Wigneshweraraj, S. R., Chaney, M. K., Ishihama, A., and Buck, M. (2001) *J. Mol. Biol.* **306**, 681–701
42. Pérez-Martín, J., and de Lorenzo, V. (1996) *J. Mol. Biol.* **258**, 562–574
43. Marschall, C., Labrousse, V., Kreimer, M., Weichart, D., Kolb, A., and Hengge-Aronis, R. (1998) *J. Mol. Biol.* **276**, 339–353
44. Blaszczak, A., Zylicz, M., Georgopoulos, C., and Liberek, K. (1995) *EMBO J.* **14**, 5085–5093
45. Huang, X., and Helmann, J. D. (1998) *J. Mol. Biol.* **279**, 165–173
46. Pribnow, D. (1975) *Proc. Natl. Acad. Sci. U.S.A.* **72**, 784–788
47. Lewin, B. (2000) *Genes VII*, Oxford Univ. Press, New York
48. Bown, J. A., Owens, J. T., Meares, C. F., Fujita, N., Ishihama, A., Busby, S. J. W., and Minchin, S. D. (1999) *J. Biol. Chem.* **274**, 2263–2270
49. Blattner, F. R., Plunkett, G., III, Bloch, C. A., Perna, N. T., Burland, V., Riley, M., Collado-Vides, J., Glasner, J. D., Rode, C. K., Mayhew, G. F., Gregor, J., Davis, N. W., Kirkpatrick, H. A., Goeden, M. A., Rose, D. J., Mau, B., and Shao, Y. (1997) *Science* **277**, 1453–1462
50. Severinov, K., Mustaev, A., Severinova, E., Kozlov, M., Darst, S. A., and Goldfarb, A. (1995) *J. Biol. Chem.* **270**, 29428–29432
51. Yang, X., and Price, C. W. (1995) *J. Biol. Chem.* **270**, 23930–23933
52. Traviglia, S. L., Datwyler, S. A., and Meares, C. F. (1999) *Biochemistry* **38**, 4259–4265
53. Ross, W., Gosink, K. K., Salomon, J., Igarashi, K., Zou, C., Ishihama, A., Severinov, K., and Gourse, R. L. (1993) *Science* **262**, 1407–1413
54. Ho Jeon, Y., Negishi, T., Shirakawa, M., Yamazaki, T., Fujita, N., Ishihama, A., and Kyogoku, Y. (1995) *Science* **270**, 1495–1497
54a. Zhang, G., and Darst, S. A. (1998) *Science* **281**, 262–266
54b. Wada, T., Yamazaki, T., and Kyogoku, Y. (2000) *J. Biol. Chem.* **275**, 16057–16063
54c. Calles, B., Monsalve, M., Rojo, F., and Salas, M. (2001) *J. Mol. Biol.* **307**, 487–497
54d. Benoff, B., Yang, H., Lawson, C. L., Parkinson, G., Liu, J., Blatter, E., Ebright, Y. W., Berman, H. M., and Ebright, R. H. (2002) *Science* **297**, 1562–1566
55. Bokal, A. J., IV, Ross, W., and Gourse, R. L. (1995) *J. Mol. Biol.* **245**, 197–207
56. Muskhelishvili, G., Buckle, M., Heumann, H., Kahmann, R., and Travers, A. A. (1997) *EMBO J.* **16**, 3655–3665
57. Travers, A., and Muskhelishvili, G. (1998) *J. Mol. Biol.* **279**, 1027–1043
58. Pan, C. Q., Finkel, S. E., Cramton, S. E., Feng, J.-A., Sigman, D. S., and Johnson, R. C. (1996) *J. Mol. Biol.* **264**, 675–695
59. Lamond, A. I. (1985) *Trends Biochem. Sci.* **10**, 271–274
59a. Pemberton, I. K., Muskhelishvili, G., Travers, A. A., and Buckle, M. (2000) *J. Mol. Biol.* **299**, 859–864
60. Cashel, M., and Rudd, K. E. (1987) in *Escherichia coli and Salmonella typhimurium* (Neidhardt, F. C., ed), pp. 1410–1438, Am. Soc. for Microbiology, Washington, DC
61. Baracchi, E., and Bremer, H. (1988) *J. Biol. Chem.* **263**, 2597–2602
62. Choy, H. E. (2000) *J. Biol. Chem.* **275**, 6783–6789
62a. Barker, M. M., Gaal, T., and Gourse, R. L. (2001) *J. Mol. Biol.* **305**, 689–702
63. Yamagishi, M., Cole, J. R., Nomura, M., Studier, F. W., and Dunn, J. J. (1987) *J. Biol. Chem.* **262**, 3940–3943
64. Singer, P. T., and Wu, C.-W. (1988) *J. Biol. Chem.* **263**, 4208–4214
65. Gelles, J., and Landick, R. (1998) *Cell* **93**, 13–16
66. Bustamante, C., Guthold, M., Zhu, X., and Yang, G. (1999) *J. Biol. Chem.* **274**, 16665–16668
67. Lefèvre, J.-F., Lane, A. N., and Jardetzky, O. (1988) *Biochemistry* **27**, 1086–1094
68. Ellinger, T., Behnke, D., Knaus, R., Bujard, H., and Gralla, J. D. (1994) *J. Mol. Biol.* **239**, 466–475
69. von Hippel, P. H. (1998) *Science* **281**, 660–665
70. Levin, J. R., Blake, J. J., Ganunis, R. A., and Tullius, T. D. (2000) *J. Biol. Chem.* **275**, 6885–6893
71. Yager, T. D., and von Hippel, P. H. (1987) in *Escherichia coli and Salmonella typhimurium* (Neidhardt, F. C., ed), pp. 1241–1275, Am. Soc. for Microbiology, Washington, DC
72. Barile, M. F., Razin, S., Tully, J. G., and Whitcomb, R. F., eds. (1979) *The Mycoplasmas*, Vol. I, Academic Press, New York
73. Harris, E. H. (1989) *The Chlamydomonas Sourcebook*, Academic Press, San Diego, California
74. Milan, S., D'Ari, L., and Chamberlin, M. J. (1999) *Biochemistry* **38**, 218–225
75. Davenport, R. J., Wuite, G. J. L., Landick, R., and Bustamante, C. (2000) *Science* **287**, 2497–2500
76. Ellinger, T., Behnke, D., Bujard, H., and Gralla, J. D. (1994) *J. Mol. Biol.* **239**, 455–465
76a. Lee, K.-B., Wang, D., Lippard, S. J., and Sharp, P. A. (2002) *Proc. Natl. Acad. Sci. U.S.A.* **99**, 4239–4244
77. Selby, C. P., and Sancar, A. (1993) *Science* **260**, 53–58
78. Viswanathan, A., You, H. J., and Doetsch, P. W. (1999) *Science* **284**, 159–162
79. Nudler, E. (1999) *J. Mol. Biol.* **288**, 1–12
80. Erie, D. A., Hajiseyedjavadi, O., Young, M. C., and von Hippel, P. H. (1993) *Science* **262**, 867–873
81. Frank-Kamenetskii, M. (1989) *Nature (London)* **337**, 206
82. Krasilnikov, A. S., Podtelezhnikov, A., Vologodskii, A., and Mirkin, S. M. (1999) *J. Mol. Biol.* **292**, 1149–1160
83. Wu, H. (1988) *Cell* **53**, 433–440
84. Giaever, G. N., and Wang, J. C. (1988) *Cell* **55**, 849–856
85. Miller, O. L., Jr. (1973) *Sci. Am.* **228**(March), 34–42
86. Weickert, M. J., and Adhya, S. (1992) *J. Biol. Chem.* **267**, 15869–15874
87. Takeda, Y., Kim, J. G., Caday, C. G., Steers, E., Jr., Ohlendorf, D. H., Anderson, W. F., and Matthews, B. W. (1986) *J. Biol. Chem.* **261**, 8608–8616
88. Matthews, B. W., Ohlendorf, D. H., Anderson, W. F., Fisher, R. G., and Takedo, Y. (1983) *Trends Biochem. Sci.* **8**, 25–29
89. Jana, R., Hazbun, T. R., Fields, J. D., and Mossing, M. C. (1998) *Biochemistry* **37**, 6446–6455
90. Pabo, C. O., and Lewis, M. (1982) *Nature (London)* **298**, 443–447
90a. Bell, C. E., and Lewis, M. (2001) *J. Mol. Biol.* **314**, 1127–1136
91. Jordan, S. R., and Pabo, C. O. (1988) *Science* **242**, 893–899
92. Aggarwal, A. K., Rodgers, D. W., Drottar, M., Ptashne, M., and Harrison, S. C. (1988) *Science* **242**, 899–907
93. Wolberger, C., Dong, Y., Ptashne, M., and Harrison, S. C. (1988) *Nature (London)* **335**, 789–795
94. Padmanabhan, S., Jiménez, M. A., González, C., Sanz, J. M., Giménez-Gallego, G., and Rico, M. (1997) *Biochemistry* **36**, 6424–6436
95. Wharton, R. P., and Ptashne, M. (1985) *Nature (London)* **316**, 601–605
96. Wharton, R. P., and Ptashne, M. (1986) *Trends Biochem. Sci.* **11**, 71–73
97. Sauer, R. T., Milla, M. E., Waldburger, C. D., Brown, B. M., and Schildbach, J. F. (1996) *FASEB J.* **10**, 42–48
98. Nooren, I. M. A., Rietveld, A. W. M., Melacini, G., Sauer, R. T., Kaptein, R., and Boelens, R. (1999) *Biochemistry* **38**, 6035–6042
99. He, Y.-y, Stockley, P. G., and Gold, L. (1996) *J. Mol. Biol.* **255**, 55–66
100. Dai, X., Kloster, M., and Rothman-Denes, L. B. (1998) *J. Mol. Biol.* **283**, 43–58
101. Wang, J. C., Barkley, M. D., and Bourgeois, S. (1974) *Nature (London)* **251**, 247–249
102. Horwitz, M. S. Z., and Loeb, L. A. (1988) *Science* **241**, 703–705
103. Arndt, K. T., Boschelli, F., Cook, J., Takeda, Y., Tecza, E., and Lu, P. (1983) *J. Biol. Chem.* **258**, 4177–4183
104. Metzler, W. J., and Lu, P. (1989) *J. Mol. Biol.* **205**, 149–164
105. Borowiec, J. A., and Gralla, J. D. (1986) *Biochemistry* **25**, 5051–5057
106. Ho, Y.-S., Wulff, D. L., and Rosenberg, M. (1983) *Nature (London)* **304**, 703–708
107. Brennan, R. G., Vasu, S., Matthews, B. W., and Otsuka, A. J. (1989) *J. Biol. Chem.* **264**, 5

References

108. Eisenstein, E., and Beckett, D. (1999) *Biochemistry* **38**, 13077–13084
109. Streaker, E. D., and Beckett, D. (1999) *J. Mol. Biol.* **292**, 619–632
109a. Weaver, L. H., Kwon, K., Beckett, D., and Matthews, B. W. (2001) *Protein Sci.* **10**, 2618–2622
110. Dahl, M. K., Degenkolb, J., and Hillen, W. (1994) *J. Mol. Biol.* **243**, 413–424
111. Yoshida, K.-I., Shibayama, T., Aoyama, D., and Fujita, Y. (1999) *J. Mol. Biol.* **285**, 917–929
112. Hinrichs, W., Kisker, C., Düvel, M., Müller, A., Tovar, K., Hillen, W., and Saenger, W. (1994) *Science* **264**, 418–420
113. Kisker, C., Hinrichs, W., Tovar, K., Hillen, W., and Saenger, W. (1995) *J. Mol. Biol.* **247**, 260–280
113a. Schubert, P., Schnappinger, D., Pfleiderer, K., and Hillen, W. (2001) *Biochemistry* **40**, 3257–3263
113b. Lathe, W. C., III, Snel, B., and Bork, P. (2000) *Trends Biochem. Sci.* **25**, 474–479
114. Cohen, G. N. (1986) in *Regulation of Gene Expression* (Booth, I. R., and Higgins, C. F., eds), pp. 1–20, Cambridge Univ. Press, London
115. Cornish, E. C., Argyropoulos, V. P., Pittard, J., and Davidson, B. E. (1986) *J. Biol. Chem.* **261**, 403–410
116. Weiss, D. L., Johnson, D. I., Weith, H. L., and Somerville, R. L. (1986) *J. Biol. Chem.* **261**, 9966–9971
117. Choi, K. Y., Lu, F., and Zalkin, H. (1994) *J. Biol. Chem.* **269**, 24066–24072
118. Xu, H., Moraitis, M., Reedstrom, R. J., and Matthews, K. S. (1998) *J. Biol. Chem.* **273**, 8958–8964
119. Glasfeld, A., Koehler, A. N., Schumacher, M. A., and Brennan, R. G. (1999) *J. Mol. Biol.* **291**, 347–361
120. Charlier, D., Roovers, M., Van Vliet, F., Boyen, A., Cunin, R., Nakamura, Y., Glansdorff, N., and Piérard, A. (1992) *J. Mol. Biol.* **226**, 367–386
121. Van Duyne, G. D., Ghosh, G., Maas, W. K., and Sigler, P. B. (1996) *J. Mol. Biol.* **256**, 377–391
122. Wang, H., Glansdorff, N., and Charlier, D. (1998) *J. Mol. Biol.* **277**, 805–824
123. Ni, J., Sakanyan, V., Charlier, D., Glansdorff, N., and Van Duyne, G. D. (1999) *Nature Struct. Biol.* **6**, 427–432
124. Escolar, L., Pérez-Martín, J., and de Lorenzo, V. (1998) *J. Mol. Biol.* **283**, 537–547
125. White, A., Ding, X., vanderSpek, J. C., Murphy, J. R., and Ringe, D. (1998) *Nature (London)* **394**, 502–506
126. Pohl, E., Holmes, R. K., and Hol, W. G. J. (1998) *J. Biol. Chem.* **273**, 22420–22427
127. Pohl, E., Holmes, R. K., and Hol, W. G. J. (1999) *J. Mol. Biol.* **285**, 1145–1156
127a. Gourley, D. G., Schüttelkopf, A. W., Anderson, L. A., Price, N. C., Boxer, D. H., and Hunter, W. N. (2001) *J. Biol. Chem.* **276**, 20641–20647
128. Postma, P. W. (1986) in *Regulation of Gene Expression* (Booth, I. R., and Higgins, C. F., eds), Cambridge Univ. Press, London
129. Busby, S., and Ebright, R. H. (1999) *J. Mol. Biol.* **293**, 199–213
130. Schultz, S. C., Shields, G. C., and Steitz, T. A. (1991) *Science* **253**, 1001–1007
131. Passner, J. M., and Steitz, T. A. (1997) *Proc. Natl. Acad. Sci. U.S.A.* **94**, 2843–2847
132. Parkinson, G., Wilson, C., Gunasekera, A., Ebright, Y. W., Ebright, R. E., and Berman, H. M. (1996) *J. Mol. Biol.* **260**, 395–408
133. Gunasekera, A., Ebright, Y. W., and Ebright, R. H. (1992) *J. Biol. Chem.* **267**, 14713–14720
134. Vossen, K. M., Stickle, D. F., and Fried, M. G. (1996) *J. Mol. Biol.* **255**, 44–54
135. Dalma-Weiszhausz, D. D., and Brenowitz, M. (1996) *Biochemistry* **35**, 3735–3745
136. Roy, S., Garges, S., and Adhya, S. (1998) *J. Biol. Chem.* **273**, 14059–14062
137. Danot, O., Vidal-Ingigliardi, D., and Raibaud, O. (1996) *J. Mol. Biol.* **262**, 1–11
138. Richet, E., and Sogaard-Andersen, L. (1994) *EMBO J.* **13**, 4558–4567
139. Ziegelhoffer, E. C., and Kiley, P. J. (1995) *J. Mol. Biol.* **245**, 351–361
140. Lazazzera, B. A., Beinert, H., Khoroshilova, N., Kennedy, M. C., and Kiley, P. J. (1996) *J. Biol. Chem.* **271**, 2762–2768
141. Popescu, C. V., Bates, D. M., Beinert, H., Münck, E., and Kiley, P. J. (1998) *Proc. Natl. Acad. Sci. U.S.A.* **95**, 13431–13435
142. Shelver, D., Thorsteinsson, M. V., Kerby, R. L., Chung, S.-Y., Roberts, G. P., Reynolds, M. F., Parks, R. B., and Burstyn, J. N. (1999) *Biochemistry* **38**, 2669–2678
142a. Thorsteinsson, M. V., Kerby, R. L., Youn, H., Conrad, M., Serate, J., Staples, C. R., and Roberts, G. P. (2001) *J. Biol. Chem.* **276**, 26807–26813
143. Greene, J. R., Morrissey, L. M., Foster, L. M., and Geiduschek, E. P. (1986) *J. Biol. Chem.* **261**, 12820–12827
144. Takahashi, K., Kawazoe, Y., Sakumi, K., Nakabeppu, Y., and Sekiguchi, M. (1988) *J. Biol. Chem.* **263**, 13490–13492
145. Lobell, R. B., and Schleif, R. F. (1990) *Science* **250**, 528–532
146. Zhang, X., Reeder, T., and Schleif, R. (1996) *J. Mol. Biol.* **258**, 14–24
147. Niland, P., Hühne, R., and Müller-Hill, B. (1996) *J. Mol. Biol.* **264**, 667–674
147a. Wu, M., and Schleif, R. (2001) *J. Mol. Biol.* **307**, 1001–1009
147b. LaRonde-LeBlanc, N., and Wolberger, C. (2000) *Biochemistry* **39**, 11593–11601
148. Echols, H. (1990) *J. Biol. Chem.* **265**, 14697–14700
149. Rippe, K., von Hippel, P. H., and Langowski, J. (1995) *Trends Biochem. Sci.* **20**, 500–506
150. Blackwood, E. M., and Kadonaga, J. T. (1998) *Science* **281**, 60–63
151. Magasanik, B. (1988) *Trends Biochem. Sci.* **13**, 475–479
152. Rippe, K., Guthold, M., von Hippel, P. H., and Bustamante, C. (1997) *J. Mol. Biol.* **270**, 125–138
153. Flashner, Y., Weiss, D. S., Keener, J., and Kustu, S. (1995) *J. Mol. Biol.* **249**, 700–713
153a. Schulz, A., Langowski, J., and Rippe, K. (2000) *J. Mol. Biol.* **300**, 709–725
154. Brahms, G., Brahms, S., and Magasanik, B. (1995) *J. Mol. Biol.* **246**, 35–42
155. Cullen, P. J., Bowman, W. C., Hartnett, D.-F., Reilly, S. C., and Kranz, R. G. (1998) *J. Mol. Biol.* **278**, 903–914
156. Dworkin, J., Jovanovic, G., and Model, P. (1997) *J. Mol. Biol.* **273**, 377–388
157. Solà, M., Gomis-Rüth, F. X., Serrano, L., González, A., and Coll, M. (1999) *J. Mol. Biol.* **285**, 675–687
158. Makino, K., Amemura, M., Kawamoto, T., Kimura, S., Shinagawa, H., Nakata, A., and Suzuki, M. (1996) *J. Mol. Biol.* **259**, 15–26
159. Shi, L., Liu, W., and Hulett, F. M. (1999) *Biochemistry* **38**, 10119–10125
159a. Georgellis, D., Kwon, O., and Lin, E. C. C. (2001) *Science* **292**, 2314–2316
160. Strohmaier, H., Noiges, R., Kotschan, S., Sawers, G., Högenauer, G., Zechner, E. L., and Koraimann, G. (1998) *J. Mol. Biol.* **277**, 309–316
161. Miyatake, H., Mukai, M., Adachi, S.-i, Nakamura, H., Tamura, K., Iizuka, T., Shiro, Y., Strange, R. W., and Hasnain, S. S. (1999) *J. Biol. Chem.* **274**, 23176–23184
161a. Rusnak, F., and Reiter, T. (2000) *Trends Biochem. Sci.* **25**, 527–529
162. Li, J., Kustu, S., and Stewart, V. (1994) *J. Mol. Biol.* **241**, 150–165
163. Asayama, M., Yamamoto, A., and Kobayashi, Y. (1995) *J. Mol. Biol.* **250**, 11–23
164. Madhusudan, Zapf, J., Hoch, J. A., Whiteley, J. M., Xuong, N. H., and Varughese, K. I. (1997) *Biochemistry* **36**, 12739–12745
165. Eguchi, Y., Itoh, T., and Tomizawa, J.-i. (1991) *Ann. Rev. Biochem.* **60**, 631–652
166. Weintraub, H. M. (1990) *Sci. Am.* **262**(Jan), 40–46
166a. Nellen, W., and Lichtenstein, C. (1993) *Trends Biochem. Sci.* **18**, 419–423
166b. Delihas, N., and Forst, S. (2001) *J. Mol. Biol.* **313**, 1–12
167. Green, P. J., Pines, O., and Inouye, M. (1986) *Ann. Rev. Biochem.* **55**, 569–597
168. Cesareni, G., and Banner, D. W. (1985) *Trends Biochem. Sci.* **10**, 303–306
169. Asano, K., Niimi, T., Yokoyama, S., and Mizobuchi, K. (1998) *J. Biol. Chem.* **273**, 11826–11838
170. Ikenaka, K., Ramakrishnan, G., Inouye, M., Tsung, K., and Inouye, M. (1986) *J. Biol. Chem.* **261**, 9316–9320
171. Andersen, J., Forst, S. A., Zhao, K., Inouye, M., and Delihas, N. (1989) *J. Biol. Chem.* **264**, 17961–17970
171a. Matzke, M., Matzke, A. J. M., and Kooter, J. M. (2001) *Science* **293**, 1080–1083
172. Madison-Antenucci, S., Sabatini, R. S., Pollard, V. W., and Hajduk, S. L. (1998) *EMBO J.* **17**, 6368–6376
173. Nicoloso, M., Qu, L.-H., Michot, B., and Bachellerie, J.-P. (1996) *J. Mol. Biol.* **260**, 178–195
174. Lafontaine, D. L. J., and Tollervey, D. (1998) *Trends Biochem. Sci.* **23**, 383–388
174a. Gerdes, K., Nielsen, A., Thorsted, P., and Wagner, E. G. H. (1992) *J. Mol. Biol.* **226**, 637–649
175. Cohen, J. S., and Hogan, M. E. (1994) *Sci. Am.* **271**(Jun), 76–82
176. Askari, F. K., and McDonnell, W. M. (1996) *N. Engl. J. Med.* **334**, 316–318
176a. Sullenger, B. A., and Gilboa, E. (2002) *Nature (London)* **418**, 252–258
176b. Gitlin, L., Karelsky, S., and Andino, R. (2002) *Nature (London)* **418**, 430–434
177. Wittung-Stafshede, P. (1998) *Science* **281**, 657–658
178. Wagner, R. W. (1994) *Nature (London)* **372**, 333–335
179. Tereshko, V., Gryaznov, S., and Egli, M. (1998) *J. Am. Chem. Soc.* **120**, 269–283
180. Aramini, J. M., Kalisch, B. W., Pon, R. T., van de Sande, J. H., and Germann, M. W. (1996) *Biochemistry* **35**, 9355–9365
181. Mujeeb, A., Reynolds, M. A., and James, T. L. (1997) *Biochemistry* **36**, 2371–2379
182. Branch, A. D. (1998) *Trends Biochem. Sci.* **23**, 45–50
182a. Braasch, D. A., and Corey, D. R. (2002) *Biochemistry* **41**, 4503–4510
183. Platt, T. (1986) *Ann. Rev. Biochem.* **55**, 339–372
184. Yanofsky, C. (1981) *Nature (London)* **289**, 751–758
185. Yanofsky, C. (1988) *J. Biol. Chem.* **263**, 609–612
186. Kuroda, M. I., and Yanofsky, C. (1984) *J. Biol. Chem.* **259**, 12838–12843
187. Baumann, C., Otridge, J., and Gollnick, P. (1996) *J. Biol. Chem.* **271**, 12269–12274
187a. Losick, R., and Sonenshein, A. L. (2001) *Science* **293**, 2018–2019
187b. Valbuzzi, A., and Yanofsky, C. (2001) *Science* **293**, 2057–2059
187c. Yakhnin, A. V., and Babitzke, P. (2002) *Proc. Natl. Acad. Sci. U.S.A.* **99**, 11067–11072

References

188. Antson, A. A., Dodson, E. J., Dodson, G., Greaves, R. B., Chen, X.-p, and Gollnick, P. (1999) *Nature (London)* **401**, 235 – 242
189. Jensen, K. F., Bonekamp, F., and Poulsen, P. (1986) *Trends Biochem. Sci.* **11**, 362 – 365
190. Hauser, C. A., Sharp, J. A., Hatfield, L. K., and Hatfield, G. W. (1985) *J. Biol. Chem.* **260**, 1765 – 1770
191. Watson, M. D. (1981) *Trends Biochem. Sci.* **6**, 180 – 182
192. Stitt, B. L. (1988) *J. Biol. Chem.* **263**, 11130 – 11137
193. Bear, D. G., and Peabody, D. S. (1988) *Trends Biochem. Sci.* **13**, 343 – 347
194. Nehrke, K. W., and Platt, T. (1994) *J. Mol. Biol.* **243**, 830 – 839
195. Liu, K., and Hanna, M. M. (1995) *J. Mol. Biol.* **247**, 547 – 558
196. Greenblatt, J. (1991) *Trends Biochem. Sci.* **16**, 408 – 411
196a. Toulokhonov, I., Artsimovitch, I., and Landick, R. (2001) *Science* **292**, 730 – 733
196b. Gopal, B., Haire, L. F., Gamblin, S. J., Dodson, E. J., Lane, A. N., Papavinasasundaram, K. G., Colston, M. J., and Dodson, G. (2001) *J. Mol. Biol.* **314**, 1087 – 1095
196c. Zhou, Y., Mah, T.-F., Yu, Y.-T. N., Mogridge, J., Olson, E. R., Greenblatt, J., and Friedman, D. I. (2001) *J. Mol. Biol.* **310**, 33 – 49
197. Zhu, A. Q., and von Hippel, P. H. (1998) *Biochemistry* **37**, 11202 – 11214
198. Kim, D.-E., and Patel, S. S. (1999) *J. Biol. Chem.* **274**, 32667 – 32671
198a. Vincent, F., Openshaw, M., Trautwein, M., Gaskell, S. J., Kohn, H., and Widger, W. R. (2000) *Biochemistry* **39**, 9077 – 9083
198b. Yu, X., Horiguchi, T., Shigesada, K., and Egelman, E. H. (2000) *J. Mol. Biol.* **299**, 1279 – 1287
199. Greenblatt, J., Nodwell, J. R., and Mason, S. W. (1993) *Nature (London)* **364**, 401 – 406
200. O'Hara, B. P., Norman, R. A., Wan, P. T. C., Roe, S. M., Barrett, T. E., Drew, R. E., and Pearl, L. H. (1999) *EMBO J.* **18**, 5175 – 5186
200a. Gopal, B., Papavinasasundaram, K. G., Dodson, G., Colston, M. J., Major, S. A., and Lane, A. N. (2001) *Biochemistry* **40**, 920 – 928
200b. Carlomagno, M. S., and Nappo, A. (2001) *J. Mol. Biol.* **309**, 19 – 28
201. Sozhamannan, S., and Stitt, B. L. (1997) *J. Mol. Biol.* **268**, 689 – 703
202. Dunn, J. J., and Studier, F. W. (1983) *J. Mol. Biol.* **166**, 477 – 535
203. Daniels, C. J., Gupta, R., and Doolittle, W. F. (1985) *J. Biol. Chem.* **260**, 3132 – 3134
204. Lykke-Andersen, J., Aagaard, C., Semionenkov, M., and Garrett, R. A. (1997) *Trends Biochem. Sci.* **22**, 326 – 331
205. Abelson, J., Trotta, C. R., and Li, H. (1998) *J. Biol. Chem.* **273**, 12685 – 12688
206. Starzyk, R. M. (1986) *Trends Biochem. Sci.* **11**, 60
207. Sollner-Webb, B., and Mougey, E. B. (1991) *Trends Biochem. Sci.* **16**, 58 – 62
208. Li, Z., Pandit, S., and Deutscher, M. P. (1999) *EMBO J.* **18**, 2878 – 2885
208a. Nomura, M. (1999) *Proc. Natl. Acad. Sci. U.S.A.* **96**, 1820 – 1822
209. Perry, R. P. (1976) *Ann. Rev. Biochem.* **45**, 605 – 629
210. King, T. C., Sirdeshmukh, R., and Schlessinger, D. (1984) *Proc. Natl. Acad. Sci. U.S.A.* **81**, 185 – 188
211. Denoya, C., and Dubnau, D. (1989) *J. Biol. Chem.* **264**, 2615 – 2624
212. Tscherne, J. S., Nurse, K., Popienick, P., Michel, H., Sochacki, M., and Ofengand, J. (1999) *Biochemistry* **38**, 1884 – 1892
213. Stahl, D. A., Pace, B., Marsh, T., and Pace, N. R. (1984) *J. Biol. Chem.* **259**, 11448 – 11453
214. Raychaudhuri, S., Niu, L., Conrad, J., Lane, B. G., and Ofengand, J. (1999) *J. Biol. Chem.* **274**, 18880 – 18886
215. Niu, L., and Ofengand, J. (1999) *Biochemistry* **38**, 629 – 635
216. Ramamurthy, V., Swann, S. L., Spedaliere, C. J., and Mueller, E. G. (1999) *Biochemistry* **38**, 13106 – 13111
217. Gu, X., Liu, Y., and Santi, D. V. (1999) *Proc. Natl. Acad. Sci. U.S.A.* **96**, 14270 – 14275
218. Conrad, J., Sun, D., Englund, N., and Ofengand, J. (1998) *J. Biol. Chem.* **273**, 18562 – 18566
218a. Spedaliere, C. J., Hamilton, C. S., and Mueller, E. G. (2000) *Biochemistry* **39**, 9459 – 9465
218b. Arluison, V., Buckle, M., and Grosjean, H. (1999) *J. Mol. Biol.* **289**, 491 – 502
219. Kiss-László, Z., Henry, Y., and Kiss, T. (1998) *EMBO J.* **17**, 797 – 807
220. Segal, D. M., and Eichler, D. C. (1991) *J. Biol. Chem.* **266**, 24385 – 24389
221. Frank, D. N., and Pace, N. R. (1998) *Ann. Rev. Biochem.* **67**, 153 – 180
222. Christian, E. L., and Harris, M. E. (1999) *Biochemistry* **38**, 12629 – 12638
222a. Zuleeg, T., Hansen, A., Pfeiffer, T., Schübel, H., Kreutzer, R., Hartmann, R. K., and Limmer, S. (2001) *Biochemistry* **40**, 3363 – 3369
222b. Cole, K. B., and Dorit, R. L. (2001) *J. Mol. Biol.* **307**, 1181 – 1193
222c. Mikkelsen, N. E., Brännvall, M., Virtanen, A., and Kirsebom, L. A. (1999) *Proc. Natl. Acad. Sci. U.S.A.* **96**, 6155 – 6160
223. Misra, T. K., and Apirion, D. (1978) *J. Biol. Chem.* **253**, 5594 – 5599
224. Nashimoto, M., Tamura, M., and Kaspar, R. L. (1999) *J. Mol. Biol.* **287**, 727 – 740
225. Deutscher, M. P., and Marlor, C. W. (1985) *J. Biol. Chem.* **260**, 7067 – 7071
226. Robertson, H. D., Altman, S., and Smith, J. D. (1972) *J. Biol. Chem.* **247**, 5243 – 5251
227. Reilly, R. M., and RajBhandary, U. L. (1986) *J. Biol. Chem.* **261**, 2928 – 2935
228. Khorana, H. G., and 19 other authors. (1976) *J. Biol. Chem.* **251**, 565 – 570
229. Ramamoorthy, B., Lees, R. G., Kleid, D. G., and Khorana, H. G. (1976) *J. Biol. Chem.* **251**, 676 – 694
230. Sekiya, T., Contreras, R., Kupper, H., Landy, A., and Khorana, H. G. (1976) *J. Biol. Chem.* **251**, 5124 – 5140
231. Gommers-Ampt, J. H., and Borst, P. (1995) *FASEB J.* **9**, 1034 – 1042
232. Söll, D., and RajBhandary, U. L., eds. (1995) *tRNA Structure, Biosynthesis, and Function*, Am. Soc. for Microbiology, Washington, D.C.
233. Deutscher, M. P., Marlor, C. W., and Zaniewski, R. (1985) *Proc. Natl. Acad. Sci. U.S.A.* **82**, 6429 – 6430
234. Yue, D., Weiner, A. M., and Maizels, N. (1998) *J. Biol. Chem.* **273**, 29693 – 29700
235. Kline, L. K., and Söll, D. (1982) in *The Enzymes*, Vol. 15 (Boyer, P. D., ed), pp. 567 – __, Academic Press, New York
236. Greenberg, R., and Dudock, B. (1980) *J. Biol. Chem.* **255**, 8296 – 8302
237. Kwong, L. K., Moore, V. G., and Kaiser, I. I. (1977) *J. Biol. Chem.* **252**, 6310 – 6315
238. Becker, H. F., Motorin, Y., Sissler, M., Florentz, C., and Grosjean, H. (1997) *J. Mol. Biol.* **274**, 505 – 518
239. Kambampati, R., and Lauhon, C. T. (2000) *J. Biol. Chem.* **275**, 10727 – 10730
240. Kaiser, J. T., Clausen, T., Bourenkow, G. P., Bartunik, H.-D., Steinbacher, S., and Huber, R. (2000) *J. Mol. Biol.* **297**, 451 – 464
241. Schaeffer, K. P., Altman, S., and Soll, D. (1973) *Proc. Natl. Acad. Sci. U.S.A.* **70**, 3626 – 3630
242. Reyniers, J. P., Pleasants, J. R., Wostmann, B. S., Katze, J. R., and Farkas, W. R. (1981) *J. Biol. Chem.* **256**, 11591 – 11594
243. Katze, J. R., Gündüz, U., Smith, D. L., Cheng, C. S., and McCloskey, J. A. (1984) *Biochemistry* **23**, 1171 – 1176
244. Slany, R. K., Bösl.M, Crain, P. F., and Kersten, H. (1993) *Biochemistry* **32**, 7811 – 7817
245. Romier, C., Reuter, K., Suck, D., and Ficner, R. (1996) *Biochemistry* **35**, 15734 – 15739
246. Curnow, A. W., and Garcia, G. A. (1995) *J. Biol. Chem.* **270**, 17264 – 17267
246a. Kittendorf, J. D., Barcomb, L. M., Nonekowski, S. T., and Garcia, G. A. (2001) *Biochemistry* **40**, 14123 – 14133
247. Phillipson, D. W., Edmonds, C. G., Crain, P. F., Smith, D. L., Davis, D. R., and McCloskey, J. A. (1987) *J. Biol. Chem.* **262**, 3462 – 3471
248. Kolesky, S., Ouhammouch, M., Brody, E. N., and Geiduschek, E. P. (1999) *J. Mol. Biol.* **291**, 267 – 281
249. Gopal, V., Brieba, L. G., Guajardo, R., McAllister, W. T., and Sousa, R. (1999) *J. Mol. Biol.* **290**, 411 – 431
249a. Liu, C., and Martin, C. T. (2001) *J. Mol. Biol.* **308**, 465 – 475
249b. Severinov, K. (2001) *Proc. Natl. Acad. Sci. U.S.A.* **98**, 5 – 7
250. Gottesman, M. (1999) *J. Mol. Biol.* **293**, 177 – 180
251. Gottesman, S. (1988) in *Biochemistry*, 2nd ed. (Zubay, G., ed), pp. 1000 – 1007, Macmillan, New York
252. Johnson, A. D., Poteete, A. R., Lauer, G., Sauer, R. T., Ackers, G. K., and Ptashne, M. (1981) *Nature (London)* **294**, 217 – 223
253. Ptashne, M., Johnson, A. D., and Pabo, C. O. (1982) *Sci. Am.* **247**(Nov), 128 – 140
254. Albright, R. A., and Matthews, B. W. (1998) *Proc. Natl. Acad. Sci. U.S.A.* **95**, 3431 – 3436
255. Honigman, A., Hu, S.-L., Chase, R., and Szbalski, W. (1976) *Nature (London)* **262**, 112 – 116
256. Shimatake, H., and Rosenberg, M. (1981) *Nature (London)* **292**, 128 – 132
257. Maurizi, M. R. (1987) *J. Biol. Chem.* **262**, 2696 – 2793
258. Kastelein, R. A., Remaut, E., Fiers, W., and van Duin, J. (1982) *Nature (London)* **295**, 35 – 41
259. Voyles, B. A. (1993) *The Biology of Viruses*, Mosby, St. Louis, Missouri
260. Brown, D., and Gold, L. (1996) *Proc. Natl. Acad. Sci. U.S.A.* **93**, 11558 – 11562
261. Schuppli, D., Georgijevic, J., and Weber, H. (2000) *J. Mol. Biol.* **295**, 149 – 154
262. Blumenthal, T., and Carmichael, G. G. (1979) *Ann. Rev. Biochem.* **48**, 525 – 548
263. Holland, J., Spindler, K., Horodyski, F., Grabau, E., Nichol, S., and VandePol, S. (1982) *Science* **215**, 1577 – 1585
264. Mills, D. R., Kramer, F. R., Dobkin, C., Nishihara, T., and Spiegelman, S. (1975) *Proc. Natl. Acad. Sci. U.S.A.* **72**, 4252 – 4256
264a. Weinzierl, R. O. J. (1999) *Mechanisms of Gene Expression.Structure,Function and Evolution of the Basal Transcriptional Machinery*, Imperial College Press,
265. Langer, D., Hain, J., Thuriaux, P., and Zillig, W. (1995) *Proc. Natl. Acad. Sci. U.S.A.* **92**, 5768 – 5772
266. Bell, S. D., Kosa, P. L., Sigler, P. B., and Jackson, S. P. (1999) *Proc. Natl. Acad. Sci. U.S.A.* **96**, 13662 – 13667
266a. Bell, S. D., and Jackson, S. P. (2000) *J. Biol. Chem.* **275**, 31624 – 31629
266b. Thomsen, J., De Biase, A., Kaczanowski, S., Macario, A. J. L., Thomm, M., Zielenkiewicz, P., MacColl, R., and de Macario, E. C. (2001) *J. Mol. Biol.* **309**, 589 – 603
267. Krüger, K., Hermann, T., Armbruster, V., and Pfeifer, F. (1998) *J. Mol. Biol.* **279**, 761 – 771

References

268. Kyrpides, N. C., and Ouzounis, C. A. (1999) *Proc. Natl. Acad. Sci. U.S.A.* **96**, 8545–8550

268a. Lewis, P. J., Thaker, S. D., and Errington, J. (2000) *EMBO J.* **19**, 710–718

268b. Iborra, F. J., Jackson, D. A., and Cook, P. R. (2001) *Science* **293**, 1139–1142

268c. von Mering, C., and Bork, P. (2002) *Nature (London)* **417**, 797–798

269. Jackson, D. A., Pombo, A., and Iborra, F. (2000) *FASEB J.* **14**, 242–254

270. Workman, J. L., and Kingston, R. E. (1998) *Ann. Rev. Biochem.* **67**, 545–579

271. Hagmann, M. (1999) *Science* **285**, 1200–1203

271a. Jenuwein, T., and Allis, C. D. (2001) *Science* **293**, 1074–1080

271b. Bird, A. (2001) *Science* **294**, 2113–2115

271c. Gamble, M. J., and Freedman, L. P. (2002) *Trends Biochem. Sci.* **27**, 165–167

271d. Burke, L. J., and Baniahmad, A. (2000) *FASEB J.* **14**, 1876–1888

272. Kornberg, R. D. (1999) *Trends Biochem. Sci.* **24**, M46–M49

273. Carey, M., and Smale, S. T. (2000) *Transcriptional Regulation in Eukaryotes*, Cold Spring Harbor Lab. Press, Cold Spring Harbor, New York

273a. Steidl, S., Hynes, M. J., and Brakhage, A. A. (2001) *J. Mol. Biol.* **306**, 643–653

273b. Niehof, M., Kubicka, S., Zender, L., Manns, M. P., and Trautwein, C. (2001) *J. Mol. Biol.* **309**, 855–868

274. Vandromme, M., Gauthier-Rouvière, C., Lamb, N., and Fernandez, A. (1996) *Trends Biochem. Sci.* **21**, 59–64

275. Pennisi, E. (2000) *Science* **288**, 1372–1373

275a. Lo, W.-S., Duggan, L., Emre, N. C. T., Belotserkovskya, R., Lane, W. S., Shiekhattar, R., and Berger, S. L. (2001) *Science* **293**, 1142–1146

275b. Wang, H., Huang, Z.-Q., Xia, L., Feng, Q., Erdjument-Bromage, H., Strahl, B. D., Briggs, S. D., Allis, C. D., Wong, J., Tempst, P., and Zhang, Y. (2001) *Science* **293**, 853–857

275c. Nakayama, J.-i, Rice, J. C., Strahl, B. D., Allis, C. D., and Grewal, S. I. S. (2001) *Science* **292**, 110–113

275d. Jenuwein, T. (2002) *Science* **297**, 2215–2218

275e. Volpe, T. A., Kidner, C., Hall, I. M., Teng, G., Grewal, S. I. S., and Martienssen, R. A. (2002) *Science* **297**, 1833–1837

276. Chandler, S. P., Guschin, D., Landsberger, N., and Wolffe, A. P. (1999) *Biochemistry* **38**, 7008–7018

276a. Burnett, E., Christensen, J., and Tattersall, P. (2001) *J. Mol. Biol.* **314**, 1029–1039

277. Ng, H. H., and Bird, A. (2000) *Trends Biochem. Sci.* **25**, 121–126

278. Alland, L., Muhle, R., Hou, H., Jr., Potes, J., Chin, L., Schreiber-Agus, N., and DePinho, R. A. (1997) *Nature (London)* **387**, 49–55

278a. Arévalo-Rodríguez, M., Cardenas, M. E., Wu, X., Hanes, S. D., and Heitman, J. (2000) *EMBO J.* **19**, 3739–3749

279. Heinzel, T., Lavinsky, R. M., Mullen, T.-M., Söderström, M., Laherty, C. D., Torchia, J., Yang, W.-M., Brard, G., Ngo, S. D., Davie, J. R., Seto, E., Eisenman, R. N., Rose, D. W., Glass, C. K., and Rosenfeld, M. G. (1997) *Nature (London)* **387**, 43–48

279a. Lechner, T., Lusser, A., Pipal, A., Brosch, G., Loidl, A., Goralik-Schramel, M., Sendra, R., Wegener, S., Walton, J. D., and Loidl, P. (2000) *Biochemistry* **39**, 1683–1692

279b. Bernstein, B. E., Tong, J. K., and Schreiber, S. L. (2000) *Proc. Natl. Acad. Sci. U.S.A.* **97**, 13708–13713

279c. Johnson, C. A., Padget, K., Austin, C. A., and Turner, B. M. (2001) *J. Biol. Chem.* **276**, 4539–4542

280. Bernhard, D., Ausserlechner, M. J., Tonko, M., Löffler, M., Hartmann, B. L., Csordas, A., and Kofler, R. (1999) *FASEB J.* **13**, 1991–2001

280a. Gross, D. S. (2001) *Trends Biochem. Sci.* **26**, 685–686

280b. Smith, J. S., and Boeke, J. D. (2001) *Science* **291**, 608–609

280c. Tanny, J. C., and Moazed, D. (2001) *Proc. Natl. Acad. Sci. U.S.A.* **98**, 415–420

280d. Sauve, A. A., Celic, I., Avalos, J., Deng, H., Boeke, J. D., and Schramm, V. L. (2001) *Biochemistry* **40**, 15456–15463

280e. Ghosh, D., Gerasimova, T. I., and Corces, V. G. (2001) *EMBO J.* **20**, 2518–2527

280f. Donze, D., and Kamakaka, R. T. (2001) *EMBO J.* **20**, 520–531

280g. Bell, A. C., West, A. G., and Felsenfeld, G. (2001) *Science* **291**, 447–450

280h. Tijsterman, M., Ketting, R. F., Okihara, K. L., Sijen, T., and Plasterk, R. H. A. (2002) *Science* **295**, 694–697

280i. Ahlquist, P. (2002) *Science* **296**, 1270–1273

280j. Jorgensen, R. A. (2002) *Proc. Natl. Acad. Sci. U.S.A.* **99**, 11561–11563

281. Kouzarides, T. (2000) *EMBO J.* **19**, 1176–1179

282. Brown, C. E., Lechner, T., Howe, L., and Workman, J. L. (2000) *Trends Biochem. Sci.* **25**, 15–19

283. Trievel, R. C., Rojas, J. R., Sterner, D. E., Venkataramani, R. N., Wang, L., Zhou, J., Allis, C. D., Berger, S. L., and Marmorstein, R. (1999) *Proc. Natl. Acad. Sci. U.S.A.* **96**, 8931–8936

283a. Sterner, D. E., Belotserkovskaya, R., and Berger, S. L. (2002) *Proc. Natl. Acad. Sci. U.S.A.* **99**, 11622–11627

283b. Hughes, J. D., Estep, P. W., Tavazoie, S., and Church, G. M. (2000) *J. Mol. Biol.* **296**, 1205–1214

283c. Chen, J., Rappsilber, J., Chiang, Y.-C., Russell, P., Mann, M., and Denis, C. L. (2001) *J. Mol. Biol.* **314**, 683–694

284. Bochar, D. A., Savard, J., Wang, W., Lafleur, D. W., Moore, P., Côté, J., and Shiekhattar, R. (2000) *Proc. Natl. Acad. Sci. U.S.A.* **97**, 1038–1043

284a. Fry, C. J., and Peterson, C. L. (2002) *Science* **295**, 1847–1848

284b. Aalfs, J. D., and Kingston, R. E. (2000) *Trends Biochem. Sci.* **25**, 548–555

284c. Becker, P. B., and Hörz, W. (2002) *Ann. Rev. Biochem.* **71**, 247–273

285. Kwon, H., Imbalzano, A. N., Khavarl, P. A., Kingston, R. E., and Green, M. R. (1994) *Nature (London)* **370**, 477–481

286. Richard-Foy, H. (1994) *Nature (London)* **370**, 417–418

287. Gavin, I. M., and Simpson, R. T. (1997) *EMBO J.* **16**, 6263–6271

287a. Xue, Y., Canman, J. C., Lee, C. S., Nie, Z., Yang, D., Moreno, G. T., Young, M. K., Salmon, E. D., and Wang, W. (2000) *Proc. Natl. Acad. Sci. U.S.A.* **97**, 13015–13020

287b. Olave, I. A., Reck-Peterson, S. L., and Crabtree, G. R. (2002) *Ann. Rev. Biochem.* **71**, 755–781

287c. Asturias, F. J., Chung, W.-H., Kornberg, R. D., and Lorch, Y. (2002) *Proc. Natl. Acad. Sci. U.S.A.* **99**, 13477–13480

288. de la Cruz, J., Kressler, D., and Linder, P. (1999) *Trends Biochem. Sci.* **24**, 192–198

288a. Linder, P., Tanner, N. K., and Banroques, J. (2001) *Trends Biochem. Sci.* **26**, 339–341

288b. Henn, A., Medalia, O., Shi, S.-P., Steinberg, M., Franceschi, F., and Sagi, I. (2001) *Proc. Natl. Acad. Sci. U.S.A.* **98**, 5007–5012

288c. Story, R. M., Li, H., and Abelson, J. N. (2001) *Proc. Natl. Acad. Sci. U.S.A.* **98**, 1465–1470

289. Wu, X., Wilcox, C. B., Devasahayam, G., Hackett, R. L., Arévalo-Rodríguez, M., Cardenas, M. E., Heitman, J., and Hanes, S. D. (2000) *EMBO J.* **19**, 3727–3738

289a. Conaway, R. C., Brower, C. S., and Conaway, J. W. (2002) *Science* **296**, 1254–1258

289b. Zheng, N., Schulman, B. A., Song, L., Miller, J. J., Jeffrey, P. D., Wang, P., Chu, C., Koepp, D. M., Elledge, S. J., Pagano, M., Conaway, R. C., Conaway, J. W., Harper, J. W., and Pavietich, N. P. (2002) *Nature (London)* **416**, 703–709

289c. Ottosen, S., Herrera, F. J., and Triezenberg, S. J. (2002) *Science* **296**, 479–481

289d. Gonzalez, F., Delahodde, A., Kodadek, T., and Johnston, S. A. (2002) *Science* **296**, 548–550

289e. Tsuge, T., Matsui, M., and Wei, N. (2001) *J. Mol. Biol.* **305**, 1–9

289f. Goodson, M. L., Hong, Y., Rogers, R., Matunis, M. J., Park-Sarge, O.-K., and Sarge, K. D. (2001) *J. Biol. Chem.* **276**, 18513–18518

289g. Sampson, D. A., Wang, M., and Matunis, M. J. (2001) *J. Biol. Chem.* **276**, 21664–21669

289h. Kito, K., Yeh, E. T. H., and Kamitani, T. (2001) *J. Biol. Chem.* **276**, 20603–20609

289i. Marx, J. (2001) *Science* **292**, 838–839

289j. Buchberger, A., Howard, M. J., Proctor, M., and Bycroft, M. (2001) *J. Mol. Biol.* **307**, 17–24

290. Cramer, P., Bushnell, D. A., Fu, J., Gnatt, A. L., Maier-Davis, B., Thompson, N. E., Burgess, R. R., Edwards, A. M., David, P. R., and Kornberg, R. D. (2000) *Science* **288**, 640–649

291. Kim, W.-Y., and Dahmus, M. E. (1989) *J. Biol. Chem.* **264**, 3169–3176

292. Geiduschek, E. P., and Tocchini-Valentini, G. P. (1988) *Ann. Rev. Biochem.* **57**, 873–914

293. Shaw, P. J., Highett, M. I., Beven, A. F., and Jordan, E. G. (1995) *EMBO J.* **14**, 2896–2906

294. Seither, P., Iben, S., and Grummt, I. (1998) *J. Mol. Biol.* **275**, 43–53

295. Kelly, J. L., and Lehman, I. R. (1986) *J. Biol. Chem.* **261**, 10340–10347

295a. Masters, B. S., Stohl, L. L., and Clayton, D. A. (1987) *Cell* **51**, 89–99

296. Hanley–Bowdoin, L., and Chua, N.-H. (1987) *Trends Biochem. Sci.* **12**, 67–70

297. Myer, V. E., and Young, R. A. (1998) *J. Biol. Chem.* **273**, 27757–27760

298. Fu, J., Gerstein, M., David, P. R., Gnatt, A. L., Bushnell, D. A., Edwards, A. M., and Kornberg, R. D. (1998) *J. Mol. Biol.* **280**, 317–322

299. Jensen, G. J., Meredith, G., Bushnell, D. A., and Kornberg, R. D. (1998) *EMBO J.* **17**, 2353–2358

299a. Todone, F., Weinzierl, R. O. J., Brick, P., and Onesti, S. (2000) *Proc. Natl. Acad. Sci. U.S.A.* **97**, 6306–6310

299b. Cramer, P., Bushnell, D. A., and Kornberg, R. D. (2001) *Science* **292**, 1863–1876

299c. Gnatt, A. L., Cramer, P., Fu, J., Bushnell, D. A., and Kornberg, R. D. (2001) *Science* **292**, 1876–1882

300. Simanek, E. E., Huang, D.-H., Pasternack, L., Machajewski, T. D., Seitz, O., Millar, D. S., Dyson, H. J., and Wong, C.-H. (1998) *J. Am. Chem. Soc.* **120**, 11567–11575

301. Adams, R. L. P., Knowler, J. T., and Leader, D. P. (1992) *The Biochemistry of the Nucleic Acids*, 11th ed., Chapman & Hall, London

302. Dahmus, M. E. (1996) *J. Biol. Chem.* **271**, 19009–19012

303. Corden, J. L., and Patturajan, M. (1997) *Trends Biochem. Sci.* **22**, 413–416

304. Proudfoot, N. (2000) *Trends Biochem. Sci.* **25**, 290–293

304a. Geiduschek, E. P., and Kassavetis, G. A. (2001) *J. Mol. Biol.* **310**, 1–26

304b. Kassavetis, G. A., Letts, G. A., and Geiduschek, E. P. (2001) *EMBO J.* **20**, 2823–2834

References

305. Pombo, A., Jackson, D. A., Hollinshead, M., Wang, Z., Roeder, R. G., and Cook, P. R. (1999) *EMBO J.* **18**, 2241–2253
306. Carter, K. C., Bowman, D., Carrington, W., Fogarty, K., McNeil, J. A., Fay, F. S., and Lawrence, J. B. (1993) *Science* **259**, 1330–1332
307. Nevins, J. R. (1983) *Ann. Rev. Biochem.* **52**, 441–466
308. Leff, S. E., Rosenfeld, M. G., and Evans, R. M. (1986) *Ann. Rev. Biochem.* **55**, 1091–1117
309. Ayoubi, T. A. Y., and Van de Ven, W. J. M. (1996) *FASEB J.* **10**, 453–460
310. Dantonel, J.-C., Wurtz, J.-M., Poch, O., Moras, D., and Tora, L. (1999) *Trends Biochem. Sci.* **24**, 335–339
311. Munoz-Sanjuan, I., Smallwood, P. M., and Nathans, J. (2000) *J. Biol. Chem.* **275**, 2589–2597
312. Ince, T. A., and Scotto, K. W. (1995) *J. Biol. Chem.* **270**, 30249–30252
312a. Conaway, J. W., Shilatifard, A., Dvir, A., and Conaway, R. C. (2000) *Trends Biochem. Sci.* **25**, 375–380
312b. Maniatis, T., and Reed, R. (2002) *Nature (London)* **416**, 499–506
312c. Sousa, R. (2001) *Trends Biochem. Sci.* **26**, 695–697
313. Arndt, K. T., Styles, C., and Fink, G. R. (1987) *Science* **237**, 874–880
314. Moye, W. S., and Zalkin, H. (1987) *J. Biol. Chem.* **262**, 3609–3624
315. Shuey, D. J., and Parker, C. S. (1986) *Nature (London)* **323**, 459–461
316. Prestridge, D. S. (1995) *J. Mol. Biol.* **249**, 923–932
317. Scherf, M., Klingenhoff, A., and Werner, T. (2000) *J. Mol. Biol.* **297**, 599–606
318. Kaiser, K., and Meisterernst, M. (1996) *Trends Biochem. Sci.* **21**, 342–345
318a. Tsuji, Y., Torti, S. V., and Torti, F. M. (1998) *J. Biol. Chem.* **273**, 2984–2992
319. Kadonaga, J. T., Jones, K. A., and Tjian, R. (1986) *Trends Biochem. Sci.* **11**, 20–23
320. Kadonaga, J. T., Carner, K. R., Masiarz, F. R., and Tjian, R. (1987) *Cell* **51**, 1079–1090
321. Garcia, A., Cereghini, S., and Sontag, E. (2000) *J. Biol. Chem.* **275**, 9385–9389
322. Li, B., Adams, C. C., and Workman, J. L. (1994) *J. Biol. Chem.* **269**, 7756–7763
323. Fry, C. J., and Farnham, P. J. (1999) *J. Biol. Chem.* **274**, 29583–29586
324. Daniel, S., Zhang, S., DePaoli-Roach, A. A., and Kim, K.-H. (1996) *J. Biol. Chem.* **271**, 14692–14697
324a. Johnson-Pais, T., Degnin, C., and Thayer, M. J. (2001) *Proc. Natl. Acad. Sci. U.S.A.* **98**, 2211–2216
324b. Goodwin, A. J., McInerney, J. M., Glander, M. A., Pomerantz, O., and Lowrey, C. H. (2001) *J. Biol. Chem.* **276**, 26883–26892
325. Van Dyke, M. W., Roeder, R. G., and Sawadogo, M. (1988) *Science* **241**, 1335–1338
326. Roeder, R. G. (1996) *Trends Biochem. Sci.* **21**, 327–335
327. Nikolov, D. B., Chen, H., Halay, E. D., Hoffmann, A., Roeder, R. G., and Burley, S. K. (1996) *Proc. Natl. Acad. Sci. U.S.A.* **93**, 4862–4867
328. Nikolov, D. B., and Burley, S. K. (1997) *Proc. Natl. Acad. Sci. U.S.A.* **94**, 15–22
328a. Sachs, A. B., and Buratowski, S. (1997) *Trends Biochem. Sci.* **22**, 189–192
329. Andel, F., III, Ladurner, A. G., Inouye, C., Tjian, R., and Nogales, E. (1999) *Science* **286**, 2153–2156
329a. Gangloff, Y.-G., Romier, C., Thuault, S., Werten, S., and Davidson, I. (2001) *Trends Biochem. Sci.* **26**, 250–257
330. Burley, S. K., and Roeder, R. G. (1996) *Ann. Rev. Biochem.* **65**, 769–799
331. Green, M. R. (2000) *Trends Biochem. Sci.* **25**, 59–63
332. Verrijzer, C. P., and Tjian, R. (1996) *Trends Biochem. Sci.* **21**, 338–342
332a. Metzger, D., Scheer, E., Soldatov, A., and Tora, L. (1999) *EMBO J.* **18**, 4823–4834
333. Juo, Z. S., Chiu, T. K., Leiberman, P. M., Baikalov, I., Berk, A. J., and Dickerson, R. E. (1996) *J. Mol. Biol.* **261**, 239–254
334. Parkhurst, K. M., Richards, R. M., Brenowitz, M., and Parkhurst, L. J. (1999) *J. Mol. Biol.* **289**, 1327–1341
335. Svejstrup, J. Q., Vichi, P., and Egly, J.-M. (1996) *Trends Biochem. Sci.* **21**, 346–350
335a. Holstege, F. C. P., and Young, R. A. (1999) *Proc. Natl. Acad. Sci. U.S.A.* **96**, 2–4
336. Yan, M., and Gralla, J. D. (1997) *EMBO J.* **16**, 7457–7467
336a. Mizzen, C. A., and Allis, C. D. (2000) *Science* **289**, 2290–2291
337. Nikolov, D. B., Hu, S.-H., Lin, J., Gasch, A., Hoffmann, A., Horikoshi, M., Chua, N.-H., Roeder, R. G., and Burley, S. K. (1992) *Nature (London)* **360**, 40–46
338. Nikolov, D. B., Chen, H., Halay, E. D., Usheva, A. A., Hisatake, K., Lee, D. K., Roeder, R. G., and Burley, S. K. (1995) *Nature (London)* **377**, 119–128
338a. Brivanlou, A. H., and Darnell, J. E., Jr. (2002) *Science* **295**, 813–818
338b. Smith, R. L., and Johnson, A. D. (2000) *Trends Biochem. Sci.* **25**, 325–330
338c. Dunah, A. W., Jeong, H., Griffin, A., Kim, Y.-M., Standaert, D. G., Hersch, S. M., Mouradian, M. M., Young, A. B., Tanese, N., and Krainc, D. (2002) *Science* **296**, 2238–2243
339. Marmorstein, R., Carey, M., Ptashne, M., and Harrison, S. C. (1992) *Nature (London)* **356**, 408–414
340. Ptashne, M., and Gann, A. (1997) *Nature (London)* **386**, 569–577
341. Rodgers, K. K., and Coleman, J. E. (1994) *Protein Sci.* **3**, 608–619
342. Lohr, D., Venkov, P., and Zlatanova, J. (1995) *FASEB J.* **9**, 777–787
343. Corton, J. C., Moreno, E., and Johnston, S. A. (1998) *J. Biol. Chem.* **273**, 13776–13780
344. Ruden, D. M., Ma, J., and Ptashne, M. (1988) *Proc. Natl. Acad. Sci. U.S.A.* **85**, 4262–4266
345. Sigler, P. B. (1988) *Nature (London)* **333**, 210–212
346. Ryu, S., and Tjian, R. (1999) *Proc. Natl. Acad. Sci. U.S.A.* **96**, 7137–7142
347. Zawel, L., and Reinberg, D. (1995) *Ann. Rev. Biochem.* **64**, 533–561
348. Malik, S., and Roeder, R. G. (2000) *Trends Biochem. Sci.* **25**, 277–283
349. Svejstrup, J. Q., Li, Y., Fellows, J., Gnatt, A., Bjorklund, S., and Kornberg, R. D. (1997) *Proc. Natl. Acad. Sci. U.S.A.* **94**, 6075–6078
350. Spåhr, H., Bève, J., Larsson, T., Bergström, J., Karlsson, K.-A., and Gustafsson, C. M. (2000) *J. Biol. Chem.* **275**, 1351–1356
350a. Dotson, M. R., Yuan, C. X., Roeder, R. G., Myers, L. C., Gustafsson, C. M., Jiang, Y. W., Li, Y., Kornberg, R. D., and Asturias, F. J. (2000) *Proc. Natl. Acad. Sci. U.S.A.* **97**, 14307–14310
350b. Liu, Y., Ranish, J. A., Aebersold, R., and Hahn, S. (2001) *J. Biol. Chem.* **276**, 7169–7175
351. Tan, R. C., Truong, T. N., McCammon, J. A., and Sussman, J. L. (1993) *Biochemistry* **32**, 401–403
352. Clarke, M. F., FitzGerald, P. C., Brubaker, J. M., and Simpson, R. T. (1985) *J. Biol. Chem.* **260**, 12394–12397
353. Ondek, B., Gloss, L., and Herr, W. (1988) *Nature (London)* **333**, 40–45
354. Plon, S., and Wang, J. (1986) *Cell* **45**, 575–580
355. Ptashne, M. (1986) *Nature (London)* **322**, 697–701
356. Ross, E. D., Keating, A. M., and Maher, L. J., III. (2000) *J. Mol. Biol.* **297**, 321–334
357. Udvardy, A. (1999) *EMBO J.* **18**, 1–8
357a. Struhl, K. (2001) *Science* **293**, 1054–1055
357b. Yie, J., Senger, K., and Thanos, D. (1999) *Proc. Natl. Acad. Sci. U.S.A.* **96**, 13108–13113
358. Werner, M. H., Gronenborn, A. M., and Clore, G. M. (1996) *Science* **271**, 778–784
359. Cronk, J. D., Endrizzi, J. A., and Alber, T. (1996) *Protein Sci.* **5**, 1963–1972
360. Keller, W., König, P., and Richmond, T. J. (1995) *J. Mol. Biol.* **254**, 657–667
361. Hanson, R. W. (1998) *J. Biol. Chem.* **273**, 28543
362. Gardner, K., Moore, T. C., Davis-Smyth, T., Krutzsch, H., and Levens, D. (1994) *J. Biol. Chem.* **269**, 32963–32971
363. Lekstrom-Himes, J., and Xanthopoulos, K. G. (1998) *J. Biol. Chem.* **273**, 28545–28548
364. Turner, R., and Tjian, R. (1989) *Science* **243**, 1689–1694
365. Okahata, Y., Niikura, K., Sugiura, Y., Sawada, M., and Morii, T. (1998) *Biochemistry* **37**, 5666–5672
366. Maity, S. N., and de Crombrugghe, B. (1998) *Trends Biochem. Sci.* **23**, 174–178
367. Croniger, C., Leahy, P., Reshef, L., and Hanson, R. W. (1998) *J. Biol. Chem.* **273**, 31629–31632
368. Landschulz, W. H., Johnson, P. F., and McKnight, S. L. (1988) *Science* **240**, 1759–1764
369. Ferré-D'Amaré, A. R., Prendergast, G. C., Ziff, E. B., and Burley, S. K. (1993) *Nature (London)* **363**, 38–44
370. Chen, L., Glover, J. N. M., Hogan, P. G., Rao, A., and Harrison, S. C. (1998) *Nature (London)* **392**, 42–48
371. Mackay, J. P., Shaw, G. L., and King, G. F. (1996) *Biochemistry* **35**, 4867–4877
372. Gentz, R., Rauscher, F. J., III, Abate, C., and Curran, T. (1989) *Science* **243**, 1695–1699
373. Dang, C. V., McGuire, M., Buckmire, M., and Lee, W. M. F. (1989) *Nature (London)* **337**, 664–666
374. Garlatti, M., Tchesnokov, V., Daheshia, M., Feilleux-Duché, S., Hanoune, J., Aggerbeck, M., and Barouki, R. (1993) *J. Biol. Chem.* **268**, 6567–6574
375. Umek, R. M., Friedman, A. D., and McKnight, S. L. (1991) *Science* **251**, 288–292
376. Liberati, C., di Silvio, A., Ottolenghi, S., and Mantovani, R. (1999) *J. Mol. Biol.* **285**, 1441–1455
377. Grandori, C., and Eisenman, R. N. (1997) *Trends Biochem. Sci.* **22**, 177–181
378. Noguchi, K., Kitanaka, C., Yamana, H., Kokubu, A., Mochizuki, T., and Kuchino, Y. (1999) *J. Biol. Chem.* **274**, 32580–32587
379. Grigorieva, I., Grigoriev, V. G., Rowney, M. K., and Hoover, R. G. (2000) *J. Biol. Chem.* **275**, 7343–7350
379a. Otsuki, Y., Tanaka, M., Kamo, T., Kitanaka, C., Kuchino, Y. and Sugimura, H. (2002) *J. Biol. Chem.* In press 206733200
380. Lalli, E., and Sassone-Corsi, P. (1994) *J. Biol. Chem.* **269**, 17359–17362
381. Shaywitz, A. J., and Greenberg, M. E. (1999) *Ann. Rev. Biochem.* **68**, 821–861
382. De Cesare, D., Fimia, G. M., and Sassone-Corsi, P. (1999) *Trends Biochem. Sci.* **24**, 281–284
382a. Goren, I., Tavor, E., Goldblum, A., and Honigman, A. (2001) *J. Mol. Biol.* **313**, 695–709
382b. Nagadoi, A., Nakazawa, K.-i, Uda, H., Okuno, K., Maekawa, T., Ishii, S., and Nishimura, Y. (1999) *J. Mol. Biol.* **287**, 593–607
382c. Yeagley, D., Agati, J. M., and Quinn, P. G. (1998) *J. Biol. Chem.* **273**, 18743–18750

References

382d. Vo, N., and Goodman, R. H. (2001) *J. Biol. Chem.* **276**, 13505–13508
383. Xu, W., Chen, H., Du, K., Asahara, H., Tini, M., Emerson, B. M., Montminy, M., and Evans, R. M. (2001) *Science* **294**, 2507–2511
383a. Martínez-Balbás, M. A., Bauer, U.-M., Nielsen, S., Brehm, A., and Kouzarides, T. (2000) *EMBO J.* **19**, 662–671
383b. Classon, M., Salama, S., Gorka, C., Mulloy, R., Braun, P., and Harlow, E. (2000) *Proc. Natl. Acad. Sci. U.S.A.* **97**, 10820–10825
383c. Ogawa, H., Ishiguro, K.-i, Gaubatz, S., Livingston, D. M., and Nakatani, Y. (2002) *Science* **296**, 1132–1136
383d. La Thangue, N. B. (2002) *Science* **296**, 1034–1035
383e. Chin, L., Pomerantz, J., and DePinho, R. A. (1998) *Trends Biochem. Sci.* **23**, 291–296
383f. Chang, L., and Karin, M. (2001) *Nature (London)* **410**, 37–40
383g. Weston, C. R., Lambright, D. G., and Davis, R. J. (2002) *Science* **296**, 2345, 2347
383h. Cantley, L. C. (2001) *Science* **292**, 2019–2021
383i. Janknecht, R., and Hunter, T. (1999) *Science* **284**, 443–444
383j. Horvath, C. M. (2000) *Trends Biochem. Sci.* **25**, 496–502
383k. ten Dijke, P., Miyazono, K., and Helden, C.-H. (2000) *Trends Biochem. Sci.* **25**, 64–70
383l. Wu, G., Chen, Y.-G., Ozdamar, B., Gyuricza, C. A., Chong, P. A., Wrana, J. L., Massagué, J., and Shi, Y. (2000) *Science* **287**, 92–97
384. Towle, H. C. (1995) *J. Biol. Chem.* **270**, 23235–23238
384a. Osborne, T. F. (2000) *J. Biol. Chem.* **275**, 32379–32382
384b. Duan, R., Xie, W., Burghardt, R. C., and Safe, S. (2001) *J. Biol. Chem.* **276**, 11590–11598
384c. Hassler, M., and Richmond, T. J. (2001) *EMBO J.* **20**, 3018–3028
385. Treisman, R. (1992) *Trends Biochem. Sci.* **17**, 423–426
386. Ling, Y., West, A. G., Roberts, E. C., Lakey, J. H., and Sharrocks, A. D. (1998) *J. Biol. Chem.* **273**, 10506–10514
387. Iyer, V. R., Eisen, M. B., Ross, D. T., Schuler, G., Moore, T., Lee, J. C. F., Trent, J. M., Staudt, L. M., Hudson, J., Jr., Boguski, M. S., Lashkari, D., Shalon, D., Botstein, D., and Brown, P. O. (1999) *Science* **283**, 83–87
388. Kerrigan, L. A., Croston, G. E., Lira, L. M., and Kadonaga, J. T. (1991) *J. Biol. Chem.* **266**, 574–582
388a. Bieker, J. J. (2001) *J. Biol. Chem.* **276**, 34355–34358
389. Mackay, J. P., and Crossley, M. (1998) *Trends Biochem. Sci.* **23**, 1–4
390. Turner, J., and Crossley, M. (1999) *Trends Biochem. Sci.* **24**, 236–241
391. Clarke, N. D., and Berg, J. M. (1998) *Science* **282**, 2018–2022
392. Naltner, A., Ghaffari, M., Whitsett, J. A., and Yan, C. (2000) *J. Biol. Chem.* **275**, 56–62
393. Pizzi, S., Dieci, G., Frigeri, P., Piccoli, G., Stocchi, V., and Ottonello, S. (1999) *J. Biol. Chem.* **274**, 2539–2548
394. Laity, J. H., Dyson, H. J., and Wright, P. E. (2000) *J. Mol. Biol.* **295**, 719–727
395. Klug, A., and Rhodes, D. (1987) *Trends Biochem. Sci.* **12**, 464–469
396. Nolte, R. T., Conlin, R. M., Harrison, S. C., and Brown, R. S. (1998) *Proc. Natl. Acad. Sci. U.S.A.* **95**, 2938–2943
397. Muro-Pastor, M. I., Gonzalez, R., Strauss, J., Narendja, F., and Scazzocchio, C. (1999) *EMBO J.* **18**, 1584–1597
397a. Molkentin, J. D. (2000) *J. Biol. Chem.* **275**, 38949–38952
398. Omichinski, J. G., Clore, G. M., Schaad, O., Felsenfeld, G., Trainor, C., Appella, E., Stahl, S. J., and Gronenborn, A. M. (1993) *Science* **261**, 438–446
399. Haas, H., Zadra, I., Stöffler, G., and Angermayr, K. (1999) *J. Biol. Chem.* **274**, 4613–4619
400. Van Winkle, L. J. (1999) *Biomembrane Transport*, Academic Press, San Diego, California
401. Zeng, J., Vallee, B. L., and Kägi, J. H. R. (1991) *Proc. Natl. Acad. Sci. U.S.A.* **88**, 9984–9988
402. Lawson, C. L., and Carey, J. (1993) *Nature (London)* **366**, 178–182
403. Marsden, I., Jin, C., and Liao, X. (1998) *J. Mol. Biol.* **278**, 293–299
404. Clark, K. L., Halay, E. D., Lai, E., and Burley, S. K. (1993) *Nature (London)* **364**, 412–420
405. Pio, F., Kodandapani, R., Ni, C.-Z., Shepard, W., Klemsz, M., McKercher, S. R., Maki, R. A., and Ely, K. R. (1996) *J. Biol. Chem.* **271**, 23329–23337
406. Wasylyk, B., Hagman, J., and Gutierrez-Hartmann, A. (1998) *Trends Biochem. Sci.* **23**, 213–216
406a. Alvarez, B., Martínez-A, C., Burgering, B. M. T., and Carrera, A. C. (2001) *Nature (London)* **413**, 744–747
407. Tisné, C., Delepierre, M., and Hartmann, B. (1999) *J. Mol. Biol.* **293**, 139–150
408. Barnes, P. J., and Karin, M. (1997) *N. Engl. J. Med.* **336**, 1066–1071
408a. Nourbakhsh, M., Kälble, S., Dörrie, A., Hauser, H., Resch, K., and Kracht, M. (2001) *J. Biol. Chem.* **276**, 4501–4508
408b. Tam, W. F., and Sen, R. (2001) *J. Biol. Chem.* **276**, 7701–7704
408c. Schmitz, M. L., Bacher, S., and Kracht, M. (2001) *Trends Biochem. Sci.* **26**, 186–190
409. Malek, S., Huxford, T., and Ghosh, G. (1998) *J. Biol. Chem.* **273**, 25427–25435
410. Chen, F. E., Huang, D.-B., Chen, Y.-Q., and Ghosh, G. (1998) *Nature (London)* **391**, 410–413
411. Shlyakhtenko, L. S., Hsieh, P., Grigoriev, M., Potaman, V. N., Sinden, R. R., and Lyubchenko, Y. L. (2000) *J. Mol. Biol.* **296**, 1169–1173
412. Murphy, F. V., IV, Sweet, R. M., and Churchill, M. E. A. (1999) *EMBO J.* **18**, 6610–6618
413. Sutrias-Grau, M., Bianchi, M. E., and Bernusés, J. (1999) *J. Biol. Chem.* **274**, 1628–1634
414. Yie, J., Merika, M., Munshi, N., Chen, G., and Thanos, D. (1999) *EMBO J.* **18**, 3074–3089
414a. Lewis, R. T., Andreucci, A., and Nikolajczyk, B. S. (2001) *J. Biol. Chem.* **276**, 9550–9557
415. Bergel, M., Herrera, J. E., Thatcher, B. J., Prymakowska-Bosak, M., Vassilev, A., Nakatani, Y., Martin, B., and Bustin, M. (2000) *J. Biol. Chem.* **275**, 11514–11520
415a. Cowell, I. G. (1994) *Trends Biochem. Sci.* **19**, 38–42
415b. Harrington, R. H., and Sharma, A. (2001) *J. Biol. Chem.* **276**, 104–113
416. Zhang, J., Ou, J., Bashmakov, Y., Horton, J. D., Brown, M. S., and Goldstein, J. L., (2001) *Proc. Natl. Acad. Sci. U.S.A.* **98**, 3756–3761
416a. Olefsky, J. M. (2001) *J. Biol. Chem.* **276**, 36863–36864
416b. Khorasanizadeh, S., and Rastinejad, F. (2001) *Trends Biochem. Sci.* **26**, 384–390
416c. Valverde, M. A., and Parker, M. G. (2002) *Trends Biochem. Sci.* **27**, 172–173
417. Vázquez, F., Rodríguez-Manzaneque, J. C., Lydon, J. P., Edwards, D. P., O'Mally, B. W., and Iruela-Arispe, M. L. (1999) *J. Biol. Chem.* **274**, 2185–2192
418. Lehmann, M., and Korge, G. (1995) *EMBO J.* **14**, 716–726
418a. Wurtz, J.-M., Guillot, B., Fagart, J., Moras, D., Tietjen, K., and Schindler, M. (2000) *Protein Sci.* **9**, 1073–1084
419. Guschin, D., Chandler, S., and Wolffe, A. P. (1998) *Biochemistry* **37**, 8629–8636
420. Baudino, T. A., Kraichely, D. M., Jefcoat, S. C., Jr., Winchester, S. K., Partridge, N. C., and MacDonald, P. N. (1998) *J. Biol. Chem.* **273**, 16434–16441
420a. Makishima, M., Lu, T. T., Xie, W., Whitfield, G. K., Domoto, H., Evans, R. M., Haussler, M. R., and Mangelsdorf, D. J. (2002) *Science* **296**, 1313–1316
421. Beato, M. (1991) *FASEB J.* **5**, 2044–2051
422. Weatherman, R. V., Fletterick, R. J., and Scanlan, T. S. (1999) *Ann. Rev.Biochem.* **68**, 559–581
423. Di Croce, L., Okret, S., Kersten, S., Gustafsson, J.-Å., Parker, M., Wahli, W., and Beato, M. (1999) *EMBO J.* **18**, 6201–6210
424. Avram, D., Fields, A., Pretty On Top, K., Nevrivy, D. J., Ishmael, J. E., and Leid, M. (2000) *J. Biol. Chem.* **275**, 10315–10322
425. Kliewer, S. A., Lehmann, J. M., and Willson, T. M. (1999) *Science* **284**, 757–760
426. Kosztin, D., Bishop, T. C., and Schulten, K. (1997) *Biophys. J.* **73**, 557–570
427. Driscoll, M. D., Sathya, G., Muyan, M., Klinge, C. M., Hilf, R., and Bambara, R. A. (1998) *J. Biol. Chem.* **273**, 29321–29330
427a. Campbell, R. A., Bhat-Nakshatri, P., Patel, N. M., Constantinidou, D., Ali, S., and Nakshatri, H. (2001) *J. Biol. Chem.* **276**, 9817–9824
427b. An, J., Tzagarakis-Foster, C., Scharaschmidt, T. C., Lomri, N., and Leitman, D. C. (2001) *J. Biol. Chem.* **276**, 17808–17814
428. Rogatsky, I., Trowbridge, J. M., and Garabedian, M. J. (1999) *J. Biol. Chem.* **274**, 22296–22302
429. Kosano, H., Stensgard, B., Charlesworth, M. C., McMahon, N., and Toft, D. (1998) *J. Biol. Chem.* **273**, 32973–32979
429a. Bennett, M. K., and Osborne, T. F. (2000) *Proc. Natl. Acad. Sci. U.S.A.* **97**, 6340–6344
429b. Luong, A., Hannah, V. C., Brown, M. S., and Goldstein, J. L. (2000) *J. Biol. Chem.* **275**, 26458–26466
430. Clarke, S. D., and Abraham, S. (1992) *FASEB J.* **6**, 3146–3152
431. Lu, C.-A., Lim, E.-K., and Yu, S.-M. (1998) *J. Biol. Chem.* **273**, 10120–10131
432. Cereghino, G. P., and Scheffler, I. E. (1996) *EMBO J.* **15**, 363–374
433. Hong, S.-P., Piper, M. D., Sinclair, D. A., and Dawes, I. W. (1999) *J. Biol. Chem.* **274**, 10523–10532
434. Svaren, J., and Hörz, W. (1997) *Trends Biochem. Sci.* **22**, 93–97
435. Shimizu, T., Toumoto, A., Ihara, K., Shimizu, M., Kyogoku, Y., Ogawa, N., Oshima, Y., and Hakoshima, T. (1997) *EMBO J.* **16**, 4689–4697
436. Addess, K. J., Basilion, J. P., Klausner, R. D., Rouault, T. A., and Pardi, A. (1997) *J. Mol. Biol.* **274**, 72–83
437. Wardrop, S. L., Watts, R. N., and Richardson, D. R. (2000) *Biochemistry* **39**, 2748–2758
438. Ke, Y., Sierzputowska-Gracz, H., Gdaniec, Z., and Theil, E. C. (2000) *Biochemistry* **39**, 6235–6242
438a. Meehan, H. A., and Connell, G. J. (2001) *J. Biol. Chem.* **276**, 14791–14796
439. Quinn, J. M., Barraco, P., Eriksson, M., and Merchant, S. (2000) *J. Biol. Chem.* **275**, 6080–6089
440. Hall, D. R., Gourley, D. G., Leonard, G. A., Duke, E. M. H., Anderson, L. A., Boxer, D. H., and Hunter, W. N. (1999) *EMBO J.* **18**, 1435–1446
441. Okamura, H., Hanaoka, S., Nagadoi, A., Makino, K., and Nishimura, Y. (2000) *J. Mol. Biol.* **295**, 1225–1236
442. Ritossa, F. (1962) *Experientia* **18**, 571–573

References

443. Schlesinger, M. J. (1990) *J. Biol. Chem.* **265**, 12111–12114
444. Ang, D., Liberek, K., Skowyra, D., Zylicz, M., and Georgopoulos, C. (1991) *J. Biol. Chem.* **266**, 24233–24236
445. Morimoto, R. I. (1993) *Science* **259**, 1409–1410
446. Giardina, C., and Lis, J. T. (1995) *J. Biol. Chem.* **270**, 10369–10372
447. Kimpel, J. A., and Key, J. L. (1985) *Trends Biochem. Sci.* **10**, 353–357
448. Craig, E. A., and Gross, C. A. (1991) *Trends Biochem. Sci.* **16**, 135–140
449. Welch, W. J. (1993) *Sci. Am.* **268**(May), 57–64
450. Kim, K. K., Kim, R., and Kim, S.-H. (1998) *Nature (London)* **394**, 595–599
451. Korber, P., Stahl, J. M., Nierhaus, K. H., and Bardwell, J. C. A. (2000) *EMBO J.* **19**, 741–748
452. Goldenberg, C. J., Luo, Y., Fenna, M., Baler, R., Weinmann, R., and Voellmy, R. (1988) *J. Biol. Chem.* **263**, 19734–19739
453. Harrison, C. J., Bohm, A. A., and Nelson, H. C. M. (1994) *Science* **263**, 224–227
454. Larson, J. S., Schuetz, T. J., and Kingston, R. E. (1995) *Biochemistry* **34**, 1902–1911
455. Wyman, C., Grotkopp, E., Bustamante, C., and Nelson, H. C. M. (1995) *EMBO J.* **14**, 117–123
456. Rouvière, P. E., De Las Penas, A., Mecsas, J., Lu, C. Z., Rudd, K. E., and Gross, C. A. (1995) *EMBO J.* **14**, 1032–1042
456a. Dartigalongue, C., Missiakas, D., and Raina, S. (2001) *J. Biol. Chem.* **276**, 20866–20875
457. Brevet, A., Plateau, P., Best-Belpomme, M., and Blanquet, S. (1985) *J. Biol. Chem.* **260**, 15566–15570
458. Wahab, S. Z., and Yang, D. C. H. (1985) *J. Biol. Chem.* **260**, 5286–5289
459. Harnett, S. P., Lowe, G., and Tansley, G. (1985) *Biochemistry* **24**, 2908–2915
460. Lee, P. C., Bochner, B. R., and Ames, B. N. (1983) *Proc. Natl. Acad. Sci. U.S.A.* **80**, 7496–7500
461. Guédon, G. F., Gilson, G. J. P., Ebel, J. P., Befort, N. M.-T., and Remy, P. M. (1986) *J. Biol. Chem.* **261**, 16459–16465
462. Kodym, R., Calkins, P., and Story, M. (1999) *J. Biol. Chem.* **274**, 5131–5137
463. Freestone, P., Nyström, T., Trinei, M., and Norris, V. (1997) *J. Mol. Biol.* **274**, 318–324
464. Discher, D. J., Bishopric, N. H., Wu, X., Peterson, C. A., and Webster, K. A. (1998) *J. Biol. Chem.* **273**, 26087–26093
465. Sundaresan, M., Yu, Z.-X., Ferrans, V. J., Irani, K., and Finkel, T. (1995) *Science* **270**, 296–299
465a. Thannickal, V. J., and Fanburg, B. L. (1995) *J. Biol. Chem.* **270**, 30334–30338
466. Burg, M. B., Kwon, E. D., and Kültz, D. (1996) *FASEB J.* **10**, 1598–1606
467. Ramagopal, S. (1987) *Proc. Natl. Acad. Sci. U.S.A.* **84**, 94–98
467a. Dagsgaard, C., Taylor, L. E., O'Brien, K. M., and Poyton, R. O. (2001) *J. Biol. Chem.* **276**, 7593–7601
467b. Ivan, M., Kondo, K., Yang, H., Kim, W., Valiando, J., Ohh, M., Salic, A., Asara, J. M., Lane, W. S., and Kaelin, W. G., Jr. (2001) *Science* **292**, 464–468
467c. Zhu, H., and Bunn, H. F. (2001) *Science* **292**, 449,451
467d. Lando, D., Peet, D. J., Whelan, D. A., Gorman, J. J., and Whitelaw, M. L. (2002) *Science* **295**, 858–861
467e. Srinivas, V., Leshchinsky, I., Sang, N., King, M. P., Minchenko, A., and Caro, J. (2001) *J. Biol. Chem.* **276**, 21995–21998
467f. Jaakkola, P., Mole, D. R., Tian, Y.-M., Wilson, M. I., Gielbert, J., Gaskell, S. J., von Kriegsheim, A., Hebestreit, H. F., Mukherji, M., Schofield, C. J., Maxwell, P. H., Pugh, C. W., and Ratcliffe, P. J. (2001) *Science* **292**, 468–472
467g. Min, J.-H., Yang, H., Ivan, M., Gertler, F., Kaelin, W. G., Jr., and Pavletich, N. P. (2002) *Science* **296**, 1886–1889
467h. Semenza, G. L. (2002) *Proc. Natl. Acad. Sci. U.S.A.* **99**, 11570–11572
467i. Hansson, L. O., Friedler, A., Freund, S., Rüdiger, S., and Fersht, A. R. (2002) *Proc. Natl. Acad. Sci. U.S.A.* **99**, 10305–10309
467j. Stebbins, C. E., Kaelin, W. G., Jr., and Pavletich, N. P. (1999) *Science* **284**, 455–461
467k. Baysal, B. E., Ferrell, R. E., Willett-Brozick, J. E., Lawrence, E. C., Myssiorek, D., Bosch, A., van der Mey, A., Taschner, P. E. M., Rubinstein, W. S., Myers, E. N., Richard, C. W., III, Cornelisse, C. J., Devilee, P., and Devlin, B. (2000) *Science* **287**, 848–851
468. Kim, M., Thum, K. E., Morishige, D. T., and Mullet, J. E. (1999) *J. Biol. Chem.* **274**, 4684–4692
469. Martínez-García, J. F., Huq, E., and Quail, P. H. (2000) *Science* **288**, 859–863
470. Wright, C. V. E., Cho, K. W. Y., Oliver, G., and De Robertis, E. M. (1989) *Trends Biochem. Sci.* **14**, 52–56
471. Scott, M. P. (1987) *Ann. Rev. Biochem.* **56**, 195–227
472. Mihara, H., and Kaiser, E. T. (1988) *Science* **242**, 925–927
473. Ford, E., and Hernandez, N. (1997) *J. Biol. Chem.* **272**, 16048–16055
474. Gstaiger, M., Georgiev, O., van Leeuwen, H., van der Vliet, P., and Schaffner, W. (1996) *EMBO J.* **15**, 2781–2790
475. Lee, L., Stollar, E., Chang, J., Grossmann, J. G., O'Brien, R., Ladbury, J., Carpenter, B., Roberts, S. and Luisi, B. (2001) *Biochemistry* **40**, 6580–6588
475a. Lundbäck, T., Chang, J.-F., Phillips, K., Luisi, B., and Ladbury, J. E. (2000) *Biochemistry* **39**, 7570–7579
476. Robertson, M. (1988) *Nature (London)* **336**, 522–524
476a. Diamond, S. E., and Gutierrez-Hartmann, A. (2000) *J. Biol. Chem.* **275**, 30977–30986
477. Tanaka, N., Kato, H., Ishikawa, Y., Hisatake, K., Tashiro, K., Kominami, R., and Muramatsu, M. (1990) *J. Biol. Chem.* **265**, 13836–13842
478. Kahl, B. F., Li, H., and Paule, M. R. (2000) *J. Mol. Biol.* **299**, 75–89
478a. Miller, G., Panov, K. I., Friedrich, J. K., Trinkle-Mulcahy, L., Lamond, A. I., and Zomerdijk, J. C. B. M. (2001) *EMBO J.* **20**, 1373–1382
479. McBryant, S. J., Kassavetis, G. A., and Gottesfeld, J. M. (1995) *J. Mol. Biol.* **250**, 315–326
480. Gunnery, S., Ma, Y., and Mathews, M. B. (1999) *J. Mol. Biol.* **286**, 745–757
481. Pieler, T., and Theunissen, O. (1993) *Trends Biochem. Sci.* **18**, 226–230
482. Miller, J., McLachlan, A. D., and Klug, A. (1985) *EMBO J.* **4**, 1609–1614
483. Veldhoen, N., You, Q., Setzer, D. R., and Romaniuk, P. J. (1994) *Biochemistry* **33**, 7568–7575
484. Gabrielsen, O. S., and Sentenac, A. (1991) *Trends Biochem. Sci.* **16**, 412–416
485. Andrau, J.-C., Sentenac, A., and Werner, M. (1999) *J. Mol. Biol.* **288**, 511–520
486. Young, L. S., Dunstan, H. M., Witte, P. R., Smith, T. P., Ottonello, S., and Sprague, K. U. (1991) *Science* **252**, 542–546
487. Choder, M., and Aloni, Y. (1988) *J. Biol. Chem.* **263**, 12994–13002
488. Lewis, J. D., and Tollervey, D. (2000) *Science* **288**, 1385–1389
489. Reines, D., Conaway, J. W., and Conaway, R. C. (1996) *Trends Biochem. Sci.* **21**, 351–355
490. Gnatt, A., Fu, J., and Kornberg, R. D. (1997) *J. Biol. Chem.* **272**, 30799–30805
491. Izban, M. G., Parsons, M. A., and Sinden, R. R. (1998) *J. Biol. Chem.* **273**, 27009–27016
492. Hemming, S. A., and Edwards, A. M. (2000) *J. Biol. Chem.* **275**, 2288–2294
493. Nelson, P. (1999) *Proc. Natl. Acad. Sci. U.S.A.* **96**, 14342–14347
494. Botuyan, M. V., Koth, C. M., Mer, G., Chakrabartty, A., Conaway, J. W., Conaway, R. C., Edwards, A. M., Arrowsmith, C. H., and Chazin, W. J. (1999) *Proc. Natl. Acad. Sci. U.S.A.* **96**, 9033–9038
495. Conaway, J. W., and Conaway, R. C. (1999) *Ann. Rev. Biochem.* **68**, 301–319
496. Iyer, V., and Struhl, K. (1996) *Proc. Natl. Acad. Sci. U.S.A.* **93**, 5208–5212
497. Guo, Z., and Sherman, F. (1996) *Trends Biochem. Sci.* **21**, 477–481
498. Wahle, E., and Keller, W. (1996) *Trends Biochem. Sci.* **21**, 247–250
499. Chanfreau, G., Noble, S. M., and Guthrie, C. (1996) *Science* **274**, 1511–1514
500. Hirose, Y., and Manley, J. L. (1997) *J. Biol. Chem.* **272**, 29636–29642
501. Ohnacker, M., Barabino, S. M. L., Preker, P. J., and Keller, W. (2000) *EMBO J.* **19**, 37–47
502. Keller, R. W., Kühn, U., Aragón, M., Bornikova, L., Wahle, E., and Bear, D. G. (2000) *J. Mol. Biol.* **297**, 569–583
503. Reeder, R. H., and Lang, W. H. (1997) *Trends Biochem. Sci.* **22**, 473–477
504. Syroid, D. E., and Capone, J. P. (1994) *J. Mol. Biol.* **244**, 482–493
504a. Ohndorf, U.-M., Steegborn, C., Knijff, R., and Sondermann, P. (2001) *J. Biol. Chem.* **276**, 27188–27196
504b. Wang, Z., Bai, L., Hsieh, Y.-J., and Roeder, R. G. (2000) *EMBO J.* **19**, 6823–6832
505. Draper, D. E. (1996) *Trends Biochem. Sci.* **21**, 145–149
506. Moore, P. B. (1999) *Ann. Rev. Biochem.* **67**, 287–300
507. Hermann, T., and Patel, D. J. (1999) *J. Mol. Biol.* **294**, 829–849
508. Pan, J., and Woodson, S. A. (1999) *J. Mol. Biol.* **294**, 955–965
508a. Zhuang, X., Bartley, L. E., Babcock, H. P., Russell, R., Ha, T., Herschlag, D., and Chu, S. (2000) *Science* **288**, 2048–2051
508b. Kent, O., Chaulk, S. G., and MacMillan, A. M. (2000) *J. Mol. Biol.* **304**, 699–705
509. Murthy, V. L., Srinivasan, R., Draper, D. E., and Rose, G. D. (1999) *J. Mol. Biol.* **291**, 313–327
510. Michel, F., and Westhof, E. (1996) *Science* **273**, 1676–1677
511. Gultyaev, A. P., van Batenburg, F. H. D., and Pleij, C. W. A. (1995) *J. Mol. Biol.* **250**, 37–51
512. Narlikar, G. J., Bartley, L. E., Khosla, M., and Herschlag, D. (1999) *Biochemistry* **38**, 14192–14204
513. Sclavi, B., Sullivan, M., Chance, M. R., Brenowitz, M., and Woodson, S. A. (1998) *Science* **279**, 1940–1943
514. Silverman, S. K., and Cech, T. R. (1999) *Biochemistry* **38**, 8691–8702
515. Rivas, E., and Eddy, S. R. (1999) *J. Mol. Biol.* **285**, 2053–2068
516. Cate, J. H., Gooding, A. R., Podell, E., Zhou, K., Golden, B. L., Szewczak, A. A., Kundrot, C. E., Cech, T. R., and Doudna, J. A. (1996) *Science* **273**, 1696–1699
517. Biswas, R., Wahl, M. C., Ban, C., and Sundaralingam, M. (1997) *J. Mol. Biol.* **267**, 1149–1156
517a. Deng, J., Xiong, Y., and Sundaralingam, M. (2001) *Proc. Natl. Acad. Sci. U.S.A.* **98**, 13665–13670
517b. Snoussi, K., Nonin-Lecomte, S., and Leroy, J.-L. (2001) *J. Mol. Biol.* **309**, 139–153
517c. Schneider, C., Brandl, M., and Sühnel, J. (2001) *J. Mol. Biol.* **305**, 659–667

References

517d. Soukup, J. K., Minakawa, N., Matsuda, A., and Strobel, S. A. (2002) *Biochemistry* **41**, 10426–10438

517e. Stormo, G. D., and Ji, Y. (2001) *Proc. Natl. Acad. Sci. U.S.A.* **98**, 9465–9467

518. Costa, M., and Michel, F. (1997) *EMBO J.* **16**, 3289–3302

519. Nagai, K., Oubridge, C., Ito, N., Avis, J., and Evans, P. (1995) *Trends Biochem. Sci.* **20**, 235–240

520. Bell, S. P., Learned, R. M., Jantzen, H.–M., and Tjian, R. (1988) *Science* **241**, 1192–1197

521. Schmitz, M. L., Maier, U.-G., Brown, J. W. S., and Feix, G. (1989) *J. Biol. Chem.* **264**, 1467–1472

522. Baker, S. M., and Platt, T. (1986) *Cell* **47**, 839–840

523. Read, C., Larose, A.-M., Leblanc, B., Bannister, A. J., Firek, S., Smith, D. R., and Moss, T. (1992) *J. Biol. Chem.* **267**, 10961–10967

524. Sommerville, J. (1985) *Nature (London)* **318**, 410–411

525. Morrissey, J. P., and Tollervey, D. (1995) *Trends Biochem. Sci.* **20**, 78–82

526. Borovjagin, A. V., and Gerbi, S. A. (1999) *J. Mol. Biol.* **286**, 1347–1363

527. Wellauer, P. K., and Dawid, I. B. (1973) *Proc. Natl. Acad. Sci. U.S.A.* **70**, 2827–2831

528. Dawid, I. B., and Wellauer, P. K. (1976) *Cell* **8**, 443–448

529. Ghisolfi, L., Joseph, G., Amalric, F., and Erard, M. (1992) *J. Biol. Chem.* **267**, 2955–2959

529a. Bouvet, P., Allain, F. H.-T., Finger, L. D., Dieckmann, T., and Feigon, J. (2001) *J. Mol. Biol.* **309**, 763–775

529b. Allain, F. H.-T., Bouvet, P., Dieckmann, T., and Feigon, J. (2000) *EMBO J.* **19**, 6870–6881

530. Srivastava, M., McBride, O. W., Fleming, P. J., Pollard, H. B., and Burns, A. L. (1990) *J. Biol. Chem.* **265**, 14922–14931

531. Tycowaki, K. T., Shu, M.–D., and Steitz, J. A. (1994) *Science* **266**, 1558–1561

531a. Pederson, T. (2002) *Trends Biochem. Sci.* **27**, 111–112

532. Guthrie, C. (1986) *Trends Biochem. Sci.* **11**, 430–434

533. Birnstiel, M. L., ed. (1988) *Small Nuclear Ribonucleoprotein Particles*, Springer, Vienna/New York

534. Tarn, W.-Y., and Steitz, J. A. (1997) *Trends Biochem. Sci.* **22**, 132–137

535. Séraphin, B. (1995) *EMBO J.* **14**, 2089–2098

535a. Filipowicz, W. (2000) *Proc. Natl. Acad. Sci. U.S.A.* **97**, 14035–14037

535b. Hüttenhofer, A., Kiefmann, M., Meier-Ewert, S., O'Brien, J., Lehrach, H., Bachellerie, J.-P., and Brosius, J. (2001) *EMBO J.* **20**, 2943–2953

535c. Lowe, T. M., and Eddy, S. R. (1999) *Science* **283**, 1168–1171

535d. Xu, Y.-x, Liu, L., Lopez-Estrano, C., and Michaeli, S. (2001) *J. Biol. Chem.* **276**, 14289–14298

535e. Barneche, F., Gaspin, C., Guyot, R., and Echeverría, M. (2001) *J. Mol. Biol.* **311**, 57–73

536. Feeney, R. J., Sauterer, R. A., Feeney, J. L., and Zieve, G. W. (1989) *J. Biol. Chem.* **264**, 5776–5783

537. Hartmuth, K., Raker, V. A., Huber, J., Branlant, C., and Lührmann, R. (1999) *J. Mol. Biol.* **285**, 133–147

537a. Collins, B. M., Harrop, S. J., Kornfeld, G. D., Dawes, I. W., Curmi, P. M. G., and Mabbutt, B. C. (2001) *J. Mol. Biol.* **309**, 915–923

537b. Baulcombe, D. (2002) *Science* **297**, 2002–2003

537c. Hutvágner, G., and Zamore, P. D. (2002) *Science* **297**, 2056–2060

538. Parry, H. D., Scherly, D., and Mattaj, I. W. (1989) *Trends Biochem. Sci.* **14**, 15–19

539. Grabowski, P. J., and Sharp, P. A. (1986) *Science* **233**, 1294–1299

539a. Hannon, G. J. (2002) *Nature (London)* **418**, 244–251

540. Sollner-Webb, B., and Tower, J. (1986) *Ann. Rev. Biochem.* **55**, 801–830

541. Fournier, M. J., and Maxwell, E. S. (1993) *Trends Biochem. Sci.* **18**, 131–135

541a. Saveanu, C., Bienvenu, D., Namane, A., Gleizes, P.-E., Gas, N., Jacquier, A., and Fromont-Racine, M. (2001) *EMBO J.* **20**, 6475–6484

541b. Gadal, O., Strauss, D., Braspenning, J., Hoepfner, D., Petfalski, E., Philippsen, P., Tollervey, D., and Hurt, E. (2001) *EMBO J.* **20**, 3695–3704

541c. Jäkel, S., and Görlich, D. (1998) *EMBO J.* **17**, 4491–4502

542. Beltrame, M., and Tollervey, D. (1995) *EMBO J.* **14**, 4350–4356

543. Hughes, J. M. X. (1996) *J. Mol. Biol.* **259**, 645–654

544. Méreau, A., Fournier, R., Grégoire, A., Mougin, A., Fabrizio, P., Lührmann, R., and Branlant, C. (1997) *J. Mol. Biol.* **273**, 552–571

545. Hengst, J. A., Georgoff, I., Isom, H. C., and Jacob, S. T. (1988) *J. Biol. Chem.* **263**, 19270–19273

546. van Nues, R. W., Rientjes, J. M. J., Morré, S. A., Mollee, E., Planta, R. J., Vanema, J., and Raué, H. A. (1995) *J. Mol. Biol.* **250**, 24–36

546a. Borovjagin, A. V., and Gerbi, S. A. (2000) *J. Mol. Biol.* **300**, 57–74

546b. Peculis, B. A. (2001) *RNA* **7**, 207–219

547. Lygerou, Z., Allmang, C., Tollervey, D., and Séraphin, B. (1996) *Science* **272**, 268–270

547a. Clayton, D. A. (2001) *Nature (London)* **410**, 29, 31

548. Bachellerie, J.–P., Michot, B., Nicoloso, M., Balakin, A., Ni, J., and Fournier, M. J. (1995) *Trends Biochem. Sci.* **20**, 261–264

549. Wang, H., Boisvert, D., Kim, K. K., Kim, R., and Kim, S.-H. (2000) *EMBO J.* **19**, 317–323

550. Reddy, R., Henning, D., Das, G., Harless, M., and Wright, D. (1987) *J. Biol. Chem.* **262**, 75–81

551. Hernandez, N., and Weiner, A. M. (1986) *Cell* **47**, 249–258

552. Mimori, T., Hinterberger, M., Pettersson, I., and Steitz, J. A. (1984) *J. Biol. Chem.* **259**, 560–565

553. Rokeach, L. A., Haselby, J. A., and Hoch, S. O. (1988) *Proc. Natl. Acad. Sci. U.S.A.* **85**, 4832–4836

554. Deutscher, S. L., and Keene, J. D. (1988) *Proc. Natl. Acad. Sci. U.S.A.* **85**, 3299–3303

555. Frendeway, D., Dingermann, T., Cooley, L., and Söll, D. (1985) *J. Biol. Chem.* **260**, 449–454

556. Levinger, L., Vasisht, V., Greene, V., Bourne, R., Birk, A., and Kolla, S. (1995) *J. Biol. Chem.* **270**, 18903–18909

557. Pfeiffer, T., Tekos, A., Warnecke, J. M., Drainas, D., Engelke, D. R., Séraphin, B., and Hartmann, R. K. (2000) *J. Mol. Biol.* **298**, 559–565

558. Rossmanith, W. (1997) *J. Mol. Biol.* **265**, 365–371

558a. Doudna, J. A., and Cech, T. R. (2002) *Nature (London)* **418**, 222–228

558b. Xiao, S., Scott, F., Fierke, C. A., and Engelke, D. R. (2002) *Ann. Rev. Biochem.* **71**, 165–189

558c. Houser-Scott, F., Xiao, S., Millikin, C. E., Zengel, J. M., Lindahl, L., and Engelke, D. R. (2002) *Proc. Natl. Acad. Sci. U.S.A.* **99**, 2684–2689

559. Deutscher, M. D. (1984) *Crit. Revs. Biochem.* **17**, 45–71

559a. Li, F., Wang, J., and Steitz, T. A. (2000) *J. Mol. Biol.* **304**, 483–492

559b. Seth, M., Thurlow, D. L., and Hou, Y.-M. (2002) *Biochemistry* **41**, 4521–4532

560. Grosjean, H., Edqvist, J., Stråby, K. B., and Giegé, R. (1996) *J. Mol. Biol.* **255**, 67–85

561. Constantinesco, F., Motorin, Y., and Grosjean, H. (1999) *J. Mol. Biol.* **291**, 375–392

562. House, C. H., and Miller, S. L. (1996) *Biochemistry* **35**, 315–320

563. Lamond, A. I. (1990) *Trends Biochem. Sci.* **15**, 451–452

564. Schnierle, B. S., Gershon, P. D., and Moss, B. (1994) *J. Biol. Chem.* **269**, 20700–20706

565. Pillutla, R. C., Yue, Z., Maldonado, E., and Shatkin, A. J. (1998) *J. Biol. Chem.* **273**, 21443–21446

566. Terns, M. P., Grimm, C., Lund, E., and Dahlberg, J. E. (1995) *EMBO J.* **14**, 4860–4871

567. Shimba, S., and Reddy, R. (1994) *J. Biol. Chem.* **269**, 12419–12423

568. Booth, B. L., Jr., and Pugh, B. F. (1997) *J. Biol. Chem.* **272**, 984–991

569. Wang, S. P., Deng, L., Ho, C. K., and Shuman, S. (1997) *Proc. Natl. Acad. Sci. U.S.A.* **94**, 9573–9578

570. Kuge, H., and Richter, J. D. (1995) *EMBO J.* **14**, 6301–6310

571. Hu, G., Gershon, P. D., Hodel, A. E., and Quiocho, F. A. (1999) *Proc. Natl. Acad. Sci. U.S.A.* **96**, 7149–7154

571a. Hsu, P.-C., Hodel, M. R., Thomas, J. W., Taylor, L. J., Hagedorn, C. H., and Hodel, A. E. (2000) *Biochemistry* **39**, 13730–13736

571b. Martin, G., Keller, W., and Doublié, S. (2000) *EMBO J.* **19**, 4193–4203

571c. Hayes, R., Kudla, J., and Gruissem, W. (1999) *Trends Biochem. Sci.* **24**, 199–202

571d. Gagliardi, D., Perrin, R., Maréchal-Drouard, L., Grienenberger, J.-M., and Leaver, C. J. (2001) *J. Biol. Chem.* **276**, 43541–43547

572. Galili, G., Kawata, E. E., Smith, L. D., and Larkins, B. A. (1988) *J. Biol. Chem.* **263**, 5764–5770

573. Chen, C.-Y. A., and Shyu, A.-B. (1995) *Trends Biochem. Sci.* **20**, 465–470

574. Jacobson, A., and Peltz, S. W. (1996) *Ann. Rev. Biochem.* **65**, 693–739

575. Bashirullah, A., Halsell, S. R., Copperstock, R. L., Kloc, M., Karaiskakis, A., Fisher, W. W., Fu, W., Hamilton, J. K., Etkin, L. D., and Lipshitz, H. D. (1999) *EMBO J.* **18**, 2610–2620

575a. Matsumoto, K., Wassarman, K. M., and Wolffe, A. P. (1998) *EMBO J.* **17**, 2107–2121

575b. Bashirullah, A., Cooperstock, R. L., and Lipshitz, H. D. (2001) *Proc. Natl. Acad. Sci. U.S.A.* **98**, 7025–7028

576. Raghow, R. (1987) *Trends Biochem. Sci.* **12**, 358–360

577. Hunt, T. (1988) *Nature (London)* **334**, 567–568

578. Bouveret, E., Rigaut, G., Shevchenko, A., Wilm, M., and Séraphin, B. (2000) *EMBO J.* **19**, 1661–1671

579. Tharun, S., He, W., Mayes, A. E., Lennertz, P., Beggs, J. D., and Parker, R. (2000) *Nature (London)* **404**, 515–518

579a. Gao, M., Wilusz, C. J., Peltz, S. W., and Wilusz, J. (2001) *EMBO J.* **20**, 1134–1143

580. Phizicky, E. M., and Greer, C. L. (1993) *Trends Biochem. Sci.* **18**, 31–34

581. Cech, T. R., and Bass, B. L. (1986) *Ann. Rev. Biochem.* **55**, 599–629

582. Zaug, A. J., Grosshans, C. A., and Cech, T. R. (1988) *Biochemistry* **27**, 8924–8931

583. Saldanha, R., Mohr, G., Belfort, M., and Lambowitz, A. M. (1993) *FASEB J.* **7**, 15–24

584. Weeks, K. M., and Cech, T. R. (1995) *Biochemistry* **34**, 7728–7738

585. Ho, Y., and Waring, R. B. (1999) *J. Mol. Biol.* **292**, 987–1001

585a. Strauss-Soukup, J. K., and Strobel, S. A. (2000) *J. Mol. Biol.* **302**, 339–358

586. Bevilacqua, P. C., Sugimoto, N., and Turner, D. H. (1996) *Biochemistry* **35**, 648–658

References

587. Golden, B. L., and Cech, T. R. (1996) *Biochemistry* **35**, 3754–3763
588. Nowakowski, J., and Tinoco, I., Jr. (1996) *Biochemistry* **35**, 2577–2585
589. Michel, F., and Ferat, J.-L. (1995) *Ann. Rev. Biochem.* **64**, 435–461
590. Sharp, P. A., and Eisenberg, D. (1987) *Science* **238**, 729–730
590a. Chanfreau, G., and Jacquier, A. (1996) *EMBO J.* **15**, 3466–3476
590b. Swisher, J., Duarte, C. M., Su, L. J., and Pyle, A. M. (2001) *EMBO J.* **20**, 2051–2061
590c. Su, L. J., Qin, P. Z., Michels, W. J., and Pyle, A. M. (2001) *J. Mol. Biol.* **306**, 655–668
590d. Zhang, L., and Doudna, J. A. (2002) *Science* **295**, 2084–2088
591. Diener, T. O. (1984) *Trends Biochem. Sci.* **9**, 133–136
592. Daròs, J. A., Marcos, J. F., Hernández, C., and Flores, R. (1994) *Proc. Natl. Acad. Sci. U.S.A.* **91**, 12813–12817
593. Feldstein, P. A., Hu, Y., and Owens, R. A. (1998) *Proc. Natl. Acad. Sci. U.S.A.* **95**, 6560–6565
593a. Navarro, J.-A., and Flores, R. (2000) *EMBO J.* **19**, 2662–2670
594. Forster, A. C., Davies, C., Sheldon, C. C., Jeffries, A. C., and Symons, R. H. (1988) *Nature (London)* **334**, 265–267
595. Haseloff, J., and Gerlach, W. L. (1988) *Nature (London)* **334**, 585–591
596. Dinter-Gottlieb, G. (1986) *Proc. Natl. Acad. Sci. U.S.A.* **83**, 6250–6254
597. Greer, C. L., and Abelson, J. (1984) *Trends Biochem. Sci.* **9**, 139–141
598. Laski, F. A., Fire, A. Z., RajBhandary, U. L., and Sharp, P. A. (1983) *J. Biol. Chem.* **258**, 11974–11980
599. Phizicky, E. M., Schwartz, R. C., and Abelson, J. (1986) *J. Biol. Chem.* **261**, 2978–2986
600. McCraith, S. M., and Phizicky, E. M. (1991) *J. Biol. Chem.* **266**, 11986–11992
601. Belford, H. G., Westaway, S. K., Abelson, J., and Greer, C. L. (1993) *J. Biol. Chem.* **268**, 2444–2450
601a. Steiger, M. A., Kierzek, R., Turner, D. H., and Phizicky, E. M. (2001) *Biochemistry* **40**, 14098–14105
602. Filipowicz, W., and Gross, H. J. (1984) *Trends Biochem. Sci.* **9**, 68–71
603. Ogden, R. C., Knapp, G., Peebles, C. L., Johnson, J., and Abelson, J. (1981) *Trends Biochem. Sci.* **6**, 154–158
604. Skoglund, U., Andersson, K., Strandberg, B., and Daneholt, B. (1986) *Nature (London)* **319**, 560–564
605. Dreyfuss, G., Swanson, M. S., and Pinol-Roma, S. (1988) *Trends Biochem. Sci.* **13**, 86–91
606. Richter, J. D. (1988) *Trends Biochem. Sci.* **13**, 483–486
607. Müller, S., Wolpensinger, B., Angenitzki, M., Engel, A., Sperling, J., and Sperling, R. (1998) *J. Mol. Biol.* **283**, 383–394
608. Sharp, P. A. (1987) *Science* **235**, 766–771
609. Guthrie, C. (1991) *Science* **253**, 157–163
610. Newman, A. J. (1997) *EMBO J.* **16**, 5797–5800
611. Plessel, G., Lührmann, R., and Kastner, B. (1997) *J. Mol. Biol.* **265**, 87–94
612. Valcárcel, J., and Green, M. R. (1996) *Trends Biochem. Sci.* **21**, 296–301
613. Berget, S. M. (1995) *J. Biol. Chem.* **270**, 2411–2414
614. Oubridge, C., Ito, N., Evans, P. R., Teo, C.-H., and Nagai, K. (1994) *Nature (London)* **372**, 432–438
615. Padgett, R. A., Grabkowski, P. J., Konarska, M. M., and Sharp, P. A. (1985) *Trends Biochem. Sci.* **10**, 154–157
616. Deirdre, A., Scadden, J., and Smith, C. W. J. (1995) *EMBO J.* **14**, 3236–3246
617. Blencowe, B. J. (2000) *Trends Biochem. Sci.* **25**, 106–110
618. Reuter, K., Nottrott, S., Fabrizio, P., Lührmann, R., and Ficner, R. (1999) *J. Mol. Biol.* **294**, 515–525
619. Xie, J., Beickman, K., Otte, E., and Rymond, B. C. (1998) *EMBO J.* **17**, 2938–2946
620. Furman, E., and Glitz, D. G. (1995) *J. Biol. Chem.* **270**, 15515–15522
621. Makarov, E. M., Makarova, O. V., Achsel, T., and Lührmann, R. (2000) *J. Mol. Biol.* **298**, 567–575
622. Jiang, J., Horowitz, D. S., and Xu, R.-M. (2000) *Proc. Natl. Acad. Sci. U.S.A.* **97**, 3022–3027
623. Sontheimer, E. J., Sun, S., and Piccirilli, J. A. (1997) *Nature (London)* **388**, 801–805
624. Moore, M. J., and Sharp, P. A. (1993) *Nature (London)* **365**, 364–368
625. Tarn, W.-Y., and Steitz, J. A. (1996) *Science* **273**, 1824–1832
626. Will, C. L., Schneider, C., Reed, R., and Lührmann, R. (1999) *Science* **284**, 2003–2005
627. Bailey-Serres, J., Rochaix, J.-D., Wassenegger, M., and Filipowicz, W. (1999) *EMBO J.* **18**, 5153–5158
628. Noguchi, T., Inoue, H., and Tanaka, T. (1986) *J. Biol. Chem.* **261**, 13807–13812
629. Saitta, B., Stokes, D. G., Vissing, H., Timpl, R., and Chu, M.-L. (1990) *J. Biol. Chem.* **265**, 6473–6480
630. Magnuson, V. L., Young, M., Schattenberg, D. G., Mancini, M. A., Chen, D., Steffensen, B., and Klebe, R. J. (1991) *J. Biol. Chem.* **266**, 14654–14662
631. Nawa, H., Kotani, H., and Nakanishi, S. (1984) *Nature (London)* **312**, 729–734
632. Maeda, N., Kim, N.-S., Azen, E. A., and Smithies, O. (1985) *J. Biol. Chem.* **260**, 11123–11130
633. MacLeod, A. R., Houlker, C., Reinach, F. C., Smillie, L. B., Talbot, K., Modi, G., and Walsh, F. S. (1985) *Proc. Natl. Acad. Sci. U.S.A.* **82**, 7835–7839
634. Collins, T., Bonthron, D. T., and Orkin, S. H. (1987) *Nature (London)* **328**, 621–624
635. Hsu, T.-C., Shore, S. K., Seshsmma, T., Bagasra, O., and Walsh, P. N. (1998) *J. Biol. Chem.* **273**, 13787–13793
636. Chretien, S., Dubart, A., Beaupain, D., Raich, N., Grandchamp, B., Rosa, J., Goossens, M., and Romeo, P.-H. (1988) *Proc. Natl. Acad. Sci. U.S.A.* **85**, 6–10
637. Lamond, A. I. (1991) *Trends Biochem. Sci.* **16**, 452–453
638. Handa, N., Nureki, O., Kurimoto, K., Kim, I., Sakamoto, H., Shimura, Y., Muto, Y., and Yokoyama, S. (1999) *Nature (London)* **398**, 579–585
639. Bruzik, J. P., Van Doren, K., Hirsch, D., and Steitz, J. A. (1988) *Nature (London)* **335**, 559–562
640. Bonen, L. (1993) *FASEB J.* **7**, 40–46
641. Hannon, G. J., Maroney, P. A., Yu, Y.-T., Hannon, G. E., and Nilsen, T. W. (1992) *Science* **258**, 1775–1780
642. Davis, R. E., Hardwick, C., Tavernier, P., Hodgson, S., and Singh, H. (1995) *J. Biol. Chem.* **270**, 21813–21819
643. Ghetti, A., and Abelson, J. N. (1995) *Proc. Natl. Acad. Sci. U.S.A.* **92**, 11461–11464
644. Herrin, D. L., and Schmidt, G. W. (1988) *J. Biol. Chem.* **263**, 14601–14604
645. Eul, J., Graessmann, M., and Graessmann, A. (1995) *EMBO J.* **14**, 3226–3235
646. Vester, B., Nielsen, A. K., Hansen, L. H., and Douthwaite, S. (1998) *J. Mol. Biol.* **282**, 255–264
647. Bachellerie, J.-P., and Cavaillé, J. (1997) *Trends Biochem. Sci.* **22**, 257–261
648. Gaspin, C., Cavaillé, J., Erauso, G., and Bachellerie, J.-P. (2000) *J. Mol. Biol.* **297**, 895–906
648a. Jády, B. E., and Kiss, T. (2001) *EMBO J.* **20**, 541–551
649. Blum, B., Bakalara, N., and Simpson, L. (1990) *Cell* **60**, 189
650. Wilson, M. A., and Pohorille, A. (1996) *J. Am. Chem. Soc.* **118**, 6580–6587
651. Riley, G. R., Corell, R. A., and Stuart, K. (1994) *J. Biol. Chem.* **269**, 6101–6108
652. Hodges, P., and Scott, J. (1992) *Trends Biochem. Sci.* **17**, 77–81
652a. Keegan, L. P., Gallo, A., and O'Connell, M. A. (2000) *Science* **290**, 1707–1709
652b. Gerber, A. P., and Keller, W. (2001) *Trends Biochem. Sci.* **26**, 376–384
653. Navaratnam, N., Fujino, T., Bayliss, J., Jarmuz, A., How, A., Richardson, N., Somasekaram, A., Bhattacharya, S., Carter, C., and Scott, J. (1998) *J. Mol. Biol.* **275**, 695–714
654. Hersberger, M., and Innerarity, T. L. (1998) *J. Biol. Chem.* **273**, 9435–9442
655. Gray, M. W. (1996) *Proc. Natl. Acad. Sci. U.S.A.* **93**, 8157–8159
656. Giegé, P., and Brennicke, A. (1999) *Proc. Natl. Acad. Sci. U.S.A.* **96**, 15324–15329
657. Hirose, T., and Sugiura, M. (1997) *EMBO J.* **16**, 6804–6811
658. Paul, M. S., and Bass, B. L. (1998) *EMBO J.* **17**, 1120–1127
659. Maas, S., Melcher, T., Herb, A., Seeburg, P. H., Keller, W., Krause, S., Higuchi, M., and O'Connell, M. A. (1996) *J. Biol. Chem.* **271**, 12221–12226
660. Bass, B. L. (1997) *Trends Biochem. Sci.* **22**, 157–162
660a. Bass, B. L. (2002) *Ann. Rev. Biochem.* **71**, 817–846
661. Lehmann, K. A., and Bass, B. L. (1999) *J. Mol. Biol.* **291**, 1–13
662. Gerber, A., Grosjean, H., Melcher, T., and Keller, W. (1998) *EMBO J.* **17**, 4780–4789
663. Schwartz, T., Rould, M. A., Lowenhaupt, K., Herbert, A., and Rich, A. (1999) *Science* **284**, 1841–1845
664. Kable, M. L., Seiwert, S. D., Heidmann, S., and Stuart, K. (1996) *Science* **273**, 1189–1195
665. Kable, M. L., Heidmann, S., and Stuart, K. D. (1997) *Trends Biochem. Sci.* **22**, 162–166
666. Rusché, L. N., Cruz-Reyes, J., Piller, K. J., and Sollner-Webb, B. (1997) *EMBO J.* **16**, 4069–4081
667. Frech, G. C., Bakalara, N., Simpson, L., and Simpson, A. (1995) *EMBO J.* **14**, 178–187
667a. Müller, U. F., Lambert, L., and Göringer, H. U. (2001) *EMBO J.* **20**, 1394–1404
667b. Grams, J., McManus, M. T., and Hajduk, S. L. (2000) *EMBO J.* **19**, 5525–5532
667c. Schnaufer, A., Panigrahi, A. K., Panicucci, B., Igo, R. P., Jr., Salavati, R., and Stuart, K. (2001) *Science* **291**, 2159–2162
668. Horváth, A., Berry, E. A., and Maslov, D. A. (2000) *Science* **287**, 1639–1640
669. Wang, S. S., Mahendran, R., and Miller, D. L. (1999) *J. Biol. Chem.* **274**, 2725–2731
669a. Cheng, Y.-W., Visomirski-Robic, L. M., and Gott, J. M. (2001) *EMBO J.* **20**, 1405–1414
670. Petschek, J. P., Mermer, M. J., Scheckelhoff, M. R., Simone, A. A., and Vaughn, J. C. (1996) *J. Mol. Biol.* **259**, 885–890
671. Jacques, J.-P., Hausmann, S., and Kolakofsky, D. (1994) *EMBO J.* **13**, 5496–5503
672. Turner, P. (1985) *Nature (London)* **316**, 105–106
673. Mowry, K. L., and Steitz, J. A. (1987) *Science* **238**, 1682–1687
674. Spirin, A. S., and Ajtkhozhin, M. A. (1985) *Trends Biochem. Sci.* **10**, 162–165
675. Tiedge, H., Bloom, F. E., and Richter, D. (1999) *Science* **283**, 186–187
676. Strauss, E. (1999) *Science* **283**, 12–13
677. Tzagoloff, A., and Myers, A. M. (1986) *Ann. Rev. Biochem.* **55**, 249–285

References

678. Mueller, D. M., and Getz, G. S. (1986) *J. Biol. Chem.* **261**, 11756 – 11764
679. Hedtke, B., Börner, T., and Weihe, A. (1997) *Science* **277**, 809 – 811
680. Shadel, G. S., and Clayton, D. A. (1993) *J. Biol. Chem.* **268**, 16083 – 16086
681. Rossmanith, W., Tullo, A., Potuschak, T., Karwan, R., and Sbisà, E. (1995) *J. Biol. Chem.* **270**, 12885 – 12891
682. Mayer, M., Schiffer, S., and Marchfelder, A. (2000) *Biochemistry* **39**, 2096 – 2105
683. Yokobori, S.-i, and Pääbo, S. (1997) *J. Mol. Biol.* **265**, 95 – 99
684. Margossian, S. P., and Butow, R. A. (1996) *Trends Biochem. Sci.* **21**, 392 – 396
685. Ohyama, K., Fukazawa, H., Kohchi, T., Shirai, H., Sano, T., Sano, S., Umesono, K., Shiki, Y., Takeuchi, M., Chang, Z., Aota, S., Inokuchi, H., and Ozeki, H. (1986) *Nature (London)* **322**, 572 – 574
686. Lewin, R. (1985) *Science* **230**, 55
687. Sambrook, J. (1983) in *Biochemistry* (Zubay, G., ed), Addison-Wesley, Reading, Massachusetts (Chapter 28)
688. Freifelder, D. (1987) *Molecular Biology*, 2nd ed., Jones and Bartlett, Boston, Massachusetts (Chapter 23)
689. Wildeman, A. G. (1989) *Proc. Natl. Acad. Sci. U.S.A.* **86**, 2123 – 2127
690. Fanning, E., and Knippers, R. (1992) *Ann. Rev. Biochem.* **61**, 55 – 85
691. Nevins, J. R. (1991) *Trends Biochem. Sci.* **16**, 435 – 439
692. Swaminathan, S., and Thimmapaya, B. (1996) *J. Mol. Biol.* **258**, 736 – 746
693. Berget, S. M., Moore, C., and Sharp, P. A. (1977) *Proc. Natl. Acad. Sci. U.S.A.* **74**, 3171 – 3175
694. Bos, J. L., and van der Eb, A. J. (1985) *Trends Biochem. Sci.* **10**, 310 – 313
695. Hogle, J. M., Chow, M., and Filman, D. J. (1985) *Science* **229**, 1358 – 1365
696. Saito, T., Meshi, T., Takamatsu, N., and Okada, Y. (1987) *Proc. Natl. Acad. Sci. U.S.A.* **84**, 6074 – 6077
697. Lamb, R. A., and Choppin, P. W. (1983) *Ann. Rev. Biochem.* **52**, 467 – 506
698. Scholtissek, C., and Naylor, E. (1988) *Nature (London)* **331**, 215
699. Pekosz, A., He, B., and Lamb, R. A. (1999) *Proc. Natl. Acad. Sci. U.S.A.* **96**, 8804 – 8806
700. Ishihama, A., Mizumoto, K., Kawakami, K., Kato, A., and Honda, A. (1986) *J. Biol. Chem.* **261**, 10417 – 10421
701. Coffin, J. M., Hughes, S. H., and Varmus, H. E., eds. (1998) *Retroviruses*, Cold Spring Harbor Lab. Press, Cold Spring Harbor, New York

701a. Piot, P., Bartos, M., Ghys, P. D., Walker, N., and Schwartländer, B. (2001) *Nature (London)* **410**, 968 – 973

701b. Cohen, J. (2002) *Science* **296**, 2320 – 2324

702. Varmus, H. (1988) *Science* **240**, 1427 – 1434
703. Sugden, B. (1993) *Trends Biochem. Sci.* **18**, 233 – 235
704. Gallo, R. C., and Montagnier, L. (1988) *Sci. Am.* **259**(Oct), 41 – 48
705. Sanchez-Pescador, R., Power, M. D., Barr, P. J., Steimer, K. S., Stempien, M. M., Brown-Shimer, S. L., Gee, W. W., Renard, A., Randolph, A., Levy, J. A., Dina, D., and Luciw, P. A. (1985) *Science* **227**, 484 – 492
706. Ratner, L., Haseltine, W., Patarca, R., Livak, K. J., Starcich, B., Josephs, S. F., Doran, E. R., Rafalski, J. A., Whitehorn, E. A., Baumeister, K., Ivanoff, L., Petteway, S. R., Jr., Pearson, M. L., Lautenberger, J. A., Papas, T. S., Ghrayeb, J., Chang, N. T., Gallo, R. C., and Wong-Staal, F. (1985) *Nature (London)* **313**, 277 – 284
707. Varmus, H. E. (1985) *Nature (London)* **314**, 583 – 584
708. Vaccaro, J. A., Singh, H. A., and Anderson, K. S. (1999) *Biochemistry* **38**, 15978 – 15985
709. Gabbara, S., Davis, W. R., Hupe, L., Hupe, D., and Peliska, J. A. (1999) *Biochemistry* **38**, 13070 – 13076
710. Litvak, S., Sarih-Cottin, L., Fournier, M., Andreola, M., and Tarrago-Litvak, L. (1994) *Trends Biochem. Sci.* **19**, 114 – 118
711. Litvak, S., and Araya, A. (1982) *Trends Biochem. Sci.* **7**, 361 – 364
712. Fossé, P., Mougel, M., Keith, G., Westhof, E., Ehresmann, B., and Ehresmann, C. (1998) *J. Mol. Biol.* **275**, 731 – 746
713. Gao, H.-Q., Sarafianos, S. G., Arnold, E., and Hughes, S. H. (1999) *J. Mol. Biol.* **294**, 1097 – 1113

713a. Sevilya, Z., Loya, S., Hughes, S. H., and Hizi, A. (2001) *J. Mol. Biol.* **311**, 957 – 971

714. Yu, H., Jetzt, A. E., Ron, Y., Preston, B. D., and Dougherty, J. P. (1998) *J. Biol. Chem.* **273**, 28384 – 28391
715. Bujacz, G., Jaskólski, M., Alexandratos, J., Wlodawer, A., Merkel, G., Katz, R. A., and Skalka, A. M. (1995) *J. Mol. Biol.* **253**, 333 – 346
716. Andrake, M. D., and Skalka, A. M. (1996) *J. Biol. Chem.* **271**, 19633 – 19636
717. Greenwald, J., Le, V., Butler, S. L., Bushman, F. D., and Choe, S. (1999) *Biochemistry* **38**, 8892 – 8898
718. Maignan, S., Guilloteau, J.-P., Zhou-Liu, Q., Clément-Mella, C., and Mikol, V. (1998) *J. Mol. Biol.* **282**, 359 – 368

718a. Craigie, R. (2001) *J. Biol. Chem.* **276**, 23213 – 23216

719. Gallo, R. C. (1986) *Sci. Am.* **255**(Dec), 88 – 98
720. Gallo, R. C. (1987) *Sci. Am.* **256**(Jan), 47 – 56
721. Emerman, M., and Malim, M. H. (1998) *Science* **280**, 1880 – 1884
722. Frankel, A. D., and Young, J. A. T. (1998) *Ann. Rev. Biochem.* **67**, 1 – 25
723. Mikaélian, I., Krieg, M., Gait, M. J., and Karn, J. (1996) *J. Mol. Biol.* **257**, 246 – 264
724. Wang, Z., and Rana, T. M. (1998) *Biochemistry* **37**, 4235 – 4243
725. Ivanov, D., Kwak, Y. T., Nee, E., Guo, J., García-Martínez, L. F., and Gaynor, R. B. (1999) *J. Mol. Biol.* **288**, 41 – 56

725a. Jeang, K.-T., Xiao, H., and Rich, E. A. (1999) *J. Biol. Chem.* **274**, 28837 – 28840

726. Cullen, B. R., and Malim, M. H. (1991) *Trends Biochem. Sci.* **16**, 346 – 350
727. Charpentier, B., Stutz, F., and Rosbash, M. (1997) *J. Mol. Biol.* **266**, 950 – 962
728. Barnham, K. J., Monks, S. A., Hinds, M. G., Azad, A. A., and Norton, R. S. (1997) *Biochemistry* **36**, 5970 – 5980
729. Echarri, A., González, M. E., and Carrasco, L. (1996) *J. Mol. Biol.* **262**, 640 – 651

729a. Arold, S. T., and Baur, A. S. (2001) *Trends Biochem. Sci.* **26**, 356 – 363

729b. Pomerantz, R. J. (2002) *Nature (London)* **418**, 594 – 595

729c. Segura-Totten, M., and Wilson, K. L. (2001) *Science* **294**, 1016 – 1017

730. Greene, W. C. (1991) *N. Engl. J. Med.* **324**, 308 – 317
731. Guyadar, M., Emerman, M., Sonigo, P., Clavel, F., Montagnier, L., and Alizon, M. (1987) *Nature (London)* **326**, 662 – 669
732. Overbaugh, J., Donahue, P. R., Quackenbush, S. L., Hoover, E. A., and Mullins, J. I. (1988) *Science* **239**, 906 – 910
733. Nerenberg, M., Hinrides, S. H., Reynolds, R. K., Khoury, G., and Jay, G. (1987) *Science* **237**, 1324 – 1329
734. Haase, A. T. (1986) *Nature (London)* **322**, 130 – 136
735. Turner, B. G., and Summers, M. F. (1999) *J. Mol. Biol.* **285**, 1 – 32
736. Miedema, F., and Klein, M. R. (1996) *Science* **272**, 505 – 506

736a. Ezzell, C. (2002) *Sci. Am.* **286**(Jun), 40 – 45

737. Roberts, J. D., Bebenek, K., and Kunkel, T. A. (1988) *Science* **242**, 1171 – 1173
738. Hill, C. M., and Littman, D. R. (1996) *Nature (London)* **382**, 668 – 669
739. Cohen, J. (1995) *Science* **270**, 917 – 918
740. Kidwell, M. G., and Lisch, D. R. (1998) *Nature (London)* **393**, 22 – 23
741. Kazazian, H. H., Jr. (2000) *Science* **289**, 1152 – 1153
742. Moffat, A. S. (2000) *Science* **289**, 1455, 1457
743. Ovchinnikov, I., Rubin, A., and Swergold, G. D. (2002) *Proc. Natl. Acad. Sci. U.S.A.* **99**, 10522 – 10527
744. Perl, A., Colombo, E., Samoilova, E., Butler, M. C., and Banki, K. (2000) *J. Biol. Chem.* **275**, 7261 – 7272
745. Southworth, M. W., Adam, E., Panne, D., Byer, R., Kautz, R., and Perler, F. B. (1998) *EMBO J.* **17**, 918 – 926
746. Strauss, E. (2000) *Science* **289**, 374

Study Questions

(Additional questions are located on p. 1738.)

1. Describe the sequence of events involved in the initiation of transcription in *E. coli*. As part of your answer, describe those features that must be present in a gene for proper recognition and transcription by the RNA polymerase.
2. How does transcription in eukaryotes differ from that in prokaryotes?
3. In *E. coli* precise spacing between the conserved -35 and -10 (Pribnow) promoter elements has been found to be a critical determinant of promoter strength. What does this suggest about the interaction between RNA polymerase and these conserved sequences in the DNA?

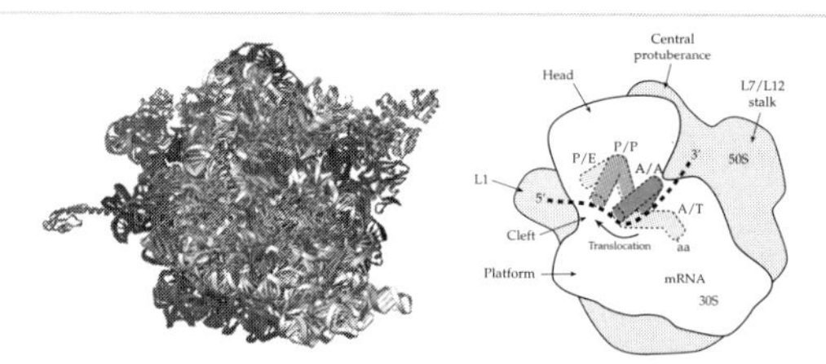

The ~15,000 ribosomes in a bacterial cell synthesize over 4000 proteins following the genetic code in messenger RNA molecules. A ribosome (left) consists of two large subunits, each composed largely of ribosomal RNA, whose folded chains can be seen. About 100 proteins are bound, largely to solvent-exposed surfaces, but with extended "tails" protruding into the ribosome. The messenger RNA (mRNA; right) moves through the ribosome between the large subunits. Amino acids, activated for reaction, are carried into the ribosomes by transfer RNAs (green) which move consecutively from A/T to A/A, A/P, P/P, and P/E sites. They insert their activated amino acids into the growing polypeptide chain in the P site of the 50S subunit. Image of ribosome[33a] courtesy of the authors.

Contents

29 Ribosomes and the Synthesis of Proteins

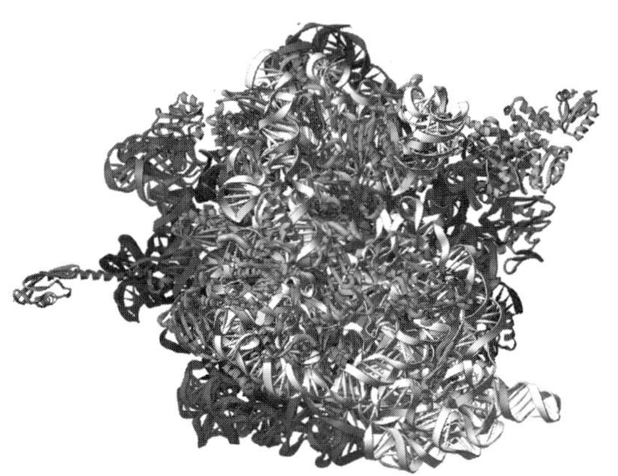

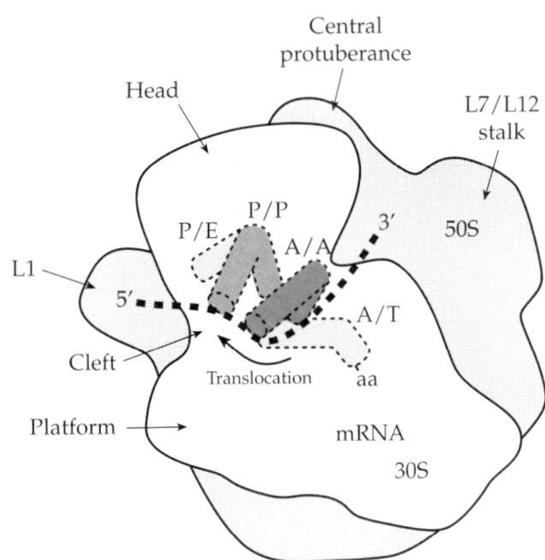

The many thousands of proteins present in each cell are made within the ribosomes, which are able to read each specific mRNA that comes their way. While ribosomes appear as little more than blurred dots in most electron micrographs, the 15,000 ribosomes of one cell of *E. coli* represent one-fourth of the total mass of the cell. Eukaryotic cells contain many times more of these little molecular machines. When ribosomes were first observed in the early 1950s,[1–3] nobody could imagine either their composition or their function. Less than 50 years later (1999) their complete three-dimensional structure was known at nearly atomic resolution, and the function of ribosomes in protein synthesis was quite well understood. However, the structure could not have been obtained without the development of a whole range of new methods.

As electron microscopy developed, the fuzzy granules in micrographs assumed well-recognized forms. By the 1970s, the approximate shapes of the two ribosomal subunits were known, locations of several ribosomal proteins had been established, and binding sites of transfer RNAs and other features of ribosomes were being mapped.[2,4–7] The resulting picture of the ribosome structure, which is shown schematically in Fig. 29-1, is quite similar to the present-day view. Later, three-dimensional images were reconstructed from electron micrographs (electron tomography),[8,9] and cryo-electron microscopy provided detailed images at a resolution of ~2 nm.[10]

In 1950, when the study of ribosomes began, no methods for determining the sequences of amino acids in proteins or of nucleotides in nucleic acids existed.[11] Sanger published the sequences of the two short chains of insulin in 1953, and the first transfer RNA sequence was published by Holley in 1965.[21] Nevertheless, by 1980 the Wittmanns and coworkers in Berlin had sequenced 53 of the *E. coli* ribosomal proteins[4,22] (Table 29-2), and the three rRNA molecules had also been sequenced.[22a,b,c] In 1950, X-ray crystallography of proteins was still in its infancy; the structure of myoglobin was not determined until 1960. Ribosomal proteins proved hard to crystallize, the first structure being solved in 1980.[23] NMR structural analysis yielded several structures including that of L30 (Fig. 3-25A). In recent years, high-resolution structures of many additional ribosomal proteins in their free forms have been established as have the structures of most of the proteins bound into ribosomes.[24]

The first crystals of bacterial ribosomes in a three-dimensional lattice suitable for study by X-ray diffraction at a resolution of ~1 nm[27] were obtained by Yonath in about 1980.[6,25,26] Now atomic structures are being established at a resolution of 0.3–0.1 nm, or less.[17–19,28–33g] However, such progress would have been impossible without information about ribosomes obtained from improved cryo-electron microscopy,[10,20,33e,34–37a] phylogenetic analysis of ribosomal RNAs,[38–39b] mutational analysis,[40,41] neutron scattering,[42–47] chemical and photochemical crosslinking,[48–53] photoaffinity labeling,[54] immunological labeling,[55] chemical footprinting,[56,57] fluorescence resonance energy transfer (FRET),[58] mass spectrometry,[59,60] and study of the effects of toxic proteins (Box 29-A)[61] and antibiotics (Box 29-B).

Why is the ribosome so large and complex? Aside from the fact that it must form the peptide linkages, it must translate the genetic code in the mRNA into the correct amino acid sequence for each of the thousands of proteins present in the cell. The process takes place

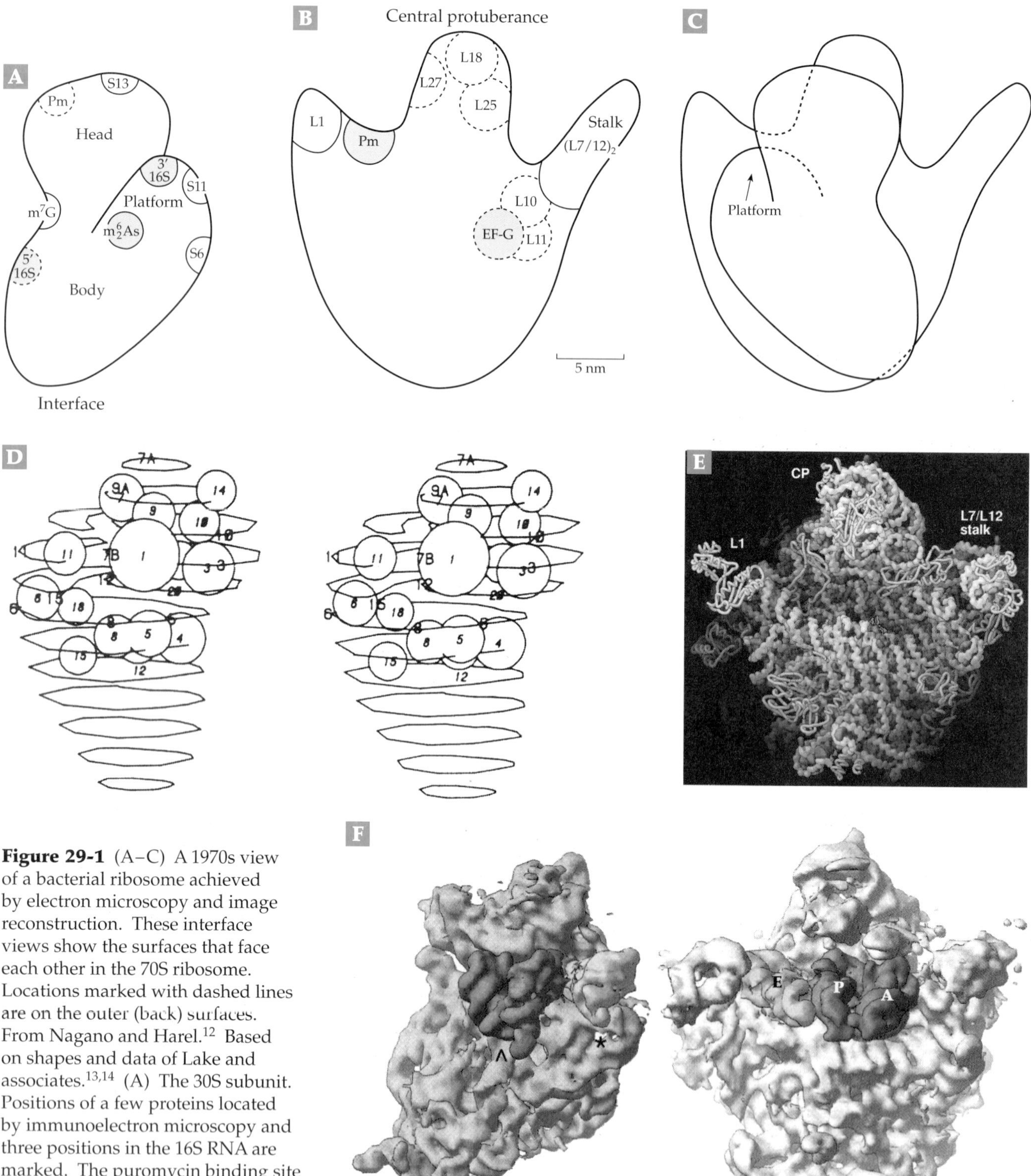

Figure 29-1 (A–C) A 1970s view of a bacterial ribosome achieved by electron microscopy and image reconstruction. These interface views show the surfaces that face each other in the 70S ribosome. Locations marked with dashed lines are on the outer (back) surfaces. From Nagano and Harel.[12] Based on shapes and data of Lake and associates.[13,14] (A) The 30S subunit. Positions of a few proteins located by immunoelectron microscopy and three positions in the 16S RNA are marked. The puromycin binding site labeled Pm was mistakenly thought to be near the peptidyltransferase center. (B) The 50S subunit. Only a few positions of many located by a variety of techniques are marked. EF-G marks the site at which elongation factor G (see Fig. 29-12) binds. A prominent feature is the "stalk", designated St in some of the drawings. It was early shown to be formed by two copies each of the nearly identical L7 and L12. The stalk is flexible and in many circumstances may be folded down or not visible as in the X-ray structures of F–H. (C) The 70S ribosome. The 50S subunit is oriented as in (A) while the 30S subunit has its outer face toward the viewer. (D) Stereoscopic view of a neutron scattering map of the 30S subunit of an *E. coli* ribosome. The proteins studied are represented as spheres with volumes corresponding to those of the anhydrous proteins. Also marked on the map are positions of proteins located by immunoelectron microscopy as mapped by Kahan *et al.*[15] Figure from Ramakrishnan *et al.*[16] Courtesy of V. Ramakrishnan. (E) Model of the 50S subunit from *Haloarcula marismortui*. From

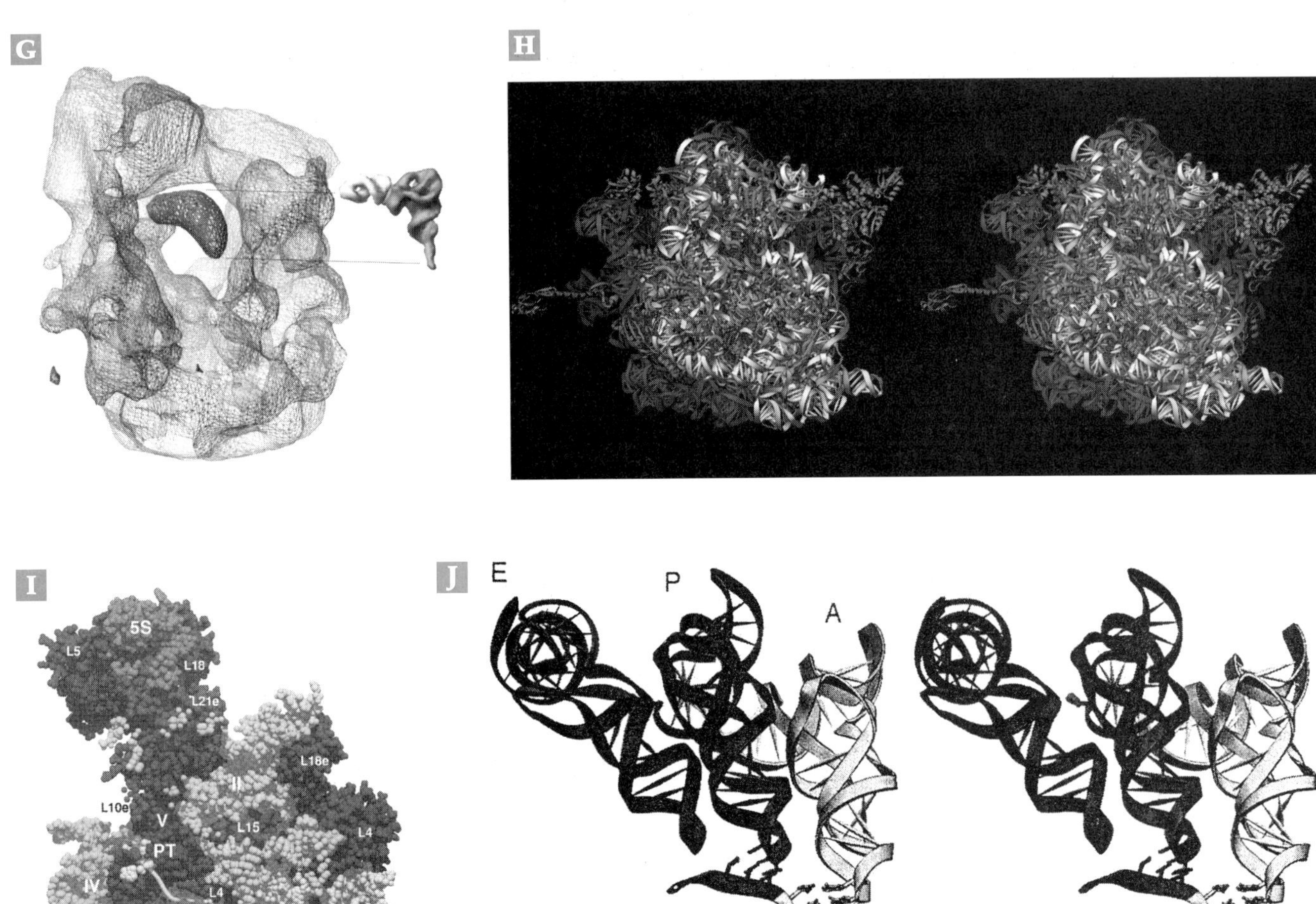

Ban *et al.*[17] Courtesy of T. A. Steitz. The peptidyltransferase center is marked by the green image of the transition state inhibitor shown in Fig. 29-13. (F) Model of three tRNAs bound to a ribosome from *Thermus thermophilus* in the A (aminoacyl), P (peptidyl), and E (exit) sites. These are based on 0.75-nm X-ray data and a number of difference electron density maps. The 3'-CCA end of the A-site tRNA is not modeled but is marked "∧". Views are left, facing the inner surface of the 30S subunit; right, facing the inner surface of the 50S subunit. (G) Schematic side view of a ribosome showing a molecule of tRNA bound in the A site between the 30S and 50S subunits. The anticodon of the tRNA is base-paired with mRNA in the "decoding site" on the 30S subunit. The 3'-CCA end with attached aminoacyl group lies in the peptidyltransferase site in the 50S subunit. (F) and (G) are courtesy of Cate *et al.*[18] (H) Stereoscopic view of a model of the 70S ribosome from *T. thermophilus*. The 30S subunit (lighter) is toward the viewer. Courtesy of Harry F. Noller and Albion Baucom. (I) Section through the 0.24 nm-resolution model of the 50S subunit shown in (E). The modeled path of the polypeptide chain through the exit tunnel is marked. Courtesy of Nissen *et al.*[19] (J) Stereo diagram of the relative orientations of the A-, P-, and E-tRNAs and mRNA showing codon–anticodon interactions and the kink between the A and P codons. (H) and (J) courtesy of Yusupov *et al.*[33a] (K) Stereoscopic view of tRNAs in the P site and in an overlapping P/E site as observed by cryo-electron microscopy at a resolution of 0.5 nm in an image of the 70S ribosome at 1.5 nm resolution. The anticodon arms are to the left. Two tRNA molecules are not present simultaneously but their images have been presented together. From Agrawal *et al.*[20] Courtesy of Rajendra Agrawal.

TABLE 29-1
The Composition of Ribosomes

Prokaryotic[a]		Eukaryotic[b]	
Component	Mass, kDa	Component	Mass, kDa
Small (30S) subunit	850	Small (40S) subunit	1440
16S RNA	500	18S RNA	700
Proteins (21)	350 (total)	Proteins (~30)	740
Large (50S) subunit	1450	Large (60S) subunit	2800
23S RNA	950	28S RNA	1700
5S RNA	40	5.8S RNA	51
		5S RNA	39
Proteins (32–34)	460	Proteins (~46)	1010
Complete (70S) ribosome	2300	Complete (80S) ribosome	4240

[a] Data from Wittmann, H. G. (1982) *Ann. Rev. Biochem.* **51**, 155–183. Based on sequences of all components. Presence of spermine, K^+, etc., may add 10%.
[b] Data from Freifelder, D. (1983) *Molecular Biology*, 2nd ed., Jones and Bartlett, Boston, Massachusetts (p. 419) and Mao, H., and Williamson, J. R. (1999) *J. Mol. Biol.* **292**, 345–349.

The pairing of codons and anticodons required for insertion of the correct amino acid into the growing polypeptide chain is often referred to as **decoding** of the gene sequence. However, an equally important part of the decoding is the attachment of the correct amino acid to its corresponding **cognate tRNA**. This occurs in the cytoplasm and also in the nucleus.[62] The base pairing of tRNAs and mRNA, which follows, occurs in the **decoding center** on the 30S ribosomal subunit. Both the A and P sites of tRNA-binding and the decoding center are formed by folds of the 16S RNA. The peptide bond formation takes place at the opposite ends of the tRNA molecules in the **peptidyltransferase center** of the 50S subunit (see Figs. 29-1, 29-14). As pointed out in Chapter 12 (top of p. 650), peptidyltransferase is a **ribozyme**. Its active site consists entirely of segments of the 23S RNA (see Fig. 29-14). Another important site is the **GTPase activating center**, at which the 23S

in several steps: (1) **Initiation** of protein synthesis in bacteria requires binding of the 30S ribosomal subunit to an mRNA molecule, location of the start signal (initiation codon and nearby Shine–Dalgarno sequence, shown in Fig. 29-2 and Eq. 29-8), and binding of the initiator tRNA carrying formylmethionine. The 30S complex must then bind to the 50S ribosomal subunit. (2) **Elongation** of the polypeptide chain in the resulting complete 70S ribosome ensues with binding of the appropriate aminoacyl tRNA to the next codon in the 5' → 3' direction. Base pairs form between the anticodon of the tRNA and the mRNA codon that lies in the **aminoacyl (A) site**; the peptide bond is then formed by the **peptidyltransferase** reaction. This reaction is followed by **translocation**, movement of the initiator tRNA into an exit site at the same time that the second tRNA (together with its mRNA codon and the attached growing peptide chain) moves into the **peptidyl (P) site**. The elongation cycle is repeated until the peptide chain is complete. (3) **Termination** of translation involves release of the completed protein and preparation of the ribosomal subunits for another cycle. The entire process is powered by the hydrolysis of ATP and GTP. The ATP is utilized in a three-step process for the activation of the amino acids, which become linked to the appropriate tRNAs (Eqs. 17-36 and 29-1).[61a] If the inorganic pyrophosphate that is formed is hydrolyzed, two molecules of ATP are required for activation of each amino acid molecule. In addition, at least two molecules of GTP are hydrolyzed to GDP and inorganic phosphate within the ribosome for each peptide linkage formed.

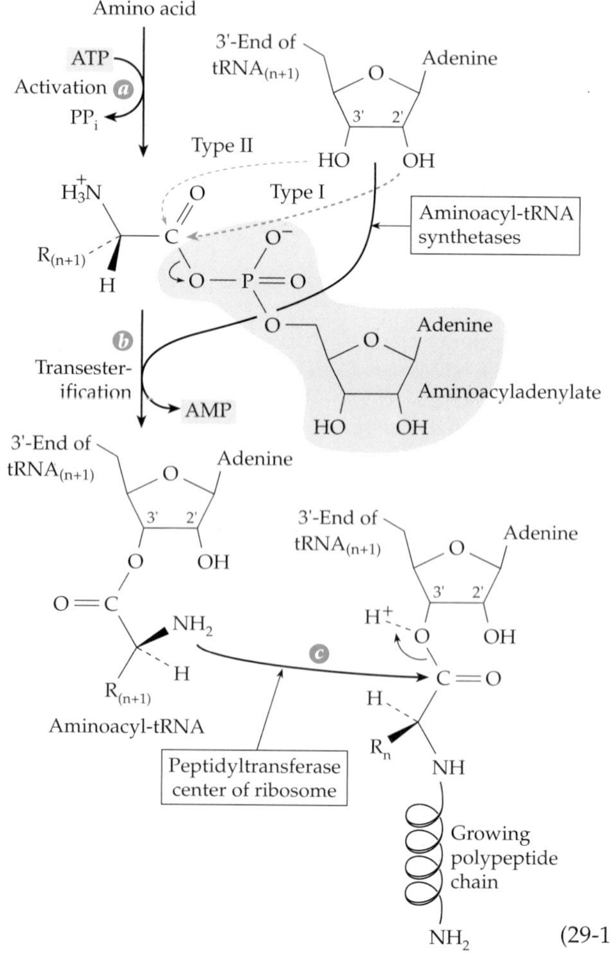

(29-1)

RNA interacts with specific G proteins known as **initiation, elongation,** and **termination factors**.

The proteins in a ribosome may help to hold the RNA into conformations that are correct for its functions. They may also catalyze conformational alterations during the various steps of the translation process. In addition, the proteins may help provide binding sites for substrate molecules and participate in regulatory activities. Both the tRNA **exit (E) site** and the **tunnel** through which the polypeptide chain leaves the ribosome are composed, in part, of ribosomal proteins.

A. The Architecture of Ribosomes

Ribosomes of *E. coli* each have a mass of ~ 2.3 x 10^6 daltons and are ~ 65% RNA and 35% protein. Ribosomes of eukaryotic organisms are larger (~4.3 x 10^6 daltons) and consist of ~50% RNA and 50% protein. Under some conditions such as a low Mg^{2+} concentration complete bacterial ribosomes, called **70S ribosomes**, dissociate into two subunits of unequal size, which are known as **30S** and **50S ribosomal subunits**. The larger 50S subunit is about twice the size of the smaller one (Table 29-1). The small 30S ribosomal subunit contains the 16S rRNA, a chain of ~1500–1700 nucleotides (nt) that, if fully extended, would stretch to a length of over 500 nm. In addition to the highly folded RNA molecule, the 30S subunit contains 21 proteins, each one unique in its amino acid composition and sequence (Table 29-2). Many of these proteins, which are designated S1, S2, S3, etc., are of relatively low molecular mass. Many are strongly basic. They contain numerous lysine and arginine residues, many of which are able to interact with RNA in the ribosome. However, neutral and acidic proteins are also present. The 50S ribosomal subunit contains the ~ 2900 nt 23S rRNA, the ~120 nt 5S RNA, and about 31–34 proteins, two of which (L7 and L12) are present as two copies each. The composition of ribosomes is variable, but most proteins are present in a strict 1:1 ratio. Others may be lacking in some of the ribosomes. Some proteins bind to the ribosomes transiently during their function in protein synthesis as do certain proteins with functions other than protein synthesis. In both subunits the RNA molecules form the internal core. Proteins are largely found on the solvent-exposed surfaces. Some of them form the stalk and other features.[17] They often have globular domains with extended tails that interact with the ribosomal RNA.

Eukaryotic ribosomes are not only larger but also (Table 29-1) contain more protein subunits than do those of bacteria: ~30 for the small subunit and 49 for the large subunit.[63] However, the number of essential proteins may be the same. Both eukaryotic ribosomal proteins and rRNA molecules are larger than those of bacteria. Bacterial ribosomes are ~22 nm in diameter and ~30 nm in the third dimension. Eukaryotic ribosomes are of the order of 1.17 times larger in linear dimensions. Ribosomes of chloroplasts resemble those of eubacteria such as *E. coli* but contain a few more proteins.[64] Mammalian mitochondrial ribosomes also resemble those of bacteria in many respects.[65] However, their RNA chains are shorter and they contain more proteins.[66,66a] The protein content is ~66% compared with ~35% for *E. coli* ribosomes.

1. Ribosomal RNA

The sequences of all three pieces of RNA in the *E. coli* ribosomes are known as are those from many other species. These include eukaryotic mitochondrial, plastid, and cytosolic rRNA. From the sequences alone, it was clear that these long molecules could fold into a complex series of hairpin loops resembling those in tRNA. For example, the 16S rRNA of *E. coli* can fold as in Fig. 29-2A and eukaryotic 18S RNA in a similar way (Fig. 29-4).[38,39,67–69] The actual secondary structures of 16S and 18S RNAs, within the folded molecules revealed by X-ray crystallography, are very similar to that shown in Fig. 29-2A. Ribosomal RNAs undergo many posttranscriptional alterations. Methylation of 2'-hydroxyls and of the nucleic acid bases as well as conversion to pseudouridines (pp. 1638–1641) predominate over 200 modifications, principally in functionally important locations that have been found in human rRNA.[69a]

Chemical modification and crosslinking. Before high-resolution X-ray data were available, two major biochemical approaches were used to deduce the secondary structures of ribosomal RNAs.[38,39] The first was the application of chemical reagents and enzymes that modify the RNA. Crosslinking reagents were used to establish pairs of nucleotides that lie close together in the three-dimensional structure. Cleavage by specific endonucleases was used to establish whether a region of the molecule is double-helical or single-stranded.[68] Nucleases were also used to clip out base-paired fragments, which were separated, denatured, and sequenced. This revealed both hairpin loops and pairings between regions that are far apart in the primary sequence. The ability of nucleic acid bases to undergo specific chemical reactions at positions not involved in base pairing was used to establish whether or not a given base was actually paired.[67,69] Thus, every position in *E. coli* 16S RNA was probed by reactions of dimethylsulfate with adenine at N1 and cytosine at N3, reaction of kethoxal (Eq. 5-16) with guanine at N1 and N2', and by reaction of a carbodiimide with uracil at N3 and with guanine at N1.[67]

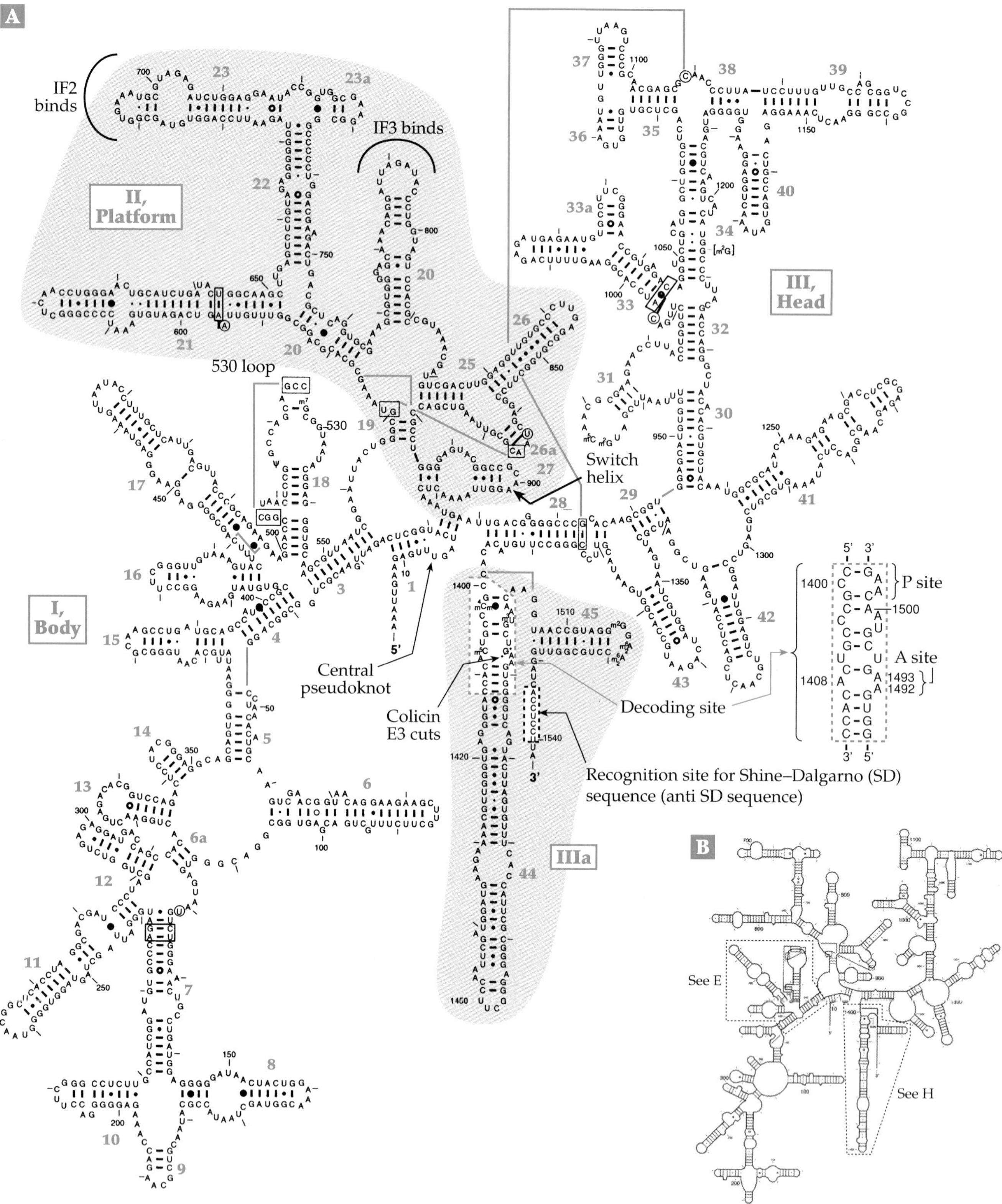

Figure 29-2 (A) Secondary structure model for the 1542-residue *E. coli* 16S rRNA based on comparative sequence analysis.[73a] Dots indicate G•U or A•G pairs; dashes indicate G•C or A•U pairs. Strongly implied tertiary interactions are shown by solid green lines. Helix numbering according to Brimacombe. Courtesy of Robin Gutell. (B) Simplified schematic drawing of type often used. (C) Positions of the A, P, and E sites on the 30S ribosomal subunit from Carter *et al.*[70] (D) Stereoscopic view of the three-dimensional fold of the 16S RNA from *Thermus thermophilus* as revealed by X-ray structural analysis at 0.3 nm resolution. Features labeled are the head (H), beak (Be), neck (N), platform (P), shoulder (Sh), spur (Sp), and body (Bo). (E-H) Selected parts of the 16S RNA. In (E) and (F) the helices are numbered as in (A). (F) and (H) are stereoscopic views. The decoding site

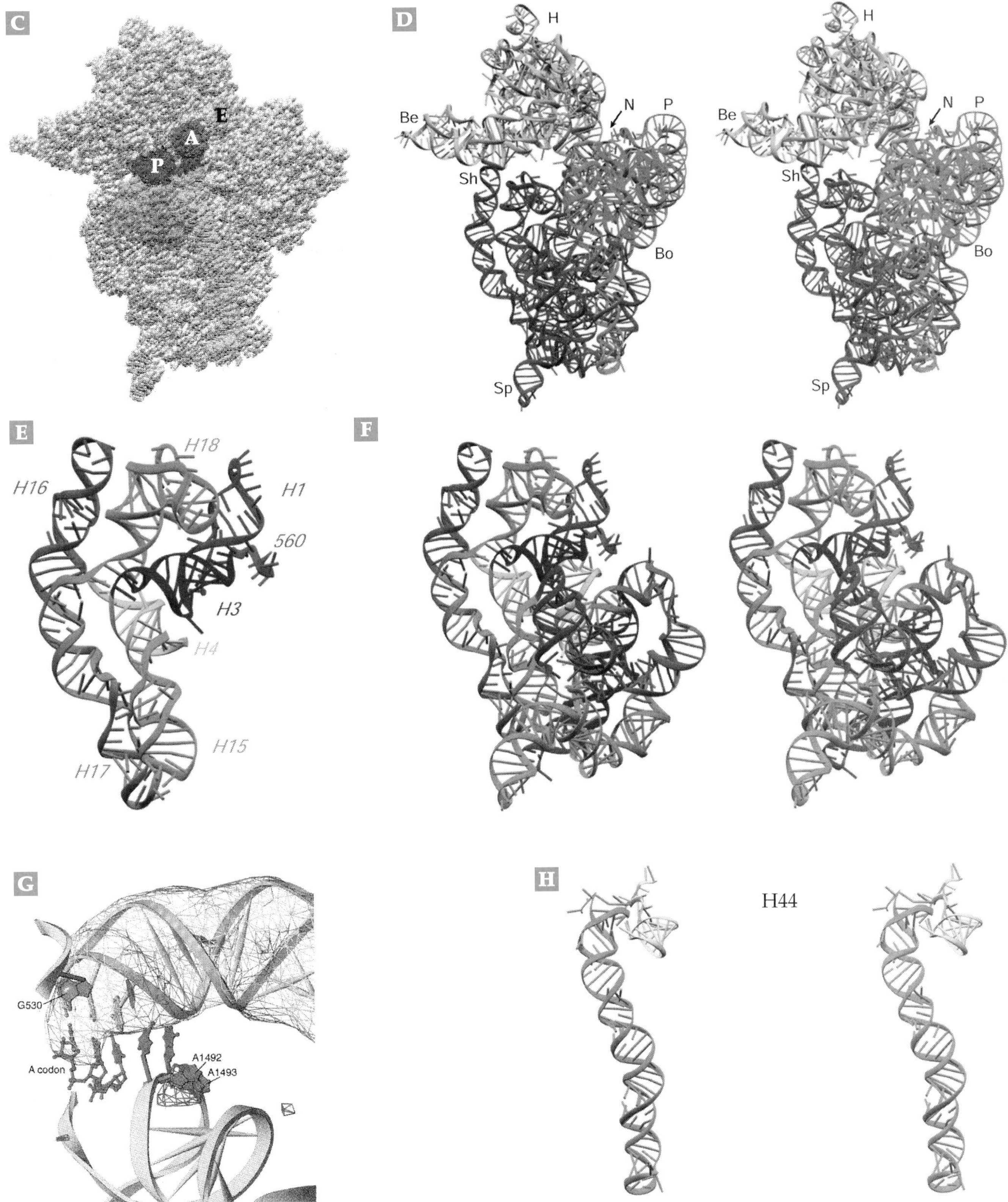

is located at the upper end of helix 44. (G), (H). In (G) the electron density difference observed upon binding of tRNA into the A site is displayed as a Fourier difference map (at 0.7-nm resolution). The molecular model of the tRNA with its anticodon paired with a codon from mRNA is superimposed. Two positions of bases A1492 and A1493 are shown as they are found in the presence and absence of paromomycin. A patch of negative density can be seen near the A1492 and A1493 labels, indicating that these groups may rearrange to interact with the minor groove of the codon-anticodon helix when the A-tRNA is bound. See also p. 1690. Courtesy of Yusupov *et al.*[33a] (D) through (F) and (H) are from Wimberly *et al.*[33] Courtesy of Venki Ramakrishnan.

Phylogenetic comparison. This technique, also called comparative sequence analysis, has proved very powerful.[38,39,39a,b,71,72] An example is illustrated in Fig. 29-3. Here a loop from 23S RNA of *E. coli* is shown and is compared with the sequence of 26S RNA from the fungus *Physarum polycephalum*.[38] Wherever the latter differs from the *E. coli* sequence, the substituted base is indicated in a box. These square boxes, which are concentrated in base-paired regions, indicate compensatory changes for which there is usually a second change that preserves base pairing in a double helical region. The studies also showed clearly that bacterial 16S RNA is homologous with eukaryotic 18S RNA,[68,69] with 17S RNA of dinoflagellates,[73] and also with 12S RNA of human mitochondria. Likewise, 23S RNA of bacteria corresponds to 28S RNA of eukaryotes.

Structural domains in 16S ribosomal RNA. Three major compact structural domains, 5', central, and 3', can be distinguished in 16S RNA.[33] An extended subdomain is also present at the 3' end. These are indicated on Fig. 29-2 as I, II, III, and IIIa. The double helical segments are also numbered. Ribosomal RNA molecules must be folded into compact forms to fit into the envelope of the ribosomal subunits. The individual structural domains form independent globular cores to which several proteins apiece are bound. Domains I, II, and III form the body, platform, and head, respectively (Fig. 29-1A). Relatively minor changes in conformation accompany the incorporation of the rRNA molecules into the ribosomes.[67] On this basis, and taking account of all available data, attempts were made for many years to predict a three-dimensional structure.[12,74–76] One of these[77] is portrayed in Fig. 5-32A as a series of cylinders representing the 45 double-helical segments suggested by the structure of Fig. 29-2A. This can be compared with the X-ray based model shown in Fig. 29-2B.

23S rRNA. The large RNA of the 50S subunit consists of six structural domains.[5] Its secondary structure is shown in Fig. 29-4. As with 16S RNA each domain is tightly folded. However, the domains are interdigitated in such a way that they form a single monolithic structural unit.[17] Nevertheless, there are distinct catalytic sites, as described in Section 4. Like proteins, which are able to undergo conformational alterations that usually involve some rearrangement in their internal hydrogen-bonding patterns, these large RNA molecules may also assume alternative conformations. Conformational changes may involve not only alternative hydrogen bonding patterns but also alternative base-pairing.[77a] Such changes may be essential to the functioning of ribosomes[86] and may also accompany maturation of pre-rRNAs.[87] Eukaryotic 28S RNAs have basically the same structures as the

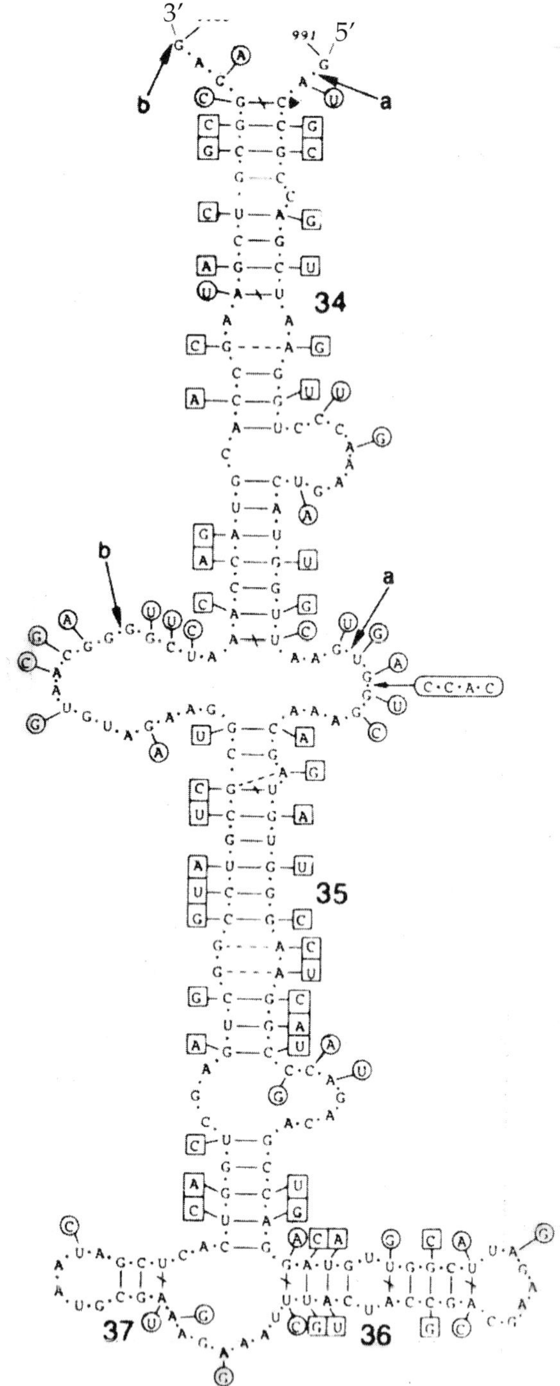

Figure 29-3 Example of phylogenetic comparisons in ribosomal RNA. The diagram shows helices 34–37 of *E. coli* 23S RNA, compared with the corresponding region of *Physarum polycephalum* 26S RNA. The diagram depicts the *E. coli* helices, with changes in the *P. polycephalum* sequences denoted by symbols in boxes on the side. Bases in square boxes are compensating; those in round boxes are mismatching or in single-stranded regions. Solid triangles denote deletions, while bases with arrows indicate insertions. Dotted lines or "crossed-out" base pairs denote modified base-pairing in *P. polycephalum*. The letters "a" and "b" indicate the termini of RNA fragments isolated as a base-paired complex. From Brimacombe.[38]

23S RNAs of bacteria but have been expanded by insertion of additional nucleotides at many places.[10,36,79]

Most of the chemical activity of ribosomes occurs in the interface between the 30S and 50S subunits. Entrance and exit tunnels for both mRNA and the aminoacylated tRNAs are formed between these subunits. The mRNA apparently moves across the platform as the tRNAs move from A to P to E sites experiencing codon selection (decoding) and peptidyltransferase activity. Many loop ends from 16S RNA interact with those of 23S RNA.[41,88]

5S rRNA. This ~120-nucleotide molecule organizes one domain of the 50S ribosomal subunit.[89] Extensive phylogenetic comparisons of 5S RNA sequences led to the secondary structure shown in Fig. 29-5.[90,91] The three-dimensional structure, as seen in a ribosome, is also shown in this figure. Study of base-pairing possibilities suggests that 5S RNA can exist in more than one conformation.[90,92] In a possible second conformation the sequence GUGUGGGG (residues 79–86) pairs in an antiparallel fashion with the sequence CCCCAUGC (residues 35–42), with loss of base pairing in stem 4 (Fig. 29-5). A structure somewhat similar to that of 5S RNA is probable for eukaryotic 5.8S RNA.[94] Nearly a thousand different prokaryotic and eukaryotic 5S RNA sequences have been compared.[95] From them **phylogenetic trees**, which suggest evolutionary pathways between species, have been constructed.[96] Sequences of 16S RNA have been used in a similar way (Fig. 1-5).[97]

2. Ribosomal Proteins

Ribosomal proteins are soluble in concentrated salt solutions. Most of them can be dissolved without damage by buffers containing 2 M LiCl and can then be separated by electrophoresis or ion exchange chromatography and gel filtration (molecular sieving).[98] Although many of them are quite insoluble and are often unstable, all ribosomal proteins of *E. coli* (Table 29-2) have been separated and sequenced, mainly by Wittmann-Liebold and coworkers.[22] The ribosomal proteins of other bacteria usually resemble those of *E. coli*.[99] The more numerous eukaryotic ribosomal proteins have also been isolated and studied individually.[100] Many of these 84 proteins appear to correspond directly in properties and functions to those of *E. coli*.[100a] As with ribosomal RNAs, the sizes of the eukaryotic proteins have been expanded.[101] Mitochondria have their own set of ribosomal proteins, which are more numerous than those of either *E. coli* or yeast.[65–66a,102–102c] Pure individual ribosomal proteins are now produced from the cloned genes as are 16S and 23S ribosomal RNAs.

Most ribosomal proteins are folded into compact forms, much of whose surfaces are accessible to added reagents. However, X-ray structures have revealed that parts of some proteins penetrate deeply into the RNA core.[17] Much of the RNA is also accessible from the outside, and the ribosome contains ~50% of its mass as internal hydration. A ribosome usually contains only one molecule of each kind of protein with the exception of proteins L7 and L12 of the large subunit. There are two of each. Sequencing of the 120-residue proteins from *E. coli* shows that L7 is

TABLE 29-2
Ribosomal Proteins from *E. coli*[a]

Proteins of 30 S Ribosomal subunits			Proteins of 30 S Ribosomal subunits		
Designation	Mass, kDa	Binding[b]	Designation	Mass, kDa	Binding[b]
S1	61.2		L1	24.6	
S2	26.6		L2	29.4	+
S3	25.8		L3	22.3	
S4	23.1	+	L4	22.1	
S5	17.5		L5	20.2	
S6	15.7		L6	18.8	+
S7					
(strain K)	19.7	+	L7	12.2	
(strain B)	17.1	+			
S8	14.0	+	L8		
S9	14.6		L9	15.5	
S10	11.7		L10	17.7	
S11	13.7		L11	14.9	
S12	13.6		L12	12.2	
S13	13.0		L13	16.0	
S14	11.1		L14	13.5	
S15	10.0	+	L15	15.0	
S16	9.2		L16	15.3	+
S17	9.6	+	L17	14.4	+
S18	8.9		L18	12.8	+
S19	10.3		L19	13.0	+
S20	9.6	+	L20	13.4	+
S21	8.4		L21	11.6	
			L22	12.2	
Total mass 350 (strain K)			L23	11.0	+
			L24	11.2	+
			L25	10.7	+
			L26 = S20	9.6	
			L27	9.0	
			L28	8.9	
			L29	7.3	
			L30	6.4	
			L31	7.0	
			L32	6.3	
			L33	6.3	
			L34	5.4	
			Total mass 460[c]		

[a] Molecular masses from Wittmann, H. G. (1982) *Ann. Rev. Biochem.* **51**, 155–183

[b] A plus sign indicates direct binding to ribosomal RNA.

[c] Four copies of L7/L12 are assumed.

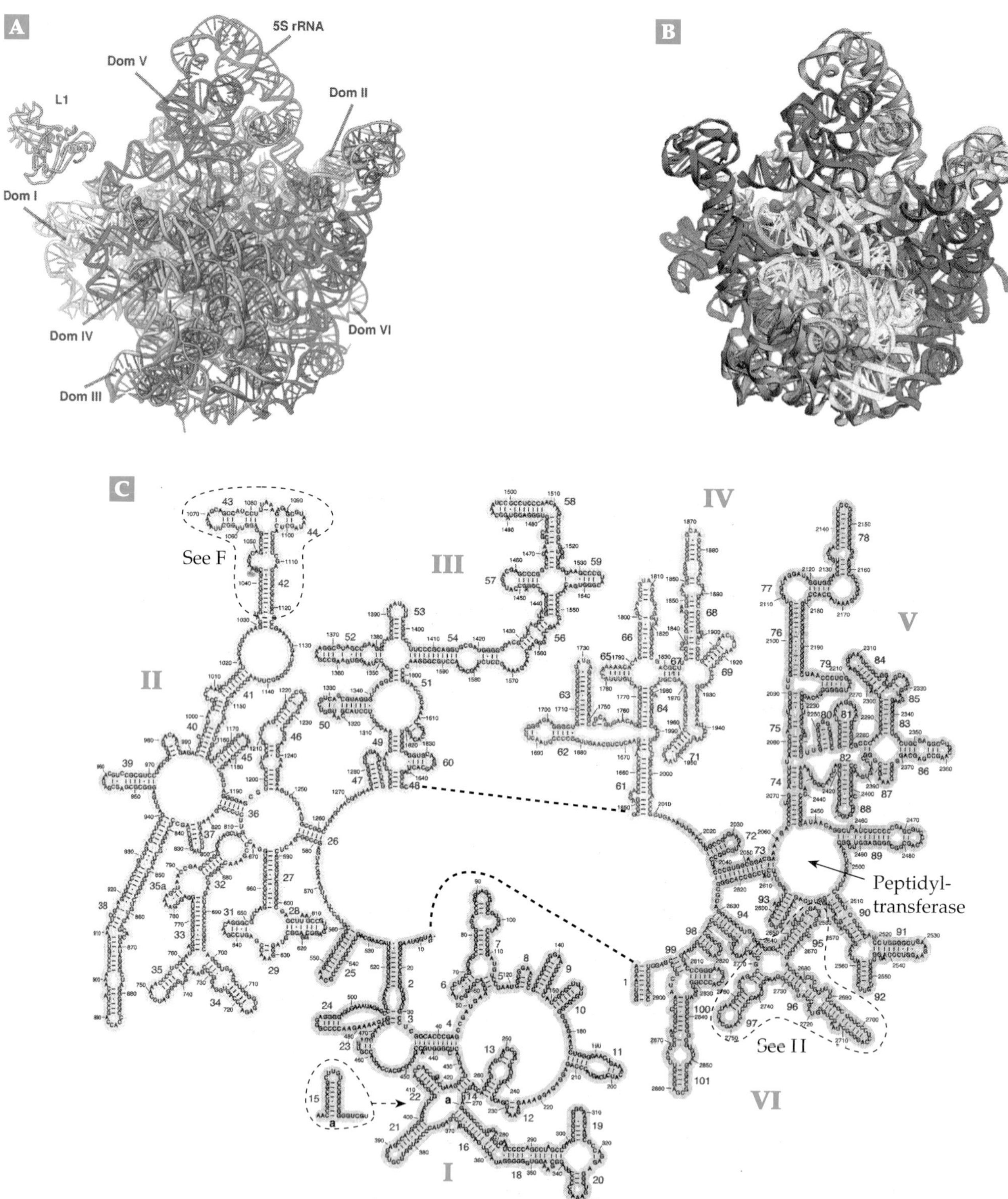

Figure 29-4 Structure of 23S–28S ribosomal RNAs. (A) The three-dimensional structure of RNA from the 50S subunit of ribosomes of *Haloarcula marismortui*. Both the 5S RNA and the six structural domains of the 23S RNA are labeled. Also shown is the backbone structure of protein L1. From Ban *et al.*[17] Courtesy of Thomas A. Steitz. (B) The corresponding structure of the 23S RNA from *Thermus thermophilus*. Courtesy of Yusupov *et al.*[33a] (C) Simplified drawing of the secondary structure of *E. coli* 23S RNA showing the six domains. The peptidyltransferase loop (see also Fig. 29-14) is labeled. This diagram is customarily presented in two halves, which are here connected by dashed lines. Stem-loop 1, which contains both residues 1 and 2000, is often shown in both halves but here only once. From Merryman *et al.*[78] Similar diagrams for *Haloarcula marismortui*[17] and for the mouse[79] reveal a largely conserved structure with nearly identical active sites. (D) Cryo-electron microscopic (Cryo-EM) reconstruction of a 50S subunit of a modified *E. coli* ribosome. The RNA has been modified genetically to have an

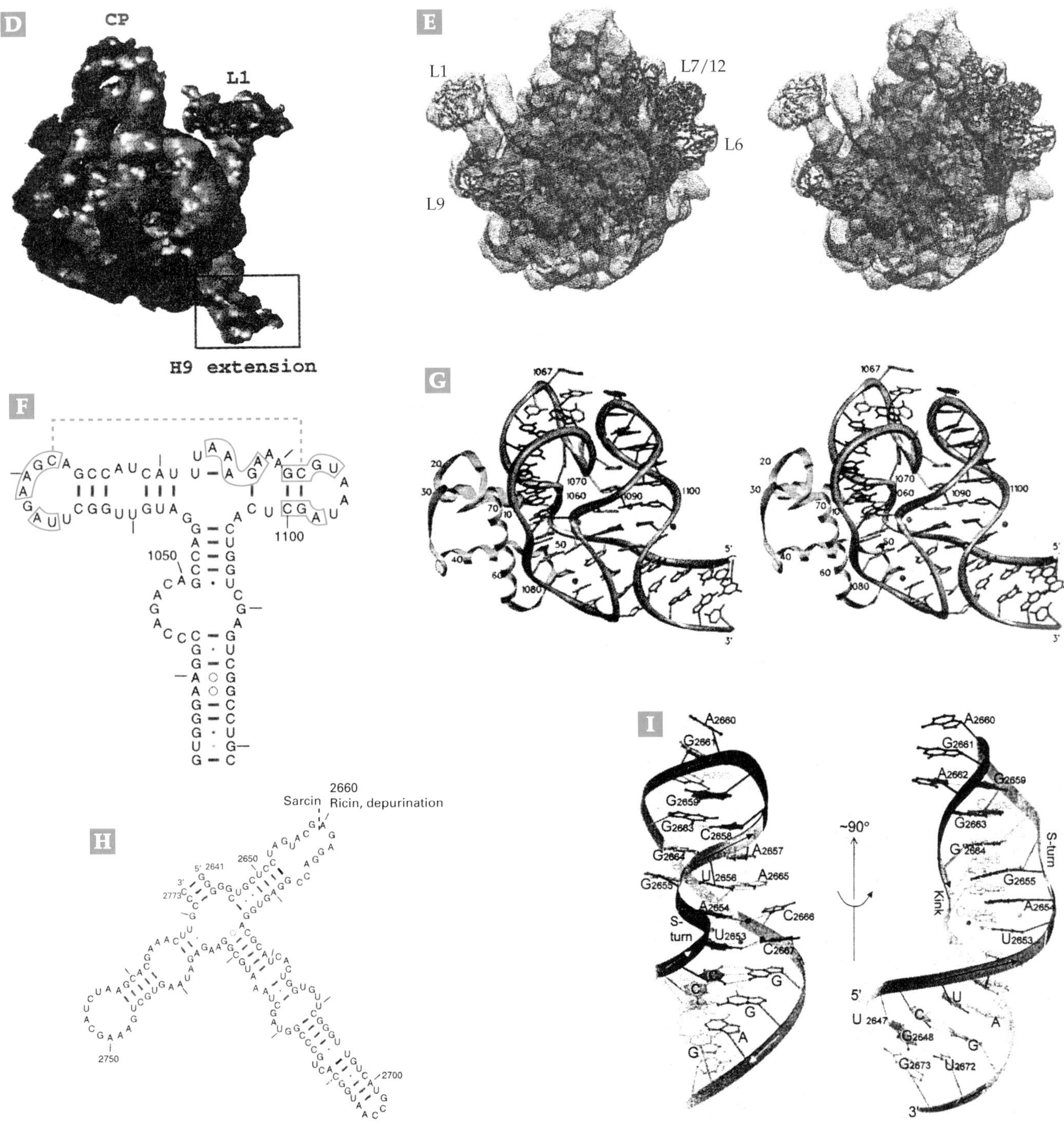

approximately 34 nt predicted extension of helix 9 of the 16S RNA (see Fig. 29-2). The helix 9 extension, clearly visible in this image, locates that helix in *E. coli* ribosomes, which have not yet given crystals satisfactory for X-ray investigation. From Matadeen *et al.*[79a] Courtesy of Richard Brimacombe. (E) Stereoscopic interface view of the 50S subunit of an *E. coli* ribosome with atomic structures of ribosomal proteins fitted to the cryo-EM density (semitransparent) of the 50S subunit. Protein structures are displayed as backbone tubes, and rRNA fragments in ball-and-stick format. Courtesy of Mueller *et al.*[37a] (F) The GTPase-activating loop of 23S RNA of *E. coli*. This loop, from domain II, binds to protein L11, which shields nucleotide A1067 from methylation and prevents the binding of the antibiotics thiostrepton (Box 29-B) and micrococcin. Green nucleotides are highly conserved in bacterial, chloroplast, and mitochondrial RNAs. The small loop (1054–1081) containing the thiostrepton-binding site is also part of the binding site for elongation factors EF-Tu and EF-G.[80,81] (G) Stereoscopic view of the 58-nucleotide loop shown in (E) with the associated protein L11. Courtesy of Conn *et al.*[82] (H) Secondary structure of the sarcin/ricin (SR domain) of the *E. coli* 23S RNA.[83–85] The site of hydrolytic cleavage by the ribonuclease sarcin (Box 29-A) is indicated as is the site of depurination catalyzed by the plant toxin ricin (Box 29-A). (I) Three-dimensional structure of the sarcin–ricin loop. The two views are from directions 90° apart. The sites of attack by ribotoxins are at the top. Courtesy of Correll *et al.*[83]

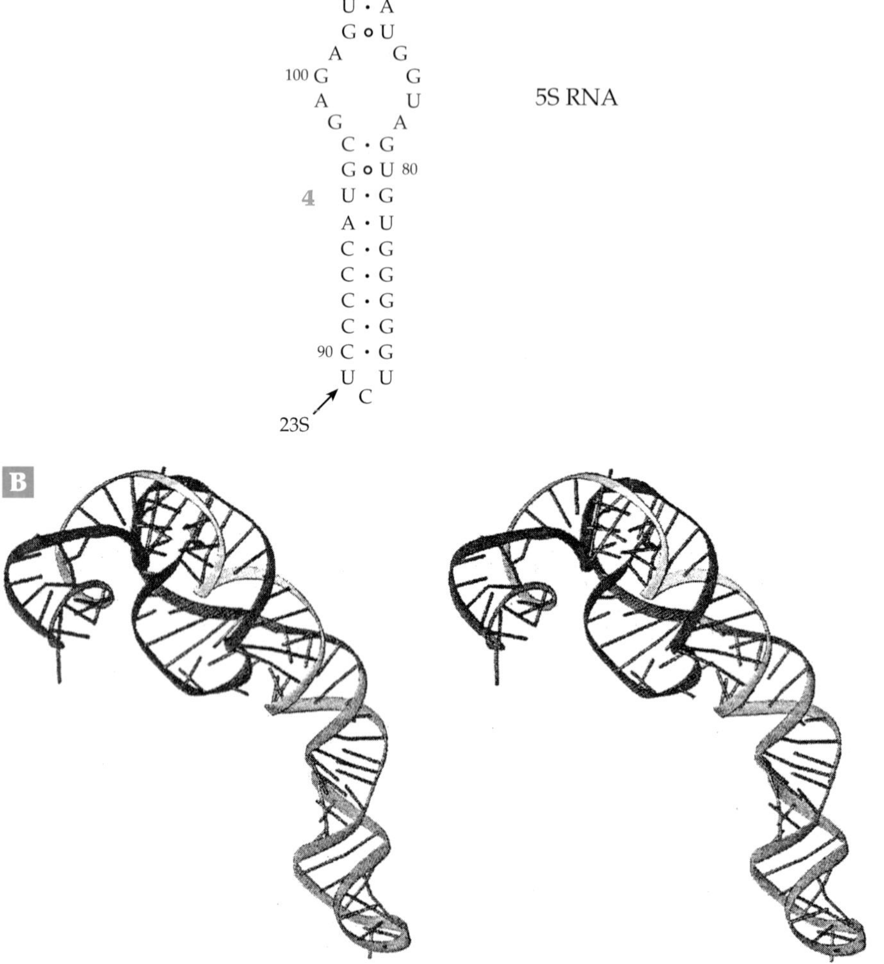

Figure 29-5 (A) Secondary structure of *E. coli* 5S RNA with five universal helical stems (labeled 1–5). This small RNA is found in the central protuberance of the 50S ribosomal subunit. See Fig 29-4A. Photocrosslinking using thiouridine-containing 5S RNA suggested a close proximity of U89 (marked by arrow) with nucleotide 2477 of the 23S RNA in the loop end of helix 89 (Fig. 29-4).[93] (B) Stereoscopic view of the 5S RNA as observed in ribosomes of *Haloarcula marismortui*. From Ban *et al.*[17] Courtesy of Thomas A. Steitz.

N-acetylated L12. Thus, the 50S ribosomal subunit is often described as containing four copies of protein L7/12. They form the flexible stalk seen in Fig. 19-1.

Most ribosomal proteins are rich in lysine and arginine and, therefore, carry a substantial net positive charge. Proteins S20, L7/12, and L10 have over 20% alanine, while L29 is almost as rich in leucine. Proteins S10, S13, L7/L12, L27, L29, and L30 are surprisingly low (<2 mol %) in aromatic amino acids. Proteins S5, S18, and L7 have acetylated N termini while L11, L3, L7/12, L11, L16, and L33 contain methylated amino acids. L11 contains nine methyl groups.[22] Protein S6 is the major phosphoprotein of eukaryotic ribosomes.[103,104] Most ribosomal proteins have no known enzymatic activity. Although often difficult to crystallize, high-resolution three-dimensional structures are known for many free ribosomal proteins.[24] Most of them have shapes resembling those previously found in globular proteins, including DNA-binding proteins. Many have extended "tails" that reach into the interior of the ribosome.[33b] A few seem to assume a defined shape only when packed into a suitable niche in the ribosome. Proteins L7/L12 and the 60-residue L30[105,105a] have similar folding patterns with 2–3 helices and a 3-strand β sheet. The structure of L30 of *E. coli* was deduced[106] by NMR methods (Fig. 3-25) and resembles that from the *Bacillus stearothermophilus* determined by X-ray diffraction.[105,107]

Many specific parts of ribosomal RNA molecules and specific proteins within the intact ribosome were located prior to the determination of high resolution crystal structures. One major approach was the use of **immunoelectron microscopy**. Antibodies to specific ribosomal proteins or to special sites in the RNA were prepared, and electron microscopy was used to map the binding sites of the antibodies on the ribosomal subunit surfaces.[108,109] In this manner, the locations of numerous proteins in both the 30S and 50S subunits were identified. A few of these are indicated in Fig. 29-1A,B.[5] In several instances more than one distinct antibody binding site was found for a given protein. Pairs of sites were sometimes 8–19 nm apart, suggesting that these proteins assumed an elongated or fibrous conformation. However, X-ray studies have established more compact structures for many of the proteins. Perhaps the ease of denaturation of the proteins led to some errors in localization with antibodies. The X-ray studies have now established exact locations for almost all of the ribosomal proteins. However, the correct identification of each protein involved extensive measurements, many of which were done prior to the availability of the X-ray structures.

A variety of crosslinking reagents have been used to locate the positions of specific proteins within ribosomes. For example, bifunctional compounds may bind covalently to two different SH groups or NH_2 groups.[110,111] Among the many crosslinked protein pairs identified in this way are S5-S8, S7-S9, S6-S18, and S13-S19.[112] Crosslinking experiments on both small and large ribosomal subunits have yielded complex distance maps that helped to establish the packing relationships.[113]

Another important approach has been to isolate ribosomal proteins from bacteria grown in D_2O and then to reconstitute ribosomal subunits with pairs of deuterated proteins. By studying **neutron scattering** the distances between the centers of mass of these pairs could be measured. By triangulation the three-dimensional relationship of the entire group of proteins could be determined. The results of such studies[43,47,114] for the 30S subunit are shown in Fig. 29-1D. Most of the results are in agreement with those obtained by other methods. Neutron scattering from the 50S subunit was investigated by using pairs of protonated proteins in a subunit consisting of otherwise deuterated components. This gives an increase in sensitivity.[47]

3. RNA–Protein Interactions and Assembly of Ribosomes

Within bacterial cells the assembly of ribosomes is coupled to rRNA synthesis and requires only 1–2 minutes.[115] In the laboratory both the 30S ribosomal subunits[116] and the 50S subunits[117–121] of *E. coli* can be completely dissociated into individual protein and RNA molecules and can be reconstituted in a functional form. This is true for both natural 16S or 23S RNA or for RNAs prepared by *in vitro* transcription. In these reassembly experiments, which were pioneered by Nomura,[116] it was found that the order of addition of the protein is important. Some proteins bind directly to ribosomal RNA. For example, S4, S7, S8, S15, S17, and S20 bind directly to 16S RNA.[31] Other proteins bind only after one or more proteins have already bound and the RNA has folded properly to form a structural core (Fig. 29-6A). Domains I, II, and III each form an independent RNA-protein assembly. The lower half of domain I of the 16S RNA, from positions ~60–300, is unreactive toward single-stranded probes and may serve as one core for assembly of the ribosome.[67] Protein S20 binds to the 240–286 stem, which is in this core. Protein S4 also binds directly to 16S RNA in the 5' domain. Proteins S8 and S15 bind in the central domain and S7, which is structurally related to the DNA-binding proteins HU and IHF (Chapter 27),[24,122] binds near the 3' end.

5′- Domain of 16S RNA. The 23-kDa protein S4, one of the largest ribosomal proteins, appears to have an important organizing role for the 5' domain.[123,124] It binds in such a way as to protect sequences 27–47 and 394–556 of the RNA (Fig. 29-2) from chemical modification. The small loop at positions 323–330 is protected in the 30S subunit, and the residues A325, A327, A379, and G331 are universally conserved.[67] The same is true of bulge loop 505–510 and the loop sequence 518–533, which contains 7-methylguanine (m^7G) at position 526. Reconstitution experiments also suggested that S16 binds to S4 as well as to S20. Some mutations in proteins S4 and S5 are associated with reduced fidelity of translation, while others lead to spectinomycin resistance.[134]

Central domain of 16S RNA. Proteins S6, S8, S15, and S18 bind to the central domain II of 16S RNA (Fig. 29-2)[31,31a,67,125–129] and organize the platform region (Fig. 29-6). Protein S8 binds with high affinity to regions 588–606 and 632–651 of helix 21 and plays a key role in ribosomal assembly.[126,130,130a] S15 protects residues 655–672 and 734–751 of helix 22. The region contains functionally important conserved loops at positions 570–571, 766–768, and 811–820 as well as many individual adenines in other locations. S15 binds not only to the 16S RNA but also to the 715 loop of 23S RNA in the large subunit and to its own mRNA.[129] A Mg^{2+}-dependent conformational change in the RNA seems to be important in the assembly of the central domain.[131] S6 and S18 bind to 16S RNA after S15 has bound (Fig. 29-6). Proteins S11 and S21 also bind after S15.[129] S11 binds to the 690 loop of the RNA, as is illustrated in Fig. 29-6B. This loop is conserved in all three phylogenetic domains. Located in the platform of the small subunit, it protrudes into the interface to interact with domain IV of the 23S RNA and is also a site of binding of initiation factor IF3.[133] The mutant A649G in 16S RNA confers resistance to **pactamycin** in *E. coli*. Protein S8 is not only an important structural protein in the central domain but also

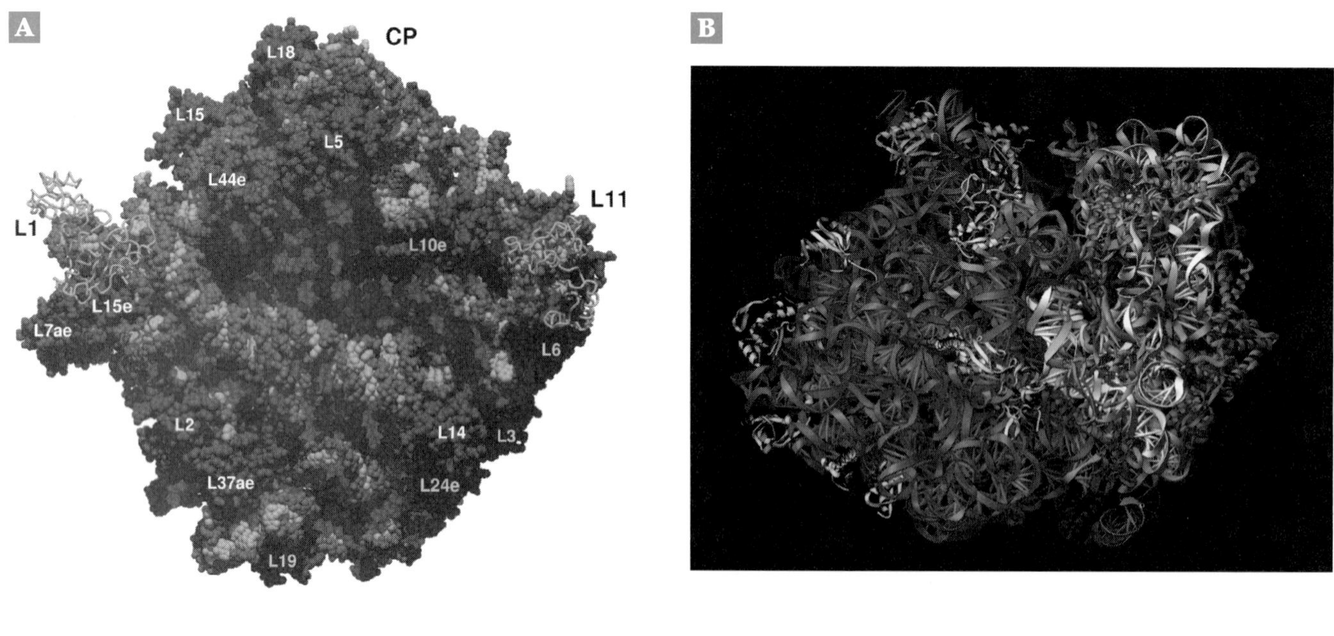

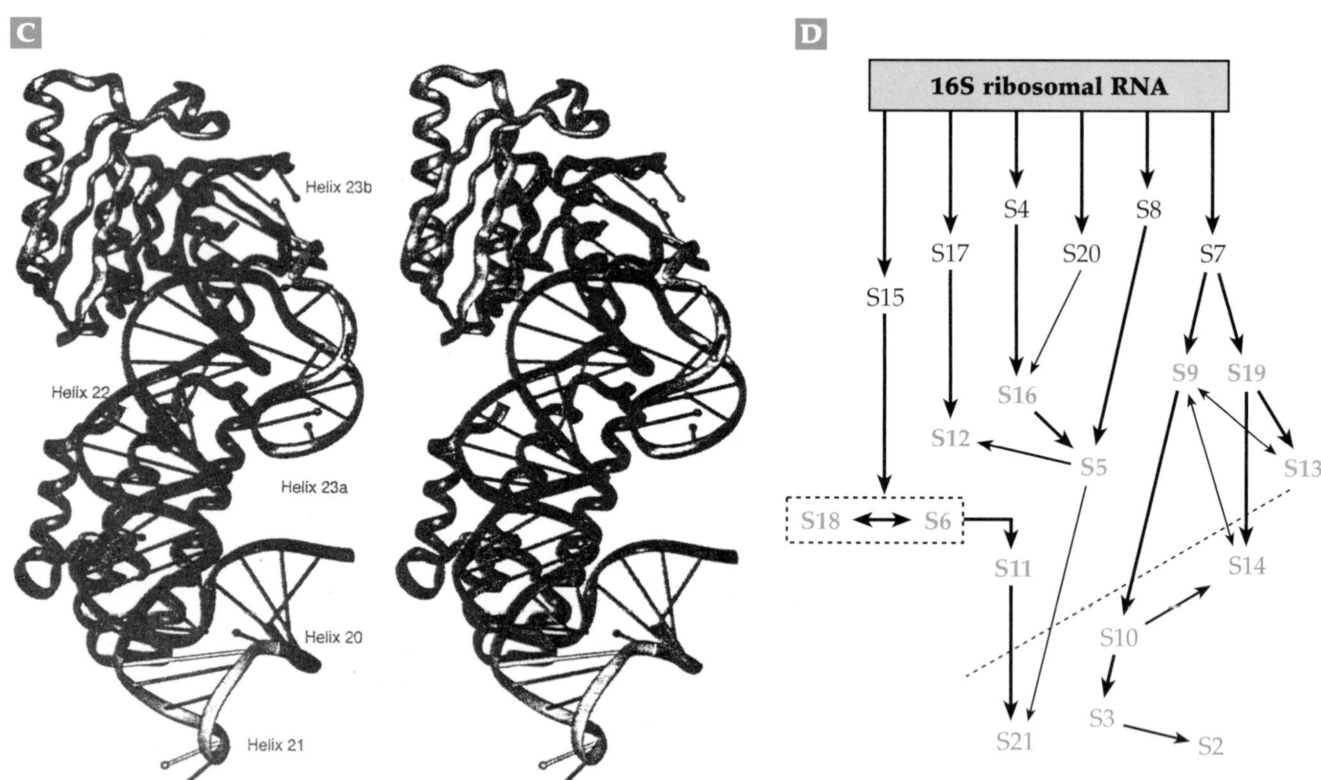

Figure 29-6 Some protein–RNA interactions within the ribosome. (A) A space-filling model of the 23S and 5S RNA with associated proteins from the ribosome of *Haloarcula marismortui*. The CCA ends of bound tRNA molecules in the A, P, and E sites are also included. The view is looking into the active site cleft. The proteins with e after the number are related to eukaryotic ribosomal proteins more closely than to those of *E. coli*.[17] Courtesy of T. A. Steitz. (B) Three-dimensional structure of a 70S ribosome from *Thermus thermophilus*. The 30S subunit is to the right of the 50S subunit. Courtesy of Yusupov *et al*.[33a] (C) Stereoscopic view of the helix 21 to helix 23b region of the 16S RNA with associated proteins S6 (upper left), S18 (upper center, front), and S15 (lower back) from *T. thermophilus*. Courtesy of Agalarov *et al*.[31] (D) Simplified *in vitro* assembly map of the central domain of the 30S bacterial ribosome. Courtesy of Gloria Culver. (E) Contacts of proteins with the central (platform) domain of the 16S RNA component. The sequence shown is that of *Thermus thermophilus*. Courtesy of Agalarov *et al*. (F) Three drawings showing alternative location of the four copies of protein L7/L12. The N-terminal and C-terminal

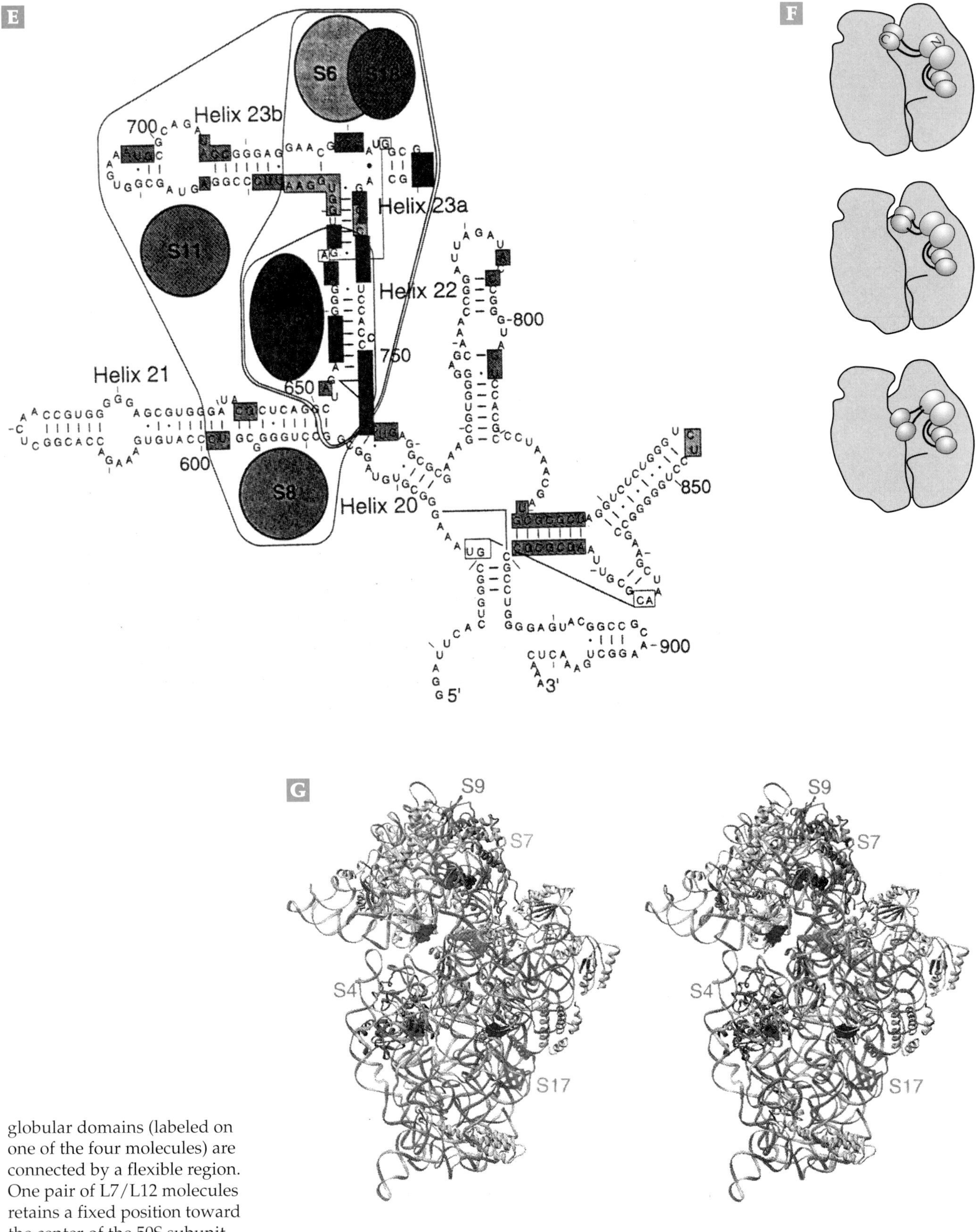

globular domains (labeled on one of the four molecules) are connected by a flexible region. One pair of L7/L12 molecules retains a fixed position toward the center of the 50S subunit but the C-terminal domains of the other pair are seen to occupy three different positions. Courtesy of Montesano-Roditis *et al.*[146a] (G) Stereoscopic view of the 30S subunit of the *T. thermophilus* ribosome with six positions at which tetracycline binds and one at which the peptide-like antibiotic **edeine** binds. From Pioletti *et a*l.[146b]

acts as a translational repressor of its own gene and of other genes of the spectinomycin-resistance operon (see Box 29-B). This operon encodes genes for ten ribosomal proteins, of both the large and small subunits.[126] S8 is a neighbor to proteins S2, S4, S5, S12, S15, and S17.

3′- Domain of 16S RNA. Domain III of 16S RNA binds proteins S2, S3, S7, S9, S10, S13, S14, and S19. Proteins S3 and S14 may be necessary for assembly of ribosomes but may no longer be needed once the 30S subunit has been correctly formed.[135] The largest of the *E. coli* ribosomal proteins is the 556-residue S1. It does not bind tightly and has sometimes been regarded as nonessential. However, mutations in the S1 gene can be lethal, and the protein seems to be essential for both initiation of translation and for elongation of polypeptide chains *in vivo*.[137] S1 behaves as an elongated molecule 22 nm in length[135] and is unusual in having an ~86-residue RNA-binding sequence repeated four times in the central and N-terminal regions. The protein possesses RNA-unwinding activity[137,138] and may employ these four motifs in unwinding mRNA as it enters the ribosome.

23S and 5S RNAs. Reconstitution of the large ribosomal subunit reveals that proteins L3 and L24 act as assembly initiators.[115,118] L1, L9, L20, and several other proteins (Table 29-2) also bind directly and independently to the 23S RNA. Assembly maps similar to that in Fig. 19-6A have been prepared for the 50S subunit.[117]

One of the most prominent features of the 50S subunit is the L1 protuberance, seen on the left side in Fig. 29-6A. This protuberance is formed almost entirely by protein L1, which is one of the largest ribosomal proteins. It binds to the 2105–2184 loop in domain V of the 23S RNA (see Fig. 29-14).[139] L1 has an important regulatory role in bacteria in which it represses translation of its own structural gene by binding to a region in its mRNA close to the Shine–Dalgarno sequence. The polygenic mRNA also carries the code for protein L11.[139] This is one of several examples of such autogenous regulation of translation of ribosomal proteins.[139a] L1 also interacts in the ribosome with the 5S rRNA.[140] The 272-residue L2 also associates directly with 23S RNA and assists in ribosome assembly.[141] Protein L2 is one of the structurally most highly conserved of the ribosomal proteins.[46,142] It binds to the 1794–1865 region of domain IV of 23S RNA. Histidine 229 of this protein may play a functional role in the ribosome. The protein is elongated, and one end contacts 16S RNA.[46,33a] Protein L9 binds to domain V of 23S RNA in the 2100–2190 region. It is an elongated molecule with two globular α/β domains separated by an α helix. This enables it to bind also to domain III, acting as a rigid strut.[143–145]

On the right side of the 50S subunit, as viewed in Fig. 29-1, is the stalk, a pentameric protein complex consisting of two L7/L12 (*E. coli*) or $(L12)_2$ dimers bound to one molecule of L10.[24,146–147] The stalk is not always seen in X-ray structures, e.g., in Fig. 29-6A, and appears to be flexible. In crosslinking experiments the N-terminal domains of L7/L12 can be linked to L10 and also to its neighbor, L11,[82,148,149] which lies in the GTPase-activating center (Fig. 29-4F) at the base of the stalk. However, the C-terminal domains can be crosslinked to three distinctly different locations: to L11 on the platform surface, to L2 and L5 near the peptidyltransferase center, and to S2, S3, and S14 of the head and neck of the 30S subunit.[146] Domain I of 23S RNA, near the 5' end, binds to protein L20.[153]

An independent and essential structural domain of the ribosome is formed around the 5S RNA.[5,108,154–156] Proteins L5, L18, and L25, whose structure is similar to that of glutaminyl-tRNA synthetase,[154] bind specifically to one loop of the 5S RNA.[156a] Furthermore, the L5-L18-L25-5S RNA complex binds the oligonucleotide TCC. This suggests an interaction between the 5S RNA and the TΨC arm of a tRNA molecule bound to the ribosome. In addition, it has been observed that L18 + either L5 or L25 cause 5S RNA to bind to 23S RNA.

Eukaryotic ribosomal proteins. The functions of the 70–80 different eukaryotic ribosomal proteins are less well known than those of *E. coli*. In eukaryotes the assembly of ribosomes begins in the nucleus with binding of proteins to the individual ribosomal RNA precursors (Chapter 28).[121,156b] Significant functional properties that are peculiar to eukaryotic ribosomal proteins include the following: S6 is the site of multiple phosphorylation reactions, which control initiation of protein synthesis.[132,132a,132b] Mammalian S3 may function in the nucleus in DNA repair.[136] Eukaryotic proteins P0, P1, and P2 are homologous to *E. coli* stalk proteins L10, L7, and L12, respectively. Higher eukaryotes possess only one type of P1 and P2,[150] but yeast,[150a,b] maize,[151] and other species have multiple forms. An L7-related protein is also required for a nucleolar function in ribosomal protein synthesis, perhaps as a component of a snoRNP complex (Chapter 28).[152] Rat liver L37 is involved in peptidyltransferase, but sequencing of the 111-residue protein reveals homology with *E. coli* L34 rather than with L16.[157] Proteins L14, L21, L24, L27, L29, and L30 bind to the 5.8S RNA of the large subunit of yeast ribosomes.[158]

Yeast protein L30, which is not homologous to any bacterial protein, controls its own synthesis by a feedback inhibition at the mRNA splicing step. L30 binds to its own pre-mRNA near the 5' splice site, blocking completion of the spliceosome assembly (Chapter 28).[159]

BOX 29-A THE DIPHTHERIA TOXIN AND OTHER RIBOSOME-INACTIVATING PROTEINS

Until a suitable vaccine was developed, an infection by *Corynebacterium diphtheriae* was one of the dread diseases of childhood. Despite the fact that the bacteria caused only superficial membranous lesions in the throat, the patient often died with evident damage to many organs. The cause is a potent heat-labile toxic protein,[a–d] which the bacterium produces when infected by a temperate bacteriophage carrying the *tox* gene and when the inorganic iron of the surroundings has been largely depleted. Diphtheria toxin is a 535-residue protein with a minimum lethal dose (LD_{50}) of only 0.16 mg kg^{-1} for the guinea pig. Tests in cell culture show that the toxin blocks incorporation of amino acids into proteins by inactivation of the eukaryotic elongation factor EF2, which is required for **translocation**, an essential step in protein synthesis in mammalian ribosomes. The toxin acts as an enzyme that transfers (with inversion at the ribose C1) an ADP-ribosyl group from NAD^+ to a side-chain ring nitrogen of the single residue of **diphthamide** in EF2. This modified histidine is found in EF2 and, apparently, in no other protein.[d]

The modified elongation factor reacts normally with GTP, but the complex so formed is unable to participate in translocation. A concentration of the toxin in the cytoplasm of 10^{-8} M is sufficient to promote the fatal reaction. The reaction with diphthamide parallels that of cholera toxin (Box 11-A).

The diphtheria toxin molecule[e,f] consists of three domains, an N-terminal catalytic (C) domain (residues 1–193), a central, largely α-helical, transmembrane (T) domain (residues 205–378), and the C-terminal receptor-binding (R) domain (residues 386–535). Before it enters a cell, the toxin molecule is "nicked" by protease activity between the catalytic and transmembrane domains, a step that is apparently necessary for binding to its receptor, which has been identified as a heparin-binding EGF-like growth factor precursor.[g] The catalytic domain (also called the A fragment) enters the cell through endocytosis from coated pits. Additional proteolytic cleavage, as well as reduction of a disulfide linkage, are required for activity.[f,h] After entering the cytosol the toxin fragment catalyzes inactivation of EF2.[f] A single molecule can kill a cell. The larger 613-residue exotoxin A of *Pseudomonas aeruginosa* catalyzes the same reaction as does diphtheria toxin. It also consists of three domains: a large β-sheet-containing N-terminal region, a central α-helical domain, and a C-terminal domain. The last contains the ADP-ribosyltransferase active site.[a,i]

What is the origin of the *tox* gene, and why is it carried by a virus? Cells do normally contain ADP-ribosyltransferases.[j] The genes for such a protein may have become incorporated into a virus and, after a period of evolution, came to specify the toxic protein.

Another family of toxins attacks ribosomes in a very different way, cleaving ribosomal RNA at specific sites. One of the best known of these is the neurotoxin from *Shigella dysentariae* (Shiga toxin). Like the cholera toxin (Box 11-A) it consists of a single catalytic A subunit and a pentameric ring of B subunits,[k] which binds to specific surface glycolipids.[l] **Verotoxin**, another poison from certain strains of *E. coli*, has a similar structure. Although they have very different effects and there is no detectable similarity in their amino acid sequences, the *structures* of the B pentamers of verotoxin and of the cholera toxin-like heat-labile enterotoxin of *E. coli* are similar.[m,n] The A subunit of Shiga toxin hydrolyzes the *N*-glycosyl linkage of adenine to the ribose ring at position 4324 of 28S ribosomal RNA.[k] A number of plants form very toxic lectins: **ricin** (from castor bean),[o] **viscumin** (from mistletoe), **modeccin**, **abrin**,[p] gelonin,[q] and **volkensin**.[o] The names are derived from the genus names of the plants. All appear to be glycoproteins consisting of two disulfide-linked chains, one of which is a lectin. The

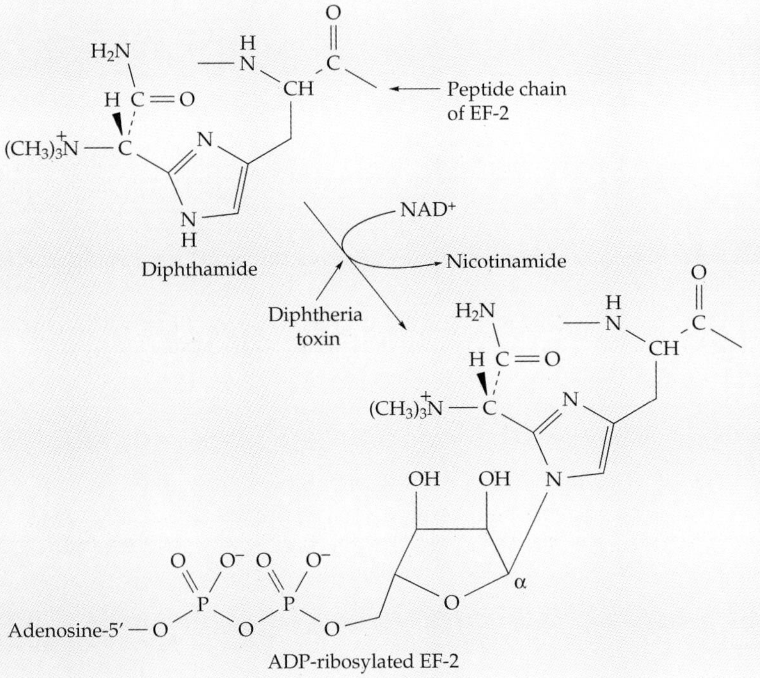

H_2N, H C=O, $(CH_3)_3\overset{+}{N}$—C, N, N H, —N H—CH—C(=O)— ← Peptide chain of EF-2

Diphthamide

Diphtheria toxin; NAD^+ → Nicotinamide

ADP-ribosylated EF-2 (Adenosine-5′—O—P(O)(O⁻)—O—P(O)(O⁻)—O—ribose (OH, OH, α)—N of diphthamide ring)

BOX 29-A THE DIPHTHERIA TOXIN AND OTHER RIBOSOME-INACTIVATING PROTEINS (continued)

lectin subunits of ricin[r] and of volkensin[s] bind to galactose residues. The A chains are cytotoxins, which enter cells and, like Shiga toxin, inactivate 60S ribosomal subunits. The 267-residue A chain of ricin is similar to that of a pokeweed viral antigen[t] and of Shiga toxin. It catalyzes the same reaction,[u] the depurination of adenosine 4324. (The pokeweed toxin also catalyzes the corresponding reaction with A2660 of the *E. coli* 23S RNA.[v]) Like the diphtheria toxin these toxic proteins bind to cell surface receptors, are taken up by endocytosis, and are transported through the Golgi to the endoplasmic reticulum. Their structures facilitate uptake but allow them to escape degradation in proteasomes.[w]

Ricin is one of the most toxic substances known. A single molecule can inactivate over 1700 ribosomes per minute and kill the cell.[u] With an LD_{50} of only 1 µg / kg of body weight for many animals, ricin has been used as a poison by assassins. Of more importance is the attempt to couple ricin and related toxins to immunoglobulins to produce **immunotoxins** that will attack cancer cells (Box 31-A). A related goal is to design a potent inhibitor that could serve as an antidote.[x] It is fortunate that most plant seeds do not contain toxins like ricin. Many plants, including such important food grains as wheat and barley, do contain ribosome-inactivating proteins similar to the A chain of ricin. However, the plants lack the B (lectin) subunits and do not enter animal cells.

A group of unusual fungal ribonucleases, which includes **α-sarcin** and **restrictocin**, are produced by *Aspergillus*. The cytotoxic nucleases enter animal cells, where they cut the 28S RNA of ribosomes, specifically on the 3' side of guanosine 4325 in the sarcin / ricin domain (see Fig. 29-4), thereby blocking protein synthesis.[u,y,z] *Staphylococcus aureus* produces a 22-kDa toxic protein thought to be responsible for **toxic shock syndrome**.[aa] Another toxic ribonuclease is **colicin E3** (Box 8-D), which cuts the 16S RNA of *E. coli* after nucleotide 1493 (see Fig. 29-1A).[bb] Colicin D stops protein synthesis by cleavage of four isoaccepting $tRNA^{Arg}$ molecules between positions 38 and 39 in the anticodon loop.[cc]

a Han, X. Y., and Galloway, D. R. (1995) *J. Biol. Chem.* **270**, 679–684
b Pappenheimer, A. M., Jr. (1977) *Ann. Rev. Biochem.* **46**, 69–94
c Pappenheimer, A. M. J. (1993) *Protein Sci.* **2**, 292–298
d Ward, W. H. J. (1987) *Trends Biochem. Sci.* **12**, 28–31
e Choe, S., Bennett, M. J., Fujii, G., Curmi, P. M. G., Kantardjieff, K. A., Collier, R. J., and Eisenberg, D. (1992) *Nature (London)* **357**, 216–222
f Weiss, M. S., Blanke, S. R., Collier, R. J., and Eisenberg, D. (1995) *Biochemistry* **34**, 773–781
g Shen, W. H., Choe, S., Eisenberg, D., and Collier, R. J. (1994) *J. Biol. Chem.* **269**, 29077–29084
h Blanke, S. R., Huang, K., Wilson, B. A., Papini, E., Covacci, A., and Collier, R. J. (1994) *Biochemistry* **33**, 5155–5161
i Douglas, C. M., and Collier, R. J. (1990) *Biochemistry* **29**, 5043–5049
j Eide, B., Gierschik, P., and Spiegel, A. (1986) *Biochemistry* **25**, 6711–6715
k Kozlov, Y. V., Chernaia, M. M., Fraser, M. E., and James, M. N. G. (1993) *J. Mol. Biol.* **232**, 704–706
l Saleh, M. T., and Gariépy, J. (1993) *Biochemistry* **32**, 918–922
m Stein, P. E., Boodhoo, A., Tyrrell, G. J., Brunton, J. L., and Read, R. J. (1992) *Nature (London)* **355**, 748–750
n Sixma, T. K., Stein, P. E., Hol, W. G. J., and Read, R. J. (1993) *Biochemistry* **32**, 191–198
o Lord, J. M., Roberts, L. M., and Robertus, J. D. (1994) *FASEB J.* **8**, 201–208
p Tahirov, T. H., Lu, T.-H., Liaw, Y.-C., Chen, Y.-L., and Lin, J.-Y. (1995) *J. Mol. Biol.* **250**, 354–367
q Hosur, M. V., Nair, B., Satyamurthy, P., Misquith, S., Surolia, A., and Kannan, K. K. (1995) *J. Mol. Biol.* **250**, 368–380
r Weston, S. A., Tucker, A. D., Thatcher, D. R., Derbyshire, D. J., and Pauptit, R. A. (1994) *J. Mol. Biol.* **244**, 410–422
s Stirpe, F., Barbieri, L., Abbondanza, A., Falasca, A. I., Brown, A. N. F., Sandvig, K., Olsnes, S., and Pihl, A. (1985) *J. Biol. Chem.* **260**, 14589–14595
t Marchant, A., and Hartley, M. R. (1995) *J. Mol. Biol.* **254**, 848–855
u Glück, A., and Wool, I. G. (1996) *J. Mol. Biol.* **256**, 838–848
v Chan, Y.-L., Sitikov, A. S., and Wool, I. G. (2000) *J. Mol. Biol.* **298**, 795–805
w Sandvig, K., and van Deurs, B. (2000) *EMBO J.* **19**, 5943–5950
x Yan, X., Hollis, T., Svinth, M., Day, P., Monzingo, A. F., Milne, G. W. A., and Robertus, J. D. (1997) *J. Mol. Biol.* **266**, 1043–1049
y Endo, Y., Chan, Y.-L., Lin, A., Tsurugi, K., and Wool, I. G. (1988) *J. Biol. Chem.* **263**, 7917–7920
z Nayak, S. K., Bagga, S., Gaur, D., Nair, D. T., Salunke, D. M., and Batra, J. K. (2001) *Biochemistry* **40**, 9115 – 9124
aa Blomster-Hautamaa, D. A., Kreiswirth, B. N., Kornblum, J. S., Novick, R. P., and Schlievert, P. M. (1986) *J. Biol. Chem.* **261**, 15783–15786
bb Lasater, L. S., Cann, P. A., and Glitz, D. G. (1989) *J. Biol. Chem.* **264**, 21798–21805
cc Tomita, K., Ogawa, T., Uozumi, T., Watanabe, K., and Masaki, H. (2000) *Proc. Natl. Acad. Sci. U.S.A.* **97**, 8278–8283

4. Locating Active Sites in Ribosomes

In early studies, antibodies against haptens covalently linked to ends of the 16S RNA were used to locate the 3' end of 16S RNA on the upper "platform" of the 30S subunit (Fig. 29-1A).[5,160] The 5' terminus was found in the lower body. Two N^6,N^6-dimethyladenosines occur at positions 1518 and 1519, about 25 residues from the 3' end of the 16S RNA (Fig. 29-2). Antibodies were used to locate this position on the lower platform. Likewise, the m^7G at position 526 lies in the "neck" as shown in Fig. 29-1A.[161] Taking into account known protein-RNA interactions, domain II of the 16S RNA was located in the "platform" on the upper left side of the "body" (viewed from the "outside" as in Fig. 29-1C), while domain III is in the "head." Recent structural studies have confirmed these biochemical localizations.[19,29,33,33a]

The tRNA-binding sites. During protein synthesis tRNA is bound sequentially in at least three places located between the 30S and 50S subunits. These are known as the **A** (**aminoacyl**), **P** (**peptidyl**), and **E** (**exit**) sites. The latter binds deacylated tRNA before it is released from the ribosome.[162,163] Because tRNA is such a large molecule, these sites have subsites in both 30S and 50S ribosomal subunits. When in the P site, a tRNA has its anticodon held firmly and base-paired with a codon in the mRNA in the decoding site of the 30S subunit. The CCA 3'-end with its attached peptidyl chain lies in the 50S subunit at the peptidyltransferase center. A "charged" aminoacyl-tRNA enters the A site, which is close to the 5S RNA in the central protuberance of the large ribosomal subunit, with its anticodon in the decoding site and its aminoacyl group at the peptidyltransferase site.

One end of the P site must be close to the 3' end of the 16S RNA near the two $m_2{}^6A$ residues (Fig. 29-1A). This conclusion, which was based on photochemical linking of a hypermodified base at position 34 (see Fig. 29-7) in $tRNA^{Val}$ with C-1400 of 16S RNA by cyclobutane dimer formation (Eq. 23-26),[164] has been confirmed by structural studies.[29,33,33d] Investigation of tRNA binding, effects of mutations in ribosomal RNA, and effects of antibiotics pointed to locations of the P and A sites in both ribosomal subunits. These have been located precisely by crystallography. See Fig. 29-1F,J; 29-2C. Residue 6530 together with nucleotides 921–927, 1390–1394, and 1491–1505 of 16S RNA participate in forming the form A and P sites in the decoding center.[33c,d,164a,b,378] The two adjacent adenine rings of A1492 and A1493 swing out from helix 44 (Fig. 29-2; see also Fig. 29-14) to form a major part of the A site. In the 50S subunit the adjacent cytosines C74 and C75 of the CCA 3′-ends of the tRNAs in the A and P sites interact respectively with G2553 of the A loop and G2252 of the P loop (Fig. 29-14B,E).[164b–e] Tetracycline (Fig. 22-7) also binds into the A site (see Box 29-B). It can be photochemically crosslinked to proteins S18 and S4.[167]

The peptidyltransferase site. The position was located by binding of derivatives of the antibiotic **puromycin** (Fig. 29-13). An arylazide derivative of puromycin was photochemically linked (Eq. 23-27) to proteins L23, L18/22, and L15; immunoelectron microscopy, using antibodies to the N^6-dimethyladenosine of puromycin,[165,166] located the binding site adjacent to the central protuberance between the 50S subunit and 30S subunit near S14.[5] 4-Thio-dT-p-C-p-puromycin was photochemically crosslinked to G2553 of the peptidyltransferase A site (see Fig. 29-14). X-ray data provided a precise structure of the peptidyltransferase site (see pp. 1702–1704).[166a] Studies of mutant ribosomes together with affinity labeling and crosslinking experiments pointed to the **peptidyltransferase loop** marked on Fig. 29-4 and further illustrated in Fig. 29-14.[164a,167a,b]

The GTPase-activating center. Also shown in Fig. 29-1C is a site for binding of the elongation factor EF-G (Section C,2). This was located, in part, because the antibiotic **thiostrepton** prevents EF-G from binding to the ribosome. Thiostrepton binds to a complex of protein L11 and a 61-fragment of the 23S RNA (positions 1052–1112; see Fig. 29-4F).[80] Another elongation factor, EF-Tu, also binds at the same site or adjacent to the EF-G site by the head of the small subunit.[168] An additional location of interest is the **polypeptide exit tunnel**, which brings the growing protein chain from the peptidyltransferase site out of the ribosome (Fig. 29-1I).[5,6,33f]

B. Transfer RNAs

The small 4S tRNA molecules have masses of ~26 kDa and consist of 75 ± 5 nucleotides (Figs. 5-30, 5-31, and 29-6). The basic structures are similar in bacteria and eukaryotic cells. The need for "adapters," to carry amino acids to the proper positions along the mRNA template, had been predicted prior to the discovery of tRNA.[169,170] It had been expected that there would be a base sequence constituting an **anticodon**, which would fit against the proper codon at some binding site on the protein-synthesizing machinery. This is just what tRNA molecules do, but their chemistry contained many surprises.

1. Structures of Transfer RNAs

The first surprise was that these molecules are much longer than seems necessary for the formation of adapters. In addition, 10–20% of their bases are modified greatly from their original form.[171] Another surprise was that the anticodons are not all made up of "standard" bases. Thus, hypoxanthine (whose nucleoside is inosine) occurs in some anticodons. Conventional "cloverleaf" representations of tRNA, which display their secondary structures, are shown in Figs. 5-30 and 29-7. However, the molecules usually have an L shape rather than a cloverleaf form (Figs. 5-31 and 29-6),[172] and the L form is essential for functioning in protein synthesis as indicated by X-ray and other data.[173] Three-dimensional structures, now determined for several different tRNAs,[174,175] are all very similar. Structures in solution are also thought to be

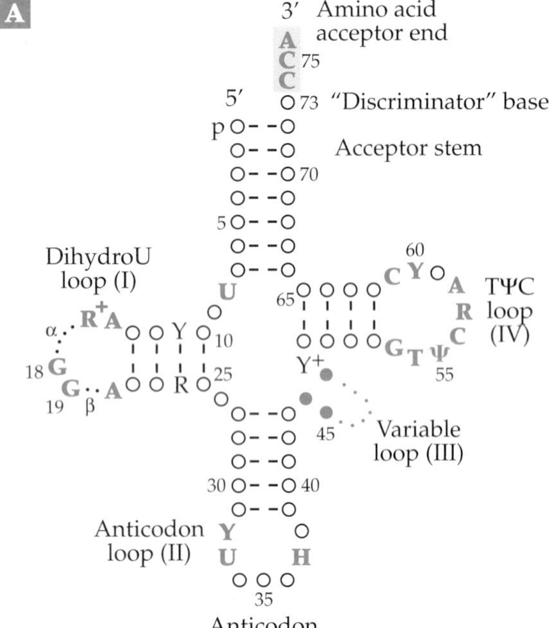

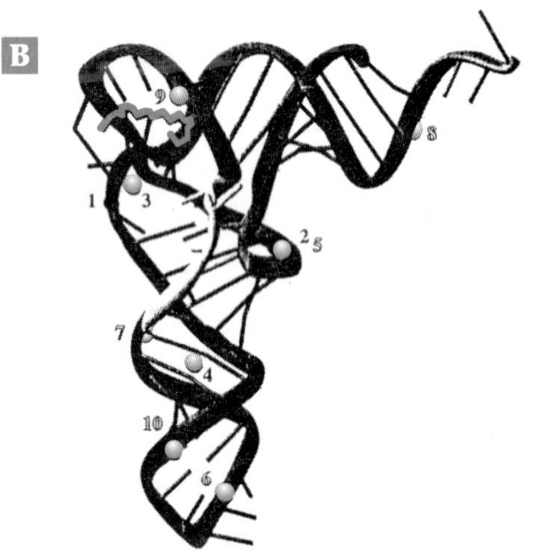

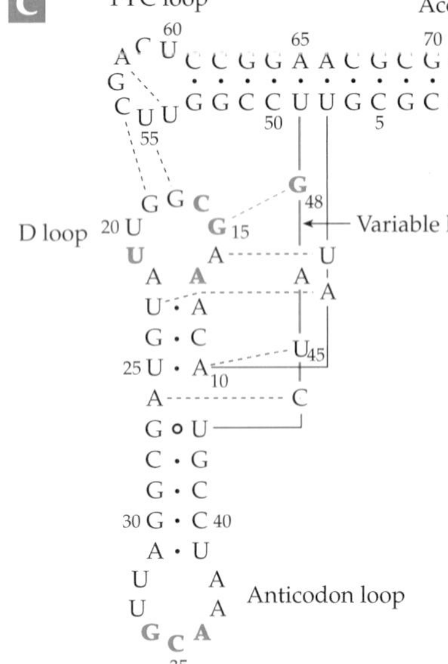

similar for the various tRNA molecules.[176,177] One of the four hydrogen-bonded "stems" of a tRNA in the cloverleaf form terminates in the universally conserved CCA-3′ **amino acid acceptor end** (Fig. 29-7), which can carry an esterified amino acid generated as in Eq. 29-1, steps *a*, *b*. The other three stems terminate in loops, which usually contain a large number of modified bases. The modifications may serve to optimize the interaction of the tRNA with other components of the protein-synthesizing machinery.[178] The **dihydroU loop** (loop I) contains 5,6-dihydrouridine in various amounts and in varying positions. The **anticodon loop** (loop II) always contains the anticodon directly opposite the amino acid acceptor end in the cloverleaf drawing. On the 5' side of the anticodon at position 33 there is almost always a U (shaded in Fig. 29-7A) preceded by another pyrimidine. A hydrogen bond from the N3 proton of U-33 and a phosphate oxygen of residue 36 stabilize the U-turn that precedes the anticodon triplet (Fig. 5-31).[177] Next to the 3' side of the anticodon there is usually a **hypermodified** base, such as N^6-(Δ^2-isopentenyl) adenosine (Fig. 5-33) or a more complex derivative.[178] The **variable loop** (loop III) can range between 5 and 21 nt in length.[177,181] The **TψC loop** (loop IV) contains the specific nucleotide sequence for which the loop is named.

Cloverleaf and L forms. Interconversion between the cloverleaf and L forms of tRNA molecules can be pictured as in Eq. 29-2. Notice that in the L

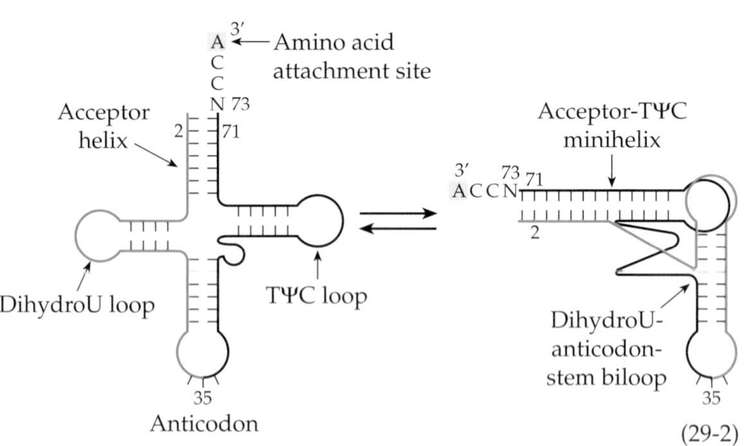

(29-2)

Figure 29-7 (A) Generalized cloverleaf diagram of all tRNA sequences except for initiator tRNAs numbered as in yeast $tRNA^{Phe}$ (Fig. 5-30). Invariant bases: A, C, G, T, U, and ψ; semivariant bases: Y (pyrimidine base), R (purine base), H (hypermodified purine base). The dotted regions (α, β, variable loop) contain different numbers of nucleotides in various tRNA sequences. See Rich.[179] (B) L form of the yeast phenylalanine-specific $tRNA^{Phe}$. The structure is the same as that in Fig. 5-31 but has recently been redetermined at a resolution of 0.20 nm.[175] The new data revealed the presence of ten bound Mg^{2+} ions (green circles) as well as bound spermine (green). (C) Schematic representation of L form of *E. coli* $tRNA^{Cys}$. Some tertiary base pairings are indicated by dashed lines. No modified bases are shown. See Hou *et al.*[180]

BOX 29-B ANTIBIOTICS THAT INHIBIT PROTEIN SYNTHESIS

Many of our most effective antibiotics act by blocking protein synthesis on ribosomes.[a–de] The usefulness of many of these remarkable drugs in human medicine depends upon the fact that they inhibit protein synthesis on bacterial 70S ribosomes but do not seriously affect eukaryotic ribosomes. Antibiotics act at a variety of sites involving both ribosomal proteins and rRNA. Some of the resulting points of inhibition of the ribosomal elongation cycle[e] are marked on Fig. 29-12. Binding sites for many antibiotics have been located by footprinting experiments, by studies of mutants, and more recently by X-ray crystallography. Exposure of bacteria to antibiotics often gives rise to antibiotic-resistant mutants. In some of these mutants a specific ribosomal protein has been altered, but in others a specific RNA base has been changed.

One of the first antibiotics to be studied was **puromycin** (Fig. 29-13), which binds to the 50S subunit and causes premature termination of peptide synthesis. A glance at its structure reveals how it can do this. It resembles in fine detail the 3' end of a tRNA molecule bearing an aminoacyl group. However, it is not an aminoacyl group, and once the growing peptide chain has been transferred onto the puromycin, further chain elongation is impossible. It was shown in 1975 that puromycin could be crosslinked to several different proteins in the ribosome by ultraviolet irradiation.[f,g] L23 of the 50S subunit and S14 of the 30S subunit were labeled most heavily. More recently puromycin, acting on 70S ribosomes, has been shown to label protein S7 and several large subunit proteins near the peptidyltransferase center in the central domain V of the 23S RNA.[h] A puromycin-derived transition state inhibitor has permitted precise identification of the peptidyltransferase site (Fig. 29-13).[i,j] In earlier work immunoelectron microscopy on *N*-bromoacetylpuromycin-labeled 50S subunits had located the site marked Pm in Fig. 29-1A. However, this is quite far from the recently determined location of the peptidyltransferase center,[k]

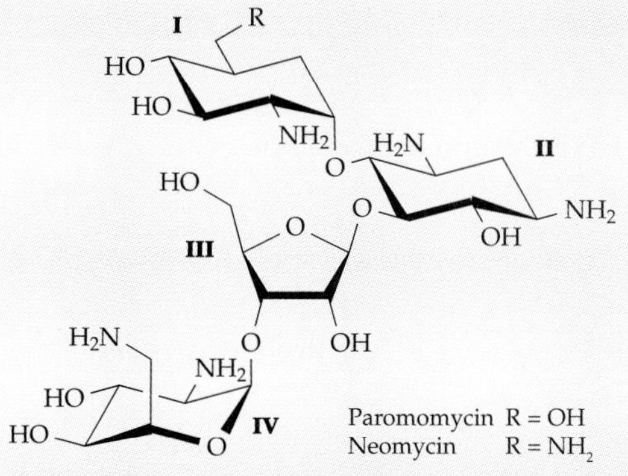

Paromomycin R = OH
Neomycin R = NH_2

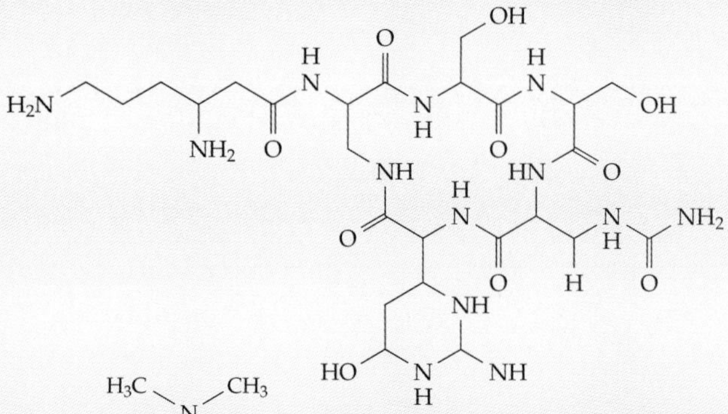

Viomycin

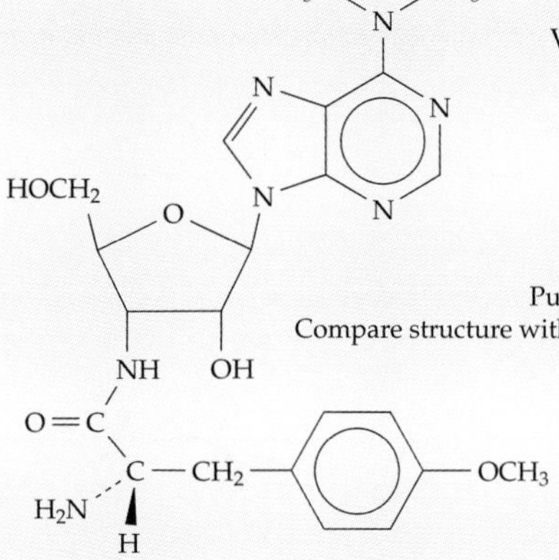

Puromycin
Compare structure with that of an aminoacyl-tRNA

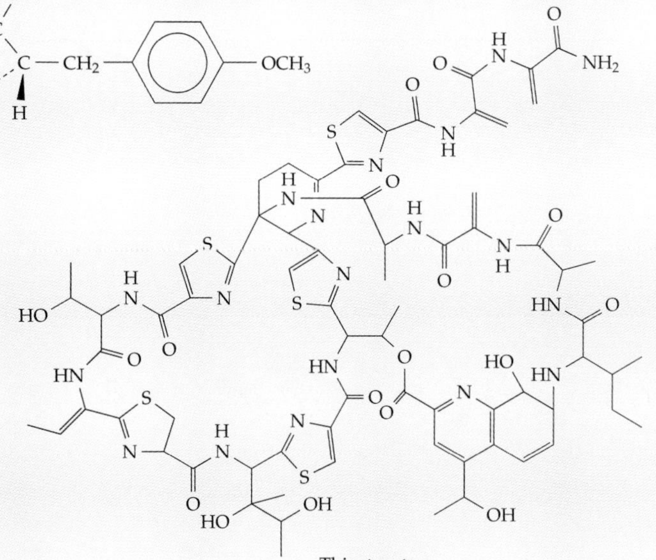

Thiostrepton

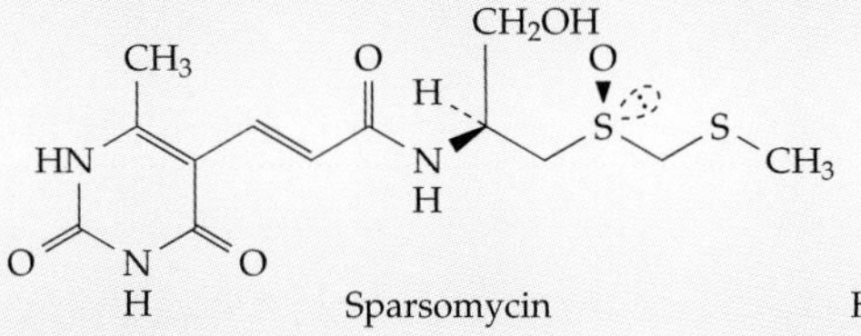

Sparsomycin

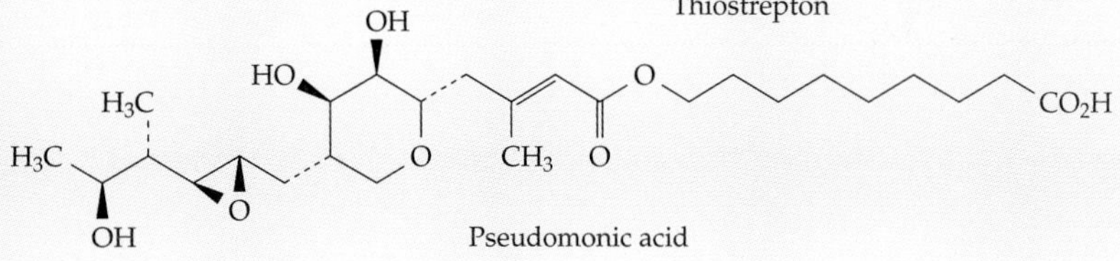

Pseudomonic acid

BOX 29-B ANTIBIOTICS THAT INHIBIT PROTEIN SYNTHESIS (continued)

illustrating the difficulty in use of active-site labeling techniques.

A large number of other antibiotics also bind in the vicinity of the peptidyltransferase center (Fig. 29-14).[l,m] Among them are macrolide antibiotics such as erythromycin (Fig. 21-1)[n] and spiramycin,[o,p] chloramphenicol (Fig. 25-10), griseofulvin (Fig. 21-10), streptogramins,[q] oxazolidinones such as linezolid,[r] sparsomycin,[s] and lincomycin. Erythromycin has been very useful in locating the peptidyltransferase center. For example, 23S RNA mutant G2057A, in which A has replaced the normal G, and mutants G2058G and G2058U are resistant to erythromycin.[l,n,t] Mutant C2518U in *Halobacterium halobium* 23S RNA (C2499 in *E. coli*) is among mutants resistant to sparsomycin.[s] Chloramphenicol not only blocks the peptidyltransferase but also causes an accumulation of the compound ppGpp (p. 1715).

The aminoglycoside antibiotics **streptomycin** (Box 20-B),[c,d,u,v] the neomycins,[w] **paromomycin** (see drawing below),[c,x–z] gentamycin,[aa] and kanamycin have one structural unit in common. They often bind to 16S ribosomal RNA in the decoding center. However, they bind in distinctly different ways. Streptomycin causes ribosomes to misread the genetic code[bb,cc] primarily at the first base of a codon. Thus, when poly(U) serves as a messenger RNA, the expected polyphenylalanine product contains 40% isoleucine. When a bacterial population is subjected to the action of any antibiotic, a few mutants are able to grow and survive in the presence of the antibiotic. Streptomycin-resistant mutants of *E. coli* arise at the very low frequency of ~10^{-12}. One of the genes affected (*rpsL*) was mapped at 72 min. Subsequently, it was shown that streptomycin binds to ribosomal protein S12, and that *rpsL* is the gene for this protein. Mutations in the universally conserved 2660 loop of 23S RNA in the sarcin/ricin domain lead to blockage of the elongation cycle. Bacteria containing both a G2661C mutation in their 23S RNA and also a streptomycin resistance mutation in protein S12 lose efficiency in the action of EF-Tu and die. However, they survive in the presence of streptomycin.[dd]

Streptomycin can also be chemically crosslinked to 16S RNA,[ee] and several aminoglycoside antibiotics including streptomycin and spectinomycin bind

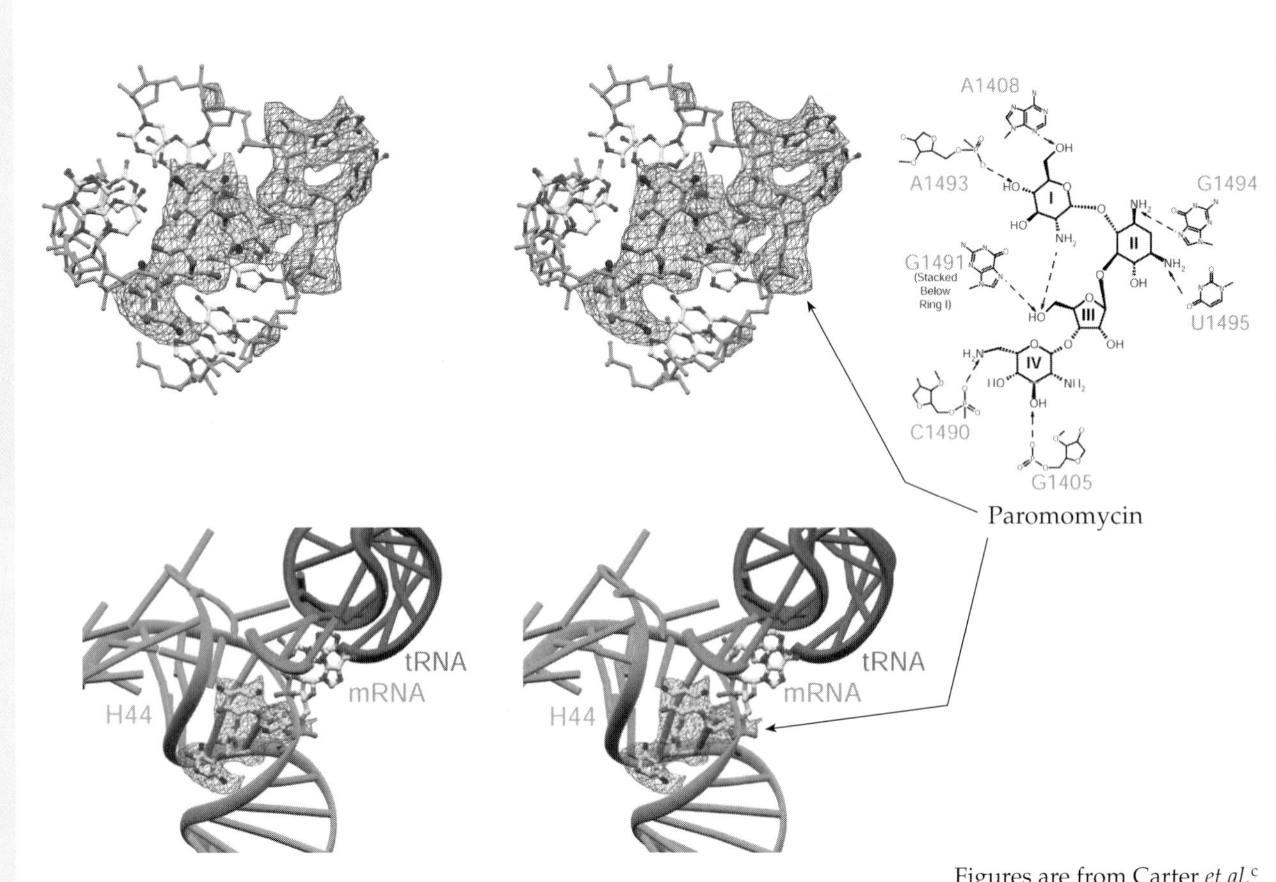

Figures are from Carter *et al.*[c]

BOX 29-B (continued)

to and, in footprinting experiments, protect specific nucleotides in 16S RNA.[ff] Streptomycin binds tightly to the upper part of helix 44 of bacterial 16S RNA (Fig. 29-2). This part of the helix is in the region that binds messenger RNA. It also contains parts of the A and P sites of the decoding region of the 30S ribosomal subunit. The same antibiotic binds less tightly to the 915 region in the center of the 16S RNA.[gg,hh] As mentioned above, some streptomycin-resistant mutants become dependent upon the antibiotic and will not grow in its absence. This streptomycin dependence sometimes results from modification in ribosomal protein S4, but the dependence can be suppressed by specific mutations in S5.[ii] It is clear that a single point mutation altering one amino acid is all that is necessary to enormously change the sensitivity of a living organism to a particular toxin, or even to make the organism dependent upon that toxin. Paromomycin also binds to the upper end of helix 44 in the major groove of the RNA (see figure in this box) and close to the streptomycin-binding site.[c,z,jj] Binding distorts the structure of the bulge loop containing adenosines A1492 and A1493, which are markers for the A site of the decoding region. Messenger RNA is also bound at this site as shown in the accompanying figure. Gentamycin also binds in the A site.[aa]

Hygromycin B also binds at the very top of helix 44 blocking the translocation step in the ribosomal cycle.[kk] **Spectinomycin** binds not only to RNA but also to protein S5, as indicated by analysis of resistant mutants. The S5 structural gene *spcA* maps at 64 min, a position in a ribosomal protein operon of the *E. coli* chromosome. The 16S RNA binding site at one end of helix 34 (with protection of G1064 and C1192) is adjacent to S5 as shown by X-ray structural analysis and directed hydroxyl radical probing (see Fig. 29-2).[c] The antibiotic also interferes with the translocation step of polypeptide elongation. **Kasugamycin** inhibits the binding of fMet-tRNA (initiation). In this case, resistant mutants appear in which it is not a protein subunit that has been modified but the 16S RNA. In resistant strains there is less methylation of adenosines 1518 and 1519 (Fig. 29-2) than in normal strains.[ll]

The **tetracyclines** (Fig. 21-10) inhibit the binding of aminoacyl-tRNA at the A site in the 30S ribosomal subunit.[kk] However, this doesn't appear to be a direct effect. Tetracyclines bind to the 16S RNA at two sites. A major site is on helix 34 near the spectinomycin site in the platform region. A second site is on helix 27, the switch helix, which plays a direct role in translocation (see Eq. 29-9).[kk] Although the basis of the inhibition is not clear, there are distinct differences in binding to the 16S bacterial and 18S eukaryotic RNAs that explain the high specificity of the antibiotic toward bacteria.

Another site of antibiotic action is the GTPase-activating center. This center contains a double hairpin structure in the 23S RNA, which binds to protein L11 and the L10•$(L12)_4$ stalk complex. Several proteins, including initiation factor IF1 and the elongation factors EF-Tu, EF-G (bacteria)/EF-2 (eukaryotes), bind to this part of the 50S ribosome (Fig. 29-4). The thiopeptide antibiotics **thiostrepton**, **micrococcin**, and siomycin[t,mm,nn] also bind in this region. Thiostrepton acts by preventing association with the ribosome of an incoming aminoacyl-tRNA as the EF-Tu•GTP complex. Its binding site is primarily in the 23S RNA, but it probably interferes with peptide elongation by interfering with a conformational change in protein L11.[t] A related cyclic peptide (GE2270A) binds to EF-Tu•GDP competing for binding of an aminoacyl-tRNA and blocking the GDP-GTP exchange.[nn] Certain mutations in the EF-Tu protein confer resistance to this antibiotic.[oo] In a similar way kirromycin prevents release of EF-Tu from the ribosome after GTP hydrolysis.[nn,pp,qq]

The binding site of initiation factor IF1 involves both the 30S and 50S ribosomal subunits. The large oligosaccharide antibiotic **evenimicin** protects a specific set of nucleotides in two loops near the peptidyltransferase center (Fig. 29-14).[rr] Erythromycin,[b] other macrolide antibiotics, cycloheximide (Fig. 21-10), and fusidic acid (p. 1266) all prevent translocation by stabilizing the pre-translocation complex.[ss,tt] Fusidic acid may bind to EF-G on the ribosome, preventing an essential conformational change in this G protein.[tt] Fusidic acid also inhibits accumulation of ppGpp. Figure 29-14 shows the locations of some mutations in *E. coli* 23S rRNA that confer resistance to erythromycins and chloramphenicol. Notice that both domains II and V are involved.[n] **Pactamycin** binds to helices 23b and 24a, a binding site for initiation factor IF3.[kk]

Many antibiotics, which inhibit protein synthesis, do not bind to ribosomes but block any of a variety of vital chemical processes needed for growth. Among them are **pseudomonic acid**, which inhibits isoleucyl-tRNA synthetase from many gram-positive bacteria.[uu,vv] **Rapamycin**, best known as an immunosuppressant (Box 9-F), inhibits phosphoinositide-3-kinase and also phosphorylation of the cap-binding protein 4G, a component of the eukaryotic initiation factor complex (Fig. 29-11).[ww] The bacterial enzyme peptide deformylase, which is absent from the human body, has been suggested as a target for design of synthetic antibiotics.[xx]

BOX 29-B ANTIBIOTICS THAT INHIBIT PROTEIN SYNTHESIS (continued)

[a] Gorini, L. (1966) *Sci. Am.* **214**(Apr), 102–109
[b] Narayanan, C. S., and Dubnau, D. (1987) *J. Biol. Chem.* **262**, 1766–1771
[c] Carter, A. P., Clemons, W. M., Brodersen, D. E., Morgan-Warren, R. J., Wimberly, B. T., and Ramakrishnan, V. (2000) *Nature (London)* **407**, 340–348
[d] Schroeder, R., Waldsich, C., and Wank, H. (2000) *EMBO J.* **19**, 1–9
[de] Schlünzen, F., Zarivach, R., Harms, J., Bashan, A., Tocilj, A., Albrecht, R., Yonath, A., and Franceschi, F. (2001) *Nature (London)* **413**, 814–821
[e] Nierhaus, K. H. (1990) *Biochemistry* **29**, 4997–5008
[f] Olson, H. M., Nicholson, A. W., Cooperman, B. S., and Glitz, D. G. (1985) *J. Biol. Chem.* **260**, 10326–10331
[g] Weitzmann, C. J., and Cooperman, B. S. (1990) *Biochemistry* **29**, 3458–3465
[h] Bischof, O., Kruft, V., and Wittmann-Liebold, B. (1994) *J. Biol. Chem.* **269**, 18315–18319
[i] Nissen, P., Hansen, J., Ban, N., Moore, P. B., and Steitz, T. A. (2000) *Science* **289**, 920–930
[j] Welch, M., Chastang, J., and Yarus, M. (1995) *Biochemistry* **34**, 385–390
[k] Lührmann, R., Bald, R., Stöffler-Meilicke, M., and Stöffler, G. (1981) *Proc. Natl. Acad. Sci. U.S.A.* **78**, 7276–7280
[l] Garrett, R. (1983) *Trends Biochem. Sci.* **8**, 189–190
[m] Rodriguez-Fonseca, C., Amils, R., and Garrett, R. A. (1995) *J. Mol. Biol.* **247**, 224–235
[n] Douthwaite, S., and Aagaard, C. (1993) *J. Mol. Biol.* **232**, 725–731
[o] Bischof, O., Urlaub, H., Kruft, V., and Wittmann-Liebold, B. (1995) *J. Biol. Chem.* **270**, 23060–23064
[p] Poulsen, S. M., Kofoed, C., and Vester, B. (2000) *J. Mol. Biol.* **304**, 471–481
[q] Porse, B. T., and Garrett, R. A. (1999) *J. Mol. Biol.* **286**, 375–387
[r] Kloss, P., Xiong, L., Shinabarger, D. L., and Mankin, A. S. (1999) *J. Mol. Biol.* **294**, 93–101
[s] Tan, G. T., DeBlasio, A., and Mankin, A. S. (1996) *J. Mol. Biol.* **261**, 222–230
[t] Porse, B. T., Leviev, I., Mankin, A. S., and Garrett, R. A. (1998) *J. Mol. Biol.* **276**, 391–404
[u] Davies, C., Bussiere, D. E., Golden, B. L., Porter, S. J., Ramakrishnan, V., and White, S. W. (1998) *J. Mol. Biol.* **279**, 873–888
[v] Recht, M. I., Douthwaite, S., and Puglisi, J. D. (1999) *EMBO J.* **18**, 3133–3138
[w] Fourmy, D., Recht, M. I., and Puglisi, J. D. (1998) *J. Mol. Biol.* **277**, 347–362
[x] Fourmy, D., Recht, M. I., Blanchard, S. C., and Puglisi, J. D. (1996) *Science* **274**, 1367–1371
[y] Recht, M. I., Douthwaite, S., Dahlquist, K. D., and Puglisi, J. D. (1999) *J. Mol. Biol.* **286**, 33–43
[z] Lynch, S. R., and Puglisi, J. D. (2001) *J. Mol. Biol.* **306**, 1037–1058
[aa] Yoshizawa, S., Fourmy, D., and Puglisi, J. D. (1998) *EMBO J.* **17**, 6437–6448
[bb] Browning, K. S., Maia, D. M., Lax, S. R., and Ravel, J. M. (1987) *J. Biol. Chem.* **262**, 538–541
[cc] Tai, P.-C., Wallace, B. J., and Davis, B. D. (1978) *Proc. Natl. Acad. Sci. U.S.A.* **75**, 275–279
[dd] Bilgin, N., and Ehrenberg, M. (1994) *J. Mol. Biol.* **235**, 813–824
[ee] Gravel, M., Melancon, P., and Brakier-Gingras, L. (1987) *Biochemistry* **26**, 6227–6232
[ff] Moazed, D., and Noller, H. F. (1987) *Nature (London)* **327**, 389–394
[gg] Pinard, R., Payant, C., Melancon, P., and Brakier-Gingras, L. (1993) *FASEB J.* **7**, 173–176
[hh] Spickler, C., Brunelle, M.-N., and Brakier-Gingras, L. (1997) *J. Mol. Biol.* **273**, 586–599
[ii] Culver, G. M., Heilek, G. M., and Noller, H. F. (1999) *J. Mol. Biol.* **286**, 355–364
[jj] VanLoock, M. S., Easterwood, T. R., and Harvey, S. C. (1999) *J. Mol. Biol.* **285**, 2069–2078
[kk] Brodersen, D. E., Clemons, W. M., Jr., Carter, A. P., Morgan-Warren, R. J., Wimberly, B. T., and Ramakrishnan, V. (2000) *Cell* **103**, 1143–1154
[ll] Vila-Sanjurjo, A., Squires, C. L., and Dahlberg, A. E. (1999) *J. Mol. Biol.* **293**, 1–8
[mm] Porse, B. T., Cundliffe, E., and Garrett, R. A. (1999) *J. Mol. Biol.* **287**, 33–45
[nn] Heffron, S. E., and Jurnak, F. (2000) *Biochemistry* **39**, 37–45
[oo] Zuurmond, A.-M., de Graff, J. M., Olsthoorn-Tieleman, L. N., van Duyl, B. Y., Mörhle, V. G., Jurnak, F., Mesters, J. R., Hilgenfeld, R., and Kraal, B. (2000) *J. Mol. Biol.* **304**, 995–1005
[pp] Mesters, J. R., Zeef, L. A. H., Hilgenfeld, R., de Graaf, J. M., Kraal, B., and Bosch, L. (1994) *EMBO J.* **13**, 4877–4885
[qq] Alexander, C., Bilgin, N., Lindschau, C., Mesters, J. R., Kraal, B., Hilgenfeld, R., Erdmann, V. A., and Lippmann, C. (1995) *J. Biol. Chem.* **270**, 14541–14547
[rr] Belova, L., Tenson, T., Xiong, L., McNicholas, P. M., and Mankin, A. S. (2001) *Proc. Natl. Acad. Sci. U.S.A.* **98**, 3726–3731
[ss] Johanson, U., AEvarsson, A., Liljas, A., and Hughes, D. (1996) *J. Mol. Biol.* **258**, 420–432
[tt] Laurberg, M., Kristensen, O., Martemyanov, K., Gudkov, A. T., Nagaev, I., Hughes, D., and Liljas, A. (2000) *J. Mol. Biol.* **303**, 593–603
[uu] Yanagisawa, T., Lee, J. T., Wu, H. C., and Kawakami, M. (1994) *J. Biol. Chem.* **269**, 24304–24309
[vv] Sassanfar, M., Kranz, J. E., Gallant, P., Schimmel, P., and Shiba, K. (1996) *Biochemistry* **35**, 9995–10003
[ww] Raught, B., Gingras, A. C., Gygi, S. P., Imataka, H., Morino, S., Gradi, A., Aebersold, R., and Sonenberg, N. (2000) *EMBO J.* **19**, 434–444
[xx] Hao, B., Gong, W., Rajagopalan, P. T. R., Zhou, Y., Pei, D., and Chan, M. K. (1999) *Biochemistry* **38**, 4712–4719

form the acceptor stem and the TψC arm form a single **acceptor-TψC-minihelix**, while the other two domains fold together to create an **anticodon-dihydrouridine** stem loop.[172] New tertiary interactions, some of which are indicated in Fig. 29-7C, are formed. Mitochondrial tRNAs of metazoa often lack some elements of the cloverleaf. An extreme example is the bovine mtRNA^{Ser}, which recognizes AGY codons and completely lacks the dihydrouridine loop. This fact suggests that the L shape of tRNAs cannot be completely invariant.[174,176] As is shown in Fig. 29-7B, divalent metal ions such as Mg^{2+} are bound at discrete sites in tRNA molecules.[171] The tertiary interactions in the "core" of the L form contains several stacked layers of base pairs and triplets (Fig. 29-7C). The top layer is usually the single base 59; below it in succession are the 15:48 pair (see Fig. 29-7), the 21:8:14 triplet, the 13:22 pair (see also p. 231), and then base pairs present in the dihydroU loop. Considerable variation is observed among the different tRNAs.[182,183] The structural features of this core may also be utilized for recognition by aminoacyl tRNA synthetases.

Initiator tRNAs. While the TψC sequence has been found in all bacterial and most eukaryotic tRNAs examined, it is replaced by UCG in eukaryotic initiator tRNAs. In these tRNAs the preceding two nucleotides, beginning in the stem of loop IV, are also conserved; the complete conserved sequence being GAUCG.[184] Other characteristics of initiator tRNAs are the absence of base-pairing between residues 1, and 72, and the presence of C rather than G at position 1, A rather than G at position 72, and CCU in place of the two dihydroU residues in loop I.[185] Initiator tRNAs of chloroplasts resemble those of bacteria,[186] whereas archaeobacteria have their own unique peculiarities.[187] These include the presence of a hypermodified base known as **archaeosine** (p. 1456) in position 15 of the dihydroU loop.[188,189]

2. Pairing of Codon and Anticodon

Accurate protein synthesis depends upon both correct charging of the tRNAs and correct recognition by an anticodon in the tRNA of the complementary codon in the mRNA. A surprise was the discovery of inosine (I) in anticodons of yeast tRNA (but not in most *E. coli* tRNAs). Another unexpected finding was that fewer than 61 kinds of tRNA exist in a given cell (61 = 64 codons minus three stop codons). Consideration of these matters led Crick, in 1966, to propose the **wobble hypothesis**.[190] According to this proposal the first two bases at the 5' end of the codon (and at the 3' end of the anticodon) must pair in the same ways as do the bases in DNA. However, the third base pair (3' end of the codon and 5' end of the anticodon) is under a less severe steric restriction. That is, there may be some "wobble." Crick suggested the accompanying rule for pairing of the third base. All of the observed

5′-Base in anticodon	Paired 3′-base in codon
G	C or U
C	G
A	U
U	A or G
I	C, A, or U

deviations from the AU, CG pairing of a Watson–Crick helix can be explained in this way. An anticodon with G at the 5' end can pair with codons with either C or U at their 3′ end. Anticodons with C or A at the 5′ end pair strictly. Anticodons ending with U at the 5′ end can pair with codons containing either A or G in the 3' position. Anticodons with I in the 5' position can recognize codons with any of the three bases in the third position. Comparison with Table 5-5 makes it immediately clear why fewer than 61 anticodons are needed. Many codons represent the same amino acid, and frequently the nature of the base in the 3' position of the codon is immaterial to the meaning of that codon. Thus, there is an economy in using less than the full array of anticodons. Crick showed that his proposal was chemically feasible if the spatial relationships for the wobble pair were allowed to vary from the usual ones in Watson–Crick base pairs. This is illustrated in Fig. 29-8 for binding of inosine to C (a normal Watson–Crick base pair) and to A and to U. Although the word wobble does not convey an exact meaning, the hypothesis has predicted many things correctly. For example, only three tRNAs are required to recognize the six serine codons. In fact, only three are found in *E. coli*.

The matter is made more complex by the fact that U34 in the first (5') anticodon position of tRNA is usually modified postranscriptionally.[190a] For

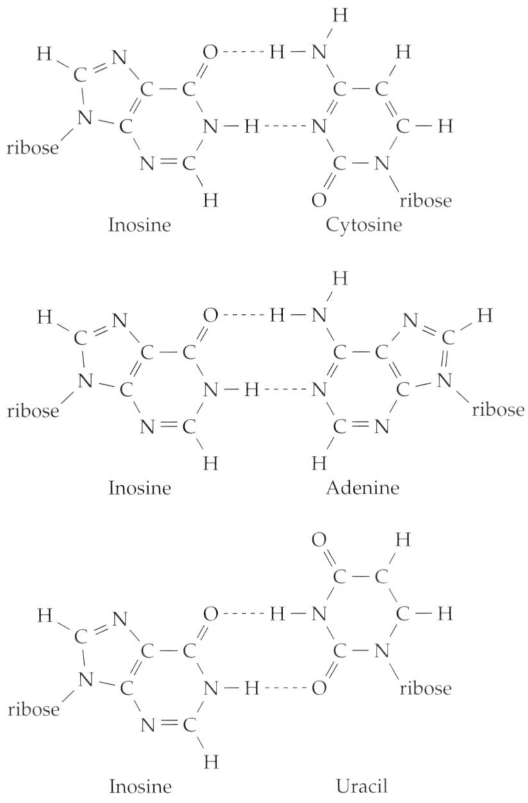

Inosine — Cytosine

Inosine — Adenine

Inosine — Uracil

Figure 29-8 Pairing of inosine with cytosine (a Watson–Crick pair) and of inosine with adenine and uracil (wobble pairs).

example, it is usually converted to the 2-thio-5-CH_3 or 2-thio-5-CH_2-NH_2^+-CH_3 derivative in anticodons recognizing A in the 3' position of a codon. For recognition of other bases the 5–OH, –OCH_3, or –OCH_2COO^- derivative is usually present. Yokoyama *et al.* attributed the selectivity to the stabilization of the C2' endo form of the ring in the former group.[191]

3. Aminoacylation of tRNAs

Discrimination between some pairs of tRNAs depends entirely on the anticodon sequence. For example, $tRNA^{Met}$ contains the anticodon CAU. That for a minor $tRNA^{Ile}$ is the same except that the cytosine has been posttranscriptionally modified by covalent linkage of a molecule of lysine via its ε-amino group to C2 of the cytosine. The latter base (**lysidine**) is correctly recognized by *E. coli* isoleucyl-tRNA synthetase; but, if the cytosine is unmodified, it is aminoacylated by methionyl-tRNA synthetase.[192] In most instances the acceptor specificity, or **tRNA identity**, is not determined solely by the anticodon sequence. Thus, when a methionine initiator tRNA was modified to contain a tryptophan anticodon, it was only partially charged with tryptophan *in vivo*. However, when A73 of the methionine tRNA was also converted to G73, only tryptophan was inserted.[193] Nucleotide 73 (Fig. 29-7) is sometimes called the **discriminator nucleotide**.[194–196] It is A in methionine and leucine tRNAs,[197] G in tryptophan tRNAs, and C in histidyl RNAs.[198] The tRNA features needed to establish its identity are sometimes referred to as its **identity-determinant set**.[196,198a,b] This includes the anticodon and other features needed for recognition by the aminoacyl-tRNA synthetases that "charge" the tRNAs with aminoacyl groups.[199–204] For example, for *E. coli* $tRNA^{Val}$ the recognition determinants are A35 and C36 of the anticodon, A73, G20, G45, and a regular A-RNA acceptor helix.[205] All known mature tRNAs contain a 3'-CCA end on which the aminoacylation occurs. Nevertheless, alterations in this sequence still allow correct aminoacylation of some tRNAs.[206]

The aminoacyl-tRNA synthetases (amino acid: tRNA ligases) join amino acids to their appropriate transfer RNA molecules for protein synthesis. They have the very important task of selecting both a specific amino acid and a specific tRNA and joining them according to Eq. 29-1.[175,195,207] These reactions represent the first step in the decoding of mRNAs. Organisms usually contain one aminoacyl-tRNA synthetase for each of the 20 amino acids. Each synthetase must select a specific amino acid and a correct tRNA for that amino acid. The same enzyme transfers an activated amino acid to all of the **isoacceptor tRNAs** specific for a given amino acid. Some aminoacyl-tRNA synthetases attach the aminoacyl group to the 2'–OH of the tRNA substrate, some to the 3'–OH. The chemical mechanism is the same in both cases.

Structures. Aminoacyl-tRNA synthetases vary in size, the subunit masses ranging from 37- to 110-kDa (329–951 residues). There are monomeric species, dimers, tetramers, and $(\alpha\beta)_2$ mixed tetramers.[207,208] Sequence comparisons, together with X-ray structural investigations, have shown that the enzymes can be classified into two groups, each containing ten enzymes.[207,209,209–209b] **Class I aminoacyl-tRNA synthetases** share two consensus sequence motifs: HIGH and KMSKS. Their ATP-binding active sites are in a Rossman fold nucleotide-binding domain (Fig. 2-13).[210] The KMSKS sequence parallels the Walker sequence found in various nucleotide-binding proteins including ATP synthase.[211–213] The actual sequences vary considerably, e.g., the KMSKS sequence for a tyrosyl-tRNA synthetase is actually KFGKT.[211]

Class II aminoacyl-tRNA synthetases contain a different set of three "signature sequences," two of which form an ATP-binding catalytic domain. The active site structure is built on an antiparallel β sheet and is surrounded by two helices (Fig. 29-9). Each class contains subgroups with inserted loops that form other domains. In the following tabulation the reference numbers refer to three-dimensional structural studies.

Class I
Glu,[209] Gln,[218–220] Arg,[221]
Tyr,[222] Trp,[212]
Ile,[223] Leu,[224] Val, Cys, Met[210,225,226]
Class II
His,[227–229] Pro, Ser,[230,231] Thr
Asp,[232,233] Asn,[234]Lys,[217,235]
Phe,[236] Ala, Gly,[204,237]

The 37-kDa 334-residue subunits of the dimeric type I tryptophanyl-tRNA synthetase[238] are the smallest known; the largest bacterial synthetase is an alanine-specific type II tetramer with 95-kDa 875-residue subunits.[239] Gene deletions show that a much smaller core, comparable in size to that of the tryptophanyl-tRNA synthetase, is needed for amino acid activation. The synthetases share little sequence homology except for a short 11-residue part of the adenylate binding site near the N terminus.[240,241] Some of the synthetases contain bound zinc ions.[225,242]

Recognition of cognate tRNAs. Many attempts have been made to learn what part or parts of tRNA molecules are involved in recognition by aminoacyl-tRNA synthetases. Nucleotide sequences of isoacceptor tRNAs have been compared. Chemically modified and fragmented tRNA molecules have been studied, and many mutant tRNAs have been made. These have often been mutants of suppressor tRNAs that place specific amino acids such as phenylalanine or alanine into a peptide at a termination codon, often the termination codon UAG (see Section C4). An alternative approach is to synthesize DNA templates, which can be transcribed *in vitro* by phage T7 RNA polymerase to give mutant tRNAs,[243] or to make such tRNAs by solid-phase chemical synthesis.[244] Although these contain no modified bases, they serve as substrates for the aminoacyl-tRNA synthetases.

The results of these efforts show that no method of tRNA recognition is universal.[244a] In some cases, e.g., for methionine- or valine-specific tRNAs, the synthetase does not aminoacylate a modified tRNA if the anticodon structure is incorrect. Although the anticodon is 7.5 nm away from the CCA end of the tRNA, the synthetases are large enzymes. Many of them are able to accommodate this large distance between a recognition site and the active site (Fig. 29-9A). For some other tRNAs the anticodon is not involved in recognition.[245] For yeast $tRNA^{Phe}$ residues in the stem of the dihydrouridine loop and at the upper end of the amino acid acceptor stem seem to be critical.[241]

For some other tRNAs only the acceptor helix is essential for recognition. Change of one base-pair, the pair G3 • U70 (a "wobble" pair) of an *E. coli* $tRNA^{Ala}$ • mRNA complex to the unnatural A3 • U70, prevents aminoacylation. Conversely, a G3 • U70 pair formed with tRNAs specific for other amino acids causes them to become substrates for the alanyl-tRNA synthetase.[241,246] Even a shortened tRNA minihelix consisting of a 7-bp acceptor stem, 6-nucleotide loop, and ACCA 3' end is a substrate for this enzyme.[247,247a] A seryl-tRNA synthetase depends upon recognition of two base pairs in the acceptor stem.[248] Synthetic DNA oligomers with sequences corresponding to those of *E. coli* $tRNA^{Phe}$ or $tRNA^{Lys}$, and with either deoxythymidine or deoxyuridine in the positions occupied by ribouridine in the tRNAs, are also substrates for the synthetases. The affinity and reaction rates are somewhat decreased, but the ribose 2'-OH is not essential for recognition.[249]

Mechanisms of reaction. Activation of an amino acid occurs by a direct in-line nucleophilic displacement by a carboxylate oxygen atom of the amino acid on the α phosphorus atom of MgATP to form the aminoacyl adenylate (Eq. 29-1, step *a*). For yeast phenylalanyl-tRNA synthetases the preferred form of MgATP appears to be the β,γ-bidentate (Λ screw sense) complex (p. 643).[250] This is followed by a second nucleophilic displacement, this one on the C=O group of the aminoacyl adenylate by the –OH group of the tRNA (Eq. 29-1, step *b*; Fig. 29-9C). A conformational change in the protein may be required to permit dissociation of the product, the aminoacyl-tRNA. In the complex of a class I synthetase with aminoacyl adenylate and tRNA the 3' CCA acceptor end of the tRNA is straight, but in a class II synthetase it is bent. In the two classes of synthetase the tRNAs approach the enzyme in a mirror-symmetric fashion. The 2'-OH of the terminal ribose is positioned to attack the carbonyl of the aminoacyl adenylate in class I enzymes, while the 3'-OH is positional for the attack in class II enzymes.[207]

The three-dimensional structure of *E. coli* glutaminyl-tRNA synthetase is shown in Fig. 29-9A. The active site lies against a β sheet in a structure similar to the nucleotide binding domain of a dehydrogenase (Fig. 2-13). The site of binding of a ATP is marked in Fig. 29-9A. The details of this binding to tyrosine-tRNA synthetase have been studied intensively.[251–255] Binding of the tRNA substrates is less well understood. A large series of mutants involving 40 basic residues were prepared by Bedouelle and Winter[256]; study of these mutant enzymes, together with computer-assisted modeling, led to a proposed structure for a transition state for a complex with tRNA as is shown in Fig. 29-9B. Kinetic studies of heterodimers prepared from mutant and normal enzyme confirm that both subunits of the dimeric enzyme interact with the tRNA.

The active site of a type II synthetase is shown in Fig. 29-9C,D.[217,217a] The expected movement of electrons in the reaction with ATP is illustrated by the green arrow in D. Both metal ions and active-site protein groups may participate as is also proposed for another type II enzyme.[229]

Some data suggested that a transient covalent linkage of tRNA to the synthetases may form through addition of a nucleophilic group of the enzyme to the 6' position of the uracil (or 4-thiouracil) present in position 8 of all tRNAs (Eq. 29-3).[257] The two isoacceptors $tRNA^{Tyr}$ species in *E. coli* contain 4-thiouracil at this position. The C=C bond in this base can be saturated by sodium borohydride reduction, which was found not only to prevent the covalent interaction with the enzyme but also to prevent aminoacylation of the tRNA. However, Eq. 29-3 probably describes a side reaction irrelevant to tRNA function.

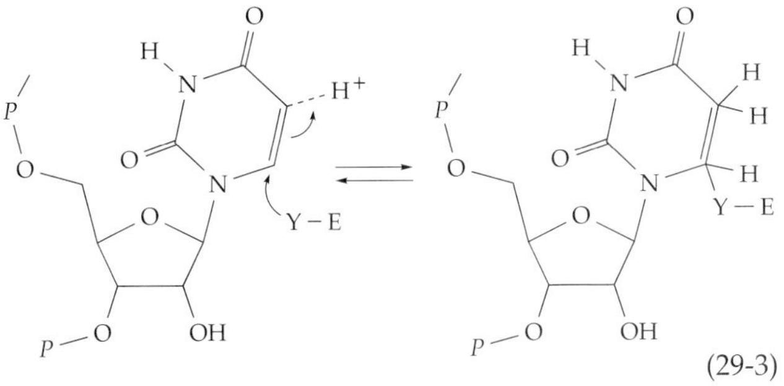

(29-3)

Correcting errors. Much attention has been devoted to "proofreading" or "editing" activities of amino acid-tRNA synthetases (see p. 482). For the majority of the enzymes proofreading is not needed.[209a] Thus, tyrosyl-tRNA synthetase (Fig. 29-9B) mistakenly chooses phenylalanine instead of tyrosine only 5 in 10^4 times, apparently a tolerable rate of error. This enzyme, as well as a tryptophan-specific synthetase,[238] depends largely upon differences in the Gibbs energy of binding to select the correct substrate. However, the discrimination between valine and isoleucine by isoleucyl-tRNA synthetase poses a more difficult problem. It is apparently solved, in part, by a "double sieve" editing mechanism,[223,224,258,259] which is described briefly on p. 482. In the first sieve competitors that are larger than the substrate or are differently shaped are excluded by steric repulsion from binding in the active site. Isoleucyl-tRNA synthetase doesn't convert leucine into an aminoacyladenylate, but it does act on the smaller valine. However, most of the resulting enzyme-bound valyl-adenylate is hydrolyzed to valine and AMP before it can be transferred to $tRNA^{Ileu}$. It shifts into an editing site, which is too small for the isoleucyl-adenylate, in effect passing through a second sieve.[259a] Some of the activated valine is transferred to tRNA and is removed in a second editing reaction.[223] Misactivation of threonine or some other amino acids by valyl-tRNA synthetase is corrected in an analogous fashion.[259b,c]

A similar editing process prevents isoleucyl-, leucyl-, and methionyl-tRNA synthetases from attaching L-homocysteine to tRNAs.[260–263] In this case, instead of hydrolysis the editing site catalyzes conversion of the homocysteinyl-adenylate into homocysteine lactone. Naturally occurring mutations in tRNA molecules can sometimes have serious consequences. For example, a human mutation is responsible for a fragile mitochondrial isoleucine tRNA and serious cardiomyopathy and opthalmophegia (see also Box 18-B).[263a]

H NH_3^+ O S

L-Homocysteine lactone

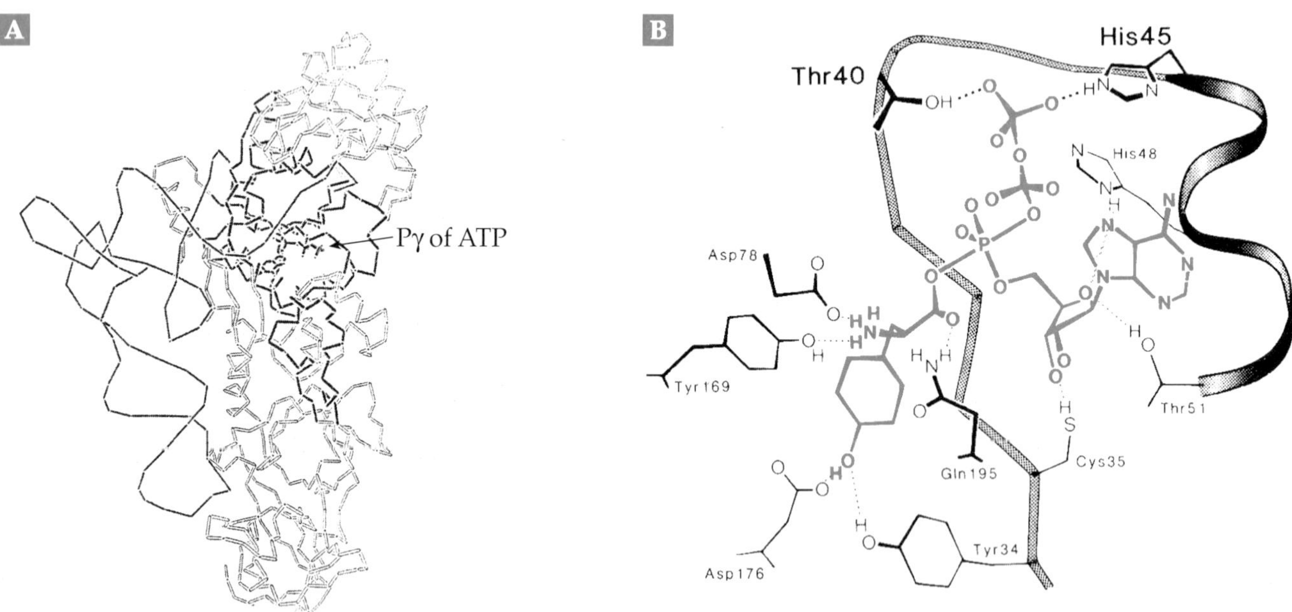

Figure 29-9 Selected views of aminoacyl-tRNA synthetase structure and action. (A) Alpha-carbon trace of the type I *E. coli* glutaminyl-tRNA synthetase. The phosphate backbone of $tRNA^{Gln}$ is shown in black; ATP is shown in the active-site cleft. The canonical dinucleotide fold domain near the N terminus is shaded. Two structural motifs (black), proposed to link the active site with regions of the protein–RNA interface involved in tRNA discrimination, are indicated. The α helix (top) connects tRNA recognition in the minor groove of the acceptor stem with binding of the ribose group of ATP. The large loop (center) connects anticodon recognition by the two β-barrel domains (bottom) with sequences flanking the MSK sequence motif, which interacts with the phosphates of ATP. From Perona *et al.*[214] Courtesy of Thomas A. Steitz. (B) The active site structure of tyrosyl-tRNA synthetase from *Bacillus stearothermophilus* with a penta-coordinate transition state structure modeled.[215] From Leatherbarrow *et al.*[216] (C) Schematic representation of the active site of the lysyl-tRNA synthetase showing potential hydrogen bonding interaction in the ternary complex with lysine and ATP. The invariant motif 2 Arg 262 plays a key role in the recognition of the lysine carboxylate and the ATP α phosphate, while the invariat motif 2 Arg 480 binds the

Proofreading involves kinetic as well as thermodynamic considerations.[264–266] The relative rates of a hydrolytic reaction and the competing activating reaction must always be considered. These ratios can be strongly affected by conformational changes, which may occur in several steps (see also Section C,2).

Essential modification reactions of aminoacyl-tRNAs. In bacteria the initiator tRNA needed to start the synthesis of a polypeptide is initially aminoacylated by methionine, but the methionyl-$tRNA^{fMet}$ must then be *N*-formylated by transfer of a formyl group from N^{10}-formyltetrahydrofolate (Fig. 15-18; Eq. 29-4).[267,268] In gram-positive bacteria and in archaea, mitochondria, and chloroplasts the glutamine-specific $tRNA^{Gln}$ is charged with glutamate to form Glu-$tRNA^{Gln}$. The latter is converted by action of an ATP-dependent amidotransferase (see Eq. 24-22) to the necessary Gln-$tRNA^{Gln}$ (Eq. 29-5).[268a] In a similar way, RNA^{Asn} in some organisms is charged with aspartate, then converted by transamidation to Asp-$tRNA^{Asn}$ (Eq. 29-6).[207,267,269] An important reaction, that occurs in all kingdoms of life, is the charging of the special $tRNA^{Sec}$ with serine and conversion of the product into selenocysteinyl-$tRNA^{Sec}$ (p. 827; Eq 29-7).[267,270]

$$tRNA^{fMet} \xrightarrow[ATP]{Met} Met\text{-}tRNA^{fMet} \xrightarrow{Methyl\text{-}THF \;\; THF} fMet\text{-}tRNA^{fMet} \quad (29\text{-}4)$$

$$tRNA^{Gln} \xrightarrow[ATP]{Glu} Glu\text{-}tRNA^{Gln} \xrightarrow[ATP]{Gln \;\; Glu} Gln\text{-}tRNA^{Gln} \quad (29\text{-}5)$$

$$tRNA^{Asn} \xrightarrow[ATP]{Asp} Asp\text{-}tRNA^{Asn} \xrightarrow[ATP]{Gln \;\; Glu} Asn\text{-}tRNA^{Asn} \quad (29\text{-}6)$$

$$tRNA^{Sec} \xrightarrow[ATP]{Ser} Ser\text{-}tRNA^{Sec} \xrightarrow[(p.\ 827)]{Se\text{-}P \;\; P_i} Selenocysteinyl\text{-}tRNA^{Sec} \quad (29\text{-}7)$$

The introduction of selenocysteine into proteins at selected stop codons using this tRNA is described in Section C,5.

Examination of the complete genome sequences of methanogens revealed an apparent lack of cysteinyl-tRNA synthetase. However, prolyl-tRNA synthetase does correctly aminoacylate the tRNAs for both proline and cysteine in these archaeobacteria.[271–272a]

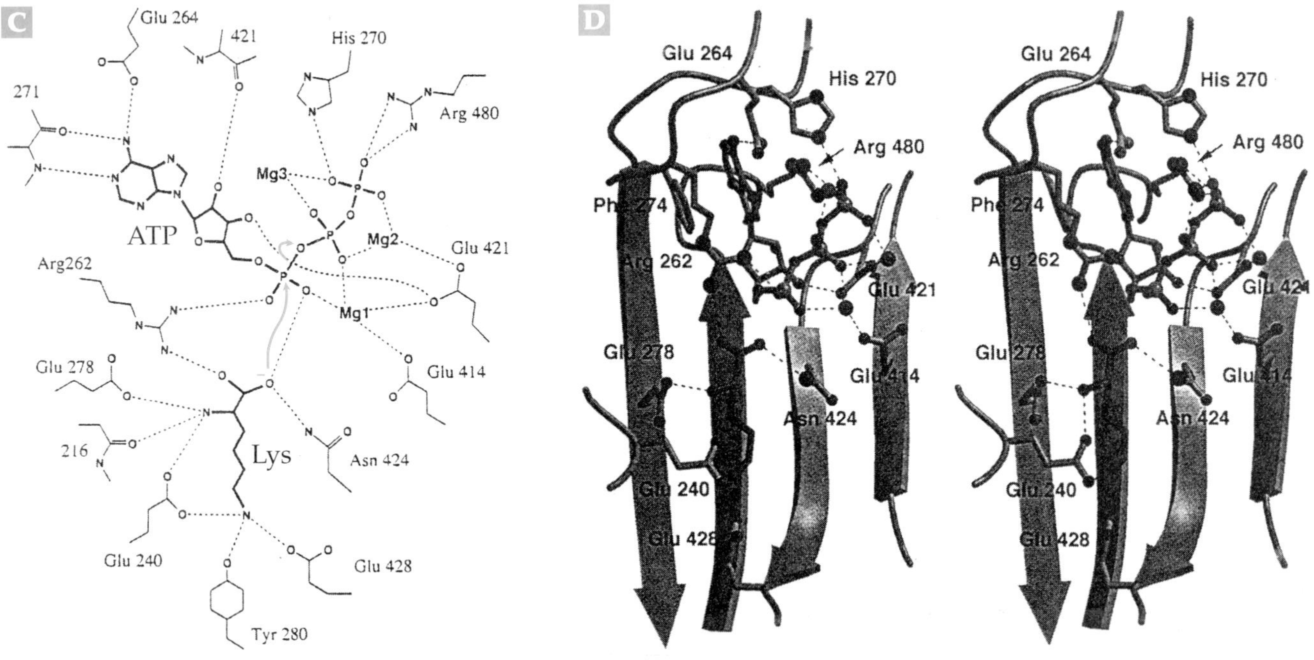

γ phosphate of the ATP. A number of conserved residues in the motif 2 loop (residues 264–271) assume an ordered conformation only upon ATP binding. The positions of the Mg^{2+} sites are indicated. (D) View of the active site of the type II lysyl-tRNA synthetase showing the conformations of the substrates lysine and ATP before the first step of the reaction takes place. The ATP molecule is located on one side of the central β sheet of the C-terminal domain, with the adenine ring sandwiched between a conserved phenylalanine (Phe 274) and the motif 3 arginine residue (Arg 480). The pyrophosphate moiety is bent toward the adenine placing the α phosphate in the correct position for nucleophilic attack of the lysine carboxylate oxygen. The hydrogen bonding and electrostatic interactions between the substrates and some of the key residues, including the invariant motif 2 arginine (Arg 262), are shown. The three Mg^{2+} ions (green) involved in catalysis are included. (C) and (D) are from Desogus *et al.*[217]

Additional functions of aminoacyl-tRNA synthetases. The primary function of these enzymes in protein synthesis is well known, but they have a whole range of other activities.[273,274] In *E. coli* the large alanyl-tRNA synthetase can repress transcription of its own gene by binding to a palindromic sequence in the control region of the gene.[275] Expression of some genes, such as that for threonyl-tRNA synthetase, is regulated at the translational level.[273,274,276] In mammalian cells the formation of the threonine-specific synthetase appears to be regulated by a phosphorylation–dephosphorylation mechanism.[277] Other synthetases participate in mitochondrial RNA splicing[278] and in aminoacylation of tRNA-like 3' ends of viral genomes (see Fig. 28-24) and of N termini of certain proteins.[279] For example, an arginyl group may be transferred onto the N terminus of a protein, marking it for rapid degradation.[280] Under conditions of apoptosis, tyrosyl tRNA synthetase is hydrolytically cleaved to form two different cytokines.[274] Phenylalanyl-tRNA synthetase is a DNA-binding protein.[281] Within the nucleus newly synthesized tRNAs are checked before being exported to the cytoplasm. Only tRNAs with mature 5' and 3' ends are exported. In both *Xenopus* oocytes[281a] and in *S. cere- visiae*[281b] the tRNAs are also tested prior to export, using aminoacyl-tRNA synthetases, to ensure that they are functional.

Many proteins have structures related to those of aminoacyl-tRNA synthetases.[282,283] For example, asparagine synthetase A functions via an aspartyl-adenylate intermediate (Chapter 24, Section B), and its structure resembls that of aspartyl-tRNA synthetase.[284] The *his G* gene of histidine biosynthesis (Fig. 25-13) encodes an ATP phosphoribosyltransferase with structural homology to the catalytic domain of histidyl-tRNA synthetase.[284] The reason is not clear, but some aminoacyl-tRNA synthetases, especially the histidyl-tRNA synthetase, are common autoantigens for the inflammatory disease **polymyositis**.[285,286]

C. Protein Synthesis: The Ribosome Cycle

Initiation (Figs. 29-10 and 29-11), elongation (Fig. 29-12), and termination are three distinct steps in the synthesis of a protein. A variety of specialized proteins are required for each stage of synthesis. Their sequential interaction with ribosomes can be viewed as a means of ensuring an orderly sequence of steps in the synthesis cycle. The rate of protein formation will depend upon the concentrations of amino acids, tRNAs, protein factors, numbers of ribosomes, and kinetic constants. The formation of specific proteins can also be inhibited by **translational repressors**, proteins that compete with ribosomes for binding to target mRNAs.[287]

1. Initiation

For most polypeptide chains initiation begins with one of the three **initiation codons**, most commonly the methionine codon AUG. When properly placed in an mRNA chain, GUG may also serve as a bacterial initiation codon. In such cases, it codes for methionine rather than for valine. Occasionally UUG, AUU, ACG, and perhaps other codons can initiate translation.[288,289] This is less frequent in eukaryotes than in bacteria. The sequence of bases preceding the initiation codon must also be important for recognition of the "start" signal.

In *E. coli* polypeptide chains are always initiated with the amino acid **N-formylmethionine**. Some bacteria can apparently live without the ability to formylate methionyl-tRNA,[290] but most eubacteria as well as mitochondria and chloroplasts use formylmethionine for initiation. In a few cases, both among bacteria and eukaryotes, initiation can sometimes occur with other amino acids.[291] The first step is the alignment of the proper initiation codon correctly on

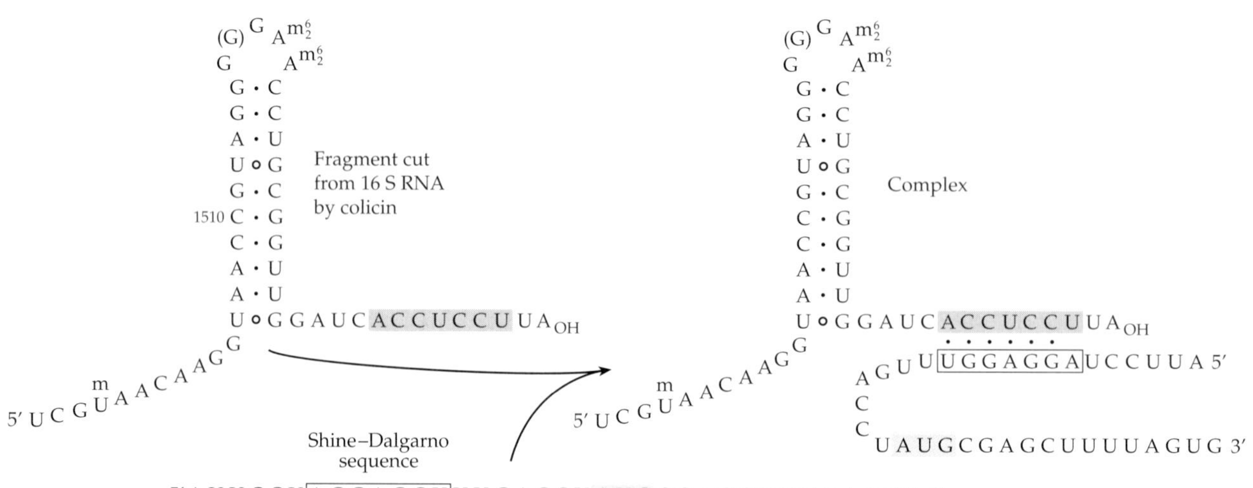

(29-8)

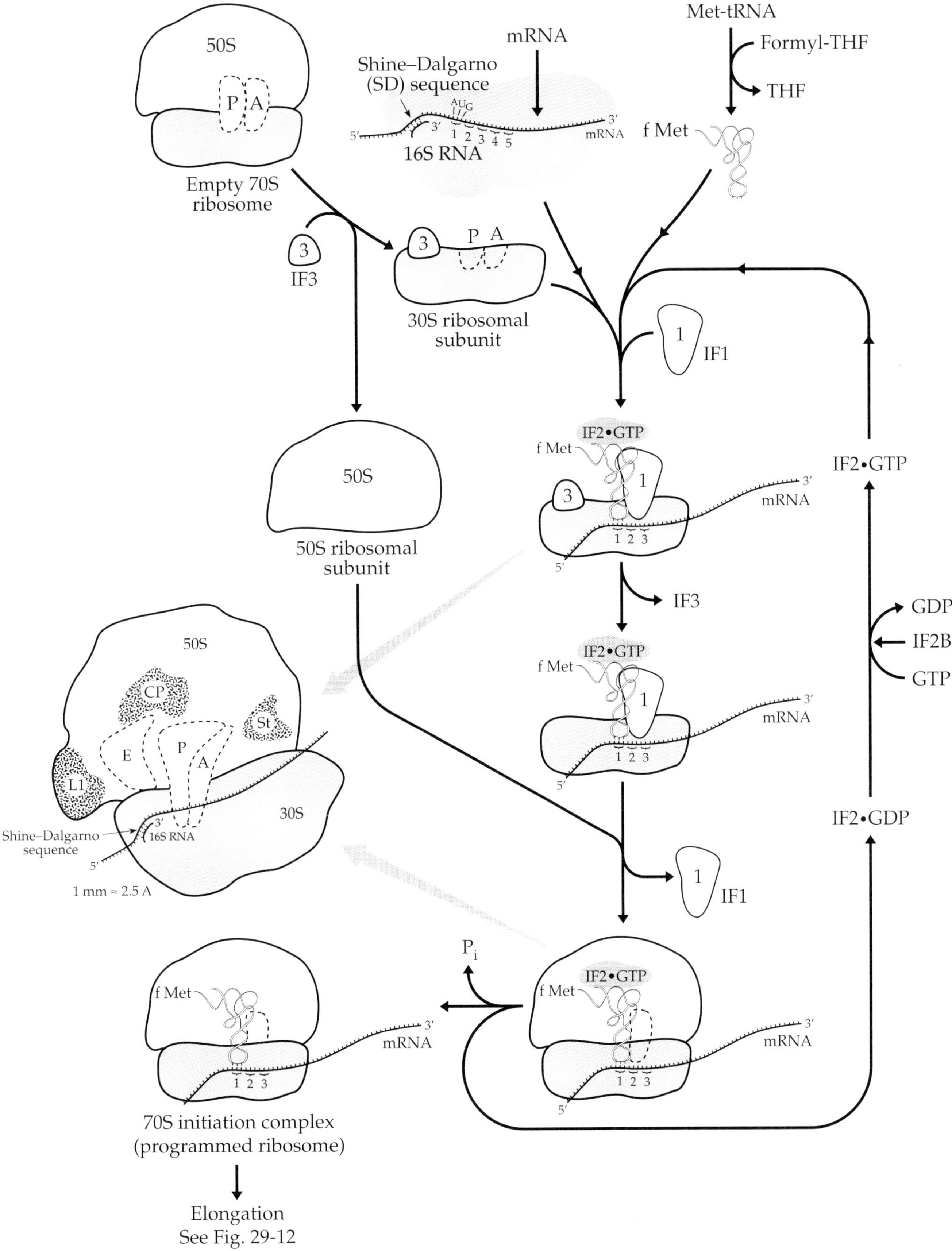

Figure 29-10 Initiation of protein synthesis on bacterial ribosomes. Images are not drawn to scale. Some details are indicated on the larger scale image at the left.

the ribosome and the binding to it of a molecule of initiator tRNA carrying *N*-formylmethionine.[268,292,293] The process by which this occurs is relatively complex, partly because it is essential for the ribosomes to distinguish the true initiation codon from the many AUG codons in internal positions in the message. In bacteria recognition of the initiation codon is assisted by base pairing between the conserved sequence ACCUCCU at positions 1534–1540 at the 3' end of the 16S RNA (Fig. 29-2A) and the complementary **Shine-Dalgarno sequence** AGGAGGU, which is found near the 5' end of most mRNA molecules.[53] This is illustrated in Eq. 29-8 for a messenger RNA in the form of the A protein initiator region from the R17 phage RNA.[294,295] Ribosomal protein S1 also seems to be required for this binding.[135,295a]

Prokaryotic initiation factors. In addition to the ribosomal proteins, the initiation factors **IF1**, **IF2**, and **IF3**, whose molecular masses are 9.5, 9.7, and 19.7 kDa, respectively,[70,296] are essential. They coordinate a sequence of reactions that begins with the dissociation of 70S ribosomes into their 30S and 50S subunits. Then, as is shown in Fig. 29-10, the mRNA, the initiator tRNA charged with formylmethionine, the three initiation factors, and the ribosomal subunits react to form 70S **programmed ribosomes**, which carry the bound mRNA and are ready to initiate protein synthesis. IF2 is a specialized G protein (Chapter 11), which binds and hydrolyzes GTP. It resembles the better known elongation factor EF-Tu (Section 2). The ~172-residue IF3 consists of two compact α/β domains linked by a flexible sequence, which may exist as an α helix.[296a–298] Its C-terminal domain binds to the central domain of the 16S RNA near nucleotides 819–859 (Fig. 29-2). When bound it protects nucleotides in the 690 loop from chemical modification[297] and induces a conformational change in the loop.[297a] Binding of IF3 prevents association of the 30S and 50S subunits, assuring the cell of a supply of free 30S subunits for translational initiation. It also promotes the binding of the other two factors: IF1 and IF2.[299] Binding of IF2, as its GTP complex, stimulates the binding of fMet-tRNA in the adjacent P site.[300] Another function of IF2, in cooperation with IF1, may be to remove peptidyl-tRNAs with short polypeptide chains under conditions in which such peptidyl-tRNAs accumulate to abnormal levels.[301] However, the order of binding, which is implied in Fig. 29-10, has been hard to establish.

IF1, which is essential to the viability of bacteria, binds and partially occludes the A site of the ribosome, preventing the initiator fMet-tRNA from incorrectly occupying the 30S A site.[70,296a,302] Binding of IF1 also causes the functionally important bases A1492 and A1493 of 16S RNA (Fig. 29-2) to be flipped out of helix 44 and to bind to pockets in IF1. This induces further long-range conformational changes.[70] It has also been hard to establish whether the charged initiator tRNA binds into the P site before or after the mRNA binds to the 30S subunit. Some evidence supports the latter possibility,[296] which is indicated in Fig. 29-10. In any case, an important step is the specific base-pairing of the initiator tRNA with the first AUG start codon. IF3 seems to be essential for this pairing process, which establishes the correct reading frame for translation of the genetic message.[303] A proofreading function at this step is often attributed to IF3.[304] Intact ribosomes bind charged tRNAs tightly in the P site. Perhaps the initial binding to the 30S subunit is loose enough to allow the mRNA, which ties itself to the 3' end of the 16S RNA via Eq. 29-3, to move back and forth until the correct reading frame is located. Then a conformational change occurs and locks the initiator tRNA in place. This change also weakens the binding of IF3, which dissociates from the complex, allowing the 50S subunits to rejoin the complex. The ribosome-binding domain of IF3 is homologous to spliceosome protein U1A (Chapter 28).[304]

The hydrolysis of GTP during initiation is essential as is shown by the fact that 5'-guanylmethylene diphosphonate, a GTP analog containing a methylene bridge between the terminal and central phosphorus atoms (see p. 558), can substitute for GTP in all steps up to and including the binding of the 50S ribosome. However, it cannot function in the final step because it cannot undergo hydrolysis. Why is GTP hydrolysis needed? It may provide energy for the conformational rearrangement of ribosomal components, or it may simply be required for release of the IF2•GDP complex. For example, IF2•GTP may bind to the ribosome with a high affinity, but IF2•GDP only weakly. Remember that G-proteins exist in at least two conformations, one stabilized by GDP and another by GTP (Chapter 11). When the hydrolysis of the bound GTP is incorporated into a reaction sequence, it provides a Gibbs energy change that may be needed to drive the reactions. In this case, it ensures that the charged initiator tRNA is firmly bound and ready to initiate translation.

Some information about spatial arrangements of the ribosomal proteins involved in initiation was provided by the fact that antibodies against proteins S19 and S21 block the formation of a complex with fMet-tRNA, while antibodies against S2, S18, and S20 block the binding of IF3. Crosslinking experiments showed that IF2 and S19 are close together and that IF3 is close to S12 (Fig. 29-1A).

Initiation of protein synthesis in eukaryotes. Most eukaryotic mRNAs have a 5' cap (p. 1642) and lack a Shine–Dalgarno sequence. Otherwise, initiation follows a pattern similar to that in bacteria but more complex.[305–308] There are at least ten eukaryotic

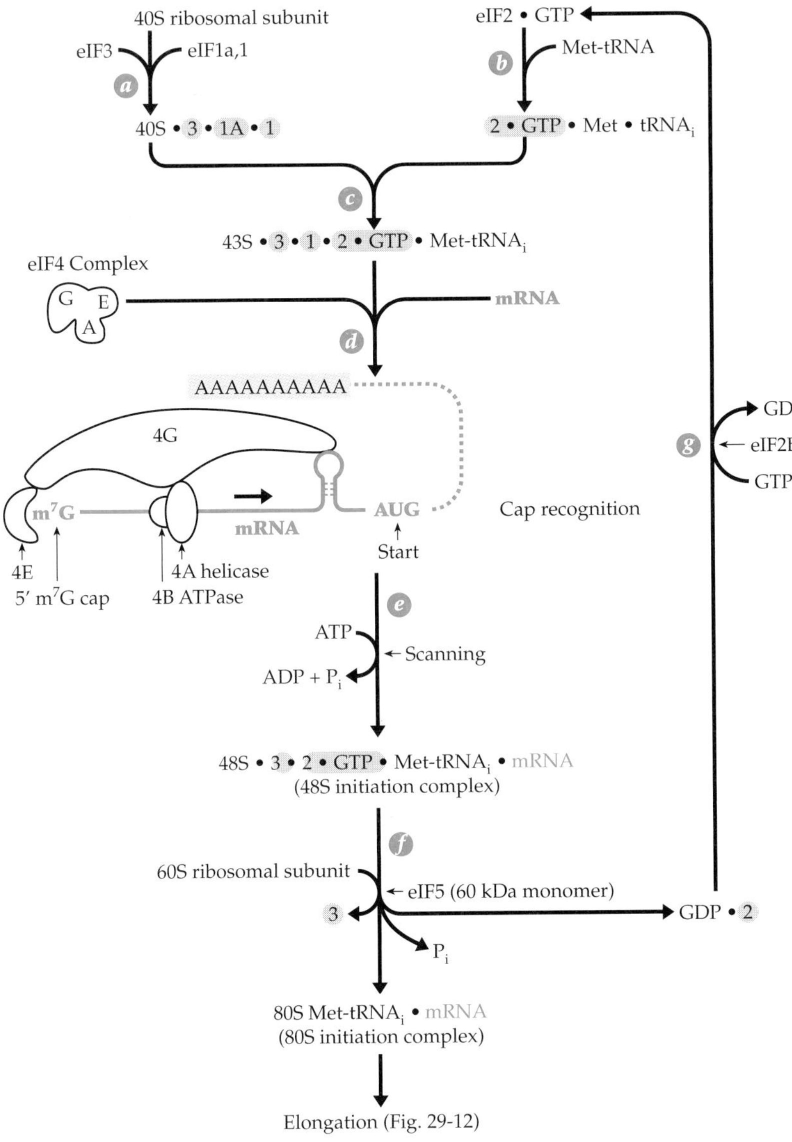

Figure 29-11 Initiation of eukaryotic protein synthesis. 1, 2, 3 = eIF1,2,3

initiation factors (eIFs), some composed of several peptides.[309] Hydrolysis of both ATP and GTP is required to form the initiation complex. Cap-binding proteins help to locate the 5' end, but the first initiation codons occur at greatly varying distances from the cap. Ribosomes apparently conduct a systematic scanning beginning at the 5' end cap of the mRNA to locate the first initiation codon.[305,310,311] Initiation of translation in eukaryotes is also often subject to controls that are more complex than those in bacteria.[308,312–314a] At least 25 polypeptides are involved.[315] Specific functions of only a few of these are described here. Functions of some are unknown or uncertain, and new proteins such as the stimulatory factor 4H[316,317] have been discovered recently.

The first initiation step is the dissociation of idle 80S ribosomes into their 40S and 60S subunits. This depends upon the ~700-kDa eIF3, a complex of 5–11 peptides of mass 30 to 170 kDa each, which binds to the 40S subunit (Fig. 29-11, step *a*).[306,311,318–320] In a separate reaction (step *b*) the charged initiator tRNA (Met-$tRNA_i$) is bound by the G protein eIF2,[321–324] an $\alpha\beta\gamma$ mixed trimer whose α subunit not only binds GTP but is also the site of regulation by a phosphorylation–dephosphorylation mechanism.[325,326] As is indicated in Fig. 29-11, GDP-containing eIF2, released from ribosomes upon formation of the initiation complex, does not bind the charged initiator tRNA. The bound GDP must first be exchanged for GTP, a process that requires a five-subunit guanine nucleotide exchange factor eIF2B (Fig. 29-11, step *g*).[327–328a] However, if the α subunit of eIF2 becomes phosphorylated, the nucleotide exchange and consequently the initiation of protein synthesis, is retarded.[329–330c] In reticulocytes protein synthesis stops rapidly if there is a deficiency of heme. This appears to occur via a **heme-sensitive eIF2α kinase** whose catalytic activity is inhibited by heme.[322,327,331] The [NAD^+]/[NADH] ratio may also be a factor in controlling the nucleotide-exchange GEF-catalyzed reaction.[332]

The ternary complex of eIF2 with GTP and Met-$tRNA_i$ binds to the 40S complex (Fig 29-11, step *c*). If the ribo-trinucleotide AUG, the initiation codon, is added to this complex, it is converted to a 43S initiation complex.[321] However, with natural mRNAs several additonal protein factors as well as ATP are needed (Fig. 29-11, step *d*). Factor 4F (eIF4F) is a large complex of several components known as factors 4A, 4B, 4E, and the large 220-kDa 4G (formerly EIF4γ or p220).[333–334a] Factor 4G is a multifunctional adapter or scaffold that apparently organizes the complex and coordinates various control mechanisms.[335–338] Factor 4E is a cap-binding protein, which recognizes and binds tightly to the ^{7}mG cap present on most mRNAs.[306,339–341a] It also binds to 4G. Factor 4A, an ATPase, acts together with 4B as an RNA helicase to unwind the mRNA and remove hairpin loops[342,343] during the scanning to locate the initiation codon (step *e*). Kozak suggested that the 43S ribosome scans from the cap at the 5' end and stops at the first initiation codon, which is usually the AUG found within the sequence (A/G)NNAUGG.

AUG codons in other positions, known as **internal ribosome entry sites** (IRES),[311,344–347] and, more rarely, non-AUG codons can also initiate translation with lower efficiency.[348] Thus, mechanisms exist for synthesis of small amounts of proteins of varying lengths and of proteins that are encoded in any one of the three reading frames.[305,349–350a] Even circular RNAs can serve as mRNAs by this mechanism.[351] It is significant that, as shown in Fig. 19-11, factor 4G, the large subunit of eIF4, also binds to the poly(A) tail present on the 3' terminus of most mRNAs. This binding, which seems to be essential for rapid initiation,[352–354] is mediated by yet another protein, the **poly(A)-binding protein**. The importance of this protein in the human body is emphasized by its identification as **ataxin-2**, the protein defective in type 2 spinocerebellar ataxia (see Table 26-4).[354] The significance to the regulation of initiation is not clear, but the poly(A) binding may favor reuse of the mRNA, which may be translated repeatedly under conditions of rapid growth.

The last initiation step (step *f*, Fig. 29-11) is the reaction of the 60S ribosomal subunit with the 48S initiation complex to form the 80S initiation complex. Initiation factors 3, 4C, the eIF2•GDP complex, and inorganic phosphate are all released in this process, which is promoted by IF5. This monomeric ~60-kDa protein[355,356] also stimulates conversion of the GTP bound to IF2 into GDP and P_i. IF5 is unique as the only known protein containing **hypusine**, N^{ε}-(4-amino-2-hydroxybutyl)lysine, a posttranslationally modified lysine. It occurs only at position 50 in the 17-kDa protein.[356–358] Hypusine is not present in eubacteria but is essential for viability of both eukaryotes and archaeobacteria[358] and is present within an invariant 12-residue sequence.

2. Elongation of Polypeptide Chains

Once the initiating fMet-tRNA of bacteria or the eukaryotic Met-$tRNA_i$ is in place in the P site of a ribosome and is paired with the initiation codon in the mRNA, peptide chain growth can commence. Amino acid residues are added in turn by insertion at the C-terminal end of the growing peptide chain. Elongation requires three processes repeated over and over until the entire peptide is formed.

1. Codon-specific binding of a charged tRNA bearing the next amino acid at the A site (decoding).
2. Formation of the peptide bond. This process transfers the growing peptide chain from the tRNA in the P site onto the aminoacyl-tRNA in the A site.
3. Translocation of the peptidyl tRNA from the A site to the P site. This process also involves movement of the used tRNA from the P site into the exit site and simultaneous movement of the mRNA to bring the next codon into place in the A site. Both the release from the A site and translocation require energy. This is provided by the hydrolysis of GTP, one molecule for each of the two processes.[359,360]

The elongation cycle for *E. coli* is shown in Fig. 29-12. That for eukaryotic ribosomes is similar except that 40S and 60S subunits are involved in formation of the complete 80S ribosome.

Codon-specific binding of an aminoacyl-tRNA (decoding). The binding of an aminoacyl-tRNA to the A site of the 70S or 80S initiation complex depends upon a protein called **elongation factor Tu** (**EF-Tu** or **eIF1** in eukaryotes), which is present as a mixed dimer with a second protein, **EF-Ts**. In *E. coli* EF-Ts is a stable 35-kDa protein, while Tu is a 43-kDa soluble protein present in a large excess over Ts. Tu is one of the most abundant soluble proteins in bacterial cells and accounts for about 5% of the total protein. Most of the tRNAs in a bacterial cell are present as complexes with Tu. Tu may also have functions other than in protein synthesis and is found associated with the plasma membrane as well as with ribosomes.

Factor Tu is a G protein. It not only carries the aminoacyl-tRNAs into the A site on ribosomes, as shown in Fig. 29-12, but also binds and hydrolyzes GTP during the elongation cycle.[361–368] Factor Ts is a nucleotide exchange factor that catalyzes the exchange of GDP bound to Tu for GTP.[369] This is shown in Fig. 29-12 (steps *a* and *b*). The GTP/GDP-binding site of EF-Tu is located in the N-terminal portion. Eukaryotic eEF-Tu is also called EF1α or EF-TA. As isolated from various sources it has a molecular mass ranging from 50 to 53 kDa. Like the bacterial counterpart, it is abundant.[368,370–373] Like bacterial EF-Tu it exists largely as a complex with a more abundant nucleotide exchange factor EF1β. The complex tends to be bound to actin filaments.[373,374] Fungal eEF-Tu contains mono-, di-, and trimethylated lysine at up to 16 positions.[371]

EF-Tu will bind to any aminoacylated tRNA other than $tRNA^{f\text{-}Met}$, the initiator tRNA[374a] (step *c*, Fig. 29-12), and carry it to the ribosome (step *d*), where it binds into the A site. There it is selected if it forms a proper base pair with the mRNA codon in the A site or is rejected if it does not. This decoding process involves both an initial step and a proofreading step. The aminoacyl-tRNA binds both to the decoding site in the 16S RNA and to the peptidyltransferase site in the 23S RNA. (See discussions on p. 1687.) The decoding site is on the platform at the upper end of helix 44 (Fig. 29-2). Nucleotide G1401 plays a crucial role.[375] When one of the isoacceptor species of *E. coli* $tRNA^{Val}$ is irradiated with ultraviolet light, the

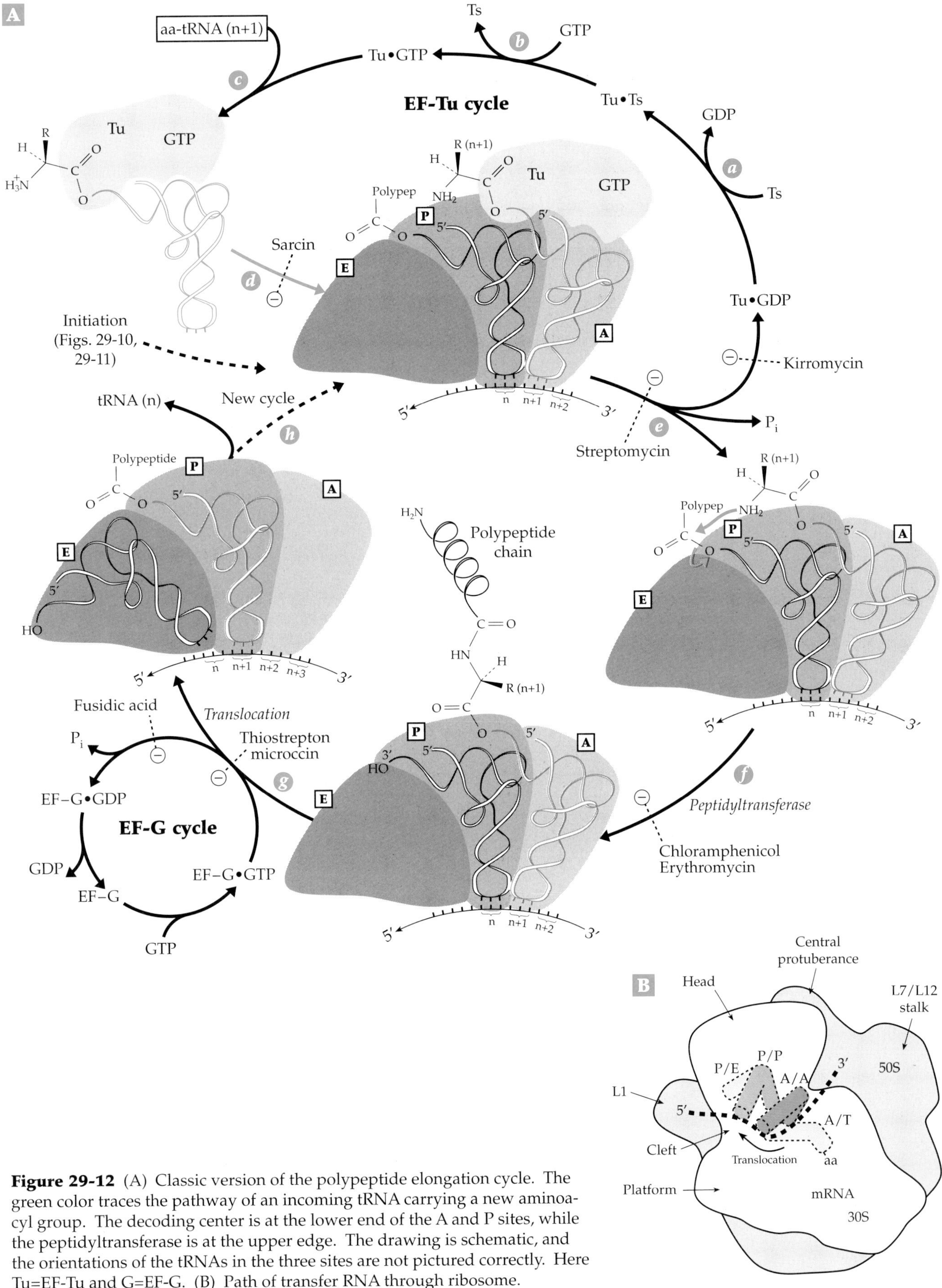

Figure 29-12 (A) Classic version of the polypeptide elongation cycle. The green color traces the pathway of an incoming tRNA carrying a new aminoacyl group. The decoding center is at the lower end of the A and P sites, while the peptidyltransferase is at the upper edge. The drawing is schematic, and the orientations of the tRNAs in the three sites are not pictured correctly. Here Tu=EF-Tu and G=EF-G. (B) Path of transfer RNA through ribosome.

5-(carboxymethoxy)uridine at position 34 in the anticodon becomes crosslinked specifically with C-1400 of the 16S rRNA in the 30S subunit.[5,376] This nucleotide lies in the deep cleft, in the decoding region, between the neck and the platform of this ribosomal subunit (Fig. 29-1). Various crosslinking and protection experiments[377,378] show that other helix 44 residues bind the tRNAs in both the P and A sites. A1492 and A1493 form part of the A site, while C1400 is in the P site.[378,379] Also strongly affecting tRNA binding and decoding is the nearby **switch helix** in the 900 region of the 16S RNA. This helix readily undergoes a shift between two hydrogen-bonded configurations (Eq. 29-9).[378,378a,380]

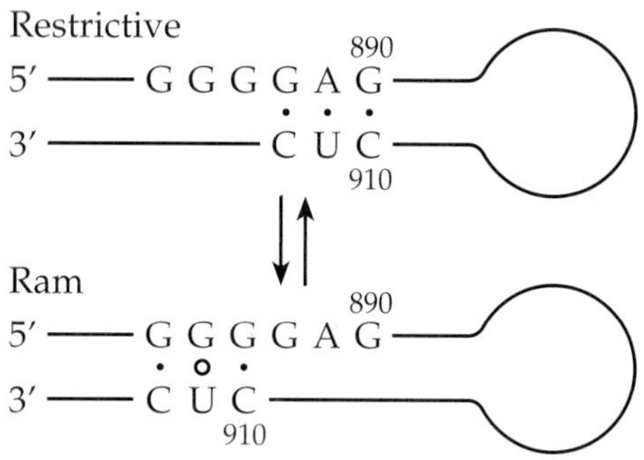

Restrictive
890
5′ — G G G G A G —
• • •
3′ — C U C —
910

⇅

Ram
890
5′ — G G G G A G —
• o •
3′ — C U C —
910

(29-9)

Judging by the effects of mutations in 16S RNA or in proteins S5 and S12 that favor one or the other conformation, the restrictive conformation gives a greater fidelity in translation than the "ram" (ribosomal ambiguity) conformation.[380] This loop is near the central pseudoknot in the 16S RNA and is involved in binding S5 and S12 as well as streptomycin (Box 29-B), all of which affect fidelity of protein synthesis. As mentioned on p. 1687 the adenine rings of residues A1492 and A1493 move out to interact with the CCA-3' ends of the tRNA (Fig. 29-14).

The location of binding of EF-Tu on ribosomes has been established directly by cryo-EM. It binds both to the L7/L12 stalk and to the body of the 50S ribosomal subunit.[381] The other end of the P site is at the peptidyltransferase locus and has been photochemically labeled by azide derivatives of aminoacyl groups bound to a tRNA.[382] The labeling is primarily in the 50S subunit of *E. coli* ribosomes and involves the central loop of domain V (residues 2043–2625) of the 23S RNA. Residues U2584 and U2585 are major sites of crosslinking (see Fig. 29-14). The presence of nearby sites of mutation leading to resistance to chloramphenicol or erythromycin[383,384] (Box 29-B) served to confirm the central loop as part of the peptidyltransferase. Domain II of 23S RNA is also involved, and there is evidence that the unique sequence UGG at positions 807–809 may also interact with the CCA end of tRNA in the P site.[361]

Bound Tu leaves a "footprint" at positions 2655 and 2661 in the sarcin/ricin loop of domain VI (Fig. 29-4H) when chemical probes are applied.[385] From a thermodynamic viewpoint, the hydrolysis of GTP to GDP and P_i during the functioning of EF-Tu is unnecessary, but it appears to drive a conformational change needed to bring the reacting groups together or as part of a proofreading mechanism.[385-385d] The hydrolysis of GTP appears to follow codon–anticodon recognition between the tRNA and mRNA in the A site (see Figs. 29-2G and 29-14). EF-Tu • GDP has a greatly reduced affinity for an aminoacyl-tRNA and dissociates, leaving the latter firmly bound into the A site.[385b–d] Simonson and Lake proposed that binding of a tRNA into the A site is *preceded* by binding into a **D site**. After the initial binding the anticodon of the tRNA wing from the D site into the A site as a result of a conformational rearrangement of the base stacking within the tRNA.[385e]

The peptidyltransferase reaction. It has been difficult to establish whether the 2' or the 3' of the terminal adenosine of tRNA carries the activated aminoacyl or peptidyl group. Rapid equilibration between the two via an orthoester may occur (Eq. 29-10), and EF-Tu of *E. coli* binds to either the 2' or 3' isomer.[386,386a] However, reaction of the 3'-aminoacyl-tRNA at the peptidyltransferase site is probable.[387,387a,b]

The peptidyltransferase reaction resembles that of the proteases (Chapter 12, pp. 649, 650), with a tetrahedrally bonded intermediate probable (Fig 29-13A). As is shown on pp. 649–650, the catalytic acid has been proposed to be the N3 atom of adenosine 2486 (2451 in *E. coli*) in the *H. marismortui* 23S RNA. This is in the central loop of domain V (Fig. 29-14). However, replacement of A2451 with G, U, or C did

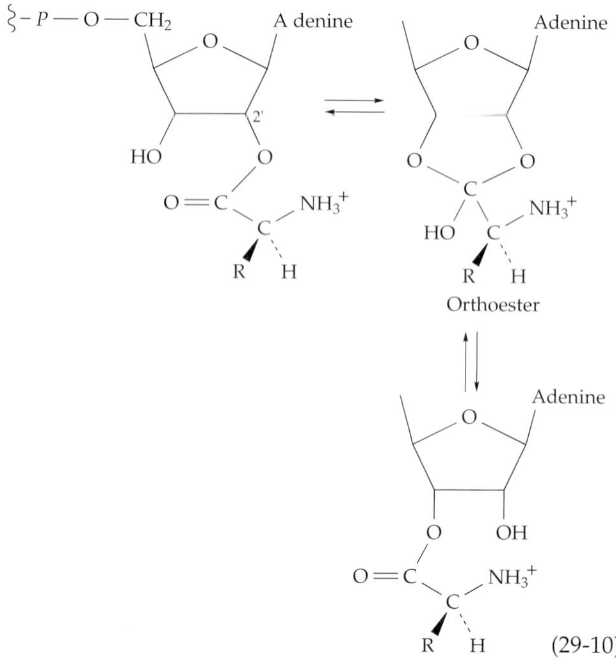

P—O—CH_2 A denine O 2' HO O O=C C NH_3^+ R H ⇌ Adenine O O O C HO C NH_3^+ R H Orthoester ⇅ Adenine O O OH O=C C NH_3^+ R H

(29-10)

not totally destroy peptidyltransferase activity in *Thermus aquaticus*, nor did replacement of the essential G2447 with A, U, or C. Polacek *et al.* suggest that the ribosome may need only to hold the reacting aminoacyl and peptidyl groups attached to tRNA in the correct positions for reaction and that no other catalysis is necessary.[387c] However, A2451G, U, or C mutant ribosomes have very low activity and organisms with such mutations are often not viable.[33f,167b] The pH rate profile for peptidyltransferase activity indicates a pK_a of ~7.5 in the RNA.[33f,387d] This is consistent with the view that A2451 may serve as a catalytic base. There has been considerable discussion about the pK_a. Can it be assigned to A2451? As mentioned on pp. 751–753, many enzymes have a broad pH region of maximum velocity over which catalytic groups of quite different microscopic pK_as (pp. 305–307) may function. For ribosomal RNAs, as for proteins, tight bonding between ionized groups in a substrate-catalyst complex may lock in an overall protonation state of the macromolecule. However, a proton may jump from one group to another within the complex (e.g., as in Fig. 29-14D) to provide a set of tautomeric species in a pH-independent equilibrium. Among these some will be on the catalytic pathway. One may arise by deprotonation of the reacting –NH_2 group of the aminoacyl-tRNA (Eq. 29-1, step *c*). Conformational changes,[387e] which may be induced by proton movements, may also be encompassed within the array of pH-independent equilibria.

A careful stereochemical analysis has led to the conclusion that for all of the different aminoacyl groups to be able to react in the same way at the peptidyltransferase site and to all generate trans amide linkages, the torsion angles ϕ and ψ of the resulting peptide must be approximately those of an α helix.[388] Thus, the peptide emerging from the ribosome exit tunnel may be largely helical.

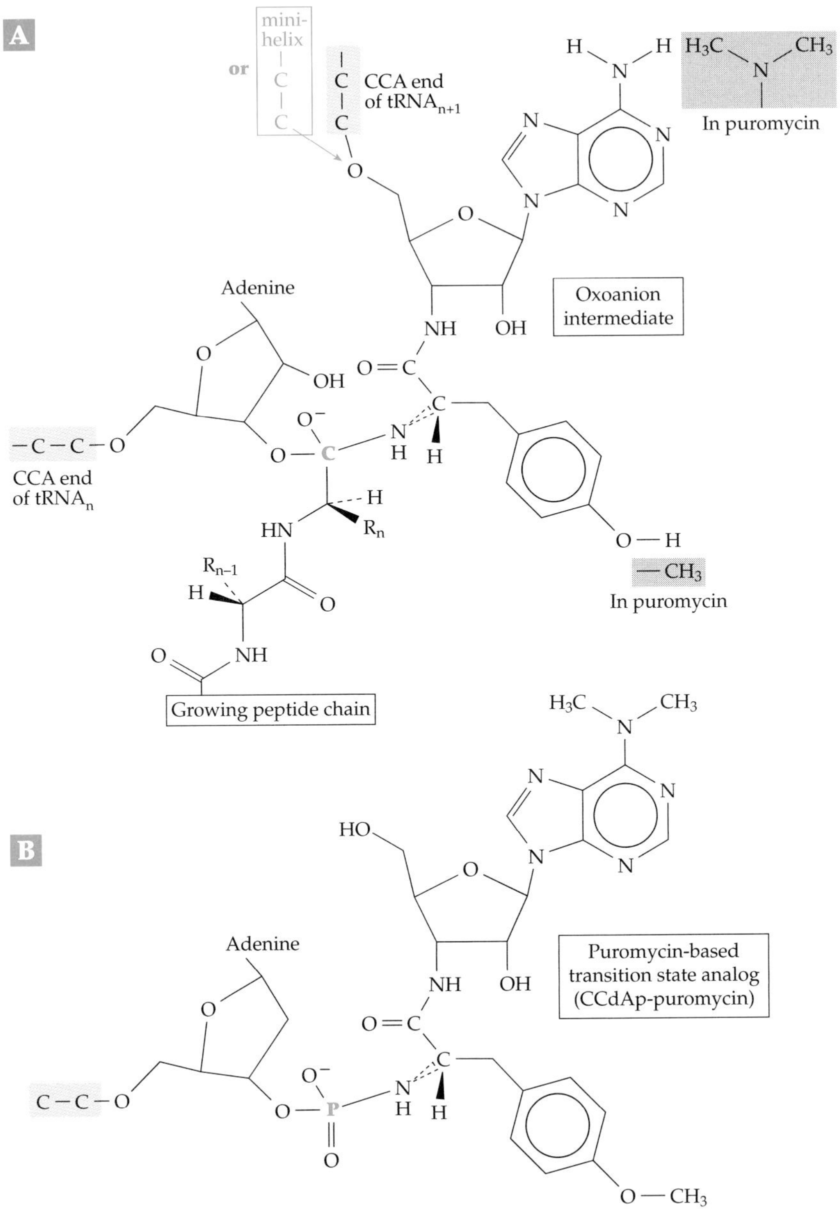

Figure 29-13 (A) Structure of expected intermediate with tetrahedral C-atom in peptidyltransferase reaction with a tRNA, with a minihelix analog, or with the antibiotic puromycin. (B) Transition-state (or bisubstrate) analog formed with puromycin and a mimic of the CCA end of a tRNA. See Box 29-B.

Elongation factor EF-G and translocation. The third step in the elongation sequence on ribosomes (Fig. 29-12, step *g*) depends upon **EF-G**, a monomeric GTP-binding protein with a sequence homologous with that of other members of the G protein family. It apparently utilizes the Gibbs energy of hydrolysis of GTP to GDP to drive translocation of the peptidyl-tRNA from the A site to the P site (Fig. 29-12) and of the previously utilized (deacylated) tRNA to the exit site. EF-G binds to the 50S ribosomal subunit at the base of the L7 / 12 stalk as indicated in Fig. 29-1.[392,393] It competes with EF-Tu, which binds in nearly the same location.[5] EF-G is a large five-domain GTPase. Domain 1 contains the GTPase site and resembles other G proteins, and domain 2 has some similarity to the

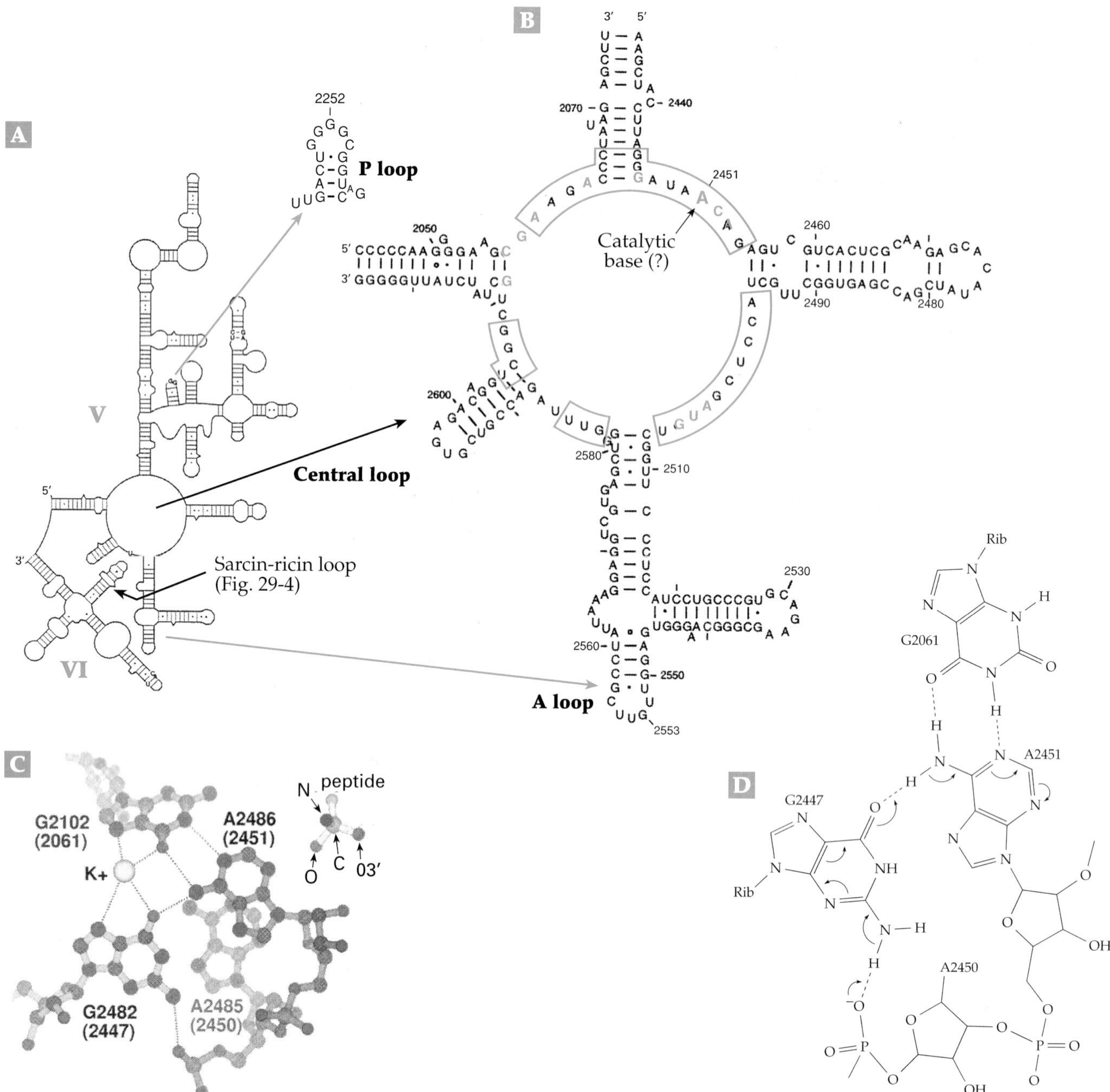

Figure 29-14 The ribosomal peptidyltransferase center in the 23S RNA. (A) Secondary structure map for domains V and VI of *E. coli* 23S RNA. From Samaha *et al.*[389] (B) Sequences of the P loop and A loops and of the central loop of domain V and of the 23S RNA of *Halobacterium halobium* with numbering according to the *E. coli* sequence. Sequences within the green boxes are highly conserved in eubacteria, chloroplast, and mitochondrial RNAs. See Kloss *et al.*[390] and Garret.[80] Sites of mutations that confer resistance to erythromycin (G2057, A2058, and C2611) and chloramphenicol (G2057, G2447, A2451, C2452, A2503, and U2504) are indicated. See Douthwaite and Aagaard.[383] (C) A skeletal representation of the peptidyltransferase active site which is depicted more completely in the stereoscopic view in (H). Included is the peptidyl group on the 2'-end of a tRNA. (D) Schematic region of domain V showing the hydrogen-bonding interactions of the catalytic base A2486 (2451, *E. coli*) with neighboring bases and also locations of nearby K^+ and Mg^{2+} ions. (C, D, G, H) are from Nissen *et al.*[19] (E) Interaction of the CCA 3'-ends of ribosome-bound tRNAs (at top) with the large ribosomal subunit. This is a cutaway view with tRNAs in the A, D, and E sites. The ribosome is sliced to show the acceptor ends of the tRNAs in the entrance to the peptide tunnel. Enlarged section shows interactions of the CCA 3'-ends in the P and A sites. The CCA end of a deacylated tRNA is shown in the P site, while a peptidyl-tRNA analog (CCA puromycin–phenylalanine–caproic acid–biotin) is shown in the A site. Bases of the 23S rRNA are numbered as in *H. marismortui*. Bases A2486 and U2620 (corresponding to A2451 and U2585) are closest to the newly formed peptide bond. From Schmeing *et al.*[33f] (F) Discrete states of the A site of the 30S subunit, as deduced from four different crystal structures. The tRNA, anticodon stem-loop (ASL), A-site mRNA codon, P-site mRNA, protein S12, and

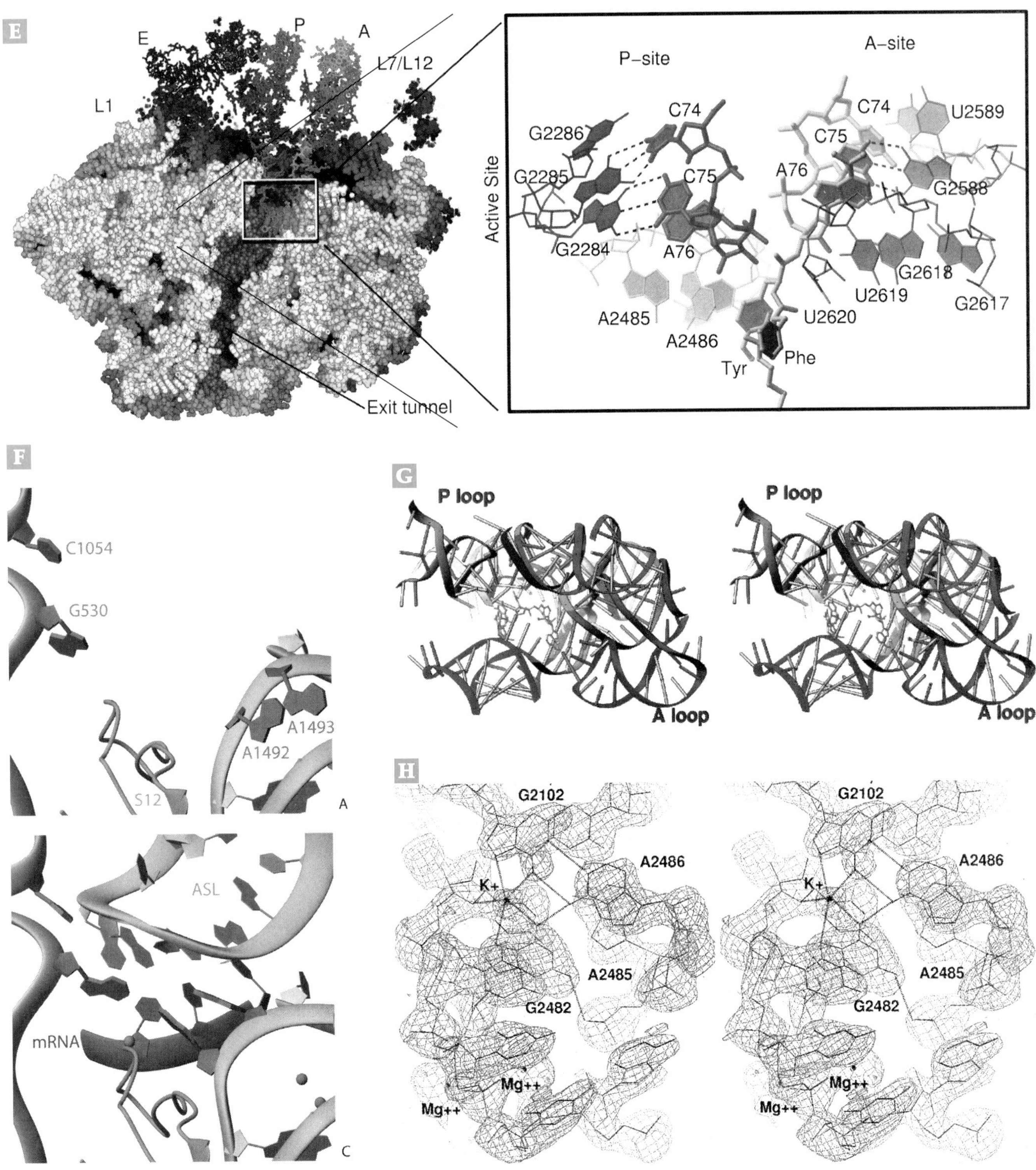

important bases involved in conformational changes are shown. A few elements of the 16S RNA such as helix 44 (lower right), 530 loop (mid and lower left), and helix 34 (upper left) are also shown. At the top, the native 30S subunit. A1492 and A1493 have been stacked in the interior of H44 and G530 is in the *syn* conformation. C1054 is shown in the upper left corner. Below, when the codon and cognate tRNA-ASL bind in the A site, A1492 and A1493 flip out to monitor the codon-anticodon interaction, and G530 switches to the *anti* conformation to interact with A1492, the anticodon in the second position, and the codon in the third. Two Mg^{2+} ions are visible near the region vacated by A1492 and A1493 in the interior of helix 44, and one is located close to the ribose of the codon in the wobble position. From Ogle *et al.*[33d] Courtesy of Venki Ramakrishnan. (G) Three-dimensional structure of the active-site region of domain V including the P loop[391] and A loop and active-site region. (H) Three-dimensional structure of the active-site region showing the hydrogen-bonding interactions of the catalytic base A 2486 (2451, *E. coli*) with neighboring bases and also locations of nearby K^+ and Mg^{2+} ions. (C, G, H) are from Nissen *et al.*[19]

corresponding domain in EF-Tu. GTP hydrolysis appears to induce within EF-G a conformational change that is coupled directly to a rapid mechanical movement within the ribosome.[394,394a] The large domain 4 protrudes from the molecule and by its shape mimics a molecule of tRNA, and the complete EF-G molecule bears a striking resemblance to an EF-Tu•aminoacyl-tRNA complex.[395,396] It leaves chemical footprints around position 1067 (the thiostrepton-binding site) in domain II of the 23S RNA and in the universally conserved loop around position 2660.[385]

Factor EF-G from eukaryotes (eEF2) is similar to the bacterial protein, but its interaction with the larger eukaryotic ribosomes seems to be more complex. For example, interaction with the ribosomal stalk is more extensive.[37] EF2 contains a single modified histidine called **diphthamide**.[397] This amino acid is not found in other proteins but is always present in eukaryotic EF2 and also in EF-G from archaeobacteria. It is the site of modification by diphtheria toxin (Box 29-A).

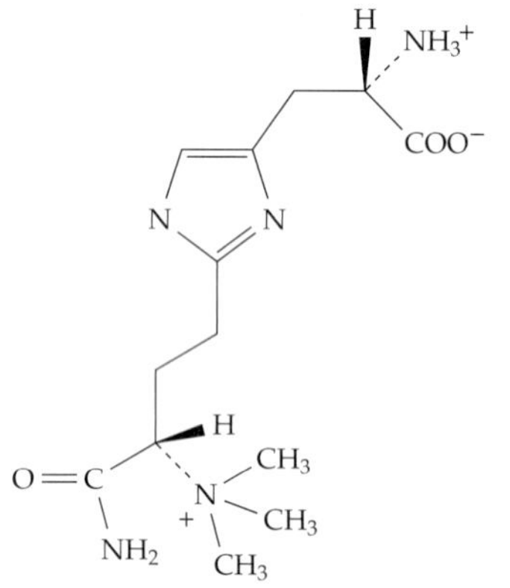

Diphthamide

The ribosomal translocation process is quite complex. As the tRNAs move from A to P to E sites on the 16S RNA platform, the mRNA must also move in discrete single-codon steps. The acceptor stems of the tRNAs in the A and P sites must react at the appropriate times in the peptidyltransferase center. Study of protection from chemical probes suggests that tRNAs sometimes lie with the anticodon loop in the A site of the small ribosomal subunit, while the acceptor stem is in the P site of the large subunit (an A/P site as illustrated in Fig. 29-12B). Each aminoacyl-tRNA enters as a complex with EF-Tu and may initially bind with its anticodon in the A site and the acceptor stem with attached EF-Tu in a transient T site, the composite state being A/T. After loss of EF-Tu the acceptor stem can move into the A site to give an A/A state. The peptidyltransferase reaction itself necessarily involves movement at the acceptor stems by 0.1 nm or more. However, additional movement of ~1 nm is needed to move the two tRNAs into states A/P and P/E, respectively. Movement of the mRNA then moves the anticodon loops of the peptidyl-tRNA into the P/P state and of the deacylated tRNA completely into the exit site (Fig. 29-12B).[86,397a] Translocation may occur at different times in the 30S and 50S subunits. The pathway of the mRNA through the ribosome is known[397b,c] and is indicated approximately in Fig. 29-12B.

EF-G seems to be the motor protein that drives translocation in the 30S subunit. When it is not attached to a ribosome, the EF-G•GTP complex is very stable, but in its functioning location GTP is hydrolyzed rapidly. This occurs prior to translocation[398,398a] and presumably causes an internal alteration in the ribosome that energizes it for the translocation step. G proteins usually undergo large changes in conformation when GTP is hydrolyzed (Chapter 11). A very large change is observed for EF-Tu, but such a change has not been seen for EF-G. However, large conformational changes in the ribosome do evidently accompany translocation.[399] The hydrolytic activity of EF-G is stimulated strongly by its binding to the L7/L12 stalk proteins.[400] Eukaryotic EF2 like EF-G binds to the stalk proteins (P1, P2, P0 complex) and also to domains of 28S rRNA equivalent to the bacterial 1067 and sarcin / ricin loops.[401] However, EF-G and EF2 are not functionally interchangeable.

Translocation occurs slowly even in the absence of GTP. However, it is greatly enhanced by GTP hydrolysis.[394a,402] Even empty ribosome, without tRNAs, but in the presence of GTP and EF-Tu and EF-G, hydrolyze GTP. The ribosome may sequentially bind EF-Tu, then EF-G, oscillating between two differing states.[403] The movement of mRNA through the ribosome has been plotted using a variety of immunochemical, crosslinking, and chemical footprinting methods.[52,404–407]

A third elongation factor, eEF3, which is an ATPase, is required by yeast and fungi.[408–410] The 1044-residue yeast protein may be required for ATP-dependent release of deacylated tRNA from the exit site.

Polyribosomes. Under suitable conditions ribosomes isolated from cells are found to sediment together in clusters, often of five or more. These **polyribosomes** (or **polysomes**), which can be seen in electron micrographs (Fig. 28-5), are held together by chains of mRNA. Polyribosomes arise because a single mRNA molecule is being translated by several ribosomes at once. As the 5' terminus of the mRNA emerges from one ribosome, it may soon combine with another and initiate translation of a second peptide chain, etc. The length of the mRNA determines how many ribosomes are likely to be associated in a polyribosome.

Rates of synthesis of ribosomes and of proteins. In a rapidly growing yeast cell with a generation time of ~100 min there are nearly 200,000 ribosomes. Almost 200 new ribosomes must be formed in one minute. Each of the 150 tandemly repeated ribosomal RNA genes must be transcribed into the 4560 nucleotides of one ribosome in less than one minute. The ~150 nuclear pores must import nearly 1000 ribosomal proteins per minute and must export ~25 ribosomal subunits per minute.[410a] The ribosomes that are formed can at 37°C add 14–17 amino acids per second to a growing polypeptide chain,[410b,c] while eukaryotic ribosomes can add 2–4 amino acids per second.[410d,e]

3. Termination of Polypeptide Synthesis

A ribosome faithfully translates the genetic message, adding amino acids to the polypeptide chain until a stop codon is reached. Then a **termination** or **release factor** acts, probably by binding directly to the stop codon on the mRNA in the A site.[411–413b] In *E. coli* termination factor **RF1**, a 47-kDa protein, recognizes UAA or UAG, while **RF2**, a very similar protein,[414,415] recognizes UAA or UGA. There are several hundred molecules per cell of these release factors. They not only recognize the stop codons but also catalyze the hydrolytic removal of the peptidyl chain from the tRNA in the ribosomal P site. They bind into the A site, where they may interact with mRNA bases in addition to those of the stop codon.[415,416] Hydrolytic release of the polypeptide chain from the tRNA in the P site may represent a change in specificity of the peptidyltransferase center induced by binding of a release factor. Genes are often terminated by a succession of two stop codons. Thus, there is a safety factor that prevents translation from continuing in case the first stop codon is missed. An example is provided by the *I* gene of the *lac* operon of *E. coli* (Fig. 28-2), which has a second stop codon in phase with the TGA codon marked in the figure and located five codons further "downstream" (to the right). A third release factor **RF3**[416] is a GTP-binding protein resembling EF-G. It is not essential to life for *E. coli*, but it accelerates the release of RF1 or RF2 and is needed for rapid growth.[417,418] Eukaryotes contain one release factor **eRF1**, which recognizes all three termination codons, and a second release factor **eRF3**, which binds and hydrolyzes GTP.[413,413b,418a,419]

Just as elongation factor EF-G mimicks the aminoacyl-tRNA•Tu complex, release factors RF1 and RF2, in their shapes, mimick molecules of tRNA.[419a] One domain of human eRF1 has an anticodon-recognition domain and a conserved GGQ sequence in a second domain, which mimicks the amino-acceptor arm of tRNA.[419,419b] Mutations in either eRF1 or eRF3 affect translational accuracy and may allow "read-through" of stop codons. In yeast (*S. cerevisiae*) a 685-residue subunit of eRF3 has an N-terminal domain, that like the human **prion protein** (Box 29-E) is capable of being transformed into a self-seeding amyloid-like conformation. In the yeast the formation of amyloid aggregate leads to depletion of the termination factor and increased readthrough.[420–422]

Recycling factors. Even though release factors remove the completed polypeptide chain, a ribosome is not ready for reuse until the deacylated tRNA in the P site is removed and the mRNA is released. This depends upon **ribosome recycling factor** (**RRF**) together with EF-G. The recycling factor is also a tRNA mimic.[419a,423-424b] It may bind into the empty A site, and in an action similar to that of the translocation step of elongation remove the P site tRNA.[417,423–425] However, probing with hydroxyl radicals indicates a different mode of binding.[425a]

4. Preventing and Correcting Errors in Translation on the Ribosome

The wrong amino acid is inserted into most positions in a protein about one time in ~10^4, a frequency[361,426,427] of ~10^{-4}. However, in *E. coli* misreading of certain codons is observed more often. For example, AAU (Asn) is read as AAA (Lys) with a frequency[428] of ~5 x 10^{-3}. Misreading also depends upon adjacent codons, i.e., the codon context.[429] Having all of the tRNAs charged with the correct amino acids, as discussed in Section B3, is a first essential for accurate translation. A second is finding the correct location of the initiation codon and binding of the aminoacyl-initiator tRNA into the P site. The decoding process by which the correct aminoacyl-tRNA is brought into the A site is still not fully understood. It has often been proposed that (as in DNA replication; Chapter 27, Section C,2) the fidelity of this process depends upon two consecutive recognition steps.[265,266] The first is the binding of the complex of EF-Tu•GTP and the charged tRNA to the ribosome. The second may be associated with the conformational change that locks the aminoacyl-tRNA into the A site and perhaps sends to the peptidyltransferase center a signal that the correct codon-anticodon pairing has been achieved. Some checking is done in the first step. For example, many of the 380 possible mischarged forms of aminoacyl-tRNAs that may have escaped previous proofreading steps (pp. 1695–1696) are rejected because they bind too loosely or too tightly to EF-Tu.[429a,b] Codon-anticodon base pairing may also be checked in the P site[429c] after translocation. The P site is buried deep in a cleft in the RNA of the large subunit. It is designed to hold the mRNAs in a kinked conformation (Fig. 29-1E) with the codons in the A and P sites oriented differently. Some tRNA residues required for high-fidelity participate in imposing this geometry. Mutants in either of the major rRNAs or in tRNAs can lead to loss of fidelity in base pairing and sometimes to excessive frame-shifting.[429c] During the proofreading process a mispaired aminoacyl-tRNA may be allowed to dissociate and be replaced by a new one. Certain mutations, such as those in ribosomal protein S12 that lead to streptomycin resistance, cause greatly increased fidelity of protein synthesis. However, these mutations slow bacterial growth,[426,430] perhaps because some misreading is necessary for synthesis of minor essential proteins.

A strictly hypothetical way in which an alteration in hydrogen bonding could be used to signal the achievement of correct base pairing is illustrated in Fig. 29-15. As indicated by the curved arrows, the approach of a negatively charged group could induce an electron pair to move from the ring nitrogen on the right side. If the pairs of hydrogen bonds were correct, a concerted flow of electrons could take place across the base pair and out into group H-Y and beyond through the postulated tautomeric chain. If the base pair were not correctly formed, the signal could not be transmitted, except during an occasional mispairing with a minor tautomer. Note that another reciprocal electron transfer in the opposite direction to that shown in the figure is also possible through the same base pair. Similar tautomeric shifts are possible for all legitimate base pairs. Initiation of a signal of the type shown could also occur by the addition of some nucleophile to a purine or pyrimidine ring, e.g., to C-6 of the cytosine ring as in Eq. 29-3. In ribosomes such electronic signals could be passed in turn through each of the base pairs involved in codon-anticodon recognition and also through other base pairs formed within loops of ribosomal RNA. If group H-Y is connected by a suitable chain of hydrogen bonds that passes through the active site of the peptidyltransferase, coupling between the recognition signal and the formation of the transition state might be accomplished. Since changes in hydrogen bonding can trigger conformational alterations, the sensing of correct hydrogen bonding could increase the rate of the peptidyltransferase reaction as has been observed experimentally.[266] Base pairing in both the A and P sites may be sensed in similar ways. The observation that rRNA residues hydrogen-bond with groups in the minor grooves of base pairs[33d] seems to be consistent with the proposal of Fig. 29-15.

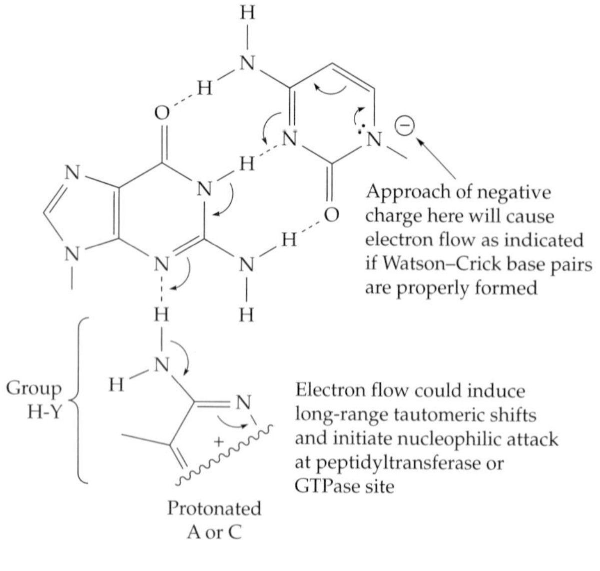

Figure 29-15 Hypothetical scheme by which an electronic signal might be sent through a base pair to initiate the peptidyltransferase reaction. See also Metzler.[431]

Codon usage. The usage of codons in specific mRNAs is not random.[432] For example, in a glyceraldehyde-3-phosphate dehydrogenase gene of yeast >96% of the 1004 codons make use of only 25 of the 61 possible coding triplets. Genes undergoing most rapid transcription are more highly biased toward these 25 than are other genes.[433] Many other evolutionary factors have affected usage. These include the need for translational accuracy.[432]

Dealing with lost peptidyl-tRNAs and broken transcripts. Many problems arise during protein synthesis. For example, a peptidyl-tRNA may become detached from a ribosome. In *E. coli* this seems to happen most frequently with peptidyl-$tRNA^{Lys}$. A 193-residue **peptidyl-tRNA hydrolase** is essential for life![434–436] It releases the tRNA for reuse, recycling all peptidyl-tRNAs other than formylmethionyl-tRNA. Perhaps the enzyme is essential because detached peptidyl-tRNAs are toxic, but it is more likely to be to avoid a shortage of free $tRNA^{Lys}$.[434]

If a ribosome starts to synthesize a protein using a broken mRNA, it will reach the end of the mRNA but will not find a termination codon. The peptidyl-tRNA will eventually fall off, but the ribosome will be stalled temporarily. Eukaryotes try to prevent this problem by sending only intact mRNAs to ribosomes,[436a,b] but bacteria have a ribosome rescue system that also tags the partially formed protein on the stalled ribosome for rapid proteolytic degradation.[437–440c] Bacteria synthesize a special 362-residue RNA that resembles a tRNA but also contains a short mRNA-like module that codes for the 11-residue peptide tag AANDENYALAA. This hybrid tRNA-mRNA, which is designated tmRNA (or *ssrA* RNA), mimicks $tRNA^{Ala}$ and is recognized and charged by alanyl-tRNA synthetase. The resulting $tmRNA^{Ala}$ binds into the A site of the stalled ribosome, undergoes the peptidyltransferase and translocation steps (Fig. 29-16). The old mRNA is released, the mRNA-like sequence of tmRNA becomes seated, and translation of the new tail sequence follows. The tail sequence is similar to C-terminal sequences that are known to mark other proteins for rapid proteolytic degradation. An associated protein SmpB is also required for functioning of the tmRNA system.[438a,b]

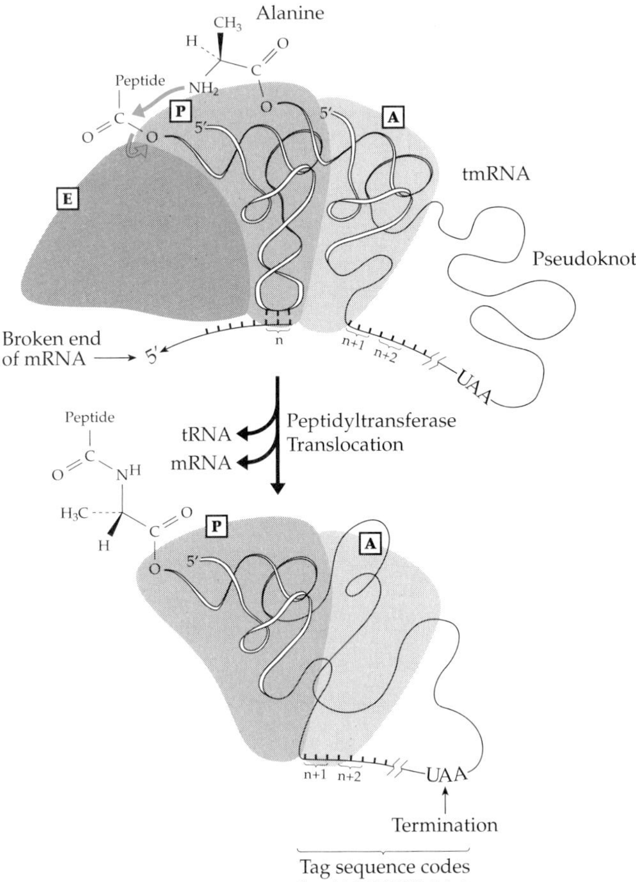

Figure 29-16 Schematic diagram of the tmRNA structure and its function in the rescue of ribosomes stalled at the end of a messenger RNA that has been broken and has lost its in-frame termination codon. After it binds into the ribosomal A site the tmRNA, which has been charged with alanine, undergoes the peptidyltransferase reaction and translocation to the P site. Then it lays down its mRNA-like coding sequence, which is used by the ribosome to add ten more amino acids to form the 11-residue C-terminal degradation signal AANDENYALAA. This induces rapid degradation of the imperfect protein that has been formed.[436a]

5. Suppressor Genes

The suppression of nonsense mutations by suppressor genes has been discussed in Chapter 26. The chemical nature of these genes was discerned, in part, from experiments involving transfer of suppressor gene *supF(su3)* into the DNA of a bacteriophage. This DNA was found to specifically hybridize with a minor transfer RNA, $tRNA_1^{Tyr}$. Subsequent investigation showed that *sup F* is a structural gene for this tRNA, and that in it the normal 5'-GUA-3' (Tyr) anticodon has been replaced with CUA. The latter can pair with the chain termination codon 5'-UAG-3' (the *amber* codon) permitting the ribosome to insert tyrosine at the site of chain termination signals introduced in *amber* mutations. It may seem puzzling that a tRNA, which prevents chain termination, does not prevent synthesis of other essential proteins within the bacterium. However, suppression is typically less than 30% efficient. Hence, many protein chains terminate normally. Since two chain termination signals are often present in a gene, most protein synthesis in the presence of the small amount of suppressor tRNA present is concluded normally. Premature chain termination caused by selected *amber* mutations will be partially inhibited, permitting the cell to make enough of the missing proteins to survive. The nucleotide sequence of a further mutated *supF* tRNA and of its longer precursor is shown in Fig. 28-10. Several other suppressor genes have also been identified as specific tRNA structural genes.[441]

A suppressor of frame-shift mutations in *Salmonella* is a tRNA containing at the anticodon position the nucleotide quartet CCCC instead of the usual CCC triplet anticodon.[442,443] It has eight unpaired bases in the anticodon loop instead of the usual seven. Other frame-shift repressor tRNAs have been identified in *E. coli*,[444] *Salmonella*, and yeast.[445] Not all suppressor genes encode tRNAs. For example, a UGA suppressor from *E. coli* is a mutant 16S rRNA from which C1054 has been deleted.[446] A general nonsense suppressor in yeast is homologous to yeast elongation factor EF-1α as well as to *E. coli* EF-Tu.[447]

Among other suppressor genes present in eukaryotic organisms[448] are mammalian genes encoding serine tRNAs that are *opal* (UGA) suppressors. These and other eukaryotic suppressor tRNAs have specific and important normal functions in cells. For example, a specific kinase phosphorylates the *opal* suppressor seryl-tRNA to its phosphoseryl derivative.[448] This suppressor tRNAs may sometimes be responsible for introducing phosphoserine at specific positions in proteins. An *opal* suppressor is also used for the introduction of selenocysteine. An *amber* suppressor is used by some methane-forming Archaea to introduce **pyrrolysine** into specific sites in methyltransferases. In pyrrolysine the epsilon amino group of lysine is joined by an amide linkage to a derivative of pyrroline-5-carboxylate (p. 1374).[448a,b]

Selenocysteine (Sec) Selenocysteine is incorporated into a small number of proteins in species from all three kingdoms of life by a suppressor $tRNA^{Sec}$ that reads certain UGA codons, which are marked as representing selenocysteine.[449,450] The selenocysteinyl-tRNA is made from a seryl-tRNA (Eq. 29-7) as described further in Chapters 16 and 24. In *E. coli* selenocysteine is present in three proteins, all formate dehydrogenases. The archaeon *Methanococcus jannaschii* contains genes for seven selenocysteine-containing proteins. Only one Sec-containing protein has been found in the nematode *Caenorhabditis elegans* and none in the yeast

Saccharomyces cerevisiae. However, there are at least 14 in the human body.[451,452] One of these, selenoprotein P, contains ten selenocysteine residues.[452,453] Products of four special genes are needed for incorporation of selenocysteine into *E. coli* proteins.[454,455] *Sel C* encodes the special *tRNA*^{Sec}, which becomes charged with selenocysteine.[456] *Sel D* encodes selenophosphate synthetase and *Sel A* selenocysteine synthetase (Eq. 29-7). *Sel B* encodes a special elongation factor, which resembles EF-Tu but has an extra domain that binds to an mRNA segment known as the **SECIS** (selenocysteine insertion sequence).[457–461a] The SECIS sequence follows the 3' end of the UGA termination codon. It is a 40-nt segment that is able to form a stem-loop structure. However, in archaea and in eukaryotes the SECIS sequence lies at the end of the selenoprotein gene in the 3' nontranslated region. It may be some distance away and may function by a foldback mechanism. It recodes the entire message, acting on any in-frame UGA codon.[461] In mammals a special SECIS-binding protein SBP2 is also required.[462]

Expanding the genetic code. Suppressor tRNAs can also be created artificially and are being used in protein engineering. *Amber*, *ochre*, or *opal* chain termination mutations can be introduced readily at many points in a protein (Chapter 26). Suppressor tRNAs can be made that will then place any one of the possible amino acids into most of the mutated positions.[463] Synthetic amino acids not normally found in proteins can also be incorporated using such tRNAs.[464–467] The TAG(UAG) *amber* stop codon is often used together with a genetically engineered tRNA. In early experiments these techniques were used to create hundreds of mutant forms of the *lac* repressor protein (see Chapter 28, p. 1606). Since then a variety of additional approaches have been explored. Transfer RNAs have been engineered to recognize four-base codons such as AGGU and CGGG.[468,469] Organisms such as *Micrococcus luteus*, in which not all of the available triplet codons have been utilized, allows development of a mutation system using an unassigned codon rather than a stop codon.[468] A general method for site-specific incorporation of any amino acid or amino acid analog requires a suppressor tRNA that is not aminoacylated by any aminoacyl-tRNA synthetase present within the host cell, and also an aminoacyl-tRNA synthetase that acts only on the suppressor tRNA and no other tRNA in the cell.[470] Several such systems are being developed.[470–472] Another idea is to utilize a 65th codon-anticodon pair, one depending upon a new synthetic nucleoside that can be incorporated into mRNA.[473]

Another possible application of suppressor genes is *in vivo* suppression of undesirable termination codons. An example comes from a β^0 thalassemia caused by mutation of lysine codon CAG to UAG. By changing the anticodon of a human tRNA^{Lys} gene to an *amber* suppressor, premature termination of globin chains was suppressed in an *in vivo* test.[474,475]

6. Read-Through Translation and Ribosomal Frameshifting

If termination codons are not recognized efficiently by termination factors, synthesis continues past the termination codons and new longer protein chains are made. This **read-through translation**[429,476] may sometimes be accidental, but it is also used by cells to form several important proteins. For example, the 14-kDa coat protein of bacteriophage Qβ is elongated by read-through during translation of the RNA about 4% of the time. This produces a 38-kDa protein known as A_1, which has an extra 200 amino acid residues at the C terminus and is essential for formation of infectious virus particles.

A similar situation is met with retroviruses (Chapter 28) whose coat protein gene *gag* is fused to the reverse transcription polyprotein precursor gene *pol* (Fig. 28-26).[477] In fact, the polyprotein gene overlaps the 3' end of the *gag* gene. Read-through translation allows synthesis of the polyprotein about 5% of the time. However, the *pol* gene is written in a different reading frame: −1 with respect to the *gag* gene. For example, in the HIV genome (Fig. 28-26) the *pol* gene begins at nucleotide 1638 in the −1 reading frame with respect to the *gag* gene. In both the Rous sarcoma virus and HIV a polyprotein that is formed contains peptide sequences from both *gag* and *pol* genes.[478,479] This fusion of the two proteins is accomplished by a frameshift, which occurs on the ribosome as it operates in the region of overlap of the *gag* and *pol* genes. This mechanism allows synthesis of relatively small amounts of the enzymes encoded by *pol* but large amounts of structural proteins encoded by *gag*.[480] Many other examples of essential ribosomal frameshifting are known.[429,477,481–483] For example, the gene for *E. coli* ribosomal release factor RF2 has a UGA termination codon at position 26, but the coding sequence for the protein continues in the +1 frame.[484]

Ribosomal frameshifting can be accounted for by more than one mechanism. It can occur when a four-base anticodon is present in a suppressor tRNA. It can result from incorrect base pairing. If a tRNA slips over by one nucleotide, a single base in the mRNA can be left unpaired with the reading frame being shifted +1. However, most frameshifts are in the –1 direction and occur at specific locations in the mRNA, i.e., they are **programmed frameshifts**.[484a] These often occur at "slippery sites"[485] including the following mRNA sequences in which three codons are marked:

$$\overbrace{\;\;X}^{1}\overbrace{X\,X\,Y}^{2}\overbrace{Y\,Y\,Z}^{3}$$

This sequence is followed closely by an element of secondary structure, most often a pseudoknot.[486–492] Eukaryotic frameshifts are almost always in the –1 direction, the exception being found in the mammalian mRNA for **antizyme**, a negative regulator of ornithine decarboxylase (Chapter 24, p. 1382).[493] The frame-shift occurs at an initially in-frame termination codon (UGA), which is followed by a pseudoknot. Most translation is terminated at this stop codon, but frameshifting, which is induced by a high polyamine concentration, allows read-through and synthesis of the antizyme protein. In rare cases frameshifting may lead to **translational bypass** of some codons on the mRNA. Such a case is found in a bacteriophage T4 mRNA for which the *E. coli* ribosomes bypass 50 nucleotides in order to complete the synthesis of a

BOX 29-C NONRIBOSOMAL PEPTIDE SYNTHESIS

Many small biologically active peptides, including hormones and some antibiotics, are synthesized on ribosomes as precursor proteins, which are cut into small pieces and may then be modified in a variety of ways. However, many other peptides including many antibiotics are made without use of ribosomes by large polyfunctional synthetases. The first of these, gramicidin S synthetase, was described by Lipmann and coworkers in 1971.[a] It is discussed on p. 994 as is the mechanism of synthesis. It is now recognized that these enzymes are modular and have much in common with fatty acid synthetases (Fig. 17-12 and p. 1186) and polyketide synthetases (Fig. 21-11). They are able to link not only the amino acids found in proteins but also modified and unusual amino acids. They may also join one or more α-hydroxy acids to a peptide to form a depsipeptide, and they may contain modules that carry out modification reactions such as methylation, acylation, or glycosylation.[b–e] Because of their modular nature they are attractive proteins for genetic engineering.[b,f,g]

Each synthetase module contains three active site domains: The **A domain** catalyzes activation of the amino acid (or hydroxyacid) by formation of an aminoacyl- or hydroxyacyl-adenylate, just as occurs with aminoacyl-tRNA synthetases. However, in three-dimensional structure the A domains do not resemble either of the classes of aminoacyl-tRNA synthetases but are similar to luciferyl adenylate (Eq. 23-46) and acyl-CoA synthetases.[h] The **T-domain** or **peptidyl carrier protein domain** resembles the acyl carrier domains of fatty acid and polyketide synthetases in containing bound phosphopantetheine (Fig. 14-1). Its –SH group, like the CCA-terminal ribosyl –OH group of a tRNA, displaces AMP, transferring the activated amino acid or hydroxy acid to the thiol sulfur of phosphopantetheine. The **C-domain** catalyzes condensation (peptidyl transfer). The first or **initiation module** lacks a C-domain, and the final **termination module** contains an extra termination domain. The process parallels that outlined in Fig. 21-11.[i]

A few of the products of nonribosomal peptide synthesis are gramicidin S (Fig. 2-4), enniatins, bacitracins, and tyrocidines (p. 994),[b,e] vancamycin (Box 20-H),[j] actinomycin (Box 28-A),[k] the siderophore yersiniabactin,[l] surfactin (Fig. 2-4),[m,n] and cyclosporin (Box 9-F).[o,p] The δ -(L-α-aminoadipoyl)-L-cysteinyl-D-valine synthetase,[q,r] which forms the precursor to penicillin and cephalosporins (Box 20-G), also belongs to this group of enzymes as do synthetases that make cyclooctadepsipeptides with antihelminthic activity[s] and many other compounds.[t–v]

[a] Lipmann, F. (1971) *Science* **173**, 875–884

[b] Mootz, H. D., Schwarzer, D., and Marahiel, M. A. (2000) *Proc. Natl. Acad. Sci. U.S.A.* **97**, 5848–5853

[c] Linne, U., and Marahiel, M. A. (2000) *Biochemistry* **39**, 10439–10447

[d] Guenzi, E., Galli, G., Grgurina, I., Pace, E., Ferranti, P., and Grandi, G. (1998) *J. Biol. Chem.* **273**, 14403–14410

[e] Belshaw, P. J., Walsh, C. T., and Stachelhaus, T. (1999) *Science* **284**, 486–489

[f] Symmank, H., Saenger, W., and Bernhard, F. (1999) *J. Biol. Chem.* **274**, 21581–21588

[g] de Ferra, F., Rodriguez, F., Tortora, O., Tosi, C., and Grandi, G. (1997) *J. Biol. Chem.* **272**, 25304–25309

[h] Pfeifer, E., Pavela-Vrancic, M., von Döhren, H., and Kleinkauf, H. (1995) *Biochemistry* **34**, 7450–7459

[i] Cane, D. E., Walsh, C. T., and Khosla, C. (1998) *Science* **282**, 63–68

[j] Trauger, J. W., and Walsh, C. T. (2000) *Proc. Natl. Acad. Sci. U.S.A.* **97**, 3112–3117

[k] Pfennig, F., Schauwecker, F., and Keller, U. (1999) *J. Biol. Chem.* **274**, 12508–12516

[l] Miller, D. A., and Walsh, C. T. (2001) *Biochemistry* **40**, 5313–5321

[m] Weinreb, P. H., Quadri, L. E. N., Walsh, C. T., and Zuber, P. (1998) *Biochemistry* **37**, 1575–1584

[n] Reuter, K., Mofid, M. R., Marahiel, M. A., and Ficner, R. (1999) *EMBO J.* **18**, 6823–6831

[o] Lawen, A., and Traber, R. (1993) *J. Biol. Chem.* **268**, 20452–20465

[p] Hoffmann, K., Schneider-Scherzer, E., Kleinkauf, H., and Zocher, R. (1994) *J. Biol. Chem.* **269**, 12710–12714

[q] Shiau, C.-Y., Byford, M. F., Aplin, R. T., Baldwin, J. E., and Schofield, C. J. (1997) *Biochemistry* **36**, 8798–8806

[r] Kallow, W., Kennedy, J., Arezi, B., Turner, G., and von Döhren, H. (2000) *J. Mol. Biol.* **297**, 395–408

[s] Weckwerth, W., Miyamoto, K., Iinuma, K., Krause, M., Glinski, M., Storm, T., Bonse, G., Kleinkauf, H., and Zocher, R. (2000) *J. Biol. Chem.* **275**, 17909–17915

[t] Milne, J. C., Roy, R. S., Eliot, A. C., Kelleher, N. L., Wokhlu, A., Nickels, B., and Walsh, C. T. (1999) *Biochemistry* **38**, 4768–4781

[u] Gaitatzis, N., Kunze, B., and Müller, R. (2001) *Proc. Natl. Acad. Sci. U.S.A.* **98**, 11136–11141

[v] Gewolb, J. (2002) *Science* **295**, 2205–2207

topoisomerase subunit.[494–495a] Ribosomal protein L9 may play a role in the bypass process.

7. RNA Viruses

The RNA-containing bacteriophages are convenient sources of relatively simple mRNA molecules, whose sequences can be studied.[496] The genetic information for these viruses is carried by RNA molecules consisting of only 3500–4500 nucleotides and which may contain only four genes (p. 247). The RNA from phages f2, R17, MS2, and the more distant Qβ have been studied intensively.[483,497]

Parts of the 2569-nucleotide sequence for the RNA of phage MS2[498] are shown in Fig. 29-17. The 5' end (upper left center) still bears the triphosphate group of the initiating GTP. Following a number of hairpin loops there is a ribosome-protected region, which begins with the initiation codon GUG for the A protein

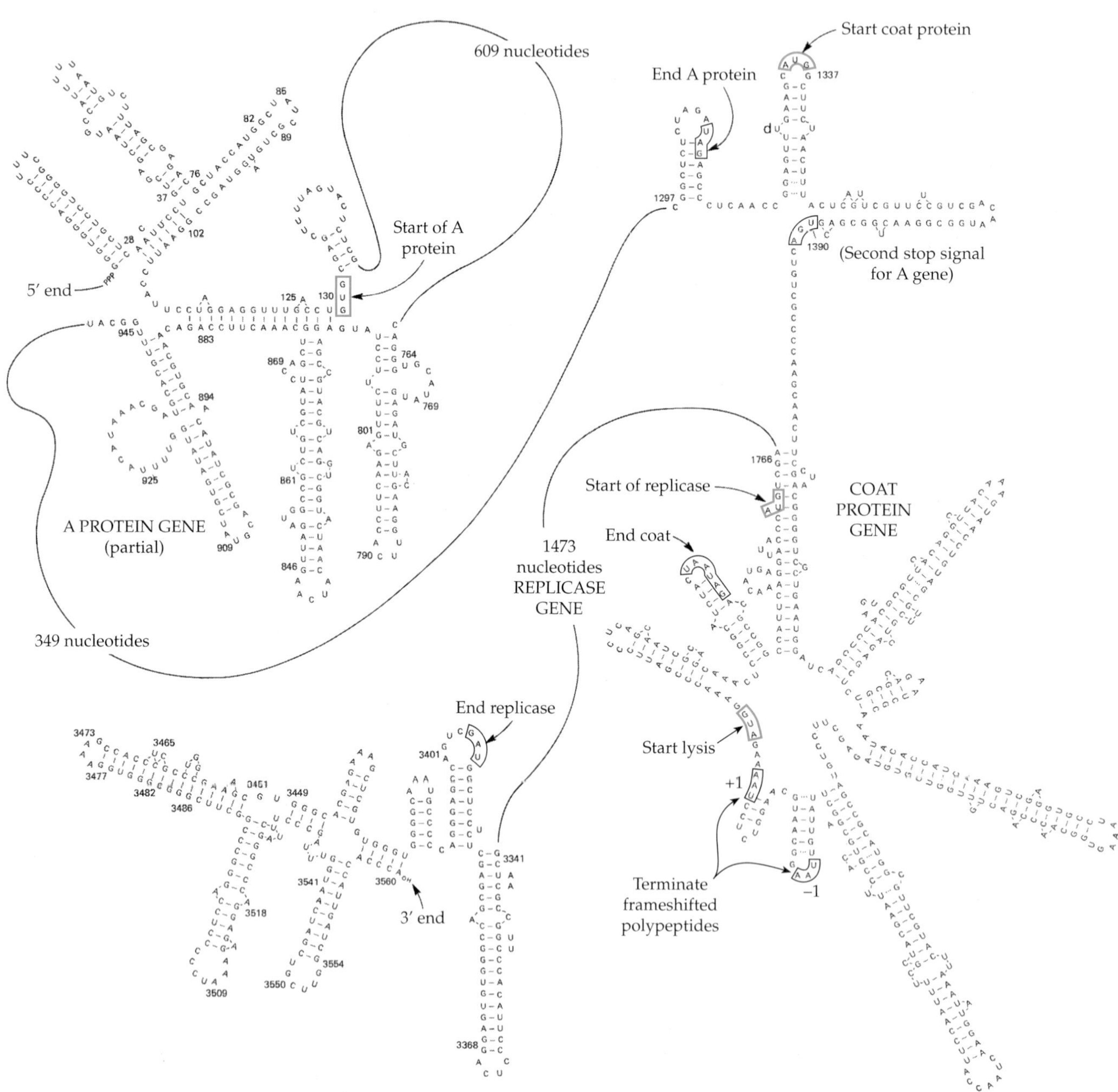

Figure 29-17 Partial sequence and secondary structure model of RNA of bacteriophage MS2. The initiation and termination codons for each of the three genes (A protein, coat protein, and replicase) are enclosed in boxes as is the second stop signal that is in-frame for the A protein gene but out-of-frame for the coat protein gene. The entire coat protein gene is shown but less than one-third of the entire sequence is given. From W. Fiers and associates.[499–501]

(enclosed in a box). Here is some of the first direct evidence ever obtained that GUG as well as AUG is a biologically important initiation codon. Following the initiation codon the nucleotide sequence codes exactly for the established amino acid sequence of the protein. The termination codon UAG is also enclosed in a box in the figure. Following this is a short intergenic region, which includes one side of a hairpin loop with the initiator codon AUG for the next gene at the end. The nucleotide sequence following this codes exactly for the experimentally established sequence end of the coat protein.[499] One other feature of the sequence shown is the UGA termination codon in a box shortly after the beginning of the coat protein gene (at position 1390). This termination signal is out of phase with the initiator codon AUG; hence, it does not represent a termination point for the coat gene. However, it is in phase with the UAG termination codon for the A protein. In the presence of various host amber (UAG) suppressor genes, the A protein is elongated and terminated at this UGA codon.

The coat gene, containing only 390 nucleotides, is shown in its entirety. The secondary structure proposed resembles a flower.[499] The gene ends with a double stop signal UAAUAG. Following an intergenic sequence of 36 nucleotides the long replicase gene starts with an AUG codon. It ends at position 3395 leaving an untranslated segment of 174 nucleotides at the 3' end.

Initially it was thought that MS2 RNA contained only the three genes mentioned in the preceding paragraphs, but later it was found to have an additional gene required for lysis of the host cell.[483] The initiation codon for this gene begins at position 1678 (Fig. 29-17) in the +1 reading frame. There is a UAA stop just two codons before this in the same frame and another UAA stop codon in the −1 frame beginning at position 1652. As a result of these stop codons any reading frameshift during synthesis of the coat protein yields mistranslated proteins that are terminated at these codons. There is no Shine–Dalgarno sequence to bind ribosomes for initiation in this region, but because the initiation codon for the lysis (L) protein is nearby, reinitiation occurs and the L protein is made in the relatively small amounts needed. This arrangement permits efficient use of the RNA by making use of overlapping genes. It also ensures that enough coat protein has been synthesized to make new virus particles before the L protein accumulation causes lysis.[483]

Many viral RNAs that are formed within eukaryotic cells lack a 5' cap. They depend upon internal ribosomal entry sites (IRESs). This has been studied most with picorna viruses.[338,502,503] These viruses not only initiate translation at discrete sites in uncapped RNA but carry out a proteolytic cleavage of initiation factor 4G (Fig. 29-11), which seems to be necessary for initiation of viral-RNA translation.[338,504] The IRES region of hepatitis C viral RNA contains a complex pseudoknotted secondary structure that is necessary for initiation.[346,505,506] Cryo-EM reveals a pronounced change in the 40S ribosomal subunit structure when the viral IRES binds.[346] Some RNA viruses of plants have complex secondary structures in the untranslated 3' region that promote efficient initiation of translation.[506a]

8. Other Functions of Ribosomes

In addition to making proteins, ribosomes also participate in regulatory mechanisms that influence the entire cell. One such mechanism is seen in the **stringent response**.[507–510] Many amino acid-requiring auxotrophs of *E. coli* and other bacteria, when deprived of an essential amino acid, respond by decreasing their production of ribosomal RNA, ribosomal proteins, purine nucleoside triphosphates, lipids, and other essential materials. However, mutations in the gene *rel* (relaxed) lead to continued production of rRNA even in the absence of an essential amino acid. (The stringent response is "relaxed.") It was observed that the **guanosine polyphosphates ppGpp** and **pppGpp**, originally termed MS or "magic spot" compounds, accumulate in stringent (rel^+) strains to a concentration of ~1 mM but not in relaxed (rel^-) strains. Guanosine polyphosphates are synthesized on the ribosomes by transfer of a pyrophospho group from ATP (Eq. 29-11):

$$\text{ATP} + \text{GDP(GTP)} \longrightarrow \text{ppGpp(pppGpp)} + \text{AMP} \quad (29\text{-}11)$$

The reaction is catalyzed by the 84-kDa ppGpp synthetase (**stringent factor**), which is encoded by the *rel* gene and is present only in stringent strains.[511–512a] It binds to ribosomes and becomes active only if mRNA is bound to the ribosomes and if codon-selected uncharged tRNA is present in the A sites. A second ppGpp synthetase (PSII) is encoded by gene **spoT**, which also codes for a ppGpp hydrolase.[510] The presence of an uncharged tRNA in the ribosomal A site is expected during amino acid starvation. The stringent factor competes with elongation factor EF-G for its ribosomal site.[513]

The most important effect of accumulating ppGpp may be to bind to an allosteric site on RNA polymerase.[509] The ppGpp-polymerase complex appears to be inefficient in initiating transcription of genes for rRNA, other stable RNAs, and ribosomal proteins. However, it stimulates expression of various amino acid biosynthetic genes and catabolic genes, perhaps via the "discriminator sequence" (Chapter 28; p. 1608). This is not the only effect of ppGpp. The fidelity of translation is decreased when amino acid concentrations fall and

BOX 29-D PROTEIN SPLICING, INTEINS, AND HOMING ENDONUCLEASES

Like self-splicing RNAs, which excise introns from their chains, a few proteins are able to splice out segments of their own chains as **inteins**. The surrounding protein sequences can be referred to as **exteins**. Over 100 self-splicing proteins are known. They are found in all kingdoms of life.[a–d] The inteins, which are excised, are typically 50 kDa in size but range from ~360 to over 500 residues. The mechanism of splicing is related to the chemistry of pyruvoyl enzyme activation (Eq. 14-41), succinimide formation from asparagine residues (Eq. 2-24), and protein carboxymethylation (Box 12-A). The intein always contains serine or cysteine in its N-terminal (1)-position and asparagine in its C-terminal position. The latter is always followed by cysteine, serine, or threonine in the N-terminal

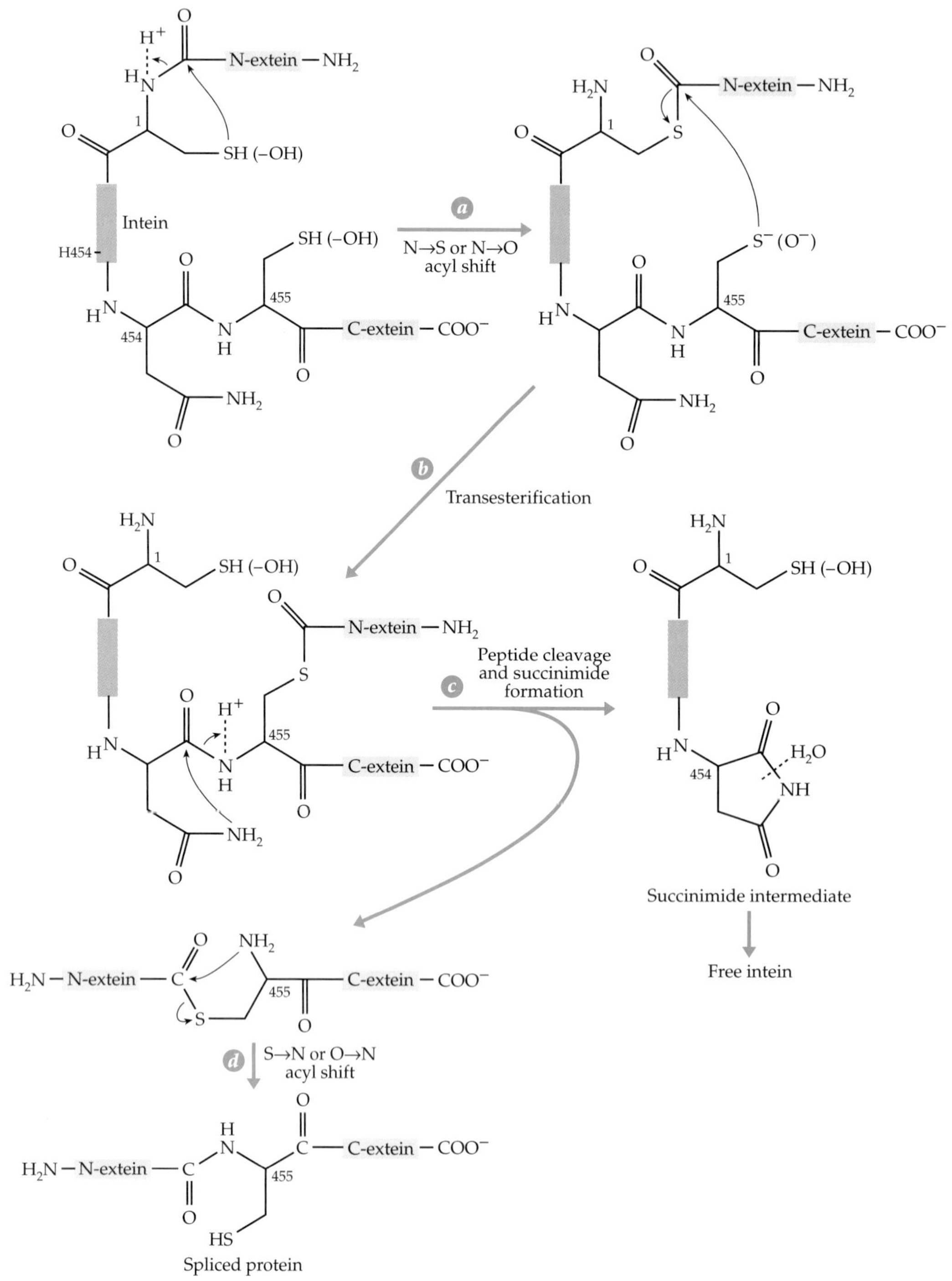

BOX 29-D (continued)

position of the C-extein. The penultinate residue in the intein is usually (~90 %) histidine, which is thought to play a catalytic role. Other residues in the catalytic domains, which form the ends of the inteins, may also participate in catalysis.

One of the first inteins discovered was found in the 119-kDa precursor to a subunit of a vacuolar ATPase of yeast.[a,c] In this 50-kDa intein Thr 72, His 75, and His 197 may have catalytic functions.[d] The intein is spliced out to form the 69-kDa subunit. The splicing mechanism, which is illustrated for this intein, is shown in the accompanying equations.[b,d–g] Step *a* is an N→S or N→O acyl shift. This is followed by transesterification (step *b*) which involves either thioesters (as illustrated) or oxygen esters. Formation of a succinmide intermediate (step *c*) releases the intein and the spliced protein. The latter must undergo an S→N or O→N acyl shift (step *d*), and the succinimide in the extein must be hydrolyzed to complete the process.

Why do cells ever splice proteins? It isn't clear. However, a curious fact is that many inteins are **homing endonucleases**.[h–k] The genes for these nucleases are often present in introns in mRNA, and the homing endonuclease often cuts DNA in such a way as to initiate movement of its own gene (Chapter 27). The endonuclease itself is found in the center of the intein between the two end domains, which contain the catalytic centers for the splicing reaction.

A few cases are known in which proteins undergo *trans* splicing. For example, the *dnaE* gene of *Synechocystis*, which codes for DNA polymerase III, is actually two partial genes that are 745 kb apart and on opposite strands of the DNA. One of the partial genes codes for a protein containing the N-terminal splice site for an intein, and the other gene codes for a polypeptide containing the C-terminal splice site. Evidently the two splicing domains associate and then catalyze the splicing sequence in the usual way. Split inteins have become very useful in protein engineering because they can be used to join various polypeptide sequences.[k–m] They have also provided an efficient system for purification of specific proteins.[b,n]

[a] Cooper, A. A., and Stevens, T. H. (1995) *Trends Biochem. Sci.* **20**, 351–356

[b] Chong, S., Shao, Y., Paulus, H., Benner, J., Perler, F. B., and Xu, M.-Q. (1996) *J. Biol. Chem.* **271**, 22159–22168

[c] Chong, S., Williams, K. S., Wotkowicz, C., and Xu, M.-Q. (1998) *J. Biol. Chem.* **273**, 10567–10577

[d] Poland, B. W., Xu, M.-Q., and Quiocho, F. A. (2000) *J. Biol. Chem.* **275**, 16408–16413

[e] Clarke, N. D. (1994) *Proc. Natl. Acad. Sci. U.S.A.* **91**, 11084–11088

[f] Xu, M.-Q., and Perler, F. B. (1996) *EMBO J.* **15**, 5146–5153

[g] Shao, Y., Xu, M.-Q., and Paulus, H. (1996) *Biochemistry* **35**, 3810–3815

[h] Chuprina, V. P., Heinemann, U., Nurislamov, A. A., Zielenkiewicz, P., and Dickerson, R. E. (1991) *Proc. Natl. Acad. Sci. U.S.A.* **88**, 593–597

[i] Gimble, F. S., and Wang, J. (1996) *J. Mol. Biol.* **263**, 163 – 180

[j] Pietrokovski, S. (1998) *Protein Sci.* **7**, 64–71

[k] Perler, F. B. (1999) *Trends Biochem. Sci.* **24**, 209–211

[l] Martin, D. D., Xu, M.-Q., and Evans, T. C., Jr. (2001) *Biochemistry* **40**, 1393–1402

[m] Otomo, T., Ito, N., Kyogoku, Y., and Yamazaki, T. (1999) *Biochemistry* **38**, 16040–16044

[n] Evans, T. C., Jr., Martin, D., Kolly, R., Panne, D., Sun, L., Ghosh, I., Chen, L., Benner, J., Liu, X.-Q., and Xu, M.-Q. (2000) *J. Biol. Chem.* **275**, 9091–9094

ribosomal action slows. However, ppGpp apparently binds to the ribosome and slows the binding of the aminoacyl-tRNA•EF-Tu•GTP complex. This allows more time for rejection of mispaired tRNAs and increases the fidelity of translation.[514] Under conditions of nutrient starvation the accumulating ppGpp may promote enzymatic degradation of unneeded proteins[512a,b] and may also induce programmed cell death.[515]

Another "alarmone" that regulates both transcription and DNA replication and other cell functions is diadenosine tetraphosphate (Ap_4A). Effects of Ap_4A and related compounds have been discussed in Chapter 28 (p. 1635). These compounds affect many biological events including replication, growth, and differentiation.[516] However, the synthesis of Ap_4A is a reaction not of ribosomes but of an aminoacyl-tRNA synthetase. An enzyme-bound aminoacyl adenylate carries out adenylylation of ATP rather than aminoacylation of tRNA, especially when Zn^{2+} is present. Ap_4A is abundant in blood platelets, where it is stored in dense granules.[517] Both Ap_2A and AP_3A accumulate as gran- ules in myocardial tissues,[518] and Ap_5A and Ap_6A are also present in adrenal chromaffin cells, in blood plate- lets, and in synaptic vesicles.[519] These compounds are catabolized by hydrolases or in lower eukaryotes by phosphorylases. For example, Ap_4A may be converted into ATP + AMP or converted into ATP and ADP (Eq. 29-12).[516,520]

$$\text{A—}P\text{—}P\text{—}P\text{—}P\text{—A} \xrightarrow[\text{H}_2\text{O}\ \ \text{AMP}]{\text{P}_i\ \ \text{ADP}} \text{ATP} \quad (29\text{-}12)$$

A quite different role of ribosomes is to regulate the life span of certain mRNA molecules. The best studied example is the mRNA for the microtubule

BOX 29-E PRIONS AND AMYLOID DISEASES

The mysterious prions (proteinaceous infective agents), which are described briefly on p. 248, are under intensive investigation. Prion diseases affect fewer than one in 100,000 persons, but there is fear of a possible epidemic. Furthermore, there is a close relationship of prions to a large family of **amyloid diseases**. The most frequent of these is Alzheimer disease, which is estimated to affect one-third of people over 85 years of age in the United States.[a,b]

Prion diseases include **scrapie** of sheep and goats, **bovine spongiform encephalopathy** (**BSE** or mad cow disease), **chronic wasting disease** (**CWD**) of deer and elk, and the human diseases **kuru**, **Creutzfeldt-Jakob disease** (**CJD**), **Gerstmann-Sträussler-Scheinker syndrome** (**GSS**),[c] and **fatal familial insomnia** (**FFI**).[a,d–f] The diseases have a variety of symptoms that include dementia, ataxia (loss of muscular coordination), insomnia, and behavioral problems. All involve some loss of neurons, which may or may not be indicated by a sponge-like appearance of the brain. A characteristic feature of prion disease is the appearance of **amyloid (starch-like) plaques**, which consist of fibrils of insoluble protein.[a] Exhaustive attempts failed to identify a virus particle or an associated DNA or RNA. On this basis, Stanley Prusiner suggested that the diseases are transmitted by pure proteins.[a] All of the diseases seem to involve the same protein, which is known as the **prion protein** (**PrP**). It is encoded by a single-copy gene on human chromosome 20.[g] The amino acid sequence of the C-terminal region of PrP is highly conserved among all animals. However, there are more than 20 known human genetic variants, and a second prion protein has been found in mice.[h] The function of the normal cellular prion protein (**PrP^C**) is unknown, but it appears to be a copper ion carrier, which may be essential to proper synaptic function.[i,j] "Knockout mice" lacking PrP are resistant to prion disease[j,k] but may not be completely healthy.

If it were not for the diseases, PrP^C might be viewed as just another cell surface glycoprotein. Determination of its three-dimensional structure has been difficult, but use of NMR spectroscopy and modeling has given a nearly complete picture, which is shown in the accompanying drawing.[d,l–n] The 250-residue (~220 residues after removal of N- and C-terminal signal sequences) has a long N-terminal tail, a glycosylated globular domain, and a C terminus that is anchored in the outer membrane of neurons by a glycosylphosphatidylinositol (GPI) anchor similar to that shown in Fig. 8-13. The globular domain contains three α helices, a small β sheet, and two glycosylation sites. These last carry

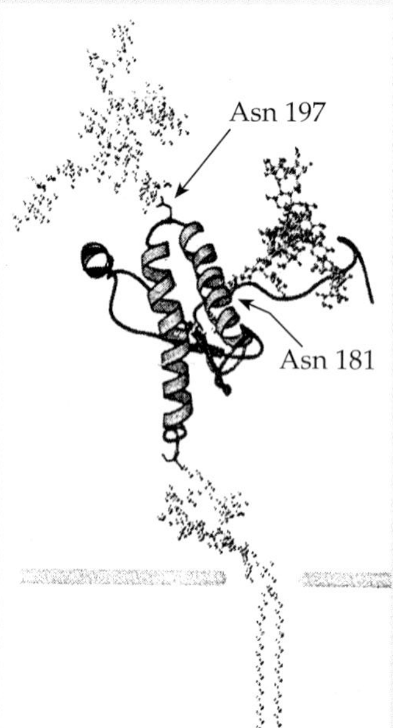

typical N-linked, branched sialic acid-containing oligosaccharides with a total of 52 or more sugar residues. The N-terminal 120 amino acid residue "tail" appears to be largely unstructured. However, it contains five octapeptide repeats with the consensus sequence PHGGGWGQ, each able to bind one Cu^{2+} or Mn^{2+} ion.[d,i,o,p]

How can this ordinarily harmless protein become a killer? The prion is a 20- to 30-kDa hydrophobic particle, which is thought to arise from PrP^C by a conformational alteration in which the α helices are largely changed into a β structure. The new conformer is often designated PrP^{Sc} or PrP-res. The latter abbreviation arises from the fact that native PrP^C can be completely hydrolyzed to small fragments by proteinase K, but PrP^{Sc} contains a 142-residue extremely resistant core (residues 90–231), which is not hydrolyzed and is over 80% β sheet.[q] Evidently the PrP^{Sc} form is able to associate to form a "seed" that, when conditions are favorable, can induce the conformational change in other molecules spreading the PrP^{Sc} form throughout the brain and even into tissues of the immune system.[r] With prion diseases and other amyloid diseases the body may be able to fight off the process by normal proteolytic turnover of the prion protein.

About 85% of all cases of prion disease are **sporadic CJD**. These are thought to arise by spontaneous conversion of PrP^C to PrP^{Sc}. Inherited (familial) forms of CJD, GSS, and FFI are also known. A series of point mutations as well as expansion of the octapeptide repeats[s] account for the various diseases, which have an autosomal dominant inheritance. At least 23 pathogenic mutations have been

BOX 29-E (continued)

reported.[t,u] The point mutations occur at several locations, some of them adjacent to the glycosylation sites.[m] These mutant proteins may be more readily converted to the less soluble PrP^{Sc} type structure, initiating the disease process. However, a mutant with a stop codon (TAG) in the place of the tyrosine 145 codon loses its C-terminal anchor and is degraded rapidly in the proteasomal pathway.[t]

The infectious forms of prion diseases are more puzzling. They account for less than one percent of all cases. Attention was first focused on kuru, a disease of the Fore people of New Guinea. In earlier times they practiced a ritualistic cannibalism of brain tissue that apparently propagated the disease, which is now nearly extinct. Of present concern are over 100 cases of a "new variant" form of CJD, some involving teenage persons and young adults, which have been reported in Europe.[u,v] This disease may have originated in sheep, then jumped to cattle, where it was spread by the ingestion of prion-contaminated meat and bone meal.[a] In addition, more than 120 cases of CJD have arisen from injection of prion-contaminated human growth hormone. Other cases have been traced to contaminated surgical instruments, to tissue grafts, and to use of contaminated human pituitary gonadotrophin.[a]

A hard-to-understand aspect of the "protein-only" theory of prion diseases is the existence of various "strains" of prion proteins. These do not involve differences in amino acid sequence but differences in the conformations of the PrP^{Sc} forms and in the glycosylation patterns.[d,m,w] How can there be several different conformations of the same protein, all of which seed the conversion of normal PrP into differing insoluble forms? In spite of this puzzle, support for the explanation of strain differences comes from a yeast prion system, which involves transcription termination factor eRF3.[x–z] In this system, which involves a prion whose insoluble form can be redissolved by guanidine hydrochloride,[aa] differing strains have also been described.[y,bb,cc] Nevertheless, the presence of the various strains of animal prions, as well as observed vaccination of inbred mice against specific strains,[dd] may be more readily understood if the disease is transmitted by an unidentified virus rather than by a pure protein.[r,u,ee,ff] In fact, the diseases have not been successfully transmitted by truly virus-free proteins synthesized from recombinant DNA.[ee]

What are the prospects for a cure for prion diseases? Several compounds show some effect in slowing accumulation of amyloid plaques,[d,v,gg,hh] but suitable drugs have not been developed. Prevention is the best cure, but more needs to be known about the basic biology of the disease transmission before effective strategies for prevention can be developed.[u]

What is the nature of the insoluble forms of the prion protein? They are hard to study because of the extreme insolubility, but the conversion of α helix to β sheet seems to be fundamental to the process and has been confirmed for the yeast prion by X-ray diffraction.[ii] It has been known since the 1950s that many soluble α-helix-rich proteins can be transformed easily into a fibrillar form in which the polypeptide chains are thought to form a β sheet. The chains are probably folded into hairpin loops that form an antiparallel β sheet (see Fig. 2-11).[jj–ll] For example, by heating at pH 2 insulin can be converted to fibrils, whose polarized infrared spectrum (Fig. 23-3A) indicates a **cross-β structure** with strands lying perpendicular to the fibril axis.[jj,mm] Many other proteins are also able to undergo similar transformation. Most biophysical evidence is consistent with the cross-β structure for the fibrils, which typically have diameters of 7–12 nm.[ii,ll,nn] These may be formed by association of thinner 2 to 5 nm fibrils.[oo] However, β-helical structures have been proposed for some amyloid fibrils[pp] and polyproline II helices for others.[qq]

A wide range of human diseases involving amyloid deposits are known. These include not only the prion diseases and the neurodegenerative diseases, Alzheimer, Parkinson, and the polyglutamine repeat diseases (Table 26-4)[rr,ss] but also **systemic amyloidoses**.[tt] Among the latter are deposits of transthyretin,[uu] the 37-residue **amylin** that develops in the β cells of the pancrease in type II diabetes,[vv] mutant forms of lysozyme,[ww] and of β2 microglobulin,[xx] and gelsolin.[yy] A serum protein amyloid P, a calcium-binding protein, is usually also a component of amyloid deposits.[zz]

---

[a] Prusiner, S. B. (2001) *N. Engl. J. Med.* **344**, 1516–1526
[b] Manuelidis, L., Fritch, W., and Xi, Y.-G. (1997) *Science* **277**, 94–98
[c] Tagliavini, F., Lievans, P. M.-J., Tranchant, C., Warter, J.-M., Mohr, M., Giaccone, G., Perini, F., Rossi, G., Salmona, M., Piccardo, P., Ghetti, B., Beavis, R. C., Bugiani, O., Frangione, B., and Prelli, F. (2001) *J. Biol. Chem.* **276**, 6009–6015
[d] Caughey, B. (2001) *Trends Biochem. Sci.* **26**, 235–242
[e] Prusiner, S. B., ed. (1999) *Prion Biology and Diseases*, Cold Spring Harbor Lab. Press, Cold Spring Harbor, New York
[f] Prusiner, S. B. (1996) *Trends Biochem. Sci.* **21**, 482–487
[g] Prusiner, S. B. (1995) *Proc. Natl. Acad. Sci. U.S.A.* **95**, 13363–13383
[h] Moore, R. C., and 20 other authors. (1999) *J. Mol. Biol.* **292**, 797–817

BOX 29-E PRIONS AND AMYLOID DISEASES (continued)

[i] Kramer, M. L., Kratzin, H. D., Schmidt, B., Römer, A., Windl, O., Liemann, S., Hornemann, S., and Kretzschmar, H. (2001) *J. Biol. Chem.* **276**, 16711–16719

[j] Yokoyama, T., Kimura, K. M., Ushiki, Y., Yamada, S., Morooka, A., Nakashiba, T., Sassa, T., and Itohara, S. (2001) *J. Biol. Chem.* **276**, 11265–11271

[k] Weissmann, C., and Aguzzi, A. (1999) *Science* **286**, 914–915

[l] Zahn, R., Liu, A., Lührs, T., Riek, R., von Schroetter, C., Garcia, F. L., Billeter, M., Calzolai, L., Wider, G., and Wüthrich, K. (2000) *Proc. Natl. Acad. Sci. U.S.A.* **97**, 145–150

[m] Rudd, P. M., Wormald, M. R., Wing, D. R., Prusiner, S. B., and Dwek, R. A. (2001) *Biochemistry* **40**, 3759–3766

[n] Viles, J. H., Donne, D., Kroon, G., Prusiner, S. B., Cohen, F. E., Dyson, H. J., and Wright, P. E. (2001) *Biochemistry* **40**, 2743–2753

[o] Prince, R. C., and Gunson, D. E. (1998) *Trends Biochem. Sci.* **23**, 197–198

[p] Brown, D. R., Hafiz, F., Glasssmith, L. L., Wong, B.-S., Jones, I. M., Clive, C., and Haswell, S. J. (2000) *EMBO J.* **19**, 1180–1186

[q] Baskakov, I. V., Aagaard, C., Mehlhorn, I., Wille, H., Groth, D., Baldwin, M. A., Prusiner, S. B., and Cohen, F. E. (2000) *Biochemistry* **39**, 2792–2804

[r] Balter, M. (1999) *Science* **286**, 660–662

[s] Narwa, R., and Harris, D. A. (1999) *Biochemistry* **38**, 8770–8777

[t] Zanusso, G., Petersen, R. B., Jin, T., Jing, Y., Kanoush, R., Ferrari, S., Gambetti, P., and Singh, N. (1999) *J. Biol. Chem.* **274**, 23396–23404

[u] Almond, J., and Pattison, J. (1997) *Nature (London)* **389**, 437–438

[v] Thompson, C. (2001) *Nature (London)* **409**, 660–661

[w] Hill, A. F., Desbruslais, M., Joiner, S., Sidle, K. C. L., Gowland, I., Collinge, J., Doey, L. J., and Lantos, P. (1997) *Nature (London)* **389**, 448–450

[x] Patino, M. M., Liu, J.-J., Glover, J. R., and Lindquist, S. (1996) *Science* **273**, 622–626

[y] Liebman, S. W., and Derkatch, I. L. (1999) *J. Biol. Chem.* **274**, 1181–1184

[z] Zhou, P., Derkatch, I. L., Uptain, S. M., Patino, M. M., Lindquist, S., and Liebman, S. W. (1999) *EMBO J.* **18**, 1182–1191

[aa] Wickner, R. B., Edskes, H. K., Maddelein, M.-L., Taylor, K. L., and Moriyama, H. (1999) *J. Biol. Chem.* **274**, 555–558

[bb] Sparrer, H. E., Santoso, A., Szoka, F. C., Jr., and Weissman, J. S. (2000) *Science* **289**, 595–599

[cc] Chien, P., and Weissman, J. S. (2001) *Nature (London)* **410**, 223–227

[dd] Manuelidis, L. (1998) *Proc. Natl. Acad. Sci. U.S.A.* **95**, 2520–2525

[ee] Manuelidis, L., Sklaviadis, T., Akowitz, A., and Fritch, W. (1995) *Proc. Natl. Acad. Sci. U.S.A.* **92**, 5124–5128

[ff] Manuelidis, L. (1997) *Annales De L'Institut Pasteur* **8**, 311–326

[gg] Tagliavini, F., and 20 other authors. (1997) *Science* **276**, 1119–1122

[hh] Supattapone, S., Nguyen, H.-O. B., Cohen, F. E., Prusiner, S. B., and Scott, M. R. (1999) *Proc. Natl. Acad. Sci. U.S.A.* **96**, 14529–14534

[ii] Balbirnie, M., Grothe, R., and Eisenberg, D. S. (2001) *Proc. Natl. Acad. Sci. U.S.A.* **98**, 2375–2380

[jj] Burke, M. J., and Rougvie, M. A. (1972) *Biochemistry* **11**, 2435–2439

[kk] Bradbury, E. M., Brown, L., Downie, A. R., Elliott, A., Fraser, R. D. B., Hanby, W. E., and Macdonald, T. R. R. (1960) *J. Mol. Biol.* **2**, 276–286

[ll] Sunde, M., Serpell, L. C., Bartlam, M., Fraser, P. E., Pepys, M. B., and Blake, C. C. F. (1997) *J. Mol. Biol.* **273**, 729–739

[mm] Bouchard, M., Zurdo, J., Nettleton, E. J., Dobson, C. M., and Robinson, C. V. (2000) *Protein Sci.* **9**, 1960–1967

[nn] Hörnberg, A., Eneqvist, T., Olofsson, A., Lundgren, E., and Sauer-Eriksson, A. E. (2000) *J. Mol. Biol.* **302**, 649–669

[oo] Chiti, F., Taddei, N., Bucciantini, M., White, P., Ramponi, G., and Dobson, C. M. (2000) *EMBO J.* **19**, 1441–1449

[pp] Lazo, N. D., and Downing, D. T. (1998) *Biochemistry* **37**, 1731–1736

[qq] Blanch, E. W., Morozova-Roche, L. A., Cochran, D. A. E., Doig, A. J., Hecht, L., and Barron, L. D. (2000) *J. Mol. Biol.* **301**, 553–563

[rr] Koo, E. H., Lansbury, P. T., Jr., and Kelly, J. W. (1999) *Proc. Natl. Acad. Sci. U.S.A.* **96**, 9989–9990

[ss] Kaytor, M. D., and Warren, S. T. (1999) *J. Biol. Chem.* **274**, 37507–37510

[tt] Hardy, J., and Gwinn-Hardy, K. (1998) *Science* **282**, 1075–1079

[uu] Sebastiao, M. P., Lamzin, V., Saraiva, M. J., and Damas, A. M. (2001) *J. Mol. Biol.* **306**, 733–744

[vv] Moriarty, D. F., and Raleigh, D. P. (1999) *Biochemistry* **38**, 1811–1818

[ww] Booth, D. R., Sunde, M., Bellotti, V., Robinson, C. V., Hutchinson, W. L., Fraser, P. E., Hawkins, P. N., Dobson, C. M., Radford, S. E., Blake, C. C. F., and Pepys, M. B. (1997) *Nature (London)* **385**, 787–793

[xx] Chiti, F., Mangione, P., Andreola, A., Giorgetti, S., Stefani, M., Dobson, C. M., Bellotti, V., and Taddei, N. (2001) *J. Mol. Biol.* **307**, 379–391

[yy] Robinson, R. C., Choe, S., and Burtnick, L. D. (2001) *Proc. Natl. Acad. Sci. U.S.A.* **98**, 2117–2118

[zz] Lambowitz, A. M., and Perlman, P. S. (1990) *Trends Biochem. Sci.* **15**, 440–444

proteins tubulin.[521–523] Accumulating α and β tubulin subunits act in a feedback loop to induce the degradation of their mRNA. They do this by binding to the N-terminal sequence MREI of β tubulin as it is emerging from ribosomes. This binding allosterically activates an endonuclease that cuts the polysome-bound mRNA. A similar mechanism for tubulin mRNA may involve the MREC N-terminal sequence of that protein.

An unexpected finding was that **phosphatidylserine synthase** of *E. coli* is tightly bound to ribosomes.[524] This enzyme, which incorporates serine into phospholipids according to step *h* of Fig. 21-4, is responsible for synthesis of the principal membrane lipid of *E. coli*. Its localization on ribosomes may be linked to the joint regulation of the synthesis of proteins and lipids.

D. Processing, Secretion, and Turnover of Proteins

The concepts of processing and turnover of proteins have been introduced in Chapter 10, and many details have been presented in other chapters. However, as we complete our discussion of protein synthesis, it is appropriate to discuss processing further. As polypeptide chains leave the ribosomes via the exit channels, they may follow several different paths. They may enter the cytosol and fold quickly into a compact form. This may require only a few seconds, whereas the translation process in the ribosome may take many seconds. The folding will therefore be **cotranslational**.[525] Depending upon the N-terminal signal peptide the protein may later unfold and pass through a membrane pore or **translocon** into the endoplasmic reticulum (ER), a mitochondrion, chloroplast, or peroxisome. Wherever it is, it will be crowded together with thousands of other proteins. It will interact with many of these, and evolution will have enabled some of these to become chaperones (discussed in Chapter 10).[526]

A single rapidly growing cell of *E. coli* may contain ~2.3 x 10^6 soluble polypeptide chains of ~2600 different types with an average length of ~317 residues and average mass of ~35 kDa.[527] These are chaperoned in a variety of ways,[528,529] not only in the cytoplasm but in the periplasm (see p. 364).[530,531] The three chaperones **trigger factor** (TF), DnaK, and GroEL participate in folding newly formed proteins.[525a,b] TF is a prolyl-*cis-trans* isomerase (Box 9-F), which associates with the large ribosomal subunit with a 1:1 stoichiometry.[525b–d] DnaK and related chaperones hold and protect newly formed polypeptides in extended conformations, while the GroES-GroEL chaperonins assist folding within their internal cavities (Box 7-A).[525b,e–g] **Prefoldin**, a 90-kDa complex,[525b] has a special function in chaperoning microtubule subunits. A Type II chaperonin also assists the folding of actin and tubulin.[525h] Chaperones assist not only in the folding of proteins but also in translocation into the ER (e.g., by the Hsp 70 homolog BiP), and into mitochondria and other organelles.[532,533] **Co-chaperones** are additional proteins that act as selective agents to direct proteins to a particular chaperone. For example, the DnaJ protein is a scanning factor for the Hsp 70 chaperone DnaK. As is described on p. 518, it catalyzes ATP-dependent association of a substrate with the binding cavity of DnaK.[533] A chaperone whose function has long remained elusive is an abundant secreted glycoprotein known as **clusterin.**[534] It seems to have a protective function in protection against stress.

How are the possible choices for newly formed proteins made? Much seems to depend upon the amino acid sequences at the ends of the polypeptide chains. As they emerge from a ribosome, some N-terminal signal sequences bind to recognition proteins. One such protein labels the ends of proteins destined for secretion into the vesicles of the ER. This protein ensures that the protein end binds to the **signal recognition particle** (SRP), enters a translocon pore, and undergoes cotranslational passage into the periplasmic space in bacteria or the ER in eukaryotes. Cotranslational modification reactions also occur both in the cytosol and in the ER vesicles. These too influence the choice of destinations as do additional signal sequences in the polypeptide chains. Proteins may be directed to the various organelles, to residence in membranes,[535] or to secretion into the external medium. It was somewhat surprising to discover that under some circumstances most newly synthesized proteins are degraded in proteasomes.[536] Cotranslational degradation of proteins with imperfect ends may account for some of this.[537–539] In addition, imperfect proteins that are retained in the ER may be sent back into the cytoplasm for degradation in proteasomes.[540]

1. Cotranslational and Posttranslational Processing

The modifications that lead to the presence of fully functional proteins in their proper locations begin while peptide chains are still emerging from the ribosomes.[541] In bacteria and in eukaryotic organelles the first of these modifications is hydrolytic removal of the *N*-formyl group by an Fe^{2+}-dependent **deformylase** leaving the N-terminal methionine.[542–543] Deformylases are present in eukaryotes as well as in bacteria, making the deformylase a less attractive target for antibiotic design than has sometimes been proposed.[544,544a] When the chain is only 20–30 residues long, the terminal methionine that remains after deformylation may be removed by a ribosome-associated methionine aminopeptidase.[545] The methionine is usually removed if followed by P, G, A, S, or T and is retained if followed by K, R, L, I, F, or N. With other amino acids removal is variable.[546] A ribosome-bound *N*-acetyltransferase may acetylate the N terminus either before or after removal of Met.[547] Approximate rules for eukaryotic cells are[541]:

N-terminal D, E, N	Acetylation without removal of Met
N-terminal P, V, C	Removal of Met; no acetylation
N-terminal G, A, S, T	Removal of Met followed by acetylation
Other N termini	No modification

An example is provided by actin, which contains acetyl-Met-Asp, acetyl-Met-Gln or acetyl-Met-Cys-Asp at the N terminus immediately after synthesis. Then, within ~15 min the acetyl-Met is cleaved off, and the next terminal residue is acetylated.[548] *N*-Acylation of nascent peptides by fatty acyl groups can also occur cotranslationally. For example, 14-carbon myristoyl groups are added in amide linkage to the N-terminal glycines of many cellular and virally encoded proteins.[535,549,550] This may take place on the ribosomes,[551] but it is often not clear whether the modification is cotranslational or posttranslational. The same may be said of many other "posttranslational" alterations, many of which may begin on a nascent polypeptide chain. Fatty acyl groups (mainly palmitoyl) may form thioester linkages with cysteine side chains.[552] This often occurs near the C terminus (see p. 559). Other C-terminal modifications include prenylation (p. 559)[553] and attachment to diacylglycerols via thioester linkages to cysteine (pp. 402, 428)[553a] or to glycosylphosphatidylinositol glycan anchors (Fig. 8-13; Eq. 20-23).

The addition of an N-terminal myristoyl group to a protein causes a relatively permanent alteration as does methylation of histidine, lysine, or arginine side chains.[554,555] So do hydroxylation, vitamin K-dependent carboxylation (Eq. 15-55), and many other alterations. In contrast, glycosylation, phosphorylation, and sulfation produce reversible alterations. Sometimes, as in the conversion of proenzymes to active enzymes, a modification step is used to generate a catalytic activity. In other instances, as in the processing of glycoproteins in the Golgi, the major function of the modification reaction seems to be one of directing a protein to the correct intracellular location.

2. Forming Functional Proteins

Proenzymes and other precursor proteins are often almost totally inactive until they are activated by some alteration that occurs when they reach their destination in a cell or in the body. Cleavage of the polypeptide chains of proenzymes, covalent attachment of coenzymes,[556,557] oxidation (Eq. 16-57) or halogenation (Eq. 25-6) of tyrosine or tryptophan[558] side chains, and oxidation of cysteine in the sequence CTPSR to formylglycine (in sulfatase formation; Eq. 12-44)[559] are only a few of many modifications needed to form functional proteins. Sometimes, as in activation of chymotrypsinogen, a single simple modification creates the active protein. In other cases modification may be quite complex. For example, although many polypeptide antibiotics are formed by nonribosomal synthesis (Box 29-C), some are created on ribosomes and may require extensive subsequent alteration. An example is **microcin**, a 69-residue peptide antibiotic formed by some strains of *E. coli*. Eight Gly-Cys and Gly-Ser pairs in a pre-microcin chain are cyclized to thiazole and oxazole rings. Then the 69-residue antibiotic is cut out from the precursor and secreted into the medium.[560,561] The 22-residue antibiotic **epidermin** is one of a family of **lantibiotics** that contain lanthionine as a characteristic component. Biosynthesis involves dehydration of serine and threonine residues, sulfide (thioether) bridge formation, oxidative decarboxylation, and removal of a leader peptide.[562]

3. Translocation through Membranes

The processes by which proteins are selected for secretion into the periplasmic space of bacteria or into the vesicles of the endoplasmic reticulum of eukaryotic cells are similar and have been discussed in Chapter 10 (pp. 519–521). However, some details are still being worked out. The first step in translocation is binding of the N terminus of a protein that is emerging from a ribosome to the **signal recognition particle**.[563,564] The core of this particle has a universally conserved structure consisting of two proteins and an RNA molecule.[565–571b]

Bacteria	Eukaryotes
4.5S RNA (~114 nt)	7S RNA (~295 nt)
Protein Ffh	Protein SRP54
Protein Ftsy (SRP receptor)[572]	Protein SRα

All of these proteins are GTPases, and in eukaryotes SRα is associated with a third protein SRβ, which is also a GTPase.[573] Either protein Ffh or SRP54 recognizes the N terminus of the protein that is to be translocated, chaperoning it to the receptor Ftsy[574] or SRα, where it may be anchored to the translocating pore (**translocon**). Hydrolysis of GTP by both proteins accompanies the recognition process. The eukaryotic SRP is more complex, containing six proteins and a larger RNA than in bacteria.[568,575–576] One domain of the 7S RNA is homologous to the bacterial 4.5S RNA, while an additional domain is closely related in its sequence to that of the highly repetitive *Alu* sequences in DNA (Fig. 27-9). However, the significance of this similarity is unclear.

In eukaryotic cells binding of SRP54 induces a transient retardation of translocation, an **elongation arrest**, while the SRP complex binds to its receptor SRα. This 70-kDa peripheral membrane protein is tightly associated with the 30-kDa integral membrane protein SRβ. Binding to this receptor leads the nascent polypeptide chain from the ribosome directly into the Sec61 translocon,[33f,576a] which consists of α, β, and γ subunits and a central aqueous pore. The ribosome apparently also becomes bonded firmly to the translocon until synthesis of the polypeptide chain has been completed.[573,577,578] The translocon complex also contains additional components[563,577,579] including the **leader peptidase** (signal peptidase, p. 620)[563,580–582a] and the **oligosaccharyltransferase** of ER membranes (Eq. 20-21).[563,583] The latter transfers an oligosaccharide from a lipid carrier onto certain asparagine side chains of polypeptides entering the ER. This and other glycosylation reactions help to keep the polypeptide moving to its correct destination, whether it be in some membrane surface, a lysosome or other organelle, or a secretion vesicle (Chapter 20). Furthermore, some proteins are translocated by a mechanism that doesn't depend upon SRP but utilizes a different complex, which consists in yeast of proteins Sec62, Sec63, Sec71, Sec72, and Kar2 (the chaperone BiP present in the ER lumen).[577]

Translocation of most bacterial proteins occurs posttranslationally rather than cotranslationally.[584] After recognition by SRP the polypeptide chains are transferred to chaperone complexes.[585] Some proteins are escorted to the folding compartment of the GroEL-GroES chaperonin (Box 7-A). Those that are to be secreted are often chaperoned by the protein **SecB**.[586] Genetic analysis shows that for *E. coli* the secretion of many proteins requires the products of genes *secA, secB, secD, secE, secF, secG*, and *secY*. Gene *secA* encodes a 92-kDa cytoplasmic ATPase protein (**SecA**), SecB is a 64-kDa homotetramic chaperone that prevents folding of preproteins prior to export, and SecY is a 42-kDa integral membrane protein.[587,588]

A complex of the three transmembrane proteins **SecYEG** forms the translocation channel in *E. coli* membranes. **SecY** and **SecE** are essential components in most bacteria and are homologous to components of the eukaryotic **Sec61p** translocon complex.[589] From their sequences SecY, SecE, and SecG are predicted to have ten, three, and two transmembrane sequences, respectively. Additional accessory proteins in the complex are designated SecD, SecF, and yajC. Their functions are uncertain, and they are not essential for transport. The driving force for translocation is provided by the peripheral ATPase Sec A.[588,590] For many proteins the signal sequence, which is usually positively charged, stays on the negatively charged cytoplasmic surface of the membrane, while SecA in an ATP-dependent process pushes a loop of protein through the membrane. Hydrolysis of a second ATP molecule seems to be required to release SecA, allowing it to reload with ATP and to assist the next 20–30 residue polypeptide sequence to be translocated.[591,592] Surprisingly, SecA is also an ATP-dependent RNA helicase.[590] The significance of its apparent ability to translocate along either an RNA or a polypeptide chain is unclear. The protonmotive force provided by the membrane potential is another important factor in the translocation of many proteins.[591,593–596] Yet another factor is the lipid composition of the membrane. Non-bilayer lipids seem to be required for efficient transport.[597]

As is mentioned in Chapter 10, bacteria have additional mechanisms of polypeptide transport. A recently recognized Sec-independent pathway is used by *E. coli* and many other bacteria to secrete proteins that contain the twin-arginine motif RRX$\phi\phi$, where ϕ is a hydrophobic amino acid, in their N-terminal signal sequences. Proteins encoded by genes *tatABC* are required by this **Tat pathway**.[594,598,599] Related pathways have been identified both in mitochondria and in chloroplasts. Small peptides may pass out of the periplasmic space through the porins in the outer membrane of gram-negative bacteria. However, larger proteins require conduit molecules such as the **TolC** channel-tunnel, which directly connects an inner membrane translocon with a channel in the outer membrane of *E. coli* cells.[600,601]

Eukaryotic cells also have additional transport mechanisms. One of these is an ABC transporter (p. 417) known as the **transporter associated with antigen processing** (**TAP**). It carries small polypeptides generated by proteasomes from the cytosol into the ER for export and binding to **MHC Class I** molecules and subsequent presentation to the immune system (Fig. 31-15).[540,602]

4. Translocation into Organelles

Most of the proteins of mitochondria are encoded in nuclear DNA and are synthesized on cytoplasmic ribosomes. Mitochondria do not utilize proteins homologous to those of the bacterial Sec system but have their own set of transport proteins.[603–605] These proteins, which include an outer membrane complex (Tom) and an inner membrane complex (Tim), are discussed in Chapter 18 (see Fig. 18-4). Perhaps these specialized mitochondrial proteins are needed because transport into the mitochondrial matrix is in an opposite direction to the transport out through bacterial membranes.

The transport of proteins into chloroplasts also occurs by more than one mechanism. An SRP-dependent pathway may be needed only for insertion of proteins into membranes.[594] Other proteins, among which are the 23-kDa and 16-kDa photosystem II proteins (Chapter 23), enter by a pathway related to the Tat pathway of bacteria. In thylakoids this pathway is directly dependent upon the large pH difference (Δ pH) across the thylakoid membrane. In contrast to the bacterial Sec pathway, the Δ pH pathway seems to be able to transport completely folded proteins.

Proteins destined for peroxisomes have their own targeting signals. One of these (**PTS1**) is the sequence SKL at the C terminus. A second signal (**PTS2**) is an N-terminal nonapeptide (R / K)(L / V / I) X_5(H /Q) (L / A).[606,607]

5. Membrane Proteins

Some proteins enter membranes immediately after synthesis. The translocon channel is not required. However, in *E. coli* an additional protein **YidC** is needed.[603] Homologs of this protein are found in mitochondria (**Oxa1** protein) and in thylakoid membranes of chloroplasts (**Alb3** protein).[608] These proteins may function in cotranslational insertion. If a protein carries a positively charged N-terminal region, it will tend to stick to the negatively charged cytoplasmic surface of a cell membrane. This "positive inside" rule (p. 401)[609] is strong for bacterial proteins but somewhat weaker for eukaryotic cells. A second topological rule is that hydrophobic segments of proteins will be attracted to membrane surfaces and can enter the membrane (perhaps via translocon pores) as loops (Fig. 29-18).[596] Passage of the loop out through the membrane will be facilitated if negatively charged groups are present in the loop and are acted upon by the membrane potential (pp. 401,402).[595,610,611] If the entire polypeptide chain follows the loop out through the membrane, the protein will be anchored to the inside of the cell membrane with its C terminus outside. On the other hand, if the C terminus also has a positively charged cluster nearby, a membrane associated **leader peptidase** (or signal peptidase) may cut the loop past the signal sequence on the outside of the membrane leaving the bulk of the protein with its N terminus outside (Fig. 29-18B). How are **polytopic** integral membrane proteins with multiple cytoplasmic and external loops formed? Hydrophobic signal sequences are not always at the N terminus of a polypeptide chain. Suitable **internal signal sequences** may be found in the sequences that form the transmembrane helices, e.g., those present in the many 7-helix receptors found in a membrane. This suggests the possibility that successive loops may be translocated. If the N terminus is allowed to pass through a translocon in one of the steps, the topology of Fig. 11-6 or of Fig. 23-41 will result.[611]

Genetic methods have also been applied to study the insertion of coat subunits of phage M13 into the plasma membrane of *E. coli*.[612–614] The subunits are stored in the plasma membrane waiting to form a cylindrical shell about a viral DNA molecule as it is extruded from the bacterium.[612] The rod-like subunits (Fig. 7-7) have their N termini in the periplasmic space and their C termini in the cytoplasm. Each end carries a cluster of electrically charged residues, mostly negative at the N terminus and positive at the C terminus. Insertion into the membrane occurs only if the membrane has its normal membrane potential with a positive external surface charge and a negative internal charge, complementary to the charges on the coat subunit. Insertion does not occur unless the leader peptide with its positively charged N-terminal cluster and the C-terminal positive cluster are both present.[610,615] This suggests insertion by the loop mechanism of Fig. 29-18A. Genetic studies of the *E. coli* leader peptidase revealed that this protein also has an internal signal sequence, which becomes inserted into the membrane and which is not cleaved.[612,616,617] The final orientation of the mature enzyme is indicated in Fig. 29-18C.

Targeting of proteins to specialized domains of a membrane are less well understood. These include caveolae and lipid rafts, domains that are high in cholesterol and sphingolipids and which function in endocytosis and in cell signaling. A recent proposal is that proteins with hydrophobic surfaces needed in these domains become coated with a lipid "shell" before entering the membrane.[617a]

6. Secretion of Proteins

Cells continuously secrete materials via small cytoplasmic vesicles, which in eukaryotes arise largely from the Golgi apparatus (pp. 425–427; Fig. 20-8). The vesicles of this **constitutive pathway** may have diameters of ~50 nm. They carry phospholipids, proteins, and other constituents for incorporation into the plasma membrane of the cell.[618,619] In addition, there are

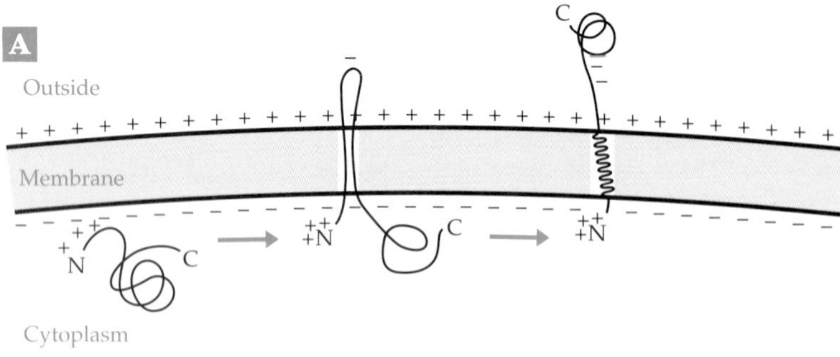

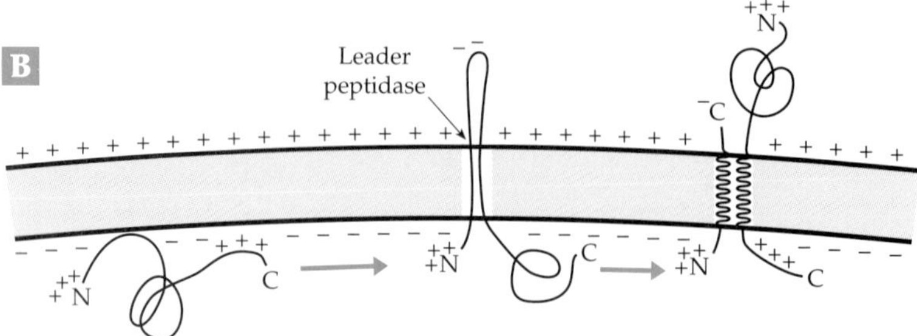

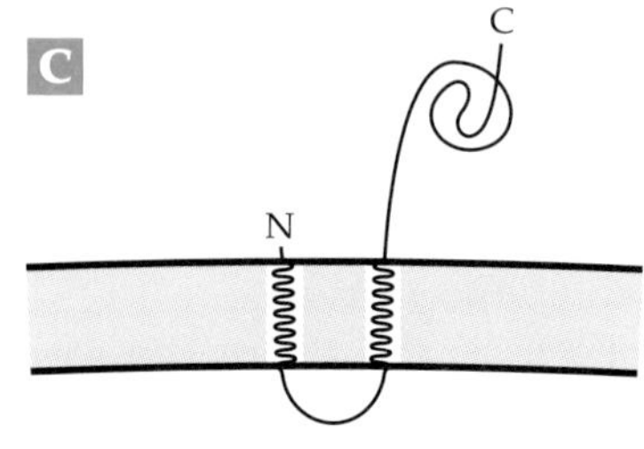

Figure 29-18 (A) Proposed mechanism for insertion of a loop of polypeptide chain through a translocon pore in a membrane with the positively charged N terminus anchored to the negatively charged inner membrane surface. (B) Cleavage of a polypeptide loop formed as in (A) by a leader peptidase to give a polypeptide chain anchored by a positively charged cluster near its C terminus. (C) Membrane topology of the *E. coli* leader peptidase. The active site is in the periplasmic domain. See Tschantz *et al.*[580]

BOX 29-F THE YEAST TWO-HYBRID SYSTEM FOR IDENTIFYING PROTEIN–PROTEIN INTERACTIONS

Many techniques including ultracentrifugation, chemical crosslinking, and X-ray crystallography are used to identify interactions between proteins. However, for study of the entire proteome new approaches are needed. One new technique that has already been widely applied is the yeast two-hybrid system.[a–f] In its original form[a,b] a transcriptional activator is utilized together with a reporter gene, e.g., a fused *GAL1– lac2* gene that when transcribed yields β-galactosidase. This enzyme then cleaves a chromogenic substrate (see p. 1494) to give a blue color. The 881-residue transcription factor GAL4 (p. 1630) is often used in two-hybrid systems. It binds upstream of the TATA sequence in the promoter regions of genes that code for enzymes of galactose catabolism. The N-terminal **DNA-binding domain** of *GAL4* binds to a specific 17-bp palindromic sequence in the DNA, while the acidic C-terminal **activation domain** activates transcription by interacting with the RNA polymerase complex bound to the promoter (see figure). The GAL4 activator seems to be quite flexible and is able to activate transcription, even if the distance from its binding site on DNA to the transcription initiation site is varied considerably. The two-hybrid system was constructed by cloning separately the pieces of DNA that code for residues 1–147 of the DNA-binding domain of the GAL4 protein (DBD in the figure) and for residues 768–881 of the activation domain (AD). The intervening nucleotides coding for the remaining 620 residues of GAL4 are discarded. The two-hybrid system tests whether a protein X, sometimes called the **bait**, binds or otherwise interacts strongly with another protein (Y, often called the **prey**) or with a series of other proteins (Y_1, $Y_2 \cdots Y_n$). To carry out the test the gene for protein X is fused with that for the DNA-binding domain of GAL4. When expressed in a living yeast cell the hybrid protein DBD–X will be formed. The gene fusion must be in-frame to ensure a correct structure for the X portion. Likewise, genes for protein Y, or for a series Y_1, $Y_2 \cdots Y_n$, are fused in-frame to the gene fragment carrying the GAL4 activation domain. Y–AD hybrids will be formed. The test is made using a strain of yeast in which the GAL4 gene has been replaced with a hybrid *GAL1– lacZ* reporter gene. If both DBD–X and Y–AD are present, and if they interact strongly (bind tightly), transcription of the reporter gene will be activated, and a blue colony will grow from the yeast cell. It is useful to create the hybrids with protein X, or proteins X_1 $X_2 \cdots$ (the baits) in a haploid strain of one of the two yeast mating types *MATa* or *MATα* (pp. 20, 1574) and the proteins Y (the prey) in a strain of the other mating type. Mating of the two strains will produce diploid cells that express both the DBD–X and Y–AD hybrids.

Many variants of the two-hybrid system have been devised.[d] For example, a green fluorescent protein reporter can be used.[g] Because significant biological protein–protein interactions often require that three or more proteins interact,[d] hybrid systems involving more than two proteins have been developed. Two-hybrid systems for bacteria have also been devised.[h] A virtue of the two-hybrid methods is that they work with undenatured, if not totally natural, proteins. This is in contrast to widely used methods that involve separation of denatured proteins on gels or columns.

The most popular two-hybrid systems utilize microarrays.[d,f] In the simplest approach hybrid DBD–X is tested against a library of Y–ADs

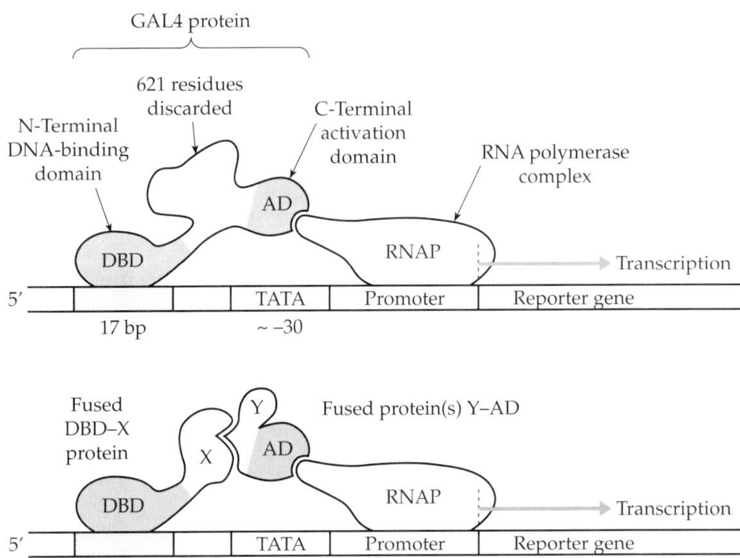

Top: The yeast GAL4 protein interacts with the RNA polymerase complex to activate transcription of a suitable reporter gene. Bottom: Two hybrid proteins, one containing the DNA-binding domain of GAL4, fused to protein X and the other containing protein Y fused to the activation domain of GAL4, are present in a cell. If X and Y bind strongly to each other, activation domain AD will be held close to the RNA polymerase and will activate transcription.

BOX 29-F THE YEAST TWO-HYBRID SYSTEM FOR IDENTIFYING PROTEIN–PROTEIN INTERACTIONS (continued)

prepared by random cleavage of DNA of known sequence. For example, the entire genome of the gastric pathogenic bacterium *Helicobacter pylori* was cut into ~1000 nucleotide pieces. These were cloned into plasmids in *E. coli*, then into yeast. Over 10 million *E. coli* clones provided a final two million independent yeast colonies, which carried the Y–ADs (prey). The genomic DNA fragments were also used to prepare 285 DBD–X (baits) from 261 genes. In a series of two-hybrid screening tests more than 1200 different protein–protein interactions connecting 47% of the proteome were detected.[i]

The complete yeast (*S. cerevisiae*) has been probed using at least two large-scale two-hybrid investigations. Uetz *et al.*[j] generated a large set of ~6000 genetically engineered yeast colonies, each one expressing just one of the possible Y–AD hybrid proteins (prey) derived from the ~6000 gene products identified in the yeast genome. These strains were distributed into microtiter plates and were individually crossed with 192 strains of yeast, each of which expressed a single BDB–X hybrid. This simple automated array screening identified 281 interacting pairs. In a second approach, the cells producing the Y–AD prey hybrids were mixed to give a single library. This was then screened against nearly all of the possible DBD–X hybrids in a large-scale automated procedure. The two approaches together detected 957 probable interactions involving 1004 different proteins.[j] In an independent study, using similar approaches but different cloning vehicles, Ito *et al.*[k,l] identified 4549 two-hybrid interactions among 3278 proteins. Of these 841 interactions were judged to be most relevant (core). Surprisingly, only 135 were identical to those found by Uetz *et al.* The significance and possible reasons for this disparity have been discussed.[f,j,l]

Interpretation of results of these studies is still difficult. Results of two-hybrid methods become more useful if they can be coordinated with other approaches. For example, computational methods can predict interactions from genome sequences alone.[m,n,o] More than 45,000 interactions have been predicted among yeast proteins.[m] Reliable identification of such motifs as DNA-binding domains and Ca^{2+}- binding domains can complement two-hybrid analysis.[n] The yeast genome is predicted to contain 162 coiled-coil sequences and at least 213 unique interactions between them.[o] Examination of sequences of protein families in the Protein Data Bank (PDB) led to prediction of 8151 interactions of 664 types between protein families in yeast.[p]

Improved experimental procedures of other types can also complement two-hybrid methods. Among these are formaldehyde crosslinking with immunoprecipitation,[q] methods that couple mass spectrometry and crosslinking,[r] and detection of intermolecular nuclear Overhauser enhancements in protein–protein complexes.[s] Phage display methods (see Fig. 3-16) have been developed as another method of detecting protein–protein interactions[e] as has fluorescence resonance energy transfer (FRET; p. 1291).[d] Evanescent wave methods, e.g., surface plasmon resonance (Box 3-F) are increasingly being used to quantify protein–protein interactions. These may be combined with single-hybrid methods in inexpensive and rapid micro-devices.[d]

---

[a] Fields, S., and Song, O.-K. (1989) *Nature (London)* **340**, 245–246
[b] Chien, C.-t, Bartel, P. L., Sternglanz, R., and Fields, S. (1991) *Proc. Natl. Acad. Sci. U.S.A.* **88**, 9578–9582
[c] Finley, R. L., Jr., and Brent, R. (1994) *Proc. Natl. Acad. Sci. U.S.A.* **91**, 12980–12984
[d] Mendelsohn, A. R., and Brent, R. (1999) *Science* **284**, 1948–1950
[e] Allen, J. B., Walberg, M. W., Edwards, M. C., and Elledge, S. J. (1995) *Trends Biochem. Sci.* **20**, 511–516
[f] Oliver, S. (2000) *Nature (London)* **403**, 601–603
[g] Shioda, T., Andriole, S., Yahata, T., and Isselbacher, K. J. (2000) *Proc. Natl. Acad. Sci. U.S.A.* **97**, 5220–5224
[h] Joung, J. K., Ramm, E. I., and Pabo, C. O. (2000) *Proc. Natl. Acad. Sci. U.S.A.* **97**, 7382–7387
[i] Rain, J.-C., and 12 other authors. (2001) *Nature (London)* **409**, 211–215
[j] Uetz, P., and 19 other authors. (2000) *Nature (London)* **403**, 623–627
[k] Ito, T., Tashiro, K., Muta, S., Ozawa, R., Chiba, T., Nishizawa, M., Yamamoto, K., Kuhara, S., and Sakaki, Y. (2000) *Proc. Natl. Acad. Sci. U.S.A.* **97**, 1143–1147
[l] Ito, T., Chiba, T., Ozawa, R., Yoshida, M., Hattori, M., and Sakaki, Y. (2001) *Proc. Natl. Acad. Sci. U.S.A.* **98**, 4569–4574
[m] Marcotte, E. M., Pellegrini, M., Ng, H.-L., Rice, D. W., Yeates, T. O., and Eisenberg, D. (1999) *Science* **285**, 751–753
[n] Gallet, X., Charloteaux, B., Thomas, A., and Brasseur, R. (2000) *J. Mol. Biol.* **302**, 917–926
[o] Newman, J. R. S., Wolf, E., and Kim, P. S. (2000) *Proc. Natl. Acad. Sci. U.S.A.* **97**, 13203–13208
[p] Park, J., Lappe, M., and Teichmann, S. A. (2001) *J. Mol. Biol.* **307**, 929–938
[q] Orlando, V. (2000) *Trends Biochem. Sci.* **25**, 99–104
[r] Bennett, K. L., Kussmann, M., Björk, P., Godzwon, M., Mikkelsen, M., Sorensen, P., and Roepstorff, P. (2000) *Protein Sci.* **9**, 1503–1518
[s] Clore, G. M. (2000) *Proc. Natl. Acad. Sci. U.S.A.* **97**, 9021–9025

regulated pathways for storage and release by exocytosis of hormones, neuropeptides, and neurotransmitters. The last are secreted from both small (~ 50 nm diam.) synaptic vesicles and larger **dense core vesicles** (>100 nm diam.),[620,621] which are discussed in Chapter 30, Section B,8. In every case specialized proteins (discussed on pp. 427 and 521 and in Chapters 20 and 30) are involved.[622–624]

7. Protein Folding

Before a protein can function its polypeptide chain must fold into its own native tertiary structure.[624a] This folding is influenced by many surrounding proteins, by the state of glycosylation of side chains,[625] and by other posttranslational modifications, by the presence of cis amide linkages in unfolded or folded forms (pp. 82, 83; Box 9-F),[626] and by possibility for formation of disulfide bridges (pp. 521, 522).[626a–628] The prediction of the folding pattern of proteins from the amino acid sequence remains a major goal of protein chemistry. In principle, a protein fold can be predicted from the DNA sequence of the genes, with proper allowance for effects of posttranscriptional modification. This goal once seemed intractable, but two things have provided new hope. (1) The speed and power of computers is still increasing. This not only has allowed more rapid calculations but also has led to improvement in experimental methods. (2) Methods for studying folding, which include mass spectroscopy,[629] NMR spectroscopy,[630–633] and optical methods,[634,635] have become more rapid and more sensitive.[636] As a consequence, we have an abundance of new data.

Anfinsen, in 1963, proposed that the three-dimensional structure of a protein is in its lowest Gibbs energy state when present in its natural environment.[637,638] However, there is a problem with this suggestion (the "Levinthal paradox"; p. 82). Even if a polypeptide chain occupies only two of the lowest energy regions of the Ramachandran diagram (Fig. 2-9), a 100-residue protein would have ~10^{30} possible conformations. If a folding protein checked all of these conformations at a realistic rate of ~10^{11} s^{-1}, it would take ~10^{11} years to fold. Furthermore, in a test tube of protein, which would contain at most 10^{18} molecules, each of the molecules would probably have a different conformation.[639] In fact, most proteins fold reliably to the same final structure in less than a second, and some in a millisecond.[639–643] It is also true that proteins, under altered solvent conditions, can misfold into totally "incorrect" structures.[644] Most can assume an amyloid structure under some conditions (Box 29-E). One clear conclusion is that folding is not totally random but follows a **folding pathway,** which is dictated by the sequence. Nevertheless, experimental data indicate that there is an ensemble of related structures at each stage of folding.[631]

We cannot answer the question posed by Anfinsen's hypothesis. Does the native state have a minimum value of the Gibbs energy? Nevertheless, it is observed that proteins usually behave as if folded, unfolded forms are in a true thermodynamic equilibrium, and that this equilibrium is attained rapidly. The difference ΔG between a folded and a denatured protein is only 21–63 kJ mol^{-1}, which shows that folded proteins are only marginally more stable than are unfolded polypeptide chains.[645] The value of ΔG of unfolding as a function of temperature T is given by Eq. 29-13, where $\Delta H(T)$ and ΔC_p are the changes in enthalpy and heat capacity upon unfolding.[645,646]

$$\Delta G\,(T) = \Delta H_m\,(1 - T/T_m) - \Delta C_p\,[(T_m - T) - T\ln(T/T_m)] \quad (29\text{-}13)$$

ΔH_m is the enthalpy change at T_m, the midpoint of the thermal unfolding curve (the "melting" temperature). The temperature of maximum stability T_s occurs when $\Delta S = 0$ (Eq. 29-14). T_s is usually between –10°C and

$$T_s = T_m \exp(-\Delta H_m / [T_m \Delta C_p]) \quad (29\text{-}14)$$

35°C. For staphylococcal nuclease (Fig. 12-29) T_s = 18°C and T_m = –19°C and +57°C, i.e., *the protein is denatured by either cooling below 18°C or heating above 57°C*, a behavior that is common for many proteins. Cold denaturation is observed whenever the unfolded state has a higher heat capacity than does the folded state.[647]

Can we predict the Gibbs energy of unfolding from the protein sequence? To do this it is necessary to utilize experimental data on known proteins to obtain a series of terms (Eq. 29-15) that can be summed to give the $\Delta G°_u$, the standard Gibbs energy of unfolding.

$$\Delta G°_u = \Delta G°_{charge} + \Delta G°_{hyd} + \Delta G°_{conf} + \Delta G°_{vW} + \Delta G°_{Hbond} \quad (29\text{-}15)$$

The terms refer to summations of all of the charge-charge, hydrophobic, configurational, van der Waals, and hydrogen-bonding interactions between both main-chain and side-chain atoms.[638,646,648] Such computations are formidable and are uncertain, especially for electrostatic (charge-charge and hydrogen-bonding) interactions.[649–652] Both the folded and denatured state must be considered,[653] as must the heat capacities[654] and configurational entropies.[655]

While we tend to think of proteins as having fixed structures, conformational changes are basic to life. Many proteins are very flexible and in part disordered.[655a] At the same time proteins can be misfolded leading to amyloid formation (Box 29-E) and to other diseases.[655b–e]

8. Completing the Cycle: Proteolytic Degradation of Proteins

Like all other body constituents proteins must undergo breakdown as well as synthesis. Regular turnover of all proteins is essential, and defects in the process may lead to amyloid deposits (Box 29-E). Turnover and degradation of proteins depends upon a variety of proteases, many of which are discussed in Chapter 12, Section C. Because of the changing needs of body cells, specific proteins turn over at widely varying rates. Some enzymes, hormones, and regulatory proteins have half-lives of only a few minutes while others may function for months or years. How can the regulation of the breakdown of thousands of different proteins be controlled? The answer seems to lie in the amino acid sequence. Just as the sequence determines the location that a protein occupies in the cell and its folding pattern, it also determines the turnover rate.

Much of the breakdown takes place in the cytosol in proteasomes (Box 7-A) and is controlled by the ubiquitin system (Box 10-C), which selects the proteins for degradation; control of the system is quite complex.[656–661] One aspect depends upon the N-terminal amino acid of the substrate protein. Defective proteins often have N-terminal destabilizing amino acids such as phenylalanine, leucine, aspartic acid, lysine, and arginine (p. 527). If an internal lysine is also present, the protein may be conjugated with ubiquitin and degraded rapidly. Many metabolic processes, such as the cell cycle (Fig. 11-15), are controlled by protein degradation.[538,662] In some cases an arginine residue is transferred onto the N terminus of a protein by an **arginyl-tRNA protein transferase**. This creates a better substrate for ubiquitination and rapid degradation.[280,663] In other cases proteolytic cleavage uncovers an arginine or other destabilizing residue and speeds hydrolysis.[538] While ubiquitination often initiates the degradation of proteins, it also helps to direct proteins to specific locations within a cell.[663a] Rapid degradation of a ubiquitinated protein may require hydrolytic **deubiquitination** by a metalloprotease, which is a subunit of the 26S proteasome lid (Box 7-A). This allows the ubiquitin to be recycled and also directs the deubiquitinated protein into the proteasome.[663b]

As mentioned on p. 1854, an important function of proteasomes is formation of short antigenic peptides for use by the immune system.[664] Inhibition of proteasome activity reduces or prevents antigen presentation (Chapter 31).[665,666] In this immune surveillance system mature proteins of host cells are cut up and checked for self-identity. The checking also includes the rapidly degraded imperfect proteins and foreign proteins from invading organisms or viruses.[667,667a]

Lysosomes, which contain more than 50 proteases, lipases, glycosidases, and other hydrolases, also play a major role in protein degradation.[668] Their importance is emphasized by the range of lysosomal deficiency diseases (Table 20-1).[668] Lysosomes also function in the process of **autophagy**, by which cells can sacrifice a whole section, organelles and all, by walling off a large vesicle or **autophagosome** and fusing it with a lysosome.[669] In such a way a tadpole can resorb its tail while becoming a frog. We have now come full circle: our proteins have been converted back to the amino acids and other small molecules derived from them. The amino acids can be reutilized or can be catabolized, depending upon the needs of the organism.

E. Proteomics

The vast amount of data on protein structures and improved methods of predicting structures[670–672] have led to development of new areas of science variously designated as genomics, proteomics, transcriptomics,[673,674] and bioinformatics.[675–677] These fields encompass all of the methods for sequence determination, observation of gene expression, protein synthesis by cells, and mathematical analysis of resulting data. Proteomics includes new approaches to polypeptide separation[678] and identification,[678,679] sequencing at the attomole level,[680–681] and comparison of sequences between species.[682–684] Protein separation by liquid chromatography,[685] capillary electrophoresis,[686] or two-dimensional gel electrophoresis[687,688] can be followed by mass spectrometry of intact proteins or of proteolytic fragments.[689,690] Microarrays on proteome chips can be used to observe production of thousands of proteins simultaneously.[691–694] Structural genomics centers have been established for rapid determination of protein structures using NMR[695,696] and X-ray methods.[697–699] If each center determines 200 or more structures per year there will soon be 16,000 new structures, enough to allow us to predict much about all the rest.[699] Then we can study all the important remaining details for millions and millions of proteins.

Current efforts to understand the structures, conformational movements, and functions of these molecules range from the classification of nearly 10,000 different protein folds[700] to investigation of the dynamics of single protein molecules.[701] Well known motifs such as β sheets and α helices are studied with the goal of more accurate predictions of structure and better understanding of interactions between proteins in solution and in membranes. For example, one natural topology is the β barrel, which may contain 8 to 22 strands (e.g., see Fig. 8-20). These cylindrical proteins are abundant in outer membranes of gram-negative bacteria.[702–704] The partial electrical changes at the edges of the β sheets (see Fig. 2-11,B) may interact to help stabilize the barrels. In contrast, soluble β barrel-containing proteins are designed to avoid edge-to-edge interactions, which could cause aggregation of the proteins.[704] A recently discovered membrane-protein motif is an α **barrel**, which is composed of 12 α helices stacked side-by-side with side-chain groups fitting together in a knobs-in-holes fashion.[705] An example is the TolC protein of *E. coli*. A trimer of 428-residue subunits forms a long cylinder, which is a 12-stranded β barrel at one end and an α barrel at the other.[705] More common are transmembrane α helices, many of which are present in 7-helix receptors (e.g., Fig. 11-6). Relatively accurate prediction methods are now available for these structures,[706,707] but there are still uncertainties about mechanisms of transmission of signals across the membranes.

Predictions of structures of more complex proteins from their amino acid sequences presents a major challenge.[708] Assignment of domains within the protein is a first step.[709–712] Regions of probable helix or β-strand structure can be recognized but it is difficult to predict the exact lengths of the helices and the structures of connecting loops and strands. These depend upon many factors including the possible formation of ion pairs[713] and of locks at the ends of strands created by van der Waals interactions.[714] There are also circular proteins.[715] Composite structures such as those of silks (Box 2-B) have surprising properties. Both silkworm,[716,717] and spider silks[718,719] undergo marked changes in properties upon spinning of the random coil forms of the proteins found in silk glands into the drawn fibers.

Whether we discuss silk, proteins embedded in membranes, or soluble complexes of cytosolic proteins, we must ask questions about interactions. A first step is to identify interactions[720–730] among proteins either *in vitro* or in living cells.[731] Proteomic methods, which include the yeast two-hybrid method (Box 29-F), are widely used for this purpose. It is possible to identify large sets of interacting proteins, to identify disease states, to observe effects of drugs, and to compare metabolism among species.

References

1. Nomura, M. (1990) *Trends Biochem. Sci.* **15**, 244–247
2. Lake, J. A. (1981) *Sci. Am.* **245**(Aug), 84–97
3. Schachman, H. D., Pardee, A. B., and Stanier, R. Y. (1952) *Arch. Biochem. Biophys.* **38**, 245–260
4. Wittmann, H. G. (1983) *Ann. Rev. Biochem.* **52**, 35–65
5. Prince, J. B., Gutell, R. R., and Garrett, R. A. (1983) *Trends Biochem. Sci.* **8**, 359–363
6. Yonath, A., Leonard, K. R., and Wittmann, H. G. (1987) *Science* **236**, 813–816
7. Lake, J. A. (1985) *Ann. Rev. Biochem.* **54**, 507–530
8. Frank, J., Verschoor, A., Wagenknecht, T., Radermacher, M., and Carazo, J.-M. (1988) *Trends Biochem. Sci.* **13**, 123–127
9. Baumeister, W., and Steven, A. C. (2000) *Trends Biochem. Sci.* **25**, 624–631
10. Morgan, D. G., Ménétret, J.-F., Radermacher, M., Neuhof, A., Akey, I. V., Rapoport, T. A., and Akey, C. W. (2000) *J. Mol. Biol.* **301**, 301–321
11. De Chadarevian, S. (1999) *Trends Biochem. Sci.* **24**, 203–206
12. Nagano, K., and Harel, M. (1987) *Prog. Biophys. and Mol. Biol.* **48**, 67–101
13. Lake, J. A. (1977) *Proc. Natl. Acad. Sci. U.S.A.* **74**, 1903–1907
14. Lake, J. A., and Strycharz, W. A. (1981) *J. Mol. Biol.* **153**, 979–992
15. Kahan, L., Winkelmann, D. A., and Lake, J. A. (1981) *J. Mol. Biol.* **145**, 193–214
16. Ramakrishnan, V., Capel, M., Kjeldgaard, M., Engelman, D. M., and Moore, P. B. (1984) *J. Mol. Biol.* **174**, 265–284
17. Ban, N., Nissen, P., Hansen, J., Moore, P. B., and Steitz, T. A. (2000) *Science* **289**, 905–920
18. Cate, J. H., Yusupov, M. M., Yusupova, G. Z., Earnest, T. N., and Noller, H. F. (1999) *Science* **285**, 2095–2104
19. Nissen, P., Hansen, J., Ban, N., Moore, P. B., and Steitz, T. A. (2000) *Science* **289**, 920–930
20. Agrawal, R. K., Penczek, P., Grassucci, R. A., Burkhardt, N., Nierhaus, K. H., and Frank, J. (1999) *J. Biol. Chem.* **274**, 8723–8729
21. Holley, R. W., Apgar, J., Everett, G. A., Madison, J. T., Marquisee, M., Merrill, S. H., Penswick, J. R., and Zamir, A. (1965) *Science* **147**, 1462–1465
22. Wittmann, H. G. (1982) *Ann. Rev. Biochem.* **51**, 155–183
22a. Brownlee, G. G., Sanger, F., and Burell, B. G. (1967) *Nature (London)* **215**, 735–736
22b. Brosius, J., Palmer, M. L., Kennedy, P. J., and Noller, H. F. (1978) *Proc. Natl. Acad. Sci. U.S.A.* **75**, 4801–4805
22c. Brosius, J., Dull, T. J., and Noller, H. F. (1980) *Proc. Natl. Acad. Sci. U.S.A.* **77**, 201–204
23. Leijonmarck, M., Eriksson, S., and Liljas, A. (1980) *Nature (London)* **286**, 824–826
24. Ramakrishnan, V., and White, S. W. (1998) *Trends Biochem. Sci.* **23**, 208–212
25. Yonath, A. (1984) *Trends Biochem. Sci.* **9**, 227–230
26. Yonath, A., and Wittmann, H. G. (1989) *Trends Biochem. Sci.* **14**, 329–335
27. Yonath, A., Glotz, C., Gewitz, H. S., Bartels, K. S., von Böhlen, K., Makowski, I., and Wittman, H. G. (1988) *J. Mol. Biol.* **203**, 831–834
28. Moore, P. B. (2001) *Biochemistry* **40**, 3243–3250
29. Schluenzen, F., Tociij, A., Zarivach, R., Harms, J., Gluehmann, M., Janell, D., Bashan, A., Bartels, H., Agmon, I., Franceschi, F., and Yonath, A. (2000) *Cell* **102**, 615–623
30. Culver, G. M., Cate, J. H., Yusupova, G. Z., Yusupov, M. M., and Noller, H. F. (1999) *Science* **285**, 2133–2135
31. Agalarov, S. C., Prasad, G. S., Funke, P. M., Stout, C. D., and Williamson, J. R. (2000) *Science* **288**, 107–112
31a. Recht, M. I., and Williamson, J. R. (2001) *J. Mol. Biol.* **313**, 35–48
32. Pennisi, E. (1999) *Science* **285**, 2048–2051
33. Wimberly, B. T., Brodersen, D. E., Clemons, W. M., Jr., Morgan-Warren, R. J., Carter, A. P., Vonrhein, C., Hartsch, T., and Ramakrishnan, V. (2000) *Nature (London)* **407**, 327–339
33a. Yusupov, M. M., Yusupova, G. Z., Baucom, A., Lieberman, K., Earnest, T. N., Cate, J. H. D., and Noller, H. F. (2001) *Science* **292**, 883–896
33b. Ramakrishnan, V., and Moore, P. B. (2001) *Curr. Opin. in Struct.Biol.* **11**, 144–154
33c. Nissen, P., Ippolito, J. A., Ban, N., Moore, P. B., and Steitz, T. A. (2001) *Proc. Natl. Acad. Sci. U.S.A.* **98**, 4899–4903
33d. Ogle, J. M., Brodersen, D. E., Clemons, W. M. J., Tarry, M. J., Carter, A. P., and Ramakrishnan, V. (2001) *Science* **292**, 897–902
33e. Mathews, M. B., and Pe'ery, T. (2001) *Trends Biochem. Sci.* **26**, 585–587
33f. Moore, P. B., and Steitz, T. A. (2002) *Nature (London)* **418**, 229–235
33g. Pioletti, M., Schlünzen, F., Harms, J., Zarivach, R., Glühmann, M., Avila, H., Bashan, A., Bartels, H., Auerbach, T., Jacobi, C., Hartsch, T., Yonath, A., and Franceschi, F. (2001) *EMBO J.* **20**, 1829–1839
34. Malhotra, A., Penczek, P., Agrawal, R. K., Gabashvili, I. S., Grassucci, R. A., Jünemann, R., Burkhardt, N., Nierhaus, K. H., and Frank, J. (1998) *J. Mol. Biol.* **280**, 103–116
35. Liljas, A. (1999) *Science* **285**, 2077–2078
36. Dube, P., Bacher, G., Stark, H., Mueller, F., Zemlin, F., van Heel, M., and Brimacombe, R. (1998) *J. Mol. Biol.* **279**, 403–421
37. Gomez-Lorenzo, M. G., Spahn, C. M. T., Agrawal, R. K., Grassucci, R. A., Penczek, P., Chakraburtty, K., Ballesta, J. P. G., Lavandera, J. L., Garcia-Bustos, J. F., and Frank, J. (2000) *EMBO J.* **19**, 2710–2718
37a. Mueller, F., Sommer, I., Baranov, P., Matadeen, R., Stoldt, M., Wöhnert, J., Görlach, M., van Heel, M., and Brimacombe, R. (2000) *J. Mol. Biol.* **298**, 35–59
38. Brimacombe, R. (1984) *Trends Biochem. Sci.* **9**, 273–277
39. Noller, H. F. (1984) *Ann. Rev. Biochem.* **53**, 119–162
39a. Gutell, R. R., Noller, H. F., and Woese, C. R. (1986) *EMBO J.* **5**, 1111–1113
39b. Pace, N. R., Olsen, G. J., and Woese, C. R. (1986) *Cell* **45**, 325–326

References

40. Dragon, F., Payant, C., and Brakier-Gingras, L. (1994) *J. Mol. Biol.* **244**, 74–85
41. Lee, K., Varma, S., SantaLucia, J., Jr., and Cunningham, P. R. (1997) *J. Mol. Biol.* **269**, 732–743
42. Capel, M. S., Kjeldgaard, M., Engelman, D. M., and Moore, P. B. (1988) *J. Mol. Biol.* **200**, 65–87
43. Moore, P. B. (1988) *Nature (London)* **331**, 223–227
44. Svergun, D. I., Koch, M. H. J., Pedersen, J. S., and Serdyuk, I. N. (1994) *J. Mol. Biol.* **240**, 78–86
45. Wadzack, J., Burkhardt, N., Jünemann, R., Diedrich, G., Nierhaus, K. H., Frank, J., Penczek, P., Meerwinck, W., Schmitt, M., Willumeit, R., and Stuhrmann, H. B. (1997) *J. Mol. Biol.* **266**, 343–356
46. Willumeit, R., Forthmann, S., Beckmann, J., Diedrich, G., Ratering, R., Stuhrmann, H. B., and Nierhaus, K. H. (2001) *J. Mol. Biol.* **305**, 167–177
47. Svergun, D. I., and Nierhaus, K. H. (2000) *J. Biol. Chem.* **275**, 14432–14439
48. Herwig, S., Kruft, V., Eckart, K., and Wittmann-Liebold, B. (1993) *J. Biol. Chem.* **268**, 4643–4650
49. Urlaub, H., Kruft, V., Bischof, O., Müller, E.-C., and Wittmann-Liebold, B. (1995) *EMBO J.* **14**, 4578–4588
50. Wang, R., Alexander, R. W., VanLoock, M., Vladimirov, S., Bukhtiyarov, Y., Harvey, S. C., and Cooperman, B. S. (1999) *J. Mol. Biol.* **286**, 521–540
51. Culver, G. M., Heilek, G. M., and Noller, H. F. (1999) *J. Mol. Biol.* **286**, 355–364
52. Wower, J., Kirillov, S. V., Wower, I. K., Guven, S., Hixson, S. S., and Zimmermann, R. A. (2000) *J. Biol. Chem.* **275**, 37887–37894
53. Shapkina, T. G., Dolan, M. A., Babin, P., and Wollenzien, P. (2000) *J. Mol. Biol.* **299**, 615–628
54. Juzumiene, D. I., Shapkina, T. G., and Wollenzien, P. (1995) *J. Biol. Chem.* **270**, 12794–12800
55. Montesano-Roditis, L., Glitz, D. G., Perrault, A. R., and Cooperman, B. S. (1997) *J. Biol. Chem.* **272**, 8695–8703
56. Laughrea, M., and Tam, J. (1992) *Biochemistry* **31**, 12035–12041
57. Hüttenhofer, A., and Noller, H. F. (1994) *EMBO J.* **13**, 3892–3901
58. Czworkowski, J., Odom, O. W., and Hardesty, B. (1991) *Biochemistry* **30**, 4821–4830
59. Urlaub, H., Thiede, B., Müller, E.-C., Brimacombe, R., and Wittmann-Liebold, B. (1997) *J. Biol. Chem.* **272**, 14547–14555
60. Benjamin, D. R., Robinson, C. V., Hendrick, J. P., Hartl, F. U., and Dobson, C. M. (1998) *Proc. Natl. Acad. Sci. U.S.A.* **95**, 7391–7395
61. Glück, A., and Wool, I. G. (1996) *J. Mol. Biol.* **256**, 838–848
61a. Zamecnik, P. (1984) *Trends Biochem. Sci.* **9**, 464–466
62. Schimmel, P., and Wang, C.-C. (1999) *Trends Biochem. Sci.* **24**, 127–128
63. Freifelder, D. (1983) *Molecular Biology*, 2nd ed., Jones and Bartlett, Boston, Massachusetts (p. 419)
64. Yamaguchi, K., and Subramanian, A. R. (2000) *J. Biol. Chem.* **275**, 28466–28482
65. O'Brien, T. W., Fiesler, S. E., Denslow, N. D., Thiede, B., Wittmann-Liebold, B., Mougey, E. B., Sylvester, J. E., and Graack, H.-R. (1999) *J. Biol. Chem.* **274**, 36043–36051
66. Koc, E. C., Burkhart, W., Blackburn, K., Koc, H., Moseley, A., and Spremulli, L. L. (2001) *Protein Sci.* **10**, 471–481
66a. Koc, E. C., Burkhart, W., Blackburn, K., Moyer, M. B., Schlatzer, D. M., Moseley, A., and Spremulli, L. L. (2001) *J. Biol. Chem.* **276**, 43958–43969
67. Moazed, D., Stern, S., and Noller, H. F. (1986) *J. Mol. Biol.* **187**, 399–416
68. Choi, Y. C. (1985) *J. Biol. Chem.* **260**, 12769–12772
69. Rairkar, A., Rubino, H. M., and Lockard, R. E. (1988) *Biochemistry* **27**, 582–592
69a. Decatur, W. A., and Fournier, M. J. (2002) *Trends Biochem. Sci.* **27**, 344 – 351
70. Carter, A. P., Clemons, W. M., Jr., Brodersen, D. E., Morgan-Warren, R. J., Hartsch, T., Wimberly, B. T., and Ramakrishnan, V. (2001) *Science* **291**, 498–501
71. Leontis, N. B., and Westhof, E. (1998) *J. Mol. Biol.* **283**, 571–583
72. Schnare, M. N., Damberger, S. H., Gray, M. W., and Gutell, R. R. (1996) *J. Mol. Biol.* **256**, 701–719
73. Herzog, M., and Maroteaux, L. (1986) *Proc. Natl. Acad. Sci. U.S.A.* **83**, 8644–8648
73a. Gutell, R. R. (1994) *Nucleic Acids Res.* **22**, 3502–3507
74. Brimacombe, R., Atmadja, J., Stiege, W., and Schüler, D. (1988) *J. Mol. Biol.* **199**, 115–136
75. Malhotra, A., and Harvey, S. C. (1994) *J. Mol. Biol.* **240**, 308–340
76. Stern, S., Weiser, B., and Noller, H. F. (1988) *J. Biochem. Biophys. Methods* **204**, 447–481
77. Mueller, F., and Brimacombe, R. (1997) *J. Mol. Biol.* **271**, 524–544
77a. Lodmell, J. S., and Dahlberg, A. E. (1997) *Science* **277**, 1262–1267
78. Merryman, C., Moazed, D., Daubresse, G., and Noller, H. F. (1999) *J. Mol. Biol.* **285**, 107–113
79. Larsson, S. L., and Nygård, O. (2001) *Biochemistry* **40**, 3222–3231
79a. Matadeen, R., Sergiev, P., Leonov, A., Pape, T., van der Sluis, E., Mueller, F., Osswald, M., von Knoblauch, K., Brimacombe, R., Bogdanov, A., van Heel, M., and Dontsova, O. (2001) *J. Mol. Biol.* **307**, 1341–1349
80. Garrett, R. (1983) *Trends Biochem. Sci.* **8**, 189–190
81. Porse, B. T., Cundliffe, E., and Garrett, R. A. (1999) *J. Mol. Biol.* **287**, 33–45
82. Conn, G. L., Draper, D. E., Lattman, E. E., and Gittis, A. G. (1999) *Science* **284**, 1171–1174
83. Correll, C. C., Wool, I. G., and Munishkin, A. (1999) *J. Mol. Biol.* **292**, 275–287
84. Macbeth, M. R., and Wool, I. G. (1999) *J. Mol. Biol.* **285**, 965–975
85. Chan, Y.-L., Sitikov, A. S., and Wool, I. G. (2000) *J. Mol. Biol.* **298**, 795–805
86. Wilson, K. S., and Noller, H. F. (1998) *Cell* **92**, 337–349
87. Klein, B. K., Staden, A., and Schlessinger, D. (1985) *J. Biol. Chem.* **260**, 8114–8120
88. Yusupov, M. M., Yusupova, G. Z., Baucom, A., Lieberman, K., Earnest, T. N., Cate, J. H. D., and Noller, H. F. (2001) *Science* **1**, 1–20
89. Khaitovich, P., and Mankin, A. S. (1999) *J. Mol. Biol.* **291**, 1025–1034
90. Christensen, A., Mathiesen, M., Peattie, D., and Garrett, R. A. (1985) *Biochemistry* **24**, 2284–2291
91. Romby, P., Westhof, E., Toukifimpa, R., Mache, R., Ebel, J.-P., Ehresmann, C., and Ehresmann, B. (1988) *Biochemistry* **27**, 4721–4730
92. Leontis, N. B., and Moore, P. B. (1986) *Biochemistry* **25**, 3916–3925
93. Dontsova, O., Tishkov, V., Dokudovskaya, S., Bogdanov, A., Döring, T., Rinke-Appel, J., Thamm, S., Greuer, B., and Brimacombe, R. (1994) *Proc. Natl. Acad. Sci. U.S.A.* **91**, 4125–4129
94. Lee, K. M., and Marshall, A. G. (1986) *Biochemistry* **25**, 8245–8252
95. Funari, S. S., Rapp, G., Perbandt, M., Dierks, K., Vallazza, M., Betzel, C., Erdmann, V. A., and Svergun, D. I. (2000) *J. Biol. Chem.* **275**, 31283–31288
96. Garrett, R. A. (1985) *Nature (London)* **318**, 233–235
97. Olsen, G. J., and Woese, C. R. (1993) *FASEB J.* **7**, 113–123
98. Dijk, J., and Littlechild, J. A. (1979) *Methods Enzymol.* **59**, 481–502
99. Kruft, V., and Wittmann-Liebold, B. (1991) *Biochemistry* **30**, 11781–11787
100. Bielka, H., ed. (1982) *The Eukaryotic Ribosome*, Springer, Berlin
100a. Wool, I. G., Chan, Y. L., and Gluck, A. (1995) *Biochem. Cell Biol.* **73**, 933–947
101. van Beekvelt, C. A., Kooi, E. A., de Graaff-Vincent, M., van't Riet, J., Venema, J., and Raué, H. A. (2000) *J. Mol. Biol.* **296**, 7–17
102. Goldschmidt-Reisin, S., Kitakawa, M., Herfurth, E., Wittmann-Liebold, B., Grohmann, L., and Graack, H.-R. (1998) *J. Biol. Chem.* **273**, 34828–34836
102a. Saveanu, C., Fromont-Racine, M., Harington, A., Ricard, F., Namane, A., and Jacquier, A. (2001) *J. Biol. Chem.* **276**, 15861–15867
102b. Graack, H.-R., Bryant, M. L., and O'Brien, T. W. (1999) *Biochemistry* **38**, 16569–16577
102c. Suzuki, T., Terasaki, M., Takemoto-Hori, C., Hanada, T., Ueda, T., Wada, A., and Watanabe, K. (2001) *J. Biol. Chem.* **276**, 33181–33195
103. Chan, Y.-L., and Wool, I. G. (1988) *J. Biol. Chem.* **263**, 2891–2896
104. Andres, J. L., Johansen, J. W., and Maller, J. L. (1987) *J. Biol. Chem.* **262**, 14389–14393
105. Wilson, K. S., Appelt, K., Badger, J., Tanaka, I., and White, S. W. (1986) *Proc. Natl. Acad. Sci. U.S.A.* **83**, 7251–7255
105a. Leijonmarck, M., Appelt, K., Badger, J., Liljas, A., Wilson, K. S., and White, S. W. (1988) *Proteins* **3**, 243–251
106. Stern, S., Wilson, R. C., and Noller, H. F. (1986) *J. Mol. Biol.* **192**, 101–110
107. van de Ven, F. J. M., and Hilbers, C. W. (1986) *J. Mol. Biol.* **192**, 389–417
108. Nag, B., Tewari, D. S., Sommer, A., Olson, H. M., Glitz, D. G., and Traut, R. R. (1987) *J. Biol. Chem.* **262**, 9681–9687
109. Wittmann, H.-G. (1976) *Eur. J. Biochem.* **61**, 1–13
110. Brockmöller, J., and Kamp, R. M. (1988) *Biochemistry* **27**, 3372–3381
111. Xiang, R. H., and Lee, J. C. (1989) *J. Biol. Chem.* **264**, 10542–10546
112. Pohl, T., and Wittmann-Liebold, B. (1988) *J. Biol. Chem.* **263**, 4293–4301
113. Yeh, Y.-C., Traut, R. R., and Lee, J. C. (1986) *J. Biol. Chem.* **261**, 14148–14153
114. Capel, M. S., Engelman, D. M., Freeborn, B. R., Kjeldgaard, M., Langer, J. A., Ramakrishnan, V., Schindler, D. G., Schneider, D. K., Schoenborn, B. P., Sillers, I.-Y., Yabuki, S., and Moore, P. B. (1987) *Science* **238**, 1403–1406
115. Liiv, A., Tenson, T., and Remme, J. (1996) *J. Mol. Biol.* **263**, 396–410
116. Nomura, M. (1973) *Science* **179**, 864–873
117. Herold, M., and Nierhaus, K. H. (1987) *J. Biol. Chem.* **262**, 8826–8833
118. Liiv, A., and Remme, J. (1998) *J. Mol. Biol.* **276**, 537–545
119. Green, R., and Noller, H. F. (1999) *Biochemistry* **38**, 1772–1779
120. Khaitovich, P., Tenson, T., Kloss, P., and Mankin, A. S. (1999) *Biochemistry* **38**, 1780–1788
121. Michael, W. M., and Dreyfuss, G. (1996) *J. Biol. Chem.* **271**, 11571–11574
122. Wimberly, B. T., White, S. W., and Ramakrishnan, V. (1997) *Structure* **5**, 1187–1198

References

123. Davies, C., Gerstner, R. B., Draper, D. E., Ramakrishnan, V., and White, S. W. (1998) *EMBO J.* **17**, 4545–4558
124. Markus, M. A., Gerstner, R. B., Draper, D. E., and Torchia, D. A. (1998) *EMBO J.* **17**, 4559–4571
125. Agalarov, S. C., Zheleznyakova, E. N., Selivanova, O. M., Zheleznaya, L. A., Matvienko, N. I., Vasiliev, V. D., and Spirin, A. S. (1998) *Proc. Natl. Acad. Sci. U.S.A.* **95**, 999–1003
126. Nevskaya, N., Tishchenko, S., Nikulin, A., Al-Karadaghi, S., Liljas, A., Ehresmann, B., Ehresmann, C., Garber, M., and Nikonov, S. (1998) *J. Mol. Biol.* **279**, 233–244
127. Kean, J. M., and Draper, D. E. (1985) *Biochemistry* **24**, 5052–5061
128. Svensson, P., Changchien, L., Craven, G. R., and Noller, H. F. (1988) *J. Mol. Biol.* **200**, 301–308
129. Serganov, A., Bénard, L., Portier, C., Ennifar, E., Garber, M., Ehresmann, B., and Ehresmann, C. (2001) *J. Mol. Biol.* **305**, 785–803
130. Kalurachchi, K., and Nikonowicz, E. P. (1998) *J. Mol. Biol.* **280**, 639–654
130a. Tishchenko, S., Nikulin, A., Fomenkova, N., Nevskaya, N., Nikonov, O., Dumas, P., Moine, H., Ehresmann, B., Ehresmann, C., Piendl, W., Lamzin, V., Garber, M., and Nikonov, S. (2001) *J. Mol. Biol.* **311**, 311–324
131. Orr, J. W., Hagerman, P. J., and Williamson, J. R. (1998) *J. Mol. Biol.* **275**, 453–464
132. Hei, Y.-j, Pelech, S. L., Chen, X., Diamond, J., and McNeill, J. H. (1994) *J. Biol. Chem.* **269**, 7816–7823
132a. Martin, K. A., Schalm, S. S., Richardson, C., Romanelli, A., Keon, K. L., and Blenis, J. (2001) *J. Biol. Chem.* **276**, 7884–7891
132b. Martin, K. A., Schalm, S. S., Romanelli, A., Keon, K. L., and Blenis, J. (2001) *J. Biol. Chem.* **276**, 7892–7898
133. Morosyuk, S. V., Cunningham, P. R., and SantaLucia, J., Jr. (2001) *J. Mol. Biol.* **307**, 197–211
134. Davies, C., Bussiere, D. E., Golden, B. L., Porter, S. J., Ramakrishnan, V., and White, S. W. (1998) *J. Mol. Biol.* **279**, 873–888
135. Subramanian, A. R. (1984) *Trends Biochem. Sci.* **9**, 491–494
136. Kim, J., Chubatsu, L. S., Admon, A., Stahl, J., Fellous, R., and Linn, S. (1995) *J. Biol. Chem.* **270**, 13620–13629
137. Sorensen, M. A., Fricke, J., and Pedersen, S. (1998) *J. Mol. Biol.* **280**, 561–569
138. Laughrea, M., and Tam, J. (1991) *Biochemistry* **30**, 11412–11420
139. Nikonov, S., Nevskaya, N., Eliseikina, I., Fomenkova, N., Nikulin, A., Ossina, N., Garber, M., Jonsson, B.-H., Briand, C., Al-Karadaghi, S., Svensson, A., AEvarsson, A., and Liljas, A. (1996) *EMBO J.* **15**, 1350–1359
139a. Zengel, J. M., and Lindahl, L. (1994) *Prog. Nucleic Acid Res. Mol. Biol.* **47**, 331–370
140. Yeh, L.-C. C., and Lee, J. C. (1995) *J. Mol. Biol.* **246**, 295–307
141. Nag, B., Tewari, D. S., Etchison, J. R., Sommer, A., and Traut, R. R. (1986) *J. Biol. Chem.* **261**, 13892–13897
142. Nakagawa, A., Nakashima, T., Taniguchi, M., Hosaka, H., Kimura, M., and Tanaka, I. (1999) *EMBO J.* **18**, 1459–1467
143. Hoffman, D. W., Cameron, C. S., Davies, C., White, S. W., and Ramakrishnan, V. (1996) *J. Mol. Biol.* **264**, 1058–1071
144. Kuhlman, B., Yang, H. Y., Boice, J. A., Fairman, R., and Raleigh, D. P. (1997) *J. Mol. Biol.* **270**, 640–647
145. Lieberman, K. R., Firpo, M. A., Herr, A. J., Nguyenle, T., Atkins, J. F., Gesteland, R. F., and Noller, H. F. (2000) *J. Mol. Biol.* **297**, 1129–1143
146. Dey, D., Bochkariov, D. E., Jokhadze, G. G., and Traut, R. R. (1998) *J. Biol. Chem.* **273**, 1670–1676
146a. Montesano-Roditis, L., Glitz, D. G., Traut, R. R., and Stewart, P. L. (2001) *J. Biol. Chem.* **276**, 14117–14123

146a. Montesano-Roditis, L., Glitz, D. G., Traut, R. R., and Stewart, P. L. (2001) *J. Biol. Chem.* **276**, 14117–14123

146b. Pioletti, M., Schlünzen, F., Harms, J., Zarivach, R., Glühmann, M., Avila, H., Bashan, A., Bartels, H., Auerbach, T., Jacobi, C., Hartsch, T., Yonath, A., and Franceschi, F. (2001) *EMBO J.* **20**, 1829–1839
147. Wahl, M. C., Bourenkov, G. P., Bartunik, H. D., and Huber, R. (2000) *EMBO J.* **19**, 174–186
148. Porse, B. T., Leviev, I., Mankin, A. S., and Garrett, R. A. (1998) *J. Mol. Biol.* **276**, 391–404
149. Holmberg, L., and Noller, H. F. (1999) *J. Mol. Biol.* **289**, 223–233
150. Gonzalo, P., Lavergne, J.-P., and Reboud, J.-P. (2001) *J. Biol. Chem.* **276**, 19762–19769
150a. Zurdo, J., González, C., Sanz, J. M., Rico, M., Remacha, M., and Ballesta, J. P. G. (2000) *Biochemistry* **39**, 8935–8943
150b. Guarinos, E., Remacha, M., and Ballesta, J. P. G. (2001) *J. Biol. Chem.* **276**, 32474–32479
151. Szick-Miranda, K., and Bailey-Serres, J. (2001) *J. Biol. Chem.* **276**, 10921–10928
152. Dunbar, D. A., Gragon, F., Lee, S. J., and Baserga, S. J. (2000) *Proc. Natl. Acad. Sci. U.S.A.* **97**, 13027–13032
153. Rohl, R., and Nierhaus, K. H. (1982) *Proc. Natl. Acad. Sci. U.S.A.* **79**, 729–733
154. Stoldt, M., Wöhnert, J., Görlach, M., and Brown, L. R. (1998) *EMBO J.* **17**, 6377–6384
155. Lu, M., and Steitz, T. A. (2000) *Proc. Natl. Acad. Sci. U.S.A.* **97**, 2023–2028
156. Chan, Y. L., Lin, A., McNally, J., and Wool, I. G. (1987) *J. Biol. Chem.* **262**, 12879–12886
156a. Stoldt, M., Wöhnert, J., Ohlenschläger, O., Görlach, M., and Brown, L. R. (1999) *EMBO J.* **18**, 6508–6521
156b. Johnson, A. W., Lund, E., and Dahlberg, J. (2002) *Trends Biochem. Sci.* **27**, 580 – 585
157. Lin, A., McNally, J., and Wool, I. G. (1983) *J. Biol. Chem.* **258**, 10664–10671
158. Lee, J. C., Henry, B., and Yeh, Y.-C. (1983) *J. Biol. Chem.* **258**, 854–858
159. Mao, H., and Williamson, J. R. (1999) *J. Mol. Biol.* **292**, 345–359
160. Stoffler-Meilicke, M., Stoffler, G., Odom, O. W., Zinn, A., Kramer, G., and Hardesty, B. (1981) *Proc. Natl. Acad. Sci. U.S.A.* **78**, 5538–5542
161. Trempe, M. R., Ohgi, K., and Glitz, D. G. (1982) *J. Biol. Chem.* **257**, 9822–9829
162. Hausner, T.-P., Geigenmüller, U., and Nierhaus, K. N. (1988) *J. Biol. Chem.* **263**, 13103–13111
163. Prince, J. B., and Garrett, R. A. (1982) *Trends Biochem. Sci.* **7**, 79
164. Denman, R., Nègre, D., Cunningham, P. R., Nurse, K., Colgan, J., Weitzmann, C., and Ofengand, J. (1989) *Biochemistry* **28**, 1012–1019
164a. Noller, H. F. (1993) *FASEB J.* **7**, 87–89
164b. Vila-Sanjurjo, A., and Dahlberg, A. E. (2001) *J. Mol. Biol.* **308**, 457–463
164c. Samaha, R. R., Green, R., and Noller, H. F. (1995) *Nature (London)* **377**, 309–314
164d. Porse, B. T., Thi-Ngoc, H. P., and Garrett, R. A. (1996) *J. Mol. Biol.* **264**, 472–483
164e. Green, R., Samaha, R. R., and Noller, H. F. (1997) *J. Mol. Biol.* **266**, 40–50
165. Lührmann, R., Bald, R., Stöffler-Meilicke, M., and Stöffler, G. (1981) *Proc. Natl. Acad. Sci. U.S.A.* **78**, 7276–7280
166. Olson, H. M., Nicholson, A. W., Cooperman, B. S., and Glitz, D. G. (1985) *J. Biol. Chem.* **260**, 10326–10331
166a. Green, R., Switzer, C., and Noller, H. F. (1998) *Science* **280**, 286–289
167. Goldman, R. A., Hasan, T., Hall, C. C., Strycharz, W. A., and Cooperman, B. S. (1983) *Biochemistry* **22**, 359–368
167a. Porse, B. T., and Garrett, R. A. (1995) *J. Mol. Biol.* **249**, 1–10
167b. Thompson, J., Kim, D. F., O'Connor, M., Lieberman, K. R., Bayfield, M. A., Gregory, S. T., Green, R., Noller, H. F., and Dahlberg, A. E. (2001) *Proc. Natl. Acad. Sci. U.S.A.* **98**, 9002–9007
168. Langer, J. A., and Lake, J. A. (1986) *J. Mol. Biol.* **187**, 617–621
169. Ibba, M., Becker, H. D., Stathopoulos, C., Tumbula, D. L., and Söll, D. (2000) *Trends Biochem. Sci.* **25**, 311–316
170. Weisblum, B. (1999) *Trends Biochem. Sci.* **24**, 247–250
171. Yue, D., Kintanar, A., and Horowitz, J. (1994) *Biochemistry* **33**, 8905–8911
172. Schimmel, P. (1996) *Proc. Natl. Acad. Sci. U.S.A.* **93**, 4521–4522
173. Senger, B., Aphasizhev, R., Walter, P., and Fasiolo, F. (1995) *J. Mol. Biol.* **249**, 45–58
174. Frazer-Abel, A. A., and Hagerman, P. J. (1999) *J. Mol. Biol.* **285**, 581–593
175. Jovine, L., Djordjevic, S., and Rhodes, D. (2000) *J. Mol. Biol.* **301**, 401–414
176. Auffinger, P., and Westhof, E. (1999) *J. Mol. Biol.* **292**, 467–483
177. Perreau, V. M., Keith, G., Holmes, W. M., Przykorska, A., Santos, M. A. S., and Tuite, M. F. (1999) *J. Mol. Biol.* **293**, 1039–1053
178. Wilson, R. K., and Roe, B. A. (1989) *Proc. Natl. Acad. Sci. U.S.A.* **86**, 409–413
179. Rich, A. (1978) *Trends Biochem. Sci.* **3**, 34–37
180. Hou, Y.-M., Westhof, E., and Giegé, R. (1993) *Proc. Natl. Acad. Sci. U.S.A.* **90**, 6776–6780
181. Himeno, H., Yoshida, S., Soma, A., and Nishikawa, K. (1997) *J. Mol. Biol.* **268**, 704–711
182. Christian, T., Lipman, R. S. A., Evilia, C., and Hou, Y.-M. (2000) *J. Mol. Biol.* **303**, 503–514
183. Hamann, C. S., and Hou, Y.-M. (2000) *J. Mol. Biol.* **295**, 777–789
184. Drabkin, H. J., and RajBhandary, U. L. (1985) *J. Biol. Chem.* **260**, 5580–5587
185. Woo, N. H., Roe, B. A., and Rich, A. (1980) *Nature (London)* **286**, 346–351
186. Calagan, J. L., Pirtle, R. M., Pirtle, I. L., Kashdan, M. A., Vreman, H. J., and Dudock, B. S. (1980) *J. Biol. Chem.* **255**, 9981–9984
187. Kuchino, Y., Ihara, M., Yabusaki, Y., and Nishimura, S. (1982) *Nature (London)* **298**, 684–685
188. Bai, Y., Fox, D. T., Lacy, J. A., Van Lanen, S. G., and Iwata-Reuyl, D. (2000) *J. Biol. Chem.* **275**, 28731–28738
189. Watanabe, M., Nameki, N., Matsuo-Takasaki, M., Nishimura, S., and Okada, N. (2001) *J. Biol. Chem.* **276**, 2387–2394
190. Crick, F. H. C. (1966) *J. Mol. Biol.* **19**, 548–555
190a. Inagaki, Y., Kojima, A., Bessho, Y., Hori, H., Ohama, T., and Osawa, S. (1995) *J. Mol. Biol.* **251**, 486–492
191. Booth, V., Keizer, D. W., Kamphuis, M. B., Clark-Lewis, I., and Sykes, B. D. (2002) *Biochemistry* **41**, 10418 – 10425
192. Auld, D. S., and Schimmel, P. (1995) *Science* **267**, 1994–1996
193. Pak, M., Willis, I. M., and Schulman, L. H. (1994) *J. Biol. Chem.* **269**, 2277–2282
194. Li, S., Pelka, H., and Schulman, L. H. (1993) *J. Biol. Chem.* **268**, 18335–18339
195. Hong, K.-W., Ibba, M., Weygand-Durasevic, I., Rogers, M. J., Thomann, H.-U., and Söll, D. (1996) *EMBO J.* **15**, 1983–1991

References

196. Nureki, O., Niimi, T., Muramatsu, T., Kanno, H., Kohno, T., Florentz, C., Giegé, R., and Yokoyama, S. (1994) *J. Mol. Biol.* **236**, 710–724
197. Tocchini-Valentini, G., Saks, M. E., and Abelson, J. (2000) *J. Mol. Biol.* **298**, 779–793
198. Yan, W., and Francklyn, C. (1994) *J. Biol. Chem.* **269**, 10022–10027
198a. Shimada, A., Nureki, O., Goto, M., Takahashi, S., and Yokoyama, S. (2001) *Proc. Natl. Acad. Sci. U.S.A.* **98**, 13537–13542
198b. Xu, F., Jiang, G., Li, W., He, X., Jin, Y., and Wang, D. (2002) *Biochemistry* **41**, 8087–8092
199. Normanly, J., and Abelson, J. (1989) *Ann. Rev. Biochem.* **58**, 1029–1049
199a. Sampson, J. R., DiRenzo, A. B., Behlen, L. S., and Uhlenbeck, O. C. (1989) *Science* **243**, 1363–1366
200. Saks, M. E., Sampson, J. R., and Abelson, J. N. (1994) *Science* **263**, 191–197
201. McClain, W. H. (1993) *FASEB J.* **7**, 72–78
202. Ribas de Pouplana, L., and Schimmel, P. (2001) *J. Biol. Chem.* **276**, 6881–6884
203. Sherman, J. M., Thomann, H.-U., and Söll, D. (1996) *J. Mol. Biol.* **256**, 818–828
204. Arnez, J. G., Dock-Bregeon, A.-C., and Moras, D. (1999) *J. Mol. Biol.* **286**, 1449–1459
205. Horowitz, J., Chu, W.-C., Derrick, W. B., Liu, J. C.-H., Liu, M., and Yue, D. (1999) *Biochemistry* **38**, 7737–7746
206. Liu, M., and Horowitz, J. (1994) *Proc. Natl. Acad. Sci. U.S.A.* **91**, 10389–10393
207. Arnez, J. G., and Moras, D. (1997) *Trends Biochem. Sci.* **22**, 211–216
208. Schimmel, P. (1991) *Trends Biochem. Sci.* **16**, 1–3
209. Nureki, O., Vassylyev, D. G., Katayanagi, K., Shimizu, T., Sekine, S.-i, Kigawa, T., Miyazawa, T., Yokoyama, S., and Morikawa, K. (1995) *Science* **267**, 1958–1965
209a. Liu, J., Ibba, M., Hong, K.-W., and Söll, D. (1998) *Biochemistry* **37**, 9836–9842
209b. Ribas de Pouplana, L., and Schimmel, P. (2001) *Trends Biochem. Sci.* **26**, 591–596
210. Zelwer, C., Risler, J. L., and Brunie, S. (1982) *J. Mol. Biol.* **155**, 63–81
211. Xin, Y., Li, W., and First, E. A. (2000) *Biochemistry* **39**, 340–347
212. Ilyin, V. A., Temple, B., Hu, M., Li, G., Yin, Y., Vachette, P., and Carter, C. W., Jr. (2000) *Protein Sci.* **9**, 218–231
213. First, E. A., and Fersht, A. R. (1995) *Biochemistry* **34**, 5030–5043
214. Perona, J. J., Rould, M. A., and Steitz, T. A. (1993) *Biochemistry* **32**, 8758–8771
215. Fersht, A. R., Leatherbarrow, R. J., and Wells, T. N. C. (1986) *Trends Biochem. Sci.* **11**, 321–325
216. Leatherbarrow, R. J., Fersht, A. R., and Winter, G. (1985) *Proc. Natl. Acad. Sci. U.S.A.* **82**, 7840–7844
217. Desogus, G., Todone, F., Brick, P., and Onesti, S. (2000) *Biochemistry* **39**, 8418–8425
217a. Onesti, S., Desogus, G., Brevet, A., Chen, J., Plateau, P., Blanquet, S., and Brick, P. (2000) *Biochemistry* **39**, 12853–12861
218. Rould, M. A., Perona, J. J., Söll, D., and Steitz, T. A. (1989) *Science* **246**, 1135–1142
219. Rould, M. A., Perona, J. J., and Steitz, T. A. (1991) *Nature (London)* **352**, 213–218
220. Arnez, J. G., and Steitz, T. A. (1994) *Biochemistry* **33**, 7560–7567
221. Cavarelli, J., Delagoutte, B., Eriani, G., Gangloff, J., and Moras, D. (1998) *EMBO J.* **17**, 5438–5448
222. Brick, P., Bhat, T. N., and Blow, D. M. (1988) *J. Mol. Biol.* **208**, 83–98
223. Nureki, O., Vassylyev, D. G., Tateno, M., Shimada, A., Nakama, T., Fukai, S., Konno, M., Hendrickson, T. L., Schimmel, P., and Yokoyama, S. (1998) *Science* **280**, 578–582
224. Cusack, S., Yaremchuk, A., and Tukalo, M. (2000) *EMBO J.* **19**, 2351–2361
225. Fourmy, D., Mechulam, Y., and Blanquet, S. (1995) *Biochemistry* **34**, 15681–15688
226. Mechulam, Y., Schmitt, E., Maveyraud, L., Zelwer, C., Nureki, O., Yokoyama, S., Konno, M., and Blanquet, S. (1999) *J. Mol. Biol.* **294**, 1287–1297
227. Arnez, J. G., Harris, D. C., Mitschler, A., Rees, B., Francklyn, C. S., and Moras, D. (1995) *EMBO J.* **14**, 4143–4155
228. Qiu, X., Janson, C. A., Blackburn, M. N., Chhohan, I. K., Hibbs, M., and Abdel-Meguid, S. S. (1999) *Biochemistry* **38**, 12296–12304
229. Arnez, J. G., Augustine, J. G., Moras, D., and Francklyn, C. S. (1997) *Proc. Natl. Acad. Sci. U.S.A.* **94**, 7144–7149
230. Biou, V., Yaremchuk, A., Tukalo, M., and Cusack, S. (1994) *Science* **263**, 1404–1410
231. Cusack, S., Yaremchuk, A., and Tukalo, M. (1996) *EMBO J.* **15**, 2834–2842
232. Eiler, S., Dock-Bregeon, A.-C., Moulinier, L., Thierry, J.-C., and Moras, D. (1999) *EMBO J.* **18**, 6532–6541
233. Archontis, G., Simonson, T., and Karplus, M. (2001) *J. Mol. Biol.* **306**, 307–327
234. Berthet-Colominas, C., Seignovert, L., Härtlein, M., Grotli, M., Cusack, S., and Leberman, R. (1998) *EMBO J.* **17**, 2947–2960
235. Commans, S., Lazard, M., Delort, F., Blanquet, S., and Plateau, P. (1998) *J. Mol. Biol.* **278**, 801–813
236. Reshetnikova, L., Moor, N., Lavrik, O., and Vassylyev, D. G. (1999) *J. Mol. Biol.* **287**, 555–568
237. Logan, D. T., Mazauric, M.-H., Kern, D., and Moras, D. (1995) *EMBO J.* **14**, 4156–4167
238. Xu, Z.-J., Love, M. L., Ma, L. Y. Y., Blum, M., Bronskill, P. M., Bernstein, J., Grey, A. A., Hofmann, T., Camerman, N., and Wong, J. T.-F. (1989) *J. Biol. Chem.* **264**, 4304–4311
239. Jasin, M., Regan, L., and Schimmel, P. (1983) *Nature (London)* **306**, 441–447
240. Webster, T. A., Lathrop, R. H., and Smith, T. F. (1987) *Biochemistry* **26**, 6950–6957
241. Schimmel, P. (1989) *Biochemistry* **28**, 2747–2759
242. Liu, J., Lin, S.-x, Blochet, J.-E., Pézolet, M., and Lapointe, J. (1993) *Biochemistry* **32**, 11390–11396
243. Sampson, J., and Uhlenbeck, O. C. (1988) *Proc. Natl. Acad. Sci. U.S.A.* **85**, 1033–1037
244. Ogilvie, K. K., Usman, N., Nicoghosian, K., and Cedergren, R. J. (1988) *Proc. Natl. Acad. Sci. U.S.A.* **85**, 5764–5768
244a. Hendrickson, T. L. (2001) *Proc. Natl. Acad. Sci. U.S.A.* **98**, 13473–13475
245. Alexander, R. W., and Schimmel, P. (1999) *Biochemistry* **38**, 16359–16365
246. McClain, W. H., Chen, Y.-M., Foss, K., and Schneider, J. (1988) *Science* **242**, 1681–1684
247. Francklyn, C., and Schimmel, P. (1989) *Nature (London)* **337**, 478–481
247a. Sardesai, N. Y., Green, R., and Schimmel, P. (1999) *Biochemistry* **38**, 12080–12088
248. Rogers, M. J., and Söll, D. (1988) *Proc. Natl. Acad. Sci. U.S.A.* **85**, 6627–6631
249. Khan, A. S., and Roe, B. A. (1988) *Science* **241**, 74–79
250. Connolly, B. A., Von der Haar, F., and Eckstein, F. (1980) *J. Biol. Chem.* **255**, 11301–11307
251. Fersht, A. R. (1987) *Biochemistry* **26**, 8031–8037
252. Leatherbarrow, R. J., and Fersht, A. R. (1987) *Biochemistry* **26**, 8524–8528
253. Fersht, A. R. (1988) *Biochemistry* **27**, 1577–1580
254. Fersht, A. (1985) *Enzyme Structure and Mechanism*, 2nd ed., Freeman, New York
255. Xin, Y., Li, W., and First, E. A. (2000) *J. Mol. Biol.* **303**, 299–310
256. Bedouelle, H., and Winter, G. (1986) *Nature (London)* **320**, 371–373
257. Starzyk, R. M., Koontz, S. W., and Schimmel, P. (1982) *Nature (London)* **298**, 136–140
258. Fersht, A. R. (1977) *Enzyme Structure and Mechanism*, 1st ed., Freeman, San Francisco, California (p. 283)
259. Fersht, A. R. (1998) *Science* **280**, 541
259a. Mursinna, R. S., Lincecum, T. L., Jr., and Martinis, S. A. (2001) *Biochemistry* **40**, 5376–5381
259b. Tardif, K. D., Liu, M., Vitseva, O., Hou, Y.-M., and Horowitz, J. (2001) *Biochemistry* **40**, 8118–8125
259c. Tardif, K. D., and Horowitz, J. (2002) *Nucleic Acids Res.* **30**, 2538–2545
260. Gao, W., Goldman, E., and Jakubowski, H. (1994) *Biochemistry* **33**, 11528–11535
261. Jakubowski, H. (1995) *J. Biol. Chem.* **270**, 17672–17673
262. Jakubowski, H. (1996) *Biochemistry* **35**, 8252–8259
263. Serre, L., Verdon, G., Choinowski, T., Hervouet, N., Risler, J.-L., and Zelwer, C. (2001) *J. Mol. Biol.* **306**, 863–876
263a. Kelley, S. O., Steinberg, S. V., and Schimmel, P. (2001) *J. Biol. Chem.* **276**, 10607–10611
264. Yarus, M. (1992) *Trends Biochem. Sci.* **17**, 171–174
265. Farabaugh, P. J., and Björk, G. R. (1999) *EMBO J.* **18**, 1427–1434
266. Rodnina, M. V., and Wintermeyer, W. (2001) *Trends Biochem. Sci.* **26**, 124–130
267. Ibba, M., Curnow, A. W., and Söll, D. (1997) *Trends Biochem. Sci.* **22**, 39–42
268. Wallis, N. G., Dardel, F., and Blanquet, S. (1995) *Biochemistry* **34**, 7668–7677
268a. Horiuchi, K. Y., Harpel, M. R., Shen, L., Luo, Y., Rogers, K. C., and Copeland, R. A. (2001) *Biochemistry* **40**, 6450–6457
269. Curnow, A. W., Hong,, K-w., Yuan, R., Kim, S.-i, Martins, O., Winkler, W., Henkin, T. M., and Söll, D. (1997) *Proc. Natl. Acad. Sci. U.S.A.* **94**, 11819–11826
270. Böck, A., Forchhammer, K., Heider, J., and Baron, C. (1991) *Trends Biochem. Sci.* **16**, 463–467
271. Yarus, M. (2000) *Science* **287**, 440–441
272. Stathopoulos, C., Li, T., Longman, R., Vothknecht, U. C., Becker, H. D., Ibba, M., and Söll, D. (2000) *Science* **287**, 479–482
272a. Stathopoulos, C., Kim, W., Li, T., Anderson, I., Deutsch, B., Palioura, S., Whitman, W., and Söll, D. (2001) *Proc. Natl. Acad. Sci. U.S.A.* **98**, 14292–14297
273. Wool, I. G. (1996) *Trends Biochem. Sci.* **21**, 164–165
274. Martinis, S. A., Plateau, P., Cavarelli, J., and Florentz, C. (1999) *EMBO J.* **18**, 4591–4596
275. Putney, S. D., and Schimmel, P. (1981) *Nature (London)* **291**, 632–635
276. Romby, P., Caillet, J., Ebel, C., Sacerdot, C., Graffe, M., Eyermann, F., Brunel, C., Moine, H., Ehresmann, C., Ehresmann, B., and Springer, M. (1996) *EMBO J.* **15**, 5976–5987
277. Gerken, S. C., and Arfin, S. M. (1984) *J. Biol. Chem.* **259**, 11160–11161
278. Lambowitz, A. M., and Perlman, P. S. (1990) *Trends Biochem. Sci.* **15**, 440–444
279. Abramochkin, G., and Shrader, T. E. (1996) *J. Biol. Chem.* **271**, 22901–22907
280. Li, J., and Pickart, C. M. (1995) *Biochemistry* **34**, 139–147
281. Dou, X., Limmer, S., and Kreutzer, R. (2001) *J. Mol. Biol.* **305**, 451–458
281a. Lund, E., and Dahlberg, J. E. (1998) *Science* **282**, 2003–2004
281b. Sarkar, S., Azad, A. K., and Hopper, A. K. (1999) *Proc. Natl. Acad. Sci. U.S.A.* **96**, 14366–14371

References

282. Schimmel, P., and Ribas De Pouplana, L. (2000) *Trends Biochem. Sci.* **25**, 207–209
283. Kisselev, L., Frolova, L., and Haenni, A.-L. (1993) *Trends Biochem. Sci.* **18**, 263–267
284. Sissler, M., Delorme, C., Bond, J., Ehrlich, S. D., Renault, P., and Francklyn, C. (1999) *Proc. Natl. Acad. Sci. U.S.A.* **96**, 8985–8990
285. Raben, N., Nichols, R., Dohlman, J., McPhie, P., Sridhar, V., Hyde, C., Leff, R., and Plotz, P. (1994) *J. Biol. Chem.* **269**, 24277–24283
286. Ge, Q., Trieu, E. P., and Targoff, I. N. (1994) *J. Biol. Chem.* **269**, 28790–28797
287. Winter, R. B., Morrissey, L., Gauss, P., Gold, L., Hsu, T., and Karam, J. (1987) *Proc. Natl. Acad. Sci. U.S.A.* **84**, 7822–7826
288. Peabody, D. S. (1989) *J. Biol. Chem.* **264**, 5031–5035
289. Gupta, K. C., and Patwardhan, S. (1988) *J. Biol. Chem.* **263**, 8553–8556
290. Newton, D. T., Creuzenet, C., and Mangroo, D. (1999) *J. Biol. Chem.* **274**, 22143–22146
291. RajBhandary, U. L. (2000) *Proc. Natl. Acad. Sci. U.S.A.* **97**, 1325–1327
292. Li, S., Kumar, N. V., Varshney, U., and RajBhandary, U. L. (1996) *J. Biol. Chem.* **271**, 1022–1028
293. Takeuchi, N., Kawakami, M., Omori, A., Ueda, T., Spremulli, L. L., and Watanabe, K. (1998) *J. Biol. Chem.* **273**, 15085–15090
294. Shine, J., and Dalgarno, L. (1975) *Nature (London)* **254**, 34–38
295. Steitz, J. A., and Jakes, K. (1975) *Proc. Natl. Acad. Sci. U.S.A.* **72**, 4734–4738
295a. van Dieijen, G., Zipori, P., van Prooijen, W., and van Duin, J. (1978) *Eur. J. Biochem.* **90**, 571–580
296. Wu, X.-Q., Iyengar, P., and RajBhandary, U. L. (1996) *EMBO J.* **15**, 4734–4739
296a. Roll-Mecak, A., Shin, B.-S., Dever, T. E., and Burley, S. K. (2001) *Trends Biochem. Sci.* **26**, 705–709
297. Biou, V., Shu, F., and Ramakrishnan, V. (1995) *EMBO J.* **14**, 4056–4064
297a. Morosyuk, S. V., Cunningham, P. R., SantaLucia, J., Jr. (2001) *J. Mol. Biol.* **307**, 197–211
298. Moreau, M., de Cock, E., Fortier, P.-L., Garcia, C., Albaret, zC., Blanquet, S., Lallemand, J.-Y., and Dardel, F. (1997) *J. Mol. Biol.* **266**, 15–22
299. Yu, N.-J., and Spremulli, L. L. (1997) *Biochemistry* **36**, 14827–14835
300. La Teana, A., Pon, C. L., and Gualerzi, C. O. (1996) *J. Mol. Biol.* **256**, 667–675
301. Karimi, R., Pavlov, M. Y., Heurgué-Hamard, V., Buckingham, R. H., and Ehrenberg, M. (1998) *J. Mol. Biol.* **281**, 241–252
302. Dahlquist, K. D., and Puglisi, J. D. (2000) *J. Mol. Biol.* **299**, 1–15
303. Meinnel, T., Sacerdot, C., Graffe, M., Blanquet, S., and Springer, M. (1999) *J. Mol. Biol.* **290**, 825–837
304. Garcia, C., Fortier, P.-L., Blanquet, S., Lallemand, J.-Y., and Dardel, F. (1995) *J. Mol. Biol.* **254**, 247–259
305. Kozak, M. (1983) *Microbiol. Rev.* **47**, 1–45
306. Moldave, R. (1985) *Ann. Rev. Biochem.* **54**, 1109–1149
307. Lorsch, J. R., and Herschlag, D. (1999) *EMBO J.* **18**, 6705–6717
308. Rhoads, R. E. (1999) *J. Biol. Chem.* **274**, 30337–30340
309. Rhoads, R. E. (1993) *J. Biol. Chem.* **268**, 3017–3020
310. Browning, K. S., Maia, D. M., Lax, S. R., and Ravel, J. M. (1987) *J. Biol. Chem.* **262**, 538–541
311. Pestova, T. V., and Hellen, C. U. T. (1999) *Trends Biochem. Sci.* **24**, 85–87
312. Altmann, M., and Trachsel, H. (1993) *Trends Biochem. Sci.* **18**, 429–432
313. Sonenberg, N., Hershey, J. W. B., and Mathews, M. B., eds. (2000) *Translational Control of Gene Expression*, Cold Spring Harbor Lab. Press, Plainview, New York
314. Carey, M., and Smale, S. T. (2000) *Transcriptional Regulation in Eukaryotes: Concepts, Strategies, and Techniques*, Cold Spring Harbor Lab. Press, Plainview, New York
314a. Pestova, T. V., Kolupaeva, V. G., Lomakin, I. B., Pilipenko, E. V., Shatsky, I. N., Agol, V. I., and Hellen, C. U. T. (2001) *Proc. Natl. Acad. Sci. U.S.A.* **98**, 7029–7036
315. Jackson, R. J. (1998) *Nature (London)* **394**, 829–831
316. Richter-Cook, N. J., Dever, T. E., Hensold, J. O., and Merrick, W. C. (1998) *J. Biol. Chem.* **273**, 7579–7587
317. Richter, N. J., Rogers, G. W., Jr., Hensold, J. O., and Merrick, W. C. (1999) *J. Biol. Chem.* **274**, 35415–35424
318. Melander, Y., Holmberg, L., and Nygård, O. (1997) *J. Biol. Chem.* **272**, 3254–3258
319. Korneeva, N. L., Lamphear, B. J., Hennigan, F. L. C., and Rhoads, R. E. (2000) *J. Biol. Chem.* **275**, 41369–41376
319a. Browning, K. S., Gallie, D. R., Hershey, J. W. B., Hinnebusch, A. G., Maitra, U., Merrick, W. C., and Norbury, C. (2001) *Trends Biochem. Sci.* **26**, 284
320. Valásek, L., Phan, L., Schoenfeld, L. W., Valásková, V., and Hinnebusch, A. G. (2001) *EMBO J.* **20**, 891–904
321. Raychaudhuri, P., and Maitra, U. (1986) *J. Biol. Chem.* **261**, 7723–7728
322. Proud, C. G. (1986) *Trends Biochem. Sci.* **11**, 73–77
323. Gil, J., Esteban, M., and Roth, D. (2000) *Biochemistry* **39**, 7521–7530
324. Sudhakar, A., Krishnamoorthy, T., Jain, A., Chatterjee, U., Hasnain, S. E., Kaufman, R. J., and Ramaiah, K. V. A. (1999) *Biochemistry* **38**, 15398–15405
325. Kimball, S. R., Fabian, J. R., Pavitt, G. D., Hinnebusch, A. G., and Jefferson, L. S. (1998) *J. Biol. Chem.* **273**, 12841–12845
326. De Haro, C., Méndez, R., and Santoyo, J. (1996) *FASEB J.* **10**, 1378–1387
327. Rose, D. W., Welch, W. J., Kramer, G., and Hardesty, B. (1989) *J. Biol. Chem.* **264**, 6239–6244
328. Williams, D. D., Pavitt, G. D., and Proud, C. G. (2001) *J. Biol. Chem.* **276**, 3733–3742
328a. Asano, K., Krishnamoorthy, T., Phan, L., Pavitt, G. D., Hinnebusch, A. G. (1999) *EMBO J.* **18**, 1673 – 1688
329. Mueller, P. P., Grueter, P., Hinnebusch, A. G., and Trachsel, H. (1998) *J. Biol. Chem.* **273**, 32870–32877
330. Qiu, H., Dong, J., Hu, C., Francklyn, C. S., and Hinnebusch, A. G. (2001) *EMBO J.* **20**, 1425–1438
330a. Wang, X., Paulin, F. E. M., Campbell, L. E., Gomez, E., O'Brien, K., Morrice, N., and Proud, C. G. (2001) *EMBO J.* **20**, 4349–4359
330b. Wang, X., Li, W., Williams, M., Terada, N., Alessi, D. R., and Proud, C. G. (2001) *EMBO J.* **20**, 4370–4379
330c. Cho, S., and Hoffman, D. W. (2002) *Biochemistry* **41**, 5730–5742
331. Chen, J.-J., and London, I. M. (1995) *Trends Biochem. Sci.* **20**, 105–108
332. Dholakia, J. N., Mueser, T. C., Woodley, C. L., Parkhurst, L. J., and Wahba, A. J. (1986) *Proc. Natl. Acad. Sci. U.S.A.* **83**, 6746–6750
333. Gingras, A.-C., Raught, B., and Sonenberg, N. (1999) *Ann. Rev. Biochem.* **68**, 913–963
334. Herbert, T. P., Kilhams, G. R., Batty, I. H., and Proud, C. G. (2000) *J. Biol. Chem.* **275**, 11249–11256
334a. Karim, M. M., Hughes, J. M. X., Warwicker, J., Scheper, G. C., Proud, C. G., and McCarthy, J. E. G. (2001) *J. Biol. Chem.* **276**, 20750–20757
335. Dever, T. E. (1999) *Trends Biochem. Sci.* **24**, 398–403
336. De Gregorio, E., Preiss, T., and Hentze, M. W. (1999) *EMBO J.* **18**, 4865–4874
337. Korneeva, N. L., Lamphear, B. J., Hennigan, F. L. C., Merrick, W. C., and Rhoads, R. E. (2001) *J. Biol. Chem.* **276**, 2872–2879
338. Lamphear, B. J., Kirchweger, R., Skern, T., and Rhoads, R. E. (1995) *J. Biol. Chem.* **270**, 21975–21983
339. McCubbin, W. D., Edery, I., Altmann, M., Sonenberg, N., and Kay, C. M. (1988) *J. Biol. Chem.* **263**, 17663–17671
340. Keiper, B. D., Lamphear, B. J., Deshpande, A. M., Jankowska-Anyszka, M., Aamodt, E. J., Blumenthal, T., and Rhoads, R. E. (2000) *J. Biol. Chem.* **275**, 10590–10596
341. Rao, G. N. (2000) *J. Biol. Chem.* **275**, 16993–16999
341a. Niedzwiecka, A., Stepinski, J., Darzynkiewicz, E., Sonenberg, N., and Stolarski, R. (2002) *Biochemistry* **41**, 12140–12148
342. Naranda, T., Strong, W. B., Menaya, J., Fabbri, B. J., and Hershey, J. W. B. (1994) *J. Biol. Chem.* **269**, 14465–14472
343. Dominguez, D., Altmann, M., Benz, J., Baumann, U., and Trachsel, H. (1999) *J. Biol. Chem.* **274**, 26720–26726
344. Chappell, S. A., Edelman, G. M., and Mauro, V. P. (2000) *Proc. Natl. Acad. Sci. U.S.A.* **97**, 1536–1541
345. Hentze, M. W. (1997) *Science* **275**, 500–501
345a. Fernandez, J., Yaman, I., Mishra, R., Merrick, W. C., Snider, M. D., Lamers, W. H., and Hatzoglou, M. (2001) *J. Biol. Chem.* **276**, 12285–12291
346. Spahn, C. M. T., Kieft, J. S., Grassucci, R. A., Penczek, P. A., Zhou, K., Doudna, J. A., and Frank, J. (2001) *Science* **291**, 1959–1962
347. Hudder, A., and Werner, R. (2000) *J. Biol. Chem.* **275**, 34586–34591
348. Kozak, M. (1997) *EMBO J.* **16**, 2482–2492
349. Geballe, A. P., and Morris, D. R. (1994) *Trends Biochem. Sci.* **19**, 159–164
350. Cigan, A. M., Feng, L., and Donahue, T. F. (1988) *Science* **242**, 93–97
350a. Galy, B. (2001) *Trends Biochem. Sci.* **26**, 220
351. Chen, C.-y, and Sarnow, P. (1995) *Science* **268**, 415–417
352. Otero, L. J., Ashe, M. P., and Sachs, A. B. (1999) *EMBO J.* **18**, 3153–3163
353. Le, H., Browning, K. S., and Gallie, D. R. (2000) *J. Biol. Chem.* **275**, 17452–17462
354. Kozlov, G., Trempe, J.-F., Khaleghpour, K., Kahvejian, A., Ekiel, I., and Gehring, K. (2001) *Proc. Natl. Acad. Sci. U.S.A.* **98**, 4409–4413
355. Das, S., Maiti, T., Das, K., and Maitra, U. (1997) *J. Biol. Chem.* **272**, 31712–31718
356. Joe, Y. A., and Park, M. H. (1994) *J. Biol. Chem.* **269**, 25916–25921
357. Park, M. H., Wolff, E. C., and Folk, J. E. (1993) *Trends Biochem. Sci.* **18**, 475–479
358. Joao, H. C., Csonga, R., Klier, H., Koettnitz, K., Auer, M., and Eder, J. (1995) *Biochemistry* **34**, 14703–14711
359. Chetverin, A. B., and Spirin, A. S. (1982) *Biochim. Biophys. Acta.* **683**, 153–179
360. Gnirke, A., Geigenmüller, U., Rheinberger, H.-J., and Nierhaus, K. H. (1989) *J. Biol. Chem.* **264**, 7291–7301
361. Thompson, R. C. (1988) *Trends Biochem. Sci.* **13**, 91–93
362. Abel, K., and Jurnak, F. (1996) *Structure* **4**, 229–238
363. Jurnak, F. (1985) *Science* **230**, 32–36

References

364. Sprinzl, M. (1994) *Trends Biochem. Sci.* **19**, 245–250
365. Krásny, L., Mesters, J. R., Tieleman, L. N., Kraal, B., Fucík, V., Hilgenfeld, R., and Jonák, J. (1998) *J. Mol. Biol.* **283**, 371–381
366. Song, H., Parsons, M. R., Rowsell, S., Leonard, G., and Phillips, S. E. V. (1999) *J. Mol. Biol.* **285**, 1245–1256
367. Krab, I. M., and Parmeggiani, A. (1999) *Biochemistry* **38**, 13035–13041
368. Andersen, G. R., Thirup, S., Spremulli, L. L., and Nyborg, J. (2000) *J. Mol. Biol.* **297**, 421–436
369. Jiang, Y., Nock, S., Nesper, M., Sprinzl, M., and Sigler, P. B. (1996) *Biochemistry* **35**, 10269–10278
370. Uetsuki, T., Naito, A., Nagata, S., and Kaziro, Y. (1989) *J. Biol. Chem.* **264**, 5791–5798
371. Riis, B., Rattan, S. I. S., Clark, B. F. C., and Merrick, W. C. (1990) *Trends Biochem. Sci.* **15**, 420–424
372. Cavallius, J., and Merrick, W. C. (1998) *J. Biol. Chem.* **273**, 28752–28758
373. Edmonds, B. T., Bell, A., Wyckoff, J., Condeelis, J., and Leyh, T. S. (1998) *J. Biol. Chem.* **273**, 10288–10295
374. Condeelis, J. (1995) *Trends Biochem. Sci.* **20**, 169–170
374a. Pleiss, J. A., and Uhlenbeck, O. C. (2001) *J. Mol. Biol.* **308**, 895–905
375. Cunningham, P. R., Nurse, K., Weitzmann, C. J., Nègre, D., and Ofengand, J. (1992) *Biochemistry* **31**, 7629–7637
376. Gornicki, P., Nurse, K., Hellmann, W., Boublik, M., and Ofengand, J. (1984) *J. Biol. Chem.* **259**, 10493–10498
377. Weller, J., and Hill, W. E. (1994) *J. Biol. Chem.* **269**, 19369–19374
378. Lynch, S. R., and Puglisi, J. D. (2001) *J. Mol. Biol.* **306**, 1023–1035
378a. Ericson, G., Minchew, P., and Wollenzien, P. (1995) *J. Mol. Biol.* **250**, 407–419
379. VanLoock, M. S., Easterwood, T. R., and Harvey, S. C. (1999) *J. Mol. Biol.* **285**, 2069–2078
380. Lodmell, J. S., and Dahlberg, A. E. (1997) *Science* **277**, 1262–1267
381. Stark, H., Rodnina, M. V., Rinke-Appel, J., Brimacombe, R., Wintermeyer, W., and van Heel, M. (1997) *Nature (London)* **389**, 403–406
382. Hall, C. C., Smith, J. E., and Cooperman, B. S. (1985) *Biochemistry* **24**, 5702–5711
383. Douthwaite, S., and Aagaard, C. (1993) *J. Mol. Biol.* **232**, 725–731
384. Villsen, I. D., Vester, B., and Douthwaite, S. (1999) *J. Mol. Biol.* **286**, 365–374
385. Moazed, D., Robertson, J. M., and Noller, H. F. (1988) *Nature (London)* **334**, 362–364
385a. Rodnina, M. V., Pape, T., Fricke, R., Kuhn, L., and Wintermeyer, W. (1996) *J. Biol. Chem.* **271**, 646–652
385b. Piepenburg, O., Pape, T., Pleiss, J. A., Wintermeyer, W., Uhlenbeck, O. C., and Rodnina, M. V. (2000) *Biochemistry* **39**, 1734–1738
385c. Knudsen, C., Wieden, H.-J., and Rodnina, M. V. (2001) *J. Biol. Chem.* **276**, 22183–22190
385d. Vogeley, L., Palm, G. J., Mesters, J. R., and Hilgenfeld, R. (2001) *J. Biol. Chem.* **276**, 17149–17155
385e. Simonson, A. B., and Lake, J. A. (2002) *Nature (London)* **416**, 281–285
386. Hecht, S. M., Tan, K. H., Chinault, A. C., and Arcari, P. (1977) *Proc. Natl. Acad. Sci. U.S.A.* **74**, 437–441
386a. Sprinzl, M., Kucharzewski, M., Hobbs, J. B., and Cramer, F. (1977) *Eur. J. Biochem.* **78**, 55–61
387. Taiji, M., Yokoyama, S., and Miyazawa, T. (1985) *Biochemistry* **24**, 5776–5780
387a. Wagner, T., and Sprinzl, M. (1983) *Biochemistry* **22**, 94–98
387b. Limmer, St., Vogtherr, M., Nawrot, B., Hillenbrand, R., and Sprinzl, M. (1997) *Angew. Chem. Int. Ed. Engl.* **36**, 2485–2489
387c. Polacek, N., Gaynor, M., Yassin, A., and Mankin, A. S. (2001) *Nature (London)* **411**, 498–501
387d. Katunin, V. I., Muth, G. W., Strobel, S. A., Wintermeyer, W., and Rodnina, M. V. (2002) *Molecular Cell* **10**, 339–346
387e. Bayfield, M. A., Dahlberg, A. E., Schulmeister, U., Dorner, S., and Barta, A. (2001) *Proc. Natl. Acad. Sci. U.S.A.* **98**, 10096–10101
388. Lim, V. I., and Spirin, A. S. (1986) *J. Mol. Biol.* **188**, 565–577
389. Samaha, R. R., Green, R., and Noller, H. F. (1995) *Nature (London)* **377**, 309–314
390. Kloss, P., Xiong, L., Shinabarger, D. L., and Mankin, A. S. (1999) *J. Mol. Biol.* **294**, 93–101
391. Gregory, S. T., and Dahlberg, A. E. (1999) *J. Mol. Biol.* **285**, 1475–1483
392. Agrawal, R. K., Penczek, P., Grassucci, R. A., and Frank, J. (1998) *Proc. Natl. Acad. Sci. U.S.A.* **95**, 6134–6138
393. Wilson, K. S., and Noller, H. F. (1998) *Cell* **92**, 131–139
394. Savelsbergh, A., Matassova, N. B., Rodnina, M. V., and Wintermeyer, W. (2000) *J. Mol. Biol.* **300**, 951–961
394a. Katunin, V. I., Savelsbergh, A., Rodnina, M. V., and Wintermeyer, W. (2002) *Biochemistry* **41**, 12806–12812
395. Nissen, P., Kjeldgaard, M., Thirup, S., Polekhina, G., Reshetnikova, L., Clark, B. F. C., and Nyborg, J. (1995) *Science* **270**, 1464–1472
396. Nyborg, J., Nissen, P., Kjeldgaard, M., Thirup, S., Polekhina, G., Clark, B. F. C., and Reshetnikova, L. (1996) *Trends Biochem. Sci.* **21**, 81–82
397. Van Ness, B. G., Howard, J. B., and Bodley, J. W. (1980) *J. Biol. Chem.* **255**, 10710–10716
397a. Nierhaus, K. H., Wadzack, J., Burkhardt, N., Jünemann, R., Meerwinck, W., Willumeit, R., and Stuhrmann, H. B. (1998) *Proc. Natl. Acad. Sci. U.S.A.* **95**, 945–950
397b. Frank, J., and Agrawal, R. K. (1998) *Biophys. J.* **74**, 589–594
397c. Culver, G. M. (2001) *Structure* **9**, 751–758
398. Rodnina, M. V., Savelsbergh, A., Katunin, V. I., and Wintermeyer, W. (1997) *Nature (London)* **385**, 37–41
398a. Agrawal, R. K., Linde, J., Sengupta, J., Nierhaus, K. H., and Frank, J. (2001) *J. Mol. Biol.* **311**, 777–787
399. Gabashvili, I. S., Agrawal, R. K., Grassucci, R., Squires, C. L., Dahlberg, A. E., and Frank, J. (1999) *EMBO J.* **18**, 6501–6507
400. Mohr, D., Wintermeyer, W., and Rodnina, M. V. (2000) *EMBO J.* **19**, 3458–3464
401. Uchiumi, T., Hori, K., Nomura, T., and Hachimori, A. (1999) *J. Biol. Chem.* **274**, 27578–27582
402. VanLoock, M. S., Agrawal, R. K., Gabashvili, I. S., Qi, L., Frank, J., and Harvey, S. C. (2000) *J. Mol. Biol.* **304**, 507–515
403. Mesters, J. R., Potapov, A. P., de Graaf, J. M., and Kraal, B. (1994) *J. Mol. Biol.* **242**, 644–654
404. Montesano-Roditis, L., and Glitz, D. G. (1994) *J. Biol. Chem.* **269**, 6458–6470
405. Bhangu, R., and Wollenzien, P. (1992) *Biochemistry* **31**, 5937–5944
406. Graifer, D. M., Juzumiene, D. I., Karpova, G. G., and Wollenzien, P. (1994) *Biochemistry* **33**, 6201–6206
407. Sergiev, P. V., Lavrik, I. N., Wlasoff, V. A., Dokudovskaya, S. S., Dontsova, O. A., Bogdanov, A. A., and Brimacombe, R. (1997) *RNA* **3**, 464–475
408. Qin, S., Moldave, K., and McLaughlin, C. S. (1987) *J. Biol. Chem.* **262**, 7802–7807
409. Triana-Alonso, F. J., Chakraburtty, K., and Nierhaus, K. H. (1995) *J. Biol. Chem.* **270**, 20473–20478
410. Kambampati, R., Pellegrino, C., Paiva, A., Huang, L., Mende-Mueller, L., and Chakraburtty, K. (2000) *J. Biol. Chem.* **275**, 16963–16968
410a. Warner, J. R. (1999) *Trends Biochem. Sci.* **24**, 437–440
410b. Schleif, R., Hess, W., Finkelstein, S., and Ellis, D. (1973) *J. Bacteriol.* **115**, 9–14
410c. Andersson, D. I., Bohman, K., Isaksson, L. A., and Kurland, C. G. (1982) *Mol. Gen. Genet.* **187**, 467–472
410d. Dintzis, H. M. (1961) *Proc. Natl. Acad. Sci. U.S.A.* **47**, 247–261
410e. Goustin, A. S., and Wilt, F. H. (1982) *Biochim. Biophys. Acta.* **699**, 22–27
411. Tate, W. P., and Brown, C. M. (1992) *Biochemistry* **31**, 2443–2450
412. Caskey, C. T. (1980) *Trends Biochem. Sci.* **5**, 234–237
413. Stansfield, I., Jones, K. M., and Tuite, M. F. (1995) *Trends Biochem. Sci.* **20**, 489–491
413a. Wilson, K. S., Ito, K., Noller, H. F., and Nakamura, Y. (2000) *Nature Struct. Biol.* **7**, 866–870
413b. Kisselev, L. L., and Buckingham, R. H. (2000) *Trends Biochem. Sci.* **25**, 561–566
414. Tate, W. P., Hornig, H., and Luhrmann, R. (1983) *J. Biol. Chem.* **258**, 10360–10365
415. Poole, E. S., Brown, C. M., and Tate, W. P. (1995) *EMBO J.* **14**, 151–158
416. Zhang, S., Rydén-Aulin, M., and Isaksson, L. A. (1998) *J. Mol. Biol.* **284**, 1243–1246
417. Pavlov, M. Y., Freistroffer, D. V., MacDougall, J., Buckingham, R. H., and Ehrenberg, M. (1997) *EMBO J.* **16**, 4134–4141
418. Pavlov, M. Y., Freistroffer, D. V., Heurguè-Hamard, V., Buckingham, R. H., and Ehrenberg, M. (1997) *J. Mol. Biol.* **273**, 389–401
418a. Ito, K., Frolova, L., Seit-Nebi, A., Karamyshev, A., Kisselev, L., and Nakamura, Y. (2002) *Proc. Natl. Acad. Sci. U.S.A.* **99**, 8494–8499
419. Velichutina, I. V., Hong, J. Y., Mesecar, A. D., Chernoff, Y. O., and Liebman, S. W. (2001) *J. Mol. Biol.* **305**, 715–727
419a. Nakamura, Y., Ito, K., and Ehrenberg, M. (2000) *Cell* **101**, 349–352
419b. Song, H., Mugnier, P., Das, A. K., Webb, H. M., Evans, D. R., Tuite, M. F., Hemmings, B. A., and Barford, D. (2000) *Cell* **100**, 311–321
420. Wickner, R. B., Edskes, H. K., Maddelein, M.-L., Taylor, K. L., and Moriyama, H. (1999) *J. Biol. Chem.* **274**, 555–558
421. Sparrer, H. E., Santoso, A., Szoka, F. C., Jr., and Weissman, J. S. (2000) *Science* **289**, 595–599
422. King, C.-Y. (2001) *J. Mol. Biol.* **307**, 1247–1260
423. Selmer, M., Al-Karadaghi, S., Hirokawa, G., Kaji, A., and Liljas, A. (1999) *Science* **286**, 2349–2352
424. Yoshida, T., Uchiyama, S., Nakano, H., Kashimori, H., Kijima, H., Ohshima, T., Saihara, Y., Ishino, T., Shimahara, H., Yoshida, T., Yokose, K., Ohkubo, T., Kaji, A., and Kobayashi, Y. (2001) *Biochemistry* **40**, 2387–2396
424a. Kim, K. K., Min, K., and Suh, S. W. (2000) *EMBO J.* **19**, 2362–2370
424b. Karimi, R., Pavlov, M. Y., Buckingham, R. H., and Ehrenberg, M. (1999) *Mol. Cell. Biol.* **3**, 601–609
425. Rolland, N., Janosi, L., Block, M. A., Shuda, M., Teyssier, E., Miège, C., Chéniclet, C., Carde, J.-P., Kaji, A., and Joyard, J. (1999) *Proc. Natl. Acad. Sci. U.S.A.* **96**, 5464–5469
425a. Lancaster, L., Kiel, M. C., Kaji, A., and Noller, H. F. (2002) *Cell* **111**, 129–140

References

426. Gast, F.-U., Peters, F., and Pingoud, A. (1987) *J. Biol. Chem.* **262**, 11920–11926
427. Ibba, M., and Söll, D. (1999) *Science* **286**, 1893–1897
428. Parker, J., Johnston, T. C., Borgia, P. T., Holtz, G., Remaut, E., and Fiers, W. (1983) *J. Biol. Chem.* **258**, 10007–10012
429. Precup, J., and Parker, J. (1987) *J. Biol. Chem.* **262**, 11351–11355
429a. LaRiviere, F. J., Wolfson, A. D., and Uhlenbeck, O. C. (2001) *Science* **294**, 165–168
429b. Ibba, M. (2001) *Science* **294**, 70–71
429c. Stahl, G., McCarty, G. P., and Farabaug, P. J. (2002) *Trends Biochem. Sci.* **27**, 178–183
430. Kurland, C. G. (1987) *Trends Biochem. Sci.* **12**, 169–171
431. Metzler, D. E. (1977) *Biochemistry; The Chemical Reactions of Living Cells*, Academic Press, New York (pp. 914–915)
432. Oresic, M., and Shalloway, D. (1998) *J. Mol. Biol.* **281**, 31–48
433. Bennetzen, J. L., and Hall, B. D. (1982) *J. Biol. Chem.* **257**, 3026–3031
434. Heurgué-Hamard, V., Mora, L., Guarneros, G., and Buckingham, R. H. (1996) *EMBO J.* **15**, 2826–2833
435. Schmitt, E., Mechulam, Y., Fromant, M., Plateau, P., and Blanquet, S. (1997) *EMBO J.* **16**, 4760–4769
436. Fromant, M., Plateau, P., Schmitt, E., Mechulam, Y., and Blanquet, S. (1999) *Biochemistry* **38**, 4982–4987
436a. van Hoof, A., Frischmeyer, P. A., Dietz, H. C., and Parker, R. (2002) *Science* **295**, 2262–2264
436b. Maquat, L. E. (2002) *Science* **295**, 2221–2222
437. Keiler, K. C., Waller, P. R. H., and Sauer, R. T. (1996) *Science* **271**, 990–994
438. Karzai, A. W., and Sauer, R. T. (2001) *Proc. Natl. Acad. Sci. U.S.A.* **98**, 3040–3044
438a. Roche, E. D., and Sauer, R. T. (2001) *J. Biol. Chem.* **276**, 28509–28515
438b. Gillet, R., and Felden, B. (2001) *EMBO J.* **20**, 2966–2976
439. Nameki, N., Tadaki, T., Muto, A., and Himeno, H. (1999) *J. Mol. Biol.* **289**, 1–7
440. Barends, S., Wower, J., and Kraal, B. (2000) *Biochemistry* **39**, 2652–2658
440a. Barends, S., Karzai, A. W., Sauer, R. T., and Wower, J. (2001) *J. Mol. Biol.* **314**, 9–21
440b. Zwieb, C., Guven, S. A., Wower, I. K., and Wower, J. (2001) *Biochemistry* **40**, 9587–9595
440c. Stagg, S. M., Frazer-Abel, A. A., Hagerman, P. J., and Harvey, S. C. (2001) *J. Mol. Biol.* **309**, 727–735
441. Littauer, U. Z., and Inouye, H. (1973) *Ann. Rev. Biochem.* **42**, 439–470
442. Riddle, D. L., and Carbon, J. (1973) *Nature New Biol.* **242**, 230–234
443. Magliery, T. J., Anderson, J. C., and Schultz, P. G. (2001) *J. Mol. Biol.* **307**, 755–769
444. Murgola, E. J., Prather, N. E., Mims, B. H., Pagel, F. T., and Hijazi, K. A. (1983) *Proc. Natl. Acad. Sci. U.S.A.* **80**, 4936–4939
445. Bossi, L., and Smith, D. M. (1984) *Proc. Natl. Acad. Sci. U.S.A.* **81**, 6105–6109
446. Murgola, E. J., Hijazi, K. A., Göringer, H. U., and Dahlberg, A. E. (1988) *Proc. Natl. Acad. Sci. U.S.A.* **85**, 4162–4165
447. Surguchov, A. P. (1988) *Trends Biochem. Sci.* **13**, 120–123
448. Hatfield, D. (1985) *Trends Biochem. Sci.* **10**, 201–204
448a. Srinivasan, G., James, C. M., and Krzycki, J. A. (2002) *Science* **296**, 1459 – 1462
448b. Hao, B., Gong, W., Ferguson, T. K., James, C. M., Krzycki, J. A., and Chan, M. K. (2002) *Science* **296**, 1462 – 1466
449. Engelberg-Kulka, H., and Schoulaker-Schwarz, R. (1988) *Trends Biochem. Sci.* **13**, 419–421
450. Stadtman, T. C. (1996) *Ann. Rev. Biochem.* **65**, 83–100
451. Kryukov, G. V., Kryukov, V. M., and Gladyshev, V. N. (1999) *J. Biol. Chem.* **274**, 33888–33897
452. Lescure, A., Gautheret, D., Carbon, P., and Krol, A. (1999) *J. Biol. Chem.* **274**, 38147–38154
453. Saito, Y., Hayashi, T., Tanaka, A., Watanabe, Y., Suzuki, M., Saito, E., and Takahashi, K. (1999) *J. Biol. Chem.* **274**, 2866–2871
454. Low, S. C., and Berry, M. J. (1996) *Trends Biochem. Sci.* **21**, 203–208
455. Suppmann, S., Persson, B. C., and Böck, A. (1999) *EMBO J.* **18**, 2284–2293
456. Ioudovitch, A., and Steinberg, S. V. (1999) *J. Mol. Biol.* **290**, 365–371
457. Forchhammer, K., Leinfelder, W., and Böck, A. (1989) *Nature (London)* **342**, 453–456
458. Kromayer, M., Wilting, R., Tormay, P., and Böck, A. (1996) *J. Mol. Biol.* **262**, 413–420
459. Liu, Z., Reches, M., and Engelberg-Kulka, H. (1999) *J. Mol. Biol.* **294**, 1073–1086
460. Rother, M., Wilting, R., Commans, S., and Böck, A. (2000) *J. Mol. Biol.* **299**, 351–358
461. Low, S. C., Grundner-Culemann, E., Harney, J. W., and Berry, M. J. (2000) *EMBO J.* **19**, 6882–6890
461a. Nasim, M. T., Jaenecke, S., Belduz, A., Kollmus, H., Flohé, L., and McCarthy, J. E. G. (2000) *J. Biol. Chem.* **275**, 14846 – 14852
462. Copeland, P. R., Fletcher, J. E., Carlson, B. A., Hatfield, D. L., and Driscoll, D. M. (2000) *EMBO J.* **19**, 306–314
463. Normanly, J., Masson, J.-M., Kleina, L. G., Abelson, J., and Miller, J. H. (1986) *Proc. Natl. Acad. Sci. U.S.A.* **83**, 6548–6552
464. Noren, C. J., Anthony-Cahill, S. J., Griffith, M. C., and Schultz, P. G. (1989) *Science* **244**, 182–188
465. Anthony-Cahill, S. J., Griffith, M. C., Noren, C. J., Suich, D. J., and Schultz, P. G. (1989) *Trends Biochem. Sci.* **14**, 400–403
466. Mendel, D., Ellman, J. A., Chang, Z., Veenstra, D. L., Kollman, P. A., and Schultz, P. G. (1992) *Science* **266**, 1798–1802
467. Chung, H.-H., Benson, D. R., and Schultz, P. G. (1993) *Science* **259**, 806–809
468. Moore, B., Persson, B. C., Nelson, C. C., Gesteland, R. F., and Atkins, J. F. (2000) *J. Mol. Biol.* **298**, 195–209
469. Hohsaka, T., Ashizuka, Y., Sasaki, H., Murakami, H., and Sisido, M. (1999) *J. Am. Chem. Soc.* **121**, 12194–12195
470. Kowal, A. K., Köhrer, C., and RajBhandary, U. L. (2001) *Proc. Natl. Acad. Sci. U.S.A.* **98**, 2268–2273
471. Wang, L., Brock, A., Herberich, B., and Schultz, P. G. (2001) *Science* **292**, 498–500
472. Döring, V., Mootz, H. D., Nangle, L. A., Hendrickson, T. L., de Crécy-Lagard, V., Schimmel, P., and Marlière, P. (2001) *Science* **292**, 501–504
473. Bain, J. D., Switzer, C., Chamberlin, A. R., and Benner, S. A. (1992) *Nature (London)* **356**, 537–539
474. Temple, G. F., Dozy, A. M., Roy, K. L., and Kan, Y. W. (1982) *Nature (London)* **296**, 537–540
475. Ho, Y.-S., and Kan, Y. W. (1987) *Proc. Natl. Acad. Sci. U.S.A.* **84**, 2185–2188
476. Ryoji, M., Hsia, K., and Kaji, A. (1983) *Trends Biochem. Sci.* **8**, 88–90
477. Hatfield, D., and Oroszlan, S. (1990) *Trends Biochem. Sci.* **15**, 186–190
478. Jacks, T., and Varmus, H. E. (1985) *Science* **230**, 1237–1242
479. Jacks, T., Power, M. D., Masiarz, F. R., Luciw, P. A., Barr, P. J., and Varmus, H. E. (1988) *Nature (London)* **331**, 280–283
480. Varmus, H. (1988) *Science* **240**, 1427–1434
481. Alam, S. L., Atkins, J. F., and Gesteland, R. F. (1999) *Proc. Natl. Acad. Sci. U.S.A.* **96**, 14177–14179
482. Gesteland, R. F., and Atkins, J. F. (1996) *Ann. Rev. Biochem.* **65**, 741–768
483. Kastelein, R. A., Remaut, E., Fiers, W., and van Duin, J. (1982) *Nature (London)* **295**, 35–41
484. Dayhuff, T. J., Atkins, J. F., and Gesteland, R. F. (1986) *J. Biol. Chem.* **261**, 7491–7500
484a. Harger, J. W., Meskauskas, A., and Dinman, J. D. (2002) *Trends Biochem. Sci.* **27**, 448 – 454
485. Spanjaard, R. A., and van Duin, J. (1988) *Proc. Natl. Acad. Sci. U.S.A.* **85**, 7967–7971
486. Klovins, J., and van Duin, J. (1999) *J. Mol. Biol.* **294**, 875–884
487. Barak, Z., Lindsley, D., and Gallant, J. (1996) *J. Mol. Biol.* **256**, 676–684
487a. Chamorro, M., Parkin, N., and Varmus, H. E. (1992) *Proc. Natl. Acad. Sci. U.S.A.* **89**, 713–717
487b. Chen, X., Chamorro, M., Lee, S. I., Shen, L. X., Hines, J. V., Tinoco, I., Jr., and Varmus, H. E. (1995) *EMBO J.* **14**, 842–852
487c. Chen, X., Kang, H., Shen, L. X., Chamorro, M., Varmus, H. E., and Tinoco, I., Jr. (1996) *J. Mol. Biol.* **260**, 479–483
488. Giedroc, D. P., Theimer, C. A., and Nixon, P. L. (2000) *J. Mol. Biol.* **298**, 167–185
489. Marczinke, B., Fisher, R., Vidakovic, M., Bloys, A. J., and Brierley, I. (1998) *J. Mol. Biol.* **284**, 205–225
490. Marczinke, B., Hagervall, T., and Brierley, I. (2000) *J. Mol. Biol.* **295**, 179–191
491. Liphardt, J., Napthine, S., Kontos, H., and Brierley, I. (1999) *J. Mol. Biol.* **288**, 321–335
492. Kim, Y.-G., Su, L., Maas, S., O'Neill, A., and Rich, A. (1999) *Proc. Natl. Acad. Sci. U.S.A.* **96**, 14234–14239
493. Matsufuji, S., Matsufuji, T., Wills, N. M., Gesteland, R. F., and Atkins, J. F. (1996) *EMBO J.* **15**, 1360–1370
494. Adamski, F. M., Atkins, J. F., and Gesteland, R. F. (1996) *J. Mol. Biol.* **261**, 357–371
495. Herr, A. J., Gesteland, R. F., and Atkins, J. F. (2000) *EMBO J.* **19**, 2671–2680
495a. Herr, A. J., Wills, N. M., Nelson, C. C., Gesteland, R. F., and Atkins, J. F. (2001) *J. Mol. Biol.* **311**, 445–452
496. Weissmann, C., Billeter, M. A., Goodman, H. M., Hindley, J., and Weber, H. (1973) *Ann. Rev. Biochem.* **42**, 303–328
497. Beekwilder, J., Nieuwenhuizen, R., Poot, R., and van Duin, J. (1996) *J. Mol. Biol.* **256**, 8–19
498. Fiers, W., Contreras, R., Duerinck, F., Haegeman, G., Iserentant, D., Merregaert, J., Min Jou, W., Molemans, F., Raeymaekers, A., Van den Berghe, A., Volckaert, G., and Ysebaert, M. (1976) *Nature (London)* **260**, 500–507
499. Min Jou, W., Haegeman, G., Ysebaert, M., and Fiers, W. (1972) *Nature (London)* **237**, 82–88
500. Fiers, W., Contreras, R., Duerinck, F., Haegeman, G. H., Iserentant, D., Merregaert, J., Min Jou, W., Molemans, F., Raeymaekers, A., Van den Berghe, A., Volckaert, C., and Ysebaert, M. (1976) *Nature (London)* **260**, 500–507
501. Fiers, W., Contreras, R., Duerinck, F., Haegeman, G., Merregaert, J., Min Jou, W., Raeymakers, A., Volckaert, G., Ysebaert, M., Van de Kerckhove, J., Nolf, F., and Van Montagu, M. (1975) *Nature (London)* **256**, 273–278
502. Jackson, R. J., Howell, M. T., and Kaminski, A. (1990) *Trends Biochem. Sci.* **15**, 477–483
503. Pilipenko, E. V., Gmyl, A. P., Maslova, S. V., Belov, G. A., Sinyakov, A. N., Huang, M., Brown, T. D. K., and Agol, V. I. (1994) *J. Mol. Biol.* **241**, 398–414
504. Gan, W., and Rhoads, R. E. (1996) *J. Biol. Chem.* **271**, 623–626

References

505. Kieft, J. S., Zhou, K., Jubin, R., Murray, M. G., Lau, J. Y. N., and Doudna, J. A. (1999) *J. Mol. Biol.* **292**, 513–529
506. Lott, W. B., Takyar, S. S., Tuppen, J., Crawford, D. H. G., Harrison, M., Sloots, T. P., and Gowans, E. J. (2001) *Proc. Natl. Acad. Sci. U.S.A.* **98**, 4916–4921
506a. Guo, L., Allen, E., and Miller, W. A. (2000) *RNA* **6**, 1808–1820
507. Kaplan, S., Atherly, A. G., and Barrett, A. (1973) *Proc. Natl. Acad. Sci. U.S.A.* **70**, 689–692
508. Cozzone, A. J. (1981) *Trends Biochem. Sci.* **6**, 108–110
509. Baracchini, E., and Bremer, H. (1988) *J. Biol. Chem.* **263**, 2597–2602
510. Murray, K. D., and Bremer, H. (1996) *J. Mol. Biol.* **259**, 41–57
511. Metzger, S., Dror,, IB, Aizenman, E., Schreiber, G., Toone, M., Friesen, J. D., Cashel, M., and Glaser, G. (1988) *J. Biol. Chem.* **263**, 15699–15704
512. Hernandez, V. J., and Cashel, M. (1995) *J. Mol. Biol.* **252**, 536–549
512a. Barker, M. M., Gaal, T., Josaitis, C. A., and Gourse, R. L. (2001) *J. Mol. Biol.* **305**, 673–688
512b. Gottesman, S., and Maurizi, M. R. (2001) *Science* **293**, 614 – 615
513. Wagner, E. G. H., and Kurland, C. G. (1980) *Biochemistry* **19**, 1234–1240
514. Nègre, D., Cortay, J.-C., Donini, P., and Cozzone, A. J. (1989) *Biochemistry* **28**, 1814–1819
515. Marianovsky, I., Aizenman, E., Engelberg-Kulka, H., and Glaser, G. (2001) *J. Biol. Chem.* **276**, 5975–5984
516. Swarbrick, J. D., Bashtannyk, T., Maksel, D., Zhang, X.-R., Blackburn, G. M., Gayler, K. R., and Gooley, P. R. (2000) *J. Mol. Biol.* **302**, 1165–1177
517. Kim, B. K., Zamecnik, P., Taylor, G., Guo, M. J., and Blackburn, G. M. (1992) *Proc. Natl. Acad. Sci. U.S.A.* **89**, 11056–11058
518. Luo, J., Jankowski, J., Knobloch, M., van der Giet, M., Gardanis, K., Russ, T., Vahlensieck, U., Neumann, J., Schmitz, W., Tepel, M., Deng, M. C., Zidek, W., and Schlüter, H. (1999) *FASEB J.* **13**, 695–705
519. Ingram, S. W., Stratemann, S. A., and Barnes, L. D. (1999) *Biochemistry* **38**, 3649–3655
520. Cartwright, J. L., and McLennan, A. G. (1999) *J. Biol. Chem.* **274**, 8604–8610
521. Yen, T. J., Machlin, P. S., and Cleveland, D. W. (1988) *Nature (London)* **334**, 580–585
522. Hunt, T. (1988) *Nature (London)* **334**, 567–568
523. Cleveland, D. W. (1988) *Trends Biochem. Sci.* **13**, 339–343
524. Louie, K., and Dowhan, W. (1980) *J. Biol. Chem.* **255**, 1124–1127
525. Fedorov, A. N., and Baldwin, T. O. (1997) *J. Biol. Chem.* **272**, 32715–32718
525a. Patzelt, H., Rüdiger, S., Brehmer, D., Kramer, G., Vorderwülbecke, S., Schaffitzel, E., Waitz, A., Hesterkamp, T., Dong, L., Schneider-Mergener, J., Bukau, B., and Deuerling, E. (2001) *Proc. Natl. Acad. Sci. U.S.A.* **98**, 14244–14249
525b. Hartl, F. U., and Hayer-Hartl, M. (2002) *Science* **295**, 1852–1858
525c. Stoller, G., Rücknagel, K. P., Nierhaus, K. H., Schmid, F. X., Fischer, G., and Rahfeld, J.-U. (1995) *EMBO J.* **14**, 4939–4948
525d. Maier, R., Scholz, C., and Schmid, F. X. (2001) *J. Mol. Biol.* **314**, 1181–1190
525e. Richardson, A., Landry, S. J., and Georgopoulos, C. (1998) *Trends Biochem. Sci.* **23**, 138–143
525f. Keskin, O., Bahar, I., Flatow, D., Covell, D. G., and Jernigan, R. L. (2002) *Biochemistry* **41**, 491–501
525g. Flaux, J., Bertelsen, E. B., Horwich, A. L., and Wüthrich, K. (2002) *Nature (London)* **418**, 207–211
525h. Llorca, O., Martín-Benito, J., Grantham, J., Ritco-Vonsovici, M., Willison, K. R., Carrascosa, J. L., and Valpuesta, J. M. (2001) *EMBO J.* **20**, 4065–4075
526. Ruddon, R. W., and Bedows, E. (1997) *J. Biol. Chem.* **272**, 3125–3128
527. Ellis, R. J. (2000) *Trends Biochem. Sci.* **25**, 210–212
528. Sakikawa, C., Taguchi, H., Makino, Y., and Yoshida, M. (1999) *J. Biol. Chem.* **274**, 21251–21256
529. Ma, J., Sigler, P. B., Xu, Z., and Karplus, M. (2000) *J. Mol. Biol.* **302**, 303–313
530. Schäfer, U., Beck, K., and Müller, M. (1999) *J. Biol. Chem.* **274**, 24567–24574
531. Jones, C. H., Danese, P. N., Pinkner, J. S., Silhavy, T. J., and Hultgren, S. J. (1997) *EMBO J.* **16**, 6394–6406
532. Frydman, J., and Höhfeld, J. (1997) *Trends Biochem. Sci.* **22**, 87–92
533. Rüdiger, S., Schneider-Mergener, J., and Bukau, B. (2001) *EMBO J.* **20**, 1042–1050
534. Wilson, M. R., and Easterbrook-Smith, S. B. (2000) *Trends Biochem. Sci.* **25**, 95–98
535. Zha, J., Weiler, S., Oh, K. J., Wei, M. C., and Korsmeyer, S. J. (2000) *Science* **290**, 1761–1765
536. Schubert, U., Antón, L. C., Gibbs, J., Norbury, C. C., Yewdell, J. W., and Bennink, J. R. (2000) *Nature (London)* **404**, 770–774
537. Turner, G. C., and Varshavsky, A. (2000) *Science* **289**, 2117–2120
538. Scheffner, M., and Whitaker, N. J. (2001) *Nature (London)* **410**, 882–883
539. Wickner, S., Maurizi, M. R., and Gottesman, S. (1999) *Science* **286**, 1888–1893
540. Bonifacino, J. S. (1996) *Nature (London)* **384**, 405–406
541. Arfin, S. M., and Bradshaw, R. A. (1988) *Biochemistry* **27**, 7979–7984
542. Rajagopalan, P. T. R., Grimme, S., and Pei, D. (2000) *Biochemistry* **39**, 779–790
542a. Deng, H., Callender, R., Zhu, J., Nguyen, K. T., and Pei, D. (2002) *Biochemistry* **41**, 10563–10569
543. Solbiati, J., Chapman-Smith, A., Miller, J. L., Miller, C. G., and Cronan, J. E., JR. (1999) *J. Mol. Biol.* **290**, 607–614
544. Giglione, C., Serero, A., Pierre, M., Boisson, B., and Meinnel, T. (2000) *EMBO J.* **19**, 5916–5929
544a. Serero, A., Giglione, C., and Meinnel, T. (2001) *J. Mol. Biol.* **314**, 695–708
545. Boissel, J.-P., Kasper, T. J., and Bunn, H. F. (1988) *J. Biol. Chem.* **263**, 8443–8449
546. Ben-Bassat, A., and Bauer, K. (1987) *Nature (London)* **326**, 315
547. Polevoda, B., and Sherman, F. (2000) *J. Biol. Chem.* **275**, 36479–36482
548. Strauch, A. R., and Rubenstein, P. A. (1984) *J. Biol. Chem.* **259**, 7224–7229
549. Farazi, T. A., Waksman, G., and Gordon, J. I. (2001) *Biochemistry* **40**, 6335–6343
550. Utsumi, T., Sato, M., Nakano, K., Takemura, D., Iwata, H., and Ishisaka, R. (2001) *J. Biol. Chem.* **276**, 10505–10513
551. Wilcox, C., Hu, J.-S., and Olson, E. N. (1987) *Science* **238**, 1275–1278
552. Towler, D. A., Gordon, J. I., Adams, S. P., and Glaser, L. (1988) *Ann. Rev. Biochem.* **57**, 69–99
553. Wu, Z., Demma, M., Strickland, C. L., Radisky, E. S., Poulter, C. D., Le, H. V., and Windsor, W. T. (1999) *Biochemistry* **38**, 11239–11249
553a. Matsuyama, S., Yokota, N., and Tokuda, H. (1997) *EMBO J.* **16**, 6947 – 6955
554. Aletta, J. M., Cimato, T. R., and Ettinger, M. J. (1998) *Trends Biochem. Sci.* **23**, 89–91
555. Zobel-Thropp, P., Gary, J. D., and Clarke, S. (1998) *J. Biol. Chem.* **273**, 29283–29286
556. Chapman-Smith, A., and Cronan, J. E., Jr. (1999) *Trends Biochem. Sci.* **24**, 359–363
557. Campeau, E., and Gravel, R. A. (2001) *J. Biol. Chem.* **276**, 12310–12316
558. Craig, A. G., Jimenez, E. C., Dykert, J., Nielsen, D. B., Gulyas, J., Abogadie, F. C., Porter, J., Rivier, J. E., Cruz, L. J., Olivera, B. M., and McIntosh, J. M. (1997) *J. Biol. Chem.* **272**, 4689–4698
559. Dierks, T., Lecca, M. R., Schlotterhose, P., Schmidt, B., and von Figura, K. (1999) *EMBO J.* **18**, 2084–2091
560. Stubbe, J. (1996) *Science* **274**, 1152–1153
561. Kelleher, N. L., Hendrickson, C. L., and Walsh, C. T. (1999) *Biochemistry* **38**, 15623–15630
562. Kupke, T., Kempter, C., Gnau, V., Jung, G., and Götz, F. (1994) *J. Biol. Chem.* **269**, 5653–5659
563. Rapoport, T. A. (1992) *Science* **258**, 931–935
564. Corsi, A. K., and Schekman, R. (1996) *J. Biol. Chem.* **271**, 30299–30302
565. Freymann, D. M., Keenan, R. J., Stroud, R. M., and Walter, P. (1997) *Nature (London)* **385**, 361–364
566. Montoya, G., Svensson, C., Luirink, J., and Sinning, I. (1997) *Nature (London)* **385**, 365–368
567. Peluso, P., Herschlag, D., Nock, S., Freymann, D. M., Johnson, A. E., and Walter, P. (2000) *Science* **288**, 1640–1643
568. Batey, R. T., Rambo, R. P., Lucast, L., Rha, B., and Doudna, J. A. (2000) *Science* **287**, 1232–1239
569. Batey, R. T., Sagar, M. B., and Doudna, J. A. (2001) *J. Mol. Biol.* **307**, 229–246
570. Diener, J. L., and Wilson, C. (2000) *Biochemistry* **39**, 12862–12874
571. Walter, P., Keenan, R., and Schmitz, U. (2000) *Science* **287**, 1212–1213
571a. Peluso, P., Shan, S.-o, Nock, S., Herschlag, D., and Walter, P. (2001) *Biochemistry* **40**, 15224–15233
571b. Pool, M. R., Stumm, J., Fulga, T. A., Sinning, I., and Dobberstein, B. (2002) *Science* **297**, 1345–1348
572. de Leeuw, E., te Kaat, K., Moser, C., Menestrina, G., Demel, R., de Kruijff, B., Oudega, B., Luirink, J., and Sinning, I. (2000) *EMBO J.* **19**, 531–541
573. Fulga, T. A., Sinning, I., Dobberstein, B., and Pool, M. R. (2001) *EMBO J.* **20**, 2338–2347
574. Bibi, E., Herskovits, A. A., Bochkareva, E. S., and Zelazny, A. (2001) *Trends Biochem. Sci.* **26**, 15–16
575. Lütcke, H. (1995) *Eur. J. Biochem.* **228**, 531–550
575a. Huang, Q., Abdulrahman, S., Yin, J., and Zwieb, C. (2002) *Biochemistry* **41**, 11362 – 11371
575b. Hainzl, T., Huang, S., and Sauer-Eriksson, A. E. (2002) *Nature (London)* **417**, 767 – 771
576. Weichenrieder, O., Wild, K., Strub, K., and Cusack, S. (2000) *Nature (London)* **408**, 167–173
576a. Beckman, R. (2001) *Cell* **107**, 361–72
577. Young, B. P., Craven, R. A., Reid, P. J., Willer, M., and Stirling, C. J. (2001) *EMBO J.* **20**, 262–271
578. Römisch, K. (2001) *Trends Biochem. Sci.* **26**, 13
579. Beswick, V., Baleux, F., Huynh-Dinh, T., Képès, F., Neumann, J.-M., and Sanson, A. (1996) *Biochemistry* **35**, 14717–14724
580. Tschantz, W. R., Paetzel, M., Cao, G., Suciu, D., Inouye, M., and Dalbey, R. E. (1995) *Biochemistry* **34**, 3935–3941
581. Chatterjee, S., Suciu, D., Dalbey, R. E., Kahn, P. C., and Inouye, M. (1995) *J. Mol. Biol.* **245**, 311–314

References

582. VanValkenburgh, C., Chen, X., Mullins, C., Fang, H., and Green, N. (1999) *J. Biol. Chem.* **274**, 11519–11525

582a. Weihofen, A., Binns, K., Lemberg, M. K., Ashman, K., and Martoglio, B. (2002) *Science* **296**, 2215 – 2218

583. Suzuki, T., Yan, Q., and Lennarz, W. J. (1998) *J. Biol. Chem.* **273**, 10083–10086

584. Netzer, W. J., and Hartl, F. U. (1997) *Nature (London)* **388**, 343–349

585. Wickner, W., and Leonard, M. R. (1996) *J. Biol. Chem.* **271**, 29514–29516

586. Kim, J., Miller, A., Wang, L., Müller, J. P., and Kendall, D. A. (2001) *Biochemistry* **40**, 3674–3680

587. Verner, K., and Schatz, G. (1988) *Science* **241**, 1307–1313

588. van der Does, C., Manting, E. H., Kaufmann, A., Lutz, M., and Driessen, A. J. M. (1998) *Biochemistry* **37**, 201–210

589. Collinson, I., Breyton, C., Duong, F., Tziatzios, C., Schubert, D., Or, E., Rapoport, T., and Kühlbrandt, W. (2001) *EMBO J.* **20**, 2462–2471

590. Sianidis, G., Karamanou, S., Vrontou, E., Boulias, K., Repanas, K., Kyrpides, N., Politou, A. S., and Economou, A. (2001) *EMBO J.* **20**, 961–970

591. Kluger, R., and Smyth, T. (1981) *J. Am. Chem. Soc.* **103**, 1216–1218

592. Yahr, T. L., and Wickner, W. T. (2000) *EMBO J.* **19**, 4393–4401

593. Driessen, A. J. M. (1992) *Trends Biochem. Sci.* **17**, 219–223

594. Dalbey, R. E., and Robinson, C. (1999) *Trends Biochem. Sci.* **24**, 17–22

595. Delgado-Partin, V. M., and Dalbey, R. E. (1998) *J. Biol. Chem.* **273**, 9927–9934

596. Gafvelin, G., Sakaguchi, M., Andersson, H., and von Heijne, G. (1997) *J. Biol. Chem.* **272**, 6119–6127

597. Rietveld, A. G., Koorengevel, M. C., and de Kruijff, B. (1995) *EMBO J.* **14**, 5506–5513

598. Jongbloed, J. D. H., Martin, U., Antelmann, H., Hecker, M., Tjalsma, H., Venema, G., Bron, S., van Dijl, J. M., and Müller, J. (2000) *J. Biol. Chem.* **275**, 41350–41357

599. Sambasivarao, D., Turner, R. J., Simala-Grant, J. L., Shaw, G., Hu, J., and Weiner, J. H. (2000) *J. Biol. Chem.* **275**, 22526–22531

600. Koronakis, V., Sharff, A., Koronakis, E., Luisi, B., and Hughes, C. (2000) *Nature (London)* **405**, 914–919

601. Buchanan, S. K. (2001) *Trends Biochem. Sci.* **26**, 3–6

602. Reits, E. A. J., Vos, J. C., Grommé, M., and Neefjes, J. (2000) *Nature (London)* **404**, 774–778

603. Samuelson, J. C., Chen, M., Jiang, F., Möller, I., Wiedmann, M., Kuhn, A., Phillips, G. J., and Dalbey, R. E. (2000) *Nature (London)* **406**, 637–640

604. Tokatlidis, K., and Schatz, G. (1999) *J. Biol. Chem.* **274**, 35285–35288

605. Schatz, G. (1998) *Nature (London)* **395**, 439–440

606. McNew, J. A., and Goodman, J. M. (1996) *Trends Biochem. Sci.* **21**, 54–58

607. Subramani, S. (1996) *J. Biol. Chem.* **271**, 32483–32486

608. Hell, K., Neupert, W., and Stuart, R. A. (2001) *EMBO J.* **20**, 1281–1288

609. von Heijne, G. (1989) *Nature (London)* **341**, 456–458

610. Kuhn, A., Wickner, W., and Kreil, G. (1986) *Nature (London)* **322**, 335–339

611. Bibi, E. (1998) *Trends Biochem. Sci.* **23**, 51–55

612. Dalbey, R. E., Kahn, A., and Wickner, W. (1987) *J. Biol. Chem.* **262**, 13241–13245

613. Yamane, K., Ichehara, S., and Mizushima, S. (1987) *J. Biol. Chem.* **263**, 2358–2362

614. Meijer, A. B., Spruijt, R. B., Wolfs, C. J. A. M., and Hemminga, M. A. (2001) *Biochemistry* **40**, 5081–5086

615. Soekarjo, M., Eisenhawer, M., Kuhn, A., and Vogel, H. (1996) *Biochemistry* **35**, 1232–1241

616. Moore, K. E., and Miura, S. (1988) *J. Biol. Chem.* **263**, 11575–11583

617. von Heijne, G., Wickner, W., and Dalbey, R. E. (1988) *Proc. Natl. Acad. Sci. U.S.A.* **85**, 3363–3366

617a. Anderson, R. G. W., and Jacobson, K. (2002) *Science* **296**, 1821– 1825

618. Thomas-Reetz, A. C., and De Camilli, P. (1994) *FASEB J.* **8**, 209–216

619. Rothman, J. E., and Wieland, F. T. (1996) *Science* **272**, 227–234

620. Nicholls, D. G. (1994) *Proteins, Transmitters and Synapses*, Blackwell Scientific Publications, Oxford

621. Bean, A. J., Zhang, X., and Hökfelt, T. (1994) *FASEB J.* **8**, 630–638

622. Huttner, W. B., Gerdes, H.-H., and Rosa, P. (1991) *Trends Biochem. Sci.* **16**, 27–30

623. Guo, W., Grant, A., and Novick, P. (1999) *J. Biol. Chem.* **274**, 23558–23564

624. Peters, C., Bayer, M. J., Bühler, S., Andersen, J. S., Mann, M., and Mayer, A. (2001) *Nature (London)* **409**, 581–588

624a. Pain, R., ed. (2000) *Mechanisms of Protein Folding*, 2nd ed., Oxford Univ. Press, London

625. Holst, B., Bruun, A. W., Kielland-Brandt, M. C., and Winther, J. R. (1996) *EMBO J.* **15**, 3538–3546

626. Pirkl, F., and Buchner, J. (2001) *J. Mol. Biol.* **308**, 795–806

626a. Gilbert, H. F. (1997) *J. Biol. Chem.* **272**, 29399 – 29402

627. Qiao, Z.-S., Guo, Z.-Y., and Feng, Y.-M. (2001) *Biochemistry* **40**, 2662–2668

628. Tu, B. P., Ho-Schleyer, S. C., Travers, K. J., and Weissman, J. S. (2000) *Science* **290**, 1571–1573

629. Miranker, A., Robinson, C. V., Radford, S. E., and Dobson, C. M. (1996) *FASEB J.* **10**, 93–101

630. Balbach, J., Steegborn, C., Schindler, T., and Schmid, F. X. (1999) *J. Mol. Biol.* **285**, 829–842

631. Arrington, C. B., Teesch, L. M., and Robertson, A. D. (1999) *J. Mol. Biol.* **285**, 1265–1275

632. Rumbley, J., Hoang, L., Mayne, L., and Englander, S. W. (2001) *Proc. Natl. Acad. Sci. U.S.A.* **98**, 105–112

633. Steegborn, C., Schneider-Hassloff, H., Zeeb, M., and Balbach, J. (2000) *Biochemistry* **39**, 7910–7919

634. Arai, M., Ikura, T., Semisotnov, G. V., Kihara, H., Amemiya, Y., and Kuwajima, K. (1998) *J. Mol. Biol.* **275**, 149–162

635. Panick, G., Malessa, R., Winter, R., Rapp, G., Frye, K. J., and Royer, C. A. (1998) *J. Mol. Biol.* **275**, 389–402

636. Radford, S. E. (2000) *Trends Biochem. Sci.* **25**, 611–618

637. Anfinsen, C. B. (1973) *Science* **181**, 223–230

638. Dill, K. A. (1990) *Biochemistry* **29**, 7133–7155

639. Dinner, A. R., Sali, A., Smith, L. J., Dobson, C. M., and Karplus, M. (2000) *Trends Biochem. Sci.* **25**, 331–339

640. Baker, D. (2000) *Nature (London)* **405**, 39–42

641. Clarke, D. T., Doig, A. J., Stapley, B. J., and Jones, G. R. (1999) *Proc. Natl. Acad. Sci. U.S.A.* **96**, 7232–7237

642. Nölting, B., Golbik, R., Neira, J. L., Soler-Gonzalez, A. S., Schreiber, G., and Fersht, A. R. (1997) *Proc. Natl. Acad. Sci. U.S.A.* **94**, 826–830

643. Zhou, Y., and Karplus, M. (1999) *J. Mol. Biol.* **293**, 917–951

644. Damaschun, G., Damaschun, H., Gast, K., and Zirwer, D. (1999) *J. Mol. Biol.* **291**, 715–725

645. Pace, C. N. (1990) *Trends Biochem. Sci.* **15**, 14–17

646. Rees, D. C., and Robertson, A. D. (2001) *Protein Sci.* **10**, 1187– 1194

647. Andersen, N. H., Cort, J. R., Liu, Z., Sjoberg, S. J., and Tong, H. (1996) *J. Am. Chem. Soc.* **118**, 10309–10310

648. Swint-Kruse, L., and Robertson, A. D. (1995) *Biochemistry* **34**, 4724–4732

649. Yang, A.-S., and Honig, B. (1993) *J. Mol. Biol.* **231**, 459–474

650. Yang, A.-S., and Honig, B. (1994) *J. Mol. Biol.* **237**, 602–614

651. Pace, C. N. (2001) *Biochemistry* **40**, 310–313

652. Sippl, M. J. (1996) *J. Mol. Biol.* **260**, 644–648

653. Shortle, D. (1996) *FASEB J.* **10**, 27–34

654. Häckel, M., Hinz, H.-J., and Hedwig, G. R. (1999) *J. Mol. Biol.* **291**, 197–213

655. Makhatadze, G. I., and Privalov, P. L. (1996) *Protein Sci.* **5**, 507–510

655a. Dunker, A. K., Brown, C. J., Lawson, J. D., Iakoucheva, L. M., and Obradovic, Z. (2002) *Biochemistry* **41**, 6575 – 6582

655b. Dobson, C. M. (1999) *Trends Biochem. Sci.* **24**, 329 – 332

655c. Stevens, F. J., Pokkuluri, P. R., and Schiffer, M. (2000) *Biochemistry* **39**, 15291 – 15296

655d. Ellis, R. J., and Pinheiro, T. J. T. (2002) *Nature (London)* **416**, 483 – 484

655e. Oxenoid, K., Sönnichsen, F. D., and Sanders, C. R. (2001) *Biochemistry* **40**, 5111 – 5118

656. Huang, L., Kinnucan, E., Wang, G., Beaudenon, S., Howley, P. M., Huibregtse, J. M., and Pavletich, N. P. (1999) *Science* **286**, 1321–1326

657. Reyes, J. C. (2001) *Trends Biochem. Sci.* **26**, 18–20

658. Johnston, S. C., Riddle, S. M., Cohen, R. E., and Hill, C. P. (1999) *EMBO J.* **18**, 3877–3887

659. Whitby, F. G., Masters, E. I., Kramer, L., Knowlton, J. R., Yao, Y., Wang, C. C., and Hill, C. P. (2000) *Nature (London)* **408**, 115–120

660. Joazeiro, C. A. P., and Hunter, T. (2000) *Science* **289**, 2061–2062

661. Satoh, K., Sasajima, H., Nyoumura, K.-i, Yokosawa, H., and Sawada, H. (2001) *Biochemistry* **40**, 314–319

662. Byrd, C., Turner, G. C., and Varshavsky, A. (1998) *EMBO J.* **17**, 269–277

663. Davydov, I. V., and Varchavsky, A. (2000) *J. Biol. Chem.* **275**, 22931–22941

663a. Marx, J. (2002) *Science* **297**, 1792 – 1794

663b. Verma, R., Aravind, L., Oania, R., McDonald, W. H., Yates, J. R., III, Koonin, E. V., and Deshaies, R. J. (2002) *Science* **298**, 611 – 615

664. Kuttler, C., Nussbaum, A. K., Dick, T. P., Rammensee, H.-G., Schild, H., and Hadeler, K.-P. (2000) *J. Mol. Biol.* **298**, 417–429

665. Fineschi, B., and Miller, J. (1997) *Trends Biochem. Sci.* **22**, 377–382

666. Cascio, P., Hilton, C., Kisselev, A. F., Rock, K. L., and Goldberg, A. L. (2001) *EMBO J.* **20**, 2357–2366

667. Schild, H., and Rammensee, H.-G. (2000) *Nature (London)* **404**, 709–710

667a. Cyr, D. M., Höhfeld, J., and Patterson, C. (2002) *Trends Biochem. Sci.* **27**, 368 – 375

668. Dell'Angelica, E. C., Mullins, C., Caplan, S., and Bonifacino, J. S. (2000) *FASEB J.* **14**, 1265–1278

669. Jentsch, S., and Ulrich, H. D. (1998) *Nature (London)* **395**, 321–323

670. Wolf, Y. I., Grishin, N. V., and Koonin, E. V. (2000) *J. Mol. Biol.* **299**, 897–905

671. Simons, K. T., Strauss, C., and Baker, D. (2001) *J. Mol. Biol.* **306**, 1191–1199

672. Takada, S. (1999) *Proc. Natl. Acad. Sci. U.S.A.* **96**, 11698–11700

673. Abbott, A. (1999) *Nature (London)* **402**, 715–720

References

674. Pradet-Balade, B., Boulmé, F., Beug, H., Müllner, E. W., and Garcia-Sanz, J. A. (2001) *Trends Biochem. Sci.* **26**, 225–229
675. Attwood, T. K., and Parry-Smith, D. J. (1999) *Introduction to Bioinformatics*, Longman, Green, New York
676. Spengler, S. J. (2000) *Science* **287**, 1221–1223
677. Attwood, T. K. (2000) *Science* **290**, 471–473
678. Gygi, S. P., Corthals, G. L., Zhang, Y., Rochon, Y., and Aebersold, R. (2000) *Proc. Natl. Acad. Sci. U.S.A.* **97**, 9390–9395
679. MacBeath, G., and Schreiber, S. L. (2000) *Science* **289**, 1760–1763
680. Horn, D. M., Zubarev, R. A., and McLafferty, F. W. (2000) *Proc. Natl. Acad. Sci. U.S.A.* **97**, 10313–10317
680a. Rajagopol, I., and Ahern, K. (2001) *Science* **294**, 2571 – 2573
681. Miyashita, M., Presley, J. M., Buchholz, B. A., Lam, K. S., Lee, Y. M., and Vogel, J. S. (2001) *Proc. Natl. Acad. Sci. U.S.A.* **98**, 4403–4408
682. Mann, M., and Pandey, A. (2001) *Trends Biochem. Sci.* **26**, 54–61
683. Baldauf, S. L., Roger, A. J., Wenk-Siefert, I., and Doolittle, W. F. (2000) *Science* **290**, 972–977
684. Gerstein, M. (1997) *J. Mol. Biol.* **274**, 562–576
685. Gómez, S. M., Nishio, J. N., Faull, K. F., and Whitelegge, J. P. (2002) *Molecular & Cellular Proteomics* **1**, 46 – 59
686. Michels, D. A., Hu, S., Schoenherr, R. M., Eggertson, M. J., and Dovichi, N. J. (2002) *Molecular & Cellular Proteomics* **1**, 69 – 74
687. Smolka, M., Zhou, H., and Aebersold, R. (2002) *Molecular & Cellular Proteomics* **1**, 19 – 29
688. Herbert, B. R., Harry, J. L., Packer, N. H., Gooley, A. A., Pedersen, S. K., and Williams, K. L. (2001) *Trends in Biotechnology* **19**, S3 – S9
689. Ge, Y., Lawhorn, B. G., ElNaggar, M., Strauss, E., Park, J.-H., Begley, T. P., and McLafferty, F. W. (2002) *J. Am. Chem. Soc.* **124**, 672 – 678
690. Griffin, T. J., and Aebersold, R. (2001) *J. Biol. Chem.* **276**, 45497 – 45500
691. Zhu, H., Bilgin, M., Bangham, R., Hall, D., Casamayor, A., Bertone, P., Lan, N., Jansen, R., Bidlingmaier, S., Houfek, T., Mitchell, T., Miller, P., Dean, R. A., Gerstein, M., and Snyder, M. (2001) *Science* **293**, 2101 – 2105
692. Kodadek, T. (2002) *Trends Biochem. Sci.* **27**, 295 – 300
693. Smith, G. P., Patel, S. U., Windass, J. D., Thornton, J. M., Winter, G., and Griffiths, A. D. (1998) *J. Mol. Biol.* **277**, 317 – 332
694. Blagoev, B., Pandey, A. (2001) *Trends Biochem. Sci.* **26**, 639 – 641
695. Yee, A., and 23 other authors. (2002) *Proc. Natl. Acad. Sci. U.S.A.* **99**, 1825 – 1830
696. Riek, R., Fiauz, J., Bertelsen, E. B., Horwich, A. L., and Wüthrich, K. (2002) *J. Am. Chem. Soc.* **124**, 12144 – 12153
697. Tugarinov, V., Muhandiram, R., Ayed, A., and Kay, L. E. (2002) *J. Am. Chem. Soc.* **124**, 10025 – 10035
698. Jhoti, H. (2001) *Trends in Biotechnology* **19**, S67 – S71
699. Thornton, J. (2001) *Trends Biochem. Sci.* **26**, 88–89
700. Koonin, E. V., Wolf, Y. I., and Karev, G. P. (2002) *Nature (London)* **420**, 218 – 223
701. Mallis, R. J., Brazin, K., Jez, J. M., Wilson, E. K., Dieckmann, G. R., Robic, S., and Harrahy, J. (2001) *Trends Biochem. Sci.* **26**, 642 – 643
702. Zhang, C., and Kim, S.-H. (2000) *J. Mol. Biol.* **299**, 1075 – 1089
703. Wimley, W. C. (2002) ***Protein Sci.*** 11, 301 – 312
704. Richardson, J. S., and Richardson, D. C. (2002) *Proc. Natl. Acad. Sci. U.S.A.* **99**, 2754 – 2759
705. Calladine, C. R., Sharff, A., and Luisi, B. (2001) *J. Mol. Biol.* **305**, 603 – 618
706. Chen, C. P., Kernytsky, A., and Rost, B. (2002) *Protein Sci.* **11**, 2774 – 2791
707. Chen, C. P., and Rost, B. (2002) *Protein Sci.* **11**, 2766 – 2773
708. Baker, D., and Sali, A. (2001) *Science* **294**, 93 – 96
709. Marsden, R. L., McGuffin, L. J., and Jones, D. T. (2002) *Protein Sci.* **11**, 2814 – 2824
710. Jiang, W. Baker, M. L., Ludtke, S. J., and Chiu, W. (2001) *J. Mol. Biol.* **308**, 1033 – 1044
711. Taylor, W. R. (2002) *Nature (London)* **416**, 657 – 660
712. Apic, G., Gough, J., and Teichmann, S. A. (2001) *J. Mol. Biol.* **310**, 311 – 325
713. Lee, K. K., Fitch, C. A., and Garcia-Moreno, E. B. (2002) *Protein Sci.* **11**, 1004 – 1016
714. Berezovsky, I. N., and Trifonov, E. N. (2002) *J. Mol. Biol.* **307**, 1419 – 1426
715. Trabi, M., and Craik, D. J. (2002) *Trends Biochem. Sci.* **27**, 132 – 138
716. Shao, Z., and Vollrath, F. (2002) *Nature (London)* **418**, 741
717. Asakura, T., Ashida, J., Yamane, T., Kameda, T., Nakazawa, Y., Ohgo, K., and Komatsu, K. (2001) *J. Mol. Biol.* **306**, 291 – 305
718. van Beek, J. D., Hess, S., Vollrath, F., and Meier, B. H. (2002) *Proc. Natl. Acad. Sci. U.S.A.* **99**, 10266 – 10271
719. Vollrath, F., and Knight, D. P. (2001) *Nature (London)* **410**, 541 – 548
720. Sprinzak, E., and Margalit, H. (2001) *J. Mol. Biol.* **311**, 681 – 692
721. von Mering, C., Krause, R., Snel, B., Cornell, M., Oliver, S. G., Fields, S., and Bork, P. (2002) *Nature (London)* **417**, 399 – 403
722. Gavin, A.-C., and 37 other authors. (2002) *Nature (London)* **415**, 141 – 147
723. Stagljar, I., and Fields, S. (2002) *Trends Biochem. Sci.* **27**, 559 – 563
724. Tong, A. H. Y., Drees, B., Nardelli, G., Bader, G. D., Brannetti, B., Castagnoli, L., Evangelista, M., Ferracuti, S., Nelson, B., Paoluzi, S., Quondam, M., Zucconi, A., Hogue, C. W. V., Fields, S., Boone, C., and Cesareni, G. (2002) *Science* **295**, 321 – 324
725. Cornish-Bowden, A., and Cárdenas, M. L. (2001) *Trends Biochem. Sci.* **26**, 463 – 465
726. Grant, S. G. N., and Husi, H. (2001) *Trends in Biotechnology* **19**, S49 – S54
727. Natsume, T., Nakayama, H., and Isobe, T. (2001) *Trends in Biotechnology* **19**, S28 – S33
728. Ray, P., Pimenta, H., Paulmurugan, R., Berger, F., Phelps, M. E. Iyer, M., and Gambhir, S. S. (2002) *Proc. Natl. Acad. Sci. U.S.A.* **99**, 3105 – 3110
729. Florens, L., and 17 other authors. (2002) *Nature (London)* **419**, 520 – 526
730. Koller, A., Washburn, M. P., Lange, B. M., Andon, N. L., Deciu, C., Haynes, P. A., Hays, L. Schieltz, D., Ulaszek, R., Wei, J., Wolters, D., and Yates, J. R., III. (2002) *Proc. Natl. Acad. Sci. U.S.A.* **99**, 11969 – 11974
731. Remy, I., and Michnick, S. W. (2001) *Proc. Natl. Acad. Sci. U.S.A.* **98**, 7678 – 7683

Study Questions for chapters 28 and 29

1. Describe the role of sigma factors (σ) in transcription by prokaryotic RNA polymerases. What is the effect of the release of σ from the holoenzyme once transcription has been initiated. How would a mutation that prevents a σ factor from dissociating from core RNA polymerase affect the rate of transcription?

2. Explain how histidine biosynthesis is controlled in *E. coli*, a bacterium that has no *his* repressor.

3. One mechanism of transcriptional control in prokaryotes, especially of several operons controlling the biosynthesis of amino acids, is **attenuation**. Briefly describe the mechanism of attenuation. How does the supply of amino acid in the cell affect the process?

4. Is transcription attenuation likely to be an important mechanism of transcriptional regulation in eukaryotic cells?

5. Discuss two main DNA-recognition motifs found in eukaryotic transcription factors. Describe their structures, indicate how they bind to DNA, and discuss how each specifically recognizes its DNA binding site.

Study Questions

6. How can a DNA enhancer sequence located as many as several thousand base pairs from a gene transcription start site influence transcription even if its orientation is reversed?

7. Some eukaryotic DNA viruses code for two or more mRNA transcripts of differing lengths from the same region on the DNA. Suggest an explanation. How do you expect the two translation products of these mRNAs to differ?

8. High salt concentrations weaken the interaction of histones with DNA but have little effect on the binding of many regulatory proteins. Explain this observation in terms of the modes of interaction of the two types of protein.

9. Discuss the changes that must be made in a typical eukaryotic structural gene to allow its protein product to be synthesized in bacteria.

10. List the different types of covalent modification that may be made to tRNA. To ribosomal RNA. To messenger RNA.

11. List various small RNAs and their functions within cells.

12. Some amino acids utilize only one codon of the 64 in the genetic code. Other amino acids use as many as six codons (Tables 5-5, 5-6). What advantages to a cell is provided by utilization of several codons for a single amino acid?

13. In what ways is the genetic code not quite "universal?" What is meant by "editing" of mRNA?

14. Why is it necessary to have "adapters" in the form of tRNAs to read the genetic code during translation?

15. Most nonsense suppressor genes are mutants of tRNA genes. In view of this fact, how can cells survive the presence of such mutations?

16. Explain how the protein synthesizing machinery is able to differentiate the initiation AUG codon from an internal AUG (methionine) codon in prokaryotes. How is this accomplished in eukaryotes?

17. The amino acid sequence of a mature protein sometimes differs from that deduced from the DNA nucleotide sequence of the structural gene for that protein. Discuss three ways by which this may occur.

18. Write out in detail, using structural formulas, the chemical mechanism of synthesis of an aminoacyl-tRNA and of incorporation of the aminoacyl group into a peptide chain being formed by a ribosome.

19. a) Calculate the minimum number of ATP equivalents consumed in the biosynthesis of a 300-amino acid *E. coli* protein, having the N-terminal sequence Ala-Ser-Val-Tyr, from the free amino acids.

    b) Much of this energy involves hydrolysis of GTP. What is the role of this GTP hydrolysis in protein synthesis?

20. How do the polypeptide products produced in the presence of (a) puromycin and (b) streptomycin differ from polypeptides synthesized in the absence of these inhibitors? Explain your answer.

21. What is the significance to protein synthesis of each of the following?
    Shine–Dalgarno sequence
    Signal recognition particle
    proteasome

22. How can useful antibiotics that act on ribosomes kill bacteria but not people?

23. Compare termination of translation in bacteria and in eukaryotes.

24. List some types of error that are likely to be made during protein synthesis. What mechanisms have cells developed to deal with these?

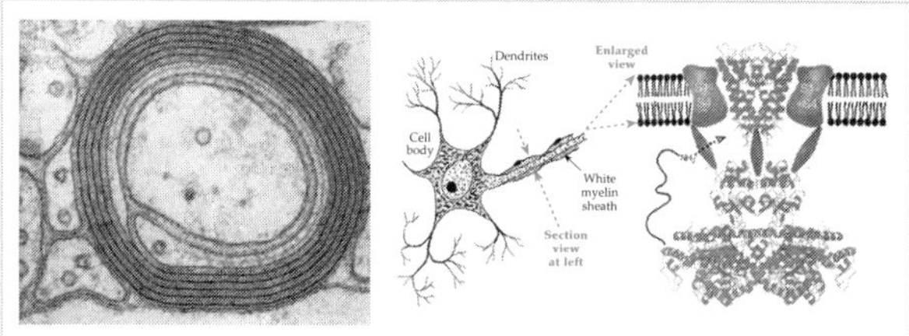

Center: Diagram of the cell body of a neuron with dendrites and a short section of its long myelinated axon (see Fig. 30-8). Left: Electron micrograph of a thin section through an axon showing the myelin sheath formed by the wrapping of the plasma membrane of a neuroglial cell around the axon (see p. 390 and Fig. 30-9). Right: Model of a voltage-regulated K^+ channel in the cell membrane of an axon. The pore, which is formed from four α-subunits, is represented by that of the bacterial pore shown in Fig. 8-21. Also shown are an inner cytoplasmic activation gate consisting of four β-subunits, which are proposed to form ball-and-chain devices that can close the pores in response to voltage changes. From Zhou *et al.* See Fig. 30-18.

Contents

Chemical Communication Between Cells

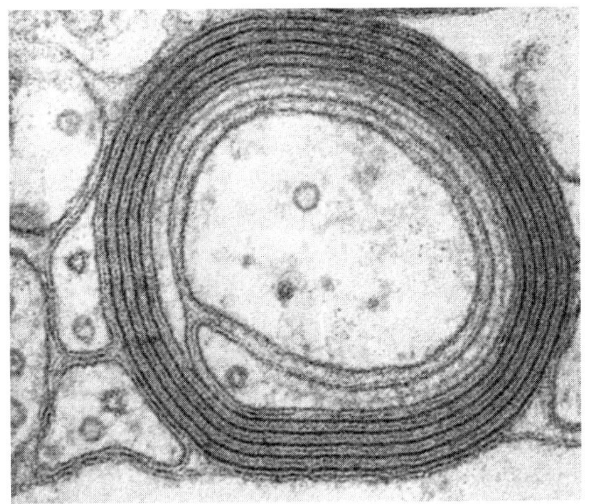

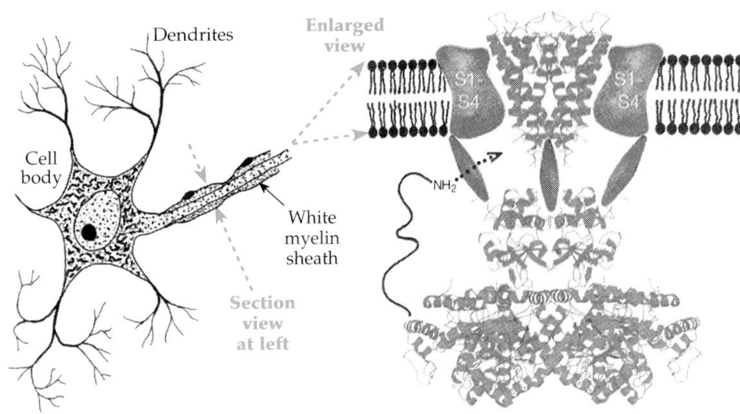

The regulation of growth and metabolism of complex multicellular organisms depends heavily upon chemical messages sent between cells. This includes secretion of hormones into the circulatory system,[1–3] chemical transfer of information through communicating cell junctions, and passage of signals between neurons in the brain. This chapter deals with these matters and also with communication between organisms, i.e., with the biochemistry of ecological relationships. Embryonic development and differentiation of tissues also require communication between cells as does the functioning of the immune system. These topics are considered in Chapters 31 and 32.

A. The Hormones

The term hormone has traditionally been applied to substances synthesized in and secreted by one tissue and which act to influence distant target organs or tissues. However, many peptide hormones also act as neurotransmitters, passing across very short gaps between cells. In addition, many chemical messengers, including the peptide growth factors, act more locally. Looking at lower invertebrates as simple as *Hydra*, we find peptides resembling our own hormones and neurotransmitters. These are secreted by neuroendocrine cells of *Hydra* and diffuse throughout the body. In higher animals hormones regulate the concentrations of nutrients such as glucose and of ions such as Ca^{2+} and phosphate in the blood. They control the volume and osmotic pressure of body fluids, as well as digestion, growth, reproduction, and responses to stress.

1. Receptors, Feedback Loops, and Cascades

Every hormone must have one or more receptors, most of which are proteins. These may be found embedded in the outer surface of the plasma membrane, in the cytoplasm, or in the cell nucleus. Binding of a hormone to its receptor often elicits both a rapid response and a slower one. For example, we have seen that glucagon, adrenaline, and vasopressin bind to cell surface receptors and promote the synthesis of cyclic AMP (Fig. 11-4). The cAMP induces rapid chemical modifications of many proteins. Some of these may diffuse into the nucleus and affect transcription of genes, a slower response. Insulin (Chapter 11, Section G) also exerts both rapid and slower responses.

Receptor types. Many different kinds of protein can serve as hormone receptors. Some of these are discussed in Chapter 11. The most abundant are the G protein-coupled 7-helix receptors[4–5c] such as that of a β adrenergic receptor pictured in Fig. 11-6. Glucagon, adrenaline, ACTH, and gastrin are a few of the hormones that bind to receptors of this type. Similar receptors respond to light (rhodopsin; Chapter 23) and over 1000 different 7-helix receptors respond to smell and taste. The G proteins and their controlling cycles, Eq. 11-10,[5–7] have also been considered in Chapter 11. The reality of the dissociation and reassociation of the α and βγ subunits in response to binding of a hormone has been demonstrated in living cells by the use of fluorescence resonance energy transfer (FRET).[8] Not all receptors activate G proteins. One large group of membrane-associated receptors have single transmembrane helices but require dimerization to be effective.

The bacterial chemoreceptor (Figs. 11-8 and 19-5) has a very small ligand-binding domain and a larger internal domain that activates a histidine kinase. Many growth-factor receptors, including the insulin receptor (Figs. 11-11, 11-12), have internal domains with protein tyrosine kinase activity.

In contrast, steroid hormones, thyroxine, and retinoids bind to internal receptors. In 1968, Gorski *et al.*[9] and Jensen *et al.*[10,11] proposed independently that steroid hormone receptors in the cytoplasm bind incoming steroid molecules and after an "activation" step carry the hormone into the nucleus, where the hormone-receptor complex would bind at many sites in the chromatin inducing transcription of selected genes.[12] Doubt has been cast on the assumption that the steroid hormone receptors must bind hormone initially in the cytoplasm. However, the role of steroid receptors in regulating transcription is well established (see discussion in Chapter 22, Section E,5; Chapter 28, Section C,6).

Feedback loops. Maintenance of a steady state within an organism depends upon numerous negative-feedback loops. Hormones assist in adjusting reaction rates to maintain a steady state when conditions are changed. For example, blood glucose rises after a meal. This increase is sensed in the pancreatic beta cells (pp. 998, 999), which release insulin. The released insulin promotes uptake of glucose by cells and its conversion into glycogen and lipid stores. When the glucose level falls, inhibitory mechanisms that decrease insulin release are allowed to operate.

Similar regulatory loops can be traced for nearly all hormones. Sometimes they involve several stages and involve sensing devices in the central nervous system. In such cases neural impulses stimulate the **hypothalamus** of the brain (Fig. 30-1) to release **neurohormones**, which travel to the anterior lobe of the pituitary gland. The pituitary, in turn, releases hormones such as **corticotropin** (adrenocorticotropic hormone, **ACTH**), which stimulate the adrenal cortex to release its hormones. The latter exerts feedback inhibition upon the hypothalamus to decrease the secretion of ACTH by the pituitary. Steroids also participate in feedback loops to the hypothalamus.[13] Using ^{3}H-labeled hormones or fluorescent analogs, it has been possible to locate specific brain cells sensitive to a given hormone by autoradiography.[14]

A characteristic of hormonal effects is that they are seldom unique, and are often balanced by counteracting effects of other hormones. For example, both glucagon and adrenaline promote the release of glucose from liver glycogen into the bloodstream. The glucocorticoids stimulate the rate of production of glucose from other body constituents (Chapter 11). Growth hormone tends to increase glucose levels by inhibiting utilization of sugar by tissues. On the other hand, insulin acts to promote uptake of glucose by tissues and a more efficient utilization. The thyroid

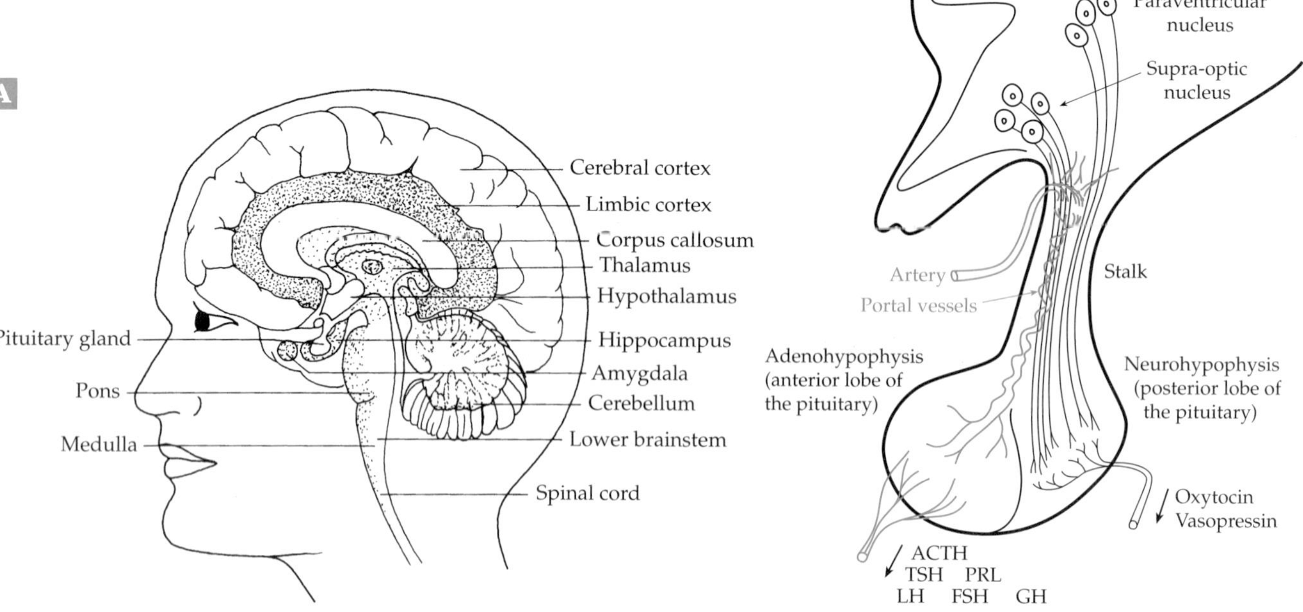

Figure 30-1 (A) Median sagittal section of the human brain. From Maya Pines.[15] (B) Drawing illustrating the synthesis of peptide hormones in the hypothalamus and transport via portal blood vessels into the anterior lobe of the pituitary gland or via nerve tracts into the posterior lobe.[16]

hormone increases the overall rate of metabolism of cells and also tends to promote a decrease in blood glucose.

Signaling cascades. As we have seen in Chapter 11, hormones frequently elicit the synthesis of second messengers such as cAMP, inositol phosphates, or diacyglycerol. This not only provides amplification of the initial hormonal signal but also allows a single hormone to control a "domain" of many metabolic processes. Each of these processes, in turn, can influence others. Many processes are affected by several different hormones and by more than one second messenger. Since we are far from knowing how many hormones exist and how many second messengers are released, the network of regulatory interactions within cells may be one of overwhelming complexity. An abbreviated version of a mitogen-activated kinase (MAP kinase) cascade[4,17] is shown in Fig. 11-13. These cascades are not only initiated but also are propagated by a series of phosphorylation reactions catalyzed by more than 1000 protein kinases encoded by the human genome.[18] Together with more than 500 protein phosphatases, which are often joined together as a bifunctional protein (p. 545),[19,20] they form a complex branching network of interactions.[21,22] These help to control responses not only to hormones but also to varying metabolite concentrations and physical stimuli.

Second messengers as hormones. The same compounds that serve as intracellular second messengers sometimes act as hormones. Tomkins suggested[23] that cAMP and some other small molecules serve as "symbols" indicating a metabolic need. For example, in bacteria ppGpp (p. 1715) serves as a symbol of nitrogen or amino acid deficiency. In cells ranging from those of bacteria to animals, cAMP is a symbol for carbon-source starvation. In *E. coli* cAMP levels increase during carbon-source starvation and stimulate the initiation of transcription of many bacterial operons (Chapter 28). In *Dictyostelium discoideum* (Box 11-C) cAMP is released by cells, when substrate depletion occurs. In this instance, the cyclic nucleotide acts as a hormone transmitting a signal to other cells.

Whereas cAMP is sometimes used by lower organisms as a hormone, its metabolic lability makes it unsuitable for higher animals. Thus, in our bodies the hormones glucagon and adrenaline carry a message to cell surfaces, where binding to receptors stimulates cAMP production. This, in turn, leads to mobilization of metabolic stores such as those of glycogen and triglycerides, just as if these cells had also been subjected to acute starvation. Tomkins suggested that hormones are produced by "sensor" cells in direct contact with environmental signals and travel to and activate more sequestered "responder" cells. The picture can be generalized further by realizing that neurotransmitters are largely derivatives of amino acids. These amino acids may have originally served as intracellular symbols reflecting changes in environmental amino acid concentration but were later utilized in short-range intercellular communication within the nervous system.

The vertebrate hormones. The principal established vertebrate hormones are listed in Tables 30-1 and 30-2. Also given are references to other parts of the text, where specific hormones are discussed. The hormones can be divided into four groups on the basis of chemical structure: (1) peptides and proteins, (2) derivatives of the aromatic amino acids, (3) steroids and prostaglandins, and (4) volatile compounds such as NO and CO. The most numerous are the peptide hormones, many of which also act as neurotransmitters. Peptide hormones, e.g., those with insulin-like effects, function in all phyla of the metazoa, and hormone-like molecules are found in bacteria.[24]

2. Hormones of the Pituitary Gland (Hypophysis) and Hypothalamus

Connected to the brain by a stalk (Fig. 30-1), the pituitary gland releases at least ten peptide or protein hormones that regulate the activity of other **endocrine** (hormone-producing) glands in distant parts of the body. The pituitary is composed of several distinct parts: the anterior lobe (**adenohypophysis**), a thin intermediate portion (**pars intermedia**), and a posterior lobe (**neurohypophysis**). Each has its own characteristic endocrine functions.

The anterior lobe of the pituitary secretes a series of ten or more peptide hormones ranging in size from the ~20-residue β-melanotropin to the ~200-residue growth hormone (somatotropin). Several of these contain a common heptapeptide unit, which is marked in green in the following structure:

Ac•Ser•Tyr•Ser•Met•Glu•His•Phe•Arg•Trp•Gly•Lys•Pro•ValNH_2

Structure of α-melanotropin from pig, beef, and horse

Not only this heptapeptide but also the entire amino acid sequence of **α-melanotropin** is found within the sequence of **corticotropin** (Fig. 30-2), which has an additional 29 amino acids at the C-terminal end.[25] The same heptapeptide was also found in the **lipotropins**. The explanation is, in part, that several of these hormones arise from a single 31-kDa precursor protein called **prepro-opiomelanocortin**.[25,26] It contains an N-terminal signal sequence that is removed shortly after synthesis, as well as pairs of adjacent basic residues (Arg-Arg, Arg-Lys, Lys-Arg and Lys-Lys) at a number of places (Fig. 30-2). After removal of the signal sequence, further cleavage is thought to occur within the secretory vesicles by proteases, which cut either on the carboxyl side of these basic pairs or between them.[25,27,28] The same precursor is made in both anterior and intermediate lobes and is rapidly cut to ACTH, β-lipotropin, and an N-terminal part. In the intermediate lobe the ACTH is then cleaved at the Lys-Lys and Arg-Arg pairs to form αMSH and another peptide called corticotropinlike intermediate lobe peptide (CLIP). Beta lipotropin is degraded rapidly in the intermediate lobe and more slowly in the anterior lobe to γ-lipotropin and the opioid peptide β-endorphin

TABLE 30-1
Peptide and Protein Hormones of Vertebrates

Source and name of hormone	No. residues	Principal site of action	References Chapter, Section
A. Pituitary gland (hypophysis)			
1. Adenohypophysis (anterior portion)			
Corticotropin (ACTH)[a]	39	Adrenal cortex, adipose tissue	Fig. 2-4
β-Melanotropin (β melanocyte-stimulating hormone, β-MSH)[a]	18–22	Skin	Fig. 30-2
β-Lipotropin (β-LPH)[a]	91	Precursor of β-MSH and β-endorphin	Fig. 30-2
γ-Lipotropin (γ-LPH)[a]	58	Precursor of β-MSH	Fig. 30-2
β Endorphin[a]	31	Brain	
Somatotropin (growth hormone, GH)	~200	All tissues	
Prolactin (mammotropin)	~200		
Thyrotropin (thyroid-stimulating hormone, TSH)[b]		Thyroid	Ch 25, B2
Follitropin (follicle-stimulating hormone, FH)[b]		Ovaries, testes	
Lutropin (luteinizing hormone, ICSH or LH)[b]		Ovaries, testes	
2. Pars intermedia (intermediate portion)			
α-Melanotropin (α-melanocyte-stimulating hormone, αMSH)[a]	13	Skin	Fig. 30-2, pp. 1742, 1748
3. Neurohypophysis (posterior portion)			
Oxytocin (ocytocin)	9	Uterus, mammary glands	Fig. 2-4
Vasopressin (antidiuretic hormone)	9		Fig. 2-4
B. Pancreas			
Insulin	51	All cells	Fig. 7-17, Ch 11, G
Glucagon	29	Liver, adipose tissue	Ch 11, D
C. Ovary (corpus luteum)			
Relaxin	—	Pelvic ligaments	p. 1746
D. Thyroid			Ch 25, B2
Calcitonin (thyrocalcitonin)	32	Bones, kidney	Box 6-D
E. Parathyroid			
Parathyrin (parathyroid hormone)	84	Bones, kidney	Box 6-D
F. Kidney			
Erythropoietin		Bone marrow	
Renin (an enzyme)		Adrenal cortex	p. 621; Box 22-D

[a] Arise by cleavage of pro-opiomelanocortin.
[b] Related two-subunit (αβ) proteins with a common β subunit for these three hormones, for human chorionic gonadotropin (hCGH), and for mitogen-regulated protein (proliferin).

(Section B,10). Precursor proteins have been identified for many other peptide hormones, even those with very short chains.[29–30a] Proteolytic cleavages and other processing reactions occur within the secretory pathways of organisms from yeast to humans.[30b,c]

Many pituitary hormones have a pyroglutamate (5-oxoproline) residue at the N terminus (e.g., see Fig. 2-4). This presumably arises by attack of the terminal $-NH_2$ group on the amide carbon of an N-terminal glutamine side chain with displacement of NH_3 (Eq. 10-10).[29] The C terminus is often an amide of the carboxyl group with ammonia, which usually arises from a peptide chain containing one additional glycine residue at the C terminus (Eq. 10-11). The processing of peptide hormones doesn't end with their synthesis. They are usually degraded quickly or are converted into derivatives with weaker hormonal activity.

Pituitary growth hormone and related hormones. The pituitary growth hormone (**somatotropin**)[31] and **prolactin** are 22- to 23-kDa proteins, which share homology also with human **placental lactogen**, a lactogenic hormone secreted by the placenta,[32] and with a growth factor called **mitogen-regulated protein** (or proliferin).[33–35] The polypeptide chain of the 191-residue porcine[36] and human[37] growth hormone folds into an antiparallel four helix bundle (similar to that in Fig. 2-22) but with two long irregular connecting strands. The high degree of homology among somatotropins of many other species indicates a near identity of three-dimensional structures. However, biological function is species-specific. Humans and monkeys respond only to growth hormone from primates. The interaction of an aspartate side chain at position 171 with arginine 43 of the receptor protein may account for some of this specificity.[38] The receptor is a member of a large superfamily of receptor proteins with single transmembrane helices and extracellular domains similar to that of tissue factor (Fig. 12-18) and also, in some respects, to immunoglobulin domains.[39,40] Receptors for growth hormones bind in specific ways to two molecules of receptor protein.[41,42]

Human growth hormone produced in bacteria is used to help very short children to grow. Bovine growth hormone produced in bacteria is used to increase milk production from cows.[43] However, this use may be damaging to the cows.[44] Some humans produce too much growth hormone, often as a result of tumors in the pituitary. The resulting condition of **acromegaly**[45] causes excessive bone growth and many other problems. Growth hormone has a broad range of other effects, e.g., mimicking the action of insulin.[46]

Lactogenic hormones also have 4-helix bundle structures, and the prolactin receptor structure resembles that of growth hormone as well as those of a large cytokine family.[47,48] Prolactin affects placental development during pregnancy. However, during the latter half of pregnancy the placenta has a dominant endocrine effect, synthesizing both progesterone and the placental lactogen.[32,49]

The pituitary glycoprotein hormones. The thyroid-stimulating hormone **thyrotropin** (TSH), together with **folitropin** (FH) and **lutropin** (LH; Table 30-1), form a family of related ~28-kDa dimeric glycoproteins in which each subunit has a three-loop structure stabilized by a characteristic "cystine knot."[50] Also included in the family is the placental **chorionic gonadotropin**,[51,52] which is found only in human beings and a few other species. LH has a central role in promoting both spermatogenesis and ovulation by stimulating synthesis of steroid hormones in the testes and ovary, respectively.[53] Human chorionic gonadotropin (hCG) is also essential for maintenance of pregnancy and acts by stimulating the ovaries to secrete required steroid hormones.

All of these glycoprotein hormones are αβ dimers, and within a single species the subunits of TSH, FH, LH, and CG are identical. However, the β subunits are all different.[54,55] In the human there are at least six genes or pseudogenes for the hCG β chain in a cluster that also contains a single LH β chain gene.[54] The hormones undergo glycosylation and sulfation in the Golgi before secretion.[56] The hormones bind to 7-helix receptors, which are coupled to formation of cAMP or inositol trisphosphate.[57] Mutations in LH may cause male infertility,[58] while mutations in the corresponding receptor may cause male precocious puberty.[59]

Hypothalamic releasing hormones. As was mentioned previously, the anterior lobe of the pituitary

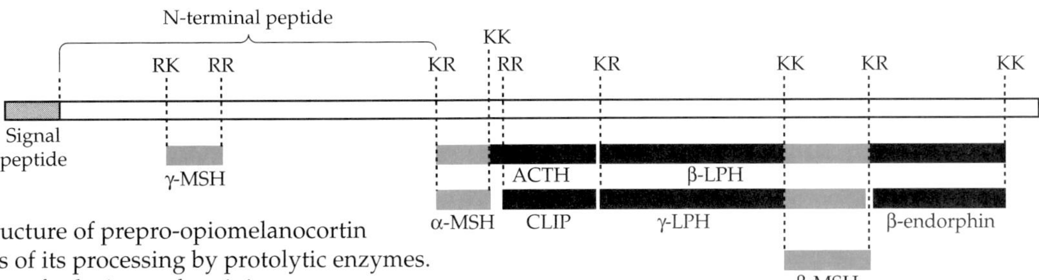

Figure 30-2 Schematic structure of prepro-opiomelanocortin and of some of the products of its processing by protolytic enzymes. The abbreviations K and R are for lysine and arginine.

releases its hormones in response to at least nine neurohormones known as **releasing hormones** or releasing factors.[16,60,61] They are secreted in minute quantities by the hypothalamus into a special portal vein that carries them directly to the pituitary where they exert their effects (Fig. 30-1B). As is indicated in Fig. 2-4 and in Table 30-3, several releasing factors are small peptides, but others are quite large. **Thyrotropin-releasing hormone** (THR, thyroliberin)[62] is a tripeptide; but human **growth hormone-releasing hormone** (somatoliberin) is a 44-residue peptide. Both are synthesized as larger proteins, which are cleaved and processed to form the mature C-amidated hormones.[63,64] **Corticotropin-releasing hormone** (CRH; CRF; corticoliberin), the 41-residue ACTH-releasing factor, is also cut from the much larger prepro-CRF.[65,66] Release of both LH and FSH is stimulated by a single **gonadotropin-releasing hormone** (GnRH).[5a,10,67,68] The releasing factors bind to 7-helix G-protein coupled receptor.[68] Both the releasing factor and gonadotropin are released into the appropriate parts of the bloodstream in a pulsatile fashion emphasizing the neural origin of their release.[69,70]

The hypothalamus also synthesizes **release-inhibiting factors**.[67] One of these, **somatostatin** (Table 30-3), inhibits release of somatotropin, thus counteracting the effect of the growth hormone releasing hormone. Somatostatin acts both in the pituitary and also in the pancreas, where it inhibits the release of both insulin and glucagon.[71,72] The result is a lowering of blood glucose. This suggested a new approach to the treatment of diabetes. However, because of the many other effects of somatostatin[73] and its rapid degradation it has not been useful clinically. Nevertheless, hundreds of analogs of somatostatin have been synthesized, some of which may be of practical

TABLE 30-2
Nonpeptide Vertebrate Hormones

Type, source, and name of hormone	Principal site of action	References Chapter, Section
A. Amino acid derivatives		
1. Thyroid		Ch 25,B,2
Thyroxine and triiodothyronine	Most cells	
2. Adrenal medulla		Ch 11
Adrenaline, noradrenaline (epinephrine, norepinephrine)	Most cells	
3. Pineal gland		
Melatonin	Melanophores	Fig. 25-12
4. Nerves and other cells		
Serotonin (5-hydroxytryptamine)	Arterioles, central nervous system	Fig. 25-12
B. Steroids and prostaglandins		Ch 22, F
1. Testes		
Testosterone	Most cells	Fig. 22-11
2. Ovaries		
Estrogen (estradiol-17β)	Most cells	Fig. 22-11
3. Corpus luteum		
Progesterone	Uterus, mammary glands	Ch 12; Ch 28
4. Adrenal cortex		Fig. 22-11
Corticosterone, cortisol	Most cells	Fig. 22-11
Aldosterone	Kidney	Fig. 22-11
5. Various tissues		
Prostaglandins	Smooth muscle	Ch 21, D
C. Volatile hormones		Fig. 21-8
1. Nitric oxide, NO	Endothelium, brain	
2. Carbon monoxide, CO	Brain	

value.[67] The biological activity of somatostatin resides largely in the sequence FYKT at positions 6–10, a sequence that is thought to form a beta turn (Fig. 30-3). Much of the rest of the molecule can be left off and the disulfide bridge moved up as far as positions 6–11 with retention of high potency. Human somatostatin is synthesized initially as a 116-residue precursor.[74]

A 56-residue peptide, which is formed from the 10-kDa precursor to GnRH, inhibits secretion of prolactin.[75] Inhibition of FSH release is accomplished by feedback inhibition. Hormones known as **inhibins** are produced in the gonads and act to inhibit release of FSH from the pituitary.[76]

Vasopressin and oxytocin. In contrast to the large peptide hormones made in the anterior lobe of the pituitary are **vasopressin** and **oxytocin**, which are secreted from the neurohypophysis, the posterior lobe.[60] The neurohypophysis consists of neural tissue, whose secretions are directly controlled by the central nervous system. In fact, the cell bodies of the secretory neurons are located in specific nuclei of the hypothalamus (Fig. 30-1B). About 4000 vasopressin-secretory neurons and a similar number of oxytocin neurons are present in the neurohypophysis of the rat.[16] Vasopressin is a major regulator of blood volume and pressure,[77] and its secretion is influenced by stress. It increases the water permeability of the kidney collecting duct cells by inducing translocation of aquaporin proteins from intracellular storage vesicles into the apical plasma membrane.[78] Vasopressin binds to

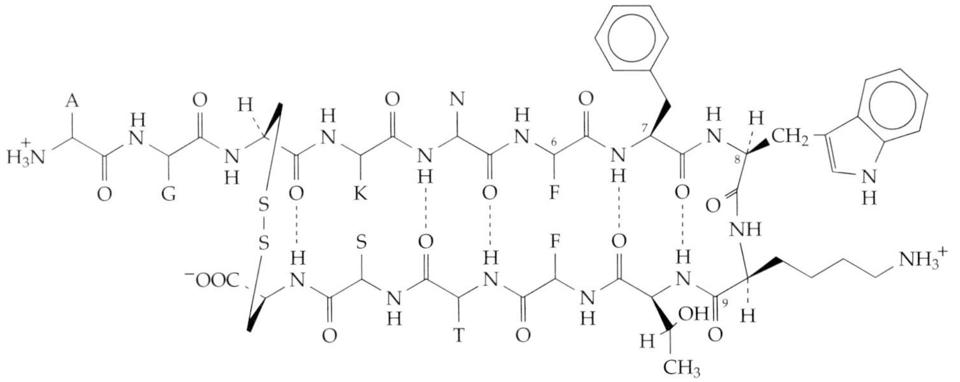

Figure 30-3 Possible secondary structure of somatostatin with a beta turn at residues 7–10 and a disulfide bond between positions 3 and 14. The true conformation would have the plane of the beta sheet puckered and twisted.

TABLE 30-3
Releasing and Inhibiting Hormones from the Hypothalamus

Name	Number of amino acid residues	Sequence[a]
Thyrotropin-releasing hormone (thyroliberin, TRH)	3	*p*EHP–NH_2
Gonadotropin-releasing hormone (GnRH, LH- and FSH-releasing hormone)	10	*p*EHWSYGLRPG–NH_2
GH-releasing hormone (somatoliberin)	44	
Corticotropin-releasing hormone (corticoliberin, CRH)	41	
MSH-releasing factor (melanoliberin)[b]	5	CYIQNC (S–S between the two C)
Somatostatin (GH-release-inhibiting hormone)	14	AGCKNFFWKTFTSC (S–S between C3 and C14)
MSH-release-inhibiting factor[b]	3	PLG
Prolactin-releasing factor		
Dopamine (prolactin-release-inhibiting factor)		

[a] Standard one-letter abbreviations are used. *p*E is pyroglutamyl (5-oxoprolyl) and –NH_2 at the right indicates a C-terminal *carboxamide*.
[b] Ring and tail fragments of oxytocin.

G-protein-coupled receptors.[79] A defect in the type 2 vasopressin receptor leads to the condition of nephrogenic **diabetes insipidus** in which the body fails to concentrate the urine.[77,80] Oxytocin acts on smooth muscles of the uterus during childbirth and triggers the release of milk from the mammary glands.[81] The latter response is partially controlled by the suckling of the infant, which induces the nervous system to release oxytocin into the bloodstream.

Hormones related to oxytocin and vasopressin occur in most vertebrates, the compound **vasotocin** shown in Fig. 30-4 being the most common. Substitution of phenylalanine for isoleucine at position 3 gives arginine vasopressin, the vasopressin found in our bodies. Structure of oxytocin and related hormones[82] are also shown in Fig. 30-4. Like somatostatin, vasopressin and oxytocin may also form antiparallel pleated sheet structures with β turns. The structural requirements for hormone activity have been studied intensively. Both the macrocyclic hexapeptide ring and the tripeptide side chains are necessary for maximal activity.[83]

The gene for arginine vasopressin is that of a 166-residue precursor protein carrying a 19-residue signal sequence at the N terminus.[84] This sequence is followed by that of vasopressin, then after a GKR linker by the 95-residue **neurophysin II**. Finally, after one additional arginine there is a 39-residue glycopeptide. Oxytocin originates in a parallel way from its own precursor.[85]

The 93- to 95-residue neurophysins act as carriers for vasopressin and ocytocin, forming specific complexes with them. Neurophysins contain 14 cysteine residues, which form seven disulfide bonds. There is a striking similarity in sequence between the neurophysins, snake venom toxins, a wheat germ lectin (agglutinin), a ragweed pollen allergen, and a small plant protein called hevein. On the basis of the alignment of cysteine residues, Drenth proposed[86] that all of these proteins have a disulfide-linked core whose structure is shown in Fig. 30-16.

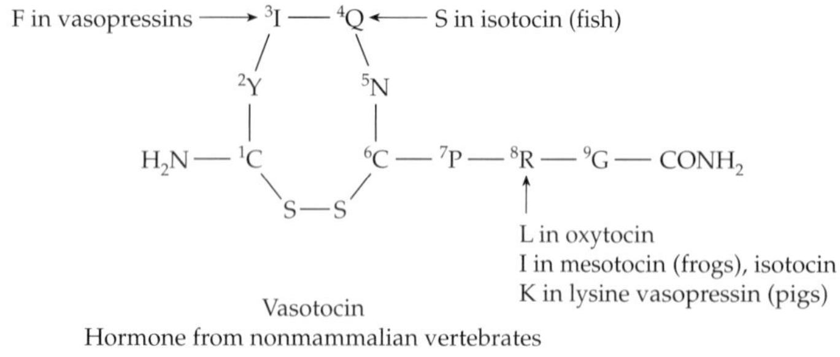

F in vasopressins → ³I — ⁴Q ← S in isotocin (fish)

²Y   ⁵N

H_2N — ¹C   ⁶C — ⁷P — ⁸R — ⁹G — $CONH_2$

S—S

L in oxytocin
I in mesotocin (frogs), isotocin
K in lysine vasopressin (pigs)

Vasotocin
Hormone from nonmammalian vertebrates

Figure 30-4 Structure of the nonmammalian hormone vasotocin and of related hormones including ocytocin and vasopressin.

Melanocortins. The melanocortin peptides, which are derived from pro-opiomelanocortin as indicated in Fig. 30-2, are formed in varying amounts in the pituitary, in two brain nuclei, and in some peripheral tissues.[25] In some animals α-MSH arises primarily in an intermediate part of the pituitary. The hormone has a direct effect on the melanocytes (Box 8-F) causing darkening of the skin. In addition, the various melanocortin peptides (ACTH, α-, β-, and γ-MSH) bind to five different types of receptors. These have been linked to the control of energy homeostasis, appetite, and obesity in both mice and humans.[87–89] The sequence His-Phe-Arg-Trp (of the green-shaded sequence in Fig. 30-2) is essential for binding. In keratinocytes α-MSH may form a 1:1 complex with tetrahydrobiopterin,[90] the coenzyme for tyrosine hydroxylase, and a regulator of tyrosinase, an essential enzyme for melanin formation (Fig. 25-6).

3. Pancreatic and Related Hormones

The functions of the 51-residue insulin (Figs. 7-17 and 7-18) are discussed in Chapter 11. Its actions begin early in life. Mammalian preimplantation blastocysts already show a response to insulin. The glucose transporter GLUT1 is present at the earliest stages; synthesis of GLUT2 and GLUT3 (p. 416) begins at the eight-cell stage. However, the insulin-regulated transporter GLUT4 is not present in the blastocyst. A newly discovered insulin-regulated GLUT8 may function during preimplantation development.[91] Although the secretion of insulin is of primary importance to the regulation of the glucose concentration in mammals, it is still not clearly understood.[92] The beta cells have insulin receptors and other components of the insulin signaling system such as the insulin receptor substrates IRS-1 and IRS-2 and phosphatidylinositol 3-kinase (PI3-K). The sensing of glucose by the beta cells is also not yet well understood.[93] This lack of knowledge has made it difficult to improve the treatment of diabetes. A new approach is to engineer non-beta cells to secrete a steady supply of insulin. Such a possibility has been demonstrated in mice using gut K-cells.[94]

The **insulinlike growth factors** (IGF-I and IGF-II) are produced in many different tissues and promote growth of other cells (see Section 6). **Relaxin**,[94a] which is produced in the corpus luteum of ovaries during pregnancy, is responsible for inducing widening of the birth canal during the late stages of pregnancy and inhibits contraction of uterine muscle, perhaps by decreasing the activity of the kinase that phosphorylates the 20-kDa light chains of myosin.[95] Relaxin is found throughout the animal kingdom, even in the protozoan *Tetrahymena*.[96] Its structure is apparently identical in pigs, whales, and in a primitive tunicate.[97] In human males relaxin is apparently produced in the prostate, where it may function as a sperm motility factor.[97] Relaxin, IGF-I, and IGF-II are all structurally homologous to proinsulin and contain the characteristic 3-disulfide structure of insulin. The IGF-I receptor structure resembles that of the insulin receptor (Fig. 11-11) and also that of the epidermal growth factor (EGF) receptor.[98]

Glucagon belongs to a family that also includes the gastrointestinal hormones **secretin**, **gastrointestinal inhibitory peptide** (**GIP**), **vasoactive intestinal peptide** (**VIP**), and **glicentin** (Table 30-4). The function of glucagon in regulation of the blood glucose level is considered in Chapter 17, but the hormone may have other effects. A complex processing pathway converts 14- to 16-kDa preproglucagons into the active hormone.[99,100] Proglucagon is processed to glucagon in the pancreas, but in the endocrine L cells of intestinal mucosa it yields glicentin, a polypeptide containing the entire glucagon sequence, and other products.[101–102a] Glucagon receptors generate both cAMP and Ca^{2+} as second messengers.[103]

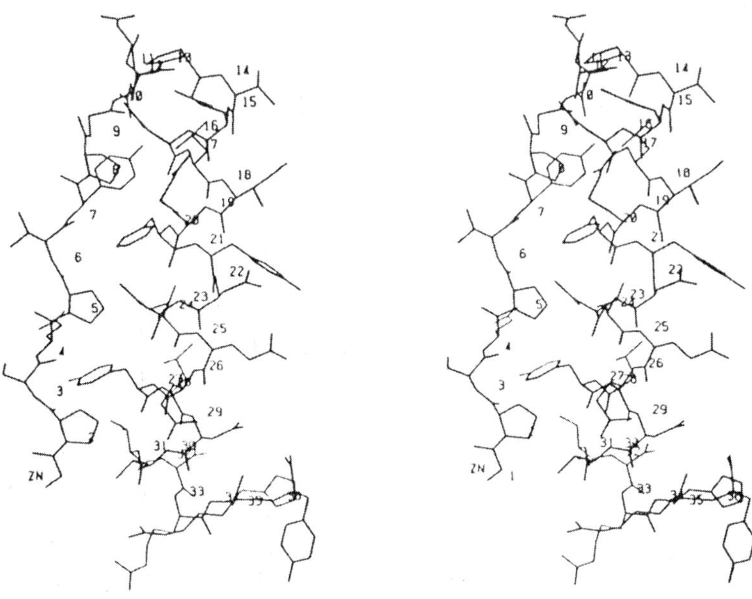

Figure 30-5 Structure of the avian pancreatic polypeptide, a small globular protein. From Blundell *et al.*[107]

The 27-residue secretin stimulates secretion of bicarbonate into the pancreatic juice and inhibits gastric secretion of acid. The 28-residue VIP is found throughout the gastrointestinal tract of mammals and birds as well as in the brain and the lungs. It is a potent vasodilator and may be the major relaxant of pulmonary smooth muscle.[104] It has been reported totally absent from lungs of asthma patients.[105] The gastrointestinal inhibitory peptide (GIP) is larger than VIP but also has a close homology with glucagon (Table 30-4).[106]

The 36-residue **pancreatic polypeptide** is a hormone of uncertain functions. The crystalline polypeptide has at the N terminus an 8-residue collagenlike helix that lies parallel to a C-terminal α helix (Fig. 30-5). The overall shape resembles that of both insulin and glucagon.[107,108] This PP-fold includes also neuropeptide Y, which is considered in the next section,[109] and neuropeptide YY.[110]

4. Gastrointestinal and Brain Peptides

The largest endocrine gland in the body is the gastointestinal tract, which produces a profusion of peptide hormones, many of which are also found in the brain.[111,112] Indeed, a majority of the known vertebrate peptide hormones occur in the brain.[82,112,113] For example, glucagon has been found in the brainstem and hypothalamus.[114] Many of these peptides, or closely related ones, are also found in lower invertebrates.[115] For example, the 10-residue *Hydra* **head activator** is also present in mammalian brain.[115,116] The concentrations of these peptides in the brain is very low (10^{-12} 10^{-15} M).

Gastrin is produced in the lower portion of the stomach and regulates the secretion of acid as well as growth of the gastrointestinal mucosa.[112,117] It may also function as an Fe^{3+} carrier.[117a] The shorter gastrin 17 as well as the longer 34-residue gastrin 34 are both active as is a synthetic pentapeptide with the hormone's C-terminal sequence.[13] The family of **pancreozymin-cholecystokinins** (CCK) are 8- to 58-residue peptides produced in the upper intestinal tract. They have a 4-residue amidated C-terminal sequence in common with gastrin. This tetrapeptide has some biological activity, but eight residues are required for full activity as is conversion of the tyrosine at position seven from the C terminus to an *O*-sulfate ester. Both gastrin and CCK molecules are partially converted to sulfate esters.[118,119] Most regions of the brain contain CCK peptides in amounts exceeding those of other neurotransmitters. These arise (in pigs) from a 114-residue preproCCK.[120–122] The sulfated insect neuropeptide, **leukosulfakinin** (Table 30-4), is homologous to gastrin and CCK.[123] Another hormone **gastrotropin**[124] is produced by cells of the intestinal mucosa in the distal ileum and stimulates gastric secretion.

Motilin, a 22-residue intestinal neuropeptide, stimulates motor activity of the gastrointestinal tract.[125,126] **Bombesin** was first isolated from frog skin but probably also functions in both the intestinal tract and the brain. It has a powerful hypothermic effect.[127,128] A mammalian homolog of bombesin, the 27-residue **gastrin-releasing peptide** (GRP), is found throughout the gastrointestinal and pulmonary tracts as well as the central nervous system.[120,129] Bombesin-like material, possibly GRP, is produced by some cancers and may serve as an autocrine growth factor.[130] The 29-residue **galanin** was originally isolated from procine intestine but is found throughout the central nervous system. It may function as a neurotransmitter or modulator.[131] The 15-residue **guanylin** is an important regulator of epithelial transport in the intestine and probably in other tissues.[132,133] The active hormone, which is cut from the 99-residue proguanylin, contains two disulfide bonds. Guanylin receptors activate guanylate cyclase with production of cyclic GMP, which functions in the regulation of intestinal fluid and electrolyte absorption. The 18- or 19-residue heat stable enterotoxins of some strains of *E. coli* bind to and activate the guanylin receptors. The resulting overproduction of cGMP causes severe diarrhea (see also Box 11-A).[133]

Neuropeptides Y (NPY) and YY are 36-residue amidated peptides that are members of the pancreatic polypeptide (PP) family (Fig. 30-5). NPY is produced both in the peripheral nervous system and in the brain,[110,134] where it is one of the most abundant neuropeptides. Another member of the PP family is **seminalplasmin**, a regulator of calcium ion transport in bovine sperm.[135] NPY is best known for its stimulation of appetite. It also inhibits anxiety and increases memory retention. It has a vasoconstrictive effect on blood vessels, participating in cardiovascular regulation.[136,137] Peptide YY is formed in endocrine cells of the intestine, while NPY is formed in neurons of the parasympathetic system.[138] Both participate in regulation of fluid and electrolyte secretion. Both are found in other vertebrate species.[139]

NPY is one of the most important of several regulators of feeding behavior of animals. PYY_{3-36}, another member of the neuropeptide Y family, suppresses appetite by antagonizing the action of NPY.[139a,b] A large variety of hormonal effects seem to be involved in control of appetite.[139b,140,141] There are both short-term and long-term mechanisms. For example, when introduced into the gut of rats prior to feeding, CCK and various other gastrointestinal peptides decrease the amount of food eaten.[140]

Much attention has been focused on the 146-residue cytokine **leptin**, a hormone produced by adipose tissue.[141–144] Leptin, which is sometimes described as the antiobesity hormone, was recognized by mutations of the *obese* gene (OB) or of the OB receptor in genetically obese mice. When food is scarce, the fat cells shrink and decrease their secretion of leptin. The decrease is sensed by receptors in the hypothalamus, which signal for increased NPY secretion and decreased secretion of αMSH. NPY increases appetite, while αMSH has an opposing role of blocking feelings of hunger .[144–145a] Nevertheless, there are doubts that leptin's primary role is control of obesity.[146,147]

The 13-residue **neurotensin** was first isolated from the hypothalamus but is more abundant in cells of the ileum.[13] It induces gut contraction, lowers blood pressure, and has a variety of other effects.[127,148] **Substance P** (SP; Table 30-4) has been regarded as a possible neurotransmitter for some time[127] but is also found in the digestive tract. It is the most abundant of a family of five neurokinins (or tachykinins). Others include neurokinin A (substance K), neurokinin B, neuropeptide K, and neuropeptide γ. They have a common C-terminal sequence $FXGLM\text{-}NH_2$.[149,150] Substance P is thought to be involved in the perception of pain, and mice lacking a substance P receptor appear to have reduced sensitivity to pain.[151,152] Substance P as well as the related substance K are derived from two large precursor proteins, which appear to arise as a result of alternative modes of splicing of mRNA.[29]

5. Other Mammalian Peptide Hormones

The action of the 32-residue thyroid hormone **calcitonin**[153] has been described in Box 22-C. This calcium-regulatory hormone is produced in the thyroid C cells from a precursor having an extra 82 residues at the N terminus and 16 residues at the C terminus. The same gene gives rise in neural tissues to a neuropeptide, possibly a neurotransmitter, called **calcitonin gene-related polypeptide** (CGRPP).[29,154] The 84-residue **parathyrin** (parathyroid hormone) is present in secretion granules as a 90-residue prohormone containing six extra residues at the N terminus. The primary biosynthetic product **preproparathyrin** contains an additional 25 residues.[155] An N-terminal 34-residue fragment of the hormone, when injected subcutaneously daily, causes an increase in bone density in persons with osteoporosis.[156–157] The hormone acts via a G-protein-coupled receptor in bone and kidney (see Box 22-C).[158,159] A calcium ion receptor, which binds Ca^{2+} cooperatively, acts as a sensor that regulates release of the parathyroid hormone to regulate the serum Ca^{2+} concentration.[160] A 141-residue **parathyroid hormone-related protein** has an N-terminal sequence homologous with that of parathyroid hormone, eight of the first 13 residues being identical. It is secreted by a variety of cells and serves as a growth factor.[161]

TABLE 30-4
Some Pancreatic and Gastrointestinal Hormones and Neurohormones and Their Sequences

Name and source	No. of residues	Sequence[a]
Glucagon	29	HSQGTFTSDYSKYLDSRRAQDFVQWLMNT
Secretin (pancreas)	27	HSDGTFTSELSRLRDSARLQRLLQGLV–NH_2
Vasoactive intestinal peptide	28	HSDAVFTDNYTRLRKQMARKKYLNSILN–NH_2
Gastrointestinal inhibitory peptide (GIP)	43	YAEGTFISDYSIAMDKIRQQDFVNWLLAQ–Q^{43}
Glicentin	100	A proglucagon containing the entire glucagon sequence in residues 64–92
Pancreatic polypeptide	36	
Neuropeptides Y (NPY) and YY	36	
Gastrin (stomach)		
Gastrin-17	17	pEGPWLEEEEEAYGWMDF–NH_2[b]
Cholecystokinin, CCK or pancreozymin (gallbladder, pancreas), many forms exist		
CCK 58	58	
CCK 8	8	DYMGWMDF–NH_2
Motilin (porcine)	23	FVPIFTYGELQRMQEKERNKGQ
Bombesin	14	pEQRLGNQWAVGHLM–NH_2
Gastrin-releasing peptide	27	
Galanin	29	
Guanylin	15	PNTCEICAYAACTGC (disulfide bonds: S–S, S–S)
Neurotensin	13	pELYENKPRRPYIL
Substance P	11	RPKPQQFFGLM–NH_2
Physaelemin (frog skin)	11	pEADPNKFYGLM
Neurophysins	93 – 95	
β-Endorphin	31	YGGFMTSEKSQTPLVTLFKNAIIKNAHKKGQ
Dynorphin	17	YGGFLRRIRPKLKWDNQ
Met-enkephalin	5	YGGFM
Leu-eukephalin	5	YGGFL
Angiotensin II	8	DRVYIHPF
Bradykinin (BK)	9	RPPGFSPFR
Lys-bradykinin (kallidin)	10	KRPPGFSPFR
Sleep peptide	9	WAGGDASGE
Atrial natriuretic hormone	28	
Chemotactic factors		
for neutrophils	3	f–MLF
for phagocytes	4	TKPR
Speract	10	GFDLNGGGVG

[a] Standard one-letter abbreviations; pE, 5-oxoprolyl; f–, formyl; –NH_2, C-terminal carboxamide.
[b] Y-12 may be sulfated.

Endogenous opioid peptides. Extensive processing is also involved in formation of analgesic opioid peptides, which are present naturally in the brain (see also Section B). The formation of β-endorphin in the hypothalamus from prepro-opiomelanocortin (Fig. 30-2) has already been mentioned. Prior to the discovery of β-endorphin, the pentapeptides **Met-enkephalin** and **Leu-enkephalin** (Table 30-4) were discovered and were found to compete with opiate drugs for receptors in the brain. The larger β-endorphin, which contains the Met-enkephalin sequence at its N terminus, is a far more potent opiate antagonist than are the enkephalins. Since the Met-enkephalin sequence within β-endorphin is not flanked by basic residues, it apparently is normally not released. Two other recently discovered brain peptides are **endomorphin-1** (YPWF–NH_2) and **endomorphin-2** (YPFF–NH_2). They are also potent agonists for the opioid receptors, especially the μ receptor (see Section B,10).[161a,161b]

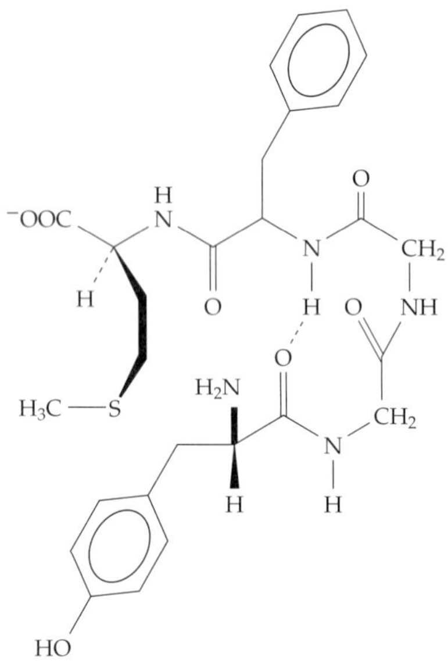

Tyr-Gly-Gly-Phe-Met
(Met-enkephalin)

Both Met-enkephalin and Leu-enkephalin have their own pro- and prepro forms.[29] Bovine **preproenkephalin A** is a 268-residue protein containing a 20-residue signal sequence and four sequences of Met-enkephalin and one of Leu-enkephalin, each flanked by pairs of basic residues. There are also Met-enkephalin-Arg-Gly-Leu (YGGFMRGL) and Met-enkephalin-Arg-Phe sequences. Not all of these are cut out cleanly, and other peptides such as Met-enkephalin-Arg-Arg-Val-NH_2 are also found in brain. **Proenkephalin B** contains three copies of Leu-enkephalin contained within longer peptides. One of these, **β-dynorphin** (Table 30-4), is also a potent opioid compound. The enkephalins are thought to act as neurotransmitters, which are rapidly degraded after their release by two or three membrane-bound peptidases.[29,162] Attempts are being made to design inhibitors that might inactivate these enzymes allowing buildup of enkephalin concentrations with a resultant analgesic effect.

ATP, ADP, and adenosine. Usually regarded as a strictly intracellular compound, ATP is also released into extracellular space. There the ATP, as well as ADP and adenosine, have a variety of local hormonal functions. ATP receptors are found in many tissues and are present in some nerve synapses.[163–166] ATP is one of the substances that induces sensations of pain.[164,167,168] It may affect secretion of saliva,[163] signal a full urinary bladder, induce a feeling of warmth,[167] and have functions in the immune system, in platelet clotting,[166] and as a neurotransmitter. Adenosine has been recognized for many years as an extracellular signaling molecule, a local hormone that can arise by breakdown of ATP or by secretion from cells.[169,170] At least four types of receptor are present in the human body.[170–174] Adenosine is thought to modulate neural responses in many tissues. It may be involved in sleep,[172] in regulation of serotonin transport,[171] and in control of appetite.[171,174] Extracellular ADP appears to have a role in controlling bone osteoclasts (p. 441).[175]

Kinins. These hormones are small peptides that induce contraction of smooth muscles, lower blood pressure (Box 22-D), and increase vascular permeability.[176] They also have a function in contact-activated blood coagulation. The most important human kinins are the nonapeptide **bradykinin**[177,178] and the related decapeptide **lysine-bradykinin** (Table 30-4). Other forms such as Met-Lys-bradykinin and Ile-Ser-bradykinin (T-kinin) are also known. The precursors to the kinins, the **kininogens**,[176] are cleaved by the protease **kallikrein** (Fig. 12-17) or by kallikreinlike enzymes to form the kinins. Kinins are suspected of being important producers of pain in inflammatory conditions such as arthritis.[176a]

Endothelins. Endothelial cells of blood vessels produce **endothelins** that cause vascular smooth muscle contraction and a rise in blood pressure.[179–183] Three human genes code for the closely related endothelins-1, -2, and -3. A 203-residue preproendothelin-1 is processed to form the 39-residue prohormone called **big endothelin-1**. Some of this peptide is secreted and circulates in plasma, where it may have various hormonal functions. Cleavage of the prohormone by a cellular metalloprotease yields endothelin-1, a 21-residue peptide held in a looped configuration by two disulfide bridges. It is homologous to a group of neurotoxins that includes the α-scorpion toxins and ω-conotoxin.[181] These toxins act on voltage-dependent ion channels. Endothelin-2 is produced largely in the kidneys and intestine, while endothelin-3 is found in

Endothelin-1

high concentrations in the brain. Type A endothelin receptors are 7-helix G-protein-coupled proteins, which activate phospholipase C with generation of inositol 1,4,5-trisphosphate and diacylglycerol (Ins-P_3; Figs. 11-4, 11-9). The Ins-P_3 causes release of Ca^{2+}, while diacylglycerol mediates mitogenic responses.

Opposing the effects of the endothelins, which act slowly, is a fast-acting endothelium-derived **relaxing factor**, which has been identified as nitric oxide, NO. It is discussed in Chapter 18, Section F, and in Section 7 of this chapter. Also affecting blood pressure is the potent vasorelaxant **atrial natriuretic factor**. This 28-residue peptide, which is discussed in Box 22-D, is produced by the cardiac atria and stimulates the excretion of Na^+ and of water by the kidneys.[184] It also promotes hydrolysis of lipids within human adipocytes.[185]

Peptides as attractants. Small peptides as well as larger polypeptides serve to attract cells within the human body and other multicelled organisms. Both unicellular and multicellular organisms also use peptides as pheromones. The human immune system depends upon hormonelike **chemotactic factors**. Neutrophils are attracted by such peptides as formyl-Met-Leu-Phe,[186–188] which have a bacterial origin, while the basic tetrapeptide Thr-Lys-Pro-Arg activates the phagocytic polymorphonuclear leukocytes and macrophages.[189] Larger 8- to 10-kDa proteins known as **chemokines** (chemotactic cytokines) attract leukocytes to sites of inflammation (Fig. 30-6).[190–192] Some proteins serve as pheromones. Examples range from the 40-residue mating pheromones of protozoa of the genus *Euplotes*[193,194] to the 17-kDa sex pheromone of the female hamster.[195] The decapeptide **speract** (Table 30-4) is produced by sea urchin eggs and stimulates the respiration of spermatozoa.[196] Similar factors probably function in fertilization of human ova.

6. Protein Growth Factors and Cytokines

The pituitary growth hormone is only one of a large family of protein growth factors that are secreted by cells and which promote the growth of other cells.[200] Many of the growth factors are also described as **cytokines**, local protein hormones that conduct cell-to-cell communication to regulate growth, development, and differentiation.[197,201–203] Among the first growth factors to be recognized were the **insulinlike growth factors** (IGF or somatomedins), mitogenic peptides isolated from plasma. They share some of the metabolic effects of insulin but are less active. On the other hand, they are much more active than insulin[204] in promoting cell growth and proliferation of cells.[205] The abundant IGF-I (somatomedin C), a 70-residue single-chain basic peptide with a sequence and three-dimensional structure homologous to that of proinsulin,[206,207] is considered a major mediator of the action of the pituitary growth hormone (GH, somatotropin). Studies in cell culture suggest that GH may induce differentiation of cells, and that IGF-I may then cause a rapid proliferation of the newly differentiated cells.[208] The homologous 67-residue IGF-II may have a similar function in fetal development.[209] The cell surface receptor for IGF-I is similar to the insulin receptor, but IGF-II receptor is structurally different. It is a monomeric 250-kDa protein; and although it is a substrate for a tyrosine kinase, it has no kinase activity of its own.[210]

The 53-residue **epidermal growth factor** (EGF or **urogastrone**) is found in human urine and in very high concentration in the submaxillary salivary glands of male mice. Like the pancreas these glands contain both endocrine and exocrine tissues. EGF is synthesized in mice as a 1217-residue precursor, which contains not only the EGF sequence but also seven other related sequences.[211] Related growth factors include transforming growth factor-α (TGF-α), neuregulins,[212–214] betacelulin, and epiregulin, all of which promote growth of epithelial cells and are involved in wound healing.[215] The EGF and related peptide chains are each crosslinked by three disulfide bridges. The three-dimensional structure of EGF, deduced from NMR measurements, contains largely β structure and loops and is organized into two domains in a "mitten shape."[216] The receptor for EGF is a 1186-residue transmembrane glycoprotein. Its relationship to cellular oncogene c-*erbB* has been discussed in Chapter 11, Section H. The extracellular glycosylated N-terminal region of the receptor contains the EGF-binding site.[217–220] It also contains two cysteine-rich repeat sequences homologous to one of those in the insulin receptor A chain (Fig. 11-11). The cytoplasmic C-terminal part of the EGF receptor contains a 250-residue tyrosine-specific protein kinase sequence. Following dimerization the EGF receptor phosphorylates tyrosine residues in various proteins including itself (autophosphorylation).[212,219,221] The receptor is also phosphorylated on Thr 654 and other residues through the action of the Ca^{2+}- and phospholipid-dependent diacylglycerol-activated **protein kinase C** (Fig. 11-9).[222] Serines 1002, 1046, and 1047 may also become phosphorylated, perhaps resulting in desensitization of the receptor.[223]

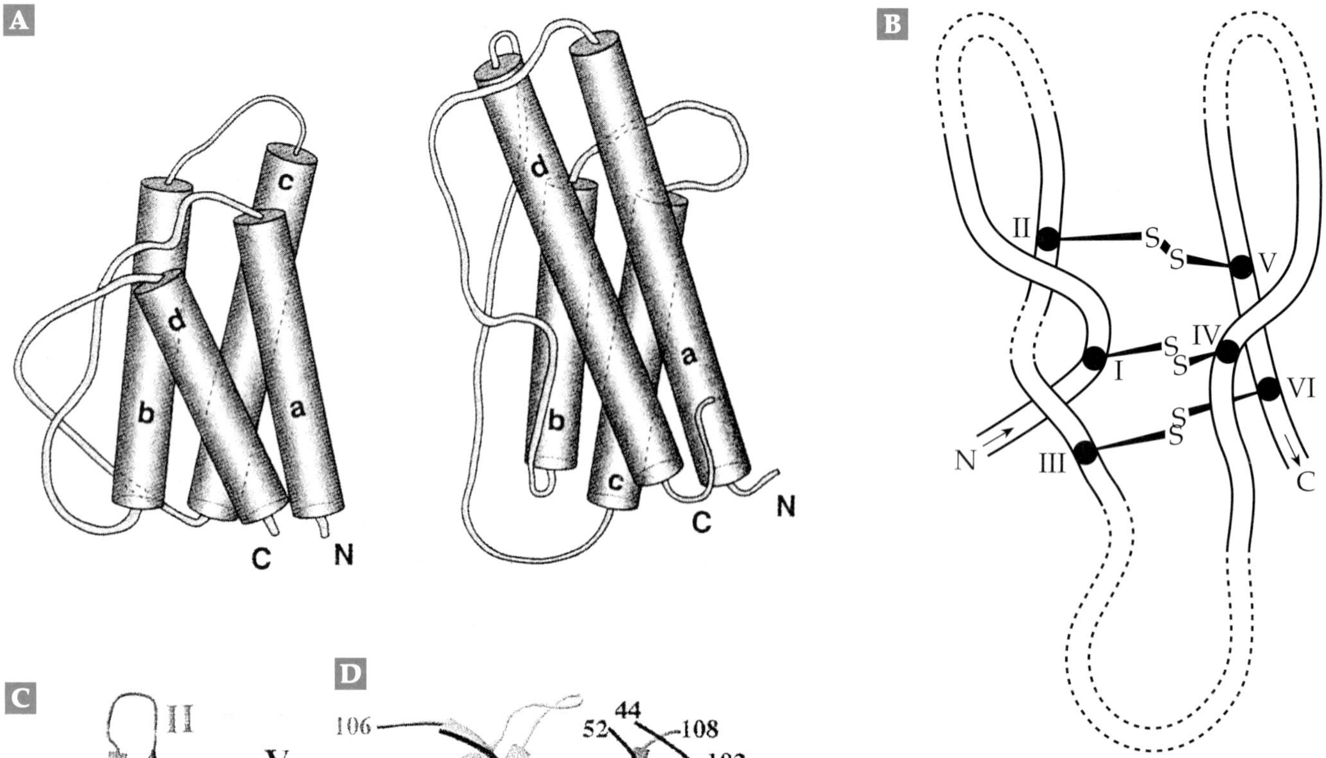

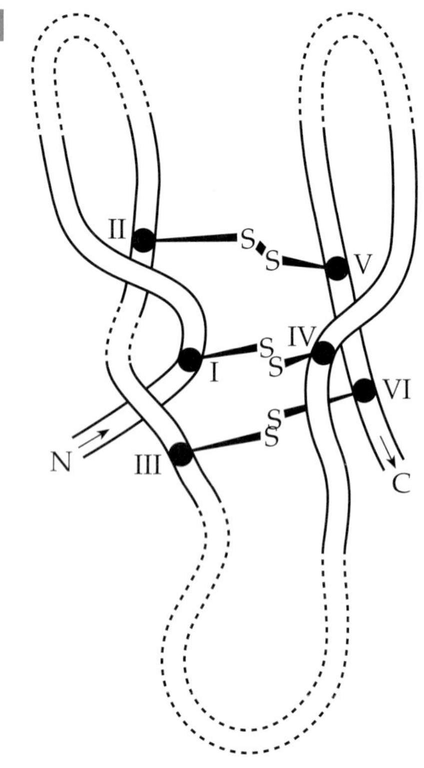

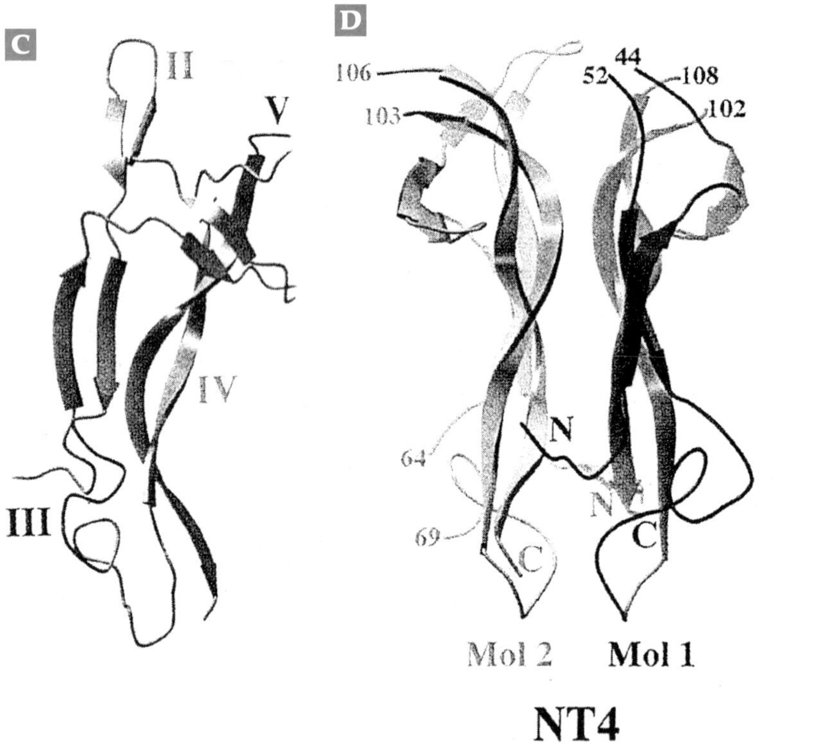

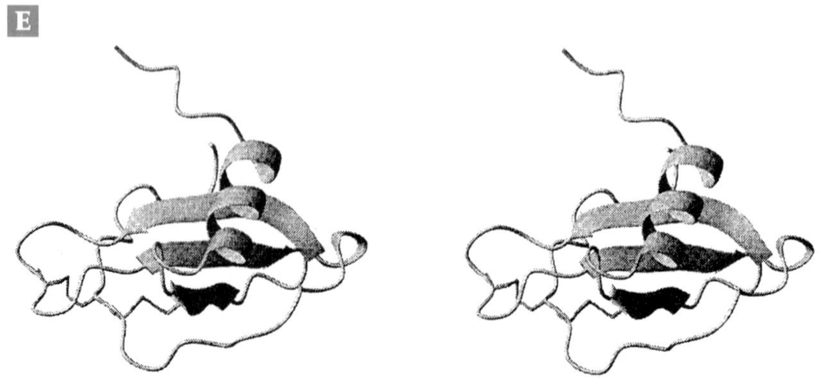

Figure 30-6 Structures of some cytokines and other growth factor proteins. (A) Schematic drawing of representative structures of "short" and "long" helical cytokines. Note the difference in the topology of the connection between helices A and B in the two models. The short cytokines (left) include interleukins-2 and -4. Human growth hormone has a long cytokine structure. Both groups include various colony-stimulating factors. (B) Schematic representation of the disulfide knot topology of cystine knot cytokines. The cysteine residues are numbered I to VI in order of occurrence. See Davies and Wlodawer.[197] (C) Ribbon drawing of a monomer of nerve growth factor (NGF). (D) Ribbon drawing of a dimer of the closely related neurotrophin 4 (NT4). (C) and (D) are from Robinson *et al.*[198] (E) Stereoscopic ribbon drawing of the chemokine **eotaxin-3**. From Ye *et al.*[199]

Binding of EGF to its receptor produces within minutes an increased transcription rate for the prolactin gene and other nearby genes.[221] The urinary form of EGF, urogastrone, is an inhibitor of ulcer formation. It is found in relatively large amounts in the urine of pregnant women (who tend not to develop ulcers).

Platelet-derived growth factor (PDGF)[224] is released from the α-granules of blood platelets during clot formation and is thought to stimulate the growth and mitosis in fibroblasts that is necessary for wound healing.[225] It consists of two chains, A and B. The 31-kDa precursor of the A chain is encoded by the cellular oncogene *c-sis* (p. 571).[224] The **PDGF receptor** is another transmembrane glycoprotein with a C-terminal tyrosine kinase domain. However, its construction differs from that of the insulin or EGF receptors. The external part of the single-chain receptor appears to contain five immunoglobulinlike domains (see Figs. 2-16 and 12-18).[226] Binding of PDGF to the receptor causes responses within minutes.[217,226] These include activation of the tyrosine kinase, hydrolysis of phosphatidylinositides, increases in the levels of cAMP and of Ca^{2+}, and increased transcription of a group of genes. The last include the proto-oncogenes c-*myc* and c-*fos*, which encode proteins that regulate transcription. The PDGF receptor is part of a recognized autocrine stimulatory loop in cells infected with a virus carrying the v-*sis* oncogene.[217] The oncogene product resembles PDGF and binds to the PDGF receptors of the cell producing the v-*sis* product. In this way the cancer cell stimulates its own growth.

Transformation of kidney fibroblasts into cancer-like cells can be induced by the concerted action of PDGF, an analog of EGF, and the **transforming growth factor** (TGF-β).[225,227,228] The latter is one of 30 or more related growth factors that have numerous functions in normal tissues.[229] Platelets produce a relatively large amount of the 25-kDa TGF-β, and it too may be involved in wound healing. TGF-α is a smaller protein with a structure resembling that of EGF.[227,228] While TGF-β inhibits epidermal cell growth, TGF-α stimulates growth. It is found in elevated levels in the skin lesions of **psoriasis** (Box 8-F) and may be the cause of the excessive epithelial growth in that disease.[230]

There are at least nine **fibroblast growth factors** (FGFs). Originally found in brain, they act on many cells including the endothelial cells that line blood vessels.[231] Basic FGF[232] and acidic FGFs[233] have homologous sequences[234] and are also related to the lymphokine interleukin-1. **Vascular endothelial cell growth factor** (VEGF), which is similar to PDGF, is essential for maintenance of the endothelium. The FGFs and VEGF as well as TFG are potent **angiogenic factors** needed for growth of blood vessels.[235–238] These proteins are important not only to normal blood vessels but also to invasive tumors that must develop blood vessels in order to grow. Excessive production of angiogenic factors may also be a factor in eye diseases including the retinal deterioration caused by diabetes.[239] Another protein, **angiogenin**, is a ribonuclease,[240] which is discussed on p. 648.

There are four closely related transmembrane FGF receptors and subforms that arise by alternative mRNA splicing.[241–243] The receptor structures include three external immunoglobulinlike domains and an internal tyrosine kinase domain at the C terminus. Mutations in FGF receptors are associated with a variety of skeletal defects and other hereditary problems.[241,244] For example, the Gly380Arg substitution in the transmembrane segment of FGF receptor 3 is the major cause of **human dwarfism** (achondroplasia).[245] The fibroblast growth factors, as well as other proteins such as the IGFs, HGF, and TGF-β, bind not only to their receptors but also to heparan and heparin. This binding appears to be a major factor in controlling the availability of the growth factors.[242,246,247]

The **nerve growth factor** (NGF) was identified over 40 years ago by Rita Levi-Montalcini[248] on the basis of its activity in promoting the profuse outgrowth of neurites from embryonic neurons (Fig. 30-7). The 118-residue monomer consists largely of three β-hairpin loops, which are held together by three disulfide bridges that form a "cystine knot." The C15–C80 disulfide passes through a ring formed by the C58–C108 and C68–C110 disulfide bridges (Fig. 30-6B,C).[249] A similar folding pattern and disulfide core are found in TGF-β2 and also in several other **neurotrophins**, growth factors involved in the development and survival of neurons (Fig. 30-6D).[198,250–251a] NGF may also have a more general function in promoting tissue repair.[252] Like EGF nerve growth factor is most abundant in the submaxillary glands of male mice. Larger oligomers containing bound Zn^{2+} are present in mouse submaxillary glands. Two different receptor proteins, one of which is a tyrosine kinase, are present on many cell surfaces.[198,251,253] The glial cells, which lie between the neurons, have their own growth factors.[254]

Bone formation and resorption are influenced by several protein factors. For example, IGF-I stimulates formation of bone, but EGF promotes breakdown.[256] Additional **bone-derived growth factors** and **morphogenetic factors** also have been described.[256,257] A **cartilage-inducing** factor has been identified as TGF-β.[258]

A group of glycoproteins function as hematopoietic growth regulators in the development of blood cells.[259–264] The 166-residue cytokine **erythropoietin** is the primary regulator of red blood cell formation in mammals.[260,264,265] At least four glycoprotein **colony-stimulating factors** (CSF) promote proliferation of granulocytes and macrophages.[259,266,267] The lymphocyte-produced **lymphokines** include the **interleukins** and other proteins. Two species of

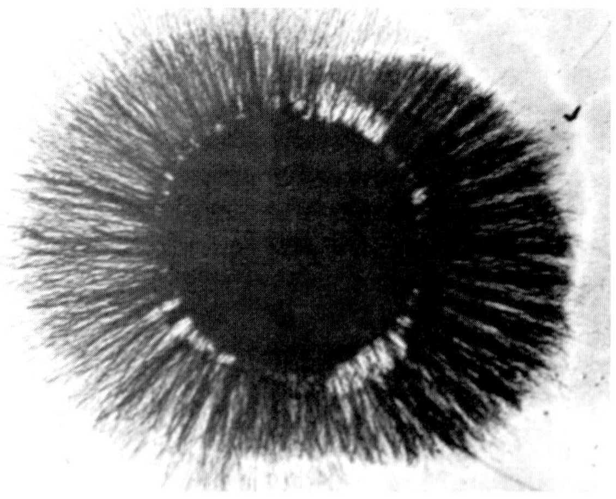

Figure 30-7 Effect of one ng of nerve growth factor in promoting the production of neurites in a chick embryonic sensory ganglion. From Frazier *et al.*[255]

interleukin-1 (IL-1) serve as mediators of inflammation.[267,268] They induce proliferation of T lymphocytes and fibroblasts, bone resorption, release of acute phase proteins (Section C), breakdown of cartilage, and fever.

Interleukin-2 (T-cell growth factor; Fig. 30-6A) is secreted by some activated T-lymphocytes. This 133-residue largely helical protein is involved in generation of cytotoxic T-cells, stimulation of interferon release, and of release of a B-cell growth factor.[269] Considerable excitement has accompanied the possibility of activating lymphocytes with IL-2 produced from cloned genes in bacteria to increase their ability to kill cancer cells. However, IL-2 is toxic, and this is limiting its use. See also Chapter 31, Section C.

IL-3 is one of the colony-stimulating factors, which stimulates the growth of many types of blood cells.[270] Other lymphokines include one derived from T helper cells, which activates resting T lymphocytes thus amplifying an immune response. Others (Chapter 31) are **α-interferon** and the neurotrophic factor (autocrine motility factor) **neuroleukin**.[271] It acts in monomeric form, but as a dimer it seems to be identical to the enzyme phosphoglucose isomerase.[272] While most hormones regulating growth and differentiation seem to be large peptides or proteins, **bursin**, which induces differentiation of lymphocytes, is the amidated tripeptide Lys-His-Gly-NH_2.[273] The corresponding differentiation hormone for T lymphocytes is the 49-residue **thymopoietin**, a hormone of the thymus gland.[274]

Tumor necrosis factor (TNF, also called cachetin) is a 157-residue hormone secreted by macrophages. It is a mediator of inflammatory responses including fever, shock, and **cachexia**, the wasting of the body during chronic diseases including cancer. TNF was isolated as the causative agent of cachexia and also as a factor produced in acute bacterial infections, which sometimes caused death of tumor cells and spontaneous recovery from cancer. In the latter case, it is the lipopolysaccharide (Fig. 8-30) and other bacterial endotoxins that induce the release of TNF by macrophages. Its extreme toxicity has prevented immediate harnessing of the tumor-killing potential of TNF. One function of TNF is regulation of transcription factor NF-κB (Fig. 5-40) in neutrophils and macrophages,[275] a key part of the inflammatory response. TNF also mediates programmed cell death (apoptosis)[276] and has been linked to obesity-induced insulin resistance.[277] The cell surface TNF receptors have a variety of modular structures consisting of various disulfide-linked subdomains.[276]

This long list of vertebrate peptide growth and regulatory hormones is not complete. The biological actions of these hormones are also complex. Growth factors usually have pleiotropic effects, which may involve many tissues as well as many regulatory systems. Are there any simplifying generalizations? Loret *et al.*[278] point out that some growth factors such as IFG-1 and EGF are ubiquitous, affecting virtually all tissues. Others, such as PDGF and thrombin (Fig. 12-17), are more localized in their effects. Some, such as the lymphokines, are more specialized. For one group of hormone receptors the effects are mediated by tyrosine kinases and internalization of the receptors. Another group of receptors activate G proteins and, in turn, adenylate cyclase or phospholipase C. The regulatory domains of the various receptors overlap, a property that allows different tissues to respond differently to hormonal stimuli. The result is the network of interactions that makes the body so sensitive and responsive.

7. Nonpeptide Mammalian Hormones

Most nonpeptide hormones have been considered in other places in the book as indicated in Table 30-2. Because of their importance in the brain adrenaline, noradrenaline, serotonin, and melatonin are also dealt with in Section B,9 of this chapter.

The volatile hormones nitric oxide (NO) and carbon monoxide (CO). The free radical molecule nitric oxide, commonly abbreviated as either •NO or simply NO, is formed by hydroxylation of guanidine groups of arginine (Eq. 18-65). First recognized as the **endothelium-derived relaxing factor**,[279] NO has received increasing attention because of its involvement in a broad range of physiological processes. These include regulation of blood pressure through effects on smooth muscles of the vascular endothelium, regulation of several aspects of the innate immune system (Chapter 31), and neurotransmitter functions both in the brain and in the peripheral nervous system.[279–281] Roles for NO in bacteria, other microorganisms, and plants have also been discovered.[281,282] These often involve regulation of transcription.[283] As mentioned in Chapter 18, Section F,2, many of the effects of NO are a result of activation of soluble guanylate cyclase (p. 561).[283–285a] In the endothelium other hormones, such as the endothelins (p. 1750), atrial natriuretic factor, and bradykinin (Box 22-D), cooperate in the regulation of NO synthase.[279] Neuronal NO synthase functions in the brain in olfaction and in formation of memory. In the peripheral system it mediates penile erection[284,285–286a] and plays a variety of roles in the enteric nervous system.[287] Neuronal NO synthase is often localized to synaptic regions by binding to tissue-specific proteins.[288] NO may also regulate cellular respiration by inhibition of cytochrome *c* oxidase.[285]

In high enough concentrations NO is toxic. It is formed of phagocytic cells and utilized in the killing of ingested pathogens.[279] It also contributes to the inflammatory response of tissues.[289,290] Even the firefly's flash is triggered by a pulse of NO.[291] The dangerous **stonefish**, whose sting causes death within six hours, apparently utilizes NO to kill its victim. A 148-kDa lethal protein (stonustoxin) in its venom induces rapid formation of NO, which causes a fatal drop in blood pressure.[291a]

Like carbon monoxide, NO binds tightly to many metal centers within a cell.[292,293] This has added greatly to the problem of understanding the mechanisms of its action. NO also reacts rapidly with thiol groups of proteins and of small molecules such as glutathione.[294,295] Because of its importance in the regulation of blood pressure, reactions of NO with hemoglobin and the related myoglobin have been studied intensively.[296–300a] NO binds to hemoglobin 1000 times more tightly than either O_2 or CO, preferentially occupying the hemes of the α subunits.[298] Because there is so much hemoglobin in red blood cells, at most one NO per hemoglobin molecule can react. This allows as much as one NO to be carried to tissues along with three O_2 molecules. If the NO could be released in the capillaries, it would activate guanylate cyclase. The resulting cGMP would induce relaxation of smooth muscles and reduce blood pressure.[279] However, tight bonding of NO to deoxyhemoglobin would prevent this release. A plausible possibility (with experimental support) is that NO is not bound to Fe but to the SH group of the conserved cysteine 93 of a β subunit of hemoglobin as SNO (*S*-nitrosothiol) hemoglobin. The NO may bind initially to the iron atom of an α subunit, but then be transferred to the nearby β Cys 93 (p. 359) to form the SNO-Hb.[296] NO may then move from SNO-Hb to thiol groups in the tissues. Recent evidence suggests that the transfer occurs first to an SH group in the anion exchange **AE1** (p. 420).[299] An alternative explanation, which does not involve SNO-Hb, is that hemoglobin Fe-NO is converted to **nitrite** via oxidation of the iron to form a methemoglobin subunit (Eq. 30-1), and that it is nitrite which serves as the endothelial relaxing agent.[300]

$$\underset{\substack{\text{Oxyhemoglobin}\\ \text{or}\\ \text{Mb-Fe(II)}\cdot O_2\\ \text{Oxymyoglobin}}}{\text{Hb-Fe(II)}\cdot O_2} \xrightarrow{NO \;\; NO_3^-} \underset{\substack{\text{Methemoglobin}\\ \text{or}\\ \text{Mb-Fe(III)}^+\\ \text{Metmyoglobin}}}{\text{Hb-Fe(III)}^+} \qquad (30\text{-}1)$$

The function of the monomeric myoglobin has often been assumed to be participation in facilitated diffusion of O_2. Although this is an important function under some circumstances, an additional role for myoglobin may be to scavenge NO, via Eq. 30-1, preventing its buildup to dangerous levels. The metmyoglobin produced can be reduced by methemoglobin reductase.[301,302] A different situation is met by the parasitic nematode *Ascaris*, whose hemoglobin binds O_2 so tightly that it can't serve as an O_2 carrier. It may serve as an **NO-activated deoxygenase**, again using Eq. 30-1 to remove O_2, which can be toxic to the nematode.[303] Free myoglobin can also react with NO to form heme-NO and heme-nitroxyl complexes.[304,305]

Like NO, CO also binds tightly to heme iron and is able to activate guanylate cyclase.[306] CO is formed in the human body by the action of heme oxygenases (Fig. 24-24). Synthesis of heme oxygenase-1 (HO-1) in smooth muscle is induced by a low oxygen tension (hypoxia). The resulting elevated level of CO not only may produce increased vasodilation, but also may inhibit synthesis of vascular smooth muscle cells.[307] Heme oxygenase-2 (HO-2) is found in the brain, where it is colocalized with soluble guanylate cyclase.[308] Some other organisms have a more active CO metabolism. The CO oxidation system of *Rhodospirillum rubrum* is activated by a CO-sensing heme protein, which acts as a transcriptional regulator. The CO binds to the heme iron, apparently inducing a conformational change that allows the protein to bind to its target DNA sequence.[309,310]

Hormonal lipids. We have already considered a number of hormones that are not water-soluble but may have to be transported by carrier proteins to their sites of action. These include retinoic acid (Box 22-A), metabolites of vitamin D (Box 22-C), and the platelet-activating factor (Box 8-A). The last functions in the brain[311] as well as in blood. Hormonal lipids also include the prostaglandins (Fig. 21-7), leukotrienes, and lipoxins (Fig. 21-8). These are products of the eicosenoid cascade or network, which is activated by receptors linked to phospholipase C (Fig. 11-9). Ceramide formed by hydrolysis of sphingomyelin initiates additional responses.[312,313] Sphingolipids may also be important mediators of apoptosis.[314]

The sleep-inducing oleamide (p. 382) modulates signaling by serotonin-dependent and Gaba-dependent neurons and blocks gap junction signaling in brain glial cells.[315–318] Oleamide is one of a family of fatty acid amides found in human plasma. One of these, found also in the brain, is **anandamide** (arachidonoyl-ethanolamide). It is an endogenous activator of the brain cannabinoid receptors.[318–320] The 22-carbon **erucamide** (*cis*-13-docosenamide) stimulates growth of blood vessels.[316] The fatty acid amides are apparently synthesized from corresponding acylglycines (Eq. 30-2) by the action of the peptidylglycine α-amidating enzyme using the mechanism of Eq. 10-11. See also pp. 1792, 1793.

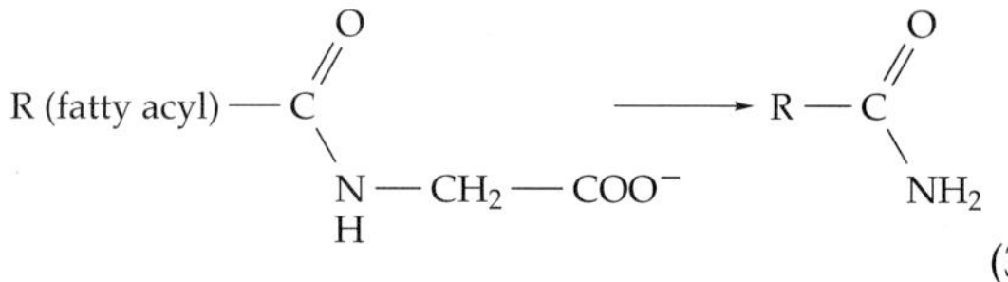

$$R\text{(fatty acyl)}-C(=O)-NH-CH_2-COO^- \longrightarrow R-C(=O)-NH_2 \quad (30\text{-}2)$$

The fatty acid amides are destroyed by an integral membrane protein, a **fatty acid amide hydrolase**.[321,322]

8. Nonvertebrate Hormones and Pheromones

The chemical signals that are passed between bacteria and other microbial cells often resemble hormones of vertebrates. Thus, some bacteria secrete peptide mating pheromones. The sequence of an octapeptide[270] of this type from *Streptococcus faecalis* is given in Table 30-5. Many bacteria utilize "quorum signaling." They do not secrete signaling molecules until they sense that there are enough of them to be effective if they act in unison. Then they all secrete an inducer. The best-known example is the induction of bioluminescence of *Vibrio fischeri* (Eq. 23-49). Long-chain fatty acyl derivatives of L-homoserine lactone act as secreted inducers.[323,324] A lactonase hydrolytically inactivates the inducer to avoid excessive accumulation.

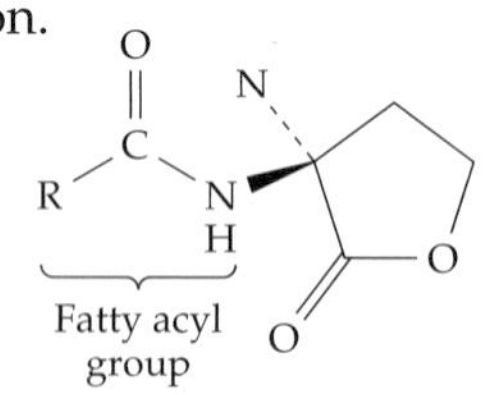

Fatty acyl group

N-Acylhomoserine lactones

Depending upon the types of bacteria and the specific response a variety of different fatty acyl groups may be present in the inducer. Other responses include the formation of bacterial film (biofilms) on a surface and release of virulence factors that induce attack on a host.

Sexual conjugation in yeast is also induced by pheromones (mating factors).[325–327] Yeast cells of mating type **a** synthesize the 12-residue mating factor **a** which contains a C-terminal cysteine methyl ester *S*-alkylated with a *trans,trans*-farnesyl group (Table 30-5). Cells of type **α** synthesize a 13-residue factor **α**.[327a] Cells are attracted to the pheromone produced by cells of the opposite type. The **tremerogens**, sex hormones of certain basidiomycetes, have related structures (Table 30-5).[328]

Peptides are not the only fungal hormones. The water mold *Blastocladiella* releases a **zoospore maintenance factor**, a cyclic phosphate derivative of 5'-phosphoribosyl-5-aminoimidazolecarboxamide.[329] It is similar to the succinocarboxamide, which is an

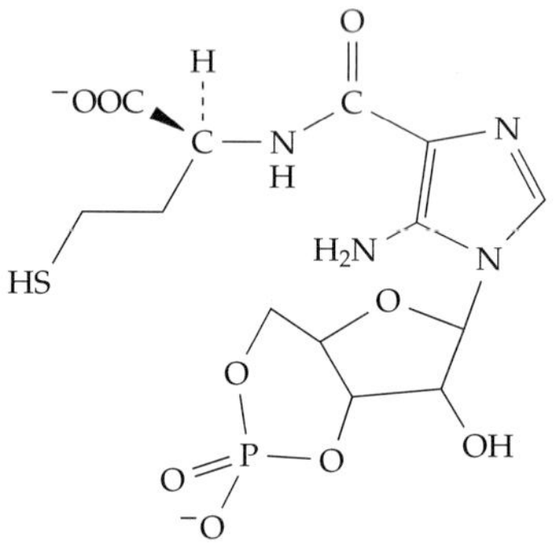

Proposed structure of zoospore maintenance factor of *Blastocladiella*.

intermediate in *de novo* synthesis of purines (Fig. 25-15). Substitution of homocysteine for L-aspartate in step *g* of that sequence could generate a precursor to the zoospore maintenance factor.

Male sperm cells of the alga *Chlamydomonas allensworthii* (Fig. 1-11) are attracted to female gametes by a pentosylated isoprenoid quinone.[330]

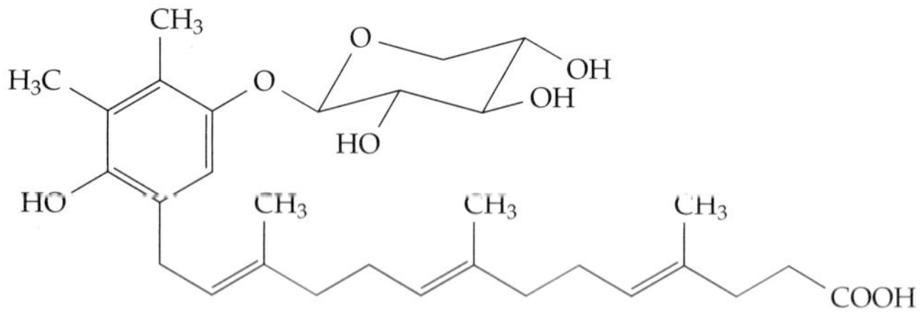

Compound 1

Neurohormones of invertebrate animals. The Cnidaria (coelenterates) have the simplest known nervous system. Simple amidated tetrapeptides, some of which are also found in molluscs, are among their neurotransmitters.[331] EGRFamide and L-3- phenyllactyl-LRNamide are found in some sea anemones.[332] When a hydra (Fig. 1-13) is cut into two pieces, one containing a head and one a foot, each piece reforms the missing end. The decapeptide **head activator** (Table 30-5) diffuses upward from the foot end and induces formation of a head. Similarly, a hormone produced by head cells may induce growth of a new foot end. Glutathione, flowing out from the hydra's prey after wounding by a nematocyst, is the feeding attractant for *Hydra vulgaris*.[282,333] Termination of the response is dependent upon nitric oxide in this primitive invertebrate.

Certain large anemones enter into a symbiotic relationship with fishes, which recognize chemical signals from the anemones and are also chemically protected from the anemones' stings. One of the fish attractants, **amphikuemin**,[334] is effective at a concentration of 10^{-10} M.

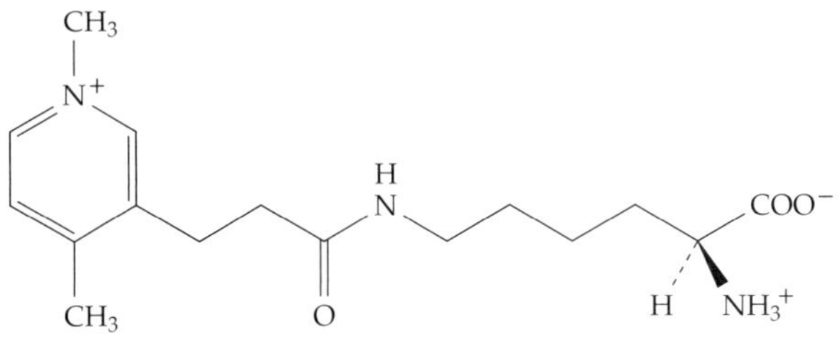

Amphikuemin, a powerful fish attractant

A variety of peptide neurohormones are produced by molluscs. Among these are the sea snail *Aplysia*, which is studied because of its simple nervous system and giant neurons.[335,336] Proteolytic processing of precursors within single cells often yields neurohormones specific to those cells. The structures of two **small cardioactive peptides** secreted from single neurons[337] are shown in Table 30-5. A 4.4-kDa egg-laying hormone is formed in at least eight processing steps.[338–341] Among the peptides identified in the freshwater snail *Lymnaea* are many that are characteristic of mammalian pituitary gland, pancreas, brain, and intestinal tract. These include TRH, ACTH, αMSH, arginine vasopressin, ocytocin, calcitonin, gastrin, gastrointestinal peptide, glucogen, insulin, Met-enkephalin, pancreatic poly- peptide, secretin, somatostatin, substance P, and vasoactive intestinal peptide. Also present are FMRF amide and arginine vasotocin.[342]

The cardioacceleratory peptide FMRFamide (Table 30-5), which was discovered in 1977, was the first in a large series of related neuropeptides that are found in organisms ranging from the nematode *Caenorhobditis*

TABLE 30-5
Some Microbial and Invertebrate Peptide Hormones

Source name	Number of residues	Sequence[a]
Streptococcus faecalis sex hormone	8	FLVMFLSG
Yeast mating factor a	12	YIIKGV(L)FWDPAC(farnesyl[b])–OCH_3
Tremerogen A-10	10	EHDPSAPGNGYC(farnesyl[b])–OCH_3
Hydra head activator	10	pEPPGGSKVIF
Antho-RF amide (sea anemone)	4	pEGRF–NH_2
Small cardiovascular peptides (*Aplysia*)		
SCP-A	11	ARPGYLAFPRM–NH_2
SCP-B	9	MNYLAFPRM–NH_2
FMRFamide (coelenterates, molluscs)	4	FMRF–NH_2
(octopus)	7	YGGFMRF–NH_2
Shrimp blanching hormone	8	pELNFSPGW–NH_2
Fidler crab pigment-dispersing hormone	18	NSELINSILGLPKVMNDA–NH_2
Proctolin (cockroaches)	5	RYLPT
Myotropic neuropeptide (cockroaches)	11	EQFEDYGHMRF–NH_2 (Y: sulfate ester in leukosulfakinin)
Adipokinetic hormone (locust)	10	PELNFTPNWGT–NH_2
Crustacean cardioactive peptide	9	EPFCNAFTGC–NH_2 (C–S–S–C)

[a] One-letter abbreviations, pE, 5-oxoproline; –NH_2, C-terminal carboxamide.
[b] Alkylated on S.

elegans to vertebrate animals. At least 18 genes in *C. elegans* encode 53 distinct FMRFamide-related peptides. Disruption of one of these genes causes hyperactivity, uncoordination, and other behavioral difficulties in the nematodes.[343] Several groups of FMRFamide-related peptides have been found in *Drosophilia*[344] and other insects.[345] One group has the C-terminal FLRFamide and another HMRFamide. Among the latter are sulfate esters such as the cockroach neuropeptide shown in Table 30-5. The sequence of the sea urchin sperm chemoattractant **speract** (sperm attractant peptide-1; SAP-1) is shown in Table 30-4. This is one of a family of egg-associated peptides that stimulate sperm metabolism and mobility. The DNA sequence that codes for speract predicts formation of a 296-residue protein that contains four speract sequences plus six related decapeptide sequences, each separated by a single lysine residue.[346] Many other SAP peptides, some containing the unusual amino acid *o*-bromo-L-phenylalanine,[347] are formed.

Hormones of insects and crustaceans. Peptide neurohormones of insect brains[348] include the pentapeptide **proctolin** (Table 30-5), which was first iso-lated from the cockroach and has since been found in crustaceans and in mammalian brain. It has been traced to specific insect neurons.[349] A nonapeptide neurohormone from the shore crab does not resemble any other known vertebrate or invertebrate hormone.[350]

The prawn *Pandalus borealis* changes its body color by means of movable pigment granules. The neurosecretory octapeptide **blanching hormone**[351] (Table 30-5) controls the process. This is a member of a larger family of peptide hormones with related functions such as the 18-residue **pigment-dispersing hormone** of the fiddler crab[352] and similar hormones in insects.[353]

One of the insect neurohormones, the **activation hormone**, controls the secretion of the corpora allata, paired glands that synthesize the **juvenile hormone** (Fig. 22-4) in insect larvae. While the structure of the juvenile hormone varies somewhat with species, it is usually a polyprenyl ester. A specific binding protein provides the hormone with protection from degradative enzymes. However, in the tobacco hornworm an esterase, able to hydrolyze the protein-bound juvenile hormone, is produced at the start of pupal differentiation.[354] The exact mechanism of action of juvenile hormones has been difficult to determine. However, it affects polyamine synthesis.[355,356]

The **mandibular organ-inhibiting hormone** of the crab *Cancer pagurus* produces two neurohormones that inhibit the secretion of methyl farnesoate, which is thought to function as the juvenile hormone in crustaceans.[357]

A related role of the insect corpora allata is to store and release the **prothoracicotropic hormone**, a peptide neurohormone formed in a single neurosecretory cell of the brain.[358,359] The steroid hormone **ecdysone** (Fig. 22-12) is secreted by the insect's prothoracic gland. Also known as the **molting hormone**, ecdysone is required for the periodic replacement of the exoskeleton of the larvae.[359a] It induces molting in crayfish and other arthropods and appears to be needed by such members of lower phyla as schistosomes and nematodes. It also controls the biting behavior of mosquitos.[360] In addition to α-ecdysone, the 20- and 26-hydroxyecdysones and 20,26-dihydroxyecdysone have been identified in insects.[348] It has been suggested that different ecdysones may function at different stages of insect development.

Ecdysone stimulates the synthesis of RNA in tissues. Visual demonstration of the effect is provided by its action on polytene chromosomes of fly larvae (Fig. 26-14).[361] Fifteen minutes after the application of ecdysone, a puff is induced on one band of the chromosome; a second puff forms at a later time while a preexisting puff diminishes. Thus, like steroid hormones in mammals, ecdysone appears to have a direct controlling effect on transcription. The cuticle-shedding process (**ecdysis**) is initiated by the brain peptide **eclosian**. However, the brain may be responding to the **ecdysis-triggering hormone**, a peptide that is secreted by a series of epitracheal glands located in various segments of the body.[362]

Adipokinetic hormones control metabolism of insects during long-distance flight.[359,363] In the migratory locust these hormones consist of a pair of related octapeptides and a decapeptide (Table 30-5). The hormones stimulate triacylglycerol lipase in the insects' fat bodies, induce release of carbohydrates from body stores, and affect many other aspects of metabolism.[363] Insects also have hormones of the insulin family, proteins consisting of disulfide-linked A and B chains as in insulin. The silkworm *Bombyx mori* has 38 genes for the insulinlike **bombyxins**, which are synthesized in the brain.[364]

Insects produce many different types of sex attractant pheromones (e.g., see p. 382). By 1995 more than 300 structures had been determined for pheromones from >1600 insect species.[365]

9. Plant Hormones

Plants possess a kind of circulatory system by which fluids are transported from the roots upward in the xylem and downward from the leaves through the phloem. Many compounds are carried between cells in this manner, while others are transported across cell membranes and against concentration gradients by active transport. A number of compounds that move between cells in either of these two manners have been classified as hormones.[366–369] The major plant hormones consist of five compounds or groups of compounds: **auxins** (p. 1446), **gibberellins** (Eq. 22-5), **cytokinins** (Fig. 5-33), **abscisic acid** (Fig. 22-4), and **ethylene** (Fig. 24-16). A number of other plant regulators, some involved in defensive reactions, are sometimes also described as hormone. These include the **brassinosterols** (Fig. 22-9) and related compounds,[366,367,369,370] **jasmonic acid** (Eq. 21-18), **salicylic acid** (Chapter 25, Section B,7), bacterially produced **lipooligosaccharides** such as the NOD factors (Box 20-E), and polypeptides such as **systemin**. The sugars glucose and sucrose have hormonelike functions as does light, which controls many plant functions.[368]

Plant hormones have multiple and overlapping functions, which are exerted predominantly by repression of gene expression. This makes it difficult to discuss their functions briefly. Most studied are the auxins, of which the principal member is **indole-3-acetic acid** (Fig. 25-12). This compound, whose biosynthesis is discussed on p. 1446, has been implicated as a controlling agent for cell division and cell elongation. In this capacity auxin influences a great variety of plant processes. Produced principally by growing shoots, auxin diffuses down the stem aided by special **efflux carriers**[371,371a] and inhibits the growth of lateral buds. However, the hormone stimulates the growth of stems, thus establishing the apical dominance of the tip of a plant. Other hormones also have an influence. Auxin is well established as the controlling agent in phototropism, the tendency of a plant to bend toward the sun.

A sensitive test for auxin, which is dependent on the bending of the coleoptile of *Avena sativa* (the common oat) in response to the hormone, allows detection of as little as three pmol of auxin. Using this assay, it was shown that auxin is transported laterally away from the illuminated side of plants, causing the darker side to elongate more rapidly. Both membrane-associated and soluble binding sites, which may represent natural auxin receptors, have been identified, and auxin response elements have been located in DNA. The membrane-bound receptors may regulate an ATP-dependent proton pump, while the soluble receptors may act to regulate gene transcription.

The gibberellins together with brassinosterols are active in helping to determine the *form* of plants. There are 66 different known gibberellins, which are synthesized (Eq. 22-5) in mature leaves and are transported downward. Their very active effect in stimulating RNA synthesis in dwarf varieties of vegetables suggests that gibberellins also serve as gene activators to promote RNA synthesis. A possible function in the **geotropic** response of plant roots is suggested by the presence of higher concentrations of glibberellins in the upper half of horizontal roots than in the lower half.[372] On the other hand, auxin has long been known to have a higher concentration on the lower side of the root,[371a] and it has been assumed to inhibit elongation (in contrast to stimulation in stems). Because they are structurally and in physical properties somewhat similar to sterols, gibberellins have been assumed to act by a steroid hormonelike mechanism. The brassinosterols, which are true sterols, may also be expected to bind to soluble receptors and to regulate transcription.

The cytokinins are isopentenyladenosine derivatives (Fig. 5-33), which may be hydroxylated or substituted in the 2 position by a methylthio group.[373,374] Cytokinins are synthesized in roots and translocated upward to other parts of the plant. They may originate in part from degradation of cytokinin-containing tRNA, but there is evidence that they may also be synthesized independently.[374–376] The N^6-isopentenyladenine in tRNAs is generated by transfer of the isopentenyl group from Δ^2-isopentenyl pyrophosphate. The role of cytokinins may be at the level of gene transcription, but it has been difficult to identify signaling sequences.[368] Their hormonal influence on plants appears to be independent of their function in tRNA. The most striking effect of cytokinins in solution is on differentiation of plant cells (Chapter 32).

The terpenoid abscisic acid (Fig. 22-4) is synthesized by degradation of a carotenoid precursor. It is formed by plants in response to *stress* from low temperature, high salinity, or drought. Abscisic acid appears to block the growth-promoting effects of hormones such as the gibberellins and cytokinins. It is sometimes regarded as a general gene repressor, which prepares plants for dormancy. Synthesis of abscisic acid occurs in response to the short day and long night pattern of the fall. It often opposes the action of gibberellins. The signaling pathway for abscisic acid apparently involves release of Ca^{2+} ions and formation of cyclic ADP-ribose (cADPR; p. 564). Among other effects this induces the closing of stomata in leaf surfaces.[368]

Ethylene not only hastens the ripening of fruit but also tends to promote senescence in all parts of plants. Its signaling mechanisms are the best-known for any plant hormone.[368,369] The synthesis and action of ethylene are discussed in Chapter 24, Section D,4. Ethylene is metabolized slowly in plants by oxidation to ethylene oxide. The latter is hydrolyzed to form ethylene glycol, which is metabolized further to CO_2 (Eq. 30-3).

$$CH_2{=}CH_2 \xrightarrow{O_2} \underset{\text{(epoxide, O bridging)}}{CH_2{-}CH_2} \xrightarrow{H_2O} HO{-}CH_2{-}CH_2{-}OH \longrightarrow CO_2 \quad (30\text{-}3)$$

A postulated flowering hormone, **florigen**, has not been isolated.[366] Flowering seems to be controlled by a variety of different hormonal effects.[369] Jasmonic acid (jasmonate) and salicylic acid act as plant defense signals.[377,378] Salicylic acid activates a large number of transcription factors, which induce resistance to a variety of pathogenic organisms, a response referred to as **systemic acquired resistance**. See also Chapter 31, Section G. Among other compounds synthesized as part of the systemic acquired resistance of plants are proteins known as **systemins**.[382] Initially discovered in tomatoes, systemins have been discovered in more than 100 other species of plants. Jasmonic acid emitted from tomato plants acts as a pheromone that attracts wasps to attack caterpillars that feed on the tomato plants.[379] In fact, wounding by herbivores may stimulate emission of a variety of volatile compounds that may attract predators to the attacking herbivores.[380,381]

Many other compounds influence plant growth. Among them are the vitamins, thiamine, pyridoxine, and nicotinic acid, which are synthesized in the leaves and transported downward to the roots. Since they promote growth of roots, they are sometimes referred to as **root growth hormones**. However, they are nutrients universally needed by cells. Various compounds secreted by other organisms can either stimulate or inhibit growth of a given plant. Some are powerful toxins. Others, such as the previously mentioned NOD factors, and evidently also the riboflavin degradation product **lumichrome** (Box 15-B), are beneficial.[383] These **plant growth regulators** may be produced by other plants, by microorganisms, or by fungi.[384,385] Much use is made in agriculture of synthetic growth regulators.

10. Secretion of Hormones

In Chapter 11 the effects of binding of hormones to cell surface receptors have been emphasized. Equally important are the mechanisms that control the secretion of hormones. The topic of exocytosis has been considered briefly in Chapter 8, Section C,6 and aspects of the Golgi in Fig. 20-8 and associated text. Both hormones and neurotransmitters are secreted by exocytosis of vesicles. Cells have two pathways for secretion.[386,387] The **constitutive pathway** is utilized for continuous secretion of membrane constituents, enzymes, growth factors, viral proteins, and components of the extracellular matrix. This pathway carries small vesicles that originate in the trans-Golgi network (TGN; Fig. 20-8). The **regulated pathway** is utilized for secretion of hormones and neurotransmitters in response to chemical, electrical, or other stimuli.

Many neurotransmitters are packaged into **small synaptic vesicles** ~50 nm in diameter. These may originate from large endosomes rather than from the Golgi. They are usually recycled and refilled repeatedly.[386] Secretion of hormones, and of some neurotransmitters, occurs via **large dense-core vesicles** of ~100 nm diameter. These originate from the TGN and are not recycled. They are prominent in chromaffin cells and other cells that secrete large amounts of a signaling molecule. Secretion of hormones and that of neurotransmitters have several common features. Indeed, hormones of the hypothalamus, neurohypophysis, and the adrenal medulla are secreted by specialized neurons. However, while hormones are often carried in the bloodstream, neurotransmitters are most often secreted into the very small volume of a single synapse. The exocytosis must occur very rapidly from a small number of SSVs.

A common feature, and also a puzzle, of vesicular signaling is the nearly universal response to calcium ions. Exocytosis is usually triggered by a rise in the concentration of Ca^{2+}, and most receptor signaling also leads to an increase in cytosolic Ca^{2+}.[388–391] The puzzle lies in the ability of cells to use a common mechanism for so many specific purposes. This topic is considered further in Section B,8. There are also many other factors that can control exocytosis. Recent evidence suggests that NO may play a role.[392]

B. Neurochemistry

The nervous system, a network of neurons in active communication, reaches its ultimate development in the 1.5 kg human brain.[149,393–396] Many invertebrates, such as leeches,[396a] crayfish, insects, and snails, have brains containing no more than 10^4 to 10^5 neurons,[396b,397,398] but the human brain contains ~10^{11}. Each of these neurons interconnects through **synapses** with hundreds or thousands of other neurons. The number of connections is estimated to be as many as 60,000 with each Purkinje cell of the human cerebellum. There may be many more than 10^{14} synapses in the human brain.[399,400]

In addition to neurons, the brain contains 5–10 times as many **glial** cells of several types. The neuroglia occupy 40% of the volume of brain and spinal cord in the human. Some glial cells seem to bridge the space between neurons and bloodcarrying capillaries. Others synthesize myelin. Some are very irregular in shape.

1. The Anatomy and Functions of Neurons

Although neurons have many shapes and forms, a common pattern is evident.[400a] At one end of the elongated cell (Fig. 30-8) is a series of **dendrites**, thin fibers often less than 1 μm in diameter. The ends of the dendrites form synapses with other neurons and act as receivers of incoming messages. Additional messages come into synapses on the **cell body**, while the **axon** serves as the output end of the cell. The axon, a long fiber of diameter 1–20 μm, is also branched. As a consequence, the nervous system contains both highly branching and highly converging pathways. Many of the axons are wrapped in a myelin sheath (Fig. 30-9; pp. 390 and 1767).

The ends of the fine nerve fibers are thickened to form the **synaptic knobs**, which make synaptic contacts with dendrites on cell bodies of other neurons. In most instances the arrival of a nerve signal at the **presynaptic** end of a neuron causes the release of a transmitter substance (neurohormone). The transmitter passes across the 10–50 nm (typically 20 nm) **synaptic cleft** between the two cells and induces a change in the electrical potential of the **postsynaptic** membrane of the next neuron (Fig. 30-10).[149,401] Excitatory transmitters usually cause **depolarization** of the membrane. By this we mean that the membrane potential, which in a resting neuron is –50 to –70 mv (Chapter 8), falls to nearly zero often as a consequence of an increased permeability to Na^+ and a resultant inflow of sodium ions. The resulting **postsynaptic potential** (really a drop in the potential difference) is propagated to the cell body and axon and under appropriate circumstances may initiate an **action potential**. This is a narrow spikelike region of depolarization that travels down the axon at a constant velocity and with undiminished intensity (Fig. 30-11).

A characteristic of many neurons is an *all-or-none response* or firing. An action potential passes down the axon only if there is sufficient depolarization. In general, a stimulus must reach a neuron through *more than one synapse* before the neuron will fire. Furthermore, neurons are often *inhibitory*, releasing transmitters that counter the excitatory synapses and tend to prevent firing. Inhibition is important in damping out small excitations; thus sharpening the response of the nervous system toward strong stimuli. Another characteristic of basic importance to the operation of the brain is that neurons fire at longer or shorter intervals depending upon the strength and duration of the stimulus. The stronger the stimulus to a given neuron, the more rapid the train of spikes that passes down the axon. Thus, the brain functions to a large extent in

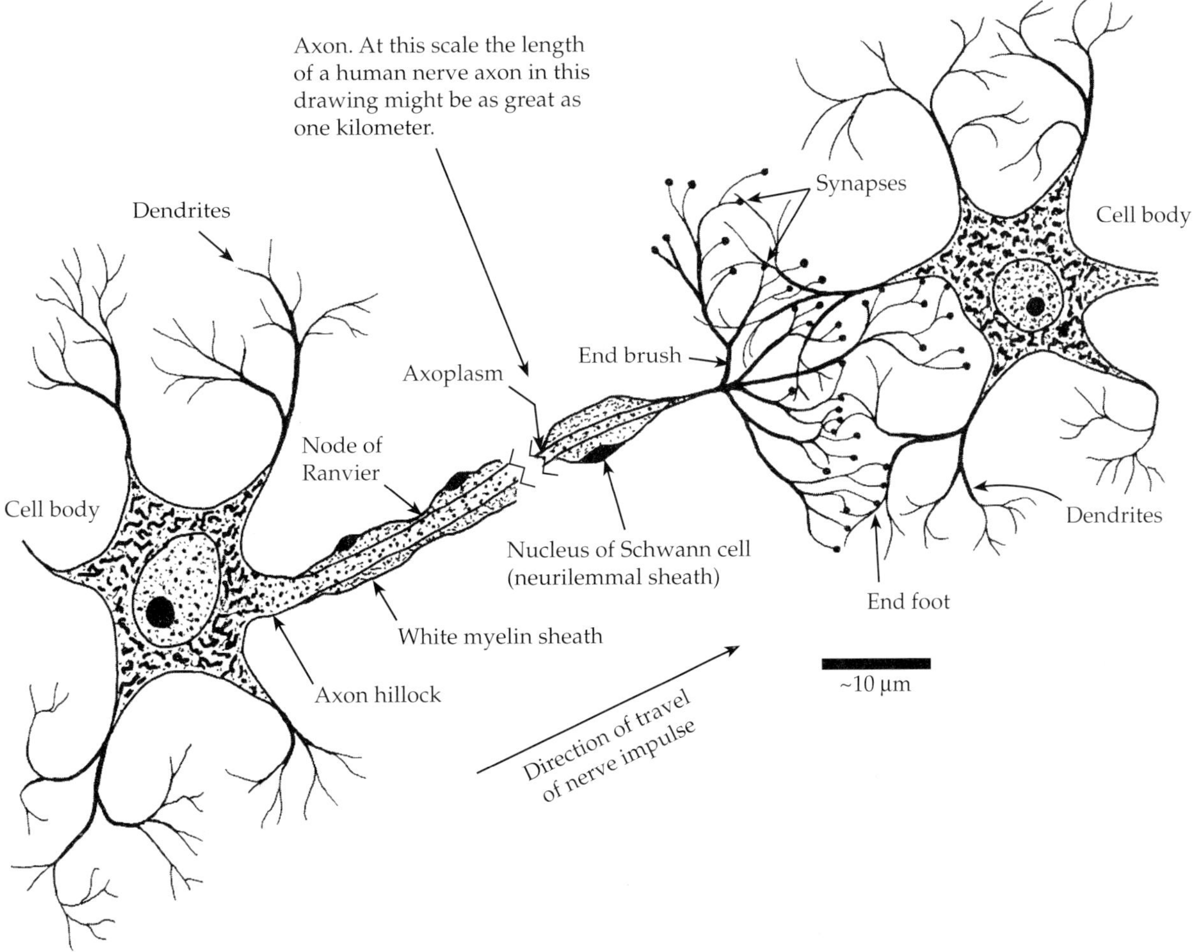

Figure 30-8 Schematic drawing of a neuron (after Brand and Westfall,[401] p. 1192).

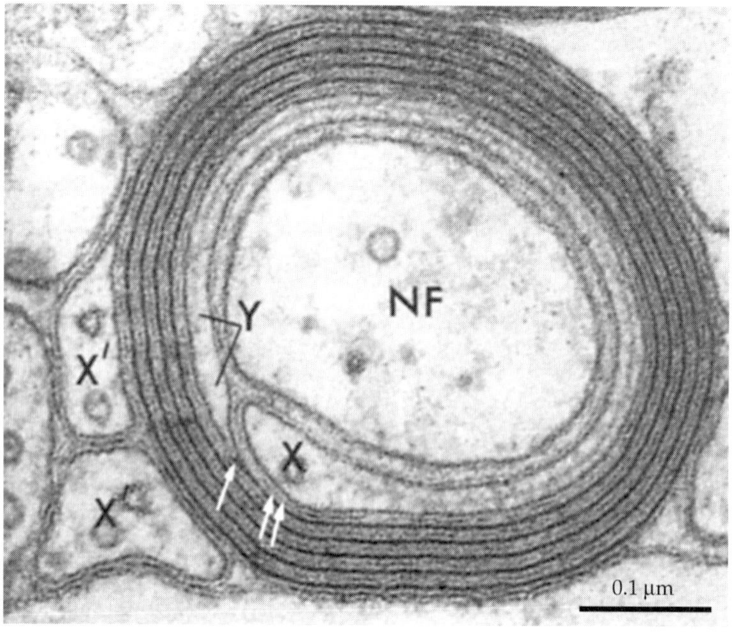

Figure 30-9 Micrograph of a section through an axon of a neuron from rat brain. The structure of the myelin sheath can be seen clearly. The growing lip of cytoplasm (X) from a neuroglial cell is advancing around the axon process (NF) and insinuating itself into the space between the plasma membrane of the axon and the membrane that limits the thin layer of cytoplasm (Y) left behind by the growing lip during its previous turn. This cytoplasmic layer disappears as the inner leaflets of its plasma membrane fuse to form the major dense line of the myelin sheath. This process is occurring at the point indicated by the single arrow. The outer leaflet of the plasma membrane surrounding the lip fuses with its own outer leaflet laid down on the previous turn. The two outer leaflets thus give rise to the less dense intermediate line of the sheath (double arrow). The cell body from which the investing cytoplasmic sheet originated cannot be seen in this micrograph, but cytoplasm within the lateral margins of the sheet does appear (X′). The micrograph, by A. Hirano and H. M. Dembritzer, originally appeared in *J. Cell Biol.* **34**, 555 (1967), where a more complete explanation of myelin sheath formation is provided. Figure copied from Porter and Bonneville.[402] Courtesy of Mary Bonneville.

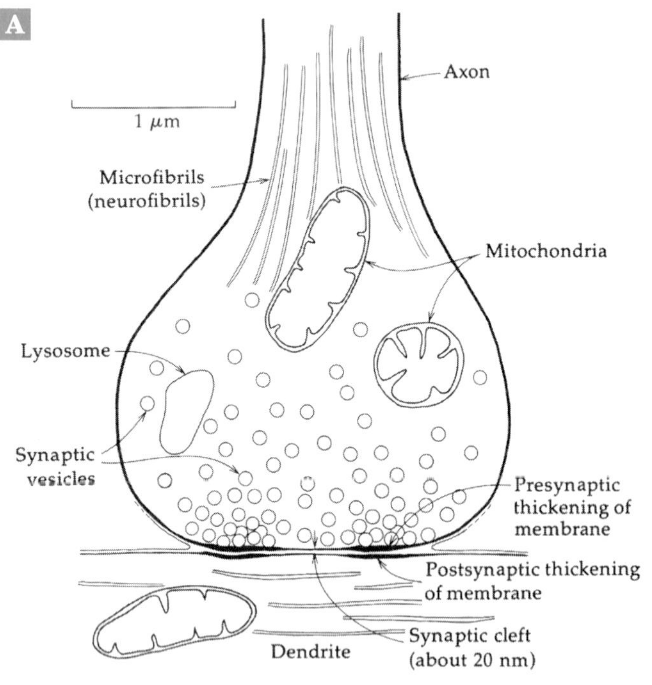

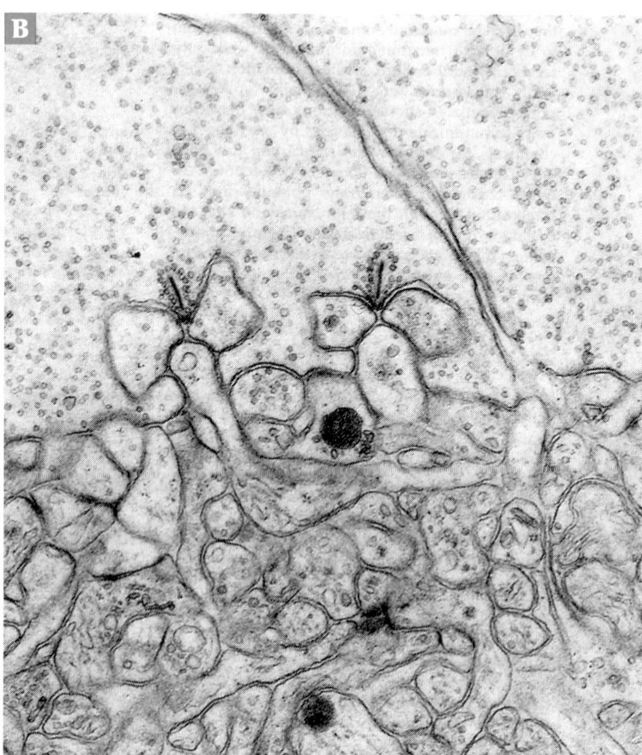

Figure 30-10 (A) Schematic drawing of a synapse. (B) Electron micrograph showing the synaptic junctions in the basal part (pedicle) of a retinal cone cell of a monkey.[403] Each pedicle contains synaptic contacts with 12 triads, each made up of processes from a bipolar cell center that carries the principal output signal and processes from two horizontal cells that also synapse with other cones. A ribbon structure within the pedicle is characteristic of these synapses. Note the numerous synaptic vesicles in the pedicle, some arranged around the ribbon, the synaptic clefts, and the characteristic thickening of the membranes surrounding the cleft (below the ribbons). Micrograph courtesy of John Dowling.

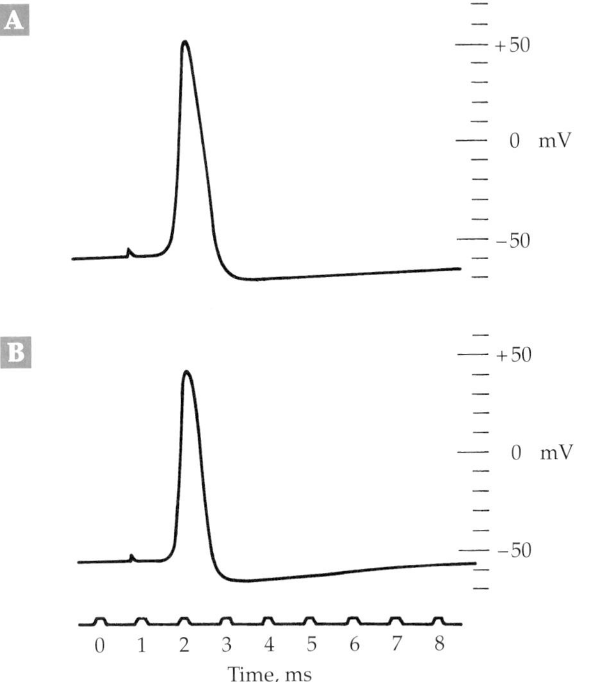

Figure 30-11 (A) Action potential recorded with internal electrode from extruded axon filled with potassium sulfate (16°C). (B) Action potential of an intact axon, with same amplification and time scale (18°C). The voltage scale gives the potential of the internal electrode relative to its potential in the external solution with no correction for junction potential. From A. Hodgkin, *Conduction of Nervous Impulses*, 1964. Courtesy of Charles C. Thomas, Publisher, Springfield, Illinois.

decoding trains of impulses. The frequency of the impulses from neurons varies from a few per second to a maximum of about 200 s^{-1} in most nerves (up to 1600 s^{-1} in the Renshaw cells of the spinal cord). The maximum frequency is dictated by the refractory period of ~1 ms (Section B,3).

Although the concepts of neuronal function outlined in the preceding paragraphs have been accepted for many years, more recent discoveries require that they be modified somewhat. Dendrites seem to be able to transmit information as well as to receive it. Furthermore, while information is certainly transmitted long distances by spike action potentials, shorter neurons and dendrites may communicate extensively by exchange of chemicals through low resistance gap junctions, also called **electric synapses** (Chapter 1). Small changes in membrane potential transmitted through these junctions may alter the behavior of adjacent neurons. Chemical transmitters do not always have an electrical effect on postsynaptic neurons but may influence metabolism or gene transcription.

2. Organization of the Brain

The anatomy of the brain is quite complex, and only a few terms will be defined here. The **cerebrum**, which is made up of two hemispheres, accounts for the largest part of the brain. The deeply folded outermost layer, the **cerebral cortex**, consists of **gray matter**, a mass of cell bodies, and fine unmyelinated nerve fibers. Beneath this lies a layer of **white matter** made up of myelin-covered axons connecting the cerebral cortex with other parts of the brain. The two cerebral hemispheres are connected by the **corpus callosum**, a band of ~2×10^8 nerve fibers. Remarkably, these fibers can be completely severed with a relatively minimal disruption of the nervous system. In the past the corpus callosum was sometimes cut to control almost incessant epileptic seizures that could not be prevented by drugs. The "split-brain" patients suffered relatively little disability as long as both eyes functioned normally. Studies of these patients provided some insights into the differing functions of the two hemispheres of the cerebrum.[395]

Deeper in the cerebrum lie the **basal ganglia,** which include the caudate, lenticular, and amygdaloid nuclei. The lenticular nuclei are further divided into putamen (an outer portion) and the globus pallidus. The putamen and caudate nuclei together are known as the **striatum** (Fig. 30-12). The lower lying subthalamic nuclei and substantia nigra are sometimes also included in the basal ganglia.

The outer parts of the cerebrum, including the basal ganglia, make up the telencephalon. Deep in the center of the brain is the diencephalon consisting of the **thalamus** (actually two thalami), **hypothalamus**, **hypophysis** (Figs. 30-1, 30-13), and other attached regions. A major structure at the back of the brain is the **cerebellum**. Like the cerebrum, its cortex is highly folded. The 30 billion neurons of the cerebellum are organized in a highly regular fashion.[393,404] The interconnections of the seven types of neurons present in this part of the brain have been worked out in fine detail.

The basal part of the brain or **brain stem** consists of the medulla oblongata and the pons. While the bulk of the tissue consists of myelinated nerve tracts passing into the spinal cord, synaptic regions such as the olivary nucleus are also present.

The brain, which must function in a chemically stable environment, is protected by a tough outer covering, the **arachnoid membrane**, and by the **blood–brain barrier**[406,407] and the **blood–cerebrospinal barrier**. Both of these barriers consist of tight junctions similar to those seen in Fig. 1-15A. They are formed between the endothelial cells of the cerebral capillaries and between the epithelial cells that surround the capillaries of the **choroid plexus**. The choroid plexus consists of capillary beds around portions

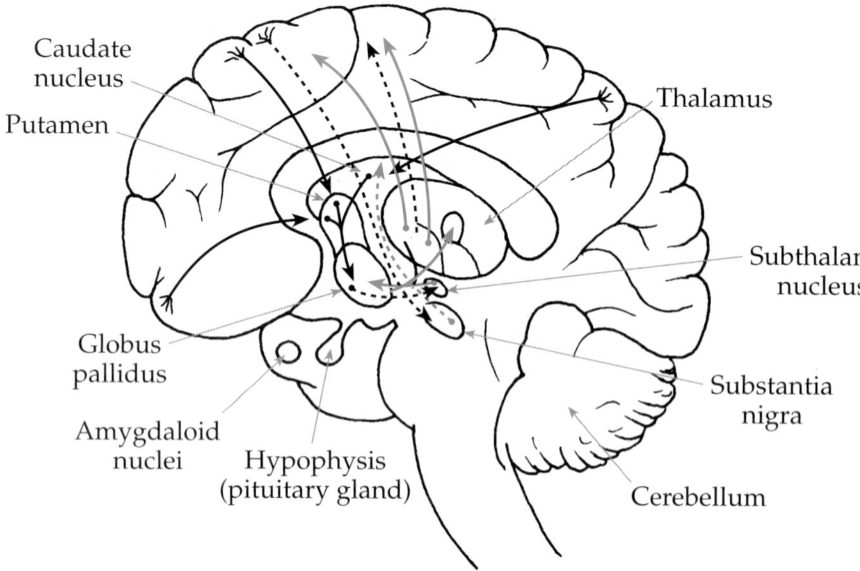

Figure 30-12 Diagram illustrating some of the major interconnections of the "extrapyramidal system" of the brain. Arrows indicate major direction of projections. The nigrostriatal (substantia nigra to striatum) and related neuronal pathways are indicated with dashed lines. After Noback and Demarest,[405] pp. 182 and 183.

of the fluid-filled **ventricles** deep in the interior of the brain. They serve as a kind of "kidney" for the brain assisting in bringing nutrients in from the blood and helping to keep dangerous compounds out.[406]

3. Neuronal Pathways and Systems

Consider a message originating with a nerve receptor in the skin or in another sense organ. A nerve signal passes via a **sensory neuron (afferent fiber)** upward toward the brain. It may pass through two or more synapses (often through one in the spinal cord and one in the thalamus) finally reaching a spot in the sensory region of the cerebral cortex. From there the signal in modified form spreads through the **interneurons** of virtually the entire cortex. In each synapse, as well as in the cortex, the impulse excites inhibitory fibers that dampen impulses flowing through adjacent fibers. Likewise, if a given impulse is not strong enough, it will itself be inhibited before reaching the cortex. Among the important sensory neurons are those from the seven million cone cells and 100 million rod cells of the eye. The nerve signals pass out of the retina by way of a million axons from retinal ganglion cells reaching, among other parts of the brain, the **visual cortex** (Fig. 30-14).[408]

The neuronal events that occur within the cerebral cortex are extraordinarily complex and little understood.[409] In what way the brain is able to initiate voluntary movement of muscles is obscure. However, it is established that the signals that travel out of the brain down the **efferent fibers** to the muscles arise from large **motor neurons** of the **motor cortex**,[410] a region that extends in a band across the brain and adjacent to the sensory cortex (Fig. 30-14). The axons of these cells form the **pyramidal tract** that carries impulses downward to synapses in the spinal cord and from there to the **neuromuscular junctions**. These are specialized synapses at which acetycholine is released, carrying the signal to the muscle fibers themselves. Passing over the cell surface and into the

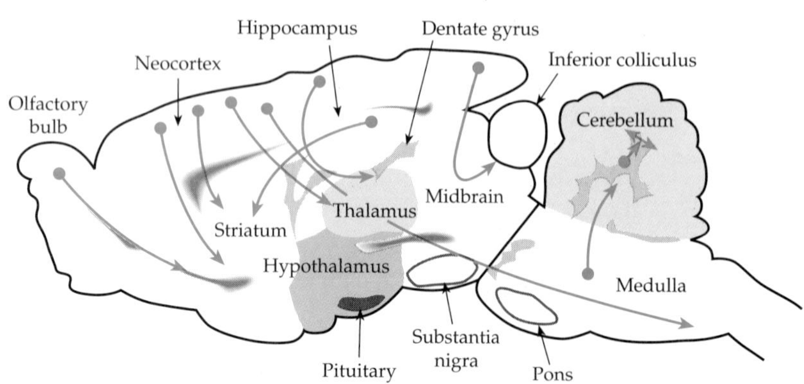

Figure 30-13 Section through a rat brain. This brain, which has been very widely used in neurochemical studies, appears superficially to be quite different from the human brain (Fig. 30-1), which is characterized by its large cerebral cortex. However, basic pathways are the same. Some major pathways for glutamate-secreting (glutamatergic) neurons are marked by arrows. Most of these originate in the neocortex (outer layers of the cerebral cortex) and the hippocampus. From Nicholls.[149] Courtesy of David G. Nicholls.

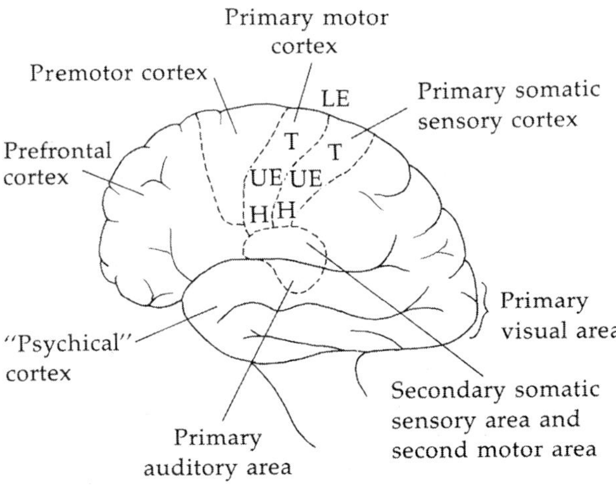

Figure 30-14 The location of several functional areas of the cerebral cortex. The representation of body parts on the primary motor and somatic sensory cortices include the head (H), upper extremity (UE), trunk (T), and lower extremity (LE). After Noback and Demarest,[405] p. 193.

T tubules (Chapter 19, Section B,4; Fig. 19-21), a wave of depolarization initiates the release of calcium and muscular contraction.

At the same time that the motor neurons send signals to the muscles, branches travel into other parts of the brain including the olivary nuclei, which send neurons into the cerebellum. The cerebellum acts as a kind of computer needed for fine tuning of the impulses to the muscles. Injury to the cerebellum leads to difficulty in finely coordinated motions. Input to the Purkinje cells arises from the climbing fibers, which originate in the inferior olive of the brain stem. Each climbing fiber activates a single Purkinje cell, but the dendrites of each Purkinje cell also form as many as 200,000 different synapses with parallel fibers that run across the cortex of the cerebellum (Fig. 30-15). The parallel fibers receive input from many sources via a complex series of mossy fibers and granule cells and influence the firing of the Punkinje cells. The output from the Purkinje cells is entirely inhibitory. It is transmitted via synapses in the cerebellar nuclei to neurons that lead back to the cerebral cortex, into the thalamus, and down the spinal cord.[411] The pathway to the cortex completes an inhibitory feedback loop, of which there are many in the nervous system. For details see Llinás[404] and Nicholls.[149]

In addition to the **somatic motor system** that operates the voluntary (striated) muscles via the pyramidal tract, there is the **autonomic system,** which controls the involuntary (smooth) muscles, glands, heartbeat, blood pressure, and body temperature. This system has its origins in both the cerebral cortex and hypothalamus. It is subdivided into two systems, the **sympathetic** and **parasympathetic** systems, which are anatomically distinct. The sympathetic system is geared to the fight and fright reactions. Its **postganglionic fibers** (those below the ganglia in the spinal cord) liberate norephinephrine (noradrenaline) and include the adrenal medulla, which consists of specialized neurons, the **chromaffin** cells. The parasympathetic system has to do more with homeostasis and maintenance of body systems. Biochemically it is characterized by the release of acetylcholine as a transmitter substance.

The hypothalamus, a four gram portion of the brain, receives a great deal of biochemical attention because of its function in the autonomic nervous system, in homeostasis, and in endocrine secretion. Its liberation of neurohormones that stimulate the hypophysis has already been considered in Section A,3. The hypothalamus is also involved in the regulation of the body temperature, of water balance, and possibly of glucose concentration.

Two other systems of importance in the brain are the **reticular system** and the **limbic system**. The former is the mediator of the sleep–wake cycle and is responsible for characteristic waves in the electroencephalogram. The limbic system is the mediator of **affect** or mood and of **instincts**. It is anatomically complex with centers in the amygdala, other subcortical nuclei, and the limbic lobe of the cortex. The limbic cortex forms a ring lying largely within the longitudinal fissure between the two hemispheres. It includes the olfactory cortex, the **hippocampus**, a region associated with formation of conscious memories, and other evolutionarily older regions of the cerebral cortex. Within the limbic lobe are the **pleasure centers**. When electrodes are implanted in these regions, animals will repeatedly push levers that are designed to electrically stimulate these centers. There are also **punishing centers,** whose stimulation causes animals to avoid further stimulation.

4. The Propagation of Nerve Impulses

Although the chemical basis of the conduction of nerve impulses via an action potential is not entirely clear, the electrical events have been described with precision. If the permeability of a membrane toward sodium ions is increased in a local region, sodium ions flow through the membrane into the cell neutralizing the negative charge inside and depolarizing the membrane. Such depolarization leads to propagation of an electrical signal of diminishing intensity over the surface of the membrane in a manner analogous to the flow of electrical current along a coaxial cable. It is thought that local increases in Na^+ permeability of the plasma membrane often trigger nerve impulses. Other

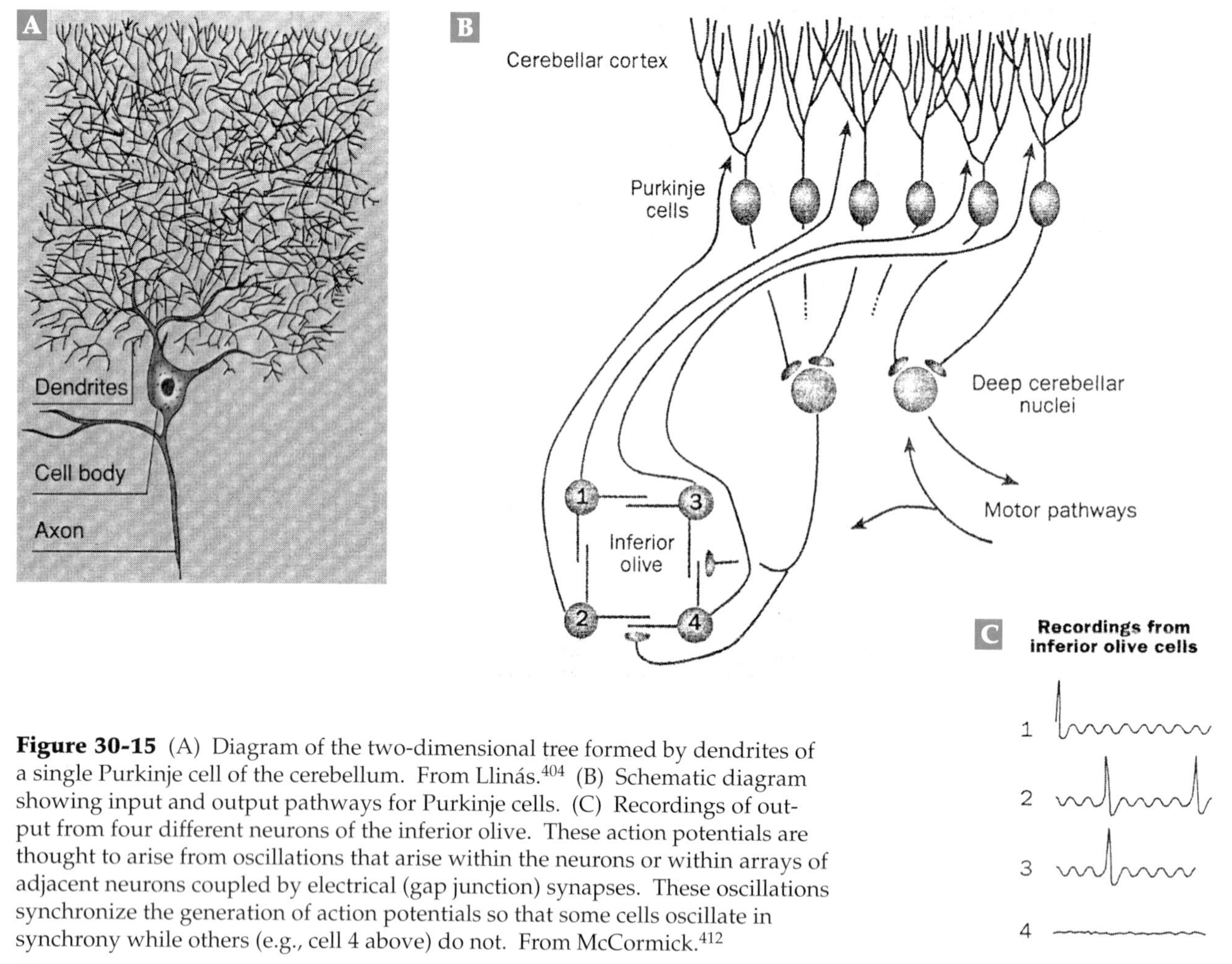

Figure 30-15 (A) Diagram of the two-dimensional tree formed by dendrites of a single Purkinje cell of the cerebellum. From Llinás.[404] (B) Schematic diagram showing input and output pathways for Purkinje cells. (C) Recordings of output from four different neurons of the inferior olive. These action potentials are thought to arise from oscillations that arise within the neurons or within arrays of adjacent neurons coupled by electrical (gap junction) synapses. These oscillations synchronize the generation of action potentials so that some cells oscillate in synchrony while others (e.g., cell 4 above) do not. From McCormick.[412]

ions such as Ca^{2+} may also play a role. While the kind of passive transmission of electrical signals that results from a local depolarization of the membrane is suitable for very short nerve cells, it cannot be used to send signals for long distances. Most nerve axons employ the more efficient action potential. This is an impulse that passes along the axon and for a short fraction of a second (~0.5 ms in mammalian nerves) changes the membrane potential in the characteristic way shown in Fig. 30-11. Initially, the negative potential of 50–70 mV drops rapidly to zero and then becomes positive by as much as 40–50 mV, after which it returns to the resting potential. The remarkable thing about the action potential is that it is propagated down the axons at velocities of 1–100 m / s without loss of intensity.

To establish the chemical basis of the action potential, A. L. Hodgkin and A. F. Huxley in the 1950s devised the **voltage clamp**, a sophisticated device by which the transmembrane current can be measured while using a feedback mechanism to fix the membrane potential at a preselected value.[413–417] Using the voltage clamp the membrane conductance could be measured as a function of the membrane potential and of time. It was found that immediately after a decrease in membrane potential was imposed with the voltage clamp, the permeability of the membrane toward sodium ions rose rapidly. Since an increased sodium ion permeability automatically leads to depolarization in an adjacent region of the membrane, a self-propagating wave is established and moves down the axon. The voltage clamp studies also revealed that after a fraction of a millisecond the permeability to potassium ions also increases. At the same time the sodium ion permeability decreases again, and the normal membrane potential is soon reestablished. However, during an **absolute refractory period** of ~0.5 ms no other nerve impulse can be passed. The sequence of events during passage of the nerve impulse can be described as the opening of sodium channels followed by the opening of potassium channels, and then by a closing of the channels in the same sequence. The results of these investigations led Hodgkin and Huxley to propose equations that quantitatively describe the action potential and that predict the observed conduction velocities and other features of nerve impulses.

A special feature of nerves that are designed to transmit impulses very rapidly is the presence of the wrapping of **myelin** (Fig. 30-9). As can be seen in this figure, the extracellular surfaces of the consecutive wraps bind tightly together, and the cytoplasm of the cell interior is squeezed out to form the compact myelin sheath.[418] Mutations in the integral membrane proteolipid protein (p. 401) are associated with a variety of defects in myelin formation. Some of these are severe, for example, leading to loosely wrapped myelin.[419,420] The proteolipid protein is encoded by an X-linked gene. The most abundant protein in peripheral nerve myelin is the integral membrane **peripheral myelin glycoprotein P_0**. It is encoded by an autosomal gene for which 29 known defects account for a variety of human diseases,[421–422a] including an autoimmune inner ear disease.[423] The extracellular domain of P_0, like many other cell adhesion molecules (p. 407), has a structure related to that of immunoglobulins. Four molecules of P_0, each of which carries a single immunoglobulin domain, associate via these domains in a kind of square donut that protrudes from the outer cell surface. There it can interact with four similar donuts from the apposed cell surface, zipping up the cell–cell interface by a kind of Velcro action.[422,424,425] Protein P_0 accounts for 50% of the total protein of peripheral myelin, but the **myelin basic protein**, which constitutes 20% of the total protein, is also essential.[426] This protein exists as a variety of forms that arise from differential splicing of its mRNA and extensive posttranslational modification. Deimination of argine side chains to form citrulline residues has been associated with development of the autoimmune disease **multiple sclerosis**.[427,428] **Peripheral myosin protein 22** is a 160-residue polypeptide with four membrane-spanning helices. It accounts for 2–5% of the myelin protein and is the site of defects that cause the demyelinating **Charcot–Marie–Tooth disease** and other serious human diseases.[428a,b]

The axon is effectively insulated from the surrounding medium by the myelin sheets except for special regions, the **nodes of Ranvier**, which lie at 1- to 2-mm intervals along the nerve. The nerve impulse in effect jumps from one nerve to the next. This **saltatory conduction** occurs much more rapidly (up to 100 m / s) than conduction in unmyelinated axons. It depends upon Na^+ and K^+ channels that are concentrated in the nodes of Ranvier.

5. Ion Conducting Channels

What is known about the channels through which Na^+ and K^+ flow during nerve excitation? That the channels for the two ions are separate was shown by the fact that **tetrodotoxin** (found in the puffer fish)[429,429a] and **saxitoxin** of dinoflagellates, as well as scorpion toxins (see Fig. 30-16), exert their toxic action by blocking the Na^+ channels while having no effect upon conductance for K^+. At the same time the K^+ channels can be blocked by certain quaternary ammonium salts. Since the binding constants for the toxins are high ($K_f \sim 3 \times 10^8 \ M^{-1}$ for tetrodotoxin), it is possible to titrate the sodium channels. The number is usually quite small, about 10–400 Na^+ channels / μm^2 of surface[430] (the same surface area contains 2×10^6 phopholipid molecules). However, membranes in the nodes of Ranvier of mammalian nerve fibers[431] contain ~12,000 channels / μm^2. Note that the ion channels described here are not the same as those in the ion pump, i.e., the Na^+,K^+-ATPase (Fig. 8-25). In some neurons the number of conduction channels for Na^+ appears to be ten times less than the number of pumping channels, i.e., of Na^+,K^+-ATPase.[432,432a]

Since the number of ion-conducting channels is small, the rate of sodium passage through the open channels must be extremely rapid and has been estimated as ~10^8 ions / s.[433] This is within an order of magnitude of the diffusion-limited rate (Eq. 9-30). On this basis it is clear that the channels cannot act by means of ionophoric carriers but form pores that can be opened and closed (**gated**) in response to changes in the membrane potential. They are **voltage-sensitive ion channels**.[433,434] The channels are selective for specific ions and the selectivity parallels that of sites in some cation-exchange resins such as those containing carboxylate groups. This suggested that the inside surface of the channel might contain one or more carboxylate groups from protein side chains as well as other polar groups. A Na^+ ion approaching the channel entrance might exchange some of its hydration sphere for ligands from the channel surface. The differing affinity of the "ion exchange" sites for various cations could ensure that it is predominately Na^+ that passes through the channel. Anions could be excluded by electrostatic repulsion. Recent structural studies have allowed these speculations to be replaced with experimental findings as described in the following paragraphs. They have revealed that the selectivity mechanism are similar for Na^+ and Ca^{2+} channels.

The sodium ion channel of the electric eel. Making use of the binding of radioactively labeled specific toxins to identify them, the subunits of the sodium channel proteins were purified from several sources including the electrical tissue of the electric eel *Electrophorus electricus*,[437–439] heart and skeletal muscle, and brain.[440–441b] In all cases a large ~260-kDa glycoprotein, which may be 30% carbohydrate, is present. The saxitoxin-binding protein from rat brain has two additional 33–36 kDa subunits with a stoichiometry of $\alpha\beta_1\beta_2$. The *Electrophorus* α subunit consists of 1820 residues,[437] while rat brain contains α proteins of 2009

A

Tetrodotoxin of the puffer fish and certain newts

B

Saxitoxin of *Gonyaulax*, a dinoflagellate whose toxin accumulates in shellfish. As little as 0.3 mg of toxin can be fatal.

C

Structure of erabutoxin b. Many snake venom neurotoxins have similar structures with conserved disulfide bridge positions. See Hatanaka, H. *et. al.*[513]

D

Brevetoxin A

E

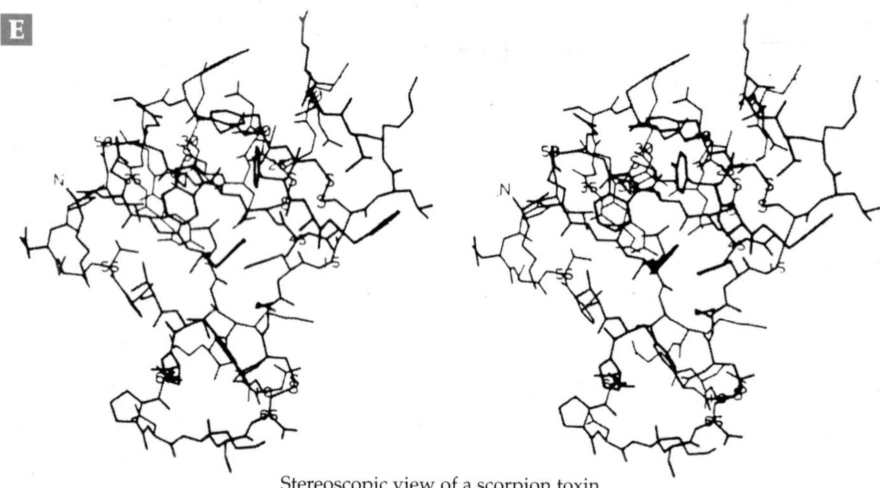

Stereoscopic view of a scorpion toxin

F

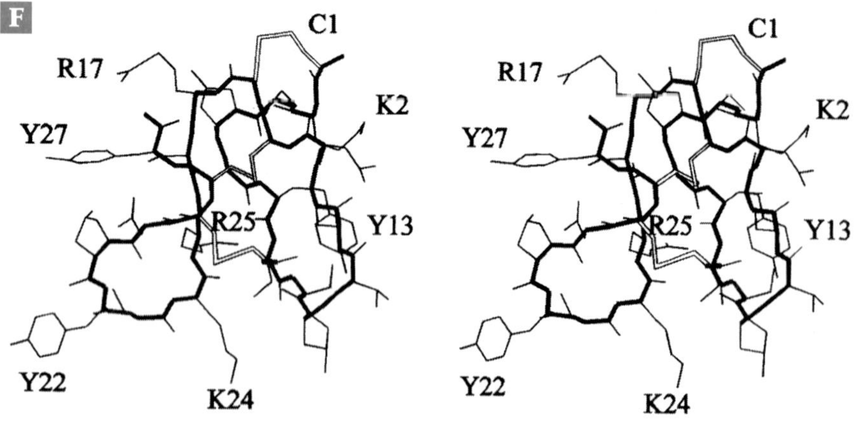

ω Conotoxin, blocks presynaptic voltage-regulated Ca^{2+} channels

Figure 30-16 Structures of some neurotoxins that affect ion channels. Other neurotoxins include the Na^+, K^+-ATPase inhibitor ouabain (Fig. 22-12), batrachotoxin (Fig. 22-12), and picrotoxin (Fig. 22-4). The structure of a scorpion toxin is from Almassy *et al.*,[494a] that of ω conotoxin is from Pallaghy *et al.*,[435] and that of brevetoxin is redrawn after Shimizu *et al.*[436]

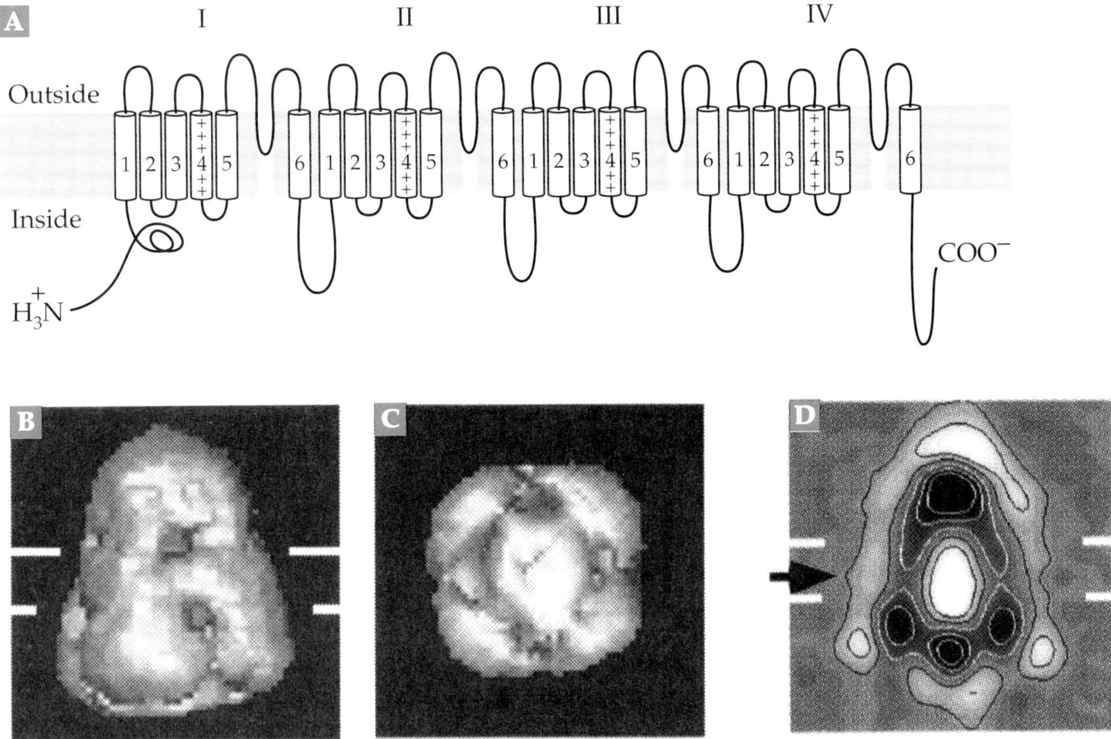

Figure 30-17 (A) Two-dimensional map of the ~260-kDa α subunit of the voltage-gated Na^+ channel from the electric eel *Electrophorus electris*.[438,441] (B) Image of the sodium channel protein obtained by cryo-electron microscopy and image analysis at 1.9 nm resolution. In this side view the protein appears to be bell-shaped with a height of ~13.5 nm, a square bottom (cytoplasmic surface) ~10 nm on a side, and a hemispherical top with a diameter of ~6.5 nm. (C) Bottom view of the protein. (D) Axial section which cuts the bottom, as viewed in (C), approximately along a diagonal. From Sato *et al.*[438] Notice the cavities (dark) and domain structures (light). The black arrow marks a constriction between upper (extracelllar) and lower (cytoplasmic) cavities. White lines indicate approximate position of the lipid bilayer. From Sato *et al.*[438] Courtesy of Chikara Sato.

and 2005 residues, respectively, for Na^+ channels designated I and II.[441] In fact, mammals contain ten distinct Na^+ channel genes.[442] In every case the channel proteins contain four consecutive homologous sequences of about 300 residues apiece. Within these the hydro-pathy plots (see Fig. 2-30) suggest that each homology region forms six membrane-spanning helices as shown in Fig. 30-17A.[441] The four sets may then fold together into a square arrangement that provides a pore somewhat familar to that of the voltage-gated K^+ channel (see Fig. 30-18). The three-dimensional structure of the sodium channel protein, based on cry-electron microscopy, appears to be complex. The central channel may resemble that of Fig. 30-18, but there also seem to be smaller peripheral channels (Fig. 30-17).[438] Bacteria also contain Na^+ channels but they are tetramers of smaller subunits, resembling in this respect bacterial K^+ channels (Figs. 30-18).[442a,b]

How do the "gates" to ion channels open? Presumably some part of the channel protein senses the change in potential and undergoes an appropriate alteration in conformation that opens the gate.[434] The current carried by the ions flowing out through a small number or even a single channel can be measured with tiny **patch electrodes** having openings ~1 μm^2 in area. These are pressed against the nerve membrane, where they form a tight seal. With such a small patch of membrane surface the electrical noise level is low, and it is possible to measure the conductance of the pore.[149,443] From such measurements it was found that a single pore can allow $>10^8$ ions to pass through in one second. Another thing that is apparently measured with patch electrodes is a small **gating current**, which precedes the opening of the channels by ~0.1 ms. This has been interpreted as a flow of ~6 charges across the membrane or the movement of a larger number of dipoles needed to open the gate. One possibility is that a loss of the electrical field from the surface charges on the bilayer induces a rearrangement of charges on protein side chains within the bilayer or induces changes in interactions between two or more dipoles. Such changes could trigger conformation alterations within the proteins, allowing the channel to switch from open to closed.

Recordings with single channels indicate that after a sodium channel is open for a random length of time it spontaneously closes and passes into a third state, an "inactive" state from which it cannot reopen during the refractory period. After the membrane is repolarized it can function again.[444a]

Calcium ion channels. Immediately after the Na^+ pores open as a result of membrane depolarization, voltage-sensitive Ca^{2+} channels also open. These allow a rapid influx of Ca^{2+}, which can trigger many processes including the secretion of neurotransmitters within the synapses.[434,444] There are several types of voltage-sensitive Ca^{2+} channels.[444a,b] The most abundant type are specifically inhibited by dihydropyridines and are called **dihydropyridine-sensitive** or L-type channels.[434,445–445b] They are most numerous in the transverse tubular membranes of skeletal muscle where they appear to form a complex with the very large calcium release channels, the **ryanodine receptors** (Fig. 19-21 and associated discussion).[446,446a] These channels appear to have a structure similar to that of the Na^+ channels.[434] Calcium channels are also discussed on p. 422 and on pp. 1114-1115. Calcium ions play a central role in cell signaling and there are a large number of different calcium channels in bacteria, plants, and animals. Many of these are coupled to specific receptors.[445b,447] Some are involved in controlling intracellular stores.[447–449] Some release Ca^{2+} in response to mechanical movement and function in feeling, hearing, maintaining balance, and cardiovascular regulation. Plants sense wind and gravity, and microorganisms sense changes in osmotic pressure with the aid of these channel proteins.[450–452]

Potassium ion channels. Several types of K^+-selective cation channels have been recognized on the basis of electrophysiological and pharmacological studies.[149] More recently, the cloning of channel genes has permitted the study of the proteins by X-ray crystallography. The first structure determined[452a] was that of the *Streptococcus lividans* K^+ channel (designated KcsA; Fig. 8-21) There are three large structural families of K^+ channel proteins.[453–455a] One group consists of voltage-regulated (K_v) channels, such as those involved in the action potential of neurons. Like the *S. lividans* channel, they are tetramers whose predicted structure contains six transmembrane helices per subunit with a pore-forming loop (P region) between helices 5 and 6. This is just what is seen in the *S. lividans* channel and in one-fourth of the much larger Na^+ channel protein (Fig. 30-17A). Furthermore, all known potassium channels, from bacteria to human beings, have the conserved sequence GYGD in the C-terminal half of the P region.[453] A great variety of K_v channels are known. There are ~70 genes for these channels in the *Caenorhabditis elegans* genome.[456] One of the first K_v channel genes to be cloned was from a *Drosophilia* mutant known for its neurological defect as *shaker*. Its structure (Fig. 30-18), which is based in part on modeling from the KcsA channel, has the ion selective filter with the conserved sequence **TVGYG** in the expected location. At the cytoplasmic end of the pore is an additional structure not found in the KcsA channel. This is the **inactivation gate**, so called because it accomplished the rapid self-inactivation of the K^+ channels during passage of the action potential (Fig. 30-18A). This is one of the factors necessary for recovery and repolarization of the axon membrane. The inactivation gate is composed of N-terminal ~130 residue "T1" domains of the α subunits together with parts of the β subunits, which are associated as a tetramer beneath the channel in the cytoplasm (Fig. 30-18B). Various experimental data including mutational analysis suggest that small ball-like domains at the N termini of the β subunits block the channel.[456–458] Zhou *et al.* propose that the N termini unfold into an extended conformation, passing through "windows" between the T1 domains and the channel and allowing the $-NH_3^+$ ends to bind into the central cavity in the channel.[459] The same site can be blocked by well-known quaternary amine inhibitors such as tetraethylammonium, tetrabutylammonium ions, or tetrabutylantimony, an analog used for X-ray crystallography.

The T1 domain of the channel not only participates in control of the ion flux but also stabilizes the pore complex.[459a] Among the various K^+, Ca^{2+}, and Na^+ channels the regulatory β subunits are quite variable in their structures and mechanisms of gating.[459b] Some β subunits have bound NADH. A speculative possibility (p. 737) is that the rapid interconversion of the positively charged thiazolium ion and negatively charged thiolate ion forms of thiamin (Eq. 7-19) plays some role in nerve conduction, e.g., voltage sensing.

Some questions about ion channels have been hard to answer. For example, how are small cations allowed to flow rapidly through a very small opening in a 2–3 nm thick nonpolar core of a membrane?[460,461] From basic electrostatic principles ΔG for transfer of an ion to the center of a membrane has been estimated as ~160 kJ/mol, a high thermodynamic barrier to transport. A solution to this problem apparently lies partly in the fact that at the center of the lipid bilayer the ion channel contains a cavity large enough (~0.5 nm diameter) to hold about 50 water molecules. Cations tend to enter this cavity, and X-ray studies have shown that the electron-dense Rb^+ does occupy the cavity. A second stabilizing factor is that four helices have their negative (C-terminal) ends pointing toward the cavity. Although the electrostatic effect of these helix dipoles (Fig. 2-20A) might be regarded as negligible, computations indicate that within the low dielectric bilayer the stabilizing effect of the helices becomes significant.[460,461]

How are the pores in these channels opened and closed? Different channels are gated in different ways.

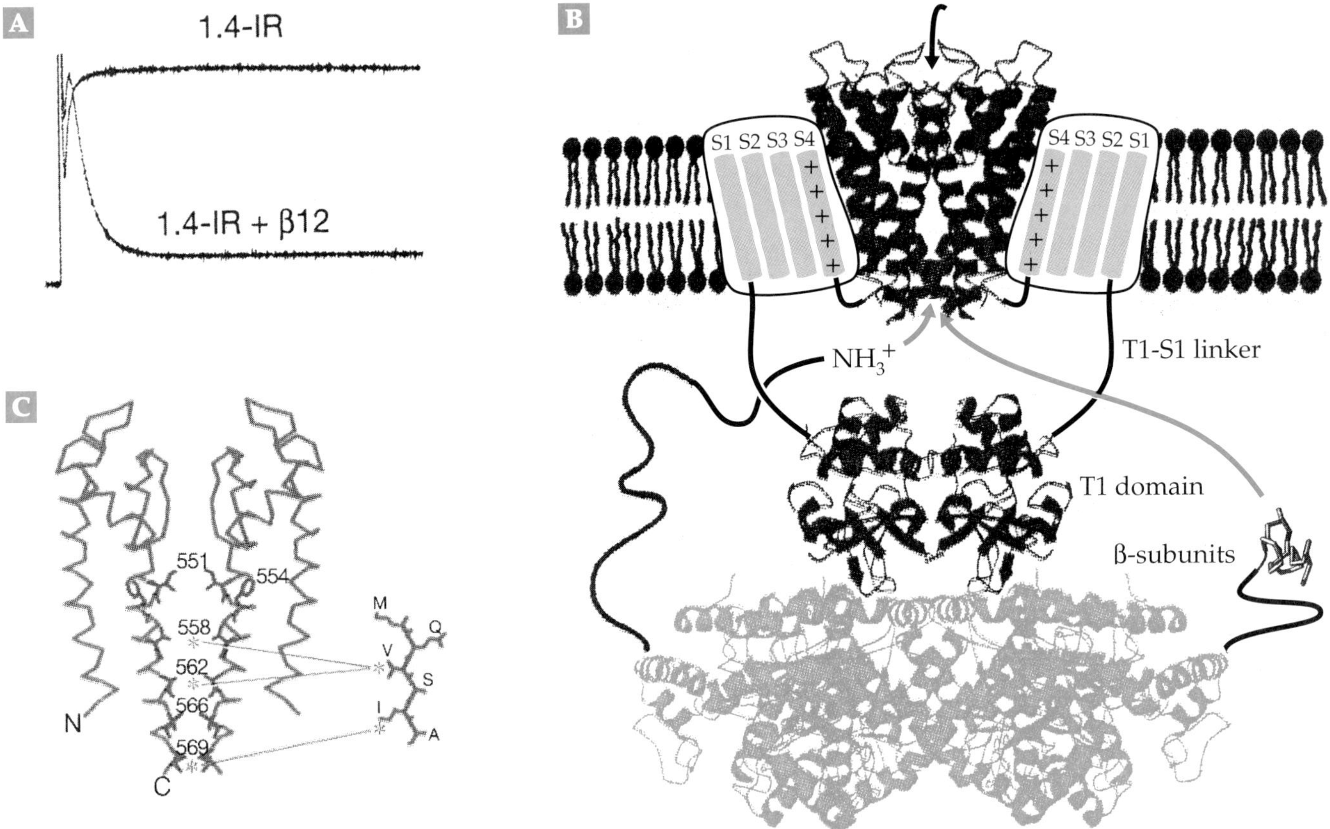

Figure 30-18 (A) K^+ currents recorded from *Xenopus laevus* oocytes carrying cloned genes of *Drosophila* shaker K^+ channels under two-electrode voltage-clamp conditions. Trace 1.4-IR was obtained from a cell expressing channels that lack the inactivation gate. Trace 1.4-IR + β_{12}, obtained from a cell expressing β subunits as well, shows rapid self-inactivation. (From Zhou *et al.*[459]) (B) Composite model of a voltage-dependent K^+ channel. The pore structure in the α subunit is represented by the KcsA channel (Fig. 8-21). The structure of the T1-β complex is from Gulbis *et al.*[458a] The drawing is modified from that of Zhou.[459] (C) Ball-and-stick view of the selectivity filter showing positions of four bound K^+ ions. Two of the four TVGYG peptide strands of the conduction pore are shown. Courtesy of Roderick MacKinnon.

The KcsA channel, which is mostly closed at neutral pH, responds by opening at a low external pH.[462] Using methods of spin labeling and EPR spectroscopy, Perozo *et al.* found small translational and rotational movements of the helices that form the pore (Fig. 30-18). These may alter the diameter of the pore, opening or closing it.[463] How do the electrostatic sensors control the process? The details are uncertain, but the sensor is thought to lie in a conserved sequence of arginine and lysine residues interspersed with hydrophobic amino acids in transmembrane helix 4 of the channel protein (Fig. 30-18; see also Fig. 30-17).[456] How do potassium pores select K^+ over Na^+ or Ca^{2+}? One factor is that Na^+ is more heavily hydrated than K^+ (p. 311). This allows K^+ to pass through the channel more readily than Na^+.[464] Potassium ions travel through the 1.2-nm-long selectivity filter at a rate of $\sim 10^8\ s^{-1}$ in consecutive steps of dehydration, movement, and rehydration occurring in ~10 ns.[464a–d] The process is catalyzed by polypeptides and may depend upon competition between a state in which a ring of four hydrogen-bonded peptide groups is formed and a state in which the four carbonyl groups coordinate a K^+ ion.[464d]

Belonging to the same structural group as the K_v channels are Ca^{2+}-regulated K^+ channels[465,466] Some bacterial channels are controlled by binding of Ca^{2+} ions to a "gating ring" on the intracellular membrane surface.[466a] A mammalian channel is controlled by a complex of calmodulin with the intracellular end of the α subunits of the channel[466b] and others.[453,467] A second large group of K^+ channels, containing seven subfamilies, are the **inward rectifying** (Kir) channels.[455,468] They are tetramers of 360- to 500-residue polypeptide chains, each chain forming two transmembrane helices with a P region between them.[453,469] These channels support a large conductance when K^+ ions flow out from a cell but only a small conductance when they flow in.[470] Kir channels are subject to a variety of controls, which include effects of pH.[471,472]

Some are inhibited by ATP,[473–474b] and others by eicosanoids[475] or inositol hexaphosphate.[476] Some of the ATP-sensitive channels contain an ABC transporter subunit and are binding sites for sulfonylureas and other drugs. See discussion on p. 421. A number of human disorders in Kir channels have been identified.[468] The human Kir channels participate in regulation of resting membrane potentials in K^+ homeostasis, control of heart rate, and hormone secretion.[468] A third group of K^+ channels are dimeric, but each subunit contains two tandem P regions and 4–8 transmembrane helices.[455]

Chloride channels and the ionic environment of neurons. All cells contain voltage-gated chloride channels, which are encoded by the *Clc* genes mentioned on pp. 420, 421.[477,477a] Recently crystal structures[477a–c] have revealed chloride channels formed in single polypeptide chains arranged as dimers. The selectivity filter involves stabilization by the positive ends of α-helix dipoles. The importance of the corresponding proteins to the human body is shown by the existence of several specific diseases arising from mutations in their genes (p. 420).[478,479] A calcium-regulated Cl^- channel is also present[480] as is the ATP-gated CFTR channel (Box 26-A).[480a] In addition, other ligand-gated Cl^- channels, such as γ-aminobutyrate receptor channels (Section B,9), are found in the central nervous system.[481] A glutamate-gated chloride channel in invertebrate organisms is the site of action of the antihelminthic and insecticidal compound **ivermectin**.[481a]

The significance of ion channels can be better appreciated by considering the ionic environment of nerve axons.[149] Mammalian neurons have roughly the following millimolar concentrations of ions in the cytosol and in the external medium. (The concentration gradients for the much-studied squid axon are substantially higher.[149,482]) The membrane potentials that could arise from each one of these concentrations, according to Eq. 8-2, are also given.[149] In a resting

	Cytosol	Extracellular	E_m (mV)
K^+	150	5.5	–90
Na^+	15	150	+60
Ca^{2+}	10^{-4}	1.5	+270
Cl^-	9	12.5	–70

neuron the K^+ potential dominates with an observed membrane potential of ~–80 mV. Some K^+ channels are open and the K^+ and Cl^- concentrations are nearly in Donnan equilibrium across the membrane. The Na^+ and Ca^{2+} channels are closed, and the sodium and calcium pumps keep the internal concentrations of these ions low.

When an action potential is propagated, a wave of depolarization moves along the axon, changing the membrane potential suddenly to a less negative value. When it reaches ~50 mV the Na^+ channels open, allowing sodium ions to flow into the cell causing further propagation of the wave of depolarization. After ~1–2 ms the Na^+ channels begin to deactivate. At the same time the slower K^+ channels open allowing potassium ions to flow out and to repolarize the membrane, the membrane potential sometimes transiently reaching more negative values (hyperpolarization) than the ~80 mV resting potential. Action of the Na^+,K^+-ATPase then restores the original state. The finely tuned properties and sequential opening and closing of the channel proteins are essential to the conduction of nerve impulses.

The existence of voltage-gated ion channels in bilayers are not limited to nerve membranes. They are present to some extent in all cell membranes. Even the paramecium has at least seven kinds of Na^+, K^+, and Ca^{2+} channels.[483] Channels may also be formed by many peptide antibiotics. Among them are the human defensins (Chapter 31) and the ~20-residue **alamethicin**. Six to eleven of the mostly helical monomers of that antibiotic assemble to form a single voltage-dependent channel.[484,484a] The bacterial toxin colicin E1 (Box 8-D) forms voltage-dependent channels within bacterial membranes.[485]

Receptor-associated ion channels. Many neurotransmitters, including acetylcholine and glutamate, act to open ion channels that are part of the receptor protein or of a tight complex of proteins.[149,486] Such **ionotropic receptors** are responsible for most rapid neuronal action. For example, binding of acetylcholine to its receptor in the neuromuscular junction causes the release of Ca^{2+} ions from the exterior into the muscle fibers. Binding of glutamate to its ionotropic receptor in a synaptic ending of a dendrite causes an influx of ions into the cytoplasm, initiating an action potential in the dendrite. In most instances the properties of the receptor channel favor the rapid flow of Ca^{2+} ions into the cytoplasm.

Many other receptors are 7-helix transmembrane proteins, which activate guanine nucleotide G proteins (Chapter 11, Section D, 3). The G proteins couple some receptors directly to Ca^{2+} channels; they couple other receptors to adenylate cyclase and cyclic AMP-activated channels and yet others via phospholipase C to K^+ channels and indirectly to Ca^{2+} channels (Fig. 30-19). All of these G protein coupled receptors are referred to as **metabotropic receptors**. A single synapse often contains both ionotropic receptors and metabotropic receptors. The ionotropic receptors induce a rapid (<1 ms) response, while the metabotropic receptors act more slowly. However, in most cases the final effect is the release of calcium ions into the cytoplasm

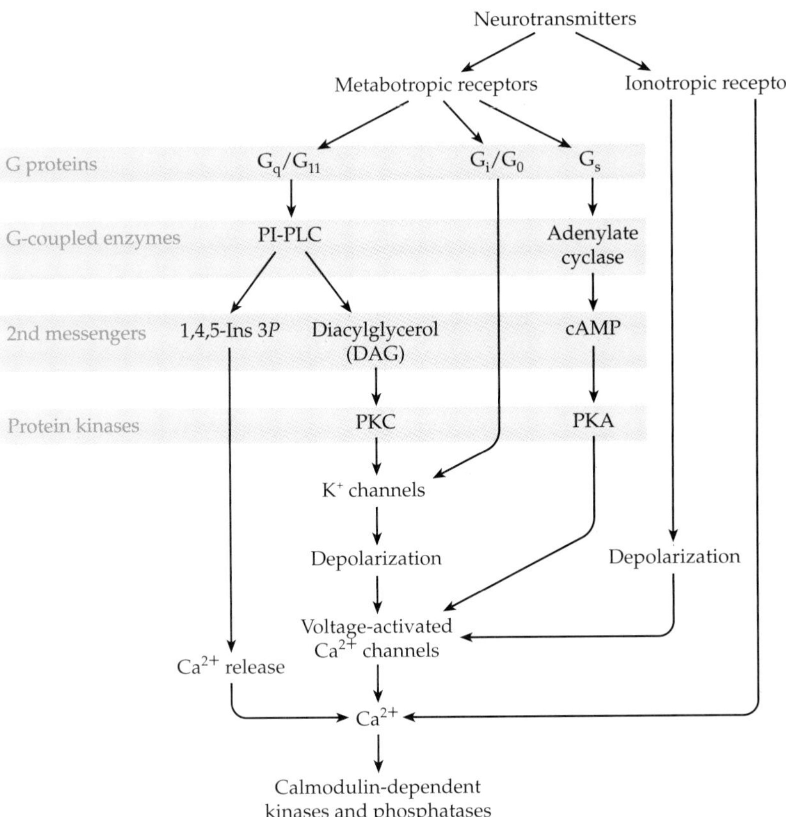

Figure 30-19 Major signaling pathways from metabotropic and ionotropic receptors in neurons. Various G proteins control the signaling from mutabotropic receptors using phosphatidylinisitol-specific phospholipase C (PI-PLC) and adenylate cyclase or acting directly on K^+ ion channels. Adapted from Fig. 5.1 of Nicholls' *Proteins, Transmitters, and Synapses.*[149]

(Fig. 30-19). The rapid response may be initiation of an action potential, while the slow response may be activation of calmodulin-dependent kinases and phosphatases.[149]

6. A Plethora of Neurotoxins

Bacteria, protozoa, and venomous animals synthesize numerous toxins that are used to kill their prey or to defend themselves. Sea anemones, jellyfish, cone snails, insects, spiders, scorpions, and snakes all make potent and highly specific neurotoxins. Plants form a host of alkaloids and other specialized products, some of which are specifically neurotoxic and able to deter predators. More than 500 species of marine cone snails of the genus *Conus* synthesize a vast array of polypeptide toxins (**conotoxins**),[487–489] some with unusual posttranslational modifications.[490,491] The slow-moving snails are voracious predators that use their toxins, which they inject with a disposible harpoon-like tooth,[492] to paralyze fish, molluscs, or worms.[493]

The targets for natural biological toxins include ion channels and receptors for transmitters. At least four parts of the voltage-gated sodium channels are binding sites for extremely toxic natural products.[494–499] **Tetrodotoxin** (Fig. 30-16),[496,497] which is found in the puffer fish, certain newts,[429a] and venom of the blue-ringed octopus, and also the shellfish poison **saxitoxin** (Fig. 30-16) block the entry of sodium ions into the channels.[498] **Bactrachotoxin** (Fig. 22-12) and related lipophilic compounds such as **veratridine** *increase* sodium permeability by blocking the channels permanently open. **Pyrethroid insecticides** (p. 1237) prolong the time that the sodium channels stay open after excitation. Some **scorpion toxins** (Fig. 30-16),[494,499] which all have a hydrophobic core made from a short α helix and a three-strand antiparallel β sheet,[500–502] and **sea anemone toxins**[495,503–505] also stabilize the open conformation of the Na^+ channels. Other smaller ~4-kDa scorpion toxins block K^+ or Cl^- channels or other receptors.[500,506,507] Some are most toxic to insects and others to mammals.[500] Although their three-dimensional structures resemble those of scorpion toxins, the amino acid sequences of anemone toxins show no homology.[505] The most potent poison produced by the red tide organism, the dinoflagellate *Gymnodinium breve* (Fig. 1-9), is **brevetoxin A** (Fig. 30-16).[436,508] It selectively opens one class of sodium channels.[495]

Venoms of **cobras**, **sea snakes**, and pit vipers contain several 6- to 7-kDa proteins that bind to acetylcholine receptors (Fig. 30-23) of the postsynaptic neurons, preventing binding of the neurotransmitter and opening of the ion channels.[509,510] All of these toxins contain four disulfide bridges and share with certain plant proteins a folding pattern that has been called the **toxin-agglutinin fold**[511,512] (Fig. 30-16). These toxins include **erabutoxin a** (Fig. 30-16) from a sea snake[513,514] as well as the 74-residue toxin **bungaratoxin a** (from the banded Krait). This toxin, which has been used to titrate acetylcholine receptors in neuromuscular junctions, is a member of the *long neurotoxin* group, which contains 71–74 residues and five disulfide bonds.[515] Other *short neurotoxins* are 60–62 residues in length with four disulfide bridges.[516] Cobra toxins contain both neurotoxins and **cardiotoxins**, which have somewhat similar structures but quite different modes of action.[517,518] In contrast, **crotoxin** from the venom of a South American rattlesnake[510,510a] and **β-bungarotoxin**[519] consist of 13-kDa phospholipases A_2 complexed with smaller 7.5-kDa proteins. They act at the presynaptic membranes of selected neurons by blocking neurotransmitter release.[520]

The seven types of **botulinum toxin**[521–523a] and the **tetanus toxin**[524] are the most neurotoxic substances known. Only 10^8 molecules are sufficient to kill a mouse. Both toxins are zinc proteases, which block presynaptic transmitter release by cleaving specific synaptic vesicles proteins (see p. 1780 and Fig. 30-20).[522,523,525–528] They bind initially to ganglioside in the neuromuscular junction, one subunit then being internalized as with the diphtheria toxin (Box 29-A). Botulinum toxins specifically enter motor neurons,[521,528a] while tetanus toxin is taken up via synaptic vesicle endocytosis[529] by both peripheral and central neurons. Retrograde axonal transport carries the toxin into the central nervous system and across synaptic clefts into cholinergic interneurons, which are poisoned.

The black widow spider produces the 130-kDa **α-latrotoxin,** which causes massive release of acetylcholine, norepinephrine, dopamine, and GABA from synaptosomal endings.[530,531] The small **anatoxin-a** or "very fast death factor" (Fig. 30-22), which is synthesized by various cyanobacteria, antagonizes both muscarinic and nicotinic acetylcholine receptors.[532] Cone snails synthesize mixtures of the 13- to 17-residue conotoxins (Fig. 30-16).[493] They cause rapid paralysis of fish permitting the snails to prey on the much faster fish. They bind to a variety of targets, which include Na^+, K^+, and Ca^{2+} channels,[435,492] and acetylcholine,[533,534] and glutamate[490] receptors. One of the toxins is a 17-residue peptide containing five residues of γ-carboxyglutamate and is also notable for the fact that intercerebral injection of less than one microgram of the toxin induces a prolonged sleeplike state in mice.[490,493] The venom of *Conus geographicus* is so toxic that two-thirds of human stinging cases are fatal.

The most deadly nonproteinaceous toxin known, **palytoxin**, is also the most complex structure ever established without the aid of X-ray crystallography.[535,536] It is produced by marine zoanthids of the genus *Palythoa* and has the molecular formula $C_{129}H_{223}N_3O_{54}$.

7. Neuronal Metabolism

The brain has a very high rate of metabolism. Although accounting for only 1/50 of the body mass its utilization of energy amounts to 1/5 of the basal metabolism. This is ~20 watts and is nearly constant day and night. It reflects the unusually active metabolism of neurons, a major part of which can be attributed to the sodium–potassium ion pumps in the membranes and to the maintenance of the excitable state.[536a] The source of energy for these processes is the ATP that is utilized to drive the ion pumps and thereby to maintain the membrane potential needed to drive the action potentials. The ATP is formed largely by oxidative metabolism of glucose and, to a lesser extent, of acetoacetate. The large surface area of the axons as well as the frequency with which they transmit nerve impulses accounts for the high rate of metabolism.

Another factor peculiar to neurons doubtless contributes also to their rapid metabolism. The nucleus and most of the ribosomes are found in the cell body. Although few ribosomes are seen in axons and dendrites[536b], many proteins are needed in high concentrations within the axons and synaptic endings. Among these are enzymes catalyzing synthesis and catabolism of neurotransmitters and membrane proteins. If an axon is cut, the separated synaptic endings soon atrophy, an observation that long ago suggested that essential materials, which may include mRNAs,[537] might flow from the cell body. It has now been established experimentally that many materials do move at the rate of 0.3–3 mm / day from the cell body down the axon.[538] More remarkable is **fast axonal transport** by which proteins and other materials move at rates of up to 5 μm / s (0.4 m / day). This transport is specifically blocked by vinblastine (Box 7-D) and batrachotoxin (Fig. 22-12). As has been pointed out in Chapter 20, an ATP-hydrolyzing protein chemically related to the myosin heads functions together with microtubules to provide a kind of miniature railway that moves materials along the microtubules. Transport is sometimes in the opposite direction, i.e., from the synaptic endings to the cell body. This **retrograde axonal transport** may be of importance in altering neuronal properties in response to electrical activity at synaptic endings. It also provides a means of recycling materials originally sent in the other direction.

Brain cells appear to transcribe an unusually large fraction of the genome.[539–541] About 20% of the DNA of human brain was found to hybridize with mRNA formed by brain cells. In other tissues about half this amount of DNA appears to be transcribed. A related observation that seems surprising is the absence of common electrophoretic variants of enzymes in the brain.[539] However, brain cells synthesize specialized isoforms of many proteins, e.g., of the G proteins (p. 558), the cytoskeletal protein 4.1 (Fig. 8-14),[542] and transglutaminase.[543] Unusual lipids, such as the cationic acetal of a galactosylcerebroside shown above,[544] are also formed. Adult rat brain contains about 30,000 different kinds of polyadenylated messenger RNA,[540] much of which lacks the poly(A) tail.[513] Many of these mRNAs contain a specific 82-nucleotide sequence within at least one of their introns. Sutcliffe *et al.* suggest that this is an **identifier sequence** instructing brain cells to express these genes.[540,545] However, the sequence is also found in genes transcribed in other tissues, and its significance is not clear.[546,547]

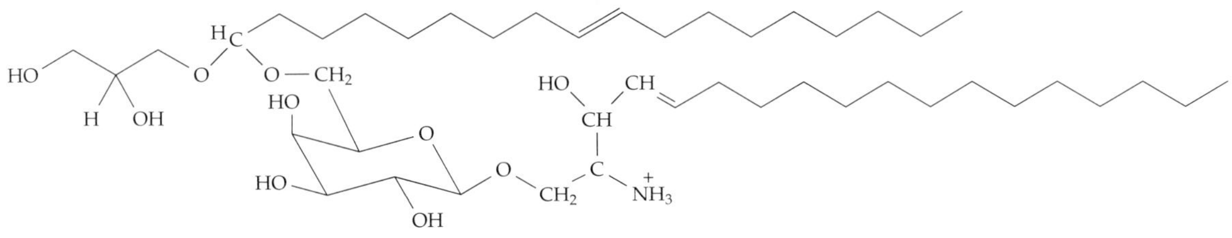

8. Synapses and Gap Junctions

Like the micro-transistors in a computer chip, synapses are the devices by which the brain operates. Synapses process and integrate information from many input channels, send signals on to other neurons, and store information. The information is not stored in digital form, but as chemical alterations in the synapses themselves.[482,548,549] Synapses are formed when axons, growing in response to a chemical trail, reach their destinations and send out branches, each with a bulbous terminal knob (**bouton**). When these boutons meet receptive regions on dendrites of another axon, synapses are formed.[550] The synapse is a very firm connection with a thin, tight synaptic cleft through which signaling takes place. It is surrounded in part by astrocytes or other glial cells (Fig. 30-20A,C).

With the advent of electron microscopy, the fine structure of synaptic contacts became evident. The synaptic knobs were often found to contain vesicles of ~30–80 nm diameter, which were later shown by chemical analysis and staining procedures to contain the neurotransmitters (Fig. 30-10). In the case of the acetylcholine-releasing synapses (**cholinergic synapses**) each 80-nm vesicle contains ~40,000 molecules of acetylcholine,[551] the concentration in the vesicle being of the order of 0.5 M. To show that the acetylcholine released at a synapse stimulated the postsynaptic membrane to initiate an impulse, the technique of **electrophoretic injection** or **microiontophoresis** was developed.[552] By using ultramicrocapillaries a small pulse of current, e.g., 3×10^8 amp for 1 ms, can be used to inject electrically a compound directly into a synaptic cleft. The results may be observed with separate recording electrodes, one of which is inserted into an axon or a muscle fiber. By this means it was shown that amounts of acetylcholine comparable to those released at the large synapses of the neuromuscular junction do cause muscles to contract.

How does the release of neurotransmitter occur? That the release is "quantal," i.e., involving the entire content of a vesicle, was established from the observation of **miniature end-plate potentials**. These are fluctuations in the postsynaptic potential observed under conditions of weak stimulation of the presynaptic neuron. They reflect the randon release of neurotransmitter from individual vesicles.[553] Normally, a strong impulse will release on the order of 100–200 quanta of transmitter, enough to initiate an action potential in the postsynaptic neuron.

A synaptic vesicle cycle. The number of synaptic vesicles in a single synapse in the brain varies from fewer than 100 to several hundred. In specialized synapses there may be thousands. However, at any moment only a fraction of the total are in the "active zone," often aligned along the presynaptic membrane (Fig. 30-20A) or in specialized ribbons such as those in Fig. 30-10B. The vesicles are normally reused repeatedly, undergoing a cycle of filling with neurotransmitter, translocation to the active zone, ATP-dependent priming, exocytosis with release of the neurotransmitter into the synaptic cleft, coating with clathrin, endocytosis, and acidification as outlined in Fig. 30-20B.[554–557] The entire cycle may be completed within 40–60 s to avoid depletion of active vesicles.[558,559] A key event in the cycle is the arrival of an action potential at the presynaptic neuron end.

The accompanying depolarization of the membrane at the synaptic ending permits a rapid inflow of calcium ions through a voltage-gated calcium channel.[444,560] Within less than 0.1 ms the transient increase in intracellular [Ca^{2+}] triggers the release of the contents of the vesicles. About four calcium ions are needed to release one clathrin-coated vesicle (Fig. 30-20A,B). The membrane fusion required for transmitter release involves cytoskeletal proteins of the synaptic endings as well as specialized proteins that are present in the membranes of the synaptic vesicles (Table 30-6). In fact, every step in the cycle depends upon specialized proteins.[387]

Synaptic vesicles can be isolated in large quantities. Their composition is well known, and the proteins have been studied intensively. Indeed, much of what we know about exocytosis and vescular transport has been learned from investigation of synaptic vesicles.[554,561,562] A small synaptic vesicle of 35 nm diameter will contain ~10,000 phospholipid molecules in its membrane and only about 200 protein molecules, at least one of which must be a 13-subunit vacuolar type proton pump (Fig. 18-14). This pump acidifies the vacuole, allowing uptake of a neurotransmitter. Although many different proteins may be found in synaptic membranes, only about 15, which are listed in Table 30-6, are found in all synaptic vesicles and appear essential to function.

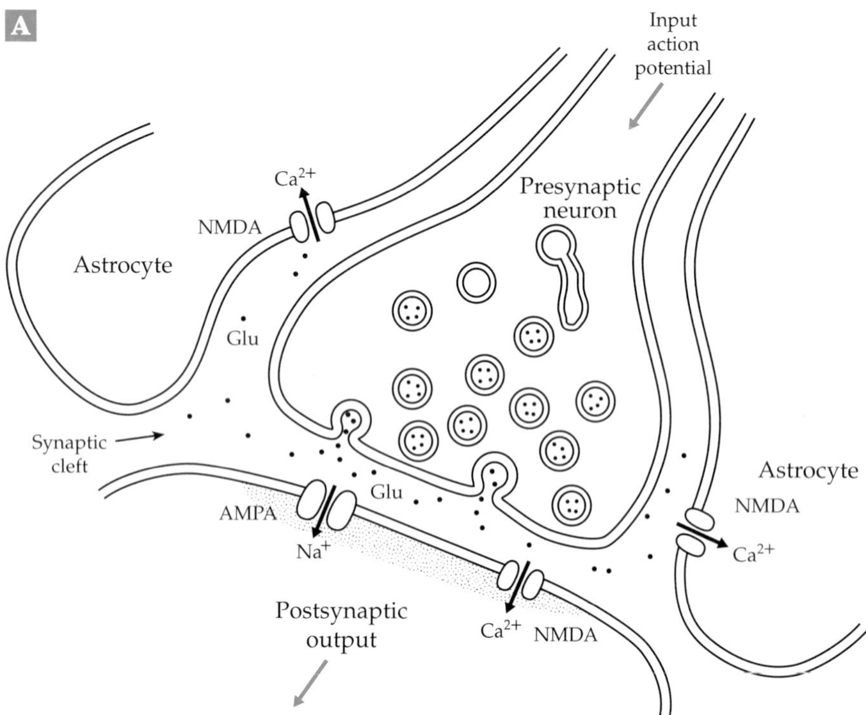

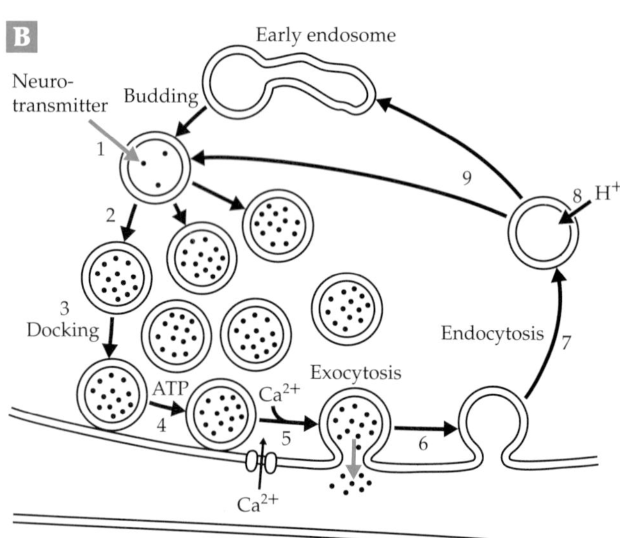

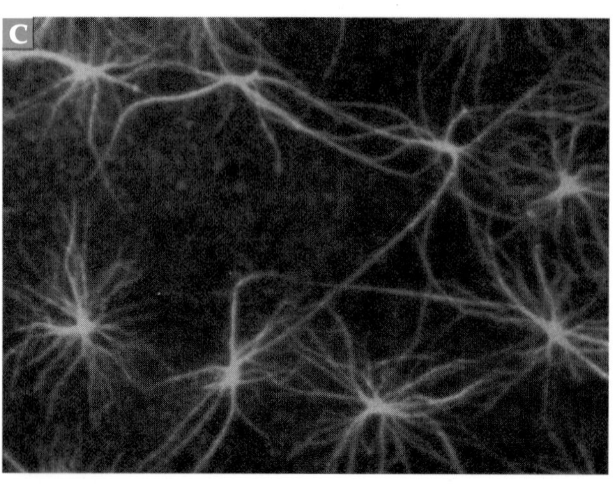

Figure 30-20 (A) Schematic drawing of a fast glutamatergic synapse. An action potential arrives at the synapse, depolarizing the presynaptic membrane and allowing calcium ions to enter the cytoplasm via voltage-gated Ca^{2+} channels. The Ca^{2+} ions induce exocytosis of small synaptic vesicles from the "active zone" near the membrane, releasing glutamate into the synaptic cleft. After diffusing rapidly across the narrow ~50-nm synaptic gap the glutamate binds to its receptors on the ending on a dendrite from a second (postsynaptic) neuron. Glutamatergic synapses usually have two types of receptor, NMDA and AMPA (see Fig. 30-24 and text). Both are ligand-gated ion channels, which release Ca^{2+} and Na^+ into the cytosol of the postsynaptic ending depolarizing its membrane and possibly initiating an action potential. (B) The synaptic vesicle cycle. The synaptic vesicles, which are formed by budding from an early endosome, are filled with neurotransmitter (1). The filled vesicles are then transported to the active zone near the presynaptic membrane (2), are "docked" on the membrane surface (3), and undergo ATP-dependent priming (4). Binding of four Ca^{2+} ions induces exocytosis and rapid release of the neurotransmitter (5). The empty vesicles receive a clathrin coat (6) and undergo endocytosis (7) and uptake of protons (8) to acidify the content in preparation for a second round of neurotransmitter uptake. Alternatively the vesicle can fuse with an endosome as part of the cycle. After Südhof and Scheller.[554] (C) Small section of brain stained to reveal the astrocytes whose extensions form synapses not only with neurons, as in (A), but also with capillary blood vessels.[149] From Kimelberg and Norenberg.[564] Micrograph from Andreas Karschin, Heinz Wässle, and Jutta Schnitzer. (D) Illustration of some proteins essential to the synaptic vesicle cycle. Several are integral membrane proteins. Synaptotagmins contain Ca^{2+}-binding domains and may serve as calcium sensors. The vesicle is portrayed as if docked to the presynaptic membrane by interaction of the SNARE proteins synaptobrevin, syntaxin, and synaptotagmin. The 4-helix bundle is as portrayed by Sutton *et al.*[563]

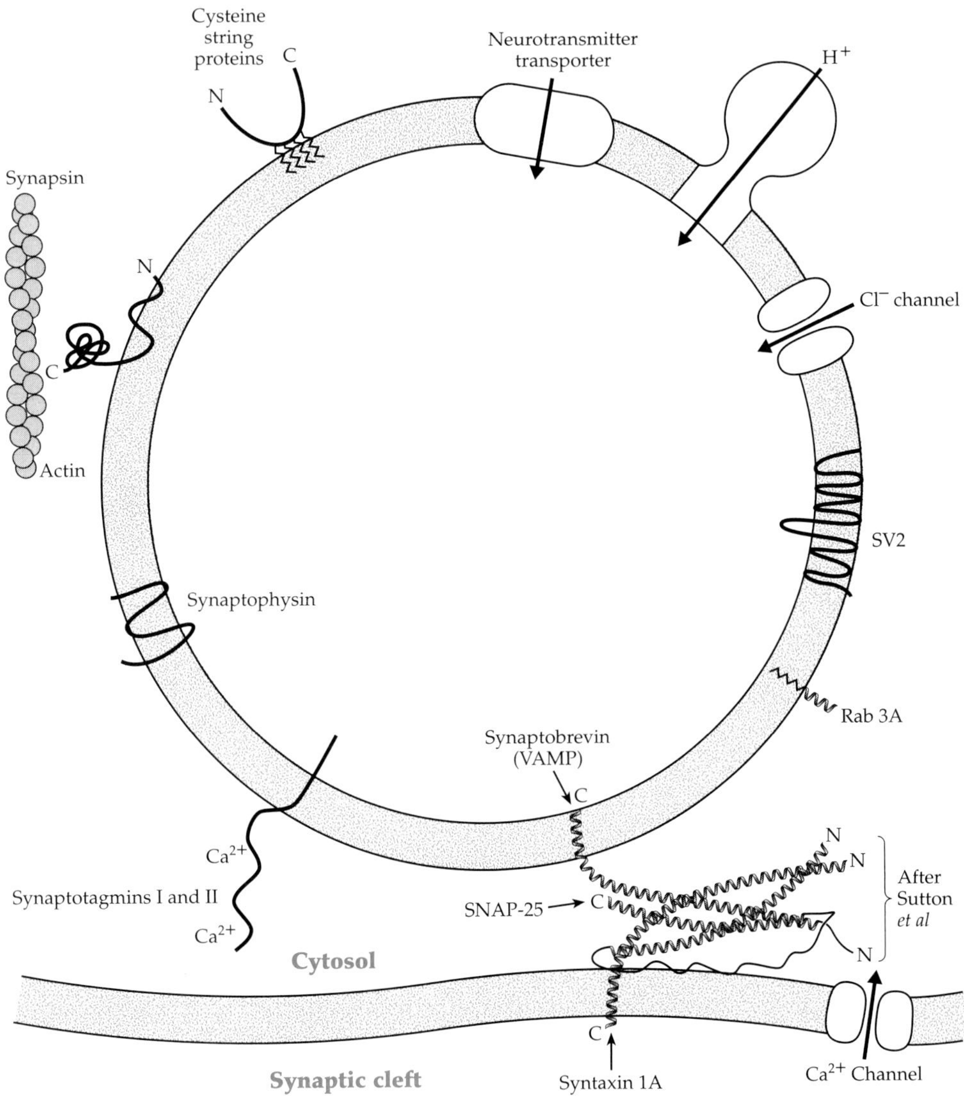

The synaptic vesicles, which are formed by budding from early endosomes, take up neurotransmitters using one of the transporters (step 1 in Fig. 30-20B). Transmitter uptake is G-protein dependent[565] and is driven by the proton electrochemical gradient generated by a vacuolar type (V-type) ATPase (Chapter 18).[149,566] The filled vesicles move into the active zone where they undergo an ATP-dependent priming of uncertain nature.[555,567] Exocytosis (step 5 in Fig. 30-20B) requires membrane fusion, and it is possible that partial fusion occurs during the priming steps. Priming is also thought to involve interaction between vesicle-associated v-SNARES and synaptic membrane-associated t-SNARES (p. 521).[556,563] A major v-SNARE has been identified as **synaptobrevin**, which is also known as **VAMP** (vesicle-associated membrane protein).[563,568,568a] The C-terminal-anchored synaptobrevin is inserted into the plasma membrane of neuronal and neuroendocrine cells prior to endocytosis and budding of the synaptic vesicles.[568] The target t-SNARES have been identified as the synaptic plasma membrane proteins **syntaxin**[568b] and **SNAP-25**.[569–573] Syntaxin is an integral membrane protein, whereas SNAP-25 is anchored by palmitoylation.[571] These proteins bind together to form a synaptobrevin•syntaxin•SNAP-25 complex, which forms a four-helix bundle as shown in Fig. 30-20. Synaptobrevin and syntaxin each contribute one helix, while SNAP-25 provides two; all four have a mutually parallel orientation.[563,574,574a] The helix bundle is so tight that it has a high melting temperature and is resistant to proteolytic cleavage. Nevertheless, the helical domains of both synaptobrevin and syntaxin are sites of very specific cleavage by the zinc proteases of tetanus and botulinin toxins.[527,563,570] Cutting of the protein chains by these toxins prevents proper formation of the four-helix bundle and prevents release of neurotransmitter. It is thought that the complex, which probably forms at several points on the periphery of the docked synaptic vesicle, is essential for membrane fusion.

Other proteins are also needed. All cell fusion processes seem to require regulatory proteins that are essential to neurotransmission in the nematode *C. elegans*. Two of these are encoded by the nematode genes *unc*-13 and *unc*-18. The corresponding mammalian proteins **munc-13** and **munc-18** interact with syntaxin and are essential for exocytosis of synaptic vesicles.[572,575] An ATPase is also needed for correct functioning of the SNARE complex[574] as are other additional proteins.[570]

Details of the control of exocytosis are also uncertain. **Synaptotagmin I**, which contains two Ca^{2+}-binding domains, is probably the sensor that detects the rapid influx of Ca^{2+} that initiates exocytosis.[576–578b] It binds several Ca^{2+} ions via a β-sandwich motif that contains five aspartate side chains at its tip. This motif is conserved in a large family of synaptotagmins. A possibility is that Ca^{2+}–synaptotagmin complexes may self-associate to form a protein ring around the site where the fusion pore forms.[576] Synaptotagmin I also interacts with both syntaxin and with **neurexins**, proteins related to laminin (Fig. 8-33) and present in numerous variant forms in nerve endings. Neurexins are also targets for the α-lathrotoxin of the black widow spider.[531,579] Other proteins that may participate in membrane fusion include the unique **cysteine string proteins**, which in *Drosophila* contain 13 cysteine residues, 11 of which are palmitoylated.[580,581] Nitric

TABLE 30-6
Some Proteins Important to the Formation and Functioning of Synaptic Vesicles[a]

1. Synaptic vesicle proteins	
Synapsins Ia, Ib, IIa, IIb	Peripheral, abundant
Rab3, rabphilin	Rab 3 has lipid anchor
Cysteine string proteins (CSP)	Ca^{2+}-binding
Synaptotagmins	Single transmembrane helix; Ca^{2+} receptor N terminus in vesicle
Synaptobrevins (VAMPs)[b]	SNARE proteins, C termini in vesicle
Synaptophysins, synaptogyrin	Integral membrane protein
SV2 A, B, C	Integral membrane protein, Cl^- transporter
SCAMPS 1 and 4	Integral membrane protein
SVOP	Integral membrane protein
Vacuolar H^+ pump	13 subunits
Cytochrome 561	H^+ generator
Neurotransmitter transporters	For acetylcholine, glutamate, GABA/glycine, catecholamines, ATP
Ancillary transporters	Zn^{2+}, Cl^-
2. Presynaptic membrane proteins	
Syntaxin [b]	t-SNARE
SNAP-25[b]	t-SNARE
Munc-13	
Ca^{2+} channel	
Agrin	
Neurexin	
Actin and microtubules	In dendrites
3. Postsynaptic specializations	
Receptors	e.g., NMDA, AMPA

[a] Based on data of Südhof and Scheller: Südhof, T. C., and Scheller, R. H. (2000) in *Synapses* (Cowan, W. M., Südhof, T. C., and Stevens, C. F., eds), pp. 177–215, Johns Hopkins Univ. Press, Baltimore, Maryland and Südhof, T. C. (1995) *Nature (London)* **375**, 645–653.

[b] Targets for clostridial toxins, tetanus, botulinin.

oxide NO may be involved in a late stage of exocytosis,[392] and phospholipase D1 may also be required.[582]

Presynaptic nerve terminals may contain as few as a hundred vesicles, which must be recycled rapidly after exocytosis in order to allow for repetitive firing.[558,559,583] Several proteins are needed for endocytosis (step 7 in Fig. 30-20). These include **endophilin I**,[584] the vesicle transport ATPase **NSF**,[574] GTPases,[565] and the soluble NSF attachment protein α-SNAP (which is not related to SNAP-25).[585]

Functions of some other abundant proteins of synaptic vesicles have not yet been accurately defined. The **synapsins** are abundant peripheral membrane ATP-binding proteins with multiple phosphorylation sites and variable C-terminal domains that interact with cytoskeletal proteins such as actin microtubules, microfilaments, and spectrin.[554,561,586,587] Another abundant protein is **synaptophysin**, an integral membrane protein found in all synaptic vesicles.[554,561,588] Other proteins are discussed by Südhof and Scheller.[554] The small G protein **rab 3** together with the Ca^{2+}-binding protein **rabphilin** participate in a G-protein cycle that helps to drive exocytosis.[554] Synaptotagmin, as well as clathrin assembly proteins bind inositol hexaphosphate ($InsP_6$; Fig. 11-9), which undergoes active turnover in synapses. This suggests a role for $InsP_6$ in the endocytosis steps of the synaptic vesicle cycle.[589] The brain is rich in zinc ions. Much of the Zn^{2+} is bound into zinc finger domains of transcriptional regulators, but much is also present in a relatively free form within synapses of the hippocampus, cerebral cortex, and other regions.[590,591] Zinc ions may function as a neuromodulator in glutamatergic synapses.[591]

What does a neurotransmitter do at the postsynaptic membrane? In the case of acetylcholine in neuromuscular junctions the principal action appears to be one of opening sodium channels and thereby depolarizing the postsynaptic membrane. If enough nerve impulses arrive, an action potential will be initiated in the postsynaptic neuron. In other cases, the first response may be activation of a protein kinase either directly or by opening a channel for Ca^{2+}, which indirectly regulates protein kinases and phosphatases.[592] Thus, a complex cascade may be activated. See also Fig. 30-19.

The postsynaptic nerve ending, which is usually the tip of an axonal dendrite, has its own set of proteins, which varies to some extent with the nature of the neurotransmitter. In excitatory cells the plasma membrane of the postsynaptic neuron is thickened to ~30–40 nm to form the "**postsynaptic density**," a disc-like structure of clustered receptors of two types, which extends ~ 30 nm into the cytosol.[593,594] Only single receptor channels are indicated in Fig. 30-20, but many receptors are present in the clusters[594,595] as are other specialized proteins. One of these, designated PSD-95, was found to associate with the NMDA receptor using the yeast two-hybrid system (Box 29-F).[594] Neuronal nitric oxide synthase may also be present.

The large neuromuscular junctions, which contain clusters of acetylcholine receptors, have wider synaptic clefts (> 40 nm), which contain basal lamina, a dense network of collagen fibrils together with the heparan sulfate proteoglycan **agrin** (p. 437). Agrin activates a muscle-specific kinase MusK, which phosphorylates the acetylcholine receptors inducing clustering of the receptors together with other proteins embedded in the plasma membrane and binding to the cytosolic protein **rapsyn** (see Fig. 30-23B).[596,597] Agrin is also a component of **immunological synapses,** which are important in lymphocyte development (Chapter 31).[596,598,599] The neuromuscular junction is formed between two cell types, a neuron and a muscle myotube. Both contribute proteins, which include a muscle-specific laminin.[600]

Astrocytes and other glia. Although the glial cells greatly outnumber neurons, they were long regarded simply as glue, as implied by the name glia. We know now that the several types of glial cells have functions in many different aspects of brain chemistry.[149,564,601–605] The oligodendrocytes generate myelin sheaths around many brain neurons. Macrophages that invade the brain differentiate into **microglia** that serve as part of the innate immune system (Chapter 31). **Bergmann glia** of the cerebellum help guide axons during brain development. The astrocytes have many processes, which not only contact synapses directly (Fig. 30-20A,D) but also form contacts with capillary blood vessels. They often contain receptor ion channels of the same types as are found in postsynaptic membranes (see Fig. 30-20A) and respond to Ca^{2+} influx as do neurons.[602–603a] Glia often take up neurotransmitters and ions from synapses in order to prepare for consecutive nerve impulses. Glia may also control the number of synapses formed,[604–604b] and they may have other roles in brain development. For example, an iodothyronine deiodinase (Eq. 15-60) is expressed primarily in neonatal brain, where it supplies thyroid hormone essential to brain development.[605]

Gap junctions in synapses. Not all neurons communicate via chemical synapses. Gap junctions, which are found in both neurons, astrocytes, and other cells, serve as **electrical synapses**. Thus, heart cells are all electrically coupled together by gap junctions.[606 - 608] Gap junctions are formed with the aid of hexameric **connexons,** which are present in each of the opposed membranes and are aligned one with the other (Fig. 1-15F,G).[607,609,610] There may be thousands of connexons in a single gap junction, which resemble ion channels in appearance but contain pores ~1.5 nm in diameter. They are formed from 26- to 43- kDa

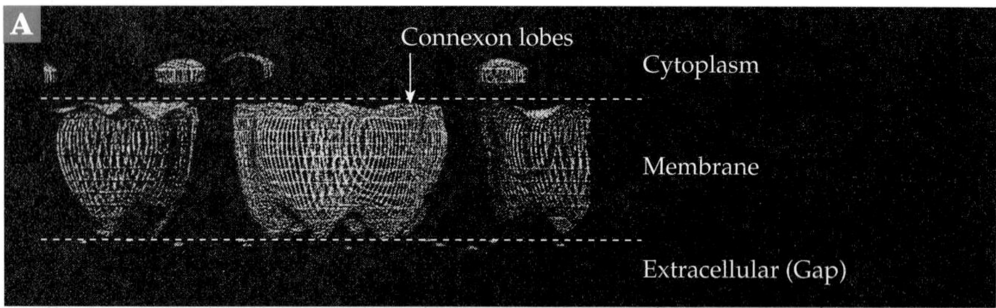

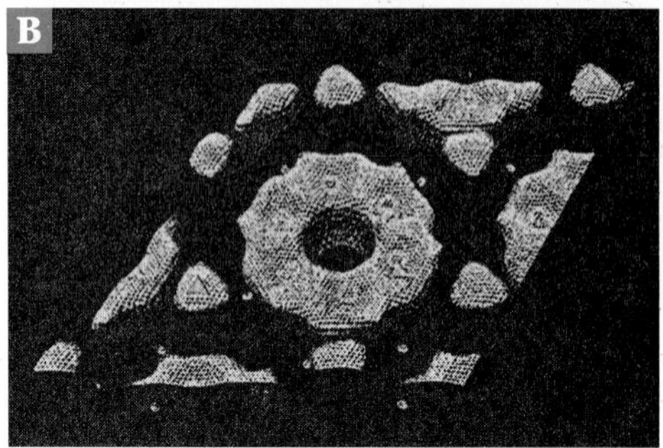

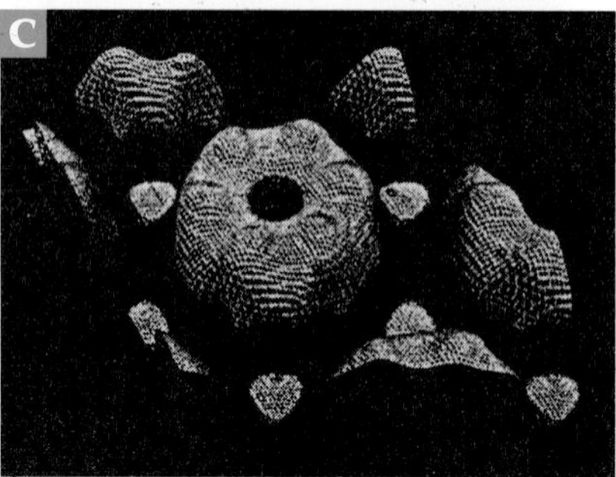

Figure 30-21 Images of gap junction connexins obtained by electron crystallographic methods at a resolution of 1.6 nm. (A) Cross-section. The thinkness of the (43 x 6) kDa hexameric connexin is 5.0 mm. (B) View of the connexin from the cytoplasmic side. (C) View from the extracellular side. From Perkins, Goodenough, and Sosinsky.[609] Courtesy of Guy Perkins.

protein subunits of the multigene family of **connexins**.[610–611a] Each gap junction consists of a pair of hexameric rings of connexins (Fig. 30-21), one ring from each of the two juxtaposed membrane surfaces.[609] Defects in connexins cause inherited deafness, neuropathy, malignancy, and cataract formation.[612–613a] The connexin subunits each contain four transmembrane helices and are related structurally to the peripheral myosin protein 22, the myelin proteolipid (p. 1767), and the protein stargazin (p. 1901), which is involved in synapse formation in the brain.[428a]

Another type of channel has been recognized quite recently. An ion channel, which regulates Mg^{2+} ion transport in kidney tubules, forms within the tight junctions that seal the extracellular space between cells (Fig. 1-15B). A protein **paracellin** forms channels through the tight junction protein complexes that surround the cells.[614,615]

9. Neurotransmitters

Studies of neuromuscular junctions of the autonomic nervous system as early as 1904 led to the suggestion that adrenaline might be released at the nerve endings. Later it was shown that, while adrenaline does serve as a transmitter at neuromuscular junctions in amphibians, it is primarily a hormone in mammals. Nevertheless, it was through this proposal that the concept of chemical communication in synapses was formulated. By 1921, it was shown that acetylcholine is released at nerve endings of the parasympathetic system, and it later became clear the motor nerve endings of the somatic system also release acetylcholine.

Acetylcholine is an established neurotransmitter because it meets five important criteria: (1) a synthetic mechanism exists within the presynaptic neuron; (2) a mechanism of storage (in vesicles) is evident; (3) the transmitter is released in proportion to the strength of the stimulus (frequency of firing); (4) postsynaptic action of the transmitter has been demonstrated directly by microiontophoresis; and (5) an efficient means for inactivation of the transmitter is present. The same five criteria must be met by other compounds if they are to be considered as transmitters.

At present, in addition to acetylcholine, glutamate, and γ-aminobutyrate (GABA), glycine, noradrenaline (norepinephrine), and dopamine and 5-hydroxytryptamine (serotonin) are regarded as established transmitters. Other probable (**putative**) or possible **candidate transmitters** are also known. Aspartate, taurine, and a large number of peptides (Tables 30-1, 30-4) are under consideration.

Some transmitters, including noradrenaline, dopamine, serotonin, and various neuropeptides, are sometimes called **neuromodulators** rather than neurotransmitters. These compounds may not initiate a nerve impulse but may act on adenylate cyclase to increase or decrease cAMP levels and protein kinase activity. They may also diffuse through the extracellular space to influence a region of the brain greater than a single synaptic cleft. However, the distinction between transmitters and modulators is not exact.

For many years it was assumed that a single neuron released only a single transmitter. We know now that this is incorrect.[616] For example, enzymes in neuromuscular junctions synthesize not only acetylcholine but also catecholamines, taurine, and GABA.[617] Some synapses in the central nervous system release both glycine and GABA.[618]

Acetylcholine in a skewed conformation

L(+)-Muscarine, a cholinergic agonist from the mushroom *Amanita muscaria*

Acetylcholine in extended conformation

Atropine

Decamethonium, a synthetic drug with potent curare-like activity

Physostigmine, an acetylcholinesterase inhibitor widely used in treatment of glaucoma. Compare with other carbamate esters (Box 12-E)

D-Tubocurarine, the principal ingredient of South American arrow poisons. Blocks cholinergic receptors in skeletal muscle

Nicotine (protonated)

Tetramethylammonium ion

Anatoxin-a (very fast death factor) from the cyanobacterium *Anabaena*

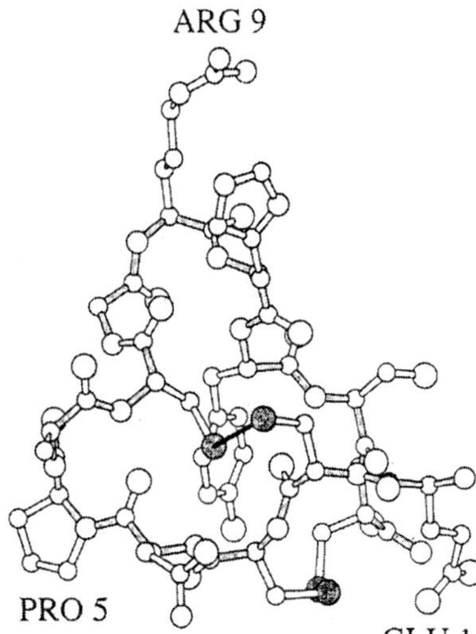

Figure 30-22 Some inhibitors of cholinergic synapses. The structure of conotoxin GI is from Guddat *et al.*[533] Courtesy of A. B. Admundson.

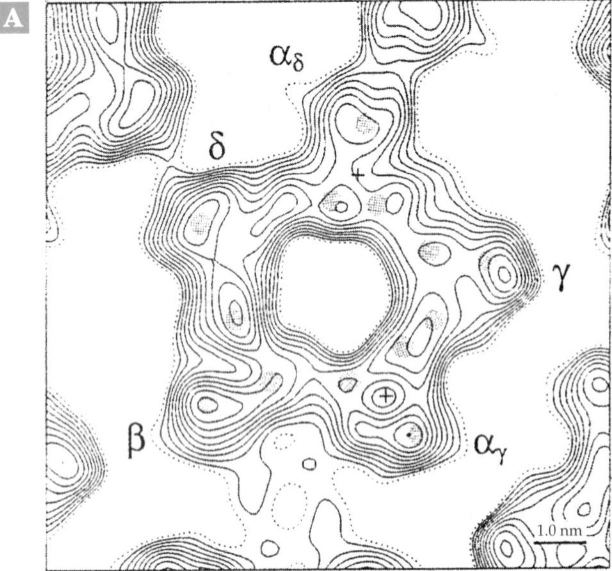

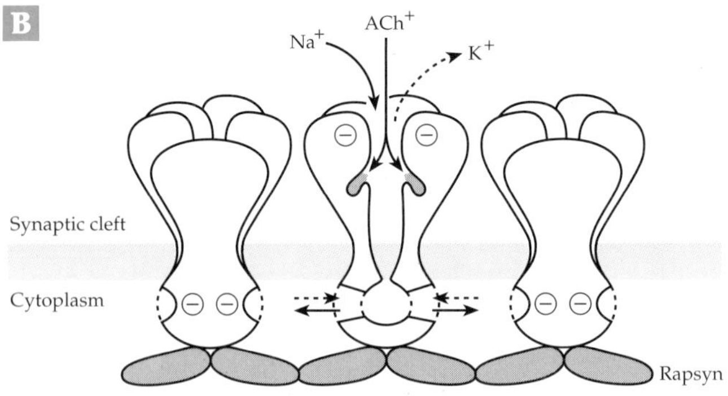

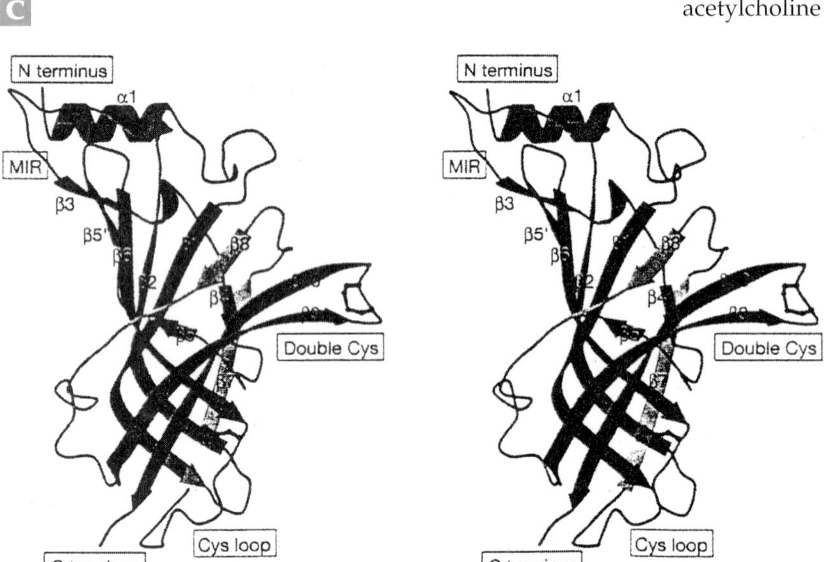

Figure 30-23 The nicotinic acetylcholine receptor from the *Torpedo* ray. (A) The mouth of the receptor channel viewed from the synaptic cleft based on reconstruction from cryoEm images. Addition of acetylcholine, which binds to the two α subunits, induces small rotations in the five subunits of the $\alpha_2\beta\gamma\delta$ complex causing the channel to open. From Unwin.[640] (B) Architecture of the subsynaptic membrane and the acetylcholine receptor. The binding of acetylcholine and the movement of cations through the open channel is illustrated. Cations that leave the cytoplasm may be filtered through narrow openings that lead into the central channel, which is formed by transmembrane helices. Negatively charged amino acid residues may help exclude anions from the region of the pore. From Miyazawa *et al.*[624] (A) and (B) Courtesy of Nigel Unwin. (C) Stereoscopic ribbon drawing of one subunit of a pentameric acetylcholine-binding protein, which mimics the receptor structure. Disulfide bonds are shown in a ball-and-stick form. The N terminus in a receptor would point toward the synaptic cleft and the C terminus would continue at the bottom into the transmembrane helix. Courtesy of Brejc *et al.*[627]

Cholinergic receptors and their agonists and antagonists. Among the acetylcholine-releasing (cholinergic) neurons are the motor neurons that form synapses at neuromuscular junctions, the preganglionic neurons of the entire autonomic system, and the postganglionic neurons of the parasympathetic system. There are also many cholinergic synapses within the brain. In contrast, in insects neuromuscular transmission is mediated by glutamate while acetylcholine is the principal neurotransmitter in the central nervous system.[619]

Important in the study of neurotransmitters is the identification of specific agonists, which mimic the action of a transmitter, and of antagonists, which block the action of the transmitter. Two groups of compounds influence acetylcholine-secreting neurons, leading to the classification of these neurons either as **muscarinic** (activated by muscarine; Fig. 30-22) or **nicotinic** (stimulated by nicotine). The muscarinic receptors, which are found in many autonomic neurons, are specifically inhibited by **atropine** and **decamethonium** (Fig. 30-22). The nicotinic synapses occur in ganglia and skeletal muscle. They are inhibited by curare and its active ingredient **D-tubocurarine** (Fig. 30-22) and by the protein snake venom **α-bungarotoxin**. This toxin has been used to titrate the number of acetylcholine receptors in the motor end plate of the rat diaphragm. About 4 x 10^7 receptors per end plate (or 13,000 / μm^2) were found.[620]

Nicotinic receptors (nAChRs; Fig. 30-23) of the type found in neuromuscular junctions are most frequently isolated from the electric organs of the electric eel *Electrophorus* or from electric fish of the genus *Torpedo*. They have been studied more intensively than any other receptor.[621–626a] They contain four kinds of subunit with a stoichiometry $\alpha_2\beta\gamma\sigma$ and molecular masses of 39, 48, 58, and 64 kDa respectively. The amino acid sequences of the four proteins contain homologous regions, some of which are thought to represent membrane-spanning segments of the peptides. These receptors are ligand-gated ion channels and are closely similar to $GABA_A$ and $GABA_C$ receptors, to glycine receptors, and to 5-hydroxytryptamine (serotonin) receptors of the 5-HT_3 type. Parts of their amino acid sequences are also homologous to those of both the voltage-gated Na^+ channels and gap junctions,[433,627] suggesting that the transmembrane domain may resemble that of Fig. 30-18.[628] However, notice the difference in symmetry. Acetylcholine binds to the two α subunits (Fig. 30-23). Neurotoxins may bind at several sites.[629] Some indication of the possible function of the various subunits comes from studies of the neuromuscular junction in which the different subunits are degraded at different rates with half-lives of from one to ten days. During development fetal ε subunits are replaced by adult γ subunits. Perhaps more rapid changes in receptor composition are sometimes needed.[630]

Similar nAChRs are also found in the brain.[621,631,632] However, they are not identical but have at least 17 differing amino acid sequences (α1–α10, β1–β4, γ, δ, and ε). The neuromuscular junction receptor (muscle type) from fish is described as $(\alpha1)_2 \bullet \beta1 \bullet \gamma / \varepsilon \bullet \delta$.[626] The brain contains homopentamers of subunits α7, α8, and α9 as well as various heteropentamers. The various forms possess different affinities for acetylcholine and for antagonists such as nicotine.[633,634] In the brain the highest affinity for nicotine is shown by an $\alpha4\beta2$ form, which represents over 80% of the nAChR in mammalian brain.[634,635] Knockout mice in which the β2 subunit gene has been deleted lose their sensitivity to nicotine.

Conductance measurements showed that the nicotinic receptors contain channels permeable to Na^+ and other cations and that they are acetylcholine-gated ion channels. Construction of a three-dimensional image from electron micrographs at various angles of tilt shows a tube with approximate pentagonal symmetry and a narrow channel through the center (Fig. 30-23).[622,624,636] Acetylcholine binds to sites on the two α subunits ~3 nm away from the ion channel. An allosteric change opens the channel, allowing cations (largely Na^+) to flow out, depolarizing the membrane. There are at least four structural states in the channel opening-and-closing cycle.[637,638] The three-dimensional structure has been modeled using an acetylcholine-binding protein of known structure from a snail[626,627,639] as a mimic of the cytoplasmic nicotine-binding domain of the receptor. The structure of one subunit of the binding protein is shown in Fig. 30-23C. This protein, which is secreted into synapses by glial cells, may provide a buffering action by binding the acetylcholine. Although the most rapid effect of acetylcholine binding to the nicotinic receptor is depolarization of the postsynaptic membrane, other slower effects follow. Thus, protein kinases are activated and phosphorylate the receptor as well as other proteins.[641]

After a pulse of transmitter is released, it must be removed or inactivated quickly to prepare the synapse for arrival of a new nerve impulse. This is accomplished in two ways in cholinergic synapses. The first is via hydrolytic destruction by acetylcholinesterase[642–645] (pp. 634–637; Eq. 12-25). This esterase and the related butyrylcholinesterase[646] are present in the synaptic membrane itself. The second mechanism is energy-dependent transport of acetylcholine into the neuron for reuse. Since much of the transmitter is hydrolyzed, new acetylcholine is synthesized by transfer of an acetyl group of acetyl-CoA to choline.[647]

In the central nervous system muscarinic acetylcholine receptors are more abundant than nicotinic receptors. They consist of single-chain proteins of mass ~70 kDa. They are not ion channels but are 7-helix receptors homologous in sequence with β-adrenergic receptors (Fig. 11-6) and with rhodopsin.[648] Five different subtypes (M1–M5) have been characterized. The M1, M3, and M5 receptors are coupled to the G_q/G_{11} family of G proteins (pp. 557–558), and M2 and M4 are coupled to G_i/G_o proteins.[649–651] Their effects are slower and of longer duration than those of the nicotinic receptors. It has been difficult to assign functions to the individual types. Most regions of the brain contain more than one type, but they are thought to be involved in locomotion, learning, memory, thermoregulation, and cardiac and pulmonary functions. Many drugs, some of which are used in treatment of Parkinson and Alzheimer diseases, epilepsy, and asthma, affect muscarinic receptors. The M2 receptors predominate in the heart where they help to regulate the beating frequency and atrial contractility. Sudden infant death may sometimes result from a defect in muscarinic receptors.[652] Knockout mice lacking M2 receptors also have problems with movement control, body temperature, and pain responses.[651] Mice lacking M3 receptors are lean with very low levels of serum leptin and insulin.[653] Many of the muscarinic receptors activate adenylate cyclase, while others are coupled to the phosphoinositide cascade. Some indirectly activate K^+ channels.[654] Muscarinic receptors are also studied in insects, but it is difficult to correlate the insect and mammalian receptors.[655]

Amino acids as neurotransmitters. The concentrations of **glutamate** and of its decarboxylation product **γ-aminobutyrate** (GABA) are high in all regions of the brain. The two compounds are generated sequentially in the γ-aminobutyrate shunt, a pathway that accounts for a quantitatively significant part of the total metabolism of the brain (Fig. 17-5). Because they are present in all parts of the brain in high concentrations, there was initially reluctance to accept glutamate and GABA as neurotransmitters. However, it is now accepted that L-glutamate is the major excitatory transmitter in the central nervous system.[656–658] It seems to be responsible for nearly all of the very fast acting nerve impulses in the brain. At the same time GABA is recognized as the most important inhibitory transmitter. The role of glutamate as an excitatory transmitter was first established for the neuromuscular junction of arthropods.[659] Although it is a constituent of all animal tissues, the concentration of glutamate is much higher in brain than in other tissues, and it is higher in neurons than in glia. Microiontophoretic application of either glutamate or aspartate to the brain cortex leads to very strong excitatory responses.

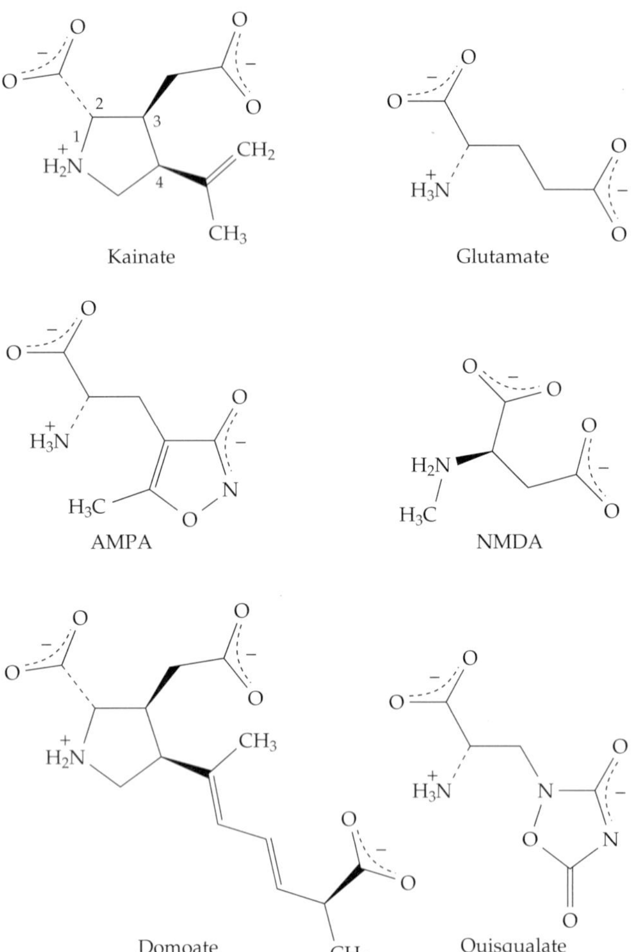

Figure 30-24 Chemical structures of some agonists of ionotropic glutamate receptors (iGluR).

Three subtypes of ionotropic glutamate receptors (iGluR) are named for the specific agonists **α-amino-3-hydroxy-5-methyl-4-isoxazolepropionic acid** (AMPA), ***N*-methyl-D-aspartate** (NMDA), and **kainate**. The receptors resemble the acetylcholine receptor in containing a cation channel.[149,660–662] In addition, there are 7-helix **metabotropic glutamate receptors,** which are coupled to G proteins.[663,664] The AMPA receptors were in the recent past called **quisqualate** receptors, because they are also activated by the agonist with that name. The toxic domoate (Fig. 30-24) also binds to kainate receptors. Both domoic acid and kainic acid are terrible convulsant toxins. They are formed by two different red algae. Domoic acid accumulates in contaminated mussels and causes shellfish poisoning. The ionotropic glutamate receptors, which may be stimulated by either glutamate or aspartate, are directly linked to the opening of cation channels. Their activation may also induce the inositol phosphate cascade and slower Ca^{2+}-dependent changes. A peculiarity of the high-conductance NMDA channels is that they are blocked by Mg^{2+} in a voltage-dependent manner. They do not open unless the frequency of nerve impulses is high or some other factor causes membrane depolarization.[656]

The AMPA receptors, which are thought to be the predominant mediators of fast excitatory transmission in the brain,[665] are oligomers (probably tetramers[666,666a]) of 950- to 1500-residue protein subunits. These subunits have large N-terminal domains in the synaptic cleft. There are probably three transmembrane helices and a membrane-associated loop similar to those depicted in Fig. 30-17A. A long C-terminal tail protrudes into the cytosol, while a large loop between transmembrane regions extends from the outer membrane surface, joining with the N-terminal domain to form the ligand-binding site, the structure of which resembles those of bacterial periplasmic binding proteins.[661,665,667] Four related AMP receptors, designated GluR1, 2, 3, and 4, have been identified. Related kainate receptors, whose properties overlap those of AMPA receptors, are designated GluR5, 6, and 7.[662] Although AMPA receptors are essential for fast signal transmission they lose sensitivity rapidly (on a millisecond time scale) as a result of conformational alterations.[667a] Many factors, including inhibition by polyamines,[667b] affect these receptors. However, brief high-frequency activation of some AMP receptors leads to a long-lasting increase in efficiency, termed LTP, which is important to learning (see p. 1801).[666a]

The NMDA receptors are heterooligomers with two type of subunits. The NR1 (or ζ) subunits exist as a series of at least eight splice variants. The NR2A, B, C, and D (ε series) are encoded by four different genes.[668,669] NR1 is regarded as the principal subunit ard NR2 as a regulatory subunit. As with the AMPA receptors[670] the oligomeric NMDA receptors are anchored at appropriate locations in the postsynaptic membrane by scaffolding proteins containing PDZ domains (Table 7-3).[671] The C-terminal domains of the ε subunits are unusually long and participate in anchoring. NMDA receptors are found not only in neurons but also in astrocytes (Fig. 30-20), where they are thought to have important signaling functions.[672,673] These include regulation of Ca^{2+} flow, in part via gap junctions.[603a]

The N-terminal domain of the NR1 subunit of the NMDA receptor contains a glycine-binding site.[674] Full activity of the receptor requires a **coagonist** bound in this site. Surprisingly, **D-serine** seems to be the normal coagonist, at least in some sites.[675,676] This newly recognized neurotransmitter is synthesized from L-serine by a pyridoxal phosphate-dependent recemase and is destroyed by the flavoprotein D-amino acid oxidase. Associated with NMDA receptors are clusters of **ephrin receptors**, proteins that bind the glycosylphosphatidylinositol (GPI)-anchored proteins known as ephrins in presynaptic membranes. Binding of ephrins to their postsynaptic receptors activates tyrosine kinases and enhances the influx of Ca^{2+} ions.[676a,b]

Specific inhibitors of NMDA channels include a 27-residue "spasmotic" conotoxin,[490] 2-amino-4-phosphonobutyrate, related longer chain aminophosphonates, and the following potent anticonvulsant drug, which is able to penetrate the blood–brain barrier.[677]

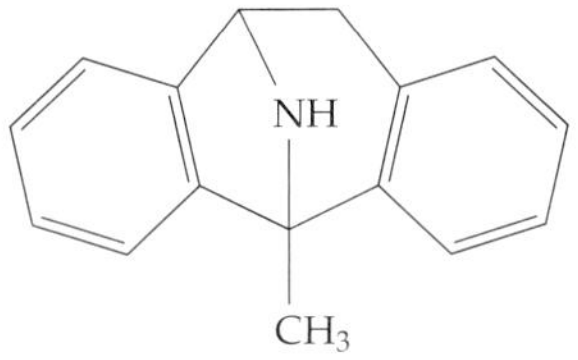

NH

CH_3

(+)5-Methyl-10,11-dihydro-5*H*-dibenzo[*a,d*]cyclohepten-5,10-imine

Metabotropic glutamate receptors have been classified into eight types (mGluRs1–8).[678–680a] Group I (mGluRs1–5) are selectively activated by 3,5-dihydroxyphenylglycine; Group II (mGluR2 and mGluR3) are activated by L-2-(carboxycyclopropyl)glycine; and Group III (mRluR4 and mGluR 6–8) are activated by L-2-aminophosphonobutyrate. They are all 7-helix G-protein-coupled receptors with external ligand-binding domains that resemble those of bacterial periplasmic binding proteins.[680] Splice variants for at least mGlnR1 are known.[678] Metabotropic glutamate receptors are neuromodulary but nevertheless play essential roles in the cerebellum and other parts of the brain. For example, mice deficient in the mGluR1 protein have severe problems with motor coordination and learning.[681,682] Metabotropic glutamate receptors may participate in calcium sensing and signaling.[683,684]

Synaptosomal particles have a high-affinity proton-dependent uptake system for glutamate.[685] Glutamate and aspartate may also be taken up from the synaptic cleft by neurons or by glial cells, which then transfer the glutamate into neurons for reuse.[686,687] Five distinct mammalian transporter genes have been cloned.[688] They are driven by concentration gradients of Na^+ and K^+ across the membrane.[689,690] However, some serve as glutamate-gated chloride ion channels.[691,691a]

Excitotoxicity. As essential as glutamate is for brain function it is toxic in excess. Excessive stimulation of the NMDA receptors, which occurs during convulsions, strokes, or traumatic injury and which can accompany anoxia or hypoglycemia, causes neuronal death.[660,692–694] Blocking these receptors with the above-mentioned anticonvulsant drug or aminophosphonates has a remarkable protective effect against the neurotoxicity of the accumulating glutamate.[658,677] Vitamin E and **tocotrienols** (Fig. 15-24) may also be protective.[695]

The inhibitory neurotransmitter gamma-aminobutyrate (GABA). Glutamate, aspartate, and cysteic acid are all potent excitors, but their decarboxylation products γ-aminobutyrate (**GABA**), β-alanine, and taurine are inhibitors as is also glycine. Of these GABA is the most important.[696] Its concentration in the brain is high and varies at least threefold in different parts of the brain. It is hardly present elsewhere in the body. GABA and GABA-binding sites are found in 30–50% of the nerve endings. The function as an inhibitory transmitter has also been demonstrated in inhibitory neurons present in the peripheral nervous system of arthropods. Virtually every neuron in the brain is to some extent subject to inhibition by GABA.[697,698] Glial cells also have GABA receptors.

The receptors for GABA are divided into type A, which are blocked by **bicuculline**,[699] and type B, which are stimulated by **baclofen** (Fig. 30-25).[698] The $GABA_A$ receptors are the major sites of fast synaptic inhibition in the central nervous system.[700] They are structurally related to the nicotinic acetylcholine, glycine, and serotonin type 3 ($5\text{-}HT_3$) receptors. Cloning has revealed 16 different mammalian subunits: α1–α4, β1–β3, γ1–γ3, δ, ε, π, and Φ.[701–704a] The oligomeric receptors are ligand-gated chloride ion channels[481,705] as are also glycine receptors. These receptors are clustered in synaptic membranes, apparently anchored in part by their β subunits[706] and scaffold proteins such as the microtubule-binding **gephyrin** (from the Greek word for bridge)[701,707] and a small ~14-kDa GABA receptor-associated protein.[708] A novel serine protein kinase is also associated with GABA receptors.[703]

Whereas excitatory transmitters lead to depolarization of the postsynaptic membrane, inhibitory transmitters cause **hyperpolarization**, apparently by increasing the conductance of K^+ and Cl^-. The result is

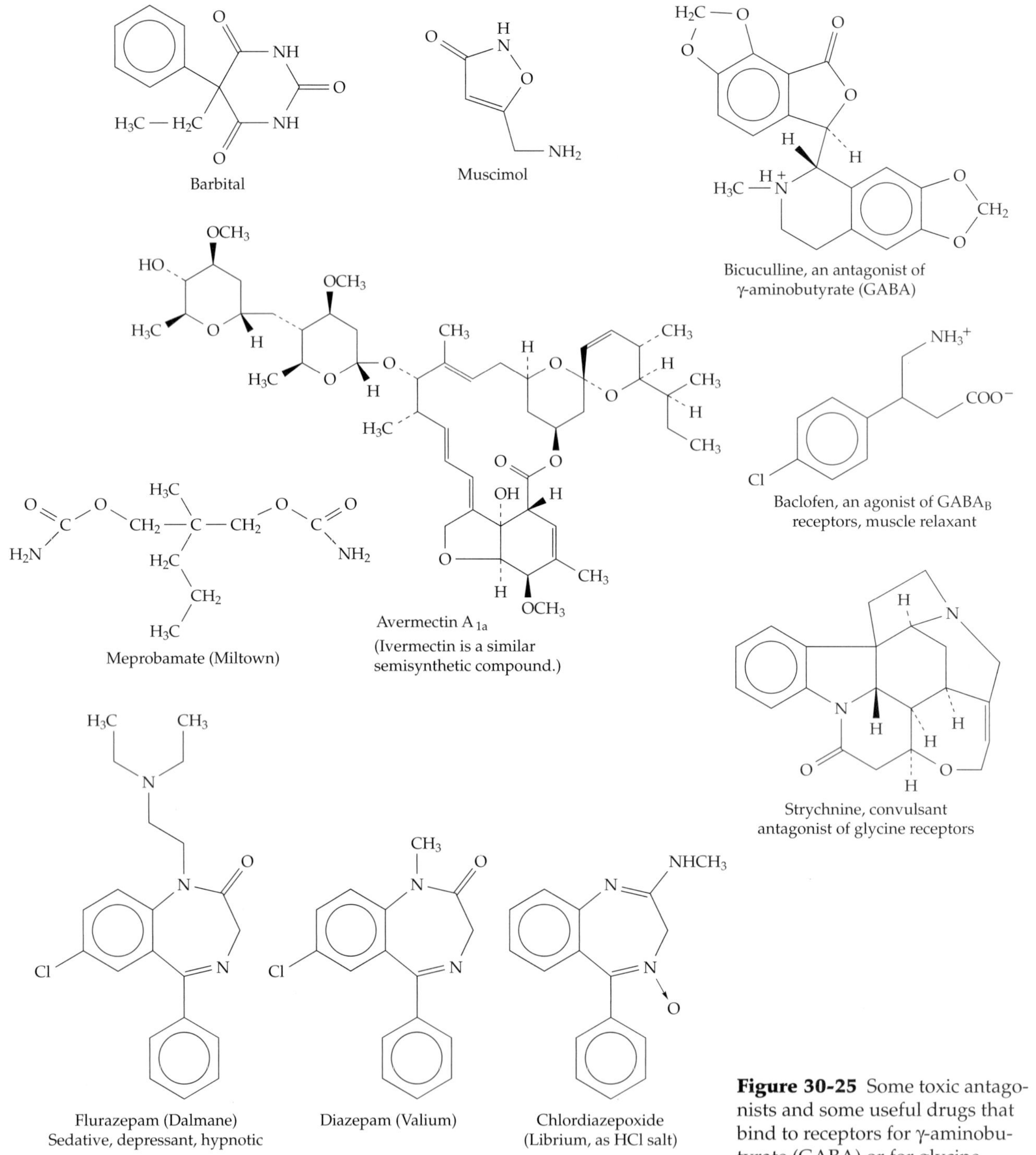

Barbital

Muscimol

Bicuculline, an antagonist of γ-aminobutyrate (GABA)

Avermectin A_{1a}
(Ivermectin is a similar semisynthetic compound.)

Baclofen, an agonist of $GABA_B$ receptors, muscle relaxant

Meprobamate (Miltown)

Strychnine, convulsant antagonist of glycine receptors

Flurazepam (Dalmane)
Sedative, depressant, hypnotic

Diazepam (Valium)

Chlordiazepoxide
(Librium, as HCl salt)

Figure 30-25 Some toxic antagonists and some useful drugs that bind to receptors for γ-aminobutyrate (GABA) or for glycine.

that it is more difficult to excite the postsynaptic membrane in the presence of, than in the absence of, these transmitters. GABA-dependent interneurons also contain the calcium-binding **parvalbumin** (Fig. 6-7), which suggests that a Ca^{2+}-dependent process is involved.[709]

The $GABA_B$ receptors resemble metabotropic glutamate receptors.[710,711] They are 7-helix G-protein coupled proteins, which activate adenylate cyclase. They tend to dimerize, and maximum activity is observed for heterodimers of $GABA_B1$ and $GABA_B2$ receptors.[712,713] They are often coupled to inward rectifying K^+ channels.[714]

The GABA receptors provide binding sites for a great variety of toxins and drugs.[481] These include barbiturates, anesthetics, antianxiety drugs, and the insecticides such as toxaphine, cyclodienes, and pyrethroids.[481] **Diazepam**, **chlordiazepoxide**, and flurazepam[700,702,715–717] (Fig. 30-25) are antianxiety drugs and muscle relaxants, which, during the 1970s, were the most frequently prescribed drugs in the United States.[716] Binding of benzodiazepines to GABA receptor-chloride channels enhances the effect of GABA. The drugs induce relaxation but can interfere with memory, reduce concentration, and cause physical clumsiness. They may also intensify the effects of alcohol and can be addictive.[718]

Specific antagonists for $GABA_A$ receptors include the alkaloid convulsants bicuculline (Fig. 30-25)[699] and **picrotoxin** (Fig. 22-4) and the convulsant terpenoid compound **thujone** (Fig. 22-3), which is present in the wormwood plant *Artemesia absinthium*. Thujone is present in the liqueur absinthe, which was the national drink of France in the late 19th century but, because of its toxicity, has been illegal in most countries since ~1915.[719]

GABA enters synaptic vesicles via a vesicular GABA transporter, an integral membrane protein whose gene has been found in *Caenorhabditis elegans*.[720] Termination of GABA neurotransmission is accomplished by rapid Na^+-dependent uptake into neurons for reuse and uptake into glial cells.[721,722] Excess GABA is continuously oxidized to succinic semialdehyde by GABA aminotransferase[723] in the GABA cycle of Fig. 17-4. Notice the manner in which this cycle incorporates synthesis of both of the neurotransmitters glutamate and GABA. Glutamine also functions in neurons, perhaps serving as a buffer for glutamate.

The hereditary triple-repeat disease Huntington's chorea (**Huntington disease**), with an incidence of 5–10 per 100,000 persons, affects principally persons of age over 40 and is associated with a deficiency of GABA in basal ganglia.[724] The cortex is also affected. Severe neurologic symptoms arise as a result of premature death of neurons in the basal ganglia. Convulsions may also arise because of a deficiency of GABA in the brain.

Glycine. Glycine appears to be the most important neuroinhibitor in the spinal cord and brainstem. It is present at concentration of 3–5 mM in the spinal cord and in the medulla but is low in the cerebral cortex. **Strychnine** (Fig. 30-25) is a specific antagonist of glycine receptors in spinal synapses.[725] Ivermectin (Fig. 30-25) also blocks glycine Cl^- channels.[726] A mutant mouse called *spastic* is deficient in glycine receptor function. A small dose of strychnine produces an effect on a normal mouse that resembles the effect of this mutation.[727,728] A similar disorder affects some Hereford calves.[729] Strychnine-binding studies have suggested a deficit of glycine receptors in human spasticity and in the loss of motor control associated with **Parkinson disease** and **amyotrophic lateral sclerosis**.[725] A human **startle disease**, which causes an exaggerated muscular response to unexpected stimuli, also results from reduced glycinergic neurotransmission.[730]

Most glycine receptors are Cl^- ion channels that open in response to transmitter binding.[725] The strychnine-binding subunit shows significant homology with the nAChR proteins,[725] and the overall structures resemble those of GABA receptors and of nAChRs.[731,732,732a] Human α1–α4 and β subunits have been identified.[733,734] Two integral membrane glycine transporters are known.[735–737]

Anesthetics. Several types of neurotransmitter receptors provide binding sites for anesthetics. Some anesthetics are molecules of moderate size, e.g., **barbiturate** derivatives, while others, such as **diethyl ether** or **halothane** ($CF_3CHClBr$), are very small. The latter is one of the most widely used inhalation anesthetics. Both Mg^{2+} and Mn^{2+} are also powerful CNS depressants and can cause general anesthesia. It has often been proposed that the effectiveness of anesthetics is related to solubility in lipids, but it has been difficult to pinpoint a site of action. Now it is clear that specific synaptic proteins often provide the binding sites for anesthetics. Important among these are the glycine receptors.[715,738,739] GABA receptors[740,740a] and kainate glutamate receptors may also bind anesthetics.[741]

Adrenergic synapses: the catecholamines. The three closely related tyrosine metabolites, **dopamine**, **noradrenaline**, and **adrenaline**, known collectively as catecholamines, are important products of neuronal metabolism.[149,393] Dopamine and noradrenaline serve as neurotransmitters. Catecholamine-containing neurons are found throughout the brain, including the cortex and cerebellum regions. Very large dopamine-containing neurons are present in the brains of gastropod molluscs.[742] In the human brain a prominent series of dopamine neurons run from the substantia nigra to the caudate nuclei and putamen of the striatum, the **nigrostriatal** pathway (Fig. 30-12).[149,743,743a] In many invertebrates **octopamine**,[744–746] which is synthesized via tyramine (Fig. 30-26), apparently functions in place of noradrenaline. Note the precursor–product relationship between dopamine, noradrenaline, and adrenaline. The synthetic pathways to these neurotransmitters involve decarboxylation and hydroxylation, types of reaction important in formation of other transmitters as well. The most important process for terminating the action of released catecholamine transmitters is reuptake by the neurons. High-affinity uptake systems transport the catecholamine molecules back into the neurons and then into the synaptic vesicles. The uptake is specifically blocked by the drug **reserpine** (Fig. 25-12).[746a] The dopamine transporter is a major binding site for cocaine (see Fig. 30-28).[747–751] Catecholamine transmitters are catabolized by two enzymes. One is the

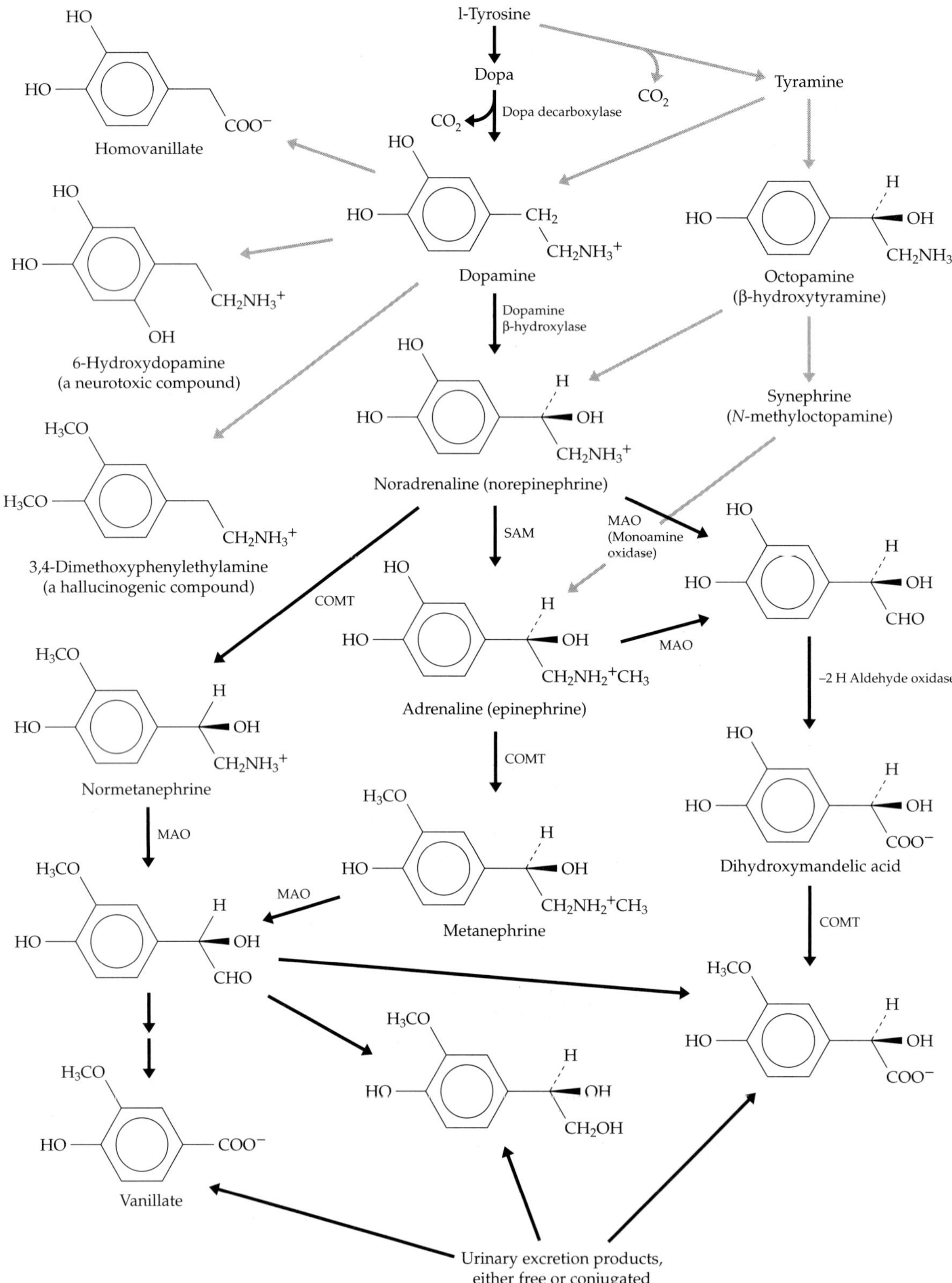

Figure 30-26 Some pathways of metabolism of the catecholamines. See also Fig. 25-5.

flavoprotein **monoamine oxidase** (**MAO**; Chapter 16), an enzyme present within the mitochondria of neurons as well as in other cells in all parts of the body.[752] The second enzyme is **catechol-*O*-methyltransferase** (**COMT**; Eq. 12-3), which is found in postsynaptic membranes as well as in liver, kidney, and other tissues. It apparently provides the principal means of inactivating circulating catecholamines. In a process that occurs in all organisms sulfo groups are transferred from PAPS (Eq. 17-38) onto hydroxyl groups of catecholamines, steroid compounds, and proteins (p. 659). Sulfation of catecholamines is relatively specific to humans.[752a,b]

Both adrenaline and noradrenaline stimulate smooth muscles throughout the body and have a hypertensive effect. Their postsynaptic receptors are 7-helix transmembrane proteins (Fig. 11-6). A comparison of the effects of various analogs led to the classification of these receptors into classes α, α_2, β, and β_2, which are discussed briefly on pp. 553–555. The α receptors, which are structurally closely related to rhodopsin,[753,754] are coupled via Gq / 11 proteins to a phosphoinositide-activated phospholipase C (Figs. 11-9, 30-19).[755] They usually provoke an excitatory response. However, in intestinal smooth muscles they are inhibitory. Adrenaline is usually more active at α receptors than is noradrenaline. A specific antagonist is **dibenane** (Fig. 30-27). The β receptors usually induce muscular relaxation but cause myocardial stimulation. Noradrenaline is usually more active than adrenaline. In most cases the β receptors of the postsynaptic membrane respond to the neurotransmitter by causing a hyperpolarization of the cell membrane and inhibition of nerve impulses. A specific antagonist is **propranolol** (Fig. 30-27). The β receptors are coupled via proteins of the G_s family (pp. 557, 558). The β2 receptors have received special attention because of their importance to heart and pulmonary functions. Both heart failure and asthma are associated with poor β2 receptor function.[756,757] The β2 receptors affect many other processes including insulin action.[758] Intense efforts are being made to understand them at the structural level.[757,759,760] Of special interest are the mechanisms by which receptors are densensitized after passage of impulses, a process that often involves multiple phosphorylation reactions[761]

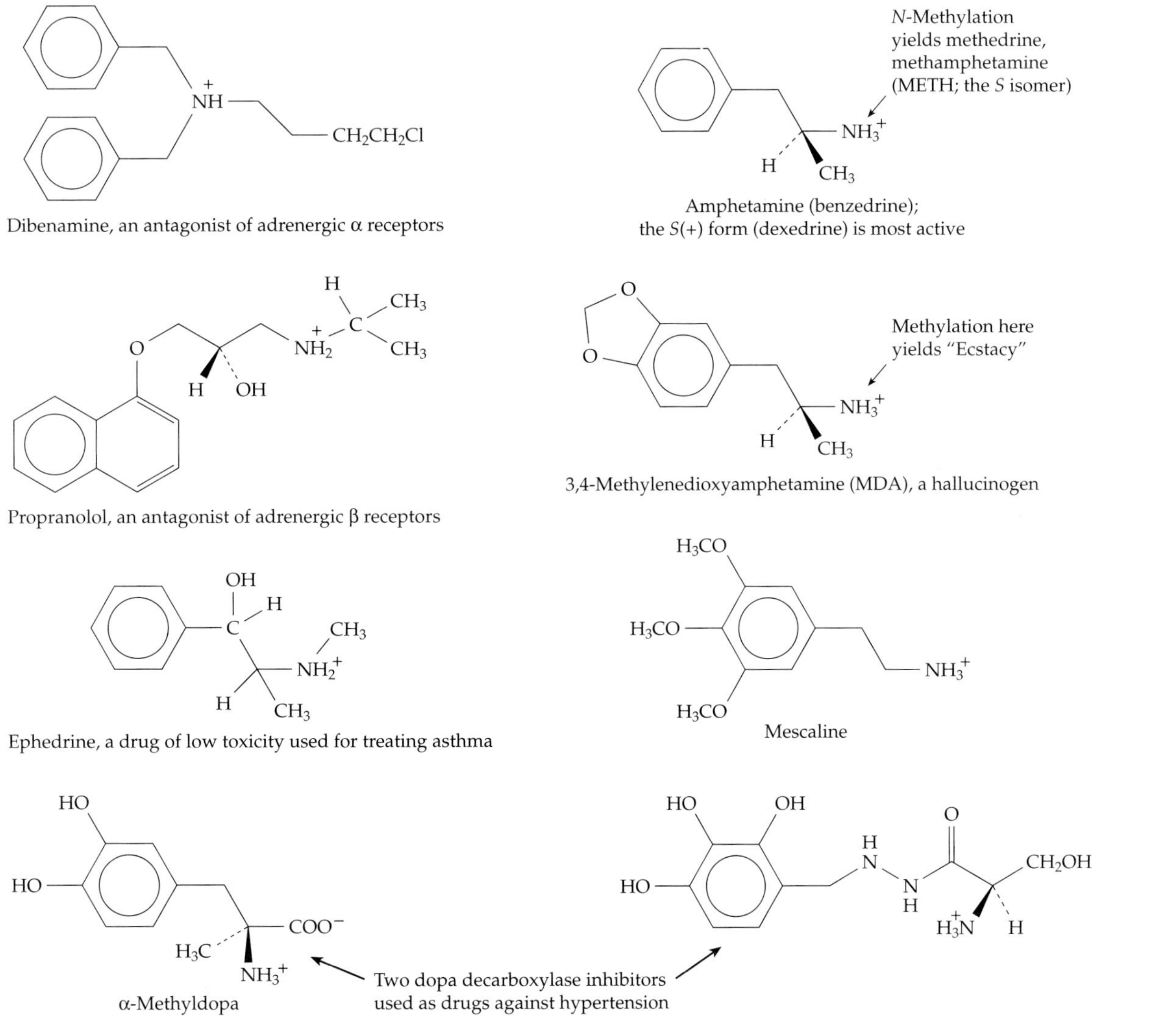

NH^+ CH_2CH_2Cl

Dibenamine, an antagonist of adrenergic α receptors

N-Methylation yields methedrine, methamphetamine (METH; the *S* isomer)

NH_3^+ H CH_3

Amphetamine (benzedrine); the *S*(+) form (dexedrine) is most active

O H OH NH_2^+ H C CH_3 CH_3

Propranolol, an antagonist of adrenergic β receptors

O O Methylation here yields "Ecstacy" NH_3^+ H CH_3

3,4-Methylenedioxyamphetamine (MDA), a hallucinogen

OH H C CH_3 NH_2^+ H CH_3

Ephedrine, a drug of low toxicity used for treating asthma

H_3CO H_3CO H_3CO NH_3^+

Mescaline

HO HO COO^- H_3C NH_3^+

α-Methyldopa

Two dopa decarboxylase inhibitors used as drugs against hypertension

HO OH HO H N N H O CH_2OH H_3N^+ H

Figure 30-27 Some agonists and antagonists of adrenergic synapses (shown as cations in most cases).

as well as interaction with **arrestin** (Fig. 23-43) and receptor internalization.[762]

Attention has been focused on dopamine because of its relationship to neurological diseases and to addiction (discussed in Section 10). Dopamine receptors constitute a large family, which are classified into two main subfamilies. The D_1 subfamily consists of D_{1A} and D_{1B} (D_5) receptors and the D_2 subfamily of D_2, D_3, and D_4 receptors.[763,764] The D_1 receptors, which are prominent in the prefrontal cortex and also in the striatum, are more abundant than the D_2 receptors, which are also present in the striatum and the pituitary and are targets for antipsychotic drugs such as **haloperidol** (Fig. 30-33).[765] The recently discovered and less numerous D_3 receptors are present in only a few regions of the brain. However, a deficiency of D_3 receptors may also be involved in addiction, schizophrenia, and Parkinson disease.[766,767]

The role of the catecholamines as transmitters in the sympathetic nervous system and in the peripheral ganglia has been well established, but the function in the central nervous system is less clear. Catecholamines are present in varying quantities throughout the brain, and histochemical techniques[149,768] have made it possible to visualize both dopamine and noradrenaline-containing neurons by the green fluorescence produced from reaction with formaldehyde or glyoxylate.[769] The reactions are presumably analogous to those in Fig. 25-10. Another method for tracing dopamine receptors in the central nervous sytem is through labeling with specific antibodies to dopamine-β-hydroxylase (Eq. 18-53), the enzyme that converts dopamine to noradrenaline, to tyrosine hydroxylase, or to other specific neuronal enzymes.[770]

Parkinson disease. Neurons of the nigrostriatal pathway degenerate in Parkinson disease, a condition accompanied by severe tremors and rigidity. The significance of dopamine was illustrated by the finding that the precursor amino acid **L-dopa** caused dramatic improvement in many persons with Parkinson disease.[771] Dopamine and other catecholamines do not cross the blood–brain barrier but L-dopa does. This leads to an increase in the dopamine level in the basal ganglia of the brain, which apparently compensates for the deficiency resulting from the neuronal degeneration.

In 1982 a number of young people in California injected themselves with an illegally manufactured opiate drug that was subsequently found contaminated with *N*-methyl-4-phenyltetrahydropyridine (MPTP). Within a few days they developed irreversible symptoms of Parkinson disease. Subsequent investigation revealed that MPTP itself is not toxic but that it is oxidized by monoamine oxidase B (MAO-B) to the corresponding pyridinium derivative MPP^+ (Eq. 30-4). It is this pyridinium derivative, or perhaps related free radicals, that is toxic.[772] MPP^+ is readily taken up by mitochondria and is apparently concentrated in the mitochondria of the nigrostriatal cells to a toxic level.[773] The MAO inhibitor pargyline (Fig. 30-33) interferes with the oxidation of Eq. 30-4 and prevents development of Parkinson disease in squirrel monkeys exposed to MPTP.[774] These results suggested

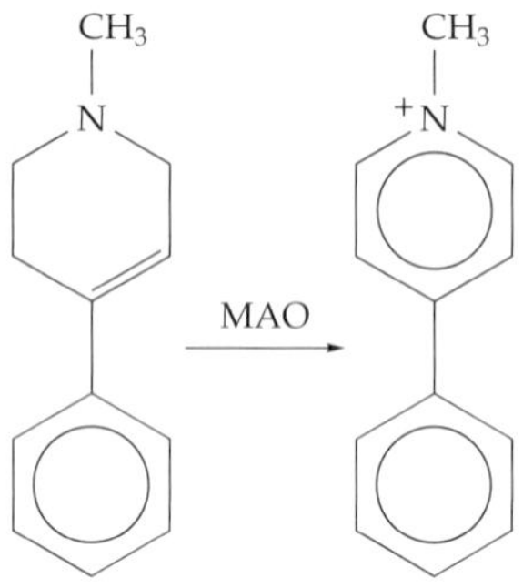

CH_3 N MAO CH_3 ^+N

MPTP MPP^+ (30-4)

possible environmental causes for Parkinson disease and also a new approach to treatment.[775,776] MPP^+ has been marketed as a herbicide, and it has a close structural relationship to another herbicide, **paraquat**.

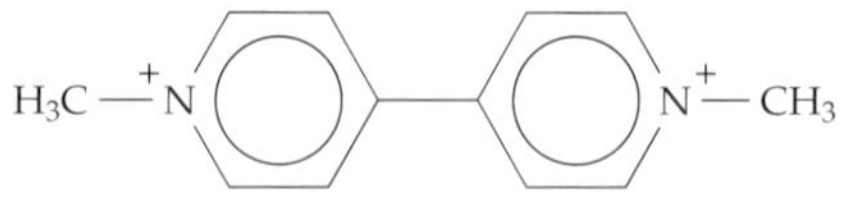

$H_3C—{}^+N$ $^+N—CH_3$

Paraquat

Many food constituents including peppermint, spearmint, and tea contain 4-phenylpyridine, another close relative.[775] While administration of L-dopa to replace the deficit in the basal ganglia seemed the ideal treatment for Parkinson disease, mental deterioration is not stopped, and for some patients the drug loses its effectiveness in about three years. Based on the new information about MPTP, treatment with extra vitamin E as an antioxidant along with an MAO inhibitor is being tested as a way to prevent further damage from environmental toxins.[776]

Serotonin and melatonin. The indolealkyl amine serotonin (5-hydroxytryptamine, 5-HT; Fig. 30-28), is found in all mammalian brains and in invertebrates as well. Its distribution in the brain is limited, serotonin-containing neurons being found in the raphe nuclei of the brainstem from which they ascend into the brain and down the spinal cord. Serotonin-containing neurons have been traced within brains of snails using ^3H-labeled serotonin.[777] Studies with these simpler brains have revealed both inhibitory and excitatory responses to these neurons. Serotonin-accumulating neurons are also found in the retina[778] and are widely distributed in the peripheral nervous

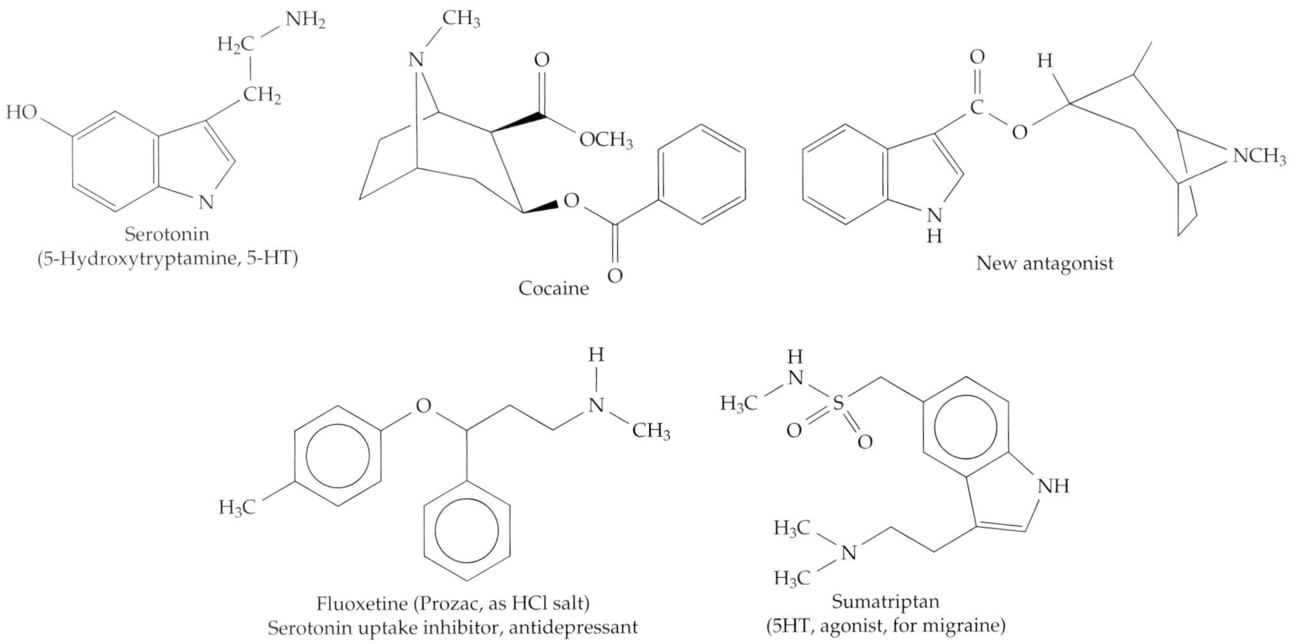

Serotonin
(5-Hydroxytryptamine, 5-HT)

Cocaine

New antagonist

Fluoxetine (Prozac, as HCl salt)
Serotonin uptake inhibitor, antidepressant

Sumatriptan
(5HT, agonist, for migraine)

Figure 30-28 Serotonin (5-hydroxytryptamine) and some drugs that affect receptors and transporters.

system.[779] Serotonin-containing granules are present in blood platelets.[780]

Serotonin appears to be involved in activation of pain fibers, when tissues are injured. **Cocaine** (Fig. 30-28) is a powerful pain killer and a weak antagonist of responses to serotonin, a fact that has led to the synthesis of new antagonists, such as the one in Fig. 30-28 whose structure encompasses that of both cocaine and serotonin.[779,781] It is active at a concentration as low as 10^{-14} M and is among the most potent known drugs of any type.

Serotonin is synthesized via tryptophan and 5-hydroxytryptophan with decarboxylation of the latter (Fig. 25-12). Within the **pineal body** of the brain and in the retina, serotonin is acetylated to *N*-acetylserotonin,[782,783] which is then *O*-methylated to **melatonin**, the pineal hormone (Fig. 25-12). A specific inhibitor of serotonin synthesis is *p*-chlorophenylalanine, and studies with this and other inhibitors suggest that serotonin is required for sleep.[784]

At least 14 distinct types of serotonin receptors (5-HT_{1A}, 5-HT_{1B}, 5-HT_{2A}, etc.) have been identified.[785,786] They are present in the heart, in gastrointestinal tissues, adrenal and other glands,[787] and bone[788] as well as in the brain. Drugs, such as **sumatriptan** (Fig. 30-28), which activate serotonin receptors are important in the treatment of **migraine**. This common disorder of serotonin metabolism is characterized by severe or moderately severe headache and a variety of other symptoms, which are frequently preceded by a visual aura.[789] Serotonin is removed from synapses via a transporter, which also contains the binding site of the widely used antidepressant **Prozac** (Fig. 30-28) and related drugs.[790–791b]

Serotonin and melatonin are evidently involved in maintenance of the 24-h circadian rhythm of the body (see Section 13).[792,792a] Melatonin regulates the sexual cycle in photoperiodic animals and influences the onset of puberty.[792–794] The serotonin content of the brain is influenced by the diet, being higher after a meal rich in carbohydrates. Serotonin may serve as a chemical message sent from one set of neurons to the rest of the brain, reporting on the nature of dietary intake.[784] Melatonin, which can readily form free radicals, may function as part of the body's antioxidant system.[795,796]

Other neurotransmitters. The abundant glutamate, GABA, and glycine are major neurotransmitters. Do other amino acids also function in the brain? Roles for L-aspartate and D-serine (p. 1785) have been identified, but it is very difficult either to discover or to disprove a neurotransmitter function for other amino acids. It is even more difficult for small amounts of various amines and small peptides that are present in the brain. **Taurine** (Fig. 24-25) is one of the most abundant free amino acids in animals and meets several criteria for consideration as both an inhibitory and an excitatory transmitter.[797,798] However, its function is still uncertain (see Chapter 24). **Homocysteic acid**, formed by oxidation of homocysteine, is a powerful neuroexcitatory substance, but its concentration in the brain is very low.[149] D-Aspartate is also present at high concentrations in the cerebellum, pituitary, pineal gland, and adrenal chromaffin cells. It appears to be a modulator of melatonin synthesis.[799,800]

Receptors for **histamine**, which probably acts as a neuromodulator,[801] occur in the brain.[802] Histamine is formed by decarboxylation of histidine (p. 745)[803] and is inactivated by histidine *N*-methyltransferase. Histamine is best known for its presence in mast cells,[804] components of the immune system that release histamine during inflammatory and allergic reactions (Chapter 31). However, histaminergic neurons of the hypothalamus extend throughout the whole forebrain,[805] and specific receptors have been found both in the brain and in peripheral tissues.[806] Several other amines that are formed by decarboxylation of amino acids are present in trace amounts but may have important functions, some of which may be related to psychiatric disorders. These include tyramine (from tyrosine), β-phenylethylamine (from phenylalanine), and tryptamine (from tryptophan). As previously mentioned, octopamine is also present in trace amounts in mammalian brains.[807]

ATP, **ADP**, and **adenosine** are among the purines that are present in some synapses and activate a variety of receptors. Adenosine receptors are blocked specifically by methylated xanthines such as caffeine (Fig. 25-18) and theophylline.[808–808b] A drug almost 10^5 times as potent as theophylline is 1,3-dipropyl-8-(2-amino-4-chlorophenyl)xanthine.[809] Adenosine receptors, which are present in large numbers in the hippocampus,[149] form functional complexes with metabotropic glutamate receptors.[678] Adenosine

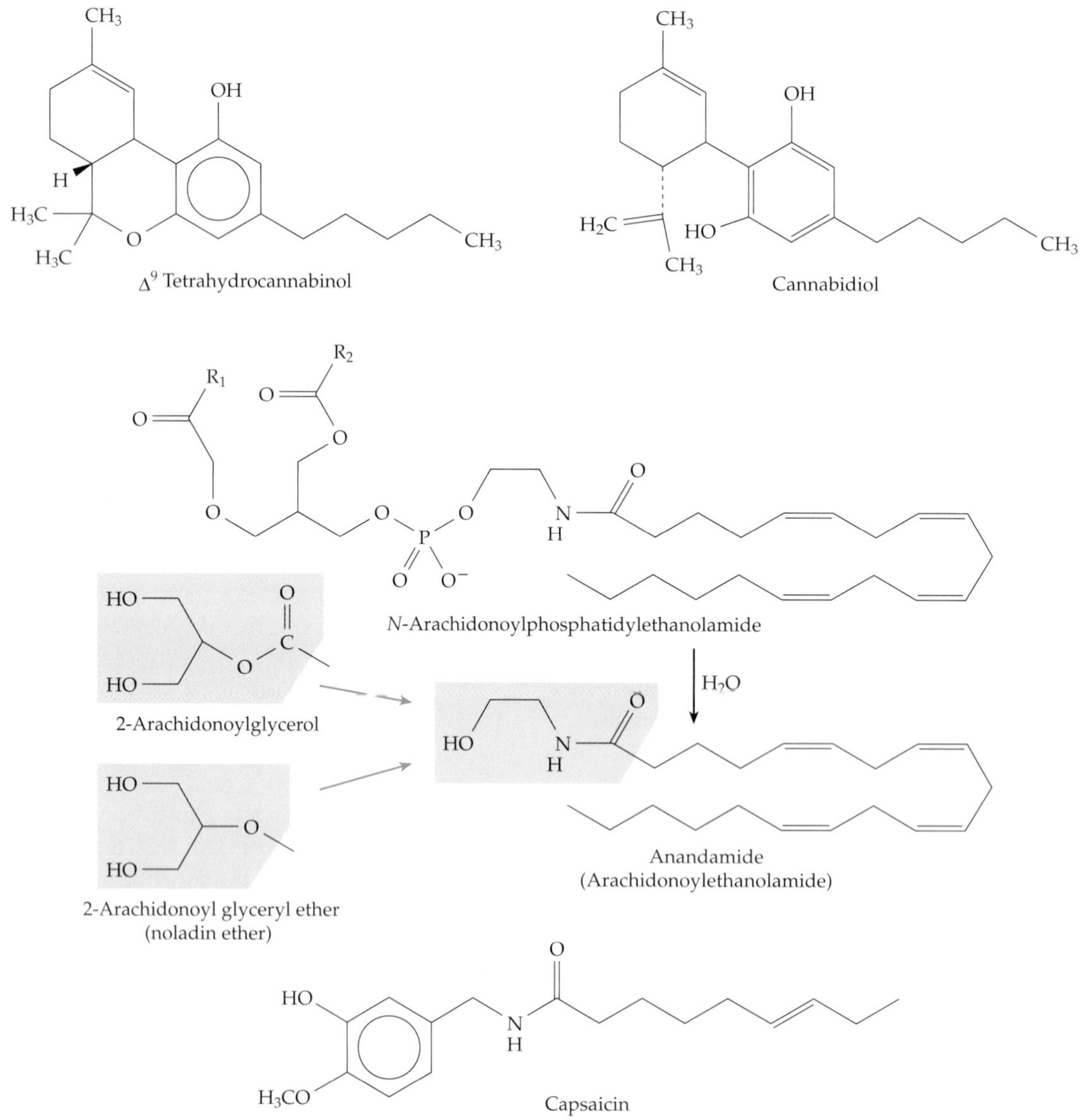

Δ⁹ Tetrahydrocannabinol

Cannabidiol

N-Arachidonoylphosphatidylethanolamide

2-Arachidonoylglycerol

Anandamide (Arachidonoylethanolamide)

2-Arachidonoyl glyceryl ether (noladin ether)

Capsaicin

Figure 30-29 Structures of the active components of cannabis, tetrahydrocannabinol, and cannabidiol, and structures of endogenous cannabinoids and of the vanilloid lipid capsaicin.

usually has a depressive effect. Craving for chocolate is often attributed to the methylxanthines present, but it may be a result of anandamide and related compounds.[810]

The occurrence of a variety of **neuropeptides** in the brain has been discussed in Section A. The first of these to be discovered[811] was the 11-residue **substance P** (Table 30-4), which was isolated in 1931. Like other neuropeptides it may function either as a transmitter or neuromodulator or perhaps both. Substance P, as well as many other neuropeptides, has been localized to specific neurons. Along with somatostatin, CCK, and enkephalins, it is found in high concentrations in the basal ganglia. Enkephalin and substance P are also found in specific neural elements in the visual system of lobsters.[812] In some cases a neuron contains both synaptic vesicles containing a major neurotransmitter and also vesicles containing a peptide or other cotransmittor. The peptide pituitary hormones ACTH, MSH, and vasopressin as well as the hypothalamic neurohormones may have effects on learning and behavior.[813]

Lipid mediators in the brain. The brain is rich in phospholipids, glycolipids, and long-chain unsaturated fatty acids. Many signaling functions seem likely, and some are discussed in Section A,7. Prostaglandin D_2 is a major prostanoid in the brain, which induces both hypothermia and sleep.[814] As mentioned in Section A,7, **oleamide** also induces sleep, perhaps by modulating the effects of 5-HT receptors.[815,816] **Anandamide** is a lipid derived by hydrolysis of the unusual phospholipid *N*-arachidonoylphosphatidylethanolamide. This is one of a recently discovered series of amides, esters, and ethers derived from arachidonic acid (Fig. 30-29).[816a,b] They have been identified as endogenous ligands of the abundant **cannabinoid receptors**.[816] The latter were identified as binding sites of Δ^9-tetrahydrocannabinol and cannabadiol (Fig. 30-29), both of which are constituents of **marijuana**. Anandamide was the first of the endogenous cannabinoids to be isolated.[817] However, the monoglyceride **2-arachidonoylglycerol** (Fig. 30-29) is much more abundant in brain and also activates cannabinoid receptors.[818] It arises by hydrolysis of a diglyceride.[819,820] Recently 2-arachidonoyl glyceryl ether (**noladin ether**; Fig. 30-29) has been identified as another endogenous agonist of the CB_1 cannabinoid receptors.[821] A possible alternative pathway for anandamide synthesis is via an energy-dependent coupling of arachidonic acid with ethanolamine.[822,823] The two known types of cannabinoid receptors are both 7-helix proteins coupled by G_i or G_o proteins to adenylate cyclase and to Ca^{2+} and K^+ channels.[824,825] The CB_1 receptors are found largely in the brain and are responsible for the psychoactive effects of cannabis, while the CB_2 receptors are more widely distributed. They seem to have a special role in cells of the immune system, e.g., in macrophages and B cells.[818,820,825–828] Palmitoylethanolanide has been proposed as an additional endogenous ligand for CB_2 receptors.[820,829] Cannabinoid receptors of invertebrate immune system cells and of human monocytes have been found coupled to NO release.[830]

Cannabinoid receptors are present at extremely high levels in the basal ganglia of the brain,[831,832] but they do not appear to be essential. Knockout mice lacking the CB_1 receptors appear normal in most respects. However, they do not respond to cannabinoid drugs and, curiously, do not become addicted to morphine as normal mice and have less severe withdrawal symptoms than normal after morphine addiction.[826] The CB_1 receptors in the basal ganglia modulate GABA neurons that have outputs to the substantia nigra and the globus pallidus (Fig. 30-30B). The nigrostriatal neurons also secrete substance P and dynorphin, while those extending to the globus pallidus generally contain enkephalin as a cotransmitter.[832] These interconnections affect the dopaminergic neurons. Cannabinoids also have pain supressing and neuro-protective effects. They may have many possible medicinal uses, which are being explored.[833–837]

The endogenous cannabinoid compounds are lipids and are not stored in synaptic vesicles but are presumably released by enzymatic action following passage of a nerve impulse. Recent evidence suggests that the endocannabinoids are released at a postsynaptic membrane and then diffuse back to a presynaptic surface and outward to other cell surfaces where they affect signaling.[838–840] This **retrograde signaling** in synapses of the hippocampus is thought to be involved in **long-term potentiation** (**LTP**), the changes in synaptic properties that occur during learning and in the formation of memories (Section 12). A monoglyceride lipase participates in inactivation of endocannabinoids.[840a] Anandamide is also a substrate for cyclooxygenase-2 (Eq. 21-16), whose action may lead to formation of additional immunomodulatory compounds.[841,842] Long-chain relatives of arachidonic acid such as docosohexaenoic acid (DHA; Box 21-B) are especially high in brain lipids.[843,843a]

Nitric oxide and carbon monoxide. The gaseous molecules NO and CO have both been found in the brain, and neuronal NO synthase (nNOS or NOS I) has been studied intensively.[844–847] NO synthases and the functions of NO and CO are discussed in Section A7 and in Chapter 18. Complexity in understanding the role of NO in the brain arises from the fact that different isoenzyme forms of NO synthase occur in three different types of cell: nNOS in neurons, iNOS from microglial immune system cells, and eNOS from endothelial cells of capillary blood vessels.[846] All three types of cells are so tightly intermingled in the brain that it is hard to interpret observed experimental effects. Elevated Ca^{2+} concentrations that can arise from stimulation of NMDA receptors in the hippocampus seem particularly effective in activating the calmodulin-dependent nNOS. This suggests that, like the endogenous cannabinoids, NO may be a retrograde messenger in LTP.[149] The possibility that CO may function in a similar way also remains uncertain, as does any pathway for metabolism of CO. Certainly NO and CO generated in the brain will have some effects that arise from their very tight binding to heme groups. An example is the observed inhibition of dopamine β-hydroxylase by N_2O_3 with a resulting decrease in noradrenaline synthesis.[848]

10. Some Addictive, Psychotropic, and Toxic Drugs

Humans have a long history of use of stimulant and mind-altering substances. Tea, coffee, alcohol, tobacco, opium, cocaine, marijuana, and a host of modern synthetic compounds have been used as stimulants, as medications, and for pleasurable experiences.[849–851] Many are also addictive and sometimes lethal. Stimulant drugs such as nicotine, cocaine, methamphetamine (METH), and other amphetamines (Fig. 30-27)[849,852–854b] can give users feelings of increased energy, well-being, and self-confidence. Nicotine enhances fast excitatory transmission[855] and may sharpen memory.[856] However, all are acutely toxic and are highly addictive. Amphetamines and cocaine act directly to increase the brain dopamine level causing euphoria. However, in response the dopamine receptors rapidly decrease their sensitivity. This leads to mental depression and the desire for more drug. Nicotine appears to indirectly affect the same dopamine neurons.[857] The wisdom and ethics of giving hypoactive children the addictive stimulant **methylphenidate** (Ritalin; see Fig. 30-33) have been questioned.[858,859] The depressive drugs, including **morphine** and other narcotics (Fig. 30-30), barbiturates (Fig. 30-22), and ethanol, are all strongly addic-

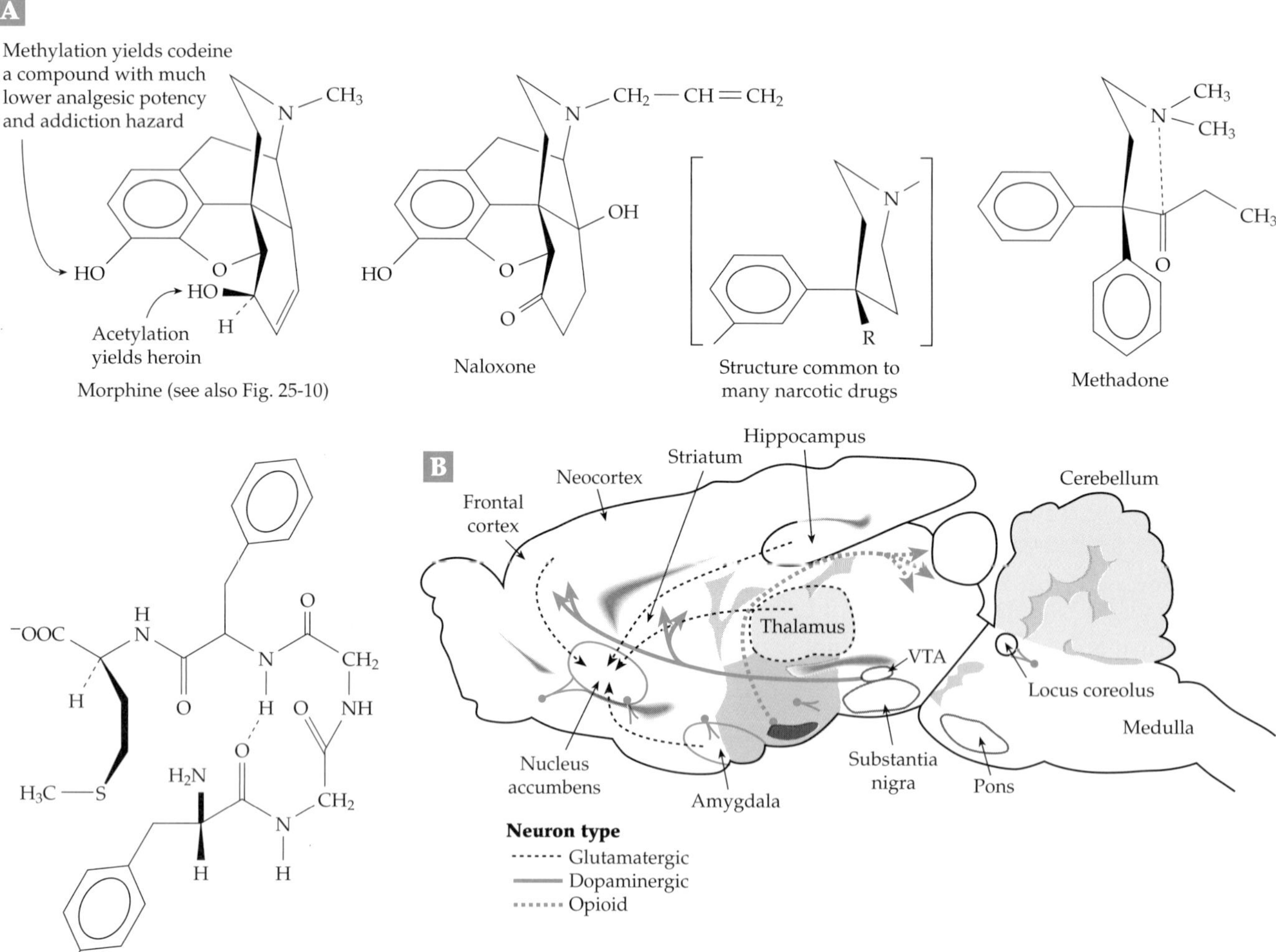

Figure 30-30 (A) The structures of morphine and of some analogs including the brain peptide Met-enkephalin. Also shown is a structure common to many narcotic drugs. (B) Diagram of a rat brain as shown in Fig. 30-13 with some aspects of the mesolimbic dopamine system emphasized. See Shulteis and Koob.[869]

tive for susceptible individuals. The phenomenon is most striking in the case of the opiates. Addiction leads to physical dependence, a situation in which painful withdrawal symptoms occur in the absence of the drug. At the same time a striking tolerance to the drug is developed. The addicted individual can survive what would otherwise be a fatal dose without ill effect. Aside from the pathological hunger for the drug, an addict can function normally in almost every respect.[860] Dependence develops only from frequent doses of drug over a long period of time and is not observed with cocaine or amphetamines.[861] Marijuana is only mildly addictive, according to some data about the same as caffeine.[862] However, this conclusion is controversial.

Opioid receptors. Direct binding of highly radioactive opiates has permitted localization of specific opiate receptors of several types.[863–866] The three major types (μ, δ, κ) are all 7-helix receptors coupled to adenylate cyclase, K^+ and Ca^{2+} channels, and the MAP kinase cascade.[866] The μ receptors bind morphine most tightly.[867,867a] These receptors are found in various cortical and subcortical regions of the brain. Most narcotics are polycyclic in nature and share the grouping indicated in Fig. 30-30. However, the flexible molecule **methadone** binds to the same receptors.[868] Among antagonists that block the euphoric effects of opiates the most effective is **naloxone** (Fig. 30-30).

What is the natural function of opiate receptors? Opiates are the most powerful **analgesic agents** known. The existence of the **enkephalins**, **endorphins**, and **endomorphins** (Section A,5; Table 30-4) in the brain suggests that opiate drugs mimic the normal action of these peptides, which may function in controlling pain. Although opiates are powerful drugs, their efficiency in diminishing pain is directly related to their addiction potential. To date, it has not been possible to design a nonaddictive analgesic drug of the potency of morphine.

Addiction seems to follow compensatory changes in the receptor–agonist system that result from the occupation of the receptor sites by the drug. For example, studies of opiate receptors indicate that morphine acts in an inhibitory fashion, lowering the internal level of cAMP.[861,870] The neuron then compensates by increasing the number or activity of adenylate cyclase molecules restoring the internal cAMP level. This leads to dependence upon morphine because in its absence the cAMP level rises too high. The increased number of adenylate cyclase molecules and associated receptors also accounts for the observed tolerance. It is now clear that this adaptation is complex. The properties of many synapses in various parts of the brain are altered by phosphorylation or dephosphorylation or other reactions of receptors and other synaptic proteins. Some changes are rapid, but others are slower and involve alterations in transcriptional patterns within neurons. These changes occur in three different neuronal systems: (1) physical control systems, in which changes lead to physical dependence; (2) motivational control systems; and (3) associative memory systems.[861]

The **mesolimbic dopamine system** is thought to be involved either directly or indirectly in addiction to many drugs. The dopaminergic neurons of this system have cell bodies in the **ventral tegmental area** (VTA) of the brain (Fig. 30-30B) and extend into the **nucleus accumbens**, a region at the base of the striatum that is thought to provide the "rewarding effects," i.e., pleasure from drugs such as cocaine or amphetamines. There is direct experimental support for this conclusion.[871] Less certain is the proposal that opiates and other depressive drugs indirectly cause a similar effect in the nucleus accumbens.[869,870,872] A more recent view is to regard addiction as an aberrant form of learning.[861,871,873] This concept is applicable also to "behavioral addictions."[873a]

Ethanol. As with morphine addiction, tolerance to alcohol is developed, and a lack of ethanol produces withdrawal symptoms. The principal route of metabolism of ethanol (both ingested and the small amount of endogenous alcohol) is believed to be oxidation in the liver to the chemically reactive acetaldehyde (p. 774),[874,875] which is further oxidized to acetate. Some theories of alcoholism assume that addiction, and possibly also the euphoric feeling experienced by some drinkers, results from a metabolite of ethanol in the brain. For example, acetaldehyde could form alkaloids (Eq. 30-5).[876]

$$\text{Dopamine} \xrightarrow[2[H]]{H_3C-CHO} \text{(6,7-dihydroxy-1-methyl-1,2,3,4-tetrahydroisoquinoline: HO, HO, NH, C, H, } CH_3\text{)} \quad (30\text{-}5)$$

In fact, small amounts of morphine, 6-acetylmorphine, codeine, and thebaine, all opiate compounds, have been found in mammalian brain[877,878] and have presumably arisen by the same pathway observed in plants (Fig. 25-10). However, there is no cross reactivity between morphine and alcohol in addicted mice,[879] and acetaldehyde is probably not the addictive agent. Acetaldehyde is very reactive and may be responsible for much of the damage caused by ethanol.[880] At a blood ethanol concentration of 20 mm a person is legally intoxicated, and large amounts of acetaldehyde may be formed and react with many amines, nucleotides, proteins, etc. Ethanol blocks glutamatergic NMDA receptors and activates GABA receptors.[740a] These effects may be involved in the neurodegeneration of fetal alcohol syndrome.[881,882] Ecitotoxicity may also be a factor in alcohol damage.[692] Alcoholic liver disease may involve malnutrition as well as direct damage.[883]

Experiments with mice and rats have established a genetic propensity toward addiction to alcohol. Animals from some strains shun alcohol and become addicted only if force-fed for prolonged periods. Others, which may have low levels of neuropeptide Y in the brain,[884] accept the alcohol readily and become addicted quickly. That a similar situation holds for humans is quite possible. However, a specific "alcoholism gene" has not been found.

Psychotropic or mind-changing drugs. Hallucinogenic compounds have long been a source of special fascination to many people. The presence of the indole ring in the powerful hallucinogen **lysergic acid diethylamide** (LSD; Fig. 25-12) suggests that this compound may mimic the action of serotonin. However, other experiments suggest antagonism of dopamine receptors in the striatum.[885] Other hallucinogens include 3,4-methylenedioxyamphetamine (MDA; Fig. 30-27),[886] a compound that damages serotonergic neurons, its derivative "ecstasy" (Fig. 30-27),[886a] mescaline (Fig. 30-27), and phencyclidine (angel dust), a compound introduced in the late 1950s as a general anesthetic. Unfortunately, it produces a long-lasting condition resembling schizophrenia.[887] A common site of action for a large variety of hallucinogens has been suggested.[888]

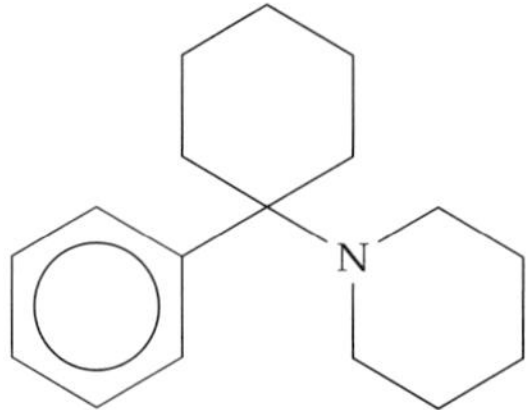

Phencyclidine (Angel dust)
a discredited anesthetic

Neurotoxins produced by the body. Some normal body constituents are neurotoxic in excess. These incluse **quinolinic acid** (Fig. 25-11),[889] **3-hydroxykynurenine** (Fig. 25-11; p. 1444),[890] and homocysteine.[891] Elevated levels of homocysteine are also associated with vascular disease and stroke (Chapter 24). 3-Hydroxykynurenine is a precursor to ommochrome pigments of insects and an intermediate in conversion of tryptophan into the nicotinamide ring of NAD in humans (Fig. 25-11). 6-Hydroxydopamine (Fig. 30-26), which may be formed in the body, is severely toxic to catecholaminergic neurons.[892]

Other neurotoxins can be formed from environmental pollutants. The solvent 1,4-butanediol is converted to **γ-hydroxybutyrate**, which is also a drug of abuse.[893] Many compounds in commercial use have not been adequately tested as neurotoxins.[894]

11. The Senses: Sight, Smell, Taste, Hearing, Touch, and Others

Our brains receive a continuous stream of impulses from receptors that sense light, taste and odor molecules, sound waves, touch, pain, gravitational pull, etc. Of these receptors those of vision, which are discussed in Chapter 23, may be the best known. The photoreceptors consist of rhodopsin and related 7-helix proteins embedded in membranes of the rod and cone cells (Fig. 23-40). A complex series of control mechanisms, some of which are outlined in Fig. 23-43, permit enormous amplification of the initial signal generated by a G protein and a cGMP-gated ion channel. The array of rods and cones in the retina send messages via the optic nerve to the **visual cortex**, an area of ~15 cm^2 on the cerebral cortex surface at the back of the brain.[149,895] The visual cortex is divided into two halves, but curiously, the right eye sends its signals to the left brain and vice versa. The image viewed by the retina can be mapped to the visual cortex. There it may reside in the form of chemical alterations in the ~40,000 neurons thought to be present in the visual cortex[409,896–898] for a short time until it is stored in short-term working memory locations.

Receptors for the other senses, like those for sight, also consist of clusters, often in regular arrays, of 7-helix receptors. Most of these are also G protein-coupled ion channels that are controlled by cAMP or cGMP.[899]

Odor. Even bacteria possess something akin to our ability to taste and smell. As is discussed in Chapter 19, Section A, many bacteria are attracted to L-serine or D-ribose and are repelled by phenol. Receptor proteins in the plasma membrane are involved in sensing these compounds and in allowing bacteria to move toward food and away from danger. Many other examples of chemotaxis are known among the lower invertebrates such as *Euglena*. Chemoreceptors in *Hydra* sense glutathione that flows from the broken tissue of their prey and control the animal's feeding behavior. Related organisms respond to proline. Asparagine induces the bending of the tentacles of the sea anemone *Anthopleura*, while glutathione induces swallowing.[900] Salmon return to their home streams using a memory of specific odors.[901]

Thoughout the animal kingdom the sense of smell is essential for survival. Perhaps it is not surprising that from the nematode *C. elegans* to human beings there is a largely conserved mechanism for sensing odors.[902] A large array of 7-helix G protein-coupled olfactory receptors embedded in an epithelial membrane carry signals directly into the nervous system.

In *C. elegans*, which has only 302 neurons, there are 32 chemosensory neurons and more than 100 genes for 7-helix receptors that are expressed in these neurons.[903–905] The fruit fly *Drosophila melanogaster* has at least 59 genes for olfactory receptors.[906,906a] Zebrafish and catfish have ~100.[905,907] Mice and rats have ~1000 olfactory receptor genes and human beings at least 500, which account for about 1–4 % of the genome.[905,908,909] In higher animals most receptors are coupled via G proteins, adenylate cyclase, and cAMP to ion channels in the membrane.[905,908] Insects utilize both cAMP and Ins3-*P* in their chemosensory receptors.[906] The signaling pathways parallel those of the visual receptors (Figs. 23-40, 23-43), which, however, utilize cGMP. Each gene is thought to give rise to a receptor of a specific **type** able to respond to specific structural features in an odor molecule.

Human olfactory cells are located in the **olfactory epithelium** on the upper surface of the back portion of the nasal cavity. They are neurons with chemosensory cilia similar to the rods and cones of the retina (Fig. 23-40). The cilia, which can be detached and isolated from the olfactory epithelium, contain the odorant-stimulated G-protein-dependent adenylate cyclase.[910,911] There are ~10 million receptor cells of at least 500–1000 different types. The 10 million axons form bundles of ~5000 axons each and pass through small perforations in the skull directly into the **olfactory bulb** (at the front of the brain before the pituitary, Fig. 30-13), a distance of 3–4 cm. The cortex of the olfactory bulb is lined with ~1800 **glomeruli**. Each glomerulus is a bundle, ~0.1–0.2 mm in diameter, of synaptic endings of the neurosensory nerves coming from the olfactory epithelium with dendrites of neurons that run to the **olfactory cortex** and other regions of the brain.[909] Each sensory receptor sends signals to a single glomerulus, but the glomerulus receives signals from 500 or more sensory neurons, which are not all of the same types. The glomerular cortex of the mouse is divided into four zones, each of which contains only some of the types of receptor. It seems that the cortex contains a crude "map" that relates position to the type of smell.[912] The neural processing involved in the discrimination of odors is not yet clear.[912,913] Interneurons of the olfactory bulb are unusual, being continuously discarded and replaced by new neurons that arise from neural stem cells.[908,914] This process seems to be essential for odor discrimination but not for the sensitivity of odor detection.

Most mammals have a second olfactory apparatus, the **vomeronasal organ** (VNO) or "sexual nose," which is located on the lower surface of the nasal cavity. It is a fluid-filled cavity containing chemosensory receptors through which nasal fluid is literally pumped, when the animal seeks to maximize the sensitivity of detection.[908,915] The VNO is especially important to reproduction, defense, and food-seeking. A specialized set of olfactory sensory neurons that project to atypical glomeruli in the olfactory bulb utilize cGMP signaling and may also function in reproductive behavior.[916]

The olfactory epithelia are bathed in an aqueous mucus through which odorant molecules must pass. A number of specialized proteins, including **odorant-binding proteins,** are secreted in this fluid.[917–919] Many odorant-binding proteins are **lipocalins** (Box 21-A) and presumably assist in transporting lipophilic odorant molecules to the olfactory receptors. They tend to have a low specificity for the odorant and a weak binding affinity, properties that are consistent with this function. Pheromone-binding lipocalins encoded by ~30 genes are also found in rodent urine,[920] where they play a similar role. In contrast, the pheromone-binding proteins of some male moths are largely α helical.[920a] Although pheromones are not as important to human physiology, axillary odors from both males and females do apparently carry chemical signals. One well-established effect is the synchronization of menstrual cycles of women living in the same house or dormitory. Alipoprotein D apparently serves as a binding protein that carries odorant precursors that are acted on by bacteria to produce the pheromones.[921]

Virtually all people lack the ability to detect some specific odors. A striking example of such an **anosmia** is the inability to smell the volatile steroid **androstenone** (5α-androst-16-en-3-one), a constituent of perspiration, of some pork products, truffles, and celery.[922]

Taste. Less is known about the biochemistry of taste. The taste that we perceive is affected by odor, temperature, and physical contact. However, five primary tastes are recognized.[923,923a]

Salty: apparently perceived by an ion-channel-linked receptor
Sour: apparently linked to an H^+ channel
Bitter: perceived by bitter-sweet G protein-coupled receptors
Sweet: also perceived by bitter-sweet receptors
Umami: a recently recognized taste, that of glutamate

An experimental difficulty lies in the fact that there are only a few thousand taste buds in the tongue, with only 50–100 cells in a bud. They age rapidly, having a lifespan of only about ten days.[924] There may be only 30,000–50,000 hard-to-isolate taste receptor cells on the tongue's surface.[923] However, very recently published reports describe a large family of bitter and sweet receptors in mice and humans[924–928] and in *Drosophila*.[929,930] The sweet-sour receptors are thought to activate a G protein called **gustducin**,[931,932] which plays a role similar to that of transducin in vision and

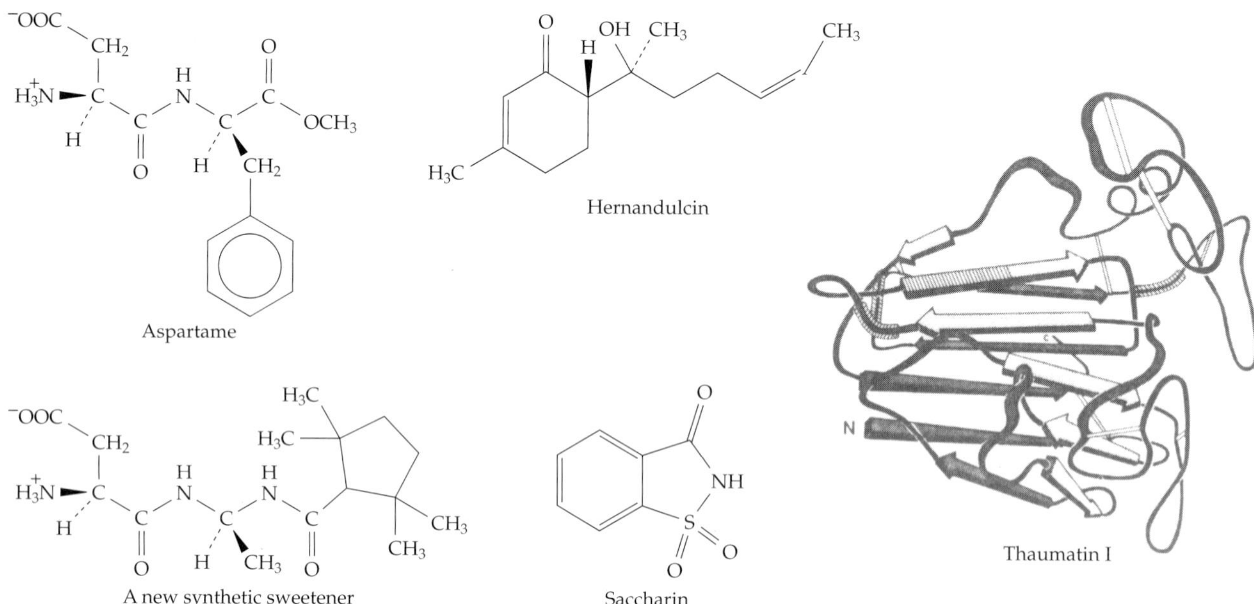

Figure 30-31 Structures of some very sweet compounds. The backbone structure of the protein thaumatin I is included. The main body of this structure consists of two β sheets forming a flattened β barrel. β Strands in the top sheet are shaded light, and those in the bottom sheet are darker. Open bars represent disulfide bonds, and the regions with sequences homologous to monellin are indicated by the hatched marks. From de Vos *et al.*[940]

also activates ion channels.[933] Like the odor receptors, taste buds are also bathed in a special fluid. The **von Ebner's glands** in the tongue contain binding proteins,[934,935] at least some of which are lipocalins.

The relationship between structures and sweet taste in humans has been investigated intensively, but no simple rules have been discovered (see Robyt[936] for a discussion). Sucrose is usually perceived as very sweet. D-Fructose and D-xylose are nearly twice as sweet, but D-glucose is less sweet than sucrose. D-Galactose is usually perceived as not sweet and D-mannose as bitter. Many sucrose derivatives, in which hydroxyl groups have been replaced with Cl or other halogen, are very sweet. One tetrachloro derivative of this type is 7000 times sweeter than sucrose.[936] Some especially sweet materials are depicted in Fig. 30-31. These include peptide derivatives[937,938] such as Asp-Phe-OCH_3 (**aspartame**), the sesquiterpene **hernandulcin**,[939] and **chemostimulatory proteins**. Among these are some of the sweetest substances known, the 207-residue **thaumatins**[940,941] and **monellin**[942,943] present in certain tropical berries. Thaumatins are ~3000 times sweeter than sucrose on a weight basis and 10^5 times sweeter on a molar basis. Thus, sucrose tastes sweet at a concentration of 10^{-3} M or higher but thaumatin[944] at 3×10^{-8} M.

The proteins **miraculin** and **circulin** from tropical fruits modify taste. Acids taste sweet rather than sour after the tongue has been treated with either protein.[945,946] Exposure of the tongue to artichokes often makes water taste sweet.[947] Thus, the response of taste receptors can be temporarily altered by binding of other substances, perhaps at adjacent sites on a receptor.

Pain. Receptors for pain (**nociceptors**) are spread over the body in nerve endings found in the skin, muscle, joints, and internal organs. There are several types of receptors, most of which are present in excitatory glutamatergic neurons.[948] Some release substance P. Some activate tyrosine kinases and others ATP-gated ion channels. Some pain receptors are also activated by intense heat or pressure or by irritant compounds. Among the latter is capsaicin (Fig. 30-29), the active ingredient in chili peppers, and an ultrapotent compound, **resiniferatoxin**. Both capsaicin, which is 10,000 times more potent than jalapeño peppers,[949] and resiniferatoxin, which is 20-fold more potent than capsaicin, bind to **vanilloid receptors**. These are ligand-gated ion channels related to the Shaker K^+ channel (Fig. 30-18). They are nonselective but with a high permeability to Ca^{2+} and are members of the **transient receptor potential** (TRP) family.[948,950–954] Pain seems to stimulate an increase in anandamide (Fig. 30-29), which has an analgesic effect. Nevertheless, anandamide and *N*-vanillyloleamide activate capsaicin receptors.[955] Because the activated receptors become desensitized rapidly, capsaicin has been used in a paradoxical manner as an analgesic agent.[951] Sensing of temperature changes also depends upon TRP channels.[955a,b]

Mechanoreceptors. The transduction of mechanical force into a chemical signal provides the basis for the senses of touch and hearing. Plants detect wind and gravitational force,[956] and many organisms, even bacteria, respond to changes in osmotic pressure using mechanoreceptors.[957] One of the best known mechanoreceptors is from *Mycobacterium tuberculosis*. It is a homopentamer whose three-dimensional structure[450–452,958] resembles that of the nicotinic acetylcholine receptor (Fig. 30-23). A second type of mechanoreceptor is found in the inner membranes of *E. coli* and in plasma membranes of many other bacteria, archaea, and some eukaryotes.[957a] These receptors, which are also sensitive to voltage changes, are heptamers of a 282-residue protein that forms a symmetric ion channel in the center. There is also a large cytoplasmic domain consisting largely of β structure.[957b] How do such receptors sense mechanical stress? In bacteria they respond to stretch in the membrane induced by an increase in osmotic pressure. One suggestion is that the membrane expansion pulls apart the radially symmetric ion channel in the receptor.[958,958a] In higher organisms transmembrane adhesion receptors and their linkage to the internal cytoskeleton provide a framework for detection of mechanical forces and linkage to mechanoreceptors.[956,957,959]

Hearing. Movement of the **stereocilia** of the hair cells of the inner ear activates mechanoreceptors. Each stereocilium contains a core of crosslinked actin filaments, and tens to hundreds of these cilia are connected in hair bundles, which move in response to arrival of sound waves of appropriate frequencies. The movement of the stereocilia induces the opening of receptor ion channels in the hair cell membrane allowing K^+ and other ions to flow inward.[960] The matter is much more complex than this because of the tuning and amplification mechanisms in the cochlea of the inner ear.[960–962] These mechanisms allow receptors in hair cells to respond to very weak vibrations of specific frequencies. Both mechanical and biochemical mechanisms are involved. A number of specific proteins participate. Among these is a motor protein called **prestin**, which seems to be involved in the rapid changes in length and stiffness of some hair cells in the cochlea.[962,963]

Other sets of hair cells are formed in specialized parts of the inner ear.[964] The three semicircular canals detect angular acceleration in three directions, while the sac-like utricle and saccule detect linear acceleration including gravitational attraction. These two organs each contain a patch of hair cells whose tips project into a gelatinous layer, which is overlain by a field of small crystals of calcium carbonate. These little stones (**otoliths**) provide an inertial mass, which resists movement causing the hair cell tips to bend and activate mechanoreceptors to send information about balance and orientation to the brain.

While discussing vibrations we may ask whether 60 cycle electromagnetic field fluctuations caused by electrical power transmission can affect the human body? Considerable effort has been expended in addressing this question. The tentative conclusion is that such effects, if they exist, are extremely difficult to detect. However, the possibility has not been disproven.[965]

12. The Chemistry of Learning, Memory, and Thinking

What is known about the chemistry underlying memory, thinking, and the generation of the stream of consciousness within the brain? Nerve impulses originating in sensory receptors are sent to several regions of the brain, among which are the sensory regions of the cerebral cortex (Fig. 30-14). Memory also depends upon other regions of the brain including the hippocampus, amygdala, and cerebellum (Fig. 30-1). Learning, remembering, and thinking all require transfers of information between various neurons and between different parts of the brain. These tranfers may perhaps be coordinated via endogenous electrical rhythms (brain waves).

Memory stystems. Memories exist in several forms and are found in various regions that are reached by several pathways.[966–967a] Two major forms of memory are:

Explicit (declarative, episodic):
Conscious recall of facts and events involving people, places, and things
Implicit (associative):
Nonconscious recall of motor skills, conditioned responses, etc.

Explicit memory depends upon the **temporal lobe** of the midbrain, an area that includes the hippocampus and the nearby subiculum and entorhinal cortex.[966,968–971] Implicit associative learning and memory involve the cerebellum, amygdala, and other regions.[972,972a]

Both types of memory possess both **short-term** and **long-term** components. Short-term memory lasts only minutes to hours, but long-term memory lasts days, weeks, and sometimes a lifetime. The difference between the two is clearly seen in individuals who have damage to the hippocampus and impairment of short-term memory. A blow to the head may cause total loss of short-term memory of associations (**amnesia**).[969,973,974] Some persons with damage to the hippocampus may never regain their temporary memory, but long-term memories are intact, and new long-term memories may still be formed. An increasingly important tool for study of memory is brain imaging using

PET or **fMRI** (Box 30-A). These tools have become rapid and sensitive with the ability to observe regions of the brain that become activated by visual, auditory, or other stimuli.[897,898]

The brain often needs to store information for a short period of time. For example, one can recall many details of a visual image after closing one's eyes or shifting one's gaze. The sensory images may be stored in **working memory**.[896,975] Similarly, if one mentally multiplies two 2-digit numbers the partial product obtained by multiplying the two right most digits is temporarily stored in working memory until the next arithmetic operation is completed, etc. PET and fMRI tomography indicates that regions in the prefrontal cortex may be involved.[976,977]

Some short-term memory appears to be stored by neurons that continue to fire after a stimulus has stopped. It has been proposed that such memory consists of reverberations of electrical activity in loops of coupled axons.[978,979]

Implicit memory can be studied in animals. Much has been learned from the large marine snails *Aplysia* and *Hermissendra* whose simple nervous systems and large neurons have been investigated for over 40 years.[967a,980–984] The basic chemical mechanism associated with learning in these creatures seems to be similar to those in our own brain. Olfactory memory can be studied in *Drosophila*, even though the organization of the fly's brain differs from ours.[985–987b]

To be useful for more than a few minutes stored information must be transferred from the temporary to more permanent forms. We know that even temporary memory depends upon chemical changes in synapses. Long-term memory involves both stable chemical changes and also changes in the physical connections between neurons. Before discussing these changes let us consider briefly the waves of nerve impulses that drive the necessary alterations.

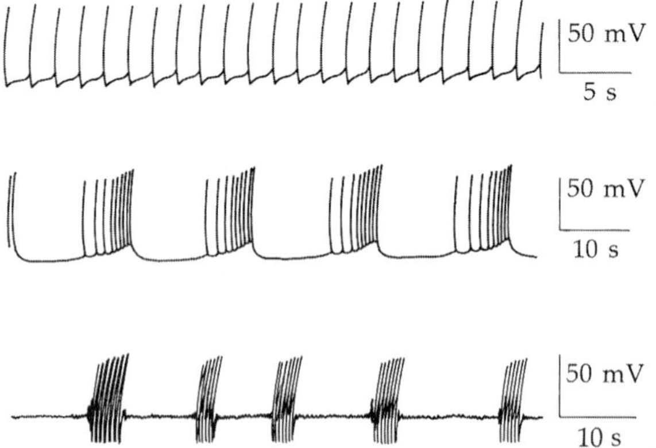

Figure 30-32 Intracellular recordings from isolated neurons of the mollusc *Aplysia*: (A) beating pacemaker, (B) bursting pacemaker, and (C) oscillating pacemaker. From Chen *et al.*[1002]

Brain rhythms. A live brain displays characteristic oscillatory activity. Using electrodes placed on the scalp of an awake but relaxed individual, a rhythmic change in the recorded voltage with a frequency of ~10 Hz (**alpha waves**) can be detected. Such **electroencephalograms** (EEGs) contain other rhythms at ~5–6 (theta), ~40 (gamma), and ~200 (high frequency) Hz.[978,988,989] More recent studies employ microelectrodes placed on individual neurons. Some cells generate spikes at frequencies as high as 800 Hz. However, the significance is uncertain.[990] The 40-Hz frequency, which is prominent in the hippocampus, has aroused the most interest[991–995] because of its probable relationship to learning and memory. Psychophysical experiments have suggested that humans can store only 7 ± 2 items, such as digits in a telephone number. To remember more digits usually requires a conscious effort to place them in longer-term memory. One proposal is that the seven items in temporary memory are stored as 40-Hz oscillations and that ~7 such items can be stored within a single 5-Hz theta oscillation.[978,979] Thinking rates have also been estimated as 7±2 thoughts per second. This is also the same as the syllable rate in speech. This allows us to speak at the same rate that we think.[996] Stuttering may be a result of lack of synchronization of thinking and speaking.

Individual cells or groups of cells are able to initiate rhythms.[997] Examples are provided by the slower Ca^{2+} oscillations shown in Box 6-D,[998–1001] by the periodic release (at ~155 intervals) of cAMP by cells of *Dictyostelium* (p. 20), and by the 24-h circadian cycle observed for virtually all living cells (Section 13). In simple invertebrates the source of neural rhythms appears to reside in **pacemaker neurons** that fire spontaneously at regular intervals. Their cell membranes apparently undergo a cyclic series of changes in ionic permeabilities sufficient to initiate action potentials. Three types of pacemaker output from molluscan neurons[1002] are illustrated in Fig. 30-32. In lobsters three-neuron **pacemaker groups** provide a pyloric rhythm. In these groups the oscillation period of a pacemaker neuron is adjusted from its intrinsic value by feedback through inhibitory and electrical connections to the other cells.[1003] Electrical coupling seems to be basic to oscillatory cell networks.[1004] Individual neurons or small groups of neurons in our own bodies act as pacemakers for the heartbeat rhythm.[705] The slow 3- to 8-Hz rhythm observed in the EEG apparently originates in pacemaker bursts from the basal ganglia.[993] Rhythms from these endogenous pacemakers may combine with pulses from sensory neurons to evoke **conscious thought**. However, the basis of consciousness is still poorly understood.[1005,1006]

As can be seen from Fig. 30-32, neurons send "trains" of spikes down their axons. These form synapses with dendrites, usually on dendritic spikes, of a postsynaptic cell.[593,1007–1009] However, each such cell typically receives input from thousands of other neurons. At any moment most of these are probably "silent," but others are sending trains of impulses. Among the important questions are "How does the postsynaptic neuron know whether to fire or not?" and "What kinds of information, if any, are encoded in the trains of impulses both in the presynaptic inputs and in the output of the postsynaptic neuron?"[1010,1011] Part of the answer to the first question is probably that firing occurs if two or more input impulses arrive synchronously,[1010,1012–1014] and if there are not too many inhibitory impulses that damp the response. In the hippocampus a network of neurons electrically coupled via gap junctions may be synchronized to the theta and gamma brain rhythms by high-frequency (150–200 Hz) oscillations.[988] See also Fig. 30-15.

Chemical changes in synapses. It has long been recognized that the synapses are the probable sites of alterations that lead to memory, whether long-term or short-term. Study of individual synapses has demonstrated the phenomena of **potentiation** (facilitation) and **depression** (habituation). Potentiation refers to the fact that a second impulse will often be transmitted through a synapse more effectively than the first, while depression refers to a decreased response to repeated stimuli. Memory may consist of potentiation and depression at specific synapses. The underlying chemical changes in the synapses are referred to collectively as **synaptic plasticity**. Chemical changes associated with short-term memory are often transient. Those associated with long-term memory are described as **long-term potentiation** (**LTP**) and **long-term depression** (**LTD**).

Many experimental results have confirmed the chemical basis of memory. For example, learning is facilitated by administration to animals of small doses of strychnine.[1015] Puromycin and other inhibitors of protein synthesis disrupt the transfer of information into long-term memory. They are especially effective during the first hour after the initial learning event.[1016] Increased synthesis both of mRNA and of proteins within the cell bodies of neurons is observed.

Short-term memory is not affected by inhibitors of protein synthesis, but alteration of synaptic proteins and membranes may be induced by covalent modification of existing macromolecules.[1016] One way in which this happens has been described for *Aplysia*. As the snail learns a simple gill-withdrawal reflex, the duration of the action potentials in sensory neurons is increased, and there is a greater release of transmitters. This change comes about because stimulation of the sensory neuron causes simultaneous activation of interneurons that synapse with the sensory neurons. The interneurons release the neuromodulator serotonin, which binds to receptors in the membrane of the sensory neurons. This activates adenylate cyclase, which in turn activates a protein kinase that phosphorylates a class of open K^+ channels. Phosphorylation causes the channels to close with a consequent strengthening of the action potential. Thus serotonin brings about presynaptic facilitation.[1017] The peptide FMRF amide (Table 30-5) has the opposite effect. It causes hyperpolarization and a decrease in the duration of the action potential. It also binds to a receptor on the neuronal membrane, and presumably via a different second messenger than cAMP causes the K^+ channels to stay open a longer fraction of the time.

Evidence that LTP is essential to learning in rats was provided by the observation that the antagonist 2-aminophosphonovalerate, which blocks the NMDA class of glutamate receptors (Fig. 30-20A), impairs both LTP and learning.[1018] The potentiation is thought to result, in part, from Ca^{2+} influx through the nonselective NMDA cation channels. The increased intracellular calcium may then induce phosphorylation of various proteins with associated long-lasting changes in the postsynaptic endings.[1019] A large amount of evidence favors this interpretation of LTP.[1020] However, it is a great oversimplification. Most studies of LTP in mammals (usually rodents) have focused on the CA1 region of the hippocampus and nearby brain regions. The excitatory axons in this organ are largely glutamatergic, and, as shown in Fig. 30-20A, the postsynaptic (dendritic) membranes contain both fast AMPA receptors and the slower NMDA receptors. Both are ionotropic. The AMPA receptor channels allow mainly K^+ and Na^+ to pass and are responsible for most of the nerve transmission. However, the NMDA receptors have an important controlling influence.

A generally accepted theory is that no LTP arises unless both the presynaptic and postsynaptic neurons are activated. This can happen if a presynaptic action potential activates many AMPA receptors in a synapse allowing enough flow of $Na^+ + K^+$ to depolarize the postsynaptic membrane and possibly to initiate an action potential in the postsynaptic neuron. (However, many factors, probably including influences from neighboring neurons,[1021] will affect this outcome.) The NMDA receptors are usually blocked by extracellular Mg^{2+} ions, and their ion channels remain closed. However, when the postsynaptic membrane becomes depolarized, the Mg^{2+} dissociates, and if the NMDA receptors are also occupied by glutamate, their channels will open, permitting Ca^{2+} to enter the neuron (Fig. 30-20A). This not only enhances the probability of developing a postsynaptic action potential but is also the trigger for LTP.[1022] Ca^{2+} ions have a variety of effects, one of which is to bind to calmodulin. This activates a calcium–calmodulin-dependent protein kinase, which phosphorylates postsynaptic structural and signaling proteins to increase the synapse strength.[641,1023–1025]

Modifications in existing proteins, such as are induced by Ca^{2+} and calmodulin, can provide LTP for a few hours, but other mechanisms must provide for longer-term effects. These require transcription of genes and protein synthesis, processes that occur in the cell bodies of neurons and may depend upon axonal transport mechanisms.[1026,1027] Among the experimentally observed results of LTP are ultrastructural changes in synapses and in dendrites.[1009,1028] Long-term memory is also thought to involve changes in the neocortex. Again, NMDA receptor activation seems to be involved.[1029,1030]

LTP has been demonstrated experimentally, but does it really influence memory? Evidence that it does has been provided by clever experiments with transgenic mice. Using the Cre recombinase (Chapter 27) the NR2B subunit of the NMDA receptor was overexpressed in the hippocampus[1031] and in the forebrain of mice.[1032] This was expected to provide better synaptic strengthening than for receptors with the similar NR2A subunit. It was found experimentally that these transgenic mice were more intelligent than normal mice.

LTP is also thought to affect the presynaptic as well as the postsynaptic neuron. One way in which this may happen is for a **retrograde messenger** to pass across the synapse and induce alterations in the presynaptic cell. One proposed retrograde messenger is nitric oxide, **NO**.[1033,1034] Neuronal NO synthase (nNOS; NOS1) contains a calmodulin binding site and is activated by Ca^{2+}. However, other substances as simple as K^+ might also be the messenger.

Long-term depression (LTD) is the *loss* of synaptic strength after passage of an impulse. There is evidence that during brain activity, including that in the hippocampus, both LTP and LTD are essential.[1035,1036] LTD may depend upon cyclic ADP-ribose (p. 564).[1037] LTD, like LTP, may also spread via retrograde signaling.[1038]

Does learning affect a few specific neurons or a large number of neurons? The rate of glucose utilization in different parts of the brain can be estimated from the rate at which labeled 2-deoxyglucose is taken up (see Box 30-A). From changes in this rate (obtained using ^{14}C-labeling) in brains of split-brain cats performing visual tasks it was estimated that from 10^{10} to 10^{11} neurons are activated.[1039] This supports the idea that memory is distributed over a large area of the brain, just as information about an image is stored in all parts of a hologram.

An alternative to the idea that synaptic potentiation and depression provide the chemical basis for learning is **molecular coding**. Thus, it was reported that a 15-amino acid peptide isolated from rats trained to avoid the dark carries behavioral information. When this peptide was injected into brains of untrained rats, they also avoided the dark.[1040] This was one of several reports of transfer of learned behavior through chemical substances extracted from the brain. These ideas are hard to accept in the light of our present knowledge of the brain. However, in view of the large number of different neuropeptides known (Table 30-4) the possibility that some aspects of long-term memory may be associated with transcription of specific amino acid sequences within specific neurons should perhaps still be considered.

The complexity of the brain. A major obstacle to our understanding of the human brain is its enormous complexity.[400] This problem can be appreciated if we consider the small nematode *Caenorhabditis elegans*. All of the synaptic connections among its 302 neurons had been mapped by 1986.[400,1041,1042] There are 5000 chemical synapses and 600 electrical (gap junction) connections. There are 80 different types of K^+-selective channels, 90 types of ligand-gated receptors, and ~1000 G-protein-linked receptors. Twenty-six of the neurons are GABAergic and are involved in three distinct behavioral motions that involve muscular contractions. Despite intensive efforts the system has been hard to understand. The brain of the macaque monkey has been described in great detail.[409] Over half of its cerebral cortex is devoted to vision, and this can be subdivided into 20 functional areas. However, the human brain, with its extremely large cerebral cortex, cannot be compared accurately with the monkey brain. Whereas anatomical studies are done on postmortem human tissues, *in vivo* studies rely largely on fMRI and PET imaging (Box 30-A). The resolution of these images is now less than 1 mm, but 1 mm^3 of human visual cortex contains more than 40,000 neurons![409] The microcircuits in the neocortex are still largely unknown and the tissue is of "apparently impenetrable complexity." There may be several hundred different classes of neocortical neurons.[1043] The tissue is rich in GABAergic interneurons.[1044,1045] Some fast-spiking GABAergic neurons are also connected by electrical synapses and may be involved in detecting and promoting synchronous activity.[1046]

Intelligence. We must all agree that there is such a thing as intelligence, but can it be measured? In 1904, Spearman proposed the existence of a general intelligence factor g that could be measured as the IQ (intelligence quotient). Since then various tests have been devised that attempt to measure IQ.[1047,1048] Most recently use of PET scan data has indicated that various types of analytical analysis lead to brain activity in the lateral frontal cortex in one or both cerebral hemispheres[1049] suggesting that this is a region important to

IQ. A question that has been raised is whether analytical intelligence, creative intelligence, and practical intelligence are correlated?[1048]

Is intelligence hereditary? Both logic and observation say that heredity must be a major factor. However, it is hard to know how to measure the hereditary component.[1050–1052] Also hard to understand is why IQ scores have been increasing about one standard deviation unit per generation.[1047] Is this really true?[843] Does environment also influence IQ? The fact that new hippocampal nerve cells are formed continuously provides one mechanism by which learning, nutrition, and other influences may alter intelligence.

A difficult-to-explain aspect of the brain is the existence of rare **savants**, persons with amazing mental abilites in music, art, or computation but who are unable to communicate (autistic) and mentally retarded.[1053–1055] One boy at age four could play Mozart piano sonatas flawlessly after a single hearing. A three-year old girl drew horses with lifelike perspective from memory but was unable to communicate. Some mathematical savants can instantly state the day of the week for any arbitrary date such as June 12, 1929; others rapidly identify prime numbers. They evidently use the same strategies as mathematically trained persons. Do we all have these abilities but can't have access to them? How can we explain the fact that rarely a blow to the head will convert a person into a savant?

Behavior. It may seem impossible to interpret complex behavioral patterns at the molecular level. However, the genetics of behavior is a well recognized field of investigation, and some behavioral traits have been linked to single genes. If a gene can be located cloning, sequencing, and biochemical studies may follow quickly. The behavioral genetics of lower organisms, e.g., of *Drosophila*, have provided many insights.[1056] Recently, however, the mouse has become a major object of behavioral studies.[1057] Its genome is well known, and a very large number of mutations have been mapped. The ability to prepare **knockout mice** (p. 1501) and to carry out gene transfer experiments on such animals makes them very attractive for study.

Some behavioral traits are based on simple alterations, often defects, in motor skills. For example, the following traits in mutant mice have been traced to specific brain structures and often to specific biochemical alterations.

Staggerer	Purkinje cell defect
Vibrator	Phosphatidylinositol transfer protein gene
Tottering	Mutation in voltage-gated Ca^{2+} channel
Lurcher	Abnormality in cerebellum
Weaver	Gly → Ser mutation in K^+ channel

Knockout mice lacking oxytocin or vasopressin have altered social behavior toward other mice. Those lacking galanin seem less intelligent than normal mice, as if they had Alzheimer disease. Mice lacking neuronal NO synthase became aggressive.[1057] Human personality,[1058] language abilities,[1059] and sexual behavior all have a genetic component. However, claims that a "gay gene" has been found are not generally accepted.[1060]

13. Circadian Cycles and Sleep

In mammals an approximately 24-h (**circadian**) rhythm controls behavior and affects many physiological functions. As previously mentioned (p. 1800), the brain has its own rhythms, which originate with pacemaker neurons. The heart beats with another neurally established rhythm. The circadian rhythm has a much longer period and, therefore, seems more mysterious. It is observable, even with single cells and for virtually all organisms.[1061–1062a] In most instances the cycle becomes synchronized with the daily light–dark cycle with the aid of suitable light-absorbing pigments often crytochromes (see Chapter 23, Section I,1). However, the cycle can be observed in various ways under conditions of constant light intensity and temperature. For example, the unicellular marine alga *Gonyaulax* undergoes dramatic circadian changes in the intensity of its bioluminescence. Over one 10-day period the luminescence peaked every 22.99 ± 0.01 hours.[1063,1064] It is more difficult to measure the period for human beings (see Czeisler *et al.*[1065] for a discussion), but under suitable conditions during which time cues were missing a precise period of 24.18 hours was observed for the level of melatonin in the blood, the body temperature, and other quantities.[1065] From cyanobacteria,[1066,1067] fungi (*Neurospora*),[1068] insects (*Drosophila*),[1069–1071b] and frogs[1072] to mice and people,[1073,1074] the circadian cycle affects the organism's chemistry and behavior. Green plants likewise observe a circadian cycle.[1075]

The cycle is thought to originate in feedback loops that control transcription of a small set of genes. In *Drosophila* the set includes seven genes: *period* (*per*), *timeless* (*tim*), *clock* (*clk*), *cycle* (*cyc*), *double-time* (*dbt*), *vrille*, and *cryptochrome* (*cry*).[1070] Many corresponding genes have been found in mammals. For example, the mouse NPAS2 is a close relative of the *Drosophila* CLOCK protein, and the period proteins PER1 and PER2 and the cryptochromes CRY1 and CRY2 are also related to the *Drosophila* proteins.[1076,1077] In *Drosophila* the heterodimers PER•TIM and CYC•CLK are thought to serve as DNA-binding transcription factors that repress transcription of their own genes when they reach a high enough concentration in the nucleus.[1073,1076,1077] Because some time is required for transcription and

BOX 30-A POSITRON EMISSION TOMOGRAPHY (PET), FUNCTIONAL MAGNETIC RESONANCE (FMRI), AND OTHER IMAGING TECHNIQUES

In the widely used technique of transmission computerized tomography (CT) an image of a slice through the body of a patient is obtained using X-rays. An X-ray source moves in a ring around the patient while detectors measure the intensity of the transmitted radiation and send it in digital form to a computer. which generates the desired image. A chemically more sophisticated view of the body can be obtained by positron emission tomography (PET). This technique makes use of a metabolite or drug labeled with a short-lived radioisotope that decays by emission of positrons (antielectrons). Among these are ^{11}C, ^{13}N, ^{15}O, and ^{18}F with half lives of 20 min, 10 min, 2 min, and 110 min, respectively. The isotopes are produced in a cyclotron, and are rapidly introduced into suitable compounds, which can be injected into a bloodstream.[a–d] An emitted positron travels only a few millimeters before undergoing annihilation with an electron to produce two high-energy (50 keV) photons (γ-rays) that travel in opposite directions and are detected by an array of scintillation detectors.

$$\beta^+ + e^- \rightarrow 2\,hv$$

Present-day PET technology allows images to be formed in a few seconds, and in some cases in a fraction of a second. Among the useful compounds for PET imaging is [^{18}F]2-fluoro-2-deoxy-D-glucose (see figure). This compound, which contains the longer-lived ^{18}F, is phosphorylated by hexokinase, and the resulting phosphate ester is effectively trapped in the brain. 3-Deoxy-3-fluoro-D-glucose is another useful tracer.[e] One of the most useful PET measurements has been blood-flow monitored by ^{15}O-containing H_2O, which is administered into a vein in the arm.[b] The ^{15}O has a half-life of only two minutes and is almost completely gone in ten minutes. However, very low doses of radioactivity are used, and several images can be obtained before the radioactivity has decayed. A common practice is to subtract images obtained after the isotope has decayed from those obtained at various times while it was still present. The technique is also useful for study of the binding and transport of hormones,[f] other metabolites, drugs, and other inhibitors.[c]

The NMR technique **magnetic resonance imaging** (**MRI**),[g–i] so called to avoid the word nuclear, is rapidly displacing many applications of PET scanning. MRI uses proton NMR spectroscopy to generate very sharp images based largely on the water present in tissues. These images can be made

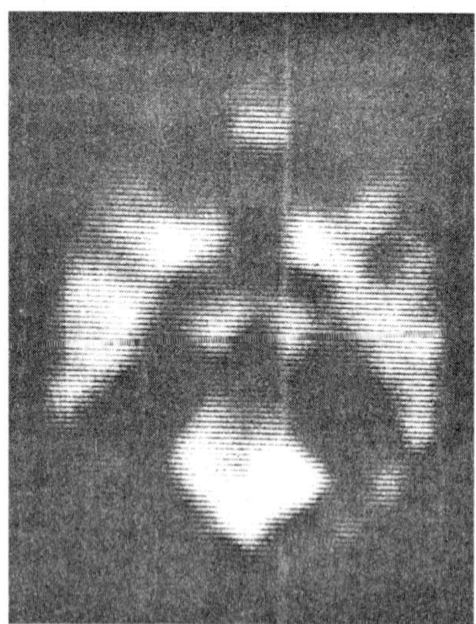

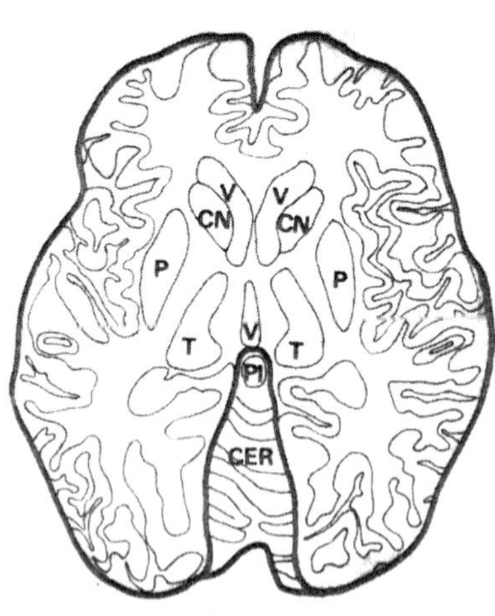

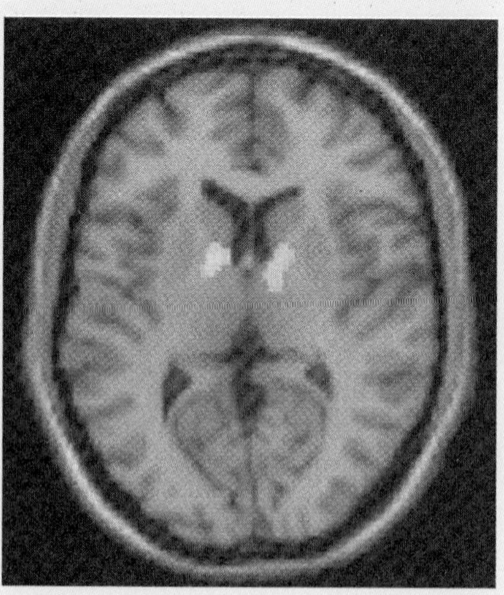

Left: PET image of a human brain obtained using 2-[^{18}F]fluoro-2-deoxyglucose. This tomographic brain slice at the level of the basal ganglia shows the cortical gray matter and subcortical white matter. As marked on the drawing on the right: V, ventricles; CN, caudate nucleus; P, putamen; T, thalamus; PI, pineal gland; CER, cerebellum. From Rottenberg and Cooper.[p] Right: fMRI image illustrating modulation of neural activity in the ventral striatum, an area of the brain associated with reward, when eye contact was made with an attractive face. The activation map shown is derived in a complex manner and is based on recorded brain activity of persons viewing images of a series of faces. It portrays the differences in neural activity when viewing images of attractive faces of either sex with the eye gaze directed at the subject and with the eyes averted. See the report of Kampe *et al.* for details.[q]

BOX 30-A (continued)

to depend upon variations in T_1 and T_2 (Chapter 3) as well as upon differences in the water content. The first MRI scans required 20 minutes, but the use of more powerful magnets and more sensitive instruments has reduced the acquisition times in ultrafast MRI to ~0.1 s. The decay of the NMR signal from a single RF pulse is observed at several different times.[h] The dynamics of blood flow and neural activity can be followed. Every technique has disadvantages as well as advantages. MRI does not use radioisotopes, but overheating of the brain must be carefully avoided. In addition, patients may suffer from uncomfortably loud noises generated by rapidly changing magnetic gradients.[h] As with PET scans isotopic tracers may be used. However, most MRI scanning is done with ^1H from the solvent water. As with PET, MRI is often used to measure blood flow but with an indirect method. The Fe of deoxyhemoglobin (Hb) is paramagnetic, but upon oxygenation to HbO_2 it becomes diamagnetic (pp. 850–851), and the ^1H signal of the solvent H_2O becomes sharper. In metabolically active regions of the brain the demand for oxygenated blood is greatly increased. Perhaps surprisingly, the ratio $[HbO_2]/[Hb]$ is greater in these areas than in less active areas where a greater fraction of the hemoglobin remains unoxygenated.[g] However, the exact interpretation of the ultrafast MRI images is uncertain. In **functional MRI (fMRI)**, differences in images acquired after some physiological change are recorded. For example, after a visual or other sensory stimulus a change in the MRI image of some region of the cortex will be observed (see figure).[i–l] The technique is allowing many deductions about learning, memory, and communication pathways in the brain[j,m] and is being used to investigate many aspects of brain disease.

Another brain imaging technique is **magnetoencephalography (MEG)**.[c,n] It has been uniquely valuable in mapping the sensory regions of the human cerebral cortex. Looking ahead, optical methods, which include use of infrared radiation, are also under development.[o] They may not be adequate for study of the human brain but can be used for smaller animals, for studies of embryonic development, etc.

[a] Ter-Pogossian, M. M., Raichle, M. E., and Sobel, B. E. (1980) *Sci. Am.* **243**(Oct), 171–181

[b] Raichle, M. E. (1994) *Sci. Am.* **270** (Apr), 58–64

[c] Volkow, N. D., Rosen, B., and Farde, L. (1997) *Proc. Natl. Acad. Sci. U.S.A.* **94**, 2787–2788

[d] Phelps, M. E. (2000) *Proc. Natl. Acad. Sci. U.S.A.* **97**, 9226–9233

[e] Berkowitz, B. A., Moriyama, T., Fales, H. M., Byrd, R. A., and Balaban, R. S. (1990) *J. Biol. Chem.* **265**, 12417–12423

[f] Berman, K. F., Schmidt, P. J., Rubinow, D. R., Danaceau, M. A., Van Horn, J. D., Esposito, G., Ostrem, J. L., and Weinberger, D. R. (1997) *Proc. Natl. Acad. Sci. U.S.A.* **94**, 8836–8841

[g] Shulman, R. G., Rothman, D. L., and Blamire, A. M. (1994) *Trends Biochem. Sci.* **19**, 522–526

[h] McKinstry, R. C., and Feinberg, D. A. (1998) *Science* **279**, 1965–1966

[i] Disbrow, E. A., Slutsky, D. A., Roberts, T. P. L., and Krubitzer, L. A. (2000) *Proc. Natl. Acad. Sci. U.S.A.* **97**, 9718–9723

[j] Ungerleider, L. G. (1995) *Science* **270**, 769–775

[k] McCarthy, G., Blamire, A. M., Puce, A., Nobre, A. C., Bloch, G., Hyder, F., Goldman-Rakic, P., and Shulman, R. G. (1994) *Proc. Natl. Acad. Sci. U.S.A.* **91**, 8690–8694

[l] Chen, W., Zhu, X.-H., Thulborn, K. R., and Ugurbil, K. (1999) *Proc. Natl. Acad. Sci. U.S.A.* **96**, 2430–2434

[m] Wagner, A. D., Schacter, D. L., Rotte, M., Koutstaal, W., Maril, A., Dale, A. M., Rosen, B. R., and Buckner, R. L. (1998) *Science* **281**, 1188–1191

[n] Yang, T. T., Gallen, C. C., Schwartz, B. J., and Bloom, F. E. (1993) *Proc. Natl. Acad. Sci. U.S.A.* **90**, 3098–3102

[o] Lok, C. (2001) *Nature (London)* **412**, 372–374

[p] Rottenberg, D. A., and Cooper, A. J. L. (1981) *Trends Biochem. Sci.* **6**, 120–122

[q] Kampe, K. K. W., Frith, C. D., Dolan, R. J., and Frith, U. (2001) *Nature (London)* **413**, 589

protein synthesis, this inhibitory feedback can lead to oscillations in the concentrations of the circadian clock proteins. Proteosomal degradation of the TIM protein may also be a factor.[1071] The need for proteins encoded by other genes indicates that the matter is more complex. Individual cells or individual tissues, e.g., mammalian retinas,[1074,1078] may independently set up circadian cycles. However, these normally become **entrained** by the daylight cycle and are reset daily as discussed in Chapter 23, Section I,1. Other factors such as temperature, activity, and food may also affect the resetting. One factor, which may be influenced by food, is the $NAD^+/NADH$ and $NADP^+/NADPH$ ratios within cells.[1077] The circadian cycles for mammalian tissues are synchronized by a **master clock** that originates in neural tissues[1062a] and specifically in a region of the hypothalamus containing the **suprachiasmatic nuclei**.[1079–1080b]

The pineal gland appears also to play a role in maintaining the mammalian circadian cycle.[1081–1083] The concentration of the pineal hormone melatonin (Fig. 27-11) as well as its precursor *N*-acetylserotonin and the enzyme serotonin *N*-acetyltransferase (Eq. 30-4) all fluctuate far more than do the concentrations of other metabolites during the 24-h cycle. These metabolites increase over 10-fold concentration at night and decrease by day. During the daytime the serotonin *N*-acyltransferase, which forms the precursor, is rapidly and apparently irreversibly inactivated, perhaps through a disulfide exchange reaction.[1081] Bright light will reset the circadian cycle[1084] keeping it approximately (circa) daily. The effect of light is apparently a result of signals sent to the hypothalamus from the optic nerves. In chickens and in lower vertebrates the pineal gland may directly sense light passing through the skull.

The circadian cycle is not the only timing device used by animals. A short-term interval timer helps male doves to know how long to sit on a nest[1085] and helps all of us in timing everyday tasks.

We spend a third of our lives asleep, but our understanding of sleep from a molecular viewpoint is minimal. Sleep is essential for the life of mammals, which die if completely deprived of sleep. It has been shown that during prolonged sleep deprivation sleep-inducing material accumulates in the brain. One such substance, isolated from human urine, appears to be a peptide containing glutamate, alanine, diaminopimedic acid, and muramic acid.[1086] Thus, it resembles a fragment of bacterial peptidoglycan. Prostaglandin D_2 also induces sleep.[1087] Hayashi proposed that a balance between this substance and prostaglandin E_2, which induces wakefulness, is in part responsible for the sleep–wake cycle.[1087] More recently oleic acid amide (**oleamide**; p. 382) was identified as a sleep-inducing compound. A fact observed by everyone is that the longer one is awake the higher the probability of going to sleep. The accumulation of sleep inducers is part of a homeostatic mechanism. On the other hand, the circadian cycle probably provides the signal to awake and tends to consolidate our sleep into the characteristic 8-hour period.[1088] The melatonin level, which drops in daylight, plays a role.[1089] Release of adrenocorticotropin (ACTH) one hour before waking may also be important.[1090]

During much of the night's sleep the EEG is characterized by the slow 5- to 6-Hz waves.[989,993] However, after ~90 min there is an ~10-min period of **rapid eye movement** (**REM**) **sleep** during which the EEG resembles that of an awake person and dreaming occurs. The closed eyes move rapidly in unison, breathing is irregular, and the heart rate increases. Motor neurons are inhibited allowing only minimal body movement. Three more periods of slow-wave sleep, each shorter than the preceding one, are followed by REM sleep. The REM sleep periods become successively longer. The fourth period lasts 20–30 min and is followed by awakening. All placental and marsupial mammals follow a similar sleep pattern and all dream.[989] The importance of dreaming is not obvious[1088a] but is often thought of as a reprocessing of memory, a means of ridding the mind of unneeded memories, a process of **unlearning**. However, this is uncertain as is the relationship of sleep to learning and memory.[1088a,b]

A number of disorders of sleep are known. Among these is **narcolepsy**, uncontrollable, sudden daytime sleepiness. It affects 1 in 2000 individuals.[1088] The same occurs in dogs.[1091] After a 10-year effort at great expense the narcolepsy gene of dogs (*canarc*-1) was located by positional cloning.[1088,1092] The corresponding human (and rat) gene was independently discovered by other investigators. It encodes a receptor for neuropeptides produced by the hypothalamus and named **hypocretins** or **orexins** for their stimulation of appetite. It seems probable that the hypocretin/orexin neuropeptides are involved in promoting wakefulness. Another sleep disorder is **familial advanced sleep phase syndrome**. Persons with this trait are "morning larks" who tend to fall asleep at ~7:30 p.m. and awake suddenly at ~4:30 a.m., about four hours in advance of a typical sleep period. A missence Ser→Gly mutation in the human period gene (hper2) has been found.[1093]

Some mammals hibernate. Special blood proteins that induce hibernation apparently control the process.[1094]

14. Mental Illness

Whereas many metabolic defects affect only a small number of individuals, emotional illnesses including depression, **schizophrenia**, and other **affective disorders** at one time or another afflict a large fraction of the population. Autism affects thousands of children.[1055] Parkinson disease and **Alzheimer disease** are just two of a number of degenerative neural diseases attacking older people. Less commonly, young persons contact **multiple sclerosis** and **muscular dystrophy,** which is often a disease of neuromuscular junctions.

Depression. Depression is our most common mental problem. One in four women and one in ten men will have a major depression during their lifetime.[1095] More than 15 million people in the United States are affected by severe depression in any given year and more than 30,000 may commit suicide.[1096,1097] Worldwide psychiatric problems, mostly depression, account for 28% of all disabilities.[1098] The **biogenic amine hypothesis** states that depression results from the depletion of neurotransmitters in the areas of the brain involved in sleep, arousal, appetite, sex drive, and psychomotor activity. An excess of transmitters is proposed to give rise to the manic phase of the bipolar (manic–depressive) cycle that is sometimes observed. In support of this hypothesis is the observation that administration of reserpine precipitates depression, which may be serious in 15–20% of hypertensive patients receiving the drug. Similar effects are observed with the dopa decarboxylase inhibitor **α-methyldopa**

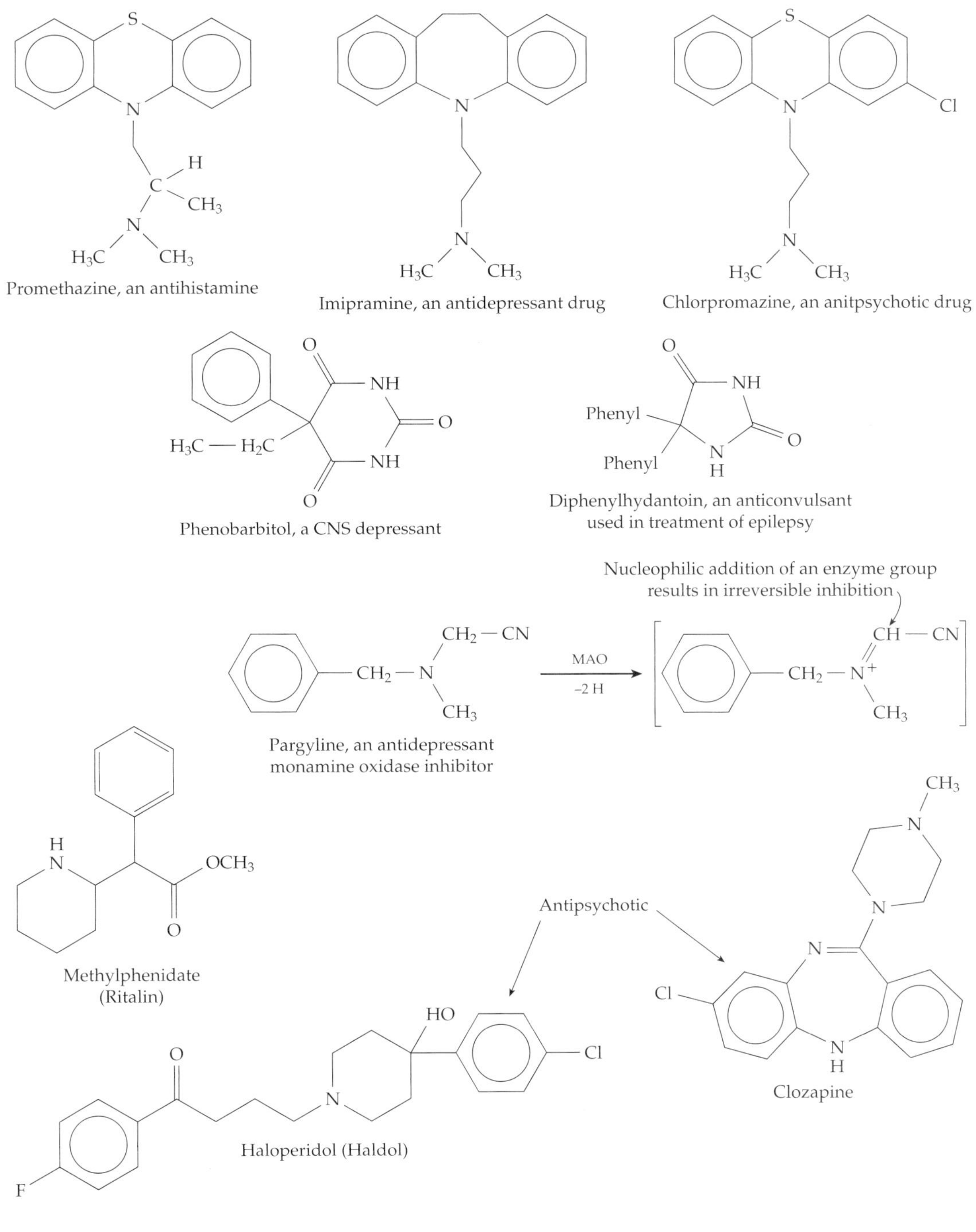

Promethazine, an antihistamine

Imipramine, an antidepressant drug

Chlorpromazine, an anitpsychotic drug

Phenobarbitol, a CNS depressant

Diphenylhydantoin, an anticonvulsant used in treatment of epilepsy

Nucleophilic addition of an enzyme group results in irreversible inhibition

Pargyline, an antidepressant monamine oxidase inhibitor

Methylphenidate (Ritalin)

Antipsychotic

Haloperidol (Haldol)

Clozapine

Figure 30-33 Some drugs used to treat psychiatric disorders. See also Figs. 30-25 and 30-28.

(Fig. 30-27). The fact that L-tryptophan has some antidepressant activity, but L-dopa does not, was one clue that a low concentration of serotonin (5-hydroxytryptamine) might be responsible for depression. Excessive formation of histamine[1099] and decreased formation of tyramine and octopamine[1100] have also been suggested as causes of depression.

Strong support for the biogenic amine theory of depression is provided by the powerful antidepressant effect of inhibitors of monoamine oxidase. An example is **pargyline** (Fig. 30-33), which forms a covalent adduct with the flavin of MAO.[1101] Although effective, this drug is somewhat dangerous. Because their monoamine oxidase activity is so low, patients taking pargyline have been killed by ingesting compounds such as tyramine, which occurs in cheese. Less easy to understand but clinically more important are tricyclic antidepressants such as **imipramine** (Fig. 30-33),[746a] whose antidepressant action was discovered accidentally. Notice the close similarity to chlorpromazine but the greater flexibility of the central ring.[1102] Imipramine was found to block transporters of both noradrenaline and serotonin. In 1986, the less toxic serotonin reuptake inhibitor fluoxetine (Prozac; Fig. 30-28) was introduced and is now used by many millions of people.[1103,1104] Nevertheless, its mode of action is not entirely clear. For example, it blocks nicotinic acetylcholine receptors and may have many other effects. Interestingly, depression sometimes responds to a placebo just as well as to an antidepressant drug.[1105] In addition to newer drugs related to Prozac, antagonists of substance P are also effective antidepressants.[1106] MRI images of brains of depressed patients show that hippocampal volume has decreased and suggest that formation of new neurons is inhibited.[1106a] Antidepressants seem to stimulate growth of new cells as does exercise, which also has an antidepressant effect.[1107] Dietary treatment can also help.[843] Among older people depression may be caused by deficiency of vitamin B_{12} and can be treated by injection of the vitamin.[1108] An **anxiety peptide** that may be the natural ligand for benzodiazepine receptors has been reported.[1109]

Another recognized type of depression is **seasonal affective disorder** (**SAD**). People in far northern or southern latitudes develop this condition in the winter, apparently from lack of sunshine needed to lower the melatonin level in the morning (see Section 13). Light therapy is beneficial.[1110] Persons with the SAD syndrome also tend to crave carbohydrates and to stay in bed for 9–10 hours.

An effective treatment for **bipolar disorder** (manic–depressive illness) is the administration of lithium salts.[445,1111–1113a] Inhibition of the hydrolysis of inositol phosphate by Li^+ (Fig. 11-9) may be related to its therapeutic effect. Reduced phosphatidylinositol turnover may dampen responses to neurotransmitters.[1114] Li^+ may affect gene expression in neuropeptide-secreting neurons.[1115] Bipolar disorder apparently has more than one cause. There are strong indications of genetic susceptibility,[1116] and genes that increase susceptibility have been located on chromosomes 4, 12, 13, 18, 21, and X.[1117]

Schizophrenia. Among the most baffling of mental illnesses are the group of diseases known as schizophrenia. They involve thought disorder, disturbance of the affect, and withdrawal from interactions with other people. Hallucinations and paranoid feelings are common.[813] In some cases a striking loss of gray matter in some areas of the brain is revealed by MRI scans.[1117a,b] The schizophrenias are of varying degrees of severity and shade continuously into the affective or mood disorders, which include manic-depression and depression. As many as one person in a hundred is affected by schizophrenia.[1118–1119a] There is a complex genetic susceptibility.[1119,1120] One theory about the persistence of the genes favoring schizophrenia is that they are also associated with creativity.[1121]

A revolution in the treatment of the schizophrenias, as well as in thinking about mental illnesses, took place following the synthesis, in 1950, of the antipsychotic drug **chlorpromazine** (Fig. 30-33). At about the same time the effect of the *Rauwolfia* alkaloid reserpine (Fig. 25-12) in calming mentally disturbed persons was rediscovered. The Indian plant *Rauwolfia* had been used for centuries in Hindu medicine for the same purpose. The tricyclic phenothiazines such as promethazine (Fig. 30-33) earlier had been found to have powerful **antihistamine** activity. It was the search for better antihistamine drugs that led to the synthesis of chlorpromazine.[1122] As many as 250 million people throughout the world were treated with chlorpromazine and related drugs in the 20 years following its discovery before newer and safer drugs (e.g., **clozapine**; Fig. 30-33) were developed.[1123] What does chlorpromazine do? A possible clue comes from the fact that it sometimes induces serious "extrapyramidal" side effects including tremors and other symptoms of Parkinson disease. This suggested that chlorpromazine may block dopamine receptors in the corpus striatum, thereby precipitating a functional deficiency of dopamine.[1124] If so, it is possible that schizophrenia may result from an overactivity of dopamine neurons, perhaps including some of the same neurons that are hypoactive in Parkinson disease. Supporting this view is the observation that amphetamines (Fig. 30-27), which may substitute for dopamine, worsen the symptoms of schizophrenia and in very high doses induce striking schizophrenialike symptoms in normal individuals.

A stereotyped compulsive behavior is induced both in humans and in laboratory animals by amphetamines. This provided the basis for a method that has been used to measure the action of drugs on amphetamine-sensitive centers of the brain. A lesion in the nigrostriatal bundle on one side of a rat brain was made by injection of a neurotoxic compound such as 6-hydroxydopamine. This caused degeneration of dopamine-containing neurons on one side of the brain. When rats that had been injured in this way were given amphetamines, they developed a compulsive rotational behavior. Administration of chlorpromazine and several other antipsychotic drugs neutralized this behavior and in direct proportion to the efficacy in clinical use, an observation that also supports the theory that schizophrenia involves overactivity of dopamine neurons.

If schizophrenia results from an elevated dopamine content of the brain, the fault may lie with either an oversupply or a reduced rate of metabolism of dopamine. The possibilities of reduced activity of monoamine oxidase or of dopamine β-hydroxylase have both been suggested. The plasma level of the dopamine metabolite **homovanillic acid** (Fig. 30-26) is elevated in schizophrenia and is correlated with the severity of the illness,[1125] suggesting the hypothesis of a decreased rate of metabolism. Possible defects in dopamine receptors may be at fault.[1126]

Chlorpromazine may also act on brain cholinergic neurons.[1127] Blockage of muscarinic acetylcholine receptors in the brain by belladonna alkaloids such as atropine (Fig. 30-22) has often been used in treatment of Parkinson disease. Apparently antagonizing acetylcholine action is to some extent functionally equivalent to increasing dopamine concentrations. There is evidence that suggests a role for cholecystokinin (CCK) in development of schizophrenia. CCK-containing neurons interact with dopaminergic neurons in the midbrain.[1128] GABA neurons in the prefrontal cortex may be faulty.[1129] Excessive glutamate may also induce schizophrenia. The schizophrenia-like symptoms induced by phenylcyclidine (p. 1796) are eased by antagonists of metabotropic glutamate receptors. This suggests another possible therapy.[1130] Among other suggested causes of schizophrenia are dysregulation by **retinoids**[1131] and action of retroviruses.[1132] Demyelination in portions of the prefrontal cortex may disrupt neural connectivity.[1132a] Recent genetic evidence points to a possible defect in proline dehydrogenase which reduces Δ^1-pyrroline-5-carboxylate to L-proline (Fig. 24-9).[1132b]

Numerous theories of mental illness have embodied proposals that a toxic metabolite is produced in abnormal quantities. An example is **6-hydroxydopamine** (Fig. 30-26), which is known to damage dopamine-containing neurons.[1133] Overactive methylation of catecholamines has also been suggested as a cause of mental disorders.[1134] The hallucinogen **3,4-dimethoxyphenylethylamine** (Fig. 30-26) has been identified in urine during acute schizophrenic attacks, but the variability is so high that no definite conclusion has been reached. *N*-Methylation of serotonin yields **bufotenin** (*N*-methylserotonin) and ***N*-dimethylserotonin**, known hallucinogenic agents. Enzymatic synthesis of the latter by human brain and other tissues has been demonstrated,[1135] and administration of tryptophan and methionine to schizophrenic patients exacerbates their illness.

Another theory of mental illness postulates endogenous alkaloid formation. Aldehydes formed by oxidation of catecholamines as well as formaldehyde and acetaldehyde are present in tissues in small amounts. Condensation with amines could generate Schiff bases and alkaloids as in Fig. 25-10. This "plant chemistry" is spontaneous and can apparently take place in the brain, where it may have a potent effect.

H_3C-N^+ … N^+-CH_3

Paraquat

Incubation of tryptamine derivatives with 5-methyltetrahydrofolic acid and an enzyme preparation from brain gives **tryptolines**. Dopamine and its derivatives form related tetrahydroisoquinolines such as the product that arises from reaction with acetaldehyde (see Eq. 30-5). This product has been found in elevated amounts in alcoholics (who synthesize excess acetaldehyde), in phenylketonurics, and in L-dopa-treated patients with Parkinson disease.[1136]

Epilepsy. The brain disorders known as epilepsies affect 1–2 % of the population worldwide. Characteristic of epilepsies are recurrent **seizures**, sudden brief changes in behavior caused by the simultaneous, disordered firing of large numbers of neurons in the brain. Many seizures are thought to be initiated in specific areas of the cerebral cortex. For example, seizure-induced firing of neurons in the thumb area of the right motor cortex will be accompanied by rhythmic jerking in the left thumb. More than 40 different types of epilepsy are known.[1137–1138a]

GABA is the principal inhibitor neurotransmitter, and one cause of epilepsy may be a deficiency in GABA formation from glutamate. The brain contains two isoforms of glutamate decarboxylase, designated GAD65 and GAD67, in accordance with their molecular masses in kDa. They are encoded by separate genes.[1139,1140] GAD67 is formed mainly in cell bodies of neurons, binds its cofactor PLP tightly, and is essential to survival of young mice. GAD65 is associated mainly with nerve termini, where it is anchored, apparently by association with other proteins to the membranes of synaptic vesicles.[1141] It binds PLP weakly. Some convulsive agents such as 1,1-dimethylhydrazine are thought to act by interfering with PLP-dependent enzymes (Box 14-C) among which is GAD.[1140,1142,1143] Convulsions are one of the most striking symptoms of a severe vitamin B_6 deficiency. A zinc deficiency can also cause convulsions, apparently because pyridoxine kinase is a Zn•ATP-requiring protein and the rate of synthesis of PLP is too slow to supply apo-GAD with the PLP needed for GABA synthesis. The PLP in GAD65 undergoes rapid substrate-dependent transamination to pyridoxamine phosphate (see Chapter 14), which must be replaced by new PLP.[1143,1144]

Epilepsy may arise also from defects in a GABA transporter[1145] or receptor.[1146] One form of epilepsy is a triple-repeat disease of cystatin B (Table 26-4). Mutation in potassium channels,[1147] glutamate receptors,[1148] absence of neuropeptide Y,[1149] and absence of L-isoaspartyl / D-aspartyl *O*-methyltransferase (Box 12-A)[1150] have all been associated with epilepsy.

Neurodegenerative diseases. As many as 5% of persons of age 65 and 20% of those of age 80 are afflicted with the progressive senile dementia known as **Alzheimer disease**. The condition is characterized by a gradual loss of memory and of the abilities to speak, think, or take care of one's self. Histologically Alzheimer disease is marked by the accumulation within neurons of **paired helical filaments**. These filaments, of 10 nm diameter, twist about each other to form a helix with an 80-nm pitch. The helices aggregate to create **neurofibrillary tangles**. The tangles are composed largely of a highly phosphorylated form of the microtubule-associated protein **tau** (p. 372)[1151–1155a,b] together with phosphorylated neurofilaments, apolipoprotein E (p. 1183), and other materials. The tangles are found in the cell bodies, axons, and dendrites of neurons in the hippocampus, amygdala, cerebral cortex, and other areas of the brain. Tangles may also be present in Parkinson disease, in the nearly extinct **Guam disease**,[1156,1157] and in some types of prion disease.[1158] Outside of the diseased neurons are numerous, spherical **amyloid plaques**. Their principal component is a 40- to 43-residue fragment called **amyloid β-protein** (**Aβ**), which appears to be toxic to neurons. Aβ is cut from a larger **amyloid precursor protein** (APP).[1154,1159–1160a] The APP gene is a member of a family of 16 related genes found in many organisms including nematodes, flies, and mammals. In humans the APP gene is found on chromosome 21, the chromosome that is present in three copies in **Down syndrome**. People with Down syndrome who live into their late thirties or beyond develop Alzheimer disease,[1153] presumably from excessive synthesis of APP. Both APP and its cleavage product Aβ are formed by nonneuronal cells throughout the body. However, the Aβ plaques form only in the brain, and the APP gene is essential for life. The rare **familial British dementia** resembles Alzheimer disease in producing amyloid plaques and neurofibrillary tangles. They appear principally in the cerebellum and arise from a different precursor protein.[1161,1162]

There are many other neurodegenerative diseases, some with a high incidence, and others rare. They include **Parkinson disease** (p. 1790), **Huntington disease** (Table 26-4), **spinal muscular atrophy** (SMA; a leading hereditary cause of infant mortality),[1162a,b] amyotrophic lateral sclerosis (**ALS**), prion diseases (Box 29-E), **ataxias**, and other diseases caused by triple-repeat DNA sequences (Table 26-4) and X-linked adrenoleukodystrophy (ALD; p. 945).[1163] In the last, membrane function is disrupted. Although these diseases arise from a variety of causes many of them have in common amyloidosis, the deposition of insoluble proteins in or around neurons.[1163a]

Parkinson disease, some cases of Alzheimer disease, and some types of prion disease are accompanied by the presence of **Lewy bodies** within the cytoplasm of neurons and also in nearby glia. These deposits consist largely of a dense core of fibrils of **α-synuclein**, a small 140-residue protein abundant in various parts of the brain.[1164,1164a,b] Mutations in the α-synuclein gene are associated with autosomal-dominant inheritance of early-onset Parkinson disease.[1165,1166] Just as tau tends to be associated with microtubules, α-synuclein may function in cooperation with microfilaments.[1158] Studies of an autosomal-recessive form of inherited juvenile Parkinson disease led to mutations in a large (>1 Mbp) gene on chromosome 6. It encodes the 465-residue **Parkin.**[1167,1168] Parkin is an E3 ubiquitin ligase (p. 524), which ubiquitinates α-synuclein.[1168,1169] This finding suggests that abnormally slow degradation of synuclein may be an important cause of Parkinson disease.

One of the triple-repeat polyglutamine diseases discussed in Chapter 26 (Table 26-4) is Huntington disease. The defective **Huntingtin** is a cytosolic protein that normally protects neurons but fails when the polygultamine sequence becomes too long.[1170,1171] Neurons of the cerebral cortex and striatum die, apparently by apoptosis.[1172] Huntingtin interacts with p53, with a CREB-binding protein, and with an EGF receptor suggesting that it functions in regulation of transcription.[1172–1173a] One of the genes whose transcription is regulated is that of the neurotrophin known as **brain-derived neurotrophic factor** (**BDNF**).[1170]

Many approaches have been taken in therapy of Parkinson disease. As mentioned on p. 1790 enhancing dopamine production by administration of L-dopa or by use of MAO inhibitors is a standard treattment. Experimental gene therapy with a glial cell line-derived neurotrophic factor also appears promising.[1174,1175]

Another aspect of neurodegeneration involves oxidative damage. A clue comes from amyotrophic lateral sclerosis (**ALS**), which struck down the New York Yankees baseball player Lou Gehrig, after he had started 2130 consecutive games over a 15-year period. ALS (Lou Gehrig disease) is the most prevalent of more than 70 diseases that cause loss of motor neurons.[1176] As pointed out on p.1075, the cause of a rare hereditary form of ALS is a defect in superoxide dismutase, which appears to promote excessive formation of free radicals.[1177] However, this interpretation is uncertain.[1178] Parkinson disease induced by the compound MPTP (Eq. 30-4) may also arise as a result of free radical damage.[1179] Among possible effects, MPTP may induce apoptosis.[1180] Both oxidative damage and apoptosis may be factors in Alzheimer and other neurodegenerative diseases as well.[1181,1182]

In every disease in which an abnormal protein is found there must be pathways of processing the protein to generate its functional form and pathways for degradation. These pathways are being investi gated for all of the neurodegenerative diseases and none more intensively than for Alzheimer disease. The amyloid precursor protein APP is an integral membrane glycoprotein with a large ~687-residue extracellular N-terminal portion, which resembles a cell surface receptor. It contains both a protease inhibitor-like domain and a zinc-binding region, which can be phosphor-ylated. It binds heparin and collagen as well as other proteins.[1183,1184] Rare mutations that cause early-onset familial Alzheimer disease are found in the APP gene. Some of these mutations alter the regulation of pre-mRNA splicing. Splicing generates eight different APP isoforms vaying in length from 677 to 770 residues. The properties of the isoforms vary. For example, if the 18 residues of exon 15 are spliced out a new motif for posttranslational modification is created by fusion of exons 14 and 16. The newly created sequence ENEGSG is recognized by a xylosyltransferase, which initiates formation of the terminal unit for glycosaminoglycan formation. The resulting proteoglycan is known as **appican** (p. 1154).[1185] The precursor protein APP is transported down axons to the nerve endings and is proteolytically cleaved to form the insoluble amyloid deposits. Alzheimer disease may occur when there is excessively rapid proteolysis of the precursor or if there is a failure to metabolize the amyloid protein.[1186] The folding and glycosylation reactions of APP occur in the ER and the Golgi, but the major problem in Alzheimer disease appears to be in subsequent proteolytic processing, some of which may occur as the APP is being transported through the Golgi to the cytoplasmic membrane. The protein may be cleaved at three sites by enzymes known as α-, β-, and γ-**secretases** as is indicated in Fig. 30-34, in which the protein is represented as an unfolded "stick."[1154,1187–1189a] Most of the mutations in APP that cause Alzheimer disease are near these three cleavage sites. As indicated in Fig. 30-34, cutting at the α site liberates into the extracellular space the large N-terminal portion as a soluble protein called APPsα.[1190,1191] It is thought to have a protective effect on neurons. If this cleavage occurs the protein is not cut at the β site and fragment Aβ is not formed. However, if cleavage by β-secretase (beta-site APP-cleaving enzyme or BACE) occurs first and is accompanied by or followed by cleavage at the γ-site, Aβ is liberated (Fig. 30-34). The β-secretase is an integral membrane protein, which carries a pepsinlike domain in its lumenal (or extracellular) part.[1192–1194] Since the Aβ peptide in an aggregated form appears to be toxic to neurons, a logical therapy for Alzheimer disease may be to block either the β- or γ-secretase.[1195,1196]

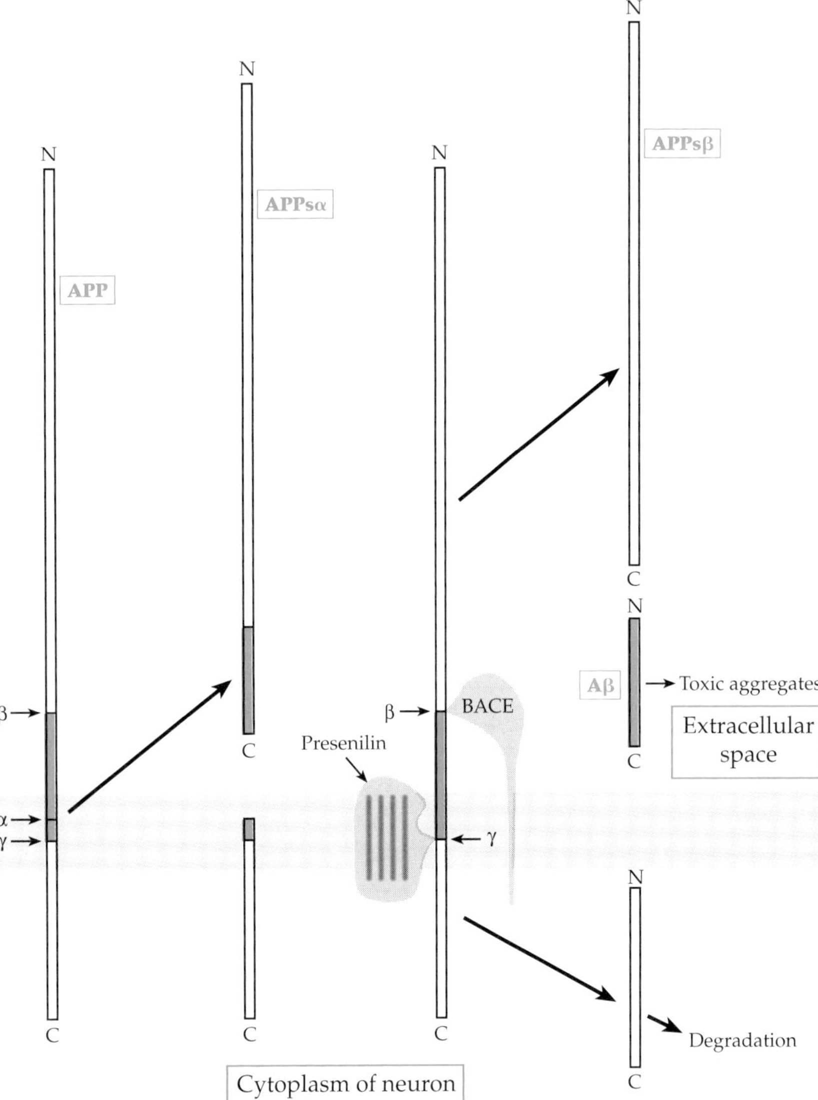

Figure 30-34 Cleavage of the amyloid precursor protein APP with liberation of amyloid Aβ protein. The proteins are represented as sticks (not to scale) but in reality contain both intracellular and extracellular globular domains.

The γ-secretase has been difficult to locate but has been identified as a result of other rare familial forms of Alzheimer disease. These are caused by mutations in genes for proteins known as **presenilin-1** (on chromosome 14) and **presenilin-2** (on chromosome 1).[1188,1197,1198] The presenilins are integral membrane proteins with multiple transmembrane helices.

They have been regarded as regulators of γ-secretase, but there is much evidence that the presenilin molecules may be cleaved proteolytically and that the C- and N-terminal domains formed in this way may associate to form an unusual aspartyl protease. It, too, is a target for inhibitors.[1198] The picture is made more complex by the fact that presenilins form complexes with other proteins. These include a newly discovered protein **nicastrin**,[1199–1200a] a large 709-residue transmembrane glycoprotein. Nicastrin not only seems to modulate presenilin action but also participates in an important developmental process via the highly conserved **Notch pathway** (Chapter 32).[1200] Many other proteins are found in the amyloid plaques of Alzheimer disease. Among them are acetylcholinesterase,[1201] proteoglycans,[1202] hydroxyacyl-CoA dehydrogenase,[1203] GM1 ganglioside,[1204] apolipoprotein A-1,[1187] and lithostatine.[1205]

What are the possible adverse consequences of accumulation of the Aβ protein? It may cause inflammation by activation of **microglia**,[1157] which may cause damage by release of NO.[1206] Aβ may induce death of neurons by apoptosis.[1201,1207–1209] A defect in protesomal degradation may be a factor.[1208] Both Aβ and the prion protein may promote oxidative damage. The brain derives most of its energy from oxidative metabolism, a major source of damaging radicals. Mitochondria are found in dendrites as well as cell bodies.[1210] Methionine residues in glycine-rich parts of the Aβ and prion proteins are suspected as centers of free radical formation.[1202,1211]

Both amyloid plaques and the tangles of protein tau-containing paired helical filaments are typically present in Alzheimer disease. Which comes first? Some hereditary neurodegenerative diseases are known in which tau filaments are present in neurons and sometimes also in glia.[1203,1212,1213] Since mutations in tau don't lead to Alzheimer disease whereas mutations affecting APP do, it is often assumed that the primary defect in the disease is with APP and that accumulating Aβ induces the observed changes in tau. However, this is by no means certain. Six different isoforms of tau (the longest with 441 residues) are created by alternative splicing of the mRNA. During its normal functioning tau is phosphorylated and carries an average of 2–3 phospho groups. In Alzheimer disease the level of tau is greatly increased (4- to 8-fold) and the molecules carry 5–9 phospho groups.[1204,1213a] It is this hyperphosphorylated tau that forms the paired helical filaments and tau tangles, which appear to clog the slender neurons.

What does tau do normally? Although it has been studied for many years, its exact functions are elusive. However, the role of the microtubules in axonal transport is well established. The tau isoforms may play a functional role in this process. The hyperphosphorylated tau of Alzheimer disease doesn't promote proper assembly of microtubules and may interfere with axonal transport of materials along the microtubules (see p. 1119).[1214,1215] Alzheimer disease may reflect an imbalance between the phosphorylation and dephosphorylation processes. Another possible problem with tau may be slow isomerization of prolyl linkages because of a deficiency of a prolyl *cis–trans* isomerase (Box 9-F).[1216]

Until 1993 **apolipoprotein E** was best known for its central role in plasma lipoproteins and cholesterol transport (Fig. 21-1). However, one of the three common alleles of the apoE gene confers a significant risk of development of Alzheimer disease.[1217,1218] A high blood cholesterol level is also correlated with increased risk.[1219,1220] Membrane abnormalities in mitochondria have been associated with Alzheimer disease.[1221] Also related to membranes and lipid metabolism, **vitamin E** appears to combat Alzheimer disease.[843,1218]

Environmental and nutritional factors may also affect the development of Alzheimer disease and other mental illness. Aluminum frequently accumulates in the neurons containing neurofibrillary tangles.[1206,1222] Copper and zinc ions can cause the amyloid Aβ to aggregate. However, Zn^{2+} may actually protect against neurotoxicity.[1223] The amino acid β-*N*-methylamino-L-alanine, a constituent of the toxic seeds of a type of palm (*Cycas circinalis* L.), may have induced both ALS and Guam disease, a condition resembling Parkinson disease, in a population in Guam that traditionally used these seeds as food.[1157]

Can neurodegenerative diseases be prevented or delayed? Much evidence suggests that the answer is yes. Rare early onset forms pose a special problem, but for most of us maintaining an active life style, using our minds, and choosing a good diet with adequate amounts of vitamins and essential ω3 fatty acids (Box 21-B) may be very helpful.[843] New methods of treatment are being tested. Antiinflammatory drugs are helpful,[1218,1218a] and even vaccination against Aβ and other amyloid proteins appears possible.[1224] Is it possible that antibodies and phagocytic cells can clear the cobwebs from our brains?

References

1. Norman, A., and Litwack, G. (1987) *Hormones*, Harcourt Brace Jovanovich, Orlando, Florida
2. Baulieu, E.-E., and Kelly, P. A., eds. (1990) *Hormones*, Hermann, Paris, France
3. Weintraub, B. D., ed. (1994) *Molecular Endocrinology*, Raven Press, New York
4. Gutkind, J. S. (1998) *J. Biol. Chem.* **273**, 1839–1842
5. Lefkowitz, R. J. (1998) *J. Biol. Chem.* **273**, 18677–18680

5a. Benard, O., Naor, Z., and Seger, R. (2001) *J. Biol. Chem.* **276**, 4554–4563

5b. Rockman, H. A., Koch, W. J., and Lefkowitz, R. J. (2002) *Nature (London)* **415**, 206–212

5c. Neves, S. R., Ram, P. T., and Iyengar, R. (2002) *Science* **296**, 1636–1639

6. Iiri, T., Farfel, Z., and Bourne, H. R. (1998) *Nature (London)* **394**, 35–38
7. Zheng, B., De Vries, L., and Farquhar, M. G. (1999) *Trends Biochem. Sci.* **24**, 411–414
8. Janetopoulos, C., Jin, T., and Devreotes, P. (2001) *Science* **291**, 2408–2411
9. Gorski, J., Toft, D., Shyamala, G., Smith, D., and Notides, A. (1968) *Rec. Progr. in Horm. Res.* **24**, 45–80
10. Jensen, E. V., Suzuki, T., Kawashima, T., Stumpf, W. E., Jungblut, P. W., and DeSombre, E. R. (1968) *Proc. Natl. Acad. Sci. U.S.A.* **59**, 632–638
11. Sluyser, M., ed. (1985) *Interaction of Steroid Hormone Receptors with DNA*, Ellis Horwood, Chichester
12. O'Malley, B. W., and Schrader, W. T. (1976) *Sci. Am.* **234**(Feb), 32–43
13. Wallis, M., Howell, S. L., and Taylor, K. W. (1985) *The Biochemistry of the Polypeptide Hormones*, Wiley, Chichester
14. McEwen, B. S. (1976) *Sci. Am.* **235**(Jul), 48–58
15. Pines, M. (1975) *Saturday Rev.* **Aug. 9**, 14
16. Guillemin, R., and Burgus, R. (1972) *Sci. Am.* **227**(Nov), 24–33
17. Sharrocks, A. D., Yang, S.-H., and Galanis, A. (2000) *Trends Biochem. Sci.* **25**, 448–453
18. Cohen, P. (2000) *Trends Biochem. Sci.* **25**, 596–601
19. Hafen, E. (1998) *Science* **280**, 1212–1213
20. Peles, E., Schlessinger, J., and Grumet, M. (1998) *Trends Biochem. Sci.* **23**, 121–124
21. Weng, G., Bhalla, U. S., and Lyengar, R. (1999) *Science* **284**, 92–96
22. Roberts, C. J., Nelson, B., Marton, M. J., Stoughton, R., Meyer, M. R., Bennett, H. A., He, Y. D., Dai, H., Walker, W. L., Hughes, T. R., Tyers, M., Boone, C., and Friend, S. H. (2000) *Science* **287**, 873–880
23. Tomkins, G. M. (1975) *Science* **189**, 760–763
24. Lenard, J. (1992) *Trends Biochem. Sci.* **17**, 147–150
25. Oosterom, J., Nijenhuis, W. A. J., Schaaper, W. M. M., Slootstra, J., Meloen, R. H., Gispen, W. H. H., Burbach, J. P. H., and Adan, R. A. H. (1999) *J. Biol. Chem.* **274**, 16853–16860
26. Muller, L., Zhu, P., Juliano, M. A., Juliano, L., and Lindberg, I. (1999) *J. Biol. Chem.* **274**, 21471–21477
27. Jutras, I., Seidah, N. G., Reudelhuber, T. L., and Brechler, V. (1997) *J. Biol. Chem.* **272**, 15184–15188
28. Feliciangeli, S., Kitabgi, P., and Bidard, J.-N. (2001) *J. Biol. Chem.* **276**, 6140–6150
29. Lynch, D. R., and Snyder, S. H. (1986) *Ann. Rev. Biochem.* **55**, 773–799

29a. Bateman, R. C., Jr., Temple, J. S., Misquitta, S. A., and Booth, R. E. (2001) *Biochemistry* **40**, 11246–11250

30. Loh, Y. P., Parish, D. C., and Tuteja, R. (1985) *J. Biol. Chem.* **260**, 7194–7205

30a. Yasothornsrikul, S., Aaron, W., Toneff, T., and Hook, V. Y. H. (1999) *Biochemistry* **38**, 7421–7430

30b. Jutras, I., Seidah, N. G., and Reudelhuber, T. L. (2000) *J. Biol. Chem.* **275**, 40337–40343

30c. Rockwell, N. C., and Fuller, R. S. (2001) *Biochemistry* **40**, 3657–3665

31. Paladini, A. C., Pena, C., and Retegui, L. A. (1979) *Trends Biochem. Sci.* **4**, 256–260
32. Lowman, H. B., Cunningham, B. C., and Wells, J. A. (1991) *J. Biol. Chem.* **266**, 10982–10988
33. Nilsen-Hamilton, M., Shapiro, J. M., Massoglia, S. L., and Hamilton, R. T. (1980) *Cell* **20**, 19–28
34. Parfett, C. L. J., Hamilton, R. T., Howell, B. W., Edwards, D. R., Nilsen-Hamilton, M., and Denhardt, D. T. (1985) *Mol. Cell. Biol.* **5**, 3289–3292
35. Lee, S., and Nathans, D. (1988) *J. Biol. Chem.* **263**, 3521–3527
36. Abdul-Meguid, S. S., Shieh, H., Smith, W. W., Dayringer, H. E., Violang, B. N., and Bentle, L. A. (1987) *Proc. Natl. Acad. Sci. U.S.A.* **84**, 6434–6437
37. Ultsch, M. H., Somers, W., Kossiakoff, A. A., and de Vos, A. M. (1994) *J. Mol. Biol.* **236**, 286–299
38. Souza, S. C., Frick, G. P., Wang, X., Kopchick, J. J., Lobo, R. B., and Goodman, H. M. (1995) *Proc. Natl. Acad. Sci. U.S.A.* **92**, 959–963
39. Sundström, M., Lundqvist, T., Rödin, J., Giebel, L. B., Milligan, D., and Norstedt, G. (1996) *J. Biol. Chem.* **271**, 32197–32203
40. de Vos, A. M., Ultsch, M., and Kossiakoff, A. A. (1992) *Science* **255**, 306–312
41. Wells, J. A. (1996) *Proc. Natl. Acad. Sci. U.S.A.* **93**, 1–6
42. Clackson, T., Ultsch, M. H., Wells, J. A., and de Vos, A. M. (1998) *J. Mol. Biol.* **277**, 1111–1128
43. Talmadge, K., Vamvakopoulos, N. C., and Fiddes, J. C. (1984) *Nature (London)* **307**, 37–40
44. Millstone, E., Brunner, E., and White, I. (1994) *Nature (London)* **371**, 647–648
45. Melmed, S. (1990) *N. Engl. J. Med.* **322**, 966–977
46. Souza, S. C., Frick, G. P., Yip, R., Lobo, R. B., Tai, L.-R., and Goodman, H. M. (1994) *J. Biol. Chem.* **269**, 30085–30088
47. Gertler, A., Grosclaude, J., Strasburger, C. J., Nir, S., and Djiane, J. (1996) *J. Biol. Chem.* **271**, 24482–24491
48. Herman, A., Helman, D., Livnah, O., and Gertler, A. (1999) *J. Biol. Chem.* **274**, 7631–7639
49. Cross, J. C., Werb, Z., and Fisher, S. J. (1994) *Science* **266**, 1508–1517
50. Darling, R. J., Wilken, J. A., Miller-Lindholm, A. K., Urlacher, T. M., Ruddon, R. W., Sherman, S. A., and Bedows, E. (2001) *J. Biol. Chem.* **276**, 10692–10699
51. Weller, C. T., Lustbader, J., Seshadri, K., Brown, J. M., Chadwick, C. A., Kolthoff, C. E., Ramnarain, S., Pollak, S., Canfield, R., and Homans, S. W. (1996) *Biochemistry* **35**, 8815–8823
52. Moyle, W. R., Campbell, R. K., Rao, S. N. V., Ayad, N. G., Bernard, M. P., Han, Y., and Wang, Y. (1995) *J. Biol. Chem.* **270**, 20020–20031
53. Seger, R., Hanoch, T., Rosenberg, R., Dantes, A., Merz, W. E., Strauss, J. F., III, and Amsterdam, A. (2001) *J. Biol. Chem.* **276**, 13957–13964
54. Policastro, P. F., Daniels-McQueen, S., Carle, G., and Boime, I. (1986) *J. Biol. Chem.* **261**, 5907–5916
55. Sairam, M. R. (1989) *FASEB J.* **3**, 1915–1926
56. Kaetzel, D. M., and Nilson, J. H. (1988) *J. Biol. Chem.* **263**, 6344–6351
57. Fernandez, L. M., and Puett, D. (1996) *J. Biol. Chem.* **271**, 925–930
58. Weiss, J., Axelrod, L., Whitcomb, R. W., Harris, P. E., Crowley, W. F., and Jameson, J. L. (1992) *N. Engl. J. Med.* **326**, 179–183
59. Laue, L., Chan, W.-Y., Hsueh, A. J. W., Kudo, M., Hsu, S. Y., Wu, S.-M., Blomberg, L., and Cutler, G. B., Jr. (1995) *Proc. Natl. Acad. Sci. U.S.A.* **92**, 1906–1910
60. Schally, A. V., Arimura, A., and Kastin, A. J. (1973) *Science* **179**, 341–350
61. Ling, N., Zeytin, F., Böhlen, P., Esch, F., Brazeau, P., Wehrenberg, W. B., Baird, A., and Guillemin, R. (1985) *Ann. Rev. Biochem.* **54**, 403–423
62. Laakkonen, L. J., Guarnieri, F., Perlman, J. H., Gershengorn, M. C., and Osman, R. (1996) *Biochemistry* **35**, 7651–7663
63. Bulant, M., Delfour, A., Vaudry, H., and Nicolas, P. (1988) *J. Biol. Chem.* **263**, 17189–17196
64. Perez de la Cruz, I., and Nillni, E. A. (1996) *J. Biol. Chem.* **271**, 22736–22745
65. Qi, L. J., Leung, A. T., Xiong, Y., Marx, K. A., and Abou-Samra, A.-B. (1997) *Biochemistry* **36**, 12442–12448
66. Taylor, A. L., and Fishman, L. M. (1988) *N. Engl. J. Med.* **319**, 213–222
67. Strulovici, B., Tahilramani, R., and Nestor, J. J., Jr. (1987) *Biochemistry* **26**, 6005–6011
68. Flanagan, C. A., Rodic, V., Konvicka, K., Yuen, T., Chi, L., Rivier, J. E., Millar, R. P., Weinstein, H., and Sealfon, S. C. (2000) *Biochemistry* **39**, 8133–8141
69. Marshall, J. C., and Kelch, R. P. (1986) *N. Engl. J. Med.* **315**, 1459–1468
70. Pincus, S. M., Mulligan, T., Iranmanesh, A., Gheorghiu, S., Godschalk, M., and Veldhuis, J. D. (1996) *Proc. Natl. Acad. Sci. U.S.A.* **93**, 14100–14105
71. Koch, B. D., Dorflinger, L. J., and Schonbrunn, A. (1985) *J. Biol. Chem.* **260**, 13138–13145
72. Reichlin, S. (1983) *N. Engl. J. Med.* **309**, 1556–1564
73. Pfeiffer, M., Koch, T., Schröder, H., Klutzny, M., Kirscht, S., Kreienkamp, H.-J., Höllt, V., and Schulz, S. (2001) *J. Biol. Chem.* **276**, 14027–14036
74. Cai, R.-Z., Szoke, B., Lu, R., Fu, D., Redding, T. W., and Schally, A. V. (1986) *Proc. Natl. Acad. Sci. U.S.A.* **83**, 1896–1900
75. Nikolics, K., Mason, A. J., Szönyi, E., Ramachandran, J., and Seeburg, P. H. (1985) *Nature (London)* **316**, 511–517
76. Tsonis, C. G., and Sharpe, R. M. (1986) *Nature (London)* **321**, 724–725
77. Sharif, M., and Hanley, M. R. (1992) *Nature (London)* **357**, 279–280
78. Nielsen, S., Chou, C.-L., Marples, D., Christensen, E. I., Kishore, B. K., and Knepper, M. A. (1995) *Proc. Natl. Acad. Sci. U.S.A.* **92**, 1013–1017
79. Innamorati, G., Le Gouill, C., Balamotis, M., and Birnbaumer, M. (2001) *J. Biol. Chem.* **276**, 13096–13103
80. Rosenthal, W., Antaramian, A., Gilbert, S., and Birnbaumer, M. (1993) *J. Biol. Chem.* **268**, 13030–13033
81. Nishimori, K., Young, L. J., Guo, Q., Wang, Z., Insel, T. R., and Matzuk, M. M. (1996) *Proc. Natl. Acad. Sci. U.S.A.* **93**, 11699–11704
82. Bloom, F. E. (1981) *Sci. Am.* **245**(Oct), 148–168
83. Larive, C. K., Guerra, L., and Rabenstein, D. L. (1992) *J. Am. Chem. Soc.* **114**, 7331–7337
84. Land, H., Schutz, G., Schmale, H., and Richter, D. (1982) *Nature (London)* **295**, 299–303
85. Brownstein, M. J., Russell, J. T., and Gainer, H. (1980) *Science* **207**, 373–378
86. Drenth, J. (1981) *J. Biol. Chem.* **256**, 2601–2602
87. Kiefer, L. L., Veal, J. M., Mountjoy, K. G., and Wilkison, W. O. (1998) *Biochemistry* **37**, 991–997
88. Haskell-Luevano, C., Cone, R. D., Monck, E. K., and Wan, Y.-P. (2001) *Biochemistry* **40**, 6164–6179
89. Seeley, R. J., Yagaloff, K. A., Fisher, S. L., Burn, P., Thiele, T. E., van Dijk, G., Baskin, D. G., and Schwartz, M. W. (1997) *Nature (London)* **390**, 349
90. Moore, J., Wood, J. M., and Schallreuter, K. U. (1999) *Biochemistry* **38**, 15317–15324

References

91. Carayannopoulos, M. O., Chi, M. M.-Y., Cui, Y., Pingsterhaus, J. M., McKnight, R. A., Mueckler, M., Devaskar, S. U., and Moley, K. H. (2000) *Proc. Natl. Acad. Sci. U.S.A.* **97**, 7313–7318
92. Aspinwall, C. A., Qian, W.-J., Roper, M. G., Kulkarni, R. N., Kahn, C. R., and Kennedy, R. T. (2000) *J. Biol. Chem.* **275**, 22331–22338
93. Rolland, F., Winderickx, J., and Thevelein, J. M. (2001) *Trends Biochem. Sci.* **26**, 310–317
94. Cheung, A. T., Dayanandan, B., Lewis, J. T., Korbutt, G. S., Rajotte, R. V., Bryer-Ash, M., Boylan, M. O., Wolfe, M. M., and Kieffer, T. J. (2000) *Science* **290**, 1959–1962
94a. Hsu, S. Y., Nakabayashi, K., Nishi, S., Kumagai, J., Kudo, M., Sherwood, O. D., and Hsueh, A. J. W. (2002) *Science* **295**, 671–674
95. Nishikori, K., Weisbrodt, N. W., Sherwood, O. D., and Sanborn, B. M. (1983) *J. Biol. Chem.* **258**, 2468–2474
96. Schwabe, C., LeRoith, D., Thompson, R. P., Shiloach, J., and Roth, J. (1983) *J. Biol. Chem.* **258**, 2778–2781
97. Georges, D., and Schwabe, C. (1999) *FASEB J.* **13**, 1269–1275
98. Garrett, T. P. J., McKern, N. M., Lou, M., Frenkel, M. J., Bentley, J. D., Lovrecz, G. O., Elleman, T. C., Cosgrove, L. J., and Ward, C. W. (1998) *Nature (London)* **394**, 395–399
99. Andrews, P. C., Hawke, D. H., Lee, T. D., Legesse, K., Noe, B. D., and Shively, J. E. (1986) *J. Biol. Chem.* **261**, 8128–8133
100. Rouillé, Y., Martin, S., and Steiner, D. F. (1995) *J. Biol. Chem.* **270**, 26488–26496
101. Holst, J. J. (1980) *Biochem. J.* **187**, 337–343
102. Rouillé, Y., Kantengwa, S., Irminger, J.-C., and Halban, P. A. (1997) *J. Biol. Chem.* **272**, 32810–32816
102a. Lovshin, J., Estall, J., Yusta, B., Brown, T. J., and Drucker, D. J. (2001) *J. Biol. Chem.* **276**, 21489–21499
103. Jelinek, L. J., Lok, S., Rosenberg, G. B., Smith, R. A., Grant, F. J., Biggs, S., Bensch, P. A., Kuijper, J. L., Sheppard, P. O., Sprecher, C. A., O'Hara, P. J., Foster, D., Walker, K. M., Chen, L. H. J., McKernan, P. A., and Kindsvogel, W. (1993) *Science* **259**, 1614–1616
104. Patthi, S., Simerson, S., and Velicelebi, G. (1988) *J. Biol. Chem.* **263**, 19363–19369
105. Ollerenshaw, S., Jarvis, D., Woolcock, A., Sullivan, C., and Scheibner, T. (1989) *N. Engl. J. Med.* **320**, 1244–1248
106. Miyawaki, K., and 16 other authors. (1999) *Proc. Natl. Acad. Sci. U.S.A.* **96**, 14843–14847
107. Blundell, T. L., Pitts, J. E., Tickle, I. J., Wood, S P., and Wu, C.-W. (1981) *Proc. Natl. Acad. Sci. U.S.A.* **78**, 4175–4179
108. Tonan, K., Kawata, Y., and Hamaguchi, K. (1990) *Biochemistry* **29**, 4424–4429
109. Bjornholm, B., Jorgensen, F. S., and Schwartz, T. W. (1993) *Biochemistry* **32**, 2954–2959
110. Cabrele, C., Langer, M., Bader, R., Wieland, H. A., Doods, H. N., Zerbe, O., and Beck-Sickinger, A. G. (2000) *J. Biol. Chem.* **275**, 36043–36048
111. Turner, A. J., ed. (1987) *Neuropeptides and Their Peptidases*, Ellis Horwood, Chichester
112. Uvnäs-Moberg, K. (1989) *Sci. Am.* **261**(Jul), 78–83
113. Krieger, D. T. (1983) *Science* **222**, 975–985
114. Drucker, D. J., and Asa, S. (1988) *J. Biol. Chem.* **263**, 13475–13478
115. Hoffman, J., and Porchet, M., eds. (1984) *Biosynthesis, Metabolism and Mode of Action of Invertebrate Hormones*, Springer, Berlin
116. Schaller, H. C., and Bodemüller, H. (1981) *Proc. Natl. Acad. Sci. U.S.A.* **78**, 7000–7004
117. Kowalski-Chauvel, A., Pradayrol, L., Vaysse, N., and Seva, C. (1996) *J. Biol. Chem.* **271**, 26356–26361
117a. Pannequin, J., Barnham, K. J., Hollande, F., Shulkes, A., Norton, R. S., and Baldwin, G. S. (2002) *J. Biol. Chem.* **277**, 48602–48609
118. Gigoux, V., Escrieut, C., Fehrentz, J.-A., Poirot, S., Maigret, B., Moroder, L., Gully, D., Martinez, J., Vaysse, N., and Fourmy, D. (1999) *J. Biol. Chem.* **274**, 20457–20464
119. Ding, X.-Q., Dolu, V., Hadac, E. M., Holicky, E. L., Pinon, D. I., Lybrand, T. P., and Miller, L. J. (2001) *J. Biol. Chem.* **276**, 4236–4244
120. Rehfeld, J. F. (1986) *J. Biol. Chem.* **261**, 5841–5847
121. Giragossian, C., and Mierke, D. F. (2001) *Biochemistry* **40**, 3804–3809
122. Anders, J., Blüggel, M., Meyer, H. E., Kühne, R., ter Laak, A. M., Kojro, E., and Fahrenholz, F. (1999) *Biochemistry* **38**, 6043–6055
123. Nachman, R. J., Holman, G. M., Haddon, W. F., and Ling, N. (1986) *Science* **234**, 71–73
124. Walz, D. A., Wider, M. D., Snow, J. W., Dass, C., and Desiderio, D. M. (1988) *J. Biol. Chem.* **263**, 14189–14195
125. Edmondson, S., Khan, N., Shriver, J., Zdunek, J., and Gräslund, A. (1991) *Biochemistry* **30**, 11271–11279
126. Feighner, S. D., Tan, C. P., McKee, K. K., Palyha, O. C., Hreniuk, D. L., Pong, S.-S., Austin, C. P., Figueroa, D., MacNeil, D., Cascieri, M. A., Nargund, R., Bakshi, R., Abramovitz, M., Stocco, R., Kargman, S., O'Neill, G., Van Der Ploeg, L. H. T., Evans, J., Patchett, A. A., Smith, R. G., and Howard, A. D. (1999) *Science* **284**, 2184–2188
127. Brown, M., Rivier, J., and Vale, W. (1977) *Science* **196**, 998–1000
128. Erne, D., and Schwyzer, R. (1987) *Biochemistry* **26**, 6316–6319
129. Nagalla, S. R., Gibson, B. W., Tang, D., Reeve, J. R., Jr., and Spindel, E. R. (1992) *J. Biol. Chem.* **267**, 6916–6922
130. Takuwa, N., Takuwa, Y., Bollag, W. E., and Rasmussen, H. (1987) *J. Biol. Chem.* **262**, 182–188
131. Xu, Z.-Q., Shi, T.-J., and Hökfelt, T. (1996) *Proc. Natl. Acad. Sci. U.S.A.* **93**, 14901–14905
132. Forte, L. R., and Currie, M. G. (1995) *FASEB J.* **9**, 643–650
133. Skelton, N. J., Garcia, K. C., Goeddel, D. V., Quan, C., and Burnier, J. P. (1994) *Biochemistry* **33**, 13581–13592
134. Bader, R., Bettio, A., Beck-Sickinger, A. G., and Zerbe, O. (2001) *J. Mol. Biol.* **305**, 307–329
135. Herzog, H., Hort, Y., Schneider, R., and Shine, J. (1995) *Proc. Natl. Acad. Sci. U.S.A.* **92**, 594–598
136. Pellieux, C., Sauthier, T., Domenighetti, A., Marsh, D. J., Palmiter, R. D., Brunner, H.-R., and Pedrazzini, T. (2000) *Proc. Natl. Acad. Sci. U.S.A.* **97**, 1595–1600
137. Larhammar, D., Blomqvist, A. G., Yee, F., Jazin, E., Yoo, H., and Wahlestedt, C. (1992) *J. Biol. Chem.* **267**, 10935–10938
138. Voisin, T., Couvineau, A., Rouyer-Fessard, C., and Laburthe, M. (1991) *J. Biol. Chem.* **266**, 10762–10767
139. Blomqvist, A. G., Söderberg, C., Lundell, I., Milner, R. J., and Larhammar, D. (1992) *Proc. Natl. Acad. Sci. U.S.A.* **89**, 2350–2354
139a. Batterham, R. L., Cowley, M. A., Small, C. J., Herzog, H., Cohen, M. A., Dakin, C. L., Wren, A. M., Brynes, A. E., Low, M. J., Ghatel, M. A., Cone, R. D., and Bloom, S. R. (2002) *Nature (London)* **418**, 650–654
139b. Schwartz, M. W., and Morton, G. J. (2002) *Nature (London)* **418**, 595–597
140. Woods, S. C., Seeley, R. J., Porte, D., Jr., and Schwartz, M. W. (1998) *Science* **280**, 1378–1383
141. Friedman, J. M., and Halaas, J. L. (1998) *Nature (London)* **395**, 763–770
142. Leibel, R. L., Chung, W. K., and Chua, S. C., Jr. (1997) *J. Biol. Chem.* **272**, 31937–31940
143. Zhang, F., Basinski, M. B., Beals, J. M., Briggs, S. L., Churgay, L. M., Clawson, D. K., DiMarchi, R. D., Furman, T. C., Hale, J. E., Hsiung, H. M., Schoner, B. E., Smith, D. P., Zhang, X. Y., Wery, J.-P., and Schevitz, R. W. (1997) *Nature (London)* **387**, 207–209
144. Chicurel, M. (2000) *Nature (London)* **404**, 538–540
145. Cowley, M. A., Smart, J. L., Rubinstein, M., Cerdán, M. G., Diano, S., Horvath, T. L., Cone, R. D., and Low, M. J. (2001) *Nature (London)* **411**, 480–484
145a. Bednarek, M. A., Hreniuk, D. L., Tan, C., Palyha, O. C., MacNeil, D. J., Van der Ploeg, L. H. Y., Howard, A. D., and Feighner, S. D. (2002) *Biochemistry* **41**, 6383–6390
146. Flier, J. S. (1997) *Proc. Natl. Acad. Sci. U.S.A.* **94**, 4242–4245
147. Unger, R. H., Zhou, Y.-T., and Orci, L. (1999) *Proc. Natl. Acad. Sci. U.S.A.* **96**, 2327–2332
148. Wahlestedt, C. (1998) *Science* **281**, 1624–1625
149. Nicholls, D. G. (1994) *Proteins, Transmitters and Synapses*, Blackwell Scientific Publications, Oxford
150. Li, Y.-M., Marnerakis, M., Stimson, E. R., and Maggio, J. E. (1995) *J. Biol. Chem.* **270**, 1213–1220
151. Iversen, L. (1998) *Nature (London)* **392**, 334–335
152. De Felipe, C., Herrero, J. F., O'Brien, J. A., Palmer, J. A., Doyle, C. A., Smith, A. J. H., Laird, J. M. A., Belmonte, C., Cervero, F., and Hunt, S. P. (1998) *Nature (London)* **392**, 394–397
153. Motta, A., Temussi, P. A., Wünsch, E., and Bovermann, G. (1991) *Biochemistry* **30**, 2364–2371
154. Rosenfeld, M. G., Amara, S. G., and Evans, R. M. (1984) *Science* **225**, 1315–1320
155. Habener, J. E., Potts, J. T., Jr., and Rich, A. (1976) *J. Biol. Chem.* **251**, 3893–3899
156. Jin, L., Briggs, S. L., Chandrasekhar, S., Chirgadze, N. Y., Clawson, D. K., Schevitz, R. W., Smiley, D. L., Tashjian, A. H., and Zhang, F. (2000) *J. Biol. Chem.* **275**, 27238–27244
156a. Shimizu, M., Potts, J. T., Jr., and Gardella, T. J. (2000) *J. Biol. Chem.* **275**, 21836–21843
157. Neer, R. M., and 11 other authors. (2001) *N. Engl. J. Med.* **344**, 1434–1441
158. Greenberg, Z., Bisello, A., Mierke, D. F., Rosenblatt, M., and Chorev, M. (2000) *Biochemistry* **39**, 8142–8152
159. Rölz, C., Pellegrini, M., and Mierke, D. F. (1999) *Biochemistry* **38**, 6397–6405
160. Ruat, M., Snowman, A. M., Hester, L. D., and Snyder, S. H. (1996) *J. Biol. Chem.* **271**, 5972–5975
161. Mierke, D. F., Maretto, S., Schievano, E., DeLuca, D., Bisello, A., Mammi, S., Rosenblatt, M., Peggion, E., and Chorev, M. (1997) *Biochemistry* **36**, 10372–10383
161a. Julius, D. (1997) *Nature (London)* **386**, 442
161b. Fiori, S., Renner, C., Cramer, J., Pegoraro, S., and Moroder, L. (1999) *J. Mol. Biol.* **291**, 163–175
162. Dickenson, A. H. (1986) *Nature (London)* **320**, 681–682
163. Tenneti, L., Gibbons, S. J., and Talamo, B. R. (1998) *J. Biol. Chem.* **273**, 26799–26808
164. Cockayne, D. A., Hamilton, S. G., Zhu, Q.-M., Dunn, P. M., Zhong, Y., Novakovic, S., Malmberg, A. B., Cain, G., Berson, A., Kassotakis, L., Hedley, L., Lachnit, W. G., Burnstock, G., McMahon, S. B., and Ford, A. P. D. W. (2000) *Nature (London)* **407**, 1011–1015
165. Surprenant, A., and Evans, R. J. (1993) *Nature (London)* **362**, 211–212
166. Grinthal, A., and Guidotti, G. (2000) *Biochemistry* **39**, 9–16
167. Cook, S. P., and McCleskey, E. W. (2000) *Nature (London)* **407**, 951–952

References

168. Goettsch, S., and Bayer, P. (2001) *Trends Biochem. Sci.* **26**, 12
169. Newby, A. C. (1984) *Trends Biochem. Sci.* **9**, 42–44
170. Huang, N.-K., Lin, Y.-W., Huang, C.-L., Messing, R. O., and Chern, Y. (2001) *J. Biol. Chem.* **276**, 13838–13846
171. Miller, K. J., and Hoffman, B. J. (1994) *J. Biol. Chem.* **269**, 27351–27356
172. Satoh, S., Matsumura, H., Suzuki, F., and Hayaishi, O. (1996) *Proc. Natl. Acad. Sci. U.S.A.* **93**, 5980–5984
173. Porkka-Heiskanen, T., Strecker, R. E., Thakkar, M., Bjorkum, A. A., Greene, R. W., and McCarley, R. W. (1997) *Science* **276**, 1265–1268
174. LaNoue, K. F., and Martin, L. F. (1994) *FASEB J.* **8**, 72–80
175. Hoebertz, A., Meghji, S., Burnstock, G., and Arnett, T. R. (2001) *FASEB J.* **15**, 1139–1148
176. Müller-Esterl, W., Iwanaga, S., and Nakanishi, S. (1986) *Trends Biochem. Sci.* **11**, 336–339
176a. Chuang, H.-h, Prescott, E. D., Kong, H., Shields, S., Jordt, S.-E., Basbaum, A. I., Chao, M. V., and Julius, D. (2001) *Nature (London)* **411**, 957–962
177. Steranka, L. R., Farmer, S. G., and Burch, R. M. (1989) *FASEB J.* **3**, 2019–2025
178. Jarnagin, K., Bhakta, S., Zuppan, P., Yee, C., Ho, T., Phan, T., Tahilramani, R., Pease, J. H. B., Miller, A., and Freedman, R. (1996) *J. Biol. Chem.* **271**, 28277–28286
179. Yanagisawa, M., Kurihara, H., Kimura, S., Tomobe, Y., Kobayashi, M., Mitsui, Y., Yazaki, Y., Goto, K., and Masaki, T. (1988) *Nature (London)* **332**, 411–415
180. Takuwa, N., Takuwa, Y., Yanagisawa, M., Yamashita, K., and Masaki, T. (1989) *J. Biol. Chem.* **264**, 7856–7861
181. Levin, E. R. (1995) *N. Engl. J. Med.* **333**, 356–363
182. Sokolovsky, M. (1991) *Trends Biochem. Sci.* **16**, 261–264
183. Pollock, D. M., Keith, T. L., and Highsmith, R. F. (1995) *FASEB J.* **9**, 1196–1204
184. Inagami, T. (1989) *J. Biol. Chem.* **264**, 3043–3046
185. Sengenès, C., Berlan, M., De Glisezinski, I., Lafontan, M., and Galitzky, J. (2000) *FASEB J.* **14**, 1345–1351
186. Volpi, M., Naccache, P. H., Molski, T. F. P., Shefcyk, J., Huang, C.-K., Marsh, M. L., Munoz, J., Becker, E. L., and Sha'afi, R. I. (1985) *Proc. Natl. Acad. Sci. U.S.A.* **82**, 2708–2712
187. Mills, J. S., Miettinen, H. M., Barnidge, D., Vlases, M. J., Wimer-Mackin, S., Dratz, E. A., Sunner, J., and Jesaitis, A. J. (1998) *J. Biol. Chem.* **273**, 10428–10435
188. Miettinen, H. M., Gripentrog, J. M., Mason, M. M., and Jesaitis, A. J. (1999) *J. Biol. Chem.* **274**, 27934–27942
189. Tzehoval, E., Segal, S., Stabinsky, Y., Fridkin, M., Spirer, Z., and Feldman, M. (1978) *Proc. Natl. Acad. Sci. U.S.A.* **75**, 3400–3404
190. Handel, T. M., and Domaille, P. J. (1996) *Biochemistry* **35**, 6569–6584
191. Clore, G. M., and Gronenborn, A. M. (1995) *FASEB J.* **9**, 57–62
192. Sticht, H., Escher, S. E., Schweimer, K., Forssmann, W.-G., Rösch, P., and Adermann, K. (1999) *Biochemistry* **38**, 5995–6002
193. Raffioni, S., Miceli, C., Vallesi, A., Chowdhury, S. K., Chait, B. T., Luporini, P., and Bradshaw, R. A. (1992) *Proc. Natl. Acad. Sci. U.S.A.* **89**, 2071–2075
194. Anderson, D. H., Weiss, M. S., and Eisenberg, D. (1997) *J. Mol. Biol.* **273**, 479–500
195. Vincent, F., Löbel, D., Brown, K., Spinelli, S., Grote, P., Breer, H., Cambillau, C., and Tegoni, M. (2001) *J. Mol. Biol.* **305**, 459–469
196. Dangott, L. J., and Garbers, D. L. (1984) *J. Biol. Chem.* **259**, 13712–13716
197. Davies, D. R., and Wlodawer, A. (1995) *FASEB J.* **9**, 50–56
198. Robinson, R. C., Radziejewski, C., Spraggon, G., Greenwald, J., Kostura, M. R., Burtnick, L. D., Stuart, D. I., Choe, S., and Jones, E. Y. (1999) *Protein Sci.* **8**, 2589–2597
199. Ye, J., Mayer, K. L., Mayer, M. R., and Stone, M. J. (2001) *Biochemistry* **40**, 7820–7831
200. Heath, J. K. (1994) *Growth Factors*, IRL Press, Oxford
201. Taniguchi, T. (1995) *Science* **268**, 251–255
202. Wells, J. A., ed. (1998) *Advances in Protein Chemistry, Cytokines*, Vol. 52, Academic Press, San Diego, California
203. Oppenheim, J. J., Feldmann, M., Durum, S. K., Hirano, T., Vilcek, J., and Nicola, N. A., eds. (2000) *Cytokine Reference*, Academic Press, San Diego
204. Harada, S., Smith, R. M., Smith, J. A., White, M. F., and Jarett, L. (1996) *J. Biol. Chem.* **271**, 30222–30226
205. Jackson, J. G., White, M. F., and Yee, D. (1998) *J. Biol. Chem.* **273**, 9994–10003
206. Rotwein, P., Pollock, K. M., Didier, D. K., and Krivi, G. G. (1986) *J. Biol. Chem.* **261**, 4828–4832
207. Dubaquié, Y., and Lowman, H. B. (1999) *Biochemistry* **38**, 6386–6396
208. Zezulak, K. M., and Green, H. (1986) *Science* **233**, 551–553
209. Frunzio, R., Chiariotti, L., Brown, A. L., Graham, D. E., Rechler, M. M., and Bruni, C. B. (1986) *J. Biol. Chem.* **261**, 17138–17149
210. Corvera, S., Whitehead, R. E., Mottola, C., and Czech, M. P. (1986) *J. Biol. Chem.* **261**, 7675–7679
211. Urdea, M. S., Merryweather, J. P., Mullenbach, G. T., Coit, D., Heberlein, U., Valenzuela, P., and Barr, P. J. (1983) *Proc. Natl. Acad. Sci. U.S.A.* **80**, 7461–7465
212. Abe, Y., Odaka, M., Inagaki, F., Lax, I., Schlessinger, J., and Kohda, D. (1998) *J. Biol. Chem.* **273**, 11150–11157
213. Marchionni, M. A. (1995) *Nature (London)* **378**, 334–335
214. Chang, H., Riese, D. J., II, Gilbert, W., Stern, D. F., and McMahan, U. J. (1997) *Nature (London)* **387**, 509–512
215. Strachan, L., Murison, J. G., Prestidge, R. L., Sleeman, M. A., Watson, J. D., and Kumble, K. D. (2001) *J. Biol. Chem.* **276**, 18265–18271
216. Kohda, D., and Imagaki, F. (1988) *J. Biochem.* **103**, 554–571
217. Yarden, Y., and Ullrich, A. (1988) *Ann. Rev. Biochem.* **57**, 443–478
218. Kumagai, T., Davis, J. G., Horie, T., O'Rourke, D. M., and Greene, M. I. (2001) *Proc. Natl. Acad. Sci. U.S.A.* **98**, 5526–5531
219. Schlessinger, J. (1988) *Trends Biochem. Sci.* **13**, 443–447
220. Morrison, P., Chung, K.-C., and Rosner, M. R. (1996) *Biochemistry* **35**, 14618–14624
221. Lin, C. R., Chen, W. S., Kruiger, W., Stolarsky, L. S., Weber, W., Evans, R. M., Verma, I. M., Gill, G. N., and Rosenfeld, M. G. (1984) *Science* **224**, 843–848
222. Hunter, T., Ling, N., and Cooper, J. A. (1984) *Nature (London)* **311**, 480–483
223. Kuppuswamy, D., Dalton, M., and Pike, L. J. (1993) *J. Biol. Chem.* **268**, 19134–19142
224. Betsholtz, C., Johnsson, A., Heldin, C.-H., Westermark, B., Lind, P., Urdea, M. S., Eddy, R., Shows, T. B., Philpott, K., Mellor, A. L., Knott, T. J., and Scott, J. (1986) *Nature (London)* **320**, 695–699
225. Assoian, R. K., Grotendorst, G. R., Miller, D. M., and Sporn, M. B. (1984) *Nature (London)* **309**, 804–806
226. Williams, L. T. (1989) *Science* **243**, 1564–1570
227. Massagué, J. (1985) *Trends Biochem. Sci.* **10**, 237–240
228. Kohda, D., Shimada, I., Miyake, T., Fuwa, T., and Inagaki, F. (1989) *Biochemistry* **28**, 953–958
229. Piek, E., Heldin, C.-H., and ten Dijke, P. (1999) *FASEB J.* **13**, 2105–2124
230. Elder, J. T., Fisher, G. J., Lindquist, P. B., Bennett, G. L., Pittelkow, M. R., Coffey, R. J., Jr., Ellingsworth, L., Derynck, R., and Voorhees, J. J. (1989) *Science* **243**, 811–814
231. Thomas, K. A., and Gimenez-Gallego, G. (1986) *Trends Biochem. Sci.* **11**, 81–84
232. Riboni, L., Viani, P., Bassi, R., Giussani, P., and Tettamanti, G. (2001) *J. Biol. Chem.* **276**, 12797–12804
233. Blaber, M., DiSalvo, J., and Thomas, K. A. (1996) *Biochemistry* **35**, 2086–2094
234. Thomas, K. A. (1988) *Trends Biochem. Sci.* **13**, 327–328
235. Schreiber, A. B., Winkler, M. E., and Derynck, R. (1986) *Science* **232**, 1250–1253
236. Folkman, J., and Klagsbrun, M. (1987) *Science* **235**, 442–447
237. Thomas, K. A. (1996) *J. Biol. Chem.* **271**, 603–606
238. Radisavljevic, Z., Avraham, H., and Avraham, S. (2000) *J. Biol. Chem.* **275**, 20770–20774
239. Antonetti, D. A., Barber, A. J., Hollinger, L. A., Wolpert, E. B., and Gardner, T. W. (1999) *J. Biol. Chem.* **274**, 23463–23467
240. Leonidas, D. D., Shapiro, R., Allen, S. C., Subbarao, G. V., Veluraja, K., and Acharya, K. R. (1999) *J. Mol. Biol.* **285**, 1209–1233
241. Burke, D., Wilkes, D., Blundell, T. L., and Malcolm, S. (1998) *Trends Biochem. Sci.* **23**, 59–62
242. Richard, C., Liuzzo, J. P., and Moscatelli, D. (1995) *J. Biol. Chem.* **270**, 24188–24196
243. Pellegrini, L., Burke, D. F., von Delft, F., Mulloy, B., and Blundell, T. L. (2000) *Nature (London)* **407**, 1029–1034
244. Ibrahimi, O. A., Eliseenkova, A. V., Plotnikov, A. N., Yu, K., Ornitz, D. M., and Mohammadi, M. (2001) *Proc. Natl. Acad. Sci. U.S.A.* **98**, 7182–7187
245. Webster, M. K., and Donoghue, D. J. (1996) *EMBO J.* **15**, 520–527
246. Faham, S., Hileman, R. E., Fromm, J. R., Linhardt, R. J., and Rees, D. C. (1996) *Science* **271**, 1116–1120
247. Taipale, J., and Keski-Oja, J. (1997) *FASEB J.* **11**, 51–59
248. Levi-Montalcini, R. (1987) *Science* **237**, 1154–1162
249. Bradshaw, R. A., Blundell, T. L., Lapatto, R., McDonald, N. Q., and Murray-Rust, J. (1993) *Trends Biochem. Sci.* **18**, 48–52
250. Thoenen, H. (1995) *Science* **270**, 593–598
251. Wiesmann, C., Ultsch, M. H., Bass, S. H., and de Vos, A. M. (1999) *Nature (London)* **401**, 184–188
251a. Tartaglia, N., Du, J., Tyler, W. J., Neale, E., Pozzo-Miller, L., and Lu, B. (2001) *J. Biol. Chem.* **276**, 37585–37593
252. Micera, A., Vigneti, E., Pickholtz, D., Reich, R., Pappo, O., Bonini, S., Maquart, F. X., Aloe, L., and Levi-Schaffer, F. (2001) *Proc. Natl. Acad. Sci. U.S.A.* **98**, 6162–6167
253. Wooten, M. W., Seibenhener, M. L., Mamidipudi, V., Diaz-Meco, M. T., Barker, P. A., and Moscat, J. (2001) *J. Biol. Chem.* **276**, 7709–7712
254. Brockes, J. P., Lemke, G. E., and Balzer, D. R., Jr. (1980) *J. Biol. Chem.* **255**, 8374–8377
255. Frazier, W. A., Ohléndorf, C. E., Boyd, L. F., Aloe, L., Johnson, E. M., Ferrendelli, J. A., and Bradshaw, R. A. (1973) *Proc. Natl. Acad. Sci. U.S.A.* **70**, 2448–2452
256. Simpson, E. (1984) *Trends Biochem. Sci.* **9**, 527–530

References

257. Wang, E. A., Rosen, V., Cordes, P., Hewick, R. M., Kriz, M. J., Luxenberg, D. P., Sibley, B. S., and Wozney, J. M. (1988) *Proc. Natl. Acad. Sci. U.S.A.* **85**, 9484–9488
258. Seyedin, S. M., Thompson, A. Y., Bentz, H., Rosen, D. M., McPherson, J. M., Conti, A., Siegel, N. R., Galluppi, G. R., and Piez, K. A. (1986) *J. Biol. Chem.* **261**, 5693–5695
259. Whetton, A. D., and Dexter, T. M. (1986) *Trends Biochem. Sci.* **11**, 207–211
260. Lai, P.-H., Everett, R., Wang, F.-F., Arakawa, T., and Goldwasser, E. (1986) *J. Biol. Chem.* **261**, 3116–3121
261. Goldberg, M. A., Dunning, S. P., and Bunn, H. F. (1988) *Science* **242**, 1412–1415
262. D'Andrea, A. D. (1994) *N. Engl. J. Med.* **330**, 839–845
263. Mufson, R. A. (1997) *FASEB J.* **11**, 37–44
264. Livnah, O., Stura, E. A., Johnson, D. L., Middleton, S. A., Mulcahy, L. S., Wrighton, N. C., Dower, W. J., Jolliffe, L. K., and Wilson, I. A. (1996) *Science* **273**, 464–471
265. Wells, J. A. (1996) *Science* **273**, 449–450
266. Kawasaki, E. S., Ladner, M. B., Wang, A. M., Van Arsdell, J., Warren, M. K., Coyne, M. Y., Schweickart, V. L., Lee, M.-T., Wilson, K. J., Boosman, A., Stanley, E. R., Ralph, P., and Mark, D. F. (1985) *Science* **230**, 291–296
267. Taylor, E. W., Fear, A. L., Bohm, A., Kim, S.-H., and Koths, K. (1994) *J. Biol. Chem.* **269**, 31171–31177
268. Dinarello, C. A., and Wolff, S. M. (1993) *N. Engl. J. Med.* **328**, 106–112
269. Cohen, F. E., Kosen, P. A., Kuntz, I. D., Epstein, L. B., Ciardelli, T. L., and Smith, K. A. (1986) *Science* **234**, 349–355
270. Clark-Lewis, I., Aebersold, R., Ziltener, H., Schrader, J. W., Hood, L. E., and Kent, S. B. H. (1986) *Science* **231**, 134–139
271. Gurney, M. E., Heinrich, S. P., Lee, M. R., and Yin, H.-S. (1986) *Science* **234**, 566–574
272. Sun, Y.-J., Chou, C.-C., Chen, W.-S., Wu, R.-T., Meng, M., and Hsiao, C.-D. (1999) *Proc. Natl. Acad. Sci. U.S.A.* **96**, 5412–5417
273. Audhya, T., Kroon, D., Heavner, G., Viamontes, G., and Goldstein, G. (1986) *Science* **231**, 997–999
274. Audhya, T., Schlesinger, D. H., and Goldstein, G. (1981) *Biochemistry* **20**, 6195–6200
275. Vancurova, I., Miskolci, V., and Davidson, D. (2001) *J. Biol. Chem.* **276**, 19746–19752
276. Naismith, J. H., and Sprang, S. R. (1998) *Trends Biochem. Sci.* **23**, 74–79
277. Kanety, H., Feinstein, R., Papa, M. Z., Hemi, R., and Karasik, A. (1995) *J. Biol. Chem.* **270**, 23780–23784
278. Loret, C., Sensenbremmer, M., and Labourdette, G. (1989) *J. Biol. Chem.* **264**, 8319–8327
279. Moncada, S., and Higgs, A. (1993) *N. Engl. J. Med.* **329**, 2002–2010
280. Mayer, B., and Hemmens, B. (1997) *Trends Biochem. Sci.* **22**, 477–481
281. Durner, J., Gow, A. J., Stamler, J. S., and Glazebrook, J. (1999) *Proc. Natl. Acad. Sci. U.S.A.* **96**, 14206–14207
282. Colasanti, M., Lauro, G. M., and Venturini, G. (1995) *Nature (London)* **374**, 505
283. Lawson, D. M., Stevenson, C. E. M., Andrew, C. R., and Eady, R. R. (2000) *EMBO J.* **19**, 5661–5671
284. Champion, H. C., Bivalacqua, T. J., Hyman, A. L., Ignarro, L. J., Hellstrom, W. J. G., and Kadowitz, P. J. (1999) *Proc. Natl. Acad. Sci. U.S.A.* **96**, 11648–11652
285. Clementi, E., Brown, G. C., Foxwell, N., and Moncada, S. (1999) *Proc. Natl. Acad. Sci. U.S.A.* **96**, 1559–1562
285a. Corbin, J. D., and Francis, S. H. (1999) *J. Biol. Chem.* **274**, 13729–13732
286. Kim, N. N., Cox, J. D., Baggio, R. F., Emig, F. A., Mistry, S. K., Harper, S. L., Speicher, D. W., Morris, S. M., Jr., Ash, D. E., Traish, A., and Christianson, D. W. (2001) *Biochemistry* **40**, 2678–2688
286a. Lue, T. F. (2000) *N. Engl. J. Med.* **342**, 1802–1805
287. Xue, L., Farrugia, G., Miller, S. M., Ferris, C. D., Snyder, S. H., and Szurszewski, J. H. (2000) *Proc. Natl. Acad. Sci. U.S.A.* **97**, 1851–1855
288. Hillier, B. J., Christopherson, K. S., Prehoda, K. E., Bredt, D. S., and Lim, W. A. (1999) *Science* **284**, 812–815
289. Denlinger, L. C., Fisette, P. L., Garis, K. A., Kwon, G., Vazquez-Torres, A., Simon, A. D., Nguyen, B., Proctor, R. A., Bertics, P. J., and Corbett, J. A. (1996) *J. Biol. Chem.* **271**, 337–342
290. Hall, J. P., Merithew, E., and Davis, R. J. (2000) *Proc. Natl. Acad. Sci. U.S.A.* **97**, 14022–14024
291. Trimmer, B. A., Aprille, J. R., Dudzinski, D. M., Lagace, C. J., Lewis, S. M., Michel, T., Qazi, S., and Zayas, R. M. (2001) *Science* **292**, 2486–2488
291a Yew, W. S., Kolatkar, P. R., Kuhn, P., and Khoo, H. E. (1999) *J Struct. Biol.* **128**, 216–218
292. Inoue, K., Akaike, T., Miyamoto, Y., Okamoto, T., Sawa, T., Otagiri, M., Suzuki, S., Yoshimura, T., and Maeda, H. (1999) *J. Biol. Chem.* **274**, 27069–27075
293. Andersen, J. F., Champagne, D. E., Weichsel, A., Ribeiro, J. M. C., Balfour, C. A., Dress, V., and Montfort, W. R. (1997) *Biochemistry* **36**, 4423–4428
294. Spencer, N. Y., Zeng, H., Patel, R. P., and Hogg, N. (2000) *J. Biol. Chem.* **275**, 36562–36567
295. Mayer, B., Pfeiffer, S., Schrammel, A., Koesling, D., Schmidt, K., and Brunner, F. (1998) *J. Biol. Chem.* **273**, 3264–3270
296. Gross, S. S., and Lane, P. (1999) *Proc. Natl. Acad. Sci. U.S.A.* **96**, 9967–9969
297. Herold, S., Exner, M., and Nauser, T. (2001) *Biochemistry* **40**, 3385–3395
298. Nagatomo, S., Nagai, M., Tsuneshige, A., Yonetani, T., and Kitagawa, T. (1999) *Biochemistry* **38**, 9659–9666
299. Pawloski, J. R., Hess, D. T., and Stamler, J. S. (2001) *Nature (London)* **409**, 622–626
300. Gladwin, M. T., Shelhamer, J. H., Schechter, A. N., Pease-Fye, M. E., Waclawiw, M. A., Panza, J. A., Ognibene, F. P., and Cannon, R. O., III. (2000) *Proc. Natl. Acad. Sci. U.S.A.* **97**, 11482–11487
300a. Ruschitzka, F. T., Wenger, R. H., Stallmach, T., Quaschning, T., Wit, C., Wagner, K., Labugger, R., Kelm, M., Noll, G., Rülicke, T., Shaw, S., Lindberg, R. L. P., Rodenwaldt, B., Lutz, H., Bauer, C., Lüscher, T. F., and Gassmann, M. (2000) *Proc. Natl. Acad. Sci. U.S.A.* **97**, 11609–11613
301. Brunori, M. (2001) *Trends Biochem. Sci.* **26**, 209–210
302. Flögel, U., Merx, M. W., Gödecke, A., Decking, U. K. M., and Schrader, J. (2001) *Proc. Natl. Acad. Sci. U.S.A.* **98**, 735–740
303. Minning, D. M., Gow, A. J., Bonaventura, J., Braun, R., Dewhirst, M., Goldberg, D. E., and Stamler, J. S. (1999) *Nature (London)* **401**, 497–502
304. Lin, R., and Farmer, P. J. (2000) *J. Am. Chem. Soc.* **122**, 2393–2394
305. Witting, P. K., Douglas, D. J., and Mauk, A. G. (2001) *J. Biol. Chem.* **276**, 3991–3998
306. Friebe, A., Schultz, G., and Koesling, D. (1996) *EMBO J.* **15**, 6863–6868
307. Morita, T., Mitsialis, S. A., Koike, H., Liu, Y., and Kourembanas, S. (1997) *J. Biol. Chem.* **272**, 32804–32809
308. Arakawa, T., and Timasheff, S. N. (1982) *Biochemistry* **21**, 6545–6552
309. Uchida, T., Ishikawa, H., Ishimori, K., Morishima, I., Nakajima, H., Aono, S., Mizutani, Y., and Kitagawa, T. (2000) *Biochemistry* **39**, 12747–12752
310. Nakajima, H., Honma, Y., Tawara, T., Kato, T., Park, S.-Y., Miyatake, H., Shiro, Y., and Aono, S. (2001) *J. Biol. Chem.* **276**, 7055–7061
311. Ho, Y. S., Swenson, L., Dwrewenda, U., Serre, L., Wei, Y., Dauter, Z., Hattori, M., Adachi, T., Aoki, J., Arai, H., Inoue, K., and Derewenda, Z. S. (1997) *Nature (London)* **385**, 89–93
312. Serhan, C. N., Haeggström, J. Z., and Leslie, C. C. (1996) *FASEB J.* **10**, 1147–1158
313. Ruvolo, P. P., Gao, F., Blalock, W. L., Deng, X., and May, W. S. (2001) *J. Biol. Chem.* **276**, 11754–11758
314. Lehtonen, J. Y. A., Horiuchi, M., Daviet, L., Akishita, M., and Dzau, V. J. (1999) *J. Biol. Chem.* **274**, 16901–16906
315. Cravatt, B. F., Prospero-Garcia, O., Siuzdak, G., Gilula, N. B., Henriksen, S. J., Boger, D. L., and Lerner, R. A. (1995) *Science* **268**, 1506–1509
316. Wilcox, B. J., Ritenour-Rogers, K. J., Asser, A. S., Baumgart, L. E., Baumgart, M. A., Boger, D. L., DeBlassio, J. L., deLong, M. A., Glufke, U., Henz, M. E., King, L., III, Merkler, K. A., Patterson, J. E., Robleski, J. J., Vederas, J. C., and Merkler, D. J. (1999) *Biochemistry* **38**, 3235–3245
317. Boger, D. L., Patterson, J. E., Guan, X., Cravatt, B. F., Lerner, R. A., and Gilula, N. B. (1998) *Proc. Natl. Acad. Sci. U.S.A.* **95**, 4810–4815
318. Patricelli, M. P., Lovato, M. A., and Cravatt, B. F. (1999) *Biochemistry* **38**, 9804–9812
319. Maccarrone, M., Bari, M., Lorenzon, T., Bisogno, T., Di Marzo, V., and Finazzi-Agró, A. (2000) *J. Biol. Chem.* **275**, 13484–13492
320. Zygmunt, P. M., Petersson, J., Andersson, D. A., Chaunag, H.-H., Sorgard, M., Di Marzo, V., Julius, D., and Högestätt, E. D. (1999) *Nature (London)* **400**, 452–456
321. Patricelli, M. P., Lashuel, H. A., Giang, D. K., Kelly, J. W., and Cravatt, B. F. (1998) *Biochemistry* **37**, 15177–15787
322. Patterson, J. E., Ollmann, I. R., Cravatt, B. F., Boger, D. L., Wong, C.-H., and Lerner, R. A. (1996) *J. Am. Chem. Soc.* **118**, 5938–5945
323. Strauss, E. (1999) *Science* **284**, 1302–1304
324. Dong, Y.-H., Wang, L.-H., Xu, J.-L., Zhang, H.-B., Zhang, X.-F., and Zhang, L.-H. (2001) *Nature (London)* **411**, 813–817
325. Anderegg, R. J., Betz, R., Carr, S. A., Crabb, J. W., and Duntze, W. (1987) *J. Biol. Chem.* **263**, 18236–18240
326. Kurjan, J. (1992) *Ann. Rev. Biochem.* **61**, 1097–1129
327. Song, J., and Dohlman, H. G. (1996) *Biochemistry* **35**, 14806–14817
327a. Mentesana, P. E., and Konopka, J. B. (2001) *Biochemistry* **40**, 9685–9694
328. Sakagami, Y., Yoshida, M., Isogai, A., and Suzuki, A. (1981) *Science* **212**, 1525–1527
329. Gottschalk, W. K., and Sonneborn, D. R. (1985) *J. Biol. Chem.* **260**, 6592–6599
330. Starr, R. C., Marner, F. J., and Jaenicke, L. (1995) *Proc. Natl. Acad. Sci. U.S.A.* **92**, 641–645
331. Grimmelikhuijzen, C. J. P., and Graff, D. (1986) *Proc. Natl. Acad. Sci. U.S.A.* **83**, 9817–9821
332. Grimmelikhuijzen, C. J. P., Rinehart, K. L., Jacob, E., Graff, D., Reinscheid, R. K., Nothacker, H.-P., and Staley, A. L. (1990) *Proc. Natl. Acad. Sci. U.S.A.* **87**, 5410–5414
333. Bellis, S. L., Kass-Simon, G., and Rhoads, D. E. (1992) *Biochemistry* **31**, 9838–9843
334. Murata, M., Miyagawa-Kohshima, K., Nakanishi, K., and Naya, Y. (1986) *Science* **234**, 585–587

References

335. Scheller, R. H., Kaldany, R.-R., Kreiner, T., Mahon, A. C., Nambu, J. R., Schaefer, M., and Taussig, R. (1984) *Science* **225**, 1300–1308
336. Chiu, D. T., and Zare, R. N. (1998) *Proc. Natl. Acad. Sci. U.S.A.* **95**, 3338–3340
337. Lloyd, P. E., Schacher, S., Kupfermann, I., and Weiss, K. R. (1986) *Proc. Natl. Acad. Sci. U.S.A.* **83**, 9794–9798
338. Berry, R. W. (1981) *Biochemistry* **20**, 6200–6205
339. Kaldany, R.-R. J., Campanelli, J. T., Makk, G., Evans, C. J., and Scheller, R. H. (1986) *J. Biol. Chem.* **261**, 5751–5757
340. Newcomb, R., Fisher, J. M., and Scheller, R. H. (1988) *J. Biol. Chem.* **263**, 12514–12521
341. Garden, R. W., Shippy, S. A., Li, L., Moroz, T. P., and Sweedler, J. V. (1998) *Proc. Natl. Acad. Sci. U.S.A.* **95**, 3972–3977
342. Joose, J. (1984) in *Biosynthesis, Metabolism and Mode of Action of Invertebrate Hormones* (Hoffmann, J., and Porchet, M., eds), pp. 17–35, Springer, Berlin
343. Nelson, L. S., Rosoff, M. L., and Li, C. (1998) *Science* **281**, 1686–1690
344. Duve, H., Johnsen, A. H., Sewell, J. C., Scott, A. G., Orchard, I., Rehfeld, J. F., and Thorpe, A. (1992) *Proc. Natl. Acad. Sci. U.S.A.* **89**, 2326–2330
345. Lange, A. B., Orchard, I., Wang, Z., and Nachman, R. J. (1995) *Proc. Natl. Acad. Sci. U.S.A.* **92**, 9250–9253
346. Ramarao, C. S., Burks, D. J., and Garbers, D. L. (1990) *Biochemistry* **29**, 3383–3388
347. Yoshino, K.-i, Takao, T., Suhara, M., Kitai, T., Hori, H., Nomura, K., Yamaguchi, M., Shimonishi, Y., and Suzuki, N. (1991) *Biochemistry* **30**, 6203–6209
348. Downer, R. G. H., and Laufer, H., eds. (1983) *Endocrinology of Insects*, Liss, New York
349. O'Shea, M., Adams, M. E., and Bishop, C. A. (1982) *Fed. Proc.* **41**, 2940–2947
350. Stangier, J., Hilbich, C., Beyreuther, K., and Keller, R. (1987) *Proc. Natl. Acad. Sci. U.S.A.* **84**, 575–579
351. Fernlund, P., and Josefsson, L. (1972) *Science* **177**, 173–175
352. Rao, K. R., Riehm, J. P., Zahnow, C. A., Kleinholz, L. H., Tarr, G. E., Johnson, L., Norton, S., Landau, M., Semmes, O. J., Sattelberg, R. M., Jorenby, W. H., and Hintz, M. F. (1985) *Proc. Natl. Acad. Sci. U.S.A.* **82**, 5319–5322
353. Rao, K. R., Mohrherr, C. J., Riehm, J. P., Zahnow, C. A., Norton, S., Johnson, L., and Tarr, G. E. (1987) *J. Biol. Chem.* **262**, 2672–2675
354. Sanburg, L. L., Kramer, K. J., Kezdy, F. J., Law, J. H., and Oberlander, H. (1975) *Nature (London)* **253**, 266–267
355. Touhara, K., Lerro, K. A., Bonning, B. C., Hammock, B. D., and Prestwich, G. D. (1993) *Biochemistry* **32**, 2068–2075
356. Cayre, M., Strambi, C., Charpin, P., Augier, R., and Strambi, A. (1997) *Proc. Natl. Acad. Sci. U.S.A.* **94**, 8238–8242
357. Wainwright, G., Webster, S. G., Wilkinson, M. C., Chung, J. S., and Rees, H. H. (1996) *J. Biol. Chem.* **271**, 12749–12754
358. Agui, N., Bollenbacher, W. E., Granger, N. A., and Gilbert, L. I. (1980) *Nature (London)* **285**, 669–670
359. Coudron, T. A., Law, J. H., and Koeppe, J. K. (1981) *Trends Biochem. Sci.* **6**, 248–252
359a. Takeuchi, H., Chen, J.-H., O'Reilly, D. R., Turner, P. C., and Rees, H. H. (2001) *J. Biol. Chem.* **276**, 26819–26828
360. Beach, R. (1979) *Science* **205**, 829–831
361. Pongs, O. (1985) in *Interaction of Steroid Hormone Receptors with DNA* (Sluyser, M., ed), pp. 226–240, Ellis Horwood, Chichester
362. Zitnan, D., Kingan, T. G., Hermesman, J. L., and Adams, M. E. (1996) *Science* **271**, 88–91
363. Oudejans, R. C. H. M., Vroemen, S. F., Jansen, R. F. R., and Van der Horst, D. J. (1996) *Proc. Natl. Acad. Sci. U.S.A.* **93**, 8654–8659
364. Kondo, H., Ino, M., Suzuki, A., Ishizaki, H., and Iwami, M. (1996) *J. Mol. Biol.* **259**, 926–937
365. Roelofs, W. L. (1995) *Proc. Natl. Acad. Sci. U.S.A.* **92**, 44–49
366. Takahashi, N., ed. (1986) *Chemistry of Plant Hormones*, CRC Press, Boca Raton, Florida
367. Crozier, A., and Hillman, J. R., eds. (1984) *The Biosynthesis and Metabolism of Plant Hormones*, Cambridge Univ. Press, London
368. Moller, S. G., and Chua, N.-H. (1999) *J. Mol. Biol.* **293**, 219–234
369. Buchanan, B. B., Gruissem, W., and Jones, R. L. (2000) *Biochemistry and Molecular Biology of Plants*, American Society of Plant Physiologists, Rockville, Maryland
370. Busby, S., and Ebright, R. H. (1999) *J. Mol. Biol.* **293**, 199–213
371. Steinmann, T., Geldner, N., Grebe, M., Mangold, S., Jackson, C. L., Paris, S., Gälweiler, L., Palme, K., and Jürgens, G. (1999) *Science* **286**, 316–318
371a. Friml, J., Wisniewska, J., Benková, E., Mendgen, K., and Palme, K. (2002) *Nature (London)* **415**, 806–809
372. El-Antably, H. M. M., and Larsen, P. (1974) *Nature (London)* **250**, 76–77
373. Shaw, G. (1994) in *Cytokinins. Chemistry, Activity and Function* (Mok, D. W. S., and Mok, M. C., eds), CRC Press, Boca Raton, Florida
374. Martin, R. C., Mok, M. C., Habben, J. E., and Mok, D. W. S. (2001) *Proc. Natl. Acad. Sci. U.S.A.* **98**, 5922–5926
375. Koshimiza, K., and Iwamura, H. (1986) *Chemistry of Plant Hormones*, CRC Press, Boca Raton, Florida (pp. 153–199)
376. Hall, M. A. (1986) *Hormones, Receptors and Cellular Interactions in Plants*, Cambridge Univ. Press, London (pp. 69–89)
377. Menke, F. L. H., Champion, A., Kijne, J. W., and Memelink, J. (1999) *EMBO J.* **18**, 4455–4463
378. Niggeweg, R., Thurow, C., Kegler, C., and Gatz, C. (2000) *J. Biol. Chem.* **275**, 19897–19905
379. Whitfield, J. (2001) *Nature (London)* **410**, 736–737
380. Kessler, A., and Baldwin, I. T. (2001) *Science* **291**, 2141–2144
381. Farmer, E. E. (2001) *Nature (London)* **411**, 854–856
382. Pearce, G., Moura, D. S., Stratmann, J., and Ryan, C. A. (2001) *Nature (London)* **411**, 817–820
383. Phillips, D. A., Joseph, C. M., Yang, G.-P., Martinez-Romero, E., Sanborn, J. R., and Volpin, H. (1999) *Proc. Natl. Acad. Sci. U.S.A.* **96**, 12275–12280
384. Wain, R. L. (1977) *Chem. Soc. Rev.* **6**, 261–276
385. Nickell, L. G. (1978) *Chem. Eng. News* **Oct. 9**, 18–34
386. Thomas-Reetz, A. C., and De Camilli, P. (1994) *FASEB J.* **8**, 209–216
387. Bean, A. J., Zhang, X., and Hökfelt, T. (1994) *FASEB J.* **8**, 630–638
388. Berridge, M. J., Bootman, M. D., and Lipp, P. (1998) *Nature (London)* **395**, 645–648
389. Cancela, J. M., Gerasimenko, O. V., Gerasimenko, J. V., Tepikin, A. V., and Petersen, O. H. (2000) *EMBO J.* **19**, 2549–2557
390. Putney, J. W., Jr. (2001) *Nature (London)* **410**, 648–649
391. Ma, H.-T., Patterson, R. L., van Rossum, D. B., Birnbaumer, L., Mikoshiba, K., and Gill, D. L. (2000) *Science* **287**, 1647–1651
392. Machado, J. D., Segura, F., Brioso, M. A., and Borges, R. (2000) *J. Biol. Chem.* **275**, 20274–20279
393. Siegel, G. J., Albers, R. W., Agronoff, B. W., and Katzman, R. (1981) *Basic Neurochemistry*, 3rd ed., Little, Brown, Boston, Massachusetts
394. entire issue. (1979) *Sci. Am.* **241**(Sept)
395. Eccles, J. C. (1973) *The Understanding of the Brain*, McGraw-Hill, New York
396. McCool, B. A., Plonk, S. G., Martin, P. R., and Singleton, C. K. (1993) *J. Biol. Chem.* **268**, 1397–1404
396a. Johansen, K. M., and Johansen, J. (1995) *J. Neurobiol.* **27**, 227–239
396b. Nicholls, J. G., and Van Essen, D. (1974) *Sci. Am.* **230**(Jan), 38–48
397. Nicholls, J. G., and Van Essen, D. (1974) *Sci. Am.* **230**(Jan), 38–48
398. McKay, R. D. G., Hockfield, S., Johansen, J., Thompson, I., and Frederiksen, K. (1983) *Science* **222**, 788–799
399. Barnes, D. M. (1986) *Science* **233**, 155–156
400. Koch, C., and Laurent, G. (1999) *Science* **284**, 96–98
400a. Kandel, E. R., and Squire, L. R. (2000) *Science* **290**, 1113–1120
401. Brand, E. D., and Westfall, T. C. (1970) in *Medicinal Chemistry*, 3rd ed. (Burger, A., ed), pp. 1190 –1234, Wiley (Interscience), New York (Part II)
402. Porter, K. R., and Bonneville, M. A. (1973) *Fine Structure of Cells and Tissues*, 4th ed., Lea & Febiger, Philadelphia, Pennsylvania
403. Dowling, J. E. (1965) *Science* **147**, 57–59
404. Llinás, R. R. (1975) *Sci. Am.* **232**(Jan), 56–71
405. Noback, C. R., and Demarest, R. J. (1972) *The Nervous System; Introduction and Review*, McGraw–Hill, New York
406. Spector, R., and Johanson, C. E. (1989) *Sci. Am.* **261**(Nov), 68–74
407. Boado, R., Li, J. Y., Nagaya, M., Zhang, C., and Pardridge, W. M. (1999) *Proc. Natl. Acad. Sci. U.S.A.* **96**, 12079–12084
408. Glickstein, M., and Gibson, A. R. (1976) *Sci. Am.* **235**(Nov), 90–98
409. Crick, F., and Jones, E. (1993) *Nature (London)* **361**, 109–110
410. Barinaga, M. (1995) *Science* **268**, 1696–1698
411. Vranesic, I., Iijima, T., Ichikawa, M., Matsumoto, G., and Knöpfel, T. (1994) *Proc. Natl. Acad. Sci. U.S.A.* **91**, 13014–13017
412. McCormick, D. A. (1995) *Nature (London)* **374**, 412–413
413. Aidley, D. J. (1971) *The Physiology of Excitable Cells*, Cambridge Univ. Press, London and New York
414. Adelman, W. J., Jr., ed. (1971) *Biophysics and Physiology of Excitable Membranes*, Van Nostrand–Reinhold, New York
415. Nystrom, R. A. (1973) *Membrane Physiology*, Prentice-Hall, Englewood Cliffs, New Jersey
416. Black, S. (1973) *Adv. Enzymol.* **38**, 193–234
417. Hodgkin, A. L. (1964) *The Conduction of the Nervous Impulse*, Thomas, Springfield, Illinois
418. Hirano, A., and Dembitzer, H. M. A. (1967) *J. Cell Biol.* **34**, 555
419. Weimbs, T., and Stoffel, W. (1994) *Biochemistry* **33**, 10408–10415
420. Griffiths, I., Klugmann, M., Anderson, T., Yool, D., Thomson, C., Schwab, M. H., Schneider, A., Zimmermann, F., McCulloch, M., Nadon, N., and Nave, K.-A. (1998) *Science* **280**, 1610–1613
421. Su, Y., Brooks, D. G., Li, L., Lepercq, J., Trofatter, J. A., Ravetch, J. V., and Lebo, R. V. (1993) *Proc. Natl. Acad. Sci. U.S.A.* **90**, 10856–10860
422. Barinaga, M. (1996) *Science* **273**, 1657–1658
422a. Gallego, R. G., Blanco, J. L. J., Thijssen-van Zuylen, C. W. E. M., Gotfredsen, C. H., Voshol, H., Duus, J. O., Schachner, M., and Vliegenthart, J. F. G. (2001) *J. Biol. Chem.* **276**, 30834–30844

References

423. Cao, M. Y., Dupriez, V. J., Rider, M. H., Deggouj, N., Gersdorff, M. C. H., Rousseau, G. G., and Tomasi, J.-P. (1996) *FASEB J.* **10**, 1635–1640
424. Shapiro, L., Doyle, J. P., Hensley, P., Colman, D. R., and Hendrickson, W. (1996) *Neuron* **17**, 435–440
425. Lemke, G. (1996) *Nature (London)* **383**, 395–396
426. Beniac, D. R., Luckevich, M. D., Czarnota, G. J., Tompkins, T. A., Ridsdale, R. A., Ottensmeyer, F. P., Moscarello, M. A., and Harauz, G. (1997) *J. Biol. Chem.* **272**, 4261–4268
427. Pritzker, L. B., Joshi, S., Harauz, G., and Moscarello, M. A. (2000) *Biochemistry* **39**, 5382–5388
428. Li, Y., Li, H., Martin, R., and Mariuzza, R. A. (2000) *J. Mol. Biol.* **304**, 177–188
428a. Sanders, C. R., Ismail-Beigi, F., and McEnery, M. W. (2001) *Biochemistry* **40**, 9453–9459
428b. Notterpek, L., Roux, K. J., Amici, S. A., Yazdanpour, A., Rahner, C., and Fletcher, B. S. (2001) *Proc. Natl. Acad. Sci. U.S.A.* **98**, 14404–14409
429. Fuhrman, F. A. (1967) *Sci. Am.* **217**(Aug), 61–71
429a. Huey, R. B., and Moody, W. J. (2002) *Science* **297**, 1289–1290
430. Spiro, T. G. (1974) *Acc. Chem. Res.* **7**, 339–344
431. Ritchie, J. M., and Bogart, R. B. (1977) *Proc. Natl. Acad. Sci. U.S.A.* **74**, 211–215
432. Ritchie, J. M. (1973) *Prog. Biophys. and Mol. Biol.* **26**, 149–187
432a. Kaplan, J. H. (2002) *Ann. Rev. Biochem.* **71**, 511–535
433. Catterall, W. A. (1986) *Ann. Rev. Biochem.* **55**, 953–985
434. Catterall, W. A. (1988) *Science* **242**, 50–61
435. Pallaghy, P. K., Duggan, B. M., Pennington, M. W., and Norton, R. S. (1993) *J. Mol. Biol.* **234**, 405–420
436. Shimizu, Y., Chou, H.-N., Bando, H., Van Duyne, G., and Clardy, J. C. (1986) *J. Am. Chem. Soc.* **108**, 514–515
437. Noda, M., Shimizu, S., Tanabe, T., Takai, T., Kayano, T., Ikeda, T., Takahashi, H., Nakayama, H., Kanaoka, Y., Minamino, N., Kangawa, K., Matsuo, H., Raftery, M. A., Hirose, T., Inayama, S., Hayashida, H., Miyata, T., and Numa, S. (1984) *Nature (London)* **312**, 121–127
438. Sato, C., Ueno, Y., Asai, K., Takahashi, K., Sato, M., Engel, A., and Fujiyoshi, Y. (2001) *Nature (London)* **409**, 1047–1051
439. Catterall, W. A. (2001) *Nature (London)* **409**, 988–991
440. Messner, D. J., and Catterall, W. A. (1986) *J. Biol. Chem.* **261**, 211–215
441. Noda, M., Ikeda, T., Kayano, T., Suzuki, H., Takeshima, H., Kurasaki, M., Takahashi, H., and Numa, S. (1986) *Nature (London)* **320**, 188–192
441a. Hilber, K., Sandtner, W., Kudlacek, O., Glaaser, I. W., Weisz, E., Kyle, J. W., French, R. J., Fozzard, H. A., Dudley, S. C., and Todt, H. (2001) *J. Biol. Chem.* **276**, 27831–27839
441b. Tan, H. L., Bink-Boelkens, M. T. E., Bezzina, C. R., Viswanathan, P. C., Beaufort-Krol, G. C. M., van Tintelen, P. J., van den Berg, M. P., Wilde, A. A. M., and Balser, J. R. (2001) *Nature (London)* **409**, 1043–1047
442. Lopreato, G. F., Lu, Y., Southwell, A., Atkinson, N. S., Hillis, D. M., Wilcox, T. P., and Zakon, H. H. (2001) *Proc. Natl. Acad. Sci. U.S.A.* **98**, 7588–7592
442a. Ren, D., Navarro, B., Xu, H., Yue, L., Shi, Q., and Clapham, D. E. (2001) *Science* **294**, 2372–2375
442b. Catterall, W. A. (2001) *Science* **294**, 2306–2308
443. Reichardt, L. F. (1984) *Trends Biochem. Sci.* **9**, 173–176
444. Akerman, K. E. O., and Nicholls, D. G. (1983) *Trends Biochem. Sci.* **8**, 63–64
444a. Kinoshita, M., Nukada, T., Asano, T., Mori, Y., Akaike, A., Satoh, M., and Kaneko, S. (2001) *J. Biol. Chem.* **276**, 28731–28738
444b. Saegusa, H., Kurihara, T., Zong, S., Kazuno, A.-a, Matsuda, Y., Nonaka, T., Han, W., Toriyama, H., and Tanabe, T. (2001) *EMBO J.* **20**, 2349–2356
445. Ellis, S. B., Williams, M. E., Ways, N. R., Brenner, R., Sharp, A. H., Leung, A. T., Campbell, K. P., McKenna, E., Koch, W. J., Hui, A., Schwartz, A., and Harpold, M. M. (1988) *Science* **241**, 1661–1664
445a. Serysheva, I. I., Ludtke, S. J., Baker, M. R., Chiu, W., and Hamilton, S. L. (2002) *Proc. Natl. Acad. Sci. U.S.A.* **99**, 10370–10375
445b. Yang, S.-N., Yu, J., Mayr, G. W., Hofmann, F., Larsson, O., and Berggren, P.-O. (2001) *FASEB J.* **15**, 1753–1763
446. Sharp, A. H., and Campbell, K. P. (1989) *J. Biol. Chem.* **264**, 2816–2825
446a. Zhao, M., Li, P., Li, X., Zhang, L., Winkfein, R. J., and Chen, S. R. W. (1999) *J. Biol. Chem.* **274**, 25971–25974
447. Putney, J. W., Jr. (1999) *Proc. Natl. Acad. Sci. U.S.A.* **96**, 14669–14671
448. Zhao, X.-S., Shin, D. M., Liu, L. H., Shull, G. E., and Muallem, S. (2001) *EMBO J.* **20**, 2680–2689
449. Churchill, G. C., and Galione, A. (2001) *EMBO J.* **20**, 2666–2671
450. Blount, P., Sukharev, S. I., Moe, P. C., Schroeder, M. J., Guy, H. R., and Kung, C. (1996) *EMBO J.* **15**, 4798–4805
451. Oakley, A. J., Martinac, B., and Wilce, M. C. J. (1999) *Protein Sci.* **8**, 1915–1921
452. Chang, G., Spencer, R. H., Lee, A. T., Barclay, M. T., and Rees, D. C. (1998) *Science* **282**, 2220–2226
452a. Abbott, A. (2002) *Nature (London)* **418**, 268–269
453. Li, H. L., Sui, H. X., Ghanshani, S., Lee, S., Walian, P. J., Wu, C. L., Chandy, K. G., and Jap, B. K. (1998) *J. Mol. Biol.* **282**, 211–216
454. Caprini, M., Ferroni, S., Planells-Cases, R., Rueda, J., Rapisarda, C., Ferrer-Montiel, A., and Montal, M. (2001) *J. Biol. Chem.* **276**, 21070–21076
455. Katz, A. M. (1993) *N. Engl. J. Med.* **328**, 1244–1251
455a. Yellen, G. (2002) *Nature (London)* **419**, 35–42
456. Choe, S., Kreusch, A., and Pfaffinger, P. J. (1999) *Trends Biochem. Sci.* **24**, 345–349
457. Jing, J., Chikvashvili, D., Dinger-Lahat, D., Thornhill, W. B., Reuveny, E., and Lotan, I. (1999) *EMBO J.* **18**, 1245–1256
458. Aldrich, R. W. (2001) *Nature (London)* **411**, 643–644
458a. Gulbis, J. M., Zhou, M., Mann, S., and MacKinnon, R. (2000) *Science* 289, 123–127
459. Zhou, M., Morais-Cabral, J. H., Mann, S., and MacKinnon, R. (2001) *Nature (London)* **411**, 657–661
459a. Strang, C., Cushman, S. J., DeRubeis, D., Peterson, D., and Pfaffinger, P. J. (2001) *J. Biol. Chem.* **276**, 28493–28502
459b. Hanlon, M. R., and Wallace, B. A. (2002) *Biochemistry* **41**, 2886–2894
460. Roux, B., Bernéche, S., and Im, W. (2000) *Biochemistry* **39**, 13295–13306
461. Roux, B., and MacKinnon, R. (1999) *Science* **285**, 100–102
462. Zagrovic, B., and Aldrich, R. (1999) *Science* **285**, 59–61
463. Perozo, E., Cortes, D. M., and Cuello, L. G. (1999) *Science* **285**, 73–78
464. Guidoni, L., Torre, V., and Carloni, P. (1999) *Biochemistry* **38**, 8599–8604
464a. Zhou, Y., Morais-Cabral, J. H., Kaufman, A., and MacKinnon, R. (2001) *Nature (London)* **414**, 43–48
464b. Morais-Cabral, J. H., Zhou, Y., and MacKinnon, R. (2001) *Nature (London)* **414**, 37–42
464c. Bernèche, S., and Roux, B. (2001) *Nature (London)* **414**, 73–76
464d. Rivas, J. C. M., Schwalbe, H., and Lippard, S. J. (2001) *Proc. Natl. Acad. Sci. U.S.A.* **98**, 9478–9483
465. Brenner, R., Peréz, G. J., Bonev, A. D., Eckman, D. M., Kosek, J. C., Wiler, S. W., Patterson, A. J., Nelson, M. T., and Aldrich, R. W. (2000) *Nature (London)* **407**, 870–871
466. Tian, L., Duncan, R. R., Hammond, M. S. L., Coghill, L. S., Wen, H., Rusinova, R., Clark, A. G., Levitan, I. B., and Shipston, M. J. (2001) *J. Biol. Chem.* **276**, 7717–7720
466a. Jiang, Y., Lee, A., Chen, J., Cadene, M., Chait, B. T., and MacKinnon, R. (2002) *Nature (London)* **417**, 515–522
466b. Schumacher, M. A., Rivard, A. F., Bächinger, H. P., and Adelman, J. P. (2001) *Nature (London)* **410**, 1120–1124
467. Wang, H.-S., Pan, Z., Shi, W., Brown, B. S., Wymore, R. S., Cohen, I. S., Dixon, J. E., and McKinnon, D. (1998) *Science* **282**, 1890–1893
468. Abraham, M. R., Jahangir, A., Alekseev, A. E., and Terzic, A. (1999) *FASEB J.* **13**, 1901–1910
469. Maingret, F., Patel, A. J., Lesage, F., Lazdunski, M., and Honoré, E. (1999) *J. Biol. Chem.* **274**, 26691–26696
470. Lu, Z., and MacKinnon, R. (1995) *Biochemistry* **34**, 13133–13138
471. Qu, Z., Yang, Z., Cui, N., Zhu, G., Liu, C., Xu, H., Chanchevalap, S., Shen, W., Wu, J., Li, Y., and Jiang, C. (2000) *J. Biol. Chem.* **275**, 31573–31580
472. Repunte, V. P., Nakamura, H., Fujita, A., Horio, Y., Findlay, I., Pott, L., and Kurachi, Y. (1999) *EMBO J.* **18**, 3317–3324
473. Tucker, S. J., Gribble, F. M., Proks, P., Trapp, S., Ryder, T. J., Haug, T., Reimman, F., and Ashcroft, F. M. (1998) *EMBO J.* **17**, 3290–3296
474. Carrasco, A. J., Dzeja, P. P., Alekseev, A. E., Pucar, D., Zingman, L. V., Abraham, M. R., Hodgson, D., Bienengraeber, M., Puceat, M., Janssen, E., Wieringa, B., and Terzic, A. (2001) *Proc. Natl. Acad. Sci. U.S.A.* **98**, 7623–7628
474a. Loussouarn, G., Pike, L. J., Ashcroft, F. M., Makhina, E. N., and Nichols, C. G. (2001) *J. Biol. Chem.* **276**, 29098–29103
474b. Moreau, C., Jacquet, H., Prost, A.-L., D'hahan, N., and Vivaudou, M. (2000) *EMBO J.* **19**, 6644–6651
475. Rogalski, S. L., and Chavkin, C. (2001) *J. Biol. Chem.* **276**, 14855–14860
476. Lemtiri-Chlieh, F., MacRobbie, E. A. C., and Brearley, C. A. (2000) *Proc. Natl. Acad. Sci. U.S.A.* **97**, 8687–8692
477. Petalcorin, M. I. R., Oka, T., Koga, M., Ogura, K.-i, Wada, Y., Ohshima, Y., and Futai, M. (1999) *J. Mol. Biol.* **294**, 347–355
477a. Harrop, S. J., DeMaere, M. Z., Fairlie, W. D., Reztsova, T., Valenzuela, S. M., Mazzanti, M., Tonini, R., Qiu, M. R., Jankova, L., Warton, K., Bauskin, A. R., Wu, W. M., Pankhurst, S., Campbell, T. J., Breit, S. N., and Curmi, P. M. G. (2001) *J. Biol. Chem.* **276**, 44993–45000
477b. Dutzler, R., Campbell, E. B., Cadene, M., Chait, B. T., and MacKinnon, R. (2002) *Nature (London)* **415**, 287–294
477c. Jentsch, T. J. (2002) *Nature (London)* **415**, 276–277
478. George, A. L., Jr. (1998) *Proc. Natl. Acad. Sci. U.S.A.* **95**, 7843–7845
479. Cooper, E. C., and Jan, L. Y. (1999) *Proc. Natl. Acad. Sci. U.S.A.* **96**, 4759–4766
480. Ho, M. W. Y., Kaetzel, M. A., Armstrong, D. L., and Shears, S. B. (2001) *J. Biol. Chem.* **276**, 18673–18680
480a. Zou, X., and Hwang, T.-C. (2001) *Biochemistry* **40**, 5579–5586
481. Eldefrawi, A. T., and Eldefrawi, M. E. (1987) *FASEB J.* **1**, 262–271

References

481a. Smith, M. M., Warren, V. A., Thomas, B. S., Brochu, R. M., Ertel, E. A., Rohrer, S., Schaeffer, J., Schmatz, D., Petuch, B. R., Tang, Y. S., Meinke, P. T., Kaczorowski, G. J., and Cohen, C. J. (2000) *Biochemistry* **39**, 5543–5554

482. Llinas, R. R. (2000) *The Squid Synapse: A Model for Chemical Transmission*, Oxford Univ. Press, London

483. Saimi, Y., Hinrichsen, R. D., Forte, M., and Kung, C. (1983) *Proc. Natl. Acad. Sci. U.S.A.* **80**, 5112–5116

484. Woolley, G. A., and Wallace, B. A. (1993) *Biochemistry* **32**, 9819–9825

484a. Cafiso, D. S. (1994) *Annu. Rev. Biophys. Biomol. Struct.* **23**, 141–165

485. Cleveland, MvB., Slatin, S., Finkelstein, A., and Levinthal, C. (1983) *Proc. Natl. Acad. Sci. U.S.A.* **80**, 3706–3710

486. Nicoll, R. A. (1988) *Science* **241**, 545–551

487. Olivera, B. M., Rivier, J., Clark, C., Ramilo, C. A., Corpuz, G. P., Abogadie, F. C., Mena, E. E., Woodward, S. R., Hillyard, D. R., and Cruz, L. J. (1990) *Science* **249**, 257–263

488. McIntosh, J. M., Santos, A. D., and Olivera, B. M. (1999) *Ann. Rev. Biochem.* **68**, 59–88

489. Fainzilber, M., Nakamura, T., Lodder, J. C., Zlotkin, E., Kits, K. S., and Burlingame, A. L. (1998) *Biochemistry* **37**, 1470–1477

490. Lirazan, M. B., Hooper, D., Corpuz, G. P., Ramilo, C. A., Bandyopadhyay, P., Cruz, L. J., and Olivera, B. M. (2000) *Biochemistry* **39**, 1583–1588

491. Rigby, A. C., Lucas-Meunier, E., Kalume, D. E., Czerwiec, E., Hambe, B., Dahlqvist, I., Fossier, P., Baux, G., Roepstorff, P., Baleja, J. D., Furie, B. C., Furie, B., and Stenflo, J. (1999) *Proc. Natl. Acad. Sci. U.S.A.* **96**, 5758–5763

492. Shon, K.-J., Grilley, M. M., Marsh, M., Yoshikami, D., Hall, A. R., Kurz, B., Gray, W. R., Imperial, J. S., Hillyard, D. R., and Olivera, B. M. (1995) *Biochemistry* **34**, 4913–4918

493. McIntosh, J. M., Olivera, B. M., Cruz, L. J., and Gray, W. R. (1984) *J. Biol. Chem.* **259**, 14343–14346

494. Jover, E., Bablito, J., and Couraud, F. (1984) *Biochemistry* **23**, 1147–1152

494a. Almassy, R. J., Fontecilla-Camps, J. C., Suddath, F. L., Bugg, C. E. (1983) *J. Mol. Biol.* **170**, 497–527

495. Schweitz, H., Bidard, J.-N., Frelin, C., Pauron, D., Vijverberg, H. P. M., Mahasneh, D. M., and Lazdunski, M. (1985) *Biochemistry* **24**, 3554–3561

496. Shimizu, Y., Hsu, C.-P., and Genenah, A. (1981) *J. Am. Chem. Soc.* **103**, 605–609

497. Kao, C. Y., and Levinson, S. R. (1986) *Ann. N.Y. Acad. Sci.* **479**, entire volume

498. Morabito, M. A., and Moczydlowski, E. (1994) *Proc. Natl. Acad. Sci. U.S.A.* **91**, 2478–2482

499. Ovchinnikov, Y. A., and Grishin, E. V. (1982) *Trends Biochem. Sci.* **7**, 26–28

500. Jablonsky, M. J., Jackson, P. L., and Krishna, N. R. (2001) *Biochemistry* **40**, 8273–8282

501. Tugarinov, V., Kustanovich, I., Zilberberg, N., Gurevitz, M., and Anglister, J. (1997) *Biochemistry* **36**, 2414–2424

502. Gurevitz, M., Gordon, D., Ben-Natan, S., Turkov, M., and Froy, O. (2001) *FASEB J.* **15**, 1201–1205

503. Fontecilla-Camps, J. C., Almassy, R. J., Ealick, S. E., Suddath, F. L., Watt, D. D., Feldmann, R. J., and Bugg, C. E. (1981) *Trends Biochem. Sci.* **6**, 291–296

504. Fogh, R. H., Kem, W. R., and Norton, R. S. (1990) *J. Biol. Chem.* **265**, 13016–13028

505. Loret, E. P., Menendez Soto del Valle, R., Mansuelle, P., Sampieri, F., and Rochat, H. (1994) *J. Biol. Chem.* **269**, 16785–16788

506. Ellis, K. C., Tenenholz, T. C., Jerng, H., Hayhurst, M., Dudlak, C. S., Gilly, W. F., Blaustein, M. P., and Weber, D. J. (2001) *Biochemistry* **40**, 5942–5953

507. Delepierre, M., Prochnicka-Chalufour, A., Boisbouvier, J., and Possani, L. D. (1999) *Biochemistry* **38**, 16756–16765

508. Baden, D. G. (1989) *FASEB J.* **3**, 1807–1817

509. Duffton, M. J., and Hider, R. C. (1980) *Trends Biochem. Sci.* **5**, 53–56

510. Achari, A., Radvanyi, F. R., Scott, D., Bon, C., and Sigler, P. B. (1985) *J. Biol. Chem.* **260**, 9385–9387

510a. Délot, E., and Bon, C. (1993) *Biochemistry* **32**, 10708–10713

511. Juillerat, M. A., Schwendimann, B., Hauert, J., Fulpius, B. W., and Bargetzi, J. P. (1982) *J. Biol. Chem.* **257**, 2901–2907

512. Drenth, J., Low, B. W., Richardson, J. S., and Wright, C. S. (1980) *J. Biol. Chem.* **255**, 2652–2655

513. Hatanaka, H., Oka, M., Kohda, D., Tate, S-i, Suda, A., Tamiya, N., and Inagaki, F. (1994) *J. Mol. Biol.* **240**, 155–166

514. Corfield, P. W. R., Lee, T.-J., and Low, B. W. (1989) *J. Biol. Chem.* **264**, 9239–9242

515. Basus, V. J., Billeter, M., Love, R. A., Stroud, R. M., and Kuntz, I. D. (1988) *Biochemistry* **27**, 2763–2771

516. Ruoppolo, M., Moutiez, M., Mazzeo, M. F., Pucci, P., Ménez, A., Marino, G., and Quéméneur, E. (1998) *Biochemistry* **37**, 16060–16068

517. Bilwes, A., Rees, B., Moras, D., Ménez, R., and Ménez, A. (1994) *J. Mol. Biol.* **239**, 122–136

518. Sun, Y.-J., Wu, W.-g, Chiang, C.-M., Hsin, A.-Y., and Hsiao, C.-D. (1997) *Biochemistry* **36**, 2403–2413

519. Rehm, H., and Betz, H. (1982) *J. Biol. Chem.* **257**, 10015–10022

520. Montecucco, C., and Rossetto, O. (2000) *Trends Biochem. Sci.* **25**, 266–270

521. Montecucco, C. (1986) *Trends Biochem. Sci.* **11**, 314–317

522. Montecucco, C., and Schiavo, G. (1993) *Trends Biochem. Sci.* **18**, 324–327

523. Lacy, D. B., and Stevens, R. C. (1999) *J. Mol. Biol.* **291**, 1091–1104

523a. Eswaramoorthy, S., Kumaran, D., and Swaminathan, S. (2002) *Biochemistry* **41**, 9795–9802

524. Eisel, U., Jarausch, W., Goretzki, K., Henschen, A., Engels, J., Weller, U., Hudel, M., Habermann, E., and Niemann, H. (1986) *EMBO J.* **5**, 2495–2502

525. Fu, F.-N., Lomneth, R. B., Cai, S., and Singh, B. R. (1998) *Biochemistry* **37**, 5267–5278

526. Keller, J. E., and Neale, E. A. (2001) *J. Biol. Chem.* **276**, 13476–13482

527. Turton, K., Chaddock, J. A., and Acharya, K. R. (2002) *Trends Biochem. Sci.* **27**, 552–558

528. Blasi, J., Chapman, E. R., Link, E., Binz, T., Yamasaki, S., De Camilli, P., Südhof, T. C., Niemann, H., and Jahn, R. (1993) *Nature (London)* **365**, 160–163

528a. Li, Y., Foran, P., Lawrence, G., Mohammed, N., Chan-Kwo-Chion, C.-K.-N., Lisk, G., Aoki, R., and Dolly, O. (2001) *J. Biol. Chem.* **276**, 31394–31401

529. Matteoli, M., Verderio, C., Rossetto, O., Iezzi, N., Coco, S., Schiavo, G., and Montecucco, C. (1996) *Proc. Natl. Acad. Sci. U.S.A.* **93**, 13310–13315

530. Nicholls, D. G., Rugolo, M., Scott, I. G., and Meldolesi, J. (1982) *Proc. Natl. Acad. Sci. U.S.A.* **79**, 7924–7928

531. Khvotchev, M., and Südhof, T. C. (2000) *EMBO J.* **19**, 3250–3262

532. Tufariello, J. J., Meckler, H., Pushpananda, K., and Senartne, A. (1984) *J. Am. Chem. Soc.* **106**, 7979–7980

533. Guddat, L. W., Martin, J. A., Shan, L., Edmundson, A. B., and Gray, W. R. (1996) *Biochemistry* **35**, 11329–11335

534. Hu, S.-H., Loughnan, M., Miller, R., Weeks, C. M., Blessing, R. H., Alewood, P. F., Lewis, R. J., and Martin, J. L. (1998) *Biochemistry* **37**, 11425–11433

535. Klein, L. L., McWhorter, W. W., Jr., Ko, S. S., Pfaff, K.-P., and Kishi, Y. (1982) *J. Am. Chem. Soc.* **104**, 7362–7364

536. Shimizu, Y. (1983) *Nature (London)* **302**, 212

536a. Raichle, M. E., and Gusnard, D. A. (2002) *Proc. Natl. Acad. Sci. U.S.A.* **99**, 10237–10239

536b. Job, C., and Eberwine, J. (2001) *Proc. Natl. Acad. Sci. U.S.A.* **98**, 13037–13042

537. Mohr, E., Prakash, N., Vieluf, K., Fuhrmann, C., Buck, F., and Richter, D. (2001) *Proc. Natl. Acad. Sci. U.S.A.* **98**, 7072–7079

538. Terada, S., Nakata, T., Peterson, A. C., and Hirokawa, N. (1996) *Science* **273**, 784–788

539. Caplan, R., Cheung, S. C.-Y., and Omenn, G. S. (1974) *J. Neurochemistry* **22**, 517–520

540. Sutcliffe, J. G., and Milner, R. J. (1984) *Trends Biochem. Sci.* **9**, 95–99

541. Geschwind, D. H. (2000) *Proc. Natl. Acad. Sci. U.S.A.* **97**, 10676–10678

542. Parra, M., Gascard, P., Walensky, L. D., Gimm, J. A., Blackshaw, S., Chan, N., Takakuwa, Y., Berger, T., Lee, G., Chasis, J. A., Snyder, S. H., Mohandas, N., and Conboy, J. G. (2000) *J. Biol. Chem.* **275**, 3247–3255

543. Kim, S.-Y., Grant, P., Lee, J.-H., Pant, H. C., and Steinert, P. M. (1999) *J. Biol. Chem.* **274**, 30715–30721

544. Hikita, T., Tadano-Aritomi, K., Iida-Tanaka, N., Anand, J. K., Ishizuka, I., and Hakomori, S.-i. (2001) *J. Biol. Chem.* **276**, 23084–23091

545. Sutcliffe, J. G., Milner, R. J., Gottesfeld, J. M., and Reynolds, W. (1984) *Science* **225**, 1308–1315

546. Owens, G. P., Chaudhari, N., and Hahn, W. E. (1985) *Science* **229**, 1263–1265

547. Lone, Y.-C., Simon, M.-P., Kahn, A., and Marie, J. (1986) *J. Biol. Chem.* **261**, 1499–1502

548. Eccles, J. (1965) *Sci. Am.* **212**(Jan), 56–66

549. Cowan, W. M., Südhof, T. C., and Stevens, C. F., eds. (2001) *Synapses*, The Johns Hopkins Univ. Press, Baltimore

550. Kalil, R. E. (1989) *Sci. Am.* **261**(Dec), 76–85

551. Hall, Z. W. (1972) *Ann. Rev. Biochem.* **41**, 925–952

552. Krnjevic, K. (1971) *Methods Neurochem.* **1**, 129–172

553. Katz, B. (1971) *Science* **173**, 123–126

554. Südhof, T. C., and Scheller, R. H. (2000) in *Synapses* (Cowan, W. M., Südhof, T. C., and Stevens, C. F., eds), pp. 177–215, Johns Hopkins Univ. Press, Baltimore, Maryland

555. Südhof, T. C. (1995) *Nature (London)* **375**, 645–653

556. Bajjalieh, S. M., and Scheller, R. H. (1995) *J. Biol. Chem.* **270**, 1971–1974

557. Bennett, M. K., and Scheller, R. H. (1994) *Ann. Rev. Biochem.* **63**, 63–100

558. Klingauf, J., Kavalali, E. T., and Tsien, R. W. (1998) *Nature (London)* **394**, 581–585

559. Wang, L.-Y., and Kaczmarek, L. K. (1998) *Nature (London)* **394**, 384–388

560. Zucker, R. S., and Landó, L. (1986) *Science* **231**, 574–579

561. Greengard, P., Valtorta, F., Czernik, A. J., and Benfenati, F. (1993) *Science* **259**, 780–785

562. Jahn, R., and Südhof, T. C. (1999) *Ann. Rev. Biochem.* **68**, 863–911

563. Sutton, R. B., Fasshauer, D., Jahn, R., and Brunger, A. T. (1998) *Nature (London)* **395**, 347–353

564. Kimelberg, H. K., and Norenberg, M. D. (1989) *Sci. Am.* **260**(Apr), 66–76

References

565. Takahashi, T., Hori, T., Kajikawa, Y., and Tsujimoto, T. (2000) *Science* **289**, 460–463
566. Erickson, J. D., Varoqui, H., Schäfer, M. K.-H., Modi, W., Diebler, M.-F., Weihe, E., Rand, J., Eiden, L. E., Bonner, T. I., and Usdin, T. B. (1994) *J. Biol. Chem.* **269**, 21929–21932
567. Hartwell, L. H., and Kastan, M. B. (1994) *Science* **266**, 1821–1828
568. Whitley, P., Grahn, E., Kutay, U., Rapoport, T. A., and von Heijne, G. (1996) *J. Biol. Chem.* **271**, 7583–7586
568a. Schoch, S., Deák, F., Königstorfer, A., Mozhayeva, M., Sara, Y., Südhof, T. C., and Kavalali, E. T. (2001) *Science* **294**, 1117–1122
568b. Misura, K. M. S., Bock, J. B., Gonzalez, L. C., Jr., Scheller, R. H., and Weis, W. I. (2002) *Proc. Natl. Acad. Sci. U.S.A.* **99**, 9184–9189
569. Weis, W. I., and Scheller, R. H. (1998) *Nature (London)* **395**, 328–329
570. Tsujimoto, S., and Bean, A. J. (2000) *J. Biol. Chem.* **275**, 2938–2942
571. Vogel, K., Cabaniols, J.-P., and Roche, P. A. (2000) *J. Biol. Chem.* **275**, 2959–2965
572. Betz, A., Okamoto, M., Benseler, F., and Brose, N. (1997) *J. Biol. Chem.* **272**, 2520–2526
572a. Augustin, I., Rosenmund, C., Südhof, T. C., and Brose, N. (1999) *Nature (London)* **400**, 457–460
573. Puffer, E. B., Lomneth, R. B., Sarkar, H. K., and Singh, B. R. (2001) *Biochemistry* **40**, 9374–9378
574. May, A. P., Whiteheart, S. W., and Weis, W. I. (2001) *J. Biol. Chem.* **276**, 21991–21994
574a. Kim, C. S., Kweon, D.-H., and Shin, Y.-K. (2002) *Biochemistry* **41**, 10928–10933
575. Dulubova, I., Sugita, S., Hill, S., Hosaka, M., Fernandez, I., Südhof, T. C., and Rizo, J. (1999) *EMBO J.* **18**, 4372–4382
576. Ubach, J., Lao, Y., Fernandez, I., Arac, D., Südhof, T. C., and Rizo, J. (2001) *Biochemistry* **40**, 5854–5860
577. Ubach, J., Zhang, X., Shao, X., Südhof, T. C., and Rizo, J. (1998) *EMBO J.* **17**, 3921–3930
578. Fukuda, M., Kojima, T., and Mikoshiba, K. (1996) *J. Biol. Chem.* **271**, 8430–8434
578a. Fernández-Chacón, R., Königstorfer, A., Gerber, S. H., Garcia, J., Matos, M. F., Stevens, C. F., Brose, N., Rizo, J., Rosenmund, C., and Südhof, T. C. (2001) *Nature (London)* **410**, 41–49
578b. Coppola, T., Magnin-Lüthi, S., Perret-Menoud, V., Gattesco, S., Schiavo, G., and Regazzi, R. (2001) *J. Biol. Chem.* **276**, 32756–32762
579. Ushkaryov, Y. A., Petrenko, A. G., Geppert, M., and Südhof, T. C. (1992) *Science* **257**, 50–56
580. Gundersen, C. B., Mastrogiacomo, A., Faull, K., and Umbach, J. A. (1994) *J. Biol. Chem.* **269**, 19197–19199
581. Chamberlain, L. H., and Burgoyne, R. D. (1996) *J. Biol. Chem.* **271**, 7320–7323
582. Vitale, N., Caumont, A.-S., Chasserot-Golaz, S., Du, G., Wu, S., Sciorra, V. A., Morris, A. J., Frohman, M. A., and Bader, M.-F. (2001) *EMBO J.* **20**, 2424–2434
583. Ryan, T. A., Smith, S. J., and Reuter, H. (1996) *Proc. Natl. Acad. Sci. U.S.A.* **93**, 5567–5571
584. Schmidt, A., Wolde, M., Thiele, C., Fest, W., Kratzin, H., Podtelejnikov, A. V., Witke, W., Huttner, W. B., and Söling, H.-D. (1999) *Nature (London)* **401**, 133–141
585. Hayashi, T., Yamasaki, S., Nauenburg, S., Binz, T., and Niemann, H. (1995) *EMBO J.* **14**, 2317–2325
586. Südhof, T. C., Czernik, A. J., Kao, H.-T., Takei, K., Johnston, P. A., Horiuchi, A., Kanazir, S. D., Wagner, M. A., Perin, M. S., De Camilli, P., and Greengard, P. (1989) *Science* **245**, 1474–1480
587. Hosaka, M., and Südhof, T. C. (1999) *J. Biol. Chem.* **274**, 16747–16753
588. Johnston, P. A., and Südhof, T. C. (1990) *J. Biol. Chem.* **265**, 8869–8873
589. Hilton, J. M., Plomann, M., Ritter, B., Modregger, J., Freeman, H. N., Falck, J. R., Krishna, U. M., and Tobin, A. B. (2001) *J. Biol. Chem.* **276**, 16341–16347
590. Huang, E. P. (1997) *Proc. Natl. Acad. Sci. U.S.A.* **94**, 13386–13387
591. Palmiter, R. D., Cole, T. B., Quaife, C. J., and Findley, S. D. (1996) *Proc. Natl. Acad. Sci. U.S.A.* **93**, 14934–14939
592. Hemmings, H. C., Jr., Nairn, A. C., McGuinness, T. L., Huganir, R. L., and Greengard, P. (1989) *FASEB J.* **3**, 1583–1592
593. Kennedy, M. B. (2000) *Science* **290**, 750–754
594. Sheng, M. (2001) *Proc. Natl. Acad. Sci. U.S.A.* **98**, 7058–7061
595. Malenka, R. C., and Nicoll, R. A. (1998) *Nature (London)* **396**, 414–415
596. Trautmann, A., and Vivier, E. (2001) *Science* **292**, 1667–1668
597. Lin, W., Burgess, R. W., Dominguez, B., Pfaff, S. L., Sanes, J. R., and Lee, K.-F. (2001) *Nature (London)* **410**, 1057–1064
598. Khan, A. A., Bose, C., Yam, L. S., Soloski, M. J., and Rupp, F. (2001) *Science* **292**, 1681–1686
599. Qi, S. Y., Groves, J. T., and Chakraborty, A. K. (2001) *Proc. Natl. Acad. Sci. U.S.A.* **98**, 6548–6553
600. Hall, Z. W. (1995) *Science* **269**, 362–363
601. Travis, J. (1994) *Science* **266**, 970–972
602. Temburni, M. K., and Jacob, M. H. (2001) *Proc. Natl. Acad. Sci. U.S.A.* **98**, 3631–3632
603. Lino, M., Goto, K., Kakegawa, W., Okado, H., Sudo, M., Ishiuchi, S., Miwa, A., Yakaysau, Y., Saito, I., Tsuzuki, K., and Ozawa, S. (2001) *Science* **292**, 926–927
603a. Kast, B. (2001) *Nature (London)* **412**, 674–676
604. Ullian, E. M., Sapperstein, S. K., Christopherson, K. S., and Barres, B. A. (2001) *Science* **291**, 657–661
604a. Song, H., Stevens, C. F., and Gage, F. H. (2002) *Nature (London)* **417**, 39–44
604b. Svendsen, C. N. (2002) *Nature (London)* **417**, 29–32
605. Guadaño-Ferraz, A., Obregón, M. J., St. Germain, D. L., and Bernal, J. (1997) *Proc. Natl. Acad. Sci. U.S.A.* **94**, 10391–10396
606. Veenstra, R. D., and DeHaan, R. L. (1986) *Science* **233**, 972–974
607. Revel, J. P., Yancey, S. B., and Nicholson, B. J. (1986) *Trends Biochem. Sci.* **11**, 375–377
608. Zimmer, D. B., Green, C. R., Evans, W. H., and Gilula, N. B. (1987) *J. Biol. Chem.* **262**, 7751–7763
609. Perkins, G., Goodenough, D., and Sosinsky, G. (1997) *Biophys. J.* **72**, 533–544
610. Sosinsky, G. (1995) *Proc. Natl. Acad. Sci. U.S.A.* **92**, 9210–9214
611. Goodenough, D. A., Goliger, J. A., and Paul, D. L. (1996) *Ann. Rev. Biochem.* **65**, 475–502
611a. Unger, V. M., Kumar, N. M., Gilula, N. B., and Yeager, M. (1999) *Science* **283**, 1176–1180
612. Bergoffen, J., Scherer, S. S., Wang, S., Scott, M. O., Bone, L. J., Paul, D. L., Chen, K., Lensch, M. W., Chance, P. F., and Fischbeck, K. H. (1993) *Science* **262**, 2039–2042
613. George, C. H., Kendall, J. M., Campbell, A. K., and Evans, W. H. (1998) *J. Biol. Chem.* **273**, 29822–29829
613a. del Castillo, I., Villamar, M., Moreno-Pelayo, M. A., del Castillo, F. J., Alvarez, A., Telleria, D., Menendez, I., and Moreno, F. (2002) *N. Engl. J. Med.* **346**, 243–249
614. Wong, V., and Goodenough, D. A. (1999) *Science* **285**, 62
615. Simon, D. B., Lu, Y., Choate, K. A., Velazquez, H., Al-Sabban, E., Praga, M., Casari, G., Bettinelli, A., Colussi, G., Rodriguez-Soriano, J., McCredie, D., Milford, D., Sanjad, S., and Lifton, R. P. (1999) *Science* **285**, 103–106
616. Osborne, N. N. (1977) *Nature (London)* **270**, 622–623
617. Chan-Palay, V., Engel, A. G., Wu, J.-Y., and Palay, S. L. (1982) *Proc. Natl. Acad. Sci. U.S.A.* **79**, 7027–7030
618. Geha, R. S., and Rosen, F. S. (1994) *N. Engl. J. Med.* **330**, 1008–1009
619. Hermsen, B., Stetzer, E., Thees, R., Heiermann, R., Schrattenholz, A., Ebbinghaus, U., Kretschmer, A., Methfessel, C., Reinhardt, S., and Maelicke, A. (1998) *J. Biol. Chem.* **273**, 18394–18404
620. Fambrough, D. M., and Hartzell, H. C. (1972) *Science* **176**, 189–191
621. Changeux, J.-P. (1993) *Sci. Am.* **269**(Nov), 58–62
622. Brisson, A., and Unwin, P. N. T. (1985) *Nature (London)* **315**, 474–477
623. Sakmann, B., Methfessel, C., Mishina, M., Takahashi, T., Takai, T., Kurasaki, M., Fukuda, K., and Numa, S. (1985) *Nature (London)* **318**, 538–543
624. Miyazawa, A., Fujiyoshi, Y., Stowell, M., and Unwin, N. (1999) *J. Mol. Biol.* **288**, 765–786
625. Arias, H. R. (1998) *Biochim. Biophys. Acta.* **1376**, 173–220
626. Grutter, T., and Changeux, J.-P. (2001) *Trends Biochem. Sci.* **26**, 459–463
626a. Bezakova, G., Rabben, I., Sefland, I., Fumagalli, G., and Lømo, T. (2001) *Proc. Natl. Acad. Sci. U.S.A.* **98**, 9924–9929
627. Brejc, K., van Dijk, W. J., Klaassen, R. V., Schuurmans, M., van der Oost, J., Smit, A. B., and Sixma, T. K. (2001) *Nature (London)* **411**, 269–276
628. Barrantes, F. J., Antollini, S. S., Blanton, M. P., and Prieto, M. (2000) *J. Biol. Chem.* **275**, 37333–37339
629. Osaka, H., Malany, S., Molles, B. E., Sine, S. M., and Taylor, P. (2000) *J. Biol. Chem.* **275**, 5478–5484
630. Salpeter, M. M. (1999) *Science* **286**, 424–425
631. Boulter, J., Evans, K., Goldman, D., Martin, G., Treco, D., Heinemann, S., and Patrick, J. (1986) *Nature (London)* **319**, 368–374
632. Vicente-Agullo, F., Rovira, J. C., Sala, S., Sala, F., Rodriguez-Ferrer, C., Campos-Caro, A., Criado, M., and Ballesta, J. J. (2001) *Biochemistry* **40**, 8300–8306
633. Osaka, H., Sugiyama, N., and Taylor, P. (1998) *J. Biol. Chem.* **273**, 12758–12765
634. Labarca, C., Schwarz, J., Deshpande, P., Schwarz, S., Nowak, M. W., Fonck, C., Nashmi, R., Kofuji, P., Dang, H., Shi, W., Fidan, M., Khakh, B. S., Chen, Z., Bowers, B. J., Boulter, J., Wehner, J. M., and Lester, H. A. (2001) *Proc. Natl. Acad. Sci. U.S.A.* **98**, 2786–2791
635. Jeanclos, E. M., Lin, L., Treuil, M. W., Rao, J., DeCoster, M. A., and Anand, R. (2001) *J. Biol. Chem.* **276**, 28281–28290
636. Tierney, M. L., and Unwin, N. (2000) *J. Mol. Biol.* **303**, 185–196
637. Avizonis, D. Z., Farr-Jones, S., Kosen, P. A., and Basus, V. J. (1996) *J. Am. Chem. Soc.* **118**, 13031–13039
638. Grutter, T., Goeldner, M., and Kotzyba-Hibert, F. (1999) *Biochemistry* **38**, 7476–7484
639. Dougherty, D. A., and Lester, H. A. (2001) *Nature (London)* **411**, 252–255
640. Unwin, N. (1995) *Nature (London)* **373**, 37–43
641. Swope, S. L., Moss, S. J., Blackstone, C. D., and Huganir, R. L. (1992) *FASEB J.* **6**, 2514–2523
642. Hass, R., Marshall, T. L., and Rosenberry, T. L. (1988) *Biochemistry* **27**, 6453–6457
643. Arnon, R., Silman, I., and Tarrab-Hazdai, R. (1999) *Protein Sci.* **8**, 2553–2561
644. Perrier, A. L., Cousin, X., Boschetti, N., Haas, R., Chatel, J.-M., Bon, S., Roberts, W. L., Pickett, S. R., Massoulié, J., Rosenberry, T. L., and Krejci, E. (2000) *J. Biol. Chem.* **275**, 34260–34265
645. Simon, S., Krejci, E., and Massoulié, J. (1998) *EMBO J.* **17**, 6178–6187

References

646. Kaplan, D., Ordentlich, A., Barak, D., Ariel, N., Kronman, C., Velan, B., and Shafferman, A. (2001) *Biochemistry* **40**, 7433–7445
647. Blusztajn, J. K., and Wurtman, R. J. (1983) *Science* **221**, 614–620
648. Kubo, T., Fukuda, K., Mikami, A., Maeda, A., Takahashi, H., Mishina, M., Haga, T., Haga, K., Ichiyama, A., Kangawa, K., Kojima, M., Matsuo, H., Hirose, T., and Numa, S. (1986) *Nature (London)* **323**, 411–416
649. Gainetdinov, R. R., and Caron, M. G. (1999) *Proc. Natl. Acad. Sci. U.S.A.* **96**, 12222–12223
650. Nathanson, N. M. (2000) *Proc. Natl. Acad. Sci. U.S.A.* **97**, 6245–6247
651. Gomeza, J., Shannon, H., Kostenis, E., Felder, C., Zhang, L., Brodkin, J., Grinberg, A., Sheng, H., and Wess, J. (1999) *Proc. Natl. Acad. Sci. U.S.A.* **96**, 1692–1697
652. Kinney, H. C., Filiano, J. J., Sleeper, L. A., Mandell, F., Valdes-Dapena, M., and White, W. F. (1995) *Science* **269**, 1446–1449
653. Yamada, M., Miyakawa, T., Duttaroy, A., Yamanaka, A., Moriguchi, T., Makita, R., Ogawa, M., Chou, C. J., Xia, B., Crawley, J. N., Felder, C. C., Deng, C.-X., and Wess, J. (2001) *Nature (London)* **410**, 207–212
654. Hill, J. J., and Peralta, E. G. (2001) *J. Biol. Chem.* **276**, 5505–5510
655. Heinrich, R., Wenzel, B., and Elsner, N. (2001) *Proc. Natl. Acad. Sci. U.S.A.* **98**, 9919–9923
656. Baudry, M. (1986) in *Excitatory Amino Acids* (Roberts, P. J., Storm-Mathisen, J., and Bradford, H. F., eds), pp. 301–321, Macmillian, London
657. Sladeczek, F., Pin, J.-P., Récasens, M., Bockaert, J., and Weiss, S. (1985) *Nature (London)* **317**, 717–719
658. Wieloch, T. (1985) *Science* **230**, 681–683
659. Johnson, J. L. (1972) *Brain Res.* **37**, 1–19
660. Choi, D. W. (1992) *Science* **258**, 241–243
661. Armstrong, N., Sun, Y., Chen, G.-Q., and Gouaux, E. (1998) *Nature (London)* **395**, 913–917
662. Bortolotto, Z. A., Clarke, V. R. J., Delany, C. M., Parry, M. C., Smolders, I., Vignes, M., Ho, K. H., Miu, P., Brinton, B. T., Fantaske, R., Ogden, A., Gates, M., Ornstein, P. L., Lodge, D., Bleakman, D., and Collingridge, G. L. (1999) *Nature (London)* **402**, 297–301
663. Nakanishi, S., and Masu, M. (1994) *Annu. Rev. Biophys. Biomol. Struct.* **23**, 319–348
664. Nakanishi, S. (1992) *Science* **258**, 597–603
665. Abele, R., Keinänen, K., and Madden, D. R. (2000) *J. Biol. Chem.* **275**, 21355–21363
666. Rosenmund, C., Stern-Bach, Y., and Stevens, C. F. (1998) *Science* **280**, 1596–1599

666a. Zamanillo, D., Sprengel, R., Hvalby, O., Jensen, V., Burnashev, N., Rozov, A., Kaiser, K. M. M., Köster, H. J., Borchardt, T., Worley, P., Lübke, J., Frotscher, M., Kelly, P. H., Sommer, B., Andersen, P., Seeburg, P. H., and Sakmann, B. (1999) *Science* **284**, 1805–1811

667. Jayaraman, V., Keesey, R., and Madden, D. R. (2000) *Biochemistry* **39**, 8693–8697

667a. Sun, Y., Olson, R., Horning, M., Armstrong, N., Mayer, M., and Gouaux, E. (2002) *Nature (London)* **417**, 245–253

667b. Rozov, A., and Burnashev, N. (1999) *Nature (London)* **401**, 594–598

668. Meddows, E., Le Bourdellés, B., Grimwood, S., Wafford, K., Sandhu, S., Whiting, P., and McIlhinney, R. A. J. (2001) *J. Biol. Chem.* **276**, 18795–18803
669. Nakazawa, T., Komai, S., Tezuka, T., Hisatsune, C., Umemori, H., Semba, K., Mishina, M., Manabe, T., and Yamamoto, T. (2001) *J. Biol. Chem.* **276**, 693–699
670. Wong, R. W. C., and Hirokawa, N. (2001) *Trends Biochem. Sci.* **26**, 410–411
671. Sheng, M. (1997) *Nature (London)* **386**, 221–223
672. Schipke, C. G., Ohlemeyer, C., Matyash, M., Nolte, C., Kettenmann, H., and Kirchhoff, F. (2001) *FASEB J.* **15**, 1270–1272
673. Hansson, E., and Rönnbäck, L. (1995) *FASEB J.* **9**, 343–350
674. Ivanovic, A., Reiländer, H., Laube, B., and Kuhse, J. (1998) *J. Biol. Chem.* **273**, 19933–19937
675. Mothet, J.-P., Parent, A. T., Wolosker, H., Brady, R. O., Jr., Linden, D. J., Ferris, C. D., Rogawski, M. A., and Snyder, S. H. (2000) *Proc. Natl. Acad. Sci. U.S.A.* **97**, 4926–4931
676. Snyder, S. H., and Ferris, C. D. (2001) in *Synapses* (Cowan, W. M., Südhof, T. C., and Stevens, C. F., eds), pp. 651–680, Johns Hopkins Univ. Press, Baltimore

676a. Takasu, M. A., Dalva, M. B., Zigmond, R. E., and Greenberg, M. E. (2002) *Science* **295**, 491–495

676b. Ghosh, A. (2002) *Science* **295**, 449–451

677. Wong, E. H. F., Kemp, J. A., Priestley, T., Knight, A. R., Woodruff, G. N., and Iversen, L. L. (1986) *Proc. Natl. Acad. Sci. U.S.A.* **83**, 7104–7108
678. Ciruela, F., Escriche, M., Burgueño, J., Angulo, E., Casadó, V., Soloviev, M. M., Canela, E. I., Mallol, J., Chan, W.-Y., Lluis, C., McIlhinney, R. A. J., and Franco, R. (2001) *J. Biol. Chem.* **276**, 18345–18351
679. Herrero, I., Miras-Portugal, and Sánchez-Prieto, J. (1998) *J. Biol. Chem.* **273**, 1951–1958
680. Kunishima, N., Shimada, Y., Tsuji, Y., Sato, T., Yamamoto, M., Kumaska, T., Nakaniashi, S., Jingami, H., and Morikawa, K. (2000) *Nature (London)* **407**, 971–977

680a. Tsuchiya, D., Kunishima, N., Kamiya, N., Jingami, H., and Morikawa, K. (2002) *Proc. Natl. Acad. Sci. U.S.A.* **99**, 2660–2665

681. Conquet, F., Bashir, Z. I., Davies, C. H., Daniel, H., Ferraguti, F., Bordi, F., Franz-Bacon, K., Reggiani, A., Matarese, V., Condé, F., Collingridge, G. L., and Crépel, F. (1994) *Nature (London)* **372**, 237–247
682. Ichise, T., Kano, M., Hashimoto, K., Yanagihara, D., Nakao, K., Shigemoto, R., Katsuki, M., and Aiba, A. (2000) *Science* **288**, 1832–1835
683. Kubo, Y., Miyashita, T., and Murata, Y. (1998) *Science* **279**, 1722–1725
684. Kawabata, S., Kohara, A., Tsutsumi, R., Itahana, H., Hayashibe, S., Yamaguchi, T., and Okada, M. (1998) *J. Biol. Chem.* **273**, 17381–17385
685. Bellocchio, E. E., Reimer, R. J., Fremeau, R. T., Jr., and Edwards, R. H. (2000) *Science* **289**, 957–960
686. Barbour, B., Brew, H., and Attwell, D. (1988) *Nature (London)* **335**, 433–435
687. Mitrovic, A. D., Amara, S. G., Johnston, G. A. R., and Vandenberg, R. J. (1998) *J. Biol. Chem.* **273**, 14698–14706
688. Grewer, C., Watzke, N., Wiessner, M., and Rauen, T. (2000) *Proc. Natl. Acad. Sci. U.S.A.* **97**, 9706–9711
689. Eskandari, S., Kreman, M., Kavanaugh, M. P., Wright, E. M., and Zampighi, G. A. (2000) *Proc. Natl. Acad. Sci. U.S.A.* **97**, 8641–8646
690. Grewer, C., Mobarekeh, S. A. M., Watzke, N., Rauen, T., and Schaper, K. (2001) *Biochemistry* **40**, 232–240
691. Smith, M. M., Warren, V. A., Thomas, B. S., Brochu, R. M., Ertel, E. A., Rohrer, S., Schaeffer, J., Schmatz, D., Petuch, B. R., Tang, Y. S., Meinke, P. T., Kaczorowski, G. J., and Cohen, C. J. (2000) *Biochemistry* **39**, 5543–5554

691a. Slotboom, D. J., Konings, W. N., and Lolkema, J. S. (2001) *Trends Biochem. Sci.* **26**, 534–539

692. Cavalheiro, E. A., and Olney, J. W. (2001) *Proc. Natl. Acad. Sci. U.S.A.* **98**, 5947–5948
693. Mukherjee, P. K., DeCoster, M. A., Campbell, F. Z., Davis, R. J., and Bazan, N. G. (1999) *J. Biol. Chem.* **274**, 6493–6498
694. Lipton, S. A., and Rosenberg, P. A. (1994) *N. Engl. J. Med.* **330**, 613–621
695. Sen, C. K., Khanna, S., Roy, S., and Packer, L. (2000) *J. Biol. Chem.* **275**, 13049–13055
696. Olsen, R. W., Wong, E. H. F., Stauber, G. B., and King, R. G. (1984) *Fed. Proc.* **43**, 2773–2778
697. Dingledine, R., Myers, S. J., and Nicholas, R. A. (1990) *FASEB J.* **4**, 2636–2645
698. Burt, D. R., and Kamatchi, G. L. (1991) *FASEB J.* **5**, 2916–2923
699. Gilardi, R. D. (1973) *Nature (London)* **245**, 86–88
700. Rudolph, U., Crestani, F., Benke, D., Brünig, I., Benson, J. A., Fritschy, J.-M., Martin, J. R., Bluethmann, H., and Möhler, H. (1999) *Nature (London)* **401**, 796–800
701. Schaerer, M. T., Kannenberg, K., Hunziker, P., Baumann, S. W., and Sigel, E. (2001) *J. Biol. Chem.* **276**, 26597–26604
702. Renard, S., Olivier, A., Granger, P., Avenet, P., Graham, D., Sevrin, M., George, P., and Besnard, F. (1999) *J. Biol. Chem.* **274**, 13370–13374
703. Kannenberg, K., Schaerer, M. T., Fuchs, K., Sieghart, W., and Sigel, E. (1999) *J. Biol. Chem.* **274**, 21257–21264
704. O'Shea, S. M., and Harrison, N. L. (2000) *J. Biol. Chem.* **275**, 22764–22768

704a. Cromer, B. A., Morton, C. J., and Parker, M. W. (2002) *Trends Biochem. Sci.* **27**, 280–287

705. Jayaraman, V., Thiran, S., and Hess, G. P. (1999) *Biochemistry* **38**, 11372–11378
706. Connolly, C. N., Wooltorton, J. R. A., Smart, T. G., and Moss, S. J. (1996) *Proc. Natl. Acad. Sci. U.S.A.* **93**, 9899–9904
707. Kennedy, M. B. (2000) *Proc. Natl. Acad. Sci. U.S.A.* **97**, 11135–11136
708. Chen, L., Wang, H., Vicini, S., and Olsen, R. W. (2000) *Proc. Natl. Acad. Sci. U.S.A.* **97**, 11557–11562
709. Celio, M. R. (1986) *Science* **231**, 995–997
710. Kaupmann, K., Huggel, K., Heid, J., Flor, P. J., Bischoff, S., Mickel, S. J., McMaster, G., Angst, C., Bittiger, H., Froestl, W., and Bettler, B. (1997) *Nature (London)* **386**, 239–246
711. Galvez, T., Prézeau, L., Milioti, G., Franek, M., Joly, C., Froestl, W., Bettler, B., Bertrand, H.-O., Blahos, J., and Pin, J.-P. (2000) *J. Biol. Chem.* **275**, 41166–41174
712. Kuner, R., Köhr, G., Grünewald, S., Eisenhardt, G., Bach, A., and Kornau, H.-C. (1999) *Science* **283**, 74–77
713. Galvez, T., Duthey, B., Kniazeff, J., Blahos, J., Rovelli, G., Bettler, B., Prézeau, L., and Pin, J.-P. (2001) *EMBO J.* **20**, 2152–2159
714. Slesinger, P. A., Stoffel, M., Jan, Y. N., and Jan, L. Y. (1997) *Proc. Natl. Acad. Sci. U.S.A.* **94**, 12210–12217
715. Yamakura, T., Mihic, S. J., and Harris, R. A. (1999) *J. Biol. Chem.* **274**, 23006–23012
716. Tallman, J. F., Paul, S. M., Skolnick, P., and Gallager, D. W. (1980) *Science* **207**, 274–284
717. Löw, K., Crestani, F., Keist, R., Benke, D., Brünig, I., Benson, J. A., Fritschy, J.-M., Rülicke, T., Bluethmann, H., Möhler, H., and Rudolph, U. (2000) *Science* **290**, 131–134
718. Wisden, W., and Stephens, D. N. (1999) *Nature (London)* **401**, 751–752
719. Olsen, R. W. (2000) *Proc. Natl. Acad. Sci. U.S.A.* **97**, 4417–4418
720. McIntire, S. L., Reimer, R. J., Schuske, K., Edwards, R. H., and Jorgensen, E. M. (1997) *Nature (London)* **389**, 870–876
721. Tamura, S., Nelson, H., Tamura, A., and Nelson, N. (1995) *J. Biol. Chem.* **270**, 28712–28715
722. Bismuth, Y., Kavanaugh, M. P., and Kanner, B. I. (1997) *J. Biol. Chem.* **272**, 16096–16102
723. Storici, P., Capitani, G., De Biase, D., Moser, M., John, R. A., Jansonius, J. N., and Schirmer, T. (1999) *Biochemistry* **38**, 8628–8634
724. Gray, T. M., and Matthews, B. W. (1984) *J. Mol. Biol.* **175**, 75–81

References

725. Grenningloh, G., Rienitz, A., Schmitt, B., Methfessel, C., Zensen, M., Beyreuther, K., Gundelfinger, E. D., and Betz, H. (1987) *Nature (London)* **328**, 215–220
726. Shan, Q., Haddrill, J. L., and Lynch, J. W. (2001) *J. Biol. Chem.* **276**, 12556–12564
727. White, W. F., and Heller, A. H. (1982) *Nature (London)* **298**, 655–657
728. Becker, C.-M. (1990) *FASEB J.* **4**, 2767–2774
729. Gundlach, A. L. (1990) *FASEB J.* **4**, 2761–2766
730. Rajendra, S., Lynch, J. W., Pierce, K. D., French, C. R., Barry, P. H., and Schofield, P. R. (1994) *J. Biol. Chem.* **269**, 18739–18742
731. Lynch, J. W., Rajendra, S., Pierce, K. D., Handford, C. A., Barry, P. H., and Schofield, P. R. (1997) *EMBO J.* **16**, 110–120
732. Griffon, N., Büttner, C., Nicke, A., Kuhse, J., Schmalzing, G., and Betz, H. (1999) *EMBO J.* **18**, 4711–4721
732a. Leite, J. F., and Cascio, M. (2002) *Biochemistry* **41**, 6140–6148
733. Matzenbach, B., Maulet, Y., Sefton, L., Courtier, B., Avner, P., Guénet, J.-L., and Betz, H. (1994) *J. Biol. Chem.* **269**, 2607–2612
734. García-Alcocer, G., García-Colunga, J., Martínez-Torres, A., and Miledi, R. (2001) *Proc. Natl. Acad. Sci. U.S.A.* **98**, 2781–2785
735. Martínez-Maza, R., Poyatos, I., López-Corcuera, B., Núñez, E., Giménez, C., Zafra, F., and Aragón, C. (2001) *J. Biol. Chem.* **276**, 2168–2173
736. Horiuchi, M., Nicke, A., Gomeza, J., Aschrafi, A., Schmalzing, G., and Betz, H. (2001) *Proc. Natl. Acad. Sci. U.S.A.* **98**, 1448–1453
737. Geerlings, A., Núnez, E., López-Corcuera, B., and Aragón, C. (2001) *J. Biol. Chem.* **276**, 17584–17590
738. Franks, N. P., and Lieb, W. R. (1997) *Nature (London)* **389**, 334–335
739. Beckstead, M. J., Phelan, R., and Mihic, S. J. (2001) *J. Biol. Chem.* **276**, 24959–24964
740. Davies, P. A., Hanna, M. C., Hales, T. G., and Kirkness, E. F. (1997) *Nature (London)* **385**, 820–823
740a. Mihic, S. J., Ye, Q., Wick, M. J., Koltchine, V. V., Krasowski, M. D., Finn, S. E., Mascia, M. P., Valenzuela, C. F., Hanson, K. K., Greenblatt, E. P., Harris, R. A., and Harrison, N. L. (1997) *Nature (London)* **389**, 385–389
741. Minami, K., Wick, M. J., Stern-Bach, Y., Dildy-Mayfield, J. E., Brozowski, S. J., Gonzales, E. L., Trudell, J. R., and Harris, R. A. (1998) *J. Biol. Chem.* **273**, 8248–8255
742. Berry, M. S., and Cottrell, G. A. (1973) *Nature New Biol.* **242**, 250–253
743. Palacios, J. M. (1986) *Nature (London)* **323**, 205
743a. Nishi, A., Bibb, J. A., Snyder, G. L., Higashi, H., Nairn, A. C., and Greengard, P. (2000) *Proc. Natl. Acad. Sci. U.S.A.* **97**, 12840–12845
744. Axelrod, J., and Saavedra, J. M. (1977) *Nature (London)* **265**, 501–504
745. Gerhardt, C. C., Lodder, H. C., Vincent, M., Bakker, R. A., Planta, R. J., Vreugdenhil, E., Kits, K. S., and van Heerikhuizen, H. (1997) *J. Biol. Chem.* **272**, 6201–6207
746. Chang, D.-J., Li, X.-C., Lee, Y.-S., Kim, H.-K., Kim, U. S., Cho, N. J., Lo, X., Weiss, K. R., Kandel, E. R., and Kaang, B.-K. (2000) *Proc. Natl. Acad. Sci. U.S.A.* **97**, 1829–1834
746a. Carlsson, A. (2001) *Science* **294**, 1021–1024
747. Balter, M. (1996) *Science* **271**, 909
748. White, F. J. (1998) *Nature (London)* **393**, 118–119
749. Daniels, G. M., and Amara, S. G. (1999) *J. Biol. Chem.* **274**, 35794–35801
750. Lin, Z., Itokawa, M., and Uhl, G. R. (2000) *FASEB J.* **14**, 715–728
751. Reith, M. E. A., Berfield, J. L., Wang, L. C., Ferrer, J. V., and Javitch, J. A. (2001) *J. Biol. Chem.* **276**, 29012–29018
752. Mitchell, D. J., Nikolic, D., Rivera, E., Sablin, S. O., Choi, S., van Breemen, R. B., Singer, T. P., and Silverman, R. B. (2001) *Biochemistry* **40**, 5447–5456
752a. Dajani, R., Cleasby, A., Neu, M., Wonacott, A. J., Jhoti, H., Hood, A. M., Modi, S., Hersey, A., Taskinen, J., Cooke, R. M., Manchee, G. R., and Coughtrie, M. W. H. (1999) *J. Biol. Chem.* **274**, 37862–37868
752b. Bidwell, L. M., McManus, M. E., Gaedigk, A., Kakuta, Y., Negishi, M., Pedersen, L., and Martin, J. L. (1999) *J. Mol. Biol.* **293**, 521–530
753. Chen, S., Xu, M., Lin, F., Lee, D., Riek, P., and Graham, R. M. (1999) *J. Biol. Chem.* **274**, 16320–16330
754. Salminen, T., Varis, M., Nyrönen, T., Pihlavisto, M., Hoffrén, A.-M., Lönnberg, T., Marjamäki, A., Frang, H., Savola, J.-M., Scheinin, M., and Johnson, M. S. (1999) *J. Biol. Chem.* **274**, 23405–23413
755. Wu, D., Jiang, H., and Simon, M. I. (1995) *J. Biol. Chem.* **270**, 9828–9832
756. Turki, J., Lorenz, J. N., Green, S. A., Donnelly, E. T., Jacinto, M., and Liggett, S. B. (1996) *Proc. Natl. Acad. Sci. U.S.A.* **93**, 10483–10488
757. Small, K. M., Brown, K. M., Forbes, S. L., and Liggett, S. B. (2001) *J. Biol. Chem.* **276**, 31596–31601
758. Wang, H.-y, Doronin, S., and Malbon, C. C. (2000) *J. Biol. Chem.* **275**, 36086–36093
759. Ghanouni, P., Steenhuis, J. J., Farrens, D. L., and Kobilka, B. K. (2001) *Proc. Natl. Acad. Sci. U.S.A.* **98**, 5997–6002
760. Liapakis, G., Ballesteros, J. A., Papachristou, S., Chan, W. C., Chen, X., and Javitch, J. A. (2000) *J. Biol. Chem.* **275**, 37779–37788
761. Choi, D.-J., Koch, W. J., Hunter, J. J., and Rockman, H. A. (1997) *J. Biol. Chem.* **272**, 17223–17229
762. Shiina, T., Kawasaki, A., Nagao, T., and Kurose, H. (2000) *J. Biol. Chem.* **275**, 29082–29090
763. Lee, S.-H., Minowa, M. T., and Mouradian, M. M. (1996) *J. Biol. Chem.* **271**, 25292–25299
764. Demchyshyn, L. L., McConkey, F., and Niznik, H. B. (2000) *J. Biol. Chem.* **275**, 23446–23455
765. Castner, S. A., Williams, G. V., and Goldman-Rakic, P. S. (2000) *Science* **287**, 2020–2022
766. Guillin, O., Diaz, J., Carroll, P., Griffon, N., Schwartz, J.-C., and Sokoloff, P. (2001) *Nature (London)* **411**, 86–89
767. White, F. J. (2001) *Nature (London)* **411**, 35–37
768. Iversen, L. L. (1974) *Nature (London)* **250**, 700–701
769. Kater, S. B., and Nicholson, C. (1973) *Intracellular Staining in Neurobiology*, Springer-Verlag., New York
770. Hökfelt, T., Johansson, O., and Goldstein, M. (1984) *Science* **225**, 1326–1334
771. Antelman, S. M., and Caggiula, A. R. (1977) *Science* **195**, 646–653
772. Javitch, J. A., Uhl, G. R., and Snyder, S. H. (1984) *Proc. Natl. Acad. Sci. U.S.A.* **81**, 4591–4595
773. Ramsay, R. R., and Singer, T. P. (1986) *J. Biol. Chem.* **261**, 7585–7587
774. Langston, J. W., Irwin, I., Langston, E. B., and Forno, L. S. (1984) *Science* **225**, 1480–1482
775. Snyder, S. H., and D'Amato, R. J. (1985) *Nature (London)* **317**, 198–199
776. Lewin, R. (1985) *Science* **230**, 527–528
777. Pentreath, V. W., and Cottrell, G. A. (1974) *Nature (London)* **250**, 655–658
778. Vaney, D. I. (1986) *Science* **233**, 444–446
779. Richardson, B. P., Engel, G., Donatsch, P., and Stadler, P. A. (1985) *Nature (London)* **316**, 126–131
780. Johnson, R. G., and Scarpa, A. (1981) *J. Biol. Chem.* **256**, 11966–11969
781. Iversen, L. L. (1985) *Nature (London)* **316**, 107–108
782. Chong, N. W., Bernard, M., and Klein, D. C. (2000) *J. Biol. Chem.* **275**, 32991–32998
783. Ganguly, S., Gastel, J. A., Weller, J. L., Schwartz, C., Jaffe, H., Namboodiri, M. A. A., Coon, S. L., Hickman, A. B., Rollag, M., Obsil, T., Beauverger, P., Ferry, G., Boutin, J. A., and Klein, D. C. (2001) *Proc. Natl. Acad. Sci. U.S.A.* **98**, 8083–8088
784. Kolata, G. B. (1976) *Science* **192**, 41–42
785. Schmidt, A. W., and Peroutka, S. J. (1989) *FASEB J.* **3**, 2242–2249
786. Julius, D. (1998) *Proc. Natl. Acad. Sci. U.S.A.* **95**, 15153–15154
787. Hegde, S. S., and Eglen, R. M. (1996) *FASEB J.* **10**, 1398–1407
788. Westbroek, I., van der Plas, A., de Rooij, K. E., Klein-Nulend, J., and Nijweide, P. J. (2001) *J. Biol. Chem.* **276**, 28961–28968
789. The Subcutaneous Sumatriptan International Study Group. (1991) *N. Engl. J. Med.* **325**, 316–321
790. Tang, H., Braun, T. F., and Blair, D. F. (1996) *J. Mol. Biol.* **261**, 209–221
791. Penado, K. M. Y., Rudnick, G., and Stephan, M. M. (1998) *J. Biol. Chem.* **273**, 28098–28106
791a. Scanlon, S. M., Williams, D. C., and Schloss, P. (2001) *Biochemistry* **40**, 10507–10513
791b. Ni, Y. G., Chen, J.-G., Androutsellis-Theotokis, A., Huang, C.-J., Moczydlowski, E., and Rudnick, G. (2001) *J. Biol. Chem.* **276**, 30942–30947
792. Brzezinski, A. (1997) *N. Engl. J. Med.* **336**, 186–195
792a. Roy, D., and Belsham, D. D. (2002) *J. Biol. Chem.* **277**, 251–258
793. Tamarkin, L., Baird, C. J., and Almeida, O. F. X. (1985) *Science* **227**, 714–720
794. Dubocovich, M. L. (1988) *FASEB J.* **2**, 2765–2773
795. Turjanski, A. G., Leonik, F., Estrin, D. A., Rosenstein, R. E., and Doctorovich, F. (2000) *J. Am. Chem. Soc.* **122**, 10468–10469
796. Martín, M., Macías, M., Escames, G., León, J., and Acuña-Castroviejo, D. (2000) *FASEB J.* **14**, 1677–1679
797. Chan-Palay, V., Lin, C.-T., Palay, S., Yamamoto, M., and Wu, J.-Y. (1982) *Proc. Natl. Acad. Sci. U.S.A.* **79**, 2695–2699
798. McBride, W. J., and Frederickson, R. C. A. (1980) *Fed. Proc.* **39**, 2701–2705
799. Schnell, M. J., Cooper, O. B., and Snyder, S. H. (1997) *Proc. Natl. Acad. Sci. U.S.A.* **94**, 2013–2018
800. Nakatsuka, S., Hayashi, M., Muroyama, A., Otsuka, M., Kozaki, S., Yamada, H., and Moriyama, Y. (2001) *J. Biol. Chem.* **276**, 26589–26596
801. Bowsher, R. R., Verburg, K. M., and Henry, D. P. (1983) *J. Biol. Chem.* **258**, 12215–12220
802. Steinberg, G. H., Kandel, M., Kandel, S. I., and Wells, J. W. (1985) *Biochemistry* **24**, 6107–6115
803. Tanaka, S., Nemoto, K.-i, Yamamura, E., and Ichikawa, A. (1998) *J. Biol. Chem.* **273**, 8177–8182
804. Chuang, W.-L., Christ, M. D., Peng, J., and Rabenstein, D. L. (2000) *Biochemistry* **39**, 3542–3555
805. Chiang, P. K., and Sacktor, B. (1975) *J. Biol. Chem.* **250**, 3399–3408
806. Arrang, J.-M., Garbarg, M., Lancelot, J.-C., Lecomte, J.-M., Pollard, H., Robba, M., Schunack, W., and Schwartz, J.-C. (1987) *Nature (London)* **327**, 117–123
807. Borowsky, B., and 16 other authors. (2001) *Proc. Natl. Acad. Sci. U.S.A.* **98**, 8966–8971
808. Snyder, S. H., Katims, J. J., Annau, Z., Bruns, R. F., and Daly, J. W. (1981) *Proc. Natl. Acad. Sci. U.S.A.* **78**, 3260–3264

References

808a. Lindskog, M., Svenningsson, P., Pozzi, L., Kim, Y., Fienberg, A. A., Bibb, J. A., Fredholm, B. B., Nairn, A. C., Greengard, P., and Fisone, G. (2002) *Nature (London)* **418**, 774–778

808b. Vaugeois, J.-M. (2002) *Nature (London)* **418**, 734–736

809. Bruns, R. F., Daly, J. W., and Snyder, S. H. (1983) *Proc. Natl. Acad. Sci. U.S.A.* **80**, 2077–2080

810. di Tomaso, E., Beltramo, M., and Piomelli, D. (1996) *Nature (London)* **382**, 677–678

811. Jan, L. (1983) *Science* **220**, 64–65

812. Mancillas, J. R., McGinty, J. F., Selverston, A. I., Karten, H., and Bloom, F. E. (1981) *Nature (London)* **293**, 576–578

813. Snyder, S. H., Banerjee, S. P., Yamamura, H. I., and Greenberg, D. (1974) *Science* **184**, 1243–1253

814. Urade, Y., Fujimoto, N., and Hayaishi, O. (1985) *J. Biol. Chem.* **260**, 12410–12415

815. Huidobro-Toro, J. P., and Harris, R. A. (1996) *Proc. Natl. Acad. Sci. U.S.A.* **93**, 8078–8082

816. Boger, D. L., Patterson, J. E., and Jin, Q. (1998) *Proc. Natl. Acad. Sci. U.S.A.* **95**, 4102–4107

816a. Huang, S. M., Bisogno, T., Petros, T. J., Chang, S. Y., Zavitsanos, P. A., Zipkin, R. E., Sivakumar, R., Coop, A., Maeda, D. Y., De Petrocellis, L., Burstein, S., Di Marzo, V., and Walker, J. M. (2001) *J. Biol. Chem.* **276**, 42639–42644

816b. Wilson, R. I., and Nicoll, R. A. (2002) *Science* **296**, 678–682

817. Devane, W. A., Hanus, L., Breuer, A., Pertwee, R. G., Stevenson, L. A., Griffin, G., Gibson, D., Mandelbaum, A., Etinger, A., and Mechoulam, R. (1992) *Science* **258**, 1946–1949

818. Sugiura, T., Kodaka, T., Nakane, S., Miyashita, T., Kondo, S., Suhara, Y., Takayama, H., Waku, K., Seki, C., Baba, N., and Ishima, Y. (1999) *J. Biol. Chem.* **274**, 2794–2801

819. Kozak, K. R., Rowlinson, S. W., and Marnett, L. J. (2000) *J. Biol. Chem.* **275**, 33744–33749

820. Bisogno, T., Maurelli, S., Melck, D., De Petrocellis, L., and Di Marzo, V. (1997) *J. Biol. Chem.* **272**, 3315–3323

821. Hanus, L., Abu-Lafi, S., Fride, E., Breuer, A., Vogel, Z., Shalev, D. E., Kustanovich, I., and Mechoulam, R. (2001) *Proc. Natl. Acad. Sci. U.S.A.* **98**, 3662–3665

822. Kruszka, K. K., and Gross, R. W. (1994) *J. Biol. Chem.* **269**, 14345–14348

823. Iversen, L. (1994) *Nature (London)* **372**, 619

824. Bayewitch, M., Rhee, M.-H., Avidor-Reiss, T., Breuer, A., Mechoulam, R., and Vogel, Z. (1996) *J. Biol. Chem.* **271**, 9902–9905

825. Bouaboula, M., Dussossoy, D., and Casellas, P. (1999) *J. Biol. Chem.* **274**, 20397–20405

826. Ledent, C., Valverde, O., Cossu, G., Petitet, F., Aubert, J.-F., Beslot, F., Böhme, G. A., Imperato, A., Pedrazzini, T., Roques, B. P., Vassart, G., Fratta, W., and Parmentier, M. (1999) *Science* **283**, 401–404

827. Xie, X.-Q., Melvin, L. S., and Makriyannis, A. (1996) *J. Biol. Chem.* **271**, 10640–10647

828. Hanus, L., Breuer, A., Tchilibon, S., Shiloah, S., Goldenberg, D., Horowitz, M., Pertwee, R. G., Ross, R. A., Mechoulam, R., and Fride, E. (1999) *Proc. Natl. Acad. Sci. U.S.A.* **96**, 14228–14233

829. Calignano, A., La Rana, G., Giuffrida, A., and Piomelli, D. (1998) *Nature (London)* **394**, 277–281

830. Stefano, G. B., Liu, Y., and Goligorsky, M. S. (1996) *J. Biol. Chem.* **271**, 19238–19242

831. Iversen, L. L. (2000) *The Science of Marijuana*, Oxford Univ. Press, Oxford

832. Steiner, H., Bonner, T. I., Zimmer, A. M., Kitai, S. T., and Zimmer, A. (1999) *Proc. Natl. Acad. Sci. U.S.A.* **96**, 5786–5790

833. Meng, I. D., Manning, B. H., Martin, W. J., and Fields, H. L. (1998) *Nature (London)* **395**, 381–383

833a. Panikashvili, D., Simeonidou, C., Ben-Shabat, S., Hanus, L., Breuer, A., Mechoulam, R., and Shohami, E. (2001) *Nature (London)* **413**, 527–531

834. Straus, S. E. (2000) *Proc. Natl. Acad. Sci. U.S.A.* **97**, 9363–9364

835. Baker, D., Pryce, G., Croxford, J. L., Brown, P., Pertwee, R. G., Huffman, J. W., and Layward, L. (2000) *Nature (London)* **404**, 84–87

836. editorial. (2001) *Nature (London)* **410**, 613

837. De Petrocellis, L., Melck, D., Palmisano, A., Bisogno, T., Laezza, C., Bifulco, M., and Di Marzo, V. (1998) *Proc. Natl. Acad. Sci. U.S.A.* **95**, 8375–8380

838. Wilson, R. I., and Nicoll, R. A. (2001) *Nature (London)* **410**, 588–592

839. Christie, M. J., and Vaughan, C. W. (2001) *Nature (London)* **410**, 527–530

840. Barinaga, M. (2001) *Science* **291**, 2530–2531

840a. Dinh, T. P., Carpenter, D., Leslie, F. M., Freund, T. F., Katona, I., Sensi, S. L., Kathuria, S., and Piomelli, D. (2002) *Proc. Natl. Acad. Sci. U.S.A.* **99**, 10819–10824

841. Yu, M., Ives, D., and Ramesha, C. S. (1997) *J. Biol. Chem.* **272**, 21181–21186

842. Kozak, K. R., Prusakiewicz, J. J., Rowlinson, S. W., Schneider, C., and Marnett, L. J. (2001) *J. Biol. Chem.* **276**, 30072–30077

843. Carper, J. (2000) *Your Miracle Brain*, Harper Collins Publ., New York

843a. Kitajka, K., Puskás, L. G., Zvara, A., Hackler, L., Jr., Barceló-Coblijn, G., Yeo, Y. K., and Farkas, T. (2002) *Proc. Natl. Acad. Sci. U.S.A.* **99**, 2619–2624

844. Stuehr, D., Pou, S., and Rosen, G. M. (2001) *J. Biol. Chem.* **276**, 14533–14536

845. Wolff, D. J., Mialkowski, K., Richardson, C. F., and Wilson, S. R. (2001) *Biochemistry* **40**, 37–45

846. Pou, S., Keaton, L., Surichamorn, W., and Rosen, G. M. (1999) *J. Biol. Chem.* **274**, 9573–9580

847. Tochio, H., Zhang, Q., Mandal, P., Li, M., and Zhang, M. (1999) *Nature Struct. Biol.* **6**, 417–421

848. Zhou, X., Espey, M. G., Chen, J. X., Hofseth, L. J., Miranda, K. M., Hussain, S. P., Wink, D. A., and Harris, C. C. (2000) *J. Biol. Chem.* **275**, 21241–21246

849. Cho, A. K. (1990) *Science* **249**, 631–634

850. Musto, D. F. (1991) *Sci. Am.* **265**(Jul), 40–47

851. Courtwright, D. T. (2001) *Forces of Habit Drugs and the Making of the Modern World*, Harvard Univ. Press, Cambridge, Massachusetts

852. Jayanthi, S., Deng, X., Bordelon, M., McCoy, M. T., and Cadt, J. L. (2001) *FASEB J.* **15**, 1745–1752

853. Billman, G. E. (1990) *FASEB J.* **4**, 2469–2475

854. Leshner, A. I. (1996) *N. Engl. J. Med.* **335**, 128–129

854a. Regan, C. (2001) *Intoxicating Minds*, Weidenfeld & Nicolson,

854b. Helmuth, L. (2001) *Science* **294**, 983–984

855. McGehee, D. S., Heath, M. J. S., Gelber, S., Devay, P., and Role, L. W. (1995) *Science* **269**, 1692–1696

856. McGehee, D. S., and Role, L. W. (1996) *Nature (London)* **383**, 670–671

857. Pich, E. M., Pagliusi, S. R., Tessari, M., Talabot-Ayer, D., Hooft van Huijsduijnen, R., and Chiamulera, C. (1997) *Science* **275**, 83–86

858. Breggin, P. R., and Baughman, F. A., Jr. (2001) *Science* **291**, 595

859. Marshall, E. (2000) *Science* **290**, 1280–1282

860. Dole, V. P. (1970) *Ann. Rev. Biochem.* **39**, 821–840

861. Hyman, S. E. (1996) *Science* **273**, 611–612

862. Grinspoon, L., Bakalar, J. B., Zimmer, L., and Morgan, J. P. (1997) *Science* **277**, 748

863. Knapp, R. J., Malatynska, E., Collins, N., Fang, L., Wang, J. Y., Hruby, V. J., Roeske, W. R., and Yamamura, H. I. (1995) *FASEB J.* **9**, 516–525

864. Pak, Y., O'Dowd, B. F., Wang, J. B., and George, S. R. (1999) *J. Biol. Chem.* **274**, 27610–27616

865. Befort, K., Zilliox, C., Filliol, D., Yue, S., and Kieffer, B. L. (1999) *J. Biol. Chem.* **274**, 18574–18581

866. Xu, W., Li, J., Chen, C., Huang, P., Weinstein, H., Javitch, J. A., Shi, L., de Riel, J. K., and Liu-Chen, L.-Y. (2001) *Biochemistry* **40**, 8018–8029

867. Zubieta, J.-K., Smith, Y. R., Bueller, J. A., Xu, Y., Kilbourn, M. R., Jewett, D. M., Meyer, C. R., Koeppe, R. A., and Stohler, C. S. (2001) *Science* **293**, 311–315

867a. Zhang, L., DeHaven, R. N., and Goodman, M. (2002) *Biochemistry* **41**, 61–68

868. Bürgi, H. B., Dunitz, J. D., and Shefter, E. (1973) *Nature (London), New Biology* **244**, 186–188

869. Schulteis, G., and Koob, G. (1994) *Nature (London)* **371**, 108–109

870. Nestler, E. J., and Aghajanian, G. K. (1997) *Science* **278**, 58–63

871. Robbins, T. W., and Everitt, B. J. (1999) *Nature (London)* **398**, 567–570

872. Wickelgren, I. (1997) *Science* **278**, 36–37

873. Nestler, E. J. (2001) *Science* **292**, 2266–2267

873a. Holden, C. (2001) *Science* **294**, 980–982

874. Lieber, C. S. (1988) *N. Engl. J. Med.* **319**, 1639–1650

875. Hoek, J. B., and Taraschi, T. F. (1988) *Trends Biochem. Sci.* **13**, 269–274

876. Pari, K., Sundari, C. S., Chandani, S., and Balasubramanian, D. (2000) *J. Biol. Chem.* **275**, 2455–2462

877. Kodaira, H., Lisek, C. A., Jardine, I., Arimura, A., and Spector, S. (1989) *Proc. Natl. Acad. Sci. U.S.A.* **86**, 716–719

878. Weitz, C. J., Lowney, L. I., Faull, K. F., and Feistner, G. (1988) *Proc. Natl. Acad. Sci. U.S.A.* **85**, 5335–5338

879. Goldstein, A., and Judson, B. A. (1971) *Science* **172**, 290–292

880. Braun, K. P., Cody, R. B., Jr., Jones, D. R., and Peterson, C. M. (1995) *J. Biol. Chem.* **270**, 11263–11266

881. Kumari, M. (2001) *J. Biol. Chem.* **276**, 29764–29771

882. Ikonomidou, C., Bittigau, P., Ishimaru, M. J., Wozniak, D. F., Koch, C., Genz, K., Price, M. T., Stefovska, V., Hörster, F., Tenkova, T., Dikranian, K., and Olney, J. W. (2000) *Science* **287**, 1056–1060

883. Tsukamoto, H., and Lu, S. C. (2001) *FASEB J.* **15**, 1335–1349

884. Thiele, T. E., Marsh, D. J., Marie, L. S., Bernstein, I. L., and Palmiter, R. D. (1998) *Nature (London)* **396**, 366–369

885. von Hungen, K., Roberts, S., and Hill, D. F. (1974) *Nature (London)* **252**, 588–589

886. Ricaurte, G., Bryan, G., Strauss, L., Seiden, L., and Schuster, C. (1985) *Science* **229**, 986–988

886a. Simantov, R., and Tauber, M. (1997) *FASEB J.* **11**, 141–146

887. Vincent, J. P., Kartalovski, B., Geneste, P., Kamenka, J. M., and Lazdunski, M. (1979) *Proc. Natl. Acad. Sci. U.S.A.* **76**, 4678–4682

888. Keup, W., ed. (1970) *Origin and Mechanisms of Hallucinations*, Plenum, New York

889. Holzman, D. (1993) *Science* **259**, 25–26

890. Aquilina, J. A., Carver, J. A., and Truscott, R. J. W. (1999) *Biochemistry* **38**, 11455–11464

891. Lipton, S. A., Kim, W.-K., Choi, Y.-B., Kumar, S., D'Emilia, D. M., Rayudu, P. V., Arnelle, D. R., and Stamler, J. S. (1997) *Proc. Natl. Acad. Sci. U.S.A.* **94**, 5923–5928

892. Borisenko, G. G., Kagan, V. E., Hsia, C. J. C., and Schor, N. F. (2000) *Biochemistry* **39**, 3392–3400

893. Zvosec, D. L., Smith, S. W., McCutcheon, J. R., Spillane, J., Hall, B. J., and Peacock, E. A. (2001) *N. Engl. J. Med.* **344**, 87–94

894. Williams, S. (1990) *Science* **248**, 958

References

895. Glickstein, M. (1988) *Sci. Am.* **259**(Sep), 118–127
896. Supèr, H., Spekreijse, H., and Lamme, V. A. F. (2001) *Science* **293**, 120–124
897. Miyashita, Y. (1995) *Science* **268**, 1719–1720
898. Brewer, J. B., Zhao, Z., Desmond, J. E., Glover, G. H., and Gabrieli, J. D. E. (1998) *Science* **281**, 1185–1187
899. Gold, G. H., and Pugh, E. N., Jr. (1997) *Nature (London)* **385**, 677–679
900. Lindstedt, K. J. (1971) *Science* **173**, 333–334
901. Nevitt, G. A., Dittman, A. H., Quinn, T. P., and Moody, W. J., Jr. (1994) *Proc. Natl. Acad. Sci. U.S.A.* **91**, 4288–4292
902. Krieger, J., and Breer, H. (1999) *Science* **286**, 720–723
903. Thomas, J. H. (1994) *Science* **264**, 1698–1699
904. Zhang, Y., Chou, J. H., Bradley, J., Bargmann, C. I., and Zinn, K. (1997) *Proc. Natl. Acad. Sci. U.S.A.* **94**, 12162–12167
905. Mombaerts, P. (1999) *Science* **286**, 707–711
906. Störtkuhl, K. F., and Kettler, R. (2001) *Proc. Natl. Acad. Sci. U.S.A.* **98**, 9381–9385
906a. Pennisi, E. (1999) *Science* **283**, 1239
907. Yoshihara, Y., Nagao, H., and Mori, K. (2001) *Science* **291**, 835,837
908. Axel, R. (1995) *Sci. Am.* **273**(Oct), 154–159
909. Firestein, S. (2001) *Nature (London)* **413**, 211–218
910. Snyder, S. H., Sklar, P. B., and Pevsner, J. (1988) *J. Biol. Chem.* **263**, 13971–13974
911. Sklar, P. B., Anholt, R. R. H., and Snyder, S. H. (1986) *J. Biol. Chem.* **261**, 15538–15543
912. Mori, K., Nagao, H., and Yoshihara, Y. (1999) *Science* **286**, 711–715
913. Floriano, W. B., Vaidehi, N., Goddard, W. A., III, Singer, M. S., and Shepherd, G. M. (2000) *Proc. Natl. Acad. Sci. U.S.A.* **97**, 10712–10716
914. Gheusi, G., Cremer, H., McLean, H., Chazal, G., Vincent, J.-D., and Lledo, P.-M. (2000) *Proc. Natl. Acad. Sci. U.S.A.* **97**, 1823–1828
915. Keverne, E. B. (1999) *Science* **286**, 716–720
916. Meyer, M. R., Angele, A., Kremmer, E., Kaupp, U. B., and Müller, F. (2000) *Proc. Natl. Acad. Sci. U.S.A.* **97**, 10595–10600
917. Bal, R. S., and Anholt, R. R. H. (1993) *Biochemistry* **32**, 1047–1053
918. Spinelli, S., Ramoni, R., Grolli, S., Bonicel, J., Cambillau, C., and Tagoni, M. (1998) *Biochemistry* **37**, 7913–7918
919. Vincent, F., Spinelli, S., Ramoni, R., Grolli, S., Pelosi, P., Cambillau, C., and Tegoni, M. (2000) *J. Mol. Biol.* **300**, 127–139
920. Timm, D. E., Baker, L. J., Mueller, H., Zidek, L., and Novotny, M. V. (2001) *Protein Sci.* **10**, 997–1004
920a. Horst, R., Damberger, F., Luginbühl, P., Güntert, P., Peng, G., Nikonova, L., Leal, W. S., and Wüthrich, K. (2001) *Proc. Natl. Acad. Sci. U.S.A.* **98**, 14374–14379
921. Zeng, C., Spielman, A. I., Vowels, B. R., Leyden, J. J., Biemann, K., and Preti, G. (1996) *Proc. Natl. Acad. Sci. U.S.A.* **93**, 6626–6630
922. Wysocki, C. J., Dorries, K. M., and Beauchamp, G. K. (1989) *Proc. Natl. Acad. Sci. U.S.A.* **86**, 7976–7978
923. Firestein, S. (2000) *Nature (London)* **404**, 552–553
923a. Margolskee, R. F. (2002) *J. Biol. Chem.* **277**, 1–4
924. Lindemann, B. (2001) *Nature (London)* **413**, 219–225
925. Matsunami, H., Montmayeur, J.-P., and Buck, L. B. (2000) *Nature (London)* **404**, 601–604
926. Adler, E., Hoon, M. A., Mueller, K. L., Chandrashekar, J., Ryba, N. J. P., and Zuker, C. S. (2000) *Cell* **100**, 693–702
927. Chandrashekar, J. (2000) *Cell* **100**, 703–711
928. Caicedo, A., and Roper, S. D. (2001) *Science* **291**, 1557–1560
929. Clyne, P. J., Warr, C. G., and Carlson, J. R. (2000) *Science* **287**, 1830–1834
930. Ishimoto, H., Matsumoto, A., and Tanimura, T. (2000) *Science* **289**, 116–119
931. McLaughlin, S. K., McKinnon, P. J., and Margolskee, R. F. (1992) *Nature (London)* **357**, 563–569
932. Ruiz-Avila, L., Wong, G. T., Damak, S., and Margolskee, R. F. (2001) *Proc. Natl. Acad. Sci. U.S.A.* **98**, 8868–8873
933. Misaka, T., Kusakabe, Y., Emori, Y., Gonoi, T., Arai, S., and Abe, K. (1997) *J. Biol. Chem.* **272**, 22623–22629
934. Li, X.-J., and Snyder, S. H. (1995) *J. Biol. Chem.* **270**, 17674–17679
935. Gachon, A. M. F. (1993) *Trends Biochem. Sci.* **18**, 206–207
936. Robyt, J. F. (1998) *Essentials of Carbohydrate Chemistry*, Springer, New York
937. Stegink, L. D., and Filer, L. J., Jr. (1984) *Aspartame*, Dekker, New York
938. Fuller, W. D., Goodman, M., and Verlander, M. S. (1985) *J. Am. Chem. Soc.* **107**, 5821–5822
939. Compadre, C. M., Pezzuto, J. M., Kinghorn, A. D., and Kamath, S. K. (1985) *Science* **227**, 417–418
940. de Vos, A. M., Hatada, M., Van Der Wel, H., Krabbendam, H., Peerdeman, A. F., and Kim, S.-H. (1985) *Proc. Natl. Acad. Sci. U.S.A.* **82**, 1406–1409
941. Kim, S.-H., de Vos, A., and Ogata, C. (1988) *Trends Biochem. Sci.* **13**, 13–15
942. Somoza, J. R., Jiang, F., Tong, L., Kang, C.-H., Cho, J. M., and Kim, S.-H. (1993) *J. Mol. Biol.* **234**, 390–404
943. Spadaccini, R., Crescenzi, O., Tancredi, T., De Casamassimi, N., Saviano, G., Scognamiglio, R., Di Donato, A., and Temussi, P. A. (2001) *J. Mol. Biol.* **305**, 505–514
944. Van Der Wel, H. (1980) *Trends Biochem. Sci.* **5**, 122–123
945. Theerasilp, S., and Kurihara, Y. (1988) *J. Biol. Chem.* **263**, 11536–11539
946. Harada, S., Otani, H., Maeda, S., Kai, Y., Kasai, N., and Kurihara, Y. (1994) *J. Mol. Biol.* **238**, 286–287
947. Bartoshuk, L. M., Lee, C.-H., and Scarpellino, R. (1972) *Science* **178**, 988–990
948. Julius, D., and Basbaum, A. I. (2001) *Nature (London)* **413**, 203–210
949. Clapham, D. E. (1997) *Nature (London)* **389**, 783–784
950. Vogel, G. (2000) *Science* **288**, 241–242
951. Kedei, N., Szabo, T., Lile, J. D., Treanor, J. J., Olah, Z., Iadarola, M. J., and Blumberg, P. M. (2001) *J. Biol. Chem.* **276**, 28613–28619
952. Caterina, M. J., Schumacher, M. A., Tominaga, M., Rosen, T. A., Levine, J. D., and Julius, D. (1997) *Nature (London)* **389**, 816–824
953. Levitan, I. B., and Cibulsky, S. M. (2001) *Science* **293**, 1270–1271
954. Cahalan, M. D. (2001) *Nature (London)* **411**, 542–544
955. Boger, D. L., Sato, H., Lerner, A. E., Hedrick, M. P., Fecik, R. A., Miyauchi, H., Wilkie, G. D., Austin, B. J., Patricelli, M. P., and Cravatt, B. F. (2000) *Proc. Natl. Acad. Sci. U.S.A.* **97**, 5044–5049
955a. McKemy, D. D., Neuhausser, W. M., and Julius, D. (2002) *Nature (London)* **416**, 52–58
955b. Smith, G. D., Gunthorpe, M. J., Kelsell, R. E., Hayes, P. D., Reilly, P., Facer, P., Wright, J. E., Jerman, J. C., Walhin, J.-P., Ooi, L., Egerton, J., Charles, K. J., Smart, D., Randall, A. D., Anand, P., and Davis, J. B. (2002) *Nature (London)* **418**, 186–190
956. Ingber, D. (1999) *FASEB J.* **13**, S3–S15
957. Gillespie, P. G., and Walker, R. G. (2001) *Nature (London)* **413**, 194–202
957a. Bezanilla, F., and Perozo, E. (2002) *Science* **298**, 1562–1563
957b. Bass, R. B., Strop, P., Barclay, M., and Rees, D. C. (2002) *Science* **298**, 1582–1587
958. Sukharev, S., Betanzos, M., Chiang, C.-S., and Guy, H. R. (2001) *Nature (London)* **409**, 720–724
958a. Perozo, E., Cortes, D. M., Sompornpisut, P., Kloda, A., and Martinac, B. (2002) *Nature (London)* **418**, 942–948
959. Watson, P. A. (1991) *FASEB J.* **5**, 2013–2019
960. Steel, K. P. (1999) *Science* **285**, 1363–1364
961. Cho, A. (2000) *Science* **288**, 1954–1955
962. Zheng, J., Shen, W., He, D. Z. Z., Long, K. B., Madison, L. D., and Dallos, P. (2000) *Nature (London)* **405**, 149–155
963. Oliver, D., He, D. Z. Z., Klöcker, N., Ludwig, J., Schulte, U., Waldegger, S., Ruppersberg, J. P., Dallos, P., and Fakler, B. (2001) *Science* **292**, 2340–2343
964. Parker, D. E. (1980) *Sci. Am.* **243**(Nov), 118–132
965. Lacy-Hulbert, A., Metcalfe, J. C., and Hesketh, R. (1998) *FASEB J.* **12**, 395–420
966. Bailey, C. H., Bartsch, D., and Kandel, E. R. (1996) *Proc. Natl. Acad. Sci. U.S.A.* **93**, 13445–13452
967. Thompson, R. F., and Kim, J. J. (1996) *Proc. Natl. Acad. Sci. U.S.A.* **93**, 13438–13444
967a. Kandel, E. R. (2001) *Science* **294**, 1030–1038
968. Squire, L. R., and Zola-Morgan, S. (1991) *Science* **253**, 1380–1386
969. Henke, K., Weber, B., Kneifel, S., Wieser, H. G., and Buck, A. (1999) *Proc. Natl. Acad. Sci. U.S.A.* **96**, 5884–5889
970. Biegler, R., McGregor, A., Krebs, J. R., and Healy, S. D. (2001) *Proc. Natl. Acad. Sci. U.S.A.* **98**, 6941–6944
971. Frey, S., and Petrides, M. (2000) *Proc. Natl. Acad. Sci. U.S.A.* **97**, 8723–8727
972. McGaugh, J. L., Cahill, L., and Roozendaal, B. (1996) *Proc. Natl. Acad. Sci. U.S.A.* **93**, 13508–13514
972a. Nakazawa, K., Quirk, M. C., Chitwood, R. A., Watanabe, M., Yeckel, M. F., Sun, L. D., Kato, A., Carr, C. A., Johnston, D., Wilson, M. A., and Tonegawa, S. (2002) *Science* **297**, 211–218
973. Shallice, T., Fletcher, P., Frith, C. D., Grasby, P., Frackowiak, R. S. J., and Dolan, R. J. (1994) *Nature (London)* **368**, 633–635
974. Teng, E., and Squire, L. R. (1999) *Nature (London)* **400**, 675–677
975. Jiang, Y., Haxby, J. V., Martin, A., Ungerleider, L. G., and Parasuraman, R. (2000) *Science* **287**, 643–646
976. Smith, E. E., Geva, A., Jonides, J., Miller, A., Reuter-Lorenz, P., and Koeppe, R. A. (2001) *Proc. Natl. Acad. Sci. U.S.A.* **98**, 2095–2100
977. Courtney, S. M., Ungerleider, L. G., Keil, K., and Haxby, J. V. (1997) *Nature (London)* **386**, 608–611
978. Lisman, J. (1998) *Nature (London)* **394**, 132–133
979. Lisman, J. E., and Idiart, M. A. P. (1995) *Science* **267**, 1512–1514
980. Alkon, D. L. (1983) *Sci. Am.* **249**(Jul), 70–84
981. Bailey, C. H., Chen, M., Keller, F., and Kandel, E. R. (1992) *Science* **256**, 645–649
982. Johnston, D. (1997) *Science* **278**, 401–402
983. Sherff, C. M., and Carew, T. J. (1999) *Science* **285**, 1911–1914
984. Levenson, J., Endo, S., Kategaya, L. S., Fernandez, R. I., Brabham, D. G., Chin, J., Byrne, J. H., and Eskin, A. (2000) *Proc. Natl. Acad. Sci. U.S.A.* **97**, 12858–12863
985. Menzel, R., and Müller, U. (2001) *Nature (London)* **411**, 433–434
986. Zars, T., Fischer, M., Schulz, R., and Heisenberg, M. (2000) *Science* **288**, 672–675
987. Grotewiel, M. S., Beck, C. D. O., Wu, K. H., Zhu, X.-R., and Davis, R. L. (1998) *Nature (London)* **391**, 455–460
987a. Pascual, A., and Préat, T. (2001) *Science* **294**, 1115–1117
987b. Perez-Orive, J., Mazor, O., Turner, G. C., Cassenaer, S., Wilson, R. I., and Laurent, G. (2002) *Science* **297**, 359–365

References

988. Draguhn, A., Traub, R. D., Schmitz, D., and Jefferys, J. G. R. (1998) *Nature (London)* **394**, 189–192
989. Winson, J. (1990) *Sci. Am.* **263**(Nov), 86–96
990. Schechter, B. (1996) *Science* **274**, 339–340
991. Joliot, M., Ribary, U., and Llinás, R. (1994) *Proc. Natl. Acad. Sci. U.S.A.* **91**, 11748–11751
992. Fisahn, A., Pike, F. G., Buhl, E. H., and Paulsen, O. (1998) *Nature (London)* **394**, 186–189
993. Wichmann, T., and DeLong, M. R. (1999) *Nature (London)* **400**, 621–622
994. Plenz, D., and Kital, S. T. (1999) *Nature (London)* **400**, 677–682
995. Whittington, M. A., Traub, R. D., and Jefferys, J. G. R. (1995) *Nature (London)* **373**, 612–615
996. Perkins, W. H. (2001) *Science* **294**, 786
997. Goldbeter, A. (1996) *Biochemical Oscillations and Cellular Rhythms: The Molecular Basis of Periodic and Chaotic Behavior*, Cambridge Univ. Press, London and New York
998. Vergara, L. A., Stojikovic, S. S., and Rojas, E. (1995) *Biophys. J.* **69**, 1606–1614
999. Thomas, A. P., Bird, G. S. J., Hajnóczky, G., Robb-Gaspers, L. D., and Putney, J. W., Jr. (1996) *FASEB J.* **10**, 1505–1517
1000. Hoyer, J., Köhler, R., and Distler, A. (1998) *FASEB J.* **12**, 359–366
1001. Marchant, J. S., and Parker, I. (2001) *EMBO J.* **20**, 65–76
1002. Chen, C. F., von Baumgarten, R., and Takeda, R. (1971) *Nature New Biol.* **233**, 27–29
1003. Nadim, F., Manor, Y., Kopell, N., and Marder, E. (1999) *Proc. Natl. Acad. Sci. U.S.A.* **96**, 8206–8211
1004. Loewenstein, Y., Yarom, Y., and Sompolinsky, H. (2001) *Proc. Natl. Acad. Sci. U.S.A.* **98**, 8095–8100
1005. Posner, M. I. (1994) *Proc. Natl. Acad. Sci. U.S.A.* **91**, 7398–7403
1006. Crick, F., and Koch, C. (1995) *Nature (London)* **375**, 121–123
1007. Koch, C., Zador, A., and Brown, T. H. (1992) *Science* **256**, 973–974
1008. Häusser, M., Spruston, N., and Stuart, G. J. (2000) *Science* **290**, 739–744
1009. Kaech, S., Parmar, H., Roelandse, M., Bornmann, C., and Matus, A. (2001) *Proc. Natl. Acad. Sci. U.S.A.* **98**, 7086–7092
1010. Gerstner, W., Kreiter, A. K., Markram, H., and Herz, A. V. M. (1997) *Proc. Natl. Acad. Sci. U.S.A.* **94**, 12740–12741
1011. Fairhall, A. L., Lewen, G. D., Bialek, W., and de Ruyter van Steveninck, R. R. (2001) *Nature (London)* **412**, 787–792
1012. Salinas, E., and Romo, R. (2000) *Nature (London)* **404**, 131–133
1013. MacLeod, K., and Laurent, G. (1996) *Science* **274**, 976–979
1014. Barinaga, M. (1998) *Science* **280**, 376–378
1015. Alpern, H. P., and Crabbe, J. C. (1972) *Science* **177**, 722–724
1016. Goelet, P., Castellucci, V. F., Schacher, S., and Kandel, E. R. (1986) *Nature (London)* **322**, 419–422
1017. Belardetti, F., Kandel, E. R., and Siegelbaum, S. A. (1987) *Nature (London)* **325**, 153–156
1018. Morris, R. G. M., Anderson, E., Lynch, G. S., and Baudry, M. (1986) *Nature (London)* **319**, 774–776
1019. Kennedy, M. B. (1988) *Nature (London)* **335**, 770–772
1020. Bliss, T. V. P., and Collingridge, G. L. (1993) *Nature (London)* **361**, 31–39
1021. Schuman, E. M., and Madison, D. V. (1994) *Science* **263**, 532–536
1022. Malenka, R. C., and Nicoll, R. A. (1999) *Science* **285**, 1870–1874
1023. Shen, K., and Meyer, T. (1999) *Science* **284**, 162–166
1024. Silva, A. J., Paylor, R., Wehner, J. M., and Tonegawa, S. (1992) *Science* **257**, 206–211
1025. Dosemeci, A., Tao-Cheng, J.-H., Vinade, L., Winters, C. A., Pozzo-Miller, L., and Reese, T. S. (2001) *Proc. Natl. Acad. Sci. U.S.A.* **98**, 10428–10432
1026. Frey, U., and Morris, R. G. M. (1997) *Nature (London)* **385**, 533–536
1027. Barinaga, M. (2000) *Science* **290**, 736–738
1028. Buchs, P.-A., and Muller, D. (1996) *Proc. Natl. Acad. Sci. U.S.A.* **93**, 8040–8045
1029. Bear, M. F. (1996) *Proc. Natl. Acad. Sci. U.S.A.* **93**, 13453–13459
1030. Rioult-Pedotti, M.-S., Friedman, D., and Donoghue, J. P. (2000) *Science* **290**, 533–536
1031. Tsien, J. Z. (2000) *Sci. Am.* **282**(Apr), 62–68
1032. Tang, Y.-P., Shimizu, E., Dube, G. R., Rampon, C., Kerchner, G. A., Zhuo, M., Liu, G., and Tsien, J. Z. (1999) *Nature (London)* **401**, 63–69
1033. Schuman, E. M., and Madison, D. V. (1991) *Science* **254**, 1503–1506
1034. Montague, P. R., Gancayco, C. D., Winn, M. J., Marchase, R. B., and Friedlander, M. J. (1994) *Science* **263**, 973–977
1035. Bear, M. F. (1999) *Proc. Natl. Acad. Sci. U.S.A.* **96**, 9457–9458
1036. Mulkey, R. M., Herron, C. E., and Malenka, R. C. (1993) *Science* **261**, 1051–1055
1037. Reyes-Harde, M., Empson, R., Potter, B. V. L., Galione, A., and Stanton, P. K. (1999) *Proc. Natl. Acad. Sci. U.S.A.* **96**, 4061–4066
1038. Nicoll, R. A., and Malenka, R. C. (1997) *Nature (London)* **388**, 427–428
1039. John, E. R., Tang, Y., Brill, A. B., Young, R., and Ono, K. (1986) *Science* **233**, 1167–1175
1040. Ungar, G. (1972) *Naturwissenschaften* **59**, 85–91
1041. White, J. G., Southgate, E., Thomson, J. N., and Brenner, S. (1986) *Phil. Trans. Roy. Soc. London B* **314**, 1–
1042. Herman, R. K. (1993) *Nature (London)* **364**, 282–283
1043. Kozloski, J., Hamzei-Sichani, F., and Yuste, R. (2001) *Science* **293**, 868–872
1044. Gupta, A., Wang, Y., and Markram, H. (2000) *Science* **287**, 273–278
1045. Steriade, M. (2001) *Proc. Natl. Acad. Sci. U.S.A.* **98**, 3625–3627
1046. Galarreta, M., and Hestrin, S. (2001) *Science* **292**, 2295–2299
1047. Mackintosh, N. J. (1998) *IQ and Human Intelligence*, Oxford Univ. Press, Oxford
1048. Sternberg, R. J. (2000) *Science* **289**, 399–401
1049. Duncan, J., Seitz, R. J., Kolodny, J., Bor, D., Herzog, H., Ahmed, A., Newell, F. N., and Emslie, H. (2000) *Science* **289**, 457–460
1050. Devlin, B., Daniels, M., and Roeder, K. (1997) *Nature (London)* **388**, 468–471
1051. Herrnstein, R. J., and Murray, C. (1994) *The Bell Curve: Intelligence and Class Structure in American Life*, Free Press, New York
1052. McClearn, G. E., Johansson, B., Berg, S., Pedersen, N. L., Ahern, F., Petrill, S. A., and Plomin, R. (1997) *Science* **276**, 1560–1563
1053. Hermelin, B. (2001) *Bright Splinters of the Mind: A Personal Story of Research with Autistic Savants*, Jessica Kingsley Publishers, London
1054. Snyder, A. (2001) *Nature (London)* **413**, 251–252
1055. Stokstad, E. (2001) *Science* **294**, 34–37
1056. Greenspan, R. J. (1995) *Sci. Am.* **272**(Apr), 72–78
1057. Pfaff, D. (2001) *Proc. Natl. Acad. Sci. U.S.A.* **98**, 5957–5960
1058. Bouchard, T. J., Jr. (1994) *Science* **264**, 1700–1701
1059. Pinker, S. (2001) *Nature (London)* **413**, 465–466
1060. Wickelgren, I. (1999) *Science* **284**, 571
1061. Young, M. W. (2000) *Sci. Am.* **282**(Mar), 64–71
1062. Sassone-Corsi, P. (1998) *Nature (London)* **392**, 871–874
1062a. Reppert, S. M., and Weaver, D. R. (2002) *Nature (London)* **418**, 935–941
1063. Morse, D. S., Fritz, L., and Hastings, J. W. (1990) *Trends Biochem. Sci.* **15**, 262–265
1064. Roenneberg, T., and Rehman, J. (1996) *FASEB J.* **10**, 1443–1447
1065. Czeisler, C. A., Duffy, J. F., Shanahan, T. L., Brown, E. N., Mitchell, J. F., Rimmer, D. W., Ronda, J. M., Silva, E. J., Allan, J. S., Emens, J. S., Dijk, D.-J., and Kronauer, R. E. (1999) *Science* **284**, 2177–2181
1066. Kondo, T., Mori, T., Lebedeva, N. V., Aoki, S., Ishiura, M., and Golden, S. S. (1997) *Science* **275**, 224–227
1067. Iwasaki, H., Taniguchi, Y., Ishiura, M., and Kondo, T. (1999) *EMBO J.* **18**, 1137–1145
1068. Crosthwaite, S. K., Dunlap, J. C., and Loros, J. J. (1997) *Science* **276**, 763–769
1069. Lee, C., Parikh, V., Itsukaichi, T., Bae, K., and Edery, I. (1996) *Science* **271**, 1740–1744
1070. Young, M. W. (2000) *Science* **288**, 451–453
1071. Naidoo, N., Song, W., Hunter–Ensor, M., and Sehgal, A. (1999) *Science* **285**, 1737–1741
1071a. Williams, J. A., Su, H. S., Bernards, A., Field, J., and Sehgal, A. (2001) *Science* **293**, 2251–2256
1071b. Panda, S., Hogenesch, J. B., and Kay, S. A. (2002) *Nature (London)* **417**, 329–335
1072. Green, C. B., and Besharse, J. C. (1996) *Proc. Natl. Acad. Sci. U.S.A.* **93**, 14884–14888
1073. Okamura, H., Miyake, S., Sumi, Y., Yamaguchi, S., Yasui, A., Muijtjens, M., Hoeijmakers, J. H. J., and van der Horst, G. T. J. (1999) *Science* **286**, 2531–2534
1074. Gekakis, N., Staknis, D., Nguyen, H. B., Davis, F. C., Wilsbacher, L. D., King, D. P., Takahashi, J. S., and Weitz, C. J. (1998) *Science* **280**, 1564–1569
1075. Suárez-López, P., Wheatley, K., Robson, F., Onouchi, H., Valverde, F., and Coupland, G. (2001) *Nature (London)* **410**, 1116–1120
1076. Schibler, U., Ripperger, J. A., and Brown, S. A. (2001) *Science* **293**, 437–438
1077. Rutter, J., Reick, M., Wu, L. C., and McKnight, S. L. (2001) *Science* **293**, 510–514
1078. Tosini, G., and Menaker, M. (1996) *Science* **272**, 419–421
1079. Morell, V. (1995) *Science* **272**, 349
1080. Turek, F. W. (1981) *Nature (London)* **292**, 289–290
1080a. Pennartz, C. M. A., de Jeu, M. T. G., Bos, N. P. A., Schaap, J., and Geurtsen, A. M. S. (2002) *Nature (London)* **416**, 286–290
1080b. Cheng, M. Y., Bullock, C. M., Li, C., Lee, A. G., Bermak, J. C., Belluzzi, J., Weaver, D. R., Leslie, F. M., and Zhou, Q.-Y. (2002) *Nature (London)* **417**, 405–410
1081. Klein, D. C., and Namboodiri, M. A. A. (1982) *Trends Biochem. Sci.* **7**, 98–102
1082. Reppert, S. M., Weaver, D. R., Rivkees, S. A., and Stopa, E. G. (1988) *Science* **242**, 78–81
1083. Baler, R., and Klein, D. C. (1995) *J. Biol. Chem.* **270**, 27319–27325
1084. Czeisler, C. A., Allan, J. S., Strogatz, S. H., Ronda, J. M., Sánchez, R., Riós, C. D., Freitag, W. O., Richardson, G. S., and Kronauer, R. E. (1986) *Science* **233**, 667–671
1085. Morell, V. (1996) *Science* **271**, 905–906
1086. Krueger, J. M., Pappenheimer, J. R., and Karnovsky, M. L. (1982) *J. Biol. Chem.* **257**, 1664–1669
1087. Hayaishi, O. (1988) *J. Biol. Chem.* **263**, 14593–14596
1088. Takahashi, J. S. (1999) *Science* **285**, 2076–2077
1088a. Siegel, J. M. (2001) *Science* **294**, 1058–1063
1088b. Maquet, P. (2001) *Science* **294**, 1048–1052
1089. Roush, W. (1995) *Science* **269**, 1220–1221

References

1090. Born, J., Hansen, K., Marshall, L., Mölle, M., and Fehm, H. L. (1999) *Nature (London)* **397**, 29–30
1091. Mefford, I. N., Baker, T. L., Boehme, R., Foutz, A. S., Ciaranello, R. D., Barchas, J. D., and Dement, W. C. (1983) *Science* **220**, 629–632
1092. Chicurel, M. (2000) *Nature (London)* **407**, 554–556
1093. Toh, K. L., Jones, C. R., He, Y., Eide, E. J., Hinz, W. A., Virshup, D. M., Ptácek, L. J., and Fu, Y.-H. (2001) *Science* **291**, 1040–1043
1094. Kondo, N., and Kondo, J. (1992) *J. Biol. Chem.* **267**, 473–478
1095. Holden, C. (1986) *Science* **233**, 723–726
1096. Holden, C. (1991) *Science* **254**, 1450–1452
1097. Stinson, S. C. (1990) *Chem. Eng. News* **Oct 15**, 33–68
1098. Holden, C. (2000) *Science* **288**, 39–40
1099. Kanof, P. D., and Greengard, P. (1978) *Nature (London)* **272**, 329–333
1100. Sandler, M., Ruthven, C. R. J., Goodwin, B. L., Reynolds, G. P., Rao, V. A. R., and Cooper, A. (1979) *Nature (London)* **278**, 357–358
1101. Maycock, A. L., Abeles, R. H., Salach, J. I., and Singer, T. P. (1976) *Biochemistry* **15**, 114–125
1102. Post, M. L., Kennard, O., and Horn, A. S. (1974) *Nature (London)* **252**, 492–495
1103. Barondes, S. H. (1994) *Science* **263**, 1102–1103
1104. García-Colunga, J., Awad, J. N., and Miledi, R. (1997) *Proc. Natl. Acad. Sci. U.S.A.* **94**, 2041–2044
1105. Enserink, M. (1999) *Science* **284**, 238–240
1106. Kramer, M. S., and 26 other authors. (1998) *Science* **281**, 1640–1645
1106a. Sapolsky, R. M. (2001) *Proc. Natl. Acad. Sci. U.S.A.* **98**, 12320–12322
1107. Vogel, G. (2000) *Science* **290**, 258–259
1108. Lindenbaum, J., Healton, E. B., Savage, D. G., Brust, J. C. M., Garrett, T. J., Podell, E. R., Marcell, P. D., Stabler, S. P., and Allen, R. H. (1988) *N. Engl. J. Med.* **318**, 1720–1728
1109. Marx, J. L. (1985) *Science* **227**, 934
1110. Wurtman, R. J., and Wurtman, J. J. (1989) *Sci. Am.* **260**(Jan), 68–75
1111. Segal, D. S., Callaghan, M., and Mandell, A. J. (1975) *Nature (London)* **254**, 58–59
1112. Price, L. H., and Heninger, G. R. (1994) *N. Engl. J. Med.* **331**, 591–598
1113. Dixon, J. F., Los, G. V., and Hokin, L. E. (1994) *Proc. Natl. Acad. Sci. U.S.A.* **91**, 8358–8362
1113a. Dichtl, B., Stevens, A., and Tollervey, D. (1997) *EMBO J.* **16**, 7184–7195
1114. Menkes, H. A., Baraban, J. M., Freed, A. N., and Snyder, S. H. (1986) *Proc. Natl. Acad. Sci. U.S.A.* **83**, 5727–5730
1115. Dubner, P. R., Tischler, A. S., Lee, Y. C., Bloom, S. R., and Donahue, S. R. (1988) *J. Biol. Chem.* **263**, 13983–13986
1116. Morell, V. (1996) *Science* **272**, 31–32
1117. Kelsoe, J. R., Spence, M. A., Loetscher, E., Foguet, M., Sadovnick, A. D., Remick, R. A., Flodman, P., Khristich, J., Mroczkowski-Parker, Z., Brown, J. L., Masser, D., Ungerleider, S., Rapaport, M. H., Wishart, W. L., and Luebbert, H. (2001) *Proc. Natl. Acad. Sci. U.S.A.* **98**, 585–590
1117a. Thompson, P. M., Vidal, C., Giedd, J. N., Gochman, P., Blumenthal, J., Nicolson, R., Toga, A. W., and Rapoport, J. L. (2001) *Proc. Natl. Acad. Sci. U.S.A.* **98**, 11650–11655
1117b. Cannon, T. D., Thompson, P. M., van Erp, T. G. M., Toga, A. W., Poutanen, V.-P., Huttunen, M., Lonnqvist, J., Standerskjold-Nordenstam, C.-G., Narr, K. L., Khaledy, M., Zoumalan, C. I., Dail, R., and Kaprio, J. (2002) *Proc. Natl. Acad. Sci. U.S.A.* **99**, 3228–3233
1118. Barnes, D. H. (1987) *Science* **235**, 430–433
1119. Brzustowicz, L. M., Hodgkinson, K. A., Chow, E. W. C., Honer, W. G., and Bassett, A. S. (2000) *Science* **288**, 678–682
1119a. Sawa, A., and Snyder, S. H. (2002) *Science* **296**, 692–695
1120. Egan, M. F., Goldberg, T. E., Kolachana, B. S., Callicott, J. H., Mazzanti, C. M., Straub, R. E., Goldman, D., and Weinberger, D. R. (2001) *Proc. Natl. Acad. Sci. U.S.A.* **98**, 6917–6922
1121. Nettle, D. (2001) *Nature (London)* **412**, 119
1122. Zirkle, C. L., and Kaiser, C. (1970) in *Medicinal Chemistry*, 3rd ed., Vol. Part II (Burger, A., ed), pp. 1410–1469, Wiley (Interscience), New York
1123. Meltzer, H. Y., Park, S., and Kessler, R. (1999) *Proc. Natl. Acad. Sci. U.S.A.* **96**, 13591–13593
1124. Seeman, P., and Lee, T. (1975) *Science* **188**, 1217–1219
1125. Davis, K. L., Davidson, M., Mohs, R. C., Kendler, K. S., Davis, B. M., Johns, C. A., DeNigris, Y., and Horvath, T. B. (1985) *Science* **227**, 1601–1602
1126. Taubes, G. (1994) *Science* **265**, 1034–1035
1127. Hornykiewicz, O. (1982) *Nature (London)* **299**, 484–486
1128. Wang, R. Y., and Schoenfeld, R., eds. (1987) *Cholecystokinin Antagonists*, Liss, New York
1129. Woo, T.-U., Whitehead, R. E., Melchitzky, D. S., and Lewis, D. A. (1998) *Proc. Natl. Acad. Sci. U.S.A.* **95**, 5341–5346
1130. Moghaddam, B., and Adams, B. W. (1998) *Science* **281**, 1349–1351
1131. Goodman, A. B. (1998) *Proc. Natl. Acad. Sci. U.S.A.* **95**, 7240–7244
1132. Lewis, D. A. (2001) *Proc. Natl. Acad. Sci. U.S.A.* **98**, 4293–4294
1132a. Hakak, Y., Walker, J. R., Li, C., Wong, W. H., Davis, K. L., Buxbaum, J. D., Haroutunian, V., and Fienberg, A. A. (2001) *Proc. Natl. Acad. Sci. U.S.A.* **98**, 4746–4751
1132b. Chakravarti, A. (2002) *Proc. Natl. Acad. Sci. U.S.A.* **99**, 4755–4756
1133. Stein, L., and Wise, C. D. (1971) *Science* **171**, 1032–1036
1134. Friedhoff, A. J. (1973) in *Biological Psychiatry* (Mendels, J., ed), pp. 113–129, Wiley, New York
1135. Wyatt, R. J., Erdelyi, E., Do Amaral, J. R., Elliott, G. R., Renson, J., and Barchas, J. D. (1975) *Science* **187**, 853–855
1136. Shen, R.-S., Smith, R. V., Davis, P. J., Brubaker, A., and Abell, C. W. (1982) *J. Biol. Chem.* **257**, 7294–7297
1137. Dichter, M. A., and Ayala, G. F. (1987) *Science* **237**, 157–163
1138. McNamara, J. O. (1999) *Nature (London)* **399**, A15–A22
1138a. Browne, T. R., and Holmes, G. L. (2001) *N. Engl. J. Med.* **344**, 1145–1151
1139. Kash, S. F., Tecott, L. H., Hodge, C., and Baekkeskov, S. (1999) *Proc. Natl. Acad. Sci. U.S.A.* **96**, 1698–1703
1140. Kanaani, J., Lissin, D., Kash, S. F., and Baekkeskov, S. (1999) *J. Biol. Chem.* **274**, 37200–37209
1141. Hsu, C.-C., Davis, K. M., Jin, H., Foos, T., Floor, E., Chen, W., Tyburski, J. B., Yang, C.-Y., Schloss, J. V., and Wu, J.-Y. (2000) *J. Biol. Chem.* **275**, 20822–20828
1142. Erlander, M. G., Tillakaratne, N. J. K., Feldblum, S., Patel, N., and Tobin, A. J. (1991) *Neuron* **7**, 91–100
1143. Qu, K., Martin, D. L., and Lawrence, C. E. (1998) *Protein Sci.* **7**, 1092–1105
1144. Porter, T. G., and Martin, D. L. (1988) *Biochim. Biophys. Acta.* **874**, 235–244
1145. During, M. J., Ryder, K. M., and Spencer, D. D. (1995) *Nature (London)* **376**, 174–177
1146. Rice, A., Rafiq, A., Shapiro, S. M., Jakoi, E. R., Coulter, D. A., and DeLorenzo, R. J. (1996) *Proc. Natl. Acad. Sci. U.S.A.* **93**, 9665–9669
1147. Dedek, K., Kunath, B., Kananura, C., Reuner, U., Jentsch, T. J., and Steinlein, O. K. (2001) *Proc. Natl. Acad. Sci. U.S.A.* **98**, 12272–12277
1148. Brusa, R., Zimmermann, F., Koh, D.-S., Feldmeyer, D., Gass, P., Seeburg, P. H., and Sprengel, R. (1995) *Science* **270**, 1677–1680
1149. Erickson, J. C., Clegg, K. E., and Palmiter, R. D. (1996) *Nature (London)* **381**, 415–418
1150. Kim, E., Lowenson, J. D., Clarke, S., and Young, S. G. (1999) *J. Biol. Chem.* **274**, 20671–20678
1151. Mandelkow, E.-M., and Mandelkow, E. (1993) *Trends Biochem. Sci.* **18**, 480–483
1152. Huang, Y., Liu, X. Q., Wyss-Coray, T., Brecht, W. J., Sanan, D. A., and Mahley, R. W. (2001) *Proc. Natl. Acad. Sci. U.S.A.* **98**, 8838–8843
1153. Ashall, F., and Goate, A. M. (1994) *Trends Biochem. Sci.* **19**, 42–46
1154. Selkoe, D. J. (1996) *J. Biol. Chem.* **271**, 18295–18298
1155. von Bergen, M., Friedhoff, P., Biernat, J., Heberle, J., Mandelkow, E.-M., and Mandelkow, E. (2000) *Proc. Natl. Acad. Sci. U.S.A.* **97**, 5129–5134
1155a. Götz, J., Chen, F., van Dorpe, J., and Nitsch, R. M. (2001) *Science* **293**, 1491–1495
1155b. Lewis, J., Dickson, D. W., Lin, W.-L., Chisholm, L., Corral, A., Jones, G., Yen, S.-H., Sahara, N., Skipper, L., Yager, D., Eckman, C., Hardy, J., Hutton, M., and McGowan, E. (2001) *Science* **293**, 1487–1491
1156. Stone, R. (1993) *Science* **261**, 424–426
1157. Spencer, P. S., Nunn, P. B., Hugon, J., Ludolph, A. C., Ross, S. M., Roy, D. N., and Robertson, R. C. (1987) *Science* **237**, 517–522
1158. Hardy, J., and Gwinn–Hardy, K. (1998) *Science* **282**, 1075–1079
1159. Selkoe, D. J. (1991) *Sci. Am.* **265**(Nov), 68–71
1160. Suzuki, T., Ando, K., Isohara, T., Oishi, M., Lim, G. S., Satoh, Y., Wasco, W., Tanzi, R. E., Nairn, A. C., Greengard, P., Gandy, S. E., and Kirino, Y. (1997) *Biochemistry* **36**, 4643–4649
1160a. Hardy, J., and Selkoe, D. J. (2002) *Science* **297**, 353–356
1161. El-Agnaf, O. M. A., Sheridan, J. M., Sidera, C., Siligardi, G., Hussain, R., Haris, P. I., and Austen, B. M. (2001) *Biochemistry* **40**, 3449–3457
1162. El-Agnaf, O. M. A., Nagala, S., Patel, B. P., and Austen, B. M. (2001) *J. Mol. Biol.* **310**, 157–168
1162a. Paushkin, S., Charroux, B., Abel, L., Perkinson, R. A., Pellizzoni, L., and Dreyfuss, G. (2000) *J. Biol. Chem.* **275**, 23841–23846
1162b. Chang, J. G., Hsieh-Li, H.-M., Jong, Y.-J., Wang, N. M., Tsai, C.-H., and Li, H. (2001) *Proc. Natl. Acad. Sci. U.S.A.* **98**, 9808–9813
1163. Min, K.-T., and Benzer, S. (1999) *Science* **284**, 1985–1988
1163a. Taylor, J. P., Hardy, J., and Fischbeck, K. H. (2002) *Science* **296**, 1991–1995
1164. Masliah, E., Rockenstein, E., Veinbergs, I., Mallory, M., Hashimoto, M., Takeda, A., Sagara, Y., Sisk, A., and Mucke, L. (2000) *Science* **287**, 1265–1269
1164a. Manning, B., AB, McCormack, A. L., Li, J., Uversky, V. N., Fink, A. L., and Di Monte, D. A. (2002) *J. Biol. Chem.* **277**, 1641–1644
1164b. Auluck, P. K., Chan, H. Y. E., Trojanowski, J. Q., Lee, V. M.-Y., and Bonini, N. M. (2002) *Science* **295**, 865–868
1165. Li, J., Uversky, V. N., and Fink, A. L. (2001) *Biochemistry* **40**, 11604–11613
1166. Sung, J. Y., Kim, J., Paik, S. R., Park, J. H., Ahn, Y. S., and Chung, K. C. (2001) *J. Biol. Chem.* **276**, 27441–27448
1167. Kitada, T., Asakawa, S., Hattori, N., Matsumine, H., Yamamura, Y., Minoshima, S., Yokochi, M., Mizuno, Y., and Shimizu, N. (1998) *Nature (London)* **392**, 605–608

References

1168. Kahle, P. J., Leimer, U., and Haass, C. (2000) *Trends Biochem. Sci.* **25**, 524–527

1169. Shimura, H., Schlossmacher, M. G., Hattori, N., Frosch, M. P., Trockenbacher, A., Schneider, R., Mizuno, Y., Kosik, K. S., and Selkoe, D. J. (2001) *Science* **293**, 263–269

1170. Trottier, Y., and Mandel, J. L. (2001) *Science* **293**, 445–446

1171. Zuccato, C., Ciammola, A., Rigamonti, D., Leavitt, B. R., Goffredo, D., Conti, L., MacDonald, M. E., Friedlander, R. M., Silani, V., Hayden, M. R., Timmusk, T., Sipione, S., and Cattaneo, E. (2001) *Science* **293**, 493–498

1172. Steffan, J. S., Kazantsev, A., Spasic-Boskovic, O., Greenwald, M., Zhu, Y.-Z., Gohler, H., Wanker, E. E., Bates, G. P., Housman, D. E., and Thompson, L. M. (2000) *Proc. Natl. Acad. Sci. U.S.A.* **97**, 6763–6768

1173. Liu, Y. F., Deth, R. C., and Devys, D. (1997) *J. Biol. Chem.* **272**, 8121–8124

1173a. Freiman, R. N., and Tjian, R. (2002) *Science* **296**, 2149–2150

1174. Choi-Lundberg, D. L., Lin, Q., Chang, Y.-N., Chiang, Y. L., Hay, C. M., Mohajeri, H., Davidson, B. L., and Bohn, M. C. (1997) *Science* **275**, 838–841

1175. Olson, L. (2000) *Science* **290**, 721–724

1176. Brady, S. (1995) *Nature (London)* **375**, 12–13

1177. Liu, H., Zhu, H., Eggers, D. K., Nersissian, A. M., Faull, K. F., Goto, J. J., Ai, J., Sanders-Loehr, J., Gralla, E. B., and Valentine, J. S. (2000) *Biochemistry* **39**, 8125–8132

1178. Singh, R. J., Karoui, H., Gunther, M. R., Beckman, J. S., Mason, R. P., and Kalyanaraman, B. (1998) *Proc. Natl. Acad. Sci. U.S.A.* **95**, 6675–6680

1179. Muralikrishnan, D., and Mohanakumar, K. P. (1998) *FASEB J.* **12**, 905–912

1180. Mochizuki, H., Hayakawa, H., Migita, M., Shibata, M., Tanaka, R., Suzuki, A., Shimo-Nakanishi, Y., Urabe, T., Yamada, M., Tamayose, K., Shimada, T., Miura, M., and Mizuno, Y. (2001) *Proc. Natl. Acad. Sci. U.S.A.* **98**, 10918–10923

1181. Giasson, B. I., Duda, J. E., Murray, I. V. J., Chen, Q., Souza, J. M., Hurtig, H. I., Ischiropoulos, H., Trojanowski, J. Q., and Lee, V. M.-Y. (2000) *Science* **290**, 985–989

1182. Utton, M. A., Gibb, G. M., Burdett, I. D. J., Anderton, B. H., and Vandecandelaere, A. (2001) *J. Biol. Chem.* **276**, 34288–34297

1183. Scheuermann, S., Hambsch, B., Hesse, L., Stumm, J., Schmidt, C., Beher, D., Bayer, T. A., Beyreuther, K., and Multhaup, G. (2001) *J. Biol. Chem.* **276**, 33923–33929

1184. Minopoli, G., de Candia, P., Bonetti, A., Faraonio, R., Zambrano, N., and Russo, T. (2001) *J. Biol. Chem.* **276**, 6545–6550

1185. Bergsdorf, C., Paliga, K., Kreger, S., Masters, C. L., and Beyreuther, K. (2000) *J. Biol. Chem.* **275**, 2046–2056

1186. Iwata, N., Tsubuki, S., Takaki, Y., Shirotani, K., Lu, B., Gerard, N. P., Gerard, C., Hama, E., Lee, H.-J., and Saido, T. C. (2001) *Science* **292**, 1550–1552

1187. Selkoe, D. J. (1999) *Nature (London)* **399**, A23–A31

1188. Wolfe, M. S., and Haass, C. (2001) *J. Biol. Chem.* **276**, 5413–5416

1189. Hong, L., Koelsch, G., Lin, X., Wu, S., Terzyan, S., Ghosh, A. K., Zhang, X. C., and Tang, J. (2000) *Science* **290**, 150–153

1189a. Esler, W. P., and Wolfe, M. S. (2001) *Science* **293**, 1449–1454

1190. Haass, C., Hung, A. Y., Selkoe, D. J., and Teplow, D. B. (1994) *J. Biol. Chem.* **269**, 17741–17748

1191. Tomita, S., Kirino, Y., and Suzuki, T. (1998) *J. Biol. Chem.* **273**, 19304–19310

1192. Yan, R., and 14 other authors. (1999) *Nature (London)* **402**, 533–537

1193. Vassar, R., and 23 other authors. (1999) *Science* **286**, 735–741

1194. Creemers, J. W. M., Dominguez, D. I., Plets, E., Serneels, L., Taylor, N. A., Multhaup, G., Craessaerts, K., Annaert, W., and De Strooper, B. (2001) *J. Biol. Chem.* **276**, 4211–4217

1195. Pennisi, E. (1999) *Science* **286**, 650–651

1196. Steinhilb, M. L., Turner, R. S., and Gaut, J. R. (2001) *J. Biol. Chem.* **276**, 4476–4484

1197. Haass, C., and De Strooper, B. (1999) *Science* **286**, 916–919

1198. Zhang, L., Song, L., Terracina, G., Liu, Y., Pramanik, B., and Parker, E. (2001) *Biochemistry* **40**, 5049–5055

1199. Fagan, R., Swindells, M., Overington, J., and Weir, M. (2001) *Trends Biochem. Sci.* **26**, 213–214

1200. Xia, X., Qian, S., Soriano, S., Wu, Y., Fletcher, A. M., Wang, X.-J., Koo, E. H., Wu, X., and Zheng, H. (2001) *Proc. Natl. Acad. Sci. U.S.A.* **98**, 10863–10868

1200a. Esler, W. P., Kimberly, W. T., Ostaszewski, B. L., Ye, W., Diehl, T. S., Selkoe, D. J., and Wolfe, M. S. (2002) *Proc. Natl. Acad. Sci. U.S.A.* **99**, 2720–2725

1201. Hashimoto, Y., Niikura, T., Ito, Y., and Nishimoto, I. (2000) *J. Biol. Chem.* **275**, 34541–34551

1202. Varadarajan, S., Kanski, J., Aksenova, M., Lauderback, C., and Butterfield, D. A. (2001) *J. Am. Chem. Soc.* **123**, 5625–5631

1203. Evans, D. B., Rank, K. B., Bhattacharya, K., Thomsen, D. R., Gurney, M. E., and Sharma, S. K. (2000) *J. Biol. Chem.* **275**, 24977–24983

1204. del C. Alonso, A., Zaidi, T., Novak, M., Grundke-Iqbal, I., and Iqbal, K. (2001) *Proc. Natl. Acad. Sci. U.S.A.* **98**, 6923–6928

1205. Grégoire, C., Marco, S., Thimonier, J., Duplan, L., Laurine, E., Chauvin, J.-P., Michel, B., Peyrot, V., and Verdier, J.-M. (2001) *EMBO J.* **20**, 3313–3321

1206. Perl, D. P., and Brody, A. R. (1980) *Science* **208**, 297–299

1207. Zhou, Y., Morais-Cabral, J. H., Kaufman, A., and MacKinnon, R. (2001) *Nature (London)* **414**, 43–48

1208. Lam, Y. A., Pickart, C. M., Alban, A., Landon, M., Jamieson, C., Ramage, R., Mayer, R. J., and Layfield, R. (2000) *Proc. Natl. Acad. Sci. U.S.A.* **97**, 9902–9906

1209. Marx, J. (2001) *Science* **293**, 2192–2194

1210. Coyle, J. T., and Puttfarcken, P. (1993) *Science* **262**, 689–695

1211. Rauk, A., Armstrong, D. A., and Fairlie, D. P. (2000) *J. Am. Chem. Soc.* **122**, 9761–9767

1212. Spillantini, M. G., Goedert, M., Crowther, R. A., Murrell, J. R., Farlow, M. R., and Ghetti, B. (1997) *Proc. Natl. Acad. Sci. U.S.A.* **94**, 4113–4118

1213. Pérez, M., Arrasate, M., de Garcini, E. M., Munoz, V., and Avila, J. (2001) *Biochemistry* **40**, 5983–5991

1213a. Alonso, AdC., Zaidi, T., Novak, M., Barra, H. S., Grundke-Iqbal, I., and Iqbal, K. (2001) *J. Biol. Chem.* **276**, 37967–37973

1214. Goldstein, L. S. B. (2001) *Science* **291**, 2102–2103

1215. Planel, E., Yasutake, K., Fujita, S. C., and Ishiguro, K. (2001) *J. Biol. Chem.* **276**, 34298–34306

1216. Lu, P.-J., Wulf, G., Zhou, X. Z., Davies, P., and Lu, K. P. (1999) *Nature (London)* **399**, 784–788

1217. Weisgraber, K. H., and Mahley, R. W. (1996) *FASEB J.* **10**, 1485–1494

1218. Reiman, E. M., Caselli, R. J., Chen, K., Alexander, G. E., Bandy, D., and Frost, J. (2001) *Proc. Natl. Acad. Sci. U.S.A.* **98**, 3334–3339

1218a. Helmuth, L. (2002) *Science* **297**, 1260–1262

1219. Wolozin, B. (2001) *Proc. Natl. Acad. Sci. U.S.A.* **98**, 5371–5373

1220. Yip, C. M., Elton, E. A., Darabie, A. A., Morrison, M. R., and McLaurin, J. (2001) *J. Mol. Biol.* **311**, 723–734

1221. Farber, S. A., Slack, B. E., and Blusztajn, J. K. (2000) *FASEB J.* **14**, 2198–2206

1222. Abdel-Ghany, M., El-Sebae, A. K., and Shalloway, D. (1993) *J. Biol. Chem.* **268**, 11976–11981

1223. Yoshiike, Y., Tanemura, K., Murayama, O., Akagi, T., Murayama, M., Sato, S., Sun, X., Tanaka, N., and Takashima, A. (2001) *J. Biol. Chem.* **276**, 32293–32299

1224. Lee, VM-Y (2001) *Proc. Natl. Acad. Sci. U.S.A.* **98**, 8931–8932

1225. Watanabe, C. M. H., Wolffram, S., Ader, P., Rimbach, G., Packer, L., Maguire, J. J., Schultz, P. G., and Gohil, K. (2001) *Proc. Natl. Acad. Sci. U.S.A.* **98**, 6577–6580

Study Questions

1. Compare the sensing of signals and responses to signals in liver, muscle, or other tissue with signaling in the brain.

2. Compare signaling by ionotropic receptors, metabotropic receptors and gap junctions.

3. List major neurotransmitters in the brain. In what other locations do these compounds function?

4. Compare addiction to gambling with addiction to cocaine. Are they similar on a biochemical basis? What about addiction to Internet games, chatting, pornography, compulsive overeating, etc.?[873a]

5. Can extracts of leaves of *Gingko biloba* counteract age-related neurological disorders?[1225]

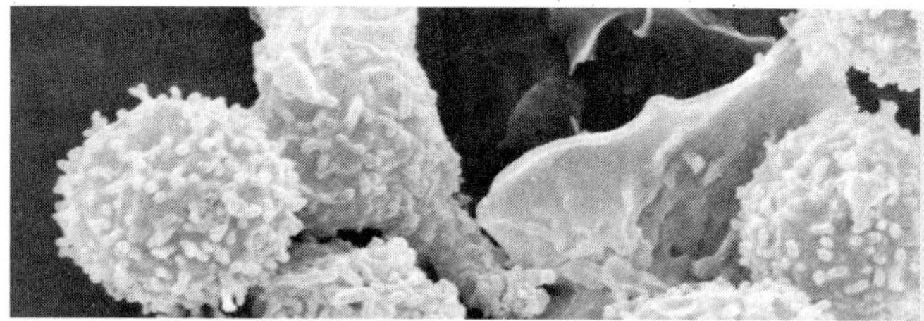

The large flat cell, a portion of which is seen here, is a macrophage which has ingested bacterial proteins and is displaying peptide fragments on its surface. Some of the small, spherical T lymphocytes (T cells) seen here, interact with the macrophage, recognize an antigen, and respond by becoming helper T cells. They can then stimulate B lymphocytes (B cells) to multiply and produce antibodies. See Fig. 31-11 for an enlarged view. Scanning electron micrograph courtesy of Morton H. Nielsen and Ole Werdelin, University of Copenhagen.

Contents

Biochemical Defense Mechanisms

31

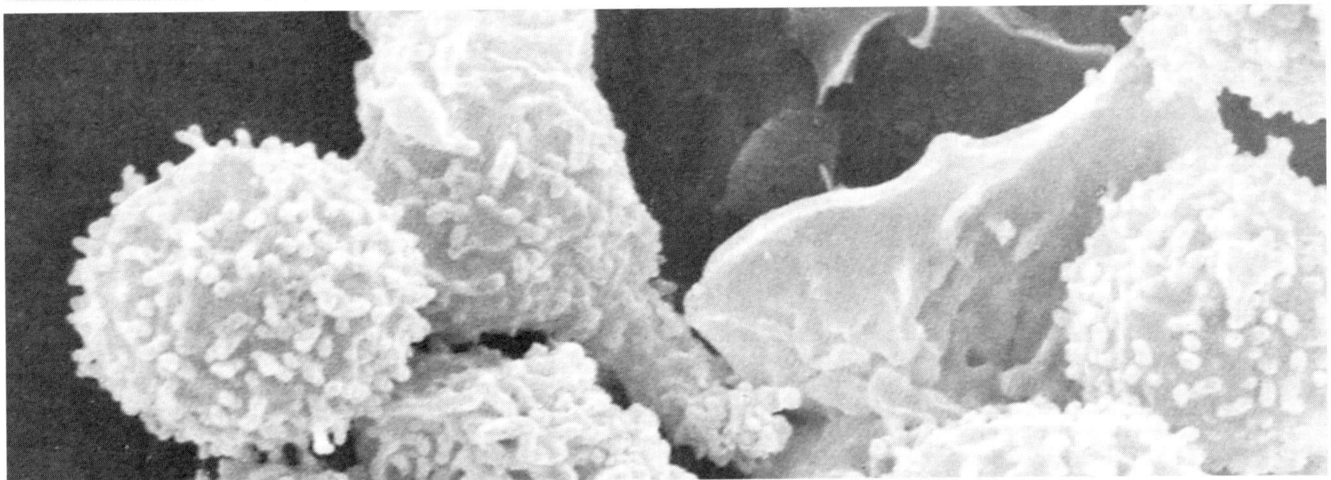

Our bodies are under constant attack by viruses, bacteria, protozoa, and metazoan parasites. Persons born without an immune system adequate to fight off these invaders die very quickly unless heroic measures are taken. We have learned to cooperate with our immune systems by immunizing ourselves against some bacteria and viruses. At other times we may fight a stubborn battle with our own defense systems against allergic reactions and a variety of autoimmune responses.[1]

A. Locations and Organization of the Immune System

The immune system has many components, many of which are dispersed throughout the human body (Fig. 31-1). Lymphocytes, which are a foundation of the immune system, constitute little more than 1% of the body's mass. However, this represents ~10^{12} cells of several types, about 10 times more than there are neurons in the brain. These 10^{12} cells make antibodies and T-cell receptors, both of which are thought to have ~10^{15} different peptide sequences.[2] And this is only the beginning of the complexities.[3–8]

Immune responses have often been described in terms of **humoral** and **cellular** components. The humoral response involves the small circulating **B lymphocytes** (B cells), the **antibodies** (immunoglobulins), and proteins of the **complement** system. The cellular response is mediated by another group of small lymphocytes, the **T lymphocytes** (T cells). They resemble B cells in appearance but have quite different functions. However, newer knowledge has provided a somewhat different description of the body's defense systems, which can be classified into three levels. (1) The skin and internal mucous membranes, which are resistant to infection and have antibacterial properties, provide the first level of defense.[2,9] (2) A fast-acting **innate immune system** can respond within a few minutes to breaches in the barriers provided by the tough outer skin and the glycoproteins of mucous surfaces and provides a second level.[2,10–14] (3) A slower **adaptive** (acquired) part of the immune system leads to synthesis of antibodies and to long-term immunity, providing the third level (Table 31-1). Both B and T lymphocytes together with **antigen-presenting cells** (**APCs**) are necessary for the selection and development of immunoglobulin structures appropriate for attack on an invading organism.

The innate immune system utilizes **phagocytic cells** including neutrophils, monocytes, and macrophages[15] to ingest and kill invading organisms. Basophils, mast cells,[16] eosinophils, and other cells release inflammatory mediators, which attract additional lymphocytes and affect their development.[17] Specialized T lymphocytes called **natural killer (NK) cells**[18] may also attack foreign cells (Table 31-2). The innate system is ancient and has apparently evolved to recognize molecular structures that are foreign to the host but are characteristic of pathogens. These structures, which are described as **pathogen-associated molecular patterns** (**PAMPs**), include those of lipopolysaccharides of bacterial cell walls (Fig. 8-30), mannans (p. 175),[19] other carbohydrates of surface layers,[20] oxidized phosphatidylcholines,[20a] bacterial flagellins,[21] various posttranscriptionally modified proteins, teichoic acids (p. 431), etc. However, they do not include patterns characteristic of host cells. Those have been avoided during evolution of the system. Although

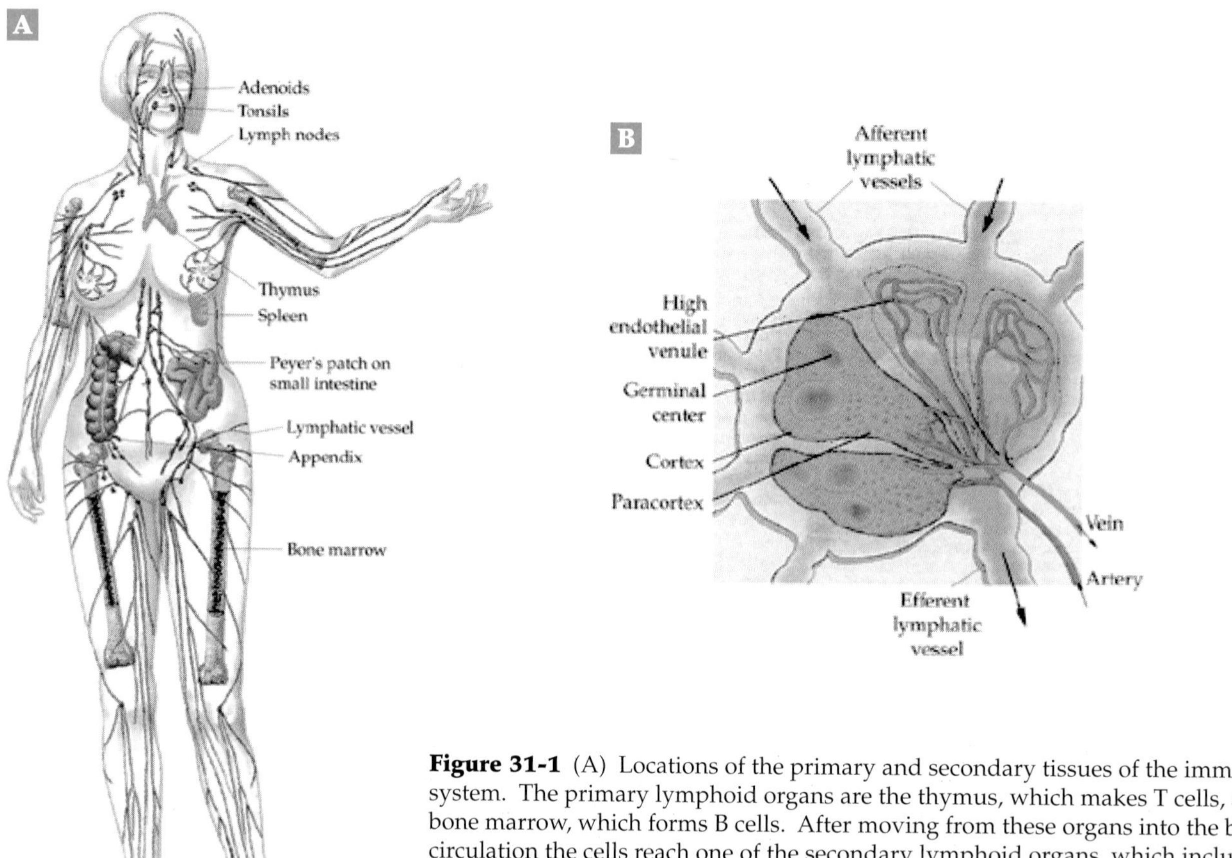

Figure 31-1 (A) Locations of the primary and secondary tissues of the immune system. The primary lymphoid organs are the thymus, which makes T cells, and the bone marrow, which forms B cells. After moving from these organs into the blood circulation the cells reach one of the secondary lymphoid organs, which include lymph nodes, spleen, tonsils, and Peyer's patches on the small intestine. Immature dendritic cells are found in body tissues including skin and mucous membranes and respond to foreign proteins by inducing attack by T lyphocytes and antibody formation by B cells. (B) Schematic drawing of a lymph node. From Nossal.[1] Courtesy of Gustav J. V. Nossal.

T lymphocytes are major mediators of the innate immune response they are under control of the **dendritic cells** (**DCs**), which are found in "immature" forms in tissues throughout the body.[9,22–26a]

The immature DCs are phagocytic cells that act as "immunological sensors." They recognize various PAMPs, which act as **danger signals**,[11b] using what are known as **toll-like receptors** (TLRs).[23,24,27,27a,b] They are also the most active APCs. Their proteasomes cleave proteins, both of the host and of invading organisms, into short peptides. These peptide fragments are displayed on the APC surfaces for recognition by T lymphocytes and for activation of adaptive immune responses. Some "autoreactive" B cells are also part of the innate immune system[28,29] as are the IgA antibodies present in mucous membranes.[30] The innate immune system of insects resembles that of vertebrates. The toll-like receptors of the latter are named for their resemblance to the Toll receptors of *Drosophila,* which are utilized in resistance to fungi. In both mammals and insects the innate immune system activates responses via the transcription factor NF-κB. However, many details of the signaling pathways differ.[30a–c]

The innate system also provides for synthesis of small antibiotic peptides called **defensins**[12,31–33] as well as larger proteins. Some of these proteins constitute the **complement system**, while others are described as **acute-phase reactants**. Some defensins are also **cytokines**, which attract lymphocytes.

The innate system is of special importance during early infancy. Prior to birth and for at least 4–12 months after birth a child's immune system is poorly developed. It may not become fully competent until age ~5.[34,35] During the prenatal period maternal antibodies are transferred to the child. IgG crosses the placenta and enters the fetal circulation. Breast milk provides IgA, which remains largely in the child's gut, as well as other protective proteins. UNICEF and the World Health Organization recommend breast-feeding to two years or beyond.[34]

While the innate immune system provides for immediate and direct attack on invaders, it also provides information to the slower adaptive system. Genes both for immunoglobulins and for the T-cell receptors of the adaptive system undergo extensive rearrangement during development of an individual.

TABLE 31-1
The Two Major Branches of the Immune System

	Innate (natural)	Adaptive (acquired)
Cells	Dendritic cells Phagocytic cells (neutrophils, monocytes, macrophages) Cells that release inflammatory mediators (basophils, mast cells, eosinophils) Natural killer (NK) cells	Dendritic cells B lymphocytes (B cells) T lymphocytes (T cells) Other antigen-presenting cells, e.g., macrophages
Molecular components	Antibacterial peptides (defensins, complement, acute-phase proteins)	Immunoglobulins
Receptor genes	Fixed in genome	Complement proteins Encoded in gene segments; rearrangement necessary
Recognition	Conserved molecular patterns	Small molecular groups (epitopes)
Immunogenic memory	Absent	Present
Self–nonself discrimination	Perfect	Imperfect
Action time	Immediate	Delayed

This provides potential defensive proteins directed at almost every imaginable invader. It also ensures that every individual has a set of proteins that labels its own cells as "self," and that virtually every individual on earth has cell surface proteins different from those of every other person. In both the innate and adaptive responses the immune system must carefully distinguish "self" from "nonself."[36,37] In the innate system this discrimination developed during evolution of the host and its pathogens. In the adaptive system it depends upon interaction of the T cells with surface molecules, primarily those of the **major histocompatibility complex** (MHC).

Another basic characteristic of immune responses is the development of **immunologic memory**.[38–40] This is exemplified by the fact that vaccination can sometimes impart immunity for a person's lifetime. If a foreign protein is injected into an animal, after a lag period of 2–5 days the animal will synthesize antibodies against this foreign antigen. This is called a **primary adaptive immune response**. If after a few days or weeks a second injection of the same protein is made, a much more rapid synthesis of additional antibodies occurs. This **secondary immune response** may take place within hours and will last longer than the primary response. It is a manifestation of immunologic memory.

1. Development of Lymphocytes and Other Specialized Cells

Both the B cells and T cells arise in the fetal liver or bone marrow (Fig. 31-1) from pluripotent stem cells. In birds the B cells develop in a special organ, the bursa of Fabricius. Mammalian B cells complete their differentiation into mature **lymphocytes** within the bone marrow. However, the T cells must travel to the **thymus,** where they complete their maturation. The T lymphocytes include the previously mentioned NK cells as well as the somewhat similar **cytolytic T cells** and **immunoregulatory T cells**. The latter are further characterized as **helper T cells**[41] or **suppressor T cells**. The adaptive response requires cooperation of helper T cells in many instances. The mature B and T cells leave the bone marrow and thymus, which are known as the **primary lymphoid tissues,** and enter the blood circulation. Following "homing" signals[42] they take up residence in a variety of locations in the lymph nodes, spleen, adenoids, tonsils, and Peyer's patches. The last are small clusters of lymphoid cells in the wall of the intestine. All of these tissues, which are referred to as **secondary lymphoid tissues**, are the sites in which the adaptive immune system is developed.

An important component of the immune system that was neglected until recently is located in the mucous membranes and the skin.[9,30,43,43a] The mucosal surfaces of airways and the gastrointestinal tract provide the point of entry for many diseases. Both internal and external body surfaces are protected by dendritic cells, whose immature forms in skin are called Langerhans cells. See figure in Box 8-F.[22,26a]

Eight different types of cells of the immune system (Table 31-2) develop by differentiation of pluripotent stem cells[44–48] as indicated in Fig. 31-2. Dendritic cells[26a] (which are not shown in this figure) may be formed from monocytes but may also arise by other routes.[22] The development of the various cells takes place under the influence of a number of **hemopoictic regulators**. Among these are the protein hormone **erythropoietin** and various interleukins and colony-stimulating factors.[44,45,49]

TABLE 31-2
Cells of the Immune System[a]

Type	Functions
B lymphocytes	
Plasma cells	Antibody synthesis
Memory B cells	Immunologic memory
T lymphocytes	
Cytolytic	Destroy infected and malignant cells
Helper cells	
Type 1 (T_H1)	Participate in activation of B cells
Type 2 (T_H2)	
Memory T cells	Immunologic memory
Natural killer (NK) cells	Destroy infected and malignant cells with pore-forming protein perforin and cytotoxic granules
Dendritic cells	
Interdigitating	Antigen recognition and processing
Follicular	Antigen presentation
Microglia	Defensive network in brain[b]
Cells that release inflammatory mediators	
Mast cells and basophils	Posses high-affinity receptors for IgE May secrete histamine, prostaglandins, leukotrienes Important in allergies
Eosinophils	Weakly phagocytic, secrete cationic proteins, reactive reduced oxygen species, leukotrienes, prostaglandins, cytokinins
Phagocytic cells	
Neutrophils	Acute inflammatory response
Monocytes	
Macrophages	Carry receptors for carbohydrates not normally exposed on surfaces of cells in vertebrates e.g., mannose; kill engulfed organisms with $^{\bullet}O_2^-$, HOCl, NO, cationic proteins and peptides, lysozyme Antigen processing and presentation

[a] General reference: Delves, P. J., and Roitt, I. M. (2000) *N. Engl. J. Med.* **343**, 37–49.
[b] Streit, W. J., and Kincaid-Colton, C. A. (1995) *Sci. Am.* **273** (Nov), 54–61.

2. Triggering an Immune Response

When a foreign antigen enters the body the B cells, with receptors of appropriate specificity and present in the lymph nodes, are stimulated to divide repeatedly and to produce a large clone of **plasma cells**. These contain a highly developed ER and actively synthesize and secrete immunoglobulins. One activated B cell may produce 10 million antibody molecules per hour.[1] This B-cell response occurs within a network of **follicular dendritic cells** in the **germinal centers** of the lymph nodes (Fig. 31-1B).[2,50,51] The antibodies are "adaptor" molecules, which bind to antigenic proteins often on surfaces of invading microorganisms. Another part of the antibody binds to one of several **effectors** systems. These immobilize microorganisms, induce phagocytosis, activate the complement system, carry antibodies across placental membranes, etc.[52,53]

Induction of a T-cell response is more complex and is very demanding.[22,47,54] Antigenic peptide frag-

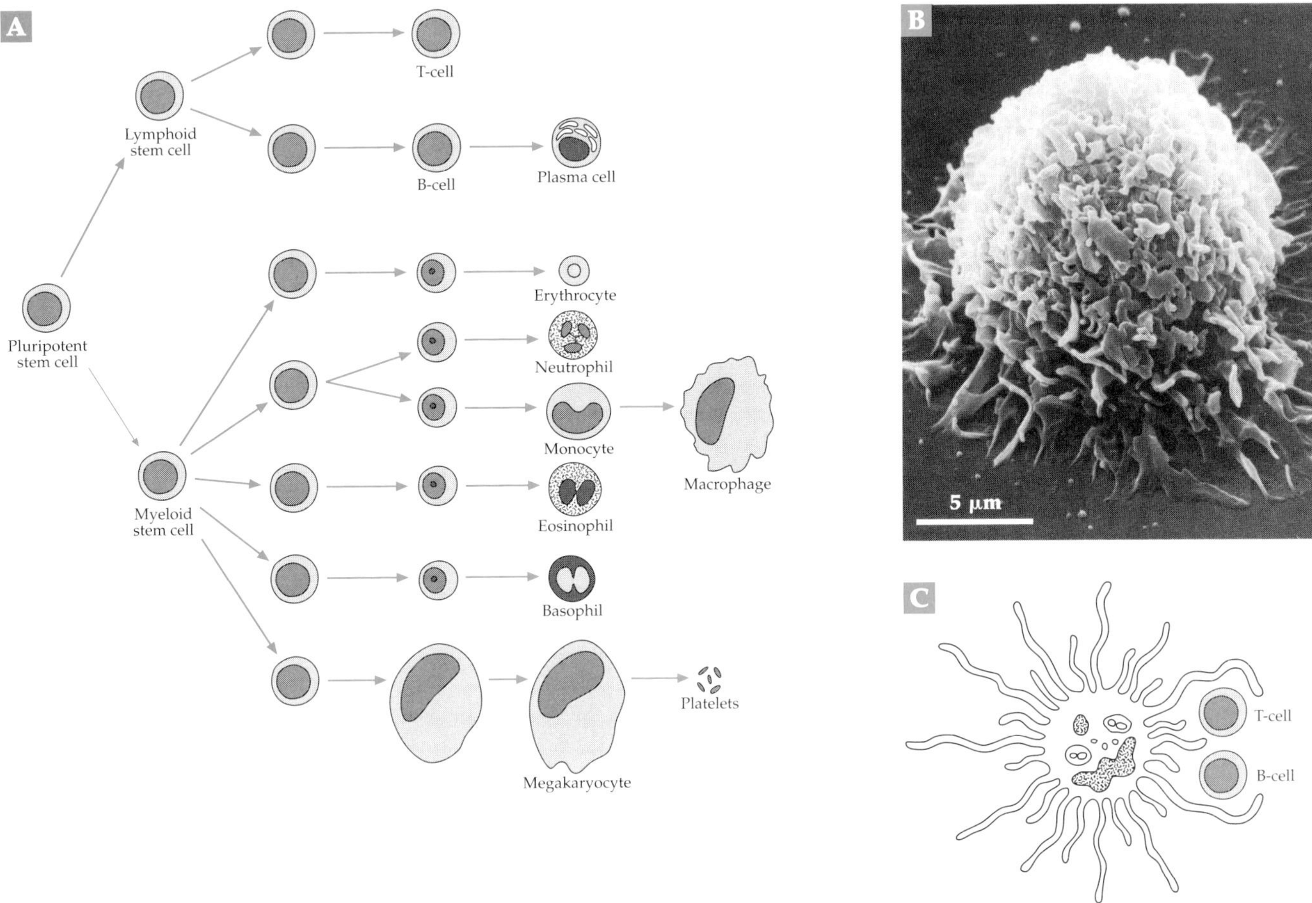

Figure 31-2 (A) Development of eight types of blood cells including those of the immune system from pluripotent (multipotential) stem cells. The cells develop under the influence of a variety of protein growth factors. Some steps, e.g., maturation of B cells, involve complex rearrangements in the DNA of the cell. (B) Scanning electron micrograph of a macrophage, a large motile cell that plays a key role in the immune system. It moves by means of its surface "ruffles." It actively phagocytoses both pathogens and waste materials and is also one of the cells that releases the hormones known as colony-stimulating factors. Micrograph courtesy of Shirley G. Quam. Drawings courtesy of David W. Golde and Judith C. Gasson.[44] (C) Schematic drawing of a dendritic cell. Redrawn from Banchereau and Steinman.[22] Both macrophages and dendritic cells present antigens for recognition by T cells and synthesize cytokines, which affect lymphocyte development.

ments from infected or malignant cells anywhere in the body must be recognized by an appropriate receptor on a T cell that is circulating in the blood. Only a few T cells with receptors for any given antigen specificity exist.[54,55] The foreign antigen fragment must bind to a protein of the **major histocompatibility complex** (**MHC**) while present within a dendritic cell or other APC. The resulting MHC•antigen complexes pass through the ER and the Golgi to the outer cell surface in a rather complex process. Recognition of the antigen that is "presented" in this manner on the APC surface is accomplished with the aid of ~10^{15} different receptor proteins (**T-cell receptors**) on the T-cell surfaces. When a T-cell receptor is occupied by an HMC•antigen complex of appropriate specificity the T cell is activated to participate in adaptive immunity. However, some T cells, notably γδ T cells, like antibodies bind to antigen directly.[56,56a,b] The recognition process occurs in an **immunological synapse**, which has elements of similarity to neurological synapses.[54,57–58a]

B. The Immunoglobulins (Antibodies)

1. Molecular Structures

There are five classes of antibodies or immunoglobulins.[59,60] The first three, IgG, IgM, and IgA, are quantitatively the most significant, but IgD and IgE are also important. For example, the content of IgE is elevated in allergic persons. The basic structure of all of the immunoglobulins is that of a quasi-symmetric dimer composed of a pair of light chains and a pair of heavy chains whose lengths vary among the different

Symbol	Mass (kDa)	Formula
IgG	150	$\kappa_2\gamma_2$ or $\lambda_2\gamma_2$
IgM	950	$(\kappa_2\mu_2)_{2.5}$ or $(\lambda_2\mu_2)_{2.5}$
IgA	320	$(\kappa_2\alpha_2)_n$ or $(\lambda_2\alpha_2)_n$
IgD	180	$\kappa_2\delta_2$ or $\lambda_2\delta_2$
IgE	190	$\kappa_2\varepsilon_2$ or $\lambda_2\varepsilon_2$

classes of immunoglobulins. Two classes of **light chains,** κ and λ, are found in human antibodies. The **heavy chains** are designated γ, μ, α, δ, and ε (see accompanying tabulation). Both IgM and IgA contain an additional J chain.

Treatment with mercaptoethanol splits the disulfide linkages holding the chains together, permitting preparation of monomeric light and heavy chains. When peptide chains of the immunoglobulins were hydrolyzed enzymatically, the resulting peptide fragments were found to be extremely heterogeneous. They were mixtures of many different kinds of peptides. The result was not unexpected, for it had long been recognized that the body contains millions of different antibodies, with binding sites specific for different antigenic determinants. It had been unclear how different binding sites could be formed, but the heterogeneity in amino acid sequence suggested the correct answer: Each antibody has its own sequence.

Progress toward understanding of the detailed structure of antibodies came when it was recognized that patients with tumors of the lymphatic system, e.g., the bone marrow tumors **multiple myeloma**, produced tremendous quantities of homogeneous immunoglobulins or parts thereof. Similar tumors were soon discovered in mice and provide a ready source of experimental material. The **Bence-Jones proteins** that are secreted in the urine of myeloma patients were found to be light chains of immunoglobulins. Sequence determinations showed that each Bence-Jones protein was homogeneous, even though no two patients secreted the same protein.[61,62] Later, intact myeloma globulins and macroglobulins (IgM) of a homogeneous kind were also obtained.

The first complete amino acid sequence of an IgG molecule was announced in 1969.[63] The protein contained 446 amino acids in each heavy chain and 214 in each light chain. The longer heavy chains of IgM molecules contain 576 amino acids.[64] In all of the immunoglobulins the heavy and light chains are held together by disulfide linkages, and the chains are folded into loops to form compact domains. The IgM molecule is polymerized through additional disulfide linkages to form a pentamer readily visible with the electron microscope (Fig. 31-3). The heavy chains also carry oligosaccharide units. In IgM there are five of these, as indicated in Fig. 31-4A. They contain mannose and *N*-acetylglucosamine units linked to asparagine. Other immunoglobulins (IgA, IgE, and IgG)

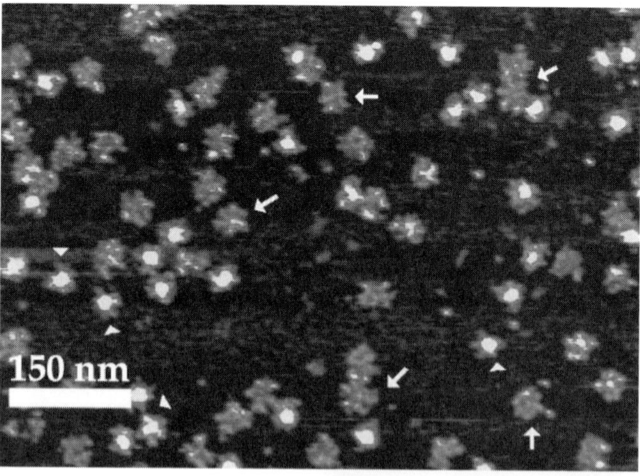

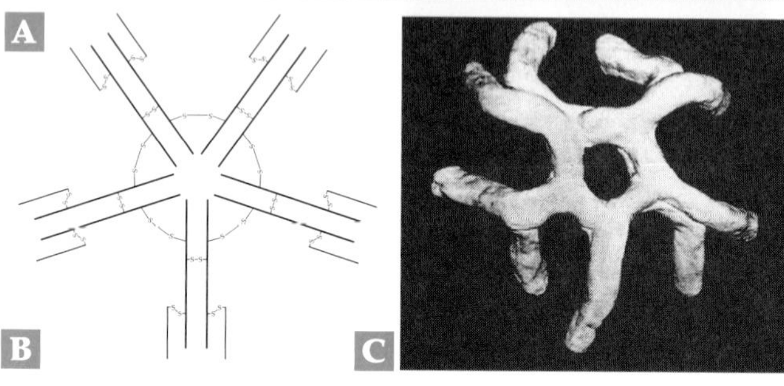

Figure 31-3 (A) Cryo atomic force (AFM) micrograph of molecules of the human immunoglobulin IgM. Courtesy of Zhifeng Shao, University of Virginia. (B) Schematic diagram. One-fifth of this structure is shown in greater detail in Fig. 31-4A. (C) Model based on earlier electron microscopic images. From Feinstein and Munn. [64c]

contain fucose, galactose, and *N*-acetylneuraminic acid as well. In fact, almost all of the most important macromolecules that participate in innate and adaptive immune responses are glycoproteins.[64b]

Digestion of an intact molecule of IgG with papain cleaves both heavy chains in the hinge region near the interchain disulfide bridge. This splits the molecule into three parts; two **Fab** (antibody-binding) **fragments**, each containing the N-terminal end of a heavy chain together with a linked light chain, and an **Fc fragment**. Even before it was known that IgG could be split into two Fab fragments, the antibody was known to be divalent, i.e., capable of binding with two different antigens (Fig. 31-4). The shape and overall structure of IgG molecules have been verified by electron microscopy and numerous X-ray diffraction studies.

Sequence determinations showed that in some regions of immunoglobulin molecules there is extreme variation in the amino acid sequence between one homogeneous antibody and the next; other regions have a constant sequence. The molecule can also be divided into domains. The **variable regions**, which occupy the N-terminal ends of the chains, are designated

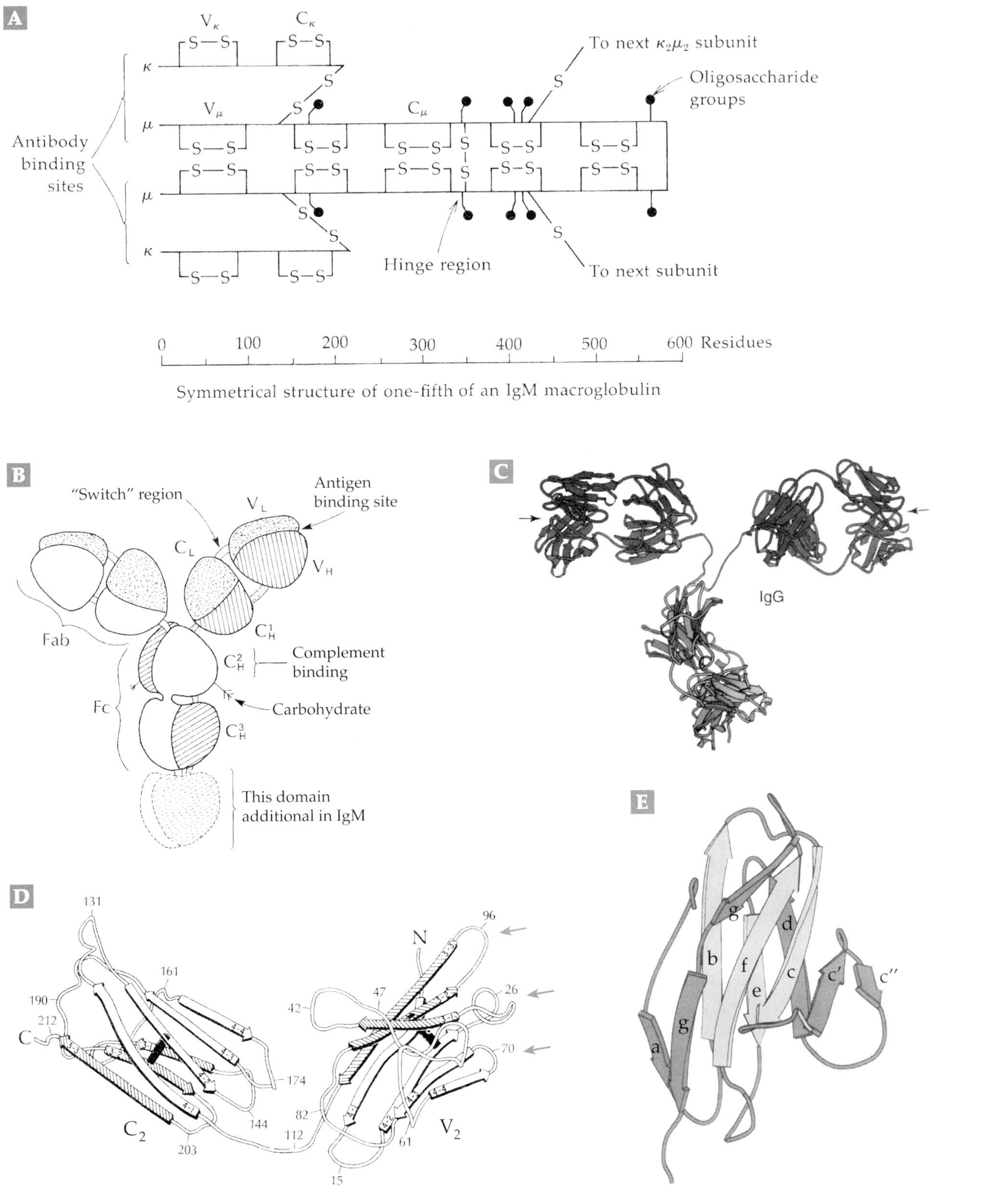

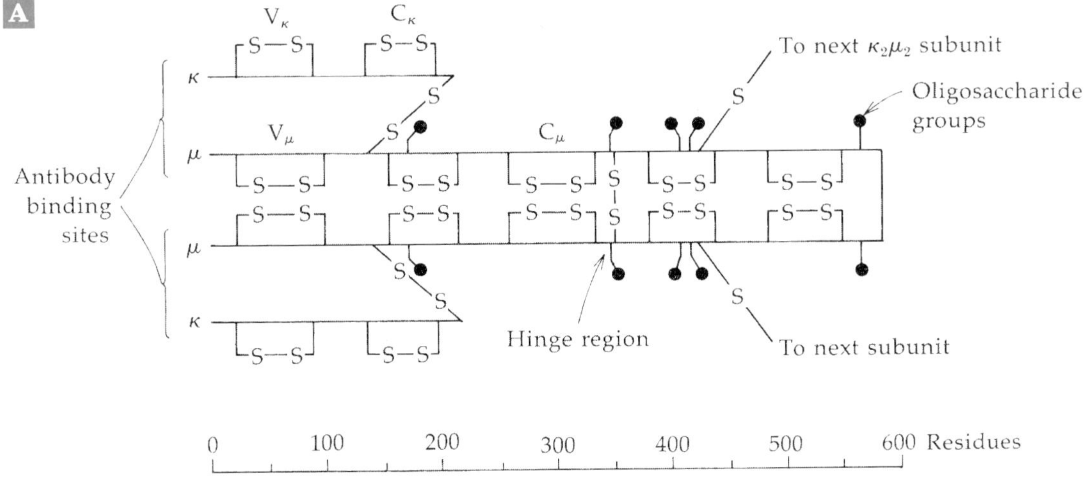

Figure 31-4 Schematic structure of one-fifth of an IgM molecule. From Putnam *et al.*[64] (A) Covalent structure. (B) Schematic three-dimensional representation. (C) Ribbon diagram of an IgG molecule. From Cochran *et al.*[64a] (D) Folding patterns of one chain in a constant and a variable domain of a Bence-Jones protein. From Schiffer *et al.*[66] Green arrows indicate hypervariable regions. (E) MolScript drawing of the common core structure of Ig-like domains. The lighter shaded strands (b, c, e, f) form the core common to all Ig-like domains, which is surrounded by structurally more varied additional strands (darker). The front sheet has up to five strands (a, f, c, e, c″) and the back sheet up to four (a, b, e, d). Strand c″ is very flexible and is not always a part of the β sheet. From Bork, Holm, and Sander.[65] See also Fig. 2-16.

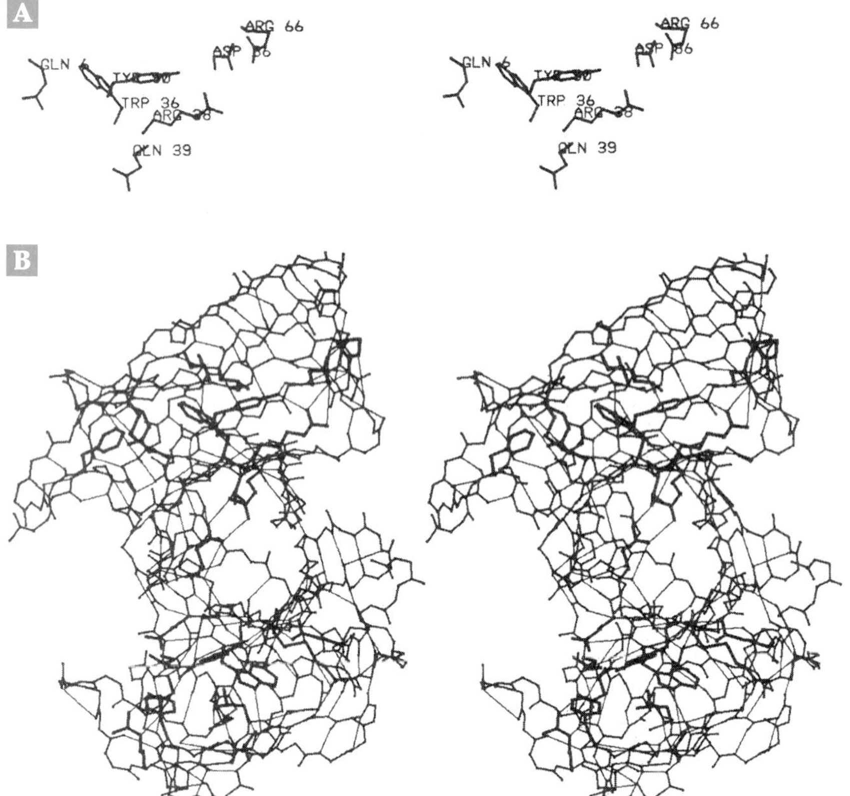

Figure 31-5 The extensive conserved hydrogen-bonding pattern in an immunoglobulin variable domain provided by polar residues buried inside the V_L and V_H domains. (A) To facilitate orientation, prominent side chains are displayed here and identified by names and numbers in the same orientation as in (B). (B) Polypeptide chain backbones of both domains are denoted by heavy lines and hydrogen bonds by light lines. In addition to the regular interbackbone hydrogen-bonding network characteristic of antiparallel β-sheets, there are hydrogen bonds provided by side-chain atoms. Note the two hydrogen bonds of Gln-38 (V_L) and Gln-39 (V_H) that span the domain–domain interface. From Novotny and Haber.[75]

V_L and V_H for the light and heavy chains, respectively. The **constant regions** are $C_L$and C_H. Examination of the C_H region showed that much of the sequence is repeated after ~110 residues. In the IgG molecule the constant region of the heavy chains is made up of three such homologous domains (C_H1, C_H2, and C_H3). A fourth C_H domain is present in IgM. These facts suggest that duplication of a smaller gene coding for about 110 amino acids took place in the evolution of the immunoglobulins. Within the variable regions of immunoglobulin chains are **hypervariable regions** that form the antigen binding sites. These regions are located at the ends of the Fab fragments and involve both the light and heavy chains.[66a]

Within all of the domains each of the two peptide chains is folded in a similar way. Seven extended lengths of chain form two mostly antiparallel sheets between which hydrophobic side chains are packed. The overall size of the unit is ~4.0 x 2.5 x 2.5 nm. An S–S bridge links the two sheets in the center of each domain. The folding patterns in the variable domains are somewhat more complex. Different domains are linked by segments of extended peptide chain known as hinge or switch regions.[67–69] These impart a segmental mobility, which seems to be important for functioning of the molecules.[70]

The exact mode of binding to Fab fragments has been established for several specific **haptens**. Haptens are small molecules having the binding properties of antigenic determinants but unable by themselves to induce formation of antibodies when injected into animals. Binding of the hapten phosphocholine to one Fab fragment and of vitamin K to another[67] involves the hypervariable regions of both the heavy and light chains. The same is true for the binding of lysozyme[71–73] and a bacterial oligosaccharide.[74] The binding sites for the haptens or for the antigenic determinants on larger antigens are largely within the nine-stranded elliptical barrel or β sandwich formed from the two β sheets (Fig. 31-5).[69,75] Four strands come from the V_L domain and five from the V_H domain. The barrel forms the bottom and sides of the antibody-binding site, which can also be viewed as consisting of six separate loops of peptide chain.[76] IgA molecules are similar to those of IgG but have structurally different hinge regions as well as an extra 18-residue tailpiece at the C terminus of each heavy chain.[77]

As shown in Fig. 31-5, which provides a three-dimensional view of the variable domain of an Fab fragment of an immunoglobulin, there is a conserved hydrogen-bonded network even in this region. There are also "framework residues," which are highly conserved.[78] The antibody-binding site is provided largely by the three hypervariable regions present in each of the V_L and V_H domains. These are usually referred to in current literature as **complementarity-determining regions** (**CDRs**).[78–81] Each pair of heavy and light chains is held together by a conserved disulfide bridge.[82] Three-dimensional structures of a substantial number of different Fab fragments have provided precise knowledge about the antibody-binding cavities and about the forces involved in binding.[78,79,81,83] Among the established structures are those of Fab fragments specific for the following antigens: the haptens *p*-azophenylarsonate[84] and phencyclidine (p. 1798),[85] a sweet-tasting hapten,[86] triple-stranded DNA,[87] a DNA photoproduct,[88] creatine kinase,[89] staphylococcal nuclease,[90] an HIV capsid protein,[91] and an EGF receptor.[92] Structures are also known for single-domain antibodies from camels and llamas. These antibodies are naturally lacking in heavy chains but have single chains that fold back to mimic two-chain Fab fragments.[93,94] Similar single-chain antibody domains have also been created artifically.[95]

Not all proteins bind to antibodies in the usual binding cavity. **Protein G**, a cell surface protein from *Streptococcus* bonds to IgG molecules from many different species. Its binding site is on the outer surface of the heavy chain C_H1 domain.[96]

2. Antigenicity

Antibodies often bind haptens or complete antigens very tightly. The association constants K_f observed for monoclonal antibodies (Box 31-A) range from 10^6 to 10^{12} M^{-1}.[97] However, most natural antibodies have a lower affinity for their antigens. When protein antigens are denatured, the binding constants often decrease by 10^{-4} to 10^{-5}. This suggested that only antigenic determinants of relatively rigid structures serve as good antigens. However, when the reaction of antibodies with proteins of well-established three-dimensional structure were studied, it was found that the best antigenic determinants are those with some segmental mobility.[98,99] Furthermore, while some small peptides are good antigens, peptides are most highly antigenic when they can readily fold into a bend or other definite conformation.[100] Good antigenicity apparently requires some segmental flexibility as well as a definite conformation for the antigenic determinant.

3. Responses to Antibody Binding

Both B cells and T cells circulate throughout the body, spending only about 30 min during each cycle. They may meet and bind to an antigen in one of several different places.[50] Lymphocytes, which encounter blood-borne pathogens, usually initiate an immune response in the spleen. Responses to microorganisms in tissues are usually generated in lymph nodes. Ingested pathogens activate lymphocytes in specialized epithelial **microfold (M) cells** from which the antigen is transported to the Peyer's patches. Responses to inhaled or intranasal pathogens arise in the tonsils and adenoids. In every case one major aspect of the immune response results from binding of antibodies to antigens.

Antibodies by themselves do not destroy bacteria or viruses, but they induce responses that do. One immediate effect of antibodies is to remove offending materials or cells from circulation. When multivalent antibodies each combine with two different cells **agglutination** occurs. The agglutinated cells or multicellular organisms can then be destroyed by phagocytes. The coating of a cell surface by IgG is one form of a process called **opsonization**, a process that marks the cell as foreign and a target for phagocytosis.[3,53] Antibody–antigen interactions trigger several other responses as well. One of these results from the binding of protein **C1q**, a component of complement. Complement consists of a series of blood proteins that is poised to respond and to *complement* the action of antibodies in a variety of ways that are described in Section C,2 (see also Figs. 31-8 and 31-9). It has been established that it is the C_H2 domain of the Fc region of IgG that binds to C1q.[101] The binding occurs only after antigen (but not a small hapten) binds to the immunoglobulin.

Complement C1q is only one of several types of **Fc receptor**.[53,102,103] Others are involved in antigenic stimulation of B and T lymphocytes, macrophages, polymorphonuclear lymphocytes, and mast cells. Binding of the antibody–antigen complex to the receptors on phagocytic cells induces phagocytosis and release of oxygen metabolites, leukotrienes, prostaglandins, and other mediators of inflammation. The Fc domain mediates the uptake of antibodies from the mother's milk by young rats.[104,105] It also is the binding site of antibodies to **protein A**, a constituent of the cell wall of *Staphylococcus aureus*,[106] which is also widely used as a tool in immunological studies (Box 31-C). The neonatal Fc receptor, which is related structurally to Class I MHC antigens (Section D,5), is one of three major types of Fc receptor. The other two are the receptors for Fcγ (of IgG) and Fcε (of IgE). They (like their ligands) are members of the immunoglobulin superfamily. An exception is FcεRII (also called CD23), which resembles a C-type lectin. Some Fc receptors, e.g., FcγRI (CD64) and FcεRI, have a high affinity for their ligands with K_d ~10^{-8} to 10^{-10} M. Others, such as FcγRII (CD32) and FcγRIII (CD16), have lower affinities with K_d ~10^{-5} to 10^{-7} M.[103] Three-dimensional structures of several Fc receptor fragments, some in complexes with Fc fragments (Fig. 31-6), are known.[53,103,107–109] These include both IgG and IgE receptors.

It may be worthwhile to recall that many quite different proteins are members of the immunoglobulin structural family (Fig. 2-16). These include proteins

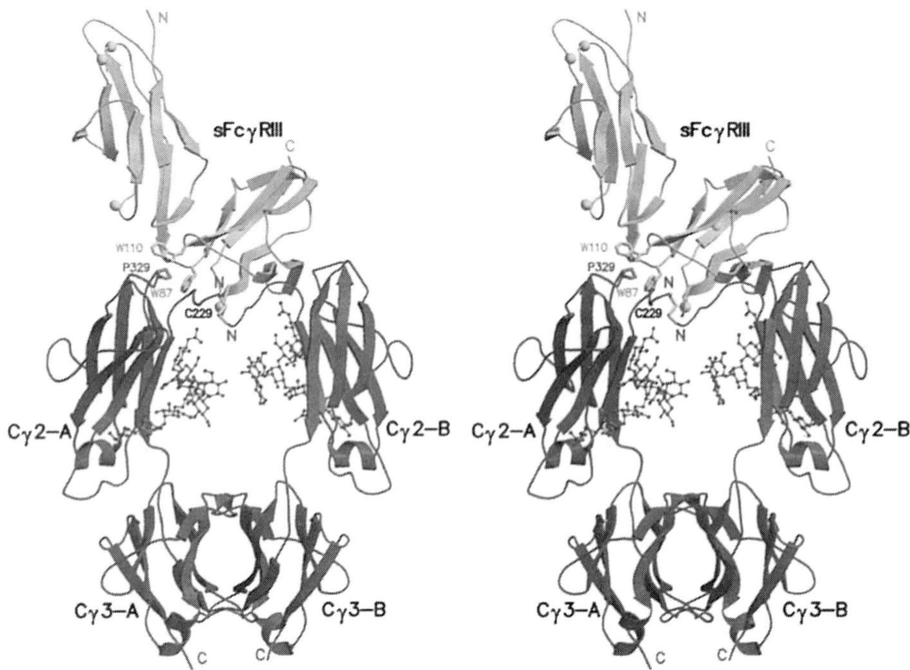

Figure 31-6 Three-dimensional ribbon representation of the structure of a complex of a soluble Fc fragment of a human IgG1 molecule. Pro 329 of the IgG and Trp 87 and Trp 110 of the Fc-receptor fragment form a "proline sandwich," which is shown in ball-and-stick form. The oligosaccharide attached to the Fc fragment of the antibody and the disulfide bridge between the two Cys 229 residues (at the N termini of the C2 domains of the heavy γ chains) are also shown. The small spheres on the Fc receptor fragment are potential sites for *N*-glycosylation. From Sondermann *et al.*[107] Courtesy of Uwe Jacob.

encoded by 64 genes of *E. elegans*, an organism that doesn't form antibodies.[110]

4. Clonal Expansion of B Cells; Plasma Cells

The immunoglobulins are synthesized both by plasma cells and by their precursors, the B cells. Each B lymphocyte makes antibodies of specific sequence in two forms, secreted and membrane-bound or **antigen receptor**. Mature "virgin" B cells, which are responsible for the primary immune response, make largely monomeric IgM and some IgD.[111] It isn't clear why IgD should be the predominant surface immunoglobulin on most B cells.[112] The secreted and membrane-bound (receptor) antibodies differ in their C-terminal sequences but are otherwise the same. A B cell responds to the binding of an antigen with a shape complementary to that of its IgM antigen receptor by multiplying and differentiating. Some of the progeny B cells start to divide and begin to differentiate into clones of plasma cells that secrete IgG or into cells of the gut that secrete IgA. Some of the B cells give rise to **memory cells**, long-lived lymphocytes that can be triggered into rapid proliferation many years later if the same antigen is encountered. The B cells also undergo a shift to secretion of pentameric IgM rather than to synthesis of membrane-bound antibody.

5. Help from T Cells

The maturation of B cells is a complex process that requires the cooperation of helper T cells. The B cells must process some antigen and present the peptide fragments for recognition by the MHC complex and T-cell receptors as described in Section D. If the antigen is recognized as foreign, the differentiation goes forward with the activated T cells secreting lymphokines that promote B-cell growth.

Before activated lymphocytes start to divide, interesting surface phenomena occur. If fluorescent antigens are allowed to bind to a lymphocyte, the cell surface is seen to be relatively evenly covered with the antibody–antigen complexes. Then after a short time the antibodies aggregate to form "patches" and begin to migrate to one side of the cell, where they eventually form a "cap." At still longer times the cap material is engulfed by the lymphocytes. Perhaps this phenomenon simply reflects the oriented flow of liquid within membranes (Chapter 8). On the other hand, the membrane-bound immunoglobulins, like other cell surface receptors, are integral membrane proteins whose cytoplasmic C termini may be attached on the inside of the cell to the cytoskeleton, which may control the capping process. The binding of lectins sometimes triggers lymphocytes into antibody synthesis, but it is not clear how the binding of a lectin to a carbohydrate receptor can have the same effect as binding of an antigen to a surface IgM.[113]

C. Some Specialized Proteins of the Immune System

The immune response depends not only upon recognition of foreign antigens but also upon an extensive signaling network and upon a series of specialized

BOX 31-A MONOCLONAL ANTIBODIES

A mouse may make over 10 million different antibodies. Because of this heterogeneity it was impossible to learn antibody structures until the discovery of the myeloma proteins (Fig. 31-4D). These were produced in the bone marrow by clones of specific immunoglobulin-forming malignant cells. However, it was still not possible to obtain homogeneous antibodies to any desired antigenic determinant. The discovery of a method of forming such **monoclonal antibodies** by Milstein and Köhler[a–c] in 1975 provided a new tool with many biochemical and medical applications.[d–f] What Milstein and Kohler did was to immunize mice against an antigen of interest. They then fused B cells from the spleen of the immunized mouse with cultured myeloma cells. The resulting **hybridomas** grow vigorously and produce antibodies of the type dictated by the B cells. Since each hybridoma cell is derived from a single B cell, it makes a single kind of antibody. By plating out and selecting clones of hybridoma cells it is often possible to find a monoclonal antibody that binds well to a specific antigenic determinant. The hybridoma can be cultured indefinitely, producing its monoclonal antibody in any desired quantity.

A major application of monoclonal antibodies is in clinical assays for drugs, bacterial and viral products, tumor antigens, hormones, and other circulating proteins. Their use in conjunction with immunoassays (Box 31-C) has provided increased specificity and sensitivity. Another major application is to observe binding of antibodies to specific proteins by electron microscopy. The location of specific receptor proteins can be established[g–j] as can the locations of ribosomal proteins and many other cellular components (Fig. 29-1). Monoclonal antibodies to acetylcholine receptors have been shown to induce symptoms of myasthenia gravis (Box 31-D), supporting the autoimmune origin of that disease.[h] Monoclonal antibodies specific for such a small hapten as mercuric ion have been isolated.[k]

Several problems have limited the wider use of monoclonal antibodies created by the hydridoma method. The antibodies are those of a mouse and are antigenic to humans.[f,l–n] This long prevented many medicinal uses. Years of effort have gone into attempts to "humanize" the antibodies. One approach is to introduce human immunoglobin genes into mice. Another is to use recombinant DNA techniques to clone genes for immunoglobulin fragments and to introduce these into cells of *E. coli* in which additional genetic diversity in the antibodies arises.[l,o] Selection of antibody fragments is often accomplished using bacteriophage display systems (Fig. 3-16).[f,l] After selection gene fragments can be reassembled into a final form. Recently, using cloning of large pieces of the several Mbp of human immunoglobin genes into yeast artificial chromosomes (p. 1497), it has been possible to prepare purely human monoclonal antibodies.[f,n]

Many attempts have been made to link monoclonal antibodies specific for antigenic determinants on cancer cells to protein toxins such as ricin (Box 29-A). It is hoped that this may provide an effective way of carrying toxins into cancer cells.[f,p–r] Therapeutic human monoclonal antibodies are already in use as antirejection drugs for kidney transplantation, for treatment of rheumatoid arthritis, Crohn disease, and for some types of cancer.[f]

---

[a] Milstein, C. (1980) *Sci. Am.* **243**(Oct), 66–74
[b] Milstein, C. (1986) *Science* **231**, 1261–1268
[c] Kohler, G. (1986) *Science* **233**, 1281–1286
[d] Yelton, D. E., and Scharff, M. D. (1981) *Ann. Rev. Biochem.* **50**, 657–680
[e] Birch, J. R., and Lennox, E. S., eds. (1994) *Monoclonal Antibodies*, Wiley-Liss, New York
[f] Ezzell, C. (2001) *Sci. Am.* **285**(Oct), 36–41
[g] Greaves, M. F., ed. (1984) *Monoclonal Antibodies of Receptors: Probes for Receptor Structure*, Chapman & Hall, London
[h] Tzartos, S. J. (1984) *Trends Biochem. Sci.* **9**, 63–67
[i] Harlow, E., and Lane, D. (1999) *Using Antibodies: A Laboratory Manual*, Cold Spring Harbor Lab. Press, Cold Spring Harbor, New York
[j] Goldman, R. D. (2000) *Trends Biochem. Sci.* **25**, 593–595
[k] Wylie, D. E., Lu, D., Carlson, L. D., Carlson, R., Babacan, K. F., Schuster, S. M., and Wagner, F. W. (1992) *Proc. Natl. Acad. Sci. U.S.A.* **89**, 4104–4108
[l] Marks, C., and Marks, J. D. (1996) *N. Engl. J. Med.* **335**, 730–733
[m] Neuberger, M. S. (1985) *Trends Biochem. Sci.* **10**, 347–349
[n] Neuberger, M., and Brüggemann, M. (1997) *Nature (London)* **386**, 25–26
[o] Plückthun, A. (1990) *Nature (London)* **347**, 497–498
[p] Collier, R. J., and Kaplan, D. A. (1984) *Sci. Am.* **251**(Jul), 56–64
[q] Pastan, I., and FitzGerald, D. (1991) *Science* **254**, 1173–1177
[r] Oeltmann, T. N., and Frankel, A. E. (1991) *Trends Biochem. Sci.* **5**, 2334–2337

antibiotics, toxins, and hormones. Some of these, such as the defensins, are major, rapid-acting components of the innate system. Both the innate and adaptive systems utilize the complement proteins, and both employ numerous cytokines and other signaling proteins. Plants, and perhaps also other organisms, employ gene silencing by small RNA molecules as part of their defense against viruses.[113a,b]

BOX 31-B CATALYTIC ANTIBODIES

Both enzymes and antibodies are proteins. Antibodies consist of subunits with multiple domains, just as do some enzymes. Both enzymes and antibodies have binding sites for small molecules between domains or subunits. In view of such similarities it isn't surprising that some antibodies have catalytic properties. The possibility was suggested in 1969 by Jencks.[a] He also proposed that injection of a mouse with a hapten, that resembled a transition state for an enzyme, might induce formation of antibodies complementary to the transition-state structure. These might be catalytic. By the early 1980s such antibodies were discovered.[b–d] Some of the first catalytic antibodies (also dubbed **abzymes**) had esterase activity. The haptens used to induce antibody formation were phosphonates such as the following.[e,f]

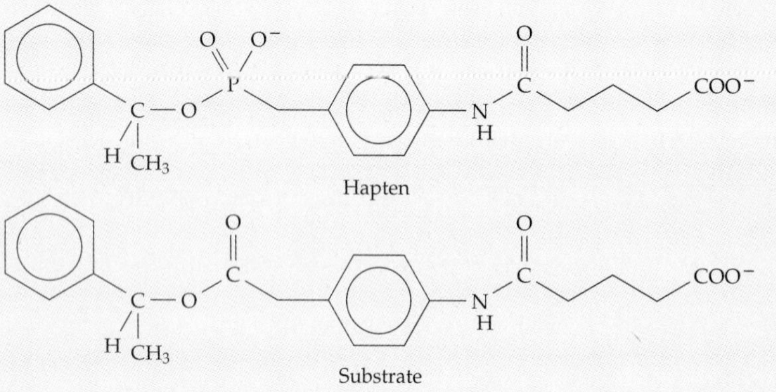

Hapten

Substrate

Using the transition-state analog shown on p. 485 a catalytic antibody with chorismate mutase activity was isolated.[g] Many antibodies catalyzing additional reactions have also been found. Although they are usually less active than natural enzymes, in some cases they approach enzymatic rates. Furthermore, they may catalyze reactions for which no known enzymes exist.[h]

Catalytic antibodies, like enzymes, must be isolated and purified to homogeneity before they can be studied. Initially this was done by using the hybridoma technique for isolation of monoclonal antibodies (Box 31-A). After induction of antibody formation by injecting a selected hapten into a mouse, large numbers of monoclonal antibodies had to be tested for catalytic activity. Even if several thousand different monoclonal antibodies were tested, only a few with catalytic properties could be found.[i] Newer methods have incorporated recombinant DNA techniques (Box 31-A) and use of combinatorial libraries and phage display.[j–m] Incorporation of acidic or basic groups into the haptens used to induce antibody formation may yield antibodies capable of mimicking the acid–base catalysis employed by natural enzymes.[n,o]

A sample of the types of reaction for which catalytic antibodies have been discovered or designed include the following: ester hydrolysis,[e] transesterification,[p] amide hydrolysis,[q] serine protease-like hydrolysis,[r] elimination,[h,s] aldol cleavage,[t] decarboxylation,[u,v] deiodination by a selenium-containing antibody,[w] pericyclic rearrangements,[g,x] and the Diels–Alder reaction.[y,z] Like natural enzymes catalytic enzymes can be mutated and engineered and can be used to study fundamental aspects of catalysis.[aa] Fluorescent probes incorporate near active sites may provide information about mechanisms or may signal information of diagnostic significance.[bb]

In science we must always expect the unexpected. Do antibodies all catalyze the reaction of singlet molecular oxygen $^1O_2$* with H_2O to form H_2O_3 and H_2O_2? How?[cc]

1. Defensins and Other Antibacterial Polypeptides

Only higher vertebrates have an adaptive immune system with circulating antibodies. However, from bacteria to higher plants and human beings all of us utilize defensive polypeptides for protection. More than 500 have been identified.[114] Many have a broad specificity, attacking both bacteria and other pathogens. Among these peptides are more than 200 bacterially produced antibiotics such as gramicidin, tyrocidines, and colicins (Boxes 20-G, 8-D). More recently discovered are the 37- to 59-residue **bacteriocins**, formed by lactic acid bacteria.[115] Like colicin E1 (Box 8-D) and alamethicin (p. 1774) they disrupt cytoplasmic membranes of some other groups of bacteria.

Helicobacter pylori, which is associated with stomach ulcers, forms a 38-residue antibiotic that may help protect infected persons from other bacteria.[116] This peptide forms a simple two-helix structure and is one of a large number of simple helical antimicrobial polypeptides 40 residues or less in length. Among them

BOX 31-B (continued)

[a] Jencks, W. P. (1969) *Catalysis in Chemistry and Enzymology*, McGraw-Hill, New York (p. 288)
[b] Lerner, R. A., and Tramontano, A. (1987) *Trends Biochem. Sci.* **12**, 427–430
[c] Benkovic, S. J. (1992) *Ann. Rev. Biochem.* **61**, 29–54
[d] Lerner, R. A., Benkovic, S. J., and Schultz, P. G. (1991) *Science* **252**, 659–667
[e] Wedemayer, G. J., Wang, L. H., Patten, P. A., Schultz, P. G., and Stevens, R. C. (1997) *J. Mol. Biol.* **268**, 390–400
[f] Janda, K. D., Benkovic, S. J., and Lerner, R. A. (1989) *Science* **244**, 437–440
[g] Haynes, M. R., Stura, E. A., Hilvert, D., and Wilson, I. A. (1994) *Science* **263**, 646–652
[h] Larsen, N. A., Heine, A., Crane, L., Cravatt, B. F., Lerner, R. A., and Wilson, I. A. (2001) *J. Mol. Biol.* **314**, 93–102
[i] Tawfik, D. S., Zemel, R. R., Arad-Yellin, R., Green, B. S., and Eshhar, Z. (1990) *Biochemistry* **29**, 9916–9921
[j] Posner, B., Smiley, J., Lee, I., and Benkovic, S. (1994) *Trends Biochem. Sci.* **19**, 145–150
[k] Janda, K. D., Lo, L.-C., Lo, C.-H. L., Sim, M.-M., Wang, R., Wong, C.-H., and Lerner, R. A. (1997) *Science* **275**, 945–948
[l] Gao, C., Lavey, B. J., Lo, C.-H. L., Datta, A., Wentworth, P., Jr., and Janda, K. D. (1998) *J. Am. Chem. Soc.* **120**, 2211–2217
[m] Baca, M., Scanlan, T. S., Stephenson, R. C., and Wells, J. A. (1997) *Proc. Natl. Acad. Sci. U.S.A.* **94**, 10063–10068
[n] Kemp, D. S. (1995) *Nature (London)* **373**, 196–197
[o] Thorn, S. N., Daniels, R. G., Auditor, M.-T. M., and Hilvert, D. (1995) *Nature (London)* **373**, 228–230
[p] Wirsching, P., Ashley, J. A., Benkovic, S. J., Janda, K. D., and Lerner, R. A. (1991) *Science* **252**, 680–685
[q] Thayer, M. M., Olender, E. H., Arvai, A. S., Koike, C. K., Canestrelli, I. L., Stewart, J. D., Benkovic, S. J., Getzoff, E. D., and Roberts, V. A. (1999) *J. Mol. Biol.* **291**, 329–345
[r] Zhou, G. W., Guo, J., Huang, W., Fletterick, R. J., and Scanlan, T. S. (1994) *Science* **265**, 1059–1064
[s] Romesberg, F. E., Flanagan, M. E., Uno, T., and Schultz, P. G. (1998) *J. Am. Chem. Soc.* **120**, 5160–5167
[t] Karlstrom, A., Zhong, G., Rader, C., Larsen, N. A., Heine, A., Fuller, R., List, B., Tanaka, F., Wilson, I. A., Barbas, C. F., III, and Lerner, R. A. (2000) *Proc. Natl. Acad. Sci. U.S.A.* **97**, 3878–3883
[u] Hotta, K., Lange, H., Tantillo, D. J., Houk, K. N., Hilvert, D., and Wilson, I. A. (2000) *J. Mol. Biol.* **302**, 1213–1225
[v] Smiley, J. A., and Benkovic, S. J. (1994) *Proc. Natl. Acad. Sci. U.S.A.* **91**, 8319–8323
[w] Lian, G., Ding, L., Chen, M., Liu, Z., Zhao, D., and Ni, J. (2001) *J. Biol. Chem.* **276**, 28037–28041
[x] Driggers, E. M., Cho, H. S., Liu, C. W., Katzka, C. P., Braisted, A. C., Ulrich, H. D., Wemmer, D. E., and Schultz, P. G. (1998) *J. Am. Chem. Soc.* **120**, 1945–1958
[y] Romesberg, F. E., Spiller, B., Schultz, P. G., and Stevens, R. C. (1998) *Science* **279**, 1929–1933
[z] Heine, A., Stura, E. A., Yli-Kauhaluoma, J. T., Gao, C., Deng, Q., Beno, B. R., Houk, K. N., Janda, K. D., and Wilson, I. A. (1998) *Science* **279**, 1934–1940
[aa] Romesberg, F. E., Santarsiero, B. D., Spiller, B., Yin, J., Barnes, D., Schultz, P. G., and Stevens, R. C. (1998) *Biochemistry* **37**, 14404–14409
[bb] Simeonov, A., and 14 other authors. (2000) *Science* **290**, 307–313
[cc] Wentworth, P., Jr., Jones, L. H., Wentworth, A. D., Zhu, X., Larsen, N. A., Wilson, I. A., Xu, X., Goddard, W. A., III, Janda, K. D., Eschenmoser, A., and Lerner, R. A. (2001) *Science* **293**, 1806–1811

are the **cecropins** of insects and **magainins** and **buforins** of amphibians.[117,117a] Many of these kill by disrupting membranes or by forming pores in membranes. However, others enter bacteria and disrupt functions of nucleic acids, enzymes, etc.[117,118] Many antibacterial peptides have been isolated from insects,[12,119] scorpions,[120] spiders and horseshoe crabs,[121,122] and amphibians.[123] All of these organisms lack adaptive immunity but have strong innate immunity.

The human body is protected by two groups of defensins formed in the skin, in mucous membranes, in secretions of neutrophils, and other phagocytic cells. The α-defensins (Fig. 31-7) are 29–35 residues in length and are active against both gram-positive and gram-negative bacteria as well as fungi and enveloped viruses including HIV.[12,124–125a] The β-defensins are mainly active against gram-negative bacteria and yeast. They also possess immunostimulatory activity that is important in activating the adaptive immune response.[32,125,126] Various tissue-specific defensins have been discovered.[127] Peptides of the **trefoil family** protect the gastrointestinal tract.[128–130] Eosinophils,[131] leukocytes, and neutrophils make additional protective proteins. One leukocyte defensin is a macrocyclic peptide, whose gene may have arisen by fusion of two segments encoding nonapeptide segments of α-defensins.[132] Neutrophils form, in addition to defensins, α-helical peptides called **cathelicidins**, which protect skin from invasive bacterial infection.[133,133a] Their synthesis is greatly increased after wounding. They may be among the proteins whose absence after severe burning is likely to be fatal.

Both α and β-defensins consist largely of β strands (Fig. 31-7) and are linked by three disulfide bridges. Some scorpion and insect defensins resemble scorpion toxins (Fig. 30-16) and have four S–S bridges. Fungi and green plants[135,136] also form antimicrobial peptides. A 30-residue fungal protein is highly knotted and contains four S–S bridges.[137] Some polypeptides from the oleander and related plants are 29- to 31-residue macrocyclic structures with two S–S bridges in a **cysteine knot** structure[138–140] (Fig. 31-7). They are exceptionally stable and protease-resistant and may have defensive activity against insects. Defensins are small polypeptides, but larger proteins are also part of the innate defense system. For example, a

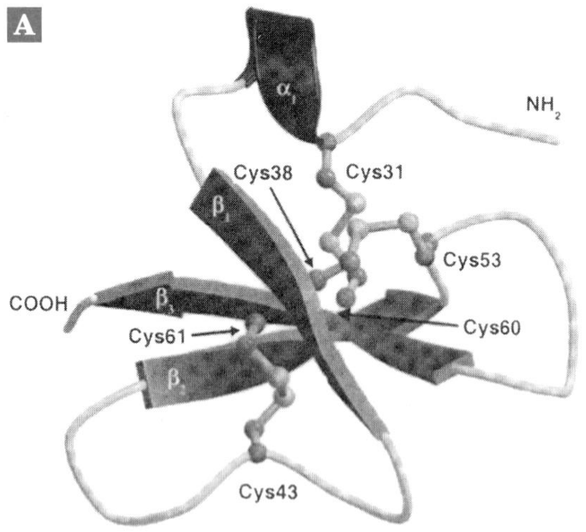

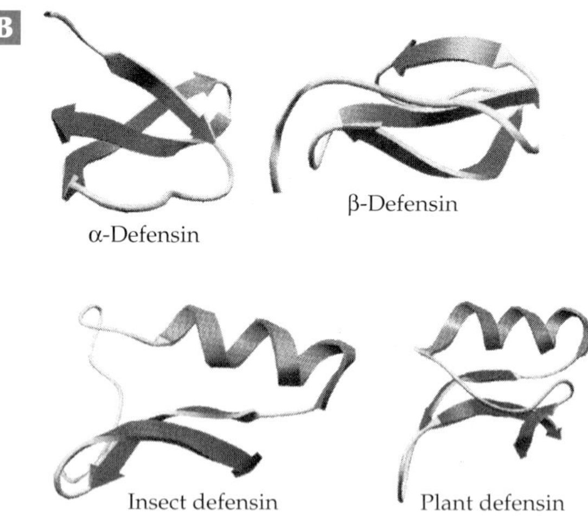

Figure 31-7 Ribbon structures of some defensins. (A) Structure of a human β-defensin showing the three disulfide bonds. From Bauer *et al.*[134] Courtesy of Heinrich Sticht. (B) Comparison of the folding patterns of four types of defensins. Mammalian α- and β-defensins are all β sheets with somewhat different arrangements of disulfide bridges. Insect and plant defensins have an α helix joined to the β sheet. Mammalian and insect defensins have three disulfide bridges, while plant defensins have four. From Hoffmann *et al.*[12] Courtesy of Jules A. Hoffmann.

93-residue protein from onion seeds resembles plant in lipid-transfer proteins.[135] Some frog skins contain a 60-residue trypsin inhibitor.[141] Ribosome-inactivating proteins are well known (Box 29-A).

2. Complement

Complement is a group of more than 30 proteins found in blood serum, which are activated in a cascade mechanism when antibody and antigen combine[10,111,142–148] (Eq. 31-1). This **classical pathway** for activation of complement is outlined in Fig. 31-8. The proteins involved in the cascade are designated C1 to C9. Many of them undergo proteolytic cleavage, the

$$\begin{array}{l}\text{Antigen + antibody} \\ \quad\downarrow \\ \text{Complex} \xrightarrow{\text{Complement proteins}} \text{Cell lysis, phagocytosis, etc.}\end{array} \qquad (31\text{-}1)$$

cleavage products being designated by a or b, e.g., C3a and C3b. The b fragment is usually the larger of the two. There is also an **alternative pathway** that is part of the innate system. It is activated by such foreign surfaces as lipopolysaccharides of bacterial cell walls. Its special proteins are called **factors**, e.g., factor B, factor D. A third pathway, the **lectin pathway**, is activated by microbial surface mannans, which bind to a serum **mannan-binding lectin** (**MBL**). This protein, a so-called defense collagen, resembles protein C1q (next paragraph).[149] It activates two associated serine proteases (MASP1 and MASP2), which are able to cause cleavage of proteins C4 and C2 and possibly C3 in the classical pathway (Fig. 31-8). The ultimate effects of the action of complement include destruction of cells by lysis and activation of leukocytes, which engulf foreign cells by phagocytosis. Complement also induces the release of **chemotactic factors** that attract polymorphonuclear leukocytes and monocytes to the site involved.[150]

The classical pathway begins with the **recognition component C1** of complement. This is a complex of three proteins, C1q, C1r, and C1s. Proteins C1r and C1s form a mixed tetramer Clr_2s_2, while C1q binds to the C_H2 domain of "activated" antibodies, that is, with immunoglobulins that have combined with an antigen. It takes at least a dimer or larger aggregate of IgG to activate C1q, whereas a single molecule of the naturally pentameric IgM suffices. The mechanism by which this activation occurs is uncertain. Perhaps a change of conformation within the immunoglobulin acompanies antibody binding and is responsible for generation of a binding site for C1q. It may seem strange that haptens cannot cause complement binding, and that they do not cause detectable conformational alterations in Fab. Only multivalent antigens able to bind to more than one antibody induce complement binding. However, as we have learned in recent years, many biological responses involve transient assembly of large aggregates of different protein components. In this context, the requirement for two or more antibody molecules doesn't seem so strange.

The 400-kDa C1q consists of a central portion of diameter 3–6 nm and length 10–12 nm to which are attached six very thin connecting strands. These are

~14.5 nm long and ~1.5 nm in diameter and terminate ~135-residue globular ends of ~6 nm diameter,[151–153] which are thought to be the sites of combination with the immunoglobulins. The thin connecting strands have, for most of their length, a collagenlike structure with a high content of hydroxyproline and hydroxylysine. The latter is glycosylated by glucosylgalactosyl disaccharides as in collagen itself (pp. 181, 432, 433). The reason for this unusual structure is not obvious. We do know that the binding of antigens activates the complement-binding regions of antibodies, and that the activated antibodies then bind C1q. This binding in some manner activates C1q, which in turn activates C1r subunits of the Clr_2s_2 tetramer.[154] The latter is thought to bind at the center of C1q, while the antibodies bind at the outer ends. We don't know how the activation message is carried from the outer arms to the center. C1q is a member of a group of collagen-like proteins that includes protein MBL (also designated MBP) and surfactant protein A (SP-A; p. 436).[155]

Activated C1r (often designated $\overline{C1r}$ but here and in Fig. 31-8 as C1**r**) is one of five different serine proteases involved in activation of complement.[156] The substrate for the trypsinlike C1**r** is C1s, a proenzyme which is converted by the action of C1**r** into another trypsinlike serine protease C1**s**.[157–160] Through a rather elaborate cascade mechanism, depicted in Fig. 31-8, the important proenzyme C2 is activated.[161] Its active form C2**a** is a serine protease, which cleaves proteins C3 and C5 to the active forms C3b and C5b. Protein C4 is also cleaved to C4b by activated C1. C4 and C3 are also activated, and protein C5 is cleaved

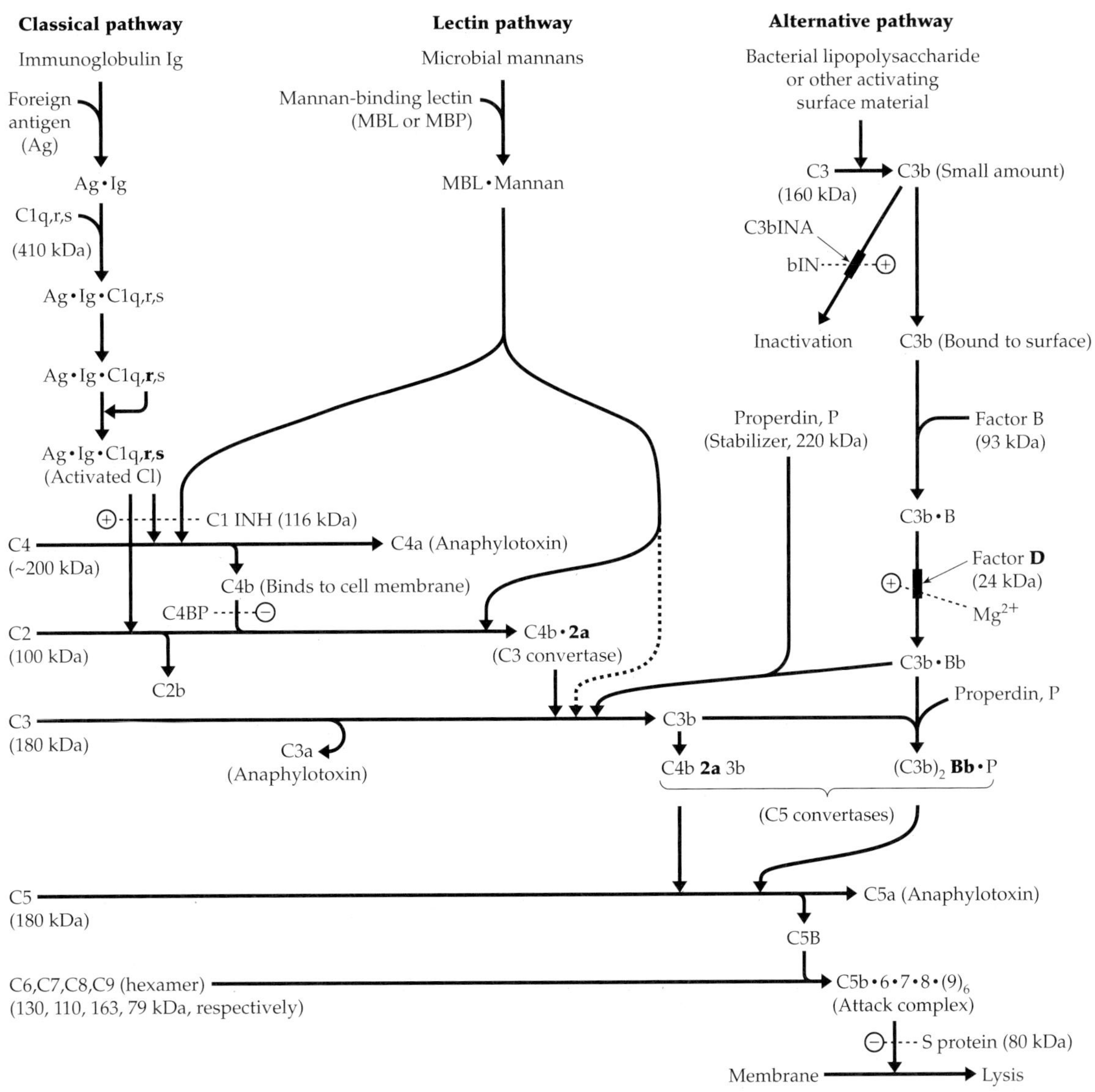

Figure 31-8 Pathways for activation of the complement system. Active proteases are designated by abbreviations in boldface.

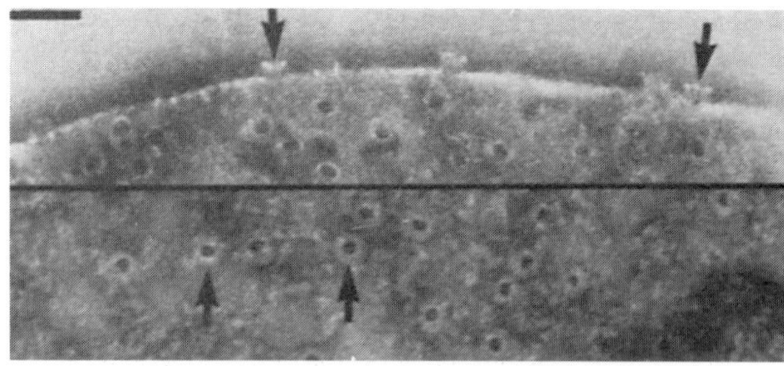

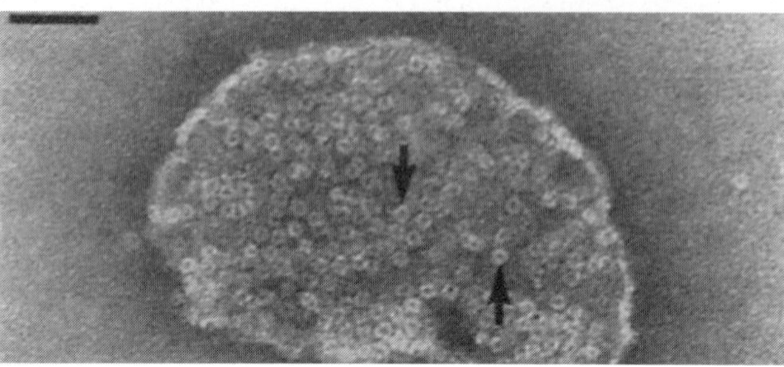

Figure 31-9 Electron micrograph of a negatively stained sheep erythrocyte lysed with human complement. The cylindrical "attack complex" embedded in the membrane is seen in the upper left frame in side projection and in the lower frames in axial projection. The top views are of a proteolytically "stripped" ghost, the lower view from a freshly lysed ghost. The inner diameter of the cylinders is 10 nm; scale bars 50 nm. From Bhakdi and Tranum-Jensen.[169]

to C5b. The 200-kDa C4 consists of three chains, all derived by proteolytic cleavage of a precursor. It is also glycosylated and sulfated. Both C4 and C3 contain internal thiol ester linkages and act as "molecular mousetraps" (Box 12-D).[162] They react to fix the proteins covalently to the assembling complement complex.[163] Protein C5b interacts with C6, C7, C8; and six molecules of C9 to generate an "attack complex," which inserts donutlike rings into the cell membrane being attacked (Fig. 31-9).[164,164a] Although there has been some uncertainty about the mechanism of lysis, it seems likely that it is at least partly a result of loss of ions through the holes in the donut.

In the alternative pathway of activation a small amount of C3b is formed and becomes bound to the cell surface. This binds another proenzyme factor B,[165] which is converted by protease factor D[166] to active protease **Bb**. The latter in its complex with C3b is the enzyme that cleaves C3 in large amounts and permits a rapid formation of more **Bb** and also of the complex $(C3b)_2$ **Bb•**P, which attacks C5 (Fig. 31-8). These complexes are stabilized by the abundant serum protein **properdin** (P).[167]

Other components of complement are the plasma C1 protease inhibitor,[168] which prevents accidental activation of the system, and protease C3bINA, which inactivates C3b. The latter depends upon accessory protein bIN. Another component, serum carboxypeptidase B (SCPB), inactivates anaphylotoxins C3a, C4a, and C5a. These small ~80-residue pieces have a variety of powerful biological activities.[170–174] They are chemotactic factors for leukocytes and induce release of histamine from mast cells. In excess, they can cause anaphylaxis; hence their rapid degradation is essential. An excess of C5a may be present both in asthma and in rheumatoid arthritis.

The ninth component of complement, C9, is a 70-kDa 537-residue water-soluble glycoprotein, which contains a hydrophobic domain that aggregates to form the ion conducting channels.[175,176] Proteins that closely resemble C9 and are called **perforins** or **cytolysins** are found in cytoplasmic granules of cytotoxic T-lymphocytes and natural killer cells. These ~66-kDa proteins are assembled into rings similar to those formed by C9 and may be involved in the killing action of these cells (Fig. 31-9).[177–179] Certain pathogenic amebas, which may cause a fatal infection, also utilize a similar pore-forming protein.[180]

Every regulatory system in the body must be prevented from overactivity or activity that is unnecessarily prolonged. This can help us understand that, just as with blood clotting (Fig. 12-17), a network of regulatory factors controls the complement system. Among these are an inhibitory C4b-binding protein (C4BP),[181] which acts to prevent excessive formation of the C4b•C2**a** complex (Fig. 31-8). Complement **cofactor I** is a serine protease that cleaves both C3b and C4b into smaller pieces in the presence of **cofactor H**[181a] or of C4BP. Its absence leads to excessive consumption of C3 and recurrent pyogenic infections.[182] The **membrane cofactor protein** (MCP) stimulates this action of cofactor I in inhibiting attacks of complement on the cells that carry MCP.[183] Acting in the opposite direction is **complement receptor 2** (CR2 or CD21), which acts as a receptor for proteolytic fragment C3d. This fragment binds CR2-bearing cells to the B cell receptor, amplifying the B cell response to foreign antigens.[184]

Complement is involved not only in attacking foreign cells but in inflammation. Unfortunately, this is sometimes accompanied by serious problems. Human diseases in which complement is thought to be involved include glomerulonephritis, rheumatoid arthritis, myasthenia gravis, and lupus erythematosus.

3. Cytokines, Interferons, and the Acute-Phase Response

The body responds in many ways to infection, injury, or cancer. These include the secretion of cytokines, interferons, and proteins of the acute phase response. These proteins, many of which are quite small, are involved in communication with other cells, often with specialized cells of the immune system. **Cytokines**, some of which are considered in Chapter 30 (Section A,6; Fig. 30-6), are small hormonelike molecules. They may stimulate, inhibit, or exhibit other effects on cells of the immune system. They often have pleiotropic effects, not acting in the same way on all types of cells.[185] The cytokines known as **interleukins** (IL-1, etc.) are produced by leukocytes. **Lymphokines** are formed by lymphocytes and **monokines** by monocytes. Based on their functions there are four categories of cytokines.[111]

(1) Mediation of natural immunity: type I interferons, IL-1, IL-6, and more than 40 **chemokines** (small highly basic chemotactic proteins).
(2) Regulation of lymphocytes, activation, growth, and differentiation of B and T cells: IL-2, IL-4, IL-21, TGF-β.
(3) Regulation of immune-mediated inflammation: Interferon-γ, tumor necrosis factor (TNF), IL-5, IL-10, IL-12, and migration inhibition factor (MIF).
(4) Stimulation of hematopoeiesis (IL-3, IL-7), colony-stimulating factors (CSF; see also Chapter 32).

Cytokines all function using a group of transmembrane receptors embedded in the plasma membranes of target cells. The receptors have no tyrosine kinase activity but associate with and activate kinases known as **Janus kinases (JAKs)**. These kinases phosphorylate tyrosine side chains in their receptors, and the phosphorylated receptors activate transcription factors of the **STAT** (signal transducer–activators of transcription) group.[186–195] The specificity of cytokine action results from a combination of receptor recognition and recognition of the various STAT molecules by different JAKs.[111] Cytokines have a variety of structures. Many are helix bundles or have β sheet structures (Fig. 30-6).

Interferons. The interferons (IFNs),[196,197] which were discovered in 1957, are proteins secreted by leukocytes, fibroblasts, and activated lymphocytes. They inhibit replication of viruses as well as the growth of host cells and also have antitumor activity. Interferons are classified as α (from leukocytes), β (from fibroblasts), and γ (from lymphocytes). According to their affinities for the two types of known interferon receptors, interferons IFN-α, IFN-β, and the less well known IFN-ω and IFN-τ are designated type I,[198–201] while interferon γ (IFN-γ) is type II. At least 15 homologous 166-residue human α interferons are known.

The binding of interferons to their receptors induces a rapid increase in the transcription of particular genes and synthesis of corresponding proteins.[196,202] One of the proteins induced is a **double-stranded RNA-activated 2'-5'A synthase**, which polymerizes ATP to a series of 2'-5' linked oligonucleotides containing triphosphates at the 5' termini.[202–204] Double-stranded RNA is uncommon except in replicating viruses, and it is thought that the activation by dsRNA is related to establishment of an antiviral state. Another interferon-induced enzyme is the small subunit of eukaryotic protein synthesis initiation factor eIF-2. This is converted to an inactive phosphorylated form by a dsRNA-dependent protein kinase[205] (Fig. 31-10). The protein kinase also appears to be an interferon-induced protein[206] as is the oligo(2'-5' A)-activated RNAse indicated in Fig. 31-10.[207] Interferons have effects other than inducing the antiviral state. Thus, human IFN-β_2 is identical to a B-cell differentiation factor.[208] Both IFN-α and IFN-β have antigrowth activity and are currently in use for treatment of some forms of cancer as well as for viral infections.[209]

Interleukin-1 (Il-1) plays a key role in the body's response to microbes and to tissue injury.[210,211] It actually consists of three similar proteins, **Il-1α**, **IL-1β**, and **Il-1 receptor antagonist**. The first two are the active cytokines with a wide range of effects among

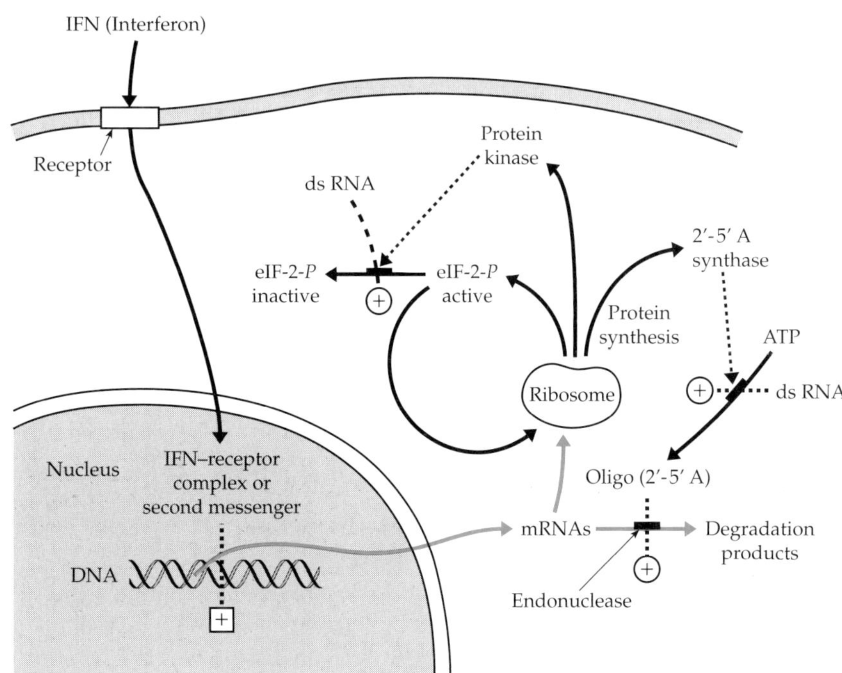

Figure 31-10 Some responses to the binding of an interferon to a cell surface receptor.

which are induction of inflammation and pain.[212] Il-1β is thought to be most active in promoting inflammation but only after it is cleaved by **interleukin-1β-converting enzyme** (see p. 619).[213] Blocking of Il-1 receptors provides a potential new method for control of pain.[214] **Interleukin-6** (IL-6) is also needed for an optimal immune system. Its effects overlap those of Il-1, and it has a potent activity in inducing the acute-phase response.[215] Like Il-1, it is a four-helix cytokine.

Also required by the immune response are the numerous **chemokines**. Chemoattractant molecules provide concentration gradients that direct the movement of B and T cells and other leukocytes.[216–219] Chemokines bind to seven-helix receptors, often

BOX 31-C IMMUNOASSAYS

Among the important techniques that have permitted rapid progress in studies of hormone action is the use of specific antibodies formed against hormones, hormone-protein conjugates, or other molecules.[a–c] The first of these techniques to come into general use was the **radioimmunoassay** (RIA),[d–f] which was devised by Yalow and Berson.[d] In one form of RIA various amounts of a sample containing an unknown quantity of hormone, e.g., insulin, are placed in a series of tubes. Additional tubes containing known amounts of the hormone are also prepared. Then a standard quantity of radiolabeled hormone (often iodinated with a γ emitter such as ^{125}I) is added to each tube together with a standard quantity of the specific antibody to the hormone. The solution is incubated for minutes or hours to obtain equilibrium between hormone (the antigen) and antibody-hormone complex. The antibody-hormone complex is then separated, e.g., by gel filtration or ammonium sulfate precipitation, and the radioactivity of the complex is measured. In the tubes containing higher concentrations of hormone, the labeled hormone has been diluted more, and the amount bound to antibody is less than in tubes with lower concentrations of hormone. The tubes of known concentrations are used to construct a standard curve from which the unknown concentrations can be read. As little as a femptomole of hormone (i.e., the amount present in 1 ml of a 10^{-12} M solution can be detected).[f] Methods are available for virtually every pure hormone.[c]

The RIA methods were made more convenient by adsorbing either the antibody of antigen to the plastic surface of a tube or depression plate. This facilitates separation of the antibody-ligand complex and washing. A variety of other immunoassays techniques have been devised. For example, in **enzyme-linked immunoabsorbent assays** (ELISA),[c] the amount of adsorbed antibody-ligand complex is measured by treating the washed surface with a second antibody, which is directed against the first. The second antibody is linked covalently to an enzyme, whose activity can then be measured by a suitable colorimetric procedure. The reactions involved are as follows.[g,h]

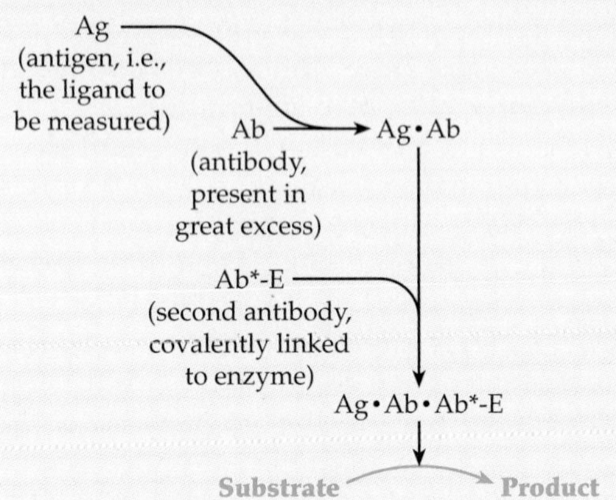

Variations, which avoid the use of radioisotopes, are replacing RIA. Some utilize stable isotopes. However, ^{14}C at such low levels that there is no radioactive waste can be coupled with accelerator mass spectrometry to provide very sensitive immunoassays.[i] A great variety of other procedures are available. Some involve coupling to antibodies that carry fluorescent labels. Many are now automated. Often protein A from *Staphylococcus aureus* is utilized in various ways that take advantage of its ability to bind to the Fc portion of IgG from virtually all mammals. For example, it may fix antibodies to a surface or to a label.[j]

[a] Price, C. P., and Newman, D. J., eds. (1991) *Principles and Practice of Immunoassay*, Stockton Press, New York
[b] Lindbladh, C., Mosbach, K., and Bülow, L. (1993) *Trends Biochem. Sci.* **18**, 279–283
[c] Crowther, J. R. (1995) *ELISA: Theory and Practice*, Humana Press, Totowa, New Jersey
[d] Yalow, R. S. (1978) *Science* **200**, 1236–1245
[e] Brooker, B., Terasake, W. L., and Price, M. G. (1976) *Science* **194**, 270–276
[f] Jaffe, B. M., and Behrmann, H. R., eds. (1974) *Methods of Hormone Radioimmunoassay*, Academic Press, New York
[g] van Vunakis, H. and Langone, J. J.(1980) *Methods Enzymol.* **70**, entire volume
[h] Langone, J. J., and van Vunakis, H..eds. (1983) *Methods of Enzymology* **92**, entire volume
[i] Shan, G., Huang, W., Gee, S. J., Buchholz, B. A., Vogel, J. S., and Hammock, B. D. (2000) *Proc. Natl. Acad. Sci. U.S.A.* **97**, 2445–2449
[j] Surolia, A., Pain, D., and Khan, M. I. (1982) *Trends Biochem. Sci.* **7**, 74–76

without rigid specificies. However, some such as **eotaxin** (Fig. 30-6E) are more specific. Eotaxin attracts primarily eosinophils and basophils during allergic reactions.[220] Interleukin-8 (IL-8) is a proinflammatory cytokine and a powerful attractant for neutrophils. Neutrophils are attracted into affected tissues, where they undergo a respiratory burst and generate toxic compounds from O_2 (pp. 1072–1074).[221–222c]

The second group of cytokines regulate B and T lymphocytes. Among them interleukin-2 (IL-2) stands out as the major promoter of growth and differentiation of T cells. It was the first hormone of the immune system to be recognized.[223] Both Il-2 and Il-4 have short four-helix structures (Fig. 30-6A). Il-2 is synthesized by activated T cells and binds to a multisubunit receptor complex. The latter associates with tyrosine kinases of both the Src family (p. 572) and with Janus kinases and also activates phosphatidylinositol 3-kinase.[224–227] Il-2 promotes growth and differentiation, and clonal expansion of T cells, a key aspect of the cellular immune system. It also acts as an immunomodulator of B cells, macrophages, and NK cells. Considerable excitement has accompanied the possibility of activating lymphocytes with IL-2 produced from cloned genes in bacteria to increase their ability to kill cancer cells. However, IL-2 is toxic, and this is limiting its use.

Interleukin-4 (IL-4), cooperating with IL-21,[227a] stimulates growth of activated B cells, T lymphocytes, and mast cells, induces formation of cytotoxic $CD8^+$ T cells, and enhances formation of IgG.[228,229] The **transforming growth factor-β** (TGF-β) is another cytokine that modulates the development of the immune system. It affects a very broad range of tissues and is discussed in Chapter 32.

Inflammatory influences. Inflammation (p. 1211), which usually accompanies infection and can also arise from allergic responses, is affected by many substances.[229a–e] These include chemotactic factors that attract neutrophils and monocytes[222b,229d] and the adhesion molecules that assist in the movement of lymphocytes.[229e,f] Some epithelial tissues, such as the mucosal surfaces of the gastrointestinal tract, are maintained in a continuous very low level of inflammation. This reflects the balance between activation of the immune system and inhibition of the system by signals from microorganisms both pathogenic and commensal.[229g,h]

The third group of cytokines[229b,c] are among the molecules that regulate inflammation. One of these is **interferon-γ** (IFN-γ). Like the type I interferons it induces an antiviral state. However, its most important biological function lies in modulation of the immune system. It induces synthesis of both class I and class II (HLA-DR) antigens (see Section D), activates macrophages, and regulates synthesis and activity of other lymphokines.[230–231b] One mechanism of immunomodulation may involve induction of an aminopeptidase that participates in "trimming" of antigenic peptides that participate in antigen presentation (Section D,6). This may directly alter the T-cell response.[232] IFN-γ has a major effect on skin cell-mediated immune responses.[233] INF-γ exists in solution as a symmetric dimer, which binds to two molecules of its major surface receptor.[231,234] The antiviral activity of IFN-γ is largely a result of induction of large amounts of **guanylate-binding proteins**, large 60- to 100-kDa GTPases.[235,236]

Occasionally a well-established cancer regresses and disappears spontaneously. In the late 1800s it was observed that this sometimes happened, when a person had a concurrent severe bacterial infection. W. B. Coley pursued this lead for many years, treating cancer patients with extracts of killed bacteria, which, although highly toxic were safer than live bacteria. In the laboratory filtrates from cultures of gram-negative bacteria were shown to kill some experimental mouse tumors. The active ingredient was identified as a highly toxic and pyrogenic lipopolysaccharide (Fig. 8-28; Chapter 20, Section E).[237,238] This lipopolysaccharide has a powerful activating effect on macrophages. More recently it was found that the activated macrophages produce a protein known as **tumor necrosis factor** (TNF-α) that can destroy tumor cells and also acts together with interferons in inducing resistance to viruses.[239,240] A similar **lymphotoxin** (TNF-β) is secreted by lymphocytes.[241] Although highly toxic there has been hope of obtaining engineered forms of these proteins more specifically toxic to tumors.

TNF-α is identical to **cachetin**, a protein that suppresses completely the lipoprotein lipase of adipose tissue and is believed to be responsible for **cachexia**, a condition of general ill health, malnutrition, weight loss, and wasting of muscle that accompanies cancer and other chronic diseases. Nevertheless, TNF-α may be overproduced in obesity as well. It has been suggested that abnormal production of TNF-α may induce cachexia while abnormal action of the cytokine may cause obesity.[233] Some TNF receptors have "death domains" and trigger apoptosis, while other receptors promote proliferation and differentiation via transcription factor NF-κB.[242]

Other cytokines with lymphocyte-regulatory functions are IL-5, IL-10, IL-12, and the **macrophage migration-inhibition factor** (MIF). IL-10 is secreted by B cells, T cells, keratinocytes, monocytes, and macrophages. It suppresses synthesis of many cytokines but stimulates growth and activity of activated B cells.[243] IL-12 is formed by monocytes, macrophages, neutrophils, and dendritic cells. It activates T cells and NK cells, is a very potent stimulator of INF-γ formation, and also inhibits angiogenesis in tumor cells. It stimulates defenses against a wide range of infectious diseases caused by bacteria, fungi, protozoa, and worms.[244,245] The 115-residue MIF is formed not only by the immune system but also by many other tissues. The first lymphokine to be discovered, MIF, inhibits migration of macrophages and is also a mediator of toxic shock.[246,247] MIF is also an enzyme, a **phenylpyruvate tautomerase** (p. 692).[247]

The fourth group of cytokines are involved in hematopoiesis and control the developmental steps portrayed in Fig. 31-2. They are discussed in Chapter 32.

The **acute-phase response** consists of increased production of a group of plasma proteins in response to tissue injury or inflammation.[229c] Important acute-phase reactants are the **C-reactive protein**,[229a,248,249] **serum amyloid A**,[250,251] **haptoglobin**, **hemopexin**, **α1-acid glycoprotein**,[252] and **α2-macroglobulin**.[253] The C-reactive protein precipitates pneumoccocal polysaccharides in the presence of Ca^{2+}. It is present in primitive invertebrates and may serve as a rudimentary immunoglobulin.[248] Serum amyloid A is one of the apolipoproteins associated with high-density lipoproteins (Chapter 21, Section A). Its concentration may increase as much as 1000-fold during the acute-phase response, and during prolonged stress it may precipitate as extracellular amyloid fibers. This secondary amyloidosis is sometimes a severe pathological problem. The level of the general protease inhibitor α_2-macroglobulin can increase several hundredfold.

D. Organizing the Immune Response

A person's immune system must be able to respond to a large variety of foreign antigens without reacting against the individual's own tissues. The huge variety of antibodies that can be formed arise from the existence of B cells with millions of different sequences in their antibody genes. When an immune response occurs only a few B cells are stimulated to proliferate, and it is these selected clones that provide the needed specific antibodies and memory cells. However, it is not immediately obvious how we avoid a disastrous attack of the immune system triggered by the many antigenic determinants (**epitopes**) present in our own cell surfaces and macromolecules. Part of the answer is that the immune system "learns" early in life what is self and what is nonself. Thus, while foreign tissues cannot usually be grafted without rejection, cells of two immunologically incompatible embryos can be mixed at a very early stage of development, and an animal tolerant to both types of cell will develop.

A full understanding of self-discrimination is not yet available.[253a] The adaptive system, as generally understood, is outlined in the following pages. A current view of the innate system is presented by Medzhitov and Janeway.[11a] However, an alternative description, the **Danger model**, is being developed by Matzinger. Her view is that the immune system is designed not so much to recognize *nonself* as to send *alarm signals* from injured tissues.[11b] Most of the basic mechanisms of the adaptive immune system are not in dispute, but many hard-to-explain phenomena remain uncertain.

1. Coreceptors and the B-Cell Response

Early in life most B cells that would produce antibodies directed against a person's own tissues (autoreactive B cells) are eliminated or altered to reduce their reactivity.[254] When functional B-cell receptors do bind an antigen, the B cell will not be activated unless **coreceptors** also bind to the antigen-bearing particle. The transmembrane glycoproteins known as CD22, CD21, CD72, and FcγRIIb are among the many coreceptor molecules. Coreceptors often induce tyrosine phosphorylation of internal receptor domains and attract other molecules to form a signaling complex that may release cytokines.[255–256] The coreceptors ensure that an immune response doesn't take place without at least two signals. They also help to localize the immune response.

Among the most important factors in B-cell activation are the effects of T cells. B cells can independently mount an attack using IgMs against surface antigens. However, B-cell responses to many antigens, e.g., those present on flagella or inserted into membranes, are also dependent upon assistance from helper T cells (T_H cells).[257] These cells also have a major role in determining the longer term fate of B cells. Upon activation B cells may survive or die via apoptosis. They may proliferate (clonal expansion) and differentiate into plasma cells or may become unreactive (**anergy**). They may become long-lived memory cells.

2. The Leukocyte Differentiation Antigens

Before discussing T-cell responses it seems appropriate to mention the nomenclature of molecules (largely glycoproteins) that have been recognized as antigens present on leukocyte surfaces. These same molecules are found on other cells, but the designation of the antigens by a **cluster of differentiation** number, such as CD1, CD4, or CD8, has provided a convenient way of distinguishing different types of leukocytes.[258–261] For example, helper T_H cells are usually $CD4^+$, carrying predominantly CD4. Cytotoxic T cells are predominantly $CD8^+$.[262,263] Both CD4 and CD8 consist largely of Ig-like domains. CD4 is a 55-kDa transmembrane protein with tyrosine kinase activity.[264] It is a monomer containing four Ig-like domains, but CD8 is a disulfide-linked αβ dimer.[263]

Not all CD molecules are related to IgG. Proteins are often designated by a specific name followed by a CD number, e.g., FcγRII / CD32, ICAM-1 (CD54).

3. Functions of T Cells

T cells carry the responsibility of identifying antigens as foreign or as belonging to self. They do this in immunological synapses (Fig. 31-11) in conjunction with the major histocompatibility complex MHC (Section 5). T cells circulate through the body searching for antigens that indicate danger to the body. To avoid being swept through the bloodstream too rapidly and to be able to enter the lymphoid organs lymphocytes form tethers with adhesion molecules such as the **selectins** (p. 188).[266] They then roll more slowly to their destination. Within the lymphoid tissues the T cells may form synapses with activated B cells, dendritic cells, and macrophages. Within these cells proteosomes generate a stream of peptide fragments, some of which arise from phagocytosed pathogens. These foreign peptide fragments are displayed on the cell surfaces as complexes with type II proteins of the major histocompatibility complex (MHC; see Fig. 31-13). The complexes are checked by $CD4^+$ T cells, some of whose **T-cell receptors** (TCRs) will probably be complementary to the surfaces of the complex of the class II MHC protein and the foreign peptide. The T cell will recognize two things about this complex: the MHC protein is of *self* origin but the antigen is *foreign*. The CD4 on the T cell must also bind to the MHC on the surface of the antigen-presenting cell. Other costimulatory interactions may be needed as well.[50,267] Both $CD4^+$ and $CD8^+$ T cells tend to bind to oligomeric **activation clusters** of receptors within the immunological synapses.[267a] Other proteins also participate in assembly of these activation complexes.[267b] Of

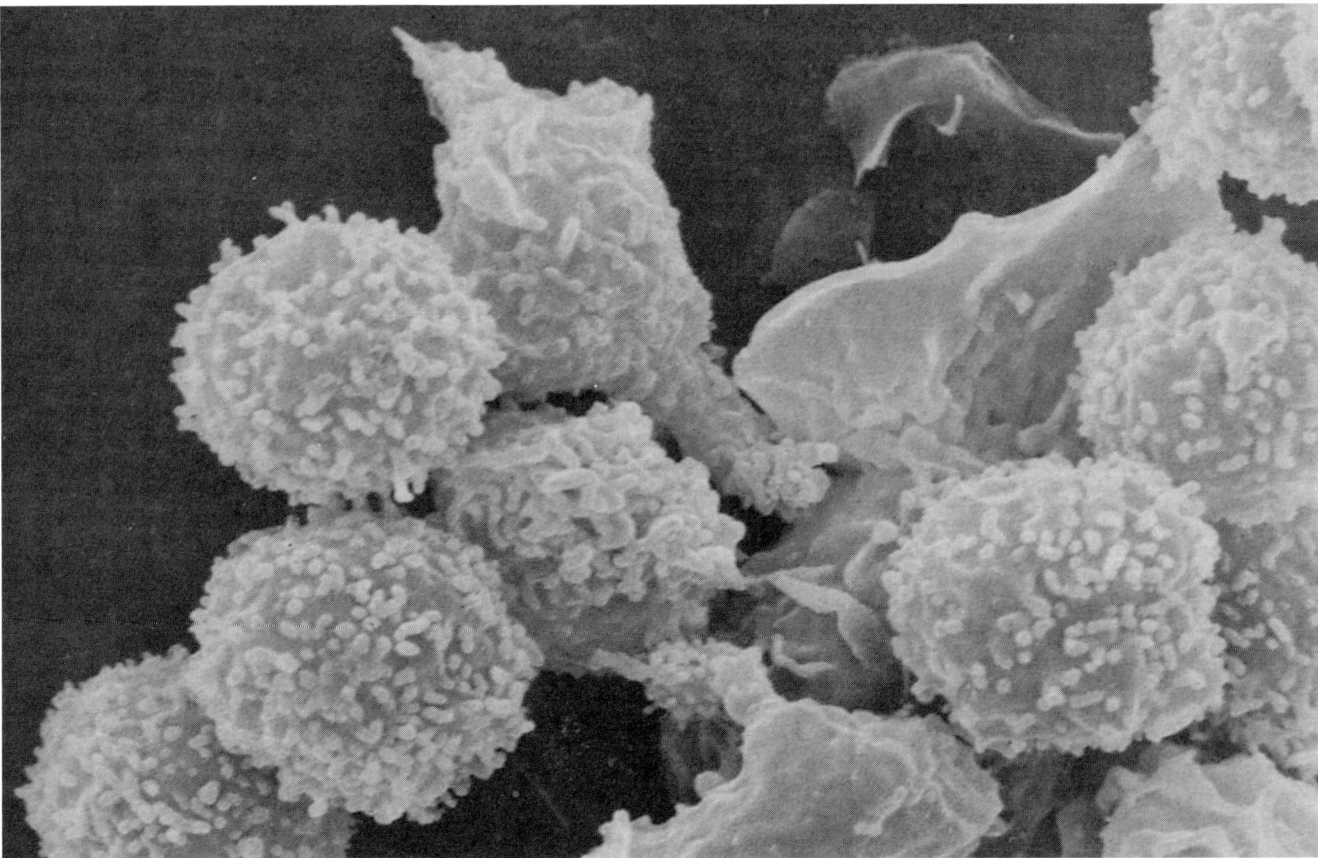

Figure 31-11 The T lymphocytes seen here are forming synapses with the large flat macrophage in the center. The macrophage is displaying antigenic peptide fragments bound to molecules of the major histocompatibility complex (MHC). Most T cells carry their own specific type of receptor. If it is complementary to a displayed antigen fragment it will bind, and the T cell will respond. Depending upon what other coreceptors are activated, it may become a T_H helper cell, or a cytotoxic T cell, or it may become inactive. Micrograph from Grey *et al.*[265] Scanning electron micrograph courtesy of Morton H. Nielsen and Ole Werdelin, University of Copenhagen.

particular interest is the recognition of an antigenic peptide produced by a B cell. The B cell has probably already recognized and phagocytized a foreign protein and is displaying peptides from that protein on its MHC I molecules. Recognition of this peptide complex by a $CD4^+$ T cell will stimulate the cell to become a T_H helper cell, which will in turn stimulate the B cell to proliferate and differentiate into a clone of as many antibody-forming plasma cells. The essential nature of the costimulation by CD4 is emphasized by the fact that infection by HIV-1, which is mediated by CD4, leads to loss of CD4 from plasma membranes and to the weakening of the immune response toward various pathogens that is characteristic of AIDS.[268–270]

In a similar manner $CD8^+$ T cells recognize peptide fragments displayed on MHC class I molecules. These fragments arise via a somewhat different pathway that forms fragments of viral proteins or proteins of other intracellular pathogens. Recognition by a $CD8^+$ T cell usually converts it into a **cytotoxic (killer) T cell**, which will kill the infected cell.[270a,b] This type of immune reaction was first recognized by the phenomena of **delayed hypersensitivity** and of **transplantation immunity**, i.e., the rejection of transplanted tissues. Both phenomena are caused by cytotoxic T cells. In delayed hypersensitivity they appear to be confused and to attack host cells.

Some very hydrophobic antigens are presented by neither a class I nor a class II MHC molecule but by members of the CD1 family, leukocyte surface proteins that are not encoded in the MHC gene region.[266,271,272]

4. Natural Killer Cells

An additional line of defense is provided by **natural killer cells** (NK cells), a type of circulating lymphoid cell able to kill cancer cells, to participate in antiviral defenses, and to help control immune responses.[273–276] NK cells, which utilize their own signaling pathways, are also able to use MHC class I molecules to recognize and to spare the lives of normal, healthy cells.[277,277a,b] Partial deprivation of a night's sleep can reduce NK cell activity, damaging the cellular immune response.[278]

5. Identifying Self: The Major Histocompatibility Complex

Proteins encoded by a single cluster of genes are known as the **major histocompatibility complex** (MHC).[279] These proteins, which are essential to T-cell function, were first recognized as the primary determinants of the compatibility of grafted tissues with the host's immune system.[280] A lack of histocompatibility can be disastrous. Not only are grafted tissues rejected but T lympocytes from the grafted tissues sometimes proliferate, attack, and kill the host. The MHC of mice is usually referred to as the **H-2 complex**[281,282] and that of humans as **leukocyte locus A (HLA)**.[283,284] Although the MHC is the most important determinant of histocompatibility, differences in other genes may also lead to a slow rejection of transplanted tissues. Since there are many different MHC genes, transplantation is successful only within inbred lines.

Some of the MHC genes have a large number (50–100) of alleles. So great is this genetic polymophism that it is extremely unlikely that two individuals will have an identical set of histocompatibility genes. The MHC (HLA) genes are located in a 2-centimorgan (~3.6 kb) region of the short arm of human chromosome 6 (Fig. 31-12)[284,285] and on chromosome 17 of mice. These genes are of at least three classes. **Class I genes** (called HLA-A, HLA-B, HLA-C, HLA-E, HLA-F, and HLA-G in humans; see Fig. 31-12) encode the major transplantation antigens, which are found on the surfaces of nearly all cells of the body. **Class II genes** encode proteins found largely on the membranes of B lymphocytes, macrophages, and dendritic cells. They are designated HLA-DP, HLA-DN, HLA-DM, HLA-DQ, HLA-DR, and HLA-DO.[286] **Class III genes** encode several components of the complement system. Many other genes and pseudogenes are interspersed with those of the MHC.

All type I MHC molecules are integral membrane glycoproteins each of which is composed of a 45-kDa heavy chain of about 350 residues together with a noncovalently linked 17-kDa light chain. The genetic variation occurs in the heavy chain between residues 43 and 195 in the human proteins. This chain appears

Figure 31-12 (Opposite page) Arrangement of genes of the human major histocompatibility complex (MHC). Left: Banding pattern of a stained chromosome 6 with the MHC region marked. Center and right: locations of all genes and pseudogenes in this region. The MHC molecules can be divided into three classes on the basis of their structure and function. The class I antigens constitute a single class structurally but fall into two functional groups. The first of these contains the "classical" class I antigens, first discovered as the transplantation antigens and now known to function as target antigens in the recognition and destruction of virus-infected cells by cytotoxic T lymphocytes. They are expressed on virtually all somatic cells. The class II antigens are expressed largely on B lymphocytes and macrophages of the immune system and are essential for presenting antigen to the helper and suppressor T cells that regulate the immune response. Many class III products are components of the complement system. These maps are based on serological and biochemical data, together with complete sequences. From the MHC sequencing consortium.[284]

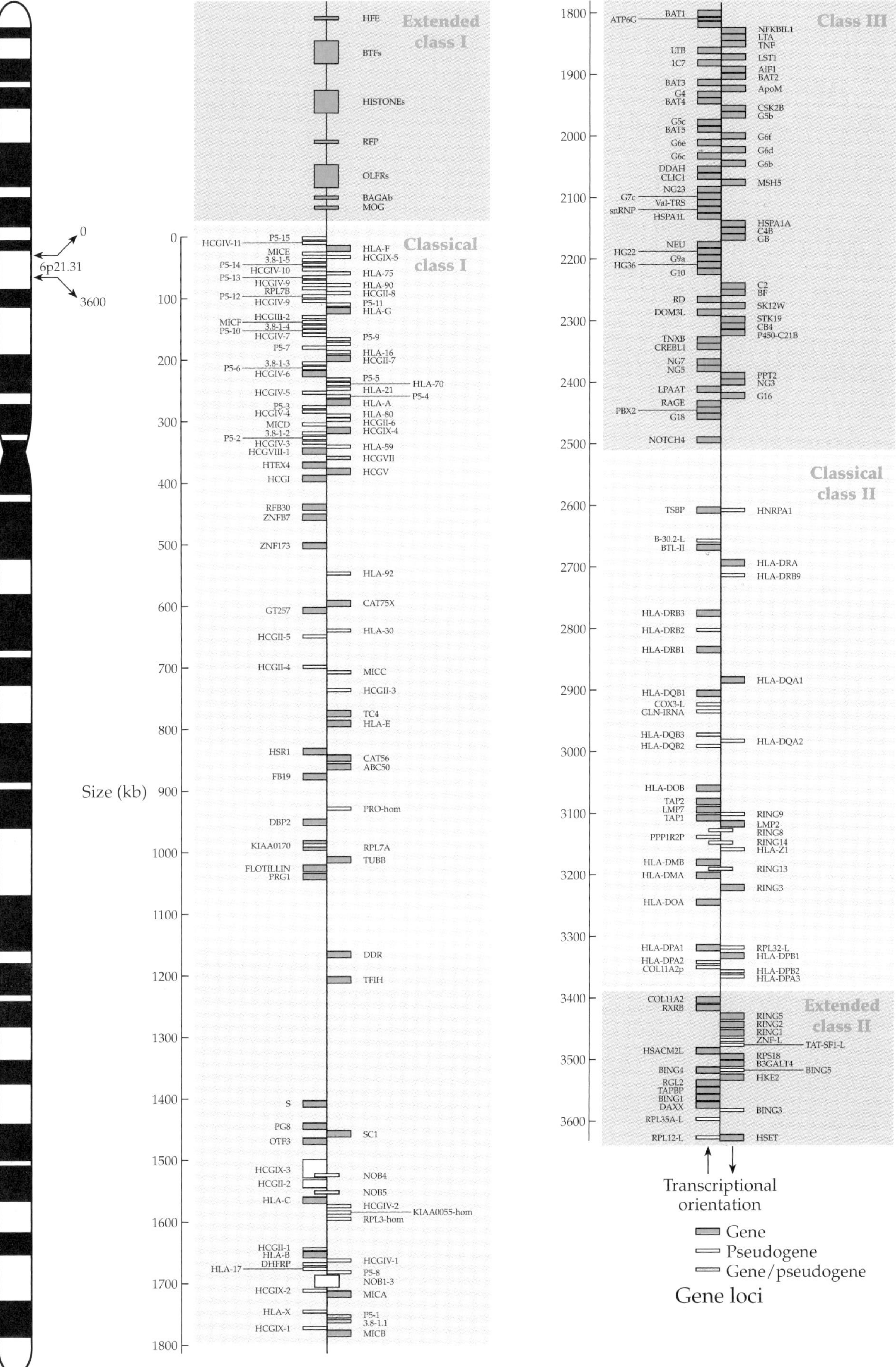

0
6p21.31
3600
Size (kb)
Extended class I
HFE
BTFs
HISTONEs
RFP
OLFRs
BAGAb
MOG
Classical class I
P5-15
HCGIV-11
MICE
3.8-1-5
P5-14
HCGIV-10
P5-13
HCGIV-9
RPL7B
P5-12
HCGIV-9
HCGIII-2
MICF
3.8-1-4
P5-10
HCGIV-7
P5-7
3.8-1-3
P5-6
HCGIV-6
HCGIV-5
P5-3
HCGIV-4
MICD
3.8-1-2
P5-2
HCGIV-3
HCGVIII-1
HTEX4
HCGI
RFB30
ZNFB7
ZNF173
GT257
HCGII-5
HCGII-4
HSR1
FB19
DBP2
KIAA0170
FLOTILLIN
PRG1
S
PG8
OTF3
HCGIX-3
HCGII-2
HLA-C
HCGII-1
HLA-B
DHFRP
HLA-17
HCGIX-2
HLA-X
HCGIX-1
HLA-F
HCGIX-5
HLA-75
HLA-90
HCGII-8
P5-11
HLA-G
P5-9
HLA-16
HCGII-7
P5-5
HLA-70
HLA-21
P5-4
HLA-A
HLA-80
HCGII-6
HCGIX-4
HLA-59
HCGVII
HCGV
HLA-92
CAT75X
HLA-30
MICC
HCGII-3
TC4
HLA-E
CAT56
ABC50
PRO-hom
RPL7A
TUBB
DDR
TFIH
SC1
NOB4
NOB5
HCGIV-2
KIAA0055-hom
RPL3-hom
HCGIV-1
P5-8
NOB1-3
MICA
P5-1
3.8-1.1
MICB
Class III
BAT1
ATP6G
NFKBIL1
LTA
TNF
LTB
1C7
LST1
AIF1
BAT2
BAT3
ApoM
G4
BAT4
CSK2B
G5b
G5c
BA15
G6f
G6e
G6d
G6c
G6b
DDAH
CLIC1
MSH5
NG23
G7c
Val-TRS
snRNP
HSPA1L
HSPA1A
C4B
GB
NEU
HG22
G9a
HG36
G10
C2
BF
RD
SK12W
DOM3L
STK19
CB4
P450-C21B
TNXB
CREBL1
NG7
NG5
PPT2
NG3
LPAAT
G16
RAGE
PBX2
G18
NOTCH4
Classical class II
TSBP
HNRPA1
B-30.2-L
BTL-II
HLA-DRA
HLA-DRB9
HLA-DRB3
HLA-DRB2
HLA-DRB1
HLA-DQA1
HLA-DQB1
COX3-L
GLN-IRNA
HLA-DQB3
HLA-DQB2
HLA-DQA2
HLA-DOB
TAP2
LMP7
TAP1
RING9
LMP2
RING8
PPP1R2P
RING14
HLA-Z1
HLA-DMB
RING13
HLA-DMA
RING3
HLA-DOA
HLA-DPA1
RPL32-L
HLA-DPB1
HLA-DPA2
COL11A2p
HLA-DPB2
HLA-DPA3
Extended class II
COL11A2
RXRB
RING5
RING2
RING1
ZNF-L
TAT-SF1-L
HSACM2L
RPS18
B3GALT4
BING4
BING5
HKE2
RGL2
TAPBP
BING1
DAXX
BING3
RPL35A-L
RPL12-L
HSET
Transcriptional orientation
Gene
Pseudogene
Gene/pseudogene
Gene loci

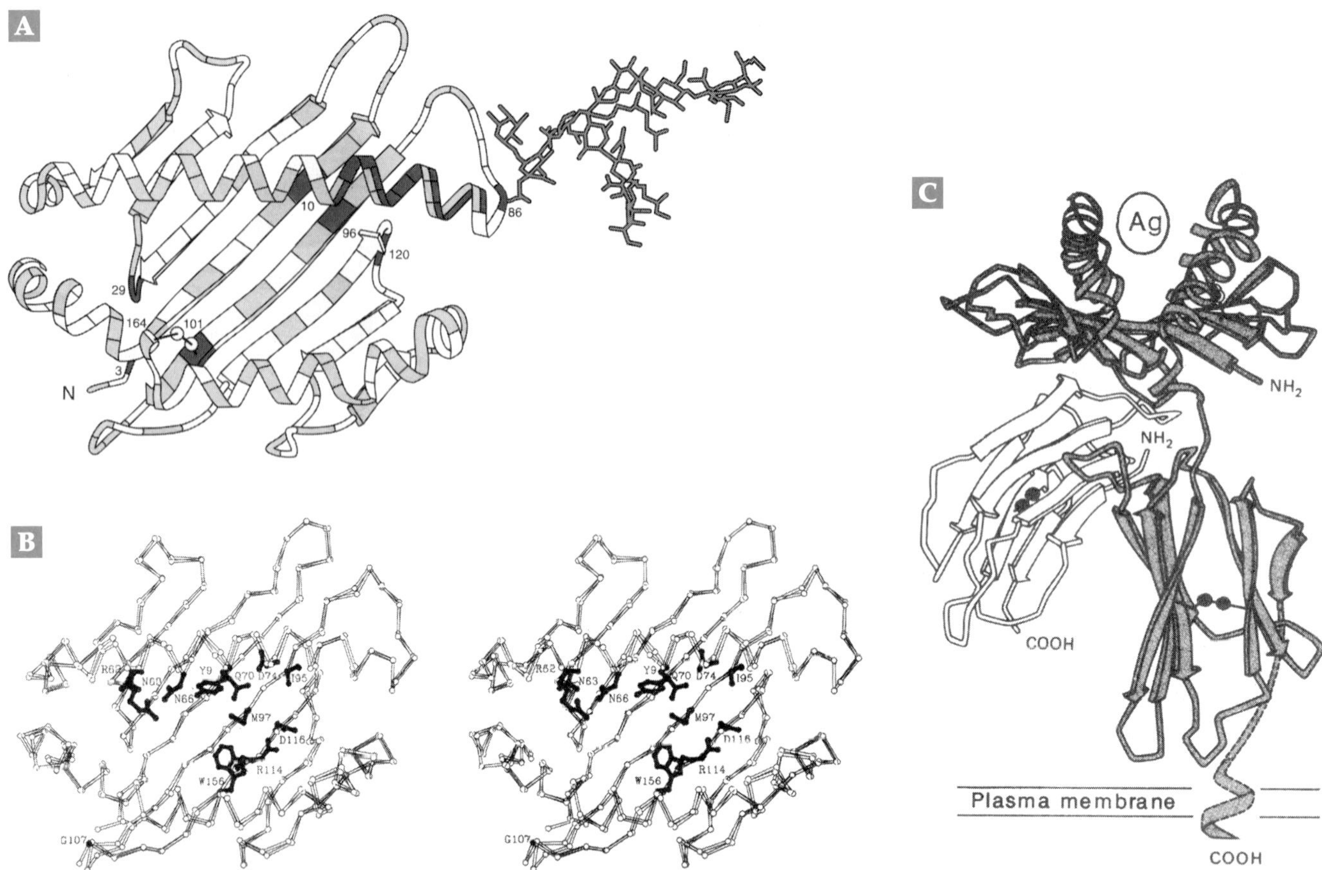

Figure 31-13 The structure of Class I MHC molecules. (A) The specificity pocket in the N-terminal part of the ~360-residue α chain. The numbered residues are invariant in all of the ~20 different Class I molecules. An oligosaccharide is shown on the invariant Asn 86. (B) A stereoscopic view of a similar MHC molecule showing some of the polar residues that protrude into the peptide-binding groove and may form hydrogen bonds with the peptide. From Garrett *et al.*[291] Courtesy of Don C. Wiley. (C) Side view of a complete MHC molecule with an antigenic peptide (Ag) bound into the peptide-binding groove. The C terminus of the long ~360-residue α chain is in the cytoplasm of the displaying cell. The small 99-residue β chain (unshaded) is a molecule of β microglobulin, which is also a constituent of blood plasma.[292] Courtesy of Peter Parham.

to consist of three ~90-residue domains protruding from the outside of the cell, about 25 residues embedded in the membrane, and a short C-terminal tail in the cytoplasm (Fig. 31-13).[287] The light chain has an invariant composition and is identical to the plasma protein β_2-microglobulin, whose gene is located on a different chromosome. Its structure closely resembles that of a single immunoglobulin domain.[288] The MHC Class II antigens (Fig. 31-14) are also αβ dimers, the α chains being 34-kDa glycoproteins and the 28-kDa β chains being larger than in the type I antigen.[283,289] While the MHC of humans and mice have been studied the most, all vertebrates possess similar self-identification systems. Although both the sequences and the folding patterns of the MHC antigens (Figs. 31-13, 31-14) are somewhat similar to those of immunoglobulins, there are many differences. Furthermore, the cause and significance of the polymorphism is quite different in the two cases. Each individual has millions of antibodies with different variable regions but only one set of HLA antigens, which are largely the same on germ, embryonic cells, and adult cells.

Serological tests allow tissue types to be defined by the HLA genes.[290] Thus the commonest HLA type in Caucasian populations is HLA-Al / B8 / Dw3, whereas A1 / B17 is common among Asian Indians. In every case subtypes can be defined, and this fact together with the polymorphism in other genes leads to a unique HLA type for nearly every individual. As is indicated in Fig. 31-12, complete nucleotide sequences are known for typical alleles of all of these genes.[284] It is of medical interest that the susceptibility of an individual to many degenerative diseases is determined in part by the HLA type.[289,295] Thus among patients with a kind of arthritis, **ankylosing spondylitis** that affects 1 or 2 per 1000 men of Caucasian origin, 96% have the HLA-B27 antigen. Of patients with **celiac disease**, a type of intolerance to gluten, 60% have the HLA-B8 antigen. Persons with

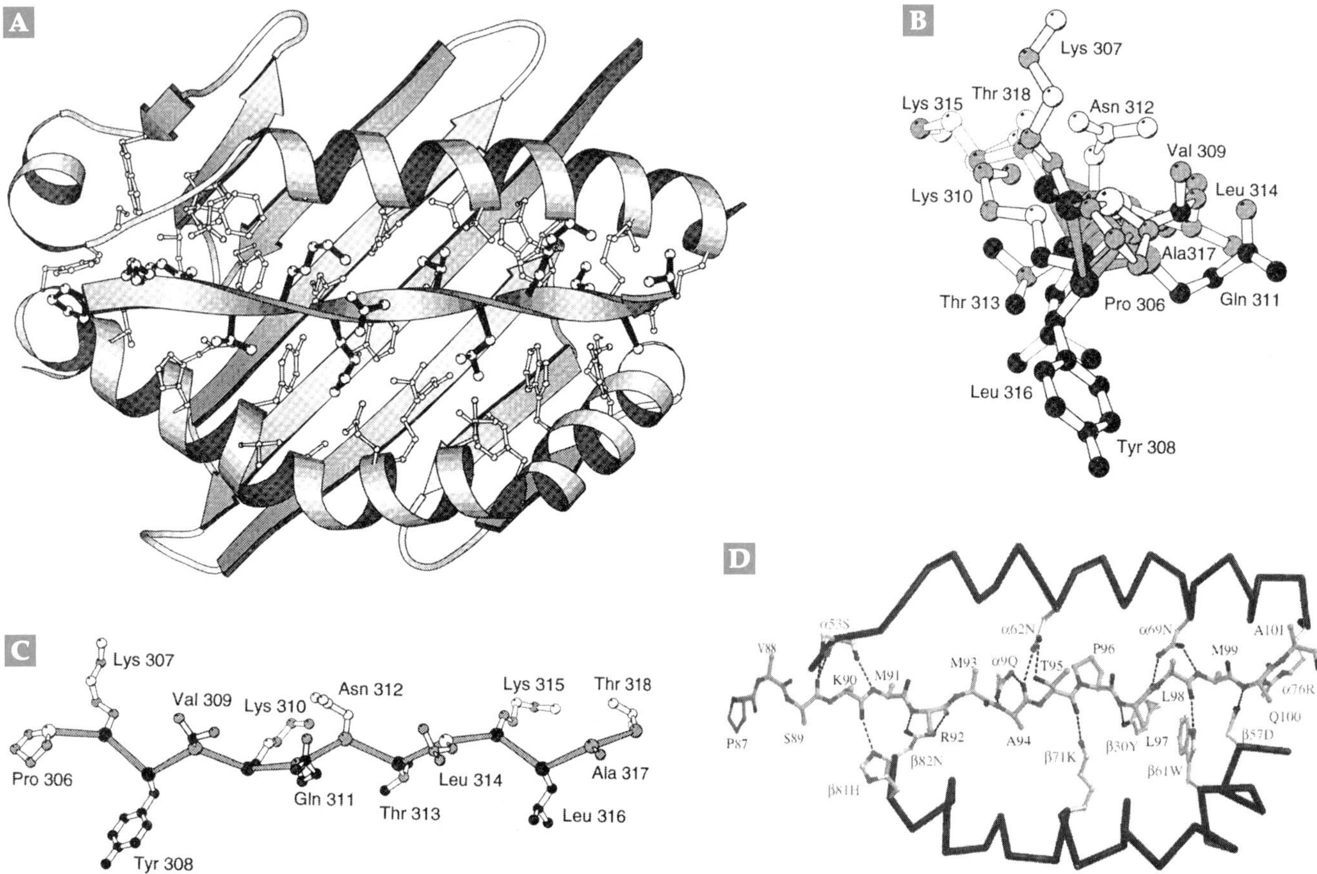

Figure 31-14 Illustration of the binding of a short peptide fragment (central ribbon) into the antigen-binding groove of an MHC type II molecule. (A) The binding groove of molecule HLA-DR1 with a bound peptide (HA) derived from an influenza virus. (B) End view of the bound peptide. (C) Side view of the same peptide. From Stern *et al.*[293] (D) A similar HLA molecule (HLA-DR3) with the peptide CLIP (class II associated invariant chain peptide) bound into the antigen-binding groove. The binding is almost identical to that in (A). Notice the specific hydrogen bonding to side chains of the HLA molecule. From Ghosh *et al.*[294] Courtesy of Don C. Wiley.

HLA-Bw17 and B13 have an increased susceptibility to **psoriasis** and those with HLA-DRw4 an increased tendency to develop **rheumatoid arthritis**. Susceptibility to **autoimmune insulin-dependent (Type I) diabetes** is strongly correlated with the presence of the neutral residues Ala, Val, or Ser at position 57 of the HLA-DQ β chain.[296,297] However, aspartate in position 57 protects against the disease. It may prevent development of a dangerous autoantibody to this cell surface protein. The presence in populations of both humans and apes of a balanced polymorphism among the residues Ala, Val, Ser, and Asp at this position suggests an essential evolutionary origin to this disease susceptibility.[297] HLA-B53 protects against severe malaria in Africa. Other diseases with a strong association with HLA type include multiple sclerosis, **Crohn disease** (inflammatory bowel disease),[298] and several diseases induced by infections with viruses, bacteria, trypanosomes, etc. For example, arthritis can follow infection by *Salmonella*. This suggests that the killer T cells can be confused when stimulated by foreign antigens, which are too closely related to the HLA antigens of the host.

6. Antigen Presentation and MHC Restriction

T cells usually do not respond to intact antigens on cell surfaces but only to partially degraded antigens. Antigen-presenting cells (APCs) of various types must process the antigen through endocytosis and partial digestion before the foreign antigen can bind as an MHC complex to a T-cell receptor. Apparently the processed foreign antigen must lie in the binding site of an MHC Class I or Class II chain (Figs. 31-13C, 31-14) before the T cell will recognize the complex and respond.

Antigen processing begins with cytosolic proteasomes that are present in all cells (Box 7-A; Chapter 29, Section D,8). They cleave proteins of the cell and of intracellular parasites into short peptide fragments,[55,299–301] which may need to be trimmed to shorter 8- to 11-residue peptides suitable for binding into the groove in an MHC Class I molecule.[302,303] The peptides are carried into the ER with the aid of the **TAP** (transporter associated with antigen processing) complex,[304–305a] which is discussed briefly in Chapter 29, Section D,3. Chaperones, such as hsp70, may also participate in the transport. In contrast, MHC Class II proteins receive their antigenic peptides via an endosomal–lysosomal pathway. Proteins from phagocytized pathogens are cleaved by proteases in an endosome or lysosome into fragments that tend to be longer (13–25 residues) than the 9- to 11-residue peptides generated by proteasomes.[299,306]

Peptides bind into the groove in a Class I MHC molecule in a manner similar to that illustrated in Fig. 31-14 for a Class II MHC molecule. However, in the Class II complex the longer peptides extend from the two ends of the binding groove. The peptide, which assumes a polyproline II helical conformation, is held by hydrogen bonds from the Class II MHC molecule to the peptide backbone.[307–311] A single peptide may shift and bind in a different register with the possibility of being recognized by a different T cell receptor, when it is displayed.[308] In contrast, MHC Class I molecules bind best to 8- or 9-residue peptides, which are held by an array of hydrogen bonds to the $-NH_3^+$ and $-COO^-$ termini (not shown in Fig. 31-13, which displays the empty binding groove). The central part of this groove contains a deep pocket, which together with smaller pockets near the ends provides specificity.[312–318] However, there is a puzzle. Because of the great genetic variability in the MHC genes there will be great differences (polymorphism) in the shapes of the binding pockets among different people. However, an individual has at most six different kinds of MHC molecules. Yet, a single MHC molecule has been estimated to be able to bind more than 10,000 different peptides.[314] Essential to this process is a final trimming of the peptides at their N termini to provide better fits.[318a]

Peptides are loaded onto Class I MHC molecules, while they are together in the ER. They move as tightly bound complexes through the Golgi and into the external plasma membrane, where they remain tethered via the MHC molecule (Fig. 31-13C). In contrast, binding to Class II MHC molecules occurs in the endosomes or lysosomes. The process is somewhat complex. Class II MHC molecules are chaperoned from the cytosol into late endosomal/lysosomal organelles with their antigen-binding grooves occupied by a peptide fragment known as the Class II-associated invariant chain peptide (CLIP; Fig. 31-14D). This is cut from the end of an invariant chain known as Ii.[319–321]

The MHC proteins HLA-DO and HLA-DM (Fig. 31-12) are resident in the lysosome-like organelles and chaperone the class II molecules, until they are ready for loading with peptides. HLA-DM assists in removal of the CLIP peptide when loading occurs.[322] An asparaginyl endopeptidase may also be required[323] as well as a disulfide reductase.[324] After loading the Class II MHC•peptide complexes, like the Class I MHC•peptide complexes, are exported to the plasma membrane. There they may be recognized by a T cell, which utilizes its T-cell receptor to recognize an antigen and its CD4 or CD8 proteins to distinguish Class I from Class II MHC complexes. The binding of CD4 and CD8 to their cognate MHC molecules has also been described at the molecular level.[310,325] As mentioned previously, some hydrophobic antigens are presented by CD1 molecules. They also have an MHC-like fold with a large hydrophobic binding groove.[272]

An interesting approach to the treatment of autoimmune diseases is design of peptide mimics that bind into the antigen-binding groove of specific MHC proteins. For example, a protease-resistant pyrrolinone–peptide hybrid has been designed to bind to the rheumatoid-arthritis-associated HLA-DR1.[326]

An important distinction between B- and T-cell responses is that T cells recognize a foreign antigen only when associated with an MHC antigen of the same type as is carried by the T cell. This "MHC restriction" limits the actions of T lymphocytes. The function of cytotoxic T cells appears to be primarily one of killing virus-infected cells and perhaps cancer cells. MHC restriction ensures that the T cell is attached by its rereceptor to a Class I MHC molecule belonging to self. The dual recognition ensures that the cell probably contains a foreign antigen and should be killed. The Class I MHC antigens are found on almost all body cells. Therefore, cytotoxic $CD8^+$ T cells can attack infected cells of all types. They may kill their target cells by injecting their cell membranes with the complement C9-like cytolysins.

Regulatory $CD4^+$ T cells recognize the Class II MHC molecules, which are found primarily on lymphocytes. Thus, the attention of regulatory T cells is directed towards other lymphocytes. In this case, too, MHC restriction enables helper T cells to recognize B lymphocytes as self. If foreign antigen is present so that B lymphocytes have been activated by the binding of a foreign antigen, they will be stimulated by the T_H cells to proliferate and make and release immunoglobulins. This is accomplished in part by secretion of the lymphokine interleukin-2 (Fig. 30-6) and B-cell growth factors. Some T cells become suppressor T cells.

7. T-Cell Receptors

T cells both mediate the recognition of self and also participate in the immunologic response to foreign antigens. Their surface receptors function much like the immunoglobulins that are attached to the surfaces of B cells. The T-cell receptors are αβ disulfide-linked heterodimeric glycoproteins (Fig. 31-15) consisting of 40- to 45-kDa α subunits and 42- to 44-kDa β subunits.[267,327–330a] They are associated in the T-cell membrane with a larger complex called CD3, which contains additional 26- to 28-kDa γ, δ, ε, and ζ chains.[267,331,332] The polypeptides of the CD3 complex have C-terminal cytoplasmic tails that contain tyrosine residues within several immune system tyrosine-based activation motifs (ITAMs; Fig. 31-15). As is illustrated in this figure, the activating antigen is cradled in the binding groove of an MHC molecule attached to the APC (top) with some side-chain and backbone atoms of the peptide available for bonding to the T-cell receptor.[333,334] Notice that in the synapse the T-cell receptor also makes direct contact with the MHC molecule (Fig. 31-15).

Signaling by an activated T-cell receptor is quite complex. The ITAMs are sites of tyrosine phosphorylation by kinases of the Src family.[335,336] Another tyrosine kinase, Zap-70 (Fig. 31-15), associates with the C-terminal tails of the disulfide-linked dimer of subunits ζ. It recognizes the phosphotyrosines groups via its SH2 domains (see Fig. 7-30). Zap-70 appears to act synergistically with the Src kinases.[332,337,338] The nature of the APC is also of importance. For example, some dendritic cells secrete Il-12, which favors formation of T_H1 helper cells. A second type of dendritic cell induces formation of T_H2 helper cells.[338a] Dendritic cells may also control growth and proliferation of T cells by regulating the availability of cysteine, which is a nutritional essential for lymphocytes.[338b] Other effects may result from endocytosis by T cells of occupied T-cell receptor•MHC complexes.[338c]

A second type of T-cell receptor, the γδ receptor, is carried by a small subgroup of T cells. It may have a distinct role in generating an immune response to certain microorganisms including *Mycobacterium tuberculosis*.[339–342a] Like immunoglobulins, T-cell receptors have a great variety of amino acid sequences. They have C-terminal constant domains and N-terminal variable and hypervariable regions. Thus, T cells can bind to a variety of foreign antigens. However, except as a result of autoimmune diseases, they do not attack cells recognized as self unless these cells are infected with a virus or for some other reason carry foreign surface antigens.

Certain bacterial immunostimulatory molecules known as **superantigens** are able to stimulate MHC Class II molecules to activate large numbers of T cells without any assistance of an antigenic peptide, sometimes with disastrous results.[343,344] Superantigens are a group of related proteins that includes enterotoxins from species of *Staphylococcus*[344–346] and *Streptococcus*,[347] a staphylococcal exfoliative toxin,[348] and **toxic shock syndrome** toxins.[349] Superantigen molecules don't occupy the peptide-binding groove in the MHC molecule but bind as intact proteins at an external site. They also bind to the variable region of the TCR in the MHC-TCR complex. Attempts are being made to design decoy molecules that prevent binding of a particular superantigen.[350]

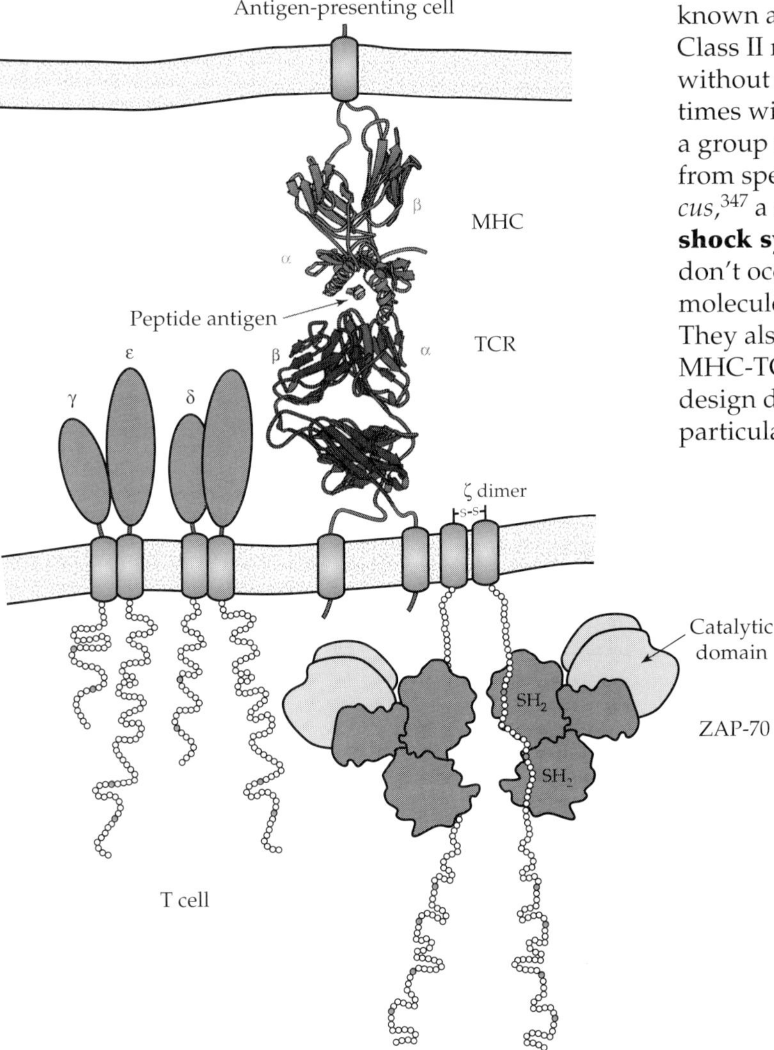

Figure 31-15 Interaction between an MHC•peptide complex on an antigen-presenting cell (APC) with a T-cell antigen receptor (TCR) that is attached to the plasma membrane of a T cell. Structures of the α- and β-subunits of the MHC molecule and of the T-cell receptor are based on crystallographic data. The detailed structures of the disulfide-linked σε and γε modules of the T-cell receptor are not shown. The dimeric $ζ_2$ subunit has large cytoplasmic domains that are thought to be involved in signaling the protein ZAP-70 (zeta-associated protein of M_r 70,000). The σ, ε, and γ subunits may also undergo phosphorylation of their tyrosyl groups (gray spheres), which are found in immune system tyrosine-based activation motifs (ITAMS). Drawing modified from those of Cochran *et al.*[267] and Hatada *et al.*[332]

8. Self-Tolerance

The immune system is flexible enough and powerful enough to protect us from a great variety of dangers, even from viruses and organisms that may be entirely new. However, it is hard to understand how the immune system completely avoids fatal damage to our own bodies. The answer is complex. It has baffled generations of investigators[350a,b] and is still not fully understood.

Since the discovery of vaccination in 1796 immunology has claimed the attention of many scientists. However, it was not until 1891 that the German bacteriologist Emil von Behring proposed the term antibody for the protective materials in blood.[351] By about 1900 Paul Ehrlich and Svante Tiselius, who wrote the first immunochemistry book,[352] initiated serious investigations. Ehrlich proposed that binding of an antigen to a surface receptor would induce the cell to make additional identical surface receptors, which would be released to become antibodies.[351,353] The concept was correct, but it would be many years before knowledge of the structures and biosynthesis of antibodies became available.

Nevertheless, immunological tolerance interested Ehrlich and other immunologists.[353a] One proposal, offered by Niels Jerne in 1974, was that self-tolerance depends upon **immunological networks**.[354–356] Consider a lymphocyte bearing a bound immunoglobulin receptor or a bound T cell receptor. It will be specific for some epitope E. The receptors are shown in Fig. 31-16 as having V-shaped antigen-binding sites. Jerne pointed out that the variable region of this receptor will itself carry epitopes that can be recognized by other appropriate antibodies. These epitopes on the receptor are called **idiotopes** and as a group define the **idiotype** of that receptor or immunoglobulin.[356a] There will usually be other lymphocytes with receptors that recognize the idiotype of the first lymphocyte (see Fig. 31-16). These constitute an **antiidiotype set**. In addition there will be lymphocytes, whose idiotopes resemble those of epitope E and which will therefore be recognized by the first lymphocyte as foreign. These lymphocytes constitute an **internal image set**. There will be other sets of lymphocytes that recognize the lymphocytes of the antiidiotype set or of the internal image set. Thus, there will be an elaborate network of clones of interacting lymphocytes. When an immune response occurs many members of this network will respond. A B lymphocyte will recognize a particular antigen and gives rise to a clone of plasma cells making antibodies against that antigen. The body will then make new antibodies against the first antibodies formed, etc. A whole segment of the network will respond in this fashion. Jerne suggested that the overall effect would be to limit and suppress the immune response. Anti-idiotype antibodies as well as anti-anti-idiotypic antibodies have been prepared[356] and have been used in studies of receptors.[357] However, Jerne's theory is generally regarded as incorrect or at least a great oversimplification.[358]

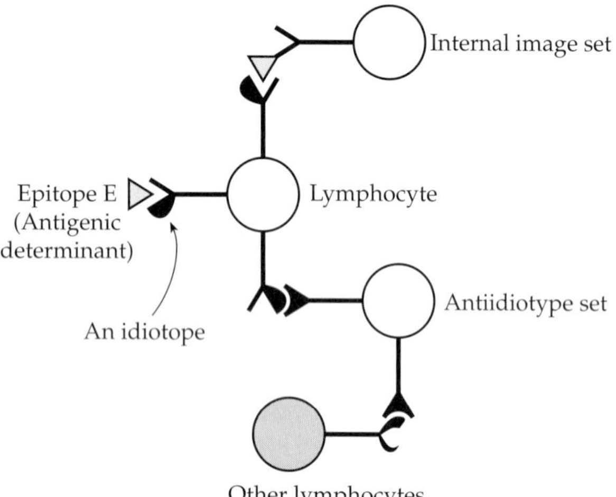

Figure 31-16 Schematic depiction of lymphocyte receptors forming anti-idiotype and internal image sets as proposed by Jerne.

A process of **clonal selection**[359] is now thought to be basic to self-tolerance. Credit for the theory, developed in the 1950s, is usually given to F. M. Burnet. However, Ehrlich, Jerne, and David Talmage were also prominent contributors.[37,351,359,360] An essential postulate of the clonal selection theory is that each B lymphocyte is predetermined to make antibodies of only a single specificity. The mechanism of **allelic exclusion**,[359] which makes this possible, is described in Section E. Clonal selection can occur because the B cells carry their antibodies as surface receptors. Binding of an antigen provides a signal for clonal expansion. However, during development in the thymus the progenitor B cells carrying self-reactive antibodies are killed by apoptosis. Later, peripheral B lymphocytes also undergo selection by a complex network of signaling and apoptosis.[361–363] In a similar manner excess lymphocytes that build up during an immune response must be removed.[364]

The learning of self by T lymphocytes also happens in the thymus early in development, during the first three weeks of life in mice.[365,366] This again involves selection against potentially autoreactive lymphocytes carrying idiotypes that are also present on the body's own tissues and which have a high affinity for self peptide•MHC complexes. However, T cells with a weaker affinity for a self MHC molecule but a high potential affinity for a nonself peptide are allowed to develop.[366,367] Only about 1% of the lymphocytes that develop in the thymus emerge as mature T cells.[368,369] Others appear to be killed (clonal deletion) or to become unreactive toward antigen.[353a] The latter enter a state referred to as **anergy**.[370–373]

As with every aspect of metabolism, homeostasis is essential to the immune system, which must be able to both grow and shrink rapidly.[373a–c] Antigens, cytokines, apoptosis-inducing signals, immune inhibitor receptors,[229h] and receptor tyrosine kinases[373d] all participate in preserving the delicate balance that is required.

9. Immunologic Memory and Vaccination

In 1796, Edward Jenner carried out the first human vaccination. Attempting to protect a teenaged boy from disfigurement and possible death from smallpox, he vaccinated him with material from a cowpox lesion on the hand of a milkmaid. (She had contracted the disease from a cow named Blossom, whose hide hangs in St. George's Hospital in London.)[353,374,375] Six weeks later he inoculated the boy with virulent smallpox. Fortunately, the boy didn't contract the disease. Today vaccination is in use for more than 70 bacteria, viruses, parasites, and fungi, and the results have been impressive.[376] Poliomyelitis has been almost eliminated.[377] Smallpox has not been seen for many years, and the decreases in diphtheria, measles, mumps, whooping cough, and rubella have been impressive.

A nagging question is "How long will vaccination last?" One unplanned experiment resulted from two epidemics of measles in the remote Faroe Islands. The first outbreak was in 1781 after which the islands remained free of measles for 65 years. The second outbreak in 1846 affected 75–95 % of the population, but according to a physician who investigated the epidemic not a single one of the many aged people living who had had measles in 1781 contracted the disease a second time.[38] This bit of history confirms that immunological memory is sometimes very long-lived (although it doesn't prove that the smallpox vaccination older people received is still good!).[377a]

Vaccines have been prepared traditionally by use of viruses or organisms killed by compounds such as formaldehyde or by attenuated viruses or live organisms. These are selected for a low degree of virulence after repeated passages through live animals or cell cultures.[376] Newer methods utilize purified viral proteins, bacterial capsular polysaccharides, or DNA.[378–379a] In the future edible vaccines may be produced in plants.[380] Nevertheless, it is often very difficult to devise effective vaccines. In spite of 80 years of effort better vaccines against tuberculosis are needed.[381,382] All efforts to produce an AIDS vaccine have failed.[375,379] A satisfactory vaccine must activate both B cells and T cells. Activation of the latter may be especially difficult. Continuous development of new strains of bacteria is a problem for vaccination against tuberculosis and has been an insurmountable barrier to vaccination against AIDS. However, in the latter case it may be possible to use vaccination to prevent an HIV infection from progressing to AIDS with complete destruction of $CD4^+$ T cells.[379] As the immune system and also diseases become better understood, it is possible that new strategies for induction of specific cytolytic T cells can be devised,[383] e.g., for AIDS[384] and even for cancer.[385] A current obstacle to development of new vaccines is that pharmaceutical companies view vaccines as unprofitable.[385a]

We need a better understanding of how memory B and T cells are formed and selected for long-term survival.[386] After differentiation and selection in the bone marrow, B cells move to the spleen. These **transitional B cells** within germinal centers undergo further selection to become mature B cells and B memory cells.[256,387,388] Interactions with cytokines and with coreceptors play important roles. A subset of B lymphocytes may remain in the germinal centers, serving as a kind of stem cell providing new memory B cells continuously.[39] **Naive T cells**, which have not yet encountered antigen, travel throughout the body but apparently don't enter nonlymphoid tissues. However, after being presented with antigen by dendritic cells within lymph nodes, some T cells move to the skin and other peripheral locations.[389,390] After a pathogen is destroyed most of these T cells die, but a few remain as long-lived memory cells. These are able to respond to a second encounter with a pathogen.[40,383,391] Apparently continuous new exposure to antigen is not needed for long-term immunity, but the slowly dividing $CD8^+$ memory T cells may require continuous stimulation by Il-15 to counteract inhibition by Il-2.[392]

E. The Rearranging Genes for Immunoglobulins and T-Cell Receptors

An impressive example of the kind of permanent changes that can occur in the genome of specialized cells is provided by the genes of immunoglobulins and T-cell receptors. B lymphocytes make tens of millions of antibodies of differing sequence, and T lymphocytes make a similar number of different T-cell receptors. This diversity is established in major part in the DNA but also by alternative splicing and editing of RNA.

1. Rearrangements of Germline DNA

Each ~110-residue domain of an immunoglobulin is encoded by a single exon, but the exon for the amino-terminal or variable domain is assembled from two or three small genes (or segments) selected from a large family of such genes present in the germ cells and in the lymphoid progenitor cells.[393–396] Within the

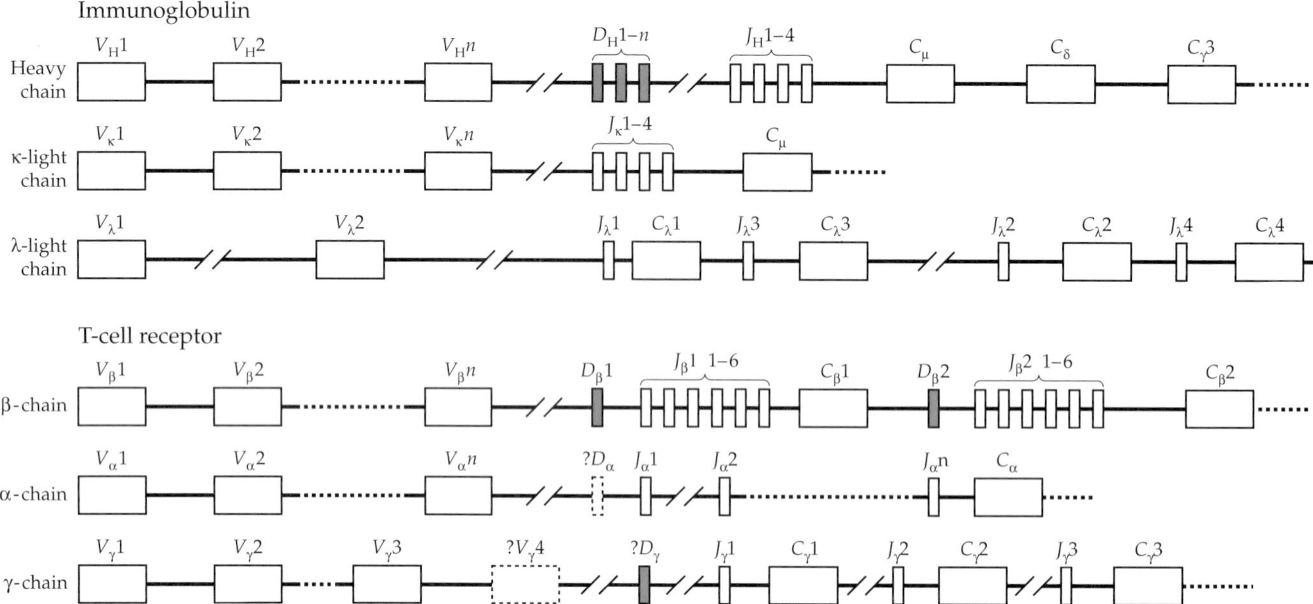

Figure 31-17 Organization of the immunoglobulin and T-cell receptor gene families of the mouse. The human γ-gene pool is larger as is the λ light-chain immunoglobin gene pool. All six gene pools contain separate gene segments encoding the variable and constant regions of the antigen receptors of lymphocytes. In the course of lymphocyte development, one of the V segments is juxtaposed by chromosomal rearrangement with one of the J segments and, where applicable, a D segment to form a complete variable-region gene. Each V segment has two regions of hypervariability, which are known in the case of the immunoglobulins to contribute to the antigen-binding site in the folded molecule. A third hypervariable region, which also contributes to the antigen-binding site, is generated by the junction of the V segment with the J or the D and J segment(s) at recombination. There is more germline diversity in the T-cell receptor than in the immunoglobulin gene pools in the J segments. The β-gene pool is organized in a way that also allows more combinatorial diversity. The β-gene pool contains fewer J regions than the gene pool, but it has two distinctive features that allow for exceptional diversification during somatic rearrangement. First, the 'rules' for recombination allow in principle the joining of both D to D segments and V segments directly to J segments, neither of which is possible in the immunoglobulin heavy-chain pool. The D segments of the β genes can be read in any of the three possible reading frames, so that varying the site of the V – D junction alone can make a substantial contribution to the diversity of the third hypervariable region. From Robertson.[397]

V region the three short hypervariable segments alternate with four framework segments that have a more nearly constant structure. The V region of the light-chain genes (of either the κ or λ type) is put together in part from a V gene that encodes an approximately 95-residue sequence making up the first three framework regions plus two hypervariable regions and part of the third hypervariable segment. There are ~100 different V_κ genes and ~30 V_λ genes[398–400] in the light-chain family. The similar arrangement of genes in the mouse is indicated in Fig. 31-17. These V genes are spaced at intervals of 14–30 kb within the DNA.

The rest of the third hypervariable region and the fourth framework region of the κ and λ chains are encoded by short **J (joining) genes** that specify ~15 residues. There are about five J_κ and three J_λ genes in the mouse. The first mechanism for creating antibody diversity lies in the large number of V genes, which are especially diverse in their hypervariable regions. The second mechanism is the joining of any one of these V genes with any one of the J genes, the joining taking place within the third hypervariable region.

There are also at least six different constant (C_λ) genes in the human genome.

The heavy-chain genes are more complex. There are over 200 V_H genes located at 14- to 16-kbp intervals, which are followed by ten ~15–17 bp **D (diversity) genes** and 4–6 J_H genes. During the differentiation of the lymphoid stem cells pre-pro-B cells can be identified in which the V, D, and J genes are still separate. Later there are pro-B cells containing joined DJ_H segments, then pre-B cells with joined VDJ_H segments, then B cells with a VJ_L (either κ or λ) segment also joined. The overall process of gene rearrangement, mRNA splicing, and immunoglobulin synthesis is outlined in Fig. 31-18. The joining of the gene segments occurs by recombinational mechanisms that involve 7-bp and 9-bp recognition sequences known as **recombination signal sequences** (RSSs)[401–402b]:

heptamer	nonamer
5′-CACTGTG	5′-GGTTTTTGT
3′-GTGACAC	3′-CCAAAAACA

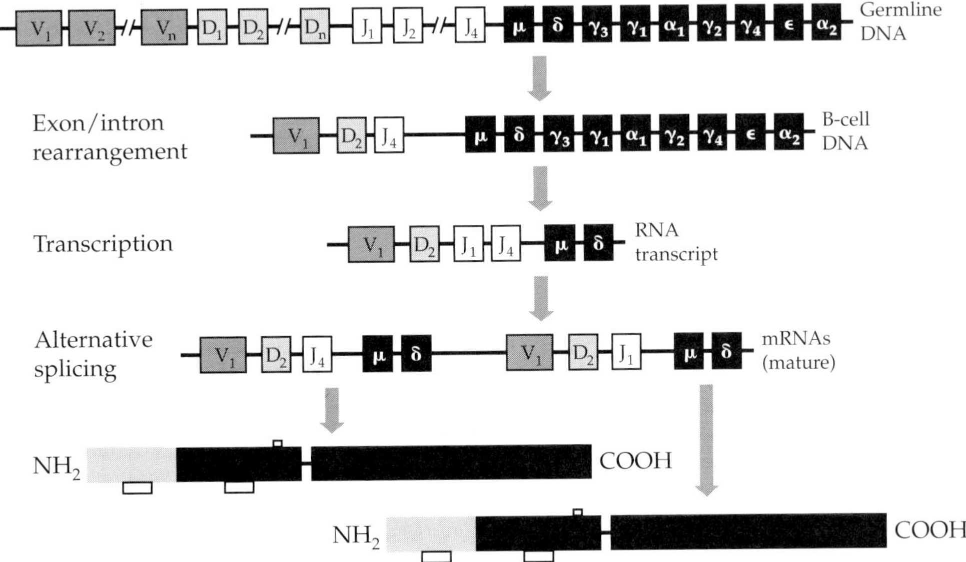

Figure 31-18 Human immunoglobulin heavy-chain gene structure and gene processing. Exons of the heavy-chain genes that encode the variable regions of the immunoglobulin molecule are labeled V_1, V_2, ... V_n. Selection from these V exons during embryonic development produces the unique sequences of each B-cell clone. The germline genes for the immunoglobulin heavy chains also contain diversity exons, labeled D_1, D_2, ... D_n. Recombination between the V and D regions occurs more frequently than that between the V and J exons in the light-chain exons. Introns between the V and D and between the D and J exons contain signal sequences that regulate synthesis of the Rag1/Rag2 recombinase. This enzyme is responsible for the efficient recombination that gives rise to the epitope-specific B-cell clones with their individual Ig genes. The heavy-chain genes contain exons that encode all of the isotype heavy chains. Class switching, i.e., the change in chain expression that occurs during antibody synthesis after B-cell activation, results from alternative splicing between the J exons and the exons for the various heavy-chain isotypes. Redrawn from Bhagavan.[403] Courtesy of N. V. Bhagavan.

The heptamers have twofold rotational symmetry. The recognition sequences adjacent to one or both sides of the coding segments are separated by 12- or 23-bp spacers, and an empirical rule states that joining can occur only when one pair of recognition sequences are separated by 12 bp and the other by 23. Recombination involves cleavage and rejoining of the DNA, a process also used for reorganization of the T-cell receptor genes. The cleavage is catalyzed by a complex of two proteins RAG1 and RAG2, which are encoded by the adjacent **recombination activating genes** *RAG1* and *RAG2*.[402–406] Cleavage is assisted by HMG chromatin proteins (Chapter 27, Section A,4) and by the level of histone acetylation in an associated enhancer.[407] In the presence of a divalent metal such as Mg^{2+} or Mn^{2+} a single-stranded nick is made between the 3'-end of the DNA coding region and the heptamer sequence. The released 3'–OH group then attacks the phosphodiester bond on the opposite strand, cleaving it in a transesterification reaction. This leaves the coding sequences capped by a hairpin end and a blunt cut end on the RSS.[408] The cut ends are apparently held until synapsis with a nonamer RSS further in the 3' direction and rejoining can occur.[409] Rejoining requires DNA ligase IV.[410] The process resembles that of transposition (Chapter 27, Section D,4).[411,412] The details are still uncertain.[408,413,413a]

The recombinational events are not entirely regular.[401,404,414,415] There are excisions of pieces of DNA, and additional nucleotides may be inserted randomly through the action of **terminal deoxynucleotidyl transferase**.[416,417] All of these mechanisms lead to additional diversity. After the rearrangements are completed, the fused VDJ and VJ gene segments are close enough to suitable constant-region (C) sequences that when the genes are transcribed the intervening sequences in the RNA are spliced out to yield the mature mRNAs for light κ and λ chains and H chains.

2. Somatic Hypermutation and Affinity Maturation

Yet another factor that introduces immunoglobulin diversity is the occurrence of somatic mutation at unusually high rates (**hypermutation**) in the hypervariable regions. Hypermutation consists largely of point mutations in the V-region gene segment and occurs at a rate 10^5- to 10^6-fold higher than for the rest of the genome.[51,418–423] It occurs after B cells have been presented with antigen by dendritic cells in the germinal centers in a process called affinity maturation. The result is generation of an enlarged repetoire of B cells, some of which synthesize antibodies with an increased affinity for the antigen. These will be selected for clonal expansion.[51] Further genetic alteration occurs by **gene conversion** (see p. 1566), which involves copying from a homologous segment of DNA, perhaps from a nearby pseudogene.[423a,b] Some experiments suggested that RNA editing (p. 1642) may also contribute to antibody diversity.[29,424,424a] However, it appears that the observed deamination of cytosine to uracil rings occurs at the DNA level and is initiated by an **activation-induced cytosine deaminase**.[424b] The generation of uracil, which may be removed from the DNA by uracil-DNA glycosylase (p. 1579), apparently triggers both somatic hypermutation and gene conversion. Both processes also depend upon DNA repair via homologous recombination or nonhomologous end joining (p. 1581). The error-prone DNA polymerase τ is also needed.[424c] Class-switching recombination, discussed in the next section, is apparently also initiated by the activation-induced deaminase.[424d] These are affected by **RNA editing** (Chapter 28).[79,424] **Receptor editing**, gene rearrangements that occur in the peripheral immune system, also contributes to affinity maturation.[424,425]

It is important that the genes as finally assembled maintain correct reading frames so that a potentially useful antibody can be made.[426] Although lymphocytes are diploid and therefore contain two sets of immunoglobulin genes, a single cell produces only one kind of immunoglobin (**allelic exclusion**). When the genetic rearrangements produce a light chain able to combine with a heavy chain to form a functional immunoglobulin, a signal may be sent that stops the rearrangements in the other chromosome.[427] However, recent findings suggest that one allele may be marked for inactivation early in development, just as one X chromosome becomes inactive in females (Chapter 32).[428]

3. Immunoglobulin Class (Isotype) Switching

A newly matured B cell produces initially IgM bound to its surfaces. The difference between the C_H domains of bound and secreted antibodies of a given type seems to lie in alternative splicing of the mRNA. More mysterious are the consecutive switches from IgM to other types in the following sequence:

$$IgM \rightarrow IgG \rightarrow IgE \rightarrow IgA$$

This is the same sequence in which the C_H genes lie (Fig. 31-18). However, the C_δ gene, which codes for IgD, is not utilized in this sequence. A newly matured B cell transcribes the C_μ gene to give IgM. Later, the class switch occurs, apparently by a looping-out recombinational mechanism, allowing a C_γ gene (for IgG) to be expressed.[429–433] The switches are mediated by tandemly repeated DNA sequences in 1- to 10-kbp switch regions and are controlled in part by cytokines. Other complexities are involved in synthesis of the J chain of IgM[434] and in synthesis of IgD. Although the latter is a major surface immunoglobulin on B lymphocytes, its exact functions have been hard to understand. The δ exon of its heavy chain gene is joined to a J exon by alternative splicing of the mRNA (Fig. 31-18).[112] In a similar manner, the difference in the surface-bound and soluble forms of IgM arises by alternative mRNA splicing.[403]

The T-cell receptor gene families are also indicated in Fig. 31-17. Their development is remarkably similar to that of the immunoglobulin genes and involves most of the same mechanism of diversification[365,397,435] with the exception of somatic hypermutation. The same recombinase may cut the DNA to initiate rearrangements of all of these gene families.[436] A single T-cell precursor may give rise to 1000 or more clones with unique β-chain sequences.[437] With a total of ~10^6 different β chains there are potentially ~10^{15} unique T-cell receptor structures that could arise from the 42V and 61J segments of the α-chain gene and the 47V, 2D, and 13J segments of the human β-chain gene.[438] Allelic exclusion is observed, as with the immunoglobulins.[439]

F. Disorders of the Immune System

Many things can go wrong with a system as complex as the human immune system. In immunodeficiency disease some component is missing or has been inactivated. In autoimmune diseases the immune system attacks some component of the body. Of the known problems none is more common than **allergy**,[440–442] which may be described as the inappropriate activation of the immune system by environmental antigens (**allergens**).

1. Allergy

One in 10 persons, ~22 million people, in the United States have allergies. Ten million of these suffer from the nasal discomfort of "hay fever" and six million from the more serious **asthma**. Substantial numbers of people in the United States die of allergic reactions to insect stings (more than 30 per year) or to injections of penicillin (300 per year in 1970). Foods, drugs, pollens, mold spores, mites in house dust, and even heat or cold can evoke serious allergic reactions. Among these **eczema** (atopic dermatitis) is very common. A major cause of allergic reactions has been traced to molecules of immunoglobulins IgE, which bind to the **basophils** in the blood and to the related **mast cells** of tissues. Binding of an antigen to these IgE molecules activates them. These activated antibodies bind (as in Fig. 31-6) to the α subunits of the $F_{c\varepsilon}RI$, a transmembrane receptor on basophil or mast cell surfaces.[108,109,443] If two or more IgE molecules bind to a mast cell, they may aggregate and activate the mast cell to release its histamine-containing granules.[444,445] The granules also release cytokines and arachidonate, which is converted primarily into prostaglandin D_2 (Fig. 21-7) and into products of the 5-lipoxygenase pathway (Fig. 21-8). The products include the chemotactic leukotriene B_4 and leukotrienes C_4 and D_4. The latter two constitute the slow-reacting substance of anaphylaxis (Fig. 21-8). The result is a rapid inflammatory response with dilation of blood vessels, increased vascular permeability, infiltration of leukocytes, and destruction of tissues.

What is the normal function of IgE and the mast cells? These cells are located at places where parasites might enter tissues. IgE is involved in killing of schistosomes, and elevated IgE levels are seen in patients infected with various parasites. The killing of schistosomes seems to be mediated by blood platelets as well as by neutrophils and eosinophils with the help of mast cells.[446,447] Allergic persons often have an IgE level over ten times normal. This high level makes the individual especially sensitive to IgE-mediated reactions, a condition called **atopy** (meaning "strange disease").[448,448a] Allergies may also be accompanied by increased B cell levels.[442] This can sometimes be responsible for the sudden and sometimes fatal systemic reaction of **anaphylaxis**. T-cell responses may also cause anaphylaxis.

Most allergy-inducing antigens are proteins, but proteins vary widely in their antigenicity. Only a few natural proteins are major allergens. Many of these are relatively small, with molecular masses of 5–50 kDa. Most are soluble, and some are glycoproteins.[449] Mites are the closest animals to human life and carry allergens that are among the most important causes of asthma and allergic dermatitis. The allergens are 125- to 129-kDa proteins crosslinked by three disulfide bridges.[450,451] Cockroaches[449,452] and other insects also form many allergens. Among them are the hemoglobins of small flies of the chironomid family.[453] Studies of the latter suggest that antigenicity may arise both from flexible regions and also from the presence of a preponderance of amino acids with polar side chains.

Proteins of cat saliva dry and flake off as dander, which contains major indoor allergens linked to asthma. Dogs, horses, cattle, and other animals also provide several allergens,[454] some of which are lipocalins (Box 21-A).[455] Other close-to-home allergens are provided by fungi that live on skin or nails.[456]

Plants provide a host of allergens. Major allergens are found in pollen of rye-grass,[457] of many other grasses, of ragweed,[458] and of olive trees.[459] Natural rubber latex would appear to be a harmless high polymer, but it contains antigenic proteins, which have been blamed for 1100 anaphylactic attacks with at least 15 deaths between 1988 and 1992.[460,461]

Food is a major source of allergens, which are often overlooked. Food allergies may be hard to diagnose and symptoms such as headache, diarrhea, itching, and asthma may be attributed to other causes. However, the occasional rapid death from anaphylactic shock, e.g., from exposure to peanuts,[461a] is a reminder that unrecognized food allergies exist. About 100–200 persons die annually of food allergies. About 90% of recognized food allergies involve milk, eggs, fish, crustacea, peanuts, tree nuts, soybeans, and wheat.[462] There are usually only a few allergenic proteins in any one food. Many of these are resistant to digestion in the stomach. Some but not all are compact proteins with multiple disulfide crosslinkages. However, no structural generalization can be applied to all food allergens.

The increasing use of genetic engineering of foods poses both risks and hopes. The accidental incorporation of an allergenic protein into a plant or animal product can make a food previously safe for a person deadly. For this reason, attempts are now made to identify likely allergens and to avoid transferring their genes. However, this process can't be completely reliable, partly because we each have our own personal immune system. This is one reason for requiring accurate labeling of foods. On the positive side recognized major allergens can probably be eliminated by genetic engineering. The muscle protein tropomyosin (Fig. 19-9) is a well-known allergen whose allergenicity varies among different sources. Tropomyosin from beef, pork, and chicken is usually not highly allergenic, but that from shrimp often is.[462] Perhaps safer shrimp can be created.

Allergies are treated in various ways, often with antihistamines or corticosteroid ointments. Injection of adrenaline is an emergency treatment of anaphylaxis or asthma. Also important is **specific immunotherapy**, better known as "allergy shots."[441] Small amounts of allergen in increasing amounts are injected subcutaneously at intervals to desensitize the patient. Use of purer antigens, which may be engineered to decrease their antigenicity, may provide advances in this technique.[463]

Asthma is one of the most common chronic diseases in industrialized countries, affecting 10% or more of young children in some countries. An atopic disease with high IgE levels, asthma is induced by small particles of antigen, which are able to penetrate deep into the lungs.[464] About 80% of asthmatic children are allergic to house mites.[450] Animal danders are another major cause. The incidence of asthma appears to be increasing in many modern societies, but

BOX 31-D MYASTHENIA GRAVIS

One of the best-understood autoimmune diseases is myasthenia gravis, a condition associated with a decrease in the number of functional postsynaptic nicotinic acetylcholine receptors (Fig. 30-23) in neuromuscular junctions.[a–e] The resulting extreme muscular weakness can be fatal. Myasthenia gravis is not rare and affects about one in 10,000 people.[c] An interesting treatment consists of the administration of physostigmine, diisopropylphosphofluoridate (Chapter 12, Section C,1), or other acetylcholinesterase inhibitors (Box 12-E). These very toxic compounds, when administered in controlled amounts, permit accumulation of higher acetylcholine concentration with a resultant activation of muscular contraction. The same compounds are widely used in the treatment of glaucoma.

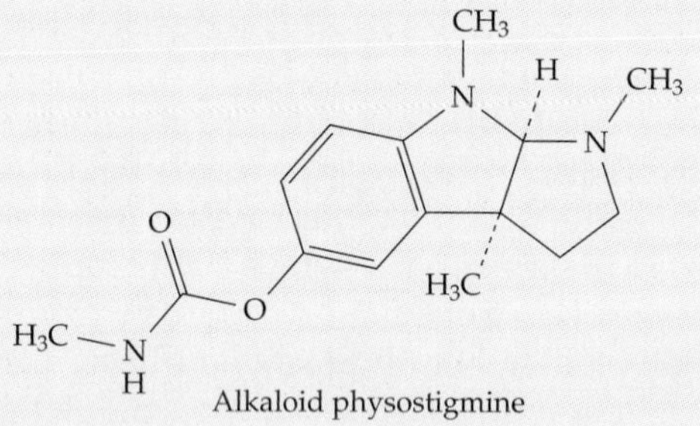

Alkaloid physostigmine

More than 90% of patients with myasthenia gravis have circulating antibodies directed against a subunit of the acetylcholine receptor.[f] Immunosuppressive drugs and steroids help to cut down on these autoantibodies, and many patients are benefitted by removal of the thymus. Newer approaches involve specific immunotherapy aimed at increasing tolerance to either T cells or to B cells.[c,d] For example, oral ingestion of purified acetylcholine receptors to desensitize the body's response or inhibition of production of Il-2.

A possible cause for the production of the damaging antibodies may be the sharing of common antigenic determinants between the receptor protein and surface proteins of bacteria such as *E. coli*.[f]

[a] Fuchs, S. (1980) *Trends Biochem. Sci.* **5**, 259–262
[b] Tzartos, S. J. (1984) *Trends Biochem. Sci.* **9**, 63–67
[c] Drachman, D. B. (1994) *N. Engl. J. Med.* **330**, 1797–1808
[d] Steinman, L., and Mantegazza, R. (1990) *FASEB J.* **4**, 2726–2731
[e] Barnes, D. M. (1986) *Science* **232**, 160–161
[f] Stefansson, K., Dieperink, M. E., Richman, D. P., Gomez, C. M., and Marton, L. S. (1985) *N. Engl. J. Med.* **312**, 221–225

the reasons are unclear,[465,466] and the increase doesn't appear to be linked to air pollution. A chronic atopic condition usually precedes an acute attack of asthma. In addition to IgE and eosinophils there are excessive numbers of neutrophils with high-affinity IgE receptors in the airway tissues.[467] Prostaglandin D_2 released from mast cells may play a role in triggering an attack.[468] Cytokines,[469] nitric oxide,[447] and nerve growth factor[470] may also participate in the response. The presence of a high concentration of glutathione and of glutathione peroxidase, whose concentration increases in asthmatic lungs, may reflect the action of the antioxidant system in combatting inflammation.[471] The surfactant proteins SP-A and SP-D (Box 8-B) are Ca^{2+}-dependent lectins, which serve as regulators of the innate immune response. Their concentrations also increase in asthma.[472,473]

Treatment of asthma has depended upon inhaled glucocorticoids, quick-acting bronchodilators usually β-adrenergic antagonists, and long-acting beta agonists such as theophyllins and leukotriene antagonists.[464]

2. Autoimmune Diseases

There are numerous **autoimmune diseases** in which the body makes antibodies against its own cells (Table 31-3).[363,474–476] In **myasthenia gravis** (Box 31-D) antibodies attack the acetylcholine receptors in postsynaptic membranes.[477] In **Graves disease** aberrant antibodies are directed against receptors for thyrotropin. They have a stimulatory rather than an inhibitory effect and cause hyperthyroidism.[478] **Childhood onset (Type I) diabetes** results from destruction of insulin-secreting cells by an autoimmune reaction triggered by environmental factors in genetically susceptible persons (Box 17-G).[479,480] The principal autoantigen appears to be the 65-kDa form of glutamate decarboxylase (GAD).[481–483] While GAD has an essential function in the formation of γ-aminobutyrate in the brain, its role in the pancreatic islets is not clear. What is established is the presence of specific GAD65-stimulated T cells in diabetic individuals. The stimulating autoantibodies, which may appear in the blood years before diabetes is evident, carry HLA-DR4 type surface MHC antigens.[484,485] After the activated T cells kill enough of the pancreatic β cells, diabetes appears.[486]

Myasthenia gravis, Graves disease, and type I diabetes are organ-specific autoimmune diseases. Another group of autoimmune diseases are systemic, affecting many tissues. For example, in the severe **systemic lupus erythematosus** there are often antibodies against the victim's own DNA.[487–488a] The antibodies may then attack any tissue, e.g., the red blood cells. Antibodies against a variety of other nuclear constitutents such as histones,[207] ribosomal protein L7,[489] ubiquitin,[490] enzymes, cardiolipin,[491] and small nuclear RNAs[492] are also made. The primary defect appears to be an intolerance to chromatin and in particular to nucleosomes.[493] Antibodies to nucleolar and other components of nuclei are also present in progressive systemic sclerosis (**scleroderma**).[494,495] In **rheumatoid arthritis**, a chronic inflammation of joints, the serum and joint fluids contain abnormal complexes, which appear to consist entirely of immunoglobulins. They may be antibody – antiidiotype antibody complexes.[496,496a] Immunization of animals with type II collagen induces a very similar arthritis,[496,497] but this collagen probably doesn't supply the offending human antigen. Susceptibility to rheumatoid arthritis is linked to HLA-DR4 class II MHC genes. Molecules of the class II DR4 subtypes may associate with antigenic peptides of uncertain origin to induce a T-cell response. $CD4^+$ T cells are thought to drive inflammation in the disease.[497] Monocytes are attracted from the blood and become inflammatory macrophages.[498]

Another autoimmune disease in which antibodies attack collagen is **Goodpasture disease**. It is mediated by B cells, which form antibodies directed at the N-terminal domain of the α3 chain of collagen IV (pp. 435–438).[499–501] The antibodies attack the glomerular basement membranes causing a rapidly progressing glomerulonephritis and also lung hemorrhages. Primary **glomerulonephritis**, a major kidney disease, may be caused by a cross-reaction between the membrane of streptococci and the glomerular basement membranes.

In **Sjögren syndrome** autoantibodies are directed against α fodrin (p. 405).[502] In primary **biliary cirrhosis** they are directed at mitochondria and specificially to a pyruvate dehydrogenase subunit (Fig. 15-14).[503] In the inflammatory muscle disease **polymyocystitis** autoantibodies are often directed against cytoplasmic proteins including aminoacyl-tRNA synthetases.[504] In the rare **paroxysmal cold hemoglobinuria** autoantibodies attack red blood cell membranes only when the temperature of an extremity is lowered. **Paroxysmal nocturnal hemoglobinuria**, a serious complement-mediated condition, results from deficiency in the complement decay accelerating factor. This is a result of a defect in the PGI tail on this factor.[505]

Celiac disease (celiac sprue) is an allergic inflammatory condition caused by poorly digested proline-rich sequences of wheat gluten and related proteins (p. 74). The disease is usually not recognized, but it may occur in 3% or more of the United States population. A T-cell response that causes destruction of the smaller intestinal mucosa, celiac disease is characterized by malabsorption and diarrhea.[505a–c] It can cause death by starvation. A primary target of the autoantibodies is a transglutaminase.[505c,d]

Most cells of the immune system are ordinarily kept apart from those of the nervous system by means of the blood–brain barrier. However, allergic encephalomyelitis, in which T cells attack the myelin sheath of brain neurons, can easily be induced in mice.[506] A similar autoimmune process is thought to be involved in human **multiple sclerosis** (see Chapter 30, pp. 1769, 1808, and Fig. 30-9).[507,508] High levels of circulating IgM are found in some demyelinating diseases of peripheral neurons.[508] In **Rasmussen's encephalitis**, which causes brain inflammation and epilepsy, serum antibodies attack a glutamate receptor subunit **GluR3**.[509]

The causes of autoimmune disease doubtless lie largely in the difficulty of developing a repetoire of immunoglobulin-forming B cells and of T-cell receptors that will always reliably distinguish self from a foreign antigen. The problem can lie either with B-cell recognition or with the T-cell receptors. Extensive medical use is made of **immunosuppressants** in treatment of persistent allergic reactions, autoimmune problems, and rejection of transplanted tissues. Among these compounds are the steroidal

TABLE 31-3
Some Autoimmune Diseases

Addison disease	Adrenal glands
Ankylosing spondilitis	
Celiac disease	Upper intestines
Crohn disease	Intestines
Diabetes, type I	Pancreatic islets
Glomerulonephritis	Kidney
Goodpasture disease	Kidney
Graves disease	Thyroid gland
Guillain – Barré syndrome	Gangloisides
Multiple sclerosis (MS)	Peripheral myelin
Myasthenia gravis	
Paroxysmal cold hemoglobinuria	Red blood cells
Primary biliary cirrhosis	
Psoriasis	Skin
Polymyocystitis	
Rasmussen encephalomyelitis	Cerebral cortex
Rheumatoid arthritis	Joints
Scleroderma	Skin
Sjögren syndrome	
Systemic lupus erythematosis	Many tissues
Thyroiditis	Thyroid gland
Ureitis	

BOX 31-E EVADING THE IMMUNE SYSTEM

Parasitic species always have a problem with the antibodies and killer T cells of their hosts, and the chemical makeup of the external coats of parasites tends to reflect this fact.[a] An example is provided by **trypanosomes,** which cause sleeping sickness and which make much of Africa unsuitable for cattle grazing.[b–d] Trypanosomes in the bloodstream evade the immune system by covering the outer surface of their plasma membrane, flagella and all, with a dense 12- to 15-mm thick monolayer of an ~60-kDa **variable surface glycoprotein**.[e,f] The glycoprotein molecules are anchored in the cell membrane by C-terminal glycosylphosphatidylinositol (GPI) anchors (Fig. 8-13).[g] The glycoprotein layer protects the parasite but is soon attacked by the host's immune system. However, the parasite has perhaps 1000 different genes for the variable surface protein, and every ten days or so new clones of trypanosomes appear with new coats that the immune system is not prepared to attack. To accomplish this cells occasionally copy one of the previously unused variable surface glycoprotein genes and place it into a new location in the genome, where it is expressed.[h–j]

Parasitic **nematodes** shed the outer layers of their external cuticle and like trypanosomes reveal a new layer with different antigenic proteins.[k] *Giardia* protects itself in a similar fashion.[l] **Schistosomes**, tiny parasitic flatworms, evade a host's immune system by shedding complex glycoproteins from specialized double outer membranes.[m] Antigenic determinants including MHC antigens characteristic of the mouse have been identified in the membrane of schistosomes from infected mice. Thus, one aspect of the parasite's defense may be to hide behind surface recognition markers stolen from its host.[n,o] Schistosomes also secrete the peptide Thr-Lys-Pro, which inhibits macrophages, as well as a small molecule that inhibits T lymphocytes.[o]

The malaria parasite *Plasmodium* has a complex life cycle with several forms and spends much of its life hiding within red blood cells.[p] It may also suppress the immune system. The unicellular sporozoites, which are injected into the bloodstream by mosquitos, are protected by an external coat protein that is unusual in containing many short repeated sequences. For example, that of *P. falciparum*, which causes the most deadly form of malaria, contains the sequence Asn-Ala-Asn-Pro repeated 37 times.[q] These coat proteins undergo unusually rapid evolution, which makes the preparation of vaccines difficult.[r]

Trypanosomes, schistosomes, and malaria parasites still represent major health problems. Malaria kills two to four million persons a year and endangers almost a third of the world's population. It has been impossible to produce suitable vaccines for any of these parasites. However, the cloning of genes for individual parasite proteins has given hope that effective vaccines can be devised.[e,o–v] One problem is the lack of interest in financing the effort.[w]

Many other protozoan parasites and bacteria invade cells and take up residence in macrophages.[x] These include species of *Salmonella*, *Legionella*, and *Mycobacterium*.[y] Bacteria often employ structural mimicry to gain access to a cell,[z] e.g., by mimicking the type III secretion system (p. 520).[aa] Some bacteria have developed defenses against reactive oxygen species, allowing them to evade the action of phagocytes.[bb] *Borrelia burgdorferi*, the Lyme disease spirochete, synthesizes an unusual single-layer β-sheet outer surface protein,[cc] which becomes coated with complement protein H. This may protect the bacteria and allow them to live for a long time within cells.[dd]

Even the lowly **influenza virus** finds a way around our immunity so that it can strike us repeatedly. As this virus matures, it acquires a lipid membrane by budding from the host cell. Two virally encoded proteins are present in the membrane. One is a trimeric hemagglutinin, which forms small 7.6-nm spikes that protrude from the virus surface.[ee,ff] The hemagglutinin monomer is a 550-residue peptide containing four antigenic regions. The RNA genome of the virus undergoes rapid mutation (Chapter 28, Section E,2). At least one amino acid substitution was found in each antigenic region, when hemagglutinins from influenza viruses causing epidemics in 1972 and 1975 were compared with the strain that caused a worldwide epidemic in 1968. Recently the type A influenza virus that caused the 1918–1919 pandemic, the greatest acute plague of the 20th century, has been "resurrected" and investigated using viral RNA from three victims.[gg–ii] The globular part of the hemagglutinin appears to have come from a pig and the "stalk" from a human lineage. The virus takes advantage of the pool of virus in swine, humans, and birds to vary its structure and create new strains. The reason for the deadly nature of the 1918–1919 strain, which killed 20–40 million people, an unusually large number of whom were young, previously healthy adults, is not clear.

Viruses use a large variety of mechanisms to evade cellular defense mechanisms. Almost every aspect of the innate or adaptive immune systems provides some opportunity for evasion.[jj] The rapid

BOX 31-E (continued)

mutation rate in a population of virus particles contributes greatly to this ability, allowing chronic infections such as those of hepatitis C[kk] or delayed catastrophic infections such as those of HIV.

[a] Bloom, B. R. (1979) *Nature (London)* **279**, 21–26
[b] Englund, P. T., Hajduk, S. L., and Marini, J. C. (1982) *Ann. Rev. Biochem.* **51**, 695–726
[c] Rice-Ficht, A. C., Chen, K. K., and Donelson, J. E. (1981) *Nature (London)* **294**, 53–57
[d] Muñoz-Jordán, J. L., Davies, K. P., and Cross, G. A. M. (1996) *Science* **272**, 1795–1797
[e] Metcalf, P., Blum, M., Freymann, D., Turner, M., and Wiley, D. C. (1987) *Nature (London)* **325**, 84–86
[f] Gardiner, P. R., Pearson, T. W., Clarke, M. W., and Mutharia, L. M. (1987) *Science* **235**, 774–777
[g] Ferguson, M. A. J., Haldar, K., and Cross, J. A. M. (1985) *J. Biol. Chem.* **260**, 4963–4968
[h] Raibaud, A., Gaillard, C., Longacre, S., Hibner, V., Buck, G., Bernardi, G., and Eisen, H. (1983) *Proc. Natl. Acad. Sci. U.S.A.* **80**, 4306–4310
[i] Borst, P., and Greaves, D. R. (1987) *Science* **235**, 658–667
[j] Navarro, M., and Gull, K. (2001) *Nature (London)* **414**, 759–763
[k] Philipp, M., Parkhouse, R. M. E., and Ogilvie, B. M. (1980) *Nature (London)* **287**, 538–540
[l] Papanastasiou, P., Hiltpold, A., Bommeli, C., and Köhler, P. (1996) *Biochemistry* **35**, 10143–10148
[m] Cummings, R. D., and Nyame, A. K. (1996) *FASEB J.* **10**, 838–848
[n] Simpson, A. J. G., and Cioli, D. (1982) *Nature (London)* **296**, 285–287
[o] Kolata, G. (1987) *Science* **227**, 285–287
[p] Kolata, G. (1984) *Science* **226**, 679–682
[q] Young, K. J. F., Hockmeyer, W. T., Gross, M., Ballou, W. R., Wirtz, R. A., Trosper, J. H., Beaudoin, R. L., Hollingdale, M. R., Miller, L. H., Diggs, C. L., and Rosenberg, M. (1985) *Science* **228**, 958–962
[r] de la Cruz, V. F., Lal, A. A., Welsh, J. A., and McCutchan, T. F. (1987) *J. Biol. Chem.* **262**, 6464–6467
[s] Peterson, D. S., Wrightsman, R. A., and Manning, J. E. (1986) *Nature (London)* **322**, 566–568
[t] Balloul, J. M., Sondermeyer, P., Dreyer, D., Capron, M., Grzych, J. M., Pierce, R. J., Cavillo, D., Lecocq, J. P., and Capron, A. (1987) *Nature (London)* **326**, 149–153
[u] Capron, A., DeSaint, J. P., Capron, M., Ouma, J. H., and Butterworth, A. E. (1987) *Science* **238**, 1065–1072
[v] Marshall, E. (1997) *Science* **275**, 299
[w] Enserink, M. (2000) *Science* **287**, 1956–1958
[x] Small, P. L. C., Ramakrishnan, L., and Falkow, S. (1994) *Science* **263**, 637–639
[y] Schorey, J. S., Carroll, M. C., and Brown, E. J. (1997) *Science* **277**, 1091–1093
[z] Vallance, B. A., and Finlay, B. B. (2000) *Proc. Natl. Acad. Sci. U.S.A.* **97**, 8799–8806
[aa] Stebbins, C. E., and Galán, J. E. (2001) *Nature (London)* **412**, 701–705
[bb] Hassett, D. J., and Cohen, M. S. (1989) *FASEB J.* **3**, 2574–2582
[cc] Huang, X., Nakagawa, T., Tamura, A., Link, K., Koide, A., and Koide, S. (2001) *J. Mol. Biol.* **308**, 367–375
[dd] Hellwage, J., Meri, T., Heikkila, T., Alitalo, A., Panelius, J., Lahdenne, P., Seppälä, I. J. T., and Meri, S. (2001) *J. Biol. Chem.* **276**, 8427–8435
[ee] Wilson, I. A., Skehel, J. J., and Wiley, D. C. (1981) *Nature (London)* **289**, 366–373
[ff] Chen, J., Skehel, J. J., and Wiley, D. C. (1999) *Proc. Natl. Acad. Sci. U.S.A.* **96**, 8967–8972
[gg] Lederberg, J. (2001) *Proc. Natl. Acad. Sci. U.S.A.* **98**, 2115–2116
[hh] Basler, C. F., Reid, A. H., Dybing, J. K., Janczewski, T. A., Fanning, T. G., Zheng, H., Salvatore, M., Perdue, M. L., Swayne, D. E., Garcîa-Sastre, A., Palese, P., and Taubenberger, J. K. (2001) *Proc. Natl. Acad. Sci. U.S.A.* **98**, 2746–2751
[ii] Gibbs, M. J., Armstrong, J. S., and Gibbs, A. J. (2001) *Science* **293**, 1842–1845
[jj] Ploegh, H. L. (1998) *Science* **280**, 248–253
[kk] Farci, P., Shimoda, A., Coiana, A., Diaz, G., Peddis, G., Melpolder, J. C., Strazzera, A., Chien, D. Y., Munoz, S. J., Balestrieri, A., Purcell, R. H., and Alter, H. J. (2000) *Science* **288**, 339–344

anti-inflammatory agents such as prednisone and the folate antagonist methotrexate. A new vision of the possibilities for immunosuppression came, however, with the discovery of **cyclosporin A** (Box 9-F). This fungal metabolite inhibits lymphokine formation by helper T cells. It alleviates rejection of grafted tissues and prevents graft-versus-host disease. The use of cyclosporin and FK506 (Box 9-F) has permitted organ transplantation to the extent that by 1987 surgeons had transplanted in one year 1200 livers, 1500 hearts, and 9000 kidneys with one-year survival rates of 80% for hearts and over 90% for kidneys.[510] As mentioned in Box 9-F, cyclosporins bind very tightly to **cyclophilins**,[511] which have peptidylprolyl *cis – trans* isomerase activity and perhaps other independent functions.[512]

3. Immunodeficiencies

More than 95 different problems of impaired immunity have been identified. They affect about 1 in 10,000 persons born. The defects may involve T cells, B cells, NK cells, or phagocytic cells.[513,514] There may be problems in lymphocyte development.[515] Complement proteins may be lacking,[161,513,516] or their regulation may be faulty.[517] The immune system has specific "blind spots" and fails to recognize all dangerous foreign antigens.[518] In the fatal **X-linked immunoproliferative syndrome**,[519,520] the immune system of susceptible males does not respond to the Epstein – Barr virus-induced mononucleosis by killing the persistently lymphoblastoid cells characteristic of that disease. Some individuals are born with **severe combined immunodeficiency** disease. This condition was made well known to the public by the plight of David, the "bubble boy," who lived 12 years in protective sterile rooms and a plastic bubble-like "space suit."[521] The condition is often caused by a defect in the interleukin-2 receptor, but there are a variety of other causes. About 15% of cases arise from a defect

BOX 31-F AN INSECTICIDAL PROTEIN

During sporulation the bacterium *Bacillus thuringiensis* forms within its own cells large protein crystals, which are highly toxic to some insect larvae. The crystals account for 20–30% of the dry weight of the bacterial spores and contain more than one toxin[a] and, curiously, a 20-kbp piece of DNA.[b] Dusting of garden plants with dried spores from these bacteria has become a popular and effective way of combating cabbage worms and other insects. The toxic protein from one strain of bacteria is encoded by a 4222-bp gene.[c] The corresponding 133-kDa 1176-residue polypeptide protoxin undergoes glycosylation and perhaps other modifications, presumably prior to crystallization. After ingestion by susceptible insect larvae (largely Lepidoptera) the protein is cleaved to form a smaller ~65-kDa protease-resistant core, which is the active toxin. Other strains of bacteria produce toxins specific for Diptera or Coleoptera.[d–f]

X-ray crystallography of the 65-kDa form reveal a three-domain structure. The central domain varies among different strains and is probably involved in recognition and in binding to cell surface receptors.[e–h] The toxin binds to a receptor, apparently an aminopeptidase N,[i] after which the toxin is rapidly inserted into the membrane forming a 1- to 2-nm diameter pore. This leads to cell death.[j]

Because the toxins appear to be harmless to human beings and higher animals the toxin genes have been transferred into various other bacteria, which are symbiotic with plants and into plants themselves. Toxin genes in suitably modified form (Chapter 27) were first transferred into bacteria that live naturally in association with roots of *Zea mays* and into tobacco and tomato plants. The new host organisms expressed the toxin genes and protected the plants from damage by caterpillars.[k,l] Since then the toxin genes have been transferred into many crop plants, which are widely planted.

Two problems must be considered. Insects do develop resistance to the Bt toxin.[m] This problem can be combated by protein engineering[n] and by location of new sources of toxins.[o,p] A second problem deals with the environmental impact.[q] Will Bt toxin kill desirable insects? Will the gene be transferred in nature to other species and into the environment? The latter may seem unlikely, but as the toxins are applied to fight soil organisms such as nematodes, transfer into organisms of the largely unstudied soil ecosphere may pose problems.

[a] Donovan, W. P., Dankocsik, C. C., Gilbert, M. P., Gawron-Burke, M. C., Groat, R. G., and Carlton, B. C. (1988) *J. Biol. Chem.* **263**, 561–567

[b] Clairmont, F. R., Milne, R. E., Pham, V. T., Carrière, M. B., and Kaplan, H. (1998) *J. Biol. Chem.* **273**, 9292–9296

[c] Schnepf, H. E., Wong, H. C., and Whiteley, H. R. (1985) *J. Biol. Chem.* **260**, 6264–6272

[d] Sekar, V., Thompson, D. V., Maroney, M. J., Bookland, R. G., and Adang, M. J. (1987) *Proc. Natl. Acad. Sci. U.S.A.* **84**, 7036–7040

[e] Garfield, J. L., and Stout, C. D. (1988) *J. Biol. Chem.* **263**, 11800–11801

[f] Li, J., Koni, P. A., and Ellar, D. J. (1996) *J. Mol. Biol.* **257**, 129–152

[g] Li, J., Carroll, J., and Ellar, D. J. (1991) *Nature (London)* **353**, 815–821

[h] Grochulski, P., Masson, L., Borisova, S., Pusztai-Carey, M., Schwartz, J.-L., Brousseau, R., and Cygler, M. (1995) *J. Mol. Biol.* **254**, 447–464

[i] Gill, S. S., Cowles, E. A., and Francis, V. (1995) *J. Biol. Chem.* **270**, 27277–27282

[j] Burton, S. L., Ellar, D. J., Li, J., and Derbyshire, D. J. (1999) *J. Mol. Biol.* **287**, 1011–1022

[k] Vaeck, M., Reynaerts, A., Hofte, H., Jansens, S., De Beuckeleer, M., Dean, C., Zabeau, M., Van Montagu, M., and Leemans, J. (1987) *Nature (London)* **328**, 33–37

[l] Baum, R. M. (1987) *Chem. Eng. News* **65** **Aug 10**, 9–14

[m] Oppert, B., Kramer, K. J., Beeman, R. W., Johnson, D., and McGaughey, W. H. (1997) *J. Biol. Chem.* **272**, 23473–23476

[n] Rajamohan, F., Alzate, O., Cotrill, J. A., Curtiss, A., and Dean, D. H. (1996) *Proc. Natl. Acad. Sci. U.S.A.* **93**, 14338–14343

[o] Estruch, J. J., Warren, G. W., Mullins, M. A., Nye, G. J., Craig, J. A., and Koziel, M. G. (1996) *Proc. Natl. Acad. Sci. U.S.A.* **93**, 5389–5394

[p] Guo, L., Fatig, R. O., III, Orr, G. L., Schafer, B. W., Strickland, J. A., Sukhapinda, K., Woodsworth, A. T., and Petell, J. K. (1999) *J. Biol. Chem.* **274**, 9836–9842

[q] Pimentel, D. S., and Raven, P. H. (2000) *Proc. Natl. Acad. Sci. U.S.A.* **97**, 8198–8199

in adenosine deaminase, an enzyme of the purine salvage pathway (Fig. 25-17). As mentioned in Chapter 25, genetic therapy for this condition is being used. However, the most reliable treatment for these immunodeficiencies seems to be bone marrow transplantation. By 2000 more than 375 patients worldwide had received this treatment with up to 95% chance of survival.[513] Virus-induced immunodeficiency is the prime characteristic of HIV infection and **AIDS**.[522,523] Both the amounts of autoantibodies and of amyloid deposits increase with age, and immune complex disease is suspected of being a cause of aging.

4. Cancers of the Immune System

A major function of the immune system is thought to be destruction of cancer cells. In this case altered cell surface carbohydrates or proteins elicit an antibody response with destruction of the offending cells.

That this process works imperfectly may explain why the incidence of cancer increases with age and also why the concentration of autoantibodies increases. The immune system is also susceptible to cancers, which include multiple myeloma, leukemias, and lymphomas. Some of these, such as Burkitt's lymphoma, involve rearrangement of chromosome segments that carry immunoglobulin genes.[524,525] These may result from errors in the gene rearrangements involved in the development of lymphocytes.

G. Defense Mechanisms of Plants

Plants make many compounds that repel or poison animals that eat them. Among such compounds are alkaloids, terpenes, calcium oxalate, fluoroacetate, cyanogenic glycosides, and phenolic compounds.[526] The chewing of insects or other wounding of plant tissues releases phenolic glycosides and other reactive compounds from vesicles. Some of these compounds, which are often referred to as **phytoalexins** (see also Box 20-E),[527] are repellent to predators, have antimicrobial activity, and / or participate in chemical crosslinking and strengthening of the plant cell wall.[526,528] Some are protease inhibitors that interfere with a predator's nutrition.[529] Some released compounds attract insects that may assist in defense by feeding on predator eggs or by attracting wasps that deposit eggs in predator larvae.[530] These can all be regarded as part of an innate defense system that in some respects resembles our own innate immune system. For example, plant defensins (Fig. 31-7), most of which are directed against fungi,[531,532] resemble those in our tissues.

A system of **receptor-mediated surveillance**, part of the innate system, triggers both immediate **local responses** and secondary immunity throughout the plant.[530,533,534] Immediate responses include **programmed cell death** (called the **hypersensitive response**[535]), tissue reinforcement, and production of antimicrobial metabolites. Secondary responses, known as **systemic acquired resistance**, develop immunity throughout the plant. The surveillance system of some plants consists of a series of receptors known as **resistance (R) proteins**, which recognize signaling molecules produced by pathogens.[535a,b] The R proteins are thought of as being paired with **avirulence (Avr) proteins** of the pathogen. If the resistance protein is missing, the plant will be susceptible to attack by the pathogen. The pathogen's Avr protein is thought to be part of the chemical attack on the plant, apparently assisting in the invasion. However, if the *Avr* gene has been lost or is mutated, the R protein won't detect the invasion, and the pathogen may have increased virulence. What are the characteristics of the Avr proteins? They are often small and may be crosslinked by S–S bridges. They may be taken up by plant cells via receptors that resemble the type III translocation system of bacteria (p. 520).[534] Similar small protein **elicitors** are released directly by wounding even in plants that do not have paired *R–Avr* genes.[529,536]

The R proteins, which act as receptors for Avr, and other elicitor proteins, are usually leucine-rich-repeat proteins with a characteristic nucleotide binding site attached (NB-LRR proteins).[534,537] Like other cell surface receptors they participate in signaling and utilize both ion channels and Ser/Thr protein kinases.[538] The *Arabidopsis* genome contains ~150 sequences that may represent NB-LRR receptors.[530]

What do these receptors do? Like other cell membrane receptors they may induce both rapid and slower responses. The rapid responses may result from transmembrane flow of ions, just as in neurotransmitter action (Figs. 30-19, 30-20). The first response observed is an oxidative burst, which within minutes generates reduced oxygen intermediates (ROIs; pp. 1072–1074).[533,539–543] These compounds may participate in crosslinking and lignification of cell walls. Together with nitric oxide (NO) and endogenous salicylic acid (Chapter 25, Section B,7),[533,544,545] they promote transcription of defense-related genes and participate in the hypersensitive response. A second pathway, utilized against some pathogens especially those that kill plants to obtain nutrients,[533] involves production of jasmonic acid (Eq. 21-18 and associated text)[545a] and ethylene (Fig. 24-16).[529,546]

Plants also have mechanisms for minimizing the damage from the over 500 known viruses. These don't often kill plants but can cause great damage. There are interferon-like responses[547] and gene-silencing mechanisms.[548] The latter often involve synthesis of dsRNA, cleavage by the enzyme Dicer, and interference with transcription as described on p. 1640.[113a,b,549,550] This defensive reaction can spread between cells and throughout a plant, apparently by transport of RNA through plasmadesmata and the phloem.[550]

References

1. Nossal, G. J. V. (1993) *Sci. Am.* **269** (Sep), 53–62
2. Delves, P. J., and Roitt, I. M. (2000) *N. Engl. J. Med.* **343**, 37–49
3. Bhagovan, N. V. (2001) *Medical Biochemistry*, 4th ed., Harcourt/Academic Press, San Diego, California
4. Cruse, J. M., and Lewis, R. E. (1997) *Atlas of Immunology*, CRC Press, Boca Raton, Florida
5. Van Oss, C. J., and Van Regenmortel, M. H. V. (1994) *Immunochemistry*, Dekker, New York
6. Kuby, J. (1992) *Immunology*, Freeman, New York
7. Male, D., Champion, B., Cooke, A., and Owen, M. (1991) *Advanced Immunology*, 2nd ed., Raven Press, New York
8. Hames, B. D., and Glover, D. M. (1988) *Molecular Immunology*, IRL Press, Oxford
9. Edelson, R. L., and Fink, J. M. (1985) *Sci. Am.* **252**(Jun), 46–53
10. Janeway, C. A., Jr. (1993) *Sci. Am.* **269**(Sep), 73–89
11. Medzhitov, R., and Janeway, C., Jr. (2000) *N. Engl. J. Med.* **343**, 338–344

11a. Medzhitov, R., and Janeway, C. A., Jr. (2002) *Science* **296**, 298 – 300

11b. Matzinger, P. (2002) *Science* **296**, 301 – 305

12. Hoffmann, J. A., Kafatos, F. C., Janeway, C. A., Jr., and Ezekowitz, R. A. B. (1999) *Science* **284**, 1313–1318
13. Brown, P. (2001) *Nature (London)* **410**, 1018–1020
14. Janeway, C. A., Jr. (2001) *Proc. Natl. Acad. Sci. U.S.A.* **98**, 7461–7468
15. Paul, W. E. (1993) *Sci. Am.* **269**(Sep), 91–97
16. Marone, G., Lichtenstein, L. M., and Galli, S. J., eds. (2000) *Mast Cells and Basophils*, Academic Press, San Diego, California
17. Metcalf, D. (1991) *Science* **254**, 529–533
18. Boyington, J. C., Motyka, S. A., Schuck, P., Brooks, A. G., and Sun, P. D. (2000) *Nature (London)* **405**, 537–543
19. Sheriff, S., Chang, C. Y. Y., and Ezekowitz, R. A. B. (1994) *Nature Struct. Biol.* **1**, 789–794
20. Riddihough, G. (1994) *Nature (London)* **372**, 114

20a. Hazen, S. L., and Chisolm, G. M. (2002) *Proc. Natl. Acad. Sci. U.S.A.* **99**, 12515 – 12517

21. Hayashi, F., Smith, K. D., Ozinsky, A., Hawn, T. R., Yi, E. C., Goodlett, D. R., Eng, J. K., Akira, S., Underhill, D. M., and Aderem, A. (2001) *Nature (London)* **410**, 1099–1103
22. Banchereau, J., and Steinman, R. M. (1998) *Nature (London)* **392**, 245–252
23. Pulendran, B., Palucka, K., and Banchereau, J. (2001) *Science* **293**, 253–256
24. Thoma-Uszynski, S., Stenger, S., Takeuchi, O., Ochoa, M. T., Engele, M., Sieling, P. A., Barnes, P. F., Röllinghoff, M., Bölcskei, P. L., Wagner, M., Akira, S., Norgard, M. V., Belisle, J. T., Godowski, P. J., Bloom, B. R., and Modlin, R. L. (2001) *Science* **291**, 1544–1547
25. Huang, Q., Liu, D., Majewski, P., Schulte, L. C., Korn, J. M., Young, R. A., Lander, E. S., and Hacohen, N. (2001) *Science* **294**, 870–874
26. Streit, W. J., and Kincaid-Colton, C. A. (1995) *Sci. Am.* **273**(Nov), 54–61

26a. Banchereau, J. (2002) *Sci. Am.* **287** (Nov), 52 – 59

27. Re, F., and Strominger, J. L. (2001) *J. Biol. Chem.* **276**, 37692–37699

27a. Janssens, S., and Beyaert, R. (2002) *Trends Biochem. Sci.* **27**, 474 – 482

27b. Mak, T. W., and Yeh, W.-C (2002) *Nature (London)* **418**, 835 – 836

28. Boissy, G., O'Donohue, M., Gaudemer, O., Perez, V., Pernollet, J.-C., and Brunie, S. (1999) *Protein Sci.* **8**, 1191–1199
29. Fagarasan, S., and Honjo, T. (2000) *Science* **290**, 89–92
30. Macpherson, A. J., Gatto, D., Sainsbury, E., Harriman, G. R., Hengartner, H., and Zinkernagel, R. M. (2000) *Science* **288**, 2222–2226

30a. Khush, R. S., Leulier, F., and Lemaitre, B. (2002) *Science* **296**, 273 – 275

30b. Choe, K.-M, Werner, T., Stöven, S., Hultmark, D., and Anderson, K. V. (2002) *Science* **296**, 359 – 362

30c. Christophides, G. K., and 34 other authors. (2002) *Science* **298**, 159 – 165

31. Hancock, R. E. W., and Scott, M. G. (2000) *Proc. Natl. Acad. Sci. U.S.A.* **97**, 8856–8861
32. Sawai, M. V., Jia, H. P., Liu, L., Aseyev, V., Wiencek, J. M., McCray, P. B., Jr., Ganz, T., Kearney, W. R., and Tack, B. F. (2001) *Biochemistry* **40**, 3810–3816
33. Ganz, T. (1999) *Science* **286**, 420–421
34. Newman, J. (1995) *Sci. Am.* **273**(Dec), 76–79
35. Zinkernagel, R. M. (2001) *N. Engl. J. Med.* **345**, 1331–1335
36. von Boehmer, H., and Kisielow, P. (1991) *Sci. Am.* **265**(Oct), 74–81
37. Nossal, G. J. V. (2001) *Nature (London)* **412**, 685–686
38. Ahmed, R., and Gray, D. (1996) *Science* **272**, 54–60
39. Fearon, D. T., Manders, P., and Wagner, S. D. (2001) *Science* **293**, 248–250
40. Mackay, C. R., and von Andrian, U. H. (2001) *Science* **291**, 2323–2324
41. O'Garra, A. (2000) *Nature (London)* **404**, 719–720
42. Butcher, E. C., and Picker, L. J. (1996) *Science* **272**, 60–66
43. Service, R. F. (1994) *Science* **265**, 1522–1524

43a. Hayday, A., and Viney, J. L. (2000) *Science* **290**, 97 – 100

44. Golde, D. W., and Gasson, J. C. (1988) *Sci. Am.* **259**(July), 62–70
45. Socolovsky, M., Lodish, H. F., and Daley, G. Q. (1998) *Proc. Natl. Acad. Sci. U.S.A.* **95**, 6573–6575
46. Phillips, R. L., Ernst, R. E., Brunk, B., Ivanova, N., Mahan, M. A., Deanehan, J. K., Moore, K. A., Overton, G. C., and Lemischka, I. R. (2000) *Science* **288**, 1635–1640
47. Weissman, I. L., and Cooper, M. D. (1993) *Sci. Am.* **269**(Sep), 65–71
48. Akashi, K., Traver, D., Miyamoto, T., and Weissman, I. L. (2000) *Nature (London)* **404**, 193–197
49. Metcalf, D. (1992) *Trends Biochem. Sci.* **17**, 286–289
50. Delves, P. J., and Roitt, I. M. (2000) *N. Engl. J. Med.* **343**, 108–117
51. Liu, Y.-J. (1997) *Science* **278**, 238–239
52. Burton, D. R. (1990) *Trends Biochem. Sci.* **15**, 64–69
53. Sondermann, P., Kaiser, J., and Jacob, U. (2001) *J. Mol. Biol.* **309**, 737–749
54. Grakoui, A., Bromley, S. K., Sumen, C., Davis, M. M., Shaw, A. S., Allen, P. M., and Dustin, M. L. (1999) *Science* **285**, 221–227
55. Fineschi, B., and Miller, J. (1997) *Trends Biochem. Sci.* **22**, 377–382
56. Williams, N. (1998) *Science* **280**, 198–200

56a. Chen, Y., Chou, K., Fuchs, E., Havran, W. L., and Boismenu, R. (2002) *Proc. Natl. Acad. Sci. U.S.A.* **99**, 14338 – 14343

56b. Jameson, J. Ugarte, K., Chen, N., Yachi, P., Fuchs, E., Boismenu, R., and Havran, W. L. (2002) *Science* **296**, 747 – 749

57. Qi, S. Y., Groves, J. T., and Chakraborty, A. K. (2001) *Proc. Natl. Acad. Sci. U.S.A.* **98**, 6548–6553

57a. Dustin, M. L., and Colman, D. R. (2002) *Science* **298**, 785 – 789

57b. Lee, K.-H., Holdorf, A. D., Dustin, M. L., Chen, A. C., Allen, P. M., and Shaw, A. S. (2002) *Science* **295**, 1539 – 1542

58. Trautmann, A., and Vivier, E. (2001) *Science* **292**, 1667–1668

58a. Khan, A. A., Bose, C., Yam, L. S., Soloski, M. J., and Rupp, F. (2001) *Science* **292**, 1681 – 1686

59. Edelman, G. M. (1973) *Science* **180**, 830–840
60. Porter, R. R. (1973) *Science* **180**, 713–716
61. Edmundson, A. B., Ely, K. R., Abola, E. E., Schiffer, M., and Panagiotopoulos, N. (1975) *Biochemistry* **14**, 3953–3961
62. Putnam, F. W. (1993) *Protein Sci.* **2**, 1536–1542
63. Verfuèse, D. (1972) *Science* **178**, 384–385
64. Putnam, F. W., Florent, G., Paul, C., Shinoda, T., and Shimizu, A. (1973) *Science* **182**, 287–291

64a. Cochran, J. R., Aivazian, D., Cameron, T. O., and Lawrence, L. J. (2001) *Trends Biochem. Sci.* **26**, 304-310

64b. Rudd, P. M., Elliott, T., Cresswell, P., Wilson, I. A., and Dwek, R. A. (2001) *Science* **292**, 2370 – 2376

64c. Feinstein, A., and Munn, E. A. (1969) *Nature (London)* **224**, 1307 – 1309

65. Bork, P., Holm, L., and Sander, C. (1994) *J. Mol. Biol.* **242**, 309–320
66. Schiffer, M., Girling, R. L., Ely, K. R., and Edmundson, A. B. (1973) *Biochemistry* **12**, 4620–4631

66a. Chothia, C., Lesk, A. M., Tramontano, A., Levitt, M., Smith-Gill, S. J., Air, G., Sheriff, S., Padlan, E. A., Davies, D., Tulip, W. R., Colman, P. M., Spinelli, S., Alzari, P. M., and Poljak, R. J. (1989) *Nature (London)* **342**, 877 – 883

67. Capra, J. D., and Edmundson, A. B. (1977) *Sci. Am.* **236**(Jan), 50–59
68. Amzel, L. M., and Poljak, R. J. (1979) *Ann. Rev. Biochem.* **48**, 961–997
69. Davies, D. R., Sheriff, S., and Padlan, E. A. (1988) *J. Biol. Chem.* **263**, 10541–10544
70. Lesk, A. M., and Chothia, C. (1988) *Nature (London)* **335**, 188–190
71. Amit, A. G., Mariuzza, R. A., Phillips, S. E. V., and Poljak, R. J. (1986) *Science* **233**, 747–753
72. Lescar, J., Pellegrini, M., Souchon, H., Tello, D., Poljak, R. J., Peterson, N., Greene, M., and Alzari, P. M. (1995) *J. Biol. Chem.* **270**, 18067–18076
73. Sundberg, E. J., Urrutia, M., Braden, B. C., Isern, J., Tsuchiya, D., Fields, B. A., Malchiodi, E. L., Tormo, J., Schwarz, F. P., and Mariuzza, R. A. (2000) *Biochemistry* **39**, 15375–15387
74. Cygler, M., Rose, D. R., and Bundle, D. R. (1991) *Science* **253**, 442–445
75. Novotny, J., and Haber, E. (1985) *Proc. Natl. Acad. Sci. U.S.A.* **82**, 4592–4596
76. Bruccoleri, R. E., Haber, E., and Novotny, J. (1988) *Nature (London)* **335**, 564–568
77. Boehm, M. K., Woof, J. M., Kerr, M. A., and Perkins, S. J. (1999) *J. Mol. Biol.* **286**, 1421–1447
78. Jung, S., Spinelli, S., Schimmele, B., Honegger, A., Pugliese, L., Cambillau, C., and Plückthun, A. (2001) *J. Mol. Biol.* **309**, 701–716
79. Braden, B. C., and Poljak, R. J. (1995) *FASEB J.* **9**, 9–16
80. Padlan, E. A., Abergel, C., and Tipper, J. P. (1995) *FASEB J.* **9**, 133–139
81. Chatellier, J., Van Regenmortel, M. H. V., Vernet, T., and Altschuh, D. (1996) *J. Mol. Biol.* **264**, 1–6
82. Smith, D. K., and Xue, H. (1997) *J. Mol. Biol.* **274**, 530–545
83. Vargas-Madrazo, E., Lara-Ochoa, F., and Almagro, J. C. (1995) *J. Mol. Biol.* **254**, 497–504

References

84. Strong, R. K., Campbell, R., Rose, D. R., Petsko, G. A., Sharon, J., and Margolies, M. N. (1991) *Biochemistry* **30**, 3739–3748
85. Lim, K., Owens, S. M., Arnold, L., Sacchattini, J. C., and Linthicum, D. S. (1998) *J. Biol. Chem.* **273**, 28576–28582
86. Guddat, L. W., Shan, L., Broomell, C., Ramsland, P. A., Fan, Z.-c, Anchin, J. M., Linthicum, D. S., and Edmundson, A. B. (2000) *J. Mol. Biol.* **302**, 853–872
87. Mol, C. D., Muir, A. K. S., Cygler, M., Lee, J. S., and Anderson, W. F. (1994) *J. Biol. Chem.* **269**, 3615–3622
88. Yokoyama, H., Mizutani, R., Satow, Y., Komatsu, Y., Ohtsuka, E., and Nikaido, O. (2000) *J. Mol. Biol.* **299**, 711–723
89. Augustine, J. G., de la Calle, A., Knarr, G., Buchner, J., and Frederick, C. A. (2001) *J. Biol. Chem.* **276**, 3287–3294
90. Bossart-Whitaker, P., Chang, C. Y., Novotny, J., Benjamin, D. C., and Sheriff, S. (1995) *J. Mol. Biol.* **253**, 559–575
91. Berthet-Colominas, C., Monaco, S., Novelli, A., Sibai, G., Mallet, F., and Cusack, S. (1999) *EMBO J.* **18**, 1124–1136
92. Landry, R. C., Klimowicz, A. C., Lavictoire, S. J., Borisova, S., Kottachchi, D. T., Lorimer, I. A. J., and Evans, S. V. (2001) *J. Mol. Biol.* **308**, 883–893
93. Muyldermans, S., Cambillau, C., and Wyns, L. (2001) *Trends Biochem. Sci.* **26**, 230–235
94. Conrath, K. E., Lauwereys, M., Wyns, L., and Muyldermans, S. (2001) *J. Biol. Chem.* **276**, 7346–7350
95. Zdanov, A., Li, Y., Bundle, D. R., Deng, S.-J., MacKenzie, C. R., Narang, S. A., Young, N. M., and Cygler, M. (1994) *Proc. Natl. Acad. Sci. U.S.A.* **91**, 6423–6427
96. Derrick, J. P., and Wigley, D. B. (1992) *Nature (London)* **359**, 752–754
97. Todd, P. E. E., East, I. J., and Leach, S. J. (1982) *Trends Biochem. Sci.* **7**, 212–216
98. Westhof, E., Altschuh, D., Moras, D., Bloomer, A. C., Mondragon, A., Klug, A., and Van Regenmortel, M. H. V. (1984) *Nature (London)* **311**, 123–126
99. Tainer, J. A., Getzoff, E. D., Alexander, H., Houghten, R. A., Olson, A. J., Lerner, R. A., and Hendrickson, W. A. (1984) *Nature (London)* **312**, 127–134
100. Dyson, H. J., Cross, K. J., Houghten, R. A., Wilson, I. A., Wright, P. E., and Lerner, R. A. (1985) *Nature (London)* **318**, 480–483
101. Low, T. L. K., Liu, Y.-S. V., and Putnam, F. W. (1976) *Science* **191**, 390–392
102. Blank, U., Ra, C., Miller, L., White, K., Metzger, H., and Kinet, J.-P. (1989) *Nature (London)* **337**, 187–189
103. Radaev, S., Motyka, S., Fridman, W.-H., Sautes-Fridman, C., and Sun, P. D. (2001) *J. Biol. Chem.* **276**, 16469–16477
104. Burmeister, W. P., Huber, A. H., and Bjorkman, P. J. (1994) *Nature (London)* **372**, 379–383
105. Weng, Z., Gulukota, K., Vaughn, D. E., Bjorkman, P. J., and DeLisi, C. (1998) *J. Mol. Biol.* **282**, 217–225
106. Deisenhofer, J. (1981) *Biochemistry* **20**, 2361–2370
107. Sondermann, P., Huber, R., Oosthuizen, V., and Jacob, U. (2000) *Nature (London)* **406**, 267–273
108. Garman, S. C., Wurzburg, B. A., Tarchevskaya, S. S., Kinet, J.-P., and Jardetzky, T. S. (2000) *Nature (London)* **406**, 259–266
109. Garman, S. C., Sechi, S., Kinet, J.-P., and Jardetzky, T. S. (2001) *J. Mol. Biol.* **311**, 1049–1062
110. Teichmann, S. A., and Chothia, C. (2000) *J. Mol. Biol.* **296**, 1367–1383
111. Bhagavan, N. V. (2002) *Medical Biochemistry*, 4th ed., Harcourt/Academic Press, San Diego, California (pp. 803–837)
112. Blattner, F. R., and Tucker, P. W. (1984) *Nature (London)* **307**, 417–422
113. Edelman, G. M. (1976) *Science* **192**, 218–226
113a. Klahre, U., Crété, P., Leuenberger, S. A., Iglesias, V. A., and Meins, F., Jr. (2002) *Proc. Natl. Acad. Sci. U.S.A.* **99**, 11981 – 11986
113b. Zamore, P. D. (2002) *Science* **296**, 1265 – 1269
114. Zhang, L., Benz, R., and Hancock, R. E. W. (1999) *Biochemistry* **38**, 8102–8111
115. Wang, Y., Henz, M. E., Gallagher, N. L. F., Chai, S., Gibbs, A. C., Yan, L. Z., Stiles, M. E., Wishart, D. S., and Vederas, J. C. (1999) *Biochemistry* **38**, 15438–15447
116. Pütsep, K., Brändén, C.-I., Boman, H. G., and Normark, S. (1999) *Nature (London)* **398**, 671–672
117. Park, C. B., Yi, K.-S., Matsuzaki, K., Kim, M. S., and Kim, S. C. (2000) *Proc. Natl. Acad. Sci. U.S.A.* **97**, 8245–8250
117a. Kobayashi, S., Hirakura, Y., and Matsuzaki, K. (2001) *Biochemistry* **40**, 14330 – 14335
118. Kragol, G., Lovas, S., Varadi, G., Condie, B. A., Hoffmann, R., and Otvos, L., Jr. (2001) *Biochemistry* **40**, 3016–3026
119. Gura, T. (2001) *Science* **291**, 2068–2071
120. Ehret-Sabatier, L., Loew, D., Goyffon, M., Fehlbaum, P., Hoffmann, J. A., van Dorsselaer, A., and Bulet, P. (1996) *J. Biol. Chem.* **271**, 29537–29544
121. Osaki, T., Omotezako, M., Nagayama, R., Hirata, M., Iwanaga, S., Kasahara, J., Hattori, J., Ito, I., Sugiyama, H., and Kawabata, S.-i. (1999) *J. Biol. Chem.* **274**, 26172–26178
122. Silva, P. I., Jr., Daffre, S., and Bulet, P. (2000) *J. Biol. Chem.* **275**, 33464–33470
123. Sai, K. P., Jagannadham, M. V., Vairamani, M., Raju, N. P., Devi, A. S., Nagaraj, R., and Sitaram, N. (2001) *J. Biol. Chem.* **276**, 2701–2707
124. Lohner, K., Latal, A., Lehrer, R. I., and Ganz, T. (1997) *Biochemistry* **36**, 1525–1531
125. Hoover, D. M., Chertov, O., and Lubkowski, J. (2001) *J. Biol. Chem.* **276**, 39021–39026
125a. Ganz, T. (2002) *Science* **298**, 977 – 979
126. Jia, H. P., Wowk, S. A., Schutte, B. C., Lee, S. K., Vivado, A., Tack, B. F., Bevins, C. L., and McCray, P. B., Jr. (2000) *J. Biol. Chem.* **275**, 33314–33320
127. Park, C. H., Valore, E. V., Waring, A. J., and Ganz, T. (2001) *J. Biol. Chem.* **276**, 7806–7810
128. Suemori, S., Lynch-Devaney, K., and Podolsky, D. K. (1991) *Proc. Natl. Acad. Sci. U.S.A.* **88**, 11017–11021
129. Mashimo, H., Wu, D.-C., Podolsky, D. K., and Fishman, M. C. (1996) *Science* **274**, 262–265
130. Plaut, A. G. (1997) *N. Engl. J. Med.* **336**, 506–507
131. Boix, E., Leonidas, D. D., Nikolovski, Z., Nogués, M. V., Cuchillo, C. M., and Acharya, K. R. (1999) *Biochemistry* **38**, 16794–16801
132. Tang, Y.-Q., Yuan, J., Ösapsy, G., Ösapay, K., Tran, D., Miller, C. J., Ouellette, A. J., and Selsted, M. E. (1999) *Science* **286**, 498–502
133. Nizet, V., Ohtake, T., Lauth, X., Trowbridge, J., Rudisill, J., Dorschner, R. A., Pestonjamasp, V., Piraino, J., Huttner, K., and Gallo, R. L. (2001) *Nature (London)* **414**, 454–457
133a. Nagpal, S., Kaur, K. J., Jain, D., and Salunke, D. M. (2002) *Protein Sci.*, **11**, 2158 – 2167
134. Bauer, F., Schweimer, K., Klüver, E., Conejo-Garcia, J.-R., Forssmann, W.-G., Rösch, P., Adermann, K., and Sticht, H. (2001) *Protein Sci.* **10**, 2470–2479
135. Tassin, S., Broekaert, W. F., Marion, D., Acland, D. P., Ptak, M., Vovelle, F., and Sodano, P. (1998) *Biochemistry* **37**, 3623–3637
136. Gao, G.-H., Liu, W., Dai, J.-X., Wang, J.-F., Hu, Z., Zhang, Y., and Wang, D.-C. (2001) *Biochemistry* **40**, 10973–10978
137. Qi, J., Wu, J., Somkuti, G. A., and Watson, J. T. (2001) *Biochemistry* **40**, 4531–4538
138. Tam, J. P., Lu, Y.-A., Yang, J.-L., and Chiu, K.-W. (1999) *Proc. Natl. Acad. Sci. U.S.A.* **96**, 8913–8918
139. Jennings, C., West, J., Waine, C., Craik, D., and Anderson, M. (2001) *Proc. Natl. Acad. Sci. U.S.A.* **98**, 10614–10619
140. Craik, D. J., Daly, N. L., Bond, T., and Waine, C. (1999) *J. Mol. Biol.* **294**, 1327–1336
141. Rosengren, K. J., Daly, N. L., Scanlon, M. J., and Craik, D. J. (2001) *Biochemistry* **40**, 4601–4609
142. Porter, R. R., and Reid, K. B. M. (1978) *Nature (London)* **275**, 699–704
143. Law, S. K. A., and Reid, K. B. M. (1988) *Complement*, IRL Press, Oxford
144. Reid, K. B. M., and Porter, R. R. (1981) *Ann. Rev. Biochem.* **50**, 433–464
145. Müller-Eberhard, H. J. (1988) *Ann. Rev. Biochem.* **57**, 321–347
146. Winkelstein, J. A., Sullivan, K. E., and Colten, H. R. (1995) in *The Metabolic and Molecular Bases of Inherited Disease*, 7th ed., Vol. 1 (Scriver, C. R., Beaudet, A. L., Sly, W. S., and Valle, D., eds), pp. 3911–3941, McGraw-Hill, New York
147. Walport, M. J. (2001) *N. Engl. J. Med.* **344**, 1058–1066
148. Walport, M. J. (2001) *N. Engl. J. Med.* **344**, 1140–1144
149. Arora, M., Munoz, E., and Tenner, A. J. (2001) *J. Biol. Chem.* **276**, 43087–43094
150. Gilbert, T. L., Bennett, T. A., Maestas, D. C., Cimino, D. F., and Prossnitz, E. R. (2001) *Biochemistry* **40**, 3467–3475
151. Perkins, S. J. (1985) *Biochem. J.* **228**, 13–26
152. Hanson, D. C., Siegel, R. C., and Schumaker, V. N. (1985) *J. Biol. Chem.* **260**, 3576 –3583
153. Marqués, G., Antón, L. C., Barrio, E., Sánchez, A., Ruiz, S., Gavilanes, F., and Vivanco, F. (1993) *J. Biol. Chem.* **268**, 10393–10402
154. Lacroix, M., Rossi, V., Gaboriaud, C., Chevallier, S., Jaquinod, M., Thielens, N. M., Gagnon, J., and Arlaud, G. J. (1997) *Biochemistry* **36**, 6270–6282
155. Hoppe, H.-J., and Reid, K. B. M. (1994) *Protein Sci.* **3**, 1143–1158
156. Leytus, S. P., Kurachi, K., Sakariassen, K. S., and Davie, E. W. (1986) *Biochemistry* **25**, 4855–4863
157. Gaboriaud, C., Rossi, V., Bally, I., Arlaud, G. J., and Fontecilla-Camps, J. C. (2000) *EMBO J.* **19**, 1755–1765
158. Luo, C., Thielens, N. M., Gagnon, J., Gal, P., Sarvari, M., Tseng, Y., Tosi, M., Zavodszky, P., Arlaud, G. J., and Schumaker, V. N. (1992) *Biochemistry* **31**, 4254–4262
159. Rossi, V., Gaboriaud, C., Lacroix, M., Ulrich, J., Fontecilla-Camps, J. C., Gagnon, J., and Arlaud, G. J. (1995) *Biochemistry* **34**, 7311–7321
160. Ruiz, S., Henschen-Edman, A. H., and Tenner, A. J. (1995) *J. Biol. Chem.* **270**, 30627–30634
161. Johnson, C. A., Densen, P., Hurford, R. K., Jr., Colten, H. R., and Wetsel, R. A. (1992) *J. Biol. Chem.* **267**, 9347–9353
162. Hortin, G. L., Farries, T. C., Graham, J. P., and Atkinson, J. P. (1989) *Proc. Natl. Acad. Sci. U.S.A.* **86**, 1338–1342
163. Sahu, A., and Pangburn, M. K. (1994) *J. Biol. Chem.* **269**, 28997–29002

References

164. Plumb, M. E., and Sodetz, J. M. (2000) *Biochemistry* **39**, 13078–13083
164a. Scibek, J. J., Plumb, M. E., and Sodetz, J. M. (2002) *Biochemistry* **41**, 14546 – 14551
165. Hourcade, D. E., Wagner, L. M., and Oglesby, T. J. (1995) *J. Biol. Chem.* **270**, 19716–19722
166. Narayana, S. V. L., Carson, M., El-Kabbani, O., Kilpatrick, J. M., Moore, D., Chen, X., Bugg, C. E., Volanakis, J. E., and DeLucas, L. J. (1994) *J. Mol. Biol.* **235**, 695–708
167. Smith, C. A., Pangburn, M. K., Vogel, C.-W., and Müller-Eberhard, H. J. (1984) *J. Biol. Chem.* **259**, 4582–4588
168. Skriver, K., Radziejewska, E., Silbermann, J. A., Donaldson, V. H., and Bock, S. C. (1989) *J. Biol. Chem.* **264**, 3066–3071
169. Bhakdi, S., and Tranum-Jensen, J. (1983) *Trends Biochem. Sci.* **8**, 134–136
170. Gerard, C., Showell, H. J., Hoeprich, P. D., Jr., Huglis, T. E., and Stimler, N. P. (1985) *J. Biol. Chem.* **260**, 2613–2616
171. Johnson, R. J., and Chenoweth, D. E. (1985) *J. Biol. Chem.* **260**, 10339–10345
172. Humbles, A. A., Lu, B., Nilsson, C. A., Lilly, C., Israel, E., Fujiwara, Y., Gerard, N. P., and Gerard, C. (2000) *Nature (London)* **406**, 998–1001
173. Cain, S. A., Coughlan, T., and Monk, P. N. (2001) *Biochemistry* **40**, 14047–14052
174. Chen, Z., Zhang, X., Gonnella, N. C., Pellas, T. C., Boyar, W. C., and Ni, F. (1998) *J. Biol. Chem.* **273**, 10411–10419
175. Young, J. D.-E., Cohn, Z. A., and Podack, E. R. (1986) *Science* **233**, 184–190
176. Persechini, P. M., Ojcius, D. M., Adeodato, S. C., Notaroberto, P. C., Daniel, C. B., and Young, J. D.-E. (1992) *Biochemistry* **31**, 5017–5021
177. Lichtenheld, M. G., Olsen, K. J., Lu, P., Lowrey, D. M., Hameed, A., Hengartner, H., and Podack, E. R. (1988) *Nature (London)* **335**, 448–451
178. Young, J. D.-E., Cohen, Z. A., and Podack, E. R. (1986) *Science* **233**, 184–190
179. Stenger, S., Hanson, D. A., Teitelbaum, R., Dewan, P., Niazi, K. R., Froelich, C. J., Ganz, T., Thoma-Uszynski, S., Melián, A., Bogdan, C., Porcelli, S. A., Bloom, B. R., Krensky, A. M., and Modlin, R. L. (1998) *Science* **282**, 121–125
180. Young, J. D.-E., and Lowrey, D. M. (1989) *J. Biol. Chem.* **264**, 1077–1083
181. Blom, A. M., Kask, L., and Dahlbäck, B. (2001) *J. Biol. Chem.* **276**, 27136–27144
181a. Aslam, M., and Perkins, S. J. (2001) *J. Mol. Biol.* **309**, 1117 – 1138
182. Chamberlain, D., Ullman, C. G., and Perkins, S. J. (1998) *Biochemistry* **37**, 13918–13929
183. Liszewski, M. K., Tedja, I., and Atkinson, J. P. (1994) *J. Biol. Chem.* **269**, 10776–10779
184. Szakonyi, G., Guthridge, J. M., Li, D., Young, K., Holers, V. M., and Chen, X. S. (2001) *Science* **292**, 1725–1728
185. Oppenheim, J. J., Feldmann, M., Durum, S. K., Hirano, T., Vilcek, J., and Nicola, N. A., eds. (2000) *Cytokine Reference*, Academic Press, San Diego
186. Schindler, C., and Darnell, J. E., Jr. (1995) *Ann. Rev. Biochem.* **64**, 621–651
187. Darnell, J. E., Jr. (1997) *Science* **277**, 1630–1635
188. Beuvink, I., Hess, D., Flotow, H., Hofsteenge, J., Groner, B., and Hynes, N. E. (2000) *J. Biol. Chem.* **275**, 10247–10255
189. Becker, S., Groner, B., and Müller, C. W. (1998) *Nature (London)* **394**, 145–151
190. Williams, J. G. (1999) *Trends Biochem. Sci.* **24**, 333–334
191. Su, L., and David, M. (2000) *J. Biol. Chem.* **275**, 21661–21666
192. Naka, T., Fujimoto, M., and Kishimoto, T. (1999) *Trends Biochem. Sci.* **24**, 394–398
193. Ihle, J. N. (1995) *Nature (London)* **377**, 591–594
194. Ihle, J. N., Witthuhn, B. A., Quelle, F. W., Yamamoto, K., Thierfelder, W. E., Kreider, B., and Silvennoinen, O. (1994) *Trends Biochem. Sci.* **19**, 222–227
195. Chow, D.-c, He, X.-L., Snow, A. L., Rose-John, S., and Garcia, K. C. (2001) *Science* **291**, 2150–2155
196. Revel, M., and Chebath, J. (1986) *Trends Biochem. Sci.* **11**, 166–170
197. Johnson, H. M., Bazer, F. W., Szente, B. E., and Jarpe, M. A. (1994) *Sci. Am.* **270**(May), 68–75
198. Runkel, L., deDios, C., Karpusas, M., Betzenhauser, M., Muldowney, C., Zafari, M., Benjamin, C. D., Miller, S., Hochman, P. S., and Whitty, A. (2000) *Biochemistry* **39**, 2538–2551
199. Li, J., and Roberts, R. M. (1994) *J. Biol. Chem.* **269**, 24826–24833
200. Klaus, W., Gsell, B., Labhardt, A. M., Wipf, B., and Senn, H. (1997) *J. Mol. Biol.* **274**, 661–675
201. Rani, M. R. S., Asthagiri, A. R., Singh, A., Sizemore, N., Sathe, S. S., Li, X., DiDonato, J. D., Stark, G. R., and Ransohoff, R. M. (2001) *J. Biol. Chem.* **276**, 44365–44368
202. Eppstein, D. A., Schryver, B. B., and Marsh, Y. V. (1986) *J. Biol. Chem* **261**, 5999–6003
203. Sarkar, S. N., Ghosh, A., Wang, H.-W., Sung, S.-S., and Sen, G. C. (1999) *J. Biol. Chem.* **274**, 25535–25542
204. Rebouillat, D., Hovnanian, A., Marié, I., and Hovanessian, A. G. (1999) *J. Biol. Chem.* **274**, 1557–1565
205. Koromilas, A. E., Roy, S., Barber, G. N., Katze, M. G., and Sonenberg, N. (1992) *Science* **257**, 1685–1689
206. Bischoff, J. R., and Samuel, C. E. (1985) *J. Biol. Chem.* **260**, 8237–8239
207. Hardin, J. A., and Thomas, J. O. (1983) *Proc. Natl. Acad. Sci. U.S.A.* **80**, 7410–7414
208. Sehgal, P. B., May, L. T., Tamm, I., and Vilcek, J. (1987) *Science* **235**, 731–732
209. Gamero, A. M., and Larner, A. C. (2001) *J. Biol. Chem.* **276**, 13547–13553
210. Dinarello, C. A. (1988) *FASEB J.* **2**, 108–115
211. Dinarello, C. A., and Wolff, S. M. (1993) *N. Engl. J. Med.* **328**, 106–112
212. Samad, T. A., Moore, K. A., Sapirstein, A., Billet, S., Allchorne, A., Poole, S., Bonventre, J. V., and Woolf, C. J. (2001) *Nature (London)* **410**, 471–475
213. Yamin, T.-T., Ayala, J. M., and Miller, D. K. (1996) *J. Biol. Chem.* **271**, 13273–13282
214. Vigers, G. P. A., Dripps, D. J., Edwards, C. K., III, and Brandhuber, B. J. (2000) *J. Biol. Chem.* **275**, 36927–36933
215. Simpson, R. J., Hammacher, A., Smith, D. K., Matthews, J. M., and Ward, L. D. (1997) *Protein Sci.* **6**, 929–955
216. Cyster, J. G. (1999) *Science* **286**, 2098–2102
217. Baggiolini, M. (1998) *Nature (London)* **392**, 565–568
218. Mellado, M., Rodríguez-Frade, J. M., Vila-Coro, A. J., Fernández, S., de Ana, A. M., Jones, D. R., Torán, J. L., and Martínez-A, C. (2001) *EMBO J.* **20**, 2497–2507
219. Shao, W., Jerva, L. F., West, J., Lolis, E., and Schweitzer, B. I. (1998) *Biochemistry* **37**, 8303–8313
220. Martinelli, R., Sabroe, I., LaRosa, G., Williams, T. J., and Pease, J. E. (2001) *J. Biol. Chem.* **276**, 42957–42964
221. Ben-Baruch, A., Bengali, K. M., Biragyn, A., Johnston, J. J., Wang, J.-M., Kim, J., Chuntharapai, A., Michiel, D. F., Oppenheim, J. J., and Kelvin, D. J. (1995) *J. Biol. Chem.* **270**, 9121–9128
222. Ben-Baruch, A., Michiel, D. F., and Oppenheim, J. J. (1995) *J. Biol. Chem.* **270**, 11703–11706
222a. Roos, D., and Winterbourn, C. C. (2002) *Science* **296**, 669 – 671
222b. Malkowski, M. G., Wu, J. Y., Lazar, J. B., Johnson, P. H., and Edwards, B. F. P. (1995) *J. Biol. Chem.* **270**, 7077 – 7087
222c. Masuda, M., Suzuki, T., Friesen, M. D., Ravanat, J.-L., Cadet, J., Pignatelli, B., Nishino, H., and Ohshima, H. (2001) *J. Biol. Chem.* **276**, 40486 – 40496
223. Smith, K. A. (1990) *Sci. Am.* **262**(Mar), 50–57
224. Waldmann, T. A. (1991) *J. Biol. Chem.* **266**, 2681–2684
225. Reif, K., Burgering, B. M. T., and Cantrell, D. A. (1997) *J. Biol. Chem.* **272**, 14426–14433
226. Fukushima, K., and Yamashita, K. (2001) *J. Biol. Chem.* **276**, 7351–7356
227. Gadina, M., Sudarashan, C., Visconti, R., Zhou, Y.-J., Gu, H., Neel, B. G., and O'Shea, J. J. (2000) *J. Biol. Chem.* **275**, 26959–26966
227a. Ozaki, K., Spolski, R., Feng, C. G., Qi, C.-F., Cheng, J., Sher, A., Morse, H. C., III, Liu, C., Schwartzberg, P. L., and Leonard, W. J. (2002) *Science* **298**, 1630 – 1634
228. Powers, R., Garrett, D. S., March, C. J., Frieden, E. A., Gronenborn, A. M., and Clore, G. M. (1992) *Science* **256**, 1673–1677
229. Walter, M. R., Cook, W. J., Zhao, B. G., Cameron, R. P., Jr., Ealick, S. E., Walter, R. L., Jr., Reichert, P., Nagabhushan, T. L., Trotta, P. P., and Bugg, C. E. (1992) *J. Biol. Chem.* **267**, 20371–20376
229a. Taubes, G. (2002) *Science* **296**, 242 – 245
229b. Langen, R. C. J., Schols, A. M. W. J., Kelders, M. C. J. M., Wouters, E. F. M., and Janssen-Heininger, Y. M. W. (2001) *FASEB J.* **15**, 1169 – 1180
229c. Poli, V. (1998) *J. Biol. Chem.* **273**, 29279 – 29282
229d. Dekker, L. V., and Segal, A. W. (2000) *Science* **287**, 982 – 985
229e. Frenette, P. S., and Wagner, D. D. (1996) *N. Engl. J. Med.* **335**, 43 – 45
229f. Albelda, S. M., Smith, C. W., and Ward, P. A. (1994) *FASEB J.* **8**, 504 – 512
229g. Xavier, R. J., and Podolsky, D. K. (2000) *Science* **289**, 1483 – 1484
229h. Ravetch, J. V., and Lanier, L. L. (2000) *Science* **290**, 84 –89
230. Samudzi, C. T., Burton, L. E., and Rubin, J. R. (1991) *J. Biol. Chem.* **266**, 21791–21797
231. Walter, M. R., Windsor, W. T., Nagabhushan, T. L., Lundell, D. J., Lunn, C. A., Zauodny, P. J., and Narula, S. K. (1995) *Nature (London)* **376**, 230–235
231a. Samuel, C. E. (2002) *Proc. Natl. Acad. Sci. U.S.A.* **99**, 11555 – 11557
231b. Nguyen, K. B., Watford, W. T., Salomon, R., Hofmann, S. R., Pien, G. C., Morinobu, A., Gadina, M., O'Shea, J. J., and Biron, C. A. (2002) *Science* **297**, 2063 –2066
232. Beninga, J., Rock, K. L., and Goldberg, A. L. (1998) *J. Biol. Chem.* **273**, 18734–18742
233. Dhabhar, F. S., Satoskar, A. R., Bluethmann, H., David, J. R., and McEwen, B. S. (2000) *Proc. Natl. Acad. Sci. U.S.A.* **97**, 2846–2851
234. Bazan, J. F. (1995) *Nature (London)* **376**, 217–218
235. Prakash, B., Renault, L., Praefcke, G. J. K., Herrmann, C., and Wittinghofer, A. (2000) *EMBO J.* **19**, 4555–4564
236. Prakash, B., Praefcke, G. J. K., Renault, L., Wittinghofer, A., and Herrmann, C. (2000) *Nature (London)* **403**, 567–571
237. Old, L. J. (1985) *Science* **230**, 630 – 632
238. Argilés, J. M., López-Soriano, J., Busquets, S., and López-Soriano, F. J. (1997) *FASEB J.* **11**, 743–751

References

239. Beutler, B., and Cerami, A. (1988) *Ann. Rev. Biochem.* **57**, 505–518
240. Eck, M. J., Beutler, B., Kuo, G., Merryweather, J. P., and Sprang, S. R. (1988) *J. Biol. Chem.* **263**, 12816–12819
241. Gray, P. W., Aggarwal, B. B., Benton, C. V., Bringman, T. S., Henzel, W. J., Jarrett, J. A., Leung, D. W., Moffat, B., Ng, P., Svedersky, L. P., Palladino, M. A., and Nedwin, G. E. (1984) *Nature (London)* **312**, 721–724
242. Laâbi, Y., and Strasser, A. (2000) *Science* **289**, 883–884
243. Zdanov, A., Schalk-Hihi, C., and Wlodawer, A. (1996) *Protein Sci.* **5**, 1955–1962
244. Scott, P. (1993) *Science* **260**, 496–497
245. Hall, S. S. (1995) *Science* **268**, 1432–1434
246. Sun, H.-W., Bernhagen, J., Bucala, R., and Lolis, E. (1996) *Proc. Natl. Acad. Sci. U.S.A.* **93**, 5191–5196
247. Taylor, A. B., Johnson, W. H., Jr., Czerwinski, R. M., Li, H.-S., Hackert, M. L., and Whitman, C. P. (1999) *Biochemistry* **38**, 7444–7452
248. Nguyen, N. Y., Suzuki, A., Boykins, R. A., and Liu, T.-Y. (1986) *J. Biol. Chem.* **261**, 10456–10465
249. Kirkpatrick, L. L., Matzuk, M. M., Dodds, D. C., and Perin, M. S. (2000) *J. Biol. Chem.* **275**, 17786–17792
250. Coetzee, G. A., Strachan, A. F., van der Westhuyzen, D. R., Hoppe, H. C., Jeenah, M. S., and de Beer, F. C. (1986) *J. Biomol. Struct. Dyn.* **261**, 9644–9651
251. Lowell, C. A., Potter, D. A., Stearman, R. S., and Morrow, J. F. (1986) *J. Biol. Chem.* **261**, 8442–8452
252. Hochepied, T., Van Molle, W., Berger, F. G., Baumann, H., and Libert, C. (2000) *J. Biol. Chem.* **275**, 14903–14909
253. Gehring, M. R., Shiels, B. R., Northemann, W., de Bruijn, M. H. L., Kan, C.-C., Chain, A. C., Noonan, D. J., and Fey, G. H. (1987) *J. Biol. Chem.* **262**, 446–454
253a. Simpson, S. J., and Hines, P. J. (2002) *Science* **296**, 297
254. Levine, M. H., Haberman, A. M., Sant'Angelo, D. B., Hannum, L. G., Cancro, M. P., Janeway, C. A., Jr., and Shlomchik, M. J. (2000) *Proc. Natl. Acad. Sci. U.S.A.* **97**, 2743–2748
255. Otipoby, K. L., Draves, K. E., and Clark, E. A. (2001) *J. Biol. Chem.* **276**, 44315–44322
255a. Malissen, B. (1999) *Science* **285**, 207 – 208
256. Fischer, M. B., Goerg, S., Shen, L., Prodeus, A. P., Goodnow, C. C., Kelsoe, G., and Carroll, M. C. (1998) *Science* **280**, 582–585
257. Zinkernagel, R. M., and Hengartner, H. (2001) *Science* **293**, 251–253
258. Shaw, S. (1989) *Nature (London)* **338**, 539 – 540
259. Porcelli, S., Brenner, M. B., Greestein, J. L., Balk, S. P., Terhorst, C., and Bleicher, P. A. (1989) *Nature (London)* **341**, 447–450
260. Feizi, T. (1991) *Trends Biochem. Sci.* **16**, 84–86
261. Barclay, A. N., Birkeland, M. L., Brown, M. H., Beyers, A. D., Davis, S. J., Somoza, C., and Williams, A. F. (1993) *The Leucocyte Antigens Facts Book*, Academic Press, San Diego, California
262. Brady, R. L., Dodson, E. J., Dodson, G. G., Lange, G., Davis, S. J., Williams, A. F., and Barclay, A. N. (1993) *Science* **260**, 979–983
263. Leahy, D. J. (1995) *FASEB J.* **9**, 17–25
264. Veillette, A., Bookman, M. A., Horak, E. M., Samelson, L. E., and Bolen, J. B. (1989) *Nature (London)* **338**, 257–259
265. Grey, H. M., Sette, A., and Buus, S. (1989) *Sci. Am.* **261**(Nov), 56–64
266. von Andrian, U. H., and Mackay, C. R. (2000) *N. Engl. J. Med.* **343**, 1020–1034
267. Cochran, J. R., Aivazian, D., Cameron, T. O., and Lawrence, L. J. (2001) *Trends Biochem. Sci.* **26**, 304–310
267a. Potter, T. A., Grebe, K., Freiberg, B., and Kupfer, A. (2001) *Proc. Natl. Acad. Sci. U.S.A.* **98**, 12624 – 12629
267b. Chen, M., Stafford, W. F., Diedrich, G., Khan, A., and Bouvier, M. (2002) *Biochemistry* **41**, 14539 – 14545
268. Geleziunas, R., Bour, S., and Wainberg, M. A. (1994) *FASEB J.* **8**, 593–600
269. Kwong, P. D., Wyatt, R., Robinson, J., Sweet, R. W., Sodroski, J., and Hendrickson, W. A. (1998) *Nature (London)* **393**, 648–650
270. McMichael, A. J., and Rowland-Jones, S. L. (2001) *Nature (London)* **410**, 980–987
270a. Hill, A. V. S. (1999) *Nature (London)* **398**, 668 – 669
270b. Welsh, R. M. (2001) *Nature (London)* **411**, 541 – 542
271. Sieling, P. A., Chatterjee, D., Porcelli, S. A., Prigozy, T. I., Mazzaccaro, R. J., Soriano, T., Bloom, B. R., Brenner, M. B., Kronenberg, M., Brennan, P. J., and Modlin, R. L. (1995) *Science* **269**, 227–230
272. Zeng, Z.-H., Castaño, A. R., Segelke, B. W., Stura, E. A., Peterson, P. A., and Wilson, I. A. (1997) *Science* **277**, 339–345
273. Gumperz, J. E., and Parham, P. (1995) *Nature (London)* **378**, 245–248
274. Maghazachi, A. A., and Al-Aoukaty, A. (1998) *FASEB J.* **12**, 913–924
275. Colonna, M. (1998) *Nature (London)* **391**, 642–643
276. McVicar, D. W., Taylor, L. S., Gosselin, P., Willette-Brown, J., Mikhael, A. I., Geahlen, R. L., Nakamura, M. C., Linnemeyer, P., Seaman, W. E., Anderson, S. K., Ortaldo, J. R., and Mason, L. H. (1998) *J. Biol. Chem.* **273**, 32934–32942
277. Kärre, K., and Schneider, G. (2000) *Nature (London)* **405**, 527–528
277a. MacDonald, H. R. (2002) *Science* **296**, 481 – 482
277b. Vivier, E., and Biron, C. A. (2002) *Science* **296**, 1248 – 1249
278. Irwin, M., McClintick, J., Costlow, C., Fortner, M., White, J., and Gillin, J. C. (1996) *FASEB J.* **10**, 643–653
279. Steinmetz, M., and Hood, L. (1983) *Science* **222**, 727–733
280. Lu, C. Y., Khair-El-Din, T. A., Dawidson, I. A., Butler, T. M., Brasky, K. M., Vazquez, M. A., and Sicher, S. C. (1994) *FASEB J.* **8**, 1122–1130
281. Steinmetz, M. (1984) *Trends Biochem. Sci.* **9**, 224–226
282. Flavell, R. A., Allen, H., Burkly, L. C., Sherman, D. H., Waneck, G. L., and Widera, G. (1986) *Science* **233**, 437–443
283. Kappes, D., and Strominger, J. L. (1988) *Ann. Rev. Biochem.* **57**, 991–1028
284. MHC sequencing consortium. (1999) *Nature (London)* **401**, 921–923
285. Klein, J., and Sato, A. (2000) *N. Engl. J. Med.* **343**, 702–709
286. Radley, E., Alderton, R. P., Kelly, A., Trowsdale, J., and Beck, S. (1994) *J. Biol. Chem.* **269**, 18834–18838
287. Nathenson, S. G., Uehara, H., Ewenstein, B. M., Kindt, T. J., and Coligan, J. E. (1981) *Ann. Rev. Biochem.* **50**, 1025–1052
288. Åkerström, B., and Lögdberg, L. (1990) *Trends Biochem. Sci.* **15**, 240–243
289. Glassy, M. C. (1982) *Trends Biochem. Sci.* **7**, 286–288
290. Arguello, R., Avakian, H., Goldman, J. M., and Madrigal, J. A. (1996) *Proc. Natl. Acad. Sci. U.S.A.* **93**, 10961–10965
291. Garrett, T. P. J., Saper, M. A., Bjorkman, P. J., Strominger, J. L., and Wiley, D. C. (1989) *Nature (London)* **342**, 692–696
292. Parham, P. (1989) *Nature (London)* **340**, 426–428
293. Stern, L. J., Brown, J. H., Jardetzky, T. S., Gorga, J. C., Urban, R. G., Strominger, J. L., and Wiley, D. C. (1994) *Nature (London)* **368**, 215–221
294. Ghosh, P., Amaya, M., Mellins, E., and Wiley, D. C. (1995) *Nature (London)* **378**, 457–462
295. Zinkernagel, R. M. (1979) *Ann. Rev. Microbiol.* **33**, 201–213
296. Morel, P. A., Dorman, J. S., Todd, J. A., McDevitt, H. O., and Trucco, M. (1988) *Proc. Natl. Acad. Sci. U.S.A.* **85**, 8111–8115
297. Erlich, H. A. (1989) *Nature (London)* **337**, 415
298. Forcione, D. G., Sands, B., Isselbacher, K. J., Rustgi, A., Podolsky, D. K., and Pillai, S. (1996) *Proc. Natl. Acad. Sci. U.S.A.* **93**, 5094–5098
299. Goldberg, A. L., and Rock, K. L. (1992) *Nature (London)* **357**, 375–379
300. Kuttler, C., Nussbaum, A. K., Dick, T. P., Rammensee, H.-G., Schild, H., and Hadeler, K.-P. (2000) *J. Mol. Biol.* **298**, 417–429
301. Shimbara, N., Ogawa, K., Hidaka, Y., Nakajima, H., Yamasaki, N., Niwa, S.-i, Tanahashi, N., and Tanaka, K. (1998) *J. Biol. Chem.* **273**, 23062–23071
302. Heemels, M.-T., and Ploegh, H. (1995) *Ann. Rev. Biochem.* **64**, 463–491
303. Cascio, P., Hilton, C., Kisselev, A. F., Rock, K. L., and Goldberg, A. L. (2001) *EMBO J.* **20**, 2357–2366
304. de la Salle, H., and 16 other authors. (1994) *Science* **265**, 237–241
305. Binder, R. J., Blachere, N. E., and Srivastava, P. K. (2001) *J. Biol. Chem.* **276**, 17163–17171
305a. van Endert, P. M., Saveanu, L., Hewitt, E. W., and Lehner, P. J. (2002) *Trends Biochem. Sci.* **27**, 454 – 461
306. Vogt, A. B., and Kropshofer, H. (1999) *Trends Biochem. Sci.* **24**, 150–154
307. Fremont, D. H., Hendrickson, W. A., Marrack, P., and Kappler, J. (1996) *Science* **272**, 1001–1004
308. McFarland, B. J., Sant, A. J., Lybrand, T. P., and Beeson, C. (1999) *Biochemistry* **38**, 16663–16670
309. Jardetzky, T. S., Brown, J. H., Gorga, J. C., Stern, L. J., Urban, R. G., Strominger, J. L., and Wiley, D. C. (1996) *Proc. Natl. Acad. Sci. U.S.A.* **93**, 734–738
310. Wang, J.-h, Meijers, R., Xiong, Y., Liu, J.-h, Sakihama, T., Zhang, R., Joachimiak, A., and Reinherz, E. L. (2001) *Proc. Natl. Acad. Sci. U.S.A.* **98**, 10799–10804
311. Brown, J. H., Jardetzky, T. S., Gorga, J. C., Stern, L. J., Urban, R. G., Strominger, J. L., and Wiley, D. C. (1993) *Nature (London)* **364**, 33–39
312. Matsumura, M., Fremont, D. H., Peterson, P. A., and Wilson, I. A. (1992) *Science* **257**, 927–934
313. Fahnestock, M. L., Johnson, J. L., Feldman, R. M. R., Tsomides, T. J., Mayer, J., Narhi, L. O., and Bjorkman, P. J. (1994) *Biochemistry* **33**, 8149–8158
314. Young, A. C. M., Nathenson, S. G., and Sacchettini, J. C. (1995) *FASEB J.* **9**, 26–36
315. Udaka, K. (1996) *Trends Biochem. Sci.* **21**, 7–11
316. Parham, P., and Ohta, T. (1996) *Science* **272**, 67–74
317. Zhang, C., Anderson, A., and DeLisi, C. (1998) *J. Mol. Biol.* **281**, 929–947
318. Parham, P. (1996) *Trends Biochem. Sci.* **21**, 427–433
318a. Rammensee, H.-G. (2002) *Nature (London)* **419**, 443 – 445

References

319. Wilson, I. A. (1996) *Science* **272**, 973–974
320. Liang, M. N., Lee, C., Xia, Y., and McConnell, H. M. (1996) *Biochemistry* **35**, 14734–14742
321. Bénaroch, P., Yilla, M., Raposo, G., Ito, K., Miwa, K., Geuze, H. J., and Ploegh, H. L. (1995) *EMBO J.* **14**, 37–49
322. Kropshofer, H., Vogt, A. B., Thery, C., Armandola, E. A., Li, B.-C., Moldenhauer, G., Amigorena, S., and Hämmerling, G. J. (1998) *EMBO J.* **17**, 2971–2981
323. Bogyo, M., and Ploegh, H. L. (1998) *Nature (London)* **396**, 625–627
324. Watts, C. (2001) *Science* **294**, 1294–1295
325. Gao, G. F., Tormo, J., Gerth, U. C., Wyer, J. R., McMichael, A. J., Stuart, D. I., Bell, J. I., Jones, E. Y., and Jakobsen, B. K. (1997) *Nature (London)* **387**, 630–634
326. Smith, A. B., III, Benowitz, A. B., Sprengeler, P. A., Barbosa, J., Guzman, M. C., Hirschmann, R., Schweiger, E. J., Bolin, D. R., Nagy, Z., Campbell, R. M., Cox, D. C., and Olson, G. L. (1999) *J. Am. Chem. Soc.* **121**, 9286–9298
327. Saito, H., Kranz, D. M., Takagaki, Y., Hayday, A. C., Eisen, H. N., and Tonegawa, S. (1984) *Nature (London)* **309**, 757–762
328. Arden, B., Klotz, J. L., Siu, G., and Hood, L. E. (1985) *Nature (London)* **316**, 783–787
329. Garcia, K. C., Degano, M., Stanfield, R. L., Brunmark, A., Jackson, M. R., Peterson, P. A., Teyton, L., and Wilson, I. A. (1996) *Science* **274**, 209–219
330. Swan, K. A., Alberola-lla, J., Gross, J. A., Appleby, M. W., Forbush, K. A., Thomas, J. F., and Perlmutter, R. M. (1995) *EMBO J.* **14**, 276–285
330a. Rudolph, M. G., Huang, M., Teyton, L., and Wilson, I. A. (2001) *J. Mol. Biol.* **314**, 1 – 8
331. Abraham, R. T., Karnitz, L. M., Secrist, J. P., and Leibson, P. J. (1992) *Trends Biochem. Sci.* **17**, 434–438
332. Hatada, M. H., Lu, X., Laird, E. R., Green, J., Morgenstern, J. P., Lou, M., Marr, C. S., Phillips, T. B., Ram, M. K., Theriault, K., Zoller, M. J., and Karas, J. L. (1995) *Nature (London)* **377**, 32–38
333. Reinherz, E. L., Tan, K., Tang, L., Kern, P., Liu, J.-h, Xiong, Y., Hussey, R. E., Smolyar, A., Hare, B., Zhang, R., Joachimiak, A., Chang, H.-C., Wagner, G., and Wang, J.-h. (1999) *Science* **286**, 1913–1921
334. Hennecke, J., Carfi, A., and Wiley, D. C. (2000) *EMBO J.* **19**, 5611–5624
335. Germain, R. N. (2001) *J. Biol. Chem.* **276**, 35223–35226
336. Lang, P., Stolpa, J. C., Freiberg, B. A., Crawford, F., Kappler, J., Kupfer, A., and Cambier, J. C. (2001) *Science* **291**, 1537–1540
337. Howe, L. R., and Weiss, A. (1995) *Trends Biochem. Sci.* **20**, 59–64
338. LoGrasso, P. V., Hawkins, J., Frank, L. J., Wisniewski, D., and Marcy, A. (1996) *Proc. Natl. Acad. Sci. U.S.A.* **93**, 12165–12170
338a. Bottomly, K. (1999) *Science* **283**, 1124 – 1125
338b. Edinger, A. L., and Thompson, C. B. (2002) *Proc. Natl. Acad. Sci. U.S.A.* **99**, 1107 – 1109
338c. Hudrisier, D., and Bongrand, P. (2002) *FASEB J.* **16**, 477 – 486
339. Janis, E. M., Kaufmann, S. H. E., Schwartz, R. H., and Pardoll, D. M. (1989) *Science* **244**, 713–716
340. Steinle, A., Groh, V., and Spies, T. (1998) *Proc. Natl. Acad. Sci. U.S.A.* **95**, 12510–12515
341. Wingren, C., Crowley, M. P., Degano, M., Chien, Y.-h, and Wilson, I. A. (2000) *Science* **287**, 310–314
342. Allison, T. J., Winter, C. C., Fournié, J.-J., Bonneville, M., and Garboczi, D. N. (2001) *Nature (London)* **411**, 820–823
342a. Belmant, C., Espinosa, E., Poupot, R., Peyrat, M.-A., Guiraud, M., Poquet, Y., Bonneville, M., and Fournié, J.-J. (1999) J. Biol. Chem. **274**, 32079 – 32084
343. Johnson, H. M., Russell, J. K., and Pontzer, C. H. (1992) *Sci. Am.* **266**(April), 92–101
344. Jardetzky, T. S., Brown, J. H., Gorga, J. C., Stern, L. J., Urban, R. G., Chi, Y., Stauffacher, C., Strominger, J. L., and Wiley, D. C. (1994) *Nature (London)* **368**, 711–718
345. Abrahmsén, L., Dohlsten, M., Segrén, S., Björk, P., Jonsson, E., and Kalland, T. (1995) *EMBO J.* **14**, 2978–2986
346. Håkansson, M., Petersson, K., Nilsson, H., Forsberg, G., Björk, P., Antonsson, P., and Svensson, L. A. (2000) *J. Mol. Biol.* **302**, 527–537
347. Arcus, V. L., Proft, T., Sigrell, J. A., Baker, H. M., Fraser, J. D., and Baker, E. N. (2000) *J. Mol. Biol.* **299**, 157–168
348. Vath, G. M., Earhart, C. A., Monie, D. D., Iandolo, J. J., Schlievert, P. M., and Ohlendorf, D. H. (1999) *Biochemistry* **38**, 10239–10246
349. Papageorgiou, A. C., Brehm, R. D., Leonidas, D. D., Tranter, H. S., and Acharaya, K. R. (1996) *J. Mol. Biol.* **260**, 553–569
350. Lehnert, N. M., Allen, D. L., Allen, B. L., Catasti, P., Shiflett, P. R., Chen, M., Lehnert, B. E., and Gupta, G. (2001) *Biochemistry* **40**, 4222–4228
350a. Mackay, I. R. (1999) *Science* **284**, 269 – 270
350b. Podulsky, S. H., and Tauber, A. I. (1999) *The Generation of Diversity Clonal Selection*, Harvard Univ. Press, Cambridge, Massachusetts
351. Ada, G. L., and Nossal, G. (1987) *Sci. Am.* **257**(Aug), 62–69
352. Arrhenius, S. (1907) *Immunochemistry*, Macmillan Co., New York
353. Haurowitz, F. (1979) *Trends Biochem. Sci.* **4**, N 268–N 270
353a. Steinman, R. M., and Nussenzweig, M. C. (2002) *Proc. Natl. Acad. Sci. U.S.A.* **99**, 351 – 358
354. Howard, J. C. (1985) *Nature (London)* **314**, 494–495
355. Jerne, N. K. (1985) *Science* **229**, 1057– 1059
356. Ban, N., Day, J., Wang, X., Ferrone, S., and McPherson, A. (1996) *J. Mol. Biol.* **255**, 617–627
356a. Greenspan, N. S., and Bona, C. A. (1993) *FASEB J.* **7**, 437 – 444
357. Vaux, D., Tooze, J., and Fuller, S. (1990) *Nature (London)* **345**, 495–502
358. Marshall, E. (1996) *Science* **273**, 174–175
359. Rajewsky, K. (1996) *Nature (London)* **381**, 751–758
360. Forsdyke, D. R. (1995) *FASEB J.* **9**, 164–166
361. Goodnow, C. C., Crosbie, J., Jorgensen, H., Brink, R. A., and Basten, A. (1989) *Nature (London)* **342**, 385–391
362. Boise, L. H., and Thompson, C. B. (1996) *Science* **274**, 67–68
363. Kamradt, T., and Mitchison, N. A. (2001) *N. Engl. J. Med.* **344**, 655–664
364. Van Parijs, L., and Abbas, A. K. (1998) *Science* **280**, 243 – 248
365. Crabtree, G. R. (1989) *Science* **243**, 355–361
366. Mondino, A., Khoruts, A., and Jenkins, M. K. (1996) *Proc. Natl. Acad. Sci. U.S.A.* **93**, 2245–2252
367. Barton, G. M., and Rudensky, A. Y. (1999) *Science* **283**, 67–70
368. Marrack, P. (1987) *Science* **235**, 1311–1313
369. Janeway, C. A., Jr. (1988) *Nature (London)* **335**, 208–210
370. Ramsdell, F., and Fowlkes, B. J. (1990) *Science* **248**, 1342–1348
371. Schwartz, R. H. (1993) *Sci. Am.* **269**(Aug), 62–71
372. Akdis, C. A., and Blaser, K. (1999) *FASEB J.* **13**, 603–609
373. Weiner, H. L. (1994) *Proc. Natl. Acad. Sci. U.S.A.* **91**, 10762–10765
373a. Schwartzberg, P. L. (2001) *Science* **293**, 228 – 229
373b. Petty, H. R., and Kindzelskii, A. L. (2001) *Proc. Natl. Acad. Sci. U.S.A.* **98**, 3145 – 3149
373c. Germain, R. N. (2001) *Science* **293**, 240 – 245
373d. Lu, Q., and Lemke, G. (2001) *Science* **293**, 306 – 311
374. Rosen, F. S., and Mackay, I. (2001) *N. Engl. J. Med.* **345**, 1343–1344
375. Nabel, G. J. (2001) *Nature (London)* **410**, 1002–1007
376. Ada, G. (2001) *N. Engl. J. Med.* **345**, 1042–1053
377. Clarke, T. (2001) *Nature (London)* **409**, 278–280
377a. Cohen, J. (2001) *Science* **294**, 985
378. Cohen, A. D., Boyer, J. D., and Weiner, D. B. (1998) *FASEB J.* **12**, 1611–1626
379. Shen, X., and Siliciano, R. F. (2000) *Science* **290**, 463–465
379a. Modlin, R. L. (2000) *Nature (London)* **408**, 659 – 660
380. Langridge, W. H. R. (2000) *Sci. Am.* **283**(Sep), 66–71
381. Behr, M. A., Wilson, M. A., Gill, W. P., Salamon, H., Schoolnik, G. K., Rane, S., and Small, P. M. (1999) *Science* **284**, 1520–1523
382. Enserink, M. (2001) *Science* **293**, 234–235
383. Sprent, J., and Tough, D. F. (2001) *Science* **293**, 245–248
384. Barouch, D. H., and 28 other authors. (2000) *Science* **290**, 486–492
385. Riddell, S. R. (2001) *Proc. Natl. Acad. Sci. U.S.A.* **98**, 8933–8935
385a. Rappuoli, R., Miller, H. I., and Falkow, S. (2002) *Science* **297**, 937, 939
386. Simpson, S., and Marshall, E. (2001) *Science* **293**, 233
387. Waldschmidt, T. J., and Noelle, R. J. (2001) *Science* **293**, 2012–2013
388. Martin, S., and Goodnow, C. (2000) *Nature (London)* **407**, 576–577
389. Hayday, A., and Viney, J. L. (2000) *Science* **290**, 97–100
390. Morales, J., Homey, B., Vicari, A. P., Hudak, S., Oldham, E., Hedrick, J., Orozco, R., Copeland, N. G., Jenkins, N. A., McEvoy, L. M., and Zlotnik, A. (1999) *Proc. Natl. Acad. Sci. U.S.A.* **96**, 14470–14475
391. Lanzavecchia, A., and Sallusto, F. (2000) *Science* **290**, 92–97
392. Ku, C. C., Murakami, M., Sakamoto, A., Kappler, J., and Marrack, P. (2000) *Science* **288**, 675–678
393. Tonegawa, S. (1983) *Nature (London)* **302**, 575–581
394. Marx, J. L. (1987) *Science* **238**, 484–485
395. Honjo, T., and Habu, S. (1985) *Ann. Rev. Biochem.* **54**, 803–830
396. Kodaira, M., Kinashi, T., Umemura, I., Matsuda, F., Noma, T., Ono, Y., and Honjo, T. (1986) *J. Mol. Biol.* **190**, 529–541
397. Robertson, M. (1985) *Nature (London)* **317**, 768–771
398. Williams, S. C., Frippiat, J.-P., Tomlinson, I. M., Ignatovich, O., Lefranc, M.-P., and Winter, G. (1996) *J. Mol. Biol.* **264**, 220–232
399. Ignatovich, O., Tomlinson, I. M., Popov, A. V., Brüggemann, M., and Winter, G. (1999) *J. Mol. Biol.* **294**, 457–465
400. Tomlinson, I. M., Cox, J. P. L., Gherardi, E., Lesk, A. M., and Chothia, C. (1995) *EMBO J.* **14**, 4628–4638
401. Halligan, B. D., and Desiderio, S. V. (1987) *Proc. Natl. Acad. Sci. U.S.A.* **84**, 7019–7023
402. Oettinger, M. A., Schatz, D. G., Gorka, C., and Baltimore, D. (1990) *Science* **248**, 1517–1523

References

402a. Agrawal, A., Eastman, Q. M., and Schatz, D. G. (1998) *Nature (London)* **394**, 744 – 751
402b. Gellert, M. (2002) *Ann. Rev. Biochem.* **71**, 101 – 132
403. Bhagavan, N. V. (2002) *Medical Biochemistry*, 4th ed., Harcourt/Academic Press, San Diego, California
404. Hope, T. J., Aguilera, R. J., Minie, M. E., and Sakano, H. (1986) *Science* **231**, 1141 – 1145
405. Roman, C. A. J., and Baltimore, D. (1996) *Proc. Natl. Acad. Sci. U.S.A.* **93**, 2333 – 2338
406. Lin, W.-C., and Desiderio, S. (1993) *Science* **260**, 953 – 959
407. McMurry, M. T., and Krangel, M. S. (2000) *Science* **287**, 495 – 498
408. Jones, J. M., and Gellert, M. (2001) *Proc. Natl. Acad. Sci. U.S.A.* **98**, 12926 – 12931
409. Tevelev, A., and Schatz, D. G. (2000) *J. Biol. Chem.* **275**, 8341 – 8348
410. Frank, K. M., Sekiguchi, J. M., Seidl, K. J., Swat, W., Rathbun, G. A., Cheng, H.-L., Davidson, L., Kangaloo, L., and Alt, F. W. (1998) *Nature (London)* **396**, 173 – 177
411. Chu, G. (1997) *J. Biol. Chem.* **272**, 24097 – 24100
412. Plasterk, R. (1998) *Nature (London)* **394**, 718 – 719
413. Cherry, S. R., Beard, C., Jaenisch, R., and Baltimore, D. (2000) *Proc. Natl. Acad. Sci. U.S.A.* **97**, 8467 – 8472
413a. Fugmann, S. D. (2002) *Nature (London)* **416**, 691 – 694
414. Baltimore, D. (1986) *Nature (London)* **319**, 12 – 13
415. Raghavan, S. C., Kirsch, I. R., and Lieber, M. R. (2001) *J. Biol. Chem.* **276**, 29126 – 29133
416. Pandey, V. N., and Modak, M. J. (1989) *J. Biol. Chem.* **264**, 867 – 871
417. Gough, N. (1983) *Trends Biochem. Sci.* **8**, 227 – 228
418. Rada, C., and Milstein, C. (2001) *EMBO J.* **20**, 4570 – 4576
419. Milstein, C., Neuberger, M. S., and Staden, R. (1998) *Proc. Natl. Acad. Sci. U.S.A.* **95**, 8791 – 8794
420. Tomlinson, I. M., Walter, G., Jones, P. T., Dear, P. H., Sonnhammer, E. L. L., and Winter, G. (1996) *J. Mol. Biol.* **256**, 813 – 817
421. Kelsoe, G. (1998) *Proc. Natl. Acad. Sci. U.S.A.* **95**, 6576 – 6577
422. Papavasiliou, F. N., and Schatz, D. G. (2000) *Nature (London)* **408**, 216 – 221
423. Foster, P. L. (1999) *Proc. Natl. Acad. Sci. U.S.A.* **96**, 7617 – 7618
423a. Arakawa, H. Hauschild, J., and Buerstedde, J.-M. (2002) *Science* **295**, 1301 – 1306
423b. Fugmann, S. D., and Schatz, D. G. (2002) *Science* **295**, 1244 – 1245
424. de Wildt, R. M. T., Hoet, R. M. A., van Venrooij, W. J., Tomlinson, I. M., and Winter, G. (1999) *J. Mol. Biol.* **285**, 895 – 901
424a. Neuberger, M. S., and Scott, J. (2000) *Science* **289**, 1705 – 1706
424b. Di Noia, J., and Neuberger, M. S. (2002) *Nature (London)* **419**, 43 – 48
424c. Faili, A., Aoufouchi, S., Flatter, E., Guéranger, Q., Reynaud, C.-A., and Weill, J.-C. (2002) *Nature (London)* **419**, 944 – 947
424d. Petersen, S., and 15 other authors. (2001) *Nature (London)* **414**, 660 – 665
425. Casellas, R., Shih, T.-A. Y., Kleinewietfeld, M., Rakonjac, J., Nemazee, D., Rajewsky, K., and Nussenzweig, M. C. (2001) *Science* **291**, 1541 – 1544
426. Reth, M., Gehrmann, P., Petrac, E., and Wiese, P. (1986) *Nature (London)* **322**, 840 – 842
427. Ritchie, K. A., Brinster, R. L., and Storb, U. (1984) *Nature (London)* **312**, 517 – 520
428. Mostoslavsky, R., Singh, N., Tenzen, T., Goldmit, M., Gabay, C., Elizur, S., Qi, P., Reubinoff, B. E., Chess, A., Cedar, H., and Bergman, Y. (2001) *Nature (London)* **414**, 221 – 225
429. Noma, Y., Sideras, P., Naito, T., Bergstedt-Lindquist, S., Azuma, C., Severinson, E., Tanabe, T., Kinashi, T., Matsuda, F., Yaoita, Y., and Honjo, T. (1986) *Nature (London)* **319**, 640 – 646
430. Geha, R. S., and Rosen, F. S. (1994) *N. Engl. J. Med.* **330**, 1008 – 1009
431. Fukita, Y., Mizuta, T.-R., Shirozu, M., Ozawa, K., Shimizu, A., and Honjo, T. (1993) *J. Biol. Chem.* **268**, 17463 – 17470
432. Stavnezer, J. (2000) *Science* **288**, 984 – 985
433. Levitzki, A., and Gazit, A. (1995) *Science* **267**, 1782 – 1787
434. Wallin, J. J., Rinkenberger, J. L., Rao, S., Gackstetter, E. R., Koshland, M. E., and Zwollo, P. (1999) *J. Biol. Chem.* **274**, 15959 – 15965
435. Goldwrth, A. W., and Bevan, M. J. (1999) *Nature (London)* **402**, 255 – 262
436. Yancopoulos, G. D., Blackwell, T. K., Suh, H., Hood, L., and Alt, F. W. (1986) *Cell* **44**, 251 – 259
437. Bousso, P., Wahn, V., Douagi, I., Horneff, G., Pannetier, C., Le Deist, F., Zepp, F., Niehues, T., Kourilsky, P., Fischer, A., and de Saint Basile, G. (2000) *Proc. Natl. Acad. Sci. U.S.A.* **97**, 274 – 278
438. Arstila, T. P., Casrouge, A., Baron, V., Even, J., Kanellopoulos, J., and Kourilsky, P. (1999) *Science* **286**, 958 – 961
439. Davodeau, F., Difilippantonio, M., Roldan, E., Malissen, M., Casanova, J.-L., Couedel, C., Morcet, J.-F., Merkenschlager, M., Nussenzweig, A., Bonneville, M., and Malissen, B. (2001) *EMBO J.* **20**, 4717 – 4729
440. Buisseret, P. D. (1982) *Sci. Am.* **247**(Aug), 86 – 95
441. Lichtenstein, L. M. (1993) *Sci. Am.* **269**(Sep), 117 – 124
442. Kay, A. B. (2001) *N. Engl. J. Med.* **344**, 30 – 37
443. Cook, J. P. D., Henry, A. J., McDonnell, J. M., Owens, R. J., Sutton, B. J., and Gould, H. J. (1997) *Biochemistry* **36**, 15579 – 15588
444. Corry, D. B., and Kheradmand, F. (1999) *Nature (London)* **402, Supp.**, B18 – B23
445. Turner, H., and Kinet, J.-P. (1999) *Nature (London)* **402, Supp.**, B24 – B30
446. Joseph, M., Auriault, C., Capron, A., Vorng, H., and Viens, P. (1983) *Nature (London)* **303**, 810 – 812
447. Dweik, R. A., Comhair, S. A. A., Gaston, B., Thunnissen, F. B. J. M., Farver, C., Thomassen, M. J., Kavuru, M., Hammel, J., Abu-Soud, H. M., and Erzurum, S. C. (2001) *Proc. Natl. Acad. Sci. U.S.A.* **98**, 2622 – 2627
448. Cookson, W. (1999) *Nature (London)* **402, Supp.**, B5 – B11
448a. Yazdanbakhsh, M., Kremsner, P. G., and van Ree, R. (2002) *Science* **296**, 490 – 494
449. Pomés, A., Melén, E., Vailes, L. D., Retief, J. D., Arruda, L. K., and Chapman, M. D. (1998) *J. Biol. Chem.* **273**, 30801 – 30807
450. Ichikawa, S., Hatanaka, H., Yuuki, T., Iwamoto, N., Kojima, S., Nishiyama, C., Ogura, K., Okumura, Y., and Inagaki, F. (1998) *J. Biol. Chem.* **273**, 356 – 360
451. Mueller, G. A., Smith, A. M., Chapman, M. D., Rule, G. S., and Benjamin, D. C. (2001) *J. Biol. Chem.* **276**, 9359 – 9365
452. Rosenstreich, D. L., Eggleston, P., Kattan, M., Baker, D., Slavin, R. G., Gergen, P., Mitchell, H., McNiff-Mortimer, K., Lynn, H., Ownby, D., and Malveaux, F. (1997) *N. Engl. J. Med.* **336**, 1356 – 1363
453. Baur, X., Aschauer, H., Mazur, G., Dewair, M., Prelicz, H., and Steigemann, W. (1986) *Science* **233**, 351 – 354
454. Lascombe, M.-B., Grégoire, C., Poncet, P., Tavares, G. A., Rosinski-Chupin, I., Rabillon, J., Goubran-Botros, H., Mazié, J.-C., David, B., and Alzari, P. M. (2000) *J. Biol. Chem.* **275**, 21572 – 21577
455. Rouvinen, J., Rautiainen, J., Virtanen, T., Zeiler, T., Kauppinen, J., Taivainen, A., and Mäntyjärvi, R. (1999) *J. Biol. Chem.* **274**, 2337 – 2343
456. Woodfolk, J. A., Wheatley, L. M., Piyasena, R. V., Benjamin, D. C., and Platts-Mills, T. A. E. (1998) *J. Biol. Chem.* **273**, 29489 – 29496
457. Ansari, A. A., Shenbagamurthi, P., and Marsh, D. G. (1989) *J. Biol. Chem.* **264**, 11181 – 11185
458. Metzler, W. J., Valentine, K., Roebber, M., Marsh, D. G., and Mueller, L. (1992) *Biochemistry* **31**, 8697 – 8705
459. Villalba, M., Batanero, E., Monsalve, R. I., González de la Peña, M. A., and Lahoz, C. (1994) *J. Biol. Chem.* **269**, 15217 – 15222
460. Slater, J. E., Vedvick, T., Arthur-Smith, A., Trybul, D. E., and Kekwick, R. G. O. (1996) *J. Biol. Chem.* **271**, 25394 – 25399
461. Akasawa, A., Hsieh, L.-S., Martin, B. M., Liu, T., and Lin, Y. (1996) *J. Biol. Chem.* **271**, 25389 – 25393
461a. Sampson, H. A. (2002) *N. Engl. J. Med.* **346**, 1294 – 1299
462. Wilkinson, S. L. (1998) *Chem. Eng. News* **Sep 7**, 38 – 40
463. Ferreira, F., Ebner, C., Kramer, B., Casari, G., Briza, P., Kungl, A. J., Grimm, R., Jahn-Schmid, B., Breiteneder, H., Kraft, D., Breitenbach, M., Rheinberger, H.-J., and Scheiner, O. (1998) *FASEB J.* **12**, 231 – 242
464. Busse, W. W., and Lemanske, R. F., Jr. (2001) *N. Engl. J. Med.* **344**, 350 – 361
465. Cookson, W. O. C. M., and Moffatt, M. F. (1997) *Science* **275**, 41 – 42
466. Barnes, P. J. (1991) *Trends Biochem. Sci.* **16**, 365 – 369
467. Gounni, A. S., Lamkhioued, B., Koussih, L., Ra, C., Renzi, P. M., and Hamid, Q. (2001) *FASEB J.* **15**, 940 – 949
468. Matsuoka, T., and 17 other authors. (2000) *Science* **287**, 2013 – 2017
469. Grünig, G., Warnock, M., Wakil, A. E., Venkayya, R., Brombacher, F., Rennick, D. M., Sheppard, D., Mohrs, M., Donaldson, D. D., Locksley, R. M., and Corry, D. B. (1998) *Science* **282**, 2261 – 2263
470. Bonini, S., Lambiase, A., Bonini, S., Angelucci, F., Magrini, L., Manni, L., and Aloe, L. (1996) *Proc. Natl. Acad. Sci. U.S.A.* **93**, 10955 – 10960
471. Comhair, S. A. A., Bhathena, P. R., Farver, C., Thunnissen, F. B. J. M., and Erzurum, S. C. (2001) *FASEB J.* **15**, 70 – 78
472. Mishra, A., Weaver, T. E., Beck, D. C., and Rothenberg, M. E. (2001) *J. Biol. Chem.* **276**, 8453 – 8459
473. Khubchandani, K. R., and Snyder, J. M. (2001) *FASEB J.* **15**, 59 – 69
474. Davidson, A., and Diamond, B. (2001) *N. Engl. J. Med.* **345**, 340 – 350
475. Ridgway, W. M., Fassò, M., and Fathman, C. G. (1999) *Science* **284**, 749 – 751
476. Mitchison, N. A., and Wedderburn, L. R. (2000) *Proc. Natl. Acad. Sci. U.S.A.* **97**, 8750 – 8751
477. Rose, N. R. (1981) *Sci. Am.* **244**, 80 – 98
478. Chen, C.-R., Tanaka, K., Chazenbalk, G. D., McLachlan, S. M., and Rapoport, B. (2001) *J. Biol. Chem.* **276**, 14767 – 14772

References

479. Corper, A. L., Stratmann, T., Apostolopoulos, V., Scott, C. A., Garcia, K. C., Kang, A. S., Wilson, I. A., and Teyton, L. (2000) *Science* **288**, 505–511
480. Wilson, S. B., Kent, S. C., Horton, H. F., Hill, A. A., Bollyky, P. L., Hafler, D. A., Strominger, J. L., and Byrne, M. C. (2000) *Proc. Natl. Acad. Sci. U.S.A.* **97**, 7411–7416
481. Schwartz, H. L., Chandonia, J.-M., Kash, S. F., Kanaani, J., Tunnell, E., Domingo, A., Cohen, F. E., Banga, J. P., Madec, A.-M., Richter, W., and Baekkeskov, S. (1999) *J. Mol. Biol.* **287**, 983–999
482. Yoon, J.-W., Yoon, C.-S., Lim, H.-W., Huang, Q. Q., Kang, Y., Pyun, K. H., Hirasawa, K., Sherwin, R. S., and Jun, H.-S. (1999) *Science* **284**, 1183–1187
483. von Boehmer, H., and Sarukhan, A. (1999) *Science* **284**, 1135–1137
484. Gianani, R., and Sarvetnick, N. (1996) *Proc. Natl. Acad. Sci. U.S.A.* **93**, 2257–2259
485. Nepom, G. T., Lippolis, J. D., White, F. M., Masewicz, S., Marto, J. A., Herman, A., Luckey, C. J., Falk, B., Shabanowitz, J., Hunt, D. F., Engelhard, V. H., and Nepom, B. S. (2001) *Proc. Natl. Acad. Sci. U.S.A.* **98**, 1763–1768
486. Mathis, D., Vence, L., and Benoist, C. (2001) *Nature (London)* **414**, 792–798
487. Kumar, S., Kalsi, J., Latchman, D. S., Pearl, L. H., and Isenberg, D. A. (2001) *J. Mol. Biol.* **308**, 527–539
488. Stevens, S. Y., and Glick, G. D. (1999) *Biochemistry* **38**, 560–568

488a. Marshall, E. (2002) *Science* **296**, 689 – 691

489. Witte, S., Neumann, F., Krawinkel, U., and Przybylski, M. (1996) *J. Biol. Chem.* **271**, 18171–18175
490. Muller, S., Briand, J.-P., and Van Regenmortel, M. H. V. (1988) *Proc. Natl. Acad. Sci. U.S.A.* **85**, 8176–8180
491. Pereira, B., Benedict, C. R., Le, A., Shapiro, S. S., and Thiagarajan, P. (1998) *Biochemistry* **37**, 1430–1437
492. Wieben, E. D., Rohleder, A. M., Nenninger, J. M., and Pederson, T. (1985) *Proc. Natl. Acad. Sci. U.S.A.* **82**, 7914–7918
493. Morel, L., Blenman, K. R., Croker, B. P., and Wakeland, E. K. (2001) *Proc. Natl. Acad. Sci. U.S.A.* **98**, 1787–1792
494. Reddy, R., Tan, E. M., Henning, E., Nohga, K., and Busch, H. (1983) *J. Biol. Chem.* **258**, 1383–1386
495. Bolívar, J., Guelman, S., Iglesias, C., Ortíz, M., and Valdivia, M. M. (1998) *J. Biol. Chem.* **273**, 17122–17127
496. Schultz, G. E., and Schirmer, R. H. (1979) *Principles of Protein Structure*, Springer-Verlag, New York

496a. Vinuesa, C. G., and Goodnow, C. C. (2002) *Nature (London)* **416**, 595 – 598

497. Kotzin, B. L., Falta, M. T., Crawford, F., Rosloniec, E. F., Bill, J., Marrack, P., and Kappler, J. (2000) *Proc. Natl. Acad. Sci. U.S.A.* **97**, 291–296
498. Nishiura, H., Shibuya, Y., Matsubara, S., Tanase, S., Kambara, T., and Yamamoto, T. (1996) *J. Biol. Chem.* **271**, 878–882
499. Hellmark, T., Burkhardt, H., and Wieslander, J. (1999) *J. Biol. Chem.* **274**, 24862–25868
500. Netzer.K-O, Leinonen, A., Boutaud, A., Borza, D.-B., Todd, P., Gunwar, S., Langeveld, J. P. M., and Hudson, B. G. (1999) *J. Biol. Chem.* **274**, 11267–11274
501. Gunnarsson, A., Hellmark, T., and Wieslander, J. (2000) *J. Biol. Chem.* **275**, 30844–30848
502. Haneji, N., Nakamura, T., Takio, K., Yanagi, K., Higashiyama, H., Saito, I., Noji, S., Sugino, H., and Hayashi, Y. (1997) *Science* **276**, 604–607
503. Yip, T.-T., Van de Water, J., Gershwin, M. E., Coppel, R. L., and Hutchens, T. W. (1996) *J. Biol. Chem.* **271**, 32825–32833
504. Dang, C. V., Tan, E. M., and Traugh, J. A. (1988) *FASEB J.* **2**, 2376–2379
505. Miyata, T., Yamada, N., Iida, Y., Nishimura, J., Takeda, J., Kitani, T., and Kinoshita, T. (1994) *N. Engl. J. Med.* **330**, 249–255

505a. Farrell, R. J., and Kelly, C. P. (2002) *N. Engl. J. Med.* **346**, 180 – 188

505b. Shan, L., Molberg, O., Parrot, I., Hausch, F., Filiz, F., Gray, G. M., Sollid, L. M., and Khosla, C. (2002) *Science* **297**, 2275 – 2279

505c. Schuppan, D., and Hahn, E. G. (2002) *Science* **297**, 2218 – 2220

505d. Fesus, L., and Piacentini, M. (2002) *Trends Biochem. Sci.* **27**, 534 – 539

506. Shields, D. C., Tyor, W. R., Deibler, G. E., Hogan, E. L., and Banik, N. L. (1998) *Proc. Natl. Acad. Sci. U.S.A.* **95**, 5768–5772
507. Noseworthy, J. H., Lucchinetti, C., Rodriguez, M., and Weinshenker, B. G. (2000) *N. Engl. J. Med.* **343**, 938–952
508. Li, Y., Li, H., Martin, R., and Mariuzza, R. A. (2000) *J. Mol. Biol.* **304**, 177–188
509. Rogers, S. W., Andrews, P. I., Gahring, L. C., Whisenand, T., Cauley, K., Crain, B., Hughes, T. E., Heinemann, S. F., and McNamara, J. O. (1994) *Science* **265**, 648–651
510. Byrne, G. (1988) *Science* **242**, 198
511. Kallen, J., Mikol, V., Taylor, P., and Walkinshaw, M. D. (1998) *J. Mol. Biol.* **283**, 435–449
512. Montague, J. W., Hughes, F. M., Jr., and Cidlowski, J. A. (1997) *J. Biol. Chem.* **272**, 6677–6684
513. Mackay, I. R., and Rosen, F. S. (2000) *N. Engl. J. Med.* **343**, 1313–1324
514. Greenberg, P. D., and Riddell, S. R. (1999) *Science* **285**, 546–551
515. Fischer, A., and Malissen, B. (1998) *Science* **280**, 237–243
516. Singer, L., Whitehead, W. T., Akama, H., Katz, Y., Fishelson, Z., and Wetsel, R. A. (1994) *J. Biol. Chem.* **269**, 28494–28499
517. Xu, C., Mao, D., Holers, V. M., Palanca, B., Cheng, A. M., and Molina, H. (2000) *Science* **287**, 498–501
518. Vidovic, D., and Matzinger, P. (1988) *Nature (London)* **336**, 222–225
519. Skare, J. C., Milunsky, A., Byron, K. S., and Sullivan, J. L. (1987) *Proc. Natl. Acad. Sci. U.S.A.* **84**, 2015–2018
520. Sayos, J., Wu, C., Morra, M., Wang, N., Zhang, X., Allen, D., van Schaik, S., Notarangelo, L., Geha, R., Roncarolo, M. G., Oettgen, H., De Vries, J. E., Aversa, G., and Terhorst, C. (1998) *Nature (London)* **395**, 462–469
521. Rennie, J. (1993) *Sci. Am.* **268**(Jun), 34–35
522. Nowak, M. A., and McMichael, A. J. (1995) *Sci. Am.* **273**(Aug), 58–65
523. McCune, J. M. (2001) *Nature (London)* **410**, 974–979
524. Magrath, I., Erikson, J., Whang-Peng, J., Sieverts, H., Armstrong, G., Benjamin, D., Triche, T., Alabaster, O., and Croce, C. M. (1983) *Science* **222**, 1094–1098
525. Tsujimoto, Y., Gorham, J., Cossman, J., Jaffe, E., and Croce, C. M. (1985) *Science* **229**, 1390–1393
526. Dixon, R. A. (2001) *Nature (London)* **411**, 843–847
527. Currier, W. W. (1981) *Trends Biochem. Sci.* **6**, 191–194
528. Konno, K., Hirayama, C., Yasui, H., and Nakamura, M. (1999) *Proc. Natl. Acad. Sci. U.S.A.* **96**, 9159–9164
529. Bergey, D. R., Howe, G. A., and Ryan, C. A. (1996) *Proc. Natl. Acad. Sci. U.S.A.* **93**, 12053–12058
530. Dangl, J. L., and Jones, J. D. G. (2001) *Nature (London)* **411**, 826–833
531. De Samblanx, G. W., Goderis, I. J., Thevissen, K., Raemaekers, R., Fant, F., Borremans, F., Acland, D. P., Osborn, R. W., Patel, S., and Broekaert, W. F. (1997) *J. Biol. Chem.* **272**, 1171–1179
532. Fant, F., Vranken, W., Broekaert, W., and Borremans, F. (1998) *J. Mol. Biol.* **279**, 257–270
533. McDowell, J. M., and Dangl, J. L. (2000) *Trends Biochem. Sci.* **25**, 79–82
534. Parker, J. E., and Coleman, M. J. (1997) *Trends Biochem. Sci.* **22**, 291–296
535. Lam, E., Kato, N., and Lawton, M. (2001) *Nature (London)* **411**, 848–853

535a. Austin, M. J., Muskett, P., Kahn, K., Feys, B. J., Jones, J. D. G., and Parker, J. E. (2002) *Science* **295**, 2077 – 2080

535b. Nishimura, M., and Somerville, S. (2002) *Science* **295**, 2032 – 2033

536. Fefeu, S., Bouaziz, S., Huet, J.-C., Pernollet, J.-C., and Guittet, E. (1997) *Protein Sci.* **6**, 2279–2284
537. Van Der Biezen, E. A., and Jones, J. D. G. (1998) *Trends Biochem. Sci.* **23**, 454–456
538. Tang, X., Frederick, R. D., Zhou, J., Halterman, D. A., Jia, Y., and Martin, G. B. (1996) *Science* **274**, 2060–2063
539. Chandra, S., and Low, P. S. (1995) *Proc. Natl. Acad. Sci. U.S.A.* **92**, 4120–4123
540. Legendre, L., Yueh, Y. G., Crain, R., Haddock, N., Heinstein, P. F., and Low, P. S. (1993) *J. Biol. Chem.* **268**, 24559–24563
541. Jabs, T., Tschöpe, M., Colling, C., Hahlbrock, K., and Scheel, D. (1997) *Proc. Natl. Acad. Sci. U.S.A.* **94**, 4800–4805
542. Chandra, S., and Low, P. S. (1997) *J. Biol. Chem.* **272**, 28274–28280
543. Jennings, D. B., Ehrenshaft, M., Pharr, D. M., and Williamson, J. D. (1998) *Proc. Natl. Acad. Sci. U.S.A.* **95**, 15129–15133
544. Wildermuth, M. C., Dewdney, J., Wu, G., and Ausubel, F. M. (2001) *Nature (London)* **414**, 562–565
545. Klessig, D. F., Durner, J., Noad, R., Navarre, D. A., Wendehenne, D., Kumar, D., Zhou, J. M., Shah, J., Zhang, S., Kachroo, P., Trifa, Y., Pontier, D., Lam, E., and Silva, H. (2000) *Proc. Natl. Acad. Sci. U.S.A.* **97**, 8849–8855

545a. Howe, G. A. (2001) *Proc. Natl. Acad. Sci. U.S.A.* **98**, 12317 – 12319

546. Schenk, P. M., Kazan, K., Wilson, I., Anderson, J. P., Richmond, T., Somerville, S. C., and Manners, J. M. (2000) *Proc. Natl. Acad. Sci. U.S.A.* **97**, 11655–11660
547. Sela, I. (1981) *Trends Biochem. Sci.* **6**, 31–33
548. Waterhouse, P. M., Wang, M.-B., and Lough, T. (2001) *Nature (London)* **411**, 834–842
549. Marx, J. (2000) *Science* **288**, 1370 – 1372
550. Jorgensen, R. A. (2002) *Proc. Natl. Acad. Sci. U.S.A.* **99**, 11561 – 11563

Study Questions

1. Describe major aspects of the vertebrate innate and adaptive immune system. In what ways do they cooperate?

2. Describe briefly the functions of each of the folowing: antibodies, defensins, cytokines, complement, MHC proteins, B cells, T cells, dendritic cells, monocytes, macrophages, and neutrophils.

3. Why do antibodies produced using a native protein tend to bind only weakly to the corresponding denatured protein?

4. Discuss the topic of self-identity.

5. What are autoimmune diseases? How does the body avoid most autoimmune diseases?

6. List some methods by which viruses, bacteria, protozoa, and pathogenic fungi gain access to cells or to tissues.

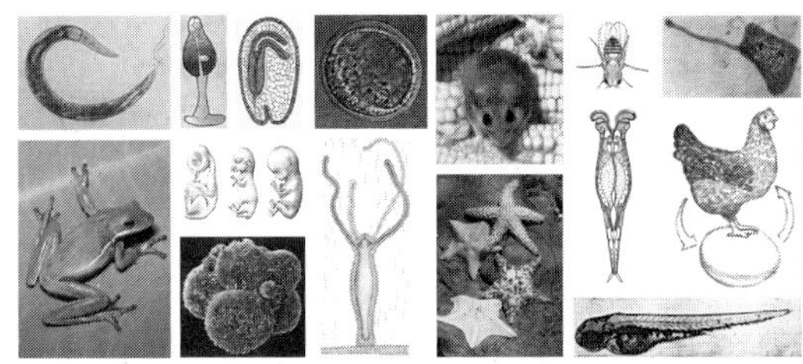

Single cells develop into an astonishing variety of different species, all of which find their niches in the ecosystem. Whether a rectagular bacterium, a plant, a frog, or a human being, the size, shape, the body construction and metabolic pathways are established by the sequential expression of the organism's genes. Recent investigations have confirmed many similarities among major families of proteins from virtually all species. The same studies also emphasize the profound genetic differences between species. Understanding these differences, as well as the interrelationships among species, provides a continuing challenge to biochemists and biologists. Such understanding may even be essential to the survival of the human species in a changing environment.

Contents

Growth and Development

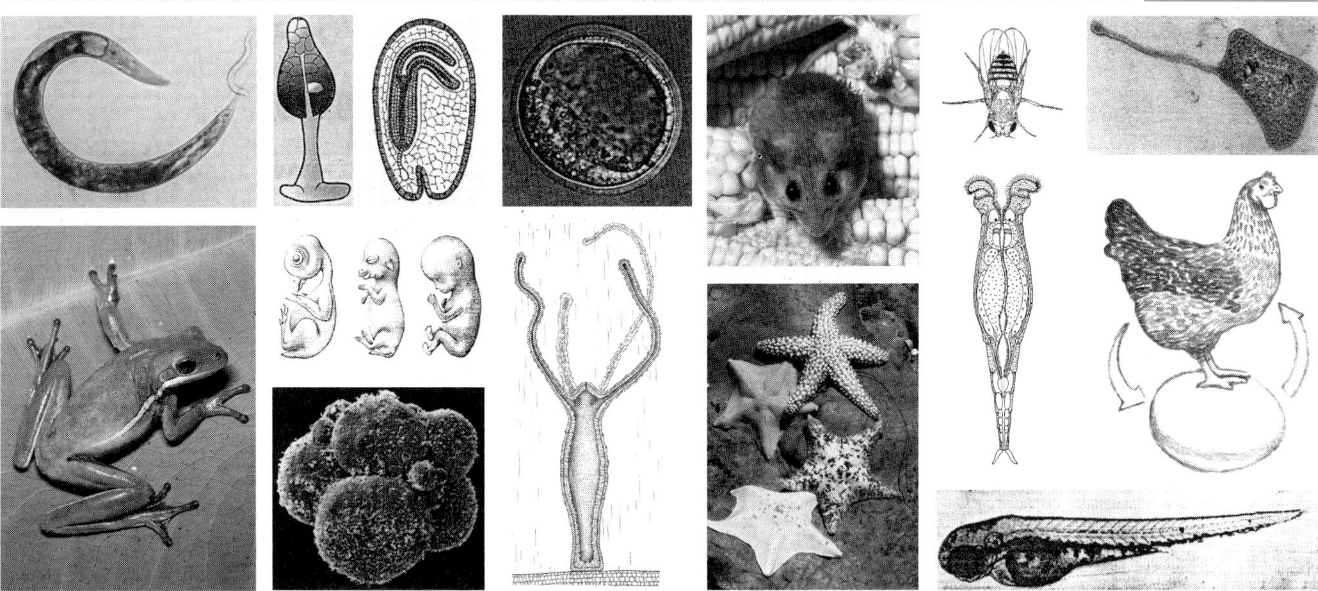

One of the most fascinating of all biological phenomena is the development of an animal from a fertilized egg. From the early embryonic cells, which appear to be much alike, there arise during the course of a very few cell divisions differentiated organs and tissues such as liver, brain, kidney, muscle, skin, and red blood cells. The biochemical properties of differentiated cells are often highly specialized. Red blood cells make hemoglobin, while muscle cells make large amounts of myosin and actin. The endocrine cells of the pancreas make insulin or glucagon, while the exocrine cells form the digestive enzymes that are secreted into the intestinal tract.

Looking more broadly at the biological world we see many additional specialized features at every level of observation. Developmental patterns differ for every organism. Specialized organs abound. The structures of proteins, lipids, carbohydrates, and nucleic acids all vary, and every species has its own metabolic peculiarities. Even among bacteria we find extreme variation. Furthermore, many unicellular eukaryotic organisms undergo complex development within a single cell. The topics of this chapter are too complex for any detailed discussion. We will examine some aspects of growth and development for a few organisms and will ask whether there are simplifying generalizations. Comprehensive textbooks are available.[1,2]

Enough has been learned about development to make it clear that the DNA contains genetically coded **developmental programs**, which are followed by cells.[3] However, both transcription and translation are controlled by many chemical signals, which influence the execution of the genetic program. Such signals may arise from within a single cell, from the external environment, and from adjacent cells. The tight control is reflected by the fact that in most human tissues at any stage of development no more than about 10% of the total genes are transcribed at any one time. Chemical analysis makes it clear that most specialized cells contain a full complement of DNA but 90% of the genes are turned off.

A. Basic Concepts and Molecular Essentials

Listed in Table 32-1 are some essential aspects of growth and differentiation. Some of them are obvious. **Cohesion** between molecules provides the basis of specificity. **Receptors**, whether they be enzymes, hormone receptors, or receptors for chemotaxis, are essential. They are usually activated by a **conformational change** that accompanies binding. A cell must have receptors that can respond to a variety of **signals**, which may come from within the cell, from the external medium, or from neighboring cells. The receptor–signal pairs are essential to **local control**, which provides the basis for all of development.[4]

To have any kind of spatial differentiation a cell must develop **directionality** (polarity).[5] This permits **asymmetric cell division**[5] and development of poles in a developing cell or embryo. **Adhesion** molecules hold cells together, allowing a cell to have a fixed position relative to other cells. Development of directionality and of developmental **patterns**[6–8] are dependent upon **gradients** of concentration, of foods, heat, light, gravitation, etc., that can be detected by receptors. Gradients of compounds called **morphogens**[9,10] help to provide a **positional identity** to cells.[10–12] **Movement** of molecules of organelles and of intact cells is also essential. In multicellular organisms cells often migrate to new locations by following chemotactic signals.

Growth of individual cells enlarges them and often leads to **cell division**. The **cell cycle** describes this process with emphasis on replication of DNA. **Homeostasis** encompasses adaptation to altered nutrient or other environmental variables and to all processes which influence a cell to change. It provides for defensive reactions to many types of stress.[12a,b]

TABLE 32-1
Some Essentials for Growth and Differentiation

Cohesion of molecules, utilizing specific, hydrogen-bonding, and complementary surface shapes

Recognition, and conformational changes
- **Receptors** and **signals**
- **Local controls**

Polarity (directionality)
- Asymmetric cell division
- Poles

Gradients that can be sensed by receptors
- Food, physical qualities, morphogens

Adhesion between cells is required to hold an organism together and also, together with morphogen gradients, to provide a **positional identity**

Movement of molecules, organelles, and cells

Growth to enlarge cell size and numbers of cells
- **A cell cycle** for **replication of DNA** and **cell division**

Homeostasis to permit adaptation to changes in nutrient concentrations, stress

A developmental program, which is encoded in the genome
- Implementing this program usually requires many preformed compounds
- **Alternative developmental programs** often provide flexibility to an organism

Stem cells of **totipotent**, **pluripotent,** or multipotent nature supply new germ cells and other cells for multicellular organisms when needed

Programmed cell death (**apoptosis**) is part of many developmental programs and provides for removal of unneeded cells without inflammation

1. DNA and Developmental Programs

The genetic developmental program of an organism is encoded in the DNA. The expression of the program is implemented initially at the transcriptional level by a host of transcription factors that act at appropriate times and in appropriate places. Their action usually requires the presence of many preformed compounds. **Alternative developmental programs** may be used by unicellular organisms to adapt to new environmental conditions or by multicellular organisms to differentiate tissues.[6]

Programmed alterations and rearrangements of DNA. Part of the developmental program may involve a temporary or permanent change in the DNA. One of the simplest of these changes makes use of the **transposable recombinational switch** in which a small piece of DNA is present in either of the two possible orientations. (See Chapter 27, Section D,3.) An example, illustrated by Eqs. 27-15 and 27-16, is the variation in "phase" of the flagella produced by *Salmonella*. A somewhat different example is provided by the unicellular yeast *Saccharomyces cerevisiae*, which changes the mating type of its haploid forms in a highly regulated pattern.[13,14] The **a** mating type is expressed constitutively, but this is frequently switched to the **α** type, which produces a different mating pheromone (see Table 30-5) and responds to a pheromone from **a**-type cells. The change occurs through the transposition of different "cassettes" of DNA from "silent sites" into an expression site.[14a,b] The cassettes contain several genes, which are copied into the expression site called MAT. This site always contains genes of either **a** or **α** type. However, both **a** and **α** genes are present in other storage locations. When the mating type is switched, a copy is made of one of the stored cassettes and is placed into the MAT locus replacing the cassette already present. The MAT α genes encode two regulatory proteins, α1 and α2. Protein α1 is a positive regulator of the α-cell-specific genes, while protein α2 is a repressor of the a-cell-specific genes.[15–17] A similar mechanism appears to be employed by trypanosomes in changing their variable cell surface proteins (Box 31-E).[18]

Inactivation of genes and imprinting. Under some circumstances a chromosome or part of a chromosome is permanently inactivated but remains within the cell as compactly folded heterochromatin. Heterochromatin often consists of reiterated sequences of unknown function, but it may also contain groups of inactivated genes. The most impressive case is the total inactivation of one of the two X chromosomes in cells of female mammals.[19–21] The entire chromosome appears as heterochromatin. The inactivation occurs early in embryonic development and is random with respect to the two X chromosomes. In some cells the maternal X chromosome, in others the paternal X chromosome, becomes inactive. However, upon further cell divisions the same chromosome in each clone remains inactive. As a consequence of the random inactivation, the female body is a mosaic with respect to genes in the X chromosomes. Formation of heterochromatin in X chromosomes is described in Section C,1. It depends upon **epigenetic markers**, which consist, at least in part, of methylated cytosines in 5'-CpG-3' sequences. These CG pairs are palindromic and can be methylated on both strands (see Eq. 27-2 and Fig. 32-3).

Selective inactivation of genes by methylation accounts for **genomic imprinting**, which occurs in mammals and which marks a gene of either maternal or paternal origin for silencing.[22–24] About 50 imprinted genes have been identified in mice and in humans.[23,25] Imprinting, known as **epigenetic inheritance**, also appears to arise in part by methylation of cytosines in CpG sequences. Four m^5C-methyltransferase genes have been identified in the human genome.[26] One of these presumably provides the initial methylation of the cytosine in one strand of the DNA. A second methyltransferase (apparently encoded by gene *Dnmt1*)[27] methylates the second strand and serves as the **maintenance methylase** that preserves the methylation pattern upon DNA replication (see Eq. 27-2). The faithful maintenance of imprinting has been demonstrated experimentally for genes of mice formed by nuclear transplantation, i.e., mice grown from somatic cell nuclei transplanted into ova.[28] Imprinted genes usually occur in clusters, which also contain imprinted genes for noncoding RNA molecules. The transcription of those genes often is correlated with repression (**silencing**) of protein-coding genes on the same chromosome. This observation is consistent with evidence that genes may be silenced by the binding of small RNA molecules to complementary sequences in the DNA or of mRNAs. The mechanism may be limited to **imprint control regions** or silencer units in the DNA, while the genes to be controlled are set off by **insulator elements** (see also pp. 1882, 1894 and Chapter 28, Section C,1). The significance of genomic imprinting is not clear.[29,30] However, it is very important to nuclear transplantation because the methylation state of imprinted genes is normally reset (**epigenetic reprogramming**) before embryonic development begins.[23,25,27,31–33] See Section 5. Methylation of DNA is not essential to the life of filamentous fungi. Nevertheless, *Neurospora* does have a DNA methyltransferase, which methylates about 1.5% of the cytosines present.

Loss of DNA. While some genes are selectively inactivated, others may be irreversibly lost during development. The extreme case is that of the human red blood cell from which the entire nucleus is expelled. Loss of DNA may result from recombinational events. For example, crossing-over between sister chromatids during mitosis has been demonstrated for some cells and chromosomes. This does not alter the genetics of the progeny cells if equal amounts of genetic material are exchanged. However, if two or more similar base sequences occur in tandem in a DNA molecule, unequal crossing-over (Chapter 27, Section D,2) can occur with the loss of genetic material from a chromosome of one of the progeny cells. This may be a deliberately programmed route of differentiation for some cells.

Alternatively, loss of DNA from a chromosome may occur through a **looping-out excision** mechanism. Like the excision of a prophage from the chromosome of *E. coli* (Chapter 27, Section D,3) this loss of genes occurs at specific sites in the DNA. The best known case is the joining of gene segments during the differentiation of lymphocytes (Fig. 31-18). The extent to which similar changes occur during terminal differentiation of other tissues or in nonmammalian species is uncertain.

Amplification of DNA of chromosomes. During formation of oocytes parts of the DNA are "amplified" by repeated replication. This provides a way for the ovum to accumulate ribosomal RNA and various proteins in large amounts. Similarly, genes for two abundant proteins of the egg shell or **chorion** of insects are amplified. Bidirectional replication initiated at discrete positions yields an "onion skin" structure containing many copies of an ~90-kb sequence containing the two genes. The polyploidy observed in some highly specialized cells such as the Purkinje cells

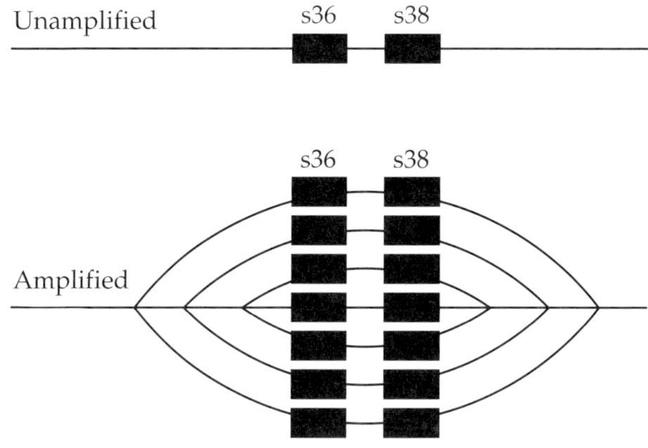

of the cerebellum and of many cells of Diptera larva (Chapter 26, Section F,3) represent another way of amplifying genes. Polyploid cells of animals generally represent a terminal stage of differentiation and do not divide. They tend to contain their full complement of genes in each copy of their DNA, but most genes are not expressed.

2. Receptors and Signals

Receptors and their ligands are numerous, varied, and essential to all forms of life. Cell-surface receptors on bacteria detect feeding attractants as well as dangerous molecules. From bacteria to humans seven-helix receptors function to detect light, odors, hormones, and other molecules. The numbers of different receptors are impressive. For example, the tiny nematode *C. elegans* has 650 seven-helix transmembrane receptors and 411 protein kinases, many of which may be associated with receptors.[34] Our bodies have thousands.

Every ligand that binds to a receptor is a signal of some kind. How many signaling molecules are there, and what are their structures? The number of proteins and small peptides affecting growth and development of cells may be enormous. Many of these have been considered in Chapter 30. In addition, the catecholamines, serotonin, histamine, and even bicarbonate ions may act as local hormones.[35] For example, HCO_3^- determines whether a thin-walled sporangium or a thick-walled heat-resistant sporangium will appear in the phycomycete water mold *Blastocladiella*. We know that a very large number of different proteins and small peptides are secreted from cells, many in very small amounts.[36] Some protein morphogens act at concentrations so low, e.g., 50 pM, that it has been hard to detect them.[10] T-cell receptors respond to extremely small amounts of short peptides that stream outward from many cells bound to surface MHC molecules. Do some of these same peptides act as signals between other cells during development?

Small RNA molecules are now being found to function in many ways within cells,[37–39] They may also act as messengers within cells and between cells. The small 22 nt RNAs transcribed from genes *lin-4* and *let-7* of the nematode *C. elegans* control key developmental decisions.[38b,d] In green plants small RNAs move between cells and throughout the plants to trigger defensive responses (p. 1869).[38c] Much of the RNA transcribed from genes of any eukaryotic cell lacks any known function. Since evolution tends to act on all of the molecular constituents of a cell, this vast amount of RNA may have acquired vital roles in the control of metabolism and growth.[38e] A poorly understood intracellular structure, the **ribonucleoprotein vault** also remains a biochemical mystery. Differentiated cells may contain 10^4 and embryonic cells over 10^7 of these ~42 x 75 nm hollow objects. Vaults have internal cavities large enough to hold two intact ribosomes. Some vaults appear to be empty, but others contain materials, a fact that suggests some role in transport or storage.[38f]

Homeostasis. A large fraction of the receptors and signaling system of cells is devoted to the maintenance of a constant internal environment. This homeostasis is essential if a cell is to respond to changes in external conditions without disastrous consequences.[12a,b] Some special aspects of these processes are discussed in Chapters 11, 17, and 28. Within mammalian cells the hypothalamus, pituitary, and adrenals have a primary responsibility for maintaining homeostasis.[40]

Transcriptional control in differentiation. Development of an organism depends upon an orderly sequence of transcription of genes. Some genes are transcribed in germ cells, others within cells of an early embryo, and others later. As the embryo develops sequential waves of synthesis of the needed protein are observed.[3,41–43] These are controlled by the actions of numerous transcription factors that act at a variety of **cis-regulatory modules** (CRMs) associated with promoter regions[42,44–46] and also by controls on translation of mRNAs,[47] and by negative feedback loops.[48] The latter may involve hormones, morphogenic proteins, coregulatory proteins of various types, small RNAs, etc.[49] More than 2000 transcription factors are encoded in the human genome. Most are positive-acting, i.e., they *promote* transcription. Several families are specifically involved in regulation of development (see Table 28-2). These include the Zn^{2+}-containing GATA-1 (p. 1634), which regulates globin synthesis, embryonic factors of *Drosophila* (e.g., bicoid), vertebrate homeotic genes (Hox clusters), Pit1, the muscle-specific helix–loop–helix proteins MyoD and Myf5, and several forkhead (winged helix) proteins.[3] A single CRM can bind many different regulatory molecules and single regulators can bind to a variety of CRMs. During development of the simple embryo of the sea urchin *Strongylocentratus purpuratus* a network of 40 highly regulated genes is needed to coordinate growth and development with the production of the proteins needed at each step.[41,43]

Part of the control of differentiation lies in the interaction of proteins that regulate transcription with metabolites and hormones. For example, substrate depletion not only decreases growth rate of bacteria but also alters gene transcription. This occurs in *E. coli* as a result of a rise in internal cAMP concentration. The presence of an alternative energy source such as lactose induces changes in gene transcription (Chapter 28, Section A,1). Such **physiological modulation** of a developmental pattern can also be seen in higher organisms.

More striking is the fact that environmental signals can trigger a cell to switch to an alternative developmental program by which enough new genes are activated to rebuild the cell into a new form. An example is spore formation, a process that occurs with some bacteria when external conditions become unfavorable for vegetative growth (Section B,1). Alternative developmental programs are also evident in eukaryotic organisms that undergo metamorphosis, and they may be important to development. Perhaps persistent states of repression of groups of genes can be passed through several generations of cells until a specific chemical signal triggers the unwrapping of the appropriate nucleosomes and transcription of formerly inactive genes.

3. Adhesion, Cell–Cell Recognition, and Cell Migration

Development of multicellular organisms depends upon both adhesion and on recognition of a correct interaction. Like enzyme–substrate, receptor–hormone, and antibody–antigen binding these interactions of macromolecules on cell surfaces often show a high degree of specificity. They may be accompanied by conformational changes and may trigger signaling cascades. We have already discussed some of these interactions, for example, the binding of a molecule of IgG attached to a surface antigen to protein C1q of the complement system (Fig. 31-8) and the binding of an MHC–antigen complex to a T-cell receptor (Fig. 31-15).

There are many other cell–surface **adhesins**, several of which have been discussed on pp. 402–409. Among them are proteins that contain immunoglobulin-like domains and numerous **glycoproteins**. An example of the latter is the binding of a type of pili found in pathogenic strains of *E. coli* to epithelial cells of the urinary tract. The pilin subunits (Fig. 7-9), like lectins (Box 4-C), bind specifically to the disaccharide group Gal*p*α1→4Gal. A lectinlike protein specific for *N*-acetylglucosamine rings is involved in invasion of erythrocytes by the malaria parasite *Plasmodium*. The unicellular alga *Chlamydomonas* (Fig. 1-11) produces sexual gametes of two mating types. When mixed together, gametes of opposite mating types, prior to fusion, adhere to each other via **agglutinins** present on their flagella. The agglutinins are glycoproteins rich in hydroxyproline, serine, glycine, arabinose, and galactose.[50] As mentioned on p. 29, colored cells of different strains of the marine sponge *Microciona prolifera* find others of the same strain using highly specific proteoglycan-like aggregation factors.[51–53] These compounds are highly polymorphic, and it has been suggested that they are part of a primitive immune system. The aggregation reaction requires calcium ions. In our own bodies Ca^{2+}-dependent lectins, the **selectins** (p. 187, 188), bind leukocytes and help to guide them to their sites of action.

Other adhesins include the **integrins**,[53a] cellular adhesion molecules (**CAMs**), **cadherins**,[53a–c] and **fibronectin** (Fig. 8-19). These are also discussed on pp. 402–409. The CAMs (Fig. 8-18A),[54,55] which are members of the immunoglobulin-like protein family, are glycoproteins bearing large 2,8-linked sialic acid polymers.[56–58] They promote Ca^{2+}-dependent aggregation. However, the effect of NCAM, which is widely distributed in a developmentally regulated fashion, can be antiadhesive if long chains of sialic acid are present. NCAM appears to play a role in remodeling and repair of tissues. Adhesion of molecules within cell membranes and the binding of substances to membrane surfaces provides another driving force in development. Within membranes molecules spontaneously sort themselves into lipid **microdomains**, often called **lipid rafts**.[58a] Related to lipid rafts are caveolae (p. 426). These little craters arise in cholesterol-rich microdomains. They often contain the protein **caveolin** as well as glycosphingolipids and GPI-tailed proteins (Fig. 8-13).[58b] ATP-dependent linking reactions may also occur to provide more permanent bonding. Membrane-associated molecules, in turn, become centers for attachment of cytoskeletal proteins and other protein complexes. As with the cytosol and extracellular fluids homeostatic mechanisms act to provide a relatively constant membrane–lipid environment.[58c]

Several types of cell junctions are associated with adhesion and participate in intercellular communication (Fig. 1-15).[59] The cadherins are transmembrane proteins with large extracellular domains (Fig. 8-18B). They are prominent components of adherens junctions[59–61a] in which they join the exterior surfaces of pairs of cells in a zipper-like manner. Another protein, **β-catenin**, links the short C-terminal tails of cadherin through a-catenin subunits to the actin cytoskeleton.[60] In desmosomal junctions other specialized proteins including desmoglein have functions similar to that of cadherins.[59] Tight junctions, from zebrafish to humans, depend upon a complex of several proteins including those of the claudin family.[62] Significantly, the cohesive powers of some adhesins, e.g., of cadherin, are altered during development. Cadherin E is nonadhesive in a four-cell mouse embryo but becomes adhesive after the eight-cell stage.[59] It is obvious that many other changes in intercellular adhesion must also occur during growth and development.

The integrins (see also p. 405) comprise a large family of adhesive receptors that are found in animals from sponges to humans.[63–65] They have both adhesive and signaling functions. Both subunits of their αβ heterodimeric structures[64] have single transmembrane helices and short C-terminal cytoplasmic tails. The β1 subunit tails interact with cytoplasmic proteins. The distribution of integrins varies among cell types. Human leukocytes contain alpha subunits of types αd, αl, αm, and αx with molecular masses of 150–180 kDa. Two ~95-kDa beta subunits (β1 and β2) are present. However, T lymphocytes express β1, β2, and β7 integrins. Other patterns are observed for other leukocytes,[66] in skin,[67] and in other tissues.[68] Integrin molecules tend to aggregate into clusters, which are found together with other proteins, at the ends of actin stress fibers (p. 370).[63] The largest of these clusters are known as focal adhesions. Signals may be sent through integrins in either of the two directions.[63] The extracellular domains of integrins interact with a variety of proteins of the extracellular matrix. These include fibronectin, fibrinogen, vitronectin, collagen, and entactin.[63,69] Other large cell surface adhesins include laminin and osteopontin (Chapter 8), thrombospondin, von Willebrand factor, and related proteins.[70] These adhesins appear to depend upon the sequence Arg-Gly-Asp (RGD), which binds noncovalently to integrins, which act as cell-surface receptors.[71,71a] See also Chapter 12, Section C,9.

The functioning of the complex network of integrins, adhesins, and other components of the extracellular matrix is not understood in detail. One fundamental question is how the strength of adhesion can vary with time and stage of development. Roseman postulated an association of an oligosaccharide chain of a glycoprotein attached to one cell with a specific glycosyltransferase of another cell.[72] The specific interaction would hold cells together, but addition of another glycosyl unit to the oligosaccharide by the transferase would alter the surface properties of the cell carrying the glycoprotein. This, in turn, could cause disaggregation of the cells. Glycosyltransferases can be found on the outer surfaces of cell membranes, and Roseman's proposal may correctly describe one aspect of cell adhesion.

Other molecules that are abundant on cell surfaces include heparan sulfate proteoglycans. Although they have often been regarded as providing a nonspecific "extracellular fly paper," recent evidence from studies of development in *Drosophila* suggest specific and important functions in signaling and in developmental patterning.[73] Both hyaluronan and chitin also have been proposed to play an important role in vertebrate development.[74,75] Proteoglycans of plant cell surfaces, as well as the hydroxyproline-rich proteins of cell walls, may function in plant development.[76]

Movement of cells from one location to another is essential to embryonic development as well as to wound repair and to the immune response. Many brain structures are composed exclusively of immigrant cells.[77] These **cell migrations** depend upon the cytoskeletal actin filaments, integrins, and focal adhesions.[78,79] Chemotactic signals are also required.[80] A great complexity of underlying chemistry is being elucidated.[79,81] See also Chapter 19, Section C.

4. Polarity, Asymmetric Cell Division, and Morphogens

Cells of *E. coli* usually divide exactly in the center to form two seemingly identical cells.[82–84a] However, under the right conditions some bacteria, e.g., *Caulobacter crescentus* and *Bacillus subtilis*, undergo asymmetric division to form two different types of cells (Fig. 32-1).[83] There is clearly an axial polarity. This polarity is evident even in *E. coli*, which has flagella streaming out at one end and its chemoreceptor-bearing "nose" at the other (Fig. 32-1).[85] Axial polarity is also obvious in other bacteria (Fig. 19-1).

Polarity is evident in eukaryotic cells from protists to higher organisms.[85a,b] Cells of the yeast *S. cerevisiae*, whether haploid or diploid, divide in an asymmetric way by budding.[5,7] Among body cells of higher animals those of the epithelium are among the most polarized (Fig. 32-2; see also Fig. 1-6, Box 8-F). Polarity is always present in ova of eukaryotes, but the ova may initially be radially symmetric.[86–88] The **anterior-posterior** axis, which is formed first, establishes a head-to-tail direction.[89] In bacteria this major axis is determined as perpendicular to the division plane. In the tiny worm *C. elegans* the anterior–posterior axis of an ovum is determined by the position of entrance of the sperm. This marks the posterior end.[90] In higher animals the axis, which is also known as the **animal–vegetal axis**, is established by uneven distribution of materials that include mRNAs and proteins in the unfertilized ovum. During embryonic development of bilaterial species two other axes, the **dorsal–ventral** and **right–left** axes, are also developed and help to establish the body plan. Thoughout development polarized movements allow cells to intercalate between one another to help shape the body.[90a]

Early studies of simple organisms such as Hydra (Fig. 1-13) and planaria (flatworms, Fig. 1-14A) showed that distinct chemical differences can be detected along the anterior–posterior axis. These organisms can be cut into pieces, many of which can regenerate complete individuals.[91,92] Regions near the head regenerate most readily. These and other observations led to the concept of gradients of diffusible **morphogens**, or form-giving molecules.[9,10] In *Drosophila* eggs an mRNA specified by the gene *bicoid* is localized at the anterior pole. The translation product, the bicoid protein, diffuses through the embryo, which in *Drosophila* lacks cell walls at this stage (see also Section C,4).[9,93–95] Bicoid is a transcription factor and also one of a number of established morphogens. Many other morphogens are members of the **TGF-β**

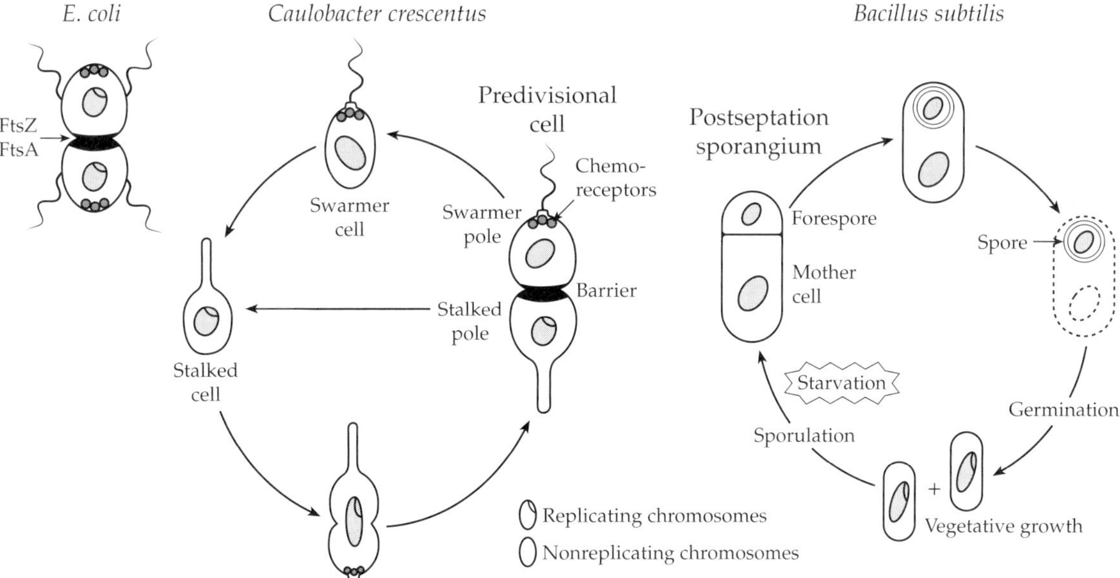

Figure 32-1 Comparison of cell division in three species of bacteria. *Escherchia coli* divides symmetrically after forming a septum in a plane marked by a ring of FtsZ (tubulin-like) and other cell division proteins. *Caulobacter crescentus* divides asymmetrically to give one flagellated swarmer cell and one stalked cell. *Bacillus subtilis*, under starvation conditions, divide to form a mother cell and a forespore. The latter is engulfed by the mother cell, which promotes its conversion to a resistant spore. From Shapiro and Losick.[83] Courtesy of L. Shapiro.

(transforming growth factor beta) family. Among them are proteins that establish the dorsal–ventral axis (Section C,4)[86,96–98] and also bone morphogenic proteins (p. 443). Retinoids also appear to act as morphogens.

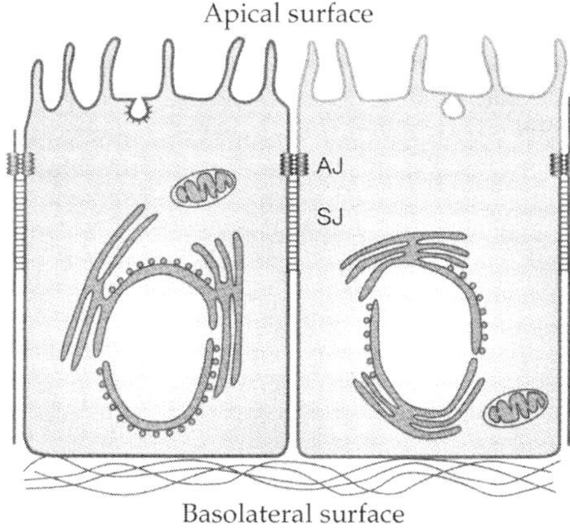

Figure 32-2 A pair of epithelial cells of *Drosophila*. The apical surface (top), e.g., of epithelial cells of the gut, faces the external surface, while the basolateral surface (bottom) binds to a basal membrane. Adheren junctions (AJ) and septate junctions (SJ) are shown between the cells. From Peifer and Tepass.[84] Drawing by S. Whitfield.

5. Totipotency and Stem Cells

The cambium layer of plant stems (Fig. 1-16) differentiates continuously to form phloem on the outside of the cambium and xylem on the inside. At the same time, cambium cells are retained. Thus, at each cell division one daughter cell becomes a differentiated cell, while another remains the less differentiated cambium. This pattern of continuous differentiation from a line of **stem cells** with constant properties is found in animals as well as in plants. In the differentiation of cambium it appears that chemical signals obtained from the surrounding cells on either the inside or the outside of the cambium layer determine whether the differentiated cell becomes phloem or xylem. Sucrose, auxin, and cytokinins are all involved.

Cloning. Asexual propagation (cloning) of plants ordinarily occurs by virture of the ability of embryonic meristematic tissue to differentiate into roots and shoots. If isolated phloem cells or other more differentiated cells are cultured, the result is often the formation of a **callus**, a dedifferentiated mass of cells somewhat reminiscent of embryonic cells. Under proper conditions, e.g., in a coconut milk culture and in the presence of the correct auxin-to-cytokinin ratio, some carrot root phloem cells revert to embryonic cells and develop into intact plants.[99] This experiment provided proof that the differentiated carrot phloem cells contained a complete genome for the plant. Its nucleus is **totipotent**, able to generate all cell types. However, the experiment cannot be done easily with most plants, and dedifferentiation is not always automatic. It does occurs often enough to establish the totipotency of the nucleus of many differentiated cells.

Similar considerations apply to animal cells. In the earliest stages differentiation is readily reversible. Later it becomes difficult to convert a differentiated animal cell into one resembling an embryonic cell. However, Gurdon demonstrated that this is sometimes possible. Nuclei from cells of intestinal epithelia and other tissues were substituted, by transplantation, for the nuclei of egg cells. The process is called **nuclear transplantation**[100–101a] or nuclear transfer cloning.[102] The resulting eggs in some cases grew into adult toads. Thus, the full genetic information of the toad was present in the differentiated cells.[103,104] However, it was not possible to accomplish this result with nuclei of neurons, which may have undergone irreversible differentiation. More recently mammalian nuclei have been utilized in the same way to create the famous sheep Dolly as well as mice, calves, pigs, and kittens.[105–108] These animals are commonly said to have been **cloned**, a term that has long been used to denote asexual propagation, e.g., in a colony of dividing cells or in propagation of plants by grafting.

In nuclear transplantation it is the DNA that is hoped to be the same in every individual in a clone. However, the ovum used for the transplantation contains mitochondria. Some mitochondria may also accompany the nucleus during the transfer. If the donors of the ovum and of the nucleus are different individuals the offspring will be mitochondrial hybrids.[105] In addition, there are questions about the methylation state of DNA in the donated nucleus and about the age and health of the donated mitochondria. That these questions are significant is emphasized by a bit of 3000-year-old knowledge from mule breeders: a mare crossed with a donkey yields a mule but a stallion crossed with a donkey yields a hinny, which has shorter ears, a thicker mane and tail, and stronger legs than does a mule.[109] There are worries because Dolly and many other animals produced by nuclear transplantation have not been completely healthy.[107b,110] Is something missing from the transplanted DNA or does it carry something extra, such as methyl groups? Recently it has been recognized that incorrect epigenetic marking of cytosine in CpG pairs that control maternally imprinted genes, especially those on chromosomes 11 or 15, may cause death of embryos or devastating human diseases.[25,111] An important related question for those wishing to clone an animal by nuclear transplantation is "Should the cell that donates the nucleus be in the G_1 state of the cell cycle (Fig. 11-15) or the G_0 or paused state that precedes G_1?"[105] See also Chapter 27, Section B,6.

Stem cells. For many years it has been appreciated that, as shown in Fig. 31-2, both erythrocytes and other blood cells arise throughout life from self-renewing stem cells in the bone marrow.[13,112] Stem cells are also needed for renewal of bones, muscle, skin, neurons, etc. Stem cells appear to be present only in small numbers and in well-protected special **niches** in the body.[53c,113–115] They are able to live throughout an individual's lifetime, dividing quite rarely and always producing one or more highly differentiated cells as well as a new stem cell.[113] A fertilized egg (zygote) is totipotent, able to generate all the cells of an animal including those of the placenta and other tissues that are not part of the embryo. However, the most capable stem cells are **pluripotent**, able to form more than one type of specialized cell.[113,116] Mammalian pluripotent stem cells include tumor cells, **embryonic stem cells**, derived from preimplantation embryos, and **embryonic germ cells**, derived from the primordial germ cells of the postimplantation embryo.[117] These germ cells are not only totipotent but, with good luck, may be immortal.[118]

Recent results indicate that adult-derived somatic cell nuclei may still retain full pluripotency.[107a,117a] Some confusion has arisen because of the discovery that stem cells may sometimes fuse with differentiated cells.[118a,b] It is only recently that it has been possible to locate and to cultivate human stem cells. These cells, which may be recovered from both embryonic and mature tissues, include the blood-cell-forming **hematopoietic** stem cells, fetal **neuronal** stem cells, **melanocyte** stem cells, and **mesenchymal** stem cells (or marrow stromal cells). The last give rise to muscle, bone, cartilage, and tendons.[119–121a] Most stem cells may arise late in development and function principally in tissue renewal.[122] Among the most abundant are those of epthelial tissues, whose cells provide 60% of differentiated tissue types in the mammalian body.[123,124] Epidermal stem cells must provide for regular replacement of the outer skin surface (Box 8-F) but must also provide cells for rapid repair of wounds[125] The exact locations of epidermal skin cells have been difficult to find. The cells appear to be well-protected in areas deep in the skin. Some are located in hair follicles.[120a,125] Stem cells of plants are present in specialized structures called **meristems**. A seedling typically has two meristems, at the tips of the shoot and root, respectively.[126] See Fig. 32-8B.

Cloning of human stem cells is of great medical interest because of the possibility of replacing defective cells or tissues. Tissue engineering may supply urgently needed differentiated cells for replacement purposes[107,127–128a] and may eventually lead to replacement organs.[129–131] These efforts must be pursued with caution, but most researchers see a bright future for cloning of tissue cells.[128,132–134] At the same time there is nearly universal agreement that nuclear transplantation cloning of human beings should not be attempted.[132,135] One key objection is the near certainty that many seriously defective human embryos would be created.

How can we explain a pattern in which one daughter cell resulting from division of a stem cell undergoes differentiation while the other remains a stem cell? A hypothetical way in which this might happen is illustrated in Fig. 32-3. It depends upon methylation of CpG sequences in DNA (Eq. 27-2). A methyltransferase, E_1 in Fig. 32-3, would modify a site in the DNA that controls differentiation by methylating a base in one of the two strands of a palindromic sequence such as a CpG pair (Eq. 27-2). The maintenance methylase E_2 would further methylate the corresponding cytosine in the palindrome in the second strand. During tissue growth replication would produce a hemimethylated CpG in each daughter cell. These would also be methylated by E_2 (upper part of Fig. 32-3) to complete the replication process. In contrast, a stem cell would need to mark one DNA strand, e.g., by action of a third methyltransferase (E_3) or some other type of DNA-modifying enzyme. As depicted in Fig. 32-3, E_3 would add a methyl group to DNA on some location outside of the palindromic region. Replication in this case would yield one hemimethylated palindome, which would allow one daughter cell to follow the normal growth and replication cycle. However, the other daughter cell would carry the mark designating it as a stem cell. The presence of both E_1 and E_2 in cells would lead to the continuous differentiation of the modified cells from unmodified ones, the situation found in stem line cells. A "maintenance methylase" with the properties of E_2 has been identified (Section A,1).

One question that has been asked is whether the 200 cell types of the vertebrate body all arise as a result of chemical interactions between cells and hormones and other external signals? Alternatively, does a **developmental clock** count the number of cell divisions and at the appropriate time turn off one set of genes and turn on another?[136] Methylases as well as other enzymes might modify DNA at specific times during development. For example, a hydrolase might deaminate the adenine in an AT pair to inosine. Upon replication and cell division one daughter cell would receive an unaltered DNA molecule, but the other would contain in place of the AT base pair an IC pair. Following a second replication, a GC pair would be formed (Eq. 27-20) resulting in AT to GC mutation at a specific site in the DNA of some of the daughter cells. Such a simple change, occurring in response to an enzyme formed at a certain stage of development, could alter the expression of genes in a cell. Schemes involving palindromic sites and modification enzymes that could turn off specific genes after a given number of cell replications have been suggested.[136] In fact, there seems to be little support for such mechanisms.

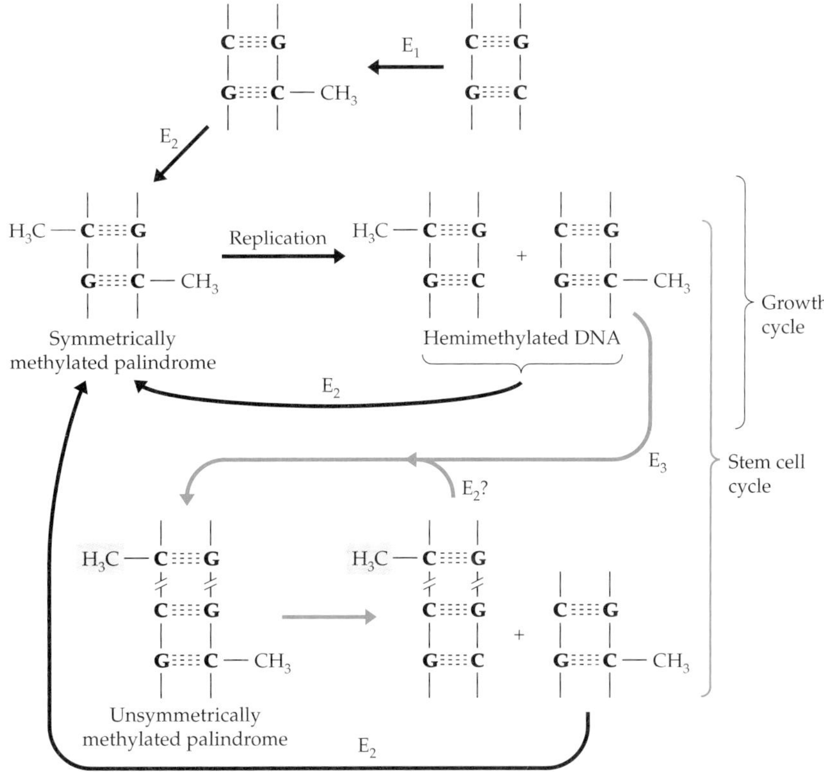

Figure 32-3 Hypothetical way of controlling stem cell replication by methylation or other marking system. Methyltransferases E_1 and E_2 methylate the cytosine in a 5'-CpG-3' or other palindromic sequence. In freely replicating cells these two enzymes keep the CG sequences methylated on both DNA strands. In stem cells another enzyme, perhaps a third methylase (E_3), marks a location outside the palindromic DNA on one strand ( ). Replication leaves the mark in the duplex, which is retained in a stem cell. The other strand will yield a hemimethylated duplex allowing the cell to follow the normal growth and replication pattern. Based on proposals of Holliday and Pugh.[136]

Transcriptional controls (Section 2) and the sensing of a cell's location may provide adequate control.

If methylation and possibly other covalent modifications of DNA occur, how can one explain the totipotency observed for some nuclei of differentiated cells? During development of the ovum and the sperm there appears to be a "resetting" of the developmental clocks that led to differentiation. At this time all of the methyl groups on the CpG pairs of imprinted genes are removed (Section C,2).[25,137] The mechanism is uncertain. Perhaps marking of newly replicated DNA stops. For example, in the case of the methylated DNA of Fig. 32-3 if E_1 and E_2 were absent in the cytoplasm of the ovum, no further methylation would occur during subsequent cell divisions. However, there may be active enzyme-catalyzed demethylation (see p.1541). Other mechanisms of gene silencing are also known (pp. 1881,1894).[138]

6. Apoptosis or How the Tadpole Eats Its Tail

Observers have long been fascinated by the rapid resorption of a tadpole's tail as it turns into a frog or toad. The process, designated by the Greek word apoptosis (whose "pop" is pronounced),[139] or as **programmed cell death**, plays a major role in many aspects of development in nearly all organisms.[140–142] For example, during human development about one-half of all the neurons generated die.[143,144] Unneeded lymphocytes, some of which produce antibodies or T-cell receptors directed against a person's self, are also killed. Cells may die accidentally from injuries. In many cases the resulting death occurs by **necrosis** rather than apoptosis. Necrotic death is accompanied by swelling and bursting of the cell and a subsequent inflammatory response.[145,146] In contrast, cells dying by apoptosis shrink, break into fragments, and are rapidly eaten by surrounding cells.[146a,b] There is no inflammation. Because of this it has been difficult to determine the extent to which apoptosis contributes to normal development. Apoptosis is also distinguished from **autophagy**, which is intracellular turnover under starvation conditions. Cells may need to scavenge unneeded proteins and organelles, recycling them within the cell.[147]

Cells damaged by disease, e.g., dopaminergic neurons in Parkinson disease, may die by apoptosis.[147a,b] A second form of self-destruction occurs when an axon is cut.[147b] Failure of the elaborate network of mechanism for repair of DNA and maintenance of the genome normally leads to apoptosis. In cancer essential steps in the apoptosis pathway are often inactivated.[147c]

Our view of apoptosis (outlined in Fig. 32-4) changed with the tracing of the origins and fates of all of the ~1000 cells of the nematode *C. elegans*.[13] During development of the adult worm just 131 specific cells die by apoptosis. Studies of mutant worms revealed mutations in several cell-death (*ced*) genes. Three proteins, encoded by genes *ced-3*, *ced-4*, and *egl-1*, are essential for apoptosis.[140,148] Somewhat surprisingly worms with defective Ced proteins are apparently healthy, even though they have 131 extra cells. On the other hand, in the fruit fly *Drosophila* mutations in similar death genes are sometimes fatal.[140]

The nucleotide sequence of the *ced-3* gene revealed that the Ced3 protein is closely related to the **interleukin-converting enzyme ICE**,[149–152] which is discussed on p. 619. ICE is a member of the **caspase family** of thiol proteases (p. 619). At least 14 different caspases are found in the human body. Some of them

TABLE 32-2
Some Components of Apoptotic Systems

Apaf1	Mammalian homolog of Ced4; component of apoptosome
Apoptosome	Cytosolic complex: Apaf1•caspase-9•cytochrome *c*
Bcl2	Mammalian homolog of Ced9 protein of *C. elegans;* inhibitor of apoptosis
Bcl-2 family	Group of regulators of apoptosis, both inhibitory and stimulatory (Bad, Bax, Bik, etc.)
CARD	
CD95 (AP-1, Fas)	One of the most studied death receptors
Ced3	*C. elegans* thiol protease, related to mammalian caspase9
Ced4	Activator of Ced3, related to mammalian Apaf1
Ced9	Inhibitor of Ced3 and related caspases
DD	Death domain of a death receptor
DED	Death effector domain
DISC	Death-inducing signaling complex, formed in plasma membrane
FADD	Fas-associated death domain, an adapter protein
ICE	Interleukin-converting enzyme, structurally related to Ced 3
TNF	Tumor necrosis factor (a family of cytokines secreted by macrophages)
TNFR	Receptors for a TNF family member

function in apoptosis[152a] and others in maturation of pro-inflammatory cytokines.[152–155] Most exist as pro-enzymes, which must be activated by proteolysis.[156] The mammalian homology of Ced3 is caspase 9.[153,157] The Ced4 protein of *C. elegans* is an activator for Ced3. Its mammalian counterpart is called **Apaf1** (apoptotic protease-activating factor 1).[153,158] Protein **Ced9** is an inhibitor of apoptosis, which probably protects the worm from erroneous deaths.[159] Its mammalian equivalents are proteins of the *Bcl-2* gene family.[142,152a]

It is well established that caspases participate in the final stages of apoptosis (Fig. 32-4), but what initiates the process? There appear to be many ways in which apoptosis can be triggered. If every cell has a proper location in the body, which is determined by signals from adjacent cells, what will happen if the cell becomes detached? There is evidence that such detachment with the loss of survival signals causes apoptosis.[53a,152b] Cell damage is also a major trigger. In other cases the cell is "instructed" to die. An example is the death of unneeded lymphocytes, one of many cellular processes induced by cytokines of the tumor necrosis factor (TNF) family. To allow for this process cells have surface receptors of the TNF receptor (TNFR) superfamily.[159a] Some TNFRs are **death receptors,** which are called by many names.[142,160–163] One of the best known is CD95[164a] (also called Fas[164] or Apo1). CD95 is involved in death of mature T lymphocytes at the end of an immune response and also in the killing of virus-infected cells and cancer cells by cytotoxic T cells or NK cells.

Members of the TNF family that activate CD95 (CD95 ligands or CD95Ls) are trimers. They bind to the cysteine-rich external domains of the transmembrane CD95 molecules inducing them to aggregate (Fig. 32-4). The cytosolic portion of each of these death receptors contain a **death domain** (DD). The bundle of aggregated receptors also bind to an adapter protein such as the Fas-associated death domain (FADD).[156] It is one of many proteins involved in apoptosis whose structures are known.[157,163] The FADD molecule contains a **death effector domain** (DED), which associates with a similar domain in the proenzyme procaspase 8. A rather large membrane-associated molecular complex, the **death-inducing signaling complex** (DISC; Fig. 32-4), is assembled in this way.[161,165] Oligomerization of the procaspase domain causes activation via self-cleavage to give active

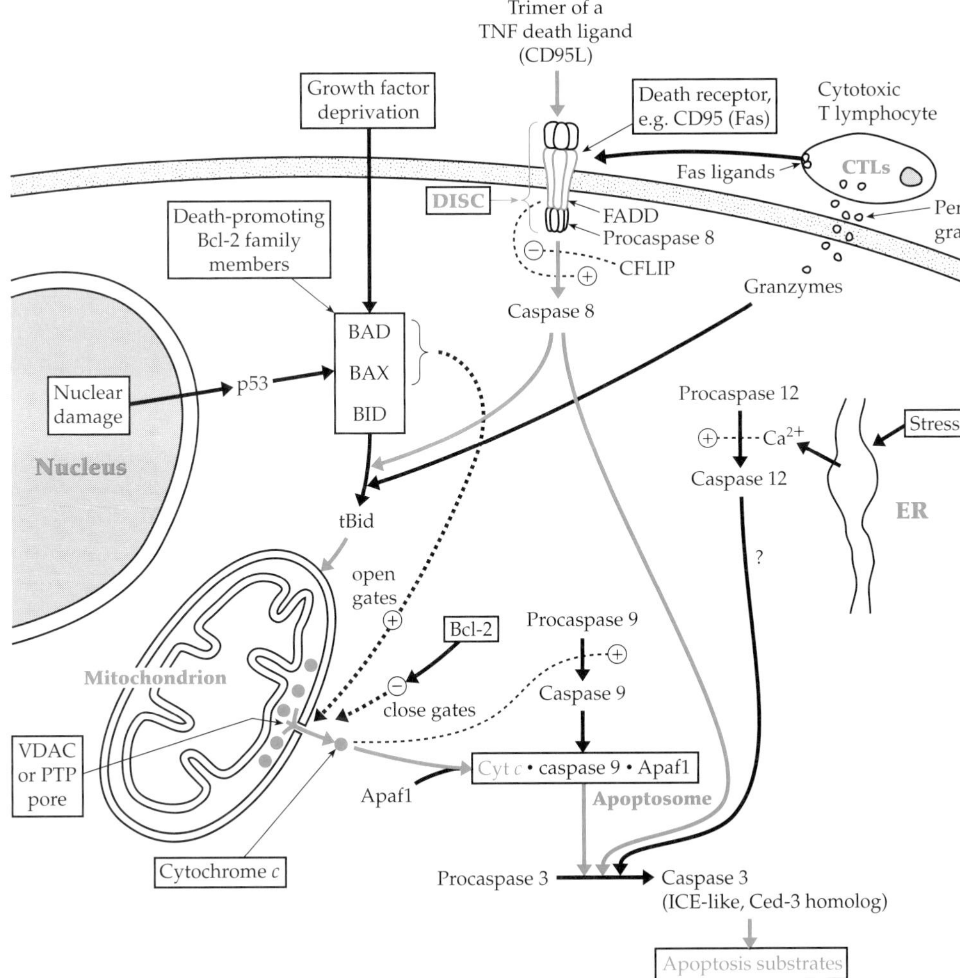

Figure 32-4 Sketch illustrating only a few of the many aspects of apoptosis in a mammalian cell. Emphasis here is on the death receptor pathways and cytochrome *c*-activated apoptosis. A third pathway is initiated by stress in endoplasmic reticulum membranes. In addition, attack by cytolytic T cells sometimes causes apoptosis by action of a granzyme on protein Bid or via a death receptor. Objects in scheme are not drawn to a single scale.

caspase 8, which initiates the apoptotic response. Other "upstream" caspases (caspases 2, 9, and 10) also participate in initiation of apoptosis. In contrast, caspases 3, 6, and 7, the "executioner caspases," are thought to participate directly in demolition of the cell.[154] A caspase-activated DNase also participates by degrading DNA.[166,167]

A second major pathway of activation of apoptosis depends upon mitochondria. Various stresses such as lack of needed growth factors, exposure to ultraviolet light, or other apoptosis-inducing signals apparently open pores or gates in mitochondrial membranes allowing materials that would promote apoptosis to flow out into the cytoplasm and stimulate the effector caspases.[167a] This possibility was supported by the discovery that cytochrome *c* stimulates apoptosis.[168] Cytochrome *c* is a small protein, which is present in inner mitochondrial membranes in a 1:1 ratio with other electron carriers. It is more mobile and less tightly bound than the other components. It carries electrons from complex III to cytochrome *c* oxidase within the intermembrane space (Fig. 18-5). As a result of apoptotic stimuli cytochrome *c* rapidly flows out of the intermembrane space into the cytoplasm both interfering with respiration and triggering other changes in the cell.[169–172] The outflow of cytochrome *c* may occur via the mitochondrial porin VDAC (p. 1047) or under some circumstances via the mitochondrial permeability transition pore (PTP; p. 1049).[173] Within the cytosol the escaped cytochrome *c*, together with caspase 9 and Apaf1, forms a large multimeric complex (cyt*c* • caspase 9 • Apaf1) called an **apoptosome**. The apoptosome catalyzes activation of caspase 3, initiating the caspase cascade.

Control of the gates or pores by which cytochrome *c* escapes from mitochondria is poorly understood.[171,171a,173] Whereas in *C. elegans* a single protein Ced9 has been identified as an inhibitor of apoptosis, vertebrate animals have a large family of proteins that are related to the Ced9 homolog Bcl-2.[157,174] Of these Bcl-2 and Bcl-x_L *inhibit* the flow of cytochrome *c* out of mitochondria, but several other members of the family, e.g., Bad, Bid, Bik, and Bax, *promote* apoptosis.[152a,175,176] Bad carries a signal that indicates a lack of growth factor stimulation. Bid carries a death message from QD95R and other death receptors. Bax carries a signal from p53 (Fig. 11-15) indicating unacceptable DNA damage. However, a truncated form of Bax may prevent apoptosis of neurons.[177] In every one of these pathways there are many complexities. In one of the best known pathways Bid is cleaved by caspase 8 to form a 15-kDa fragment t-Bid that becomes an integral membrane protein in the outer membrane of mitochondria. There it promotes the release of cytochrome *c* (Fig. 32-4).[178]

A quite different source of apoptotic signals are ER membranes, which respond to stress by releasing Ca^{2+} ions that activate caspase 12 (Fig. 32-4).[179] Yet another type of apoptosis is sometimes induced by granzyme B (p. 610), which is released from cytolytic T cells.[180–182]

B. Differentiation in Prokaryotic Cells and in Simple Eukaryotes

Every species undergoes developmental changes. Only a few of these will be considered here briefly.

1. Bacteria

Although they are usually regarded as unicellular, some bacteria develop more than one type of cell,[83] and some even form colonies with filamentous growth[183,184] or other distinct morphology.[185] Many bacteria alter their development in response to changes in environment. For example, unfavorable conditions lead bacteria such as *Bacillus subtilis* to form compact endospores inside the vegetative cells.[83,184] Many other bacteria including *E. coli* divide symmetrically. This fact also poses a question. How does a cell locate its center and divide? The answer is only partially understood. In all kinds of bacteria a protein known as **FtsZ** (filamentation temperature-sensitive protein Z), a GTP-binding protein homologous to eukaryotic tubulins (Fig. 7-33), is essential. Prior to division FtsZ accumulates as a **septal ring** at the center of the *E. coli* cell. Contraction of the ring is thought to be an essential step in cell division.[186–188] The FtsZ ring nucleates a growing complex of eight additional proteins known as FtsA, T, K, L, N, Q, W, and ZipA. While ZipA is not highly conserved among bacteria, in *E. coli* it is the first protein to add to the FtsZ ring.[187,189] ZipA is somewhat related to eukaryotic actin. Another group of proteins is also needed for location of the midcell plane. These are known as MinC, MinD, and MinE. A MinC•MinD complex inhibits potential binding sites for FtsZ. MinD is an ATPase, which structurally resembles the Fe protein of nitrogenase (Fig. 24-2) and appears to propel the MinC•MinD complex in an oscillatory fashion from pole-to-pole.[190–193] This behavior is not understood, but in some manner the 10-kDa MinE is able to overcome the inhibition and bind to FtsZ initiating division. Division in *E. coli* follows DNA replication by a constant time period (20 min at 37°C). The timing apparently depends upon diadenosine 5'-tetraphosphate (Ap_4A), which acts as a signal to couple division to replication.[194]

Bacteria with stalks. *Caulobacter crescentis* spreads to new areas while retaining a presence at home. As is illustrated in Fig. 32-1, asymmetric division produces two distinctly different cells. One, like the maternal cell, has a stalk, while the other has a rotary flagellum with which it travels. In time the motile "swarmer" cell sheds its flagellum and undergoes metamorphoses into a stalked cell. What controls this process? There is apparently a two-component system similar to the one that controls flagellar movement (Fig. 19-5). Two sensor histidine kinases phosphorylate a central response regulator **CtrA**.[195,195a,b] This represses initiation of replication in the swarmer cells as well as transcription of the cell division gene *ftsZ*. As mentioned on p.1094 changes in DNA methylation may also occur. Flagellar biosynthesis and other steps of differentiation then occur in the swarmer cells but not in the stalked cells in which normal growth and replication take place.[196–201] Action of proteases is also essential.[202]

Sporulation. Bacteria of the genera *Bacillus* and *Clostridium* form metabolically inert spores when deprived of adequate nutrients (Fig. 32-1).[83,203,204] Bacterial spores are remarkably resistant to heat and can survive boiling water for prolonged periods. Their metabolic rate is essentially zero, but they can be revived and can grow even after many years. For example, bacteria have been grown from a 118-year old can of meat. Some data suggest that spores can survive for no more than ~1000 years, but recent reports, not yet fully verified, say that spores have survived when embedded in salt crystals for 250 million years.[203,205] At the onset of sporulation the synthesis of ribosomal RNA is turned off completely, and new classes of mRNA are made. More than 50 genetic loci are affected by mutations that cause spore formation. As was mentioned in Chapter 28, one or more specific forms of the 70-kDa σ subunit of RNA polymerase (Chapter 28, Section A,2) are produced and direct the initiation of the new mRNA molecules encoding new proteins.[204,206–209] Prior to asymmetric cell division the first of the new σ factors, σF, is formed together with two regulatory proteins, Spo0A and Spo0B. Spo0B, a protein kinase, phosphorylates Spo0A, inactivating it.[210] SpoAB also forms an inactive complex with σF. After asymmetric cell division σF remains inactive in the mother cell but is released in the prespore by action of Spo0E. This is a membrane-bound phosphatase, which dephosphorylates Spo0A-*P*, allowing it to form a complex with Spo0B with release of σF. Another protein, SpoIIIE, appears to direct one copy of the replicated DNA into the forespore.[211] The σF factor then directs the transcription of genes in the forespore. In contrast, σE is produced only in the mother cell.

One of the most striking metabolic changes in metabolism during sporulation is the accumulation of large amounts of dipicolinic acid (Fig. 24-14). This requires the appearance of at least one new enzyme. In addition, as the spores develop the bacteria take up large amounts of Ca^{2+} and substantial concentrations of Mn^{2+} and other metal ions. In many bacteria 3-L-sulfolactic acid is also formed.

$$^{-}OOC - C(H)(OH) - CH_2SO_3^{-}$$

3-L-Sulfolactate

These components account for the following percentages of the total dry weight of spores of *B. subtilis*: dipicolinic acid, 10%; sulfolactic acid, 3–6%; Ca^{2+}, 3%; and Mn^{2+}, 0.3%. It is often suggested that the dipicolinic acid and other ions protect the proteins from denaturation. However, the heat resistance may arise from the maintenance of the core of the spore in a highly dehydrated state.[212] When conditions become appropriate for growth again, the spore germinates, and the bacterium again follows the cell growth and division program.

More complex alternative developmental programs are followed by colonial forms of bacteria such as the myxobacteria. The life cycle involves aggregation of cells and formation of fruiting bodies as well as sporulation.[185]

Signaling among bacteria. Even bacteria respond to signals from other bacteria. Individuals of a single species often react by secreting pheromones called **autoinducers** using a process called **quorum sensing**. Among the responses are swarming of cells, emission of light by luminous bacteria, synthesis of antibiotics, and formation of biofilms. As mention on p.1758, autoinducers used by gram-negative bacteria are often *N*-acetylhomoserine lactones.[213–215] A furanosyl borate diester (see Box 11-F) may be a more nearly universal autoinducer.[216] Programmed cell death can also be observed among bacterial populations.[217]

2. Yeasts

The budding yeast *Saccharomyces cerevisiae* and the fission yeast *Schizosaccharomyces pombe* are the best known fungi. Although they usually grow as individual cells, they can grow with a filamentous form under some conditions.[218] Other yeast, notably *Candida albicans*, are important pathogens and can also grow in either yeast or pseudohyphal filamentous forms.[219,220] Like *E. coli*, *S. pombe* undergoes symmetric cell division. However, the strong β(1→3) linked glycans with their β1→6 crosslinkages, mannose polysaccharides, and chitin provide a cell wall very different from those of bacteria.[221] Cells of *S. pombe* grow mainly at their tips and begin early in mitosis to form a ring of actomyosin and other proteins at the center. This corresponds to the mammalian contractile ring. At the end of anaphase the ring contracts, and the septum that separates the two cells develops.[222–223a] In both types of yeast the septum is rich in chitin, which is secreted from the cell membrane or from vesicles known as chitosomes. After the septum is fully formed and thickened, a chitinase partially hydrolyzes the chitin releasing the cells. In *S. cerevisiae* the cell division is asymmetric.[5,221] The position of the bud seems to be directed by the actin cytoskeleton.

All fungi form spores (gametes) during their haploid stage, which follows meiosis (Fig. 1-10). The transcriptional program for *S. cerevisiae* involves at least four sets of genes, which are transcribed consecutively. During spore formation the mRNA levels of more than 1000 of the ~6200 protein-encoding genes are changed. About 50% are elevated and ~50% are depressed.[224] The mating type changes in the haploid state have been mentioned on p. 1574. Similar mating and sporulation pathways are observed for *S. pombe*.[225]

3. The Cellular Slime Molds

The life cycle of *Dictyostelium discoideum* is described briefly in Box 11-C. About 10^5 individual amebas aggregate to form a moving "slug" in response to the chemoattractant cAMP. Some other species of *Dictyostelium* are attracted to a folic acid derivative or to the ethyl ester of *N*-propionyl-γ-L-glutamyl-L-ornithine-δ-lactam.[226]

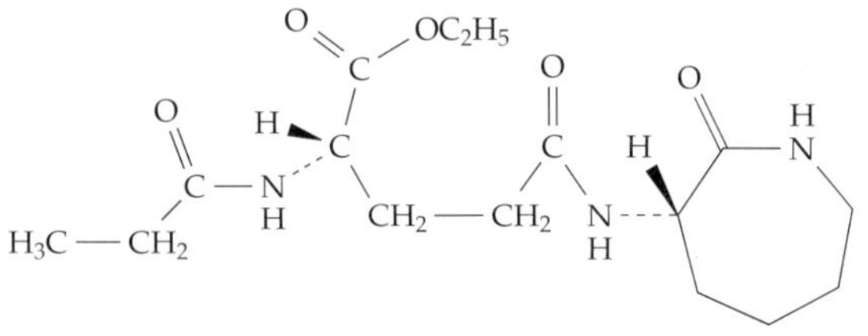

N-Propionyl-γ-L-glutamyl-L-ornithine-δ-lactam

In all cases the cells also utilize cAMP as an internal second messenger. For *D. discoideum* the components of the chemotactic-aggregation system include a 41-kDa cAMP receptor on the outside, adenylate cyclase, an extracellular diesterase that specifically hydrolyzes the cAMP to AMP, and a diesterase inhibitor protein.[35,227–230] The inhibitor keeps the phosphodiesterase largely inactive initially, but when cAMP concentrations build up synthesis of the inhibitor is repressed and the cAMP is hydrolyzed, a necessary condition for retaining sensitivity of the receptors for the arriving pulses of cAMP.

The slug of aggregated amebas continues to move and to undergo differentiation into two cell types: about 80% of the cells become pre-spores and the remaining 20%, which are at the "head" of the slug, become pre-stalk cells. The front-to-back gradient of cAMP within the moving aggregate seems to be involved in differentiation. However, another "differentiation-inducing factor" as well as NH_3 may be involved in the formation of stalk.[230] As the aggregate forms, the cells become cohesive, an 80-kDa surface glycoprotein being involved.[231] Later other adhesive **discoidins**, 24- to 27-kDa RGD-containing galactose-binding lectins,[232] also participate in holding the colony together. Some cells begin to produce cellulose. Trehalose is also formed and is stored in the spores. New enzymes have to be made to synthesize these materials. An alternative developmental pattern for some strains of *Dictyostelium* is formation of macrocysts between cells of two different mating types. A diffusible 12-kDa inducing factor appears to be released by cells of one strain.[233]

4. The Hydra

A well-fed hydra (Fig. 1-13) appears immortal. Its body cells are sloughed off and replaced at a steady rate so that within a month or so its body has been completely renewed.[35] The hydra contains only ten cell types. These include two kinds of stem cells that give rise to the ectodermal and endodermal cells of the body wall as well as small **interstitial stem cells** (Fig. 1-13) that differentiate nerve cells, germ cells, and the nematocytes or stinging cells. Of the ~10^5 cells in a hydra about 3600 are interstitial stem cells. Each day they generate 400 nerve cells and 1800 nematocyte precursor cells as well as 3500 new interstitial cells. The nematocyte precursors move up the body of the hydra and take up residence in the tentacles. Their movement is thought to be guided by chemotaxis. The head activator peptide (Table 30-5), which was identified following isolation from 3 x 10^6 hydras (3 kg),[234] diffuses from the foot end of the animal forming a gradient. A foot activator may diffuse from the opposite end. The interstitial stem cells of hydra also give rise to clones that develop into the gametes. Female hydra always develop female gametes, but stem cells of male hydra give rise to both male and female gametes.[235] This sex switching is reminiscent of the mating type variation of yeast.

5. Cell-Constant Animals

While the hydra is almost immortal as a result of the continuous differentiation of its stem cell lines, other small invertebrates follow a very different course of development. Both the rotifers and the annelid worms (Fig. 1-14) tend to have a constant number of cells in the adult body. The entire developmental program is specified genetically in strict detail.

The one millimeter long adult nematode *Caenorhabditis elegans* contains only 959 somatic cells. The lineal descent of all of these has been traced.[236–240] The development follows an almost exactly defined pathway with 113 programmed cell deaths during formation of the 558-cell newly hatched larva. In addition, each adult worm contains 302 neurons that make about 8000 synapses. This little nematode also has an alternative developmental pathway. The larvae shed their cuticles in four consecutive molts. If the food supply is inadequate, they enter a persistent nonfeeding state in which they may survive for months and are able to resume development when conditions are appropriate.[241]

C. Development of Animal Embryos

The shapes and body plans of animals vary enormously. Consequently, the study of embryonic development of sea urchins, insects, frogs, chickens, mice, and humans might appear to lead to quite unrelated conclusions. However, there are many similarities as well as variety.

1. Germ Cells and Gametes

Throughout the animal kingdom from protozoa to human beings sexual reproduction predominates. It is true that there are about 1000 species that reproduce asexually.[242,243] Among them are ~350 species of all-female rotifers[242] and even a species of tiny mites, all of which are haploid females.[244] Nevertheless, sex seems to have conferred some advantage on most species. There are two thories that may explain this: (1) Sex brings different combinations of genes together, allowing especially favorable combinations to survive, when changing conditions make life difficult.[242] (2) Sex helps to remove deleterious mutations from a population.[243] A large fraction of human fetuses (at least 10–25%) contain an "incorrect" number of chromosomes and as many as 20% of oocytes are defective. In contrast only 3–4% of sperm are chromosomally abnormal. Female meiosis I appears to be highly error-prone.[243a] Abnormal fertilized eggs or embryos are eliminated later in development.

Sex determination. The sex of an individual is determined by the chromosomes. In humans and other mammals presence or absence of a Y chromosome determines the sex. However, in many organisms including *C. elegans* and *Drosophila* this is not true. Although *Drosophila* males like human males are XY, it is the ratio of the number of X chromosomes to the number of sets of autosome (A) that determines the sex. This is also true for *C. elegans*, which has no Y chromosomes.[244,245] Apparently because of the differing ratios of X:A in the two sexes, organisms utilize a variety of **dosage compensation** methods. In cells of human females only one X chromosome is active. In *Drosophila* the rate of expression of genes from the X chromosome is roughly doubled in males.[245–248] In *C. elegans* the expression from both X chromosomes of the (hermaphroditic) female is roughly halved.[245,246,249,250] The biochemistry underlying these processes is quite complex.

The mammalian Y chromosome. The basic plan of the gonads prior to differentiation is female. However, if a Y chromosome is present (or if genes from a Y chromosome have been translocated to other locations) testes develop and begin to secrete androgen as early as the 60th day of gestation. A male-specific DNA sequence, ***SRY*** (sex determining region Y), constituting the gene for the **testes-determining factor**, is located in the small arm of the Y chromosome (Fig. 32-5).[251–253] A small pseudoautosomal segment at the end of the short arm of the Y chromosome carries other genes and undergoes crossing-over during meiosis.[254,255] The *SRY* gene lies between this and the centromere. The SRY protein is a member of the HMGA subgroup of **HMG** DNA-binding proteins (p. 1535).[256,257] It binds tightly to the sequence AACAA(A/T)(G/C) broadening the minor groove of the B-DNA and bending the DNA by more than 70°.[252,258,259]

Both SRY and the related SOX proteins are critical developmental regulators.[260] In early fetal life the mammalian embryo contains an indifferent gonad, able to differentiate into either a testis or ovary. Adjacent to the gonad are two simple ducts, the Müllerian (female) and Wolffian (male).[261] In the male SRY acts in the developing gonads to induce differentiation into the Sertoli cells of the testis. In the mouse the *Sry* gene is active for only a brief period about ten days after fertilization. During that period cells of the genital ridge start to differentiate. In the absence of protein SRY they develop into the female follicle (granulosa) cells but in the presence of SRY into Sertoli cells.[252,262] This is, in part, a result of production of the **Müllerian inhibitory substance** (MIS), which induces regression of the Müllerian duct, and later production of testosterone. MIS is a glycoprotein of the TGF-β family. Binding of SRY to a site in the Mis gene promoter appears to be involved in activation of the Mis gene.[252] Recent evidence points to a role for both SRY and SOX proteins in pre-mRNA splicing.[260] At least 25 other genes are also involved in spermatogenesis in the mouse.[263] Many of these testis-specific genes have completely unmethylated CpG sequences.[264,265] For example, a cAMP-responsive element present in a promoter sequence for a testis-specific subunit of pyruvate dehydrogenase must be demethylated for

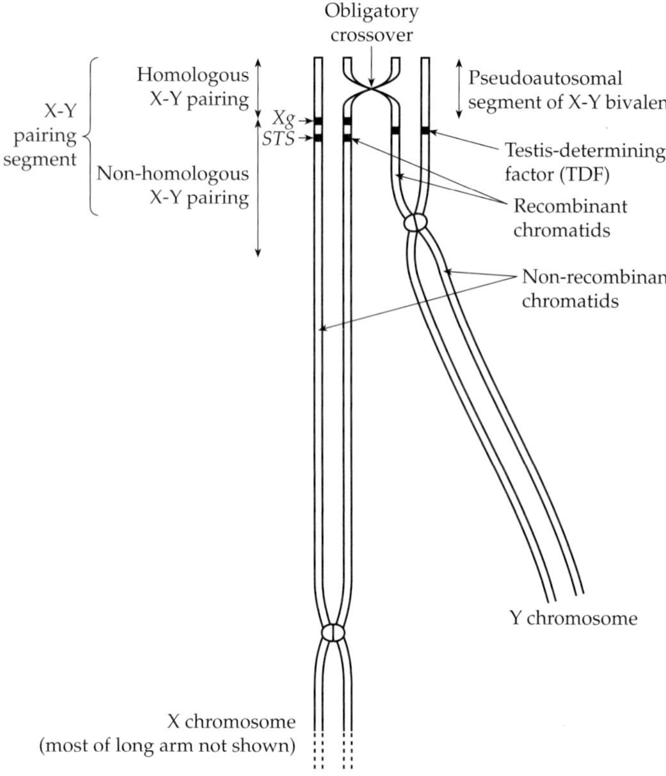

Figure 32-5 Schematic diagram showing crossing-over between the human X and Y chromosome. The pseudoautosomal segment is that part of the X-Y bivalent where there can be X-Y exchange by crossing over. X-Y homology in this segment is maintained by, and may be necessary for, this crossing over. There is always one "obligatory" X-Y crossover, whose position varies. The length of the X-Y pairing segment varies with meiotic stage and can extend well beyond the pseudoautosomal segment into the Y long arm. Much (perhaps all) of the synaptonemal complex formed outside the pseudoautosomal segment represents non-homologous pairing. From Burgoyne.[254]

transcription to occur. The developing germ cells interact with the surrounding Sertoli cells at every stage both through direct cell-cell contacts and via secreted signals.[266] Interstitial cells of the testis differentiate into Leydig cells, which secrete testosterone, promoting development of the Wolffian duct.[261] A small population of germ-line stem cells provide for continuing spermatogenesis. In *Drosophila* their self-renewal depends upon signals from special hub cells.[115,267] Other proteins needed for normal male development include the X-linked androgen receptor, whose absence causes testicular feminization, and dihydrotestosterone reductase (Chapter 22).

The development of spermatozoa is unlike that of somatic cells. Extensive reorganization of chromatin occurs under the direction of cis-regulatory elements that are controlled by cells of the testis.[267a] Among specialized proteins that are synthesized is a testis-specific polyadenylate polymerase.[267b] During the remodeling histones are replaced by arginine- and cysteine-rich protamines.[267a–c] In mammals this occurs in two stages. Small intermediate proteins (TP1 and TP2) replace histones in the first stage and are displaced by protamines in the second.[267a,d] Sulfolipids, which are also present in myelin, are essential to spermatogenesis.[267e] Both the sphingolipid 3-sulfogalactosylceramide and **seminolipid**, a sulfate ester of monogalastosylalkylacylglycerol (structure on p. 387) are present in large amounts. However, their functions are not clear.

Selenium plays a special role in development and protection of spermatozoa (Chapter 15). The selenoprotein **phospholipid hydroperoxide glutathione peroxidase** (PHGPx; Eq. 15-58, Table 15-4) has a high activity in the testis and in spermatids. However, in mature spermatozoa it forms an enzymatically inactive oxidatively crosslinked capsular material around the midpiece of the cell perhaps providing mechanical stability.[268] A similar 34-kDa selenoprotein is present in sperm nuclei and may be essential for condensation of DNA.[269] Sperm tails contain specialized cytoskeletal proteins which form "outer dense fibers."[270] In contrast to mammalian spermatozoa, nematode sperm move by ameboid motility that depends upon a specialized actin-like molecule.[271] Sperm cells are unusually rich in polyamines, most of which are bound to RNA and DNA (Chapter 24).

The X chromosomes. The phenomenon of X chromosome inactivation in mammalian female cells is closely related to imprinting, which has been discussed in Section 1. The inactivation process is quite complex. It involves methylation of 5'-CpG-3' sequences of DNA, as is described in Chapter 27, Section B,6. It also depends upon an **inactivation center**, the *Xist* gene, which is expressed only from the inactivated X_i chromosome, whose *Xist* DNA is unmethylated. On the X_a chromosome this DNA is methylated, and the gene is silent.[272] The *Xist* transcript is a long RNA that may bind to and coat much of the X_i

DNA.[19,21,273–275] The associated chromatin is enriched in a variant of histone H2A and is underacetylated on the tails of histones H2A, H3, and H4 (Fig. 27-4). Also noteworthy is the fact that not all genes on X_i are inactivated. As many as 19% escape this control.[276] Another gene *Tsix*, which is adjacent to *Xist* in the DNA, is also involved. *Tsix* encodes an RNA that is *antisense* to the *Xist* transcript[21,275,277] but is transcribed from the active X chromosome X_a. One hypothesis is that the *Xist* transcript causes X inactivation and that the *Tsix* transcript acts in an opposite way to favor activation of the chromosome. A transcription factor known as **CTCF** has been identified as a possible regulator of the inactivation process.[21,277] This is a trans-acting factor that is encoded on a chromosome other than X or Y. The process also depends upon methylation of histone H3.[278] Methylation of H3 may also be a factor in gene silencing in other organisms.[279] CTCF also regulates a number of other genes, e.g., those of the globin gene cluster (Fig. 27-10). It binds to 60-bp sequences, perhaps in enhancer elements.

2. Development of the Ovum

In the early mammalian female embryo the absence of the Müllerian inhibitory substance MIS permits continuing development of the Müllerian duct, while the absence of testosterone permits the Wolffian duct to degenerate. However, positive developmental signals are also required. Among these is the protein **Wnt-4**, a member of a large family of locally acting signal molecules (Section 4). Wnt-4 may be needed both for oocyte development and for further suppression of male development.[261]

The earliest studies of oocyte development were done with sea urchins (often *S. purpuratus*) and with amphibians (often the South African clawed toad *Xenopus laevis*) whose eggs are as much as 1000 times larger than those of mammals.[280] However, despite the differences in size, modes of fertilization, and ovary development, oocytes of nematodes, sea urchins, frogs, insects, and mammals have much in common. Oocytes of *C. elegans* and of most other animals undergo a temporary arrest in development at the prophase stage of the first meiotic division (Fig. 26-12).[13,281–283a] At this stage oocytes transcribe many genes. In some species chromosomes may develop a "lampbrush" appearance (Fig. 27-6) as a result of the transcriptional activity. Many mRNA molecules are stored in the expanding cytoplasm. Proteins are also synthesized and stored.[284] Among these are specialized proteins of yolk granules and proteins used to construct an outer egg coat. Surrounding **follicle cells** also contribute nutrients to the oocyte.[284a] In insects, whose early embryonic development has some special characteristics, 16 surrounding **nurse cells** are connected to the oocyte by cytoplasmic bridges.[285,286]

Oocytes may remain in arrest at the beginning of meiosis for prolonged periods before continuing through the **maturation** stage to form an ovum (egg). Women and other female mammals are born with thousands of oocytes, but only a few at a time develop into eggs. Maturation is often delayed until sexual maturity, when it is stimulated by hormones.[13] In *X. laevis* progesterone stimulates maturation.[282,287] In *C. elegans* and many other animals a signal from a sperm cell is needed to induce maturation.[281,283] Maturation of the oocyte is often arrested again, this time at metaphase of the second meiotic division. Transcription is halted, and protein synthesis is slowed. Fertilization then induces rapid completion of meiosis. Penetration of the sperm leads to "activation" of the egg and completion of meiosis. In lower organisms activation can often be carried out by chemical or physical treatment in the absence of a sperm cell, with formation of parthenogenetic offspring.

3. Fertilization

Fertilization of the egg is a biochemically complex process.[35,288] It involves recognition of sperm and egg, often in a species-specific manner.[289] The jelly layer around sea urchin eggs contains peptides such as the **sperm-activating peptide** (speract; Table 30-4), which stimulates increased respiration and motility of the sperm cells.[290] Additional chemotactic peptides may also be released from the jelly layer of invertebrate eggs. Chemoattractants for vertebrate eggs are less well known, but a 21-kDa sperm attractant protein from *X. laevis* egg jelly has been characterized.[291] Both in sea urchins and in mammals the jelly layer, which is called the **zona pellucida**, contains sperm cell receptors.[288,292–295] These are glycoproteins that interact with proteins (spermadhesins)[296] of the sperm cell membrane. One of these is the integrin-associated CD9, an integral membrane protein.[297,298] Penetration of the sperm through the zona pellucida often involves a large specialized secretory vesicle, the **acrosome,** as well as the enzyme hyaluronidase.[298a] In some species the acrosome releases a large amount of monomeric G actin, which polymerizes suddenly into a tube of polymeric F actin, which in some way assists the penetration of sperm.[35] In the horseshoe crab *Limulus polyphemus* the acrosome in an unactivated sperm cell contains a twisted bundle of as many as 120 cross-linked actin filaments. When the sperm is activated by contact with the jelly coat of the egg, the acrosome straightens into a 50-µm-long crystalline bundle, which is driven into the egg coat.[299] Of importance to all types of sperm cells are proteases and other materials that are also released from the acrosome and which help to etch a hole that allows the sperm to enter the

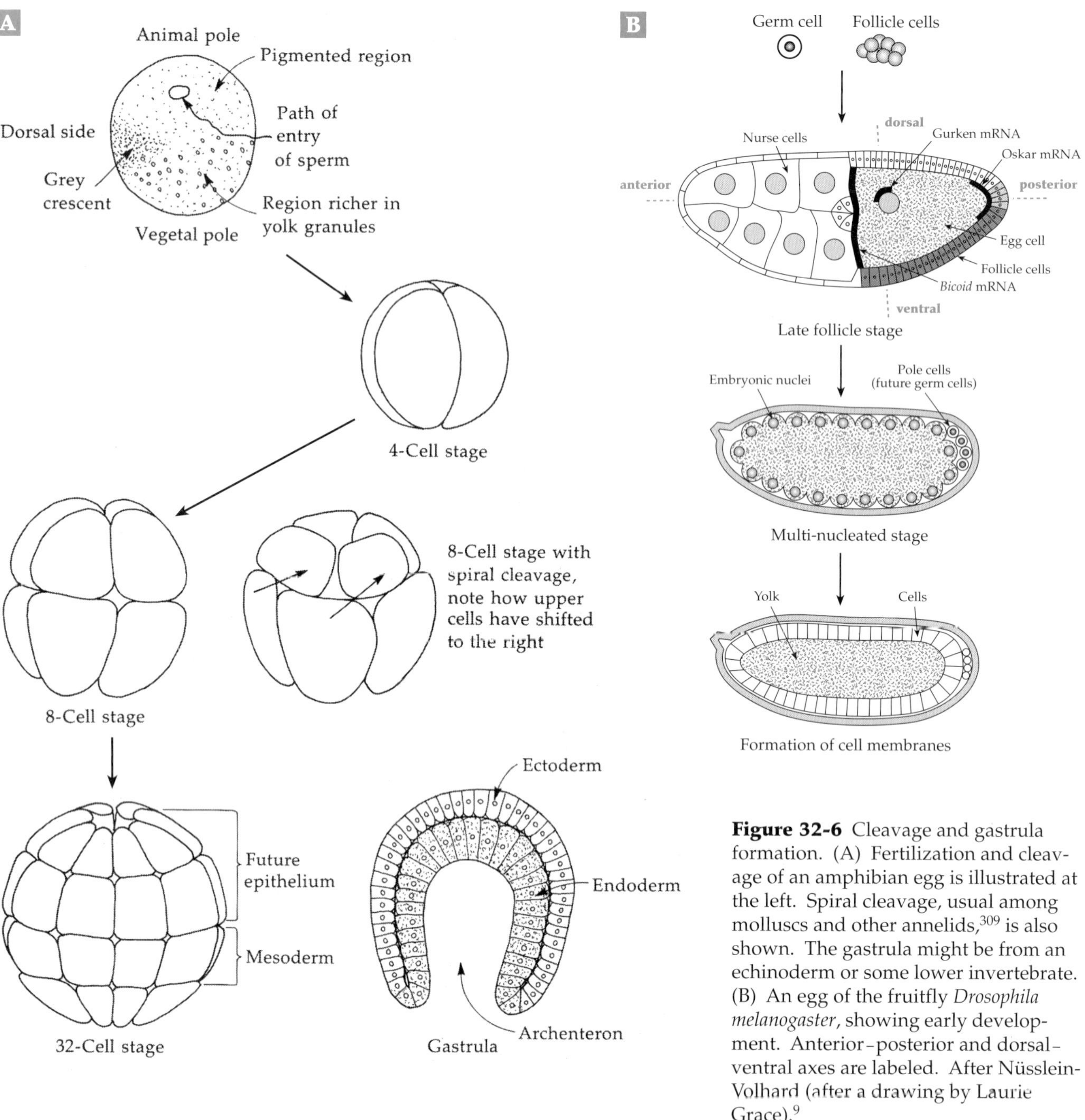

Figure 32-6 Cleavage and gastrula formation. (A) Fertilization and cleavage of an amphibian egg is illustrated at the left. Spiral cleavage, usual among molluscs and other annelids,[309] is also shown. The gastrula might be from an echinoderm or some lower invertebrate. (B) An egg of the fruitfly *Drosophila melanogaster*, showing early development. Anterior–posterior and dorsal–ventral axes are labeled. After Nüsslein-Volhard (after a drawing by Laurie Grace).[9]

ovum.[300,301] The acrosome reaction also activates the egg, a process that may depend upon NO.[302] In the sea urchin the acrosome contains a large amount of the protein **bindin**, which mediates a species-specific adhesion of the gametes and presumably fusion with the egg membrane.[302a]

Fusion of the sperm membrane with that of the ovum causes a rapid depolarization of the membrane of the ovum and an influx of calcium ions.[303,304] This causes an immediate block to the entrance of any other sperm cells in most species. It also causes the fusion of the membranes of **cortical granules** (several thousand in a mouse ovum) with the cell membrane and release of their contents.[305] The material released to the outside of the ovum includes various proteins and enzymes such as a peroxidase that catalyzes crosslinking of tyrosine side chains (Chapter 25) and hardens the material immediately around the ovum into a tough **fertilization membrane**. Within the ovum a respiratory burst resembling that of activated neutrophils (p. 1074) provides H_2O_2 to the peroxidase.

Fertilization also induces completion of meiosis

and formation of a one-cell embryo containing a maternal pronucleus contributed by the egg and a paternal pronucleus derived from the sperm. Each pronucleus undergoes DNA replication and then enters the first mitosis, which yields a two-cell embryo containing one diploid zygotic nucleus per cell.[283,306] Under the influence of the cyclin-dependent kinase cdc2 (see Eq. 26-3) and a hyperphosphorylated form of the protein **nucleoplasmin**, the chromatin of the compact sperm nucleus undergoes decondensation. The sperm basic proteins that coated the DNA are replaced by histones H_2A and H_2B.[307]

Although an amphibian egg is nearly spherical, there is already a strong polarity. The nucleus lies nearer to the **animal pole** than to the **vegetal pole**, which in many eggs is rich in yolk granules. In eggs of amphibians the animal pole is highly pigmented, but the vegetal pole is less so. On one side above the equator, there is a gray crescent. In some animals this marking appears on the opposite side of the egg from the point of sperm cell entry (Fig. 32-6). The gray crescent marks the future back (dorsal) side of the organism and the opposite part of the cell, the future ventral side.[280] The point of sperm entry also marks the ventral side for the mouse, a fact that suggests that the plan of development of mammalian embryos may be basically the same as that of frogs.[308] However, it is the internal components of the cell that actually determine the cell's axes. The cytoplasm of the mature ovum contains an unequal distribution of many materials with a well-developed bilateral symmetry. That this distribution is important is seen from the fact that centrifugation of eggs prior to fertilization often leads to formation of abnormal embryos because of displacement of preformed ribosomes and other materials. It is probably gradients in the concentrations of dormant mRNAs[284] and other metabolites that lead to uneven growth of cells and to the indentation of cells at the vegetal pole, a process that initiates the formation of the endodermal layer of the gastrula (Fig. 32-6). In insects the polarity of the developing ovum (the oocyte) is established by the cytoplasmic bridges from surrounding nurse cells, which are asymmetrically arranged (Fig. 32-6B).[285]

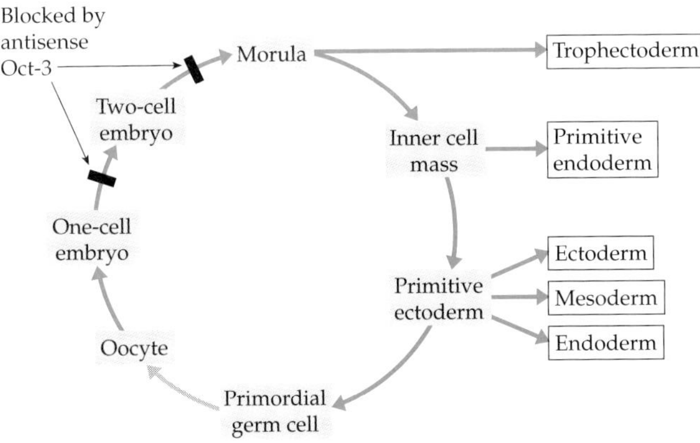

Figure 32-7 Expression pattern of Oct-3 mRNA during mouse development. The green boxes indicate those stages in which Oct-3 is expressed, while the white boxes at the right indicate all types in which little or no Oct-3 mRNA is formed. After Rosner *et al.*[314]

4. Embryonic Development

The fertilized (activated) ovum rapidly undergoes several mitotic divisions, known as **cleavage**, during which no overall growth occurs. The number of cells increases and the DNA replicates at each division, but the overall size of the resulting cell cluster is the same as that of the original ovum (Fig. 32-6). Further development leads quickly to a stage in which a layer of cells (called blastomeres at this stage) surrounds an internal cavity forming a **blastula**. In the sea urchin the blastula, which is released from its protective fertilization membrane ~11h after fertilization,[310a] consists of a single layer of cells In frogs and many other organisms there are two or more layers. In *X. laevis* about 4000 cells are formed in eight hours.[307,310] In mammals a solid cell mass (**morula**) forms first and is later transformed to a **blastocyst**, a hollow ball with an internal cavity.

Early mammalian development has been hard to study because of two facts: the ovum is very small, and a first priority is development of the placenta and of the layers of tissues that surround the embryo.[308,311,312] This occurs in humans within the first week after fertilization. Both the trophectoderm and cells of the inner cell mass (Fig. 32-7) contribute to the extra embryonic tissues.

Development of a mouse beyond the one-cell stage is dependent upon a regulatory gene of the *Oct* family (see p. 1631). The **Oct-3/4** protein, which binds specifically to the DNA motif 5'-ATTTGGAT, consists of two domains, both of which are essential for tight binding to this sequence. One domain is a 75-residue **POU domain** that consists of a helix-turn-helix motif with an amino acid sequence that is highly conserved among mammalian **P**it and **O**ct regulatory proteins as well as some **U**nc proteins of *C. elegans*.[313] The second domain is a 60-residue homeodomain (see p.1900).[313,314] Oct-3/4 appears to be essential not only for cleavage of a one-cell egg but also for progression from a two-cell to a four-cell egg (Fig. 32-7) and also in other embryonic cells.[315] Up to the two-cell stage very little transcription of zygotic genes is observed but further development requires zygotic genes and by the 8-cell stage protein synthesis from maternal mRNAs ends.[283] Oct-3/4 continues to be synthesized in the embryo and is necessary for establishment of pluripotent stem cells in the embryo.[315]

Development in *Drosophila* and other insects follows a somewhat different pathway, as is indicated in Fig. 32-6B. The egg, which is surrounded by follicle cells and 16 nurse cells, does not divide. However, its nucleus divides repeatedly, about once every nine minutes, to form ~6000 nuclei. Only then do separating membranes form to give individual cells.[9,285,316,317] During the first two hours the nuclei form a **syncytium**, in which they are embedded in a common cytoplasm, that allows free diffusion of signaling compounds. At first the nuclei are in the center, but later most migrate to the periphery and form a single layer of cells comparable to the blastoderm of amphibian cells. A few nuclei remain in the central cavity to become yolk cells, and some at the posterior pole become separated into pole cells.

The next stage in embryo formation, which occurs universally, is the invagination of the blastula at the vegetal pole to form a **gastrula**. At this stage the embryo has distinct ectoderm and endoderm cell layers. The cavity, formed in the gastrulation process and connecting to the outside, is referred to as the **archenteron** and is the forerunner of the gastrointestinal tract or enteron. Gastrula formation is more complex in the frog embryo and still more so in the human embryo. In all but the most primitive of animals a third layer of **mesosomal** cells is formed between the endoderm and ectoderm. These three **germ layers** differentiate further as follows. The **ectoderm** yields the skin and nervous sytem; the **mesoderm** the skeleton, muscles, connective tissues, and circulatory system; and the **endoderm** the digestive tract, lungs, and other internal organs and germ cells.

Organ development occurs largely by infoldings of cells from the endoderm and ectoderm. These infoldings appear to be induced by chemical substances secreted by cells of an adjacent germ layer. Thus, ectodermal cells form the **neural plate**, the prospective brain, and spinal cord in response to induction by mesodermal cells lying beneath the neural plate area. The mammary glands also arise from interactions of mesodermal and ectodermal cells, while the formation of the pancreas, liver, and lungs depends upon interactions of groups of cells from endoderm and mesoderm. Because of their transparency zebrafish are especially useful for study of organ formation.[318–321]

During the early stages of embryo growth, development seems to be directed largely by the polarity and gradients of the large amounts of mRNA, yolk and other constituents, which form a prepattern in the ovum.[321a] However, even at very early stages signaling from the nucleus of the ovum to surrounding cells is a necessary part of establishing the cell axes and developmental pattern.

The anterior–posterior axis. Only recently has it become possible to identify some of the specific mRNAs, the signaling molecules, and receptors that are involved in establishing the principal axes of the ovum (Fig. 32-6). Even for this aspect of development the genetic program is very complex, there are many uncertainties, and the details go far beyond the scope of this book. Much of the most important work has been done with *Drosophila* for which numerous mutants have been identified and characterized. Many names of genes and of proteins are derived from a description of the phenotype of a mutant fly. These same names are often used for the corresponding genes or proteins in other organisms. However, a protein may be known by more than one name, depending upon the species.

A *Drosophila* mother deposits mRNA for ~80% of all of her genes in the egg, but not all of the encoded information is used. Because genetic experiments can be done so readily with *Drosophila*, it is possible to ask what **maternal effects** do come from this mRNA and what effects come from the genes of the zygote.[9,88,316] For example, the *bicoid* mRNA that accumulates at the anterior end of the *Drosophila* egg is produced by nurse cells and is transported into the ovum. If the mother has defective *bicoid* genes the eggs die. A normal bicoid gene present in the father does not prevent death. Another maternal effect protein is encoded by *nanos*, whose mRNA accumulates at the posterior pole of the ovum. The maternal *torso* gene, which acts on follicle cells, is also needed for the anterior-posterior axis formation.[316] About 30 of these maternal effect genes are active in organzing the pattern of the embryo.

The *Drosophila bicoid* mRNA forms an anterior-posterior (A-P) concentration gradient, which controls early development along the axis. Another protein called **Staufen**, which forms a complex with the *bicoid* mRNA, assists in moving the RNA along on microtubules to form the gradient.[321b] The microtubules, together with associated transport proteins, are also essential to the prepatterning of the ovum.[9,88,285,322,323] Bicoid protein is a transcription factor that is synthesized at the sites of its mRNA accumulation. It is absent from the eggs when they are laid but soon appears.[316] Bicoid binds to the CRMs of target genes and activates them. Cooperative binding of multiple copies of the transcription factor may be necessary to provide the observed sensitivity to concentration. A transcription factor may bind at many sites on the DNA of the zygote or of nurse cells. For example, a gene called *even-skipped* (*eve*) is expressed in seven stripes in the blastoderm embryo. The enhancer that controls *eve's* second stripe contains at least five binding sites for Bicoid as well as others for the Hunchback, Giant, and Krüppel proteins.[44]

A different gradient along the A-P axis is formed by the **Nanos** protein, whose mRNA localizes in the cytoplasm of the posterior part of the ovum. Like the *bicoid* mRNA, *nanos* mRNA forms a complex with the Staufen protein. Messenger RNA transcribed from a gene called *oskar* is also necessary for development of the posterior region of the ovum. An additional gene, called *gurken*, is also involved in establishing the A-P axis. The nucleus of the ovum secretes *gurken* mRNA (Fig. 32-6), which is translated to the protein **Gurken**, a TGF-α-like protein that carries signals to follicle cells. They, in turn, help to organize the microtubules in the ovum. The nucleus of the ovum moves, sending the Gurken signal first to the posterior pole cells, then to a position that marks the dorsal side of the ovum. Thus, it participates in establishing both the A-P and dorsal-ventral axes.[324]

The animal-vegetal axis of sea urchin eggs is established during oogenesis, but the mechanisms are still unclear. After fertilization a distinct pattern along the axis is established by the 60-cell stage. Signaling from the vegetal pole appears to play an important role.[89] In the presence of lithium chloride, which is known to affect inositol triphosphate ($InsP_3$) metabolism (Fig. 11-9), embryos develop an excessive fraction of endoderm and mesoderm (tissues of the vegetal half) at the expense of ectoderm (tissues of the animal half).[89,325] Some data suggest that this effect implicates the **Wnt signaling pathway** and the protein **β-catenin**. This is the same protein that links C-terminal tails of cadherin to the actin cytoskeleton.[60] Its cytoskeletal and signaling functions seem to be controlled separately.

In the frog *Xenopus* the ovum accumulates an RNA (Vg1 mRNA) that encodes a growth factor of the TGF-β family (Chapter 30, Section A,6).[280] Initially present throughout the ovum, it moves and forms a gradient of concentration that is highest at the vegetal pole.

The dorsal–ventral (D-V) and right–left axes. Establishment of the D-V axis in *Drosophila* requires the participation of several genes. As mentioned above, *gurken* affects follicle cells. These cells cooperate with the ovum to set up a concentration gradient of the maternal gene **Dorsal**, the protein encoded by the *dorsal* gene.[9,280,326–328] Dorsal is a transcription factor related to NF-κB (Fig. 5-40). In the early embryo of the fly it forms a gradient in which it remains largely in the cytoplasm on the dorsal side but is mostly taken up into the nuclei on the ventral side. There it binds to a series of complex enhancers, each 300–1000 kbp in length, which act along the D-V axis. The enhancers interpret the concentration of Dorsal at five different threshold levels.[328]

The establishment of a second axis in vertebrate embryos is a complex process, which depends upon prior formation of a mesoderm layer. Development of mesodermal cells (not shown in the more primitive gastrula illustrated in Fig. 32-6) is induced by diffusion of a growth factor from the vegetal pole.[329–331] **Activin**, a member of the TGF-β family, has been proposed as the natural inducer of mesoderm. More recently **nodal**, a different TGF-β-like protein, has been suggested.[332] The fact that induction can occur through thin (20 μm) filters without any cell-cell contact indicates that specific chemical agents are responsible. Induction of the mesodermal layer in *X. laevis* appears to be an effect of an epidermal growth-factor-like protein. Additional factors are also needed to establish the D-V axis in vertebrates[326] in which the dorsal side is homologous to the ventral region in *Drosophila*.

Expression of another set of genes establishes the **right-left** axis and characteristic asymmetries of the body.[333–336] In the chick activin 2β, also a member of the TGF-β family, as well as Nodal and **Sonic hedgehog** (Shh) participate in control. The gene *Pitx*2 is a downstream transcription target for this signaling cascade in vertebrates.

Spemann's organizer. In 1924, the German physiologist Hans Spemann with Hilde Mangold transplanted a small piece of tissue from the dorsal lip of a newt blastula to a site on the ventral side of an early blastula of a differently colored species. The embryo developed a small secondary embryo, most of the tissues of which came from the host, not from the transplanted piece. It was concluded from this famous experiment that the transplanted vegetal tissues had supplied a diffusible inducer.[13,14,337–339] This morphogen caused the cells of the ventral surface of the blastula to secrete other morphogens. The signaling center in this part of the blastula surface is known as **Spemann's organizer**, or simply the organizer. It utilizes more than one inducer and a complex set of signaling interactions.[340] The cascade that induces formation of the organizer involves the Wnt–β-catenin and TGF-β pathways and the transcription factor **Smad4**.[41,341,342] The organizer secretes protein such as **noggin**, **folistatin**, and **chordin**, as well as nodal and other members of the TGF-β family.[332,343,344] They establish the D-V axis and also direct the development of the head and the initial patterning of the central nervous system. Noggin seems to be a neural inducer.[338]

Patterns, signaling pathways, and homeotic genes. While gradients in *Drosophila* eggs establish the anterior-posterior axis, products of other genes specify the developmental fates of cells of specific lineages and of cells found in particular spatial domains. Many *Drosophila* genes are needed to establish patterns, e.g., spacings of similar elements such as hairs, components of compound eyes, and whorls of plants.[345] Insects are organized into a pattern of consecutive segments from head to tail along the A-P axis.[13,14] In *Drosophila* there are typically 17 of these segments, some carrying appendages such as antenna, legs, and wings. Each segment develops under a different set of influences from neighboring cells. Development is controlled by ~30 **segmentation genes**, which determine the number of segments and their internal organization. Of these genes one set of at least six **gap genes**, among them *hunchback* (Hb) and *Krüppel*, are expressed first. Both of these genes encode zinc-finger transcription factors (see Fig. 5-38). The Hunchback protein acts mainly on head parts and upper thorax, while Krüppel influences development of the thorax. Also among the segmentation genes are eight "pair-rule" genes and at least 16 segment polarity genes. The interactions of the products of these genes creates a prepattern that provides positional information and guides further development.[346] Somewhat similar to the segmentation of the insect body is the development of skeletal muscle from a series of mammalian embryonic blocks known as somites.[13,347,348] In *Drosophila* signaling pathways involving the secreted proteins EGF, Decapentaplegic (Dpp),[349] Wingless (Wg), Hedgehog (Hh),[349,350] and Notch are used repeatedly to provide positional information. The names of these proteins describe effects of mutations on the limbs of the insect, but the proteins have a much broader significance. They represent an evolutionarily conserved set of intercellular signaling pathways.[346] Wingless is the first member of the previously mentioned Wnt family. Proteins of the Hh family, including the vertebrate Shh, control a large variety of processes that include development both of limbs and of the nervous system.[351–355] The Hh and Shh proteins all carry a molecule of cholesterol covalently bound in ester linkage to the C termini of the biologically active N-terminal domains of these proteins.[350,353] Defects in Shh signaling causes some human birth defects, and it is possible that drugs that inhibit cholesterol synthesis may have adverse effects on embryonic development.[352,355]

The *Drosophila* Notch 1 is a 300-kDa integral membrane protein that contains 36 EGF-like repeats. Its activation by proteolysis produces a 200-kDa N-terminal portion and a 120-kDa C-terminal fragment, which contains a transmembrane domain. The small intracellular domain of this fragment is then released by protease action and travels to the nucleus where it activates several target genes.[356–358] The Notch signaling pathway is conserved in all metazoans and influences many interactions that control cell fate during development.[359] The proteolytic cleavages of the Notch protein parallel those of both the ErbB-4 growth factor receptor (see Table 11-3) and the amyloid precursor protein APP that is pictured in Fig. 30-34. The same type of protease (γ-secretase or presenilin in the human brain) cleaves all three of these proteins.[359–361] Actions of Notch are modulated by posttranslational alteration, e.g., glycosylation by a fucosyltransferase encoded by gene *fringe*.[362,363] A homolog of *fringe*, called *lunatic fringe*, encodes an essential component of somite formation in the mouse.[348,364]

Functioning together with the signaling pathways are **selector genes** that determine which specific pathway is to be influenced. For example, eyes, antennae, legs, or wings of a fly may be selected.[346] *Antennapedia* (*Antp*) is one of these genes. Several of the *Drosophila* selector genes as well as some pair-rule genes[365] are also known as **homeotic genes** (see also Chapter 28, Section C,6). Homeotic genes were first recognized by the fact that mutation causes conversion of one segment of an insect's body into the homologous tissues of another segment.[366,367] For example, a mutation in *antennapedia* (*ante*) changes the antenna into a leg. Similar genes are also active in vertebrates, e.g., in the development of the chick embryo limb bud, the very tip (the last 20 cell diameters of length) contains cells that differentiate into the various elements of the limb in a relatively autonomous manner. If this *progress zone* from one limb bud is grafted onto the end of another limb bud, the bones and cartilaginous elements of the limb are repeated. Both the number and morphology of fingers and toes are determined by homeotic genes[367a] as is the formation of sphincters in the developing gut.[367b] Homeotic genes (also known as *Hox* genes) contain a conserved sequence of 180 bp that specifies a 60-residue protein **homeodomain** (also known as a homeobox). The homeodomain folds into a helix–turn–helix motif characteristic of many transcriptional regulators (Figs. 5-35, 28-3, p. 1631).[365,367–371] *Hox* genes are found among all forms of life. Hundreds have been described.[372] They include genes for the previously mentioned transcription factors of the Oct and Pit families as well as for the yeast mating type proteins MATa1 and MATα2 (p. 1880).[17] The *Drosophila* genome contains eight *Hox* genes, while the human genome has ~40, which are organized into Hox clusters.[373] Hox genes are also abundant in plants.[345,372,374]

Despite intense interest the role of homeodomain proteins in development is not well understood. The highly conserved motif binds to DNA at many places in the genome. Current thinking is that homeodomain proteins interact with other regulatory proteins, and that various combinations of these proteins provide the information needed to direct development.[375]

D. Specialized Tissues and Organs

Here are a few details about development of mammalian tisssues. We'll begin with the blood and connective tissues, which arise from embryonic mesodermal cells.

1. Blood Cells and the Circulatory System

Every second of life a human must produce about 2.5 million red blood cells, about 2 million granulocytes, and many lymphocytes as well as other less numerous leukocytes. All of these arise from **multipotential stem cells** found in the bone marrow.[376–379] Each of these stem cells divides to form one daughter stem cell and one **progenitor cell**.[380] The progenitor cells are also stem cells but have differentiated into **myeloid**,[381] **erythroid**, and **lymphoid**[382,383] cells. These differentiate further as is indicated in Fig. 31-2. Mature blood cells of most types have short lifespans and must be regenerated from stem cells continuously.

At all stages the differentiation process is regulated by the microenvironment which is rich in specific protein growth factors, several of which have been discussed in Chapter 30. Among the 20 interleukins, three stimulate growth of both multipotential progenitor cells and erythroid progenitor cells.[380,384] The acetylated tetrapeptide Ac-Ser-Asp-Lys-Pro inhibits stem cell proliferation. Granulocyte-macrophage colony-stimulating factors stimulate proliferation of both granulocytes and macrophages. The kidney cytokine **erythropoietin**, a 30.4-kDa glycoprotein,[385] is a primary regulator of red blood cell formation. Its action on a differentiated stem cell initiates massive hemoglobin synthesis and terminal differentiation of the erythrocyte. **Thrombopoietin** promotes formation of megakaryocytes and also their maturation and release of platelets to the blood.[386] **Thymopoietin** promotes early T-cell differentiation. Activated macrophages secrete interleukin-1, which stimulates maturation and proliferation of B lymphocytes. Interleukin-2 (T-cell growth factor; Fig. 30-8) is produced by activated T lymphocytes and is needed by T lymphocytes for long-term helper and cytotoxic functions. Differentiation of the stem cells into the erythroid lineage requires transcription factor GATA-1, development into erythrocytes requires GATA-2, while development into T lymphocytes requires GATA-3.[387]

Globin genes. The genes that encode the human globins from which hemoglobin is formed are found in two clusters, the α-like genes on chromosome 16 and the β-like on chromosome 11. They are developmentally regulated, different genes in the clusters being active at different stages of development. Mammalian hemoglobins (Chapter 7) each contain two α chains or two related ζ chains and two other chains, β, γ, δ, or ε. Adult hemoglobin is mainly $\alpha_2\beta_2$ but contains small amounts of $\alpha_2\delta_2$. In early embryos the hemoglobin is $\zeta_2\varepsilon_2$, but during the second to sixth weeks of embryonic life the two fetal hemoglobin chains γ^G and γ^A replace the ε chains. The switch from fetal to adult hemoglobin begins a few weeks before birth and is complete by about ten weeks after birth. The β-like gene cluster contains five genes encoding globins ε, γ^G, γ^A, δ and β (Fig. 27-10). Each gene consists of three exons separated by two introns and has rather similar control signals. These include CACCC at ~–100, CACA at ~– 92, CCAAT at ~– 75, and ATAAAA (TATA sequence) at ~– 30, as well as AATAAA (cleavage and polyadenylation). The CCAAT sequence appears twice in the γ^A promoter. A variety of transcription factors and chromatin modifiers influence the expression of these genes.[388,389]

Only a small amount of fetal hemoglobin ($\alpha_2\gamma_2$) is produced after infancy. There are two genes, γ^G and γ^A, for the β-like chains of fetal hemoglobin. A few adults make large amounts of fetal hemoglobin and this hereditary **persistence of fetal hemoglobin** has survival value for persons carrying thalassemia genes. This condition may result from a single base change in the CCAAT sequence found upstream (~75 bp) of the globin genes. Many other genes are also preceded by the same sequence, which in extended form is often TTGGPyCAAT. In one individual with persistence of fetal hemoglobin the first G in this sequence was replaced by A in one of the two CCAAT sequences present in the γ^A gene.

Thalassemias. In these important hemoglobin diseases the α or β chain either is absent or is present in far less than stoichiometric amounts.[14,390] About 40 point mutations in the β globin gene have been described among patients with β thalassemia, in which β chains are missing (β^0 thalassemia) or are present in reduced amounts (β^+ thalassemia). These mutations sometimes occur in control regions at the 5' end of the gene. For example, a change of the TATA sequence from ATAAAA to ATACAA causes decreased transcriptional efficiency and β^+ thalassemia. Other mutations result in abnormal splicing or in instability of the β globin. Deletions may result in the complete absence of the gene or in a frameshift that results in nonfunctional globin and β^0 thalassemia.

How do embryonic cells choose to transcribe only the embryonic globin genes? The decisions to switch from embryonic to fetal and from fetal to adult at appropriate times appear to be controlled by interactions with appropriately expressed transcription factors. Especially impressive is the total and permanent cessation of transcription of the embryonic globin ε gene at about the seventh week of gestation. Transcription of the β globin gene cluster is controlled by a powerful enhancer, the **locus control region** ~6–22 kbp upstream of the promoter.[391] In addition, ~270 bp upstream of the ε gene promoter is a **silencer**, a DNA sequence to which inhibitory proteins may bind and, in cooperation with the enhancer, may completely silence the ε gene while allowing transcription of other genes in the cluster.[392,393] The ζ gene, in the α globin cluster, is silenced by the binding of an NF-κB transcription factor to a 108-bp segment of DNA located 1.2 kbp to the 3' end of the gene.[394] The globin genes are shielded from action of nearby enhancers by **insulators**, DNA sequences that often contain CpG islands.[395]

Blood vessels: vasculogenesis and angiogenesis. Early in development of an embryo a network of blood vessels is formed from mesenchymal progenitor cells (vasculogenesis).[396–398] Later, in either the embryo or the adult, new capillaries are formed (angiogenesis). As the organism develops these new vessels are "pruned," and the vascular bed is remodeled to a tree-like form with vessels of both large and small diameter.[399,400] A key activator of angiogenesis is the vascular endothelial growth factor **VEGF**.[401,401a] However, a number of other proteins including **endoglin**, a TGF-β binding protein,[402] the clotting factor thrombin,[400,403] and the ribonuclease **angiogenin** (p. 648) exert their influences.[404] There are diseases that result from defective angiogenesis. However, a major interest in this process arises because of the essential role of angiogenesis in the growth of cancer[405] and as a complication of inflammatory processes. Angiogenesis in the retina is a major cause of blindness resulting from diabetes mellitus or from macular degeneration.[405,406] There are natural antagonists of angiogenesis,[407–409] and efforts are being made to utilize them in therapy. A related project is development of completely tissue-engineered blood vessels for surgical use.[410]

2. Cartilage, Tendons, Bone, Muscle, and Fat

Mesenchymal cells differentiate into cartilage, bone, muscle, adipose, and other connective tissues.[411] **Chondrocytes** synthesize the variety of collagens (pp. 431–426) that are needed for synthesis of cartilage and other connective tissues. The 32 or more genes encoding the polypeptides needed for synthesis of the 19 types of vertebrate collagens (Table 8-4) are developmentally regulated in a complex manner.[411] Their promoters contain TATA and CCAAT sequences as well as other presumed regulatory codes. The first intron of several collagen genes has also been identified as a control region containing enhancers.[412] The elastic fibers (p. 436) owe their properties in part to elastin. The control region of elastin genes lacks the TATA sequence but has SP1 binding sites. As with many other mammalian genes, a diversity of protein products, many in small amounts, are made by alternative splicing.

Collagen fibrils provide the scaffolding for formation of bone, whose composition is considered on pp. 440–443. Bone develops under the influence of **bone morphogenic proteins** BMP-2 to BMP-7.[413–415] Most of these are cytokines of the TGF-β family. Noggin (p. 1899) antagonizes the action of the BMPs.[344] A characteristic of bone is rapid remodeling (p. 441) by which ten per cent of skeletal bone is replaced every year. The balance between action of the bone-forming osteoblasts and the bone-resorbing osteoblasts is regulated by surface proteins responding to **c-Fos** and to **interferon-β**.[416]

Muscle, whose structure and function are discussed in Chapter 19, develops in response to four members of the myoD family. These include myoD, **myogenin**, **myf5**, and **MRF4**.[417–419] All are muscle-specific transcription factors of the basic helix-loop-helix class. An unusual aspect of muscle development is formation of multinucleate **myotubes** (muscle fibers; p. 1096).[420] Apoptosis plays an important role in muscle development and can present significant complications in damaged cardiac muscle.[421] Defects in several developmental control genes are responsible for congenital heart diseases.[422]

3. Epithelia

Epithelial tissues, which line both internal and external surfaces, arise from all three cell layers of the blastula. The epidermis (Box 8-F) arises from ectoderm, while the lining of the digestive tract is formed by endodermal cells. Mesoderm provides the linings of blood vessels. About 60% of differentiated tissue types in the mammalian body are epithelia.[123] Stem cells or progenitor cells are present and provide for renewal.[123,124] While epidermal stem cells are located in deep layers of the skin, **keratinocytes** are readily cultured *in vitro* and can give rise to fully differentiated multilayered skin.[423] Development appears to require transcription factors related to **Oct-2**[424] as well as **p63**, a homolog of the tumor suppressor p53.[425,426] Mice deficient in the aspartyl protease **presenilin 1** (which is defective in some forms of Alzheimer disease, Chapter 30) develop characteristic epidermal skin tumors. The β-catenin–Wnt signaling pathway (p. 1899) seems to be involved.[427] The gastrointestinal endoderm develops its highly convoluted surface under some control by the Notch signaling pathway.[428,428a] Endothelial progenitor cells (**angioblasts**) are responsive to many signaling molecules[429] including thrombin.[430] The *Drosophila* eye develops from the epithelium, again through signaling via Notch and other morphogens.[430a] Of outstanding importance to epithelial cells in general is their ability to form complex communicating junctions.[430b]

4. The Nervous System

Development of the vertebrate central nervous systems is initiated during gastrulation through an interaction between the dorsal ectoderm and an infolding of the dorsal mesoderm. Several different diffusible inducers are involved. These include noggin,[344] folistatin, and other members of the TGF-β family as well as thyroid hormones, basic fibroblast growth factor (bFGF), and sonic hedgehog.[329] The nervous system develops over a period of a few days with differentiation of precise numbers of neurons, astrocytes, and oligodendrites in successive waves. The order in which various cell types arise is determined by the order in which transcription factors such as Hunchback, Krüppel, and others are expressed.[431] Multipotential neural stem cells provide the new cells that are required.[432,433] Neural tissue from a region called the **neural plate** develops into a neural fold. The latter is closed to form the **neural tube**,[434] within which the notochord, the precursor to the spine, as well as the neurons, glia, and other cells grow. The **neural crest** forms as an outgrowth from the dorsal surface of the neural tube under the influence of inducers of the Wnt and BMP families.[434a] Cells migrate from this crest to form the peripheral nervous system, melanocytes, and cranial cartilage.[434b] The pituitary, a central component of the neuroendocrine system, develops from tissues from the midline part of the anterior neural ridge.[435]

The **floor plate**, which develops along the midline of the vental surface of the neural tube, is the source of Sonic hedgehog,[436,437] **netrin-1**, and other secreted molecules.[436] Some of these participate in **axon guidance** by which the growing tips (**growth cones**) of axons are able to connect to the correct "targets."[437a] For example, every visual receptor cell in the retina must send its signal to the correct locations in the visual cortex of the brain.[438] How can this be accomplished? Over a century ago Ramon y Cajal proposed that chemoattraction, analogous to chemotaxis of bacteria, might be involved.[439,440] A hundred years later Tessier-Lavigne and coworkers isolated the first of these attractants, **netrin-1** and **netrin-2,** from 25,000 pulverized chick brains.[441,442] Of these two closely related 75- and 78-kDa molecules netrin-1 is produced only in the floor-plate cells. Like the less well understood nerve growth factor (Fig. 30-7), netrins induce outgrowths of neurites and also are chemoattractants for nerve growth cones.[443] The netrin receptor is known as **DCC** (Deleted in Colorectal Cancer).[444] A nematode protein UNC-6 is related to the netrins.[445]

Growth cones are subject to both chemoattractant and chemorepellent effects of guidance molecules and also to attraction or repulsion resulting from cell–cell-contacts. To complicate the picture further, the netrins and also the brain-derived neurotrophic factor BDNF (see Fig. 30-6D) may first attract, and then after a period of adaptation or desensitization, repel a growth cone.[443,444,446–448] Consecutive phases of desensitization and resensitization may result in a zig–zag path of growth. The netrins were recognized first by observing growth of **commissural neurons**. These neurons originate within the spinal cord on one side or the other or the midline. They grow down toward the floor plate attracted by the netrin-1 or **BDNF** produced there. The neurons then cross the midline before turning and growing toward the brain. After crossing the midline the growth cones become insensitive to netrin but are repelled by a molecule (first recognized in *Drosophila*) called **slit**. Its receptor is appropriately named roundabout (**Robo**).[449] A similar receptor in zebrafish (called **astray**) is required for retinal axon guidance.[449] An important aspect of neuron guidance is apoptosis induced by misdirected growth.[450] The Notch receptor apparently participates in this decision in the mammalian CNS.[357] Positive signals for axon growth often involve the MAP kinase pathways, while inhibition may involve INS-2*P*–Ca^{2+} signaling.[451]

Chemoattractants that function in development of the cerebral cortex include several **semaphorins**.[445,452,453] A separate family of attractant and repellant compounds, the semaphorins have been identified in insects, chickens, and mammals. They play a role in regulation of communication between neurons. Because of the complexity of the brain, the study of growth of neurons with the brain is difficult. The $>10^{12}$ neurons each contact, on the average, 100 different cells. Some insight comes from mutant mice with names such as *reeler*, *scrambler*, *stargazer*, and *Yotari* (Japanese for tottering).[454] The single defective gene in these mice can be identified and studied. For example, *reeler* mice are defective in **reelin**, a large glycoprotein of the extracellular matrix (ECM).[366,379,455,456] The *reeler* phenotype can also result from mutation of the gene *disabled*-1, which encodes a cytosolic tyrosine kinase. Other mutations in mice implicate **VLDL** and **apoE** receptors (Chapter 21) in these developmental abnormalities.[379,455]

The *stargazer* mutant mouse is ataxic and epileptic. It lacks functional **AMPA receptors** (Fig. 30-1), which apparently are not delivered successfully to the synapses in the cerebellum in which they function.[380,386] Mutation of a transmembrane protein **stargazin**, which may interact with the AMP receptor, causes the symptoms.[457,458] **NMDA receptors** (Fig. 30-20) are involved in synapse formation in the brain. Filopodial extensions on dendrites, triggered by electrical activity, are essential for synapse formation,[459] which occurs rapidly.[459a] Activation of NMDA receptors is apparently also necessary.[379,460] Without this stimulation the excitatory glutamatergic neurons of the developing brain undergo apoptosis.

Why do neurons grow in the embryo but not in most parts of the adult CNS? Two proteins called **Nogo** were isolated from bovine brain. Their sequences were utilized in identifying the *Nogo* gene and three human isoforms of the protein.[389,461] The large 250-kDa Nogo-A is present both in myelin and in the endoplasmic reticulum. Both Nogo-A and the diffusible 35-kDa Nogo-B are inhibitory of neurite outgrowth.[462] This effect, as well as the crowding of regenerating neurons by the chondroitin and other matrix components,[396] may provide obstacles to nerve regeneration.[463]

E. Development of Green Plants

Green plants may have diverged from a common ancestor with animals ~1.6 billion (1.6×10^9) years ago. How do the genomes of present-day plants and animals compare? There are many similarities in basic metabolism. These arise from the intrinsic chemical properties and reactivities of cellular components and from the coevolution of plants and animals. Plants and animals also utilize similar structures and similar control of chromatin. However, in the control of development there are great differences.[464] For example, the *Arabidopsis* genome contains no relative of the *Drosophila* Gurken, no receptor tyrosine kinases, no relatives of transcription factor NF-κB. However, there are similarities in parallel pathways utilized by plants and animals.

The structures and life cycle of angiosperms[465] are described briefly on pp. 29–30. The alternating haploid (n, **gametophytic**) and diploid ($2n$, **sporophytic**) phases of the life cycle[466,467] are diagrammed in Fig. 32-8A. Following flowering a diploid **mother cell** within the **ovule** undergoes meiosis to form four haploid **megaspores**. After mitosis a single egg cell is formed. Within pollen sacs in the anthers of each mother undergoes meiosis to yield four haploid **microspores**. Following mitosis these develop into pollen grains each of which contains two sperm cells as well as a vegetative nucleus. After falling upon the stigma surface and growth of the pollen tube, one of the sperm cells fuses with the egg to give the diploid zygote. The other sperm unites with the specialized diploid **central cell** in the ovule to form a triploid ($3n$) **endosperm nucleus**, which develops into the **endosperm**, the food storage tissue of the seed. Endosperm contains two tissues, a starchy inner layer and a protein- and oil-rich outer layer.[468]

Variations of the life cycle occur. For example, a process called **apomixis** leads to asexual formation of seed.[469] In many plants, including maize, separate flowers form the ovule and the pollen. This is one mechanism for avoiding inbreeding.[470] In many plants systems of **self-incompatibility** have evolved.[471] In some, e.g., *Arabidopsis* and other crucifers, pollen germination is disrupted unless it falls on a stigma possessing a different allele-specific receptor. In other cases development of the pollen tube is disrupted at a later stage. In maize and in more than 150 other species a mitochondrial and therefore maternally inherited trait prevents formation of a functional pollen.[472,473] Male-sterile plants, which carry this trait, are very useful in plant breeding. However, a near disaster occurred in the United States in 1970 when the fungal disease Southern corn leaf blight attacked the male-sterile maize that had been used for production of 85% of commercial hybrid maize. The mitochondrial defect in pollen formation also resulted in an increased sensitivity to the fungal toxin. The target of the toxin is a 115-residue pore-forming polypeptide in the inner mitochondrial membrane.[474] The male sterility can be reversed if the plant carries two nuclear **restorer genes**.[472] One of these encodes an aldehyde dehydrogenase, but its mechanism of action is unclear.

The plant embryo is a juvenile form, the seedling. In *Arabidopsis* the zygote, which is surrounded by maternal diploid tissue, divides asymmetrically. The resulting apical and basal cells (Fig. 32-8B) differ in several ways. The small cytoplasm-rich apical cell is partitioned into eight proembryo cells by two rounds of vertical division and one horizontal division. The larger basal cell contains a vacuole and divides repeatedly horizontally to give 7–9 aligned cells. Only the uppermost of these becomes a part of the embryo. The others form an extra-embryonic suspensor (Fig. 32-8B).[475] The apical part of the embryo develops the shoot meristem and the central part the radial pattern of tissue layers characteristic of plants. The root meristem develops from the basal portion of the embryo. Movements of proteins that provide positional cues are involved in the development of the embryo.[476,477] Early embryonic and endosperm development is largely under maternal control. Most paternal genes may be initially silent.[478]

Many angiosperms develop **fruit** from tissues of the ovary (Fig. 32-8A). The development and ripening of fruit is also complex and highly regulated.[479,480]

Formation of seeds is a slow process. For example, in wheat the mature embryo, which consists of ~10^5 cells, develops over a seven week period. Seeds may live from a few years to 1000 years or more.[481] Subsequent germination of the seed into a seedling requires only two days.[482] The very dry embryo is converted into a highly hydrated plant whose further growth requires uptake of very large amounts of water. Many plants also synthesize large amounts of oxalic acid. This may arise from ascorbate (p. 1135) or via oxidation of glycine (Fig. 24-20). One of the earliest mRNAs to appear during seed germination encodes a 125-kDa glycoprotein called **germin**. This protein, which exists as multiple isoforms, is a copper-dependent oxalate oxidase (Eq. 32-1) which generates hydrogen peroxide. The latter is probably needed to

$$C_2O_4^{2-} + 2\,H^+ + O_2 \rightarrow 2\,CO_2 + H_2O_2 \quad (32\text{-}1)$$

crosslink cell-wall polymers. Germin may also be useful to plants in defense against oxalate-forming fungi.[482]

The rapid vegative growth, which includes development of shoots, leaves, and flowers, is controlled by a variety of transcription factors.[483] Among these are homeodomain proteins that control differentiation of meristem cells.[484–486] The induction of flowering is especially complex, involved day length, light quality, and effects of gibberellins.[486] At the ends of their lives plant cells die slowly from **senescence**. In this process many materials are recycled for use by new cells. Other plant cells die via the **hypersensitive response**, a form of programmed cell death.[486a]

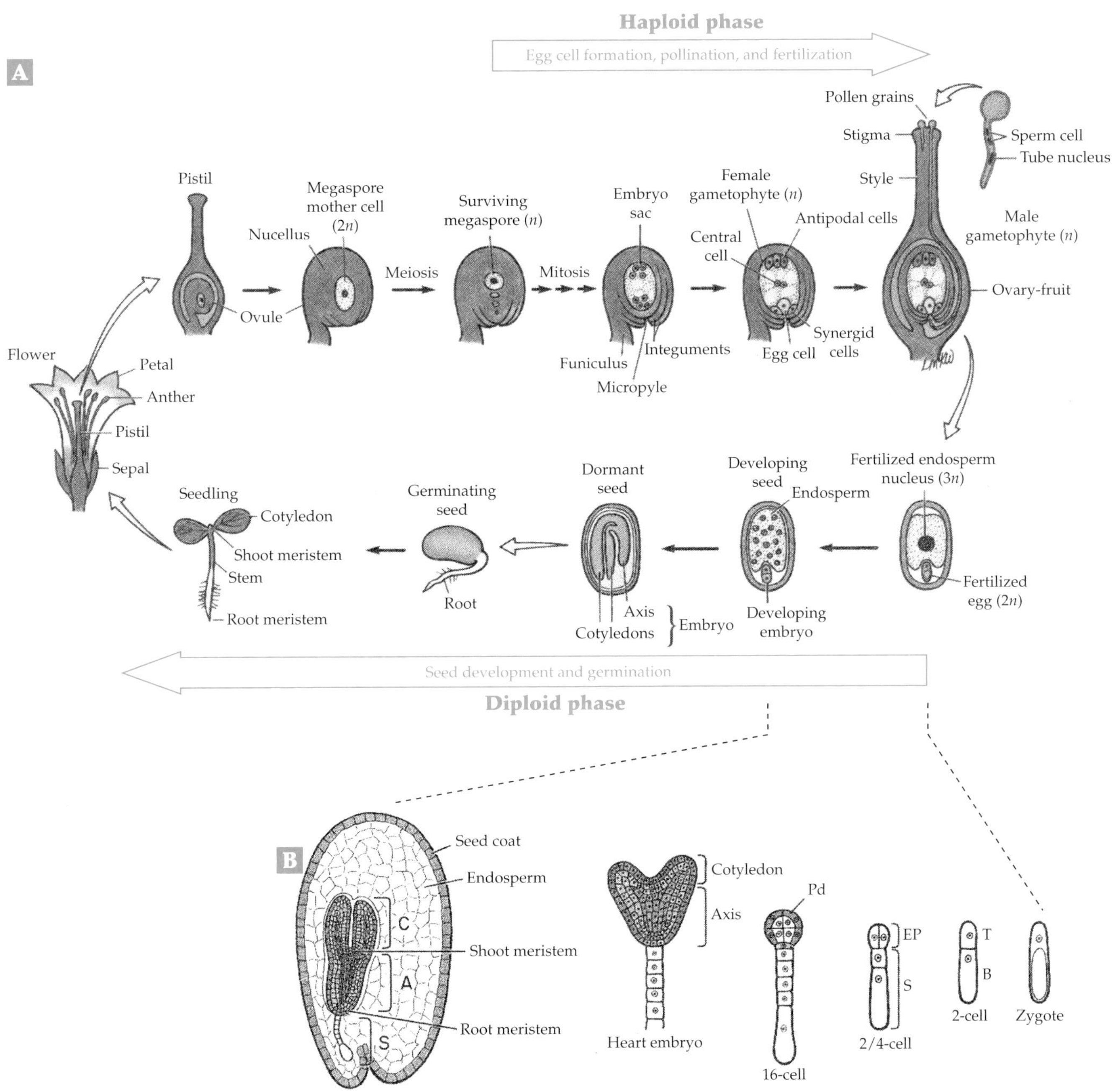

Figure 32-8 (A) The life cycle of a flowering plant with emphasis on egg-cell formation and seed development. (B) Some further details of embryo development. T, terminal cell; B, basal cell; C, cotyledon; A, axis; SC, seed coat; En, endosperm; EP, embryo proper; S, suspensor; SM, shoot meristem; Pd, protoderm; RM, root meristem. From Goldberg *et al.*[466] with modification.

F. Aging

Why do we age? This question is often asked but the answers are not simple. Do our tissues deteriorate with age as a result of damage to proteins? From an accumulation of mutations in our DNA? From attacks by free radicals? From loss of hormone receptors? From misregulation of mitosis?[487] From loss of telomeres on the ends of replicating DNA (Chapter 27, Section C,8)?[488,489] From an internal genetic program that dictates our life span? All of these possibilities may be partially correct. The simple answer is that "we just wear out." However, different parts wear out at different rates and in different ways.

Perhaps we should be amazed that the human body can live for an average of 75–80 years.[490,491] If we all avoided accidents and could cure all recognized diseases we might live an average of ~90 years.[492] A few very healthy people live for 100 years or more, ~20 years longer than average.[493] The maximum lifespan at present seems to be ~114 years. Long life tends to run in families, indicating a genetic component that can be identified.[493] However, this component is relatively small.[494]

Why do small rodents live only 2–3 years while we often live nearly 100 years?[494,495] Is it because their rate of metabolism is high? But bats have a comparable metabolic rate to mice, yet live ten times longer.[496] Nematodes live only ~20 days and fruitflies ~10 days. At the other extreme fish and some reptiles continue to grow throughout their lifetimes, surviving even longer than mammals.[495] Except for their germ cells nematodes, rotifers, and many insects have no dividing cells in their adult bodies. Their lifespan is presumably determined by the loss of cells through injury or death. In contrast, some simple animals, such as *Hydra*, other coelenterates, and flatworms maintain a pool of pluripotent stem cells that, except for accidental death, seems to make them immortal.[495]

Considerations such as these have helped move contemporary thinking toward an evolutionary view.[488,489,495–499] If the mortality of an animal in the wild (**extrinsic mortality**) is high it will evolve to have rapid development, good reproductive ability, and a short lifetime. If the extrinsic mortality is low the lifetime will be long. Such animals will require development of good protective functions including a highly developed brain.

Many factors must affect aging. It is generally agreed that one of these is the deleterious effects of free radicals derived from oxygen[500–503] (see pp. 1074, 1075). The lowered turnover rate of aging tissues may allow the damage to become lethal. According to this theory we might anticipate that free radical scavengers such as vitamin E could prolong life as might a restriction in food consumption.[504] For example, decreased fat intake might cut down on production of malondialdehyde (p. 1205) and lipid peroxides that may be especially damaging to cell membranes. The life of rodents can be prolonged substantially by a semi-starvation diet. Although there are uncertainties, a convincing case can be made for humans to keep food intake to a minimum and to eat foods rich in antioxidants and other nutrients.[491,504a]

For many years after techniques of cell culture had been developed it was commonly believed that cells in tissue culture were potentially immortal.[505] Challenging this idea, Medvedev[506] and others proposed that cells are internally programmed for a certain lifetime. This might explain why we have short-lived mammals as well as long-lived mammals. Support for the idea was supplied by Hayflick,[507,508] who observed that animal cells in culture have a limited potential for doubling. For example, normal human diploid embryonic fibroblasts grow in culture and double their number aproximately 50 ± 10 times. Regardless of cultural conditions, the cells die after this number of doublings. Cells taken from older humans undergo a smaller number of doublings before dying, as do those taken from shorter lived animals such as the mouse (14–28 doublings).[508] These experiments suggested that there is an internal program by which cells are scheduled to die from **replicative senescence**.[509–511] Malignant transformation overrides this program and transformed cells appear "immortal."[512] However, unlike fibroblasts some glial stem cells have been identified as possibly having unlimited proliferative capacity. These include cultured rat oligodendrocyte[509] and Schwann cells.[513] These results suggest that replicative senescence may not be inevitable.

As mentioned in Chapter 27 (p. 1568), "erosion" of the telomere ends on chromosomes is thought to be a major cause of cell senescence. Old cells have little or no telomerase. However, most human cancer cells, as well as those from immortalized cell cultures, do synthesize telomerase and maintain telomeres of adequate length.[514,515] Inhibition of telomerase activity in immortalized cells causes telomere shortening and cell death.[516] A second pathway for telomere maintenance is based upon homologous recombination.[517] Experimental elongation of telomeres extends the lifespan of cells in culture.[518] Furthermore, apparently healthy calves have been produced by nuclear transfer cloning from senescent fibroblast cells for which four or fewer cell doublings were expected to be possible. The cells of the cloned animals had the capacity for 90 or more cell doublings.[519]

What information about aging can we obtain by study of the "model" organisms *S. cerevisiae*, *C. elegans*, *D. melanogaster*, and the mouse? In every case a variety of mutations may shorten or lengthen the lifespan. In every case dietary energy restriction can lengthen life. Yeast cells grown on 0.5% glucose instead of 2%

glucose may undergo ~25% more cell doublings before a culture becomes senescent. However, mutant strains with a defect in the *SIR2* gene have a shortened lifespan, which is not increased by caloric restriction.[520–522] *SIR2* encodes Sir2p, an NAD^+-dependent histone deacetylase (p. 1626).[522,523] It is likely that caloric restriction causes the yeast to switch from anaerobic fermentation to oxidative metabolism. The resulting increase in the $[NAD^+] / [NADH]$ ratio activates Sir2p, thereby altering chromatin and silencing a group of genes. Mechanisms by which this shift in metabolism decreases deleterious mutations, even though respiration is increased, are probably complex.[523]

Nematodes (*C. elegans*) usually live about three weeks but the simultaneous presence of life-extending mutations in two different groups of genes lengthen the lifespan as much as fivefold.[524] One of the genes is the maternal-effect Clock gene *Clk*-1. It has been found to code for a hydroxylase involved in the final step of synthesis of ubiquinone-9 (p. 1429, Fig. 25-4).[525] The lifetime of wild-type nematodes is lengthened by ~60% by elimination of ubiquinone from the diet.[526] The second group of genes that affect longevity regulate an insulin-like signaling system.[527–529] In every case metabolism is slowed, an effect which may reduce the rate of harmful mutations. However, mutant animals may not be healthy. Some of these mutations induce formation of long-lived "dauer" larvae, providing a means for the larvae to survive for up to two months during periods of starvation.[527,530] Others affect sensory cilia.[531] Mutation of a cytosolic catalase gene *reduces* the lifespan.[530] The nematode's lifetime is also affected negatively by its own germ cells, perhaps via a steroid hormone.[532] The heat shock proteins (p. 1630), by chaperoning newly synthesized proteins and preventing aggregation, also increase life span both in *C. elegans* and *Drosophila*.[532a,b]

Some mutants of *D. melanogaster* with extended lifespans have a defective insulin / IGF signaling pathway.[526,533] The **methuselah** mutant, whose lifespan is 35% greater than average, appears to involve a G-protein-coupled transmembrane receptor.[534] Mutation of an insulin receptor homolog extends lifespan, apparently by causing a juvenile hormone deficiency.[535] *Drosophila* lifespan is also lengthened by mutation of a transmembrane dicarboxylate transporter[536] or by overexpression of a protein repair carboxyl methyltransferase (p. 594).[537]

Some mutant mice have extended lifespans. The Ames dwarf mouse has a mutation in $p66^{shc}$, a cell-surface protein that contains both Src-homology and collagen-homology domains. It lives almost one-third longer than do wild-type mice.[538] Mice deficient in methionine sulfoxide reductase have a reduced lifespan[539] but fruit flies with overexpressed activity of the enzyme are more resistant than wild-type flies to oxidative damage.[540]

In humans 100 or more years in age some mitochondrial mutations are associated with good health and longevity.[541] Dietary factors doubtless play a role. For example, supplementation of rats' diet with lipoic acid improved mitochondrial function and increased the metabolic rate of old animals.[542]

A number of genetic **progeroid diseases** result in premature aging.[543,544] Several of these arise from deficiencies in repair of DNA (Box 27-A). Among them are some types of cancer, and **Werner syndrome**, which arises from a defect in a 1432-residue protein with a central domain homologous to the RecQ family of DNA helicases (p. 1550).[545] Defects in other RecQ homologs cause **Hutchinson-Gilford progeria** as well as the Bloom syndrome (see Box 27-A).[546] Yet another DNA helicase, a subunit of transcription factor II (TFIIh, p. 1628), is defective in trichothiodystrophy (TTD, see Box 27-A).[547,548] Another gene which helps to prevent aging is *KLOTHO*. First identified in mice, it encodes a transmembrane protein that has sequence similarities to β-glucosidases.[549,550] Some mice with mutations in the tumor suppressor **p53** (Box 11-D) have enhanced resistance to tumors but age rapidly.[551,552]

Aging seems to be inevitably linked to an increase in the incidence of cancer. This uncontrolled growth of cells appears to be allowed by the stepwise accumulation of mutations that affect growth, differentiation, and survival.[553] Several aspects of cancer are discussed in other chapters of this book (see Box 11-D). However, the topic is so complex and research so active that it is hard to give even a thumbnail sketch of more recent discoveries.

Much effort is being dedicated to identifying the many signaling pathways that control growth, the mechanisms that cells employ to recognize problems in the control of growth, and the means by which cells can correct the problems or undergo apoptosis and avoid cancer.[553–558] Some of the complexity arises because of the large number of signaling pathways in which mutations may produce activated proto-onco-genes or faulty tumor suppressors. A large network of these suppressors is present in human cells.[553,559] Among the relevant signaling pathways are the following:

RAS – RAF – ERK (Fig. 11-12)[553–555]
p53[556–558]
the PtdIns 3-kinase – PKB/Akt pathway (Fig 11-9), which is opposed by PTEN[560,561]
EFG receptor (EFGR) signaling
Wnt-Catenin signaling[559,562]
E. Cadherin[559,563]

The importance of oncogenes and tumor suppressors has been demonstrated by conversion of human cells in culture into tumor cells in vitro.[564] Introduction of an activated *ras* gene, an SV40 viral protein that inhibits formation of both p53 and the Rb gene (Fig. 11-15), and an active telomerase gene sufficed. However, there is some doubt about the relevance of this work to human cancer.

Most cancerous cells have extra chromosomes. The karyotype (p. 1472) is rarely normal.[565] This and other evidence suggest that **genomic instability** may be the major cause of cancer.[566,567] In healthy cells stalled RNA polymerase is removed by transcription-coupled repair and lesions in DNA are either repaired (Chapter 27) or the cell undergoes apoptosis. Telomere dysfunction is also a factor.[567] The two breast cancer susceptibility genes *BRCA1* and *BRCA2* are apparently responsible for about half of all hereditary breast and ovarian cancers.[568,569] Protein BRCA1 is an 1863-residue nuclear protein, which is thought to function in transcription. However, recent evidence indicates that BRAC2 is directly involved in repair of double-strand breaks in DNA by homologous recombination.[569,570] Other data implicate the Neu-Ras pathway, proto-oncogenes *c-myc* and *Wnt-1*, and cyclin D1 in breast cancer.[571]

Yet another aspect of cancer is the **aberrant glycosylation** observed for many proteins.[572] The state of glycosylation of cell-surface proteins is one of many factors that affect metastasis, which is critical to growth of tumors.[573–575] The recognition that causes of cancer are numerous has led to a new large-scale project to identify as many cancer-associated mutations as possible within the entire human genome. One early success from this effort is identification of mutations in the gene *BRAF*, one of the three human *RAF* genes. These mutations are present in 15% of human malignant melanomas.[555,576]

G. Ecological Matters (Author's Personal Postscript)

The final section of this chapter deals with interactions among different species. As humans, beset by problems arising from our inability to communicate with other humans, we may feel that ecological relationships are relatively unimportant. However, any careful look at what can be regarded as an extension of metabolic cycles into the biosphere should convince us of the significance of this aspect of biochemistry.

Recall that the original development of eukaryotic creatures may have started with a symbiotic relationship between two prokaryotes and that symbiosis between algae and nonphotosynthetic organisms may have led to development of higher plants. Associations between species are still important today. For example, the bacteria in the protozoa of the digestive tract of ruminant animals are essential to production of meat. Our own bodies play host to bacteria, fungi, and other organisms with whom we have to try to maintain friendly relations. We depend upon antibiotics produced by bacteria or by fungi to fight our bacterial infections. Plants provide both essential nutrients and oxygen. Our environment has been created in large part by other living forms that coexist with us and which are subject to ecological checks and balances. It is therefore important that we learn more about the effects of one group of organisms on another and also about the effects of human activities on plants and animals of all degrees of complexity. This includes the poorly understood world of soil microorganisms. The consequences of environmental pollution, of depletion of atmospheric ozone or other alterations that affect the radiant energy reaching us, and of the availability to humans of excessive amounts of energy must all be considered. Just as a steady state within cells is often essential to the life of organisms, maintenance of a steady state in the chemical cycles of the biosphere may also be a necessity.

Biochemists and molecular biologists are being called upon to play an increasing role in medicine, agriculture, and industry. As such, they must be prepared to help in the making of decisions that may affect the future of life on earth. Biochemical approaches will be required to cope with many important problems. Among these are the long-term effects of the growing number of synthetic compounds in the environment, problems of antibiotic resistance, and effects of bioengineering of plants, fishes, and other organisms in the biosphere. Some of these scientific and ethical questions have been discussed in Chapter 26, and more are considered in the Study Questions that follow in this chapter.

Despite attempts to ignore it, we cannot avoid facing the war problem. The possibility of virtually total destruction of the more complex forms of life by genetic damage from radiation is real. That we have lived with nuclear weapons as long as we have is encouraging but continuing threats to use them as a last resort may bring eventual castastrophe. A race to put weapons into space might result in having computers decide to fight a war in which all people could be destroyed, but one computer might win! Perhaps biochemists, who understand the technical problems of radiation damage and mutation, have a special obligation to point out the hazard to others.

Just as threatening is the possibility of biochemical warfare, e.g., the use of artificial viruses. Biological weapons have been little used because of their lack of discrimination between friend and foe. However, our increasing knowledge of molecular biology makes possible insidious attacks on a population of unvaccinated persons. Since biochemical work does not require elaborate facilities, the development of biological weapons can be carried on by small groups in a clandestine manner. The recent assembly of a viable polio virus from oligonucleotides purchased from a commercial supplier emphasizes the ease with which virus warfare might be launched. Finding a way to protect ourselves may be more difficult.

Should we really worry about such matters? Since biochemistry is unable to ascribe any purpose to life, shouldn't we scientists stick to science? Science is amoral, isn't it? And besides, won't society do just what it wants to regardless of our opinions? Questions like these will always be with us, but most of the best scientists in the world seem to act with a great deal of responsibility. Not only do they want the pleasure and excitement of discovery and recognition for their work, but also they want a world for their children and grandchildren. They tend to feel compassion for other human beings. Many of them will give as a principal motivation for becoming biochemists the desire to contribute to the understanding of living things for the purpose of improving health, medical care, nutrition, etc. Most of them would not like to see the evolution of human beings ended through a disaster with nuclear or biological weapons or by irreversible pollution of land and sea. It will be a strange irony if we use our marvelous inquisitive, ingenous, inventive, and compassionate brains, the pinnacle of biological evolution, to destroy our environment and ourselves.

At a conference in Berkeley in 1971,[577] Joshua Lederberg, discoverer of genetic recombination in bacteria, talked about these matters. Lederberg asked if fairness and objectivity are possible outside the laboratory. He thought so. He pointed out that the nations of the world agreed to stop production of biological weapons and that genuine steps had been taken to decrease some of the hazards facing us. Nevertheless, progress is slow. Some insist on inspection for violation of agreements. But how can one inspect thoroughly enough? Lederberg suggested that the only possible form of control is now evolving. It must come from scientists themselves who must step out of their roles as "pure" scientists and accept the responsibility of preventing foolish uses of new biological discoveries. It may seem impossible that there could be a scientific community which could be counted on always to act in a responsible way, but it may be the only way that the human beings can survive for long on this planet. Lederberg believes it possible (and so do I).

If this book has helped to bring to the reader some awareness of the knowledge and power of molecular biology, I hope that these final words may lead the reader to heed the advice of Professor Lederberg. I sincerely hope that all the young people now studying biochemistry and modern biology will commit themselves to using the fantastic new knowledge available to us for the betterment of mankind and to proceeding with caution and responsibility as they move into positions of influence in the scientific community.

References

1. Gilbert, S. F., and Raunio, A. M., eds. (1997) *Embryology: Constructing the Organism*, Sinauer Assoc., Stamford, Connecticut
2. Wolpert, L., Beddington, R., Jessell, T., Lawrence, P., Meyerowitz, E., and Smith, J. (2002) *Principles of Development*, 2nd ed., Current Biology/Oxford Univ. Press, London
3. Brivanlou, A. H., and Darnell, J. E., Jr. (2002) *Science* **295**, 813–818
4. Britten, R. J. (1998) *Proc. Natl. Acad. Sci. U.S.A.* **95**, 9372–9377
5. Jan, Y. N., and Jan, L. Y. (1998) *Nature (London)* **392**, 775–778
6. Wolpert, L. (1994) *Science* **266**, 571–572
7. Shulman, J. M., and Johnston, D. S. (1999) *Trends Biochem. Sci.* **24**, M60–M64
8. Meyerowitz, E. M. (1999) *Trends Biochem. Sci.* **24**, M65–M68
9. Nüsslein-Volhard, C. (1996) *Sci. Am.* **275**(Aug), 54–61
10. Gurdon, J. B., and Bourillot, P.-Y. (2001) *Nature (London)* **413**, 797–803
11. Tsonis, P. A. (1987) *Trends Biochem. Sci.* **12**, 249
12. Basler, K. (2000) *EMBO J.* **19**, 1169–1175

12a. Hochachka, P. W., and Somero, G. N. (2002) *Biochemical Adaptation: Mechanism and Process in Physiological Evolution*, Oxford Univ. Press, New York

12b. Johnson, G. L., and Lapadat, R. (2002) *Science* **298**, 1911 – 1912

13. Alberts, B., Bray, D., Lewis, J., Raff, M., Roberts, K., and Watson, J. D. (1994) *Molecular Biology of the Cell*, 3rd ed., Garland, New York
14. Lewin, B. (2000) *Genes VII*, Oxford Univ. Press, New York

14a. Foss, M., McNally, F. J., Laurenson, P., and Rine, J. (1993) *Science* **262**, 1838 – 1844

14b. Arcangioli, B. (2000) *EMBO Reports* **1**, 145 – 150

15. Caldwell, G. A., Wang, S.-H., Xue, C.-B., Jiang, Y., Lu, H.-F., Naider, F., and Becker, J. M. (1994) *J. Biol. Chem.* **269**, 19817–19826
16. Tan, S., and Richmond, T. J. (1998) *Nature (London)* **391**, 660–666
17. Li, T., Stark, M. R., Johnson, A. D., and Wolberger, C. (1995) *Science* **270**, 262–269
18. Acosta-Serrano, A., Vassella, E., Liniger, M., Renggli, C. K., Brun, R., Roditi, I., and Englund, P. T. (2001) *Proc. Natl. Acad. Sci. U.S.A.* **98**, 1513–1518
19. Penny, G. D., Kay, G. F., Sheardown, S. A., Rastan, S., and Brockdorff, N. (1996) *Nature (London)* **379**, 131–137
20. Park, Y., and Kuroda, M. I. (2001) *Science* **293**, 1083–1085
21. Chao, W., Huynh, K. D., Spencer, R. J., Davidow, L. S., and Lee, J. T. (2002) *Science* **295**, 345–347
22. Ferguson-Smith, A. C., and Surani, M. A. (2001) *Science* **293**, 1086–1089
23. Surani, M. A. (2001) *Nature (London)* **414**, 122–128
24. Reik, W., and Murrell, A. (2000) *Nature (London)* **405**, 408–409
25. Surani, M. A. (2002) *Nature (London)* **416**, 491–493
26. Lorincz, M. C., and Groudine, M. (2001) *Proc. Natl. Acad. Sci. U.S.A.* **98**, 10034–10036
27. Reik, W., Dean, W., and Walter, J. (2001) *Science* **293**, 1089–1093
28. Inoue, K., Kohda, T., Lee, J., Ogonuki, N., Mochida, K., Noguchi, Y., Tanemura, K., Kaneko-Ishino, T., Ishino, F., and Ogura, A. (2002) *Science* **295**, 297
29. Burns, J. L., Jackson, D. A., and Hassan, A. B. (2001) *FASEB J.* **15**, 1694–1703
30. Jones, P. A., and Takai, D. (2001) *Science* **293**, 1068–1070
31. Rideout, W. M., III, Eggan, K., and Jaenisch, R. (2001) *Science* **293**, 1093–1098
32. Kang, Y.-K., Koo, D.-B., Park, J. S., Choi, Y.-H., Kim, H.-N., Chang, W.-K., Lee, K.-K., and Han, Y.-M. (2001) *J. Biol. Chem.* **276**, 39980–39984
33. Dean, W., Santos, F., Stojkovic, M., Zakhartchenko, V., Walter, J., Wolf, E., and Reik, W. (2001) *Proc. Natl. Acad. Sci. U.S.A.* **98**, 13734–13738

References

34. Plowman, G. D., Sudarsanam, S., Bingham, J., Whyte, D., and Hunter, T. (1999) *Proc. Natl. Acad. Sci. U.S.A.* **96**, 13603–13610
35. Loomis, W. E. (1986) *Developmental Biology*, Macmillan, New York
36. Schrader, M., and Schulz-Knappe, P. (2001) *Trends in Biotechnology* **19**, S55–S60
37. Hernandez, N. (2001) *J. Biol. Chem.* **276**, 26733–26736
38. Hutvágner, G., McLachlan, J., Pasquinelli, A. E., Bálint, E., Tuschl, T., and Zamore, P. D. (2001) *Science* **293**, 834–838
38a. Storz, G. (2002) *Science* **296**, 1260 – 1262
38b. Lee, R. C., and Ambros, V. (2001) *Science* **294**, 862 – 864
38c. Caplen, N. J., Parrish, S., Imani, F., Fire, A., and Morgan, R. A. (2001) *Proc. Natl. Acad. Sci. U.S.A.* **98**, 9742 – 9747
38d. Ambros, V. (2001) *Science* **293**, 811– 813
38e. Dennis, C. (2002) *Nature (London)* **418**, 122 – 124
38f. Suprenant, K. A. (2002) *Biochemistry* **41**, 14447– 14454
39. Lagos–Quintana, M., Rauhut, R., Lendeckel, W., and Tuschl, T. (2001) *Science* **294**, 853–858
40. Hu, R.-M., and 28 other authors. (2000) *Proc. Natl. Acad. Sci. U.S.A.* **97**, 9543–9548
41. Davidson, E. H., and 24 other authors. (2002) *Science* **295**, 1669–1678
42. Michelson, A. M. (2002) *Proc. Natl. Acad. Sci. U.S.A.* **99**, 546–548
43. Yuh, C.-H., Bolouri, H., and Davidson, E. H. (1998) *Science* **279**, 1896–1902
44. Berman, B. P., Nibu, Y., Pfeiffer, B. D., Tomancak, P., Celniker, S. E., Levine, M., Rubin, G. M., and Eisen, M. B. (2002) *Proc. Natl. Acad. Sci. U.S.A.* **99**, 757–762
45. Roush, W. (1996) *Science* **272**, 652–653
46. Veenstra, G. J. C., and Wolffe, A. P. (2001) *Trends Biochem. Sci.* **26**, 665–671
47. Richter, J. D., and Theurkauf, W. E. (2001) *Science* **293**, 60–62
48. Freeman, M. (2000) *Nature (London)* **408**, 313–319
49. Mannervik, M., Nibu, Y., Zhang, H., and Levine, M. (1999) *Science* **284**, 606–609
50. Musgrave, A., and van den Ende, H. (1987) *Trends Biochem. Sci.* **12**, 470–473
51. Misevic, G. N., and Burger, M. M. (1990) *J. Biol. Chem.* **265**, 20577–20584
52. Spillmann, D., Hård, K., Thomas-Oates, J., Vliegenthart, J. F. G., Misevic, G., Burger, M. M., and Finne, J. (1993) *J. Biol. Chem.* **268**, 13378–13387
53. Fernàndez-Busquets, X., Gerosa, D., Hess, D., and Burger, M. M. (1998) *J. Biol. Chem.* **273**, 29545–29553
53a. Yamada, K. M., and Clark, K. (2002) *Nature (London)* **419**, 790 – 791
53b. Boggon, T. J., Murray, J., Chappuis-Flament, S., Wong, E., Gumbiner, B. M., and Shapiro, L. (2002) *Science* **296**, 1308 – 1313
53c. Song, X., and Xie, T. (2002) *Proc. Natl. Acad. Sci. U.S.A.* **99**, 14813 – 14818
54. Edelman, G. M. (1984) *Sci. Am.* **250**(Apr), 118–129
55. Edelman, G. M. (1985) *Ann. Rev. Biochem.* **54**, 135–169
56. Kiss, J. Z., Wang, C., Olive, S., Rougon, G., Lang, J., Baetens, D., Harry, D., and Pralong, W. F. (1994) *EMBO J.* **13**, 5284–5292
57. Nelson, R. W., Bates, P. A., and Rutishauser, U. (1995) *J. Biol. Chem.* **270**, 17171–17179
58. Su, X.-D., Gastinel, L. N., Vaughn, D. E., Faye, I., Poon, P., and Bjorkman, P. J. (1998) *Science* **281**, 991–995
58a. van Meer, G. (2002) *Science* **296**, 855 – 857
58b. Shin, J.-S., and Abraham, S. N. (2001) *Science* **293**, 1447– 1448
58c. Nohturfft, A., and Losick, R. (2002) *Science* **296**, 857– 858
59. Collins, J. E., and Fleming, T. P. (1995) *Trends Biochem. Sci.* **20**, 307–312
60. Daniels, D. L., Spink, K. E., and Weis, W. I. (2001) *Trends Biochem. Sci.* **26**, 672–678
61. Wu, Q., and Maniatis, T. (2000) *Proc. Natl. Acad. Sci. U.S.A.* **97**, 3124–3129
61a. Boggon, T. J., Murray, J., Chappuis-Flament, S., Wong, E., Gumbiner, B. M., and Shapiro, L. (2002) *Science* **296**, 1308–1313
62. Kollmar, R., Nakamura, S. K., Kappler, J. A., and Hudspeth, A. J. (2001) *Proc. Natl. Acad. Sci. U.S.A.* **98**, 10196–10201
63. Giancotti, F. G., and Ruoslahti, E. (1999) *Science* **285**, 1028–1032
64. Couzin, J. (2001) *Science* **293**, 1743–1746
65. Calderwood, D. A., Shattil, S. J., and Ginsberg, M. H. (2000) *J. Biol. Chem.* **275**, 22607–22610
66. Harris, E. S., McIntyre, T. M., Prescott, S. M., and Zimmerman, G. A. (2000) *J. Biol. Chem.* **275**, 23409–23412
67. Brakebusch, C., Grose, R., Quondamatteo, F., Ramirez, A., Jorcano, J. L., Pirro, A., Svensson, M., Herken, R., Sasaki, T., Timpl, R., Werner, S., and Fässler, R. (2000) *EMBO J.* **19**, 3990–4003
68. Friedlander, M., Theesfeld, C. L., Sugita, M., Fruttiger, M., Thomas, M. A., Chang, S., and Cheresh, D. A. (1996) *Proc. Natl. Acad. Sci. U.S.A.* **93**, 9764–9769
69. Edwards, S. W. (1995) *Trends Biochem. Sci.* **20**, 362–367
70. Perez-Vilar, J., and Hill, R. L. (1997) *J. Biol. Chem.* **272**, 33410–33415
71. D'Souza, S. E., Ginsberg, M. H., Burke, T. A., Lam, S. C.-T., and Plow, E. F. (1988) *Science* **242**, 91–93
71a. Xiong, J.-P., Stehle, T., Zhang, R., Joachimiak, A., Frech, M., Goodman, S. L., and Arnaout, M. A. (2002) *Science* **296**, 151–155
72. Roth, S., McGuire, E. J., and Roseman, S. (1971) *J. Cell Biol.* **51**, 536–547
73. Perrimon, N., and Bernfield, M. (2000) *Nature (London)* **404**, 725–728
74. Zhuo, L., Yoneda, M., Zhao, M., Yingsung, W., Yoshida, N., Kitagawa, Y., Kawamura, K., Suzuki, T., and Kimata, K. (2001) *J. Biol. Chem.* **276**, 7693–7696
75. Varki, A. (1996) *Proc. Natl. Acad. Sci. U.S.A.* **93**, 4523–4525
76. Knox, J. P. (1995) *FASEB J.* **9**, 1004–1012
77. Rakic, P. (1999) *Nature (London)* **400**, 315–316
78. Horwitz, A. R., and Parsons, J. T. (1999) *Science* **286**, 1102–1103
79. Chicurel, M. (2002) *Science* **295**, 606–609
80. Condliffe, A. M., and Hawkins, P. T. (2000) *Nature (London)* **404**, 135, 137
81. Ng, T., Parsons, M., Hughes, W. E., Monypenny, J., Zicha, D., Gautreau, A., Arpin, M., Gschmeissner, S., Verveer, P. J., Bastiaens, P. I. H., and Parker, P. J. (2001) *EMBO J.* **20**, 2723–2741
82. D'Ari, R., and Bouloc, P. (1990) *Trends Biochem. Sci.* **15**, 191–194
83. Shapiro, L., and Losick, R. (1997) *Science* **276**, 712–718
84. Peifer, M., and Tepass, U. (2000) *Nature (London)* **403**, 611–612
84a. Shapiro, L., McAdams, H. H., and Losick, R. (2002) *Science* **298**, 1942 – 1946
85. Losick, R., and Shapiro, L. (1993) *Science* **262**, 1227–1228
85a. Bourne, H. R., and Weiner, O. (2002) *Nature (London)* **419**, 21
85b. Pellettieri, J., and Seydoux, G. (2002) *Science* **298**, 1946 – 1950
86. Dawid, I. B. (1994) *J. Biol. Chem.* **269**, 6259–6262
87. Rodriguez-Boulan, E., and Nelson, W. J. (1989) *Science* **245**, 718–725
88. St Johnston, D. (2001) *EMBO J.* **20**, 6169–6179
89. Wikramanayake, A. H., Huang, L., and Klein, W. H. (1998) *Proc. Natl. Acad. Sci. U.S.A.* **95**, 9343–9348
90. Wallenfang, M. R., and Seydoux, G. (2000) *Nature (London)* **408**, 89–92
90a. Keller, R. (2002) *Science* **298**, 1950 – 1954
91. Martelly, I., and Franquinet, R. (1984) *Trends Biochem. Sci.* **9**, 468–471
92. Slack, J. M. W. (1987) *Trends Biochem. Sci.* **12**, 200–204
93. Patel, N. H., and Lall, S. (2002) *Nature (London)* **415**, 748–749
94. Houchmandzadeh, B., Wieschaus, E., and Leibler, S. (2002) *Nature (London)* **415**, 798–802
95. Wimmer, E. A., Carleton, A., Harjes, P., Turner, T., and Desplan, C. (2000) *Science* **287**, 2476–2479
96. Vincent, S., and Perrimon, N. (2001) *Nature (London)* **411**, 533–536
97. Sampath, K., Rubinstein, A. L., Cheng, A. M. S., Liang, J. O., Fekany, K., Solnica-Krezel, L., Korzh, V., Halpern, M. E., and Wright, C. V. E. (1998) *Nature (London)* **395**, 185–189
98. Chen, Y., and Schier, A. F. (2001) *Nature (London)* **411**, 607–610
99. Pasternak, C. A. (1970) *Biochemistry of Differentiation*, Wiley (Interscience), New York
100. Vogelstein, B., Alberts, B., and Shine, K. (2002) *Science* **295**, 1237
101. Shin, T., Kraemer, D., Pryor, J., Liu, L., Rugila, J., Howe, L., Buck, S., Murphy, K., Lyons, L., and Westusin, M. (2002) *Nature (London)* **415**, 859
101a. Rossant, J. (2002) *Nature (London)* **415**, 967, 969
102. Lai, L., Kolber-Simonds, D., Park, K.-W., Cheong, H.-T., Greenstein, J. L., Im, G.-S., Samuel, M., Bonk, A., Rieke, A., Day, B. N., Murphy, C. N., Carter, D. B., Hawley, R. J., and Prather, R. S. (2002) *Science* **295**, 1089–1092
103. Gurdon, J. B. (1974) *Nature (London)* **248**, 772–776
104. De Robertis, E. M., and Gurdon, J. B. (1979) *Sci. Am.* **241**(Dec), 74–82
105. Anderson, G. B., and Seidel, G. E. (1998) *Science* **280**, 1400–1401
106. Aldhous, P. (2000) *Nature (London)* **405**, 610–612
107. Gurdon, J. B., and Colman, A. (1999) *Nature (London)* **402**, 743–746
107a. Hochedlinger, K., and Jaenisch, R. (2002) *Nature (London)* **415**, 1035 – 1038
107b. Wilmut, I., Beaujean, N., de Sousa, P. A., Dinnyes, A., King, T. J., Paterson, L. A., Wells, D. N., and Young, L. E. (2002) *Nature (London)* **419**, 583 – 586
108. Kubota, C., Yamakurchi, H., Todoroki, J., Mizoshita, K., Tabara, N., Barber, M., and Yang, X. (2000) *Proc. Natl. Acad. Sci. U.S.A.* **97**, 990–995
109. Pennisi, E. (2001) *Science* **293**, 1064–1067
110. De Bie, I., Savaria, D., Roebroek, A. J. M., Day, R., Lazure, C., Van de Ven, W. J. M., and Seidah, N. G. (1995) *J. Biol. Chem.* **270**, 1020–1028
111. Judson, H., Hayward, B. E., Sheridan, E., and Bonthron, D. T. (2002) *Nature (London)* **416**, 539–542
112. Golde, D. W. (1991) *Sci. Am.* **265**(Dec), 86–93
113. Watt, F. M., and Hogan, B. L. M. (2000) *Science* **287**, 1427–1430
114. Spradling, A., Drummond-Barbosa, D., and Kai, T. (2001) *Nature (London)* **414**, 98–104
115. Kiger, A. A., Jones, D. L., Schulz, C., Rogers, M. B., and Fuller, M. T. (2001) *Science* **294**, 2542–2545
116. McKay, R. (2000) *Nature (London)* **406**, 361–364

References

117. Donovan, P. J., and Gearhart, J. (2001) *Nature (London)* **414**, 92–97
117a. Wakayama, T., Tabar, V., Rodriguez, I., Perry, A. C. F., Studer, L., and Mombaerts, P. (2001) *Science* **292**, 740 – 743
118. Ahmed, S., and Hodgkin, J. (2000) *Nature (London)* **403**, 159–164
118a. Wurmser, A. E., and Gage, F. H. (2002) *Nature (London)* **416**, 485 – 487
118b. Blau, H. M. (2002) *Nature (London)* **419**, 437
119. Vogel, G. (1999) *Science* **283**, 1432–1434
119a. Ivanova, N. B., Dimos, J. T., Schaniel, C., Hackney, J. A., Moore, K. A., and Lemischka, I. R. (2002) *Science* **298**, 601– 604
119b. Zhang, Z., Zhang, R., Joachimiak, A., Schlessinger, J., and Kong, X.-P. (2000) *Proc. Natl. Acad. Sci. U.S.A.* **97**, 7732 – 7737
120. Vogel, G. (2001) *Science* **292**, 1820–1822
120a. Nishimura, E. K., Jordan, S. A., Oshima, H., Yoshida, H., Osawa, M., Moriyama, M. Jackson, I. J., Barrandon, Y., Miyachi, Y., and Nishikawa, S.-i. (2002) *Nature (London)* **416**, 854 – 860
121. Colter, D. C., Sekiya, I., and Prockop, D. J. (2001) *Proc. Natl. Acad. Sci. U.S.A.* **98**, 7841–7845
121a. Jiang, Y., Jahagirdar, B. N., Reinhard, R. L., Schwartz, R. E., Keene, C. D., Ortiz-Gonzalez, X. R., Reyes, M., Lenvik, T., Lund, T., Blackstad, M., Du, J., Aldrich, S., Lisberg, A., Low, W. C., Largaespada, D. A., and Verfaillie, C. M. (2002) *Nature (London)* **418**, 41 – 49
122. van der Kooy, D., and Weiss, S. (2000) *Science* **287**, 1439–1441
123. Slack, J. M. W. (2000) *Science* **287**, 1431 – 1433
124. Ghazizadeh, S., and Taichman, L. B. (2001) *EMBO J.* **20**, 1215–1222
125. Lavker, R. M., and Sun, T.-T. (2000) *Proc. Natl. Acad. Sci. U.S.A.* **97**, 13473–13475
126. Weigel, D., and Jürgens, G. (2002) *Nature (London)* **415**, 751–754
127. Langer, R., and Vacanti, J. P. (1993) *Science* **260**, 920–926
128. Solter, D., and Gearhart, J. (1999) *Science* **283**, 1468–1470
128a. Lumelsky, N., Blondel, O., Laeng, P., Velasco, I., Ravin, R., and McKay, R. (2001) *Science* **292**, 1389 – 1394
129. Langer, R., and Vacanti, J. P. (1995) *Sci. Am.* **273**(Sep), 130–133
130. Ferber, D. (1999) *Science* **284**, 422–425
131. Aldhous, P. (2001) *Nature (London)* **410**, 622–625
132. McLaren, A. (2000) *Science* **288**, 1775–1780
133. Lovell-Badge, R. (2001) *Nature (London)* **414**, 88–91
134. Weissman, I. L. (2000) *Science* **287**, 1442–1446
135. Jaenisch, R., and Wilmut, I. (2001) *Science* **291**, 2552
136. Holliday, R., and Pugh, J. E. (1975) *Science* **187**, 226–232
137. Thomassin, H., Flavin, M., Espinás, M.-L., and Grange, T. (2001) *EMBO J.* **20**, 1974–1983
138. Clerc, P., and Avner, P. (2000) *Science* **290**, 1518–1519
139. Georgatsos, J. G. (1995) *Nature (London)* **375**, 100
140. Raff, M. (1998) *Nature (London)* **396**, 119–122
141. Aravind, L., Dixit, V. M., and Koonin, E. V. (2001) *Science* **291**, 1279–1284
142. Ashkenazi, A., and Dixit, V. M. (1998) *Science* **281**, 1305–1308
143. White, K., Grether, M. E., Abrams, J. M., Young, L., Farrell, K., and Steller, H. (1994) *Science* **264**, 677–683
144. Yuan, J., and Yankner, B. A. (2000) *Nature (London)* **407**, 802–809
145. Kim, Y.-M., Talanian, R. V., and Billiar, T. R. (1997) *J. Biol. Chem.* **272**, 31138–31148
146. Bonfoco, E., Krainc, D., Ankarcrona, M., Nicotera, P., and Lipton, S. A. (1995) *Proc. Natl. Acad. Sci. U.S.A.* **92**, 7162–7166
146a. Green, D. R., and Beere, H. M. (2002) *Nature (London)* **405**, 28 – 29
146b. Savill, J., and Fadok, V. (2000) *Nature (London)* **407**, 784 – 788
147. Klionsky, D. J., and Emr, S. D. (2000) *Science* **290**, 1717–1721
147a. Xia, X. G., Harding, T., Weller, M., Bieneman, A., Uney, J. B., and Schulz, J. B. (2001) *Proc. Natl. Acad. Sci. U.S.A.* **98**, 10433 – 10438
147b. Raff, M. C., Whitmore, A. V., and Finn, J. T. (2002) *Science* **296**, 868 – 871
147c. Jones, P. A. (2001) *Nature (London)* **409**, 141–144
148. Meier, P., Finch, A., and Evan, G. (2000) *Nature (London)* **407**, 796–801
149. Härtel, U., Eckel, E., Koch, J., Fuchs, G., Linder, D., and Buckel, W. (1993) *Arch Microbiol* **159**, 174–181
150. Kumar, S. (1995) *Trends Biochem. Sci.* **20**, 198–202
151. Haecker, G., and Vaux, D. L. (1994) *Trends Biochem. Sci.* **19**, 99–100
152. Wolf, B. B., and Green, D. R. (1999) *J. Biol. Chem.* **274**, 20049–20052
152a. Adams, J. M., and Cory, S. (2001) *Trends Biochem. Sci.* **26**, 61– 66
152b. Ruoslahti, E., and Reed, J. (1999) *Nature (London)* **397**, 479 – 480
153. Qin, H., Srinivasula, S. M., Wu, G., Fernandes-Alnemri, T., Alnemri, E. S., and Shi, Y. (1999) *Nature (London)* **399**, 549–557
154. Slee, E. A., Adrain, C., and Martin, S. J. (2001) *J. Biol. Chem.* **276**, 7320–7326
155. Earnshaw, W. C., Martins, L. M., and Kaufmann, S. H. (1999) *Ann. Rev. Biochem.* **68**, 383–424
156. Kumar, S., and Colussi, P. A. (1999) *Trends Biochem. Sci.* **24**, 1–4
157. Liang, H., and Fesik, S. W. (1997) *J. Mol. Biol.* **274**, 291–302
158. Chu, Z.-L., Pio, F., Xie, Z., Welsh, K., Krajewska, M., Krajewski, S., Godzik, A., and Reed, J. C. (2001) *J. Biol. Chem.* **276**, 9239–9245
159. Hengartner, M. O., Ellis, R. E., and Horvitz, H. R. (1992) *Nature (London)* **356**, 494–499
159a. Chen, G., and Goeddel, D. V. (2002) *Science* **296**, 1634 – 1635
160. Aravind, L., Dixit, V. M., and Koonin, E. V. (1999) *Trends Biochem. Sci.* **24**, 47–53
161. Walczak, H., and Sprick, M. R. (2001) *Trends Biochem. Sci.* **26**, 452–453
162. Jeong, E.-J., Bang, S., Lee, T. H., Park, Y. I., Sim, W.-S., and Kim, K.-S. (1999) *J. Biol. Chem.* **274**, 16337–16342
163. Weber, C. H., and Vincenz, C. (2001) *Trends Biochem. Sci.* **26**, 475–481
164. Nagata, S., and Golstein, P. (1995) *Science* **267**, 1449–1456
164a. Krammer, P. H. (2000) *Nature (London)* **407**, 789 – 795
165. Imai, Y., Kimura, T., Murakami, A., Yajima, N., Sakamaki, K., and Yonehara, S. (1999) *Nature (London)* **398**, 777–785
166. Enari, M., Sakahira, H., Yokoyama, H., Okawa, K., Iwamatsu, A., and Nagata, S. (1998) *Nature (London)* **391**, 43–50
167. Hengartner, M. O. (2000) *Nature (London)* **407**, 770–776
167a. Lassus, P., Opitz-Araya, X., and Lazebnik, Y. (2002) *Science* **297**, 1352 – 1354
168. Liu, X., Kim, C. N., Yang, J., Jemmerson, R., and Wang, X. (1996) *Cell* **86**, 147–157
169. Mootha, V. K., Wei, M. C., Buttle, K. F., Scorrano, L., Panoutsakopoulou, V., Mannella, C. A., and Korsmeyer, S. J. (2001) *EMBO J.* **20**, 661–671
170. Bernardi, P., Petronilli, V., Di Lisa, F., and Forte, M. (2001) *Trends Biochem. Sci.* **26**, 112–117
171. Adrain, C., and Martin, S. J. (2001) *Trends Biochem. Sci.* **26**, 390–397
171a. Marsden, V. S., O'Conner, L., O'Reilly, L. A., Silke, J., Metcalf, D., Ekert, P. G., Huang, D. C. S., Cecconi, F., Kuida, K., Tomaselll, K. J., Roy, S., Nicholson, D. W., Vaux, D. L., Bouillet, P., Adams, J. M., and Strasser, A. (2002) *Nature (London)* **419**, 634 – 637
172. Finkel, E. (2001) *Science* **292**, 624–626
173. Qin, Z.-H., Wang, Y., Kikly, K. K., Sapp, E., Kegel, K. B., Aronin, N., and DiFiglia, M. (2001) *J. Biol. Chem.* **276**, 8079–8086
174. Adams, J. M., and Cory, S. (1998) *Science* **281**, 1322–1326
175. Pawlowski, J., and Kraft, A. S. (2000) *Proc. Natl. Acad. Sci. U.S.A.* **97**, 529–531
176. Wei, M. C., Zong, W.-X., Cheng, E. H.-Y., Lindsten, T., Panoutsakopoulou, V., Ross, A. J., Roth, K. A., MacGregor, G. R., Thompson, C. B., and Korsmeyer, S. J. (2001) *Science* **292**, 727–730
177. Pozniak, C. D., Radinovic, S., Yang, A., McKeon, F., Kaplan, D. R., and Miller, F. D. (2000) *Science* **289**, 304–306
178. Gross, A., Yin, X.-M., Wang, K., Wei, M. C., Jockel, J., Milliman, C., Erdjument-Bromage, H., Tempst, P., and Korsmeyer, S. J. (1999) *J. Biol. Chem.* **274**, 1156–1163
179. Mehmet, H. (2000) *Nature (London)* **403**, 29–30
180. Podack, E. R. (1999) *Proc. Natl. Acad. Sci. U.S.A.* **96**, 8312–8314
181. Alimonti, J. B., Shi, L., Baijal, P. K., and Greenberg, A. H. (2001) *J. Biol. Chem.* **276**, 6974–6982
182. Pinkoski, M. J., Waterhouse, N. J., Heibein, J. A., Wolf, B. B., Kuwana, T., Goldstein, J. C., Newmeyer, D. D., Bleackley, R. C., and Green, D. R. (2001) *J. Biol. Chem.* **276**, 12060–12067
183. Schauer, A., Ranes, M., Santamaria, R., Guijarro, J., Lawlor, E., Mendez, C., Chater, K., and Losick, R. (1988) *Science* **240**, 768–772
184. Youngman, P., Zuber, P., Perkins, J. B., Sandman, K., Igo, M., and Losick, R. (1985) *Science* **229**, 285–291
185. Wireman, J. W., and Dworkin, M. (1975) *Science* **189**, 516–523
186. Sossong, T. M., Jr., Brigham-Burke, M. R., Hensley, P., and Pearce, K. H., Jr. (1999) *Biochemistry* **38**, 14843–14850
187. Mosyak, L., Zhang, Y., Glasfeld, E., Haney, S., Stahl, M., Seehra, J., and Somers, W. S. (2000) *EMBO J.* **19**, 3179–3191
188. Romberg, L., Simon, M., and Erickson, H. P. (2001) *J. Biol. Chem.* **276**, 11743–11753
189. van den Ent, F., and Löwe, J. (2000) *EMBO J.* **19**, 5300–5307
190. Cordell, S. C., Anderson, R. E., and Löwe, J. (2001) *EMBO J.* **20**, 2454–2461
191. Hayashi, I., Oyama, T., and Morikawa, K. (2001) *EMBO J.* **20**, 1819–1828
192. Jacobs, C., and Shapiro, L. (1999) *Proc. Natl. Acad. Sci. U.S.A.* **96**, 5891–5893
193. RayChaudhuri, D., Gordon, G. S., and Wright, A. (2001) *Proc. Natl. Acad. Sci. U.S.A.* **98**, 1332–1334
194. Nishimura, A. (1998) *Trends Biochem. Sci.* **23**, 157–159
195. Domian, I. J., Reisenauer, A., and Shapiro, L. (1999) *Proc. Natl. Acad. Sci. U.S.A.* **96**, 6648–6653
195a. Ouimet, M.-C., and Marczynski, G. T. (2000) *J. Mol. Biol.* **302**, 761–775
195b. Laub, M. T., Chen, S. L., Shapiro, L., and McAdams, H. H. (2002) *Proc. Natl. Acad. Sci. U.S.A.* **99**, 4632–4637
196. Jenal, U., and Shapiro, L. (1996) *EMBO J.* **15**, 2393–2406

References

197. Brun, Y. V., Marczynski, G., and Shapiro, L. (1994) *Ann. Rev. Biochem.* **63**, 419–450
198. Hecht, G. B., Lane, T., Ohta, N., Sommer, J. M., and Newton, A. (1995) *EMBO J.* **14**, 3915–3924
199. Quon, K. C., Yang, B., Domian, I. J., Shapiro, L., and Marczynski, G. T. (1998) *Proc. Natl. Acad. Sci. U.S.A.* **95**, 120–125
200. Wortinger, M., Sackett, M. J., and Brun, Y. V. (2000) *EMBO J.* **19**, 4503–4512
201. Laub, M. T., McAdams, H. H., Feldblyum, T., Fraser, C. M., and Shapiro, L. (2000) *Science* **290**, 2144–2148
202. Jenal, U., and Fuchs, T. (1998) *EMBO J.* **17**, 5658–5669
203. Ponnuraj, K., Rowland, S., Nessi, C., Setlow, P., and Jedrzejas, M. J. (2000) *J. Mol. Biol.* **300**, 1–10
204. Sharp, M. D., and Pogliano, K. (1999) *Proc. Natl. Acad. Sci. U.S.A.* **96**, 14553–14558
205. Parkes, R. J. (2000) *Nature (London)* **407**, 844–845
206. Arigoni, F., Pogliano, K., Webb, C. D., Stragier, P., and Losick, R. (1995) *Science* **270**, 637–640
207. Lucet, I., Feucht, A., Yudkin, M. D., and Errington, J. (2000) *EMBO J.* **19**, 1467–1475
208. Ducros, V. M.-A., Lewis, R. J., Verma, C. S., Dodson, E. J., Leonard, G., Turkenburg, J. P., Murshudov, G. N., Wilkinson, A. J., and Brannigan, J. A. (2001) *J. Mol. Biol.* **306**, 759–771
209. Shazand, K., Frandsen, N., and Stragier, P. (1995) *EMBO J.* **14**, 1439–1445
210. Lewis, R. J., Brannigan, J. A., Muchová, K., Barák, I., and Wilkinson, A. J. (1999) *J. Mol. Biol.* **294**, 9–15
211. Bath, J., Wu, L. J., Errington, J., and Wang, J. C. (2000) *Science* **290**, 995–997
212. Gould, G. W., and Dring, G. J. (1975) *Nature (London)* **258**, 402–405
213. Dong, Y.-H., Xu, J.-L., Li, X.-Z., and Zhang, L.-H. (2000) *Proc. Natl. Acad. Sci. U.S.A.* **97**, 3526–3531
213a. Zhang, R.-g, Pappas, T., Brace, J. L., Miller, P. C., Oulmassov, T., Molyneaux, J. M., Anderson, J. C., Bashkin, J. K., Winans, S. C., and Joachimiak, A. (2002) *Nature (London)* **417**, 917–974
214. Fuqua, C., and Greenberg, E. P. (1998) *Proc. Natl. Acad. Sci. U.S.A.* **95**, 6571–6572
215. Daniels, R., De Vos, D. E., Desair, J., Raedschelders, G., Luyten, E., Rosemeyer, V., Verreth, C., Schoeters, E., Vanderleyden, J., and Michiels, J. (2002) *J. Biol. Chem.* **277**, 462–468
216. Chen, X., Schauder, S., Potier, N., Van Dorsselaer, A., Pelczer, I., Bassler, B. L., and Hughson, F. M. (2002) *Nature (London)* **415**, 545–549
217. Yarmolinsky, M. B. (1995) *Science* **267**, 836–837
218. Hoffman, M. (1992) *Science* **255**, 1510–1511
219. Magee, P. T. (1997) *Science* **277**, 52–53
220. Herskowitz, I. (1992) *Nature (London)* **357**, 190–191
221. Cabib, E., Roh, D.-H., Schmidt, M., Crotti, L. B., and Varma, A. (2001) *J. Biol. Chem.* **276**, 19679–19682
222. Utzig, S., Fankhauser, C., and Simanis, V. (2000) *J. Mol. Biol.* **302**, 751–759
223. Desautels, M., Den Haese, J. P., Slupsky, C. M., McIntosh, L. P., and Hemmingsen, S. M. (2001) *J. Biol. Chem.* **276**, 5932–5942
223a. Pelham, R. J., Jr., and Chang, F. (2002) *Nature (London)* **419**, 82–86
224. Chu, S., DeRisi, J., Eisen, M., Mulholland, J., Botstein, D., Brown, P. O., and Herskowitz, I. (1998) *Science* **282**, 699–705
225. Obara, T., Nakafuku, M., Yamamoto, M., and Kaziro, Y. (1991) *Proc. Natl. Acad. Sci. U.S.A.* **88**, 5877–5881
226. Shimomura, O., Suthers, H. L. B., and Bonner, J. T. (1982) *Proc. Natl. Acad. Sci. U.S.A.* **79**, 7376–7379
227. Rutherford, C. L., Taylor, R. D., Merkle, R. K., and Frame, L. T. (1982) *Trends Biochem. Sci.* **7**, 108–111
228. Chiew, Y. Y., Reimers, J. M., and Wright, B. E. (1985) *J. Biol. Chem.* **260**, 15325–15331
229. Rutherford, C. L., and Brown, S. S. (1983) *Biochemistry* **22**, 1251–1258
230. Morrissey, J. H. (1983) *Nature (London)* **303**, 203–204
231. Siu, C.-H., Lam, T. Y., and Choi, A. H. C. (1985) *J. Biol. Chem.* **260**, 16030–16036
232. Berger, E. A., and Armant, D. R. (1982) *Proc. Natl. Acad. Sci. U.S.A.* **79**, 2162–2166
233. O'Day, D. H., and Lewis, K. E. (1975) *Nature (London)* **254**, 431–432
234. Schaller, H. C., and Bodenmüller, H. (1981) *Proc. Natl. Acad. Sci. U.S.A.* **78**, 7000–7004
235. Bosch, T. C. G., and David, C. N. (1986) *Proc. Natl. Acad. Sci. U.S.A.* **83**, 9478–9482
236. Marx, J. L. (1984) *Science* **225**, 40–42
237. Kenyon, C. J. (1983) *Trends Biochem. Sci.* **8**, 349–351
238. Wood, W. B., ed. (1988) *The Nematode Caenorhabditis elegans*, Cold Spring Harbor Lab. Press, Cold Spring Harbor, New York
239. Kenyon, C. (1988) *Science* **240**, 1448–1453
240. Roberts, L. (1990) *Science* **248**, 1310–1313
241. Riddle, D. L., Swanson, M. M., and Albert, P. S. (1981) *Nature (London)* **290**, 668–671
242. Wuethrich, B. (1998) *Science* **281**, 1980–1982
243. Zeyl, C., and Bell, G. (1997) *Nature (London)* **388**, 465–468
243a. Hunt, P. A., and Hassold, T. J. (2002) *Science* **296**, 2181–2183
244. Weeks, A. R., Marec, F., and Breeuwer, J. A. J. (2001) *Science* **292**, 2479–2482
245. Parkhurst, S. M., and Meneely, P. M. (1994) *Science* **264**, 924–932
246. Williams, N. (1995) *Science* **269**, 1826–1827
247. Lee, A. L., Volkman, B. F., Robertson, S. A., Rudner, D. Z., Barbash, D. A., Cline, T. W., Kanaar, R., Rio, D. C., and Wemmer, D. E. (1997) *Biochemistry* **36**, 14306–14317
248. Scott, M. J., Pan, L. L., Cleland, S. B., Knox, A. L., and Heinrich, J. (2000) *EMBO J.* **19**, 144–155
249. Kuroda, M. I., and Kelley, R. L. (1999) *Science* **284**, 1787–1788
250. Carmi, I., Kopczynski, J. B., and Meyer, B. J. (1998) *Nature (London)* **396**, 168–173
251. Sinclair, A. H., Berta, P., Palmer, M. S., Hawkins, J. R., Griffiths, B. L., Smith, M. J., Foster, J. W., Frischauf, A.-M., Lovell-Badge, R., and Goodfellow, P. N. (1990) *Nature (London)* **346**, 240–244
252. Werner, M. H., Huth, J. R., Gronenborn, A. M., and Clore, G. M. (1996) *Trends Biochem. Sci.* **21**, 302–308
253. Werner, M. H., Huth, J. R., Gronenborn, A. M., and Clore, G. M. (1995) *Cell* **81**, 705–714
254. Burgoyne, P. S. (1986) *Nature (London)* **319**, 258–259
255. Ellis, N. A., Goodfellow, P. J., Pym, B., Smith, M., Palmer, M., Frischauf, A.-M., and Goodfellow, P. N. (1989) *Nature (London)* **337**, 81–84
256. Thomas, J. O., and Travers, A. A. (2001) *Trends Biochem. Sci.* **26**, 167–174
257. Bustin, M. (2001) *Trends Biochem. Sci.* **26**, 431–437
258. Haqq, C. M., King, C.-Y., Ukiyama, E., Falsafi, S., Haqq, T. N., Donahoe, P. K., and Weiss, M. A. (1994) *Science* **266**, 1494–1499
259. Benevides, J. M., Chan, G., Lu, X.-J., Olson, W. K., Weiss, M. A., and Thomas, G. J., Jr. (2000) *Biochemistry* **39**, 537–547
260. Ohe, K., Lalli, E., and Sassone-Corsi, P. (2002) *Proc. Natl. Acad. Sci. U.S.A.* **99**, 1146–1151
261. Vainio, S., Heikkilä, M., Kispert, A., Chin, N., and McMahon, A. P. (1999) *Nature (London)* **397**, 405–409
262. Swain, A., Narvaez, V., Burgoyne, P., Camerino, G., and Lovell-Badge, R. (1998) *Nature (London)* **391**, 761–767
263. Hurst, L. D. (2001) *Nature (London)* **411**, 149–150
264. Ariel, M., McCarrey, J., and Cedar, H. (1991) *Proc. Natl. Acad. Sci. U.S.A.* **88**, 2317–2321
265. Iannello, R. C., Gould, J. A., Young, J. C., Giudice, A., Medcalf, R., and Kola, I. (2000) *J. Biol. Chem.* **275**, 19603–19608
266. Akama, T. O., Nakagawa, H., Sugihara, K., Narisawa, S., Ohyama, C., Nishimura, S.-I., O'Brien, D. A., Moremen, K. W., Millán, J. L., and Fukuda, M. N. (2002) *Science* **295**, 124–127
267. Tulina, N., and Matunis, E. (2001) *Science* **294**, 2546–2549
267a. Sassone-Corsi, P. (2002) *Science* **296**, 2176–2178
267b. Kashiwabara, S.-i, Noguchi, J., Zhuang, T., Ohmura, K., Honda, A., Sugiura, S., Miyamoto, K., Takahashi, S., Inoue, K., Ogura, A., and Baba, T. (2002) *Science* **298**, 1999–2002
267c. Càceres, C., Giménez-Bonafé, P., Ribes, E., Wouters-Tyrou, D., Martinage, A., Kouach, M., Sautière, P., Muller, S., Palau, J., Subirana, J. A., Cornudella, L., and Chiva, M. (1999) *J. Biol. Chem.* **274**, 649–656
267d. Kundu, T. K., and Rao, M. R. S. (1996) *Biochemistry* **35**, 15626–15632
267e. Honke, K., Hirahara, Y., Dupree, J., Suzuki, K., Popko, B., Fukushima, K., Fukushima, J., Nagasawa, T., Yoshida, N., Wada, Y., and Taniguchi, N. (2002) *Proc. Natl. Acad. Sci. U.S.A.* **99**, 4227–4232
268. Ursini, F., Helm, S., Kiess, M., Maiorino, M., Roveri, A., Wissing, J., and Flohé, L. (1999) *Science* **285**, 1393–1396
269. Pfeifer, H., Conrad, M., Roethlein, D., Kyriakopoulos, A., Brielmeier, M., Bornkamm, G. W., and Behne, D. (2001) *FASEB J.* **15**, 1236–1238
270. Shao, X., Tarnasky, H. A., Schalles, U., Oko, R., and van der Hoorn, F. A. (1997) *J. Biol. Chem.* **272**, 6105–6113
271. Haaf, A., Butler, P. J. G., Kent, H. M., Fearnley, I. M., Roberts, T. M., Neuhaus, D., and Stewart, M. (1996) *J. Mol. Biol.* **260**, 251–260
272. Eggan, K., Akutsu, H., Hochedlinger, K., Rideout, W., III, Yanagimachi, R., and Jaenisch, R. (2000) *Science* **290**, 1578–1581
273. O'Neill, L. P., Keohane, A. M., Lavender, J. S., McCabe, V., Heard, E., Avner, P., Brockdorff, N., and Turner, B. M. (1999) *EMBO J.* **18**, 2897–2907
274. Gilbert, S. L., and Sharp, P. A. (1999) *Proc. Natl. Acad. Sci. U.S.A.* **96**, 13825–13830
275. Willard, H. F., and Carrel, L. (2001) *Proc. Natl. Acad. Sci. U.S.A.* **98**, 10025–10027
276. Disteche, C. M. (1999) *Proc. Natl. Acad. Sci. U.S.A.* **96**, 14180–14182
277. Percec, I., and Bartolomei, M. S. (2002) *Science* **295**, 287–288
278. Tamaru, H., and Selker, E. U. (2001) *Nature (London)* **414**, 277–283
279. Rice, J. C., and Allis, C. D. (2001) *Nature (London)* **414**, 258–259
280. Melton, D. A. (1991) *Science* **252**, 234–241
281. Miller, M. A., Nguyen, V. Q., Lee, M.-H., Kosinski, M., Schedl, T., Caprioli, R. M., and Greenstein, D. (2001) *Science* **291**, 2144–2147

References

282. Ferrell, J. E., Jr., and Machleder, E. M. (1998) *Science* **280**, 895–898
283. Nothias, J.-Y., Majumder, S., Kaneko, K. J., and DePamphilis, M. L. (1995) *J. Biol. Chem.* **270**, 22077–22080
283a. Mehlmann, L. M., Jones, T. L. Z., and Jaffe, L. A. (2002) *Science* **297**, 1343 – 1345
284. Kuge, H., and Richter, J. D. (1995) *EMBO J.* **14**, 6301–6310
284a. Matzuk, M. M., Burns, K. H., Viveiros, M. M., and Eppig, J. J. (2002) *Science* **296**, 2178 – 2180
285. Cooley, L., and Theurkauf, W. E. (1994) *Science* **266**, 590–596
286. Godt, D., and Tepass, U. (1998) *Nature (London)* **395**, 387–391
287. Cau, J., Faure, S., Vigneron, S., Labbé, J. C., Delsert, C., and Morin, N. (2000) *J. Biol. Chem.* **275**, 2367–2375
288. Wassarman, P. M. (1988) *Sci. Am.* **259**(Dec), 78–84
289. Vacquier, V. D. (1998) *Science* **281**, 1995–1998
290. Babcock, D. F., Bosma, M. M., Battaglia, D. E., and Darszon, A. (1992) *Proc. Natl. Acad. Sci. U.S.A.* **89**, 6001–6005
291. Olson, J. H., Xiang, X., Ziegert, T., Kittelson, A., Rawls, A., Bieber, A. L., and Chandler, D. E. (2001) *Proc. Natl. Acad. Sci. U.S.A.* **98**, 11205–11210
292. Ohlendieck, K., and Lennarz, W. J. (1995) *Trends Biochem. Sci.* **20**, 29–32
293. Vilela-Silva, A.-C. E. S., Castro, M. O., Valente, A.-P., Biermann, C. H., and Mourao, P. A. S. (2002) *J. Biol. Chem.* **277**, 379–387
294. Litscher, E. S., Juntunen, K., Seppo, A., Penttilä, L., Niemelä, R., Renkonen, O., and Wassarman, P. M. (1995) *Biochemistry* **34**, 4662–4669
295. Johnston, D. S., Wright, W. W., Shaper, J. H., Hokke, C. H., Van den Eijnden, D. H., and Joziasse, D. H. (1998) *J. Biol. Chem.* **273**, 1888–1895
296. Varela, P. F., Romero, A., Sanz, L., Romao, M. J., Töpfer-Petersen, E., and Calvete, J. J. (1997) *J. Mol. Biol.* **274**, 635–649
297. Chen, M. S., Tung, K. S. K., Coonrod, S. A., Takahashi, Y., Bigler, D., Chang, A., Yamashita, Y., Kincade, P. W., Herr, J. C., and White, J. M. (1999) *Proc. Natl. Acad. Sci. U.S.A.* **96**, 11830–11835
298. Miyado, K., Yamada, G., Yamada, S., Hasuwa, H., Nakamura, Y., Ryu, F., Suzuki, K., Kosai, K., Inoue, K., Ogura, A., Okabe, M., and Mekada, E. (2000) *Science* **287**, 321–324
298a. Primakoff, P., and Myles, D. G. (2002) *Science* **296**, 2183 – 2185
299. Sherman, M. B., Jakana, J., Sun, S., Matsudaira, P., Chiu, W., and Schmid, M. F. (1999) *J. Mol. Biol.* **294**, 139–149
300. Yamagata, K., Murayama, K., Okabe, M., Toshimori, K., Nakanishi, T., Kashiwabara, S.-i, and Baba, T. (1998) *J. Biol. Chem.* **273**, 10470–10474
301. Mengerink, K. J., Moy, G. W., and Vacquier, V. D. (2002) *J. Biol. Chem.* **277**, 943–948
302. Kuo, R. C., Baxter, G. T., Thompson, S. H., Stricker, S. A., Patton, C., Bonaventura, J., and Epel, D. (2000) *Nature (London)* **406**, 633–636
302a. Glaser, R. W., Grüne, M., Wandelt, C., and Ulrich, A. S. (1999) *Biochemistry* **38**, 2560 – 2569
303. Parrington, J., Swann, K., Shevchenko, V. I., Sesay, A. K., and Lai, F. A. (1996) *Nature (London)* **379**, 364–368
304. Fontanilla, R. A., and Nuccitelli, R. (1998) *Biophys. J.* **75**, 2079 – 2087
305. Shapiro, B. M. (1991) *Science* **252**, 533–536
306. Lawinger, P., Rastelli, L., Zhao, Z., and Majumder, S. (1999) *J. Biol. Chem.* **274**, 8002–8011
307. Leno, G. H., Mills, A. D., Philpott, A., and Laskey, R. A. (1996) *J. Biol. Chem.* **271**, 7253–7256
308. Pedersen, R. A. (2001) *Nature (London)* **409**, 473–474
309. Shankland, M., and Seaver, E. C. (2000) *Proc. Natl. Acad. Sci. U.S.A.* **97**, 4434–4437
310. Blow, J. J. (2001) *EMBO J.* **20**, 3293–3297
310a. Nomura, K., Shimizu, T., Kinoh, H., Sendai, Y., Inomata, M., and Suzuki, N. (1997) *Biochemistry* **36**, 7225 – 7238
311. Cross, J. C., Werb, Z., and Fisher, S. J. (1994) *Science* **266**, 1508–1517
312. Beddington, R. (1998) *Nature (London)* **395**, 641–643
313. He, X., Treacy, M. N., Simmons, D. M., Ingraham, H. A., Swanson, L. W., and Rosenfeld, M. G. (1989) *Nature (London)* **340**, 35–42
314. Rosner, M. H., Vigano, M. A., Rigby, P. W. J., Arnheiter, H., and Staudt, L. M. (1991) *Science* **253**, 144–145
315. Barnea, E., and Bergman, Y. (2000) *J. Biol. Chem.* **275**, 6608–6619
316. Lawrence, P. (1992) *The Making of a Fly: The Genetics of Animal Design*, Blackwell Scientific Publ., Oxford
317. Leptin, M. (1999) *EMBO J.* **18**, 3187–3192
318. Roush, W. (1996) *Science* **274**, 1608–1609
319. Fishman, M. C. (2001) *Science* **294**, 1290–1291
320. Thisse, C., and Zon, L. I. (2002) *Science* **295**, 457–462
321. Farber, S. A., Pack, M., Ho, S.-Y., Johnson, I. D., Wagner, D. S., Dosch, R., Mullins, M. C., Hendrickson, H. S., Hendrickson, E. K., and Halpern, M. E. (2001) *Science* **292**, 1385–1388
321a. Pearson, H. (2002) *Nature (London)* **418**, 14 – 15
321b. Brendza, R. P., Serbus, L. R., Duffy, J. B., and Saxton, W. M. (2000) *Science* **298**, 2120 – 2122
322. Bullock, S. L., and Ish-Horowicz, D. (2001) *Nature (London)* **414**, 611–616
323. Gergely, F., Kidd, D., Jeffers, K., Wakefield, J. G., and Raff, J. W. (2000) *EMBO J.* **19**, 241–252
324. Anderson, K. (1995) *Science* **269**, 2189–2190
325. McCoon, P. E., Angerer, R. C., and Angerer, L. M. (1996) *J. Biol. Chem.* **271**, 20119–20125
326. De Robertis, E. M., and Sasai, Y. (1996) *Nature (London)* **380**, 37–40
327. Dissing, M., Giordano, H., and DeLotto, R. (2001) *EMBO J.* **20**, 2387–2393
328. Markstein, M., Markstein, P., Markstein, V., and Levine, M. S. (2002) *Proc. Natl. Acad. Sci. U.S.A.* **99**, 763–768
329. Kessler, D. S., and Melton, D. A. (1994) *Science* **266**, 596–604
330. Dawid, I. B., and Sargent, T. D. (1988) *Science* **240**, 1443–1448
331. Hemmati-Brivanlou, A., and Melton, D. A. (1992) *Nature (London)* **359**, 609–614
332. Schier, A. F., and Shen, M. M. (2000) *Nature (London)* **403**, 385–389
333. Ewing, T. (1993) *Science* **260**, 624– 625
334. Rodriguez Esteban, C., Capdevila, J., Economides, A. N., Pascual, J., Ortiz, A., and Izpisúa, B., JC. (1999) *Nature (London)* **401**, 243–251
335. Ryan, A. K., Blumberg, B., Rodriguez-Esteban, C., Yonei-Tamura, S., Tamura, K., Tsukui, T., de la Pena, J., Sabbagh, W., Greenwald, J., Choe, S., Norris, D. P., Robertson, E. J., Evans, R. M., Rosenfeld, M. G., and Belmonte, J. C. I. (1998) *Nature (London)* **394**, 545–551
336. Mochizuki, T., Saijoh, Y., Tsuchiya, K., Shirayoshi, Y., Takai, S., Taya, C., Yonekawa, H., Yamada, K., Nihei, H., Nakatsuji, N., Overbeek, P. A., Hamada, H., and Yokoyama, T. (1998) *Nature (London)* **395**, 177– 181
337. Witkowski, J. (1985) *Trends Biochem. Sci.* **10**, 379–381
338. Lamb, T. M., Knecht, A. K., Smith, W. C., Stachel, S. E., Economides, A. N., Stahl, N., Yancopolous, G. D., and Harland, R. M. (1993) *Science* **262**, 713–718
339. Bouwmeester, T., Kim, S.-H., Sasai, Y., Lu, B., and De Robertis, E. M. (1996) *Nature (London)* **382**, 595–601
340. Ruiz i Altaba, A. (1998) *Nature (London)* **391**, 748–749
341. Nishita, M., Hashimoto, M. K., Ogata, S., Laurent, M. N., Ueno, N., Shibuya, H., and Cho, K. W. Y. (2000) *Nature (London)* **403**, 781–785
342. Peifer, M., and Polakis, P. (2000) *Science* **287**, 1606–1609
343. Feldman, B., Gates, M. A., Egan, E. S., Dougan, S. T., Rennebeck, G., Sirotkin, H. I., Schier, A. F., and Talbot, W. S. (1998) *Nature (London)* **395**, 181–185
344. Brunet, L. J., McMahon, J. A., McMahon, A. P., and Harland, R. M. (1998) *Science* **280**, 1455–1457
345. Goodrich, J., Puangsomlee, P., Martin, M., Long, D., Meyerowitz, E. M., and Coupland, G. (1997) *Nature (London)* **386**, 44–51
346. Affolter, M., and Mann, R. (2001) *Science* **292**, 1080–1081
347. Cossu, G., and Borello, U. (1999) *EMBO J.* **18**, 6867–6872
348. Zhang, N., and Gridley, T. (1998) *Nature (London)* **394**, 374–377
349. Basler, K., and Struhl, G. (1994) *Nature (London)* **368**, 208–214
350. Ingham, P. W. (1998) *EMBO J.* **17**, 3505–3511
351. Blair, S. S. (1995) *Nature (London)* **373**, 656–657
352. Chiang, C., Litingtung, Y., Lee, E., Young, K. E., Corden, J. L., Westphal, H., and Beachy, P. A. (1996) *Nature (London)* **383**, 407–413
353. Martin, G. (1996) *Science* **274**, 203–204
354. Strauss, E. (1998) *Science* **280**, 1528–1529
355. Tozawa, R.-i, Ishibashi, S., Osuga, J.-i, Yagyu, H., Oka, T., Chen, Z., Ohashi, K., Perrey, S., Shionoiri, F., Yahagi, N., Harada, K., Gotoda, T., Yazaki, Y., and Yamada, N. (1999) *J. Biol. Chem.* **274**, 30843–30848
356. Kimble, J., Henderson, S., and Crittenden, S. (1998) *Trends Biochem. Sci.* **23**, 353–361
357. Chenn, A., and Walsh, C. A. (1999) *Science* **286**, 689–690
358. Gupta-Rossi, N., Bail, O. L., Gonen, H., Brou, C., Logeat, F., Six, E., Ciechanover, A., and Israël, A. (2001) *J. Biol. Chem.* **276**, 34371–34378
359. Goutte, C., Tsunozaki, M., Hale, V. A., and Priess, J. R. (2002) *Proc. Natl. Acad. Sci. U.S.A.* **99**, 775–779
360. Heldin, C.-H., and Ericsson, J. (2001) *Science* **294**, 2111–2113
361. Gu, Y., Misonou, H., Sato, T., Dohmae, N., Takio, K., and Ihara, Y. (2001) *J. Biol. Chem.* **276**, 35235–35238
362. Brückner, K., Perez, L., Clausen, H., and Cohen, S. (2000) *Nature (London)* **406**, 411– 415
363. Moloney, D. J., Panin, V. M., Johnston, S. H., Chen, J., Shao, L., Wilson, R., Wang, Y., Stanley, P., Irvine, K. D., Haltiwanger, R. S., and Vogt, T. F. (2000) *Nature (London)* **406**, 369– 375
364. Evrard, Y. A., Lun, Y., Aulehla, A., Gan, L., and Johnson, R. L. (1998) *Nature (London)* **394**, 377–381
365. Hirsch, J. A., and Aggarwal, A. K. (1995) *EMBO J.* **14**, 6280–6291
366. De Robertis, E. M., Oliver, G., and Wright, C. V. E. (1990) *Sci. Am.* **263**(Jul), 46–52
367. McGinnis, W., and Kuziora, M. (1994) *Sci. Am.* **270**(Feb), 58–66

References

367a. Kmita, M., Fraudeau, N., Hérault, Y., and Duboule, D. (2002) *Nature (London)* **420**, 145 – 150

367b. Zákány, J., and Duboule, D. (1999) *Nature (London)* **401**, 761

368. Ades, S. E., and Sauer, R. T. (1995) *Biochemistry* **34**, 14601 – 14608

369. Fraenkel, E., Rould, M. A., Chambers, K. A., and Pabo, C. O. (1998) *J. Mol. Biol.* **284**, 351 – 361

370. Carr, A., and Biggin, M. D. (1999) *EMBO J.* **18**, 1598 – 1608

371. Jabet, C., Gitti, R., Summers, M. F., and Wolberger, C. (1999) *J. Mol. Biol.* **291**, 521 – 530

372. Ippel, H., Larsson, G., Behravan, G., Zdunek, J., Lundqvist, M., Schleucher, J., Lycksell, P.-O., and Wijmenga, S. (1998) *J. Mol. Biol.* **288**, 689 – 703

373. Duboule, D. (2000) *Nature (London)* **403**, 607 – 609

374. Capili, A. D., Schultz, D. C., Rauscher, F. J., III, and Borden, K. L. B. (2001) *EMBO J.* **20**, 165 – 177

375. Wray, G. A. (2001) *Science* **292**, 2256 – 2257

376. Vaziri, H., Dragowska, W., Allsopp, R. C., Thomas, T. E., Harley, C. B., and Lansdorp, P. M. (1994) *Proc. Natl. Acad. Sci. U.S.A.* **91**, 9857 – 9860

377. Orkin, S. H. (1995) *J. Biol. Chem.* **270**, 4955 – 4958

378. Ziegler, B. L., Valtieri, M., Porada, G. A., Maria, R. D., Müller, R., Masella, B., Gabbianelli, M., Casella, I., Pelosi, E., Bock, T., Zanjani, E. D., and Peschle, C. (1999) *Science* **285**, 1553 – 1558

379. Kaufman, D. S., Hanson, E. T., Lewis, R. L., Auerbach, R., and Thomson, J. A. (2001) *Proc. Natl. Acad. Sci. U.S.A.* **98**, 10716 – 10721

380. Nony, P., Hannon, R., Gould, H., and Felsenfeld, G. (1998) *J. Biol. Chem.* **273**, 32910 – 32919

381. Akashi, K., Traver, D., Miyamoto, T., and Weissman, I. L. (2000) *Nature (London)* **404**, 193 – 197

382. Nutt, S. L., Heavey, B., Rolink, A. G., and Busslinger, M. (1999) *Nature (London)* **401**, 556 – 562

383. Natarajan, K., Sawicki, M. W., Margulies, D. H., and Mariuzza, R. A. (2000) *Biochemistry* **39**, 14779 – 14786

384. Metcalf, D. (1992) *Trends Biochem. Sci.* **17**, 286 – 289

385. Wen, D., Boissel, J.-P., Showers, M., Ruch, B. C., and Bunn, H. F. (1994) *J. Biol. Chem.* **269**, 22839 – 22846

386. Metcalf, D. (1994) *Nature (London)* **369**, 519 – 520

387. Lebestky, T., Chang, T., Hartenstein, V., and Banerjee, U. (2000) *Science* **288**, 146 – 149

388. McMorrow, T., van den Wijngaard, A., Wollenschlaeger, A., van de Corput, M., Monkhorst, K., Trimborn, T., Fraser, P., van Lohuizen, M., Jenuwein, T., Djabali, M., Philipsen, S., Grosveld, F., and Milot, E. (2000) *EMBO J.* **19**, 4986 – 4996

389. Razin, S. V., Loudinkova, E. S., and Scherrer, K. (2000) *J. Mol. Biol.* **299**, 845 – 852

390. Weatherall, D. J., Clegg, J. B., Higgs, D. R., and Wood, W. G. (1995) in *The Metabolic and Molecular Bases of Inherited Disease*, 7th ed., Vol. 1 (Scriver, C. R., Beaudet, A. L., Sly, W. S., and Valle, D., eds), pp. 3417 – 3484, McGraw-Hill, New York

391. Elnitski, L., Li, J., Noguchi, C. T., Miller, W., and Hardison, R. (2001) *J. Biol. Chem.* **276**, 6289 – 6298

392. Li, Q., Blau, C. A., Clegg, C. H., Rohde, A., and Stamatoyannopoulos, G. (1998) *J. Biol. Chem.* **273**, 17361 – 17367

393. Li, J., Noguchi, C. T., Miller, W., Hardison, R., and Schechter, A. N. (1998) *J. Biol. Chem.* **273**, 10202 – 10209

394. Wang, Z., and Liebhaber, S. A. (1999) *EMBO J.* **18**, 2218 – 2228

395. Chung, J. H., Bell, A. C., and Felsenfeld, G. (1997) *Proc. Natl. Acad. Sci. U.S.A.* **94**, 575 – 580

396. Hanahan, D. (1997) *Science* **277**, 48 – 50

397. Yancopoulos, G. D., Davis, S., Gale, N. W., Rudge, J. S., Wiegand, S. J., and Holash, J. (2000) *Nature (London)* **407**, 242 – 248

398. Carmeliet, P. (2000) *Nature (London)* **408**, 43,45

399. Risau, W. (1997) *Nature (London)* **386**, 671 – 674

400. Browder, T., Folkman, J., and Pirie-Shepherd, S. (2000) *J. Biol. Chem.* **275**, 1521 – 1524

401. Tao, Q., Backer, M. V., Backer, J. M., and Terman, B. I. (2001) *J. Biol. Chem.* **276**, 21916 – 21923

401a. Funamoto, M., Fujio, Y., Kunisada, K., Negoro, S., Tone, E., Osugi, T., Hirota, H., Izumi, M., Yoshizaki, K., Walsh, K., Kishimoto, T., and Yamauchi-Takihara, K. (2000) *J. Biol. Chem.* **275**, 10561 – 10566

402. Li, D. Y., Sorensen, L. K., Brooke, B. S., Urness, L. D., Davis, E. C., Taylor, D. G., Boak, B. B., and Wendel, D. P. (1999) *Science* **284**, 1534 – 1537

403. Carmeliet, P. (2001) *Science* **293**, 1602 – 1604

404. Leonidas, D. D., Shapiro, R., Subbarao, G. V., Russo, A., and Acharya, K. R. (2002) *Biochemistry* **41**, 2552 – 2562

405. Carmeliet, P., and Jain, R. K. (2000) *Nature (London)* **407**, 249 – 257

406. Otani, A., Slike, B. M., Dorrell, M. I., Hood, J., Kinder, K., Ewalt, K. L., Cheresh, D., Schimmel, P., and Friedlander, M. (2002) *Proc. Natl. Acad. Sci. U.S.A.* **99**, 178 – 183

407. Maisonpierre, P. C., Suri, C., Jones, P. F., Bartunkova, S., Wiegand, S. J., Radziejewski, C., Compton, D., McClain, J., Aldrich, T. H., Papadopoulos, N., Daly, T. J., Davis, S., Sato, T. N., and Yancopoulos, G. D. (1997) *Science* **277**, 55 – 60

408. Dawson, D. W., Volpert, O. V., Gillis, P., Crawford, S. E., Xu, H.-J., Benedict, W., and Bouck, N. P. (1999) *Science* **285**, 245 – 248

409. Hohenester, E., Sasaki, T., Olsen, B. R., and Timpl, R. (1998) *EMBO J.* **17**, 1656 – 1664

410. L'Heureux, N., Pâquet, S., Labbé, R., Germain, L., and Auger, F. A. (1998) *FASEB J.* **12**, 47 – 56

411. Alliston, T., Choy, L., Ducy, P., Karsenty, G., and Derynck, R. (2001) *EMBO J.* **20**, 2254 – 2272

412. Liu, Y., Li, H., Tanaka, K., Tsumaki, N., and Yamada, Y. (2000) *J. Biol. Chem.* **275**, 12712 – 12718

413. ten Dijke, P., Yamashita, H., Sampath, T. K., Reddi, A. H., Estevez, M., Riddle, D. L., Ichijo, H., Heldin, C.-H., and Miyazono, K. (1994) *J. Biol. Chem.* **269**, 16985 – 16988

414. Scheufler, C., Sebald, W., and Hülsmeyer, M. (1999) *J. Mol. Biol.* **287**, 103 – 115

415. Harland, R. M. (2001) *Nature (London)* **410**, 423 – 424

416. Alliston, T., and Derynck, R. (2002) *Nature (London)* **416**, 686 – 687

417. Hasty, P., Bradley, A., Morris, J. H., Edmondson, D. G., Venuti, J. M., Olson, E. N., and Klein, W. H. (1993) *Nature (London)* **364**, 501 – 506

418. Zhang, J.-M., Zhao, X., Wei, Q., and Paterson, B. M. (1999) *EMBO J.* **18**, 6983 – 6993

419. Charbonnier, P., Gaspera, B. D., Armand, A.-S., Van der Laarse, W. J., Launay, T., Becker, C., Gallien, C.-L., and Chanoine, C. (2002) *J. Biol. Chem.* **277**, 1139 – 1147

420. Miller, J. B. (1995) *Nature (London)* **377**, 575 – 576

421. Stephanou, A., Brar, B. K., Scarabelli, T. M., Jonassen, A. K., Yellon, D. M., Marber, M. S., Knight, R. A., and Latchman, D. S. (2000) *J. Biol. Chem.* **275**, 10002 – 10008

422. Srivastava, D., and Olson, E. N. (2000) *Nature (London)* **407**, 221 – 226

423. Andreadis, S. T., Hamoen, K. E., Yarmush, M. L., and Morgan, J. R. (2001) *FASEB J.* **15**, 898 – 906

424. Andersen, B., Schonemann, M. D., Flynn, S. E., Pearse, R. V., II, Singh, H., and Rosenfeld, M. G. (1993) *Science* **260**, 78 – 82

425. Mills, A. A., Zheng, B., Wang, X.-J., Vogel, H., Roop, D. R., and Bradley, A. (1999) *Nature (London)* **398**, 708 – 713

426. Yang, A., Schweitzer, R., Sun, D., Kaghad, M., Walker, N., Bronson, R. T., Tabin, C., Sharpe, A., Caput, D., Crum, C., and McKeon, F. (1999) *Nature (London)* **398**, 714 – 718

427. Hartmann, D. (2001) *Proc. Natl. Acad. Sci. U.S.A.* **98**, 10522 – 10523

428. van den Brink, G. R., de Santa Barbara, P., and Roberts, D. J. (2001) *Science* **294**, 2115 – 2116

428a. Peifer, M. (2002) *Nature (London)* **410**, 274 – 277

429. Risau, W. (1995) *FASEB J.* **9**, 926 – 933

430. Griffin, C. T., Srinivasan, Y., Zheng, Y.-W., Huang, W., and Coughlin, S. R. (2001) *Science* **293**, 1666 – 1670

430a. Cooper, M. T. D., and Bray, S. J. (1999) *Nature (London)* **397**, 526 – 530

430b. Knust, E., and Bossinger, O. (2002) *Science* **298**, 1955–1959

431. Livesey, R., and Cepko, C. (2001) *Nature (London)* **413**, 471,473

432. McKay, R. (1997) *Science* **276**, 66 – 71

433. Ourednik, V., Ourednik, J., Flax, J. D., Zawada, W. M., Hutt, C., Yang, C., Park, K. I., Kim, S. U., Sidman, R. L., Freed, C. R., and Snyder, E. Y. (2001) *Science* **293**, 1820 – 1824

434. Tanabe, Y., and Jessell, T. M. (1996) *Science* **274**, 1115 – 1123

434a. García-Castro, M. I., Marcelle, C., and Bronner-Fraser, M. (2002) *Science* **297**, 848 – 851

434b. Hatten, M. E. (2002) *Science* **297**, 1660 – 1663

435. Scully, K. M., and Rosenfeld, M. G. (2002) *Science* **295**, 2231 – 2235

436. Dodd, J., Jessell, T. M., and Placzek, M. (1998) *Science* **282**, 1654 – 1657

437. Jeong, J., and McMahon, A. P. (2001) *Nature (London)* **412**, 136 – 137

437a. Dickson, B. J. (2002) *Science* **298**, 1959 – 1964

438. Shatz, C. J. (1992) *Sci. Am.* **267**(Sep), 61 – 67

439. Ramón y Cajal, S. (1892) *La Cellle* **9**, 119

440. Baier, H., and Bonhoeffer, F. (1994) *Science* **265**, 1541 – 1542

441. Serafini, T., Kennedy, T. E., Galko, M. J., Mirzayan, C., Jessell, T. M., and Tessier-Lavigne, M. (1994) *Cell* **78**, 409 – 424

442. Kennedy, T. E., Serafini, T., de la Torre, J. R., and Tessier-Lavigne, M. (1994) *Cell* **78**, 425 – 435

443. Tessier-Lavigne, M., and Goodman, C. S. (1996) *Science* **274**, 1123 – 1133

444. Stein, E., and Tessier-Lavigne, M. (2001) *Science* **291**, 1928 – 1938

445. Marx, J. (1995) *Science* **268**, 971 – 973

446. Ming, G.-L., Wong, S. T., Henley, J., Yuan, X.-b, Song, H.-j, Spitzer, N. C., and Poo, M.-m. (2002) *Nature (London)* **417**, 411 – 418

447. Tear, G. (2001) *Nature (London)* **409**, 472 – 473

448. Dickson, B. J. (2001) *Science* **291**, 1910 – 1911

449. Fricke, C., Lee, J.-S., Geiger-Rudolph, S., Bonhoeffer, F., and Chien, C.-B. (2001) *Science* **292**, 507 – 510

450. Wang, H., and Tessier-Lavigne, M. (1999) *Nature (London)* **401**, 765 – 769

451. Takei, K., Shin, R.-M., Inoue, T., Kato, K., and Mikoshiba, K. (1998) *Science* **282**, 1705 – 1708

452. Polleux, F., Morrow, T., and Ghosh, A. (2000) *Nature (London)* **404**, 567 – 573

References

453. Marín, O., Yaron, A., Bagri, A., Tessier-Lavigne, M., and Rubenstein, J. L. R. (2001) *Science* **293**, 872–875
454. Goffinet, A. M. (1997) *Nature (London)* **389**, 668–669
455. Bar, I., and Goffinet, A. M. (1999) *Nature (London)* **399**, 645–646
456. Yip, J. W., Yip, Y. P. L., Nakajima, K., and Capriotti, C. (2000) *Proc. Natl. Acad. Sci. U.S.A.* **97**, 8612–8616
457. Chen, L., Chetkovich, D. M., Petralia, R. S., Sweeney, N. T., Kawasaki, Y., Wenthold, R. J., Bredt, D. S., and Nicoll, R. A. (2000) *Nature (London)* **408**, 936–943
458. Nakagawa, T., and Sheng, M. (2000) *Science* **290**, 2270–2271
459. Smith, S. J. (1999) *Science* **283**, 1860–1861
459a. Cohen-Cory, S. (2002) *Science* **298**, 770–776
460. Ikonomidou, C., Bosch, F., Miksa, M., Bittigau, P., Vöckler, J., Dikranian, K., Tenkova, T. I., Stefovska, V., Turski, L., and Olney, J. W. (1999) *Science* **283**, 70–74
461. Goldberg, J. L., and Barres, B. A. (2000) *Nature (London)* **403**, 369–370
462. Prinjha, R., Moore, S. E., Vinson, M., Blake, S., Morrow, R., Christie, G., Michalovich, D., Simmons, D. L., and Walsh, F. S. (2000) *Nature (London)* **403**, 33–384
463. Olson, L. (2002) *Nature (London)* **416**, 589–590
464. Meyerowitz, E. M. (2002) *Science* **295**, 1482–1485
465. Buchanan, B. B., Gruissem, W., and Jones, R. L., eds. (2000) *Biochemistry and Molecular Biology of Plants*, American Society of Plant Physiologists, Rockville, Maryland
466. Goldberg, R. B., de Paiva, G., and Yadegari, R. (1994) *Science* **266**, 605–614
467. Bewley, J. D., Hempel, F. D., McCormick, S., and Zambryski, P. (2000) in *Biochemistry and Molecular Biology of Plants* (Buchanan, B., Gruissem, W., and Jones, R., eds), American Society of Plant Physiologists, Rockville, Maryland
468. Thompson, R. D. (2000) *Nature (London)* **408**, 39,41
469. Vielle Calzada, J.-P., Crane, C. F., and Stelly, D. M. (1996) *Science* **274**, 1322–1323
470. Dellaporta, S. L., and Calderon-Urrea, A. (1994) *Science* **266**, 1501–1505
471. Nasrallah, J. B. (2002) *Science* **296**, 305–308
472. Cui, X., Wise, R. P., and Schnable, P. S. (1996) *Science* **272**, 1334–1336
473. Lichtenstein, C. (1990) *Trends Biochem. Sci.* **15**, 453–454
474. Levings, C. S., III. (1996) *Science* **272**, 1279–1280
475. Jürgens, G. (2001) *EMBO J.* **20**, 3609–3616
476. van den Berg, C., Willemsen, V., Hage, W., Weisbeek, P., and Scheres, B. (1995) *Nature (London)* **378**, 62–65
477. Hake, S. (2001) *Nature (London)* **413**, 261–264
478. Vielle-Calzada, J.-P., Baskar, R., and Grossniklaus, U. (2000) *Nature (London)* **404**, 91–94
479. Ferrándiz, C., Pelaz, S., and Yanofsky, M. F. (1999) *Ann. Rev. Biochem.* **68**, 321–354
480. Vrebalov, J., Ruezinsky, D., Padmanabhan, V., White, R., Medrano, D., Drake, R., Schuch, W., and Giovannoni, J. (2002) *Science* **296**, 343–346
481. Brown, K. (2001) *Science* **291**, 1884–1885
482. Lane, B. G. (1994) *FASEB J.* **8**, 294–301
483. Berardini, T. Z., Bollman, K., Sun, H., and Poethig, R. S. (2001) *Science* **291**, 2405–2407
484. Jürgens, G. (1997) *Nature (London)* **386**, 17
485. Timmermans, M. C. P., Hudson, A., Becraft, P. W., and Nelson, T. (1999) *Science* **284**, 151–153
486. Simpson, G. G., and Dean, C. (2002) *Science* **296**, 285–289
486a Dangl, J. L., Dietrich, R. A., and Thomas, H. (2000) in *Biochemistry & Molecular Biology of Plants* (Buchanan, B., Gruissem, W., and Jones, R., eds), pp. 1044–1100, American Society of Plant Physiologists, Rockville, Maryland
487. Ly, D. H., Lockhart, D. J., Lerner, R. A., and Schultz, P. G. (2000) *Science* **287**, 2486–2492
488. Shay, J. W., and Wright, W. E. (2001) *Science* **291**, 839–840
489. Takahashi, Y., Kuro-o, M., and Ishikawa, F. (2000) *Proc. Natl. Acad. Sci. U.S.A.* **97**, 12407–12408
490. Olshansky, S. J., Carnes, B. A., and Désesquelles, A. (2001) *Science* **291**, 1491–1492
491. Weindruch, R. (1996) *Sci. Am.* **274**(Jan), 46–52
492. Hayflick, L. (2000) *Nature (London)* **408**, 267–269
493. Puca, A. A., Daly, M. J., Brewster, S. J., Matise, T. C., Barrett, J., Shea-Drinkwater, M., Kang, S., Joyce, E., Nicoli, J., Benson, E., Kunkel, L. M., and Perls, T. (2001) *Proc. Natl. Acad. Sci. U.S.A.* **98**, 10505–10508
494. Finch, C. E., and Tanzi, R. E. (1997) *Science* **278**, 407–411
495. Holliday, R. (2001) *Trends Biochem. Sci.* **26**, 68–71
496. Guarente, L., and Kenyon, C. (2000) *Nature (London)* **408**, 255–262
497. Kirkwood, T. B. L., and Austad, S. N. (2000) *Nature (London)* **408**, 233–238
498. Rusting, R. L. (1992) *Sci. Am.* **267**(Dec), 130–141
499. Stearns, S. C., Ackermann, M., Doebeli, M., and Kaiser, M. (2000) *Proc. Natl. Acad. Sci. U.S.A.* **97**, 3309–3313
500. Raha, S., and Robinson, B. H. (2000) *Trends Biochem. Sci.* **25**, 502–508
501. Finkel, T., and Holbrook, N. J. (2000) *Nature (London)* **408**, 239–247
502. Hamilton, M. L., Van Remmen, H., Drake, J. A., Yang, H., Guo, Z. M., Kewitt, K., Walter, C. A., and Richardson, A. (2001) *Proc. Natl. Acad. Sci. U.S.A.* **98**, 10469–10474
503. Zou, S., Meadows, S., Sharp, L., Jan, L. Y., and Jan, Y. N. (2000) *Proc. Natl. Acad. Sci. U.S.A.* **97**, 13726–13731
504. Murray, C. A., and Lynch, M. A. (1998) *J. Biol. Chem.* **273**, 12161–12168
504a. Roth, G. S., Lane, M. A., Ingram, D. K., Mattison, J. A., Elahi, D., Tobin, J. D., Muller, D., and Metter, E. J. (2002) *Science* **297**, 811
505. Witkowski, J. (1985) *Trends Biochem. Sci.* **10**, 258–260
506. Medvedev, Z. A. (1972) *Exptl. Gerontol.* **1**, 227–238
507. Hayflick, L., and Moorhead, P. S. (1961) *Exp. Cell. Res.* **25**, 585–621
508. Hayflick, L. (1980) *Sci. Am.* **242**, 58–65
509. Tang, D. G., Tokumoto, Y. M., Apperly, J. A., Lloyd, A. C., and Raff, M. C. (2001) *Science* **291**, 868–871
510. Yang, J., Chang, E., Cherry, A. M., Bangs, C. D., Oei, Y., Bodnar, A., Bronstein, A., Chiu, C.-P., and Herron, G. S. (1999) *J. Biol. Chem.* **274**, 26141–26148
511. Romanov, S. R., Kozakiewicz, B. K., Hoist, C. R., Stampfer, M. R., Haupt, L. M., and Tisty, T. D. (2001) *Nature (London)* **409**, 633–637
512. Freshney, R. I., and Freshney, M. G., eds. (1996) *Culture of Immortalized Cells*, Wiley Liss, New York
513. Mathon, N. F., Malcolm, D. S., Harrisingh, M. C., Cheng, L., and Lloyd, A. C. (2001) *Science* **291**, 872–875
514. de Lange, T., and DePinho, R. A. (1999) *Science* **283**, 947–949
515. Bryan, T. M., Englezou, A., Gupta, J., Bacchetti, S., and Reddel, R. R. (1995) *EMBO J.* **14**, 4240–4248
516. Herbert, B. S., Pitts, A. E., Baker, S. I., Hamilton, S. E., Wright, W. E., Shay, J. W., and Corey, D. R. (1999) *Proc. Natl. Acad. Sci. U.S.A.* **96**, 14276–14281
517. Kucherlapati, R., and DePinho, R. A. (2001) *Nature (London)* **411**, 647–648
518. Wright, W. E., Brasiskyte, D., Piatyszek, M. A., and Shay, J. W. (1996) *EMBO J.* **15**, 1734–1741
519. Lanza, R. P., Cibelli, J. B., Blackwell, C., Cristofalo, V. J., Francis, M. K., Baerlocher, G. M., Mak, J., Schertzer, M., Chavez, E. A., Sawyer, N., Lansdorp, P. M., and West, M. D. (2000) *Science* **288**, 665–669
520. Lin, S.-J., Defossez, P.-A., and Guarente, L. (2000) *Science* **289**, 2126–2128
521. Schweitzer, B. I., Dicker, A. P., and Bertino, J. R. (1990) *FASEB J.* **4**, 2441–2452
522. Lin, S.-J., Kaeberiein, M., Andalis, A. A., Sturtz, L. A., Defossez, P.-A., Culotta, V. C., Fink, G. R., and Guarente, L. (2002) *Nature (London)* **418**, 344–348
523. Campisi, J. (2000) *Science* **289**, 2062–2063
524. Lakowski, B., and Hekimi, S. (1996) *Science* **272**, 1010–1013
525. Jonassen, T., Larsen, P. L., and Clarke, C. F. (2001) *Proc. Natl. Acad. Sci. U.S.A.* **98**, 421–426
526. Larsen, P. L., and Clarke, C. F. (2002) *Science* **295**, 120–123
527. Kimura, K. D., Tissenbaum, H. A., Liu, Y., and Ruvkun, G. (1997) *Science* **277**, 942–946
528. Lin, K., Dorman, J. B., Rodan, A., and Kenyon, C. (1997) *Science* **278**, 1319–1322
529. Strauss, E. (2001) *Science* **292**, 41,43
530. Taub, J., Lau, J. F., Ma, C., Hahn, J. H., Hoque, R., Rothblatt, J., and Chalfie, M. (1999) *Nature (London)* **399**, 162–166
531. Apfeld, J., and Kenyon, C. (1999) *Nature (London)* **402**, 804–807
532. Arantes-Oliveira, N., Apfeld, J., Dillin, A., and Kenyon, C. (2002) *Science* **295**, 502–505
532a. Garigan, D., Hsu, A. L., Fraser, A. G., Kamath, R. S., Ahringer, J., and Kenyon, C. (2002) *Genetics* **161**, 1101–1112
532b. Tower, J. (2000) *Mech Ageing Dev* **118**, 1–14
533. Clancy, D. J., Gems, D., Hafen, E., Leevers, S. J., and Partridge, L. (2002) *Science* **296**, 319
534. West, A. P., Jr., Llamas, L. L., Snow, P. M., Benzer, S., and Bjorkman, P. J. (2001) *Proc. Natl. Acad. Sci. U.S.A.* **98**, 3744–3749
535. Tatar, M., Kopelman, A., Epstein, D., Tu, M.-P., Yin, C.-M., and Garofalo, R. S. (2001) *Science* **292**, 107–110
536. Rogina, B., Reenan, R. A., Nilsen, S. P., and Helfand, S. L. (2000) *Science* **290**, 2137–2140
537. Chavous, D. A., Jackson, F. R., and O'Connor, C. M. (2001) *Proc. Natl. Acad. Sci. U.S.A.* **98**, 14814–14818
538. Guarente, L. (1999) *Nature (London)* **402**, 243,245
539. Moskovitz, J., Bar-Noy, S., Williams, W. M., Requena, J., Berlett, B. S., and Stadtman, E. R. (2001) *Proc. Natl. Acad. Sci. U.S.A.* **98**, 12920–12925
540. Ruan, H., Tang, X. D., Chen, M.-L., Joiner, M. A., Sun, G., Brot, N., Weissbach, H., Heinemann, S. H., Iverson, L., Wu, C.-F., and Hoshi, T. (2002) *Proc. Natl. Acad. Sci. U.S.A.* **99**, 2748–2753
541. De Benedictis, G., Rose, G., Carrieri, G., De Luca, M., Falcone, E., Passarino, G., Bonafé, M., Monti, D., Baggio, G., Bertolini, S., Mari, D., Mattace, R., and Franceschi, C. (1999) *FASEB J.* **13**, 1532–1536
542. Hagen, T. M., Ingersoll, R. T., Lykkesfeldt, J., Liu, J., Wehr, C. M., Vinarsky, V., Bartholomew, J. C., and Ames, B. N. (1999) *FASEB J.* **13**, 411–418
543. Martin, G. M., and Oshima, J. (2000) *Nature (London)* **408**, 263–266

544. Marx, J. (2000) *Science* **287**, 2390
545. Kamath-Loeb, A. S., Johansson, E., Burgers, P. M. J., and Loeb, L. A. (2000) *Proc. Natl. Acad. Sci. U.S.A.* **97**, 4603–4608
546. Kusano, K., Johnson-Schlitz, D. M., and Engels, W. R. (2001) *Science* **291**, 2600–2602
547. de Boer, J., Andressoo, J. O., de Wit, J., Huijmans, J., Beems, R. B., van Steeg, H., Weeda, G., van der Horst, G. T. J., van Leeuwen, W., Themmen, A. P. N., Meradji, M., and Hoeijmakers, J. H. J. (2002) *Science* **296**, 1276–1279
548. Hasty, P., and Vijg, J. (2002) *Science* **296**, 1250–1251
549. Kuro-o, M., Matsumura, Y., Aizawa, H., Kawaguchi, H., Suga, T., Utsugi, T., Ohyama, Y., Kurabayashi, M., Kaname, T., Kume, E., Iwasaki, H., Iida, A., Shiraki-Iida, T., Nishikawa, S., Nagai, R., and Nabeshima, Y.-i. (1997) *Nature (London)* **390**, 45–51
550. Arking, D. E., Krebsova, A., Macek, M., Sr., Macek, M., Jr., Arking, A., Mian, I. S., Fried, L., Hamosh, A., Dey, S., McIntosh, I., and Dietz, H. C. (2002) *Proc. Natl. Acad. Sci. U.S.A.* **99**, 856–861
551. Tyner, S. D., Venkatachalam, S., Choi, J., Jones, S., Ghebranious, N., Igelmann, H., Lu, X., Soron, G., Cooper, B., Brayton, C., Park, S. H., Thompson, T., Karsenty, G., Bradley, A., and Donehower, L. A. (2002) *Nature (London)* **415**, 45–53
552. Strauss, E. (2002) *Science* **295**, 28–29
553. McCormick, F. (1999) *Trends Biochem. Sci.* **24**, M53–M56
554. Aksan, I., and Stinson, J. A. (2002) *Trends Biochem. Sci.* **27**, 387–389
555. Pollock, P. M., and Meltzer, P. S. (2002) *Nature (London)* **417**, 906–907
556. Ryan, K. M., and Vousden, K. H. (2002) *Nature (London)* **419**, 795, 797
557. Vogelstein, B., Lane, D., and Levine, A. J. (2000) *Nature (London)* **408**, 307–310
558. Evan, G. I., and Vousden, K. H. (2001) *Nature (London)* **411**, 342–348
559. Massagué, J., and Serrano, M. (2000) *EMBO Reports* **1**, 115–119
560. Brazil, D. P., and Hemmings, B. A. (2001) *Trends Biochem. Sci.* **26**, 657–664
561. Mayo, L. D., and Donner, D. B. (2002) *Trends Biochem. Sci.* **27**, 462–467
562. Taipale, J., and Beachy, P. A. (2001) *Nature (London)* **411**, 349–354
563. Huntsman, D. G., Carneiro, F., Lewis, F. R., MacLeod, P. M., Hayashi, A., Monaghan, K. G., Maung, R., Seruca, R., Jackson, C. E., and Caldas, C. (2001) *N. Engl. J. Med.* **344**, 1904–1909
564. Weitzman, J. B., and Yaniv, M. (1999) *Nature (London)* **400**, 401–402
565. Marx, J. (2002) *Science* **297**, 544–546
566. Hoeijmakers, J. H. J. (2001) *Nature (London)* **411**, 366–374
567. Maser, R. S., and DePinho, R. A. (2002) *Science* **297**, 565–569
568. Monteiro, A. N. A. (2000) *Trends Biochem. Sci.* **25**, 469–474
569. Wilson, J. H., and Elledge, S. J. (2002) *Science* **297**, 1822–1823
570. Yang, H., Jeffrey, P. D., Miller, J., Kinnucan, E., Sun, Y., Thomä, N. H., Zheng, N., Chen, P.-L., Lee, W.-H., and Pavletich, N. P. (2002) *Science* **297**, 1837–1848
571. Yu, Q., Geng, Y., and Sicinski, P. (2001) *Nature (London)* **411**, 1017–1021
572. Hakomori, S. (2002) *Proc. Natl. Acad. Sci. U.S.A.* **99**, 10231–10233
573. Bernards, R., and Weinberg, R. A. (2002) *Nature (London)* **418**, 823
574. Varambally, S., Dhanasekaran, S. M., Zhou, M., Barrette, T. R., Kumar-Sinha, C., Sanda, M. G., Ghosh, D., Pienta, K. J., Sewalt, R. G. A. B., Otte, A. P., Rubin, M. A., and Chinnalyan, A. M. (2002) *Nature (London)* **419**, 624–629
575. Müller, A., Homey, B., Soto, H., Ge, N., Catron, D., Buchanan, M. E., McClanahan, T., Murphy, E., Yuan, W., Wagner, S. N., Barrera, J. L., Mohar, A., Verástegui, E., and Zlotnik, A. (2001) *Nature (London)* **410**, 50–56
576. Davies, H., and 51 other authors. (2002) *Nature (London)* **417**, 949–954
577. Barinaga, M. (2000) *Science* **287**, 1584–1585
578. Raven, P. H. (2002) *Science* **297**, 954–958
579. Myers, N. (2001) *Nature (London)* **410**, 631–632
580. Serageldin, I. (2002) *Science* **296**, 54–58
581. Vitousek, P. M., Mooney, H. A., Lubchenco, J., and Melillo, J. M. (1997) *Science* **277**, 494–499
582. Rojstaczer, S., Sterling, S. M., and Moore, N. J. (2001) *Science* **294**, 2549–2552
583. Rosenzweig, M. L. (2001) *Proc. Natl. Acad. Sci. U.S.A.* **98**, 5404–5410
584. Krebs, J. R., Wilson, J. D., Bradbury, R. B., and Siriwardena, G. M. (1999) *Nature (London)* **400**, 6111–6112
585. Pimm, S. L., and 32 other authors. (2001) *Science* **293**, 2207–2208
586. Schiermeier, Q. (2002) *Nature (London)* **419**, 662–665
587. Naylor, R. L., Goldburg, R. J., Primavera, J. H., Kautsky, N., Beveridge, M. C. M., Clay, J., Folke, C., Lubchenco, J., Mooney, H., and Troell, M. (2000) *Nature (London)* **405**, 1017–1024
588. Ziman, J. (1996) *Nature (London)* **382**, 751–754

Study Questions

1. Discuss the roles of the following proteins in development: receptors, transcription factors, protein kinases, histones, DNA methylases, adhesion molecules, ubiquitin. How do small RNA molecules participate in development?

2. Are all body cells totipotent?

3. Discuss the roles of apoptosis in various groups of organisms.

4. Compare signaling between bacteria and other unicellular organisms with signaling in higher eukaryotes.

5. Are human beings the most highly developed organisms? If so, in what ways? Has evolution of humans stopped or will it continue? Will it be upward?

6. Is it important for the world to achieve a sustainable state in which the population is constant and the environment stable?[578] How will the world support a projected increase in population from the present 6 billion to 9 billion in 50 years?[579,580]

7. How seriously is the earth's ecosystem dominated by human activity?[581,582] Human activities have greatly reduced the amount of area available to wild species. Will the ensuing extinction of many organisms impoverish future diversity?[583–585] Can the world's fisheries become sustainable?[586,587]

8. Is science losing its objectivity because of an emphasis on monetary gain rather than on meeting social needs?[588]

Complete Index to Volumes 1 and 2

Page numbers set in **boldface** refer to major discussions.
The symbol s after a page number refers to a chemical structure.

A

C

E

G

H

I

J

K

L

N

O

P

Q

R

S

T

U

V

W

X

Y

Z